COLLEGE ALGEBRA

Graphs and Models

John W. Coburn
St. Louis Community College at Florissant Valley

J.D. Herdlick
St. Louis Community College at Meramec-Kirkwood

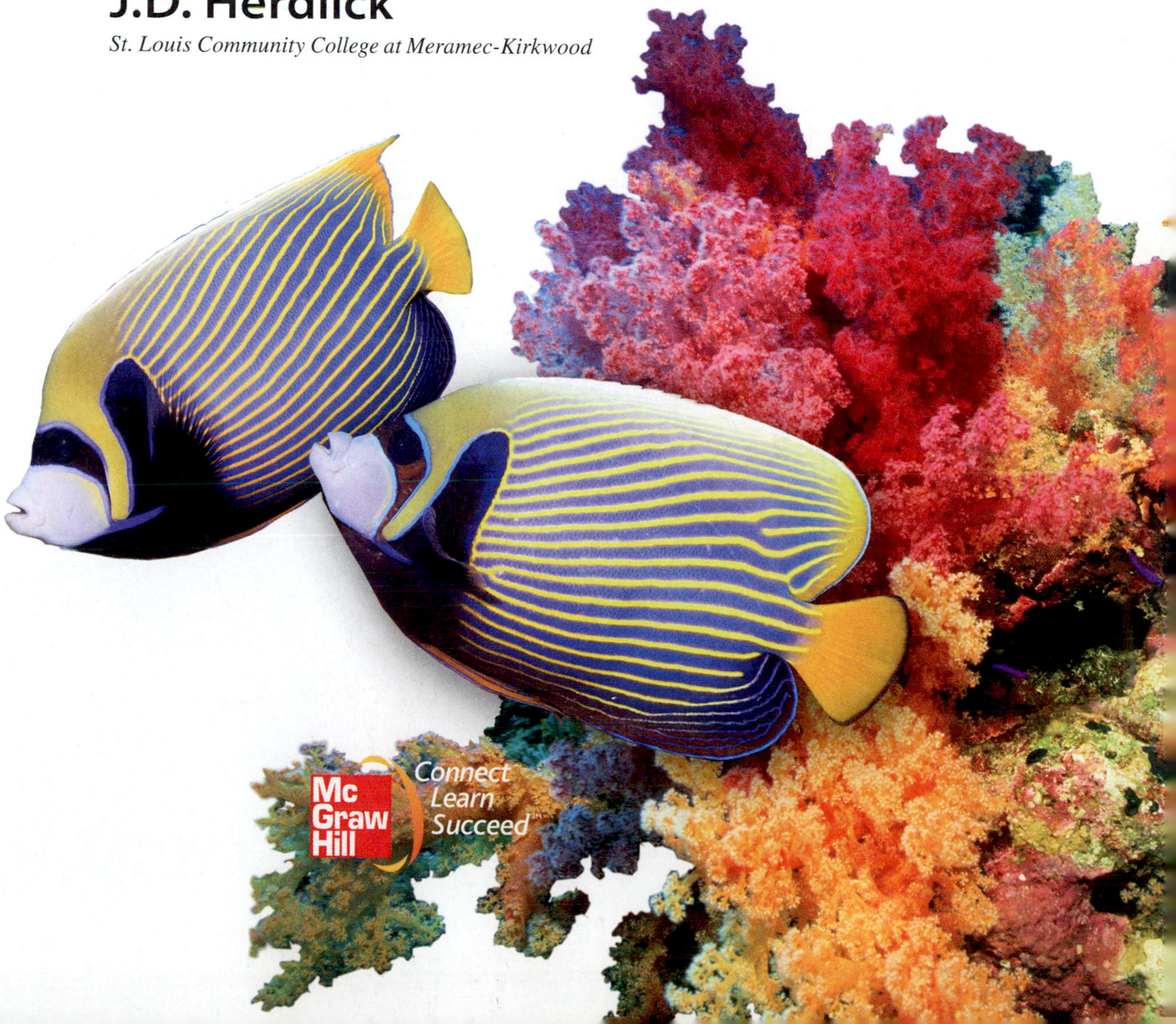

Mc Graw Hill

Connect
Learn
Succeed™

COLLEGE ALGEBRA: GRAPHS AND MODELS

Published by McGraw-Hill, a business unit of The McGraw-Hill Companies, Inc., 1221 Avenue of the
Americas, New York, NY 10020. Copyright © 2012 by The McGraw-Hill Companies, Inc. All rights reserved.
No part of this publication may be reproduced or distributed in any form or by any means, or stored in
a database or retrieval system, without the prior written consent of The McGraw-Hill Companies, Inc.,
including, but not limited to, in any network or other electronic storage or transmission, or broadcast for
distance learning.

Some ancillaries, including electronic and print components, may not be available to customers outside the
United States.

This book is printed on acid-free paper.

1 2 3 4 5 6 7 8 9 0 DOW/DOW 1 0 9 8 7 6 5 4 3 2 1

ISBN 978–0–07–351954–8
MHID 0–07–351954–5

ISBN 978–0–07–723057–9 (Annotated Instructor's Edition)
MHID 0–07–723057–4

Vice President, Editor-in-Chief: *Marty Lange*
Vice President, EDP: *Kimberly Meriwether David*
Senior Director of Development: *Kristine Tibbetts*
Editorial Director: *Stewart K. Mattson*
Sponsoring Editor: *John R. Osgood*
Developmental Editor: *Eve L. Lipton*
Marketing Manager: *Kevin M. Ernzen*
Senior Project Manager: *Vicki Krug*

Buyer II: *Sherry L. Kane*
Senior Media Project Manager: *Sandra M. Schnee*
Senior Designer: *Laurie B. Janssen*
Cover Image: © *Georgette Douwma and Sami Sarkis / Gettyimages*
Senior Photo Research Coordinator: *John C. Leland*
Compositor: *Aptara, Inc.*
Typeface: *10.5/12 Times Roman*
Printer: *R. R. Donnelley*

All credits appearing on page or at the end of the book are considered to be an extension of the copyright page.

Library of Congress Cataloging-in-Publication Data

Coburn, John W.
    College algebra : graphs and models / John W. Coburn, J.D. Herdlick.
        p. cm.
    Includes index.
    ISBN 978–0–07–351954–8 — ISBN 0–07–351954–5 (hard copy : alk. paper) 1. Algebra—
Textbooks. 2. Algebra—Graphic methods—Textbooks. I. Herdlick, John D. II. Title.
    QA154.3.C5953 2012
    512.9—dc22
                                    2010035347

www.mhhe.com

Brief Contents

About the Authors

John Coburn

John Coburn grew up in the Hawaiian Islands, the seventh of sixteen children. He received his Associate of Arts degree in 1977 from Windward Community College, where he graduated with honors. In 1979 he earned a Bachelor's Degree in Education from the University of Hawaii. After working in the business world for a number of years, he returned to teaching, accepting a position in high school mathematics where he was recognized as Teacher of the Year (1987). Soon afterward, the decision was made to seek a Master's Degree, which he received two years later from the University of Oklahoma. John is now a full professor at the Florissant Valley campus of St. Louis Community College. During his tenure there he has received numerous nominations as an outstanding teacher by the local chapter of Phi Theta Kappa, two nominations to *Who's Who Among America's Teachers,* and was recognized as Post Secondary Teacher of the Year in 2004 by the Mathematics Educators of Greater St. Louis (MEGSL). He has made numerous presentations and local, state, and national conferences on a wide variety of topics and maintains memberships in several mathematics organizations. Some of John's other interests include body surfing, snorkeling, and beach combing whenever he gets the chance. He is also an avid gamer, enjoying numerous board, card, and party games. His other loves include his family, music, athletics, composition, and the wild outdoors.

J.D. Herdlick

J.D. Herdlick was born and raised in St. Louis, Missouri, very near the Mississippi river. In 1992, he received his bachelor's degree in mathematics from Santa Clara University (Santa Clara, California). After completing his master's in mathematics at Washington University (St. Louis, Missouri) in 1994, he felt called to serve as both a campus minister and an aid worker for a number of years in the United States and Honduras. He later returned to education and spent one year teaching high school mathematics, followed by an appointment at Washington University as visiting lecturer, a position he held until 2006. Simultaneously teaching as an adjunct professor at the Meramec campus of St. Louis Community College, he eventually joined the department full time in 2001. While at Santa Clara University, he became a member of the honorary societies Phi Beta Kappa, Pi Mu Epsilon, and Sigma Xi under the tutelage of David Logothetti, Gerald Alexanderson, and Paul Halmos. In addition to the Dean's Award for Teaching Excellence at Washington University, J.D. has received numerous awards and accolades for his teaching at St. Louis Community College. Outside of the office and classroom, he is likely to be found in the water, on the water, and sometimes above the water, as a passionate wakeboarder and kiteboarder. It is here, in the water and wind, that he finds his inspiration for writing. J.D. and his family currently split their time between the United States and Argentina.

Dedication

With boundless gratitude, we dedicate this work to the special people in our lives. To our children, whom we hope were joyfully oblivious to the time, sacrifice, and perseverance required; and to our wives, who were well acquainted with every minute of it.

▼ Formulas from Analytical Geometry: $P_1 \rightarrow (x_1, y_1)$, $P_2 \rightarrow (x_2, y_2)$

Distance between P_1 and P_2

$$d = \sqrt{(x_2 - x_1)^2 + (y_2 - y_1)^2}$$

Slope of Line Containing P_1 and P_2

$$m = \frac{\Delta y}{\Delta x} = \frac{y_2 - y_1}{x_2 - x_1}$$

Equation of Line Containing P_1 and P_2

Point-Slope Form

$$y - y_1 = m(x - x_1)$$

Equation of Line Containing P_1 and P_2

Slope-Intercept Form (slope m, y-intercept b)

$$y = mx + b, \text{ where } b = y_1 - mx_1$$

Parallel Lines

Slopes Are Equal: $m_1 = m_2$

Perpendicular Lines

Slopes Have a Product of -1: $m_1 m_2 = -1$

Intersecting Lines

Slopes Are Unequal: $m_1 \neq m_2$

Dependent (Coincident) Lines

Slopes and y-Intercepts Are Equal: $m_1 = m_2$, $b_1 = b_2$

▼ Logarithms and Logarithmic Properties

$$y = \log_b x \Leftrightarrow b^y = x \qquad \log_b b = 1 \qquad \log_b 1 = 0$$

$$\log_b b^x = x \qquad b^{\log_b x} = x \qquad \log_c x = \frac{\log_b x}{\log_b c}$$

$$\log_b MN = \log_b M + \log_b N \qquad \log_b \frac{M}{N} = \log_b M - \log_b N \qquad \log_b M^P = P \cdot \log_b M$$

▼ Applications of Exponentials and Logarithms

$A \rightarrow$ amount accumulated $P \rightarrow$ initial deposit, $p \rightarrow$ periodic payment $n \rightarrow$ compounding periods/year

$r \rightarrow$ interest rate per year $R \rightarrow$ interest rate per time period $\left(\dfrac{r}{n}\right)$ $t \rightarrow$ time in years

Interest Compounded n Times per Year

$$A = P\left(1 + \frac{r}{n}\right)^{nt}$$

Interest Compounded Continuously

$$A = Pe^{rt}$$

Accumulated Value of an Annuity

$$A = \frac{P}{R}\left[(1 + R)^{nt} - 1\right]$$

Payments Required to Accumulate Amount A

$$P = \frac{AR}{(1 + R)^{nt} - 1}$$

▼ Sequences and Series:

$a_1 \rightarrow$ 1st term, $a_n \rightarrow n$th term, $S_n \rightarrow$ sum of n terms, $d \rightarrow$ common difference, $r \rightarrow$ common ratio

Arithmetic Sequences

$$a_1, a_2 = a_1 + d, a_3 = a_1 + 2d, \ldots, a_n = a_1 + (n - 1)d$$

$$S_n = \frac{n}{2}(a_1 + a_n)$$

$$S_n = \frac{n}{2}[2a_1 + (n - 1)d]$$

Geometric Sequences

$$a_1, a_2 = a_1 r, a_3 = a_1 r^2, \ldots, a_n = a_1 r^{n-1}$$

$$S_n = \frac{a_1 - a_1 r^n}{1 - r}$$

$$S_\infty = \frac{a_1}{1 - r}; \; |r| < 1$$

▼ Binomial Theorem

$$(a + b)^n = \binom{n}{0}a^n b^0 + \binom{n}{1}a^{n-1}b^1 + \binom{n}{2}a^{n-2}b^2 + \cdots + \binom{n}{n-1}a^1 b^{n-1} + \binom{n}{n}a^0 b^n$$

$$n! = n(n - 1)(n - 2)\cdots(3)(2)(1) \qquad \binom{n}{k} = \frac{n!}{k!(n - k)!}; \qquad 0! = 1$$

▼ The Toolbox and Other Functions

linear

$y = mx + b$

$(0, b)$

$m < 0, b > 0$

linear

$y = mx + b$

$(0, b)$

$m > 0, b > 0$

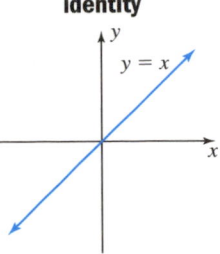

identity

$y = x$

$m = 1, b = 0$

constant

$y = b$

$m = 0, b > 0$

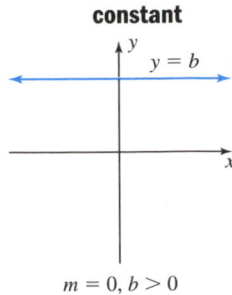

absolute value

$y = |x|$

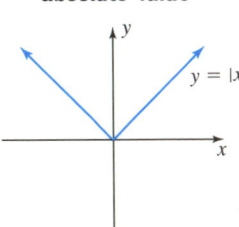

squaring

$y = x^2$

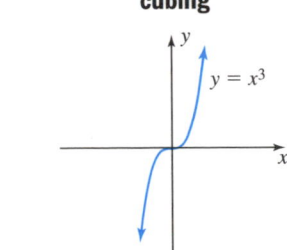

cubing

$y = x^3$

square root

$y = \sqrt{x}$

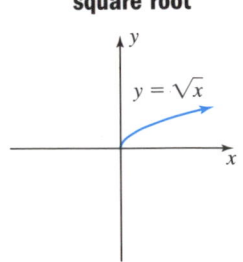

cube root

$y = \sqrt[3]{x}$

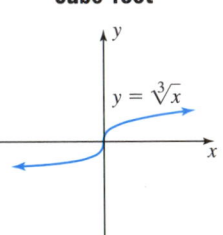

floor function

$y = \lfloor x \rfloor$

1 2

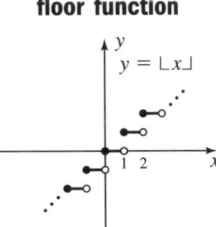

reciprocal

$y = \frac{1}{x}$

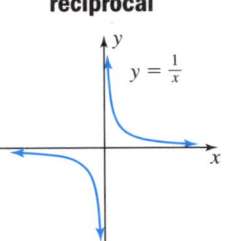

reciprocal quadratic

$y = \frac{1}{x^2}$

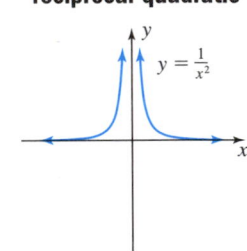

exponential

$y = b^x$

1

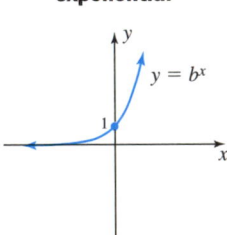

exponential

$y = b^{-x}$

1

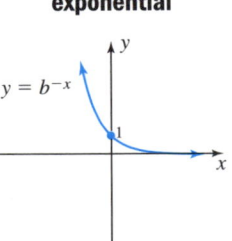

logarithmic

$y = \log_b x$

1

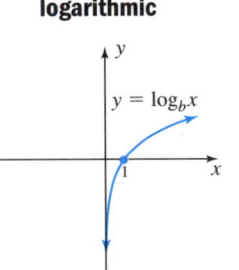

logistic

$y = \dfrac{c}{1 + ae^{-bx}}$

$y = c$

$\left(0, \dfrac{c}{1 + a}\right)$

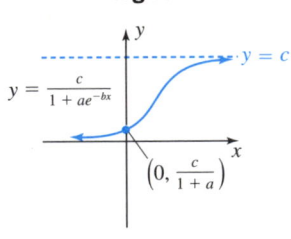

▼ Transformations of Basic Graphs

Given Function

$y = f(x)$

Transformation of Given Function

$y = af(x \pm h) \pm k$

vertical reflections
vertical stretches/compressions

horizontal shift h units,
opposite direction of sign

vertical shift k units,
same direction as sign

▼ Average Rate of Change of $f(x)$

For linear function models, the average rate of change on the interval $[x_1, x_2]$ is constant, and given by the slope formula: $\dfrac{\Delta y}{\Delta x} = \dfrac{y_2 - y_1}{x_2 - x_1}$. The average rate of change for other function models is nonconstant. By writing the slope formula in function form using $y_1 = f(x_1)$ and $y_2 = f(x_2)$, we can compute the average rate of change of other functions on this interval:

$$\frac{\Delta y}{\Delta x} = \frac{f(x_2) - f(x_1)}{x_2 - x_1}$$

About the Cover

Most coral reefs in the world are 7000–9000 years old, but new reefs can fully develop in as few as 20 years. In addition to being home to over 4000 species of tropical or reef fish, coral reefs are immensely beneficial to humans and must be carefully preserved. They buffer coastal regions from strong waves and storms, provide millions of people with food and jobs, and prompt advances in modern medicine.

Similar to the ancient reefs, a course in College Algebra is based on thousands of years of mathematical curiosity, insight, and wisdom. In this one short course, we study a wealth of important concepts that have taken centuries to mature. Just as the variety of fish in the sea rely on the coral reefs to survive, students in a College Algebra course rely on mastery of this bedrock of concepts to successfully pursue more advanced courses, as well as their career goals.

From the Authors

In the last two decades, mathematics education has seen some enormous changes. From the introduction of graphing calculators and the advent of the Internet, to online homework and visual supplements we could only dream about decades ago, the changes have been unrelenting. Together, John Coburn and J.D. Herdlick share a combined 40 years of experience teaching college algebra with graphing calculators and other technologies, and have developed a wealth of firsthand experience related to the endeavor.

In the Coburn/Herdlick Graphs and Models text, we have combined the conversational style and the wealth of applications that our texts are known for, with this depth of experience. As one of our primary goals, we set out to help students think visually, to a point where they see functions like $f(x) = x^2 - 4x$ as one of a family of graphs, with attributes that immediately lead to a discussion of maximums and minimums, end-behavior, zeroes, solutions to inequalities, the nature of the roots, and the application of these attributes in context—instead of merely an equation that must be solved by factoring or by interpreting a graph on the screen of a calculator. And while graphing calculators may relieve some computational drudgery, we believe our text offers much more than a simple side-by-side comparison of algebraic methods versus graphical methods, with the calculator playing a more significant role than simply checking answers to work done manually. Graphing calculators are used to work and investigate far beyond what's possible with paper and pencil, with the technology used to solve more true-to-life equations, engage more applications, and explore more substantial questions of interest. In the end we believe you'll see this text is built on strong fundamentals, yet one that offers a visual and dynamic excursion that accentuates the organizational planning and problem solving acumen that students will use in all areas of their lives. To this end we offer the Coburn/Herdlick Graphs and Models text as an ideal tool for the teaching and learning of mathematics. —John Coburn and J.D. Herdlick

Making Connections . . .

College Algebra tends to be a challenging course for many students. They may not see the connections that College Algebra has to their life or why it is so critical that they succeed in this course. Others may enter into this course underprepared or improperly placed and with very little motivation.

Instructors are faced with several challenges as well. They are given the task of improving pass rates and student retention while ensuring the students are adequately prepared for more advanced courses, as a College Algebra course attracts a very diverse audience, with a wide variety of career goals and a large range of prerequisite skills.

The goal of this textbook series is to provide both students and instructors with tools to address these challenges, so that both can experience greater success in College Algebra. For instance, the comprehensive exercise sets have a range of difficulty that provides very strong support for weaker students, while advanced students are challenged to reach even further. The rest of this preface further explains the tools that John Coburn, J.D. Herdlick, and McGraw-Hill have developed and how they can be used to *connect* students to College Algebra and *connect* instructors to their students.

The Coburn/Herdlick College Algebra Series provides you with strong tools to achieve better outcomes in your College Algebra course as follows:

▶ *Making Connections Visually, Symbolically, Numerically, and Verbally*

▶ *Better Student Preparedness Through Superior Course Management*

▶ *Increased Student Engagement*

▶ *Solid Skill Development*

▶ *Strong Mathematical Connections*

▶ Making Connections Visually, Symbolically, Numerically, and Verbally

In writing their Graphs and Models series, the Coburn/Herdlick team took great care to help students think visually by relating a basic graph to an algebraic equation at every opportunity. This empowers students to see the "Why?" behind many algebraic rules and properties, and offers solid preparation for the connections they'll need to make in future courses which often depend on these visual skills.

▶ Better Student Preparedness Through Superior Course Management

McGraw-Hill is proud to offer instructors a choice of course management options to accompany Coburn/ Herdlick. If you prefer to assign text-specific problems in a brand new, robust online homework system that contains stepped out and guided solutions for all questions, Connect Math Hosted by ALEKS may be for you. Or perhaps you prefer the diagnostic nature and artificial intelligence engine that is the driving force behind our ALEKS 360 Course product, a true online learning environment, which has been expanded to contain hundreds of new College Algebra & Precalculus topics. We encourage you to take a closer look at each product on preface pages **x** through **xiii** and to consult your McGraw-Hill sales representative to setup a demonstration.

▶ Increased Student Engagement

There are many texts that claim they "engage" students, but only the Coburn Series has carefully studied and implemented features and options that make it truly possible. From the on-line support, to the textbook design and a wealth of quality applications, students will remain engaged throughout their studies.

▶ Solid Skill Development

The Coburn/Herdlick series intentionally relates the examples to the exercise sets so there is a strong connection between what students are learning while working through the examples in each section and the homework exercises that they complete. This development of strong mechanical skills is followed closely by a careful development of problem solving skills, with the use of interesting and engaging applications that have been carefully chosen with regard to difficulty and the skills currently under study.

There is also an abundance of exercise types to choose from to ensure that homework challenges a wide variety of skills. Furthermore, John and J.D. reconnect students to earlier chapter material with Mid-Chapter Checks; students have praised these exercises for helping them understand what key concepts require additional practice.

▶ Strong Mathematical Connections

John Coburn and J.D. Herdlick's experience in the classroom and their strong connections to how students comprehend the material are evident in their writing style. This is demonstrated by the way they provide a tight weave from topic to topic and foster an environment that doesn't just focus on procedures but illustrates the big picture, which is something that so often is sacrificed in this course. Moreover, they employ a clear and supportive writing style, providing the students with a tool they can depend on when the teacher is not available, when they miss a day of class, or simply when working on their own.

Making Connections...

Visually, Symbolically, Numerically, and Verbally

"It is widely known that for students to grow stronger algebraically, the concrete and numeric experiences from their past must give way to more symbolic representations. In this transition from numeric, to symbolic, to algebraic thinking, the importance of visual connections and verbal connections is too often overlooked. To reach a deep understanding of rich concepts or subtle ideas, students must develop the ability to mentally "see" and discuss the concept or idea using the terms and names needed to describe it accurately. Only then can they begin seeing the connections that exist between each new concept, and concepts that are already known. A large part of this involves helping our students to begin thinking visually, to a point where they're able to see functions like $f(x) = x^2 - 4x$ as only one of a large family of functions, with graphical attributes that immediately lead to a discussion of maximums and minimums, end-behavior, zeroes, solutions to inequalities, the nature of the roots, and the application of these attributes in context. And while it's important for students to see that zeroes are x-intercepts and x-intercepts are zeroes, and that the intersection of two graphs provides a simultaneous solution to the equations forming these graphs, these should not remain the sole focus of the tool. Graphing calculators allow explorations, investigations, connections, and visualizations far beyond what's possible with paper and pencil, and we should use the technology to aid the development of these mental-visual skills, in addition to solving more true-to-life equations, engaging more applications, and exploring the more substantial questions involving real data, domain and range, anticipated graphical behavior, additional uses of lists and tables, and other questions of interest. We believe this text offers instructors the tools they need to be successful in these endeavors."

—The Authors

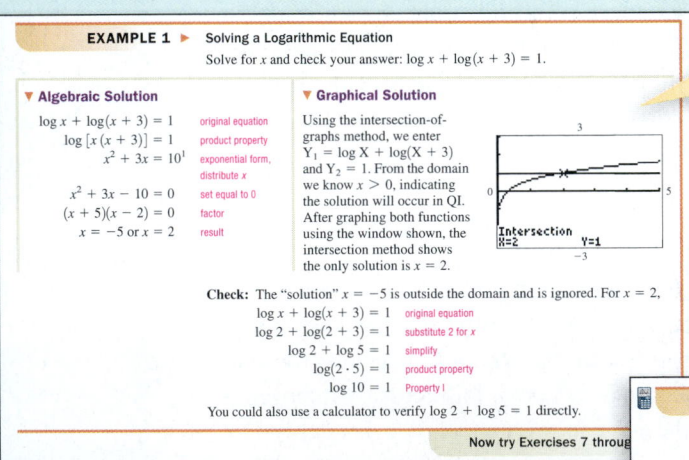

EXAMPLE 1 ▶ Solving a Logarithmic Equation

Solve for x and check your answer: $\log x + \log(x + 3) = 1$.

▼ **Algebraic Solution**

$\log x + \log(x + 3) = 1$ original equation
$\log[x(x + 3)] = 1$ product property
$x^2 + 3x = 10^1$ exponential form, distribute x
$x^2 + 3x - 10 = 0$ set equal to 0
$(x + 5)(x - 2) = 0$ factor
$x = -5$ or $x = 2$ result

▼ **Graphical Solution**

Using the intersection-of-graphs method, we enter $Y_1 = \log X + \log(X + 3)$ and $Y_2 = 1$. From the domain we know $x > 0$, indicating the solution will occur in QI. After graphing both functions using the window shown, the intersection method shows the only solution is $x = 2$.

Check: The "solution" $x = -5$ is outside the domain and is ignored. For $x = 2$,
$\log x + \log(x + 3) = 1$ original equation
$\log 2 + \log(2 + 3) = 1$ substitute 2 for x
$\log 2 + \log 5 = 1$ simplify
$\log(2 \cdot 5) = 1$ product property
$\log 10 = 1$ Property I

You could also use a calculator to verify $\log 2 + \log 5 = 1$ directly.

Now try Exercises 7 throug[h]

❝I think there is a good balance between technology and paper/pencil techniques. I particularly like how the technology portion does not take the place of paper/pencil, but instead supplements it. I think a lot of departments will like that.**❞**

—Daniel Brock, Arkansas State University-Beebe

▶ **Graphical Examples** show students how the calculator can be used to supplement their understanding of a problem.

❝I have certainly found the Coburn/Herdlick's Precalculus: Graphs and Models textbook the best approach ever to the teaching of Precalculus with the inclusion of graphing calculator.**❞**

—Alvio Dominguez, Miami-Dade College-Wolfson

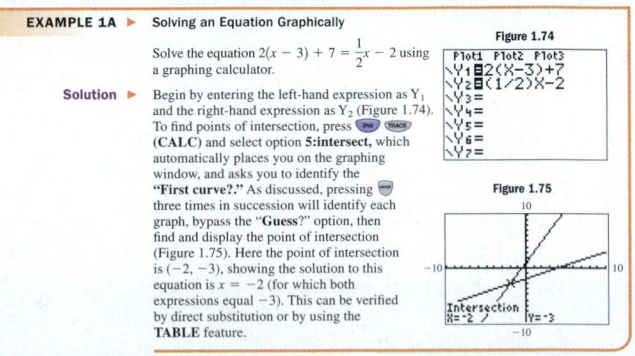

EXAMPLE 1A ▶ Solving an Equation Graphically

Solve the equation $2(x - 3) + 7 = \frac{1}{2}x - 2$ using a graphing calculator.

Solution ▶ Begin by entering the left-hand expression as Y_1 and the right-hand expression as Y_2 (Figure 1.74). To find points of intersection, press ⬛ ⬛ (**CALC**) and select option **5:intersect**, which automatically places you on the graphing window, and asks you to identify the **"First curve?."** As discussed, pressing ⬛ three times in succession will identify each graph, bypass the "Guess?" option, then find and display the point of intersection (Figure 1.75). Here the point of intersection is $(-2, -3)$, showing the solution to this equation is $x = -2$ (for which both expressions equal -3). This can be verified by direct substitution or by using the **TABLE** feature.

► **Calculator Explanations** incorporate the calculator without sacrificing.

Most graphing calculators are programmed to work with imaginary and complex numbers, though for some models the calculator must be placed in complex number mode. After pressing the ⬛ MODE key (located to the right of the ⬛ option key), the screen shown in Figure 3.2 appears and we use the arrow keys to access "$a + bi$" and active this mode (by pressing ⬛). Once active, we can validate our previous statements about imaginary numbers (Figure 3.3), as well as verify our previous calculations like those in Examples 3(a), 3(d), and 4(a) (Figure 3.4). Note the imaginary unit i is the ⬛ option for the decimal point.

Figure 3.2

```
NORMAL  SCI  ENG
FLOAT  0123456789
RADIAN  DEGREE
FUNC  PAR  POL  SEQ
CONNECTED  DOT
SEQUENTIAL  SIMUL
REAL  a+bi  re^θi
FULL  HORIZ  G-T
SET CLOCK 09/06/09 18:04
```

Figure 3.3

```
√(-1)
              i
i²
             -1
√(-16)
             4i
```

Figure 3.4

```
2+√(-49)
            2+7i
(4+3√(-25))/20
           .2+.75i
(2+3i)+(-5+2i)
           -3+5i
```

> **"The technology (graphing calculator) explanations and illustrations are superb. The level of detail is valuable; even an experienced user (myself) learned some new techniques and "tricks" in reading through the text. The text frequently references use of the calculator—yet without sacrificing rigor or mathematical integrity."**
> —*Light Bryant, Arizona Western College*

To help illustrate the Intermediate Value Theorem, many graphing calculators offer a useful feature called *split screen viewing,* that enables us to view a table of values and the graph of a function at the same time. To illustrate, enter the function $y = x^3 - 9x + 6$ (from Example 6) as Y_1 on the ⬛ Y= screen, then set the viewing window as shown in Figure 4.4. Set your table in **AUTO** mode with ΔTbl = 1, then press the ⬛ MODE key (see Figure 4.4A) and notice the second-to-last entry on this screen reads: **Full** for full screen viewing, **Horiz** for splitting the screen horizontally with the graph above a reduced home screen, and **G-T,** which represents **Graph-Table** and splits the screen vertically. In the **G-T** mode, the graph appears on the left and the table of values on the right. Navigate the cursor to the **G-T** mode and press ⬛. Pressing the ⬛ GRAPH key at this point should give you a screen similar to Figure 4.5. Scrolling downward shows the function also changes sign between $x = 2$ and $x = 3$. For more on this idea, **see Exercises 31 and 32.**

As a final note, while the intermediate value theorem is a powerful yet simple tool, it must be used with care. For example, given $p(x) = -x^4 + 10x^2 - 5$, $p(-1) > 0$ and $p(1) > 0$, seeming to indicate that no zeroes exist in the interval $(-1, 1)$. Actually there are two zeroes, as seen in Figure 4.6.

Figure 4.4A

☑ **B.** You've just seen how we can use the intermediate value theorem to identify intervals containing a polynomial zero

Figure 4.5

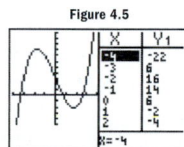

Figure 4.6

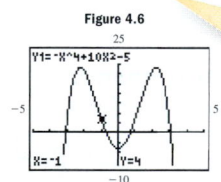

> **"The authors give very good uses of the calculator in every section. I have been using TI calculators for 15 years and I learned a few new tricks while reading this book."**
> —*George Hurlburt, Corning Community College*

► **Technology Applications** show students how technology can be used to help apply lessons from the classroom to real life.

> **"I think that the graphing examples, explanations, and problems are perfect for the average college algebra student who has never touched a graphing calculator. ..... I think this book would be great to actually have in front of the students."**
> —*Dale Duke, Oklahoma City Community College*

📱 Use Newton's law of cooling to complete Exercises 75 and 76: $T(x) = T_R + (T_0 - T_R)e^{kx}$.

75. Cold party drinks: Janae was late getting ready for the party, and the liters of soft drinks she bought were still at room temperature (73°F) with guests due to arrive in 15 min. If she puts these in her freezer at −10°F, will the drinks be cold enough (35°F) for her guests? Assume $k \approx -0.031$.

76. Warm party drinks: Newton's law of cooling applies equally well if the "cooling is negative," meaning the object is taken from a colder medium and placed in a warmer one. If a can of soft drink is taken from a 35°F cooler and placed in a room where the temperature is 75°F, how long will it take the drink to warm to 65°F? Assume $k \approx -0.031$.

📱 **Photochromatic sunglasses:** Sunglasses that darken in sunlight (photochromatic sunglasses) contain millions of molecules of a substance known as *silver halide.* The molecules are transparent indoors in the absence of ultraviolet (UV) light. Outdoors, UV light from the sun causes the molecules to change shape, darkening the lenses in response to the intensity of the UV light. For certain lenses, the function $T(x) = 0.85^x$ models the transparency of the lenses (as a percentage) based on a UV index x. Find the transparency (to the nearest percent), if the lenses are exposed to

77. sunlight with a UV index of 7 (a high exposure).

78. sunlight with a UV index of 5.5 (a moderate exposure).

80. Use a trial-and-error process and a graphing calculator to determine the UV index when the lenses are 50% transparent.

📱 **Modeling inflation:** Assuming the rate of inflation is 5% per year, the predicted price of an item can be modeled by the function $P(t) = P_0(1.05)^t$, where P_0 represents the initial price of the item and t is in years. Use this information to solve Exercises 81 and 82.

81. What will the price of a new car be in the year 2015, if it cost $20,000 in the year 2010?

82. What will the price of a gallon of milk be in the year 2015, if it cost $3.95 in the year 2010? Round to the nearest cent.

📱 **Modeling radioactive decay:** The half-life of a radioactive substance is the time required for half an initial amount of the substance to disappear through decay. The amount of the substance remaining is given by the formula $Q(t) = Q_0(\frac{1}{2})^{\frac{t}{h}}$, where h is the half-life, t represents the elapsed time, and $Q(t)$ represents the amount that remains (t and h must have the same unit of time). Use this information to solve Exercises 83 and 84.

83. Some isotopes of the substance known as thorium have a half-life of only 8 min. (a) If 64 grams are initially present, how many grams (g) of the substance remain after 24 min? (b) How many minutes until only 1 gram (g) of the substance remains?

MATH

Hosted by **ALEKS Corp.**

Connect Math Hosted by ALEKS Corporation is an exciting, new assignment and assessment platform combining the strengths of McGraw-Hill Higher Education and ALEKS Corporation. Connect Math Hosted by ALEKS is the first platform on the market to combine an artificially-intelligent, diagnostic assessment with an intuitive ehomework platform designed to meet your needs.

Connect Math Hosted by ALEKS Corporation is the culmination of a one-of-a-kind market development process involving math full-time and adjunct Math faculty at every step of the process. This process enables us to provide you with a solution that best meets your needs.

Connect Math Hosted by ALEKS Corporation is built by Math educators for Math educators!

1 *Your students want a well-organized homepage where key information is easily viewable.*

Modern Student Homepage

▶ This homepage provides a dashboard for students to immediately view their assignments, grades, and announcements for their course. (Assignments include HW, quizzes, and tests.)

▶ Students can access their assignments through the course Calendar to stay up-to-date and organized for their class.

Modern, intuitive, and simple interface.

2 *You want a way to identify the strengths and weaknesses of your class at the beginning of the term rather than after the first exam.*

Integrated ALEKS® Assessment

▶ This artificially-intelligent (AI), diagnostic assessment identifies precisely what a student knows and is ready to learn next.

▶ Detailed assessment reports provide instructors with specific information about where students are struggling most.

▶ This AI-driven assessment is the only one of its kind in an online homework platform.

Recommended to be used as the first assignment in any course.

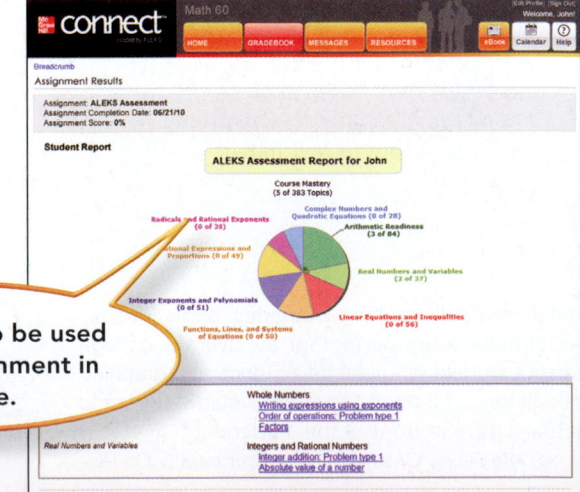

ALEKS is a registered trademark of ALEKS Corporation.

Built by Math Educators
for Math Educators

3 *Your students want an assignment page that is easy to use and includes lots of extra help resources.*

Efficient Assignment Navigation

▶ Students have access to immediate feedback and help while working through assignments.

▶ Students have direct access to a media-rich eBook for easy referencing.

▶ Students can view detailed, step-by-step solutions written by instructors who teach the course, providing a unique solution to each and every exercise.

Students can easily monitor and track their progress on a given assignment.

4 *You want a more intuitive and efficient assignment creation process because of your busy schedule.*

Assignment Creation Process

▶ Instructors can select textbook-specific questions organized by chapter, section, and objective.

▶ Drag-and-drop functionality makes creating an assignment quick and easy.

▶ Instructors can preview their assignments for efficient editing.

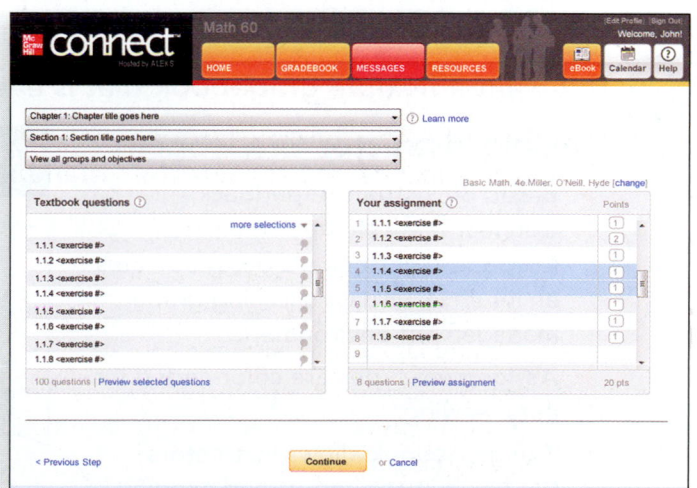

Connect
Learn
Succeed™

www.connectmath.com

CONNECT | MATH

Hosted by **ALEKS Corp.**

5 *Your students want an interactive eBook with rich functionality integrated into the product.*

 McGraw Hill **CONNECT** plus+ | MATH

Hosted by **ALEKS Corp.**

Integrated Media-Rich eBook

▶ A Web-optimized eBook is seamlessly integrated within ConnectPlus Math Hosted by ALEKS Corp for ease of use.

▶ Students can access videos, images, and other media in context within each chapter or subject area to enhance their learning experience.

▶ Students can highlight, take notes, or even access shared instructor highlights/notes to learn the course material.

▶ The integrated eBook provides students with a cost-saving alternative to traditional textbooks.

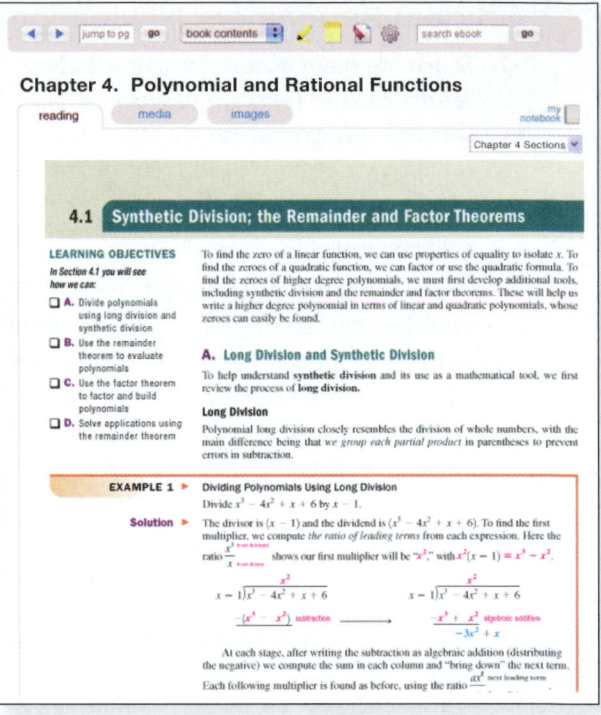

6 *You want a flexible gradebook that is easy to use.*

Flexible Instructor Gradebook

▶ Based on instructor feedback, Connect Math Hosted by ALEKS Corp's straightforward design creates an intuitive, visually pleasing grade management environment.

▶ Assignment types are color-coded for easy viewing.

▶ The gradebook allows instructors the flexibility to import and export additional grades.

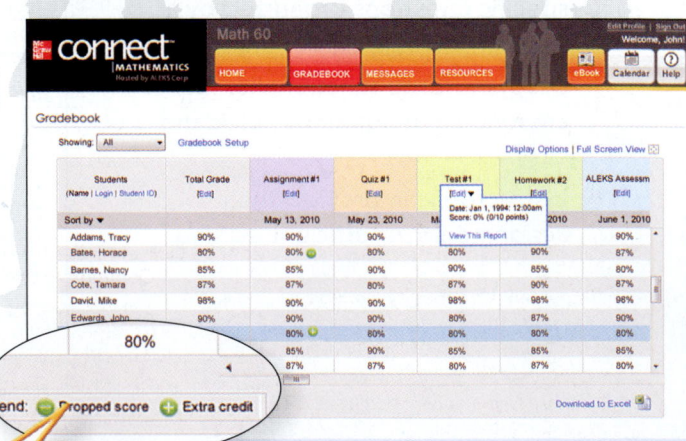

Instructors have the ability to drop grades as well as assign extra credit.

Built by Math Educators for Math Educators

 7 *You want algorithmic content that was developed by math faculty to ensure the content is pedagogically sound and accurate.*

Digital Content Development Story

The development of McGraw-Hill's Connect Math Hosted by ALEKS Corp. content involved collaboration between McGraw-Hill, experienced instructors, and ALEKS, a company known for its high-quality digital content. The result of this process, outlined below, is accurate content created with your students in mind. It is available in a simple-to-use interface with all the functionality tools needed to manage your course.

1. McGraw-Hill selected experienced instructors to work as Digital Contributors.
2. The Digital Contributors selected the textbook exercises to be included in the algorithmic content to ensure appropriate coverage of the textbook content.
3. The Digital Contributors created detailed, stepped-out solutions for use in the Guided Solution and Show Me features.
4. The Digital Contributors provided detailed instructions for authoring the algorithm specific to each exercise to maintain the original intent and integrity of each unique exercise.
5. Each algorithm was reviewed by the Contributor, went through a detailed quality control process by ALEKS Corporation, and was copyedited prior to being posted live.

Connect Math Hosted by ALEKS Corp.
Built by Math Educators for Math Educators

Lead Digital Contributors

Tim Chappell
Metropolitan Community College, Penn Valley

Jeremy Coffelt
Blinn College

Nancy Ikeda
Fullerton College

Amy Naughten

Digital Contributors

Al Bluman, *Community College of Allegheny County*
John Coburn, *St. Louis Community College, Florissant Valley*
Vanessa Coffelt, *Blinn College*
Donna Gerken, *Miami-Dade College*
Kimberly Graham
J.D. Herdlick, *St. Louis Community College, Meramec*

Vickie Flanders, *Baton Rouge Community College*
Nic LaHue, *Metropolitan Community College, Penn Valley*
Nicole Lloyd, *Lansing Community College*
Jackie Miller, *The Ohio State University*
Anne Marie Mosher, *St. Louis Community College, Florissant Valley*
Reva Narasimhan, *Kean University*
David Ray, *University of Tennessee, Martin*

Kristin Stoley, *Blinn College*
Stephen Toner, *Victor Valley College*
Paul Vroman, *St. Louis Community College, Florissant Valley*
Michelle Whitmer, *Lansing Community College*

ALEKS® College Algebra

Enhanced Course Coverage Enables Seamless Integration with Textbooks and Syllabi

ALEKS College Algebra features hundreds of new course topics to provide comprehensive course coverage, and ALEKS AI-2, the next generation intelligence engine to dramatically improve student learning outcomes. This enhanced ALEKS course product allows for better curriculum coverage and seamless textbook integration to help students succeed in mathematics, while allowing instructors to customize course content to align with their course syllabi.

ALEKS is a Web-based program that uses artificial intelligence and adaptive questioning to assess precisely a student's knowledge in College Algebra and provide personalized instruction on the exact topics the student is most **ready to learn.** By providing individualized assessment and learning, ALEKS helps students to master course content quickly and easily.

Topics Added For Comprehensive Coverage:

ALEKS College Algebra includes hundreds of new topics for comprehensive coverage of course material. To view College Algebra course content in more detail, please visit:

www.aleks.com/highered/math/course_products

> **The ALEKS Pie** summarizes a student's current knowledge of course material and provides an individualized learning path with topics each student is most ready to learn.

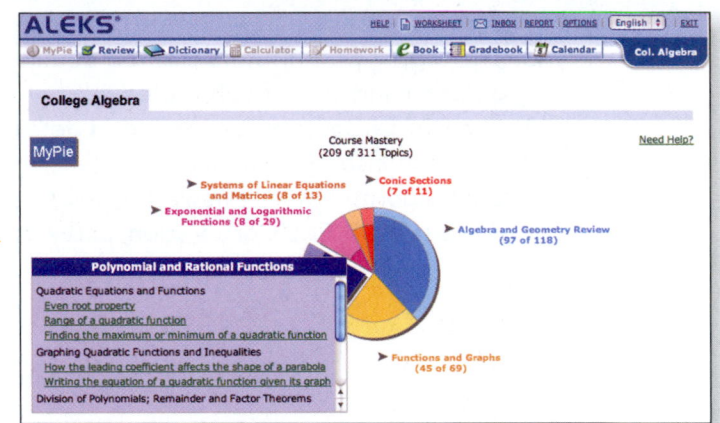

Robust Graphing Features:

ALEKS College Algebra provides more graphing coverage and includes a built-in graphing calculator, an adaptive, open-response environment, and realistic answer input tools to ensure student mastery.

> **The ALEKS Graphing Calculator** is accessible via the Student Module and can be turned on or off by the instructor.

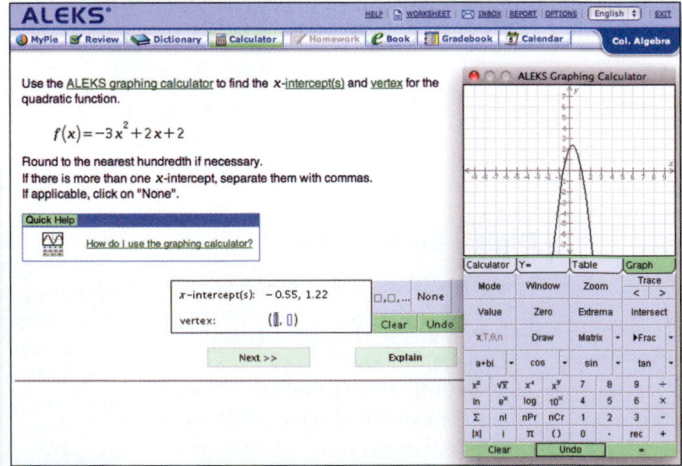

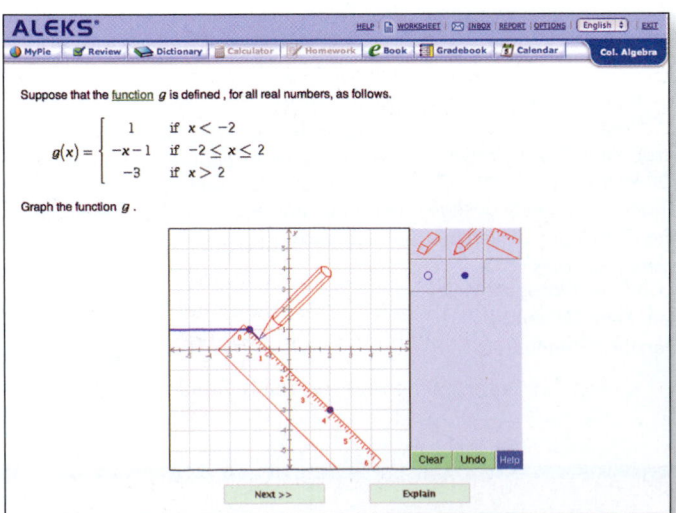

◀ **Realistic Input Tools** provide an adaptive, open-response environment that avoids multiple-choice questions and ensures student mastery.

ALEKS® New Instructor Module Features for College Algebra

Help Students Achieve Success While Saving Instructor Time

ALEKS includes an **Instructor Module** with powerful, assignment-driven features and extensive content flexibility to simplify course management so instructors spend less time with administrative tasks and more time directing student learning. The ALEKS Instructor Module also includes two new features that further simplify course management and provide content flexibility: **Partial Credit on Assignments** and **Supplementary Textbook Integration Topic Coverage.**

Partial Credit On Assignments:

With the addition of many more multipart questions to ALEKS College Algebra, instructors now have the option to have ALEKS automatically assign partial credit to students' responses on multipart questions in an ALEKS Homework, Test, or Quiz. Instructors can also manually adjust scores.

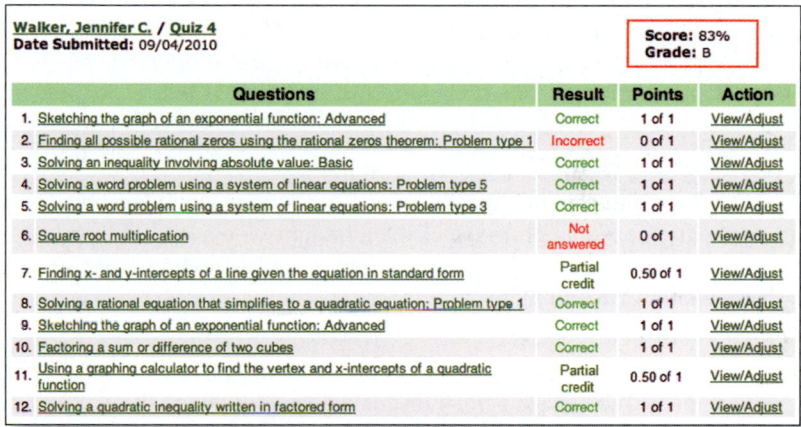

Walker, Jennifer C. / Quiz 4
Date Submitted: 09/04/2010

Score: 83%
Grade: B

Questions	Result	Points	Action
1. Sketching the graph of an exponential function: Advanced	Correct	1 of 1	View/Adjust
2. Finding all possible rational zeros using the rational zeros theorem: Problem type 1	Incorrect	0 of 1	View/Adjust
3. Solving an inequality involving absolute value: Basic	Correct	1 of 1	View/Adjust
4. Solving a word problem using a system of linear equations: Problem type 5	Correct	1 of 1	View/Adjust
5. Solving a word problem using a system of linear equations: Problem type 3	Correct	1 of 1	View/Adjust
6. Square root multiplication	Not answered	0 of 1	View/Adjust
7. Finding x- and y-intercepts of a line given the equation in standard form	Partial credit	0.50 of 1	View/Adjust
8. Solving a rational equation that simplifies to a quadratic equation: Problem type 1	Correct	1 of 1	View/Adjust
9. Sketching the graph of an exponential function: Advanced	Correct	1 of 1	View/Adjust
10. Factoring a sum or difference of two cubes	Correct	1 of 1	View/Adjust
11. Using a graphing calculator to find the vertex and x-intercepts of a quadratic function	Partial credit	0.50 of 1	View/Adjust
12. Solving a quadratic inequality written in factored form	Correct	1 of 1	View/Adjust

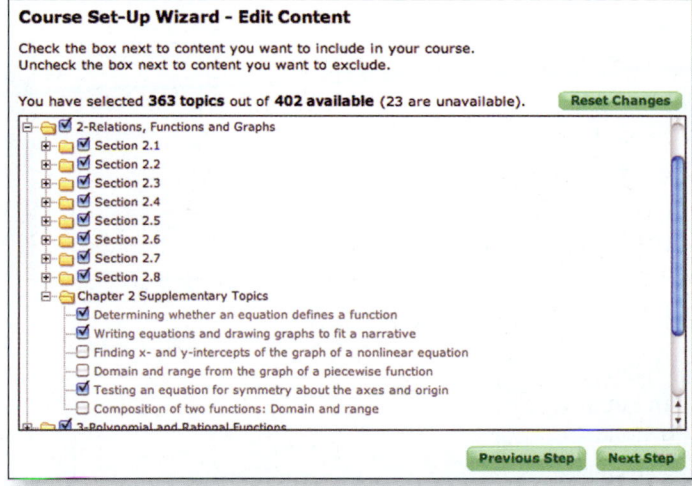

Course Set-Up Wizard - Edit Content

Check the box next to content you want to include in your course.
Uncheck the box next to content you want to exclude.

You have selected **363 topics** out of **402 available** (23 are unavailable).   Reset Changes

- 2-Relations, Functions and Graphs
  - Section 2.1
  - Section 2.2
  - Section 2.3
  - Section 2.4
  - Section 2.5
  - Section 2.6
  - Section 2.7
  - Section 2.8
  - Chapter 2 Supplementary Topics
    - Determining whether an equation defines a function
    - Writing equations and drawing graphs to fit a narrative
    - Finding x- and y-intercepts of the graph of a nonlinear equation
    - Domain and range from the graph of a piecewise function
    - Testing an equation for symmetry about the axes and origin
    - Composition of two functions: Domain and range
  - 3-Polynomial and Rational Functions

Previous Step   Next Step

Supplementary Textbook Integration Topic Coverage:

Instructors have access to ALL course topics in ALEKS College Algebra, and can include supplementary course topics even if they are not specifically tied to an integrated textbook's table of contents.

To learn more about how other instructors have successfully implemented ALEKS, please visit:
www.aleks.com/highered/math/implementations

"Overall, both students and I have been very pleased with ALEKS. Students like the flexibility it offers them. I like that students are working where they need to be and can spend as much time reviewing as they need. . . . Students have made such comments as 'I never liked math in high school but this is kind of fun,' or 'I never understood this in high school but now I do.'"

—*Linda Flanery, Instructor, Sisseton Wahpeton College*

For more information about ALEKS, please visit: www.aleks.com/highered/math

ALEKS is a registered trademark of ALEKS Corporation.

Increased Student Engagement...

Through Meaningful Applications

Making mathematics meaningful requires that students experience a connection between the mathematics they study, and its impact on the world they live in. This text is the result of a powerful commitment to provide applications of the highest quality, and greatest interest, having close ties to the examples, and with carefully monitored levels of difficulty. We particularly made an effort to supply these in sufficient quantity to be used for in-class illustrations, included as homework assignments, and employed in the construction of quizzes and tests, without exhausting their supply prematurely. Many applications were born of our own diverse experiences, with others coming from a curious, even visionary folly that enables one to seize on the everyday events of life, and to see the mathematics in the background. These were supported by a substantial library of reference and research tools, with an eye toward history, current events, and modern trends. —The Authors

▶ **Chapter Openers** highlight Chapter Connections, an interesting application exercise from the chapter, and provide a list of other real-world connections to give context for students who wonder how math relates to them.

> ❝I think the book has very modern applications and quite a few of them. The calculator instructions are very well done.❞
> —*Nezam Iraniparast, Western Kentucky University*

▶ **Examples** throughout the text feature word problems, providing students with a starting point for how to solve these types of problems in their exercise sets.

> ❝The students always want to know 'When am I ever going to have to use algebra anyway?' Now it will not be hard for them to see for themselves some REAL ways.❞ —*Sally Haas, Angelina College*

▶ **Application Exercises** at the end of each section are the hallmark of the Coburn series. Never contrived, always creative, and born out of the author's life and experiences, each application tells a story and appeals to a variety of teaching styles, disciplines, backgrounds, and interests. The authors have ensured that the applications reflect the most common majors of college algebra students.

> ❝The amount of technology is great, as are the applications. The quality of the applications is better than my current text.❞
> —*Daniel Russow, Arizona Western College–Yuma*

▶ **Math in Action Applets,** located online, enable students to work collaboratively as they manipulate applets that apply mathematical concepts in real-world contexts.

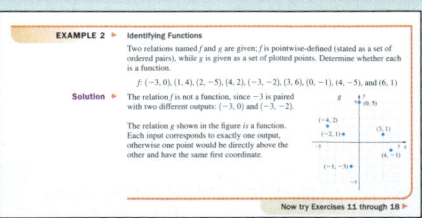

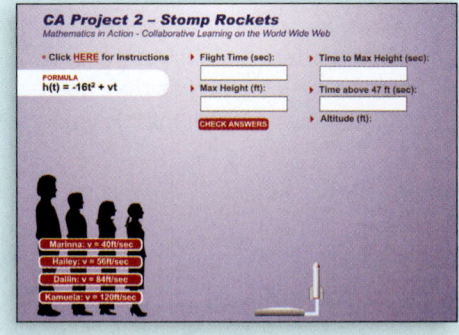

Through Timely Examples

In mathematics, it would be difficult to overstate the importance of examples that set the stage for learning. Not a few educational experiences have faltered due to an example that was too difficult, a poor fit, out of sequence, or had a distracting result. In this series, a careful and deliberate effort was made to select examples that were timely and clear, with a direct focus on the concept or skill at hand. Everywhere possible, they were further designed to link previous concepts to current ideas, and to lay the groundwork for concepts to come. As a trained educator knows, the best time to answer a question is often before it's ever asked, and a timely sequence of carefully constructed examples can go a long way in this regard, making each new idea simply the next logical, even anticipated step. When successful, the mathematical maturity of a student grows in unnoticed increments, as though it was just supposed to be that way.   —The Authors

▶ **Side by side graphical and algebraic solutions** illustrate the difference between problem-solving methods, emphasize the connections between algebraic and graphical information, and enable students to understand why one method might be preferable to another for any given problem.

▶ **Titles** have been added to examples to highlight relevant learning objectives and reinforce the importance of speaking mathematically using vocabulary.

▶ **Annotations** located to the right of the solution sequence help the student recognize which property or procedure is being applied.

▶ **"Now Try"** boxes immediately following examples guide students to specific matched exercises at the end of the section, helping them identify exactly which homework problems coincide with each discussed concept.

EXAMPLE 8 ▶ Solving a Quadratic Inequality

Solve the inequality $-x^2 + 6x \leq 9$.

Analytical Solution ▶ Begin by writing the inequality in standard form: $-x^2 + 6x - 9 \leq 0$. Note this is equivalent to $g(x) \leq 0$ for $g(x) = -x^2 + 6x - 9$. Since $a < 0$, the graph of g will open downward. The factored form is $g(x) = -(x - 3)^2$, showing 3 is a zero and a repeated root. Using the x-axis, we plot the point $(3, 0)$ and visualize a parabola opening downward through this point.

Figure 3.29 shows the graph is *below* the x-axis (outputs are negative) for *all values* of x except $x = 3$. But since this is a *less than or equal to* inequality, the solution is $x \in \mathbb{R}$.

WORTHY OF NOTE
Since $x = 3$ was a zero of multiplicity 2, the graph "bounced off" the x-axis at this point, with no change of sign for g. The graph is entirely below the x-axis, except at the vertex $(3, 0)$.

Figure 3.29

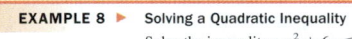

$a < 0$

Graphical Solution ▶ The complete graph of g shown in Figure 3.30 confirms the analytical solution (using the zeroes method). For the intervals of the domain shown in red: $(-\infty, 3) \cup (3, \infty)$, the graph of g is below the x-axis $[g(x) < 0]$. The point $(3, 0)$ is *on* the x-axis $[g(3) = 0]$. As with the analytical solution, the solution to this "less than or equal to" inequality is all real numbers. A calculator check of the original inequality is shown in Figure 3.31.

Figure 3.30

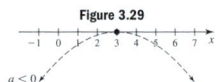

$g(x)$

Figure 3.31

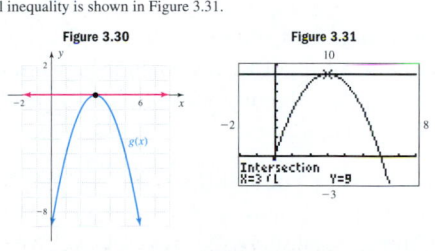

Now try Exercises 121 through 132 ▶

xvii

Solid Skill Development...

Through Exercises

Mid-Chapter Checks

Mid-Chapter Checks provide students with a good stopping place to assess their knowledge before moving on to the second half of the chapter.

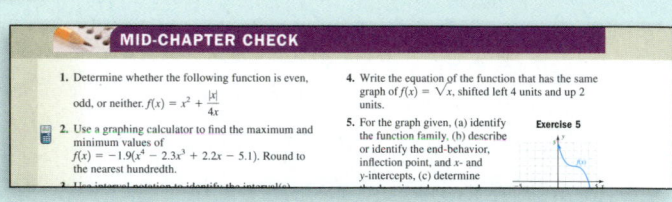

End-of-Section Exercise Sets

▶ **Concepts and Vocabulary** exercises to help students recall and retain important terms.

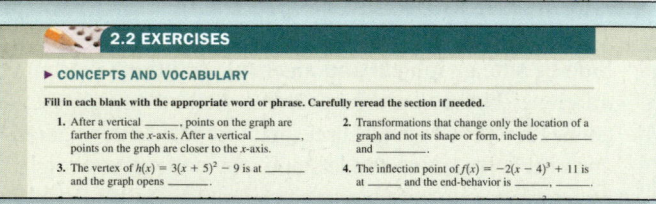

▶ **Developing Your Skills** exercises to provide practice of relevant concepts just learned with increasing levels of difficulty.

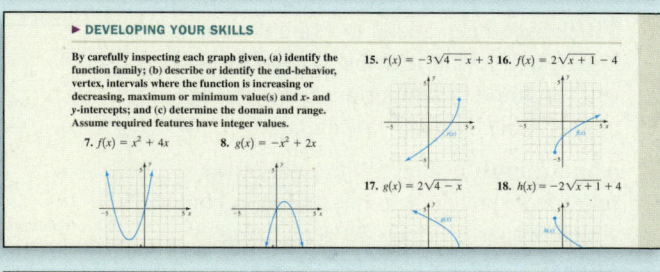

> "The sections in the assignments headed working with formulas and applications bring forward some interesting ideas and problems that are more in depth. These would help hold the students' interest in the topic."
>
> —Sherri Rankin, Huchinson Community College

▶ **Working with Formulas** exercises to demonstrate contextual applications of well-known formulas.

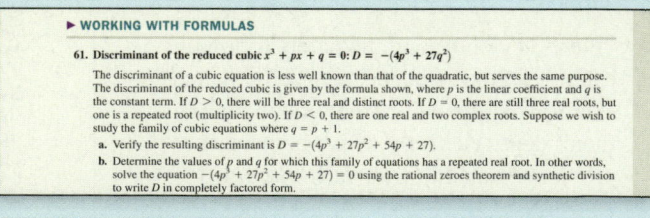

▶ **Extending the Concept** exercises that require communication of topics, synthesis of related concepts, and the use of higher-order thinking skills.

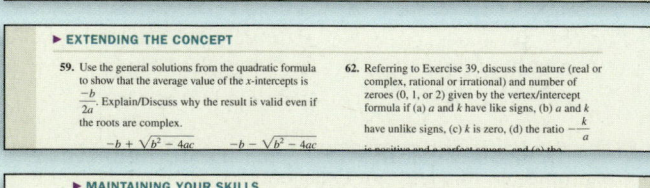

▶ **Maintaining Your Skills** exercises that address skills from previous sections to help students retain previously learning knowledge.

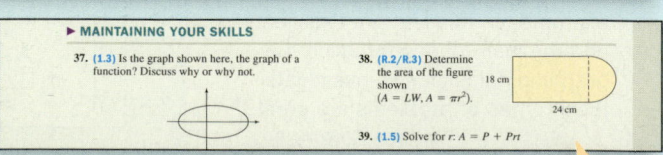

> "The exercise sets are plentiful. I like having many to choose from when assigning homework. When there are only one or two exercises of a particular type, it's hard for the students to get the practice they need."
>
> —Sarah Jackson, Pratt Community College

> "There seems to be a good selection of easy, moderate, and difficult problems in the exercises."
>
> —Ed Gallo, Sinclair Community College

End-of-Chapter Review Material

Exercises located at the end of the chapter provide students with the tools they need to prepare for a quiz or test. Each chapter features the following:

▶ **Making Connections** matching exercises are groups of problems where students must identify graphs based on an equation or description. This feature helps students make the connection between graphical and algebraic information while it enhances students' ability to read and interpret graphical data.

▶ **Chapter Summary and Concept Reviews** that present key concepts with corresponding exercises by section in a format easily used by students.

> **"** The problem sets are really magnificent. I deeply enjoy and appreciate the many problems that incorporate telescopes, astronomy, reflector design, nuclear cooling tower profiles, charged particle trajectories, and other such examples from science, technology, and engineering. **"** *—Light Bryant, Arizona Western College*

▶ **Practice Tests** that give students the opportunity to check their knowledge and prepare for classroom quizzes, tests, and other assessments.

▶ **Cumulative Reviews** that are presented at the end of each chapter help students retain previously learned skills and concepts by revisiting important ideas from earlier chapters (starting with Chapter 2).

▶ **Graphing Calculator** icons appear next to exercises where important concepts can be supported by the use of graphing technology.

> **"** Not only was the algebra rigorously treated, but it was reinforced throughout the chapters with the Mid-Chapter Check and the Chapter Review and Tests. **"** *—Mark Crawford, Waubonsee Community College*

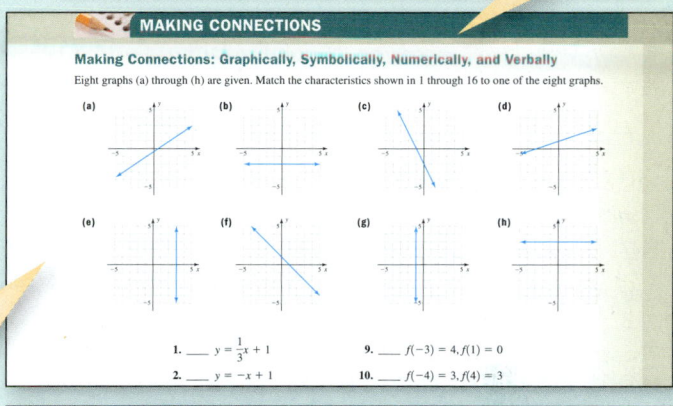

> **"** The authors give very good uses of the calculator in every section. I have been using TI calculators for 15 years and I learned a few new tricks while reading this book. **"** *—George Hurlburt, Corning Community College*

Homework Selection Guide

A list of suggested homework exercises has been provided for each section of the text (Annotated Instructor's Edition only). This feature may prove especially useful for departments that encourage consistency among many sections, or those having a large adjunct population. The feature was also designed as a convenience to instructors, enabling them to develop an inventory of exercises that is more in tune with the course as they like to teach it. The guide provides prescreened and preselected assignments at four different levels: *Core, Standard, Extended,* and *In Depth.*

- **Core:** These assignments go right to the heart of the material, offering a minimal selection of exercises that cover the primary concepts and solution strategies of the section, along with a small selection of the best applications.

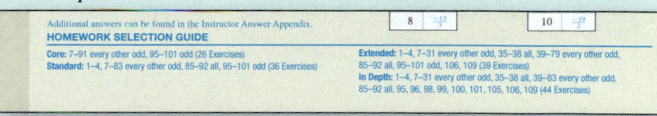

- **Standard:** The assignments at this level include the *Core* exercises, while providing for additional practice without excessive drill. A wider assortment of the possible variations on a theme are included, as well as a greater variety of applications.

- **Extended:** Assignments from the *Extended* category expand on the *Standard* exercises to include more applications, as well as some conceptual or theory-based questions. Exercises may include selected items from the *Concepts and Vocabulary, Working with Formulas,* and the *Extending the Concept* categories of the exercise sets.

- **In Depth:** The *In Depth* assignments represent a more comprehensive look at the material from each section, while attempting to keep the assignment manageable for students. These include a selection of the most popular and highest-quality exercises from each category of the exercise set, with an additional emphasis on *Maintaining Your Skills.*

Strong Mathmatical Connections...

Through a Conversational Writing Style

While examples and applications are arguably the most prominent features of a mathematics text, it's the readability and writing style of the authors that bind them together. It may be true that some students don't read the text, and that others open the text only when looking for an example similar to the exercise they're working on. But when they do and for those students that do (read the text), it's important they have a text that "speaks to them," relating concepts in a form and at a level they understand and can relate to. We feel the writing style of this text will help draw students in and keep their interest, becoming a positive experience and bringing them back a second and third time, until it becomes habitual. At this point students might begin to see the true value of their text, as it becomes a resource for learning on equal footing with any other form of supplemental instruction. This text represents our best efforts in this direction.   —The Authors

Conversational Writing Style

John and J.D.'s experience in the classroom and their strong connections to how students comprehend the material are evident in their writing style. They use a conversational and supportive writing style, providing the students with a tool they can depend on when the teacher is not available, when they miss a day of class, or simply when working on their own. The effort they have put into the writing is representative of John Coburn's unofficial mantra: "If you want more students to reach the top, you gotta put a few more rungs on the ladder."

> "Coburn strikes a good balance between providing all of the important information necessary for a certain topic without going too deep."
> —Barry Monk, Macon State College

> "I think the authors have done an excellent job of interweaving the formal explanations with the 'plain talk' descriptions, illustrating with meaningful examples and applications."
> —Ken Gamber, Hutchinson Community College

Through Student Involvement

How do you design a student-friendly textbook? We decided to get students involved by hosting two separate focus groups. During these sessions we asked students to advise us on how they use their books, what

pedagogical elements are useful, which elements are distracting and not useful, as well as general feedback on page layout. During this process there were times when we thought, "Now why hasn't anyone ever thought of that before?" Clearly these student focus groups were invaluable. Taking direct student feedback and incorporating what is feasible and doesn't detract from instructor use of the text is the best way to design a truly student-friendly text. The next two pages will highlight what we learned from students so you can see for yourself how their feedback played an important role in the development of the Coburn/Herdlick series.

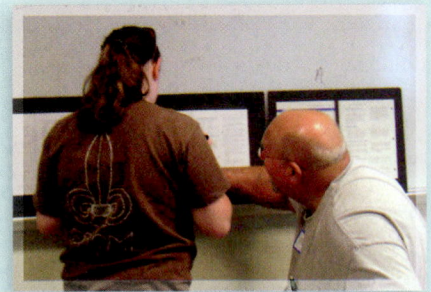

Students said that **Learning Objectives** should clearly define the goals of each section.

5.2 Exponential Functions

LEARNING OBJECTIVES

In Section 5.2 you will see how we can:

- **A.** Evaluate an exponential function
- **B.** Graph general exponential functions
- **C.** Graph base-*e* exponential functions
- **D.** Solve exponential equations and applications

Demographics is the statistical study of human populations. In this section, we introduce the family of *exponential functions*, which are widely used to model population growth or decline with additional applications in science, engineering, and many other fields. As with other functions, we begin with a study of the graph and its characteristics.

A. Evaluating Exponential Functions

In the boomtowns of the old west, it was not uncommon for a town to double in size every year (at least for a time) as the lure of gold drew more and more people westward. When this type of growth is modeled using mathematics, exponents play a lead role. Suppose the town of Goldsboro

Examples are "boxed" so students can clearly see where they begin and end.

Examples are called out in the margins so they are easy for students to spot.

EXAMPLE 4 ▶ Graphing Exponential Functions Using Transformations

Graph $F(x) = 2^{x-1} + 2$ using transformations of the basic function $f(x) = 2^x$ (not by simply plotting points). Clearly state what transformations are applied.

Solution ▶ The graph of F is that of the basic function $f(x) = 2^x$ with a horizontal shift 1 unit right and a vertical shift 2 units up. With this in mind the horizontal asymptote also shifts from $y = 0$ to $y = 2$ and $(0, 1)$ shifts to $(1, 3)$. The y-intercept of F is at $(0, 2.5)$:

$F(x) = 2^x$ is shifted 1 unit right 2 units up. $(3, 6)$ $(0, 2.5)$ $(1, 3)$ $y = 2$

$$F(0) = 2^{(0)-1} + 2$$
$$= 2^{-1} + 2$$
$$= \frac{1}{2} + 2$$
$$= 2.5$$

✓ **B.** You've just seen how we can graph general exponential functions

To help sketch a more accurate graph, the point $(3, 6)$ can be used: $F(3) = 6$.

Now try Exercises 15 through 30 ▶

Students asked for **Check Points** throughout each section to alert them when a specific learning objective has been covered and to reinforce the use of correct mathematical terms.

Students told us they liked when the examples were linked to the exercises.

Described by students as one of the most useful features in a math text, **Caution Boxes** signal a student to stop and take note in order to avoid mistakes in problem solving.

⚠ **CAUTION** ▶ For equations like those in Example 1, be careful not to treat the absolute value bars as simple grouping symbols. The equation $-5(x - 7) + 2 = -13$ has only the solution $x = 10$, and "misses" the second solution since it yields $x - 7 = 3$ in simplified form. The equation $-5|x - 7| + 2 = -13$ simplifies to $|x - 7| = 3$ and there are actually two solutions. Also note that $-5|x - 7| \neq |-5x + 35|$!

Students told us that the color red should only be used for things that are really important. Also, anything significant should be included in the body of the text; marginal readings imply optional.

Students told us that directions should be in bold so they are easily distinguishable from the problems.

Students said having a lot of icons was confusing. The graphing calculator is the only icon used in the exercise sets; no unnecessary icons are used.

▶ **APPLICATIONS**

Use the information given to build a linear equation model, then use the equation to respond. For exercises 71 to 74, develop both an algebraic and a graphical solution.

71. Business depreciation: A business purchases a copier for $8500 and anticipates it will depreciate in value $1250 per year.
  a. What is the copier's value after 4 yr of use?
  b. How many years will it take for this copier's value to decrease to $2250?

72. Baseball card value: After purchasing an autographed baseball card for $85, its value increases by $1.50 per year.
  a. What is the card's value 7 yr after purchase?
  b. How many years will it take for this card's value to reach $100?

74. Gas mileage: When empty, a large dump-truck gets about 15 mi per gallon. It is estimated that for each 3 tons of cargo it hauls, gas mileage decreases by $\frac{3}{4}$ mi per gallon.
  a. If 10 tons of cargo is being carried, what is the truck's mileage?
  b. If the truck's mileage is down to 10 mi per gallon, how much weight is it carrying?

75. Parallel/nonparallel roads: Aberville is 38 mi north and 12 mi west of Boschertown, with a straight "farm and machinery" road (FM 1960) connecting the two cities. In the next county, Crownsburg is 30 mi north and 9.5 mi west of Dower, and these cities are likewise connected by a straight road (FM 830). If the two roads continued indefinitely in both directions, would they intersect at some point?

Because students spend a lot of time in the exercise section of a text, they said that a white background is hard on their eyes...so we used a soft, off-white color for the background.

Coburn's Precalculus Series

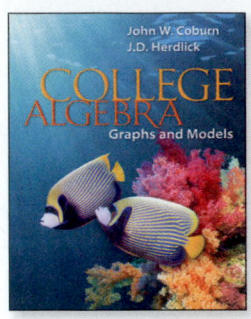

College Algebra: Graphs & Models, First Edition

A Review of Basic Concepts and Skills ◆ Functions and Graphs ◆ Relations; More on Functions ◆ Quadratic Functions and Operations on Functions ◆ Polynomial and Rational Functions ◆ Exponential and Logarithmic Functions ◆ Systems of Equations and Inequalities ◆ Matrices and Matrix Applications ◆ Analytic Geometry and the Conic Sections ◆ Additional Topics in Algebra

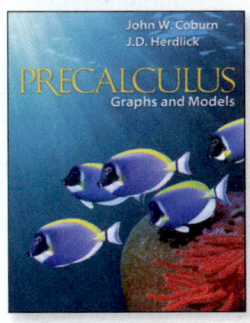

Precalculus: Graphs & Models, First Edition

Functions and Graphs ◆ Relations; More on Functions ◆ Quadratic Functions and Operations on Functions ◆ Polynomial and Rational Functions ◆ Exponential and Logarithmic Functions ◆ Introduction to Trigonometry ◆ trigonometric Identities, Inverses, and Equations ◆ Applications of Trigonometry ◆ Systems of Equations and Inequalities; Matrices ◆ Analytic Geometry; Polar and parametric Equations ◆ Sequences, Series, Counting, and Probability ◆ Bridges to Calculus—An Introduction to Limits

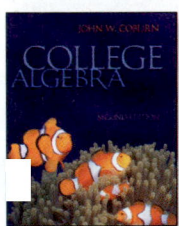

College Algebra
Second Edition
Review ◆ Equations and Inequalities ◆ Relations, Functions, and Graphs ◆ Polynomial and Rational Functions ◆ Exponential and Logarithmic Functions ◆ Systems of Equations and Inequalities ◆ Matrices ◆ Geometry and Conic Sections ◆ Additional Topics in Algebra
MHID 0-07-351941-3, ISBN 978-0-07-351941-8

College Algebra Essentials
Second Edition
Review ◆ Equations and Inequalities ◆ Relations, Functions, and Graphs ◆ Polynomial and Rational Functions ◆ Exponential and Logarithmic Functions ◆ Systems of Equations and Inequalities
MHID 0-07-351968-5, ISBN 978-0-07-351968-5

Algebra and Trigonometry
Second Edition
Review ◆ Equations and Inequalities ◆ Relations, Functions, and Graphs ◆ Polynomial and Rational Functions ◆ Exponential and Logarithmic Functions ◆ Trigonometric Functions ◆ Trigonometric Identities, Inverses, and Equations ◆ Applications of Trigonometry ◆ Systems of Equations and Inequalities ◆ Matrices ◆ Geometry and Conic Sections ◆ Additional Topics in Algebra
MHID 0-07-351952-9, ISBN 978-0-07-351952-4

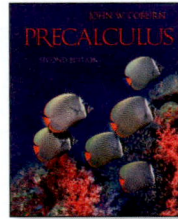

Precalculus
Second Edition
Equations and Inequalities ◆ Relations, Functions, and Graphs ◆ Polynomial and Rational Functions ◆ Exponential and Logarithmic Functions ◆ Trigonometric Functions ◆ Trigonometric Identities, Inverses, and Equations ◆ Applications of Trigonometry ◆ Systems of Equations and Inequalities ◆ Matrices ◆ Geometry and Conic Sections ◆ Additional Topics in Algebra ◆ Limits
MHID 0-07-351942-1, ISBN 978-0-07-351942-5

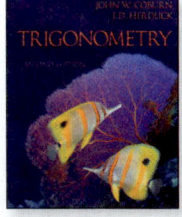

Trigonometry
Second Edition
Introduction to Trigonometry ◆ Right Triangles and Static Trigonometry ◆ Radian Measure and Dynamic Trigonometry ◆ Trigonometric Graphs and Models ◆ Trigonometric Identities ◆ Inverse Functions and Trigonometric Equations ◆ Applications of Trigonometry ◆ Trigonometric Connections to Algebra
MHID 0-07-351948-0, ISBN 978-0-07-351948-7

Making Connections...

Through 360° Development

McGraw-Hill's 360° Development Process is an ongoing, never-ending, market-oriented approach to building accurate and innovative print and digital products. It is dedicated to continual large-scale and incremental improvement driven by multiple customer feedback loops and checkpoints. This process is initiated during the early planning stages of our new products, intensifies during the development and production stages, and then begins again on publication, in anticipation of the next edition.

A key principle in the development of any mathematics text is its ability to adapt to teaching specifications in a universal way. The only way to do so is by contacting those universal voices—and learning from their suggestions. We are confident that our book has the most current content the industry has to offer, thus pushing our desire for accuracy to the highest standard possible. In order to accomplish this, we have moved through an arduous road to production. Extensive and open-minded advice is critical in the production of a superior text.

By investing in this extensive endeavor, McGraw-Hill delivers to you a product suite that has been created, refined, tested, and validated to be a successful tool in your course.

Student Focus Groups

Two student focus groups were held at Illinois State University and Southeastern Louisiana University to engage students in the development process and provide feedback as to how the design of a textbook impacts homework and study habits in the College Algebra, Precalculus, and Trigonometry course areas.

Francisco Arceo, *Illinois State University*
Candace Banos, *Southeastern Louisiana University*
Dave Cepko, *Illinois State University*
Andrea Connell, *Illinois State University*
Nicholas Curtis, *Southeastern Louisiana University*
M. D. "Boots" Feltenberger, *Southeastern Louisiana University*
Regina Foreman, *Southeastern Louisiana University*
Ashley Lae, *Southeastern Louisiana University*
Brian Lau, *Illinois State University*
Daniel Nathan Mielneczek, *Illinois State University*
Mingaile Orakauskaite, *Illinois State University*
Todd Michael Rapnikas, *Illinois State University*
Bethany Rollet, *Illinois State University*
Teddy Schrishuhn, *Illinois State University*

Josh Schultz, *Illinois State University*
Jessica Smith, *Southeastern Louisiana University*
Andy Thurman, *Illinois State University*
Ashley Youngblood, *Southeastern Louisiana University*

Digital Contributors

Jeremy Coffelt, *Blinn College*
Vanessa Coffelt, *Blinn College*
Vickie Flanders, *Baton Rouge Community College*
Anne Marie Mosher, *Saint Louis Community College-Florissant Valley*

Kristen Stoley, *Blinn College*
David Ray, *University of Tennessee-Martin*
Stephen Toner, *Victor Valley Community College*
Paul Vroman, *Saint Louis Community College-Florissant Valley*

Special Thanks

Sherry Meier, *Illinois State University*
Rebecca Muller, *Southeastern Louisiana University*
Anne Schmidt, *Illinois State University*

Making Connections...

Developmental Editing

The manuscript has been impacted by numerous developmental reviewers who edited for clarity and consistency. Efforts resulted in cutting length from the manuscript, while retaining a conversational and casual narrative style. Editorial work also ensured the positive visual impact of art and photo placement.

Chapter Reviews and Manuscript Reviews

Teachers and academics from across the country reviewed the current edition text, the proposed table of contents, and first-draft manuscript to give feedback on reworked narrative, design changes, pedagogical enhancements, and organizational changes. This feedback was summarized by the book team and used to guide the direction of the second-draft manuscript.

Betty Anderson, *Howard Community College*
David Bosworth, *Hutchinson Community College*
Daniel Brock, *Arkansas State University-Beebe*
Barry Brunson, *Western Kentucky University*
Light Bryant, *Arizona Western College*
Brenda Burns-Williams, *North Carolina State University-Raleigh*
Charles Cooper, *Hutchinson Community College*
Mark Crawford, *Waubonsee Community College*
Joseph Demaio, *Kennesaw State University*
Alvio Dominguez, *Miami-Dade College-Wolfson*
Dale Duke, *Oklahoma City Community College*
Frank Edwards, *Southeastern Louisiana University*
Caleb Emmons, *Pacific University*
Mike Everett, *Santa Ana College*
Maggie Flint, *Northeast State Technical Community College*
Ed Gallo, *Sinclair Community College*
Ken Gamber, *Hutchinson Community College*
David Gurney, *Southeastern Louisiana University*

Sally Haas, *Angelina College*
Ben Hill, *Lane Community College*
Jody Hinson, *Cape Fear Community College*
Lynda Hollingsworth, *Northwest Missouri State University*
George Hurlburt, *Corning Community College*
Sarah Jackson, *Pratt Community College*
Laud Kwaku, *Owens Community College*
Kathryn Lavelle, *Westchester Community College*
Joseph Lloyd Harris, *Gulf Coast Community College*
Austin Lovenstein, *Pulaski Technical College*
Rodolfo Maglio, *Northeastern Illinois University*
Barry Monk, *Macon State College*
Camille Moreno, *Cosumnes River College*
Anne Marie Mosher, *Saint Louis Community College-Florissant Valley*
Lilia Orlova, *Nassau Community College*
Susan Pfeifer, *Butler Community College*
Sherri Rankin, *Hutchinson Community College*
Daniel Russow, *Arizona Western College-Yuma*
Rose Shirey, *College of the Mainland*
Joy Shurley, *Abraham Baldwin Agricultural College*
Sean Simpson, *Westchester Community College*
Pam Stogsdill, *Bossier Parish Community College*
Allison Sutton, *Austin Community College*
Linda Tremer, *Three Rivers Community Collge*
Dahlia Vu, *Santa Ana College*
Jackie Wing, *Angelina College*

Acknowledgments

We first want to express a deep appreciation for the guidance, comments, and suggestions offered by all reviewers of the manuscript. We have once again found their collegial exchange of ideas and experience very refreshing and instructive, and always helping to create a better learning tool for our students.

Vicki Krug has continued to display an uncanny ability to bring innumerable pieces from all directions into a unified whole, in addition to providing spiritual support during some extremely trying times; Patricia Steele's skill as a copy editor is as sharp as ever, and her attention to detail continues to pay great dividends; which helps pay the debt we owe Katie White, Michelle Flomenhoft, Christina Lane, and Eve Lipton for their useful suggestions, infinite patience, tireless efforts, and art-counting eyes, which helped in bringing the manuscript to completion. We must also thank John Osgood for his ready wit, creative energies, and ability to step into the flow without missing a beat; Laurie Janssen and our magnificent design team, and Dawn Bercier whose influence on this project remains strong although she has moved on, as it was her indefatigable spirit that kept the ship on course through trial and tempest, and her ski-jumper's vision that brought J.D. on board. In truth, our hats are off to all the fine people at McGraw-Hill for their continuing support and belief in this series. A final word of thanks must go to Rick Armstrong, whose depth of knowledge, experience, and mathematical connections seems endless; Anne Marie Mosher for her contributions to various features of the text, Mitch Levy for his consultation on the exercise sets, Stephen Toner for his work on the videos, Jon Booze and his team for their work on the test bank, Cindy Trimble for her invaluable ability to catch what everyone else misses; and to Rick Pescarino, Kelly Ballard, John Elliot, Jim Frost, Barb Kurt, Lillian Seese, Nate Wilson, and all of our colleagues at St. Louis Community College, whose friendship, encouragement, and love of mathematics makes going to work each day a joy.

Making Connections...

Through Supplements

*All online supplements are available through the book's website: www.mhhe.com/coburn.

Instructor Supplements

- **Computerized Test Bank Online:** Utilizing Brownstone Diploma® algorithm-based testing software enables users to create customized exams quickly.
- **Instructor's Solutions Manual:** Provides comprehensive, worked-out solutions to all exercises in the text.
- **Annotated Instructor's Edition:** Contains all answers to exercises in the text, which are printed in a second color, adjacent to corresponding exercises, for ease of use by the instructor.

Student Supplements

- **Student Solutions Manual** provides comprehensive, worked-out solutions to all of the odd-numbered exercises.
- **Graphing Calculator Manual** includes detailed instructions for using different calculator models to solve problems throughout the text. Written by the authors to accompany their text, it is designed to match and supplement the text.
- **Videos**
  - Interactive video lectures are provided for each section in the text, which explain to the students how to do key problem types, as well as highlighting common mistakes to avoid.
  - Exercise videos provide step-by-step instruction for the key exercises which students will most wish to see worked out.
  - Graphing calculator videos help students master the most essential calculator skills used in the college algebra course.
  - The videos are closed-captioned for the hearing impaired, subtitled in Spanish, and meet the Americans with Disabilities Act Standards for Accessible Design.

Connect Math™ Hosted by ALEKS®

www.connectmath.com

Connect Math Hosted by ALEKS is an exciting, new assessment and assignment platform combining the strengths of McGraw-Hill Higher Education and ALEKS Corporation. Connect Math Hosted by ALEKS is the first platform on the market to combine an artificial-intelligent, diagnostic assessment with an intuitive ehomework platform designed to meet your needs.

Connect Hosted by ALEKS is the culmination of a one-of-a-kind market development process involving math full-time faculty members and adjuncts at every step of the process. This process enables us to provide you with an end product that better meets your needs.

Connect Math Hosted by ALEKS is built by mathematicians educators for mathematicians educators!

ALEKS® www.aleks.com

ALEKS (**A**ssessment and **LE**arning in **K**nowledge **S**paces) is a dynamic online learning system for mathematics education, available over the Web 24/7. ALEKS assesses students, accurately determines their knowledge, and then guides them to the material that they are most ready to learn. With a variety of reports, Textbook Integration Plus, quizzes, and homework assignment capabilities, ALEKS offers flexibility and ease of use for instructors.

- ALEKS uses artificial intelligence to determine exactly what each student knows and is ready to learn. ALEKS remediates student gaps and provides highly efficient learning and improved learning outcomes
- ALEKS is a comprehensive curriculum that aligns with syllabi or specified textbooks. Used in conjunction with McGraw-Hill texts, students also receive links to text-specific videos, multimedia tutorials, and textbook pages.
- ALEKS offers a dynamic classroom management system that enables instructors to monitor and direct student progress toward mastery of course objectives.

ALEKS Prep/Remediation:

- Helps instructors meet the challenge of remediating underprepared or improperly placed students.
- Assesses students on their prerequisite knowledge needed for the course they are entering (i.e., Calculus students are tested on Precalculus knowledge) and prescribes a unique and efficient learning path specifically to address their strengths and weaknesses.
- Students can address prerequisite knowledge gaps outside of class freeing the instructor to use class time pursuing course outcomes.

Making Connections...

Do More

McGraw-Hill Higher Education and Blackboard® have teamed up.

Blackboard, the Web-based course-management system, has partnered with McGraw-Hill to better allow students and faculty to use online materials and activities to complement face-to-face teaching. Blackboard features exciting social learning and teaching tools that foster more logical, visually impactful and active learning opportunities for students. You'll transform your closed-door classrooms into communities where students remain connected to their educational experience 24 hours a day.

This partnership allows you and your students access to McGraw-Hill's Connect™ and Create™ right from within your Blackboard course—all with one single sign-on.

Not only do you get single sign-on with Connect and Create, you also get deep integration of McGraw-Hill content and content engines right in Blackboard. Whether you're choosing a book for your course or building Connect assignments, all the tools you need are right where you want them—inside of Blackboard.

Gradebooks are now seamless. When a student completes an integrated Connect assignment, the grade for that assignment automatically (and instantly) feeds your Blackboard grade center.

McGraw-Hill and Blackboard can now offer you easy access to industry leading technology and content, whether your campus hosts it, or we do. Be sure to ask your local McGraw-Hill representative for details.

TEGRITY—tegritycampus.mhhe.com

McGraw-Hill Tegrity Campus™ is a service that makes class time available all the time by automatically capturing every lecture in a searchable format for students to review when they study and complete assignments. With a simple one-click start and stop process, you capture all computer screens and corresponding audio. Students replay any part of any class with easy-to-use browser-based viewing on a PC or Mac.

Educators know that the more students can see, hear, and experience class resources, the better they learn. With Tegrity, students quickly recall key moments by using Tegrity's unique search feature. This search helps students efficiently find what they need, when they need it across an entire semester of class recordings. Help turn all your students' study time into learning moments immediately supported by your lecture.

To learn more about Tegrity watch a 2-minute Flash demo at **tegritycampus.mhhe.com.**

Electronic Books:

If you or your students are ready for an alternative version of the traditional textbook, McGraw-Hill eBooks offer a cheaper and eco-friendly alternative to traditional textbooks. By purchasing eBooks from McGraw-Hill, students can save as much as 50% on selected titles delivered on the most advanced eBook platform available. Contact your McGraw-Hill sales representative to discuss eBook packaging options.

Create:

Craft your teaching resources to match the way you teach! With McGraw-Hill Create, **www.mcgrawhillcreate.com,** you can easily rearrange chapters, combine material from other content sources, and quickly upload content you have written like your course syllabus or teaching notes. Find the content you need in Create by searching through thousands of leading McGraw-Hill textbooks. Arrange your book to fit your teaching style. Create even allows you to personalize your book's appearance by selecting the cover and adding your name, school, and course information. Order a Create book and you'll receive a complimentary print review copy in 3–5 business days or a complimentary electronic review copy (eComp) via email in minutes.

Go to **www.mcgrawhillcreate.com** today and register to experience how McGraw-Hill Create empowers you to teach *your* students *your* way.

Contents

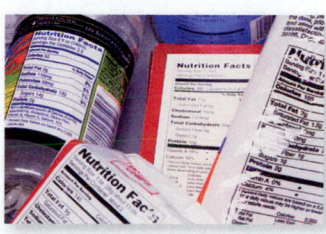

Index of Applications

A Review of Basic Concepts and Skills

CHAPTER CONNECTIONS

One of the primary goals of a college algebra course is to develop the mathematical tools necessary to model, explain, and understand the world around us. Speaking very broadly, this understanding gives us a better perspective of "where we've been" and "where we're going." For example, you're likely aware that in 2009, the federal minimum wage got a nice boost. The ability to model increases in the minimum wage over time can give us information of what this wage might become in future years, or remind us of what the wage was long ago. This application is explored in Exercise 96 of Section R.1.

Check out these other real-world connections:

1

LEARNING OBJECTIVES

In Section R.1 you will review how to:

☐ **A.** Identify terms, coefficients, and expressions

☐ **B.** Create mathematical models

☐ **C.** Evaluate algebraic expressions

☐ **D.** Identify and use properties of real numbers

☐ **E.** Simplify algebraic expressions

To effectively use mathematics as a problem-solving tool, we must develop the ability to translate written or verbal information into a mathematical model. After obtaining a model, many applications require that you work effectively with algebraic terms and expressions. The basic ideas involved are reviewed here.

A. Terms, Coefficients, and Algebraic Expressions

An **algebraic term** is a *collection of factors* that may include numbers, variables, or expressions within parentheses. Here are some examples:

(a) 3   (b) $-6P$ (c) $5xy$ (d) $-8n^2$ (e) n (f) $2(x + 3)$

If a term consists of a single nonvariable number, it is called a **constant** term. In (a), 3 is a constant term. Any term that contains a variable is called a **variable term.** We call the constant factor of a term the **numerical coefficient** or simply the **coefficient.** The coefficients for (a), (b), (c), and (d) are 3, -6, 5, and -8, respectively. In (e), the coefficient of n is 1, since $1 \cdot n = 1n = n$. The term in (f) has two factors as written, 2 and $(x + 3)$. The coefficient is 2.

An **algebraic expression** can be a single term or a sum or difference of terms. To avoid confusion when identifying the coefficient of each term, the expression can be rewritten using algebraic addition if desired: $A - B = A + (-B)$. For instance, $4 - 3x = 4 + (-3x)$ shows the coefficient of x is -3. To identify the coefficient of a rational term, it sometimes helps to **decompose** the term, rewriting it using a unit fraction as in $\frac{n-2}{5} = \frac{1}{5}(n - 2)$ and $\frac{x}{2} = \frac{1}{2}x$.

EXAMPLE 1 ▶ **Identifying Terms and Coefficients**

State the number of terms in each expression as given, then identify the coefficient of each term.

a. $2x - 5y$ **b.** $\dfrac{x + 3}{7} - 2x$ **c.** $-(x - 12)$ **d.** $-2x^2 - x + 5$

Solution ▶ We can begin by rewriting each subtraction using algebraic addition.

Rewritten:	**a.** $2x + (-5y)$	**b.** $\frac{1}{7}(x + 3) + (-2x)$	**c.** $-1(x - 12)$	**d.** $-2x^2 + (-1x) + 5$
Number of terms:	two	two	one	three
Coefficient(s):	2 and -5	$\frac{1}{7}$ and -2	-1	$-2, -1,$ and 5

☑ **A.** You've just seen how we can identify terms, coefficients, and expressions

Now try Exercises 7 through 14 ▶

B. Translating Written or Verbal Information into a Mathematical Model

The key to solving many applied problems is finding an algebraic expression that accurately models relationships described in context. First, we assign a variable to represent an unknown quantity, then build related expressions using words from the English language that suggest mathematical operations. Variables that remind us of what they represent are often used in the modeling process, such as D = RT for Distance equals Rate times Time. These are often called **descriptive variables.** Capital letters are also used due to their widespread appearance in other fields.

EXAMPLE 2 ▶ **Translating English Phrases into Algebraic Expressions**

Assign a variable to the unknown number, then translate each phrase into an algebraic expression.

- **a.** twice a number, increased by five
- **b.** eleven less than eight times the width
- **c.** ten less than triple the payment
- **d.** two hundred fifty dollars more than double the amount

Solution ▶

a. Let n represent the number. Then $2n$ represents twice the number, and $2n + 5$ represents twice the number, increased by five.

b. Let W represent the width. Then $8W$ represents eight times the width, and $8W - 11$ represents 11 less than eight times the width.

c. Let p represent the payment. Then $3p$ represents triple the payment, and $3p - 10$ represents 10 less than triple the payment.

d. Let A represent the amount in dollars. Then $2A$ represents double the amount, and $2A + 250$ represents 250 dollars more than double the amount.

Now try Exercises 15 through 28 ▶

Identifying and translating such phrases *when they occur in context* is an important problem-solving skill. Note how this is done in Example 3.

EXAMPLE 3 ▶ **Creating a Mathematical Model**

The cost for a rental car is $35 plus 15 cents per mile. Express the cost of renting a car in terms of the number of miles driven.

Solution ▶

Let m represent the number of miles driven. Then $0.15m$ represents the cost for each mile and $C = 35 + 0.15m$ represents the total cost for renting the car.

☑ **B.** You've just seen how we can create mathematical models

Now try Exercises 29 through 40 ▶

C. Evaluating Algebraic Expressions

We often need to **evaluate** expressions to investigate patterns and note relationships.

Evaluating a Mathematical Expression

1. Replace each variable with open parentheses ( ).
2. Substitute the values given for each variable.
3. Simplify using the order of operations.

In this process, it's best to use a **vertical format,** with the original expression written first, the substitutions shown next, followed by the simplified forms and the final result. The numbers substituted or "plugged into" the expression are often called the **input values,** with the result called the **output** value.

EXAMPLE 4 ▶   Evaluating an Algebraic Expression

Evaluate the expression $x^3 - 2x^2 + 5$ for $x = -3$.

Solution ▶   For $x = -3$: $x^3 - 2x^2 + 5 = ()^3 - 2()^2 + 5$ replace variables with open parentheses

$= (-3)^3 - 2(-3)^2 + 5$ substitute -3 for x

$= -27 - 2(9) + 5$ simplify: $(-3)^3 = -27, (-3)^2 = 9$

$= -27 - 18 + 5$ simplify: $2(9) = 18$

$= -40$ result

When the input is -3, the output is -40.

> **WORTHY OF NOTE**
>
> In Example 4, note the importance of the first step in the evaluation process: *replace each variable with open parentheses.* Skipping this step could easily lead to confusion as we try to evaluate the squared term, since $-3^2 = -9$, while $(-3)^2 = 9$. **Also see Exercises 55 and 56.**

Now try Exercises 41 through 60 ▶

If the same expression is evaluated repeatedly, results are often collected and analyzed in a table of values, as shown in Example 5. As a practical matter, the substitutions and simplifications are often done mentally or on scratch paper, with the table showing only the input and output values.

EXAMPLE 5 ▶   Evaluating an Algebraic Expression

Evaluate $x^2 - 2x - 3$ to complete the table shown. Which input value(s) of x cause the expression to have an output of 0?

Solution ▶

Input x	Output $x^2 - 2x - 3$
-2	$(-2)^2 - 2(-2) - 3 = 5$
-1	0
0	-3
1	-4
2	-3
3	0
4	5

The expression has an output of 0 when $x = -1$ and $x = 3$.

Now try Exercises 61 through 66 ▶

Figure R.1

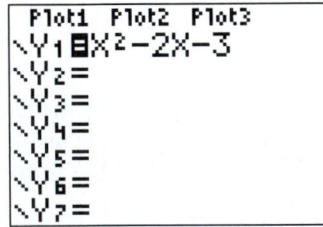

Graphing calculators provide an efficient means of evaluating many expressions. After entering the expression on the ⬭Y= screen (Figure R.1), we can set up the table using the keystrokes ⬭2nd ⬭WINDOW **(TBLSET).** For this exercise, we'll put the table in the "**Indpnt: Auto Ask**" mode, which will have the calculator "automatically" generate the input and output values. In this mode, we can tell the calculator where to start the inputs (we chose **TblStart** $= -2$), and have the calculator produce the input values using any increment desired (we choose Δ**Tbl** $= 1$), as shown in Figure R.2. We access the completed table using ⬭2nd ⬭GRAPH **(TABLE),** and the result for Example 5 is shown in Figure R.3.

For exercises that combine the skills from Examples 3 through 5, **see Exercises 91 to 98.**

☑ **C.** You've just seen how we can evaluate algebraic expressions

Figure R.2

Figure R.3

D. Properties of Real Numbers

While the phrase, "an unknown number times five," is accurately modeled by the expression $n5$ for some number n, in algebra we prefer to have numerical coefficients precede variable factors. When we reorder the *factors* as $5n$, we are using the **commutative property of multiplication**. A reordering of *terms* involves the **commutative property of addition.**

The Commutative Properties

Given that a and b represent real numbers:

ADDITION: $a + b = b + a$ MULTIPLICATION: $a \cdot b = b \cdot a$

Terms can be combined in   Factors can be multiplied in
any order without changing   any order without changing
the sum.   the product.

Each property can be extended to include any number of terms or factors. While the commutative property implies a *reordering* or *movement* of terms (to commute implies back-and-forth movement), the **associative property** implies a *regrouping* or reassociation of terms. For example, the sum $(\frac{3}{4} + \frac{3}{5}) + \frac{2}{5}$ is easier to compute if we regroup the addends as $\frac{3}{4} + (\frac{3}{5} + \frac{2}{5})$. This illustrates the **associative property of addition.** Multiplication is also associative.

The Associative Properties

Given that a, b, and c represent real numbers:

ADDITION: MULTIPLICATION:

$$(a + b) + c = a + (b + c)$$ $$(a \cdot b) \cdot c = a \cdot (b \cdot c)$$

Terms can be regrouped.   Factors can be regrouped.

EXAMPLE 6 ▶ **Simplifying Expressions Using Properties of Real Numbers**

Use the commutative and associative properties to simplify each calculation.

a. $\frac{3}{8} - 19 + \frac{5}{8}$ **b.** $[-2.5 \cdot (-1.2)] \cdot 10$

Solution ▶

a. $\frac{3}{8} - 19 + \frac{5}{8} = -19 + \frac{3}{8} + \frac{5}{8}$ commutative property (order changes)

$\qquad = -19 + (\frac{3}{8} + \frac{5}{8})$ associative property (grouping changes)

$\qquad = -19 + 1$ simplify

$\qquad = -18$ result

b. $[-2.5 \cdot (-1.2)] \cdot 10 = -2.5 \cdot [(-1.2) \cdot 10]$ associative property (grouping changes)

$\qquad = -2.5 \cdot (-12)$ simplify

$\qquad = 30$ result

Now try Exercises 67 and 68 ▶

WORTHY OF NOTE

Is subtraction commutative? Consider a situation involving money. If you had $100, you could easily buy an item costing $20: $100 − $20 leaves you with $80. But if you had $20, could you buy an item costing $100? Obviously $100 − $20 is not the same as $20 − $100. Subtraction is *not* commutative. Likewise, 100 ÷ 20 is not the same as 20 ÷ 100, and division is *not* commutative.

For any real number x, $x + 0 = x$ and 0 is called the **additive identity** since the original number was returned or "identified." Similarly, 1 is called the **multiplicative identity** since $1 \cdot x = x$. The identity properties are used extensively in the process of solving equations.

The Additive and Multiplicative Identities

Given that x is a real number,

$$x + 0 = x$$

Zero is the identity
for addition.

$$1 \cdot x = x$$

One is the identity
for multiplication.

For any real number x, there is a real number $-x$ such that $x + (-x) = 0$. The number $-x$ is called the **additive inverse** of x, since their sum results in the additive identity. Similarly, the **multiplicative inverse** of any nonzero number x is $\frac{1}{x}$, since $x \cdot \frac{1}{x} = 1$ (the multiplicative identity). This property can also be stated as $\frac{p}{q} \cdot \frac{q}{p} = 1$ ($p, q \neq 0$) for any rational number $\frac{p}{q}$. Note that $\frac{p}{q}$ and $\frac{q}{p}$ are **reciprocals.**

The Additive and Multiplicative Inverses

Given that p, q, and x represent real numbers ($p, q \neq 0$):

$$x + (-x) = 0$$

x and $-x$ are
additive inverses.

$$\frac{p}{q} \cdot \frac{q}{p} = 1$$

$\frac{p}{q}$ and $\frac{q}{p}$ are
multiplicative inverses.

EXAMPLE 7 ▶ **Determining Additive and Multiplicative Inverses**

Replace the box to create a true statement:

a. $\boxed{} \cdot \dfrac{-3}{5}x = 1 \cdot x$ **b.** $x + 4.7 + \boxed{} = x$

Solution ▶ **a.** $\boxed{} = \dfrac{5}{-3}$, since $\dfrac{5}{-3} \cdot \dfrac{-3}{5} = 1$

b. $\boxed{} = -4.7$, since $4.7 + (-4.7) = 0$

Now try Exercises 69 and 70 ▶

Note that if no coefficient is indicated, it is assumed to be 1, as in $x = 1x$, $(x^2 + 3x) = 1(x^2 + 3x)$, and $-(x^3 - 5x^2) = -1(x^3 - 5x^2)$.

The **distributive property of multiplication over addition** is widely used in a study of algebra, because it enables us to rewrite a product as an equivalent sum and vice versa.

The Distributive Property of Multiplication over Addition

Given that a, b, and c represent real numbers:

$$a(b + c) = ab + ac$$

A factor outside a sum can be
distributed to each addend in
the sum.

$$ab + ac = a(b + c)$$

A factor common to each addend
in a sum can be "undistributed"
and written outside a group.

EXAMPLE 8 ▶ **Simplifying Expressions Using the Distributive Property**

Apply the distributive property as appropriate. Simplify if possible.

a. $7(p + 5.2)$ **b.** $-(2.5 - x)$ **c.** $7x^3 - x^3$ **d.** $\dfrac{5}{2}n + \dfrac{1}{2}n$

Solution ▶ **a.** $7(p + 5.2) = 7p + 7(5.2)$ **b.** $-(2.5 - x) = -1(2.5 - x)$
$\qquad\qquad\quad = 7p + 36.4$ $\qquad\qquad\qquad = -1(2.5) - (-1)(x)$
$\qquad\qquad\qquad\qquad\qquad\qquad\qquad\qquad\qquad = -2.5 + x$

c. $7x^3 - x^3 = 7x^3 - 1x^3$ **d.** $\dfrac{5}{2}n + \dfrac{1}{2}n = \left(\dfrac{5}{2} + \dfrac{1}{2}\right)n$
$\qquad\qquad\quad = (7 - 1)x^3$ $\qquad\qquad\qquad\quad = \left(\dfrac{6}{2}\right)n$
$\qquad\qquad\quad = 6x^3$ $\qquad\qquad\qquad\quad = 3n$

> **WORTHY OF NOTE**
>
> From Example 8b we learn that a negative sign outside a group changes the sign of all terms within the group: $-(2.5 - x) = -2.5 + x$.

☑ **D.** You've just seen how we can identify and use properties of real numbers

Now try Exercises 71 through 78 ▶

E. Simplifying Algebraic Expressions

Two terms are **like terms** only if they have the *same variable factors* (the coefficient is not used to identify like terms). For instance, $3x^2$ and $-\frac{1}{7}x^2$ are like terms, while $5x^3$ and $5x^2$ are not. We simplify expressions by **combining like terms** using the distributive property, along with the commutative and associative properties. Many times the distributive property is used to eliminate grouping symbols *and* combine like terms within the same expression.

EXAMPLE 9 ▶ **Simplifying an Algebraic Expression**

Simplify the expression completely: $7(2p^2 + 1) - (p^2 + 3)$.

Solution ▶ $7(2p^2 + 1) - 1(p^2 + 3)$ *original expression; note coefficient of* -1
$\quad = 14p^2 + 7 - 1p^2 - 3$ *distributive property*
$\quad = (14p^2 - 1p^2) + (7 - 3)$ *commutative and associative properties (collect like terms)*
$\quad = (14 - 1)p^2 + 4$ *distributive property*
$\quad = 13p^2 + 4$ *result*

Now try Exercises 79 through 88 ▶

The steps for simplifying an algebraic expression are summarized here:

> **To Simplify an Expression**
>
> 1. Eliminate parentheses by applying the distributive property.
> 2. Use the commutative and associative properties to group like terms.
> 3. Use the distributive property to combine like terms.

☑ **E.** You've just seen how we can simplify algebraic expressions

As you practice with these ideas, many of the steps will become more automatic. At some point, the distributive property, the commutative and associative properties, as well as the use of algebraic addition will all be performed mentally.

R.1 EXERCISES

▶ CONCEPTS AND VOCABULARY

Fill in each blank with the appropriate word or phrase. Carefully reread the section, if necessary.

1. A term consisting of a single number is called a(n) ___constant___ term.

2. A term containing a variable is called a(n) ___variable___ term.

3. The constant factor in a variable term is called the ___coefficient___.

4. When $3 \cdot 14 \cdot \frac{2}{3}$ is written as $3 \cdot \frac{2}{3} \cdot 14$, the ___commutative___ property has been used.

5. Discuss/Explain why the additive inverse of -5 is 5, while the multiplicative inverse of -5 is $-\frac{1}{5}$. Answers will vary.

6. Discuss/Explain how we can rewrite the sum $3x + 6y$ as a product, and the product $2(x + 7)$ as a sum. Answers will vary.

▶ DEVELOPING YOUR SKILLS

Identify the number of terms in each expression and the coefficient of each term.

7. $3x - 5y$ two; 3 and -5 8. $-2a - 3b$ two; -2 and -3

9. $2x + \dfrac{x + 3}{4}$ two; 2 and $\frac{1}{4}$ 10. $\dfrac{n - 5}{3} + 7n$ two; $\frac{1}{3}$ and 7

11. $-2x^2 + x - 5$ three; -2, 1, and -5 12. $3n^2 + n - 7$ three; 3, 1, and -7

13. $-(x + 5)$ one; -1 14. $-(n - 3)$ one; -1

Translate each phrase into an algebraic expression.

15. seven fewer than a number $n - 7$

16. a number decreased by six $n - 6$

17. the sum of a number and four $n + 4$

18. a number increased by nine $n + 9$

19. the difference between a number and five is squared $(n - 5)^2$

20. the sum of a number and two is cubed $(n + 2)^3$

21. thirteen less than twice a number $2n - 13$

22. five less than double a number $2n - 5$

23. a number squared plus the number doubled $n^2 + 2n$

24. a number cubed less the number tripled $n^3 - 3n$

25. five fewer than two-thirds of a number $\frac{2}{3}n - 5$

26. fourteen more than one-half of a number $\frac{1}{2}n + 14$

27. three times the sum of a number and five, decreased by seven $3(n + 5) - 7$

28. five times the difference of a number and two, increased by six $5(n - 2) + 6$

Additional answers can be found in the Instructor Answer Appendix.

Create a mathematical model using descriptive variables.

29. The length of the rectangle is three meters less than twice the width.

30. The height of the triangle is six centimeters less than three times the base.

31. The speed of the car was fifteen miles per hour more than the speed of the bus.

32. It took Romulus three minutes more time than Remus to finish the race.

33. **Hovering altitude:** The helicopter was hovering 150 ft above the top of the building. Express the altitude of the helicopter in terms of the building's height. $h = b + 150$

34. **Stacks on a cruise liner:** The smoke stacks of the luxury liner cleared the bridge by 25 ft as it passed beneath it. Express the height of the stacks in terms of the bridge's height. $S = b - 25$

35. **Dimensions of a city park:** The length of a rectangular city park is 20 m more than twice its width. Express the length of the park in terms of the width. $L = 2W + 20$

HOMEWORK SELECTION GUIDE

Core: 7–87 every other odd, 91–97 odd, 99 (26 Exercises)
Standard: 1–4, 7–87 every other odd, 89, 91–97 odd, 99 (31 Exercises)

Extended: 1–4, 7–87 every other odd, 89, 90, 91–97 odd, 99, 100 (33 Exercises)
In Depth: 1–6, 7–87 every other odd, 89, 90, 91–98 all, 99, 100 (39 Exercises)

36. Dimensions of a parking lot: In order to meet the city code while using the available space, a contractor planned to construct a parking lot with a length that was 50 ft less than three times its width. Express the length of the lot in terms of the width. $L = 3W - 50$

37. Cost of milk: In 2010, a gallon of milk cost two and one-half times what it did in 1990. Express the cost of a gallon of milk in 2010 in terms of the 1990 cost. $M = 2.5N$

38. Cost of gas: In 2010, a gallon of gasoline cost two and one-half times what it did in 1990. Express the cost of a gallon of gas in 2010 in terms of the 1990 cost. $G = 2.5N$

39. Pest control: In her pest control business, Judy charges $50 per call plus $12.50 per gallon of insecticide for the control of spiders and certain insects. Express the total charge in terms of the number of gallons of insecticide used. $T = 12.50g + 50$

40. Computer repairs: As his reputation and referral business grew, Keith began to charge $75 per service call plus an hourly rate of $50 for the repair and maintenance of home computers. Express the cost of a service call in terms of the number of hours spent on the call. $C = 50H + 75$

Evaluate each algebraic expression given $x = 2$ and $y = -3$.

41. $4x - 2y$ 14
42. $5x - 3y$ 19
43. $-2x^2 + 3y^2$ 19
44. $-5x^2 + 4y^2$ 16
45. $2y^2 + 5y - 3$ 0
46. $3x^2 + 2x - 5$ 11
47. $-2(3y + 1)$ 16
48. $-3(2y + 5)$ 3
49. $3x^2y$ -36
50. $6xy^2$ 108
51. $(-3x)^2 - 4xy - y^2$ 51
52. $(-2x)^2 - 5xy - y^2$ 37
53. $\frac{1}{2}x - \frac{1}{3}y$ 2
54. $\frac{2}{3}x - \frac{1}{2}y$ $\frac{17}{6}$
55. $(3x - 2y)^2$ 144
56. $(2x - 3y)^2$ 169
57. $\dfrac{-12y + 5}{-3x + 1}$ $-\frac{41}{5}$
58. $\dfrac{12x + (-3)}{-3y + 1}$ $\frac{21}{10}$
59. $\sqrt{-12y} \cdot 4$ 24
60. $7 \cdot \sqrt{-27y}$ 63

Evaluate each expression for integers from -3 to 3 inclusive. Verify results using a graphing calculator. What input(s) give an output of zero?

61. $x^2 - 3x - 4$
62. $x^2 - 2x - 3$
63. $-3(1 - x) - 6$
64. $5(3 - x) - 10$
65. $x^3 - 6x + 4$
66. $x^3 + 5x + 18$

Rewrite each expression using the given property and simplify if possible.

67. Commutative property of addition
 a. $-5 + 7$ $7 + (-5) = 2$ **b.** $-2 + n$ $n + (-2)$
 c. $-4.2 + a + 13.6$ $a + (-4.2) + 13.6 = a + 9.4$ **d.** $7 + x - 7$ $x + 7 - 7 = x$

68. Associative property of multiplication
 a. $2 \cdot (3 \cdot 6)$ $(2 \cdot 3) \cdot 6 = 36$ **b.** $3 \cdot (4 \cdot b)$ $(3 \cdot 4) \cdot b = 12b$
 c. $-1.5 \cdot (6 \cdot a)$ $(-1.5 \cdot 6) \cdot a = -9a$ **d.** $-6 \cdot (-\frac{5}{6} \cdot x)$ $(-6 \cdot -\frac{5}{6}) \cdot x = 5x$

Replace the box so that a true statement results.

69. a. $x + (-3.2) + \boxed{} = x$ 3.2

 b. $n - \frac{5}{6} + \boxed{\frac{5}{6}} = n$

70. a. $\boxed{\frac{3}{2}} \cdot \frac{2}{3}x = 1x$

 b. $\boxed{\frac{-3}{1}} \cdot \dfrac{n}{-3} = 1n$

Apply the distributive property and simplify if possible.

71. $-5(x - 2.6)$ $-5x + 13$
72. $-12(v - 3.2)$ $-12v + 38.4$
73. $\frac{2}{3}(-\frac{1}{5}p + 9)$ $-\frac{2}{15}p + 6$
74. $\frac{5}{6}(-\frac{2}{15}q + 24)$ $-\frac{1}{9}q + 20$
75. $3a + (-5a)$ $-2a$
76. $13m + (-5m)$ $8m$
77. $\frac{2}{3}x + \frac{3}{4}x$ $\frac{17}{12}x$
78. $\frac{5}{12}y - \frac{3}{8}y$ $\frac{1}{24}y$

Simplify by removing all grouping symbols (as needed) and combining like terms.

79. $3(a^2 + 3a) - (5a^2 + 7a)$ $-2a^2 + 2a$
80. $2(b^2 + 5b) - (6b^2 + 9b)$ $-4b^2 + b$
81. $x^2 - (3x - 5x^2)$ $6x^2 - 3x$
82. $n^2 - (5n - 4n^2)$ $5n^2 - 5n$
83. $(3a + 2b - 5c) - (a - b - 7c)$ $2a + 3b + 2c$
84. $(x - 4y + 8z) - (8x - 5y - 2z)$ $-7x + y + 10z$
85. $\frac{3}{5}(5n - 4) + \frac{5}{8}(n + 16)$ $\frac{29}{8}n + \frac{38}{5}$
86. $\frac{2}{3}(2x - 9) + \frac{3}{4}(x + 12)$ $\frac{25}{12}x + 3$
87. $(3a^2 - 5a + 7) + 2(2a^2 - 4a - 6)$ $7a^2 - 13a - 5$
88. $2(3m^2 + 2m - 7) - (m^2 - 5m + 4)$ $5m^2 + 9m - 18$

Additional answers can be found in the Instructor Answer Appendix.

▶ WORKING WITH FORMULAS

89. Electrical resistance: $R = \dfrac{kL}{d^2}$

The electrical resistance in a wire depends on the length and diameter of the wire. This resistance can be modeled by the formula shown, where R is the resistance in ohms, L is the length in feet, and d is the diameter of the wire in inches. Find the resistance if $k = 0.000025$, $d = 0.015$ in., and $L = 90$ ft   10 ohms

90. Volume and pressure: $P = \dfrac{k}{V}$

If temperature remains constant, the pressure of a gas held in a closed container is related to the volume of gas by the formula shown, where P is the pressure in pounds per square inch, V is the volume of gas in cubic inches, and k is a constant that depends on given conditions. Find the pressure exerted by the gas if $k = 440,310$ and $V = 22,580$ in^3.   19.5 psi

▶ APPLICATIONS

Translate each key phrase into an algebraic expression, then evaluate as indicated.

91. Cruising speed: A turbo-prop airliner has a cruising speed that is one-half the cruising speed of a 767 jet aircraft. (a) Express the speed of the turbo-prop in terms of the speed of the jet, and (b) determine the speed of the airliner if the cruising speed of the jet is 550 mph.   **a.** $t = \frac{1}{2}j$ **b.** $t = 275$ mph

92. Softball toss: Macklyn can throw a softball two-thirds as far as her father. (a) Express the distance that Macklyn can throw a softball in terms of the distance her father can throw. (b) If her father can throw the ball 210 ft, how far can Macklyn throw the ball?   **a.** $m = \frac{2}{3}d$ **b.** $m = 140$ ft

93. Dimensions of a lawn: The length of a rectangular lawn is 3 ft more than twice its width. (a) Express the length of the lawn in terms of the width. (b) If the width is 52 ft, what is the length?
a. $L = 2W + 3$ **b.** 107 ft

94. Pitch of a roof: To obtain the proper pitch, the crossbeam for a roof truss must be 2 ft less than three-halves the rafter. (a) Express the length of the crossbeam in terms of the rafter. (b) If the rafter is 18 ft, how long is the crossbeam?
a. $c = \frac{3}{2}r - 2$ **b.** $c = 25$ ft

95. Postage costs: In 2009, a first class stamp cost 29¢ more than it did in 1978. Express the cost of a 2009 stamp in terms of the 1978 cost. If a stamp cost 15¢ in 1978, what was the cost in 2009? $t = c + 29$; 44¢

96. Minimum wage: In 2009, the federal minimum wage was $4.95 per hour more than it was in 1976. Express the 2009 wage in terms of the 1976 wage. If the hourly wage in 1976 was $2.30, what was it in 2009? $t = w + 4.95$; $7.25

97. Repair costs: The TV repair shop charges a flat fee of $43.50 to come to your house and $25 per hour for labor. Express the cost of repairing a TV in terms of the time it takes to repair it. If the repair took 1.5 hr, what was the total cost? $C = 25t + 43.50$; $81

98. Repair costs: At the local car dealership, shop charges are $79.50 to diagnose the problem and $85 per shop hour for labor. Express the cost of a repair in terms of the labor involved. If a repair takes 3.5 hr, how much will it cost? $C = 85t + 79.50$; $377.00

▶ EXTENDING THE CONCEPT

99. If C must be a positive odd integer and D must be a negative even integer, then $C^2 + D^2$ must be a:
  a. positive odd integer.   **a.** positive odd integer
  b. positive even integer.
  c. negative odd integer.
  d. negative even integer.
  e. cannot be determined.

100. Historically, several attempts have been made to create metric time using factors of 10, but our current system won out. If 1 day was 10 metric hours, 1 metric hour was 10 metric minutes, and 1 metric minute was 10 metric seconds, what time would it really be if a metric clock read 4:3:5? Assume that each new day starts at midnight.
10:26:24 A.M.

In this section, we review basic exponential properties and operations on polynomials. Although there are five to eight exponential properties (depending on how you count them), all can be traced back to the basic definition involving repeated multiplication.

A. The Properties of Exponents

An exponent is a superscript number or letter occurring to the upper right of a base number, and indicates how many times the base occurs as a factor. For $b \cdot b \cdot b = b^3$, we say b^3 is written in *exponential form.* In some cases, we may refer to b^3 as an **exponential term.**

Exponential Notation

For any positive integer n,

$$b^n = \underbrace{b \cdot b \cdot b \cdot \ldots \cdot b}_{n \text{ times}} \quad \text{and} \quad \underbrace{b \cdot b \cdot b \cdot \ldots \cdot b}_{n \text{ times}} = b^n$$

The Product and Power Properties

There are two properties that follow immediately from this definition. When b^3 is multiplied by b^2, we have an uninterrupted string of five factors: $b^3 \cdot b^2 = (b \cdot b \cdot b) \cdot (b \cdot b)$, which can then be written as b^5. This is an example of the **product property of exponents.**

Product Property of Exponents

For any base b and positive integers m and n:

$$b^m \cdot b^n = b^{m+n}$$

WORTHY OF NOTE

In this statement of the product property and the exponential properties that follow, it is assumed that for any expression of the form 0^m, $m > 0$ (hence $0^m = 0$).

In words, the property says, *to multiply exponential terms with the **same base,** keep the common base and add the exponents.* A special application of the product property uses repeated factors of the *same* exponential term, as in $(x^2)^3$. Using the product property, we have $(x^2)(x^2)(x^2) = x^6$. Notice the same result can be found more quickly by multiplying the inner exponent by the outer exponent: $(x^2)^3 = x^{2 \cdot 3} = x^6$. We generalize this idea to state the **power property of exponents.** In words the property says, *to raise an exponential term to a power, keep the same base and multiply the exponents.*

Power Property of Exponents

For any base b and positive integers m and n:

$$(b^m)^n = b^{m \cdot n}$$

EXAMPLE 1 ▶   **Multiplying Terms Using Exponential Properties**

Compute each product.

a. $-4x^3 \cdot \frac{1}{2}x^2$ **b.** $(p^3)^2 \cdot (p^4)^5$

Solution ▶   **a.** $-4x^3 \cdot \frac{1}{2}x^2 = (-4 \cdot \frac{1}{2})(x^3 \cdot x^2)$ commutative and associative properties

$= (-2)(x^{3+2})$ simplify; product property

$= -2x^5$ result

b. $(p^3)^2 \cdot (p^4)^5 = p^{3 \cdot 2} \cdot p^{4 \cdot 5}$ power property

$= p^6 \cdot p^{20}$ simplify

$= p^{6+20}$ product property

$= p^{26}$ result

Now try Exercises 7 through 12 ▶

The power property can easily be extended to include more than one factor within the parentheses. This application of the power property is sometimes called the **product to a power property** and can be extended to include any number of factors. We can also raise a quotient of exponential terms to a power. The result is called the **quotient to a power property.** In words the properties say, to raise a product or quotient of exponential terms to a power, *multiply every exponent inside* the parentheses *by the exponent outside* the parentheses.

Product to a Power Property

For any bases a and b, and positive integers m, n, and p:

$$(a^m b^n)^p = a^{mp} \cdot b^{np}$$

Quotient to a Power Property

For any bases a and $b \neq 0$, and positive integers m, n, and p:

$$\left(\frac{a^m}{b^n}\right)^p = \frac{a^{mp}}{b^{np}}$$

EXAMPLE 2 ▶ **Simplifying Terms Using the Power Properties**

Simplify using the power property (if possible):

a. $(-3a)^2$ **b.** $-3a^2$ **c.** $\left(\dfrac{-5a^3}{2b}\right)^2$

Solution ▶

a. $(-3a)^2 = (-3)^2 \cdot (a^1)^2$ **b.** $-3a^2$ is in simplified form
$= 9a^2$

c. $\left(\dfrac{-5a^3}{2b}\right)^2 = \dfrac{(-5)^2(a^3)^2}{2^2 b^2}$

$= \dfrac{25a^6}{4b^2}$

WORTHY OF NOTE

Regarding Examples 2a and 2b, note the difference between the expressions $(-3a)^2 = (-3 \cdot a)^2$ and $-3a^2 = -3 \cdot a^2$. In the first, the exponent acts on both the negative 3 *and* the a; in the second, the exponent acts on only the a and there is no "product to a power."

Now try Exercises 13 through 24 ▶

Applications of exponents sometimes involve linking one exponential term with another using a substitution. The result is then simplified using exponential properties.

EXAMPLE 3 ▶ **Applying the Power Property after a Substitution**

The formula for the volume of a cube is $V = S^3$, where S is the length of one edge. If the length of each edge is $2x^2$:

a. Find a formula for volume in terms of x.
b. Find the volume if $x = 2$.

Solution ▶

a. $V = S^3$
       substitute $2x^2$ for S
   $= (2x^2)^3$
   $= 8x^6$

b. For $V = 8x^6$,
   $V = 8(2)^6$ substitute 2 for x
   $= 8 \cdot 64$ or 512 $(2)^6 = 64$
   The volume of the cube would be 512 units^3.

Now try Exercises 25 and 26 ▶

The Quotient Property of Exponents

By combining exponential notation and the property $\frac{x}{x} = 1$ for $x \neq 0$, we note a pattern that helps to simplify a *quotient* of exponential terms. For $\frac{x^5}{x^2} = \frac{x \cdot x \cdot x \cdot x \cdot x}{x \cdot x} = x^3$, the exponent of the final result appears to be the *difference between the exponent in the numerator and the exponent in the denominator*. This seems reasonable since the subtraction would indicate a removal of the factors that reduce to 1. Regardless of how many factors are used, we can generalize the idea and state the **quotient property of exponents.** In words the property says, to divide two exponential terms with the same base, *keep the common base and subtract the exponent of the denominator from the exponent of the numerator.*

> **Quotient Property of Exponents**
>
> For any base $b \neq 0$ and positive integers m and n:
>
> $$\frac{b^m}{b^n} = b^{m-n}$$

Zero and Negative Numbers as Exponents

If the exponent of the denominator is *greater* than the exponent in the numerator, the quotient property yields a negative exponent: $\frac{x^2}{x^5} = x^{2-5} = x^{-3}$. To help understand what a negative exponent *means,* let's look at the expanded form of the expression: $\frac{x^2}{x^5} = \frac{x \cdot x^1}{x \cdot x \cdot x \cdot x \cdot x} = \frac{1}{x^3}$. A negative exponent can literally be interpreted as "write the factors as a reciprocal." A good way to remember this is

$$2^{-3} \quad \frac{2^{-3}}{1} = \frac{1}{2^3} = \frac{1}{8}$$

three factors of 2 written as a reciprocal

Since the result would be similar regardless of the base used, we can generalize this idea and state the **property of negative exponents.**

> **Property of Negative Exponents**
>
> For any base $b \neq 0$ and integer n:
>
> $$\frac{b^{-n}}{1} = \frac{1}{b^n} \qquad \frac{1}{b^{-n}} = \frac{b^n}{1} \qquad \left(\frac{a}{b}\right)^{-n} = \left(\frac{b}{a}\right)^n ; a \neq 0$$

WORTHY OF NOTE

The use of zero as an exponent should not strike you as strange or odd; it's simply a way of saying that *no factors of the base remain,* since all terms have been reduced to 1.

For $\frac{2^3}{2^3}$, we have $\frac{8}{8} = 1$, or

$$\frac{\overset{1}{\cancel{2}} \cdot \overset{1}{\cancel{2}} \cdot \overset{1}{\cancel{2}}}{2 \cdot 2 \cdot 2} = 1, \text{ or } 2^{3-3} = 2^0 = 1.$$

Finally, when we consider that $\frac{x^3}{x^3} = 1$ by division, and $\frac{x^3}{x^3} = x^{3-3} = x^0$ using the quotient property, we conclude that $x^0 = 1$ as long as $x \neq 0$. We can also generalize this observation and state the meaning of zero as an exponent. In words the property says, *any nonzero quantity raised to an exponent of zero is equal to 1.*

> **Zero Exponent Property**
>
> For any base $b \neq 0$:
>
> $$b^0 = 1$$

EXAMPLE 4 ▶   Simplifying Expressions Using Exponential Properties

Simplify using exponential properties. Answer using positive exponents only.

a. $\left(\dfrac{2a^3}{b^2}\right)^{-2}$
  b. $(3hk^{-2})^3(6h^{-2}k^{-3})^{-2}$

c. $(3x)^0 + 3x^0 + 3^{-2}$
  d. $\dfrac{(-2m^2n^3)^5}{(4mn^2)^3}$

Solution ▶   **a.** $\left(\dfrac{2a^3}{b^2}\right)^{-2} = \left(\dfrac{b^2}{2a^3}\right)^2$ property of negative exponents

$$= \dfrac{(b^2)^2}{2^2(a^3)^2}$$ power properties

$$= \dfrac{b^4}{4a^6}$$ result

b. $(3hk^{-2})^3(6h^{-2}k^{-3})^{-2} = 3^3h^3(k^{-2})^3 \cdot 6^{-2}(h^{-2})^{-2}(k^{-3})^{-2}$ power property

$$= 3^3h^3k^{-6} \cdot 6^{-2}h^4k^6$$ simplify

$$= 3^3 \cdot 6^{-2} \cdot h^{3+4} \cdot k^{-6+6}$$ product property

$$= \dfrac{27h^7k^0}{36}$$ simplify $\left(6^{-2} = \dfrac{1}{6^2} = \dfrac{1}{36}\right)$

$$= \dfrac{3h^7}{4}$$ result $(k^0 = 1)$

c. $(3x)^0 + 3x^0 + 3^{-2} = 1 + 3(1) + \dfrac{1}{3^2}$ zero exponent property; property of negative exponents

$$= 4 + \dfrac{1}{9}$$ simplify: $(3x)^0 = 1, 3x^0 = 3 \cdot 1 = 3$

$$= 4\dfrac{1}{9} = \dfrac{37}{9}$$ result

d. $\dfrac{(-2m^2n^3)^5}{(4mn^2)^3} = \dfrac{(-2)^5(m^2)^5(n^3)^5}{4^3m^3(n^2)^3}$ power property

$$= \dfrac{-32m^{10}n^{15}}{64m^3n^6}$$ simplify

$$= \dfrac{-1m^7n^9}{2}$$ quotient property

$$= -\dfrac{m^7n^9}{2}$$ result

Now try Exercises 27 through 66 ▶

WORTHY OF NOTE

Notice in Example 4(c), we have $(3x)^0 = (3 \cdot x)^0 = 1$, while $3x^0 = 3 \cdot x^0 = 3(1)$. This is another example of operations and grouping symbols working together: $(3x)^0 = 1$ because any *quantity* to the zero power is 1. However, for $3x^0$ there are no grouping symbols, so the exponent 0 acts only on the x and not the 3: $3x^0 = 3 \cdot x^0 = 3(1) = 3$.

Summary of Exponential Properties

For real numbers a and b, and integers m, n, p (excluding 0 raised to a nonpositive power)

Product property: $\qquad b^m \cdot b^n = b^{m+n}$

Power property: $\qquad (b^m)^n = b^{m \cdot n}$

Product to a power: $\qquad (a^m b^n)^p = a^{mp} \cdot b^{np}$

Quotient to a power: $\qquad \left(\dfrac{a^m}{b^n}\right)^p = \dfrac{a^{mp}}{b^{np}}(b \neq 0)$

Quotient property: $\qquad \dfrac{b^m}{b^n} = b^{m-n}(b \neq 0)$

Zero exponents: $\qquad b^0 = 1(b \neq 0)$

Negative exponents: $\qquad \dfrac{b^{-n}}{1} = \dfrac{1}{b^n}, \dfrac{1}{b^{-n}} = b^n, \left(\dfrac{a}{b}\right)^{-n} = \left(\dfrac{b}{a}\right)^n (a, b \neq 0)$

☑ **A.** You've just seen how we can apply properties of exponents

B. Exponents and Scientific Notation

In many technical and scientific applications, we encounter numbers that are either extremely large or very, very small. For example, the mass of the Moon is over 73 quintillion kilograms (73 followed by 18 zeroes), while the constant for universal gravitation contains 10 zeroes before the first nonzero digit. When computing with numbers of this magnitude, scientific notation has a distinct advantage over the common decimal notation (base-10 place values).

WORTHY OF NOTE

Recall that multiplying by 10's (or multiplying by 10^k, $k > 0$) shifts the decimal point to the right k places, making the number larger. Dividing by 10's (or multiplying by 10^{-k}, $k > 0$) shifts the decimal point to the left k places, making the number smaller.

Scientific Notation

A non-zero number written in scientific notation has the form

$$N \times 10^k$$

where $1 \leq |N| < 10$ and k is an integer.

To convert a number from decimal notation into scientific notation, we begin by placing the decimal point to the immediate right of the first nonzero digit (creating a number less than 10 but greater than or equal to 1) and multiplying by 10^k. Then we determine the power of 10 (the value of k) needed to ensure that the two forms are equivalent. When writing large or small numbers in scientific notation, we sometimes round the value of N to two or three decimal places.

EXAMPLE 5 ▶   **Converting from Decimal Notation to Scientific Notation**

The mass of the Moon is about 73,000,000,000,000,000,000 kg. Write this number in scientific notation.

Solution ▶   Place decimal to the right of first nonzero digit (7) and multiply by 10^k.

$$73{,}000{,}000{,}000{,}000{,}000{,}000 = 7.3 \times 10^k$$

To return the decimal to its original position would require 19 shifts to the *right*, so k must be *positive* 19.

$$73{,}000{,}000{,}000{,}000{,}000{,}000 = 7.3 \times 10^{19}$$

The mass of the Moon is 7.3×10^{19} kg.

Now try Exercises 67 and 68 ▶

Converting a number from scientific notation to decimal notation is simply an application of multiplication or division with powers of 10.

EXAMPLE 6 ▶ **Converting from Scientific Notation to Decimal Notation**

The constant of gravitation is 6.67×10^{-11}. Write this number in common decimal form.

Solution ▶ Since the exponent is *negative* 11, shift the decimal 11 *places to the left,* using placeholder zeroes as needed to return the decimal to its original position:

$$6.67 \times 10^{-11} = 0.000\,000\,000\,066\,7$$

Now try Exercises 69 and 70 ▶

Computations that involve scientific notation typically use real number properties and the properties of exponents.

EXAMPLE 7 ▶ **Storage Space on a Hard Drive**

A typical 320-gigabyte portable hard drive can hold about 340,000,000,000 bytes of information. A 2-hr DVD movie can take up as much as 8,000,000,000 bytes of storage space. Find the number of movies (to the nearest whole movie) that can be stored on this hard drive.

Solution ▶ Using the ideas from Example 5, the hard drive holds 3.4×10^{11} bytes, while the DVD requires 8.0×10^9 bytes. Divide to find the number of DVDs the hard drive will hold.

$$\frac{3.4 \times 10^{11}}{8.0 \times 10^9} = \frac{3.4}{8.0} \times \frac{10^{11}}{10^9} \quad \text{rewrite the expression}$$
$$= 0.425 \times 10^2 \quad \text{divide; subtract exponents}$$
$$= 42.5 \quad \text{result}$$

```
(3.4*10^11)/(8.0
*10^9)
               42.5
```

The drive will hold approximately 42 DVD movies. A calculator check is shown in the figure.

☑ **B.** You've just seen how we can perform operations in scientific notation

Now try Exercises 71 and 72 ▶

C. Identifying and Classifying Polynomial Expressions

A **monomial** is a term using *only whole number exponents* on variables, with no variables in the denominator. One important characteristic of a monomial is its **degree.** For a monomial in one variable, the degree is the same as the exponent *on the variable.* The degree of a monomial in two or more variables is the sum of exponents occurring on variable factors. A **polynomial** is a monomial or any sum or difference of monomial terms. For instance, $\frac{1}{2}x^2 - 5x + 6$ is a polynomial, while $3n^{-2} + 2n - 7$ is not (the exponent -2 is not a whole number). Identifying polynomials is an important skill because they represent a very different kind of real-world model than nonpolynomials. In addition, there are different **families of polynomials,** with each family having different characteristics. We classify polynomials according to their *degree* and *number of terms.* The **degree of a polynomial** in one variable is the largest exponent occurring on the variable. The degree of a polynomial in more than one variable is the largest sum of exponents in any one term. A polynomial with two terms is called a **binomial** (*bi* means two) and a polynomial with three terms is called a **trinomial** (*tri* means three). There are special names for polynomials with four or more terms, but for these, we simply use the general name *polynomial* (*poly* means many).

EXAMPLE 8 ▶ **Classifying and Describing Polynomials**

For each expression:

a. Classify as a monomial, binomial, trinomial, or polynomial.

b. State the degree of the polynomial.

c. Name the coefficient of each term.

Solution ▶

Expression	Classification	Degree	Coefficients
$5x^2y - 2xy$	binomial	three	$5, -2$
$x^2 - 0.81$	binomial	two	$1, -0.81$
$z^3 - 3z^2 + 9z - 27$	polynomial (four terms)	three	$1, -3, 9, -27$
$\frac{-3}{4}x + 5$	binomial	one	$\frac{-3}{4}, 5$
$2x^2 + x - 3$	trinomial	two	$2, 1, -3$

Now try Exercises 73 through 78 ▶

A polynomial expression is in **standard form** when the terms of the polynomial are written in *descending order of degree,* beginning with the highest-degree term. The coefficient of the highest-degree term is called the **leading coefficient.**

EXAMPLE 9 ▶ **Writing Polynomials in Standard Form**

Write each polynomial in standard form, then identify the leading coefficient.

Solution ▶

Polynomial	Standard Form	Leading Coefficient
$9 - x^2$	$-x^2 + 9$	-1
$5z + 7z^2 + 3z^3 - 27$	$3z^3 + 7z^2 + 5z - 27$	3
$2 + \left(\frac{-3}{4}\right)x$	$\frac{-3}{4}x + 2$	$\frac{-3}{4}$
$-3 + 2x^2 + x$	$2x^2 + x - 3$	2

☑ **C.** You've just seen how we can identify and classify polynomial expressions

Now try Exercises 79 through 84 ▶

D. Adding and Subtracting Polynomials

Adding polynomials simply involves using the distributive, commutative, and associative properties to combine like terms (at this point, the properties are usually applied mentally). As with real numbers, the subtraction of polynomials involves adding the opposite of the second polynomial using algebraic addition. This can be viewed as distributing -1 to the second polynomial and combining like terms.

EXAMPLE 10 ▶ **Adding and Subtracting Polynomials**

Perform the indicated operations:

$(0.7n^3 + 4n^2 + 8) + (0.5n^3 - n^2 - 6n) - (3n^2 + 7n - 10)$.

Solution ▶ $0.7n^3 + 4n^2 + 8 + 0.5n^3 - n^2 - 6n - 3n^2 - 7n + 10$ eliminate parentheses (distributive property)

$= 0.7n^3 + 0.5n^3 + 4n^2 - 1n^2 - 3n^2 - 6n - 7n + 8 + 10$ use properties to collect like terms

$= 1.2n^3 - 13n + 18$ combine like terms

Now try Exercises 85 through 90 ▶

Sometimes it's easier to add or subtract polynomials using a vertical format and aligning like terms. Note the use of a placeholder zero in Example 11.

EXAMPLE 11 ▶ **Subtracting Polynomials Using a Vertical Format**

Compute the difference of $x^3 - 5x + 9$ and $x^3 + 3x^2 + 2x - 8$ using a vertical format.

Solution ▶
$$
\begin{array}{ll}
x^3 + \mathbf{0}x^2 - 5x + 9 & x^3 + \mathbf{0}x^2 - 5x + 9 \\
\underline{-(x^3 + 3x^2 + 2x - 8)} \longrightarrow & \underline{-x^3 - 3x^2 - 2x + 8} \\
& -3x^2 - 7x + 17
\end{array}
$$

☑ **D.** You've just seen how we can add and subtract polynomials

The difference is $-3x^2 - 7x + 17$.

Now try Exercises 91 and 92 ▶

E. The Product of Two Polynomials

Monomial Times Monomial

The simplest case of polynomial multiplication is the product of monomials shown in Example 1a. These were computed using exponential properties and the properties of real numbers.

Monomial Times Polynomial

To compute the product of a monomial and a polynomial, we use the distributive property.

EXAMPLE 12 ▶ **Multiplying a Monomial by a Polynomial**

Find the product: $-2a^2(a^2 - 2a + 1)$.

Solution ▶
$$
\begin{aligned}
-2a^2(a^2 - 2a + 1) &= -2a^2(a^2) - (-2a^2)(2a^1) + (-2a^2)(1) \quad \text{distribute} \\
&= -2a^4 + 4a^3 - 2a^2 \quad \text{simplify}
\end{aligned}
$$

Now try Exercises 93 and 94 ▶

Binomial Times Polynomial

For products involving binomials, we still use a version of the distributive property—this time to distribute one polynomial to each term of the other polynomial factor. Note the distribution can be performed either from the left or from the right.

EXAMPLE 13 ▶ **Multiplying a Binomial by a Polynomial**

Multiply as indicated:

a. $(2z + 1)(z - 2)$ b. $(2v - 3)(4v^2 + 6v + 9)$

Solution ▶
$$
\begin{aligned}
\text{a. } (2z + 1)(z - 2) &= 2z(z - 2) + 1(z - 2) \quad &&\text{distribute to every term in the first binomial} \\
&= 2z^2 - 4z + 1z - 2 \quad &&\text{eliminate parentheses (distribute again)} \\
&= 2z^2 - 3z - 2 \quad &&\text{simplify}
\end{aligned}
$$

$$
\begin{aligned}
\text{b. } (2v - 3)(4v^2 + 6v + 9) &= 2v(4v^2 + 6v + 9) - 3(4v^2 + 6v + 9) \quad &&\text{distribute} \\
&= 8v^3 + 12v^2 + 18v - 12v^2 - 18v - 27 \quad &&\text{simplify} \\
&= 8v^3 - 27 \quad &&\text{combine like terms}
\end{aligned}
$$

Now try Exercises 95 through 100 ▶

The F-O-I-L Method

By observing the product of two binomials in Example 13(a), we note a pattern that can make the process more efficient. The product of two binomials can quickly be computed using the **F**irst, **O**uter, **I**nner, **L**ast (**FOIL**) method, an acronym giving the respective position of each term in a product of binomials in relation to the other terms. We illustrate here using the product $(2x - 1)(3x + 2)$.

> **WORTHY OF NOTE**
>
> Consider the product $(x + 3)(x + 2)$ in the context of *area*. If we view $x + 3$ as the length of a rectangle (an unknown length plus 3 units), and $x + 2$ as its width (the same unknown length plus 2 units), a diagram of the total area would look like the following, with the result $x^2 + 5x + 6$ clearly visible.
>
>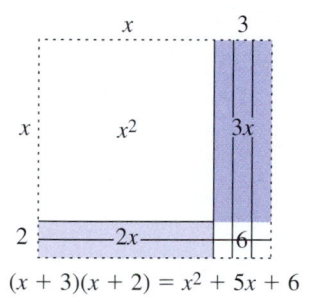
>
> $(x + 3)(x + 2) = x^2 + 5x + 6$

The F-O-I-L Method for Multiplying Binomials

$$(2x - 1)(3x + 2)$$

First, Outer, Inner, Last

$$6x^2 + 4x - 3x - 2$$
$$\underline{\text{First}} \quad \underline{\text{Outer}} \quad \underline{\text{Inner}} \quad \underline{\text{Last}}$$

Combine like terms
$$6x^2 + x - 2$$

The first term of the result will always be the product of the first terms from each binomial, and the last term of the result is the product of their last terms. We also note that here, the middle term is found by adding the *outermost product* with the *innermost product*. As you practice with the F-O-I-L process, much of the work can be done mentally and you can often compute the entire product without writing anything down except the answer.

EXAMPLE 14 ▶ **Multiplying Binomials Using F-O-I-L**

Compute each product mentally:

a. $(5n - 1)(n + 2)$

b. $(2b + 3)(5b - 6)$

Solution ▶

a. $(5n - 1)(n + 2)$: $5n^2 + 9n - 2$

$10n + (-1n) = 9n$

product of first two terms   sum of outer and inner products   product of last two terms

b. $(2b + 3)(5b - 6)$: $10b^2 + 3b - 18$

$-12b + 15b = 3b$

product of first two terms   sum of outer and inner products   product of last two terms

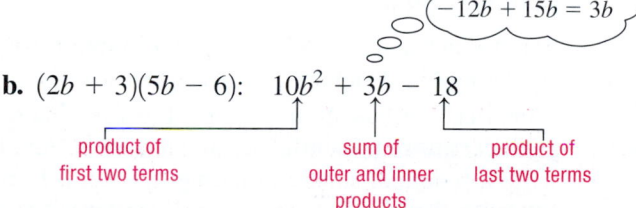

☑ **E.** You've just seen how we can compute the product of two polynomials

Now try Exercises 101 through 116 ▶

F. Special Polynomial Products

Certain polynomial products are considered "special" for two reasons: (1) the product follows a predictable pattern, and (2) the result can be used to simplify expressions, graph functions, solve equations, and/or develop other skills.

Binomial Conjugates

Expressions like $x + 7$ and $x - 7$ are called **binomial conjugates.** For any given binomial, its conjugate is found by using the same two terms with the opposite sign between them. Example 15 shows that when we multiply a binomial and its conjugate, the "outers" and "inners" sum to zero and the result is a **difference of two squares.**

EXAMPLE 15 ▶   **Multiplying Binomial Conjugates**

Compute each product mentally:

a. $(x + 7)(x - 7)$ **b.** $(2x - 5y)(2x + 5y)$ **c.** $\left(x + \dfrac{2}{5}\right)\left(x - \dfrac{2}{5}\right)$

$$-7x + 7x = 0x$$

Solution ▶   **a.** $(x + 7)(x - 7) = x^2 - 49$ difference of squares $(x)^2 - (7)^2$

$$10xy + (-10xy) = 0xy$$

b. $(2x - 5y)(2x + 5y) = 4x^2 - 25y^2$ difference of squares: $(2x)^2 - (5y)^2$

$$-\frac{2}{5}x + \frac{2}{5}x = 0$$

c. $\left(x + \dfrac{2}{5}\right)\left(x - \dfrac{2}{5}\right) = x^2 - \dfrac{4}{25}$ difference of squares: $x^2 - \left(\dfrac{2}{5}\right)^2$

Now try Exercises 117 through 124 ▶

In summary, we have the following.

The Product of a Binomial and Its Conjugate

Given any expression that can be written in the form $A + B$, the conjugate of the expression is $A - B$ and their product is a difference of two squares:

$$(A + B)(A - B) = A^2 - B^2$$

Binomial Squares

Expressions like $(x + 7)^2$ are called **binomial squares** and are useful for solving many equations and sketching a number of basic graphs. Note $(x + 7)^2 = (x + 7)(x + 7) = x^2 + 14x + 49$ using the F-O-I-L process. The expression $x^2 + 14x + 49$ is called a **perfect square trinomial** because it is the result of expanding a binomial square. If we write a binomial square in the more general form $(A + B)^2 = (A + B)(A + B)$ and compute the product, we notice a pattern that helps us write the expanded form more quickly.

$$(A + B)^2 = (A + B)(A + B) \qquad \text{repeated multiplication}$$
$$= A^2 + AB + AB + B^2 \qquad \text{F-O-I-L}$$
$$= A^2 + 2AB + B^2 \qquad \text{simplify (perfect square trinomial)}$$

The first and last terms of the trinomial are squares of the terms A and B. Also, the middle term of the trinomial is *twice the product of these two terms:* $AB + AB = 2AB$. The F-O-I-L process shows us why. Since the outer and inner products are identical, we always end up with two. A similar result holds for $(A - B)^2$ and the process can be summarized for both cases using the $\pm$ symbol.

LOOKING AHEAD

Although a binomial square can always be found using repeated factors and F-O-I-L, learning to expand them using the pattern is a valuable skill. Binomial squares occur often in a study of algebra and it helps to find the expanded form quickly.

The Square of a Binomial

Given any expression that can be written in the form $(A \pm B)^2$,

1. $(A + B)^2 = A^2 + 2AB + B^2$
2. $(A - B)^2 = A^2 - 2AB + B^2$

⚠ CAUTION ▶ Note the square of a binomial always results in a trinomial (three terms). In particular, $(A + B)^2 \neq A^2 + B^2$.

EXAMPLE 16 ▶ Find each binomial square without using F-O-I-L:

  a. $(a + 9)^2$ **b.** $(3x - 5)^2$ **c.** $(3 + \sqrt{x})^2$

Solution ▶   **a.** $(a + 9)^2 = a^2 + 2(a \cdot 9) + 9^2$ $(A + B)^2 = A^2 + 2AB + B^2$
                    $= a^2 + 18a + 81$ simplify

  b. $(3x - 5)^2 = (3x)^2 - 2(3x \cdot 5) + 5^2$ $(A - B)^2 = A^2 - 2AB + B^2$
                    $= 9x^2 - 30x + 25$ simplify

  c. $(3 + \sqrt{x})^2 = 9 + 2(3 \cdot \sqrt{x}) + (\sqrt{x})^2$ $(A + B)^2 = A^2 + 2AB + B^2$
                    $= 9 + 6\sqrt{x} + x$ simplify

Now try Exercises 125 through 136 ▶

☑ **F.** You've just seen how we can compute special products: binomial conjugates and binomial squares

With practice, you will be able to go directly from the binomial square to the resulting trinomial.

R.2 EXERCISES

▶ CONCEPTS AND VOCABULARY

Fill in each blank with the appropriate word or phrase. Carefully reread the section, if necessary.

1. The equation $(x^2)^3 = x^6$ is an example of the ___power___ property of exponents.

2. The equation $(x^3)^{-2} = \dfrac{1}{x^6}$ is an example of the property of ___negative___ exponents.

3. The sum of the "outers" and "inners" for $(2x + 5)^2$ is ___20x___, while the sum of the outers and inners for $(2x + 5)(2x - 5)$ is ___0___.

4. The expression $2x^2 - 3x - 10$ can be classified as a ___trinomial___ of degree ___2___, with a leading coefficient of ___2___.

5. Discuss/Explain why one of the following expressions can be simplified further, while the other cannot: (a) $-7n^4 + 3n^2$; (b) $-7n^4 \cdot 3n^2$.
   a. cannot be simplified, unlike terms b. can be simplified, like bases

6. Discuss/Explain why the degree of $2x^2y^3$ is greater than the degree of $2x^2 + y^3$. Include additional examples for contrast and comparison.
   Answers will vary, degree 5 and degree 3

▶ DEVELOPING YOUR SKILLS

Determine each product using the product and/or power properties.

7. $\dfrac{2}{3}n^2 \cdot 21n^5$ $14n^7$

8. $24g^5 \cdot \dfrac{3}{8}g^9$ $9g^{14}$

9. $(-6p^2q)(2p^3q^3)$ $-12p^5q^4$

10. $(-1.5vy^2)(-8v^4y)$ $12v^5y^3$

11. $(a^2)^4 \cdot (a^3)^2 \cdot b^2 \cdot b^5$ $a^{14}b^7$

12. $d^2 \cdot d^4 \cdot (c^5)^2 \cdot (c^3)^2$ d^6c^{16}

HOMEWORK SELECTION GUIDE

Core: 7–135 every other odd, 137, 139–144 (40 Exercises)
Standard: 1–4, 7–135 every other odd, 137, 139–144 (44 Exercises)

Extended: 1–4, 7–135 every other odd, 137–145 (46 Exercises)
In Depth: 1–6, 7–135 every other odd, 137–146 (49 Exercises)

Simplify using the product to a power property.

13. $(6pq^2)^3$ $216p^3q^6$

14. $(-3p^2q)^2$ $9p^4q^2$

15. $(3.2hk^2)^3$ $32.768h^3k^6$

16. $(-2.5h^5k)^2$ $6.25h^{10}k^2$

17. $\left(\dfrac{p}{2q}\right)^2$ $\dfrac{p^2}{4q^2}$

18. $\left(\dfrac{b}{3a}\right)^3$ $\dfrac{b^3}{27a^3}$

19. $(-0.7c^4)^2(10c^3d^2)^2$ $49c^{14}d^4$

20. $(-2.5a^3)^2(3a^2b^2)^3$ $168.75a^{12}b^6$

21. $(\tfrac{3}{4}x^3y)^2$ $\tfrac{9}{16}x^6y^2$

22. $(\tfrac{4}{5}x^3)^2$ $\tfrac{16}{25}x^6$

23. $(-\tfrac{3}{8}x)^2(16xy^2)$ $\tfrac{9}{4}x^3y^2$

24. $(\tfrac{2}{3}m^2n)^2 \cdot (\tfrac{1}{2}mn^2)$ $\tfrac{2}{9}m^5n^4$

25. **Volume of a cube:** The formula for the volume of a cube is $V = S^3$, where S is the length of one edge. If the length of each edge is $3x^2$,

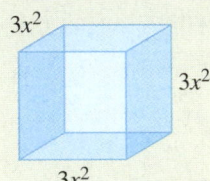

$3x^2$

$3x^2$

$3x^2$

   a. Find a formula for volume in terms of the variable x. $V = 27x^6$

   b. Find the volume of the cube if $x = 2$. 1728 units^3

26. **Area of a circle:** The formula for the area of a circle is $A = \pi r^2$, where r is the length of the radius. If the radius is given as $5x^3$,

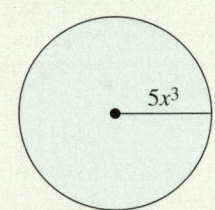

$5x^3$

   a. Find a formula for area in terms of the variable x. $A = 25\pi x^6$

   b. Find the area of the circle if $x = 2$. 1600π units^2

Simplify using the quotient property or the property of negative exponents. Write answers using positive exponents only.

27. $\dfrac{-6w^5}{-2w^2}$ $3w^3$

28. $\dfrac{8z^7}{16z^5}$ $\dfrac{z^2}{2}$

29. $\dfrac{-12a^3b^5}{4a^2b^4}$ $-3ab$

30. $\dfrac{5m^3n^5}{10mn^2}$ $\dfrac{m^2n^3}{2}$

31. $(\tfrac{2}{3})^{-3}$ $\dfrac{27}{8}$

32. $(\tfrac{5}{6})^{-1}$ $\dfrac{6}{5}$

33. $\dfrac{2}{h^{-3}}$ $2h^3$

34. $\dfrac{3}{m^{-2}}$ $3m^2$

35. $(-2)^{-3}$ $\dfrac{-1}{8}$

36. $(-4)^{-2}$ $\dfrac{1}{16}$

37. $(\tfrac{-1}{2})^{-3}$ -8

38. $(\tfrac{-2}{3})^{-2}$ $\dfrac{9}{4}$

Simplify each expression using the quotient to a power property.

39. $\left(\dfrac{2p^4}{q^3}\right)^2$ $\dfrac{4p^8}{q^6}$

40. $\left(\dfrac{-5v^4}{7w^3}\right)^2$ $\dfrac{25v^8}{49w^6}$

41. $\left(\dfrac{0.2x^2}{0.3y^3}\right)^3$ $\dfrac{8x^6}{27y^9}$

42. $\left(\dfrac{-0.5a^3}{0.4b^2}\right)^2$ $\dfrac{25a^6}{16b^4}$

43. $\left(\dfrac{5m^2n^3}{2r^4}\right)^2$ $\dfrac{25m^4n^6}{4r^8}$

44. $\left(\dfrac{4p^3}{3x^2y}\right)^3$ $\dfrac{64p^9}{27x^6y^3}$

45. $\left(\dfrac{5p^2q^3r^4}{-2pq^2r^4}\right)^2$ $\dfrac{25p^2q^2}{4}$

46. $\left(\dfrac{9p^3q^2r^3}{12p^5qr^2}\right)^3$ $\dfrac{27q^3r^3}{64p^6}$

Use properties of exponents to simplify the following. Write the answer using positive exponents only.

47. $\dfrac{9p^6q^4}{-12p^4q^6}$ $\dfrac{3p^2}{-4q^2}$

48. $\dfrac{5m^5n^2}{10m^5n}$ $\dfrac{n}{2}$

49. $\dfrac{20h^{-2}}{12h^5}$ $\dfrac{5}{3h^7}$

50. $\dfrac{5k^3}{20k^{-2}}$ $\dfrac{k^5}{4}$

51. $\dfrac{(a^2)^3}{a^4 \cdot a^5}$ $\dfrac{1}{a^3}$

52. $\dfrac{(5^3)^4}{5^9}$ 5^3 or 125

53. $\left(\dfrac{a^{-3} \cdot b}{c^{-2}}\right)^{-4}$ $\dfrac{a^{12}}{b^4c^8}$

54. $\dfrac{(p^{-4}q^8)^2}{p^5q^{-2}}$ $\dfrac{q^{18}}{p^{13}}$

55. $\dfrac{-6(2x^{-3})^2}{10x^{-2}}$ $\dfrac{-12}{5x^4}$

56. $\dfrac{18n^{-3}}{-8(3n^{-2})^3}$ $\dfrac{n^3}{-12}$

57. $\dfrac{14a^{-3}bc^0}{-7(3a^2b^{-2}c)^3}$ $\dfrac{-2b^7}{27a^9c^3}$

58. $\dfrac{-3(2x^3y^{-4}z)^2}{18x^{-2}yz^0}$ $\dfrac{2x^8z^2}{-3y^9}$

59. $4^0 + 5^0$ 2

60. $(-3)^0 + (-7)^0$ 2

61. $2^{-1} + 5^{-1}$ $\dfrac{7}{10}$

62. $4^{-1} + 8^{-1}$ $\dfrac{3}{8}$

63. $3^0 + 3^{-1} + 3^{-2}$ $\dfrac{13}{9}$

64. $2^{-2} + 2^{-1} + 2^0$ $\dfrac{7}{4}$

65. $-5x^0 + (-5x)^0$ -4

66. $-2n^0 + (-2n)^0$ -1

Convert the following numbers to scientific notation.

67. In mid-2009, the U.S. Census Bureau estimated the world population at nearly 6,770,000,000 people. 6.77×10^9

68. The mass of a proton is generally given as 0.000 000 000 000 000 000 000 000 001 670 kg. 1.67×10^{-27}

Convert the following numbers to decimal notation.

69. The smallest microprocessors in common use measure 6.5×10^{-9} m across. 0.000 000 006 5

70. In 2009, the estimated net worth of Bill Gates, the founder of Microsoft, was 5.8×10^{10} dollars. $58,000,000,000

Compute using scientific notation. Show all work.

71. The average distance between the Earth and the planet Jupiter is 465,000,000 mi. How many hours would it take a satellite to reach the planet if it traveled an average speed of 17,500 mi per hour? How many days? Round to the nearest whole. 26,571 hrs; 1107 days

72. In fiscal terms, a nation's debt-per-capita is the ratio of its total debt to its total population. In the year 2009, the total U.S. debt was estimated at $11,300,000,000,000, while the population was estimated at 305,000,000. What was the U.S. debt-per-capita ratio for 2009? Round to the nearest whole dollar. $37,049

Identify each expression as a polynomial or nonpolynomial (if a nonpolynomial, state why); classify each as a monomial, binomial, trinomial, or none of these; and state the degree of the polynomial.

73. $-35w^3 + 2w^2 + (-12w) + 14$

74. $-2x^3 + \frac{2}{3}x^2 - 12x + 1.2$

75. $5n^{-2} + 4n + \sqrt{17}$ 76. $\frac{4}{r^3} + 2.7r^2 + r + 1$

77. $p^3 - \frac{2}{5}$ 78. $q^3 + 2q^{-2} - 5q$

Write each polynomial in standard form and name the leading coefficient.

79. $7w + 8.2 - w^3 - 3w^2$ $-w^3 - 3w^2 + 7w + 8.2; -1$

80. $-2k^2 - 12 - k$ $-2k^2 - k - 12; -2$

81. $c^3 + 6 + 2c^2 - 3c$ $c^3 + 2c^2 - 3c + 6; 1$

82. $-3v^3 + 14 + 2v^2 + (-12v)$ $-3v^3 + 2v^2 - 12v + 14; -3$

83. $12 - \frac{2}{3}x^2$ $\frac{-2}{3}x^2 + 12; \frac{-2}{3}$

84. $8 + 2n^2 + 7n$ $2n^2 + 7n + 8; 2$

Find the indicated sum or difference.

85. $(3p^3 - 4p^2 + 2p - 7) + (p^2 - 2p - 5)$ $3p^3 - 3p^2 - 12$

86. $(5q^2 - 3q + 4) + (-3q^2 + 3q - 4)$ $2q^2$

87. $(5.75b^2 + 2.6b - 1.9) + (2.1b^2 - 3.2b)$ $7.85b^2 - 0.6b - 1.9$

88. $(0.4n^2 + 5n - 0.5) + (0.3n^2 - 2n + 0.75)$ $0.7n^2 + 3n + 0.25$

89. $(\frac{3}{4}x^2 - 5x + 2) - (\frac{1}{2}x^2 + 3x - 4)$ $\frac{1}{4}x^2 - 8x + 6$

90. $(\frac{5}{9}n^2 + 4n - \frac{1}{2}) - (\frac{2}{3}n^2 - 2n + \frac{3}{4})$ $-\frac{1}{9}n^2 + 6n - \frac{5}{4}$

91. Subtract $q^5 + 2q^4 + q^2 + 2q$ from $q^6 + 2q^5 + q^4 + 2q^3$ using a vertical format. $q^6 + q^5 - q^4 + 2q^3 - q^2 - 2q$

92. Find $x^4 + 2x^3 + x^2 + 2x$ decreased by $x^4 - 3x^3 + 4x^2 - 3x$ using a vertical format. $5x^3 - 3x^2 + 5x$

Compute each product.

93. $-3x(x^2 - x - 6)$ $-3x^3 + 3x^2 + 18x$

94. $-2v^2(v^2 + 2v - 15)$ $-2v^4 - 4v^3 + 30v^2$

95. $(3r - 5)(r - 2)$ $3r^2 - 11r + 10$

96. $(s - 3)(5s + 4)$ $5s^2 - 11s - 12$

97. $(x - 3)(x^2 + 3x + 9)$ $x^3 - 27$

98. $(z + 5)(z^2 - 5z + 25)$ $z^3 + 125$

99. $(b^2 - 3b - 28)(b + 2)$ $b^3 - b^2 - 34b - 56$

100. $(2h^2 - 3h + 8)(h - 1)$ $2h^3 - 5h^2 + 11h - 8$

101. $(7v - 4)(3v - 5)$ 102. $(6w - 1)(2w + 5)$

103. $(3 - m)(3 + m)$ 104. $(5 + n)(5 - n)$

105. $(p - 2.5)(p + 3.6)$ 106. $(q - 4.9)(q + 1.2)$

107. $(x + \frac{1}{2})(x + \frac{1}{4})$ 108. $(z + \frac{1}{3})(z + \frac{5}{6})$

109. $(m + \frac{3}{4})(m - \frac{3}{4})$ 110. $(n - \frac{2}{5})(n + \frac{2}{5})$

111. $(3x - 2y)(2x + 5y)$ 112. $(6a + b)(a + 3b)$

113. $(4c + d)(3c + 5d)$ 114. $(5x + 3y)(2x - 3y)$

115. $(2x^2 + 5)(x^2 - 3)$ 116. $(3y^2 - 2)(2y^2 + 1)$

For each binomial, determine its conjugate and find the product of the binomial with its conjugate.

117. $4m - 3$ $4m + 3; 16m^2 - 9$ 118. $6n + 5$ $6n - 5; 36n^2 - 25$

119. $7x - 10$ $7x + 10; 49x^2 - 100$ 120. $c + 3$ $c - 3; c^2 - 9$

121. $6 + 5k$ $6 - 5k; 36 - 25k^2$ 122. $11 - 3r$ $11 + 3r; 121 - 9r^2$

123. $x + \sqrt{6}$ $x - \sqrt{6}; x^2 - 6$ 124. $p - \sqrt{2}$ $p + \sqrt{2}; p^2 - 2$

Find each binomial square.

125. $(x + 4)^2$ $x^2 + 8x + 16$ 126. $(a - 3)^2$ $a^2 - 6a + 9$

127. $(4g + 3)^2$ $16g^2 + 24g + 9$ 128. $(5x - 3)^2$ $25x^2 - 30x + 9$

129. $(4p - 3q)^2$ $16p^2 - 24pq + 9q^2$ 130. $(5c + 6d)^2$ $25c^2 + 60cd + 36d^2$

131. $(4 - \sqrt{x})^2$ $16 - 8\sqrt{x} + x$ 132. $(\sqrt{x} + 7)^2$ $x + 14\sqrt{x} + 49$

Compute each product.

133. $(x - 3)(y + 2)$ $xy + 2x - 3y - 6$

134. $(a + 3)(b - 5)$ $ab - 5a + 3b - 15$

135. $(k - 5)(k + 6)(k + 2)$ $k^3 + 3k^2 - 28k - 60$

136. $(a + 6)(a - 1)(a + 5)$ $a^3 + 10a^2 + 19a - 30$

▶ WORKING WITH FORMULAS

137. **Medication in the bloodstream:** $M = 0.5t^4 + 3t^3 - 97t^2 + 348t$

If 400 mg of a pain medication are taken orally, the number of milligrams in the bloodstream is modeled by the formula shown, where M is the number of milligrams and t is the time in hours, $0 \le t < 5$. Construct a table of values for $t = 1$ through 5, then answer the following.

a. How many milligrams are in the bloodstream after 2 hr? After 3 hr?   340 mg, 292.5 mg

b. Based on part a, would you expect the number of milligrams in the bloodstream after 4 hr to be less or more? Why?   Less, amount is decreasing.

c. Approximately how many hours until the medication wears off (the number of milligrams in the bloodstream is 0)?   after 5 hr

138. Amount of a mortgage payment: $M = \dfrac{A\left(\dfrac{r}{12}\right)\left(1 + \dfrac{r}{12}\right)^n}{\left(1 + \dfrac{r}{12}\right)^n - 1}$

The monthly mortgage payment required to pay off (or amortize) a loan is given by the formula shown, where M is the monthly payment, A is the original amount of the loan, r is the annual interest rate, and n is the term of the loan in months. Find the monthly payment (to the nearest cent) required to purchase a $198,000 home, if the interest rate is 6.5% and the home is financed over 30 yr. $1251.49

▶ **APPLICATIONS**

139. Attraction between particles: In electrical theory, the force of attraction between two particles P and Q with opposite charges is modeled by $F = \dfrac{kPQ}{d^2}$, where d is the distance between them and k is a constant that depends on certain conditions. This is known as Coulomb's law. Rewrite the formula using a negative exponent. $F = kPQd^{-2}$

140. Intensity of light: The intensity of illumination from a light source depends on the distance from the source according to $I = \dfrac{k}{d^2}$, where I is the intensity measured in footcandles, d is the distance from the source in feet, and k is a constant that depends on the conditions. Rewrite the formula using a negative exponent. $I = kd^{-2}$

141. Rewriting an expression: In advanced mathematics, negative exponents are widely used because they are easier to work with than rational expressions. Rewrite the expression $\dfrac{5}{x^3} + \dfrac{3}{x^2} + \dfrac{2}{x^1} + 4$ using negative exponents. $5x^{-3} + 3x^{-2} + 2x^{-1} + 4$

142. Swimming pool hours: A swimming pool opens at 8 A.M. and closes at 6 P.M. In summertime, the

number of people in the pool at any time can be approximated by the formula $S(t) = -t^2 + 10t$, where S is the number of swimmers and t is the number of hours the pool has been open (8 A.M.: $t = 0$, 9 A.M.: $t = 1$, 10 A.M.: $t = 2$, etc.).

  a. How many swimmers are in the pool at 6 P.M.? Why?   0, pool closes at 6 P.M.

  b. Between what times would you expect the largest number of swimmers?   between noon and 2 P.M.

  c. Approximately how many swimmers are in the pool at 3 P.M.?   21

  d. Create a table of values for $t = 1, 2, 3, 4, \ldots$ and check your answer to part b.

 143. Maximizing revenue: A sporting goods store finds that if they price their video games at $20, they make 200 sales per day. For each decrease of $1, 20 additional video games are sold. This means the store's revenue can be modeled by the formula $R = (20 - 1x)(200 + 20x)$, where x is the number of $1 decreases. Multiply out the binomials and use a table of values to determine what price will give the most revenue. $15

 144. Maximizing revenue: Due to past experience, a jeweler knows that if they price jade rings at $60, they will sell 120 each day. For each decrease of $2, five additional sales will be made. This means the jeweler's revenue can be modeled by the formula $R = (60 - 2x)(120 + 5x)$, where x is the number of $2 decreases. Multiply out the binomials and use a table of values to determine what price will give the most revenue. $54

▶ **EXTENDING THE CONCEPT**

145. If $(3x^2 + kx + 1) - (kx^2 + 5x - 7) + (2x^2 - 4x - k) = -x^2 - 3x + 2$, what is the value of k?   6

146. If $\left(2x + \dfrac{1}{2x}\right)^2 = 5$, then the expression $4x^2 + \dfrac{1}{4x^2}$ is equal to what number?   3

R.3 Solving Linear Equations and Inequalities

LEARNING OBJECTIVES

In Section R.3 you will review how to:

- ☐ **A.** Solve linear equations using properties of equality
- ☐ **B.** Recognize equations that are identities or contradictions
- ☐ **C.** Solve linear inequalities
- ☐ **D.** Solve compound inequalities
- ☐ **E.** Solve basic applications of linear equations and inequalities
- ☐ **F.** Solve applications of basic geometry

In a study of algebra, you will encounter many **families of equations,** or groups of equations that share common characteristics. Of interest to us here is the family of **linear equations in one variable,** a study that lays the foundation for understanding more advanced families. This section will also lay the foundation for solving a formula for a specified variable, a practice widely used in science, business, industry, and research.

A. Solving Linear Equations Using Properties of Equality

An **equation** is *a statement that two expressions are equal.* From the expressions $3(x - 1) + x$ and $-x + 7$, we can form the equation

$$3(x - 1) + x = -x + 7,$$

which is a **linear equation in one variable** (the exponent on any variable is a 1). To solve an equation, we attempt to find a specific input or x-value that will make the equation true, meaning the left-hand expression will be equal to the right. Using Table R.1, we find that $3(x - 1) + x = -x + 7$ is a true equation when x is replaced by 2, and is a false equation otherwise. Replacement values that make the equation true are called **solutions** or **roots** of the equation.

Table R.1

x	$3(x - 1) + x$	$-x + 7$
-2	-11	9
-1	-7	8
0	-3	7
1	1	6
2	**5**	**5**
3	9	4
4	13	3

⚠ **CAUTION** ▶ From Section R.1, an algebraic *expression* is a sum or difference of algebraic terms. Algebraic expressions can be simplified, evaluated or written in an equivalent form, but cannot be "*solved*," since we're not seeking a specific value of the unknown.

Solving equations using a table is too time consuming to be practical. Instead we attempt to write a sequence of **equivalent equations,** each one simpler than the one before, until we reach a point where the solution is obvious. Equivalent equations are those that have the same solution set, and can be obtained by using the distributive property to simplify the expressions on each side of the equation. The additive and multiplicative properties of equality are then used to obtain an equation of the form $x = $ constant.

The Additive Property of Equality	**The Multiplicative Property of Equality**
If A, B, and C represent algebraic expressions and $A = B$,	If A, B, and C represent algebraic expressions and $A = B$,
then $A + C = B + C$	then $AC = BC$ and $\dfrac{A}{C} = \dfrac{B}{C}, (C \neq 0)$

In words, the additive property says that like quantities, numbers, or terms can be added to both sides of an equation. A similar statement can be made for the multiplicative property. These properties are combined into a general guide for solving linear equations, which you've likely encountered in your previous studies. Note that not all steps in the guide are required to solve every equation.

> **Guide to Solving Linear Equations in One Variable**
>
> - Eliminate parentheses using the distributive property, then combine any like terms.
> - Use the additive property of equality to write the equation with all variable terms on one side, and all constants on the other. Simplify each side.
> - Use the multiplicative property of equality to obtain an equation of the form $x =$ constant.
> - For applications, answer in a complete sentence and include any units of measure indicated.

For our first example, we'll use the equation $3(x - 1) + x = -x + 7$ from our initial discussion.

EXAMPLE 1 ▶ **Solving a Linear Equation Using Properties of Equality**

Solve for x: $3(x - 1) + x = -x + 7$.

Solution ▶

$$3(x - 1) + x = -x + 7 \quad \text{original equation}$$
$$3x - 3 + x = -x + 7 \quad \text{distributive property}$$
$$4x - 3 = -x + 7 \quad \text{combine like terms}$$
$$5x - 3 = 7 \quad \text{add } x \text{ to both sides (additive property of equality)}$$
$$5x = 10 \quad \text{add 3 to both sides (additive property of equality)}$$
$$x = 2 \quad \text{multiply both sides by } \tfrac{1}{5} \text{ or divide both sides by 5 (multiplicative property of equality)}$$

As we noted in Table R.1, the solution is $x = 2$.

Now try Exercises 7 through 12 ▶

To check a solution by substitution means we substitute the solution back into the original equation (this is sometimes called **back-substitution**), and verify the left-hand side is equal to the right. For Example 1 we have:

$$3(x - 1) + x = -x + 7 \quad \text{original equation}$$
$$3(2 - 1) + 2 = -2 + 7 \quad \text{substitute 2 for } x$$
$$3(1) + 2 = 5 \quad \text{simplify}$$
$$5 = 5 ✓ \quad \text{solution checks}$$

If any coefficients in an equation are fractional, multiply both sides by the least common denominator (LCD) to *clear the fractions*. Since any decimal number can be written in fraction form, the same idea can be applied to decimal coefficients.

EXAMPLE 2 ▶ **Solving a Linear Equation with Fractional Coefficients**

Solve for n: $\frac{1}{4}(n + 8) - 2 = \frac{1}{2}(n - 6)$.

Solution ▶

$$\frac{1}{4}(n + 8) - 2 = \frac{1}{2}(n - 6) \quad \text{original equation}$$
$$\frac{1}{4}n + 2 - 2 = \frac{1}{2}n - 3 \quad \text{distributive property}$$
$$\frac{1}{4}n = \frac{1}{2}n - 3 \quad \text{combine like terms}$$
$$4(\tfrac{1}{4}n) = 4(\tfrac{1}{2}n - 3) \quad \text{multiply both sides by LCD} = 4$$
$$n = 2n - 12 \quad \text{distributive property}$$
$$-n = -12 \quad \text{subtract } 2n$$
$$n = 12 \quad \text{multiply by } -1$$

☑ **A.** You've just seen how we can solve linear equations using properties of equality

Verify the solution is $n = 12$ using back-substitution.

Now try Exercises 13 through 30 ▶

B. Identities and Contradictions

Example 1 illustrates what is called a **conditional equation,** since the equation is true for $x = 2$, but false for all other values of x. The equation in Example 2 is also conditional. An **identity** is an equation that is *always true,* no matter what value is substituted for the variable. For instance, $2(x + 3) = 2x + 6$ is an identity with a solution set of all real numbers, written as $\{x|x \in \mathbb{R}\}$, or $x \in (-\infty, \infty)$ in interval notation. **Contradictions** are equations that are *never true,* no matter what real number is substituted for the variable. The equations $x - 3 = x + 1$ and $-3 = 1$ are contradictions. To state the solution set for a contradiction, we use the symbol "$\varnothing$" (the null set) or "{ }" (the empty set). Recognizing these special equations will prevent some surprise and indecision in later chapters.

EXAMPLE 3 ▶   Solving Equations (Special Cases)

Solve each equation and state the solution set.

    a. $2(x - 4) + 10x = 8 + 4(3x + 1)$ **b.** $8x - (6 - 10x) = 24 + 6(3x - 5)$

Solution ▶

    a. $2(x - 4) + 10x = 8 + 4(3x + 1)$ original equation

        $2x - 8 + 10x = 8 + 12x + 4$ distributive property

            $12x - 8 = 12x + 12$ combine like terms

                 $-8 = 12$ subtract 12x; contradiction

Since -8 is never equal to 12, the original equation is a contradiction. The solution set is empty: { }

    b. $8x - (6 - 10x) = 24 + 6(3x - 5)$ original equation

        $8x - 6 + 10x = 24 + 18x - 30$ distributive property

           $18x - 6 = 18x - 6$ combine like terms

                $-6 = -6$ subtract 18x; identity

The result shows that the original equation is an identity, with an infinite number of solutions: $\{x \mid x \in \mathbb{R}\}$. You may recall this notation is read, "the set of all numbers x, such that x is a real number."

Now try Exercises 31 through 36 ▶

In Example 3(a), our attempt to solve for x ended with all variables being eliminated, leaving an equation that is *always false*—a contradiction (-8 is never equal to 12). There is nothing wrong with the solution process, the result is simply telling us the original equation has *no solution.* In Example 3(b), all variables were again eliminated but the end result was *always true*—an identity (-6 is always equal to -6). Once again we've done nothing wrong mathematically, the result is just telling us that the original equation will be true no matter what value of x we use for an input.

☑ **B. You've just seen how we can recognize equations that are identities or contradictions**

C. Solving Linear Inequalities

A linear *inequality* resembles a linear *equality* in many respects:

	Linear Inequality	**Related Linear Equation**
(1)	$x < 3$	$x = 3$
(2)	$\frac{3}{8}p - 2 \geq -12$	$\frac{3}{8}p - 2 = -12$

    A linear inequality in one variable is one that can be written in the form $ax + b < c$, where a, b, and $c \in \mathbb{R}$ and $a \neq 0$. This definition and the following properties also apply when other inequality symbols are used. Solutions to simple inequalities are easy to spot. For instance, $x = -2$ is a solution to $x < 3$ since $-2 < 3$. For more involved inequalities we use the **additive property of inequality** and the **multiplicative property of**

inequality. Similar to solving equations, we solve inequalities by isolating the variable on one side to obtain a solution form such as *variable < number*.

The Additive Property of Inequality

If A, B, and C represent algebraic expressions and $A < B$,

$$\text{then } A + C < B + C$$

Like quantities (numbers or terms) can be added to both sides of an inequality.

While there is little difference between the additive property of *equality* and the additive property of *inequality,* there is an *important difference* between the multiplicative property of *equality* and the multiplicative property of *inequality.* To illustrate, we begin with $-2 < 5$. Multiplying both sides by positive three yields $-6 < 15$, a true inequality. But notice what happens when we **multiply both sides by negative three:**

$$-2 < 5 \qquad \text{\color{magenta}original inequality}$$
$$-2(-3) < 5(-3) \qquad \text{\color{magenta}multiply by negative three}$$
$$6 < -15 \qquad \text{\color{magenta}false}$$

The result is a *false* inequality, because 6 is *to the right* of -15 on the number line. Multiplying (or dividing) an inequality by a negative quantity *reverses the order relationship between two quantities* (we say it changes the *sense* of the inequality). We must compensate for this by reversing the inequality symbol.

$$6 > -15 \qquad \text{\color{magenta}change direction of symbol to maintain a true statement}$$

For this reason, the multiplicative property of inequality is stated in two parts.

The Multiplicative Property of Inequality

If A, B, and C represent algebraic expressions and $A < B$, $\qquad$ then $AC < BC$	If A, B, and C represent algebraic expressions and $A < B$, $\qquad$ then $AC > BC$
if C is a *positive quantity* (inequality symbol remains the same).	if C is a *negative quantity* (inequality symbol must be reversed).

EXAMPLE 4 ▶ **Solving an Inequality**

Solve the inequality, then graph the solution set and write it in interval notation: $\frac{-2}{3}x + \frac{1}{2} \le \frac{5}{6}$.

Solution ▶

$$\frac{-2}{3}x + \frac{1}{2} \le \frac{5}{6} \qquad \text{\color{magenta}original inequality}$$

$$6\left(\frac{-2}{3}x + \frac{1}{2}\right) \le (6)\frac{5}{6} \qquad \text{\color{magenta}clear fractions (multiply by LCD)}$$

$$-4x + 3 \le 5 \qquad \text{\color{magenta}simplify}$$

$$-4x \le 2 \qquad \text{\color{magenta}subtract 3}$$

$$\frac{-4x}{-4} \ge \frac{2}{-4} \qquad \text{\color{magenta}divide by } -4, \text{ reverse inequality sign}$$

$$x \ge -\frac{1}{2} \qquad \text{\color{magenta}result}$$

• Graph:

• Interval notation: $x \in \left[-\frac{1}{2}, \infty\right)$

WORTHY OF NOTE

As an alternative to multiplying or dividing by a negative value, the additive property of inequality can be used to ensure the variable term will be positive. From Example 4, the inequality $-4x \le 2$ can be written as $-2 \le 4x$ by adding $4x$ to both sides and subtracting 2 from both sides. This gives the solution $-\frac{1}{2} \le x$, which is equivalent to $x \ge -\frac{1}{2}$.

Now try Exercises 37 through 46 ▶

To check a linear inequality, you often have an infinite number of choices—any number from the solution set/interval. If a test value from the solution interval results in a true inequality, all numbers in the interval are solutions. For Example 4, using $x = 0$ results in the true statement $\frac{1}{2} \leq \frac{5}{6}$ ✓.

Some inequalities have all real numbers as the solution set: $\{x | x \in \mathbb{R}\}$, while other inequalities have no solutions, with the answer given as the empty set: { }.

EXAMPLE 5 ▶ **Solving Inequalities**

Solve the inequality and write the solution in set notation:

    a. $7 - (3x + 5) \geq 2(x - 4) - 5x$ **b.** $3(x + 4) - 5 < 2(x - 3) + x$

Solution ▶

a. $7 - (3x + 5) \geq 2(x - 4) - 5x$ original inequality

     $7 - 3x - 5 \geq 2x - 8 - 5x$ distributive property

        $-3x + 2 \geq -3x - 8$ combine like terms

              $2 \geq -8$ add 3x

Since the resulting statement is always true, the original inequality is true for all real numbers. The solution is all real numbers $\mathbb{R}$.

b. $3(x + 4) - 5 < 2(x - 3) + x$ original inequality

     $3x + 12 - 5 < 2x - 6 + x$ distribute

        $3x + 7 < 3x - 6$ combine like terms

           $7 < -6$ subtract 3x

Since the resulting statement is always false, the original inequality is false for all real numbers. The solution is { }.

☑ **C.** You've just seen how we can solve linear inequalities

Now try Exercises 47 through 52 ▶

D. Solving Compound Inequalities

In some applications of inequalities, we must consider more than one solution interval. These are called **compound inequalities,** and these require us to take a close look at the operations of **union** "∪" and **intersection** "∩". The intersection of two sets A and B, written $A \cap B$, is the set of all elements *common to both sets*. The union of two sets A and B, written $A \cup B$, is the set of all elements *that are in either set*. When stating the union of two sets, repetitions are unnecessary.

EXAMPLE 6 ▶ **Finding the Union and Intersection of Two Sets**

For set $A = \{-2, -1, 0, 1, 2, 3\}$ and set $B = \{1, 2, 3, 4, 5\}$, determine $A \cap B$ and $A \cup B$.

Solution ▶

$A \cap B$ is the set of all elements in *both A and B*:
$A \cap B = \{1, 2, 3\}$.

$A \cup B$ is the set of all elements in *either A or B*:
$A \cup B = \{-2, -1, 0, 1, 2, 3, 4, 5\}$.

Now try Exercises 53 through 58 ▶

WORTHY OF NOTE

For the long term, it may help to rephrase the distinction as follows. The intersection is a *selection* of elements that are common to two sets, while the union is a *collection* of the elements from two sets (with no repetitions).

Notice the intersection of two sets is described using the word "and," while the union of two sets is described using the word "or." When compound inequalities are formed using these words, the solution is modeled after the ideas from Example 6. If "and" is used, the solutions must satisfy *both* inequalities. If "or" is used, the solutions can satisfy *either* inequality.

EXAMPLE 7 ▶ **Solving a Compound Inequality**

Solve the compound inequality, then write the solution in interval notation:
$-3x - 1 < -4$ **or** $4x + 3 < -6$.

Solution ▶ Begin with the statement as given:

$$-3x - 1 < -4 \quad \text{or} \quad 4x + 3 < -6 \qquad \text{original statement}$$
$$-3x < -3 \quad \text{or} \quad 4x < -9 \qquad \text{isolate variable term}$$
$$x > 1 \quad \text{or} \quad x < -\frac{9}{4} \qquad \text{solve for } x, \text{ reverse first inequality symbol}$$

The solution $x > 1$ **or** $x < -\frac{9}{4}$ is better understood by graphing each interval separately, *then selecting both intervals (the union).*

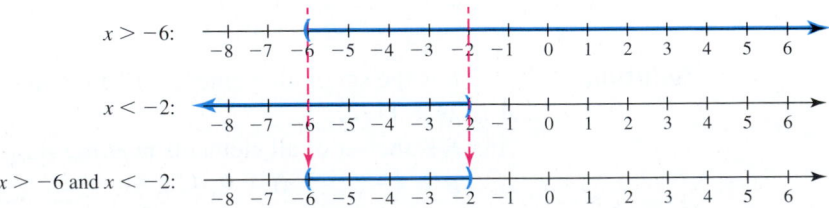

Interval notation: $x \in \left(-\infty, -\frac{9}{4}\right) \cup (1, \infty)$.

Now try Exercises 59 and 60 ▶

WORTHY OF NOTE

The graphs from Example 7 clearly show the solution consists of two disjoint (disconnected) intervals. This is reflected in the "or" statement: $x < -\frac{9}{4}$ or $x > 1$, and in the interval notation. Also, note the solution $x < -\frac{9}{4}$ or $x > 1$ is not equivalent to $-\frac{9}{4} > x > 1$, as there is no single number that is both greater than 1 and less than $-\frac{9}{4}$ at the same time.

EXAMPLE 8 ▶ **Solving a Compound Inequality**

Solve the compound inequality, then write the solution in interval notation:
$3x + 5 > -13$ **and** $3x + 5 < -1$.

Solution ▶ Begin with the statement as given:

$$3x + 5 > -13 \quad \text{and} \quad 3x + 5 < -1 \qquad \text{original statement}$$
$$3x > -18 \quad \text{and} \quad 3x < -6 \qquad \text{subtract five}$$
$$x > -6 \quad \text{and} \quad x < -2 \qquad \text{divide by 3}$$

The solution $x > -6$ **and** $x < -2$ can best be understood by graphing each interval separately, then *noting where they intersect.*

Interval notation: $x \in (-6, -2)$.

Now try Exercises 61 through 72 ▶

WORTHY OF NOTE

The inequality $a < b$ (a is less than b) can equivalently be written as $b > a$ (b is greater than a). In Example 8, the solution is read, "$x > -6$ and $x < -2$," but if we rewrite the first inequality as $-6 < x$ (with the "arrowhead" still pointing at -6), we have $-6 < x$ and $x < -2$ and can clearly see that x must be in the single interval between -6 and -2.

The solution from Example 8 consists of the single interval $(-6, -2)$, indicating the original inequality could actually be *joined* and written as $-6 < x < -2$, called a **joint inequality.** We solve joint inequalities in much the same way as linear inequalities, but must remember they *have three parts (left, middle, and right).* This means operations must be applied to *all three parts* in each step of the solution process, to obtain a solution form such as *smaller number* $< x <$ *larger number.* The same ideas apply when other inequality symbols are used.

EXAMPLE 9 ▶ **Solving a Joint Inequality**

Solve the joint inequality, then graph the solution set and write it in interval notation: $1 > \dfrac{2x + 5}{-3} \geq -6$.

Solution ▶ $1 > \dfrac{2x + 5}{-3} \geq -6$ original inequality

$-3 < 2x + 5 \leq 18$ multiply all parts by -3; reverse the inequality symbols

$-8 < 2x \leq 13$ subtract 5 from all parts

$-4 < x \leq \dfrac{13}{2}$ divide all parts by 2

• Graph:

• Interval notation: $x \in \left(-4, \dfrac{13}{2}\right]$

☑ **D.** You've just seen how we can solve compound inequalities

Now try Exercises 73 through 78 ▶

E. Solving Basic Applications of Linear Equations and Inequalities

Applications of linear equations and inequalities come in many forms. In most cases, you are asked to translate written relationships or information given verbally into an equation using words or phrases that indicate mathematical operations or relationships. Here, we'll practice this skill using ideas that were introduced in Section R.1, where we translated English phrases into mathematical expressions. Very soon these skills will be applied in much more significant ways.

EXAMPLE 10 ▶ **Translating Written Information into an Equation**

Translate the following relationships into equations, then solve:

In an effort to lower the outstanding balance on her credit card, Laura paid $10 less than triple her normal payment. If she sent the credit card company $350.75, how much was her normal payment? (See Section R.1, Examples 2 and 3.)

Solution ▶   Let p represent her normal payment. Then "triple her normal payment" would be $3p$, and "ten less than triple" would be $3p - 10$. Since "she sent the company $350.75," we have

$$3p - 10 = 350.75 \quad \text{equation form}$$
$$3p = 360.75 \quad \text{add 10}$$
$$p = 120.25 \quad \text{divide by 3}$$

Laura's normal payment is $120.25 per month.
A calculator check is shown in the figure.

```
120.25→X
            120.25
3X-10
            350.75
```

> **Now try Exercises 83 through 90 ▶**

Inequalities are widely used to help gather information, and to make comparisons that will lead to informed decisions.

EXAMPLE 11 ▶   **Using an Inequality to Compute Desired Test Scores**

Justin earned scores of 78, 72, and 86 on the first three out of four exams. What score must he earn on the fourth exam to have an average of at least 80?

Solution ▶   The current scores are 78, 72, and 86. An average of *at least* 80 means $A \geq 80$. In organized form:

Test 1	Test 2	Test 3	Test 4	Computed Average	Minimum
78	72	86	x	$\dfrac{78 + 72 + 86 + x}{4}$	80

Let x represent Justin's score on the fourth exam, then $\dfrac{78 + 72 + 86 + x}{4}$ represents his average score.

$$\frac{78 + 72 + 86 + x}{4} \geq 80 \quad \text{average must be greater than or equal to 80}$$
$$78 + 72 + 86 + x \geq 320 \quad \text{multiply by 4}$$
$$236 + x \geq 320 \quad \text{simplify}$$
$$x \geq 84 \quad \text{solve for } x \text{ (subtract 236)}$$

Justin must score at least an 84 on the last exam to earn an 80 average.

☑ **E.** You've just seen how we can solve applications of linear equations and inequalities

> **Now try Exercises 91 through 100 ▶**

F. Solving Applications of Basic Geometry

As your translation skills grow, your ability to solve a wider range of more significant applications will grow as well. In many cases, the applications will involve some basic geometry and the most often used figures and formulas appear here. For a more complete review of geometry, see *Appendix II Geometry Review with Unit Conversions,* which is posted online at www.mhhe.com/coburn.

Perimeter and Area

Perimeter is a measure of the distance around a two dimensional figure. As this is a linear measure, results are stated in linear units as in centimeters (cm), feet (ft), kilometers (km), miles (mi), and so on. If no unit is specified, simply write the result as x *units*. Area is a measure of the surface of a two dimensional figure, with results stated in square units as in x *units*^{2}. Some of the most common formulas involving perimeter and area are given in Table R.2A.

Table R.2A

	Definition and Diagram	Perimeter Formula (linear units or *units*)	Area Formula (square units or *units²*)
Triangle	a three-sided polygon s_1 s_2 s_3 h b	$P = s_1 + s_2 + s_3$	$A = \dfrac{1}{2}bh$
Rectangle	a quadrilateral with four right angles and opposite sides parallel W L	$P = 2L + 2W$	$A = LW$
Square	a rectangle with four equal sides S	$P = 4S$	$A = S^2$
Trapezoid	a quadrilateral with one pair of parallel sides (called bases b_1 and b_2) s_2 s_1 s_3 s_4 b_1 h b_2	sum of all sides $P = s_1 + s_2 + s_3 + s_4$	$A = \dfrac{h}{2}(b_1 + b_2)$
Circle	the set of all points lying in a plane that are an equal distance (called the radius r) from a given point (called the center C). r C	$C = 2\pi r$ or $C = \pi d$	$A = \pi r^2$

If an exercise or application uses a formula, begin by *stating the formula first*. Using the formula as a template for the values substituted will help to prevent many careless errors.

EXAMPLE 12A ▶   Computing the Area of a Trapezoidal Window

A basement window is shaped like an isosceles trapezoid (base angles equal, nonparallel sides equal in length), with a height of 10 in. and bases of 1.5 ft and 2 ft. What is the area of the glass in the window?

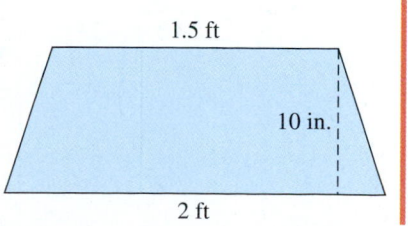
1.5 ft
10 in.
2 ft

Solution ▶ Before applying the area formula, all measures must use the same unit. In inches, we have 1.5 ft = 18 in. and 2 ft = 24 in.

$$A = \frac{h}{2}(b_1 + b_2) \qquad \text{given formula}$$

$$A = \frac{10 \text{ in.}}{2}(18 \text{ in.} + 24 \text{ in.}) \qquad \text{substitute 10 for } h, \text{ 18 for } b_1, \text{ and 24 for } b_2$$

$$A = (5 \text{ in.})(42 \text{ in.}) \qquad \text{simplify}$$

$$A = 210 \text{ in}^2 \qquad \text{result}$$

The area of the glass in the window is 210 in^2.

Now try Exercises 101 and 102 ▶

Volume

Volume is a measure of the amount of space occupied by a three dimensional object and is measured in **cubic units.** Some of the more common formulas are given in Table R.2B.

Table R.2B

	Definition and Diagram		Volume Formula (cubic units or *units*^{3})
Rectangular solid	a six-sided, solid figure with opposite faces congruent and adjacent faces meeting at right angles		$V = LWH$
Cube	a rectangular solid with six congruent, square faces		$V = S^3$
Sphere	the set of all points in space, an equal distance (called the radius) from a given point (called the center)		$V = \frac{4}{3}\pi r^3$
Right circular cylinder	union of all line segments connecting two congruent circles in parallel planes, meeting each at a right angle		$V = \pi r^2 h$
Right circular cone	union of all line segments connecting a given point (vertex) to a given circle (base) and whose altitude meets the center of the base at a right angle		$V = \frac{1}{3}\pi r^2 h$
Right pyramid	union of all line segments connecting a given point (vertex) to a given square (base) and whose altitude meets the center of the base at a right angle		$V = \frac{1}{3}s^2 h$

EXAMPLE 12B ▶   **Computing the Volume of a Composite Figure**

Sand at a cement factory is being dumped from a conveyor belt into a pile shaped like a right circular cone atop a right circular cylinder (see figure). How many cubic feet of sand are there at the moment the cone is 6 ft high with a diameter of 10 ft?

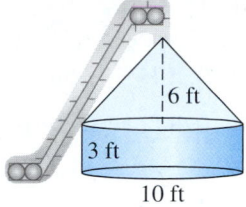

Solution ▶   Total Volume = volume of cylinder + volume of cone   verbal model

$$V = \pi r^2 h_1 + \frac{1}{3}\pi r^2 h_2$$ formula model (note $h_1 \neq h_2$)

$$= \pi(5)^2(3) + \frac{1}{3}\pi(5)^2(6)$$ substitute 5 for r, 3 for h_1, and 6 for h_2

$$= 75\pi + 50\pi$$ simplify

$$= 125\pi$$ result (exact form)

☑ **F.** You've just seen how we can solve applications of basic geometry

There are about 392.7 ft^3 of sand in the pile.

Now try Exercises 103 and 104 ▶

R.3 EXERCISES

▶ CONCEPTS AND VOCABULARY

Fill in each blank with the appropriate word or phrase. Carefully reread the section, if necessary.

1. A(n) ___identity___ is an equation that is always true, regardless of the ___unknown___ value while a(n) ___contradiction___ is an equation that is always false, regardless of the ___unknown___ value.

2. For inequalities, the three ways of writing a solution set are ___set___ notation, a number line graph, and ___interval___ notation.

3. The mathematical sentence $3x + 5 < 7$ is a(n) ___simple___ inequality, while $-2 < 3x + 5 < 7$ is a(n) ___compound___ inequality.

4. The ___intersection___ of sets A and B is written $A \cap B$. The ___union___ of sets A and B is written $A \cup B$.

5. Discuss/Explain the similarities and differences between the properties of equality for equations and those for inequalities.   Answers will vary.

6. Discuss/Explain the use of the words "and" and "or" in the statement of compound inequalities. Include a few examples to illustrate.   Answers will vary.

▶ DEVELOPING YOUR SKILLS

Solve each equation. Check your answer by substitution.

7. $4x + 3(x - 2) = 18 - x$ $x = 3$

8. $15 - 2x = -4(x + 1) + 9$ $x = -5$

9. $21 - (2v + 17) = -7 - 3v$ $v = -11$

10. $-12 - 5w = -9 - (6w + 7)$ $w = -4$

11. $8 - (3b + 5) = -5 + 2(b + 1)$ $b = \frac{6}{5}$

12. $2a + 4(a - 1) = 3 - (2a + 1)$ $a = \frac{3}{4}$

Solve each equation.

13. $\frac{1}{5}(b + 10) - 7 = \frac{1}{3}(b - 9)$ $b = -15$

14. $\frac{1}{6}(n - 12) = \frac{1}{4}(n + 8) - 2$ $n = -24$

15. $\frac{2}{3}(m + 6) = \frac{-1}{2}$ $m = -\frac{27}{4}$

16. $\frac{4}{5}(n - 10) = \frac{-8}{9}$ $n = \frac{80}{9}$

17. $\frac{1}{2}x + 5 = \frac{1}{3}x + 7$ $x = 12$

18. $-4 + \frac{2}{3}y = \frac{1}{2}y - 5$ $y = -6$

19. $\dfrac{x + 3}{5} + \dfrac{x}{3} = 7$ $x = 12$

20. $\dfrac{z - 4}{6} - 2 = \dfrac{z}{2}$ $z = -8$

HOMEWORK SELECTION GUIDE

Core: 7–75 every other odd, 79, 81, 83–103 every other odd (26 Exercises)
Standard: 1–4, 7–75 every other odd, 79, 81, 83–103 every other odd, 105, 109–116 (39 Exercises)

Extended: 1–4, 7–75 every other odd, 79, 81, 83–103 odd, 105, 109–116 (44 Exercises)
In Depth: 1–6, 7–75 every other odd, 79–81, 83–103 odd, 105, 108, 109–116 (49 Exercises)

21. $15 = -6 - \dfrac{3p}{8}$ **22.** $-15 - \dfrac{2q}{9} = -21$ $q = 27$
$p = -56$

23. $0.2(24 - 7.5a) - 6.1 = 4.1$ $a = -3.6$

24. $0.4(17 - 4.25b) - 3.15 = 4.16$ $b = -0.3$

25. $6.2v - (2.1v - 5) = 1.1 - 3.7v$ $v = -0.5$

26. $7.9 - 2.6w = 1.5w - (9.1 + 2.1w)$ $w = 8.5$

27. $\dfrac{n}{2} + \dfrac{n}{5} = \dfrac{2}{3}$ $n = \dfrac{20}{21}$

28. $\dfrac{m}{3} - \dfrac{2}{5} = \dfrac{m}{4}$ $m = \dfrac{24}{5}$

29. $3p - \dfrac{p}{4} - 5 = \dfrac{p}{6} - 2p + 6$ $p = \dfrac{12}{5}$

30. $\dfrac{q}{6} + 1 - 3q = 2 - 4q + \dfrac{q}{8}$ $q = \dfrac{24}{25}$

Identify the following equations as an identity, a contradiction, or a conditional equation, then state the solution.

31. $-3(4z + 5) = -15z - 20 + 3z$ contradiction; { }

32. $5x - 9 - 2 = -5(2 - x) - 1$ identity; $\{x|x \in \mathbb{R}\}$

33. $8 - 8(3n + 5) = -5 + 6(1 + n)$ conditional; $n = -\dfrac{11}{10}$

34. $2a + 4(a - 1) = 1 + 3(2a + 1)$ contradiction; { }

35. $-4(4x + 5) = -6 - 2(8x + 7)$ identity; $\{x|x \in \mathbb{R}\}$

36. $-(5x - 3) + 2x = 11 - 4(x + 2)$ conditional; $x = 0$

Write the solution set illustrated on each graph in set notation and interval notation.

37. $\{x|x \ge -2\}; [-2, \infty)$

38. $\{x|x < 1\}; (-\infty, 1)$

39. $\{x|-2 \le x \le 1\}; [-2, 1]$

40. $\{x|-2 \le x < 3\}; [-2, 3)$

Solve the inequality and write the solution in set notation. Then graph the solution and write it in interval notation.

41. $5a - 11 \ge 2a - 5$

42. $-8n + 5 > -2n - 12$

43. $2(n + 3) - 4 \le 5n - 1$

44. $-5(x + 2) - 3 < 3x + 11$

45. $\dfrac{3x}{8} + \dfrac{x}{4} < -4$ **46.** $\dfrac{2y}{5} + \dfrac{y}{10} < -2$

Additional answers can be found in the Instructor Answer Appendix.

Solve each inequality and write the solution in set notation.

47. $7 - 2(x + 3) \ge 4x - 6(x - 3)$ { }

48. $-3 - 6(x - 5) \le 2(7 - 3x) + 1$ { }

49. $4(3x - 5) + 18 < 2(5x + 1) + 2x$ $\{x|x \in \mathbb{R}\}$

50. $8 - (6 + 5m) > -9m - (3 - 4m)$ $\{m|m \in \mathbb{R}\}$

51. $-6(p - 1) + 2p \le -2(2p - 3)$ $\{p|p \in \mathbb{R}\}$

52. $9(w - 1) - 3w \ge -2(5 - 3w) + 1$ $\{w|w \in \mathbb{R}\}$

Determine the intersection and union of sets A, B, C, and D as indicated, given $A = \{-3, -2, -1, 0, 1, 2, 3\}$, $B = \{2, 4, 6, 8\}$, $C = \{-4, -2, 0, 2, 4\}$, and $D = \{4, 5, 6, 7\}$.

53. $A \cap B$ and $A \cup B$
$\{2\}; \{-3, -2, -1, 0, 1, 2, 3, 4, 6, 8\}$
54. $A \cap C$ and $A \cup C$
$\{-2, 0, 2\}; \{-4, -3, -2, -1, 0, 1, 2, 3, 4\}$

55. $A \cap D$ and $A \cup D$ **56.** $B \cap C$ and $B \cup C$
$\{ \}; \{-3, -2, -1, 0, 1, 2, 3, 4, 5, 6, 7\}$ $\{2, 4\}; \{-4, -2, 0, 2, 4, 6, 8\}$

57. $B \cap D$ and $B \cup D$ **58.** $C \cap D$ and $C \cup D$
$\{4, 6\}; \{2, 4, 5, 6, 7, 8\}$ $\{4\}; \{-4, -2, 0, 2, 4, 5, 6, 7\}$

Express the compound inequalities graphically and in interval notation.

59. $x < -2$ or $x > 1$ **60.** $x < -5$ or $x > 5$

61. $x < 5$ and $x \ge -2$ **62.** $x \ge -4$ and $x < 3$

63. $x \ge 3$ and $x \le 1$ **64.** $x \ge -5$ and $x \le -7$
no solution   no solution

Solve the compound inequalities and graph the solution set.

65. $4(x - 1) \le 20$ or $x + 6 > 9$

66. $-3(x + 2) > 15$ or $x - 3 \le -1$

67. $-2x - 7 \le 3$ and $2x \le 0$

68. $-3x + 5 \le 17$ and $5x \le 0$

69. $\dfrac{3}{5}x + \dfrac{1}{2} > \dfrac{3}{10}$ and $-4x > 1$

70. $\dfrac{2}{3}x - \dfrac{5}{6} \le 0$ and $-3x < -2$

71. $\dfrac{3x}{8} + \dfrac{x}{4} < -3$ or $x + 1 > -5$

72. $\dfrac{2x}{5} + \dfrac{x}{10} < -2$ or $x - 3 > 2$

73. $-3 \le 2x + 5 < 7$

74. $2 < 3x - 4 \le 19$

75. $-0.5 \le 0.3 - x \le 1.7$

76. $-8.2 < 1.4 - x < -0.9$

77. $-7 < -\tfrac{3}{4}x - 1 \le 11$

78. $-21 \le -\tfrac{2}{3}x + 9 < 7$

▶ WORKING WITH FORMULAS

79. Euler's Polyhedron Formula: $V + F - E = 2$

Discovered by Leonhard Euler in 1752, this simple but powerful formula states that in any regular polyhedron, the number of vertices V and faces F is always two more than the number of edges E. (a) Verify the formula for a simple cube. (b) Verify the formula for the octahedron shown in the figure. (c) If a dodecahedron has 12 faces and 30 edges, how many vertices does it have?

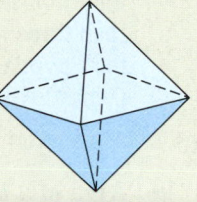

a. $8 + 6 - 12 = 2$✓   b. $6 + 8 - 12 = 2$✓   c. 20 vertices

80. Area of a Regular Polygon: $A = \dfrac{1}{2}ap$

The area of any regular polygon can be found using the formula shown, where a is the apothem of the polygon (perpendicular distance from center to any edge), and p is the perimeter.

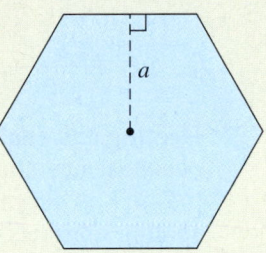

(a) Verify the formula using a square with sides of length 6 cm. (b) If the hexagon shown has an area of 259.8 cm² with sides 10 cm in length, what is the length a of the apothem?

a. $A = 36$ cm², $36 = \frac{1}{2}(3)(24)$ ✓   b. 8.66 cm

81. Body mass index: $B = \dfrac{704W}{H^2}$

The U.S. government publishes a body mass index formula to help people consider the risk of heart disease. An index "B" of 27 or more means that a person is at risk. Here W represents weight in pounds and H represents height in inches. If your height is 5′8″ what range of weights will help ensure you remain safe from the risk of heart disease? $W < 177.34$ lb

Source: www.surgeongeneral.gov/topics.

82. Lift capacity: $75S + 125B \le 750$

The capacity in pounds of the lift used by a roofing company to place roofing shingles and buckets of roofing nails on rooftops is modeled by the formula shown, where S represents packs of shingles and B represents buckets of nails. Use the formula to find (a) the largest number of shingle packs that can be lifted, (b) the largest number of nail buckets that can be lifted, and (c) the largest number of shingle packs that can be lifted along with three nail buckets.   a. 10   b. 6   c. 5

▶ APPLICATIONS

Write an equation to model the given information and solve.

83. Celebrity Travel: To avoid paparazzi and overzealous fans, the arrival gates of planes carrying celebrities are often kept secret until the last possible moment. While awaiting the arrival of Angelina Jolie, a large crowd of fans and photographers had gathered at Terminal A, Gate 18. However, the number of fans waiting at Gate 32 was *twice that number increased by 5*. If there were 73 fans at Gate 32, how many were waiting at Gate 18? (See Section R.1, Example 2a.) $2f + 5 = 73$; 34 fans

84. Famous Architecture: The *Hall of Mirrors* is the central gallery of the Palace of Versailles and is one of the most famous rooms in the world. The length of this hall is *11 m less than 8 times the width*. If the hall is 73 m long, what is its width? (See Section R.1, Example 2b.) $73 = 8W - 11$; 10.5 m

85. Dietary Goals: At the picnic, Mike abandoned his diet and consumed 13 calories more than twice the number of calories he normally allots for lunch. If he consumed 1467 calories, how many calories are normally allotted for lunch? $2c + 13 = 1467$; 727 cal

86. Marathon Training: While training for the Chicago marathon, Christina's longest run of the week was 5 mi less than double the shortest. If the longest run was 11.2 mi, how long was the shortest? $2s - 5 = 11.2$; 8.1 mi

87. Actor's Ages: At the time of this writing, actor Will Smith (*Enemy of the State*, *Seven Pounds*, others), was 1 yr older than two-thirds the age of Samuel Jackson (*The Negotiator*, *Die Hard III*, others). If Will Smith was 41 at this time, how old was Samuel Jackson? $41 = \frac{2}{3}S + 1$; 60 yr

88. Football versus Fútbol: The area of a regulation field for American football is about 410 square meters (m²) less than three-fifths of an Olympic-sized soccer field. If an American football field covers 5350 m², what is the area of an Olympic soccer field? $5350 = \frac{3}{5}S - 410$; 9600 m²

89. Forensic Studies: In forensic studies, skeletal remains are analyzed to determine the height, gender, race, age, and other characteristics of the decedent. For instance, the height of a male individual is approximated as 34 in. more than three and one-third times the length of the radial bone. If a live individual is 74 in. tall, how long is his radial bone? $74 = \frac{10}{3}r + 34$; 12 in.

90. Famous Waterways: The Suez Canal and the Panama Canal are two of the most important waterways in the world, saving ships thousands of miles as they journey from port to destination. The length of the Suez Canal is 39 kilometers (km) less than three times the length of the Panama Canal. If the Egyptian canal is 192 km long, how long is the Central American canal? $192 = 3P - 39$; 77 km

Write an inequality to model the given information and solve.

91. Exam scores: Jacques is going to college on an academic scholarship that requires him to maintain at least a 75% average in all of his classes. So far he has scored 82%, 76%, 65%, and 71% on four exams. What scores are possible on his last exam that will enable him to keep his scholarship?
$\frac{82 + 76 + 65 + 71 + x}{5} \geq 75$, at least 81%

92. Timed trials: In the first three trials of the 100-m butterfly, Johann had times of 50.2, 49.8, and 50.9 sec. How fast must he swim the final timed trial to have an average time of at most 50 sec?
$\frac{50.2 + 49.8 + 50.9 + x}{4} \leq 50$, less than 49.1 sec

93. Checking account balance: If the average daily balance in a certain checking account drops below $1000, the bank charges the customer a $7.50 service fee. The table gives the daily balance for one customer. What must the daily balance be for Friday to avoid a service charge?

Weekday	Balance
Monday	$1125
Tuesday	$850
Wednesday	$625
Thursday	$400

$\frac{1125 + 850 + 625 + 400 + x}{5} \geq 1000$, at least $2000

94. Average weight: In the National Football League, many consider an offensive line to be "small" if the average weight of the five down linemen is less than 325 lb. Using the table, what must the weight of the right tackle be so that the line will not be considered small?

Lineman	Weight
Left tackle	318 lb
Left guard	322 lb
Center	326 lb
Right guard	315 lb
Right tackle	?

$\frac{318 + 322 + 326 + 315 + x}{5} \geq 325$, at least 344 lb

95. Area of a rectangle: Given the rectangle shown, what is the range of values for the width, in order to keep the area less than 150m²?

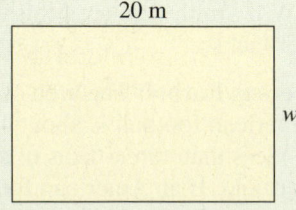

20 m

w

$0 < 20w < 150, 0 < w < 7.5$ m

96. Area of a triangle: Using the triangle shown, find the height that will guarantee an area equal to or greater than 48 in². $\frac{1}{2}(12h) \geq 48$, $h \geq 8$ in.

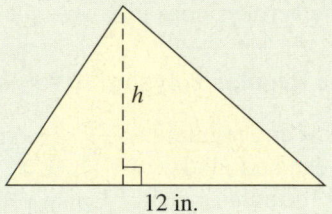

h

12 in.

97. Heating and cooling subsidies: As long as the outside temperature is over 45°F and less than 85°F ($45 < F < 85$), the city does not issue heating or cooling subsidies for low-income families. What is the corresponding range of Celsius temperatures C? Recall that $F = \frac{9}{5}C + 32$.
$45 < \frac{9}{5}C + 32 < 85$, $7.2° < C < 29.4°$

98. U.S. and European shoe sizes: To convert a European male shoe size "E" to an American male shoe size "A," the formula $A = 0.76E - 23$ can be used. Lillian has five sons in the U.S. military, with shoe sizes ranging from size 9 to size 14 ($9 \leq A \leq 14$). What is the corresponding range of European sizes? Round to the nearest half-size.
$9 \leq 0.76E - 23 \leq 14$, $42 \leq E \leq 48.5$

99. Power tool rentals: Sunshine Equipment Co. rents its power tools for a $20 fee, plus $4.50/hr. Kealoha's Rentals offers the same tools for an $11 fee plus $6.00/hr. How many hours h must a tool be rented to make the cost at Sunshine a better deal? $S = 4.5h + 20$, $K = 6h + 11$, more than 6 hr

100. Moving van rentals: Stringer Truck Rentals will rent a moving van for $15.75/day plus $0.35 per mile. Bertz Van Rentals will rent the same van for $25/day plus $0.30 per mile. How many miles m must the van be driven to make the cost at Bertz a better deal? $D = 0.35m + 15.75$, $B = 0.30m + 25$, more than 185 mi

101. Cost of drywall: After the studs are up, the wall shown in the figure must be covered in drywall. (a) How many square feet of drywall are needed? (b) If drywall is sold only in 4-ft by 8-ft sheets, approximately how many sheets are required for this job?
a. 216.5 ft² **b.** 7 sheets will be needed

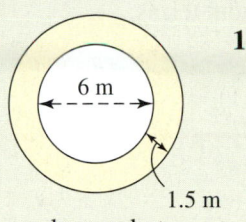

102. Paving a walkway: Current plans call for building a circular fountain 6 m in diameter with a circular walkway around it that is 1.5 m wide. (a) What is the approximate area of the walkway? (b) If the concrete for the walkway is to be 6 cm deep, what volume of cement must be used (1 cm = 0.01 m)?
a. 35.34 m² **b.** 2.12 m³

103. Trophy bases: The base of a new trophy has the form of a cylinder sitting atop a rectangular solid. If the base is to be cast in a special aluminum, determine the volume of aluminum to be used.   about 337.4 in³

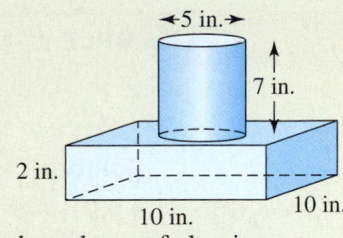

104. Grain storage: The dimensions of a grain silo are shown in the figure. If the maximum storage capacity of the silo is 95% of the total volume of the silo, how many cubic meters of corn can be stored?
153.9π m³ or about 483.5 m³

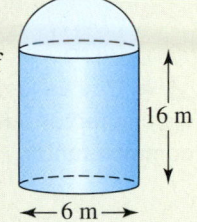

▶ **EXTENDING THE CONCEPT**

105. Solve for x: $-3(4x^2 + 5x - 2) + 7x = 6(4 - x - 2x^2) - 19$ $x = \frac{1}{2}$

106. Solve for n: $5\{3 - [4 - 2(5 - 9n)]\} + 15 = -6\{5 + 2[n - 10(9 + n)]\}$ $n = -5$

107. Use your local library, the Internet, or another resource to find the highest and lowest point on each of the seven continents. Express the range of altitudes for each continent as a joint inequality. Which continent has the greatest range?   Answers may vary.

108. The sum of two consecutive even integers is greater than or equal to 12 and less than or equal to 22. List all possible values for the two integers.
6 and 8; 8 and 10; 10 and 12

Place the correct inequality symbol in the blank to make the statement true.

109. If $m > 0$ and $n < 0$, then mn ___<___ 0.

110. If $m > n$ and $p > 0$, then mp ___>___ np.

111. If $m < n$ and $p > 0$, then mp ___<___ np.

112. If $m \leq n$ and $p < 0$, then mp ___≥___ np.

113. If $m > n$, then $-m$ ___<___ $-n$.

114. If $0 < m < n$, then $\frac{1}{m}$ ___>___ $\frac{1}{n}$.

115. If $m > 0$ and $n < 0$, then m^2 ___>___ n.

116. If $m < 0$, then m^3 ___<___ 0.

R.4 Factoring Polynomials and Solving Polynomial Equations by Factoring

LEARNING OBJECTIVES

In Section R.4 you will review:

☐ **A.** Factoring out the greatest common factor

☐ **B.** Common binomial factors and factoring by grouping

☐ **C.** Factoring quadratic polynomials

☐ **D.** Factoring special forms and quadratic forms

☐ **E.** Solving Polynomial Equations by Factoring

It is often said that knowing which tool to use is just as important as knowing how to use the tool. In this section, we review the tools needed to factor an expression, an important part of solving polynomial equations. This section will also help us decide which factoring tool is appropriate when many different factorable expressions are presented.

A. The Greatest Common Factor

To **factor** an expression means to *rewrite the expression as an equivalent product*. The distributive property is an example of factoring in action. To factor $2x^2 + 6x$, we might first rewrite each term using the common factor $2x$: $2x^2 + 6x = 2x \cdot x + 2x \cdot 3$, then apply the distributive property to obtain $2x(x + 3)$. We commonly say that we have *factored out 2x*. The **greatest common factor** (or GCF) is the largest factor common to *all* terms in the polynomial.

EXAMPLE 1 ▶   **Factoring Polynomials**

Factor each polynomial:

   a. $12x^2 + 18xy - 30y$ **b.** $x^5 + x^2$

Solution ▶   **a.** 6 is common to all three terms:

$$12x^2 + 18xy - 30y \qquad \text{\textcolor{magenta}{mentally: } } 6 \cdot 2x^2 + 6 \cdot 3xy - 6 \cdot 5y$$
$$= 6(2x^2 + 3xy - 5y)$$

   b. x^2 is common to both terms:

$$x^5 + x^2 \qquad \text{\textcolor{magenta}{mentally: } } x^2 \cdot x^3 + x^2 \cdot 1$$
$$= x^2(x^3 + 1)$$

☑ **A.** You've just seen how we can factor out the greatest common factor

Now try Exercises 7 and 8 ▶

B. Common Binomial Factors and Factoring by Grouping

If the terms of a polynomial have a **common** *binomial* **factor,** it can also be factored out using the distributive property.

EXAMPLE 2 ▶   **Factoring Out a Common Binomial Factor**

Factor:

   a. $(x + 3)x^2 + (x + 3)5$ **b.** $x^2(x - 2) - 3(x - 2)$

Solution ▶   **a.** $(x + 3)x^2 + (x + 3)5$ **b.** $x^2(x - 2) - 3(x - 2)$
$$= (x + 3)(x^2 + 5) \qquad\qquad\qquad = (x - 2)(x^2 - 3)$$

Now try Exercises 9 and 10 ▶

One application of removing a binomial factor involves **factoring by grouping.** At first glance, the expression $x^3 + 2x^2 + 3x + 6$ appears unfactorable. But by grouping the terms (applying the associative property), we can remove a monomial factor from each subgroup, which then reveals a common binomial factor.

$$\underline{x^3 + 2x^2} + \underline{3x + 6} = x^2(x + 2) + 3(x + 2)$$
$$= (x + 2)(x^2 + 3)$$

This grouping of terms must take into account any sign changes and common factors, as seen in Example 3. Also, it will be helpful to note that a general four-term polynomial $A + B + C + D$ is factorable by grouping only if $AD = BC$.

EXAMPLE 3 ▶   **Factoring by Grouping**

Factor $3t^3 + 15t^2 - 6t - 30$.

Solution ▶   Notice that all four terms have a common factor of 3. Begin by factoring it out.

$$3t^3 + 15t^2 - 6t - 30 \qquad \text{\textcolor{magenta}{original polynomial}}$$
$$= 3(t^3 + 5t^2 - 2t - 10) \qquad \text{\textcolor{magenta}{factor out 3}}$$
$$= 3(\underline{t^3 + 5t^2} - \underline{2t - 10}) \qquad \text{\textcolor{magenta}{group remaining terms}}$$
$$= 3[t^2(t + 5) - 2(t + 5)] \qquad \text{\textcolor{magenta}{factor common } monomial}$$
$$= 3(t + 5)(t^2 - 2) \qquad \text{\textcolor{magenta}{factor common } binomial}$$

Now try Exercises 11 and 12 ▶

☑ **B. You've just seen how we can factor common binomial factors and factor by grouping**

When asked to factor an expression, first look for common factors. The resulting expression will be easier to work with and help ensure the final answer is written in **completely factored form.** If a four-term polynomial cannot be factored as written, try rearranging the terms to find a combination that enables factoring by grouping.

C. Factoring Quadratic Polynomials

A quadratic polynomial is one that can be written in the form $ax^2 + bx + c$, where $a, b, c \in \mathbb{R}$ and $a \neq 0$. One common form of factoring involves quadratic trinomials such as $x^2 + 7x + 10$ and $2x^2 - 13x + 15$. While we know $(x + 5)(x + 2) = x^2 + 7x + 10$ and $(2x - 3)(x - 5) = 2x^2 - 13x + 15$ using F-O-I-L, how can we factor these trinomials without seeing the original expression in advance? First, it helps to place the trinomials in two families—those with a leading coefficient of 1 and those with a leading coefficient other than 1.

$ax^2 + bx + c$, where $a = 1$

When $a = 1$, the only factor pair for x^2 (other than $1 \cdot x^2$) is $x \cdot x$ and the first term in each binomial will be x: $(x \quad)(x \quad)$. The following observation helps guide us to the complete factorization. Consider the product $(x + b)(x + a)$:

$$(x + b)(x + a) = x^2 + ax + bx + ab \quad \text{F-O-I-L}$$

$$= x^2 + (a + b)x + ab \quad \text{distributive property}$$

Note the last term is the product ab (the *lasts*), while the coefficient of the middle term is $a + b$ (the sum of the *outers* and *inners*). Since the last term of $x^2 - 8x + 7$ is 7 and the coefficient of the middle term is -8, we are seeking two numbers with a product of positive 7 and a sum of negative 8. The numbers are -7 and -1, so the factored form is $(x - 7)(x - 1)$. It is also helpful to note that if the constant term is positive, the binomials will have *like* signs, since only *the product of like signs is positive.* If the constant term is negative, the binomials will have *unlike* signs, since only *the product of unlike signs is negative.* This means we can use the sign of the linear term (the term with degree 1) to guide our choice of factors.

> **Factoring Trinomials with a Leading Coefficient of 1**
>
> If the constant term is positive, the binomials will have *like* signs:
>
> $$(x + \quad)(x + \quad) \text{ or } (x - \quad)(x - \quad),$$
>
> to match the sign of the linear (middle) term.
>
> If the constant term is negative, the binomials will have *unlike* signs:
>
> $$(x + \quad)(x - \quad),$$
>
> with the larger factor placed in the binomial whose sign *matches* the linear (middle) term.

EXAMPLE 4 ▶ **Factoring Trinomials**

Factor these expressions:

  a. $-x^2 + 11x - 24$ **b.** $x^2 - 10 - 3x$

Solution ▶   **a.** First rewrite the trinomial in standard form as $-1(x^2 - 11x + 24)$. For $x^2 - 11x + 24$, the constant term is positive so the binomials will have like signs. Since the linear term is negative,

$$-1(x^2 - 11x + 24) = -1(x - \quad)(x - \quad) \quad \text{like signs, both negative}$$

$$= -1(x - 8)(x - 3) \quad (-8)(-3) = 24; -8 + (-3) = -11$$

b. First rewrite the trinomial in standard form as $x^2 - 3x - 10$. The constant term is negative so the binomials will have unlike signs. Since the linear term is negative,

$$x^2 - 3x - 10 = (x + \quad)(x - \quad) \qquad \text{unlike signs, one positive and one negative}$$
$$\qquad\qquad\qquad = (x + 2)(x - 5) \qquad \begin{array}{l}5 > 2, 5 \text{ is placed in the second binomial;} \\ (2)(-5) = -10; 2 + (-5) = -3\end{array}$$

> **Now try Exercises 13 and 14** ▶

Sometimes we encounter **prime polynomials,** or polynomials that cannot be factored. For $x^2 + 9x + 15$, the factor pairs of 15 are $1 \cdot 15$ and $3 \cdot 5$, with neither pair having a sum of $+9$. We conclude that $x^2 + 9x + 15$ is prime.

$ax^2 + bx + c$, where $a \neq 1$

If the leading coefficient is not one, the possible combinations of outers and inners are more numerous. Furthermore, the sum of the outer and inner products will change depending on the position of the possible factors. Note that $(2x + 3)(x + 9) = 2x^2 + 21x + 27$ and $(2x + 9)(x + 3) = 2x^2 + 15x + 27$ result in a different middle term, even though identical numbers were used.

To factor $2x^2 - 13x + 15$, note the constant term is positive so the binomials *must have like signs*. The negative linear term indicates these signs will be negative. We then list possible factors for the first and last terms of each binomial, then sum the outer and inner products.

Possible First and Last Terms for $2x^2$ and 15	Sum of Outers and Inners
1. $(2x - 1)(x - 15)$	$-30x - 1x = -31x$
2. $(2x - 15)(x - 1)$	$-2x - 15x = -17x$
3. $(2x - 3)(x - 5)$	$-10x - 3x = -13x$ ←
4. $(2x - 5)(x - 3)$	$-6x - 5x = -11x$

As you can see, only possibility 3 yields a linear term of $-13x$, and the correct factorization is then $(2x - 3)(x - 5)$. With practice, this **trial-and-error** process can be completed very quickly.

If the constant term is negative, the number of possibilities can be reduced by finding a factor pair with a sum *or* difference equal to the *absolute value* of the linear coefficient, as we can then arrange the sign in each binomial to obtain the needed result as shown in Example 5.

> **WORTHY OF NOTE**
>
> The number of trials needed to factor a polynomial can also be reduced by noting that the two terms in any binomial cannot share a common factor (all common factors are removed in a preliminary step).

EXAMPLE 5 ▶ **Factoring a Trinomial Using Trial and Error**

Factor $6z^2 - 11z - 35$.

Solution ▶ Note the constant term is negative (binomials will have unlike signs) and $|-11| = 11$. The factors of 35 are $1 \cdot 35$ and $5 \cdot 7$. Two possible first terms are: $(6z \quad)(z \quad)$ and $(3z \quad)(2z \quad)$, and we begin with 5 and 7 as factors of 35.

$(6z \quad)(z \quad)$	Outer and Inner Products		$(3z \quad)(2z \quad)$	Outer and Inner Products	
	Sum	Difference		Sum	Difference
1. $(6z \quad 5)(z \quad 7)$	$42z + 5z$ $47z$	$42z - 5z$ $37z$	3. $(3z \quad 5)(2z \quad 7)$	$21z + 10z$ $31z$	$21z - 10z$ $11z$
2. $(6z \quad 7)(z \quad 5)$	$30z + 7z$ $37z$	$30z - 7z$ $23z$	4. $(3z \quad 7)(2z \quad 5)$	$15z + 14z$ $29z$	$15z - 14z$ $1z$

Since possibility 3 yields a linear term of $11z$, we need not consider other factors of 35 and write the factored form as $6z^2 - 11z - 35 = (3z \quad 5)(2z \quad 7)$. The signs can then be arranged to obtain a middle term of $-11z$: $(3z + 5)(2z - 7)$, $-21z + 10z = -11z$ ✓.

☑ **C.** You've just seen how we can factor quadratic polynomials

Now try Exercises 15 and 16 ▶

D. Factoring Special Forms and Quadratic Forms

Next we consider methods to factor each of the special products we encountered in Section R.2.

The Difference of Two Squares

> **WORTHY OF NOTE**
>
> In an attempt to factor a *sum* of two perfect squares, say $v^2 + 49$, let's list all possible binomial factors. These are (1) $(v + 7)(v + 7)$, (2) $(v - 7)(v - 7)$, and (3) $(v + 7)(v - 7)$. Note that (1) and (2) are the binomial squares $(v + 7)^2$ and $(v - 7)^2$, with each product resulting in a "middle" term, whereas (3) is a binomial times its conjugate, resulting in a *difference* of squares: $v^2 - 49$. With all possibilities exhausted, we conclude that *the sum of two squares is prime!*

Multiplying and factoring are inverse processes. Since $(x - 7)(x + 7) = x^2 - 49$, we know that $x^2 - 49 = (x - 7)(x + 7)$. In words, *the difference of two squares will factor into a binomial and its conjugate*. To find the terms of the factored form, rewrite each term in the original expression as a square: $(\quad)^2$.

> **Factoring the Difference of Two Perfect Squares**
>
> Given any expression that can be written in the form $A^2 - B^2$,
> $$A^2 - B^2 = (A + B)(A - B)$$

Note that the *sum* of two perfect squares $A^2 + B^2$ *cannot be factored* using real numbers (the expression is prime). As a reminder, always check for a common factor first and be sure to write all results in completely factored form. See Example 6(c).

EXAMPLE 6 ▶ **Factoring the Difference of Two Perfect Squares**

Factor each expression completely.

    a. $4w^2 - 81$ **b.** $v^2 + 49$ **c.** $-3n^2 + 48$ **d.** $z^4 - \frac{1}{81}$ **e.** $x^2 - 7$

Solution ▶ **a.** $4w^2 - 81 = (2w)^2 - 9^2$ write as a difference of squares

                  $= (2w + 9)(2w - 9)$ $A^2 - B^2 = (A + B)(A - B)$

b. $v^2 + 49$ is prime.

c. $-3n^2 + 48 = -3(n^2 - 16)$ factor out -3

               $= -3[n^2 - (4)^2]$ write as a difference of squares

               $= -3(n + 4)(n - 4)$ $A^2 - B^2 = (A + B)(A - B)$

d. $z^4 - \frac{1}{81} = (z^2)^2 - (\frac{1}{9})^2$ write as a difference of squares

            $= (z^2 + \frac{1}{9})(z^2 - \frac{1}{9})$ $A^2 - B^2 = (A + B)(A - B)$

            $= (z^2 + \frac{1}{9})[z^2 - (\frac{1}{3})^2]$ write as a difference of squares ($z^2 + \frac{1}{9}$ is prime)

            $= (z^2 + \frac{1}{9})(z + \frac{1}{3})(z - \frac{1}{3})$ result

e. $x^2 - 7 = (x)^2 - (\sqrt{7})^2$ write as a difference of squares

          $= (x + \sqrt{7})(x - \sqrt{7})$ $A^2 - B^2 = (A + B)(A - B)$

Now try Exercises 17 and 18 ▶

Perfect Square Trinomials

Since $(x + 7)^2 = x^2 + 14x + 49$, we know that $x^2 + 14x + 49 = (x + 7)^2$. In words, *a perfect square trinomial will factor into a binomial square*. To use this idea effectively, we must learn to *identify* perfect square trinomials. Note that the first and last terms of $x^2 + 14x + 49$ are *the squares* of x and 7, and the middle term is *twice the product of these two terms*: $2(7x) = 14x$. These are the characteristics of a perfect square trinomial.

Factoring Perfect Square Trinomials

Given any expression that can be written in the form $A^2 \pm 2AB + B^2$,

    1. $A^2 + 2AB + B^2 = (A + B)^2$

    2. $A^2 - 2AB + B^2 = (A - B)^2$

EXAMPLE 7 ▶ **Factoring a Perfect Square Trinomial**

Factor $12m^3 - 12m^2 + 3m$.

Solution ▶ $12m^3 - 12m^2 + 3m$ check for common factors: GCF = $3m$

    $= 3m(4m^2 - 4m + 1)$ factor out $3m$

For the remaining trinomial $4m^2 - 4m + 1 \dots$

1. Are the first and last terms perfect squares?

$$4m^2 = (2m)^2 \text{ and } 1 = (1)^2 ✓ \text{ Yes.}$$

2. Is the linear term twice the product of $2m$ and 1?

$$2 \cdot 2m \cdot 1 = 4m ✓ \text{ Yes.}$$

Factor as a binomial square: $4m^2 - 4m + 1 = (2m - 1)^2$

This shows $12m^3 - 12m^2 + 3m = 3m(2m - 1)^2$.

Now try Exercises 19 and 20 ▶

⚠ **CAUTION** ▶ As shown in Example 7, be sure to include the GCF in your final answer. It is a common error to "leave the GCF behind."

In actual practice, these calculations can be performed mentally, making the process much more efficient.

Sum or Difference of Two Perfect Cubes

Recall that the *difference* of two perfect squares is factorable, but the *sum* of two perfect squares is prime. In contrast, *both the sum and difference of two perfect* **cubes** *are factorable.* For either $A^3 + B^3$ or $A^3 - B^3$ we have the following:

1. Each will factor into the product of a binomial and a trinomial: $(\quad)(\quad\quad)$
                                               binomial   trinomial

2. The terms of the binomial are the quantities being cubed: $(A \quad B)(\quad\quad)$

3. The terms of the trinomial are the square of A, the product AB, and the square of B, respectively: $(A \quad B)(A^2 \quad AB \quad B^2)$

4. The binomial takes the same sign as the original expression $(A \pm B)(A^2 \quad AB \quad B^2)$

5. The middle term of the trinomial takes the opposite sign of the original expression (the last term is always positive): $(A \pm B)(A^2 \mp AB + B^2)$

Factoring the Sum or Difference of Two Perfect Cubes: $A^3 \pm B^3$

    1. $A^3 + B^3 = (A + B)(A^2 - AB + B^2)$

    2. $A^3 - B^3 = (A - B)(A^2 + AB + B^2)$

EXAMPLE 8 ▶   **Factoring the Sum and Difference of Two Perfect Cubes**

Factor completely:

a. $x^3 + 125$ **b.** $-5m^3n + 40n^4$

Solution ▶   **a.** $x^3 + 125 = x^3 + 5^3$ write terms as perfect cubes

Use $A^3 + B^3 = (A + B)(A^2 - AB + B^2)$ factoring template

$x^3 + 5^3 = (x + 5)(x^2 - 5x + 25)$ $A \rightarrow x$ and $B \rightarrow 5$

b. $-5m^3n + 40n^4 = -5n(m^3 - 8n^3)$ check for common factors (GCF = $-5n$)

$= -5n[m^3 - (2n)^3]$ write terms as perfect cubes

Use $A^3 - B^3 = (A - B)(A^2 + AB + B^2)$ factoring template

$m^3 - (2n)^3 = (m - 2n)[m^2 + m(2n) + (2n)^2]$ $A \rightarrow m$ and $B \rightarrow 2n$

$= (m - 2n)(m^2 + 2mn + 4n^2)$ simplify

$\Rightarrow -5m^3n + 40n^4 = -5n(m - 2n)(m^2 + 2mn + 4n^2).$ factored form

The results for parts (a) and (b) can be checked using multiplication.

Now try Exercises 21 and 22 ▶

Quadratic Forms and *u*-Substitution

For any quadratic expression $ax^2 + bx + c$ in standard form, the degree of the leading term is twice the degree of the middle term. Generally, a trinomial is in **quadratic form** if it can be written as $a(\_\_)^2 + b(\_\_) + c$, where the parentheses "hold" the same factors. The equation $x^4 - 13x^2 + 36 = 0$ is in quadratic form since $(x^2)^2 - 13(x^2) + 36 = 0$. In many cases, we can factor these expressions using a **placeholder substitution** that transforms them into a more recognizable form. In a study of algebra, the letter "*u*" often plays this role. If we let *u* represent x^2, the expression $(x^2)^2 - 13(x^2) + 36$ becomes $u^2 - 13u + 36$, which can be factored into $(u - 9)(u - 4)$. After "unsubstituting" (replace *u* with x^2), we have $(x^2 - 9)(x^2 - 4) = (x + 3)(x - 3)(x + 2)(x - 2)$.

EXAMPLE 9 ▶   **Factoring a Quadratic Form**

Write in completely factored form: $(x^2 - 2x)^2 - 2(x^2 - 2x) - 3$.

Solution ▶   Expanding the binomials would produce a fourth-degree polynomial that would be very difficult to factor. Instead we note the expression is in *quadratic form*. Letting *u* represent $x^2 - 2x$ (the variable part of the "middle" term), $(x^2 - 2x)^2 - 2(x^2 - 2x) - 3$ becomes $u^2 - 2u - 3$.

$$u^2 - 2u - 3 = (u - 3)(u + 1) \quad \text{factor}$$

To finish up, write the expression in terms of *x*, substituting $x^2 - 2x$ for *u*.

$$= (x^2 - 2x - 3)(x^2 - 2x + 1) \quad \text{substitute } x^2 - 2x \text{ for } u$$

The resulting trinomials can be further factored.

$$= (x - 3)(x + 1)(x - 1)^2 \quad x^2 - 2x + 1 = (x - 1)^2$$

Now try Exercises 23 and 24 ▶

It is well known that information is retained longer and used more effectively when it's placed in an organized form. The "factoring flowchart" provided in Figure R.4 offers a streamlined and systematic approach to factoring and the concepts involved. However, with some practice the process tends to "flow" more naturally than following a chart, with many of the decisions becoming automatic.

☑ **D.** You've just seen how we can factor special forms and quadratic forms

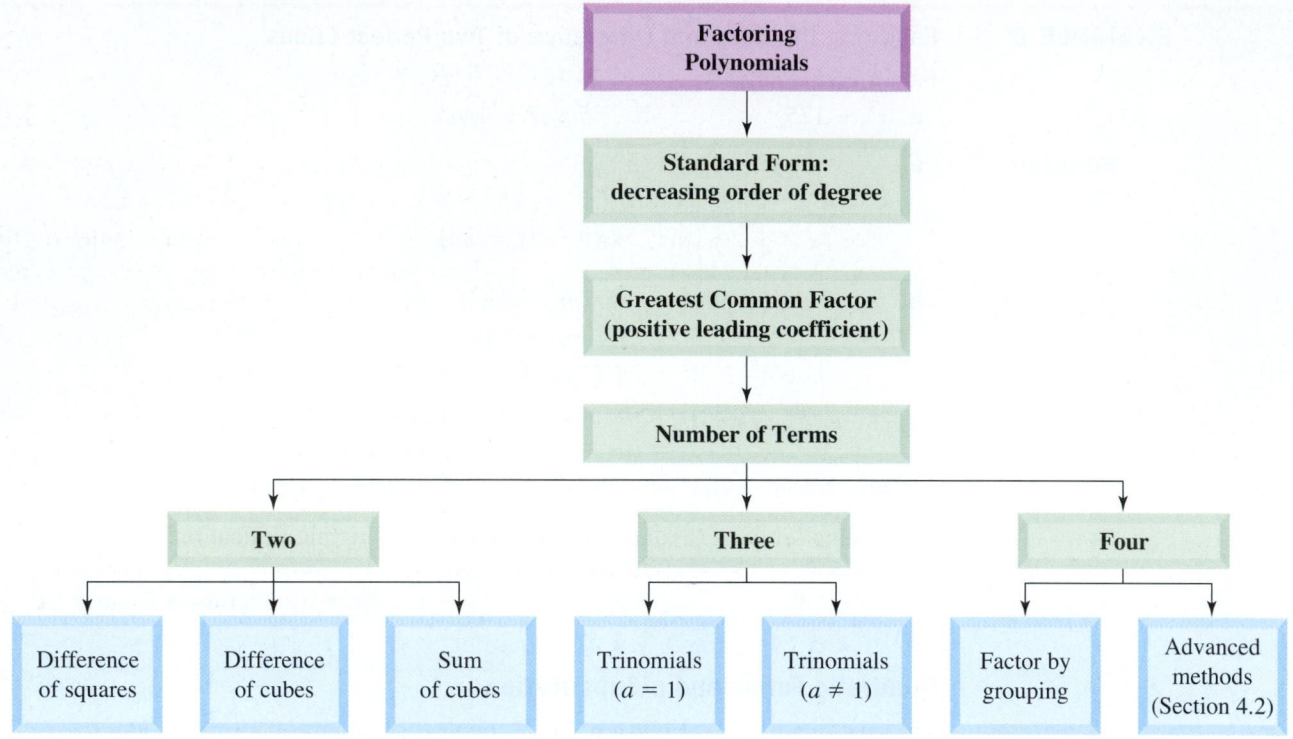

Figure R.4

• Can any result be factored further?   • Polynomials that cannot be factored are said to be *prime*.

For additional practice with these ideas, **see Exercises 25 through 52.**

E. Polynomial Equations and the Zero Product Property

The ability to solve linear and quadratic equations is the foundation on which a large percentage of our future studies are built. Both are closely linked to the solution of other equation types, as well as to the graphs of these equations.

In standard form, linear and quadratic equations have a known number of terms, so we commonly represent their coefficients using the early letters of the alphabet, as in $ax^2 + bx + c = 0$. However, these equations belong to the larger family of **polynomial equations.** To write a general polynomial, where the number of terms is unknown, we often represent the coefficients using subscripts on a single variable, such as a_1, a_2, a_3, and so on.

Polynomial Equations

A *polynomial equation of degree n* is one of the form

$$a_n x^n + a_{n-1} x^{n-1} + \cdots + a_1 x^1 + a_0 = 0$$

where $a_n, a_{n-1}, \ldots, a_1, a_0$ are real numbers and $a_n \neq 0$.

As a prelude to solving polynomial equations of higher degree, we'll first look at quadratic equations and the *zero product property.* As before, a **quadratic equation** is one that can be written in the form $ax^2 + bx + c = 0$, where a, b, and c are real numbers and $a \neq 0$. As written, the equation is in **standard form,** meaning the terms are in decreasing order of degree and the equation is set equal to zero.

Quadratic Equations

A quadratic equation can be written in the form

$$ax^2 + bx + c = 0,$$

with $a, b, c \in \mathbb{R}$, and $a \neq 0$.

Notice that a is the leading coefficient, b is the coefficient of the linear (first degree) term, and c is a constant. All quadratic equations have degree two, but can have one, two, or three terms. The equation $n^2 - 81 = 0$ is a quadratic equation with two terms, where $a = 1$, $b = 0$, and $c = -81$.

EXAMPLE 10 ▶ **Determining Whether an Equation Is Quadratic**

State whether the given equation is quadratic. If yes, identify coefficients a, b, and c.

a. $2x^2 - 18 = 0$ **b.** $z - 12 - 3z^2 = 0$ **c.** $\dfrac{-3}{4}x + 5 = 0$

d. $z^3 - 2z^2 + 7z = 8$ **e.** $0.8x^2 = 0$

Solution ▶

	Standard Form	Quadratic	Coefficients
a.	$2x^2 - 18 = 0$	yes, deg 2	$a = 2$ $b = 0$ $c = -18$
b.	$-3z^2 + z - 12 = 0$	yes, deg 2	$a = -3$ $b = 1$ $c = -12$
c.	$\dfrac{-3}{4}x + 5 = 0$	no, deg 1	(linear equation)
d.	$z^3 - 2z^2 + 7z - 8 = 0$	no, deg 3	(cubic equation)
e.	$0.8x^2 = 0$	yes, deg 2	$a = 0.8$ $b = 0$ $c = 0$

WORTHY OF NOTE

The word *quadratic* comes from the Latin word *quadratum,* meaning square. The word historically refers to the "four sidedness" of a square, but mathematically to the *area* of a square. Hence its application to polynomials of the form $ax^2 + bx + c$, where the variable of the leading term is *squared.*

Now try Exercises 53 through 64 ▶

With quadratic and other polynomial equations, we generally cannot isolate the variable on one side using only properties of equality, because the variable is raised to different powers. Instead we attempt to solve the equation by factoring and applying the **zero product property.**

Zero Product Property

If A and B represent real numbers or real-valued expressions

and $A \cdot B = 0$,

then $A = 0$ or $B = 0$.

In words, the property says, *If the product of any two (or more) factors is equal to zero, then at least one of the factors must be equal to zero.* We can use this property to solve higher degree equations after rewriting them in terms of equations with lesser degree. As with linear equations, values that make the original equation true are called *solutions* or *roots* of the equation.

EXAMPLE 11 ▶ **Solving Equations Using the Zero Product Property**

Solve by writing the equations in factored form and applying the zero product property.

a. $3x^2 = 5x$ **b.** $-5x + 2x^2 = 3$ **c.** $4x^2 = 12x - 9$

Solution ▶ **a.**

$$3x^2 = 5x \qquad \text{given equation}$$
$$3x^2 - 5x = 0 \qquad \text{standard form}$$
$$x(3x - 5) = 0 \qquad \text{factor}$$
$$x = 0 \quad \text{or} \quad 3x - 5 = 0 \qquad \text{set factors equal to zero (zero product property)}$$
$$x = 0 \quad \text{or} \qquad x = \frac{5}{3} \qquad \text{result}$$

b.

$$-5x + 2x^2 = 3 \qquad \text{given equation}$$
$$2x^2 - 5x - 3 = 0 \qquad \text{standard form}$$
$$(2x + 1)(x - 3) = 0 \qquad \text{factor}$$
$$2x + 1 = 0 \quad \text{or} \quad x - 3 = 0 \qquad \text{set factors equal to zero (zero product property)}$$
$$x = -\frac{1}{2} \quad \text{or} \qquad x = 3 \qquad \text{result}$$

c.

$$4x^2 = 12x - 9 \qquad \text{given equation}$$
$$4x^2 - 12x + 9 = 0 \qquad \text{standard form}$$
$$(2x - 3)(2x - 3) = 0 \qquad \text{factor}$$
$$2x - 3 = 0 \quad \text{or} \quad 2x - 3 = 0 \qquad \text{set factors equal to zero (zero product property)}$$
$$x = \frac{3}{2} \quad \text{or} \qquad x = \frac{3}{2} \qquad \text{result}$$

This equation has only the solution $x = \frac{3}{2}$, which we call a *repeated root*.

Now try Exercises 65 through 88 ▶

⚠ **CAUTION** ▶ Consider the equation $x^2 - 2x - 3 = 12$. While the left-hand side is factorable, the result is $(x - 3)(x + 1) = 12$ and finding a solution becomes a "guessing game" because the equation is not set equal to zero. If you *misapply* the zero factor property and say that $x - 3 = 12$ or $x + 1 = 12$, the "solutions" are $x = 15$ or $x = 11$, which are both incorrect! After subtracting 12 from both sides, $x^2 - 2x - 3 = 12$ becomes $x^2 - 2x - 15 = 0$ giving $(x - 5)(x + 3) = 0$ with solutions $x = 5$ or $x = -3$.

EXAMPLE 12 ▶ **Solving Polynomials by Factoring**

Solve by factoring: $4x^3 - 40x = 6x^2$.

Solution ▶

$$4x^3 - 40x = 6x^2 \qquad \text{given equation}$$
$$4x^3 - 6x^2 - 40x = 0 \qquad \text{standard form}$$
$$2x(2x^2 - 3x - 20) = 0 \qquad \text{common factor is } 2x$$
$$2x(2x + 5)(x - 4) = 0 \qquad \text{factored form}$$
$$2x = 0 \quad \text{or} \quad 2x + 5 = 0 \quad \text{or} \quad x - 4 = 0 \qquad \text{zero product property}$$
$$x = 0 \quad \text{or} \qquad x = \frac{-5}{2} \quad \text{or} \qquad x = 4 \qquad \text{result—solve for } x$$

Substituting these values into the original equation verifies they are solutions.

Now try Exercises 89 through 92 ▶

Example 12 reminds us that in the process of factoring polynomials, there may be a common monomial factor. This factor is also set equal to zero in the solution process (if the monomial is a constant, no solution is generated).

EXAMPLE 13 ▶   **Solving Higher Degree Equations**

Solve each equation by factoring.

a. $x^3 - 4x + 20 = 5x^2$ **b.** $x^4 - 10x^2 + 9 = 0$

Solution ▶   **a.**

$$x^3 - 4x + 20 = 5x^2 \qquad \text{original equation}$$

$$\underline{x^3 - 5x^2} \ \underline{- 4x + 20} = 0 \qquad \text{standard form; factor by grouping}$$

$$x^2(x - 5) - 4(x - 5) = 0 \qquad \text{remove common factors from each group}$$

$$(x - 5)(x^2 - 4) = 0 \qquad \text{factor common binomial}$$

$$(x - 5)(x + 2)(x - 2) = 0 \qquad \text{factored form}$$

$$x - 5 = 0 \quad \text{or} \quad x + 2 = 0 \quad \text{or} \quad x - 2 = 0 \qquad \text{zero product property}$$

$$x = 5 \quad \text{or} \qquad x = -2 \quad \text{or} \qquad x = 2 \qquad \text{solve}$$

The solutions are $x = 5$, $x = -2$, and $x = 2$.

b. The equation appears to be in *quadratic form* and we begin by substituting u for x^2 and u^2 for x^4.

$$x^4 - 10x^2 + 9 = 0 \qquad \text{original equation}$$

$$u^2 - 10u + 9 = 0 \qquad \text{substitute } u \text{ for } x^2 \text{ and } u^2 \text{ for } x^4$$

$$(u - 9)(u - 1) = 0 \qquad \text{factored form}$$

$$u - 9 = 0 \quad \text{or} \quad u - 1 = 0 \qquad \text{zero product property}$$

$$x^2 - 9 = 0 \quad \text{or} \quad x^2 - 1 = 0 \qquad \text{substitute } x^2 \text{ for } u$$

$$(x + 3)(x - 3) = 0 \quad \text{or} \quad (x + 1)(x - 1) = 0 \qquad \text{factor}$$

$$x = -3 \text{ or } x = 3 \quad \text{or} \qquad x = -1 \text{ or } x = 1 \qquad \text{zero product property}$$

The solutions are $x = -3$, $x = 3$, $x = -1$, and $x = 1$.

Now try Exercises 93 through 100 ▶

☑ **E.** You've just seen how we can solve polynomial equations by factoring

In Examples 12 and 13, we were able to solve higher degree polynomial equations by "breaking them down" into linear and quadratic forms. This basic idea can be applied to other kinds of equations as well.

R.4 EXERCISES

▶ **CONCEPTS AND VOCABULARY**

Fill in each blank with the appropriate word or phrase. Carefully reread the section, if necessary.

1. To factor an expression means to rewrite the expression as an equivalent __product__.

2. If a polynomial will not factor, it is said to be a(n) __prime__ polynomial.

3. The difference of two perfect squares always factors into the product of a(n) __binomial__ and its __conjugate__.

4. The expression $x^2 + 6x + 9$ is said to be a(n) __perfect__ __square__ trinomial, since its factored form is a perfect (binomial) square.

5. Discuss/Explain why $4x^2 - 36 = (2x - 6)(2x + 6)$ is not written in completely factored form, then rewrite it so it is factored completely.   Answers will vary.

6. Discuss/Explain why $a^3 + b^3$ is factorable, but $a^2 + b^2$ is not. Demonstrate by writing $x^3 + 64$ in factored form, and by exhausting all possibilities for $x^2 + 64$ to show it is prime.   Answers will vary.

HOMEWORK SELECTION GUIDE

Core: 7–99 every other odd, 101, 103–109 odd, 111 (30 Exercises)
Standard: 1–4, 7–99 every other odd, 101, 103–109 odd, 111, 112 (35 Exercises)

Extended: 1–4, 7–99 every other odd, 101–107, 109, 111–113 (39 Exercises)
In Depth: 1–6, 7–99 every other odd, 101–120 (50 Exercises)

▶ **DEVELOPING YOUR SKILLS**

Factor each expression using the method indicated.

Greatest Common Factor

7. a. $-17x^2 + 51$
$-17(x^2 - 3)$
b. $21b^3 - 14b^2 + 56b$
$7b(3b^2 - 2b + 8)$
c. $-3a^4 + 9a^2 - 6a^3$
$-3a^2(a^2 + 2a - 3)$

8. a. $-13n^2 - 52$
$-13(n^2 + 4)$
b. $9p^2 + 27p^3 - 18p^4$
$-9p^2(2p^2 - 3p - 1)$
c. $-6g^5 + 12g^4 - 9g^3$
$-3g^3(2g^2 - 4g + 3)$

Common Binomial Factor

9. a. $2a(a + 2) + 3(a + 2)$ $(a + 2)(2a + 3)$
b. $(b^2 + 3)3b + (b^2 + 3)2$ $(b^2 + 3)(3b + 2)$
c. $4m(n + 7) - 11(n + 7)$ $(n + 7)(4m - 11)$

10. a. $5x(x - 3) - 2(x - 3)$ $(x - 3)(5x - 2)$
b. $(v - 5)2v + (v - 5)3$ $(v - 5)(2v + 3)$
c. $3p(q^2 + 5) + 7(q^2 + 5)$ $(q^2 + 5)(3p + 7)$

Grouping

11. a. $9q^3 + 6q^2 + 15q + 10$ $(3q + 2)(3q^2 + 5)$
b. $h^5 - 12h^4 - 3h + 36$ $(h - 12)(h^4 - 3)$
c. $k^5 - 7k^3 - 5k^2 + 35$ $(k^2 - 7)(k^3 - 5)$

12. a. $6h^3 - 9h^2 - 2h + 3$ $(2h - 3)(3h^2 - 1)$
b. $4k^3 + 6k^2 - 2k - 3$ $(2k + 3)(2k^2 - 1)$
c. $3x^2 - xy - 6x + 2y$ $(3x - y)(x - 2)$

Trinomial Factoring where $|a| = 1$

13. a. $-p^2 + 5p + 14$
$-1(p - 7)(p + 2)$
b. $q^2 - 4q + 12$
prime
c. $n^2 + 20 - 9n$
$(n - 4)(n - 5)$

14. a. $-m^2 + 13m - 42$
$-1(m - 6)(m - 7)$
b. $x^2 + 12 + 13x$
$(x + 12)(x + 1)$
c. $v^2 + 10v + 15$
prime

Trinomial Factoring where $a \neq 1$

15. a. $3p^2 - 13p - 10$
$(3p + 2)(p - 5)$
b. $4q^2 + 7q - 15$
$(4q - 5)(q + 3)$
c. $10u^2 - 19u - 15$
$(5u + 3)(2u - 5)$

16. a. $6v^2 + v - 35$
$(2v + 5)(3v - 7)$
b. $20x^2 + 53x + 18$
$(4x + 9)(5x + 2)$
c. $15z^2 - 22z - 48$
$(3z - 8)(5z + 6)$

Difference of Perfect Squares

17. a. $4s^2 - 25$
$(2s + 5)(2s - 5)$
b. $9x^2 - 49$
$(3x + 7)(3x - 7)$
c. $50x^2 - 72$
$2(5x + 6)(5x - 6)$
d. $121h^2 - 144$
$(11h + 12)(11h - 12)$
e. $b^2 - 5$
$(b + \sqrt{5})(b - \sqrt{5})$

18. a. $9v^2 - \frac{1}{25}$
$(3v + \frac{1}{5})(3v - \frac{1}{5})$
b. $25w^2 - \frac{1}{49}$
$(5w + \frac{1}{7})(5w - \frac{1}{7})$
c. $v^4 - 1$
$(v^2 + 1)(v + 1)(v - 1)$
d. $16z^4 - 81$
$(4z^2 + 9)(2z + 3)(2z - 3)$
e. $x^2 - 17$
$(x + \sqrt{17})(x - \sqrt{17})$

Perfect Square Trinomials

19. a. $a^2 - 6a + 9$ $(a - 3)^2$ **b.** $b^2 + 10b + 25$ $(b + 5)^2$
c. $4m^2 - 20m + 25$
$(2m - 5)^2$
d. $9n^2 - 42n + 49$ $(3n - 7)^2$

20. a. $x^2 + 12x + 36$ $(x + 6)^2$ **b.** $z^2 - 18z + 81$ $(z - 9)^2$
c. $25p^2 - 60p + 36$
$(5p - 6)^2$
d. $16q^2 + 40q + 25$
$(4q + 5)^2$

Sum/Difference of Perfect Cubes

21. a. $8p^3 - 27$
$(2p - 3)(4p^2 + 6p + 9)$
b. $m^3 + \frac{1}{8}$ $(m + \frac{1}{2})(m^2 - \frac{1}{2}m + \frac{1}{4})$
c. $g^3 - 0.027$
$(g - 0.3)(g^2 + 0.3g + 0.09)$
d. $-2t^4 + 54t$
$-2t(t - 3)(t^2 + 3t + 9)$

22. a. $27q^3 - 125$
$(3q - 5)(9q^2 + 15q + 25)$
b. $n^3 + \frac{8}{27}$ $(n + \frac{2}{3})(n^2 - \frac{2}{3}n + \frac{4}{9})$
c. $b^3 - 0.125$
$(b - 0.5)(b^2 + 0.5b + 0.25)$
d. $3r^4 - 24r$
$3r(r - 2)(r^2 + 2r + 4)$

u-Substitution

23. a. $x^4 - 10x^2 + 9$
$(x + 3)(x - 3)(x + 1)(x - 1)$
b. $x^4 + 13x^2 + 36$
$(x^2 + 9)(x^2 + 4)$
c. $x^6 - 7x^3 - 8$
$(x - 2)(x^2 + 2x + 4)(x + 1)(x^2 - x + 1)$

24. a. $x^6 - 26x^3 - 27$ $(x - 3)(x^2 + 3x + 9)(x + 1)(x^2 - x + 1)$
b. $3(n + 5)^2 + 2(n + 5) - 21$ $(3n + 8)(n + 8)$
c. $2(z + 3)^2 + 3(z + 3) - 54$ $(2z - 3)(z + 9)$

25. Completely factor each of the following (recall that "1" is its own perfect square and perfect cube).
a. $n^2 - 1$ $(n + 1)(n - 1)$ **b.** $n^3 - 1$ $(n - 1)(n^2 + n + 1)$
c. $n^3 + 1$
$(n + 1)(n^2 - n + 1)$
d. $28x^3 - 7x$
$7x(2x + 1)(2x - 1)$

26. Carefully factor each of the following trinomials, if possible. Note differences and similarities.
a. $x^2 - x + 6$ prime
b. $x^2 + x - 6$ $(x + 3)(x - 2)$
c. $x^2 + x + 6$ prime
d. $x^2 - x - 6$ $(x - 3)(x + 2)$
e. $x^2 - 5x + 6$
$(x - 2)(x - 3)$
f. $x^2 + 5x - 6$
$(x + 6)(x - 1)$

Factor each expression completely, if possible. Rewrite the expression in standard form (factor out "−1" if needed) and factor out the GCF if one exists. If you believe the expression will not factor, write "prime."

27. $a^2 + 7a + 10$
$(a + 5)(a + 2)$
28. $b^2 + 9b + 20$
$(b + 5)(b + 4)$
29. $2x^2 - 24x + 40$
$2(x - 2)(x - 10)$
30. $10z^2 - 140z + 450$
$10(z - 9)(z - 5)$
31. $64 - 9m^2$
$-1(3m + 8)(3m - 8)$
32. $25 - 16n^2$
$-1(4n + 5)(4n - 5)$
33. $-9r + r^2 + 18$
$(r - 3)(r - 6)$
34. $28 + s^2 - 11s$
$(s - 7)(s - 4)$
35. $2h^2 + 7h + 6$
$(2h + 3)(h + 2)$
36. $3k^2 + 10k + 8$
$(3k + 4)(k + 2)$
37. $9k^2 - 24k + 16$
$(3k - 4)^2$
38. $4p^2 - 20p + 25$
$(2p - 5)^2$
39. $-6x^3 + 39x^2 - 63x$
$-3x(2x - 7)(x - 3)$
40. $-28z^3 + 16z^2 + 80z$
$-4z(7z + 10)(z - 2)$
41. $12m^2 - 40m + 4m^3$
$4m(m + 5)(m - 2)$
42. $-30n - 4n^2 + 2n^3$
$2n(n - 5)(n + 3)$
43. $a^2 - 7a - 60$
$(a + 5)(a - 12)$
44. $b^2 - 9b - 36$
$(b - 12)(b + 3)$
45. $8x^3 - 125$
$(2x - 5)(4x^2 + 10x + 25)$
46. $27r^3 + 64$
$(3r + 4)(9r^2 - 12r + 16)$

47. $m^2 + 9m - 24$ prime   **48.** $n^2 - 14n - 36$ prime

49. $x^3 - 5x^2 - 9x + 45$
$(x - 5)(x + 3)(x - 3)$
50. $x^3 + 3x^2 - 4x - 12$
$(x + 3)(x + 2)(x - 2)$

51. Match each expression with the description that fits *best*.

    __H__ **a.** prime polynomial

    __E__ **b.** standard trinomial $a = 1$

    __C__ **c.** perfect square trinomial

    __F__ **d.** difference of cubes

    __B__ **e.** binomial square

    __A__ **f.** sum of cubes

    __I__ **g.** binomial conjugates

    __D__ **h.** difference of squares

    __G__ **i.** standard trinomial $a \neq 1$

A. $x^3 + 27$ **B.** $(x + 3)^2$

C. $x^2 - 10x + 25$ **D.** $x^2 - 144$

E. $x^2 - 3x - 10$ **F.** $8s^3 - 125t^3$

G. $2x^2 - x - 3$ **H.** $x^2 + 9$

I. $(x - 7)$ and $(x + 7)$

52. Match each polynomial to its factored form. Two of them are prime.

    __C__ **a.** $4x^2 - 9$

    __D__ **b.** $4x^2 - 28x + 49$

    __A__ **c.** $x^3 - 125$

    __H__ **d.** $8x^3 + 27$

    __I__ **e.** $x^2 - 3x - 10$

    __E__ **f.** $x^2 + 3x + 10$

    __B__ **g.** $2x^2 - x - 3$

    __G__ **h.** $2x^2 + x - 3$

    __F__ **i.** $x^2 + 25$

A. $(x - 5)(x^2 + 5x + 25)$

B. $(2x - 3)(x + 1)$ **C.** $(2x + 3)(2x - 3)$

D. $(2x - 7)^2$ **E.** prime trinomial

F. prime binomial   **G.** $(2x + 3)(x - 1)$

H. $(2x + 3)(4x^2 - 6x + 9)$

I. $(x - 5)(x + 2)$

Determine whether each equation is quadratic. If so, identify the coefficients a, b, and c. If not, discuss why.

53. $2x - 15 - x^2 = 0$
$a = -1; b = 2; c = -15$
54. $21 + x^2 - 4x = 0$
$a = 1; b = -4; c = 21$

55. $\frac{2}{3}x - 7 = 0$
not quadratic $(a = 0)$
56. $12 - 4x = 9$
not quadratic $(a = 0)$

57. $\frac{1}{4}x^2 = 6x$
$a = \frac{1}{4}; b = -6; c = 0$
58. $0.5x = 0.25x^2$
$a = 0.25; b = -0.5; c = 0$

59. $2x^2 + 7 = 0$
$a = 2; b = 0; c = 7$
60. $5 = -4x^2$
$a = -4; b = 0; c = -5$

61. $-3x^2 + 9x - 5 + 2x^3 = 0$
not quadratic (degree 3)

62. $z^2 - 6z + 9 - z^3 = 0$ not quadratic (degree 3)

63. $(x - 1)^2 + (x - 1) + 4 = 9$ $a = 1; b = -1; c = -5$

64. $(x + 5)^2 - (x + 5) + 4 = 17$ $a = 1; b = 9; c = 7$

Solve using the zero factor property. Be sure each equation is in standard form and factor out any common factors before attempting to solve. Check all answers in the original equation.

65. $x^2 - 15 = 2x$
$x = 5$ or $x = -3$
66. $z^2 - 10z = -21$
$z = 7$ or $z = 3$

67. $m^2 = 8m - 16$
$m = 4$
68. $-10n = n^2 + 25$
$n = -5$

69. $5p^2 - 10p = 0$
$p = 0$ or $p = 2$
70. $6q^2 - 18q = 0$
$q = 0$ or $q = 3$

71. $-14h^2 = 7h$
$h = 0$ or $h = \frac{-1}{2}$
72. $9w = -6w^2$
$w = 0$ or $w = \frac{-3}{2}$

73. $a^2 - 17 = -8$
$a = 3$ or $a = -3$
74. $b^2 + 8 = 12$
$b = 2$ or $b = -2$

75. $g^2 + 18g + 70 = -11$ $g = -9$

76. $h^2 + 14h - 2 = -51$ $h = -7$

77. $m^3 + 5m^2 - 9m - 45 = 0$ $m = -5$ or $m = -3$ or $m = 3$

78. $n^3 - 3n^2 - 4n + 12 = 0$ $n = 3$ or $n = 2$ or $n = -2$

79. $(c - 12)c - 15 = 30$ $c = -3$ or $c = 15$

80. $(d - 10)d + 10 = -6$ $d = 8$ or $d = 2$

81. $9 + (r - 5)r = 33$ $r = 8$ or $r = -3$

82. $7 + (s - 4)s = 28$ $s = 7$ or $s = -3$

83. $(t + 4)(t + 7) = 54$ $t = -13$ or $t = 2$

84. $(g + 17)(g - 2) = 20$ $g = 3$ or $g = -18$

85. $2x^2 - 4x - 30 = 0$ $x = 5$ or $x = -3$

86. $-3z^2 + 12z + 36 = 0$ $z = 6$ or $z = -2$

87. $2w^2 - 5w = 3$ $w = -\frac{1}{2}$ or $w = 3$

88. $-3v^2 = -v - 2$ $v = \frac{-2}{3}$ or $v = 1$

89. $22x = x^3 - 9x^2$ $x = -2, x = 0, x = 11$

90. $x^3 = 13x^2 - 42x$ $x = 0, x = 6, x = 7$

91. $3x^3 = -7x^2 + 6x$ $x = -3, x = 0, x = \frac{2}{3}$

92. $7x^2 + 15x = 2x^3$ $x = -\frac{3}{2}, x = 0, x = 5$

93. $p^3 + 7p^2 - 63 = 9p$ $p = -7, p = -3, p = 3$

94. $q^3 - 4q + 24 = 6q^2$ $q = -2, q = 2, q = 6$

95. $x^3 - 25x = 2x^2 - 50$ $x = -5, x = 2, x = 5$

96. $3c^2 + c = c^3 + 3$ $c = -1, c = 1, c = 3$

97. $x^4 - 29x^2 + 100 = 0$ $x = -5, x = -2, x = 2, x = 5$

98. $z^4 - 20z^2 + 64 = 0$ $z = -4, z = -2, z = 2, z = 4$

99. $(b^2 - 3b)^2 - 14(b^2 - 3b) + 40 = 0$
$b = -2, b = -1, b = 4, b = 5$

100. $(d^2 - d)^2 - 8(d^2 - d) + 12 = 0$
$d = -2, d = -1, d = 2, d = 3$

▶ WORKING WITH FORMULAS

101. Surface area of a cylinder: $2\pi r^2 + 2\pi rh$

The surface area of a cylinder is given by the formula shown, where h is the height of the cylinder and r is the radius. Factor out the GCF and use the result to find the surface area of a cylinder where $r = 35$ cm and $h = 65$ cm. Answer in exact form and in approximate form rounded to the nearest whole number. $2\pi r(r + h), 7000\pi \text{ cm}^2; 21{,}991 \text{ cm}^2$

102. Volume of a cylindrical shell: $\pi R^2 h - \pi r^2 h$

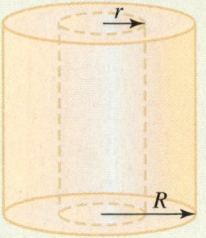

The volume of a cylindrical shell (a larger cylinder with a smaller cylinder removed) can be found using the formula shown, where R is the radius of the larger cylinder and r is the radius of the smaller. Factor the expression completely and use the result to find the volume of a shell where $R = 9$ cm, $r = 3$ cm, and $h = 10$ cm. Answer in exact form and in approximate form rounded to the nearest whole number. $\pi h(R + r)(R - r); 720\pi, 2262 \text{ cm}^3$

▶ APPLICATIONS

In many cases, factoring an expression can make it easier to evaluate as in the following applications.

103. Conical shells: The volume of a conical shell (like the shell of an ice cream cone) is given by the formula $V = \dfrac{1}{3}\pi R^2 h - \dfrac{1}{3}\pi r^2 h$, where R is the outer radius and r is the inner radius of the cone. Write the formula in completely factored form, then find the volume of a shell when $R = 5.1$ cm, $r = 4.9$ cm, and $h = 9$ cm. Answer in exact form and in approximate form rounded to the nearest tenth. $V = \frac{1}{3}\pi h(R + r)(R - r); 6\pi \text{ cm}^3; 18.8 \text{ cm}^3$

104. Spherical shells: The volume of a spherical shell (like the outer shell of a cherry cordial) is given by the formula $V = \frac{4}{3}\pi R^3 - \frac{4}{3}\pi r^3$, where R is the outer radius and r is the inner radius of the shell. Write the right-hand side in completely factored form, then find the volume of a shell where $R = 1.8$ cm and $r = 1.5$ cm. Answer in exact form and in approximate form rounded to the nearest tenth.

105. Volume of a box: The volume of a rectangular box x inches in height is given by the relationship $V = x^3 + 8x^2 + 15x$. Factor the right-hand side to determine: (a) The number of inches that the width exceeds the height, (b) the number of inches the length exceeds the height, and (c) the volume given the height is 2 ft.

106. Shipping textbooks: A publisher ships paperback books stacked x copies high in a box. The total number of books shipped per box is given by the relationship $B = x^3 - 13x^2 + 42x$. Factor the

right-hand side to determine (a) how many more or fewer books fit the width of the box (than the height), (b) how many more or fewer books fit the length of the box (than the height), and (c) the number of books shipped per box if they are stacked 10 high in the box.

107. Space-Time relationships: Due to the work of Albert Einstein and other physicists who labored on space-time relationships, it is known that the faster an object moves the shorter it appears to become. This phenomenon is modeled by the

Lorentz transformation $L = L_0\sqrt{1 - \left(\dfrac{v}{c}\right)^2}$,

where L_0 is the length of the object at rest, L is the relative length when the object is moving at velocity v, and c is the speed of light. Factor the radicand and use the result to determine the relative length of a 12-in. ruler if it is shot past a stationary observer at 0.75 times the speed of light ($v = 0.75c$).

108. Tubular fluid flow: As a fluid flows through a tube, it is flowing faster at the center of the tube than at the sides, where the tube exerts a backward drag. **Poiseuille's law** gives the velocity of the flow at any point of the cross section: $v = \dfrac{G}{4\eta}(R^2 - r^2)$,

where R is the inner radius of the tube, r is the distance from the center of the tube to a point in the flow, G represents what is called the pressure gradient, and η is a constant that depends on the viscosity of the fluid. Factor the right-hand side and find v given $R = 0.5$ cm, $r = 0.3$ cm, $G = 15$, and $\eta = 0.25$.

Solve by factoring.

109. Envelope sizes: Large mailing envelopes often come in standard sizes, with 5- by 7-in. and 9- by 12-in. envelopes being the most common. The next larger size envelope has an area of 143 in^2, with a length that is 2 in. longer than the width. What are the dimensions of the larger envelope? *11 in. by 13 in.*

110. Paper sizes: Letter size paper is 8.5 in. by 11 in. Legal size paper is $8\frac{1}{2}$ in. by 14 in. The next larger (common) size of paper has an area of 187 in^2, with

a length that is 6 in. longer than the width. What are the dimensions of the Ledger size paper? *11 in. by 17 in.*

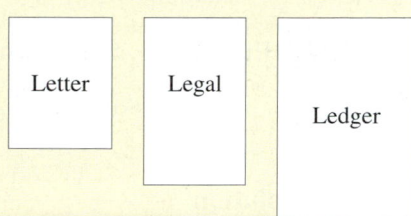

▶ EXTENDING THE CONCEPT

111. Factor out a constant that leaves integer coefficients for each term:

   a. $\frac{1}{2}x^4 + \frac{1}{8}x^3 - \frac{3}{4}x^2 + 4$ $\frac{1}{8}(4x^4 + x^3 - 6x^2 + 32)$

   b. $\frac{2}{3}b^5 - \frac{1}{6}b^3 + \frac{4}{9}b^2 - 1$ $\frac{1}{18}(12b^5 - 3b^3 + 8b^2 - 18)$

112. If $x = 2$ is substituted into $2x^3 + hx + 8$, the result is zero. What is the value of h? -12

113. Factor the expression: $192x^3 - 164x^2 - 270x$.
   $2x(16x - 27)(6x + 5)$

114. As an alternative to evaluating polynomials by direct substitution, **nested factoring** can be used. The method has the advantage of using only products and sums — no powers. For $P = x^3 + 3x^2 + 1x + 5$, we begin by grouping all variable terms and factoring x: $P = [x^3 + 3x^2 + 1x] + 5 =$

$x[x^2 + 3x + 1] + 5$. Then we group the inner terms with x and factor again:
$P = x[x^2 + 3x + 1] + 5 = x[x(x + 3) + 1] + 5$.
The expression can now be evaluated using any input and the order of operations. If $x = 2$, we quickly find that $P = 27$. Use this method to evaluate $H = x^3 + 2x^2 + 5x - 9$ for $x = -3$. $H = -33$

Factor each expression completely.

115. $x^4 - 81$
   $(x + 3)(x - 3)(x^2 + 9)$

116. $16n^4 - 1$
   $(2n + 1)(2n - 1)(4n^2 + 1)$

117. $p^6 - 1$
   $(p + 1)(p^2 - p + 1)(p - 1)(p^2 + p + 1)$

118. $m^6 - 64$
   $(m + 2)(m^2 - 2m + 4)(m - 2)(m^2 + 2m + 4)$

119. $q^4 - 28q^2 + 75$
   $(q + 5)(q - 5)(q + \sqrt{3})(q - \sqrt{3})$

120. $a^4 - 18a^2 + 32$
   $(a + 4)(a - 4)(a + \sqrt{2})(a - \sqrt{2})$

R.5   Rational Expressions and Equations

LEARNING OBJECTIVES

In Section R.5 you will review how to:

- ☐ **A.** Write a rational expression in simplest form
- ☐ **B.** Multiply and divide rational expressions
- ☐ **C.** Add and subtract rational expressions
- ☐ **D.** Simplify compound fractions
- ☐ **E.** Solve rational equations

A rational number is one that can be written as the quotient of two integers. Similarly, a *rational expression* is one that can be written as the quotient of two polynomials. We can apply the skills developed in a study of fractions (how to reduce, add, subtract, multiply, and divide) to **rational expressions,** sometimes called **algebraic fractions.**

A. Writing a Rational Expression in Simplest Form

A rational expression is in **simplest form** when the numerator and denominator have no common factors (other than 1). After factoring the numerator and denominator, we apply the **fundamental property of rational expressions.**

Fundamental Property of Rational Expressions

If P, Q, and R are polynomials, with $Q, R \neq 0$,

$$(1) \frac{P \cdot R}{Q \cdot R} = \frac{P}{Q} \quad \text{and} \quad (2) \frac{P}{Q} = \frac{P \cdot R}{Q \cdot R}$$

In words, the property says (1) a rational expression can be simplified by canceling common factors in the numerator and denominator, and (2) an equivalent expression can be formed by multiplying numerator and denominator by the same nonzero polynomial.

EXAMPLE 1 ▶ **Simplifying a Rational Expression**

Write the expression in simplest form: $\dfrac{x^2 - 1}{x^2 - 3x + 2}$.

Solution ▶
$$\frac{x^2 - 1}{x^2 - 3x + 2} = \frac{(x - 1)(x + 1)}{(x - 1)(x - 2)}$$ factor numerator and denominator

$$= \frac{\cancel{(x - 1)}(x + 1)}{\cancel{(x - 1)}(x - 2)}$$ common factors reduce to 1

$$= \frac{x + 1}{x - 2}$$ simplest form

Now try Exercises 7 through 10 ▶

WORTHY OF NOTE

If we view a and b as two points on the number line, we note that they are the same distance apart, regardless of the order they are subtracted. This tells us the numerator and denominator will have the same absolute value but be opposite in sign, giving a value of -1 (check using a few test values).

When simplifying rational expressions, we sometimes encounter expressions of the form $\dfrac{a - b}{b - a}$. If we factor -1 from the numerator, we see that $\dfrac{a - b}{b - a} = \dfrac{-1\cancel{(b - a)}}{\cancel{b - a}} = -1$.

⚠ **CAUTION** ▶ When reducing rational numbers or expressions, only common *factors* can be reduced.

It is incorrect to reduce (or divide out) individual terms: $\dfrac{-6 + 4\sqrt{3}}{2} \neq -3 + 4\sqrt{3}$, and

$\dfrac{x + 1}{x + 2} \neq \dfrac{1}{2}$ (except for $x = 0$)

Note that after simplifying an expression, we are actually saying the resulting (simpler) expression is equivalent to the original expression for all values where both are defined. The first expression is not defined when $x = 1$ or $x = 2$, the second when $x = 2$ (since the denominators would be zero). The calculator screens shown in Figure R.5 help to illustrate this fact, and it appears that we would very much prefer to be working with the simpler expression!

Figure R.5

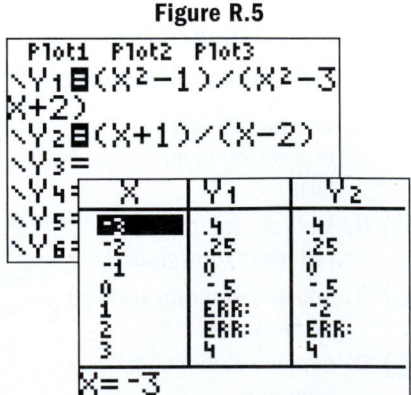

EXAMPLE 2 ▶ **Simplifying a Rational Expression**

Write the expression in simplest form: $\dfrac{6 - 2x}{x^2 - 9}$.

Solution ▶ $\dfrac{6 - 2x}{x^2 - 9} = \dfrac{2(3 - x)}{(x - 3)(x + 3)}$ factor numerator and denominator

$= \dfrac{(2)(-1)}{x + 3}$ reduce: $\dfrac{3 - x}{x - 3} = -1$

$= \dfrac{-2}{x + 3}$ simplest form

☑ **A.** You've just seen how we can write a rational expression in simplest form

Now try Exercises 11 through 16 ▶

B. Multiplication and Division of Rational Expressions

Operations on rational expressions use the factoring skills reviewed earlier, along with much of what we know about rational numbers.

> **Multiplying Rational Expressions**
>
> Given that P, Q, R, and S are polynomials with $Q, S \neq 0$,
>
> $$\frac{P}{Q} \cdot \frac{R}{S} = \frac{PR}{QS}$$
>
> 1. Factor all numerators and denominators completely.
> 2. Reduce common factors.
> 3. Multiply numerator × numerator and denominator × denominator.

EXAMPLE 3 ▶ **Multiplying Rational Expressions**

Compute the product: $\dfrac{2a + 2}{3a - 3a^2} \cdot \dfrac{3a^2 - a - 2}{9a^2 - 4}$.

Solution ▶ $\dfrac{2a + 2}{3a - 3a^2} \cdot \dfrac{3a^2 - a - 2}{9a^2 - 4} = \dfrac{2(a + 1)}{3a(1 - a)} \cdot \dfrac{(3a + 2)(a - 1)}{(3a - 2)(3a + 2)}$ factor

$= \dfrac{2(a + 1)}{3a\cancel{(1 - a)}} \cdot \dfrac{\cancel{(3a + 2)}\cancel{(a - 1)}}{(3a - 2)\cancel{(3a + 2)}}$ reduce: $\dfrac{a - 1}{1 - a} = -1$

$= \dfrac{-2(a + 1)}{3a(3a - 2)}$ simplest form

Now try Exercises 17 through 20 ▶

To divide fractions, we multiply the first expression by the *reciprocal of the second* (we sometimes say, "invert the divisor and multiply"). The quotient of two rational expressions is computed in the same way.

Dividing Rational Expressions

Given that P, Q, R, and S are polynomials with $Q, R, S \neq 0$,

$$\frac{P}{Q} \div \frac{R}{S} = \frac{P}{Q} \cdot \frac{S}{R} = \frac{PS}{QR}$$

Invert the divisor and multiply.

EXAMPLE 4 ▶   **Dividing Rational Expressions**

Compute the quotient $\dfrac{4m^3 - 12m^2 + 9m}{m^2 - 49} \div \dfrac{10m^2 - 15m}{m^2 + 4m - 21}$.

Solution ▶ $\dfrac{4m^3 - 12m^2 + 9m}{m^2 - 49} \div \dfrac{10m^2 - 15m}{m^2 + 4m - 21}$

$= \dfrac{4m^3 - 12m^2 + 9m}{m^2 - 49} \cdot \dfrac{m^2 + 4m - 21}{10m^2 - 15m}$ invert and multiply

$= \dfrac{m(4m^2 - 12m + 9)}{(m + 7)(m - 7)} \cdot \dfrac{(m + 7)(m - 3)}{5m(2m - 3)}$ factor

$= \dfrac{\overset{1}{\cancel{m}}(2m - 3)\overset{1}{\cancel{(2m - 3)}}}{\cancel{(m + 7)}(m - 7)} \cdot \dfrac{\overset{1}{\cancel{(m + 7)}}(m - 3)}{5\cancel{m}\cancel{(2m - 3)}}$ factor and reduce

$= \dfrac{(2m - 3)(m - 3)}{5(m - 7)}$ lowest terms

Note that we sometimes refer to simplest form as *lowest terms*.

Now try Exercises 21 through 42 ▶

⚠ **CAUTION** ▶   For products like $\dfrac{(w + 7)(w - 7)}{(w - 7)(w - 2)} \cdot \dfrac{w - 2}{w + 7}$, it is a common mistake to think that all factors "cancel," leaving an answer of zero. Actually, all factors *reduce to 1,* and the result is a value of 1 for all inputs where the product is defined.

$$\frac{\overset{1}{\cancel{(w + 7)}}\overset{1}{\cancel{(w - 7)}}}{\cancel{(w - 7)}\cancel{(w - 2)}} \cdot \frac{\overset{1}{\cancel{w - 2}}}{\cancel{w + 7}} = 1$$

☑ **B.** You've just seen how we can multiply and divide rational expressions

C. Addition and Subtraction of Rational Expressions

Recall that the addition and subtraction of *fractions* requires finding the lowest common denominator (LCD) and building equivalent fractions. The sum or difference of the numerators is then placed over this denominator. The procedure for the addition and subtraction of *rational expressions* is very much the same. Note that the LCD can also be described as the least common multiple (LCM) of all denominators.

Addition and Subtraction of Rational Expressions

1. Find the LCD of all rational expressions.
2. Build equivalent expressions using the LCD.
3. Add or subtract numerators as indicated.
4. Write the result in lowest terms.

EXAMPLE 5 ▶ **Adding and Subtracting Rational Expressions**

Compute as indicated:

a. $\dfrac{7}{10x} + \dfrac{3}{25x^2}$ b. $\dfrac{10x}{x^2 - 9} - \dfrac{5}{x - 3}$

Solution ▶

a. The LCM for $10x$ and $25x^2$ is $50x^2$.   find the LCD

$$\dfrac{7}{10x} + \dfrac{3}{25x^2} = \dfrac{7}{10x} \cdot \dfrac{5x}{5x} + \dfrac{3}{25x^2} \cdot \dfrac{2}{2}$$ write equivalent expressions

$$= \dfrac{35x}{50x^2} + \dfrac{6}{50x^2}$$ simplify

$$= \dfrac{35x + 6}{50x^2}$$ add the numerators and write the result over the LCD

The result is in simplest form.

b. The LCM for $x^2 - 9$ and $x - 3$ is $(x - 3)(x + 3)$.   find the LCD

$$\dfrac{10x}{x^2 - 9} - \dfrac{5}{x - 3} = \dfrac{10x}{(x - 3)(x + 3)} - \dfrac{5}{x - 3} \cdot \dfrac{x + 3}{x + 3}$$ write equivalent expressions

$$= \dfrac{10x - 5(x + 3)}{(x - 3)(x + 3)}$$ subtract numerators, write the result over the LCD

$$= \dfrac{10x - 5x - 15}{(x - 3)(x + 3)}$$ distribute

$$= \dfrac{5x - 15}{(x - 3)(x + 3)}$$ combine like terms

$$= \dfrac{5\overset{1}{(x - 3)}}{(x - 3)(x + 3)} = \dfrac{5}{x + 3}$$ factor and reduce

Now try Exercises 43 through 48 ▶

EXAMPLE 6 ▶ **Adding and Subtracting Rational Expressions**

Perform the operations indicated:

a. $\dfrac{5}{n + 2} - \dfrac{n - 3}{n^2 - 4}$ b. $\dfrac{b^2}{4a^2} - \dfrac{c}{a}$

Solution ▶

a. The LCM for $n + 2$ and $n^2 - 4$ is $(n + 2)(n - 2)$.

$$\dfrac{5}{n + 2} - \dfrac{n - 3}{n^2 - 4} = \dfrac{5}{(n + 2)} \cdot \dfrac{n - 2}{n - 2} - \dfrac{n - 3}{(n + 2)(n - 2)}$$ write equivalent expressions

$$= \dfrac{5(n - 2) - (n - 3)}{(n + 2)(n - 2)}$$ subtract numerators, write the result over the LCD

$$= \dfrac{5n - 10 - n + 3}{(n + 2)(n - 2)}$$ distribute

$$= \dfrac{4n - 7}{(n + 2)(n - 2)}$$ result

b. The LCM for a and $4a^2$ is $4a^2$: $\dfrac{b^2}{4a^2} - \dfrac{c}{a} = \dfrac{b^2}{4a^2} - \dfrac{c}{a} \cdot \dfrac{4a}{4a}$ write equivalent expressions

$$= \dfrac{b^2}{4a^2} - \dfrac{4ac}{4a^2}$$ simplify

$$= \dfrac{b^2 - 4ac}{4a^2}$$ subtract numerators, write the result over the LCD

Now try Exercises 49 through 64 ▶

⚠ **CAUTION ▶** When the second term in a subtraction has a binomial numerator as in Example 6a, be sure the subtraction *is applied to both terms*. It is a common error to write

$$\dfrac{5(n-2)}{(n+2)(n-2)} - \dfrac{n-3}{(n+2)(n-2)} = \dfrac{5n-10 \; \boxed{-\; n-3}}{(n+2)(n-2)}\;\text{✗}$$ in which the subtraction is applied

to the first term only. This is incorrect!

☑ **C.** You've just seen how we can add and subtract rational expressions

D. Simplifying Compound Fractions

Rational expressions whose numerator or denominator contain a fraction are called **compound fractions.** The expression $\dfrac{\dfrac{2}{3m} - \dfrac{3}{2}}{\dfrac{3}{4m} - \dfrac{1}{3m^2}}$ is a compound fraction with a

numerator of $\dfrac{2}{3m} - \dfrac{3}{2}$ and a denominator of $\dfrac{3}{4m} - \dfrac{1}{3m^2}$. The two methods commonly used to simplify compound fractions are summarized in the following boxes.

Simplifying Compound Fractions (Method I)

1. Add/subtract fractions in the numerator, writing them as a single expression.
2. Add/subtract fractions in the denominator, also writing them as a single expression.
3. Multiply the numerator by the reciprocal of the denominator and simplify if possible.

Simplifying Compound Fractions (Method II)

1. Find the LCD of all fractions in the numerator and denominator.
2. Multiply the numerator and denominator by this LCD and simplify.
3. Simplify further if possible.

Method II is illustrated in Example 7.

EXAMPLE 7 ▶   **Simplifying a Compound Fraction**

Simplify the compound fraction:

$$\dfrac{\dfrac{2}{3m} - \dfrac{3}{2}}{\dfrac{3}{4m} - \dfrac{1}{3m^2}}$$

Solution ▶ The LCD for all fractions is $12m^2$.

$$\frac{\dfrac{2}{3m} - \dfrac{3}{2}}{\dfrac{3}{4m} - \dfrac{1}{3m^2}} = \frac{\left(\dfrac{2}{3m} - \dfrac{3}{2}\right)\left(\dfrac{12m^2}{1}\right)}{\left(\dfrac{3}{4m} - \dfrac{1}{3m^2}\right)\left(\dfrac{12m^2}{1}\right)}$$

multiply numerator and denominator by $12m^2 = \dfrac{12m^2}{1}$

$$= \frac{\left(\dfrac{2}{3m}\right)\left(\dfrac{12m^2}{1}\right) - \left(\dfrac{3}{2}\right)\left(\dfrac{12m^2}{1}\right)}{\left(\dfrac{3}{4m}\right)\left(\dfrac{12m^2}{1}\right) - \left(\dfrac{1}{3m^2}\right)\left(\dfrac{12m^2}{1}\right)}$$

distribute

$$= \frac{8m - 18m^2}{9m - 4}$$

simplify

$$= \frac{2m(4 \overset{-1}{-} 9m)}{9m - 4} = -2m$$

factor and write in lowest terms

☑ **D.** You've just seen how we can simplify compound fractions

Now try Exercises 65 through 74 ▶

E. Solving Rational Equations

In Section R.3 we solved linear equations using basic properties of equality. If any equation contained fractional terms, we "cleared the fractions" using the least common denominator (LCD). We can also use this idea to solve **rational equations,** or equations that contain rational *expressions.*

> **Solving Rational Equations**
>
> 1. Identify and exclude any values that cause a zero denominator.
> 2. Multiply both sides by the LCD and simplify (this will eliminate all denominators).
> 3. Solve the resulting equation.
> 4. Check all solutions in the original equation.

EXAMPLE 8 ▶ **Solving a Rational Equation**

Solve for m: $\dfrac{2}{m} - \dfrac{1}{m - 1} = \dfrac{4}{m^2 - m}$.

Solution ▶ Since $m^2 - m = m(m - 1)$, the LCD is $m(m - 1)$, where $m \neq 0$ and $m \neq 1$.

$$m(m - 1)\left(\frac{2}{m} - \frac{1}{m - 1}\right) = m(m - 1)\left[\frac{4}{m(m - 1)}\right]$$

multiply by LCD

$$\frac{m(m - 1)}{1}\left(\frac{2}{m}\right) - \frac{m(m - 1)}{1}\left(\frac{1}{m - 1}\right) = \frac{m(m - 1)}{1}\left(\frac{4}{m(m - 1)}\right)$$

distribute and simplify

$$2(m - 1) - m = 4$$

denominators are eliminated

$$2m - 2 - m = 4$$

distribute

$$m = 6$$

solve for m

Checking by substitution we have:

$$\frac{2}{m} - \frac{1}{m-1} = \frac{4}{m^2 - m} \qquad \text{original equation}$$

$$\frac{2}{(6)} - \frac{1}{(6)-1} = \frac{4}{(6)^2 - (6)} \qquad \text{substitute 6 for } m$$

$$\frac{1}{3} - \frac{1}{5} = \frac{4}{30} \qquad \text{simplify}$$

$$\frac{5}{15} - \frac{3}{15} = \frac{2}{15} \qquad \text{common denominator}$$

$$\frac{2}{15} = \frac{2}{15} \checkmark \qquad \text{result}$$

```
6→X
                          6
(2/X)-1/(X-1)
              .133333333
4/(X²-X)
              .133333333
```

A calculator check is shown in the figure.

Now try Exercises 75 through 80 ▶

Multiplying both sides of an equation by a variable sometimes introduces a solution that satisfies the *resulting equation,* but not the original equation—the one we're trying to solve. Such "solutions" are called **extraneous roots** and illustrate the need to check all apparent solutions in the original equation. In the case of rational equations, we are particularly aware that any value that causes a zero denominator is outside the domain and cannot be a solution.

EXAMPLE 9 ▶ **Solving a Rational Equation**

Solve: $x + \dfrac{12}{x-3} = 1 + \dfrac{4x}{x-3}$.

Solution ▶ The LCD is $x - 3$, where $x \neq 3$.

$$(x-3)\left(x + \frac{12}{x-3}\right) = (x-3)\left(1 + \frac{4x}{x-3}\right) \qquad \text{multiply both sides by LCD}$$

$$(x-3)x + \left(\frac{\cancel{x-3}}{1}\right)\left(\frac{12}{\cancel{x-3}}\right) = (x-3)(1) + \left(\frac{\cancel{x-3}}{1}\right)\left(\frac{4x}{\cancel{x-3}}\right) \qquad \text{distribute and simplify}$$

$$x^2 - 3x + 12 = x - 3 + 4x \qquad \text{denominators are eliminated}$$

$$x^2 - 8x + 15 = 0 \qquad \text{set equation equal to zero}$$

$$(x-3)(x-5) = 0 \qquad \text{factor}$$

$$x = 3 \quad \text{or} \quad x = 5 \qquad \text{zero factor property}$$

Checking shows $x = 3$ is an extraneous root, and $x = 5$ is the only valid solution.

☑ **E.** You've just seen how we can solve rational equations

Now try Exercises 81 through 86 ▶

R.5 EXERCISES

▶ CONCEPTS AND VOCABULARY

Fill in each blank with the appropriate word or phrase. Carefully reread the section, if necessary.

1. In simplest form, $(a - b)/(a - b)$ is equal to _1_, while $(a - b)/(b - a)$ is equal to _−1_.

2. A rational expression is in _simplest_ _form_ when the numerator and denominator have no common factors, other than _1_.

3. As with numeric fractions, algebraic fractions require a _common_ _denominator_ for addition and subtraction.

4. Since $x^2 + 9$ is prime, the expression $(x^2 + 9)/(x + 3)$ is already written in _simplest_ _form_.
   (or lowest terms)

State T or F and discuss/explain your response.

5. $\dfrac{x}{x + 3} - \dfrac{x + 1}{x + 3} = \dfrac{1}{x + 3}$ F; numerator should be −1

6. $\dfrac{\cancel{(x + 3)}\cancel{(x - 2)}}{\cancel{(x - 2)}\cancel{(x + 3)}} = 0$ F; result is 1

▶ DEVELOPING YOUR SKILLS

Reduce to lowest terms.

7. **a.** $\dfrac{a - 7}{-3a + 21}$ $-\dfrac{1}{3}$ **b.** $\dfrac{2x + 6}{4x^2 - 8x}$ $\dfrac{x + 3}{2x(x - 2)}$

8. **a.** $\dfrac{x - 4}{-7x + 28}$ $-\dfrac{1}{7}$ **b.** $\dfrac{3x - 18}{6x^2 - 12x}$ $\dfrac{x - 6}{2x(x - 2)}$

9. **a.** $\dfrac{x^2 - 5x - 14}{x^2 + 6x - 7}$ simplified   **b.** $\dfrac{a^2 + 3a - 28}{a^2 - 49}$ $\dfrac{a - 4}{a - 7}$

10. **a.** $\dfrac{r^2 + 3r - 10}{r^2 + r - 6}$ $\dfrac{r + 5}{r + 3}$ **b.** $\dfrac{m^2 + 3m - 4}{m^2 - 4m}$ simplified

11. **a.** $\dfrac{x - 7}{7 - x}$ -1 **b.** $\dfrac{5 - x}{x - 5}$ -1

12. **a.** $\dfrac{v^2 - 3v - 28}{49 - v^2}$ $\dfrac{-1(v + 4)}{v + 7}$ **b.** $\dfrac{u^2 - 10u + 25}{25 - u^2}$ $\dfrac{-1(u - 5)}{u + 5}$

13. **a.** $\dfrac{-12a^3 b^5}{4a^2 b^{-4}}$ $-3ab^9$ **b.** $\dfrac{7x + 21}{63}$ $\dfrac{x + 3}{9}$

   c. $\dfrac{y^2 - 9}{3 - y}$ $-1(y + 3)$ **d.** $\dfrac{m^3 n - m^3}{m^4 - m^4 n}$ $\dfrac{-1}{m}$

14. **a.** $\dfrac{5m^{-3} n^5}{-10mn^2}$ $\dfrac{1n^3}{-2m^4}$ **b.** $\dfrac{-5v + 20}{25}$ $\dfrac{-v + 4}{5}$

   c. $\dfrac{n^2 - 4}{2 - n}$ $-1(n + 2)$ **d.** $\dfrac{w^4 - w^4 v}{w^3 v - w^3}$ $-w$

15. **a.** $\dfrac{2n^3 + n^2 - 3n}{n^3 - n^2}$ $\dfrac{2n + 3}{n}$ **b.** $\dfrac{6x^2 + x - 15}{4x^2 - 9}$ $\dfrac{3x + 5}{2x + 3}$

   c. $\dfrac{x^3 + 8}{x^2 - 2x + 4}$ $x + 2$ **d.** $\dfrac{mn^2 + n^2 - 4m - 4}{mn + n + 2m + 2}$ $n - 2$

16. **a.** $\dfrac{x^3 + 4x^2 - 5x}{x^3 - x}$ $\dfrac{x + 5}{x + 5}$ **b.** $\dfrac{5p^2 - 14p - 3}{5p^2 + 11p + 2}$ $\dfrac{p - 3}{p + 2}$

   c. $\dfrac{12y^2 - 13y + 3}{27y^3 - 14y - 3}$ $\dfrac{x + 1}{9y^2 + 3y + 1}$ **d.** $\dfrac{ax^2 - 5x^2 - 3a + 15}{ax - 5x + 5a - 25}$ $\dfrac{x^2 - 3}{x + 5}$

Compute as indicated. Write final results in lowest terms.

17. $\dfrac{a^2 - 4a + 4}{a^2 - 9} \cdot \dfrac{a^2 - 2a - 3}{a^2 - 4}$ $\dfrac{(a - 2)(a + 1)}{(a + 3)(a + 2)}$

18. $\dfrac{b^2 + 5b - 24}{b^2 - 6b + 9} \cdot \dfrac{b}{b^2 - 64}$ $\dfrac{b}{(b - 3)(b - 8)}$

19. $\dfrac{x^2 - 7x - 18}{x^2 - 6x - 27} \cdot \dfrac{2x^2 + 7x + 3}{2x^2 + 5x + 2}$ 1

20. $\dfrac{6v^2 + 23v + 21}{4v^2 - 4v - 15} \cdot \dfrac{4v^2 - 25}{3v + 7}$ $2v + 5$

21. $\dfrac{p^3 - 64}{p^3 - p^2} \div \dfrac{p^2 + 4p + 16}{p^2 - 5p + 4}$ $\dfrac{(p - 4)^2}{p^2}$

22. $\dfrac{a^2 + 3a - 28}{a^2 + 5a - 14} \div \dfrac{a^3 - 4a^2}{a^3 - 8}$ $\dfrac{a^2 + 2a + 4}{a^2}$

23. $\dfrac{3x - 9}{4x + 12} \div \dfrac{3 - x}{5x + 15}$ $\dfrac{-15}{4}$

24. $\dfrac{5b - 10}{7b - 28} \div \dfrac{2 - b}{5b - 20}$ $\dfrac{-25}{7}$

25. $\dfrac{a^2 + a}{a^2 - 3a} \cdot \dfrac{3a - 9}{2a + 2}$ $\dfrac{3}{2}$

HOMEWORK SELECTION GUIDE

Core: 7–83 every other odd, 87, 89–93 odd (25 Exercises)
Standard: 1–4, 7–83 every other odd, 87, 88, 89–95 odd (31 Exercises)

Extended: 1–4, 7–83 every other odd, 85–96 (36 Exercises)
In Depth: 1–6, 7–83 every other odd, 84–97 (40 Exercises)

26. $\dfrac{p^2-36}{2p} \cdot \dfrac{4p^2}{2p^2+12p}$ $\quad p-6$

27. $\dfrac{8}{a^2-25} \cdot (a^2-2a-35)$ $\quad \dfrac{8(a-7)}{a-5}$

28. $(m^2-16) \cdot \dfrac{m^2-5m}{m^2-m-20}$ $\quad m(m-4)$

29. $\dfrac{xy-3x+2y-6}{x^2-3x-10} \div \dfrac{xy-3x}{xy-5y}$ $\quad \dfrac{y}{x}$

30. $\dfrac{2a-ab+7b-14}{b^2-14b+49} \div \dfrac{ab-2a}{ab-7a}$ $\quad \dfrac{7-a}{b-7}$

31. $\dfrac{m^2+2m-8}{m^2-2m} \div \dfrac{m^2-16}{m^2}$ $\quad \dfrac{m}{m-4}$

32. $\dfrac{18-6x}{x^2-25} \div \dfrac{2x^2-18}{x^3-2x^2-25x+50}$ $\quad \dfrac{-3(x-2)}{x+3}$

33. $\dfrac{y+3}{3y^2+9y} \cdot \dfrac{y^2+7y+12}{y^2-16} \div \dfrac{y^2+4y}{y^2-4y}$ $\quad \dfrac{y+3}{3y(y+4)}$

34. $\dfrac{x^2+4x-5}{x^2-5x-14} \div \dfrac{x^2-1}{x^2-4} \cdot \dfrac{x+1}{x+5}$ $\quad \dfrac{x-2}{x-7}$

35. $\dfrac{x^2-0.49}{x^2+0.5x-0.14} \div \dfrac{x^2-x+0.21}{x^2-0.09}$ $\quad \dfrac{x+0.3}{x-0.2}$

36. $\dfrac{x^2-0.25}{x^2+0.1x-0.2} \div \dfrac{x^2-0.8x+0.15}{x^2-0.16}$ $\quad \dfrac{x+0.4}{x-0.3}$

37. $\dfrac{n^2-\frac{4}{9}}{n^2-\frac{13}{15}n+\frac{2}{15}} \div \dfrac{n^2+\frac{4}{3}n+\frac{4}{9}}{n^2-\frac{1}{25}}$ $\quad \dfrac{n+\frac{1}{5}}{n+\frac{2}{3}}$

38. $\dfrac{q^2-\frac{9}{25}}{q^2-\frac{1}{10}q-\frac{3}{10}} \div \dfrac{q^2+\frac{17}{20}q+\frac{3}{20}}{q^2-\frac{1}{16}}$ $\quad \dfrac{q-\frac{1}{4}}{q+\frac{1}{2}}$

39. $\dfrac{3a^3-24a^2-12a+96}{a^2-11a+24} \div \dfrac{6a^2-24}{3a^3-81}$ $\quad \dfrac{3(a^2+3a+9)}{2}$

40. $\dfrac{p^3+p^2-49p-49}{p^2+6p-7} \div \dfrac{p^2+p+1}{p^3-1}$ $\quad (p+1)(p-7)$

41. $\dfrac{4n^2-1}{12n^2-5n-3} \cdot \dfrac{6n^2+5n+1}{2n^2+n} \cdot \dfrac{12n^2-17n+6}{6n^2-7n+2}$ $\quad \dfrac{2n+1}{n}$

42. $\left(\dfrac{4x^2-25}{x^2-11x+30} \div \dfrac{2x^2-x-15}{x^2-9x+18}\right) \cdot \dfrac{4x^2+25x-21}{12x^2-5x-3}$

$\dfrac{(2x-5)(x+7)}{(x-5)(3x+1)}$

Compute as indicated. Write answers in lowest terms [recall that $a-b=-1(b-a)$].

43. $\dfrac{3}{8x^2}+\dfrac{5}{2x}$ $\quad \dfrac{3+20x}{8x^2}$

44. $\dfrac{15}{16y}-\dfrac{7}{2y^2}$ $\quad \dfrac{15y-56}{16y^2}$

45. $\dfrac{7}{4x^2y^3}-\dfrac{1}{8xy^4}$ $\quad \dfrac{14y-x}{8x^2y^4}$

46. $\dfrac{3}{6a^3b}+\dfrac{5}{9ab^3}$ $\quad \dfrac{9b^2+10a^2}{18a^3b^3}$

47. $\dfrac{4p}{p^2-36}-\dfrac{2}{p-6}$ $\quad \dfrac{2}{p+6}$

48. $\dfrac{3q}{q^2-49}-\dfrac{3}{2q-14}$ $\quad \dfrac{3}{2(q+7)}$

49. $\dfrac{m}{m^2-16}+\dfrac{4}{4-m}$ $\quad \dfrac{-3m-16}{(m+4)(m-4)}$

50. $\dfrac{2}{4-p^2}+\dfrac{p}{p-2}$ $\quad \dfrac{2-2p-p^2}{(2+p)(2-p)}$

51. $\dfrac{2}{m-7}-5$ $\quad \dfrac{-5m+37}{m-7}$

52. $\dfrac{4}{x-1}-9$ $\quad \dfrac{-9x+13}{x-1}$

53. $\dfrac{y+1}{y^2+y-30}-\dfrac{2}{y+6}$ $\quad \dfrac{-y+11}{(y+6)(y-5)}$

54. $\dfrac{4n}{n^2-5n}-\dfrac{3}{4n-20}$ $\quad \dfrac{13}{4(n-5)}$

55. $\dfrac{1}{a+4}+\dfrac{a}{a^2-a-20}$ $\quad \dfrac{2a-5}{(a+4)(a-5)}$

56. $\dfrac{2x-1}{x^2+3x-4}-\dfrac{x-5}{x^2+3x-4}$ $\quad \dfrac{1}{x-1}$

57. $\dfrac{3y-4}{y^2+2y+1}-\dfrac{2y-5}{y^2+2y+1}$ $\quad \dfrac{1}{y+1}$

58. $\dfrac{-2}{3a+12}-\dfrac{7}{a^2+4a}$ $\quad \dfrac{-2a-21}{3a(a+4)}$

59. $\dfrac{2}{m^2-9}+\dfrac{m-5}{m^2+6m+9}$ $\quad \dfrac{m^2-6m+21}{(m+3)^2(m-3)}$

60. $\dfrac{m+2}{m^2-25}-\dfrac{m+6}{m^2-10m+25}$ $\quad \dfrac{-2(7m+20)}{(m-5)^2(m+5)}$

61. $\dfrac{y+2}{5y^2+11y+2}+\dfrac{5}{y^2+y-6}$ $\quad \dfrac{y^2+26y-1}{(5y+1)(y+3)(y-2)}$

62. $\dfrac{m-4}{3m^2-11m+6}+\dfrac{m}{2m^2-m-15}$

$\dfrac{5(m^2-m-4)}{(3m-2)(m-3)(2m+5)}$

Write each term as a rational expression. Then compute the sum or difference indicated.

63. a. $p^{-2}-5p^{-1}$ $\quad \dfrac{1}{p^2}-\dfrac{5}{p}; \dfrac{1-5p}{p^2}$ **b.** $x^{-2}+2x^{-3}$ $\quad \dfrac{1}{x^2}+\dfrac{2}{x^3}; \dfrac{x+2}{x^3}$

64. a. $3a^{-1}+(2a)^{-1}$ **b.** $2y^{-1}-(3y)^{-1}$

$\dfrac{3}{a}+\dfrac{1}{2a}; \dfrac{7}{2a}$ $\qquad\qquad \dfrac{2}{y}-\dfrac{1}{3y}; \dfrac{5}{3y}$

Simplify each compound fraction. Use either method.

65. $\dfrac{\dfrac{5}{a} - \dfrac{1}{4}}{\dfrac{25}{a^2} - \dfrac{1}{16}}$ $\dfrac{4a}{a + 20}$ 66. $\dfrac{\dfrac{8}{x^3} - \dfrac{1}{27}}{\dfrac{2}{x} - \dfrac{1}{3}}$ $\dfrac{x^2 + 6x + 36}{9x^2}$

67. $\dfrac{p + \dfrac{1}{p - 2}}{1 + \dfrac{1}{p - 2}}$ $p - 1$ 68. $\dfrac{1 + \dfrac{3}{y - 6}}{y + \dfrac{9}{y - 6}}$ $\dfrac{1}{y - 3}$

69. $\dfrac{\dfrac{2}{3 - x} + \dfrac{3}{x - 3}}{\dfrac{4}{x} + \dfrac{5}{x - 3}}$ $\dfrac{x}{3(3x - 4)}$ 70. $\dfrac{\dfrac{1}{y - 5} - \dfrac{2}{5 - y}}{\dfrac{3}{y - 5} - \dfrac{2}{y}}$ $\dfrac{3y}{y + 10}$

71. $\dfrac{\dfrac{2}{y^2 - y - 20}}{\dfrac{3}{y + 4} - \dfrac{4}{y - 5}}$ $\dfrac{-2}{y + 31}$ 72. $\dfrac{\dfrac{2}{x^2 - 3x - 10}}{\dfrac{6}{x + 2} - \dfrac{4}{x - 5}}$ $\dfrac{1}{x - 19}$

Rewrite each expression as a compound fraction. Then simplify using either method.

73. a. $\dfrac{1 + 3m^{-1}}{1 - 3m^{-1}}$ $\dfrac{1 + \dfrac{3}{m}}{1 - \dfrac{3}{m}}; \dfrac{m + 3}{m - 3}$ b. $\dfrac{1 + 2x^{-2}}{1 - 2x^{-2}}$ $\dfrac{1 + \dfrac{2}{x^2}}{1 - \dfrac{2}{x^2}}; \dfrac{x^2 + 2}{x^2 - 2}$

74. a. $\dfrac{4 - 9a^{-2}}{3a^{-2}}$ $\dfrac{4 - \dfrac{9}{a^2}}{\dfrac{3}{a^2}}; \dfrac{4a^2 - 9}{3}$ b. $\dfrac{3 + 2n^{-1}}{5n^{-2}}$ $\dfrac{3 + \dfrac{2}{n}}{\dfrac{5}{n^2}}; \dfrac{3n^2 + 2n}{5}$

Solve each equation. Identify any extraneous roots.

75. $\dfrac{2}{x} + \dfrac{1}{x + 1} = \dfrac{5}{x^2 + x}$ $x = 1$

76. $\dfrac{3}{m + 3} - \dfrac{5}{m^2 + 3m} = \dfrac{1}{m}$ $m = 4$

77. $\dfrac{21}{a + 2} = \dfrac{3}{a - 1}$ $a = \dfrac{3}{2}$

78. $\dfrac{4}{2y - 3} = \dfrac{7}{3y - 5}$ $y = \dfrac{1}{2}$

79. $\dfrac{1}{3y} - \dfrac{1}{4y} = \dfrac{1}{y^2}$ $y = 12$ 80. $\dfrac{3}{5x} - \dfrac{1}{2x} = \dfrac{1}{x^2}$ $x = 10$

81. $x + \dfrac{14}{x - 7} = 1 + \dfrac{2x}{x - 7}$ $x = 3; x = 7$ is extraneous

82. $\dfrac{10}{x - 5} + x = 1 + \dfrac{2x}{x - 5}$ $x = 3; x = 5$ is extraneous

83. $\dfrac{6}{n + 3} + \dfrac{20}{n^2 + n - 6} = \dfrac{5}{n - 2}$ $n = 7$

84. $\dfrac{7}{p + 2} - \dfrac{1}{p^2 + 5p + 6} = -\dfrac{2}{p + 3}$ $p = -\dfrac{8}{3}$

85. $\dfrac{a}{2a + 1} - \dfrac{2a^2 + 5}{2a^2 - 5a - 3} = \dfrac{3}{a - 3}$ $a = -1, a = -8$

86. $\dfrac{-18}{6n^2 - n - 1} + \dfrac{3n}{2n - 1} = \dfrac{4n}{3n + 1}$ $n = -9, n = 2$

▶ WORKING WITH FORMULAS

 87. **Cost to seize illegal drugs:** $C = \dfrac{450P}{100 - P}$

The cost C, in millions of dollars, for a government to find and seize $P\%$ ($0 \le P < 100$) of a certain illegal drug is modeled by the rational equation shown. Complete the table (round to the nearest dollar) and answer the following questions.

a. What is the cost of seizing 40% of the drugs? Estimate the cost at 85%.
$300 million $2550 million
b. Why does cost increase dramatically the closer you get to 100%?
It would require many resources.
c. Will 100% of the drugs ever be seized?   No

P	$\dfrac{450P}{100 - P}$
40	300
60	675
80	1800
90	4050
93	5979
95	8550
98	22050
100	ERROR

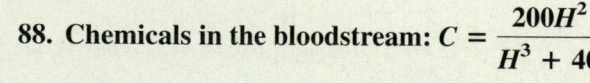 88. **Chemicals in the bloodstream:** $C = \dfrac{200H^2}{H^3 + 40}$

Rational equations are often used to model chemical concentrations in the bloodstream. The percent concentration C of a certain drug H hours after injection into muscle tissue can be modeled by the equation shown ($H \ge 0$). Complete the table (round to the nearest tenth of a percent) and answer the following questions.

a. What is the percent concentration of the drug 3 hr after injection?   26.9%
b. Why is the concentration virtually equal at $H = 4$ and $H = 5$?
It reaches maximum concentration between these hours.
c. Why does the concentration begin to decrease?
The drug is absorbed into the body (it begins to wear off).
d. How long will it take for the concentration to become less than 10%?   20 hr

H	$\dfrac{200H^2}{H^3 + 40}$
0	0
1	4.9
2	16.7
3	26.9
4	30.8
5	30.3
6	28.1
7	25.6

▶ APPLICATIONS

89. Stock prices: When a hot new stock hits the market, its price will often rise dramatically and then taper off over time. The equation $P = \dfrac{50(7d^2 + 10)}{d^3 + 50}$ models the price of stock XYZ d days after it has "hit the market." (a) Create a table of values showing the price of the stock for the first 10 days (rounded to the nearest dollar) and comment on what you notice. (b) Find the opening price of the stock. (c) Does the stock ever return to its original price?

90. Population growth: The Department of Wildlife introduces 60 elk into a new game reserve. It is projected that the size of the herd will grow according to the equation $N = \dfrac{10(6 + 3t)}{1 + 0.05t}$, where N is the number of elk and t is the time in years. (a) Approximate the population of elk after 14 yr. (b) If recent counts find 225 elk, approximately how many years have passed?
a. $N = 282$ **b.** $t = 8.8$; just less than 9 yr

91. Typing speed: The number of words per minute that a beginner can type is approximated by the equation $N = \dfrac{60t - 120}{t}$, where N is the number

of words per minute after t weeks, $3 < t < 12$. Use a table to determine how many weeks it takes for a student to be typing an average of forty-five words per minute. $t = 8$ weeks

92. Memory retention: A group of students is asked to memorize 50 Russian words that are unfamiliar to them. The number N of these words that the average student remembers D days later is modeled by the equation $N = \dfrac{5D + 35}{D} (D \geq 1)$. How many words are remembered after (a) 1 day? (b) 5 days? (c) 12 days? (d) 35 days? (e) 100 days? According to this model, is there a certain number of words that the average student never forgets? yes How many?   5 words
a. 40 words   **b.** 12 words   **c.** 8 words   **d.** 6 words   **e.** 5 words

93. Pollution removal: For a steel mill, the cost C (in millions of dollars) to remove toxins from the resulting sludge is given by $C = \dfrac{22P}{100 - P}$, where P is the percent of the toxins removed. What percent can be removed if the mill spends $88,000,000 on the cleanup? Round to tenths of a percent. $P \approx 80\%$

▶ EXTENDING THE CONCEPT

94. One of these expressions is *not* equal to the others. Identify which and explain why.

a. $\dfrac{20n}{10n}$

b. $20 \cdot n \div 10 \cdot n$
b. $20 \cdot n \div 10 \cdot n = 2n^2$, all others equal 2

c. $20n \cdot \dfrac{1}{10n}$

d. $\dfrac{20}{10} \cdot \dfrac{n}{n}$

95. The average of A and B is x. The average of C, D, and E is y. The average of A, B, C, D, and E is: b

a. $\dfrac{3x + 2y}{5}$

b. $\dfrac{2x + 3y}{5}$

c. $\dfrac{2(x + y)}{5}$

d. $\dfrac{3(x + y)}{5}$

96. Given the rational numbers $\dfrac{2}{5}$ and $\dfrac{3}{4}$, what is the reciprocal of the sum of their reciprocals? Given that $\dfrac{a}{b}$ and $\dfrac{c}{d}$ are *any* two numbers—what is the reciprocal of the sum of their reciprocals? $\dfrac{6}{23}$; $\dfrac{ac}{ad + bc}$

Radicals, Rational Exponents, and Radical Equations

LEARNING OBJECTIVES

In Section R.6 you will review how to:

☐ **A.** Simplify radical expressions of the form $\sqrt[n]{a^n}$

☐ **B.** Rewrite and simplify radical expressions using rational exponents

☐ **C.** Use properties of radicals to simplify radical expressions

☐ **D.** Add and subtract radical expressions

☐ **E.** Multiply and divide radical expressions; write a radical expression in simplest form

☐ **F.** Solve equations and use formulas involving radicals

Square roots and cube roots come from a much larger family called **radical expressions**. Expressions containing radicals can be found in virtually every field of mathematical study, and are an invaluable tool for modeling many real-world phenomena.

A. Simplifying Radical Expressions of the Form $\sqrt[n]{a^n}$

In previous coursework, you likely noted that $\sqrt{a} = b$ only if $b^2 = a$. This definition cannot be applied to expressions like $\sqrt{-16}$, since there is no number b such that $b^2 = -16$. In other words, the expression $\sqrt{a}$ represents a real number only if $a \geq 0$ (for a full review of the real numbers and other sets of numbers, see Appendix I at www.mhhe.com/coburn). Of particular interest to us now is an inverse operation for a^2. In other words, what operation can be applied to a^2 to return a? Consider the following.

EXAMPLE 1 ▶ Evaluating a Radical Expression

Evaluate $\sqrt{a^2}$ for the values given:

  a. $a = 3$ **b.** $a = 5$ **c.** $a = -6$

Solution ▶   **a.** $\sqrt{3^2} = \sqrt{9}$ **b.** $\sqrt{5^2} = \sqrt{25}$ **c.** $\sqrt{(-6)^2} = \sqrt{36}$

              $= 3$ $= 5$ $= 6$

                                       Now try Exercises 7 and 8 ▶

The pattern seemed to indicate that $\sqrt{a^2} = a$ and that our search for an inverse operation was complete—until Example 1(c), where we found that $\sqrt{(-6)^2} \neq -6$. Using the absolute value concept, we can "repair" this apparent discrepancy and state a general rule for simplifying these expressions: $\sqrt{a^2} = |a|$. For expressions like $\sqrt{49x^2}$ and $\sqrt{y^6}$, the radicands can be rewritten as perfect squares and simplified in the same manner: $\sqrt{49x^2} = \sqrt{(7x)^2} = 7|x|$ and $\sqrt{y^6} = \sqrt{(y^3)^2} = |y^3|$.

The Square Root of a^2: $\sqrt{a^2}$

For any real number a,
$$\sqrt{a^2} = |a|.$$

EXAMPLE 2 ▶ Simplifying Square Root Expressions

Simplify each expression.

  a. $\sqrt{169x^2}$ **b.** $\sqrt{x^2 - 10x + 25}$

Solution ▶   **a.** $\sqrt{169x^2} = |13x|$

                 $= 13|x|$ since *x* could be negative

    b. $\sqrt{x^2 - 10x + 25} = \sqrt{(x - 5)^2}$

                     $= |x - 5|$ since *x* − 5 could be negative

                                       Now try Exercises 9 and 10 ▶

> ⚠ **CAUTION** ▶ In Section R.2, we noted that $(A + B)^2 \neq A^2 + B^2$, indicating that you cannot square the individual terms in a sum (the square of a binomial results in a perfect square trinomial). In a similar way, $\sqrt{A^2 + B^2} \neq A + B$, and you cannot take the square root of individual terms. There is a big difference between the expressions $\sqrt{A^2 + B^2}$ and $\sqrt{(A + B)^2} = |A + B|$. Try evaluating each when $A = 3$ and $B = 4$.

To investigate expressions like $\sqrt[3]{x^3}$, note the radicand in both $\sqrt[3]{8}$ and $\sqrt[3]{-64}$ can be written as a perfect cube. From our earlier definition of cube roots we know $\sqrt[3]{8} = \sqrt[3]{(2)^3} = 2$, $\sqrt[3]{-64} = \sqrt[3]{(-4)^3} = -4$, and that every real number has only one real cube root. For this reason, absolute value notation is not used or needed when taking cube roots.

The Cube Root of a^3: $\sqrt[3]{a^3}$

For any real number a,
$$\sqrt[3]{a^3} = a.$$

EXAMPLE 3 ▶ **Simplifying Cube Root Expressions**

Simplify each expression.

  a. $\sqrt[3]{-27x^3}$ **b.** $\sqrt[3]{-64n^6}$

Solution ▶   **a.** $\sqrt[3]{-27x^3} = \sqrt[3]{(-3x)^3}$ **b.** $\sqrt[3]{-64n^6} = \sqrt[3]{(-4n^2)^3}$
$$= -3x \qquad\qquad\qquad = -4n^2$$

Now try Exercises 11 and 12 ▶

We can extend these ideas to fourth roots, fifth roots, and so on. For example, the fifth root of a is b only if $b^5 = a$. In symbols, $\sqrt[5]{a} = b$ implies $b^5 = a$. Since an odd number of negative factors is always negative: $(-2)^5 = -32$, and an even number of negative factors is always positive: $(-2)^4 = 16$, we must take the index into account when evaluating expressions like $\sqrt[n]{a^n}$. If n is even and the radicand is unknown, absolute value notation must be used.

WORTHY OF NOTE

Just as $\sqrt[2]{-16}$ is not a real number, $\sqrt[4]{-16}$ and $\sqrt[6]{-16}$ do not represent real numbers. An even number of repeated factors is always positive!

The nth Root of a^n: $\sqrt[n]{a^n}$

For any real number a,
  1. $\sqrt[n]{a^n} = |a|$ when n is even.   **2.** $\sqrt[n]{a^n} = a$ when n is odd.

EXAMPLE 4 ▶ **Simplifying Radical Expressions**

Simplify each expression.

  a. $\sqrt[4]{81}$ **b.** $\sqrt[4]{-81}$ **c.** $\sqrt[5]{32}$ **d.** $\sqrt[5]{-32}$
  e. $\sqrt[4]{16m^4}$ **f.** $\sqrt[5]{32p^5}$ **g.** $\sqrt[6]{(m + 5)^6}$ **h.** $\sqrt[7]{(x - 2)^7}$

Solution ▶   **a.** $\sqrt[4]{81} = 3$ **b.** $\sqrt[4]{-81}$ is not a real number
  c. $\sqrt[5]{32} = 2$ **d.** $\sqrt[5]{-32} = -2$
  e. $\sqrt[4]{16m^4} = \sqrt[4]{(2m)^4}$ **f.** $\sqrt[5]{32p^5} = \sqrt[5]{(2p)^5}$
$$= |2m| \text{ or } 2|m| \qquad\qquad = 2p$$
  g. $\sqrt[6]{(m + 5)^6} = |m + 5|$ **h.** $\sqrt[7]{(x - 2)^7} = x - 2$

✓ **A.** You've just seen how we can simplify radical expressions of the form $\sqrt[n]{a^n}$

Now try Exercises 13 and 14 ▶

B. Radical Expressions and Rational Exponents

As an alternative to radical notation, a rational (fractional) exponent can be used, along with the power property of exponents. For $\sqrt[3]{a^3} = a$, notice that an exponent of one-third can replace the cube root notation and produce the same result: $\sqrt[3]{a^3} = (a^3)^{\frac{1}{3}} = a^{\frac{3}{3}} = a$. In the same way, an exponent of one-half can replace the square root notation: $\sqrt{a^2} = (a^2)^{\frac{1}{2}} = a^{\frac{2}{2}} = |a|$. In general, we have the following:

Rational Exponents

If a is a real number and n is an integer greater than 1,
$$\text{then } \sqrt[n]{a} = \sqrt[n]{a^1} = a^{\frac{1}{n}}$$
provided $\sqrt[n]{a}$ represents a real number.

EXAMPLE 5 ▶ **Simplifying Radical Expressions Using Rational Exponents**

Simplify by rewriting each radicand as a perfect nth power and converting to rational exponent notation.

a. $\sqrt[3]{-125}$ **b.** $-\sqrt[4]{16x^{20}}$ **c.** $\sqrt[4]{-81}$ **d.** $\sqrt[3]{\dfrac{8w^3}{27}}$

Solution ▶
a. $\sqrt[3]{-125} = \sqrt[3]{(-5)^3}$ **b.** $-\sqrt[4]{16x^{20}} = -\sqrt[4]{(2x^5)^4}$
$= [(-5)^3]^{\frac{1}{3}}$ $= -[(2x^5)^4]^{\frac{1}{4}}$
$= (-5)^{\frac{3}{3}}$ $= -|2x^5|$
$= -5$ $= -2|x|^5$

c. $\sqrt[4]{-81} = (-81)^{\frac{1}{4}}$ **d.** $\sqrt[3]{\dfrac{8w^3}{27}} = \sqrt[3]{\left(\dfrac{2w}{3}\right)^3}$
    is not a real number $= \left[\left(\dfrac{2w}{3}\right)^3\right]^{\frac{1}{3}} = \dfrac{2w}{3}$

Now try Exercises 15 and 16 ▶

When a rational exponent is used, as in $\sqrt[n]{a} = \sqrt[n]{a^1} = a^{\frac{1}{n}}$, the **denominator** of the exponent represents the index number, while the **numerator** of the exponent represents the original power on a. *This is true even when the exponent on a is something other than one!* In other words, the radical expression $\sqrt[4]{16^3}$ can be rewritten as $(16^3)^{\frac{1}{4}} = (16^{\frac{3}{1}})^{\frac{1}{4}}$ or $16^{\frac{3}{4}}$. This is further illustrated in Figure R.6 where we see the rational exponent has the form, "power over root." To evaluate this expression without the aid of a calculator, we use the commutative property to rewrite $(16^{\frac{3}{1}})^{\frac{1}{4}}$ as $(16^{\frac{1}{4}})^{\frac{3}{1}}$ and begin with the fourth root of 16: $(16^{\frac{1}{4}})^{\frac{3}{1}} = 2^3 = 8$.

Figure R.6

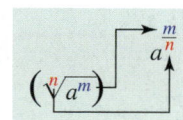

In general, if m and n have no common factors (other than 1) the expression $a^{\frac{m}{n}}$ can be interpreted in the following two ways.

Rational Exponents

If $\frac{m}{n}$ is a rational number expressed in lowest terms with $n \geq 2$, then
$$(1)\ a^{\frac{m}{n}} = (\sqrt[n]{a})^m \qquad \text{or} \qquad (2)\ a^{\frac{m}{n}} = \sqrt[n]{a^m}$$
(compute $\sqrt[n]{a}$, then take the mth power),   (compute a^m, then take the nth root), provided $\sqrt[n]{a}$ represents a real number.

WORTHY OF NOTE

Any rational number can be decomposed into the product of a unit fraction and an integer: $\dfrac{m}{n} = \dfrac{1}{n} \cdot m$.

Expressions with rational exponents are generally easier to evaluate if we compute the root first, then apply the exponent. Computing the root first also helps us determine whether or not an expression represents a real number.

EXAMPLE 6 ▶ **Simplifying Expressions with Rational Exponents**

Simplify each expression, if possible.

    a. $-49^{\frac{3}{2}}$ **b.** $(-49)^{\frac{3}{2}}$ **c.** $(-8)^{\frac{2}{3}}$ **d.** $-8^{-\frac{2}{3}}$

Solution ▶

a. $-49^{\frac{3}{2}} = -(49^{\frac{1}{2}})^3$
$\qquad\quad = -(\sqrt{49})^3$
$\qquad\quad = -(7)^3$ or -343

b. $(-49)^{\frac{3}{2}} = [(-49)^{\frac{1}{2}}]^3,$
$\qquad\qquad = (\sqrt{-49})^3$
$\qquad$ not a real number

c. $(-8)^{\frac{2}{3}} = [(-8)^{\frac{1}{3}}]^2$
$\qquad\quad = (\sqrt[3]{-8})^2$
$\qquad\quad = (-2)^2$ or 4

d. $-8^{-\frac{2}{3}} = -(8^{\frac{1}{3}})^{-2}$
$\qquad\quad = -(\sqrt[3]{8})^{-2}$
$\qquad\quad = -2^{-2}$ or $-\dfrac{1}{4}$

> **WORTHY OF NOTE**
>
> While the expression $(-8)^{\frac{1}{3}} = \sqrt[3]{-8}$ represents the real number -2, the expression $(-8)^{\frac{2}{6}} = (\sqrt[6]{-8})^2$ is not a real number, even though $\dfrac{1}{3} = \dfrac{2}{6}$. Note that the second exponent is not in lowest terms.

☑ **B.** You've just seen how we can rewrite and simplify radical expressions using rational exponents

Now try Exercises 17 through 22 ▶

C. Using Properties of Radicals to Simplify Radical Expressions

The properties used to simplify radical expressions are closely connected to the properties of exponents. For instance, the product to a power property holds even when n is a rational number. This means $(xy)^{\frac{1}{2}} = x^{\frac{1}{2}}y^{\frac{1}{2}}$ and $(4 \cdot 25)^{\frac{1}{2}} = 4^{\frac{1}{2}} \cdot 25^{\frac{1}{2}}$. When the second statement is expressed in radical form, we have $\sqrt{4 \cdot 25} = \sqrt{4} \cdot \sqrt{25}$, with both forms having a value of 10. This suggests the **product property of radicals,** which can be extended to include cube roots, fourth roots, and so on.

> **Product Property of Radicals**
>
> If $\sqrt[n]{A}$ and $\sqrt[n]{B}$ represent real-valued expressions, then
>
> $$\sqrt[n]{AB} = \sqrt[n]{A} \cdot \sqrt[n]{B} \qquad \text{and} \qquad \sqrt[n]{A} \cdot \sqrt[n]{B} = \sqrt[n]{AB}.$$

⚠ **CAUTION** ▶ Note that this property applies only to a *product* of two terms, not to a sum or difference. In other words, while $\sqrt{9x^2} = |3x|$, $\sqrt{9 + x^2} \neq |3 + x|$!

One application of the product property is to simplify radical expressions. In general, the expression $\sqrt[n]{a}$ is in simplified form if a has no factors (other than 1) that are perfect nth roots.

EXAMPLE 7 ▶ **Simplifying Radical Expressions**

Write each expression in simplest form using the product property.

    a. $\sqrt{18}$ **b.** $5\sqrt[3]{125x^4}$ **c.** $\dfrac{-4 + \sqrt{20}}{2}$ **d.** $1.2\sqrt[3]{16n^4}\,\sqrt[3]{4n^5}$

Solution ▶

a. $\sqrt{18} = \sqrt{9 \cdot 2}$
$\qquad\quad = \sqrt{9}\sqrt{2}$
$\qquad\quad = 3\sqrt{2}$

b. $5\sqrt[3]{125x^4} = 5 \cdot \sqrt[3]{125 \cdot x^4}$
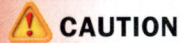 *These steps can be done mentally* $\begin{cases} = 5 \cdot \sqrt[3]{125} \cdot \sqrt[3]{x^3} \cdot \sqrt[3]{x^1} \\ = 5 \cdot 5 \cdot x \cdot \sqrt[3]{x} \end{cases}$
$\qquad\qquad\qquad = 25x\sqrt[3]{x}$

WORTHY OF NOTE

For expressions like those in Example 7(c), students must resist the "temptation" to reduce individual terms as in
$$\frac{-4 + \sqrt{20}}{2} \neq -2 + \sqrt{20}.$$
Remember, only *factors* can be reduced.

c. $\dfrac{-4 + \sqrt{20}}{2} = \dfrac{-4 + \sqrt{4 \cdot 5}}{2}$

$\qquad\qquad\quad = \dfrac{-4 + 2\sqrt{5}}{2}$

$\qquad\qquad\quad = \dfrac{\overset{1}{2}(-2 + \sqrt{5})}{\cancel{2}}$

$\qquad\qquad\quad = -2 + \sqrt{5}$

d. $1.2\sqrt[3]{16n^4}\sqrt[3]{4n^5} = 1.2\sqrt[3]{64 \cdot n^9}$

$\qquad\qquad\qquad\quad = 1.2\sqrt[3]{64}\sqrt[3]{n^9}$

$\qquad\qquad\qquad\quad = 1.2\sqrt[3]{64}\sqrt[3]{(n^3)^3}$

$\qquad\qquad\qquad\quad = 1.2(4)n^3$

$\qquad\qquad\qquad\quad = 4.8n^3$

Now try Exercises 23 through 26 ▶

When radicals are *combined* using the product property, the result may contain a perfect *n*th root, which should be simplified. Note that the *index numbers must be the same* in order to use this property.

The **quotient property of radicals** can also be established using exponential properties. The fact that $\dfrac{\sqrt{100}}{\sqrt{25}} = \sqrt{\dfrac{100}{25}} = 2$ suggests the following:

Quotient Property of Radicals

If $\sqrt[n]{A}$ and $\sqrt[n]{B}$ represent real-valued expressions with $B \neq 0$, then

$$\sqrt[n]{\dfrac{A}{B}} = \dfrac{\sqrt[n]{A}}{\sqrt[n]{B}} \quad \text{and} \quad \dfrac{\sqrt[n]{A}}{\sqrt[n]{B}} = \sqrt[n]{\dfrac{A}{B}}.$$

Many times the product and quotient properties must work together to simplify a radical expression, as shown in Example 8A.

EXAMPLE 8A ▶ **Simplifying Radical Expressions**

Simplify each expression:

a. $\dfrac{\sqrt{18a^5}}{\sqrt{2a}}$

b. $\sqrt[3]{\dfrac{81}{125x^3}}$

Solution ▶   **a.** $\dfrac{\sqrt{18a^5}}{\sqrt{2a}} = \sqrt{\dfrac{18a^5}{2a}}$

$\qquad\qquad\qquad\quad = \sqrt{9a^4}$

$\qquad\qquad\qquad\quad = 3a^2$

b. $\sqrt[3]{\dfrac{81}{125x^3}} = \dfrac{\sqrt[3]{81}}{\sqrt[3]{125x^3}}$

$\qquad\qquad\qquad = \dfrac{\sqrt[3]{27 \cdot 3}}{5x}$

$\qquad\qquad\qquad = \dfrac{3\sqrt[3]{3}}{5x}$

Radical expressions can also be simplified using rational exponents.

EXAMPLE 8B ▶ **Using Rational Exponents to Simplify Radical Expressions**

Simplify using rational exponents:

a. $\sqrt{36p^4q^5}$ **b.** $v\sqrt[3]{v^4}$ **c.** $\sqrt[3]{m}\sqrt{m}$

Solution ▶   **a.** $\sqrt{36p^4q^5} = (36p^4q^5)^{\frac{1}{2}}$

$\qquad\qquad\qquad = 36^{\frac{1}{2}}p^{\frac{4}{2}}q^{\frac{5}{2}}$

$\qquad\qquad\qquad = 6p^2q^{(\frac{4}{2} + \frac{1}{2})}$

$\qquad\qquad\qquad = 6p^2q^2q^{\frac{1}{2}}$

$\qquad\qquad\qquad = 6p^2q^2\sqrt{q}$

b. $v\sqrt[3]{v^4} = v^1 \cdot v^{\frac{4}{3}}$

$\qquad\qquad = v^{\frac{3}{3}} \cdot v^{\frac{4}{3}}$

$\qquad\qquad = v^{\frac{7}{3}}$

$\qquad\qquad = v^{\frac{6}{3}}v^{\frac{1}{3}}$

$\qquad\qquad = v^2\sqrt[3]{v}$

c. $\sqrt[3]{m}\sqrt{m} = m^{\frac{1}{3}}m^{\frac{1}{2}}$

$\qquad\qquad = m^{\frac{1}{3} + \frac{1}{2}}$

$\qquad\qquad = m^{\frac{5}{6}}$

$\qquad\qquad = \sqrt[6]{m^5}$

☑ **C.** You've just seen how we can use properties of radicals to simplify radical expressions

Now try Exercises 27 through 30 ▶

D. Addition and Subtraction of Radical Expressions

Since $3x$ and $5x$ are like terms, we know $3x + 5x = 8x$. If $x = \sqrt[3]{7}$, the sum becomes $3\sqrt[3]{7} + 5\sqrt[3]{7} = 8\sqrt[3]{7}$, illustrating how *like* radical expressions can be combined. Like radicals are those that have *the same index and radicand*. In some cases, we can identify like radicals only after radical terms have been simplified.

EXAMPLE 9 ▶ **Adding and Subtracting Radical Expressions**

Simplify and combine (if possible).

    a. $\sqrt{45} + 2\sqrt{20}$ **b.** $\sqrt[3]{16x^5} - x\sqrt[3]{54x^2}$

Solution ▶   **a.** $\sqrt{45} + 2\sqrt{20} = 3\sqrt{5} + 2(2\sqrt{5})$ simplify radicals: $\sqrt{45} = \sqrt{9 \cdot 5}$; $\sqrt{20} = \sqrt{4 \cdot 5}$

                     $= 3\sqrt{5} + 4\sqrt{5}$ like radicals

                     $= 7\sqrt{5}$ result

    b. $\sqrt[3]{16x^5} - x\sqrt[3]{54x^2} = \sqrt[3]{8 \cdot 2 \cdot x^3 \cdot x^2} - x\sqrt[3]{27 \cdot 2 \cdot x^2}$

                        $= 2x\sqrt[3]{2x^2} - 3x\sqrt[3]{2x^2}$ simplify radicals

                        $= -x\sqrt[3]{2x^2}$ result

☑ **D.** You've just seen how we can add and subtract radical expressions

Now try Exercises 31 through 34 ▶

E. Multiplication and Division of Radical Expressions; Radical Expressions in Simplest Form

Multiplying radical expressions is simply an extension of our earlier work. The multiplication can take various forms, from the distributive property to any of the special products reviewed in Section R.2. For instance, $(A \pm B)^2 = A^2 \pm 2AB + B^2$, even if A or B is a radical term.

EXAMPLE 10 ▶ **Multiplying Radical Expressions**

Compute each product and simplify.

    a. $5\sqrt{3}(\sqrt{6} - 4\sqrt{3})$ **b.** $(2\sqrt{2} + 6\sqrt{3})(3\sqrt{10} + \sqrt{15})$

    c. $(x + \sqrt{7})(x - \sqrt{7})$ **d.** $(3 - \sqrt{2})^2$

Solution ▶   **a.** $5\sqrt{3}(\sqrt{6} - 4\sqrt{3}) = 5\sqrt{18} - 20(\sqrt{3})^2$ distribute

                        $= 5(3)\sqrt{2} - (20)(3)$ simplify: $\sqrt{18} = 3\sqrt{2}$, $(\sqrt{3})^2 = 3$

                        $= 15\sqrt{2} - 60$ result

    b. $(2\sqrt{2} + 6\sqrt{3})(3\sqrt{10} + \sqrt{15}) = 6\sqrt{20} + 2\sqrt{30} + 18\sqrt{30} + 6\sqrt{45}$ F-O-I-L

                        $= 12\sqrt{5} + 20\sqrt{30} + 18\sqrt{5}$ extract roots *and* simplify

                        $= 30\sqrt{5} + 20\sqrt{30}$ result

    c. $(x + \sqrt{7})(x - \sqrt{7}) = x^2 - (\sqrt{7})^2$ $(A + B)(A - B) = A^2 - B^2$

                      $= x^2 - 7$ result

    d. $(3 - \sqrt{2})^2 = (3)^2 - 2(3)(\sqrt{2}) + (\sqrt{2})^2$ $(A - B)^2 = A^2 - 2AB + B^2$

                     $= 9 - 6\sqrt{2} + 2$ simplify each term

                     $= 11 - 6\sqrt{2}$ result

LOOKING AHEAD

Notice that the answer for Example 10c contains no radical terms, since the outer and inner products sum to zero. This result will be used to simplify certain radical expressions in this section and later in Chapter 2.

Now try Exercises 35 through 38 ▶

One application of products and powers of radical expressions is to evaluate certain quadratic expressions, as illustrated in Example 11.

EXAMPLE 11 ▶ **Evaluating a Quadratic Expression**

Show that when $x^2 - 4x + 1$ is evaluated at $x = 2 + \sqrt{3}$, the result is zero.

Solution ▶

```
2+√(3)→X
         3.732050808
X²-4X+1
              0
```

$$x^2 - 4x + 1 \qquad \text{original expression}$$
$$(2 + \sqrt{3})^2 - 4(2 + \sqrt{3}) + 1 \qquad \text{substitute } 2 + \sqrt{3} \text{ for } x$$
$$4 + 4\sqrt{3} + 3 - 8 - 4\sqrt{3} + 1 \qquad \text{square binomial; distribute}$$
$$(4 + 3 - 8 + 1) + (4\sqrt{3} - 4\sqrt{3}) \qquad \text{commutative and associative properties}$$
$$0 \checkmark$$

A calculator check is shown in the figure.

Now try Exercises 39 through 42 ▶

When we applied the quotient property in Example 8A, we obtained a denominator free of radicals. Sometimes the denominator is not automatically free of radicals, and the need to write radical expressions in *simplest form* comes into play. This process is called **rationalizing the denominator.**

> **Radical Expressions in Simplest Form**
>
> A radical expression is in simplest form if:
> 1. The radicand has no perfect *n*th root factors.
> 2. The radicand contains no fractions.
> 3. No radicals occur in a denominator.

As with other types of simplification, the desired form can be achieved in various ways. If the denominator is a single radical term, we multiply the numerator and denominator by the factors required to eliminate the radical in the denominator [see Examples 12(a) and 12(b)]. If the radicand is a rational expression, it is generally easier to build an equivalent fraction *within the radical* having perfect *n*th root factors in the denominator [see Example 12(c)].

EXAMPLE 12 ▶ **Simplifying Radical Expressions**

Simplify by rationalizing the denominator. Assume $a, x \neq 0$.

a. $\dfrac{2}{5\sqrt{3}}$ **b.** $\dfrac{-7}{\sqrt[3]{x}}$

Solution ▶ **a.** $\dfrac{2}{5\sqrt{3}} = \dfrac{2}{5\sqrt{3}} \cdot \dfrac{\sqrt{3}}{\sqrt{3}}$ — multiply numerator and denominator by $\sqrt{3}$

$$= \dfrac{2\sqrt{3}}{5(\sqrt{3})^2} = \dfrac{2\sqrt{3}}{15} \qquad \text{simplify—denominator is now rational}$$

b. $\dfrac{-7}{\sqrt[3]{x}} = \dfrac{-7(\sqrt[3]{x})(\sqrt[3]{x})}{\sqrt[3]{x}(\sqrt[3]{x})(\sqrt[3]{x})}$ — multiply using two additional factors of $\sqrt[3]{x}$

$$= \dfrac{-7\sqrt[3]{x^2}}{\sqrt[3]{x^3}} \qquad \text{product property}$$

$$= \dfrac{-7\sqrt[3]{x^2}}{x} \qquad \sqrt[3]{x^3} = x$$

Now try Exercises 43 and 44 ▶

In some applications, the denominator may be a sum or difference containing a radical term. In this case, the methods from Example 12 are ineffective, and instead we multiply by a conjugate since $(A + B)(A - B) = A^2 - B^2$. If either A or B is a square root, the result will be a denominator free of radicals.

EXAMPLE 13 ▶ **Simplifying Radical Expressions Using a Conjugate**

Simplify the expression by rationalizing the denominator. Write the answer in exact form and approximate form rounded to three decimal places. $\dfrac{2 + \sqrt{3}}{\sqrt{6} - \sqrt{2}}$

Solution ▶ $\dfrac{2 + \sqrt{3}}{\sqrt{6} - \sqrt{2}} = \dfrac{2 + \sqrt{3}}{\sqrt{6} - \sqrt{2}} \cdot \dfrac{\sqrt{6} + \sqrt{2}}{\sqrt{6} + \sqrt{2}}$ multiply by the conjugate of the denominator

$\qquad = \dfrac{2\sqrt{6} + 2\sqrt{2} + \sqrt{18} + \sqrt{6}}{(\sqrt{6})^2 - (\sqrt{2})^2}$ FOIL

difference of squares

$\qquad = \dfrac{3\sqrt{6} + 2\sqrt{2} + 3\sqrt{2}}{6 - 2}$ simplify

$\qquad = \dfrac{3\sqrt{6} + 5\sqrt{2}}{4}$ exact form

$\qquad \approx 3.605$ approximate form

☑ **E.** You've just seen how we can multiply and divide radical expressions and write a radical expression in simplest form

Now try Exercises 45 through 48 ▶

F. Equations and Formulas Involving Radicals

A **radical equation** is any equation that contains terms with a variable in the radicand. To solve a radical equation, we attempt to isolate a radical term on one side, then apply the appropriate nth power to free up the radicand and solve for the unknown. This is an application of the **power property of equality.**

The Power Property of Equality

If $\sqrt[n]{u}$ and v are real-valued expressions and $\sqrt[n]{u} = v$,

$$\text{then } (\sqrt[n]{u})^n = v^n$$
$$u = v^n$$

for n an integer, $n \geq 2$.

Raising both sides of an equation to an *even* power can also introduce a false solution (extraneous root). Note that by inspection, the equation $x - 2 = \sqrt{x}$ has only the solution $x = 4$. But the equation $(x - 2)^2 = x$ (obtained by squaring both sides) has both $x = 4$ *and* $x = 1$ as solutions, yet $x = 1$ does not satisfy the original equation. This means we should *check all solutions of an equation where an even power is applied.*

EXAMPLE 14 ▶ **Solving Radical Equations**

Solve each radical equation:

a. $\sqrt{3x - 2} + 12 = x + 10$ **b.** $2\sqrt[3]{x - 5} + 4 = 0$

Solution ▶

a.

$\sqrt{3x - 2} + 12 = x + 10$	original equation
$\sqrt{3x - 2} = x - 2$	isolate radical term (subtract 12)
$(\sqrt{3x - 2})^2 = (x - 2)^2$	apply power property, power is even
$3x - 2 = x^2 - 4x + 4$	simplify, square binomial
$0 = x^2 - 7x + 6$	set equal to zero
$0 = (x - 6)(x - 1)$	factor
$x - 6 = 0$ or $x - 1 = 0$	apply zero product property
$x = 6$ or $x = 1$	result, check for extraneous roots

Check ▶ $x = 6$:

$$\sqrt{3(6) - 2} + 12 = (6) + 10$$
$$\sqrt{16} + 12 = 16$$
$$16 = 16 \checkmark$$

Check ▶ $x = 1$:

$$\sqrt{3(1) - 2} + 12 = (1) + 10$$
$$\sqrt{1} + 12 = 11$$
$$13 = 11\textcolor{red}{\times}$$

```
6→X
                    6
√(3X-2)+12
                   16
X+10
                   16
```

```
1→X
                    1
√(3X-2)+12
                   13
X+10
                   11
```

The only solution is $x = 6$; $x = 1$ is extraneous. A calculator check is shown in the figures.

b.

$2\sqrt[3]{x - 5} + 4 = 0$	original equation
$\sqrt[3]{x - 5} = -2$	isolate radical term (subtract 4, divide by 2)
$(\sqrt[3]{x - 5})^3 = (-2)^3$	apply power property, power is odd
$x - 5 = -8$	simplify: $\sqrt[3]{x - 5})^3 = x - 5$
$x = -3$	solve

Substituting -3 for x in the original equation verifies it is a solution.

Now try Exercises 49 through 52 ▶

Sometimes squaring both sides of an equation still results in an equation with a radical term, but often there is *one fewer* than before. In this case, we simply repeat the process, as indicated by the flowchart in Figure R.7.

EXAMPLE 15 ▶ **Solving Radical Equations**

Solve the equation: $\sqrt{x + 15} - \sqrt{x + 3} = 2$.

Solution ▶

$\sqrt{x + 15} - \sqrt{x + 3} = 2$	original equation
$\sqrt{x + 15} = \sqrt{x + 3} + 2$	isolate one radical
$(\sqrt{x + 15})^2 = (\sqrt{x + 3} + 2)^2$	power property
$x + 15 = (x + 3) + 4\sqrt{x + 3} + 4$	$(A + B)^2$; $A = \sqrt{x + 3}, B = 2$
$x + 15 = x + 4\sqrt{x + 3} + 7$	simplify
$8 = 4\sqrt{x + 3}$	isolate radical
$2 = \sqrt{x + 3}$	divide by four
$4 = x + 3$	power property
$1 = x$	possible solution

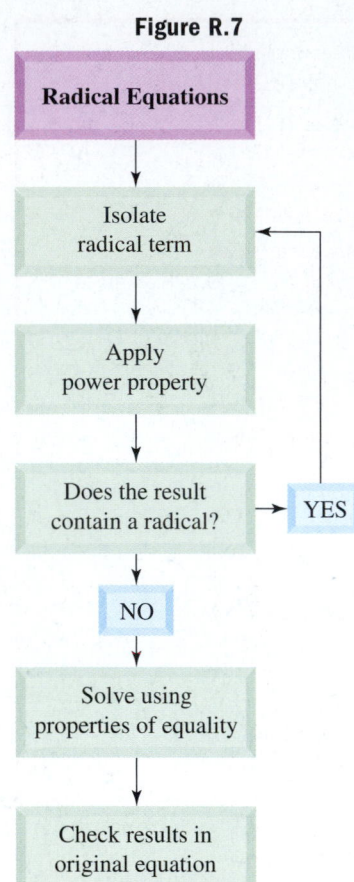

Figure R.7

Radical Equations

Isolate radical term

Apply power property

Does the result contain a radical? → YES

NO

Solve using properties of equality

Check results in original equation

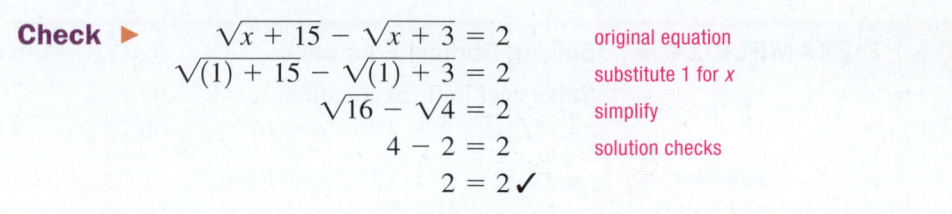

Check ▶

$$\sqrt{x + 15} - \sqrt{x + 3} = 2 \qquad \text{original equation}$$
$$\sqrt{(1) + 15} - \sqrt{(1) + 3} = 2 \qquad \text{substitute 1 for } x$$
$$\sqrt{16} - \sqrt{4} = 2 \qquad \text{simplify}$$
$$4 - 2 = 2 \qquad \text{solution checks}$$
$$2 = 2 \checkmark$$

Now try Exercises 53 and 54 ▶

Since rational exponents are so closely related to radicals, the solution process for each is very similar. The goal is still to "undo" the radical (rational exponent) and solve for the unknown.

Power Property of Equality

For real-valued expressions u and v, with positive integers m, n, and $\frac{m}{n}$ in lowest terms:

If m is odd	If m is even
and $u^{\frac{m}{n}} = v$,	and $u^{\frac{m}{n}} = v \, (v > 0)$,
then $(u^{\frac{m}{n}})^{\frac{n}{m}} = v^{\frac{n}{m}}$	then $(u^{\frac{m}{n}})^{\frac{n}{m}} = \pm v^{\frac{n}{m}}$
$u = v^{\frac{n}{m}}$	$u = \pm v^{\frac{n}{m}}$

The power property of equality basically says that if certain conditions are satisfied, both sides of an equation can be raised to any needed power.

EXAMPLE 16 ▶   **Solving Equations with Rational Exponents**

Solve each equation:

a. $3(x + 1)^{\frac{3}{4}} - 9 = 15$ **b.** $(x - 3)^{\frac{2}{3}} = 4$

Solution ▶

a. $3(x + 1)^{\frac{3}{4}} - 9 = 15$ original equation; $\frac{m}{n} = \frac{3}{4}$

$\qquad (x + 1)^{\frac{3}{4}} = 8$ isolate variable term (add 9, divide by 3)

$\qquad [(x + 1)^{\frac{3}{4}}]^{\frac{4}{3}} = 8^{\frac{4}{3}}$ apply power property, note m is odd

$\qquad x + 1 = 16$ simplify $[8^{\frac{4}{3}} = (8^{\frac{1}{3}})^4 = 16]$

$\qquad x = 15$ result

Check ▶

$3(15 + 1)^{\frac{3}{4}} - 9 = 15$ substitute 15 for x in the original equation

$3(16^{\frac{1}{4}})^3 - 9 = 15$ simplify, rewrite exponent

$3(2)^3 - 9 = 15$ $\sqrt[4]{16} = 2$

$3(8) - 9 = 15$ $2^3 = 8$

$15 = 15 \checkmark$ solution checks

b. $(x - 3)^{\frac{2}{3}} = 4$ original equation; $\frac{m}{n} = \frac{2}{3}$

$\qquad [(x - 3)^{\frac{2}{3}}]^{\frac{3}{2}} = \pm 4^{\frac{3}{2}}$ apply power property, note m is even

$\qquad x - 3 = \pm 8$ simplify $[4^{\frac{3}{2}} = (4^{\frac{1}{2}})^3 = 8]$

$\qquad x = 3 \pm 8$ result

The solutions are $3 + 8 = 11$ and $3 - 8 = -5$. Verify by checking both in the original equation.

Now try Exercises 55 through 58 ▶

> ⚠️ **CAUTION** ▶ As you continue solving equations with radicals and rational exponents, be careful not to arbitrarily place the "±" sign in front of terms *given* in radical form. The expression $\sqrt{18}$ indicates the positive square root of 18, where $\sqrt{18} = 3\sqrt{2}$. The equation $x^2 = 18$ becomes $x = \pm\sqrt{18}$ after applying the power property, with solutions $x = \pm 3\sqrt{2}$ ($x = -3\sqrt{2}, x = 3\sqrt{2}$), since the square of either number produces 18.

In Section R.4, we used a technique called *u-substitution* to factor expressions in quadratic form. The following equations are also in quadratic form since the degree of the leading term is twice the degree of the middle term: $x^{\frac{2}{3}} - 3x^{\frac{1}{3}} - 10 = 0$ and $(x^2 + x)^2 - 8(x^2 + x) + 12 = 0$. The first equation and its solution appear in Example 17.

EXAMPLE 17 ▶ **Solving Equations in Quadratic Form**

Solve using a *u*-substitution: $x^{\frac{2}{3}} - 3x^{\frac{1}{3}} - 10 = 0$.

Solution ▶ This equation is in quadratic form since it can be rewritten as: $(x^{\frac{1}{3}})^2 - 3(x^{\frac{1}{3}})^1 - 10 = 0$, where the degree of leading term is twice that of second term. If we let $u = x^{\frac{1}{3}}$, then $u^2 = x^{\frac{2}{3}}$ and the equation becomes $u^2 - 3u^1 - 10 = 0$, which is factorable.

$$(u - 5)(u + 2) = 0 \qquad \text{factor}$$

$u = 5$	or	$u = -2$	solution in terms of u
$x^{\frac{1}{3}} = 5$	or	$x^{\frac{1}{3}} = -2$	resubstitute $x^{\frac{1}{3}}$ for u
$(x^{\frac{1}{3}})^3 = 5^3$	or	$(x^{\frac{1}{3}})^3 = (-2)^3$	cube both sides: $\frac{1}{3}(3) = 1$
$x = 125$	or	$x = -8$	solve for x

Both solutions check.

Now try Exercises 59 and 60 ▶

Figure R.8

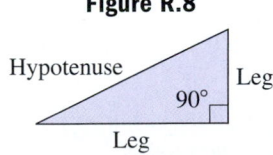

A right triangle is one that has a 90° angle (see Figure R.8). The longest side (opposite the right angle) is called the **hypotenuse,** while the other two sides are simply called "legs." The **Pythagorean theorem** is a formula that says if you add the square of each leg, the result will be equal to the square of the hypotenuse. Furthermore, we note the converse of this theorem is also true.

> **Pythagorean Theorem**
>
> 1. For any right triangle with legs a and b and hypotenuse c, $a^2 + b^2 = c^2$
> 2. For any triangle with sides a, b, and c, if $a^2 + b^2 = c^2$, then the triangle is a right triangle.

Figure R.9

A geometric interpretation of the theorem is given in Figure R.9, which shows $3^2 + 4^2 = 5^2$.

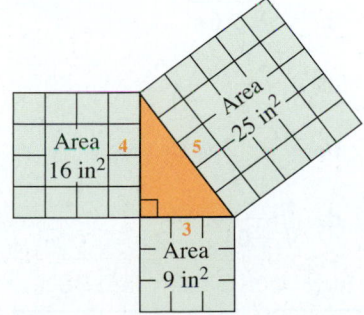

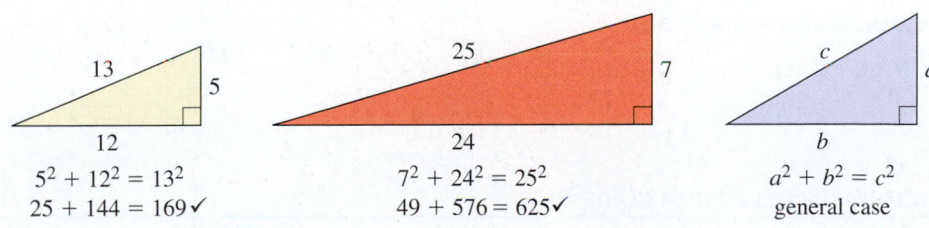

$5^2 + 12^2 = 13^2$
$25 + 144 = 169 ✓$

$7^2 + 24^2 = 25^2$
$49 + 576 = 625 ✓$

$a^2 + b^2 = c^2$
general case

EXAMPLE 18 ▶ **Applying the Pythagorean Theorem**

An extension ladder is placed 9 ft from the base of a building in an effort to reach a third-story window that is 27 ft high. What is the minimum length of the ladder required? Answer in exact form using radicals, and in approximate form by rounding to one decimal place.

 **Solution** ▶ We can assume the building makes a 90° angle with the ground, and use the Pythagorean theorem to find the required length. Let c represent this length.

$$c^2 = a^2 + b^2 \qquad \text{Pythagorean theorem}$$
$$c^2 = (9)^2 + (27)^2 \qquad \text{substitute 9 for } a \text{ and 27 for } b$$
$$c^2 = 81 + 729 \qquad 9^2 = 81, 27^2 = 729$$
$$c^2 = 810 \qquad \text{add}$$
$$c = \sqrt{810} \qquad \text{definition of square root; } c > 0$$
$$c = 9\sqrt{10} \qquad \text{exact form: } \sqrt{810} = \sqrt{81 \cdot 10} = 9\sqrt{10}$$
$$c \approx 28.5 \text{ ft} \qquad \text{approximate form}$$

The ladder must be at least 28.5 ft tall.

c 27 ft

9 ft

☑ **F.** You've just seen how we can solve equations and use formulas involving radicals

Now try Exercises 63 and 64 ▶

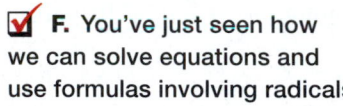 **R.6 EXERCISES**

▶ **CONCEPTS AND VOCABULARY**

Fill in each blank with the appropriate word or phrase. Carefully reread the section, if necessary.

1. $\sqrt[n]{a^n} = |a|$ if $n > 0$ is a(n) ___even___ integer.

2. The conjugate of $2 - \sqrt{3}$ is ___$2 + \sqrt{3}$___.

3. By decomposing the rational exponent, we can rewrite $16^{\frac{3}{4}}$ as $(16^{\frac{2}{7}})^?$. $(16^{\frac{1}{4}})^3$

4. $(x^{\frac{3}{2}})^{\frac{2}{3}} = x^{\frac{3}{2} \cdot \frac{2}{3}} = x^1$ is an example of the ___power___ property of exponents.

5. Discuss/Explain what it means when we say an expression like $\sqrt{A}$ has been written in simplest form.   Answers will vary.

6. Discuss/Explain why it would be easier to simplify the expression given using rational exponents rather than radicals. $\dfrac{x^{\frac{1}{2}}}{x^{\frac{1}{3}}}$ Answers will vary.

▶ **DEVELOPING YOUR SKILLS**

Evaluate the expression $\sqrt{x^2}$ for the values given.

7. a. $x = 9$ $|9| = 9$ b. $x = -10$ $|-10| = 10$

8. a. $x = 7$ $|7| = 7$ b. $x = -8$ $|-8| = 8$

Simplify each expression, assuming that variables can represent any real number.

9. a. $\sqrt{49p^2}$ $7|p|$ b. $\sqrt{(x-3)^2}$ $|x - 3|$

    c. $\sqrt{81m^4}$ $9m^2$ d. $\sqrt{x^2 - 6x + 9}$ $|x - 3|$

10. a. $\sqrt{25n^2}$ $5|n|$ b. $\sqrt{(y+2)^2}$ $|y + 2|$

    c. $\sqrt{v^{10}}$ $|v^5|$ d. $\sqrt{4a^2 + 12a + 9}$ $|2a + 3|$

11. a. $\sqrt[3]{64}$ 4   b. $\sqrt[3]{-216x^3}$ $-6x$

    c. $\sqrt[3]{216z^{12}}$ $6z^4$ d. $\sqrt[3]{\dfrac{v^3}{-8}}$ $\dfrac{v}{-2}$

12. a. $\sqrt[3]{-8}$ -2 b. $\sqrt[3]{-125p^3}$ $-5p$

    c. $\sqrt[3]{27q^9}$ $3q^3$ d. $\sqrt[3]{\dfrac{w^3}{-64}}$ $\dfrac{w}{-4}$

HOMEWORK SELECTION GUIDE

Core: 7–19 every other odd, 21, 23–59 every other odd, 61, 63–71 odd, 73 (23 Exercises)
Standard: 1–4, 7–19 every other odd, 21, 23–59 every other odd, 61, 63–68, 69, 71, 73, 76 (30 Exercises)

Extended: 1–4, 7–19 every other odd, 21, 23–59 every other odd, 61–69, 71, 73–76 (33 Exercises)
In Depth: 1–6, 7–19 every other odd, 21, 23–59 every other odd, 61–80 (41 Exercises)

13. a. $\sqrt[6]{64}$ _2_   **b.** $\sqrt[6]{-64}$ _not a real number_

c. $\sqrt[5]{243x^{10}}$ _$3x^2$_   **d.** $\sqrt[5]{-243x^5}$ _$-3x$_

e. $\sqrt[5]{(k-3)^5}$ _$k-3$_   **f.** $\sqrt[6]{(h+2)^6}$ _$|h+2|$_

14. a. $\sqrt[4]{81}$ _3_   **b.** $\sqrt[4]{-81}$ _not a real number_

c. $\sqrt[5]{1024z^{15}}$ _$4z^3$_   **d.** $\sqrt[5]{-1024z^{20}}$ _$-4z^4$_

e. $\sqrt[5]{(q-9)^5}$ _$q-9$_ **f.** $\sqrt[6]{(p+4)^6}$ _$|p+4|$_

15. a. $\sqrt[3]{-125}$ _-5_   **b.** $-\sqrt[4]{81n^{12}}$ _$-3|n^3|$_

c. $\sqrt{-36}$ _not real number_   **d.** $\sqrt{\dfrac{49v^{10}}{36}}$ _$\dfrac{7|v^5|}{6}$_

16. a. $\sqrt[3]{-216}$ _-6_   **b.** $-\sqrt[4]{16m^{24}}$ _$-2m^6$_

c. $\sqrt{-121}$ _not real number_   **d.** $\sqrt{\dfrac{25x^6}{4}}$ _$\dfrac{5|x^3|}{2}$_

17. a. $8^{\frac{2}{3}}$ _4_   **b.** $\left(\dfrac{16}{25}\right)^{\frac{3}{2}}$ _$\dfrac{64}{125}$_

c. $\left(\dfrac{4}{25}\right)^{-\frac{3}{2}}$ _$\dfrac{125}{8}$_   **d.** $\left(\dfrac{-27p^6}{8q^3}\right)^{\frac{2}{3}}$ _$\dfrac{9p^4}{4q^2}$_

18. a. $9^{\frac{3}{2}}$ _27_   **b.** $\left(\dfrac{4}{9}\right)^{\frac{3}{2}}$ _$\dfrac{8}{27}$_

c. $\left(\dfrac{16}{81}\right)^{-\frac{3}{4}}$ _$\dfrac{27}{8}$_   **d.** $\left(\dfrac{-125v^9}{27w^6}\right)^{\frac{2}{3}}$ _$\dfrac{25v^6}{9w^4}$_

19. a. $-144^{\frac{3}{2}}$ _-1728_   **b.** $\left(-\dfrac{4}{25}\right)^{\frac{3}{2}}$ _not a real number_

c. $(-27)^{-\frac{2}{3}}$ _$\dfrac{1}{9}$_   **d.** $-\left(\dfrac{27x^3}{64}\right)^{-\frac{4}{3}}$ _$\dfrac{-256}{81x^4}$_

20. a. $-100^{\frac{3}{2}}$ _-1000_   **b.** $\left(-\dfrac{49}{36}\right)^{\frac{3}{2}}$ _not a real number_

c. $(-125)^{-\frac{2}{3}}$ _$\dfrac{1}{25}$_   **d.** $-\left(\dfrac{x^9}{8}\right)^{-\frac{4}{3}}$ _$\dfrac{-16}{x^{12}}$_

Use properties of exponents to simplify. Answer in exponential form without negative exponents.

21. a. $(2n^2p^{-\frac{2}{5}})^5$ _$\dfrac{32n^{10}}{p^2}$_   **b.** $\left(\dfrac{8y^{\frac{3}{4}}}{64y^{\frac{3}{2}}}\right)^{\frac{1}{3}}$ _$\dfrac{1}{2y^{\frac{1}{4}}}$_

22. a. $\left(\dfrac{24x^{\frac{3}{8}}}{4x^{\frac{1}{2}}}\right)^2$ _$\dfrac{36}{x^{\frac{1}{4}}}$_   **b.** $(2x^{-\frac{1}{4}}y^{\frac{3}{4}})^4$ _$\dfrac{16y^3}{x}$_

Simplify each expression. Assume all variables represent nonnegative real numbers.

23. a. $\sqrt{18m^2}$ _$3m\sqrt{2}$_   **b.** $-2\sqrt[3]{-125p^3q^7}$ _$10pq^2\sqrt[3]{q}$_

c. $\dfrac{3}{8}\sqrt[3]{64m^3n^5}$ _$\dfrac{3}{2}mn\sqrt[3]{n^2}$_ **d.** $\sqrt{32p^3q^6}$ _$4pq^3\sqrt{2p}$_

e. $\dfrac{-6+\sqrt{28}}{2}$ _$-3+\sqrt{7}$_   **f.** $\dfrac{27-\sqrt{72}}{6}$ _$\dfrac{9}{2}-\sqrt{2}$_

24. a. $\sqrt{8x^6}$ _$2x^3\sqrt{2}$_   **b.** $3\sqrt[3]{128a^4b^2}$ _$12a\sqrt[3]{2ab^2}$_

c. $\dfrac{2}{9}\sqrt[3]{27a^2b^6}$ _$\dfrac{2}{3}b^2\sqrt[3]{a^2}$_ **d.** $\sqrt{54m^6n^8}$ _$3m^3n^4\sqrt{6}$_

e. $\dfrac{12-\sqrt{48}}{8}$ _$\dfrac{3}{2}-\dfrac{\sqrt{3}}{2}$_ **f.** $\dfrac{-20+\sqrt{32}}{4}$ _$-5+\sqrt{2}$_

25. a. $2.5\sqrt{18a}\sqrt{2a^3}$ _$\dfrac{15a^2}{}$_   **b.** $-\dfrac{2}{3}\sqrt{3b}\sqrt{12b^2}$ _$-4b\sqrt{b}$_

c. $\sqrt{\dfrac{x^3y}{3}}\sqrt{\dfrac{4x^5y}{12y}}$ _$\dfrac{x^4\sqrt{y}}{3}$_ **d.** $\sqrt[3]{9v^2u}\sqrt[3]{3u^5v^2}$ _$3u^2v\sqrt[3]{v}$_

26. a. $5.1\sqrt{2p}\sqrt{32p^5}$ _$40.8\,p^3$_   **b.** $-\dfrac{4}{5}\sqrt{5q}\sqrt{20q^3}$ _$-8q^2$_

c. $\sqrt{\dfrac{ab^2}{3}}\sqrt{\dfrac{25ab^4}{27}}$ _$\dfrac{5}{9}ab^3$_ **d.** $\sqrt[3]{5cd^2}\sqrt[3]{25cd}$ _$5d\sqrt[3]{c^2}$_

27. a. $\dfrac{\sqrt{8m^5}}{\sqrt{2m}}$ _$2m^2$_   **b.** $\dfrac{\sqrt[3]{108n^4}}{\sqrt[3]{4n}}$ _$3n$_

c. $\sqrt{\dfrac{45}{16x^2}}$ _$\dfrac{3\sqrt{5}}{4x}$_   **d.** $12\sqrt[3]{\dfrac{81}{8z^9}}$ _$\dfrac{18\sqrt[3]{3}}{z^3}$_

28. a. $\dfrac{\sqrt{27y^7}}{\sqrt{3y}}$ _$3y^3$_   **b.** $\dfrac{\sqrt[3]{72b^5}}{\sqrt[3]{3b^2}}$ _$2b\sqrt[3]{3}$_

c. $\sqrt{\dfrac{20}{4x^4}}$ _$\dfrac{\sqrt{5}}{x^2}$_   **d.** $-9\sqrt[3]{\dfrac{125}{27x^6}}$ _$\dfrac{-15}{x^2}$_

29. a. $\sqrt[5]{32x^{10}y^{15}}$ _$2x^2y^3$_ **b.** $x\sqrt[4]{x^5}$ _$x^2\sqrt[4]{x}$_

c. $\sqrt[4]{\sqrt[3]{b}}$ _$\sqrt[12]{b}$_   **d.** $\dfrac{\sqrt[3]{6}}{\sqrt{6}}$ _$\dfrac{1}{\sqrt[6]{6}}=\dfrac{\sqrt[6]{6^5}}{6}$_

e. $\sqrt{b}\sqrt[4]{b}$ _$b^{\frac{3}{4}}$_

30. a. $\sqrt[4]{81a^{12}b^{16}}$ _$3a^3b^4$_ **b.** $a\sqrt[5]{a^6}$ _$a^2\sqrt[5]{a}$_

c. $\sqrt{\sqrt[4]{a}}$ _$\sqrt[8]{a}$_   **d.** $\dfrac{\sqrt[3]{3}}{\sqrt[4]{3}}$ _$\sqrt[12]{3}$_

e. $\sqrt[3]{c}\sqrt[4]{c}$ _$c^{\frac{7}{12}}$_

Simplify and add (if possible).

31. a. $12\sqrt{72}-9\sqrt{98}$ _$9\sqrt{2}$_

b. $8\sqrt{48}-3\sqrt{108}$ _$14\sqrt{3}$_

c. $7\sqrt{18m}-\sqrt{50m}$ _$16\sqrt{2m}$_

d. $2\sqrt{28p}-3\sqrt{63p}$ _$-5\sqrt{7p}$_

32. a. $-3\sqrt{80}+2\sqrt{125}$ _$-2\sqrt{5}$_

b. $5\sqrt{12}+2\sqrt{27}$ _$16\sqrt{3}$_

c. $3\sqrt{12x}-5\sqrt{75x}$ _$-19\sqrt{3x}$_

d. $3\sqrt{40q}+9\sqrt{10q}$ _$15\sqrt{10q}$_

33. a. $3x\sqrt[3]{54x}-5\sqrt[3]{16x^4}$ _$-x\sqrt[3]{2x}$_

b. $\sqrt{4}+\sqrt{3x}-\sqrt{12x}+\sqrt{45}$ _$2-\sqrt{3x}+3\sqrt{5}$_

c. $\sqrt{72x^3}+\sqrt{50}-\sqrt{7x}+\sqrt{27}$ _$6x\sqrt{2x}+5\sqrt{2}-\sqrt{7x}+3\sqrt{3}$_

34. a. $5\sqrt[3]{54m^3}-2m\sqrt[3]{16m^3}$ _$15m\sqrt[3]{2}-4m^2\sqrt[3]{2}$_

b. $\sqrt{10b}+\sqrt{200b}-\sqrt{20}+\sqrt{40}$ _$\sqrt{10b}-10\sqrt{2b}-2\sqrt{5}+2\sqrt{10}$_

c. $\sqrt{75r^3}+\sqrt{32}-\sqrt{27r}+\sqrt{38}$ _$5r\sqrt{3r}+4\sqrt{2}-3\sqrt{3r}+\sqrt{38}$_

Compute each product and simplify the result.

35. a. $(7\sqrt{2})^2$ 98   **b.** $\sqrt{3}(\sqrt{5} + \sqrt{7})$ $\sqrt{15} + \sqrt{21}$

c. $(n + \sqrt{5})(n - \sqrt{5})$ $n^2 - 5$ **d.** $(6 - \sqrt{3})^2$ $39 - 12\sqrt{3}$

36. a. $(0.3\sqrt{5})^2$ 0.45   **b.** $\sqrt{5}(\sqrt{6} - \sqrt{2})$ $\sqrt{30} - \sqrt{10}$

c. $(4 + \sqrt{3})(4 - \sqrt{3})$ 13   **d.** $(2 + \sqrt{5})^2$ $9 + 4\sqrt{5}$

37. a. $(3 + 2\sqrt{7})(3 - 2\sqrt{7})$ -19

b. $(\sqrt{5} - \sqrt{14})(\sqrt{2} + \sqrt{13})$ $\sqrt{10} + \sqrt{65} - 2\sqrt{7} - \sqrt{182}$

c. $(2\sqrt{2} + 6\sqrt{6})(3\sqrt{10} + \sqrt{7})$
$12\sqrt{5} + 2\sqrt{14} + 36\sqrt{15} + 6\sqrt{42}$

38. a. $(5 + 4\sqrt{10})(1 - 2\sqrt{10})$ $-75 - 6\sqrt{10}$

b. $(\sqrt{3} + \sqrt{2})(\sqrt{10} + \sqrt{11})$ $\sqrt{30} + \sqrt{33} + 2\sqrt{5} + \sqrt{22}$

c. $(3\sqrt{5} + 4\sqrt{2})(\sqrt{15} + \sqrt{6})$ $23\sqrt{3} + 7\sqrt{30}$

Use a substitution to verify the solutions to the quadratic equation given. Verify results using a calculator.

39. $x^2 - 4x + 1 = 0$ Verified

a. $x = 2 + \sqrt{3}$ **b.** $x = 2 - \sqrt{3}$

40. $x^2 - 10x + 18 = 0$ Verified

a. $x = 5 - \sqrt{7}$ **b.** $x = 5 + \sqrt{7}$

41. $x^2 + 2x - 9 = 0$ Verified

a. $x = -1 + \sqrt{10}$ **b.** $x = -1 - \sqrt{10}$

42. $x^2 - 14x + 29 = 0$ Verified

a. $x = 7 - 2\sqrt{5}$ **b.** $x = 7 + 2\sqrt{5}$

Rationalize each expression by building perfect nth root factors for each denominator. Assume all variables represent positive quantities.

43. a. $\dfrac{3}{\sqrt{12}}$ $\dfrac{\sqrt{3}}{2}$ **b.** $\sqrt{\dfrac{20}{27x^3}}$ $\dfrac{2\sqrt{15x}}{9x^2}$

c. $\sqrt{\dfrac{27}{50b}}$ $\dfrac{3\sqrt{6b}}{10b}$ **d.** $\sqrt[3]{\dfrac{1}{4p}}$ $\dfrac{\sqrt[3]{2p^2}}{2p}$

e. $\dfrac{5}{\sqrt[3]{a}}$ $\dfrac{5\sqrt[3]{a^2}}{a}$

44. a. $\dfrac{-4}{\sqrt{20}}$ $\dfrac{-2\sqrt{5}}{5}$ **b.** $\sqrt{\dfrac{125}{12n^3}}$ $\dfrac{5\sqrt{15n}}{6n^2}$

c. $\sqrt{\dfrac{5}{12x}}$ $\dfrac{\sqrt{15x}}{6x}$ **d.** $\sqrt[3]{\dfrac{3}{2m^2}}$ $\dfrac{\sqrt[3]{12m}}{2m}$

e. $\dfrac{-8}{3\sqrt[3]{5}}$ $\dfrac{-8\sqrt[3]{25}}{15}$

Simplify the following expressions by rationalizing the denominators. Where possible, state results in exact form and approximate form, rounded to hundredths.

45. a. $\dfrac{8}{3 + \sqrt{11}}$ **b.** $\dfrac{6}{\sqrt{x} - \sqrt{2}}$ $\dfrac{6\sqrt{x} + 6\sqrt{2}}{x - 2}$
$-12 + 4\sqrt{11}$; 1.27

Additional answers can be found in the Instructor Answer Appendix.

46. a. $\dfrac{7}{\sqrt{7} + 3}\dfrac{7\sqrt{7} - 21}{-2}$; 1.24   **b.** $\dfrac{12}{\sqrt{x} + \sqrt{3}}$ $\dfrac{12\sqrt{x} - 12\sqrt{3}}{x - 3}$

47. a. $\dfrac{\sqrt{10} - 3}{\sqrt{3} + \sqrt{2}}$
$\sqrt{30} - 2\sqrt{5} - 3\sqrt{3} + 3\sqrt{2}$; 0.05

b. $\dfrac{7 + \sqrt{6}}{3 - 3\sqrt{2}}$ $\dfrac{7 + 7\sqrt{2} + \sqrt{6} + 2\sqrt{3}}{-3}$; -7.60

48. a. $\dfrac{1 + \sqrt{2}}{\sqrt{6} + \sqrt{14}}$
$\dfrac{\sqrt{6} - \sqrt{14} + 2\sqrt{3} - 2\sqrt{7}}{-8}$; 0.39

b. $\dfrac{1 + \sqrt{6}}{5 + 2\sqrt{3}}$ $\dfrac{5 - 2\sqrt{3} + 5\sqrt{6} - 6\sqrt{2}}{13}$; 0.41

Solve each equation and check your solutions by substitution. Identify any extraneous roots.

49. a. $-3\sqrt{3x - 5} = -9$ $x = \frac{14}{3}$

b. $x = \sqrt{3x + 1} + 3$ $x = 8, x = 1$ is extraneous

50. a. $-2\sqrt{4x - 1} = -10$ $x = \frac{13}{2}$

b. $-5 = \sqrt{5x - 1} - x$ $x = 13, x = 2$ is extraneous

51. a. $2 = \sqrt[3]{3m - 1}$ $m = 3$

b. $2\sqrt[3]{7 - 3x} - 3 = -7$ $x = 5$

c. $\dfrac{\sqrt[3]{2m + 3}}{-5} + 2 = 3$ $m = -64$

d. $\sqrt[3]{2x - 9} = \sqrt[3]{3x + 7}$ $x = -16$

52. a. $-3 = \sqrt[3]{5p + 2}$ $p = -\frac{29}{5}$

b. $3\sqrt[3]{3 - 4x} - 7 = -4$ $x = \frac{1}{2}$

c. $\dfrac{\sqrt[3]{6x - 7}}{4} - 5 = -6$ $x = \frac{-19}{2}$

d. $3\sqrt[3]{x + 3} = 2\sqrt[3]{2x + 17}$ $x = 5$

53. a. $\sqrt{x - 9} + \sqrt{x} = 9$ $x = 25$

b. $x = 3 + \sqrt{23 - x}$ $x = 7; x = -2$ is extraneous

c. $\sqrt{x - 2} - \sqrt{2x} = -2$ $x = 2, x = 18$

d. $\sqrt{12x + 9} - \sqrt{24x} = -3$ $x = 6; x = 0$ is extraneous

54. a. $\sqrt{x + 7} - \sqrt{x} = 1$ $x = 9$

b. $\sqrt{2x + 31} + x = 2$ $x = -3; x = 9$ is extraneous

c. $\sqrt{3x} = \sqrt{x - 3} + 3$ $x = 3, x = 12$

d. $\sqrt{3x + 4} - \sqrt{7x} = -2$ $x = 7; x = 0$ is extraneous

Write the equation in simplified form, then solve. Check all answers by substitution.

55. a. $x^{\frac{3}{5}} + 17 = 9$ $x = -32$

b. $-2x^{\frac{3}{4}} + 47 = -7$ $x = 81$

56. a. $0.\overline{3}x^{\frac{5}{2}} - 39 = 42$ $x = 9$

b. $0.\overline{5}x^{\frac{5}{3}} + 92 = -43$ $x = -27$

57. a. $2(x + 5)^{\frac{2}{3}} - 11 = 7$ $x = -32, x = 22$

b. $-3(x - 2)^{\frac{4}{5}} + 29 = -19$ $x = -30, x = 34$

58. a. $3x^{\frac{1}{3}} - 10 = \sqrt[3]{x}$ $x = 125$

b. $2\sqrt[5]{x^2} - 4 = x^{\frac{2}{5}}$ $x = 32$

59. $x^{\frac{2}{3}} - 2x^{\frac{1}{3}} - 15 = 0$ $x = -27, x = 125$

60. $x^3 - 9x^{\frac{3}{2}} = -8$ $x = 1, x = 4$

▶ WORKING WITH FORMULAS

61. Fish length to weight relationship: $L = 1.13(W)^{\frac{1}{3}}$

The length to weight relationship of a female Pacific halibut can be approximated by the formula shown, where W is the weight in pounds and L is the length in feet. A fisherman lands a halibut that weighs 400 lb. Approximate the length of the fish (round to two decimal places).   8.33 ft

62. Timing a falling object: $t = \dfrac{\sqrt{s}}{4}$

The time it takes an object to fall a certain distance is given by the formula shown, where t is the time in seconds and s is the distance the object has fallen. Approximate the time it takes an object to hit the ground, if it is dropped from the top of a building that is 80 ft in height (round to hundredths).   2.24 sec

▶ APPLICATIONS

63. Length of a cable: A radio tower is secured by cables that are anchored in the ground 8 m from its base. If the cables are attached to the tower 24 m above the ground, what is the length of each cable? Answer in (a) exact form using radicals, and (b) approximate form by rounding to one decimal place.   **a.** $8\sqrt{10}\,m$
b. about 25.3 m

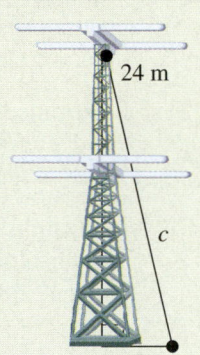

64. Height of a kite: Benjamin Franklin is flying his kite in a storm once again. John Adams has walked to a position directly under the kite and is 75 ft from Ben. If the kite is 50 ft above John Adams' head, how much string S has Ben let out? Answer in (a) exact form using radicals, and (b) approximate form by rounding to one decimal place.   **a.** $25\sqrt{13}$ ft   **b.** about 90.1 ft

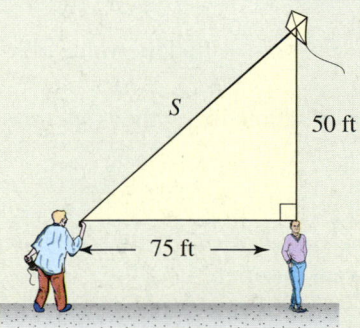

The time T (in days) required for a planet to make one revolution around the sun is modeled by the function $T = 0.407R^{\frac{3}{2}}$, where R is the maximum radius of the planet's orbit (in millions of miles). This is known as *Kepler's third law of planetary motion*. Use the equation given to approximate the number of days required for one complete orbit of each planet, given its maximum orbital radius.

65. a. Earth: 93 million mi   365.02 days
   b. Mars: 142 million mi   688.69 days
   c. Mercury: 36 million mi   87.91 days

66. a. Venus: 67 million mi   223.21 days
   b. Jupiter: 480 million mi   4280.12 days
   c. Saturn: 890 million mi   10,806.36 days

67. Accident investigation: After an accident, police officers will try to determine the approximate velocity V that a car was traveling using the formula $V = 2\sqrt{6L}$, where L is the length of the skid marks in feet and V is the velocity in miles per hour. (a) If the skid marks were 54 ft long, how fast was the car traveling? (b) Approximate the speed of the car if the skid marks were 90 ft long.
   a. 36 mph   **b.** 46.5 mph

68. Wind-powered energy: If a wind-powered generator is delivering P units of power, the velocity V of the wind (in miles per hour) can be determined using $V = \sqrt[3]{\dfrac{P}{k}}$, where k is a constant that depends on the size and efficiency of the generator. Rationalize the radical expression and use the new version to find the velocity of the wind if $k = 0.004$ and the generator is putting out 13.5 units of power.   15 mph

69. Surface area: The lateral surface area (surface area excluding the base) S of a cone is given by the formula $S = \pi r \sqrt{r^2 + h^2}$, where r is the radius of the base and h is the height of the cone. Find the lateral surface area of a cone that has a radius of 6 m and a height of 10 m. Answer in simplest form. $12\pi\sqrt{34} \approx 219.82 \text{ m}^2$

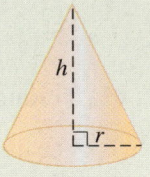

70. Surface area: The lateral surface area S of a frustum (a truncated cone) is given by the formula $S = \pi(a + b)\sqrt{h^2 + (b - a)^2}$, where a is the radius of the upper base, b is the radius of the lower base, and h is the height. Find the surface area of a frustum where $a = 6$ m, $b = 8$ m, and $h = 10$ m. Answer in simplest form. $28\pi\sqrt{26} \approx 448.53 \text{ m}^2$

71. Planetary motion: The time T (in days) for a planet to make one revolution (elliptical orbit) around the sun is modeled by $T = 0.407R^{\frac{3}{2}}$, where R is the maximum radius of the planet's orbit in millions of miles (*Kepler's third law of planetary motion*). Use the equation to approximate the maximum radius of each orbit, given the number of days it takes for one revolution. (See Exercises 65 and 66.)

   a. Mercury: 88 days   36 million mi
   b. Venus: 225 days   67 million mi
   c. Earth: 365 days   93 million mi
   d. Mars: 687 days   142 million mi
   e. Jupiter: 4333 days   484 million mi
   f. Saturn: 10,759 days   887 million mi

72. Wind-powered energy: If a wind-powered generator is delivering P units of power, the velocity V of the wind (in miles per hour) can be determined using $V = \sqrt[3]{\dfrac{P}{k}}$, where k is a constant that depends on the size and efficiency of the generator. Given $k = 0.004$, approximately how many units of power are being delivered if the wind is blowing at 27 miles per hour? (See Exercise 68.) $P \approx 78.7 \text{ units}$

▶ EXTENDING THE CONCEPT

The expression $x^2 - 7$ is not factorable using *integer values*. But the expression *can be written* in the form $x^2 - (\sqrt{7})^2$, enabling us to factor it as a "binomial" and its conjugate: $(x + \sqrt{7})(x - \sqrt{7})$. Use this idea to factor the following expressions.

73. a. $x^2 - 5$ $(x + \sqrt{5})(x - \sqrt{5})$
   b. $n^2 - 19$ $(n + \sqrt{19})(n - \sqrt{19})$

74. a. $4v^2 - 11$ $(2v + \sqrt{11})(2v - \sqrt{11})$
   b. $9w^2 - 17$ $(3w + \sqrt{17})(3w - \sqrt{17})$

75. The following terms form a pattern that continues until the sixth term is found:
$\sqrt{3x} + \sqrt{9x} + \sqrt{27x} + \ldots$ (a) Compute the sum of all six terms; (b) develop a system (investigate the pattern further) that will enable you to find the sum of 12 such terms *without actually writing out the terms.*   **a.** $13\sqrt{3x} + 39\sqrt{x}$ **b.** Answers will vary.

76. Find a quick way to simplify the expression without the aid of a calculator.   3

$$\left(\left(\left(\left(\left(3^{\frac{5}{6}}\right)^{\frac{3}{2}}\right)^{\frac{4}{5}}\right)^{\frac{3}{4}}\right)^{\frac{2}{5}}\right)^{\frac{10}{3}}$$

77. If $(x^{\frac{1}{2}} + x^{-\frac{1}{2}})^2 = \dfrac{9}{2}$, find the value of $x^{\frac{1}{2}} + x^{-\frac{1}{2}}$. $\dfrac{3\sqrt{2}}{2}$

78. Rewrite by rationalizing the *numerator:*

$$\frac{\sqrt{x + h} - \sqrt{x}}{h} \qquad \frac{1}{\sqrt{x + h} + \sqrt{x}}$$

Determine the values of x for which each expression represents a real number.

79. $\dfrac{\sqrt{x - 1}}{x^2 - 4}$ $x \in [1, 2) \cup (2, \infty)$

80. $\dfrac{x^2 - 4}{\sqrt{x - 1}}$ $x \in (1, \infty)$

OVERVIEW OF CHAPTER R

Prerequisite Definitions, Properties, Formulas, and Relationships

Notation and Relations

concept	notation	description	example
• Set notation:	$\{members\}$	braces enclose the members of a set	set of even whole numbers $A = \{0, 2, 4, 6, 8, \ldots\}$
• Is an element of	$\in$	indicates membership in a set	$14 \in A$
• Empty set	$\varnothing$ or $\{\ \}$	a set having no elements	odd numbers in A
• Is a proper subset of	$\subset$	indicates the elements of one set are entirely contained in another	$S = \{0, 6, 12, 18, 24, \ldots\}$ $S \subset A$
• Defining a set	$\{x \mid x \ldots\}$	the set of all x, *such that $x \ldots$*	$S = \{x \mid x = 6n \text{ for } n \in \mathbb{W}\}$

Sets of Numbers

- Natural: $\mathbb{N} = \{1, 2, 3, 4, \ldots\}$

- Whole: $\mathbb{W} = \{0, 1, 2, 3, \ldots\}$

- Integers: $\mathbb{Z} = \{\ldots, -3, -2, -1, 0, 1, 2, 3, \ldots\}$

- Rational: $\mathbb{Q} = \left\{ \dfrac{p}{q}, \text{ where } p, q \in \mathbb{Z}; q \neq 0 \right\}$

- Irrational: $\mathbb{H} = \{$numbers with a nonterminating, nonrepeating decimal form$\}$

- Real: $\mathbb{R} = \{$all rational and irrational numbers$\}$

Absolute Value of a Number

- The distance between a number n and zero (always positive)

$$|n| = \begin{cases} n & \text{if } n \geq 0 \\ -n & \text{if } n < 0 \end{cases}$$

Distance between numbers *a* and *b* on the number line

$$|a - b| \text{ or } |b - a|$$

For a complete review of these ideas, go to www.mhhe.com/coburn.

R.1 Properties of Real Numbers: For real numbers a, b, and c,

Commutative Property
- Addition: $a + b = b + a$
- Multiplication: $a \cdot b = b \cdot a$

Associative Property
- Addition: $(a + b) + c = a + (b + c)$
- Multiplication: $(a \cdot b) \cdot c = a \cdot (b \cdot c)$

Identities
- Additive: $0 + a = a$
- Multiplicative: $1 \cdot a = a$

Inverses
- Additive: $a + (-a) = 0$
- Multiplicative: $\dfrac{p}{q} \cdot \dfrac{q}{p} = 1; p, q \neq 0$

R.2 Properties of Exponents: For real numbers a and b, and integers m, n, and p (excluding 0 raised to a nonpositive power),

- Product property: $b^m \cdot b^n = b^{m+n}$
- Product to a power: $(a^m b^n)^p = a^{mp} \cdot b^{np}$
- Quotient property: $\dfrac{b^m}{b^n} = b^{m-n} \ (b \neq 0)$

- Negative exponents: $b^{-n} = \dfrac{1}{b^n}; \left(\dfrac{a}{b}\right)^{-n} = \left(\dfrac{b}{a}\right)^n$
  $(a, b \neq 0)$

- Power property: $(b^m)^n = b^{mn}$
- Quotient to a power: $\left(\dfrac{a^m}{b^n}\right)^p = \dfrac{a^{mp}}{b^{np}} \ (b \neq 0)$
- Zero exponents: $b^0 = 1 \ (b \neq 0)$
- Scientific notation: $N \times 10^k; 1 \leq |N| < 10, k \in \mathbb{Z}$

R.2 Polynomials

- A polynomial is a sum or difference of monomial terms
- Polynomials are classified as a monomial, binomial, trinomial, or polynomial, depending on the number of terms
- The degree of a polynomial in one variable is the same as the largest exponent occuring on the variable in any term
- A polynomial expression is in standard form when written with the terms in descending order of degree

R.3 Solving Linear Equations and Inequalities

- **Properties of Equality:**

additive property

If A and B are algebraic expressions, where $A = B$, then $A + C = B + C$.

[C can be positive or negative]

multiplicative property

If A and B are algebraic expressions, where $A = B$, then $A \cdot C = B \cdot C$

and $\dfrac{A}{C} = \dfrac{B}{C}; C \neq 0$

- A linear equation in one variable is one that can be written in the form $ax + b = c$, where the exponent on the variable is a 1.
- To solve a linear equation, we attempt to isolate the term containing the variable using the additive property, then solve for the variable using the multiplicative property.
- An equation can be an identity (always true), a contradiction (never true) or conditional (true or false depending on the input value[s]).
- If an equation contains fractions, multiplying both sides by the least common denominator of all fractions will "clear the denominators" and reduce the amount of work required to solve.
- Inequalities are solved using properties similar to those used for solving equations. The one exception is when multiplying or dividing by a negative quantity, as the inequality symbol *must then be reversed* to maintain the truth of the resulting statement.
- Solutions to an inequality can be given using a simple inequality, graphed on a number line, stated in set notation, or stated using interval notation.
- Given two sets A and B: A intersect B $(A \cap B)$ is the set of elements *shared* by both A **and** B (elements common to both sets). A union B $(A \cup B)$ is the set of elements *in either* A **or** B (elements are combined to form a larger set).
- Compound inequalities are formed using the conjunction "and" or the conjunction "or." The result can be either a joint inequality as in $-3 < x \leq 5$, or a disjoint inequality, $x < -2$ or $x > 7$.

R.4 Special Factorizations

- $A^2 - B^2 = (A + B)(A - B)$
- $A^3 - B^3 = (A - B)(A^2 + AB + B^2)$
- $A^2 \pm 2AB + B^2 = (A \pm B)^2$
- $A^3 + B^3 = (A + B)(A^2 - AB + B^2)$
- Certain equations of higher degree can be solved using factoring skills and the zero product property.

R.5 Rational Expressions: For polynomials P, Q, R, and S with no denominator of zero,

- Lowest terms: $\dfrac{P \cdot R}{Q \cdot R} = \dfrac{P}{Q}$
- Equivalence: $\dfrac{P}{Q} = \dfrac{P \cdot R}{Q \cdot R}$
- Multiplication: $\dfrac{P}{Q} \cdot \dfrac{R}{S} = \dfrac{P \cdot R}{Q \cdot S} = \dfrac{PR}{QS}$
- Division: $\dfrac{P}{Q} \div \dfrac{R}{S} = \dfrac{P}{Q} \cdot \dfrac{S}{R} = \dfrac{PS}{QR}$
- Addition: $\dfrac{P}{R} + \dfrac{Q}{R} = \dfrac{P + Q}{R}$
- Subtraction: $\dfrac{P}{R} - \dfrac{Q}{R} = \dfrac{P - Q}{R}$

- Addition/subtraction with unlike denominators:
  1. Find the LCD of all rational expressions.
  2. Build equivalent expressions using LCD.
  3. Add/subtract numerators as indicated.
  4. Write the result in lowest terms.
- To solve rational equations, first clear denominators using the LCD, noting values that must be excluded.
- Multiplying an equation by a variable quantity sometimes introduces extraneous solutions. Check all results in the original equation.

R.6 Properties of Radicals

- $\sqrt{a}$ is a real number only for $a \geq 0$
- $\sqrt[n]{a} = b$, only if $b^n = a$

- For any real number a, $\sqrt[n]{a^n} = |a|$ when n is even

- If a is a real number and n is an integer greater than 1, then $\sqrt[n]{a} = a^{\frac{1}{n}}$ provided $\sqrt[n]{a}$ represents a real number

- If $\sqrt[n]{A}$ and $\sqrt[n]{B}$ represent real numbers, $\sqrt[n]{AB} = \sqrt[n]{A} \cdot \sqrt[n]{B}$

- $\sqrt{a} = b$, only if $b^2 = a$
- If n is even, $\sqrt[n]{a}$ represents a real number only if $a \geq 0$

- For any real number a, $\sqrt[n]{a^n} = a$ when n is odd

- If $\frac{m}{n}$ is a rational number written in lowest terms with $n \geq 2$, then $a^{\frac{m}{n}} = (\sqrt[n]{a})^m$ and $a^{\frac{m}{n}} = \sqrt[n]{a^m}$ provided $\sqrt[n]{a}$ represents a real number.

- If $\sqrt[n]{A}$ and $\sqrt[n]{B}$ represent real numbers and $B \neq 0$, $\sqrt[n]{\dfrac{A}{B}} = \dfrac{\sqrt[n]{A}}{\sqrt[n]{B}}$

- A radical expression is in simplest form when:
  1. the radicand has no factors that are perfect nth roots,
  2. the radicand contains no fractions, and
  3. no radicals occur in a denominator.

- To solve radical equations, isolate the radical on one side, then apply the appropriate "nth power" to free up the radicand. Repeat the process if needed. See flowchart on page 74.
- For equations with a rational exponent $\frac{m}{n}$, isolate the variable term and raise both sides to the $\frac{n}{m}$ power. If m is even, there will be two real solutions.

R.6 Pythagorean Theorem

- For any right triangle with legs a and b and hypotenuse c: $a^2 + b^2 = c^2$.

- For any triangle with sides a, b, and c, if $a^2 + b^2 = c^2$, then the triangle is a right triangle.

PRACTICE TEST

1. State true or false. If false, state why.

   a. $(3 + 4)^2 = 25$ **b.** $\dfrac{7}{0} = 0$ False; undefined

   False; parentheses first

   c. $x - 3 = -3 + x$ True   **d.** $-2(x - 3) = -2x - 3$

   False; $-2x + 6$

2. State the value of each expression.

   a. $\sqrt{121}$ 11   **b.** $\sqrt[3]{-125}$ -5

   c. $\sqrt{-36}$ not a real number   **d.** $\sqrt{400}$ 20

3. Evaluate each expression:

   a. $\dfrac{7}{8} - \left(-\dfrac{1}{4}\right)$ $\frac{9}{8}$ **b.** $-\dfrac{1}{3} - \dfrac{5}{6}$ $\frac{-7}{6}$

   c. $-0.7 + 1.2$ 0.5   **d.** $1.3 + (-5.9)$ -4.6

4. Evaluate each expression:

   a. $(-4)\left(-2\dfrac{1}{3}\right)$ $\frac{28}{3}$ **b.** $(-0.6)(-1.5)$ 0.9

   c. $\dfrac{-2.8}{-0.7}$ 4   **d.** $4.2 \div (-0.6)$ -7

   5. Evaluate using a calculator: $2000\left(1 + \dfrac{0.08}{12}\right)^{12 \cdot 10}$

   ≈ 4439.28

6. State the value of each expression, if possible.

   a. $0 \div 6$ 0   **b.** $6 \div 0$ undefined

7. State the number of terms in each expression and identify the coefficient of each.

   a. $-2v^2 + 6v + 5$ **b.** $\dfrac{c + 2}{3} + c$ $2; \frac{1}{3}, 1$

   3; $-2, 6, 5$

8. Evaluate each expression given $x = -0.5$ and $y = -2$. Round to hundredths as needed.

   a. $2x - 3y^2$ -13 **b.** $\sqrt{2} - x(4 - x^2) + \dfrac{y}{x}$

   ≈ 7.29

9. Translate each phrase into an algebraic expression.

   a. Nine less than twice a number is subtracted from the number cubed. $x^3 - (2x - 9)$

   b. Three times the square of half a number is subtracted from twice the number. $2n - 3\left(\dfrac{n}{2}\right)^2$

10. Create a mathematical model using descriptive variables.

    a. The radius of the planet Jupiter is approximately 119 mi less than 11 times the radius of the Earth. Express the radius of Jupiter in terms of the Earth's radius.   Let r represent Earth's radius. Then $11r - 119$ represents Jupiter's radius.

    b. Last year, Video Venue Inc. earned $1.2 million more than four times what it earned this year. Express last year's earnings of Video Venue Inc. in terms of this year's earnings.   Let e represent this year's earnings. Then $4e + 1.2$ million represents last year's earnings.

11. Simplify by combining like terms.

a. $8v^2 + 4v - 7 + v^2 - v$ $9v^2 + 3v - 7$

b. $-4(3b - 2) + 5b$ $-7b + 8$

c. $4x - (x - 2x^2) + x(3 - x)$ $x^2 + 6x$

12. Factor each expression completely.

a. $9x^2 - 16$ $(3x + 4)(3x - 4)$ **b.** $4v^3 - 12v^2 + 9v$ $v(2v - 3)^2$

c. $x^3 + 5x^2 - 9x - 45$ $(x + 5)(x + 3)(x - 3)$

13. Simplify using the properties of exponents.

a. $\dfrac{5}{b^{-3}}$ $5b^3$ **b.** $(-2a^3)^2(a^2b^4)^3$ $4a^{12}b^{12}$

c. $\left(\dfrac{m^2}{2n}\right)^3$ $\dfrac{m^6}{8n^3}$ **d.** $\left(\dfrac{5p^2q^3r^4}{-2pq^2r^4}\right)^2$ $\dfrac{25}{4}p^2q^2$

14. Simplify using the properties of exponents.

a. $\dfrac{-12a^3b^5}{3a^2b^4}$ $-4ab$

b. $(3.2 \times 10^{-17}) \times (2.0 \times 10^{15})$ $6.4 \times 10^{-2} = 0.064$

c. $\left(\dfrac{a^{-3} \cdot b}{c^{-2}}\right)^{-4}$ $\dfrac{a^{12}}{b^4c^8}$ **d.** $-7x^0 + (-7x)^0$ -6

15. Compute each product.

a. $(3x^2 + 5y)(3x^2 - 5y)$ $9x^4 - 25y^2$

b. $(2a + 3b)^2$ $4a^2 + 12ab + 9b^2$

16. Add or subtract as indicated. $7a^4 - 5a^3 + 8a^2 - 3a - 18$

a. $(-5a^3 + 4a^2 - 3) + (7a^4 + 4a^2 - 3a - 15)$

b. $(2x^2 + 4x - 9) - (7x^4 - 2x^2 - x - 9)$

$-7x^4 + 4x^2 + 5x$

Simplify or compute as indicated.

17. a. $\dfrac{x - 5}{5 - x}$ -1 **b.** $\dfrac{4 - n^2}{n^2 - 4n + 4}$ $\dfrac{2 + n}{2 - n}$

c. $\dfrac{x^3 - 27}{x^2 + 3x + 9}$ $x - 3$ **d.** $\dfrac{3x^2 - 13x - 10}{9x^2 - 4} \cdot \dfrac{x - 5}{3x - 2}$

e. $\dfrac{x^2 - 25}{3x^2 - 11x - 4} \div \dfrac{x^2 + x - 20}{x^2 - 8x + 16}$ $\dfrac{x - 5}{3x + 1}$

f. $\dfrac{m + 3}{m^2 + m - 12} - \dfrac{2}{5(m + 4)}$ $\dfrac{3(m + 7)}{5(m + 4)(m - 3)}$

18. a. $\sqrt{(x + 11)^2}$ $|x + 11|$ **b.** $\sqrt[3]{\dfrac{-8}{27v^3}}$ $\dfrac{-2}{3v}$

c. $\left(\dfrac{25}{16}\right)^{-\frac{3}{2}}$ $\dfrac{64}{125}$ **d.** $\dfrac{-4 + \sqrt{32}}{8}$ $-\dfrac{1}{2} + \dfrac{\sqrt{2}}{2}$

e. $7\sqrt{40} - \sqrt{90}$ $11\sqrt{10}$ **f.** $(x + \sqrt{5})(x - \sqrt{5})$ $x^2 - 5$

g. $\sqrt{\dfrac{2}{5x}}$ $\dfrac{\sqrt{10x}}{5|x|}$ **h.** $\dfrac{8}{\sqrt{6} - \sqrt{2}}$ $2(\sqrt{6} + \sqrt{2})$

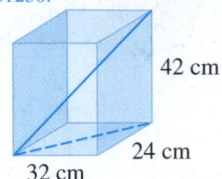

19. Maximizing revenue: Due to past experience, the manager of a video store knows that if a popular video game is priced at $30, the store will sell 40 each day. For each decrease of $0.50, one additional sale will be made. The formula for the store's revenue is then $R = (30 - 0.5x)(40 + x)$, where x represents the number of times the price is decreased. Multiply the binomials and use a table of values to determine (a) the number of 50¢ decreases that will give the most revenue and (b) the maximum amount of revenue. $-0.5x^2 + 10x + 1200$;   **a.** 10 decreases of 0.50 or $5.00   **b.** Maximum revenue is $1250.

20. Diagonal of a rectangular prism: Use the Pythagorean theorem to determine the length of the diagonal of the rectangular prism shown in the figure. (*Hint:* First find the diagonal of the base.)   58 cm

42 cm

24 cm

32 cm

21. Solve each linear equation.

a. $-2b + 7 = -5$ $b = 6$ **b.** $3(2n - 6) + 1 = 7$ $n = 4$

c. $4m - 5 = 11m + 2$ $m = -1$ **d.** $\dfrac{1}{2}x + \dfrac{2}{3} = \dfrac{3}{4}$ $x = \dfrac{1}{6}$

e. $-8(3p + 5) - 9 = 6(3 - 4p)$ $\{ \}$ (contradiction)

f. $-\dfrac{g}{6} = 3 - \dfrac{1}{2} - \dfrac{5g}{12}$ $g = 10$

22. If one-fourth of the sum of a number and twelve is added to three, the result is sixteen. Find an equation model for this statement, then use it to find the number. $3 + \frac{1}{4}(n + 12) = 16$, the number is 40

23. Solve each polynomial equation by factoring.

a. $x^3 - 7x^2 = 4x - 28$ $x = -2, x = 2, x = 7$

b. $-7r^3 + 21r^2 + 28r = 0$ $r = 0, r = -1, r = 4$

c. $g^4 - 10g^2 + 9 = 0$ $g = -3, g = -1, g = 1, g = 3$

24. Solve each rational equation.

a. $\dfrac{3}{5x} + \dfrac{7}{10} = \dfrac{1}{4x}$ $x = -\dfrac{1}{2}$

b. $\dfrac{3h}{h + 3} - \dfrac{7}{h^2 + 3h} = \dfrac{1}{h}$ $h = -\dfrac{5}{3}, h = 2$

c. $\dfrac{n}{n + 2} - \dfrac{3}{n - 4} = \dfrac{4n + 20}{n^2 - 2n - 8}$

$n = 13$ (-2 is extraneous)

25. Solve each radical equation.

a. $\dfrac{\sqrt{x^2 + 7}}{2} + 3 = 5$ $x = -3, x = 3$

b. $3\sqrt{x + 4} = x + 4$ $x = -4, x = 5$

c. $\sqrt{3x + 4} = 2 - \sqrt{x + 2}$ $x = -1$ (7 is extraneous)

1

Relations, Functions, and Graphs

CHAPTER OUTLINE

CHAPTER CONNECTIONS

Viewing relations and functions in terms of an equation, a table of values, and the related graph, often brings a clearer understanding of the relationships involved. For instance, while many business are aware that internet use is increasing with time, they are very interested in the *rate of growth*, in order to prepare and develop related goods and services. This application appears as Exercise 109 in Section 1.4.

Check out these other real-world connections

LEARNING OBJECTIVES

In Section 1.1 you will see how we can:

☐ **A.** Express a relation in mapping notation and ordered pair form

☐ **B.** Graph a relation

☐ **C.** Graph relations on a calculator

☐ **D.** Develop the equation and graph of a circle using the distance and midpoint formulas

In everyday life, we encounter a large variety of relationships. For instance, the time it takes us to get to work is related to our average speed; the monthly cost of heating a home is related to the average outdoor temperature; and in many cases, the amount of our charitable giving is related to changes in the cost of living. In each case we say that a relation exists between the two quantities.

A. Relations, Mapping Notation, and Ordered Pairs

In the most general sense, a **relation** is simply a correspondence between two sets. Relations can be represented in many different ways and may even be very "unmathematical," like the one shown in Figure 1.1 between a set of people and the set of their corresponding birthdays. If *P* represents the set of people and *B* represents the set of birthdays, we say that elements of *P* correspond to elements of *B*, or the birthday relation maps elements of *P* to elements of *B*. Using what is called **mapping notation,** we might simply write $P \rightarrow B$. From a purely practical standpoint, we note that while it is possible for two different people to share the same birthday, it is quite impossible for the same person to have two different birthdays. Later, this observation will help us mark the difference between a relation and special kind of relation called a function.

Figure 1.1

P		*B*
Missy		April 12
Jeff		Nov 11
Angie		Sept 10
Megan		Nov 28
Mackenzie		May 7
Michael		April 14
Mitchell		

The bar graph in Figure 1.2 is also an example of a relation. In the graph, each year is related to annual consumer spending per person on cable and satellite television. As an alternative to mapping or a bar graph, this relation could also be represented using ordered pairs. For example, the ordered pair (5, 234) would indicate that in 2005, spending per person on cable and satellite TV in the United States averaged $234. When a relation is represented using ordered pairs, we say the relation is **pointwise-defined.**

Over a long period of time, we could collect many ordered pairs of the form (*t*, *s*), where consumer spending *s depends* on the time *t*. For this reason we often call the second coordinate of an ordered pair (in this case *s*) the **dependent variable,** with the first coordinate designated as the **independent variable.** The set of all first coordinates is called the **domain** of the relation. The set of all second coordinates is called the **range.**

Figure 1.2

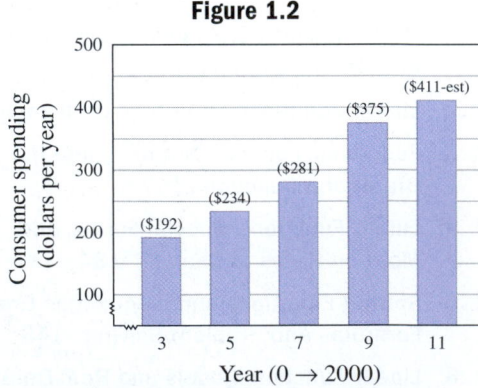

Consumer spending (dollars per year)

($192) ($234) ($281) ($375) ($411-est)

Year (0 → 2000): 3, 5, 7, 9, 11

Source: 2009 Statistical Abstract of the United States, Table 1089 (some figures are estimates)

EXAMPLE 1 ▶ **Expressing a Relation as a Mapping and as a Pointwise-Defined Relation**

Represent the relation from Figure 1.2 in mapping notation and as a pointwise-defined relation, then state its domain and range.

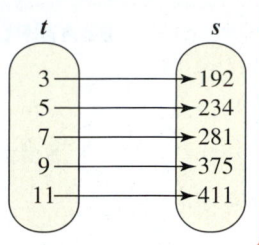

Solution ▶ Let t represent the year and s represent consumer spending. The mapping $t \rightarrow s$ gives the diagram shown. As a pointwise-defined relation we have (3, 192), (5, 234), (7, 281), (9, 375), and (11, 411). The domain is the set {3, 5, 7, 9, 11}; the range is {192, 234, 281, 375, 411}.

☑ **A.** You've just seen how we can express a relation in mapping notation and ordered pair form

Now try Exercises 7 through 12 ▶

For more on this relation, **see Exercise 93.**

B. The Graph of a Relation

Table 1.1 $y = x - 1$

x	y
-4	-5
-2	-3
0	-1
2	1
4	3

Table 1.2 $x = |y|$

x	y
2	-2
1	-1
0	0
1	1
2	2

Relations can also be stated in **equation form.** The equation $y = x - 1$ expresses a relation where each y-value is one less than the corresponding x-value (see Table 1.1). The equation $x = |y|$ expresses a relation where each x-value corresponds to the absolute value of y (see Table 1.2). In each case, the relation is the set of all ordered pairs (x, y) that create a true statement when substituted, and a few ordered pair solutions are shown in the tables for each equation.

Relations can be expressed graphically using a **rectangular coordinate system.** It consists of a horizontal number line (the x-axis) and a vertical number line (the y-axis) intersecting at their zero marks. The point of intersection is called the *origin.* The x- and y-axes create a flat, two-dimensional surface called the xy**-plane** and divide the plane into four regions called **quadrants.** These are labeled using a capital "Q" (for quadrant) and the Roman numerals I through IV, beginning in the upper right and moving counterclockwise (Figure 1.3). The **grid lines** shown denote the integer values on each axis and further divide the plane into a **coordinate grid,** where every point in the plane corresponds to an ordered pair. Since a point at the origin has not moved along either axis, it has coordinates $(0, 0)$. To plot a point (x, y) means we place a dot at its location in the xy-plane. A few of the ordered pairs from $y = x - 1$ are plotted in Figure 1.4, where a noticeable pattern emerges—the points seem to lie along a straight line.

If a relation is pointwise-defined, the graph of the relation is simply the plotted points. The graph of a relation *in equation form,* such as $y = x - 1$, is the set of *all* ordered pairs (x, y) that are solutions (make the equation true).

Figure 1.3

Figure 1.4

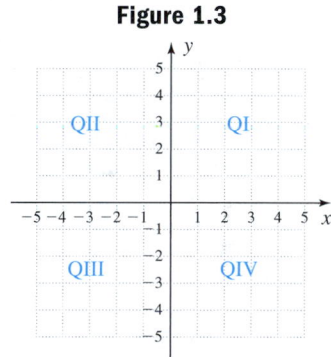

Solutions to an Equation in Two Variables

1. If substituting $x = a$ and $y = b$ results in a true equation, the ordered pair (a, b) is a solution and on the graph of the relation.

2. If the ordered pair (a, b) is on the graph of a relation, it is a solution (substituting $x = a$ and $y = b$ will result in a true equation).

We generally use only a few select points to determine the shape of a graph, then draw a straight line or smooth curve through these points, as indicated by any patterns formed.

EXAMPLE 2 ▶ **Graphing Relations**

Graph the relations $y = x - 1$ and $x = |y|$ using the ordered pairs given in Tables 1.1 and 1.2.

Solution ▶ For $y = x - 1$, we plot the points then connect them with a straight line (Figure 1.5). For $x = |y|$, the plotted points form a V-shaped graph made up of two half lines (Figure 1.6).

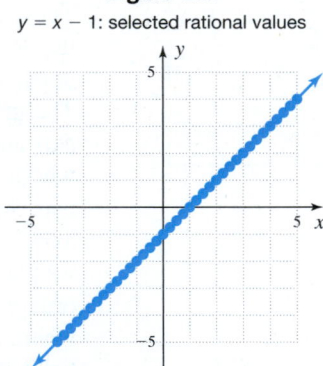

Figure 1.5

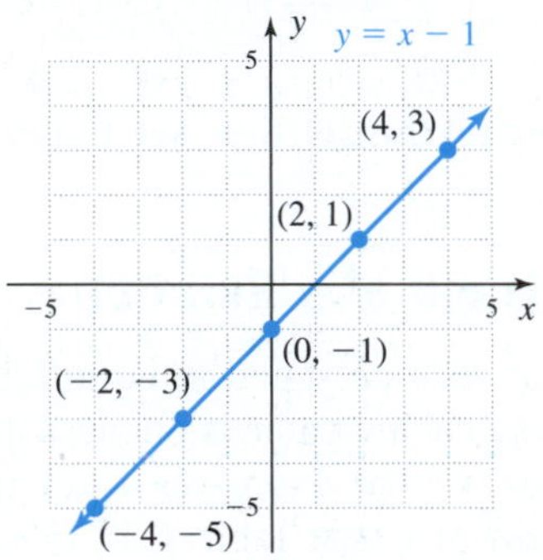

$y = x - 1$

(4, 3)
(2, 1)
(0, −1)
(−2, −3)
(−4, −5)

Figure 1.6

$x = |y|$

(2, 2)
(0, 0)
(2, −2)

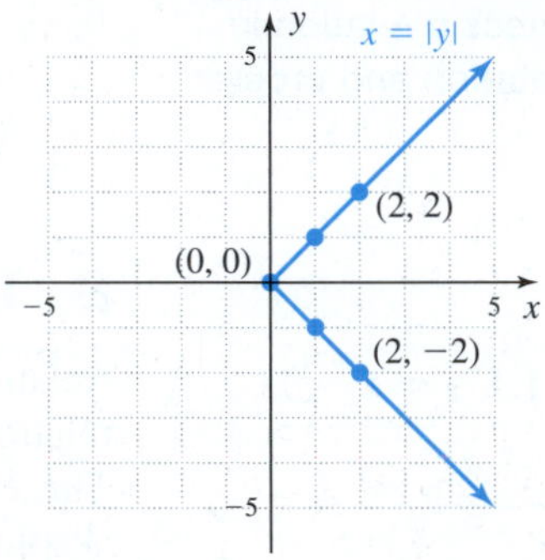

Now try Exercises 13 through 16 ▶

WORTHY OF NOTE

As the graphs in Example 2 indicate, arrowheads are used where appropriate to indicate the infinite extension of a graph.

While we used only a few points to graph the relations in Example 2, they are actually made up of an *infinite number of ordered pairs* that satisfy each equation, including those that might be rational or irrational. This understanding is an important part of reading and interpreting graphs, and is illustrated for you in Figures 1.7 through 1.10.

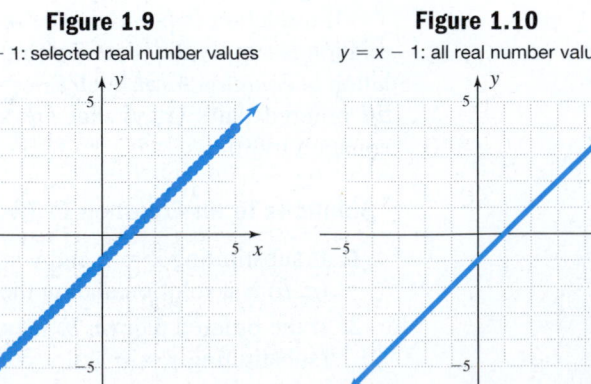

Figure 1.7

$y = x - 1$: selected integer values

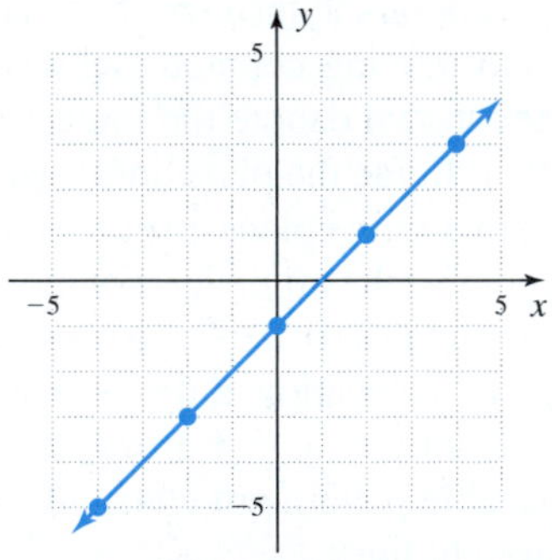

Figure 1.8

$y = x - 1$: selected rational values

Figure 1.9

$y = x - 1$: selected real number values

Figure 1.10

$y = x - 1$: all real number values

Since there are an infinite number of ordered pairs forming the graph of $y = x - 1$, the domain cannot be given in list form. Here we note x can be any real number and write $D: x \in \mathbb{R}$. Likewise, y can be any real number and for the range we have $R: y \in \mathbb{R}$. All of these points together make these graphs **continuous,** which for our purposes means you can draw the entire graph without lifting your pencil from the paper.

Actually, a majority of graphs cannot be drawn using only a straight line or directed line segments. In these cases, we rely on a "sufficient number" of points to outline the basic shape of the graph, then connect the points with a smooth curve. As your experience with graphing increases, this "sufficient number of points" tends to get smaller as you learn to anticipate what the graph of a given relation should look like. In particular, for the linear graph in Figure 1.5 we notice that both the x- and y-variables have an *implied exponent of 1.* This is in fact a characteristic of linear equations and graphs. In Example 3 we'll notice that if the exponent on one of the variables is 2 (either x or y is *squared*) while the other exponent is 1, the result is a graph called a **parabola.** If the x-term is squared (Example 3a) the parabola is oriented vertically, as in Figure 1.11, and its highest or lowest point is called the **vertex.** If the y-term is squared (Example 3c), the parabola is oriented horizontally, as in Figure 1.13, and the leftmost or rightmost point is the vertex. The graphs and equations of other relations likewise have certain identifying characteristics. **See Exercises 85 through 92.**

EXAMPLE 3 ▶ **Graphing Relations**

Graph the following relations by completing the tables given. Then use the graph to state the domain and range of the relation.

a. $y = x^2 - 2x$ **b.** $y = \sqrt{9 - x^2}$ **c.** $x = y^2$

Solution ▶ For each relation, we use each x-input in turn to determine the related y-output(s), if they exist. Results can be entered in a table and the ordered pairs used to assist in drawing a complete graph.

a. $y = x^2 - 2x$ **Figure 1.11**

x	y	(x, y) Ordered Pairs
-4	24	$(-4, 24)$
-3	15	$(-3, 15)$
-2	8	$(-2, 8)$
-1	3	$(-1, 3)$
0	0	$(0, 0)$
1	-1	$(1, -1)$
2	0	$(2, 0)$
3	3	$(3, 3)$
4	8	$(4, 8)$

The resulting **vertical parabola** is shown in Figure 1.11. Although $(-4, 24)$ and $(-3, 15)$ cannot be plotted here, the arrowheads indicate an infinite extension of the graph, which will include these points. This "infinite extension" in the *upward* direction shows there is no largest y-value (the graph becomes infinitely "tall"). Since the smallest possible y-value is -1 [from the vertex $(1, -1)$], the range is $y \geq -1$. However, this extension also continues forever in the *outward* direction as well (the graph gets wider and wider). This means the x-value of all possible ordered pairs could vary from negative to positive infinity, and the domain is all real numbers. We then have $D: x \in \mathbb{R}$ and $R: y \geq -1$.

b. $y = \sqrt{9 - x^2}$

x	y	(x, y) Ordered Pairs
-4	not real	—
-3	0	$(-3, 0)$
-2	$\sqrt{5}$	$(-2, \sqrt{5})$
-1	$2\sqrt{2}$	$(-1, 2\sqrt{2})$
0	3	$(0, 3)$
1	$2\sqrt{2}$	$(1, 2\sqrt{2})$
2	$\sqrt{5}$	$(2, \sqrt{5})$
3	0	$(3, 0)$
4	not real	—

Figure 1.12

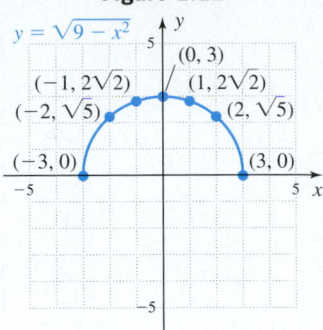

The result is the graph of a **semicircle** (Figure 1.12). The points with irrational coordinates were graphed by estimating their location. Note that when $x < -3$ or $x > 3$, the relation $y = \sqrt{9 - x^2}$ does not represent a real number and no points can be graphed. Also note that no arrowheads are used since the graph terminates at $(-3, 0)$ and $(3, 0)$. These observations and the graph itself show that for this relation, $D: -3 \leq x \leq 3$, and $R: 0 \leq x \leq 3$.

c. Similar to $x = |y|$, the relation $x = y^2$ is defined only for $x \geq 0$ since y^2 is always nonnegative ($-1 = y^2$ has no real solutions). In addition, we reason that each positive x-value will correspond to two y-values. For example, given $x = 4$, $(4, -2)$ and $(4, 2)$ are both solutions to $x = y^2$.

$x = y^2$

x	y	(x, y) Ordered Pairs
-2	not real	—
-1	not real	—
0	0	$(0, 0)$
1	$-1, 1$	$(1, -1)$ and $(1, 1)$
2	$-\sqrt{2}, \sqrt{2}$	$(2, -\sqrt{2})$ and $(2, \sqrt{2})$
3	$-\sqrt{3}, \sqrt{3}$	$(3, -\sqrt{3})$ and $(3, \sqrt{3})$
4	$-2, 2$	$(4, -2)$ and $(4, 2)$

Figure 1.13

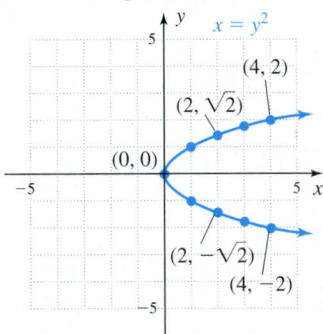

This relation is a horizontal parabola, with a vertex at $(0, 0)$ (Figure 1.13). The graph begins at $x = 0$ and extends infinitely to the right, showing the domain is $x \geq 0$. Similar to Example 3a, this "infinite extension" also extends in both the *upward* and *downward* directions and the *y-value* of all possible ordered pairs could vary from negative to positive infinity. We then have $D: x \geq 0$ and $R: y \in \mathbb{R}$.

✓ **B.** You've just seen how we can graph relations

Now try Exercises 17 through 24 ▶

C. Graphing Relations on a Calculator

For relations given in equation form, the **TABLE** feature of a graphing calculator can be used to compute ordered pairs, and the (GRAPH) feature to draw the related graph. To use these features, we first solve the equation for the variable y (write y in terms of x), then enter the right-hand expression on the calculator's (Y=) (equation editor) screen.

We can then select either the GRAPH feature, or set-up, create, and use the **TABLE** feature. We'll illustrate here using the relation $-2x + y = 3$.

1. Solve for y in terms of x.

   $-2x + y = 3$ given equation

   $y = 2x + 3$ add $2x$ to each side

2. Enter the equation.

   Press the Y= key to access the equation editor, then enter $2x + 3$ as Y_1 (see Figure 1.14). The calculator automatically highlights the equal sign, showing that equation Y_1 is now active. If there are other equations on the screen, you can either CLEAR them or deactivate them by moving the cursor to overlay the equal sign and pressing ENTER.

3. Use the **TABLE** or GRAPH.

   To set up the table, we use the keystrokes 2nd WINDOW (**TBLSET**). For this exercise, we'll put the table in the "**Indpnt: Auto Ask**" mode, which will have the calculator automatically generate the input and output values. In this mode, we can tell the calculator where to start the inputs (we chose **TblStart = −3**), and have the calculator produce the input values using any increment desired (we choose **ΔTbl = 1**). See Figure 1.15A. Access the table using 2nd GRAPH (**TABLE**), and the table resulting from this setup is shown in Figure 1.15B. Notice that all ordered pairs satisfy the equation $y = 2x + 3$, or "y is twice x increased by 3."

Figure 1.14

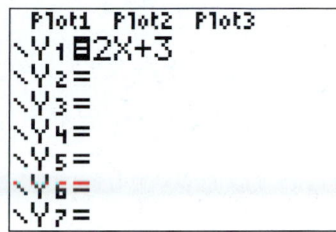

Figure 1.15A

Figure 1.15B

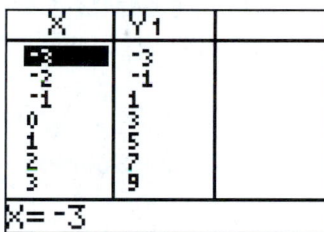

Since much of our graphical work is centered at $(0, 0)$ on the coordinate grid, the calculator's default settings for the standard viewing WINDOW are $[-10, 10]$ for both x and y (Figure 1.16). The **Xscl** and **Yscl** values give the scale used on each axis, and indicate here that each "tick mark" will be 1 unit apart. To graph the line in this window, we can use the ZOOM key and select **6:ZStandard** (Figure 1.17), which resets the window to these default settings and automatically graphs the line (Figure 1.18).

Figure 1.16

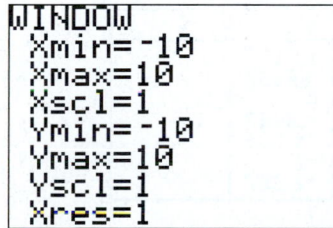

Figure 1.17

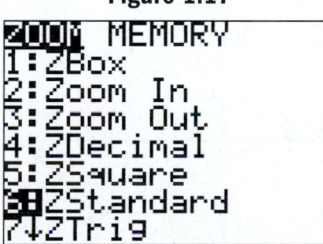

Figure 1.18

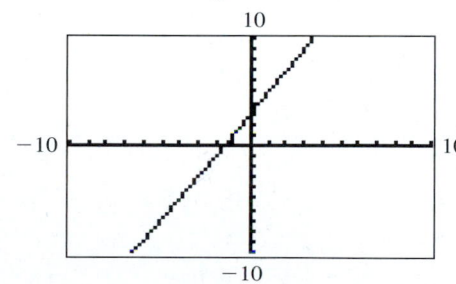

In addition to using the calculator's **TABLE** feature to find ordered pairs for a given graph, we can also use the calculator's TRACE feature. As the name implies, this feature allows us to "trace" along the graph by moving a cursor to the left ◄ and right ► using the arrow keys. The calculator displays the coordinates of the cursor's location each time it moves. After pressing the TRACE key, the marker appears automatically and as you move it to the left or right, the current coordinates are shown at the bottom of the screen (Figure 1.19). While not very "pretty," $(-0.8510638, 1.2978723)$ is a point on this line (rounded to seven decimal places) and satisfies its equation. The calculator is displaying these decimal values because the viewing screen is exactly 95 pixels wide, 47 pixels to the left of the y-axis, and 47 pixels to the right. This means that each time you press the left or right arrow, the x-value changes by 1/47th—which is not a nice round number. To have the calculator TRACE through "friendlier" values, we can use the ZOOM **4:ZDecimal** feature, which sets **Xmin** $= -4.7$ and **Xmax** $= 4.7$, or ZOOM **8:ZInteger,** which sets **Xmin** $= -47$ and **Xmax** $= 47$. Let's use the ZOOM **4:ZDecimal** option here, noting the calculator automatically regraphs the line. Pressing the TRACE key once again and moving the marker shows that more "friendly" ordered pair solutions are displayed (Figure 1.20). Other methods for finding a friendly window are discussed later in this section.

Figure 1.19

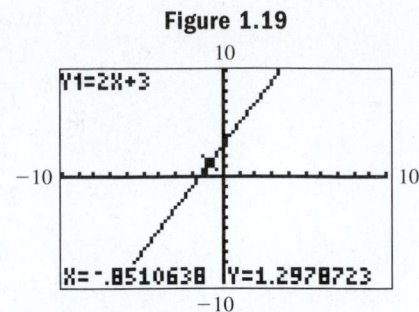

Figure 1.20

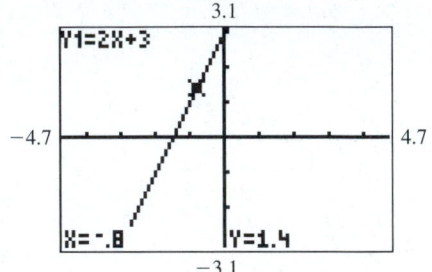

EXAMPLE 4 ▶ **Graphing a Relation Using Technology**

Use a calculator to graph $2x + 3y = -6$. Then use the **TABLE** feature to determine the value of y when $x = 0$, and the value of x when $y = 0$. Write each result in ordered pair form.

Solution ▶ We begin by solving the equation for y, so we can enter it on the Y= screen.

$$2x + 3y = -6 \qquad \text{given equation}$$
$$3y = -2x - 6 \qquad \text{subtract } 2x \text{ (isolate the } y\text{-term)}$$
$$y = \frac{-2}{3}x - 2 \qquad \text{divide by 3}$$

Entering $y = \dfrac{-2}{3}x - 2$ on the Y= screen and using ZOOM **6:ZStandard** produces the graph shown. Using the **TABLE** and scrolling as needed, shows that when $x = 0$, $y = -2$, and when $y = 0$, $x = -3$. As ordered pairs we have $(0, -2)$ and $(-3, 0)$.

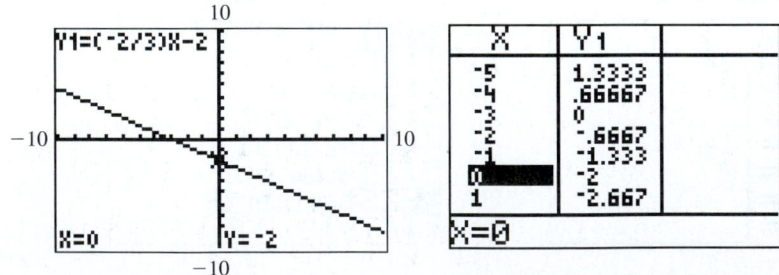

☑ **C.** You've just seen how we can graph a relation using a calculator

Now try Exercises 25 through 28 ▶

D. The Equation and Graph of a Circle

Using the midpoint and distance formulas, we can develop the equation of another important relation, that of a circle. As the name suggests, the **midpoint of a line segment** is located halfway between the endpoints. On a standard number line, the midpoint of the line segment with endpoints 1 and 5 is 3, but more important, note that 3 is the

average distance (from zero) of 1 unit and 5 units: $\dfrac{1+5}{2} = \dfrac{6}{2} = 3$. This observation can be extended to find the midpoint between any two points (x_1, y_1) and (x_2, y_2) in the xy-plane. We simply find the average distance between the x-coordinates and the average distance between the y-coordinates.

The Midpoint Formula

Given any line segment with endpoints $P_1 = (x_1, y_1)$ and $P_2 = (x_2, y_2)$, the midpoint M is given by

$$M: \left(\frac{x_1 + x_2}{2}, \frac{y_1 + y_2}{2} \right)$$

The midpoint formula can be used in many different ways. Here we'll use it to find the coordinates of the center of a circle.

EXAMPLE 5 ▶ **Using the Midpoint Formula**

The diameter of a circle has endpoints at $P_1 = (-3, -2)$ and $P_2 = (5, 4)$. Use the midpoint formula to find the coordinates of the center, then plot this point.

Solution ▶ Midpoint: $\left(\dfrac{x_1 + x_2}{2}, \dfrac{y_1 + y_2}{2} \right)$

$$M: \left(\frac{-3+5}{2}, \frac{-2+4}{2} \right)$$

$$M: \left(\frac{2}{2}, \frac{2}{2} \right) = (1, 1)$$

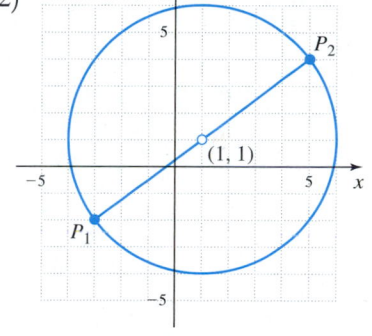

The center is at $(1, 1)$, which we graph directly on the diameter as shown.

Now try Exercises 29 through 38 ▶

Figure 1.21

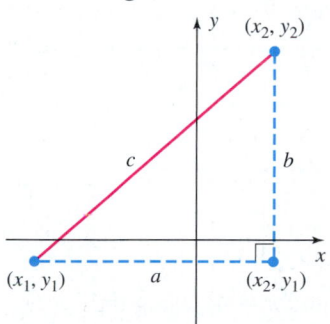

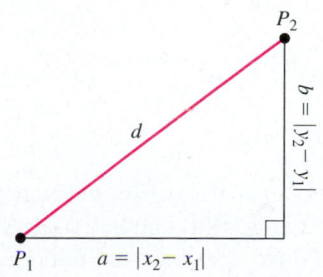

The Distance Formula

In addition to a line segment's midpoint, we are often interested in the *length* of the segment. For any two points (x_1, y_1) and (x_2, y_2) not lying on a horizontal or vertical line, a right triangle can be formed as in Figure 1.21. Regardless of the triangle's orientation, the length of side a (the horizontal segment or base of the triangle) will have length $|x_2 - x_1|$ units, with side b (the vertical segment or height) having length $|y_2 - y_1|$ units. From the Pythagorean theorem (Section R.6), we see that $c^2 = a^2 + b^2$ corresponds to $c^2 = (|x_2 - x_1|)^2 + (|y_2 - y_1|)^2$. By taking the square root of both sides we obtain the length of the hypotenuse, *which is identical to the distance between these two points:* $c = \sqrt{(x_2 - x_1)^2 + (y_2 - y_1)^2}$. The result is called the **distance formula,** although it's most often written using d for distance, rather than c. Note the absolute value bars are dropped from the formula, since the square of any quantity is always nonnegative. This also means that *either* point can be used as the initial point in the computation.

The Distance Formula

Given any two points $P_1 = (x_1, y_1)$ and $P_2 = (x_2, y_2)$, the straight line distance d between them is

$$d = \sqrt{(x_2 - x_1)^2 + (y_2 - y_1)^2}$$

EXAMPLE 6 ▶ **Using the Distance Formula**

Use the distance formula to find the diameter of the circle from Example 5.

Solution ▶ For $(x_1, y_1) = (-3, -2)$ and $(x_2, y_2) = (5, 4)$, the distance formula gives

$$d = \sqrt{(x_2 - x_1)^2 + (y_2 - y_1)^2}$$
$$= \sqrt{[5 - (-3)]^2 + [4 - (-2)]^2}$$
$$= \sqrt{8^2 + 6^2}$$
$$= \sqrt{100} = 10$$

The diameter of the circle is 10 units long.

Now try Exercises 39 through 48 ▶

A circle can be defined as the set of all points in a plane that are a *fixed distance* called the **radius,** from a *fixed point* called the **center.** Since the definition involves *distance,* we can construct the general equation of a circle using the distance formula. Assume the center has coordinates (h, k), and let (x, y) represent any point on the graph. The distance between these points is equal to the radius r, and the distance formula yields: $\sqrt{(x - h)^2 + (y - k)^2} = r$. Squaring both sides gives the equation of a circle in **standard form:** $(x - h)^2 + (y - k)^2 = r^2$.

The Equation of a Circle

A circle of radius r with center at (h, k) has the equation

$$(x - h)^2 + (y - k)^2 = r^2$$

If $h = 0$ and $k = 0$, the circle is centered at $(0, 0)$ and the graph is a **central circle** with equation $x^2 + y^2 = r^2$. At other values for h or k, the center is at (h, k) with no change in the radius. Note that an open dot is used for the center, as it's actually a point of reference and not a part of the graph.

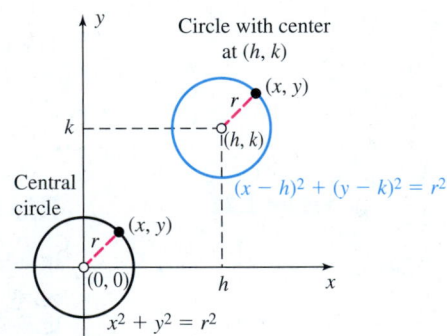

EXAMPLE 7 ▶ **Finding the Equation of a Circle in Standard Form**

Find the equation of a circle with center $(0, -1)$ and radius 4.

Solution ▶ Since the center is at $(0, -1)$ we have $h = 0$, $k = -1$, and $r = 4$. Using the standard form $(x - h)^2 + (y - k)^2 = r^2$ we obtain

$$(x - 0)^2 + [y - (-1)]^2 = 4^2 \quad \text{substitute 0 for } h, -1 \text{ for } k, \text{ and 4 for } r$$
$$x^2 + (y + 1)^2 = 16 \quad \text{simplify}$$

The graph of $x^2 + (y + 1)^2 = 16$ is shown in the figure.

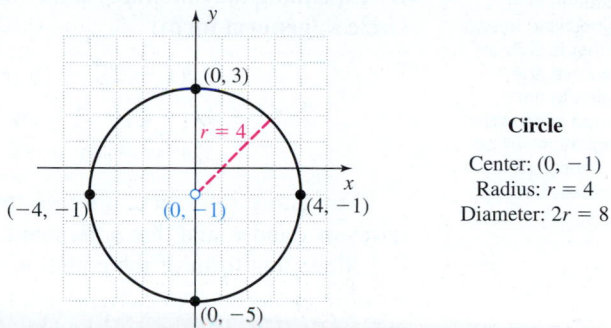

Circle

Center: $(0, -1)$
Radius: $r = 4$
Diameter: $2r = 8$

Now try Exercises 49 through 66 ▶

The graph of a circle can be obtained by first identifying the coordinates of the center and the length of the radius from the equation in standard form. After plotting the center point, we count a distance of r units left and right of center in the horizontal direction, and up and down from center in the vertical direction, obtaining four points on the circle. Neatly graph a circle containing these four points.

EXAMPLE 8 ▶ **Graphing a Circle**

Graph the circle represented by $(x - 2)^2 + (y + 3)^2 = 12$. Clearly label the center and radius.

Solution ▶ Comparing the given equation with the standard form, we find the center is at $(2, -3)$ and the radius is $r = 2\sqrt{3} \approx 3.5$.

$$(x - h)^2 + (y - k)^2 = r^2 \qquad \text{standard form}$$
$$\downarrow \qquad \quad \downarrow \qquad \quad \downarrow$$
$$(x - 2)^2 + (y + 3)^2 = 12 \qquad \text{given equation}$$
$$-h = -2 \qquad -k = 3 \qquad r^2 = 12$$
$$h = 2 \qquad \quad k = -3 \qquad r = \sqrt{12} = 2\sqrt{3} \quad \text{radius must be positive}$$
$$\approx 3.5$$

Plot the center $(2, -3)$ and count approximately 3.5 units in the horizontal and vertical directions. Complete the circle by freehand drawing or using a compass. The graph shown is obtained.

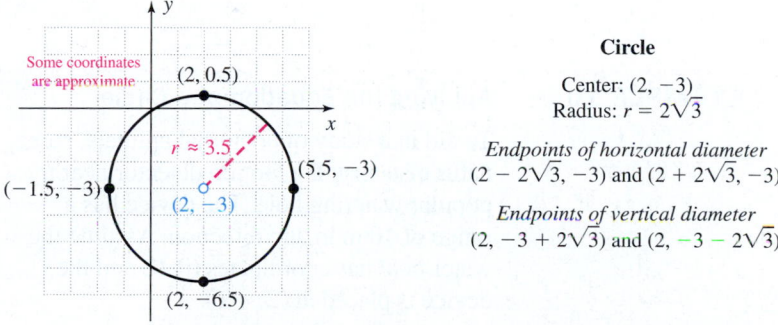

Circle

Center: $(2, -3)$
Radius: $r = 2\sqrt{3}$

Endpoints of horizontal diameter
$(2 - 2\sqrt{3}, -3)$ and $(2 + 2\sqrt{3}, -3)$

Endpoints of vertical diameter
$(2, -3 + 2\sqrt{3})$ and $(2, -3 - 2\sqrt{3})$

Now try Exercises 67 through 72 ▶

WORTHY OF NOTE

After writing the equation in standard form, it is possible to end up with a constant that is zero or negative. In the first case, the graph is a single point. In the second case, no graph is possible since roots of the equation will be complex numbers. These are called *degenerate cases*. **See Exercise 105.**

In Example 8, note the equation is composed of binomial squares in both x and y. By expanding the binomials and collecting like terms, we can write the equation of the circle in **general form:**

$$(x - 2)^2 + (y + 3)^2 = 12 \qquad \text{standard form}$$
$$x^2 - 4x + 4 + y^2 + 6y + 9 = 12 \qquad \text{expand binomials}$$
$$x^2 + y^2 - 4x + 6y + 1 = 0 \qquad \text{combine like terms—general form}$$

For future reference, observe the general form contains a *sum* of second-degree terms in x and y, and that *both terms have the same coefficient* (in this case, "1").

Since this form of the equation was derived by squaring binomials, it seems reasonable to assume we can go back to the standard form by creating binomial squares in x and y. This is accomplished by *completing the square*.

EXAMPLE 9 ▶ **Finding the Center and Radius of a Circle**

Find the center and radius of the circle with equation $x^2 + y^2 + 2x - 4y - 4 = 0$. Then sketch its graph and label the center and radius.

Solution ▶ To find the center and radius, we complete the square in both x and y.

$$x^2 + y^2 + 2x - 4y - 4 = 0 \qquad \text{given equation}$$
$$(x^2 + 2x + \underline{\quad}) + (y^2 - 4y + \underline{\quad}) = 4 \qquad \text{group } x\text{-terms and } y\text{-terms; add 4}$$
$$(x^2 + 2x + 1) + (y^2 - 4y + 4) = 4 + 1 + 4 \qquad \text{complete each binomial square}$$
$$\underbrace{\qquad\qquad}_{\text{adds 1 to left side}} \quad \underbrace{\qquad\qquad}_{\text{adds 4 to left side}} \qquad \underbrace{\qquad}_{\text{add } 1 + 4 \text{ to right side}}$$
$$(x + 1)^2 + (y - 2)^2 = 9 \qquad \text{factor and simplify}$$

The center is at $(-1, 2)$ and the radius is $r = \sqrt{9} = 3$.

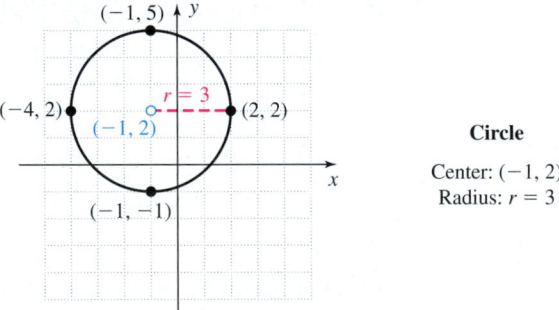

Circle

Center: $(-1, 2)$
Radius: $r = 3$

Now try Exercises 73 through 84 ▶

EXAMPLE 10 ▶ **Applying the Equation of a Circle**

To aid in a study of nocturnal animals, some naturalists install a motion detector near a popular watering hole. The device has a range of 10 m in any direction. Assume the water hole has coordinates $(0, 0)$ and the device is placed at $(2, -1)$.

a. Write the equation of the circle that models the maximum effective range of the device.

b. Use the distance formula to determine if the device will detect a badger that is approaching the water and is now at coordinates $(11, -5)$.

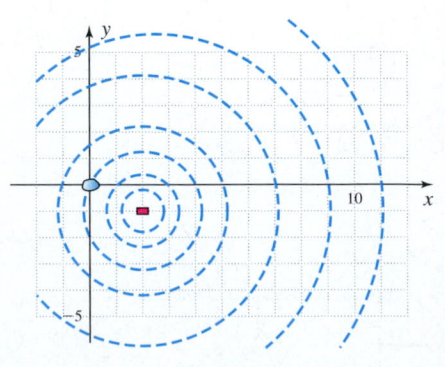

Solution ▶ **a.** Since the device is at $(2, -1)$ and the radius (or reach) of detection is 10 m, any movement in the interior of the circle defined by $(x - 2)^2 + (y + 1)^2 = 10^2$ will be detected.

b. Using the points $(2, -1)$ and $(11, -5)$ in the distance formula yields:

$$d = \sqrt{(x_2 - x_1)^2 + (y_2 - y_1)^2} \qquad \text{distance formula}$$
$$= \sqrt{(11 - 2)^2 + [-5 - (-1)]^2} \qquad \text{substitute given values}$$
$$= \sqrt{9^2 + (-4)^2} \qquad \text{simplify}$$
$$= \sqrt{81 + 16} \qquad \text{compute squares}$$
$$= \sqrt{97} \approx 9.85 \qquad \text{result}$$

Since $9.85 < 10$, the badger is within range of the device and will be detected.

Now try Exercises 95 through 100 ▶

When using a graphing calculator to study circles, it's important to note that most standard viewing windows have the x- and y-values preset at $[-10, 10]$ even though the calculator screen is not square. This tends to compress the y-values and give a skewed image of the graph. If the circle appears oval in shape, use ⓩⓞⓞⓜ **5:ZSquare** to obtain the correct perspective. Graphing calculators can produce the graph of a circle in various ways, and the choice of method simply depends on what you'd like to accomplish. To simply view the graph or compare two circular graphs, the **DRAW** command is used. From the home screen press: ⓶ⓝⓓ ⓟⓡⓖⓜ **(DRAW) 9:Circle(**. This generates the **"Circle("** command, with the left parentheses indicating we need to supply three inputs, separated by commas. These inputs are the x-coordinate of the center, the y-coordinate of the center, and the radius of the circle. For the circle defined by the equation $(x - 3)^2 + (y + 2)^2 = 49$, we know the center is at $(3, -2)$ and the radius is 7 units. The resulting command and graph are shown in Figures 1.22 and 1.23.

While the **DRAW** command will graph any circle, we are unable to use the ⓣⓡⓐⓒⓔ or **CALC** commands to interact with the graph. To make these features available, we must first solve for x in terms of y, as we did previously (the **1:ClrDraw** command is used to clear the graph). Consider the *relation* $x^2 + y^2 = 25$, which we know is the equation of a circle centered at $(0, 0)$ with radius $r = 5$.

$$x^2 + y^2 = 25 \qquad \text{original equation}$$
$$y^2 = 25 - x^2 \qquad \text{isolate } y^2$$
$$y = \pm\sqrt{25 - x^2} \qquad \text{solve for } y$$

Note that we can separate this result into two parts, enabling the calculator to graph $Y_1 = \sqrt{25 - x^2}$ (giving the "upper half" of the circle), and $Y_2 = -\sqrt{25 - x^2}$ (giving the "lower half"). Enter these on the ⓨ⁼ screen (note that $Y_2 = -Y_1$ can be used instead of reentering the entire expression: ⓥⓐⓡⓢ ⓥ ⓔⓝⓣⓔⓡ). If we graph Y_1 and Y_2 on the standard screen, the result appears more oval than circular (Figure 1.24). Using the ⓩⓞⓞⓜ **5:ZSquare** option, the tick marks become equally spaced on both axes (Figure 1.25).

Figure 1.22, 1.23

Figure 1.24

Figure 1.25

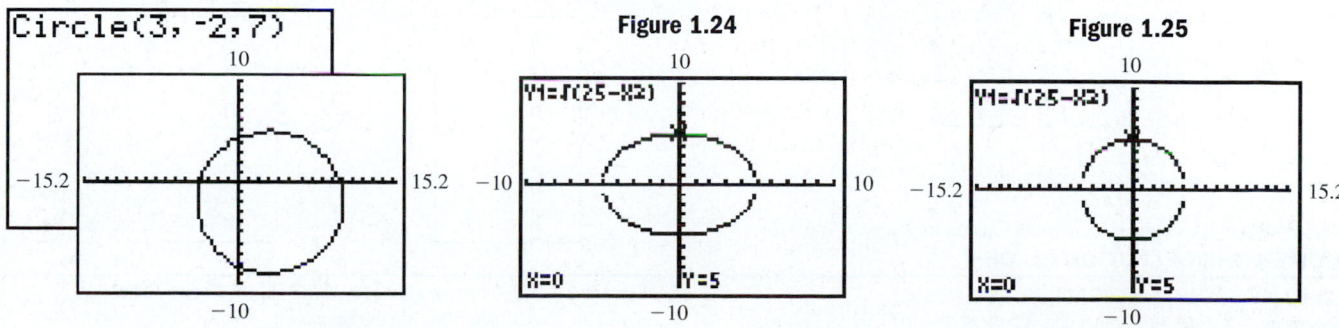

 **D.** You've just seen how we can develop the equation and graph of a circle using the distance and midpoint formulas

Although it is a much improved graph, the circle does not appear "closed" as the calculator lacks sufficient pixels to show the proper curvature. A second alternative is to manually set a "friendly" window. Using Xmin = −9.4, Xmax = 9.4, Ymin = −6.2, and Ymax = 6.2 will generate a better graph due to the number of pixels available. Note that we can jump between the upper and lower halves of the circle using the up ⌃ or down ⌄ arrows. **See Exercises 101 and 102.**

1.1 EXERCISES

▶ CONCEPTS AND VOCABULARY

Fill in each blank with the appropriate word or phrase. Carefully reread the section if needed.

1. If a relation is defined by a set of ordered pairs, the domain is the set of all ____first____ components, the range is the set of all ___second___ components.

2. For the equation $y = x + 5$ and the ordered pair (x, y), x is referred to as the input or ___independent___ variable, while y is called the ___output___ or dependent variable.

3. A circle is defined as the set of all points that are an equal distance, called the ____radius____, from a given point, called the ____center____.

4. For $x^2 + y^2 = 25$, the center of the circle is at ___(0, 0)___ and the length of the radius is ___5___ units. The graph is called a ___central___ circle.

5. Discuss/Explain how to find the center and radius of the circle defined by the equation $x^2 + y^2 - 6x = 7$. How would this circle differ from the one defined by $x^2 + y^2 - 6y = 7$?   Answers will vary.

6. In Example 3(b) we graphed the semicircle defined by $y = \sqrt{9 - x^2}$. Discuss how you would obtain the equation of the full circle from this equation, and how the two equations are related.   Answers will vary.

▶ DEVELOPING YOUR SKILLS

Represent each relation in mapping notation, then state the domain and range.

7.
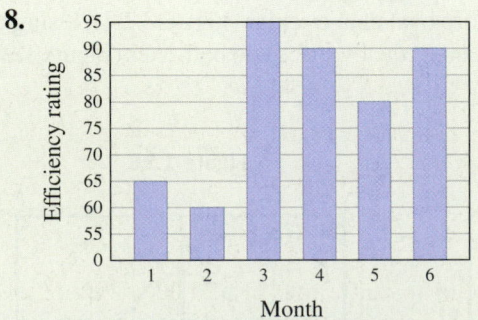
Year in college

8.

State the domain and range of each pointwise-defined relation.

9. {(1, 2), (3, 4), (5, 6), (7, 8), (9, 10)}
   $D = \{1, 3, 5, 7, 9\}; R = \{2, 4, 6, 8, 10\}$

10. {(−2, 4), (−3, −5), (−1, 3), (4, −5), (2, −3)}
   $D = \{-2, -3, -1, 4, 2\}; R = \{4, -5, 3, -3\}$

11. {(4, 0), (−1, 5), (2, 4), (4, 2), (−3, 3)}
   $D = \{4, -1, 2, -3\}; R = \{0, 5, 4, 2, 3\}$

12. {(−1, 1), (0, 4), (2, −5), (−3, 4), (2, 3)}
   $D = \{-1, 0, 2, -3\}; R = \{1, 4, -5, 3\}$

Complete each table using the given equation. For Exercises 15, 16, 21, and 22, each input may correspond to two outputs (be sure to find both if they exist). Use these points to graph the relation. For Exercises 17 through 24, also state the domain and range.

13. $y = -\dfrac{2}{3}x + 1$

x	y
−6	5
−3	3
0	1
3	−1
6	−3
8	$\frac{-13}{3}$

14. $y = -\dfrac{5}{4}x + 3$

x	y
−8	13
−4	8
0	3
4	−2
8	−7
10	$\frac{-19}{2}$

Additional answers can be found in the Instructor Answer Appendix.

HOMEWORK SELECTION GUIDE

Core: 7–91 every other odd, 95–101 odd (26 Exercises)
Standard: 1–4, 7–83 every other odd, 85–92 all, 95–101 odd (36 Exercises)

Extended: 1–4, 7–31 every other odd, 35–38 all, 39–79 every other odd, 85–92 all, 95–101 odd, 106, 109 (39 Exercises)
In Depth: 1–4, 7–31 every other odd, 35–38 all, 39–83 every other odd, 85–92 all, 95, 96, 98, 99, 100, 101, 105, 106, 109 (44 Exercises)

15. $x + 2 = |y|$

x	y
−2	0
0	2, −2
1	3, −3
3	5, −5
6	8, −8
7	9, −9

16. $|y + 1| = x$

x	y
0	−1
1	0, −2
3	2, −4
5	4, −6
6	5, −7
7	6, −8

17. $y = x^2 - 1$

x	y
−3	8
−2	3
0	−1
2	3
3	8
4	15

18. $y = -x^2 + 3$

x	y
−2	−1
−1	2
0	3
1	2
2	−1
3	−6

19. $y = \sqrt{25 - x^2}$

x	y
−4	3
−3	4
0	5
2	$\sqrt{21}$
3	4
4	3

20. $y = \sqrt{169 - x^2}$

x	y
−12	5
−5	12
0	13
3	$4\sqrt{10}$
5	12
12	5

21. $x - 1 = y^2$

x	y
10	3, −3
5	2, −2
4	$\sqrt{3}, -\sqrt{3}$
2	1, −1
1.25	0.5, −0.5
1	0

22. $y^2 - 2 = x$

x	y
−2	0
−1	1, −1
0	$\sqrt{2}, -\sqrt{2}$
1	$\sqrt{3}, -\sqrt{3}$
2	2, −2
7	3, −3

23. $y = \sqrt[3]{x + 1}$

x	y
−9	−2
−2	−1
−1	0
0	1
4	$\sqrt[3]{5}$
7	2

24. $y = (x - 1)^3$

x	y
−2	−27
−1	−8
0	−1
1	0
2	1
3	8

Use a graphing calculator to graph the following relations. Then use the TABLE feature to determine the value of y when x = 0, and the value(s) of x when y = 0, and write the results in ordered pair form.

25. $-2x + 5y = 10$

26. $x + 2y = 6$

27. $y = x^2 - 4x$

28. $y + x^2 = 2x + 3$

Find the midpoint of each segment with the given endpoints.

29. $(1, 8), (5, -6)$ (3, 1)

30. $(5, 6), (6, -8)$ $(\frac{11}{2}, -1)$

31. $(-4.5, 9.2), (3.1, -9.8)$ $(-0.7, -0.3)$

32. $(5.2, 7.1), (6.3, -7.1)$ $(5.75, 0)$

33. $\left(\frac{1}{5}, -\frac{2}{3}\right), \left(-\frac{1}{10}, \frac{3}{4}\right)$ $(\frac{1}{20}, \frac{1}{24})$

34. $\left(-\frac{3}{4}, -\frac{1}{3}\right), \left(\frac{3}{8}, \frac{5}{6}\right)$ $(-\frac{3}{16}, \frac{1}{4})$

Find the midpoint of each segment.

35.

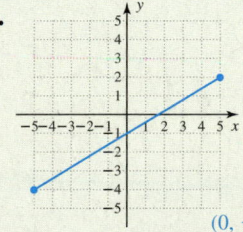

(0, −1)

36.

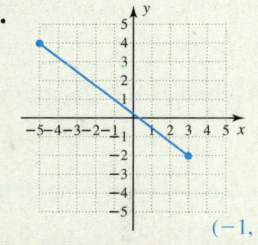

(−1, 1)

Find the center of each circle with the diameter shown.

37.

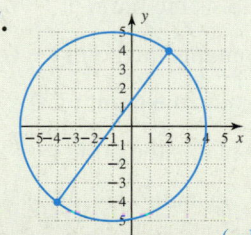

(−1, 0)

38.

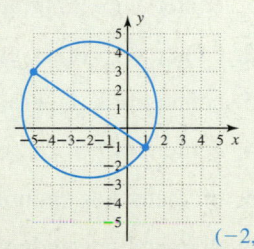

(−2, 1)

39. Use the distance formula to find the length of the line segment in Exercise 35. $2\sqrt{34}$

40. Use the distance formula to find the length of the line segment in Exercise 36.   10

41. Use the distance formula to find the length of the diameter of the circle in Exercise 37.   10

42. Use the distance formula to find the length of the diameter of the circle in Exercise 38. $2\sqrt{13}$

Additional answers can be found in the Instructor Answer Appendix.

In Exercises 43 to 48, three points that form the vertices of a triangle are given. Use the distance formula to determine if any of the triangles are right triangles (the three sides satisfy the Pythagorean Theorem $a^2 + b^2 = c^2$).

43. $(-3, 7), (2, 2), (5, 5)$ *right triangle*

44. $(7, 0), (-1, 0), (7, 4)$ *right triangle*

45. $(-4, 3), (-7, -1), (3, -2)$ *not a right triangle*

46. $(5, 2), (0, -3), (4, -4)$ *not a right triangle*

47. $(-3, 2), (-1, 5), (-6, 4)$ *right triangle*

48. $(0, 0), (-5, 2), (2, -5)$ *not a right triangle*

Find the equation of a circle satisfying the conditions given, then sketch its graph.

49. center $(0, 0)$, radius 3

50. center $(0, 0)$, radius 6

51. center $(5, 0)$, radius $\sqrt{3}$

52. center $(0, 4)$, radius $\sqrt{5}$

53. center $(4, -3)$, radius 2

54. center $(3, -8)$, radius 9

55. center $(-7, -4)$, radius $\sqrt{7}$

56. center $(-2, -5)$, radius $\sqrt{6}$

57. center $(1, -2)$, diameter 6

58. center $(-2, 3)$, diameter 10

59. center $(4, 5)$, diameter $4\sqrt{3}$

60. center $(5, 1)$, diameter $4\sqrt{5}$

61. center at $(7, 1)$, graph contains the point $(1, -7)$

62. center at $(-8, 3)$, graph contains the point $(-3, 15)$

63. center at $(3, 4)$, graph contains the point $(7, 9)$

64. center at $(-5, 2)$, graph contains the point $(-1, 3)$

65. diameter has endpoints $(5, 1)$ and $(5, 7)$

66. diameter has endpoints $(2, 3)$ and $(8, 3)$

Identify the center and radius of each circle, then graph. Also state the domain and range of the relation.

67. $(x - 2)^2 + (y - 3)^2 = 4$

68. $(x - 5)^2 + (y - 1)^2 = 9$

69. $(x + 1)^2 + (y - 2)^2 = 12$

70. $(x - 7)^2 + (y + 4)^2 = 20$

71. $(x + 4)^2 + y^2 = 81$

72. $x^2 + (y - 3)^2 = 49$

Write each equation in standard form to find the center and radius of the circle. Then sketch the graph.

73. $x^2 + y^2 - 10x - 12y + 4 = 0$

74. $x^2 + y^2 + 6x - 8y - 6 = 0$

75. $x^2 + y^2 - 10x + 4y + 4 = 0$

76. $x^2 + y^2 + 6x + 4y + 12 = 0$

77. $x^2 + y^2 + 6y - 5 = 0$

78. $x^2 + y^2 - 8x + 12 = 0$

79. $x^2 + y^2 + 4x + 10y + 18 = 0$

80. $x^2 + y^2 - 8x - 14y - 47 = 0$

81. $x^2 + y^2 + 14x + 12 = 0$

82. $x^2 + y^2 - 22y - 5 = 0$

83. $2x^2 + 2y^2 - 12x + 20y + 4 = 0$

84. $3x^2 + 3y^2 - 24x + 18y + 3 = 0$

In this section we looked at characteristics of equations that generated linear graphs, and graphs of parabolas and circles. Use this information and ordered pairs of your choosing to match the eight graphs given with their corresponding equation (two of the equations given have no matching graph).

a. $y = x^2 - 6x$ **b.** $x^2 + (y - 3)^2 = 36$

c. $x^2 + y = 9$ **d.** $3x - 4y = 12$

e. $y = \dfrac{-3}{2}x + 4$ **f.** $(x - 1)^2 + (y + 2)^2 = 49$

g. $(x - 3)^2 + y^2 = 16$ **h.** $(x - 1)^2 + (y + 2)^2 = 9$

i. $4x - 3y = 12$ **j.** $6x + y = x^2 + 9$

85.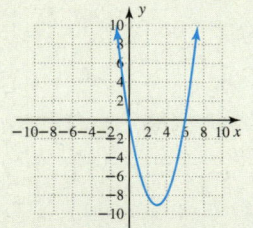

a. $y = x^2 - 6x$

86.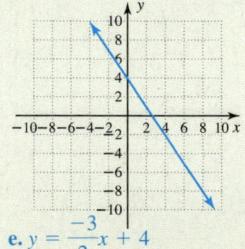

e. $y = \dfrac{-3}{2}x + 4$

87.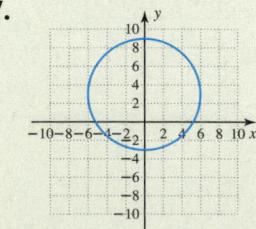

b. $y = x^2 + (y - 3)^2 = 36$

88.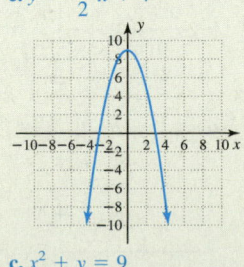

c. $x^2 + y = 9$

Additional answers can be found in the Instructor Answer Appendix.

89.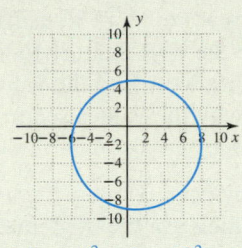

90.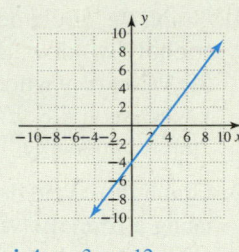

91.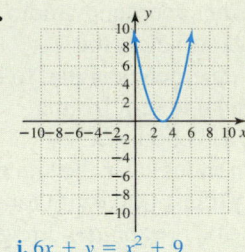

92.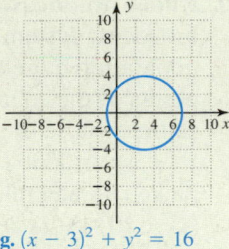

f. $(x - 1)^2 + (y + 2)^2 = 49$ **i.** $4x - 3y = 12$ **j.** $6x + y = x^2 + 9$ **g.** $(x - 3)^2 + y^2 = 16$

▶ WORKING WITH FORMULAS

93. Spending on Cable and Satellite TV:
$s = 29t + 96$

The data from Example 1 is closely modeled by the formula shown, where t represents the year ($t = 0$ corresponds to the year 2000) and s represents the average amount spent per person, per year in the United States. (a) List five ordered pairs for this relation using $t = 3, 5, 7, 9, 11$. Does the model give a good approximation of the actual data? (b) According to the model, what will be the average amount spent on cable and satellite TV in the year 2013? (c) According to the model, in what year will annual spending surpass \$500? (d) Use the table to graph this relation by hand.
a. (3, 183), (5, 241), (7, 299), (9, 357), (11, 415); yes   **b.** \$473
c. 2014

94. Radius of a circumscribed circle: $r = \sqrt{\dfrac{A}{2}}$

The radius r of a circle circumscribed around a square is found by using the formula given, where A is the area of the square. Solve the formula for A and use the result to find the area of the square shown. $A = 2r^2$, 50 units^2

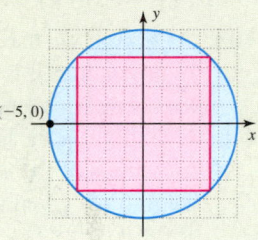

▶ APPLICATIONS

95. Radar detection: A luxury liner is located at map coordinates (5, 12) and has a radar system with a range of 25 nautical miles in any direction. (a) Write the equation of the circle that models the range of the ship's radar, and (b) Use the distance formula to determine if the radar can pick up the liner's sister ship located at coordinates (15, 36).
a. $(x - 5)^2 + (y - 12)^2 = 625$ **b.** no

96. Earthquake range: The epicenter (point of origin) of a large earthquake was located at map coordinates (3, 7), with the quake being felt up to 12 mi away. (a) Write the equation of the circle that models the range of the earthquake's effect. (b) Use the distance formula to determine if a person living at coordinates (13, 1) would have felt the quake.   **a.** $(x - 3)^2 + (y - 7)^2 = 144$ **b.** yes

97. Inscribed circle: Find the equation for both the red and blue circles, then find the area of the region shaded in blue.
Red: $(x - 2)^2 + (y - 2)^2 = 4$;
Blue: $(x - 2)^2 + y^2 = 16$;
Area blue $= 12\pi$ units^2

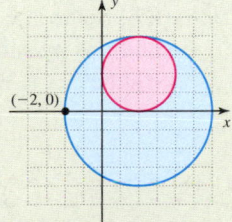

98. Inscribed triangle: The area of an equilateral triangle inscribed in a circle is given by the formula $A = \dfrac{3\sqrt{3}}{4}r^2$, where r is the radius of the circle. Find the area of the equilateral triangle shown. $\dfrac{75\sqrt{3}}{4}$ units^2

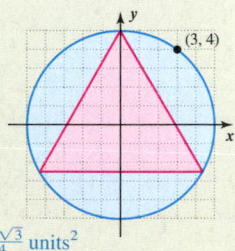

99. Radio broadcast range: Two radio stations may not use the same frequency if their broadcast areas *overlap*. Suppose station KXRQ has a broadcast area bounded by $x^2 + y^2 + 8x - 6y = 0$ and WLRT has a broadcast area bounded by $x^2 + y^2 - 10x + 4y = 0$. Graph the circle representing each broadcast area on the same grid to determine if both stations may broadcast on the same frequency.

Additional answers can be found in the Instructor Answer Appendix.

100. Radio broadcast range: The emergency radio broadcast system is designed to alert the population by relaying an emergency signal to all points of the country. A signal is sent from a station whose broadcast area is bounded by $x^2 + y^2 = 2500$ (x and y in miles) and the signal is picked up and relayed by a transmitter with range $(x - 20)^2 + (y - 30)^2 = 900$. Graph the circle representing each broadcast area on the same grid to determine the greatest distance from the original station that this signal can be received. Be sure to scale the axes appropriately.

 101. Graph the circle defined by $x^2 + y^2 = 36$ using a friendly window, then use the ⟨TRACE⟩ feature to find the value of y when $x = 3.6$. Now find the value of y when $x = 4.8$. Explain why the values seem "interchangeable." $y = \pm 4.8; y = \pm 3.6$, Answers will vary.

 102. Graph the circle defined by $(x - 3)^2 + y^2 = 16$ using a friendly window, then use the ⟨TRACE⟩ feature to find the value of the y-intercepts. Show you get the same intercept by computation. $(0, \pm 2.6457513), (0, \pm \sqrt{7})$

▶ **EXTENDING THE CONCEPT**

103. Although we use the word "domain" extensively in mathematics, it is also commonly seen in literature and heard in everyday conversation. Using a college-level dictionary, look up and write out the various meanings of the word, noting how closely the definitions given are related to its mathematical use. Answers will vary.

104. Consider the following statement, then determine whether it is true or false and discuss why. *A graph will exhibit some form of symmetry if, given a point that is h units from the x-axis, k units from the y-axis, and d units from the origin, there is a second point on the graph that is a like distance from the origin and each axis.* Statement is true; answers will vary.

105. When completing the square to find the center and radius of a circle, we sometimes encounter a value for r^2 that is negative or zero. These are called **degenerate cases.** If $r^2 < 0$, no circle is possible, while if $r^2 = 0$, the "graph" of the circle is simply the point (h, k). Find the center and radius of the following circles (if possible).
   a. $x^2 + y^2 - 12x + 4y + 40 = 0$ center: (6, −2); $r = 0$ (degenerate case)
   b. $x^2 + y^2 - 2x - 8y - 8 = 0$ center: (1, 4); $r = 5$
   c. $x^2 + y^2 - 6x - 10y + 35 = 0$ $r^2 = -1$; degenerate case

▶ **MAINTAINING YOUR SKILLS**

106. (R.2) Evaluate/Simplify the following expressions.
   a. $\dfrac{x^2 x^5}{x^3}$ x^4

   b. $3^3 + 3^2 + 3^1 + 3^0 + 3^{-1}$ $40\frac{1}{3}$

   c. $125^{-\frac{1}{3}}$ $\frac{1}{5}$

   d. $27^{\frac{2}{3}}$ 9

   e. $(2m^3 n)^2$ $4m^6 n^2$

   f. $(5x)^0 + 5x^0$ 6

107. (R.3) Solve the following equation.
$$\frac{x}{3} + \frac{1}{4} = \frac{5}{6}$$ $x = \frac{7}{4}$

108. (R.4) Solve $x^2 - 27 = 6x$ by factoring. $x = -3, 9$

109. (R.6) Solve $1 - \sqrt{n + 3} = -n$ and check solutions by substitution. If a solution is extraneous, so state. $n = 1$ is a solution, $n = -2$ is extraneous

1.2 Linear Equations and Rates of Change

LEARNING OBJECTIVES

In Section 1.2 you will see how we can:

☐ **A.** Graph linear equations using the intercept method

☐ **B.** Find the slope of a line and interpret it as a rate of change

☐ **C.** Graph horizontal and vertical lines

☐ **D.** Identify parallel and perpendicular lines

☐ **E.** Apply linear equations in context

In preparation for sketching graphs of other equations, we'll first look more closely at the characteristics of linear graphs. While linear graphs are fairly simple models, they have many substantive and meaningful applications. For instance, most of us are aware that satellite and cable TV have been increasing in popularity since they were first introduced. A close look at Figure 1.2 from Section 1.1 reveals that spending on these forms of entertainment increased from $192 per person per year in 2003 to $281 in 2007 (Figure 1.26). From an investor's or a producer's point of view, there is a very high interest in the questions, "How fast are sales increasing? Can this relationship be modeled mathematically to help predict sales in future years?" Answers to these and other questions are precisely what our study in this section is all about.

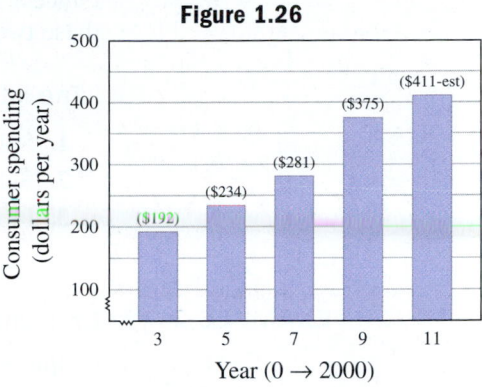

Figure 1.26

Source: 2009 Statistical Abstract of the United States, Table 1089 (some figures are estimates)

A. The Graph of a Linear Equation

A linear equation can be identified using these three tests:

1. the exponent on any variable is one,
2. no variable occurs in a denominator, and
3. no two variables are multiplied together.

The equation $3y = 9$ is a linear equation in one variable, while $2x + 3y = 12$ and $y = -\frac{2}{3}x + 4$ are linear equations in two variables. In general, we have the following definition:

> **Linear Equations**
>
> A linear equation is one that can be written in the form
>
> $$ax + by = c$$
>
> where a, b, and c are real numbers, with a and b not simultaneously equal to zero.

As in Section 1.1, the most basic method for graphing a line is to simply plot a few points, then draw a straight line through the points.

EXAMPLE 1 ▶ Graphing a Linear Equation in Two Variables

Graph the equation $3x + 2y = 4$ by plotting points.

Solution ▶ Selecting $x = -2$, $x = 0$, $x = 1$, and $x = 4$ as inputs, we compute the related outputs and enter the ordered pairs in a table. The result is

WORTHY OF NOTE

If you cannot draw a straight line through the plotted points, a computational error has been made. All points satisfying a linear equation *lie on a straight line.*

x input	y output	(x, y) ordered pairs
−2	5	$(-2, 5)$
0	2	$(0, 2)$
1	$\frac{1}{2}$	$(1, \frac{1}{2})$
4	−4	$(4, -4)$

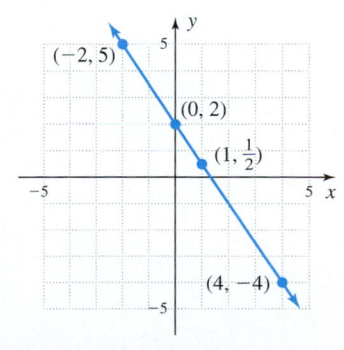

Now try Exercises 7 through 12 ▶

Notice that the line in Example 1 crosses the y-axis at $(0, 2)$, and this point is called the **y-intercept** of the line. In general, y-intercepts have the form $(0, y)$. Although difficult to see graphically, substituting 0 for y and solving for x shows this line crosses the x-axis at $(\frac{4}{3}, 0)$ and this point is called the **x-intercept.** In general, x-intercepts have the form $(x, 0)$. The x- and y-intercepts are usually easier to calculate than other points (since $y = 0$ or $x = 0$, respectively) and we often graph linear equations using only these two points. This is called the **intercept method** for graphing linear equations.

The Intercept Method

1. Substitute 0 for x and solve for y. This will give the y-intercept $(0, y)$.
2. Substitute 0 for y and solve for x. This will give the x-intercept $(x, 0)$.
3. Plot the intercepts and use them to graph a straight line.

EXAMPLE 2 ▶   Graphing Lines Using the Intercept Method

Graph $3x + 2y = 9$ using the intercept method.

Solution ▶   Substitute 0 for x (y-intercept)   Substitute 0 for y (x-intercept)

$$3(0) + 2y = 9 \qquad\qquad\qquad 3x + 2(0) = 9$$
$$2y = 9 \qquad\qquad\qquad\qquad 3x = 9$$
$$y = \frac{9}{2} \qquad\qquad\qquad\qquad x = 3$$
$$\left(0, \frac{9}{2}\right) \qquad\qquad\qquad\qquad (3, 0)$$

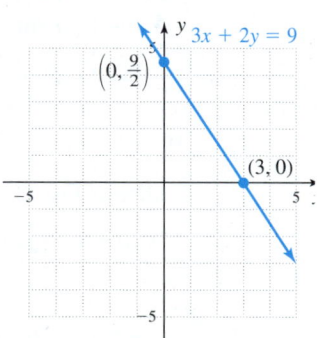

A. You've just seen how we can graph linear equations using the intercept method

Now try Exercises 13 through 32 ▶

B. The Slope of a Line and Rates of Change

After the x- and y-intercepts, we next consider the **slope of a line.** We see applications of this concept in many diverse areas, including the *grade* of a highway (trucking), the *pitch* of a roof (carpentry), the *climb* of an airplane (flying), the *drainage* of a field (landscaping), and the *slope* of a mountain (parks and recreation). While the general concept is an intuitive one, we seek to quantify the concept (assign it a numeric value) for purposes of comparison and decision making. In each of the preceding examples (grade, pitch, climb, etc.), slope is a measure of "steepness," as defined by the ratio $\frac{\text{vertical change}}{\text{horizontal change}}$. Using a line segment through arbitrary points $P_1 = (x_1, y_1)$ and $P_2 = (x_2, y_2)$, we can create the right triangle shown in Figure 1.27 to help us quantify this relationship. The figure illustrates that the **vertical change** or the

Figure 1.27

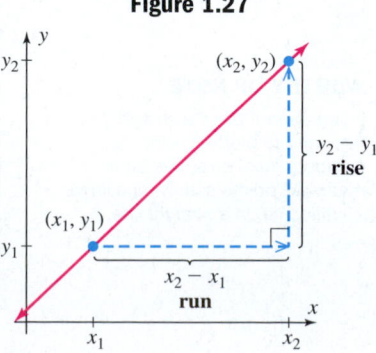

change in y (also called the **rise**) is simply the difference in y-coordinates: $y_2 - y_1$. The **horizontal change** or **change in x** (also called the **run**) is the difference in x-coordinates: $x_2 - x_1$. In algebra, we typically use the letter "m" to represent slope, giving $m = \frac{y_2 - y_1}{x_2 - x_1}$ as the $\frac{\text{change in } y}{\text{change in } x}$. The result is called the **slope formula.**

WORTHY OF NOTE

While the original reason that "m" was chosen for slope is uncertain, some have speculated that it was because in French, the verb for "to climb" is *monter*. Others say it could be due to the "*modulus* of slope," the word *modulus* meaning a numeric measure of a given property, in this case the inclination of a line.

The Slope Formula

Given two points $P_1 = (x_1, y_1)$ and $P_2 = (x_2, y_2)$, the slope of the nonvertical line through P_1 and P_2 is

$$m = \frac{y_2 - y_1}{x_2 - x_1}$$

where $x_2 \neq x_1$.

Actually, the slope value does much more than quantify the slope of a line, it expresses a **rate of change** between the quantities measured along each axis. In applications of slope, the ratio $\frac{\text{change in } y}{\text{change in } x}$ is symbolized as $\frac{\Delta y}{\Delta x}$. The symbol Δ is the Greek letter **delta** and has come to represent a change in some quantity, and the notation $m = \frac{\Delta y}{\Delta x}$ is read, "slope is equal to the *change in y* over the *change in x*." Interpreting slope as a rate of change has many significant applications in college algebra and beyond.

EXAMPLE 3 ▶   **Using the Slope Formula**

Find the slope of the line through the given points, then use $m = \dfrac{\Delta y}{\Delta x}$ to find an additional point on the line.

a. $(2, 1)$ and $(8, 4)$ **b.** $(-2, 6)$ and $(4, 2)$

Solution ▶   **a.** For $P_1 = (2, 1)$ and $P_2 = (8, 4)$,   **b.** For $P_1 = (-2, 6)$ and $P_2 = (4, 2)$,

$$m = \frac{y_2 - y_1}{x_2 - x_1} \qquad\qquad\qquad m = \frac{y_2 - y_1}{x_2 - x_1}$$

$$= \frac{4 - 1}{8 - 2} \qquad\qquad\qquad\qquad = \frac{2 - 6}{4 - (-2)}$$

$$= \frac{3}{6} = \frac{1}{2} \qquad\qquad\qquad\qquad = \frac{-4}{6} = \frac{-2}{3}$$

The slope of this line is $\frac{1}{2}$. | The slope of this line is $\frac{-2}{3}$.

Using $\dfrac{\Delta y}{\Delta x} = \dfrac{1}{2}$, we note that y increases 1 unit (the y-value is positive), as x increases 2 units. Since $(8, 4)$ is known to be on the line, the point $(8 + 2, 4 + 1) = (10, 5)$ must also be on the line.

Using $\dfrac{\Delta y}{\Delta x} = \dfrac{-2}{3}$, we note that y decreases 2 units (the y-value is negative), as x increases 3 units. Since $(4, 2)$ is known to be on the line, the point $(4 + 3, 2 - 2) = (7, 0)$ must also be on the line.

Now try Exercises 33 through 40 ▶

> ⚠️ **CAUTION** ▶ When using the slope formula, try to avoid these common errors.
>
> 1. The order that the x- and y-coordinates are subtracted must be consistent, since $\frac{y_2 - y_1}{x_2 - x_1} \neq \frac{y_2 - y_1}{x_1 - x_2}$.
>
> 2. The vertical change (involving the y-values) always occurs in the numerator: $\frac{y_2 - y_1}{x_2 - x_1} \neq \frac{x_2 - x_1}{y_2 - y_1}$.
>
> 3. When x_1 or y_1 is negative, use parentheses when substituting into the formula to prevent confusing the negative sign with the subtraction operation.

EXAMPLE 4 ▶ **Interpreting the Slope Formula as a Rate of Change**

Jimmy works on the assembly line for an auto parts remanufacturing company. By 9:00 A.M. his group has assembled 29 carburetors. By 12:00 noon, they have completed 87 carburetors. Assuming the relationship is linear, find the slope of the line and discuss its meaning in this context.

Solution ▶ First write the information as ordered pairs using c to represent the carburetors assembled and t to represent time. This gives $(t_1, c_1) = (9, 29)$ and $(t_2, c_2) = (12, 87)$. The slope formula then gives:

$$\frac{\Delta c}{\Delta t} = \frac{c_2 - c_1}{t_2 - t_1} = \frac{87 - 29}{12 - 9}$$
$$= \frac{58}{3} \text{ or } 19.\overline{3}$$

Here the slope ratio measures $\frac{\text{carburetors assembled}}{\text{hours}}$, and we see that Jimmy's group can assemble 58 carburetors every 3 hr, or about $19\frac{1}{3}$ carburetors per hour.

WORTHY OF NOTE

Actually, the assignment of (t_1, c_1) to (9, 29) and (t_2, c_2) to (12, 87) was arbitrary. The slope ratio will be the same *as long as the order of subtraction is the same*. In other words, if we reverse this assignment and use $(t_1, c_1) = (12, 87)$ and $(t_2, c_2) = (9, 29)$, we have

$m = \frac{29 - 87}{9 - 12} = \frac{-58}{-3} = \frac{58}{3}$.

Now try Exercises 41 through 44 ▶

Positive and Negative Slope

If you've ever traveled by air, you've likely heard the announcement, "Ladies and gentlemen, please return to your seats and fasten your seat belts as we begin our descent." For a time, the descent of the airplane follows a linear path, but the *slope of the line is negative* since the altitude of the plane is decreasing. Positive and negative slopes, as well as the rate of change they represent, are important characteristics of linear graphs. In Example 3(a), the slope was a positive number ($m > 0$) and the line will slope upward from left to right since the y-values are increasing. If $m < 0$ as in Example 3(b), the slope of the line is negative and the line slopes downward as you move left to right since y-values are decreasing.

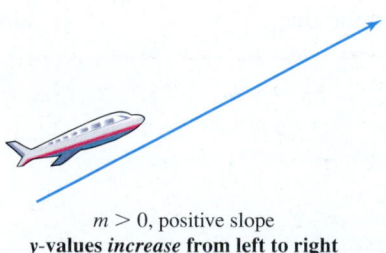

$m > 0$, positive slope
y-values *increase* from left to right

$m < 0$, negative slope
y-values *decrease* from left to right

EXAMPLE 5 ▶ **Applying Slope as a Rate of Change in Altitude**

At a horizontal distance of 10 mi after take-off, an airline pilot receives instructions to decrease altitude from their current level of 20,000 ft. A short time later, they are 17.5 mi from the airport at an altitude of 10,000 ft. Find the slope ratio for the descent of the plane and discuss its meaning in this context. Recall that 1 mi = 5280 ft.

Solution ▶ Let a represent the altitude of the plane and d its horizontal distance from the airport. Converting all measures to feet, we have $(d_1, a_1) = (52{,}800, 20{,}000)$ and $(d_2, a_2) = (92{,}400, 10{,}000)$, giving

$$\frac{\Delta a}{\Delta d} = \frac{a_2 - a_1}{d_2 - d_1} = \frac{10{,}000 - 20{,}000}{92{,}400 - 52{,}800}$$

$$= \frac{-10{,}000}{39{,}600} = \frac{-25}{99}$$

Since this slope ratio measures $\frac{\Delta \text{altitude}}{\Delta \text{distance}}$, we note the plane is decreasing 25 ft in altitude for every 99 ft it travels horizontally.

Now try Exercises 45 through 48 ▶

☑ **B.** You've just seen how we can find the slope of a line and interpret it as a rate of change

C. Horizontal Lines and Vertical Lines

Horizontal and vertical lines have a number of important applications, from finding the boundaries of a given graph (the domain and range), to performing certain tests on nonlinear graphs. To better understand them, consider that in *one dimension*, the graph of $x = 2$ is a single point (Figure 1.28), indicating a location on the number line 2 units from zero in the positive direction. In *two dimensions*, the equation $x = 2$ represents **all points** with an x-coordinate of 2. A few of these are graphed in Figure 1.29, but since there are an infinite number, we end up with a solid *vertical line* whose equation is $x = 2$ (Figure 1.30).

Figure 1.28

Figure 1.29   **Figure 1.30**

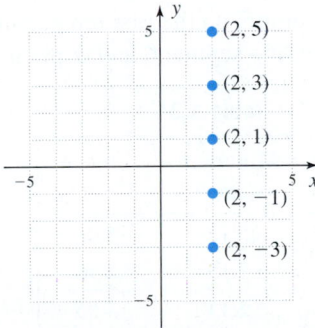   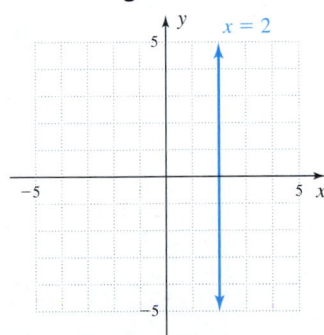

WORTHY OF NOTE

If we write the equation $x = 2$ in the form $ax + by = c$, the equation becomes $x + 0y = 2$, since the original equation has no y-variable. Notice that regardless of the value chosen for y, x will always be 2 and we end up with the set of ordered pairs $(2, y)$, which gives us a vertical line.

The same idea can be applied to horizontal lines. In *two dimensions,* the equation $y = 4$ represents *all points* with a y-coordinate of positive 4, and there are an infinite number of these as well. The result is a solid horizontal line whose equation is $y = 4$. **See Exercises 49 through 54.**

Horizontal Lines	Vertical Lines
The equation of a horizontal line is	The equation of a vertical line is
$y = k$	$x = h$
where $(0, k)$ is the y-intercept.	where $(h, 0)$ is the x-intercept.

So far, the slope formula has only been applied to lines that were nonhorizontal or nonvertical. So what *is* the slope of a horizontal line? On an intuitive level, we expect that a perfectly level highway would have an incline or slope of zero. In general, for any two points on a horizontal line, $y_2 = y_1$ and $y_2 - y_1 = 0$, giving a slope of $m = \frac{0}{x_2 - x_1} = 0$. For any two points on a vertical line, $x_2 = x_1$ and $x_2 - x_1 = 0$, making the slope ratio undefined: $m = \frac{y_2 - y_1}{0}$ (see Figures 1.31 and 1.32).

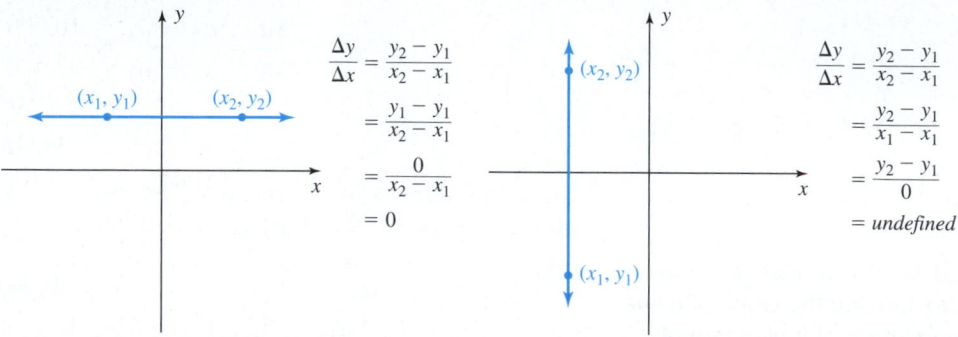

Figure 1.31

For any horizontal line, $y_2 = y_1$

$$\frac{\Delta y}{\Delta x} = \frac{y_2 - y_1}{x_2 - x_1}$$
$$= \frac{y_1 - y_1}{x_2 - x_1}$$
$$= \frac{0}{x_2 - x_1}$$
$$= 0$$

Figure 1.32

For any vertical line, $x_2 = x_1$

$$\frac{\Delta y}{\Delta x} = \frac{y_2 - y_1}{x_2 - x_1}$$
$$= \frac{y_2 - y_1}{x_1 - x_1}$$
$$= \frac{y_2 - y_1}{0}$$
$$= undefined$$

The Slope of a Horizontal Line	The Slope of a Vertical Line
The slope of any horizontal line is zero.	The slope of any vertical line is undefined.

EXAMPLE 6 ▶ **Calculating Slopes**

The federal minimum wage remained constant from 1997 through 2006. However, the buying power (in 1996 dollars) of these wage earners fell each year due to inflation (see Table 1.3). This decrease in buying power is approximated by the red line shown.

a. Using the data or graph, find the slope of the line segment representing the minimum wage.

b. Select two points on the line representing buying power to approximate the slope of the line segment, and explain what it means in this context.

Table 1.3

Time t (years)	Minimum wage w	Buying power p
1997	5.15	5.03
1998	5.15	4.96
1999	5.15	4.85
2000	5.15	4.69
2001	5.15	4.56
2002	5.15	4.49
2003	5.15	4.39
2004	5.15	4.28
2005	5.15	4.14
2006	5.15	4.04

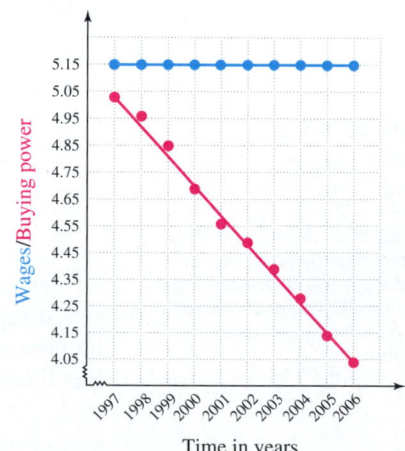

Solution ▶

a. Since the minimum wage did not increase or decrease from 1997 to 2006, the line segment has slope $m = 0$.

b. The points (1997, 5.03) and (2006, 4.04) from the table appear to be on or close to the line drawn. For buying power p and time t, the slope formula yields:

$$\frac{\Delta p}{\Delta t} = \frac{p_2 - p_1}{t_2 - t_1}$$

$$= \frac{4.04 - 5.03}{2006 - 1997}$$

$$= \frac{-0.99}{9} = \frac{-0.11}{1}$$

> **WORTHY OF NOTE**
>
> In the context of lines, try to avoid saying that a horizontal line has "no slope," since it's unclear whether a slope of zero or an undefined slope is intended.

The buying power of a minimum wage worker decreased by 11¢ per year during this time period.

☑ **C.** You've just seen how we can graph horizontal and vertical lines

Now try Exercises 55 and 56 ▶

D. Parallel and Perpendicular Lines

Two lines in the same plane that never intersect are called **parallel lines.** When we place these lines on the coordinate grid, we find that "never intersect" is equivalent to saying "the lines have equal slopes but different y-intercepts." In Figure 1.33, notice the rise and run of each line is identical, and that by counting $\frac{\Delta y}{\Delta x}$ both lines have slope $m = \frac{3}{4}$.

Figure 1.33

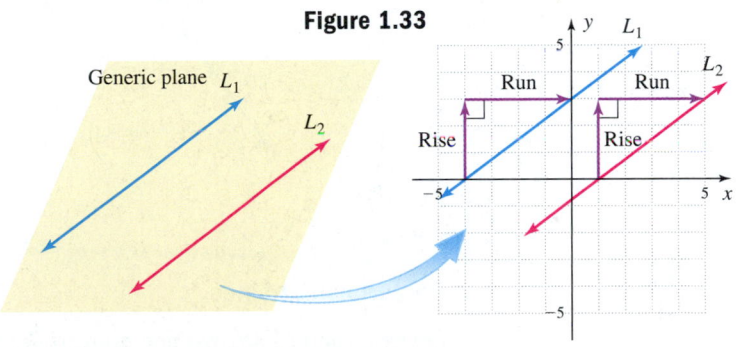

Coordinate plane

> **Parallel Lines**
>
> Given L_1 and L_2 are distinct, nonvertical lines with slopes of m_1 and m_2, respectively.
>
> **1.** If $m_1 = m_2$, then L_1 is parallel to L_2.
> **2.** If L_1 is parallel to L_2, then $m_1 = m_2$.
>
> In symbols, we write $L_1 \| L_2$.
> *Any two vertical lines (undefined slope) are parallel.*

EXAMPLE 7A ▶ **Determining Whether Two Lines Are Parallel**

Teladango Park has been mapped out on a rectangular coordinate system, with a ranger station at (0, 0). Brendan and Kapi are at coordinates $(-24, -18)$ and have set a direct course for the pond at (11, 10). Caden and Kymani are at $(-27, 1)$ and are heading straight to the lookout tower at $(-2, 21)$. Are they hiking on parallel or nonparallel courses?

Solution ▶   To respond, we compute the slope of each trek across the park.

For Brendan and Kapi: For Caden and Kymani:

$$m = \frac{y_2 - y_1}{x_2 - x_1}$$ $$m = \frac{y_2 - y_1}{x_2 - x_1}$$

$$= \frac{10 - (-18)}{11 - (-24)}$$ $$= \frac{21 - 1}{-2 - (-27)}$$

$$= \frac{28}{35} = \frac{4}{5}$$ $$= \frac{20}{25} = \frac{4}{5}$$

Since the slopes are equal, the two groups are hiking on parallel courses.

Two lines in the same plane that intersect at right angles are called **perpendicular lines.** Using the coordinate grid, we note that *intersect at right angles* suggests that *their slopes are negative reciprocals.* While certainly not a proof, notice in Figure 1.34, the ratio $\frac{\text{rise}}{\text{run}}$ for L$_1$ is $\frac{4}{3}$ and the ratio $\frac{\text{rise}}{\text{run}}$ for L$_2$ is $\frac{-3}{4}$. Alternatively, we can say their **slopes have a product of −1,** since $m_1 \cdot m_2 = -1$ implies $m_1 = -\frac{1}{m_2}$.

Figure 1.34

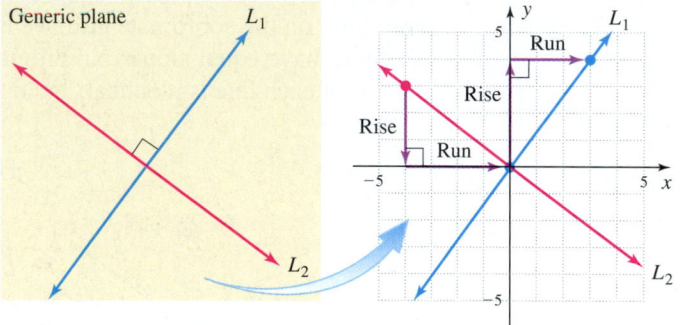

Coordinate plane

Perpendicular Lines

Given L_1 and L_2 are distinct, nonvertical lines with slopes of m_1 and m_2, respectively.

1. If $m_1 \cdot m_2 = -1$, then L_1 is perpendicular to L_2.
2. If L_1 is perpendicular to L_2, then $m_1 \cdot m_2 = -1$.

In symbols we write $L_1 \perp L_2$.
*Any vertical line (undefined slope) is perpendicular
to any horizontal line (slope m = 0).*

We can easily find the slope of a line perpendicular to a second line whose slope is known or can be found—just find the reciprocal and make it negative. For a line with slope $m_1 = -\frac{3}{7}$, any line perpendicular to it will have a slope of $m_2 = \frac{7}{3}$. For $m_1 = -5$, the slope of any line perpendicular would be $m_2 = \frac{1}{5}$.

EXAMPLE 7B ▶ **Determining Whether Two Lines Are Perpendicular**

The three points $P_1 = (5, 1)$, $P_2 = (3, -2)$, and $P_3 = (-3, 2)$ form the vertices of a triangle. Use these points to draw the triangle, then use the slope formula to determine if they form a *right* triangle.

Solution ▶ For a right triangle to be formed, two of the lines through these points must be perpendicular (forming a right angle). From Figure 1.35, it *appears* a right triangle is formed, but we must *verify* that two of the sides are actually perpendicular. Using the slope formula, we have:

Figure 1.35

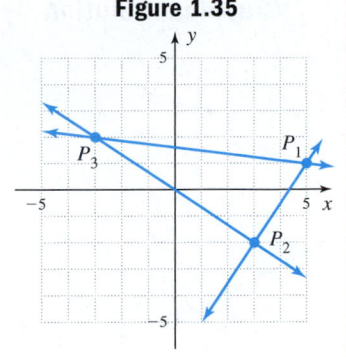

For P_1 and P_2

$$m_1 = \frac{-2 - 1}{3 - 5}$$

$$= \frac{-3}{-2} = \frac{3}{2}$$

For P_1 and P_3

$$m_2 = \frac{2 - 1}{-3 - 5}$$

$$= \frac{1}{-8}$$

For P_2 and P_3

$$m_3 = \frac{2 - (-2)}{-3 - 3}$$

$$= \frac{4}{-6} = \frac{2}{-3}$$

Since $m_1 \cdot m_3 = -1$, the triangle has a right angle and must be a right triangle.

☑ **D.** You've just seen how we can identify parallel and perpendicular lines

Now try Exercises 57 through 68 ▶

E. Applications of Linear Equations

The graph of a linear equation can be used to help solve many applied problems. If the numbers you're working with are either very small or very large, **scale the axes** appropriately. This can be done by letting each tic mark represent a smaller or larger unit so the data points given will fit on the grid. Also, many applications use only nonnegative values and although points with negative coordinates may be used to graph a line, only ordered pairs in QI can be meaningfully interpreted.

EXAMPLE 8 ▶ **Applying a Linear Equation Model—Commission Sales**

Use the information given to create a linear equation model in two variables, then graph the line and answer the question posed:

A salesperson gets a daily $20 meal allowance plus $7.50 for every item she sells. How many sales are needed for a daily income of $125?

Verify your answer by graphing the line on a calculator and using the TRACE feature.

Algebraic Solution ▶ Let x represent sales and y represent income. This gives

verbal model: Daily income (y) equals $7.5 per sale (x) + $20 for meals

equation model: $y = 7.5x + 20$

Using $x = 0$ and $x = 10$, we find $(0, 20)$ and $(10, 95)$ are points on this line and these are used to sketch the graph. From the graph, it appears that 14 sales are needed to generate a daily income of $125.00.

Since daily income is given as $125, we substitute 125 for y and solve for x.

$$125 = 7.5x + 20 \qquad \text{substitute 125 for } y$$

$$105 = 7.5x \qquad \text{subtract 20}$$

$$14 = x \qquad \text{divide by 7.5}$$

Graphical Solution ▶ Begin by entering the equation $y = 7.5x + 20$ on the ⬭Y= screen, recognizing that in this context, both the input and output values must be positive. Reasoning the 10 sales will net $95 (less than $125) and 20 sales will net $170 (more than $125), we set the viewing ⬭WINDOW as shown in Figure 1.36. We can then ⬭GRAPH the equation and use the ⬭TRACE feature to estimate the number of sales needed. The result shows that income is close to $125 when x is close to 14 (Figure 1.37). In addition to letting us trace along a graph, the ⬭TRACE option enables us to evaluate the equation at specific points. Simply entering the number "14" causes the calculator to accept 14 as the desired input (Figure 1.38), and after pressing ⬭ENTER, it verifies that (14, 125) is indeed a point on the graph (Figure 1.39).

Figure 1.36

```
WINDOW
 Xmin=0
 Xmax=20
 Xscl=2
 Ymin=0
 Ymax=200
 Yscl=20
 Xres=1
```

Figure 1.37

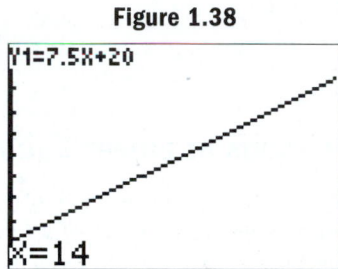

Figure 1.38

Figure 1.39

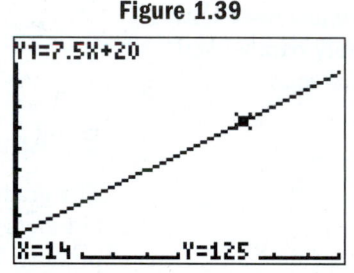

 **E.** You've just seen how we can apply linear equations in context

Now try Exercises 71 through 80 ▶

1.2 EXERCISES

▶ CONCEPTS AND VOCABULARY

Fill in each blank with the appropriate word or phrase. Carefully reread the section if needed.

1. To find the x-intercept of a line, substitute ___0___ for y and solve for x. To find the y-intercept, substitute ___0___ for x and solve for y.

2. The slope formula is $m = \dfrac{y_2 - y_1}{x_2 - x_1} = \dfrac{\Delta y}{\Delta x}$, and indicates a rate of change between the x- and y-variables.

3. If $m < 0$, the slope of the line is __negative__ and the line slopes __downward__ from left to right.

4. The slope of a horizontal line is __zero__, the slope of a vertical line is __undefined__, and the slopes of two parallel lines are __equal__.

5. Discuss/Explain If $m_1 = 2.1$ and $m_2 = 2.01$, will the lines intersect? If $m_1 = \frac{2}{3}$ and $m_2 = -\frac{2}{3}$, are the lines perpendicular?   yes   no
$m_1 \neq m_2$ $m_1 \cdot m_2 \neq -1$

6. Discuss/Explain the relationship between the slope formula, the Pythagorean theorem, and the distance formula. Include several illustrations.
Answers will vary.

HOMEWORK SELECTION GUIDE

Core: 7–67 every other odd, 69, 71–79 odd (21 Exercises)
Standard: 1–4, 7–67 every other odd, 69, 71–79 odd, 81 (26 Exercises)

Extended: 1–4, 7–67 every other odd, 69, 70, 71, 73, 75–79, 81, 82, 86, 87 (81 Exercises)
In Depth: 1–6, 7–67 every other odd, 69, 70, 71–79, 81, 82, 83, 84–87 (39 Exercises)

► DEVELOPING YOUR SKILLS

Create a table of values for each equation and sketch the graph.

7. $2x + 3y = 6$

x	y
-6	6
-3	4
0	2
3	0

8. $-3x + 5y = 10$

x	y
-5	-1
0	2
2	3.2
5	5

9. $y = \dfrac{3}{2}x + 4$

x	y
-2	1
0	4
2	7
4	10

10. $y = \dfrac{5}{3}x - 3$

x	y
-6	-13
-3	-8
0	-3
3	2

11. If you completed Exercise 9, verify that $(-3, -0.5)$ and $(\frac{1}{2}, \frac{19}{4})$ also satisfy the equation given. Do these points appear to be on the graph you sketched?

12. If you completed Exercise 10, verify that $(-1.5, -5.5)$ and $(\frac{11}{2}, \frac{37}{6})$ also satisfy the equation given. Do these points appear to be on the graph you sketched?

Graph the following equations using the intercept method. Plot a third point as a check.

13. $3x + y = 6$ **14.** $-2x + y = 12$

15. $5y - x = 5$ **16.** $-4y + x = 8$

17. $-5x + 2y = 6$ **18.** $3y + 4x = 9$

19. $2x - 5y = 4$ **20.** $-6x + 4y = 8$

21. $2x + 3y = -12$ **22.** $-3x - 2y = 6$

23. $y = -\dfrac{1}{2}x$ **24.** $y = \dfrac{2}{3}x$

25. $y - 25 = 50x$ **26.** $y + 30 = 60x$

27. $y = -\dfrac{2}{5}x - 2$ **28.** $y = \dfrac{3}{4}x + 2$

29. $2y - 3x = 0$ **30.** $y + 3x = 0$

31. $3y + 4x = 12$ **32.** $-2x + 5y = 8$

Compute the slope of the line through the given points, then graph the line and use $m = \dfrac{\Delta y}{\Delta x}$ to find two additional points on the line. Answers may vary.

33. $(3, 5), (4, 6)$ **34.** $(-2, 3), (5, 8)$

Additional answers can be found in the Instructor Answer Appendix.

35. $(10, 3), (4, -5)$ **36.** $(-3, -1), (0, 7)$

37. $(1, -8), (-3, 7)$ **38.** $(-5, 5), (0, -5)$

39. $(-3, 6), (4, 2)$ **40.** $(-2, -4), (-3, -1)$

41. The graph shown models the relationship between the cost of a new home and the size of the home in square feet. (a) Determine the slope of the line and interpret what the slope ratio means in this context and (b) estimate the cost of a 3000 ft² home.
a. $m = 125$, cost increased \$125,000 per 1000 sq ft   **b.** \$375,000

Exercise 41

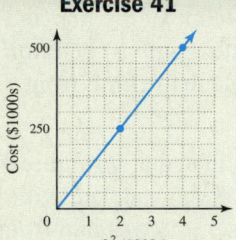

Exercise 42

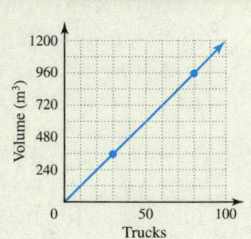

42. The graph shown models the relationship between the volume of garbage that is dumped in a landfill and the number of commercial garbage trucks that enter the site. (a) Determine the slope of the line and interpret what the slope ratio means in this context and (b) estimate the number of trucks entering the site daily if 1000 m³ of garbage is dumped per day.
a. $m = 12$, 12 m³ dumped per garbage truck   **b.** 83 trucks

43. The graph shown models the relationship between the distance of an aircraft carrier from its home port and the number of hours since departure. (a) Determine the slope of the line and interpret what the slope ratio means in this context and (b) estimate the distance from port after 8.25 hours.
a. $m = 22.5$, distance increases 22.5 mph   **b.** about 186 mi

Exercise 43

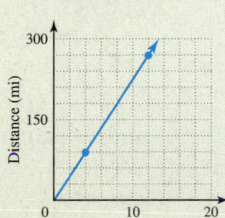

Exercise 44

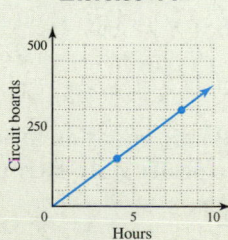

44. The graph shown models the relationship between the number of circuit boards that have been assembled at a factory and the number of hours since starting time. (a) Determine the slope of the line and interpret what the slope ratio means in this context and (b) estimate how many hours the factory has been running if 225 circuit boards have been assembled.
a. $m = 37.5$, 37.5 circuit boards are assembled per hour   **b.** 6 hr

45. Height and weight: While there are many exceptions, numerous studies have shown a close relationship between an average height and average weight. Suppose a person 70 in. tall weighs 165 lb, while a person 64 in. tall weighs 142 lb. Assuming the relationship is linear, (a) find the slope of the line and discuss its meaning in this context and (b) determine how many pounds are added for each inch of height.

a. $m = \frac{23}{6}$, a person weighs 23 lb more for each additional 6 in. in height   **b.** ≈ 3.8

46. Rate of climb: Shortly after takeoff, a plane increases altitude at a constant (linear) rate. In 5 min the altitude is 10,000 ft. Fifteen minutes after takeoff, the plane has reached its cruising altitude of 32,000 ft. (a) Find the slope of the line and discuss its meaning in this context and (b) determine how long it takes the plane to climb from 12,200 ft to 25,400 ft.

47. Sewer line slope: Fascinated at how quickly the plumber was working, Ryan watched with great interest as the new sewer line was laid from the house to the main line, a distance of 48 ft. At the edge of the house, the sewer line was 6 in. under ground. If the plumber tied in to the main line at a depth of 18 in., what is the slope of the (sewer) line? What does this slope indicate?

48. Slope (pitch) of a roof: A contractor goes to a lumber yard to purchase some trusses (the triangular frames) for the roof of a house. Many sizes are available, so the contractor takes some measurements to ensure the roof will have the desired slope. In one case, the height of the truss (base to ridge) was 4 ft, with a width of 24 ft (eave to eave). Find the slope of the roof if these trusses are used. What does this slope indicate?

Graph each line using two or three ordered pairs that satisfy the equation.

49. $x = -3$ **50.** $y = 4$

51. $x = 2$ **52.** $y = -2$

Write the equation for each line L_1 and L_2 shown. Specifically state their point of intersection.

53.   **54.**

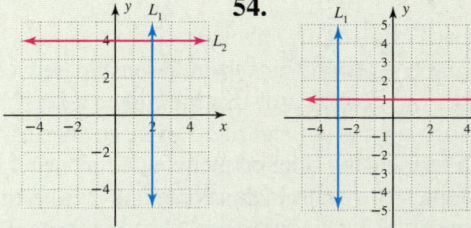

55. Supreme Court justices: The table given shows the total number of justices j sitting on the Supreme Court of the United States for selected time periods t (in decades), along with the number of nonmale, nonwhite justices n for the same years. (a) Use the data to graph the linear relationship between t and j, then determine the slope of the line and discuss its meaning in this context. (b) Use the data to graph the linear relationship between t and n, then determine the slope of the line and discuss its meaning.

Exercise 55

Time t (1960 → 0)	Justices j	Nonwhite, nonmale n
0	9	0
10	9	1
20	9	2
30	9	3
40	9	4
50	9	5

56. Boiling temperature: The table shown gives the boiling temperature t of water as related to the altitude h. Use the data to graph the linear relationship between h and t, then determine the slope of the line and discuss its meaning in this context.

Exercise 56

Altitude h (ft)	Boiling Temperature t (°F)
0	212.0
1000	210.2
2000	208.4
3000	206.6
4000	204.8
5000	203.0
6000	201.2

Two points on L_1 and two points on L_2 are given. Use the slope formula to determine if lines L_1 and L_2 are parallel, perpendicular, or neither.

57. L_1: $(-2, 0)$ and $(0, 6)$ **58.** L_1: $(1, 10)$ and $(-1, 7)$
L_2: $(1, 8)$ and $(0, 5)$ L_2: $(0, 3)$ and $(1, 5)$
parallel   neither

59. L_1: $(-3, -4)$ and $(0, 1)$ **60.** L_1: $(6, 2)$ and $(8, -2)$
L_2: $(0, 0)$ and $(-4, 4)$ L_2: $(5, 1)$ and $(3, 0)$
neither   perpendicular

61. L_1: $(6, 3)$ and $(8, 7)$ **62.** L_1: $(-5, -1)$ and $(4, 4)$
L_2: $(7, 2)$ and $(6, 0)$ L_2: $(4, -7)$ and $(8, 10)$
parallel   neither

In Exercises 63 to 68, three points that form the vertices of a triangle are given. Use the points to draw the triangle, then use the slope formula to determine if any of the triangles are right triangles. Also see Exercises 43–48 in Section 1.1.

63. $(-3, 7)$, $(2, 2)$, $(5, 5)$ right triangle

64. $(7, 0)$, $(-1, 0)$, $(7, 4)$ right triangle

65. $(-4, 3)$, $(-7, -1)$, $(3, -2)$ not a right triangle

66. $(5, 2)$, $(0, -3)$, $(4, -4)$ not a right triangle

67. $(-3, 2)$, $(-1, 5)$, $(-6, 4)$ right triangle

68. $(0, 0)$, $(-5, 2)$, $(2, -5)$ not a right triangle

▶ WORKING WITH FORMULAS

69. Human life expectancy: $L = 0.15T + 73.7$

In the United States, the average life expectancy has been steadily increasing over the years due to better living conditions and improved medical care. This relationship is modeled by the formula shown, where L is the average life expectancy and T is number of years since 1980. (a) What was the life expectancy in the year 2010? (b) In what year will average life expectancy reach 79 yr?

a. 78.2 yr   **b.** 2015

70. Interest earnings: $100I = 35,000T$

If $5000 dollars is invested in an account paying 7% simple interest, the amount of interest earned is given by the formula shown, where I is the interest and T is the time in years. Begin by solving the formula for I. (a) How much interest is earned in 5 yr? (b) How much is earned in 10 yr? (c) Use the two points (5 yr, interest) and (10 yr, interest) to calculate the slope of this line. What do you notice? $I = 350T$;   **a.** $1750   **b.** $3500   **c.** $350 Interest increases $350 per year

▶ APPLICATIONS

Use the information given to build a linear equation model, then use the equation to respond. For exercises 71 to 74, develop both an algebraic and a graphical solution.

71. Business depreciation: A business purchases a copier for $8500 and anticipates it will depreciate in value $1250 per year.

  a. What is the copier's value after 4 yr of use?

  b. How many years will it take for this copier's value to decrease to $2250? $v = -1250t + 8500$ **a.** $3500   **b.** 5 yr

72. Baseball card value: After purchasing an autographed baseball card for $85, its value increases by $1.50 per year.

  a. What is the card's value 7 yr after purchase?

  b. How many years will it take for this card's value to reach $100? $v = 1.5t + 85$ **a.** $95.50   **b.** 10 yr

73. Water level: During a long drought, the water level in a local lake decreased at a rate of 3 in. per month. The water level before the drought was 300 in.

  a. What was the water level after 9 months of drought?

  b. How many months will it take for the water level to decrease to 20 ft? $h = -3t + 300$ **a.** 273 in.   **b.** 20 months

74. Gas mileage: When empty, a large dump-truck gets about 15 mi per gallon. It is estimated that for each 3 tons of cargo it hauls, gas mileage decreases by $\frac{3}{4}$ mi per gallon.

  a. If 10 tons of cargo is being carried, what is the truck's mileage?

  b. If the truck's mileage is down to 10 mi per gallon, how much weight is it carrying? $m = -0.25w + 15$ **a.** 12.5 mpg   **b.** 20 tons

75. Parallel/nonparallel roads: Aberville is 38 mi north and 12 mi west of Boschertown, with a straight "farm and machinery" road (FM 1960) connecting the two cities. In the next county, Crownsburg is 30 mi north and 9.5 mi west of Dower, and these cities are likewise connected by a straight road (FM 830). If the two roads continued indefinitely in both directions, would they intersect at some point?   Yes they will meet, the two roads are not parallel: $\frac{38}{12} \neq \frac{30}{9.5}$.

76. Perpendicular/nonperpendicular course headings: Two shrimp trawlers depart Charleston Harbor at the same time. One heads for the shrimping grounds located 12 mi north and 3 mi east of the harbor. The other heads for a point 2 mi south and 8 mi east of the harbor. Assuming the harbor is at (0, 0), are the routes of the trawlers perpendicular? If so, how far apart are the boats when they reach their destinations (to the nearest one-tenth mi)?   Yes, the routes are perpendicular: $\frac{12}{3} \cdot \frac{-2}{8} = -1$; 14.9 mi.

77. Cost of college: For the years 2000 to 2008, the cost of tuition and fees per semester (in constant dollars) at a public 4-yr college can be approximated by the equation $y = 386x + 3500$, where y represents the cost in dollars and $x = 0$ represents the year 2000. Use the equation to find: (a) the cost of tuition and fees in 2010 and (b) the year this cost will exceed $9000.   **a.** $7360   **b.** 2015

Source: The College Board

78. Female physicians: In 1960 only about 7% of physicians were female. Soon after, this percentage began to grow dramatically. For the years 1990 to 2000, the percentage of physicians that were female can be approximated by the equation $y = 0.6x + 18.1$, where y represents the percentage (as a whole number) and $x = 0$ represents the year 1990. Use the equation to find: (a) the percentage of physicians that were female in 2000 and (b) the projected year this percentage would have exceeded 30%.   **a.** 24.1   **b.** 2010

Source: American Journal of Public Health

79. Decrease in smokers: For the years 1990 to 2000, the percentage of the U.S. adult population who were smokers can be approximated by the equation $y = \frac{-13}{25}x + 28.7$, where y represents the percentage of smokers (as a whole number) and $x = 0$ represents 1990. Use the equation to find: (a) the percentage of adults who smoked in the year 2005 and (b) the year the percentage of smokers is projected to fall below 15%.   **a.** 21%   **b.** 2016

Source: WebMD

80. Temperature and cricket chirps: Biologists have found a strong relationship between temperature and the number of times a cricket chirps. This is modeled by the equation $T = \frac{1}{4}N + 40$, where N is the number of times the cricket chirps per minute and T is the temperature in Fahrenheit. Use the equation to find: (a) the outdoor temperature if the cricket is chirping 48 times per minute and (b) the number of times a cricket chirps if the temperature is 70°.   **a.** 52°F   **b.** 120 chirps per minute

▶ EXTENDING THE CONCEPT

81. If the lines $4y + 2x = -5$ and $3y + ax = -2$ are perpendicular, what is the value of a? $a = -6$

82. Let $m_1, m_2, m_3,$ and m_4 be the slopes of lines $L_1, L_2, L_3,$ and L_4, respectively. Which of the following statements is true?

  a. $m_4 < m_1 < m_3 < m_2$

  b. $m_3 < m_2 < m_4 < m_1$

  c. $m_3 < m_4 < m_2 < m_1$

  d. $m_1 < m_3 < m_4 < m_2$

  e. $m_1 < m_4 < m_3 < m_2$ e

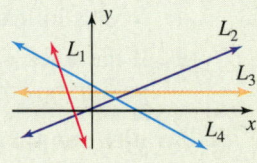

83. An *arithmetic sequence* is a sequence of numbers where each successive term is found by adding a fixed constant, called the common difference d, to the preceding term. For instance 3, 7, 11, 15, . . . is an arithmetic sequence with $d = 4$. The formula for the "nth term" t_n of an arithmetic sequence is a linear equation of the form $t_n = t_1 + (n - 1)d$, where d is the common difference and t_1 is the first term of the sequence. Use the equation to find the term specified for each sequence.

  a. 2, 9, 16, 23, 30, . . . ; 21st term   142

  b. 7, 4, 1, −2, −5, . . . ; 31st term   −83

  c. 5.10, 5.25, 5.40, 5.55, . . . ; 27th term   9

  d. $\frac{3}{2}, \frac{9}{4}, 3, \frac{15}{4}, \frac{9}{2}, \ldots$; 17th term $\frac{27}{2}$

▶ MAINTAINING YOUR SKILLS

84. (1.1) Name the center and radius of the circle defined by $(x - 3)^2 + (y + 4)^2 = 169$
$(3, -4), r = 13$

85. (R.6) Compute the sum and product indicated:

  a. $\sqrt{20} + 3\sqrt{45} - \sqrt{5}$ $10\sqrt{5}$

  b. $(3 + \sqrt{5})(3 - \sqrt{5})$ 4

86. (R.4) Solve the equation by factoring, then check the result(s) using substitution:

$12x^2 - 44x - 45 = 0$ $x = -\frac{5}{6}, \frac{9}{2}$

87. (R.5) Factor the following polynomials completely:

  a. $x^3 - 3x^2 - 4x + 12$ $(x - 3)(x + 2)(x - 2)$

  b. $x^2 - 23x - 24$ $(x - 24)(x + 1)$

  c. $x^3 - 125$ $(x - 5)(x^2 + 5x + 25)$

LEARNING OBJECTIVES

In Section 1.3 you will see how we can:

☐ **A.** Distinguish the graph of a function from that of a relation

☐ **B.** Determine the domain and range of a function

☐ **C.** Use function notation and evaluate functions

☐ **D.** Read and interpret information given graphically

In this section we introduce one of the most central ideas in mathematics—the concept of a function. Functions can model the cause-and-effect relationship that is so important to using mathematics as a decision-making tool. In addition, the study will help to unify and expand on many ideas that are already familiar.

A. Functions and Relations

There is a special type of relation that merits further attention. A **function** is a relation where each element of the domain corresponds to exactly one element of the range. In other words, for each first coordinate or input value, there is only one possible second coordinate or output.

> **Functions**
>
> A *function* is a relation that pairs each element from the *domain*
> with exactly one element from the *range*.

If the relation is defined by a mapping, we need only check that each element of the domain is mapped to exactly one element of the range. This is indeed the case for the mapping $P \rightarrow B$ from Figure 1.1 (page 2), where we saw that each person corresponded to only one birthday, and that it was impossible for one person to be born on two different days. For the relation $x = |y|$ shown in Figure 1.6 (page 4), each element of the domain except zero is paired with *more than one* element of the range. The relation $x = |y|$ is *not* a function.

EXAMPLE 1 ▶ **Determining Whether a Relation is a Function**

Three different relations are given in mapping notation below. Determine whether each relation is a function.

a.   b.   c.

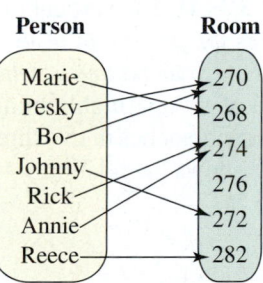
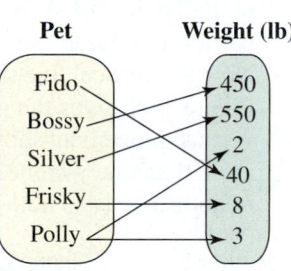
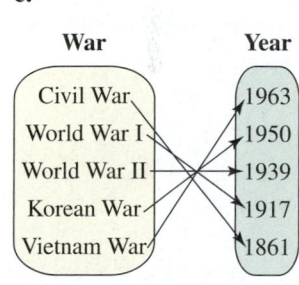

Solution ▶ Relation (a) is a function, since each person corresponds to exactly one room. This relation pairs math professors with their respective office numbers. Notice that while two people can be in one office, it is impossible for one person to physically be in two different offices.

Relation (b) is not a function, since we cannot tell whether Polly the Parrot weighs 2 lb or 3 lb (one element of the domain is mapped to two elements of the range).

Relation (c) is a function, where each major war is paired with the year it began.

Now try Exercises 7 through 10 ▶

If the relation is pointwise-defined or given as a set of individual and distinct plotted points, we need only check that no two points have the same first coordinate with a different second coordinate. This gives rise to an alternative definition for a function.

Functions (Alternate Definition)

A function is a set of ordered pairs (x, y), in which each first component is paired with only one second component.

EXAMPLE 2 ▶ **Identifying Functions**

Two relations named f and g are given; f is pointwise-defined (stated as a set of ordered pairs), while g is given as a set of plotted points. Determine whether each is a function.

$$f: (-3, 0), (1, 4), (2, -5), (4, 2), (-3, -2), (3, 6), (0, -1), (4, -5), \text{ and } (6, 1)$$

Solution ▶ The relation f is not a function, since -3 is paired with two different outputs: $(-3, 0)$ and $(-3, -2)$.

The relation g shown in the figure *is* a function. Each input corresponds to exactly one output, otherwise one point would be directly above the other and have the same first coordinate.

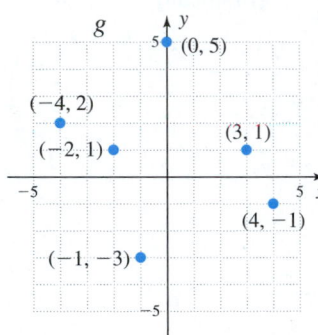

Now try Exercises 11 through 18 ▶

The graphs of $y = x - 1$ and $x = |y|$ from Section 1.1 offer additional insight into the definition of a function. Figure 1.40 shows the line $y = x - 1$ with emphasis on the plotted points $(4, 3)$ and $(-3, -4)$. The vertical movement shown from the x-axis to a point on the graph illustrates *the pairing of a given x-value with one related y-value.* Note the vertical line shows *only one related y-value* ($x = 4$ is paired with only $y = 3$). Figure 1.41 gives the graph of $x = |y|$, highlighting the points $(4, 4)$ and $(4, -4)$. The vertical movement shown here branches in two directions, associating one x-value with more than one y-value. This shows the relation $y = x - 1$ is also a function, while the relation $x = |y|$ is not.

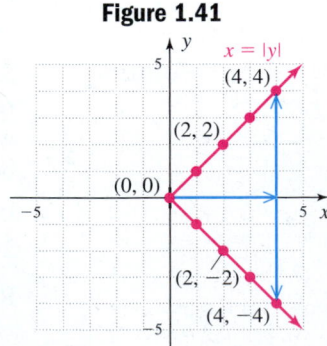

Figure 1.40   Figure 1.41

This "vertical connection" of a location on the x-axis to a point on the graph can be generalized into a **vertical line test** for functions.

> **Vertical Line Test**
>
> A given graph is the graph of a function, if and only if every vertical line intersects the graph in at most one point.

Applying the test to the graph in Figure 1.40 helps to illustrate that the graph of any nonvertical line must be the graph of a function, as is the graph of any pointwise-defined relation where no x-coordinate is repeated. Compare the relations f and g from Example 2.

EXAMPLE 3 ▶ **Using the Vertical Line Test**

Use the vertical line test to determine if any of the relations shown (from Section 1.1) are functions.

Solution ▶ Visualize a vertical line on each coordinate grid (shown in solid blue), then mentally shift the line to the left and right as shown in Figures 1.42, 1.43, and 1.44 (dashed lines). In Figures 1.42 and 1.43, every vertical line intersects the graph only once, indicating both $y = x^2 - 2x$ and $y = \sqrt{9 - x^2}$ are functions. In Figure 1.44, a vertical line intersects the graph twice for any $x > 0$ [for instance, both (4, 2) and (4, −2) are on the graph]. The relation $x = y^2$ is not a function.

Figure 1.42

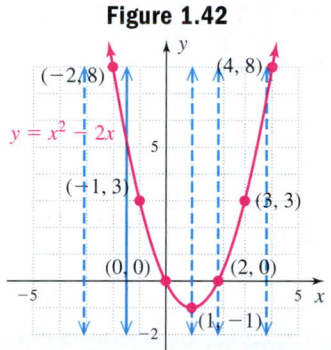

Figure 1.43

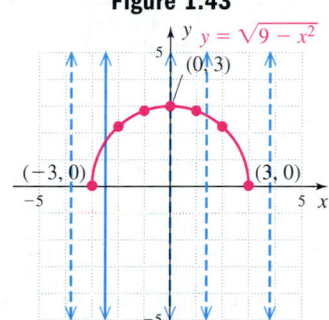

Figure 1.44

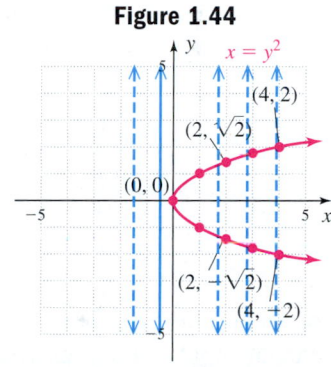

Now try Exercises 19 through 30 ▶

EXAMPLE 4 ▶ **Using the Vertical Line Test**

Use a table of values to graph the relations defined by
a. $y = |x|$ **b.** $y = \sqrt{x}$,
then use the vertical line test to determine whether each relation is a function.

Solution ▶ **a.** For $y = |x|$, using input values from $x = -4$ to $x = 4$ produces the following table and graph (Figure 1.45). Note the result is a V-shaped graph that "opens upward." The point (0, 0) of this absolute value graph is called the **vertex.** Since any vertical line will intersect the graph in at most one point, this is the graph of a function.

$y = |x|$

x	$y = |x|$
-4	4
-3	3
-2	2
-1	1
0	0
1	1
2	2
3	3
4	4

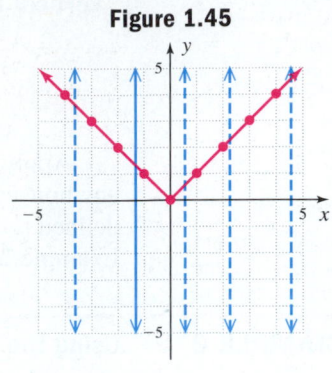

Figure 1.45

b. For $y = \sqrt{x}$, values less than zero do not produce a real number, so our graph actually begins at (0, 0) (see Figure 1.46). Completing the table for nonnegative values produces the graph shown, which appears to rise to the right and remains in the first quadrant. Since any vertical line will intersect this graph in at most one place, $y = \sqrt{x}$ is also a function.

Figure 1.46

$y = \sqrt{x}$

x	$y = \sqrt{x}$
0	0
1	1
2	$\sqrt{2} \approx 1.4$
3	$\sqrt{3} \approx 1.7$
4	2

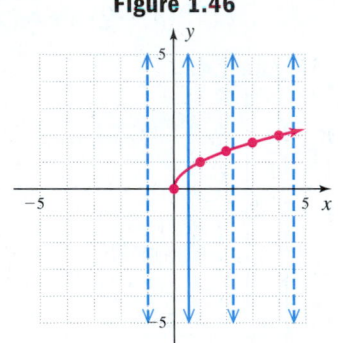

☑ **A.** You've just seen how we can distinguish the graph of a function from that of a relation

Now try Exercises 31 through 34 ▶

B. The Domain and Range of a Function

Vertical Boundary Lines and the Domain

In addition to its use as a graphical test for functions, a vertical line can help determine the domain of a function from its graph. For the graph of $y = \sqrt{x}$ (Figure 1.46), a vertical line will not intersect the graph until $x = 0$, and then will intersect the graph for all values $x \geq 0$ (showing the function is defined for these values). These **vertical boundary lines** indicate the domain is $x \geq 0$.

Instead of using a simple inequality to write the domain and range, we will often use (1) a form of **set notation,** (2) a **number line** graph, or (3) **interval notation.** Interval notation is a symbolic way of indicating a selected interval of the real numbers. When a number acts as the **boundary point** for an interval (also called an **endpoint**), we use a left bracket "[" or a right bracket "]" to indicate **inclusion** of the endpoint. If the boundary point is **not included,** we use a left parenthesis "(" or right parenthesis ")."

WORTHY OF NOTE

On a number line, some texts will use an open dot "○" to mark the location of an endpoint that is not included, and a closed dot "●" for an included endpoint.

EXAMPLE 5 ▶ **Using Notation to State the Domain and Range**

Model the given phrase using the correct inequality symbol. Then state the result in set notation, graphically, and in interval notation: "The set of real numbers greater than or equal to 1."

Solution ▶ Let n represent the number: $n \geq 1$.

WORTHY OF NOTE

Since infinity is really a *concept* and not a number, it is *never included* (using a bracket) as an endpoint for an interval.

- Set notation: $\{n | n \geq 1\}$
- Graph:

- Interval notation: $n \in [1, \infty)$

Now try Exercises 35 through 50 ▶

The "$\in$" symbol says the number n is *an element of the set or interval* given. The "∞" symbol represents positive infinity and indicates the interval continues forever to the right. Note that the endpoints of an interval must occur in the same order as on the number line *(smaller value on the left; larger value on the right)*.

A short summary of other possibilities is given here for any real number x. Many variations are possible.

Conditions ($a < b$)	Set Notation	Number Line	Interval Notation
x is greater than k	$\{x | x > k\}$		$x \in (k, \infty)$
x is less than or equal to k	$\{x | x \leq k\}$		$x \in (-\infty, k]$
x is less than b and greater than a	$\{x | a < x < b\}$		$x \in (a, b)$
x is less than b and greater than or equal to a	$\{x | a \leq x < b\}$		$x \in [a, b)$
x is less than a or x is greater than b	$\{x | x < a$ or $x > b\}$		$x \in (-\infty, a) \cup (b, \infty)$

For the graph of $y = |x|$ (Figure 1.45), a vertical line will intersect the graph (or its infinite extension) for *all values* of x, and the domain is $x \in (-\infty, \infty)$. Using vertical lines in this way also affirms the domain of $y = x - 1$ (Section 1.1, Figure 1.5) is $x \in (-\infty, \infty)$ while the domain of the relation $x = |y|$ (Section 1.1, Figure 1.6) is $x \in [0, \infty)$.

Range and Horizontal Boundary Lines

The range of a relation can be found using a **horizontal "boundary line,"** since it will associate a value on the y-axis with a point on the graph (if it exists). Simply visualize a horizontal line and move the line up or down until you determine the graph will always intersect the line, or will no longer intersect the line. This will give you the boundaries of the range. Mentally applying this idea to the graph of $y = \sqrt{x}$ (Figure 1.46) shows the range is $y \in [0, \infty)$. Although shaped very differently, a horizontal boundary line shows the range of $y = |x|$ (Figure 1.45) is also $y \in [0, \infty)$.

EXAMPLE 6 ▶ **Determining the Domain and Range of a Function**

Use a table of values to graph the functions defined by

  a. $y = x^2$ **b.** $y = \sqrt[3]{x}$

Then use boundary lines to determine the domain and range of each.

Solution ▶   **a.** For $y = x^2$, it seems convenient to use inputs from $x = -3$ to $x = 3$, producing the following table and graph. Note the result is a basic parabola that "opens upward" (both ends point in the positive y direction), with a vertex at $(0, 0)$. Figure 1.47 shows a vertical line will intersect the graph or its extension anywhere it is placed. The domain is $x \in (-\infty, \infty)$. Figure 1.48 shows a horizontal line will intersect the graph only for values of y that are greater than or equal to 0. The range is $y \in [0, \infty)$.

Squaring Function

x	$y = x^2$
-3	9
-2	4
-1	1
0	0
1	1
2	4
3	9

Figure 1.47

Figure 1.48

b. For $y = \sqrt[3]{x}$, we select points that are perfect cubes where possible, then a few others to round out the graph. The resulting table and graph are shown. Notice there is a "pivot point" at $(0, 0)$ called a **point of inflection,** and the ends of the graph point in opposite directions. Figure 1.49 shows a vertical line will intersect the graph or its extension anywhere it is placed. Figure 1.50 shows a horizontal line will likewise always intersect the graph. The domain is $x \in (-\infty, \infty)$, and the range is $y \in (-\infty, \infty)$.

Cube Root Function

x	$y = \sqrt[3]{x}$
-8	-2
-4	≈ -1.6
-1	-1
0	0
1	1
4	≈ 1.6
8	2

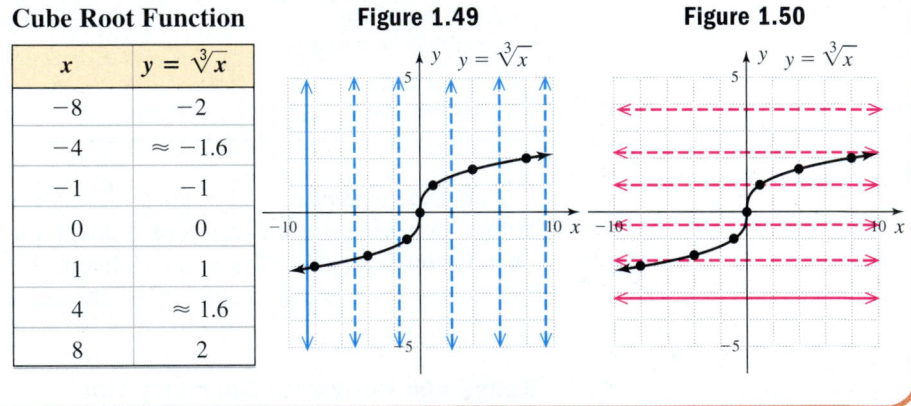

Figure 1.49

Figure 1.50

Now try Exercises 51 through 62 ▶

Implied Domains

When stated in equation form, the domain of a function is implicitly given by the expression used to define it, since the expression will dictate what input values are allowed. The **implied domain** is the set of all real numbers for which the function represents a real number. If the function involves a rational expression, the domain will exclude any input that causes a denominator of zero, since division by zero is undefined. If the function involves a square root expression, the domain will exclude inputs that create a negative radicand, since $\sqrt{A}$ represents a real number only when $A \geq 0$.

EXAMPLE 7 ▶ **Determining Implied Domains**

State the domain of each function using interval notation.

a. $y = \dfrac{3}{x + 2}$ b. $y = \sqrt{2x + 3}$

c. $y = \dfrac{x - 1}{x^2 - 9}$ d. $y = x^2 - 5x + 7$

Solution ▶

Figure 1.51

$Y_1 = (X - 1)/(X^2 - 9)$

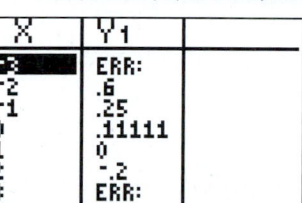

X	Y1
-3	ERR:
-2	.6
-1	.25
0	.11111
1	0
2	-.2
3	ERR:

X = -3

a. By inspection, we note an x-value of -2 results in a zero denominator and must be excluded. The domain is $x \in (-\infty, -2) \cup (-2, \infty)$.

b. Since the radicand must be nonnegative, we solve the inequality $2x + 3 \geq 0$, giving $x \geq \frac{-3}{2}$. The domain is $x \in [\frac{-3}{2}, \infty)$.

c. To prevent division by zero, inputs of -3 and 3 must be excluded (set $x^2 - 9 = 0$ and solve by factoring). The domain is $x \in (-\infty, -3) \cup (-3, 3) \cup (3, \infty)$. Note that $x = 1$ *is in the domain* since $\frac{0}{-8} = 0$ is defined. See Figure 1.51, where $Y_1 = \dfrac{X - 1}{X^2 - 9}$.

d. Since squaring a number and multiplying a number by a constant are defined for all real numbers, the domain is $x \in (-\infty, \infty)$.

Now try Exercises 63 through 80 ▶

EXAMPLE 8 ▶ **Determining Implied Domains**

Determine the domain of each function:

a. $y = \sqrt{\dfrac{7}{x + 3}}$ b. $y = \dfrac{2x}{\sqrt{4x + 5}}$

Solution ▶

a. For $y = \sqrt{\dfrac{7}{x + 3}}$, we must have $\dfrac{7}{x + 3} \geq 0$ (for the radicand) **and** $x + 3 \neq 0$ (for the denominator). Since the numerator is *always* positive, we need $x + 3 > 0$, which gives $x > -3$. The domain is $x \in (-3, \infty)$.

b. For $y = \dfrac{2x}{\sqrt{4x + 5}}$, we must have $4x + 5 \geq 0$ **and** $\sqrt{4x + 5} \neq 0$. This shows we need $4x + 5 > 0$, so $x > \frac{-5}{4}$. The domain is $x \in (\frac{-5}{4}, \infty)$.

☑ **B.** You've just seen how we can determine the domain and range of a function

Now try Exercises 81 through 96 ▶

C. Function Notation

Figure 1.52

Input x →

f

Sequence of operations on x as defined by f

→ $y = f(x)$

Output

In our study of functions, you've likely noticed that the relationship between input and output values is an important one. To highlight this fact, think of a function as a simple machine, which can *process inputs* using a stated sequence of operations, then deliver a single output. The inputs are x-values, a program we'll name f performs the operations on x, and y is the resulting output (see Figure 1.52). Once again we see that "the value of y depends on the value of x," or simply "y is a function of x." Notationally, we write "y is a function of x" as $y = f(x)$ using **function notation.** You are already familiar with letting a variable represent a number. Here we do something quite different, as the letter f is used to represent *a sequence of operations to be performed on x.* Consider the function $y = \dfrac{x}{2} + 1$, which we'll now write as $f(x) = \dfrac{x}{2} + 1$ [since $y = f(x)$].

In words the function says, "divide inputs by 2, then add 1." To evaluate the function at $x = 4$ (Figure 1.53) we have:

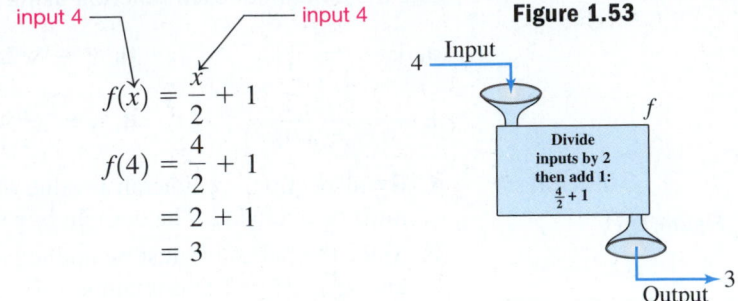

Figure 1.53

$$f(x) = \frac{x}{2} + 1$$
$$f(4) = \frac{4}{2} + 1$$
$$= 2 + 1$$
$$= 3$$

Function notation enables us to summarize the three most important aspects of a function using a single expression, as shown in Figure 1.54.

Figure 1.54

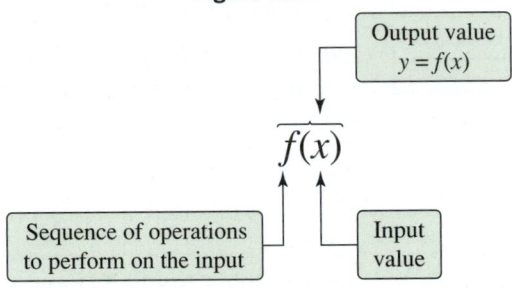

Instead of saying, ". . . when $x = 4$, the value of the function is 3," we simply say "f of 4 is 3," or write $f(4) = 3$. Note that the ordered pair $(4, 3)$ is equivalent to $(4, f(4))$.

> ⚠️ **CAUTION** ▶ Although $f(x)$ is the favored notation for a "function of x," other letters can also be used. For example, $g(x)$ and $h(x)$ also denote functions of x, where g and h represent different sequences of operations on the x-inputs. It is also important to remember that these represent *function values* and not the product of two variables: $f(x) \neq f \cdot (x)$.

EXAMPLE 9 ▶ **Evaluating a Function**

Given $f(x) = -2x^2 + 4x$, find

a. $f(-2)$ **b.** $f\left(\frac{7}{2}\right)$ **c.** $f(2a)$ **d.** $f(a + 1)$

Solution ▶

a. $f(x) = -2x^2 + 4x$
$f(-2) = -2(-2)^2 + 4(-2)$
$= -8 + (-8) = -16$

b. $f(x) = -2x^2 + 4x$
$f\left(\frac{7}{2}\right) = -2\left(\frac{7}{2}\right)^2 + 4\left(\frac{7}{2}\right)$
$= \frac{-49}{2} + 14 = \frac{-21}{2}$ or -10.5

c. $f(x) = -2x^2 + 4x$
$f(2a) = -2(2a)^2 + 4(2a)$
$= -2(4a^2) + 8a$
$= -8a^2 + 8a$

d. $f(x) = -2x^2 + 4x$
$f(a + 1) = -2(a + 1)^2 + 4(a + 1)$
$= -2(a^2 + 2a + 1) + 4a + 4$
$= -2a^2 - 4a - 2 + 4a + 4$
$= -2a^2 + 2$

Now try Exercises 87 through 102 ▶

Figure 1.55

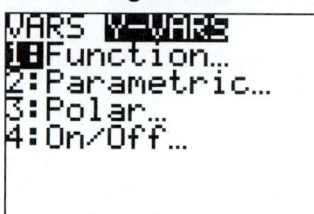

A graphing calculator can evaluate the function $Y_1 = -2X^2 + 4X$ using the **TABLE** feature, the (TRACE) feature, or function notation (on the home screen). The first two have been illustrated previously. To use function notation, we access the function names using the (VARS) key and right arrow to select **Y-VARS** (Figure 1.55). The **1:Function** option is the default, so pressing (ENTER) will enable us to make our choice (Figure 1.56). In this case, we selected **1:Y$_1$**, which the calculator then places on the home screen, enabling us to enclose the desired input value in parentheses (function notation). Pressing ENTER completes the evaluation (Figure 1.57), which verifies the result from Example 9(b).

Figure 1.56

Figure 1.57

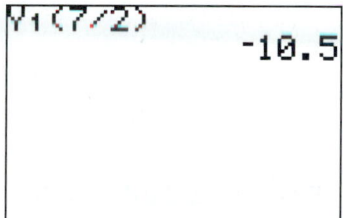

✓ **C.** You've just seen how we can use function notation and evaluate functions

D. Reading and Interpreting Information Given Graphically

Graphs are an important part of studying functions, and learning to read and interpret them correctly is a high priority. A graph highlights and emphasizes the all-important input/output relationship that defines a function. In this study, we hope to firmly establish that the following statements are synonymous:

1. $f(-2) = 5$
2. $(-2, f(-2)) = (-2, 5)$
3. $(-2, 5)$ is on the graph of f, and
4. when $x = -2, f(x) = 5$

EXAMPLE 10A ▶ **Reading a Graph**

For the functions f and g whose graphs are shown in Figures 1.58 and 1.59

   a. State the domain of the function.

   b. Evaluate the function at $x = 2$.

   c. Determine the value(s) of x for which $y = 3$.

   d. State the range of the function.

Figure 1.58

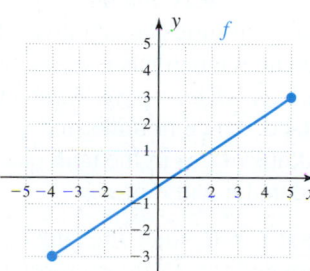

Figure 1.59

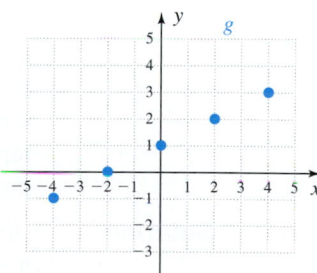

Solution ▶ For f,

   a. The graph is a continuous line segment with endpoints at $(-4, -3)$ and $(5, 3)$, so we state the domain in interval notation. Using a vertical boundary line we note the smallest input is -4 and the largest is 5. The domain is $x \in [-4, 5]$.

b. The graph shows an input of $x = 2$ corresponds to $y = 1$: $f(2) = 1$ since $(2, 1)$ is a point on the graph.

c. For $f(x) = 3$ (or $y = 3$) the input value must be $x = 5$ since $(5, 3)$ is the point on the graph.

d. Using a horizontal boundary line, the smallest output value is -3 and the largest is 3. The range is $y \in [-3, 3]$.

For g,

a. Since g is given as a set of plotted points, we state the domain as the set of first coordinates: D: $\{-4, -2, 0, 2, 4\}$.

b. An input of $x = 2$ corresponds to $y = 2$: $g(2) = 2$ since $(2, 2)$ is on the graph.

c. For $g(x) = 3$ (or $y = 3$) the input value must be $x = 4$, since $(4, 3)$ is a point on the graph.

d. The range is the set of all second coordinates: R: $\{-1, 0, 1, 2, 3\}$.

EXAMPLE 10B ▶   **Reading a Graph**

Use the graph of $f(x)$ given to answer the following questions:

a. What is the value of $f(-2)$?

b. What value(s) of x satisfy $f(x) = 1$?

Solution ▶   **a.** The notation $f(-2)$ says to find the value of the function f when $x = -2$. Expressed graphically, we go to $x = -2$ and locate the corresponding point on the graph (blue arrows). Here we find that $f(-2) = 4$.

b. For $f(x) = 1$, we're looking for x-inputs that result in an output of $y = 1$ [since $y = f(x)$]. From the graph, we note there are two points with a y-coordinate of 1, namely, $(-3, 1)$ and $(0, 1)$. This shows $f(-3) = 1$, $f(0) = 1$, and the required x-values are $x = -3$ and $x = 0$.

Now try Exercises 103 through 108 ▶

In many applications involving functions, the domain and range can be determined by the context or situation given.

EXAMPLE 11 ▶   **Determining the Domain and Range from the Context**

Paul's 2009 Voyager has a 20-gal tank and gets 18 mpg. The number of miles he can drive (his range) depends on how much gas is in the tank. As a function we have $M(g) = 18g$, where $M(g)$ represents the total distance in miles and g represents the gallons of gas in the tank (see graph). Find the domain and range.

Solution ▶   Begin evaluating at $x = 0$, since the tank cannot hold less than zero gallons. With an empty tank, the (minimum) range is $M(0) = 18(0)$ or 0 miles. On a full tank, the maximum range is $M(20) = 18(20)$ or 360 miles. As shown in the graph, the domain is $g \in [0, 20]$ and the corresponding range is $M(g) \in [0, 360]$.

☑ **D.** You've just seen how we can read and interpret information given graphically

Now try Exercises 112 through 119 ▶

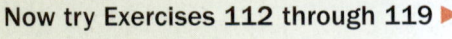

1.3 EXERCISES

► CONCEPTS AND VOCABULARY

Fill in each blank with the appropriate word or phrase. Carefully reread the section if needed.

1. If a relation is given in ordered pair form, we state the domain by listing all of the ___first___ coordinates in a set.

2. A relation is a function if each element of the ___domain___ is paired with ___exactly___ ___one___ element of the range.

3. The set of output values for a function is called the ___range___ of the function.

4. Write using function notation: The function f evaluated at 3 is negative 5: ___$f(3) = -5$___

5. Discuss/Explain why the relation $y = x^2$ is a function, while the relation $x = y^2$ is not. Justify your response using graphs, ordered pairs, and so on.　*Answers will vary.*

6. Discuss/Explain the process of finding the domain and range of a function given its graph, using vertical and horizontal boundary lines. Include a few illustrative examples.　*Answers will vary.*

► DEVELOPING YOUR SKILLS

Determine whether the mappings shown represent functions or nonfunctions. If a nonfunction, explain how the definition of a function is violated.

7.　　**Woman**　　　**Country**　　*function*

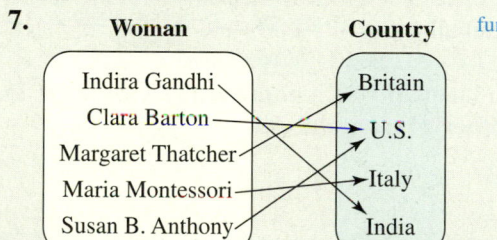

8.　　**Book**　　　**Author**　　*function*

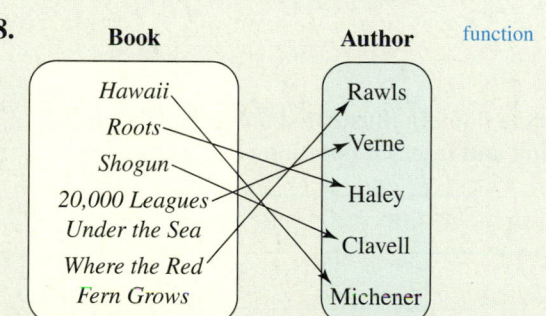

9.　**Basketball star**　**Reported height**　*Not a function. The Shaq is paired with two heights.*

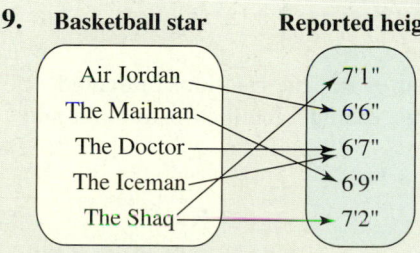

10.　**Country**　　　**Language**　*Not a function. Canada is paired with 2 languages and Brazil is paired with 2.*

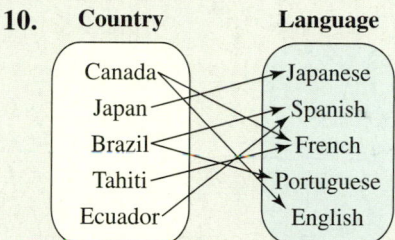

Determine whether the relations indicated represent functions or nonfunctions. If the relation is a nonfunction, explain how the definition of a function is violated.

11. $(-3, 0)$, $(1, 4)$, $(2, -5)$, $(4, 2)$, $(-5, 6)$, $(3, 6)$, $(0, -1)$, $(4, -5)$, and $(6, 1)$
Not a function; 4 is paired with 2 and -5.

12. $(-7, -5)$, $(-5, 3)$, $(4, 0)$, $(-3, -5)$, $(1, -6)$, $(0, 9)$, $(2, -8)$, $(3, -2)$, and $(-5, 7)$
Not a function; -5 is paired with 3 and 7.

13. $(9, -10)$, $(-7, 6)$, $(6, -10)$, $(4, -1)$, $(2, -2)$, $(1, 8)$, $(0, -2)$, $(-2, -7)$, and $(-6, 4)$　*function*

14. $(1, -81)$, $(-2, 64)$, $(-3, 49)$, $(5, -36)$, $(-8, 25)$, $(13, -16)$, $(-21, 9)$, $(34, -4)$, and $(-55, 1)$
function

15.　　　　　　　　　　　16.

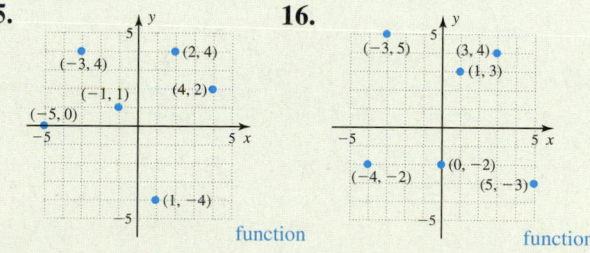

function　　　　　　　　　*function*

HOMEWORK SELECTION GUIDE

Core: 7–91 every other odd, 105, 113, 115, 117 (27 Exercises)
Standard: 1–4, 7–91 every other odd, 105, 109, 113–119 odd, 120 (34 Exercises)

Extended: 1–4, 7–107 every other odd, 109, 110, 113–119 odd, 120, 122, 125, 127 (40 Exercises)
In Depth: 1–6, 7–107 every other odd, 109–111, 113–119 odd, 120, 122, 123–127 (46 Exercises)

17. 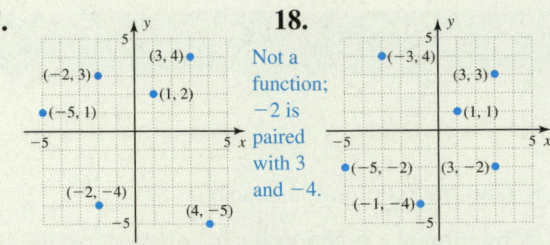 **18.** Not a function; −2 is paired with 3 and −4.

Not a function; 3 is paired with 3 and −2.

Determine whether or not the relations given represent a function. If not, explain how the definition of a function is violated.

19. 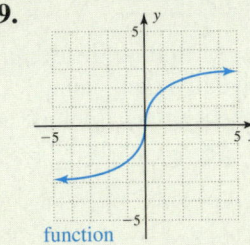 **20.**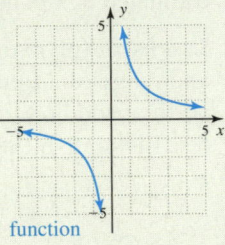

function   function

21. 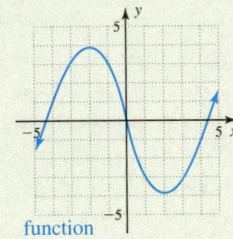 **22.**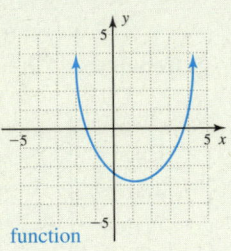

function   function

23. 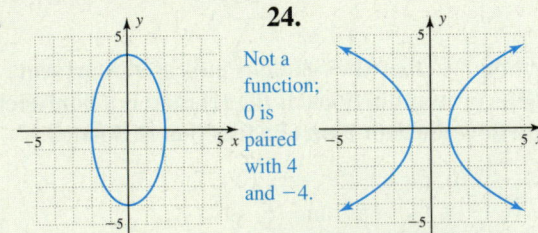 **24.** Not a function; 0 is paired with 4 and −4.

Not a function; 2 is paired with −2.3 and 2.3.

25. 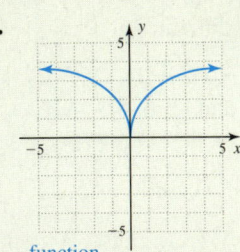 **26.**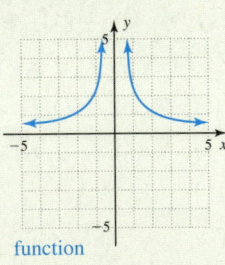

function   function

27. 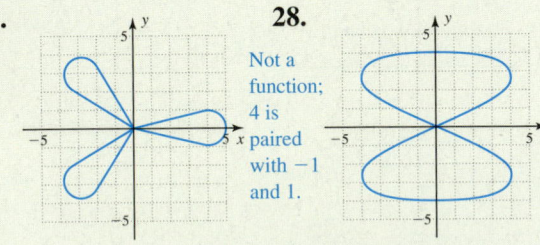 **28.** Not a function; 4 is paired with −1 and 1.

Not a function; 0 is paired with 4 and −4.

29. 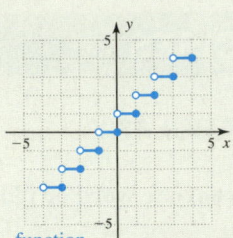 **30.**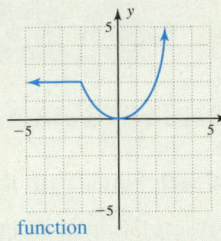

function   function

Graph each relation using a table, then use the vertical line test to determine if the relation is a function.

31. $y = x$ **32.** $y = \sqrt[3]{x}$

33. $y = (x + 2)^2$ **34.** $x = |y - 2|$

Use an inequality to write a mathematical model for each statement, then write the relation in interval notation.

35. To qualify for a secretarial position, a person must type at least 45 words per minute. $w \geq 45;\ w \in [45, \infty)$

36. The balance in a checking account must remain above \$1000 or a fee is charged.
$b > 1000;\ b \in (1000, \infty)$

37. To bake properly, a turkey must be kept between the temperatures of 250° and 450°.
$250 < T < 450;\ T \in (250, 450)$

38. To fly effectively, the airliner must cruise at or between altitudes of 30,000 and 35,000 ft.
$30{,}000 \leq a \leq 35{,}000;\ a \in [30{,}000, 35{,}000]$

Graph each inequality on a number line, then write the relation in interval notation.

39. $p < 3$ **40.** $x > -2$

41. $m \leq 5$ **42.** $n \geq -4$

43. $x \neq 1$ **44.** $x \neq -3$

45. $5 > x > 2$ **46.** $-3 < p \leq 4$

Write the domain illustrated on each graph in set notation and interval notation.

47. $\{x | x \geq -2\}; [-2, \infty)$

48. $\{x | x < 1\}; (-\infty, 1)$

49. $\{x | -2 \leq x \leq 1\}; [-2, 1]$

50. $\{x | -2 \leq x < 3\}; [-2, 3)$

Determine whether or not the relations indicated represent functions, then determine the domain and range of each.

51. 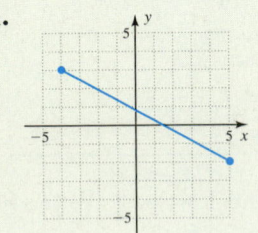 **52.**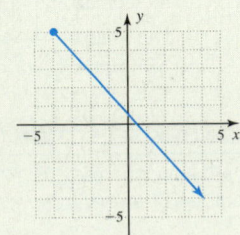

Additional answers can be found in the Instructor Answer Appendix.

53.
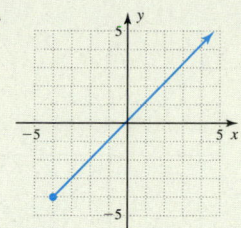

54.
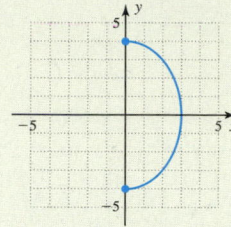

55.
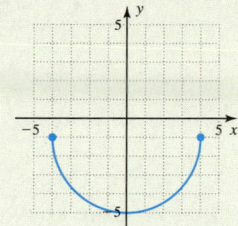

56.
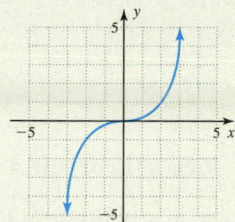

57.
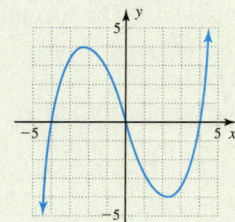

58.
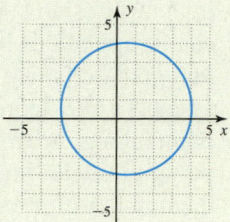

59.
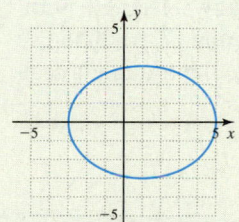

60.
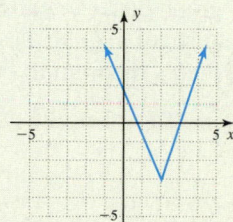

61.
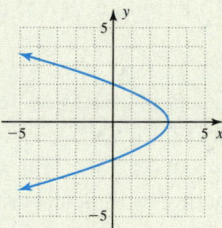

62.
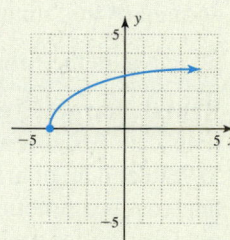

Determine the domain of the following functions, and write your response in interval notation.

63. $f(x) = \dfrac{3}{x - 5}$　　　**64.** $g(x) = \dfrac{-2}{3 + x}$

65. $h(a) = \sqrt{3a + 5}$　　**66.** $p(a) = \sqrt{5a - 2}$

67. $v(x) = \dfrac{x + 2}{x^2 - 25}$　　**68.** $w(x) = \dfrac{x - 4}{x^2 - 49}$

69. $u = \dfrac{v - 5}{v^2 - 18}$　　　**70.** $p = \dfrac{q + 7}{q^2 - 12}$

71. $y = \dfrac{17}{25}x + 123$　　**72.** $y = \dfrac{11}{19}x - 89$

73. $m = n^2 - 3n - 10$　　**74.** $s = t^2 - 3t - 10$

75. $y = 2|x| + 1$　　　**76.** $y = |x - 2| + 3$

77. $y_1 = \dfrac{x}{x^2 - 3x - 10}$　　**78.** $y_2 = \dfrac{x - 4}{x^2 + 2x - 15}$

79. $y = \dfrac{\sqrt{x - 2}}{2x - 5}$　　**80.** $y = \dfrac{\sqrt{x + 1}}{3x + 2}$

81. $h(x) = \dfrac{-2}{\sqrt{x + 4}}$　　**82.** $f(x) = \sqrt{\dfrac{5}{x - 2}}$

83. $g(x) = \sqrt{\dfrac{-4}{3 - x}}$　　**84.** $p(x) = \dfrac{-7}{\sqrt{5 - x}}$

85. $r(x) = \dfrac{2x - 1}{\sqrt{3x - 7}}$　　**86.** $s(x) = \dfrac{x^2 - 4}{\sqrt{11 - 2x}}$

$\qquad\qquad x \in (\frac{7}{3}, \infty)$　　　　　$x \in (-\infty, \frac{11}{2})$

For Exercises 87 through 102, determine the value of $f(-6), f(\frac{3}{2}), f(2c),$ and $f(c + 1)$, then simplify. Verify results using a graphing calculator where possible.

87. $f(x) = \dfrac{1}{2}x + 3$　　**88.** $f(x) = \dfrac{2}{3}x - 5$

89. $f(x) = 3x^2 - 4x$　　**90.** $f(x) = 2x^2 + 3x$

Determine the value of $h(3), h(-\frac{2}{3}), h(3a),$ and $h(a - 2)$, then simplify.

91. $h(x) = \dfrac{3}{x}$　　　**92.** $h(x) = \dfrac{2}{x^2}$

93. $h(x) = \dfrac{5|x|}{x}$　　**94.** $h(x) = \dfrac{4|x|}{x}$

Determine the value of $g(4), g(\frac{3}{2}), g(2c),$ and $g(c + 3)$, then simplify.

95. $g(r) = 2\pi r$　　　**96.** $g(r) = 2\pi rh$

97. $g(r) = \pi r^2$　　　**98.** $g(r) = \pi r^2 h$

Determine the value of $p(5), p(\frac{3}{2}), p(3a),$ and $p(a - 1)$, then simplify.

99. $p(x) = \sqrt{2x + 3}$　　**100.** $p(x) = \sqrt{4x - 1}$

101. $p(x) = \dfrac{3x^2 - 5}{x^2}$　　**102.** $p(x) = \dfrac{2x^2 + 3}{x^2}$

Use the graph of each function given to (a) state the domain, (b) state the range, (c) evaluate $f(2)$, and (d) find the value(s) x for which $f(x) = k$ (k a constant). Assume all results are integer-valued.

103. $k = 4$　　　　　**104.** $k = 3$

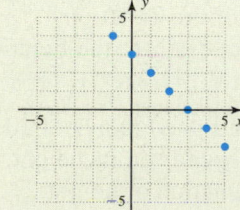　　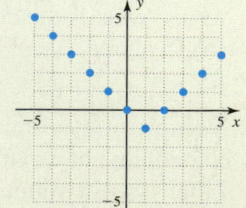

105. $k = 1$ 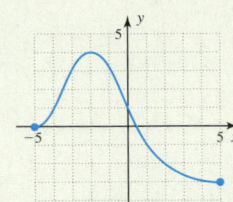   **106.** $k = -3$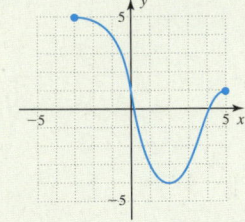

107. $k = 2$ 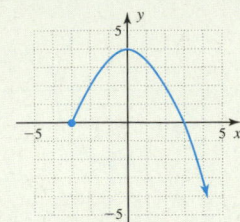   **108.** $k = -1$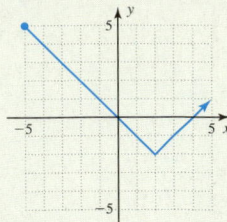

► **WORKING WITH FORMULAS**

109. Ideal weight for males: $W(H) = \dfrac{9}{2}H - 151$

The ideal weight for an adult male can be modeled by the function shown, where W is his weight in pounds and H is his height in inches. (a) Find the ideal weight for a male who is 75 in. tall. (b) If I am 72 in. tall and weigh 210 lb, how much weight should I lose?   **a.** 186.5 lb   **b.** 37 lb

110. Celsius to Fahrenheit conversions: $C = \dfrac{5}{9}(F - 32)$

The relationship between Fahrenheit degrees and degrees Celsius is modeled by the function shown. (a) What is the Celsius temperature if °F = 41? (b) Use the formula to solve for F in terms of C, then substitute the result from part (a). What do you notice?   **a.** 5°C   **b.** They give the same result.

111. Pick's theorem: $A = \dfrac{1}{2}B + I - 1$

Pick's theorem is an interesting yet little known formula for computing the area of a polygon drawn in the Cartesian coordinate system. The formula can be applied as long as the vertices of the polygon are lattice points (both x and y are integers). If B represents the number of lattice points lying directly on the boundary of the polygon (including the vertices), and I represents the number of points in the interior, the area of the polygon is given by the formula shown. Use some graph paper to carefully draw a triangle with vertices at $(-3, 1)$, $(3, 9)$, and $(7, 6)$, then use Pick's theorem to compute the triangle's area.
$A = \dfrac{1}{2}(8) + 22 - 1 = 25$ units^2

► **APPLICATIONS**

112. Gas mileage: John's old '87 LeBaron has a 15-gal gas tank and gets 23 mpg. The number of miles he can drive is a function of how much gas is in the tank. (a) Write this relationship in equation form and (b) determine the domain and range of the function in this context.
a. $N(g) = 23g$ **b.** $g \in [0, 15]$; $N \in [0, 345]$

113. Gas mileage: Jackie has a gas-powered model boat with a 5-oz gas tank. The boat will run for 2.5 min on each ounce. The number of minutes she can operate the boat is a function of how much gas is in the tank. (a) Write this relationship in equation form and (b) determine the domain and range of the function in this context.
a. $N(g) = 2.5g$ **b.** $g \in [0, 5]$; $N \in [0, 12.5]$

114. Volume of a cube: The volume of a cube depends on the length of the sides. In other words, volume is a function of the sides: $V(s) = s^3$. (a) In practical terms, what is the domain of this function? (b) Evaluate $V(6.25)$ and (c) evaluate the function for $s = 2x^2$.   **a.** $[0, \infty)$ **b.** about 244 units^3 **c.** $8x^6$ units^3

115. Volume of a cylinder: For a fixed radius of 10 cm, the volume of a cylinder depends on its height. In other words, volume is a function of height:

$V(h) = 100\pi h$. (a) In practical terms, what is the domain of this function? (b) Evaluate $V(7.5)$ and (c) evaluate the function for $h = \dfrac{8}{\pi}$.
a. $[0, \infty)$ **b.** about 2356 units^3 **c.** 800 units^3

116. Rental charges: Temporary Transportation Inc. rents cars (local rentals only) for a flat fee of $19.50 and an hourly charge of $12.50. This means that cost is a function of the hours the car is rented plus the flat fee. (a) Write this relationship in equation form; (b) find the cost if the car is rented for 3.5 hr; (c) determine how long the car was rented if the bill came to $119.75; and (d) determine the domain and range of the function in this context, if your budget limits you to paying a maximum of $150 for the rental.   **a.** $c(t) = 12.50t + 19.50$ **b.** $63.25
c. about 8 hr   **d.** $t \in [0, 10.44]$; $c \in [0, 150]$

117. Cost of a service call: Paul's Plumbing charges a flat fee of $50 per service call plus an hourly rate of $42.50. This means that cost is a function of the hours the job takes to complete plus the flat fee. (a) Write this relationship in equation form; (b) find the cost of a service call that takes $2\frac{1}{2}$ hr; (c) find the number of hours the job took if the charge came to $262.50; and (d) determine the domain and range of the function in this context, if your insurance company has agreed to pay for all charges over $500 for the service call.

118. Predicting tides: The graph shown approximates the height of the tides at Fair Haven, New Brunswick, for a 12-hr period. (a) Is this the graph of a function? Why? (b) Approximately what time did high tide occur? (c) How high is the tide at 6 P.M.? (d) What time(s) will the tide be 2.5 m?

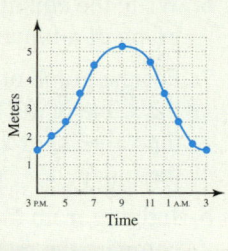

119. Predicting tides: The graph shown approximates the height of the tides at Apia, Western Samoa, for a 12-hr period. (a) Is this the graph of a function? Why? (b) Approximately what time did low tide occur? (c) How high is the tide at 2 A.M.? (d) What time(s) will the tide be 0.7 m?

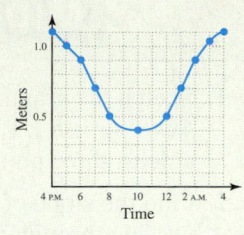

▶ **EXTENDING THE CONCEPT**

120. A father challenges his son to a 400-m race, depicted in the graph shown here.

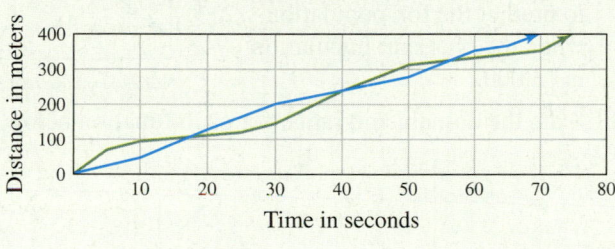

Father: ——   Son: ——

   a. Who won and what was the approximate winning time?   *Father, 70 sec*

   b. Approximately how many meters behind was the second place finisher?   *50 m*

   c. Estimate the number of seconds the father was in the lead in this race.   *≈40 sec*

   d. How many times during the race were the father and son tied?   *3*

121. Sketch the graph of $f(x) = x$, then discuss how you could use this graph to obtain the graph of $F(x) = |x|$ without computing additional points. What would the graph of $g(x) = \dfrac{|x|}{x}$ look like?

122. Sketch the graph of $f(x) = x^2 - 4$, then discuss how you could use this graph to obtain the graph of $F(x) = |x^2 - 4|$ without computing additional points. Determine what the graph of $g(x) = \dfrac{|x^2 - 4|}{x^2 - 4}$ would look like.

123. If the equation of a function is given, the domain is implicitly defined by input values that generate real-valued outputs. But unless the graph is given or can be easily sketched, we must attempt to find the range analytically *by solving for x in terms of y*. We should note that sometimes this is an easy task, while at other times it is virtually impossible and we must rely on other methods. For the following functions, determine the implicit domain and find the range by solving for x in terms of y.

   a. $y = \frac{x-3}{x+2}$ **b.** $y = x^2 - 3$

▶ **MAINTAINING YOUR SKILLS**

124. (1.1) Find the equation of a circle whose center is $(4, -1)$ with a radius of 5. Then graph the circle.

125. (R.6) Compute the sum and product indicated:

   a. $\sqrt{24} + 6\sqrt{54} - \sqrt{6}$ *19$\sqrt{6}$*

   b. $(2 + \sqrt{3})(2 - \sqrt{3})$ *1*

126. (R.4) Solve the equation by factoring, then check the result(s) using substitution: $3x^2 - 4x = 7$.
$x = \frac{7}{3}, -1$

127. (R.4) Factor the following polynomials completely:

   a. $x^3 - 3x^2 - 25x + 75$ *$(x-3)(x-5)(x+5)$*

   b. $2x^2 - 13x - 24$ *$(2x+3)(x-8)$*

   c. $8x^3 - 125$ *$(2x-5)(4x^2+10x+25)$*

MID-CHAPTER CHECK

1. Sketch the graph of the line $4x - 3y = 12$. Plot and label at least three points.

2. Find the slope of the line passing through the given points: $(-3, 8)$ and $(4, -10)$. $m = \frac{-18}{7}$

3. In 2009, Data.com lost $2 million. In 2010, they lost $0.5 million. Will the slope of the line through these points be positive or negative? Why? Calculate the slope. Were you correct? Write the slope as a unit rate and explain what it means in this context.

4. To earn some spending money, Sahara takes a job in a ski shop working primarily with her specialty—snowboards. She is paid a monthly salary of $950 pus a commission of $7.50 for each snowboard she sells. (a) Write a function that models her monthly earnings E. (b) Use a graphing calculator to determine her income if she sells 20, 30, or 40 snowboards in one month. (c) Use the results of parts a and b to set an appropriate viewing window and graph the line. (d) Use the TRACE feature to determine the number of snowboards that must be sold for Sahara's monthly income to top $1300.

5. Write the equation for line L_1 shown. Is this the graph of a function? Discuss why or why not.

Exercises 5 and 6

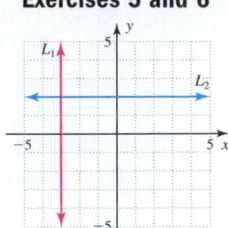

Exercises 7 and 8

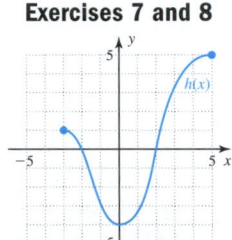

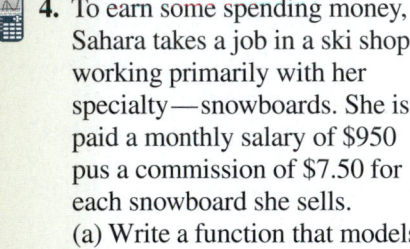

6. Write the equation for line L_2 shown. Is this the graph of a function? Discuss why or why not.

7. For the graph of function $h(x)$ shown, (a) determine the value of $h(2)$; (b) state the domain; (c) determine the value(s) of x for which $h(x) = -3$; and (d) state the range.   **a.** 0   **b.** $x \in [-3, 5]$ **c.** -1 and 1   **d.** $y \in [-4, 5]$

8. Judging from the appearance of the graph alone, compare the rate of change (slope) from $x = 1$ to $x = 2$ to the rate of change from $x = 4$ to $x = 5$. Which rate of change is larger? How is that demonstrated graphically?

9. Compute the slope of the line shown, and explain what it means as a rate of change in this context. Then use the slope to predict the fox population when the pheasant population is 13,000.

Exercise 9

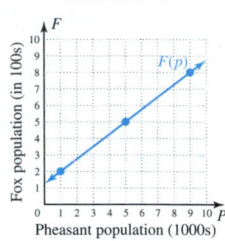

10. State the domain and range for each function below.

a.
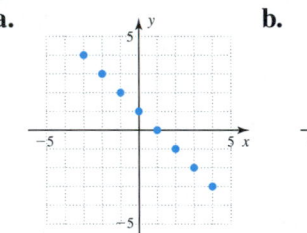

b.
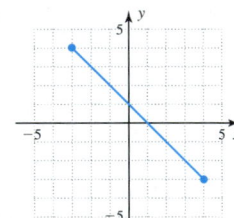

c.
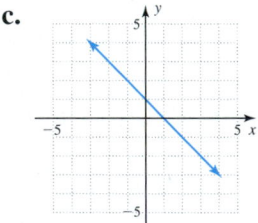

REINFORCING BASIC CONCEPTS

Finding the Domain and Range of a Relation from Its Graph

The concepts of domain and range are an important and fundamental part of working with relations and functions. In this chapter, we learned to determine the domain of any relation from its graph using a "vertical boundary line," and the range by using a "horizontal boundary line." These approaches to finding the domain and range can be combined into a single step by envisioning a rectangle drawn around or about the graph. If the entire graph can be "bounded" within the rectangle, the domain and range can be based on the rectangle's related length and width. If it's impossible to bound the graph in a particular direction, the related x- or y-values continue infinitely. Consider the graph in Figure 1.60. This is the graph of an ellipse (Section 8.2), and a rectangle that bounds the graph in all directions is shown in Figure 1.61.

Figure 1.60

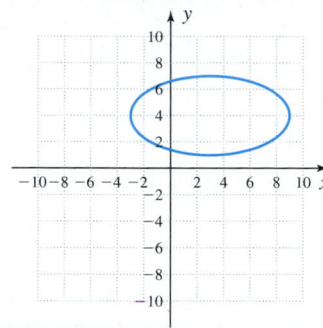

Figure 1.61

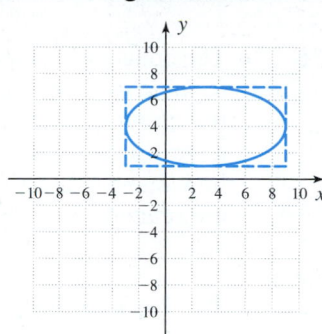

The rectangle extends from $x = -3$ to $x = 9$ in the horizontal direction, and from $y = 1$ to $y = 7$ in the vertical direction. The domain of this relation is $x \in [-3, 9]$ and the range is $y \in [1, 7]$.

The graph in Figure 1.62 is a parabola, and no matter how large we draw the rectangle, an infinite extension of the graph will extend beyond its boundaries in the left and right directions, and in the upward direction (Figure 1.63).

Figure 1.62

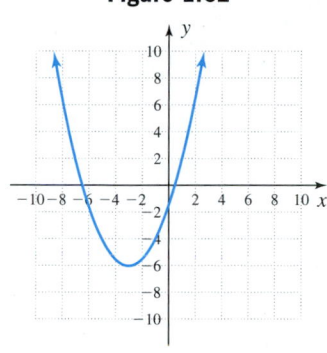

Figure 1.63

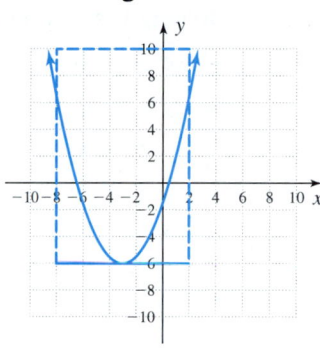

The domain of this relation is $x \in (-\infty, \infty)$ and the range is $y \in [-6, \infty)$.

Finally, the graph in Figure 1.64 is the graph of a square root function, and a rectangle can be drawn that bounds the graph below and to the left, but not above or to the right (Figure 1.65).

Figure 1.64

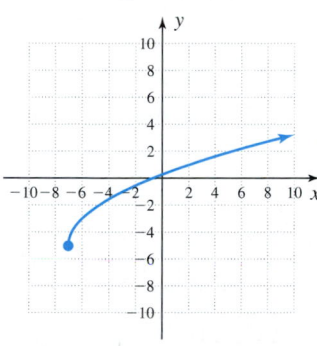

Figure 1.65

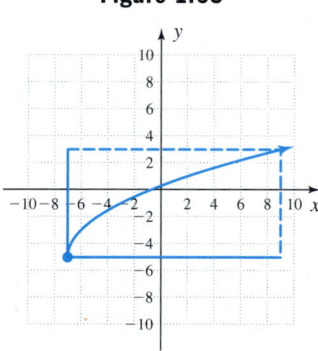

The domain of this relation is $x \in [-7, \infty)$ and the range is $y \in [-5, \infty)$.

Use this approach to find the domain and range of the following relations and functions.

Exercise 1:

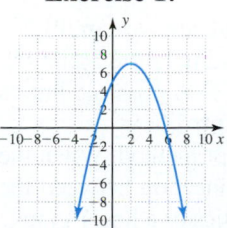

$D: x \in (-\infty, \infty), R: y \in (-\infty, 7)$

Exercise 2:

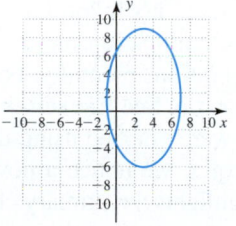

$D: x \in [-1, 7], R: y \in [-6, 9]$

Exercise 3:

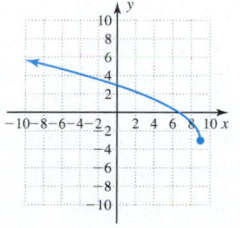

$D: x \in (-\infty, 9], R: y \in [-3, \infty)$

Exercise 4:

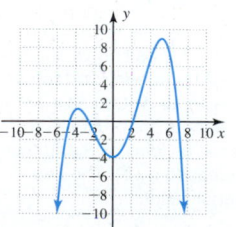

$D: x \in (-\infty, \infty), R: y \in (-\infty, 9]$

In Section 1.4 you will see how we can:

- ☐ **A.** Write a linear equation in slope-intercept form and function form
- ☐ **B.** Use slope-intercept form to graph linear equations
- ☐ **C.** Write a linear equation in point-slope form
- ☐ **D.** Apply the slope-intercept form and point-slope form in context

The concept of slope is an important part of mathematics, because it gives us a way to measure and compare change. The value of an automobile changes with time, the circumference of a circle increases as the radius increases, and the tension in a spring grows the more it is stretched. The real world is filled with examples of how one change affects another, and slope helps us understand how these changes are related.

A. Linear Equations, Slope-Intercept Form and Function Form

In Section 1.2, we learned that a linear equation is one that can be written in the form $ax + by = c$. Solving for y in a linear equation offers distinct advantages to understanding linear graphs and their applications.

EXAMPLE 1 ▶ **Solving for y in a Linear Equation**

Solve $2y - 6x = 4$ for y, then evaluate at $x = 4$, $x = 0$, and $x = -\frac{1}{3}$.

Solution ▶
$$2y - 6x = 4 \qquad \text{given equation}$$
$$2y = 6x + 4 \qquad \text{add } 6x$$
$$y = 3x + 2 \qquad \text{divide by 2}$$

Since the coefficients are integers, evaluate the function mentally. Inputs are multiplied by 3, then increased by 2, yielding the ordered pairs $(4, 14)$, $(0, 2)$, and $\left(-\frac{1}{3}, 1\right)$.

Now try Exercises 7 through 12 ▶

This form of the equation (where y has been written in terms of x) enables us to quickly identify what operations are performed on x in order to obtain y. Once again, for $y = 3x + 2$: *multiply inputs by 3, then add 2.*

EXAMPLE 2 ▶ **Solving for y in a Linear Equation**

Solve the linear equation $3y - 2x = 6$ for y, then identify the new coefficient of x and the constant term.

Solution ▶
$$3y - 2x = 6 \qquad \text{given equation}$$
$$3y = 2x + 6 \qquad \text{add } 2x$$
$$y = \frac{2}{3}x + 2 \qquad \text{divide by 3}$$

The coefficient of x is $\frac{2}{3}$ and the constant term is 2.

Now try Exercises 13 through 18 ▶

WORTHY OF NOTE

In Example 2, the final form can be written $y = \frac{2}{3}x + 2$ as shown (inputs are multiplied by two-thirds, then increased by 2), or written as $y = \frac{2x}{3} + 2$ (inputs are multiplied by two, the result divided by 3 and this amount increased by 2). The two forms are equivalent.

When the coefficient of x is rational, it's helpful to select inputs that are multiples of the denominator if the context or application requires us to evaluate the equation. This enables us to perform most operations mentally. For $y = \frac{2}{3}x + 2$, possible inputs might be $x = -9, -6, 0, 3, 6$, and so on. **See Exercises 19 through 24.**

In Section 1.2, linear equations were graphed using the intercept method. When the equation is written with y in terms of x, we notice a powerful connection between the graph and its equation—one that highlights the primary characteristics of a linear graph.

EXAMPLE 3 ▶ **Noting Relationships between an Equation and Its Graph**

Find the intercepts of $4x + 5y = -20$ and use them to graph the line. Then,

  a. Use the intercepts to calculate the slope of the line, then identify the y-intercept.

  b. Write the equation with y in terms of x and compare the calculated slope and y-intercept to the equation in this form. Comment on what you notice.

Solution ▶ Substituting 0 for x in $4x + 5y = -20$, we find the y-intercept is $(0, -4)$. Substituting 0 for y gives an x-intercept of $(-5, 0)$. The graph is displayed here.

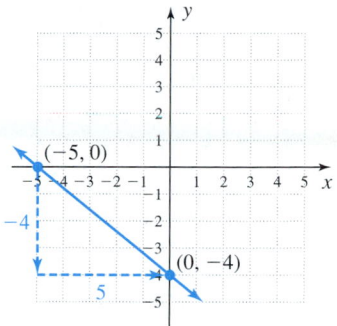

  a. The y-intercept is $(0, -4)$ and by calculation or counting $\dfrac{\Delta y}{\Delta x}$, the slope is $m = \frac{-4}{5}$ [from the intercept $(-5, 0)$ we count down 4, giving $\Delta y = -4$, and right 5, giving $\Delta x = 5$, to arrive at the intercept $(0, -4)$].

  b. Solving for y:

$$4x + 5y = -20 \qquad \text{given equation}$$
$$5y = -4x - 20 \qquad \text{subtract } 4x$$
$$y = \frac{-4}{5}x - 4 \qquad \text{divide by 5}$$

The slope value seems to be the coefficient of x, while the y-intercept is the constant term.

Now try Exercises 25 through 30 ▶

After solving a linear equation for y, an input of $x = 0$ causes the "x-term" to become zero, so the y-intercept automatically involves the constant term. As Example 3 illustrates, we can also identify the slope of the line—it is the coefficient of x. In general, a linear equation of the form $y = mx + b$ is said to be in **slope-intercept form,** since the slope of the line is m and the y-intercept is $(0, b)$.

Slope-Intercept Form

For a nonvertical line whose equation is $y = \boldsymbol{mx + b}$,
the slope of the line is $\boldsymbol{m}$ and the y-intercept is $(0, \boldsymbol{b})$.

Solving a linear equation for y in terms of x is sometimes called writing the equation in **function form,** as this form clearly highlights what operations are performed on the input value in order to obtain the output (see Example 1). In other words, this form plainly shows that "y depends on x," or "y is a function of x," and that the equations $y = mx + b$ and $f(x) = mx + b$ are equivalent.

Linear Functions

A linear function is one of the form

$$f(x) = mx + b,$$

where m and b are real numbers.

Note that if $m = 0$, the result is a **constant function** $f(x) = b$. If $m = 1$ and $b = 0$, the result is $f(x) = x$, called the **identity function.**

EXAMPLE 4 ▶ **Finding the Function Form of a Linear Equation**

Write each equation in both slope-intercept form and function form. Then identify the slope and y-intercept of the line.

    a. $3x - 2y = 9$ **b.** $y + x = 5$ **c.** $2y = x$

Solution ▶

a. $3x - 2y = 9$ **b.** $y + x = 5$ **c.** $2y = x$

$$-2y = -3x + 9 \qquad\qquad y = -x + 5 \qquad\qquad y = \frac{x}{2}$$

$$y = \frac{3}{2}x - \frac{9}{2} \qquad\qquad y = -1x + 5 \qquad\qquad y = \frac{1}{2}x$$

$$f(x) = \frac{3}{2}x - \frac{9}{2} \qquad\qquad f(x) = -1x + 5 \qquad\qquad f(x) = \frac{1}{2}x$$

$$m = \frac{3}{2}, b = -\frac{9}{2} \qquad\qquad m = -1, b = 5 \qquad\qquad m = \frac{1}{2}, b = 0$$

$$y\text{-intercept}\left(0, -\frac{9}{2}\right) \qquad y\text{-intercept }(0, 5) \qquad y\text{-intercept }(0, 0)$$

☑ **A.** You've just seen how we can write a linear equation in slope-intercept form and function form

Now try Exercises 31 through 38 ▶

Note that we can analytically develop the slope-intercept form of a line using the slope formula. Figure 1.66 shows the graph of a general line through the point (x, y) with a y-intercept of $(0, b)$. Using these points in the slope formula, we have

Figure 1.66

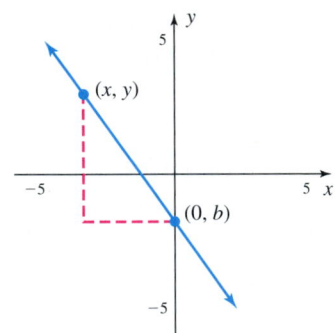

$$\frac{y_2 - y_1}{x_2 - x_1} = m \qquad \text{slope formula}$$

$$\frac{y - b}{x - 0} = m \qquad \text{substitute: } (0, b) \text{ for } (x_1, y_1), (x, y) \text{ for } (x_2, y_2)$$

$$\frac{y - b}{x} = m \qquad \text{simplify}$$

$$y - b = mx \qquad \text{multiply by } x$$

$$y = mx + b \qquad \text{add } b \text{ to both sides}$$

This approach confirms the relationship between the graphical characteristics of a line and its slope-intercept form. Specifically, for any linear equation written in the form $y = mx + b$, the slope must be m and the y-intercept is $(0, b)$.

B. Slope-Intercept Form and the Graph of a Line

If the slope and y-intercept of a linear equation are known or can be found, we can construct its equation by substituting these values directly into the slope-intercept form $y = mx + b$.

EXAMPLE 5 ▶ **Finding the Equation of a Line from Its Graph**

Find the slope-intercept equation of the line shown.

Solution ▶ Using $(-3, -2)$ and $(-1, 2)$ in the slope formula, or by simply counting $\dfrac{\Delta y}{\Delta x}$, the slope is $m = \frac{4}{2}$ or $\frac{2}{1}$.
By inspection we see the y-intercept is $(0, 4)$.
Substituting $\frac{2}{1}$ for m and 4 for b in the slope-intercept form we obtain the equation $y = 2x + 4$.

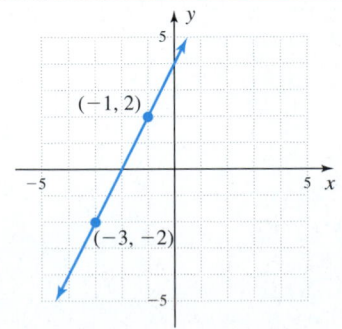

Now try Exercises 39 through 44 ▶

Actually, if the slope is known and we have *any* point (x, y) on the line, we can still construct the equation since the given point *must satisfy the equation of the line*. In this case, we're treating $y = mx + b$ as a simple formula, solving for b after substituting known values for m, x, and y.

EXAMPLE 6 ▶ **Using $y = mx + b$ as a Formula**

Find the slope-intercept equation of a line that has slope $m = \frac{4}{5}$ and contains $(-5, 2)$. Verify results on a graphing calculator.

Solution ▶ Use $y = mx + b$ as a "formula," with $m = \frac{4}{5}$, $x = -5$, and $y = 2$.

$y = mx + b$ slope-intercept form

$2 = \frac{4}{5}(-5) + b$ substitute $\frac{4}{5}$ for m, -5 for x, and 2 for y

$2 = -4 + b$ simplify

$6 = b$ solve for b

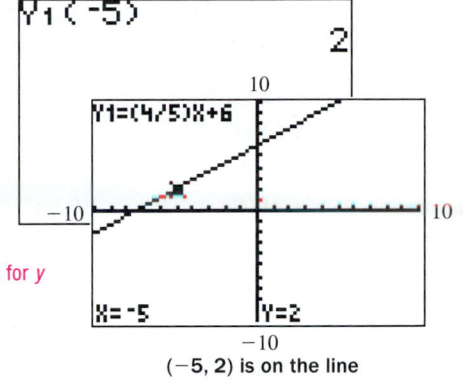

$(-5, 2)$ is on the line

The equation of the line is $y = \frac{4}{5}x + 6$. After entering the equation on the ⟨Y=⟩ screen of a graphing calculator, we can evaluate $x = -5$ on the home screen, or use the ⟨TRACE⟩ feature. See the figures provided.

Now try Exercises 45 through 50 ▶

Writing a linear equation in slope-intercept form enables us to draw its graph with a minimum of effort, since we can easily locate the y-intercept and a second point using the rate of change $\frac{\Delta y}{\Delta x}$. For instance, $\frac{\Delta y}{\Delta x} = \frac{-2}{3}$ indicates that counting down 2 and right 3 from a known point will locate another point on this line.

EXAMPLE 7 ▶ **Graphing a Line Using Slope-Intercept Form and the Rate of Change**

Write $3y - 5x = 9$ in slope-intercept form, then graph the line using the y-intercept and the rate of change (slope).

Solution ▶ $3y - 5x = 9$ given equation

$ 3y = 5x + 9$ isolate y term

$ y = \frac{5}{3}x + 3$ divide by 3

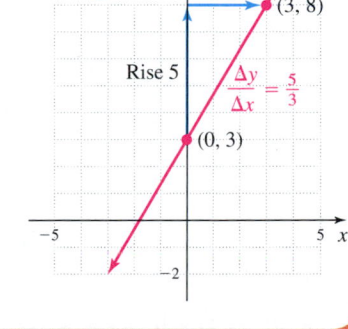

The slope is $m = \frac{5}{3}$ and the y-intercept is $(0, 3)$.

Plot the y-intercept, then use $\frac{\Delta y}{\Delta x} = \frac{5}{3}$ (up 5 and right 3—shown in blue) to find another point on the line (shown in red). Finish by drawing a line through these points.

WORTHY OF NOTE

Noting the fraction $\frac{5}{3}$ is equal to $\frac{-5}{-3}$, we could also begin at $(0, 3)$ and count $\frac{\Delta y}{\Delta x} = \frac{-5}{-3}$ (down 5 and left 3) to find an additional point on the line: $(-3, -2)$. Also, for any negative slope $\frac{\Delta y}{\Delta x} = -\frac{a}{b}$, note $\frac{a}{b} = \frac{-a}{b} = \frac{a}{-b}$.

Now try Exercises 51 through 62 ▶

For a discussion of what graphing method might be most efficient for a given linear equation, **see Exercises 103 and 114.**

Parallel and Perpendicular Lines

From Section 1.2 we know parallel lines have equal slopes: $m_1 = m_2$, and perpendicular lines have slopes with a product of -1: $m_1 \cdot m_2 = -1$ or $m_1 = -\dfrac{1}{m_2}$. In some applications, we need to find the equation of a second line parallel or perpendicular to a given line, through a given point. Using the slope-intercept form makes this a simple four-step process.

Finding the Equation of a Line Parallel or Perpendicular to a Given Line

1. Identify the slope m_1 of the given line.
2. Find the slope m_2 of the new line using the parallel or perpendicular relationship.
3. Use m_2 with the point (x, y) in the "formula" $y = mx + b$ and solve for b.
4. The desired equation will be $y = m_2x + b$.

EXAMPLE 8 ▶ **Finding the Equation of a Parallel Line**

Find the slope-intercept equation of a line that goes through $(-6, -1)$ and is parallel to $2x + 3y = 6$.

Solution ▶ Begin by writing the equation in slope-intercept form to identify the slope.

$$2x + 3y = 6 \qquad \text{given line}$$
$$3y = -2x + 6 \qquad \text{isolate } y\text{-term}$$
$$y = \tfrac{-2}{3}x + 2 \qquad \text{result}$$

The original line has slope $m_1 = \tfrac{-2}{3}$ and this will also be the slope of any line parallel to it. Using $m_2 = \tfrac{-2}{3}$ with $(x, y) \to (-6, -1)$ we have

$$y = mx + b \qquad \text{slope-intercept form}$$
$$-1 = \frac{-2}{3}(-6) + b \qquad \text{substitute } \tfrac{-2}{3} \text{ for } m, -6 \text{ for } x, \text{ and } -1 \text{ for } y$$
$$-1 = 4 + b \qquad \text{simplify}$$
$$-5 = b \qquad \text{solve for } b$$

The equation of the new line is $y = \tfrac{-2}{3}x - 5$.

Now try Exercises 63 through 76 ▶

Graphing the lines from Example 8 as Y_1 and Y_2 on a graphing calculator, we note the lines do appear to be parallel (they actually *must* be since they have identical slopes). Using the ZOOM **8:ZInteger** feature of the calculator, we can quickly verify that Y_2 indeed contains the point $(-6, -1)$.

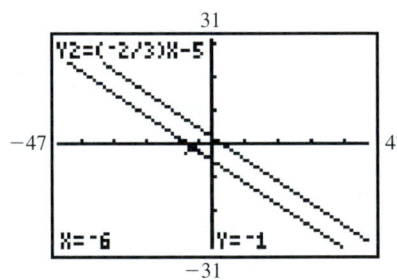

For any nonlinear graph, a straight line drawn through two points on the graph is called a **secant line.** The slope of a secant line, and lines parallel and perpendicular to this line, play fundamental roles in the further development of the rate-of-change concept.

EXAMPLE 9 ▶ **Finding Equations for Parallel and Perpendicular Lines**

A secant line is drawn using the points $(-4, 0)$ and $(2, -2)$ on the graph of the function shown. Find the equation of a line that is

a. parallel to the secant line through $(-1, -4)$.
b. perpendicular to the secant line through $(-1, -4)$.

Solution ▶ Either by using the slope formula or counting $\dfrac{\Delta y}{\Delta x}$, we find the secant line has slope

$$m = \frac{-2}{6} = \frac{-1}{3}.$$

a. For the parallel line through $(-1, -4)$, $m_2 = \dfrac{-1}{3}$.

$$y = mx + b \qquad \text{slope-intercept form}$$

$$-4 = \frac{-1}{3}(-1) + b \qquad \begin{array}{l}\text{substitute } \frac{-1}{3} \text{ for } m,\\ -1 \text{ for } x, \text{ and } -4 \text{ for } y\end{array}$$

$$-\frac{12}{3} = \frac{1}{3} + b \qquad \text{simplify } \left(-4 = -\frac{12}{3}\right)$$

$$-\frac{13}{3} = b \qquad \text{result}$$

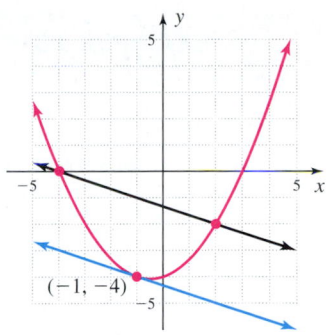

The equation of the parallel line (in blue) is $y = \dfrac{-1}{3}x - \dfrac{13}{3}$.

b. For the perpendicular line through $(-1, -4)$, $m_2 = 3$.

$$y = mx + b \qquad \text{slope-intercept form}$$

$$-4 = 3(-1) + b \qquad \text{substitute 3 for } m, -1 \text{ for } x, \text{ and } -4 \text{ for } y$$

$$-4 = -3 + b \qquad \text{simplify}$$

$$-1 = b \qquad \text{result}$$

The equation of the perpendicular line (in yellow) is $y = 3x - 1$.

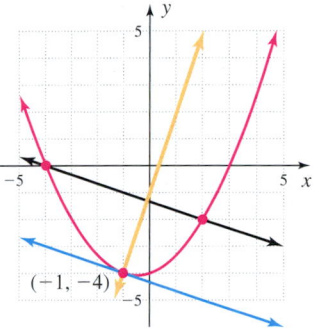

☑ **B.** You've just seen how we can use the slope-intercept form to graph linear equations

Now try Exercises 77 through 82 ▶

C. Linear Equations in Point-Slope Form

As an alternative to using $y = mx + b$, we can find the equation of the line using the slope formula $\dfrac{y_2 - y_1}{x_2 - x_1} = m$, and the fact that *the slope of a line is constant.* For a given slope m, we can let (x_1, y_1) represent a *given* point on the line and (x, y) represent *any other point* on the line, and the formula becomes $\dfrac{y - y_1}{x - x_1} = m$. Isolating the "$y$" terms on one side gives a new form for the equation of a line, called the **point-slope form:**

$$\frac{y - y_1}{x - x_1} = m \qquad \text{slope formula}$$

$$\frac{(\cancel{x - x_1})}{1}\left(\frac{y - y_1}{\cancel{x - x_1}}\right) = m(x - x_1) \qquad \text{multiply both sides by } (x - x_1)$$

$$y - y_1 = m(x - x_1) \qquad \text{simplify} \rightarrow \text{point-slope form}$$

The Point-Slope Form of a Linear Equation

For a nonvertical line whose equation is $y - y_1 = m(x - x_1)$, the slope of the line is m and (x_1, y_1) is a point on the line.

While using $y = mx + b$ (as in Example 6) may appear to be easier, both the slope-intercept form and point-slope form have their own advantages and it will help to be familiar with both.

EXAMPLE 10 ▶ **Using $y - y_1 = m(x - x_1)$ as a Formula**

Find the equation of the line in point-slope form, if $m = \frac{2}{3}$ and $(-3, -3)$ is on the line. Then graph the line.

Solution ▶

$y - y_1 = m(x - x_1)$ point-slope form

$y - (-3) = \dfrac{2}{3}[x - (-3)]$ substitute $\frac{2}{3}$ for m; $(-3, -3)$ for (x_1, y_1)

$y + 3 = \dfrac{2}{3}(x + 3)$ simplify, point-slope form

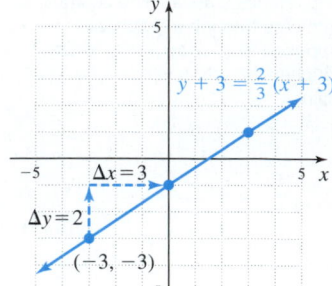

To graph the line, plot $(-3, -3)$ and use $\dfrac{\Delta y}{\Delta x} = \dfrac{2}{3}$ to find additional points on the line.

☑ **C.** You've just seen how we can write a linear equation in point-slope form

Now try Exercises 83 through 94 ▶

D. Applications of Linear Equations

As a mathematical tool, linear equations rank among the most common, powerful, and versatile. In all cases, it's important to remember that slope represents a *rate of change*. The notation $m = \dfrac{\Delta y}{\Delta x}$ literally means the quantity measured along the y-axis, is changing with respect to changes in the quantity measured along the x-axis.

EXAMPLE 11 ▶ **Relating Temperature to Altitude**

In meteorological studies, atmospheric temperature depends on the altitude according to the formula $T(h) = -3.5h + 58.5$, where $T(h)$ represents the approximate Fahrenheit temperature at height h (in thousands of feet, $0 \le h \le 36$).

  a. Interpret the meaning of the slope in this context.
  b. Determine the temperature at an altitude of 12,000 ft.
  c. If the temperature is $-8°F$ what is the approximate altitude?

Algebraic Solution ▶

  a. Notice that h is the input variable and T is the output. This shows $\dfrac{\Delta T}{\Delta h} = \dfrac{-3.5}{1}$, meaning the temperature drops $3.5°F$ for every 1000-ft increase in altitude.

  b. Since height is in thousands, use $h = 12$.

$T(h) = -3.5h + 58.5$ original formula
$T(12) = -3.5(12) + 58.5$ substitute 12 for h
$= 16.5$ result

Technology Solution ▶

```
Y1(12)
              16.5
```

At a height of 12,000 ft, the temperature is about $16.5°$.

c. Replacing $T(h)$ with -8 and solving gives

Algebraic Solution ▶

$$T(h) = -3.5h + 58.5 \quad \text{original formula}$$
$$-8 = -3.5h + 58.5 \quad \text{substitute } -8 \text{ for } T(h)$$
$$-66.5 = -3.5h \quad \text{subtract 58.5}$$
$$19 = h \quad \text{divide by } -3.5$$

The temperature is about $-8°$F at a height of $19 \times 1000 = 19{,}000$ ft.

Graphical Solution ▶

Since we're given $0 \le h \le 36$, we can set Xmin $= 0$ and Xmax $= 40$. At ground level ($x = 0$), the formula gives a temperature of $58.5°$, while at $h = 36$, we have $T(36) = -67.5$. This shows appropriate settings for the range would be Ymin $= -50$ and Ymax $= 50$ (see figure). After setting $Y_1 = -3.5X + 58.5$, we press ⟨TRACE⟩ and move the cursor until we find an output value near -8, which occurs when X is near 19. To check, we input 19 for x and the calculator displays an output of -8, which corresponds with the algebraic result (at 19,000 ft, the temperature is $-8°$F).

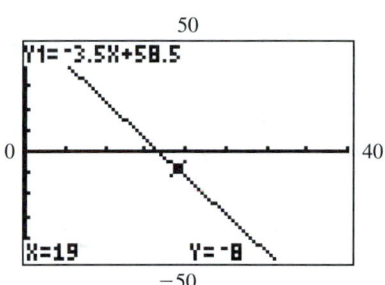

Now try Exercises 105 and 106 ▶

In many applications, outputs that are integer or rational values are rare, making it difficult to use the ⟨TRACE⟩ feature alone to find an exact solution. In the Section 1.5, we'll develop additional ways that graphs and technology can be used to solve equations.

In some applications, the relationship is known to be linear but only a few points on the line are given. In this case, we can use two of the known data points to calculate the slope, then the point-slope form to find an equation model. One such application is *linear depreciation,* as when a government allows businesses to depreciate vehicles and equipment over time (the less a piece of equipment is worth, the less you pay in taxes).

EXAMPLE 12A ▶ **Using Point-Slope Form to Find a Function Model**

Five years after purchase, the auditor of a newspaper company estimates the value of their printing press is $60,000. Eight years after its purchase, the value of the press had depreciated to $42,000. Find a linear equation that models this depreciation and discuss the slope and y-intercept in context.

Solution ▶

Since the value of the press depends on time, the ordered pairs have the form (time, value) or (t, v) where *time* is the input, and *value* is the output. This means the ordered pairs are $(5, 60{,}000)$ and $(8, 42{,}000)$.

$$m = \frac{v_2 - v_1}{t_2 - t_1} \quad \text{slope formula}$$
$$= \frac{42{,}000 - 60{,}000}{8 - 5} \quad (t_1, v_1) = (5, 60{,}000); (t_2, v_2) = (8, 42{,}000)$$
$$= \frac{-18{,}000}{3} = \frac{-6000}{1} \quad \text{simplify and reduce}$$

The slope of the line is $\dfrac{\Delta\text{value}}{\Delta\text{time}} = \dfrac{-6000}{1}$, indicating the printing press loses $6000 in value with each passing year.

$$
\begin{aligned}
v - v_1 &= m(t - t_1) && \text{point-slope form} \\
v - 60{,}000 &= -6000(t - 5) && \text{substitute } -6000 \text{ for } m; \text{ (5, 60,000) for } (t_1, v_1) \\
v - 60{,}000 &= -6000t + 30{,}000 && \text{simplify} \\
v &= -6000t + 90{,}000 && \text{solve for } v
\end{aligned}
$$

The depreciation equation is $v(t) = -6000t + 90{,}000$. The v-intercept (0, 90,000) indicates the original value (cost) of the equipment was $90,000.

Once the depreciation equation is found, it represents the (time, value) relationship for all future (and intermediate) ages of the press. In other words, we can now predict the value of the press for any given year. However, note that some equation models are valid for only a set period of time, and each model should be used with care.

EXAMPLE 12B ▶ **Using a Function Model to Gather Information**

From Example 12A,

  a. How much will the press be worth after 11 yr?
  b. How many years until the value of the equipment is $9000?
  c. Is this function model valid for $t = 18$ yr (why or why not)?

Solution ▶ **a.** Find the value v when $t = 11$:

$$
\begin{aligned}
v(t) &= -6000t + 90{,}000 && \text{equation model} \\
v(11) &= -6000(11) + 90{,}000 && \text{substitute 11 for } t \\
&= 24{,}000 && \text{result (11, 24,000)}
\end{aligned}
$$

After 11 yr, the printing press will only be worth $24,000.

  b. "... value is $9000" means $v(t) = 9000$:

$$
\begin{aligned}
v(t) &= 9000 && \text{value at time } t \\
-6000t + 90{,}000 &= 9000 && \text{substitute } -6000t + 90{,}000 \text{ for } v(t) \\
-6000t &= -81{,}000 && \text{subtract 90,000} \\
t &= 13.5 && \text{divide by } -6000
\end{aligned}
$$

After 13.5 yr, the printing press will be worth $9000.

  c. Since substituting 18 for t gives a negative quantity, the function model is not valid for $t = 18$. In the current context, the model is only valid while $v \geq 0$ and solving $-6000t + 90{,}000 \geq 0$ shows the domain of the function in this context is $t \in [0, 15]$.

☑ **D. You've just seen how we can apply the slope-intercept form and point-slope form in context**

Now try Exercises 107 through 112 ▶

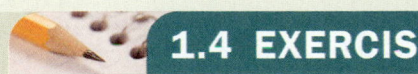

1.4 EXERCISES

▶ CONCEPTS AND VOCABULARY

Fill in each blank with the appropriate word or phrase. Carefully reread the section if needed.

1. For the equation $y = \dfrac{-7}{4}x + 3$, the slope is ___$\frac{-7}{4}$___ and the y-intercept is ___(0, 3)___.

2. The notation $\dfrac{\Delta \text{cost}}{\Delta \text{time}}$ indicates the ___cost___ is changing in response to changes in ___time___.

3. Line 1 has a slope of -0.4. The slope of any line perpendicular to line 1 is ___2.5___.

4. The equation $y - y_1 = m(x - x_1)$ is called the ___point-slope___ form of a line.

5. Discuss/Explain how to graph a line using only the slope and a point on the line (no equations).
Answers will vary.

6. Given $m = -\frac{3}{5}$ and $(-5, 6)$ is on the line. Compare and contrast finding the equation of the line using $y = mx + b$ versus $y - y_1 = m(x - x_1)$.
Answers will vary.

▶ DEVELOPING YOUR SKILLS

Solve each equation for y and evaluate the result using $x = -5$, $x = -2$, $x = 0$, $x = 1$, and $x = 3$.

7. $4x + 5y = 10$
8. $3y - 2x = 9$
9. $-0.4x + 0.2y = 1.4$
10. $-0.2x + 0.7y = -2.1$
11. $\frac{1}{3}x + \frac{1}{5}y = -1$
12. $\frac{1}{7}y - \frac{1}{3}x = 2$

For each equation, solve for y and identify the new coefficient of x and new constant term.

13. $6x - 3y = 9$　　$y = 2x - 3: 2, -3$
14. $9y - 4x = 18$　　$y = \frac{4}{9}x + 2: \frac{4}{9}, 2$
15. $-0.5x - 0.3y = 2.1$　　$y = \frac{-5}{3}x - 7: \frac{-5}{3}, -7$
16. $-0.7x + 0.6y = -2.4$　　$y = \frac{7}{6}x - 4: \frac{7}{6}, -4$
17. $\frac{5}{6}x + \frac{1}{7}y = -\frac{4}{7}$　　$y = \frac{-35}{6}x - 4: \frac{-35}{6}, -4$
18. $\frac{7}{12}y - \frac{4}{15}x = \frac{7}{6}$　　$y = \frac{16}{35}x + 2: \frac{16}{35}, 2$

Evaluate each equation by selecting three inputs that will result in integer values. Then graph each line.

19. $y = -\frac{4}{3}x + 5$
20. $y = \frac{5}{4}x + 1$
21. $y = -\frac{3}{2}x - 2$
22. $y = \frac{2}{5}x - 3$
23. $y = -\frac{1}{6}x + 4$
24. $y = -\frac{1}{3}x + 3$

Find the x- and y-intercepts for each line, then (a) use these two points to calculate the slope of the line, (b) write the equation with y in terms of x (solve for y) and (c) compare the calculated slope and y-intercept to the equation from part (b). Comment on what you notice.

25. $3x + 4y = 12$
26. $3y - 2x = -6$
27. $2x - 5y = 10$
28. $2x + 3y = 9$
29. $4x - 5y = -15$
30. $5y + 6x = -25$

Write each equation in slope-intercept form (solve for y) and function form, then identify the slope and y-intercept.

31. $2x + 3y = 6$
32. $4y - 3x = 12$
33. $5x + 4y = 20$
34. $y + 2x = 4$
35. $x = 3y$
36. $2x = -5y$
37. $3x + 4y - 12 = 0$
38. $5y - 3x + 20 = 0$

For Exercises 39 to 50, use the slope-intercept form to state the equation of each line. Verify your solutions to Exercises 45 to 47 using a graphing calculator.

39. $y = \frac{2}{3}x + 1$

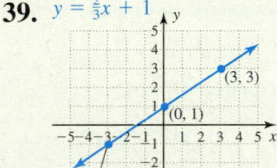

40. $y = \frac{-2}{5}x + 3$

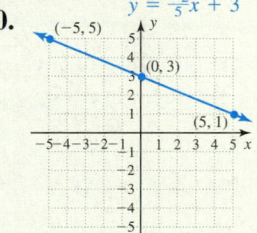

41. $y = 3x + 3$

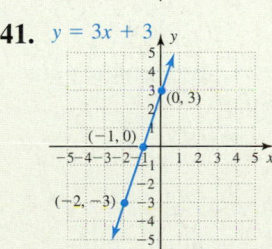

42. $m = -2$; y-intercept $(0, -3)$　　$y = -2x - 3$

43. $m = 3$; y-intercept $(0, 2)$　　$y = 3x + 2$

44. $m = \frac{-3}{2}$; y-intercept $(0, -4)$　　$y = \frac{-3}{2}x - 4$

45. $m = -4$; $(-3, 2)$ is on the line　　$y = -4x - 10$

Additional answers can be found in the Instructor Answer Appendix.

HOMEWORK SELECTION GUIDE

Core: 7–99 every other odd, 105–111 odd (28 Exercises)
Standard: 1–4, 7–99 every other odd, 105–111 odd, 115 (33 Exercises)

Extended: 1–4, 7–99 every other odd, 103, 105–111 odd, 114, 115–119 odd (37 Exercises)
In Depth: 1–6, 7–99 every other odd, 103–112 all, 114–119 all (46 Exercises)

46. $m = 2$; $(5, -3)$ is on the line $y = 2x - 13$

47. $m = \frac{-3}{2}$; $(-4, 7)$ is on the line $y = \frac{-3}{2}x + 1$

48. 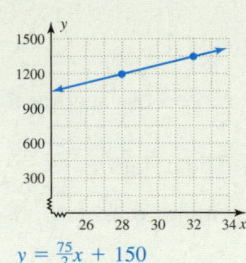   **49.**

$y = 250x + 500$ $y = \frac{75}{2}x + 150$

50.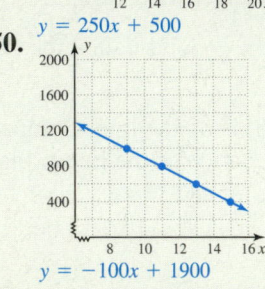

$y = -100x + 1900$

Write each equation in slope-intercept form, then use the rate of change (slope) and y-intercept to graph the line.

51. $3x + 5y = 20$ **52.** $2y - x = 4$

53. $2x - 3y = 15$ **54.** $-3x + 2y = 4$

Graph each linear equation using the y-intercept and rate of change (slope) determined from each equation.

55. $y = \frac{2}{3}x + 3$ **56.** $y = \frac{5}{2}x - 1$

57. $y = \frac{-1}{3}x + 2$ **58.** $y = \frac{-4}{5}x + 2$

59. $y = 2x - 5$ **60.** $y = -3x + 4$

61. $y = \frac{1}{2}x - 3$ **62.** $y = \frac{-3}{2}x + 2$

Find the equation of the line using the information given. Write answers in slope-intercept form.

63. parallel to $2x - 5y = 10$, through the point $(-5, 2)$ $y = \frac{2}{5}x + 4$

64. parallel to $6x + 9y = 27$, through the point $(-3, -5)$ $y = \frac{-2}{3}x - 7$

65. perpendicular to $5y - 3x = 9$, through the point $(6, -3)$ $y = \frac{-5}{3}x + 7$

66. perpendicular to $x - 4y = 7$, through the point $(-5, 3)$ $y = -4x - 17$

67. parallel to $12x + 5y = 65$, through the point $(-2, -1)$ $y = \frac{-12}{5}x - \frac{29}{5}$

68. parallel to $15y - 8x = 50$, through the point $(3, -4)$ $y = \frac{8}{15}x - \frac{28}{5}$

69. parallel to $y = -3$, through the point $(2, 5)$ $y = 5$

70. perpendicular to $y = -3$ through the point $(2, 5)$

$x = 2$

Additional answers can be found in the Instructor Answer Appendix.

Write the equations in slope-intercept form and state whether the lines are parallel, perpendicular, or neither.

71. $4y - 5x = 8$ **72.** $3y - 2x = 6$
  $5y + 4x = -15$ $2x + 3y = 6$
  perpendicular   parallel

73. $2x - 5y = 20$ **74.** $-4x + 6y = 12$
  $4x - 3y = 18$ $2x + 3y = 6$ neither
        neither

75. $3x + 4y = 12$ **76.** $5y = 11x + 135$
  $6x + 8y = 2$ parallel $11y + 5x = -77$
                                perpendicular

A *secant line* is one that intersects a graph at two or more points. For each graph given, find an equation of the line (a) parallel and (b) perpendicular to the secant line, through the point indicated.

77. 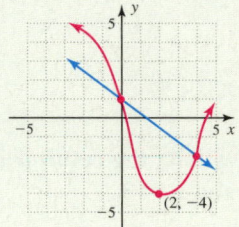   **78.**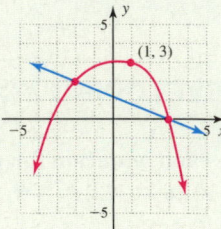

79. 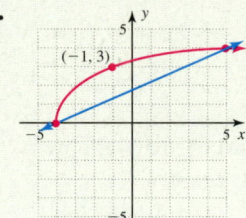   **80.**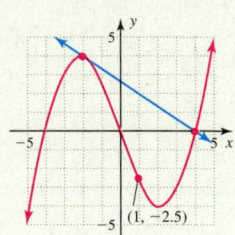

81. 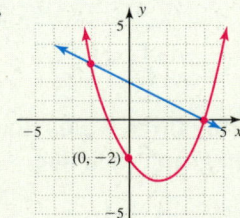   **82.**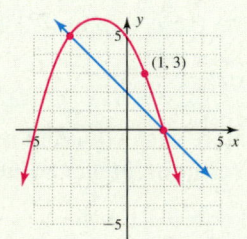

Find the equation of the line in point-slope form, then graph the line.

83. $m = 2$; $P_1 = (2, -5)$

84. $m = -1$; $P_1 = (2, -3)$

85. $P_1 = (3, -4)$, $P_2 = (11, -1)$

86. $P_1 = (-1, 6)$, $P_2 = (5, 1)$

87. $m = 0.5$; $P_1 = (1.8, -3.1)$

88. $m = 1.5$; $P_1 = (-0.75, -0.125)$

Find the equation of the line in point-slope form, and state the meaning of the slope in context—what information is the slope giving us?

89.

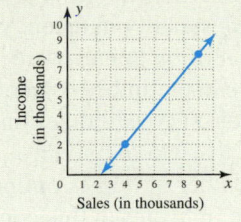

90.

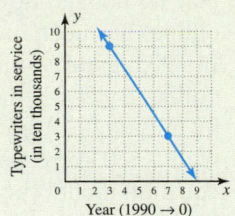

91.

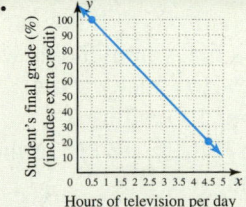

92.

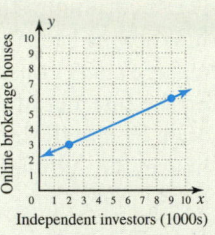

93.

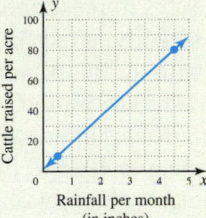

94.

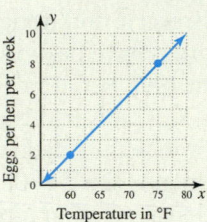

Using the concept of slope, match each description with the graph that best illustrates it. Assume time is scaled on the horizontal axes, and height, speed, or distance from the origin (as the case may be) is scaled on the vertical axis.

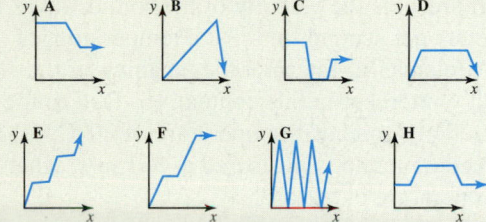

95. While driving today, I got stopped by a state trooper. After she warned me to slow down, I continued on my way.   C

96. After hitting the ball, I began trotting around the bases shouting, "Ooh, ooh, ooh!" When I saw it wasn't a home run, I began sprinting.   F

97. At first I ran at a steady pace, then I got tired and walked the rest of the way.   A

98. While on my daily walk, I had to run for a while when I was chased by a stray dog.   H

99. I climbed up a tree, then I jumped out.   B

100. I steadily swam laps at the pool yesterday.   G

101. I walked toward the candy machine, stared at it for a while then changed my mind and walked back. D

102. For practice, the girls' track team did a series of 25-m sprints, with a brief rest in between.   E

▶ WORKING WITH FORMULAS

103. General linear equation: $ax + by = c$

The general equation of a line is shown here, where a, b, and c are real numbers, with a and b not simultaneously zero. Solve the equation for y and note the slope (coefficient of x) and y-intercept (constant term). Use these to find the slope and y-intercept of the following lines, without solving for y or computing points.

  a. $3x + 4y = 8$ **b.** $2x + 5y = -15$
  c. $5x - 6y = -12$ **d.** $3y - 5x = 9$

104. Intercept-Intercept form of a linear equation: $\dfrac{x}{h} + \dfrac{y}{k} = 1$

The x- and y-intercepts of a line can also be found by writing the equation in the form shown (with the equation set equal to 1). The x-intercept will be $(h, 0)$ and the y-intercept will be $(0, k)$. Find the x- and y-intercepts of the following lines using this method. How is the slope of each line related to the values of h and k?

  a. $2x + 5y = 10$ **b.** $3x - 4y = -12$
  c. $5x + 4y = 8$

▶ APPLICATIONS

105. Speed of sound: The speed of sound as it travels through the air depends on the temperature of the air according to the function $V = \frac{3}{5}T + 331$, where V represents the velocity of the sound waves in meters per second (m/s), at a temperature of $T°$ Celsius. (a) Interpret the meaning of the slope and y-intercept in this context. (b) Determine the speed of sound at a temperature of 20°C. (c) If the speed of sound is measured at 361 m/s, what is the temperature of the air?

106. Acceleration: A driver going down a straight highway is traveling 60 ft/sec (about 41 mph) on cruise control, when he begins accelerating at a rate of 5.2 ft/sec². The final velocity of the car is given by $V = \frac{26}{5}t + 60$, where V is the velocity at time t. (a) Interpret the meaning of the slope and y-intercept in this context. (b) Determine the velocity of the car after 9.4 seconds. (c) If the car is traveling at 100 ft/sec, for how long did it accelerate?

107. Investing in coins: The purchase of a "collector's item" is often made in hopes the item will increase in value. In 1998, Mark purchased a 1909-S VDB Lincoln Cent (in fair condition) for $150. By the year 2004, its value had grown to $190. (a) Use the relation (time since purchase, value) with $t = 0$ corresponding to 1998 to find a linear equation modeling the value of the coin. (b) Discuss what the slope and y-intercept indicate in this context. (c) How much was the penny worth in 2009? (d) How many years after purchase will the penny's value exceed $250? (e) If the penny is now worth $170, how many years has Mark owned the penny?

108. Depreciation: Once a piece of equipment is put into service, its value begins to depreciate. A business purchases some computer equipment for $18,500. At the end of a 2-yr period, the value of the equipment has decreased to $11,500. (a) Use the relation (time since purchase, value) to find a linear equation modeling the value of the equipment. (b) Discuss what the slope and y-intercept indicate in this context. (c) What is the equipment's value after 4 yr? (d) How many years after purchase will the value decrease to $6000? (e) Generally, companies will sell used equipment while it still has value and use the funds to purchase new equipment. According to the function, how many years will it take this equipment to depreciate in value to $1000?

109. Internet connections: The number of households that are hooked up to the Internet (homes that are online) has been increasing steadily in recent years. In 1995, approximately 9 million homes were online. By 2001 this figure had climbed to about 51 million. (a) Use the relation (year, homes online) with $t = 0$ corresponding to 1995 to find an equation model for the number of homes online. (b) Discuss what the slope indicates in this context. (c) According to this model, in what year did the first homes begin to come online? (d) If the rate of change stays constant, how many households were on the Internet in 2006? (e) How many years after 1995 will there be over 100 million households connected? (f) If there are 115 million households connected, what year is it?

Source: 2004 Statistical Abstract of the United States, Table 965

110. Prescription drugs: Retail sales of prescription drugs have been increasing steadily in recent years. In 1995, retail sales hit $72 billion. By the year 2000, sales had grown to about $146 billion. (a) Use the relation (year, retail sales of prescription drugs) with $t = 0$ corresponding to 1995 to find a linear equation modeling the growth of retail sales. (b) Discuss what the slope indicates in this context. (c) According to this model, in what year will sales reach $250 billion? (d) According to the model, what was the value of retail prescription drug sales in 2005? (e) How many years after 1995 will retail sales exceed $279 billion? (f) If yearly sales totaled $294 billion, what year is it?

Source: 2004 Statistical Abstract of the United States, Table 122

111. Prison population: In 1990, the number of persons sentenced and serving time in state and federal institutions was approximately 740,000. By the year 2000, this figure had grown to nearly 1,320,000. (a) Find a linear function with $t = 0$ corresponding to 1990 that models this data, (b) discuss the slope ratio in context, and (c) use the equation to estimate the prison population in 2010 if this trend continues.

Source: Bureau of Justice Statistics at www.ojp.usdoj.gov/bjs

112. Eating out: In 1990, Americans bought an average of 143 meals per year at restaurants. This phenomenon continued to grow in popularity and in the year 2000, the average reached 170 meals per year. (a) Find a linear function with $t = 0$ corresponding to 1990 that models this growth, (b) discuss the slope ratio in context, and (c) use the equation to estimate the average number of times an American will eat at a restaurant in 2010 if the trend continues.

Source: The NPD Group, Inc., National Eating Trends, 2002

▶ EXTENDING THE CONCEPT

113. Locate and read the following article. Then turn in a one-page summary. "Linear Function Saves Carpenter's Time," Richard Crouse, *Mathematics Teacher,* Volume 83, Number 5, May 1990: pp. 400–401.   Answers will vary.

114. The general form of a linear equation is $ax + by = c$, where a and b are not simultaneously zero. (a) Find the x- and y-intercepts using the general form (substitute 0 for x, then 0 for y). Based on what you see, when does the intercept method work most efficiently? (b) Find the slope and y-intercept using the general form (solve for y). Based on what you see, when does the slope-intercept method work most efficiently?   **a.** $(\frac{c}{a}, 0)$ and $(0, \frac{c}{b})$; when a and b are factors of c **b.** $m = -\frac{a}{b}$; y-int: $(0, \frac{c}{b})$; when b is a factor of c

115. Match the correct graph to the conditions stated for m and b. There are more choices than graphs.
   a. $m < 0, b < 0$ **b.** $m > 0, b < 0$
   c. $m < 0, b > 0$ **d.** $m > 0, b > 0$
   e. $m = 0, b > 0$ **f.** $m < 0, b = 0$
   g. $m > 0, b = 0$ **h.** $m = 0, b < 0$

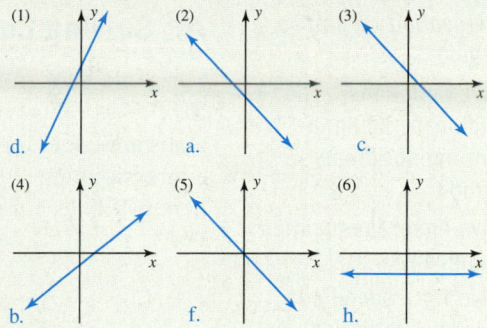

▶ MAINTAINING YOUR SKILLS

116. **(1.3)** Determine the domain:

   a. $y = \sqrt{2x - 5}$ $x \in [\frac{5}{2}, \infty)$

   b. $y = \dfrac{5}{2x - 5}$ $x \neq \frac{5}{2}$

117. **(R.6)** Simply without the use of a calculator.

   a. $27^{\frac{2}{3}}$ 9   **b.** $\sqrt{81x^2}$ $9|x|$

118. **(R.3)** Three equations follow. One is an identity, another is a contradiction, and a third has a solution. State which is which.

$2(x - 5) + 13 - 1 = 9 - 7 + 2x$ identity

$2(x - 4) + 13 - 1 = 9 + 7 - 2x$ has a solution ($x = 3$)

$2(x - 5) + 13 - 1 = 9 + 7 + 2x$ contradiction

119. **(R.2)** Compute the area of the circular sidewalk shown here $(A = \pi r^2)$. Use your calculator's value of π and round the answer (only) to hundredths.

113.10 yd²

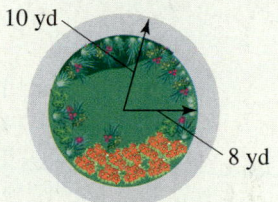

10 yd

8 yd

1.5 Solving Equations and Inequalities Graphically; Formulas and Problem Solving

In this section, we'll build on many of the ideas developed in Section R.3 (Solving Linear Equations and Inequalities), as we learn to manipulate formulas and employ certain problem-solving strategies. We will also extend our understanding of graphical solutions to a point where they can be applied to virtually any family of functions.

A. Solving Equations Graphically Using the *Intersect Method*

For some background on why a graphical solution is effective, consider the equation $2x - 9 = -3(x - 1) - 2$. By definition, an **equation** is a statement that two expressions are equal for some value of the variable (Section R.3). To highlight this fact, the expressions $2x - 9$ and $-3(x - 1) - 2$ are evaluated independently for selected integers in Tables 1.4 and 1.5.

Table 1.4

x	$2x - 9$
-3	-15
-2	-13
-1	-11
0	-9
1	-7
2	-5
3	-3

Table 1.5

x	$-3(x - 1) - 2$
-3	10
-2	7
-1	4
0	1
1	-2
2	-5
3	-8

Note the two expressions are equal (the equation is true) only when the input is $x = 2$. Solving equations graphically is a simple extension of this observation. By treating the expression on the left as the independent function Y_1, we have $Y_1 = 2X - 9$ and the related linear graph will contain *all ordered pairs shown* in Table 1.4 (see Figure 1.67). Doing the same for the right-hand expression yields $Y_2 = -3(X - 1) - 2$, and its related graph will likewise contain all ordered pairs shown in the Table 1.5 (see Figure 1.68).

$$\underbrace{2x - 9}_{Y_1} = \underbrace{-3(x - 1) - 2}_{Y_2}$$

The solution is then found where $Y_1 = Y_2$, or in other words, at the point where these two lines intersect (if it exists). See Figure 1.69.

Most graphing calculators have an **intersect** feature that can quickly find the point(s) where two graphs intersect. On many calculators, we access this ability using the sequence **2nd** **TRACE** (**CALC**) and selecting option **5:intersect** (Figure 1.70).

Figure 1.67

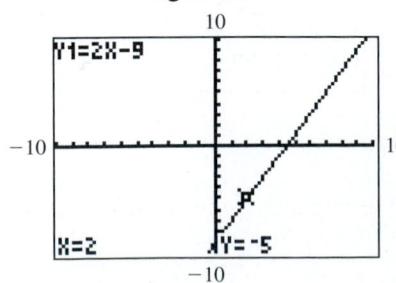

Figure 1.68

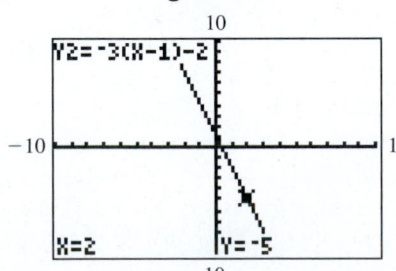

Figure 1.69

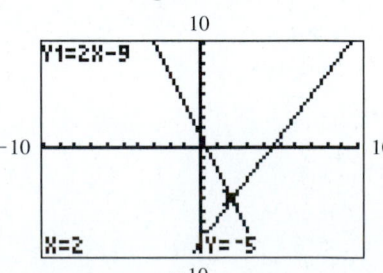

Figure 1.70

Figure 1.71

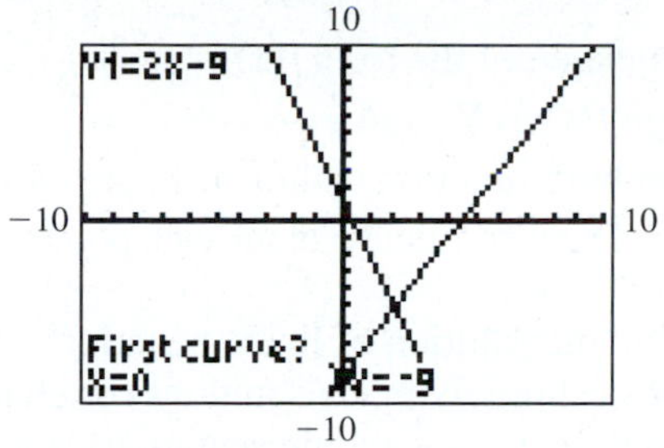

Figure 1.72

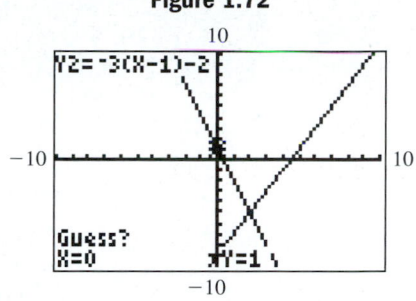

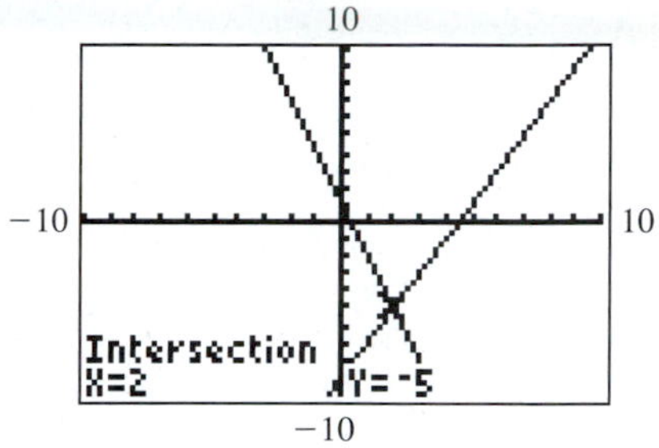

Because the calculator can work with up to 10 expressions at once, it will ask you to identify each graph you want to work with—even when there are only two. A marker is displayed on each graph in turn, and named in the upper left corner of the window (Figure 1.71). You can select a graph by pressing [ENTER], or bypass a graph by pressing one of the arrow keys. For situations involving multiple graphs or multiple solutions, the calculator offers a "**GUESS?**" option that enables you to specify the approximate location of the solution you're interested in (Figure 1.72). For now, we'll simply press [ENTER] two times in succession to identify each graph, and a third time to bypass the "**GUESS?**" option. The calculator then finds and displays the point of intersection (Figure 1.73). Be sure to check the settings on your viewing window before you begin, and if the point of intersection is not visible, try [ZOOM] **3:Zoom Out** or other window-resizing features to help locate it.

Figure 1.73

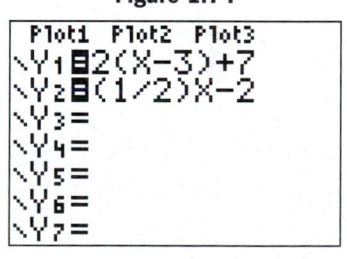

EXAMPLE 1A ▶ Solving an Equation Graphically

Solve the equation $2(x - 3) + 7 = \frac{1}{2}x - 2$ using a graphing calculator.

Solution ▶ Begin by entering the left-hand expression as Y_1 and the right-hand expression as Y_2 (Figure 1.74). To find points of intersection, press [2nd] [TRACE] **(CALC)** and select option **5:intersect,** which automatically places you on the graphing window, and asks you to identify the "**First curve?**" As discussed, pressing [ENTER] three times in succession will identify each graph, bypass the "**Guess?**" option, then find and display the point of intersection (Figure 1.75). Here the point of intersection is $(-2, -3)$, showing the solution to this equation is $x = -2$ (for which both expressions equal -3). This can be verified by direct substitution or by using the **TABLE** feature.

Figure 1.74

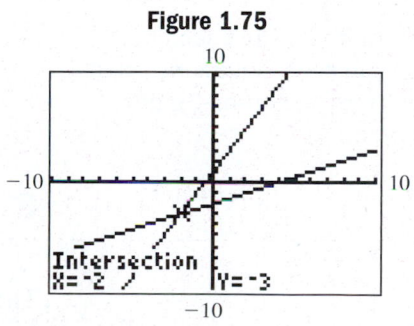

Figure 1.75

This method of solving equations is called the **Intersection-of-Graphs** method, and can be applied to many different equation types.

> **Intersection-of-Graphs Method for Solving Equations**
>
> For any equation of the form $f(x) = g(x)$,
>
> **1.** Assign $f(x)$ as Y_1 and $g(x)$ as Y_2.
>
> **2.** Graph both function and identify any point(s) of intersection, if they exist. The x-coordinate of all such points is a solution to the equation.

Recall that in the solution of linear equations, we sometimes encounter equations that are identities (infinitely many solutions) or contradictions (no solutions). These possibilities also have graphical representations, and appear as coincident lines and parallel lines respectively. These possibilities are illustrated in Figure 1.76.

Figure 1.76

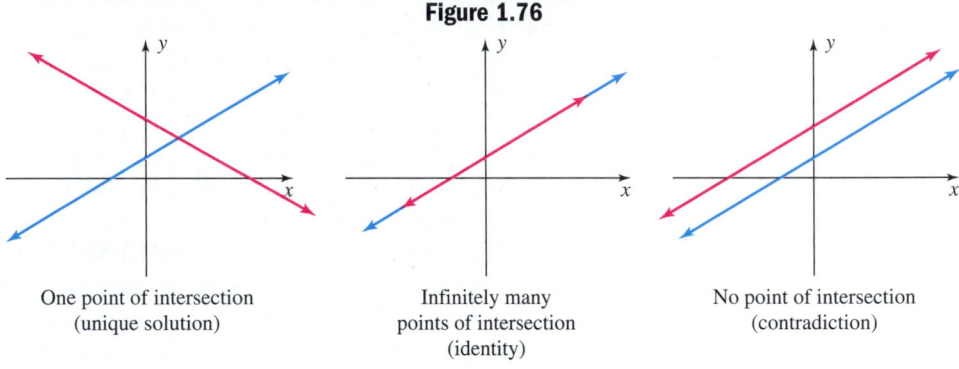

One point of intersection (unique solution)   Infinitely many points of intersection (identity)   No point of intersection (contradiction)

EXAMPLE 1B ▶   **Solving an Equation Graphically**

Solve $0.75x + 2 = 0.5(1 + 1.5x) - 3$ using a graphing calculator.

Solution ▶   With $0.75X + 2$ as Y_1 and $0.5(1 + 1.5X) - 3$ as Y_2, we use the [2nd] [TRACE] **(CALC)** option and select **5:intersect.** The graphs appear to be parallel lines (Figure 1.77), and after pressing [ENTER] three times we obtain the error message shown (Figure 1.78), confirming there are no solutions.

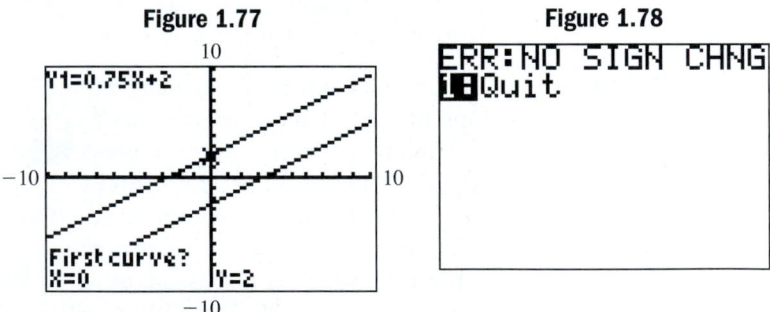

Figure 1.77   **Figure 1.78**

✓ **A.** You've just seen how we can solve equations using the Intersection-of-Graphs method

Now try Exercises 7 through 16 ▶

B. Solving Equations Graphically Using the *x-Intercept/Zeroes Method*

The intersection-of-graphs method works extremely well when the graphs of $f(x)$ and $g(x)$ (Y_1 and Y_2) are simple and "well-behaved." Later in this course, we encounter a number of graphs that are more complex, and it will help to develop alternative methods for solving graphically. Recall that two equations are equivalent if they have the same solution set. For instance, the equations $2x = 6$ and $2x - 6 = 0$ are equivalent (since $x = 3$ is a solution to both), as are $3x - 1 = x + 5$ and $2x - 6 = 0$ (since

$x = 3$ is a solution to both). Applying the intersection-of-graphs method to the last two equivalent equations, gives

$$\underbrace{3x - 1}_{Y_1} = \underbrace{x + 5}_{Y_2} \qquad \text{and} \qquad \underbrace{2x - 6}_{Y_1} = \underbrace{0}_{Y_2}$$

WORTHY OF NOTE

The *intersection-of-graphs method* focuses on a point of intersection (a, b), and names $x = a$ as the solution. The *zeroes method* focuses on the input $x = r$, for which the output is 0 $[Y_1(r) = 0]$. All such values r are called the **zeroes** of the function. Much more will be said about functions and their zeroes in later chapters.

The intersection method will work equally well in both cases, but the equation on the right has *only one variable expression,* and will produce a single (visible) graph (since $Y_2 = 0$ is simply the x-axis). Note that here we seek an input value that will result in an output of 0. In other words, all solutions will have the form $(x, 0)$, *which is the x-intercept* of the graph. For this reason, the method is alternatively called the **zeroes method** or the **x-intercept method.** The method employs the approach shown above, in which the equation $f(x) = g(x)$ is rewritten as $f(x) - g(x) = 0$, with $f(x) - g(x)$ assigned as Y_1.

Zeroes/x-Intercept Method for Solving Equations

For any equation of the form $f(x) = g(x)$,
1. Rewrite the equation as $f(x) - g(x) = 0$.
2. Assign $f(x) - g(x)$ as Y_1.
3. Graph the resulting function and identify any x-intercepts, if they exist.

     Any x-intercept(s) of the graph will be a solution to the equation.

To locate the zero (x-intercept) for $2x - 6 = 0$ on a graphing calculator, enter $2X - 6$ for Y_1 and use the **2:zero** option found on the same menu as the **5:intercept** option (Figure 1.79). Since some equations have more than one zero, the **2:zero** option will ask you to "narrow down" the interval it should search, even though there is only one zero here. It does this by asking for a **"Left Bound?"**, a **"Right Bound?"**, and a **"GUESS?"** (the **Guess?** option can once again be bypassed). You can enter these bounds by tracing along the graph or by inputting a chosen value, then pressing [ENTER] (note how the calculator posts a marker at each bound). Figure 1.80 shows we entered $x = 0$ as the left bound and $x = 4$ as the right, and the calculator will search for the x-intercept in this interval (note that in general, the cursor will be either above or below the x-axis for the left bound, but must be on *the opposite side* of the x-axis for the right bound). Pressing [ENTER] once more bypasses the **Guess?** option and locates the x-intercept at $(3, 0)$. The solution is $x = 3$ (Figure 1.81).

Figure 1.79

```
CALCULATE
1:value
2:zero
3:minimum
4:maximum
5:intersect
6:dy/dx
7:∫f(x)dx
```

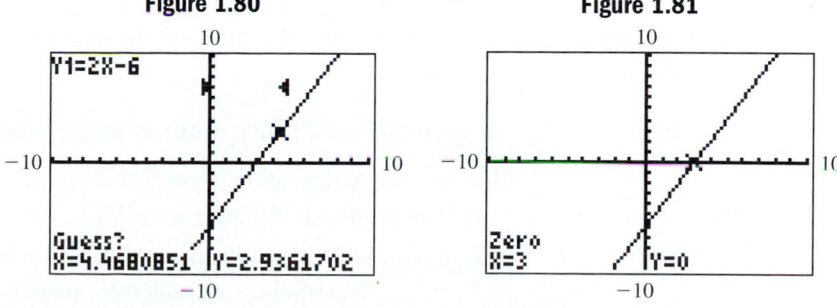

Figure 1.80

Figure 1.81

EXAMPLE 2 ▶ **Solving an Equation Using the Zeroes Method**

Solve the equation $-4(x + 3) - 6 = 2x + 3$ using the zeroes method.

Solution ▶ As given, we have $f(x) = -4(x + 3) - 6$ and $g(x) = 2x + 3$. Rewriting the equation as $f(x) - g(x) = 0$ gives $-4(x + 3) - 6 - (2x + 3) = 0$, where the expression for $g(x)$ is parenthesized to ensure the equations remain equivalent. Entering $-4(X + 3) - 6 - (2X + 3)$ as Y_1 and pressing **2nd** **TRACE** **(CALC) 2:zero** produces the screen shown in Figure 1.82, with the calculator requesting a left bound. We can input any x-value that is obviously to the left of the x-intercept, or move the cursor to any position left of the x-intercept and press **ENTER** (we input $x = -4$, see Figure 1.83). The calculator then asks for a right bound and as before we can input any x-value obviously to the right, or simply move the cursor to any location on the *opposite side* of the x-axis and press **ENTER** (we chose $x = -2$, see Figure 1.84). After bypassing the **Guess?** option (press **ENTER** once again), the calculator locates the x-intercept at $(-3.5, 0)$, and the solution to the original equation is $x = -3.5$ (Figure 1.85).

Figure 1.82

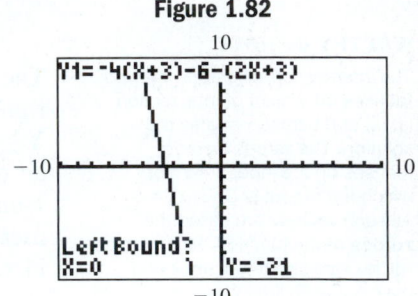

Figure 1.83

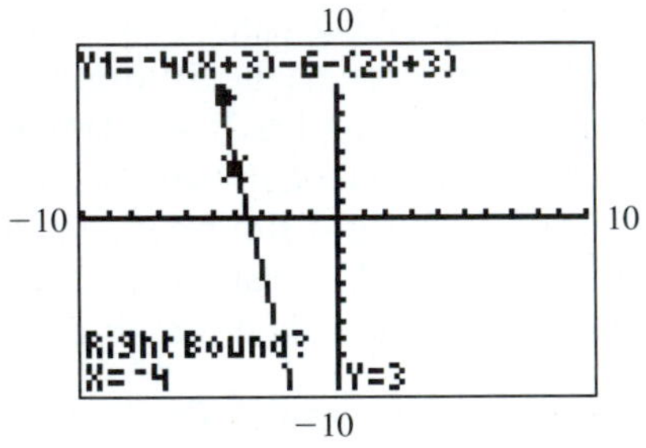

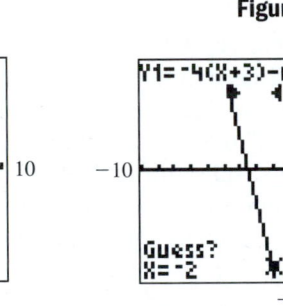

Figure 1.84

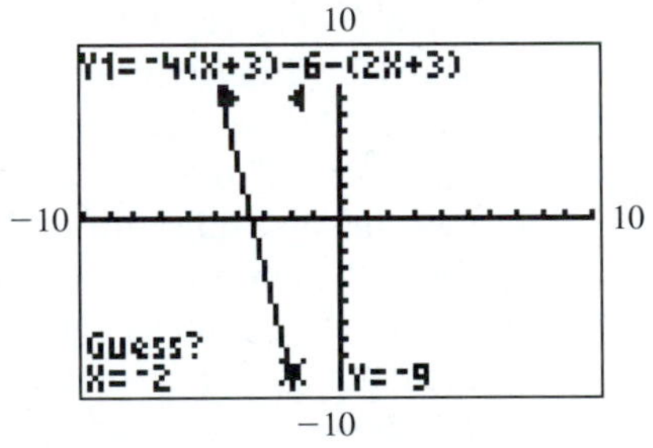

Figure 1.85

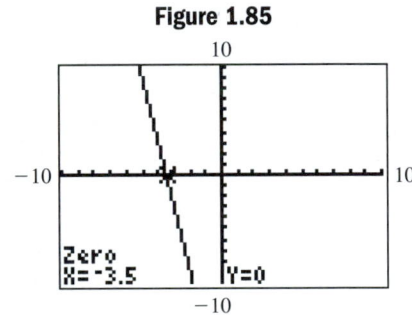

☑ **B.** You've just seen how we can solve equations using the x-intercept/zeroes method

Now try Exercises 17 through 26 ▶

C. Solving Linear Inequalities Graphically

The intersection-of-graphs method can also be applied to solve linear inequalities. The point of intersection simply becomes one of the boundary points for the solution interval, and is included or excluded depending on the inequality given. For the inequality $f(x) > g(x)$ written as $Y_1 > Y_2$, it becomes clear the inequality is true for all inputs x where the outputs for Y_1 are greater than the outputs for Y_2, meaning *the graph of $f(x)$ is above the graph of $g(x)$*. A similar statement can be made for $f(x) < g(x)$ written as $Y_1 < Y_2$.

Intersection-of-Graphs Method for Solving Inequalities

For any inequality of the form $f(x) > g(x)$,
1. Assign $f(x)$ as Y_1 and $g(x)$ as Y_2.
2. Graph both functions and identify any point(s) of intersection, if they exist.
   The solution set is all real numbers x for which the graph of Y_1 is *above* the graph of Y_2.

For strict inequalities, the boundary of the solution interval is not included. A similar process is used for the inequalities $f(x) \geq g(x), f(x) < g(x)$, and $f(x) \leq g(x)$. Note that we can actually draw the graphs of Y_1 and Y_2 differently (one more bold than the other) to clearly tell them apart in the viewing window. This is done on the [Y=] screen, by moving the cursor to the far left of the current function and pressing [ENTER] until a bold line appears. From the default setting, pressing [ENTER] one time produces this result.

EXAMPLE 3 ▶ **Solving an Inequality Using the Intersection-of-Graphs Method**

Solve $0.5(3 - x) - 5 \leq 2x + 4$ using the intersection-of-graphs method.

Figure 1.86

Solution ▶ To assist with the clarity of the solution, we set the calculator to graph Y_2 using a bolder line than Y_1 (Figure 1.86). With $0.5(3 - X) - 5$ as Y_1 and $2X + 4$ as Y_2, we use the [2nd] [TRACE] (**CALC**) option and select **5:intersect.** Pressing [ENTER] three times serves to identify both graphs, bypass the **"Guess?"** option, and display the point of intersection $(-3, -2)$ (Figure 1.87). Since the graph of Y_1 is *below* the graph of Y_2 $(Y_1 \leq Y_2)$ for all values of x to the right of $(-3, -2)$, $x = -3$ is the left boundary, with the interval extending to positive infinity. Due to the less than *or equal to* inequality, we *include* $x = -3$ and the solution interval is $x \in [-3, \infty)$.

```
Plot1  Plot2  Plot3
\Y1 ◼0.5(3-X)-5
\Y2 ◼2X+4
\Y3 =
\Y4 =
\Y5 =
\Y6 =
\Y7 =
```

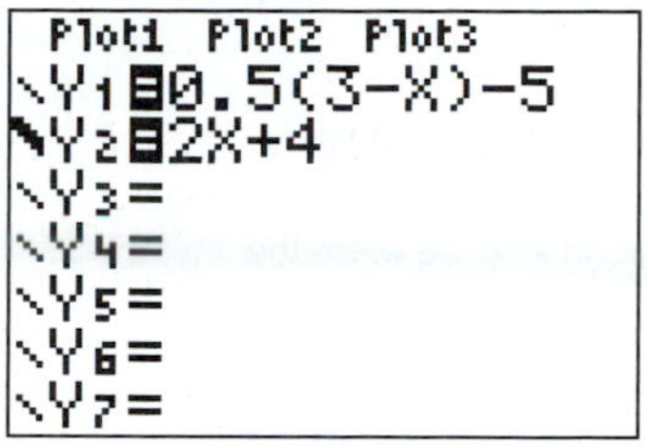

Figure 1.87

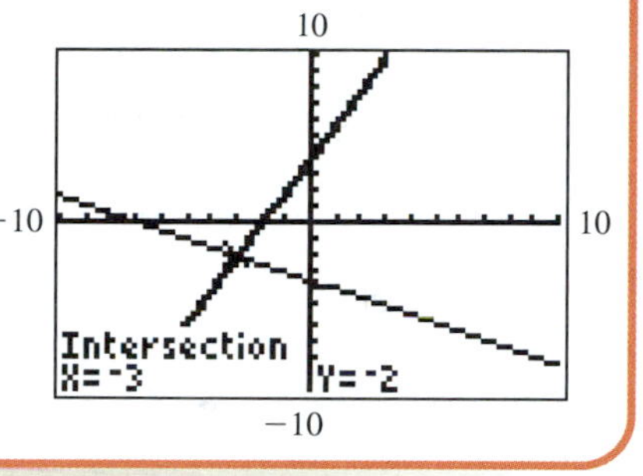

☑ **C.** You've just seen how we can solve linear inequalities graphically

Now try Exercises 27 through 36 ▶

D. Solving for a Specified Variable in Literal Equations

A **formula** is an equation that models a known relationship between two or more quantities. A **literal equation** is simply one that has two or more variables. Formulas are a type of literal equation, but not every literal equation is a formula. For example, the formula $A = P + PRT$ models the growth of money in an account earning simple interest, where A represents the total amount accumulated, P is the initial deposit, R is the annual interest rate, and T is the number of years the money is left on deposit. To *describe* $A = P + PRT$, we might say the formula has been "solved for A" or that "A is written in terms of P, R, and T." In some cases, before using a formula it may be convenient to solve for one of the other variables, say P. In this case, P is called the **object variable.**

EXAMPLE 4 ▶ **Solving for Specified Variable**

Given $A = P + PRT$, write P in terms of A, R, and T (solve for P).

Solution ▶ Since the object variable occurs in more than one term, we first apply the distributive property.

$$A = P + PRT \qquad \text{focus on } P\text{—the object variable}$$

$$A = P(1 + RT) \qquad \text{factor out } P$$

$$\frac{A}{1 + RT} = \frac{P(1 + RT)}{(1 + RT)} \qquad \text{solve for } P \text{ [divide by } (1 + RT)\text{]}$$

$$\frac{A}{1 + RT} = P \qquad \text{result}$$

Now try Exercises 37 through 48 ▶

We solve literal equations for a specified variable using the same methods we used for other equations and formulas. Remember that it's good practice to *focus on the object variable* to help guide you through the solution process, as again shown in Example 5.

EXAMPLE 5 ▶ **Solving for a Specified Variable**

Given $2x + 3y = 15$, write y in terms of x (solve for y).

Solution ▶

$2x + 3y = 15$ focus on the object variable

$\quad\quad 3y = -2x + 15$ subtract $2x$ (isolate y-term)

$\quad \frac{1}{3}(3y) = \frac{1}{3}(-2x + 15)$ multiply by $\frac{1}{3}$ (solve for y)

$\quad\quad\quad y = \frac{-2}{3}x + 5$ simplify and distribute

WORTHY OF NOTE

In Example 5, notice that in the second step we wrote the subtraction of $2x$ as $-2x + 15$ instead of $15 - 2x$. For reasons that will become clearer as we continue our study, we generally write variable terms before constant terms.

Now try Exercises 49 through 54 ▶

Literal Equations and General Solutions

Solving literal equations for a specified variable can help us develop the general solution for an entire family of equations. This is demonstrated here for the family of linear equations written in the form $ax + b = c$. A side-by-side comparison with a specific linear equation demonstrates that identical ideas are used.

Specific Equation		**Literal Equation**
$2x + 3 = 15$	focus on object variable	$ax + b = c$
$2x = 15 - 3$	subtract constant	$ax = c - b$
$x = \dfrac{15 - 3}{2}$	divide by coefficient	$x = \dfrac{c - b}{a}$

Of course the solution on the left would be written as $x = 6$ and checked in the original equation. On the right we now have a general formula for all equations of the form $ax + b = c$.

EXAMPLE 6 ▶ **Solving Equations of the Form $ax + b = c$ Using a General Formula**

Solve $6x - 1 = -25$ using the formula just developed, and check your solution in the original equation.

Solution ▶ For this equation, $a = 6$, $b = -1$, and $c = -25$, giving

$$x = \frac{c - b}{a}$$

$$= \frac{-25 - (-1)}{6}$$ Check: $6x - 1 = -25$

$$= \frac{-24}{6}$$ $6(-4) - 1 = -25$

$$= -4$$ $-24 - 1 = -25$

$$-25 = -25 \checkmark$$

Now try Exercises 55 through 60 ▶

☑ **D.** You've just seen how we can solve for a specified variable in a formula or literal equation

Developing a general solution for the linear equation $ax + b = c$ seems to have little practical use. But in Section 3.2 we'll use this idea to develop a general solution for *quadratic equations,* a result with much greater significance.

E. Using a Problem-Solving Guide

Becoming a good problem solver is an evolutionary process. Over time and with continued effort, your problem-solving skills grow, as will your ability to solve a wider range of applications. Most good problem solvers develop the following characteristics:

- A positive attitude
- A mastery of basic facts
- Strong mental arithmetic skills
- Good mental-visual skills
- Good estimation skills
- A willingness to persevere

These characteristics form a solid basis for applying what we call the **Problem-Solving Guide,** which simply organizes the basic elements of good problem solving. Using this guide will help save you from two common stumbling blocks—indecision and not knowing where to start.

Problem-Solving Guide

- **Gather and organize information.**
  Read the problem several times, forming a mental picture as you read. *Highlight key phrases.* List given information, including any related formulas. *Clearly identify what you are asked to find.*
- **Make the problem visual.**
  Draw and label a diagram or create a table of values, as appropriate. This will help you see how different parts of the problem fit together.
- **Develop an equation model.**
  Assign a variable to represent what you are asked to find and build any related expressions referred to in the problem. Write an equation model based on the relationships given in the problem. *Carefully reread the problem to double-check your equation model.*
- **Use the model and given information to solve the problem.**
  Substitute given values, then simplify and solve. State the answer in sentence form, and check that the answer is reasonable. Include any units of measure indicated.

General Modeling Exercises

Translating word phrases into symbols is an important part of building equations from information given in paragraph form. Sometimes the variable *occurs more than once* in the equation, because two different items in the same exercise are related. If the relationship involves a comparison of size, we often use line segments or bar graphs to model the relative sizes.

EXAMPLE 7 ▶   **Solving an Application Using the Problem-Solving Guide**

The largest state in the United States is Alaska (AK), which covers an area that is 230 square miles (mi^2) more than 500 times that of the smallest state, Rhode Island (RI). If they have a combined area of 616,460 mi^2, how many square miles does each cover?

Solution ▶   Combined area is 616,460 mi^2, AK covers   gather and organize information
230 more than 500 times the area of RI.   highlight any key phrases

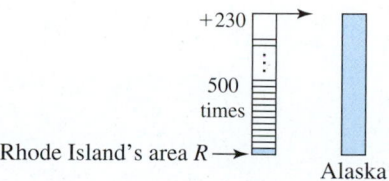

make the problem visual

Let R represent the area of Rhode Island.   assign a variable
Then $500R + 230$ represents Alaska's area.   build related expressions

Rhode Island's area + Alaska's area = Total

$$R + (500R + 230) = 616{,}460$$ write the equation model
$$501R = 616{,}230$$ combine like terms, subtract 230
$$R = 1230$$ divide by 501

Rhode Island covers an area of 1230 mi², while Alaska covers an area of $500(1230) + 230 = 615{,}230$ mi².

Now try Exercises 63 through 68 ▶

Consecutive Integer Exercises

Exercises involving **consecutive integers** offer excellent practice in assigning variables to unknown quantities, building related expressions, and the problem-solving process in general. We sometimes work with consecutive **odd** integers or consecutive **even** integers as well.

EXAMPLE 8 ▶   **Solving a Problem Involving Consecutive Odd Integers**

The sum of three consecutive *odd* integers is 69. What are the integers?

Solution ▶   The sum of three consecutive odd integers . . .   gather/organize information
highlight any key phrases

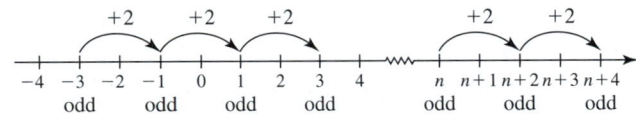

make the problem visual

Let n represent the smallest consecutive odd integer,   assign a variable
then $n + 2$ represents the second odd integer and   build related expressions
$(n + 2) + 2 = n + 4$ represents the third.

In words: first + second + third odd integer = 69   write the equation model
$$n + (n + 2) + (n + 4) = 69$$ equation model
$$3n + 6 = 69$$ combine like terms
$$3n = 63$$ subtract 6
$$n = 21$$ divide by 3

The odd integers are $n = 21$, $n + 2 = 23$, and $n + 4 = 25$.
$21 + 23 + 25 = 69$ ✓

Now try Exercises 69 through 72 ▶

WORTHY OF NOTE

The number line illustration in Example 8 shows that consecutive odd integers are *two units* apart and the related expressions were built accordingly: n, $n + 2$, $n + 4$, and so on. In particular, *we cannot use* n, $n + 1$, $n + 3$, . . . because n and $n + 1$ are *not two units apart*. If we know the exercise involves *even* integers instead, the same model is used, since even integers are also two units apart. For *consecutive* integers, the labels are n, $n + 1$, $n + 2$, and so on.

Uniform Motion (Distance, Rate, Time) Exercises

Uniform motion problems have many variations, and it's important to draw a good diagram when you get started. Recall that if speed is constant, the distance traveled is equal to the rate of speed multiplied by the time in motion: $D = RT$.

EXAMPLE 9 ▶ **Solving a Problem Involving Uniform Motion**

I live 260 mi from a popular mountain retreat. On my way there to do some mountain biking, my car had engine trouble—forcing me to bike the rest of the way. If I drove 2 hr longer than I biked and averaged 60 miles per hour driving and 10 miles per hour biking, how many hours did I spend pedaling to the resort?

Solution ▶ The sum of the two distances must be 260 mi.   gather/organize information

The **rates** are given, and the driving time is   highlight any key phrases
2 hr more than biking time.   make the problem visual

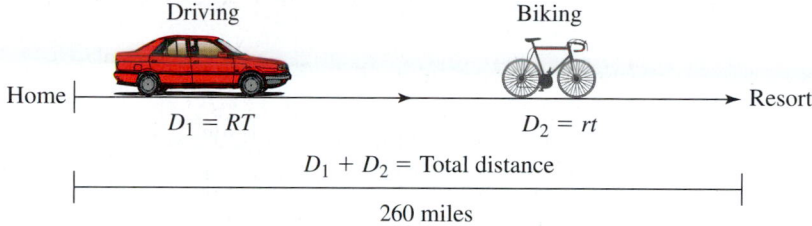

Let t represent the biking time,   assign a variable
then $T = t + 2$ represents time spent driving.   build related expressions

$$D_1 + D_2 = 260$$ write the equation model
$$RT + rt = 260$$ $RT = D_1, rt = D_2$
$$60(t + 2) + 10t = 260$$ substitute $t + 2$ for T, 60 for R, 10 for r
$$70t + 120 = 260$$ distribute and combine like terms
$$70t = 140$$ subtract 120
$$t = 2$$ divide by 70

I rode my bike for $t = 2$ hr, after driving $t + 2 = 4$ hr.

Now try Exercises 73 through 76 ▶

Exercises Involving Mixtures

Mixture problems offer another opportunity to refine our problem-solving skills while using many elements from the problem-solving guide. They also lend themselves to a very useful mental-visual image and have many practical applications.

EXAMPLE 10 ▶ **Solving an Application Involving Mixtures**

As a nasal decongestant, doctors sometimes prescribe saline solutions with a concentration between 6% and 20%. In "the old days," pharmacists had to create different mixtures, but only needed to stock these concentrations, since any percentage in between could be obtained using a mixture. An order comes in for a 15% solution. How many milliliters (mL) of the 20% solution must be mixed with 10 mL of the 6% solution to obtain the desired 15% solution? Provide both an algebraic solution and a graphical solution.

Algebraic Solution ▶   Only 6% and 20% concentrations are available;   gather/organize information
mix some 20% solution with 10 mL of the 6% solution.   highlight any key phrases
(See Figure 1.88.)

Figure 1.88

20%   6%
solution   solution

? mL   10 mL

make the problem visual

(10 + ?) mL
15% solution

Let x represent the amount of 20% solution, then   assign a variable
$10 + x$ represents the total amount of 15% solution.   build related expressions

1st quantity times its concentration		2nd quantity times its concentration		1st+2nd quantity times desired concentration	
$10(0.06)$	$+$	$x(0.2)$	$=$	$(10 + x)(0.15)$	write equation model
0.6	$+$	$0.2x$	$=$	$1.5 + 0.15x$	distribute/simplify
		$0.2x$	$=$	$0.9 + 0.15x$	subtract 0.6
		$0.05x$	$=$	0.9	subtract 0.15x
		x	$=$	18	divide by 0.05

To obtain a 15% solution, 18 mL of the 20% solution must
be mixed with 10 mL of the 6% solution.

Graphical Solution ▶   Although both methods work equally well,   **Figure 1.89**
here we elect to use the intersection-of-graphs
method and enter $10(0.06) + X(0.2)$ as Y_1 and

WORTHY OF NOTE

For mixture exercises, an estimate
assuming equal amounts of each
liquid can be helpful. For example,
assume we use 10 mL of the 6%
solution and 10 mL of the 20%
solution. The final concentration
would be halfway in between,
$\frac{6 + 20}{2} = 13\%$. This is too low a
concentration (we need a 15%
solution), so we know that more
than 10 mL of the stronger (20%)
solution must be used.

$(10 + X)(0.15)$ as Y_2. Virtually all graphical
solutions require a careful study of the context to
set the viewing window prior to graphing. If 10 mL
of liquid were used from each concentration, we
would have 20 mL of a 13% solution (see *Worthy
of Note*), so more of the stronger solution is
needed. This shows that an appropriate Xmax
might be close to 30. If all 30 mL were used,
the output would be $30(0.15) = 4.5$, so an
appropriate Ymax might be around 6 (see
Figure 1.89). Using 2nd TRACE (**CALC**)
5:Intersect and pressing ENTER three times gives
$(18, 4.2)$ as the point of intersection, showing
$x = 18$ mL of the stronger solution must be used
(Figure 1.90).

```
WINDOW
 Xmin=0
 Xmax=30
 Xscl=3
 Ymin=0
 Ymax=6
 Yscl=1
 Xres=1
```

Figure 1.90

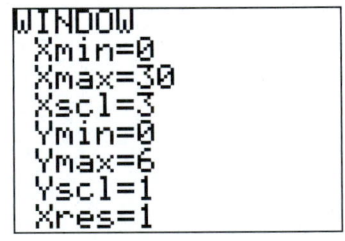

☑ **E.** You've just seen how
we can use the problem-
solving guide to solve various
problem types

Now try Exercises 77 through 84 ▶

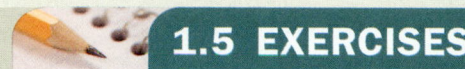

1.5 EXERCISES

▶ CONCEPTS AND VOCABULARY

Fill in each blank with the appropriate word or phrase. Carefully reread the section, if necessary.

1. When using the _intersection-of-graphs_ method, one side of an equation is entered as _Y_1_ and the other side as _Y_2_ on a graphing calculator. The _x-coordinate_ of the point of _intersection_ is the solution of the equation.

2. To solve a linear inequality using the intersection-of-graphs method, first find the point of _intersection_. The _x-coordinate_ of this point is a boundary value of the solution interval and if the inequality is not strict, this value is _included_ in the solution.

3. A(n) _literal_ equation is an equation having _two_ or more unknowns.

4. For the equation $S = 2\pi r^2 + 2\pi rh$, we can say that S is written in terms of _r_ and _h_.

5. Discuss/Explain the similarities and differences between the intersection and zeroes methods for solving equations. How can the zeroes method be applied to solving linear inequalities? Give examples in your discussion.   Answers will vary.

6. Discuss/Explain each of the four basic parts of the *problem-solving guide*. Include a solved example in your discussion.   Answers will vary.

▶ DEVELOPING YOUR SKILLS

Solve the following equations using a graphing calculator and the intersection-of-graphs method. For Exercises 7 and 8, carefully sketch the graphs you designate as Y_1 and Y_2 by hand before using your calculator.

7. $3x - 7 = -2(x + 1) + 10$

8. $-2x + 1 = 2(x - 3) - 1$

9. $0.8x + 0.4 = 0.25(2 - 0.4x) - 2.8$ $x = -3$

10. $0.5x + 2.5 = 0.75(3 - 0.2x) - 0.4$ $x = -1$

11. $x - (3x + 1) = -0.5(4x + 6) + 2$ $x \in (-\infty, \infty)$

12. $3x - (4 + x) = 0.2(6 + 10x) - 5.2$ $x \in (-\infty, \infty)$

13. $\frac{1}{3}x = \frac{-2}{3}x - 9$ $x = -9$

14. $\frac{-4}{5}x + 8 = \frac{6}{5}x$ $x = 4$

15. $\frac{1}{2}(x - 4) + 10 = x - (2 + \frac{1}{2}x)$ no solutions

16. $\frac{-1}{3}(x + 6) + 5 = -x - (6 - \frac{2}{3}x)$ no solutions

Solve the following equations using a graphing calculator and the x-intercept/zeroes method. Compare your results for Exercises 17 and 18 to those of Exercises 7 and 8.

17. $3x - 7 = -2(x + 1) + 10$ $x = 3$, answers match

18. $-2x + 1 = 2(x - 3) - 1$ $x = 2$, answers match

19. $2(3 - 2x) - 5 = -3x + 3$ $x = -2$

20. $-3(-3 - x) + 4 = 2x + 5$ $x = -8$

21. $-1.5(x + 4) + 2.5 = 3x - 3.5$ $x = 0$

22. $0.8(3x - 1) + 0.2 = -2x + 3.8$ $x = 1$

23. $2(x + 2) + 1 = x + (1 + x)$ no solutions

24. $3(2x - 1) - 1 = 2x - (1 - 4x)$ no solutions

25. $3x - (0.7x - 1.2) = 2(1.1x + 0.6) + 0.1x$
     $x \in (-\infty, \infty)$

26. $3x + 2(0.2x - 1.4) = 4(0.8x - 0.7) + 0.2x$
     $x \in (-\infty, \infty)$

Solve the following inequalities using a graphing calculator and the intersection-of-graphs method. Compare your results for Exercises 27 and 28 to those of Exercises 7 and 8.

27. $3x - 7 > -2(x + 1) + 10$
     $x \in (3, \infty)$, verified graphically in Exercise 7

28. $-2x + 1 > 2(x - 3) - 1$
     $x \in (-\infty, 2)$, verified graphically in Exercise 8

29. $2x - (3 + x) \geq -2(5 - 2x) + 7$ $x \in (-\infty, 0]$

30. $4(3x - 5) + 2 \geq 3(2 - 4x) + 24$ $x \in [2, \infty)$

31. $-0.3(x + 2) + 1.1 < 0.2x + 3$ $x \in (-5, \infty)$

32. $0.25(4 - x) + 1 < 1 - 0.5x$ $x \in (-\infty, -4)$

33. $-3(x - 1) + 1 \leq x - 4(x - 1)$ $x \in (-\infty, \infty)$

34. $1.1(2 - x) + 0.2 > 5(0.1 - 0.2x) - 0.1x$
     $x \in (-\infty, \infty)$

35. $2(1.5x - 1.1) + 0.1x \geq 4x - 0.3(3x - 4)$
     no solutions

36. $4(x - 1) - 2x + 7 < 2(x + 1.5)$
     no solutions

HOMEWORK SELECTION GUIDE

Additional answers can be found in the Instructor Answer Appendix.

Core: 7–59 every other odd, 61, 63–69 odd, 73, 75, 81, 83 (23 Exercises)
Standard: 1–4, 7–59 every other odd, 61, 63–83 odd (30 Exercises)

Extended: 1–4, 7–59 every other odd, 61, 62, 63–79 odd, 81–84 all, 86, 88, 89, 92 (37 Exercises)
In Depth: 1–6, 7–59 every other odd, 61, 62, 63–79 odd, 81–84 all, 86–88, 89–91 (42 Exercises)

Solve for the specified variable in each formula or literal equation.

37. $P = C + CM$ for C (retail) $C = \dfrac{P}{1 + M}$

38. $S = P - PD$ for P (retail) $P = \dfrac{S}{1 - D}$

39. $C = 2\pi r$ for r (geometry) $r = \dfrac{C}{2\pi}$

40. $V = LWH$ for W (geometry) $W = \dfrac{V}{LH}$

41. $\dfrac{P_1 V_1}{T_1} = \dfrac{P_2 V_2}{T_2}$ for T_2 (science) $T_2 = \dfrac{T_1 P_2 V_2}{P_1 V_1}$

42. $\dfrac{C}{P_2} = \dfrac{P_1}{d^2}$ for P_2 (communication) $P_2 = \dfrac{C d^2}{P_1}$

43. $V = \frac{4}{3}\pi r^2 h$ for h (geometry) $h = \dfrac{3V}{4\pi r^2}$

44. $V = \frac{1}{3}\pi r^2 h$ for h (geometry) $h = \dfrac{3V}{\pi r^2}$

45. $S_n = n\left(\dfrac{a_1 + a_n}{2}\right)$ for n (sequences) $n = \dfrac{2 S_n}{a_1 + a_n}$

46. $A = \dfrac{h(b_1 + b_2)}{2}$ for h (geometry) $h = \dfrac{2A}{b_1 + b_2}$

47. $S = B + \frac{1}{2}PS$ for P (geometry) $P = \dfrac{2(S - B)}{S}$

48. $s = \frac{1}{2}gt^2 + vt$ for g (physics) $g = \dfrac{2(s - vt)}{t^2}$

49. $Ax + By = C$ for y $y = \dfrac{-A}{B}x + \dfrac{C}{B}$

50. $2x + 3y = 6$ for y $y = \dfrac{-2}{3}x + 2$

51. $\frac{5}{6}x + \frac{3}{8}y = 2$ for y $y = \dfrac{-20}{9}x + \dfrac{16}{3}$

52. $\frac{2}{3}x - \frac{7}{9}y = 12$ for y $y = \dfrac{6}{7}x - \dfrac{108}{7}$

53. $y - 3 = \frac{-4}{5}(x + 10)$ for y $y = \dfrac{-4}{5}x - 5$

54. $y + 4 = \frac{-2}{15}(x + 10)$ for y $y = \dfrac{-2}{15}x - \dfrac{16}{3}$

The following equations are given in $ax + b = c$ form. Solve by identifying the value of a, b, and c, then using the formula $x = \dfrac{c - b}{a}$.

55. $3x + 2 = -19$ $a = 3; b = 2; c = -19; x = -7$

56. $7x + 5 = 47$ $a = 7; b = 5; c = 47; x = 6$

57. $-6x + 1 = 33$ $a = -6; b = 1; c = 33; x = \frac{16}{-3}$

58. $-4x + 9 = 43$ $a = -4; b = 9; c = 43; x = \frac{17}{-2}$

59. $7x - 13 = -27$ $a = 7; b = -13; c = -27; x = -2$

60. $3x - 4 = -25$ $a = 3; b = -4; c = -25; x = -7$

▶ WORKING WITH FORMULAS

61. Surface area of a cylinder: $SA = 2\pi r^2 + 2\pi rh$

The surface area of a cylinder is given by the formula shown, where h is the height of the cylinder and r is the radius of the base. Find the height of a cylinder that has a radius of 8 cm and a surface area of 1256 cm². Use $\pi \approx 3.14$. $h = 17$ cm

62. Using the equation-solving process for Exercise 61 as a model, solve the formula $SA = 2\pi r^2 + 2\pi rh$ for h.
$h = \dfrac{SA - 2\pi r^2}{2\pi r}$

▶ APPLICATIONS

Solve by building an equation model and using the problem-solving guidelines as needed. Check all answers using a graphing calculator.

General Modeling Exercises

63. Two spelunkers (cave explorers) were exploring different branches of an underground cavern. The first was able to descend 198 ft farther than twice the second. If the first spelunker descended 1218 ft, how far was the second spelunker able to descend?   510 ft

64. The area near the joining of the Tigris and Euphrates Rivers (in modern Iraq) has often been called the *Cradle of Civilization*, since the area has evidence of many ancient cultures. The length of the Euphrates River exceeds that of the Tigris by 620 mi. If they have a combined length of 2880 mi, how long is each river?   Tigris is 1130 mi; Euphrates is 1750 mi

65. U.S. postal regulations require that a package can have a maximum combined length and girth (distance around) of 108 in. A shipping carton is constructed so that it has a width of 14 in., a height of 12 in., and can be cut or folded to various lengths. What is the maximum length that can be used?   56 in

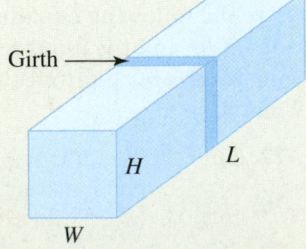

Source: www.USPS.com

66. Hi-Tech Home Improvements buys a fleet of identical trucks that cost $32,750 each. The company is allowed to depreciate the value of their trucks for tax purposes by $5250 per year. If company policies dictate that older trucks must be sold once their value declines to $6500, approximately how many years will they keep these trucks?   5 yr

67. The longest suspension bridge in the world is the Akashi Kaikyo (Japan) with a length of 6532 feet. Japan is also home to the Shimotsui Straight bridge. The Akashi Kaikyo bridge is 364 ft more than twice the length of the Shimotsui bridge. How long is the Shimotsui bridge?   3084 ft

Source: www.guinnessworldrecords.com

68. The Mars rover *Spirit* landed on January 3, 2004. Just over 1 yr later, on January 14, 2005, the *Huygens* probe landed on Titan (one of Saturn's moons). At their closest approach, the distance from the Earth to Saturn is 29 million mi more than 21 times the distance from the Earth to Mars. If the distance to Saturn is 743 million mi, what is the distance to Mars?   34 million mi

Consecutive Integer Exercises

69. Find two consecutive even integers such that the sum of twice the smaller integer plus the larger integer is one hundred forty-six.   48; 50

70. When the smaller of two consecutive integers is added to three times the larger, the result is fifty-one. Find the smaller integer.   12

71. Seven times the first of two consecutive odd integers is equal to five times the second. Find each integer.   5; 7

72. Find three consecutive even integers where the sum of triple the first and twice the second is eight more than four times the third.   20; 22; 24

Uniform Motion Exercises

73. At 9:00 A.M., Linda leaves work on a business trip, gets on the interstate, and sets her cruise control at 60 mph. At 9:30 A.M., Bruce notices she's left her briefcase and cell phone, and immediately starts after her driving 75 mph. At what time will Bruce catch up with Linda?   11: 30 A.M.

74. A plane flying at 300 mph has a 3-hr head start on a "chase plane," which has a speed of 800 mph. How far from the airport will the chase plane overtake the first plane?   1440 mi

75. Jeff had a job interview in a nearby city 72 mi away. On the first leg of the trip he drove an average of 30 mph through a long construction zone, but was able to drive 60 mph after passing through this zone. If driving time for the trip was $1\frac{1}{2}$ hr, how long was he driving in the construction zone?   36 min

76. At a high-school cross-country meet, Jared jogged 8 mph for the first part of the race, then increased his speed to 12 mph for the second part. If the race was 21 mi long and Jared finished in 2 hr, how far did he jog at the faster pace?   15 mi

Mixture Exercises

Give the total amount of the mix that results and the percent concentration or worth of the mix.

77. Two quarts of 100% orange juice are mixed with 2 quarts of water (0% juice).   4 quarts; 50% O.J.

78. Ten pints of a 40% acid are combined with 10 pints of an 80% acid.   20 pints; 60% acid

79. Eight pounds of premium coffee beans worth $2.50 per pound are mixed with 8 lb of standard beans worth $1.10 per pound.   16 lb; $1.80 lb

80. A rancher mixes 50 lb of a custom feed blend costing $1.80 per pound, with 50 lb of cheap cottonseed worth $0.60 per pound.   100 lb; $1.20 lb

Solve each application of the mixture concept.

81. To help sell more of a lower grade meat, a butcher mixes some premium ground beef worth $3.10/lb, with 8 lb of lower grade ground beef worth $2.05/lb. If the result was an intermediate grade of ground beef worth $2.68/lb, how much premium ground beef was used?   12 lb

82. Knowing that the camping/hiking season has arrived, a nutrition outlet is mixing GORP (Good Old Raisins and Peanuts) for the anticipated customers. How many pounds of peanuts worth $1.29/lb, should be mixed with 20 lb of deluxe raisins worth $1.89/lb, to obtain a mix that will sell for $1.49/lb?   40 lb

83. How many pounds of walnuts at 84¢/lb should be mixed with 20 lb of pecans at $1.20/lb to give a mixture worth $1.04/lb?   16 lb

84. How many pounds of cheese worth 81¢/lb must be mixed with 10 lb cheese worth $1.29/lb to make a mixture worth $1.11/lb?   6 lb

▶ **EXTENDING THE CONCEPT**

85. Look up and read the following article. Then turn in a one page summary. "Don't Give Up!," William H. Kraus, *Mathematics Teacher,* Volume 86, Number 2, February 1993: pages 110–112.   Answers will very

86. A chemist has four solutions of a very rare and expensive chemical that are 15% acid (cost $120 per ounce), 20% acid (cost $180 per ounce), 35% acid (cost $280 per ounce) and 45% acid (cost $359 per ounce). She requires 200 oz of a 29% acid solution. Find the combination of any two of these concentrations that will minimize the total cost of the mix. $106\frac{2}{3}$ oz of 15% acid; $93\frac{1}{3}$ oz of 45% acid

87. *P, Q, R, S, T,* and *U* represent numbers. The arrows in the figure show the sum of the two or three numbers added in the indicated direction

(Example: $Q + T = 23$). Find $P + Q + R + S + T + U.$ 69

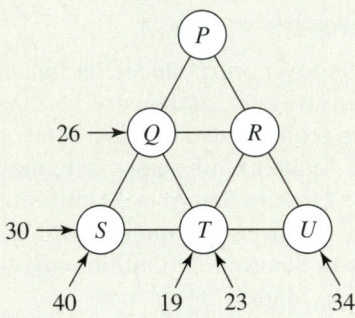

88. Given a sphere circumscribed by a cylinder, verify the volume of the sphere is $\frac{2}{3}$ that of the cylinder. Use $V = \pi r^2 h$ with $h = 2r$

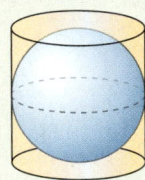

▶ **MAINTAINING YOUR SKILLS**

89. (R.5) Solve for x: $\dfrac{1}{x + 2} - \dfrac{2}{x} = \dfrac{3}{x^2 + 2x}$ $x = -7$

90. (1.4) Solve for y, then state the slope and y-intercept of the line: $-6x + 7y = 42$ $y = \frac{6}{7}x + 6, m = \frac{6}{7}, (0, 6)$

91. (R.4) Factor each expression:
   a. $4x^2 - 9$ $(2x + 3)(2x - 3)$
   b. $x^3 - 27$ $(x - 3)(x^2 + 3x + 9)$

92. (1.3) Given $g(x) = x^2 - 3x - 10$, evaluate $g(\frac{1}{3}), g(-2),$ and $g(5)$ $g(\frac{1}{3}) = -\frac{98}{9}, g(-2) = 0, g(5) = 0$

LEARNING OBJECTIVES

In Section 1.6 you will see how we can:

- [ ] **A.** Draw a scatterplot and identify positive and negative associations
- [ ] **B.** Use a scatterplot to identify linear and nonlinear associations
- [ ] **C.** Use a scatterplot to identify strong and weak correlations
- [ ] **D.** Find a linear function that models the relationships observed in a set of data
- [ ] **E.** Use linear regression to find the line of best fit

Collecting and analyzing data is a tremendously important mathematical endeavor, having applications throughout business, industry, science, and government. The link between classroom mathematics and real-world mathematics is called a **regression,** in which we attempt to find an equation that will act as a model for the raw data. In this section, we focus on situations where the data is best modeled by a linear function.

A. Scatterplots and Positive/Negative Associations

In this section, we continue our study of ordered pairs and functions, but this time using data collected from various sources or from observed real-world relationships. You can hardly pick up a newspaper or magazine without noticing it contains a large volume of data presented in graphs, charts, and tables. In addition, there are many simple experiments or activities that enable you to collect your own data. We begin analyzing the collected data using a **scatterplot,** which is simply a graph of all of the ordered pairs in a data set. Often, real data (sometimes called **raw data**) is not very "well behaved" and the points may be somewhat scattered—the reason for the name.

Positive and Negative Associations

Earlier we noted that lines with positive slope rise from left to right, while lines with negative slope fall from left to right. We can extend this idea to the data from a scatterplot. The data points in Example 1A seem to *rise* as you move from left to right, with larger input values generally resulting in larger outputs. In this case, we say there is a **positive association** between the variables. If the data seems to decrease or fall as you move left to right, we say there is a **negative association.**

EXAMPLE 1A ▶ Drawing a Scatterplot and Observing Associations

The ratio of the federal debt to the total population is known as the *per capita debt*. The per capita debt of the United States is shown in the table for the odd-numbered years from 1997 to 2007. Draw a scatterplot of the data and state whether the association is positive, negative, or cannot be determined.

Source: Data from the Bureau of Public Debt at www.publicdebt.treas.gov

Year	Per Capita Debt ($1000s)
1997	20.0
1999	20.7
2001	20.5
2003	23.3
2005	27.6
2007	30.4

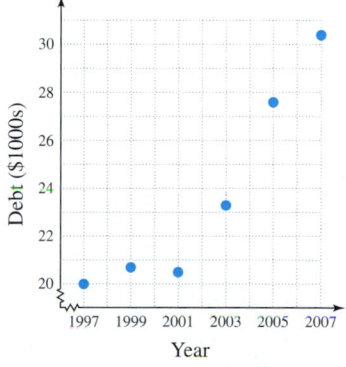

Solution ▶ Since the amount of debt depends on the year, *year* is the input *x* and *per capita debt* is the output *y*. Scale the *x*-axis from 1997 to 2007 and the *y*-axis from 20 to 30 to comfortably fit the data (the "squiggly lines," near the 20 and 1997 in the graph are used to show that some initial values have been skipped). The graph indicates a positive association between the variables, meaning the debt is generally *increasing* as time goes on.

EXAMPLE 1B ▶ Drawing a Scatterplot and Observing Associations

A cup of coffee is placed on a table and allowed to cool. The temperature of the coffee is measured every 10 min and the data are shown in the table. Draw the scatterplot and state whether the association is positive, negative, or cannot be determined.

Elapsed Time (minutes)	Temperature (°F)
0	110
10	89
20	76
30	72
40	71

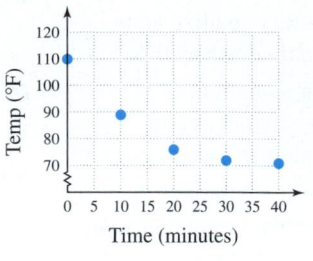

Solution ▶ Since temperature depends on cooling time, *time* is the input x and *temperature* is the output y. Scale the x-axis from 0 to 40 and the y-axis from 70 to 110 to comfortably fit the data. As you see in the figure, there is a negative association between the variables, meaning the temperature *decreases* over time.

☑ **A.** You've just seen how we can draw a scatterplot and identify positive and negative associations

Now try Exercises 7 and 8 ▶

B. Scatterplots and Linear/Nonlinear Associations

The data in Example 1A had a positive association, while the association in Example 1B was negative. But the data from these examples differ in another important way. In Example 1A, the data seem to cluster about an imaginary line. This indicates a linear equation model might be a good approximation for the data, and we say there is a **linear association** between the variables. The data in Example 1B could not accurately be modeled using a straight line, and we say the variables *time* and *cooling temperature* exhibit a **nonlinear association.**

EXAMPLE 2 ▶ Drawing a Scatterplot and Observing Associations

A college professor tracked her annual salary for 2002 to 2009 and the data are shown in the table. Draw the scatterplot and determine if there is a linear or nonlinear association between the variables. Also state whether the association is positive, negative, or cannot be determined.

Year	Salary ($1000s)
2002	30.5
2003	31
2004	32
2005	33.2
2006	35.5
2007	39.5
2008	45.5
2009	52

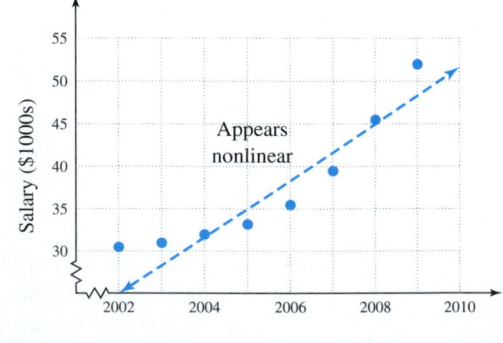

Solution ▶ Since salary earned depends on a given year, *year* is the input *x* and *salary* is the output *y*. Scale the *x*-axis from 2002 to 2010, and the *y*-axis from 30 to 55 to comfortably fit the data. A line doesn't seem to model the data very well, and the association appears to be nonlinear. The data rises from left to right, indicating a positive association between the variables. This makes good sense, since we expect our salaries to increase over time.

☑ **B.** You've just seen how we can use a scatterplot to identify linear and nonlinear associations

Now try Exercises 9 and 10 ▶

C. Identifying Strong and Weak Correlations

Using Figures 1.91 and 1.92 shown, we can make one additional observation regarding the data in a scatterplot. While both associations shown appear linear, the data in Figure 1.91 seems to cluster more tightly about an imaginary straight line than the data in Figure 1.92.

Figure 1.91

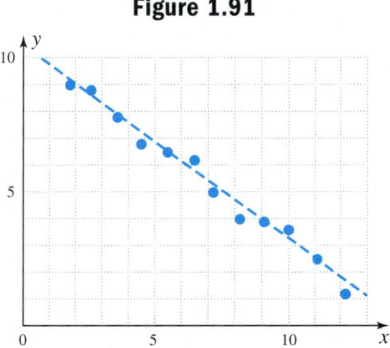

Figure 1.92

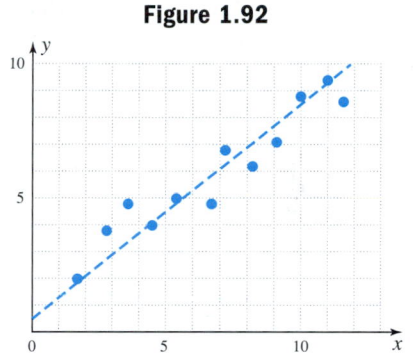

We refer to this "clustering" as the "goodness of fit," or in statistical terms, the strength of the correlation. To quantify this fit we use a measure called the correlation coefficient r, which tells whether the association is positive or negative: $r > 0$ or $r < 0$, and quantifies the strength of the association: $|r| \leq 100\%$. Actually, the coefficient is given in decimal form, making $|r| \leq 1$. If the data points form a perfectly straight line, we say the strength of the correlation is either -1 or 1, depending on the association. If the data points appear clustered about the line, but are scattered on either side of it, the strength of the correlation falls somewhere between -1 and 1, depending on how tightly/loosely they're scattered. This is summarized in Figure 1.93.

Figure 1.93

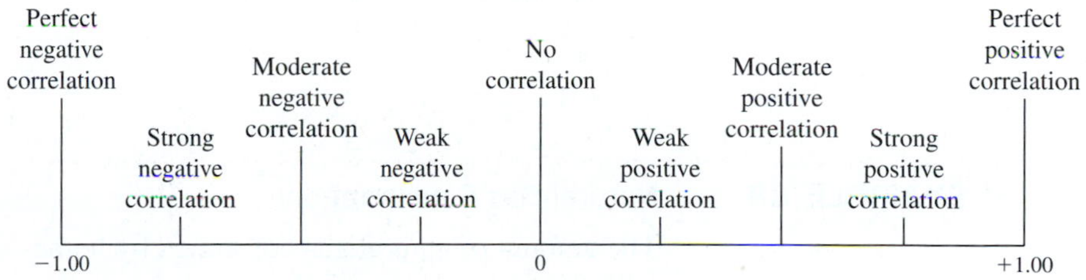

Perfect negative correlation | Moderate negative correlation | No correlation | Moderate positive correlation | Perfect positive correlation

Strong negative correlation | Weak negative correlation | Weak positive correlation | Strong positive correlation

-1.00 ———————————— 0 ———————————— $+1.00$

The following scatterplots help to further illustrate this idea. Figure 1.94 shows a linear and negative association between the value of a car and the age of a car, with a strong correlation. Figure 1.95 shows there is no apparent association between family income and the number of children, and Figure 1.96 appears to show a linear and positive association between a man's height and weight, with a weak correlation.

Figure 1.94	Figure 1.95	Figure 1.96
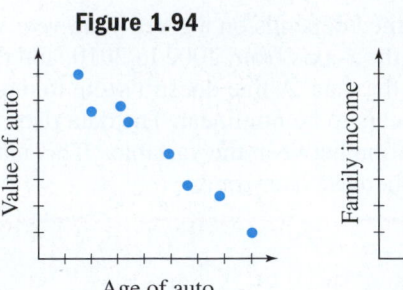	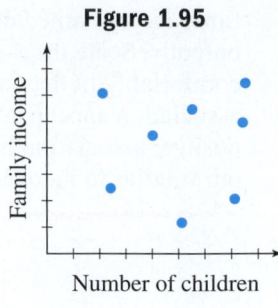	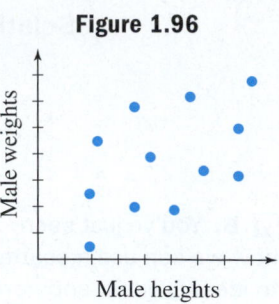

Until we develop a more accurate method of calculating a numerical value for this correlation, the best we can do are these broad generalizations: weak correlation, strong correlation, or no correlation.

EXAMPLE 3A ▶ **High School and College GPAs**

Many colleges use a student's high school GPA as a possible indication of their future college GPA. Use the data from Table 1.6 (high school/college GPA) to draw a scatterplot. Then

a. Sketch a line that seems to approximate the data, meaning it has the same general direction while passing through the observed "center" of the data.

b. State whether the association is positive, negative, or cannot be determined.

c. Decide whether the correlation is weak or strong.

Table 1.6

High School GPA	College GPA
1.8	1.8
2.2	2.3
2.8	2.5
3.2	2.9
3.4	3.6
3.8	3.9

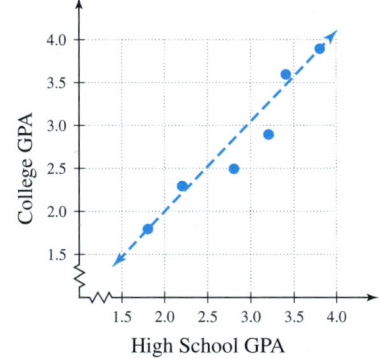

Solution ▶ **a.** A line approximating the data set as a whole is shown in the figure.

b. Since the line has positive slope, there is a positive association between a student's high school GPA and their GPA in college.

c. The correlation appears strong.

EXAMPLE 3B ▶ **Natural Gas Consumption**

The amount of natural gas consumed by homes and offices varies with the season, with the highest consumption occurring in the winter months. Use the data from Table 1.7 (outdoor temperature/gas consumed) to draw a scatterplot. Then

a. Sketch an estimated line of best fit.

b. State whether the association is positive, negative, or cannot be determined.

c. Decide whether the correlation is weak or strong.

Table 1.7

Outdoor Temperature (F°)	Gas Consumed (cubic feet)
30	800
40	620
50	570
60	400
70	290
80	220

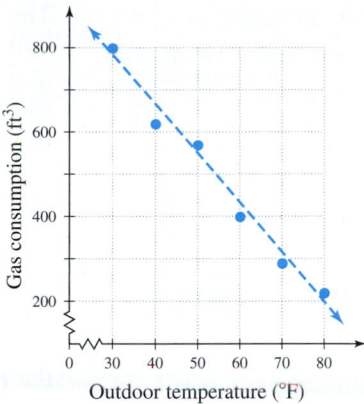

Solution ▶ a. We again use appropriate scales and sketch a line that seems to model the data (see figure).

b. There is a negative association between temperature and the amount of natural gas consumed.

c. The correlation appears to be strong.

☑ **C.** You've just seen how we can use a scatterplot to identify strong and weak correlations

Now try Exercises 11 and 12 ▶

D. Linear Functions That Model Relationships Observed in a Set of Data

Finding a **linear function model** for a set of data involves visually estimating and sketching a line that appears to best "fit" the data. This means answers will vary slightly, but a good, usable model can often be obtained. To find the function, we select two points on this imaginary line and use either the slope-intercept form or the point-slope formula to construct the function. Points on this estimated line but not actually in the data set *can still be used* to help determine the function.

EXAMPLE 4 ▶ **Finding a Linear Function to Model the Relationship Between GPAs**

Use the scatterplot from Example 3A to find a function model for the line a college might use to project an incoming student's future GPA.

Solution ▶ Any two points on or near the estimated best-fit line can be used to help determine the linear function (see the figure in Example 3A). For the slope, it's best to pick two points that are some distance apart, as this tends to improve the accuracy of the model. It appears $(1.8, 1.8)$ and $(3.8, 3.9)$ are both on the line, giving

$$m = \frac{y_2 - y_1}{x_2 - x_1} \qquad \text{slope formula}$$

$$= \frac{3.9 - 1.8}{3.8 - 1.8} \qquad \text{substitute } (x_2, y_2) \text{ for} \rightarrow (3.8, 3.9), (x_1, y_1) \text{ for} \rightarrow (1.8, 1.8)$$

$$= 1.05 \qquad \text{slope}$$

$$y - y_1 = m(x - x_1) \qquad \text{point-slope form}$$

$$y - 1.8 = 1.05(x - 1.8) \qquad \text{substitute } 1.05 \text{ for } m, (1.8, 1.8) \text{ for } (x_1, y_1)$$

$$y - 1.8 = 1.05x - 1.89 \qquad \text{distribute}$$

$$y = 1.05x - 0.09 \qquad \text{add } 1.8 \text{ (solve for } y)$$

One possible function model for this data is $f(x) = 1.05x - 0.09$. Slightly different functions may be obtained, depending on the points chosen.

Now try Exercises 13 through 22 ▶

The function from Example 4 predicts that a student with a high school GPA of 3.2 will have a college GPA of almost 3.3: $f(3.2) = 1.05(3.2) - 0.09 \approx 3.3$, yet the data gives an actual value of only 2.9. When working with data and function models, we should expect some variation when the two are compared, especially if the correlation is weak.

Applications of data analysis can be found in virtually all fields of study. In Example 5 we apply these ideas to an Olympic swimming event.

EXAMPLE 5 ▶ **Finding a Linear Function to Model the Relationship (Year, Gold Medal Times)**

The men's 400-m freestyle times (gold medal times—to the nearest second) for the 1976 through 2008 Olympics are given in Table 1.8 (1900→0). Let the year be the input x, and winning race time be the output y. Based on the data, draw a scatterplot and answer the following questions.

a. Does the association appear linear or nonlinear?

b. Is the association positive or negative?

c. Classify the correlation as weak or strong.

d. Find a function model that approximates the data, then use it to predict the winning time for the 2012 Olympics.

Table 1.8

Year (x) (1900→0)	Time (y) (sec)
76	232
80	231
84	231
88	227
92	225
96	228
100	221
104	223
108	223

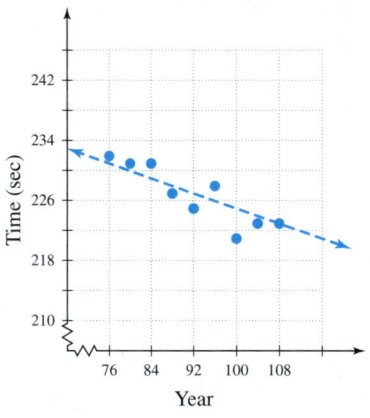

Solution ▶ Begin by choosing appropriate scales for the axes. The x-axis (year) could be scaled from 76 to 112, and the y-axis (swim time) from 210 to 246. This will allow for a "frame" around the data. After plotting the points, we obtain the scatterplot shown in the figure.

a. The association appears to be linear.

b. The association is negative, showing that finishing times tend to decrease over time.

c. There is a moderate to strong correlation.

d. The points (76, 232) and (104, 223) appear to be on a line approximating the data, and we select these to develop our equation model.

$$m = \frac{y_2 - y_1}{x_2 - x_1} \qquad \text{slope formula}$$

$$= \frac{223 - 232}{104 - 76} \qquad (x_1, y_1) \rightarrow (76, 232), (x_2, y_2) \rightarrow (104, 223)$$

$$\approx -0.32 \qquad \text{slope (rounded to tenths)}$$

$$y - 232 = -0.32(x - 76) \qquad \text{point-slope form}$$

$$y - 232 = -0.32x + 24.32 \qquad \text{distribute}$$

$$y = -0.32x + 256.32 \qquad \text{add 232 (solve for } y\text{)}$$

One model for this data is $y = -0.32x + 256.32$. Based on this model, the predicted time for the 2012 Olympics would be

$$\begin{aligned} f(x) &= -0.32x + 256.32 && \text{function model} \\ f(112) &= -0.32(112) + 256.32 && \text{substitute 112 for } x \text{ (2012)} \\ &= 220.48 && \text{result} \end{aligned}$$

In 2012 the winning time is projected to be about 220.5 sec.

Now try Exercises 23 and 24 ►

As a reminder, great care should be taken when using equation models obtained from real data. It would be foolish to assume that in the year 2700, swim times for the 400-m freestyle would be near 0 sec—even though that's what the model predicts for $x = 800$. Most function models are limited by numerous constraining factors, and data collected over a much longer period of time might even be better approximated using a nonlinear model.

☑ **D.** You've just seen how we can find a linear function that models relationships observed in a set of data

E. Linear Regression and the Line of Best Fit

There is actually a sophisticated method for calculating the equation of a line that best fits a data set, called the **regression line.** The method minimizes the vertical distance between all data points and the line itself, making it the unique **line of best fit.** Most graphing calculators have the ability to perform this calculation quickly. The process involves these steps: (1) clearing old data, (2) entering new data, (3) displaying the data, (4) calculating the regression line, and (5) displaying and using the regression line. We'll illustrate by finding the regression line for the data shown in Table 1.8 in Example 5, which gives the men's 400-m freestyle gold medal times (in seconds) for the 1976 through the 2008 Olympics, with 1900→0.

Step 1: Clear Old Data

To prepare for the new data, we first clear out any old data. Press the STAT key and select option **4:ClrList.** This places the **ClrList** command on the home screen. We tell the calculator which lists to clear by pressing 2nd 1 to indicate List1 (L1), then enter a comma using the , key, and continue entering other lists we want to clear: 2nd 2 , 2nd 3 ENTER will clear List1 (L1), List2 (L2), and List3 (L3).

Step 2: Enter New Data

Press the STAT key and select option **1:Edit.** Move the cursor to the first position of List1, and simply enter the data from the first column of Table 1.8 in order: 76 ENTER 80 ENTER 84 ENTER, and so on. Then use the right arrow ▸ to navigate to List2, and enter the data from the second column: 232 ENTER 231 ENTER 231 ENTER, and so on. When finished, you should obtain the screen shown in Figure 1.97.

Figure 1.97

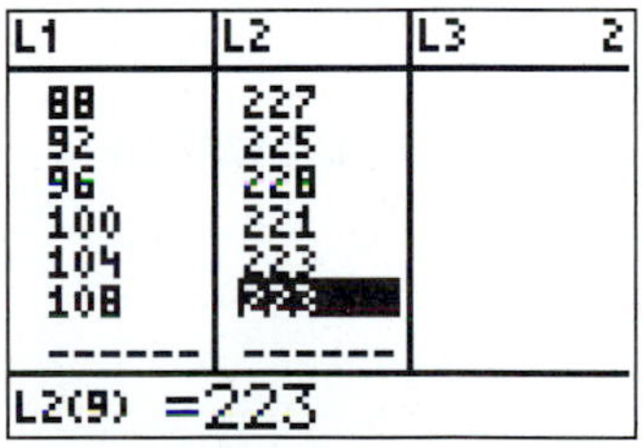
L1
L2
L3
2
88
92
96
100
104
108
227
225
228
221
223
L2(9) =223

WORTHY OF NOTE

As a rule of thumb, the tick marks for Xscl can be set by mentally estimating $\frac{|Xmax| + |Xmin|}{10}$ and using a convenient number in the neighborhood of the result (the same goes for Yscl). As an alternative to manually setting the window, the ZOOM **9:ZoomStat** feature can be used.

Step 3: Display the Data

With the data held in these lists, we can now display the related ordered pairs on the coordinate grid. First press the Y= key and CLEAR any existing equations. Then press 2nd Y= to access the "**STATPLOTS**" screen. With the cursor on **1:Plot1,** press ENTER and be sure the options shown in Figure 1.98 are highlighted. If you need to make any changes, navigate the cursor to

Figure 1.98

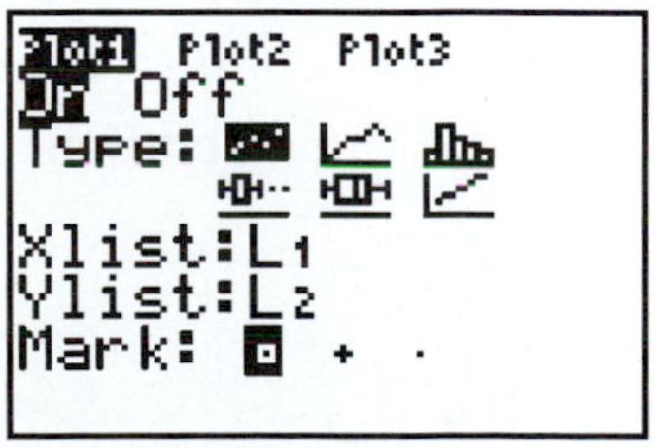
Plot1 Plot2 Plot3
On Off
Type:
Xlist:L1
Ylist:L2
Mark: + ·

the desired option and press ⏎. Note the data in L1 ranges from 76 to 108, while the data in L2 ranges from 221 to 232. This means an appropriate viewing window might be [70, 120] for the x-values, and [210, 240] for the y-values. Press the ⟨WINDOW⟩ key and set up the window accordingly. After you're finished, pressing the ⟨GRAPH⟩ key should produce the graph shown in Figure 1.99.

Figure 1.99

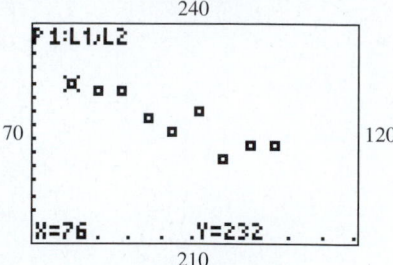

Step 4: Calculate the Regression Equation

To have the calculator compute the regression equation, press the ⟨STAT⟩ and ⟨▷⟩ keys to move the cursor over to the **CALC** options (see Figure 1.100). Since it appears the data is best modeled by a linear equation, we choose option **4:LinReg(ax + b).** Pressing the number 4 places this option on the home screen, and pressing ⏎ computes the values of a and b (the calculator automatically uses the values in L1 and L2 unless instructed otherwise). Rounded to hundredths, the linear regression model is $y = -0.33x + 257.06$ (Figure 1.101).

Figure 1.100

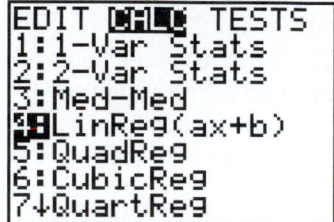

Step 5: Display and Use the Results

Although graphing calculators have the ability to paste the regression equation directly into Y_1 on the ⟨Y=⟩ screen, for now we'll enter $Y_1 = -0.33x + 257.06$ by hand. Afterward, pressing the ⟨GRAPH⟩ key will plot the data points (if Plot1 is still active) and graph the line. Your display screen should now look like the one in Figure 1.102. The regression line is the best estimator for the set of data as a whole, but there will still be some difference between the values it generates and the values from the set of raw data (the output in Figure 1.102 shows the estimated time for the 2000 Olympics was about 224 sec, when actually it was the year Ian Thorpe of Australia set a world record of 221 sec).

Figure 1.101

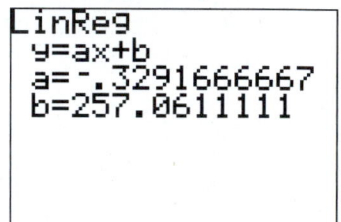

Figure 1.102

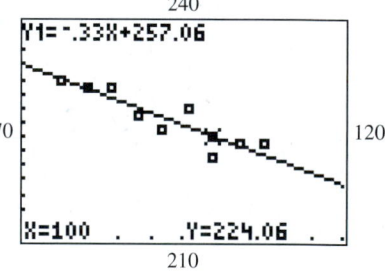

EXAMPLE 6 ▶ **Using Regression to Model Employee Performance**

Riverside Electronics reviews employee performance semiannually, and awards increases in their hourly rate of pay based on the review. The table shows Thomas' hourly wage for the last 4 yr (eight reviews). Find the regression equation for the data and use it to project his hourly wage for the year 2011, after his fourteenth review.

Review (x)	Wage (y)
(2004) 1	$9.58
2	$9.75
(2005) 3	$10.54
4	$11.41
(2006) 5	$11.60
6	$11.91
(2007) 7	$12.11
8	$13.02

Solution ▶ Following the prescribed sequence produces the equation $y = 0.48x + 9.09$. For $x = 14$ we obtain $y = 0.48(14) + 9.09$ or a wage of $15.81. According to this model, Thomas will be earning $15.81 per hour in 2011.

Now try Exercises 27 through 34 ▶

With each linear regression, the calculator can be set to compute a correlation coefficient that is a measure of how well the equation fits the data (see Subsection C). To display this "r-value" use (2nd) (0) (**CATALOG**) and activate **DiagnosticOn.** Figure 1.103 shows a scatterplot with perfect negative correlation ($r = -1$) and notice all data points are *on* the line. Figure 1.104 shows a strong positive correlation ($r \approx 0.98$) for the data from Example 6. **See Exercise 35.**

Figure 1.103

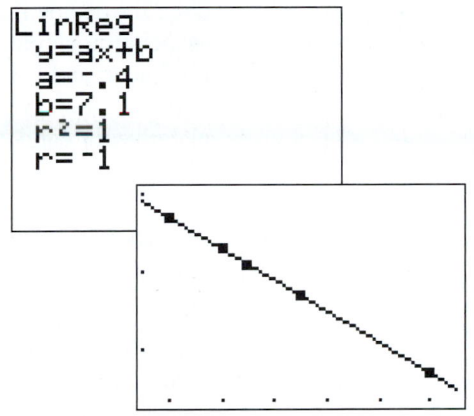

Figure 1.104

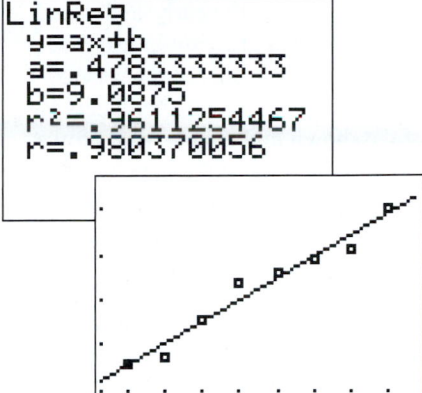

 **E.** You've just seen how we can use a linear regression to find the line of best fit

1.6 EXERCISES

▶ CONCEPTS AND VOCABULARY

Fill in each blank with the appropriate word or phrase. Carefully reread the section if needed.

1. When the ordered pairs from a set of data are plotted on a coordinate grid, the result is called a ___scatterplot___ .

2. If the data points seem to form a curved pattern or if no pattern is apparent, the data is said to have a ___nonlinear___ association.

3. If the data points seems to cluster along an imaginary line, the data is said to have a ___linear___ association.

4. If the pattern of data points seems to increase as they are viewed left to right, the data is said to have a ___positive___ association.

5. Compare/Contrast: One scatterplot is linear, with a weak and positive association. Another is linear, with a strong and negative association. Give a written description of each scatterplot.
   Answers will vary.

6. Discuss/Explain how this is possible: Working from the same scatterplot, Demetrius obtained the equation $y = -0.64x + 44$ for his equation model, while Jessie got the equation $y = -0.59x + 42$.
   Answers will vary.

HOMEWORK SELECTION GUIDE

Core: 7–33 odd (14 Exercises)
Standard: 1–4, 7–35 odd (19 Exercises)

Extended: 1–4, 7–23 odd, 25, 26, 21–29 odd, 30, 31, 33, 34, 35, 38, 40 (27 Exercises)
In Depth: 1–6, 7–23 odd, 25, 26, 21–29 odd, 30, 31, 33, 34, 35–40 (32 Exercises)

► DEVELOPING YOUR SKILLS

7. For mail with a high priority, "Express Mail" offers next day delivery by 12:00 noon to most destinations, 365 days of the year. The service was first offered by the U.S. Postal Service in the early 1980s and has been growing in use ever since. The cost of the service (in cents) for selected years is shown in the table. (a) Draw a scatterplot of the data, then (b) decide if the association is positive, negative, or cannot be determined.   **b.** positive

x	y
1981	935
1985	1075
1988	1200
1991	1395
1995	1500
1999	1575
2002	1785
2010	1830

Source: 2004 *Statistical Abstract of the United States;* USPS.com

8. After the Surgeon General's first warning in 1964, cigarette consumption began a steady decline as advertising was banned from television and radio, and public awareness of the dangers of cigarette smoking grew. The percentage of the U.S. adult population who considered themselves smokers is shown in the table for selected years. (a) Draw a scatterplot of the data, then (b) decide if the association is positive, negative, or cannot be determined.   **b.** negative

x	y
1965	42.4
1974	37.1
1979	33.5
1985	29.9
1990	25.3
1995	24.6
2000	23.1
2002	22.4
2005	16.9

Source: 1998 *Wall Street Journal Almanac* and 2009 *Statistical Abstract of the United States,* Table 1299

9. Since the 1970s women have made tremendous gains in the political arena, with more and more female candidates running for, and winning seats in the U.S. Senate and U.S. Congress. The number of women candidates for the U.S. Congress is shown in the table for selected years. (a) Draw a scatterplot of the data, (b) decide if the association is linear or nonlinear and (c) if the association is positive, negative, or cannot be determined.   **b.** linear   **c.** positive

x	y
1972	32
1978	46
1984	65
1992	106
1998	121
2004	141

Source: Center for American Women and Politics at www.cawp.rutgers.edu/Facts3.html

10. The number of shares traded on the New York Stock Exchange experienced dramatic change in the 1990s as more and more individual investors gained access to the stock market via the Internet

and online brokerage houses. The volume is shown in the table for 2002, and the odd numbered years from 1991 to 2001 (in billions of shares). (a) Draw a scatterplot of the data, (b) decide if the association is linear or nonlinear, and (c) if the association is positive, negative, or cannot be determined.   **b.** nonlinear   **c.** positive

x	y
1991	46
1993	67
1995	88
1997	134
1999	206
2001	311
2002	369

Source: 2000 and 2004 *Statistical Abstract of the United States,* Table 1202

The data sets in Exercises 11 and 12 are known to be linear.

11. The total value of the goods and services produced by a nation is called its gross domestic product or GDP. The *GDP per capita* is the ratio of the GDP for a given year to the population that year, and is one of many indicators of economic health. The GDP per capita (in $1000s) for the United States is shown in the table for selected years. (a) Draw a scatterplot using scales that appropriately fit the data, then sketch an estimated line of best fit, (b) decide if the association is positive or negative, then (c) decide whether the correlation is weak or strong.   **b.** positive   **c.** strong

x (1970 → 0)	y
0	5.1
5	7.6
10	12.3
15	17.7
20	23.3
25	27.7
30	35.0
33	37.8

Source: 2004 *Statistical Abstract of the United States,* Tables 2 and 641

12. Real estate brokers carefully track sales of new homes looking for trends in location, price, size, and other factors. The table relates the average selling price within a price range (homes in the $120,000 to $140,000 range are represented by the $130,000 figure), to the number of new homes sold by Homestead Realty in 2004. (a) Draw a scatterplot using scales that appropriately fit the data, then sketch an estimated line of best fit, (b) decide if the association is positive or negative, then (c) decide whether the correlation is weak or strong.   **b.** negative   **c.** strong

Price	Sales
130's	126
150's	95
170's	103
190's	75
210's	44
230's	59
250's	21

For the scatterplots given: (a) Arrange them in order from the weakest to the strongest correlation, (b) sketch a line that seems to approximate the data, (c) state whether the association is positive, negative, or cannot be determined, and (d) choose two points on (or near) the line and use them to approximate its slope (rounded to one decimal place).

13.

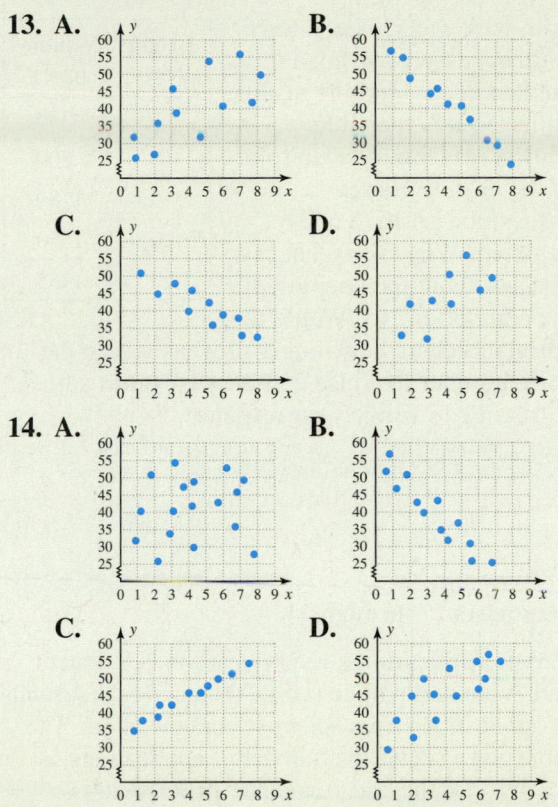

A. **B.**

C. **D.**

14. A. **B.**

C. **D.**

For the scatterplots given, (a) determine whether a linear or nonlinear model would seem more appropriate. (b) Determine if the association is positive or negative. (c) Classify the correlation as weak or strong. (d) If linear, sketch a line that seems to approximate the data and choose two points on the line and use them to approximate its slope.

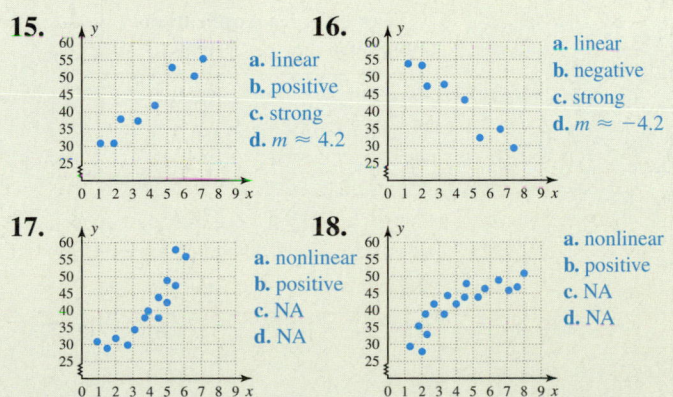

15.
a. linear
b. positive
c. strong
d. $m \approx 4.2$

16.
a. linear
b. negative
c. strong
d. $m \approx -4.2$

17.
a. nonlinear
b. positive
c. NA
d. NA

18.
a. nonlinear
b. positive
c. NA
d. NA

Additional answers can be found in the Instructor Answer Appendix.

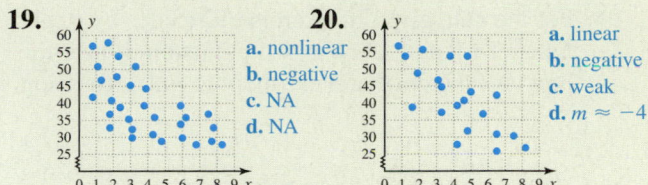

19.
a. nonlinear
b. negative
c. NA
d. NA

20.
a. linear
b. negative
c. weak
d. $m \approx -4$

21. In most areas of the country, law enforcement has become a major concern. The number of law enforcement officers employed by the federal government and having the authority to carry firearms and make arrests is shown in the table for selected years.

x (1990→0)	y (1000s)
3	68.8
6	74.5
8	83.1
10	88.5
14	93.4

(a) Draw a scatterplot using scales that appropriately fit the data and sketch an estimated line of best fit and (b) decide if the association is positive or negative. (c) Choose two points on or near the estimated line of best fit, and use them to find a function model and predict the number of federal law enforcement officers in 1995 and the projected number for 2011. Answers may vary.

Source: U.S. Bureau of Justice, Statistics at www.ojp.usdoj.gov/bjs/fedle.htm

b. positive   **c.** $f(x) = 2.4x + 62.3$, $f(5) = 74.3(74,300)$, $f(21) = 112.7 (112,700)$

22. Due to atmospheric pressure, the temperature at which water will boil varies predictably with the altitude. Using special equipment designed to duplicate atmospheric pressure, a lab experiment is set up to study this relationship for altitudes up to 8000 ft. The set of data collected is shown in the table, with the boiling temperature y in degrees Fahrenheit, depending on the altitude x in feet.

x	y
−1000	213.8
0	212.0
1000	210.2
2000	208.4
3000	206.5
4000	204.7
5000	202.9
6000	201.0
7000	199.2
8000	197.4

(a) Draw a scatterplot using scales that appropriately fit the data and sketch an estimated line of best fit, (b) decide if the association is positive or negative. (c) Choose two points on or near the estimated line of best fit, and use them to find a function model and predict the boiling point of water on the summit of Mt. Hood in Washington State (11,239 ft height), and along the shore of the Dead Sea (approximately 1312 ft below sea level). Answers may vary.   **b.** negative

c. $f(x) = -0.002x + 212.005$; $f(11,239) \approx 189.5°F$; $f(-1312) \approx 214.6°F$

23. For the data given in Exercise 11 (Gross Domestic Product per Capita), choose two points on or near the line you sketched and use them to find a function model for the data. Based on this model, what is the projected GDP per capita for the year 2010?

Using (5, 7.6) and (20, 23.3): $y \approx 1.05x + 2.37$; GDP in 2010 will be near 44,370

24. For the data given in Exercise 12 (Sales by Real Estate Brokers), choose two points on or near the line you sketched and use them to find a function model for the data. Based on this model, how many sales can be expected for homes costing $275,000? $300,000?   Using (130, 126) and (190, 75): $y = -0.85x + 236.5$; about 3 sales; no sales

▶ WORKING WITH FORMULAS

25. Circumference of a Circle: $C = 2\pi r$: The formula for the circumference of a circle can be written as a function of C in terms of r: $C(r) = 2\pi r$. (a) Set up a table of values for $r = 1$ through 6 and draw a scatterplot of the data. (b) Is the association positive or negative? Why? (c) What can you say about the strength of the correlation? (d) Sketch a line that "approximates" the data. What can you say about the slope of this line?   **b.** positive, larger radius ⇒ larger area   **c.** perfect correlation   **d.** $m = 2\pi$

26. Volume of a Cylinder: $V = \pi r^2 h$: As part of a project, students cut a long piece of PVC pipe with a diameter of 10 cm into sections that are 5, 10, 15, 20, and 25 cm long. The bottom of each is then made watertight and each section is filled to the brim with water. The volume is then measured using a flask marked in cm³ and the results collected into the table shown. (a) Draw a scatterplot of the data. (b) Is the association positive or negative? Why? (c) What can you say about the strength of the correlation? (d) Would the correlation here be stronger or weaker than the correlation in Exercise 25? Why? (e) Run a linear regression to verify your response.

Height (cm)	Volume (cm³)
5	380
10	800
15	1190
20	1550
25	1955

b. positive, greater height ⇒ greater volume   **c.** strong correlation   **d.** weaker; human error, cuts not exact, limitations on measurement   **e.** verified.

▶ APPLICATIONS

Use the regression capabilities of a graphing calculator to complete Exercises 27 through 34.

27. Height versus wingspan: Leonardo da Vinci's famous diagram is an illustration of how the human body comes in predictable proportions. One such comparison is a person's height to their wingspan (the maximum distance from the outstretched tip of one middle finger to the other). Careful measurements were taken on eight students and the set of data is shown here. Using the data, (a) draw the scatterplot; (b) determine whether the association is linear or nonlinear; (c) determine whether the association is positive or negative; and (d) find the regression equation and use it to predict the wingspan of a student with a height of 65 in.

28. Patent applications: Every year the U.S. Patent and Trademark Office (USPTO) receives thousands of applications from scientists and inventors. The table given shows the number of applications received for the odd years from 1993 to 2003 (1990 → 0). Use the data to (a) draw the scatterplot; (b) determine whether the association is linear or nonlinear; (c) determine whether the association is positive or negative; and (d) find the regression equation and use it to predict the number of applications that will be received in 2011.

Source: United States Patent and Trademark Office at www.uspto.gov/web

Height (x)	Wingspan (y)
61	60.5
61.5	62.5
54.5	54.5
73	71.5
67.5	66
51	50.75
57.5	54
52	51.5

Year (1990→0)	Applications (1000s)
3	188.0
5	236.7
7	237.0
9	278.3
11	344.7
13	355.4

b. linear   **c.** positive   **d.** $y = 17.18x + 135.94$; about 496.700

b. linear   **c.** positive   **d.** $y = 0.96x + 1.55$, 63.95 in.

Additional answers can be found in the Instructor Answer Appendix.

29. Patents issued: An increase in the number of patent applications (see Exercise 28), typically brings an increase in the number of patents issued, though many applications are denied due to improper filing, lack of scientific support, and other reasons. The table given shows the number of patents issued for the odd years from 1993 to 2003 (1999 → 0). Use the data to (a) draw the scatterplot; (b) determine whether the association is linear or nonlinear; (c) determine whether the association is positive or negative; and (d) find the regression equation and use it to predict the number of applications that will be approved in 2011. Which is increasing faster, the number of patent applications or the number of patents issued? How can you tell for sure?

Year (1990→0)	Patents (1000s)
3	107.3
5	114.2
7	122.9
9	159.2
11	187.8
13	189.6

Source: United States Patent and Trademark Office at www.uspto.gov/web
b. linear **c.** positive **d.** $y = 9.55x + 70.42$; about 271,000
The number of applications, since the line has a greater slope.

30. High jump records: In the sport of track and field, the high jumper is an unusual athlete. They seem to defy gravity as they launch their bodies over the high bar. The winning height at the summer Olympics (to the nearest unit) has steadily increased over time, as shown in the table for selected years. Using the data, (a) draw the scatterplot, (b) determine whether the association is linear or nonlinear, (c) determine whether the association is positive or negative, and (d) find the regression equation using $t = 0$ corresponding to 1900 and predict the winning height for the 2004 and 2008 Olympics. How close did the model come to the actual heights? **b.** linear **c.** positive **d.** $y = 0.21x + 73.64$, answers will vary

Source: athens2004.com

Year (1900→0)	Height in.
0	75
12	76
24	78
36	80
56	84
68	88
80	93
88	94
92	92
96	94
100	93
104	95.48
108	96.32

31. Females/males in the workforce: Over the last 4 decades, the percentage of the female population in the workforce has been increasing at a fairly steady rate. At the same time, the percentage of the male population in the workforce has been declining. The set of data is shown in the tables. Using the data, (a) draw scatterplots for both data sets, (b) determine whether the associations are linear or nonlinear, (c) determine whether the associations are positive or negative, and (d) determine if the percentage of females in the workforce is increasing faster than the percentage of males is decreasing. Discuss/Explain how you can tell for sure.

Source: 1998 *Wall Street Journal Almanac*, p. 316

Exercise 31 (women)

Year (x) (1950→0)	Percent
5	36
10	38
15	39
20	43
25	46
30	52
35	55
40	58
45	59
50	60

Exercise 31 (men)

Year (x) (1950→0)	Percent
5	85
10	83
15	81
20	80
25	78
30	77
35	76
40	76
45	75
50	73

b. women: linear **c.** positive **b.** men: linear **c.** negative **d.** yes, |slope| is greater

32. Height versus male shoe size: While it seems reasonable that taller people should have larger feet, there is actually a wide variation in the relationship between height and shoe size. The data in the table show the height (in inches) compared to the shoe size worn for a random sample of 12 male chemistry students. Using the data, (a) draw the scatterplot, (b) determine whether the association is linear or nonlinear, (c) determine whether the association is positive or negative, and (d) find the regression equation and use it to predict the shoe size of a man 80 in. tall and another that is 60 in. tall. Note that the heights of these two men fall outside of the range of our data set (see comment after Example 5 on page 168).
b. linear **c.** positive **d.** $y = 0.47x - 22.58$, 15.02(15), 5.62(5.5)

Height	Shoe Size
66	8
69	10
72	9
75	14
74	12
73	10.5
71	10
69.5	11.5
66.5	8.5
73	11
75	14
65.5	9

Additional answers can be found in the Instructor Answer Appendix.

33. Plastic money: The total amount of business transacted using credit cards has been changing rapidly over the last 15 to 20 years. The total volume (in billions of dollars) is shown in the table for selected years. (a) Use a graphing calculator to draw a scatterplot of the data and decide whether the association is linear or nonlinear. (b) Calculate a regression equation with $x = 0$ corresponding to 1990 and display the scatterplot and graph on the same screen. (c) According to the equation model, how many billions of dollars were transacted in 2003? How much will be transacted in the year 2011?

x (1990→0)	y
1	481
2	539
4	731
7	1080
8	1157
9	1291
10	1458
12	1638

Source: Statistical Abstract of the United States, various years

a. linear   **b.** $y = 108.18x + 330.20$ **c.** \$1736.54 billion; about \$2601.98 billion

34. Sales of hybrid cars: Since their mass introduction near the turn of the century, the sales of hybrid cars in the United States grew steadily until late 2007, when the price of gasoline began showing signs of weakening and eventually dipped below \$3.00/gal. Estimates for the annual sales of hybrid cars are given in the table for the years 2002 through 2009 (2000→0). (a) Use a graphing calculator to draw a scatterplot of the data and decide if the association is linear or nonlinear. (b) If linear, calculate a regression model for the data and display the scatterplot and data on the same screen. (c) Assuming that sales of hybrid cars recover, how many hybrids does the model project will be sold in the year 2012?

Year (2000→0)	Hybrid Sales (in thousands)
2	35
3	48
4	88
5	200
6	250
7	352
8	313
9	292

Source: http://www.hybridcar.com

a. (see part b), linear   **c.** 504,140 cars

▶ EXTENDING THE CONCEPT

35. It can be very misleading to rely on the correlation coefficient alone when selecting a regression model. To illustrate, (a) run a linear regression on the data set given (without doing a scatterplot), and note the strength of the correlation (the correlation coefficient). (b) Now run a quadratic regression (**STAT** CALC 5:QuadReg) and note the strength of the correlation. (c) What do you notice? What factors other than the correlation coefficient must be taken into account when choosing a form of regression?

x	y
50	67
100	125
150	145
200	275
250	370
300	550
350	600

a. $r \approx 0.9783$ **b.** $r \approx 0.9783$ **c.** they are almost identical; context, pattern of scatterplot, anticipated growth, etc.

36. In his book *Gulliver's Travels,* Jonathan Swift describes how the Lilliputians were able to measure Gulliver for new clothes, even though he was a giant compared to them. According to the text, "Then they measured my right thumb, and desired no more . . . for by mathematical computation, once around the thumb is twice around the wrist, and so on to the neck and waist." Is it true that once around the neck is twice around the waist? Find at least 10 willing subjects and take measurements of their necks and waists in millimeters. Arrange the data in ordered pair form (circumference of neck, circumference of waist). Draw the scatterplot for this data. Does the association appear to be linear? Find the equation of the best fit line for this set of data. What is the slope of this line? Is the slope near $m = 2$?
Answers will vary.

► **MAINTAINING YOUR SKILLS**

37. (1.3) Is the graph shown here, the graph of a function? Discuss why or why not.

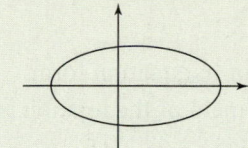

No. Except for the endpoints of the domain, one x is mapped to two y's.

38. (R.2/R.3) Determine the area of the figure shown $(A = LW, A = \pi r^2)$.

$A \approx 559.2 \text{ cm}^2$

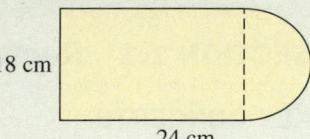

18 cm

24 cm

39. (1.5) Solve for r: $A = P + Prt$ $r = \dfrac{A - P}{Pt}$

40. (R.3) Solve for w (if possible):
$-2(6w^2 + 5) - 1 = 7w - 4(3w^2 + 1)$ $w = -1$

MAKING CONNECTIONS

Making Connections: Graphically, Symbolically, Numerically, and Verbally

Eight graphs (a) through (h) are given. Match the characteristics shown in 1 through 16 to one of the eight graphs.

(a)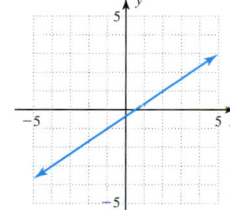

(b)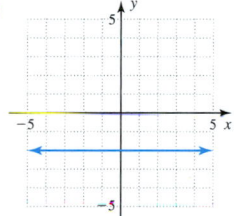

(c)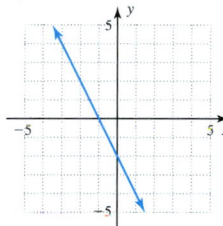

(d)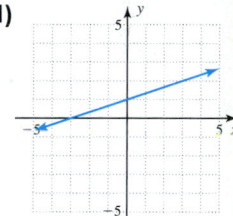

(e)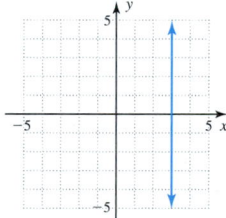

(f)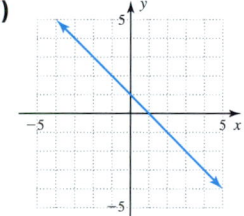

(g)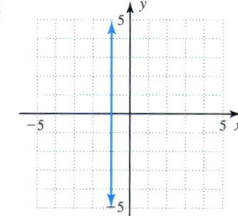

(h)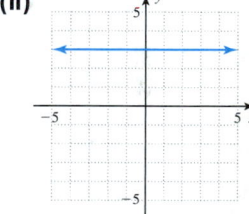

1. __d__ $y = \dfrac{1}{3}x + 1$

2. __f__ $y = -x + 1$

3. __a__ $m > 0, b < 0$

4. __g__ $x = -1$

5. __b__ $y = -2$

6. __c__ $m < 0, b < 0$

7. __c__ $m = -2$

8. __a__ $m = \dfrac{2}{3}$

9. __f__ $f(-3) = 4, f(1) = 0$

10. __h__ $f(-4) = 3, f(4) = 3$

11. __d__ $f(x) \geq 0$ for $x \in [-3, \infty)$

12. __e__ $x = 3$

13. __f__ $f(x) \leq 0$ for $x \in [1, \infty)$

14. __h__ m is zero

15. __a__ function is increasing, y-intercept is negative

16. __c__ function is decreasing, y-intercept is negative

SUMMARY AND CONCEPT REVIEW

SECTION 1.1   Rectangular Coordinates; Graphing Circles and Other Relations

KEY CONCEPTS

- A relation is a collection of ordered pairs (x, y) and can be stated as a set or in equation form.
- As a set of ordered pairs, we say the relation is pointwise-defined. The domain of the relation is the set of all first coordinates, and the range is the set of all corresponding second coordinates.
- A relation can be expressed in mapping notation $x \rightarrow y$, indicating an element from the domain is mapped to (corresponds to or is associated with) an element from the range.
- The graph of a relation in equation form is the set of all ordered pairs (x, y) that satisfy the equation. We plot a sufficient number of points and connect them with a straight line or smooth curve, depending on the pattern formed.
- The x- and y-variables of linear equations and their graphs have implied exponents of 1.
- With a relation entered on the $\boxed{\text{Y=}}$ screen, a graphing calculator can provide a table of ordered pairs and the related graph.
- The midpoint of a line segment with endpoints (x_1, y_1) and (x_2, y_2) is $\left(\dfrac{x_1 + x_2}{2}, \dfrac{y_1 + y_2}{2} \right)$.
- The distance between the points (x_1, y_1) and (x_2, y_2) is $d = \sqrt{(x_2 - x_1)^2 + (y_2 - y_1)^2}$.
- The equation of a circle centered at (h, k) with radius r is $(x - h)^2 + (y - k)^2 = r^2$.

EXERCISES

1. Represent the relation in mapping notation, then state the domain and range.

   $\{(-7, 3), (-4, -2), (5, 1), (-7, 0), (3, -2), (0, 8)\}$

2. Graph the relation $y = \sqrt{25 - x^2}$ by completing the table, then state the domain and range of the relation.

x	y
-5	0
-4	3
-2	$\sqrt{21} \approx 4.58$
0	5
2	$\sqrt{21} \approx 4.58$
4	3
5	0

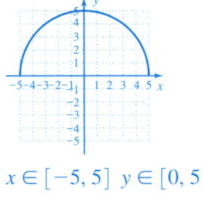

$x \in [-5, 5] \; y \in [0, 5]$

3. Use a graphing calculator to graph the relation $5x + 3y = -15$. Then use the **TABLE** feature to determine the value of y when $x = 0$, and the value(s) of x when $y = 0$, and write the results in ordered pair form.

Mr. Northeast and Mr. Southwest live in Coordinate County and are good friends. Mr. Northeast lives at *19 East and 25 North* or (19, 25), while Mr. Southwest lives at *14 West and 31 South* or $(-14, -31)$. If the streets in Coordinate County are laid out in one mile squares,

4. Use the distance formula to find how far apart they live.   65 mi

5. If they agree to meet halfway between their homes, what are the coordinates of their meeting place? $(\frac{5}{2}, -3)$

6. Sketch the graph of $x^2 + y^2 = 16$.

7. Sketch the graph of $x^2 + y^2 + 6x + 4y + 9 = 0$.

8. Find an equation of the circle whose diameter has the endpoints $(-3, 0)$ and $(0, 4)$. $(x + 1.5)^2 + (y - 2)^2 = 6.25$

SECTION 1.2   Linear Equations and Rates of Change

KEY CONCEPTS

- A linear equation can be written in the form $ax + by = c$, where a and b are not simultaneously equal to 0.

- The slope of the line through (x_1, y_1) and (x_2, y_2) is $m = \dfrac{y_2 - y_1}{x_2 - x_1}$, where $x_1 \neq x_2$.

- Other designations for slope are $m = \dfrac{\text{rise}}{\text{run}} = \dfrac{\text{change in } y}{\text{change in } x} = \dfrac{\Delta y}{\Delta x} = \dfrac{\text{vertical change}}{\text{horizontal change}}$.

- Lines with positive slope $(m > 0)$ rise from left to right; lines with negative slope $(m < 0)$ fall from left to right.

- The equation of a horizontal line is $y = k$; the slope is $m = 0$.

- The equation of a vertical line is $x = h$; the slope is undefined.

- Lines can be graphed using the intercept method. First determine $(x, 0)$ (substitute 0 for y and solve for x), then $(0, y)$ (substitute 0 for x and solve for y). Then draw a straight line through these points.

- Parallel lines have equal slopes $(m_1 = m_2)$; perpendicular lines have slopes that are negative reciprocals $\left(m_1 = -\dfrac{1}{m_2} \text{ or } m_1 \cdot m_2 = -1 \right)$.

EXERCISES

9. Plot the points and determine the slope, then use the ratio $\dfrac{\Delta y}{\Delta x} = \dfrac{\text{rise}}{\text{run}}$ to find an additional point on the line:
   a. $(-4, 3)$ and $(5, -2)$ and **b.** $(3, 4)$ and $(-6, 1)$.

10. Use the slope formula to determine if lines L_1 and L_2 are parallel, perpendicular, or neither:
    a. L_1: $(-2, 0)$ and $(0, 6)$; L_2: $(1, 8)$ and $(0, 5)$ parallel
    b. L_1: $(1, 10)$ and $(-1, 7)$: L_2: $(-2, -1)$ and $(1, -3)$ perpendicular

11. Graph each equation by plotting points: (a) $y = 3x - 2$ (b) $y = -\frac{3}{2}x + 1$.

12. Find the intercepts for each line and sketch the graph: (a) $2x + 3y = 6$ (b) $y = \frac{4}{3}x - 2$.

13. Identify each line as either horizontal, vertical, or neither, and graph each line.
    a. $x = 5$ **b.** $y = -4$ **c.** $2y + x = 5$

14. Determine if the triangle with the vertices given is a right triangle: $(-5, -4), (7, 2), (0, 16)$.   yes

15. Find the slope and y-intercept of the line shown and discuss the slope ratio in this context.

    $m = \frac{2}{3}$, y-intercept $(0, 2)$; when the rodent population increases by 3000, the hawk population increases by 200.

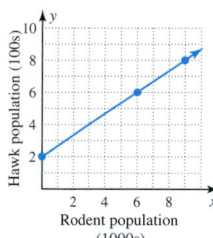

SECTION 1.3   Functions, Function Notation, and the Graph of a Function

KEY CONCEPTS

- A function is a relation, rule, or equation that pairs each element from the domain with exactly one element of the range.

- The vertical line test says that if every vertical line crosses the graph of a relation in at most one point, the relation is a function.

- The domain and range can be stated using set notation, graphed on a number line, or expressed using interval notation.

Additional answers can be found in the Instructor Answer Appendix.

- On a graph, vertical boundary lines can be used to identify the domain, or the set of "allowable inputs" for a function.
- On a graph, horizontal boundary lines can be used to identify the range, or the set of y-values (outputs) generated by the function.
- When a function is stated as an equation, the implied domain is the set of x-values that yield real number outputs.
- x-values that cause a denominator of zero or that cause the radicand of a square root expression to be negative must be excluded from the domain.
- *The phrase "y is a function of x,"* is written as $y = f(x)$. This notation enables us to summarize the three most important aspects of a function with a single expression (input, sequence of operations, output).

EXERCISES

16. State the implied domain of each function:

    a. $f(x) = \sqrt{4x + 5}$ $x \in [\frac{-5}{4}, \infty)$ **b.** $g(x) = \dfrac{x - 4}{x^2 - x - 6}$ $x \in (-\infty, -2) \cup (-2, 3) \cup (3, \infty)$

17. Determine $h(-2)$, $h(-\frac{2}{3})$, and $h(3a)$ for $h(x) = 2x^2 - 3x$. $14; \frac{26}{9}; 18a^2 - 9a$

18. Determine if the mapping given represents a function. If not, explain how the definition of a function is violated.

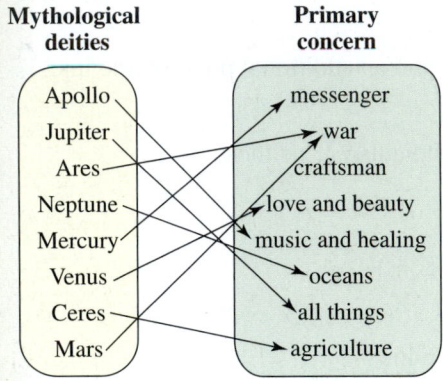

Mythological deities   **Primary concern**

Apollo
Jupiter
Ares
Neptune
Mercury
Venus
Ceres
Mars

messenger
war
craftsman
love and beauty
music and healing
oceans
all things
agriculture

It is a function.

19. For the graph of each function shown, (a) state the domain and range, (b) find the value of $f(2)$, and (c) determine the value(s) of x for which $f(x) = 1$.

  I. 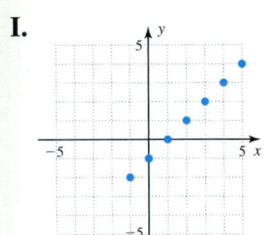   **II.** 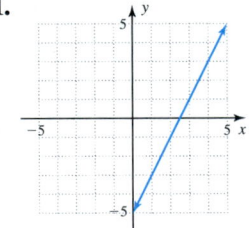   **III.**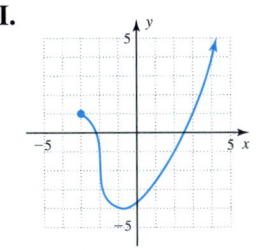

SECTION 1.4   Linear Functions, Special Forms, and More on Rates of Change

KEY CONCEPTS

- The equation of a nonvertical line in slope-intercept form is $y = mx + b$ or $f(x) = mx + b$. The slope of the line is m and the y-intercept is $(0, b)$.
- To graph a line given its equation in slope-intercept form, plot the y-intercept, then use the slope ratio $m = \dfrac{\Delta y}{\Delta x}$ to find a second point, and draw a line through these points.
- If the slope m and a point (x_1, y_1) on the line are known, the equation of the line can be written in point-slope form: $y - y_1 = m(x - x_1)$.

- A secant line is the straight line drawn through two points on a nonlinear graph.
- The notation $m = \dfrac{\Delta y}{\Delta x}$ literally means the quantity measured along the y-axis is changing with respect to changes in the quantity measured along the x-axis.

EXERCISES

20. Write each equation in slope-intercept form, then identify the slope and y-intercept.

    a. $4x + 3y - 12 = 0$ $\;y = \frac{-4}{3}x + 4,\ m = \frac{-4}{3},\ y\text{-intercept }(0, 4)$ **b.** $5x - 3y = 15$ $\;y = \frac{5}{3}x - 5,\ m = \frac{5}{3},\ y\text{-intercept }(0, -5)$

21. Graph each equation using the slope and y-intercept.

    a. $f(x) = -\frac{2}{3}x + 1$ **b.** $h(x) = \frac{5}{2}x - 3$

22. Graph the line with the given slope through the given point.

    a. $m = \frac{2}{3}; (1, 4)$ **b.** $m = -\frac{1}{2}; (-2, 3)$

23. What are the equations of the horizontal line and the vertical line passing through $(-2, 5)$? Which line is the point $(7, 5)$ on? $\;y = 5, x = -2; y = 5$

24. Find the equation of the line passing through $(1, 2)$ and $(-3, 5)$. Write your final answer in slope-intercept form.

25. Find the equation for the line that is parallel to $4x - 3y = 12$ and passes through the point $(3, 4)$. Write your final answer in slope-intercept form. $\;f(x) = \frac{4}{3}x$

Exercise 26

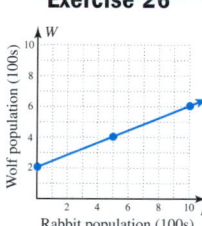

26. Determine the slope and y-intercept of the line shown. Then write the equation of the line in slope-intercept form and interpret the slope ratio $m = \dfrac{\Delta W}{\Delta R}$ in the context of this exercise. $\;m = \frac{2}{5}, y\text{-intercept }(0, 2), y = \frac{2}{5}x + 2.$ When the rabbit population increases by 500, the wolf population increases by 200.

Exercise 27

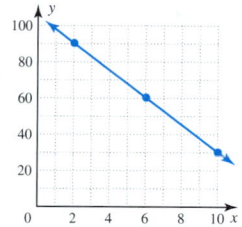

27. For the graph given, (a) find the equation of the line in point-slope form, (b) use the equation to predict the x- and y-intercepts, (c) write the equation in slope-intercept form, and (d) find y when $x = 20$, and the value of x for which $y = 15$.

SECTION 1.5 Solving Equations and Inequalities Graphically; Formulas and Problem Solving

KEY CONCEPTS

- To use the intersection-of-graphs method for solving equations, assign the left-hand expression as Y_1 and the right-hand as Y_2. The solution(s) of the original equation are the x-coordinate(s) of the point(s) of intersection of the graphs of Y_1 and Y_2.
- When an equation is written in the form $h(x) = 0$, the solutions can be found using the x-Intercept/Zeroes method. Assign $h(x)$ as Y_1, and find the x-intercepts of its graph.
- Linear inequalities can be solved by first applying the intersection-of-graphs method to identify the boundary value of the solution interval. Next, the solution is determined by a careful observation of the relative positions of the graphs (is Y_1 above or below Y_2) and the given inequality.
- To solve formulas for a specified variable, focus on the object variable and apply properties of equality to write this variable in terms of all others.
- The basic elements of good problem solving include:
  1. Gathering and organizing information
  2. Making the problem visual
  3. Developing an equation model
  4. Using the model to solve the application

For a complete review, see the problem-solving guide on page 157.

EXERCISES

28. Solve the following equation using the intersection-of-graphs method.

$3(x - 2) + 10 = 16 - 2(3 - 2x)$ $x = -6$

29. Solve the following equation using the *x*-intercept/zeroes method.

$2(x - 1) + \frac{3}{2} = 5(\frac{2}{5}x + \frac{1}{5}) - \frac{3}{2}$ $x \in (-\infty, \infty)$

30. Solve the following inequality using the intersection-of-graphs method.

$3(x + 2) - 2.2 < -2 + 4(0.2 - 0.5x)$ $x \in (-\infty, -1)$

Solve for the specified variable in each formula or literal equation.

31. $V = \pi r^2 h$ for h $h = \frac{V}{\pi r^2}$ **32.** $P = 2L + 2W$ for L $L = \frac{P - 2W}{2}$

33. $ax + b = c$ for x $x = \frac{c - b}{a}$ **34.** $2x - 3y = 6$ for y $y = \frac{2}{3}x - 2$

Use the problem-solving guidelines (page 157) to solve the following applications.

35. At a large family reunion, two kegs of lemonade are available. One is 2% sugar (too sour) and the second is 7% sugar (too sweet). How many gallons of the 2% keg, must be mixed with 12 gallons of the 7% keg to get a 5% mix?   8 gal

36. A rectangular window with a width of 3 ft and a height of 4 ft is topped by a semi-circular window. Find the total area of the window. $12 + \frac{9}{8}\pi$ ft$^2 \approx 15.5$ ft^2

37. Two cyclists start from the same location and ride in opposite directions, one riding at 15 mph and the other at 18 mph. If their radio phones have a range of 22 mi, how many minutes will they be able to communicate? $\frac{2}{3}$ hr = 40 min

SECTION 1.6   Linear Function Models and Real Data

KEY CONCEPTS

- A scatterplot is the graph of all the ordered pairs in a real data set.
- When drawing a scatterplot, be sure to scale the axes to comfortably fit the data.
- If larger inputs tend to produce larger output values, we say there is a positive association.
- If larger inputs tend to produce smaller output values, we say there is a negative association.
- If the data seem to cluster around an imaginary line, we say there is a linear association between the variables.
- If the data clearly cannot be approximated by a straight line, we say the variables exhibit a nonlinear association (or sometimes no association).
- The correlation coefficient *r* measures how tightly a set of data points cluster about an imaginary curve. The strength of the correlation is given as a value between -1 and 1. Measures close to -1 or 1 indicate a very strong correlation. Measures close to 0 indicate a very weak correlation.
- We can attempt to model linear data sets using an estimated line of best fit.
- A regression line minimizes the vertical distance between all data points and the graph itself, making it the unique line of best fit.

EXERCISES

Exercise 38

x (min study)	y (quiz score)
45	70
30	63
10	59
20	67
60	73
70	85
90	82
75	90

38. To determine the value of doing homework, a student in college algebra collects data on the time spent by classmates on their homework in preparation for a quiz. Her data is entered in the table shown. (a) Use a graphing calculator to draw a scatterplot of the data. (b) Does the association appear linear or nonlinear? (c) Is the association positive or negative?

39. If the association in Exercise 38 is linear, (a) use a graphing calculator to find a linear function that models the relation (study time, grade), then (b) graph the data and the line on the same screen. (c) Does the correlation appear weak or strong?

40. According to the function model from Exercise 39, what grade can I expect if I study for 120 minutes?

Additional answers can be found in the Instructor Answer Appendix.

PRACTICE TEST

1. Solve each equation.

a. $-\frac{2}{3}x - 5 = 7 - (x + 3)$ $x = 27$

b. $-5.7 + 3.1x = 14.5 - 4(x + 1.5)$ $x = 2$

c. $P = C + kC$; for C $C = \frac{P}{1 + k}$

d. $P = 2L + 2W$; for W $W = \frac{P - 2L}{2}$

2. How much water that is 102°F must be mixed with 25 gal of water at 91°F, so that the resulting temperature of the water will be 97°F.   30 gal

3. To make the bowling team, Jacques needs a three-game average of 160. If he bowled 141 and 162 for the first two games, what score S must be obtained in the third game so that his average is at least 160? $S \geq 177$

4. In the 2009 movie *Star Trek* (Chris Pine, Zachary Quinto, Zoy Zaldana, Eric Bana), Sulu falls off of the drill platform without a parachute, and Kirk dives off the platform to save him. To slow his fall, Zulu uses a spread-eagle tactic, while Kirk keeps his body straight and arms at his side, to maximize his falling speed. If Sulu is falling at a rate of 180 ft/sec, while Kirk is falling at 250 ft/sec, how long would it take Kirk to reach Sulu, if it took Kirk a full 2 sec to react and dive after Sulu?   about 5.1 sec

5. Two relations here are functions and two are not. Identify the nonfunctions (justify your response).

a. $x = y^2 + 2y$ **b.** $y = \sqrt{5 - 2x}$

c. $|y| + 1 = x$ **d.** $y = x^2 + 2x$

6. Determine whether the lines are parallel, perpendicular, or neither:

$L_1: 2x + 5y = -15$ and $L_2: y = \frac{2}{5}x + 7.$ neither

7. Graph the line using the slope and y-intercept: $x + 4y = 8$

8. Find the center and radius of the circle defined by $x^2 + y^2 - 4x + 6y - 3 = 0$, then sketch its graph.

9. After 2 sec, a car is traveling 20 mph. After 5 sec, its speed is 40 mph. Assuming the relationship is linear, find the velocity equation and use it to determine the speed of the car after 9 sec. $V = \frac{20}{3}t + \frac{20}{3}, 66\frac{2}{3}$ mph

10. Find the equation of the line parallel to $6x + 5y = 3$, containing the point $(2, -2)$. Answer in slope-intercept form. $y = \frac{-6}{5}x + \frac{2}{5}$

11. My partner and I are at coordinates $(-20, 15)$ on a map. If our destination is at coordinates $(35, -12)$, (a) what are the coordinates of the rest station located halfway to our destination? (b) How far away is our destination? Assume that each unit is 1 mi.   **a.** $(7.5, 1.5)$,   **b.** ≈ 61.27 mi

12. Write the equations for lines L_1 and L_2 shown.

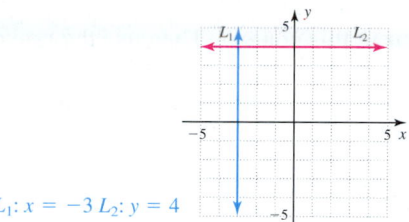

$L_1: x = -3$ $L_2: y = 4$

13. State the domain and range for the relations shown on graphs 13(a) and 13(b).

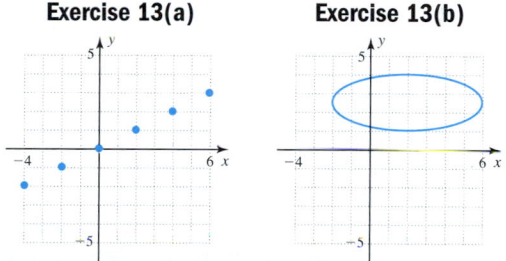

Exercise 13(a)   **Exercise 13(b)**

14. For the linear function shown,

a. Determine the value of $W(24)$ from the graph.   300

b. What input h will give an output of $W(h) = 375$?   30

c. Find a linear function for the graph. $W(h) = \frac{25}{2}h$

d. What does the slope indicate in this context?   Wages are \$12.50 per hr.

e. State the domain and range of h. $h \in [0, 40]; w \in [0, 500]$

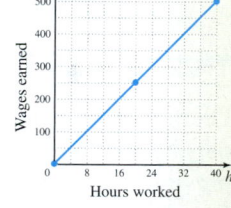

15. Given $f(x) = \dfrac{2 - x^2}{x^2}$, evaluate and simplify:

a. $f\left(\frac{2}{3}\right)$ $\frac{7}{2}$ **b.** $f(a + 3)$ $\dfrac{-a^2 - 6a - 7}{a^2 + 6a + 9}$

16. In 2007, there were 3.3 million Apple iPhones sold worldwide. By 2009, this figure had jumped to approximately 30.3 million [*Source*: http://brainstormtech.blogs. fortune.cnn.com/2009/03/12/]. Assume that for a time, this growth could be modeled by a linear function. (a) Determine the rate of change $\dfrac{\Delta \text{sales}}{\Delta \text{time}}$, and

Exercise 16

(b) interpret it in this context. Then use the rate of change to (c) approximate the number of sales in 2008, and what the projected sales would be for 2010 and 2011.

17. Solve the following equations using the *x*-intercept/zeroes method.

a. $2x + \left(4 - \dfrac{1}{3}x\right) = -(20 + x)$ $x = -9$

b. $2(0.7x - 1.3) + 2.6 = 2x - 3(0.2x - 2)$
no solution

18. Solve the following inequalities using the intersection-of-graphs method.

a. $3x - (5 - x) \geq 2(5 - x) + 3$ $x \in [3, \infty)$

b. $2(0.75x - 1) < 0.7 + 0.5(3x - 1)$ $x \in (-\infty, \infty)$

19. To study how annual rainfall affects the ability to attain certain levels of livestock production, a local university collects data on the average annual rainfall for a particular area and compares this to the average number of free-ranging cattle per acre for ranchers in that area. The data collected are shown in the table. (a) Use a graphing calculator to draw a scatterplot of the data. (b) Does the association appear linear or nonlinear? (c) Is the association positive or negative?

Exercise 19

Rainfall (in.)	Cattle per Acre
12	2
16	3
19	7
23	9
28	11
32	22
37	23
40	26

20. If the association in Exercise 19 is linear, (a) use a graphing calculator to find a linear function that models the relation (rainfall, cattle per acre), (b) use the function to find the number of cattle per acre that might be possible for an area receiving 50 in. of rainfall per year, and (c) state whether the correlation is weak or strong.   **a.** $f(x) = 0.91x - 10.78$ **b.** $f(50) \approx 35$
c. strong: $r \approx 0.974$

![pencil icon] **STRENGTHENING CORE SKILLS**

The Various Forms of a Linear Equation

Learning mathematics is very much like the construction of a skyscraper. The final height of the skyscraper ultimately depends on the strength of the foundation and quality of the frame supporting each new floor as it is built. Our previous work with linear functions and their graphs, while having a number of useful applications, is actually the foundation on which much of our future work is built. For this reason, it's important you gain a certain fluency with linear functions and relationships—even to a point where things come to you effortlessly and automatically. As noted mathematician Henri Lebesque once said, "An idea reaches its maximum level of usefulness only when you understand it so well that it seems like you have always known it. You then become incapable of seeing the idea as anything but a trivial and immediate result." These formulas and concepts, while simple, have an endless number of significant and substantial applications.

Forms and Formulas

slope formula	point-slope form	slope-intercept form	standard form
$m = \dfrac{y_2 - y_1}{x_2 - x_1}$	$y - y_1 = m(x - x_1)$	$y = mx + b$	$Ax + By = C$
given any two points on the line	given slope m and any point (x_1, y_1)	given slope m and *y*-intercept $(0, b)$	A, B, and C are integers (used in linear systems)

Characteristics of Lines

y-intercept	x-intercept	increasing	decreasing
$(0, y)$	$(x, 0)$	$m > 0$	$m < 0$
let $x = 0$, solve for y	let $y = 0$, solve for x	line slants upward from left to right	line slants downward form left to right

Relationships between Lines

intersecting	parallel	perpendicular	dependent
$m_1 \neq m_2$	$m_1 = m_2$, $b_1 \neq b_2$	$m_1 m_2 = -1$	$m_1 = m_2$, $b_1 = b_2$
lines intersect at one point	lines do not intersect	lines intersect at right angles	lines intersect at all points

Special Lines

horizontal	vertical	identity
$y = k$	$x = h$	$y = x$
horizontal line through k	vertical line through h	the input value *identifies* the output

Use the formulas and concepts reviewed here to complete the following exercises.

For the two points given: (a) compute the slope of the line through the points and state whether the line is increasing or decreasing, (b) find the equation of the line in point-slope form, then write the equation in slope-intercept form, and (c) find the x- and y-intercepts and graph the line.

Exercise 1: $P_1(0, 5)$ $P_2(6, 7)$ **Exercise 4:** $P_1(-5, -4)$ $P_2(3, 2)$

Exercise 2: $P_1(3, 2)$ $P_2(0, 9)$ **Exercise 5:** $P_1(-2, 5)$ $P_2(6, -1)$

Exercise 3: $P_1(3, 2)$ $P_2(9, 5)$ **Exercise 6:** $P_1(2, -7)$ $P_2(-8, -2)$

CALCULATOR EXPLORATION AND DISCOVERY

Evaluating Expressions and Looking for Patterns

These "explorations" are designed to explore the full potential of a graphing calculator, as well as to use this potential to investigate patterns and discover connections that might otherwise be overlooked. In this *exploration and discovery,* we point out the various ways an expression can be evaluated on a graphing calculator. Some ways seem easier, faster, and/or better than others, but each has advantages and disadvantages depending on the task at hand, and it will help to be aware of them all for future use.

One way to evaluate an expression is to use the **TABLE** feature of a graphing calculator, with the expression entered as Y_1 on the  screen. If you want the calculator to generate inputs, use the [2nd] [WINDOW] (**TBLSET**) screen to indicate a starting value (**TblStart=**) and an increment value (**ΔTbl =**), and set the calculator in **Indpnt: AUTO ASK** mode (to input specific values, the calculator should be in **Indpnt: AUTO ASK** mode). After pressing [2nd] [GRAPH] (**TABLE**), the calculator shows the corresponding input and output values.

Figure 1.105

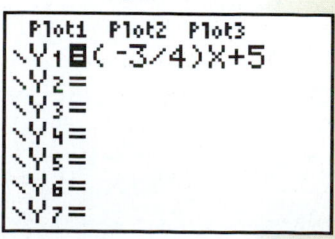

Expressions can also be evaluated on the home screen for a single value or a series of values. Enter the expression $-\frac{3}{4}x + 5$ on the  screen (see Figure 1.105) and use [2nd] [MODE] (**QUIT**) to get back to the home screen. To evaluate this expression, access Y_1 using [VARS] [▷] (**Y-VARS**), and use the first option **1:Function** [ENTER]. This brings us to a submenu where any of the equations Y_1 through Y_0 (actually Y_{10}) can be accessed. Since the default setting is the one we need (**1:Y1**), simply press [ENTER] and Y_1 appears on the home screen. To evaluate a single input, simply enclose it in parentheses. To evaluate more than one input, enter the numbers as a set of values with the set enclosed in parentheses. In Figure 1.106, Y_1 has been evaluated for $x = -4$, then simultaneously for $x = -4, -2, 0,$ and 2.

A third way to evaluate expressions is using a list, with the desired inputs entered in List 1 (L1), then List 2 (L2) defined in terms of L1. For example, $L2 = -\frac{3}{4}L1 + 5$ will return the same values for inputs of $-4, -2, 0,$ and 2 seen previously on the home screen (remember to clear the lists first). Lists are accessed by pressing **STAT** **1:Edit.** Enter the numbers $-4, -2, 0,$ and 2 in L1, then use the right arrow $\triangleright$ to move to L2. It is important to note that you *next press the up arrow key* $\triangle$ so that the cursor overlies L2. The bottom of the screen now reads "L2 = " and the calculator is waiting for us to define L2. After entering $L2 = -\frac{3}{4}L1 + 5$ (see Figure 1.107) and pressing **ENTER** we obtain the same outputs as before (see Figure 1.108). The advantage of using the "list" method is that we can *further explore or experiment with the output values* in a search for patterns.

Exercise 1: Evaluate the expression $0.2L1 + 3$ on the list screen, using consecutive integer inputs from -6 to 6 inclusive. What do you notice about the outputs?
They differ by 0.2.

Exercise 2: Evaluate the expression $\sqrt{2}L1 - \sqrt{9.1}$ on the list screen, using consecutive integer inputs from -6 to 6 inclusive. We suspect there is a pattern to the output values, but this time the pattern is very difficult to see. On the home screen, compute the difference between a few successive outputs from L2 [for example, $L2(1) - L2(2)$]. What do you notice?   They differ by $\sqrt{2} \approx 1.41$.

Figure 1.106

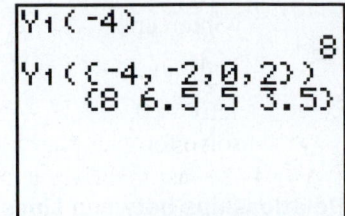

Figure 1.107

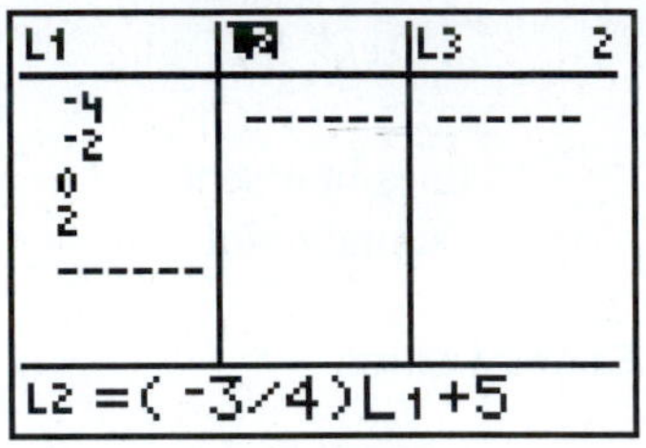

L1	▮	L3	2
-4	------	------	
-2			
0			
2			

L2 =(-3/4)L1+5

Figure 1.108

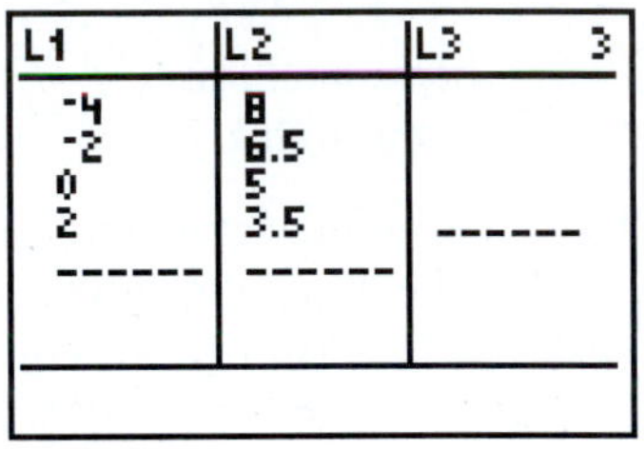

L1	L2	L3	3
-4	8		
-2	6.5		
0	5		
2	3.5	------	
------	------		

More on Functions

CHAPTER OUTLINE

CHAPTER CONNECTIONS

Viewing a function in terms of an equation, a table of values, and the related graph, often brings a clearer understanding of the relationships involved. For example, the power generated by a wind turbine is often modeled by the function $P(v) = \dfrac{8v^3}{125}$, where P is the power in watts and v is the wind velocity in miles per hour. While the formula enables us to predict the power generated for a given wind speed, the graph offers a visual representation of this relationship, where we note a rapid growth in power output as the wind speed increases. This application appears as Exercise 107 in Section 2.2.

Check out these other real-world connections:

▶ Analyzing the Path of a Projectile (Section 2.1, Exercise 57)

▶ Altitude of the Jet Stream (Section 2.3, Exercise 61)

▶ Amusement Arcades (Section 2.5, Exercise 42)

▶ Volume of Phone Calls (Section 2.6, Exercise 55)

2.1 Analyzing the Graph of a Function

In this section, we'll consolidate and refine many of the ideas we've encountered related to functions. When functions and graphs are applied as real-world models, we create numeric and visual representations that enable an informed response to questions involving *maximum* efficiency, *positive* returns, *increasing* costs, and other relationships that can have a great impact on our lives.

A. Graphs and Symmetry

While the domain and range of a function will remain dominant themes in our study, for the moment we turn our attention to other characteristics of a function's graph. We begin with the concept of symmetry.

Symmetry with Respect to the y-Axis

Figure 2.1

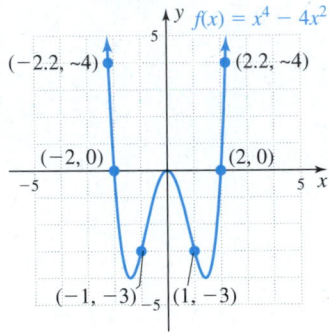

Consider the graph of $f(x) = x^4 - 4x^2$ shown in Figure 2.1, where the portion of the graph to the left of the y-axis appears to be a mirror image of the portion to the right. A function is **symmetric to the y-axis** if, given any point (x, y) on the graph, the point $(-x, y)$ is also on the graph. We note that $(-1, -3)$ is on the graph, as is $(1, -3)$, and that $(-2, 0)$ is an x-intercept of the graph, as is $(2, 0)$. Functions that are symmetric with respect to the y-axis are also known as **even functions** and in general we have:

> **Even Functions: y-Axis Symmetry**
>
> A function f is an *even function* if and only if, for each point (x, y) on the graph of f, the point $(-x, y)$ is also on the graph. In function notation
>
> $$f(-x) = f(x)$$

Symmetry can be a great help in graphing new functions, enabling us to plot fewer points and to complete the graph using properties of symmetry.

EXAMPLE 1 ▶ Graphing an Even Function Using Symmetry

a. The function $g(x)$ in Figure 2.2 (shown in solid blue) is known to be even. Draw the complete graph.

b. Show that $h(x) = x^{\frac{2}{3}}$ is an even function using the arbitrary value $x = k$ [show $h(-k) = h(k)$], then sketch the complete graph using $h(0)$, $h(1)$, $h(8)$, and y-axis symmetry.

Figure 2.2

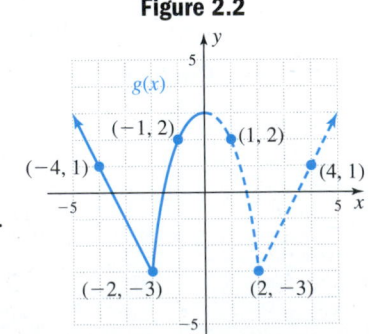

Solution ▶ **a.** To complete the graph of g (see Figure 2.2) use the points $(-4, 1)$, $(-2, -3)$, $(-1, 2)$, and y-axis symmetry to find additional points. The corresponding ordered pairs are $(4, 1)$, $(2, -3)$, and $(1, 2)$, which we use to help draw a "mirror image" of the partial graph given.

b. To prove that $h(x) = x^{\frac{2}{3}}$ is an even function, we must show $h(-k) = h(k)$ for any constant k. After writing $x^{\frac{2}{3}}$ as $[x^2]^{\frac{1}{3}}$, we have:

$h(-k) \overset{?}{=} h(k)$ first step of proof

$[(-k)^2]^{\frac{1}{3}} \overset{?}{=} [(k)^2]^{\frac{1}{3}}$ evaluate $h(-k)$ and $h(k)$

$\sqrt[3]{(-k)^2} \overset{?}{=} \sqrt[3]{(k)^2}$ radical form

$\sqrt[3]{k^2} = \sqrt[3]{k^2} \checkmark$ result: $(-k)^2 = k^2$

Using $h(0) = 0$, $h(1) = 1$, and $h(8) = 4$ with y-axis symmetry produces the graph shown in Figure 2.3.

Figure 2.3

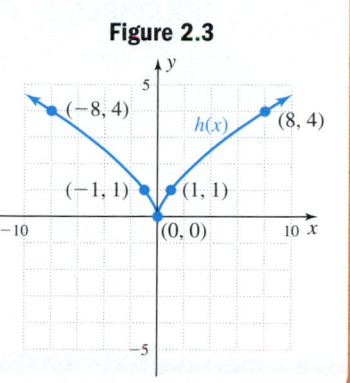

Now try Exercises 7 through 12 ▶

Symmetry with Respect to the Origin

Another common form of symmetry is known as **symmetry to the origin.** As the name implies, the graph is somehow "centered" at $(0, 0)$. This form of symmetry is easy to see for closed figures with their center at $(0, 0)$, like certain polygons, circles, and ellipses (these will exhibit both y-axis symmetry *and* symmetry with respect to the origin). Note the relation graphed in Figure 2.4 contains the points $(-3, 3)$ and $(3, -3)$, along with $(-1, -4)$ and $(1, 4)$. But the function $f(x)$ in Figure 2.5 also contains these points and is, in the same sense, symmetric to the origin (the paired points are on opposite sides of the x- and y-axes, and a like distance from the origin).

Figure 2.4

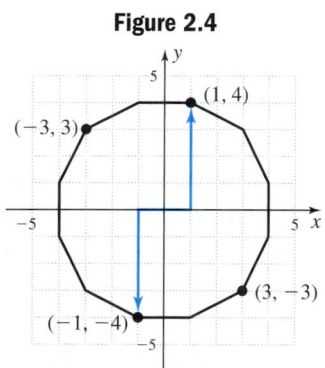

Figure 2.5

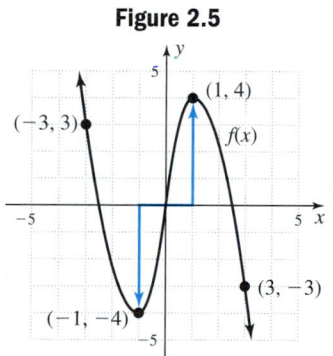

Functions symmetric to the origin are known as **odd functions** and in general we have:

Odd Functions: Symmetry About the Origin

A function f is an *odd function* if and only if, for each point (x, y) on the graph of f, the point $(-x, -y)$ is also on the graph. In function notation

$$f(-x) = -f(x)$$

EXAMPLE 2 ▶   Graphing an Odd Function Using Symmetry

a. In Figure 2.6, the function $g(x)$ given (shown in solid blue) is known to be *odd*. Draw the complete graph.

b. Show that $h(x) = x^3 - 4x$ is an odd function using the arbitrary value $x = k$ [show $h(-x) = -h(x)$], then sketch the graph using $h(-2)$, $h(-1)$, $h(0)$, and odd symmetry.

Solution ▶   **a.** To complete the graph of g, use the points $(-6, 3)$, $(-4, 0)$, and $(-2, 2)$ and odd symmetry to find additional points. The corresponding ordered pairs are $(6, -3)$, $(4, 0)$, and $(2, -2)$, which we use to help draw a "mirror image" of the partial graph given (see Figure 2.6).

Figure 2.6

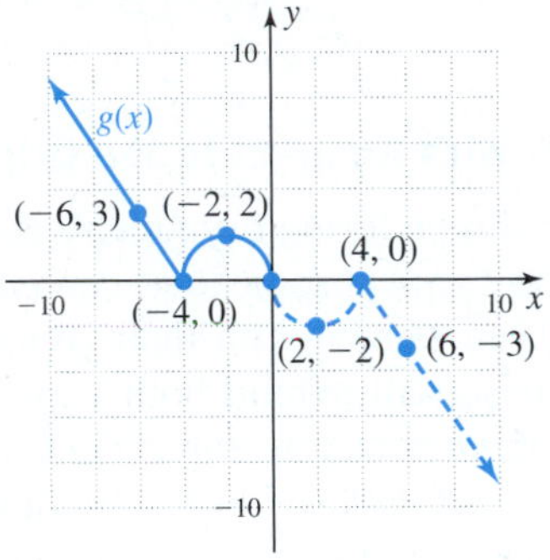

Figure 2.7

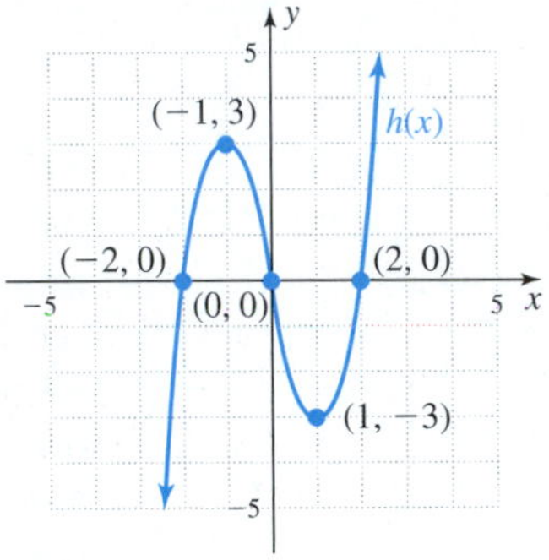

b. To prove that $h(x) = x^3 - 4x$ is an odd function, we must show that $h(-k) = -h(k)$.

$$h(-k) \overset{?}{=} -h(k)$$
$$(-k)^3 - 4(-k) \overset{?}{=} -[k^3 - 4k]$$
$$-k^3 + 4k = -k^3 + 4k \checkmark$$

Using $h(-2) = 0$, $h(-1) = 3$, and $h(0) = 0$ with symmetry about the origin produces the graph shown in Figure 2.7.

Now try Exercises 13 through 24 ▶

☑ **A.** You've just seen how we can determine whether a function is even, odd, or neither

Finally, some relations also exhibit a third form of symmetry, that of symmetry to the *x*-axis. If the graph of a circle is centered at the origin, the graph has both odd and even symmetry, and is also symmetric to the *x*-axis. Note that if a graph exhibits *x*-axis symmetry, *it cannot be the graph of a function.*

B. Intervals Where a Function Is Positive or Negative

Consider the graph of $f(x) = x^2 - 4$ shown in Figure 2.8, which has *x*-intercepts at $(-2, 0)$ and $(2, 0)$. As in Section 1.5, the *x*-intercepts have the form $(x, 0)$ and are called the **zeroes** of the function (the *x*-input causes an output of 0). Just as zero on the number line separates negative numbers from positive numbers, the zeroes of a function that crosses the *x*-axis separate *x*-intervals where a function is negative from *x*-intervals where the function is positive. Noting that outputs (*y*-values) are positive in Quadrants I and II, $f(x) > 0$ in intervals where its graph is *above the x-axis*. Conversely, $f(x) < 0$

in x-intervals where its graph is *below the x-axis*. To illustrate, compare the graph of f in Figure 2.8, with that of g in Figure 2.9.

Figure 2.8

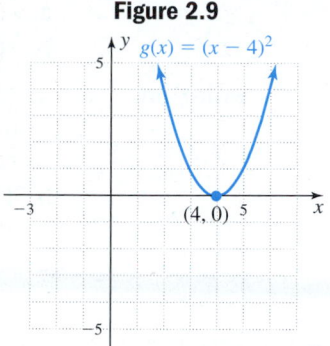

Figure 2.9

The graph of f is a parabola, with x-intercepts of $(-2, 0)$ and $(2, 0)$. Using our previous observations, we note $f(x) \geq 0$ for $x \in (-\infty, -2] \cup [2, \infty)$ since the graph is above the x-axis, and $f(x) < 0$ for $x \in (-2, 2)$. The graph of g is also a parabola, but is entirely above or on the x-axis, showing $g(x) \geq 0$ for $x \in \mathbb{R}$. The difference is that zeroes coming from factors of the form $(x - r)$ (with degree 1) allow the graph to cross the x-axis. The zeroes of f came from $(x + 2)(x - 2) = 0$. Zeroes that come from factors of the form $(x - r)^2$ (with degree 2) cause the graph to "bounce" off the x-axis (intersect without crossing) since all outputs must be nonnegative. The zero of g came from $(x - 4)^2 = 0$.

WORTHY OF NOTE

These observations form the basis for studying polynomials of higher degree in Chapter 4, where we extend the idea to factors of the form $(x - r)^n$ in a study of **roots of multiplicity**.

EXAMPLE 3 ▶ **Solving an Inequality Using a Graph**

Use the graph of $g(x) = x^3 - 2x^2 - 4x + 8$ given to solve the inequalities

a. $g(x) \geq 0$ **b.** $g(x) < 0$

Solution ▶ From the graph, the zeroes of g (x-intercepts) occur at $(-2, 0)$ and $(2, 0)$.

a. For $g(x) \geq 0$, the graph must be on or above the x-axis, meaning the solution is $x \in [-2, \infty)$.

b. For $g(x) < 0$, the graph must be below the x-axis, and the solution is $x \in (-\infty, -2)$.

As we might have anticipated from the graph, factoring by grouping gives $g(x) = (x + 2)(x - 2)^2$, with the graph crossing the x-axis at -2, and bouncing off the x-axis (intersects without crossing) at $x = 2$.

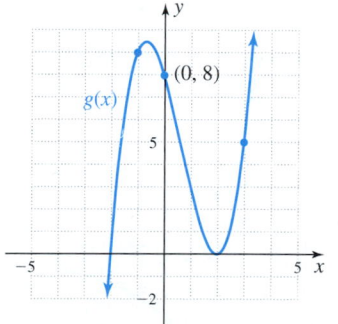

Now try Exercises 25 through 28 ▶

Even if the function is not a polynomial, the zeroes can still be used to find x-intervals where the function is positive or negative.

EXAMPLE 4 ▶ **Solving an Inequality Using a Graph**

For the graph of $r(x) = \sqrt{x+1} - 2$ shown, solve
   a. $r(x) \leq 0$
   b. $r(x) > 0$

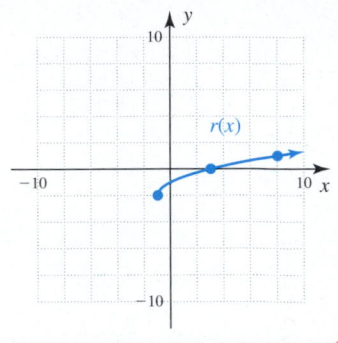

Solution ▶ **a.** The only zero of r is at $(3, 0)$. The graph is on
       or below the x-axis for $x \in [-1, 3]$, so
       $r(x) \leq 0$ in this interval.
   b. The graph is above the x-axis for $x \in (3, \infty)$,
       and $r(x) > 0$ in this interval.

Now try Exercises 29 through 32 ▶

☑ **B. You've just seen how we can determine intervals where a function is positive or negative**

This study of inequalities shows how the graphical solutions studied in Section 1.5 are easily extended to the graph of a general function. It also strengthens the foundation for the graphical solutions studied throughout this text.

C. Intervals Where a Function Is Increasing or Decreasing

In our study of linear graphs, we said a graph was increasing if it "rose" when viewed from left to right. More generally, we say the graph of a function is increasing *on a given interval* if larger and larger x-values produce larger and larger y-values. This suggests the following tests for intervals where a function is increasing or decreasing.

Increasing and Decreasing Functions

Given an interval I that is a subset of the domain, with x_1 and x_2 in I and $x_2 > x_1$,

1. A function is increasing on I if $f(x_2) > f(x_1)$ for all x_1 and x_2 in I (larger inputs produce larger outputs).

2. A function is decreasing on I if $f(x_2) < f(x_1)$ for all x_1 and x_2 in I (larger inputs produce smaller outputs).

3. A function is constant on I if $f(x_2) = f(x_1)$ for all x_1 and x_2 in I (larger inputs produce identical outputs).

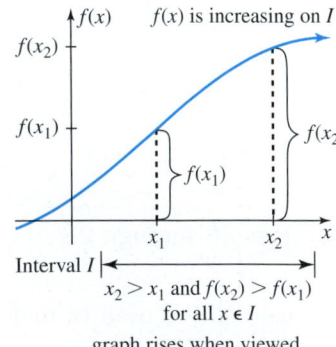

$x_2 > x_1$ and $f(x_2) > f(x_1)$
for all $x \in I$

graph rises when viewed
from left to right

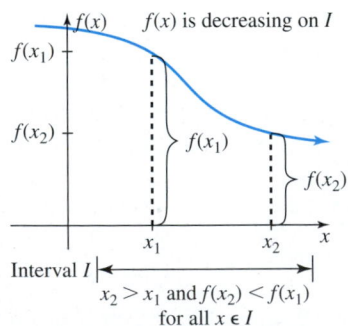

$x_2 > x_1$ and $f(x_2) < f(x_1)$
for all $x \in I$

graph falls when viewed
from left to right

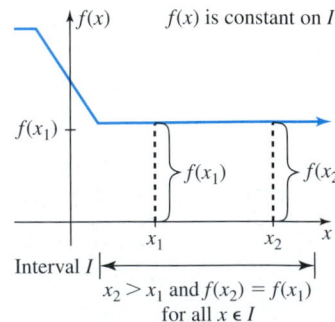

$x_2 > x_1$ and $f(x_2) = f(x_1)$
for all $x \in I$

graph is level when viewed
from left to right

Consider the graph of $f(x) = -x^2 + 4x + 5$ given in Figure 2.10. Since the parabola opens downward with the vertex at $(2, 9)$, the function must increase until it reaches this peak at $x = 2$, and decrease thereafter. Notationally we'll write this as $f(x)\uparrow$ for $x \in (-\infty, 2)$ and $f(x)\downarrow$ for $x \in (2, \infty)$. Using the interval $(-3, 2)$ shown below the figure, we see that any larger input value from the interval will indeed produce a larger output value, and $f(x)\uparrow$ on the interval. For instance,

Figure 2.10

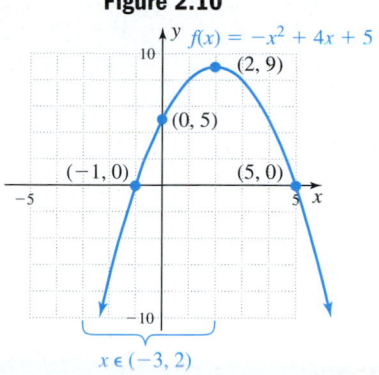

$$1 > -2 \qquad\qquad x_2 > x_1$$

$$\text{and} \qquad\qquad \text{and}$$

$$f(1) > f(-2) \qquad f(x_2) > f(x_1)$$
$$8 > -7$$

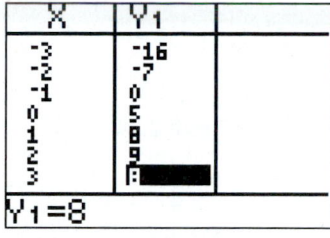

A calculator check is shown in the figure. Note the outputs are increasing until $x = 2$, then they begin decreasing.

EXAMPLE 5 ▶ **Finding Intervals Where a Function Is Increasing or Decreasing**

Use the graph of $v(x)$ given to name the interval(s) where v is increasing, decreasing, or constant.

Solution ▶ From left to right, the graph of v increases until leveling off at $(-2, 2)$, then it remains constant until reaching $(1, 2)$. The graph then increases once again until reaching a peak at $(3, 5)$ and decreases thereafter. The result is $v(x)\uparrow$ for $x \in (-\infty, -2) \cup (1, 3)$, $v(x)\downarrow$ for $x \in (3, \infty)$, and $v(x)$ is constant for $x \in (-2, 1)$.

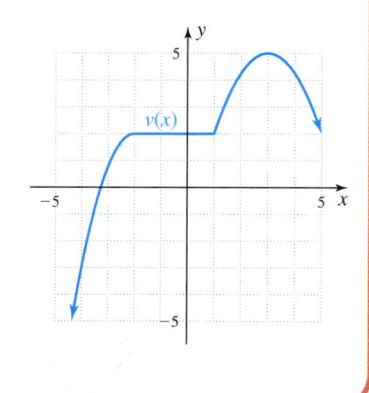

Now try Exercises 33 through 36 ▶

WORTHY OF NOTE

Questions about the behavior of a function are asked with respect to the *y* outputs: is the *function* positive, is the *function* increasing, etc. Due to the input/ output, cause/effect nature of functions, the response is given in terms of *x*, that is, what is *causing* outputs to be positive, or to be increasing.

Notice the graph of f in Figure 2.10 and the graph of v in Example 5 have something in common. It appears that both the far left and far right branches of each graph point downward (in the negative *y*-direction). We say that the **end-behavior** of both graphs is identical, which is the term used to describe what happens to a graph as $|x|$ becomes very large. For $x > 0$, we say a graph is, "up on the right" or "down on the right," depending on the direction the "end" is pointing. For $x < 0$, we say the graph is "up on the left" or "down on the left," as the case may be.

EXAMPLE 6 ▶ **Describing the End-Behavior of a Graph**

The graph of $f(x) = x^3 - 3x$ is shown. Use the graph to name intervals where f is increasing or decreasing, and comment on the end-behavior of the graph.

Solution ▶ From the graph we observe that $f(x)\uparrow$ for $x \in (-\infty, -1) \cup (1, \infty)$, and $f(x)\downarrow$ for $x \in (-1, 1)$. The end-behavior of the graph is down on the left, and up on the right (down/up).

☑ **C.** You've just seen how we can determine where a function is increasing or decreasing

Now try Exercises 37 through 40 ▶

D. Maximum and Minimum Values

The y-coordinate of the vertex of a parabola that opens downward, and the y-coordinate of "peaks" from other graphs are called **maximum values**. A **global maximum** (also called an *absolute* maximum) names the largest y-value over the entire domain. A **local maximum** (also called a *relative* maximum) gives the largest range value in a specified interval; and an **endpoint maximum** can occur at an endpoint of the domain. The same can be said for any corresponding minimum values.

We will soon develop the ability to locate maximum and minimum values for quadratic and other functions. In future courses, methods are developed to help locate maximum and minimum values for almost *any* function. For now, our work will rely chiefly on a function's graph.

EXAMPLE 7 ▶ **Analyzing Characteristics of a Graph**

Analyze the graph of function f shown in Figure 2.11. Include specific mention of

    a. domain and range,

    b. intervals where f is increasing or decreasing,

    c. maximum (max) and minimum (min) values,

    d. intervals where $f(x) \geq 0$ and $f(x) < 0$, and

    e. whether the function is even, odd, or neither.

Figure 2.11

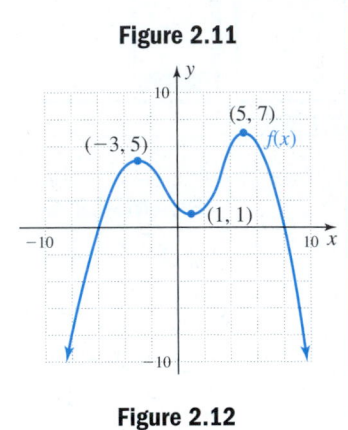

Solution ▶ **a.** Using vertical and horizontal boundary lines show the domain is $x \in \mathbb{R}$, with a range of: $y \in (-\infty, 7]$.

    b. $f(x)\uparrow$ for $x \in (-\infty, -3) \cup (1, 5)$ shown in **blue** in Figure 2.12, and $f(x)\downarrow$ for $x \in (-3, 1) \cup (5, \infty)$ as shown in **red**.

    c. From Part (b) we find that $y = 5$ at $(-3, 5)$ and $y = 7$ at $(5, 7)$ are local maximums, with a local minimum of $y = 1$ at $(1, 1)$. The point $(5, 7)$ is also a global maximum (there is no global minimum).

    d. $f(x) \geq 0$ for $x \in [-6, 8]$; $f(x) < 0$ for $x \in (-\infty, -6) \cup (8, \infty)$

    e. The function is neither even nor odd.

Figure 2.12

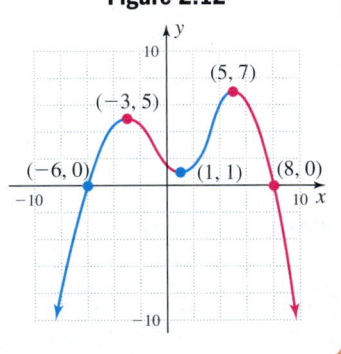

☑ **D.** You've just seen how we can identify the maximum and minimum values of a function

Now try Exercises 41 through 48 ▶

E. Locating Maximum and Minimum Values Using Technology

In Section 1.5, we used the (2nd) (TRACE) **(CALC) 2:zero** option of a graphing calculator to locate the zeroes/x-intercepts of a function. The maximum or minimum values of a function are located in much the same way. To illustrate, enter the function $y = x^3 - 3x - 2$ as Y_1 on the (Y=) screen, then graph it in the window shown, where $x \in [-4, 4]$ and $y \in [-5, 5]$. As seen in Figure 2.13, it appears a local maximum occurs at $x = -1$ and a local minimum at $x = 1$. To actually find the local maximum, we access the (2nd) (TRACE) **(CALC) 4:maximum** option, which returns you to the graph and asks for a **Left Bound?**, a **Right Bound?**, and a **Guess?** as before. Here, we entered a left bound of "-3," a right bound

Figure 2.13

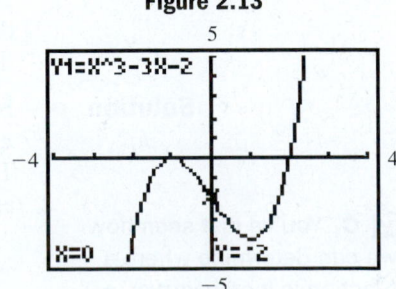

of "0" and bypassed the guess option by pressing 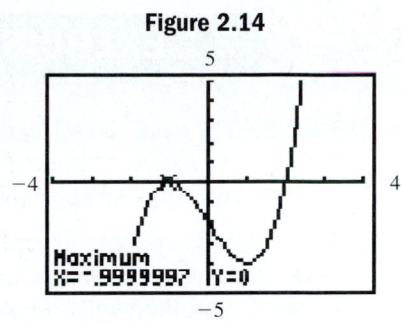 a third time (the calculator again sets the "▶" and "◀" markers to show the bounds chosen). The cursor will then be located at the local maximum in your selected interval, with the coordinates displayed at the bottom of the screen (Figure 2.14). Due to the algorithm the calculator uses to find these values, a decimal number very close to the expected value is sometimes displayed, even if the actual value is an integer (in Figure 2.14, −0.9999997 is displayed instead of −1). To check, we evaluate $f(−1)$ and find the local maximum is indeed 0.

Figure 2.14

EXAMPLE 8 ▶ Locating Local Maximum and Minimum Values on a Graphing Calculator

Find the maximum and minimum values of $f(x) = \dfrac{1}{2}(x^4 − 8x^2 + 7)$.

Solution ▶ Begin by entering $\dfrac{1}{2}(X^4 − 8X^2 + 7)$ as Y_1 on the Y= screen, and graph the function in the ZOOM **6:ZStandard** window. To locate the leftmost minimum value, we access the 2nd TRACE **(CALC) 3:minimum** option, and enter a left bound of "−4," and a right bound of "−1" (Figure 2.15). After pressing ENTER once more, the cursor is located at the minimum in the interval we selected, and we find that a local minimum of −4.5 occurs at $x = −2$ (Figure 2.16). Repeating these steps using the appropriate options shows a local maximum of $y = 3.5$ occurs at $x = 0$, and a second local minimum of $y = −4.5$ occurs at $x = 2$. Note that $y = −4.5$ is also a global minimum.

Figure 2.15

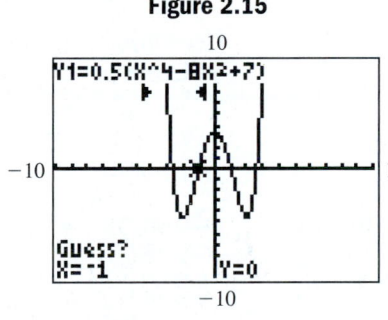

Figure 2.16

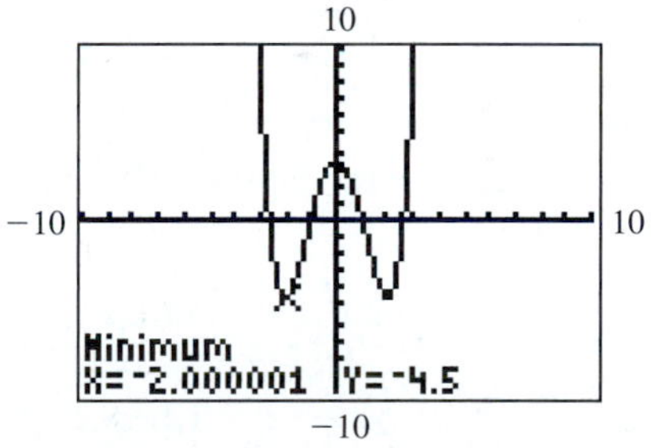

☑ **E.** You've just seen how we can locate local maximum and minimum values using a graphing calculator

Now try Exercises 49 through 54 ▶

The ideas presented here can be applied to functions of all kinds, including rational functions, piecewise-defined functions, step functions, and so on. There is a wide variety of applications in **Exercises 57 through 64.**

2.1 EXERCISES

▶ CONCEPTS AND VOCABULARY

Fill in each blank with the appropriate word or phrase. Carefully reread the section if needed.

1. The graph of a polynomial will cross through the x-axis at zeroes of __linear__ factors of degree 1, and __bounce__ off the x-axis at the zeroes from linear factors of degree 2.

2. If $f(-x) = f(x)$ for all x in the domain, we say that f is an __even__ function and symmetric to the __y__ axis. If $f(-x) = -f(x)$, the function is __odd__ and symmetric to the __origin__.

3. If $f(x_2) > f(x_1)$ for $x_1 < x_2$ for all x in a given interval, the function is __increasing__ in the interval.

4. If $f(c) \geq f(x)$ for all x in a specified interval, we say that $f(c)$ is a local __maximum__ for this interval.

5. Discuss/Explain the following statement and give an example of the conclusion it makes. "If a function f is decreasing to the left of $(c, f(c))$ and increasing to the right of $(c, f(c))$, then $f(c)$ is either a local or a global minimum."   Answers will vary.

6. Without referring to notes or textbook, list as many features/attributes as you can that are related to analyzing the graph of a function. Include details on how to locate or determine each attribute.   Answers will vary.

▶ DEVELOPING YOUR SKILLS

The following functions are known to be even. Complete each graph using symmetry.

7.
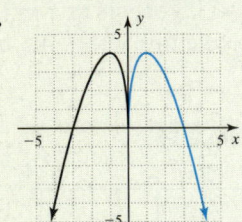

8.
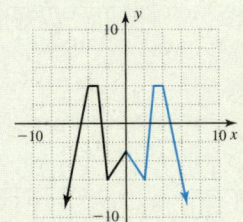

Determine whether the following functions are even: $f(-k) = f(k)$.

9. $f(x) = -7|x| + 3x^2 + 5$ even

10. $p(x) = 2x^4 - 6x + 1$ not even

11. $g(x) = \frac{1}{3}x^4 - 5x^2 + 1$ even

12. $q(x) = \frac{1}{x^2} - |x|$ even

The following functions are known to be odd. Complete each graph using symmetry.

13.
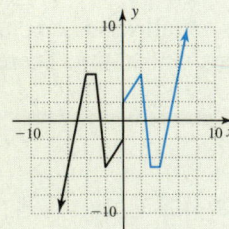

14.
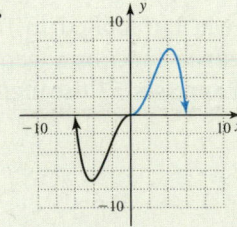

Determine whether the following functions are odd: $f(-k) = -f(k)$.

15. $f(x) = 4\sqrt[3]{x} - x$ odd

16. $g(x) = \frac{1}{2}x^3 - 6x$ odd

17. $p(x) = 3x^3 - 5x^2 + 1$ not odd

18. $q(x) = \frac{1}{x} - x$ odd

Determine whether the following functions are even, odd, or neither.

19. $w(x) = x^3 - x^2$ neither

20. $q(x) = \frac{3}{4}x^2 + 3|x|$ even

21. $p(x) = 2\sqrt[3]{x} - \frac{1}{4}x^3$ odd

22. $g(x) = x^3 + 7x$ odd

23. $v(x) = x^3 + 3|x|$ neither

24. $f(x) = x^4 + 7x^2 - 30$ even

Use the graphs given to solve the inequalities indicated. Write all answers in interval notation.

25. $f(x) = x^3 - 3x^2 - x + 3; f(x) \geq 0$

$x \in [-1, 1] \cup [3, \infty)$

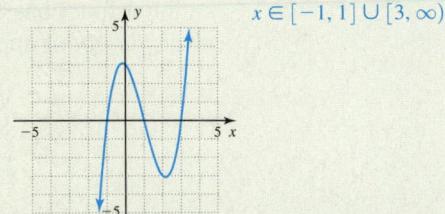

HOMEWORK SELECTION GUIDE

Core: 7–53 odd, 57, 59, 67 (27 Exercises)
Standard: 1–4, 7–53 odd, 57, 58, 59, 67, 70, 72 (34 Exercises)

Extended: 1–4, 7–53 odd, 57–60, 66, 67, 70, 72 (36 Exercises)
In Depth: 1–6, 7–53 odd, 57–60, 61, 63, 65–67, 70–73 (45 Exercises)

26. $f(x) = x^3 - 2x^2 - 4x + 8; f(x) > 0$

$x \in (-2, 2) \cup (2, \infty)$

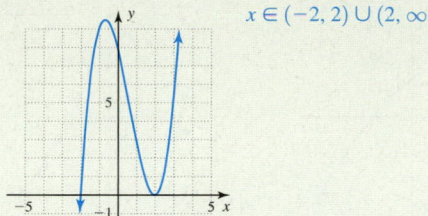

27. $f(x) = x^4 - 2x^2 + 1; f(x) > 0$

$x \in (-\infty, -1) \cup (-1, 1) \cup (1, \infty)$

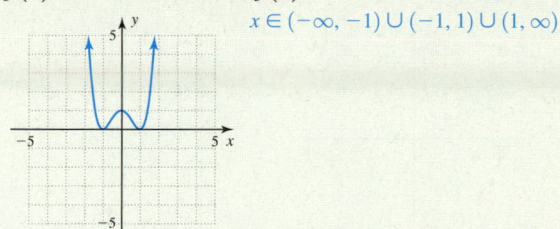

28. $f(x) = x^3 + 2x^2 - 4x - 8; f(x) \geq 0$

$x \in \{2\} \cup [2, \infty)$

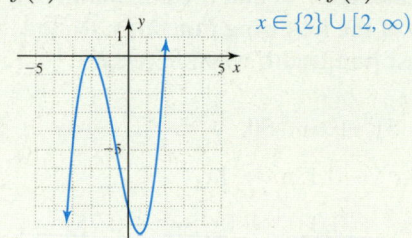

29. $p(x) = \sqrt[3]{x - 1} - 1; p(x) \geq 0$

$x \in [2, \infty)$

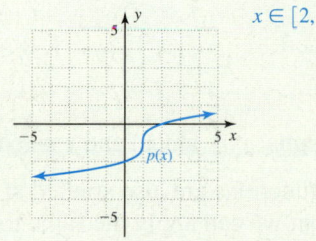

30. $q(x) = \sqrt{x + 1} - 2; q(x) > 0$

$x \in (3, \infty)$

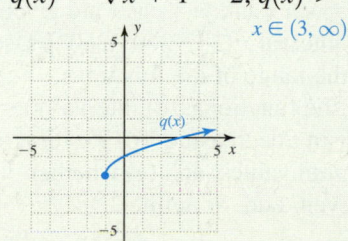

31. $f(x) = (x - 1)^3 - 1; f(x) \leq 0$

$x \in (-\infty, 2]$

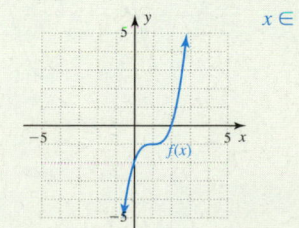

32. $g(x) = -(x + 1)^3 - 1; g(x) < 0$

$x \in (-2, \infty)$

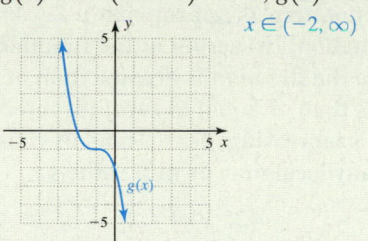

Name the interval(s) where the following functions are increasing, decreasing, or constant. Write answers using interval notation. Assume all endpoints have integer values.

33. $y = V(x)$

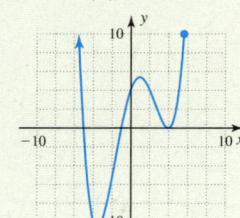

34. $y = H(x)$

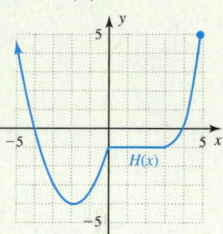

35. $y = f(x)$

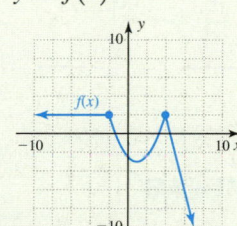

36. $y = g(x)$

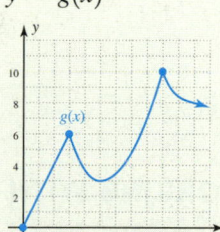

For Exercises 37 through 40, determine (a) interval(s) where the function is increasing, decreasing or constant, and (b) comment on the end-behavior.

37. $p(x) = 0.5(x + 2)^3$

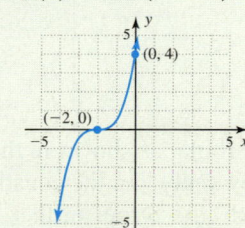

38. $q(x) = -\sqrt[3]{x + 1}$

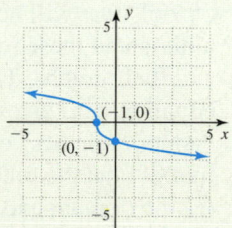

39. $y = f(x)$

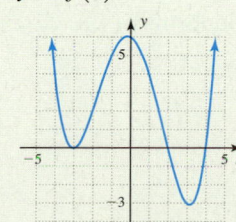

40. $y = g(x)$

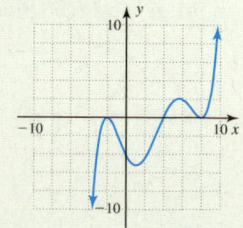

For Exercises 41 through 48, determine the following (answer in interval notation as appropriate): (a) domain and range of the function; (b) zeroes of the function; (c) interval(s) where the function is greater than or equal to zero, or less than or equal to zero; (d) interval(s) where the function is increasing, decreasing, or constant; and (e) location of any local max or min value(s).

41. $y = H(x)$

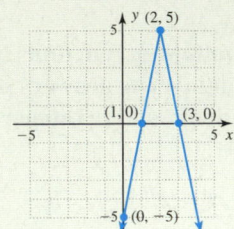

42. $y = f(x)$

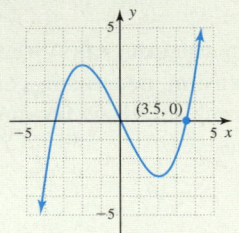

43. $y = g(x)$

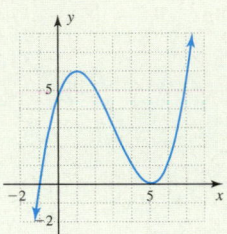

44. $y = h(x)$

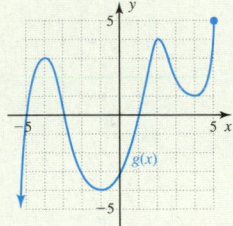

45. $y = Y_1$

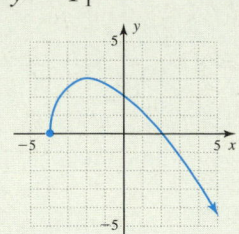

46. $y = Y_2$

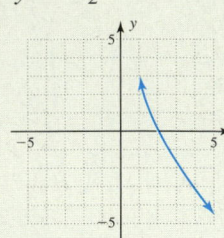

47. $p(x) = (x + 3)^3 + 1$

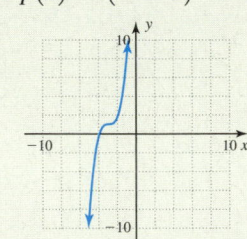

48. $q(x) = |x - 5| + 3$

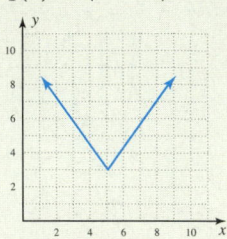

Use a graphing calculator to find the maximum and minimum values of the following functions. Round answers to nearest hundredth when necessary.

49. $y = \dfrac{3}{4}(x^3 - 5x^2 + 6x)$ **50.** $y = \dfrac{6}{5}(x^3 + 4x^2 + 3x)$

51. $y = 0.0016x^5 - 0.12x^3 + 2x$

52. $y = -0.01x^5 + 0.03x^4 + 0.25x^3 - 0.75x^2$

53. $y = x\sqrt{4 - x}$ **54.** $y = x^2\sqrt{x + 3} - 2$

▶ WORKING WITH FORMULAS

55. Conic sections—hyperbola: $y = \dfrac{1}{3}\sqrt{4x^2 - 36}$

While the conic sections are not covered in detail until later in the course, we've already developed a number of tools that will help us understand these relations and their graphs. The equation here gives the

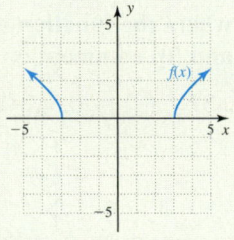

"upper branches" of a hyperbola, as shown in the figure. Find the following by analyzing the equation: (a) the domain and range; (b) the zeroes of the relation; (c) interval(s) where y is increasing or decreasing; (d) whether the relation is even, odd, or neither, and (e) solve for x in terms of y.

56. Trigonometric graphs: $y = \sin(x)$ and $y = \cos(x)$

The trigonometric functions are also studied at some future time, but we can apply the same tools to analyze the graphs of these functions as well. The graphs of $y = \sin x$ and $y = \cos x$ are given, graphed over the interval $x \in [-360°, 360°]$. Use them to find (a) the range of the functions; (b) the zeroes of the functions; (c) interval(s) where y is increasing/decreasing; (d) location of minimum/maximum values; and (e) whether each relation is even, odd, or neither.

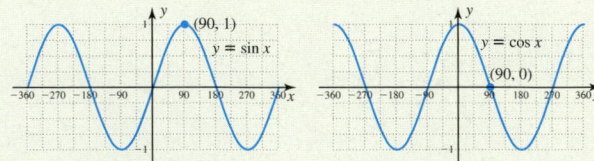

► APPLICATIONS

57. Catapults and projectiles: Catapults have a long and interesting history that dates back to ancient times, when they were used to launch javelins, rocks, and other projectiles. The diagram given illustrates the path of the projectile after release, which follows a parabolic arc. Use the graph to determine the following:

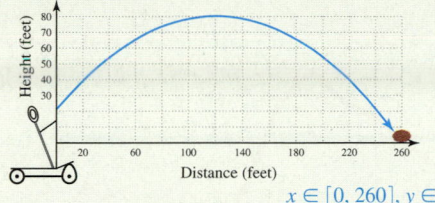

$x \in [0, 260], y \in [0, 80]$

a. State the domain and range of the projectile.

b. What is the maximum height of the projectile?   *80 ft*

c. How far from the catapult did the projectile reach its maximum height?   *120 ft*

d. Did the projectile clear the castle wall, which was 40 ft high and 210 ft away?   *yes*

e. On what interval was the height of the projectile increasing?   *(0, 120)*

f. On what interval was the height of the projectile decreasing?   *(120, 260)*

58. Profit and loss: The profit of DeBartolo Construction Inc. is illustrated by the graph shown. Use the graph to estimate the point(s) or the interval(s) for which the profit P was:

a. increasing $t \in (0, 1) \cup (3, 4) \cup (7, 10)$

b. decreasing $t \in (4, 7)$

c. constant $t \in (1, 3)$

d. a maximum   *(4, 12), (10, 16)*

e. a minimum   *(7, −4)*

f. positive $t \in (0, 6) \cup (8, 10)$

g. negative $t \in (6, 8)$

h. zero   *(6, 0) (8, 0)*

59. Functions and rational exponents: The graph of $f(x) = x^{\frac{2}{3}} - 1$ is shown. Use the graph to find:
$x \in (-\infty, \infty); y \in [-1, \infty)$

a. domain and range of the function

b. zeroes of the function   *(−1, 0), (1, 0)*

c. interval(s) where $f(x) \geq 0$ or $f(x) < 0$
$f(x) \geq 0; x \in (-\infty, -1] \cup [1, \infty); f(x) < 0; x \in (-1, 1)$

d. interval(s) where $f(x)$ is increasing, decreasing, or constant $f(x)\uparrow: x \in (0, \infty), f(x)\downarrow: x \in (-\infty, 0)$

e. location of any max or min value(s)   *min: (0, −1)*

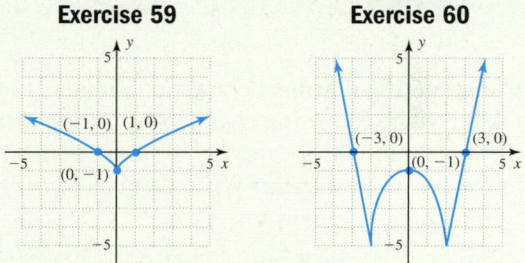

Exercise 59   Exercise 60

60. Analyzing a graph: Given $h(x) = |x^2 - 4| - 5$, whose graph is shown, use the graph to find:

a. domain and range of the function
$x \in (-\infty, \infty), y \in [-5, \infty)$

b. zeroes of the function   *(−3, 0), (3, 0)*

c. interval(s) where $h(x) \geq 0$ or $h(x) < 0$
$h(x) \geq 0; x \in (-\infty, -3] \cup [3, \infty); h(x) < 0; x \in (-3, 3)$

d. interval(s) where $f(x)$ is increasing, decreasing, or constant
$h(x)\uparrow: x \in (-2, 0) \cup (2, \infty); h(x)\downarrow: x \in (-\infty, -2) \cup (0, 2)$

e. location of any max or min value(s)
max: $y = -1$ at $x = 0$; min: $y = -5$ at $x = \pm 2$

61. Analyzing interest rates: The graph shown approximates the average annual interest rates I on 30-yr fixed mortgages, rounded to the nearest $\frac{1}{4}$%. Use the graph to estimate the following (write all answers in interval notation).

a. domain and range

b. interval(s) where $I(t)$ is increasing, decreasing, or constant

c. location of any global maximum or minimum values

d. the one-year period with the greatest rate of increase and the one-year period with the greatest rate of decrease

Source: 2009 Statistical Abstract of the United States, Table 1157

Year (1983 → 83)

Additional answers can be found in the Instructor Answer Appendix.

62. **Analyzing the surplus S:** The following graph approximates the federal surplus S of the United States. Use the graph to estimate the following. Write answers in interval notation and estimate all surplus values to the nearest $10 billion.

   a. the domain and range

   b. interval(s) where $S(t)$ is increasing, decreasing, or constant

   c. the location of any global maximum and minimum values

   d. the one-year period with the greatest rate of increase, and the one-year period with the greatest rate of decrease

   Source: 2009 Statistical Abstract of the United States, Table 451

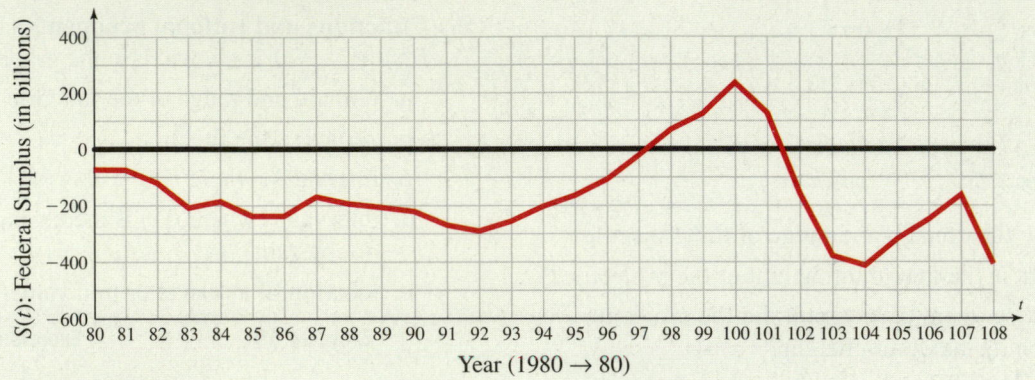

Year (1980 → 80)

63. **Constructing a graph:** Draw a continuous function f that has the following characteristics, then state the zeroes and the location of all maximum and minimum values. [*Hint:* Write them as $(c, f(c))$.]

   a. Domain: $x \in (-10, \infty)$
      Range: $y \in (-6, \infty)$

   b. $f(0) = 0; f(4) = 0$

   c. $f(x)\uparrow$ for $x \in (-10, -6) \cup (-2, 2) \cup (4, \infty)$
      $f(x)\downarrow$ for $x \in (-6, -2) \cup (2, 4)$

   d. $f(x) \geq 0$ for $x \in [-8, -4] \cup [0, \infty)$
      $f(x) < 0$ for $x \in (-\infty, -8) \cup (-4, 0)$

64. **Constructing a graph:** Draw a continuous function g that has the following characteristics, then state the zeroes and the location of all maximum and minimum values. [*Hint:* Write them as $(c, g(c))$.]

   a. Domain: $x \in (-\infty, 8]$
      Range: $y \in [-6, \infty)$

   b. $g(0) = 4.5; g(6) = 0$

   c. $g(x)\uparrow$ for $x \in (-6, 3) \cup (6, 8)$
      $g(x)\downarrow$ for $x \in (-\infty, -6) \cup (3, 6)$

   d. $g(x) \geq 0$ for $x \in (-\infty, -9] \cup [-3, 8]$
      $g(x) < 0$ for $x \in (-9, -3)$

▶ **EXTENDING THE CONCEPT**

 65. Does the function shown have a maximum value? Does it have a minimum value? Discuss/explain/justify why or why not.   no; no; Answers will vary.

Exercise 65

66. The graph drawn here depicts a 400-m race between a mother and her daughter. Analyze the graph to answer questions (a) through (f).

   a. Who wins the race, the mother or daughter?   daughter

   b. By approximately how many meters?   20 m

   c. By approximately how many seconds?   10 sec

   d. Who was leading at $t = 40$ seconds?   mother

   e. During the race, how many seconds was the daughter in the lead?   about 37 sec

   f. During the race, how many seconds was the mother in the lead?   about 28 sec

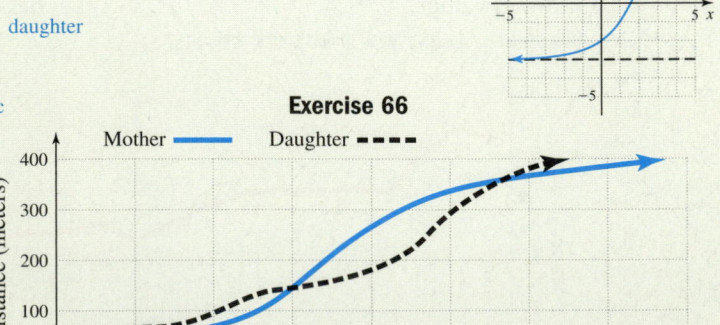

Exercise 66

67. The graph drawn here depicts the last 75 sec of the competition between Ian Thorpe (Australia) and Massimiliano Rosolino (Italy) in the men's 400-m freestyle at the 2000 Olympics, where a new Olympic record was set.

  a. Who was in the lead at 180 sec? 210 sec?   Thorpe; Rosolino

  b. In the last 50 m, how many times were they tied, and when did the ties occur?   two times, at 190 sec and 219 sec

  c. About how many seconds did Rosolino have the lead?   about 29 sec ($219 - 190 = 29$ sec)

  d. Which swimmer won the race?   Thorpe

  e. By approximately how many seconds?   about 2 sec ($223 - 221 = 2$ sec)

  f. Use the graph to approximate the new Olympic record set in the year 2000.   221 sec = 3 min 41 sec

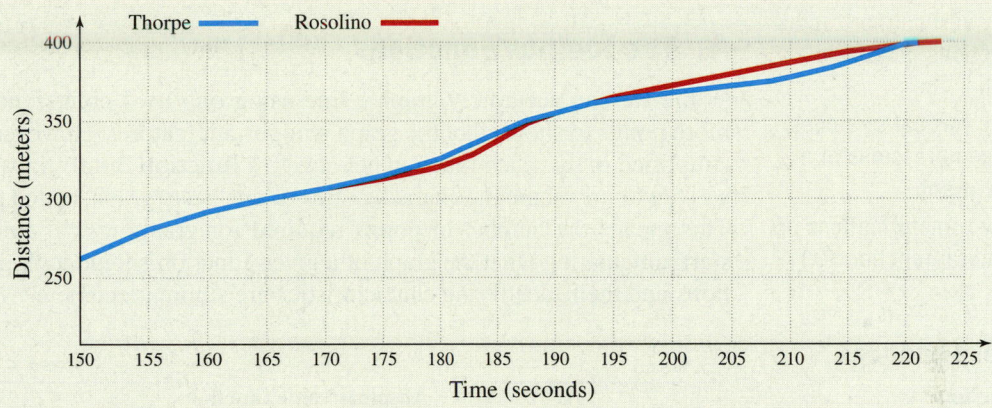

68. Draw the graph of a general function $f(x)$ that has a local *maximum* at $(a, f(a))$ and a local *minimum* at $(b, f(b))$ but with $f(a) < f(b)$.   Answers will vary.

69. Verify that $h(x) = x^{\frac{2}{3}}$ is an even function, by first rewriting h as $h(x) = (x^{\frac{1}{3}})^2$.

$$h(-k) = h(k)$$
$$[(-k)^{\frac{1}{3}}]^2 = (k^{\frac{1}{3}})^2$$
$$(-k^{\frac{1}{3}})^2 = (k^{\frac{1}{3}})^2$$
$$(k^{\frac{1}{3}})^2 = (k^{\frac{1}{3}})^2 \checkmark$$

▶ MAINTAINING YOUR SKILLS

70. **(R.4)** Solve the given quadratic equation by factoring: $x^2 - 8x - 20 = 0$. $x = -2, x = 10$

71. **(R.5)** Find the (a) sum and (b) product of the rational expressions $\dfrac{3}{x + 2}$ and $\dfrac{3}{2 - x}$.

72. **(1.4)** Write the equation of the line shown, in the form $y = mx + b$. $y = \dfrac{2}{3}x - 1$

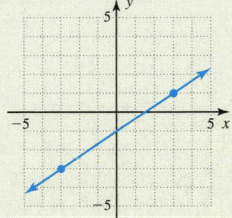

73. **(R.2)** Find the surface area and volume of the cylinder shown ($SA = 2\pi r^2 + \pi r^2 h$, $V = \pi r^2 h$).

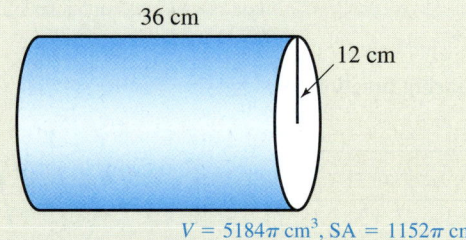

$V = 5184\pi$ cm^3, SA $= 1152\pi$ cm^2

Many applications of mathematics require that we select a function known to fit the context, or build a function model from the information supplied. So far we've looked at linear functions. Here we'll introduce the absolute value, squaring, square root, cubing, and cube root functions. Together these are the six **toolbox functions,** so called because they give us a variety of "tools" to model the real world (see Section 2.6). In the same way a study of arithmetic depends heavily on the multiplication table, a study of algebra and mathematical modeling depends (in large part) on a solid working knowledge of these functions. More will be said about each function in later sections.

A. The Toolbox Functions

While we can accurately graph a line using only two points, most functions require more points to show all of the graph's important features. However, our work is greatly simplified in that each function belongs to a **function family,** in which all graphs from a given family share the characteristics of one basic graph, called the **parent function.** This means the number of points required for graphing will quickly decrease as we start anticipating what the graph of a given function should look like. The parent functions and their identifying characteristics are summarized here.

The Toolbox Functions

Identity function

x	$f(x) = x$
−3	−3
−2	−2
−1	−1
0	0
1	1
2	2
3	3

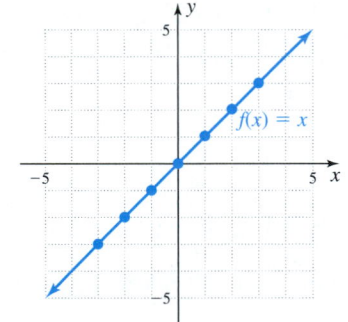

Domain: $x \in (-\infty, \infty)$, Range: $y \in (-\infty, \infty)$
Symmetry: odd
Increasing: $x \in (-\infty, \infty)$
End-behavior: down on the left/up on the right

Absolute value function

x	$f(x) = |x|$
−3	3
−2	2
−1	1
0	0
1	1
2	2
3	3

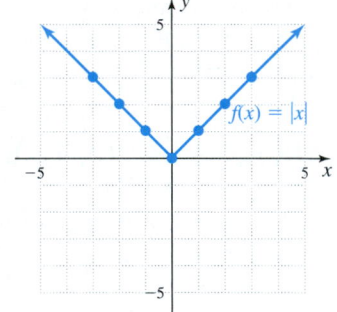

Domain: $x \in (-\infty, \infty)$, Range: $y \in [0, \infty)$
Symmetry: even
Decreasing: $x \in (-\infty, 0)$; Increasing: $x \in (0, \infty)$
End-behavior: up on the left/up on the right
Vertex at $(0, 0)$

Squaring function

x	$f(x) = x^2$
−3	9
−2	4
−1	1
0	0
1	1
2	4
3	9

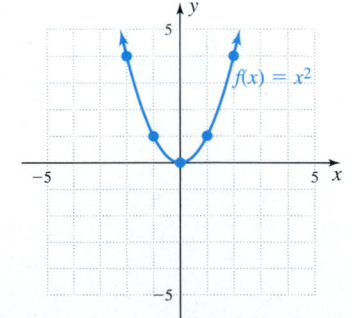

Domain: $x \in (-\infty, \infty)$, Range: $y \in [0, \infty)$
Symmetry: even
Decreasing: $x \in (-\infty, 0)$; Increasing: $x \in (0, \infty)$
End-behavior: up on the left/up on the right
Vertex at $(0, 0)$

Square root function

x	$f(x) = \sqrt{x}$
−2	—
−1	—
0	0
1	1
2	≈1.41
3	≈1.73
4	2

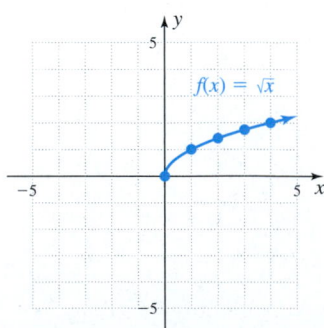

Domain: $x \in [0, \infty)$, Range: $y \in [0, \infty)$
Symmetry: neither even nor odd
Increasing: $x \in (0, \infty)$
End-behavior: up on the right
Initial point at $(0, 0)$

Cubing function

x	$f(x) = x^3$
-3	-27
-2	-8
-1	-1
0	0
1	1
2	8
3	27

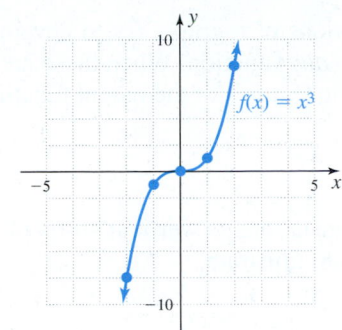

Domain: $x \in (-\infty, \infty)$, Range: $y \in (-\infty, \infty)$
Symmetry: odd
Increasing: $x \in (-\infty, \infty)$
End-behavior: down on the left/up on the right
Point of inflection at $(0, 0)$

Cube root function

x	$f(x) = \sqrt[3]{x}$
-27	-3
-8	-2
-1	-1
0	0
1	1
8	2
27	3

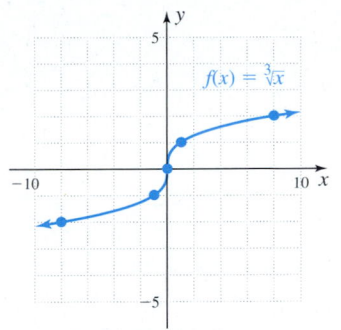

Domain: $x \in (-\infty, \infty)$, Range: $y \in (-\infty, \infty)$
Symmetry: odd
Increasing: $x \in (-\infty, \infty)$
End-behavior: down on the left/up on the right
Point of inflection at $(0, 0)$

In applications of the toolbox functions, the parent graph may be "morphed" and/or shifted from its original position, yet the graph will still retain its basic shape and features. The result is called a **transformation** of the parent graph.

EXAMPLE 1 ▶ **Identifying the Characteristics of a Transformed Graph**

The graph of $f(x) = x^2 - 2x - 3$ is given. Use the graph to identify each of the features or characteristics indicated.

  a. function family

  b. domain and range

  c. vertex

  d. max or min value(s)

  e. intervals where f is increasing or decreasing

  f. end-behavior

  g. x- and y-intercept(s)

Solution ▶   **a.** The graph is a parabola, from the squaring function family.

  b. domain: $x \in (-\infty, \infty)$; range: $y \in [-4, \infty)$

  c. vertex: $(1, -4)$

  d. minimum value $y = -4$ at $(1, -4)$

  e. decreasing: $x \in (-\infty, 1)$, increasing: $x \in (1, \infty)$

  f. end-behavior: up/up

  g. y-intercept: $(0, -3)$; x-intercepts: $(-1, 0)$ and $(3, 0)$

Now try Exercises 7 through 34 ▶

☑ **A.** You've just seen how we can identify basic characteristics of the toolbox functions

Note that for Example 1(f), we can algebraically verify the x-intercepts by substituting 0 for $f(x)$ and solving the equation by factoring. This gives $0 = (x + 1)(x - 3)$, with solutions $x = -1$ and $x = 3$. It's also worth noting that while the parabola is no longer symmetric to the y-axis, it *is* symmetric to the vertical line $x = 1$. This line is called the **axis of symmetry** for the parabola, and for a vertical parabola, it will always be a vertical line that goes through the vertex.

B. Vertical and Horizontal Shifts

As we study specific transformations of a graph, try to develop a *global view* as the transformations can be applied to *any* function. When these are applied to the toolbox functions, we rely on characteristic features of the parent function to assist in completing the transformed graph.

Vertical Translations

We'll first investigate vertical translations or vertical shifts of the toolbox functions, using the absolute value function to illustrate.

EXAMPLE 2 ▶ **Graphing Vertical Translations**

Construct a table of values for $f(x) = |x|$, $g(x) = |x| + 1$, and $h(x) = |x| - 3$ and graph the functions on the same coordinate grid. Then discuss what you observe.

Solution ▶ A table of values for all three functions is given, with the corresponding graphs shown in the figure.

x	$f(x) = |x|$	$g(x) = |x| + 1$	$h(x) = |x| - 3$
-3	3	4	0
-2	2	3	-1
-1	1	2	-2
0	0	1	-3
1	1	2	-2
2	2	3	-1
3	3	4	0

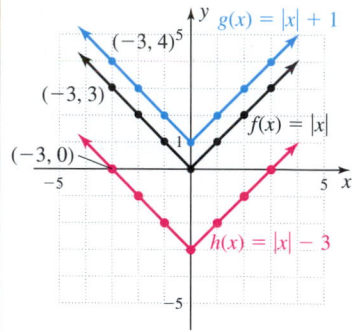

Note that outputs of $g(x)$ are one more than the outputs of $f(x)$, and that each point on the graph of f has been shifted *upward 1 unit* to form the graph of g. Similarly, each point on the graph of f has been shifted *downward 3 units* to form the graph of h, since $h(x) = f(x) - 3$.

Now try Exercises 35 through 42 ▶

We describe the transformations in Example 2 as a **vertical shift** or **vertical translation** of a basic graph. The graph of g is the graph of f *shifted up 1 unit*, and the graph of h, is the graph of f *shifted down 3 units*. In general, we have the following:

Vertical Translations of a Basic Graph

Given $k > 0$ and any function whose graph is determined by $y = f(x)$,
1. The graph of $y = f(x) + k$ is the graph of $f(x)$ shifted upward k units.
2. The graph of $y = f(x) - k$ is the graph of $f(x)$ shifted downward k units.

Graphing calculators are wonderful tools for exploring graphical transformations. To emphasize that a given graph is being shifted vertically as in Example 2, try entering $\sqrt[3]{X}$ as Y_1 on the ⬡Y= screen, then $Y_2 = Y_1 + 2$ and $Y_3 = Y_1 - 3$ (Figure 2.17 — recall the Y-variables are accessed using ⬡VARS ⬡▷ **(Y-VARS)** ⬡ENTER). Using the Y-variables in this way enables us to study identical transformations on a variety of graphs, simply by changing the function in Y_1.

Figure 2.17

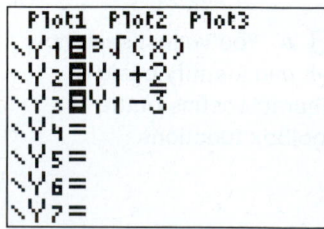

Using a window size of $x \in [-5, 5]$ and $y \in [-5, 5]$ for the cube root function, produces the graphs shown in Figure 2.18, which demonstrate the cube root graph has been shifted upward 2 units (Y_2), and downward 3 units (Y_3).

Try this exploration again using $Y_1 = \sqrt{X}$.

Figure 2.18

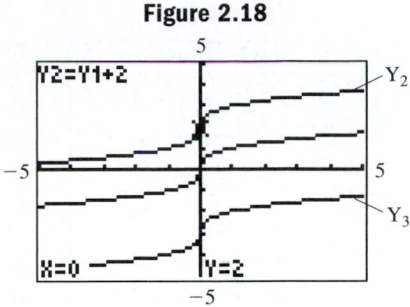

Horizontal Translations

The graph of a parent function can also be shifted left or right. This happens when we *alter the inputs to the basic function,* as opposed to adding or subtracting something to the function itself. For $Y_1 = x^2 + 2$ note that we first square inputs, then add 2, which results in a vertical shift. For $Y_2 = (x + 2)^2$, we add 2 to x *prior to squaring* and since the input values are affected, we might anticipate the graph will shift along the x-axis—horizontally.

EXAMPLE 3 ▶ **Graphing Horizontal Translations**

Construct a table of values for $f(x) = x^2$ and $g(x) = (x + 2)^2$, then graph the functions on the same grid and discuss what you observe.

Solution ▶ Both f and g belong to the quadratic family and their graphs are parabolas. A table of values is shown along with the corresponding graphs.

x	$f(x) = x^2$	$g(x) = (x + 2)^2$
-3	9	1
-2	4	0
-1	1	1
0	0	4
1	1	9
2	4	16
3	9	25

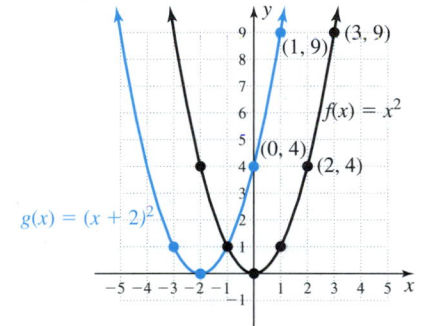

It is apparent the graphs of g and f are identical, but the graph of g has been shifted horizontally 2 units left.

Now try Exercises 43 through 46 ▶

We describe the transformation in Example 3 as a **horizontal shift** or **horizontal translation** of a basic graph. The graph of g is the graph of f, *shifted 2 units to the left.* Once again it seems reasonable that since *input* values were altered, the shift must be horizontal rather than vertical. From this example, we also learn the direction of the shift is **opposite the sign:** $y = (x + 2)^2$ is 2 units *to the left* of $y = x^2$. Although it may seem counterintuitive, the shift *opposite the sign* can be "seen" by locating the new x-intercept, which in this case is also the vertex. Substituting 0 for y gives $0 = (x + 2)^2$ with $x = -2$, as shown in the graph. In general, we have

Horizontal Translations of a Basic Graph

Given $h > 0$ and any function whose graph is determined by $y = f(x)$,

1. The graph of $y = f(x + h)$ is the graph of $f(x)$ shifted *to the left h* units.
2. The graph of $y = f(x - h)$ is the graph of $f(x)$ shifted *to the right h* units.

Figure 2.19

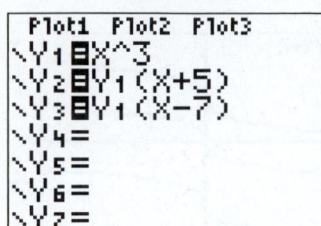

To explore horizontal translations on a graphing calculator, we input a basic function in Y_1 and indicate how we want the inputs altered in Y_2 and Y_3. Here we'll enter X^3 as Y_1 on the [Y=] screen, then $Y_2 = Y_1(X + 5)$ and $Y_3 = Y_1(X - 7)$ (Figure 2.19). Note how this duplicates the definition and notation for horizontal shifts in the orange box. Based on what we saw in Example 3, we expect the graph of $y = x^3$ will first be shifted 5 units left (Y_2), then 7 units right (Y_3). This in confirmed in Figure 2.20.

Try this exploration again using $Y_1 = \text{abs}(X)$.

Figure 2.20

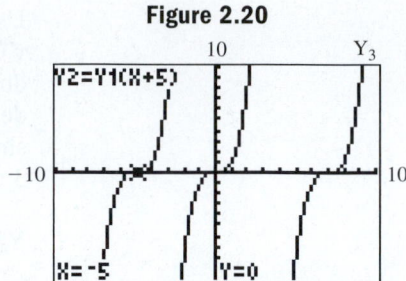

EXAMPLE 4 ▶ **Graphing Horizontal Translations**

Sketch the graphs of $g(x) = |x - 2|$ and $h(x) = \sqrt{x + 3}$ using a horizontal shift of the parent function and a few characteristic points (not a table of values).

Solution ▶ The graph of $g(x) = |x - 2|$ (Figure 2.21) is the absolute value function shifted 2 units to the right (shift the vertex and two other points from $y = |x|$). The graph of $h(x) = \sqrt{x + 3}$ (Figure 2.22) is a square root function, shifted 3 units to the left (shift the initial point and one or two points from $y = \sqrt{x}$).

Figure 2.21

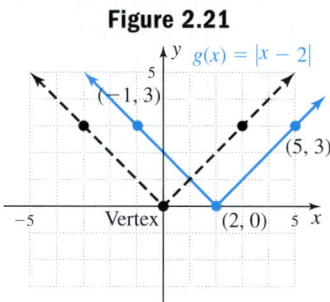

Figure 2.22

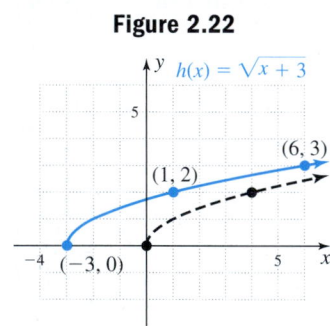

✓ **B.** You've just seen how we can perform vertical/horizontal shifts of a basic graph

Now try Exercises 47 through 50 ▶

C. Vertical and Horizontal Reflections

The next transformation we investigate is called a **vertical reflection,** in which we compare the function $Y_1 = f(x)$ with the negative of the function: $Y_2 = -f(x)$.

Vertical Reflections

EXAMPLE 5 ▶ **Graphing Vertical Reflections**

Construct a table of values for $Y_1 = x^2$ and $Y_2 = -x^2$, then graph the functions on the same grid and discuss what you observe.

Solution ▶ A table of values is given for both functions, along with the corresponding graphs.

x	$Y_1 = x^2$	$Y_2 = -x^2$
-2	4	-4
-1	1	-1
0	0	0
1	1	-1
2	4	-4

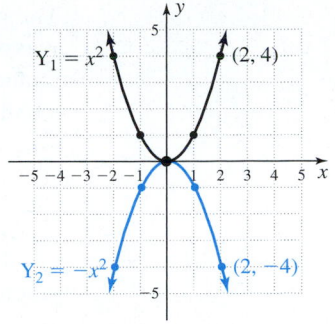

As you might have anticipated, the outputs for f and g differ only in sign. Each output is a **reflection** of the other, being an equal distance from the x-axis but on opposite sides.

Now try Exercises 51 and 52 ▶

The vertical reflection in Example 5 is called a **reflection across the x-axis.** In general,

Vertical Reflections of a Basic Graph

For any function $y = f(x)$, the graph of $y = -f(x)$
is the graph of $f(x)$ reflected across the x-axis.

Figure 2.23

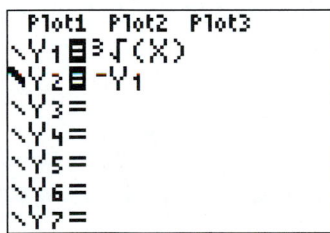

To view vertical reflections on a graphing calculator, we simply define $Y_2 = -Y_1$, as seen here using $\sqrt[3]{X}$ as Y_1 (Figure 2.23). As in Section 1.5, we can have the calculator graph Y_2 using a bolder line, to easily distinguish between the original graph and its reflection (Figure 2.24). To aid in the viewing, we have set a window size of $x \in [-5, 5]$ and $y \in [-3, 3]$.

Try this exploration again using $Y_1 = X^2 - 4$.

Figure 2.24

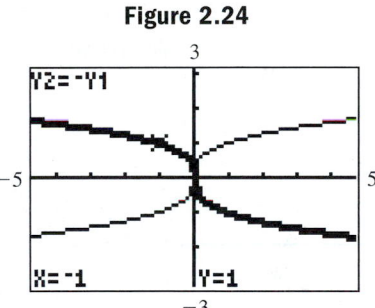

Horizontal Reflections

It's also possible for a graph to be reflected horizontally *across the y-axis.* Just as we noted that $f(x)$ versus $-f(x)$ resulted in a vertical reflection, $f(x)$ versus $f(-x)$ results in a horizontal reflection.

EXAMPLE 6 ▶   Graphing a Horizontal Reflection

Construct a table of values for $f(x) = \sqrt{x}$ and $g(x) = \sqrt{-x}$, then graph the functions on the same coordinate grid and discuss what you observe.

Solution ▶   A table of values is given here, along with the corresponding graphs.

x	$f(x) = \sqrt{x}$	$g(x) = \sqrt{-x}$
-4	not real	2
-2	not real	$\sqrt{2} \approx 1.41$
-1	not real	1
0	0	0
1	1	not real
2	$\sqrt{2} \approx 1.41$	not real
4	2	not real

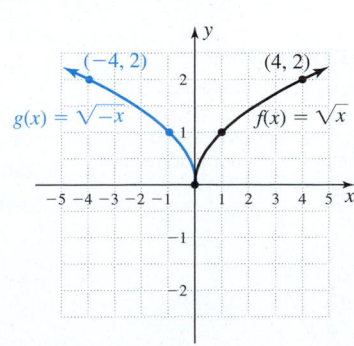

The graph of g is the same as the graph of f, but it has been reflected across the y-axis. A study of the domain shows why—f represents a real number only for nonnegative inputs, so its graph occurs to the right of the y-axis, while g represents a real number for nonpositive inputs, so its graph occurs to the left.

Now try Exercises 53 and 54 ▶

The transformation in Example 6 is called a **horizontal reflection** of a basic graph. In general,

☑ **C.** You've just seen how we can apply vertical/horizontal reflections of a basic graph

> **Horizontal Reflections of a Basic Graph**
>
> For any function $y = f(x)$, the graph of $y = f(-x)$
> is the graph of $f(x)$ reflected across the y-axis.

D. Vertically Stretching/Compressing a Basic Graph

As the words "stretching" and "compressing" imply, the graph of a basic function can also become elongated or flattened after certain transformations are applied. However, even these transformations preserve the key characteristics of the graph.

EXAMPLE 7 ▶   **Stretching and Compressing a Basic Graph**

Construct a table of values for $f(x) = x^2$, $g(x) = 3x^2$, and $h(x) = \frac{1}{3}x^2$, then graph the functions on the same grid and discuss what you observe.

Solution ▶   A table of values is given for all three functions, along with the corresponding graphs.

x	$f(x) = x^2$	$g(x) = 3x^2$	$h(x) = \frac{1}{3}x^2$
-3	9	27	3
-2	4	12	$\frac{4}{3}$
-1	1	3	$\frac{1}{3}$
0	0	0	0
1	1	3	$\frac{1}{3}$
2	4	12	$\frac{4}{3}$
3	9	27	3

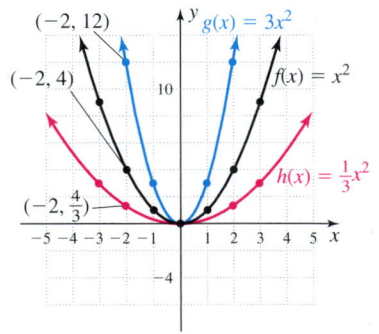

The outputs of g are triple those of f, making these outputs farther from the x-axis and *stretching* g upward (making the graph more narrow). The outputs of h are one-third those of f, and the graph of h is *compressed* downward, with its outputs closer to the x-axis (making the graph wider).

Now try Exercises 55 through 62 ▶

WORTHY OF NOTE

In a study of trigonometry, you'll find that a basic graph can also be stretched or compressed horizontally, a phenomenon known as *frequency variations*.

The transformations in Example 7 are called **vertical stretches** or **compressions** of a basic graph. Notice that while the outputs are increased or decreased by a constant factor (making the graph appear more narrow or more wide), the domain of the function remains unchanged. In general,

Stretches and Compressions of a Basic Graph

For any function $y = f(x)$, the graph of $y = af(x)$ is

    1. the graph of $f(x)$ stretched vertically if $|a| > 1$,

    2. the graph of $f(x)$ compressed vertically if $0 < |a| < 1$.

Figure 2.25

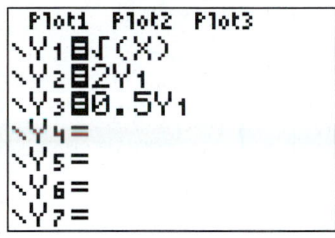

☑ **D.** You've just seen how we can apply vertical stretches and compressions of a basic graph

To use a graphing calculator in a study of stretches and compressions, we simply define Y_2 and Y_3 as constant multiples of Y_1 (Figure 2.25). As seen in Example 7, if $|a| > 1$ the graph will be stretched vertically, if $0 < |a| < 1$, the graph will be vertically compressed. This is further illustrated here using $Y_1 = \sqrt{X}$, with $Y_2 = 2Y_1$ and $Y_3 = 0.5Y_1$. Since the domain of $y = \sqrt{x}$ is restricted to nonnegative values, a window size of $x \in [0, 10]$ and $y \in [-1, 7]$ was used (Figure 2.26).

Try this exploration again using $Y_1 = \text{abs}(X) - 4$.

Figure 2.26

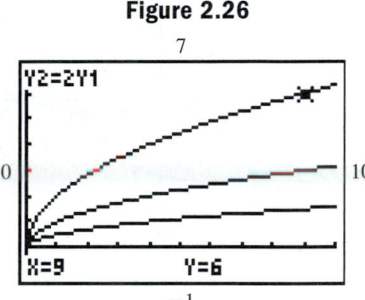

E. Transformations of a General Function

If more than one transformation is applied to a basic graph, it's helpful to use the following sequence for graphing the new function.

General Transformations of a Basic Graph

Given a function $y = f(x)$, the graph of $y = af(x \pm h) \pm k$ can be obtained by applying the following sequence of transformations:

    1. horizontal shifts   **2.** reflections   **3.** stretches/compressions   **4.** vertical shifts

We generally use a few characteristic points to track the transformations involved, then draw the transformed graph through the new location of these points.

EXAMPLE 8 ▶ **Graphing Functions Using Transformations**

Use transformations of a parent function to sketch the graphs of

    a. $g(x) = -(x + 2)^2 + 3$ **b.** $h(x) = 2\sqrt[3]{x - 2} - 1$

Solution ▶ **a.** The graph of g is a parabola, shifted left 2 units, reflected across the x-axis, and shifted up 3 units. This sequence of transformations is shown in Figures 2.27 through 2.29. Note that since the graph has been shifted 2 units left and 3 units up, the vertex of the parabola has likewise shifted from $(0, 0)$ to $(-2, 3)$.

Figure 2.27

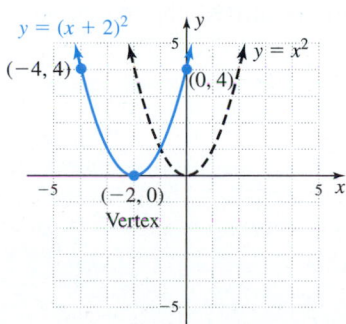

Shifted left 2 units

Figure 2.28

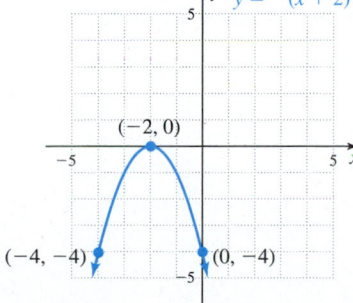

Reflected across the x-axis

Figure 2.29

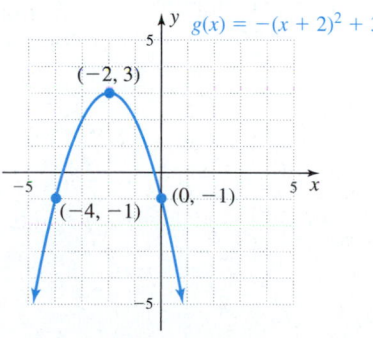

Shifted up 3 units

b. The graph of h is a cube root function, shifted right 2, stretched by a factor of 2, then shifted down 1. This sequence is shown in Figures 2.30 through 2.32 and illustrate how the inflection point has shifted from $(0, 0)$ to $(2, -1)$.

Figure 2.30

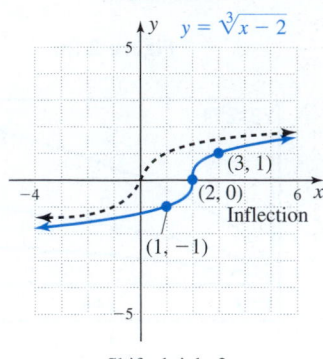

Shifted right 2

Figure 2.31

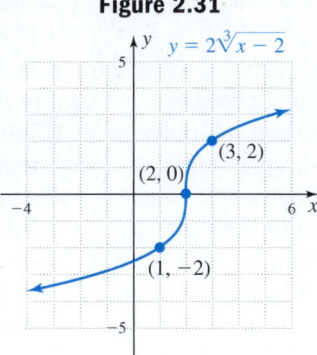

Stretched by a factor of 2

Figure 2.32

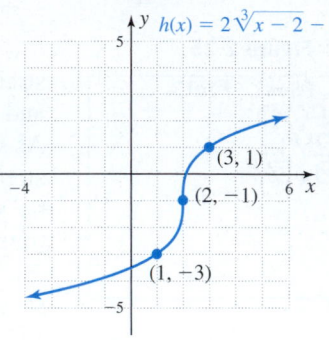

Shifted down 1

Now try Exercises 63 through 92 ▶

It's important to note that the transformations can actually be applied to *any function*, even those that are new and unfamiliar. Consider the following pattern:

Parent Function	Transformation of Parent Function
quadratic: $y = x^2$	$y = -2(x - 3)^2 + 1$
absolute value: $y = |x|$	$y = -2|x - 3| + 1$
cube root: $y = \sqrt[3]{x}$	$y = -2\sqrt[3]{x - 3} + 1$
general: $y = f(x)$	$y = -2f(x - 3) + 1$

In each case, the transformation involves a horizontal shift 3 units right, a vertical reflection, a vertical stretch, and a vertical shift up 1. Since the shifts are the same regardless of the initial function, we can generalize the results to any function $f(x)$.

General Function	Transformed Function
$y = f(x)$	$y = af(x \pm h) \pm k$

vertical reflections, vertical stretches and compressions

horizontal shift h units, opposite direction of sign

vertical shift k units, same direction as sign

WORTHY OF NOTE

Since the shape of the initial graph does not change when translations or reflections are applied, these are called **rigid transformations.** Stretches and compressions of a basic graph are called **nonrigid transformations,** as the graph is distended in some way.

Also bear in mind that the graph will be reflected across the y-axis (horizontally) if x is replaced with $-x$. This process is illustrated in Example 9 for selected transformations. Remember—if the graph of a function is shifted, the *individual points* on the graph are likewise shifted.

EXAMPLE 9 ▶ **Graphing Transformations of a General Function**

Given the graph of $f(x)$ shown in Figure 2.33, graph $g(x) = -f(x + 1) - 2$.

Solution ▶ For g, the graph of f is (1) shifted horizontally 1 unit left (Figure 2.34), (2) reflected across the x-axis (Figure 2.35), and (3) shifted vertically 2 units down (Figure 2.36). The final result is that in Figure 2.36.

Figure 2.33   **Figure 2.34**

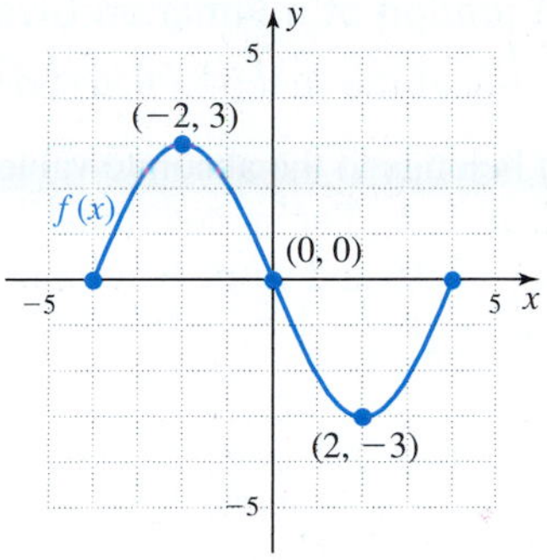

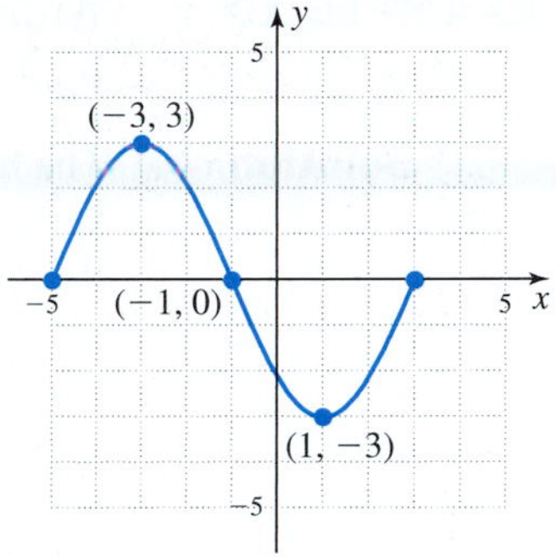

Figure 2.35   **Figure 2.36**

Now try Exercises 93 through 96 ▶

As noted in Example 9, these shifts and transformation are often combined—particularly when the toolbox functions are used as real-world models (Section 2.6). On a graphing calculator we again define Y_1 as needed, then define Y_2 as any desired combination of shifts, stretches, and/or reflections. For $Y_1 = X^2$, we'll define Y_2 as $-2 Y_1(X + 5) + 3$ (Figure 2.37), and expect that the graph of Y_2 will be that of Y_1 shifted left 5 units, reflected across the x-axis, stretched vertically, and shifted up three units. This shows the new vertex should be at $(-5, 3)$, which is confirmed in Figure 2.38 along with the other transformations.

Figure 2.38

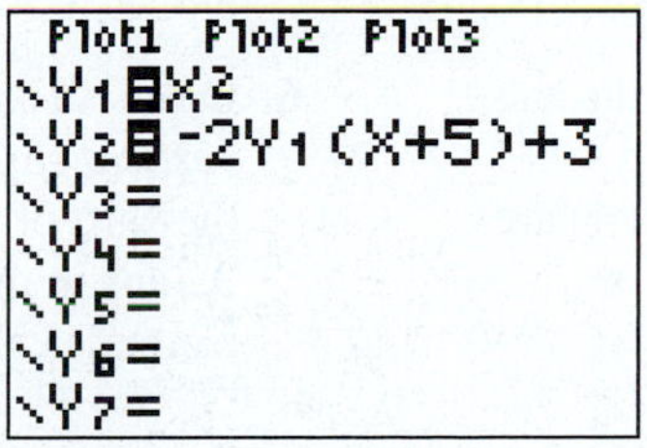

Figure 2.37

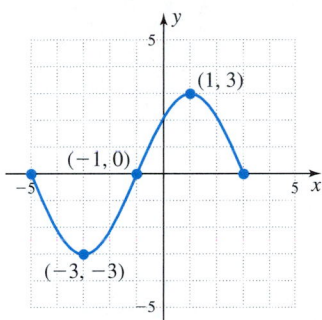

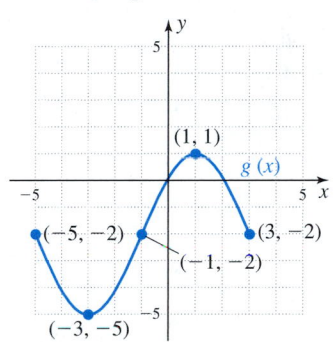

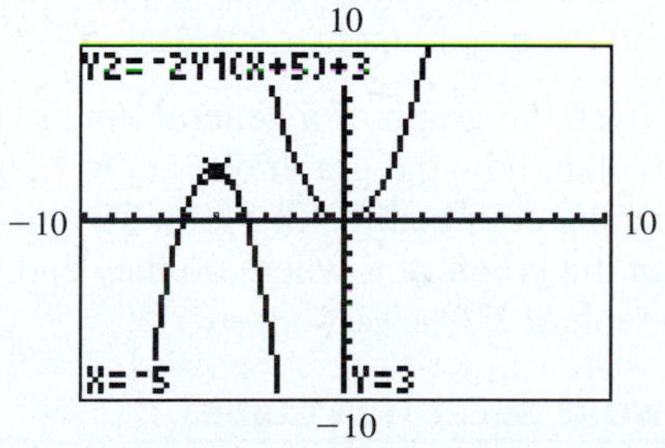

Try this exploration again using $Y_1 = \text{abs}(X)$.

Using the general equation $y = af(x \pm h) \pm k$, we can identify the vertex, initial point, or inflection point of any toolbox function and sketch its graph. Given the *graph* of a toolbox function, we can likewise identify these points and reconstruct its equation. We first identify the function family and the location (h, k) of any characteristic point. By selecting one other point (x, y) on the graph, we then use the general equation as a formula (substituting h, k, and the x- and y-values of the second point) to solve for a and complete the equation.

EXAMPLE 10 ▶ **Writing the Equation of a Function Given Its Graph**

Find the equation of the function $f(x)$ shown in the figure.

Solution ▶ The function f belongs to the absolute value family. The vertex (h, k) is at $(1, 2)$. For an additional point, choose the x-intercept $(-3, 0)$ and work as follows:

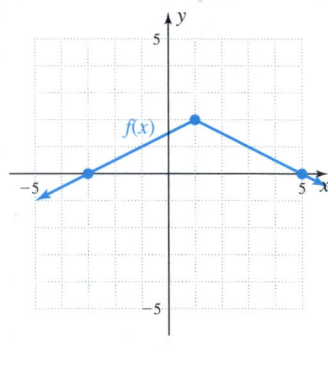

$y = a|x - h| + k$ general equation (function is shifted right and up)

$0 = a|(-3) - 1| + 2$ substitute 1 for h and 2 for k, substitute -3 for x and 0 for y

$0 = 4a + 2$ simplify

$-2 = 4a$ subtract 2

$-\dfrac{1}{2} = a$ solve for a

☑ **E.** You've just seen how we can apply transformations on a general function $f(x)$

The equation for f is $y = -\frac{1}{2}|x - 1| + 2$.

Now try Exercises 97 through 102 ▶

2.2 EXERCISES

▶ CONCEPTS AND VOCABULARY

Fill in each blank with the appropriate word or phrase. Carefully reread the section if needed.

1. After a vertical __stretch__, points on the graph are farther from the x-axis. After a vertical __compression__, points on the graph are closer to the x-axis.

2. Transformations that change only the location of a graph and not its shape or form, include __translations__ and __reflections__.

3. The vertex of $h(x) = 3(x + 5)^2 - 9$ is at __$(-5, -9)$__ and the graph opens __upward__.

4. The inflection point of $f(x) = -2(x - 4)^3 + 11$ is at __$(4, 11)$__ and the end-behavior is __up__, __down__.

5. Given the graph of a general function $f(x)$, discuss/explain how the graph of $F(x) = -2f(x + 1) - 3$ can be obtained. If $(0, 5)$, $(6, 7)$, and $(-9, -4)$ are on the graph of f, where do they end up on the graph of F? Answers will vary.

6. Discuss/Explain why the shift of $f(x) = x^2 + 3$ is a *vertical shift* of 3 units in the *positive* direction, while the shift of $g(x) = (x + 3)^2$ is a *horizontal shift* 3 units in the *negative* direction. Include several examples along with a table of values for each. Answers will vary.

HOMEWORK SELECTION GUIDE

Core: 7–99 every other odd, 105, 107, 109 (27 Exercises)
Standard: 1–4, 7–99 every other odd, 103–109 odd, 112, 115, 116 (35 Exercises)

Extended: 1–4, 7–99 every other odd, 103–109 odd, 110, 112, 113, 115, 116 (37 Exercises)
In Depth: 1–6, 7–99 every other odd, 103–117 (45 Exercises)

▶ DEVELOPING YOUR SKILLS

By carefully inspecting each graph given, (a) identify the function family; (b) describe or identify the end-behavior, vertex, intervals where the function is increasing or decreasing, maximum or minimum value(s) and *x*- and *y*-intercepts; and (c) determine the domain and range. Assume required features have integer values.

7. $f(x) = x^2 + 4x$ **8.** $g(x) = -x^2 + 2x$

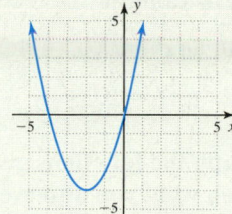   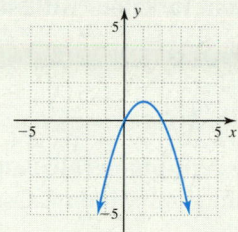

9. $p(x) = x^2 - 2x - 3$ **10.** $q(x) = -x^2 + 2x + 8$

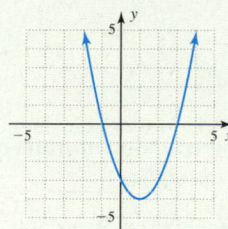   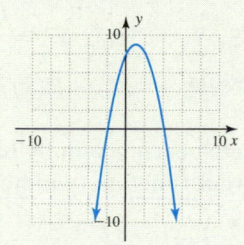

11. $f(x) = x^2 - 4x - 5$ **12.** $g(x) = x^2 + 6x + 5$

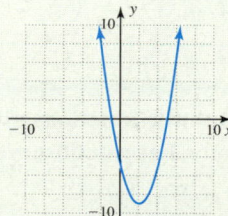   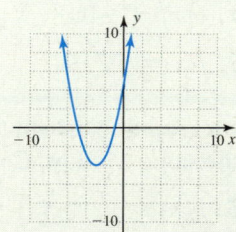

For each graph given, (a) identify the function family; (b) describe or identify the end-behavior, initial point, intervals where the function is increasing or decreasing, and *x*- and *y*-intercepts; and (c) determine the domain and range. Assume required features have integer values.

13. $p(x) = 2\sqrt{x + 4} - 2$ **14.** $q(x) = -2\sqrt{x + 4} + 2$

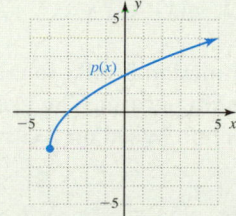   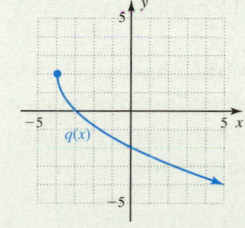

15. $r(x) = -3\sqrt{4 - x} + 3$ **16.** $f(x) = 2\sqrt{x + 1} - 4$

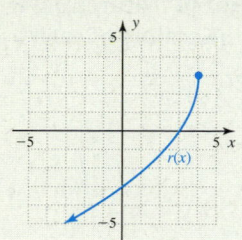   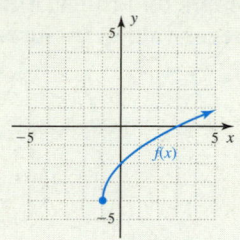

17. $g(x) = 2\sqrt{4 - x}$ **18.** $h(x) = -2\sqrt{x + 1} + 4$

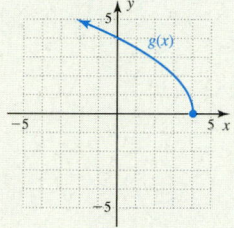   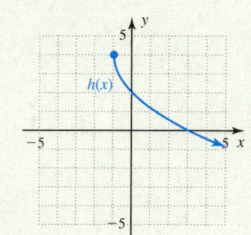

For each graph given, (a) identify the function family; (b) describe or identify the end-behavior, vertex, intervals where the function is increasing or decreasing, maximum or minimum value(s) and *x*- and *y*-intercepts; and (c) determine the domain and range. Assume required features have integer values.

19. $p(x) = 2|x + 1| - 4$ **20.** $q(x) = -3|x - 2| + 3$

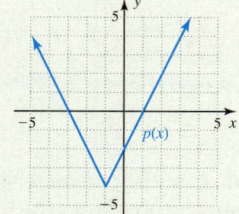   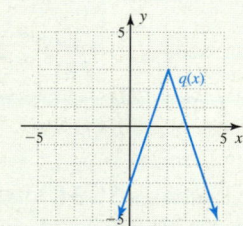

21. $r(x) = -2|x + 1| + 6$ **22.** $f(x) = 3|x - 2| - 6$

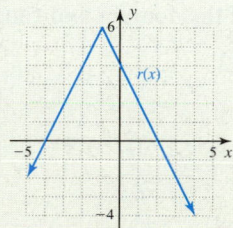   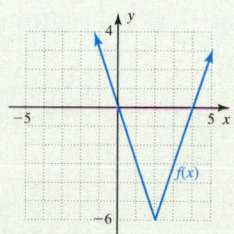

23. $g(x) = -3|x| + 6$ **24.** $h(x) = 2|x + 1|$

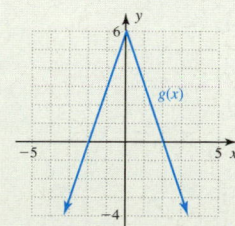   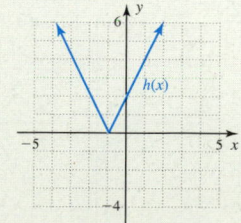

For each graph given, (a) identify the function family; (b) describe or identify the end-behavior, inflection point, and x- and y-intercepts; and (c) determine the domain and range. Assume required features have integer values. Be sure to note the scaling of each axis.

25. $f(x) = -(x - 1)^3$

26. $g(x) = (x + 1)^3$

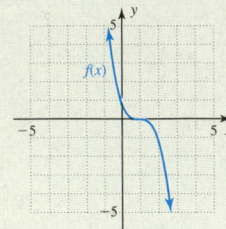
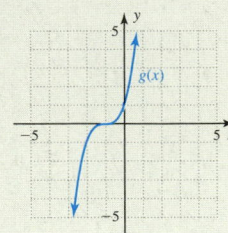

27. $h(x) = x^3 + 1$

28. $p(x) = -\sqrt[3]{x} + 1$

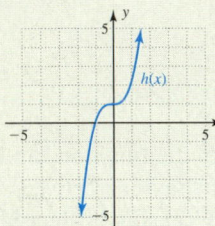
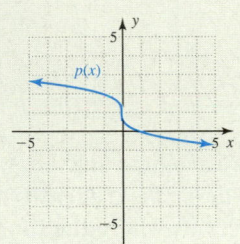

29. $q(x) = \sqrt[3]{x - 1} - 1$

30. $r(x) = -\sqrt[3]{x + 1} - 1$

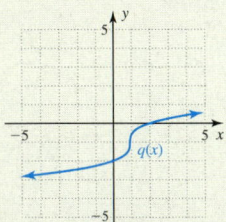
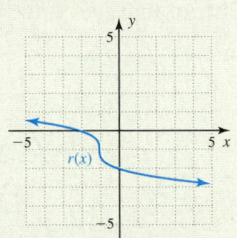

For Exercises 31–34, identify and state the characteristic features of each graph, including (as applicable) the function family, end-behavior, vertex, axis of symmetry, point of inflection, initial point, maximum and minimum value(s), x- and y-intercepts, and the domain and range.

31.

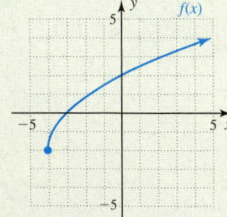

32.

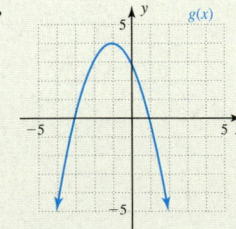

33.

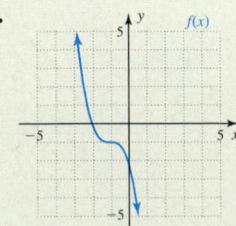

34.

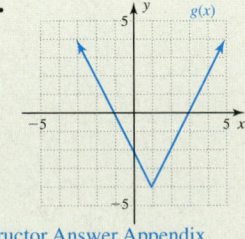

Additional answers can be found in the Instructor Answer Appendix.

Use a graphing calculator to graph the functions given in the same window. Comment on what you observe.

35. $f(x) = \sqrt{x}, \quad g(x) = \sqrt{x} + 2, \quad h(x) = \sqrt{x} - 3$

36. $f(x) = \sqrt[3]{x}, \quad g(x) = \sqrt[3]{x} - 3, \quad h(x) = \sqrt[3]{x} + 4$

37. $p(x) = |x|, \quad q(x) = |x| - 5, \quad r(x) = |x| + 2$

38. $p(x) = x^2, \quad q(x) = x^2 - 7, \quad r(x) = x^2 + 3$

Sketch each graph by hand using transformations of a parent function (without a table of values).

39. $f(x) = x^3 - 2$

40. $g(x) = \sqrt{x} - 4$

41. $h(x) = x^2 + 3$

42. $t(x) = |x| - 3$

Use a graphing calculator to graph the functions given in the same window. Comment on what you observe.

43. $p(x) = x^2, \quad q(x) = (x + 5)^2$

44. $f(x) = \sqrt{x}, \quad g(x) = \sqrt{x + 4}$

45. $Y_1 = |x|, \quad Y_2 = |x - 4|$

46. $h(x) = x^3, \quad H(x) = (x - 4)^3$

Sketch each graph by hand using transformations of a parent function (without a table of values).

47. $p(x) = (x - 3)^2$

48. $q(x) = \sqrt{x - 1}$

49. $h(x) = |x + 3|$

50. $f(x) = \sqrt[3]{x + 2}$

51. $g(x) = -|x|$

52. $j(x) = -\sqrt{x}$

53. $f(x) = \sqrt[3]{-x}$

54. $g(x) = (-x)^3$

Use a graphing calculator to graph the functions given in the same window. Comment on what you observe.

55. $p(x) = x^2, \quad q(x) = 3x^2, \quad r(x) = \frac{1}{5}x^2$

56. $f(x) = \sqrt{-x}, \quad g(x) = 4\sqrt{-x}, \quad h(x) = \frac{1}{4}\sqrt{-x}$

57. $Y_1 = |x|, \quad Y_2 = 3|x|, \quad Y_3 = \frac{1}{3}|x|$

58. $u(x) = x^3, \quad v(x) = 8x^3, \quad w(x) = \frac{1}{5}x^3$

Sketch each graph by hand using transformations of a parent function (without a table of values).

59. $f(x) = 4\sqrt[3]{x}$

60. $g(x) = -2|x|$

61. $p(x) = \frac{1}{3}x^3$

62. $q(x) = \frac{3}{4}\sqrt{x}$

Use the characteristics of each function family to match a given function to its corresponding graph. The graphs are not scaled—make your selection based on a careful comparison.

63. $f(x) = \frac{1}{2}x^3$ g

64. $f(x) = \frac{-2}{3}x + 2$ h

65. $f(x) = -(x - 3)^2 + 2$ i

66. $f(x) = -\sqrt[3]{x - 1} - 1$ d

67. $f(x) = |x + 4| + 1$ e   **68.** $f(x) = -\sqrt{x + 6}$ f

69. $f(x) = -\sqrt{x + 6} - 1$ j   **70.** $f(x) = x + 1$ k

71. $f(x) = (x - 4)^2 - 3$ l   **72.** $f(x) = |x - 2| - 5$ b

73. $f(x) = \sqrt{x + 3} - 1$ c   **74.** $f(x) = -(x + 3)^2 + 5$ a

a.

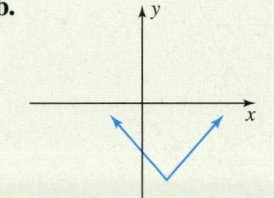

b.

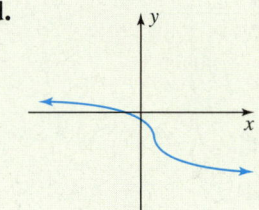

c.

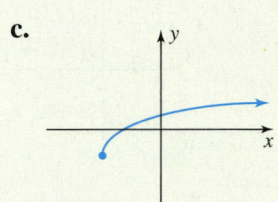

d.

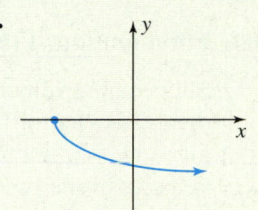

e.

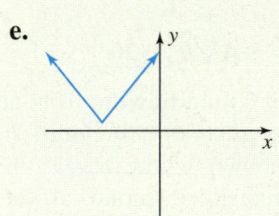

f.

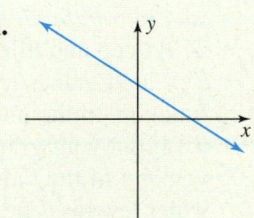

g.

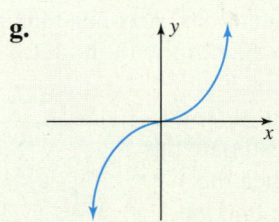

h.

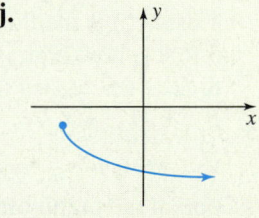

i.

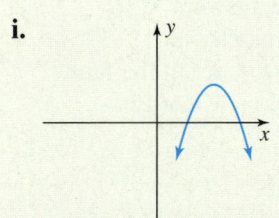

j.

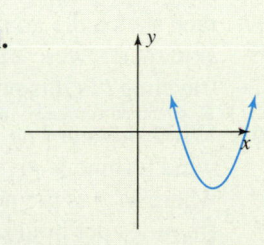

k.

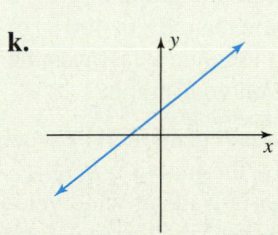

l.

Graph each function using shifts of a parent function and a few characteristic points. *Clearly state and indicate the transformations used* and identify the location of all vertices, initial points, and/or inflection points.

75. $f(x) = \sqrt{x + 2} - 1$ **76.** $g(x) = \sqrt{x - 3} + 2$

77. $h(x) = -(x + 3)^2 - 2$ **78.** $H(x) = -(x - 2)^2 + 5$

79. $p(x) = (x + 3)^3 - 1$ **80.** $q(x) = (x - 2)^3 + 1$

81. $s(x) = \sqrt[3]{x + 1} - 2$ **82.** $t(x) = \sqrt[3]{x - 3} + 1$

83. $f(x) = -|x + 3| - 2$ **84.** $g(x) = -|x - 4| - 2$

85. $h(x) = -2(x + 1)^2 - 3$ **86.** $H(x) = \frac{1}{2}|x + 2| - 3$

87. $p(x) = -\frac{1}{3}(x + 2)^3 - 1$ **88.** $q(x) = 4\sqrt[3]{x + 1} + 2$

89. $u(x) = -2\sqrt{-x - 1} + 3$ **90.** $v(x) = 3\sqrt{-x + 2} - 1$

91. $h(x) = \frac{1}{5}(x - 3)^2 + 1$ **92.** $H(x) = -2|x - 3| + 4$

Apply the transformations indicated for the graph of the general functions given.

93.

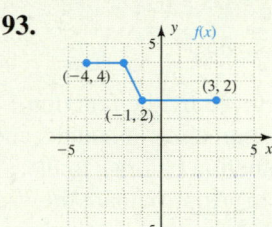

94.

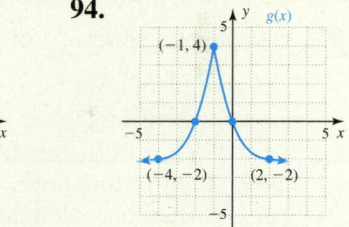

a. $f(x - 2)$

b. $-f(x) - 3$

c. $\frac{1}{2}f(x + 1)$

d. $f(-x) + 1$

a. $g(x) - 2$

b. $-g(x) + 3$

c. $2g(x + 1)$

d. $\frac{1}{2}g(x - 1) + 2$

95.

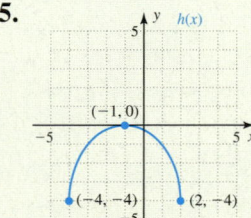

96.

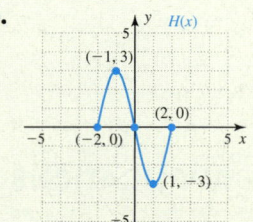

a. $h(x) + 3$

b. $-h(x - 2)$

c. $h(x - 2) - 1$

d. $\frac{1}{4}h(x) + 5$

a. $H(x - 3)$

b. $-H(x) + 1$

c. $2H(x - 3)$

d. $\frac{1}{3}H(x - 2) + 1$

Use the graph given and the points indicated to determine the equation of the function shown using the general form $y = af(x \pm h) \pm k$.

97.

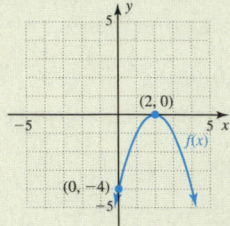

$f(x) = -(x - 2)^2$

98.

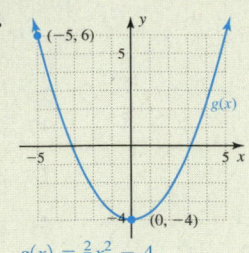

$g(x) = \frac{2}{5}x^2 - 4$

99.

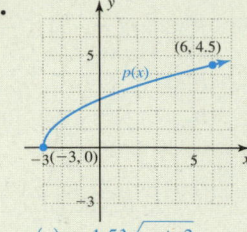

$p(x) = 1.5\sqrt{x + 3}$

100.

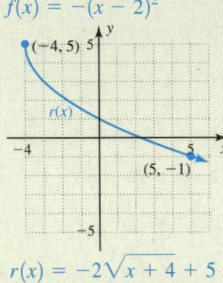

$r(x) = -2\sqrt{x + 4} + 5$

101.

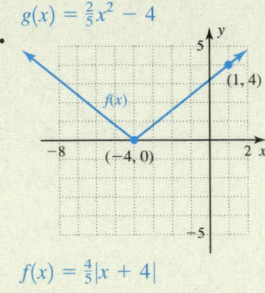

$f(x) = \frac{4}{5}|x + 4|$

102.

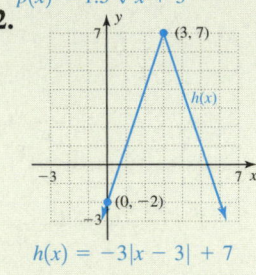

$h(x) = -3|x - 3| + 7$

▶ WORKING WITH FORMULAS

103. Volume of a sphere: $V(r) = \frac{4}{3}\pi r^3$

The volume of a sphere is given by the function shown, where $V(r)$ is the volume in cubic units and r is the radius. Note this function belongs to the *cubic family* of functions. (a) Approximate the value of $\frac{4}{3}\pi$ to one decimal place, then graph the function on the interval $[0, 3]$. (b) From your *graph*, estimate the volume of a sphere with radius 2.5 in., then compute the actual volume. Are the results close? (c) For $V = \frac{4}{3}\pi r^3$, solve for r in terms of V.   **b.** about 65 units^3, $V \approx 65.4$ units^3; yes   **c.** $r = \sqrt[3]{\frac{3}{4\pi}V}$

104. Fluid motion: $V(h) = -4\sqrt{h} + 20$

Suppose the velocity of a fluid flowing from an open tank (no top) through an opening in its side is given by the function shown, where $V(h)$ is the velocity of the fluid (in feet per second) at water height h (in feet). Note this function belongs to the *square root family* of functions. An open tank is 25 ft deep and filled to the brim with fluid. (a) Use a table of values to graph the function on the interval $[0, 25]$. (b) From your graph, estimate the velocity of the fluid when the water level is 7 ft, then find the actual velocity. Are the answers close? (c) If the fluid velocity is 5 ft/sec, how high is the water in the tank?

b. About 9 ft/sec, $V \approx 9.4$ ft/sec, yes   **c.** about 14 ft

▶ APPLICATIONS

105. Gravity, distance, time: After being released, the time it takes an object to fall x ft is given by the function $T(x) = \frac{1}{4}\sqrt{x}$, where $T(x)$ is in seconds. (a) Describe the transformation applied to obtain the graph of T from the graph of $y = \sqrt{x}$, then sketch the graph of T for $x \in [0, 100]$. (b) How long would it take an object to hit the ground if it were dropped from a height of 81 ft?
a. compressed vertically   **b.** 2.25 sec

106. Stopping distance: In certain weather conditions, accident investigators will use the function $v(x) = 4.9\sqrt{x}$ to estimate the speed of a car (in miles per hour) that has been involved in an accident, based on the length of the skid marks x (in feet). (a) Describe the transformation applied to

obtain the graph of v from the graph of $y = \sqrt{x}$, then sketch the graph of v for $x \in [0, 400]$. (b) If the skid marks were 225 ft long, how fast was the car traveling? Is this point on your graph?
a. stretched vertically   **b.** 73.5 mph; yes

107. Wind power: The power P generated by a certain wind turbine is given by the function $P(v) = \frac{8}{125}v^3$ where $P(v)$ is the power in watts at wind velocity v (in miles per hour). (a) Describe the transformation applied to obtain the graph of P from the graph of $y = v^3$, then sketch the graph of P for $v \in [0, 25]$ (scale the axes appropriately). (b) How much power is being generated when the wind is blowing at 15 mph?   **a.** compressed vertically   **b.** 216 W

Additional answers can be found in the Instructor Answer Appendix.

108. Wind power: If the power P (in watts) being generated by a wind turbine is known, the velocity of the wind can be determined using the function $v(P) = \frac{5}{2}\sqrt[3]{P}$. (a) Describe the transformation applied to obtain the graph of v from the graph of $y = \sqrt[3]{P}$, then sketch the graph of v for $P \in [0, 512]$ (scale the axes appropriately). (b) How fast is the wind blowing if 343W of power is being generated? Is this point on your graph? **a.** stretched vertically   **b.** 17.5 mph

109. Distance rolled due to gravity: The *distance* a ball rolls down an inclined plane is given by the function $d(t) = 2t^2$, where $d(t)$ represents the distance in feet after t sec. (a) Describe the transformation applied to obtain the graph of d from the graph

of $y = t^2$, then sketch the graph of d for $t \in [0, 3]$. (b) How far has the ball rolled after 2.5 sec?   **a.** vertical stretch by a factor of 2   **b.** 12.5 ft

110. Acceleration due to gravity: The *velocity* of a steel ball bearing as it rolls down an inclined plane is given by the function $v(t) = 4t$, where $v(t)$ represents the velocity in feet per second after t sec. (a) Describe the transformation applied to obtain the graph of v from the graph of $y = t$, then sketch the graph of v for $t \in [0, 3]$. (b) What is the velocity of the ball bearing after 2.5 sec? Is this point on your graph?   **a.** vertical stretch by a factor of 4   **b.** 10 ft/sec

▶ EXTENDING THE CONCEPT

111. Carefully graph the functions $f(x) = |x|$ and $g(x) = 2\sqrt{x}$ on the same coordinate grid. From the graph, in what interval is the graph of $g(x)$ *above* the graph of $f(x)$? Pick a number (call it h) from this interval and substitute it in both functions. Is $g(h) > f(h)$? In what interval is the graph of $g(x)$ below the graph of $f(x)$? Pick a number from this interval (call it k) and substitute it in both functions. Is $g(k) < f(k)$? $x \in (0, 4)$; yes, $x \in (4, \infty)$; yes

112. Sketch the graph of $f(x) = -2|x - 3| + 8$ using transformations of the parent function, then determine the area of the region in quadrant I that is beneath the graph and bounded by the vertical lines $x = 0$ and $x = 6$.   30 units^2

113. Sketch the graph of $f(x) = x^2 - 4$, then sketch the graph of $F(x) = |x^2 - 4|$ using your intuition and the meaning of absolute value (not a table of values). What happens to the graph?
Any points in Quadrants III and IV will reflect across the x-axis and move to Quadrants I and II.

▶ MAINTAINING YOUR SKILLS

114. (1.1) Find the distance between the points $(-13, 9)$ and $(7, -12)$, and the slope of the line containing these points. $d = 29$ units, $m = -\frac{21}{20}$

115. (R.2) Find the perimeter of the figure shown.
$P = (4x^2 + 16x + 14)$ units

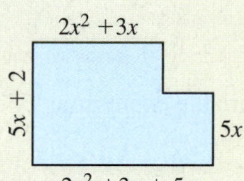

116. (1.5) Solve for x: $\frac{2}{3}x + \frac{1}{4} = \frac{1}{2}x - \frac{7}{12}$. $x = -5$

117. (2.1) Without graphing, state intervals where $f(x)\uparrow$ and $f(x)\downarrow$ for $f(x) = (x - 4)^2 + 3$.
$f(x)\uparrow: x \in (4, \infty)$; $f(x)\downarrow: x \in (-\infty, 4)$

2.3 Absolute Value Functions, Equations, and Inequalities

While the equations $x + 1 = 5$ and $|x + 1| = 5$ are similar in many respects, note the first has only the solution $x = 4$, while either $x = 4$ or $x = -6$ will satisfy the second. The fact there are two solutions shouldn't surprise us, as it's a natural result of how absolute value is defined.

A. Solving Absolute Value Equations

The absolute value of a number x can be thought of as its distance from zero on the number line, regardless of direction. This means $|x| = 4$ will have *two solutions,* since there are two numbers that are four units from zero: $x = -4$ and $x = 4$ (see Figure 2.39).

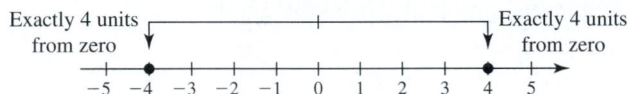

Figure 2.39

This basic idea can be extended to include situations where the quantity within absolute value bars *is an algebraic expression,* and suggests the following property.

> **Property of Absolute Value Equations**
>
> If X represents an algebraic expression and k is a positive real number,
>
> $$\text{then } |X| = k$$
>
> $$\text{implies } X = -k \text{ or } X = k$$

As the statement of this property suggests, it can only be applied *after* the absolute value expression has been isolated on one side.

EXAMPLE 1 ▶ **Solving an Absolute Value Equation**

Solve: $-5|x - 7| + 2 = -13$.

Solution ▶ Begin by isolating the absolute value expression.

$$-5|x - 7| + 2 = -13 \quad \text{original equation}$$
$$-5|x - 7| = -15 \quad \text{subtract 2}$$
$$|x - 7| = 3 \quad \text{divide by } -5 \text{ (simplified form)}$$

Now consider $x - 7$ as the variable expression "X" in the property of absolute value equations, giving

$$x - 7 = -3 \quad \text{or} \quad x - 7 = 3 \quad \text{apply the property of absolute value equations}$$
$$x = 4 \quad \text{or} \quad x = 10 \quad \text{add 7}$$

Substituting into the original equation verifies the solution set is $\{4, 10\}$.

Now try Exercises 7 through 18 ▶

⚠ **CAUTION** ▶ For equations like those in Example 1, be careful not to treat the absolute value bars as simple grouping symbols. The equation $-5(x - 7) + 2 = -13$ has only the solution $x = 10$, and "misses" the second solution since it yields $x - 7 = 3$ in simplified form. The equation $-5|x - 7| + 2 = -13$ simplifies to $|x - 7| = 3$ and there are actually *two* solutions. Also note that $-5|x - 7| \neq |-5x + 35|$!

If an equation has more than one solution as in Example 1, they cannot be simultaneously stored using the, ⬚ key to perform a calculator check (in function or "**Func**" mode, this is the variable X). While there are other ways to "get around" this (using Y_1 on the home screen, using a **TABLE** in ⬚ ASK ⬚ mode, enclosing the solutions in braces as in {4, 10}, etc.), we can also store solutions using the ⬚ keys. To illustrate, we'll place the solution $x = 4$ in storage location A, using 4 ⬚ STO▸ ⬚ ALPHA ⬚ MATH (**A**). Using this "⬚ STO▸ ⬚ ALPHA ⬚" sequence we'll next place the solution $x = 10$ in storage location B (Figure 2.40). We can then check both solutions in turn. Note that after we check the first solution, we can recall the expression using ⬚ 2nd ⬚ ENTER ⬚ and simply change the A to B (Figure 2.41).

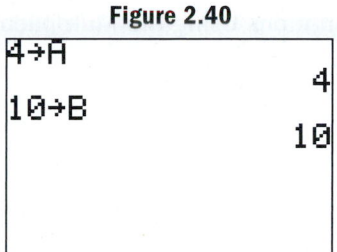

Figure 2.40

```
4→A
              4
10→B
             10
```

Figure 2.41

```
-5abs(A-7)+2
            -13
-5abs(B-7)+2
            -13
```

Absolute value equations come in many different forms. Always begin by isolating the absolute value expression, then apply the property of absolute value equations to solve.

EXAMPLE 2 ▶ **Solving an Absolute Value Equation**

Solve: $\left|5 - \dfrac{2}{3}x\right| - 9 = 8$.

Solution ▶

$$\left|5 - \frac{2}{3}x\right| - 9 = 8 \qquad\qquad \text{original equation}$$

$$\left|5 - \frac{2}{3}x\right| = 17 \qquad\qquad \text{add 9}$$

$$5 - \frac{2}{3}x = -17 \quad \text{or} \quad 5 - \frac{2}{3}x = 17 \qquad \text{apply the property of absolute value equations}$$

$$-\frac{2}{3}x = -22 \quad \text{or} \quad -\frac{2}{3}x = 12 \qquad \text{subtract 5}$$

$$x = 33 \quad \text{or} \quad x = -18 \qquad \text{multiply by } -\tfrac{3}{2}$$

Check ▶

For $x = 33$: $\left|5 - \dfrac{2}{3}(33)\right| - 9 = 8$ For $x = -18$: $\left|5 - \dfrac{2}{3}(-18)\right| - 9 = 8$

$$|5 - 2(11)| - 9 = 8 \qquad\qquad |5 - 2(-6)| - 9 = 8$$

$$|5 - 22| - 9 = 8 \qquad\qquad |5 + 12| - 9 = 8$$

$$|-17| - 9 = 8 \qquad\qquad |17| - 9 = 8$$

$$17 - 9 = 8 \qquad\qquad 17 - 9 = 8$$

$$8 = 8 ✓ \qquad\qquad 8 = 8 ✓$$

Both solutions check. The solution set is $\{-18, 33\}$.

Now try Exercises 19 through 22 ▶

WORTHY OF NOTE

As illustrated in both Examples 1 and 2, the property we use to solve absolute value equations can only be applied *after* the absolute value term has been isolated. As you will see, the same is true for the properties used to solve absolute value inequalities.

For some equations, it's helpful to apply the **multiplicative property of absolute value:**

> ### Multiplicative Property of Absolute Value
>
> If A and B represent algebraic expressions,
>
> $$\text{then } |AB| = |A||B|.$$

Note that if $A = -1$ the property says $|-1 \cdot B| = |-1| \, |B| = |B|$. More generally the property is applied where A is any constant.

EXAMPLE 3 ▶ **Solving Equations Using the Multiplicative Property of Absolute Value**

Solve: $|-2x| + 5 = 13$.

Solution ▶

$|-2x| + 5 = 13$	original equation
$|-2x| = 8$	subtract 5
$|-2||x| = 8$	apply multiplicative property of absolute value
$2|x| = 8$	simplify
$|x| = 4$	divide by 2
$x = -4 \quad \text{or} \quad x = 4$	apply property of absolute value equations

Both solutions check. The solution set is $\{-4, 4\}$.

Now try Exercises 23 and 24 ▶

In some instances, we have one absolute value quantity equal to another, as in $|A| = |B|$. From this equation, four possible solutions are immediately apparent:

$$(1)\ A = B \qquad (2)\ A = -B \qquad (3)\ -A = B \qquad (4)\ -A = -B$$

However, basic properties of equality show that equations (1) and (4) are equivalent, as are equations (2) and (3), meaning all solutions can be found using only equations (1) and (2).

EXAMPLE 4 ▶ **Solving Absolute Value Equations with Two Absolute Value Expressions**

Solve the equation $|2x + 7| = |x - 1|$.

Solution ▶ This equation has the form $|A| = |B|$, where $A = 2x + 7$ and $B = x - 1$. From our previous discussion, all solutions can be found using $A = B$ and $A = -B$.

$A = B$	solution template	$A = -B$	solution template
$2x + 7 = x - 1$	substitute	$2x + 7 = -(x - 1)$	substitute
$2x = x - 8$	subtract 7	$2x + 7 = -x + 1$	distribute
$x = -8$	subtract x	$3x = -6$	add x, subtract 7
		$x = -2$	divide by 3

The solutions are $x = -8$ and $x = -2$. Verify the solutions by substituting them into the original equation.

Now try Exercises 25 and 26 ▶

☑ **A.** You've just seen how we can solve absolute value equations

B. Solving "Less Than" Absolute Value Inequalities

Absolute value *inequalities* can be solved using the basic concept underlying the property of absolute value equalities. Whereas the equation $|x| = 4$ asks for all numbers x whose distance from zero is *equal* to 4, the inequality $|x| < 4$ asks for all numbers x whose distance from zero is *less than* 4.

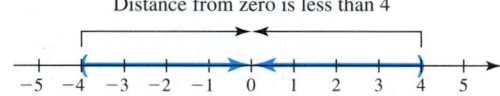

Figure 2.42

As Figure 2.42 illustrates, the solutions are $x > -4$ and $x < 4$, which can be written as the joint inequality $-4 < x < 4$. This idea can likewise be extended to include the absolute value of an algebraic expression X as follows.

> **Property I: Absolute Value Inequalities (Less Than)**
>
> If X represents an algebraic expression and k is a positive real number,
>
> $$\text{then } |X| < k$$
>
> $$\text{implies } -k < X < k$$

Property I can also be applied when the "$\leq$" symbol is used. Also notice that if $k < 0$, the solution is the empty set since the absolute value of any quantity is always positive or zero.

EXAMPLE 5 ▶ **Solving "Less Than" Absolute Value Inequalities**

Solve the inequalities:

a. $\dfrac{|3x + 2|}{4} \leq 1$ b. $|2x - 7| < -5$

Solution ▶

a.

$\dfrac{|3x + 2|}{4} \leq 1$ original inequality

$|3x + 2| \leq 4$ multiply by 4

$-4 \leq 3x + 2 \leq 4$ apply Property I

$-6 \leq 3x \leq 2$ subtract 2 from all three parts

$-2 \leq x \leq \dfrac{2}{3}$ divide all three parts by 3

The solution interval is $\left[-2, \frac{2}{3}\right]$.

b. $|2x - 7| < -5$ original inequality

Since the absolute value of any quantity is always positive or zero, the solution for this inequality is the empty set: { }.

Now try Exercises 27 through 38 ▶

As with the inequalities from Section 1.5, solutions to absolute value inequalities can be checked using a test value. For Example 5(a), substituting $x = 0$ from the solution interval yields:

$$\frac{1}{2} \leq 1 \checkmark$$

In addition to checking absolute value inequalities using a test value, the **TABLE** feature of a graphing calculator can be used, alone or in conjunction with a **relational test.** Relational tests have the calculator return a "1" if a given statement is true, and a "0" otherwise. To illustrate, consider the inequality $2|x - 3| + 1 \leq 5$. Enter the expression on the left as Y_1, recalling the **"abs("** notation is accessed in the (MATH) menu: (MATH) ▷ **(NUM)** **"1:abs("** (ENTER) (note this option gives only the left parenthesis, you must supply the right). We can then simply inspect the Y_1 column of the **TABLE** to find outputs that are less than or equal to 5. To use a relational test, we enter $Y_1 \leq 5$ as Y_2 (Figure 2.43), with the "less than or equal to" symbol accessed using (2nd) (MATH) **6:≤**. Now the calculator will automatically check the truth of the statement for any value of x (but note we are only checking integer values), and display the result in the Y_2 column of the **TABLE** (Figure 2.44). After scrolling through the table, both approaches show that $2|x - 3| + 1 \leq 5$ for $x \in [1, 5]$.

Figure 2.43

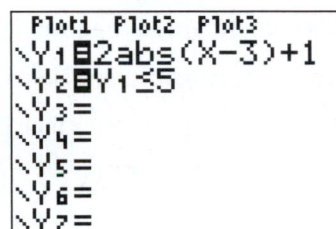

Figure 2.44

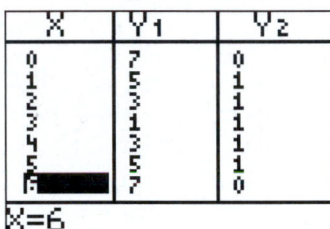

☑ **B.** You've just seen how we can solve "less than" absolute value inequalities

C. Solving "Greater Than" Absolute Value Inequalities

For "greater than" inequalities, consider $|x| > 4$. Now we're asked to find all numbers x whose distance from zero is *greater than* 4. As Figure 2.45 shows, solutions are found in the interval to the left of -4, or to the right of 4. The fact the intervals are disjoint (disconnected) is reflected in this graph, in the inequalities $x < -4$ **or** $x > 4$, as well as the interval notation $x \in (-\infty, -4) \cup (4, \infty)$.

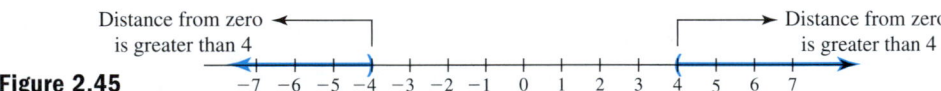

Distance from zero is greater than 4   Distance from zero is greater than 4

Figure 2.45

As before, we can extend this idea to include algebraic expressions, as follows:

Property II: Absolute Value Inequalities (Greater Than)

If X represents an algebraic expression and k is a positive real number,

$$\text{then } |X| > k$$

$$\text{implies } X < -k \quad \text{or} \quad X > k$$

EXAMPLE 6 ▶ Solving "Greater Than" Absolute Value Inequalities

Solve the inequalities:

a. $-\dfrac{1}{3}\left|3 + \dfrac{x}{2}\right| < -2$ **b.** $|5x + 2| \geq -\dfrac{3}{2}$

Solution ▶   **a.** Note the exercise is given as a *less than* inequality, but as we multiply both sides by -3, we must *reverse the inequality symbol*.

$$-\frac{1}{3}\left|3 + \frac{x}{2}\right| < -2 \qquad \text{original inequality}$$

$$\left|3 + \frac{x}{2}\right| > 6 \qquad \text{multiply by } -3, \text{ reverse the symbol}$$

$$3 + \frac{x}{2} < -6 \quad \text{or} \quad 3 + \frac{x}{2} > 6 \qquad \text{apply Property II}$$

$$\frac{x}{2} < -9 \quad \text{or} \quad \frac{x}{2} > 3 \qquad \text{subtract 3}$$

$$x < -18 \quad \text{or} \quad x > 6 \qquad \text{multiply by 2}$$

Property II yields the disjoint intervals $x \in (-\infty, -18) \cup (6, \infty)$ as the solution.

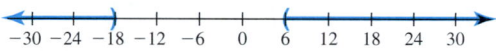

b. $|5x + 2| \geq -\dfrac{3}{2}$ original inequality

Since the absolute value of any quantity is always positive or zero, the solution for this inequality is all real numbers: $x \in \mathbb{R}$.

Now try Exercises 39 through 54 ▶

A calculator check is shown for part (a) in Figures 2.46 through 2.48.

Figure 2.46	**Figure 2.47**	**Figure 2.48**

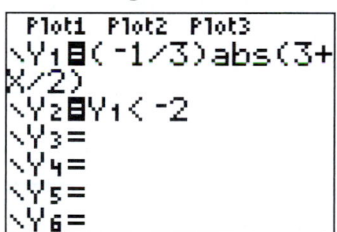

This helps to verify the solution interval is $x \in (-\infty, -18) \cup (6, \infty)$.

Due to the nature of absolute value functions, there are times when an absolute value relation cannot be satisfied. For instance the equation $|x - 4| = -2$ has no solutions, as the left-hand expression will always represent a non-negative value. The inequality $|2x + 3| < -1$ has no solutions for the same reason. On the other hand, the inequality $|9 - x| \geq 0$ is true for all real numbers, since any value substituted for x will result in a nonnegative value. We can generalize many of these special cases as follows.

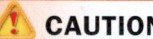 **C.** You've just seen how we can solve "greater than" absolute value inequalities

---

Absolute Value Functions—Special Cases

Given k is a positive real number and A represents an algebraic expression,

$$|A| = -k \qquad\qquad |A| < -k \qquad\qquad |A| > -k$$

has no solutions   has no solutions   is true for all real numbers

---

See **Exercises 51 through 54.**

⚠ **CAUTION** ▶   Be sure you note the difference between the individual solutions of an absolute value equation, and the solution intervals that often result from solving absolute value inequalities. The solution $\{-2, 5\}$ indicates that both $x = -2$ and $x = 5$ are solutions, while the solution $[-2, 5)$ indicates that all numbers between -2 and 5, including -2, are solutions.

D. Solving Absolute Value Equations and Inequalities Graphically

The concepts studied in Section 1.5 (solving linear equations and inequalities graphically) are easily extended to other kinds of relations. Essentially, we treat each expression forming the equation or inequality as a *separate function*, then graph both functions to find points of intersection (equations) or where one graph is above or

Figure 2.49

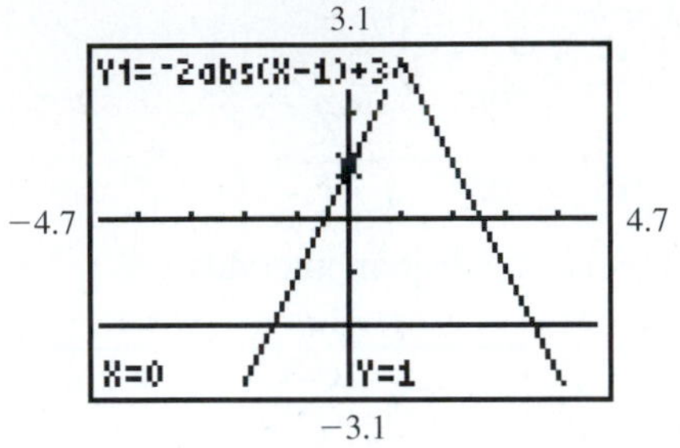

3.1

Y1=-2abs(X-1)+3

-4.7   4.7

X=0   Y=1

-3.1

Figure 2.50

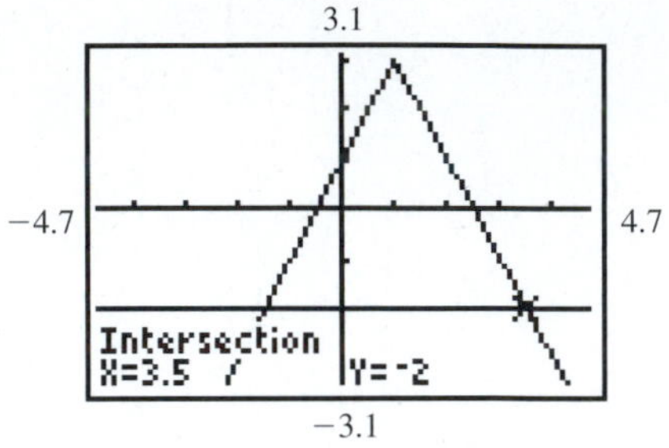

3.1

-4.7   4.7

Intersection
X=3.5   Y=-2

-3.1

below the other (inequalities). For $-2|x - 1| + 3 < -2$, enter the expression $-2|X - 1| + 3$ as Y_1 on the [Y=] screen, and -2 as Y_2. Using [ZOOM] **4:ZDecimal** produces the graph shown in Figure 2.49. Using [2nd] [TRACE] **(CALC) 5:intersect,** we find the graphs intersect at $x = -1.5$ and $x = 3.5$ (Figure 2.50), and the graph of Y_1 is above the graph of Y_2 in this interval. Since this is a "less than" inequality, the solutions are *outside* of this interval, which gives $x \in (-\infty, -1.5) \cup (3.5, \infty)$ as the solution interval. Note that the zeroes/x-intercept method could also have been used.

EXAMPLE 7 ▶ **Solving Absolute Equations and Inequalities Graphically**

For $f(x) = 2.5|x - 2| - 8$ and $g(x) = \dfrac{1}{2}x - 3$, solve

a. $f(x) = g(x)$ **b.** $f(x) \leq g(x)$ **c.** $f(x) > g(x)$

Solution ▶ **a.** With $f(x) = 2.5|x - 2| - 8$ as Y_1 and

$g(x) = \dfrac{1}{2}x - 3$ as Y_2 (set to graph in **bold**),

using [2nd] [TRACE] **(CALC) 5:intersect** shows the graphs intersect $(Y_1 = Y_2)$ at $x = 0$ and $x = 5$ (see figure). These are the solutions to $2.5|x - 2| - 8 = \dfrac{1}{2}x - 3$.

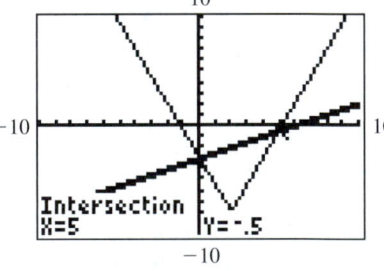

10

-10   10

Intersection
X=5   Y=-.5

-10

b. The graph of Y_1 is *below* the graph of Y_2 $(Y_1 < Y_2)$ between these points of intersection, so the solution interval for $2.5|x - 2| - 8 \leq \dfrac{1}{2}x - 3$ is $x \in [0, 5]$.

c. The graph of Y_1 is *above* the graph of Y_2 $(Y_1 > Y_2)$ outside this interval, giving a solution of $x \in (-\infty, 0) \cup (5, \infty)$ for $2.5|x - 2| - 8 > \dfrac{1}{2}x - 3$.

☑ **D.** You've just seen how we can solve absolute value equations and inequalities graphically

Now try Exercises 55 through 58 ▶

E. Applications Involving Absolute Value

Applications of absolute value often involve finding a range of values for which a given statement is true. Many times, the equation or inequality used must be modeled after a given description or from given information, as in Example 8.

EXAMPLE 8 ▶ **Solving Applications Involving Absolute Value Inequalities**

For new cars, the number of miles per gallon (mpg) a car will get is heavily dependent on whether it is used mainly for short trips and city driving, or primarily on the highway for longer trips. For a certain car, the number of miles per gallon that a driver can expect varies by no more than 6.5 mpg above or below its field tested average of 28.4 mpg. What range of mileage values can a driver expect for this car?

Solution ▶ Field tested average: 28.4 mpg   gather information
mileage varies by no more than 6.5 mpg   highlight key phrases

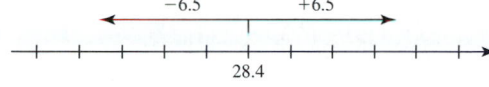

make the problem visual

Let m represent the miles per gallon a driver can expect.   assign a variable
Then the difference between m and 28.4 can be no more
than 6.5, or $|m - 28.4| \leq 6.5$.   write an equation model

$$|m - 28.4| \leq 6.5$$ equation model
$$-6.5 \leq m - 28.4 \leq 6.5$$ apply Property I
$$21.9 \leq m \leq 34.9$$ add 28.4 to all three parts

The mileage that a driver can expect ranges from a low of 21.9 mpg to a high of 34.9 mpg.

☑ **E.** You've just seen how we can solve applications involving absolute value

Now try Exercises 61 through 70 ▶

2.3 EXERCISES

▶ **CONCEPTS AND VOCABULARY**

Fill in the blank with the appropriate word or phrase. Carefully reread the section if needed.

1. When multiplying or dividing by a negative quantity, we ___reverse___ the inequality symbol to maintain a true statement.

2. To write an absolute value equation or inequality in simplified form, we ___isolate___ the absolute value expression on one side.

3. The absolute value equation $|2x + 3| = 7$ is true when $2x + 3 = $ ___-7___ or when $2x + 3 = $ ___7___.

4. The absolute value inequality $|3x - 6| < 12$ is true when $3x - 6 > $ ___-12___ and $3x - 6 < $ ___12___.

Describe the solution set for each inequality (assume $k > 0$). Justify your answer.

5. $|ax + b| < -k$ no solution; answers will vary.

6. $|ax + b| > -k$ all real numbers; answers will vary.

HOMEWORK SELECTION GUIDE

Core: 7–53 every other odd, 57, 59, 61–69 odd (20 Exercises)
Standard: 1–4, 7–53 every other odd, 57, 59, 61–69 odd, 71, 72, 75, 77 (28 Exercises)

Extended: 1–4, 7–53 every other odd, 57, 59, 60, 61–69 odd, 71–73, 75, 77 (30 Exercises)
In Depth: 1–6, 7–53 every other odd, 57, 59, 60, 61, 63, 65–77 (37 Exercises)

▶ DEVELOPING YOUR SKILLS

Solve each absolute value equation. Write the solution in set notation. For Exercises 7 to 18, verify solutions by substituting into the original equation. For Exercises 19–26 verify solutions using a calculator.

7. $2|m - 1| - 7 = 3$ $\{-4, 6\}$

8. $3|n - 5| - 14 = -2$ $\{1, 9\}$

9. $-3|x + 5| + 6 = -15$ $\{2, -12\}$

10. $-2|y + 3| - 4 = -14$ $\{-8, 2\}$

11. $2|4v + 5| - 6.5 = 10.3$ $\{-3.35, 0.85\}$

12. $7|2w + 5| + 6.3 = 11.2$ $\{-2.85, -2.15\}$

13. $-|7p - 3| + 6 = -5$ $\{-\frac{8}{7}, 2\}$

14. $-|3q + 4| + 3 = -5$ $\{-4, \frac{4}{3}\}$

15. $-2|b| - 3 = -4$ $\{-\frac{1}{2}, \frac{1}{2}\}$

16. $-3|c| - 5 = -6$ $\{-\frac{1}{3}, \frac{1}{3}\}$

17. $-2|3x| - 17 = -5$ $\{\}$

18. $-5|2y| - 14 = 6$ $\{\}$

19. $-3\left|\dfrac{w}{2} + 4\right| - 1 = -4$ $\{-10, -6\}$

20. $-2\left|3 - \dfrac{v}{3}\right| + 1 = -5$ $\{0, 18\}$

21. $8.7|p - 7.5| - 26.6 = 8.2$ $\{3.5, 11.5\}$

22. $5.3|q + 9.2| + 6.7 = 43.8$ $\{-16.2, -2.2\}$

23. $8.7|-2.5x| - 26.6 = 8.2$ $\{-1.6, 1.6\}$

24. $5.3|1.25n| + 6.7 = 43.8$ $\{-5.6, 5.6\}$

25. $|x - 2| = |3x + 4|$ $\{-3, -\frac{1}{2}\}$

26. $|2x - 1| = |x + 3|$ $\{-\frac{2}{3}, 4\}$

Solve each absolute value inequality. Write solutions in interval notation. Check solutions by back substitution, or using a calculator.

27. $3|p + 4| + 5 < 8$

28. $5|q - 2| - 7 \leq 8$

29. $-3|m| - 2 > 4$ $\{\}$

30. $-2|n| + 3 > 7$ $\{\}$

31. $|3b - 11| + 6 \leq 9$

32. $|2c + 3| - 5 < 1$

33. $|4 - 3z| + 12 < 7$

34. $|2 - 3u| + 5 \leq 4$

35. $\dfrac{|5v + 1|}{4} + 8 < 9$

36. $\dfrac{|3w - 2|}{2} + 6 < 8$

37. $\left|\dfrac{4x + 5}{3} - \dfrac{1}{2}\right| \leq \dfrac{7}{6}$

38. $\left|\dfrac{2y - 3}{4} - \dfrac{3}{8}\right| \leq \dfrac{15}{16}$

39. $|n + 3| > 7$ $(-\infty, -10) \cup (4, \infty)$

40. $|m - 1| > 5$ $(-\infty, -4) \cup (6, \infty)$

41. $-2|w| - 5 \leq -11$ $(-\infty, -3] \cup [3, \infty)$

42. $-5|v| - 3 \leq -23$ $(-\infty, -4] \cup [4, \infty)$

43. $\dfrac{|q|}{2} - \dfrac{5}{6} \geq \dfrac{1}{3}$

44. $\dfrac{|p|}{5} + \dfrac{3}{2} \geq \dfrac{9}{4}$

45. $3|5 - 7d| + 9 \geq 15$

46. $5|2c + 7| + 1 \geq 11$

47. $2 < \left|-3m + \dfrac{4}{5}\right| - \dfrac{1}{5}$ $(-\infty, -\frac{7}{15}) \cup (1, \infty)$

48. $4 \leq \left|\dfrac{5}{4} - 2n\right| - \dfrac{3}{4}$ $(-\infty, -\frac{7}{4}] \cup [3, \infty)$

49. $4|5 - 2h| - 9 > 11$ $(-\infty, 0) \cup (5, \infty)$

50. $3|7 + 2k| - 11 > 10$ $(-\infty, -7) \cup (0, \infty)$

51. $3.9|4q - 5| + 8.7 \leq -22.5$ $\{\ \}$

52. $0.9|2p + 7| + 16.11 \leq 10.89$ $\{\ \}$

53. $|4z - 9| + 6 \geq 4$ $(-\infty, \infty)$

54. $|5u - 3| + 8 > 6$ $(-\infty, \infty)$

Use the intersect command on a graphing calculator and the given functions to solve (a) $f(x) = g(x)$, **(b)** $f(x) \geq g(x)$, **and (c)** $f(x) < g(x)$.

55. $f(x) = |x - 3| + 2$, $g(x) = \frac{1}{2}x + 2$
    a. $x = 2$ and $x = 6$ **b.** $(-\infty, 2] \cup [6, \infty)$ **c.** $[2, 6]$

56. $f(x) = -|x + 2| - 1$, $g(x) = -\frac{3}{2}x - 9$
    a. $x = -4$ **b.** $[-4, \infty)$ **c.** $(-\infty, -4)$

57. $f(x) = 0.5|x + 3| + 1$, $g(x) = -2|x + 1| + 5$
    a. $x = -3$ and $x = 0.2$ **b.** $(-\infty, -3] \cup [0.2, \infty)$ **c.** $(-3, 0.2)$

58. $f(x) = 2|x - 3| + 2$, $g(x) = |x - 4| + 6$
    a. $x = -2$ and $x = 6$ **b.** $(-\infty, -2] \cup [6, \infty)$ **c.** $(-2, 6)$

Additional answers can be found in the Instructor Answer Appendix.

▶ WORKING WITH FORMULAS

59. Spring Oscillation: $|d - x| \leq L$

A weight attached to a spring hangs at rest a distance of x in. off the ground. If the weight is pulled down (stretched) a distance of L inches and released, the weight begins to bounce and its distance d off the ground must satisfy the indicated formula. (a) If x equals 4 ft and the spring is stretched 3 in. and released, solve the inequality to find what distances from the ground the weight will oscillate between. (b) Solve for x in terms of L and d.

a. $45 \leq d \leq 51$ in.   **b.** $d - L \leq x \leq d + L$

60. A "Fair" Coin: $\left| \dfrac{h - 50}{5} \right| < 1.645$

If we flipped a coin 100 times, we expect "heads" to come up about 50 times if the coin is "fair." In a study of probability, it can be shown that the number of heads h that appears in such an experiment should satisfy the given inequality to be considered "fair." (a) Solve this inequality for h. (b) If you flipped a coin 100 times and obtained 40 heads, is the coin "fair"?   **a.** $41.775 < h < 58.225$ **b.** no

▶ APPLICATIONS

Solve each application of absolute value.

61. Altitude of jet stream: To take advantage of the jet stream, an airplane must fly at a height h (in feet) that satisfies the inequality $|h - 35{,}050| \leq 2550$. Solve the inequality and determine if an altitude of 34,000 ft will place the plane in the jet stream. in feet: [32,500, 37,600]; yes

62. Quality control tests: In order to satisfy quality control, the marble columns a company produces must earn a stress test score S that satisfies the inequality $|S - 17{,}750| \leq 275$. Solve the inequality and determine if a score of 17,500 is in the passing range.   [17,475, 18,025]; yes

63. Submarine depth: The sonar operator on a submarine detects an old World War II submarine net and must decide to detour over or under the net. The computer gives him a depth model $|d - 394| - 20 > 164$, where d is the depth in feet that represents safe passage. At what depth should the submarine travel to go under or over the net? Answer using simple inequalities.   in feet: $d < 210$ or $d > 578$

64. Optimal fishing depth: When deep-sea fishing, the optimal depths d (in feet) for catching a certain type of fish satisfy the inequality $28|d - 350| - 1400 < 0$. Find the range of depths that offer the best fishing. Answer using simple inequalities.   in feet: $300 < d < 400$

For Exercises 65 through 68, (a) develop a model that uses an absolute value inequality, and (b) solve.

65. Stock value: My stock in MMM Corporation fluctuated a great deal in 2009, but never by more than $3.35 from its current value. If the stock is worth $37.58 today, what was its range in 2009?   **a.** $|s - 37.58| \leq 3.35$ **b.** [34.23, 40.93]

66. Traffic studies: On a given day, the volume of traffic at a busy intersection averages 726 cars per hour (cph). During rush hour the volume is much higher, during "off hours" much lower. Find the range of this volume if it never varies by more than 235 cph from the average.

67. Physical training for recruits: For all recruits in the 3rd Armored Battalion, the average number of sit-ups is 125. For an individual recruit, the amount varies by no more than 23 sit-ups from the battalion average. Find the range of sit-ups for this battalion.   **a.** $|s - 125| \leq 23$ **b.** [102, 148]

68. Computer consultant salaries: The national average salary for a computer consultant is $53,336. For a large computer firm, the salaries offered to their employees vary by no more than $11,994 from this national average. Find the range of salaries offered by this company.   **a.** $|s - 53{,}336| \leq 11{,}994$ **b.** [41,342, 65,330]

69. Tolerances for sport balls: According to the official rules for golf, baseball, pool, and bowling, (a) golf balls must be within 0.03 mm of $d = 42.7$ mm, (b) baseballs must be within 1.01 mm of $d = 73.78$ mm, (c) billiard balls must be within 0.127 mm of $d = 57.150$ mm, and (d) bowling balls must be within 12.05 mm of $d = 2171.05$ mm. Write each statement using an absolute value inequality, then (e) determine which sport gives the least tolerance t $\left(t = \dfrac{\text{width of interval}}{\text{average value}} \right)$ for the diameter of the ball.

a. $|d - 42.7| < 0.03$ **b.** $|d - 73.78| < 1.01$ **c.** $|d - 57.150| < 0.127$
d. $|d - 2171.05| < 12.05$ **e.** golf: $t \approx 0.0014$

Additional answers can be found in the Instructor Answer Appendix.

70. **Automated packaging:** The machines that fill boxes of breakfast cereal are programmed to fill each box within a certain tolerance. If the box is overfilled, the company loses money. If it is underfilled, it is considered unsuitable for sale.

Suppose that boxes marked "14 ounces" of cereal must be filled to within 0.1 oz. Find the acceptable range of weights for this cereal.
$|W - 14| \leq 0.1$; in ounces: [13.9, 14.1]; weight must be at least 13.9 oz, but no more than 14.1 oz.

▶ **EXTENDING THE CONCEPT**

71. Determine the value or values (if any) that will make the equation or inequality true.

   a. $|x| + x = 8$ $x = 4$ **b.** $|x - 2| \leq \dfrac{x}{2}$ $[\frac{4}{3}, 4]$

   c. $x - |x| = x + |x|$ **d.** $|x + 3| \geq 6x$ $(-\infty, \frac{3}{5}]$
   $x = 0$
   e. $|2x + 1| = x - 3$ { }

72. The equation $|5 - 2x| = |3 + 2x|$ has only one solution. Find it and explain why there is only one.
   $x = \frac{1}{2}$; answers will vary.
73. In many cases, it can be helpful to view the solutions to absolute value equations and inequalities as follows. For any algebraic expression X and positive

constant k, the equation $|X| = k$ has solutions $X = k$ and $-X = k$, since the absolute value of either quantity on the left will indeed yield the positive constant k. Likewise, $|X| < k$ has solutions $X < k$ and $-X < k$. Note the inequality symbol has not been reversed as yet, but will naturally be reversed as part of the solution process. Solve the following equations or inequalities using this idea.

   a. $|x - 3| = 5$ $\{-2, 8\}$
   b. $|x - 7| > 4$ $(-\infty, 3) \cup (11, \infty)$
   c. $3|x + 2| \leq 12$ $[-6, 2]$
   d. $-3|x - 4| + 7 = -11$ $\{-2, 10\}$

▶ **MAINTAINING YOUR SKILLS**

74. **(R.4)** Factor the expression completely:
   $18x^3 + 21x^2 - 60x.$ $3x(2x + 5)(3x - 4)$

75. **(1.5)** Solve $V^2 = \dfrac{2W}{C\rho A}$ for ρ (physics). $\rho = \frac{2W}{V^2 CA}$

76. **(R.7)** Simplify $\dfrac{-1}{3 + \sqrt{3}}$ by rationalizing the denominator. State the result in exact form and approximate form (to hundredths). $\frac{-3 + \sqrt{3}}{6} \approx -0.21$

77. **(R.3)** Solve the inequality, then write the solution set in interval notation:

   $-3(2x - 5) > 2(x + 1) - 7.$ $x \in (-\infty, \frac{5}{2})$

MID-CHAPTER CHECK

1. Determine whether the following function is even, odd, or neither. $f(x) = x^2 + \dfrac{|x|}{4x}$ neither

2. Use a graphing calculator to find the maximum and minimum values of
   $f(x) = -1.9(x^4 - 2.3x^3 + 2.2x - 5.1).$ Round to the nearest hundredth.   max: $y \approx 11.12$ at $x \approx -0.50$, $y \approx 8.55$ at $x \approx 1.47$; min: $y \approx 7.80$ at $x \approx 0.75$
3. Use interval notation to identify the interval(s) where the function from Exercise 2 is increasing, decreasing, or constant. Round to the nearest hundredth.   increasing on $(-\infty, -0.50) \cup (0.75, 1.47)$, decreasing on $(-0.50, 0.75) \cup (1.47, \infty)$

4. Write the equation of the function that has the same graph of $f(x) = \sqrt{x}$, shifted left 4 units and up 2 units. $g(x) = \sqrt{x + 4} + 2$

5. For the graph given, (a) identify the function family, (b) describe or identify the end-behavior, inflection point, and x- and y-intercepts, (c) determine the domain and range, and (d) determine the value of k if $f(k) = 2.5$. Assume required features have integer values.

Exercise 5

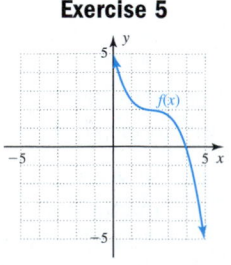

Additional answers can be found in the Instructor Answer Appendix.

6. Use a graphing calculator to graph the given functions in the same window and comment on what you observe.

$$p(x) = (x - 3)^2 \qquad q(x) = -(x - 3)^2$$
$$r(x) = -\tfrac{1}{2}(x - 3)^2$$

7. Solve the following absolute value equations. Write the solution in set notation.

a. $\dfrac{2}{3}|d - 5| + 1 = 7$ **b.** $5 - |s + 3| = \dfrac{11}{2}$

    {−4, 14}   { }

8. Solve the following absolute value inequalities. Write solutions in interval notation.

a. $3|q + 4| - 2 < 10$ **b.** $\left|\dfrac{x}{3} + 2\right| + 5 \leq 5$

    $q \in (-8, 0)$ {−6}

9. Solve the following absolute value inequalities. Write solutions in interval notation.

a. $3.1|d - 2| + 1.1 \geq 7.3$ $d \in (-\infty, 0] \cup [4, \infty)$

b. $\dfrac{|1 - y|}{3} + 2 > \dfrac{11}{2}$ $y \in \left(-\infty, -\dfrac{19}{2}\right) \cup \left(\dfrac{23}{2}, \infty\right)$

c. $-5|k - 2| + 3 < 4$ $k \in (-\infty, \infty)$

10. Kiteboarding: With the correct sized kite, a person can kiteboard when the wind is blowing at a speed w (in mph) that satisfies the inequality $|w - 17| \leq 9$. Solve the inequality and determine if a person can kiteboard with a windspeed of (a) 5 mph? (b) 12 mph? $w \in [8, 26]$; no, yes

REINFORCING BASIC CONCEPTS

Using Distance to Understand Absolute Value Equations and Inequalities

For any two numbers a and b on the number line, *the distance between a and b can be written* $|a - b|$ or $|b - a|$. In exactly the same way, the equation $|x - 3| = 4$ can be read, "the distance between 3 and an unknown number is equal to 4." The advantage of reading it in this way (instead of "the absolute value of x minus 3 is 4"), is that a much clearer *visualization* is formed, giving a constant reminder there are two solutions. In diagram form we have Figure 2.51.

Figure 2.51

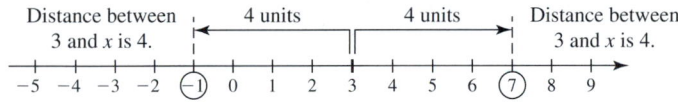

From this we note the solutions are $x = -1$ and $x = 7$.

In the case of an inequality such as $|x + 2| \leq 3$, we rewrite the inequality as $|x - (-2)| \leq 3$ and read it, "the distance between -2 and an unknown number is less than or equal to 3." With some practice, visualizing this relationship mentally enables a quick statement of the solution: $x \in [-5, 1]$. In diagram form we have Figure 2.52.

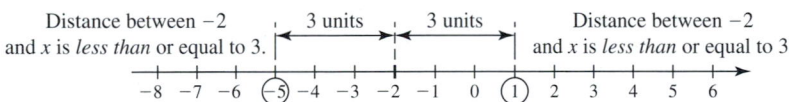
Figure 2.52

Equations and inequalities where the coefficient of x is not 1 still lend themselves to this form of conceptual understanding. For $|2x - 1| \geq 3$ we read, "the distance between 1 and twice an unknown number is greater than or equal to 3." On the number line (Figure 2.53), the number 3 units to the right of 1 is 4, and the number 3 units to the left of 1 is -2.

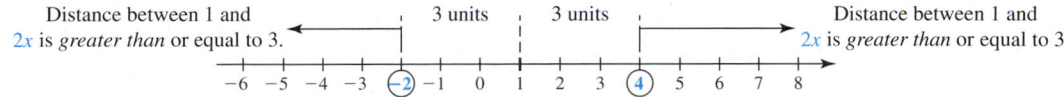
Figure 2.53

For $2x \leq -2$, $x \leq -1$, and for $2x \geq 4$, $x \geq 2$, and the solution set is $x \in (-\infty, -1] \cup [2, \infty)$.

Attempt to solve the following equations and inequalities by visualizing a number line. Check all results algebraically.

Exercise 1: $|x - 2| = 5$

    $x = -3$ or $x = 7$

Exercise 2: $|x + 1| \leq 4$

    $x \in [-5, 3]$

Exercise 3: $|2x - 3| \geq 5$

    $x \in (-\infty, -1] \cup [4, \infty)$

LEARNING OBJECTIVES

In Section 2.4 you will see how we can:

☐ **A.** Graph basic rational functions, identify vertical and horizontal asymptotes, and describe end-behavior

☐ **B.** Use transformations to graph basic rational functions and write the equation for a given graph

☐ **C.** Graph basic power functions and state their domains

☐ **D.** Solve applications involving basic rational and power functions

In this section, we introduce two new kinds of relations, **rational functions** and **power functions**. While we've already studied a variety of functions, we still lack the ability to model a large number of important situations. For example, functions that model the amount of medication remaining in the bloodstream over time, the relationship between altitude and weightlessness, and the equations modeling planetary motion come from these two families.

A. Rational Functions and Asymptotes

Just as a rational number is the ratio of two integers, a **rational function** is the ratio of two polynomials. In general,

> **Rational Functions**
>
> A rational function $V(x)$ is one of the form
>
> $$V(x) = \frac{p(x)}{d(x)},$$
>
> where p and d are polynomials and $d(x) \neq 0$.
> The domain of $V(x)$ is all real numbers, *except the zeroes of d.*

The simplest rational functions are the reciprocal function $y = \frac{1}{x}$ and the reciprocal square function $y = \frac{1}{x^2}$, as both have a constant numerator and a single term in the denominator. Since division by zero is undefined, the domain of both excludes $x = 0$. A preliminary study of these two functions will provide a strong foundation for our study of general rational functions in Chapter 4.

The Reciprocal Function: $y = \dfrac{1}{x}$

The reciprocal function takes any input (other than zero) and gives its reciprocal as the output. This means large inputs produce small outputs and vice versa. A table of values (Table 2.1) and the resulting graph (Figure 2.54) are shown.

Table 2.1

x	y	x	y
-1000	$-1/1000$	$1/1000$	1000
-5	$-1/5$	$1/3$	3
-4	$-1/4$	$1/2$	2
-3	$-1/3$	1	1
-2	$-1/2$	2	$1/2$
-1	-1	3	$1/3$
$-1/2$	-2	4	$1/4$
$-1/3$	-3	5	$1/5$
$-1/1000$	-1000	1000	$1/1000$
0	undefined		

Figure 2.54

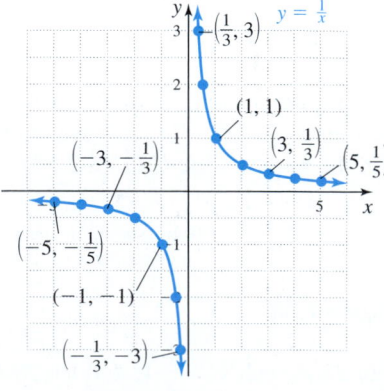

WORTHY OF NOTE

The notation used for graphical behavior always begins by describing what is happening to the x-values, and the resulting effect on the y-values. Using Figure 2.55, visualize that for a point (x, y) on the graph of $y = \frac{1}{x}$, as x gets larger, y must become smaller, particularly since their product must always be 1 ($y = \frac{1}{x} \Rightarrow xy = 1$).

Figure 2.55

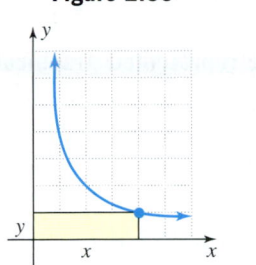

Table 2.1 and Figure 2.54 reveal some interesting features. First, the graph passes the vertical line test, verifying $y = \frac{1}{x}$ is indeed a function. Second, since division by zero is undefined, there can be no corresponding point on the graph, *creating a break at* $x = 0$. In line with our definition of rational functions, the domain is $x \in (-\infty, 0) \cup (0, \infty)$. Third, this is an odd function, with a "branch" of the graph in the first quadrant and one in the third quadrant, as the reciprocal of any input maintains its sign. Finally, we note in QI that as x becomes an infinitely large positive number, y becomes infinitely small and closer to zero. It seems convenient to symbolize this end-behavior using the following notation:

$$\text{as } x \to \infty, \qquad\qquad y \to 0$$

as x becomes an infinitely large positive number   y approaches 0

Graphically, the curve becomes very close to, or *approaches the x-axis.*

We also note that as x approaches zero from the right, y becomes an infinitely large positive number: as $x \to 0^+$, $y \to \infty$. Note a superscript $+$ or $-$ sign is used to indicate the *direction of the approach*, meaning *from the positive side* (right) or *from the negative side* (left).

EXAMPLE 1 ▶ **Describing the End-Behavior of Rational Functions**

For $y = \frac{1}{x}$ in QIII (Figure 2.54),

   a. Describe the end-behavior of the graph.

   b. Describe what happens as x approaches zero.

Solution ▶ Similar to the graph's behavior in QI, we have

   a. In words: As x becomes an infinitely large negative number, y approaches zero. In notation: As $x \to -\infty$, $y \to 0$.

   b. In words: As x approaches zero from the left, y becomes an infinitely large negative number. In notation: As $x \to 0^-$, $y \to -\infty$.

Now try Exercises 7 and 8 ▶

The Reciprocal Square Function: $y = \dfrac{1}{x^2}$

From our previous work, we anticipate this graph will also have a break at $x = 0$. But since the square of any negative number is positive, the branches of the **reciprocal square function** are both *above the x-axis.* Note the result is the graph of an even function. See Table 2.2 and Figure 2.56.

Table 2.2

x	y	x	y
-1000	$1/1{,}000{,}000$	$1/1000$	$1{,}000{,}000$
-5	$1/25$	$1/3$	9
-4	$1/16$	$1/2$	4
-3	$1/9$	1	1
-2	$1/4$	2	$1/4$
-1	1	3	$1/9$
$-1/2$	4	4	$1/16$
$-1/3$	9	5	$1/25$
$-1/1000$	$1{,}000{,}000$	1000	$1/1{,}000{,}000$
0	undefined		

Figure 2.56

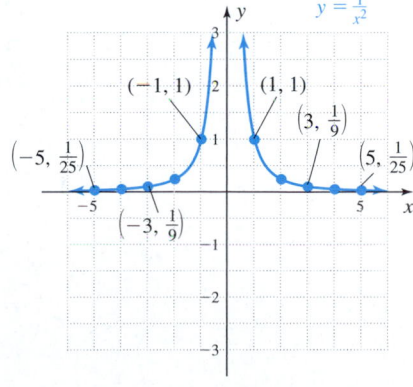

Similar to $y = \frac{1}{x}$, large positive inputs generate small, positive outputs: as $x \to \infty$, $y \to 0$. This is one indication of **asymptotic behavior** in the *horizontal* direction, and we say the line $y = 0$ (the *x*-axis) is a **horizontal asymptote** for the reciprocal and reciprocal square functions. In general,

> **Horizontal Asymptotes**
>
> Given a constant k, the line $y = k$ is a horizontal asymptote for V if, as x increases or decreases without bound, $V(x)$ approaches k:
>
> as $x \to -\infty$, $V(x) \to k$ or   as $x \to \infty$, $V(x) \to k$

As shown in Figures 2.57 and 2.58, asymptotes are represented graphically as dashed lines that seem to "guide" the branches of the graph. Figure 2.57 shows a horizontal asymptote at $y = 1$, which suggests the graph of $f(x)$ is the graph of $y = \frac{1}{x}$ shifted up 1 unit. Figure 2.58 shows a horizontal asymptote at $y = -2$, which suggests the graph of $g(x)$ is the graph of $y = \frac{1}{x^2}$ shifted down 2 units.

Figure 2.57

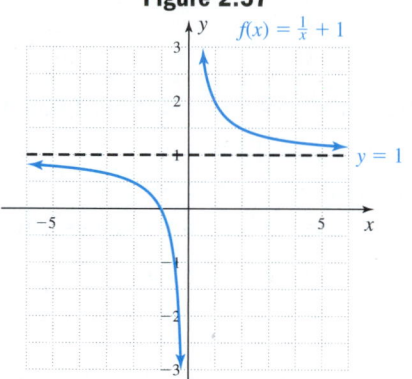

Figure 2.58

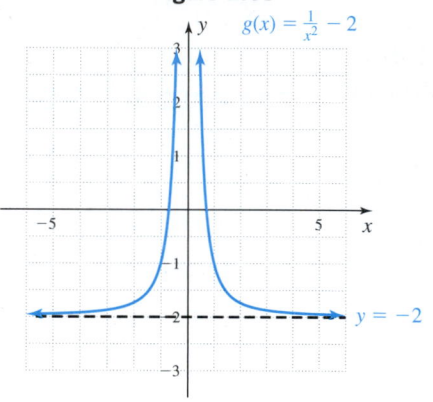

EXAMPLE 2 ▶ **Describing the End-Behavior of Rational Functions**

For the graph in Figure 2.58, use mathematical notation to
  a. Describe the end-behavior of the graph and name the horizontal asymptote.
  b. Describe what happens as x approaches zero.

Solution ▶ **a.** as $x \to -\infty$, $g(x) \to -2$, **b.** as $x \to 0^-$, $g(x) \to \infty$,
           as $x \to \infty$, $g(x) \to -2$,   as $x \to 0^+$, $g(x) \to \infty$

           $y = -2$ is a horizontal asymptote

Now try Exercises 9 and 10 ▶

While the graphical view of Example 2(a) (Figure 2.58) makes these concepts believable, a numerical view of this end-behavior can be even more compelling. Try entering $\frac{1}{x^2} - 2$ as Y_1 on the Y= screen, then go to the **TABLE** feature (TblStart = -3, ΔTbl = 1; Figure 2.59). Scrolling in either direction shows that as $|x|$ becomes very large, Y_1 becomes closer and closer to -2, but will never be equal to -2 (Figure 2.60).

Figure 2.59

X	Y₁
-3	-1.889
-2	-1.75
-1	-1
0	ERR:
1	-1
2	-1.75
3	-1.889

X=-3

Figure 2.60

X	Y₁
14	-1.995
15	-1.996
16	-1.996
17	-1.997
18	-1.997
19	-1.997
20	-1.998

X=20

From Example 2(b), we note that as x becomes *smaller and close to 0, g* becomes very large and *increases without bound.* This is one indication of asymptotic behavior in the *vertical* direction, and we say the line $x = 0$ (the y-axis) is a **vertical asymptote** for g ($x = 0$ is also a vertical asymptote for f in Figure 2.57). In general,

Vertical Asymptotes

Given a constant h, the vertical line $x = h$ is a vertical asymptote for a function V if, as x approaches h, $V(x)$ increases or decreases without bound:

$$\text{as } x \to h^+, V(x) \to \pm\infty \qquad \text{or} \qquad \text{as } x \to h^-, V(x) \to \pm\infty$$

Here is a brief summary:

Reciprocal Function	**Reciprocal Quadratic Function**
$f(x) = \dfrac{1}{x}$	$g(x) = \dfrac{1}{x^2}$
Domain: $x \in (-\infty, 0) \cup (0, \infty)$	Domain: $x \in (-\infty, 0) \cup (0, \infty)$
Range: $y \in (-\infty, 0) \cup (0, \infty)$	Range: $y \in (0, \infty)$
Horizontal asymptote: $y = 0$	Horizontal asymptote: $y = 0$
Vertical asymptote: $x = 0$	Vertical asymptote: $x = 0$

☑ **A.** You've just seen how we can graph basic rational functions, identify vertical and horizontal asymptotes, and describe end-behavior

B. Using Asymptotes to Graph Basic Rational Functions

Identifying these asymptotes is useful because the graphs of $y = \frac{1}{x}$ and $y = \frac{1}{x^2}$ can be transformed *in exactly the same way as the toolbox functions.* When their graphs shift—the vertical and horizontal asymptotes shift with them and can be used as guides to redraw the graph. In shifted form,

$$f(x) = \frac{a}{x \pm h} \pm k \text{ for the reciprocal function, and}$$

$$g(x) = \frac{a}{(x \pm h)^2} \pm k \text{ for the reciprocal square function.}$$

When horizontal and/or vertical shifts are applied to simple rational functions, we first apply them to the asymptotes, then calculate the x- and y-intercepts as before. An additional point or two can be computed as needed to round out the graph.

EXAMPLE 3 ▶ **Graphing Transformations of the Reciprocal Function**

Sketch the graph of $g(x) = \dfrac{1}{x - 2} + 1$ using transformations of the parent function.

Solution ▶ The graph of g is the same as that of $y = \dfrac{1}{x}$, but shifted 2 units right and 1 unit upward. This means the vertical asymptote is also shifted 2 units right, and the horizontal asymptote is shifted 1 unit up. The y-intercept is $g(0) = \dfrac{1}{2}$. For the x-intercept:

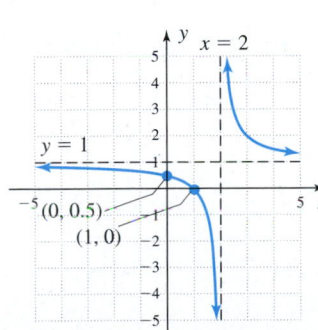

$$0 = \frac{1}{x - 2} + 1 \qquad \text{\color{red}substitute 0 for } g(x)$$

$$-1 = \frac{1}{x - 2} \qquad \text{\color{red}subtract 1}$$

$$-1(x - 2) = 1 \qquad \text{\color{red}multiply by } (x - 2)$$

$$x = 1 \qquad \text{\color{red}solve}$$

The x-intercept is $(1, 0)$. Knowing the graph is from the reciprocal function family and shifting the asymptotes and intercepts yields the graph shown.

Now try Exercises 11 through 26 ▶

These ideas can be "used in reverse" to determine the equation of a basic rational function from its given graph, as in Example 4.

EXAMPLE 4 ▶ **Writing the Equation of a Basic Rational Function, Given Its Graph**

Identify the function family for the graph given, then use the graph to write the equation of the function in "shifted form." Assume $|a| = 1$.

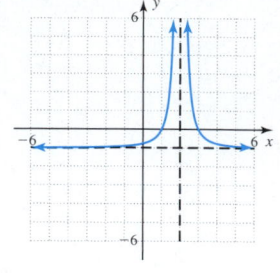

Solution ▶ The graph appears to be from the reciprocal square family, and has been shifted 2 units right (the vertical asymptote is at $x = 2$), and 1 unit down (the horizontal asymptote is at $y = -1$). From $y = \frac{1}{x^2}$, we obtain $f(x) = \frac{1}{(x - 2)^2} - 1$ as the shifted form.

Now try Exercises 27 through 38 ▶

☑ **B. You've just seen how we can use asymptotes and transformations to graph basic rational functions and write the equation for a given graph**

Using the definition of negative exponents, the basic reciprocal and reciprocal square functions can be written as $y = x^{-1}$ and $y = x^{-2}$, respectively. In this form, we note that these functions also belong to a family of functions known as the *power functions* (see Exercise 80).

C. Graphs of Basic Power Functions

Italian physicist and astronomer Galileo Galilei (1564–1642) made numerous contributions to astronomy, physics, and other fields. But perhaps he is best known for his experiments with gravity, in which he dropped objects of different weights from the Leaning Tower of Pisa. Due in large part to his work, we know that the velocity of an object after it has fallen a certain distance is $v = \sqrt{2gs}$, where g is the acceleration due to gravity (32 ft/sec^2), s is the distance in feet the object has fallen, and v is the velocity of the object in feet per second (see Exercise 71). As you will see, this is an example of a formula that uses a power function.

From previous coursework or a review of radicals and rational exponents (Section R.6), we know that $\sqrt{x}$ can be written as $x^{\frac{1}{2}}$, and $\sqrt[3]{x}$ as $x^{\frac{1}{3}}$, enabling us to write these functions in *exponential form*: $f(x) = x^{\frac{1}{2}}$ and $g(x) = x^{\frac{1}{3}}$. In this form, we see that these actually belong to a larger family of functions, where x is raised to some power, called the **power functions.**

Power Functions and Root Functions

For any constant real number p and variable x, functions of the form

$$f(x) = x^p$$

are called *power functions* in x. If p is of the form $\frac{1}{n}$ for integers $n \geq 2$, the functions

$$f(x) = x^{\frac{1}{n}} \Leftrightarrow f(x) = \sqrt[n]{x}$$

are called *root functions* in x.

The functions $y = x^2$, $y = x^{\frac{1}{4}}$, $y = x^3$, $y = \sqrt[5]{x}$, and $y = x^{\frac{3}{2}}$ are all power functions, but only $y = x^{\frac{1}{4}}$ and $y = \sqrt[5]{x}$ are also root functions. Initially we will focus on power functions where $p > 0$.

EXAMPLE 5 ▶ **Comparing the Graphs of Power Functions**

Use a graphing calculator to graph the power functions $f(x) = x^{\frac{1}{4}}$, $g(x) = x^{\frac{2}{3}}$, $h(x) = x^1$, $p(x) = x^{\frac{3}{2}}$, and $q(x) = x^2$ in the standard viewing window. Make an observation in QI regarding the effect of the exponent on each function, then discuss what the graphs of $y = x^{\frac{1}{6}}$ and $y = x^{\frac{7}{2}}$ would look like.

Solution ▶ First we enter the functions in sequence as Y_1 through Y_5 on the (Y=) screen (Figure 2.61). Using (ZOOM) **6:ZStandard** produces the graphs shown in Figure 2.62. Narrowing the window to focus on QI (Figure 2.63: $x \in [-4, 10]$, $y \in [-4, 10]$), we quickly see that for $x \geq 1$, larger values of p cause the graph of $y = x^p$ to increase at a faster rate, and smaller values at a slower rate. In other words (for $x \geq 1$), since $\frac{1}{6} < \frac{1}{4}$, the graph of $y = x^{\frac{1}{6}}$ would increase slower and appear to be "under" the graph of $Y_1 = X^{\frac{1}{4}}$.

Since $\frac{7}{2} > 2$, the graph of $y = x^{\frac{7}{2}}$ would increase faster and appear to be "more narrow" than the graph of $Y_5 = X^2$ (verify this).

Figure 2.61, 2.62

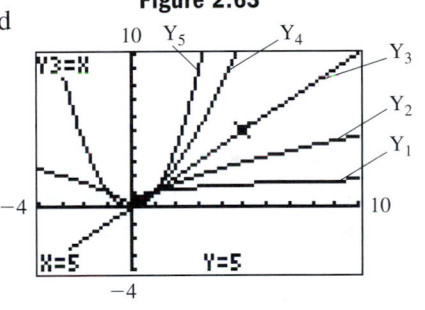

Figure 2.63

Now try Exercises 39 through 48 ▶

The Domain of a Power Function

In addition to the observations made in Example 5, we can make other important notes, particularly regarding the *domains* of power functions. When the exponent on a power function is a rational number $\frac{m}{n} > 0$ in simplest form, it appears the domain is all real numbers if n is odd, as seen in the graphs of $g(x) = x^{\frac{2}{3}}$, $h(x) = x^1 = x^{\frac{1}{1}}$, and $q(x) = x^2 = x^{\frac{2}{1}}$. If n is an even number, the domain is all nonnegative real numbers as seen in the graphs of $f(x) = x^{\frac{1}{4}}$ and $p(x) = x^{\frac{3}{2}}$. Further exploration will show that if p is irrational, as in $y = x^\pi$, the domain is also all nonnegative real numbers and we have the following:

The Domain of a Power Function

Given a power function $f(x) = x^p$ with $p > 0$.

  1. If $p = \dfrac{m}{n}$ is a rational number in simplest form,

  a. the domain of f is all real numbers if n is odd: $x \in (-\infty, \infty)$,
  b. the domain of f is all nonnegative real numbers if n is even: $x \in [0, \infty)$.
  2. If p is an irrational number, the domain of f is all nonnegative real numbers: $x \in [0, \infty)$.

Further confirmation of statement 1 can be found by recalling the graphs of $y = \sqrt{x} = x^{\frac{1}{2}}$ and $y = \sqrt[3]{x} = x^{\frac{1}{3}}$ from Section 2.2 (Figures 2.64 and 2.65).

Figure 2.64

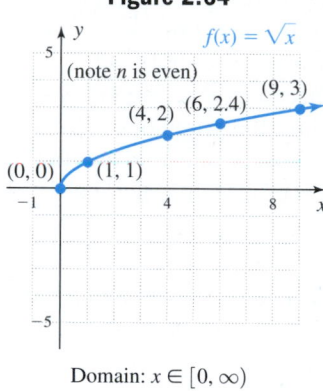

Figure 2.65

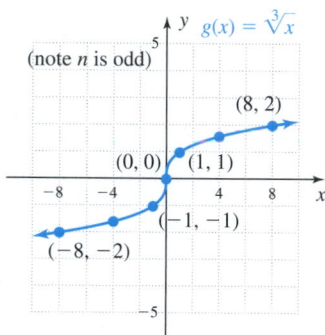

Domain: $x \in [0, \infty)$
Range: $y \in [0, \infty)$

Domain: $x \in [-\infty, \infty)$
Range: $y \in [-\infty, \infty)$

EXAMPLE 6 ▶ **Determining the Domains of Power Functions**

State the domain of the following power functions, and identity whether each is also a root function.

  a. $f(x) = x^{\frac{4}{5}}$ **b.** $g(x) = x^{\frac{1}{10}}$ **c.** $h(x) = \sqrt[8]{x}$ **d.** $q(x) = x^{\frac{2}{3}}$ **e.** $r(x) = x^{\sqrt{5}}$

Solution ▶   **a.** Since n is odd, the domain of f is all real numbers; f is not a root function.
  b. Since n is even, the domain of g is $x \in [0, \infty)$; g is a root function.
  c. In exponential form $h(x) = x^{\frac{1}{8}}$. Since n is even, the domain of h is $x \in [0, \infty)$; h is a root function.
  d. Since n is odd, the domain of q is all real numbers; q is not a root function
  e. Since p is irrational, the domain of r is $x \in [0, \infty)$; r is not a root function

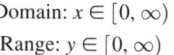

Now try Exercises 49 through 58 ▶

Transformations of Power and Root Functions

As we saw in Section 2.2 (Toolbox Functions and Transformations), the graphs of the root functions $y = \sqrt{x}$ and $y = \sqrt[3]{x}$ can be transformed using shifts, stretches, reflections, and so on. In Example 8(b) (Section 2.2) we noted the graph of $h(x) = 2\sqrt[3]{x - 2} - 1$ was the graph of $y = \sqrt[3]{x}$ shifted 2 units right, stretched by a factor of 2, and shifted 1 unit down. Graphs of other power functions can be transformed in exactly the same way.

EXAMPLE 7 ▶   **Graphing Transformations of Power Functions**

Based on our previous observations,

     a. Determine the domain of $f(x) = x^{\frac{2}{3}}$ and $g(x) = x^{\frac{3}{2}}$, then verify by graphing them on a graphing calculator.

     b. Next, discuss what the graphs of $F(x) = (x - 2)^{\frac{2}{3}} - 3$ and $G(x) = -x^{\frac{3}{2}} + 2$ will look like, then graph each on a graphing calculator to verify.

Solution ▶   **a.** Both f and g are power functions of the form $y = x^{\frac{m}{n}}$. For f, n is odd so its domain is all real numbers. For g, n is even and the domain is $x \in [0, \infty)$. Their graphs support this conclusion (Figures 2.66 and 2.67).

Figure 2.66

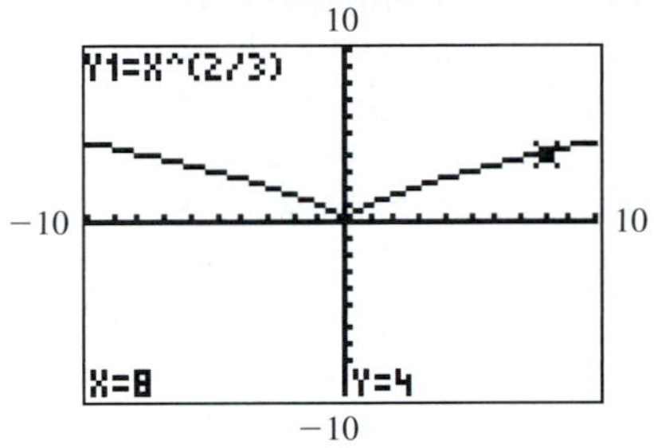

Y1=X^(2/3)

X=8   Y=4

Figure 2.67

Y1=X^(3/2)

X=4   Y=8

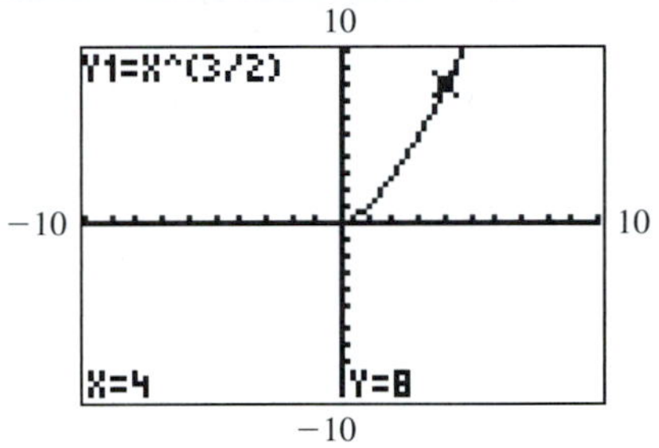

     b. The graph of F will be the same as the graph of f, but shifted two units right and three units down, moving the vertex to $(2, -3)$. The graph of G will be the same as the graph of g, but reflected across the x-axis, and shifted 2 units up (Figures 2.68 and 2.69).

Figure 2.68

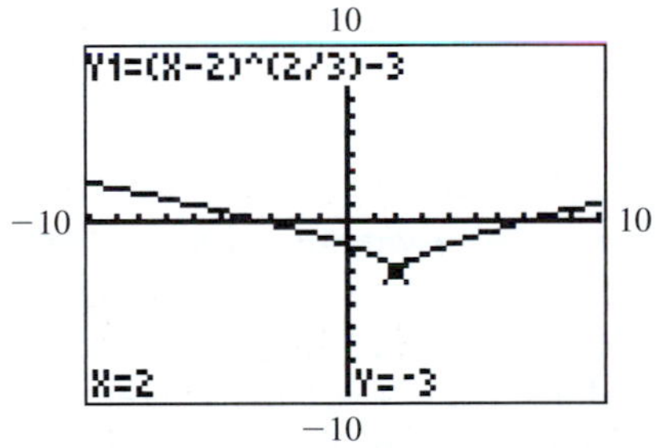

Y1=(X-2)^(2/3)-3

X=2   Y=-3

Figure 2.69

Y1=-X^(3/2)+2

X=4   Y=-6

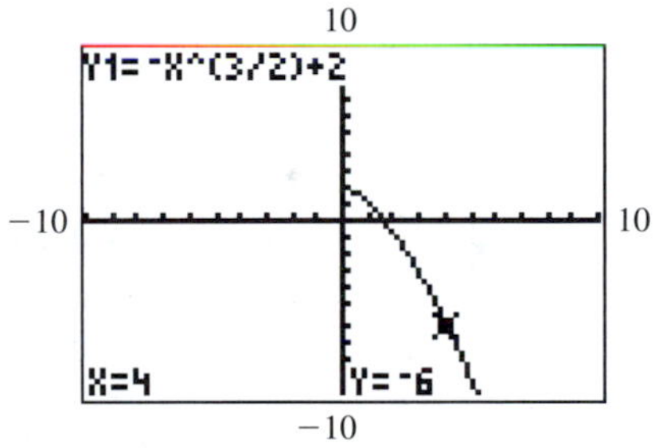

☑ **C.** You've just seen how we can graph basic power functions and state their domains

Now try Exercises 59 through 62 ▶

D. Applications of Rational and Power Functions

These new functions have a variety of interesting and significant applications in the real world. Examples 8 through 10 provide a small sample, and there are a number of additional applications in the Exercise Set. In many applications, the coefficients may be rather large, and the axes should be scaled accordingly.

EXAMPLE 8 ▶   **Modeling the Cost to Remove Waste**

For a large urban-centered county, the cost to remove chemical waste and other pollutants from a local river is given by the function $C(p) = \dfrac{-18{,}000}{p - 100} - 180$,

where $C(p)$ represents the cost (in thousands of dollars) to remove p percent of the pollutants.

a. Find the cost to remove 25%, 50%, and 75% of the pollutants and comment on the results.

b. Graph the function using an appropriate scale.

c. Use mathematical notation to state what happens as the county attempts to remove 100% of the pollutants.

Solution ▶

a. We evaluate the function as indicated, finding that $C(25) = 60$, $C(50) = 180$, and $C(75) = 540$. The cost is escalating rapidly. The change from 25% to 50% brought a $120,000 increase, but the change from 50% to 75% brought *a $360,000 increase!*

b. From the context, we need only graph the portion from $0 \leq p < 100$. For the C-intercept we substitute $p = 0$ and find $C(0) = 0$, which seems reasonable as 0% would be removed if $0 were spent. We also note there must be a vertical asymptote at $x = 100$, since this x-value causes a denominator of 0. Using this information and the points from part (a) produces the graph shown.

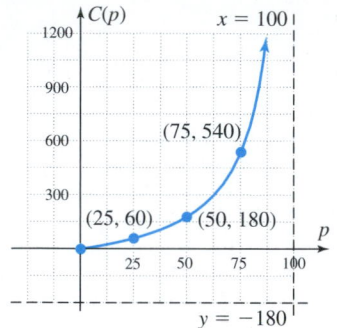

c. As the percentage of pollutants removed approaches 100%, the cost of the cleanup skyrockets. Using notation: as $p \to 100^-$, $C \to \infty$.

Now try Exercises 65 through 70 ▶

While not obvious at first, the function $C(p)$ in Example 8 is from the family of reciprocal functions $y = \dfrac{1}{x}$. A closer inspection shows it has the form $y = \dfrac{-a}{x - h} - k \to \dfrac{-18{,}000}{x - 100} - 180$, showing the graph of $y = \dfrac{1}{x}$ is shifted right 100 units, reflected across the x-axis, stretched by a factor of 18,000 and shifted 180 units down (the horizontal asymptote is $y = -180$). As sometimes occurs in real-world applications, portions of the graph were ignored due to the context. To see the full graph, we reason that the second branch occurs on the opposite side of the vertical and horizontal asymptotes, and set the window as shown in Figure 2.70. After entering $C(p)$ as Y_1 on the (Y=) screen and pressing (GRAPH), the full graph appears as shown in Figure 2.71 (for effect, the vertical and horizontal asymptotes were drawn separately using the (2nd) (PRGM) **(DRAW)** options).

Figure 2.70

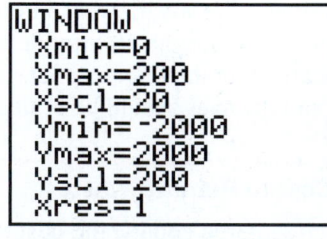

Figure 2.71

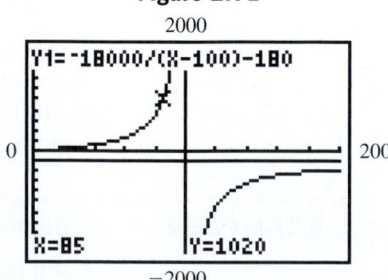

Next, we'll use a root function to model the distance to the horizon from a given height.

EXAMPLE 9 ▶ **The Distance to the Horizon**

On a clear day, the distance a person can see from a certain height (the distance to the horizon) is closely approximated by the root function $d(h) = 3.57\sqrt{h}$, where $d(h)$ represents the viewing distance (in kilometers) from a height of h meters above sea level.

   a. To the nearest kilometer, how far can a person see when standing on the observation level of the John Hancock building in Chicago, Illinois, about 335 m high?

   b. To the nearest meter, how high is the observer's eyes, if the viewing distance is 130 km?

Solution ▶ **a.** Substituting 335 for h we have

$$d(h) = 3.57\sqrt{h} \qquad \text{original function}$$
$$d(335) = 3.57\sqrt{335} \qquad \text{substitute 335 for } h$$
$$\approx 65.34 \qquad \text{result}$$

On a clear day, a person can see about 65 kilometers.

b. We substitute 130 for $d(h)$:

$$d(h) = 3.57\sqrt{h} \qquad \text{original function}$$
$$130 = 3.57\sqrt{h} \qquad \text{substitute 130 for } d(h)$$
$$36.415 \approx \sqrt{h} \qquad \text{divide by 3.57}$$
$$1326.052 \approx h \qquad \text{square both sides}$$

If the distance to the horizon is 130 km, the observer's eyes are at a height of approximately 1326 m. Check the answer to part (b) by solving graphically.

Now try Exercises 71 through 74 ▶

One area where power functions and modeling with regression are used extensively is **allometric studies.** This area of inquiry studies the relative growth of a part of an animal in relation to the growth of the whole, like the wingspan of a bird compared to its weight, or the daily food intake of a mammal or bird compared to its size.

EXAMPLE 10 ▶ **Modeling the Food Requirements of Certain Bird Species**

To study the relationship between the weight of a nonpasserine bird and its daily food intake, the data shown in the table was collected (nonpasserine: nonsinging, nonperching birds).

   a. On a graphing calculator, enter the data in L1 and L2, then set an appropriate window to view a scatterplot of the data. Does a power regression ⟨STAT⟩ **CALC, A:PwrReg** seem appropriate?

Bird	Average weight (g)	Daily food intake (g)
Common pigeon	350	25
Ring-necked duck	725	50
Ring-necked pheasant	1400	70
Canadian goose	4525	165
White swan	9075	240

b. Use a graphing calculator to find an equation model using a power regression on the data, and enter the equation in Y_1 (round values to three decimal places).

c. Use the equation to estimate the daily food intake required by a barn owl (470 g), and a gray-headed albatross (6800 g).

d. Use the intersection of graphs method to find the weight of a Great-Spotted Kiwi, given the daily food requirement is 130 g.

Solution ▶   **a.** After entering the weights in L1 and food intake in L2, we set a window that will comfortably fit the data. Using $x \in [0, 10{,}000]$ and $y \in [-30, 300]$ produces the scatterplot shown (Figure 2.72). The data does not appear linear, and based on our work in Example 5, a power function seems appropriate.

Figure 2.72

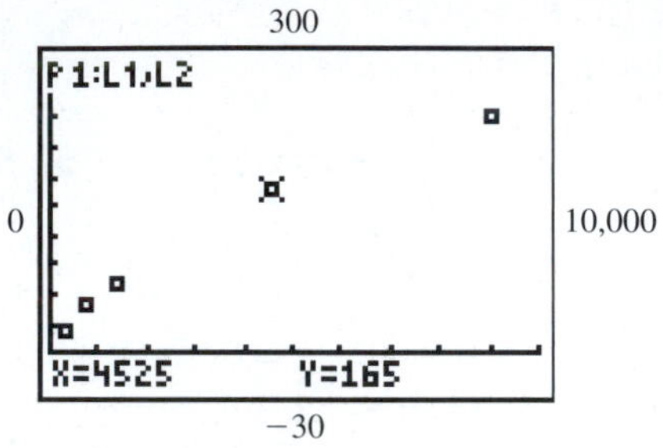

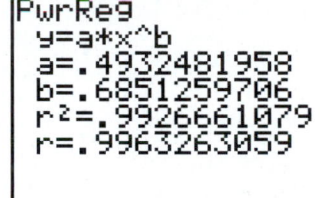

b. To access the power regression option, use **STAT** ▶ **(CALC) A:PwrReg.** To three decimal places the equation for Y_1 would be $0.493\,X^{0.685}$ (Figure 2.73).

Figure 2.73

PwrReg
y=a*x^b
a=.4932481958
b=.6851259706
r²=.9926661079
r=.9963263059

c. For the barn owl, $x = 470$ and we find the estimated food requirement is about 33.4 g per day (Figure 2.74). For the gray-headed albatross $x = 6800$ and the model estimates about 208.0 g of food daily is required.

d. Here we're given the food intake of the Great-Spotted Kiwi (the output value), and want to know what input value (weight) was used. Entering $Y_2 = 130$, we'll attempt to find where the graphs of Y_1 and Y_2 intersect (it will help to deactivate **Plot1** on the **Y=** screen, so that only the graphs of Y_1 and Y_2 appear). Using **2nd** **TRACE** **(CALC) 5:Intersect** shows the graphs intersect at about (3423.3, 130) (Figure 2.75), indicating the average weight of a Great-Spotted Kiwi is near 3423.3 g (about 7.5 lb).

Figure 2.74

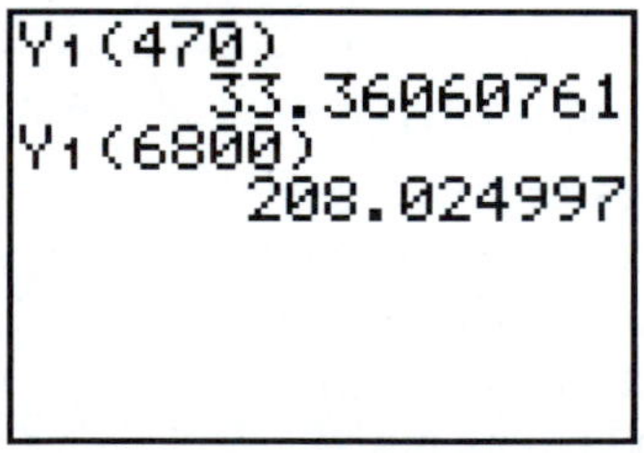

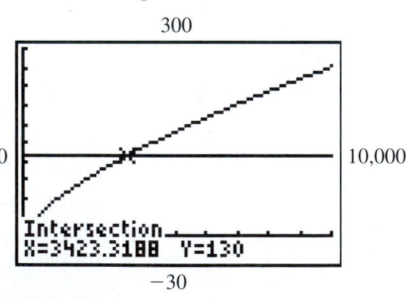

Figure 2.75

☑ **D.** You've just seen how we can solve applications involving basic rational and power functions

Now try Exercises 75 through 78 ▶

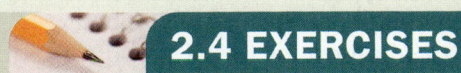

2.4 EXERCISES

► CONCEPTS AND VOCABULARY

Fill in each blank with the appropriate word or phrase. Carefully reread the section if needed.

1. Write the following in notational form. *As x becomes an infinitely large negative number, y approaches 2.* $\underline{\text{as } x \to -\infty, y \to 2}$

2. For any constant k, the notation "as $|x| \to +\infty, y \to k$" is an indication of a $\underline{\text{horizontal}}$ asymptote, while "$x \to k, |y| \to +\infty$" indicates a $\underline{\text{vertical}}$ asymptote.

3. Given the function $g(x) = \dfrac{1}{(x-3)^2} + 2$, a $\underline{\text{vertical}}$ asymptote occurs at $x = 3$ and a horizontal asymptote at $\underline{y = 2}$.

4. The graph of $Y_1 = \dfrac{1}{x}$ has branches in Quadrants I and III. The graph of $Y_2 = -\dfrac{1}{x}$ has branches in Quadrants $\underline{\text{II}}$ and $\underline{\text{IV}}$.

5. Discuss/explain how and why the range of the reciprocal function differs from the range of the reciprocal quadratic function. In the reciprocal quadratic function, all range values are positive. Answers will vary.

6. If the graphs of $Y_1 = \dfrac{1}{x}$ and $Y_2 = \dfrac{1}{x^2}$ were drawn on the same grid, where would they intersect? In what interval(s) is $Y_1 > Y_2$? $(1, 1); (1, \infty)$

► DEVELOPING YOUR SKILLS

For each graph given, (a) use mathematical notation to describe the end-behavior of each graph and (b) describe what happens as x approaches 1.

7. $V(x) = \dfrac{1}{x-1} + 2$ 8. $v(x) = \dfrac{1}{x-1} - 2$

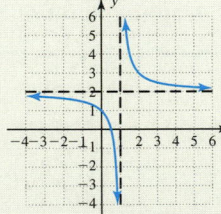   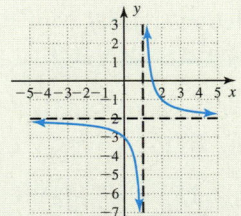

For each graph given, (a) use mathematical notation to describe the end-behavior of each graph, (b) name the horizontal asymptote, and (c) describe what happens as x approaches −2.

9. $Q(x) = \dfrac{1}{(x+2)^2} + 1$ 10. $q(x) = \dfrac{-1}{(x+2)^2} + 2$

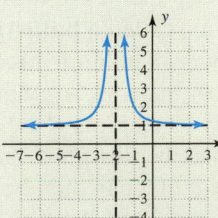   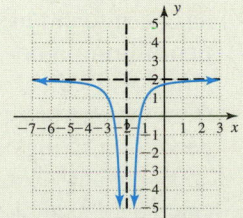

Sketch the graph of each function using transformations of the parent function (not by plotting points). Clearly state the transformations used, and label the horizontal and vertical asymptotes as well as the x- and y-intercepts (if they exist). Also state the domain and range of each function.

11. $f(x) = \dfrac{1}{x} - 1$ 12. $g(x) = \dfrac{1}{x} + 2$

13. $h(x) = \dfrac{1}{x+2}$ 14. $f(x) = \dfrac{1}{x-3}$

15. $g(x) = \dfrac{-1}{x-2}$ 16. $h(x) = \dfrac{-1}{x} - 2$

17. $f(x) = \dfrac{1}{x+2} - 1$ 18. $g(x) = \dfrac{1}{x-3} + 2$

19. $h(x) = \dfrac{1}{(x-1)^2}$ 20. $f(x) = \dfrac{1}{(x+5)^2}$

21. $g(x) = \dfrac{-1}{(x+2)^2}$ 22. $h(x) = \dfrac{-1}{x^2} - 2$

23. $f(x) = \dfrac{1}{x^2} - 2$ 24. $g(x) = \dfrac{1}{x^2} + 3$

25. $h(x) = 1 + \dfrac{1}{(x+2)^2}$ 26. $g(x) = -2 + \dfrac{1}{(x-1)^2}$

Additional answers can be found in the Instructor Answer Appendix.

HOMEWORK SELECTION GUIDE

Core: 7, 9, 11–31 every other odd, 33–61 odd, 65, 69, 71, 75, 77 (29 Exercises)
Standard: 1–4, 7, 9, 11–31 every other odd, 33–61 odd, 65, 69, 71, 75, 77, 80, 81, 84 (36 Exercises)

Extended: 1–4, 7, 9, 11–31 every other odd, 64, 65, 69, 71, 75, 77, 78, 80, 81, 84 (38 Exercises)
In Depth: 1–6, 7, 9, 11–31 every other odd, 33–61 odd, 63, 64, 65, 69, 71, 75, 77–84 (44 Exercises)

Identify the parent function for each graph given, then use the graph to construct the equation of the function in shifted form. Assume $|a| = 1$.

27.
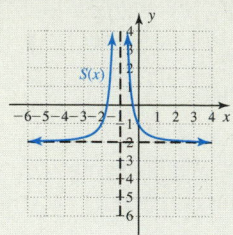

28.
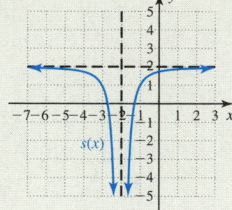

29.
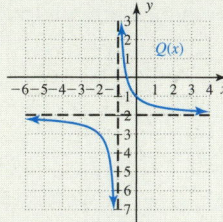

30.
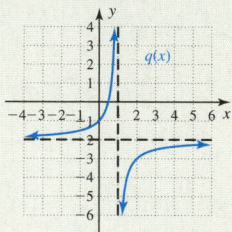

31.
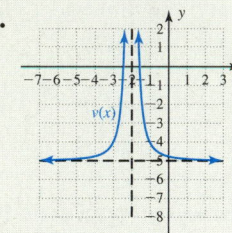

32.
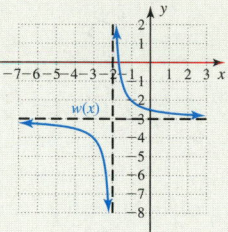

Use the graph shown to complete each statement using the direction/approach notation.

Exercises 33 through 38

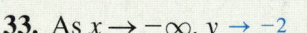

33. As $x \to -\infty$, y $\to -2$.

34. As $x \to \infty$, y $\to -2$.

35. As $x \to -1^+$, y $\to -\infty$.

36. As $x \to -1^-$, y $\to \infty$.

37. The line $x = -1$ is a vertical asymptote, since: as $x \to$ -1 , $y \to$ $\pm\infty$.

38. The line $y = -2$ is a horizontal asymptote, since: as $x \to$ $\pm\infty$, $y \to$ -2 .

For each pair of functions given, state which function increases faster for $x > 1$, then use the INTERSECT command of a graphing calculator to find where (a) $f(x) = g(x)$, (b) $f(x) > g(x)$, and (c) $f(x) < g(x)$.

39. $f(x) = x^2$, $g(x) = x^3$

40. $f(x) = x^4$, $g(x) = x^5$

41. $f(x) = x^4$, $g(x) = x^2$

42. $f(x) = x^3$, $g(x) = x^5$

43. $f(x) = x^{\frac{2}{3}}$, $g(x) = x^{\frac{4}{5}}$

44. $f(x) = x^{\frac{7}{4}}$, $g(x) = x^{\frac{3}{2}}$

45. $f(x) = \sqrt[6]{x}$, $g(x) = \sqrt[3]{x}$

46. $f(x) = \sqrt[5]{x}$, $g(x) = \sqrt[4]{x}$

47. $f(x) = \sqrt[3]{x^2}$, $g(x) = x^{\frac{3}{4}}$

48. $f(x) = x^{\frac{3}{2}}$, $g(x) = \sqrt[4]{x^3}$

State the domain of the following functions.

49. $f(x) = x^{\frac{7}{8}}$ $[0, \infty)$

50. $g(x) = x^{\frac{6}{7}}$ $(-\infty, \infty)$

51. $h(x) = x^{\frac{6}{5}}$ $(-\infty, \infty)$

52. $q(x) = x^{\frac{5}{6}}$ $[0, \infty)$

53. $r(x) = \sqrt[7]{x}$ $(-\infty, \infty)$

54. $s(x) = x^{\frac{1}{6}}$ $[0, \infty)$

Using the functions from Exercises 49–54, identify which of the following are defined and which are not. Do not use a calculator or evaluate.

55. a. $f(-2)$ **b.** $f(2)$ **c.** $g(-2)$ **d.** $g(2)$
defined: b, c, d; undefined: a

56. a. $h(0.3)$ **b.** $h(-0.3)$ **c.** $q(0.3)$ **d.** $q(-0.3)$
defined: a, b, c; undefined: d

57. a. $h(-1.2)$ **b.** $r(-7)$ **c.** $s(-\pi)$ **d.** $s(0)$
defined: a, b, d; undefined: c

58. a. $f\left(-\frac{7}{8}\right)$ **b.** $g\left(-\frac{8}{7}\right)$ **c.** $q(-1.9)$ **d.** $q(0)$
defined: b, d; undefined: a, c

Compare and discuss the graphs of the following functions. Verify your answer by graphing both on a graphing calculator.

59. $f(x) = x^{\frac{7}{8}}$; $F(x) = (x + 1)^{\frac{7}{8}} - 2$

60. $g(x) = x^{\frac{8}{7}}$; $G(x) = (x - 3)^{\frac{8}{7}} + 2$

61. $p(x) = x^{\frac{6}{5}}$; $P(x) = -(x - 2)^{\frac{6}{5}}$

62. $q(x) = x^{\frac{5}{6}}$; $Q(x) = 2x^{\frac{5}{6}} - 5$

▶ WORKING WITH FORMULAS

63. Gravitational attraction: $F = \dfrac{km_1m_2}{d^2}$

The gravitational force F between two objects with masses m_1 and m_2 depends on the distance d between them and some constant k. (a) If the masses of the two objects are constant while the distance between them gets larger and larger, what happens to F? (b) Let m_1 and m_2 equal 1 mass unit with $k = 1$ as well, and investigate using a table of values. What family does this function belong to? (c) Solve for m_2 in terms of k, m_1, d and F.

a. F becomes very small **b.** $y = \frac{1}{x^2}$ **c.** $m_2 = \frac{d^2F}{km_1}$

Additional answers can be found in the Instructor Answer Appendix.

64. Velocity of a bullet: $v = \dfrac{m + M}{m}\sqrt{2gh}$

For centuries, the velocity v of a bullet of mass m has been found using a device called a **ballistic pendulum.** In one such device, a bullet is fired into a stationary block of wood of mass M, suspended from the end of a pendulum. The height h the pendulum swings after impact is measured, and the approximate velocity of the bullet can then be calculated using $g = 9.8$ m/sec^2 (acceleration due to gravity). When a .22-caliber bullet of mass 2.6 g is fired into a wood block of mass 400 g, their combined mass swings to a height of 0.23 m. To the nearest meter per second, find the velocity of the bullet the moment it struck the wood.   329 m/s

▶ **APPLICATIONS**

65. Deer and predators: By banding deer over a period of 10 yr, a capture-and-release project determines the number of deer per square mile in the Mark Twain National Forest can be modeled by the function $D(p) = \dfrac{75}{p}$, where p is the number of predators present and D is the number of deer. Use this model to answer the following.

   a. As the number of predators increases, what will happen to the population of deer? Evaluate the function at $D(1)$, $D(3)$, and $D(5)$ to verify.   It decreases; 75, 25, 15

   b. What happens to the deer population if the number of predators becomes very large?   It approaches 0.

   c. Graph the function using an appropriate scale. Judging from the graph, use mathematical notation to describe what happens to the deer population if the number of predators becomes very small (less than 1 per square mile).   as $p \to 0$, $D \to \infty$

66. Balance of nature: A marine biology research group finds that in a certain reef area, the number of fish present depends on the number of sharks in the area. The relationship can be modeled by the function $F(s) = \dfrac{20{,}000}{s}$, where $F(s)$ is the fish population when s sharks are present.

   a. As the number of sharks increases, what will happen to the population of fish? Evaluate the function at $F(10)$, $F(50)$, and $F(200)$ to verify.   It decreases; 2000, 400, 100

   b. What happens to the fish population if the number of sharks becomes very large?   It approaches 0.

   c. Graph the function using an appropriate scale. Judging from the graph, use mathematical notation to describe what happens to the fish population if the number of sharks becomes very small.   as $s \to 0$, $F \to \infty$

67. Intensity of light: The intensity I of a light source depends on the distance of the observer from the source. If the intensity is 100 W/m² at a distance of 5 m, the relationship can be modeled by the function $I(d) = \dfrac{2500}{d^2}$. Use the model to answer the following.

   a. As the distance from the lightbulb increases, what happens to the intensity of the light? Evaluate the function at $I(5)$, $I(10)$, and $I(15)$ to verify.   It decreases; 100, 25, 11.$\overline{1}$.

   b. If the intensity is increasing, is the observer moving away or toward the light source?   toward the light source

   c. Graph the function using an appropriate scale. Judging from the graph, use mathematical notation to describe what happens to the intensity if the distance from the lightbulb becomes very small.   as $d \to 0$, $I \to \infty$

68. Electrical resistance: The resistance R (in ohms) to the flow of electricity is related to the length of the wire and its gauge (diameter in fractions of an inch). For a certain wire with fixed length, this relationship can be modeled by the function $R(d) = \dfrac{0.2}{d^2}$, where $R(d)$ represents the resistance in a wire with diameter d.

   a. As the diameter of the wire increases, what happens to the resistance? Evaluate the function at $R(0.05)$, $R(0.25)$, and $R(0.5)$ to verify.   It decreases; 80, 3.2, 0.8.

   b. If the resistance is increasing, is the diameter of the wire getting larger or smaller?   smaller

   c. Graph the function using an appropriate scale. Judging from the graph, use mathematical notation to describe what happens to the resistance in the wire as the diameter gets larger and larger.   as $d \to \infty$, $R \to 0^+$.

69. Pollutant removal: For a certain coal-burning power plant, the cost to remove pollutants from plant emissions can be modeled by $C(p) = \dfrac{-8000}{p - 100} - 80$, where $C(p)$ represents the cost (in thousands of dollars) to remove p percent of the pollutants. (a) Find the cost to remove 20%, 50%, and 80% of the pollutants, then comment on the results; (b) graph the function using an appropriate scale; and (c) use mathematical notation to state what happens if the power company attempts to remove 100% of the pollutants.

70. City-wide recycling: A large city has initiated a new recycling effort, and wants to distribute recycling bins for use in separating various recyclable materials. City planners anticipate the cost of the program can be modeled by the function $C(p) = \dfrac{-22{,}000}{p - 100} - 220$, where $C(p)$ represents the cost (in $10,000) to distribute the bins to p percent of the population. (a) Find the cost to distribute bins to 25%, 50%, and 75% of the population, then comment on the results; (b) graph the function using an appropriate scale; and (c) use mathematical notation to state what happens if the city attempts to give recycling bins to 100% of the population.

71. Hot air ballooning: If air resistance is neglected, the velocity (in ft/s) of a falling object can be closely approximated by the function $V(s) = 8\sqrt{s}$, where s is the distance the object has fallen (in feet). A balloonist suddenly finds it necessary to release some ballast in order to quickly gain altitude. (a) If she were flying at an altitude of 1000 ft, with what velocity will the ballast strike the ground? (b) If the ballast strikes the ground with a velocity of 225 ft/sec, what was the altitude of the balloon?
a. 253 ft/sec (about 172 mph) **b.** approx. 791 ft

72. River velocities: The ability of a river or stream to move sand, dirt, or other particles depends on the size of the particle and the velocity of the river. This relationship can be used to approximate the velocity (in mph) of the river using the function $V(d) = 1.77\sqrt{d}$, where d is the diameter (in inches) of the particle being moved. (a) If a creek can move a particle of diameter 0.095 in., how fast is it moving? (b) What is the largest particle that can be moved by a stream flowing 1.1 mph?
a. approx. 0.55 mph   **b.** approx. 0.39 in.

73. Shoe sizes: Although there may be some notable exceptions, the size of shoe worn by the average man is related to his height. This relationship is modeled by the function $S(h) = 0.75h^{\frac{3}{2}}$, where h is the person's height in feet and S is the U.S. shoe size. (a) Approximate Denzel Washington's shoe size given he is 6 ft, 0 in. tall. (b) Approximate Dustin Hoffman's height given his shoe size is 9.5.
a. size 11   **b.** approx. 5 ft, 5 in.

74. Whale weight: For a certain species of whale, the relationship between the length of the whale and the weight of the whale can be modeled by the function $W(l) = 0.03l^{\frac{27}{11}}$, where l is the length of the whale in meters and W is the weight of the whale in metric tons (1 metric ton $\approx$ 2205 pounds). (a) Estimate the weight of a newborn calf that is 6 m long. (b) At 81 metric tons, how long is an average adult?
a. 2.44 metric tons   **b.** 25 m

75. Gestation periods: The data shown in the table can be used to study the relationship between the weight of mammal and its length of pregnancy. Use a graphing calculator to (a) graph a scatterplot of the data and (b) find an equation model using a power regression (round to three decimal places). Use the equation to estimate (c) the length of pregnancy of a racoon (15.5 kg) and (d) the weight of a fox, given the length of pregnancy is 52 days.
b. $P = 32.251\,w^{0.246}$ **c.** about 63 days   **d.** about 6.9 kg

Mammal	Average Weight (kg)	Gestation (days)
Rat	0.4	24
Rabbit	3.5	50
Armadillo	6.0	51
Coyote	13.1	62
Dog	24.0	64

76. Bird wingspans: The data in the table explores the relationship between a bird's weight and its wingspan. Use a graphing calculator to (a) graph a scatterplot of the data and (b) find an equation model using a power regression (round to three decimal places). Use the equation to estimate (c) the wingspan of a Bald Eagle (16 lb) and (d) the weight of a Bobwhite Quail with a wingspan of 0.9 ft.
b. $S = 1.653w^{0.557}$ **c.** about 7.7 ft   **d.** about 0.3 lb

Bird	Weight (lb)	Wingspan (ft)
Golden Eagle	10.5	6.5
Horned Owl	3.1	2.6
Peregrine Falcon	3.3	4.0
Whooping Crane	17.0	7.5
Raven	1.5	2.0

77. Species-area relationship: To study the relationship between the number of species of birds on islands in the Caribbean, the data shown in the table was collected. Use a graphing calculator to (a) graph a scatterplot of the data and (b) find an equation model using a power regression (round to three decimal places). Use the equation to estimate (c) the number of species of birds on Andros (2300 mi^2) and (d) the area of Cuba, given there are 98 such species.
b. $S = 1.687a^{0.386}$ **c.** about 33 species   **d.** about 37,200 mi^2

Island	Area (mi^2)	Species
Great Inagua	600	16
Trinidad	2000	41
Puerto Rico	3400	47
Jamaica	4500	38
Hispaniola	30,000	82

78. Planetary orbits: The table shown gives the time required for the first five planets to make one complete revolution around the Sun (in years), along with the average orbital radius of the planet in astronomical units (1 AU = 92.96 million miles). Use a graphing calculator to (a) graph a scatterplot of the data and (b) find an equation model using a power regression (round to four decimal places). Use the equation to estimate (c) the average orbital radius of Saturn, given it orbits the Sun every 29.46 yr, and (d) estimate how many years it takes Uranus to orbit the Sun, given it has an average orbital radius of 19.2 AU.   **b.** $R = 1.000y^{0.6654}$ **c.** about 9.5 AU   **d.** about 84.8 yr

Planet	Years	Radius
Mercury	0.24	0.39
Venus	0.62	0.72
Earth	1.00	1.00
Mars	1.88	1.52
Jupiter	11.86	5.20

Additional answers can be found in the Instructor Answer Appendix.

▶ **EXTENDING THE CONCEPT**

79. Consider the graph of $f(x) = \dfrac{1}{x}$ once again, and the x by $f(x)$ rectangles mentioned in the Worthy of Note on page 231. Calculate the area of each rectangle formed for $x \in \{1, 2, 3, 4, 5, 6\}$. What do you notice? Repeat the exercise for $g(x) = \dfrac{1}{x^2}$ and the x by $g(x)$ rectangles. Can you detect the pattern formed here?
The area is always 1 unit²; The area is always $\frac{1}{x}$ units²

80. All of the power functions presented in this section had positive exponents, but the definition of these types of functions does allow for negative exponents as well. In addition to the reciprocal and reciprocal square functions ($y = x^{-1}$ and $y = x^{-2}$), these types of power functions have significant applications. For example, the temperature of ocean water depends on several factors, including salinity, latitude, depth, and density. However, between depths of 125 m and 2000 m, ocean temperatures are relatively predictable, as indicated by the data shown for tropical oceans in the table. Use a graphing calculator to find the power regression model and use it to estimate the water temperature at a depth of 2850 m.
$T = 222.2d^{-0.58}$, 2.2°C

Depth (meters)	Temp (°C)
125	13.0
250	9.0
500	6.0
750	5.0
1000	4.4
1250	3.8
1500	3.1
1750	2.8
2000	2.5

▶ **MAINTAINING YOUR SKILLS**

81. (1.4) Solve the equation for y, then sketch its graph using the slope/intercept method: $2x + 3y = 15$.

82. (1.3) Using a scale from 1 (lousy) to 10 (great), Charlie gave the following ratings: {(The Beatles, 9.5), (The Stones, 9.6), (The Who, 9.5), (Queen, 9.2), (The Monkees, 6.1), (CCR, 9.5), (Aerosmith, 9.2), (Lynyrd Skynyrd, 9.0), (The Eagles, 9.3), (Led
Additional answers can be found in the Instructor Answer Appendix.

Zeppelin, 9.4), (The Stones, 9.8)}. Is the relation (group, rating) as given, also a function? State why or why not.
No, The Stones are paired with two different ratings.

83. (1.5) Solve for c: $E = mc^2$. $c = \sqrt{\frac{E}{m}}$

84. (2.3) Use a graphing calculator to solve $|x - 2| + 1 \geq -2|x + 1| + 3$. $(-\infty, \infty)$

2.5   **Piecewise-Defined Functions**

LEARNING OBJECTIVES

In Section 2.5 you will see how we can:

☐ **A.** State the equation, domain, and range of a piecewise-defined function from its graph

☐ **B.** Graph functions that are piecewise-defined

☐ **C.** Solve applications involving piecewise-defined functions

Most of the functions we've studied thus far have been smooth and continuous. Although "smooth" and "continuous" are defined more formally in advanced courses, for our purposes *smooth* simply means the graph has no sharp turns or jagged edges, and *continuous* means you can draw the entire graph without lifting your pencil. In this section, we study a special class of functions, called **piecewise-defined functions**, whose graphs may be various combinations of smooth/not smooth and continuous/not continuous. The absolute value function is one example (**see Exercise 31**). Such functions have a tremendous number of applications in the real world.

A. The Domain of a Piecewise-Defined Function

For the years 1990 to 2000, the American bald eagle remained on the nation's endangered species list, although the number of breeding pairs was growing slowly. After 2000, the population of eagles grew at a much faster rate, and they were removed from the list soon afterward. From Table 2.3 and plotted points modeling this growth (see Figure 2.76), we observe that a linear model would fit the period from 1992 to 2000 very well, but a line with greater slope would be needed for the years 2000 to 2006 and (perhaps) beyond.

Figure 2.76

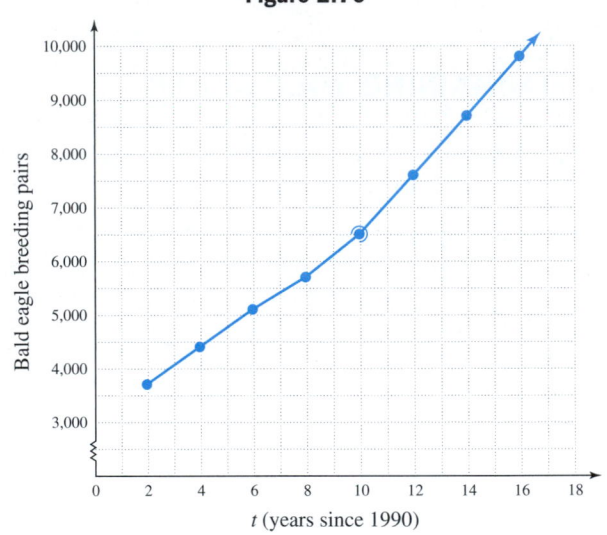

Table 2.3

Year (1990 → 0)	Bald Eagle Breeding Pairs	Year (1990 → 0)	Bald Eagle Breeding Pairs
2	3700	10	6500
4	4400	12	7600
6	5100	14	8700
8	5700	16	9800

Source: www.fws.gov/midwest/eagle/population

WORTHY OF NOTE

For the years 1992 to 2000, we can estimate the growth in breeding pairs $\frac{\Delta\text{pairs}}{\Delta\text{time}}$ using the points (2, 3700) and (10, 6500) in the slope formula. The result is $\frac{350}{1}$, or 350 pairs per year. For 2000 to 2006, using (10, 6500) and (16, 9800) shows the rate of growth is significantly larger: $\frac{\Delta\text{pairs}}{\Delta\text{years}} = \frac{550}{1}$ or 550 pairs per year.

The combination of these two lines would be a single function that modeled the population of breeding pairs from 1990 to 2006, but it would be *defined in two pieces.* This is an example of a **piecewise-defined function.**

The notation for these functions is a large "left brace" indicating the equations it groups are part of a single function. Using selected data points and techniques from Section 1.4, we find equations that could represent each piece are $p(t) = 350t + 3000$ for $0 \leq t \leq 10$ and $p(t) = 550t + 1000$ for $t > 10$, where $p(t)$ is the number of breeding pairs in year t. The complete function is then written:

function name   function pieces   domain of each piece

$$p(t) = \begin{cases} 350t + 3000, & 2 \leq t \leq 10 \\ 550t + 1000, & t > 10 \end{cases}$$

In Figure 2.76, note that we indicated the exclusion of $t = 10$ from the second piece of the function using an open half-circle.

EXAMPLE 1 ▶ **Writing the Equation and Domain of a Piecewise-Defined Function**

The linear piece of the function shown has an equation of $y = -2x + 10$. The equation of the quadratic piece is $y = -x^2 + 9x - 14$.

  a. Use the correct notation to write them as a single piecewise-defined function and state the domain of each piece by inspecting the graph.

  b. State the range of the function.

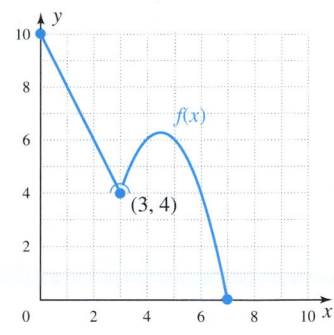

Solution ▶ **a.** From the graph we note the linear portion is defined between 0 and 3, with these endpoints included as indicated by the closed dots. The domain here is $0 \le x \le 3$. The quadratic portion begins at $x = 3$ *but does not include 3*, as indicated by the half-circle notation. The equation is

$$\underset{\text{function name}}{f(x)} = \begin{cases} \overset{\text{function pieces}}{-2x + 10,} & \overset{\text{domain}}{0 \le x \le 3} \\ -x^2 + 9x - 14, & 3 < x \le 7 \end{cases}$$

b. The largest y-value is 10 and the smallest is zero. The range is $y \in [0, 10]$.

☑ **A.** You've just seen how we can state the equation, domain, and range of a piecewise-defined function from its graph

Now try Exercises 7 and 8 ▶

Piecewise-defined functions can be composed of more than two pieces, and can involve functions of many kinds.

B. Graphing Piecewise-Defined Functions

As with other functions, piecewise-defined functions can be graphed by simply plotting points. Careful attention must be paid to the domain of each piece, both to evaluate the function correctly and to consider the inclusion/exclusion of endpoints. In addition, try to keep the transformations of a basic function in mind, as this will often help graph the function more efficiently.

EXAMPLE 2 ▶ **Graphing a Piecewise-Defined Function**

Evaluate the piecewise-defined function by noting the effective domain of each piece, then graph by plotting these points and using your knowledge of basic functions.

$$h(x) = \begin{cases} -x - 2, & -5 \le x < -1 \\ 2\sqrt{x + 1} - 1, & x \ge -1 \end{cases}$$

Solution ▶ The first piece of h is a line with negative slope, while the second is a transformed square root function. Using the endpoints of each domain specified and a few additional points, we obtain the following:

For $h(x) = -x - 2, -5 \le x < -1$,   For $h(x) = 2\sqrt{x + 1} - 1, x \ge -1$,

x	$h(x)$
-5	3
-3	1
-1	(-1)

x	$h(x)$
-1	-1
0	1
3	3

After plotting the points from the first piece, we connect them with a line segment noting the left endpoint is included, while the right endpoint is not (indicated using a semicircle around the point). Then we plot the points from the second piece and draw a square root graph, noting the left endpoint here *is* included, and the graph rises to the right. From the graph we note the complete domain of h is $x \in [-5, \infty)$, and the range is $y \in [-1, \infty)$.

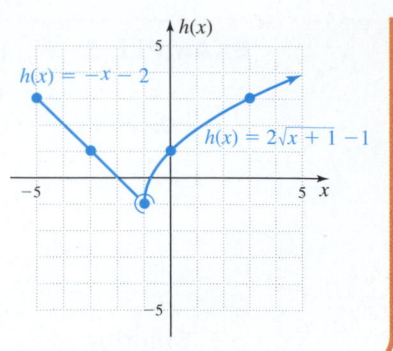

Now try Exercises 9 through 12 ▶

Most graphing calculators are able to graph piecewise-defined functions. Consider Example 3.

EXAMPLE 3 ▶ **Graphing a Piecewise-Defined Function Using Technology**

Graph the function $f(x) = \begin{cases} x + 5, & -5 \le x < 2 \\ (x - 4)^2 + 3, & x \ge 2 \end{cases}$ on a graphing calculator and evaluate $f(2)$.

Solution ▶ Both "pieces" are well known—the first is a line with slope $m = 1$ and y-intercept $(0, 5)$. The second is a parabola that opens upward, shifted 4 units to the right and 3 units up. If we attempt to graph $f(x)$ using $Y_1 = X + 5$ and $Y_2 = (X - 4)^2 + 3$ as they stand, the resulting graph may be difficult to analyze because the pieces overlap and intersect (Figure 2.77). To graph the functions we must indicate the domain for each piece, separated by a slash and enclosed in parentheses. For instance, for the first piece we enter $Y_1 = X + 5/(X \ge -5 \text{ and } X < 2)$, and for the second, $Y_2 = (X - 4)^2 + 3/(X \ge 2)$ (Figure 2.78). The slash looks like (is) the division symbol, but in this context, the calculator interprets it as a means of separating the function from the domain. The inequality symbols are accessed using the [2nd] [MATH] (**TEST**) keys. As shown for Y_1, compound inequalities must be entered in two parts, using the logical connector "and": [2nd] [MATH] [▸] (**LOGIC**) **1:and**. The graph is shown in Figure 2.79, where we see the function is linear for $x \in [-5, 2)$ and quadratic for $x \in [2, \infty)$. Using the [2nd] [GRAPH] (**TABLE**) feature reveals the calculator will give an **ERR: (ERROR)** message for inputs outside the domains of Y_1 and Y_2, and we see that f is defined for $x = 2$ only for Y_2: $f(2) = 7$ (Figure 2.80).

Figure 2.77

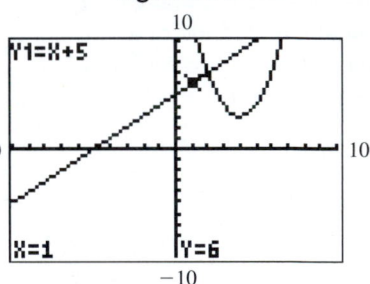

Figure 2.78

Figure 2.79

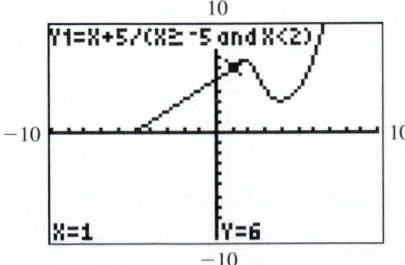

Figure 2.80

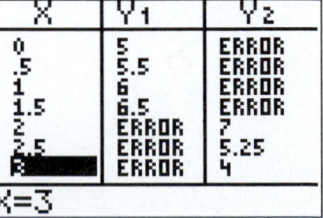

Now try Exercises 13 and 14 ▶

As an alternative to plotting points, we can graph each piece of the function using transformations of a basic graph, then erase those parts that are outside of the corresponding domain. Repeat this procedure for each piece of the function. One interesting and highly instructive aspect of these functions is the opportunity to investigate restrictions on their domain and the ranges that result.

Piecewise and Continuous Functions

EXAMPLE 4 ▶ **Graphing a Piecewise-Defined Function**

Graph the function and state its domain and range:

$$f(x) = \begin{cases} -(x-3)^2 + 12, & 0 < x \le 6 \\ 3, & x > 6 \end{cases}$$

Solution ▶ The first piece of f is a basic parabola, shifted three units right, reflected across the x-axis (opening downward), and shifted 12 units up. The vertex is at (3, 12) and the axis of symmetry is $x = 3$, producing the following graphs.

1. Graph first piece of f (Figure 2.81)

2. Erase portion outside domain. of $0 < x \le 6$ (Figure 2.82).

Figure 2.81

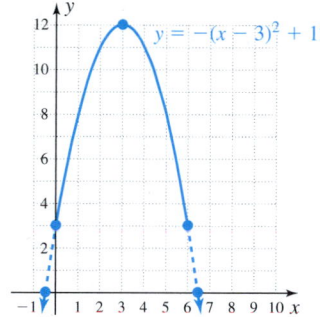

Figure 2.82

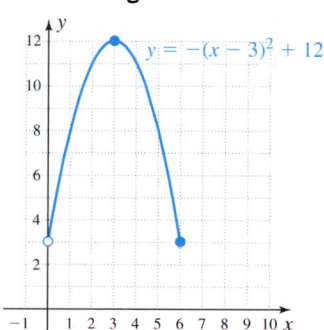

The second function is simply a horizontal line through (0, 3).

3. Graph second piece of f (Figure 2.83).

4. Erase portion outside domain of $x > 6$ (Figure 2.84).

Figure 2.83

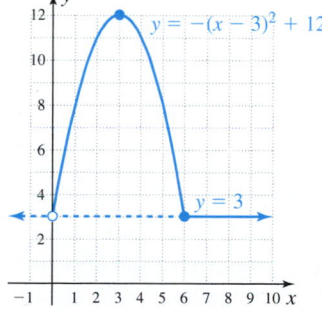

Figure 2.84

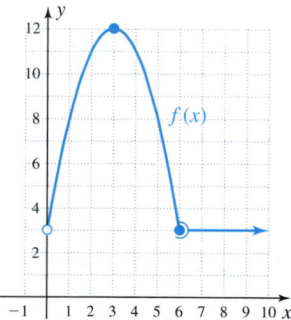

The domain of f is $x \in (0, \infty)$, and the corresponding range is $y \in [3, 12]$.

Now try Exercises 15 through 18 ▶

Piecewise and Discontinuous Functions

Notice that although the function in Example 4 was piecewise-defined, the graph was actually continuous—we could draw the entire graph without lifting our pencil. Piecewise graphs also come in the *discontinuous* variety, which makes the domain and range issues all the more important.

EXAMPLE 5 ▶ **Graphing a Discontinuous Piecewise-Defined Function**

Graph $g(x)$ and state the domain and range:

$$g(x) = \begin{cases} -\frac{1}{2}x + 6, & 0 \le x \le 4 \\ -|x - 6| + 10, & 4 < x \le 9 \end{cases}$$

Solution ▶ The first piece of g is a line, with y-intercept $(0, 6)$ and slope $\frac{\Delta y}{\Delta x} = -\frac{1}{2}$.

1. Graph first piece of g (Figure 2.85)

2. Erase portion outside domain. of $0 \le x \le 4$ (Figure 2.86).

Figure 2.85

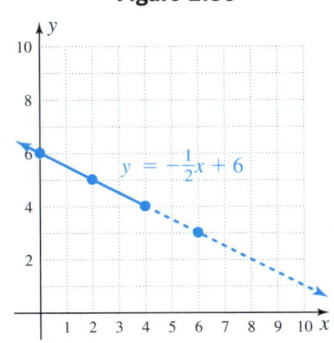

Figure 2.86

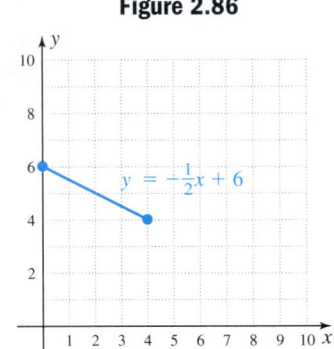

The second is an absolute value function, shifted right 6 units, reflected across the x-axis, then shifted up 10 units.

WORTHY OF NOTE

As you graph piecewise-defined functions, keep in mind that they *are* functions and the end result must pass the vertical line test. This is especially important when we are drawing each piece as a complete graph, then erasing portions outside the effective domain.

3. Graph second piece of g (Figure 2.87).

4. Erase portion outside domain of $4 < x \le 9$ (Figure 2.88).

Figure 2.87

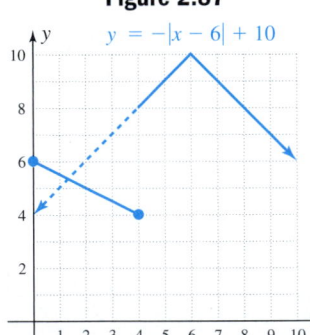

Figure 2.88

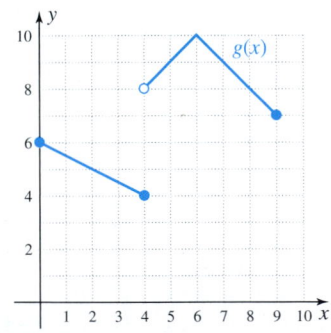

Note that the left endpoint of the absolute value portion is not included (this piece is not defined at $x = 4$), signified by the open dot. The result is a discontinuous graph, as there is no way to draw the graph other than by "jumping" the pencil from where one piece ends to where the next begins. Using a vertical boundary line, we note the domain of g includes all values between 0 and 9 inclusive: $x \in [0, 9]$. Using a horizontal boundary line shows the smallest y-value is 4 and the largest is 10, but no range values exist between 6 and 7. The range is $y \in [4, 6] \cup [7, 10]$.

Now try Exercises 19 and 20 ▶

EXAMPLE 6 ▶   **Graphing a Discontinuous Function**

The given piecewise-defined function is not continuous. Graph $h(x)$ to see why, then comment on what could be done to make it continuous.

$$h(x) = \begin{cases} \dfrac{x^2 - 4}{x - 2}, & x \neq 2 \\ 1, & x = 2 \end{cases}$$

Solution ▶   The first piece of h is unfamiliar to us, so we elect to graph it by plotting points, noting $x = 2$ is outside the domain. This produces the table shown. After connecting the points, the graph turns out to be a straight line, but with no corresponding y-value for $x = 2$. This leaves a "hole" in the graph at $(2, 4)$, as designated by the open dot (see Figure 2.89).

Figure 2.89   **Figure 2.90**

x	$h(x)$
-4	-2
-2	0
0	2
2	—
4	6

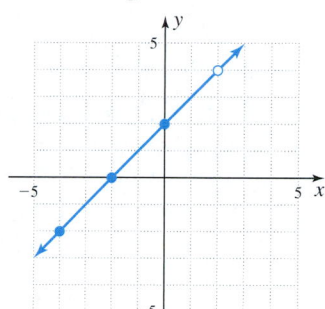
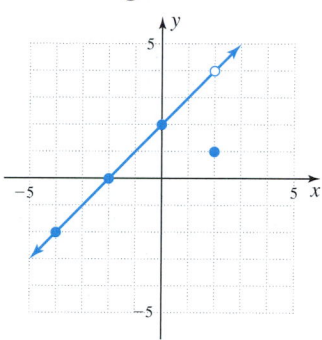

> **WORTHY OF NOTE**
>
> The discontinuity illustrated here is called a **removable discontinuity,** as the discontinuity can be removed by redefining a single point on the function. Note that after factoring the first piece, the denominator is a factor of the numerator, and writing the result in lowest terms gives $h(x) = \dfrac{(x + 2)(x - 2)}{x - 2} = x + 2, x \neq 2$. This is precisely the equation of the line in Figure 2.89 $[y = x + 2]$.

The second piece is pointwise-defined, and its graph is simply the point $(2, 1)$ shown in Figure 2.90. It's interesting to note that while the domain of h is all real numbers (h *is* defined at all points), the range is $y \in (-\infty, 4) \cup (4, \infty)$ as the function never takes on the value $y = 4$. In order for h to be continuous, we would need to redefine the second piece as $y = 4$ when $x = 2$.

Now try Exercises 21 through 26 ▶

To develop these concepts more fully, it will help to practice finding the equation of a piecewise-defined function *given its graph,* a process similar to that of Example 10 in Section 2.2.

EXAMPLE 7 ▶   **Determining the Equation of a Piecewise-Defined Function**

Determine the equation of the piecewise-defined function shown, including the domain for each piece.

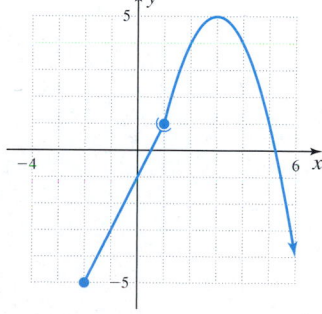

Solution ▶   By counting $\frac{\Delta y}{\Delta x}$ from $(-2, -5)$ to $(1, 1)$, we find the linear portion has slope $m = 2$, and the y-intercept must be $(0, -1)$. The equation of the line is $y = 2x - 1$. The second piece appears to be a parabola with vertex (h, k) at $(3, 5)$. Using this vertex with the point $(1, 1)$ in the general form $y = a(x - h)^2 + k$ gives

$y = a(x - h)^2 + k$	general form, parabola is shifted right and up
$1 = a(1 - 3)^2 + 5$	substitute 1 for x, 1 for y, 3 for h, 5 for k
$-4 = a(-2)^2$	simplify; subtract 5
$-4 = 4a$	$(-2)^2 = 4$
$-1 = a$	divide by 4

The equation of the parabola is $y = -(x - 3)^2 + 5$. Considering the domains shown in the figure, the equation of this piecewise-defined function must be

$$p(x) = \begin{cases} 2x - 1, & -2 \le x < 1 \\ -(x - 3)^2 + 5, & x \ge 1 \end{cases}$$

☑ **B.** You've just seen how we can graph functions that are piecewise-defined

Now try Exercises 27 through 30 ▶

C. Applications of Piecewise-Defined Functions

The number of applications for piecewise-defined functions is practically limitless. It is actually fairly rare for a single function to accurately model a situation over a long period of time. Laws change, spending habits change, and technology can bring abrupt alterations in many areas of our lives. To accurately model these changes often requires a piecewise-defined function.

EXAMPLE 8 ▶   **Modeling with a Piecewise-Defined Function**

For the first half of the twentieth century, per capita spending on police protection can be modeled by $S(t) = 0.54t + 12$, where $S(t)$ represents per capita spending on police protection in year t (1900 corresponds to year 0). After 1950, perhaps due to the growth of American cities, this spending greatly increased: $S(t) = 3.65t - 144$. Write these as a piecewise-defined function $S(t)$, state the domain for each piece, then graph the function. According to this model, how much was spent (per capita) on police protection in 2000 and 2010? How much will be spent in 2014?

Source: Data taken from the *Statistical Abstract of the United States* for various years.

Solution ▶

function name   function pieces   effective domain

$$S(t) = \begin{cases} 0.54t + 12, & 0 \le t \le 50 \\ 3.65t - 144, & t > 50 \end{cases}$$

Since both pieces are linear, we can graph each part using two points. For the first function, $S(0) = 12$ and $S(50) = 39$. For the second function $S(50) \approx 39$ and $S(80) = 148$. The graph for each piece is shown in the figure. Evaluating S at $t = 100$:

$$S(t) = 3.65t - 144$$
$$S(100) = 3.65(100) - 144$$
$$= 365 - 144$$
$$= 221$$

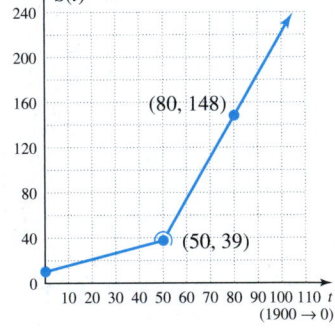

About \$221 per capita was spent on police protection in the year 2000. For 2010, the model indicates that \$257.50 per capita was spent: $S(110) = 257.5$. By 2014, this function projects the amount spent will grow to $S(114) = 272.1$ or \$272.10 per capita.

Now try Exercises 33 through 44 ▶

Step Functions

The last group of piecewise-defined functions we'll explore are the **step functions,** so called because the pieces of the function form a series of horizontal steps. These functions find frequent application in the way consumers are charged for services, and have several applications in number theory. Perhaps the most common is called the **greatest integer function,** though recently its alternative name, **floor function,** has gained popularity (see Figure 2.91). This is in large part due to an improvement in notation and as a better contrast to **ceiling functions.** The floor function of a real number x, denoted $f(x) = \lfloor x \rfloor$ or $[\![x]\!]$ (we will use the first), is the largest integer less than or equal to x. For instance, $\lfloor 5.9 \rfloor = 5$, $\lfloor 7 \rfloor = 7$, and $\lfloor -3.4 \rfloor = -4$.

In contrast, the ceiling function $C(x) = \lceil x \rceil$ is the smallest integer greater than or equal to x, meaning $\lceil 5.9 \rceil = 6$, $\lceil 7 \rceil = 7$, and $\lceil -3.4 \rceil = -3$ (see Figure 2.92). In simple terms, for any noninteger value on the number line, the floor function returns the integer to the left, while the ceiling function returns the integer to the right. A graph of each function is shown.

Figure 2.91

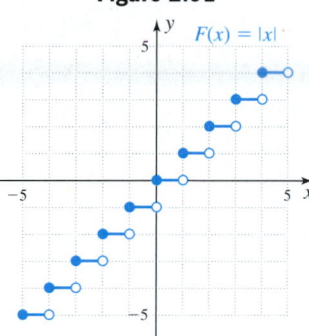

Figure 2.92

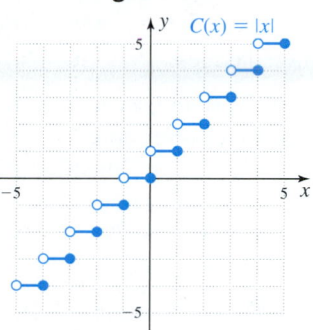

One common application of floor functions is the price of theater admission, where children 12 and under receive a discounted price. Right up until the day they're 13, they qualify for the lower price: $\lfloor 12\frac{364}{365} \rfloor = 12$. Applications of ceiling functions would include how phone companies charge for the minutes used (charging the 12-min rate for a phone call that only lasted 11.3 min: $\lceil 11.3 \rceil = 12$), and postage rates, as in Example 9.

EXAMPLE 9 ▶ **Modeling Using a Step Function**

In 2009 the first-class postage rate for large envelopes sent through the U.S. mail was 88¢ for the first ounce, then an additional 17¢ per ounce thereafter, up to 13 ounces. Graph the function and state its domain and range. Use the graph to state the cost of mailing a report weighing (a) 7.5 oz, (b) 8 oz, and (c) 8.1 oz in a large envelope.

Solution ▶ The 88¢ charge applies to letters weighing between 0 oz and 1 oz. Zero is not included since we have to mail *something,* but 1 is included since a large envelope and its contents weighing exactly one ounce still costs 88¢. The graph will be a horizontal line segment.

The function is defined for all weights between 0 and 13 oz, excluding zero and including 13: $x \in (0, 13]$. The range consists of single outputs corresponding to the step intervals: $R \in \{88, 105, 122, \ldots, 275, 292\}$.

a. The cost of mailing a 7.5-oz report is 207¢.

b. The cost of mailing an 8.0-oz report is still 207¢.

c. The cost of mailing an 8.1-oz report is $207 + 17 = 224$¢, since this brings you up to the next step.

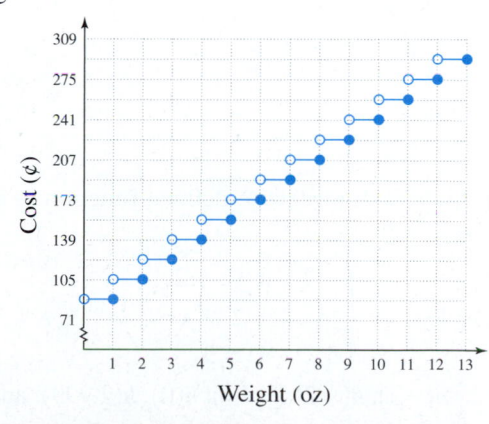

☑ **C.** You've just seen how we can solve applications involving piecewise-defined functions

Now try Exercises 45 through 48 ▶

2.5 EXERCISES

▶ CONCEPTS AND VOCABULARY

Fill in each blank with the appropriate word or phrase. Carefully reread the section if needed.

1. A function whose entire graph can be drawn without lifting your pencil is called a _continuous_ function.

2. The input values for which each part of a piecewise function is defined is the _domain_ of the function.

3. A graph is called _smooth_ if it has no sharp turns or jagged edges.

4. When graphing $2x + 3$ over a domain of $x > 0$, we leave an _open_ dot at $(0, 3)$.

5. Discuss/Explain how to determine if a piecewise-defined function is continuous, without having to graph the function. Illustrate with an example.
   Each piece must be continuous on the corresponding interval, and the function values at the endpoints of each interval must be equal. Answers will vary.

6. Discuss/Explain how it is possible for the domain of a function to be defined for all real numbers, but have a range that is defined on more than one interval. Construct an illustrative example.
   Answers will vary.

▶ DEVELOPING YOUR SKILLS

For Exercises 7 and 8, (a) use the correct notation to write them as a single piecewise-defined function and state the domain for each piece by inspecting the graph, then (b) state the range of the function.

7. $Y_1 = X^2 - 6x + 10$; $Y_2 = \frac{3}{2}X - \frac{5}{2}$

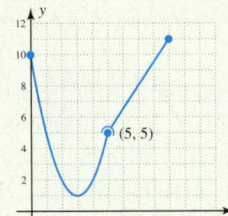

    a. $f(x) = \begin{cases} x^2 - 6x + 10 & 0 \le x \le 5 \\ \frac{3}{2}x - \frac{5}{2} & 5 < x \le 9 \end{cases}$

    b. $y \in [1, 11]$

8. $Y_1 = -1.5|X - 5| + 10$; $Y_2 = -\sqrt{X - 7} + 5$

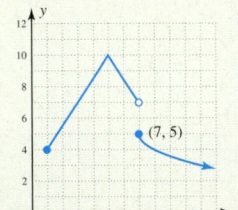

    a. $f(x) = \begin{cases} -1.5|x - 5| + 10 & 1 \le x < 7 \\ -\sqrt{x - 7} + 5 & x \ge 7 \end{cases}$

    b. $y \in (-\infty, 10]$

Evaluate each piecewise-defined function as indicated (if possible).

9. $h(x) = \begin{cases} -2 & x < -2 \\ |x| & -2 \le x < 3 \\ 5 & x \ge 3 \end{cases}$ $-2, 2, \frac{1}{2}, 0, 2.999, 5$

    $h(-5), h(-2), h(-\frac{1}{2}), h(0), h(2.999)$, and $h(3)$

Additional answers can be found in the Instructor Answer Appendix.

10. $H(x) = \begin{cases} 2x + 3 & x < 0 \\ x^2 + 1 & 0 \le x < 2 \\ 5 & x > 2 \end{cases}$ $-3, 0, 2.998, 2$, undefined, 5

    $H(-3), H(-\frac{3}{2}), H(-0.001), H(1), H(2)$, and $H(3)$

11. $p(x) = \begin{cases} 5 & x < -3 \\ x^2 - 4 & -3 \le x \le 3 \\ 2x + 1 & x > 3 \end{cases}$ $5, 5, 0, -4, 5, 11$

    $p(-5), p(-3), p(-2), p(0), p(3)$, and $p(5)$

12. $q(x) = \begin{cases} -x - 3 & x < -1 \\ 2 & -1 \le x < 2 \\ -\frac{1}{2}x^2 + 3x - 2 & x \ge 2 \end{cases}$

    $q(-3), q(-1), q(0), q(1.999), q(2)$, and $q(4)$
    $0, 2, 2, 2, 2, 2$

Graph each piecewise-defined function using a graphing calculator. Then evaluate each at $x = 2$ and $x = 0$.

13. $p(x) = \begin{cases} x + 2 & -6 \le x \le 2 \\ 2|x - 4| & x > 2 \end{cases}$ $p(2) = 4$ $p(0) = 2$

14. $q(x) = \begin{cases} \sqrt{x + 4} & -4 \le x \le 0 \\ |x - 2| & 0 < x \le 7 \end{cases}$ $q(2) = 0$ $q(0) = 2$

Graph each piecewise-defined function and state its domain and range. Use transformations of the toolbox functions where possible.

15. $g(x) = \begin{cases} -(x - 1)^2 + 5 & -2 \le x \le 4 \\ 2x - 12 & x > 4 \end{cases}$

16. $h(x) = \begin{cases} \frac{1}{2}x + 1 & x \le 0 \\ (x - 2)^2 - 3 & 0 < x \le 5 \end{cases}$

HOMEWORK SELECTION GUIDE

Core: 7–31 odd, 33, 35, 39, 43, 47 (18 Exercises)
Standard: 1–4, 7–31 odd, 33, 35, 39, 43, 47, 48, 52, 55, 56 (27 Exercises)

Extended: 1–4, 7–31 odd, 33, 34, 35, 39, 43, 45, 47, 48, 52, 55, 56 (29 Exercises)
In Depth: 1–6, 7–31 odd, 33, 34, 35, 39, 43, 45, 47, 48, 51–56 (34 Exercises)

17. $H(x) = \begin{cases} -x + 3 & x < 1 \\ -|x - 5| + 6 & 1 \le x < 9 \end{cases}$

18. $w(x) = \begin{cases} \sqrt[3]{x - 1} & x < 2 \\ (x - 3)^2 & 2 \le x \le 6 \end{cases}$

19. $f(x) = \begin{cases} -x - 3 & x < -3 \\ 9 - x^2 & -3 \le x < 2 \\ 4 & x \ge 2 \end{cases}$

20. $h(x) = \begin{cases} -\frac{1}{2}x - 1 & x < -3 \\ -|x| + 5 & -3 \le x \le 5 \\ 3\sqrt{x - 5} & x > 5 \end{cases}$

21. $p(x) = \begin{cases} \frac{1}{2}x + 1 & x \ne 4 \\ 2 & x = 4 \end{cases}$

22. $q(x) = \begin{cases} \frac{1}{2}(x - 1)^3 - 1 & x \ne 3 \\ -2 & x = 3 \end{cases}$

Each of the following functions has a removable discontinuity. Graph the first piece of each function, then find the value of c so that a continuous function results.

23. $f(x) = \begin{cases} \dfrac{x^2 - 9}{x + 3} & x \ne -3 \\ c & x = -3 \end{cases}$

24. $f(x) = \begin{cases} \dfrac{x^2 - 3x - 10}{x - 5} & x \ne 5 \\ c & x = 5 \end{cases}$

25. $f(x) = \begin{cases} \dfrac{x^3 - 1}{x - 1} & x \ne 1 \\ c & x = 1 \end{cases}$

26. $f(x) = \begin{cases} \dfrac{4x - x^3}{x + 2} & x \ne -2 \\ c & x = -2 \end{cases}$

Determine the equation of each piecewise-defined function shown, including the domain for each piece. Assume all pieces are toolbox functions.

27.

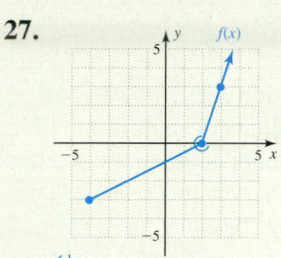

$f(x) = \begin{cases} \frac{1}{2}x - 1 & -4 \le x < 2 \\ 3x - 6 & x \ge 2 \end{cases}$

28.

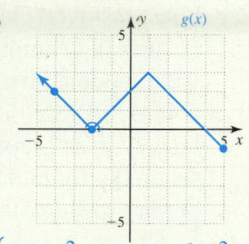

$g(x) = \begin{cases} -x - 2 & x \le -2 \\ -|x - 1| + 3 & -2 < x \le 5 \end{cases}$

29.

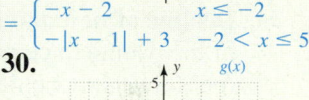

$p(x) = \begin{cases} x^2 + 2x - 3 & x \le 1 \\ x + 1 & x > 1 \end{cases}$

30.

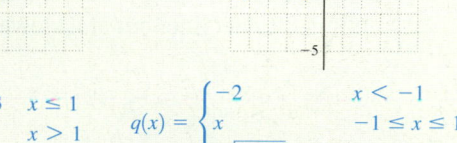

$q(x) = \begin{cases} -2 & x < -1 \\ x & -1 \le x \le 1 \\ \sqrt{x - 1} + 1 & x > 1 \end{cases}$

▶ WORKING WITH FORMULAS

31. Definition of absolute value: $|x| = \begin{cases} -x & x < 0 \\ x & x \ge 0 \end{cases}$

The absolute value function can be stated as a piecewise-defined function, a technique that is sometimes useful in graphing variations of the function or solving absolute value equations and inequalities. How does this definition ensure that the absolute value of a number is always positive? Use this definition to help sketch the graph of $f(x) = \frac{|x|}{x}$. Discuss what you notice.

32. Sand dune function:

$f(x) = \begin{cases} -|x - 2| + 1 & 1 \le x < 3 \\ -|x - 4| + 1 & 3 \le x < 5 \\ -|x - 2k| + 1 & 2k - 1 \le x < 2k + 1, \text{ for } k \in N \end{cases}$

There are a number of interesting graphs that can be created using piecewise-defined functions, and these functions have been the basis for more than one piece of modern art. (a) Use the descriptive name and the pieces given to graph the function f. Is the function accurately named? (b) Use any combination of the toolbox functions to explore your own creativity by creating a piecewise-defined function with some interesting or appealing characteristics. (c) For $y = -|x - 2| + 1$, solve for x in terms of y.   **a.** yes   **b.** Answers will vary

c. $x = 3 - y$ or $x = y + 1$

Additional answers can be found in the Instructor Answer Appendix.

► APPLICATIONS

For Exercises 33 and 34, (a) write the information given as a piecewise-defined function, and state the domain for each piece by inspecting the graph. (b) Give the range of each.

33. Results from advertising:
Due to heavy advertising, initial sales of the Lynx Digital Camera grew very rapidly, but started to decline once the advertising blitz was over. During the advertising campaign, sales were modeled by the function $S(t) = -t^2 + 6t$, where $S(t)$ represents hundreds of sales in month t. However, as Lynx Inc. had hoped, the new product secured a foothold in the market and sales leveled out at a steady 500 sales per month.

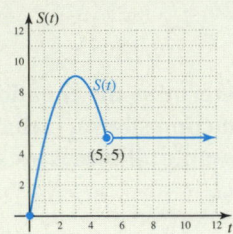

34. Decline of newspaper publishing: From the turn of the twentieth century, the number of newspapers (per thousand population) grew rapidly until the 1930s, when the growth slowed down and then declined. The years 1940 to 1946 saw a "spike" in growth, but the years 1947 to 1954 saw an almost equal decline. Since 1954 the number has continued to decline, but at a slower rate.

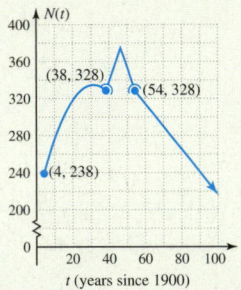

The number of papers N per thousand population for each period, respectively, can be approximated by

$N_1(t) = -0.13t^2 + 8.1t + 208$,

$N_2(t) = -5.75|t - 46| + 374$, and

$N_3(t) = -2.45t + 460$.

Source: Data from the *Statistical Abstract of the United States*, various years; data from *The First Measured Century, The AEI Press*, Caplow, Hicks, and Wattenberg, 2001.

 35. Families that own stocks: The percentage of American households that own publicly traded stocks began rising in the early 1950s, peaked in 1970, then began to decline until 1980 when there was a dramatic increase due to easy access over the Internet, an improved economy, and other factors. This phenomenon is modeled by the function $P(t)$,

where $P(t)$ represents the percentage of households owning stock in year t, with 1950 corresponding to year 0.

$$P(t) = \begin{cases} -0.03t^2 + 1.28t + 1.68 & 0 \le t \le 30 \\ 1.89t - 43.5 & t > 30 \end{cases}$$

a. According to this model, what percentage of American households held stock in the years 1955, 1965, 1975, 1985, and 1995? If this pattern continues, what percentage held stock in 2005? What percent will hold stock in 2015?

b. Why is there a discrepancy in the outputs of each piece of the function for the year 1980 ($t = 30$)? According to how the function is defined, which output should be used?

Source: 2004 *Statistical Abstract of the United States*, Table 1204; various other years.

 36. Dependence on foreign oil: America's dependency on foreign oil has always been a "hot" political topic, with the amount of imported oil fluctuating over the years due to political climate, public awareness, the economy, and other factors. The amount of crude oil imported can be approximated by the function given, where $A(t)$ represents the number of barrels imported in year t (in billions), with 1980 corresponding to year 0.

$$A(t) = \begin{cases} 0.047t^2 - 0.38t + 1.9 & 0 \le t < 8 \\ -0.075t^2 + 1.495t - 5.265 & 8 \le t \le 11 \\ 0.133t + 0.685 & t > 11 \end{cases}$$

a. Use $A(t)$ to estimate the number of barrels imported in the years 1983, 1989, 1995, and 2005. If this trend continues, how many barrels will be imported in 2015?   5.34 billion barrels

b. What was the minimum number of barrels imported between 1980 and 1988?

Source: 2004 *Statistical Abstract of the United States*, Table 897; various other years.

37. Energy rationing: In certain areas of the United States, power blackouts have forced some counties to ration electricity. Suppose the cost is $0.09 per kilowatt (kW) for the first 1000 kW a household uses. After 1000 kW, the cost increases to 0.18 per kW. (a) Write these charges for electricity in the form of a piecewise-defined function $C(h)$, where $C(h)$ is the cost for h kilowatt hours. Include the domain for each piece. Then (b) sketch the graph and determine the cost for 1200 kW.

38. Water rationing: Many southwestern states have a limited water supply, and some state governments try to control consumption by manipulating the cost of water usage. Suppose for the first 5000 gal a household uses per month, the charge is $0.05 per gallon. Once 5000 gal is used the charge doubles to $0.10 per gallon. (a) Write these charges for water usage in the form of a piecewise-defined function $C(w)$, where $C(w)$ is the cost for w gallons of water. Include the domain for each piece. Then (b) sketch the graph and determine the cost to a household that used 9500 gal of water during a very hot summer month.

39. Pricing for natural gas: A local gas company charges $0.75 per therm for natural gas, up to 25 therms. Once the 25 therms has been exceeded, the charge doubles to $1.50 per therm due to limited supply and great demand. (a) Write these charges for natural gas consumption in the form of a piecewise-defined function $C(t)$, where $C(t)$ is the charge for t therms. Include the domain for each piece. Then (b) sketch the graph and determine the cost to a household that used 45 therms during a very cold winter month.

40. Multiple births: The number of multiple births has steadily increased in the United States during the twentieth century and beyond. Between 1985 and 1995 the number of twin births could be modeled by the function $T(x) = -0.21x^2 + 6.1x + 52$, where x is the number of years since 1980 and T is in thousands. After 1995, the incidence of twins becomes more linear, with $T(x) = 4.53x + 28.3$ serving as a better model. (a) Write the piecewise-defined function modeling the incidence of twins for these years. Include the domain of each piece. Then (b) sketch the graph and use the function to estimate the incidence of twins in 1990, 2000, and 2005. If this trend continued, how many sets of twins were born in 2010?

Source: National Vital Statistics Report, Vol. 50, No. 5, February 12, 2002

41. U.S. military expenditures: Except for the year 1991 when military spending was cut drastically, the amount spent by the U.S. government on national defense and veterans' benefits rose steadily from 1980 to 1992. These expenditures can be modeled by the function $S(t) = -1.35t^2 + 31.9t + 152$, where $S(t)$ is in billions of dollars and 1980 corresponds to $t = 0$.

From 1992 to 1996 this spending declined, then began to rise in the following years. From 1992 to 2002, military-related spending can be modeled by $S(t) = 2.5t^2 - 80.6t + 950$.

Source: 2004 Statistical Abstract of the United States, Table 492

(a) Write $S(t)$ as a single piecewise-defined function. Include stating the domain for each piece. Then (b) sketch the graph and use the function to estimate the amount spent by the United States in 2005, 2008, and 2012 if this trend continues.

42. Amusement arcades: At a local amusement center, the owner has the SkeeBall machines programmed to reward very high scores. For scores of 200 or less, the function $T(x) = \frac{x}{10}$ models the number of tickets awarded (rounded to the nearest whole). For scores over 200, the number of tickets is modeled by $T(x) = 0.001x^2 - 0.3x + 40$. (a) Write these equation models of the number of tickets awarded in the form of a piecewise-defined function. Include the domain for each piece. Then (b) sketch the graph and find the number of tickets awarded to a person who scores 390 points.

43. Phone service charges: When it comes to phone service, a large number of calling plans are available. Under one plan, the first 30 min of any phone call costs only 3.3¢ per minute. The charge increases to 7¢ per minute thereafter. (a) Write this information in the form of a piecewise-defined function. Include the domain for each piece. Then (b) sketch the graph and find the cost of a 46-min phone call.

44. Overtime wages: Tara works on an assembly line, putting together computer monitors. She is paid $9.50 per hour for regular time (0, 40 hr], $14.25 for overtime (40, 48 hr], and when demand for computers is high, $19.00 for double-overtime (48, 84 hr]. (a) Write this information in the form of a simplified piecewise-defined function. Include the domain for each piece. (b) Then sketch the graph and find the gross amount of Tara's check for the week she put in 54 hr.

45. Admission prices: At Wet Willy's Water World, infants under 2 are free, then admission is charged according to age. Children 2 and older but less than 13 pay $2, teenagers 13 and older but less than 20 pay $5, adults 20 and older but less than 65 pay $7, and senior citizens 65 and older get in at the teenage rate. (a) Write this information in the form of a piecewise-defined function. Include the domain for each piece. Then (b) sketch the graph and find the cost of admission for a family of nine which includes: one grandparent (70), two adults (44/45), 3 teenagers, 2 children, and one infant.

Additional answers can be found in the Instructor Answer Appendix.

46. Demographics: One common use of the floor function $y = \lfloor x \rfloor$ is the reporting of ages. As of 2007, the record for longest living human is 122 yr, 164 days for the life of Jeanne Calment, formerly of France. While she actually lived $x = 122\frac{164}{365}$ years, ages are normally reported using the floor function, or the greatest integer number of years less than or equal to the actual age: $\lfloor 122\frac{164}{365} \rfloor = 122$ years. (a) Write a function $A(t)$ that gives a person's age, where $A(t)$ is the reported age at time t. (b) State the domain of the function (be sure to consider Madame Calment's record). Report the age of a person who has been living for (c) 36 years; (d) 36 years, 364 days; (e) 37 years; and (f) 37 years, 1 day.

47. Postage rates: The postal charge function from Example 9 is simply a transformation of the basic ceiling function $y = \lceil x \rceil$. Using the ideas from Section 2.2, (a) write the postal charges as a step function $C(w)$, where $C(w)$ is the cost of mailing a large envelope weighing w ounces, and (b) state the domain of the function. Then use the function to find the cost of mailing reports weighing: (c) 0.7 oz, (d) 5.1 oz, (e) 5.9 oz; (f) 6 oz, and (g) 6.1 oz.

48. Cell phone charges: A national cell phone company advertises that calls of 1 min or less do not count toward monthly usage. Calls lasting longer than 1 min are calculated normally using a ceiling function, meaning a call of 1 min, 1 sec will be counted as a 2-min call. Using the ideas from Section 2.2, (a) write the cell phone charges as a piecewise-defined function $C(m)$, where $C(m)$ is the cost of a call lasting m minutes, and include the domain of the function. Then (b) graph the function, and (c) use the graph or function to determine if a cell phone subscriber has exceeded the 30 free minutes granted by her calling plan for calls lasting 2 min 3 sec, 13 min 46 sec, 1 min 5 sec, 3 min 59 sec, and 8 min 2 sec. (d) What was the actual usage in minutes and seconds?

49. Combined absolute value graphs: Carefully graph the function $h(x) = |x - 2| - |x + 3|$ using a table of values over the interval $x \in [-5, 5]$. Is the function continuous? Write this function in piecewise-defined form and state the domain for each piece.

50. Combined absolute value graphs: Carefully graph the function $H(x) = |x - 2| + |x + 3|$ using a table of values over the interval $x \in [-5, 5]$. Is the function continuous? Write this function in piecewise-defined form and state the domain for each piece.

▶ **EXTENDING THE CONCEPT**

51. You've heard it said, "*any number divided by itself is one.*" Consider the functions $f(x) = \frac{x + 2}{x + 2}$, and $g(x) = \frac{|x + 2|}{x + 2}$. Are these functions continuous?

$f(x)$ has a removable discontinuity at $x = -2$;
$g(x)$ has a discontinuity at $x = -2$

52. Find a linear function $h(x)$ that will make the function shown a *continuous* function. Be sure to include its domain. $h(x) = 4x - 3, 1 \le x \le 3$

$$f(x) = \begin{cases} x^2 & x < 1 \\ h(x) & \\ 2x + 3 & x > 3 \end{cases}$$

▶ **MAINTAINING YOUR SKILLS**

53. (R.5) Solve: $\dfrac{3}{x - 2} + 1 = \dfrac{30}{x^2 - 4}$. $x = -7, x = 4$

54. (R.5) Compute the following and write the result in lowest terms:

$$\frac{x^3 + 3x^2 - 4x - 12}{x - 3} \cdot \frac{2x - 6}{x^2 + 5x + 6} \div (3x - 6)$$

55. (1.4) Find an equation of the line perpendicular to $3x + 4y = 8$, and through the point $(0, -2)$. Write the result in slope-intercept form. $y = \frac{4}{3}x - 2$

56. (R.6/1.1) For the figure shown, (a) use the Pythagorean Theorem to find the length of the missing side and (b) state the area of the triangular side.   **a.** $4\sqrt{5}$ cm   **b.** $16\sqrt{5}$ cm²

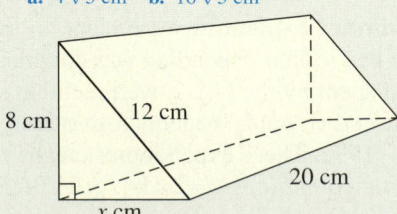

2.6 Variation: The Toolbox Functions in Action

LEARNING OBJECTIVES

In Section 2.6 you will see how we can:

- ☐ **A.** Solve direct variations
- ☐ **B.** Solve inverse variations
- ☐ **C.** Solve joint variations

A study of direct and inverse variation offers perhaps our clearest view of how mathematics is used to model real-world phenomena. While the basis of our study is elementary, involving only the toolbox functions, the applications are at the same time elegant, powerful, and far reaching. In addition, these applications unite some of the most important ideas in algebra, including functions, transformations, rates of change, and graphical analysis, to name a few.

A. Toolbox Functions and Direct Variation

If a car gets 24 miles per gallon (mpg) of gas, we could express the distance d it could travel as $d = 24g$. Table 2.4 verifies the distance traveled by the car changes in *direct* or *constant proportion* to the number of gallons used, and here we say, "distance traveled *varies directly* with gallons used." The equation $d = 24g$ is called a **direct variation,** and the coefficient 24 is called the **constant of variation.**

Using the rate of change notation, $\dfrac{\Delta \text{distance}}{\Delta \text{gallons}} = \dfrac{\Delta d}{\Delta g} = \dfrac{24}{1}$, and we note

this is actually a *linear equation* with slope $m = 24$. When working with variations, the constant k is preferred over m, and in general we have the following:

Table 2.4

g	d
1	24
2	48
3	72
4	96

Direct Variation

y varies directly with x, or *y is directly proportional to x*, if there is a nonzero constant k such that

$$y = kx.$$

k is called the *constant of variation*

EXAMPLE 1 ▶ **Writing a Variation Equation**

Write the variation equation for these statements:

   a. Wages earned varies directly with the number of hours worked.
   b. The value of an office machine varies directly with time.
   c. The circumference of a circle varies directly with the length of the diameter.

Solution ▶   **a.** **W**ages varies directly with **h**ours worked: $W = kh$
   b. The **V**alue of an office machine varies directly with **t**ime: $V = kt$
   c. The **C**ircumference varies directly with the **d**iameter: $C = kd$

Now try Exercises 7 through 10 ▶

Once we determine the relationship between two variables is a direct variation, we try to find the value of k and develop an equation model that can more generally be applied. Note that "varies directly" indicates that one value is a constant multiple of the other. In Example 1, you may have realized that if any one relationship between the variables is known, we can solve for k by substitution. For instance, if the circumference of a circle is 314 cm when the diameter is 100 cm, $C = kd$ becomes $314 = k(100)$ and division shows $k = 3.14$ (our estimate for π). The result is a formula for the circumference of *any* circle. This suggests the following procedure:

Solving Applications of Variation

1. Translate the information given into an equation model, using k as the constant of variation.
2. Substitute the first relationship (pair of values) given and solve for k.
3. Substitute this value for k in the original model to obtain the variation equation.
4. Use the variation equation to complete the application.

EXAMPLE 2 ▶ **Solving an Application of Direct Variation**

The weight of an astronaut on the surface of another planet **varies directly** with their weight on Earth. An astronaut weighing 140 lb on Earth weighs only 53.2 lb on Mars. How much would a 170-lb astronaut weigh on Mars?

Solution ▶

1. $M = kE$ "**M**ars weight varies directly with **E**arth weight"
2. $53.2 = k(140)$ substitute 53.2 for M and 140 for E
     $k = 0.38$ solve for k (constant of variation)

Substitute this value of k in the original equation to obtain the variation equation, then find the weight of a 170-lb astronaut that landed on Mars.

3. $M = 0.38E$ variation equation
4. $= 0.38(170)$ substitute 170 for E
     $= 64.6$ result

An astronaut weighing 170 lb on Earth weighs only 64.6 lb on Mars.

Now try Exercises 11 through 14 ▶

The toolbox function from Example 2 was a line with slope $k = 0.38$, or $k = \frac{19}{50}$ as a fraction in simplest form. As a rate of change, $k = \frac{\Delta M}{\Delta E} = \frac{19}{50}$, and we see that for every 50 additional pounds on Earth, the weight of an astronaut would increase by only 19 lb on Mars.

EXAMPLE 3 ▶ **Making Estimates from the Graph of a Variation**

The scientists at NASA are planning to send additional probes to the red planet (Mars), that will weigh from 250 to 450 lb. Graph the variation equation from Example 2, then *use the graph* to estimate the corresponding range of weights on Mars. Check your estimate using the variation equation.

Solution ▶

After selecting an appropriate scale, begin at $(0, 0)$ and count off the slope $k = \frac{\Delta M}{\Delta E} = \frac{19}{50}$. This gives the points $(50, 19)$, $(100, 38)$, $(200, 76)$, and so on. From the graph (see dashed arrows), it appears the weights corresponding to 250 lb and 450 lb on Earth are near 95 lb and 170 lb on Mars. Using the equation gives

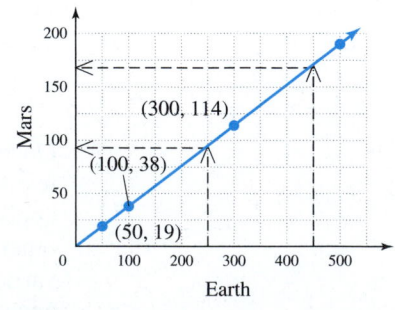

$M = 0.38E$ variation equation
$\ \ \ = 0.38(250)$ substitute 250 for E
$\ \ \ = 95,$
$M = 0.38E$ variation equation
$\ \ \ = 0.38(450)$ substitute 450 for E
$\ \ \ = 171,$ very close to our estimate from the graph.

Now try Exercises 15 and 16 ▶

When toolbox functions are used to model variations, our knowledge of their graphs and defining characteristics strengthens a contextual understanding of the application. Consider Examples 4 and 5, where the squaring function is used.

EXAMPLE 4 ▶ **Writing Variation Equations**

Write the variation equation for these statements:

a. In free fall, the distance traveled by an object varies directly with the square of the time.

b. The area of a circle varies directly with the square of its radius.

Solution ▶ **a.** Distance varies directly with the square of the time: $D = kt^2$.

b. Area varies directly with the square of the radius: $A = kr^2$.

Now try Exercises 17 through 20 ▶

Both variations in Example 4 use the squaring function, where k represents the amount of stretch or compression applied, and whether the graph will open upward or downward. However, regardless of the function used, the four-step solution process remains the same.

EXAMPLE 5 ▶ **Solving an Application of Direct Variation**

The range of a projectile varies directly with the square of its initial velocity. As part of a circus act, Bailey the Human Bullet is shot out of a cannon with an initial velocity of 80 feet per second (ft/sec), into a net 200 ft away.

a. Find the constant of variation and write the variation equation.

b. Graph the equation and *use the graph* to estimate how far away the net should be placed if initial velocity is increased to 95 ft/sec.

c. Determine the accuracy of the estimate from (b) using the variation equation.

Solution ▶ **a. 1.** $R = kv^2$ "Range varies directly with the square of the velocity"

   2. $200 = k(80)^2$ substitute 200 for *R* and 80 for *v*

       $k = 0.03125$ solve for *k* (constant of variation)

   3. $R = 0.03125v^2$ variation equation (substitute 0.03125 for *k*)

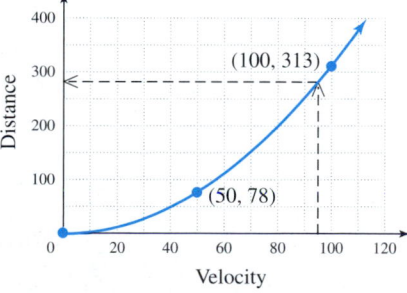

b. Since velocity and distance are positive, we again use only QI. The graph is a parabola that opens upward, with the vertex at (0, 0). Selecting velocities from 50 to 100 ft/sec, we have:

$R = 0.03125v^2$ variation equation

$\quad = 0.03125(50)^2$ substitute 50 for *v*

$\quad = 78.125$ result

Likewise substituting 100 for v gives $R = 312.5$ ft. Scaling the axes and using (0, 0), (50, 78), and (100, 313) produces the graph shown. At 95 ft/sec (dashed lines), it appears the net should be placed about 280 ft away.

c. Using the variation equation gives:

   4. $R = 0.03125v^2$ variation equation

       $\quad = 0.03125(95)^2$ substitute 95 for *v*

       $R = 282.03125$ result

Our estimate was off by about 2 ft. The net should be placed about 282 ft away.

Now try Exercises 21 through 26 ▶

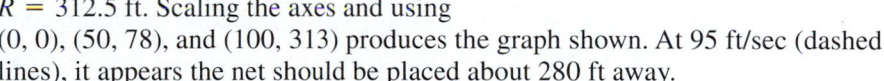

We now have a complete picture of this relationship, in which the required information can be presented graphically (Figure 2.93), numerically (Figure 2.94), verbally, and in equation form. This enables the people requiring the information, i.e., Bailey himself (for obvious reasons) and the Circus Master who is responsible, to make more informed (and safe) decisions.

Figure 2.93

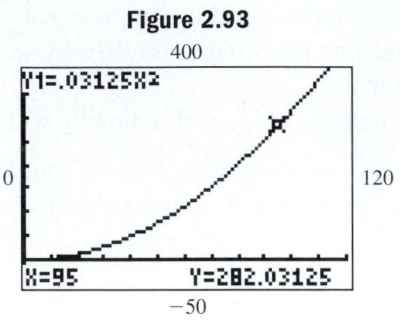

Figure 2.94

$R = 0.03125v^2$

Range R varies as the square of the velocity

✓ **A. You've just seen how we can solve direct variations**

Note: For Examples 7 and 8, the four steps of the solution process were used in sequence, but not numbered.

B. Inverse Variation

Numerous studies have been done that relate the price of a commodity to the demand—the willingness of a consumer to pay that price. For instance, if there is a sudden increase in the price of a popular tool, hardware stores know there will be a corresponding decrease in the demand for that tool. The question remains, "What is this rate of decrease?" Can it be modeled by a linear function with a negative slope? A parabola that opens downward? Some other function? Table 2.5 shows some (simulated) data regarding price versus demand. It appears that a linear function is not appropriate because the rate of change in the number of tools sold is not constant. Likewise a quadratic model seems inappropriate, since we don't expect demand to suddenly start rising again as the price continues to increase. This phenomenon is actually an example of **inverse variation,** modeled by a transformation of the reciprocal function $y = \frac{k}{x}$. We will often rewrite the equation as $y = k\left(\frac{1}{x}\right)$ to clearly see the inverse relationship. In the case at hand, we might write $D = k\left(\frac{1}{P}\right)$, where k is the constant of variation, D represents the demand for the product, and P the price of the product. In words, we say that "demand *varies inversely* as the price." In other applications of inverse variation, one quantity may vary inversely as the *square* of another (Example 6(b)), and in general we have

Table 2.5

Price (dollars)	Demand (1000s)
8	288
9	144
10	96
11	72
12	57.6

> **Inverse Variation**
>
> *y varies inversely with x,* or *y is inversely proportional to x,* if there is a nonzero constant k such that
>
> $$y = k\left(\frac{1}{x}\right).$$
>
> k is called the *constant of variation*

EXAMPLE 6 ▶ **Writing Inverse Variation Equations**

Write the variation equation for these statements:

a. In a closed container, pressure varies inversely with the volume of gas.

b. The intensity of light varies inversely with the square of the distance from the source.

Solution ▶ **a.** Pressure varies inversely with the Volume of gas: $P = k\left(\frac{1}{V}\right)$.

b. Intensity of light varies inversely with the square of the distance: $I = k\left(\frac{1}{d^2}\right)$.

Now try Exercises 27 through 30 ▶

EXAMPLE 7 ▶ **Solving an Application of Inverse Variation**

Boyle's law tells us that in a closed container with constant temperature, the volume of a gas varies inversely with the pressure applied (see illustration). Suppose the air pressure in a closed cylinder is 50 pounds per square inch (psi) when the volume of the cylinder is 60 in³.

a. Find the constant of variation and write the variation equation.

b. Use the equation to find the volume, if the pressure is increased to 150 psi.

Solution ▶ **a.** $V = k\left(\dfrac{1}{P}\right)$ "volume varies inversely with the **p**ressure"

Illustration of Boyle's Law

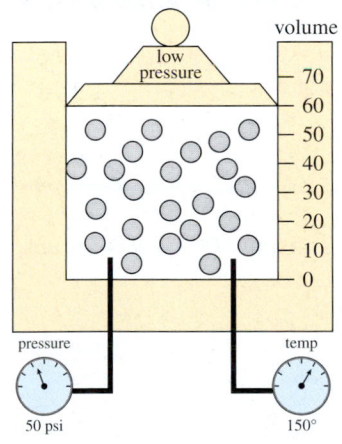

$60 = k\left(\dfrac{1}{50}\right)$ substitute 60 for V and 50 for P.

$k = 3000$ constant of variation

$V = 3000\left(\dfrac{1}{P}\right)$ variation equation (substitute 3000 for k)

b. Using the variation equation we have:

$V = 3000\left(\dfrac{1}{P}\right)$ variation equation

$\quad = 3000\left(\dfrac{1}{150}\right)$ substitute 150 for P

$\quad = 20$ result

When the pressure is increased to 150 psi, the volume decreases to 20 in³.

Now try Exercises 31 through 34 ▶

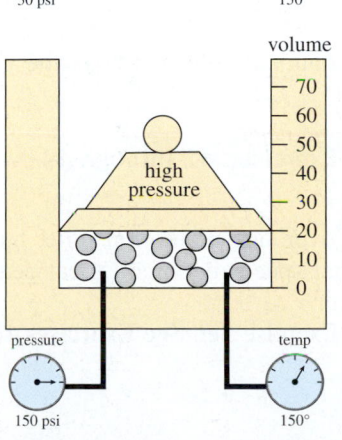

As an application of the reciprocal function, the relationship in Example 7 is easily graphed as a transformation of $y = \dfrac{1}{x}$. Using an appropriate scale and values in QI, only a vertical stretch of 3000 is required and the result is shown in Figure 2.95. As noted, when the pressure increases the volume decreases, or in notation: as $P \to \infty$, $V \to 0$. Applications of this sort can be as sophisticated as the manufacturing of industrial pumps and synthetic materials, or as simple as cooking a homemade dinner. Simply based on the equation, how much pressure is required to reduce the volume of gas to 1 in³?

Figure 2.95

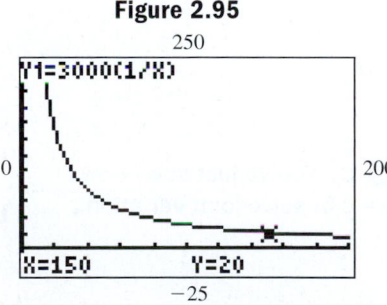

☑ **B.** You've just seen how we can solve inverse variations

C. Joint or Combined Variations

Just as some decisions might be based on many considerations, often the relationship between two variables depends on a combination of factors. Imagine a wooden plank laid across the banks of a stream for hikers to cross the streambed (see Figure 2.96). The amount of weight the plank will support depends

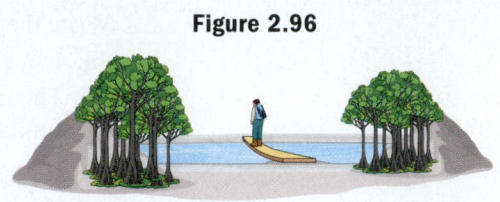

Figure 2.96

on the type of wood, the width and height of the plank's cross section, and the distance between the supported ends **(see Exercises 59 and 60).** This is an example of a **joint variation,** which can combine any number of variables in different ways. Two general possibilities are: (1) *y varies jointly with the product of x and p:* $y = kxp$; and (2) *y varies jointly with the product of x and p, and inversely with the square of q:* $y = kxp(\frac{1}{q^2})$. For practice writing joint variations as an equation model, **see Exercises 35 through 40.**

EXAMPLE 8 ▶   Solving an Application of Joint Variation

The amount of fuel used by a certain ship traveling at a uniform speed varies jointly with the distance it travels and the square of the velocity. If 200 barrels of fuel are used to travel 10 mi at 20 nautical miles per hour, how far does the ship travel on 500 barrels of fuel at 30 nautical miles per hour?

Solution ▶

$$F = kdv^2 \qquad \text{"Fuel use \textit{varies jointly} with \textbf{d}istance and \textbf{v}elocity squared"}$$
$$200 = k(10)(20)^2 \qquad \text{substitute 200 for } F, \text{10 for } d, \text{ and 20 for } v$$
$$200 = 4000k \qquad \text{simplify and solve for } k$$
$$0.05 = k \qquad \text{constant of variation}$$
$$F = 0.05dv^2 \qquad \text{equation of variation}$$

To find the distance traveled at 30 nautical miles per hour using 500 barrels of fuel, substitute 500 for F and 30 for v:

$$F = 0.05dv^2 \qquad \text{equation of variation}$$
$$500 = 0.05d(30)^2 \qquad \text{substitute 500 for } F \text{ and 30 for } v$$
$$500 = 45d \qquad \text{simplify}$$
$$11.\overline{1} = d \qquad \text{result}$$

If 500 barrels of fuel are consumed while traveling 30 nautical miles per hour, the ship covers a distance of just over 11 mi.

Now try Exercises 41 through 44 ▶

It's interesting to note that the ship covers just over one additional mile, but consumes 2.5 times the amount of fuel. The additional speed requires a great deal more fuel.

There is a variety of additional applications in the Exercise Set. **See Exercises 47 through 55.**

☑ **C. You've just seen how we can solve joint variations**

2.6 EXERCISES

▶ CONCEPTS AND VOCABULARY

Fill in each blank with the appropriate word or phrase. Carefully reread the section if needed.

1. The phrase "*y* varies directly with *x*" is written
$y = kx$, where *k* is called the ___constant___ of
variation.

2. If more than two quantities are related in a
variation equation, the result is called a ___joint___
variation.

3. For a right circular cylinder, $V = \pi r^2 h$ and we say,
the volume varies ___directly___ with the ___height___
and the ___square___ of the radius.

4. The statement "*y* varies inversely with the square
of *x*" is written ___$y = \frac{k}{x^2}$___.

5. Discuss/Explain the general procedure for solving
applications of variation. Include references to
keywords, and illustrate using an example.
Answers will vary.

6. The basic percent formula is *amount equals
percent times base*, or $A = PB$. In words, write this
out as a direct variation with *B* as the constant of
variation, then as an inverse variation with the
amount *A* as the constant of variation.
Answers will vary.

▶ DEVELOPING YOUR SKILLS

Write the variation equation for each statement.

7. distance traveled varies directly with rate of speed
$d = kr$

8. cost varies directly with the quantity purchased
$c = kq$

9. force varies directly with acceleration $F = ka$

10. length of a spring varies directly with attached
weight $l = kw$

**For Exercises 11 and 12, find the constant of variation
and write the variation equation. Then use the equation
to complete the table.**

11. *y* varies directly with *x*; $y = 0.6$ when $x = 24$.

$y = 0.025x$

x	*y*
500	12.5
650	16.25
750	18.75

12. *w* varies directly with *v*; $w = \frac{1}{3}$ when $v = 5$.

$w = \frac{1}{15}v$

v	*w*
291	19.4
327	21.8
339	22.6

13. Wages and hours worked: Wages earned varies
directly with the number of hours worked. Last
week I worked 37.5 hr and my gross pay was
$344.25. Write the variation equation and
determine how much I will gross this week if I
work 35 hr. What does the value of *k* represent in
this case? $w = 9.18h$; $321.30; the hourly wage; $k = $9.18/hr$

14. Pagecount and thickness of books: The thickness
of a paperback book varies directly as the number of
pages. A book 3.2 cm thick has 750 pages. Write the
variation equation and approximate the thickness of
Roget's 21st Century Thesaurus (paperback—2nd
edition), which has 957 pages. $T = \frac{8}{1875}P$; 4.1 cm

15. Building height and number of stairs: The
number of stairs in the stairwells of tall buildings
and other structures varies directly as the height of
the structure. The base and pedestal for the Statue
of Liberty are 47 m tall, with 192 stairs from
ground level to the observation deck at the top of
the pedestal (at the statue's feet). (a) Find the
constant of variation and write the variation
equation, (b) graph the variation equation, (c) use
the graph to estimate the number of stairs from
ground level to the observation deck in the statue's
crown 81 m above ground level, and (d) use the
equation to check this estimate. Was it close?
a. $k = \frac{192}{47}$, $S = \frac{192}{47}h$ **c.** 330 stairs   **d.** $S = 331$; yes

Additional answers can be found in the Instructor Answer Appendix.

HOMEWORK SELECTION GUIDE

Core: 7–43 odd, 47–59 odd, 64 (27 Exercises)
Standard: 1–4, 7–59 odd, 62, 65, 66 (34 Exercises)

Extended: 1–4, 7–59 odd, 60, 61, 62, 65, 66 (36 Exercises)
In Depth: 1–6, 7–45 odd, 46, 47–59 odd, 60, 61–66 (41 Exercises)

16. **Projected images:** The height of a projected image varies directly as the distance of the projector from the screen. At a distance of 48 in., the image on the screen is 16 in. high. (a) Find the constant of variation and write the variation equation, (b) graph the variation equation, (c) use the graph to estimate the height of the image if the projector is placed at a distance of 5 ft 3 in., and (d) use the equation to check this estimate. Was it close?
a. $k = \frac{1}{3}; h = \frac{1}{3}d$ **c.** 21 in. **d.** $h = 21$; yes

Write the variation equation for each statement.

17. Surface area of a cube varies directly with the square of a side. $A = kS^2$

18. Potential energy in a spring varies directly with the square of the distance the spring is compressed.
    $E = kd^2$
19. Electric power varies directly with the square of the current (amperes). $P = kc^2$

20. Manufacturing cost varies directly as the square of the number of items made. $C = km^2$

For Exercises 21 and 22, find the constant of variation and write the variation equation. Then use the equation to complete the table.

21. p varies directly with the square of q; $p = 280$ when $q = 50$ $k = 0.112; p = 0.112q^2$

q	p
45	226.8
55	338.8
70	548.8

22. n varies directly with m squared; $n = 24.75$ when $m = 30$ $k = 0.0275; n = 0.0275\,m^2$

m	n
40	44
60	99
88	212.96

For Exercises 23 to 26, supply the relationship indicated (a) in words, (b) in equation form, (c) graphically, and (d) in table form, then (e) solve the application.

23. **The Borg Collective:** The surface area of a cube varies directly as the square of one side. A cube with sides of $14\sqrt{3}$ cm has a surface area of 3528 cm². Find the surface area in square meters of the spaceships used by the Borg Collective in *Star Trek—The Next Generation,* cubical spacecraft with sides of 3036 m.

24. **Geometry and geography:** The area of an equilateral triangle varies directly as the square of one side. A triangle with sides of 50 yd has an area of 1082.5 yd². Find the area in mi² of the region bounded by straight lines connecting the cities of Cincinnati, Ohio, Washington, D.C., and Columbia, South Carolina, which are each approximately 400 mi apart.

25. **Galileo and gravity:** The distance an object falls varies directly as the square of the time it has been falling. The cannonballs dropped by Galileo from the Leaning Tower of Pisa fell about 169 ft in 3.25 sec. How long would it take a hammer, accidentally dropped from a height of 196 ft by a bridge repair crew, to splash into the water below? According to the equation, if a camera accidentally fell out of the *News 4 Eye-in-the-Sky* helicopter and hit the ground in 2.75 sec, how high was the helicopter?

26. **Soap bubble surface area:** When a child blows small soap bubbles, they come out in the form of a sphere because the surface tension in the soap seeks to minimize the surface area. The surface area of any sphere varies directly with the square of its radius. A soap bubble with a $\frac{3}{4}$ in. radius has a surface area of approximately 7.07 in². What is the radius of a seventeenth-century cannonball that has a surface area of 113.1 in²? What is the surface area of an orange with a radius of $1\frac{1}{2}$ in.?

Write the variation equation for each statement.

27. The force of gravity varies inversely as the square of the distance between objects. $F = \frac{k}{d^2}$

28. Pressure varies inversely as the area over which it is applied. $P = \frac{k}{A}$

29. The safe load of a beam supported at both ends varies inversely as its length. $S = \frac{k}{L}$

30. The intensity of sound varies inversely as the square of its distance from the source. $I = \frac{k}{d^2}$

For Exercises 31 through 34, find the constant of variation and write the variation equation. Then use the equation to complete the table or solve the application.

31. Y varies inversely as the square of Z; $Y = 1369$ when $Z = 3$

Z	Y
37	9
74	2.25
111	1

$Y = \dfrac{12{,}321}{Z^2}$

32. A varies inversely with B; $A = 2450$ when $B = 0.8$

$$A = \frac{1960}{B}$$

B	A
140	14
320	6.125
560	3.5

33. Gravitational force: The effect of Earth's gravity on an object (its weight) varies inversely as the square of its distance from the center of the planet (assume the Earth's radius is 6400 km). If the weight of an astronaut is 75 kg on Earth (when $r = 6400$), what would this weight be at an altitude of 1600 km *above the surface* of the Earth?　$w = \frac{3{,}072{,}000{,}000}{r^2}$; 48 kg

34. Popular running shoes: The demand for a popular new running shoe varies inversely with the price of the shoes. When the wholesale price is set at $45, the manufacturer ships 5500 orders per week to retail outlets. Based on this information, how many orders would be shipped per week if the wholesale price rose to $55?　$d = \frac{247{,}500}{C}$; 4500 orders

Write the variation equation for each statement.

35. Interest earned varies jointly with the rate of interest and the length of time on deposit.　$I = krt$

36. Horsepower varies jointly as the number of cylinders in the engine and the square of the cylinder's diameter.　$H = knd^2$

37. The area of a trapezoid varies jointly with its height and the sum of the bases.　$A = kh(B + b)$

38. The area of a triangle varies jointly with its base and its height.　$A = kbh$

39. The volume of metal in a circular coin varies directly with the thickness of the coin and the square of its radius.　$V = ktr^2$

40. The electrical resistance in a wire varies directly with its length and inversely as the cross-sectional area of the wire.　$R = \frac{kL}{A}$

For Exercises 41–44, find the constant of variation and write the related variation equation. Then use the equation to complete the table or solve the application.

41. C varies jointly with R and inversely with S squared, and $C = 21$ when $R = 7$ and $S = 1.5$.

$$C = \frac{6.75R}{S^2}$$

R	S	C
120	6	22.5
200	12.5	8.64
350	15	10.5

42. J varies jointly with P and inversely with the square root of Q, and $J = 19$ when $P = 4$ and $Q = 25$.

$$J = \frac{23.75P}{\sqrt{Q}}$$

P	Q	J
47.5	90.25	118.75
112	31.36	475
18.76	44.89	66.5

43. Kinetic energy: Kinetic energy (energy attributed to motion) varies jointly with the mass of the object and the square of its velocity. Assuming a unit mass of $m = 1$, an object with a velocity of 20 m per sec (m/s) has kinetic energy of 200 J. How much energy is produced if the velocity is increased to 35 m/s?　$E = 0.5mv^2$; 612.50 J

44. Safe load: The load that a horizontal beam can support varies jointly as the width of the beam, the square of its height, and inversely as the length of the beam. A beam 4 in. wide and 8 in. tall can safely support a load of 1 ton when the beam has a length of 12 ft. How much could a similar beam 10 in. tall safely support?　$S = \frac{3wh^2}{64l}$; 1.5625 tons

▶ **WORKING WITH FORMULAS**

45. Required interest rate: $R(A) = \sqrt[3]{A} - 1$

To determine the simple interest rate R that would be required for each dollar ($1) left on deposit for 3 yr to grow to an amount A, the formula $R(A) = \sqrt[3]{A} - 1$ can be applied. (a) To what function family does this formula belong? (b) Complete the table using a calculator, then use the table to estimate the interest rate required for each $1 to grow to $1.17. (c) Compare your estimate to the value you get by evaluating $R(1.17)$. (d) For $R = \sqrt[3]{A} - 1$, solve for A in terms of R.

Amount A	Rate R
1.0	0.000
1.05	0.016
1.10	0.032
1.15	0.048
1.20	0.063
1.25	0.077

a. cube root family　**b.** answers will vary　**c.** 0.054 or 5.4%　**d.** $A = (R + 1)^3$

46. Force between charged particles: $F = k\dfrac{Q_1 Q_2}{d^2}$

The force between two charged particles is given by the formula shown, where F is the force (in joules—J), Q_1 and Q_2 represent the electrical charge on each particle (in coulombs—C), and d is the distance between them (in meters). If the particles have a like charge, the force is repulsive; if the charges are unlike, the force is attractive. (a) Write the variation equation in words. (b) Solve for k and use the formula to find the electrical constant k, given $F = 0.36$J, $Q_1 = 2 \times 10^{-6}$ C, $Q_2 = 4 \times 10^{-6}$ C, and $d = 0.2$ m. Express the result in scientific notation.
a. Force varies jointly with the product of the charges and inversely as the square of the distance between them.   **b.** 1.8×10^9

► **APPLICATIONS**

Find the constant of variation "k" and write the variation equation, then use the equation to solve.

47. Cleanup time: The time required to pick up the trash along a stretch of highway varies inversely as the number of volunteers who are working. If 12 volunteers can do the cleanup in 4 hr, how many volunteers are needed to complete the cleanup in just 1.5 hr? $T = \frac{48}{V}$; 32 volunteers

48. Wind power: The wind farms in southern California contain wind generators whose power production varies directly with the cube of the wind's speed. If one such generator produces 1000 W of power in a 25 mph wind, find the power it generates in a 35 mph wind. $P = 0.064S^3$; 2744 W

49. Pull of gravity: The weight of an object on the moon varies directly with the weight of the object on Earth. A 96-kg object on Earth would weigh only 16 kg on the moon. How much would a fully suited 250-kg astronaut weigh on the moon? $M = \frac{1}{6}E$; ≈41.7 kg

50. Period of a pendulum: The time that it takes for a simple pendulum to complete one period (swing over and back) varies directly as the square root of its length. If a pendulum 20 ft long has a period of 5 sec, find the period of a pendulum 30 ft long. $T = \frac{\sqrt{5}}{2}\sqrt{L}$; ≈6.12 sec

51. Stopping distance: The stopping distance of an automobile varies directly as the square root of its speed when the brakes are applied. If a car requires 108 ft to stop from a speed of 25 mph, estimate the stopping distance if the brakes were applied when the car was traveling 45 mph. $D = 21.6\sqrt{S}$; ≈144.9 ft

52. Supply and demand: A chain of hardware stores finds that the demand for a special power tool varies inversely with the advertised price of the tool. If the price is advertised at \$85, there is a monthly demand for 10,000 units at all participating stores. Find the projected demand if the price were lowered to \$70.83. $d = \frac{850,000}{P}$; ≈12,000 units

53. Cost of copper tubing: The cost of copper tubing varies jointly with the length and the diameter of the tube. If a 36-ft spool of $\frac{1}{4}$-in.-diameter tubing

costs \$76.50, how much does a 24-ft spool of $\frac{3}{8}$-in.-diameter tubing cost? $C = 8.5LD$; \$76.50

54. Electrical resistance: The electrical resistance of a copper wire varies directly with its length and inversely with the square of the diameter of the wire. If a wire 30 m long with a diameter of 3 mm has a resistance of 25 Ω, find the resistance of a wire 40 m long with a diameter of 3.5 mm. $R = \frac{7.5L}{D^2}$; ≈24.49 Ω

55. Volume of phone calls: The number of phone calls per day between two cities varies directly as the product of their populations and inversely as the square of the distance between them. The city of Tampa, Florida (pop. 300,000), is 430 mi from the city of Atlanta, Georgia (pop. 420,000). Telecommunications experts estimate there are about 300 calls per day between the two cities. Use this information to estimate the number of daily phone calls between Amarillo, Texas (pop. 170,000), and Denver, Colorado (pop. 550,000), which are also separated by a distance of about 430 mi. Note: Population figures are for the year 2000 and rounded to the nearest ten-thousand.
Source: 2005 World Almanac, p. 626. $C \approx (4.4 \times 10^{-4})\frac{P_1 P_2}{d^2}$; about 222 calls

56. Internet commerce: The likelihood of an eBay® item being sold for its "Buy it Now®" price P, varies directly with the feedback rating of the seller, and inversely with the cube of $\frac{P}{MSRP}$, where MSRP represents the manufacturer's suggested retail price. A power eBay® seller with a feedback rating of 99.6%, knows she has a 60% likelihood of selling an item at 90% of the MSRP. What is the likelihood a seller with a 95.3% feedback rating can sell the same item at 95% of the MSRP?   about 48.8%

57. Volume of an egg: The volume of an egg laid by an average chicken varies jointly with its length and the square of its width. An egg measuring 2.50 cm wide and 3.75 cm long has a volume of 12.27 cm³. A Barret's Blue Ribbon hen can lay an egg measuring 3.10 cm wide and 4.65 cm long. (a) What is the volume of this egg? (b) As a percentage, how much greater is this volume than that of an average chicken's egg?   **a.** about 23.39 cm³   **b.** about 191%

58. Athletic performance: Researchers have estimated that a sprinter's time in the 100-m dash varies directly as the square root of her age and inversely as the number of hours spent training each week. At 20 yr old, Gail trains 10 hr per week (hr/wk) and has an average time of 11 sec. Assuming she continues to train 10 hr/wk, (a) what will her average time be at 30 yr old? (b) If she wants to keep her average time at 11 sec, how many hours per week should she train?   **a.** about 13.47 sec **b.** about 12.25 hr

59. Maximum safe load: The maximum safe load M that can be placed on a uniform horizontal beam supported at both ends varies directly as the width w and the square of the height h of the beam's cross section, and inversely as its length L

(width and height are assumed to be in inches, and length in feet). (a) Write the variation equation. (b) If a beam 18 in. wide, 2 in. high, and 8 ft long can safely support 270 lb, what is the safe load for a beam of like dimensions with a length of 12 ft?   **a.** $M = kwh^2\left(\frac{1}{L}\right)$ **b.** 180 lb

60. Maximum safe load: Suppose a 10-ft wooden beam with dimensions 4 in. by 6 in. is made from the same material as the beam in Exercise 59 (the same k value can be used). (a) What is the maximum safe load if the beam is placed so that width is 6 in. and height is 4 in.? (b) What is the maximum safe load if the beam is placed so that width is 4 in. and height is 6 in.?   **a.** $M = 288$ lb **b.** $M = 432$ lb

▶ EXTENDING THE CONCEPT

61. The gravitational force F between two celestial bodies varies jointly as the product of their masses and inversely as the square of the distance d between them. The relationship is modeled by Newton's law of universal gravitation: $F = k\frac{m_1 m_2}{d^2}$. Given that $k = 6.67 \times 10^{-11}$, what is the gravitational force exerted by a 1000-kg sphere on another identical sphere that is 10 m away?
6.67×10^{-7}

62. The intensity of light and sound both vary inversely as the square of their distance from the source.

   a. Suppose you're relaxing one evening with a copy of *Twelfth Night* (Shakespeare), and the reading light is placed 5 ft from the surface of the book. At what distance would the intensity of the light be twice as great?   about 3.5 ft

   b. *Tamino's Aria* (*The Magic Flute*—Mozart) is playing in the background, with the speakers 12 ft away. At what distance from the speakers would the intensity of sound be three times as great?   about 6.9 ft

▶ MAINTAINING YOUR SKILLS

63. (R.2) Evaluate: $\left(\dfrac{2x^4}{3x^3 y}\right)^{-2}$ $\dfrac{9y^2}{4x^2}$

64. (R.4) Solve: $x^3 + 6x^2 + 8x = 0.$ $x = 0, -4, -2$

65. (2.4) State the domains of f and g given here:

   a. $f(x) = \dfrac{x-3}{x^2 - 16}$ $x \in (-\infty, -4) \cup (-4, 4) \cup (4, \infty)$

   b. $g(x) = \dfrac{x-3}{\sqrt{x^2 - 16}}$ $x \in (-\infty, -4) \cup (4, \infty)$

66. (2.3) Graph by using transformations of the parent function and plotting a minimum number of points: $f(x) = -2|x - 3| + 5.$

MAKING CONNECTIONS

Making Connections: Graphically, Symbolically, Numerically, and Verbally

Eight graphs (a) through (h) are given. Match the characteristics shown in 1 through 16 to one of the eight graphs.

(a)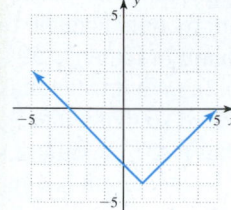

(b)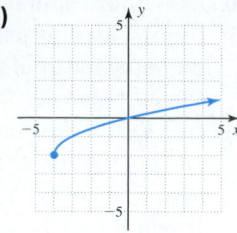

(c)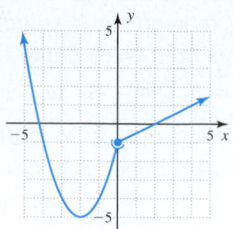

(d)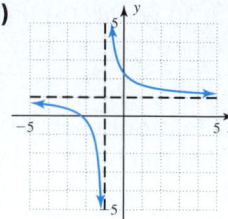

(e)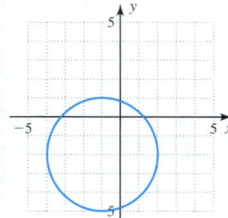

(f)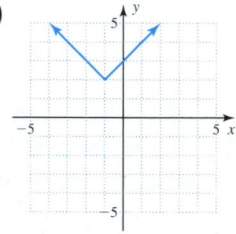

(g)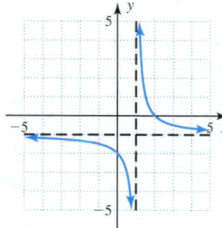

(h)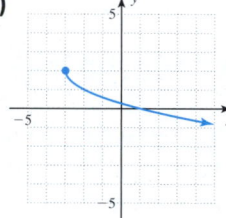

1. __g__ domain: $x \in (-\infty, 1) \cup (1, \infty)$

2. __b__ $y = \sqrt{x + 4} - 2$

3. __a__ $f(x)\uparrow$ for $x \in (1, \infty)$

4. __g__ horizontal asymptote at $y = -1$

5. __d__ $y = \dfrac{1}{x + 1} + 1$

6. __e__ domain: $x \in [-4, 2]$

7. __a__ $y = |x - 1| - 4$

8. __h__ $f(x) \le 0$ for $x \in [1, \infty)$

9. __b__ domain: $x \in [-4, \infty)$

10. __b__ $f(-3) = -1, f(5) = 1$

11. __h__ basic function is shifted 3 units left, reflected across x-axis, then shifted up 2 units

12. __f__ basic function is shifted 1 unit left, 2 units up

13. __c__ $f(-3) = -4, f(2) = 0$

14. __d__ as $x \to \infty, y \to 1$

15. __f__ $f(x) > 0$ for $x \in (-\infty, \infty)$

16. __c__ $y = \begin{cases} (x + 2)^2 - 5, & x < 0 \\ \dfrac{1}{2}x - 1, & x \ge 0 \end{cases}$

SUMMARY AND CONCEPT REVIEW

SECTION 2.1   Analyzing the Graph of a Function

KEY CONCEPTS

- A function f is even (symmetric to the y-axis), if and only if when a point (x, y) is on the graph, then $(-x, y)$ is also on the graph. In function notation: $f(-x) = f(x)$.
- A function f is odd (symmetric to the origin), if and only if when a point (x, y) is on the graph, then $(-x, -y)$ is also on the graph. In function notation: $f(-x) = -f(x)$.

Intuitive descriptions of the characteristics of a graph are given here. The formal definitions can be found within Section 2.1.

- A function is *increasing* in an interval if the graph rises from left to right (larger inputs produce larger outputs).
- A function is *decreasing* in an interval if the graph falls from left to right (larger inputs produce smaller outputs).
- A function is *positive* in an interval if the graph is above the x-axis in that interval.
- A function is *negative* in an interval if the graph is below the x-axis in that interval.
- A function is *constant* in an interval if the graph is parallel to the x-axis in that interval.
- A maximum value can be a *local* maximum, or *global* maximum. An *endpoint* maximum can occur at the endpoints of the domain. Similar statements can be made for minimum values.

EXERCISES

State the domain and range for each function $f(x)$ given. Then state the intervals where f is increasing or decreasing and intervals where f is positive or negative. Assume all endpoints have integer values.

1.

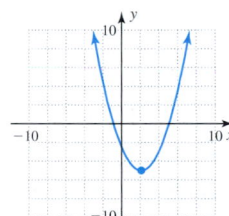

2.

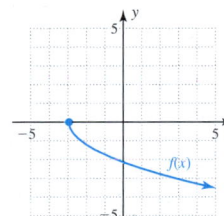

3.

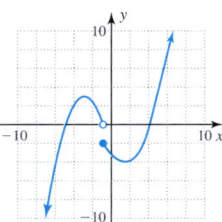

4. Determine whether the following are even $[f(-k) = f(k)]$, odd $[f(-k) = -f(k)]$, or neither.

a. $f(x) = 2x^5 - \sqrt[3]{x}$ odd   **b.** $g(x) = x^4 - \dfrac{\sqrt[3]{x}}{x}$ even

c. $p(x) = |3x| - x^3$ neither   **d.** $q(x) = \dfrac{x^2 - |x|}{x}$ odd

5. Draw the function f that has all of the following characteristics, then name the zeroes of the function and the location of all local maximum and minimum values. [*Hint:* Write them in the form $(c, f(c))$.]

a. Domain: $x \in [-6, 10)$ **b.** Range: $y \in [-8, 6)$
c. $f(0) = 0$ **d.** $f(x)\!\downarrow$ for $x \in (-6, -3) \cup (3, 7.5)$
e. $f(x)\!\uparrow$ for $x \in (-3, 3) \cup (7.5, 10)$ **f.** $f(x) < 0$ for $x \in (-6, 0) \cup (6, 9)$
g. $f(x) > 0$ for $x \in (0, 6) \cup (9, 10)$

 6. Use a graphing calculator to find the maximum and minimum values of $f(x) = 2x^5 - \sqrt[3]{x}$. Round to the nearest hundredth.   max: $y = 0.73$ at $x = -0.48$; min: $y = -0.73$ at $x = 0.48$

SECTION 2.2   The Toolbox Functions and Transformations

KEY CONCEPTS

- The *toolbox functions* and graphs commonly used in mathematics are
  - the identity function $f(x) = x$
  - square root function: $f(x) = \sqrt{x}$
  - cubing function: $f(x) = x^3$
  - squaring function: $f(x) = x^2$
  - absolute value function: $f(x) = |x|$
  - cube root function: $f(x) = \sqrt[3]{x}$
- For a basic or parent function $y = f(x)$, the general equation of the transformed function is $y = af(x \pm h) \pm k$. For any function $y = f(x)$ and $h, k > 0$,
  - the graph of $y = f(x) + k$ is the graph of $y = f(x)$ shifted upward k units
  - the graph of $y = f(x + h)$ is the graph of $y = f(x)$ shifted left h units
  - the graph of $y = -f(x)$ is the graph of $y = f(x)$ reflected across the x-axis
  - $y = af(x)$ results in a vertical stretch when $a > 1$
  - the graph of $y = f(x) - k$ is the graph of $y = f(x)$ shifted downward k units
  - the graph of $y = f(x - h)$ is the graph of $y = f(x)$ shifted right h units
  - the graph of $y = f(-x)$ is the graph of $y = f(x)$ reflected across the y-axis
  - $y = af(x)$ results in a vertical compression when $0 < a < 1$
- Transformations are applied in the following order: (1) horizontal shifts, (2) reflections, (3) stretches or compressions, and (4) vertical shifts.

EXERCISES

Identify the function family for each graph given, then (a) describe the end-behavior; (b) name the x- and y-intercepts; (c) identify the vertex, initial point, or point of inflection (as applicable); and (d) state the domain and range.

7. 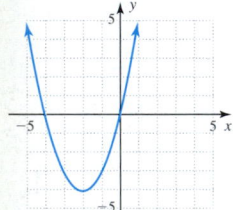   **8.** 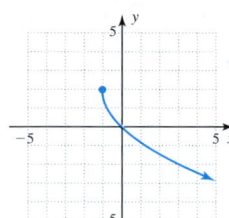   **9.**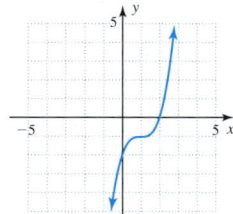

10. 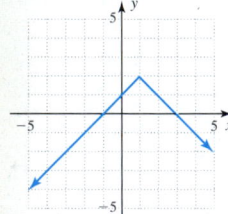   **11.**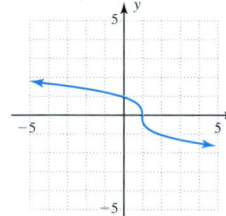

Identify each function as belonging to the linear, quadratic, square root, cubic, cube root, or absolute value family. Then sketch the graph using shifts of a parent function and a few characteristic points.

12. $f(x) = -(x + 2)^2 - 5$ **13.** $f(x) = 2|x + 3|$ **14.** $f(x) = x^3 - 1$

15. $f(x) = \sqrt{x - 5} + 2$ **16.** $f(x) = \sqrt[3]{x} + 2$

17. Apply the transformations indicated for the graph of $f(x)$ given.

  a. $f(x - 2)$
  b. $-f(x) + 4$
  c. $\frac{1}{2}f(x)$

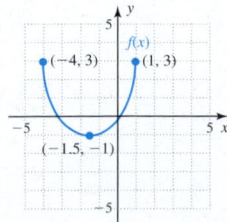

Additional answers can be found in the Instructor Answer Appendix.

SECTION 2.3   Absolute Value Functions, Equations, and Inequalities

KEY CONCEPTS

- To solve absolute value equations and inequalities, begin by writing the equation in simplified form, with the absolute value isolated on one side.
- If X and Y represent algebraic expressions and k is a nonnegative constant:
  - Absolute value equations: $|X| = k$ is equivalent to $X = -k$ or $X = k$
    $|X| = |Y|$ is equivalent to $X = Y$ or $X = -Y$
  - "Less than" inequalities: $|X| < k$ is equivalent to $-k < X < k$
  - "Greater than" inequalities: $|X| > k$ is equivalent to $X < -k$ or $X > k$
- These properties also apply when the symbols "$\leq$" or "$\geq$"are used.
- If the absolute value quantity has been isolated on the left, the solution to a less-than inequality will be a single interval, while the solution to a greater-than inequality will consist of two disjoint intervals.
- The multiplicative property states that for algebraic expressions A and B, $|AB| = |A||B|$.
- Absolute value equations and inequalities can be solved graphically using the intersect method *or* the zeroes/x-intercept method.

EXERCISES

Solve each equation or inequality. Write solutions to inequalities in interval notation.

18. $7 = |x - 3|$ $\{-4, 10\}$

19. $-2|x + 2| = -10$ $\{-7, 3\}$

20. $|-2x + 3| = 13$ $\{-5, 8\}$

21. $\dfrac{|2x + 5|}{3} + 8 = 9$ $\{-4, -1\}$

22. $-3|x + 2| - 2 < -14$ $(-\infty, -6) \cup (2, \infty)$

23. $\left|\dfrac{x}{2} - 9\right| \leq 7$ $[4, 32]$

24. $|3x + 5| = -4$ $\{\,\}$

25. $3|x + 1| < -9$ $\{\,\}$

26. $2|x + 1| > -4$ $(-\infty, \infty)$

27. $5|m - 2| - 12 \leq 8$ $[-2, 6]$

28. $\dfrac{|3x - 2|}{2} + 6 \geq 10$ $(-\infty, -2] \cup [\frac{10}{3}, \infty)$

29. Monthly rainfall received in Omaha, Nebraska, rarely varies by more than 1.7 in. from an average of 2.5 in. per month. (a) Use this information to write an absolute value inequality model, then (b) solve the inequality to find the highest and lowest amounts of monthly rainfall for this city.   **a.** $|r - 2.5| \leq 1.7$ **b.** highest: 4.2 in., lowest: 0.8 in.

SECTION 2.4   Basic Rational Functions and Power Functions; More on the Domain

KEY CONCEPTS

- A rational function is one of the form $V(x) = \dfrac{p(x)}{d(x)}$, where p and d are polynomials and $d(x) \neq 0$.

- The most basic rational functions are the reciprocal function $f(x) = \dfrac{1}{x}$ and the reciprocal square function $g(x) = \dfrac{1}{x^2}$.

- The line $y = k$ is a horizontal asymptote of V if as $|x|$ increases without bound, $V(x)$ approaches k: as $|x| \to \infty$, $V(x) \to k$.

- The line $x = h$ is a vertical asymptote of V if as x approaches h, $V(x)$ increases/decreases without bound: as $x \to h$, $|V(x)| \to \infty$.

- The reciprocal and reciprocal square functions can be transformed using the same shifts, stretches, and reflections as applied to other basic functions, with the asymptotes also shifted.

- A power function can be written in the form $f(x) = x^p$ where p is a constant real number and x is a variable.

  If $p = \dfrac{1}{n}$, where n is a natural number, $f(x) = x^{\frac{1}{n}} = \sqrt[n]{x}$ is called a root function in x.

- Given the rational exponent $\frac{m}{n}$ is in simplest form, the domain of $f(x) = x^{\frac{m}{n}}$ is $(-\infty, \infty)$ if n is odd, and $[0, \infty)$ if n is even.

EXERCISES

Sketch the graph of each function using shifts of the parent function (not by using a table of values). Find and label the
x- and y-intercepts (if they exist) and redraw the asymptotes.

30. $f(x) = \dfrac{1}{x + 2} - 1$

31. $h(x) = \dfrac{-1}{(x - 2)^2} - 3$

32. In a certain county, the cost to keep public roads free of trash is given by $C(p) = \dfrac{-7500}{p - 100} - 75$, where $C(p)$
represents the cost (thousands of dollars) to keep p percent of the trash picked up. (a) Find the cost to pick up
30%, 50%, 70%, and 90% of the trash, and comment on the results. (b) Sketch the graph using the transformation
of a toolbox function. (c) Use mathematical notation to describe what happens if the county tries to keep 100% of
the trash picked up.

33. Use a graphing calculator to graph the functions $f(x) = x^1$, $g(x) = x^{\frac{1}{2}}$, and $h(x) = x^\pi$ in the same viewing window.
What is the domain of each function?

34. The expression $T = \dfrac{2\pi}{37,840} r^{\frac{3}{2}}$ models the time T (in hr) it takes for a satellite to complete one revolution around
the Earth, where r represents the radius (in km) of the orbit measured from the center of the Earth. If the Earth
has a radius of 6370 km, (a) how long does it takes for a satellite at a height of 200 km to complete one orbit?
(b) What is the orbital height of a satellite that completes one revolution in 4 days (96 hr)?

SECTION 2.5 Piecewise-Defined Functions

KEY CONCEPTS

- Each piece of a piecewise-defined function has a domain over which that piece is defined.
- To evaluate a piecewise-defined function, identify the domain interval containing the input value, then use the
  piece of the function corresponding to this interval.
- To graph a piecewise-defined function you can plot points, or graph each piece in its entirety, then erase portions
  of the graph outside the domain indicated for each piece.
- If the graph of a function can be drawn without lifting your pencil from the paper, the function is continuous.
- A discontinuity is said to be removable if we can redefine the function to "fill the hole."
- Step functions are discontinuous and formed by a series of horizontal steps.
- The floor function $\lfloor x \rfloor$ gives the largest integer less than or equal to x.
- The ceiling function $\lceil x \rceil$ is the smallest integer greater than or equal to x.

EXERCISES

35. For the graph and functions given, (a) use the correct notation to write the relation as a
single piecewise-defined function, stating the effective domain for each piece by inspecting
the graph; and (b) state the range of the function: $Y_1 = 5$, $Y_2 = -X + 1$,
$Y_3 = 3\sqrt{X - 3} - 1$.

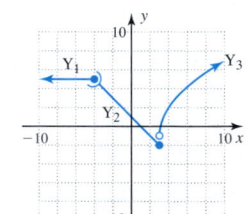

36. Use a table of values as needed to graph $h(x)$, then state its domain and range. If the
function has a pointwise (removable) **discontinuity,** state how the second piece could be
redefined so that a continuous function results.

$$h(x) = \begin{cases} \dfrac{x^2 - 2x - 15}{x + 3}, & x \neq -3 \\ -6, & x = -3 \end{cases}$$

37. Evaluate the piecewise-defined function $p(x)$: $p(-4)$, $p(-2)$, $p(2.5)$, $p(2.99)$, $p(3)$, and $p(3.5)$

$$p(x) = \begin{cases} -4, & x < -2 \\ -|x| - 2, & -2 \leq x < 3 \\ 3\sqrt{x} - 9, & x \geq 3 \end{cases}$$ $\quad -4, -4, -4.5, -4.99, 3\sqrt{3} - 9, 3\sqrt{3.5} - 9$

38. Sketch the graph of the function and state its domain and range. Use transformations of the toolbox functions where possible.

$$q(x) = \begin{cases} 2\sqrt{-x-3} - 4, & x \le -3 \\ -2|x| + 2, & -3 < x < 3 \\ 2\sqrt{x-3} - 4, & x \ge 3 \end{cases}$$

39. Many home improvement outlets now rent flatbed trucks in support of customers that purchase large items. The cost is $20 per hour for the first 2 hr, $30 for the next 2 hr, then $40 for each hour afterward. Write this information as a piecewise-defined function, then sketch its graph. What is the total cost to rent this truck for 5 hr?

SECTION 2.6   Variation: The Toolbox Functions in Action

KEY CONCEPTS

- *Direct variation:* If there is a nonzero constant k such that $y = kx$, we say, "y varies directly with x" or "y is directly proportional to x" (k is called the constant of variation).

- *Inverse variation:* If there is a nonzero constant k such that $y = k\left(\dfrac{1}{x}\right)$ we say, "y varies inversely with x" or y is inversely proportional to x.

- In some cases, direct and inverse variations work simultaneously to form a *joint variation.*
- The process for solving variation equations can be found on page 74.

EXERCISES

Find the constant of variation and write the equation model, then use this model to complete the table.

40. y varies directly as the cube root of x; $y = 52.5$ when $x = 27$. $k = 17.5; y = 17.5\sqrt[3]{x}$

x	y
216	105
0.343	12.25
729	157.5

41. z varies directly as v and inversely as the square of w; $z = 1.62$ when $w = 8$ and $v = 144$. $k = 0.72; z = \frac{0.72v}{w^2}$

v	w	z
196	7	2.88
38.75	1.25	17.856
24	0.6	48

42. Given t varies jointly with u and v, and inversely as w, if $t = 30$ when $u = 2$, $v = 3$, and $w = 5$, find t when $u = 8$, $v = 12$, and $w = 15$. $t = 160$

43. The time that it takes for a simple pendulum to complete one period (swing over and back) is directly proportional to the square root of its length. If a pendulum 16 ft long has a period of 3 sec, find the time it takes for a 36-ft pendulum to complete one period.   4.5 sec

PRACTICE TEST

1. Determine the following from the graph shown.
  a. the domain and range
  b. estimate the value of $f(-1)$
  c. interval(s) where $f(x)$ is negative or positive
  d. interval(s) where $f(x)$ is increasing, decreasing, or constant.
  e. an equation for $f(x)$

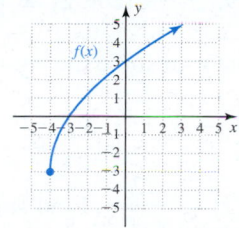

For the function $h(x)$ whose partial graph is given.

2. Complete the graph if h is known to be even.

3. Complete the graph if h is known to be odd.

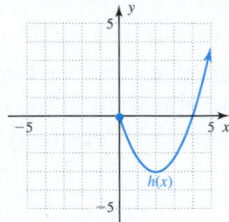

 4. Use a graphing calculator to find the maximum and minimum values of $f(x) = |x^2 + 4x - 11| - 7$. Round answers to nearest hundredth when necessary.

5. Each function graphed here is from a toolbox function family. For each graph, (a) identify the function family, (b) state the domain and range, (c) identify x- and y-intercepts, (d) discuss the end-behavior, and (e) solve the inequalities $f(x) > 0$ and $f(x) < 0$.

I.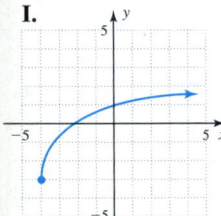

II.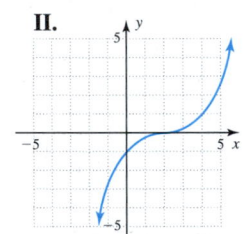

III.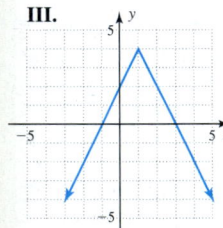

IV.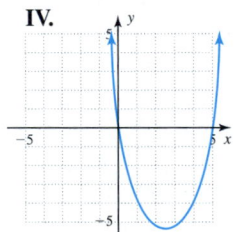

Sketch each graph using a transformation.

6. $f(x) = |x - 2| + 3$

7. $g(x) = -(x + 3)^2 - 2$

Solve each inequality. Write the solutions in interval notation.

8. $\dfrac{2}{3}|3x - 1| > 14$ **9.** $5 - 2|x + 2| \geq 1$

 10. Use a graphing calculator to solve the equation.

$$1.7|x - 0.75| + 3 = 3 - \frac{3}{5}\left|x - \frac{3}{4}\right|$$

11. Sketch the graph of $f(x) = \dfrac{2}{x + 3}$. Find and label the x- and y-intercepts, if they exist, along with all asymptotes.

 12. After the engine is cut at $t = 0$, a boat coasts for a while before stopping. The distance D it travels over t sec can be modeled by $D(t) = 60 - \dfrac{240}{(t + 2)^2}$. If the boat comes to a complete stop after traveling 59 ft, use a graphing calculator to determine the time required to stop. Round your answer to the nearest tenth. 13.5 sec

13. Find the domains of the following functions.
a. $f(x) = 2.09x^{\frac{2}{5}}$ $(-\infty, \infty)$ b. $g(x) = -4.22x^{\frac{5}{2}}$ $[0, \infty)$
c. $h(t) = 4.5t^{\pi}$ $[0, \infty)$

14. Identify the vertical and horizontal asymptotes of $g(x) = \dfrac{3}{(x + 2)^2} - 1$. VA: $x = -2$; HA: $y = -1$

 15. Using time-lapse photography, the spread of a liquid is tracked in one-fifth of a second intervals, as a small amount of liquid is dropped on a piece of fabric. A power function provides a reasonable model for the first second. Use a graphing calculator to (a) graph a scatterplot of the data and (b) find an equation model using a power regression (round to two decimal places). Use the equation to estimate (c) the size of the stain after 0.5 sec and (d) how long it will take the stain to reach a size of 15 mm.

Time (sec)	Size (mm)
0.2	0.39
0.4	1.27
0.6	3.90
0.8	10.60
1	21.50

16. The following function has two removable discontinuities. Find the values of a and b so that a continuous function results.

$$g(x) = \begin{cases} \dfrac{x^3 + x^2 - 4x - 4}{x^2 - x - 2}, & x \neq -1, 2 \\ a, & x = -1 \\ b, & x = 2 \end{cases}$$

$a = 1, b = 4$

17. The annual output of a wind turbine varies jointly with the square of the blade diameter and the cube of the average wind speed. If a 10-ft-diameter turbine in 12 mph average winds produces 2300 KWH/year, how much will a 6 ft-diameter turbine produce in 15 mph average winds? Round to the nearest KWH/year. 1617 KWH/year

18. Given $h(x) = \begin{cases} 4, & x < -2 \\ 2x, & -2 \leq x \leq 2 \\ x^2, & x > 2 \end{cases}$

a. Find $h(-3)$, $h(-2)$, and $h(\frac{5}{2})$. 4, -4, 6.25
b. Sketch the graph of h. Label important points.

19. By observing a significantly smaller object orbiting a large celestial body, astronomers can easily determine the mass of the larger. Appealing to Kepler's third law of planetary motion, we know the mass of the large body varies directly with the cube of the mean distance to the smaller and inversely with the square of its orbital period. Write the variation equation. Using the mean Earth/Sun distance of 1.496×10^8 km and the Earth's orbital period of 1 yr, the mass of the Sun has been calculated to be 1.98892×10^{30} kg. Given the orbital period of Mars is 1.88 yr, find its mean distance from the Sun.

$M = kd^3(\frac{1}{p^2})$, approx. 2.2788×10^8

20. The maximum load that can be supported by a rectangular beam varies jointly with its width and its height squared and inversely with its length. If a beam 10 ft long, 3 in. wide, and 4 in. high can support 624 lb, how many pounds could a 12-ft-long beam with the same dimensions support?   520 lb

CALCULATOR EXPLORATION AND DISCOVERY

Studying Joint Variations

Although a graphing calculator is limited to displaying the relationship between only two variables (for the most part), it has a feature that enables us to see how these two are related with respect to a third. Consider the variation equation from Example 8 in Section 2.6: $F = 0.05dv^2$. If we want to investigate the relationship between fuel consumption and velocity, we can have the calculator display multiple versions of the relationship *simultaneously for different values of d*. This is accomplished using the "{" and "}" symbols, which are [2nd] functions to the parentheses. When the calculator sees values between these grouping symbols and separated by commas, it is programmed to use each value independently of the others, graphing or evaluating the relation for each value in the set. We illustrate by graphing the relationship $f = 0.05dv^2$ for three different values of d. Enter the equation on the [Y=] screen as $Y_1 = 0.05\{10, 20, 30\}X^2$, which tells the calculator to graph the equations $Y_1 = 0.05(10)X^2$, $Y_1 = 0.05(20)X^2$, and $Y_1 = 0.05(30)X^2$ on the same grid. Note that since d is constant, each graph is a parabola. Set the viewing window using the values given in Example 8 as a guide. The result is the graph shown in Figure 2.97, where we can study the relationship between these three variables using the up [⌃] and down [⌄] arrows. From our work with the toolbox functions and transformations, we know the widest parabola used the coefficient "10," while the narrowest parabola used the coefficient "30." As shown, the graph tells us that at a speed of 15 nautical miles per hour (X = 15), it will take 112.5 barrels of fuel to travel 10 mi (the first number in the list). After pressing the [⌄] key, the cursor jumps to the second curve, which shows values of X = 15 and Y = 225. This means at 15 nautical miles per hour, it would take 225 barrels of fuel to travel 20 mi. Use these ideas to complete the following exercises:

Figure 2.97

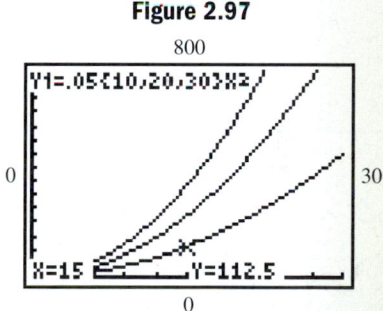

Exercise 1: The comparison of distance covered versus fuel consumption at different speeds also makes an interesting study. This time velocities are constant values and the distance varies. On the [Y=] screen, enter $Y_1 = 0.05x\{10, 20, 30\}^2$. What family of equations results? Use the up/down arrow keys for $x = 15$ (a distance of 15 mi) to find how many barrels of fuel it takes to travel 15 mi at 10 mph, 15 mi at 20 mph, and 15 mi at 30 mph. Comment on what you notice.

linear; 75; 300; 675; Fuel required increases dramatically.

Exercise 2: The maximum safe load S for a wooden horizontal plank supported at both ends varies jointly with the width W of the beam, the square of its thickness T, and inversely with its length L. A plank 10 ft long, 12 in. wide, and 1 in. thick will safely support 450 lb. Find the value of k and write the variation equation, then use the equation to explore: $k = 375, S = \frac{375 \, WT^2}{L}$

a. Safe load versus thickness for a constant width and given lengths (quadratic function). Use $w = 8$ in. and $\{8, 12, 16\}$ for L. $S = 375 \, T^2, S = 250 \, T^2, S = 187.5 \, T^2$

b. Safe load versus length for a constant width and given thickness (reciprocal functional). Use $w = 8$ in. and $\{\frac{1}{4}, \frac{1}{2}, \frac{3}{4}\}$ for thickness. $S = \frac{187.5}{L}, S = \frac{750}{L}, S = \frac{1687.5}{L}$

 **STRENGTHENING CORE SKILLS**

Variation and Power Functions: $y = kx^p$

You may have noticed that applications of power functions (Section 2.4) can also be stated as variations (Section 2.6): From the general equation shown in the title of this feature, "... y varies directly as x to the p power." Due to the nature of real data and data collection, applications of power functions based on regression yield values of k (the constant of variation) and p (the power) that cannot be written in exact form. However, "fixed" relationships modeled by power functions produce values of k and p that <u>can</u> be written in exact form. For instance, the power function that models planetary orbits states: The time T it takes a planet to complete one orbit varies directly with its orbital radius to the three-halves power: $T = kR^{\frac{3}{2}}$. Here, the power is exactly $\frac{3}{2}$ and the constant of variation turns out to be exactly 1 (also see **Section 2.4, Exercise 78**). Many times, finding this constant takes more effort, and utilizes the skills developed in this and previous chapters. Consider the following.

Illustration 1 ▶ The volume V of a sphere varies directly with its surface area S to the three-halves power. If the volume is approximately 33.51 cm^3 when the surface area is 50.30 cm^2, (a) find the constant of variation yielded by these values (round to two decimal places), and (b) find the exact constant of variation dictated by the geometry of the sphere and write the variation equation.

Solution ▶ a.

$$V = kS^{\frac{3}{2}} \qquad \text{variation equation}$$
$$33.51 = k(50.30)^{\frac{3}{2}} \qquad \text{substitute 33.51 for } V, \text{ 50.30 for } S$$
$$\frac{33.51}{(50.30)^{\frac{3}{2}}} = k \qquad \text{solve for } k$$
$$0.09 \approx k \qquad \text{approximate value for } k$$

This gives $V \approx 0.09 S^{\frac{3}{2}}$ as an approximate relationship.

b. To find the true constant of variation fixed by the nature of spheres, we begin with the same set up, but *substitute the actual formulas* for volume and surface area, then simplify.

$$V = kS^{\frac{3}{2}} \qquad \text{variation equation}$$
$$\frac{4}{3}\pi r^3 = k(4\pi r^2)^{\frac{3}{2}} \qquad \text{substitute } \frac{4}{3}\pi r^3 \text{ for } V, 4\pi r^2 \text{ for } S$$
$$\frac{4}{3}\pi r^3 = 8\pi^{\frac{3}{2}} r^3 k \qquad \text{properties of exponents: } 4^{\frac{3}{2}} = 8$$
$$\pi = 6\pi^{\frac{3}{2}} k \qquad \text{multiply by } \frac{3}{4}, \text{ divide by } r^3$$
$$\frac{\pi}{6\pi^{\frac{3}{2}}} = k \qquad \text{solve for } k$$
$$\frac{1}{6\pi^{\frac{1}{2}}} = k \qquad \text{simplify (exact form)}$$

The constant of variation for this relationship is $\dfrac{1}{6\sqrt{\pi}}$, giving a variation equation of $V = \dfrac{1}{6\sqrt{\pi}}S^{\frac{3}{2}}$. Note that $\dfrac{1}{6\sqrt{\pi}} \approx 0.09$.

Studies of this type are important, because as the radius of the sphere gets larger, so does the error generated by using an approximate value. Using a radius of $r = 18$ cm with the approximate relationship $\left[V \approx 0.09 \left(4\pi 18^2\right)^{3/2}\right]$, gives a volume of near 23,381.6 cm^3, while the volume found using the exact value for k is about 24,429.0 cm^3.

Exercise 1: Use this Strengthening Core Skills feature to find the exact constant of variation for the following relationship: The volume of a cube varies directly with its surface area to the three-halves power. $k = \dfrac{1}{6^3} \text{ or } \dfrac{1}{\sqrt{6^3}}$

CUMULATIVE REVIEW CHAPTERS R–2

1. Given $f(x) = 2x^3 + 4x^2 + 8x - 7$,
   find $f(-2)$ and $f\left(\dfrac{1}{2}\right)$. $f(-2) = -23, f\left(\dfrac{1}{2}\right) = -\dfrac{7}{4}$

2. Find the solution set for: $2 - x < 5$ **and** $3x + 2 < 8$. $-3 < x < 2$

3. The area of a circle is 69 cm². Find the circumference of the same circle.   29.45 cm

4. The surface area of a cylinder is $A = 2\pi r^2 + 2\pi rh$. Write r in terms of A and h (solve for r).

5. Solve for x: $-2(3 - x) + 5x = 4(x + 1) - 7$. $x = 1$

6. Evaluate without using a calculator: $\left(\dfrac{27}{8}\right)^{\frac{-2}{3}}$. $\frac{4}{9}$

7. Find the slope of each line:
   a. through the points: $(-4, 7)$ and $(2, 5)$. $\frac{-1}{3}$
   b. the line with equation $3x - 5y = 20$. $\frac{3}{5}$

8. Graph using transformations of a parent function.
   a. $f(x) = \sqrt{x - 2} + 3$.
   b. $f(x) = -|x + 2| - 3$.

9. Graph the line passing through $(-3, 2)$ with a slope of $m = \frac{1}{2}$, then state its equation.

10. Find (a) the length of the hypotenuse and (b) the perimeter of the triangle shown.   a. $h = 185$ cm   b. $P = 418$ cm

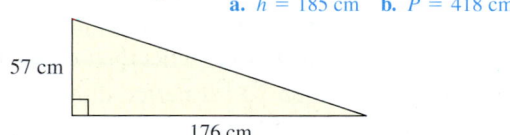

57 cm

176 cm

11. Sketch the graph of $h(x) = \dfrac{-1}{(x - 1)^2} + 3$ using a transformation of the parent function.

12. Graph by plotting the y-intercept, then counting $\dfrac{\Delta y}{\Delta x}$ to find additional points: $y = \frac{1}{3}x - 2$

13. Graph the piecewise-defined function
    $$f(x) = \begin{cases} x^2 - 4, & x < 2 \\ x - 1, & 2 \le x \le 8 \end{cases}$$
    and determine the following:
    a. the domain and range
    b. the value of $f(-3), f(-1), f(1), f(2),$ and $f(3)$
    c. the zeroes of the function
    d. interval(s) where $f(x)$ is negative/positive
    e. location of any local max/min values
    f. interval(s) where $f(x)$ is increasing/decreasing

Additional answers can be found in the Instructor Answer Appendix.

14. The graph of a function $h(x)$ is shown. (a) State the domain and estimate the range of h. (b) What are the zeroes of the function? (c) What is the value of $h(-1)$? (d) If $h(k) = 9$ what is the value of k?

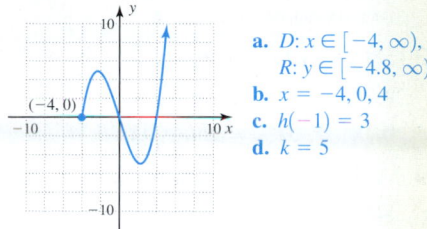

(-4, 0)

a. $D: x \in [-4, \infty)$,
   $R: y \in [-4.8, \infty)$
b. $x = -4, 0, 4$
c. $h(-1) = 3$
d. $k = 5$

15. Add the rational expressions:
    a. $\dfrac{-2}{x^2 - 3x - 10} + \dfrac{1}{x + 2}$ $\dfrac{x - 7}{(x - 5)(x + 2)}$
    b. $\dfrac{b^2}{4a^2} - \dfrac{c}{a}$ $\dfrac{b^2 - 4ac}{4a^2}$

16. Simplify the radical expressions:
    a. $\dfrac{-10 + \sqrt{72}}{4}$ $\dfrac{-5}{2} + \dfrac{3\sqrt{2}}{2}$ b. $\dfrac{1}{\sqrt{2}}$ $\dfrac{\sqrt{2}}{2}$

17. Perform the division by factoring the numerator:
    $(x^3 - 5x^2 + 2x - 10) \div (x - 5)$. $x^2 + 2$

18. Determine if the following relation is a function. If not, how is the definition of a function violated?

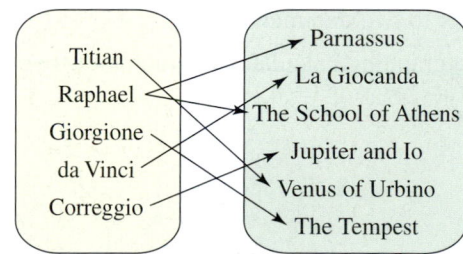

Titian
Raphael
Giorgione
da Vinci
Correggio

Parnassus
La Giocanda
The School of Athens
Jupiter and Io
Venus of Urbino
The Tempest

19. Find the center and radius of the circle defined by $x^2 + 6x + y^2 - 12y + 36 = 0$.   center $(-3, 6), r = 3$

20. The amount of pressure (in pounds per square inch—psi) felt by a professional pearl diver as she dives to harvest oysters, is 14.7 more than 0.43 times the depth of the dive (in feet). Write the equation model for this situation. If the oyster bed is at a depth of 60 ft, how much pressure is felt? $P = 0.43d + 14.7$; 40.5 psi

21. The *National Geographic Atlas of the World* is a very large, rectangular book with an almost inexhaustible panoply of information about the world we live in. The length of the front cover is 16 cm more than its width, and the area of the cover is 1457 cm². Use this information to write an equation model, then use the quadratic formula to determine the length and width of the *Atlas*. $W = 31$ cm, $L = 47$ cm

22. During a table tennis tournament, the championship game between J.W. and Mike took a dramatic and unexpected turn. At one point in the game, J.W. was losing 5–15. Facing a crushing loss, he summoned all his willpower and battled on to a 21–19 victory! Assuming the game score relationship is linear, find the slope of the line between these two scores and discuss its meaning in this context.

$m = \frac{1}{4}$; for every 1 point Mike scored, J.W. scored 4 points.

23. Solve by factoring:

a. $6x^2 - 7x = 20$ $x = \frac{-4}{3}, \frac{5}{2}$

b. $x^3 + 5x^2 - 15 = 3x$ $x = -5, -\sqrt{3}, \sqrt{3}$

24. A theorem from elementary geometry states, "*A line tangent to a circle is perpendicular to the radius at the point of tangency.*" Find the equation of the tangent line for the circle and radius shown.

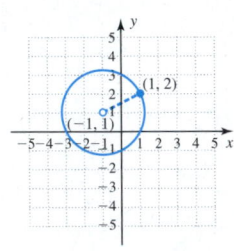

25. A triangle has its vertices at $(-4, 5)$, $(4, -1)$, and $(0, 8)$. Find the perimeter of the triangle and determine whether or not it is a *right* triangle.

📟 **Exercises 26 through 30 require the use of a graphing calculator.**

26. Use the zeroes method to solve the equation $2.7(x - 3) + 0.3 = 1.8 - 1.2(x + 4)$. Round your answer to two decimal places. $x \approx 1.23$

27. Use a graphing calculator to graph the circle defined by $(x + 2)^2 + y^2 = 4$.

28. Use a graphing calculator and the intersection-of-graphs method to solve the inequality. $|1.2(x - 0.5)| < 0.4x + 1.4$ $x \in (-0.5, 2.5)$

29. Graph the following piecewise-defined function using a graphing calculator. Then use the TRACE command to evaluate the function at $x = 1.2$.

$$f(x) = \begin{cases} -x + 2 & -5 \leq x < -2 \\ (x - 2)^2 - 3 & x \geq 1 \end{cases}$$

30. The data given shows the growth of the total U.S. National Debt (in billions) for the years 1993 to 1999 (1993 → 1), and for the years 2001 to 2007 (2001 → 1), for each set of data, enter the data into a graphing calculator, then

Year (scaled)	1993 to 1999	2001 to 2007
1	4.5	5.9
2	4.8	6.4
3	5.0	7.0
4	5.3	7.6
5	5.5	8.2
6	5.6	8.7
7	5.8	9.2

a. Set an appropriate window for viewing the scatterplots, and determine if the associations are linear or nonlinear.

b. If linear, find the regression equation for each data set and graph both (as Y_1 and Y_2) in the same window (round to two decimal places).

c. During which 7-year period did the national debt increase faster? How much faster?

Quadratic Functions and Operations on Functions

CHAPTER OUTLINE

CHAPTER CONNECTIONS

Whether it be in business, industry, animal science, space, sports, or some other sector, decision makers are preoccupied with maximum and minimum values. Questions concerning minimum cost, maximum efficiency, least amount of waste, greatest amount of area, and other similar questions require and deserve a great deal of attention. In this chapter, we'll learn to calculate some of these maximums and minimums, like the height achieved by a motocross or snowboarding athlete in the performance of jumps. This application appears as Exercise 51 in Section 3.3.

Check out these other real-world connections:

▶ Projectile Height (Section 3.2, Exercise 161)

▶ Pricing Strategies (Section 3.3, Exercise 44)

▶ Start-up Times (Section 3.4, Exercise 47)

▶ Distribution Distance (Section 3.5, Exercise 75)

3.1 Complex Numbers

For centuries, even the most prominent mathematicians refused to work with equations like $x^2 + 1 = 0$. Using the principal of square roots gave the "solutions" $x = \sqrt{-1}$ and $x = -\sqrt{-1}$, which they found baffling and mysterious, since there is no real number whose square is -1. In this section, we'll see how this dilemma was finally resolved.

A. Identifying and Simplifying Imaginary and Complex Numbers

The equation $x^2 = -1$ has no real solutions, since the square of any real number is positive. But if we apply the principle of square roots we get $x = \sqrt{-1}$ and $x = -\sqrt{-1}$, which seem to check when substituted into the original equation:

$$x^2 + 1 = 0 \qquad \text{original equation}$$

$$(1) \qquad (\sqrt{-1})^2 + 1 = 0 \qquad \text{substitute } \sqrt{-1} \text{ for } x$$

$$-1 + 1 = 0 ✓ \qquad \text{answer "checks"}$$

$$(2) \qquad (-\sqrt{-1})^2 + 1 = 0 \qquad \text{substitute } -\sqrt{-1} \text{ for } x$$

$$-1 + 1 = 0 ✓ \qquad \text{answer "checks"}$$

This observation likely played a part in prompting Renaissance mathematicians to study such numbers in greater depth, as they reasoned that while these were not *real number* solutions, they must be *solutions of a new and different kind*. Their study eventually resulted in the introduction of the set of **imaginary numbers** and the **imaginary unit i,** as follows.

Imaginary Numbers and the Imaginary Unit

- Imaginary numbers are those of the form $\sqrt{-k}$, where k is a positive real number.

- The imaginary unit i represents the number whose square is -1:

$$i = \sqrt{-1}, \text{ where } i^2 = -1$$

As a convenience to understanding and working with imaginary numbers, we rewrite them in terms of i, allowing that the product property of radicals $(\sqrt{AB} = \sqrt{A}\sqrt{B})$ still applies if *only one* of the radicands is negative. For $\sqrt{-3}$, we have $\sqrt{-1 \cdot 3} = \sqrt{-1}\sqrt{3} = i\sqrt{3}$. More generally, we simply make the following statement regarding imaginary numbers.

Rewriting Imaginary Numbers

For any positive real number k,

$$\sqrt{-k} = i\sqrt{k}.$$

For $\sqrt{-16}$ and $\sqrt{-20}$ we have:

$$\sqrt{-16} = i\sqrt{16} \qquad\qquad \sqrt{-20} = i\sqrt{20}$$
$$= i(4) \qquad\qquad\qquad = i\sqrt{4 \cdot 5}$$
$$= 4i \qquad\qquad\qquad\qquad = 2i\sqrt{5},$$

and we say the expressions have been simplified and written in terms of i. Note that for $\sqrt{-20}$ we've written the result with the unit "i" *in front of the radical* to prevent it being interpreted as being *under the radical*. In symbols, $2i\sqrt{5} = 2\sqrt{5}i \neq 2\sqrt{5i}$.

The solutions to $x^2 = -1$ also serve to illustrate that for $k > 0$, there are two solutions to $x^2 = -k$, namely, $i\sqrt{k}$ and $-i\sqrt{k}$. In other words, every negative number has two square roots, one positive and one negative. The first of these, $i\sqrt{k}$, is called the **principal square root** of $-k$.

EXAMPLE 1 ▶ **Simplifying Imaginary Numbers**

Rewrite the imaginary numbers in terms of i and simplify if possible.

    a. $\sqrt{-7}$ **b.** $\sqrt{-81}$ **c.** $\sqrt{-24}$ **d.** $-3\sqrt{-16}$

Solution ▶   **a.** $\sqrt{-7} = i\sqrt{7}$ **b.** $\sqrt{-81} = i\sqrt{81}$

                                                      $= 9i$

    c. $\sqrt{-24} = i\sqrt{24}$ **d.** $-3\sqrt{-16} = -3i\sqrt{16}$

                    $= i\sqrt{4 \cdot 6}$ $= -3i(4)$

                    $= 2i\sqrt{6}$ $= -12i$

Now try Exercises 7 through 12 ▶

EXAMPLE 2 ▶ **Writing an Expression in Terms of i**

The numbers $x = \dfrac{-6 + \sqrt{-16}}{2}$ and $x = \dfrac{-6 - \sqrt{-16}}{2}$ are not real, but are known

to be solutions of $x^2 + 6x + 13 = 0$. Simplify $\dfrac{-6 + \sqrt{-16}}{2}$.

Solution ▶ Using the i notation, we have

$$\frac{-6 + \sqrt{-16}}{2} = \frac{-6 + i\sqrt{16}}{2} \qquad \text{\color{red}write } \sqrt{-16} \text{ in } i \text{ notation}$$

$$= \frac{-6 + 4i}{2} \qquad \text{\color{red}simplify}$$

$$= \frac{2(-3 + 2i)}{2} \qquad \text{\color{red}factor numerator and reduce}$$

$$= -3 + 2i \qquad \text{\color{red}result}$$

Now try Exercises 13 through 16 ▶

WORTHY OF NOTE

The expression $\dfrac{-6 + 4i}{2}$ from the solution of Example 2 can also be simplified by rewriting it as two separate terms, then simplifying each term:

$$\frac{-6 + 4i}{2} = \frac{-6}{2} + \frac{4i}{2}$$
$$= -3 + 2i.$$

The result in Example 2 contains both a **real number part** (-3) and an **imaginary part** $(2i)$. Numbers of this type are called **complex numbers.**

> **Complex Numbers**
>
> Complex numbers are those that can be written in the form $a + bi$,
> where a and b are real numbers and $i = \sqrt{-1}$.

The expression $a + bi$ is called the **standard form** of a complex number. From this definition we note that all real numbers are also complex numbers, since $a + 0i$ is complex with $b = 0$. In addition, all imaginary numbers are complex numbers, since $0 + bi$ is a complex number with $a = 0$ (See Figure 3.1).

EXAMPLE 3 ▶ **Writing Complex Numbers in Standard Form**

Write each complex number in the form $a + bi$, and identify the values of a and b.

    a. $2 + \sqrt{-49}$ **b.** $\sqrt{-12}$ **c.** 7 **d.** $\dfrac{4 + 3\sqrt{-25}}{20}$

Solution ▶   **a.** $2 + \sqrt{-49} = 2 + i\sqrt{49}$ **b.** $\sqrt{-12} = 0 + i\sqrt{12}$

                               $= 2 + 7i$ $= 0 + 2i\sqrt{3}$

                             $a = 2, b = 7$ $a = 0, b = 2\sqrt{3}$

c. $7 = 7 + 0i$
$a = 7, b = 0$

d. $\dfrac{4 + 3\sqrt{-25}}{20} = \dfrac{4 + 3i\sqrt{25}}{20}$

$= \dfrac{4 + 15i}{20}$

$= \dfrac{1}{5} + \dfrac{3}{4}i$

$a = \dfrac{1}{5}$ or 0.2, $b = \dfrac{3}{4}$ or 0.75

Now try Exercises 17 through 24 ▶

☑ **A. You've just seen how we can identify and simplify imaginary and complex numbers**

Complex numbers complete the development of our "numerical landscape." Sets of numbers and their relationships are represented in Figure 3.1, which shows how some sets of numbers are nested within larger sets and highlights the fact that complex numbers consist of a real number part (any number within the orange rectangle), and an imaginary number part (any number within the yellow rectangle).

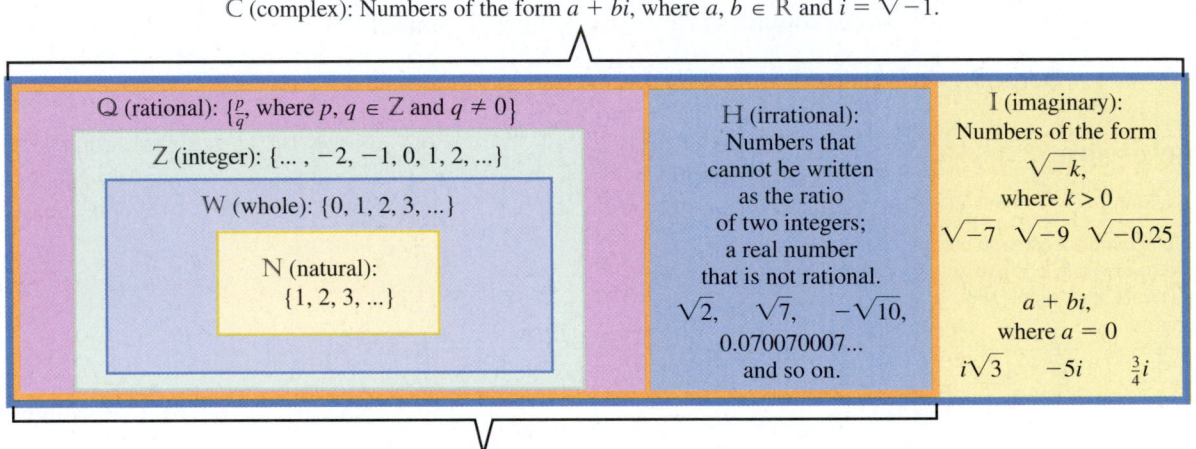

C (complex): Numbers of the form $a + bi$, where $a, b \in$ R and $i = \sqrt{-1}$.

Q (rational): $\{\frac{p}{q}$, where $p, q \in$ Z and $q \neq 0\}$

Z (integer): $\{..., -2, -1, 0, 1, 2, ...\}$

W (whole): $\{0, 1, 2, 3, ...\}$

N (natural): $\{1, 2, 3, ...\}$

H (irrational): Numbers that cannot be written as the ratio of two integers; a real number that is not rational.
$\sqrt{2}, \quad \sqrt{7}, \quad -\sqrt{10},$
0.070070007...
and so on.

I (imaginary): Numbers of the form $\sqrt{-k}$, where $k > 0$
$\sqrt{-7} \quad \sqrt{-9} \quad \sqrt{-0.25}$
$a + bi$, where $a = 0$
$i\sqrt{3} \quad -5i \quad \frac{3}{4}i$

R (real): All rational and irrational numbers

Figure 3.1

B. Adding and Subtracting Complex Numbers

The sum and difference of two polynomials is computed by identifying and combining like terms. The sum or difference of two complex numbers is computed in a similar way, by adding the real number parts from each, and the imaginary parts from each. Notice in Example 4 that the commutative, associative, and distributive properties also apply to complex numbers.

EXAMPLE 4 ▶ Adding and Subtracting Complex Numbers

Perform the indicated operation and write the result in $a + bi$ form.
a. $(2 + 3i) + (-5 + 2i)$ **b.** $(-5 - 4i) - (-2 - i\sqrt{2})$

Solution ▶

a. $(2 + 3i) + (-5 + 2i)$ original sum

$= 2 + 3i + (-5) + 2i$ distribute

$= 2 + (-5) + 3i + 2i$ commute terms

$= [2 + (-5)] + (3i + 2i)$ group like terms

$= -3 + 5i$ result

$a = -3, b = 5$

b. $(-5 - 4i) - (-2 - i\sqrt{2})$ original difference

$= -5 - 4i + 2 + i\sqrt{2}$ distribute

$= -5 + 2 + (-4i) + i\sqrt{2}$ commute terms

$= (-5 + 2) + [(-4i) + i\sqrt{2}]$ group like terms

$= -3 + (-4 + \sqrt{2})i$ result

$a = -3, b = -4 + \sqrt{2}$

Now try Exercises 25 through 30 ▶

Most graphing calculators are programmed to work with imaginary and complex numbers, though for some models the calculator must be placed in complex number mode. After pressing the [MODE] key (located to the right of the [2nd] option key), the screen shown in Figure 3.2 appears and we use the arrow keys to access "$a + bi$" and active this mode (by pressing [ENTER]). Once active, we can validate our previous statements about imaginary numbers (Figure 3.3), as well as verify our previous calculations like those in Examples 3(a), 3(d), and 4(a) (Figure 3.4). Note the imaginary unit i is the [2nd] option for the decimal point.

Figure 3.2

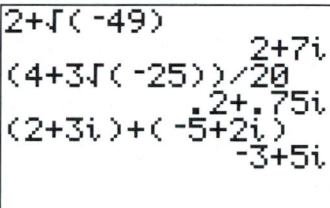

Figure 3.3

```
√(-1)
                i
i²
               -1
√(-16)
               4i
```

Figure 3.4

```
2+√(-49)
              2+7i
(4+3√(-25))/20
           .2+.75i
(2+3i)+(-5+2i)
            -3+5i
```

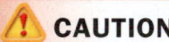 **B.** You've just seen how we can add and subtract complex numbers

C. Multiplying Complex Numbers; Powers of i

The product of two binomials is computed using the distributive property and the F-O-I-L process, then combining like terms. The product of two complex numbers is computed in a similar manner. If any result gives a factor of i^2, remember that $i^2 = -1$.

EXAMPLE 5 ▶ Multiplying Complex Numbers

Find the indicated product and write the answer in $a + bi$ form.

   a. $\sqrt{-4}\sqrt{-9}$ **b.** $\sqrt{-6}\,(2 + \sqrt{-3})$

   c. $(6 - 5i)(4 + i)$ **d.** $(2 + 3i)(2 - 3i)$

Solution ▶

a. $\sqrt{-4}\sqrt{-9} = i\sqrt{4} \cdot i\sqrt{9}$ rewrite in terms of i

     $= 2i \cdot 3i$ simplify

     $= 6i^2$ multiply

     $= -6 + 0i$ result ($i^2 = -1$)

b. $\sqrt{-6}\,(2 + \sqrt{-3}) = i\sqrt{6}(2 + i\sqrt{3})$ rewrite in terms of i

     $= 2i\sqrt{6} + i^2\sqrt{18}$ distribute

     $= 2i\sqrt{6} + (-1)\sqrt{9}\sqrt{2}$ $i^2 = -1$

     $= 2i\sqrt{6} - 3\sqrt{2}$ simplify

     $= -3\sqrt{2} + 2i\sqrt{6}$ standard form

c. $(6 - 5i)(4 + i)$

     $= (6)(4) + 6i + (-5i)(4) + (-5i)(i)$ F-O-I-L

     $= 24 + 6i + (-20i) + (-5)i^2$ $i \cdot i = i^2$

     $= 24 + 6i + (-20i) + (-5)(-1)$ $i^2 = -1$

     $= 29 - 14i$ result

d. $(2 + 3i)(2 - 3i)$

     $= (2)^2 - (3i)^2$ $(A + B)(A - B) = A^2 - B^2$

     $= 4 - 9i^2$ $(3i)^2 = 9i^2$

     $= 4 - 9(-1)$ $i^2 = -1$

     $= 13 + 0i$ result

Now try Exercises 31 through 44 ▶

⚠ **CAUTION ▶** When computing with imaginary and complex numbers, always write the square root of a negative number in terms of i before you begin, as shown in Examples 5(a) and 5(b). Otherwise we get conflicting results, since $\sqrt{-4}\,\sqrt{-9} = \sqrt{36} = 6$ if we multiply the radicands first, which is an incorrect result because the original factors were imaginary. **See Exercise 80.**

As before, computations with complex numbers are easily checked using a graphing calculator. The results from Example 5(a), 5(c), and 5(d) are verified in Figure 3.5. For Example 5(b), the coefficients are irrational numbers that run off of the screen (Figure 3.6). To check this answer, we *could* compare the decimal forms of $-3\sqrt{2}$ and $2\sqrt{6}$ to the given decimal numbers, but instead we chose to simply *check the product using division* (to check the result of a multiplication, divide by one of the factors). Dividing the result by the factor $\sqrt{-6}$, gives $2 + 1.732050808i$, which we easily recognize as the other original factor $2 + i\sqrt{3}$ ✓.

Figure 3.5

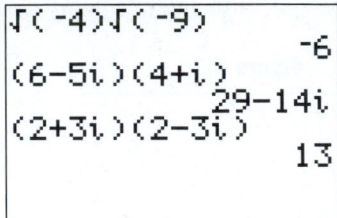

Figure 3.6

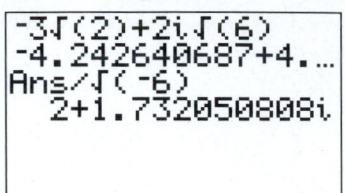

Recall that expressions $2x + 5$ and $2x - 5$ are called binomial conjugates. In the same way, $a + bi$ and $a - bi$ are called **complex conjugates.** Note from Example 5(d) that the *product* of the complex number $a + bi$ with its complex conjugate $a - bi$ is a *real number.* This relationship is useful when rationalizing expressions with a complex number in the denominator, and we generalize the result as follows:

Product of Complex Conjugates

For a complex number $a + bi$ and its conjugate $a - bi$,
their product $(a + bi)(a - bi)$ is the real number $a^2 + b^2$;
$$(a + bi)(a - bi) = a^2 + b^2$$

Showing that $(a + bi)(a - bi) = a^2 + b^2$ is left as an exercise (**see Exercise 79**), but from here on, when asked to compute the product of complex conjugates, simply refer to the formula as illustrated here: $(-3 + 5i)(-3 - 5i) = (-3)^2 + 5^2$ or 34. **See Exercises 45 through 48.**

These operations on complex numbers enable us to verify complex solutions by substitution, in the same way we verify solutions for real numbers. In Example 2 we stated that $x = -3 + 2i$ was one solution to $x^2 + 6x + 13 = 0$. This is verified here.

EXAMPLE 6 ▶   **Checking a Complex Root by Substitution**

Verify that $x = -3 + 2i$ is a solution to $x^2 + 6x + 13 = 0$.

Solution ▶

$$x^2 + 6x + 13 = 0 \quad \textcolor{magenta}{\text{original equation}}$$
$$(-3 + 2i)^2 + 6(-3 + 2i) + 13 = 0 \quad \textcolor{magenta}{\text{substitute } -3 + 2i \text{ for } x}$$
$$(-3)^2 + 2(-3)(2i) + (2i)^2 - 18 + 12i + 13 = 0 \quad \textcolor{magenta}{\text{square and distribute}}$$
$$9 - 12i + 4i^2 + 12i - 5 = 0 \quad \textcolor{magenta}{\text{simplify}}$$
$$9 + (-4) - 5 = 0 \quad \textcolor{magenta}{\text{combine terms } (12i - 12i = 0; \, i^2 = -1)}$$
$$0 = 0 ✓$$

A calculator verification is also shown.

Now try Exercises 49 through 56 ▶

EXAMPLE 7 ▶ **Checking a Complex Root by Substitution**

Show that $x = 2 - i\sqrt{3}$ is a solution of $x^2 - 4x = -7$.

Solution ▶

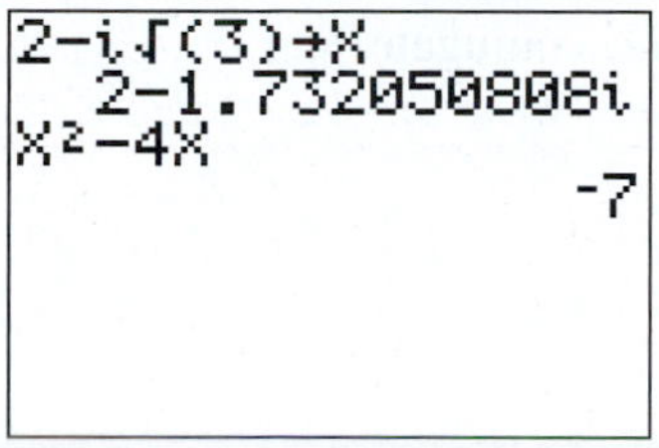

$$
\begin{array}{ll}
x^2 - 4x = -7 & \text{original equation} \\
(2 - i\sqrt{3})^2 - 4(2 - i\sqrt{3}) = -7 & \text{substitute } 2 - i\sqrt{3} \text{ for } x \\
4 - 4i\sqrt{3} + (i\sqrt{3})^2 - 8 + 4i\sqrt{3} = -7 & \text{square and distribute} \\
4 - 4i\sqrt{3} - 3 - 8 + 4i\sqrt{3} = -7 & (i\sqrt{3})^2 = -3 \\
-7 = -7\checkmark & \text{solution checks}
\end{array}
$$

A calculator verification is also shown.

```
2-i√(3)→X
      2-1.732050808i
X²-4X
              -7
```

Now try Exercises 57 through 60 ▶

The imaginary unit i has another interesting and useful property. Since $i = \sqrt{-1}$ and $i^2 = -1$, it follows that $i^3 = i^2 \cdot i = (-1)i = -i$ and $i^4 = (i^2)^2 = 1$. We can now simplify any *higher power of* i by rewriting the expression in terms of i^4, since $(i^4)^n = 1$ for any natural number n.

$$
\begin{array}{ll}
i = i & i^5 = i^4 \cdot i = i \\
i^2 = -1 & i^6 = i^4 \cdot i^2 = -1 \\
i^3 = -i & i^7 = i^4 \cdot i^3 = -i \\
i^4 = 1 & i^8 = (i^4)^2 = 1
\end{array}
$$

Notice the powers of i "cycle through" the four values i, -1, $-i$ and 1. In more advanced classes, powers of complex numbers play an important role, and next we learn to reduce higher powers using the power property of exponents and $i^4 = 1$. Essentially, we divide the exponent on i by 4, then use the remainder to compute the value of the expression. For i^{35}, $35 \div 4 = 8$ remainder 3, showing $i^{35} = (i^4)^8 \cdot i^3 = -i$.

EXAMPLE 8 ▶ **Simplifying Higher Powers of i**

Simplify:

   a. i^{22} **b.** i^{28} **c.** i^{57} **d.** i^{75}

Solution ▶

a. $22 \div 4 = 5$ remainder 2
$$
\begin{aligned}
i^{22} &= (i^4)^5 \cdot (i^2) \\
&= (1)^5(-1) \\
&= -1
\end{aligned}
$$

b. $28 \div 4 = 7$ remainder 0
$$
\begin{aligned}
i^{28} &= (i^4)^7 \\
&= (1)^7 \\
&= 1
\end{aligned}
$$

c. $57 \div 4 = 14$ remainder 1
$$
\begin{aligned}
i^{57} &= (i^4)^{14} \cdot i \\
&= (1)^{14}i \\
&= i
\end{aligned}
$$

d. $75 \div 4 = 18$ remainder 3
$$
\begin{aligned}
i^{75} &= (i^4)^{18} \cdot (i^3) \\
&= (1)^{18}(-i) \\
&= -i
\end{aligned}
$$

Now try Exercises 61 and 62 ▶

While powers of i can likewise be checked on a graphing calculator, the result must be interpreted carefully. For instance, while we know $i^{22} = -1$, the calculator returns an answer of $-1 - 2\text{E} - 13i$ (Figure 3.7). To interpret this result correctly, we identify the real number part as $a = -1$, and the imaginary part as $b = 0$, which due to limitations in the technology is *approximated by* $-2\text{E} - 13 = -2 \times 10^{-13}$ (an extremely small number).

Figure 3.7

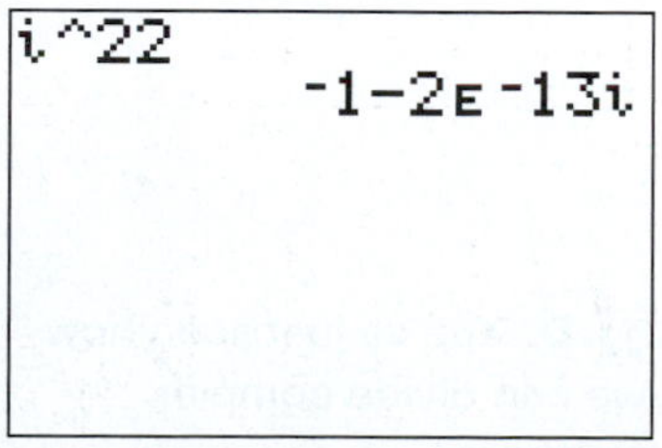

```
i^22
          -1-2E-13i
```

☑ **C.** You've just seen how we can multiply complex numbers and find powers of i

D. Division of Complex Numbers

Since $i = \sqrt{-1}$, expressions like $\dfrac{3 - i}{2 + i}$ actually have a radical in the denominator.

To divide complex numbers, we simply apply our earlier method of rationalizing denominators (Section R.6), but this time using a *complex* conjugate.

EXAMPLE 9 ▶ **Dividing Complex Numbers**

Divide and write each result in $a + bi$ form.

a. $\dfrac{2}{5 - i}$ **b.** $\dfrac{3 - i}{2 + i}$ **c.** $\dfrac{6 + \sqrt{-36}}{3 + \sqrt{-9}}$

Solution ▶

a. $\dfrac{2}{5 - i} = \dfrac{2}{5 - i} \cdot \dfrac{5 + i}{5 + i}$

$= \dfrac{2(5 + i)}{5^2 + 1^2}$

$= \dfrac{10 + 2i}{26}$

$= \dfrac{10}{26} + \dfrac{2}{26}i$

$= \dfrac{5}{13} + \dfrac{1}{13}i$

b. $\dfrac{3 - i}{2 + i} = \dfrac{3 - i}{2 + i} \cdot \dfrac{2 - i}{2 - i}$

$= \dfrac{6 - 3i - 2i + i^2}{2^2 + 1^2}$

$= \dfrac{6 - 5i + (-1)}{5}$

$= \dfrac{5 - 5i}{5} = \dfrac{5}{5} - \dfrac{5i}{5}$

$= 1 - i$

c. $\dfrac{6 + \sqrt{-36}}{3 + \sqrt{-9}} = \dfrac{6 + i\sqrt{36}}{3 + i\sqrt{9}}$ convert to *i* notation

$= \dfrac{6 + 6i}{3 + 3i}$ simplify

The expression can be further simplified by reducing common factors.

$= \dfrac{6(1 + i)}{3(1 + i)} = 2 + 0i$ factor and reduce

Now try Exercises 63 through 68 ▶

As mentioned, operations on complex numbers can be checked using inverse operations, just as we do for real numbers. To check the division from Example 9b, we multiply $1 - i$ by the divisor $2 + i$:

$$(1 - i)(2 + i) = 2 + i - 2i - i^2$$
$$= 2 - i - (-1)$$
$$= 2 - i + 1$$
$$= 3 - i \checkmark$$

Several checks are asked for in the exercises. A calculator check is shown for Example 9(a) in Figure 3.8, where we note that converting the coefficients to rational numbers (where possible): **MATH** **1: ▶Frac**, makes the result easier to understand. As you read and interpret *this* result, note the intent is $\dfrac{5}{13} + \dfrac{1}{13}i$, which must not be confused with $\dfrac{5}{13} + \dfrac{1}{13i}$. The latter is an entirely different (and incorrect) number.

Figure 3.8

```
2/(5-i)
.3846153846+.07…
Ans▶Frac
           5/13+1/13i
```

☑ **D. You've just seen how we can divide complex numbers**

3.1 EXERCISES

▶ CONCEPTS AND VOCABULARY

Fill in each blank with the appropriate word or phrase. Carefully reread the section, if necessary.

1. Given the complex number $3 + 2i$, its complex conjugate is $\underline{3 - 2i}$.

2. The product $(3 + 2i)(3 - 2i)$ gives the real number $\underline{13}$.

3. If the expression $\dfrac{4 + 6i\sqrt{2}}{2}$ is written in the standard form $a + bi$, then $a = \underline{2}$ and $b = \underline{3\sqrt{2}}$.

4. For $i = \sqrt{-1}$, $i^2 = \underline{-1}$, $i^4 = \underline{1}$, $i^6 = \underline{-1}$, and $i^8 = \underline{1}$, $i^3 = \underline{-i}$, $i^5 = \underline{i}$, $i^7 = \underline{-i}$, and $i^9 = \underline{i}$.

5. Discuss/Explain which is correct:
   a. $\sqrt{-4} \cdot \sqrt{-9} = \sqrt{(-4)(-9)} = \sqrt{36} = 6$
   b. $\sqrt{-4} \cdot \sqrt{-9} = 2i \cdot 3i = 6i^2 = -6$ (b) is correct.

6. Compare/Contrast the product $(1 + \sqrt{2})(1 - \sqrt{3})$ with the product $(1 + i\sqrt{2})(1 - i\sqrt{3})$. What is the same? What is different?   Answers will vary.

▶ DEVELOPING YOUR SKILLS

Simplify each radical (if possible). If imaginary, rewrite in terms of i and simplify.

7. a. $\sqrt{-144}$ $12i$ b. $\sqrt{-49}$ $7i$
   c. $\sqrt{27}$ $3\sqrt{3}$ d. $\sqrt{72}$ $6\sqrt{2}$

8. a. $\sqrt{-100}$ $10i$ b. $\sqrt{-169}$ $13i$
   c. $\sqrt{64}$ 8 d. $\sqrt{98}$ $7\sqrt{2}$

9. a. $-\sqrt{-18}$ $-3i\sqrt{2}$ b. $-\sqrt{-50}$ $-5i\sqrt{2}$
   c. $3\sqrt{-25}$ $15i$ d. $2\sqrt{-9}$ $6i$

10. a. $-\sqrt{-32}$ $-4i\sqrt{2}$ b. $-\sqrt{-75}$ $-5i\sqrt{3}$
    c. $3\sqrt{-144}$ $36i$ d. $2\sqrt{-81}$ $18i$

11. a. $\sqrt{-19}$ $i\sqrt{19}$ b. $\sqrt{-31}$ $i\sqrt{31}$
    c. $\sqrt{\dfrac{-12}{25}}$ $\dfrac{2\sqrt{3}}{5}i$ d. $\sqrt{\dfrac{-9}{32}}$ $\dfrac{3\sqrt{2}}{8}i$

12. a. $\sqrt{-17}$ $i\sqrt{17}$ b. $\sqrt{-53}$ $i\sqrt{53}$
    c. $\sqrt{\dfrac{-45}{36}}$ $\dfrac{\sqrt{5}}{2}i$ d. $\sqrt{\dfrac{-49}{75}}$ $\dfrac{7\sqrt{3}}{15}i$

Simplify each expression, writing the result in terms of i.

13. a. $\dfrac{2 + \sqrt{-4}}{2}$ b. $\dfrac{6 + \sqrt{-27}}{3}$

14. a. $\dfrac{16 - \sqrt{-8}}{2}$ b. $\dfrac{4 + 3\sqrt{-20}}{2}$

15. a. $\dfrac{8 + \sqrt{-16}}{2}$ b. $\dfrac{10 - \sqrt{-50}}{5}$

16. a. $\dfrac{6 - \sqrt{-72}}{4}$ b. $\dfrac{12 + \sqrt{-200}}{8}$

Additional answers can be found in the Instructor Answer Appendix.

Write each complex number in the standard form $a + bi$ and clearly identify the values of a and b.

17. a. 5 $5 + 0i; a = 5, b = 0$ b. $3i$ $0 + 3i; a = 0, b = 3$

18. a. -2 $-2 + 0i; a = -2, b = 0$ b. $-4i$ $0 - 4i; a = 0, b = -4$

19. a. $2\sqrt{-81}$ $0 + 18i; a = 0, b = 18$ b. $\dfrac{\sqrt{-32}}{8}$ $0 + \dfrac{\sqrt{2}}{2}i; a = 0, b = \dfrac{\sqrt{2}}{2}$

20. a. $-3\sqrt{-36}$ $0 + -18i; a = 0, b = -18$ b. $\dfrac{\sqrt{-75}}{15}$ $0 + \dfrac{\sqrt{3}}{3}i; a = 0, b = \dfrac{\sqrt{3}}{3}$

21. a. $4 + \sqrt{-50}$ b. $-5 + \sqrt{-27}$

22. a. $-2 + \sqrt{-48}$ b. $7 + \sqrt{-75}$

23. a. $\dfrac{14 + \sqrt{-98}}{8}$ b. $\dfrac{5 + \sqrt{-250}}{10}$

24. a. $\dfrac{21 + \sqrt{-63}}{12}$ b. $\dfrac{8 + \sqrt{-27}}{6}$

Perform the addition or subtraction. Write the result in $a + bi$ form. Check your answers using a calculator.

25. a. $(12 - \sqrt{-4}) + (7 + \sqrt{-9})$ $19 + i$
    b. $(3 + \sqrt{-25}) + (-1 - \sqrt{-81})$ $2 - 4i$
    c. $(11 + \sqrt{-108}) - (2 - \sqrt{-48})$ $9 + 10i\sqrt{3}$

26. a. $(-7 - \sqrt{-72}) + (8 + \sqrt{-50})$ $1 - i\sqrt{2}$
    b. $(\sqrt{3} + \sqrt{-2}) - (\sqrt{12} + \sqrt{-8})$ $-\sqrt{3} - i\sqrt{2}$
    c. $(\sqrt{20} - \sqrt{-3}) + (\sqrt{5} - \sqrt{-12})$ $3\sqrt{5} - 3i\sqrt{3}$

27. a. $(2 + 3i) + (-5 - i)$ $-3 + 2i$
    b. $(5 - 2i) + (3 + 2i)$ 8
    c. $(6 - 5i) - (4 + 3i)$ $2 - 8i$

HOMEWORK SELECTION GUIDE

Core: 7–59 every other odd, 61 63, 67, 73–76 (21 Exercises)
Standard: 1–4, 7–59 every other odd, 61 63, 67, 71, 73–76, 79, 84, 86 (29 Exercises)

Extended: 1–4, 7–59 every other odd, 61 63, 67, 71, 73–76, 79, 81, 84, 86 (30 Exercises)
In Depth: 1–6, 7–59 every other odd, 61 63, 67, 70, 71–78, 79, 81, 82–86 (39 Exercises)

28. a. $(-2 + 5i) + (3 - i)$ $1 + 4i$

b. $(7 - 4i) - (2 - 3i)$ $5 - i$

c. $(2.5 - 3.1i) + (4.3 + 2.4i)$ $6.8 - 0.7i$

29. a. $(3.7 + 6.1i) - (1 + 5.9i)$ $2.7 + 0.2i$

b. $\left(8 + \frac{3}{4}i\right) - \left(-7 + \frac{2}{3}i\right)$ $15 + \frac{1}{12}i$

c. $\left(-6 - \frac{5}{8}i\right) + \left(4 + \frac{1}{2}i\right)$ $-2 - \frac{1}{8}i$

30. a. $(9.4 - 8.7i) - (6.5 + 4.1i)$ $2.9 - 12.8i$

b. $\left(3 + \frac{3}{5}i\right) - \left(-11 + \frac{7}{15}i\right)$ $14 + \frac{2}{15}i$

c. $\left(-4 - \frac{5}{6}i\right) + \left(13 + \frac{3}{8}i\right)$ $9 - \frac{11}{24}i$

Multiply and write your answer in $a + bi$ form.

31. a. $5i \cdot (-3i)$ $15 + 0i$ **b.** $(4i)(-4i)$ $16 + 0i$

32. a. $3(2 - 3i)$ $6 - 9i$ **b.** $-7(3 + 5i)$ $-21 - 35i$

33. a. $-7i(5 - 3i)$ **b.** $6i(-3 + 7i)$ $-42 - 18i$
$-21 - 35i$

34. a. $(-4 - 2i)(3 + 2i)$ **b.** $(2 - 3i)(-5 + i)$
$-8 - 14i$ $-7 + 17i$

35. a. $(-3 + 2i)(2 + 3i)$ **b.** $(3 + 2i)(1 + i)$
$-12 - 5i$ $1 + 5i$

36. a. $(5 + 2i)(-7 + 3i)$ **b.** $(4 - i)(7 + 2i)$
$-41 + i$ $30 + i$

Compute the special products.

37. a. $(4 - 5i)(4 + 5i)$ 41

b. $(7 - 5i)(7 + 5i)$ 74

38. a. $(-2 - 7i)(-2 + 7i)$ 53

b. $(2 + i)(2 - i)$ 5

39. a. $(3 - i\sqrt{2})(3 + i\sqrt{2})$ 11

b. $(\frac{1}{6} + \frac{2}{3}i)(\frac{1}{6} - \frac{2}{3}i)$ $\frac{17}{36}$

40. a. $(5 + i\sqrt{3})(5 - i\sqrt{3})$ 28

b. $(\frac{1}{2} + \frac{3}{4}i)(\frac{1}{2} - \frac{3}{4}i)$ $\frac{13}{16}$

41. a. $(2 + 3i)^2$ $-5 + 12i$ **b.** $(3 - 4i)^2$ $-7 - 24i$

42. a. $(2 - i)^2$ $3 - 4i$ **b.** $(3 - i)^2$ $8 - 6i$

43. a. $(-2 + 5i)^2$ $-21 - 20i$ **b.** $(3 + i\sqrt{2})^2$ $7 + 6i\sqrt{2}$

44. a. $(-2 - 5i)^2$ $-21 + 20i$ **b.** $(2 - i\sqrt{3})^2$ $1 - 4i\sqrt{3}$

For each complex number given, name the complex conjugate and compute the product.

45. a. $4 + 5i$ $4 - 5i; 41$ **b.** $3 - i\sqrt{2}$ $3 + i\sqrt{2}; 11$

46. a. $2 - i$ $2 + i; 5$ **b.** $-1 + i\sqrt{5}$ $-1 - i\sqrt{5}; 6$

47. a. $7i$ $-7i; 49$ **b.** $\frac{1}{2} - \frac{2}{3}i$ $\frac{1}{2} + \frac{2}{3}i; \frac{25}{36}$

48. a. $-5i$ $5i; 25$ **b.** $\frac{3}{4} + \frac{1}{5}i$ $\frac{3}{4} - \frac{1}{5}i; \frac{241}{400}$

Use substitution to determine if the value shown is a solution to the given equation. Use a calculator check for Exercises 57 to 60.

49. $x^2 + 36 = 0; x = -6$ no

50. $x^2 + 16 = 0; x = -4$ no

51. $x^2 + 49 = 0; x = -7i$ yes

52. $x^2 + 25 = 0; x = -5i$ yes

53. $(x - 3)^2 = -9; x = 3 - 3i$ yes

54. $(x + 1)^2 = -4; x = -1 + 2i$ yes

55. $x^2 - 2x - 5 = 0; x = 1 - 2i$ no

56. $x^2 + 6x + 11 = 0; x = -1 - 3i$ no

57. $x^2 - 4x + 9 = 0; x = 2 + i\sqrt{5}$ yes

58. $x^2 - 2x + 4 = 0; x = 1 - i\sqrt{3}$ yes

59. Verify that $x = 1 + 4i$ is a solution to $x^2 - 2x + 17 = 0$. Then show its complex conjugate $1 - 4i$ is also a solution.   verified

60. Verify that $x = 2 - 3i\sqrt{2}$ is a solution to $x^2 - 4x + 22 = 0$. Then show its complex conjugate $2 + 3i\sqrt{2}$ is also a solution.   verified

Simplify using powers of i.

61. a. i^{48} 1 **b.** i^{26} -1 **c.** i^{39} $-i$ **d.** i^{53} i

62. a. i^{36} 1 **b.** i^{50} -1 **c.** i^{19} $-i$ **d.** i^{65} i

Divide and write your answer in $a + bi$ form. Check your answer using multiplication.

63. a. $\dfrac{-2}{\sqrt{-49}}$ $0 + \frac{2}{7}i$ **b.** $\dfrac{4}{\sqrt{-25}}$ $0 - \frac{4}{5}i$

64. a. $\dfrac{2}{1 - \sqrt{-4}}$ $\frac{2}{5} + \frac{4}{5}i$ **b.** $\dfrac{3}{2 + \sqrt{-9}}$ $\frac{6}{13} - \frac{9}{13}i$

65. a. $\dfrac{7}{3 + 2i}$ $\frac{21}{13} - \frac{14}{13}i$ **b.** $\dfrac{-5}{2 - 3i}$ $\frac{-10}{13} - \frac{15}{13}i$

66. a. $\dfrac{6}{1 + 3i}$ $\frac{3}{5} - \frac{9}{5}i$ **b.** $\dfrac{7}{7 - 2i}$ $\frac{49}{53} + \frac{14}{53}i$

67. a. $\dfrac{3 + 4i}{4i}$ $1 - \frac{3}{4}i$ **b.** $\dfrac{2 - 3i}{3i}$ $-1 - \frac{2}{3}i$

68. a. $\dfrac{-4 + 8i}{2 - 4i}$ $-2 + 0i$ **b.** $\dfrac{3 - 2i}{-6 + 4i}$ $-\frac{1}{2} + 0i$

▶ WORKING WITH FORMULAS

69. Absolute value of a complex number:
$$|a + bi| = \sqrt{a^2 + b^2}$$

The absolute value of any complex number $a + bi$ (sometimes called the *modulus* of the number) is computed by taking the square root of the sum of the squares of a and b. Find the absolute value of the given complex numbers.

    a. $|2 + 3i|$ $\sqrt{13}$ **b.** $|4 - 3i|$ 5
    c. $|3 + i\sqrt{2}|$ $\sqrt{11}$

70. Binomial cubes:
$$(A + B)^3 = A^3 + 3A^2B + 3AB^2 + B^3$$

The cube of any binomial can be found using the formula shown, where A and B are the terms of the binomial. Use the formula to compute $(1 - 2i)^3$ (note $A = 1$ and $B = -2i$). $-11 + 2i$

▶ APPLICATIONS

71. Dawn of imaginary numbers: In a day when imaginary numbers were imperfectly understood, Girolamo Cardano (1501–1576) once posed the problem, "Find two numbers that have a sum of 10 and whose product is 40." In other words, $A + B = 10$ and $AB = 40$. Although the solution is routine today, at the time the problem posed an enormous challenge. Verify that $A = 5 + i\sqrt{15}$ and $B = 5 - i\sqrt{15}$ satisfy these conditions.
$A + B = 10$ $AB = 40$

72. Verifying calculations using i: Suppose Cardano had said, "Find two numbers that have a sum of 4 and a product of 7" (see Exercise 71). Verify that $A = 2 + i\sqrt{3}$ and $B = 2 - i\sqrt{3}$ satisfy these conditions.

Although it may seem odd, complex numbers have several applications in the real world. Many of these involve a study of electrical circuits, in particular *alternating current* or AC circuits. Briefly, the components of an AC circuit are current I (in amperes), voltage V (in volts), and the impedance Z (in ohms). The impedance of an electrical circuit is a measure of the total opposition to the flow of current through the circuit and is calculated as $Z = R + iX_L - iX_C$ where R represents a pure resistance, X_C represents the capacitance, and X_L represents the inductance. Each of these is also measured in ohms (symbolized by Ω).

73. Find the impedance Z if $R = 7\ \Omega$, $X_L = 6\ \Omega$, and $X_C = 11\ \Omega$. $(7 - 5i)\ \Omega$

74. Find the impedance Z if $R = 9.2\ \Omega$, $X_L = 5.6\ \Omega$, and $X_C = 8.3\ \Omega$. $(9.2 - 2.7i)\ \Omega$

The voltage V (in volts) across any element in an AC circuit is calculated as a product of the current I and the impedance Z: $V = IZ$.

75. Find the voltage in a circuit with a current $I = 3 - 2i$ amperes and an impedance of $Z = 5 + 5i\ \Omega$. $(25 + 5i)$ V

76. Find the voltage in a circuit with a current $I = 2 - 3i$ amperes and an impedance of $Z = 4 + 2i\ \Omega$. $(14 - 8i)$ V

In an AC circuit, the total impedance (in ohms) is given by $Z = \dfrac{Z_1 Z_2}{Z_1 + Z_2}$, where Z represents the total impedance of a circuit that has Z_1 and Z_2 wired in parallel.

77. Find the total impedance Z if $Z_1 = 1 + 2i$ and $Z_2 = 3 - 2i$. $(\frac{7}{4} + i)\ \Omega$

78. Find the total impedance Z if $Z_1 = 3 - i$ and $Z_2 = 2 + i$. $(\frac{7}{5} + \frac{1}{5}i)\ \Omega$

▶ EXTENDING THE CONCEPT

79. Up to this point, we've said that expressions like $x^2 - 9$ and $p^2 - 7$ are factorable:

$$x^2 - 9 = (x + 3)(x - 3) \text{ and}$$
$$p^2 - 7 = (p + \sqrt{7})(p - \sqrt{7}),$$

while $x^2 + 9$ and $p^2 + 7$ are prime. More correctly, we should state that $x^2 + 9$ and $p^2 + 7$ are nonfactorable *using real numbers,* since they

actually *can* be factored if complex numbers are used. Specifically,

$$(x + 3i)(x - 3i) = x^2 + 9 \text{ and}$$
$$(p + i\sqrt{7})(p - i\sqrt{7}) = p^2 + 7.$$

   a. Verify that in general,
$$(a + bi)(a - bi) = a^2 + b^2.$$
$(a + bi)(a - bi) = a^2 - abi + abi - (bi)^2$
$\qquad\qquad\qquad\quad = a^2 - b^2(-1)$
$\qquad\qquad\qquad\quad = a^2 + b^2 ✓$

Use this idea to factor the following.

b. $x^2 + 36$
$(x + 6i)(x - 6i)$

c. $m^2 + 3$
$(m + i\sqrt{3})(m - i\sqrt{3})$

d. $n^2 + 12$
$(n + 2i\sqrt{3})(n - 2i\sqrt{3})$

e. $4x^2 + 49$
$(2x + 7i)(2x - 7i)$

80. In this section, we noted that the product property of radicals $\sqrt{AB} = \sqrt{A}\sqrt{B}$, can still be applied when at most one of the factors is negative. So what happens if *both* are negative? First consider the expression $\sqrt{-4 \cdot -25}$. What happens if you first multiply in the radicand, then compute the square root? Next consider the product $\sqrt{-4} \cdot \sqrt{-25}$. Rewrite each factor using the i notation, then compute the product. Do you get the same result as before? What can you say about $\sqrt{-4 \cdot -25}$ and $\sqrt{-4} \cdot \sqrt{-25}$? You get a result of 10; you get a result of -10; no, $\sqrt{-4 \cdot -25} \neq \sqrt{-4} \cdot \sqrt{-25}$.

81. Simplify the expression
$i^{17}(3 - 4i) - 3i^3(1 + 2i)^2.$ $-8 - 6i$

82. While it is a simple concept for real numbers, the square root of a complex number is much more involved due to the interplay between its real and imaginary parts. For $z = a + bi$ the square root of z can be found using the formula:
$$\sqrt{z} = \frac{\sqrt{2}}{2}(\sqrt{|z| + a} \pm i\sqrt{|z| - a}),$$ where the sign
is chosen to match the sign of b (see Exercise 69). Use the formula to find the square root of each complex number, then check by squaring.

a. $z = -7 + 24i$ $3 + 4i$
b. $z = 5 - 12i$ $3 - 2i$
c. $z = 4 + 3i$ $\frac{3\sqrt{2}}{2} + \frac{\sqrt{2}}{2}i$

▶ MAINTAINING YOUR SKILLS

83. **(1.4)** Two boats leave Nawiliwili (Kauai) at the same time, traveling in opposite directions. One travels at 15 knots (nautical miles per hour) and the other at 20 knots. How long until they are 196 mi apart?   5.6 hr (5 hr 36 min)

84. **(2.4)** State the domain of the following functions using interval notation.

a. $f(x) = x^{\frac{2}{3}}$
$x \in (-\infty, \infty)$

b. $f(x) = x^{\frac{3}{2}}$ $x \in [0, \infty)$

c. $f(x) = \dfrac{x}{x - 3}$ $x \in (-\infty, 3)\,(3, \infty)$

85. **(1.5)** John can run 10 m/sec, while Rick can only run 9 m/sec. If Rick gets a 2-sec head start, who will reach the 200-m finish line first?   John

86. **(R.4)** Factor the following expressions completely.

a. $x^4 - 16$
b. $n^3 - 27$
c. $x^3 - x^2 - x + 1$
d. $4n^2m - 12nm^2 + 9m^3$

a. $(x + 2)(x - 2)(x^2 + 4)$
b. $(n - 3)(n^2 + 3n + 9)$
c. $(x - 1)^2(x + 1)$
d. $m(2n - 3m)^2$

3.2   Solving Quadratic Equations and Inequalities

LEARNING OBJECTIVES

In Section 3.2 you will see how we can:

☐ **A.** Establish a relationship between zeroes of a quadratic function and the *x*-intercepts of its graph

☐ **B.** Solve quadratic equations using the square root property of equality

☐ **C.** Solve quadratic equations by completing the square

☐ **D.** Solve quadratic equations using the quadratic formula and the discriminant

☐ **E.** Solve quadratic inequalities

☐ **F.** Solve applications of quadratic functions and inequalities

In Section R.4 we reviewed how to solve polynomials by factoring and applying the zero factor property. While this is an extremely valuable skill, many polynomials are unfactorable, and a more general method for finding solutions is necessary. We begin our search here, with the family of quadratic polynomials (degree 2).

A. Zeroes of Quadratic Functions and *x*-Intercepts of Quadratic Graphs

Understanding quadratic equations and functions is an important step towards a more general study of polynomial functions. Due to their importance, we begin by restating their definition:

> **Quadratic Equations**
>
> A quadratic equation is one that can be written in the form
> $$ax^2 + bx + c = 0,$$
> where a, b, and c are real numbers and $a \neq 0$.

As shown, the equation is in standard form, meaning the terms are in decreasing order of degree and the equation is set equal to zero. The family of quadratic functions is similarly defined.

> ### Quadratic Functions
>
> A quadratic function is one that can be written in the form
>
> $$f(x) = ax^2 + bx + c,$$
>
> where a, b, and c are real numbers and $a \neq 0$.

In Section R.4, we noted that some quadratic equations have two real solutions, others have only one, and still others have none. When these possibilities are explored graphically, we note a clear connection between the zeroes of a quadratic function and the x-intercepts of its graph.

EXAMPLE 1 ▶   **Noting Relationships between Zeroes and x-Intercepts**

Consider the functions $f(x) = x^2 - 2x - 3$ and $g(x) = x^2 - 4x + 4$.

  a. Find the zeroes of each function algebraically.

  b. Find the x-intercepts of each function graphically.

  c. Comment on how the zeroes and x-intercepts are related.

Solution ▶   **a.** To find the zeroes algebraically, replace $f(x)$ and $g(x)$ with 0, then solve. In each case, the solutions can be found by factoring.

For $f(x)$:	$x^2 - 2x - 3 = 0$	For $g(x)$:	$x^2 - 4x + 4 = 0$
	$(x - 3)(x + 1) = 0$		$(x - 2)(x - 2) = 0$
	$x = 3$ or $x = -1$		$x = 2$ or $x = 2$

There are two real solutions,   There is only one real solution,
$x = 3$ and $x = -1$.   but it is repeated twice.

  b. To find the x-intercepts, go to the (Y=) screen and enter the first function as Y_1 with $Y_2 = 0$ (the x-axis), then graph them in the standard window. Locate the x-intercepts using the (2nd) (TRACE) **(CALC) 5:Intersect** feature.

  • For $f(x) = x^2 - 2x - 3$

Figure 3.9   **Figure 3.10**

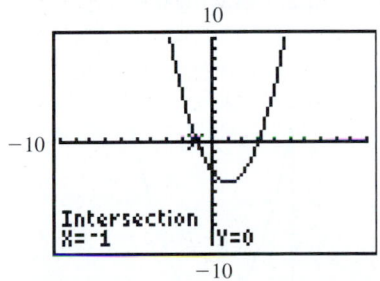

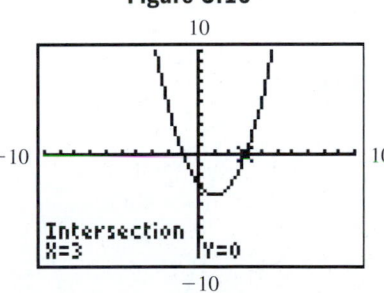

The result shows there are two x-intercepts, which occur at $x = -1$ (Figure 3.9) and $x = 3$ (Figure 3.10). These were also the real zeroes of $f(x) = x^2 - 2x - 3$.

• For $g(x) = x^2 - 4x + 4$

After entering $x^2 - 4x + 4$ as Y_1 (leaving $Y_2 = 0$) we again graph both functions in the standard window. This time we find there is only one x-intercept, located at $x = 2$ (Figure 3.11), a fact supported by the accompanying table (Figure 3.12). The graph is **tangent** to the x-axis at $x = 2$ (touching the axis at just this one point) and a closer look at g reveals why—the function is easily rewritten as $g(x) = (x - 2)^2$, producing the zero at $x = 2$ *and positive values for all other inputs!* Note that $x = 2$ was the only zero found for $g(x) = x^2 - 4x + 4$.

Figure 3.11

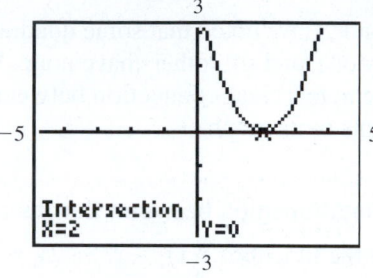

Figure 3.12

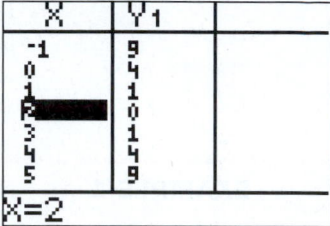

c. From parts (a) and (b), it's apparent that *the real zeroes of a function appear graphically as x-intercepts.*

Now try Exercises 7 through 20 ▶

In Section 3.1, we noted that the equation $x^2 + 1 = 0$ had no real solutions, indicating the function $f(x) = x^2 + 1$ has no real zeroes. Here we might wonder whether a graphical connection also exists for the "no real zeroes" case. The graph of $f(x) = x^2 + 1$ is given in Figure 3.13 and shows that f has no x-intercepts, further affirming the graphical connection noted in Example 1. A summary of these connections is shown in Figure 3.14 for the case where $a > 0$ (the graph opens upward). Similar statements can be made when $a < 0$ (the graph opens downward).

Figure 3.13

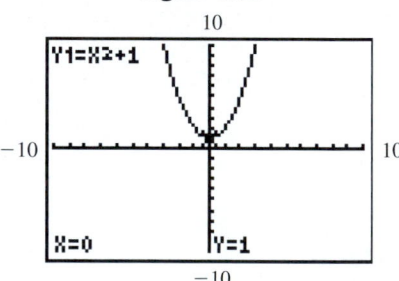

Figure 3.14

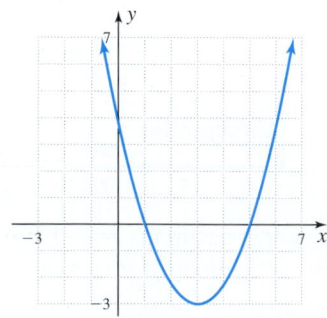

$f(x) = ax^2 + bx + c$
has two x-intercepts
and two real zeroes;
$ax^2 + bx + c = 0$
has two real solutions.

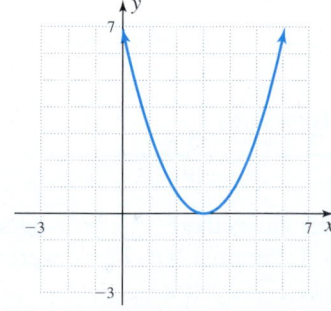

$g(x) = ax^2 + bx + c$
has one x-intercept
and one real repeated zero;
$ax^2 + bx + c = 0$
has one real repeated solution.

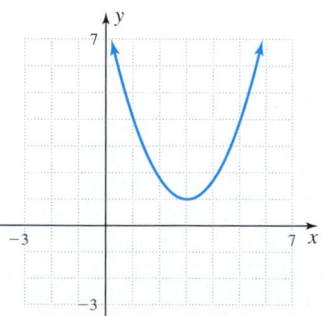

$h(x) = ax^2 + bx + c$
has no x-intercepts
and no real zeroes;
$ax^2 + bx + c = 0$
has no real solutions.

Also from our work in Example 1, we learn that the following statements are equivalent, meaning that if any one of the statements is true, then all four statements are true.

☑ **A.** You've just seen how we can establish a relationship between the zeroes of a quadratic function and the x-intercepts of its graph

For any real number r,

1. $x = r$ is a solution of $f(x) = 0$.
2. r is a zero of $f(x)$.
3. $(x - r)$ is a factor of $f(x)$.
4. $(r, 0)$ is an x-intercept of the graph of f.

B. Quadratic Equations and the Square Root Property of Equality

In Section 1.5 we solved the equation $ax + b = c$ for x to establish a general solution for equations of this form. In this section, we'll establish a general solution for $ax^2 + bx + c = 0$ using a process known as *completing the square*. To begin, we note that the equation $x^2 = 9$ can be solved by factoring. In standard form we have $x^2 - 9 = 0$ (note $b = 0$), then $(x + 3)(x - 3) = 0$. The solutions are $x = -3$ or $x = 3$, which are simply the *positive and negative square roots of 9*. This result suggests an alternative method for solving equations of the form $X^2 = k$, known as the **square root property of equality.**

Square Root Property of Equality

If X represents an algebraic expression and $X^2 = k$,

$$\text{then } X = \sqrt{k} \text{ or } X = -\sqrt{k};$$

$$\text{also written as } X = \pm\sqrt{k}$$

EXAMPLE 2 ▶ **Solving an Equation Using the Square Root Property of Equality**

Use the square root property of equality to solve each equation. Verify solutions graphically.

a. $-4x^2 + 3 = -6$ **b.** $x^2 + 12 = 0$ **c.** $(x - 5)^2 = 7$

Solution ▶   **a.** $-4x^2 + 3 = -6$ original equation

$$x^2 = \frac{9}{4}$$ subtract 3, divide by -4

$$x = \sqrt{\frac{9}{4}} \text{ or } x = -\sqrt{\frac{9}{4}}$$ square root property of equality

$$x = \frac{3}{2} \text{ or } x = -\frac{3}{2}$$ simplify radicals

WORTHY OF NOTE

In Section R.6 we noted that for any real number a, $\sqrt{a^2} = |a|$. From Example 2(a), solving the equation by taking the square root of both sides produces $\sqrt{x^2} = \sqrt{\frac{9}{4}}$. This is equivalent to $|x| = \sqrt{\frac{9}{4}}$, again showing this equation must have two solutions, $x = -\sqrt{\frac{9}{4}}$ and $x = \sqrt{\frac{9}{4}}$.

This equation has two rational solutions.

Using the Intersection Method with $Y_1 = -4x^2 + 3$ and $Y_2 = -6$, we note the graphs intersect at two points. This shows there will be two real roots, which turn out to be rational (Figures 3.15 and 3.16).

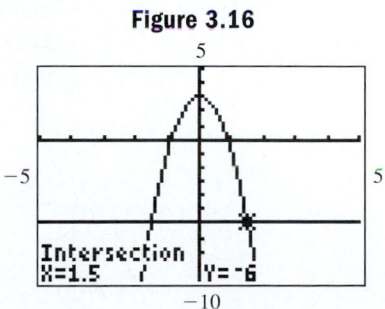

Figure 3.15

Figure 3.16

b. $x^2 + 12 = 0$ original equation

$\qquad x^2 = -12$ subtract 12

$\quad x = \sqrt{-12} \text{ or } x = -\sqrt{-12}$ square root property of equality

$\quad x = 2i\sqrt{3} \quad \text{ or } \quad x = -2i\sqrt{3}$ simplify radicals

This equation has two complex solutions.

Using the Zeroes Method with $Y_1 = x^2 + 12$ shows there are no x-intercepts, and therefore no real roots (Figure 3.17). However, the roots can still be checked on the home screen, as shown in Figure 3.18.

Figure 3.17

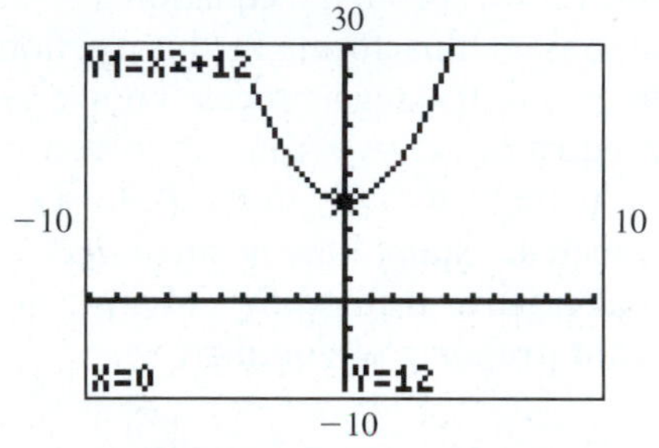

Figure 3.18

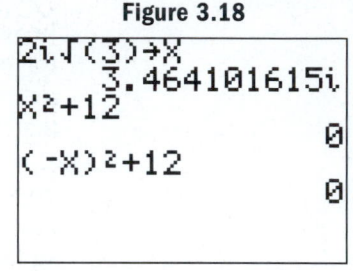

c. $(x - 5)^2 = 7$ original equation

$\quad x - 5 = \sqrt{7} \qquad \text{or} \qquad x - 5 = -\sqrt{7}$ square root property of equality

$\qquad x = 5 + \sqrt{7} \qquad\qquad\qquad x = 5 - \sqrt{7}$ solve for x

This equation has two irrational solutions.

Using the Zeroes Method with $Y_1 = (x - 5)^2 - 7$, we note the graph has two x-intercepts. This shows there will be two real roots, which turn out to be irrational (Figures 3.19 and 3.20).

Figure 3.19

Figure 3.20

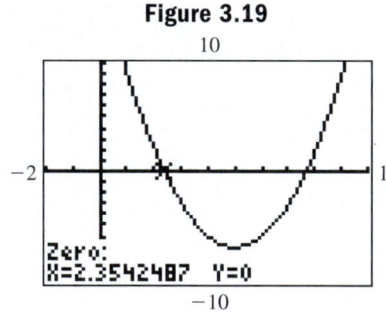 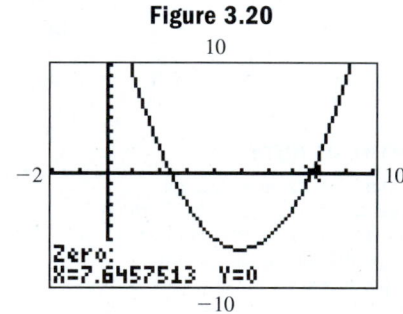

Now try Exercises 21 through 36 ▶

⚠ **CAUTION** ▶ For equations of the form $(x + d)^2 = k$ as in Example 2(c), you should resist the temptation to expand the binomial square in an attempt to simplify the equation and solve by factoring—many times the result is nonfactorable. *Any equation of the form* $(x + d)^2 = k$ can quickly be solved using the square root property of equality.

Answers written using radicals are called **exact** or **closed form** solutions. Actually checking the exact solutions is a nice application of fundamental skills. Let's check $x = 5 + \sqrt{7}$ from Example 2(c).

check:	$(x - 5)^2 = 7$	original equation
	$(5 + \sqrt{7} - 5)^2 = 7$	substitute $5 + \sqrt{7}$ for x
	$(\sqrt{7})^2 = 7$	simplify
	$7 = 7 \checkmark$	$(\sqrt{7})^2 = 7$
		result checks ($x = 5 - \sqrt{7}$ also checks)

✔ **B.** You've just seen how we can solve quadratic equations using the square root property of equality

C. Solving Quadratic Equations by Completing the Square

Again consider $(x - 5)^2 = 7$ from Example 2(c). If we had first expanded the binomial square, we would have obtained $x^2 - 10x + 25 = 7$, then $x^2 - 10x + 18 = 0$ in standard form. Note that this equation *cannot be solved by factoring*. Reversing this process leads us to a strategy for solving nonfactorable quadratic equations, by creating a *perfect square trinomial* from the quadratic and linear terms. This process is known as **completing the square.** To transform $x^2 - 10x + 18 = 0$ back into $x^2 - 10x + 25 = 7$ [which we would then rewrite as $(x - 5)^2 = 7$ and solve], we subtract 18 from both sides, then add 25:

$x^2 - 10x + 18 = 0$	
$x^2 - 10x = -18$	subtract 18
$x^2 - 10x + 25 = -18 + 25$	add 25
$(x - 5)^2 = 7$	factor, simplify

In general, after subtracting the constant term, the number that "completes the square" is computed as $\left[\frac{1}{2}(\text{coefficient of linear term})\right]^2$: $\left[\frac{1}{2}(10)\right]^2 = 25$. For additional practice finding this constant term, **see Exercises 37 through 42.**

EXAMPLE 3 ▶ **Solving a Quadratic Equation by Completing the Square**

Solve by completing the square: $x^2 + 13 = 6x$.

Solution ▶

$x^2 + 13 = 6x$	original equation
$x^2 - 6x + 13 = 0$	standard form
$x^2 - 6x + \underline{\quad} = -13 + \underline{\quad}$	subtract 13 to make room for new constant
$\left[(\frac{1}{2})(-6)\right]^2 = 9$	compute $\left[(\frac{1}{2})(\textit{linear coefficient})\right]^2$
$x^2 - 6x + 9 = -13 + 9$	add 9 to both sides (completing the square)
$(x - 3)^2 = -4$	factor and simplify
$x - 3 = \sqrt{-4} \quad$ or $\quad x - 3 = -\sqrt{-4}$	square root property of equality
$x = 3 + 2i \quad$ or $\quad x = 3 - 2i$	simplify radicals and solve for x

Now try Exercises 43 through 52 ▶

Based on our earlier work, we expect that the equation from Example 3 has no x-intercepts. Using $Y_1 = x^2 - 6x + 13$ confirms this (Figure 3.21 following), but we gain additional insight using the equation in its given form and the intersection-of-graphs method. With $Y_1 = x^2 + 13$ and $Y_2 = 6x$, we realize that if the graphs do not intersect (Figure 3.22), there can likewise be no real roots!

Figure 3.21

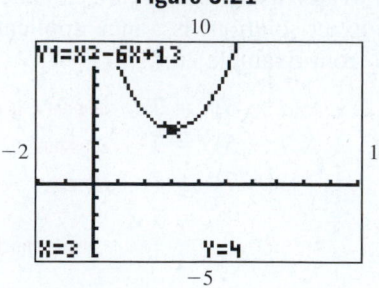

Figure 3.22

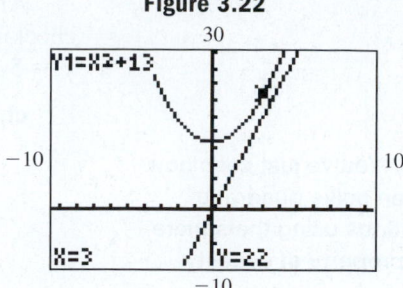

The process of completing the square can be applied to any quadratic equation with a leading coefficient of 1. If the leading coefficient is not 1, we simply divide through by a before beginning, which brings us to this summary of the process.

Completing the Square to Solve a Quadratic Equation

To solve $ax^2 + bx + c = 0$ by completing the square:

1. Subtract the constant c from both sides.
2. Divide both sides by the leading coefficient a.
3. Compute $\left[\dfrac{1}{2} \cdot \dfrac{b}{a}\right]^2 = \left(\dfrac{b}{2a}\right)^2$ and add the result to both sides.
4. Factor left-hand side as a binomial square; simplify right-hand side.
5. Solve using the square root property of equality.

EXAMPLE 4 ▶ **Solving Quadratic Equations**

Find all solutions: $-3x^2 + 1 = 4x$.

Algebraic Solution ▶ **Completing the Square**

$$-3x^2 + 1 = 4x \qquad \text{original equation}$$

$$-3x^2 - 4x + 1 = 0 \qquad \text{standard form (nonfactorable)}$$

$$-3x^2 - 4x = -1 \qquad \text{subtract 1}$$

$$x^2 + \frac{4}{3}x + \quad = \frac{1}{3} \qquad \text{divide by } -3$$

$$x^2 + \frac{4}{3}x + \frac{4}{9} = \frac{1}{3} + \frac{4}{9} \qquad \left[\left(\frac{1}{2}\right)\left(\frac{4}{3}\right)\right]^2 = \left(\frac{2}{3}\right)^2 = \frac{4}{9}; \text{ add } \frac{4}{9}$$

$$\left(x + \frac{2}{3}\right)^2 = \frac{7}{9} \qquad \text{factor and simplify } \left(\frac{1}{3} = \frac{3}{9}\right)$$

$$x + \frac{2}{3} = \sqrt{\frac{7}{9}} \quad \text{or} \quad x + \frac{2}{3} = -\sqrt{\frac{7}{9}} \qquad \text{square root property of equality}$$

$$x = -\frac{2}{3} + \frac{\sqrt{7}}{3} \quad \text{or} \quad x = -\frac{2}{3} - \frac{\sqrt{7}}{3} \qquad \text{solve for } x \text{ and simplify (exact form)}$$

$$x \approx 0.22 \quad \text{or} \quad x \approx -1.55 \qquad \text{approximate form (to hundredths)}$$

Graphical Solution ▶ **Zeroes Method**

After writing the equation in standard form, we have $-3x^2 - 4x + 1 = 0$. The graph of $Y_1 = -3x^2 - 4x + 1$ has two x-intercepts, indicating there will be two real roots. Locating the zeroes yields the decimal approximations shown and confirms our calculated results (Figures 3.23 and 3.24).

Figure 3.23

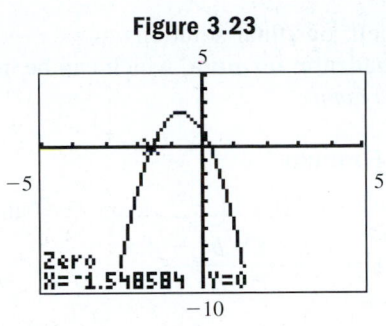

Figure 3.24

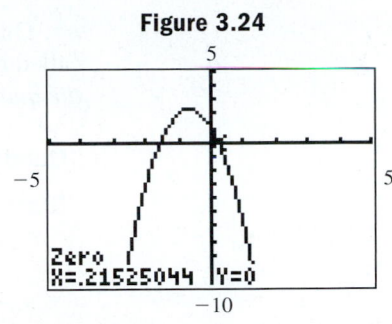

☑ **C.** You've just seen how we can solve quadratic equations by completing the square

Now try Exercises 53 through 60 ▶

⚠ **CAUTION** ▶ For many of the skills/processes needed in a study of algebra, it's actually easier to work with the fractional form of a number, rather than the decimal form. For example, computing $\left(\frac{2}{3}\right)^2$ is easier than computing $(0.\overline{6})^2$, and finding $\sqrt{\frac{9}{16}}$ is much easier than finding $\sqrt{0.5625}$.

D. The Quadratic Formula and the Discriminant

In Section 1.5 we found a general solution for the linear equation $ax + b = c$ by comparing it to $2x + 3 = 15$. Here we'll use a similar idea to find a general solution for quadratic equations. In a side-by-side format, we'll solve the equations $2x^2 + 5x + 3 = 0$ and $ax^2 + bx + c = 0$ by completing the square. Note the similarities.

$2x^2 + 5x + 3 = 0$	given equations	$ax^2 + bx + c = 0$
$2x^2 + 5x + \quad = -3$	subtract constant term	$ax^2 + bx + \quad = -c$
$x^2 + \dfrac{5}{2}x + \underline{\quad} = -\dfrac{3}{2}$	divide by lead coefficient	$x^2 + \dfrac{b}{a}x + \underline{\quad} = -\dfrac{c}{a}$
$\left[\dfrac{1}{2}\left(\dfrac{5}{2}\right)\right]^2 = \dfrac{25}{16}$	$\left[\dfrac{1}{2}(\text{linear coefficient})\right]^2$	$\left[\dfrac{1}{2}\left(\dfrac{b}{a}\right)\right]^2 = \dfrac{b^2}{4a^2}$
$x^2 + \dfrac{5}{2}x + \dfrac{25}{16} = \dfrac{25}{16} - \dfrac{3}{2}$	add to both sides	$x^2 + \dfrac{b}{a}x + \dfrac{b^2}{4a^2} = \dfrac{b^2}{4a^2} - \dfrac{c}{a}$
$\left(x + \dfrac{5}{4}\right)^2 = \dfrac{25}{16} - \dfrac{3}{2}$	left side factors as a binomial square	$\left(x + \dfrac{b}{2a}\right)^2 = \dfrac{b^2}{4a^2} - \dfrac{c}{a}$
$\left(x + \dfrac{5}{4}\right)^2 = \dfrac{25}{16} - \dfrac{24}{16}$	determine LCDs	$\left(x + \dfrac{b}{2a}\right)^2 = \dfrac{b^2}{4a^2} - \dfrac{4ac}{4a^2}$
$\left(x + \dfrac{5}{4}\right)^2 = \dfrac{1}{16}$	simplify right side	$\left(x + \dfrac{b}{2a}\right)^2 = \dfrac{b^2 - 4ac}{4a^2}$
$x + \dfrac{5}{4} = \pm\sqrt{\dfrac{1}{16}}$	square root property of equality	$x + \dfrac{b}{2a} = \pm\sqrt{\dfrac{b^2 - 4ac}{4a^2}}$
$x + \dfrac{5}{4} = \pm\dfrac{1}{4}$	simplify radicals	$x + \dfrac{b}{2a} = \pm\dfrac{\sqrt{b^2 - 4ac}}{2a}$
$x = -\dfrac{5}{4} \pm \dfrac{1}{4}$	solve for x	$x = -\dfrac{b}{2a} \pm \dfrac{\sqrt{b^2 - 4ac}}{2a}$
$x = \dfrac{-5 \pm 1}{4}$	combine terms	$x = \dfrac{-b \pm \sqrt{b^2 - 4ac}}{2a}$
$x = \dfrac{-5 + 1}{4} \quad \text{or} \quad x = \dfrac{-5 - 1}{4}$	solutions	$x = \dfrac{-b + \sqrt{b^2 - 4ac}}{2a} \quad \text{or} \quad x = \dfrac{-b - \sqrt{b^2 - 4ac}}{2a}$

On the left, our final solutions are $x = -1$ or $x = -\frac{3}{2}$. The general solution is called the **quadratic formula,** which can be used to solve *any equation belonging to the quadratic family.*

Quadratic Formula

If $ax^2 + bx + c = 0$, with a, b, and $c \in \mathbb{R}$ and $a \neq 0$, then

$$x = \frac{-b \mathbin{\color{red}{+}} \sqrt{b^2 - 4ac}}{2a} \quad \text{or} \quad x = \frac{-b \mathbin{\color{red}{-}} \sqrt{b^2 - 4ac}}{2a};$$

also written $x = \dfrac{-b \pm \sqrt{b^2 - 4ac}}{2a}$.

⚠ **CAUTION** ▶ It's very important to note the values of a, b, and c come from an equation *written in stan-dard form.* For $3x^2 - 5x = -7$, $a = 3$ and $b = -5$, but $c \neq -7$! In standard form we have $3x^2 - 5x + 7 = 0$, and note the value for use in the formula is actually $c = 7$.

EXAMPLE 5 ▶ **Solving Quadratic Equations Using the Quadratic Formula**

Solve $4x^2 + 1 = 8x$ using the quadratic formula. State the solution(s) in both exact and approximate form. Check one of the exact solutions in the original equation.

Solution ▶ Begin by writing the equation in standard form and identifying the values of a, b, and c.

$4x^2 + 1 = 8x$	original equation
$4x^2 - 8x + 1 = 0$	standard form
$a = 4, b = -8, c = 1$	
$x = \dfrac{-(-8) \pm \sqrt{(-8)^2 - 4(4)(1)}}{2(4)}$	substitute 4 for a, -8 for b, and 1 for c
$x = \dfrac{8 \pm \sqrt{64 - 16}}{8} = \dfrac{8 \pm \sqrt{48}}{8}$	simplify
$x = \dfrac{8 \pm 4\sqrt{3}}{8} = \dfrac{8}{8} \pm \dfrac{4\sqrt{3}}{8}$	simplify radical (see following CAUTION)
$x = 1 + \dfrac{\sqrt{3}}{2} \quad \text{or} \quad x = 1 - \dfrac{\sqrt{3}}{2}$	exact solutions
$x \approx 1.87 \quad \text{or} \quad x \approx 0.13$	approximate solutions

Check ▶

$4x^2 + 1 = 8x$	original equation
$4\left(1 + \dfrac{\sqrt{3}}{2}\right)^2 + 1 = 8\left(1 + \dfrac{\sqrt{3}}{2}\right)$	substitute $1 + \frac{\sqrt{3}}{2}$ for x
$4\left[1 + 2\left(\dfrac{\sqrt{3}}{2}\right) + \dfrac{3}{4}\right] + 1 = 8 + 4\sqrt{3}$	square binomial; distribute
$4 + 4\sqrt{3} + 3 + 1 = 8 + 4\sqrt{3}$	distribute
$8 + 4\sqrt{3} = 8 + 4\sqrt{3} \checkmark$	result checks

```
1+√(3)/2→X
         1.866025404
4X²+1
         14.92820323
8X
         14.92820323
```

A graphing calculator check is also shown.

Now try Exercises 61 through 90 ▶

> ⚠️ **CAUTION** ▶ For $\dfrac{8 \pm 4\sqrt{3}}{8}$, be careful not to incorrectly "cancel the eights" as in $\dfrac{\overset{1}{8} \pm 4\sqrt{3}}{\underset{1}{8}} \neq 1 \pm 4\sqrt{3}$.
>
> *No!* Use a calculator to verify that the results are not equivalent. Both terms in the numerator are divided by 8 and we must either rewrite the expression as separate terms (as above) or factor the numerator to see if the expression simplifies further:
>
> $\dfrac{8 \pm 4\sqrt{3}}{8} = \dfrac{\overset{1}{4}(2 \pm \sqrt{3})}{\underset{2}{8}} = \dfrac{2 \pm \sqrt{3}}{2}$, which is equivalent to $1 \pm \dfrac{\sqrt{3}}{2}$.

The Discriminant of the Quadratic Formula

The conclusions we reached graphically in Figure 3.14 regarding the nature of number of quadratic roots can now be seen algebraically through a closer look at the quadratic formula. For any real-valued expression X, recall that $\sqrt{X}$ represents a real number only for $X \geq 0$. Since the quadratic formula contains the radical $\sqrt{b^2 - 4ac}$, the expression $b^2 - 4ac$, called the **discriminant,** will determine the nature (real or complex) and the number of solutions to a given quadratic equation.

> **The Discriminant of the Quadratic Formula**
>
> For $f(x) = ax^2 + bx + c$, where $a, b, c \in \mathbb{R}$ and $a \neq 0$,
>
> **1.** If $b^2 - 4ac > 0$, there are two real roots
>
> **2.** If $b^2 - 4ac = 0$, there is one real (repeated) root
>
> **3.** If $b^2 - 4ac < 0$, there are two nonreal roots

These ideas are further illustrated here, by comparing the value of the discriminant with the graph of the related function shown below each calculation.

1. $f(x) = x^2 - 4x$
$a = 1, b = -4, c = 0$
$b^2 - 4ac = (-4)^2 - 4(1)(0)$
$\qquad = 16 - 0$
$\qquad = 16$
$16 > 0, f$ has two real roots

2. $g(x) = x^2 - 4x + 4$
$a = 1, b = -4, c = 4$
$b^2 - 4ac = (-4)^2 - 4(1)(4)$
$\qquad = 16 - 16$
$\qquad = 0$
$0 = 0, g$ has one real root

3. $h(x) = x^2 - 4x + 5$
$a = 1, b = -4, c = 5$
$b^2 - 4ac = (-4)^2 - 4(1)(5)$
$\qquad = 16 - 20$
$\qquad = -4$
$-4 < 0, h$ has two nonreal roots

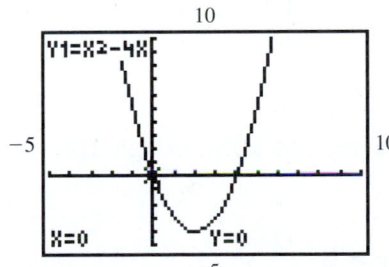
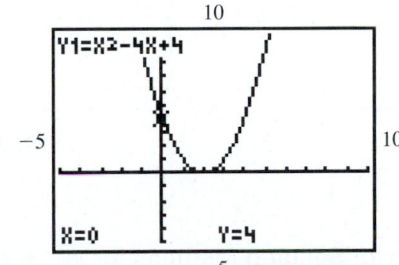
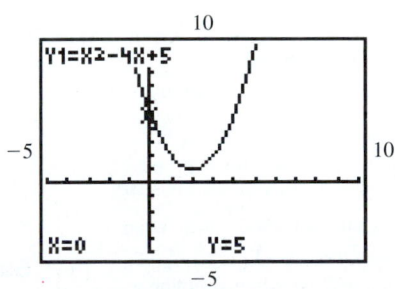

Finally, we note from (3) and the structure of the quadratic formula, that the complex solutions must occur in conjugate pairs.

> **Complex Solutions**
>
> The complex solutions of a quadratic equation with real coefficients must occur in conjugate pairs.
> If $a + bi$ is a solution, then $a - bi$ is also a solution.

Further analysis of the discriminant reveals even more concerning the nature of quadratic solutions. Namely, if a, b, and c are rational and the discriminant is

1. zero, the original equation is a perfect square trinomial.
2. a perfect square, there will be two rational roots which means the original equation can be solved by factoring.
3. not a perfect square, there will be two irrational roots.

See Exercises 91 through 102.

EXAMPLE 6 ▶ **Solving Quadratic Equations Using the Quadratic Formula**

Solve: $2x^2 - 6x + 5 = 0$.

Solution ▶ With $a = 2$, $b = -6$, and $c = 5$, the discriminant becomes $(-6)^2 - 4(2)(5) = -4$, showing there will be two complex roots. The quadratic formula then yields

$$x = \frac{-b \pm \sqrt{b^2 - 4ac}}{2a}$$ quadratic formula

$$x = \frac{-(-6) \pm \sqrt{-4}}{2(2)}$$ $b^2 - 4ac = -4$, substitute 2 for a and -6 for b

$$x = \frac{6 \pm 2i}{4}$$ simplify, write in i form

$$x = \frac{3}{2} \pm \frac{1}{2}i$$ solutions are complex conjugates

A calculator check is shown in Figures 3.25 and 3.26.

Figure 3.25, 3.26

```
1.5+0.5i→X
              1.5+.5i
2X²-6X+5
                    0
```
```
1.5-0.5i→X
              1.5-.5i
2X²-6X+5
                    0
```

Now try Exercises 103 through 108 ▶

WORTHY OF NOTE

While it's possible to solve by completing the square if $\frac{b}{a}$ is a fraction or an odd number (see Example 4), the process is usually most efficient when $\frac{b}{a}$ is an even number. This is one observation you could use when selecting a solution method.

☑ **D.** You've just seen how we can solve quadratic equations using the quadratic formula and the discriminant

Summary of Solution Methods for $ax^2 + bx + c = 0$

1. If $b = 0$, $ax^2 + c = 0$: isolate x^2 and use the square root property of equality.
2. If $c = 0$, $ax^2 + bx = 0$: factor out the GCF and use the zero product property.
3. If no coefficient is zero, you can attempt to solve by
   a. factoring   **b.** completing the square   **c.** using the quadratic formula
   d. using the intersection-of-graphs method   **e.** using the zeroes method

E. Quadratic Inequalities

The study of quadratic inequalities is simply an extension of our earlier work in analyzing functions (Section 2.1). While we've developed the ability to graph a variety of new functions, the solution set for an inequality will still be determined by analyzing the behavior of the function at its zeroes. The key idea is to recognize the following statements are synonymous:

1. $f(x) > 0$.   **2.** Outputs are positive.   **3.** The graph is *above the x-axis*.

Similar statements can be made using the other inequality symbols.

Solving a quadratic inequality only requires that we (a) locate any real zeroes of the function and (b) determine whether the graph opens upward or downward. If there are no *x*-intercepts, the graph is entirely above the *x*-axis (output values are positive), or entirely below the *x*-axis (output values are negative), making the solution either all real numbers or the empty set.

EXAMPLE 7 ▶   **Solving a Quadratic Inequality**

For $f(x) = x^2 + x - 6$, solve $f(x) > 0$.

Analytical Solution ▶   The graph of f will open upward since $a > 0$. Factoring gives $f(x) = (x + 3)(x - 2)$, with zeroes at -3 and 2. Using the *x*-axis alone (since graphing the function is not our focus), we plot $(-3, 0)$ and $(2, 0)$ and visualize a parabola opening upward through these points (Figure 3.27).

Figure 3.27

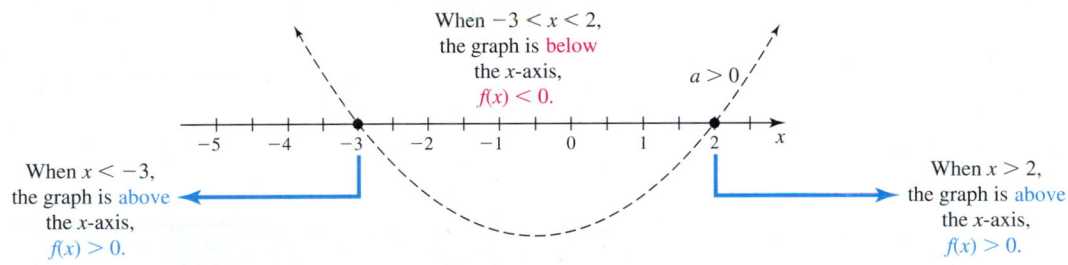

When $-3 < x < 2$, the graph is below the *x*-axis, $f(x) < 0$.

$a > 0$

When $x < -3$, the graph is above the *x*-axis, $f(x) > 0$.

When $x > 2$, the graph is above the *x*-axis, $f(x) > 0$.

The diagram clearly shows the graph is *above* the *x*-axis (outputs are positive) when $x < -3$ or when $x > 2$. The solution is $x \in (-\infty, -3) \cup (2, \infty)$.

Graphical Solution ▶   The complete graph of f shown in Figure 3.28 confirms the analytical solution. For the intervals of the domain shown in **bold** $(-\infty, -3) \cup (2, \infty)$, the graph is above the *x*-axis $(f(x) > 0)$. For the portion shown in red, the graph is below the *x*-axis $(f(x) < 0)$.

Figure 3.28

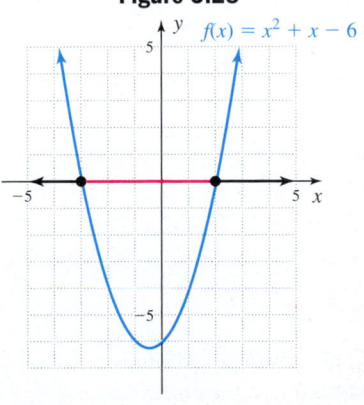

$f(x) = x^2 + x - 6$

Now try Exercises 109 through 120 ▶

When solving general inequalities, zeroes of multiplicity continue to play a role. In Example 7, the zeroes of f were both of multiplicity 1, and the graph crossed the *x*-axis at these points. In other cases, the zeroes may have even multiplicity.

EXAMPLE 8 ▶ **Solving a Quadratic Inequality**

Solve the inequality $-x^2 + 6x \leq 9$.

Analytical Solution ▶ Begin by writing the inequality in standard form: $-x^2 + 6x - 9 \leq 0$. Note this is equivalent to $g(x) \leq 0$ for $g(x) = -x^2 + 6x - 9$. Since $a < 0$, the graph of g will open downward. The factored form is $g(x) = -(x - 3)^2$, showing 3 is a zero and a repeated root. Using the x-axis, we plot the point $(3, 0)$ and visualize a parabola opening downward through this point.

Figure 3.29 shows the graph is *below* the x-axis (outputs are negative) for *all values* of x except $x = 3$. But since this is a less than *or equal to* inequality, the solution is $x \in \mathbb{R}$.

WORTHY OF NOTE

Since $x = 3$ was a zero of multiplicity 2, the graph "bounced off" the x-axis at this point, with no change of sign for g. The graph is entirely below the x-axis, except at the vertex $(3, 0)$.

Figure 3.29

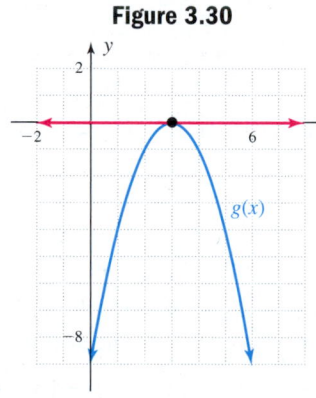

Graphical Solution ▶ The complete graph of g shown in Figure 3.30 confirms the analytical solution (using the zeroes method). For the intervals of the domain shown in red: $(-\infty, 3) \cup (3, \infty)$, the graph of g is below the x-axis $[g(x) < 0]$. The point $(3, 0)$ is *on* the x-axis $[g(3) = 0]$. As with the analytical solution, the solution to this "less than or equal to" inequality is all real numbers. A calculator check of the original inequality is shown in Figure 3.31.

Figure 3.30   **Figure 3.31**

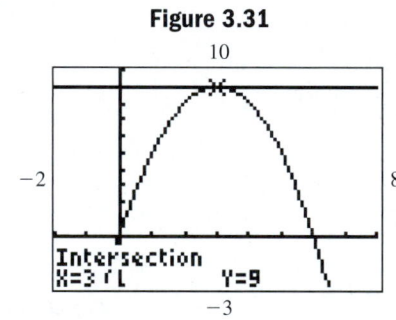

Now try Exercises 121 through 132 ▶

As an alternative to the Zeroes Method, the **Interval Test Method** can be used to solve quadratic inequalities. Using the fact that all polynomials are continuous, test values are selected from certain intervals of the domain and substituted into the original function.

Interval Test Method for Solving Inequalities

1. Find all real roots of the related equation (if they exist) and plot them on the x-axis.

2. Select any convenient test value from each interval created by the zeroes, and substitute these into the function.

3. The sign of the function at these test values will be the sign of the function for all values of x in this interval.

EXAMPLE 9 ▶ **Solving a Quadratic Inequality**

Solve the inequality $-2x^2 - 3x + 20 \leq 0$ using interval tests.

Solution ▶ To begin, we find the zeroes of $f(x) = -2x^2 - 3x + 20$ by factoring.

$$-2x^2 - 3x + 20 = 0 \qquad \text{related equation}$$
$$2x^2 + 3x - 20 = 0 \qquad \text{multiply by } -1$$
$$(2x - 5)(x + 4) = 0 \qquad \text{factored form}$$
$$2x - 5 = 0 \text{ or } x + 4 = 0 \qquad \text{zero factor property}$$
$$x = \frac{5}{2} \text{ or } x = -4 \qquad \text{solutions}$$

Plotting these intercepts creates three intervals on the x-axis (Figure 3.32).

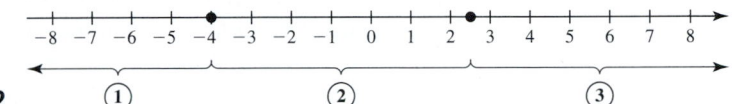

Figure 3.32 ①   ②   ③

Selecting a test value from each interval (in red) gives Figure 3.33:

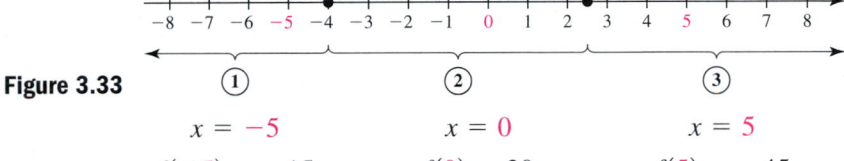

Figure 3.33   ①   ②   ③

$$\begin{array}{ccc} x = -5 & x = 0 & x = 5 \\ f(-5) = -15 & f(0) = 20 & f(5) = -45 \\ f(x) < 0 \text{ in } ① & f(x) > 0 \text{ in } ② & f(x) < 0 \text{ in } ③ \end{array}$$

WORTHY OF NOTE

When evaluating a function using the interval test method, it's usually easier to use the factored form instead of the polynomial form, since all you really need is whether the result will be positive or negative. For instance, you could likely tell $f(-5) = -[2(-5) - 5](-5 + 4)$ is going to be negative, more quickly than $f(-5) = -2(-5)^2 - 3(-5) + 20$.

The interval tests show $-2x^2 - 3x + 20 \leq 0$ for $x \in (-\infty, -4) \cup (\frac{5}{2}, \infty)$, which is supported by the graph shown in Figure 3.34.

Figure 3.34

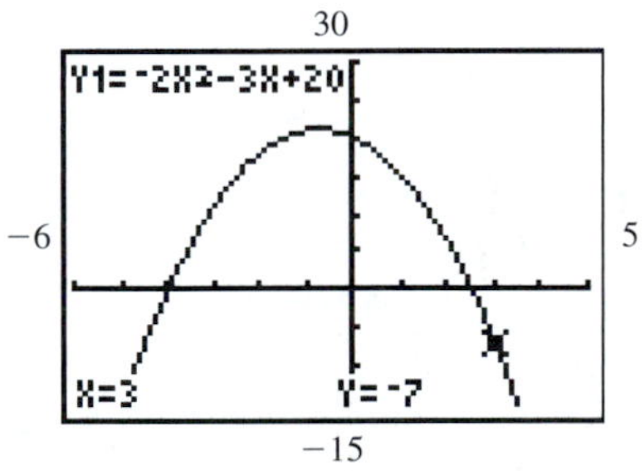

```
         30
Y1=-2X2-3X+20
-6                          5


X=3        Y=-7
        -15
```

Now try Exercises 133 through 144 ▶

The need to solve a quadratic inequality occurs in a variety of contexts. Here, the solution is used to find the domain of a radical function.

EXAMPLE 10 ▶ **Solving a Quadratic Inequality to Determine the Domain**

Find the domain of $g(x) = \sqrt{9 - x^2}$.

Solution ▶ From our earlier work, the radicand must be nonnegative and we are essentially asked to solve the inequality $9 - x^2 > 0$. The related equation is $9 - x^2 = 0$ and by inspection (or factoring), the solutions are $x = -3$ and $x = 3$. Plotting these solutions creates three intervals on the x-axis (Figure 3.35).

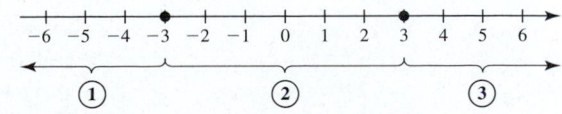

Figure 3.35   ①   ②   ③

Selecting a test value from each interval (in red) gives Figure 3.36:

Figure 3.36

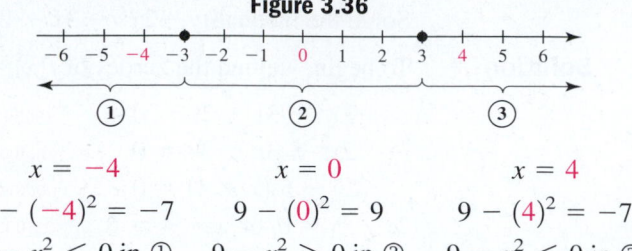

$$x = -4 \qquad\qquad x = 0 \qquad\qquad x = 4$$
$$9 - (-4)^2 = -7 \qquad 9 - (0)^2 = 9 \qquad 9 - (4)^2 = -7$$
$$9 - x^2 < 0 \text{ in } ① \qquad 9 - x^2 > 0 \text{ in } ② \qquad 9 - x^2 < 0 \text{ in } ③$$

The interval tests show that $9 - x^2 \geq 0$ for $x \in [-3, 3]$, which is the domain of $g(x) = \sqrt{9 - x^2}$. This is the same relation we graphed in Section 1.1/Example 3, a semicircle with endpoints at $x = -3$ and 3. See Figure 3.37.

Figure 3.37

Y1=√(9−X2)

X=-3   Y=0

☑ **E.** You've just seen how we can solve quadratic inequalities

Now try Exercises 145 through 154 ▶

F. Applications of Quadratic Functions and Inequalities

A projectile is any object that is thrown, shot, or *projected* upward with no sustaining source of propulsion. The height of the projectile at time t is modeled by the equation $h = -16t^2 + vt + k$, where h is the height of the object in feet, t is the elapsed time in seconds, and v is the initial velocity in feet per second. The constant k represents the initial height of the object above ground level, as when a person releases an object 5 ft above the ground in a throwing motion ($k = 5$), or when a rocket runs out of fuel at an altitude of 240 ft ($k = 240$).

EXAMPLE 11 ▶   **Solving an Application of Quadratic Equations—Rocketry**   **Figure 3.38**

A model rocketry club is testing a newly developed engine. A few seconds after liftoff, at a velocity of 160 ft/sec and a height of 240 ft, it runs out of fuel and becomes a projectile (see Figure 3.38).

   a. How high is the rocket 3 sec later?

   b. For how many seconds was the height of the rocket greater than or equal to 496 ft?

   c. How many seconds until the rocket returns to the ground?

Solution ▶   **a.** Using the information given, the function h modeling the rocket's height in the projectile phase, is $h(t) = -16t^2 + 160t + 240$. For its height at $t = 3$ we have

$$h(3) = -16(3)^2 + 160(3) + 240$$
$$= -16(9) + 480 + 240$$
$$= 576$$

Three seconds later, the rocket was at an altitude of 576 ft.

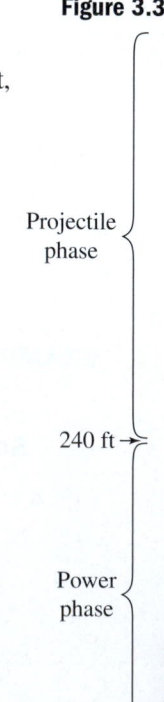

Projectile phase

240 ft →

Power phase

b. Since the height $h(t)$ must be greater than or equal to 496 ft, we use the function $h(t) = -16t^2 + 160t + 240$ to write the inequality $-16t^2 + 160t + 240 \geq 496$. In standard form, we obtain $-16t^2 + 160t - 256 \geq 0$ (subtract 496 from both sides). We begin by finding the zeroes of $-16t^2 + 160t - 256 = 0$, noting the related graph opens downward since the leading coefficient is negative.

$$-16t^2 + 160t - 256 = 0 \qquad \text{related equation}$$
$$t^2 - 10t + 16 = 0 \qquad \text{divide by } -16$$
$$(t - 2)(t - 8) = 0 \qquad \text{factor}$$
$$t - 2 = 0 \quad \text{or} \quad t - 8 = 0 \qquad \text{zero factor theorem}$$
$$t = 2 \quad \text{or} \quad t = 8 \qquad \text{result}$$

This shows the rocket is at exactly 496 ft after 2 sec (on its ascent) and after 8 sec (during its descent). We conclude the rocket's height was greater than 496 ft for $8 - 2 = 6$ sec.

c. When the rocket hits the ground, its height is $h = 0$. Substituting 0 for $h(t)$ and solving gives

$$h(t) = -16t^2 + 160t + 240 \qquad \text{original function}$$
$$0 = -16t^2 + 160t + 240 \qquad \text{substitute 0 for } h(t)$$
$$0 = t^2 - 10t - 15 \qquad \text{divide by } -16$$

The equation is nonfactorable, so we use the quadratic equation to solve, with $a = 1$, $b = -10$, and $c = -15$:

$$t = \frac{-b \pm \sqrt{b^2 - 4ac}}{2a} \qquad \text{quadratic formula}$$

$$= \frac{-(-10) \pm \sqrt{(-10)^2 - 4(1)(-15)}}{2(1)} \qquad \text{substitute 1 for } a, -10 \text{ for } b, -15 \text{ for } c$$

$$= \frac{10 \pm \sqrt{160}}{2} \qquad \text{simplify}$$

$$= \frac{10}{2} \pm \frac{4\sqrt{10}}{2} \qquad \sqrt{160} = 4\sqrt{10}$$

$$= 5 \pm 2\sqrt{10} \qquad \text{simplify}$$

Since we need the time t in seconds, we use the approximate form of the answer, obtaining $t \approx -1.32$ and $t \approx 11.32$. The rocket will return to the ground in just over 11 sec (since t represents time, the solution $t = -1.32$ does not apply). A calculator check is shown in Figure 3.39 using the Zeroes Method.

Figure 3.39

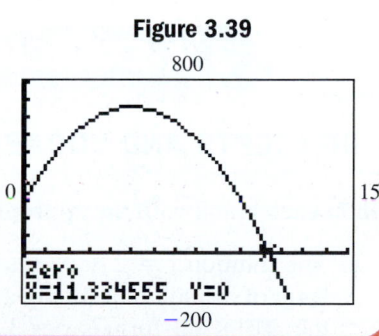

Now try Exercises 157 through 166 ▶

EXAMPLE 12 ▶ **Solving Applications of Inequalities Using the Quadratic Formula**

For the years 1995 to 2006, the amount A of annual international telephone traffic (in billions of minutes) can be modeled by $A = 0.17x^2 + 8.43x + 64.58$ where $x = 0$ represents the year 1995 [*Source:* Data from the *2009 Statistical Abstract of the United States,* Table 1344, page 846]. If this trend continues, in what year will the annual number of minutes reach or surpass 275 billion?

Analytical Solution ▶   We are essentially asked to solve the inequality $0.17x^2 + 8.43x + 64.58 \geq 275$.

$$0.17x^2 + 8.43x + 64.58 \geq 275 \qquad \text{given inequality}$$

$$0.17x^2 + 8.43x - 210.42 \geq 0 \qquad \text{subtract 275}$$

For $a = 0.17$, $b = 8.43$, and $c = -210.42$, the quadratic formula gives

$$x = \frac{-b \pm \sqrt{b^2 - 4ac}}{2a} \qquad \text{quadratic formula}$$

$$x = \frac{-8.43 \pm \sqrt{(8.43)^2 - 4(0.17)(-210.42)}}{2(0.17)} \qquad \text{substitute known values}$$

$$x = \frac{-8.43 \pm \sqrt{214.1505}}{0.34} \qquad \text{simplify}$$

$$x \approx 18.25 \quad \text{or} \quad x \approx -67.83 \qquad \text{result}$$

We disregard the negative solution (since x represents time), and find the annual number of international telephone minutes will reach or surpass 275 billion about 18 years after 1995, or in the year 2013.

Graphical Solution ▶   After entering $0.17x^2 + 8.43x + 64.58$ as Y_1, our next task is to determine an appropriate window size. With 1995 as year $x = 0$ and the data taken from year 2006 ($x = 11$) it seems that $x = -2$ to $x = 25$ would be an appropriate start for the domain (we can later adjust the window if needed). As the primary question is the year when telephone traffic surpasses 275 billion minutes, the range must include this value and $y = -100$ to $y = 350$ would be a good start (as before, the negative values were used to create a frame around the desired window). Using the Intersection-of-Graphs method, the resulting graph is shown in the figure and indicates that telephone traffic surpassed 275 billion minutes in the 18th year (2013), in line with our analytical solution.

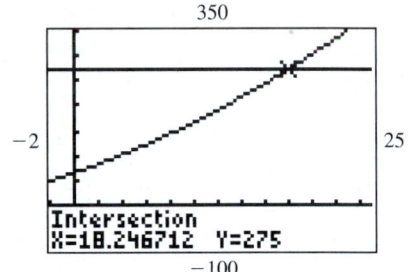

☑ **F.** You've just seen how we can solve applications of quadratic functions and inequalities

Now try Exercises 167 and 168 ▶

3.2 EXERCISES

▶ **CONCEPTS AND VOCABULARY**

Fill in each blank with the appropriate word or phrase. Carefully reread the section, if necessary.

1. The solution $x = 2 + \sqrt{3}$ is called an ___exact___ form of the solution. Using a calculator, we find the __approximate__ form is $x \approx 3.732$.

2. To solve a quadratic equation by completing the square, the coefficient of the __quadratic__ term must be a ___1___.

3. The quantity $b^2 - 4ac$ is called the __discriminant__ of the quadratic equation. If $b^2 - 4ac > 0$, there are ___2___ real roots.

4. According to the summary on page 22, the equation $4x^2 - 5x$ might best be solved by __factoring__ out the ___GCF___.

5. Discuss/Explain the relationship between solutions to $f(x) = 0$ and the x-intercepts of $y = f(x)$. Be sure to include an example.   Answers will vary.

6. Discuss/Explain why this version of the quadratic formula is incorrect:
$$x = -b \pm \frac{\sqrt{b^2 - 4ac}}{2a} \qquad \text{$-b$ must be divided by $2a$.}$$

HOMEWORK SELECTION GUIDE

Core: 7, 8, 15, 16, 29, 30, 37, 38, 47, 48, 61, 62, 73, 74, 85, 86, 109, 110, 121, 123, 151–154, 157, 159 (26 Exercises)

Standard: 1–4, 7, 8, 15, 16, 29, 30, 37, 38, 47, 48, 61, 62, 73, 74, 85, 86, 109, 110, 121, 123, 151–154, 155, 157–167 odd, 170, 177, 179 (38 Exercises)

Extended: 1–4, 7, 8, 15, 16, 29, 30, 37, 38, 47, 48, 61, 62, 73, 74, 85, 86, 109, 110, 121, 123, 151–154, 155–167 odd, 170, 171, 177, 179 (40 Exercises)

In Depth: 1–6, 7, 8, 15, 16, 29, 30, 37, 38, 47, 48, 61, 62, 73, 74, 85, 86, 91, 92, 109, 110, 121, 123, 145, 147, 151–154, 155–167 odd, 170, 171, 173, 177–180 (49 Exercises)

▶ DEVELOPING YOUR SKILLS

Six functions are shown below, along with their respective zeroes. Use these zeroes to solve the equations in Exercises 7 through 12.

$f(x) = x^2 + x - 12,$ $g(x) = x^2 - 4x - 5,$ $h(x) = -2x^2 - 4x + 6,$
$f(-4) = f(3) = 0;$ $g(-1) = g(5) = 0;$ $h(-3) = h(1) = 0;$

$j(x) = -4x^2 + 4x + 8,$ $k(x) = 2x^2 + x - 3,$ $l(x) = 3x^2 + 8x - 3,$
$j(-1) = j(2) = 0;$ $k(\frac{-3}{2}) = k(1) = 0;$ $l(-3) = l(\frac{1}{3}) = 0$

7. $x^2 + x - 12 = 0$ $x = -4$ or 3

8. $x^2 - 4x - 5 = 0$ $x = -1$ or 5

9. $2x^2 + x = 3$ $x = \frac{-3}{2}$ or 1

10. $3x^2 + 8x = 3$ $x = -3$ or $\frac{1}{3}$

11. $4x^2 = 4x + 8$ $x = -1$ or 2

12. $2x^2 + 4x = 6$ $x = -3$ or 1

Solve the following equations graphically by locating the zeroes of a related quadratic function. Round to hundredths as necessary.

13. $x^2 + 3x - 5 = 0$ $x \approx -4.19$ or 1.19

14. $x^2 - 7x - 2 = 0$ $x \approx -0.27$ or 7.27

15. $0.4x^2 - 0.6x - 2 = 0$ $x \approx -1.61$ or 3.11

16. $5x^2 + 0.2x - 1.1 = 0$ $x \approx -0.49$ or 0.45

17. $2x^2 - 3x = 6$ $x \approx -1.14$ or 2.64

18. $3x^2 + x = 3$ $x \approx -1.18$ or 0.85

19. $2.9x^2 = 1.3x - 5.7$ $\{\ \}$

20. $8.1x^2 = 5.3x - 4.2$ $\{\ \}$

Solve the following equations using the square root property of equality. Write answers in exact form and approximate form rounded to hundredths. If there are no real solutions, so state. Verify solutions graphically.

21. $m^2 = 16$

22. $p^2 = 49$

23. $y^2 - 28 = 0$

24. $m^2 - 20 = 0$

25. $p^2 + 36 = 0$

26. $n^2 + 5 = 0$

27. $x^2 = \frac{21}{16}$

28. $y^2 = \frac{13}{9}$

29. $(n - 3)^2 = 36$

30. $(p + 5)^2 = 49$

31. $(w + 5)^2 = 3$

32. $(m - 4)^2 = 5$

33. $(x - 3)^2 + 7 = 2$

34. $(m + 11)^2 + 5 = 3$

35. $(m - 2)^2 = \frac{18}{49}$

36. $(x - 5)^2 = \frac{12}{25}$

Fill in the blank so the result is a perfect square trinomial, then factor into a binomial square.

37. $x^2 + 6x +$ $9; (x + 3)^2$

38. $y^2 + 10y +$ $25; (y + 5)^2$

39. $n^2 + 3n +$ $\frac{9}{4}; (n + \frac{3}{2})^2$

40. $x^2 - 5x +$ $\frac{25}{4}; (x - \frac{5}{2})^2$

41. $p^2 + \frac{2}{3}p +$ $\frac{1}{9}; (p + \frac{1}{3})^2$

42. $x^2 - \frac{3}{2}x +$ $\frac{9}{16}; (x - \frac{3}{4})^2$

Solve by completing the square. Write your answers in both exact form and approximate form rounded to the hundredths place. If there are no real solutions, so state.

43. $x^2 + 6x = -5$ $x = -1; x = -5$

44. $m^2 + 8m = -12$ $m = -2; m = -6$

45. $p^2 - 6p + 3 = 0$

46. $n^2 = 4n + 10$

47. $p^2 + 6p = -4$

48. $x^2 - 8x - 1 = 0$

49. $m^2 + 3m = 1$

50. $n^2 + 5n - 2 = 0$

51. $n^2 = 5n + 5$

52. $w^2 - 7w + 3 = 0$

Solve the following quadratic equations (a) algebraically by completing the square and (b) graphically by using a graphing calculator and the Zeroes Method. Round answers to nearest hundredth when necessary.

53. $2x^2 = -7x + 4$

54. $3w^2 - 8w + 4 = 0$

55. $2n^2 - 3n - 9 = 0$

56. $2p^2 - 5p = 1$

57. $4p^2 - 3p - 2 = 0$

58. $3x^2 + 5x - 6 = 0$

59. $m^2 = 7m - 4$

60. $a^2 - 15 = 4a$

Solve each equation using the most efficient method: factoring, square root property of equality, or the quadratic formula. Write your answer in both exact and approximate form (rounded to hundredths). Check one of the exact solutions in the original equation.

61. $x^2 - 3x = 18$

62. $w^2 + 6w - 1 = 0$

63. $4m^2 - 25 = 0$

64. $4a^2 - 4a = 1$

65. $4n^2 - 8n - 1 = 0$

66. $2x^2 - 4x + 5 = 0$

67. $6w^2 - w = 2$

68. $3a^2 - 5a + 6 = 0$

69. $4m^2 = 12m - 15$

70. $3p^2 + p = 0$

71. $4n^2 - 9 = 0$

72. $4x^2 - x = 3$

73. $5w^2 = 6w + 8$

74. $3m^2 - 7m - 6 = 0$

75. $3a^2 - a + 2 = 0$

76. $3n^2 - 2n - 3 = 0$

77. $5p^2 = 6p + 3$

78. $2x^2 + x + 3 = 0$

79. $5w^2 - w = 1$

80. $3m^2 - 2 = 5m$

Additional answers can be found in the Instructor Answer Appendix.

81. $2a^2 + 5 = 3a$

82. $n^2 + 4n - 8 = 0$

83. $2p^2 - 4p + 11 = 0$

84. $8x^2 - 5x - 1 = 0$

85. $w^2 + \dfrac{2}{3}w = \dfrac{1}{9}$

86. $\dfrac{5}{4}m^2 - \dfrac{8}{3}m + \dfrac{1}{6} = 0$

87. $0.2a^2 + 1.2a + 0.9 = 0$

88. $-5.4n^2 + 8.1n + 9 = 0$

89. $\dfrac{2}{7}p^2 - 3 = \dfrac{8}{21}p$

90. $\dfrac{5}{9}x^2 - \dfrac{16}{15}x = \dfrac{3}{2}$

Use the discriminant to determine whether the given equation has irrational, rational, repeated, or nonreal roots. Also state whether the original equation is factorable using integers, but do not solve for x.

91. $-3x^2 + 2x + 1 = 0$

92. $2x^2 - 5x - 3 = 0$

93. $-4x + x^2 + 13 = 0$

94. $-10x + x^2 + 41 = 0$

95. $15x^2 - x - 6 = 0$

96. $10x^2 - 11x - 35 = 0$

97. $-4x^2 + 6x - 5 = 0$

98. $-5x^2 - 3 = 2x$

99. $2x^2 + 8 = -9x$

100. $x^2 + 4 = -7x$

101. $4x^2 + 12x = -9$

102. $9x^2 + 4 = 12x$

Solve the equations given. Simplify each result.

103. $-6x + 2x^2 + 5 = 0$

104. $17 + 2x^2 = 10x$

105. $5x^2 + 5 = -5x$

106. $x^2 = -2x - 19$

107. $-2x^2 = -5x + 11$

108. $4x - 3 = 5x^2$

Solve each quadratic inequality by locating the x-intercept(s) (if they exist), and noting the end-behavior of the graph. Begin by writing the inequality in function form as needed.

109. $f(x) = -x^2 + 4x; f(x) > 0$ $x \in (0, 4)$

110. $g(x) = x^2 - 5x; g(x) < 0$ $x \in (0, 5)$

111. $h(x) = x^2 + 4x - 5; h(x) \geq 0$

112. $p(x) = -x^2 + 3x + 10; p(x) \leq 0$

113. $q(x) = 2x^2 - 5x - 7; q(x) < 0$

114. $r(x) = -2x^2 - 3x + 5; r(x) > 0$

115. $7 \geq x^2$

116. $x^2 \leq 13$

117. $x^2 + 3x \leq 6$

118. $x^2 - 2 \leq 5x$

119. $3x^2 \geq -2x + 5$

120. $4x^2 \geq 3x + 7$

121. $s(x) = x^2 - 8x + 16; s(x) \geq 0$ $x \in (-\infty, \infty)$

122. $t(x) = x^2 - 6x + 9; t(x) \geq 0$ $x \in (-\infty, \infty)$

123. $r(x) = 4x^2 + 12x + 9; r(x) < 0$ { }

124. $f(x) = 9x^2 - 6x + 1; f(x) < 0$ { }

Additional answers can be found in the Instructor Answer Appendix.

125. $g(x) = -x^2 + 10x - 25; g(x) < 0$

126. $h(x) = -x^2 + 14x - 49; h(x) < 0$

127. $-x^2 > 2$ { }

128. $x^2 < -4$ { }

129. $x^2 - 2x > -5$

130. $-x^2 + 3x < 3$

131. $2x^2 \geq 6x - 9$

132. $5x^2 \geq 4x - 4$

Solve each quadratic inequality using the interval test method. Write your answer in interval notation when possible.

133. $(x + 3)(x - 1) < 0$ $(-3, 1)$

134. $(x + 2)(x - 5) < 0$ $(-2, 5)$

135. $2x^2 - x - 6 \geq 0$ $(-\infty, -\frac{3}{2}] \cup [2, \infty)$

136. $-3x^2 + x + 4 \leq 0$ $(-\infty, -1] \cup [\frac{4}{3}, \infty)$

137. $(x + 1.3)^2 > 0$ $(-\infty, -1.3) \cup (-1.3, \infty)$

138. $-(x - 2.2)^2 < 0$ $(-\infty, 2.2) \cup (2.2, \infty)$

139. $-(x - 2.9)^2 \geq 0$ $\{2.9\}$

140. $(x + 3.2)^2 \leq 0$ $\{-3.2\}$

141. $\left(x - \dfrac{3}{5}\right)^2 < 0$ { }

142. $-\left(x + \dfrac{1}{8}\right)^2 > 0$ { }

143. $x^2 + 14x + 49 \geq 0$ $(-\infty, \infty)$

144. $-x^2 + 6x - 9 \leq 0$ $(-\infty, \infty)$

Recall that for a square root expression to represent a real number, the radicand must be greater than or equal to zero. Applying this idea results in an inequality that can be solved using the skills from this section. Determine the domain of the following radical functions.

145. $h(x) = \sqrt{x^2 - 25}$ $x \in (-\infty, -5] \cup [5, \infty)$

146. $p(x) = \sqrt{25 - x^2}$ $x \in [-5, 5]$

147. $q(x) = \sqrt{x^2 - 5x}$ $x \in (-\infty, 0] \cup [5, \infty)$

148. $r(x) = \sqrt{6x - x^2}$ $x \in [0, 6]$

149. $t(x) = \sqrt{-x^2 + 3x - 4}$ { }

150. $Y_1 = \sqrt{x^2 - 6x + 9}$ $x \in (-\infty, \infty)$

Match the correct solution with the inequality and graph given.

151. $f(x) > 0$

   a. $x \in (-\infty, -2) \cup (1, \infty)$

   b. $x \in (-\infty, -2] \cup [1, \infty)$

   c. $(-2, 1)$

   d. $[-2, 1]$

   e. none of these   a

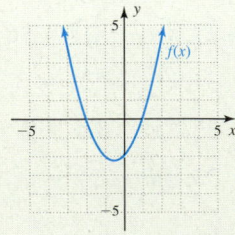

152. $g(x) \geq 0$

   a. $x \in (-\infty, -4) \cup (1, \infty)$

   b. $x \in (-\infty, -4] \cup [1, \infty)$

   c. $(-4, 1)$

   d. $[-4, 1]$

   e. none of these   d

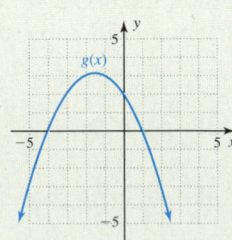

153. $r(x) \leq 0$

    a. $x \in (-\infty, 3) \cup (3, \infty)$

    b. $x \in (-\infty, \infty)$

    c. $\{3\}$

    d. $\{ \}$

    e. none of these   b

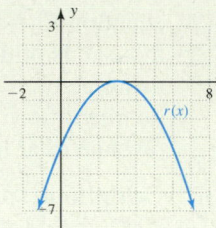

154. $s(x) < 0$

    a. $x \in (-\infty, 0) \cup (0, \infty)$

    b. $x \in (-\infty, \infty)$

    c. $\{0\}$

    d. $\{ \}$

    e. none of these   d

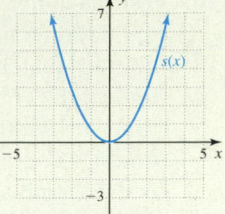

▶ WORKING WITH FORMULAS

155. Height of a projectile: $h = -16t^2 + vt$

If an object is projected vertically upward from ground level with no continuing source of propulsion, the height of the object (in feet) is modeled by the equation shown, where v is the initial velocity, and t is the time in seconds. Use the quadratic formula to solve for t in terms of v and h. (*Hint:* Set the equation equal to zero and identify the coefficients as before.)

156. Surface area of a cylinder: $A = 2\pi r^2 + 2\pi rh$

The surface area of a cylinder is given by the formula shown, where h is the height and r is the radius of the base. The equation can be considered a quadratic in the variable r. Use the quadratic formula to solve for r in terms of h and A. (*Hint:* Rewrite the equation in standard form and identify the coefficients as before.)

▶ APPLICATIONS

157. Height of a projectile: The height of an object thrown upward from the roof of a building 408 ft tall, with an initial velocity of 96 ft/sec, is given by the equation $h = -16t^2 + 96t + 408$, where h represents the height of the object after t seconds. How long will it take the object to hit the ground? Answer in exact form and decimal form rounded to the nearest hundredth.

158. Height of a projectile: The height of an object thrown upward from the floor of a canyon 106 ft deep, with an initial velocity of 120 ft/sec, is given by the equation $h = -16t^2 + 120t - 106$, where h represents the height of the object after t seconds. How long will it take the object to rise to the height of the canyon wall ($h = 0$)? Answer in exact form and decimal form rounded to hundredths.

159. Cost, revenue, and profit: The cost of raw materials to produce plastic toys is given by the cost equation $C = 2x + 35$, where x is the number of toys in hundreds. The total income (revenue) from the sale of these toys is given by $R = -x^2 + 122x - 1965$. (a) Determine the profit equation (profit = revenue − cost). (b) During the Christmas season, the owners of the company

decide to manufacture and donate as many toys as they can, without taking a loss (i.e., they break even: profit or $P = 0$). How many toys will they produce for charity?

    a. $P = -x^2 + 120x - 2000$ **b.** 10,000

160. Cost, revenue, and profit: The cost to produce bottled spring water is given by the cost equation $C = 16x + 63$, where x is the number of bottles in thousands. The total revenue from the sale of these bottles is given by the equation $R = -x^2 + 326x - 18{,}463$. (a) Determine the profit equation (profit = revenue − cost). (b) After a bad flood contaminates the drinking water of a nearby community, the owners decide to bottle and donate as many bottles of water as they can, without taking a loss (i.e., they break even: profit or $P = 0$). Approximately how many bottles will they produce for the flood victims?

    a. $P = -x^2 + 310x - 18{,}526$ **b.** 229,000

161. Height of an arrow: If an object is projected vertically upward from ground level with no continuing source of propulsion, its height (in feet) is modeled by the equation $h = -16t^2 + vt$, where v is the initial velocity and t is the time in seconds. Use the quadratic formula to solve for t, given an arrow is shot into the air with $v = 144$ ft/sec and $h = 260$ ft. See Exercise 155. $t = 2.5$ sec, 6.5 sec

162. Surface area of a cylinder: The surface area of a cylinder is given by $A = 2\pi r^2 + 2\pi rh$, where h is the height and r is the radius of the base. The equation can be considered a quadratic in the variable r. Use the quadratic formula to solve for r, given $A = 4710$ cm^2 and $h = 35$ cm. See Exercise 156. $r \approx 15$ cm

163. Tennis court dimensions: A regulation tennis court for a doubles match is laid out so that its length is 6 ft more than two times its width. The area of the doubles court is 2808 ft^2. What is the length and width of the doubles court?   36 ft, 78 ft

Exercises 163 and 164

Singles

Doubles

164. Tennis court dimensions: A regulation tennis court for a singles match is laid out so that its length is 3 ft less than three times its width. The area of the singles court is 2106 ft^2. What is the length and width of the singles court?   27 ft, 78 ft

165. Cost, revenue, and profit: The revenue for a manufacturer of microwave ovens is given by the equation $R = x(40 - \frac{1}{3}x)$, where revenue is in thousands of dollars and x thousand ovens are manufactured and sold. What is the minimum number of microwave ovens that must be sold to bring in a revenue of at least $900,000?   30,000 ovens

166. Cost, revenue, and profit: The revenue for a manufacturer of computer printers is given by the

equation $R = x(30 - 0.4x)$, where revenue is in thousands of dollars and x thousand printers are manufactured and sold. What is the minimum number of printers that must be sold to bring in a revenue of at least $440,000?   20,000 printers

167. Cell phone subscribers: For the years 1995 to 2002, the number N of cellular phone subscribers (in millions) can be modeled by the equation $N = 17.4x^2 + 36.1x + 83.3$, where $x = 0$ represents the year 1995 [*Source:* Data from the *2005 Statistical Abstract of the United States,* Table 1372, page 870]. If this trend continued, in what year did the number of subscribers reach or surpass 3750 million? $x \approx 13.5$, or the year 2008

168. U.S. international trade balance: For the years 1995 to 2003, the international trade balance B (in millions of dollars) can be approximated by the equation $B = -3.1x^2 + 4.5x - 19.9$, where $x = 0$ represents the year 1995 [*Source:* Data from the *2005 Statistical Abstract of the United States,* Table 1278, page 799]. If this trend continues, in what year will the trade balance reach a deficit of $750 million dollars or more? $x \approx 16.1$, or the year 2011

▶ EXTENDING THE CONCEPT

169. Using the discriminant: Each of the following equations can easily be solved by factoring. Using the discriminant, we can create factorable equations with identical values for b and c, but where $a \neq 1$. For instance, $x^2 - 3x - 10 = 0$ and $4x^2 - 3x - 10 = 0$ can both be solved by factoring. Find similar equations ($a \neq 1$) for the quadratics given here. (*Hint:* The discriminant $b^2 - 4ac$ must be a perfect square.)

   a. $x^2 + 6x - 16 = 0$

   b. $x^2 + 5x - 14 = 0$

   c. $x^2 - x - 6 = 0$

170. Using the discriminant: For what values of c will the equation $9x^2 - 12x + c = 0$ have

   a. no real roots

   b. one rational root

   c. two real roots

   d. two integer roots

Complex polynomials: Many techniques applied to solve polynomial equations with real coefficients can be applied to solve polynomial equations with *complex* coefficients. Here we apply the idea to carefully chosen quadratic equations, as a more general application must wait until a future course, when the square root of a complex number is fully developed. Solve each equation using the quadratic formula, noting that $\frac{1}{i} = -i$.

171. $z^2 - 3iz = -10$

172. $z^2 - 9iz = -22$

173. $4iz^2 + 5z + 6i = 0$

174. $2iz^2 - 9z + 26i = 0$

175. $0.5z^2 + (7 + i)z + (6 + 7i) = 0$

176. $0.5z^2 + (4 - 3i)z + (-9 - 12i) = 0$

► **MAINTAINING YOUR SKILLS**

177. (R.3) State the formula for the perimeter and area of each figure illustrated.

a.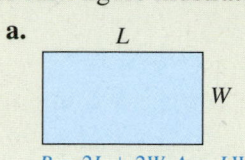

b.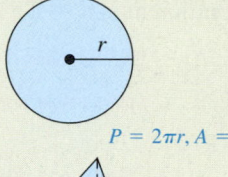

$P = 2L + 2W, A = LW$ $P = 2\pi r, A = \pi r^2$

c.   d.

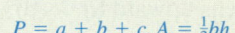

$P = c + h + b_1 + b_2, A = \frac{1}{2}h(b_1 + b_2)$ $P = a + b + c, A = \frac{1}{2}bh$

178. (R.4) Factor and solve the following equations:

    a. $x^2 - 5x - 36 = 0$

    b. $4x^2 - 25 = 0$

    c. $x^3 + 6x^2 - 4x - 24 = 0$ **a.** $x = 9$ or $x = -4$
                                        b. $x = \pm\frac{5}{2}$

179. (1.4) A total of 900 tickets were sold for a recent concert and \$25,000 was collected. If good seats were \$30 and cheap seats were \$20, how many of each type were sold?   700 \$30 tickets; 200 \$20 tickets

180. (1.5) Solve for C: $P = C + Ct$. $C = \dfrac{P}{1+t}$

3.3   **Quadratic Functions and Applications**

As our knowledge of functions grows, our ability to apply mathematics in new ways likewise grows. In this section, we'll build on the foundation laid in Section 3.2 and previous chapters, as we introduce additional tools used to apply quadratic functions effectively.

A.   **Graphing Quadratic Functions by Completing the Square**

Our earlier work suggests the graph of *any* quadratic function will be a parabola. Figure 3.40 provides a summary of the graph's characteristic features. As pictured,

1. The parabola opens upward ($y = ax^2 + bx + c, a > 0$).
2. The vertex is at (h, k) and $y = k$ is a global minimum.
3. The vertex is below the x-axis, so there are two x-intercepts.
4. The axis of symmetry contains the vertex, with equation $x = h$.
5. The y-intercept is $(0, c)$, since $f(0) = c$.

In Section 2.2, we graphed transformations of $f(x) = x^2$, using $y = a(x \pm h)^2 \pm k$. Here, we'll show that by completing the square, we can graph *any* quadratic function as a transformation of this basic graph.

When completing the square on a quadratic *equation* (Section 3.2), we applied the standard properties of equality to both sides of the equation. When completing the square on a **quadratic function,** the process is altered slightly, so that we operate on only one side.

The basic ideas are summarized here.

Figure 3.40 $f(x) = ax^2 + bx + c$

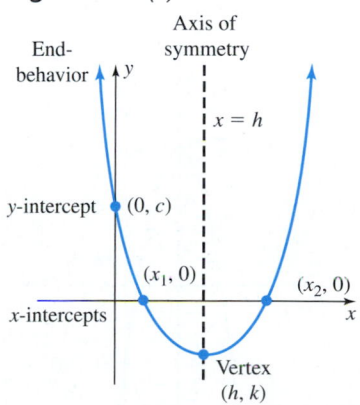

> **Graphing $f(x) = ax^2 + bx + c$ by Completing the Square**
>
> 1. Group the variable terms apart from the constant c.
> 2. Factor out the leading coefficient a from this group.
> 3. Compute $\left[\frac{1}{2}\left(\frac{b}{a}\right)\right]^2$ and add the result to the variable terms,
>   then subtract $a \cdot \left[\frac{1}{2}\left(\frac{b}{a}\right)\right]^2$ from c to maintain an equivalent expression.
> 4. Factor the grouped terms as a binomial square and combine constant terms.
> 5. Graph using transformations of $f(x) = x^2$.

EXAMPLE 1 ▶ **Graphing a Quadratic Function by Completing the Square**

Given $g(x) = x^2 - 6x + 5$, complete the square to rewrite g as a transformation of $f(x) = x^2$, then graph the function.

Solution ▶ To begin we note the leading coefficient is $a = 1$.

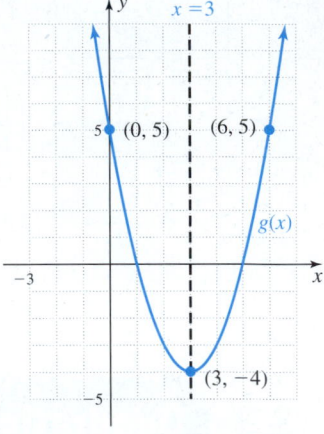

$$g(x) = x^2 - 6x + 5 \qquad \text{given function}$$
$$= 1(x^2 - 6x + \underline{\quad}) + 5 \qquad \text{group variable terms}$$
$$= 1\underbrace{(x^2 - 6x + 9)}_{\text{adds } 1 \cdot 9 = 9} - 9 + 5 \qquad \left[\left(\tfrac{1}{2}\right)(-6)\right]^2 = 9$$
$$\qquad\qquad\qquad\qquad\qquad \text{subtract } 9$$
$$= (x - 3)^2 - 4 \qquad \text{factor and simplify}$$

The graph of g is the graph of f shifted 3 units right, and 4 units down. The graph opens upward ($a > 0$) with the vertex at $(3, -4)$, and axis of symmetry $x = 3$. From the original equation we find $g(0) = 5$, giving a y-intercept of $(0, 5)$. The point $(6, 5)$ was obtained using the symmetry of the graph. The graph is shown in the figure.

Now try Exercises 7 through 10 ▶

Note that by **adding 9** and simultaneously **subtracting 9** (essentially adding "0"), we changed only the *form* of the function, not its value. In other words, the resulting expression is equivalent to the original. If the leading coefficient is not 1, we factor it out from the variable terms, but take it into account when we add the constant needed to maintain an equivalent expression (steps 2 and 3).

EXAMPLE 2 ▶ **Graphing a Quadratic Function by Completing the Square**

Given $p(x) = -2x^2 - 8x - 3$, complete the square to rewrite p as a transformation of $f(x) = x^2$, then graph the function.

Solution ▶
$$p(x) = -2x^2 - 8x - 3 \qquad \text{given function}$$
$$= (-2x^2 - 8x + \underline{\quad}) - 3 \qquad \text{group variable terms}$$
$$= -2(x^2 + 4x + \underline{\quad}) - 3 \qquad \text{factor out } a = -2 \text{ (notice sign change)}$$
$$= -2\underbrace{(x^2 + 4x + 4)}_{\text{adds } -2 \cdot 4 = -8} - (-8) - 3 \qquad \left[\left(\tfrac{1}{2}\right)(4)\right]^2 = 4$$
$$\qquad\qquad\qquad\qquad\qquad \text{subtract } -8$$
$$= -2(x + 2)^2 + 8 - 3 \qquad \text{factor trinomial, simplify}$$
$$= -2(x + 2)^2 + 5 \qquad \text{result}$$

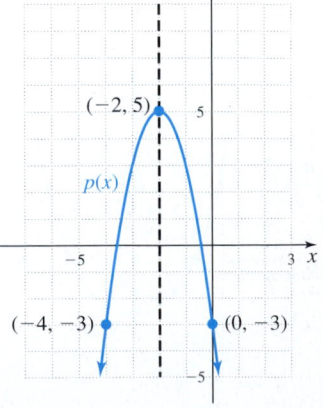

WORTHY OF NOTE

In cases like $f(x) = 3x^2 - 10x + 5$, where the linear coefficient has no integer factors of a, we factor out 3 and *simultaneously divide the linear coefficient by 3*. This yields

$$h(x) = 3\left(x^2 - \frac{10}{3}x + \underline{\quad}\right) + 5,$$

and the process continues as before: $\left[\left(\tfrac{1}{2}\right)\left(\tfrac{10}{3}\right)\right]^2 = \left(\tfrac{5}{3}\right)^2 = \tfrac{25}{9}$, and so on. For more on this idea, **see** Exercises 15 through 20.

The graph of p is a parabola, shifted 2 units left, stretched by a factor of 2, reflected across the x-axis (opens downward), and shifted up 5 units. The vertex is $(-2, 5)$, and the axis of symmetry is $x = -2$. From the original function, the y-intercept is $(0, -3)$. The point $(-4, -3)$ was obtained using the symmetry of the graph. The graph is shown in the figure.

Now try Exercises 11 through 20 ▶

☑ **A.** You've just seen how we can graph quadratic functions by completing the square

In Example 2, note that by adding 4 to the variable terms within parentheses, we actually added $-2 \cdot 4 = -8$ to the value of the function. To adjust for this we subtracted -8.

B. Graphing Quadratic Functions Using the Vertex Formula

When the process of completing the square is applied to $f(x) = ax^2 + bx + c$, we obtain a very useful result. Notice the close similarities to Example 2.

$$f(x) = ax^2 + bx + c \qquad \text{quadratic function}$$
$$= (ax^2 + bx + \underline{\quad}) + c \qquad \text{group variable terms apart from the constant } c$$
$$= a\left(x^2 + \frac{b}{a}x + \underline{\quad}\right) + c \qquad \text{factor out } a$$
$$= a\left(x^2 + \frac{b}{a}x + \frac{b^2}{4a^2}\right) - a\left(\frac{b^2}{4a^2}\right) + c \qquad \left[\left(\frac{1}{2}\right)\left(\frac{b}{a}\right)\right]^2 = \frac{b^2}{4a^2}, \text{ add within group,}$$
$$\text{subtract } a\left(\frac{b^2}{4a^2}\right)$$
$$= a\left(x + \frac{b}{2a}\right)^2 - \frac{b^2}{4a} + c \qquad \text{factor the trinomial, simplify}$$
$$= a\left(x + \frac{b}{2a}\right)^2 + \frac{4ac - b^2}{4a} \qquad \text{result}$$

By comparing this result with previous transformations, we note the x-coordinate of the vertex is $h = \dfrac{-b}{2a}$ (since the graph shifts horizontally "opposite the sign"). While we could use the expression $\dfrac{4ac - b^2}{4a}$ for k, we find it easier to substitute $\dfrac{-b}{2a}$ back into the function: $k = f\left(\dfrac{-b}{2a}\right)$. The result is called the **vertex formula.**

Vertex Formula

For the quadratic function $f(x) = ax^2 + bx + c$, the coordinates of the vertex are

$$(h, k) = \left(\frac{-b}{2a}, f\left(\frac{-b}{2a}\right)\right)$$

Since all characteristic features of the graph (end-behavior, vertex, axis of symmetry, x-intercepts, and y-intercept) can now be determined using the original equation, we'll rely on these features to sketch quadratic graphs, rather than having to complete the square.

EXAMPLE 3 ▶ **Graphing a Quadratic Function Using the Vertex Formula**

Graph $f(x) = 2x^2 + 8x + 3$ using the vertex formula and other features of a quadratic graph.

Solution ▶ The graph will open upward since $a > 0$.
The y-intercept is $(0, 3)$.
The vertex formula gives

$$h = \frac{-b}{2a} \qquad \text{x-coordinate of vertex}$$
$$= \frac{-8}{2(2)} \qquad \text{substitute 2 for } a \text{ and 8 for } b$$
$$= -2 \qquad \text{simplify}$$

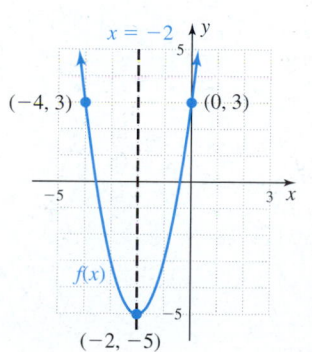

Computing $f(-2)$ to find the y-coordinate of the vertex yields

$$f(-2) = 2(-2)^2 + 8(-2) + 3 \qquad \text{substitute } -2 \text{ for } x$$
$$= 2(4) - 16 + 3 \qquad \text{multiply}$$
$$= 8 - 13 \qquad \text{simplify}$$
$$= -5 \qquad \text{result}$$

The vertex is $(-2, -5)$. The graph is shown in the figure, with the point $(-4, 3)$ obtained using symmetry.

✅ **B.** You've just seen how we can graph quadratic functions using the vertex formula

Now try Exercises 21 through 32 ▶

C. Finding the Equation of a Quadratic Function from Its Graph

While most of our emphasis so far has centered on graphing quadratic functions, it would be hard to overstate the importance of the reverse process—determining the equation of the function from its graph (as in Section 2.2). This reverse process, which began with our study of lines, will be a continuing theme each time we consider a new function.

EXAMPLE 4 ▶ **Finding the Equation of a Quadratic Function**

The graph shown is a transformation of $f(x) = x^2$. What function defines this graph?

Solution ▶ Compared to the graph of $f(x) = x^2$, the vertex has been shifted left 1 and up 2, so the function will have the form $F(x) = a(x + 1)^2 + 2$. Since the graph opens downward, we know a will be negative. As before, we select one additional point on the graph and substitute to find the value of a. Using $(x, y) \rightarrow (1, 0)$ we obtain

$$F(x) = a(x + 1)^2 + 2 \qquad \text{transformation}$$
$$0 = a(1 + 1)^2 + 2 \qquad \text{substitute 1 for } x \text{ and 0 for } F(x)$$
$$0 = 4a + 2 \qquad \text{simplify}$$
$$-2 = 4a \qquad \text{subtract 2}$$
$$-\frac{1}{2} = a \qquad \text{solve for } a$$

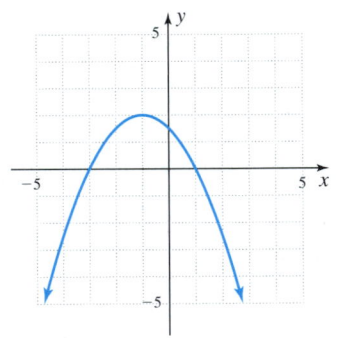

WORTHY OF NOTE

It helps to remember that any point (x, y) on the parabola can be used. To verify this, try the calculation again using $(-3, 0)$.

The equation of this function is

$$F(x) = -\frac{1}{2}(x + 1)^2 + 2.$$

✅ **C.** You've just seen how we can find the equation of a quadratic function from its graph

Now try Exercises 33 through 38 ▶

D. Quadratic Functions and Extreme Values

If $a > 0$, the parabola opens upward, and the y-coordinate of the vertex is a global minimum, the smallest value attained by the function anywhere in its domain. Conversely, if $a < 0$ the parabola opens downward and the vertex yields a global maximum. These greatest and least points are known as **extreme values** and have a number of significant applications.

Note that when graphing technology is used in a real-world context, the focus on the algebra involved is now shared with the context and meaning of the variables involved. This analysis enables us to find an appropriate viewing window, and a better interpretation of the results obtained.

EXAMPLE 5 ▶ **Applying a Quadratic Model to Manufacturing**

An airplane manufacturer can produce up to 15 planes per month. Suppose the profit made from the sale of these planes is modeled by $P(x) = -0.2x^2 + 4x - 3$, where $P(x)$ is the profit in hundred-thousands of dollars per month, and x is the number of planes sold. Based on this model,

    a. Find the y-intercept and explain what it means in this context.

    b. How many planes should be made and sold to maximize profit?

    c. What is the maximum profit?

Analytical Solution ▶

    a. $P(0) = -3$, which means the manufacturer loses \$300,000 each month if the company produces no planes.

    b. Since $a < 0$, we know the graph opens downward and has a maximum value. To find the required number of sales needed to "maximize profit," we use the vertex formula with $a = -0.2$ and $b = 4$:

$$x = \frac{-b}{2a} \qquad \text{vertex formula}$$

$$= \frac{-4}{2(-0.2)} \qquad \text{substitute } -0.2 \text{ for } a \text{ and 4 for } b$$

$$= 10 \qquad \text{result}$$

The result shows 10 planes should be sold each month for maximum profit.

    c. Evaluating $P(10)$ we find that a maximum profit of 17 "hundred-thousand dollars" will be earned (\$1,700,000).

Graphical Solution ▶

    a. Regardless of the solution method (analytical or graphical), the y-intercept will be the constant term. Here $y = -3$ indicates a loss of \$300,000 if no planes are made.

    b. and c. After entering $-0.2x^2 + 4x - 3$ as Y_1, we can use the **2nd** (**CALC**) **4:maximum** option to locate the extreme value. Before we do, we need to set a window that will enable us to view and interpret the results. First, we know that the solution must occur in QI, since both x and $P(x)$ must be positive. We're told the manufacturer can make at most 15 planes per month, so it seems reasonable to set Xmin = 0 and Xmax = 20 (to allow for a "frame" around the window). Using Ymin = 0 and leaving Ymax = 10 for now produces the graph shown in Figure 3.41. As the "top half" of the parabola is missing, we increase the maximum y-value of our window to Ymax = 20, yielding the graph in Figure 3.42. The sequence **2nd** (**CALC**) **4:maximum** (and then naming the appropriate bounds) shows the vertex occurs at (10, 17) (Figure 3.43), meaning a maximum profit of 17 × \$100,000 = \$1,700,000 will be earned when 10 planes are made and sold ($x = 10$).

Figure 3.41

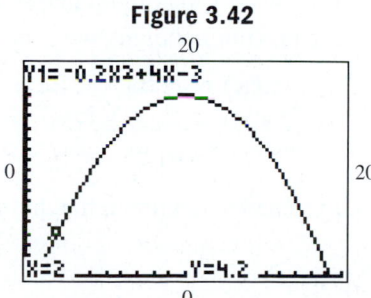

Figure 3.42

Figure 3.43

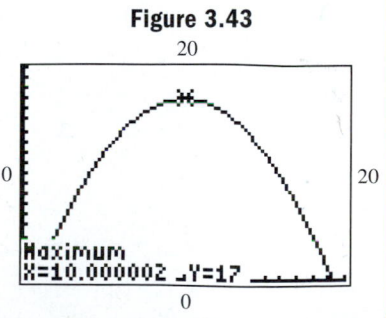

Now try Exercises 41 through 46 ▶

Recall that if the leading coefficient is positive and the vertex is below the x-axis ($k < 0$), the graph will have two x-intercepts (see Figure 3.44). If $a > 0$ and the vertex is above the x-axis ($k > 0$), the graph will not cross the x-axis (Figure 3.45). Similar statements can be made for the case where a is negative.

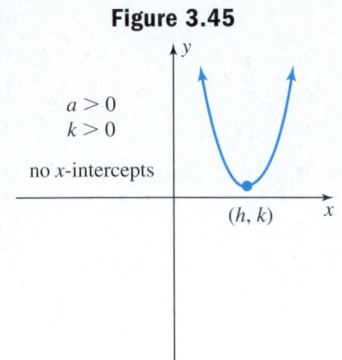

Figure 3.44   **Figure 3.45**

$a > 0$ $a > 0$
$k < 0$ $k > 0$
two x-intercepts   no x-intercepts
(h, k) (h, k)

EXAMPLE 6 ▶   **Modeling the Height of a Projectile—Football**

In the 1976 Pro Bowl, NFL punter Ray Guy of the Oakland Raiders kicked the ball so high it hit the scoreboard hanging from the roof of the New Orleans Superdome (forcing officials to raise the scoreboard from about 90 ft to 200 ft). If we assume the ball made contact with the scoreboard near the vertex of the kick, the function $h(t) = -16t^2 + 76t + 1$ is one possible model for the height of the ball, where $h(t)$ represents the height (in feet) after t sec.

a. What does the y-intercept of this function represent?

b. After how many seconds did the football reach its maximum height?

c. What was the maximum height of this kick?

d. To the nearest hundredth of a second, how long until the ball returned to the ground (what was the hang time)?

Algebraic Solution ▶   **a.** $h(0) = 1$, meaning the ball was 1 ft off the ground when Ray Guy kicked it.

b. Since $a < 0$, we know the graph opens downward and has a maximum value. To find the time needed to reach the maximum height, we use the vertex formula with $a = -16$ and $b = 76$:

$$t = \frac{-b}{2a} \qquad \text{vertex formula}$$

$$= \frac{-76}{2(-16)} \qquad \text{substitute } -16 \text{ for } a \text{ and } 76 \text{ for } b$$

$$= 2.375 \qquad \text{result}$$

The ball reached its maximum height after 2.375 sec.

c. To find the maximum height, we substitute 2.375 for t [evaluate $h(2.375)$]:

$$h(t) = -16t^2 + 76t + 1 \qquad \text{given function}$$
$$h(2.375) = -16(2.375)^2 + 76(2.375) + 1 \qquad \text{substitute 2.375 for } t$$
$$= 91.25 \qquad \text{result}$$

The ball reached a maximum height of 91.25 ft.

d. When the ball returns to the ground it has a height of 0 ft. Substituting 0 for $h(t)$ gives $0 = -16t^2 + 76t + 1$, which we solve using the quadratic formula.

$$t = \frac{-b \pm \sqrt{b^2 - 4ac}}{2a}$$ quadratic formula

$$= \frac{-76 \pm \sqrt{(76)^2 - 4(-16)(1)}}{2(-16)}$$ substitute -16 for a, 76 for b, and 1 for c

$$= \frac{-76 \pm \sqrt{5840}}{-32}$$ simplify

$$t \approx -0.013 \quad \text{or} \quad t \approx 4.763$$

The punt had a hang time of just under 5 sec.

Graphical Solution ▶

a. By inspection or using the **TABLE**, $h(0) = 1$ showing the ball was 1 ft from the ground when it was kicked.

b. and c. Here we enter $-16x^2 + 76x + 1$ as Y_1, then determine an appropriate viewing window. We are once again limited to QI, since the time t and height h must both be positive. Based on common experience (participation or observation) in throwing or kicking balls in various activities, Xmin = 0, Xmax = 8 seems to be a good place to begin for time t. For the value of Ymax (having set Ymin = 0), we decide to just "aim high" being aware of human limitations, and adjust the window afterward (it would be impossible for a human to kick a football 250 ft high!). Using some combination of the preceding we set Ymax = 120, yielding the graph in Figure 3.46. Using the sequence ⎯2nd⎯ (**CALC**) **4:maximum** with appropriate bounds shows the vertex occurs at (2.375, 91.25) (Figure 3.47), meaning a maximum height of 91.25 ft ($y = 91.25$), occurred 2.375 sec after the ball was kicked ($x = 2.375$).

Figure 3.46

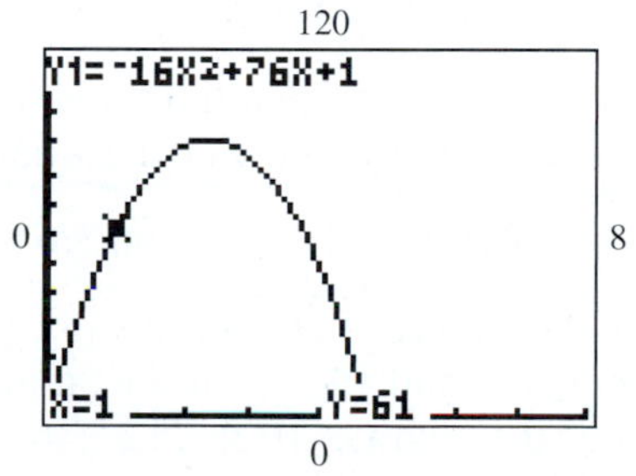

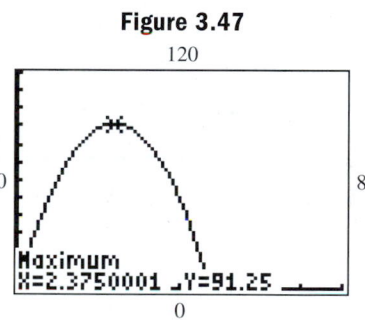

Figure 3.47

d. Before using the Zeroes Method to find the "hang time," we reset the minimum y-value to Ymin = -30, to obtain a frame that enables us to read the information at the bottom of the screen, without interfering with the graph (see *Worthy of Note*). Using the ⎯2nd⎯ (**CALC**) **2:zero** option for the rightmost x-intercept, we find the hang time is about 4.76 sec (Figure 3.48).

Figure 3.48

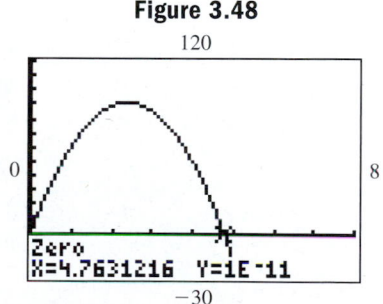

Now try Exercises 47 through 52 ▶

WORTHY OF NOTE

To find a frame of the kind mentioned in Example 6(d), we can use the formula Ymin ≈ $\frac{|\text{current Ymax}| + |\text{current Ymin}|}{4}$ as a rule of thumb.

While the calculator is a wonderful tool for removing some computational drudgery, especially when the coefficients are large, noninteger, or not very "pretty," you are likely coming to realize that it does not (cannot) replace the analytical thought required to use it effectively. Most real-world applications still require an attentive analysis of the context and the question asked, as well as a careful development of the equation model to be used. This is particularly important in applications like those in Example 7, where the original equation has more than one independent variable, and a given or known relationship must be used to eliminate one of them.

EXAMPLE 7 ▶ Due to an increase in consumer demand, a local nursery is building new chain-link pens for holding various kinds of mulch. What is the maximum total area that can be fenced off, if five open-front pens are needed (see Figure 3.49) and 112 ft of fencing is available?

Figure 3.49

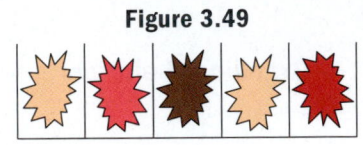

Solution ▶ We are asked to maximize the total area, an impossibility right now since the formula for area has *two independent variables*: Area = LW. Knowing that only 112 ft of fencing will be used, we observe that the finished pens will have *1 length and 6 identical widths*, giving the equation $112 = L + 6W$. This enables us to write L in terms of W (solve for L), and *substitute the result for L in $A = LW$*, giving an equation model with the single independent variable W.

$$112 = L + 6W \qquad \text{known relationship}$$
$$112 - 6W = L \qquad \text{solve for } L$$

Using $L = 112 - 6W$, we substitute for L in the formula Area = LW.

$$A = LW \qquad \text{area formula}$$
$$= (112 - 6W)W \qquad \text{substitute } 112 - 6W \text{ for } L$$
$$A(W) = 112W - 6W^2 \qquad \text{result: a function in one variable}$$

Noting the result is a quadratic function, we enter $112X - 6X^2$ as Y_1 and attempt to find an appropriate viewing window. For the maximum value at $x = -\dfrac{b}{2a}$, we substitute 112 for b and -6 for a and find this value is near 10. This shows that Xmax = 20 would work (with Xmin = 0), since the graph will be symmetric and 10 is near the "middle" (twice 10 is 20). We reason further that if the width is 10 ft, $112 - 6 \times 10 = 52$ ft is left for the length and $10 \times 52 = 520$ ft^2 would be a good estimate for Ymax (the total area). To include a frame around the viewing window, we set Ymax at 600 with Ymin = -150. Using the sequence **2nd** (**CALC**) **4:maximum** gives the result shown in Figure 3.50, which shows a vertex of $\left(9\dfrac{1}{3}, 522\dfrac{2}{3}\right)$. This indicates that a maximum area of $522\dfrac{2}{3}$ ft^2 occurs when the width of each pen is $9\dfrac{1}{3}$ ft. Note this gives a length of $L = 112 - 6\left(9\dfrac{1}{3}\right) = 56$ ft.

Figure 3.50

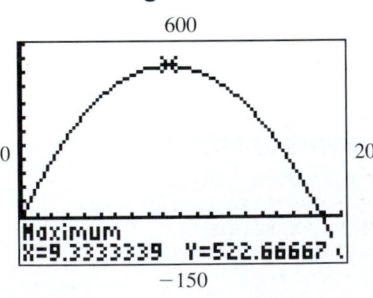

Now try Exercises 53 through 56 ▶

In a free-market economy, it is well known that if the price of an item is decreased, more people are likely to buy it. This is why stores have sales and bargain days. But if the item is sold too cheaply, revenue starts to decline because less money is coming in—even though more sales are being made. This phenomenon is analyzed in Example 8 using the standard formula for revenue: *revenue = price · number of sales* or $R = P \cdot S$.

EXAMPLE 8 ▶ **Analyzing Retail Sales Revenue**

When a popular running shoe is priced at $80, The Shoe Warehouse will sell an average of 96 pairs per week. Based on sales of similar shoes, the company believes that for each decrease of $2.50 in price, four additional pairs will be sold.

a. Construct a function that models the store's revenue at various prices. What is the maximum possible revenue under these conditions? What price should be charged to obtain this maximum revenue?

b. What is the cheapest price the manager could charge, while still bringing in a monthly revenue of at least $7000?

Solution ▶ **a.** Based on the given price and sales figures, the store has a current revenue of ($80)(96) = $7680. Using these figures, we can develop a general function model for the revenue, to show a decrease in price and an increase in sales based on the *number* of price decreases x.

$$(\$80 - 2.5x)(96 + 4x) = R(x)$$

for each price decrease of $2.50, sales increase by 4

In standard form, we have

$$7680 + 320x - 240x - 10x^2 = R(x) \qquad \text{multiply binomials}$$
$$-10x^2 + 80x + 7680 = R(x) \qquad \text{simplify}$$

To find the maximum revenue, we can enter the function as Y_1 on the ⎛Y=⎞ screen and set a window using the current revenue as a guide. We chose a window size of [0, 25] for x and [0, 9000] for y. The result is shown in Figure 3.51, where the ⎛2nd⎞ ⎛TRACE⎞ (CALC) 4:**Maximum** feature was used to locate the maximum. It appears the maximum revenue will be $7840, which occurs after $x = 4$ price decreases of $2.50. The selling price was $80 - 4(\$2.50) = \$70$.

Figure 3.51

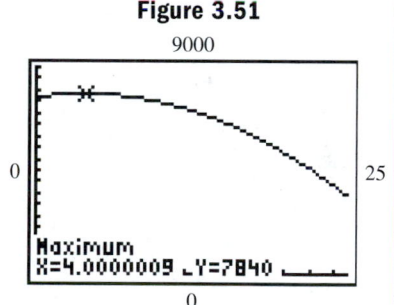

b. Since the desired revenue level is $7000, we substitute 7000 for $R(x)$.

$$-10x^2 + 80x + 7680 = 7000 \qquad \text{substitute 7000 for } R(x)$$

To solve for x, we'll use the Intersection-of-Graphs method with Y_1 as before and $Y_2 = 7000$. The result is shown in Figure 3.52 and indicates that to keep a revenue of at least $7000 per week, the price must decrease by no more than about 13.2($2.50) = $33. The selling price should be kept near 80 − 33 = $47.

Figure 3.52

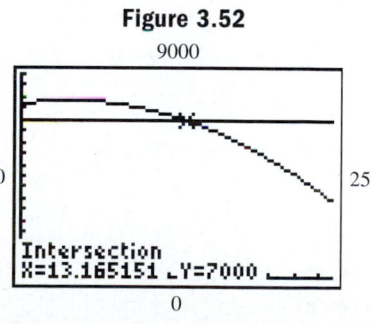

☑ **D.** You've just seen how we can solve applications involving extreme values

3.3 EXERCISES

▶ CONCEPTS AND VOCABULARY

Fill in each blank with the appropriate word or phrase. Carefully reread the section if needed.

1. Fill in the blank to complete the square, given $f(x) = -2x^2 - 10x - 7$:
$f(x) = -2(x^2 + 5x + \frac{25}{4}) - 7 + \underline{\quad \frac{25}{2} \quad}$.

2. The maximum and minimum values are called $\underline{\text{extreme}}$ values and can be found using the $\underline{\text{vertex}}$ formula.

3. To find the zeroes of $f(x) = ax^2 + bx + c$, we substitute $\underline{\quad 0 \quad}$ for $\underline{\quad f(x) \quad}$ and solve.

4. If the leading coefficient is positive and the vertex (h, k) is in Quadrant IV, the graph will have $\underline{\quad \text{two} \quad}$ x-intercepts.

5. Compare/Contrast how to complete the square on an *equation*, versus how to complete the square on a function. Use the equation $2x^2 + 6x - 3 = 0$ and the function $f(x) = 2x^2 + 6x - 3$ to illustrate.
Answers will vary.

6. Discuss/Explain why the graph of a quadratic function has no x-intercepts if a and k [vertex (h, k)] have like signs. Under what conditions will the function have a single real root?
Answers will vary; $k = 0$

▶ DEVELOPING YOUR SKILLS

Graph each function using end-behavior, intercepts, and completing the square to write the function in shifted form. Clearly state the transformations used to obtain the graph, and label the vertex and all intercepts (if they exist). If the equation is not factorable, use the quadratic formula to find the x-intercepts.

7. $f(x) = x^2 + 4x - 5$

8. $g(x) = x^2 - 6x - 7$

9. $h(x) = -x^2 + 2x + 3$

10. $H(x) = -x^2 + 8x - 7$

11. $u(x) = 3x^2 + 6x - 5$

12. $v(x) = 4x^2 - 24x + 15$

13. $f(x) = -2x^2 + 8x + 7$

14. $g(x) = -3x^2 + 12x - 7$

15. $p(x) = 2x^2 - 7x + 3$

16. $q(x) = 4x^2 - 9x + 2$

17. $f(x) = -3x^2 - 7x + 6$

18. $g(x) = -2x^2 + 9x - 7$

19. $p(x) = x^2 - 5x + 2$

20. $q(x) = x^2 + 7x + 4$

Graph each function using the vertex formula and other features of a quadratic graph. Label all important features.

21. $f(x) = x^2 + 2x - 6$

22. $g(x) = x^2 + 8x + 11$

23. $h(x) = -x^2 + 4x + 2$

24. $H(x) = -x^2 + 10x - 19$

25. $Y_1 = 0.5x^2 + 3x + 7$

26. $Y_2 = 0.2x^2 - 2x + 8$

27. $Y_1 = -2x^2 + 10x - 7$

28. $Y_2 = -2x^2 + 8x - 3$

29. $f(x) = 4x^2 - 12x + 3$

30. $g(x) = 3x^2 + 12x + 5$

31. $p(x) = \frac{1}{2}x^2 + 3x - 5$

32. $q(x) = \frac{1}{3}x^2 - 2x - 4$

State the equation of the function whose graph is shown.

33.

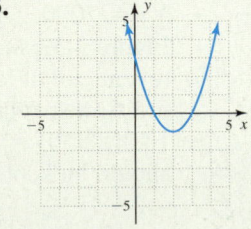

$y = 1(x - 2)^2 - 1$

34.

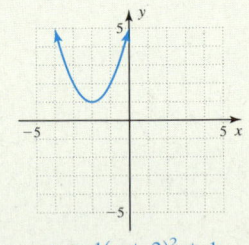

$y = 1(x + 2)^2 + 1$

HOMEWORK SELECTION GUIDE

Core: 7–37 odd, 41–57 odd (25 Exercises)
Standard: 1–4, 7–57 odd, 62, 63, 65 (33 Exercises)

Extended: 1–4, 7–53 odd, 54, 55, 57, 61, 62, 63, 65 (36 Exercises)
In Depth: 1–6, 7–39 odd, 40, 41–49 odd, 50–57, 61–66 (43 Exercises)

Additional answers can be found in the Instructor Answer Appendix.

35.

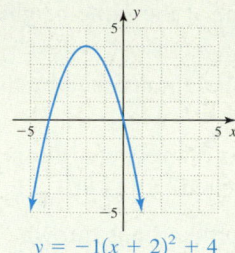

$y = -1(x + 2)^2 + 4$

36.

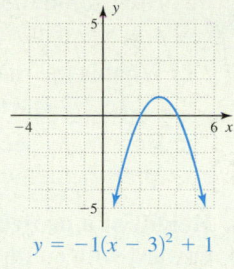

$y = -1(x - 3)^2 + 1$

37.

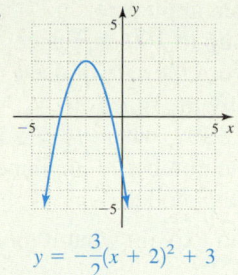

$y = -\dfrac{3}{2}(x + 2)^2 + 3$

38.

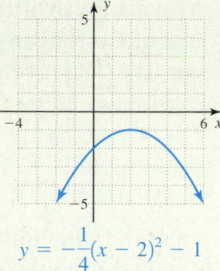

$y = -\dfrac{1}{4}(x - 2)^2 - 1$

▶ WORKING WITH FORMULAS

39. Vertex/intercept formula: $x = h \pm \sqrt{-\dfrac{k}{a}}$

As an alternative to using the quadratic formula *prior* to completing the square, the x-intercepts can easily be found using the vertex/intercept formula *after* completing the square, when the coordinates of the vertex are known. (a) Beginning with the shifted form $y = a(x - h)^2 + k$, substitute 0 for y and solve for x to derive the formula. (b) Use the formula to find all zeroes, real or complex, of the following functions.

i. $y = (x + 3)^2 - 5$ $x = -3 \pm \sqrt{5}$ **ii.** $y = -(x - 4)^2 + 3$ $x = 4 \pm \sqrt{3}$ **iii.** $y = 2(x + 4)^2 - 7$ $x = -4 \pm \dfrac{\sqrt{14}}{2}$

iv. $y = -3(x - 2)^2 + 6$ $x = 2 \pm \sqrt{2}$ **v.** $s(t) = 0.2(t + 0.7)^2 - 0.8$ $t = -2.7, t = 1.3$ **vi.** $r(t) = -0.5(t - 0.6)^2 + 2$ $t = -1.4, t = 2.6$

40. Surface area of a rectangular box with square ends: $S = 2h^2 + 4Lh$

The surface area of a rectangular box with square ends is given by the formula shown, where h is the height and width of the square ends, and L is the length of the box. (a) If L is 3 ft and the box must have a surface area of 32 ft², find the dimensions of the square ends. (b) Solve for L, then find the length if the height is 1.5 ft and surface area is 22.5 ft². **a.** 2 ft × 2 ft **b.** $L = \dfrac{S - 2h^2}{4h}$; L = 3 ft

▶ APPLICATIONS

41. Maximum profit: An automobile manufacturer can produce up to 300 cars per day. The profit made from the sale of these vehicles can be modeled by the function $P(x) = -10x^2 + 3500x - 66{,}000$, where $P(x)$ is the profit in dollars and x is the number of automobiles made and sold. Based on this model:

a. Find the y-intercept and explain what it means in this context.
(0, −66,000); when no cars are produced, there is a loss of $66,000.

b. Find the x-intercepts and explain what they mean in this context. (20, 0), (330, 0); no profit will be made if fewer than 20 or more than 330 cars are produced.

c. How many cars should be made and sold to maximize profit?   175

d. What is the maximum profit? $240,250

42. Maximum profit: The profit for a manufacturer of collectible grandfather clocks is given by the function $P(x) = -1.6x^2 + 240x - 375$, where $P(x)$ is the profit in dollars and x is the number of clocks made and sold.

a. Find the y-intercept and explain what it means in this context.
(0, −375); when no clocks are made, there is a loss of $375.

b. Find the x-intercepts and explain what they mean in this context. (1.58, 0), (148.4, 0); no profit will be made if you produce fewer than 2 or more than 148 clocks.

c. How many clocks should be made and sold to maximize profit?   75

d. What is the maximum profit? $8625

43. Optimal pricing strategy: The director of the Ferguson Valley drama club must decide what to charge for a ticket to the club's performance of *The Music Man.* If the price is set too low, the club will lose money; and if the price is too high, people won't come. From past experience she estimates that the profit P from sales (in hundreds) can be approximated by $P(x) = -x^2 + 46x - 88$, where x is the cost of a ticket and $0 \leq x \leq 50$.

a. Find the lowest cost of a ticket that would allow the club to break even. $2

b. What is the highest cost that the club can charge to break even? $44

c. If the theater were to close down before any tickets are sold, how much money would the club lose? $8800

d. How much should the club charge to maximize their profits? What is the maximum profit?
$23; $44,100

Additional answers can be found in the Instructor Answer Appendix.

44. Maximum profit: A kitchen appliance manufacturer can produce up to 200 appliances per day. The profit made from the sale of these machines can be modeled by the function $P(x) = -0.5x^2 + 175x - 3300$, where $P(x)$ is the profit in dollars, and x is the number of appliances made and sold. Based on this model,

a. Find the y-intercept and explain what it means in this context.
   (0, −3300); if no appliances are sold, the loss will be $3300.
b. Find the x-intercepts and explain what they mean in this context. (20, 0), (330, 0); if fewer than 20 or more than 330 appliances are made and sold, there will be no profit.
c. Determine the domain of the function and explain its significance. $0 \le x \le 200$; maximum capacity is 200
d. How many should be sold to maximize profit? What is the maximum profit?   175, $12,012.50

45. Cost of production: The cost of producing a plastic toy is given by the function $C(x) = 2x + 35$, where x is the number of hundreds of toys. The revenue from toy sales is given by $R(x) = -x^2 + 122x - 365$. Since profit = revenue − cost, the profit function must be $P(x) = -x^2 + 120x - 400$ (verify). How many toys sold will produce the maximum profit? What is the maximum profit?   6000; $3200

46. Cost of production: The cost to produce bottled spring water is given by $C(x) = 16x - 63$, where x is the number of thousands of bottles. The total income (revenue) from the sale of these bottles is given by the function $R(x) = -x^2 + 326x - 7463$. Since profit = revenue − cost, the profit function must be $P(x) = -x^2 + 310x - 7400$ (verify). How many bottles sold will produce the maximum profit? What is the maximum profit?   155,000; $16,625

The projectile function: $h(t) = -16t^2 + vt + k$ **applies to any object projected upward with an initial velocity** v, **from a height** k **but not to an object under propulsion (such as a rocket). Consider this situation and answer the questions that follow.**

47. Model rocketry: A member of the local rocketry club launches her latest rocket from a large field. At the moment its fuel is exhausted, the rocket has a velocity of 240 ft/sec and an altitude of 544 ft (t is in seconds).

a. Write the function that models the height of the rocket. $h(t) = -16t^2 + 240t + 544$
b. How high is the rocket at $t = 0$? If it took off from the ground, why is it this high at $t = 0$?
   544 ft; that is when the fuel is exhausted.
c. How high is the rocket 5 sec after the fuel is exhausted?   1344 ft
d. How high is the rocket 10 sec after the fuel is exhausted?   1344 ft

e. How could the rocket be at the same height at $t = 5$ and at $t = 10$?   It is coming back down.
f. What is the maximum height attained by the rocket?   1444 ft
g. How many seconds was the rocket airborne *after* its fuel was exhausted?   17 sec

48. Height of a projectile: A projectile is thrown upward with an initial velocity of 176 ft/sec. After t sec, its height $h(t)$ above the ground is given by the function $h(t) = -16t^2 + 176t$.

a. Find the projectile's height above the ground after 2 sec.   288 ft
b. Sketch the graph modeling the projectile's height.
c. What is the projectile's maximum height? What is the value of t at this height?
   484 ft; 5.5 sec
d. How many seconds after it is thrown will the projectile strike the ground?   11 sec

49. Height of a projectile: In the movie *The Court Jester* (1956; Danny Kaye, Basil Rathbone, Angela Lansbury, and Glynis Johns), a catapult is used to toss the nefarious adviser to the king into a river. Suppose the path flown by the king's adviser is modeled by the function $h(d) = -0.02d^2 + 1.64d + 14.4$, where $h(d)$ is the height of the adviser in feet at a distance of d ft from the base of the catapult.
   a. 14.4 ft   b. 41 ft   c. 48.02 ft   d. 90 ft
a. How high was the release point of this catapult?
b. How far from the catapult did the adviser reach a maximum altitude?
c. What was this maximum altitude attained by the adviser?
d. How far from the catapult did the adviser splash into the river?

50. Blanket toss competition: The fraternities at Steele Head University are participating in a blanket toss competition, an activity borrowed from the whaling villages of the Inuit Eskimos. If the person being tossed is traveling at 32 ft/sec as he is projected into the air, and the Frat members are holding the canvas blanket at a height of 5 ft,

a. Write the function that models the height at time t of the person being tossed.
   $h(t) = -16t^2 + 32t + 5$
b. How high is the person when (i) $t = 0.5$, (ii) $t = 1.5$?   (i) 17 ft (ii) 17 ft
c. From part (b) what do you know about *when* the maximum height is reached?
   it must occur between $t = 0.5$ and $t = 1.5$

d. To the nearest tenth of a second, when is the maximum height reached?   *t* = 1 sec

e. To the nearest one-half foot, what was the maximum height?   *h*(1) = 21 ft

f. To the nearest tenth of a second, how long was this person airborne?   2 sec

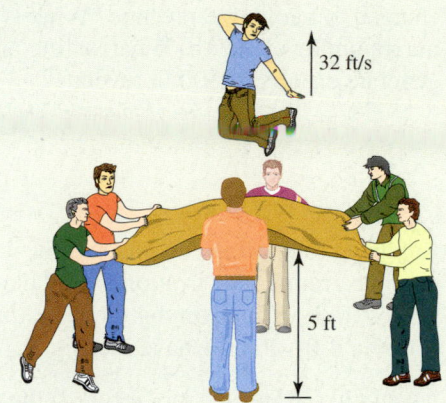

32 ft/s

5 ft

51. Motorcycle jumps: On December 31, 2007, Australian freestyle motocross legend Robbie Maddison set a world record by jumping his motorcycle 322 ft, 7½ in. During practice the day before, the prevailing wind conditions actually allowed him to jump farther. Suppose his height on one such jump is given by the equation $h = -16t^2 + 52t + 25$, where *h* represents his height (in feet) above ground level *t* sec after takeoff.

a. How high is the top of the takeoff ramp?   25 ft

b. If he touched down on the landing ramp 15 ft above ground level, how long was "Maddo" in the air?   approx. 3.43 sec

c. What would have been the daredevil's maximum height?   67.25 ft

52. SuperPipe Finals: In the Winter X Games, one of the most exciting events to watch is the SuperPipe Final. The height of a professional snowboarder during one particularly huge jump is given by the function $h = -16t^2 + 35t + 16$, where *h* represents her height (in feet) above the pipe base *t* sec after leaving the upper edge, or lip.

a. How high is the lip of the superpipe?   16 ft

b. If the snowboarder lands her trick 11 ft above the base of the pipe, how long was she in the air?   approx. 2.32 sec

c. What was the athlete's maximum height above the base of the pipe?   approx 35.1 ft

53. Fencing a backyard: Tina and Imai have just purchased a purebred German Shepherd, and need to fence in their backyard so the dog can run. What is the maximum rectangular area they can enclose with 200 ft of fencing, if (a) they use fencing material along all four sides? What are the dimensions of the rectangle? (b) What is the maximum area if they use the house as one of the sides? What are the dimensions of *this* rectangle?   **a.** 2500 ft², 50 ft × 50 ft   **b.** 5000 ft², 50 ft × 100 ft

54. Building sheep pens: It's time to drench the sheep again, so Chance and Chelsea-Lou are fencing off a large rectangular area to build some temporary holding pens. To prep the males, females, and kids, they are separated into three smaller and equal-size pens partitioned within the large rectangle. If 384 ft of fencing is available and the maximum area is desired, what will be (a) the dimensions of the larger, outer rectangle? (b) the dimensions of the smaller holding pens?   **a.** 96 ft × 48 ft   **b.** 32 ft × 48 ft

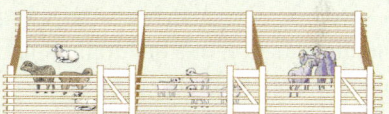

55. Building windows: Window World, Inc., is responsible for designing new windows for the expansion of the campus chapel. The current design is shown in the figure. The metal trim used to secure the perimeter of the frame is 126" long. If the maximum window area is desired (to let in the most sunlight), what will be (a) the dimensions of the rectangular portion of each window? (b) the total area of each window?

a. approx. 29.5" wide by 18.7" long   **b.** approx. 930 in² *w*

Exercise 55

w   *w*

$h = 0.866\ w$

l

56. Building windows: Before construction on the chapel expansion begins (see Exercise 55), a new design is submitted. The new design is very different (see figure), but must still have a perimeter of 126". If the maximum window area is still desired, what will be (a) the dimensions of the rectangular portion of each window? (b) the total area of each window?   **a.** approx. 35.3" wide by 17.6" long   **b.** approx. 1111.5 in²

Exercise 56

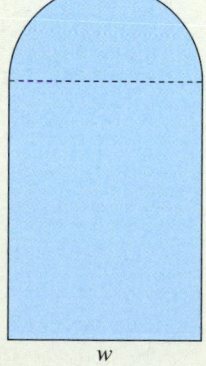

l

w

57. Maximizing soft drink revenue: A convenience store owner sells 20-oz soft drinks for $1.50 each, and sells an average of 500 per week. Using a market survey, she believes that for each $0.05 decrease in price, an additional 25 soft drinks will be sold. (a) What price should be charged to maximize revenue? What is the maximum revenue? (b) What is the lowest price the owner could charge and still bring in at least $700 per week in revenue?
a. $1.25 each, $781.25   **b.** about $0.85 each

58. Maximizing restaurant revenue: At Figaro's Pizzeria, large pizzas usually sell for $12.50 and at this price an average of 320 large pizzas are sold each weekend. Using a survey and years of experience, the manager of Figaro's believes that for each $0.25 decrease in price, 16 additional pizzas will be sold. (a) What price should be charged to maximize revenue? What is the maximum revenue? (b) What selling price(s) will generate exactly $4800 in revenue?
a. $8.75/pizza, $4900   **b.** $10.00/pizza, $7.50/pizza

▶ EXTENDING THE CONCEPT

59. Use the general solutions from the quadratic formula to show that the average value of the x-intercepts is $\dfrac{-b}{2a}$. Explain/Discuss why the result is valid even if the roots are complex.

$$x_1 = \frac{-b + \sqrt{b^2 - 4ac}}{2a}, \; x_2 = \frac{-b - \sqrt{b^2 - 4ac}}{2a}$$
Answers will vary.

60. Write the equation of a quadratic function whose x-intercepts are given by $x = 2 \pm 3i$. $f(x) = x^2 - 4x + 13$

61. Write the equation for the parabola given.

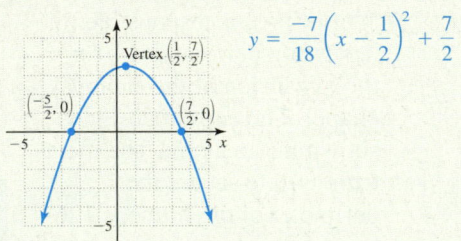

$$y = \frac{-7}{18}\left(x - \frac{1}{2}\right)^2 + \frac{7}{2}$$

62. Referring to Exercise 39, discuss the nature (real or complex, rational or irrational) and number of zeroes (0, 1, or 2) given by the vertex/intercept formula if (a) a and k have like signs, (b) a and k have unlike signs, (c) k is zero, (d) the ratio $-\dfrac{k}{a}$ is positive and a perfect square, and (e) the ratio $-\dfrac{k}{a}$ is positive and not a perfect square.

a. radicand will be negative—two complex zeroes.
b. radicand will be positive—two real zeroes.
c. radicand is zero—one real zero.
d. two real, rational zeroes.
e. two real, irrational zeroes.

▶ MAINTAINING YOUR SKILLS

63. (1.4) Identify the slope and y-intercept for $-4x + 3y = 9$. Do not graph. $m = \frac{4}{3}$, y-intercept $(0, 3)$

65. (2.2) Given $f(x) = \sqrt[3]{x} + 3$, find the equation of the function whose graph is that of $f(x)$, shifted right 2 units, reflected across the x-axis, and then down 3 units. $g(x) = -\sqrt[3]{x + 1} - 3$

64. (R.5) Multiply: $\dfrac{x^2 - 4x + 4}{x^2 + 3x - 10} \cdot \dfrac{x^2 - 25}{x^2 - 10x + 25}$ $\frac{x-2}{x-5}$

66. (3.2) Given $f(x) = 3x^2 + 7x - 6$, solve $f(x) \leq 0$ using the x-intercepts and end-behavior of f. $x \in \left[-3, \frac{2}{3}\right]$

MID-CHAPTER CHECK

1. Find the sum and product of the complex numbers $2 + 3i$ and $2 - 3i$. Comment on what you notice.
sum 4, product 13; both yield real numbers

2. Determine the quotient: $\dfrac{2 + 3i}{3 - 2i}$ i

3. Check by substitution: Is $x = 1 + 2i$ a solution to $x^2 - 2x + 5 = 0$?

4. Given $f(x)$ is a quadratic function whose graph has a vertex of $(1, -4)$ and an x-intercept of $(3, 0)$, find the two solutions of the equation $f(x) = 0$. $x = -1$ and 3

Solve the following quadratic equations using the method of your choice.

5. $3(x - 2)^2 = 5$ $x = 2 \pm \sqrt{\frac{5}{3}}$

6. $2x^2 = 7x - 4$ $x = \frac{7}{4} \pm \frac{\sqrt{17}}{4}$

 7. Solve the inequality by using a graphing calculator to locate the x-intercepts and observing the graph. Round to the nearest hundredth as necessary. $f(x) = -5.1x^2 - 3.2x + 1.9$; $f(x) \le 0$ $(-\infty, -1] \cup [0.37, \infty)$

8. Mark Twain's 1867 book, *The Celebrated Jumping Frog of Calaveras County*, has actually spawned an annual "World Championship" frog jumping contest in Calaveras County, California. As of 2009, the record for the longest triple jump (the total of three consecutive jumps) is 21 ft $5\frac{3}{4}$ in. set by Rosie the Ribeter in 1986. In a recent competition, Count Frogula felt he had beaten the record and was awaiting the judge's decision. Suppose the longest of his three jumps was modeled by the function $h(x) = -2.3x^2 + 16.5x$, where $h(x)$ represents the height of the jump in inches, x ft from the starting point. Use this function to find (a) the maximum height of the jump and (b) the length of the jump. (c) If the Count jumped this distance three times, did he beat Rosie's record?
a. about 29.6 in.   **b.** about 7.17 ft   **c.** yes; $3 \times 7.17 > 21.5$

9. Graph $g(x) = -\frac{1}{2}x^2 - 2x$ using the vertex formula and other features of a quadratic graph. Label all important features.

10. Find the equation of the quadratic function whose graph is shown.
$f(x) = \frac{4}{9}(x - 3)^2 - 2$

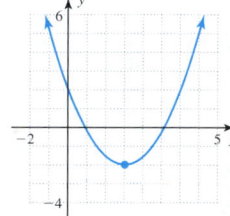

REINFORCING BASIC CONCEPTS

An Alternative Method for Checking Solutions to Quadratic Equations

To solve $x^2 - 2x - 15 = 0$ by factoring, students will often begin by looking for two numbers whose product is -15 (the constant term) and whose sum is -2 (the linear coefficient). The two numbers are -5 and 3 since $(-5)(3) = -15$ and $-5 + 3 = -2$. In factored form, we have $(x - 5)(x + 3) = 0$ with solutions $x_1 = 5$ and $x_2 = -3$. When these solutions are compared *to the original coefficients,* we can still see the sum/product relationship, but note that while $(5)(-3) = -15$ still gives the constant term, $5 + (-3) = 2$ gives the linear coefficient *with opposite sign.* Although more difficult to accomplish, this method can be applied to *any* factorable quadratic equation $ax^2 + bx + c = 0$ if we divide through by a, giving $x^2 + \dfrac{b}{a}x + \dfrac{c}{a} = 0$. For $2x^2 - x - 3 = 0$, we divide both sides by 2 and obtain $x^2 - \dfrac{1}{2}x - \dfrac{3}{2} = 0$, then look for two numbers whose product is $-\dfrac{3}{2}$ and whose sum is $-\dfrac{1}{2}$. The factors are $\left(x - \dfrac{3}{2}\right)$ and $(x + 1)$ since $\left(-\dfrac{3}{2}\right)(1) = -\dfrac{3}{2}$ and $-\dfrac{3}{2} + 1 = -\dfrac{1}{2}$, showing the solutions are $x_1 = \dfrac{3}{2}$ and $x_2 = -1$. We again note the product of the solutions is the constant $-\dfrac{3}{2} = \dfrac{c}{a}$, and the sum of the solutions is the linear coefficient *with opposite sign:* $\dfrac{1}{2} = -\dfrac{b}{a}$. No one actually promotes this method for solving trinomials where $a \ne 1$, but it does illustrate an important and useful concept:

If x_1 and x_2 are the two roots of $x^2 + \dfrac{b}{a}x + \dfrac{c}{a} = 0,$

then $x_1 x_2 = \dfrac{c}{a}$ and $x_1 + x_2 = -\dfrac{b}{a}$

Justification for this can be found by taking the product and sum of the general solutions $x_1 = \dfrac{-b}{2a} + \dfrac{\sqrt{b^2 - 4ac}}{2a}$

and $x_2 = \dfrac{-b}{2a} - \dfrac{\sqrt{b^2 - 4ac}}{2a}$. Although the computation looks impressive, the product can be computed as a binomial times its conjugate, and the radical parts add to zero for the sum, each yielding the results as already stated.

This observation provides a useful technique for checking solutions to a quadratic equation, *even those having irrational or complex roots!* Check the solutions shown in these exercises.

Exercise 1: $2x^2 - 5x - 7 = 0$

$$x_1 = \frac{7}{2}$$
$$x_2 = -1$$

Exercise 2: $2x^2 - 4x - 7 = 0$

$$x_1 = \frac{2 + 3\sqrt{2}}{2}$$
$$x_2 = \frac{2 - 3\sqrt{2}}{2}$$

Exercise 3: $x^2 - 10x + 37 = 0$

$$x_1 = 5 + 2\sqrt{3}\,i$$
$$x_2 = 5 - 2\sqrt{3}\,i$$

Exercise 4: Verify this sum/product check by computing the sum and product of the general solutions for x_1 and x_2.

3.4   Quadratic Models; More on Rates of Change

LEARNING OBJECTIVES

In Section 3.4 you will see how we can:

- ❏ **A.** Develop quadratic function models from a set of data
- ❏ **B.** Calculate the average rate of change for nonlinear functions using points from a graph
- ❏ **C.** Calculate average rates of change using the average rate of change formula

So far in our study, we've seen real data sets that were best modeled by linear functions (Section 1.6) and power functions (Section 2.4). These models enabled us to make meaningful decisions and comparisons, as well as reasonable projections for future occurrences. In this section, we explore data relationships that are best modeled by quadratic functions, and look at how the "rate of change" concept can be applied to these and other functions.

A. Quadratic Equation Models

After a set of data is collected and organized, any patterns or relationships that exist may not be readily known or easily seen. The regression model chosen for the data will depend on a number of factors, including the context of the data, any patterns noted in the scatterplot, any foreknowledge of how the data might be related, along with a careful assessment of the correlation coefficient. Most graphing calculators have the ability to find regression models for a large number of functions, and the more we are familiar with each family of functions, the better we'll be able to use the technology. Regardless of the form of regression used, the steps for inputting the data, setting a window size, viewing the scatterplot, and so on, are identical.

EXAMPLE 1 ▶   **Growth of Online Sales of Pet Supplies**

Since the year 2000, there has been a tremendous increase in the online sales of pet food, pet medications, and other pet supplies. That data shown in the table shows the amount spent (in billions of dollars) for the years indicated.

Source: 2009 Statistical Abstract of the United States, Page 646, Table 1016.

a. Input the data into a graphing calculator, set an appropriate window and view the scatterplot, then use the context and the scatterplot to decide on an appropriate form of regression.

b. Determine the regression equation, and use it to find the amount of online sales projected for the year 2012.

c. If the predicted trends continue, in what year will online sales of pet supplies surpass $25 billion dollars?

Year (2001 → 1)	Sales (billions)
1	0.8
5	4.1
6	5.6
7	7.4
8	9.1

Solution ▶

a. Begin by entering the years in L1 and the dollar amounts in L2. From the data given, a viewing window of Xmin = 0, Xmax = 15, Ymin = −2, and Ymax = 15 seems appropriate. The scatterplot in Figure 3.53 shows a definite increasing, non-linear pattern and we opt for a quadratic regression.

Figure 3.53

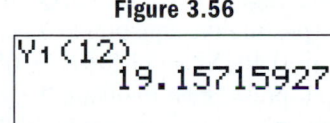

b. Using **STAT** ▶ (CALC) **5:QuadReg** places this option on the home screen. Pressing **ENTER** at this point will give us the quadratic coefficients, but we can also have the calculator paste the equation itself directly into Y_1 on the **Y=** screen, simply by appending Y_1 to the **QuadReg** command (Figure 3.54). The resulting equation is $y \approx 0.118x^2 + 0.136x + 0.537$ (to three decimal places), and the graph is shown in Figure 3.55. Using the home screen, we find that $Y_1(12) \approx 19.2$ (Figure 3.56), indicating that in 2012, about $19.2 billion is projected to be spent online for pet supplies.

c. To find the year when spending will surpass $25 billion, we set $Y_2 = 25$ and use the Intersection-of-Graphs method (be sure to increase Ymax so this graph can be seen: Ymax = 30). There result is shown in Figure 3.57 and indicates that spending will surpass $25 billion late in the year 2013.

Figure 3.54, 3.55

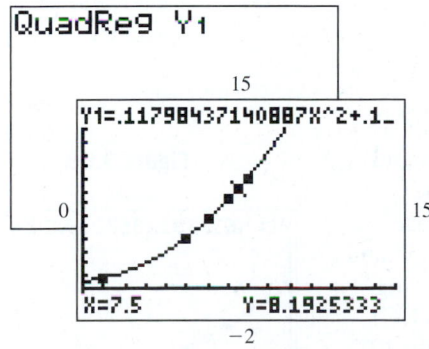

Figure 3.56

$Y_1(12)$
        19.15715927

Figure 3.57

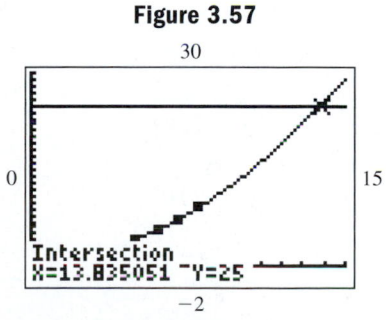

Now try Exercises 7 through 10 ▶

Applications of quadratic regression tend to focus on the defining characteristics of a quadratic graph, even if only a part of the graph is used. The data may simply indicate an increasing rate of growth beyond an initial or minimum point as in Example 1, or show values that decrease to a minimum and increase afterward. These would be data sets modeled by an equation where $a > 0$. Applications where $a < 0$ are also common.

In actual practice, data sets are often very large and need not be integer-valued. In addition, as we explore additional functions and forms of regression, the decision of which form to use must be carefully evaluated as many functions exhibit similar characteristics.

EXAMPLE 2 ▶ **Apollo 13, Retro-Rocket Option**

About 56 hr into the mission and 65,400 km from the Moon, the oxygen tanks exploded on Apollo 13. At this point, mission control explored two options: (1) firing the retro-rockets for a direct return to Earth, or (2) the so called, "sling-shot around the Moon," the option actually chosen. The data shown in the table explore a scenario where option 1 was taken, and give the distance d from the Moon, t seconds after the retro-rockets have fired.

  a. Input the data into a graphing calculator, set an appropriate window and view the scatterplot, then use the context and the scatterplot to decide on an appropriate form of regression.

  b. Determine the regression equation, and use it to find how close Apollo 13 would have come to the Moon (the minimum distance). How many seconds after the retro-rockets fired would this have occurred?

  c. How many seconds would it have taken for Apollo 13 to "turn around," and get back to the original distance of 65,400 km (where the accident occurred)?

Time (sec)	Distance from Moon (km)
0	65,400
1	65,389.0
2	65,378.8
3	65,369.6
4	65,361.4
5	65,354.0
10	65,331.2

Solution ▶

  a. After entering the data (seconds in L1 and distance in L2), we set a viewing window of Xmin = 0, Xmax = 15, and after some trial and error, Ymin = 65,300, and Ymax = 65,450. The scatterplot shows (be sure that **PLOT1** is activated on the ⬤Y= screen) a gradually decreasing, nonlinear pattern and a quadratic regression seems appropriate (Figure 3.58).

Figure 3.58

65,450

P1:L1,L2

0   15

X=5 . . . . . Y=65354 . . .

65,300

  b. We find the regression equation using **STAT** ⬤ **(CALC) 5:QuadReg** Y_1 (pasting the equation to Y_1): $Y_1 \approx 0.464X^2 - 11.519X + 65{,}400.013$. Next, press **GRAPH** to obtain the graph and scatterplot shown in Figure 3.59. The result indicates that we should increase the value of Xmax to see the full graph and to help locate the minimum value. Using Xmax = 30 and the **2nd** **TRACE** **(CALC)** **3:minimum** option shows that Apollo 13 would have come within 65,328.5 km of the Moon, after the retro-rockets had fired for 12.4 sec (Figure 3.60).

Figure 3.59

65,450

Y1=.46377912867279X^2+-_

0   15

X=7.5 . . . . Y=65339.708 .

65,300

Figure 3.60

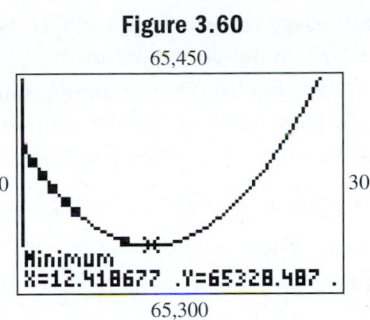

Figure 3.61

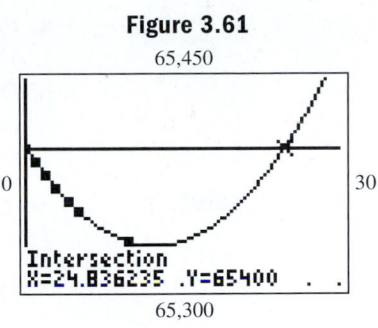

c. For the time required to return to a distance of 65,400 km, we enter $Y_2 = 65,400$ and use the (2nd) (TRACE) **(CALC) 5:intersect** option. The result shown in Figure 3.61 indicates a required time of about 25 sec. Actually, we could have reasoned that if it took 12.5 sec to reach the minimum distance, we could double this time to return to the original distance.

☑ **A. You've just seen how we can develop quadratic function models from a set of data**

Now try Exercises 11 through 14 and 45 through 48 ▶

B. Nonlinear Functions and Rates of Change

As noted in Section 1.2, one of the defining characteristics of linear functions is that their rate of change is constant: $\frac{\Delta y}{\Delta x} = \frac{y_2 - y_1}{x_2 - x_1} = m$. For nonlinear functions the rate of change is not constant, but to aid in their study and application, we use a related concept called the **average rate of change,** given by *the slope of a secant line* through points (x_1, y_1) and (x_2, y_2) on the graph.

EXAMPLE 3 ▶ Calculating Average Rates of Change

The graph shown displays the number of units shipped of vinyl records, cassette tapes, and CDs for the period 1980 to 2005.

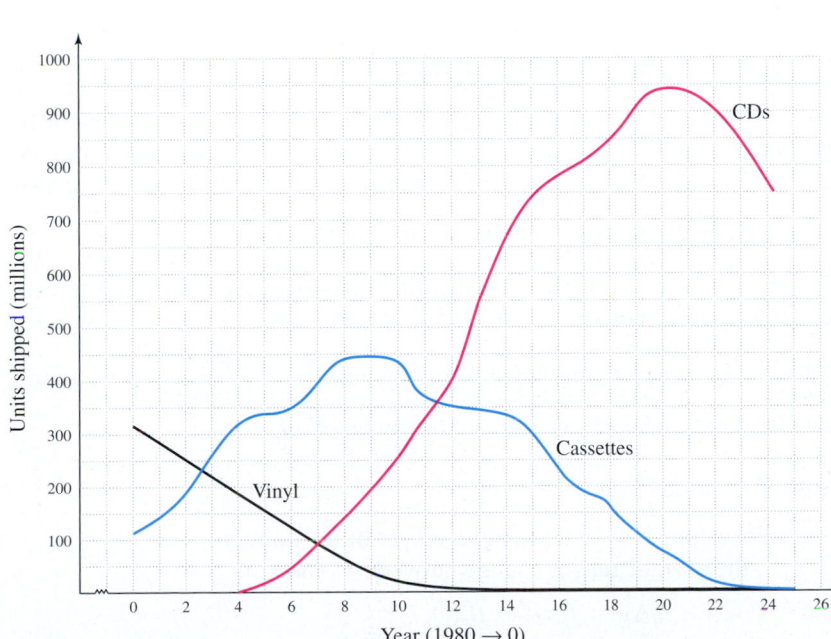

Source: Swivel.com

Units shipped in millions

Year (1980 → 0)	Vinyl	Cassette	CDs
0	323	110	0
2	244	182	0
4	205	332	6
6	125	345	53
8	72	450	150
10	12	442	287
12	2	366	408
14	2	345	662
16	3	225	779
18	3	159	847
20	2	76	942
24	1	5	767
25	1	3	705

a. Find the average rate of change in CDs shipped and in cassettes shipped from 1994 to 1998. What do you notice?

b. Does it appear that the rate of increase in CDs shipped was greater from 1986 to 1992, or from 1992 to 1996? Compute the average rate of change for each period and comment on what you find.

Solution ▶ Using 1980 as year zero (1980 → 0), we have the following:

a.

CDs	**Cassettes**

1994: (14, 662), 1998: (18, 847)   1994: (14, 345), 1998: (18, 159)

$$\frac{\Delta y}{\Delta x} = \frac{847 - 662}{18 - 14} \qquad\qquad \frac{\Delta y}{\Delta x} = \frac{159 - 345}{18 - 14}$$

$$= \frac{185}{4} \qquad\qquad\qquad = -\frac{186}{4}$$

$$= 46.25 \qquad\qquad\qquad = -46.5$$

The decrease in the number of cassettes shipped was roughly equal to the increase in the number of CDs shipped (about 46,000,000 per year).

b. From the graph, the secant line for 1992 to 1996 appears to have a greater slope.

1986–1992 CDs   **1992–1996 CDs**

1986: (6, 53), 1992: (12, 408)   1992: (12, 408), 1996: (16, 779)

$$\frac{\Delta y}{\Delta x} = \frac{408 - 53}{12 - 6} \qquad\qquad \frac{\Delta y}{\Delta x} = \frac{779 - 16}{16 - 12}$$

$$= \frac{355}{6} \qquad\qquad\qquad = \frac{371}{4}$$

$$= 59.1\overline{6} \qquad\qquad\qquad = 92.75$$

For the years 1986 to 1992, the average rate of change for CD sales was about 59.2 million per year. For the years 1992 to 1996, the average rate of change was almost 93 million per year, a significantly higher rate of growth.

☑ **B.** You've just seen how we can calculate the average rate of change for nonlinear functions using points on a graph

Now try Exercises 15 through 26, 49 and 50 ▶

C. The Average Rate of Change Formula

The importance of the rate of change concept would be hard to overstate. In many business, scientific, and economic applications, it is this attribute of a function that draws the most attention. In Example 3 we computed average rates of change by selecting two points from a graph, and computing the slope of the secant line: $m = \frac{\Delta y}{\Delta x} = \frac{y_2 - y_1}{x_2 - x_1}$.

With a simple change of notation, we can use the function's *equation* rather than relying on a graph. Note that y_2 corresponds to the function evaluated at x_2: $y_2 = f(x_2)$. Likewise, $y_1 = f(x_1)$. Substituting these into the slope formula yields $\frac{\Delta y}{\Delta x} = \frac{f(x_2) - f(x_1)}{x_2 - x_1}$, giving the average rate of change between x_1 and x_2 *for any function f.*

Average Rate of Change

For a function f and $[x_1, x_2]$ a subset of the domain, the average rate of change between x_1 and x_2 is

$$\frac{\Delta y}{\Delta x} = \frac{f(x_2) - f(x_1)}{x_2 - x_1}, x_1 \neq x_2$$

Average Rates of Change Applied to Projectile Velocity

A projectile is any object that is thrown, shot, or cast upward, with no continuing source of propulsion. The object's height (in feet) after t sec is modeled by the function $h(t) = -16t^2 + vt + k$, where v is the initial velocity of the projectile, and k is the height of the object at $t = 0$. For instance, if a soccer ball is kicked vertically upward from ground level ($k = 0$) with an initial speed of 64 ft/sec, the height of the ball t sec later is $h(t) = -16t^2 + 64t$. From Section 3.3, we recognize the graph will be a parabola and evaluating the function for $t = 0$ to 4 produces Table 3.1 and the graph shown in Figure 3.62. Experience tells us the ball is traveling at a faster rate immediately after being kicked, as compared to when it nears its maximum height where it momentarily stops, then begins its descent. In other words, the rate of change $\dfrac{\Delta \text{height}}{\Delta \text{time}}$

has a larger value at any time prior to reaching its maximum height. To quantify this we'll compute the average rate of change between (a) $t = 0.5$ and $t = 1$, and compare it to the average rates of change between, (b) $t = 1$ and $t = 1.5$, and (c) $t = 1.5$ and $t = 2$.

Table 3.1

Time in seconds	Height in feet
0	0
1	48
2	64
3	48
4	0

Figure 3.62

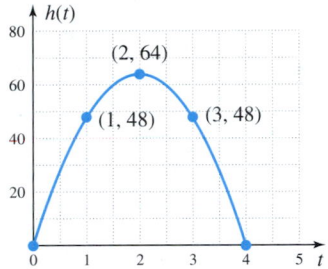

> **WORTHY OF NOTE**
> Keep in mind the graph of h represents the relationship between the soccer ball's height in feet and the elapsed time t. It does not model the actual path of the ball.

EXAMPLE 4 ▶ **Average Rates of Change Applied to Projectiles**

For the projectile function $h(t) = -16t^2 + 64t$, find the average rate of change for
  a. $t \in [0.5, 1]$.   **b.** $t \in [1, 1.5]$.   **c.** $t \in [1.5, 2.0]$.
Then graph the secant lines representing these average rates of change and comment.

Solution ▶ Using the given intervals in the formula $\dfrac{\Delta h}{\Delta t} = \dfrac{h(t_2) - h(t_1)}{t_2 - t_1}$ yields

a. $\dfrac{\Delta h}{\Delta t} = \dfrac{h(1) - h(0.5)}{1 - (0.5)}$ **b.** $\dfrac{\Delta h}{\Delta t} = \dfrac{h(1.5) - h(1)}{1.5 - 1}$ **c.** $\dfrac{\Delta h}{\Delta t} = \dfrac{h(2) - h(1.5)}{2 - 1.5}$

$\qquad = \dfrac{48 - 28}{0.5}$ $\qquad\qquad = \dfrac{60 - 48}{0.5}$ $\qquad\qquad = \dfrac{64 - 60}{0.5}$

$\qquad = 40$ $\qquad\qquad\qquad = 24$ $\qquad\qquad\qquad\quad = 8$

For $t \in [0.5, 1]$, the average rate of change is $\frac{40}{1}$, meaning the height of the ball is increasing at an average rate of 40 ft/sec. For $t \in [1, 1.5]$, the average rate of change has slowed to $\frac{24}{1}$, and the soccer ball's height is increasing at only 24 ft/sec. In the interval $[1.5, 2]$, the average rate of change has slowed to 8 ft/sec. The secant lines representing these rates of change are shown in the figure, where we note the line from the first interval (in **red**), has a much steeper slope than the line from the third interval (in **gold**).

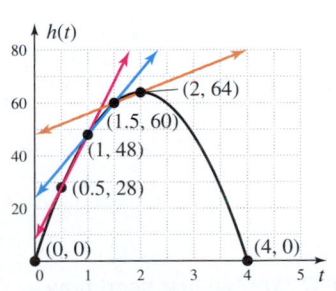

Now try Exercises 27 through 34 ▶

The calculation for average rates of change can be applied to any function $y = f(x)$ and will yield valuable information—particulary in an applied context. For practice with other functions, **see Exercises 35 to 42.**

You may have had the experience of riding in the external elevator of a modern building, with a superb view of the surrounding area as you rise from the bottom floor. For the first few floors, you note you can see much farther than from ground level. As you ride to the higher floors, you can see still farther, but not that much farther due to the curvature of the Earth. This is another example of a nonconstant rate of change.

EXAMPLE 5 ▶   **Average Rates of Change Applied to Viewing Distance**

The distance a person can see depends their elevation above level ground. On a clear day, this viewing distance can be approximated by the function $d(h) = 1.2\sqrt{h}$, where $d(h)$ represents the viewing distance (in miles) at height h (in feet) above level ground. Find the average rate of change to the nearest 100th, for

a. $h \in [9, 16]$

b. $h \in [196, 225]$

c. Graph the function along with the lines representing the average rate of change and comment on what you notice.

Solution ▶   Use the points given in the formula: $\dfrac{\Delta d}{\Delta h} = \dfrac{d(h_2) - d(h_1)}{h_2 - h_1}$

a. $\dfrac{\Delta d}{\Delta h} = \dfrac{d(16) - d(9)}{16 - 9}$ **b.** $\dfrac{\Delta d}{\Delta h} = \dfrac{d(225) - d(196)}{225 - 196}$

$= \dfrac{4.8 - 3.6}{7}$ $= \dfrac{18 - 16.8}{29}$

≈ 0.17 ≈ 0.04

c. For $h \in [9, 16]$, $\dfrac{\Delta d}{\Delta h} \approx \dfrac{0.17}{1}$, meaning the viewing distance is increasing at an average rate of 0.17 miles (about 898 feet) for each 1 ft increase in elevation. For $h \in [196, 225]$, $\dfrac{\Delta d}{\Delta h} \approx \dfrac{0.04}{1}$ and the viewing distance is increasing at a rate of only 0.04 miles (about 211 feet) for each increase of 1 ft. We'll sketch the graph using the points (0, 0), (9, 3.6), (16, 4.8), (196, 16.8) and (225, 18), along with (100, 12) and (169, 15.6) to help round out the graph (Figure 3.63).

Figure 3.63

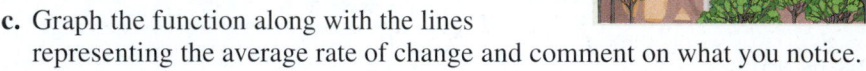

✓ **C. You've just seen how we can calculate average rates of change using the average rate of change formula**

Note the slope of the secant line through the points (9, 3.6) and (16, 4.8), has a much steeper slope than the line through (196, 16.8) and (225, 18).

Now try Exercises 51 through 56 ▶

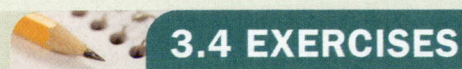

3.4 EXERCISES

▶ CONCEPTS AND VOCABULARY

Fill in each blank with the appropriate word or phrase. Carefully reread the section if needed.

1. Data that indicate a gradual increase to a maximum value, matched by a like decrease, might best be modeled using __quadratic__ regression. In the case described, the coefficient of the x^2-term will be __negative__.

2. For linear functions, the rate of change is __constant__. For __nonlinear__ functions, the rate of change is not constant.

3. The __average__ rate of change of a function can be found by calculating the __slope__ of the line that passes through two points on the graph of the function. For nonlinear functions, this is called a __secant line__.

4. The average rate of change of a function $f(x)$ between x_1 and x_2 is given by the formula $\frac{f(x_2) - f(x_1)}{x_2 - x_1}$. To avoid division by __0__, x_1 cannot equal x_2.

5. Discuss/Explain the differences between the three types of regressions we have studied so far (linear, power, and quadratic). Include examples that highlight the different behavior of each of these types of models.   *Answers will vary.*

6. Given $f(x) = 2x^2 - 12x$, compare the average rate of change near the vertex, with $\frac{\Delta y}{\Delta x}$ on either side of the vertex. Include specific intervals and values in your discussion.   *Answers will vary.*

▶ DEVELOPING YOUR SKILLS

🖩 Use a graphing calculator and the tables shown to find (a) a quadratic regression equation that models the data, (b) the output of the function for $x = 3.5$, and (c) the positive value of x where $f(x) = 39$. Round to nearest thousandths as necessary.

🖩 Use a graphing calculator and the tables shown to find (a) a quadratic regression equation $y = f(x)$ that models the data, (b) the value of $f(x)$ at the vertex, and (c) the two values of x where $f(x) = 25$. Round to nearest thousandths.

7.

x	y
1	−5.69
3	0.73
5	18.75
7	48.37
8	67.53

8.

x	y
1	4.29
2	10.72
4	37.74
5	58.33
7	113.67

11.

x	y
1	37.8
3	12.1
5	15.3
6	28.2
8	71.9

12.

x	y
2	−48.7
3	−66.2
5	−59.3
7	8.5
8	98.6

9.

x	y
1	67.8
2	63.4
3	68.5
4	67.7
5	55.1
6	51.7
7	47.3
8	31.4
9	27.2
10	9.9

10.

x	y
1	−1.0
2	−13.9
3	−14.2
4	−17.5
5	−5.1
6	3.2
7	27.2
8	42.6
9	88.4
10	105.3

13.

x	y
0	67.6
5	29.6
10	9.9
15	3.3
20	−8.2
25	−17.1
30	−7.9
35	4.6
40	6.8
45	25.7
50	54.8

14.

x	y
0	−7.6
10	30.4
20	50.1
30	56.7
40	76.5
50	71.9
60	73.6
70	55.4
80	53.2
90	19.3
100	5.2

Additional answers can be found in the Instructor Answer Appendix.

HOMEWORK SELECTION GUIDE

Core: 7–41 odd, 45–55 odd (24 Exercises)
Standard: 1–4, 7–41 odd, 43, 45–55 odd, 61, 62 (31 Exercises)

Extended: 1–4, 7–41 odd, 43, 45–55 odd, 56, 57, 61, 62 (33 Exercises)
In Depth: 1–6, 7–41 odd, 43–57, 60–63 (43 Exercises)

Using the graphs shown, find the average rate of change of *f* and *g* for the intervals specified.

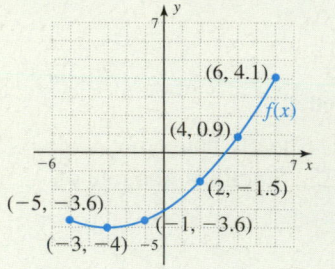

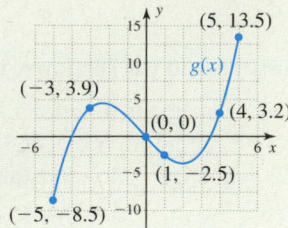

15. $f(x)$ for $x \in [-5, -1]$ 0

16. $f(x)$ for $x \in [-1, 4]$ 0.9

17. $f(x)$ for $x \in [2, 6]$ 1.4

18. $f(x)$ for $x \in [-3, 2]$ 0.5

19. $f(x)$ for $x \in [-5, -3]$ −0.2

20. $f(x)$ for $x \in [-3, -1]$ 0.2

21. $g(x)$ for $x \in [-3, 1]$ −1.6

22. $g(x)$ for $x \in [0, 1]$ −2.5

23. $g(x)$ for $x \in [-3, 4]$ −0.1

24. $g(x)$ for $x \in [0, 4]$ 0.8

25. $g(x)$ for $x \in [-3, 5]$ 1.2

26. $g(x)$ for $x \in [0, 5]$ 2.7

The functions shown are projectile equations where $h(t)$ is the height of the projectile after t sec. Calculate $\dfrac{\Delta h}{\Delta t}$ over the intervals indicated. Include units of measurement in your answer.

27. $h(t) = -16t^2 + 160t$, h in feet

    a. $[2, 5]$ 48 ft/sec   **b.** $[3, 5]$ 32 ft/sec

    c. $[4, 5]$ 16 ft/sec   **d.** $[5, 7]$ −32 ft/sec

28. $h(t) = -16t^2 + 32t$, h in feet

    a. $[0.25, 1]$ 12 ft/sec   **b.** $[0.5, 1]$ 8 ft/sec

    c. $[0.75, 1]$ 4 ft/sec   **d.** $[1, 1.5]$ −8 ft/sec

29. $h(t) = -4.9t^2 + 34.3t + 2.6$, h in meters

    a. $[1, 6]$ 0 m/sec   **b.** $[2, 5]$ 0 m/sec

    c. $[2.5, 3.5]$ 4.9 m/sec   **d.** $[3.5, 4.5]$ −4.9 m/sec

30. $h(t) = -4.9t^2 + 14.7t + 4.1$, h in meters

    a. $[1, 2]$ 0 m/sec   **b.** $[0.2, 2.8]$ 0 m/sec

    c. $[0.5, 1.5]$ 4.9 m/sec   **d.** $[1.5, 2.5]$ −4.9 m/sec

Graph the function and secant lines representing the average rates of change for the exercises given. Comment on what you notice in terms of the projectile's velocity.

31. Exercise 27.

Answers will vary.

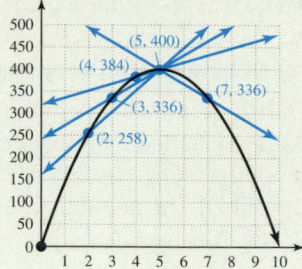

32. Exercise 28.

Answers will vary.

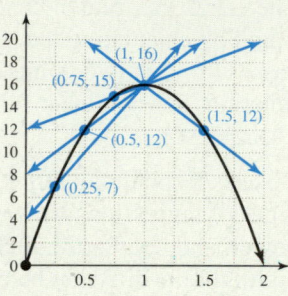

33. Exercise 29.

Answers will vary.

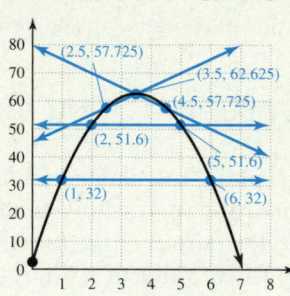

34. Exercise 30.

Answers will vary.

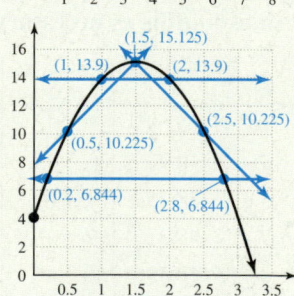

Graph each function in an appropriate window, then find the average rate of change for the interval specified. Round to hundredths as needed.

35. $y = x^3 - 8$; $[2, 5]$ $\frac{\Delta y}{\Delta x} = 39$

36. $y = \sqrt[3]{x + 5}$; $[-5, 3]$ $\frac{\Delta y}{\Delta x} = 0.25$

37. $y = 2|x + 3|$; $[-4, 0]$ $\frac{\Delta y}{\Delta x} = 1$

38. $y = |3x + 1| - 2$; $[-2, 2]$ $\frac{\Delta y}{\Delta x} = 0.5$

39. $F = 9.8m$; $[70, 100]$ $\frac{\Delta F}{\Delta m} = 9.8$

40. $\rho = 0.2m$; $[1.3, 1.5]$ $\frac{\Delta \rho}{\Delta m} = 0.2$

41. $A = \pi r^2$; $[5, 7]$ $\frac{\Delta A}{\Delta r} \approx 37.70$

42. $V = \dfrac{4}{3}\pi r^3$; $[5, 7]$ $\frac{\Delta V}{\Delta r} \approx 456.58$

▶ WORKING WITH FORMULAS

43. Height of a falling object: $h(t) = -16t^2 - v_0 t + h_0$

Neglecting air resistance, the height of an object that is *thrown* straight downward with velocity v_0 from a height of h_0 is given by the formula shown, where $h(t)$ represents the height at time t. The Earth's longest vertical drop (on land) is the Rupal Face on Nanga Parbat (Pakistan), which rises 15,000 ft above its base. From the top of this rock face, a climber's piton hammer slips from her hand and is projected downward with an initial velocity of 6 ft/sec. Determine the hammer's height after (a) $t = 5$ sec and (b) $t = 7$ sec. (c) Use the results to calculate the average rate of change over this 2-sec interval. (d) Repeat parts a, b, and c for $t = 10$ sec and $t = 12$ sec and comment.

44. The Difference Quotient: $D(x) = \dfrac{f(x + h) - f(x)}{h}$

As we'll see in Section 3.6, the difference quotient is closely related to the average rate of change.
 a. Given $f(x) = x^2$ and $h = 0.1$, evaluate $D(3)$ using the formula.
 b. Calculate the average rate of change of $f(x)$ over the interval $[3, 3.1]$ and comment on what you notice.

▶ APPLICATIONS

45. Registration for 5-km race: A local community hosts a popular 5-km race to raise money for breast cancer research. Due to certain legal restrictions, only the first 5000 registrants will be allowed to compete. The table

Day	Registration Total
1	791
2	1688
3	2407
4	3067
5	3692

shows the cumulative number of registered participants at the end of the day, for the first 5 days. (a) Use a graphing calculator to find a quadratic regression equation that models the data. Use this equation to estimate (b) the number of participants after 1 week of registration, (c) the number of days it will take for the race to fill up, and (d) the maximum number of participants that would have signed up had there been no limit. Round to the nearest hundredth when necessary.

46. Concert tickets: In San Francisco, the *Javier Mendoza Band* has scheduled a concert at Candlestick Park. Once the tickets go on sale, the band is sure to sell out this 70,000 person venue.

Week	Ticket Sales Total
1	17,751
2	31,266
3	45,311
4	54,986

The table shows the cumulative number of tickets sold each week, for the first 4 weeks. (a) Use a graphing calculator to find a quadratic regression equation that models the data. Use this equation to (b) estimate the number of tickets sold after 5 weeks, (c) estimate the number of weeks it will take for the concert to sell out, and (d) estimate the number of fans that won't get to attend the show.

47. Guided tours: A tour guide for Kalaniohana Tours noticed that for groups of two to seven people, the average time it took to organize them at the beginning of a tour actually *decreased* as the group size *increased*. For groups of eight or more,

No. of Tourists	Start-up Time (sec)
2	206
4	115
6	63
9	79
11	154
13	269

however, the logistics (and questions asked) actually caused a significant increase in the start time required. Using the given table and a graphing calculator, (a) find a quadratic regression equation that models the data. Use this equation to (b) estimate how long it would take to get a group of five tourists ready, (c) estimate the tour capacity if start-up time can be no longer than 10 min, and (d) estimate the fastest start time that could be expected. Round to the nearest hundredth as necessary.

48. Gardening: The production of a garden can be diminished not only by lack of water, but also by *over*watering. Shay has kept diligent records of her 100-ft^2 tomato garden's weekly production, as well as

Water Total (gal)	No. of Tomatoes
77	11
132	25
198	29
256	20
315	1

the amount of water it received through watering and rain. Use the given table and a graphing calculator to (a) find a quadratic regression equation that models the data. Then use this equation to (b) estimate how many tomatoes she can expect when the garden receives 156 gal of water per week, (c) estimate how much water the garden received if there were 15 tomatoes produced per week, and (d) estimate the maximum number of tomatoes she can expect from the garden in a week. Round to the nearest ten-thousandth as necessary.

49. Weight of a fetus: The growth rate of a fetus in the mother's womb (by weight in grams) is modeled by the graph shown here, beginning with the 25th week of

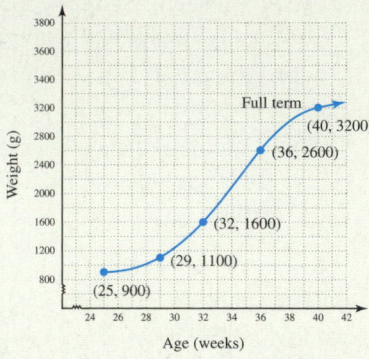

gestation. (a) Calculate the average rate of change (slope of the secant line) between the 25th week and the 29th week. Is the slope of the secant line positive or negative? Discuss what the slope means in this context. (b) Is the fetus gaining weight faster between the 25th and 29th week, or between the 32nd and 36th week? Compare the slopes of both secant lines and discuss.

50. Fertility rates: Over the years, fertility rates for women in the United States (average number of children per woman) have varied a great deal, though in the twenty-first century they've

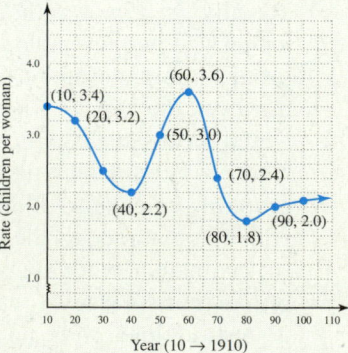

begun to level out. The graph shown models this fertility rate for most of the twentieth century. (a) Calculate the average rate of change from the years 1920 to 1940. Is the slope of the secant line positive or negative? Discuss what the slope means in this context. (b) Calculate the average rate of change from the year 1940 to 1950. Is the slope of the secant line positive or negative? Discuss what the slope means in this context. (c) Was the fertility rate increasing faster from 1940 to 1950, or from 1980 to 1990? Compare the slope of both secant lines and comment.

Source: Statistical History of the United States from Colonial Times to Present

Additional answers can be found in the Instructor Answer Appendix.

For Exercises 51 to 56, use the formula for the average rate of change $\dfrac{f(x_2) - f(x_1)}{x_2 - x_1}$.

51. Average rate of change: For $f(x) = x^3$, (a) calculate the average rate of change for the interval $x = -2$ to $x = -1$ and (b) calculate the average rate of change for the interval $x = 1$ to $x = 2$. (c) What do you notice about the answers from parts (a) and (b)? (d) Sketch the graph of this function along with the lines representing these average rates of change and comment on what you notice.

52. Average rate of change: Knowing the general shape of the graph for $f(x) = \sqrt[3]{x}$, (a) is the average rate of change greater between $x = 0$ and $x = 1$ or between $x = 7$ and $x = 8$? Why? (b) Calculate the rate of change for these intervals and verify your response. (c) Approximately how many times greater is the rate of change?
a. Between $x = 0$ and $x = 1$ **b.** 1, 0.09   **c.** 11

53. Height of an arrow: If an arrow is shot vertically from a bow with an initial speed of 192 ft/sec, the height of the arrow can be modeled by the function $h(t) = -16t^2 + 192t$,

where $h(t)$ represents the height of the arrow after t sec (assume the arrow was shot from ground level).

a. What is the arrow's height at $t = 1$ sec?   176 ft

b. What is the arrow's height at $t = 2$ sec?   320 ft

c. What is the average rate of change from $t = 1$ to $t = 2$?   144 ft/sec

d. What is the rate of change from $t = 10$ to $t = 11$? Why is it the same as (c) except for the sign?   −144 ft/sec; The arrow is going down.

54. Height of a water rocket: Although they have been around for decades, water rockets continue to be a popular toy. A plastic rocket is filled with water and then pressurized using a handheld pump. The rocket is then released and off it goes! If the rocket has an initial velocity of 96 ft/sec, the height of the rocket can be modeled by the function $h(t) = -16t^2 + 96t$, where $h(t)$ represents the height of the rocket after t sec (assume the rocket was shot from ground level).

a. Find the rocket's height at $t = 1$ and $t = 2$ sec.   80 ft, 128 ft

b. Find the rocket's height at $t = 3$ sec.   144 ft

c. Would you expect the average rate of change to be greater between $t = 1$ and $t = 2$, or between $t = 2$ and $t = 3$? Why? Between $t = 1$ and $t = 2$; the rocket is decelerating.

d. Calculate each rate of change and discuss your answer.   48 ft/sec, 16 ft/sec

55. Velocity of a falling object: The impact velocity of an object dropped from a height is modeled by $v = \sqrt{2gs}$, where v is the velocity in feet per second (ignoring air resistance), g is the acceleration due to gravity (32 ft/sec^2 near the Earth's surface), and s is the height from which the object is dropped.

a. Find the velocity at $s = 5$ ft and $s = 10$ ft.
   17.89 ft/sec; 25.30 ft/sec
b. Find the velocity at $s = 15$ ft and $s = 20$ ft.
   30.98 ft/sec; 35.78 ft/sec
c. Would you expect the average rate of change to be greater between $s = 5$ and $s = 10$, or between $s = 15$ and $s = 20$?   Between 5 and 10.
d. Calculate each rate of change and discuss your answer.   1.482 ft/sec, 0.96 ft/sec

56. Temperature drop: One day in November, the town of Coldwater was hit by a sudden winter storm that caused temperatures to plummet. During the storm, the temperature T (in degrees Fahrenheit) could be modeled by the function $T(h) = 0.8h^2 - 16h + 60$, where h is the number of hours since the storm began. Graph the function and use this information to answer the following questions.

a. What was the temperature as the storm began?
   60°
b. How many hours until the temperature dropped below zero degrees?   5 hr
c. How many hours did the temperature remain below zero?   10 hr
d. What was the coldest temperature recorded during this storm?   −20°F

▶ **EXTENDING THE CONCEPT**

57. The function $A(t) = 200t$ gives the amount of air (in cubic inches) that a compressor has pumped out after t sec. The volume of a spherical balloon being inflated by this compressor is given by $V(r) = \frac{4}{3}\pi r^3$, with the radius of the balloon modeled by $r(t) = -0.02t^2 + 0.76t + 2.26$, where $r(t)$ is the radius after t sec.

a. If the balloon pops when the radius is at its maximum, what is the maximum volume of the balloon?   about 3569 in^3
b. What amount of air (the volume) was needed to pop the balloon?   about 3800 in^3.
c. Calculate the average rates of change $\frac{\Delta r}{\Delta t}$ and $\frac{\Delta V}{\Delta r}$ during the first second of inflation and the last second of inflation. Compare the results.

58. In Exercise 44, you were provided with a formula called the difference quotient. This formula can be derived by finding the average rate of change of a function $f(x)$ over an interval $[x, x + h]$.

a. Use the definition of average rate of change to derive the formula for the difference quotient.
b. Find and simplify the difference quotient of $f(x) = x^2 + 3x$. $2x + 3 + h$

59. The floor function $f(x) = \lfloor x \rfloor$ and the ceiling function $g(x) = \lceil x \rceil$ studied in Section 2.5 can produce some interesting average rates of change. The average rate of change of these two functions over any interval 1 unit or longer must lie within what range of values? $(\frac{1}{2}, \infty)$

▶ **MAINTAINING YOUR SKILLS**

60. (1.1) Complete the squares in x and y to find the center and radius of the circle defined by $x^2 + y^2 + 6x - 8y = 0$. Then graph the circle on a graphing calculator.

61. (1.5) Solve the following inequalities graphically using the Intersection-of-Graphs method. Round to nearest hundredths when necessary.

a. $\dfrac{x-5}{x-2} < 2$ $(-\infty, -1) \cup (2, \infty)$

b. $x - \dfrac{2}{5}(3-x) > 7(0.2x + 0.5)$ { }

62. (3.3) Find an equation of the quadratic function with vertex $(-5, -3)$ and y-intercept $(0, 7)$.
   $y = 0.4(x + 5)^2 - 3$

63. (2.2/2.5) Given $f(x) = \begin{cases} |x + 3| & x < -2 \\ -\dfrac{1}{2}x & x \geq -2 \end{cases}$, find the equation of $g(x)$, given its graph is the same as $f(x)$ but translated right 3 units and reflected across the x-axis.
   $g(x) = \begin{cases} -|x| & x < 1 \\ \frac{1}{2}(x - 3) & x \geq 1 \end{cases}$

Additional answers can be found in the Instructor Answer Appendix.

3.5 The Algebra of Functions

In Section 2.2, we created new functions *graphically* by applying transformations to basic functions. In this section, we'll use two (or more) functions to create new functions *algebraically*. Previous courses often contain material on the sum, difference, product, and quotient of polynomials. Here we'll combine functions with the basic operations, noting the result is also a function that can be evaluated, graphed, and analyzed. We call these basic operations on functions the **algebra of functions.**

A. Sums and Differences of Functions

This section introduces the notation used for basic operations on functions. Here we'll note the result is also a function whose domain depends on the original functions. In general, if f and g are functions *with overlapping domains*, $f(x) + g(x) = (f + g)(x)$ and $f(x) - g(x) = (f - g)(x)$.

Sums and Differences of Functions

For functions f and g with domains P and Q respectively, the sum and difference of f and g are defined by:

	Domain of result
$(f + g)(x) = f(x) + g(x)$	$P \cap Q$
$(f - g)(x) = f(x) - g(x)$	$P \cap Q$

EXAMPLE 1A ▶ **Evaluating a Difference of Functions**

Given $f(x) = x^2 - 5x$ and $g(x) = 2x - 9$,

a. Determine the domain of $h(x) = (f - g)(x)$.　**b.** Find $h(3)$ using the definition.

Solution ▶ **a.** Since the domain of both f and g is $\mathbb{R}$, their intersection is $\mathbb{R}$, so the domain of h is also $\mathbb{R}$.

b. $h(x) = (f - g)(x)$　　　　　given difference

$\qquad = f(x) - g(x)$　　　　　by definition

$h(3) = f(3) - g(3)$　　　　　substitute 3 for *x*

$\qquad = [(3)^2 - 5(3)] - [2(3) - 9]$　　evaluate

$\qquad = [9 - 15] - [6 - 9]$　　multiply

$\qquad = -6 - [-3]$　　subtract

$\qquad = -3$　　result

If the function h is to be graphed or evaluated numerous times, it helps to compute a *new function rule* for h, rather than repeatedly apply the definition.

EXAMPLE 1B ▶ For the functions f, g, and h, as defined in Example 1A,

a. Find a new function rule for h.　**b.** Use the result to find $h(3)$.

Solution ▶ **a.** $h(x) = (f - g)(x)$　　　given difference

$\qquad = f(x) - g(x)$　　　by definition

$\qquad = (x^2 - 5x) - (2x - 9)$　　replace *f(x)* with (x^2 − 5*x*) and *g(x)* with (2*x* − 9)

$\qquad = x^2 - 7x + 9$　　distribute and combine like terms

b. $h(3) = (3)^2 - 7(3) + 9$ substitute 3 for x

$\qquad = 9 - 21 + 9$ multiply

$\qquad = -3$ result

Notice the result from Part (b) is identical to that in Example 1A.

Now try Exercises 7 through 10 ▶

⚠ **CAUTION** ▶ From Example 1A, note the importance of using grouping symbols with the algebra of functions. Without them, we could easily confuse the signs of g when computing the difference. Also, note that any operation applied to the functions f and g simply results in an *expression* representing a new function rule for h, and is not an *equation* that needs to be factored or solved.

EXAMPLE 2 ▶ **Evaluating a Sum of Functions**

For $f(x) = x^2$ and $g(x) = \sqrt{x - 2}$,

    a. Determine the domain of $h(x) = (f + g)(x)$.

    b. Find a new function rule for h.

    c. Evaluate $h(3)$.

    d. Evaluate $h(-1)$.

Solution ▶   **a.** The domain of f is $\mathbb{R}$, while the domain of g is $x \in [2, \infty)$. Since their intersection is $[2, \infty)$, this is the domain of the new function h.

    b. $h(x) = (f + g)(x)$ given sum

$\qquad\quad = f(x) + g(x)$ by definition

$\qquad\quad = x^2 + \sqrt{x - 2}$ substitute x^2 for $f(x)$ and $\sqrt{x - 2}$ for $g(x)$ (no other simplifications possible)

    c. $h(3) = (3)^2 + \sqrt{3 - 2}$ substitute 3 for x

$\qquad\quad = 10$ result

    d. $x = -1$ is outside the domain of h.

WORTHY OF NOTE

If we *did* try to evaluate $h(-1)$, the result would be $1 + \sqrt{-3}$, which is not a real number. While it's true we could write $1 + \sqrt{-3}$ as $1 + i\sqrt{3}$ and consider it an "answer," our study here focuses on real numbers and the graphs of functions in a coordinate system where x and y are both real.

Now try Exercises 11 through 14 ▶

This "intersection of domains" is illustrated in Figure 3.64.

Figure 3.64

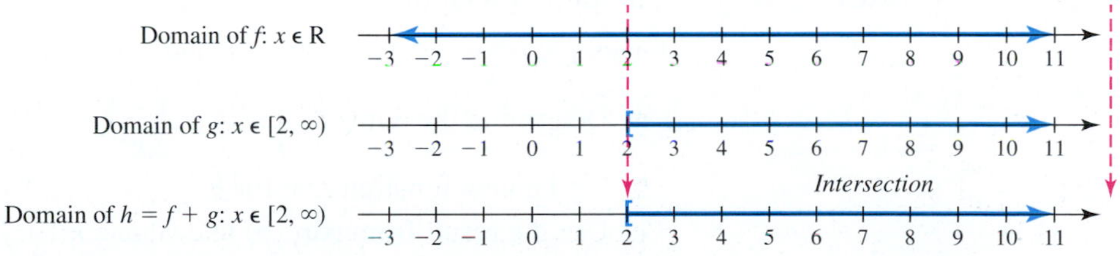

☑ **A.** You've just seen how we can compute a sum or difference of functions and determine the domain of the result

B. Products and Quotients of Functions

The product and quotient of two functions is defined in a manner similar to that for sums and differences. For example, if f and g are functions *with overlapping domains*,

$(f \cdot g)(x) = f(x) \cdot g(x)$ and $\left(\dfrac{f}{g}\right)(x) = \dfrac{f(x)}{g(x)}$. As you might expect, for quotients we must stipulate $g(x) \neq 0$.

Products and Quotients of Functions

For functions f and g with domains P and Q respectively, the product and quotient of f and g are defined by:

$$\text{Domain of result}$$

$$(f \cdot g)(x) = f(x) \cdot g(x) \qquad\qquad P \cap Q$$

$$\left(\frac{f}{g}\right)(x) = \frac{f(x)}{g(x)} \qquad\qquad P \cap Q, \text{ for all } g(x) \neq 0$$

EXAMPLE 3 ▶ **Computing a Product of Functions**

Given $f(x) = \sqrt{1 + x}$ and $g(x) = \sqrt{3 - x}$,

a. Determine the domain of $h(x) = (f \cdot g)(x)$.

b. Find a new function rule for h.

c. Use the result from part (b) to evaluate $h(2)$ and $h(4)$.

Solution ▶ **a.** The domain of f is $x \in [-1, \infty)$ and the domain of g is $x \in (-\infty, 3]$. The intersection of these domains gives $x \in [-1, 3]$, which is the domain for h.

b. $\begin{aligned} h(x) &= (f \cdot g)(x) && \text{given product} \\ &= f(x) \cdot g(x) && \text{by definition} \\ &= \sqrt{1 + x} \cdot \sqrt{3 - x} && \text{substitute } \sqrt{1+x} \text{ for } f \text{ and } \sqrt{3-x} \text{ for } g \\ &= \sqrt{3 + 2x - x^2} && \text{combine using properties of radicals} \end{aligned}$

c. $\begin{aligned} h(2) &= \sqrt{3 + 2(2) - (2)^2} && \text{substitute 2 for } x \\ &= \sqrt{3} \approx 1.732 && \text{result} \\ h(4) &= \sqrt{3 + 2(4) - (4)^2} && \text{substitute 4 for } x \\ &= \sqrt{-5} && \textit{not a real number} \end{aligned}$

The second result of Part (c) is not surprising, since $x = 4$ is not in the domain of h [meaning $h(4)$ is not defined for this function].

Now try Exercises 15 through 18 ▶

In future sections, we use polynomial division as a tool for factoring, as an aid to graphing, and to determine whether two expressions are equivalent. Understanding the notation and domain issues related to division will strengthen our ability in these areas.

EXAMPLE 4 ▶ **Computing a Quotient of Functions**

Given $f(x) = x^3 - 3x^2 + 2x - 6$ and $g(x) = x - 3$,

a. Determine the domain of $h(x) = \left(\dfrac{f}{g}\right)(x)$.

b. Find a new function rule for h.

c. Use the result from part (b) to evaluate $h(0)$.

Solution ▶ **a.** While the domain of both f and g is $\mathbb{R}$ and their intersection is also $\mathbb{R}$, we know from the definition (and past experience) *that $g(x)$ cannot be zero*. The domain of h is $x \in (-\infty, 3) \cup (3, \infty)$.

b. $\begin{aligned} h(x) &= \left(\frac{f}{g}\right)(x) && \text{given quotient} \\[2mm] &= \frac{f(x)}{g(x)} && \text{by definition} \\[2mm] &= \frac{x^3 - 3x^2 + 2x - 6}{x - 3} && \text{replace } f \text{ with } x^3 - 3x^2 + 2x - 6 \text{ and } g \text{ with } x - 3 \end{aligned}$

From our work with rational expressions in Section R.5, the expression that defines h can be simplified: $\dfrac{x^3 - 3x^2 + 2x - 6}{x - 3} = \dfrac{x^2(x - 3) + 2(x - 3)}{x - 3} =$

$\dfrac{(x^2 + 2)(x - 3)}{x - 3} = x^2 + 2$. But from the original expression, h is not defined if $g(x) = 3$, *even if the result for h is a polynomial.* In this case, we write the simplified form as $h(x) = x^2 + 2, x \neq 3$.

c. For $h(0)$ we have:

$h(0) = (0)^2 + 2$ replace *x* with 0

$h(0) = 2$

☑ **B.** You've just seen how we can compute a product or quotient of functions and determine the domain

Now try Exercises 19 through 34 ▶

For additional practice with the algebra of functions, **see Exercises 35 through 46.**

C. Graphical and Numerical Views of Operations on Functions

The algebra of functions also has an instructive *graphical interpretation,* in which values for $f(k)$ and $g(k)$ are read from a graph (k is a given constant), with operations like $(f + g)(k) = f(k) + g(k)$ then computed and lodged (**see Exercise 84**).

EXAMPLE 5 ▶   **Interpreting Operations on Functions Graphically**

Use the graph given to find the value of each expression:

a. $(f + g)(-2)$

b. $(g - f)(6)$

c. $\left(\dfrac{g}{f}\right)(8)$

d. $(f \cdot g)(4)$

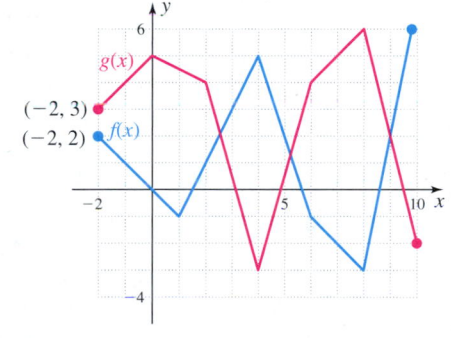

Solution ▶   Since the needed input values for this example are $x = -2$, 4, 6, and 8, we begin by reading the value of $f(x)$ and $g(x)$ at each point. From the graph, we note that $f(-2) = 2$ and $g(-2) = 3$. The other values are likewise found and appear in the table. For $(f + g)(-2)$ we have:

x	f(x)	g(x)
−2	2	3
4	5	−3
6	−1	4
8	−3	6

a. $(f + g)(-2) = f(-2) + g(-2)$ definition

$= 2 + 3$ substitute 2 for *f*(−2) and 3 for *g*(−2)

$= 5$ result

With some practice, the computations can be done mentally and we have

b. $(g - f)(6) = g(6) - f(6)$

$= 4 - (-1) = 5$

c. $\left(\dfrac{g}{f}\right)(8) = \dfrac{g(8)}{f(8)}$

$= \dfrac{6}{-3} = -2$

d. $(f \cdot g)(4) = f(4) \cdot g(4)$

$= 5(-3) = -15$

In addition to graphically viewing operations on functions *at specific points* as in Example 5, it is instructive to better understand what happens when functions are combined *for all possible x-values*. This will shed light on certain transformations of functions, as well as on the domain of the new function. For instance, consider the functions $f(x) = -x^2$ and $g(x) = 4$. The graph of *f* is a parabola opening downward with vertex (0, 0), while *g* is a horizontal line through $y = 4$. For $h(x) = f(x) + g(x)$, the notation literally says the new function is the result of adding the outputs of *f* and *g* for all values of *x* in the intersection of their domains. Using $Y_1 = -x^2$, $Y_2 = 4$, and $Y_3 = Y_1 + Y_2$, the result shown numerically in Table 3.2 is found by simply adding $Y_2 = 4$ to the output values in Y_1. The identical combination is shown graphically in Figure 3.65, where the outputs from Y_1 are all increased by 4. Note the result *h* is simply the graph *f* shifted up 4 units (a vertical shift).

Figure 3.65

Table 3.2

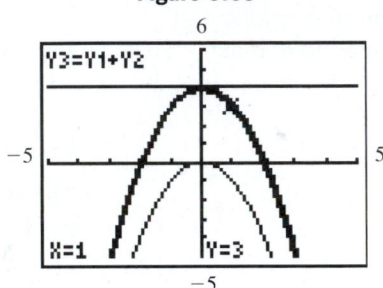

The graphical rendering also helps verify that since the domain of *f* and *g* is all real numbers, their intersection (and the domain of *h*) is all real numbers.

If the same analysis is performed on functions where one or both domains are restricted, the reason for the "intersection of domains" becomes clear. Using $f(x) = Y_1 = 2\sqrt{x + 2}$ and $g(x) = Y_2 = x - 1$, we note the domain of *f* is $x \in [-2, \infty)$ while the domain of *g* is $x \in \mathbb{R}$. For $h(x) = f(x) + g(x)$, we have $Y_3 = Y_1 + Y_2$ with the result shown numerically Table 3.3 and graphically in Figures 3.66 and 3.67 (Y_3 is the **bold** graph). The **TABLE** feature of a graphing calculator can actually be used for all three functions by activating the functions Y_1, Y_2, and Y_3, then using the right arrow to bring the column for Y_3 into view.

Table 3.3   **Figure 3.66**   **Figure 3.67**

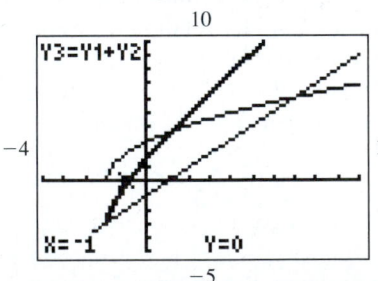 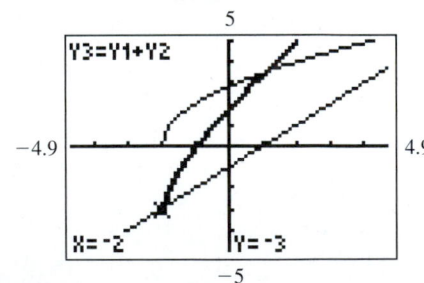

The fact that the domain of *h* is $[-2, \infty)$ is clearly seen from the table. Since Y_1 is not a real number for values less than -2, the combination $Y_1 + Y_2$ cannot produce real values unless $x \geq -2$. This is shown graphically in Figure 3.67, where we note the initial point of Y_3 (in **bold**) is the point $(-2, -3)$. Note that $[-2, \infty)$ is also the intersection of the domains for *f* and *g*.

EXAMPLE 6 ▶ In Example 3, we found that the product of the functions $f(x) = \sqrt{1+x}$ and $g(x) = \sqrt{3-x}$ was $h(x) = \sqrt{(1+x)(3-x)} = \sqrt{3+2x-x^2}$. The domain of f was stated as $[-1, \infty]$, the domain of g as $[-\infty, 3]$, and the domain of h (the intersection of domains) as $x \in [-1, 3]$. Verify these domains using a table and graph, as in Table 3.3 and Figure 3.67.

Solution ▶ Enter $\sqrt{1+x}$ as Y_1, $\sqrt{3-x}$ as Y_2, and $Y_1 Y_2$ as Y_3. The results shown in Table 3.4 (a manual table was used to show all three functions simultaneously) and the graphs in Figure 3.68 illustrate the domains of f and g stated, and show why the domain of h must be their intersection. Figure 3.69 shows the graph of Y_3 (in **bold**) on the same screen and indicates the domain of h is $[-1, 3]$.

> **WORTHY OF NOTE**
>
> In Section 1.1 we observed that the graph of $y = \sqrt{a^2 - x^2}$ was a semicircle with radius $r = a$ and center $(0, 0)$. It's interesting to note that by completing the square on the radicand of function h in Example 6, we can write it as $y = \sqrt{4 - (x-1)^2}$, which is a semicircle of radius $r = 2$, with the center shifted 1 unit right to $(1, 0)$ as shown.

Table 3.4

x	Y_1	Y_2	Y_3
-3	ERR:	2.450	ERR:
-2	ERR:	2.236	ERR:
-1	0	2	0
0	1	1.732	1.732
1	1.414	1.414	2
2	1.732	1	1.732
3	2	0	0
4	2.236	ERR:	ERR:
5	2.450	ERR:	ERR:

Figure 3.68

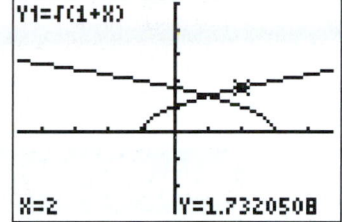

Figure 3.69

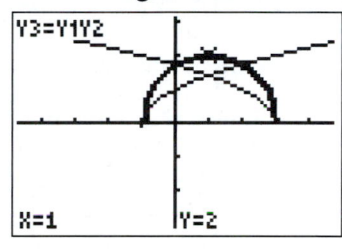

☑ **C.** You've just seen how we can interpret operations on functions graphically and numerically

Now try Exercises 57 through 68 ▶

D. Applications of the Algebra of Functions

The algebra of functions plays an important role in the business world. For example, the cost to manufacture an item, the revenue a company brings in, and the profit a company earns are all functions of the number of items made and sold. Further, we know a company "breaks even" (making $0 profit) when the difference between their revenue R and their cost C, is zero.

EXAMPLE 7 ▶ **Applying Operations on Functions in Context**
The fixed costs to publish *Relativity Made Simple* (by N.O. Way) is $2500, and the variable cost is $4.50 per book. Marketing studies indicate the best selling price for the book is $9.50 per copy.

  a. Find the cost, revenue, and profit functions for this book.

  b. Determine how many copies must be sold for the company to break even.

Solution ▶ **a.** Let x represent the number of books published and sold. The cost of publishing is $4.50 per copy, plus fixed costs (labor, storage, etc.) of $2500. The cost function is $C(x) = 4.50x + 2500$. If the company charges $9.50 per book, the revenue function will be $R(x) = 9.50x$. Since profit equals revenue minus costs,

$$
\begin{aligned}
P(x) &= R(x) - C(x) \\
&= 9.50x - (4.50x + 2500) \quad \text{\color{magenta}substitute } 9.50x \text{ for } R \text{ and } 4.50x + 2500 \text{ for } C \\
&= 9.50x - 4.50x - 2500 \quad \text{\color{magenta}distribute} \\
&= 5x - 2500 \quad \text{\color{magenta}result}
\end{aligned}
$$

The profit function is $P(x) = 5x - 2500$.

b. When a company "breaks even," the profit is zero: $P(x) = 0$.

$$P(x) = 5x - 2500 \quad \text{profit function}$$
$$0 = 5x - 2500 \quad \text{substitute 0 for } P(x)$$
$$2500 = 5x \quad \text{add 2500}$$
$$500 = x \quad \text{divide by 5}$$

 **D.** You've just seen how we can apply the algebra of functions in context

In order for the company to break even, 500 copies must be sold.

Now try Exercises 71 through 74 ▶

There are a number of additional applications in the exercise set. See **Exercises 75 through 80.**

3.5 EXERCISES

▶ CONCEPTS AND VOCABULARY

Fill in each blank with the appropriate word or phrase. Carefully reread the section, if necessary.

1. Given function f with domain A and function g with domain B, the sum $f(x) + g(x)$ can also be written $\underline{(f + g)(x)}$. The domain of the result is $\underline{A \cap B}$.

2. For the product $h(x) = f(x) \cdot g(x)$, $h(5)$ can be found by evaluating f and g then multiplying the result, or multiplying $f \cdot g$ and evaluating the result. Notationally these are written $\underline{f(5) \cdot g(5)}$ and $\underline{(f \cdot g)(5)}$.

3. When combining functions f and g using basic operations, the domain of the result is the $\underline{\text{intersection}}$ of the domains of f and g. For division, we further stipulate that $\underline{g(x)}$ cannot equal zero.

4. If $h(2) = 5$, $g(2) = 7$, and $f(2) = 9$, then $(f + g - h)(2) = \underline{11}$ and $(f - g - h)(2) = \underline{-3}$.

5. For $f(x) = 2x^3 - 50x$ and $g(x) = x - 5$, discuss/explain why the domain of $h(x) = \left(\dfrac{f}{g}\right)(x)$ must exclude $x = 5$, even though the resulting quotient is the polynomial $2x^2 + 10x$.
Answers will vary.

6. Discuss/Explain the domain of $h(x) = (f \cdot g)(x)$, given $f(x) = \sqrt{x - 3}$ and $g(x) = \sqrt{2 - x}$.
The domain of h is the empty set, since the domains of f and g do not intersect.

▶ DEVELOPING YOUR SKILLS

7. Given $f(x) = 2x^2 - x - 3$ and $g(x) = x^2 + 5x$,
(a) determine the domain for $h(x) = f(x) - g(x)$ and
(b) find $h(-2)$ using the definition.

8. Given $f(x) = 2x^2 - 18$ and $g(x) = -3x - 7$,
(a) determine the domain for $h(x) = f(x) + g(x)$
and (b) find $h(5)$ using the definition.

9. For the functions f, g, and h, as defined in Exercise 7,
(a) find a new function rule for h, and (b) use the result to find $h(-2)$. (c) How does the result compare to that of Exercise 7?

10. For the functions f, g, and h as defined in Exercise 8,
(a) find a new function rule for h, and (b) use the result to find $h(5)$. (c) How does the result compare to that in Exercise 8?

11. For $f(x) = \sqrt{x - 3}$ and $g(x) = 2x^3 - 54$,
(a) determine the domain of $h(x) = (f + g)(x)$,
(b) find a new function rule for h, and
(c) evaluate $h(4)$ and $h(2)$, if possible.

12. For $f(x) = 4x^2 - 2x + 3$ and $g(x) = \sqrt{2x - 5}$,
(a) determine the domain of $h(x) = (f - g)(x)$,
(b) find a new function rule for h, and
(c) evaluate $h(7)$ and $h(2)$, if possible.

Additional answers can be found in the Instructor Answer Appendix.

HOMEWORK SELECTION GUIDE

Core: 7–31 every other odd, 33, 35-43 every other odd, 53–67 odd, 71–79 odd
(24 Exercises)

Standard: 1–4, 7–31 every other odd, 33, 35–43 every other odd,
53–67 odd, 71–79 odd, 82, 85, 87 (31 Exercises)

Extended: 1–4, 7–31 every other odd, 33, 35–43 every other odd, 53–67 odd,
69, 70, 71–79 odd, 82, 84, 85, 87 (34 Exercises)

In Depth: 1–6, 7–31 every other odd, 33, 35–43 every other odd, 53–67 odd, 69,
70, 71–80, 82, 84, 85–87 (42 Exercises)

13. For $p(x) = \sqrt{x + 5}$ and $q(x) = \sqrt{3 - x}$,
    (a) determine the domain of $r(x) = (p + q)(x)$,
    (b) find a new function rule for r, and
    (c) evaluate $r(2)$ and $r(4)$, if possible.

14. For $p(x) = \sqrt{6 - x}$ and $q(x) = \sqrt{x + 2}$,
    (a) determine the domain of $r(x) = (p - q)(x)$,
    (b) find a new function rule for r, and
    (c) evaluate $r(-3)$ and $r(2)$, if possible.

15. For $f(x) = \sqrt{x + 4}$ and $g(x) = 2x + 3$,
    (a) determine the domain of $h(x) = (f \cdot g)(x)$,
    (b) find a new function rule for h, and
    (c) evaluate $h(-4)$ and $h(21)$, if possible.

16. For $f(x) = -3x + 5$ and $g(x) = \sqrt{x - 7}$,
    (a) determine the domain of $h(x) = (f \cdot g)(x)$,
    (b) find a new function rule for h, and
    (c) evaluate $h(8)$ and $h(11)$, if possible.

17. For $p(x) = \sqrt{x + 1}$ and $q(x) = \sqrt{7 - x}$,
    (a) determine the domain of $r(x) = (p \cdot q)(x)$,
    (b) find a new function rule for r, and
    (c) evaluate $r(15)$ and $r(3)$, if possible.

18. For $p(x) = \sqrt{4 - x}$ and $q(x) = \sqrt{x + 4}$,
    (a) determine the domain of $r(x) = (p \cdot q)(x)$,
    (b) find a new function rule for r, and
    (c) evaluate $r(-5)$ and $r(-3)$, if possible.

For the functions f and g given, (a) determine the domain of $h(x) = \left(\dfrac{f}{g}\right)(x)$ and (b) find a new function rule for h in simplified form (if possible), noting the domain restrictions alongside.

19. $f(x) = x^2 - 16$ and $g(x) = x + 4$

20. $f(x) = x^2 - 49$ and $g(x) = x - 7$

21. $f(x) = x^3 + 4x^2 - 2x - 8$ and $g(x) = x + 4$

22. $f(x) = x^3 - 5x^2 + 2x - 10$ and $g(x) = x - 5$

23. $f(x) = x^3 - 7x^2 + 6x$ and $g(x) = x - 1$

24. $f(x) = x^3 - 1$ and $g(x) = x - 1$

25. $f(x) = x + 1$ and $g(x) = x - 5$

26. $f(x) = x + 3$ and $g(x) = x - 7$

For the functions p and q given, (a) determine the domain of $r(x) = \left(\dfrac{p}{q}\right)(x)$, (b) find a new function rule for r, and (c) use it to evaluate $r(6)$ and $r(-6)$, if possible.

27. $p(x) = 2x - 3$ and $q(x) = \sqrt{-2 - x}$

28. $p(x) = 1 - x$ and $q(x) = \sqrt{3 - x}$

29. $p(x) = x - 5$ and $q(x) = \sqrt{x - 5}$

Additional answers can be found in the Instructor Answer Appendix.

30. $p(x) = x + 2$ and $q(x) = \sqrt{x + 3}$

31. $p(x) = x^2 - 36$ and $q(x) = \sqrt{2x + 13}$

32. $p(x) = x^2 - 6x$ and $q(x) = \sqrt{7 + 3x}$

For the functions f and g given, (a) find a new function rule for $h(x) = \left(\dfrac{f}{g}\right)(x)$ in simplified form. (b) If $h(x)$ were the original function, what would be its domain? (c) Since we know $h(x) = \left(\dfrac{f}{g}\right)(x) = \dfrac{f(x)}{g(x)}$, what additional restrictions exist for the domain of h?

33. $f(x) = \dfrac{6x}{x - 3}$ and $g(x) = \dfrac{3x}{x + 2}$

34. $f(x) = \dfrac{4x}{x + 1}$ and $g(x) = \dfrac{2x}{x - 2}$

For each pair of functions f and g given, determine the sum, difference, product, and quotient of f and g, and determine the domain in each case.

35. $f(x) = 2x + 3$ and $g(x) = x - 2$

36. $f(x) = x - 5$ and $g(x) = 2x - 3$

37. $f(x) = x^2 + 7$ and $g(x) = 3x - 2$

38. $f(x) = x^2 - 3x$ and $g(x) = x + 4$

39. $f(x) = x^2 + 2x - 3$ and $g(x) = x - 1$

40. $f(x) = x^2 - 2x - 15$ and $g(x) = x + 3$

41. $f(x) = 3x + 1$ and $g(x) = \sqrt{x - 3}$

42. $f(x) = x + 2$ and $g(x) = \sqrt{x + 6}$

43. $f(x) = 2x^2$ and $g(x) = \sqrt{x + 1}$

44. $f(x) = x^2 + 2$ and $g(x) = \sqrt{x - 5}$

45. $f(x) = \dfrac{2}{x - 3}$ and $g(x) = \dfrac{5}{x + 2}$

46. $f(x) = \dfrac{4}{x - 3}$ and $g(x) = \dfrac{1}{x + 5}$

47. **Reading a graph—Used vehicle sales:** The graph given shows the number of cars $C(t)$ and trucks $T(t)$ sold by Ullery Used Autos for the years 2000 to 2010. Use the graph to estimate the number of

    a. cars sold in 2005: $C(5)$

    b. trucks sold in 2008: $T(8)$

    c. vehicles sold in 2009: $C(9) + T(9)$

    d. In function notation, how would you determine how many more cars than trucks were sold in 2009? What was the actual number?

    a. 6000   **b.** 3000   **c.** 8000   **d.** $C(9) - T(9)$; 4000

Exercise 47

Ullery Used Auto Sales

48. Reading a graph—Government investment: The graph given shows a government's investment in its military $M(t)$ over time, versus its investment in public works $P(t)$, in millions of dollars. Use the graph to estimate the amount of investment in

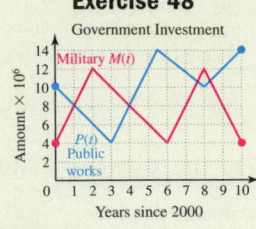

Exercise 48
Government Investment

a. the military in 2002: $M(2)$

b. public works in 2005: $P(5)$

c. public works and the military in 2009: $M(9) + P(9)$

d. In function notation, how would you determine how much more will be invested in public works than the military in 2010? What is the actual number?   **a.** $12 million   **b.** $12 million
c. $20 million   **d.** $P(10) - M(10)$; $10 million

49. Reading a graph—Space travel: The graph given shows the revenue $R(t)$ and operating costs $C(t)$ of Space Travel Resources (STR), for the years 2000 to 2010. Use the graph to find the

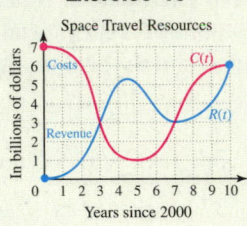

Exercise 49
Space Travel Resources

a. revenue in 2002: $R(2)$

b. costs in 2008: $C(8)$

c. years STR broke even: $R(t) = C(t)$

d. years costs exceeded revenue: $C(t) > R(t)$

e. years STR made a profit: $R(t) > C(t)$

f. For the year 2005, use function notation to write the profit equation for STR. What was their profit?

50. Reading a graph—Corporate expenditures: The graph given shows a large corporation's investment in research and development $R(t)$ over time, and the amount paid to investors as dividends $D(t)$, in billions of dollars. Use the graph to find the

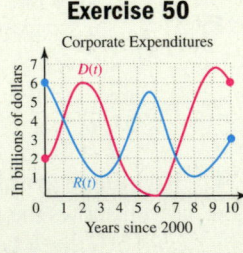

Exercise 50
Corporate Expenditures

a. dividend payments in 2002: $D(2)$

b. investment in 2006: $R(6)$

c. years where $R(t) = D(t)$

d. years where $R(t) > D(t)$

e. years where $R(t) < D(t)$

f. Use function notation to write an equation for the total expenditures of the corporation in year t. What was the total for 2010?

Additional answers can be found in the Instructor Answer Appendix.

51. Reading a graph—Operations on functions: Use the given graph to find the result of the operations indicated.

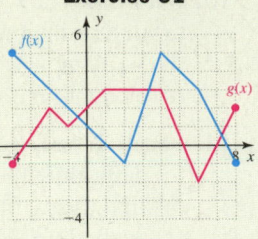

Exercise 51

Note $f(-4) = 5$, $g(-4) = -1$, and so on.

a. $(f + g)(-4)$ **b.** $(f \cdot g)(1)$

c. $(f - g)(4)$ **d.** $(f + g)(0)$

e. $\left(\dfrac{f}{g}\right)(2)$ **f.** $(f \cdot g)(-2)$

g. $(g \cdot f)(2)$ **h.** $(f - g)(-1)$

i. $(f + g)(8)$ **j.** $\left(\dfrac{f}{g}\right)(7)$

52. Reading a graph—Operations on functions: Use the given graph to find the result of the operations indicated.

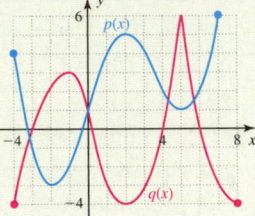

Exercise 52

Note $p(-1) = -2$, $q(5) = 6$, and so on.

a. $(p + q)(-4)$ **b.** $(p \cdot q)(1)$

c. $(p - q)(4)$ **d.** $(p + q)(0)$

e. $\left(\dfrac{p}{q}\right)(5)$ **f.** $(p \cdot q)(-2)$

g. $(q \cdot p)(2)$ **h.** $(p - q)(-1)$

i. $(p + q)(7)$ **j.** $\left(\dfrac{p}{q}\right)(6)$

Some advanced applications require that we use the algebra of functions to find a function rule for the vertical distance between two graphs. For $f(x) = 3$ and $g(x) = -2$ (two horizontal lines), we "see" this vertical distance is 5 units, or in function form: $d(x) = f(x) - g(x) = 3 - (-2) = 5$ units. However, $d(x) = f(x) - g(x)$ also serves as a *general formula* for the vertical distance between two curves (even those that are not horizontal lines), so long as $f(x) > g(x)$ in a chosen interval. Find a function rule in simplified form, for the vertical distance $h(x)$ between the graphs of f and g shown, for the interval indicated.

53. $x \in [0, 6]$

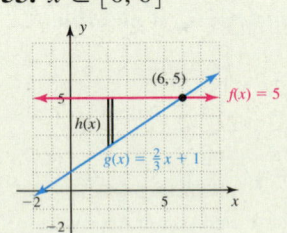

$h(x) = -\frac{2}{3}x + 4$

54. $x \in [1, 7]$

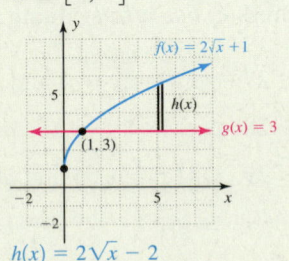

$h(x) = 2\sqrt{x} - 2$

55. $x \in [0, 4]$ **56.** $x \in [0, 5]$

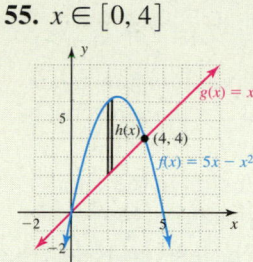

$h(x) = 4x - x^2$

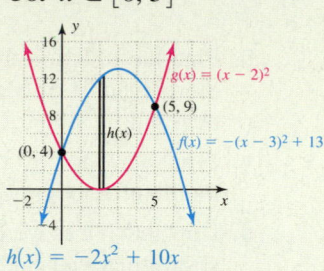

$h(x) = -2x^2 + 10x$

Use the TABLE feature of a graphing calculator to determine the domain of each newly constructed function. Write answers in interval notation. All finite endpoints of the domain intervals are integer values between -5 and 5.

57. For $Y_1 = 6 - 3x$ and $Y_2 = x - 2$,

   a. $Y_1 + Y_2$ $(-\infty, \infty)$

   b. $\dfrac{Y_1}{Y_2}$ $(-\infty, 2) \cup (2, \infty)$

58. For $Y_1 = 3x - 9$ and $Y_2 = x + 3$,

   a. $Y_1 - Y_2$ $(-\infty, \infty)$

   b. $\dfrac{Y_1}{Y_2}$ $(-\infty, -3) \cup (-3, \infty)$

59. For $Y_1 = x^2 - 1$ and $Y_2 = \sqrt{x + 1}$,

   a. $Y_1 - Y_2$ $[-1, \infty)$

   b. $\dfrac{Y_1}{Y_2}$ $(-1, \infty)$

60. For $Y_1 = 9 - x^2$ and $Y_2 = \sqrt{x - 3}$,

   a. $Y_1 + Y_2$ $[3, \infty)$

   b. $\dfrac{Y_1}{Y_2}$ $(3, \infty)$

61. For $Y_1 = \sqrt{x + 2}$ and $Y_2 = \sqrt{9 - 3x}$,

   a. $Y_1 Y_2$ $[-2, 3]$

   b. $\dfrac{Y_1}{Y_2}$ $[-2, 3)$

62. For $Y_1 = \sqrt{2x + 8}$ and $Y_2 = \sqrt{3 - x}$,

   a. $Y_1 Y_2$ $[-4, 3]$

   b. $\dfrac{Y_1}{Y_2}$ $[-4, 3)$

Use the GRAPH feature of a graphing calculator to determine the domain of each newly constructed function. Write answers in interval notation. All finite endpoints of the domain intervals are integer values between -5 and 5.

63. For $Y_1 = 3$ and $Y_2 = \sqrt{x + 4}$,

   a. $Y_1 - Y_2$ $[-4, \infty)$

   b. $Y_1 Y_2$ $[-4, \infty)$

64. For $Y_1 = x^2 - 9$ and $Y_2 = \sqrt{4 - x}$,

   a. $Y_1 + Y_2$ $(-\infty, 4]$

   b. $Y_1 Y_2$ $(-\infty, 4]$

65. For $Y_1 = \sqrt[3]{x - 1}$ and $Y_2 = x - 4$,

   a. $Y_1 + Y_2$ $(-\infty, \infty)$

   b. $\dfrac{Y_1}{Y_2}$ $(-\infty, 4) \cup (4, \infty)$

66. For $Y_1 = 4$ and $Y_2 = \sqrt[3]{x + 1}$,

   a. $Y_1 - Y_2$ $(-\infty, \infty)$

   b. $\dfrac{Y_1}{Y_2}$ $(-\infty, -1) \cup (-1, \infty)$

67. For $Y_1 = \sqrt{x + 4}$ and $Y_2 = \sqrt{6 - 2x}$,

   a. $Y_1 - Y_2$ $[-4, 3]$

   b. $\dfrac{Y_1}{Y_2}$ $[-4, 3)$

68. For $Y_1 = \sqrt{5x + 10} - 2$ and $Y_2 = -\sqrt{10 - 2x} + 2$,

   a. $Y_1 + Y_2$ $[-2, 5]$

   b. $Y_1 Y_2$ $[-2, 5]$

▶ WORKING WITH FORMULAS

69. Velocity of a falling object: $V = \dfrac{w}{k} - \dfrac{w}{k} e^{\frac{-32kt}{w}}$

When air resistance is considered, the velocity (in ft/sec) of a object weighing w lb falling for t sec can be approximated using the formula shown where k is the drag coefficient determined by the aerodynamics of the object, the density of the air, and other factors. (a) Find the velocity formula for a 16-lb bowling ball with a drag coefficient of $k = 0.04$. (b) Find two functions $f(t)$ and $g(t)$ such that $V(t) = (f - g)(t)$, and evaluate $V(1), V(2)$, and

$V(20)$. (c) Considering the calculations from Part (b), what will be the limiting or terminal velocity of the bowling ball?

70. Surface area of a cylinder: $A = 2\pi rh + 2\pi r^2$

If the height of a cylinder is fixed at 20 cm, the formula becomes $A = 40\pi r + 2\pi r^2$. Write this formula in factored form and find two functions $f(r)$ and $g(r)$ such that $A(r) = (f \cdot g)(r)$. Then find $A(5)$ by direct calculation and also by computing the product of $f(5)$ and $g(5)$, then comment on the results.
$A = 2\pi r (20 + r); f(r) = 2\pi r, g(r) = 20 + r; A(5) = 250\pi \text{ units}^2$

► APPLICATIONS

71. Boat manufacturing: Giaro Boats manufactures a popular recreational vessel, the *Revolution*. To plan for expanded production and increased labor costs, the company carefully tracks current costs and income. The fixed cost to produce this boat is $108,000 and the variable costs are $28,000 per boat. If the *Revolution* sells for $40,000, (a) find the profit function and (b) determine how many boats must be sold for the company to break even.
a. $P(x) = 12,000x - 108,000$ **b.** nine boats must be sold

72. Non-profit publications: Adobe Hope, a nonprofit agency, publishes the weekly newsletter *Community Options*. In doing so, they provide useful information to the surrounding area while giving high school dropouts valuable work experience. The fixed cost for publishing the newsletter is $900 per week, with a variable cost of $0.25 per newsletter. If the newsletter is sold for $1.50 per copy, (a) find the profit function for the newsletter, (b) determine how many newsletters must be sold to break even, and (c) determine how much money will be returned to the community if 1000 newsletters are sold (to preserve their status as a nonprofit organization). **a.** $P(x) = 1.25x - 900$
b. 720 newsletters must be sold   **c.** $350 will be returned

73. Cost, revenue, and profit: Suppose the total cost of manufacturing a certain computer component can be modeled by the function $C(n) = 0.1n^2$, where n is the number of components made and $C(n)$ is in dollars. If each component is sold at a price of $11.45, the revenue is modeled by $R(n) = 11.45n$. Use this information to complete the following.

a. Find the function that represents the total profit made from sales of the components.
$P(n) = 11.45n - 0.1n^2$

b. How much profit is earned if 12 components are made and sold? $123

c. How much profit is earned if 60 components are made and sold? $327

d. Explain why the company is making a "negative profit" after the 114th component is made and sold. $C(115) > R(115)$

74. Cost, revenue, and profit: For a certain manufacturer, revenue has been increasing but so has the cost of materials and the cost of employee benefits. Suppose revenue can be modeled by $R(t) = 10\sqrt{t}$, the cost of materials by $M(t) = 2t + 1$, and the cost of benefits by $B(t) = 0.1t^2 + 2$, where t represents the number of months since operations began and outputs are in thousands of dollars. Use this information to complete the following.

a. Find the function that represents the total manufacturing costs. $T(t) = 0.1t^2 + 2t + 3$

b. Find the function that represents how much more the cost of benefits are than the cost of materials. $D(t) = 0.1t^2 - 2t + 1$

c. What was the cost of benefits in the 10th month after operations began? $12,000

d. How much less were the benefits costs than the cost of materials in the 10th month? $9,000

e. Find the function that represents the profit earned by this company. $P(t) = 10\sqrt{t} - 0.1t^2 - 2t - 3$

f. Find the amount of profit earned in the 5th month and 10th month. Discuss each result.
approximately $6861 (profit); -$1377 (loss)

Minimizing travel distance is one of the main responsibilities of a logistics manager. From delivering pizzas on a Friday night, to building distribution centers for goods and commodities, a shorter *total* travel distance translates directly into reduced costs. On a straight highway, the distance from a distribution center at milepost x to a distribution point at milepost m_1 is $d = |x - m_1|$. Using the addition of functions, the *total* distance from a distribution center to its n distribution points is $D(x) = |x - m_1| + |x - m_2| + \cdots + |x - m_n|$.

75. Minimizing CNG distribution distance: Ruta 2 in Argentina is a well-traveled corridor for residents of Buenos Aires vacationing in Mar del Plata. Many of the cars in Argentina operate on compressed natural gas (CNG). Suppose there were CNG filling stations at kilometer-posts 30, 49, 115, 199, and 280. Using a graphing calculator, at what kilometer-post should the CNG storage tanks supplying these stations be built in order to minimize total delivery distance?   km 115

76. Minimizing H₂O distribution distance: The Trans-Australian Railway links Australia's eastern states and those of Western Australia. Most of the rail line lies in the central desert (with summer temperatures of 50°C/122°F), which is home to the world's longest stretch of dead-straight track, over 478 km/297 mi. Assume rest stops are scheduled to be built at mile-posts 25, 56, 89, 145, and 229 miles from the east. At what mile-post should the water storage tanks supplying these rest stops be built in order to minimize the total delivery distance?   mile 89

77. Minimizing CNG distribution distance: In addition to the five already mentioned in Exercise 75, another CNG station is being constructed at kilometer-post 233 on Ruta 2. With this station included, where should the CNG storage tanks be built? How does this result compare with that of Exercise 75?   Anywhere between km 115 and km 199.
Answers will vary.

78. Minimizing H₂O distribution distance: In addition to the five already mentioned in Exercise 76, another rest stop is scheduled to be built at mile-post 192 on the Trans-Australian Railway. With this rest stop included, where should the water storage tanks be built? How does this result compare with that of Exercise 76?
Anywhere between mile 89 and mile 145. Answers will vary.

 79. Measuring the depth of a well: To estimate the depth of a well, a large stone is dropped into the opening and a stop-watch is used to measure the number of seconds until a splash is heard. By solving the freefall equation $d = 16t^2$ for t, we find

the function $t_1(d) = \dfrac{\sqrt{d}}{4}$ models the time it takes

for the stone to hit the water. Since the speed of sound is known to be 1116 ft/sec, we use the distance = rate × time equation to obtain

$t_2(d) = \dfrac{d}{1116}$ as a function modeling the time it takes

the sound of the splash to reach our ears. The total time is the sum of these two functions, let's call it

 $T(d) = \dfrac{\sqrt{d}}{4} + \dfrac{d}{1116}$. (a) If the well is 230 ft deep, what is the total time? (b) If it takes 6 sec for the sound to be heard, how deep is the well?
a. 4 sec   **b.** about 494 ft

80. Predator-prey concentrations: Suppose the monthly whitetip reef shark population off the coast of Manuel Antonio National Park (Costa Rica) can be modeled by the function $s(t) = -40t^2 + 500t + 1300$ (where $t = 1$ corresponds to January). One of its favorite foods is the spiny lobster, whose monthly population in these waters might be modeled by the function $l(t) = 110t^2 - 1500t + 13{,}400$. The number of lobsters per shark in any given month can then be modeled by the quotient

$N(t) = \dfrac{110t^2 - 1500t + 13{,}400}{-40t^2 + 500t + 1300}$. (a) How many

lobsters per shark are there in February? (b) When is the number of lobsters per shark a minimum? What is this minimum?
a. about 5.07 lobsters per shark   **b.** When $t \approx 6.5$ (June) there are only 2.9 lobsters per shark.

▶ EXTENDING THE CONCEPT

 81. For a certain country, assume the function $C(x) = 0.0345x^4 - 0.8996x^3 + 7.5383x^2 - 21.7215x + 40$ approximates the number of Conservatives in the senate for the years 1995 to 2007, where $x = 0$ corresponds to 1995. The function $L(x) = -0.0345x^4 + 0.8996x^3 - 7.5383x^2 + 21.7215x + 10$ gives the number of Liberals for these years. Use this information to answer the following. (a) During what years did the Conservatives control the senate? (b) What was the greatest difference between the number of seats held by each faction in any one year? In what year did this occur? (c) What was the minimum number of seats held by the Conservatives? In what year? (d) Assuming no independent or third-party candidates are elected, what information does the function $T(x) = C(x) + L(x)$ give us? What information does $t(x) = |C(x) - L(x)|$ give us?

 82. Given $f(x) = \sqrt{1 - x}$ and $g(x) = \sqrt{x - 2}$, what can you say about the domain of $(f + g)(x)$? Enter the functions as Y_1 and Y_2 on a graphing calculator, then enter $Y_3 = Y_1 + Y_2$. See if you can determine why the calculator gives an error message for Y_3, regardless of the input.
Answers will vary.

 83. Use a graphing calculator to find the zeroes of each new function in Exercises 67 and 68, and describe their significance. **67 a.** $x = \frac{2}{3}$, where $Y_1 = Y_2$; **67 b.** $x = -4$, where $Y_1 = 0$; **68 a.** $x = 0$, where $Y_1 = -Y_2$; **68 b.** $x = -1.2$ or 3, the zeroes of Y_1 and Y_2.

Additional answers can be found in the Instructor Answer Appendix.

84. Instead of calculating the result of an operation on two functions at a *specific point* as in Exercises 47–52, we can actually *graph the function* that results from the operation. This skill, called the **addition of ordinates,** is widely applied in a study of tides and other areas. For $f(x) = (x - 3)^2 + 2$ and $g(x) = 4|x - 3| - 5$, complete a table of values like the one shown for $x \in [-2, 8]$. For the last column, remember that $(f - g)(x) = f(x) - g(x)$, and use this relation to complete the column. Finally, use the ordered pairs $(x, (f - g)(x))$ to graph the new function. Is the new function smooth? Is the new function continuous?   no, yes

Exercise 84

x	$f(x)$	$g(x)$	$(f - g)(x)$
−2	27	15	12
−1	18	11	7
0	11	7	4
1	6	3	3
2	3	−1	4
3	2	−5	7
4	3	−1	4
5	6	3	3
6	11	7	4
7	18	11	7
8	27	15	12

▶ **MAINTAINING YOUR SKILLS**

85. (R.3) Solve the equation.

$\frac{2}{7}(x - 3) - \frac{5}{3} = \frac{1}{2}x + 1$ $x = \frac{-148}{9}$

86. (2.2) Given $f(x) = 3x^2 - 2x$, find $f(x - 1)$ and describe the resulting transformation.
$f(x - 1) = 3x^2 - 8x + 5$, shift 1 unit right

87. (1.1/1.2) Given the points $(2, -5)$ and $(-3, 1)$, find (a) the standard form of the equation of the line containing these points and (b) the distance between these points.   **a.** $6x + 5y = -13$ **b.** $d = \sqrt{61}$

88. (3.4) Find the average rate of change of the function $g(x) = x^3 - 9x + 1$ over the indicated intervals.

a. $[1, 2]$ $\frac{\Delta g}{\Delta x} = -2$

b. $[1, 3]$ $\frac{\Delta g}{\Delta x} = 4$

c. $[1, 4]$ $\frac{\Delta g}{\Delta x} = 12$

3.6 The Composition of Functions and the Difference Quotient

LEARNING OBJECTIVES

In Section 3.6 you will see how we can:

☐ **A.** Compose two functions and determine the domain; decompose a function

☐ **B.** Interpret the composition of functions numerically and graphically

☐ **C.** Apply the difference quotient to find the average rate of change for nonlinear functions

☐ **D.** Apply the composition of functions and the difference quotient in context

The composition of functions gives us an efficient way to study how relationships are "linked." For example, the number of wolves w in a countywide area depends on the human population x, and a simple model might be $w(x) = 600 - 0.02x$. But the number of rodents r depends on the number of wolves, say $r(w) = 10,000 - 9.5w$, so the human population also has a measurable effect on the rodent population. In this section, we'll show that this effect is modeled by the function $r(x) = 4300 + 0.19x$.

A. The Composition of Functions

The composition of functions is best understood by studying the "input/output" nature of a function. Consider $g(x) = x^2 - 3$. For $g(x)$ we might say, "inputs are squared, then decreased by three." In diagram form we have:

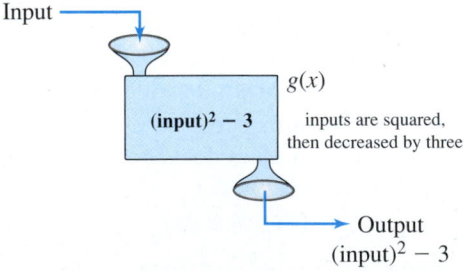

In many respects, a function box can be regarded as a very simple machine, running a simple program. It doesn't matter what the input is, this machine is going to *square the input then subtract three.*

EXAMPLE 1 ▶ **Evaluating a Function**

For $g(x) = x^2 - 3$, find

   a. $g(-5)$

   b. $g(5t)$

   c. $g(t - 4)$

Solution ▶

a.
$$g(x) = x^2 - 3 \qquad \text{original function}$$
input -5
$$g(-5) = (-5)^2 - 3 \qquad \text{square input, then subtract 3}$$
$$= 25 - 3 \qquad \text{simplify}$$
$$= 22 \qquad \text{result}$$

b.
$$g(x) = x^2 - 3 \qquad \text{original function}$$
input $5t$
$$g(5t) = (5t)^2 - 3 \qquad \text{square input, then subtract 3}$$
$$= 25t^2 - 3 \qquad \text{result}$$

c.
$$g(x) = x^2 - 3 \qquad \text{original function}$$
input $t - 4$
$$g(t - 4) = (t - 4)^2 - 3 \qquad \text{square input, then subtract 3}$$
$$= t^2 - 8t + 16 - 3 \qquad \text{expand binomial}$$
$$= t^2 - 8t + 13 \qquad \text{result}$$

Now try Exercises 7 and 8 ▶

WORTHY OF NOTE

It's important to note that t and $t - 4$ are two different, distinct values—the number represented by t, and a number four less than t. Examples would be 7 and 3, 12 and 8, as well as -10 and -14. There should be nothing awkward or unusual about evaluating $g(t)$ versus evaluating $g(t - 4)$ as in Example 1(c).

When the input value is itself a function (rather than a single number or variable), this process is called the **composition of functions.** The evaluation method is exactly the same, we are simply using a function input. Using a general function $g(x)$ and a function diagram as before, we illustrate the process in Figure 3.70.

Figure 3.70

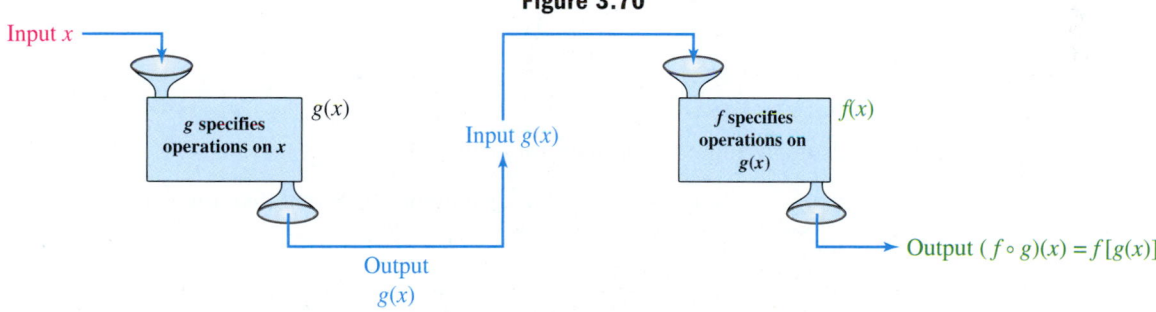

The notation used for the composition of f with g is an open dot "∘" placed between them, and is read, "f composed with g." The notation $(f \circ g)(x)$ indicates that $g(x)$ is an input for f: $(f \circ g)(x) = f[g(x)]$. If the order is reversed, as in $(g \circ f)(x)$, $f(x)$ becomes the input for g: $(g \circ f)(x) = g[f(x)]$. Figure 3.70 also helps us determine the domain of a composite function, in that the first function g can operate only if x is a valid input for g, and the second function f can operate only if $g(x)$ is a valid input for f. In other words, $(f \circ g)(x)$ is defined for *all x in the domain of g, such that $g(x)$ is in the domain of f.*

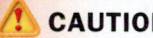 **CAUTION** ▶ Try not to confuse the new "open dot" notation for the *composition* of functions, with the multiplication dot used to indicate the *product* of two functions: $(f \cdot g)(x) = (fg)(x)$ while $(f \circ g)(x) = f[g(x)]$.

The Composition of Functions

Given two functions f and g, the composition of f with g is defined by

$$(f \circ g)(x) = f[g(x)]$$

The domain of the composition is all x in the domain of g
for which $g(x)$ is in the domain of f.

In Figure 3.71, these ideas are displayed using mapping notation, as we consider the simple case where $g(x) = x$ and $f(x) = \sqrt{x}$.

Figure 3.71

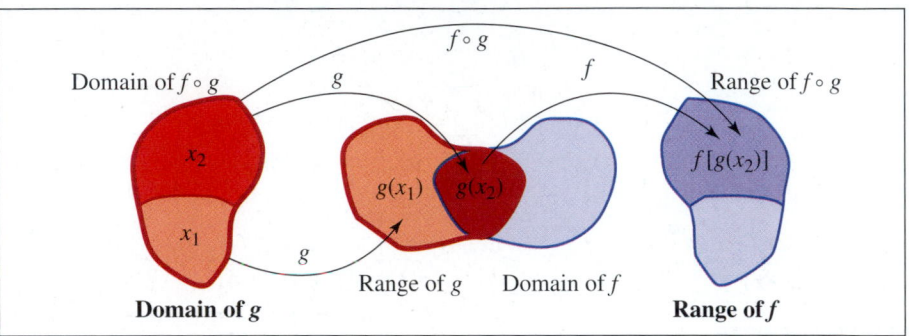

The domain of g (all real numbers) is shown within the red border, with g taking the negative inputs represented by x_1 (light red), to a like-colored portion of the range—the negative outputs $g(x_1)$. The nonnegative inputs represented by x_2 (dark red) are also mapped to a like-colored portion of the range—the nonnegative outputs $g(x_2)$. While the range of g is also all real numbers, function f can only use the nonnegative inputs represented by $g(x_2)$. This restricts the domain of $(f \circ g)(x)$ to only the inputs from g, where $g(x)$ is in the domain of f.

EXAMPLE 2 ▶   **Finding a Composition of Functions**

Given $f(x) = \sqrt{x - 4}$ and $g(x) = 3x + 2$, find
  a. $(f \circ g)(x)$
  b. $(g \circ f)(x)$

Also determine the domain for each.

Solution ▶   **a.** $f(x) = \sqrt{x - 4}$ says "decrease inputs by 4, and take the square root of the result."

$$
\begin{aligned}
(f \circ g)(x) &= f[g(x)] && \text{$g(x)$ is an input for f} \\
&= \sqrt{g(x) - 4} && \text{decrease input by 4, and take the square root of the result} \\
&= \sqrt{(3x + 2) - 4} && \text{substitute $3x + 2$ for $g(x)$} \\
&= \sqrt{3x - 2} && \text{result}
\end{aligned}
$$

While g is defined for all real numbers, $f(x)$ represents a real number only when $x \geq 4$. For $f[g(x)]$, this means we need $g(x) \geq 4$, giving $3x + 2 \geq 4$, $x \geq \frac{2}{3}$. In interval notation, the domain of $(f \circ g)(x)$ is $x \in \left[\frac{2}{3}, \infty\right)$.

  b. The function g says "inputs are multiplied by 3, then increased by 2."

$$
\begin{aligned}
(g \circ f)(x) &= g[f(x)] && \text{$f(x)$ is an input for g} \\
&= 3f(x) + 2 && \text{multiply input by 3, then increase by 2} \\
&= 3\sqrt{x - 4} + 2 && \text{substitute $\sqrt{x - 4}$ for $f(x)$}
\end{aligned}
$$

For $g[f(x)]$, g can accept any real number input, but f can supply only those where $x \geq 4$. The domain of $(g \circ f)(x)$ is $x \in [4, \infty)$.

WORTHY OF NOTE

Example 2 shows that $(f \circ g)(x)$ is generally not equal to $(g \circ f)(x)$. On those occasions when they *are* equal, the functions have a unique relationship that we'll study in Section 5.1.

Now try Exercises 9 through 18 ▶

Most graphing calculators have the ability to evaluate a composition of functions at a given point. From Example 2, enter $f(x)$ as $Y_1 = \sqrt{x-4}$ and $g(x)$ as $Y_2 = 3x + 2$. To evaluate the composition $(f \circ g)(9)$ (since we know $x = 9$ is in the domain), we can (1) return to the home screen, enter $Y_1(Y_2(9))$ and press ⬤ (Figure 3.72), or (2) enter $Y_3 = Y_1(Y_2(X))$ on the ⬤ screen and use the **TABLE** feature (Figure 3.73). **See Exercises 19 through 24.**

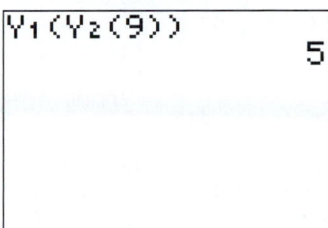

Figure 3.72　　　　　　　　**Figure 3.73**

EXAMPLE 3 ▶ **Finding a Composition of Functions**

For $f(x) = \dfrac{3x}{x-1}$ and $g(x) = \dfrac{2}{x}$, analyze the domain of

a. $(f \circ g)(x)$　　　　**b.** $(g \circ f)(x)$　　　　**c.** Find the actual compositions and comment.

Solution ▶ **a.** $(f \circ g)(x)$: For g to be defined, $x \ne 0$ is our first restriction. Once $g(x)$ is used as the input, we have $f[g(x)] = \dfrac{3g(x)}{g(x)-1}$, and additionally note that $g(x)$ cannot equal 1. This means $\dfrac{2}{x} \ne 1$, so $x \ne 2$. The domain of $f \circ g$ is all real numbers except $x = 0$ and $x = 2$.

b. $(g \circ f)(x)$: For f to be defined, $x \ne 1$ is our first restriction. Once $f(x)$ is used as the input, we have $g[f(x)] = \dfrac{2}{f(x)}$, and additionally note that $f(x)$ cannot be 0.

This means $\dfrac{3x}{x-1} \ne 0$, so $x \ne 0$. The domain of $(g \circ f)(x)$ is all real numbers except $x = 0$ and $x = 1$.

c. For $(f \circ g)(x)$:

$$f[g(x)] = \frac{3g(x)}{g(x)-1}$$ 　　　composition of *f* with *g*

$$= \frac{\left(\dfrac{3}{1}\right)\left(\dfrac{2}{x}\right)}{\left(\dfrac{2}{x}\right)-1}$$ 　　　substitute $\dfrac{2}{x}$ for *g*(x)

$$= \frac{\dfrac{6}{x}}{\dfrac{2-x}{x}} = \frac{6}{x}\cdot\frac{x}{2-x}$$ 　　　simplify denominator; invert and multiply

$$= \frac{6}{2-x}$$ 　　　result

Notice the function rule for $(f \circ g)(x)$ has an implied domain of $x \ne 2$, but does not show that g (the inner function) is undefined when $x = 0$ (see Part a). The domain of $(f \circ g)(x)$ is actually all real numbers except $x = 0$ and $x = 2$.

For $(g \circ f)(x)$ we have:

$$g[f(x)] = \frac{2}{f(x)} \qquad \text{composition of } g \text{ with } f$$

$$= \frac{2}{\dfrac{3x}{x-1}} \qquad \text{substitute } \dfrac{3x}{x-1} \text{ for } f(x)$$

$$= \frac{2}{1} \cdot \frac{x-1}{3x} \qquad \text{invert and multiply}$$

$$= \frac{2(x-1)}{3x} \qquad \text{result}$$

Similarly, the function rule for $(g \circ f)(x)$ has an implied domain of $x \neq 0$, but does not show that f (the inner function) is undefined when $x = 1$ (see Part b). The domain of $(g \circ f)(x)$ is actually all real numbers except $x = 0$ and $x = 1$.

Now try Exercises 25 through 30 ▶

As Example 3 illustrates, the domain of $h(x) = (f \circ g)(x)$ *cannot simply be taken from the new function rule for h. It* must *be determined from the functions composed to obtain h.* The graph of $(f \circ g)(x)$ is shown in Figure 3.74. We can easily see $x = 2$ is not in the domain as there a vertical asymptote at $x = 2$. The fact that $x = 0$ is also excluded is obscured by the y-axis, but the table shown in Figure 3.75 confirms that $x = 0$ is likewise not in the domain.

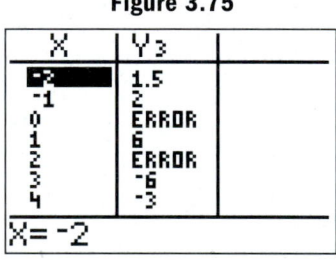

Figure 3.74   Figure 3.75

To further explore concepts related to the domain of a composition, **see Exercises 69 and 70.**

Decomposing a Composite Function

Based on Figure 3.76, would you say that the circle is inside the square or the square is inside the circle? The decomposition of a composite function is related to a similar question, as we ask ourselves what function (of the composition) is on the "inside"— the input quantity—and what function is on the "outside." For instance, consider $h(x) = \sqrt{x - 4}$, where we see that $x - 4$ is "inside" the radical. Letting $g(x) = x - 4$ and $f(x) = \sqrt{x}$, we have $h(x) = (f \circ g)(x)$ or $f[g(x)]$.

Figure 3.76

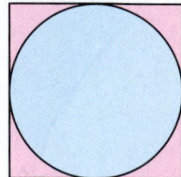

> **WORTHY OF NOTE**
>
> The decomposition of a function is not unique and can often be done in many different ways.

EXAMPLE 4 ▶ **Decomposing a Composite Function**

Given $h(x) = (\sqrt[3]{x} + 1)^2 - 3$, identify two functions f and g so that $(f \circ g)(x) = h(x)$, then check by composing the functions to obtain $h(x)$.

Solution ▶ Noting that $\sqrt[3]{x} + 1$ is inside the squaring function, we assign $g(x)$ as this inner function: $g(x) = \sqrt[3]{x} + 1$. The outer function is the squaring function decreased by 3, so $f(x) = x^2 - 3$.

Check: $(f \circ g)(x) = f[g(x)]$ *g(x) is an input for f*

$\qquad\qquad = [g(x)]^2 - 3$ *f squares inputs, then decreases the result by 3*

$\qquad\qquad = [\sqrt[3]{x} + 1]^2 - 3$ *substitute $\sqrt[3]{x}$ + 1 for g(x)*

$\qquad\qquad = h(x)$ ✓

☑ **A.** You've just seen how we can compose two functions and determine the domain, and decompose a function

Now try Exercises 31 through 34 ▶

B. A Numerical and Graphical View of the Composition of Functions

Just as with the sum, difference, product, and quotient of functions, the composition of functions can also be interpreted and understood graphically. For $(f \circ g)(x)$, once the value of $g(x)$ is known (read from the graph), the value of $(f \circ g)(x) = f[g(x)]$, can also be determined.

EXAMPLE 5 ▶ **Interpreting the Composition of Functions Numerically**

For $f(x)$ and $g(x)$ as shown, use the graph given to determine the value of each expression.

a. $f(4)$, $g(2)$, and $(f \circ g)(2)$

b. $g(6)$, $f(8)$, and $(g \circ f)(8)$

c. $(f \circ g)(8)$ and $(g \circ f)(0)$

d. $(f \circ f)(9)$ and $(g \circ g)(0)$

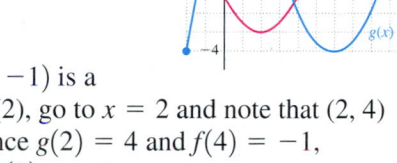

Solution ▶ **a.** For $f(4)$, we go to $x = 4$ and note that $(4, -1)$ is a point on the red graph: $f(4) = -1$. For $g(2)$, go to $x = 2$ and note that $(2, 4)$ is a point on the blue graph: $g(2) = 4$. Since $g(2) = 4$ and $f(4) = -1$, $(f \circ g)(2) = f[g(2)] = f(4) = -1$: $(f \circ g)(2) = -1$.

b. For $g(6)$, $x = 6$ and we find that $(6, -4)$ is a point on the blue graph: $g(6) = -4$. For $f(8)$, $x = 8$ and note that that $(8, 6)$ is a point on the red graph: $f(8) = 6$. Since $f(8) = 6$ and $g(6) = -4$, $(g \circ f)(8) = g[f(8)] = g(6) = -4$: $(g \circ f)(8) = -4$.

c. As illustrated in parts (a) and (b), for $(f \circ g)(8) = f[g(8)]$ we first determine $g(8)$, then substitute this value into f. From the blue graph $g(8) = -2$, and $f(-2) = 3$. For $(g \circ f)(0) = g[f(0)]$ we first determine $f(0)$, then substitute this value into g. From the red graph $f(0) = -1$, and $g(-1) = 1$.

d. To evaluate $(f \circ f)(9) = f[f(9)]$ we follow the same sequence as before. From the graph of f, $f(9) = 5$ and $f(5) = 1$, showing $(f \circ f)(9) = 1$. The same ideas are applied to $(g \circ g)(0) = g[g(0)]$. Since $g(0) = 4$ and $g(4) = -2$, $(g \circ g)(0) = -2$.

Now try Exercises 35 and 36 ▶

As with other operations on functions, a composition can be understood graphically *at specific points,* as in Example 5, or *for all allowable x-values.* Exploring the second option will help us understand certain transformations of functions, and why the domain of a composition is defined so carefully. For instance, consider the functions $f(x) = -x^2 + 3$ and $g(x) = (x - 2)$. The graph of f is a parabola opening downward with vertex $(0, 3)$. For $h(x) = (f \circ g)(x)$ we have $h(x) = -(x - 2)^2 + 3$, which we recognize as the graph of f *shifted 2 units to the right*. Using $Y_1 = -X^2 + 3$, $Y_2 = X - 2$, and $Y_3 = Y_1(Y_2(X))$, we can deactivate Y_2 so the only the graphs of Y_1 and Y_3 (in **bold**) are shown (see Figure 3.77). Here the domain of the composition was not a concern, as both f and g have domain $x \in \mathbb{R}$. But now consider $f(x) = 4\sqrt{x + 1}$, $g(x) = \dfrac{3}{2 - x}$, and $h(x) = (f \circ g)(x)$. The domain of g must exclude $x = 2$ (which is also excluded from the domain of h), while the domain of f is $x \geq -1$. But after assigning $Y_1 = 4\sqrt{X + 1}$, $Y_2 = \dfrac{3}{2 - X}$, and $Y_3 = Y_1(Y_2(X))$, the graph of Y_3 shows a noticeable gap between $x = 2$ and $x = 5$ (Figure 3.78). The reason is that $(f \circ g)(x) = f[g(x)]$ uses $g(x)$ *as the input for* f, meaning for the domain, $x \geq -1$ becomes $g(x) \geq -1 \rightarrow \dfrac{3}{2 - x} \geq -1$. Solving this inequality graphically (Figure 3.79) shows $g(x) \geq -1$ only for the intervals $x \in (-\infty, 2) \cup [5, \infty)$, leaving the gap seen in Y_3 for the interval $(2, 5]$. The domain of h is $x \in (-\infty, 2) \cup [5, \infty)$.

Figure 3.77

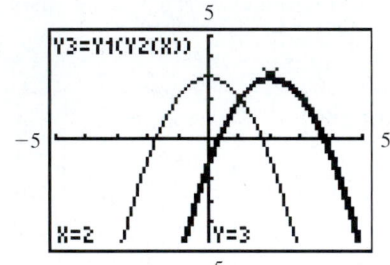

Figure 3.78

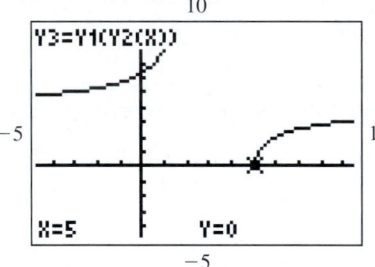

Figure 3.79

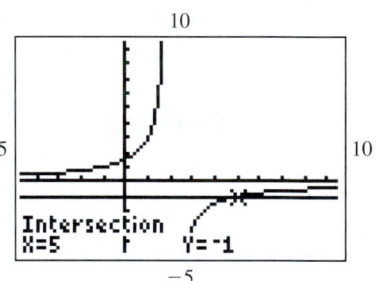

EXAMPLE 6 ▶   **Interpreting a Composition Graphically and Understanding the Domain**

Given $f(x) = 3\sqrt{1 - x}$, $g(x) = \dfrac{4}{x + 3}$,

a. State the domains of f and g.

b. Use a graphing calculator to study the graph of $h(x) = (f \circ g)(x)$.

c. Algebraically determine the domain of h, and reconcile it with the graph.

Solution ▶   **a.** The domain of g must exclude $x = -3$, and for the domain of f we must have $x \leq 1$.

b. For the graph of h, enter $Y_1 = 3\sqrt{1 - x}$, $Y_2 = \dfrac{4}{x + 3}$, and $Y_3 = Y_1(Y_2(X))$. Graphing Y_3 on the ⟨ZOOM⟩ **6:ZStandard** screen shows a gap between $x = -3$ and $x = 1$ (Figure 3.80).

c. Since the composition $(f \circ g)(x) = f[g(x)]$ uses $g(x)$ as the input for f, $x \leq 1$ (the domain for f) becomes $g(x) \leq 1 \rightarrow \dfrac{4}{x + 3} \leq 1$. Figure 3.81 shows a graphical solution to this inequality, which indicates $g(x) \leq 1$ for $x \in (-\infty, -3) \cup [1, \infty)$. The domain of h is $x \in (-\infty, -3) \cup [1, \infty)$.

Figure 3.80

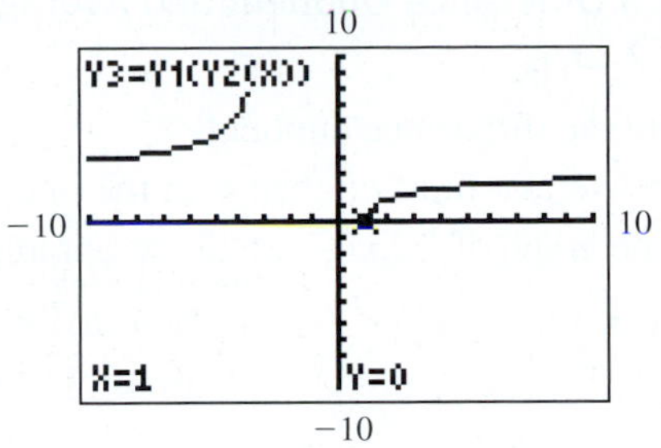

Figure 3.81

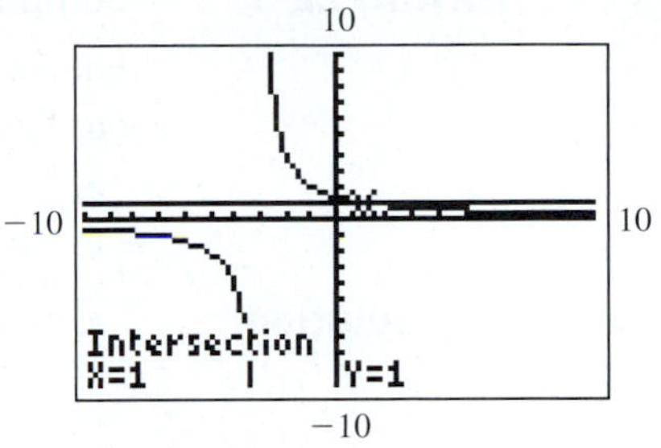

Now try Exercises 37 and 38 ▶

B. You've just seen how we can interpret the composition of functions numerically and graphically

C. Average Rates of Change and the Difference Quotient

We now return to our discussion of average rates of change, seeking to make the calculation more efficient due to (1) the large number of times the concept is applied, and (2) its value to virtually all fields of study. Our current calculation $\dfrac{\Delta y}{\Delta x} = \dfrac{f(x_2) - f(x_1)}{x_2 - x_1}$

works very well, but requires us to recalculate $\dfrac{\Delta y}{\Delta x}$ for each new interval chosen. Using a slightly different approach, we can develop a *formula* for a given function, and use it to calculate various rates of change with greater efficiency. This is done by selecting a point $x = x_1$ in the domain, and a point $x_2 = x + h$, where h is assumed to be some very small, arbitrary number. Making these substitutions in our current formula gives

$$\frac{\Delta y}{\Delta x} = \frac{f(x_2) - f(x_1)}{x_2 - x_1} \qquad \text{average rate of change}$$

$$= \frac{f(x + h) - f(x)}{(x + h) - x} \qquad \text{substitute } x + h \text{ for } x_2, x \text{ for } x_1$$

$$= \frac{f(x + h) - f(x)}{h} \qquad \text{simplify}$$

The advantage of this new formula, called the **difference quotient,** is that the result is a new function that can be evaluated repeatedly for any interval.

The Difference Quotient

For a given function $f(x)$ and constant $h \neq 0$,

$$\frac{f(x + h) - f(x)}{h}$$

is the difference quotient for f.

Note that the formula has three parts: (1) the function evaluated at $x + h$: $f(x + h)$, (2) the function f itself: $f(x)$, and (3) the constant h. The expression for $f(x + h)$ is actually a *composition of functions,* which can be evaluated and simplified prior to its use in the difference quotient.

$$\underset{(3)}{\frac{\overset{(1)}{f(x + h)} - \overset{(2)}{f(x)}}{h}}$$

EXAMPLE 7 ▶ **Computing a Difference Quotient and Average Rates of Change**

For $f(x) = x^2 - 4x$,

   a. Compute the difference quotient.

   b. Find the average rates of change in the intervals [1.9, 2.0] and [3.6, 3.7].

   c. Sketch the graph of f along with the secant lines and comment on what you notice.

Solution ▶   **a.** For $f(x) = x^2 - 4x$, $f(x + h) = (x + h)^2 - 4(x + h)$

$$= x^2 + 2xh + h^2 - 4x - 4h$$

Using this result in the difference quotient yields,

$$\frac{f(x + h) - f(x)}{h} = \frac{(x^2 + 2xh + h^2 - 4x - 4h) - (x^2 - 4x)}{h} \qquad \text{substitute into the difference quotient}$$

$$= \frac{x^2 + 2xh + h^2 - 4x - 4h - x^2 + 4x}{h} \qquad \text{eliminate parentheses}$$

$$= \frac{2xh + h^2 - 4h}{h} \qquad \text{combine like terms}$$

$$= \frac{\cancel{h}(2x + h - 4)}{\cancel{h}} \qquad \text{factor out } h$$

$$= 2x - 4 + h \qquad \text{result}$$

  b. For the interval [1.9, 2.0], $x = 1.9$ and $h = 0.1$. The slope of the secant line is

$$\frac{\Delta y}{\Delta x} = 2(1.9) - 4 + 0.1 = -0.1. \text{ For the}$$

interval [3.6, 3.7], $x = 3.6$ and $h = 0.1$. The slope of this secant line is

$$\frac{\Delta y}{\Delta x} = 2(3.6) - 4 + 0.1 = 3.3.$$

  c. After sketching the graph of f and the secant lines from each interval (see the figure), we note the slope of the first line (in red) is negative and very near zero, while the slope of the second (in blue) is positive and very steep.

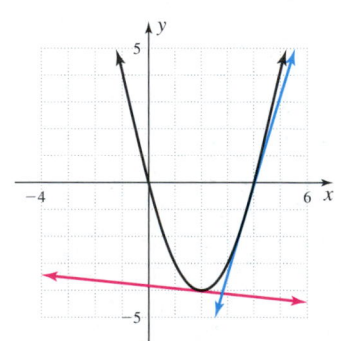

Now try Exercises 39 through 50 ▶

It is important that you see these calculations as much more than just an algebraic exercise. If the parabola were modeling the distance a rocket has traveled after the retro-rockets fired, the difference quotient could provide us with valuable information regarding the decreasing velocity of the rocket, and even help pinpoint the moment the velocity was zero.

EXAMPLE 8 ▶ Use the difference quotient to find an *average rate of change* formula for the reciprocal function $f(x) = \dfrac{1}{x}$. Then use the formula to find the average rate of change on the intervals [0.5, 0.51] and [3, 3.01]. Note $h = 0.01$ for both intervals.

Solution ▶ For $f(x) = \dfrac{1}{x}$ we have $f(x + h) = \dfrac{1}{x + h}$ and we compute as follows:

$$\frac{\Delta y}{\Delta x} = \frac{f(x + h) - f(x)}{h} \qquad \text{difference quotient}$$

$$= \frac{\dfrac{1}{(x + h)} - \dfrac{1}{x}}{h} \qquad \text{substitute } \frac{1}{x + h} \text{ for } f(x + h) \text{ and } \frac{1}{x} \text{ for } f(x)$$

$$= \frac{\dfrac{x - (x + h)}{(x + h)x}}{h} \qquad \text{common denominator}$$

$$= \frac{\dfrac{-h}{(x + h)x}}{h} \qquad \text{simplify numerator}$$

$$= \frac{-1}{(x + h)x} \qquad \text{invert and multiply}$$

This is the formula for computing rates of change given $f(x) = \dfrac{1}{x}$.

To find $\dfrac{\Delta y}{\Delta x}$ on the interval [0.5, 0.51] we have:

$$\frac{\Delta y}{\Delta x} = \frac{-1}{(0.5 + 0.01)(0.5)} \qquad \text{substitute 0.5 for } x, 0.01 \text{ for } h$$

$$\approx \frac{-4}{1} \qquad \text{result (approximate)}$$

On this interval y is decreasing by about 4 units for every 1 unit x is increasing. The slope of the secant line (in blue) through these points is $m \approx -4$. For the interval [3, 3.01] we have

$$\frac{\Delta y}{\Delta x} = \frac{-1}{(3 + 0.01)(3)} \qquad \text{substitute 3 for } x, 0.01 \text{ for } h$$

$$\approx -\frac{1}{9} \qquad \text{result (approximate)}$$

☑ **C.** You've just seen how we can apply the difference quotient to find average rates of change for nonlinear functions

On this interval, y is decreasing 1 unit for every 9 units x is increasing. The slope of the secant line (in red) through these points is $m \approx -\frac{1}{9}$.

Now try Exercises 51 through 54 ▶

In context, suppose the graph of f were modeling the declining area of the rainforest over time. The difference quotient would give us information on just how fast the destruction was taking place (and give us plenty of cause for alarm).

D. Applications of Composition and the Difference Quotient

Suppose that due to a collision, an oil tanker is spewing oil into the open ocean. The oil is spreading outward in a shape that is roughly circular, with the radius of the circle modeled by the function $r(t) = 2\sqrt{t}$, where t is the time in minutes and r is measured in feet. How could we determine the *area* of the oil slick in terms of t? As you can see, the radius depends on the time and the area depends on the radius. In diagram form we have:

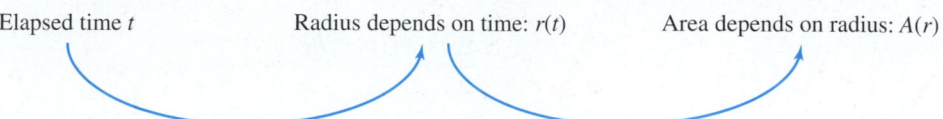

Elapsed time t Radius depends on time: $r(t)$ Area depends on radius: $A(r)$

It is possible to create a direct relationship between the elapsed time and the area of the circular spill using a composition of functions.

EXAMPLE 9 ▶ **Composition and the Area of an Oil Spill**

Given $r(t) = 2\sqrt{t}$ and $A(r) = \pi r^2$,

   a. Write A directly as a function of t by computing $(A \circ r)(t)$.

   b. Find the area of the oil spill after 30 min.

Solution ▶ **a.** The function A squares inputs, then multiplies by π.

$$
\begin{aligned}
(A \circ r)(t) &= A[r(t)] &&\text{$r(t)$ is the input for A} \\
&= [r(t)]^2 \cdot \pi &&\text{square input, multiply by π} \\
&= [2\sqrt{t}]^2 \cdot \pi &&\text{substitute $2\sqrt{t}$ for $r(t)$} \\
&= 4\pi t &&\text{result}
\end{aligned}
$$

Since the result contains no variable r, we can now compute the area of the spill directly, given the elapsed time t (in minutes): $A(t) = 4\pi t$.

   b. To find the area after 30 min, use $t = 30$.

$$
\begin{aligned}
A(t) &= 4\pi t &&\text{composite function} \\
A(30) &= 4\pi(30) &&\text{substitute 30 for t} \\
&= 120\pi &&\text{simplify} \\
&\approx 377 &&\text{result (rounded to the nearest unit)}
\end{aligned}
$$

After 30 min, the area of the spill is approximately 377 ft^2.

Now try Exercises 57 through 62 ▶

EXAMPLE 10 ▶ **Composition and Related Populations**

Using a statistical study, environmentalists find a current population of 600 wolves in the county, and believe the wolf population w will decrease as the human population x grows, according to the formula $w(x) = 600 - 0.02x$. Further, the population of rodents and small animals r depends on the number of wolves w, with the rodent population estimated by $r(w) = 10,000 - 9.5w$.

   a. Evaluate $r(w)$ for $w = 600$ to find the current rodent population, then use a composition to find a function modeling how the human population relates directly to the number of rodents.

   b. Use the function to estimate the number of rodents in the county, if the human population grows by 35,000.

Solution ▶   **a.** Since $r(600) = 4300$, there are currently 4300 rodents in the county. The number of rodents and small animals is a function of w: $r(w) = 10,000 - 9.5w$, and the number of wolves w is a function of human population x: $w(x) = 600 - 0.02x$. To find a function for r in terms of x, we use the composition $(r \circ w)(x) = r[w(x)]$.

$$r(w) = 10,000 - 9.5w \qquad \text{\textcolor{red}{r depends on w}}$$
$$r[w(x)] = 10,000 - 9.5[600 - 0.02x] \qquad \text{\textcolor{red}{compose r with $w(x)$}}$$
$$= 10,000 - 5700 + 0.19x \qquad \text{\textcolor{red}{distribute}}$$
$$r(x) = 4300 + 0.19x \qquad \text{\textcolor{red}{simplify, r depends on x}}$$

b. Evaluating $r(35,000)$ gives 10,950 rodents, showing a decline in the wolf population will eventually cause the rodent population to flourish, with an adverse effect on humans. In cases like these, a careful balance should be the goal.

Now try Exercises 63 and 64 ▶

You might be familiar with Galileo Galilei and his studies of gravity. According to popular history, he demonstrated that unequal weights will fall equal distances in equal time periods, by dropping cannonballs from the upper floors of the Leaning Tower of Pisa. Neglecting air resistance, the distance an object falls is modeled by the function $d(t) = 16t^2$, where $d(t)$ represents the distance fallen after t sec. Due to the effects of gravity, the velocity of the object increases as it falls. In other words, the velocity or the average rate of change $\dfrac{\Delta\mathbf{distance}}{\Delta\mathbf{time}}$ is a nonconstant (increasing) quantity. We can analyze this rate of change using the difference quotient.

EXAMPLE 11 ▶   **Applying the Difference Quotient in Context**

A construction worker drops a heavy wrench from atop a girder of a new skyscraper. Use the function $d(t) = 16t^2$ to

   a. Compute the distance the wrench has fallen after 2 sec and after 7 sec.

   b. Find a formula for the velocity of the wrench (average rate of change in distance per unit time).

   c. Use the formula to find the rate of change in the intervals $[2, 2.01]$ and $[7, 7.01]$.

   d. Graph the function and the secant lines representing the average rate of change. Comment on what you notice.

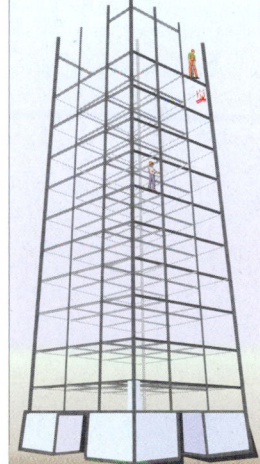

Solution ▶   **a.** Substituting $t = 2$ and $t = 7$ in the given function yields

$$d(2) = 16(2)^2 \qquad d(7) = 16(7)^2 \qquad \text{\textcolor{red}{evaluate $d(t) = 16t^2$}}$$
$$= 16(4) \qquad\qquad = 16(49) \qquad \text{\textcolor{red}{square input}}$$
$$= 64 \qquad\qquad = 784 \qquad\quad \text{\textcolor{red}{multiply}}$$

After 2 sec, the wrench has fallen 64 ft; after 7 sec, the wrench has fallen 784 ft.

b. For $d(t) = 16t^2$, $d(t + h) = 16(t + h)^2$, which we compute separately.

$$d(t + h) = 16(t + h)^2 \qquad \text{\textcolor{red}{substitute $t + h$ for t}}$$
$$= 16(t^2 + 2th + h^2) \qquad \text{\textcolor{red}{square binomial}}$$
$$= 16t^2 + 32th + 16h^2 \qquad \text{\textcolor{red}{distribute 16}}$$

Using this result in the difference quotient yields

$$\frac{d(t+h) - d(t)}{h} = \frac{(16t^2 + 32th + 16h^2) - 16t^2}{h} \quad \text{substitute into the difference quotient}$$

$$= \frac{16t^2 + 32th + 16h^2 - 16t^2}{h} \quad \text{eliminate parentheses}$$

$$= \frac{32th + 16h^2}{h} \quad \text{combine like terms}$$

$$= \frac{h(32t + 16h)}{h} \quad \text{factor out } h \text{ and simplify}$$

$$= 32t + 16h \quad \text{result}$$

For any number of seconds t and h a small increment of time thereafter, the velocity of the wrench is modeled by $\dfrac{\Delta \text{distance}}{\Delta \text{time}} = \dfrac{32t + 16h}{1}$.

c. For the interval $[t, t+h] = [2, 2.01]$, $t = 2$ and $h = 0.01$:

$$\frac{\Delta \text{distance}}{\Delta \text{time}} = \frac{32(2) + 16(0.01)}{1} \quad \text{substitute 2 for } t \text{ and 0.01 for } h$$

$$= 64 + 0.16 = 64.16$$

Two seconds after being dropped, the velocity of the wrench is close to 64.16 ft/sec (44 mph). For the interval $[t, t+h] = [7, 7.01]$, $t = 7$ and $h = 0.01$:

$$\frac{\Delta \text{distance}}{\Delta \text{time}} = \frac{32(7) + 16(0.01)}{1} \quad \text{substitute 7 for } t \text{ and 0.01 for } h$$

$$= 224 + 0.16 = 224.16$$

Seven seconds after being dropped, the velocity of the wrench is approximately 224.16 ft/sec (about 153 mph).

d.

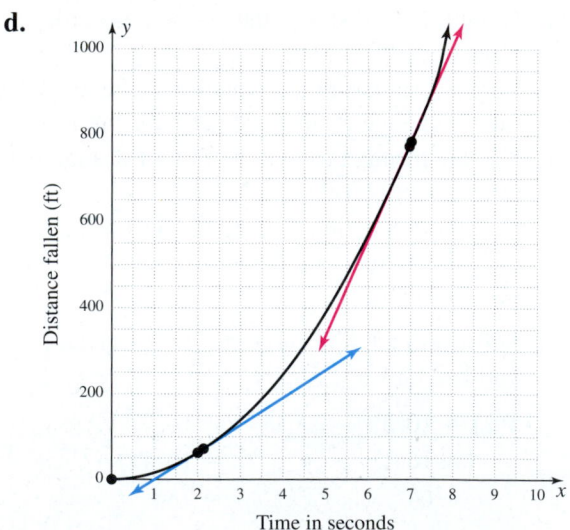

Time in seconds

The velocity increases with time, as indicated by the steepness of each secant line.

☑ **D.** You've just seen how we can apply the composition of functions and the difference quotient in context

Now try Exercises 65 through 68 ▶

3.6 EXERCISES

▶ CONCEPTS AND VOCABULARY

Fill in each blank with the appropriate word or phrase. Carefully reread the section if needed.

1. When evaluating functions, if the input value is a function itself, the process is called a <u>composition</u> of functions.

2. The notation $(f \circ g)(x)$ indicates that $g(x)$ is the input value for $f(x)$, which is written <u>$f[g(x)]$</u>.

3. For functions f and g, the domain of $(f \circ g)(x)$ is the set of all x in the <u>domain</u> of g, such that <u>$g(x)$</u> is in the domain of f.

4. The average rate of change formula becomes the <u>difference</u> quotient by substituting <u>$x + h$</u> for x_2 and <u>x</u> for x_1.

5. Discuss/Explain how and why using the the difference quotient differs from using the average rate of change formula.   *Answers will vary.*

6. Discuss/Explain how the domain of $(f \circ g)(x)$ is determined, given $f(x) = \sqrt{2x + 7}$ and $g(x) = \dfrac{2x}{x - 1}$.   *Answers will vary.*

▶ DEVELOPING YOUR SKILLS

7. Given $f(x) = x^2 - 5x - 14$, find $f(-2)$, $f(7)$, $f(2a)$, and $f(a - 2)$.

8. Given $g(x) = x^3 - 9x$, find $g(-3)$, $g(2)$, $g(3t)$, and $g(t + 1)$.

For each pair of functions below, find (a) $h(x) = (f \circ g)(x)$ and (b) $H(x) = (g \circ f)(x)$, and (c) determine the domain of each result.

9. $f(x) = \sqrt{x + 3}$ and $g(x) = 2x - 5$

10. $f(x) = x + 3$ and $g(x) = \sqrt{9 - x}$

11. $f(x) = \sqrt{x - 3}$ and $g(x) = 3x + 4$

12. $f(x) = \sqrt{x + 5}$ and $g(x) = 4x - 1$

13. $f(x) = x^2 - 3x$ and $g(x) = x + 2$

14. $f(x) = 2x^2 - 1$ and $g(x) = 3x + 2$

15. $f(x) = x^2 + x - 4$ and $g(x) = x + 3$

16. $f(x) = x^2 - 4x + 2$ and $g(x) = x - 2$

17. $f(x) = |x| - 5$ and $g(x) = -3x + 1$

18. $f(x) = |x - 2|$ and $g(x) = 3x - 5$

For the functions f and g given, $h(x) = (f \circ g)(x)$. Use a calculator to evaluate $h(-3)$, $h(\sqrt{2})$, $h\left(\dfrac{1}{2}\right)$, and $h(5)$. If an error message is received, explain why.

19. $f(x) = x^2 + 3x - 4$, $g(x) = x + 1$

20. $f(x) = -x^2 - 15x$, $g(x) = x - 2$

21. $f(x) = (x + 8)^2$, $g(x) = \dfrac{72}{x - 5}$

22. $f(x) = \dfrac{1}{x^2}$, $g(x) = \dfrac{1}{2x - 1}$

23. $f(x) = \sqrt{4 - 3x}$, $g(x) = x^2 - 9$

24. $f(x) = \dfrac{7}{x + 2}$, $g(x) = x^2 - 11$

For the functions $f(x)$ and $g(x)$ given, analyze the domain of (a) $(f \circ g)(x)$ and (b) $(g \circ f)(x)$, then (c) find the actual compositions and comment.

25. $f(x) = \dfrac{2x}{x + 3}$ and $g(x) = \dfrac{5}{x}$

26. $f(x) = \dfrac{-3}{x}$ and $g(x) = \dfrac{x}{x - 2}$

27. $f(x) = \dfrac{4}{x}$ and $g(x) = \dfrac{1}{x - 5}$

28. $f(x) = \dfrac{3}{x}$ and $g(x) = \dfrac{1}{x - 2}$

29. For $f(x) = x^2 - 8$, $g(x) = x + 2$, and $h(x) = (f \circ g)(x)$, find $h(5)$ in two ways:
    a. $(f \circ g)(5)$ 41   **b.** $f[g(5)]$ 41

30. For $p(x) = x^2 - 8$, $q(x) = x + 2$, and $H(x) = (p \circ q)(x)$, find $H(-2)$ in two ways:
    a. $(p \circ q)(-2)$ -8 **b.** $p[q(-2)]$ -8

HOMEWORK SELECTION GUIDE

Core: 7–31 odd, 35, 36, 37, 41, 47, 52, 55, 57–67 odd (19 Exercises)
Standard: 1–4, 7–31 odd, 35, 36, 37, 41, 47, 52, 55, 57–67 odd, 71, 72, 75 (26 Exercises)

Additional answers can be found in the Instructor Answer Appendix.

Extended: 1–4, 7–31 odd, 35, 36, 37, 41, 45, 47, 49, 50, 52, 55, 57–67 odd, 70, 71, 72, 75 (30 Exercises)
In Depth: 1–6, 7–31 odd, 35, 36, 37, 41, 45, 47, 49, 50, 52, 55, 57–61 odd, 62, 63–67 odd, 68, 70–75 (35 Exercises)

31. For $h(x) = (\sqrt{x - 2} + 1)^3 - 5$, find two functions f and g such that $(f \circ g)(x) = h(x)$.

32. For $H(x) = \sqrt[3]{x^2 - 5} + 2$, find two functions p and q such that $(p \circ q)(x) = H(x)$.

33. Given $f(x) = 2x - 1$, $g(x) = x^2 - 1$, and $h(x) = x + 4$, find $p(x) = f[g([h(x)])]$ and $q(x) = g[f([h(x)])]$.
$p(x) = 2(x + 4)^2 - 3, q(x) = (2x + 7)^2 - \frac{1}{2}$

34. Given $f(x) = 2x + 3$ and $g(x) = \dfrac{x - 3}{2}$, find
(a) $(f \circ f)(x)$, (b) $(g \circ g)(x)$, (c) $(f \circ g)(x)$, and
(d) $(g \circ f)(x)$.   **a.** $4x + 9$ **b.** $\frac{x - 9}{4}$ **c.** x **d.** x

35. Reading a graph: Use the given graph to find the result of the operations indicated.

Exercise 35

Note $f(-4) = 5$, $g(-4) = -1$, and so on.

a. $(f \circ g)(-4)$ 2
b. $(f \circ g)(1)$ 2
c. $(f \circ g)(4)$ 2   **d.** $(f \circ g)(0)$ -1
e. $(f \circ g)(-2)$ -1 **f.** $(g \circ f)(2)$ 1
g. $(f \circ g)(6)$ 3   **h.** $(g \circ f)(4)$ ≈ 0.5

36. Reading a graph: Use the given graph to find the result of the operations indicated.

Exercise 36

Note $p(-1) = -2$, $q(5) = 6$, and so on.

a. $(p \circ q)(-4)$ 4
b. $(p \circ q)(1)$ -1
c. $(p \circ q)(4)$ 1   **d.** $(p \circ q)(0)$ 4
e. $(p \circ q)(-2)$ 5   **f.** $(q \circ p)(2)$ 6
g. $(q \circ q)(-1)$ -3 **h.** $(p \circ p)(7)$ 2

37. Given $f(x) = 3\sqrt{x + 1}$, $g(x) = \dfrac{-3}{x + 2}$, (a) state the domain of f and g, then (b) use a graphing calculator to study the graph of $h(x) = (f \circ g)(x)$. Finally, (c) algebraically determine the domain of h, and reconcile it with the graph.
a. $f(x): x \geq -1, g(x): x \neq 2$ **c.** $h(x): x \in (-\infty, -2) \cup [1, \infty)$

38. Given $f(x) = \sqrt{x}$, $g(x) = \dfrac{1}{x^2 - 2x - 3}$, (a) state the domain of f and g, then (b) use a graphing calculator to study the graph of $h(x) = (f \circ g)(x)$. Finally, (c) algebraically determine the domain of h, and reconcile it with the graph.
a. $f(x): x \geq 0, g(x): x \neq -1, 3$ **c.** $h(x): x \in (-\infty, -1) \cup (3, \infty)$

Compute and simplify the difference quotient $\dfrac{f(x + h) - f(x)}{h}$ for each function given.

39. $f(x) = 2x - 3$ 2   **40.** $g(x) = 4x + 1$ 4
41. $j(x) = x^2 + 3$ $2x + h$ **42.** $p(x) = x^2 - 2$ $2x + h$
43. $q(x) = x^2 + 2x - 3$ $2x + 2 + h$ **44.** $r(x) = x^2 - 5x + 2$ $2x - 5 + h$
45. $f(x) = \dfrac{2}{x}$ $\frac{-2}{x(x + h)}$ **46.** $g(x) = \dfrac{-3}{x}$ $\frac{3}{x(x + h)}$

Use the difference quotient to find: **(a)** a rate of change formula for the functions given and **(b)/(c)** calculate the rate of change in the intervals shown. Then **(d)** sketch the graph of each function along with the secant lines and comment on what you notice.

47. $g(x) = x^2 + 2x$ **48.** $j(x) = x^2 - 6x$
$[-3.0, -2.9]$, $[0.50, 0.51]$ $[1.9, 2.0]$, $[5.0, 5.01]$

49. $g(x) = x^3 + 1$
$[-2.1, -2]$, $[0.40, 0.41]$

50. $v(x) = \sqrt{x}$ (*Hint*: Rationalize the numerator.)
$[1, 1.1]$, $[4, 4.1]$

Use the difference quotient to find a rate of change formula for the functions given, then calculate the rate of change for the intervals indicated. Comment on how the rate of change in each interval corresponds to the graph of the function.

51. $j(x) = \dfrac{1}{x^2}$ **52.** $f(x) = x^2 - 4x$
$[0.50, 0.51]$, $[1.50, 1.51]$ $[0.00, 0.01]$, $[3.00, 3.01]$

53. $g(x) = x^3 + 1$ **54.** $r(x) = \sqrt{x}$
$[-2.01, -2.00]$, $[0.40, 0.41]$ $[1.00, 1.01]$, $[4.00, 4.01]$

▶ **WORKING WITH FORMULAS**

55. Transformations via composition: For $f(x) = x^2 + 4x + 3$ and $g(x) = x - 2$, (a) show that $h(x) = (f \circ g)(x) = x^2 - 1$, then (b) verify the graph of h is the same as that of f, shifted 2 units to the right.

56. Compound annual growth: $A(r) = P(1 + r)^t$
The amount of money A in a savings account t yr after an initial investment of P dollars depends on the interest rate r. If \$1000 is invested for 5 yr, find $f(r)$ and $g(r)$ such that $A(r) = (f \circ g)(r)$.
$g(r) = 1 + r, f(r) = 1000r^5$; other answers possible

▶ **APPLICATIONS**

57. International shoe sizes: Peering inside her athletic shoes, Morgan notes the following shoe sizes: *US 8.5, UK 6, EUR 40*. The function that relates the U.S. sizes to the European (EUR) sizes is $g(x) = 2x + 23$, where x represents the U.S. size and $g(x)$ represents the EUR size. The function that relates European sizes to sizes in the United Kingdom (UK) is $f(x) = 0.5x - 14$, where x represents the EUR size and $f(x)$ represents the UK size. Find the function $h(x)$ that relates the U.S. measurement directly to the UK measurement by finding $h(x) = (f \circ g)(x)$. Find the UK size for a shoe that has a U.S. size of 13. $h(x) = x - 2.5; 10.5$

58. Currency conversion: On a trip to Europe, Megan had to convert American dollars to euros using the function $E(x) = 1.12x$, where x represents the number of dollars and $E(x)$ is the equivalent number of euros. Later, she converts her euros to Japanese yen using the function $Y(x) = 1061x$, where x represents the number of euros and $Y(x)$ represents the equivalent number of yen.
(a) Convert 100 U.S. dollars to euros. (b) Convert the answer from part (a) into Japanese yen.
(c) Express yen as a function of dollars by finding $M(x) = (Y \circ E)(x)$, then use $M(x)$ to convert $100 directly to yen. Do parts (b) and (c) agree?
Source: 2005 *World Almanac*, p. 231
a. 112 euros   **b.** 118,832 yen   **c.** $M(x) = 1188.32x$; yes

59. Currency conversion: While traveling in the Far East, Timi must convert U.S. dollars to Thai baht using the function $T(x) = 41.6x$, where x represents the number of dollars and $T(x)$ is the equivalent number of baht. Later she needs to convert her baht to Malaysian ringgit using the function $R(x) = 10.9x$. (a) Convert $100 to baht. (b) Convert the result from part (a) to ringgit. (c) Express ringgit as a function of dollars using $M(x) = (R \circ T)(x)$, then use $M(x)$ to convert $100 to ringgit directly. Do parts (b) and (c) agree?
Source: 2005 *World Almanac*, p. 231
a. 4160   **b.** 45,344   **c.** $M(x) = 453.44x$; yes

60. Spread of a fire: Due to a lightning strike, a forest fire begins to burn and is spreading outward in a shape that is roughly circular. The radius of the circle is modeled by the function $r(t) = 2t$, where t is the time in minutes and r is measured in meters. (a) Write a function for the area burned by the fire directly as a function of t by computing $(A \circ r)(t)$. (b) Find the area of the circular burn after 60 min.
a. $A(t) = 4\pi t^2$ **b.** $14,400\pi \, \text{m}^2$

61. Radius of a ripple: As Mark drops firecrackers into a lake one 4th of July, each "pop" caused a circular ripple that expanded with time. The radius of the circle is a function of time t. Suppose the function is $r(t) = 3t$, where t is in seconds and r is in feet. (a) Find the radius of the circle after 2 sec. (b) Find the area of the circle after 2 sec. (c) Express the area as a function of time by finding $A(t) = (A \circ r)(t)$ and use $A(t)$ to find the area of the circle after 2 sec. Do the answers agree?   **a.** 6 ft   **b.** $36\pi \, \text{ft}^2$ **c.** $A(t) = 9\pi t^2$; yes

62. Expanding supernova: The surface area of a star goes through an expansion phase prior to going *supernova*. As the star begins expanding, the radius becomes a function of time. Suppose this function is $r(t) = 1.05t$, where t is in days and $r(t)$ is in gigameters (Gm). (a) Find the radius of the star two days after the expansion phase begins. (b) Find the surface area after two days. (c) Express the surface area as a function of time by finding $h(t) = (S \circ r)(t)$, then use $h(t)$ to compute the surface area after two days directly. Do the answers agree?
a. 2.1 Gm   **b.** $17.64\pi \, \text{Gm}^2$ **c.** $h(t) = 4.41\pi t^2$; yes

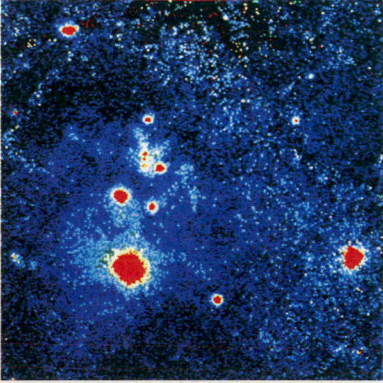

63. Composition and dependent relationships: In the wild, the balance of nature is often very fragile, with any sudden changes causing dramatic and unforeseen changes. With a huge increase in population and tourism near an African wildlife preserve, the number of lions is decreasing due to loss of habitat and a disruption in normal daily movements. This is causing a related increase in the hyena population, as the lion is one of the hyena's only natural predators. If this increase remains unchecked, animals lower in the food chain will suffer. If the lion population L depends on the increase in human population x according to the formula $L(x) = 500 - 0.015x$, and the hyena population depends on the lion population as modeled by the formula $H(L) = 650 - 0.5L$, (a) what is the current lion population ($x = 0$) and hyena population? (b) Use a composition to find a function modeling how the hyena population relates directly to the number of humans, and use the function to estimate the number of hyenas in the area if the human population grows by 16,000. (c) If the administrators of the preserve consider a population of 625 hyenas as "extremely detrimental," at what point should the human population be capped?

64. Composition and dependent relationships: The recent opening of a landfill in the area has caused the raccoon population to flourish, with an adverse effect on the number of purple martins. Wildlife specialists believe the population of martins p will decrease as the raccoon population r grows. Further, since mosquitoes are the primary diet of purple martins, the mosquito population m is likewise affected. If the first relationship is modeled by the function $p(r) = 750 - 3.75r$ and the second by $m(p) = 50,000 - 45p$, (a) what is the current number of purple martins ($r = 0$) and mosquitoes? (b) Use a composition to find a function modeling how the raccoon population relates directly to the number of mosquitoes, and use the function to estimate the number of mosquitoes in the area if the raccoon population

grows by 50. (c) If the health department considers 36,500 mosquitoes to be a "dangerous level," what increase in the raccoon population will bring this about?

65. Distance to the horizon: The distance that a person can see depends on how high they're standing above level ground. On a clear day, the distance is approximated by the function $d(h) = 1.2\sqrt{h}$, where $d(h)$ represents the viewing distance (in miles) at height h (in feet). Use the difference quotient to find the average rate of change in the intervals (a) [9, 9.01] and (b) [225, 225.01]. Then (c) graph the function along with the lines representing the average rates of change and comment on what you notice.

66. Projector lenses: A special magnifying lens is crafted and installed in an overhead projector. When the projector is x ft from the screen, the size $P(x)$ of the projected image is x^2. Use the difference quotient to find the average rate of change for $P(x) = x^2$ in the intervals (a) [1, 1.01] and (b) [4, 4.01]. Then (c) graph the function along with the lines representing the average rates of change and comment on what you notice.

67. Fortune and fame: Over the years there have been a large number of what we know as "one hit wonders," persons or groups that published a memorable or timeless song, but who were unable to repeat the feat. In some cases, their fame might be modeled by a quadratic function as their popularity rose to a maximum, then faded with time. Suppose the song *She's on Her Way* by Helyn Wheels rode to the top of the charts in January of 1988, with demand for the song modeled by $d(t) = -2t^2 + 27t$. Here, $d(t)$ represents the demand in 1000s for month t ($t = 1 \rightarrow$ Jan). (a) How many times faster was the demand growing in March (shortly after the release) than in June? Use the difference quotient and the intervals [3, 3.01] for March and [6, 6.01] for June. (b) Determine the month that demand reached its peak using a graphing calculator. (c) Was the demand increasing or decreasing in the month of August? At what rate?

68. Velocity and fuel economy: It has long been known that cars and trucks are more fuel efficient at certain speeds, which is why President Richard Nixon lowered the speed limit on all federal highways to 55 mph during the oil embargo of 1974. For heavier and less fuel-efficient vehicles, the miles per gallon for certain speeds can be modeled by the function $m(s) = -0.01s^2 + s$, where $m(s)$ represents the mileage (in miles per gallon) at speed s $(0 < s \leq 80)$. (a) Use the difference quotient to find how many times

faster fuel efficiency is growing near $s = 30$ mph than near $s = 45$ mph. Use the intervals [30, 30.1] and [45, 45.1]. (b) Use a graphing calculator to determine the speed(s) that maximizes fuel efficiency for this vehicle. How many miles per gallon are achieved? (c) Is fuel efficiency increasing or decreasing at 70 mph? At what rate?

▶ EXTENDING THE CONCEPT

69. Given $f(x) = x^3 + 2$ and $g(x) = \sqrt[3]{x - 2}$, graph each function on the same axes by plotting the points that correspond to integer inputs for $x \in [-3, 3]$. Do you notice anything? Next, find $h(x) = (f \circ g)(x)$ and $H(x) = (g \circ f)(x)$. What happened? Look closely at the functions f and g to see how they are related. Can you come up with two additional functions where the same thing occurs? Answers will vary.

70. Given $f(x) = \dfrac{1}{x^2 - 4}$, $g(x) = \sqrt{x + 1}$, and

$h(x) = (f \circ g)(x)$, (a) find the new function rule for h and (b) determine the implied domain of h. Does this *implied* domain include $x = 2$, $x = -2$, and $x = -3$ as valid inputs? (c) Determine the actual domain for $h(x) = (f \circ g)(x)$ and discuss the result.

71. Consider the functions $f(x) = \dfrac{k}{x}$ and $g(x) = \dfrac{k}{x^2}$. Both graphs appear similar in Quadrant I and both may "fit" a scatterplot fairly well, but there is a big difference between them—they decrease as x gets larger, but *they decrease at very different rates*. (a) Assume $k = 1$ and use the ideas from this section to compute the rates of change for f and g for the interval from $x = 0.5$ to $x = 0.51$. Were you surprised? (b) In the interval $x = 0.8$ to $x = 0.81$, will the rate of decrease for each function be greater or less than in the interval $x = 0.5$ to $x = 0.51$? Why?

▶ MAINTAINING YOUR SKILLS

72. (3.1) Find the sum and product of the complex numbers $2 + 3i$ and $2 - 3i$.   sum 4, product 13

73. (2.2) Draw a sketch of the functions *from memory*.
(a) $f(x) = \sqrt{x}$,   (b) $g(x) = \sqrt[3]{x}$, and
(c) $h(x) = |x|$

74. (3.2) Use the quadratic formula to solve $2x^2 - 3x + 4 = 0$. $\frac{3}{4} \pm \frac{\sqrt{23}}{4}i$

75. (1.4) Find an equation of the line perpendicular to $-2x + 3y = 9$, that also goes through the origin.
$y = -\frac{3}{2}x$

MAKING CONNECTIONS

Making Connections: Graphically, Symbolically, Numerically, and Verbally

Eight graphs (a) through (h) are given. Match the characteristics shown in 1 through 16 to one of the eight graphs.

(a) 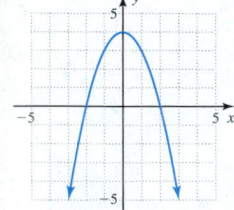   **(b)** 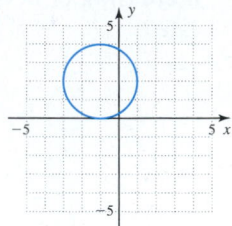   **(c)** 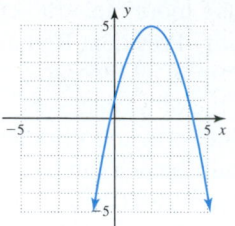   **(d)**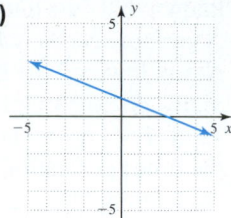

(e) 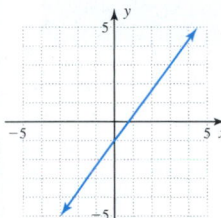   **(f)** 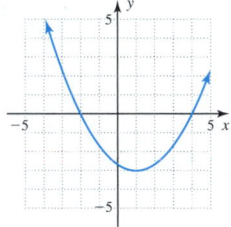   **(g)** 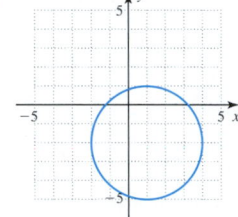   **(h)**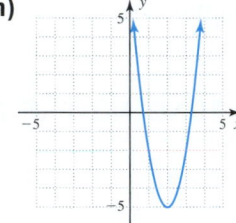

1. __b__ $(x + 1)^2 + (y - 2)^2 = 4$

2. __d__ $y = -\dfrac{2}{5}x + 1$

3. __h__ $f(x)\downarrow$ for $x \in (-\infty, 2)$, $f(x)\uparrow$ for $x \in (2, \infty)$

4. __c__ $f(x)\uparrow$ for $x \in (-\infty, 2)$, $f(x)\downarrow$ for $x \in (2, \infty)$

5. __d__ $m < 0, b > 0$

6. __e__ $m > 0, b < 0$

7. __a__ $y = -x^2 + 4$

8. __h__ $y = 3(x - 2)^2 - 5$

9. __g__ center $(1, -2)$, radius $= 3$

10. __c__ vertex $(2, 5)$, y-intercept $(0, 1)$

11. __f__ $y = \dfrac{1}{3}(x - 1)^2 - 3$

12. __g__ $(x - 1)^2 + (y + 2)^2 = 9$

13. __c__ axis of symmetry $x = 2$, opens downward

14. __e__ $4x - 3y = 3$

15. __a__ $f(x) < 0$ for $x \in (-\infty, -2) \cup (2, \infty)$, $f(x) \geq 0$ for $x \in [-2, 2]$

16. __f__ $f(-2) = 0, f(1) = -3$

SUMMARY AND CONCEPT REVIEW

SECTION 3.1   Complex Numbers

KEY CONCEPTS

- The italicized i represents the number whose square is -1. This means $i^2 = -1$ and $i = \sqrt{-1}$.
- Larger powers of i can be simplified using $i^4 = 1$.
- For $k > 0$, $\sqrt{-k} = i\sqrt{k}$ and we say the expression has been *written in terms of i*.
- The standard form of a *complex number* is $a + bi$, where a is the *real number part* and bi is the *imaginary part*.
- To add or subtract complex numbers, combine the like terms.
- For any complex number $a + bi$, its *complex conjugate* is $a - bi$.
- The *product* of a complex number and its conjugate is a real number.

- The commutative, associative, and distributive properties also apply to complex numbers and are used to perform basic operations.
- To multiply complex numbers, use the F-O-I-L method and simplify.
- To find a *quotient* of complex numbers, multiply the numerator and denominator by the conjugate of the denominator.

EXERCISES

Simplify each expression and write the result in standard form.

1. $\sqrt{-72}$ $6i\sqrt{2}$

2. $6\sqrt{-48}$ $24i\sqrt{3}$

3. $\dfrac{-10 + \sqrt{-50}}{5}$ $-2 + i\sqrt{2}$

4. $\sqrt{3}\sqrt{-6}$ $3i\sqrt{2}$

5. i^{57} i

Perform the operation indicated and write the result in standard form.

6. $(5 + 2i)^2$ $21 + 20i$

7. $\dfrac{5i}{1 - 2i}$ $-2 + i$

8. $(-3 + 5i) - (2 - 2i)$ $-5 + 7i$

9. $(2 + 3i)(2 - 3i)$ 13

10. $4i(-3 + 5i)$ $-20 - 12i$

Use substitution to show the given complex number and its conjugate are solutions to the equation shown.

11. $x^2 - 9 = -34; x = 5i$
$(5i)^2 - 9 = -34$ $(-5i)^2 - 9 = -34$
$25i^2 - 9 = -34$ $25i^2 - 9 = -34$
$-25 - 9 = -34 ✓$ $-25 - 9 = -34 ✓$

12. $x^2 - 4x + 9 = 0; x = 2 + i\sqrt{5}$
$(2 + i\sqrt{5})^2 - 4(2 + i\sqrt{5}) + 9 = 0$ $(2 - i\sqrt{5})^2 - 4(2 - i\sqrt{5}) + 9 = 0$
$4 + 4i\sqrt{5} + 5i^2 - 8 - 4i\sqrt{5} + 9 = 0$ $4 - 4i\sqrt{5} + 5i^2 - 8 + 4i\sqrt{5} + 9 = 0$
$5 + (-5) = 0 ✓$ $5 + (-5) = 0 ✓$

SECTION 3.2   Solving Quadratic Equations and Inequalities

KEY CONCEPTS

- The standard form of a quadratic equation is $ax^2 + bx + c = 0$, where a, b, and c are real numbers and $a \neq 0$. In words, we say the equation is written in decreasing order of degree and set equal to zero.
- A quadratic function is one that can be written as $f(x) = ax^2 + bx + c$, where a, b, and c are real numbers and $a \neq 0$.
- The following four statements are equivalent: (1) $x = r$ is a solution of $f(x) = 0$, (2) r is a zero of $f(x)$, (3) $(r, 0)$ is an x-intercept of $y = f(x)$, and (4) $(x - r)$ is a factor of $f(x)$.
- The square root property of equality states that if $X^2 = k$, where $k \geq 0$, then $X = \sqrt{k}$ or $X = -\sqrt{k}$.
- Quadratic equations can also be solved by *completing the square*, or using the *quadratic formula*.
- If the discriminant $b^2 - 4ac = 0$, the equation has one real (repeated) root. If $b^2 - 4ac > 0$, the equation has two real roots; and if $b^2 - 4ac < 0$, the equation has two nonreal roots.
- Quadratic inequalities can be solved using the zeroes of the function and either an understanding of quadratic graphs or mid-interval test values.

EXERCISES

13. Solve by factoring.

   a. $x^2 - 3x - 10 = 0$ **b.** $2x^2 - 50 = 0$ **c.** $3x^2 - 15 = 4x$ **d.** $x^3 - 3x^2 = 4x - 12$

14. Solve using the square root property of equality.

   a. $x^2 - 9 = 0$ **b.** $2(x - 2)^2 + 1 = 11$ **c.** $3x^2 + 15 = 0$ **d.** $-2x^2 + 4 = -46$

15. Solve by completing the square. Give real number solutions in exact and approximate form.

   a. $x^2 + 2x = 15$ **b.** $x^2 + 6x = 16$ **c.** $-4x + 2x^2 = 3$ **d.** $3x^2 - 7x = -2$

16. Solve using the quadratic formula. Give solutions in both exact and approximate form.

   a. $x^2 - 4x = -9$ **b.** $4x^2 + 7 = 12x$ **c.** $2x^2 - 6x + 5 = 0$

17. Solve by locating the x-intercepts and noting the end-behavior of the graph.

   a. $x^2 - x - 6 > 0$ **b.** $-x^2 + 1 \geq 0$ **c.** $x^2 - 2x + 2 > 0$

                                      a. $(-\infty, -2) \cup (3, \infty)$ **b.** $[-1, 1]$ **c.** $(-\infty, \infty)$

18. Solve using the interval test method.

   a. $x^2 + 3x \leq 4$ **b.** $x^2 > 20 - x$ **c.** $x^2 + 4x + 4 \leq 0$

                                        a. $[-4, 1]$ **b.** $(-\infty, -5) \cup (4, \infty)$ **c.** $\{-2\}$

Solve the following quadratic applications. For 19 and 20, recall the height of a projectile is modeled by $h = -16t^2 + v_0t + k$.

19. A projectile is fired upward from ground level with an initial velocity of 96 ft/sec. (a) To the nearest tenth of a second, how long until the object first reaches a height of 100 ft? (b) How long until the object is again at 100 ft? (c) How many seconds until it returns to the ground?   **a.** 1.3 sec   **b.** 4.7 sec   **c.** 6 sec

20. A person throws a rock upward from the top of an 80-ft cliff with an initial velocity of 64 ft/sec. (a) To the nearest tenth of a second, how long until the object is 120 ft high? (b) How long until the object is again at 120 ft? (c) How many seconds until the object hits the ground at the base of the cliff?   **a.** 0.8 sec   **b.** 3.2 sec   **c.** 5 sec

SECTION 3.3   Quadratic Functions and Applications

KEY CONCEPTS

- The graph of a quadratic function is a parabola. Parabolas have three distinctive features: (1) like end-behavior on the left and right, (2) an axis of symmetry, (3) a highest or lowest point called the vertex.
- By completing the square, $f(x) = ax^2 + bx + c$ can be written as the transformation $f(x) = a(x + h)^2 \pm k$, and graphed using transformations of $y = x^2$.
- For a quadratic function in the standard form $y = ax^2 + bx + c$,
  - End-behavior: graph opens upward if $a > 0$, opens downward if $a < 0$
  - Zeroes/x-intercepts (if they exist): substitute 0 for y and solve for x
  - y-intercept: substitute 0 for $x \rightarrow (0, c)$
  - Vertex: (h, k), where $h = \dfrac{-b}{2a}$, $k = f\left(\dfrac{-b}{2a}\right)$
  - Maximum value: If the parabola opens downward, $y = k$ is the maximum value of f.
  - Minimum value: If the parabola opens upward, $y = k$ is the minimum value of f.
  - Line of symmetry: $x = h$ is the line of symmetry. If $(h + c, y)$ is on the graph, then $(h - c, y)$ is also on the graph.

EXERCISES

Graph $p(x)$ and $f(x)$ by completing the square and using transformations of the parent function. Graph $g(x)$ and $h(x)$ using the vertex formula and y-intercept. Find the x-intercepts (if they exist) for all functions.

21. $p(x) = x^2 - 6x$ **22.** $f(x) = x^2 + 8x + 15$ **23.** $g(x) = -x^2 + 4x - 5$ **24.** $h(x) = 4x^2 - 12x + 3$

25. Height of a superball: A teenager tries to see how high she can bounce her superball by throwing it downward on her driveway. The height of the ball (in feet) at time t (in seconds) is given by $h(t) = -16t^2 + 96t$. (a) How high is the ball at $t = 0$? (b) How high is the ball after 1.5 sec? (c) How long until the ball is 135 ft high? (d) What is the maximum height attained by the ball? At what time t did this occur?   **a.** 0 ft   **b.** 108 ft   **c.** 2.25 sec   **d.** 144 ft, $t = 3$ sec

26. Theater Revenue: The manager of a large, 14-screen movie theater finds that if he charges $2.50 per person for the matinee, the average daily attendance is 4000 people. With every increase of 25 cents the attendance drops an average of 200 people. (a) What admission price will bring in a revenue of $11,250? (b) How many people will purchase tickets at this price? $3.75; 3000

SECTION 3.4   Quadratic Models; More on Rates of Change

KEY CONCEPTS

- Regardless of the form of regression chosen, obtaining a regression equation uses these five steps: (1) clear out old data, (2) enter new data, (3) set an appropriate window and display the data, (4) calculate the regression equation, and (5) display the data and equation, and once satisfied the model is appropriate, apply the result.
- The choice of a nonlinear regression model often depends on many factors, particularly the context of the data, any patterns formed by the scatterplot, some foreknowledge on how the data might be related, and/or a careful assessment of the correlation coefficient.
- Applications of quadratic regression are generally applied when a set of data indicates a gradual decrease to some minimum value, with a matching increase afterward, or a gradual increase to some maximum, with a matching decrease afterward.

- For nonlinear functions, the *average rate of change* gives an average value for how changes in the independent variable cause a change in the dependent variable within a specified interval.
- The average rate of change is given by the slope of a secant line through two points (x_1, y_1) and (x_2, y_2) on the graph, and is computed as: $\dfrac{\Delta y}{\Delta x} = \dfrac{f(x_2) - f(x_1)}{x_2 - x_1}, \; x_2 \neq x_1$.

EXERCISES

 27. While the Internet has been with us for over 20 years, its use continues to grow at a rapid pace. The data in the table gives the amount of money (in billions of dollars) consumers spent online for retail items in selected years (amounts for 2010 through 2012 are projections). Use the data and a graphing calculator to (a) draw a scatterplot and decide on an appropriate form of regression, then (b) find the regression equation and use it to estimate the amount spent by consumers in 2003, (c) the projected amount that will be spent in 2014 if this rate of growth continues, and (d) the year that $591 billion is the projected amount of retail spending over the Internet.

Year (2000 → 0)	Amount (billions)
1	31
5	84
6	108
7	128
10	267
11	301
12	335

28. The drag force on a compact car driving along the highway on a windless day, depends on a constant $\frac{k}{2}$ and the velocity of the car, where k is determined using the density of the air, the cross-sectional area of the car, and the drag coefficient of the vehicle. The data shown in the table gives the magnitude of the drag force F_d, at given velocity v. Use the data and a graphing calculator to (a) draw a scatterplot and decide on an appropriate form of regression, then (b) find a regression equation and use it to estimate the magnitude of the drag force for this car at 60 mph. Finally, (c) estimate the speed of the car if the drag force has a magnitude of 2329 units.

Velocity (mph)	F_d
10	32
30	306
50	860
70	1694

29. The graph and accompanying table show the number N of active Starbucks outlets for selected years t from 1990 to 2008. Use the graph and table to (a) find the average rate of change for the years 1994 to 1996 (the interval [4, 6]). (b) Verify that the rate of growth between the years 2000 and 2002 (the interval [10, 12]) was about 4 times greater than from 1994 to 1996. (c) Show that the average rate of change for the years 2002 to 2004 was very close to the rate of change for the years 2006 to 2008.

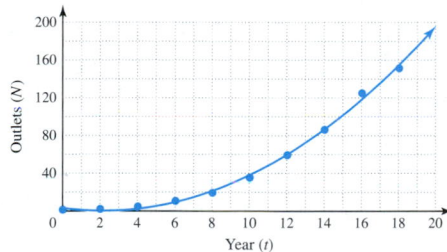

Year t (1990 → 0)	Outlets N (100s)	Year t (1990 → 0)	Outlets N (100s)
0	0.84	10	35.01
2	1.65	12	58.86
4	4.25	14	85.69
6	10.15	16	124.40
8	18.86	18	150.79

30. According to Torricelli's law for tank draining, the volume (in ft³) of a full 5 ft × 2 ft × 2 ft bathtub t sec after the plug is pulled can be modeled by the function $V(t) = (-0.2t + \sqrt{20})^2$. (a) What is the volume of the bathtub at $t = 0$ sec? (b) What is the volume of the bathtub at $t = 1$ sec? (c) What is the average rate of change from $t = 0$ to $t = 1$? (d) What is the average rate of change from $t = 20$ to $t = 21$? (e) When is the bathtub empty?

SECTION 3.5   The Algebra of Functions

KEY CONCEPTS

- The notation used to represent the basic operations on two functions is
  - $(f + g)(x) = f(x) + g(x)$
  - $(f - g)(x) = f(x) - g(x)$
  - $(f \cdot g)(x) = f(x) \cdot g(x)$
  - $\left(\dfrac{f}{g}\right)(x) = \dfrac{f(x)}{g(x)}; \; g(x) \neq 0$

- The result of these operations is a new function $h(x)$. The domain of h is the intersection of domains for f and g, excluding values that make $g(x) = 0$ for $h(x) = \left(\dfrac{f}{g}\right)(x)$.

EXERCISES

For $f(x) = x^2 + 4x$ and $g(x) = 3x - 2$, find the following:

31. $(f + g)(a)$ $a^2 + 7a - 2$

32. $(f \cdot g)(3)$ 147

33. the domain of $\left(\dfrac{f}{g}\right)(x)$
$x \in (-\infty, \frac{2}{3}) \cup (\frac{2}{3}, \infty)$

34. Use the graph given to find the value of each expression:

a. $(f + g)(-2)$ 4

b. $(g - f)(7)$ 6

c. $\left(\dfrac{g}{f}\right)(10)$ $\frac{-1}{5}$

d. $(f \cdot g)(3)$ 14

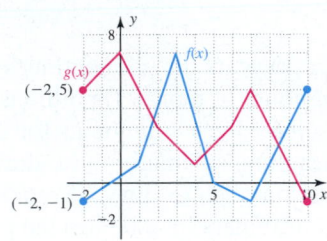

35. As the availability of free, public Wi-Fi has increased, so has the number of devices that can utilize this protocol. A new company has just released a Wi-Fi phone that provides free phone service anywhere it has access to a wireless network. The total cost for manufacturing these phones can be modeled by the function $C(n) = -0.002n^2 + 20n + 30{,}000$, where n is the number of phones made and $C(n)$ is in dollars. If each phone is sold at a price of \$84.95, the revenue is modeled by $R(n) = 84.95n$.

a. Find the function that represents the total profit made from sales of the phones.

b. How much profit is earned if 400 phones are sold? $P(x) = 84.95n - (-0.002n^2 + 20n + 30{,}000) = 0.002n^2 + 64.95n - 30{,}000$
$-\$3700$

c. How much profit is earned if 5000 phones (the production limit) are sold?   \$344,750

d. How many phone sales are necessary for the company to break even?   456

SECTION 3.6   The Composition of Functions and the Difference Quotient

KEY CONCEPTS

- The composition of two functions is written $(f \circ g)(x) = f[g(x)]$ (g is an input for f).
- The domain of $f \circ g$ is all x in the domain of g, such that $g(x)$ is in the domain of f.
- To evaluate $(f \circ g)(2)$, we find $(f \circ g)(x)$ then substitute $x = 2$. Alternatively, we can find $g(2) = k$, then find $f(k)$.
- A composite function $h(x) = (f \circ g)(x)$ can be "decomposed" into individual functions by identifying functions f and g such that $(f \circ g)(x) = h(x)$. The decomposition is not unique.
- The difference quotient for a function $f(x)$ is $\dfrac{f(x + h) - f(x)}{h}$.

EXERCISES

Given $p(x) = 4x - 3$, $q(x) = x^2 + 2x$, and $r(x) = \dfrac{x + 3}{4}$ find:

36. $(p \circ q)(x)$ $4x^2 + 8x - 3$

37. $(q \circ p)(3)$ 99

38. $(p \circ r)(x)$ and $(r \circ p)(x)$ $x; x$

For each function here, find functions $f(x)$ and $g(x)$ such that $h(x) = f[g(x)]$:

39. $h(x) = \sqrt{3x - 2} + 1$ $f(x) = \sqrt{x} + 1;\ g(x) = 3x - 2$

40. $h(x) = x^{\frac{2}{3}} - 3x^{\frac{1}{3}} - 10$ $f(x) = x^2 - 3x - 10;\ g(x) = x^{\frac{1}{3}}$

41. A stone is thrown into a pond causing a circular ripple to move outward from the point of entry. The radius of the circle is modeled by $r(t) = 2t + 3$, where t is the time in seconds. Find a function that will give the area of the circle directly as a function of time. In other words, find $A(t)$. $A(t) = \pi(2t + 3)^2$

42. Use the graph given to find the value of each expression:

    a. $(f \circ g)(-2)$　0

    b. $(g \circ f)(5)$　7

    c. $(g \circ f)(7)$　6

    d. $(g \circ f)(10)$　2

    e. $(f \circ g)(3)$　4

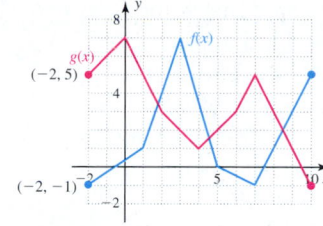

43. Use the difference quotient to find a rate of change formula for the function given, then calculate the rate of change for the interval indicated: $j(x) = x^2 - x$; $[2.00, 2.01]$.　$2x - 1 + h$; 3.01

PRACTICE TEST

1. Solve each equation or inequality.

    a. $11x - 2x^2 > 0$　$(0, \frac{11}{2})$

    b. $x^2 + 1 > 2x$　$(-\infty, 1) \cup (1, \infty)$

    c. $3x^2 \geq 2x + 21$　$(-\infty, -\frac{7}{3}] \cup [3, \infty)$

    d. $-x^2 - 2.2x \geq 3.8$　{ }

2. $x^2 + 25 = 0$　$x = \pm 5i$

3. $(x - 1)^2 + 3 = 0$　$x = 1 \pm i\sqrt{3}$

4. $3x^2 - 20x = -12$　$x = \frac{2}{3}, x = 6$

5. Due to the seasonal nature of the business, the revenue of Wet Willey's Water World can be modeled by the equation $r = -3t^2 + 42t - 135$, where t is the time in months ($t = 1$ corresponds to January) and r is the dollar revenue in thousands. (a) What month does Wet Willey's open? (b) What month does Wet Willey's close? (c) Does Wet Willey's bring in more revenue in July or August? How much more?　**a.** $t = 5$ (May)　**b.** $t = 9$ (Sept.)　**c.** July; $3000 more

6. Simplify each expression.

    a. $\dfrac{-8 + \sqrt{-20}}{6}$　$-\frac{4}{3} + \frac{\sqrt{5}}{3}i$　　**b.** i^{39}　$-i$

7. Given $x = \dfrac{1}{2} + \dfrac{\sqrt{3}}{2}i$ and $y = \dfrac{1}{2} - \dfrac{\sqrt{3}}{2}i$ find

    a. $x + y$　1　　**b.** $x - y$　$i\sqrt{3}$　　**c.** xy　1

8. Compute the quotient: $\dfrac{3i}{1 - i}$.　$-\frac{3}{2} + \frac{3}{2}i$

9. Find the product: $(3i + 5)(5 - 3i)$.　34

10. Show $x = 2 - 3i$ is a solution of $x^2 - 4x + 13 = 0$.

11. Solve by completing the square.

    a. $2x^2 - 20x + 49 = 0$　$x = 5 \pm \frac{\sqrt{2}}{2}$

    b. $2x^2 - 5x = -4$　$x = \frac{5}{4} \pm \frac{\sqrt{7}}{4}i$

12. Solve using the quadratic formula.

    a. $3x^2 + 2 = 6x$　$x = \frac{3 \pm \sqrt{3}}{3}$

    b. $x^2 = 2x - 10$　$x = 1 \pm 3i$

13. Complete the square to write each function as a transformation. Then graph each function and label the vertex and x-intercepts (if they exist).

    a. $f(x) = -x^2 + 10x - 16$

    b. $g(x) = \dfrac{1}{2}x^2 + 4x + 16$

14. The graph of a quadratic function has a vertex of $(-1, -2)$, and passes through the origin. Find the other intercept, and the equation of the graph in standard form.　$(-2, 0), y = 2x^2 + 4x$

15. Suppose the function $d(t) = t^2 - 14t$ models the depth of a scuba diver at time t, as she dives underwater from a steep shoreline, reaches a certain depth, and swims back to the surface.

    a. What is her depth after 4 sec? After 6 sec?　40 ft, 48 ft

    b. What was the maximum depth of the dive?　49 ft

    c. How many seconds was the diver beneath the surface?　14 sec

16. Homeschool education: Since the early 1980s the number of parents electing to homeschool their children has been steadily increasing. Estimates for the number of children homeschooled (in 1000s) are given in the table for selected years. (a) Use a graphing calculator to draw a scatterplot of the data and decide on an appropriate form of regression. (b) Calculate a regression equation with $x = 0$ corresponding to 1985 and display the scatterplot and graph on the same screen. (c) According to the equation model, how many children were homeschooled in 1991? If growth continues at the same rate, how many children will be homeschooled in 2010?

Year (1985 → 0)	Children (1000s)
0	183
3	225
5	301
7	470
8	588
9	735
10	800
11	920
12	1100

Source: National Home Education Research Institute

17. The graph and accompanying table show the number N of *new* books published in the United States for selected years t from 1990 to 2004. Find the average rate of change for the years (a) 1992 to 1996 and (b) 1996 to 1999. (c) In 1997, the number of new books published actually fell to 65.8 (1000s). Using this information, find the average rate of change for the years 1996 to 1997, and 1997 to 1999.

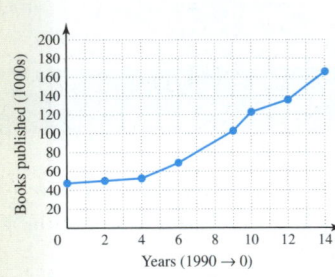

Year (1990 → 0)	Books (1000s)
0	46.7
2	49.2
4	51.7
6	68.2
9	102.0
10	122.1
12	135.1
14	165.8

18. Given $f(x) = x^2 + 2$ and $g(x) = \sqrt{3x - 1}$, determine $(f \circ g)(x)$ and its domain. $3x + 1; x \in [\frac{1}{3}, \infty)$

19. Monthly sales volume for a successful new company is modeled by $S(t) = 2t^2 - 3t$, where $S(t)$ represents sales volume in thousands in month t ($t = 0$ corresponds to January 1). (a) Would you expect the average rate of change from May to June to be greater than that from June to July? Why? (b) Calculate the rates of change in these intervals to verify your answer. (c) Calculate the difference quotient for $S(t)$ and use it to estimate the sales volume rate of change after 10, 18, and 24 months.

20. A snowball increases in size as it rolls downhill. The snowball is roughly spherical with a radius that can be modeled by the function $r(t) = \sqrt{t}$, where t is time in seconds and r is measured in inches. The volume of the snowball is given by the function $V(r) = \frac{4}{3}\pi r^3$. Use a composition to (a) write V directly as a function of t and (b) find the volume of the snowball after 9 sec.

 ## CALCULATOR EXPLORATION AND DISCOVERY

Residuals, Correlation Coefficients, and Goodness of Fit

When using technology to calculate a regression equation, we must avoid relying on the correlation coefficient as the sole indicator of how well a model fits the data. It is actually possible for a regression to have a high r-value (correlation coefficient) but fit the data very poorly. In addition, regression models are often used to predict future values, extrapolating well beyond the given set of data. Sometimes the model fails miserably when extended—even when it fits the data on the specified interval very well. This fact highlights (1) the importance of studying the behavior of the toolbox functions and other graphs, as we often need to choose between two models that seem to fit the given data; (2) the need to consider the context of the data; and (3) the need for an additional means to evaluate the "goodness of fit." For the third item, we investigate something called a **residual.** As the name implies, we are interested in the difference between the outputs generated by the equation model, and the actual data: equation value − data value = residual. Residuals that are fairly random and scattered indicate the equation model has done a good job of capturing the curvature of the data. If the residuals exhibit a detectable pattern of some sort, this often indicates trends in the data that the equation model did not account for.

Table 3.5

x	y
5	19
7.5	75
10	140
12.5	215
15	297
17.5	387
20	490

As a simplistic illustration, enter the data from Table 3.5 in L1 and L2, then graph the scatterplot. Most graphing calculators offer a window option specifically designed for scatterplots that will plot the points in an "ideal" window: **9:ZoomStat** (Figure 3.82). Upon inspection, it appears the data could be linear with positive slope, quadratic with $a > 0$, or some other toolbox function with increasing behavior. Many of these forms of regression will *give a correlation coefficient in the high 90s.* This is where an awareness of the context is important. (1) Do we expect the data to increase steadily over time (linear data)? (2) Do we expect the rate of change (the growth rate) to increase over time (quadratic data)? At what rate will values increase? (3) Do we expect the growth to increase dramatically with time (power or exponential regression)?

Figure 3.82

Running a linear regression (**LinReg L1, L2, Y₁**) gives the equation in Figure 3.83, with a very high r-value. This model appears to fit the data very well, and will reasonably approximate the data points within the interval. But could we use this model to accurately predict future values (do we expect the outputs to grow indefinitely at a linear rate)?

In some cases, the answer will be clear from the context. Other times the decision is more difficult and a study of the residuals can help. Most graphing calculators provide a residual function [2nd STAT (LIST)(▷)(OPS) 7: ΔList], but to help you understand more exactly what residuals are, for now we'll calculate them via function values. For this calculation, we go to the STAT ENTER (EDIT) screen. In the header of L3 (List3), input $Y_1(L1) - L2$ ENTER (Figure 3.84), which will evaluate the function at the input values listed in L1, compute the difference between these outputs and the data in L2, and place the results in L3. Scrolling through the residuals reveals a distinct lack of randomness, as there is a large interval where outputs are continuously positive (Figure 3.85).

Figure 3.83

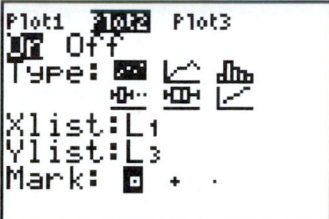

Figure 3.84

Figure 3.85

Figure 3.86

We can also analyze the residuals *graphically* by going to the 2nd Y= (STATPLOT) screen to activate **2:PLOT2**, setting it up to recognize L1 and L3 as the **XList** and **YList** respectively (Figure 3.86). After deactivating all other plots and functions, pressing ZOOM 9:**ZoomStat** gives Figure 3.87 shown, with the residuals following a definite (quadratic) pattern. Performing the same sequence of steps using quadratic regression results in a higher correlation coefficient and an increased randomness in residuals (Figures 3.88 and 3.89), with the residuals appearing to increase over time.

Figure 3.87

Figure 3.88

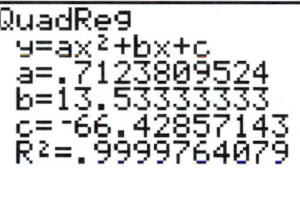

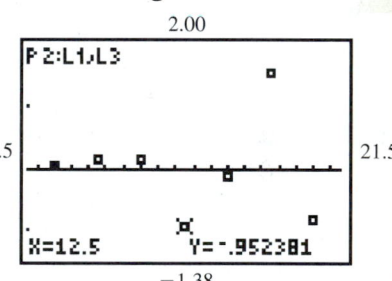

Figure 3.89

Exercise 1: As part of a science lab, students are asked to determine the relationship between the length of a pendulum and the time it takes to complete one back-and-forth cycle, called its *period*. They tie a 500-g weight to 10 different lengths of string, suspend them from a doorway, and collect the data shown in Table 3.6.

a. Use a combination of the context, the correlation coefficient, and an analysis of the residuals to determine whether a linear, quadratic, or power model is most appropriate for the data. State the *r*-value of each regression and justify your final choice of equation model.

b. According to the data, what would be the period of a pendulum with a 90-cm length? A 150-cm length?

c. If the period was 1.6 sec, how long was the pendulum? If the period were 2 sec, how long was the pendulum?

Table 3.6

Length (cm)	Time (sec)
12	0.7
20	0.9
28	1.09
36	1.24
44	1.35
52	1.55
60	1.64
68	1.73
76	1.80
84	1.95

STRENGTHENING CORE SKILLS

Base Functions and Quadratic Graphs

Certain transformations of quadratic graphs offer an intriguing alternative to graphing these functions by completing the square. In many cases, the process is less time consuming and ties together a number of basic concepts. To begin, we note that for $f(x) = ax^2 + bx + c$, $F(x) = ax^2 + bx$ is called the **base function** or *the original function less the constant term*. By comparing $f(x)$ with $F(x)$, four things are immediately apparent: (1) F and f share the same axis of symmetry since one is a vertical shift of the other; (2) the x-intercepts of F can be found by factoring; (3) the axis of symmetry is simply the average value of the x-intercepts;

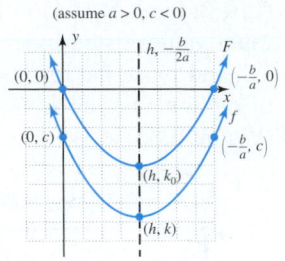

(assume $a > 0$, $c < 0$)

$h = \dfrac{x_1 + x_2}{2}$; and (4) the vertices of F and f differ only by the constant c. Consider these

vertices to be (h, k_0) and (h, k), respectively, with $k = k_0 + c$. Knowing the vertex of *any* parabola is $\left(\dfrac{-b}{2a}, f\left(\dfrac{-b}{2b}\right)\right)$, we

evaluate the base function at $h = \dfrac{x_1 + x_2}{2} = \dfrac{-b}{2a}$ and find that for the base function, $F\left(\dfrac{-b}{2a}\right) = -ah^2$:

$$F(x) = ax^2 + bx \qquad \text{original function}$$

$$F\left(-\frac{b}{2a}\right) = a\left(-\frac{b}{2a}\right)^2 + b\left(-\frac{b}{2a}\right) \qquad \text{substitute } -\frac{b}{2a} \text{ for } x$$

$$= \frac{b^2}{4a} - \frac{b^2}{2a} = \frac{b^2 - 2b^2}{4a} \text{ or } \frac{-b^2}{4a} \qquad \text{multiply and combine terms}$$

$$= \frac{-b^2}{4a} \cdot \frac{a}{a} \qquad \text{multiply by } \frac{a}{a}$$

$$= -a\left(\frac{b}{2a}\right)^2 \qquad \text{rearrange factors}$$

From $h = -\dfrac{b}{2a}$, we have $-h = \dfrac{b}{2a}$ and it follows that

$$F\left(-\frac{b}{2a}\right) = -a(-h)^2 \qquad \text{substitute } -h \text{ for } \frac{b}{2a}$$

$$= -ah^2 \qquad (-h)^2 = h^2$$

This verifies the vertex of F is (h, k_0), where $k_0 = -ah^2$.

It's significant to note that the vertex of both $F(x)$ and $f(x)$ can now be determined *using only elementary operations on the single value h*, since $k_0 = -ah^2$ and $k = k_0 + c$. By setting $y = 0$ in the quadratic equation $y = a(x - h)^2 + k$ and solving for x_1 we get the vertex/intercept formula, which can be used to find the roots of f with no further calculations:

$x = h \pm \sqrt{-\dfrac{k}{a}}$. Finally, this approach enables easy access to the exact form of the roots, even when they happen to be irrational or complex (no quadratic formula needed). Several examples follow, with the actual graphs left to the student—only the process is illustrated here.

Illustration 1 ▶ Graph $f(x) = x^2 - 10x + 17$ and locate its zeroes (if they exist).

Solution ▶ For $F(x) = x^2 - 10x$, the zeroes/x-intercepts are $(0, 0)$ and $(10, 0)$ by inspection, with $h = 5$ (halfway point) as the axis of symmetry. Noting $a = 1$ and $c = 17$, the vertex of F is at $(h, -ah^2)$ or $(5, -25)$. After adding 17 units to the y-coordinates of the points from F, we find the y-intercept for f is $(0, 17)$, its "symmetric point" is $(10, 17)$, and the vertex is at $(5, -8)$. The x-intercepts of f are $(h \pm \sqrt{-k}, 0)$ or $(5 \pm \sqrt{8}, 0)$.✓

Illustration 2 ▶ Graph $f(x) = x^2 + 7x - 15$ and locate its zeroes (if they exist).

Solution ▶ For $F(x) = x^2 + 7x$, the zeroes are $(0, 0)$ and $(-7, 0)$ by inspection, with $h = \frac{-7}{2}$ as the axis of symmetry. Noting $a = 1$, $c = -15$, and $\left(\frac{-7}{2}\right)^2 = \frac{49}{4}$ or 12.25, the vertex of F is $(-3.5, -12.25)$. After subtracting 15 units from the y-coordinates of the points from F, we find the y-intercept for f is $(0, -15)$, its "symmetric point" is $(-7, -15)$, and the vertex is at $(-3.5, -27.25)$. The x-intercepts of f are $(-3.5 \pm \sqrt{27.25}, 0)$.✓

Even when $a \neq 1$ the method lends a measure of efficiency to graphing quadratic functions, as shown in Illustration 3.

Illustration 3 ▶ Graph $f(x) = -2x^2 + 5x - 4$ and locate its zeroes (if they exist).

Solution ▶ For $F(x) = -2x^2 + 5x$, the zeroes are $(0, 0)$ and $(\frac{5}{2}, 0)$ by inspection, with $h = \frac{5}{4}$ as the halfway point and axis of symmetry. Noting $a = -2$ and $c = -4$, the vertex of F is at $(\frac{5}{4}, \frac{25}{8})$. After subtracting $4 = \frac{32}{8}$ units from the y-coordinates of the points from F, we find the y-intercept for f is $(0, -4)$, its "symmetric point" is $(\frac{5}{2}, -4)$, and the vertex is at $(\frac{5}{4}, \frac{-7}{8})$. The roots of f are $x = \frac{5}{4} \pm \sqrt{\frac{-7}{16}} = \frac{5}{4} \pm \frac{\sqrt{7}}{4}i$✓, showing the graph has no x-intercepts.

Use this method for graphing quadratic functions to sketch a complete graph of the following functions. Find and clearly indicate the axis of symmetry, vertex, x-intercept(s), and the y-intercept along with its "symmetric point."

Exercise 1: $f(x) = x^2 + 2x - 7$

Exercise 2: $g(x) = x^2 + 5x + 9$

Exercise 3: $h(x) = x^2 - 6x + 11$

Exercise 4: $H(x) = -x^2 + 10x - 17$

Exercise 5: $p(x) = 2x^2 + 12x + 21$

Exercise 6: $q(x) = 2x^2 - 7x + 8$

CUMULATIVE REVIEW CHAPTERS R–3

1. Solve for R: $\dfrac{1}{R} = \dfrac{1}{R_1} + \dfrac{1}{R_2}$ $R = \frac{R_1 R_2}{R_1 + R_2}$

2. Solve for x: $\dfrac{2}{x + 1} + 1 = \dfrac{5}{x^2 - 1}$ $x = -4$ or $x = 2$

3. Factor the expressions:

   a. $x^3 - 1$ **b.** $x^3 - 3x^2 - 4x + 12$

4. Solve using the quadratic formula. Write answers in both exact and approximate form: $2x^2 + 4x + 1 = 0$.

5. Solve the following inequality: $x + 3 < 5$ or $5 - x < 4$.   all reals

6. Name the eight toolbox functions, give their equations, then draw a sketch of each.

7. Use substitution to verify that $x = 2 - 3i$ is a solution to $x^2 - 4x + 13 = 0$.   verified

8. Given $f(x) = 3x^2 - 6x$ and $g(x) = x - 2$ find: $(f \cdot g)(x)$, $(f \div g)(x)$, and $(g \circ f)(-2)$.

9. As part of a study on traffic conditions, the mayor of a small city tracks her driving time to work each day for six months and finds a linear and increasing relationship. On day 1, her drive time was 17 min. By day 61 the drive time had increased to 28 min. Find a linear function that models the drive time and use it to estimate the drive time on day 121, if the trend continues. Explain what the slope of the line means in this context.

10. Does the relation shown represent a function? If not, discuss/explain why not.

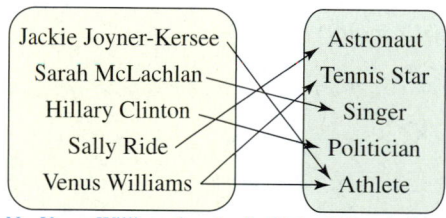

No, Venus Williams is paired with both Tennis Star and Athlete.

11. The data given shows the profit of a new company for the first 6 months of business, and is closely modeled by the function $p(m) = 1.18x^2 - 10.99x + 4.6$, where $p(m)$ is the profit earned in month m. Assuming this trend continues, use this function to find the first month a profit will be earned ($p > 0$).   Month 9

Exercise 11

Month	Profit (1000s)
1	−5
2	−13
3	−18
4	−20
5	−21
6	−19

Additional answers can be found in the Instructor Answer Appendix.

12. Graph the function $g(x) = \dfrac{-1}{(x + 2)^2} + 3$ using transformations of a basic function.

13. Given $f(x) = x^2$ and $g(x) = x^3$, use the formula for average rate of change to determine which of these functions is increasing faster in the intervals:

   a. [0.5, 0.6]   **b.** [1.5, 1.6].   **a.** $f(x)$ **b.** $g(x)$

14. Graph $f(x) = x^2 - 4x + 7$ by completing the square, then state intervals where:

   a. $f(x) \geq 0$ **b.** $f(x)\uparrow$

15. Given the graph of the general function $f(x)$ shown, graph $F(x) = -f(x + 1) + 2$.

Exercise 15

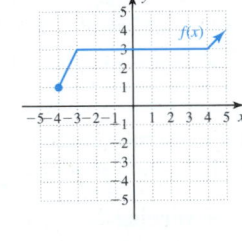

16. Graph the piecewise-defined function given:

$$f(x) = \begin{cases} -3 & x < -1 \\ x & -1 \leq x \leq 1 \\ 3x & x > 1 \end{cases}$$

17. Y varies directly with X and inversely with the square of Z. If $Y = 10$ when $X = 32$ and $Z = 4$, find X when $Z = 15$ and $Y = 1.4$. $X = 63$

18. Compute as indicated:

   a. $(2 + 5i)^2$ $-21 + 20i$ **b.** $\dfrac{1 - 2i}{1 + 2i}$ $\frac{-3}{5} - \frac{4}{5}i$

19. For $f(x)$ and $g(x)$ as shown, use the graph given to determine the value of each expression. Assume each grid line represents 1 unit.

   a. $f(4), g(2), (f \circ g)(2)$

   b. $g(4), f(8),$ and $(g \circ f)(8)$

   c. $(fg)(0)$ and $\left(\dfrac{g}{f}\right)(0)$

   d. $(f + g)(1)$ and $(g - f)(9)$

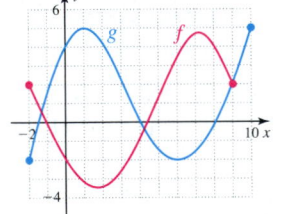

20. Solve using the quadratic formula. If solutions are complex, write them in $a + bi$ form.

   $2x^2 + 20x = -51$ $x = -5 \pm \frac{\sqrt{2}}{2}i$

Exercises 21 through 25 require the use of a graphing calculator.

21. The value $x = 2.3 - 1.4i$ is a solution to one or more of the following quadratic equations. Use your calculator to determine which one(s).

   $2x^2 - 9.2x + 14.5 = 0$ $1.1x^2 + 4.6x - 3.7 = 0$

   $1.2x^2 = 5.52x - 8.7$ $11.2x = 3.7x^2 - 2.05$

22. Use a graphing calculator to find any local maximum and minimum values of the function $g(x) = -0.2x^3 + 2.5x - 4$. Round to the nearest hundredth if necessary.

23. Use the quadratic regression feature of a graphing calculator to find the equation that contains the points $(-0.5, 8.5)$, $(1, 8.5)$, and $(1.5, 11.5)$.

24. Use a graphing calculator to find the zeroes of the function $h(x) = 4.3x - 0.9x^2$. Round to the nearest hundredth if necessary.

25. State the domain of $y = \dfrac{10x + 1}{x^2 - 0.04}$ using interval notation. Use the **TABLE** feature to verify your answer.

4

Polynomial and Rational Functions

CHAPTER OUTLINE

CHAPTER CONNECTIONS

In a study of demographics, the population density of a city and its surrounding area is measured using a unit called *people per square mile*. The population density is much greater near the city's center, and tends to decrease as you move out into suburban and rural areas. The density can be modeled using the formula $D(x) = \dfrac{ax}{x^2 + b}$, where $D(x)$ represents the density at a distance of x mi from the center of a city, and a and b are constants related to a particular city and its sprawl. Using this equation, city planners can determine how far from the city's center the population drops below a certain level, and answer other important questions to help plan for future growth. This application appears as Exercise 63 in Section 4.4.

Check out these other real-world connections:

LEARNING OBJECTIVES

In Section 4.1 you will see how we can:

☐ **A.** Divide polynomials using long division and synthetic division

☐ **B.** Use the remainder theorem to evaluate polynomials

☐ **C.** Use the factor theorem to factor and build polynomials

☐ **D.** Solve applications using the remainder theorem

To find the zero of a linear function, we can use properties of equality to isolate x. To find the zeroes of a quadratic function, we can factor or use the quadratic formula. To find the zeroes of higher degree polynomials, we must first develop additional tools, including synthetic division and the remainder and factor theorems. These will help us write a higher degree polynomial in terms of linear and quadratic polynomials, whose zeroes can easily be found.

A. Long Division and Synthetic Division

To help understand **synthetic division** and its use as a mathematical tool, we first review the process of **long division.**

Long Division

Polynomial long division closely resembles the division of whole numbers, with the main difference being that *we group each partial product in parentheses to prevent errors in subtraction.*

EXAMPLE 1 ▶ **Dividing Polynomials Using Long Division**

Divide $x^3 - 4x^2 + x + 6$ by $x - 1$.

Solution ▶ The divisor is $(x - 1)$ and the dividend is $(x^3 - 4x^2 + x + 6)$. To find the first multiplier, we compute *the ratio of leading terms* from each expression. Here the ratio $\dfrac{x^3 \text{ from dividend}}{x \text{ from divisor}}$ shows our first multiplier will be "x^2," with $x^2(x - 1) = x^3 - x^2$.

$$
\begin{array}{r}
x^2 \\
x - 1 \overline{)x^3 - 4x^2 + x + 6} \\
\underline{-(x^3 - x^2)} \text{ subtraction}
\end{array}
\qquad\longrightarrow\qquad
\begin{array}{r}
x^2 \\
x - 1 \overline{)x^3 - 4x^2 + x + 6} \\
\underline{-x^3 + x^2} \text{ algebraic addition} \\
-3x^2 + x
\end{array}
$$

At each stage, after writing the subtraction as algebraic addition (distributing the negative) we compute the sum in each column and "bring down" the next term. Each following multiplier is found as before, using the ratio $\dfrac{ax^k \text{ next leading term}}{x \text{ from divisor}}$.

$$
\begin{array}{r}
x^2 - 3x - 2 \\
x - 1 \overline{)x^3 - 4x^2 + x + 6} \\
\underline{-(x^3 - x^2)} \\
-3x^2 + x \\
\underline{-(-3x^2 + 3x)} \\
-2x + 6 \\
\underline{-(-2x + 2)} \\
4
\end{array}
$$

next multiplier: $\dfrac{-3x^2}{x} = -3x$ $\qquad$ $-3x(x - 1) = -3x^2 + 3x$, subtract $-3x^2 + 3x$

next multiplier: $\dfrac{-2x}{x} = -2$ $\qquad$ algebraic addition, bring down next term

$\qquad$ $-2(x - 1) = -2x + 2$, subtract $-2x + 2$

$\qquad$ algebraic addition, remainder is 4

The result shows $\dfrac{x^3 - 4x^2 + x + 6}{x - 1} = x^2 - 3x - 2 + \dfrac{4}{x - 1}$, or after multiplying both sides by $x - 1$, $x^3 - 4x^2 + x + 6 = (x - 1)(x^2 - 3x - 2) + 4$.

Now try Exercises 7 through 12 ▶

The process illustrated is called the **division algorithm,** and like the division of whole numbers, the final result can be checked by multiplication.

$$\text{check: } x^3 - 4x^2 + x + 6 = \overset{\text{dividend}}{} \overset{\text{divisor}}{(x-1)} \overset{\text{quotient}}{(x^2-3x-2)} + \overset{\text{remainder}}{4}$$

$$= (x^3 - 3x^2 - 2x - x^2 + 3x + 2) + 4 \qquad \text{divisor} \cdot \text{quotient}$$

$$= (x^3 - 4x^2 + x + 2) + 4 \qquad \text{combine like terms}$$

$$= x^3 - 4x^2 + x + 6 \checkmark \qquad \text{add remainder}$$

In general, the division algorithm for polynomials says

Division of Polynomials

Given polynomials $p(x)$ and $d(x) \neq 0$, there exist unique polynomials $q(x)$ and $r(x)$ such that

$$p(x) = d(x)q(x) + r(x),$$

where $r(x) = 0$ or the degree of $r(x)$ is less than the degree of $d(x)$.
Here, $d(x)$ is called the *divisor,* $q(x)$ is the *quotient,* and $r(x)$ is the *remainder.*

In other words, "a polynomial of greater degree can be divided by a polynomial of equal or lesser degree to obtain a quotient and a remainder." As with whole numbers, if the remainder is zero, the divisor is a factor of the dividend.

Synthetic Division

As the word "synthetic" implies, synthetic division not only *simulates* the long division process, but also condenses it and makes it more efficient *when the divisor is linear.* The process works by capitalizing on the repetition found in the division algorithm. First, the polynomials involved are written in decreasing order of degree, so the variable part of each term is unnecessary as we can let the *position of each coefficient* indicate the degree of the term. For the dividend from Example 1, $1 \; {-4} \; 1 \; 6$ would represent the polynomial $1x^3 - 4x^2 + 1x + 6$. Also, each stage of the algorithm involves a product of the divisor with the next multiplier, followed by a subtraction. These can likewise be computed using the coefficients only, as the degree of each term is still determined by its position. Here is the division from Example 1 in the synthetic division format. Note that we must use the *zero of the divisor* (as in $x = \frac{3}{2}$ for a divisor of $2x - 3$, or in this case, "1" from $x - 1 = 0$) and the coefficients of the dividend in the following format:

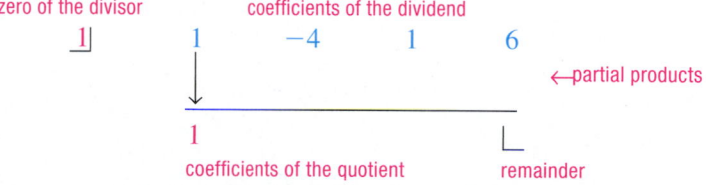

As this template indicates, the quotient and remainder will be read from the last row.

The arrow indicates we begin by "dropping the leading coefficient into place." We then multiply this coefficient by the "divisor," then place the result in the next column and add. Note that using the zero of the divisor enables us to *add in each column directly,* rather than subtracting then changing to algebraic addition as before.

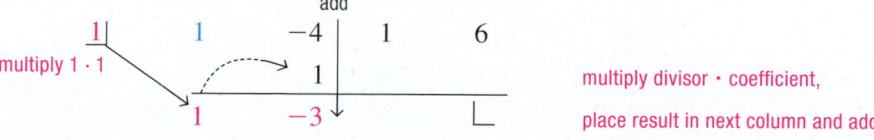

In a sense, we "multiply in the diagonal direction," and "add in the vertical direction." Repeat the process until the division is complete.

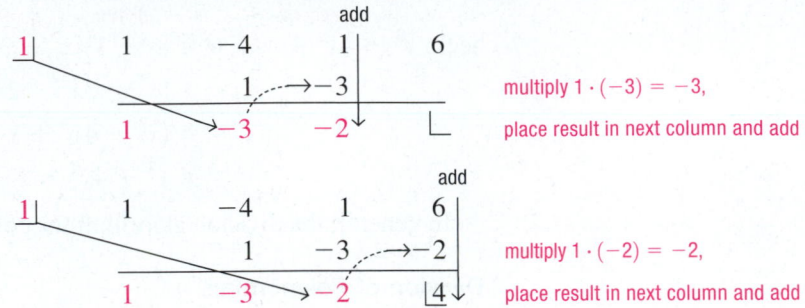

The quotient is read from the last row by noting the remainder is 4, leaving the coefficients 1 −3 −2, which translate back into the polynomial $x^2 - 3x - 2$. The final result is identical to that in Example 1, but the new process is more efficient, since all stages are actually computed on a single template as shown here:

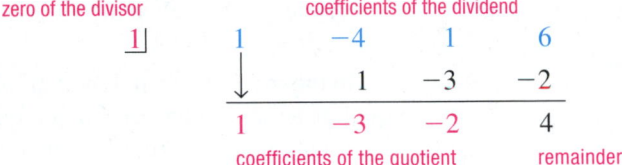

EXAMPLE 2 ▶ **Dividing Polynomials Using Synthetic Division**

Compute the quotient of $(x^3 + 3x^2 - 4x - 12)$ and $(x + 2)$, then check your answer.

Solution ▶ Using -2 as our "divisor" (from $x + 2 = 0$), we set up the synthetic division template and begin.

$$
\begin{array}{r|rrrr}
\text{use } -2 \text{ as a "divisor"} \quad -2| & 1 & 3 & -4 & -12 \\
& \downarrow & -2 & -2 & 12 \\
\hline
& 1 & 1 & -6 & 0
\end{array}
$$

drop lead coefficient into place; multiply by divisor, place result in next column and add

The result shows $\dfrac{x^3 + 3x^2 - 4x - 12}{x + 2} = x^2 + x - 6$, with no remainder.

Check ▶
$$
\begin{aligned}
x^3 + 3x^2 - 4x - 12 &= (x + 2)(x^2 + x - 6) \\
&= (x^3 + x^2 - 6x + 2x^2 + 2x - 12) \\
&= x^3 + 3x^2 - 4x - 12 \checkmark
\end{aligned}
$$

Now try Exercises 13 through 20 ▶

Note that in synthetic division, the degree of $q(x)$ will always be one less than $p(x)$, since the process requires a linear divisor (degree 1).

Since the division process is so dependent on the place value (degree) of each term, polynomials such as $2x^3 + 3x + 7$, which has no term of degree 2, must be written using a zero *placeholder*: $2x^3 + \mathbf{0}x^2 + 3x + 7$. This ensures that like place values "line up" as we carry out the division.

EXAMPLE 3 ▶   **Dividing Polynomials Using a Zero Placeholder**

Compute the quotient $\dfrac{2x^3 + 3x + 7}{x - 3}$ and check your answer.

Solution ▶   use 3 as a "divisor"

$$
\begin{array}{r|rrrr}
3 & 2 & 0 & 3 & 7 \\
  & \downarrow & 6 & 18 & 63 \\
\hline
  & 2 & 6 & 21 & 70
\end{array}
$$

note place holder **0** for "x^2" term

The result shows $\dfrac{2x^3 + 3x + 7}{x - 3} = 2x^2 + 6x + 21 + \dfrac{70}{x - 3}$. Multiplying by $x - 3$ gives

$$2x^3 + 3x + 7 = (2x^2 + 6x + 21)(x - 3) + 70$$

Check ▶

$$2x^3 + 3x + 7 = (x - 3)(2x^2 + 6x + 21) + 70$$
$$= (2x^3 + 6x^2 + 21x - 6x^2 - 18x - 63) + 70$$
$$= 2x^3 + 3x + 7 \ \checkmark$$

Now try Exercises 21 through 30 ▶

As noted earlier, for synthetic division the divisor must be a linear polynomial and the zero of this divisor is used. This means for the quotient $\dfrac{2x^3 - 3x^2 - 8x + 12}{2x - 3}$, we have $2x - 3 = 0$, and $x = \dfrac{3}{2}$ would be used for synthetic division **[see Example 6(c)]**. Finally, if the divisor is nonlinear, long division must be used.

EXAMPLE 4 ▶   **Division with a Nonlinear Divisor**

Compute the quotient: $\dfrac{2x^4 + x^3 - 7x^2 + 3}{x^2 - 2}$.

Solution ▶   Write the dividend as $2x^4 + x^3 - 7x^2 + \mathbf{0}x + 3$, and the divisor as $x^2 + \mathbf{0}x - 2$.

The quotient of leading terms gives $\dfrac{2x^4 \ \text{from dividend}}{x^2 \ \text{from divisor}} = 2x^2$ as our first multiplier.

$$
\begin{array}{r}
2x^2 + x - 3 \\
x^2 + 0x - 2 \overline{\smash{)}\ 2x^4 + x^3 - 7x^2 + 0x + 3} \\
\end{array}
$$

Multiply $2x^2(x^2 + 0x - 2)$ $-(2x^4 + 0x^3 - 4x^2)$ subtract (algebraic addition)

$\qquad\qquad\qquad\qquad x^3 - 3x^2 + 0x$ bring down next term

Multiply $x(x^2 + 0x - 2)$ $-(x^3 + 0x^2 - 2x)$ subtract (algebraic addition)

$\qquad\qquad\qquad\qquad -3x^2 + 2x + 3$ bring down next term

Multiply $-3(x^2 + 0x - 2)$ $-(-3x^2 + \mathbf{0}x + 6)$ subtract (algebraic addition)

$\qquad\qquad\qquad\qquad\qquad 2x - 3$ remainder is $2x - 3$

Since the degree of $2x - 3$ (degree 1) is less than the degree of the divisor (degree 2), the process is complete.

$$\frac{2x^4 + x^3 - 7x^2 + 3}{x^2 - 2} = (2x^2 + x - 3) + \frac{2x - 3}{x^2 - 2}$$

Now try Exercises 31 through 34 ▶

A. You've just seen how we can divide polynomials using long division and synthetic division

Note the we elected to keep the solution to Example 4 in the form $\dfrac{p(x)}{d(x)} = q(x) + \dfrac{r(x)}{d(x)}$, instead of multiplying both sides by $d(x)$.

B. The Remainder Theorem

In Example 2, we saw that $(x^3 + 3x^2 - 4x - 12) \div (x + 2) = x^2 + x - 6$, with remainder zero. Similar to whole number division, this means $x + 2$ must be a factor of $x^3 + 3x^2 - 4x - 12$, a fact made clear as we checked our answer: $x^3 + 3x^2 - 4x - 12 = (x + 2)(x^2 + x - 6)$. Now consider the functions $p(x) = x^3 + 5x^2 + 2x - 8$, $d(x) = x + 3$, and their quotient $\dfrac{p(x)}{d(x)} = \dfrac{x^3 + 5x^2 + 2x - 8}{x + 3}$. Using -3 as the divisor in synthetic division gives

$$\text{use } -3 \text{ as a "divisor"} \quad \underline{-3|} \quad \begin{array}{rrrr} 1 & 5 & 2 & -8 \\ \downarrow & -3 & -6 & 12 \\ \hline 1 & 2 & -4 & \underline{|4} \end{array}$$

This shows $x + 3$ is *not* a factor of $p(x)$, since it didn't divide evenly (the remainder is not zero). However, from the result $p(x) = (x + 3)(x^2 + 2x - 4) + 4$, we make a remarkable observation—if we evaluate $p(-3)$, *the quotient portion becomes zero*, showing $p(-3) = 4$ (the remainder).

$$\begin{aligned} p(-3) &= (-3 + 3)[(-3)^2 + 2(-3) - 4] + 4 \\ &= (\mathbf{0})(-1) + 4 \\ &= 4 \end{aligned}$$

This result can be verified by evaluating $p(-3)$ in its original form (also see Figure 4.1):

$$\begin{aligned} p(x) &= x^3 + 5x^2 + 2x - 8 \\ p(-3) &= (-3)^3 + 5(-3)^2 + 2(-3) - 8 \\ &= -27 + 45 + (-6) - 8 \\ &= 4 \end{aligned}$$

Figure 4.1

```
Plot1  Plot2  Plot3
\Y1■X^3+5X²+2X-8
\Y2=
\Y3= Y1(-3)
\Y4=              4
\Y5=
\Y6=
```

The result is no coincidence, and illustrates the conclusion of the **remainder theorem.**

> ### The Remainder Theorem
>
> If a polynomial $p(x)$ is divided by $(x - c)$ using synthetic division, the remainder is equal to $p(c)$.

This gives us a powerful tool for evaluating polynomials. Where a direct evaluation involves powers of numbers and a long series of calculations, synthetic division reduces the process to simple products and sums.

EXAMPLE 5 ▶   **Using the Remainder Theorem to Evaluate Polynomials**

Use the remainder theorem to find $p(-5)$ for $p(x) = x^4 + 3x^3 - 8x^2 + 5x - 6$. Verify the result using a substitution.

Solution ▶

$$\text{use } -5 \text{ as a "divisor"} \quad \underline{-5|} \quad \begin{array}{rrrrr} 1 & 3 & -8 & 5 & -6 \\ & -5 & 10 & -10 & 25 \\ \hline 1 & -2 & 2 & -5 & \underline{|19} \end{array}$$

The result shows $p(-5) = 19$.

Verification using algebra

$$p(-5) = (-5)^4 + 3(-5)^3 - 8(-5)^2 + 5(-5) - 6$$
$$= 625 - 375 - 200 - 25 - 6$$
$$= 625 - 606$$
$$= 19$$

Verification using technology

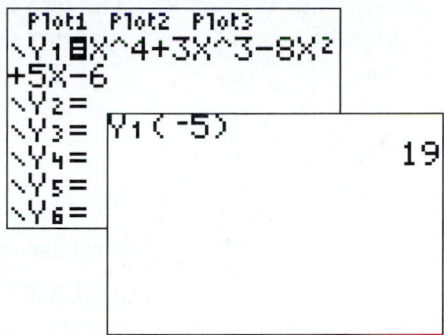

Now try Exercises 35 through 44 ▶

☑ **B.** You've just seen how we can use the remainder theorem to evaluate polynomials

Since $p(-5) = 19$, we know $(-5, 19)$ must be a point of the graph of $p(x)$. The ability to quickly evaluate polynomial functions using the remainder theorem will be used extensively in the sections that follow.

C. The Factor Theorem

As a consequence of the remainder theorem, when $p(x)$ is divided by $x - c$ and the remainder is 0, $p(c) = 0$ and c is a zero of the polynomial. The relationship between $x - c$, c, and $p(c) = 0$ are given in the **factor theorem.**

> **The Factor Theorem**
>
> For a polynomial $p(x)$,
>
> **1.** If $p(c) = 0$, then $x - c$ is a factor of $p(x)$.
> **2.** If $x - c$ is a factor of $p(x)$, then $p(c) = 0$.

The remainder and factor theorems often work together to help us find factors of higher degree polynomials.

EXAMPLE 6 ▶ **Using the Factor Theorem to Find Factors of a Polynomial**

Use the factor theorem to determine if

a. $x - 2$ **b.** $x + 1$ **c.** $3x - 2$

are factors of $p(x) = 3x^4 - 2x^3 - 21x^2 + 32x - 12$.

Solution ▶ **a.** If $x - 2$ is a factor, then $p(2)$ must be 0. Using the remainder theorem we have

$$
\begin{array}{r|rrrrr}
2 & 3 & -2 & -21 & 32 & -12 \\
  & \downarrow & 6 & 8 & -26 & 12 \\
\hline
  & 3 & 4 & -13 & 6 & 0 \\
\end{array}
$$

Since the remainder is zero, we know $p(2) = 0$ (remainder theorem) and $(x - 2)$ is a factor (factor theorem).

b. Similarly, if $x + 1$ is a factor, then $p(-1)$ must be 0.

$$
\begin{array}{r|rrrrr}
-1 & 3 & -2 & -21 & 32 & -12 \\
   & \downarrow & -3 & 5 & 16 & -48 \\
\hline
   & 3 & -5 & -16 & 48 & -60 \\
\end{array}
$$

Since the remainder is not zero, $(x + 1)$ is not a factor of p.

c. The zero of the divisor $3x - 2 = 0$ is $x = \dfrac{2}{3}$, and this value is used in the synthetic division.

$$
\begin{array}{r|rrrrr}
\frac{2}{3} & 3 & -2 & -21 & 32 & -12 \\
 & \downarrow & 2 & 0 & -14 & 12 \\
\hline
 & 3 & 0 & -21 & 18 & \underline{0}
\end{array}
$$

Since the remainder is zero, $p\!\left(\dfrac{2}{3}\right) = 0$ (remainder theorem) and $\left(x - \dfrac{2}{3}\right)$ is the related factor (factor theorem). The original factor $3x - 2$ is found by noting that the quotient polynomial $q(x) = 3x^3 - 21x + 18$ has a common factor of three, which will be factored out and applied to $x - \dfrac{2}{3}$. Starting with the partially factored form we have

$$
\begin{aligned}
p(x) &= \left(x - \frac{2}{3}\right)(3x^3 - 21x + 18) && \text{partially factored form} \\
&= \left(x - \frac{2}{3}\right)(3)(x^3 - 7x + 6) && \text{factor out 3} \\
&= (3x - 2)(x^3 - 7x + 6) && \text{multiply } 3\!\left(x - \frac{2}{3}\right)
\end{aligned}
$$

This form of simplification will always take place when the zero found using synthetic division is a fraction and the coefficients of the polynomial are integers.

Now try Exercises 45 through 56 ▶

As a final note on Example 6, there should be no hesitation to use fractions in the synthetic division process if the given polynomial has integer coefficients. The fraction will be a zero only if all values in the quotient line are integers (i.e., all of the products and sums must be integers).

EXAMPLE 7 ▶   Building a Polynomial Using the Factor Theorem

A polynomial $p(x)$ has three zeroes at $x = 3$, $\sqrt{2}$, and $-\sqrt{2}$. Use the factor theorem to find the polynomial.

Solution ▶   Using the factor theorem, the factors of $p(x)$ must be $(x - 3)$, $(x - \sqrt{2})$, and $(x + \sqrt{2})$. Computing the product will yield the polynomial.

$$
\begin{aligned}
p(x) &= (x - 3)(x - \sqrt{2})(x + \sqrt{2}) \\
&= (x - 3)(x^2 - 2) \\
&= x^3 - 3x^2 - 2x + 6
\end{aligned}
$$

Now try Exercises 57 through 64 ▶

Actually, the result obtained in Example 7 is not unique, since any polynomial of the form $a(x^3 - 3x^2 - 2x + 6)$ will also have the same three zeroes for $a \in \mathbb{R}$.

Figure 4.2 shows the graph of $Y_1 = p(x)$, as well as graph of $Y_2 = 2p(x)$. The only difference is $2p(x)$ has been vertically stretched. Likewise, the graph of $-1p(x)$ would be a vertical reflection, *but still with the same zeroes*. As in previous graph-to-equation exercises, finding a unique value for the leading coefficient a requires that we use a point (x, y) on the graph of p and substitute these values to solve for a. For Example 7, assume you were also told the graph contains the point $(1, 2)$, or $x = 1$, $p(1) = 2$. This would yield:

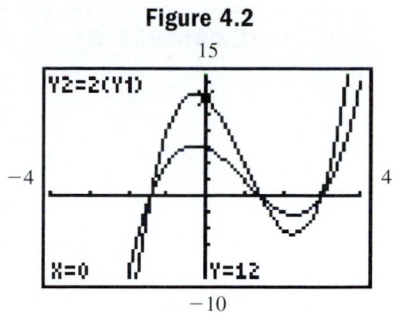

Figure 4.2

$$p(x) = a(x^3 - 3x^2 - 2x + 6) \qquad \text{original function}$$
$$p(1) = a[1^3 - 3(1)^2 - 2(1) + 6] \qquad \text{substitute 1 for } x$$
$$2 = 2a \qquad \text{substitute 2 for } p(1), \text{ simplify}$$
$$1 = a \qquad \text{result}$$

Consistent with the graph shown, if the point $(1, 4)$ were specified instead, a like calculation would show $a = 2$.

EXAMPLE 8 ▶ **Finding Zeroes Using the Factor Theorem**

Given that 2 is a zero of $p(x) = x^4 + x^3 - 10x^2 - 4x + 24$, use the factor theorem to help find all other zeroes.

Solution ▶ Using synthetic division gives:

$$
\begin{array}{r|rrrrr}
\text{use 2 as a "divisor"} \quad 2 \rfloor & 1 & 1 & -10 & -4 & 24 \\
& \downarrow & 2 & 6 & -8 & -24 \\
\hline
& 1 & 3 & -4 & -12 & \underline{|0}
\end{array}
$$

Since the remainder is zero, $(x - 2)$ is a factor and p can be written:

$$x^4 + x^3 - 10x^2 - 4x + 24 = (x - 2)(x^3 + 3x^2 - 4x - 12)$$

WORTHY OF NOTE

In Section R.4 we noted a third degree polynomial $ax^3 + bx^2 + cx + d$ is factorable if $ad = bc$. In Example 8, $1(-12) = 3(-4)$ and the polynomial is factorable.

Note the quotient polynomial can be factored by grouping to find the remaining factors of p.

$$
\begin{aligned}
x^4 + x^3 - 10x^2 - 4x + 24 &= (x - 2)(x^3 + 3x^2 - 4x - 12) &&\text{group terms (in color)}\\
&= (x - 2)[x^2(x + 3) - 4(x + 3)] &&\text{remove common factors from each group}\\
&= (x - 2)[(x + 3)(x^2 - 4)] &&\text{factor common binomial}\\
&= (x - 2)(x + 3)(x + 2)(x - 2) &&\text{factor difference of squares}\\
&= (x + 3)(x + 2)(x - 2)^2 &&\text{completely factored form}
\end{aligned}
$$

☑ **C.** You've just seen how we can use the factor theorem to factor and build polynomials

The final result shows $(x - 2)$ is actually a repeated factor, and the remaining zeroes of p are -3 and -2.

Now try Exercises 65 through 78 ▶

D. Applications

While the factor and remainder theorems are valuable tools for factoring higher degree polynomials, each has applications that extend beyond this use.

EXAMPLE 9 ▶   **Using the Remainder Theorem to Solve a Discharge Rate Application**

The *discharge rate* of a river is a measure of the river's water flow as it empties into a lake, sea, or ocean. The rate depends on many factors, but is primarily influenced by the precipitation in the surrounding area and is often seasonal. Suppose the discharge rate of the Shimote River was modeled by $D(m) = -m^4 + 22m^3 - 147m^2 + 317m + 150$, where $D(m)$ represents the discharge rate in thousands of cubic meters of water per second in month m ($m = 1 \rightarrow$ Jan).

a. What was the discharge rate in June (summer heat)?

b. Is the discharge rate higher in February (winter runoff) or October (fall rains)?

Solution ▶   a. To find the discharge rate in June, we evaluate D at $m = 6$.
Using the remainder theorem gives

$$
\begin{array}{r|rrrrr}
6| & -1 & 22 & -147 & 317 & 150 \\
 & \downarrow & -6 & 96 & -306 & 66 \\
\hline
 & -1 & 16 & -51 & 11 & |216 \\
\end{array}
$$

In June, the discharge rate is 216,000 m³/sec.

b. For the discharge rates in February ($m = 2$) and October ($m = 10$), we have

$$
\begin{array}{r|rrrrr}
2| & -1 & 22 & -147 & 317 & 150 \\
 & \downarrow & -2 & 40 & -214 & 206 \\
\hline
 & -1 & 20 & -107 & 103 & |356 \\
\end{array}
\qquad
\begin{array}{r|rrrrr}
10| & -1 & 22 & -147 & 317 & 150 \\
 & \downarrow & -10 & 120 & -270 & 470 \\
\hline
 & -1 & 12 & -27 & 47 & |620 \\
\end{array}
$$

The discharge rate during the fall rains in October is much higher: $620 > 356$.

☑ **D.** You've just seen how we can solve applications using the remainder theorem

Now try Exercises 81 through 84 ▶

4.1 EXERCISES

▶ **CONCEPTS AND VOCABULARY**

Fill in each blank with the appropriate word or phrase. Carefully reread the section if needed.

1. For __synthetic__ division, we use the __zero__ of the divisor to begin.

2. If the __remainder__ is zero after division, then the __divisor__ is a factor of the dividend.

3. If polynomial $P(x)$ is divided by a linear divisor of the form $x - c$, the remainder is identical to __$P(c)$__. This is a statement of the __remainder__ theorem.

4. If $P(c) = 0$, then __$x - c$__ must be a factor of $P(x)$. Conversely, if __$x - c$__ is a factor of $P(x)$, then $P(c) = 0$. These are statements from the __factor__ theorem.

5. Discuss/Explain how to write the quotient and remainder using the last line from a synthetic division.   Answers will vary.

6. Discuss/Explain why (a, b) is a point on the graph of P, given b was the remainder after P was divided by $x - a$ using synthetic division.   Answers will vary.

HOMEWORK SELECTION GUIDE

Core: 7–75 every other odd, 81–84 (22 Exercises)
Standard: 1–4, 7–75 every other odd, 79, 81–84, 95 (28 Exercises)
Extended: 1–4, 7–75 every other odd, 79, 81–85, 89, 92, 94, 95 (32 Exercises)
In Depth: 1–6, 7–75 every other odd, 79–85, 89, 92, 93–96 (38 Exercises)

▶ DEVELOPING YOUR SKILLS

Divide using long division. Write the result as dividend = (divisor)(quotient) + remainder.

7. $\dfrac{x^3 - 5x^2 - 4x + 23}{x - 2}$

$x^3 - 5x^2 - 4x + 23 = (x - 2)(x^2 - 3x - 10) + 3$

8. $\dfrac{x^3 + 5x^2 - 17x - 26}{x + 7}$

$x^3 + 5x^2 - 17x - 26 = (x + 7)(x^2 - 2x - 3) - 5$

9. $(2x^3 + 5x^2 + 4x + 17) \div (x + 3)$

$2x^3 + 5x^2 + 4x + 17 = (x + 3)(2x^2 - x + 7) - 4$

10. $(3x^3 + 14x^2 - 2x - 37) \div (x + 4)$

$3x^3 + 14x^2 - 2x - 37 = (x + 4)(3x^2 + 2x - 10) + 3$

11. $(x^3 - 8x^2 + 11x + 20) \div (x - 5)$

$x^3 - 8x^2 + 11x + 20 = (x - 5)(x^2 - 3x - 4) + 0$

12. $(x^3 - 5x^2 - 22x - 16) \div (x + 2)$

$x^3 - 5x^2 - 22x - 16 = (x + 2)(x^2 - 7x - 8) + 0$

Divide using synthetic division. Write answers in two ways: (a) $\frac{\text{dividend}}{\text{divisor}} = \text{quotient} + \frac{\text{remainder}}{\text{divisor}}$, and (b) dividend = (divisor)(quotient) + remainder. For Exercises 13–18, check answers using multiplication.

13. $\dfrac{2x^2 - 5x - 3}{x - 3}$ 14. $\dfrac{3x^2 + 13x - 10}{x + 5}$

15. $(x^3 - 3x^2 - 14x - 8) \div (x + 2)$

16. $(x^3 - 6x^2 - 24x - 17) \div (x + 1)$

17. $\dfrac{x^3 - 5x^2 - 4x + 23}{x - 2}$ 18. $\dfrac{x^3 + 12x^2 + 34x - 9}{x + 7}$

19. $(2x^3 - 5x^2 - 11x - 17) \div (x - 4)$

20. $(3x^3 - x^2 - 7x + 27) \div (x - 1)$

Divide using synthetic division. Note that some terms of a polynomial may be "missing." Write answers as dividend = (divisor)(quotient) + remainder.

21. $(x^3 + 5x^2 + 7) \div (x + 1)$

$x^3 + 5x^2 + 7 = (x + 1)(x^2 + 4x - 4) + 11$

22. $(x^3 - 3x^2 - 37) \div (x - 5)$

$x^3 - 3x^2 - 37 = (x - 5)(x^2 + 2x + 10) + 13$

23. $(x^3 - 13x - 12) \div (x - 4)$

$x^3 - 13x - 12 = (x - 4)(x^2 + 4x + 3) + 0$

24. $(x^3 - 7x + 6) \div (x + 3)$

$x^3 - 7x + 6 = (x + 3)(x^2 - 3x + 2) + 0$

25. $\dfrac{3x^3 - 8x + 12}{x - 1}$ 26. $\dfrac{2x^3 + 7x - 81}{x - 3}$

27. $(n^3 + 27) \div (n + 3)$

28. $(m^3 - 8) \div (m - 2)$

29. $(x^4 + 3x^3 - 16x - 8) \div (x - 2)$

30. $(x^4 + 3x^2 + 29x - 21) \div (x + 3)$

Compute each indicated quotient. Write answers in the form $\frac{\text{dividend}}{\text{divisor}} = \text{quotient} + \frac{\text{remainder}}{\text{divisor}}$.

31. $\dfrac{2x^3 + 7x^2 - x + 26}{x^2 + 3}$ 32. $\dfrac{x^4 + 3x^3 + 2x^2 - x - 5}{x^2 - 2}$

33. $\dfrac{x^4 - 5x^2 - 4x + 7}{x^2 - 1}$ 34. $\dfrac{x^4 + 2x^3 - 8x - 16}{x^2 + 5}$

Use the remainder theorem to evaluate $P(x)$ as given.

35. $P(x) = x^3 - 6x^2 + 5x + 12$

   a. $P(-2)$ -30   **b.** $P(5)$ 12

36. $P(x) = x^3 + 4x^2 - 8x - 15$

   a. $P(-2)$ 9   **b.** $P(3)$ 24

37. $P(x) = 2x^3 - x^2 - 19x + 4$

   a. $P(-3)$ -2   **b.** $P(2)$ -22

38. $P(x) = 3x^3 - 8x^2 - 14x + 9$

   a. $P(-2)$ -19   **b.** $P(4)$ 17

39. $P(x) = x^4 - 4x^2 + x + 1$

   a. $P(-2)$ -1   **b.** $P(2)$ 3

40. $P(x) = x^4 + 3x^3 - 2x - 4$

   a. $P(-2)$ -8   **b.** $P(2)$ 32

41. $P(x) = 2x^3 - 7x + 33$

   a. $P(-2)$ 31   **b.** $P(-3)$ 0

42. $P(x) = -2x^3 + 9x^2 - 11$

   a. $P(-2)$ 41   **b.** $P(-1)$ 0

43. $P(x) = 2x^3 + 3x^2 - 9x - 10$

   a. $P\left(\frac{3}{2}\right)$ -10   **b.** $P\left(-\frac{5}{2}\right)$ 0

44. $P(x) = 3x^3 + 11x^2 + 2x - 16$

   a. $P\left(\frac{1}{3}\right)$ -14   **b.** $P\left(-\frac{8}{3}\right)$ 0

Use the factor theorem to determine if the factors given are factors of $f(x)$.

45. $f(x) = x^3 - 3x^2 - 13x + 15$

   a. $(x + 3)$ yes   **b.** $(x - 5)$ yes

46. $f(x) = x^3 + 2x^2 - 11x - 12$

   a. $(x + 4)$ yes   **b.** $(x - 3)$ yes

47. $f(x) = x^3 - 6x^2 + 3x + 10$

   a. $(x + 2)$ no   **b.** $(x - 5)$ yes

48. $f(x) = x^3 + 2x^2 - 5x - 6$

    a. $(x - 2)$ yes   **b.** $(x + 4)$ no

49. $f(x) = -2x^3 - x^2 + 12x - 9$

    a. $(x + 3)$ yes   **b.** $(2x - 3)$ yes

50. $f(x) = 3x^3 - 19x^2 + 30x - 8$

    a. $(3x - 1)$ yes   **b.** $(x + 4)$ no

Use the factor theorem to show the given value is a zero of $P(x)$.

51. $P(x) = x^3 + 2x^2 - 5x - 6; x = -3$

52. $P(x) = x^3 + 3x^2 - 16x + 12; x = -6$

53. $P(x) = x^3 - 7x + 6; x = 2$

54. $P(x) = x^3 - 13x + 12; x = -4$

55. $P(x) = 9x^3 + 18x^2 - 4x - 8; x = \dfrac{2}{3}$

56. $P(x) = 5x^3 + 13x^2 - 9x - 9; x = -\dfrac{3}{5}$

A polynomial P with integer coefficients has the zeroes and degree indicated. Use the factor theorem to write the function in factored form and standard form.

57. $-2, 3, -5$; degree 3   **58.** $1, -4, 2$; degree 3

59. $-2, \sqrt{3}, -\sqrt{3}$;   **60.** $\sqrt{5}, -\sqrt{5}, 4$;
    degree 3   degree 3

61. $-5, 2\sqrt{3}, -2\sqrt{3}$;   **62.** $4, 3\sqrt{2}, -3\sqrt{2}$;
    degree 3   degree 3

63. $1, -2, \sqrt{10}, -\sqrt{10}$;   **64.** $\sqrt{7}, -\sqrt{7}, 3, -1$;
    degree 4   degree 4

In Exercises 65 through 70, a known zero of the polynomial is given. Use the factor theorem to write the polynomial in completely factored form.

65. $P(x) = x^3 - 5x^2 - 2x + 24; x = -2$
    $P(x) = (x + 2)(x - 3)(x - 4)$

66. $Q(x) = x^3 - 7x^2 + 7x + 15; x = 3$
    $Q(x) = (x + 1)(x - 5)(x - 3)$

67. $p(x) = x^4 + 2x^3 - 12x^2 - 18x + 27; x = -3$
    $p(x) = (x + 3)^2(x - 3)(x - 1)$

68. $q(x) = x^4 + 4x^3 - 6x^2 - 4x + 5; x = 1$
    $q(x) = (x - 1)^2(x + 1)(x + 5)$

69. $f(x) = 2x^3 + 11x^2 - x - 30; x = \dfrac{3}{2}$
    $f(x) = 2(x - \frac{3}{2})(x + 2)(x + 5)$

70. $g(x) = 3x^3 + 2x^2 - 75x - 50; x = -\dfrac{2}{3}$
    $g(x) = 3(x + \frac{2}{3})(x - 5)(x + 5)$

If $p(x)$ is a polynomial with rational coefficients and a leading coefficient of $a = 1$, the rational zeroes of p (if they exist) *must be factors of the constant term*. Use this property of polynomials with the factor and remainder theorems to factor each polynomial completely.

71. $p(x) = x^3 - 3x^2 - 9x + 27$ $p(x) = (x + 3)(x - 3)^2$

72. $p(x) = x^3 - 4x^2 - 16x + 64$ $p(x) = (x + 4)(x - 4)^2$

73. $p(x) = x^3 - 6x^2 + 12x - 8$ $p(x) = (x - 2)^3$

74. $p(x) = x^3 - 15x^2 + 75x - 125$ $p(x) = (x - 5)^3$

75. $p(x) = (x^2 - 6x + 9)(x^2 - 9)$ $p(x) = (x + 3)(x - 3)^3$

76. $p(x) = (x^2 - 1)(x^2 - 2x + 1)$ $p(x) = (x + 1)(x - 1)^3$

77. $p(x) = (x^3 + 4x^2 - 9x - 36)(x^2 + x - 12)$
    $p(x) = (x + 3)(x - 3)^2(x + 4)^2$

78. $p(x) = (x^3 - 3x^2 + 3x - 1)(x^2 - 3x + 2)$
    $p(x) = (x - 2)(x - 1)^4$

▶ WORKING WITH FORMULAS

Volume of an open box: $V(x) = 4x^3 - 84x^2 + 432x$

An open box is constructed by cutting square corners from a 24 in. by 18 in. sheet of cardboard and folding up the sides. Its volume is given by the formula shown, where x represents the length of the square cuts.

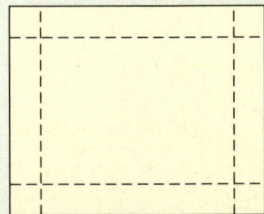

79. Given a volume of 640 in^3, use synthetic division and the remainder theorem to determine if the squares were 2-, 3-, 4-, or 5-in. squares and state the dimensions of the box. (*Hint:* Write as a function $V(x)$ and use synthetic division.)

    4-in. squares; 16 in. × 10 in. × 4 in.

80. Given the volume is 357.5 in^3, use synthetic division and the remainder theorem to determine if the squares were 5.5-, 6.5-, or 7.5-in. squares and state the dimensions of the box. (*Hint:* Write as a function $V(x)$ and use synthetic division.)

    6.5-in. squares; 11 in. × 5 in. × 6.5 in.

▶ **APPLICATIONS**

81. **Tourist population:**  During the 12 weeks of summer, the population of tourists at a popular beach resort is modeled by the polynomial $P(w) = -0.1w^4 + 2w^3 - 14w^2 + 52w + 5$, where $P(w)$ is the tourist population (in 1000s) during week w. Use the remainder theorem to help answer the following questions.

   a. Were there more tourists at the resort in week 5 ($w = 5$) or week 10? How many more tourists? *week 10, 22.5 thousand*

   b. Were more tourists at the resort one week after opening ($w = 1$) or one week before closing ($w = 11$). How many more tourists? *one week before closing, 36 thousand*

   c. The tourist population peaked (reached its highest) between weeks 7 and 10. Use the remainder theorem to determine the peak week. *week 9*

82. **Debt load:** Due to a fluctuation in tax revenues, a county government is projecting a deficit for the next 12 months, followed by a quick recovery and the repayment of all debt near the end of this period. The projected debt can be modeled by the polynomial $D(m) = 0.1m^4 - 2m^3 + 15m^2 - 64m - 3$, where $D(m)$ represents the amount of debt (in millions of dollars) in month m. Use the remainder theorem to help answer the following questions.

   a. Was the debt higher in month 5 ($m = 5$) or month 10 of this period? How much higher? *month 10, $7.5 million*

   b. Was the debt higher in the first month of this period (one month into the deficit) or after the eleventh month (one month before the expected recovery)? How much higher? *after the eleventh month, $36 million*

   c. The total debt reached its maximum between months 7 and 10. Use the remainder theorem to determine which month. *month 8*

83. **Volume of water:** The volume of water in a rectangular, inground, swimming pool is given by $V(x) = x^3 + 11x^2 + 24x$, where $V(x)$ is the volume in cubic feet when the water is x ft high. (a) Use the remainder theorem to find the volume when $x = 3$ ft. (b) If the volume is 100 ft^3 of water, what is the height x? (c) If the maximum capacity of the pool is 1000 ft^3, what is the maximum depth (to the nearest integer)? *a. 198 ft^3 b. 2 ft c. about 7 ft*

84. **Amusement park attendance:** Attendance at an amusement park depends on the weather. After opening in spring, attendance rises quickly, slows during the summer, soars in the fall, then quickly falls with the approach of winter when the park closes. The model for attendance is given by $A(m) = -\frac{1}{4}m^4 + 6m^3 - 52m^2 + 196m - 260$, where $A(m)$ represents the number of people attending in month m (in thousands). (a) Did more people go to the park in April ($m = 4$) or June ($m = 6$)? (b) In what month did maximum attendance occur? (c) When did the park close? *a. June, 4000 more b. August, about 28,000 c. October*

In these applications, synthetic division is applied in the usual way, treating k as an unknown constant.

85. Find a value of k that will make $x + 2$ a factor of $f(x) = x^3 - 3x^2 - 5x + k$. *k = 10*

86. Find a value of k that will make $x - 3$ a factor of $g(x) = x^3 + 2x^2 - 7x + k$. *k = -24*

87. For what value(s) of k will $x - 2$ be a factor of $p(x) = x^3 - 3x^2 + kx + 10$? *k = -3*

88. For what value(s) of k will $x + 5$ be a factor of $q(x) = x^3 + 6x^2 + kx + 50$? *k = 15*

▶ **EXTENDING THE CONCEPT**

89. To investigate whether the remainder and factor theorems can be applied when the coefficients or zeroes of a polynomial are complex, try using the factor theorem to find a polynomial with degree 3, whose zeroes are $x = 2i$, $x = -2i$, and $x = 3$. Then see if the result can be verified using the remainder theorem and these zeroes. What does the result suggest? **Also see Exercise 92.** *The theorems also apply to complex zeroes of polynomials.*

90. Since we use a base-10 number system, numbers like 1196 can be written in polynomial form as $p(x) = 1x^3 + 1x^2 + 9x + 6$, where $x = 10$. Divide $p(x)$ by $x + 3$ using synthetic division and write your answer as $\frac{x^3 + x^2 + 9x + 6}{x + 3} =$ quotient $+ \frac{\text{remainder}}{\text{divisor}}$. For $x = 10$, what is the value of quotient $+ \frac{\text{remainder}}{\text{divisor}}$? What is the result of dividing 1196 by $10 + 3 = 13$? What can you conclude? *$\frac{x^3 + x^2 + 9x + 6}{x + 3} = x^2 - 2x + 15 - \frac{39}{x+3}$; 92 = 92, way cool!*

91. The sum of the first n perfect cubes is given by the formula $S = \frac{1}{4}(n^4 + 2n^3 + n^2)$. Use the remainder theorem on S to find the sum of (a) the first three perfect cubes (divide by $n - 3$) and (b) the first five perfect cubes (divide by $n - 5$). Check results by adding the perfect cubes manually. To avoid working with fractions you can initially ignore the $\frac{1}{4}$ (use $n^4 + 2n^3 + n^2 + 0n + 0$), as long as you divide the remainder by 4. $S_3 = 36; S_5 = 225$

92. Though not a direct focus of this course, the remainder and factor theorems, as well as synthetic

division, *can also be applied using complex numbers.* Use the remainder theorem to show the value given is a zero of $P(x)$.

a. $P(x) = x^3 - 4x^2 + 9x - 36; x = 3i$ $P(3i) = 0$

b. $P(x) = x^4 + x^3 + 2x^2 + 4x - 8; x = -2i$
   $P(-2i) = 0$

c. $P(x) = -x^3 + x^2 - 3x - 5; x = 1 + 2i$
   $P(1 + 2i) = 0$

d. $P(x) = x^3 + 2x^2 + 16x + 32; x = -4i$
   $P(-4i) = 0$

e. $P(x) = x^4 + x^3 - 5x^2 + x - 6; x = i$ $P(i) = 0$

f. $P(x) = -x^3 + x^2 - 8x - 10; x = 1 + 3i$
   $P(1 + 3i) = 0$

▶ **MAINTAINING YOUR SKILLS**

93. (1.5) John and Rick are out orienteering. Rick finds the last marker first and is heading for the finish line, 1275 yd away. John is just seconds behind, and after locating the last marker tries to overtake Rick, who by now has a 250-yd lead. If Rick runs at 4 yd/sec and John runs at 5 yd/sec, will John catch Rick before they reach the finish line?   Yes, John wins.

94. (R.3) Solve for w:
$$-2(3w^2 + 5) + 3 = -7w + w^2 - 7 \quad w = 0, w = 1$$

95. (1.4) The profit of a small business increased linearly from \$5000 in 2005 to \$12,000 in 2010. Find a linear function $G(t)$ modeling the growth of the company's profit (let $t = 0$ correspond to 2005). $G(t) = 1400t + 5000$

96. (3.4) Given $f(x) = x^2 - 4x$, use the average rate of change formula to find $\frac{\Delta y}{\Delta x}$ in the interval $x \in [1.0, 1.1]$. -1.9

4.2 The Zeroes of Polynomial Functions

LEARNING OBJECTIVES

In Section 4.2 you will see how we can:

☐ **A.** Apply the fundamental theorem of algebra and the linear factorization theorem

☐ **B.** Use the intermediate value theorem to identify intervals containing a polynomial zero

☐ **C.** Find rational zeroes of a real polynomial function using the rational zeroes theorem

☐ **D.** Obtain more information on the zeroes of real polynomials using Descartes' rule of signs and the upper/lower bounds theorem

☐ **E.** Solve applications of polynomial functions

This section represents one of the highlights in the college algebra curriculum, because it offers a look at what many call *the big picture*. The ideas presented are the result of a cumulative knowledge base developed over a long period of time, and give a fairly comprehensive view of the study of polynomial functions.

A. The Fundamental Theorem of Algebra

From Section 3.1, we know the set of real numbers is a subset of the complex numbers. Because complex numbers are the "larger" set (containing all other number sets), properties and theorems about complex numbers are more powerful and far reaching than theorems about real numbers. In the same way, real polynomials are a subset of the complex polynomials, and the same principle applies.

Complex Polynomial Functions

A complex polynomial of degree n has the form

$$P(x) = a_n x^n + a_{n-1} x^{n-1} + \cdots + a_1 x^1 + a_0,$$

where $a_n, a_{n-1}, \cdots, a_1, a_0$ are complex numbers and $a_n \neq 0$.

Notice that real polynomials have the same form, but here $a_n, a_{n-1}, \ldots, a_1, a_0$ represent *complex numbers*. In 1799, Carl Friedrich Gauss (1777–1855) proved that *all* polynomial functions have zeroes, and that the number of zeroes is equal to the degree of the polynomial. The proof of this statement is based on a theorem that is the bedrock

WORTHY OF NOTE

Quadratic functions also belong to the larger family of **complex polynomial functions.** Since quadratics have a known number of terms, it is common to write the general form using the early letters of the alphabet: $P(x) = \mathbf{a}x^2 + \mathbf{b}x + \mathbf{c}$. For higher degree polynomials, the number of terms is unknown or unspecified, and the general form is written using subscripts on a single letter.

for a complete study of polynomial functions, and has come to be known as the **fundamental theorem of algebra.**

The Fundamental Theorem of Algebra

Every complex polynomial of degree $n \geq 1$ has at least one complex zero.

Although the statement may seem trivial, it allows us to draw two important conclusions. The first is that our search for a solution will not be fruitless or wasted—zeroes for *all* polynomial equations exist. Second, the fundamental theorem combined with the factor theorem enables us to state the **linear factorization theorem.**

The Linear Factorization Theorem

If $p(x)$ is a polynomial function of degree $n \geq 1$, then p has exactly n linear factors and can be written in the form,

$$p(x) = a(x - c_1)(x - c_2) \cdot \cdots \cdot (x - c_n)$$

where $a \neq 0$ and $c_1, c_2, \ldots, c_n$ are (not necessarily distinct) complex numbers.

In other words, every complex polynomial of degree n can be rewritten as the product of a nonzero constant and exactly n linear factors (for a proof of this theorem, see Appendix VI).

EXAMPLE 1 ▶ **Writing a Polynomial as a Product of Linear Factors**

Rewrite $P(x) = x^4 - 8x^2 - 9$ as a product of linear factors, and find its zeroes.

Solution ▶ From its given form, we know $a = 1$. Since P has degree 4, the factored form must be $P(x) = (x - c_1)(x - c_2)(x - c_3)(x - c_4)$. Noting that P is in quadratic form, we substitute u for x^2 and u^2 for x^4 and attempt to factor:

$$x^4 - 8x^2 - 9 \rightarrow u^2 - 8u - 9 \qquad \text{\textcolor{red}{substitute } } u \text{ for } x^2; u^2 \text{ for } x^4$$
$$= (u - 9)(u + 1) \qquad \text{\textcolor{red}{factor in terms of } } u$$
$$= (x^2 - 9)(x^2 + 1) \qquad \text{\textcolor{red}{rewrite in terms of } } x \text{ (substitute } x^2 \text{ for } u)$$

We know $x^2 - 9$ will factor since it is a difference of squares. From our work with complex numbers (Section 3.1), we know $(a + bi)(a - bi) = a^2 + b^2$, and the factored form of $x^2 + 1$ must be $(x + i)(x - i)$. The completely factored form is

$$P(x) = (x + 3)(x - 3)(x + i)(x - i), \text{ and}$$

the zeroes of P are -3, 3, $-i$, and i.

WORTHY OF NOTE

While polynomials with complex coefficients are not the focus of this course, interested students can investigate the wider application of these theorems by completing **Exercise 115.**

Now try Exercises 7 through 10 ▶

EXAMPLE 2 ▶ **Writing a Polynomial as a Product of Linear Factors**

Rewrite $P(x) = x^3 + 2x^2 - 4x - 8$ as a product of linear factors and find its zeroes.

Solution ▶ We observe that $a = 1$ and P has degree 3, so the factored form must be $P(x) = (x - c_1)(x - c_2)(x - c_3)$. Noting that $ad = bc$ (Section R.4, page 40), we start with factoring by grouping.

$$P(x) = x^3 + 2x^2 - 4x - 8 \qquad \text{\textcolor{red}{group terms (in color)}}$$
$$= x^2(x + 2) - 4(x + 2) \qquad \text{\textcolor{red}{remove common factors (note sign change)}}$$
$$= (x + 2)(x^2 - 4) \qquad \text{\textcolor{red}{factor common binomial}}$$
$$= (x + 2)(x + 2)(x - 2) \qquad \text{\textcolor{red}{factor difference of squares}}$$

The zeroes of P are -2, -2, and 2.

Now try Exercises 11 through 14 ▶

Note the polynomial in Example 2 has three zeroes, but the zero -2 was repeated two times. In this case we say -2 is a zero of multiplicity two, and a zero of **even multiplicity.** It is also possible for a zero to be repeated three or more times, with those repeated an odd number of times called zeroes of **odd multiplicity** [the factor $(x - 2) = (x - 2)^1$ also gives a zero of odd multiplicity]. In general, repeated factors are written in exponential form and we have

Zeroes of Multiplicity

If p is a polynomial function with degree $n \geq 1$, and $(x - c)$ occurs as a factor of p exactly m times, then c is a zero of multiplicity m.

EXAMPLE 3 ▶ **Identifying the Multiplicity of a Zero**

Factor the given function completely, writing repeated factors in exponential form. Then state the multiplicity of each zero: $P(x) = (x^2 + 8x + 16)(x^2 - x - 20)(x - 5)$

Solution ▶

$$\begin{aligned} P(x) &= (x^2 + 8x + 16)(x^2 - x - 20)(x - 5) & \text{given polynomial} \\ &= (x + 4)(x + 4)(x - 5)(x + 4)(x - 5) & \text{trinomial factoring} \\ &= (x + 4)^3(x - 5)^2 & \text{exponential form} \end{aligned}$$

For function P, -4 is a zero of multiplicity 3 (odd multiplicity), and 5 is a zero of multiplicity 2 (even multiplicity).

Now try Exercises 15 through 18 ▶

These examples help illustrate three important consequences of the linear factorization theorem. From Example 1, if the coefficients of P are real, the polynomial can be factored into linear and quadratic factors using real numbers only $[(x + 3)(x - 3)(x^2 + 1)]$, where the quadratic factors have no real zeroes. Quadratic factors of this type are said to be **irreducible.**

Corollary I: Irreducible Quadratic Factors

If p is a polynomial with real coefficients, p can be factored into a product of linear factors (which are not necessarily distinct) and irreducible quadratic factors having real coefficients.

Closely related to this corollary and our previous study of quadratic functions, complex zeroes of the irreducible factors must occur in conjugate pairs.

Corollary II: Complex Conjugates

If p is a polynomial with real coefficients, complex zeroes must occur in conjugate pairs. If $a + bi$, $b \neq 0$ is a zero, then $a - bi$ will also be a zero.

Finally, the polynomial in Example 1 has degree 4 with 4 zeroes (two real, two complex), and the polynomial in Example 2 has degree 3 with 3 zeroes (three real, one of these is a repeated root). While not shown explicitly, the polynomial in Example 3 has degree 5, and there were 5 zeroes (one repeated twice, one repeated three times). This suggests our final corollary.

Corollary III: Number of Zeroes

If p is a polynomial function with degree $n \geq 1$, then p has exactly n zeroes (real or complex), where zeroes of multiplicity m are counted m times.

These corollaries help us gain valuable information about a polynomial, when only partial information is given or known.

EXAMPLE 4 ▶ **Constructing a Polynomial from Its Zeroes**

A polynomial P of degree 3 with real coefficients has zeroes of -1 and $2 + i\sqrt{3}$. Find the polynomial (assume $a = 1$).

Solution ▶ Using the factor theorem, two of the factors are $(x + 1)$ and $x - (2 + i\sqrt{3})$. From Corollary II, $2 - i\sqrt{3}$ must also be a zero and $x - (2 - i\sqrt{3})$ is also a factor of P. This gives

$$P(x) = (x + 1)[x - (2 + i\sqrt{3})][x - (2 - i\sqrt{3})]$$
$$= (x + 1)[(x - 2) - i\sqrt{3}][(x - 2) + i\sqrt{3}] \quad \text{associative property}$$
$$= (x + 1)[(x^2 - 4x + 4) + 3] \quad (a + bi)(a - bi) = a^2 + b^2$$
$$= (x + 1)(x^2 - 4x + 7) \quad \text{simplify}$$
$$= x^3 - 3x^2 + 3x + 7 \quad \text{result}$$

> **WORTHY OF NOTE**
>
> When reconstructing a polynomial P having complex zeroes, it is often more efficient to determine the irreducible quadratic factors of P separately, as shown here. For the zeroes $2 \pm i\sqrt{3}$ we have
>
> $$x = 2 \pm i\sqrt{3}$$
> $$x - 2 = \pm i\sqrt{3}$$
> $$(x - 2)^2 = (\pm i\sqrt{3})^2$$
> $$x^2 - 4x + 4 = -3$$
> $$x^2 - 4x + 7 = 0.$$
>
> The quadratic factor is $(x^2 - 4x + 7)$.

The polynomial is $P(x) = x^3 - 3x^2 + 3x + 7$, which can be verified using the remainder theorem and any of the original zeroes. A calculator check using the zero $x = 2 + i\sqrt{3}$ is shown here.

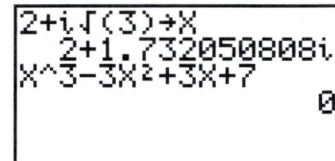
```
2+i√(3)→X
      2+1.732050808i
X^3-3X²+3X+7
                   0
```

Now try Exercises 19 through 22 ▶

EXAMPLE 5 ▶ **Building a Polynomial from Its Zeroes**

Find a fourth degree polynomial P with real coefficients, if 3 is the only real zero and $2i$ is also a zero of P.

Solution ▶ Since complex zeroes must occur in conjugate pairs, $-2i$ is also a zero, but this accounts for only three zeroes. Since P has degree 4, 3 must be a *repeated* zero, and the factors of P are $(x - 3)(x - 3)(x - 2i)(x + 2i)$.

$$P(x) = (x - 3)(x - 3)(x - 2i)(x + 2i) \quad \text{factored form}$$
$$= (x^2 - 6x + 9)(x^2 + 4) \quad \text{multiply binomials, } (a + bi)(a - bi) = a^2 + b^2$$
$$= x^4 - 6x^3 + 13x^2 - 24x + 36 \quad \text{result}$$

The polynomial is $P(x) = x^4 - 6x^3 + 13x^2 - 24x + 36$, which can be verified using a calculator or the remainder theorem and any of the original zeroes.

☑ **A.** You've just seen how we can apply the fundamental theorem of algebra and the linear factorization theorem

Now try Exercises 23 through 28 ▶

B. Real Polynomials and the Intermediate Value Theorem

The fundamental theorem of algebra is called an **existence theorem,** as it affirms the *existence* of the zeroes but does not tell us where or how to find them. Because polynomial graphs are continuous (there are no holes or breaks in the graph), the **intermediate value theorem (IVT)** can be used for this purpose.

The Intermediate Value Theorem

Given P is a polynomial with real coefficients, if $P(a)$ and $P(b)$ have opposite signs, there is *at least* one number c between a and b such that $P(c) = 0$ (see Figure 4.3).

Figure 4.3

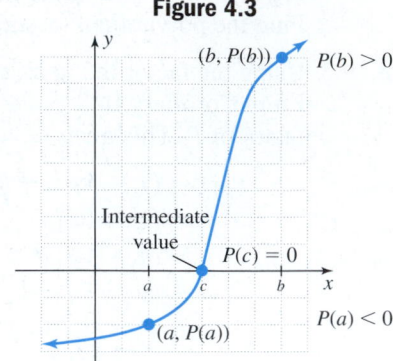

WORTHY OF NOTE

You might recall a similar idea was used in Section 2.1, where we noted the graph of $P(x)$ crosses the x-axis at the zeroes determined by linear factors, with a corresponding change of sign in the function values.

EXAMPLE 6 ▶ **Finding Zeroes Using the Intermediate Value Theorem**

Use the intermediate value theorem to show $P(x) = x^3 - 9x + 6$ has at least one zero in the interval given:

   a. $[-4, -3]$ **b.** $[0, 1]$ **c.** $[-4, 3]$

Solution ▶ **a.** Begin by evaluating P at $x = -4$ and $x = -3$.

$$P(-4) = (-4)^3 - 9(-4) + 6 \qquad P(-3) = (-3)^3 - 9(-3) + 6$$
$$= -64 + 36 + 6 \qquad\qquad = -27 + 27 + 6$$
$$= -22 \qquad\qquad\qquad = 6$$

Since $P(-4) < 0$ and $P(-3) > 0$, there must be at least one number c_1 between -4 and -3 where $P(c_1) = 0$. The graph must cross the x-axis at least once in this interval.

b. Evaluate P at $x = 0$ and $x = 1$.

$$P(0) = (0)^3 - 9(0) + 6 \qquad P(1) = (1)^3 - 9(1) + 6$$
$$= 0 - 0 + 6 \qquad\qquad = 1 - 9 + 6$$
$$= 6 \qquad\qquad\qquad = -2$$

Since $P(0) > 0$ and $P(1) < 0$, there must be at least one number c_2 between 0 and 1 where $P(c_2) = 0$.

c. In part (a) we found that $P(-4) < 0$. For $P(3)$ we have

$$P(3) = (3)^3 - 9(3) + 6$$
$$= 27 - 27 + 6$$
$$= 6$$

Since $P(-4) < 0$ and $P(3) > 0$, there must be at least one number c_3 between -4 and 3 where $P(c_3) = 0$. In fact, from parts (a) and (b) we know of two that exist, and the graph of P shown in Figure 4.4 reveals there are actually three real roots in this interval. This illustrates why the intermediate value theorem uses the phrase, "there is *at least* one number c."

Figure 4.4

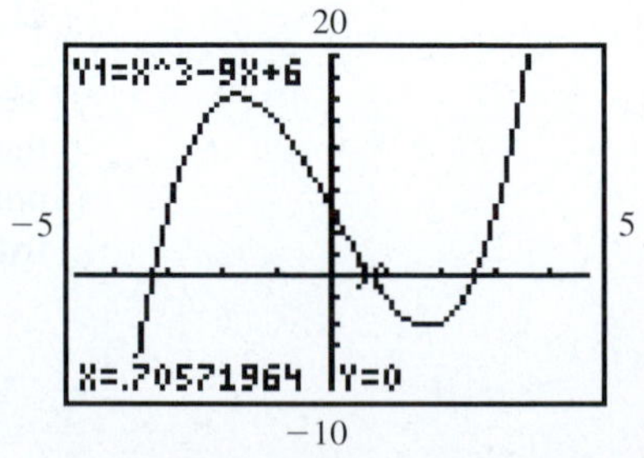

Y1=X^3-9X+6

X=.70571964 Y=0

Now try Exercises 29 and 30 ▶

Figure 4.4A

To help illustrate the Intermediate Value Theorem, many graphing calculators offer a useful feature called *split screen viewing*, that enables us to view a table of values and the graph of a function at the same time. To illustrate, enter the function $y = x^3 - 9x + 6$ (from Example 6) as Y_1 on the (Y=) screen, then set the viewing window as shown in Figure 4.4. Set your table in **AUTO** mode with $\Delta Tbl = 1$, then press the (MODE) key (see Figure 4.4A) and notice the second-to-last entry on this screen reads: **Full** for full screen viewing, **Horiz** for splitting the screen horizontally with the graph above a reduced home screen, and **G-T,** which represents **Graph-Table** and splits the screen vertically. In the **G-T** mode, the graph appears on the left and the table of values on the right. Navigate the cursor to the **G-T** mode and press (ENTER). Pressing the (GRAPH) key at this point should give you a screen similar to Figure 4.5. Scrolling downward shows the function also changes sign between $x = 2$ and $x = 3$. For more on this idea, **see Exercises 31 and 32.**

As a final note, while the intermediate value theorem is a powerful yet simple tool, it must be used with care. For example, given $p(x) = -x^4 + 10x^2 - 5$, $p(-1) > 0$ and $p(1) > 0$, seeming to indicate that no zeroes exist in the interval $(-1, 1)$. Actually, there are two zeroes, as seen in Figure 4.6.

☑ **B.** You've just seen how we can use the intermediate value theorem to identify intervals containing a polynomial zero

Figure 4.5

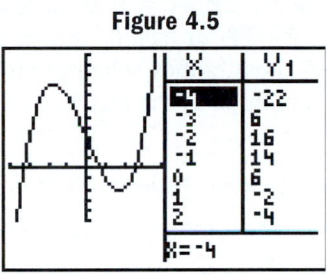

Figure 4.6

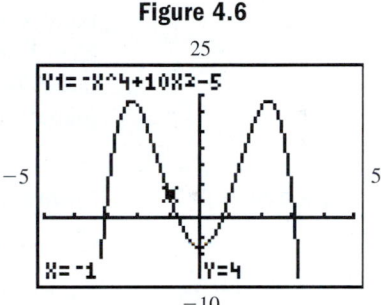

C. The Rational Zeroes Theorem

The fundamental theorem of algebra tells us that zeroes of a polynomial function *exist*. The intermediate value theorem tells us how to *locate* intervals that contain zeroes. Our next theorem gives us the information we need to actually *find* certain zeroes of a polynomial. Recall that if c is a zero of P, then $P(c) = 0$, and when $P(x)$ is divided by $x - c$ using synthetic division, the remainder is zero (from the remainder and factor theorems).

To find *divisors that give a remainder of zero,* we make the following observations. To solve $3x^2 - 11x - 20 = 0$ by factoring, a beginner might write out all possible binomial pairs where the **F**irst term in the F-O-I-L process multiplies to $3x^2$ and the **L**ast term multiplies to 20. The six possibilities are shown here:

$$(3x \quad 1)(x \quad 20) \qquad (3x \quad 20)(x \quad 1) \qquad (3x \quad 2)(x \quad 10) \qquad (3x \quad 10)(x \quad 2)$$

$$(3x \quad 4)(x \quad 5) \qquad (3x \quad 5)(x \quad 4)$$

If $3x^2 - 11x - 20$ is factorable using integers, the factors *must be somewhere in this list.* Also, the first coefficient in each binomial must be a factor of the leading coefficient, and the second coefficient must be a factor of the constant term. This means that regardless of which factored form is correct, the solution will be a rational number whose numerator comes from the factors of 20, and whose denominator comes from the factors of 3. The correct factored form is shown here, along with the solution:

$$3x^2 - 11x - 20 = 0$$
$$(3x + 4)(x - 5) = 0$$
$$3x + 4 = 0 \qquad\qquad x - 5 = 0$$

$$x = \dfrac{-4}{3} \begin{array}{l} \leftarrow \text{from the factors of 20} \\ \leftarrow \text{from the factors of 3} \end{array} \qquad\qquad x = \dfrac{5}{1} \begin{array}{l} \leftarrow \text{from the factors of 20} \\ \leftarrow \text{from the factors of 3} \end{array}$$

This same principle also applies to polynomials of higher degree, and these observations suggest the following theorem.

The Rational Zeroes Theorem

Given polynomial P with integer coefficients, and $\dfrac{p}{q}$ a rational number in lowest terms, the rational zeroes of P (if they exist) must be of the form $\dfrac{p}{q}$, where p is a factor of the constant term, and q is a factor of the leading coefficient.

Note that if the leading coefficient is 1, the possible rational zeroes are limited to factors of the constant term: $\frac{p}{1} = p$. If the leading coefficient is not "1" and the constant term has a large number of factors, the set of possible rational zeroes becomes rather large. To list these possibilities, it helps to begin with all factor *pairs* of the constant a_0, then divide each of these by the factors of a_n as shown in Example 7.

EXAMPLE 7 ▶ **Identifying the Possible Rational Zeroes of a Polynomial**

List all possible rational zeroes for the function $P(x) = 3x^4 + 14x^3 - x^2 - 42x - 24$.

Solution ▶ All rational zeroes must be of the form $\frac{p}{q}$, where p is a factor of $a_0 = -24$ and q is a factor of $a_n = 3$. The factor pairs of -24 are: $\pm 1, \pm 24, \pm 2, \pm 12, \pm 3, \pm 8, \pm 4$ and ± 6. Dividing each by ± 1 and ± 3 (the factor pairs of 3), we note division by ± 1 will not change any of the previous values, while division by ± 3 gives $\pm \frac{1}{3}, \pm \frac{2}{3}, \pm \frac{8}{3}, \pm \frac{4}{3}$ as additional possibilities. Any rational zeroes must be from the set

$$\left\{ \pm 1, \pm 24, \pm 2, \pm 12, \pm 3, \pm 8, \pm 4, \pm 6, \pm \tfrac{1}{3}, \pm \tfrac{2}{3}, \pm \tfrac{8}{3}, \pm \tfrac{4}{3} \right\}.$$

Now try Exercises 33 through 40 ▶

The actual zeroes of the function in Example 7 are $x = \sqrt{3}$, $x = -\sqrt{3}$, $x = -\frac{2}{3}$, and $x = -4$ and the graph of $P(x) = 3x^4 + 14x^3 - x^2 - 42x - 24$ is shown in Figure 4.7. Although the *rational* zeroes are indeed in the set noted, it's apparent we need a way to narrow down the number of possibilities (we don't want to try all 24 possible zeroes). If we're able to find even one factor easily, we can rewrite the polynomial using this factor and the quotient polynomial, with the hope of factoring further using trinomial factoring or factoring by grouping. Many times testing to see if 1 or -1 are zeroes will help.

Figure 4.7

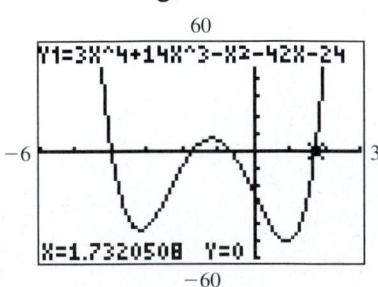

Tests to Determine If 1 or -1 is a Zero of P

For any polynomial P with real coefficients,

1. If the sum of all coefficients is zero, then 1 is a root and $(x - 1)$ is a factor.
2. After changing the sign of all terms with odd degree, if the sum of the coefficients is zero, then -1 is a root and $(x + 1)$ is a factor.

EXAMPLE 8 ▶ **Finding the Rational Zeroes of a Polynomial**

Find all rational zeroes of $P(x) = 3x^4 - x^3 - 8x^2 + 2x + 4$, and use them to write the function in completely factored form. Then use the factored form to name all zeroes of P.

Solution ▶ Instead of listing all possibilities using the rational zeroes theorem, we first test for 1 and -1, then see if we're able to complete the factorization using other means. The sum of the coefficients is: $3 - 1 - 8 + 2 + 4 = 0$, which means 1 is a zero and $x - 1$ is a factor. By changing the sign on terms of odd degree, we have $3x^4 + x^3 - 8x^2 - 2x + 4$ and $3 + 1 - 8 - 2 + 4 = -2$, showing -1 is *not* a zero. Using $x = 1$ and the factor theorem, we have

> **WORTHY OF NOTE**
>
> In the second to last line of Example 8, we factored $x^2 - 2$ as $(x + \sqrt{2})(x - \sqrt{2})$. As discussed in Section R.6, this is an application of factoring the difference of two squares: $a^2 - b^2 = (a + b)(a - b)$. By mentally rewriting $x^2 - 2$ as $x^2 - (\sqrt{2})^2$, we obtain the result shown. Also **see Exercise 113**.

use 1 as a "divisor" $\underline{1|}$ 3 -1 -8 2   4

                                    3   2 -6 -4

                         3   2 -6 -4 $\underline{|0}$

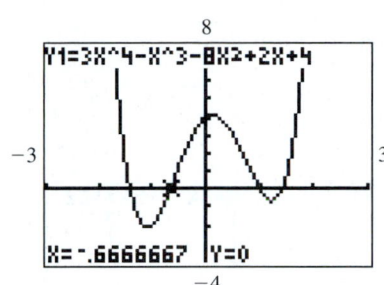

and we write P as $P(x) = (x - 1)(3x^3 + 2x^2 - 6x - 4)$. Noting the quotient polynomial can be factored by grouping ($ad = bc$), we need not continue with synthetic division or the factor theorem.

$$
\begin{aligned}
P(x) &= (x - 1)(3x^3 + 2x^2 - 6x - 4) && \text{group terms} \\
&= (x - 1)[x^2(3x + 2) - 2(3x + 2)] && \text{factor common terms} \\
&= (x - 1)(3x + 2)(x^2 - 2) && \text{factor common binomial} \\
&= (x - 1)(3x + 2)(x + \sqrt{2})(x - \sqrt{2}) && \text{completely factored form}
\end{aligned}
$$

The zeroes of P are 1, $\frac{-2}{3}$, and $\pm\sqrt{2}$. The graph of P is shown in the figure.

Now try Exercises 41 through 62 ▶

In cases where the quotient polynomial is not easily factored, we continue with synthetic division and other possible zeroes, until the remaining zeroes can be determined.

EXAMPLE 9 ▶ **Finding the Zeroes of a Polynomial**

Find all zeroes of $P(x) = x^5 - 3x^4 + 3x^3 - 5x^2 + 12$.

Solution ▶ Using the rational zeroes theorem, the possibilities are: $\{\pm 1, \pm 12, \pm 2, \pm 6, \pm 3, \pm 4\}$. The test for 1 shows 1 is not a zero. After changing the signs of all terms with odd degree, we have $-1 - 3 - 3 - 5 + 12 = 0$, and find -1 *is* a zero. Using -1 with the factor theorem, we continue our search for additional factors. Noting that P is missing a linear term, we include a placeholder zero:

use -1 as a "divisor" $\underline{-1|}$ 1 -3 3 -5 0   12   coefficients of P

                                   -1 4 -7 12 -12

                        1   -4 7 -12 12 $\underline{|0}$ coefficients of $q_1(x)$

Here the quotient polynomial $q_1(x) = x^4 - 4x^3 + 7x^2 - 12x + 12$ is not easily factored, so we next try 2, *using the quotient polynomial*:

use 2 as a "divisor" on $q_1(x)$ $\underline{2|}$ 1 -4 7 -12 12   coefficients of $q_1(x)$

                                     2   -4 6 -12

                        1   -2 3 -6 $\underline{|0}$ coefficients of $q_2(x)$

If you miss the fact that $q_2(x)$ is actually factorable ($ad = bc$), the process would continue using -2 and the current quotient.

use -2 as a "divisor" $\underline{-2|}$ $\quad 1 \quad -2 \quad 3 \quad -6$ coefficients of $q_2(x)$
$$\underline{\quad\quad\quad -2 \quad 8 \quad -22}$$
$$1 \quad -4 \quad 11 \quad \underline{|-28} \quad -2 \text{ is not a zero}$$

We find -2 is not a zero, and in fact, trying *all other* possibilities will show that *none* of them are zeroes. As there must be five zeroes, we are reminded of three things:

1. This process can only find *rational zeros* (the remaining zeroes may be irrational or complex),

2. This process cannot find irreducible quadratic factors (unless they appear as the quotient polynomial), and

3. Some of the zeroes *may have multiplicities greater than 1!*

Testing the zero 2 for a second time using $q_2(x)$ gives

use 2 as a "divisor" $\underline{2|}$ $\quad 1 \quad -2 \quad 3 \quad -6$ coefficients of $q_2(x)$
$$\underline{\quad\quad\quad 2 \quad 0 \quad 6}$$
$$1 \quad 0 \quad 3 \quad \underline{|0} \quad 2 \text{ is a } repeated \text{ zero}$$

and we see that 2 is actually a zero of multiplicity two, and the final quotient is the irreducible quadratic factor $x^2 + 3$. Using this information produces the factored form $P(x) = (x + 1)(x - 2)^2(x^2 + 3) = (x + 1)(x - 2)^2(x + i\sqrt{3})(x - i\sqrt{3})$, and the zeroes of P are $-i\sqrt{3}$, $i\sqrt{3}$, -1, and 2 with multiplicity two. The graph of P is shown in the figure.

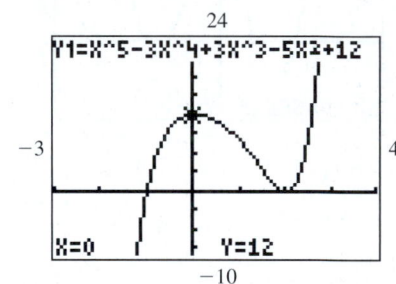

Now try Exercises 63 through 82 ▶

✓ **C.** You've just seen how we can find rational zeroes of a real polynomial function using the rational zeroes theorem

In Example 9, note that since the leading coefficient is 1, all possible rational zeroes will be integers. This means our initial search can easily be performed using the **TABLE** feature of a graphing calculator. But note that while Figure 4.8 indeed shows that $x = -1$ and $x = 2$ are zeroes, it cannot show that $x = 2$ is a repeated zero until we begin working with the quotient polynomial. In the next section, we'll learn how to locate repeated zeroes from a polynomial's graph.

Figure 4.8

D. Descartes' Rule of Signs and Upper/Lower Bounds

Testing $x = 1$ and $x = -1$ is one way to reduce the number of possible rational zeroes, but unless we're very lucky, factoring the polynomial can still be a challenge. **Descartes' rule of signs** and the **upper and lower bounds property** offer additional assistance.

> ### Descartes' Rule of Signs
>
> Given the real polynomial equation $P(x) = 0$,
>
> **1.** The number of positive real zeroes is equal to the number of variations in sign for $P(x)$, or an even number less.
>
> **2.** The number of negative real zeroes is equal to the number of variations in sign for $P(-x)$, or an even number less.

EXAMPLE 10 ▶ **Finding the Zeroes of a Polynomial**

For $P(x) = 2x^5 - 5x^4 + x^3 + x^2 - x + 6$,

a. Use the rational zeroes theorem to list all possible rational zeroes.

b. Apply Descartes' rule to count the number of possible positive, negative, and complex zeroes.

c. Use this information and the tools of this section to find all zeroes of P.

Solution ▶

a. The factors of 2 are $\{\pm 1, \pm 2\}$ and the factors of 6 are $\{\pm 1, \pm 6, \pm 2, \pm 3\}$. The possible rational zeroes for P are $\{\pm 1, \pm 6, \pm 2, \pm 3, \pm \frac{1}{2}, \pm \frac{3}{2}\}$.

b. For Descartes' rule, we organize our work in a table. Since P has degree 5, there must be a total of five zeroes. For this illustration, positive terms are in **blue** and negative terms in **red**: $P(x) = 2x^5 - 5x^4 + x^3 + x^2 - x + 6$. The terms change sign a total of four times, meaning there are four, two, or zero positive roots. For the negative roots, recall that $P(-x)$ will change the sign of *all odd-degree terms*, giving $P(-x) = -2x^5 - 5x^4 - x^3 + x^2 + x + 6$. This time there is only one sign change (from negative to positive) showing there will be exactly one negative root, a fact that is highlighted in the following table. Since there must be 5 zeroes, the number of possible complex zeroes is: none, two, or four, as shown.

possible positive zeroes	known negative zeroes	possibilities for complex roots	total number must be 5
4	1	0	5
2	1	2	5
0	1	4	5

c. Testing 1 and -1 shows $x = 1$ is not a root, but $x = -1$ *is,* and using -1 in synthetic division gives:

use -1 as a "divisor" $-1|$ 2 -5 1   1 -1 6   coefficients of $P(x)$

$\qquad\qquad\qquad\qquad\quad -2\quad 7\ -8\quad 7\ -6$

$\qquad\qquad\qquad\qquad 2\ -7\quad 8\ -7\quad 6\ \ \underline{|0}$ $q_1(x)$ is not easily factored

Since there is *only one* negative root, we need only check the remaining positive zeroes. The quotient $q_1(x)$ is not easily factored, so we continue with synthetic division using the next larger positive root, $x = 2$.

use 2 as a "divisor" $2|$ 2 -7 8 -7 6   coefficients of $q_1(x)$

$\qquad\qquad\qquad\qquad\quad 4\ -6\quad 4\ -6$

$\qquad\qquad\qquad\qquad 2\ -3\quad 2\ -3\ \ \underline{|0}$ $q_2(x)$

The partially factored form is $P(x) = (x + 1)(x - 2)(2x^3 - 3x^2 + 2x - 3)$.

WORTHY OF NOTE

As you recall from our study of quadratics, it's entirely possible for a polynomial function to have no real zeroes. Also, if the zeroes are irrational, complex, or a combination of these, they cannot be found using the rational zeroes theorem. For a look at ways to determine these zeroes, see the *Reinforcing Basic Skills* feature that follows Section 4.3.

Graphing $P(x)$ at this point (see the figure) verifies that -1 and 2 are zeroes, but also indicates there is an additional zero between 1 and 2. If it's a *rational* zero, it must be $x = \dfrac{3}{2}$ from our list of possible zeroes.

Checking $\dfrac{3}{2}$ using synthetic division and $q_2(x)$ we have

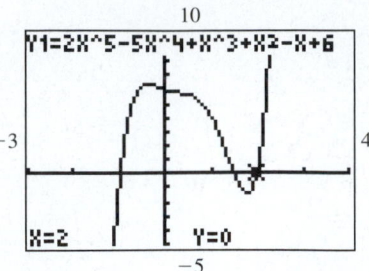

$$
\begin{array}{r|rrrr}
\frac{3}{2} & 2 & -3 & 2 & -3 \\
 & \downarrow & 3 & 0 & 3 \\
\hline
 & 2 & 0 & 2 & \underline{|0}
\end{array}
$$

Since the remainder is zero, $P\left(\dfrac{3}{2}\right) = 0$ (remainder theorem) and $\left(x - \dfrac{3}{2}\right)$ is a factor (factor theorem). Since the zero is a fraction, we'll use the ideas discussed in Section 4.1 to help write $P(x)$ in completely factored form:

$$P(x) = (x + 1)(x - 2)\left(x - \frac{3}{2}\right)(2x^2 + 2) \qquad \text{\color{magenta}partially factored form}$$

$$= (x + 1)(x - 2)\left(x - \frac{3}{2}\right)(2)(x^2 + 1) \qquad \text{\color{magenta}factor out 2}$$

$$= (x + 1)(x - 2)(2x - 3)(x^2 + 1) \qquad \text{\color{magenta}multiply } 2\left(x - \frac{3}{2}\right)$$

$$= (x + 1)(x - 2)(2x - 3)(x + i)(x - i) \qquad \text{\color{magenta}completely factored form}$$

The zeroes of P are $-1, 2, \frac{3}{2}, -i$ and i, with two positive, one negative, and two complex zeroes (row two of the table).

Now try Exercises 83 through 96 ▶

One final idea that helps reduce the number of possible zeroes is the **upper and lower bounds property**. A number b is an **upper bound** on the positive zeroes of a function if no positive zero is greater than b. In the same way, a number a is a **lower bound** on the negative zeroes if no negative zero is less than a.

> **Upper and Lower Bounds Property**
>
> Given $P(x)$ is a polynomial with real coefficients.
>
> 1. If $P(x)$ is divided by $x - b$ ($b > 0$) using synthetic division and all coefficients in the quotient row are either positive or zero, then b is an upper bound on the zeroes of P.
> 2. If $P(x)$ is divided by $x - a$ ($a < 0$) using synthetic division and all coefficients in the quotient row alternate in sign, then a is a lower bound on the zeroes of P.
>
> For both 1 and 2, zero coefficients can be either positive or negative as needed.

☑ **D. You just seen how we can obtain more information on the zeroes of real polynomials using Descartes' rule of signs and upper/lower bounds theorem**

While this test certainly helps narrow the possibilities, we gain the additional benefit of knowing the property actually places boundaries on *all* real zeroes of the polynomial, both rational and irrational. In Part (c) of Example 10, the quotient row of the first division alternates in sign, showing $x = -1$ is both a zero and a lower bound on the real zeroes of P. For more on the upper and lower bounds property, **see Exercise 111.**

E. Applications of Polynomial Functions

Polynomial functions can be very accurate models of real-world phenomena, though we often must restrict their domain, as illustrated in Example 11.

EXAMPLE 11 ▶   **Using the Remainder Theorem to Solve an Oceanography Application**

As part of an environmental study, scientists use radar to map the ocean floor from the coastline to a distance 12 mi from shore. In this study, ocean trenches appear as negative values and underwater mountains as positive values, as measured from the surrounding ocean floor. The terrain due west of a particular island can be modeled by $h(x) = x^4 - 25x^3 + 200x^2 - 560x + 384$, where $h(x)$ represents the height in feet, x mi from shore ($0 < x \leq 12$).

   a. Use the remainder theorem to find the "height of the ocean floor" 10 mi out.

   b. Use the tools developed in this section to find the number of times the ocean floor has height $h(x) = 0$ in this interval, given this occurs 12 mi out.

Solution ▶   **a.** For part (a) we simply evaluate $h(10)$ using the remainder theorem.

$$
\begin{array}{r|rrrrr}
\text{use 10 as a "divisor"} \quad 10\rfloor & 1 & -25 & 200 & -560 & 384 \quad \text{coefficients of } h(x) \\
 & & 10 & -150 & 500 & -600 \\
\hline
 & 1 & -15 & 50 & -60 & \underline{|-216} \quad \text{remainder is } -216
\end{array}
$$

Ten miles from shore, there is an ocean trench 216 ft deep.

   b. For part (b), we're given 12 is zero, so we again use the remainder theorem and work with the quotient polynomial.

$$
\begin{array}{r|rrrrr}
\text{use 12 as a "divisor"} \quad 12\rfloor & 1 & -25 & 200 & -560 & 384 \quad \text{coefficients of } h(x) \\
 & & 12 & -156 & 528 & -384 \\
\hline
 & 1 & -13 & 44 & -32 & \underline{|0} \quad q_1(x)
\end{array}
$$

The quotient is $q_1(x) = x^3 - 13x^2 + 44x - 32$. Since $a = 1$, we know the remaining rational zeroes must be factors of -32: $\{\pm 1, \pm 32, \pm 2, \pm 16, \pm 4, \pm 8\}$. Using $x = 1$ gives

$$
\begin{array}{r|rrrr}
\text{use 1 as a "divisor"} \quad 1\rfloor & 1 & -13 & 44 & -32 \quad \text{coefficients of } q_1(x) \\
 & & 1 & -12 & 32 \\
\hline
 & 1 & -12 & 32 & \underline{|0} \quad q_2(x)
\end{array}
$$

The function can now be written as $h(x) = (x - 12)(x - 1)(x^2 - 12x + 32)$ and in completely factored form $h(x) = (x - 12)(x - 1)(x - 4)(x - 8)$. The ocean floor has height zero at distances of 1, 4, 8, and 12 mi from shore.

The graph of $h(x)$ is shown in the figure. The graph shows a great deal of variation in the ocean floor, but the zeroes occurring at 1, 4, 8, and 12 mi out are clearly evident.

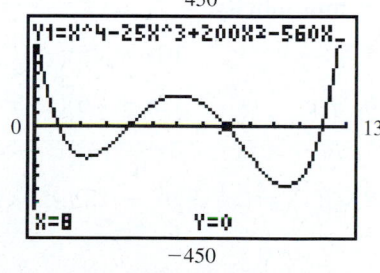

☑ **E.** You've just seen how we can solve an application of polynomial functions

Now try Exercises 99 through 110 ▶

4.2 EXERCISES

▶ CONCEPTS AND VOCABULARY

Fill in each blank with the appropriate word or phrase. Carefully reread the section if needed.

1. A complex polynomial is one where one or more <u>coefficients</u> are complex numbers.

2. A polynomial function of degree n will have exactly <u>n</u> zeroes, real or <u>complex</u>, where zeroes of multiplicity m are counted m times.

3. If $a + bi$ is a complex zero of polynomial P with real coefficients, then <u>$a - bi$</u> is also a zero.

4. According to Descartes' rule of signs, there are as many <u>positive</u> real roots as changes in sign from term to term, or an <u>even</u> number less.

5. Which of the following values is *not* a possible root of $f(x) = 6x^3 - 2x^2 + 5x - 12$:

   a. $x = \frac{4}{3}$ **b.** $x = \frac{3}{4}$ **c.** $x = \frac{1}{2}$

   Discuss/Explain why.   b; 4 is not a factor of 6

6. Discuss/Explain each of the following: (a) irreducible quadratic factors, (b) factors that are complex conjugates, (c) zeroes of multiplicity m, and (d) upper bounds on the zeroes of a polynomial.   Answers will vary.

▶ DEVELOPING YOUR SKILLS

Rewrite each polynomial as a product of linear factors, and find the zeroes of the polynomial.

7. $P(x) = x^4 + 5x^2 - 36$
   $P(x) = (x + 2)(x - 2)(x + 3i)(x - 3i)$
   $x = -2, x = 2, x = 3i, x = -3i$

8. $Q(x) = x^4 + 21x^2 - 100$
   $Q(x) = (x + 2)(x - 2)(x + 5i)(x - 5i)$
   $x = -2, x = 2, x = -5i, x = 5i$

9. $Q(x) = x^4 - 16$
   $Q(x) = (x + 2)(x - 2)(x + 2i)(x - 2i)$
   $x = -2, x = 2, x = 2i, x = -2i$

10. $P(x) = x^4 - 81$
   $P(x) = (x + 3)(x - 3)(x + 3i)(x - 3i)$
   $x = -3, x = 3, x = -3i, x = 3i$

11. $P(x) = x^3 + x^2 - x - 1$
   $P(x) = (x + 1)(x + 1)(x - 1)$
   $x = -1, x = -1, x = 1$

12. $Q(x) = x^3 - 3x^2 - 9x + 27$
   $Q(x) = (x - 3)(x + 3)(x - 3)$
   $x = 3, x = -3, x = 3$

13. $Q(x) = x^3 - 5x^2 - 25x + 125$
   $Q(x) = (x - 5)(x + 5)(x - 5)$
   $x = 5, x = -5, x = 5$

14. $P(x) = x^3 + 4x^2 - 16x - 64$
   $P(x) = (x + 4)(x + 4)(x - 4)$
   $x = -4, x = -4, x = 4$

Factor each polynomial completely. Write any repeated factors in exponential form, then name all zeroes and their multiplicity.

15. $p(x) = (x^2 - 10x + 25)(x^2 + 4x - 45)(x + 9)$

16. $q(x) = (x^2 + 12x + 36)(x^2 + 2x - 24)(x - 4)$

17. $P(x) = (x^2 - 5x - 14)(x^2 - 49)(x + 2)$

18. $Q(x) = (x^2 - 9x + 18)(x^2 - 36)(x - 3)$

Find a polynomial $P(x)$ having real coefficients, with the degree and zeroes indicated. All real zeroes are given. Assume the lead coefficient is 1. Recall $(a + bi)(a - bi) = a^2 + b^2$.

19. degree 3, $x = 3, x = 2i$ $P(x) = x^3 - 3x^2 + 4x - 12$

20. degree 3, $x = -5, x = -3i$ $P(x) = x^3 + 5x^2 + 9x + 45$

21. degree 4, $x = -1, x = 2, x = i$
   $P(x) = x^4 - x^3 - x^2 - x - 2$

22. degree 4, $x = -1, x = 3, x = -2i$
   $P(x) = x^4 - 2x^3 + x^2 - 8x - 12$

23. degree 4, $x = 3, x = 2i$
   $P(x) = x^4 - 6x^3 + 13x^2 - 24x + 36$

24. degree 4, $x = -2, x = -3i$
   $P(x) = x^4 + 4x^3 + 13x^2 + 36x + 36$

25. degree 4, $x = -1, x = 1 + 2i$
   $P(x) = x^4 + 2x^2 + 8x + 5$

26. degree 4, $x = -1, x = 1 - 3i$
   $P(x) = x^4 + 7x^2 + 18x + 10$

27. degree 4, $x = -3, x = 1 + i\sqrt{2}$
   $P(x) = x^4 + 4x^3 + 27$

28. degree 4, $x = -2, x = 1 + i\sqrt{3}$
   $P(x) = x^4 + 2x^3 + 8x + 16$

Use the intermediate value theorem to verify the given polynomial has at least one zero "c_i" in the intervals specified. Do not find the zeroes.

29. $f(x) = x^3 + 2x^2 - 8x - 5$

   a. $[-4, -3]$ yes   **b.** $[2, 3]$ yes

30. $g(x) = x^4 - 2x^2 + 6x - 3$

   a. $[-3, -2]$ yes   **b.** $[0, 1]$ yes

Additional answers can be found in the Instructor Answer Appendix.

HOMEWORK SELECTION GUIDE

Core: 7–87 every other odd, 99, 103–109 odd, 100 (27 Exercises)
Standard: 1–4, 7–87 every other odd 99, 103–109 odd, 100, 113A, 133B, 117 (34 Exercises)

Extended: 1–4, 7–87 every other odd, 97–99, 103, 105, 107–110, 113A, 113B, 115a, 117, 118 (39 Exercises)
In Depth: 1–6, 7–95 every other odd, 97–99, 103, 105–110, 113A/B, 115a/b, 117–119 (47 Exercises)

For Exercises 31 and 32, enter each function on the $Y=$ screen. Then place your graphing calculator in G-T **MODE** and set up a **TABLE** using **TblStart** $= -5$ and Δ**TBL** $= 0.1$. Use the intermediate value theorem and the resulting **GRAPH** and **TABLE** to locate intervals $[x_1, x_2]$, where $x_2 - x_1 = 0.1$, that contain zeroes of the function. Assume all real zeroes are between -5 and 5.

$[-4.7, -4.6], [-2.3, -2.2], [0.9, 1]$
31. $f(x) = x^3 + 6x^2 + 4x - 10$

32. $g(x) = 2x^4 + 3x^3 - 7x^2 - 5x + 4$
$[-2.4, -2.3], [-1.1, -1], [0.5, 0.6], [1.4, 1.5]$

List all possible rational zeroes for the polynomials given, but do not solve.

33. $f(x) = 4x^3 - 19x - 15$

34. $g(x) = 3x^3 - 2x + 20$

35. $h(x) = 2x^3 - 5x^2 - 28x + 15$

36. $H(x) = 2x^3 - 19x^2 + 37x - 14$

37. $p(x) = 6x^4 - 2x^3 + 5x^2 - 28$

38. $q(x) = 7x^4 + 6x^3 - 49x^2 + 36$

39. $Y_1 = 32t^3 - 52t^2 + 17t + 3$

40. $Y_2 = 24t^3 + 17t^2 - 13t - 6$

Use the rational zeroes theorem to write each function in factored form and find all zeroes. Note $a = 1$.

41. $f(x) = x^3 - 13x + 12$ $(x + 4)(x - 1)(x - 3), x = -4, 1, 3$

42. $g(x) = x^3 - 21x + 20$ $(x - 1)(x - 4)(x + 5), x = 1, 4, -5$

43. $h(x) = x^3 - 19x - 30$ $(x + 3)(x + 2)(x - 5), x = -3, -2, 5$

44. $H(x) = x^3 - 28x - 48$ $(x + 4)(x + 2)(x - 6), x = -4, -2, 6$

45. $p(x) = x^3 - 2x^2 - 11x + 12$
$(x + 3)(x - 1)(x - 4), x = -3, 1, 4$
46. $q(x) = x^3 - 4x^2 - 7x + 10$
$(x + 2)(x - 1)(x - 5), x = -2, 1, 5$
47. $Y_1 = x^3 - 6x^2 - x + 30$
$(x + 2)(x - 3)(x - 5), x = -2, 3, 5$
48. $Y_2 = x^3 - 4x^2 - 20x + 48$
$(x + 4)(x - 2)(x - 6), x = -4, 2, 6$
49. $Y_3 = x^4 - 15x^2 + 10x + 24$
$(x + 4)(x + 1)(x - 2)(x - 3), x = -4, -1, 2, 3$
50. $Y_4 = x^4 - 23x^2 - 18x + 40$
$(x + 4)(x + 2)(x - 1)(x - 5), x = -4, -2, 1, 5$
51. $f(x) = x^4 + 7x^3 - 7x^2 - 55x - 42$
$(x + 7)(x + 2)(x + 1)(x - 3), x = -7, -2, -1, 3$
52. $g(x) = x^4 + 4x^3 - 17x^2 - 24x + 36$
$(x + 6)(x + 2)(x - 1)(x - 3), x = -6, -2, 1, 3$

Find all rational zeroes of the functions given and use them to write the function in factored form. Use the factored form to state *all* zeroes of f. Begin by applying the tests for 1 and -1.

53. $f(x) = 4x^3 - 7x + 3$ $(2x + 3)(2x - 1)(x - 1); x = -\frac{3}{2}, \frac{1}{2}, 1$

54. $g(x) = 9x^3 - 7x - 2$
$(x - 1)(3x + 1)(3x + 2); x = 1, -\frac{1}{3}, -\frac{2}{3}$
55. $h(x) = 4x^3 + 8x^2 - 3x - 9$
$(2x + 3)^2(x - 1); x = -\frac{3}{2}, 1$

56. $H(x) = 9x^3 + 3x^2 - 8x - 4$ $(x - 1)(3x + 2)^2; x = 1, -\frac{2}{3}$

57. $Y_1 = 2x^3 - 3x^2 - 9x + 10$
$(x + 2)(x - 1)(2x - 5); x = -2, 1, \frac{5}{2}$
58. $Y_2 = 3x^3 - 14x^2 + 17x - 6$
$(3x - 2)(x - 1)(x - 3); x = \frac{2}{3}, 1, 3$
59. $p(x) = 2x^4 + 3x^3 - 9x^2 - 15x - 5$
$(x + 1)(2x + 1)(x - \sqrt{5})(x + \sqrt{5}); x = -1, -\frac{1}{2}, \sqrt{5}, -\sqrt{5}$
60. $q(x) = 3x^4 + x^3 - 11x^2 - 3x + 6$
$(x + 1)(3x - 2)(x - \sqrt{3})(x + \sqrt{3}); x = -1, \frac{2}{3}, -\sqrt{3}, \sqrt{3}$
61. $r(x) = 3x^4 + 4x^3 + 8x^2 + 16x - 16$
$(x + 2)(3x - 2)(x - 2i)(x + 2i); x = -2, \frac{2}{3}, 2i, -2i$
62. $s(x) = 2x^4 - 7x^3 + 14x^2 - 63x - 36$
$(x - 4)(2x + 1)(x - 3i)(x + 3i); x = 4, -\frac{1}{2}, 3i, -3i$

Find the zeroes of the polynomials given using any combination of the rational zeroes theorem, testing for 1 and -1, and/or the remainder and factor theorems.

63. $f(x) = 2x^4 - 9x^3 + 4x^2 + 21x - 18$ $x = 1, 2, 3, \frac{-3}{2}$

64. $g(x) = 3x^4 + 4x^3 - 21x^2 - 10x + 24$ $x = 1, 2, -3, \frac{-4}{3}$

65. $h(x) = 3x^4 + 2x^3 - 9x^2 + 4$ $x = -2, 1, \frac{-2}{3}$

66. $H(x) = 7x^4 + 6x^3 - 49x^2 + 36$ $x = -3, 2, 1, \frac{-6}{7}$

67. $p(x) = 2x^4 + 3x^3 - 24x^2 - 68x - 48$ $x = -2, \frac{-3}{2}, 4$

68. $q(x) = 3x^4 - 19x^3 + 6x^2 + 96x - 32$ $x = 4, -2, \frac{1}{3}$

69. $r(x) = 3x^4 - 20x^3 + 34x^2 + 12x - 45$ $x = 3, -1, \frac{5}{3}$

70. $s(x) = 4x^4 - 15x^3 + 9x^2 + 16x - 12$ $x = -1, 2, \frac{3}{4}$

71. $Y_1 = x^5 + 6x^2 - 49x + 42$ $x = 1, 2, -3, \pm i\sqrt{7}$

72. $Y_2 = x^5 + 2x^2 - 9x + 6$ $x = -2, 1, \pm i\sqrt{3}$

73. $P(x) = 3x^5 + x^4 + x^3 + 7x^2 - 24x + 12$
$x = -2, \frac{2}{3}, 1, \pm i\sqrt{3}$
74. $P(x) = 2x^5 - x^4 - 3x^3 + 4x^2 - 14x + 12$
$x = -2, 1, \frac{3}{2}, \pm i\sqrt{2}$
75. $Y_1 = x^4 - 5x^3 + 20x - 16$ $x = 1, 2, 4, -2$

76. $Y_2 = x^4 - 10x^3 + 90x - 81$ $x = -3, 1, 3, 9$

77. $r(x) = x^4 - x^3 - 14x^2 + 2x + 24$ $x = -3, 4, \pm\sqrt{2}$

78. $s(x) = x^4 - 3x^3 - 13x^2 + 9x + 30$ $x = -2, 5, \pm\sqrt{3}$

79. $p(x) = 2x^4 - x^3 + 3x^2 - 3x - 9$ $x = -1, \frac{3}{2}, \pm i\sqrt{3}$

80. $q(x) = 3x^4 + x^3 + 13x^2 + 5x - 10$ $x = -1, \frac{2}{3}, \pm i\sqrt{5}$

81. $f(x) = 2x^5 - 7x^4 + 13x^3 - 23x^2 + 21x - 6$
$x = \frac{1}{2}, 1, 2, \pm i\sqrt{3}$
82. $g(x) = 4x^5 + 3x^4 + 3x^3 + 11x^2 - 27x + 6$
$x = -2, 1, \frac{1}{4}, \pm i\sqrt{3}$

Gather information on each polynomial using (a) the rational zeroes theorem, (b) testing for 1 and -1, (c) applying Descartes' rule of signs, and (d) using the upper and lower bounds property. Respond explicitly to each.

83. $f(x) = x^4 - 2x^3 + 4x - 8$

84. $g(x) = x^4 + 3x^3 - 7x - 6$

85. $h(x) = x^5 + x^4 - 3x^3 + 5x + 2$

86. $H(x) = x^5 + x^4 - 2x^3 + 4x - 4$

Additional answers can be found in the Instructor Answer Appendix.

87. $p(x) = x^5 - 3x^4 + 3x^3 - 9x^2 - 4x + 12$

88. $q(x) = x^5 - 2x^4 - 8x^3 + 16x^2 + 7x - 14$

89. $r(x) = 2x^4 + 7x^2 + 11x - 20$

90. $s(x) = 3x^4 - 8x^3 - 13x - 24$

 Use Descartes' rule of signs to determine the possible combinations of real and complex zeroes for each polynomial. Then graph the function on the standard window of a graphing calculator and adjust it as needed until you're certain all real zeroes are in clear view. Use this screen and a list of the possible rational zeroes to factor the polynomial and find all zeroes (real and complex).

91. $f(x) = 4x^3 - 16x^2 - 9x + 36$
$(x - 4)(2x - 3)(2x + 3); x = 4, \frac{3}{2}, -\frac{3}{2}$

92. $g(x) = 6x^3 - 41x^2 + 26x + 24$
$(2x + 1)(3x - 4)(x - 6); x = -\frac{1}{2}, \frac{4}{3}, 6$

93. $h(x) = 6x^3 - 73x^2 + 10x + 24$
$(2x + 1)(3x - 2)(x - 12); x = -\frac{1}{2}, \frac{2}{3}, 12$

94. $H(x) = 4x^3 + 60x^2 + 53x - 42$
$(x + 14)(2x + 3)(2x - 1); x = -14, -\frac{3}{2}, \frac{1}{2}$

95. $p(x) = 4x^4 + 40x^3 - 93x^2 + 30x - 72$
$(x - 2)(x + 12)(4x^2 + 3); x = 2, -12, \frac{\sqrt{3}}{2}i, -\frac{\sqrt{3}}{2}i$

96. $q(x) = 4x^4 - 42x^3 - 70x^2 - 21x - 36$
$(2x + 3)(x - 12)(2x^2 + 1); x = -\frac{3}{2}, 12, \frac{\sqrt{2}}{2}i, -\frac{\sqrt{2}}{2}i$

▶ WORKING WITH FORMULAS

97. The absolute value of a complex number
$z = a + bi$: $|z| = \sqrt{a^2 + b^2}$

The absolute value of a complex number z, denoted $|z|$, represents the distance between the origin and the point (a, b) in the complex plane. Use the formula to find $|z|$ for the complex numbers given (also see Section 3.1, Exercise 69): (a) $3 + 4i$, (b) $-5 + 12i$, and (c) $1 + i\sqrt{3}$.   **a.** 5   **b.** 13   **c.** 2

98. The square root of $z = a + bi$:
$\sqrt{z} = \frac{\sqrt{2}}{2}\left(\sqrt{|z| + a} \pm i\sqrt{|z| - a}\right)$

The principal square root of a complex number is given by the relation shown, where $|z|$ represents the absolute value of z and the sign for the "$\pm$" is chosen to match the sign of b. Use the formula to find the square root of each complex number from Exercise 97, then check your answer by squaring the result (also see Section 3.1, Exercise 82).

$2 + i, 2 + 3i, \frac{\sqrt{6}}{2} + \frac{\sqrt{2}}{2}i$

▶ APPLICATIONS

99. Maximum and minimum values: To locate the maximum and minimum values of $F(x) = x^4 - 4x^3 - 12x^2 + 32x + 15$ requires finding the zeroes of $f(x) = 4x^3 - 12x^2 - 24x + 32$. Use the rational zeroes theorem and synthetic division to find the zeroes of f, then graph $F(x)$ on a calculator and see if the graph tends to support your calculations—do the maximum and minimum values occur at the zeroes of f?   yes

100. Graphical analysis: Use the rational zeroes theorem and synthetic division to find the zeroes of $F(x) = x^4 - 4x^3 - 12x^2 + 32x + 15$ (see Exercise 99 to verify graphically). $x = -3, 5, 1 \pm \sqrt{2}$

101. Maximum and minimum values: To locate the maximum and minimum values of $G(x) = x^4 - 6x^3 + x^2 + 24x - 20$ requires finding the zeroes of $g(x) = 4x^3 - 18x^2 + 2x + 24$. Use the rational zeroes theorem and synthetic division to find the zeroes of g, then graph $G(x)$ on a calculator and see if the graph tends to support your calculations—do the maximum and minimum values occur at the zeroes of g?   yes

102. Graphical analysis: Use the rational zeroes theorem and synthetic division to find the zeroes of $G(x) = x^4 - 6x^3 + x^2 + 24x - 20$ (see Exercise 101 to verify graphically). $x = -2, 1, 2, 5$

Geometry: The volume of a cube is $V = x \cdot x \cdot x = x^3$, where x represents the length of the edges. If a slice 1 unit thick is removed from the cube, the remaining volume is $v = x \cdot x \cdot (x - 1) = x^3 - x^2$. Use this information for Exercises 103 and 104.

103. A slice 1 unit in thickness is removed from one side of a cube. Use the rational zeroes theorem and synthetic division to find the original dimensions of the cube, if the remaining volume is (a) 48 cm³ and (b) 100 cm³.
a. 4 cm × 4 cm × 4 cm   **b.** 5 cm × 5 cm × 5 cm

104. A slice 1 unit in thickness is removed from one side of a cube, then a second slice of the same thickness is removed from a different side (not the opposite side). Use the rational zeroes theorem and synthetic division to find the original dimensions of the cube, if the remaining volume is (a) 36 cm³ and (b) 80 cm³.   **a.** 4 cm × 4 cm × 4 cm   **b.** 5 cm × 5 cm × 5 cm

Geometry: The volume of a rectangular box is $V = LWH$. For the box to satisfy certain requirements, its length must be twice the width, and its height must be two inches less than the width. Use this information for Exercises 105 and 106.

105. Use the rational zeroes theorem and synthetic division to find the dimensions of the box if it must have a volume of 150 in^3.
length 10 in., width 5 in., height 3 in.

106. Suppose the box must have a volume of 64 in^3. Use the rational zeroes theorem and synthetic division to find the dimensions required.
length 8 in., width 4 in., height 2 in.

Government deficits: Over a 14-yr period, the balance of payments (deficit versus surplus) for a certain county government was modeled by the function $f(x) = \frac{1}{4}x^4 - 6x^3 + 42x^2 - 72x - 64$, where $x = 0$ corresponds to 1990 and $f(x)$ is the deficit or surplus in tens of thousands of dollars. Use this information for Exercises 107 and 108.

107. Use the rational zeroes theorem and synthetic division to find the years when the county "broke even" (debt = surplus = 0) from 1990 to 2004. How many years did the county run a surplus during this period? 1994, 1998, 2002, about 5 yr

108. The deficit was at the $84,000 level [$f(x) = -84$], four times from 1990 to 2004. Given this occurred in 1992 and 2000 ($x = 2$ and $x = 10$), use the rational zeroes theorem, synthetic division, and the remainder theorem to find the other two years the deficit was at $84,000. 1990, 2001

109. Drag resistance on a boat: In a scientific study on the effects of drag against the hull of a sculling boat, some of the factors to consider are displacement, draft, speed, hull shape, and length, among others. If the first four are held

constant and we assume a flat, calm water surface, length becomes the sole variable (as length changes, we adjust the beam by a uniform scaling to keep a constant displacement). For a fixed sculling speed of 5.5 knots, the relationship between drag and length can be modeled by $f(x) = -0.4192x^4 + 18.9663x^3 - 319.9714x^2 + 2384.2x - 6615.8$, where $f(x)$ is the efficiency rating of a boat with length x ($8.7 < x < 13.6$). Here, $f(x) = 0$ represents an *average* efficiency rating. (a) Under these conditions, what lengths (to the nearest hundredth) will give the boat an average rating? (b) What length will maximize the efficiency of the boat? What is this rating?
a. 8.97 m, 11.29 m, 12.05 m, 12.94 m   b. 9.7 m, + 3.7

110. Comparing densities: Why is it that when you throw a rock into a lake, it sinks, while a wooden ball will float half submerged, but the bobber on your fishing line floats on the surface? It all depends on the density of the object compared to the density of water ($d = 1$). For uniformity, we'll consider spherical objects of various densities, each with a radius of 5 cm. When placed into water, the depth that the sphere will sink beneath the surface (while still floating) is modeled by the polynomial $p(x) = \frac{\pi}{3}x^3 - 5\pi x^2 + \frac{500\pi}{3}d$, where d is the density of the object and the smallest positive zero of p is the depth of the sphere below the surface (in centimeters). How far submerged is the sphere if it's made of (a) balsa wood, $d = 0.17$; (b) pine wood, $d = 0.55$; (c) ebony wood, $d = 1.12$; (d) a large bobber made of lightweight plastic, $d = 0.05$ (see figure)?
a. 2.62 cm   b. 5.33 cm   c. it sinks (no positive zeroes)
d. 1.35 cm (it floats near the surface)

► **EXTENDING THE CONCEPT**

111. In the figure, $P(x) = 0.02x^3 - 0.24x^2 - 1.04x + 2.68$ is graphed on the standard screen ($-10 \le x \le 10$), which shows two real zeroes. Since P has degree 3, there must be one more real zero but is it negative or positive? Use the upper/lower bounds property (a) to see if -10 is a lower bound and (b) to see if 10 is an upper bound. (c) Then use your calculator to find the remaining zero.
a. yes   b. no   c. about 14.88

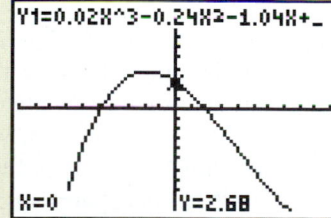

112. From Example 11, (a) what is the significance of the y-intercept? (b) If the domain were extended to include $0 < x \leq 13$, what happens when x is approximately 12.8?   **a.** The coastline is 384 ft above the ocean floor.   **b.** There is another island.

113A. It is often said that while the difference of two squares is factorable, $a^2 - b^2 = (a + b)(a - b)$, the sum of two squares is prime. To be 100% correct, we should say the sum of two squares cannot be factored *using real numbers*. If complex numbers are used, $(a^2 + b^2) = (a + bi)(a - bi)$. Use this idea to factor the following binomials.

  a. $p(x) = x^2 + 25$ **b.** $q(x) = x^2 + 9$
  $(x + 5i)(x - 5i)$ $(x + 3i)(x - 3i)$
  c. $r(x) = x^2 + 7$
  $(x + i\sqrt{7})(x - i\sqrt{7})$

113B. It is often said that while $x^2 - 16$ is factorable as a difference of squares, $a^2 - b^2 = (a + b)(a - b)$, $x^2 - 17$ is not. To be 100% correct, we should say that $x^2 - 17$ is not factorable *using integers*. Since $(\sqrt{17})^2 = 17$, it can actually be factored in the same way: $x^2 - 17 = (x + \sqrt{17})(x - \sqrt{17})$. Use this idea to solve the following equations.

  a. $x^2 - 7 = 0$ **b.** $x^2 - 12 = 0$
  $x = -\sqrt{7}, \sqrt{7}$ $x = -2\sqrt{3}, 2\sqrt{3}$
  c. $x^2 - 18 = 0$
  $x = -3\sqrt{2}, 3\sqrt{2}$

114. Every general cubic equation $aw^3 + bw^2 + cw + d = 0$ can be written in the form $x^3 + px + q = 0$ (where the squared term has been "depressed"), using the transformation

$$w = x - \frac{b}{3a}.$$ Use this transformation to solve the following equations.

  a. $w^3 - 3w^2 + 6w - 4 = 0$ $w = 1, 1 \pm i\sqrt{3}$
  b. $w^3 - 6w^2 + 21w - 26 = 0$ $w = 2, 2 \pm 3i$

Note: It is actually very rare that the transformation produces a value of $q = 0$ for the "depressed" cubic $x^3 + px + q = 0$, and general solutions must be found using what has become known as *Cardano's formula*. For a complete treatment of cubic equations and their solutions, visit our website at www.mhhe.com/coburn.

115. For each of the following complex polynomials, one of its zeroes is given. Use this zero to help write the polynomial in completely factored form. (*Hint:* Synthetic division and the quadratic formula can be applied to *all polynomials*, even those with complex coefficients.)

  a. $C(z) = z^3 + (1 - 4i)z^2 + (-6 - 4i)z + 24i$; $z = 4i$ $C(z) = (z - 4i)(z + 3)(z - 2)$
  b. $C(z) = z^3 + (5 - 9i)z^2 + (4 - 45i)z - 36i$; $z = 9i$ $C(z) = (z - 9i)(z + 4)(z + 1)$
  c. $C(z) = z^3 + (-2 - 3i)z^2 + (5 + 6i)z - 15i$; $z = 3i$ $C(z) = (z - 3i)(z - 1 - 2i)(z - 1 + 2i)$
  d. $C(z) = z^3 + (-4 - i)z^2 + (29 + 4i)z - 29i$; $z = i$ $C(z) = (z - i)(z - 2 - 5i)(z - 2 + 5i)$
  e. $C(z) = z^3 + (-2 - 6i)z^2 + (4 + 12i)z - 24i$; $z = 6i$ $C(z) = (z - 6i)(z - 1 - i\sqrt{3})(z - 1 + i\sqrt{3})$
  f. $C(z) = z^3 + (-6 + 4i)z^2 + (11 - 24i)z + 44i$; $z = -4i$ $C(z) = (z + 4i)(z - 3 - i\sqrt{2})(z - 3 + i\sqrt{2})$
  g. $C(z) = z^3 + (-2 - i)z^2 + (5 + 4i)z + (-6 + 3i)$; $z = 2 - i$ $C(z) = (z - 2 + i)(z - 3i)(z + i)$
  h. $C(z) = z^3 - 2z^2 + (19 + 6i)z + (-20 + 30i)$; $z = 2 - 3i$ $C(z) = (z - 2 + 3i)(z - 5i)(z + 2i)$

▶ MAINTAINING YOUR SKILLS

116. (2.5) Graph the piecewise-defined function and find the values of $f(-3), f(2),$ and $f(5)$.

$$f(x) = \begin{cases} 2 & x \leq -1 \\ |x - 1| & -1 < x < 5 \\ 4 & x \geq 5 \end{cases}$$

117. (3.4) For a county fair, officials need to fence off a large rectangular area, then subdivide it into three equal (rectangular) areas. If the county provides 1200 ft of fencing, (a) what dimensions will maximize the area of the larger (outer) rectangle? (b) What is the area of each smaller rectangle?
  a. $w = 150$ ft, $l = 300$ **b.** $A = 15,000$ ft^2

118. (2.1) Use the graph given to (a) state intervals where $f(x) \geq 0$, (b) locate local maximum and minimum values, and (c) state intervals where $f(x)\uparrow$ and $f(x)\downarrow$.

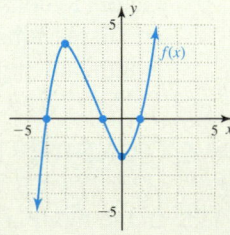

119. (2.2) Write the equation of the function shown.
  $r(x) = 2\sqrt{x + 4} - 2$

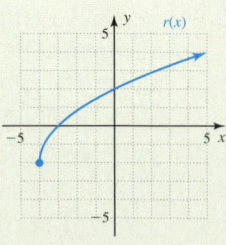

Additional answers can be found in the Instructor Answer Appendix.

LEARNING OBJECTIVES

In Section 4.3 you will see how we can:

☐ **A.** Identify the graph of a polynomial function and determine its degree

☐ **B.** Describe the end-behavior of a polynomial graph

☐ **C.** Discuss the attributes of a polynomial graph with zeroes of multiplicity

☐ **D.** Graph polynomial functions in standard form

☐ **E.** Solve applications of polynomials and polynomial modeling

As with linear and quadratic functions, understanding graphs of *polynomial* functions will help us apply them more effectively as mathematical models. Since all real polynomials can be written in terms of their linear and quadratic factors (Section 4.2), these functions provide the basis for our continuing study.

A. Identifying the Graph of a Polynomial Function

Consider the graphs of $f(x) = x + 2$ and $g(x) = (x - 1)^2$, which we know are smooth, continuous curves. The graph of f is a straight line with positive slope, that crosses the x-axis at -2. The graph of g is a parabola, opening upward, shifted 1 unit to the right, and touching the x-axis at $x = 1$. When f and g are "combined" into the single function $P(x) = (x + 2)(x - 1)^2$, the behavior of the graph at these zeroes is still evident. In Figure 4.9, the graph of P crosses the x-axis at $x = -2$, "bounces" off the x-axis at $x = 1$, and is still a smooth, continuous curve. This observation could be extended to include additional linear or quadratic factors, and helps affirm that the graph of a polynomial function is a *smooth, continuous curve*.

Figure 4.9

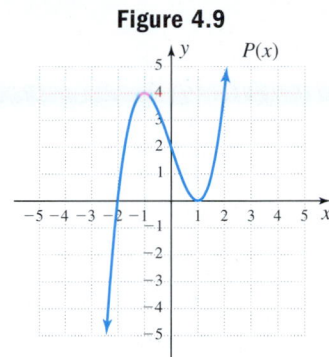

Further, after the graph of P crosses the axis at $x = -2$, it must "turn around" at some point to reach the zero at $x = 1$, then turn again as it touches the x-axis without crossing. By combining this observation with our work in Section 4.2, we can state the following:

> **Polynomial Graphs and Turning Points**
>
> **1.** If $P(x)$ is a polynomial function of degree n, then the graph of P has at most $n - 1$ turning points.
> **2.** If the graph of a function P has $n - 1$ turning points, then the degree of $P(x)$ is at least n.

While defined more precisely in a future course, we will take "smooth" to mean the graph has no sharp turns or jagged edges, and "continuous" to mean the entire graph can be drawn without lifting your pencil (Figure 4.10). In other words, a polynomial graph has none of the attributes shown in Figure 4.11.

Figure 4.10   **Figure 4.11**

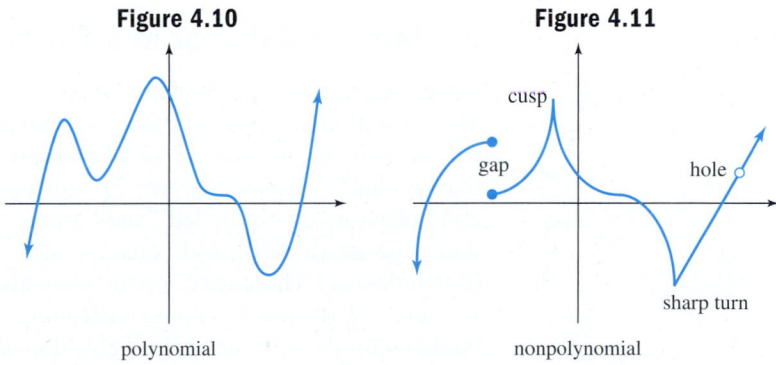

polynomial   nonpolynomial

EXAMPLE 1 ▶   Identifying Polynomial Graphs

Determine whether each graph could be the graph of a polynomial. If not, discuss why. If so, use the number of turning points and zeroes to identify the least possible degree of the function.

a.

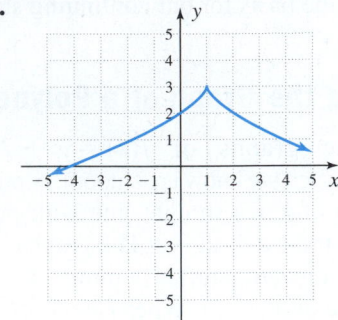

b.

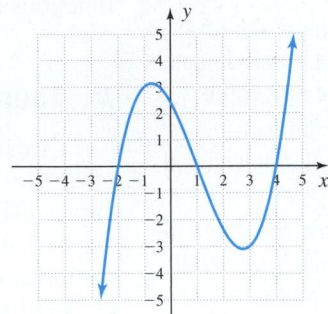

c.

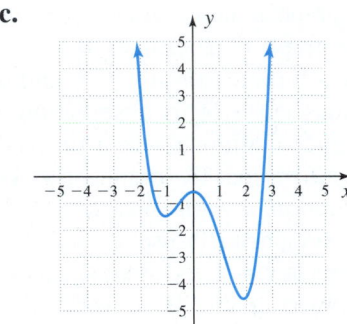

d.

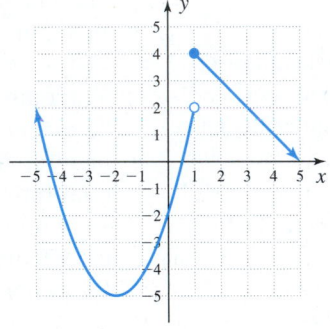

Solution ▶   **a.** This is not a polynomial graph, as it has a cusp at (1, 3). A polynomial graph is always smooth.

b. This graph is smooth and continuous, and could be that of a polynomial. With two turning points and three zeroes, the function is at least degree 3.

c. This graph is smooth and continuous, and could be that of a polynomial. With three turning points and two zeroes, the function is at least degree 4.

d. This is not a polynomial graph, as it has a gap (discontinuity) at $x = 1$. A polynomial graph is always continuous.

☑ **A.** You've just seen how we can identify the graph of a polynomial function and determine its degree

Now try Exercises 7 through 12 ▶

B. The End-Behavior of a Polynomial Graph

Once the graph of a function has "made its last turn" and crossed or touched its last real zero, it will continue to increase or decrease without bound as $|x|$ becomes large. As before, we refer to this as the **end-behavior** of the graph. In previous sections, we noted that quadratic functions (degree 2) with a positive leading coefficient ($a > 0$), had the end-behavior "up on the left" and "up on the right (up/up)." If the leading coefficient was negative ($a < 0$), end-behavior was "down on the left" and "down on the right (down/down)." These descriptions were also applied to the graph of a linear function $y = mx + b$ (degree 1). A positive leading coefficient ($m > 0$) indicates the graph will be down on the left, up on the right (down/up), and so on. All polynomial graphs exhibit some form of end-behavior, which can be likewise described.

EXAMPLE 2 ▶ **Identifying the End-Behavior of a Graph**

State the end-behavior of each graph shown:

a. $f(x) = x^3 - 4x + 1$ **b.** $g(x) = -2x^5 + 7x^3 - 4x$ **c.** $h(x) = -2x^4 + 5x^2 + x - 1$

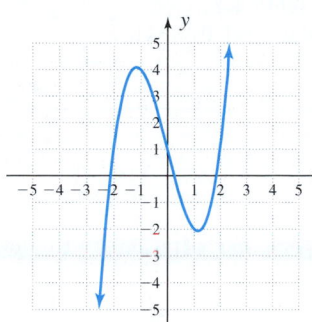 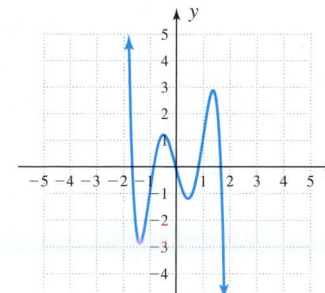 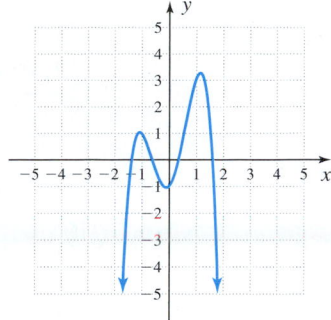

Solution ▶ **a.** down on the left, up on the right **b.** up on the left, down on the right
c. down on the left, down on the right

Now try Exercises 13 through 16 ▶

WORTHY OF NOTE

As a visual aid to end-behavior, it might help to picture a signalman using semaphore code as illustrated here. As you view the end-behavior of a polynomial graph, there is a striking resemblance.

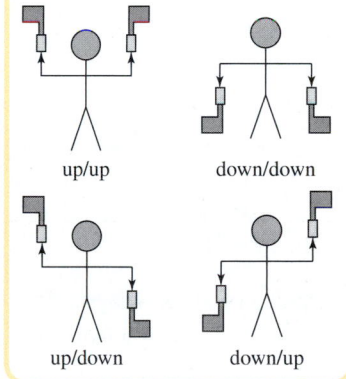

up/up   down/down

up/down   down/up

The leading term ax^n of a polynomial function is said to be the **dominant term,** because for large values of $|x|$, the value of ax^n is much larger than all other terms combined. Figure 4.12 shows a table of values for $Y_1 = 0.4x^5 - 3x^4 - 9x^3 - 15x^2 - 30$, where the leading coefficient is positive and very small, with all other coefficients negative and much larger. Initially, all outputs are negative as these terms overpower the leading term. But eventually (in this case for any integer greater than 10), the leading term will dominate all others since it becomes much more "powerful" for larger values. See Figure 4.13.

Figure 4.12

$Y_1 = 0.4x^5 - 3x^4 - 9x^3 - 15x^2 - 30$

X	Y1
0	-30
1	-56.6
2	-197.2
3	-553.8
4	-1204
5	-2155
6	-3292

X=0

Figure 4.13

X	Y1
7	-4332
8	-4779
9	-3869
10	-530
11	6673.4
12	19583
13	40496

X=13

This means that like linear and quadratic graphs, polynomial end-behavior can be predicted in advance by analyzing this term alone.

1. For ax^n when n is even, any nonzero number raised to an even power is positive, so the ends of the graph must point in the same direction. If $a > 0$, both point upward. If $a < 0$, both point downward.

2. For ax^n when n is odd, any number raised to an odd power has the same sign as the input value, so the ends of the graph must point in opposite directions. If $a > 0$, end-behavior is down on the left, up on the right. If $a < 0$, end-behavior is up on the left, down on the right.

From this we find that end-behavior depends on two things: *the degree of the function* (even or odd) and the *sign of the leading coefficient* (positive or negative). In more formal terms, this is described in terms of how the graph "behaves" for large values of x. For end-behavior that is "up on the right," we mean that as x becomes a large positive number, y becomes a large positive number. This is indicated using the notation: as $x \rightarrow \infty$, $y \rightarrow \infty$. Similar notation is used for the other possibilities. These facts are summarized in Table 4.1. The interior portion of each graph is dashed since the actual number of turning points may vary, although a polynomial of odd degree will have an even number of turning points, and a polynomial of even degree will have an odd number of turning points.

Table 4.1
Polynomial End-Behavior

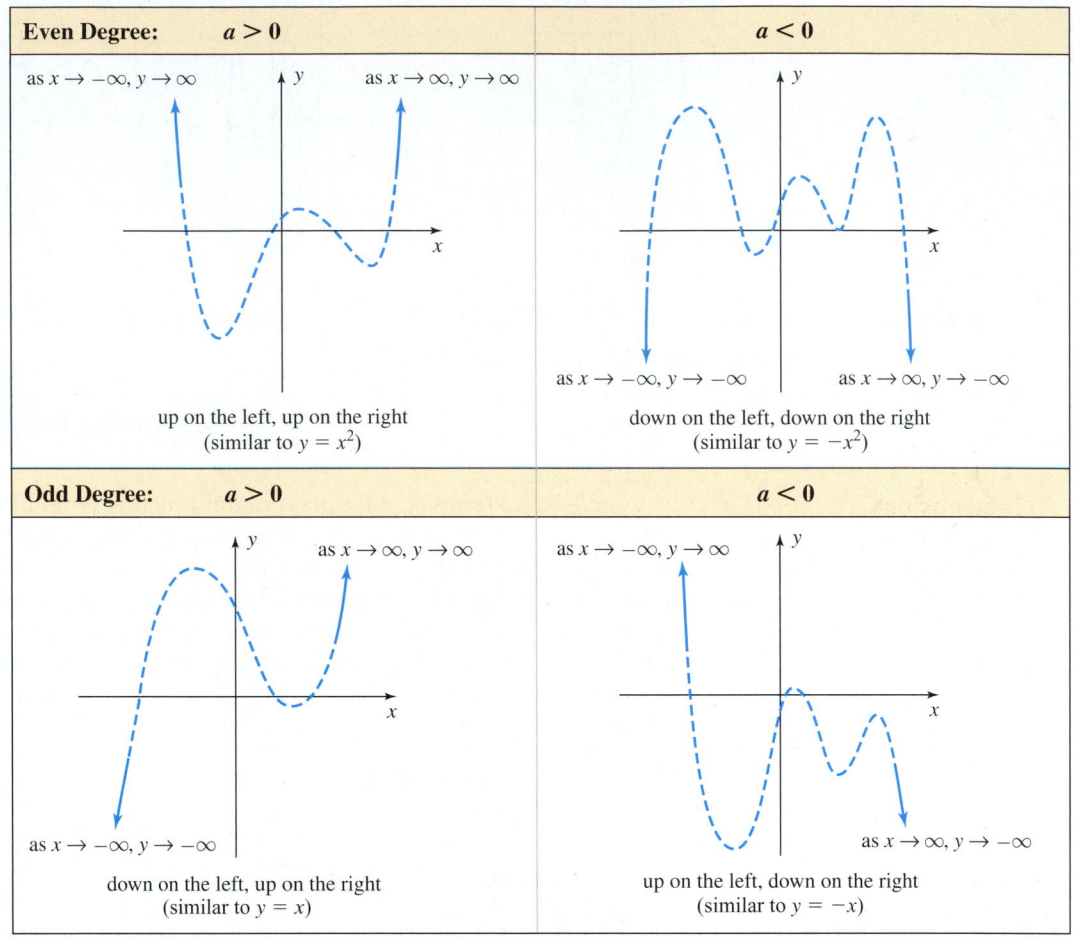

Note the end-behavior of $y = mx$ can be used as a representative of all odd degree functions, and the end-behavior of $y = ax^2$ as a representative of all even degree functions.

> **The End-Behavior of a Polynomial Graph**
>
> Given a polynomial $P(x)$ with leading term ax^n and $n \geq 1$.
> If n is **even,** ends will point in the **same direction,**
>
> **1.** for $a > 0$: up on the left, up on the right (*as with $y = x^2$*);
> $$\text{as } x \to -\infty, y \to \infty; \qquad \text{as } x \to \infty, y \to \infty$$
> **2.** for $a < 0$: down on the left, down on the right (*as with $y = -x^2$*);
> $$\text{as } x \to -\infty, y \to -\infty; \qquad \text{as } x \to \infty, y \to -\infty$$
>
> If n is **odd,** the ends will point in **opposite directions,**
>
> **1.** for $a > 0$: down on the left, up on the right (*as with $y = x$*);
> $$\text{as } x \to -\infty, y \to -\infty; \qquad \text{as } x \to \infty, y \to \infty$$
> **2.** for $a < 0$: up on the left, down on the right (*as with $y = -x$*);
> $$\text{as } x \to -\infty, y \to \infty; \qquad \text{as } x \to \infty, y \to -\infty$$

EXAMPLE 3 ▶   Identifying the End-Behavior of a Function

State the end-behavior of each function.

a. $f(x) = 0.8x^4 - 3x^3 + 0.5x^2 + 4x - 1$ **b.** $g(x) = -2x^5 + 6x^3 - 1$

Solution ▶   **a.** The function has degree 4 (even), so the ends will point in the same direction. The leading coefficient is positive, so end-behavior is up/up. See Figure 4.14.

b. The function has degree 5 (odd), so the ends will point in opposite directions. The leading coefficient is negative, so the end-behavior is up/down. See Figure 4.15.

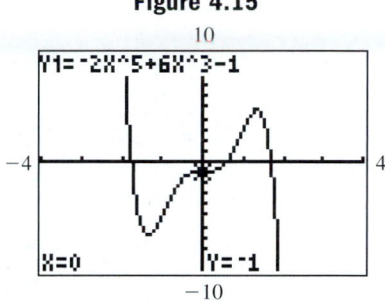

Figure 4.14

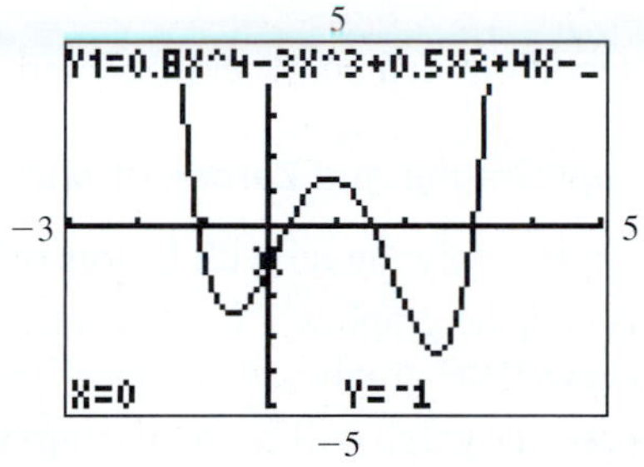

Figure 4.15

☑ **B.** You've just seen how we can describe the end-behavior of a polynomial graph

Now try Exercises 17 through 22 ▶

C. Attributes of Polynomial Graphs with Zeroes of Multiplicity

Another important aspect of polynomial functions is the behavior of a graph near its zeroes. In the simplest case, consider the functions $f(x) = x$ and $g(x) = x^3$ in Figure 4.16. Both have odd degree, like end-behavior (down/up), and a zero at $x = 0$. But the zero of f has multiplicity 1, while the zero from g has multiplicity 3. Notice the graph of g is vertically compressed near $x = 0$ and flattens out on its approach and departure from this zero.

Figure 4.16

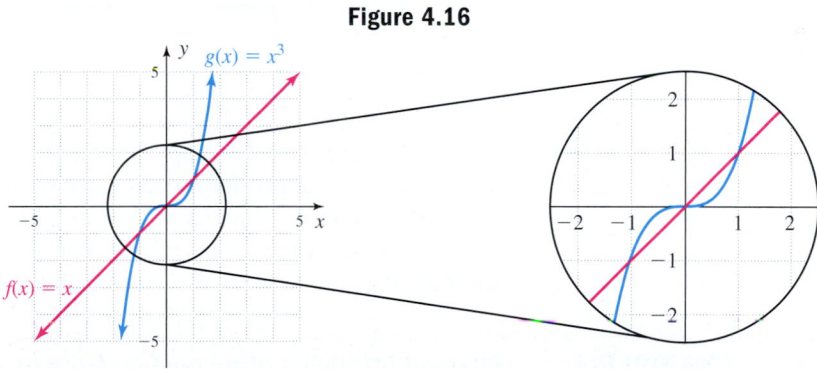

This behavior can be explained by noting that for $x = -1$ and 1, $f(x) = g(x)$. But for $|x| < 1$, the graph of g *will be closer to the x-axis* (g decreases faster than f) since the cube of a fractional number is smaller than the fraction itself. We further note that for $|x| > 1$, g increases much faster than f, and $|g(x)| > |f(x)|$. Similar observations can be made regarding $f(x) = x^2$ and $g(x) = x^4$ in Figure 4.17. Both functions have even degree, a zero at $x = 0$, and $f(x) = g(x)$ for $x = -1$ and 1. But for $|x| < 1$, the function with higher degree is once again closer to the x-axis.

Figure 4.17

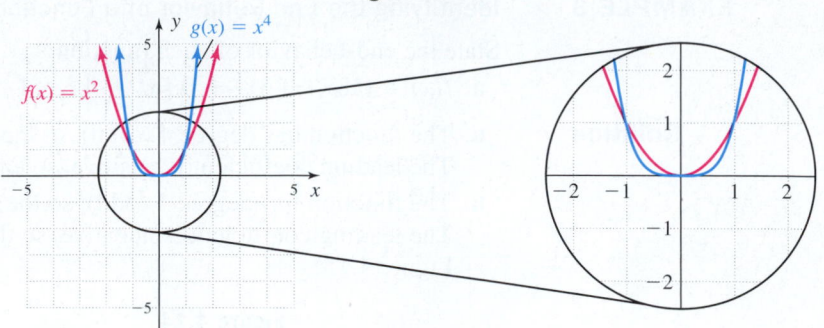

These observations can be generalized and applied to all real zeroes of a function.

> **Polynomial Graphs and Zeroes of Multiplicity**
>
> Given $P(x)$ is a polynomial with factors of the form $(x - c)^m$, with c a real number,
> * If m is odd, the graph will cross through the x-axis.
> * If m is even, the graph will "bounce" off the x-axis (touching at just one point).
>
> In each case, the graph will be more compressed (flatter) near c for larger values of m.

To illustrate, compare the graph of $P(x) = (x + 2)(x - 1)^2$ (Figure 4.18), with the graph of $p(x) = (x + 2)^3(x - 1)^4$ shown in Figure 4.19, noting the increased multiplicity of each zero.

Figure 4.18   **Figure 4.19**

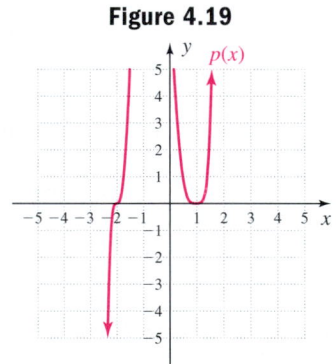

Both graphs show the expected zeroes at $x = -2$ and $x = 1$, but the graph of $p(x)$ is flatter near $x = -2$ and $x = 1$, due to the increased multiplicity of each zero. We also lose sight of the graph of $p(x)$ between $x = -2$ and $x = 0$, since the increased multiplicities produce larger values than the original grid could display.

EXAMPLE 4 ▶   **Naming Attributes of a Function from Its Graph**

The graph of a polynomial $f(x)$ is shown.

a. State whether the degree of f is even or odd.

b. Use the graph to name the zeroes of f, then state whether their multiplicity is even or odd.

c. State the minimum possible degree of f.

d. State the domain and range of f.

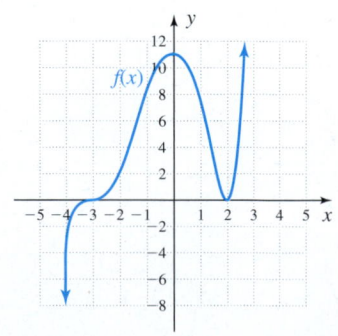

Solution ▶   **a.** Since the ends of the graph point in opposite directions, the degree of the function must be odd.

b. The graph crosses the x-axis at $x = -3$ and is compressed near -3, meaning it must have odd multiplicity with $m > 1$. The graph "bounces" off the x-axis at $x = 2$ and 2 must be a zero of even multiplicity.

c. The minimum possible degree of f is 5, as in $f(x) = a(x - 2)^2(x + 3)^3$.

d. $x \in \mathbb{R}, y \in \mathbb{R}$.

Now try Exercises 23 through 28 ▶

To find the degree of a polynomial from its factored form, add the exponents on all linear factors, then add 2 for each irreducible quadratic factor (the degree of any quadratic factor is 2). The sum gives the degree of the polynomial, from which end-behavior can be determined. To find the y-intercept, substitute 0 for x as before, noting this is equivalent to applying the exponent to the constant from each factor.

EXAMPLE 5 ▶   **Naming Attributes of a Function from Its Factored Form**

State the degree of each function, then describe the end-behavior and name the y-intercept of each graph.

    a. $f(x) = (x + 2)^3(x - 3)$ **b.** $g(x) = -(x + 2)^2(x^2 + 5)(x - 5)$

Solution ▶   **a.** The degree of f is $3 + 1 = 4$. With even degree and positive leading coefficient, end-behavior is up/up. For $f(0) = (2)^3(-3) = -24$, the y-intercept is $(0, -24)$. See Figure 4.20.

b. The degree of g is $2 + 2 + 1 = 5$. With odd degree and negative leading coefficient, end-behavior is up/down. For $g(0) = -1(2)^2(5)(-5) = 100$, the y-intercept is $(0, 100)$. See Figure 4.21.

Figure 4.20

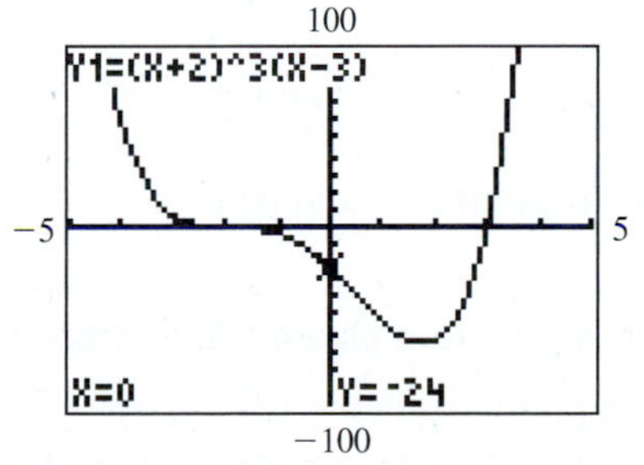

Figure 4.21

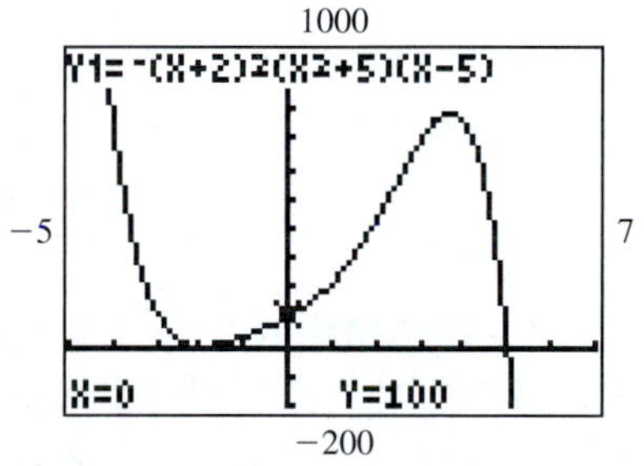

Now try Exercises 29 through 36 ▶

EXAMPLE 6 ▶   **Matching Graphs to Functions Using Zeroes of Multiplicity**

The following functions all have zeroes at $x = -2, -1$, and 1. Match each function to the corresponding graph *using its degree and the multiplicity of each zero.*

    a. $y = (x + 2)(x + 1)^2(x - 1)^3$ **b.** $y = (x + 2)(x + 1)(x - 1)^3$

    c. $y = (x + 2)^2(x + 1)^2(x - 1)^3$ **d.** $y = (x + 2)^2(x + 1)(x - 1)^3$

Solution ▶   The functions in Figures 4.22 and 4.24 must have even degree due to end-behavior, so each corresponds to (a) or (d). At $x = -1$ the graph in Figure 4.22 "crosses," while the graph in Figure 4.24 "bounces." This indicates Figure 4.22 matches equation (d), while Figure 4.24 matches equation (a).

The graphs in Figures 4.23 and 4.25 must have odd degree due to end-behavior, so each corresponds to (b) or (c). Here, one graph "bounces" at $x = -2$, while the other "crosses." The graph in Figure 4.23 matches equation (c), the graph in Figure 4.25 matches equation (b).

Figure 4.22

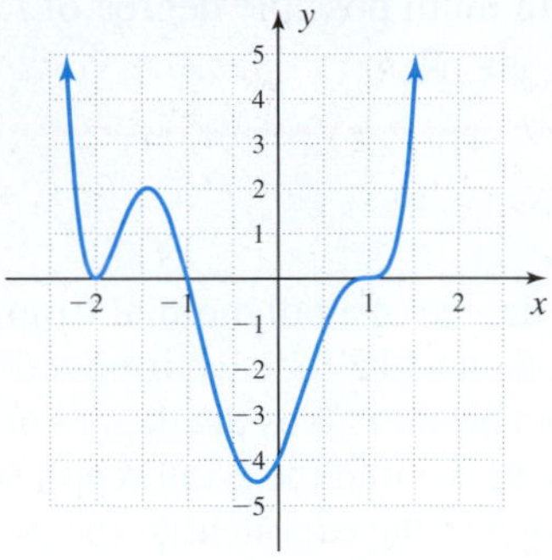

Figure 4.23

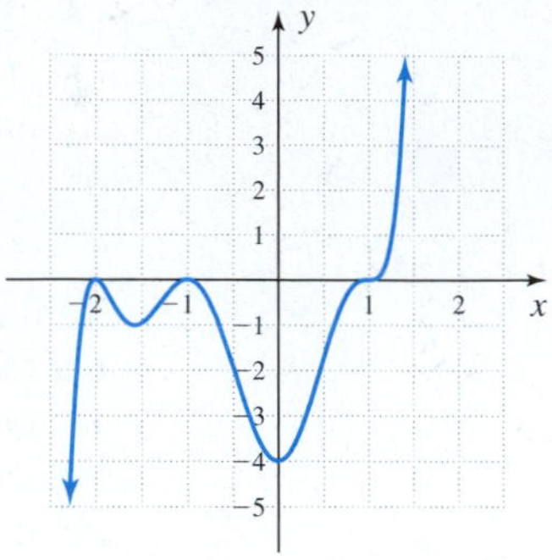

Figure 4.24

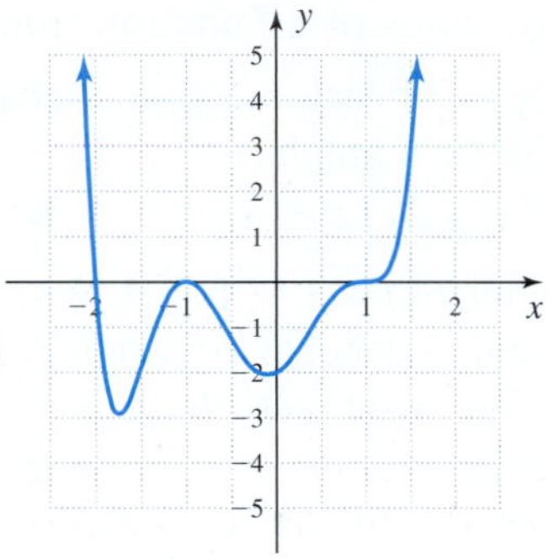

Figure 4.25

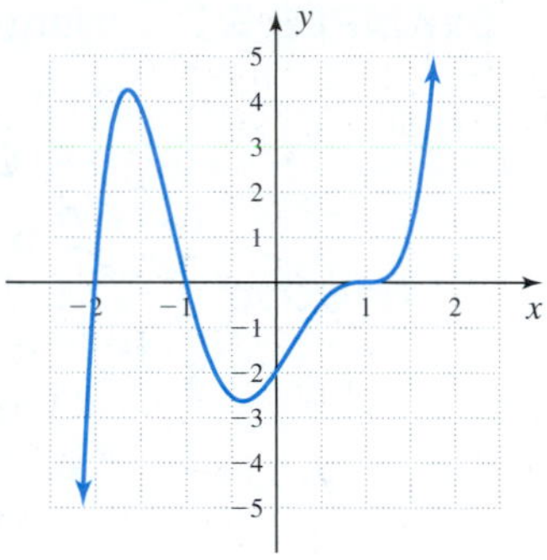

Now try Exercises 37 through 42 ▶

Using the ideas from Examples 5 and 6, we're able to draw a fairly accurate graph given the factored form of a polynomial. Convenient values between two zeroes, called **midinterval points,** should be used to help complete the graph.

EXAMPLE 7 ▶ **Graphing a Function Given the Factored Form**

Sketch the graph of $f(x) = (x - 2)(x - 1)^2(x + 1)^3$ using end-behavior; the x- and y-intercepts, and zeroes of multiplicity.

Solution ▶ Adding the exponents of each factor, we find that f is a function of degree 6 with a positive lead coefficient, so end-behavior will be up/up. Since $f(0) = -2$, the y-intercept is $(0, -2)$. The graph will bounce off the x-axis at $x = 1$ (even multiplicity), and cross the axis at $x = -1$ and 2 (odd multiplicities). The graph will "flatten out" near $x = -1$ because of its higher multiplicity. To help "round-out" the graph we evaluate f at $x = 1.5$, giving $(-0.5)(0.5)^2(2.5)^3 \approx -1.95$ (note scaling of the x- and y-axes).

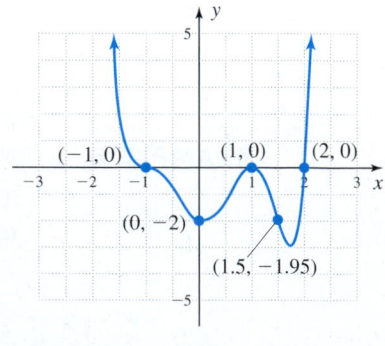

☑ **C.** You've just seen how we can discuss the attributes of a polynomial graph with zeroes of multiplicity

Now try Exercises 43 through 56 ▶

D. The Graph of a Polynomial Function

Using the cumulative observations from this and previous sections, a general strategy emerges for the graphing of polynomial functions.

Guidelines for Graphing Polynomial Functions

1. Determine the end-behavior of the graph.
2. Find the y-intercept $(0, a_0)$
3. Find the zeroes using any combination of the rational zeroes theorem, the factor and remainder theorems, tests for 1 and -1, factoring, and the quadratic formula.
4. Use the y-intercept, end-behavior, the multiplicity of each zero, and midinterval points as needed to sketch a smooth, continuous curve.

   Additional tools include (a) polynomial zeroes theorem, (b) complex conjugates theorem, (c) number of turning points, (d) Descartes' rule of signs, (e) upper and lower bounds, and (f) symmetry.

WORTHY OF NOTE

Although of somewhat limited value, symmetry (item f in the guidelines) can sometimes aid in the graphing of polynomial functions. If all terms of the function have even degree, the graph will be symmetric to the y-axis (even). If all terms have odd degree, the graph will be symmetric to the origin. Recall that a constant term has degree zero, an even number.

EXAMPLE 8 ▶

Graphing a Polynomial Function

Sketch the graph of $g(x) = -x^4 + 9x^2 - 4x - 12$.

Solution ▶

1. End-behavior: The function has degree 4 (even) with a negative leading coefficient, so end-behavior is *down on the left, down on the right.*
2. Since $g(0) = -12$, the y-intercept is $(0, -12)$.
3. Zeroes: Using the test for $x = 1$ gives $-1 + 9 - 4 - 12 = -8$, showing $x = 1$ is not a zero but $(1, -8)$ is a point on the graph. Using the test for $x = -1$ gives $-1 + 9 + 4 - 12 = 0$, so -1 is a zero and $(x + 1)$ is a factor. Using $x = -1$ with the factor theorem yields

$$
\begin{array}{r|rrrrr}
-1 & -1 & 0 & 9 & -4 & -12 \\
   &   & 1 & -1 & -8 & 12 \\
\hline
   & -1 & 1 & 8 & -12 & \underline{|0} \\
\end{array}
$$

The quotient polynomial is not easily factorable so we continue with synthetic division. Using the rational zeroes theorem, the possible rational zeroes are $\{\pm1, \pm12, \pm2, \pm6, \pm3, \pm4\}$, so we try $x = 2$.

use 2 as a "divisor" on the quotient polynomial

$$
\begin{array}{r|rrrr}
2 & -1 & 1 & 8 & -12 \\
  &   & -2 & -2 & 12 \\
\hline
  & -1 & -1 & 6 & \underline{|0} \\
\end{array}
$$

This shows $x = 2$ is a zero, $x - 2$ is a factor, and the function can now be written as

$$g(x) = (x + 1)(x - 2)(-x^2 - x + 6).$$

Factoring -1 from the trinomial gives

$$
\begin{aligned}
g(x) &= -1(x + 1)(x - 2)(x^2 + x - 6) \\
     &= -1(x + 1)(x - 2)(x + 3)(x - 2) \\
     &= -1(x + 1)(x - 2)^2(x + 3)
\end{aligned}
$$

The zeroes of g are $x = -1$ and -3, both with multiplicity 1, and $x = 2$ with multiplicity 2.

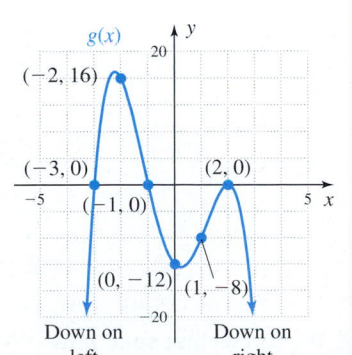

4. To help "round-out" the graph we evaluate the midinterval point $x = -2$ using the remainder theorem or the factored form of $g(x)$, which shows that $(-2, 16)$ is also a point on the graph.

$$
\begin{array}{r|rrrrr}
\text{use } -2 \text{ as a "divisor"} \quad -2 & -1 & 0 & 9 & -4 & -12 \\
& & 2 & -4 & -10 & 28 \\
\hline
& -1 & 2 & 5 & -14 & \underline{|16}
\end{array}
$$

The final result is the graph shown.

Now try Exercises 57 through 72 ▶

⚠ **CAUTION** ▶ Sometimes using a midinterval point to help draw a graph will give the illusion that a maximum or minimum value has been located. This is rarely the case, as demonstrated in the figure in Example 8, where the maximum value in Quadrant II is actually closer to $(-2.22, 16.95)$.

EXAMPLE 9 ▶ **Using the Guidelines to Sketch a Polynomial Graph**

Sketch the graph of $h(x) = x^7 - 4x^6 + 7x^5 - 12x^4 + 12x^3$.

Solution ▶

1. End-behavior: The function has degree 7 (odd) so the ends will point in opposite directions. The leading coefficient is positive and the end-behavior will be *down on the left* and *up on the right*.

2. y-intercept: Since $h(0) = 0$, the y-intercept is $(0, 0)$.

3. Zeroes: Testing 1 and -1 shows neither are zeroes but $(1, 4)$ and $(-1, -36)$ are points on the graph. Factoring out x^3 produces $h(x) = x^3(x^4 - 4x^3 + 7x^2 - 12x + 12)$, and we see that $x = 0$ is a zero of multiplicity 3. We next use synthetic division with $x = 2$ on the fourth-degree polynomial:

$$
\begin{array}{r|rrrrr}
\text{use 2 as a "divisor"} \quad 2 & 1 & -4 & 7 & -12 & 12 \\
& & 2 & -4 & 6 & -12 \\
\hline
& 1 & -2 & 3 & -6 & \underline{|0}
\end{array}
$$

This shows $x = 2$ is a zero and $x - 2$ is a factor. At this stage, it appears the quotient can be factored by grouping. From $h(x) = x^3(x - 2)(x^3 - 2x^2 + 3x - 6)$, we obtain $h(x) = x^3(x - 2)(x^2 + 3)(x - 2)$ after factoring and

$$h(x) = x^3(x - 2)^2(x^2 + 3)$$

as the completely factored form. We find that $x = 2$ is a zero of multiplicity 2, and the remaining two zeroes are complex.

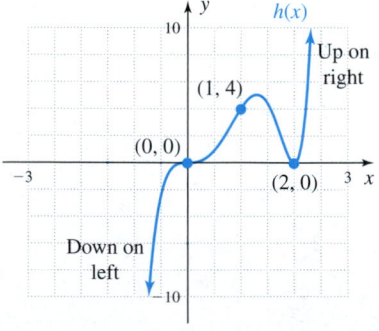

4. Using this information produces the graph shown in the figure.

Now try Exercises 73 through 76 ▶

☑ **D. You've just seen how we can graph polynomial functions in standard form**

For practice with these ideas using a graphing calculator, **see Exercises 77 through 80.** Similar to our work in previous sections, **Exercises 81 and 82** ask you to reconstruct the complete equation of a polynomial from its given graph.

E. Applications of Polynomials and Polynomial Modeling

EXAMPLE 10 ▶ **Modeling the Value of an Investment**

In the year 2000, Marc and his wife Maria decided to invest some money in precious metals. As expected, the value of the investment fluctuated over the years, sometimes being worth more than they paid, other times less. Suppose that through 2010, the gain or loss on the investment was modeled by $v(t) = t^4 - 11t^3 + 38t^2 - 40t$, where $v(t)$ represents the gain or loss (in hundreds of dollars) in year t ($t = 0 \rightarrow 2000$).

    a. Use the rational zeroes theorem to find the years when their gain/loss was zero.

    b. Sketch the graph of the function.

    c. In what years was the investment worth less than they paid?

    d. What was their gain or loss in 2010?

Solution ▶ **a.** Writing the function as $v(t) = t(t^3 - 11t^2 + 38t - 40)$, we note $t = 0$ shows no gain or loss on purchase, and attempt to find the remaining zeroes. Testing for 1 and -1 shows neither is a zero, but $(1, -12)$ and $(-1, 90)$ are points on the graph of v. Next we try $t = 2$ with the factor theorem and the cubic polynomial.

$$\underline{2|}\quad\begin{array}{rrrr} 1 & -11 & 38 & -40 \\ & 2 & -18 & 40 \\ \hline 1 & -9 & 20 & \underline{|0} \end{array}$$

> **WORTHY OF NOTE**
>
> Due to the context, the domain of $v(t)$ in Example 10 actually begins at $t = 0$, which we could designate with a point at $(0, 0)$. In addition, note there are three sign changes in the terms of $v(t)$, indicating there will be 3 or 1 positive roots (we found 3).

We find that 2 is a zero and write $v(t) = t(t - 2)(t^2 - 9t + 20)$, then factor to obtain $v(t) = t(t - 2)(t - 4)(t - 5)$. Since $v(t) = 0$ for $t = 0$, 2, 4, and 5, they "broke even" in years 2000, 2002, 2004, and 2005.

    b. With even degree and a positive leading coefficient, the end-behavior is up/up. All zeroes have multiplicity 1. As an additional midinterval point we find $v(3) = 6$:

$$\underline{3|}\quad\begin{array}{rrrrr} 1 & -11 & 38 & -40 & 0 \\ & 3 & -24 & 42 & 6 \\ \hline 1 & -8 & 14 & 2 & \underline{|6} \end{array}$$

Figure 4.26

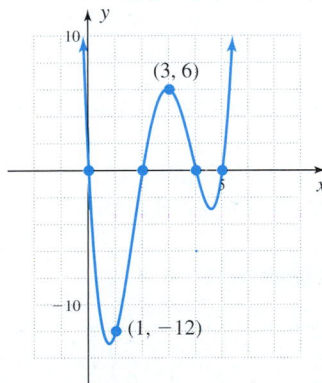

(3, 6)

(1, −12)

The graph is shown in Figure 4.26.

    c. The investment was worth less than what they paid (outputs are negative) from 2000 to 2002 and 2004 to 2005.

    d. In 2010, they were "sitting pretty," as their investment had gained $2,400.

$$\underline{10|}\quad\begin{array}{rrrrr} 1 & -11 & 38 & -40 & 0 \\ & 10 & -10 & 280 & 2400 \\ \hline 1 & -1 & 28 & 240 & \underline{|2400} \end{array}$$

Figure 4.27

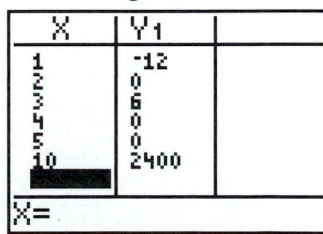

These values can also be calculated or confirmed using a graphing calculator. See Figure 4.27.

> Now try Exercises 85 through 88 ▶

As with linear and quadratic regression models, applications of other polynomial models begins with a scatterplot and a decision as to which form of regression might be appropriate. This can depend on a number of factors, such as any end-behavior that is evident, the number of apparent turning points, any anticipated behavior, and so on. However, due to the end-behavior of polynomial models, great care must be exercised when these models are used to make projections beyond the limits of the given data.

EXAMPLE 11 ▶ The Earth's atmosphere consists of several layers that are defined in terms of altitude and the characteristics of the air in each layer. In order, these are the Troposphere (0–12 km), Stratosphere (12–50 km), Mesosphere (50–80 km), and Thermosphere (80–100 km). Due to their chemical and physical characteristics, the air temperature within each layer and from layer to layer varies a great deal. The data in the table gives the temperature in °C at an altitude of h kilometers (km). Use the data to:

Altitude (km)	Temperature (°C)
0	20
4	−20
8	−45
12	−55
20	−57
30	−43
40	−16
50	−2
60	−14
70	−54
80	−91
90	−93
100	−45

a. Draw a scatterplot and decide on an appropriate form of regression, then find the regression equation.

b. Use the regression equation to find the temperature at altitudes of 32.6 km and 63.6 km.

c. As the space shuttle rockets into orbit, a temperature reading of −75°C is taken. What are the possible altitudes for the shuttle at this point?

Solution ▶ a. The scatterplot is shown in Figure 4.28. Using the characteristics exhibited (end-behavior, three turning points), it appears a quartic regression (degree 4) is appropriate and the equation is shown in Figure 4.29.

Figure 4.28

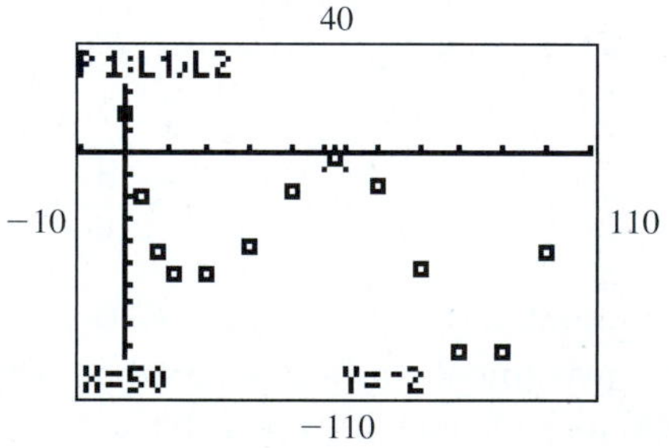

Figure 4.29

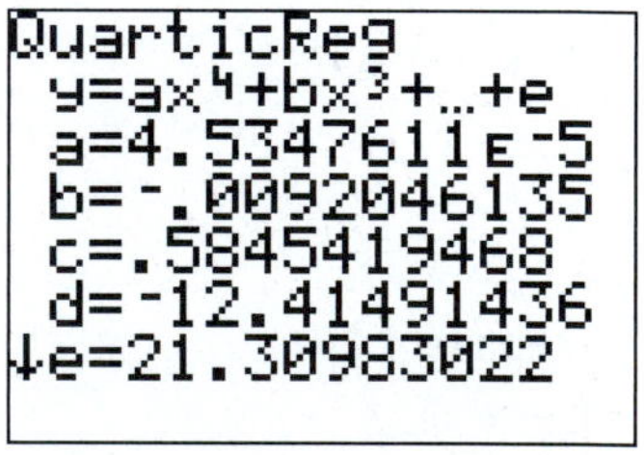

b. At altitudes of 32.6 km and 63.6 km, the temperature is very near −30.0°C (Figure 4.30).

c. Setting $Y_2 = -75$, we note the line intersects the graph in two places, one indicating an altitude of about 76.4 km, and the other an altitude of about 96.1 km (Figure 4.31).

Figure 4.30

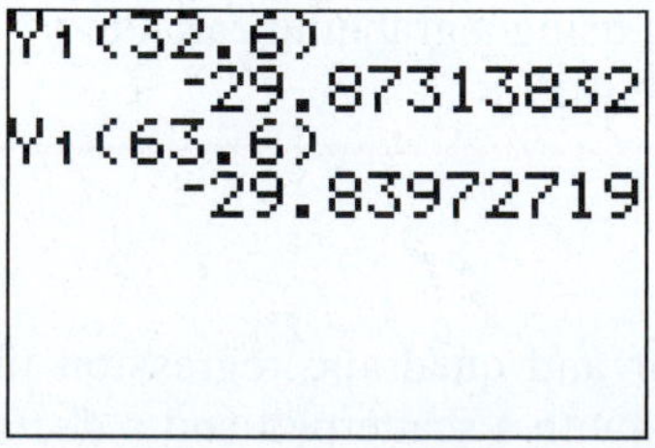

Figure 4.31

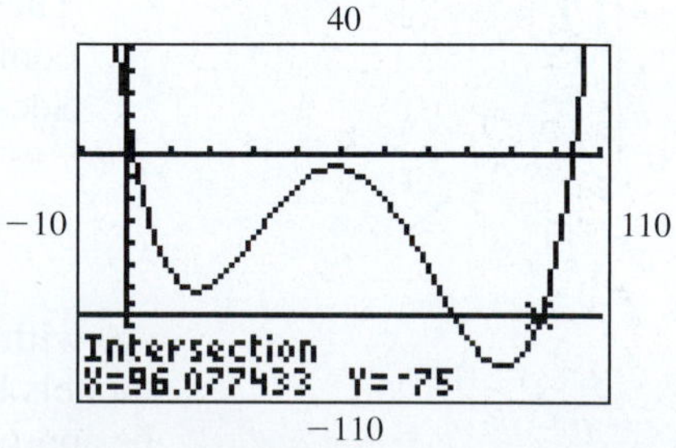

✓ **E.** You've just seen how to solve applications of polynomials and polynomial modeling

Now try Exercises 89 through 92 ▶

4.3 EXERCISES

▶ CONCEPTS AND VOCABULARY

Fill in each blank with the appropriate word or phrase. Carefully reread the section if needed.

1. For a polynomial with factors of the form $(x - c)^m$, c is called a ___zero___ of multiplicity ___m___.

2. A polynomial function of degree n has ___n___ zeroes and at most ___n − 1___ "turning points."

3. The graphs of $Y_1 = (x - 2)^2$ and $Y_2 = (x - 2)^4$ both ___bounce___ at $x = 2$, but the graph of Y_2 is ___flatter___ than the graph of Y_1 at this point.

4. Since $x^4 > 0$ for all x, the ends of its graph will always point in the ___same___ direction. Since $x^3 > 0$ when $x > 0$ and $x^3 < 0$ when $x < 0$, the ends of its graph will always point in the ___opposite___ direction.

5. In your own words, explain/discuss how to find the degree and y-intercept of a function that is given in factored form. Use $f(x) = (x + 1)^3(x - 2)(x + 4)^2$ to illustrate. Answers will vary.

6. Name all of the "tools" at your disposal that play a role in the graphing of polynomial functions. Which tools are indispensable and always used? Which tools are used only as the situation merits? Answers will vary.

▶ DEVELOPING YOUR SKILLS

Determine whether each graph is the graph of a polynomial function. If yes, state the least possible degree of the function. If no, state why.

State the end-behavior of the functions shown.

7.

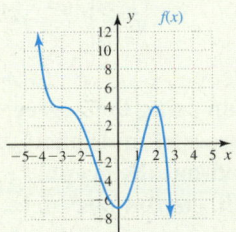

polynomial, degree 3

8.

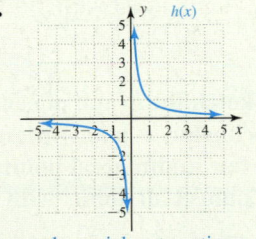

not a polynomial, not continuous

13. $f(x)$

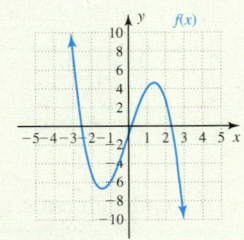

up/down

14. $g(x)$

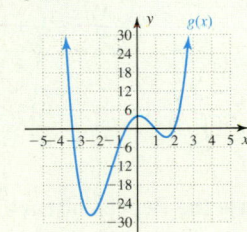

up/up

9.

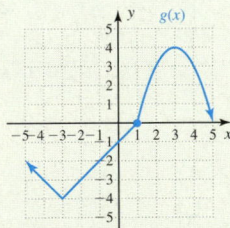

not a polynomial, sharp turns

10.

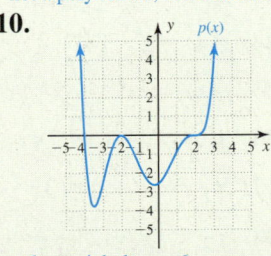

polynomial, degree 6

15. $H(x)$

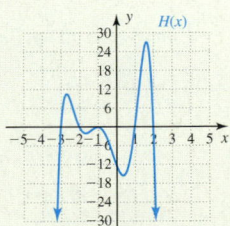

down/down

16. $h(x)$

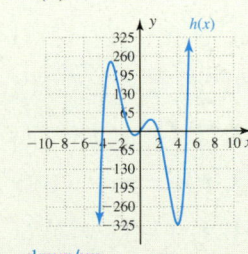

down/up

11.

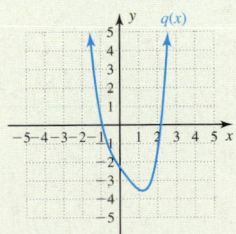

polynomial, degree 2

12.

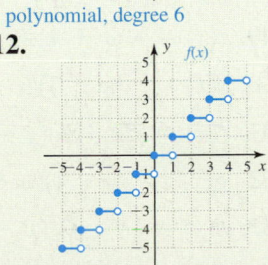

not a polynomial, not continuous

State the end-behavior and y-intercept of the functions given. Do not graph.

17. $f(x) = x^3 + 6x^2 - 5x - 2$ down/up; $(0, -2)$

18. $g(x) = x^4 - 4x^3 - 2x^2 + 16x - 12$ up/up; $(0, -12)$

19. $p(x) = -2x^4 + x^3 + 7x^2 - x - 6$ down/down; $(0, -6)$

20. $q(x) = -2x^3 - 18x^2 + 7x + 3$ up/down; $(0, 3)$

21. $Y_1 = -3x^5 + x^3 + 7x^2 - 6$ up/down; $(0, -6)$

22. $Y_2 = -x^6 - 4x^5 + 4x^3 + 16x - 12$ down/down; $(0, -12)$

HOMEWORK SELECTION GUIDE

Core: 7–79 every other odd, 81, 85–91 odd (25 Exercises)
Standard: 1–4, 7–79 every other odd, 81, 85–88, 89, 91, 93, 100 (32 Exercises)

Extended: 1–4, 7–79 every other odd, 81, 85–88, 89, 91, 93, 97, 98, 100 (34 Exercises)
In Depth: 1–6, 7–79 every other odd, 81–88, 89–97 odd, 98–101 (42 Exercises)

For each polynomial graph, (a) state whether the degree of the function is even or odd; (b) use the graph to name the zeroes of f, then state whether their multiplicity is even or odd; (c) state the minimum possible degree of f and write one possible function for f in factored form; and (d) estimate the domain and range. Assume all zeroes are real.

23.
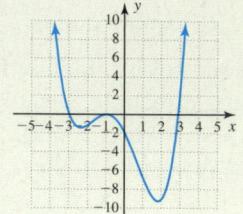

24.
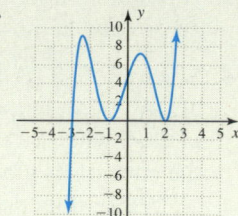

25.
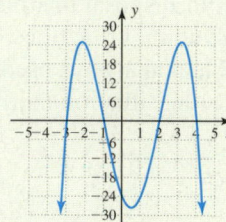

26.
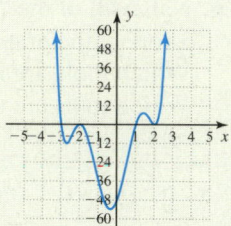

27.
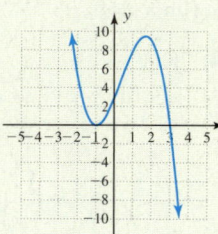

28.
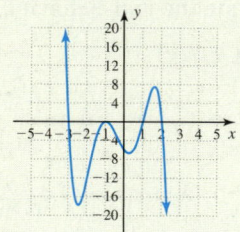

State the degree of each function, the end-behavior, and y-intercept of its graph.

29. $f(x) = (x - 3)(x + 1)^3(x - 2)^2$ degree 6; up/up; (0, −12)

30. $g(x) = (x + 2)^2(x - 4)(x + 1)$ degree 4; up/up; (0, −16)

31. $Y_1 = -(x + 1)^2(x - 2)(2x - 3)(x + 4)$
degree 5; up/down; (0, −24)

32. $Y_2 = -(x + 1)(x - 2)^3(5x - 3)$
degree 5; up/down; (0, −24)

33. $r(x) = (x^2 + 3)(x + 4)^3(x - 1)$
degree 6; up/up; (0, −192)

34. $s(x) = (x + 2)^2(x - 1)^2(x^2 + 5)$ degree 6; up/up; (0, 20)

35. $h(x) = (x^2 + 2)(x - 1)^2(1 - x)$ degree 5; up/down; (0, 2)

36. $H(x) = (x + 2)^2(2 - x)(x^2 + 4)$ degree 5; up/down; (0, 32)

Every function in Exercises 37 through 42 has the zeroes $x = -1$, $x = -3$, and $x = 2$. Match each to its corresponding graph using degree, end-behavior, and the multiplicity of each zero.

37. $f(x) = (x + 1)^2(x + 3)(x - 2)$ b

38. $F(x) = (x + 1)(x + 3)^2(x - 2)$ d

39. $g(x) = (x + 1)(x + 3)(x - 2)^3$ e

40. $G(x) = (x + 1)^3(x + 3)(x - 2)$ a

41. $Y_1 = (x + 1)^2(x + 3)(x - 2)^2$ c

42. $Y_2 = (x + 1)^3(x + 3)(x - 2)^2$ f

a.
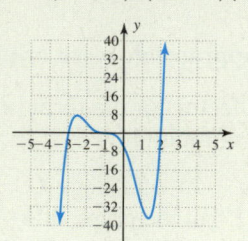

b.
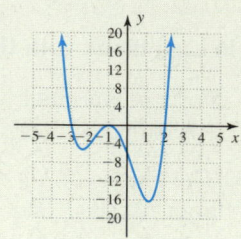

c.
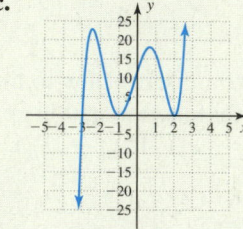

d.
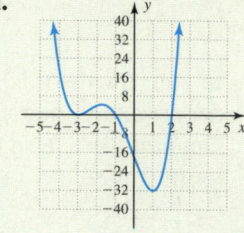

e.
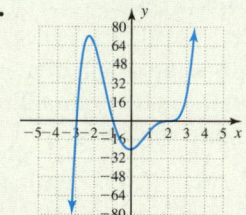

f.
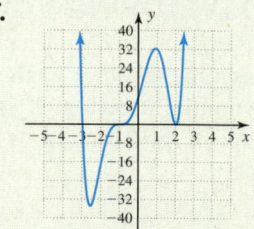

Sketch the graph of each function using the degree, end-behavior, x- and y-intercepts, zeroes of multiplicity, and a few midinterval points to round-out the graph. Connect all points with a smooth, continuous curve.

43. $f(x) = (x + 3)(x + 1)(x - 2)$

44. $g(x) = (x + 2)(x - 4)(x - 1)$

45. $p(x) = -(x + 1)^2(x - 3)$

46. $q(x) = -(x + 2)(x - 2)^2$

47. $Y_1 = (x + 1)^2(3x - 2)(x + 3)$

48. $Y_2 = (x + 2)(x - 1)^2(5x - 2)$

49. $r(x) = -(x + 1)^2(x - 2)^2(x - 1)$

50. $s(x) = -(x - 3)(x - 1)^2(x + 1)^2$

51. $f(x) = (2x + 3)(x - 1)^3$

52. $g(x) = (3x - 4)(x + 1)^3$

53. $h(x) = (x + 1)^3(x - 3)(x - 2)$

54. $H(x) = (x + 3)(x + 1)^2(x - 2)^2$

55. $Y_3 = (x + 1)^3(x - 1)^2(x - 2)$

56. $Y_4 = (x - 3)(x - 1)^3(x + 1)^2$

Use the *Guidelines for Graphing Polynomial Functions* to graph the polynomials.

57. $y = x^3 + 3x^2 - 4$

58. $y = x^3 - 13x + 12$

59. $f(x) = x^3 - 3x^2 - 6x + 8$

60. $g(x) = x^3 + 2x^2 - 5x - 6$

61. $h(x) = -x^3 - x^2 + 5x - 3$

62. $H(x) = -x^3 - x^2 + 8x + 12$

63. $p(x) = -x^4 + 10x^2 - 9$

64. $q(x) = -x^4 + 13x^2 - 36$

65. $r(x) = x^4 - 9x^2 - 4x + 12$

66. $s(x) = x^4 - 5x^3 + 20x - 16$

67. $Y_1 = x^4 - 6x^3 + 8x^2 + 6x - 9$

68. $Y_2 = x^4 - 4x^3 - 3x^2 + 10x + 8$

69. $Y_3 = 3x^4 + 2x^3 - 36x^2 + 24x + 32$

70. $Y_4 = 2x^4 - 3x^3 - 15x^2 + 32x - 12$

71. $F(x) = 2x^4 + 3x^3 - 9x^2$

72. $G(x) = 3x^4 + 2x^3 - 8x^2$

73. $f(x) = x^5 + 4x^4 - 16x^2 - 16x$

74. $g(x) = x^5 - 3x^4 + x^3 - 3x^2$

75. $h(x) = x^6 - 2x^5 - 4x^4 + 8x^3$

76. $H(x) = x^6 + 3x^5 - 4x^4$

In preparation for future course work, it becomes helpful to recognize the most common square roots in mathematics: $\sqrt{2} \approx 1.414$, $\sqrt{3} \approx 1.732$, and $\sqrt{6} \approx 2.449$. Graph the following polynomials *on a graphing calculator,* and use the calculator to locate the maximum/minimum values and all zeroes. Use the zeroes to write the polynomial in factored form, then verify the *y*-intercept from the factored form and polynomial form.

77. $h(x) = x^5 + 4x^4 - 9x - 36$

78. $H(x) = x^5 + 5x^4 - 4x - 20$

79. $f(x) = 2x^5 + 5x^4 - 10x^3 - 25x^2 + 12x + 30$

80. $g(x) = 3x^5 + 2x^4 - 24x^3 - 16x^2 + 36x + 24$

Use the graph of each function to construct its equation in factored form and in polynomial form. Be sure to check the *y*-intercept and adjust the lead coefficient if necessary.

81.

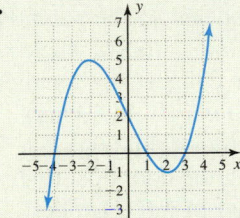

82.

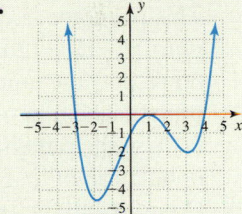

▶ WORKING WITH FORMULAS

83. Roots tests for a quartic polynomial
$ax^4 + bx^3 + cx^2 + dx + e$:

$$(r_1)^2 + (r_2)^2 + (r_3)^2 + (r_4)^2 = \frac{b^2 - 2ac}{a^2}$$

In the Chapter 3 *Reinforcing Basic Concepts* feature, we used relationships between the roots of a quadratic equation and its coefficients to verify the roots without having to substitute. Similar root/coefficient relationships exist for cubic and quartic polynomials, but the method soon becomes too time consuming (see Exercise 94). There is actually a little known formula for checking the roots of a quartic polynomial (and others) that is much more efficient. Given that r_1, r_2, r_3, and r_4 are the roots of the polynomial, the sum of the

squares of the roots must be equal to $\dfrac{b^2 - 2ac}{a^2}$.

Note that if $a = 1$, the formula reduces to $b^2 - 2c$.
(a) Use this test to verify that $x = -3$, -1, 2, and 4 are the roots of $x^4 - 2x^3 - 13x^2 + 14x + 24 = 0$, then (b) use these roots and the factored form to write the equation in polynomial form to confirm results.

84. It is worth noting that the root test in Exercise 83 still applies when the roots are irrational and/or complex. Use this test to verify that $x = -\sqrt{3}$, $\sqrt{3}$, $1 + 2i$, and $1 - 2i$ are the solutions to $x^4 - 2x^3 + 2x^2 + 6x - 15 = 0$, then use these zeroes and the factored form to write the equation in polynomial form to confirm results.

► APPLICATIONS

85. Traffic volume: Between the hours of 6:00 A.M. and 6.00 P.M., the volume of traffic at a busy intersection can be modeled by the polynomial $v(t) = -t^4 + 25t^3 - 192t^2 + 432t$, where $v(t)$ represents the number of vehicles above/below average, and t is number of hours past 6:00 A.M. (6:00 A.M. → 0). (a) Use the remainder theorem to find the volume of traffic during rush hour (8:00 A.M.), lunch time (12 noon), and the trip home (5:00 P.M.). (b) Use the rational zeroes theorem to find the times when the volume of traffic is at its average $[v(t) = 0]$. (c) Use this information to graph $v(t)$, then use the graph to estimate the maximum and minimum flow of traffic and the time at which each occurs.

86. Insect population: The population of a certain insect varies dramatically with the weather, with springlike temperatures causing a population boom and extreme weather (summer heat and winter cold) adversely affecting the population. This phenomena can be modeled by the polynomial $p(m) = -m^4 + 26m^3 - 217m^2 + 588m$, where $p(m)$ represents the number of live insects (in hundreds of thousands) in month m ($m \in (0, 1] \to$ Jan). (a) Use the remainder theorem to find the population of insects during the cool of spring (March) and the fair weather of fall (October). (b) Use the rational zeroes theorem to find the times when the population of insects becomes dormant $[p(m) = 0]$. (c) Use this information to graph $p(m)$, then use the graph to estimate the maximum and minimum population of insects, and the month at which each occurs.

87. Balance of payments: The graph shown represents the balance of payments (surplus versus deficit) for a large county over a 9-yr period. Use it to answer the following:

a. What is the minimum possible degree polynomial that can model this graph?   3

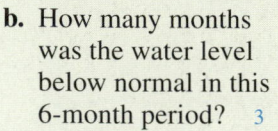

b. How many years did this county run a deficit?   5

c. Construct an equation model in factored form and in polynomial form, adjusting the lead coefficient as needed. How large was the deficit in year 8?

88. Water supply: The graph shown represents the water level in a reservoir (above and below normal) that supplies water to a metropolitan area, over a 6-month period. Use it to answer the following:

a. What is the minimum possible degree polynomial that can model this graph?   4

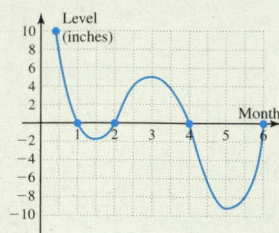

b. How many months was the water level below normal in this 6-month period?   3

c. At the beginning of this period ($m = 0$), the water level was 36 in. above normal, due to a long period of rain. Use this fact to help construct an equation model in factored form and in polynomial form, adjusting the lead coefficient as needed. Use the equation to determine the water level in months three and five.

89. In order to determine if the number of lanes on a certain highway should be increased, the flow of traffic (in vehicles per min) is carefully monitored from 6:00 A.M. to 6:00 P.M. The data collected are shown the table (6:00 A.M. corresponds to $t = 0$). (a) Draw a scatterplot and decide on an appropriate form of regression, then find the regression equation and graph the function and scatterplot on the same screen. (b) Use the regression equation and its graph to find the maximum flow of traffic for the morning and evening rush hours. (c) During what time(s) of day is the flow rate 350 vehicles per hour?

Time (6:00 A.M. → 0)	Volume (vehicles/min)
0	0
2	222
4	100
6	114
8	360
10	550
11	429

90. The Goddard Memorial Rocket Club is testing a new two-stage rocket. Using a specialized tracking device, the velocity of the rocket is monitored every second for the first 4.5 sec of flight, with the data collected in the table shown. (a) Draw a scatterplot and decide on an appropriate form of regression, then find the regression equation and graph the function and scatterplot on the same screen. (b) Use the regression equation and its graph to find how many seconds elapsed before the first stage burned out,

Time (sec)	Velocity (ft/sec)
0	0
1	441
2	484
3	459
4	696

and the rocket's velocity at this time, then (c) determine how many seconds elapsed (after liftoff) until the second stage ignited. (d) At a velocity of 1000 ft/sec, the fuel was exhausted and the return chutes deployed. How many seconds after liftoff did this occur?

 91. A posh restaurant in a thriving neighborhood opens at 10 A.M. for the lunch crowd, and closes at 9 P.M. as the dinner crowd leaves. In order to ensure that an adequate number of cooks and servers are available,

Time (10 A.M. → 0)	Customer count
0	0
2	79
4	41
6	43
9	122
11	3

their hourly customer count is monitored each day for 1 month with the data averaged and compiled in the table shown. (a) Draw a scatterplot and decide on an appropriate form of regression, then find the regression equation and graph the function and scatterplot on the same screen. (b) Use the regression equation and its graph to find what time the restaurant reaches its morning peak and its evening peak. (c) At what time is business slowest, and how many customers are in the restaurant at that time? (d) Between what times is the restaurant serving 100 customers or more?

92. Using the wind to generate power is becoming more and more prevalent. While most people are aware that a wind turbine generates more power with a stronger wind, many are not aware that the generators are built with a stall mechanism to protect the generator, blades, and infrastructure in very high winds. This affects the actual power output as the generator operates near its threshold. The power output [in watts (W)] of a certain generator is shown in the table for wind velocity v in miles per hour. (a) Draw a scatterplot and decide on an appropriate form of regression, then find the regression equation and graph the function and scatterplot on the same screen. (b) Use the regression equation and its graph to find the maximum safe power output for this generator, and the wind speed at which this occurs. (c) What is the power output in a 23 mph wind? (d) If the manufacturer stipulates that the turbine will experience automatic shutdown when power output exceeds 900 W, what is the greatest wind speed this turbine can tolerate?

Wind velocity (mph)	Power (W)
20	419
25	623
30	635
35	593
40	639

▶ EXTENDING THE CONCEPT

 93. As discussed in this section, the study of end-behavior looks at what happens to the graph of a function as $|x| \to \infty$. Notice that as $|x| \to \infty$, both $\frac{1}{x}$ and $\frac{1}{x^2}$ approach zero. This fact can be used to study the end-behavior of polynomial graphs.

a. For $f(x) = x^3 + x^2 - 3x + 6$, factoring out x^3 gives the expression

$$f(x) = x^3\left(1 + \frac{1}{x} - \frac{3}{x^2} + \frac{6}{x^3}\right).$$ What happens to the value of the expression as $x \to \infty$? As $x \to -\infty$? $f(x) \to \infty, f(x) \to -\infty$

b. Factor out x^4 from $g(x) = x^4 + 3x^3 - 4x^2 + 5x - 1$. What happens to the value of the expression as $x \to \infty$? As $x \to -\infty$? How does this affirm the end-behavior must be up/up?
$g(x) \to \infty, g(x) \to \infty; x^4 \geq 0$ for all x

94. If $u, v, w,$ and z represent the roots of the quartic polynomial $ax^4 + bx^3 + cx^2 + dx + e = 0$, then the following relationships are true:
(a) $u + v + w + z = -b$, (b) $u(v + z) + v(w + z) + w(u + z) = c$, (c) $u(vw + wz) + v(uz + wz) = -d$, and (d) $u \cdot v \cdot w \cdot z = e$. Use these tests to verify that $x = -3, -1, 2, 4$ are the solutions to $x^4 - 2x^3 - 13x^2 + 14x + 24 = 0$, then use these zeroes and the factored form to write the equation in polynomial form to confirm results.

95. For what value of c will three of the four real roots of $x^4 + 5x^3 + x^2 - 21x + c = 0$ be shared by the polynomial $x^3 + 2x^2 - 5x - 6 = 0$? $c = -18$

Show the following equations have no rational roots.

96. $x^5 - x^4 - x^3 + x^2 - 2x + 3 = 0$ verified

97. $x^5 - 2x^4 - x^3 + 2x^2 - 3x + 4 = 0$ verified

► **MAINTAINING YOUR SKILLS**

98. (3.6) Given $f(x) = x^2 - 2x$ and $g(x) = \frac{1}{x}$, find the compositions $h(x) = (f \circ g)(x)$ and $H(x) = (g \circ f)(x)$, then state the domain of each.

99. (3.1) By direct substitution, verify that $x = 1 - 2i$ is a solution to $x^2 - 2x + 5 = 0$ and name the second solution.   verified, $x = 1 + 2i$

100. (R.3/R.6) Solve each of the following equations.
  a. $-(2x + 5) - (6 - x) + 3 = x - 3(x + 2)$ $x = 2$
  b. $\sqrt{x + 1} + 3 = \sqrt{2x} + 2$ $x = 8$

  c. $\dfrac{2}{x - 3} + 5 = \dfrac{21}{x^2 - 9} + 4$ $x = 4, x = -6$

101. (1.3) Determine if the relation shown is a function. If not, explain how the definition of a function is violated.   yes

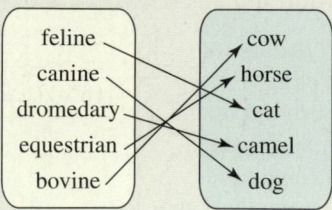

MID-CHAPTER CHECK

1. Compute $(x^3 + 8x^2 + 7x - 14) \div (x + 2)$ using long division and write the result in two ways:
  (a) dividend = (quotient)(divisor) + remainder and
  (b) $\dfrac{\text{dividend}}{\text{divisor}} = (\text{quotient}) + \dfrac{\text{remainder}}{\text{divisor}}$.

2. Given that $x - 2$ is a factor of $f(x) = 2x^4 - x^3 - 8x^2 + x + 6$, use the rational zeroes theorem to write $f(x)$ in completely factored form.

3. Use the remainder theorem to evaluate $f(-2)$, given $f(x) = -3x^4 + 7x^2 - 8x + 11$. $f(-2) = 7$

4. Use the factor theorem to find a third-degree polynomial having $x = -2$ and $x = 1 + i$ as roots.

5. Use the intermediate value theorem to show that $g(x) = x^3 - 6x - 4$ has a root in the interval $(2, 3)$.

6. Use the rational zeroes theorem, tests for -1 and 1, synthetic division, and the remainder theorem to write $f(x) = x^4 + 5x^3 - 20x - 16$ in completely factored form.

7. Find all the zeroes of h, real and complex: $h(x) = x^4 + 3x^3 + 10x^2 + 6x - 20$.

8. Sketch the graph of p using its degree, end-behavior, y-intercept, zeroes of multiplicity, and any midinterval points needed, given $p(x) = (x + 1)^2(x - 1)(x - 3)$.

9. Use the *Guidelines for Graphing* to draw the graph of $q(x) = x^3 + 5x^2 + 2x - 8$.

10. When fighter pilots train for dogfighting, a "hard-deck" is usually established below which no competitive activity can take place. The polynomial graph given shows Maverick's altitude above and below this hard-deck during a 5-sec interval.

  a. What is the minimum possible degree polynomial that could form this graph? Why?

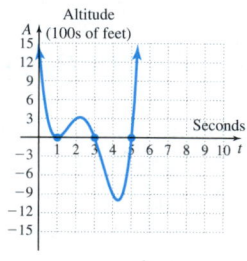

  b. How many seconds (total) was Maverick below the hard-deck for these 5 sec of the exercise?

  c. At the beginning of this time interval ($t = 0$), Maverick's altitude was 1500 ft above the hard-deck. Use this fact and the graph given to help construct an equation model in factored form and in polynomial form, adjusting the lead coefficient if needed. Use the equation to determine Maverick's altitude in relation to the hard-deck at $t = 2$ and $t = 4$.

REINFORCING BASIC CONCEPTS

Approximating Real Zeroes

Consider the equation $x^4 + x^3 + x - 6 = 0$. Using the rational zeroes theorem, the possible rational zeroes are $\{\pm 1, \pm 6, \pm 2, \pm 3\}$. The tests for 1 and -1 indicate that neither is a zero: $f(1) = -3$ and $f(-1) = -7$. Descartes' rule of signs reveals there must be one positive real zero since the coefficients of $f(x)$ change sign one time: $f(x) = x^4 + x^3 + x - 6$, and one negative real zero since $f(-x)$ also changes sign one time: $f(-x) = x^4 - x^3 - x - 6$. The remaining two zeroes must be complex. Using $x = 2$ with synthetic division shows 2 is not a zero, but the coefficients in the quotient row are all positive, so 2 is an upper bound:

$$
\begin{array}{r|rrrrr}
2 & 1 & 1 & 0 & 1 & -6 \quad \text{coefficients of } f(x)\\
  &   & 2 & 6 & 12 & 26 \\
\hline
  & 1 & 3 & 6 & 13 & \underline{|20} \quad q(x)
\end{array}
$$

Using $x = -2$ shows that -2 is a zero *and a lower bound* for all other zeroes (quotient row alternates in sign):

$$
\begin{array}{r|rrrrr}
-2 & 1 & 1 & 0 & 1 & -6 \quad \text{coefficients of } f(x)\\
   &   & -2 & 2 & -4 & 6 \\
\hline
   & 1 & -1 & 2 & -3 & \underline{|0} \quad q_1(x)
\end{array}
$$

This means the remaining real zero must be a positive irrational number less than 2 (all other possible rational zeroes were eliminated). The quotient polynomial $q_1(x) = x^3 - x^2 + 2x - 3$ is not factorable, yet we're left with the challenge of finding this final zero. While there are many advanced techniques available for approximating irrational zeroes, at this level either technology or a technique called **bisection** is commonly used. The bisection method combines the intermediate value theorem with successively smaller intervals of the input variable, to narrow down the location of the irrational zero. Although "bisection" implies halving the interval each time, any number within the interval will do. The bisection method may be most efficient using a succession of short input/output tables as shown, with the number of tables increased if greater accuracy is desired. Since $f(1) = -3$ and $f(2) = 20$, the intermediate value theorem tells us the zero must be in the interval $[1, 2]$. We begin our search here, rounding noninteger outputs to the nearest 100th. As a visual aid, positive outputs are in blue, negative outputs in red.

x	$f(x)$	Conclusion
1	-3	Zero is here, use $x = 1.25$ next
1.5	3.94	
2	20	

x	$f(x)$	Conclusion
1	-3	Zero is here, use $x = 1.30$ next
1.25	-0.36	
1.5	3.94	

x	$f(x)$	Conclusion
1.25	-0.36	Zero is here, use $x = 1.275$ next
1.30	0.35	
1.5	3.94	

A reasonable estimate for the zero appears to be $x = 1.275$. Evaluating the function at this point gives $f(1.275) \approx -0.0097$, which is very close to zero. Naturally, a closer approximation is obtained using the capabilities of a graphing calculator. To seven decimal places the zero is $x \approx 1.2756822$.

Exercise 1: Use the intermediate value theorem to show that $f(x) = x^3 - 3x + 1$ has a zero in the interval $[1, 2]$, then use bisection to locate the zero to three decimal place accuracy. 1.532

Exercise 2: The function $f(x) = x^4 + 3x - 15$ has two real zeroes in the interval $[-5, 5]$. Use the intermediate value theorem to locate the zeroes, then use bisection to find the zeroes accurate to three decimal places. -2.152, 1.765

4.4 Graphing Rational Functions

LEARNING OBJECTIVES

In Section 4.4 you will see how we can:

- ☐ **A.** Locate the vertical asymptotes and find the domain of a rational function
- ☐ **B.** Apply the concept of "roots of multiplicity" to rational functions and graphs
- ☐ **C.** Find horizontal asymptotes of rational functions
- ☐ **D.** Graph general rational functions
- ☐ **E.** Solve applications of rational functions

Our first exposure to rational functions occurred in Section 2.4, where we looked at the attributes and properties of the basic rational functions $f(x) = \frac{1}{x}$ and $g(x) = \frac{1}{x^2}$ shown in Figures 4.32 and 4.33.

Figure 4.32

$f(x) = \frac{1}{x}$

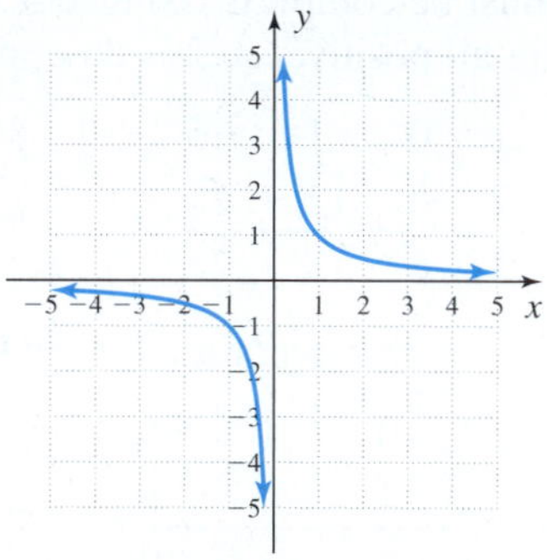

Figure 4.33

$g(x) = \frac{1}{x^2}$

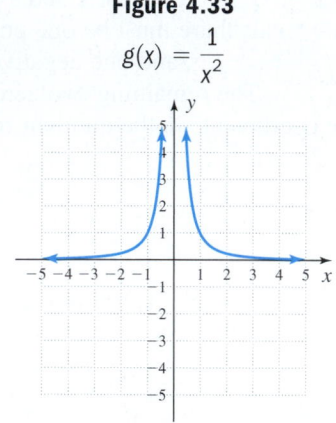

Much of what we learned about these functions can be generalized and applied to the general rational functions that follow. For convenience and emphasis, the definition of a rational function is repeated here.

Rational Functions

A rational function $V(x)$ is one of the form

$$V(x) = \frac{p(x)}{d(x)},$$

where p and d are polynomials and $d(x) \neq 0$.
The domain of $V(x)$ is all real numbers, *except the zeroes of d.*

Our study begins by taking a closer look at the zeroes of $d(x)$ that are excluded from the domain, and what happens to the graph of a rational function at or near these zeroes. These observations will form a key component of graphing general rational functions.

A. Rational Functions and Vertical Asymptotes

The graphs shown in Figures 4.34 through 4.37 illustrate that rational graphs come in many shapes, often in "pieces," and exhibit asymptotic behavior.

Figure 4.34

$g(x) = \frac{2x}{x^2 - 1}$

Figure 4.35

$w(x) = \frac{3}{x^2 + 1}$

WORTHY OF NOTE

In Section 2.5, we studied special cases of $\frac{p(x)}{d(x)}$, where p and d shared a common factor, creating a "hole" in the graph. Rational functions of this form will be investigated further in Section 4.5. In this section, we'll assume the functions are given in simplest form (the numerator and denominator have no common factors).

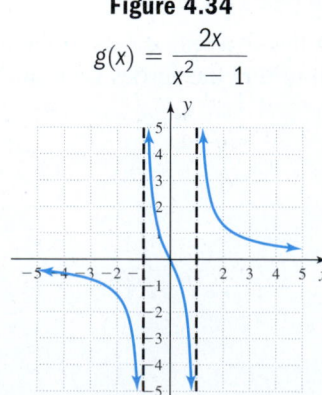

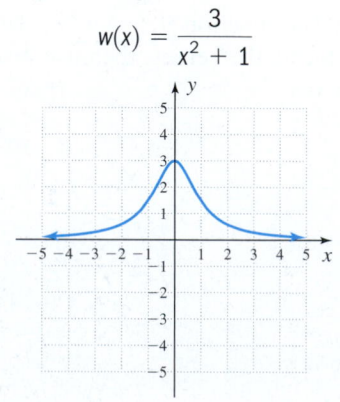

Figure 4.36

$$h(x) = \dfrac{x^2}{x^2 - 2x - 3}$$

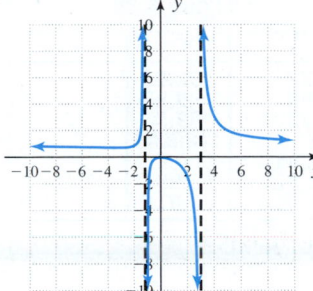

Figure 4.37

$$v(x) = \dfrac{x^2 - 4}{x + 1}$$

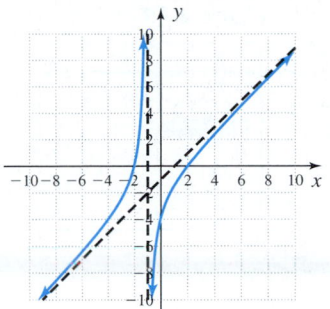

For the functions $f(x) = \frac{1}{x}$ and $g(x) = \frac{1}{x^2}$, a vertical asymptote occurred at the zero of each denominator. This actually applies to all rational functions *in simplified form*. For $V(x) = \frac{p(x)}{d(x)}$, if c is a zero of $d(x)$, the function can be evaluated at every point near c, but not *at* c. This creates a **break** or **discontinuity** in the graph of V resulting in the asymptotic behavior.

Vertical Asymptotes of a Rational Function

Given $V(x) = \dfrac{p(x)}{d(x)}$ is a rational function in simplest form,

vertical asymptotes will occur at the real zeroes of d.

Breaks created by vertical asymptotes are said to be **nonremovable,** because there is no way to repair the break, even if a piecewise-defined function were used. **See Example 5, Section 2.6.**

EXAMPLE 1 ▶   **Finding Vertical Asymptotes and the Domain of a Rational Function**

Locate the vertical asymptote(s) of each function given, then state its domain.

a. $g(x) = \dfrac{2x}{x^2 - 1}$ **b.** $w(x) = \dfrac{3}{x^2 + 1}$

c. $h(x) = \dfrac{x^2}{x^2 - 2x - 3}$ **d.** $v(x) = \dfrac{x^2 - 4}{x + 1}$

Solution ▶   **a.** Setting the denominator equal to zero gives $x^2 - 1 = 0$, so vertical asymptotes will occur at $x = -1$ and $x = 1$. The domain of g is $x \in (-\infty, -1) \cup (-1, 1) \cup (1, \infty)$. See Figure 4.34.

b. Since the equation $x^2 + 1 = 0$ has no real zeroes, there are no vertical asymptotes and the domain of w is unrestricted: $x \in \mathbb{R}$. See Figure 4.35.

c. Solving $x^2 - 2x - 3 = 0$ gives $(x + 1)(x - 3) = 0$, with solutions $x = -1$ and $x = 3$. There are vertical asymptotes at $x = -1$ and $x = 3$, and the domain of h is $x \in (-\infty, -1) \cup (-1, 3) \cup (3, \infty)$. See Figure 4.36.

d. Solving $x + 1 = 0$ gives $x = -1$, and a vertical asymptote will occur at $x = -1$. The domain of v is $x \in (-\infty, -1) \cup (-1, \infty)$. See Figure 4.37.

Now try Exercises 7 through 14 ▶

Using a **TABLE** for the function $g(x)$ from Example 1(a), we again note that we're able to evaluate rational functions for all values *near* the zeroes of the denominator, but not *at* these zeroes. See Figures 4.38 through 4.40.

Figure 4.38

X	Y₁	
-3	-.75	
-2	-1.333	
-1	ERR:	
0	0	
1	1.3333	
2	1.3333	
3	.75	

Y₁=ERR:

Figure 4.39

X	Y₁	
.97	-32.83	
.98	-49.49	
.99	-99.5	
1	ERR:	
1.01	100.5	
1.02	50.495	
1.03	33.826	

Y₁=ERR:

Figure 4.40

X	Y₁	
-1.03	-33.83	
-1.02	-50.5	
-1.01	-100.5	
-1	ERR:	
-.99	99.497	
-.98	49.495	
-.97	32.826	

Y₁=ERR:

☑ **A.** You've just seen how we can locate vertical asymptotes and find the domain of a rational function

B. Vertical Asymptotes and Multiplicities

The "cross" and "bounce" concept used for polynomial graphs can also be applied to rational graphs, particularly when viewed in terms of sign changes in the dependent variable. As you can see in Figures 4.41 to 4.43, the function $f(x) = \dfrac{1}{x + 2}$ changes sign at the asymptote $x = -2$ (negative on one side, positive on the other), and the zero of the denominator has multiplicity 1 (odd). The function $g(x) = \dfrac{1}{(x - 1)^2}$ does not change sign at the asymptote $x = 1$ (positive on both sides), and its denominator has multiplicity 2 (even). As with our earlier study of multiplicities, when these two are combined into the single function $v(x) = \dfrac{1}{(x + 2)(x - 1)^2}$, the function still changes sign at $x = -2$, and does not change sign at $x = 1$.

Figure 4.41

$$f(x) = \frac{1}{x + 2}$$

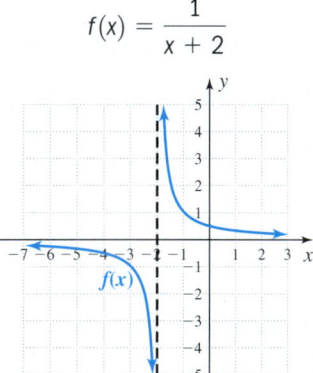

Figure 4.42

$$g(x) = \frac{1}{(x - 1)^2}$$

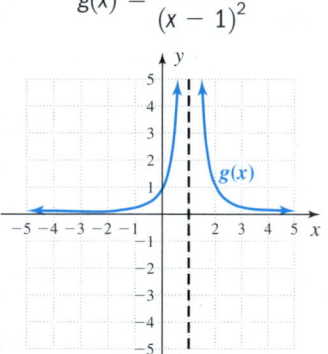

Figure 4.43

$$h(x) = \frac{1}{(x + 2)(x - 1)^2}$$

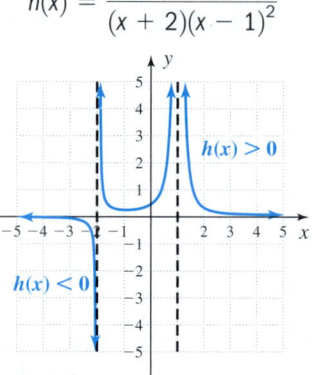

EXAMPLE 2 ▶ **Finding Sign Changes at Vertical Asymptotes**

Locate the vertical asymptotes of each function and state whether the function will change sign from one side of the asymptote(s) to the other.

a. $f(x) = \dfrac{x^2 - 4x + 4}{x^2 - 2x - 3}$ **b.** $g(x) = \dfrac{x^2 + 2}{x^2 + 2x + 1}$

Solution ▶

a. Factoring the denominator of f and setting it equal to zero gives $(x + 1)(x - 3) = 0$, and vertical asymptotes will occur at $x = -1$ and $x = 3$ (both multiplicity 1). The function will change sign at each asymptote (see Figure 4.44). Factoring the numerator gives $(x - 2)^2 = 0$, and the graph will "bounce" off the x-axis at the zero $x = 2$ (multiplicity even—no sign change).

b. Factoring the denominator of g and setting it equal to zero gives $(x + 1)^2 = 0$. There will be a vertical asymptote at $x = -1$, but the function will not change sign since it's a zero of even multiplicity (see Figure 4.45).

Figure 4.44

$$f(x) = \frac{x^2 - 4x + 4}{x^2 - 2x - 3}$$

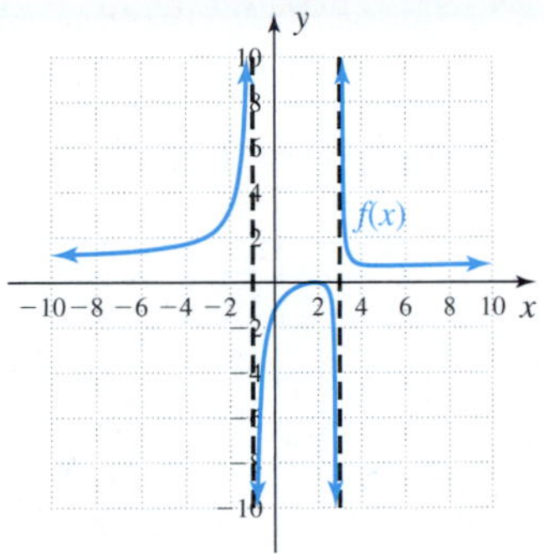

Figure 4.45

$$g(x) = \frac{x^2 + 2}{x^2 + 2x + 1}$$

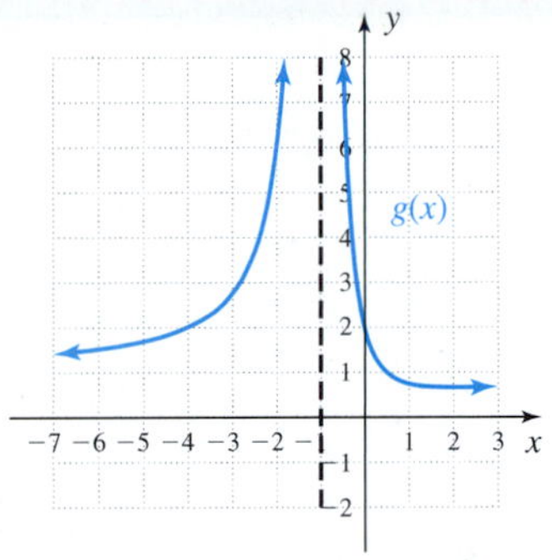

☑ **B.** You've just seen how we can apply the concept of "roots of multiplicity" to rational functions and graphs

Now try Exercises 15 through 20 ▶

C. Finding Horizontal Asymptotes

A study of horizontal asymptotes is closely related to our study of "dominant terms" in Section 4.3. Recall the highest degree term in a polynomial tends to dominate all other terms as $|x| \to \infty$. For $v(x) = \dfrac{2x^2 + 4x + 3}{x^2 + 2x + 1}$, both polynomials *have the same degree*, so $\dfrac{2x^2 + 4x + 3}{x^2 + 2x + 1} \approx \dfrac{2x^2}{x^2} = 2$ for large values of x: as $|x| \to \infty$, $y \to 2$ and $y = 2$ is a horizontal asymptote for v. When the degree of the numerator is *smaller* than the degree of the denominator, our earlier work with $y = \frac{1}{x}$ and $y = \frac{1}{x^2}$ showed there was a horizontal asymptote at $y = 0$ (the x-axis), since as $|x| \to \infty$, $y \to 0$. In general,

LOOKING AHEAD

In Section 4.5 we will explore two additional kinds of asymptotic behavior, (1) oblique (slant) asymptotes and (2) asymptotes that are nonlinear.

Horizontal Asymptotes

Given $V(x) = \dfrac{p(x)}{d(x)}$ is a rational function in lowest terms, where the leading term of p is ax^n and the leading term of d is bx^m (p has degree n, d has degree m).

    I. If $n < m$, there is a horizontal asymptote at $y = 0$ (the x-axis).

    II. If $n = m$, there is a horizontal asymptote at $y = \frac{a}{b}$.

    III. If $n > m$, the graph has no horizontal asymptote.

Finally, while the graph of a rational function can never "cross" the vertical asymptote $x = h$ (since the function simply cannot be evaluated at h), it is possible for a graph to cross the horizontal asymptote $y = k$ (some do, others do not). To find out which is the case, we set the function equal to k and solve.

EXAMPLE 3 ▶   **Locating Horizontal Asymptotes**

Locate the horizontal asymptote for each function, if one exists. Then determine if the graph will cross the asymptote.

a. $f(x) = \dfrac{3x}{x^2 + 2}$ **b.** $g(x) = \dfrac{x^2 - 4}{x^2 - 1}$ **c.** $v(x) = \dfrac{3x^2 - x - 6}{x^2 + x - 6}$

Solution ▶

a. For $f(x)$, the degree of the numerator $<$ degree of the denominator, indicating a horizontal asymptote at $y = 0$. Solving $f(x) = 0$, we find $x = 0$ is the only solution and the graph will cross the horizontal asymptote at $(0, 0)$ (see Figure 4.46).

b. For $g(x)$, the degree of the numerator and the denominator are equal. This means $g(x) \approx \dfrac{x^2}{x^2} = 1$ for large values of x, and there is a horizontal asymptote at $y = 1$. Solving $g(x) = 1$ gives

$$\dfrac{x^2 - 4}{x^2 - 1} = 1 \qquad \textcolor{magenta}{y = 1 \to \text{ horizontal asymptote}}$$

$$x^2 - 4 = x^2 - 1 \qquad \textcolor{magenta}{\text{multiply by } x^2 - 1}$$

$$-4 = -1 \qquad \textcolor{magenta}{\text{no solution}}$$

The graph will not cross the asymptote (see Figure 4.47).

c. For $v(x)$, the degree of the numerator and denominator are once again equal, so $v(x) \approx \dfrac{3x^2}{x^2} = 3$ and there is a horizontal asymptote at $y = 3$. Solving $v(x) = 3$ gives

$$\dfrac{3x^2 - x - 6}{x^2 + x - 6} = 3 \qquad \textcolor{magenta}{y = 3 \to \text{ horizontal asymptote}}$$

$$3x^2 - x - 6 = 3(x^2 + x - 6) \qquad \textcolor{magenta}{\text{multiply by } x^2 + x - 6}$$

$$3x^2 - x - 6 = 3x^2 + 3x - 18 \qquad \textcolor{magenta}{\text{distribute}}$$

$$-4x + 12 = 0 \qquad \textcolor{magenta}{\text{simplify}}$$

$$x = 3 \qquad \textcolor{magenta}{\text{result}}$$

The graph will cross its asymptote at $x = 3$ (see Figure 4.48).

Figure 4.46

$f(x) = \dfrac{3x}{x^2 + 2}$

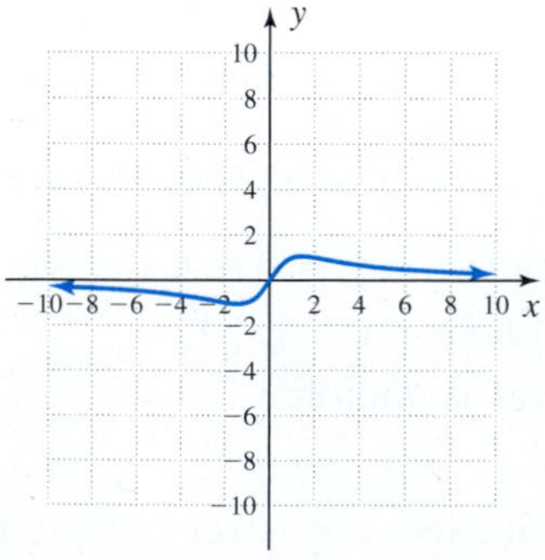

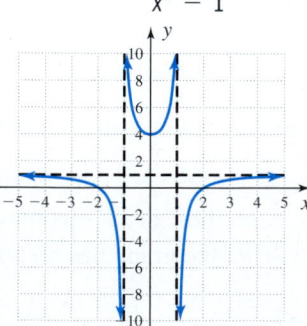

Figure 4.47

$g(x) = \dfrac{x^2 - 4}{x^2 - 1}$

Figure 4.48

$v(x) = \dfrac{3x^2 - x - 6}{x^2 + x - 6}$

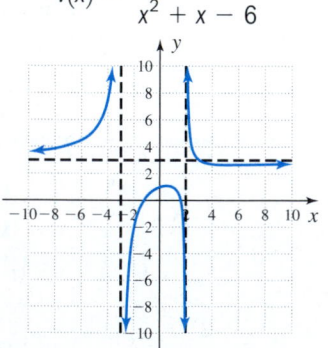

Now try Exercises 21 through 26 ▶

Using the **TABLE** feature once again gives a numerical confirmation of asymptotic behavior in the horizontal direction. For $f(x)$ from Example 3, the table verifies that larger positive values of x produce smaller and smaller values of $f(x)$: as $x \to \infty$, $f(x) \to 0$ (Figure 4.49). For $v(x)$, a table will confirm that the graph passes through the horizontal asymptote at the point $(3, 3)$ (Figure 4.50), but still approaches the asymptote $y = 3$ as x becomes very large: as $x \to \infty$, $v(x) \to 3$ (Figure 4.51).

Figure 4.49

X	Y₁
0	0
150	.02
300	.01
450	.00667
600	.005
750	.004
900	.00333

X=0

Figure 4.50

X	Y₁
0	1
.5	1.0952
1	1
1.5	.33333
2	ERR:
2.5	3.7273
3	3

Y₁=3

Figure 4.51

X	Y₁
3	3
50	2.9261
100	2.9616
150	2.974
200	2.9804
250	2.9843
400	2.9901

Y₁=2.99009938028

Finally, it's helpful to note that the location of all nonvertical asymptotes and whether or not the graph crosses through them can actually be found using division. The quotient $q(x)$ gives the equation of the asymptote, and the zeroes of the remainder $r(x)$ will indicate if and where the graph and asymptote will cross. From Example 3(c), long division gives $q(x) = 3$ and $r(x) = -4x + 12$ (verify this), showing there is a horizontal asymptote at $y = 3$, which the graph crosses at $x = 3$ since $r(3) = 0$. **See Exercises 27 through 30.**

☑ **C.** You've just seen how we can find the horizontal asymptotes of rational functions

D. The Graph of a Rational Function

Our observations to this point lead us to this general strategy for graphing rational functions. Not all graphs require every step, but together they provide an effective approach.

Guidelines for Graphing Rational Functions

Given $V(x) = \dfrac{p(x)}{d(x)}\ [d(x) \neq 0]$ is a rational function in lowest terms,

1. Find $V(0)$ (if it exists): the y-intercept

2. Find the zeroes of p (if they exist): the x-intercept(s).

3. Find the zeroes of d (if they exist): the vertical asymptotes.

4. Locate the horizontal asymptote if it exists.

5. Determine if the graph will cross the horizontal asymptote.

6. Compute any additional points needed to sketch the graph.

• Draw the asymptotes, plot the intercepts and additional points, and determine where $V(x)$ changes sign to complete the graph.

As you work to complete the graph, it helps to keep the following in mind:

• The graph *must* go through its plotted points and approach the asymptotes.
• The graph *may* cross a horizontal asymptote, but *never* a vertical asymptote.
• Function values *must* change sign at any zero of odd multiplicity.

EXAMPLE 4 ▶ **Graphing Rational Functions**

Graph each function given.

a. $f(x) = \dfrac{x^2 - x - 6}{x^2 + x - 6}$ **b.** $g(x) = \dfrac{2x^2 - 4x + 2}{x^2 - 7}$

Solution ▶ **a.** Begin by writing f in factored form: $f(x) = \dfrac{(x + 2)(x - 3)}{(x + 3)(x - 2)}$.

1. y-intercept: $f(0) = \dfrac{(2)(-3)}{(3)(-2)} = 1$, so the y-intercept is $(0, 1)$.

2. x-intercepts: Setting the numerator equal to zero gives $(x + 2)(x - 3) = 0$, showing the x-intercepts will be $(-2, 0)$ and $(3, 0)$.

3. Vertical asymptote(s): Setting the denominator equal to zero gives $(x + 3)(x - 2) = 0$, showing there will be vertical asymptotes at $x = -3$ and $x = 2$.

4. Horizontal asymptote: Since the degree of the numerator and the degree of the denominator are equal, $y = \dfrac{x^2}{x^2} = 1$ is a horizontal asymptote.

5. Solving $\dfrac{x^2 - x - 6}{x^2 + x - 6} = 1$ $f(x) = 1 \rightarrow$ horizontal asymptote

   $\qquad x^2 - x - 6 = x^2 + x - 6$ multiply by $x^2 + x - 6$

   $\qquad\qquad\qquad -2x = 0$ simplify

   $\qquad\qquad\qquad\quad x = 0$ solve

The graph will cross the horizontal asymptote at $(0, 1)$.

The information from steps 1 through 5 is shown in Figure 4.52, and indicates we have no information about the graph in the interval $(-\infty, -3)$. Since rational functions are defined for all real numbers except the zeroes of d, we know there must be a "piece" of the graph in this interval.

6. Selecting $x = -4$ to compute one additional point, we find $f(-4) = \dfrac{(-2)(-7)}{(-1)(-6)} = \dfrac{14}{6} = \dfrac{7}{3}$. The point is $\left(-4, \dfrac{7}{3}\right)$.

All factors of f are linear, so function values will alternate sign in the intervals created by x-intercepts and vertical asymptotes. The y-intercept $(0, 1)$ shows $f(x)$ is positive in the interval containing 0. To meet all necessary conditions, we complete the graph as shown in Figure 4.53.

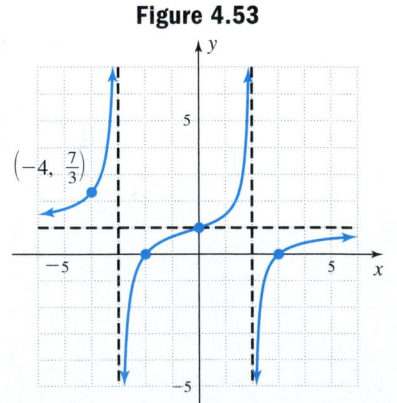

Figure 4.52   **Figure 4.53**

b. Writing g in factored form gives $g(x) = \dfrac{2(x^2 - 2x + 1)}{x^2 - 7} = \dfrac{2(x - 1)^2}{(x + \sqrt{7})(x - \sqrt{7})}$.

1. y-intercept: $g(0) = \dfrac{2(-1)^2}{(\sqrt{7})(-\sqrt{7})} = -\dfrac{2}{7}$. The y-intercept is $\left(0, -\dfrac{2}{7}\right)$.

2. x-intercept(s): Setting the numerator equal to zero gives $2(x - 1)^2 = 0$, with $x = 1$ as a zero of multiplicity 2. The x-intercept is $(1, 0)$.

3. Vertical asymptote(s): Setting the denominator equal to zero gives $(x + \sqrt{7})(x - \sqrt{7}) = 0$, showing there will be asymptotes at $x = -\sqrt{7}$ and $x = \sqrt{7}$.

4. Horizontal asymptote: The degree of the numerator is equal to the degree of denominator, so $y = \dfrac{2x^2}{x^2} = 2$ is a horizontal asymptote.

5. Solve $\dfrac{2x^2 - 4x + 2}{x^2 - 7} = 2$ $g(x) = 2 \rightarrow$ horizontal asymptote

$2x^2 - 4x + 2 = 2x^2 - 14$ multiply by $x^2 - 7$

$-4x = -16$ simplify

$x = 4$ solve

The graph will cross its horizontal asymptote at $(4, 2)$. The information from steps 1 to 5 is shown in Figure 4.54, and indicates we have no information about the graph in the interval $(-\infty, -\sqrt{7})$.

Figure 4.54

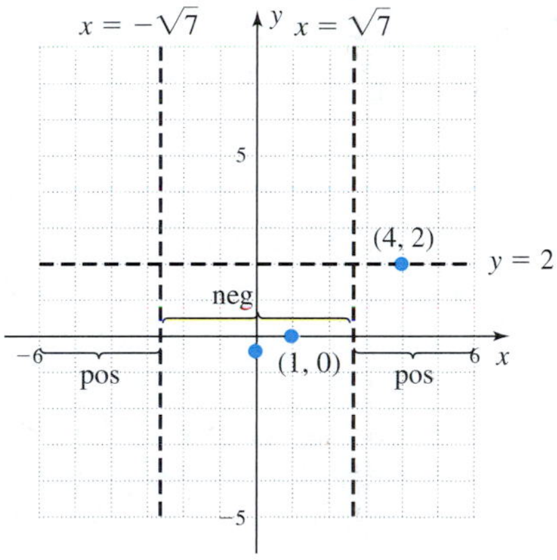

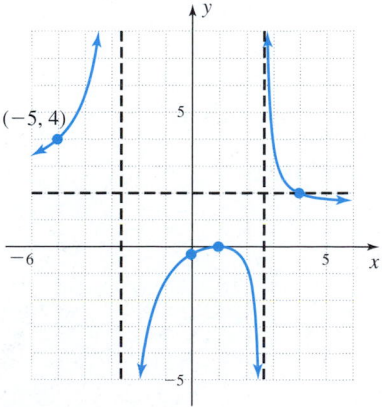

WORTHY OF NOTE

It's useful to note that the number of "pieces" forming a rational graph will always be one more than the number of vertical asymptotes. The graph of $f(x) = \dfrac{3x}{x^2 + 2}$ (Figure 4.46) has no vertical asymptotes and one piece, $y = \dfrac{1}{x}$ has one vertical asymptote and two pieces, $g(x) = \dfrac{x^2 - 4}{x^2 - 1}$ (Figure 4.47) has two vertical asymptotes and three pieces, and so on.

6. Selecting $x = -5$, $g(-5) = \dfrac{2(-5 - 1)^2}{(-5)^2 - 7}$

$= \dfrac{2(-6)^2}{25 - 7}$

$= \dfrac{2(36)}{18}$

$= 4$

The point $(-5, 4)$ is on the graph (Figure 4.55).

Since factors of the denominator have odd multiplicity, function values will alternate sign on either side of the asymptotes. The factor in the numerator has even multiplicity, so the graph will "bounce off" the x-axis at $x = 1$ (no change in sign). The y-intercept $\left(0, -\dfrac{2}{7}\right)$ shows the function is negative in the interval containing 0. This information and the completed graph are shown in Figure 4.55.

Figure 4.55

Now try Exercises 31 through 54 ▶

Examples 3 and 4 demonstrate that graphs of rational functions come in a large variety. Once the components of the graph have been found, completing the graph presents an intriguing and puzzle-like challenge as we attempt to sketch a graph that meets all conditions. As we've done with other functions, can you reverse this process? That is, given the *graph* of a rational function, can you construct its equation?

EXAMPLE 5 ▶   **Finding the Equation of a Rational Function from Its Graph**

Use the graph of $f(x)$ shown to construct its equation.

Solution ▶   The x-intercepts are $(-1, 0)$ and $(4, 0)$, so the numerator must contain the factors $(x + 1)$ and $(x - 4)$. The vertical asymptotes are $x = -2$ and $x = 3$, so the denominator must have the factors $(x + 2)$ and $(x - 3)$. So far we have:

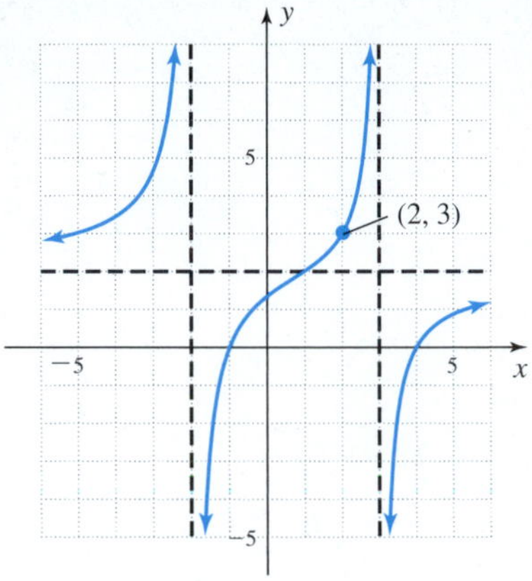

$$f(x) = \frac{a(x + 1)(x - 4)}{(x + 2)(x - 3)}$$

Since $(2, 3)$ is on the graph, we substitute 2 for x and 3 for $f(x)$ to solve for a:

$$3 = \frac{a(2 + 1)(2 - 4)}{(2 + 2)(2 - 3)}$$ substitute 3 for $f(x)$ and 2 for x

$$3 = \frac{3a}{2}$$ simplify

$$2 = a$$ solve

The result is $f(x) = \dfrac{2(x + 1)(x - 4)}{(x + 2)(x - 3)} = \dfrac{2x^2 - 6x - 8}{x^2 - x - 6}$, with a horizontal

asymptote at $y = 2$ and a y-intercept of $(0, \frac{4}{3})$, which fit the graph very well.

Now try Exercises 55 through 58 ▶

As a final note, there are many rational graphs that have x- and y-intercepts and vertical asymptotes, *but no horizontal asymptotes*. Two examples are shown in Figures 4.56 and 4.57.

Figure 4.56

$$f(x) = \frac{x^2 - 4}{x + 1}$$

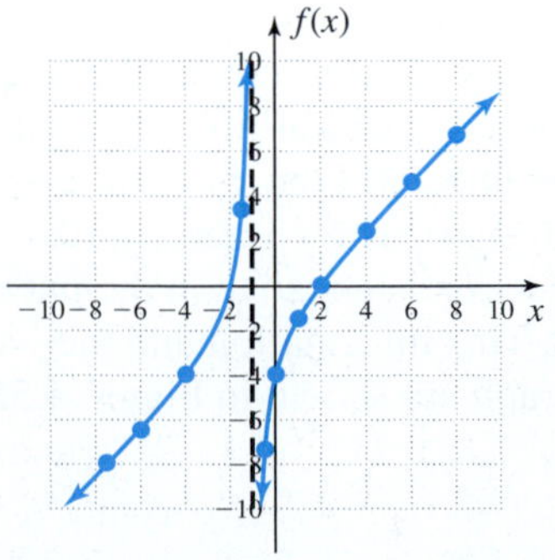

Figure 4.57

$$g(x) = \frac{x^2 - \frac{1}{4}}{x - 1}$$

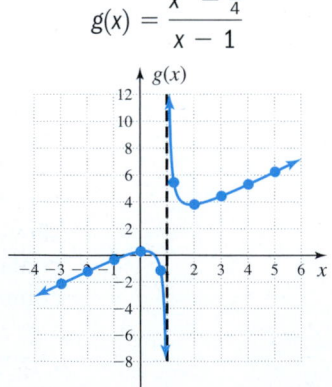

☑ **D.** You've just seen how we can graph general rational functions

Without more information regarding nonhorizontal asymptotes, sketching these graphs requires a substantial number of plotted points. Rational graphs of this type will be studied in more detail in Section 4.5, enabling us to complete each graph using far fewer points. **See Exercises 59 through 62.**

E. Applications of Rational Functions

In many applications of rational functions, the coefficients can be rather large and the graph should be scaled appropriately.

EXAMPLE 6 ▶ **Modeling the Rides at an Amusement Park**

A popular amusement park wants to add new rides and asks various contractors to submit ideas. Suppose one ride engineer offers plans for a ride that begins with a near vertical drop into a dark tunnel, quickly turns and becomes more horizontal, pops out the tunnel's end, then coasts up to the exit platform, braking 20 m from the release point. The height of a rider above ground is modeled by the function

$h(x) = \dfrac{39x^2 - 507x + 468}{3x^2 + 23x + 20}$, where $h(x)$ is the height in meters at a horizontal

distance of x meters from the release point.

  a. Graph the function for $x \in [-1, 20]$.
  b. How high is the release point for this ride?
  c. How long is the tunnel from entrance to exit?
  d. What is the height of the exit platform?

Solution ▶ **a.** Here we begin by noting the y-intercept is $h(0) = \dfrac{468}{20}$ or 23.4 m. Also, since

the degree of the numerator is equal to that of the denominator, the ratio of

leading terms $\dfrac{39x^2}{3x^2}$ indicates a horizontal

asymptote at $y = 13$. Writing $h(x)$ in

factored form gives $h(x) = \dfrac{39(x - 1)(x - 12)}{(3x + 20)(x + 1)}$,

showing the x-intercepts will be (1, 0) and
(12, 0), with vertical asymptotes at $x = -6.\overline{6}$
and $x = -1$. Computing midinterval points of
$x = 4, 8,$ and 18 gives (4, −5.85), (8, −2.76),
and (18, 2.83). Graphing the function over the
specified interval produces the graph shown
in Figure 4.58.

Figure 4.58

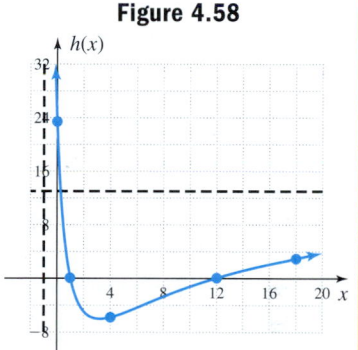

  b. From the context, the release point is at 23.4 m.
  c. The ride enters the tunnel at $x = 1$ and exits
       at $x = 12$, making the tunnel 11 m long.
  d. Since the ride begins braking at a distance of
       20 m, the platform must be $h(20) \approx 3.5$ m
       high. See Figure 4.59.

Figure 4.59

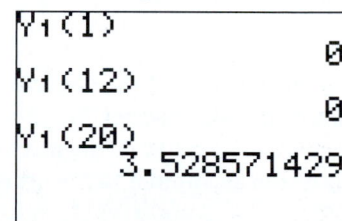

Now try Exercises 65 through 76 ▶

As a final note, the ride proposed in Example 6 was never approved due to excessive *g*-forces on the riders. Example 6 helps to illustrate that when it comes to applications of rational functions, portions of the graph may be ignored due to the context. In addition, some applications may focus on a specific attribute of the graph, such as the horizontal asymptotes in **Exercises 65, 66,** and elsewhere, or the vertical asymptotes in **Exercises 67 and 68.**

☑ **E.** You've just seen how we can solve applications of rational functions

4.4 EXERCISES

▶ CONCEPTS AND VOCABULARY

Fill in each blank with the appropriate word or phrase. Carefully reread the section if needed.

1. Write the following in direction/approach notation. *As x becomes an infinitely large negative number, y approaches 2.* as $x \to -\infty, y \to 2$

2. For any constant k, the notation as $|x| \to +\infty$, $y \to k$ is an indication of a __horizontal__ asymptote, while $x \to k$, $|y| \to +\infty$ indicates a __vertical__ asymptote.

3. Vertical asymptotes are found by setting the __denominator__ equal to zero. The x-intercepts are found by setting the __numerator__ equal to zero.

4. If the degree of the numerator is equal to the degree of the denominator, a horizontal asymptote occurs at $y = \frac{a}{b}$, where $\frac{a}{b}$ represents the ratio of the __leading__ __coefficients__.

5. Use the function $g(x) = \dfrac{3x^2 - 2x}{2x^2 - 3}$ and a table of values to discuss the concept of horizontal asymptotes. At what positive value of x is the graph of g within 0.01 of its horizontal asymptote? about $x = 98$

6. Name all of the "tools" at your disposal that play a role in the graphing of rational functions. Which tools are indispensable and always used? Which are used only as the situation merits? Answers will vary.

▶ DEVELOPING YOUR SKILLS

Give the location of the vertical asymptote(s) if they exist, and state the function's domain.

7. $f(x) = \dfrac{x + 2}{x - 3}$

8. $F(x) = \dfrac{4x}{2x - 3}$

9. $g(x) = \dfrac{3x^2}{x^2 - 9}$

10. $G(x) = \dfrac{x + 1}{9x^2 - 4}$

11. $h(x) = \dfrac{x^2 - 1}{2x^2 + 3x - 5}$

12. $H(x) = \dfrac{x - 5}{2x^2 - x - 3}$

13. $p(x) = \dfrac{2x + 3}{x^2 + x + 1}$

14. $q(x) = \dfrac{2x^3}{x^2 + 4}$

Give the location of the vertical asymptote(s) if they exist, and state whether function values will change sign (positive to negative or negative to positive) from one side of the asymptote to the other.

15. $Y_1 = \dfrac{x + 1}{x^2 - x - 6}$

16. $Y_2 = \dfrac{2x + 3}{x^2 - x - 20}$

17. $r(x) = \dfrac{x^2 + 3x - 10}{x^2 - 6x + 9}$

18. $R(x) = \dfrac{x^2 - 2x - 15}{x^2 - 4x + 4}$

19. $Y_1 = \dfrac{x}{x^3 + 2x^2 - 4x - 8}$

20. $Y_2 = \dfrac{-2x}{x^3 + x^2 - x - 1}$

For the functions given, (a) determine if a horizontal asymptote exists and (b) determine if the graph will cross the asymptote, and if so, where it crosses.

21. $Y_1 = \dfrac{2x - 3}{x^2 + 1}$

22. $Y_2 = \dfrac{4x + 3}{2x^2 + 5}$

23. $r(x) = \dfrac{4x^2 - 9}{x^2 - 3x - 18}$

24. $R(x) = \dfrac{2x^2 - x - 10}{x^2 + 5}$

25. $p(x) = \dfrac{3x^2 - 5}{x^2 - 1}$

26. $P(x) = \dfrac{3x^2 - 5x - 2}{x^2 - 4}$

Apply long division to find the quotient and remainder for each function. Use this information to determine the equation of the horizontal asymptote, and whether the graph will cross this asymptote. Verify answers by graphing the functions on a graphing calculator and locating points of intersection.

 27. $v(x) = \dfrac{8x}{x^2 + 1}$

28. $f(x) = \dfrac{4x + 8}{x^2 + 1}$

29. $g(x) = \dfrac{2x^2 - 8x}{x^2 - 4}$

30. $h(x) = \dfrac{x^2 - x - 6}{x^2 - 1}$

HOMEWORK SELECTION GUIDE

Core: 7–59 every other odd, 65–75 odd (20 Exercises)
Standard: 1–4, 7–59 every other odd, 63–75 odd, 79 (26 Exercises)

Additional answers can be found in the Instructor Answer Appendix.

Extended: 1–4, 7–55 every other odd, 57, 59, 63, 64, 65–73 odd, 74–77, 79, 81 (32 Exercises)
In Depth: 1–6, 7–55 every other odd, 57, 59, 63, 64, 65–73 odd, 74–82 (37 Exercises)

Give the location of the *x*- and *y*-intercepts (if they exist), and discuss the behavior of the function (bounce or cross) at each *x*-intercept.

31. $f(x) = \dfrac{x^2 - 3x}{x^2 - 5}$ **32.** $F(x) = \dfrac{2x - x^2}{x^2 + 2x - 3}$

33. $g(x) = \dfrac{x^2 + 3x - 4}{x^2 - 1}$ **34.** $G(x) = \dfrac{x^2 + 7x + 6}{x^2 - 2}$

35. $h(x) = \dfrac{x^3 - 6x^2 + 9x}{4 - x^2}$ **36.** $H(x) = \dfrac{4x + 4x^2 + x^3}{x^2 - 1}$

Use the *Guidelines for Graphing Rational Functions* to graph the functions given.

37. $f(x) = \dfrac{x + 3}{x - 1}$ **38.** $g(x) = \dfrac{x - 4}{x + 2}$

39. $F(x) = \dfrac{8x}{x^2 + 4}$ **40.** $G(x) = \dfrac{-12x}{x^2 + 3}$

41. $p(x) = \dfrac{-2x^2}{x^2 - 4}$ **42.** $P(x) = \dfrac{3x^2}{x^2 - 9}$

43. $q(x) = \dfrac{2x - x^2}{x^2 + 4x - 5}$ **44.** $Q(x) = \dfrac{x^2 + 3x}{x^2 - 2x - 3}$

45. $h(x) = \dfrac{-3x}{x^2 - 6x + 9}$ **46.** $H(x) = \dfrac{2x}{x^2 - 2x + 1}$

47. $Y_1 = \dfrac{x - 1}{x^2 - 3x - 4}$ **48.** $Y_2 = \dfrac{1 - x}{x^2 - 2x}$

49. $s(x) = \dfrac{4x^2}{2x^2 + 4}$ **50.** $S(x) = \dfrac{-2x^2}{x^2 + 1}$

51. $Y_1 = \dfrac{x^2 - 4}{x^2 - 1}$ **52.** $Y_2 = \dfrac{x^2 - x - 12}{x^2 + x - 12}$

53. $v(x) = \dfrac{-2x}{x^3 + 2x^2 - 4x - 8}$

54. $V(x) = \dfrac{3x}{x^3 + x^2 - x - 1}$

Use the vertical asymptotes, *x*-intercepts, and their multiplicities to construct an equation that corresponds to each graph. Be sure the *y*-intercept estimated from the graph matches the value given by your equation for *x* = 0. Check work on a graphing calculator.

55.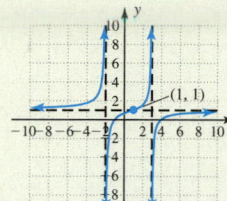

56.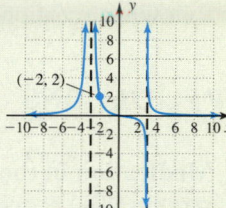

57.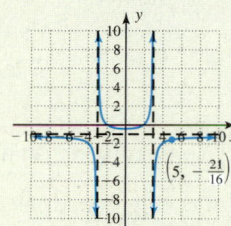

58.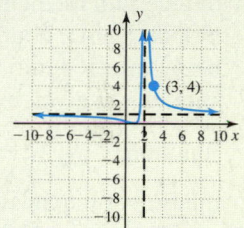

Graph the following using the *Guidelines for Graphing Rational Functions*. From the equations given, note there are no horizontal asymptotes (the degree of each numerator is greater than the degree of the denominator) and a large number of plotted points may be necessary to complete each graph.

59. $v(x) = \dfrac{x^2 - 4}{x}$ **60.** $f(x) = \dfrac{9 - x^2}{x - 1}$

61. $g(x) = \dfrac{x^2}{x - 1}$ **62.** $h(x) = \dfrac{1 - x^2}{x + 2}$

▶ WORKING WITH FORMULAS

63. Population density: $D(x) = \dfrac{ax}{x^2 + b}$

The population density of urban areas (in people per square mile) can be modeled by the formula shown, where *a* and *b* are constants related to the overall population and sprawl of the area under study, and $D(x)$ is the population density (in hundreds), *x* mi from the center of downtown.

Graph the function for *a* = 63 and *b* = 20 over the interval $x \in [0, 50]$, and then use the graph to answer the following questions.

a. What is the significance of the *horizontal asymptote* (what does it mean in this context)?

b. How far from downtown does the population density fall below 525 people per square mile? How far until the density falls below 300 people per square mile?   10 mi, 20 mi

c. Use the graph and a table to determine how far from downtown the population density reaches a maximum. What is this maximum?

Additional answers can be found in the Instructor Answer Appendix.

64. Cost of removing pollutants: $C(x) = \dfrac{kx}{100 - x}$

Some industries resist cleaner air standards because the cost of removing pollutants rises dramatically as higher standards are set. This phenomenon can be modeled by the formula given, where $C(x)$ is the cost (in thousands of dollars) of removing $x\%$ of the pollutant and k is a constant that depends on the type of pollutant and other factors.

Graph the function for $k = 250$ over the interval $x \in [0, 100]$, and then use the graph to answer the following questions.

a. What is the significance of the *vertical asymptote* (what does it mean in this context)?

b. If new laws are passed that require 80% of a pollutant to be removed, while the existing law requires only 75%, how much will the new legislation cost the company? Compare the cost of the 5% increase from 75% to 80% with the cost of the 1% increase from 90% to 91%.

c. What percent of the pollutants can be removed if the company budgets 2250 thousand dollars?

▶ **APPLICATIONS**

65. Medication in the bloodstream: The concentration C of a certain medicine in the bloodstream h hours after being injected into the shoulder is given by the

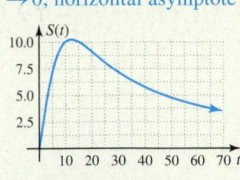

function: $C(h) = \dfrac{2h^2 + h}{h^3 + 70}$. Use the given graph of the function to answer the following questions.

a. Approximately how many hours after injection did the maximum concentration occur? What was the maximum concentration?   5 hr; about 0.28

b. Use $C(h)$ to *compute* the rate of change for the intervals $h = 8$ to $h = 10$ and $h = 20$ to $h = 22$. What do you notice?

c. Use mathematical notation to state what happens to the concentration C as the number of hours becomes infinitely large. What role does the h-axis play for this function?
   $h \to \infty, C \to 0$; horizontal asymptote

66. Supply and demand: In response to certain market demands, manufacturers will quickly get a product out on the market to take advantage of consumer interest. Once the product is released, it is not uncommon for sales to initially skyrocket, taper off, and then gradually decrease as consumer interest wanes. For a certain product, sales can be modeled by the function $S(t) = \dfrac{250t}{t^2 + 150}$, where $S(t)$ represents the daily sales (in $10,000) t days after the product has debuted. Use the given graph of the function to answer the following questions.

a. Approximately how many days after the product came out did sales reach a maximum? What was the maximum sales?

b. Use $S(t)$ to compute the rate of change for the intervals $t = 7$ to $t = 8$ and $t = 60$ to $t = 62$. What do you notice?

c. Use mathematical notation to state what happens to the daily sales S as the number of days becomes infinitely large. What role does the t-axis play for this function?

67. Cost to remove pollutants: For a certain coal-burning power plant, the cost to remove pollutants from plant emissions can be modeled by $C(p) = \dfrac{80p}{100 - p}$, where $C(p)$ represents the cost (in thousands of dollars) to remove p percent of the pollutants. (a) Find the cost to remove 20%, 50%, and 80% of the pollutants, then comment on the results; (b) graph the function using an appropriate scale; and (c) use mathematical notation to state what happens if the power company attempts to remove 100% of the pollutants.

68. Costs of recycling: A large city has initiated a new recycling effort, and wants to distribute recycling bins for use in separating various recyclable materials. City planners anticipate the cost of the program can be modeled by the function $C(p) = \dfrac{220p}{100 - p}$, where $C(p)$ represents the cost (in $10,000) to distribute the bins to p percent of the population. (a) Find the cost to distribute bins to 25%, 50%, and 75% of the population, then comment on the results; (b) graph the function using an appropriate scale; and (c) use mathematical notation to state what happens if the city attempts to give recycling bins to 100% of the population.

Memory retention: Due to their asymptotic behavior, rational functions are often used to model the mind's ability to retain information over a long period of time—the "use it or lose it" phenomenon.

69. Language retention: A large group of students is asked to memorize a list of 50 Italian words, a language that is unfamiliar to them. The group is then tested regularly to see how many of the words are retained over a period of time. The average number of words retained is modeled by the function $W(t) = \dfrac{6t + 40}{t}$, where $W(t)$ represents the number of words remembered after t days.

  a. Graph the function over the interval $t \in [1, 40]$. How many days until only half the words are remembered? How many days until only one-fifth of the words are remembered?   2; 10

  b. After 10 days, what is the average number of words retained? How many days until only 8 words can be recalled?   10; 20

  c. What is the significance of the horizontal asymptote (what does it mean in this context)?   On average, 6 words will be remembered for life.

70. Language retention: A similar study asked students to memorize 50 Hawaiian words, a language that is both unfamiliar and phonetically foreign to them (see Exercise 69). The average number of words retained is modeled by the function $W(t) = \dfrac{4t + 20}{t}$, where $W(t)$ represents the number of words after t days.

  a. Graph the function over the interval $t \in [1, 40]$. How many days until only half the words are remembered? How does this compare to Exercise 69? How many days until only one-fifth of the words are remembered?   1; lower retention; 3 days

  b. After 7 days, what is the average number of words retained? How many days until only 5 words can be recalled?   about 7; 20 days

  c. What is the significance of the horizontal asymptote (what does it mean in this context)?   On average, 4 words will be remembered for life.

Concentration and dilution: When antifreeze is mixed with water, it becomes diluted—less than 100% antifreeze. The more water added, the less concentrated the antifreeze becomes, with this process continuing until a desired concentration is met. This application and many similar to it can be modeled by rational functions.

71. Concentration of antifreeze: A 400-gal tank currently holds 40 gal of a 25% antifreeze solution. To raise the concentration of the antifreeze in the tank, x gal of a 75% antifreeze solution is pumped in.

  a. Show the formula for the resulting concentration is $C(x) = \dfrac{40 + 3x}{160 + 4x}$ after simplifying, and graph the function over the interval $x \in [0, 360]$.

  b. What is the concentration of the antifreeze in the tank after 10 gal of the new solution are added? After 120 gal have been added? How much liquid is now in the tank?   35%; 62.5%; 160 gal

  c. If the concentration level is now at 65%, how many gallons of the 75% solution have been added? How many gallons of liquid are in the tank now?   160 gal; 200 gal

  d. What is the maximum antifreeze concentration that can be attained in a tank of this size? What is the maximum concentration that can be attained in a tank of "unlimited" size?   70%; 75%

72. Concentration of sodium chloride: A sodium chloride solution has a concentration of 0.2 oz (weight) per gallon. The solution is pumped into an 800-gal tank currently holding 40 gal of pure water, at a rate of 10 gal/min.

  a. Find a function $A(t)$ modeling the amount of liquid in the tank after t min, and a function $S(t)$ for the amount of sodium chloride in the tank after t min. $A(t) = 40 + 10t$; $S(t) = 2t$

  b. The concentration $C(t)$ in ounces per gallon is measured by the ratio $\dfrac{S(t)}{A(t)}$, a rational function. Graph the function on the interval $t \in [0, 100]$. What is the concentration level (in ounces per gallon) after 6 min? After 28 min? How many gallons of liquid are in the tank at this time?

  c. If the concentration level is now 0.184 oz/gal, how long have the pumps been running? How many gallons of liquid are in the tank now?   46 min; 500 gal

  d. What is the maximum concentration that can be attained in a tank of this size? What is the maximum concentration that can be attained in a tank of "unlimited" size?   0.19 oz/gal; 0.20 oz/gal

Average cost of manufacturing an item: The cost "C" to manufacture an item depends on the relatively fixed costs "K" for remaining in business (utilities, maintenance, transportation, etc.) and the actual cost "c" of manufacturing the item (labor and materials). For x items the cost is $C(x) = K + cx$. The average cost "A" of manufacturing an item is then $A(x) = \dfrac{C(x)}{x}$.

73. Manufacturing water heaters: A company that manufactures water heaters finds their fixed costs are normally $50,000 per month, while the cost to manufacture each heater is $125. Due to factory size and the current equipment, the company can produce a maximum of 5000 water heaters per month during a good month.

a. Use the average cost function to find the average cost if 500 water heaters are manufactured each month. What is the average cost if 1000 heaters are made? $225; $175

b. What level of production will bring the average cost down to $150 per water heater?   2000 heaters

c. If the average cost is currently $137.50, how many water heaters are being produced that month?   4000 heaters

d. What's the significance of the horizontal asymptote for the average cost function (what does it mean in this context)? Will the company ever break the $130 average cost level? Why or why not?

74. Producing biodegradable disposable diapers: An enterprising company has finally developed a better disposable diaper that is biodegradable. The brand becomes wildly popular and production is soaring. The fixed cost of production is $20,000 per month, while the cost of manufacturing is $6.00 per case (48 diapers). Even while working three shifts around-the-clock, the maximum production level is 16,000 cases per month. The company figures it will be profitable if it can bring costs down to an average of $7 per case.

a. Use the average cost function to find the average cost if 2000 cases are produced each month. What is the average cost if 4000 cases are made? $16; $11

b. What level of production will bring the average cost down to $8 per case?   10,000

c. If the average cost is currently $10 per case, how many cases are being produced?   5000

d. What's the significance of the horizontal asymptote for the average cost function (what does it mean in this context)? Will the company ever reach its goal of $7/case at its maximum production? What level of production would help them meet their goal?

Additional answers can be found in the Instructor Answer Appendix.

Test averages and grade point averages: To calculate a test average we sum all test points P and divide by the number of tests N: $\dfrac{P}{N}$. To compute the score or scores needed on future tests to raise the average grade to a desired grade G, we add the number of additional tests n to the denominator, and the number of additional tests times the projected grade g on each test to the numerator: $G(n) = \dfrac{P + ng}{N + n}$. The result is a rational function with some "eye-opening" results.

75. Computing an average grade: After four tests, Bobby Lou's test average was an 84. [*Hint:* $P = 4(84) = 336$.]

a. Assume that she gets a 95 on all remaining tests ($g = 95$). Graph the resulting function on a calculator using the window $n \in [0, 20]$ and $G(n) \in [80 \text{ to } 100]$. Use the calculator to determine how many tests are required to lift her grade to a 90 under these conditions.   5

b. At some colleges, the range for an "A" grade is 93–100. How many tests would Bobby Lou have to score a 95 on, to raise her average to higher than 93? Were you surprised?   18

c. Describe the significance of the horizontal asymptote of the average grade function. Is a test average of 95 possible for her under these conditions?

d. Assume now that Bobby Lou scores 100 on all remaining tests ($g = 100$). Approximately how many more tests are required to lift her grade average to higher than 93?   6

76. Computing a GPA: At most colleges, $A \to 4$ grade points, $B \to 3$, $C \to 2$, and $D \to 1$. After taking 56 credit hours, Aurelio's GPA is 2.5. [*Hint:* In the formula given, $P = 2.5(56) = 140$.]

a. Assume Aurelio is determined to get A's (4 grade points or $g = 4$), for all remaining credit hours. Graph the resulting function on a calculator using the window $n \in [0, 60]$ and $G(n) \in [2, 4]$. Use the calculator to determine the number of credit hours required to lift his GPA to over 2.75 under these conditions.   12

b. At some colleges, scholarship money is available only to students with a 3.0 average or higher. How many (perfect 4.0) credit hours would Aurelio have to earn, to raise his GPA to 3.0 or higher? Were you surprised?   28 hr

c. Describe the significance of the horizontal asymptote of the GPA function. Is a GPA of 4.0 possible for him under these conditions?

▶ EXTENDING THE CONCEPT

77. In addition to determining *if* a function has a vertical asymptote, we are often interested in *how fast* the graph approaches the asymptote. As in previous investigations, this involves the function's rate of change over a small interval. Exercise 64 describes the rising cost of removing pollutants from the air. As noted there, the rate of increase in the cost changes as higher requirements are set. To quantify this change, we'll compute the rate of change $\dfrac{\Delta C}{\Delta x} = \dfrac{C(x_2) - C(x_1)}{x_2 - x_1}$ for $C(x) = \dfrac{250x}{100 - x}$.

  a. Find the rate of change of the function in the following intervals: 16.0, 28.7, 65.8, 277.8

$$x \in [60, 61] \quad x \in [70, 71]$$
$$x \in [80, 81] \quad x \in [90, 91]$$

  b. What do you notice? How much did the rate increase from the first interval to the second? From the second to the third? From the third to the fourth? 12.7, 37.1, 212.0

  c. Recompute parts (a) and (b) using the function $C(x) = \dfrac{350x}{100 - x}$. Comment on what you notice.
    a. 22.4, 40.2, 92.1, 388.9 **b.** 17.8, 51.9, 296.8; answers will vary.

78. Consider the function $f(x) = \dfrac{ax^2 + k}{bx^2 + h}$, where a, b, k, and h are constants and a, $b > 0$.

  a. What can you say about asymptotes and intercepts of this function if h, $k > 0$?

  b. Now assume $k < 0$ and $h > 0$. How does this affect the asymptotes? The intercepts?

  c. If $b = 1$ and $a > 1$, how does this affect the results from part (b)?

  d. How is the graph affected if $k > 0$ and $h < 0$?

  e. Find values of a, b, h, and k that create a function with a horizontal asymptote at $y = \frac{3}{2}$, x-intercepts at $(-2, 0)$ and $(2, 0)$, a y-intercept of $(0, -4)$, and no vertical asymptotes.

▶ MAINTAINING YOUR SKILLS

79. **(1.4)** Find the equation of a line that is perpendicular to $3x - 4y = 12$ and contains the point $(2, -3)$.

80. **(4.1)** Use synthetic division and the remainder theorem to find the value of $f(4)$, $f\left(\frac{3}{2}\right)$, and $f(2)$: $f(x) = 2x^3 - 7x^2 + 5x + 3$. 39, $\frac{3}{2}$, 1

81. **(3.2)** Solve the following equation using the quadratic formula, then write the equation in factored form: $12x^2 + 55x - 48 = 0$.

82. **(R.1/3.1)** Describe/Define each set of numbers: complex $\mathbb{C}$, rational $\mathbb{Q}$, and integers $\mathbb{Z}$.

Additional answers can be found in the Instructor Answer Appendix.

4.5 Additional Insights into Rational Functions

LEARNING OBJECTIVES

In Section 4.5 you will see how we can:

☐ **A.** Graph rational functions with removable discontinuities

☐ **B.** Graph rational functions with oblique or nonlinear asymptotes

☐ **C.** Solve applications involving rational functions

In Section 4.4, we studied rational functions whose graphs had horizontal and/or vertical asymptotes. In this section, we'll study functions with asymptotes that are *neither* horizontal nor vertical. In addition, we'll further explore the "break" we saw in graphs of certain piecewise-defined functions, that of a simple "hole" created when the numerator and denominator of a rational function share a common variable factor.

A. Rational Functions and Removable Discontinuities

In Example 6 of Section 2.5, we graphed the piecewise-defined function

$$h(x) = \begin{cases} \dfrac{x^2 - 4}{x - 2} & x \neq 2 \\ 1 & x = 2 \end{cases}$$. The second piece is simply the point $(2, 1)$. The first piece

is a rational function, but instead of a vertical asymptote at $x = 2$ (the zero of the denominator), its graph is actually the line $y = x + 2$ with a "hole" at $(2, 4)$ called a **removable discontinuity** (Figure 4.60). The hole occurs because the numerator and denominator of $y = \dfrac{x^2 - 4}{x - 2}$ share the common factor $(x - 2)$, and canceling these factors leaves $y = x + 2$, a continuous function. However, the *original function* is not defined at $x = 2$, so we must delete the single point at $x = 2$, $y = 4$ from the graph of the line (Figure 4.60).

Figure 4.60

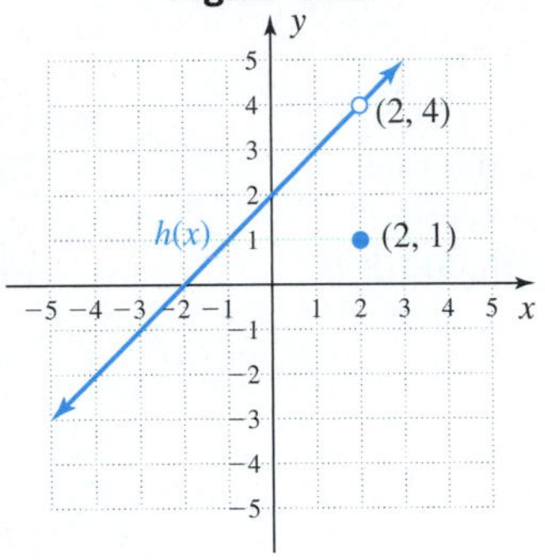

Figure 4.61

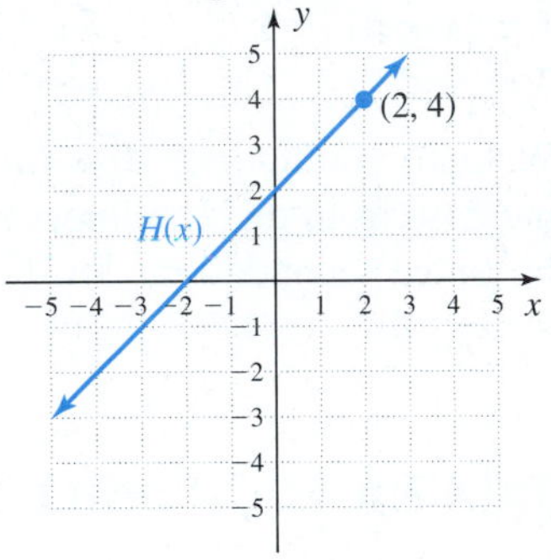

We can *remove* or fix this discontinuity by redefining the second piece as $h(x) = 4$, when $x = 2$. This would create a new and continuous function,

$$H(x) = \begin{cases} \dfrac{x^2 - 4}{x - 2} & x \neq 2 \\ 4 & x = 2 \end{cases}$$ (Figure 4.61).

It's possible for a rational graph to have more than one removable discontinuity, or to be nonlinear with a removable discontinuity. For cases where we elect to repair the break, we will adopt the convention of using the corresponding upper case letter to name the new function, as we did here.

EXAMPLE 1 ▶ **Graphing Rational Functions with Removable Discontinuities**

Graph the function $t(x) = \dfrac{x^3 + 8}{x + 2}$. If there is a removable discontinuity, repair the break using an appropriate piecewise-defined function.

Solution ▶ Note the domain of t does not include $x = -2$. We begin by factoring as before to identify zeroes and asymptotes, but find the numerator and denominator share a common factor, which we remove.

$$t(x) = \frac{x^3 + 8}{x + 2}$$
$$= \frac{(x + 2)(x^2 - 2x + 4)}{x + 2}$$
$$= x^2 - 2x + 4; \text{ where } x \neq -2$$

The graph of t will be the same as $y = x^2 - 2x + 4$ *for all values except $x = -2$.* Here we have a parabola, opening upward, with y-intercept $(0, 4)$. From the vertex formula, the x-coordinate of the vertex will be $\dfrac{-b}{2a} = \dfrac{-(-2)}{2(1)} = 1$, giving $y = 3$

after substitution. The vertex is $(1, 3)$. Evaluating $t(-1)$ we find $(-1, 7)$ is on the graph, giving the point $(3, 7)$ using the axis of symmetry. We draw a parabola through these points, noting the original function is not defined at -2, and there will be a "hole" in the graph at $(-2, y)$. The value of y is found by substituting -2 for x in the simplified form: $(-2)^2 - 2(-2) + 4 = 12$. This information produces the graph shown. We can repair the break using the function

$$T(x) = \begin{cases} \dfrac{x^3 + 8}{x + 2} & x \neq -2 \\ 12 & x = -2 \end{cases}$$

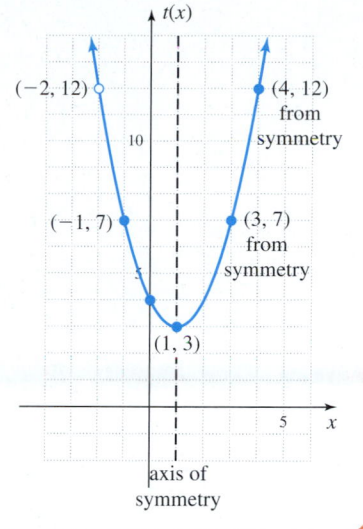

Now try Exercises 7 through 18 ▶

The ⬛TRACE feature of a graphing calculator is a wonderful tool for understanding the characteristics of a function. We'll illustrate using the function $h(x) = \dfrac{f(x)}{g(x)}$, where

$f(x) = x^3 - 2x^2 - 3x + 6$ and $g(x) = x - 2$ (similar to Example 1). Enter $\dfrac{f(x)}{g(x)}$ on

the ⬛Y= screen as Y_1 (Figure 4.62), then graph the function using ⬛ZOOM **4:ZDecimal.** Recall this will allow the calculator to trace through Δx intervals of 0.1. After pressing the ⬛TRACE key, the cursor appears on the graph at the y-intercept $(0, -3)$ and its location is displayed at the bottom of the screen. Note that there is a "hole" in the graph in the first quadrant (Figure 4.63). We can walk the cursor along the curve in either direction using the left arrow ⬛◄ and right arrow ⬛► keys to determine exactly where this hole occurs. Walking the cursor to the right, we note *that no output is displayed for x = 2.*

Figure 4.62

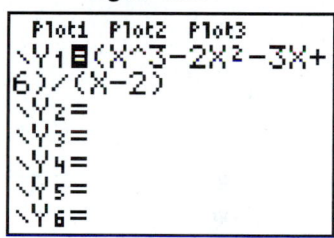

Figure 4.63

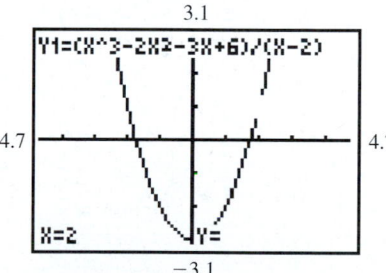

Next, simplify $\dfrac{f(x)}{g(x)}$ and enter the result as Y_2. Factoring the numerator by grouping and reducing the common factors gives $\dfrac{f(x)}{g(x)} = \dfrac{(x^2 - 3)(x - 2)}{(x - 2)}$, so $Y_2 = x^2 - 3$.

Graphing both functions reveals that they are identical, except that $Y_2 = x^2 - 3$ *covers the hole left by Y_1 using $x = 2, y = 1$.* In other words, Y_1 is equivalent to Y_2 except at $x = 2$. This can also be seen using the **TABLE** feature of a calculator, which displays an error message for Y_1 when $x = 2$ is input, but shows an output of 1 for Y_2 (Figure 4.64). The bottom line is— the domain of h is all real numbers except $x = 2$.

☑ **A.** You've just seen how we can graph rational functions with removable discontinuities

Figure 4.64

X	Y1	Y2
-2	1	1
-1	-2	-2
0	-3	-3
1	-2	-2
2	ERROR	1
3	6	6
4	13	13

X=-2

B. Rational Functions with Oblique or Nonlinear Asymptotes

In Section 4.4, we found that for $V(x) = \dfrac{p(x)}{d(x)}$, the location of nonvertical asymptotes was determined by comparing the degree of p with the degree of d. As review, for $p(x)$ with leading term ax^n and $d(x)$ with leading term bx^m,

- If $n < m$, the line $y = 0$ is a horizontal asymptote.

- If $n = m$, the line $y = \frac{a}{b}$ is a horizontal asymptote.

But what happens if the degree of the numerator is *greater than* the degree of the denominator? To investigate, consider the functions f, g, and h in Figures 4.65 to 4.68, whose only difference is the degree of the numerator.

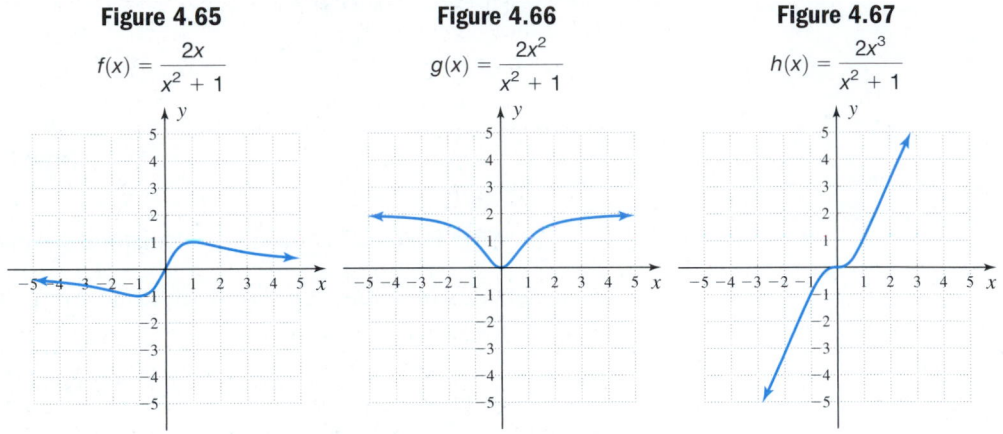

Figure 4.65

$$f(x) = \frac{2x}{x^2 + 1}$$

Figure 4.66

$$g(x) = \frac{2x^2}{x^2 + 1}$$

Figure 4.67

$$h(x) = \frac{2x^3}{x^2 + 1}$$

The graph of f has a horizontal asymptote at $y = 0$ since the denominator is of larger degree (as $|x| \to \infty$, $y \to 0$). As we might have anticipated, the horizontal asymptote for g is $y = 2$, the ratio of leading coefficients (as $|x| \to \infty$, $y \to 2$). The graph of h has no horizontal asymptote, yet appears to be asymptotic to some slanted line. The table in Figure 4.68 suggests that as $|x| \to \infty$, $y \to 2x = Y_2$. To see why, note the function $h(x) = \dfrac{2x^3}{x^2 + 1}$ can be considered an "improper fraction," similar to how we apply this designation to the fraction $\frac{3}{2}$. To write h in "proper" form, we use long division, writing the dividend as $2x^3 + 0x^2 + 0x + 0$, and the divisor as $x^2 + 0x + 1$.

Figure 4.68

$Y_2 = 2X$

X	Y1	Y2
0	0	0
25	49.92	50
50	99.96	100
75	149.97	150
100	199.98	200
125	249.98	250
150	299.99	300

X=0

The ratio $\dfrac{2x^3 \ \text{from dividend}}{x^2 \ \text{from divisor}}$ shows **$2x$** will be our first multiplier.

$$
\begin{array}{r}
2x \\
\text{divisor} \to x^2 + 0x + 1\overline{\smash{\big)}\,2x^3 + 0x^2 + 0x + 0} \\
\underline{-(2x^3 + 0x^2 + 2x)} \quad \text{multiply } 2x(x^2 + 0x + 1) \\
-2x \qquad\quad \text{subtract, next term is } 0
\end{array}
$$

The result shows $h(x) = 2x + \dfrac{-2x}{x^2 + 1}$. Note as $|x| \to \infty$, the term $\dfrac{-2x}{x^2 + 1}$ becomes very small and closer to zero, so $h(x) \approx 2x$ for large x. This is an example of an **oblique asymptote.** In general,

Oblique and Nonlinear Asymptotes

Given $V(x) = \dfrac{p(x)}{d(x)}$ is a rational function in simplest form, where the degree of p is greater than the degree of d, the graph will have an oblique or nonlinear asymptote as determined by $q(x)$, where $q(x)$ is the quotient polynomial after division.

If the denominator is a monomial, term-by-term division is the most efficient means of computing the quotient. If the denominator is not a monomial, either synthetic division or long division must be used. From our work in Section 4.4, $q(x)$ will give the location of the asymptote, and the zeroes of the remainder (if they exist) will indicate where the function crosses the asymptote.

We conclude that an oblique or slant asymptote occurs when the degree of the numerator is one more than the degree of the denominator, as that indicates $q(x)$ will be linear. A nonlinear asymptote will occur when the degree of the numerator is larger by two or more.

EXAMPLE 2 ▶ **Graphing a Rational Function with an Oblique Asymptote**

Graph the function $f(x) = \dfrac{x^2 - 1}{x}$.

Solution ▶ Using the *Guidelines*, we find $f(x) = \dfrac{(x + 1)(x - 1)}{x}$ and proceed:

1. *y-intercept:* The graph has no *y*-intercept since $f(0)$ is undefined.

2. *x-intercepts:* From $(x + 1)(x - 1) = 0$, the *x*-intercepts are $(-1, 0)$ and $(1, 0)$. Since both have multiplicity 1, the graph will cross the *x*-axis and the function will change sign at these points.

3. Vertical asymptote(s): $x = 0$ with multiplicity 1. The function will change sign at $x = 0$.

4. Horizontal/oblique asymptote: Since the degree of numerator $>$ the degree of denominator, we rewrite f using division. Using term-by-term division (the denominator is a monomial) produces $f(x) = \dfrac{x^2 - 1}{x} = \dfrac{x^2}{x} - \dfrac{1}{x} = x - \dfrac{1}{x}$. The quotient polynomial is $q(x) = x$ and the graph has the oblique asymptote $y = x$.

5. To determine if the function will cross the asymptote, we note the remainder is $r(x) = \dfrac{-1}{x}$, which has no real zeroes. The graph will not cross the oblique asymptote.

The information from steps 1 through 5 is displayed in Figure 4.69. While this is sufficient to sketch the graph, we select $x = -4$ and $x = 4$ and compute the additional points: $f(-4) = -\frac{15}{4}$ and $f(4) = \frac{15}{4}$. To meet all necessary conditions, we complete the graph as shown in Figure 4.70.

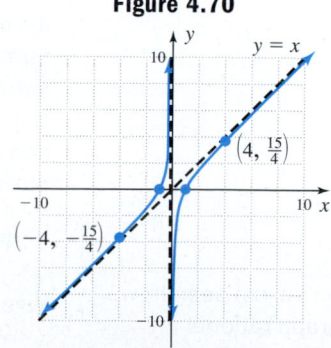

Figure 4.69

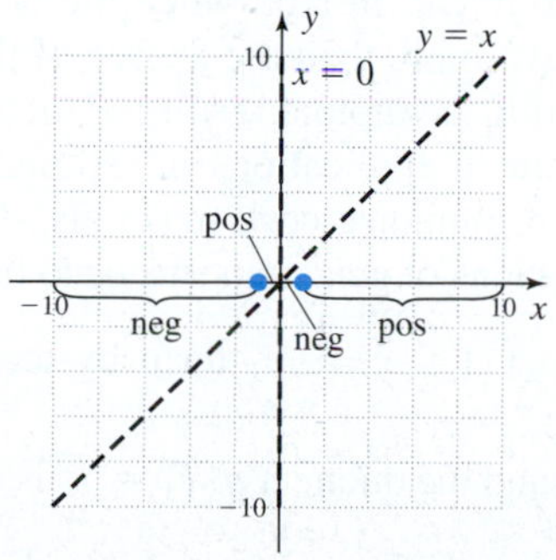

Figure 4.70

Now try Exercises 19 through 24 ▶

EXAMPLE 3 ▶ **Graphing a Rational Function with an Oblique Asymptote**

Graph the function: $h(x) = \dfrac{x^2}{x - 1}$

Solution ▶ The function is already in "factored form."

1. y-intercept: Since $h(0) = 0$, the y-intercept is $(0, 0)$.

2. x-intercept: $(0, 0)$; From, $x^2 = 0$, we have $x = 0$ with multiplicity two. The x-intercept is $(0, 0)$ and the function will not change sign here.

3. Vertical asymptote: Solving $x - 1 = 0$ gives $x = 1$ with multiplicity one. There is a vertical asymptote at $x = 1$ and the function will change sign here.

4. Horizontal/oblique asymptote: Since the degree of numerator $>$ the degree of denominator, we rewrite h using division. The denominator is linear so we use synthetic division:

use 1 as a "divisor" $\underline{1|}$ 1   0   0 coefficients of dividend

$\downarrow$ 1   1

1   1   1 quotient and remainder

Since $q(x) = x + 1$ the graph has an oblique asymptote at $y = x + 1$.

5. The remainder is $r(x) = 1$, a constant. The graph will not cross the slant asymptote.

The information gathered in steps 1 through 5 is shown Figure 4.71, and is actually sufficient to complete the graph. If you feel a little unsure about how to "puzzle" out the graph, find additional points in the first and third quadrants: $h(2) = 4$ and $h(-2) = -\frac{4}{3}$. Since the graph will "bounce" at $x = 0$ and output values must change sign at $x = 1$, all conditions are met with the graph shown in Figure 4.72.

Figure 4.71   **Figure 4.72**

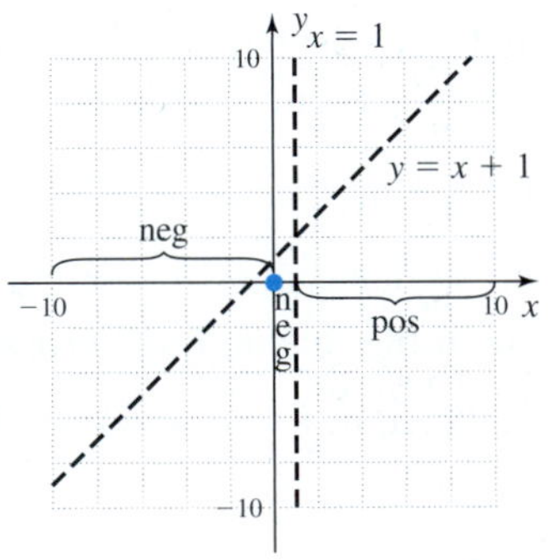

Now try Exercises 25 through 46 ▶

Finally, it would be a mistake to think that all asymptotes are linear. In fact, when the degree of the numerator is two more than the degree of the denominator, a parabolic asymptote results. Functions of this type often occur in applications of rational functions, and are used to minimize cost, materials, distances, or other considerations of great importance to business and industry. For $f(x) = \dfrac{x^4 + 1}{x^2}$, term-by-term division

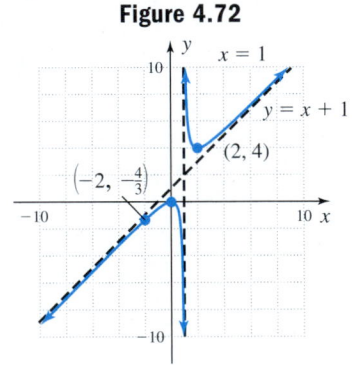

Figure 4.73

gives $x^2 + \dfrac{1}{x^2}$ and the quotient $q(x) = x^2$ is a nonlinear, parabolic asymptote (see Figure 4.73). For more on nonlinear asymptotes, **see Exercises 47 through 50.**

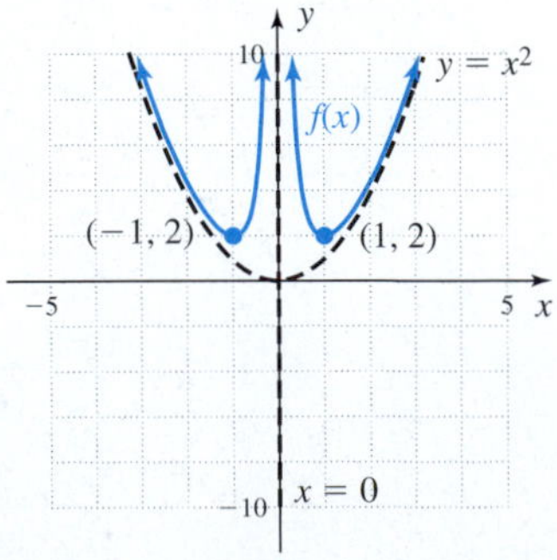

✓ **B.** You've just seen how we can graph rational functions with oblique or nonlinear asymptotes

C. Applications of Rational Functions

Rational functions have applications in a wide variety of fields, including environmental studies, manufacturing, and various branches of medicine. In most practical applications, only the values from Quadrant I have meaning since inputs and outputs must often be positive **(see Exercises 51 and 52).** Here we investigate an application involving manufacturing and average cost.

EXAMPLE 4 ▶ **Solving an Application of Rational Functions**

Suppose the cost (in thousands of dollars) of manufacturing x thousand of a given item is modeled by the function $C(x) = x^2 + 4x + 3$. The *average cost* of each item would then be expressed by

$$A(x) = \frac{x^2 + 4x + 3}{x} = \frac{\text{total cost}}{\text{number of items}}$$

a. Graph the function $A(x)$.

b. Find how many thousand items are manufactured when the average cost is $8.

c. Determine how many thousand items should be manufactured to minimize the average cost (use the graph to estimate this minimum average cost).

Solution ▶ **a.** The function is already in simplest form.

1. y-intercept: none [$A(0)$ is undefined]

2. x-intercept(s): After factoring we obtain $(x + 3)(x + 1) = 0$, and the zeroes of the numerator are $x = -1$ and $x = -3$, both with multiplicity one. The graph will cross the x-axis at each intercept.

3. Vertical asymptote: $x = 0$, multiplicity one; the function will change sign at $x = 0$.

4. Horizontal/oblique asymptote: The degree of numerator $>$ the degree of denominator, so we divide using term-by-term division:

$$\frac{x^2 + 4x + 3}{x} = \frac{x^2}{x} + \frac{4x}{x} + \frac{3}{x}$$

$$= x + 4 + \frac{3}{x}$$

The line $q(x) = x + 4$ is an oblique asymptote.

5. The remainder $\dfrac{3}{x}$ has no real zeroes, so the graph will not cross the slant asymptote.

The function changes sign at both x-intercepts and at the asymptote $x = 0$. The information from steps 1 through 5 is shown in Figure 4.74 and perhaps an additional point in Quadrant I would help to complete the graph: $A(1) = 8$. The point $(1, 8)$ is on the graph, showing A is positive in the interval containing 1 (since $y = 8$ is positive). Since output values will alternate in sign as stipulated above, all conditions are met with the graph shown in Figure 4.75.

Figure 4.74   **Figure 4.75**

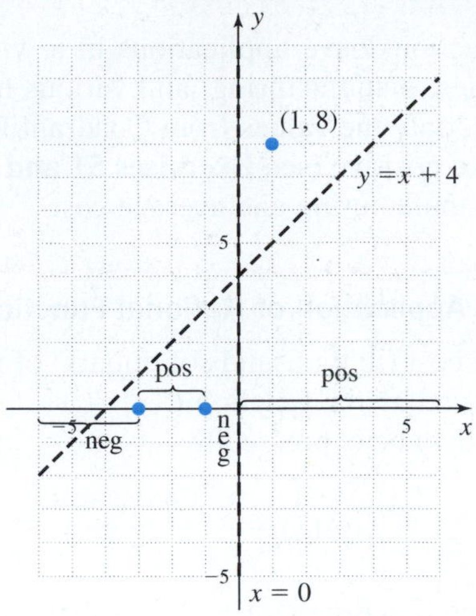

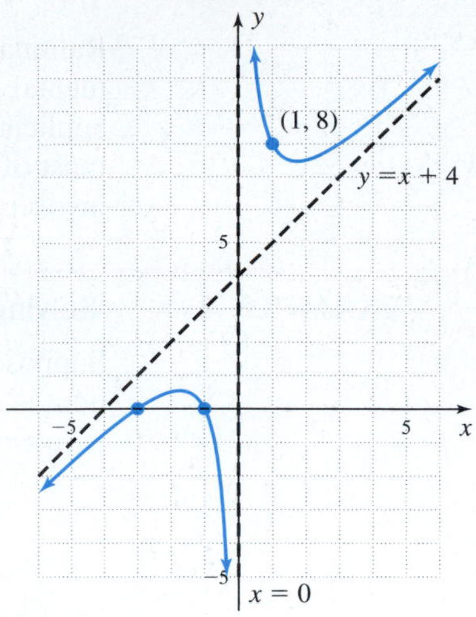

b. To find the number of items manufactured when average cost is $8, we replace $A(x)$ with 8 and solve: $\dfrac{x^2 + 4x + 3}{x} = 8$:

$$x^2 + 4x + 3 = 8x$$
$$x^2 - 4x + 3 = 0$$
$$(x - 1)(x - 3) = 0$$
$$x = 1 \quad \text{or} \quad x = 3$$

The average cost is $8 when 1000 items or 3000 items are manufactured.

c. From the graph, it appears that the minimum average cost is close to $7.50, when approximately 1500 to 1800 items are manufactured. Using a graphing calculator, we find that the minimum average cost is approximately $7.46, when about 1732 items are manufactured (Figure 4.76).

Figure 4.76

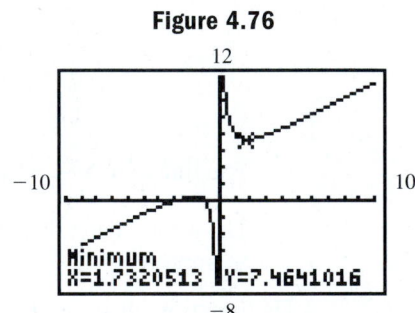

Now try Exercises 55 and 56 ▶

In some applications, the functions we use are initially defined in *two variables* rather than just one, as in $H(x, y) = (x - 50)(y - 80)$. However, in the solution process a substitution is used to rewrite the relationship as a function in one variable and we can proceed as before.

EXAMPLE 5 ▶   Using a Rational Function to Solve a Layout Application

The building codes in a new subdivision require that a rectangular home be built at least 20 ft from the street, 40 ft from the neighboring lots, and 30 ft from the rear fence line.

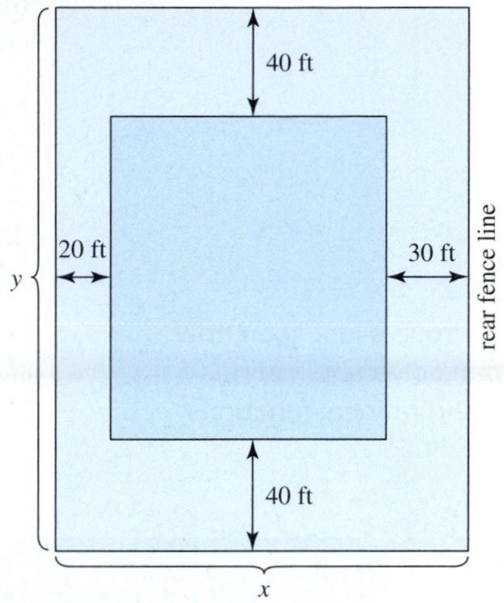

a. Find a function $A(x, y)$ for the area of the lot, and a function $H(x, y)$ for the area of the home (the inner rectangle).

b. If a new home is to have a floor area of 2000 ft², $H(x, y) = 2000$. Substitute 2000 for $H(x, y)$ and solve for y, then substitute the result in $A(x, y)$ to write the area A as a function of x alone (simplify the result).

c. Graph $A(x)$ on a calculator, using the window $X \in [-50, 150]$; $Y \in [-30{,}000, 30{,}000]$. Then graph $y = 80x + 2000$ on the same screen. How are these two graphs related?

d. Use the graph of $A(x)$ in Quadrant I to determine the minimum dimensions of a lot that satisfies the subdivision's requirements (to the nearest tenth of a foot). Also state the dimensions of the house.

Solution ▶   a. The area of the lot is simply width times length, so $A(x, y) = xy$. For the house, these dimensions are decreased by 50 ft and 80 ft respectively, so $H(x, y) = (x - 50)(y - 80)$.

b. Given $H(x, y) = 2000$ produces the equation $2000 = (x - 50)(y - 80)$, and solving for y gives

$$2000 = (x - 50)(y - 80) \quad \text{\textcolor{red}{given equation}}$$

$$\frac{2000}{x - 50} = y - 80 \quad \text{\textcolor{red}{divide by } } x - 50$$

$$\frac{2000}{x - 50} + 80 = y \quad \text{\textcolor{red}{add 80}}$$

$$\frac{2000}{x - 50} + \frac{80(x - 50)}{x - 50} = y \quad \text{\textcolor{red}{find LCD}}$$

$$\frac{80x - 2000}{x - 50} = y \quad \text{\textcolor{red}{combine terms}}$$

Substituting this expression for y in $A(x, y) = xy$ produces

$$A(x) = x\left(\frac{80x - 2000}{x - 50}\right) \quad \text{\textcolor{red}{substitute } \frac{80x - 2000}{x - 50} \text{ for } y}$$

$$= \frac{80x^2 - 2000x}{x - 50} \quad \text{\textcolor{red}{multiply}}$$

c. The graph of $Y_1 = A(x)$ appears in Figure 4.77 using the prescribed window. $Y_2 = 80x + 2000$ appears to be an oblique asymptote for A, which can be verified using synthetic division.

Figure 4.77

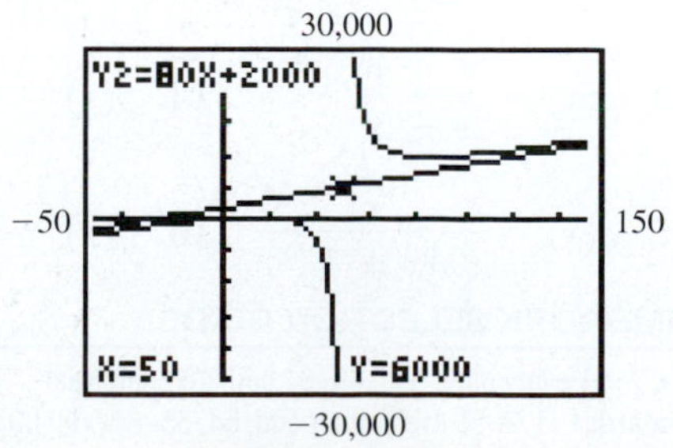

d. Using the ▣2nd▣ ▣TRACE▣ (**CALC**)
3:minimum feature of a calculator,
the minimum width is $x \approx 85.4$ ft
(see Figure 4.78). Substituting 85.4 for
x in $y = \dfrac{80x - 2000}{x - 50}$, gives the
length $y \approx 136.5$ ft. The dimensions
of the house must be
$85.4 - 50 = 35.4$ ft, by
$136.5 - 80 = 56.5$ ft.

Figure 4.78

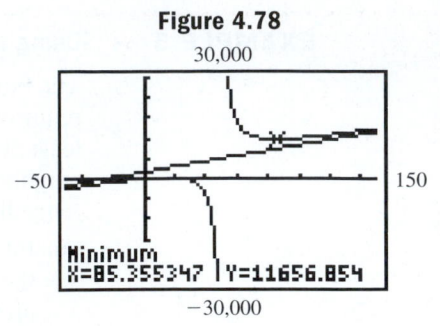

☑ **C.** You've just seen how
we can solve applications
involving rational functions

As expected, the area of the house will be $(35.4)(56.5) \approx 2000$ ft^2.

Now try Exercises 57 through 60 ▶

4.5 EXERCISES

▶ **CONCEPTS AND VOCABULARY**

Fill in each blank with the appropriate word or phrase. Carefully reread the section if needed.

1. The graph of $V(x) = \dfrac{3x^3}{x^2 + 4}$ will have a ___slant___
   asymptote, since the degree of the numerator is one
   greater than the degree of the denominator.

2. If the degree of the numerator is greater than the
   degree of the denominator, the graph will have an
   ___oblique___ or ___nonlinear___ asymptote.

3. If the degree of the numerator is ___two___ more
   than the degree of the denominator, the graph will
   have a parabolic asymptote.

4. If the denominator is a ___monomial___, use term by
   term division to find the quotient. Otherwise,
   ___synthetic___ or long division must be used.

5. Discuss/Explain how you would create a function
   with a parabolic asymptote and two vertical
   asymptotes.   Answers will vary.

6. Complete Exercise 7 in expository form. That is,
   work this exercise out completely, discussing each
   step of the process as you go.   Answers will vary.

▶ **DEVELOPING YOUR SKILLS**

**Graph each function. If there is a removable
discontinuity, repair the break using an appropriate
piecewise-defined function.**

7. $f(x) = \dfrac{x^2 - 4}{x + 2}$

8. $f(x) = \dfrac{x^2 - 9}{x + 3}$

9. $g(x) = \dfrac{x^2 - 2x - 3}{x + 1}$

10. $g(x) = \dfrac{x^2 - 3x - 10}{x - 5}$

11. $h(x) = \dfrac{3x - 2x^2}{2x - 3}$

12. $h(x) = \dfrac{4x - 5x^2}{5x - 4}$

13. $p(x) = \dfrac{x^3 - 8}{x - 2}$

14. $p(x) = \dfrac{8x^3 - 1}{2x - 1}$

15. $q(x) = \dfrac{x^3 - 7x - 6}{x + 1}$

16. $q(x) = \dfrac{x^3 - 3x + 2}{x + 2}$

17. $r(x) = \dfrac{x^3 + 3x^2 - x - 3}{x^2 + 2x - 3}$

18. $r(x) = \dfrac{x^3 - 2x^2 - 4x + 8}{x^2 - 4}$

**Graph each function using the *Guidelines for Graphing
Rational Functions,* which is simply modified to include
nonlinear asymptotes. Clearly label all intercepts and
asymptotes and any additional points used to sketch the
graph. Round to tenths as needed.**

19. $Y_1 = \dfrac{x^2 - 4}{x}$

20. $Y_2 = \dfrac{x^2 - x - 6}{x}$

21. $v(x) = \dfrac{3 - x^2}{x}$

22. $V(x) = \dfrac{7 - x^2}{x}$

HOMEWORK SELECTION GUIDE

Additional answers can be found in the Instructor Answer Appendix.

Core: 7–51 every other odd, 55–61 odd (16 Exercises)
Standard: 1–4, 7–51 every other odd, 54, 55–61 odd, 62, 67, 68 (24 Exercises)

Extended: 1–4, 7–51 every other odd, 54, 55, 57, 59, 60–63, 65, 67, 68
(27 Exercises)
In Depth: 1–6, 7–51 every other odd, 53–55, 57, 59, 60–63, 65–69 (32 Exercises)

23. $w(x) = \dfrac{x^2 + 1}{x}$ **24.** $W(x) = \dfrac{x^2 + 4}{2x}$

25. $h(x) = \dfrac{x^3 - 2x^2 + 3}{x^2}$ **26.** $H(x) = \dfrac{x^3 + x^2 - 2}{x^2}$

27. $Y_1 = \dfrac{x^3 + 3x^2 - 4}{x^2}$ **28.** $Y_2 = \dfrac{x^3 - 3x^2 + 4}{x^2}$

29. $f(x) = \dfrac{x^3 - 3x + 2}{x^2}$ **30.** $F(x) = \dfrac{x^3 - 12x - 16}{x^2}$

31. $Y_3 = \dfrac{x^3 - 5x^2 + 4}{x^2}$ **32.** $Y_4 = \dfrac{x^3 + 5x^2 - 6}{x^2}$

33. $r(x) = \dfrac{x^3 - x^2 - 4x + 4}{x^2}$

34. $R(x) = \dfrac{x^3 - 2x^2 - 9x + 18}{x^2}$

35. $g(x) = \dfrac{x^2 + 4x + 4}{x + 3}$ **36.** $G(x) = \dfrac{x^2 - 2x + 1}{x - 2}$

37. $f(x) = \dfrac{x^2 + 1}{x + 1}$ **38.** $F(x) = \dfrac{x^2 + x + 1}{x - 1}$

39. $Y_3 = \dfrac{x^2 - 4}{x + 1}$ **40.** $Y_4 = \dfrac{x^2 - x - 6}{x - 1}$

41. $v(x) = \dfrac{x^3 - 4x}{x^2 - 1}$ **42.** $V(x) = \dfrac{9x - x^3}{x^2 - 4}$

43. $w(x) = \dfrac{16x - x^3}{x^2 + 4}$ **44.** $W(x) = \dfrac{x^3 - 7x + 6}{2 + x^2}$

45. $Y_1 = \dfrac{x^3 - 3x + 2}{x^2 - 9}$ **46.** $Y_2 = \dfrac{x^3 - x^2 - 12x}{x^2 - 7}$

47. $p(x) = \dfrac{x^4 + 4}{x^2 + 1}$ **48.** $P(x) = \dfrac{x^4 - 5x^2 + 4}{x^2 + 2}$

49. $q(x) = \dfrac{10 + 9x^2 - x^4}{x^2 + 5}$ **50.** $Q(x) = \dfrac{x^4 - 2x^2 + 3}{x^2}$

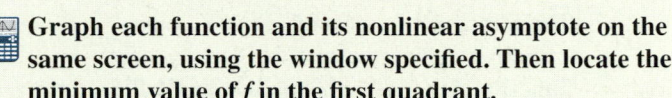 Graph each function and its nonlinear asymptote on the same screen, using the window specified. Then locate the minimum value of f in the first quadrant.

51. $f(x) = \dfrac{x^3 + 500}{x}$;

$\quad X \in [-24, 24], Y \in [-500, 500]$ 119.1

52. $f(x) = \dfrac{2\pi x^3 + 750}{x}$;

$\quad X \in [-12, 12], Y \in [-750, 750]$ 287.9

▶ **WORKING WITH FORMULAS**

 53. Area of a first quadrant triangle:

$$A(a) = \dfrac{1}{2}\left(\dfrac{ka^2}{a - h}\right)$$

The area of a right triangle in the first quadrant, formed by a line with negative slope through the point (h, k) and legs that lie along the positive axes is given by the formula shown, where a represents the x-intercept of the resulting line $(h < a)$. The area of the triangle varies with the slope of the line. Assume the line contains the point $(5, 6)$.

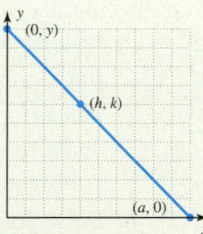

a. Find the equation of the vertical and slant asymptotes. $a = 5, y = 3a + 15$

b. Find the area of the triangle if it has an x-intercept of $(11, 0)$.   60.5

 c. Use a graphing calculator to graph the function on an appropriate window. Does the shape of the graph look familiar? Use the calculator to find the value of a that minimizes $A(a)$. That is, find the x-intercept that results in a triangle with the smallest possible area.   10

 54. Surface area of a cylinder with fixed volume:

$$S = \dfrac{2\pi r^3 + 2V}{r}$$

It's possible to construct many different cylinders that will hold a specified volume, by changing the radius and height. This is critically important to producers who want to minimize the cost of packing canned goods and marketers who want to present an attractive product. The surface area of the cylinder can be found using the formula shown, where the radius is r and $V = \pi r^2 h$ is known. Assume the fixed volume is 750 cm^3.

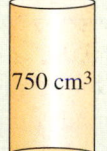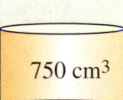
750 cm³
750 cm³

a. Find the equation of the vertical asymptote. How would you describe the nonlinear asymptote? $r = 0$, parabolic

b. If the radius of the cylinder is 2 cm, what is its surface area? ≈ 775.1 cm^2

c. Use a graphing calculator to graph the function on an appropriate window, and use it to find the value of r that minimizes $S(r)$. That is, find the radius that results in a cylinder with the smallest possible area, while still holding a volume of 750 cm^3. $r \approx 4.9$ cm

Additional answers can be found in the Instructor Answer Appendix.

▶ APPLICATIONS

Costs of manufacturing: As in Example 4, the cost $C(x)$ of manufacturing is sometimes nonlinear and can increase dramatically with each item. For the average cost function $A(x) = \dfrac{C(x)}{x}$, consider the following.

55. Assume the monthly cost of manufacturing custom-crafted storage sheds is modeled by the function $C(x) = 4x^2 + 53x + 250$.

    a. Write the average cost function and state the equation of the vertical and oblique asymptotes. $A(x) = \dfrac{4x^2 + 53x + 250}{x}; x = 0, y = 4x + 53$

    b. Enter the cost function $C(x)$ as Y_1 on a graphing calculator, and the average cost function $A(x)$ as Y_2. Using the **TABLE** feature, find the cost and average cost of making 1, 2, and 3 sheds. cost: $307, $372, $445; Avg. cost: $307, $186, $148.33

    c. Scroll down the table to where it appears that average cost is a minimum. According to the table, how many sheds should be made each month to minimize costs? What is the minimum cost? 8, $116.25

    d. Graph the average cost function and its asymptotes, using a window that shows the entire function. Use the graph to confirm the result from part (c).

56. Assume the monthly cost of manufacturing playground equipment that combines a play house, slides, and swings is modeled by the function $C(x) = 5x^2 + 94x + 576$. The company has projected that they will be profitable if they can bring their average cost down to $200 per set of playground equipment.

    a. Write the average cost function and state the equation of the vertical and oblique asymptotes.

    b. Enter the cost function $C(x)$ as Y_1 on a graphing calculator, and the average cost function $A(x)$ as Y_2. Using the TABLE feature, find the cost and average cost of making 1, 2, and 3 playground equipment combinations. Why would the average cost fall so dramatically early on?

    c. Scroll down the table to where it appears that average cost is a minimum. According to the table, how many sets of equipment should be made each month to minimize costs? What is the minimum cost? Will the company be profitable under these conditions?

    d. Graph the average cost function and its asymptotes, using a window that shows the entire function. Use the graph to confirm the result from part (c).

Additional answers can be found in the Instructor Answer Appendix.

Minimum cost of packaging: Similar to Exercise 54, manufacturers can minimize their costs by shipping merchandise in packages that use a minimum amount of material. After all, rectangular boxes come in different sizes and there are many combinations of length, width, and height that will hold a specified volume.

57. A clothing manufacturer wishes to ship lots of 12 ft^3 of clothing in boxes with square ends and rectangular sides.

    a. Find a function $S(x, y)$ for the surface area of the box, and a function $V(x, y)$ for the volume of the box. $S(x, y) = 2x^2 + 4xy; V(x, y) = x^2y$

    b. Solve for y in $V(x, y) = 12$ (volume is 12 ft^3) and use the result to write the surface area as a function $S(x)$ in terms of x alone (simplify the result). $y = \dfrac{12}{x^2}; S(x) = \dfrac{2x^3 + 48}{x}$

    c. On a graphing calculator, graph the function $S(x)$ using the window $x \in [-8, 8]$; $y \in [-100, 100]$. Then graph $y = 2x^2$ on the same screen. How are these two graphs related? $S(x)$ is asymptotic to $y = 2x^2$.

    d. Use the graph of $S(x)$ in Quadrant I to determine the dimensions that will minimize the surface area of the box, yet still hold 12 ft^3 of clothing. Clearly state the values of x and y, *in terms of feet and inches*, rounded to the nearest $\frac{1}{2}$ in. $x \approx 2$ ft 3.5 in.; $y \approx 2$ ft 3.5 in.

58. A maker of packaging materials needs to ship 36 ft^3 of foam "peanuts" to his customers across the country, using boxes with the dimensions shown.

    a. Find a function $S(x, y)$ for the surface area of the box, and a function $V(x, y)$ for the volume of the box.

    b. Solve for y in $V(x, y) = 36$ (volume is 36 ft^3), and use the result to write the surface area as a function $S(x)$ in terms of x alone (simplify the result).

    c. On a graphing calculator, graph the function $S(x)$ using the window $x \in [-10, 10]$; $y \in [-200, 200]$. Then graph $y = 2x^2 + 4x$ on the same screen. How are these two graphs related?

    d. Use the graph of $S(x)$ in Quadrant I to determine the dimensions that will minimize the surface area of the box, yet still hold the foam peanuts. Clearly state the values of x and y, *in terms of feet and inches*, rounded to the nearest $\frac{1}{2}$ in.

📠 **Printing and publishing:** In the design of magazine pages, posters, and other published materials, an effort is made to maximize the usable area of the page while maintaining an attractive border, or minimizing the page size that will hold a certain amount of print or art work.

59. An editor has a story that requires 60 in² of print. Company standards require a 1-in. border at the top and bottom of a page, and 1.25-in. borders along both sides.

    a. Find a function $A(x, y)$ for the area of the page, and a function $R(x, y)$ for the area of the inner rectangle (the printed portion).

    b. Solve for y in $R(x, y) = 60$, and use the result to write the area from part (a) as a function $A(x)$ in terms of x alone (simplify the result).

    c. On a graphing calculator, graph the function $A(x)$ using the window $x \in [-30, 30]$; $y \in [-100, 200]$. Then graph $y = 2x + 60$ on the same screen. How are these two graphs related?

    d. Use the graph of $A(x)$ in Quadrant I to determine the page of minimum size that satisfies these border requirements and holds the necessary print. Clearly state the values of x and y, rounded to the nearest hundredth of an inch.

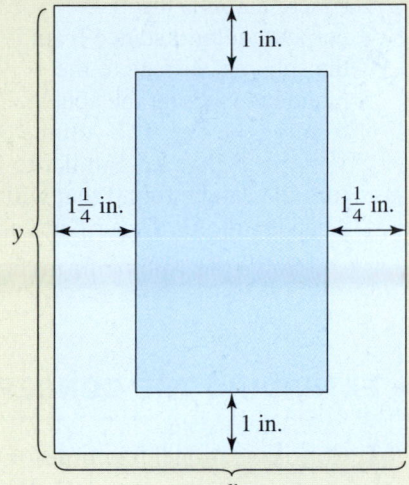

60. *The Poster Shoppe* creates posters, handbills, billboards, and other advertising for business customers. An order comes in for a poster with 500 in² of usable area, with margins of 2 in. across the top, 3 in. across the bottom, and 2.5 in. on each side.

    a. Find a function $A(x, y)$ for the area of the page, and a function $R(x, y)$ for the area of the inner rectangle (the usable area).

    b. Solve for y in $R(x, y) = 500$, and use the result to write the area from part (a) as a function $A(x)$ in terms of x alone (simplify the result).

    c. On a graphing calculator, graph $A(x)$ using the window $x \in [-100, 100]$; $y \in [-800, 1600]$. Then graph $y = 5x + 500$ on the same screen. How are these two graphs related?

    d. Use the graph of $A(x)$ in Quadrant I to determine the poster of minimum size that satisfies these border requirements and has the necessary usable area. Clearly state the values of x and y, rounded to the nearest hundredth of an inch.

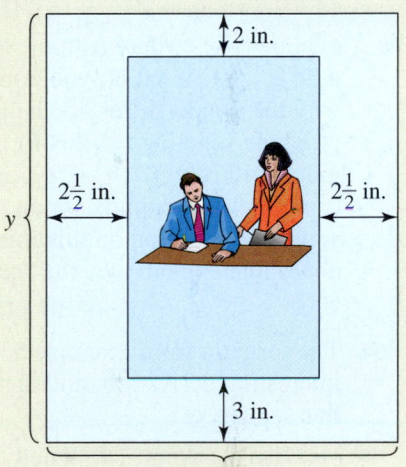

61. The formula from Exercise 54 has an interesting derivation. The volume of a cylinder is $V = \pi r^2 h$, while the surface area is given by $S = 2\pi r^2 + 2\pi r h$ (the circular top and bottom + the area of the side).

    a. Solve the volume formula for the variable h.

    b. Substitute the resulting expression for h into the surface area formula and simplify.

    c. Combine the resulting two terms using the least common denominator, and the result is the formula from Exercise 54.

    d. Assume the volume of a can must be 1200 cm³. Use a calculator to graph the function S using an appropriate window, then use it to find the radius r and height h that will result in a cylinder with the smallest possible area, while still holding a volume of 1200 cm³. What is the minimum surface area? Also see Exercise 62.

62. The surface area of a spherical cap is given by $S = 2\pi rh$, where r is the radius of the sphere and h is the perpendicular distance from the sphere's surface to the plane intersecting the sphere, forming the cap. The volume of the cap is $V = \frac{1}{3}\pi h^2(3r - h)$. Similar to Exercise 61, a formula can be found that will minimize the area of a cap that holds a specified volume.

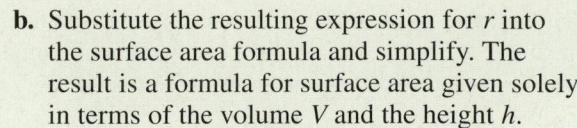

a. Solve the volume formula for the variable r.

b. Substitute the resulting expression for r into the surface area formula and simplify. The result is a formula for surface area given solely in terms of the volume V and the height h.

c. Assume the volume of the spherical cap is 500 cm³. Use a graphing calculator to graph the resulting function on an appropriate window, and use the graph to find the height h that will result in a spherical cap with the smallest possible area, while still holding a volume of 500 cm³.

d. Use this value of h and $V = 500$ cm³ to find the radius of the sphere.

▶ EXTENDING THE CONCEPT

63. Consider rational functions of the form $f(x) = \dfrac{x^2 - a}{x - b}$. Use a graphing calculator to explore cases where $a = b^2 + 1$, $a = b^2$, and $a = b^2 - 1$. What do you notice? Explain/Discuss why the graphs differ. It's helpful to note that when graphing functions of this form, the "center" of the graph will be at $(b, b^2 - a)$, and the window size can be set accordingly for an optimal view. Do some investigation on this function and determine/explain *why* the "center" of the graph is at $(b, b^2 - a)$.   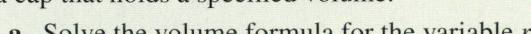 Answers will vary.

64. The formula from Exercise 53 also has an interesting derivation, and the process involves this sequence:

a. Use the points $(a, 0)$ and (h, k) to find the slope of the line, and the point-slope formula to find the equation of the line in terms of y.

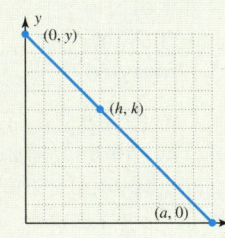

b. Use this equation to find the x- and y-intercepts of the line in terms of a, k, and h.

c. Complete the derivation using these intercepts and the triangle formula $A = \frac{1}{2}BH$.

d. If the lines goes through $(4, 4)$ the area formula becomes $A = \dfrac{1}{2}\left(\dfrac{4a^2}{a - 4}\right)$. Find the minimum value of this rational function. What can you say about the triangle with minimum area through (h, k), where $h = k$? Verify using the points $(5, 5)$, and $(6, 6)$.

65. Referring to Exercises 54 and 61, suppose that instead of a closed cylinder, with both a top and bottom, we needed to manufacture *open cylinders*, like tennis ball cans that use a lid made from a different material. Derive the formula that will minimize the surface area of an open cylinder, and use it to find the cylinder with minimum surface area that will hold 90 in³ of material.
$S = \dfrac{\pi r^3 + 2V}{r}$; $r \approx 3.1$ in., $h \approx 3$ in.

▶ MAINTAINING YOUR SKILLS

66. **(3.1)** Compute the quotient $\dfrac{5i}{1 + 2i}$, then check your answer using multiplication. $2 + i$

67. **(1.4)** Write the equation of the line in slope intercept form and state the slope and y-intercept: $-3x + 4y = -16$.

68. **(3.2)** Given $f(x) = ax^2 + bx + c$, use the discriminant to state conditions where the function will have: (a) two, real/rational roots, (b) two, real/irrational roots, (c) one real and rational root, (d) one real/irrational root, (e) one complex root, and (f) two complex roots.

Additional answers can be found in the Instructor Answer Appendix.

69. **(R.6/3.2)** For triangle ABC as shown, (a) find the perimeter; (b) find the length of $\overline{CD}$, given $(CB)^2 = \overline{AB} \cdot \overline{DB}$; (c) find the area; and (d) find the areas of the two smaller triangles.

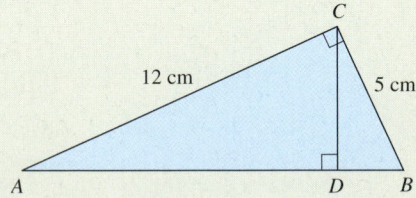

LEARNING OBJECTIVES

In Section 4.6 you will see how we can:

- ☐ **A.** Solve polynomial inequalities
- ☐ **B.** Solve rational inequalities
- ☐ **C.** Solve applications of inequalities

A. Polynomial Inequalities

Our work with quadratic inequalities (Section 3.4) transfers seamlessly to inequalities involving higher degree polynomials and the same two methods can be employed. The first involves drawing a quick sketch of the function, and using the concepts of multiplicity and end-behavior. The second involves the use of multiple interval tests, to check on the sign of the function in each interval.

Solving Inequalities Graphically

After writing the polynomial in standard form, find the zeroes, plot them on the x-axis, and determine the solution set using end-behavior and the behavior at each zero (cross—sign change; or bounce—no change in sign). In this process, any irreducible quadratic factors can be ignored, as they have no effect on the solution set. In summary,

> **Solving Polynomial Inequalities**
>
> Given $f(x)$ is a polynomial in standard form,
> 1. Write f in completely factored form.
> 2. Plot real zeroes on the x-axis, noting their multiplicity.
>   - If the multiplicity is odd the function will **change** sign.
>   - If the multiplicity is even, there will be **no change** in sign.
> 3. Use the end-behavior to determine the sign of f in the outermost intervals, then label the other intervals as $f(x) < 0$ or $f(x) > 0$ by analyzing the multiplicity of neighboring zeroes.
> 4. State the solution in interval notation.

EXAMPLE 1 ▶ **Solving a Polynomial Inequality**

Solve the inequality $x^3 - 18 < -4x^2 + 3x$.

Solution ▶ In standard form we have $x^3 + 4x^2 - 3x - 18 < 0$, which is equivalent to $f(x) < 0$ where $f(x) = x^3 + 4x^2 - 3x - 18$. The polynomial cannot be factored by grouping and testing 1 and -1 shows neither is a zero. Using $x = 2$ and synthetic division gives

use 2 as a "divisor"

$\begin{array}{r|rrrr} 2 & 1 & 4 & -3 & -18 \\ & \downarrow & 2 & 12 & 18 \\ \hline & 1 & 6 & 9 & \underline{|0} \end{array}$

with a quotient of $x^2 + 6x + 9$ and a remainder of zero.

1. The factored form is $f(x) = (x - 2)(x^2 + 6x + 9) = (x - 2)(x + 3)^2$.
2. The graph will bounce off the x-axis at $x = -3$ (f will not change sign), and cross the x-axis at $x = 2$ (f will change sign). This is illustrated in Figure 4.79, which uses open dots due to the strict inequality.

Figure 4.79

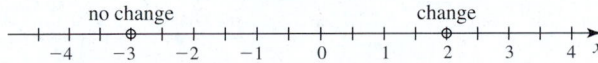

no change   change

3. The polynomial has odd degree with a positive lead coefficient, so end-behavior is down/up, which we note in the outermost intervals. Working from the left, f will not change sign at $x = -3$, showing $f(x) < 0$ in the left and middle intervals. This is supported by the y-intercept $(0, -18)$. See Figure 4.80.

Figure 4.80

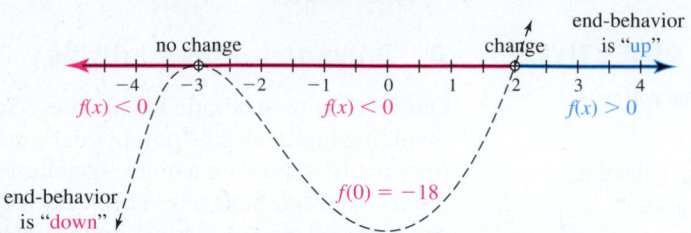

4. From the diagram, we see that $f(x) < 0$ for $x \in (-\infty, -3) \cup (-3, 2)$, which must also be the solution interval for $x^3 - 18 < -4x^2 + 3x$. The complete graph appearing in Figure 4.81 definitely shows the graph is below the x-axis $[f(x) < 0]$ from $-\infty$ to 2, except at $x = -3$ where the graph touches the x-axis without crossing.

Figure 4.81

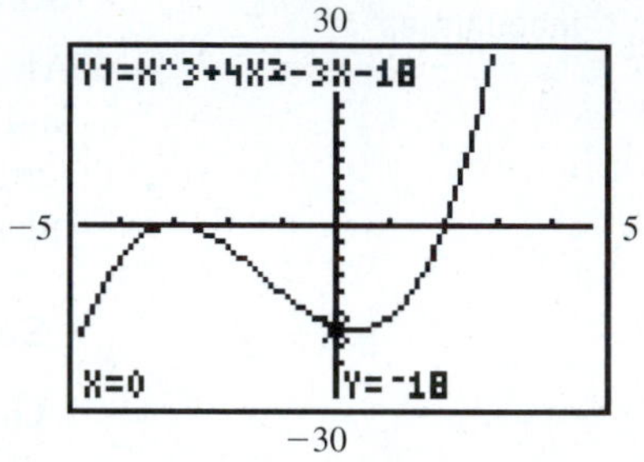

Now try Exercises 7 through 18 ▶

EXAMPLE 2 ▶ **Solving a Polynomial Inequality**

Solve the inequality $x^4 + 4x \le 9x^2 - 12$.

Solution ▶ Writing the polynomial in standard form gives $x^4 - 9x^2 + 4x + 12 \le 0$ $[f(x) \le 0]$. Testing 1 and -1 shows $x = 1$ is not a zero, but $x = -1$ is. Using synthetic division with $x = -1$ gives

use -1 as a "divisor"

$$
\begin{array}{r|rrrrr}
-1 & 1 & 0 & -9 & 4 & 12 \\
   &   & -1 & 1 & 8 & -12 \\
\hline
   & 1 & -1 & -8 & 12 & \underline{|0}
\end{array}
$$

with a quotient of $q_1(x) = x^3 - x^2 - 8x + 12$ and a remainder of zero. As $q_1(x)$ is not easily factored, we continue with synthetic division using $x = 2$.

use 2 as a "divisor"

$$
\begin{array}{r|rrrr}
2 & 1 & -1 & -8 & 12 \\
  &   & 2 & 2 & -12 \\
\hline
  & 1 & 1 & -6 & \underline{|0}
\end{array}
$$

The result is $q_2(x) = x^2 + x - 6$ with a remainder of zero.

1. The factored form is
   $f(x) = (x + 1)(x - 2)(x^2 + x - 6) = (x + 1)(x - 2)^2(x + 3)$.
2. The graph will "cross" at $x = -1$ and -3, and f will change sign. The graph will "bounce" at $x = 2$ and f will not change sign. This is illustrated in Figure 4.82 which uses closed dots since $f(x)$ *can* be equal to zero.

Figure 4.82

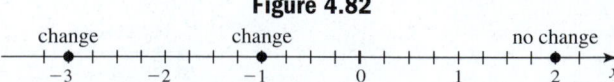

3. With even degree and positive lead coefficient, the end-behavior is up/up. Working from the leftmost interval, $f(x) > 0$, the function must change sign at $x = -3$ (going below the x-axis), and again at $x = -1$ (going above the x-axis). This is supported by the y-intercept $(0, 12)$. The graph then "bounces" at $x = 2$, remaining above the x-axis (no sign change). This produces the sketch shown in Figure 4.83.

Figure 4.83

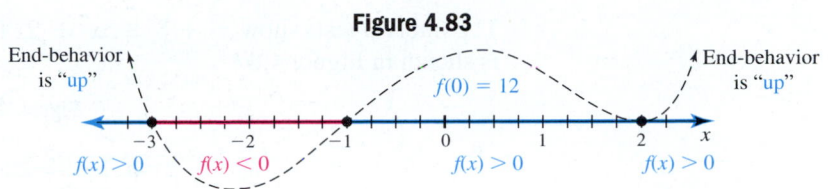

4. From the diagram, we see that
   $f(x) \leq 0$ for $x \in [-3, -1]$, and at the
   single point $x = 2$. This shows the
   solution for $x^4 + 4x \leq 9x^2 - 12$ is
   $x \in [-3, -1] \cup \{2\}$. The actual graph is
   shown in Figure 4.84. The graph is below
   or touching the x-axis from -3 to -1
   and at $x = 2$.

Figure 4.84

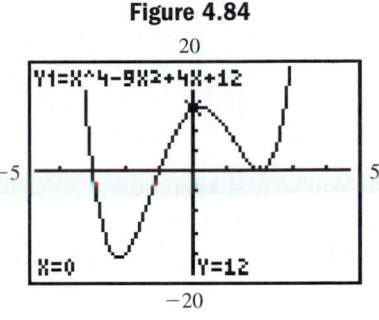

> Now try Exercises 19 through 24 ▶

Solving Function Inequalities Using Interval Tests

As an alternative to graphical analysis an **interval test method** can be used to solve polynomial (and rational) inequalities. The x-intercepts (and vertical asymptotes in the case of rational functions) are noted on the x-axis, then a test number is selected from each interval. Since polynomial and rational functions are continuous over their entire domain, the sign of the function at these test values will be the sign of the function for all values of x in the chosen interval.

EXAMPLE 3 ▶   **Solving a Polynomial Inequality**

Solve the inequality $x^3 + 8 \leq 5x^2 - 2x$.

Solution ▶   Writing the relationship in function form gives $p(x) = x^3 - 5x^2 + 2x + 8$, with
solutions needed to $p(x) \leq 0$. The tests for 1 and -1 show $x = -1$ is a root, and
using -1 with synthetic division gives

use -1 as a "divisor" $\underline{-1|}$ 1 -5 2   8
                                      $\downarrow$ -1 6 -8
                                  ─────────────────────────────────
                                      1   -6 8 $\underline{|0}$

The quotient is $q(x) = x^2 - 6x + 8$, with a remainder of 0.

The factored form is $p(x) = (x + 1)(x^2 - 6x + 8) = (x + 1)(x - 2)(x - 4)$.
The x-intercepts are $(-1, 0)$, $(2, 0)$, and $(4, 0)$. Plotting these intercepts creates four
intervals on the x-axis (Figure 4.85).

Figure 4.85

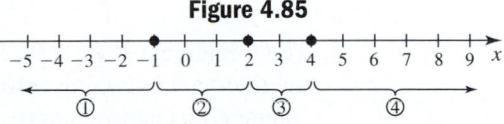

WORTHY OF NOTE

When evaluating a function using
the interval test method, it's usually
easier to use the factored form
instead of the polynomial form, since
all you really need is whether the
result will be positive or negative.
For instance, you could likely tell
$p(3) = (3 + 1)(3 - 2)(3 - 4)$ is
going to be negative, more quickly
than $p(3) = (3)^3 - 5(3)^2 + 2(3) + 8$.

Selecting a test value from each interval gives the information shown in Figure 4.86.

Figure 4.86

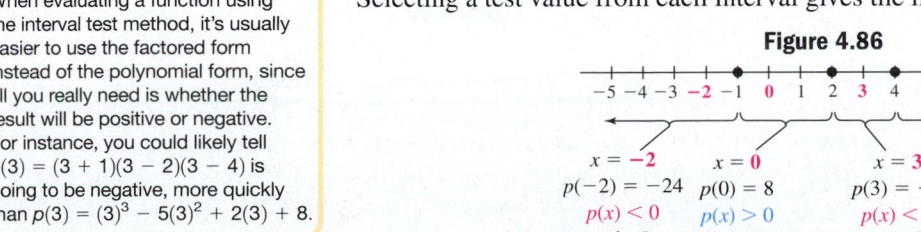

The interval tests show $x^3 + 8 \leq 5x^2 - 2x$ for $x \in (-\infty, -1] \cup [2, 4]$. The graph is shown in Figure 4.87.

Figure 4.87

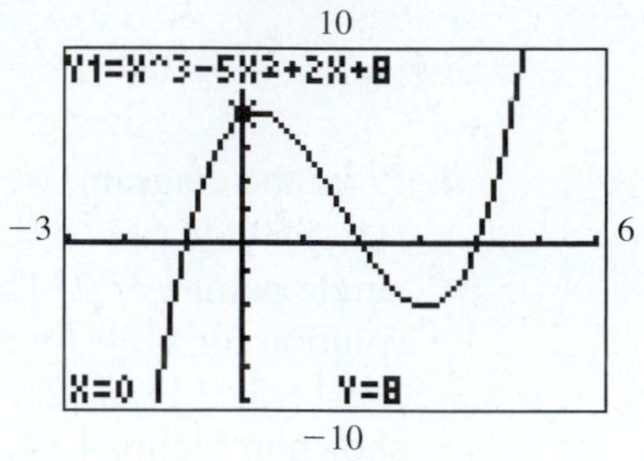

☑ A. You've just seen how we can solve polynomial inequalities

Now try Exercises 25 and 26 ▶

B. Rational Inequalities

In general, the solution process for polynomial and rational inequalities is virtually identical, once we recognize that vertical asymptotes also break the x-axis into intervals where function values may change sign. However, for rational functions it's more efficient to begin the analysis using the y-intercept or a test point, rather than end-behavior, although either will do.

EXAMPLE 4 ▶   **Solving a Rational Inequality by Analysis**

Solve $\dfrac{x^2 - 9}{x^3 - x^2 - x + 1} \leq 0$.

Solution ▶   In function form, $v(x) = \dfrac{x^2 - 9}{x^3 - x^2 - x + 1}$ and we want the solution for $v(x) \leq 0$.

The numerator and denominator are in standard form. The numerator factors easily, and the denominator can be factored by grouping.

1. The factored form is $v(x) = \dfrac{(x - 3)(x + 3)}{(x - 1)^2(x + 1)}$.

2. $v(x)$ will change sign at $x = 3, -3$, and -1 as all have odd multiplicity, but will not change sign at $x = 1$ (even multiplicity). Note that zeroes of the denominator will always be indicated by open dots (Figure 4.88) as they are excluded from any solution set.

Figure 4.88

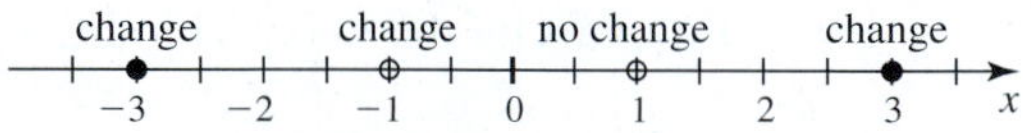

3. The y-intercept is $(0, -9)$, indicating that function values will be negative in the interval containing zero. Working outward from this interval using the "change/no change" approach, gives the solution indicated in Figure 4.89.

Figure 4.89

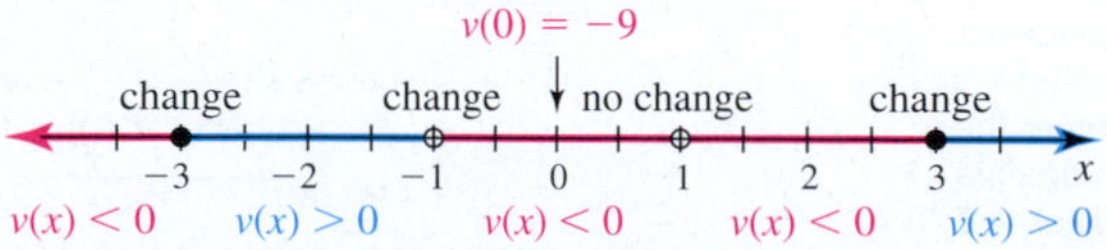

WORTHY OF NOTE

End-behavior can also be used to analyze rational inequalities, although using the *y*-intercept may be more efficient. For the function $v(x)$ from Example 5 we have
$$\frac{x^2 - 9}{x^3 - x^2 - x + 1} \approx \frac{x^2}{x^3} = \frac{1}{x} \text{ for large}$$
values of *x*, indicating $v(x) > 0$ to the far right and $v(x) < 0$ to the far left. The analysis of each interval can then begin from either side.

4. For $v(x) \leq 0$, the solution is
$$x \in (-\infty, -3] \cup (-1, 1) \cup (1, 3].$$

   Finding a window that clearly displays all features of rational function can sometimes be difficult. In these cases, we could investigate each piece separately to confirm solutions. For this example, most of the features of $v(x)$ can be seen using a window size of $X \in [-5, 5]$ and $Y \in [20, 15]$, and we note the graph in Figure 4.90 strongly tends to support our solution.

Figure 4.90

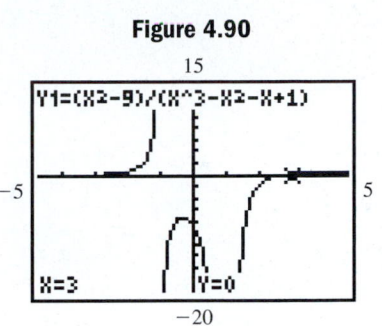

Now try Exercises 27 through 40 ▶

If the rational inequality is not given in function form or is composed of more than one term, start by writing the inequality with zero on one side, then combine terms into a single expression. This is the most efficient way of determining the zeroes of the function and the location of any vertical asymptotes.

EXAMPLE 5 ▶   Solving a Rational Inequality Using Interval Tests

Solve the inequality $\dfrac{x - 2}{x - 3} \leq \dfrac{1}{x + 3}$.

Solution ▶   Rewrite the inequality with zero on one side: $\dfrac{x - 2}{x - 3} - \dfrac{1}{x + 3} \leq 0$. This is

equivalent to $v(x) \leq 0$, where $v(x) = \dfrac{x - 2}{x - 3} - \dfrac{1}{x + 3}$. Combining the expressions

on the right, we have

$$v(x) = \frac{(x - 2)(x + 3) - 1(x - 3)}{(x + 3)(x - 3)} \qquad \text{LCD is } (x + 3)(x - 3)$$

$$= \frac{x^2 + x - 6 - x + 3}{(x + 3)(x - 3)} \qquad \text{multiply}$$

$$= \frac{x^2 - 3}{(x + 3)(x - 3)} \qquad \text{simplify}$$

1. The factored form is $v(x) = \dfrac{(x + \sqrt{3})(x - \sqrt{3})}{(x + 3)(x - 3)}$. $x^2 - k = (x + \sqrt{k})(x - \sqrt{k})$

2. The zeroes and asymptotes will occur at $x = -\sqrt{3}, \sqrt{3}, -3$, and 3, which we plot on the *x*-axis, creating five intervals (Figure 4.91).

Figure 4.91

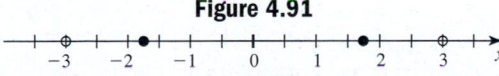

3. From left to right, we select one test value from each interval created by the zeroes and vertical asymptotes. We chose: $x = -4, -2, 0, 2$, and 4 (see Figure 4.92). The results are shown in Figure 4.93, with the conclusion noted beneath each interval.

Figure 4.92

X	Y1
-4	1.8571
-2	-.2
0	.33333
2	-.2
4	1.8571

X=

Figure 4.93

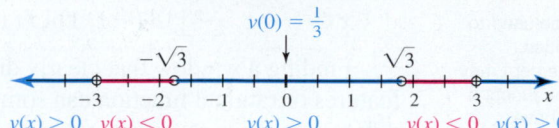

4. The solution for $\dfrac{x-2}{x-3} \le \dfrac{1}{x+3}$ is

$x \in (-3, -\sqrt{3}] \cup [\sqrt{3}, 3)$.

To check these solutions graphically,

we subtract $\dfrac{1}{x+3}$ from both sides and graph

$Y_1 = \dfrac{x-2}{x-3} - \dfrac{1}{x+3}$ to look for intervals where

Figure 4.94

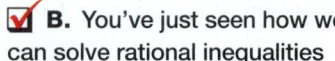

the graph is <u>below</u> the x-axis. The graph is shown in Figure 4.94 and verifies our solution.

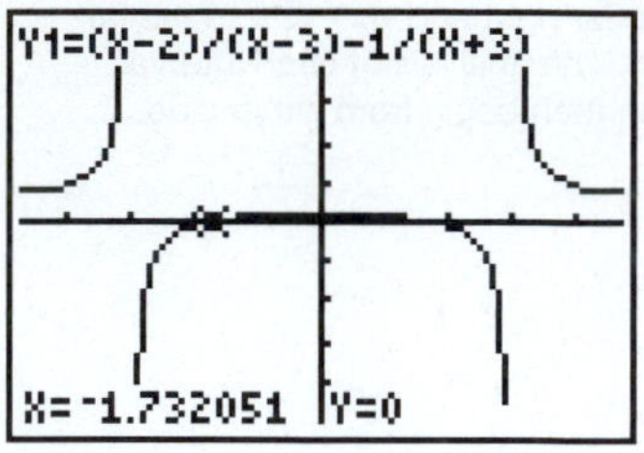

☑ **B.** You've just seen how we can solve rational inequalities

Now try Exercises 41 through 60 ▶

C. Applications of Inequalities

Applications of inequalities come in many varieties. In addition to stating the solution algebraically, these exercises often compel us to consider the *context of each application* as we state the solution set.

EXAMPLE 6 ▶ **Solving Applications of Inequalities**

The velocity of a particle (in feet per second) as it floats through air turbulence is given by $V(t) = t^5 - 10t^4 + 35t^3 - 50t^2 + 24t$, where t is the time in seconds and $0 < t < 4.5$. During what intervals of time is the particle moving in the positive direction $[V(t) > 0]$?

Solution ▶ Begin by writing V in factored form. Testing 1 and -1 shows $t = 1$ is a root. Factoring out t gives $V(t) = t(t^4 - 10t^3 + 35t^2 - 50t + 24)$, and using $t = 1$ with synthetic division yields

$$\text{use 1 as a "divisor"} \quad \underline{1|} \begin{array}{ccccc} 1 & -10 & 35 & -50 & 24 \\ \downarrow & 1 & -9 & 26 & -24 \\ \hline 1 & -9 & 26 & -24 & \underline{|0} \end{array}$$

The quotient is $q_1(t) = t^3 - 9t^2 + 26t - 24$. Using $t = 2$, we continue with the division on $q_1(t)$ which gives:

$$\text{use 2 as a "divisor"} \quad \underline{2|} \begin{array}{cccc} 1 & -9 & 26 & -24 \\ \downarrow & 2 & -14 & 24 \\ \hline 1 & -7 & 12 & \underline{|0} \end{array}$$

This shows $V(t) = t(t-1)(t-2)(t^2 - 7t + 12)$.

1. The completely factored form is $V(t) = t(t-1)(t-2)(t-3)(t-4)$.
2. All zeroes have odd multiplicity and function values will change sign.
3. With odd degree and a positive leading coefficient, end-behavior is down/up.

Function values will be negative in the far left interval and alternate in sign thereafter. The solution diagram is shown in the figure (the parentheses shown on the number line indicate the domain given).

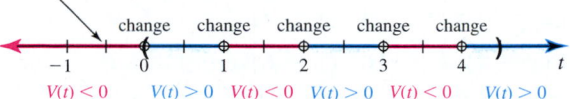

Since end-behavior is down/up, function values are negative in this interval, and will alternate thereafter.

4. For $V(t) > 0$, the solution is $t \in (0, 1) \cup (2, 3) \cup (4, 4.5)$. The particle is moving in the positive direction in these time intervals.

Now try Exercises 63 through 70 ▶

Figure 4.95

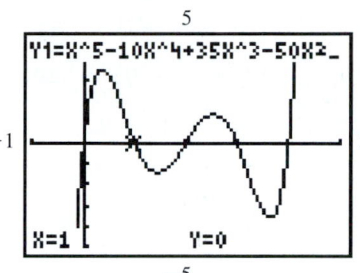

To verify the analysis in Example 6, we graph $V(t)$ using the window $X \in [-1, 5]$ and $Y \in [-5, 5]$. As Figure 4.95 shows, function values are positive (graph is above the x-axis) when $t \in (0, 1) \cup (2, 3) \cup (4, 4.5)$.

☑ **C.** You've just seen how we can solve applications of inequalities

---

4.6 EXERCISES

▶ CONCEPTS AND VOCABULARY

Fill in each blank with the appropriate word or phrase. Carefully reread the section if needed.

1. To solve a polynomial or rational inequality, begin by plotting the location of all zeroes and __vertical__ asymptotes (if they exist), then consider the __multiplicity__ of each.

2. For strict inequalities, the zeroes are __excluded__ from the solution set. For nonstrict inequalities, zeroes are __included__. The values at which vertical asymptotes occur are always __excluded__.

3. Since the graph of $g(x) = x^4 + 3$, opens upward with a vertex at $(0, 3)$ the solution set for $g(x) > 0$ is __empty__.

4. To solve a polynomial/rational inequality, it helps to find the sign of f in some interval. This can quickly be done using the __end-behavior__ or __y-intercept__ of the function.

5. Compare/Contrast the process for solving $x^2 - 3x - 4 \geq 0$ with $\dfrac{1}{x^2 - 3x - 4} \geq 0$. Are there similarities? What are the differences?
Answers will vary.

6. Compare/Contrast the process for solving $(x + 1)(x - 3)(x^2 + 1) > 0$ with $(x + 1)(x - 3) > 0$. Are there similarities? What are the differences?   Answers will vary.

HOMEWORK SELECTION GUIDE

Core: 7–55 every other odd, 57–60, 63–69 (21 Exercises)
Standard: 1–4, 7–55 every other odd, 57–60, 63–69, 72a/c, 78 (28 Exercises)

Extended: 1–4, 7–55 every other odd, 57–60, 63–69, 72a/c, 73, 78, 79 (30 Exercises)
In Depth: 1–6, 7–55 every other odd, 57–60, 63–69, 72a/c, 73, 74, 76–79 (35 Exercises)

▶ DEVELOPING YOUR SKILLS

Solve each polynomial inequality indicated using a number line and the behavior of the graph at each zero. Write all answers in interval notation.

7. $(x + 3)(x - 5) < 0$ **8.** $(x - 2)(x + 7) < 0$

9. $(x + 1)^2(x - 4) \geq 0$ **10.** $(x + 6)(x - 1)^2 \leq 0$

11. $(x + 2)^3(x - 2)^2(x - 4) \geq 0$

12. $(x - 1)^3(x + 2)^2(x - 3) \leq 0$

13. $x^2 + 4x + 1 < 0$ **14.** $x^2 - 6x + 4 > 0$

15. $x^3 + x^2 - 5x + 3 \leq 0$ $x \in (-\infty, -3] \cup \{1\}$

16. $x^3 + x^2 - 8x - 12 \geq 0$ $x \in [3, \infty) \cup \{-2\}$

17. $x^3 - 7x + 6 > 0$ **18.** $x^3 - 13x + 12 > 0$

19. $x^4 - 10x^2 > -9$ **20.** $x^4 + 36 < 13x^2$

21. $x^4 - 9x^2 > 4x - 12$

22. $x^4 - 16 > 5x^3 - 20x$

23. $x^4 - 6x^3 \leq -8x^2 - 6x + 9$ $x \in [-1, 1] \cup \{3\}$

24. $x^4 - 3x^2 + 8 \leq 4x^3 - 10x$ $x \in \{-1\} \cup [2, 4]$

Solve each inequality using the interval test method.

25. $-4x + 12 < -x^3 + 3x^2$ $x \in (-\infty, -2) \cup (2, 3)$

26. $x^3 + 8 < 5x^2 - 2x$ **27.** $\dfrac{x^2 - x - 6}{x^2 - 1} \geq 0$

28. $\dfrac{x^2 - 4x - 21}{x - 3} < 0$

Solve each rational inequality indicated using a number line and the behavior of the graph at each zero. Write all answers in interval notation.

29. $f(x) = \dfrac{x + 3}{x - 2}; f(x) \leq 0$ $x \in [-3, 2)$

30. $F(x) = \dfrac{x - 4}{x + 1}; F(x) \geq 0$ $x \in (-\infty, -1) \cup [4, \infty)$

31. $g(x) = \dfrac{x + 1}{x^2 + 4x + 4}; g(x) < 0$

$x \in (-\infty, -2) \cup (-2, -1)$

32. $G(x) = \dfrac{x - 3}{x^2 - 2x + 1}; G(x) > 0$ $x \in (3, \infty)$

33. $\dfrac{2 - x}{x^2 - x - 6} \geq 0$ **34.** $\dfrac{1 - x}{x^2 - 2x - 8} \leq 0$

35. $\dfrac{2x - x^2}{x^2 + 4x - 5} < 0$ **36.** $\dfrac{x^2 + 3x}{x^2 - 2x - 3} > 0$

37. $\dfrac{x^2 - 4}{x^3 - 13x + 12} \geq 0$ **38.** $\dfrac{x^2 + x - 6}{x^3 - 7x + 6} \leq 0$

39. $\dfrac{x^2 + 5x - 14}{x^3 + x^2 - 5x + 3} > 0$ $x \in (-7, -3) \cup (2, \infty)$

40. $\dfrac{x^2 + 2x - 8}{x^3 + 5x^2 + 3x - 9} < 0$ $x \in (-\infty, -4) \cup (1, 2)$

41. $\dfrac{2}{x - 2} \leq \dfrac{1}{x}$ **42.** $\dfrac{5}{x + 3} \geq \dfrac{3}{x}$

43. $\dfrac{x - 3}{x + 17} > \dfrac{1}{x - 1}$ **44.** $\dfrac{1}{x + 5} < \dfrac{x - 2}{x - 7}$

45. $\dfrac{x + 1}{x - 2} \geq \dfrac{x + 2}{x + 3}$ **46.** $\dfrac{x - 3}{x - 6} \leq \dfrac{x + 1}{x + 4}$

47. $\dfrac{x + 2}{x^2 + 9} > 0$ **48.** $\dfrac{x^2 + 4}{x - 3} < 0$

49. $\dfrac{x^3 + 1}{x^2 + 1} > 0$ **50.** $\dfrac{x^2 + 4}{x^3 - 8} < 0$

51. $\dfrac{x^4 - 5x^2 - 36}{x^2 - 2x + 1} > 0$ **52.** $\dfrac{x^4 - 3x^2 - 4}{x^2 - x - 20} < 0$

53. $x^2 - 2x \geq 15$ **54.** $x^2 + 3x \geq 18$

55. $x^3 \geq 9x$ **56.** $x^3 \leq 4x$

Match the correct solution with the inequality and graph given.

57. $R(x) \leq 0$

   a. $x \in (-\infty, -1) \cup (0, 2)$

   b. $x \in [0, 1] \cup (2, \infty)$

   c. $x \in [-5, -1] \cup [2, 5]$

   d. $x \in (-\infty, -1) \cup [0, 2)$

   e. none of these

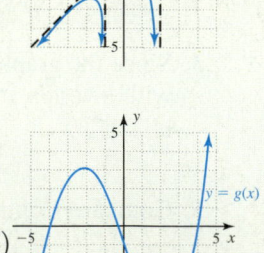

58. $g(x) \geq 0$

   a. $x \in (-4, -0.5) \cup (4, \infty)$

   b. $x \in [-0.5, 4] \cup [4, 5]$

   c. $x \in (-\infty, -4) \cup (-0.5, 4)$

   d. $x \in [-4, -0.5] \cup [4, \infty)$

   e. none of these

59. $f(x) < 0$

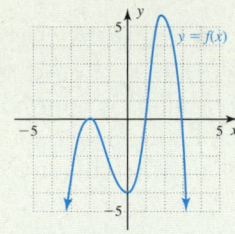

  a. $x \in (-5, -2) \cup (3, 5)$

  b. $x \in (-\infty, -2) \cup (-2, 1) \cup (3, \infty)$

  c. $x \in (-\infty, -2) \cup (3, \infty)$

  d. $x \in (-\infty, -2) \cup (-2, 1] \cup [3, \infty)$

  e. none of these

60. $r(x) \geq 0$

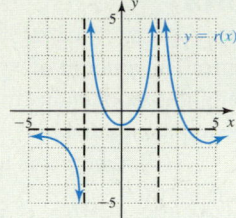

  a. $x \in (-\infty, -2) \cup [-1, 1] \cup [3, \infty)$

  b. $x \in (-2, -1] \cup [1, 2) \cup (2, 3]$

  c. $x \in (-\infty, -2) \cup (2, \infty)$

  d. $x \in (-2, -1) \cup (1, 2) \cup (2, 3]$

  e. none of these

▶ WORKING WITH FORMULAS

61. Discriminant of the reduced cubic $x^3 + px + q = 0$: $D = -(4p^3 + 27q^2)$

The discriminant of a cubic equation is less well known than that of the quadratic, but serves the same purpose. The discriminant of the reduced cubic is given by the formula shown, where p is the linear coefficient and q is the constant term. If $D > 0$, there will be three real and distinct roots. If $D = 0$, there are still three real roots, but one is a repeated root (multiplicity two). If $D < 0$, there are one real and two complex roots. Suppose we wish to study the family of cubic equations where $q = p + 1$.

  a. Verify the resulting discriminant is $D = -(4p^3 + 27p^2 + 54p + 27)$.

  b. Determine the values of p and q for which this family of equations has a repeated real root. In other words, solve the equation $-(4p^3 + 27p^2 + 54p + 27) = 0$ using the rational zeroes theorem and synthetic division to write D in completely factored form.

  c. Use the factored form from part (b) to determine the values of p and q for which this family of equations has three real and distinct roots. In other words, solve $D > 0$.

  d. Verify the results of parts (b) and (c) on a graphing calculator.

62. Coordinates for the folium of Descartes:

The interesting relation shown here is called the folium (leaf) of Descartes. The folium is most often graphed using what are called *parametric equations,* in which the coordinates a and b are expressed in terms of the parameter x ("k" is a constant that affects the size of the leaf). Since each is an individual function, the x- and y-coordinates can be investigated individually in rectangular coordinates using $F(x) = \dfrac{3x}{1 + x^3}$ and $G(x) = \dfrac{3x^2}{1 + x^3}$ (assume $k = 1$ for now).

$$\begin{cases} a = \dfrac{3kx}{1 + x^3} \\ b = \dfrac{3kx^2}{1 + x^3} \end{cases}$$

  a. Graph each function using the techniques from this chapter.

  b. Will $F(x)$ ever be equal to $G(x)$? If so, for what values of x?   yes; $x = 0$ and $x = 1$

  c. According to your graph, for what values of x will the x-*coordinate* of the folium be positive? In other words, solve $F(x) = \dfrac{3x}{1 + x^3} > 0$.

  d. For what values of x will the y-*coordinate* of the folium be positive? Solve $G(x) = \dfrac{3x^2}{1 + x^3} > 0$. $(-1, 0) \cup (0, \infty)$

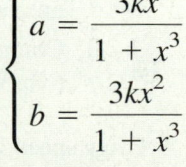
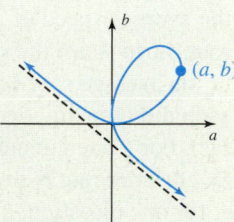

Folium of Descartes

▶ APPLICATIONS

Deflection of a beam: The amount of deflection in a rectangular wooden beam of length L ft can be approximated by $d(x) = k(x^3 - 3L^2x + 2L^3)$, where k is a constant that depends on the characteristics of the wood and the force applied, and x is the *distance from the unsupported end* of the beam $(x < L)$.

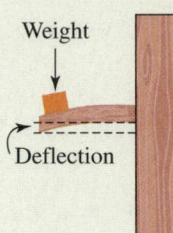

Weight

Deflection

63. Find the equation for a beam 8 ft long and use it for the following: $d(x) = k(x^3 - 192x + 1024)$

   a. For what distances x is the quantity $\dfrac{d(x)}{k}$ less than 189 units? $x \in (5, 8]$

   b. What is the amount of deflection 4 ft from the unsupported end $(x = 4)$? $320k$ units

   c. For what distances x is the quantity $\dfrac{d(x)}{k}$ greater than 475 units? $x \in [0, 3)$

   d. If safety concerns prohibit a deflection of more than $648k$ units, what is the shortest distance from the end of the beam that the force can be applied?   2 ft

64. Find the equation for a beam 9 ft long and use it for the following: $d(x) = k(x^3 - 243x + 1458)$

   a. For what distances x is the quantity $\dfrac{d(x)}{k}$ less than 216 units? $x \in (6, 9]$

   b. What is the amount of deflection 4 ft from the unsupported end $(x = 4)$? $550k$ units

   c. For what distances x is the quantity $\dfrac{d(x)}{k}$ greater than $550k$ units? $x \in [0, 4)$

    d. Compare the answer to 63b with the answer to 64b. What can you conclude?
   the longer beam gives greater deflection

Average speed for a round-trip: Surprisingly, the average speed of a round-trip is *not* the sum of the average speed in each direction divided by two. For a fixed distance D, consider rate r_1 in time t_1 for one direction, and rate r_2 in time t_2 for

the other, giving $r_1 = \dfrac{D}{t_1}$ and $r_2 = \dfrac{D}{t_2}$. The average speed for the round-trip is $R = \dfrac{2D}{t_1 + t_2}$.

Additional answers can be found in the Instructor Answer Appendix.

65. The distance from St. Louis, Missouri, to Springfield, Illinois, is approximately 80 mi. Suppose that Sione, due to the age of his vehicle, made the round-trip with an average speed of 40 mph.

   a. Use the relationships stated to verify that
   $$r_2 = \frac{20r_1}{r_1 - 20}.$$ verified

   b. Discuss the meaning of the horizontal and vertical asymptotes in this context.

   c. Verify algebraically the speed returning would be greater than the speed going for $20 < r_1 < 40$. In other words, solve the inequality $\dfrac{20r_1}{r_1 - 20} > r_1$ using the ideas from this section. $r_1 \in (20, 40)$

66. The distance from Boston, Massachusetts, to Hartford, Connecticut, is approximately 100 mi. Suppose that Stella, due to excellent driving conditions, made the round-trip with an average speed of 60 mph.

   a. Use the relationships above to verify that
   $$r_2 = \frac{30r_1}{r_1 - 30}.$$ verified

   b. Discuss the meaning of the horizontal and vertical asymptotes in this context.

   c. Verify algebraically the speed returning would be greater than the speed going for $30 < r_1 < 60$. In other words, solve the inequality $\dfrac{30r_1}{r_1 - 30} > r_1$ using the ideas from this section. $r_1 \in (30, 60)$

Electrical resistance and temperature: The amount of electrical resistance R in a medium depends on the temperature, and for certain materials can be modeled by the equation $R(t) = 0.01t^2 + 0.1t + k$, where $R(t)$ is the resistance (in ohms Ω) at temperature t $(t \geq 0°)$ in degrees Celsius, and k is the resistance at $t = 0°$C.

67. Suppose $k = 30$ for a certain medium. Write the resistance equation and use it to answer the following. $R(t) = 0.01t^2 + 0.1t + 30$

   a. For what temperatures is the resistance less than 42 Ω? $[0°, 30°)$

   b. For what temperatures is the resistance greater than 36 Ω? $(20°, \infty)$

   c. If it becomes uneconomical to run electricity through the medium for resistances greater than 60 Ω, for what temperatures should the electricity generator be shut down? $(50°, \infty)$

68. Suppose $k = 20$. Write the resistance equation and solve the following. $R(t) = 0.01t^2 + 0.1t + 20$

  a. For what temperatures is the resistance less than 26 Ω? $[0°, 20°)$

  b. For what temperatures is the resistance greater than 40 Ω? $(40°, \infty)$

  c. If it becomes uneconomical to run electricity through the medium for resistances greater than 50 Ω, for what temperatures should the electricity generator be shut down? $(50°, \infty)$

69. Sum of consecutive squares: The sum of the first n squares $1^2 + 2^2 + 3^2 + \cdots + n^2$ is given by the formula $S(n) = \dfrac{2n^3 + 3n^2 + n}{6}$. Use the equation to solve the following inequalities.

  a. For what number of consecutive squares is $S(n) \geq 30$? $n \geq 4$

b. For what number of consecutive squares is $S(n) \leq 285$? $n \leq 9$

 c. What is the maximum number of consecutive squares that can be summed without the result exceeding three digits?   13

70. Sum of consecutive cubes: The sum of the first n cubes $1^3 + 2^3 + 3^3 + \cdots + n^3$ is given by the formula $S(n) = \dfrac{n^4 + 2n^3 + n^2}{4}$. Use the equation to solve the following inequalities.

  a. For what number of consecutive cubes is $S(n) \geq 100$? $n \geq 4$

  b. For what number of consecutive cubes is $S(n) \leq 784$? $n \leq 7$

 c. What is the maximum number of consecutive cubes that can be summed without the result exceeding three digits?   7

▶ EXTENDING THE CONCEPT

71. (a) Is it possible for the solution set of a polynomial inequality to be all real numbers? If not, discuss why. If so, provide an example. (b) Is it possible for the solution set of a rational inequality to be all real numbers? If not, discuss why. If so, provide an example.

72. As in our earlier studies, if n is an even number, the expression $\sqrt[n]{A}$ represents a real number only if $A \geq 0$. Use this idea to find the domain of the following functions.

  a. $f(x) = \sqrt{2x^3 - x^2 - 16x + 15}$

  b. $g(x) = \sqrt[4]{2x^3 + x^2 - 22x + 24}$

  c. $p(x) = \sqrt[4]{\dfrac{x + 2}{x^2 - 2x - 35}}$ $x \in (-5, -2] \cup (7, \infty)$

  d. $q(x) = \sqrt{\dfrac{x^2 - 1}{x^2 - x - 6}}$
               $x \in (-\infty, -2) \cup [-1, 1] \cup (3, \infty)$

73. Find one polynomial inequality and one rational inequality that have the solution $x \in (-\infty, -2) \cup (0, 1) \cup (1, \infty)$.

74. Using the tools of calculus, it can be shown that $f(x) = x^4 - 4x^3 - 12x^2 + 32x + 39$ is increasing in the intervals where $F(x) = x^3 - 3x^2 - 6x + 8$ is positive. Solve the inequality $F(x) > 0$ using the ideas from this section, then verify $f(x)\!\uparrow$ in these intervals by graphing f on a graphing calculator and using the $\boxed{\text{TRACE}}$ feature.

75. Using the tools of calculus, it can be shown that $r(x) = \dfrac{x^2 - 3x - 4}{x - 8}$ is decreasing in the intervals where $R(x) = \dfrac{x^2 - 16x + 28}{(x - 8)^2}$ is negative. Solve the inequality $R(x) < 0$ using the ideas from this section, then verify $r(x)\!\downarrow$ in these intervals by graphing r on a graphing calculator and using the $\boxed{\text{TRACE}}$ feature. $R(x) < 0$ for $x \in (2, 8) \cup (8, 14)$

▶ MAINTAINING YOUR SKILLS

76. (2.2) Use the graph of $f(x)$ given to sketch the graph of $y = f(x + 2) - 3$.

Exercise 76

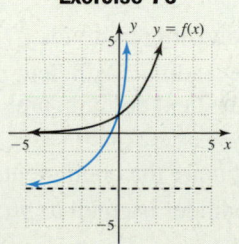

77. (2.5/4.5) Graph the function $f(x) = \dfrac{x^2 + 2x - 8}{x + 4}$. If there

is a removable discontinuity, repair the break using an appropriate piecewise-defined function.
$$F(x) = \begin{cases} f(x) & x \neq -4 \\ -6 & x = -4 \end{cases}$$

78. (R.6) Solve the equation $\dfrac{1}{2}\sqrt{16 - x} - \dfrac{x}{2} = 2$. Check solutions in the original equation. $x = 0$

79. (2.3) Solve the absolute value inequality: $-2|x - 3| + 5 \geq -7$ $x \in [-3, 9]$

Additional answers can be found in the Instructor Answer Appendix.

MAKING CONNECTIONS

Making Connections: Graphically, Symbollically, Numerically, and Verbally

Eight graphs (a) through (h) are given. Match the characteristics shown in 1 through 16 to one of the eight graphs.

(a) 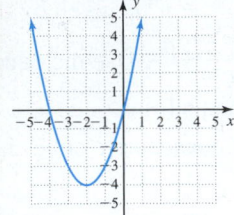 **(b)** 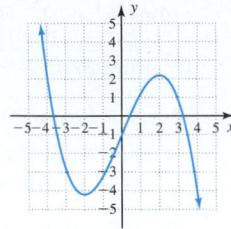 **(c)** 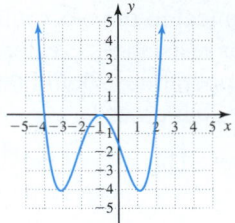 **(d)**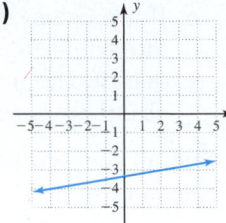

(e) 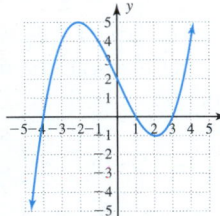 **(f)** 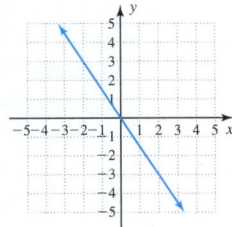 **(g)** 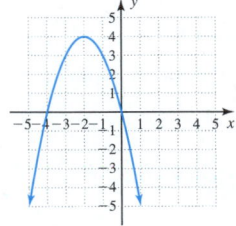 **(h)**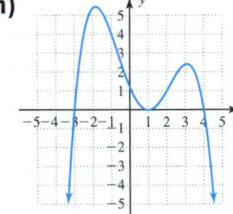

1. _e_ $\dfrac{1}{6}(x + 4)(x - 1)(x - 3)$

2. _g_ $y = -(x + 2)^2 + 4$

3. _b_ min degree 3, y-int: $(0, -1)$

4. _c_ min degree 4, one repeated root, relative max $y = 0$ at $x = -1$

5. _a_ $y = x(x - 4)$

6. _g_ $f(x) > 0$ only for $x \in (-4, 0)$

7. _h_ $f(-1) = 4, f(2) = 1$

8. _b_ $f(x)\uparrow$ only for $x \in (-2, 2)$

9. _d_ $(y + 4) = \dfrac{1}{6}(x + 4)$

10. _f_ $y = -\dfrac{3}{2}x$

11. _a_ axis of symmetry $x = -2, f(-1) = -3$

12. _h_ min degree 4, one repeated root, lead coefficient $a < 0$

13. _g_ axis of symmetry $x = -2, f(-3) = 3$

14. _e_ $f(x) > 0$ for $x \in (-4, 1) \cup (3, \infty)$

15. _c_ $f(x)\uparrow$ only for $x \in (-3, -1) \cup (1, \infty)$

16. _b_ $f(x) = 4.2$ for $x = -2$ and $x = 4$

SUMMARY AND CONCEPT REVIEW

SECTION 4.1   Synthetic Division; the Remainder and Factor Theorems

KEY CONCEPTS

- Synthetic division is an abbreviated form of long division. Only the coefficients of the dividend are used, since "standard form" ensures like place values are aligned. Zero placeholders are used for "missing" terms. The "divisor" must be linear with leading coefficient 1.
- To divide a polynomial by $x - c$, use c in the synthetic division; to divide by $x + c$, use $-c$.
- After setting up the synthetic division template, drop the leading coefficient of the dividend into place, then multiply in the diagonal direction, place the product in the next column, and add in the vertical direction, continuing to the last column.
- The final sum is the remainder r, the numbers preceding it are the coefficients of $q(x)$.
- Remainder theorem: If $p(x)$ is divided by $x - c$, the remainder is equal to $p(c)$. The theorem can be used to evaluate polynomials at $x = c$.

- Factor theorem: If $p(c) = 0$, then c is a zero of p and $(x - c)$ is a factor. Conversely, if $(x - c)$ is a factor of p, then $p(c) = 0$. The theorem can be used to factor a polynomial or build a polynomial from its zeroes.
- The remainder and factor theorems also apply when c is a complex number.

EXERCISES

Divide using long division and clearly identify the quotient and remainder:

1. $\dfrac{x^3 + 4x^2 - 5x - 6}{x - 2}$ $q(x) = x^2 + 6x + 7; r = 8$

2. $\dfrac{x^3 + 2x - 4}{x^2 - x}$ $q(x) = x + 1; r = 3x - 4$

3. Use the factor theorem to show that $x + 7$ is a factor of $2x^4 + 13x^3 - 6x^2 + 9x + 14$.

4. Complete the division and write $h(x)$ as $h(x) = d(x)q(x) + r(x)$, given $\dfrac{h(x)}{d(x)} = \dfrac{x^3 - 4x + 5}{x - 2}$.

     $x^3 - 4x + 5 = (x - 2)(x^2 + 2x) + 5$

5. Use the factor theorem to help factor $p(x) = x^3 + 2x^2 - 11x - 12$ completely. $(x + 4)(x + 1)(x - 3)$

6. Use the factor and remainder theorems to factor h, given $x = 4$ is a zero: $h(x) = x^4 - 3x^3 - 4x^2 - 2x + 8$.

         $h(x) = (x - 1)(x - 4)(x^2 + 2x + 2)$

Use the remainder theorem:

7. Show $x = \frac{1}{2}$ is a zero of V: $V(x) = 4x^3 + 8x^2 - 3x - 1$.

8. Show $x = 3i$ is a zero of W: $W(x) = x^3 - 2x^2 + 9x - 18$.

9. Find $h(-7)$ given $h(x) = x^3 + 9x^2 + 13x - 10$.

Use the factor theorem:

10. Find a degree 3 polynomial in standard form with zeroes $x = 1$, $x = -\sqrt{5}$, and $x = \sqrt{5}$. $P(x) = x^3 - x^2 - 5x + 5$

11. Find a fourth-degree polynomial in standard form with one real zero, given $x = 1$ and $x = -2i$ are zeroes.

         $C(x) = x^4 - 2x^3 + 5x^2 - 8x + 4$

12. Use synthetic division and the remainder theorem to answer: At a busy shopping mall, customers are constantly coming and going. One summer afternoon during the hours from 12 o'clock noon to 6 in the evening, the number of customers in the mall could be modeled by $C(t) = 3t^3 - 28t^2 + 66t + 35$, where $C(t)$ is the number of customers (in tens), t hours after 12 noon. (a) How many customers were in the mall at noon? (b) Were more customers in the mall at 2:00 or at 3:00 P.M.? How many more? (c) Was the mall busier at 1:00 P.M. (after lunch) or 6:00 P.M. (around dinner time)?   **a.** $C(0) = 350$ customers   **b.** more at 2 P.M., 170   **c.** busier at 1 P.M., 760 > 710

SECTION 4.2   The Zeroes of Polynomial Functions

KEY CONCEPTS

- Fundamental theorem of algebra: Every complex polynomial of degree $n \geq 1$ has at least one complex zero.
- Linear factorization theorem: Every complex polynomial of degree $n \geq 1$ has exactly n linear factors, and can be written in the form $p(x) = a(x - c_1)(x - c_2)\ldots(x - c_n)$, where $a \neq 0$ and $c_1, c_2, \ldots, c_n$ are (not necessarily distinct) complex numbers.
- For a polynomial p in factored form with repeated factors $(x - c)^m$, c is a zero of multiplicity m. If m is odd, c is a zero of odd multiplicity; if m is even, c is a zero of even multiplicity.
- Corollaries to the linear factorization theorem:

  I. If p is a polynomial with real coefficients, p can be factored into linear factors (not necessarily distinct) and irreducible quadratic factors having real coefficients.

  II. If p is a polynomial with real coefficients, the complex zeroes of p must occur in conjugate pairs. If $a + bi$ ($b \neq 0$), is a zero, then $a - bi$ is also a zero.

  III. If p is a polynomial with degree $n \geq 1$, then p will have exactly n zeroes (real or complex), where zeroes of multiplicity m are counted m times.

- Intermediate value theorem: If p is a polynomial with real coefficients where $p(a)$ and $p(b)$ have opposite signs, then there is at least one c between a and b such that $p(c) = 0$.
- Rational zeroes theorem: If a real polynomial has integer coefficients, rational zeroes must be of the form $\frac{p}{q}$, where p is a factor of the constant term and q is a factor of the leading coefficient.
- Descartes' rule of signs, upper and lower bounds property, tests for -1 and 1, and graphing technology can all be used with the rational zeroes theorem to factor, solve, and graph polynomial functions.

Additional answers can be found in the Instructor Answer Appendix.

EXERCISES

Using the tools from this section,

13. List all possible rational zeroes of $p(x) = 4x^3 - 16x^2 + 11x + 10$.

14. Find all rational zeroes of $p(x) = 4x^3 - 16x^2 + 11x + 10$.

15. Write $P(x) = 2x^3 - 3x^2 - 17x - 12$ in completely factored form.

16. Prove that $h(x) = x^4 - 7x^2 - 2x + 3$ has no rational zeroes.

 17. Identify two intervals (of those given) that contain a zero of $P(x) = x^4 - 3x^3 - 8x^2 + 12x + 6$: $[-2, -1]$, $[1, 2]$, $[2, 3]$, $[4, 5]$. Then verify your answer using a graphing calculator.

18. Discuss the number of possible positive, negative, and complex zeroes for $g(x) = x^4 + 3x^3 - 2x^2 - x - 30$. Then identify which combination is correct using a graphing calculator.

SECTION 4.3   Graphing Polynomial Functions

KEY CONCEPTS

- All polynomial graphs are smooth, continuous curves.
- A polynomial of degree n has *at most* $n - 1$ turning points. The precise location of these turning points are the local maximums or local minimums of the function.
- If the degree of a polynomial is odd, the ends of its graph will point in opposite directions (like $y = mx$). If the degree is even, the ends will point in the same direction (like $y = ax^2$). The sign of the lead coefficient determines the actual behavior.
- The "behavior" of a polynomial graph near its zeroes is determined by the multiplicity of the zero. For any factor $(x - c)^m$, the graph will "cross through" the x-axis if m is odd and "bounce off" the x-axis (touching at just one point) if m is even. The larger the value of m, the flatter (more compressed) the graph will be near c.
- To "round-out" a graph, additional *midinterval points* can be found between known zeroes.
- These ideas help to establish the *Guidelines for Graphing Polynomial Functions*. See page 419.

EXERCISES

State the degree, end-behavior, and y-intercept, but do not graph.

19. $f(x) = -3x^5 + 2x^4 + 9x - 4$ degree 5; up/down; $(0, -4)$

20. $g(x) = (x - 1)(x + 2)^2(x - 2)$ degree 4; up/up; $(0, 8)$

Graph using the *Guidelines for Graphing Polynomials*.

21. $p(x) = (x + 1)^3(x - 2)^2$

22. $q(x) = 2x^3 - 3x^2 - 9x + 10$

23. $h(x) = x^4 - 6x^3 + 8x^2 + 6x - 9$

24. For the graph of $P(x)$ shown, (a) state whether the degree of P is even or odd, (b) use the graph to locate the zeroes of P and state whether their multiplicity is even or odd, and (c) find the minimum possible degree of P and write it in factored form. Assume all zeroes are real.   a. even   b. $x = -2$, odd; $x = -1$, even; $x = 1$, odd   c. deg 6: $P(x) = (x + 2)(x + 1)^2(x - 1)^3$

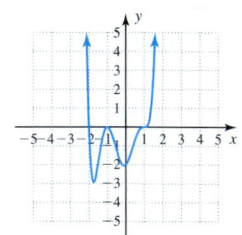

SECTION 4.4   Graphing Rational Functions

KEY CONCEPTS

- A rational function is one of the form $V(x) = \dfrac{p(x)}{d(x)}$, where p and d are polynomials and $d(x) \neq 0$.
- The domain of V is all real numbers, except the zeroes of d.
- If zero is in the domain of V, substitute 0 for x to find the y-intercept.
- The zeroes of V (if they exist), are solutions to $p(x) = 0$.
- The line $y = k$ is a horizontal asymptote of V if as $|x|$ increases without bound, $V(x)$ approaches k.
- If $\dfrac{p(x)}{d(x)}$ is in simplest form, vertical asymptotes will occur at the zeroes of d.

- The line $x = h$ is a vertical asymptote of V if as x approaches h, $V(x)$ increases/decreases without bound.
- If the degree of p is less than the degree of d, $y = 0$ (the x-axis) is a horizontal asymptote. If the degree of p is equal to the degree of d, $y = \dfrac{a}{b}$ is a horizontal asymptote, where a is the leading coefficient of p, and b is the leading coefficient of d.
- The *Guidelines for Graphing Rational Functions* can be found on page 435.

EXERCISES

25. For the function $V(x) = \dfrac{x^2 - 9}{x^2 - 3x - 4}$, state the following but do not graph: (a) domain (in set notation),
   (b) equations of the horizontal and vertical asymptotes, (c) the x- and y-intercept(s), and (d) the value of $V(1)$.

26. For $v(x) = \dfrac{(x + 1)^2}{x + 2}$, will the function change sign at $x = -1$? Will the function change sign at $x = -2$? Justify
   your responses.   No—even multiplicity; yes—odd multiplicity

Graph using the *Guidelines for Graphing Rational Functions.*

27. $v(x) = \dfrac{x^2 - 4x}{x^2 - 4}$ **28.** $t(x) = \dfrac{2x^2}{x^2 - 5}$

29. Use the vertical asymptotes, x-intercepts, and their multiplicities to construct an equation that corresponds to the given graph. Be sure the y-intercept on the graph matches the value given by your equation. Assume these features are integer-valued. Check your work on a graphing calculator. $V(x) = \dfrac{x^2 - x - 12}{x^2 - x - 6}; V(0) = 2$

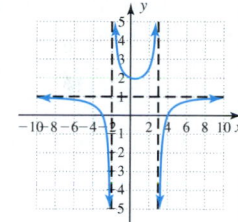

30. The average cost of producing a popular board game is given by the function
   $A(x) = \dfrac{5000 + 15x}{x}; x \geq 1000$. (a) Identify the horizontal asymptote of the function and
   explain its meaning in this context. (b) To be profitable, management believes the average cost must be below
   \$17.50. What levels of production will make the company profitable?   **a.** $y = 15$; as $|x| \to \infty$ $A(x) \to 15^+$.
   As production increases, average cost decreases and approaches 15.   **b.** $x > 2000$

SECTION 4.5   Additional Insights into Rational Functions

KEY CONCEPTS

- If $V = \dfrac{p(x)}{d(x)}$ is not in simplest form, with p and d sharing factors of the form $x - c$, the graph will have a
  removable discontinuity (a hole or gap) at $x = c$. The discontinuity can be "removed" (repaired) by redefining V
  using a piecewise-defined function.

- If $V = \dfrac{p(x)}{d(x)}$ is in simplest form, and the degree of p is greater than the degree of d, the graph will have an oblique
  or nonlinear asymptote, as determined by the quotient polynomial after division. If the degree of p is greater by 1,
  the result is a slant (oblique) asymptote. If the degree of p is greater by 2, the result is a parabolic asymptote.

EXERCISES

31. Determine if the graph of h will have a vertical asymptote or a removable discontinuity, then graph the function
   $h(x) = \dfrac{x^3 - 2x^2 - 9x + 18}{x - 2}$.

32. Sketch the graph of $h(x) = \dfrac{x^2 - 3x - 4}{x + 1}$. If there is a removable discontinuity, repair the break by redefining h
   using an appropriate piecewise-defined function.

Graph the functions using the *Guidelines for Graphing Rational Functions.*

33. $h(x) = \dfrac{x^2 - 2x}{x - 3}$ **34.** $t(x) = \dfrac{x^3 - 7x + 6}{x^2}$

Additional answers can be found in the Instructor Answer Appendix.

35. The cost to make x thousand party favors is given by $C(x) = x^2 - 2x + 6$, where $x \geq 1$ and C is in thousands of dollars. For the average cost of production $A(x) = \dfrac{x^2 - 2x + 6}{x}$, (a) graph the function, (b) use the graph to estimate the level of production that will make average cost a minimum, and (c) state the average cost of a single party favor at this level of production.   **b.** about 2450 favors   **c.** about $2.90 ea.

SECTION 4.6   Polynomial and Rational Inequalities

KEY CONCEPTS

- To solve polynomial inequalities, write $P(x)$ in factored form and note the multiplicity of each real zero.
- Plot real zeroes on a number line. The graph will cross the x-axis at zeroes of odd multiplicity (P will change sign), and bounce off the axis at zeroes of even multiplicity (P will not change sign).
- Use the end-behavior, y-intercept, or a test point to determine the sign of P in a given interval, then label all other intervals as $P(x) > 0$ or $P(x) < 0$ by analyzing the multiplicity of neighboring zeroes. Use the resulting diagram to state the solution.
- The solution process for rational inequalities and polynomial inequalities is virtually identical, considering that vertical asymptotes also create intervals where function values may change sign, depending on their multiplicity.
- Polynomial and rational inequalities can also be solved using an interval test method. Since polynomials and rational functions are continuous on their domains, the sign of the function at any one point in an interval will be the same as for all other points in that interval.

EXERCISES

Solve each inequality indicated using a number line and the behavior of the graph at each zero.

36. $x^3 + x^2 > 10x - 8$ **37.** $\dfrac{x^2 - 3x - 10}{x - 2} \geq 0$ **38.** $\dfrac{x}{x - 2} \leq \dfrac{-1}{x}$

 **PRACTICE TEST**

1. Complete the square to write each function as a transformation. Then graph each function and label the vertex and all intercepts (if they exist).

 a. $f(x) = -x^2 + 10x - 16$

 b. $g(x) = \dfrac{1}{2}x^2 + 4x + 16$

2. The graph of a quadratic function has a vertex of $(-1, -2)$, and passes through the origin. Find the other intercept, and the equation of the graph in standard form. $(-2, 0), y = 2x^2 + 4x$

3. Suppose the function $d(t) = t^2 - 14t$ models the depth of a scuba diver at time t, as she dives underwater from a steep shoreline, reaches a certain depth, and swims back to the surface.

 a. What is her depth after 4 sec? After 6 sec?   40 ft, 48 ft

 b. What was the maximum depth of the dive?   49 ft

 c. How many seconds was the diver beneath the surface?   14 sec

4. Compute the quotient using long division:

 $\dfrac{x^3 - 3x^2 + 5x - 2}{x^2 + 2x + 1}$. $x - 5 + \dfrac{14x + 3}{x^2 + 2x + 1}$

5. Find the quotient and remainder using synthetic division: $\dfrac{x^3 + 4x^2 - 5x - 20}{x + 2}$. $x^2 + 2x - 9 + \dfrac{-2}{x + 2}$

6. Use the remainder theorem to show $(x + 3)$ is a factor of $x^4 - 15x^2 - 10x + 24$.

7. Given $f(x) = 2x^3 + 4x^2 - 5x + 2$, find the value of $f(-3)$ using synthetic division and the remainder theorem. -1

8. Given $x = 2$ and $x = 3i$ are two zeroes of a real polynomial $P(x)$ with degree 3. Use the factor theorem to find $P(x)$. $P(x) = x^3 - 2x^2 + 9x - 18$

9. Factor the polynomial and state the multiplicity of each zero: $Q(x) = (x^2 - 3x + 2)(x^3 - 2x^2 - x + 2)$. $Q(x) = (x - 2)^2(x - 1)^2(x + 1)$, 2 mult 2, 1 mult 2, -1 mult 1

10. Given $C(x) = x^4 + x^3 + 7x^2 + 9x - 18$, (a) use the rational zeroes theorem to list all possible rational zeroes; (b) apply Descartes' rule of signs to count the number of possible positive, negative, and complex zeroes; and (c) use this information along with the tests for 1 and -1, synthetic division, and the factor theorem to factor C completely.

Additional answers can be found in the Instructor Answer Appendix.

11. Over a 10-yr period, the balance of payments (deficit versus surplus) for a small county was modeled by the function $f(x) = \frac{1}{2}x^3 - 7x^2 + 28x - 32$, where $x = 0$ corresponds to 2000 and $f(x)$ is the deficit or surplus in millions of dollars. (a) Use the rational roots theorem and synthetic division to find the years the county "broke even" (debt = surplus = 0) from 2000 to 2010. (b) How many years did the county run a surplus during this period? (c) What was the surplus/deficit in 2007?

12. Sketch the graph of $f(x) = (x - 3)(x + 1)^3(x + 2)^2$ using the degree, end-behavior, x- and y-intercepts, zeroes of multiplicity, and a few "midinterval" points.

13. Use the *Guidelines for Graphing Polynomials* to graph $g(x) = x^4 - 9x^2 - 4x + 12$.

14. Use the *Guidelines for Graphing Rational Functions* to graph $h(x) = \dfrac{x - 2}{x^2 - 3x - 4}$.

15. Suppose the cost of cleaning contaminated soil from a dump site is modeled by $C(x) = \dfrac{300x}{100 - x}$, where $C(x)$ is the cost (in \$1000s) to remove $x\%$ of the contaminants. Graph using $x \in [0, 100]$, and use the graph to answer the following questions.

   a. What is the significance of the *vertical asymptote* (what does it mean in this context)?

   b. If EPA regulations are changed so that 85% of the contaminants must be removed, instead of the 80% previously required, how much additional cost will the new regulations add? Compare the cost of the 5% increase from 80% to 85% with the cost of the 5% increase from 90% to 95%. What do you notice?

   c. What percent of the pollutants can be removed if the company budgets \$2,200,000?

16. Graph using the *Guidelines for Graphing Rational Functions*.

   a. $r(x) = \dfrac{x^3 - x^2 - 9x + 9}{x^2}$

   b. $R(x) = \dfrac{x^3 + 7x - 6}{x^2 - 4}$

17. Find the level of production that will minimize the average cost of an item, if production costs are modeled by $C(x) = 2x^2 + 25x + 128$, where $C(x)$ is the cost to manufacture x hundred items.   800

18. Solve each inequality

   a. $x^3 - 13x \le 12$ **b.** $\dfrac{3}{x - 2} < \dfrac{2}{x}$

19. Suppose the concentration of a chemical in the bloodstream of a large animal h hr after injection into muscle tissue is modeled by the formula
$$C(h) = \frac{2h^2 + 5h}{h^3 + 55}.$$

   a. Sketch a graph of the function for the intervals $x \in [-5, 20]$, $y \in [0, 1]$.

   b. Where is the vertical asymptote? Does it play a role in this context?

   c. What is the concentration after 2 hr? After 8 hr?

   d. How long does it take the concentration to fall below 20% $[C(h) < 0.2]$?

   e. When does the maximum concentration of the chemical occur? What is this maximum?

   f. Describe the significance of the horizontal asymptote in this context.

20. Use the vertical asymptotes, x-intercepts, and their multiplicities to construct an equation that corresponds to the given graph. Be sure the y-intercept on the graph matches the value given by your equation. Assume these features are integer-valued. Check your work on a graphing calculator. $V(x) = \dfrac{x^2 + x - 6}{x^2 - 2x - 3}$; $V(0) = 2$

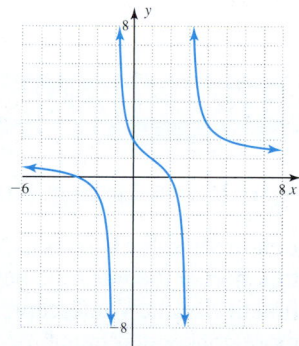

CALCULATOR EXPLORATION AND DISCOVERY

Complex Zeroes, Repeated Zeroes, and Inequalities

This *Calculator Exploration and Discovery* will explore the relationship between the solution of a polynomial (or rational) inequality and the complex zeroes and repeated zeroes of the related function. After all, if complex zeroes can never create an x-intercept, how do they affect the function? And if a zero of even multiplicity never crosses the x-axis (always bounces), can it still affect a nonstrict (*less than or equal to* or *greater than or equal to*) inequality? These are interesting and important questions, with numerous avenues of exploration. To begin, consider the function

$Y_1 = (x + 3)^2(x^3 - 1)$. In completely factored form $Y_1 = (x + 3)^2(x - 1)(x^2 + x + 1)$. This is a polynomial function of degree 5 with two real zeroes (one repeated), two complex zeroes (the quadratic factor is irreducible), and after viewing the graph on Figure 4.99, four turning points. From the graph (or by analysis), we have $Y_1 \leq 0$ for $x \leq 1$. Now let's consider $Y_2 = (x + 3)^2(x - 1)$, the same function as Y_1, less the quadratic factor. Since complex zeroes never "cross the x-axis" anyway, the removal of this factor *cannot affect the solution set of the inequality!* But how does it affect the function? Y_2 is now a function of degree three, with three real zeroes (one repeated) and only two turning points (Figure 4.100). But even so, the solution to $Y_2 \leq 0$ is the same as for $Y_1 \leq 0$: $x \leq 1$. Finally, let's look at $Y_3 = x - 1$, the same function as Y_2 but with the repeated zero removed. The key here is to notice that since $(x - 3)^2$ will be nonnegative for any value of x, it too does not change the solution set of the "less than or equal to inequality," only the shape of the graph. Y_3 is a function of degree 1, with one real zero and no turning points, *but the solution interval for $Y_3 \leq 0$ is the same solution interval as Y_2 and Y_1: $x \leq 1$* (see Figure 4.101).

Figure 4.99	Figure 4.100	Figure 4.101
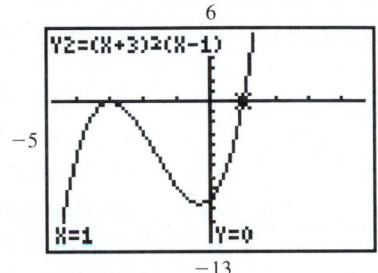		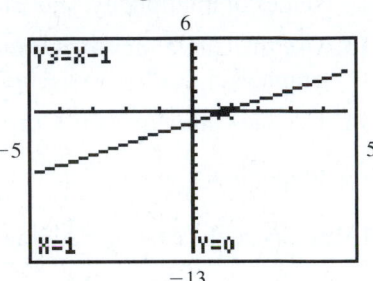

Explore these relationships further using the exercises in (A) and (B) given, using a "greater than or equal to" inequality. Begin by writing Y_1 in completely factored form.

(A) $Y_1 = (x^3 - 6x^2 + 32)(x^2 + 1)$ **(B)** $Y_1 = (x + 3)^2(x^3 - 2x^2 + x - 2)$
$\quad\ Y_2 = x^3 - 6x^2 + 32$ $\qquad\ Y_2 = (x + 3)^2(x - 2)$
$\quad\ Y_3 = x + 2$ $\qquad\ Y_3 = x - 2$

Exercise 1: Based on what you've noticed, comment on how the irreducible quadratic factors of a polynomial affect its graph. What role do they play in the solution of inequalities?   They do not affect the solution set.

Exercise 2: How do zeroes of even multiplicity affect the solution set of nonstrict inequalities (less/greater than or equal to)?   They do not affect the solution set.

 **STRENGTHENING CORE SKILLS**

Solving Inequalities Using the Push Principle

The most common method for solving polynomial inequalities involves finding the zeroes of the function and checking the sign of the function in the intervals between these zeroes. In Section 4.6, we relied on the end-behavior of the graph, the sign of the function at the y-intercept, and the multiplicity of the zeroes to determine the solution. There is a third method that is more conceptual in nature, but in many cases highly efficient. It is based on two very simple ideas, the first involving only order relations and the number line:

A. Given any number x and constant $k > 0$: $x > x - k$ and $x < x + k$.

$$x - 4 < x \qquad x < x + 3$$

This statement simply reinforces the idea that if a is left of b on the number line, then $a < b$. As shown in the diagram, $x - 4 < x$ and $x < x + 3$, from which $x - 4 < x + 3$ for any x.

B. The second idea reiterates well-known ideas regarding the multiplication of signed numbers. For any number of factors:

if there is an even number of negative factors, the result is positive;
if there is an odd number of negative factors, the result is negative.

These two ideas work together to solve inequalities using what we'll call the *push principle*. Consider the inequality $x^2 - x - 12 > 0$. The factored form is $(x - 4)(x + 3) > 0$ and we want the product of these two factors to be positive. From (A), both factors will be positive if $(x - 4)$ is positive, since it's the smaller of the two; and both factors will be negative if $x + 3 < 0$, since it's the larger. The solution set is found by solving these two simple inequalities: $x - 4 > 0$ gives $x > 4$ and $x + 3 < 0$ gives $x < -3$. If the inequality were $(x - 4)(x + 3) < 0$ instead, we require one negative factor and one positive factor. Due to order relations and the number line, the larger factor must be the positive one: $x + 3 > 0$ so $x > -3$. The smaller factor must be the negative one: $x - 4 < 0$ and $x < 4$. This gives the solution $-3 < x < 4$ as can be verified using any alternative method. Solutions to all other polynomial and rational inequalities are an extension of these two cases.

Illustration 1 ▶ Solve $x^3 - 7x + 6 < 0$ using the push principle.

Solution ▶ The polynomial can be factored using the tests for 1 and -1 and synthetic division. The factors are $(x - 2)(x - 1)(x + 3) < 0$, which we've conveniently written in increasing order. For the product of three factors to be negative we require: (1) three negative factors or (2) one negative and two positive factors. The first condition is met by simply making the largest factor negative, as it will ensure the smaller factors are also negative: $x + 3 < 0$ so $x < -3$. The second condition is met by making the smaller factor negative and the "middle" factor positive: $x - 2 < 0$ *and* $x - 1 > 0$. The second solution interval is $x < 2$ and $x > 1$, or $1 < x < 2$.

Note the push principle does not require the testing of intervals between the zeroes, nor the "cross/bounce" analysis at the zeroes and vertical asymptotes (of rational functions). In addition, irreducible quadratic factors can still be ignored as they contribute nothing to the solution of real inequalities, and factors of even multiplicity can be overlooked precisely because there is no sign change at these roots.

Illustration 2 ▶ Solve $(x^2 + 1)(x - 2)^2(x + 3) \geq 0$ using the push principle.

Solution ▶ Since the factor $(x^2 + 1)$ does not affect the solution set, this inequality will have the same solution as $(x - 2)^2(x + 3) \geq 0$. Further, since $(x - 2)^2$ will be nonnegative for all x, the original inequality *has the same solution set as* $(x + 3) \geq 0$! The solution is $x \geq -3$.

With some practice, the push principle can be a very effective tool. Use it to solve the following exercises. Check all solutions by graphing the function on a graphing calculator.

Exercise 1: $x^3 - 3x - 18 \leq 0$ $\quad x \in (-\infty, 3]$

Exercise 2: $\dfrac{x + 1}{x^2 - 4} > 0$ $\quad x \in (-2, -1) \cup (2, \infty)$

Exercise 3: $x^3 - 13x + 12 < 0$ $\quad x \in (-\infty, -4) \cup (1, 3)$

Exercise 4: $x^3 - 3x + 2 \geq 0$ $\quad x \in [-2, \infty)$

Exercise 5: $x^4 - x^2 - 12 > 0$ $\quad x \in (-\infty, -2) \cup (2, \infty)$

Exercise 6: $(x^2 + 5)(x^2 - 9)(x + 2)^2(x - 1) \geq 0$
$\quad x \in [-3, 1] \cup [3, \infty)$

CUMULATIVE REVIEW CHAPTERS R–4

1. Solve for R: $\dfrac{1}{R} = \dfrac{1}{R_1} + \dfrac{1}{R_2}$ $\quad R = \dfrac{R_1 R_2}{R_1 + R_2}$

2. Solve for x: $\dfrac{2}{x + 1} + 1 = \dfrac{5}{x^2 - 1}$ $\quad x = -4 \text{ or } x = 2$

3. Factor each expression completely:
   a. $x^3 - 1$ $\qquad$ **b.** $x^3 - 3x^2 - 4x + 12$

4. Solve using the quadratic formula. Write answers in both exact and approximate form: $2x^2 + 4x + 1 = 0$.

5. Solve the following inequality: $x + 3 < 5$ *or* $5 - x < 4$. $\quad$ all reals

6. Name the eight toolbox functions, give their equations, then draw a sketch of each.

7. Use substitution to verify that $x = 2 - 3i$ is a solution to $x^2 - 4x + 13 = 0$. $\quad$ verified

8. Solve the rational inequality:
$\dfrac{x + 4}{x - 2} < 3$. $\quad x \in (-\infty, 2) \cup (5, \infty)$

9. As part of a study on traffic conditions, the mayor of a small city tracks her driving time to work each day for six months and finds a linear and increasing relationship. On day 1, her drive time was 17 min. By day 61 the drive time had increased to 28 min. Find a linear function that models the drive time and use it to estimate the drive time on day 121, if the trend continues. Explain what the slope of the line means in this context.

10. Does the relation shown represent a function? If not, discuss/explain why not.

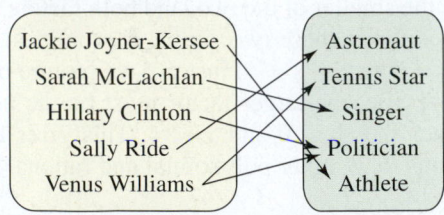

Jackie Joyner-Kersee	Astronaut
Sarah McLachlan	Tennis Star
Hillary Clinton	Singer
Sally Ride	Politician
Venus Williams	Athlete

No, Venus Williams is paired with both Tennis Star and politician.

11. The data given shows the profit of a new company for the first 6 months of business, and is closely modeled by the function $p(m) = 1.18x^2 - 10.99x + 4.6$; where $p(m)$ is the profit earned in month m. Assuming this trend continues, use this function to find the first month a profit will be earned ($p > 0$).   month 9

Exercise 11

Month	Profit (1000s)
1	−5
2	−13
3	−18
4	−20
5	−21
6	−19

12. Graph the function $g(x) = \dfrac{-1}{(x + 2)^2} + 3$ using transformations of a basic function.

13. Find $f^{-1}(x)$, given $f(x) = \sqrt[3]{2x - 3}$, then use composition to verify your inverse is correct.

14. Graph $f(x) = x^2 - 4x + 7$ by completing the square, then state intervals where:

   a. $f(x) \geq 0$ **b.** $f(x)\uparrow$

15. Given the graph of a general function $f(x)$, graph $F(x) = -f(x + 1) + 2$.

Exercise 15

16. Graph the piecewise-defined function given:

$$f(x) = \begin{cases} -3 & x < -1 \\ x & -1 \leq x \leq 1 \\ 3x & x > 1 \end{cases}$$

17. Y varies directly with X and inversely with the square of Z. If $Y = 10$ when $X = 32$ and $Z = 4$, find X when $Z = 15$ and $Y = 1.4$. $X = 63$

18. Use the rational zeroes theorem and synthetic division to find all zeroes (real and complex) of $f(x) = x^4 - 2x^2 + 16x - 15$. $x = 1, -3, 1 \pm 2i$

19. Sketch the graph of $f(x) = x^3 - 3x^2 - 6x + 8$.

20. Sketch the graph of $h(x) = \dfrac{x - 1}{x^2 - 4}$ and use the zeroes and vertical asymptotes to solve $h(x) \geq 0$.

Exercises 21 through 25 require the use of a graphing calculator.

21. Given $Y_1 = 1.47x^3 - 0.51x^2 + 1.99$, use your calculator to evaluate $Y_1(4)$, $Y_1(-3.71)$, and $Y_1(0.09)$. Round your answers to two decimal places if necessary.   87.91, −80.09, 1.99

22. Use a graphing calculator to locate the point(s) of intersection of the graph of $r(x) = \dfrac{1.12 - 3x^2}{1.5x^2 + 2.2x}$ with its horizontal asymptote. Round your answer(s) to two decimal places if necessary. $(-0.25, -2)$

23. Use a calculator in complex mode to express the following complex numbers in $a + bi$ form. Round to the nearest hundredth if necessary.

   a. $(2.4 - 1.2i)^6$ $-349.36 - 131.38i$

   b. $\dfrac{7 - 6i}{4 + 5i}$ $-\dfrac{2}{41} - \dfrac{59}{41}i$

   c. $(0.3 + 8.2i)(1.9 - 3.3i)$ $27.63 + 14.59i$

   d. i^{15} $0 - i$

24. Identify intervals containing the zeroes of $f(x) = x^4 + 3x^3 - x^2 - 7x - 3$ by using the intermediate value theorem and the table feature of your graphing calculator. Use TblStart = −5 and ΔTbl = 0.1, so none of the intervals are greater than 0.1 units wide. Assume all real zeroes are between −5 and 5. $[-2.5, -2.4], [-1.6, -1.5], [-0.6, -0.5],$ and $[1.5, 1.6]$

25. Using the calculator screen shown, identify which function (Y_2 through Y_7) performs the indicated transformation of Y_1. Verify your answers with your calculator.

```
Plot1  Plot2  Plot3
\Y1 ■√(X)
\Y2 ■Y1(-1X)
\Y3 ■Y1(X+1)
\Y4 ■Y1(X-1)
\Y5 ■-1Y1(X)
\Y6 ■Y1(X)+1
\Y7 ■Y1(X)-1
```

   a. Y_1 shifted up 1 unit Y_6

   b. Y_1 shifted down 1 unit Y_7

   c. Y_1 shifted right 1 unit Y_4

   d. Y_1 shifted left 1 unit Y_3

   e. Y_1 reflected across x-axis Y_5

   f. Y_1 reflected across y-axis Y_2

Exponential and Logarithmic Functions

CHAPTER OUTLINE

CHAPTER CONNECTIONS

The largest purchase that most individuals will make in their lifetime is that of a car or home. The monthly payment P required to amortize (pay off) the loan can be calculated using the formula

$$P = \frac{AR}{1 - (1 - R)^{-12t}}$$

where A is the amount financed; t is the time in years; and $R = \dfrac{r}{12}$, where r is the annual rate of interest. This study of exponential and logarithmic functions will help you become a more knowledgeable consumer. This application appears as Exercise 53 in Section 5.6.

Check out these other real-world connections:

▶ Effects of Inflation
(Section 5.2, Exercises 91 and 92)

▶ Intensity of Sound
(Section 5.3, Exercises 89 to 94)

▶ Proper Ventilation of a Home
(Section 5.3, Exercise 101)

▶ Freezing Time for Water Puddles
(Section 5.5, Exercise 58)

5.1 One-to-One and Inverse Functions

LEARNING OBJECTIVES

In Section 5.1 you will see how we can:

- ☐ **A.** Identify one-to-one functions
- ☐ **B.** Explore inverse functions using ordered pairs
- ☐ **C.** Find inverse functions using an algebraic method
- ☐ **D.** Graph a function and its inverse
- ☐ **E.** Solve applications of inverse functions

Consider the function $f(x) = 2x - 3$. If $f(x) = 7$, the equation becomes $2x - 3 = 7$, and the corresponding value of x can be found using *inverse operations*. In this section, we introduce the concept of an *inverse function,* which can be viewed as a formula for finding x-values that correspond to *any* given value of $f(x)$.

A. Identifying One-to-One Functions

The graphs of $y = 2x$ and $y = x^2$ are shown in Figures 5.1 and 5.2. The dashed, vertical lines clearly indicate both are functions, with each x-value corresponding to only one y. But the points on $y = 2x$ have one characteristic those from $y = x^2$ do not—*each y-value also corresponds to only one x* (for $y = x^2$, 4 corresponds to both -2 and 2). If each element from the range of a function corresponds to only one element of the domain, the function is said to be **one-to-one.**

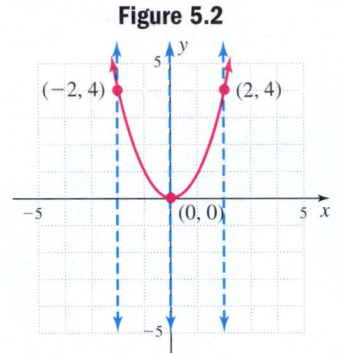

Figure 5.1

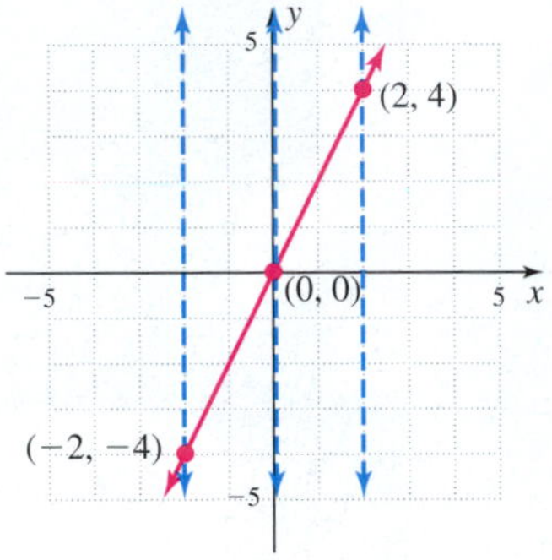

Figure 5.2

One-to-One Functions

A function f is one-to-one if every element in the range, corresponds to only one element of the domain.

In symbols, if $f(x_1) = f(x_2)$ then $x_1 = x_2$, or if $x_1 \neq x_2$, then $f(x_1) \neq f(x_2)$.

From this definition, we note the graph of a one-to-one function must not only pass a vertical line test (to show each x corresponds to only one y), but also pass a **horizontal line test** (to show each y corresponds to only one x).

Horizontal Line Test

If every horizontal line intersects the graph of a function in at most one point, the function is one-to-one.

Notice the graph of $y = 2x$ (Figure 5.3) passes the horizontal line test, while the graph of $y = x^2$ (Figure 5.4) does not.

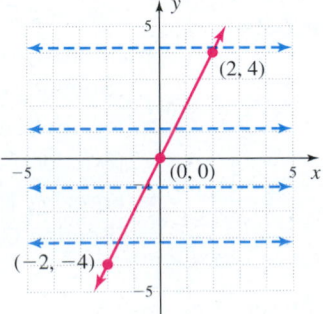

Figure 5.3

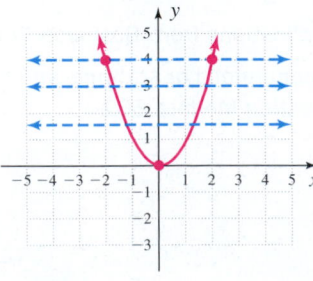

Figure 5.4

If the function is pointwise-defined or given in ordered pair form, we simply check to see that no given second coordinate is paired with more than one first coordinate.

EXAMPLE 1 ▶   Identifying One-to-One Functions

Determine whether each graph or relation shown depicts a function. If so, determine whether the function is one-to-one.

a.
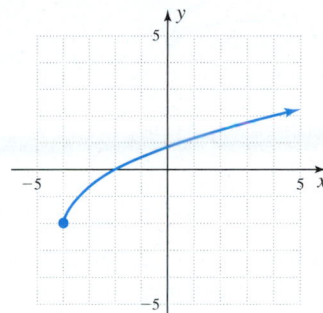

b.
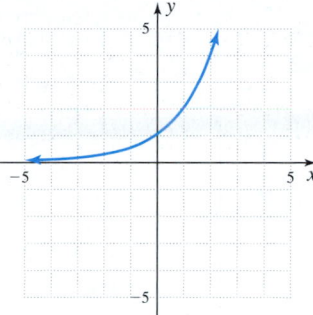

c.
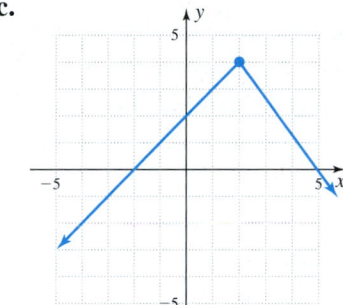

d.
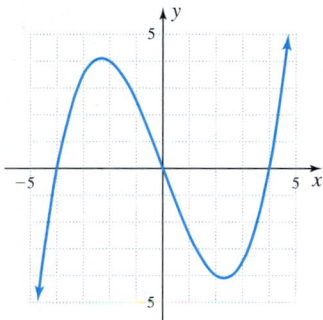

e.
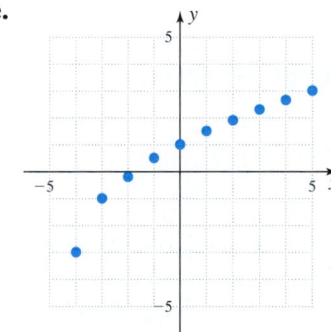

f. $\{(-4, 1), (-3, 5), (-1, 1), (-2, 3), (-3, -2), (4, 4)\}$

Solution ▶   A careful inspection shows all five graphs depict functions, since each passes the vertical line test [the relation in (f) is not a function since the input -3 corresponds to two outputs]. Only (a), (b), and (e) pass the horizontal line test and are *one-to-one* functions.

☑ **A.** You've just seen how we can identify one-to-one functions

Now try Exercises 7 through 26 ▶

B. Inverse Functions and Ordered Pairs

Consider the function $f(x) = 2x - 3$ and the solutions shown in Table 5.1. Figure 5.5 shows this function in diagram form (in blue), and illustrates that for each element of the domain, we *multiply by 2, then subtract 3*. An **inverse function** for f is one that takes the result of these operations (elements of the range), and returns the original

Table 5.1

x	$f(x)$
-3	-9
0	-3
2	1
5	7
8	13

Table 5.2

x	$F(x)$
-9	-3
-3	0
1	2
7	5
13	8

domain element. Figures 5.5 and 5.6 show that function F achieves this by "undoing" the operations in reverse order: *add 3, then divide by 2* (in red). A table of values for $F(x)$ is shown (Table 5.2).

Figure 5.5

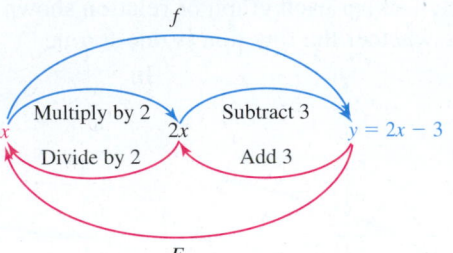

Figure 5.6

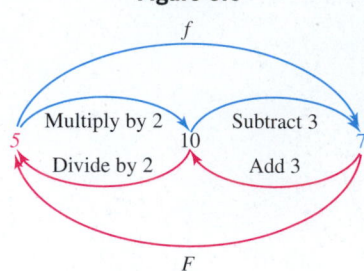

From this illustration we make the following observations regarding an inverse function, which we actually denote as $f^{-1}(x)$.

Inverse Functions

If f is a one-to-one function with ordered pairs (a, b),

1. $f^{-1}(x)$ is a one-to-one function with ordered pairs (b, a).
2. The range of f will be the domain of $f^{-1}(x)$.
3. The domain of f will be the range of $f^{-1}(x)$.

⚠️ **CAUTION** ▶ The notation $f^{-1}(x)$ is simply a way of denoting an inverse function and has nothing to do with exponential properties. In particular, $f^{-1}(x)$ does *not* mean $\frac{1}{f(x)}$.

EXAMPLE 2 ▶ **Finding the Inverse of a Function**

Find the inverse of each one-to-one function given:

a. $f(x) = \{(-4, 13), (-1, 7), (0, 5), (2, 1), (5, -5), (8, -11)\}$
b. $p(x) = -3x + 2$

Solution ▶ **a.** When a function is defined as a set of ordered pairs, the inverse function is found by simply interchanging the x- and y-coordinates:
$f^{-1}(x) = \{(13, -4), (7, -1), (5, 0), (1, 2), (-5, 5), (-11, 8)\}$.

b. Using diagrams similar to Figures 5.5 and 5.6, we reason that $p^{-1}(x)$ will subtract 2, then divide the result by -3: $p^{-1}(x) = \frac{x - 2}{-3}$. As a test, we find that $(-2, 8)$, $(0, 2)$, and $(3, -7)$ are solutions to $p(x)$, and note that $(8, -2)$, $(2, 0)$, and $(-7, 3)$ are indeed solutions to $p^{-1}(x)$.

✓ **B.** You've just seen how we can explore inverse functions using ordered pairs

Now try Exercises 27 through 38 ▶

C. Finding Inverse Functions Using an Algebraic Method

The fact that interchanging x- and y-values helps determine an inverse function can be generalized to develop an **algebraic method** for finding inverses. Instead of interchanging *specific x- and y-values*, we actually interchange the x- and y-*variables*, then solve the equation for y. The process is summarized here.

Finding an Inverse Function

1. Use y instead of $f(x)$.
2. Interchange x and y.
3. Solve the new equation for y.
4. The result gives the inverse function: substitute $f^{-1}(x)$ for y.

In this process, it might seem like we're using the *same y* to represent two different functions. To see why there is actually no contradiction, **see Exercise 97.**

EXAMPLE 3 ▶ **Finding Inverse Functions Algebraically**

State the domain and range of the function given, then use the algebraic method to find the inverse function, and state *its* domain and range.

a. $f(x) = \sqrt[3]{x + 5}$ **b.** $g(x) = \dfrac{2x}{x + 1}$

Solution ▶   **a.** $f(x) = \sqrt[3]{x + 5},\ x \in \mathbb{R},\ y \in \mathbb{R}$

$\qquad\qquad y = \sqrt[3]{x + 5}$ use y instead of $f(x)$

$\qquad\qquad x = \sqrt[3]{y + 5}$ interchange x and y

$\qquad\qquad x^3 = y + 5$ cube both sides

$\qquad\qquad x^3 - 5 = y$ solve for y

$\qquad\qquad x^3 - 5 = f^{-1}(x)$ the result is $f^{-1}(x)$

$\qquad\qquad f^{-1}(x) = x^3 - 5,\ x \in \mathbb{R},\ y \in \mathbb{R}$

  b. $g(x) = \dfrac{2x}{x + 1},\ x \neq -1,\ y \neq 2$

$\qquad\qquad y = \dfrac{2x}{x + 1}$ use y instead of $f(x)$

$\qquad\qquad x = \dfrac{2y}{y + 1}$ interchange x and y

$\qquad\qquad xy + x = 2y$ multiply by $y + 1$ and distribute

$\qquad\qquad x = 2y - xy$ gather terms with y

$\qquad\qquad x = y(2 - x)$ factor

$\qquad\qquad \dfrac{x}{2 - x} = y$ solve for y

$\qquad\qquad g^{-1}(x) = \dfrac{x}{2 - x},\ x \neq 2,\ y \neq -1$

Now try Exercises 39 through 46 ▶

In cases where a given function is *not* one-to-one, we can sometimes restrict the domain to create a function that *is,* and then determine an inverse. The restriction we use is arbitrary, and only requires that the result produce all possible range values. Most often, we simply choose a limited domain that seems convenient or reasonable.

EXAMPLE 4 ▶ **Restricting the Domain to Create a One-to-One Function**

Given $f(x) = (x - 4)^2$, restrict the domain to create a one-to-one function, then find $f^{-1}(x)$. State the domain and range of both resulting functions.

Solution ▶ The graph of f is a parabola, opening upward with the vertex at $(4, 0)$. Restricting the domain to $x \geq 4$ (see figure) leaves only the "right branch" of the parabola, creating a one-to-one function without affecting the range, $y \in [0, \infty)$. For $f(x) = (x - 4)^2$ with restricted domain $x \geq 4$, we have

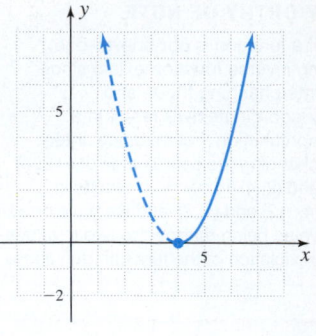

$$
\begin{array}{ll}
f(x) = (x - 4)^2 & \text{given function} \\
y = (x - 4)^2 & \text{use } y \text{ instead of } f(x) \\
x = (y - 4)^2 & \text{interchange } x \text{ and } y \\
\pm\sqrt{x} = y - 4 & \text{take square roots} \\
\sqrt{x} + 4 = y & \text{solve for } y, \text{ use } \sqrt{x} \text{ since } x \geq 4
\end{array}
$$

The result shows $f^{-1}(x) = \sqrt{x} + 4$, with domain $x \in [0, \infty)$ and range $y \in [4, \infty)$ (the domain of f becomes the range of f^{-1}, and the range of f becomes the domain of f^{-1}).

Now try Exercises 47 through 52 ▶

We can further illustrate the ideas from Example 4 using a calculator's ability to draw inverses. On the [Y=] screen, set $Y_1 = X^2$ and [GRAPH] this function on a [WINDOW] size of $[-7.5, 7.5]$ for x and $[-3, 7]$ for y. Then go to the home screen and access the **DrawInv** feature using [2nd] [PRGM] (**DRAW**) 8:DrawInv. This will place the **DrawInv** feature on the home screen, where we specify that we want the inverse of Y_1 (Figure 5.7). Pressing [ENTER] returns us to the graph, where we discover that since $y = x^2$ is *not one-to-one,* the calculator has graphed the *inverse relation* (Figure 5.8).

Figure 5.7

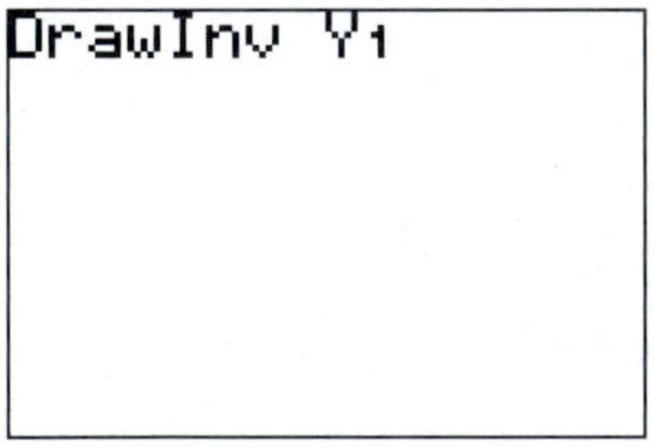

Figure 5.8

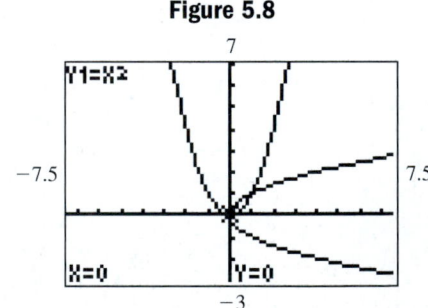

Returning to the [Y=] screen and restricting the domain of Y_1 to $x \geq 0$ (Figure 5.9), then repeating the sequence above, produces the graphs shown in Figure 5.10. Note the given function is now one-to-one, and its inverse is also now a function (and one-to one).

Figure 5.9

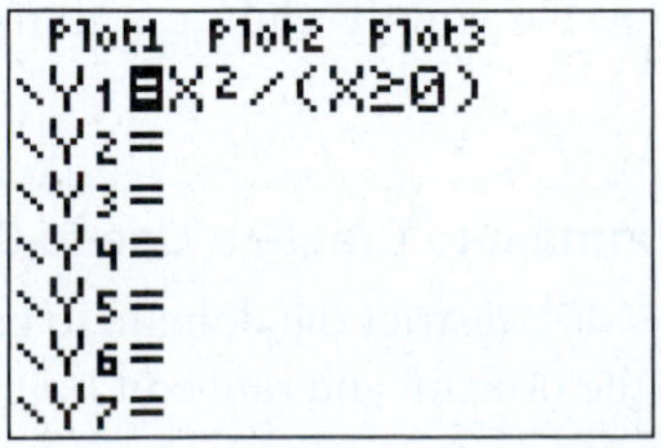

Figure 5.10

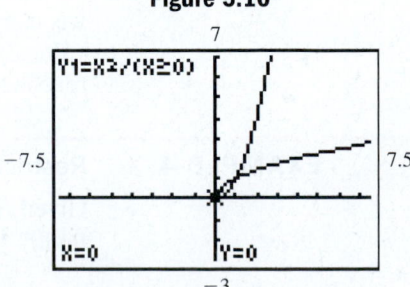

While we now have the ability to *find* the inverse of a function, we still lack a definitive method of *verifying* the inverse is correct. Actually, the diagrams in Figures 5.5 and 5.6 suggest just such a method. If we use the function f itself as an input for f^{-1}, or the function f^{-1} as an input for f, the end result should simply be x, as each function "undoes" the operations of the other. From Section 3.6 this is called a composition of functions and using the notation for composition we have,

> **Verifying Inverse Functions**
>
> If f is a one-to-one function, then the function f^{-1} exists, where
> $$(f \circ f^{-1})(x) = x \qquad \text{and} \qquad (f^{-1} \circ f)(x) = x$$

EXAMPLE 5 ▶ **Finding and Verifying an Inverse Function**

Use the algebraic method to find the inverse function for $f(x) = \sqrt{x + 2}$. Then verify the inverse you found is correct.

Solution ▶ Since the graph of f is the graph of $y = \sqrt{x}$ shifted 2 units left, we know f is one-to-one with domain $x \in [-2, \infty)$ and range $y \in [0, \infty)$. This is important since the *domain and range values will be interchanged for the inverse function*. The domain of f^{-1} will be $x \in [0, \infty)$ and its range $y \in [-2, \infty)$.

$f(x) = \sqrt{x + 2}$	given function; $x \geq -2$
$y = \sqrt{x + 2}$	use y instead of $f(x)$
$x = \sqrt{y + 2}$	interchange x and y
$x^2 = y + 2$	solve for y (square both sides)
$x^2 - 2 = y$	subtract 2
$f^{-1}(x) = x^2 - 2$	the result is $f^{-1}(x)$; D: $x \in [0, \infty)$, R: $y \in [-2, \infty)$

Verify ▶

$$
\begin{aligned}
(f \circ f^{-1})(x) &= f[f^{-1}(x)] && f^{-1}(x) \text{ is an input for } f \\
&= \sqrt{f^{-1}(x) + 2} && f \text{ adds 2 to inputs, then takes the square root} \\
&= \sqrt{(x^2 - 2) + 2} && \text{substitute } x^2 - 2 \text{ for } f^{-1}(x) \\
&= \sqrt{x^2} && \text{simplify} \\
&= x \checkmark && \text{since the domain of } f^{-1}(x) \text{ is } x \in [0, \infty)
\end{aligned}
$$

$$
\begin{aligned}
(f^{-1} \circ f)(x) &= f^{-1}[f(x)] && f(x) \text{ is an input for } f^{-1} \\
&= [f(x)]^2 - 2 && f^{-1} \text{ squares inputs, then subtracts 2} \\
&= [\sqrt{x + 2}]^2 - 2 && \text{substitute } \sqrt{x + 2} \text{ for } f(x) \\
&= x + 2 - 2 && \text{simplify} \\
&= x \checkmark && \text{result}
\end{aligned}
$$

☑ **C.** You've just seen how we can find inverse functions using an algebraic method

Now try Exercises 53 through 78 ▶

D. The Graph of a Function and Its Inverse

Graphing a function and its inverse on the same axes reveals an interesting and useful relationship—the graphs are reflections across the line $y = x$ (the identity function). Consider the function $f(x) = 2x + 3$, and its inverse $f^{-1}(x) = \dfrac{x - 3}{2} = \dfrac{1}{2}x - \dfrac{3}{2}$. In Figure 5.11, the points $(1, 5)$, $(0, 3)$, $(-\frac{3}{2}, 0)$, and $(-4, -5)$ from f (see Table 5.3) are graphed in blue, with the points $(5, 1)$, $(3, 0)$, $(0, -\frac{3}{2})$, and $(-5, -4)$ (see Table 5.4) from f^{-1} graphed in red (note the x- and y-values are reversed). Graphing both lines illustrates this symmetry (Figure 5.12).

Table 5.3

x	$f(x)$
1	5
0	3
$-\dfrac{3}{2}$	0
-4	-5

Figure 5.11

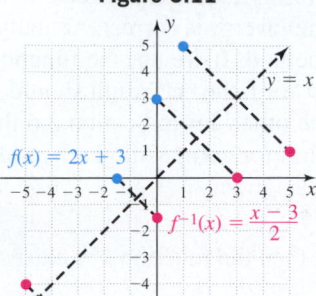

Figure 5.12

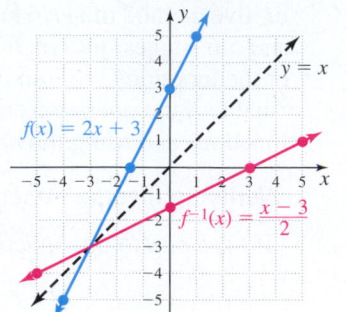

Table 5.4

x	$f^{-1}(x)$
5	1
3	0
0	$-\dfrac{3}{2}$
-5	-4

EXAMPLE 6 ▶ **Graphing a Function and Its Inverse**

Given the graph shown in Figure 5.13, draw a graph of the inverse function.

Figure 5.13

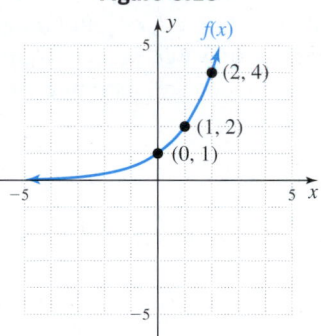

Figure 5.14

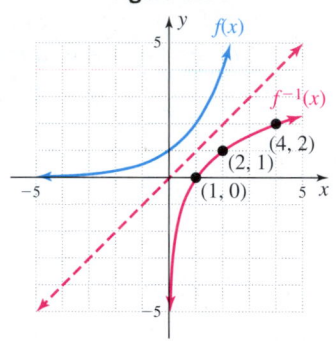

Solution ▶ From the graph, the domain of f appears to be $x \in \mathbb{R}$ and the range is $y \in (0, \infty)$. This means the domain of f^{-1} will be $x \in (0, \infty)$ and the range will be $y \in \mathbb{R}$. To sketch f^{-1}, draw the line $y = x$, interchange the x- and y-coordinates of the selected points, then plot these points and draw a smooth curve using the domain and range boundaries as a guide. The result is shown in Figure 5.14.

Now try Exercises 79 through 84 ▶

A summary of important concepts is provided here.

> **Functions and Inverse Functions**
>
> 1. If the graph of a function passes the horizontal line test, the function is one-to-one.
> 2. If a function f is one-to-one, the function f^{-1} exists.
> 3. The domain of f is the range of f^{-1}, and the range of f is the domain of f^{-1}.
> 4. For a function f and its inverse f^{-1}, $(f \circ f^{-1})(x) = x$ and $(f^{-1} \circ f)(x) = x$.
> 5. The graphs of f and f^{-1} are symmetric to the line $y = x$.

These ideas can be illustrated and reaffirmed using a graphing calculator. To begin, enter the functions $Y_1 = 2\sqrt[3]{X - 2}$ and $Y_2 = \dfrac{X^3}{8} + 2$ (which appear to be inverse functions) on the (Y=) screen, along with $Y_3 = X$. Press (ZOOM) **6:ZStandard**, then (ZOOM) **5:ZSquare** to obtain a screen that is in perspective. The graphs seem to be reflections across the line $y = x$ (Figure 5.15). To verify, we can use the **TABLE** feature. As shown in Figure 5.16, the points $(1, -2)$, $(2, 0)$ and $(3, 2)$ are on the graph of Y_1, and the points $(-2, 1)$, $(0, 2)$, and $(2, 3)$ are on the graph of Y_2. While this seems convincing (the x- and y-coordinates *are* interchanged), a graphing calculator can actually *compose the functions* to help verify the inverse relationship. After entering $Y_4 = Y_1(Y_2(X))$ and $Y_5 = Y_2(Y_1(X))$ on the Y= screen (deactivating Y_1, Y_2, and Y_3), the **TABLE** in Figure 5.17 now shows that one function indeed "undoes" the other, leaving $Y_4 = Y_5 = x$. **See Exercises 85 through 88.**

Figure 5.15

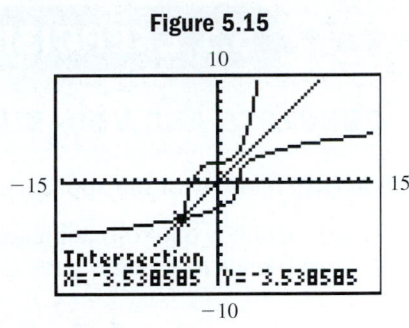

✓ D. You've just seen how we can graph a function and its inverse

Figure 5.16

X	Y1	Y2
-2	-3.175	1
-1	-2.884	1.875
0	-2.52	2
1	-2	2.125
2	0	3
3	2	5.375

X=

Figure 5.17

X	Y4	Y5
-2	-2	-2
-1	-1	-1
0	0	0
1	1	1
2	2	2
3	3	3
4	4	4

X=-2

E. Applications of Inverse Functions

Our final example illustrates one of the many ways that inverse functions can be applied.

EXAMPLE 7 ▶ **Using Volume to Understand Inverse Functions**

The volume of an equipoise cylinder (height equal to diameter) is given by $v(x) = 2\pi x^3$ (since $h = d = 2r$), where $v(x)$ represents the volume in units cubed and x represents the radius of the cylinder.

  a. Find the volume of such a cylinder if $x = 10$ ft.
  b. Find $v^{-1}(x)$, and discuss what the input and output variables represent.

Solution ▶ **a.** $v(x) = 2\pi x^3$ given function
  $v(10) = 2\pi(10)^3$ substitute 10 for *x*
  $\quad\quad = 2000\pi$ $10^3 = 1000$, exact form

With a radius of 10 ft, the volume of the cylinder would be 2000π ft^3.

  b. $v(x) = 2\pi x^3$ given function
  $\quad y = 2\pi x^3$ use *y* instead of *v(x)*
  $\quad x = 2\pi y^3$ interchange *x* and *y*
  $\quad \dfrac{x}{2\pi} = y^3$ solve for *y*
  $\quad \sqrt[3]{\dfrac{x}{2\pi}} = y$ result

The inverse function is $v^{-1}(x) = \sqrt[3]{\dfrac{x}{2\pi}}$. In this case, the input x is a given volume, the output $v^{-1}(x)$ is the radius of an equipoise cylinder that will hold this volume.

✓ E. You've just seen how we can solve applications of inverse functions

Now try Exercises 91 through 96 ▶

5.1 EXERCISES

▶ CONCEPTS AND VOCABULARY

Fill in each blank with the appropriate word or phrase. Carefully reread the section if needed.

1. A function is one-to-one if each ___second___ coordinate corresponds to exactly ___one___ first coordinate.

2. If every ___horizontal___ line intersects the graph of a function in at most ___1___ point, the function is one-to-one.

3. A certain function is defined by the ordered pairs $(-2, -11)$, $(0, -5)$, $(2, 1)$, and $(4, 19)$. The inverse function is ___$(-11, -2), (-5, 0), (1, 2), (19, 4)$___.

4. To find f^{-1} using the algebraic method, we (1) use ___y___ instead of $f(x)$, (2) ___interchange___ x and y, (3) ___solve___ for y and replace y with $f^{-1}(x)$.

5. State true or false and explain why: *To show that g is the inverse function for f, simply show that* $(f \circ g)(x) = x$. Include an example in your response.   False, answers will vary.

6. Discuss/Explain why no inverse function exists for $f(x) = (x + 3)^2$ and $g(x) = \sqrt{4 - x^2}$. How would the domain of each function have to be restricted to allow for an inverse function?   Neither function is one-to-one; one possibility is $f(x) = (x + 3)^2, x \geq -3$, and $g(x) = \sqrt{4 - x^2}, x \geq 0$.

▶ DEVELOPING YOUR SKILLS

Determine whether each graph given is the graph of a one-to-one function. If not, give examples of how the definition of one-to-oneness is violated.

7.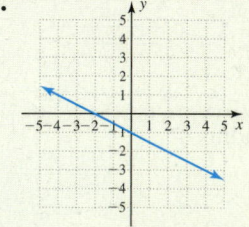

8.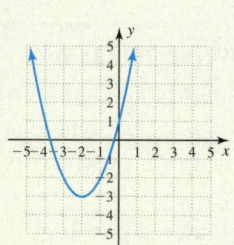

9.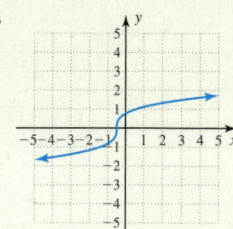

10.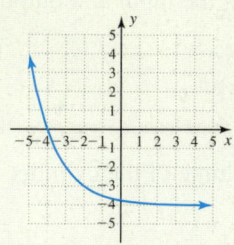

11.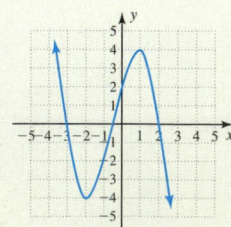

12.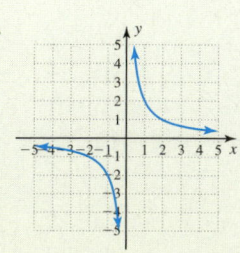

13.

14.

Determine whether the functions given are one-to-one. If not, state why.

15. $\{(-7, 4), (-1, 9), (0, 5), (-2, 1), (5, -5)\}$

16. $\{(9, 1), (-2, 7), (7, 4), (3, 9), (2, 7)\}$

17. $\{(-6, 1), (4, -9), (0, 11), (-2, 7), (-4, 5), (8, 1)\}$

18. $\{(-6, 2), (-3, 7), (8, 0), (12, -1), (2, -3), (1, 3)\}$

Determine if the functions given are one-to-one by noting the function family to which each belongs and mentally picturing the shape of the graph. If a function is not one-to-one, discuss how the definition of one-to-oneness is violated.

19. $f(x) = 3x - 5$

20. $g(x) = (x + 2)^3 - 1$

21. $h(x) = -|x - 4| + 3$

22. $p(t) = 3t^2 + 5$

23. $s(t) = \sqrt{2t - 1} + 5$

24. $r(t) = \sqrt[3]{t + 1} - 2$

25. $y = 3$

26. $y = -2x$

For Exercises 27 to 30, find the inverse function of the one-to-one functions given.

27. $f(x) = \{(-2, 1), (-1, 4), (0, 5), (2, 9), (5, 15)\}$

28. $g(x) = \{(-2, 30), (-1, 11), (0, 4), (1, 3), (2, 2)\}$

Additional answers can be found in the Instructor Answer Appendix.

HOMEWORK SELECTION GUIDE

Core: 7–87 every other odd, 89, 90, 91, 93, 95, 101, 104 (28 Exercises)
Standard: 1–4, 7–87 every other odd, 89–96, 98, 101, 104 (36 Exercises)

Extended: 1–4, 7–87 every other odd, 89–98, 101, 102, 104 (38 Exercises)
In Depth: 1–6, 7–87 every other odd, 89–98, 101–104 (41 Exercises)

29. $v(x)$ is defined by the ordered pairs shown.

X	Y₁
-4	3
-3	2
0	1
5	0
12	-1
21	-2
▓	-3

X=32

30. $w(x)$ is defined by the ordered pairs shown.

X	Y₁
-6	4
-5	2.5
-2	-2
0	-5
3	-9.5
4	-11
▓	-15.5

X=7

Find the inverse function using diagrams similar to those illustrated in Example 2. Check the result using three test points.

31. $f(x) = x + 5$ **32.** $g(x) = x - 4$

33. $p(x) = -\dfrac{4}{5}x$ **34.** $r(x) = \dfrac{3x}{4}$

35. $f(x) = 4x + 3$ **36.** $g(x) = 5x - 2$

37. $t(x) = \sqrt[3]{x - 4}$ **38.** $s(x) = \sqrt[3]{x + 2}$

State the domain and range of $f(x)$, then use the algebraic method to find the inverse function and state its domain and range. Finally, find any three ordered pairs (a, b) that satisfy f, and verify the ordered pairs (b, a) satisfy the equation for f^{-1}.

39. $f(x) = \sqrt[3]{x - 2}$ **40.** $f(x) = \sqrt[3]{x + 3}$

41. $f(x) = x^3 + 1$ **42.** $f(x) = x^3 - 2$

43. $f(x) = \dfrac{8}{x + 2}$ **44.** $f(x) = \dfrac{12}{x - 1}$

45. $f(x) = \dfrac{x}{x + 1}$ **46.** $f(x) = \dfrac{x + 2}{1 - x}$

The functions given in Exercises 47 through 52 are not one-to-one. (a) Determine a domain restriction that preserves all range values and creates a one-to-one function, then state the new domain and range. (b) State the domain and range of the inverse function and find its equation.

47. $f(x) = (x + 5)^2$ **48.** $g(x) = x^2 + 3$

49. $v(x) = \dfrac{8}{(x - 3)^2}$ **50.** $V(x) = \dfrac{4}{x^2} + 2$

51. $p(x) = (x + 4)^2 - 2$ **52.** $q(x) = \dfrac{4}{(x - 2)^2} + 1$

For each function $f(x)$ given, prove (using a composition) that $g(x) = f^{-1}(x)$.

53. $f(x) = -2x + 5$, $g(x) = \dfrac{x - 5}{-2}$

54. $f(x) = 3x - 4$, $g(x) = \dfrac{x + 4}{3}$

55. $f(x) = \sqrt[3]{x + 5}$, $g(x) = x^3 - 5$

56. $f(x) = \sqrt[3]{x - 4}$, $g(x) = x^3 + 4$

57. $f(x) = \frac{2}{3}x - 6$, $g(x) = \frac{3}{2}x + 9$

58. $f(x) = \frac{4}{5}x + 6$, $g(x) = \frac{5}{4}x - \frac{15}{2}$

59. $f(x) = x^2 - 3$; $x \geq 0$, $g(x) = \sqrt{x + 3}$

60. $f(x) = x^2 + 8$; $x \geq 0$, $g(x) = \sqrt{x - 8}$

Find the inverse of each function $f(x)$ given, then prove (by composition) your inverse function is correct. Note the domain and range of f in each case is all real numbers.

61. $f(x) = 3x - 5$ **62.** $f(x) = 5x + 4$

63. $f(x) = \dfrac{x - 5}{2}$ **64.** $f(x) = \dfrac{x + 4}{3}$

65. $f(x) = \frac{1}{2}x - 3$ **66.** $f(x) = \frac{2}{3}x + 1$

67. $f(x) = x^3 + 3$ **68.** $f(x) = x^3 - 4$

69. $f(x) = \sqrt[3]{2x + 1}$ **70.** $f(x) = \sqrt[3]{3x - 2}$

71. $f(x) = \dfrac{(x - 1)^3}{8}$ **72.** $f(x) = \dfrac{(x + 3)^3}{-27}$

The functions given in Exercises 73 through 78 are one-to-one. State the implied domain of each function given, and use these to state the domain and range of the inverse function. Then find the inverse and prove by composition that your inverse is correct.

73. $f(x) = \sqrt{3x + 2}$ **74.** $g(x) = \sqrt{2x - 5}$

75. $p(x) = 2\sqrt{x - 3}$ **76.** $q(x) = 4\sqrt{x + 1}$

77. $v(x) = x^2 + 3$; $x \geq 0$ **78.** $w(x) = x^2 - 1$; $x \geq 0$

Determine the domain and range for each one-to-one function whose graph is given, and use this information to state the domain and range of the inverse function. Then sketch in the line $y = x$, estimate the location of two or more points on the graph, and use this information to graph $f^{-1}(x)$ on the same grid.

79. 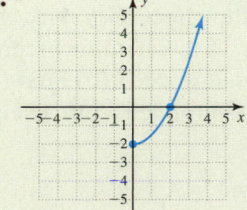   **80.**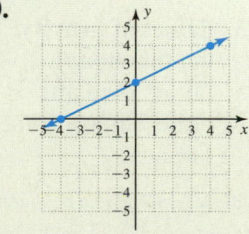

81. 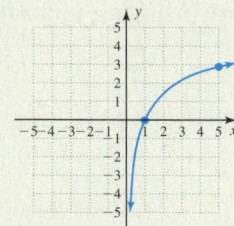   **82.** 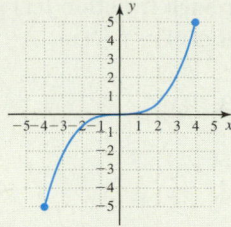   **83.** 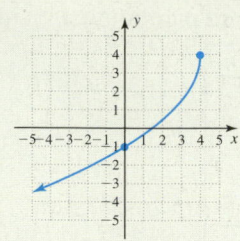   **84.**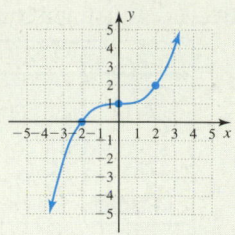

For the functions given, (a) find $f^{-1}(x)$, then use your calculator to verify they are inverses by (b) using ordered pairs, (c) composing the functions, and (d) showing their graphs are symmetric to $y = x$.

85. $f(x) = 2x + 1$

86. $g(x) = x^2 + 1; x \geq 0$

87. $h(x) = \dfrac{x}{x + 1}$

88. $j(x) = 2\sqrt{x + 9} - 6$

▶ WORKING WITH FORMULAS

89. The height of a projected image: $f(x) = \frac{1}{2}x - 8.5$

The height of an image projected on a screen is given by the formula shown, where $f(x)$ represents the actual height of the image on the projector (in centimeters) and x is the distance of the projector from the screen (in centimeters). (a) When the projector is 80 cm from the screen, how large is the image? (b) Show that the inverse function is $f^{-1}(x) = 2x + 17$, then input your answer from part (a) and comment on the result. What information does the inverse function give?

90. The radius of a sphere: $r(x) = \sqrt[3]{\dfrac{3x}{4\pi}}$

In generic form, the radius of a sphere is given by the formula shown, where $r(x)$ represents the radius and x represents the volume of the sphere in cubic units. (a) If a weather balloon that is roughly spherical holds 14,130 in^3 of helium, what is the radius of the balloon (use $\pi \approx 3.14$)? (b) Show that the inverse function is $r^{-1}(x) = \frac{4}{3}\pi x^3$, then input your answer from part (a) and comment on the result. What information does the inverse function give?

▶ APPLICATIONS

91. Temperature and altitude: The temperature (in degrees Fahrenheit) at a given altitude can be approximated by the function $f(x) = -\frac{7}{2}x + 59$, where $f(x)$ represents the temperature and x represents the altitude in thousands of feet. (a) What is the approximate temperature at an altitude of 35,000 ft (normal cruising altitude for commercial airliners)? (b) Find $f^{-1}(x)$, and state what the independent and dependent variables represent. (c) If the temperature outside a weather balloon is $-18°F$, what is the approximate altitude of the balloon?

92. Fines for speeding: In some localities, there is a set formula to determine the amount of a fine for exceeding posted speed limits. Suppose the amount of the fine for exceeding a 50 mph speed limit was given by the function $f(x) = 12x - 560 \ (x > 50)$ where $f(x)$ represents the fine in dollars for a speed of x mph. (a) What is the fine for traveling 65 mph through this speed zone? (b) Find $f^{-1}(x)$, and state what the independent and dependent variables represent. (c) If a fine of $172 were assessed, how fast was the driver going through this speed zone?

93. Effect of gravity: Due to the effect of gravity, the distance an object has fallen after being dropped is given by the function $f(x) = 16x^2; x \geq 0$, where $f(x)$ represents the distance in feet after x sec. (a) How far has the object fallen 3 sec after it has been dropped? (b) Find $f^{-1}(x)$, and state what the independent and dependent variables represent. (c) If the object is dropped from a height of 784 ft, how many seconds until it hits the ground (stops falling)?

94. Area and radius: In generic form, the area of a circle is given by $f(x) = \pi x^2$, where $f(x)$ represents the area in square units for a circle with radius x. (a) A pet dog is tethered to a stake in the backyard. If the tether is 10 ft long, how much area does the dog have to roam (use $\pi \approx 3.14$)? (b) Find $f^{-1}(x)$, and state what the independent and dependent variables represent. (c) If the owners want to allow the dog 1256 ft^2 of area to live and roam, how long a tether should be used?

95. Volume of a cone: In generic form, the volume of an equipoise cone (height equal to radius) is given by $f(x) = \frac{1}{3}\pi x^3$, where $f(x)$ represents the volume in units^3 and x represents the height of the cone. (a) Find the volume of such a cone if $r = 30$ ft (use $\pi \approx 3.14$). (b) Find $f^{-1}(x)$, and state what the independent and

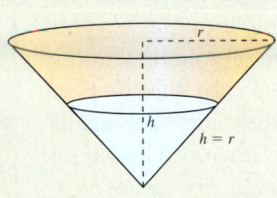

dependent variables represent. (c) If the volume of water in the cone is 763.02 ft^3, how deep is the water at its deepest point?

96. Wind power: The power delivered by a certain wind-powered generator can be modeled by the function $f(x) = \frac{x^3}{2500}$, where $f(x)$ is the horsepower (hp) delivered by the generator and x represents the speed of the wind in miles per hour. (a) Use the model to determine how much horsepower is generated by a 30 mph wind. (b) The person monitoring the output of the generators (wind generators are usually erected in large numbers) would like a function that gives the wind speed based on the horsepower readings on the gauges in the monitoring station. For this purpose, find $f^{-1}(x)$ and state what the independent and dependent variables represent. (c) If gauges show 25.6 hp is being generated, how fast is the wind blowing?

▶ **EXTENDING THE CONCEPT**

97. For a deeper understanding of the algebraic method for finding an inverse, suppose a function f is defined as $f(x):\{(x, y) \mid y = 3x - 6\}$. We can then define the inverse as $f^{-1}: \{(x, y) \mid x = 3y - 6\}$, having interchanged x and y in the equation portion. The equation for f^{-1} is not in standard form, but (x, y) still represents all ordered pairs satisfying either equation. Solving for y gives $f^{-1}:\left\{(x, y) \,\middle|\, y = \frac{x}{3} + 2\right\}$, and demonstrates the role of steps 2, 3, and 4 of the method. (a) Find five ordered pairs that satisfy the equation for f, then (b) interchange their coordinates and show they satisfy the equation for f^{-1}. *Answers will vary.*

98. By inspection, which of the following is the inverse function for $f(x) = \frac{2}{3}\left(x - \frac{1}{2}\right)^5 + \frac{4}{5}$?

a. $f^{-1}(x) = \sqrt[5]{\frac{1}{2}\left(x - \frac{2}{3}\right)} - \frac{4}{5}$

b. $f^{-1}(x) = \frac{3}{2}\sqrt[5]{(x - 2)} - \frac{5}{4}$

c. $f^{-1}(x) = \frac{3}{2}\sqrt[5]{\left(x + \frac{1}{2}\right)} - \frac{5}{4}$

d. $f^{-1}(x) = \sqrt[5]{\frac{3}{2}\left(x - \frac{4}{5}\right)} + \frac{1}{2}$

▶ **MAINTAINING YOUR SKILLS**

99. (R.3) Write as many of the following formulas as you can from memory:

a. perimeter of a rectangle
b. area of a circle
c. volume of a cylinder
d. volume of a cone
e. circumference of a circle
f. area of a triangle
g. area of a trapezoid
h. volume of a sphere
i. Pythagorean theorem

100. (3.2) Given $f(x) = x^2 - x - 2$, solve the inequality $f(x) \leq 0$ using the x-intercepts and end-behavior of the graph.

101. (3.4) For the function $y = 2\sqrt{x} + 3$, find the average rate of change between $x = 1$ and $x = 2$, and between $x = 4$ and $x = 5$. Which is greater? Why?

102. (R.4) Solve the following cubic equations by factoring:

a. $x^3 - 5x = 0$
b. $x^3 - 7x^2 - 4x + 28 = 0$
c. $x^3 - 3x^2 = 0$
d. $x^3 - 3x^2 - 4x = 0$

Additional answers can be found in the Instructor Answer Appendix.

5.2 Exponential Functions

LEARNING OBJECTIVES

In Section 5.2 you will see how we can:

- ☐ **A.** Evaluate an exponential function
- ☐ **B.** Graph general exponential functions
- ☐ **C.** Graph base-*e* exponential functions
- ☐ **D.** Solve exponential equations and applications

Demographics is the statistical study of human populations. In this section, we introduce the family of *exponential functions,* which are widely used to model population growth or decline with additional applications in science, engineering, and many other fields. As with other functions, we begin with a study of the graph and its characteristics.

A. Evaluating Exponential Functions

In the boomtowns of the old west, it was not uncommon for a town to double in size every year (at least for a time) as the lure of gold drew more and more people westward. When this type of growth is modeled using mathematics, exponents play a lead role. Suppose the town of Goldsboro had 1000 residents when gold was first discovered. After 1 yr the population doubled to **2**000 residents. The next year it doubled again to **4**000, then

again to **8**000, then to **16**,000 and so on. You probably recognize the digits in blue as powers of two (indicating the population is *doubling*), with each one multiplied by 1000 (the initial population). This suggests we can model the relationship using

$$P(x) = 1000 \cdot 2^x$$

where $P(x)$ is the population after x yr. Further, we can evaluate this function, called an **exponential function,** for *fractional parts of a year* using rational exponents. The population of Goldsboro one-and-a-half years after the gold rush was

$$P\left(\frac{3}{2}\right) = 1000 \cdot 2^{\frac{3}{2}}$$

$$= 1000 \cdot (\sqrt{2})^3$$

$$\approx 2828 \text{ people}$$

In general, exponential functions are defined as follows.

WORTHY OF NOTE

To properly understand the exponential function and its graph requires that we evaluate $f(x) = 2^x$ even when x is *irrational*. For example, what does $2^{\sqrt{5}}$ mean? While the technical details require calculus, it can be shown that successive approximations of $2^{\sqrt{5}}$ as in $2^{2.2360}$, $2^{2.23606}$, $2^{2.236067}$, . . . approach a unique real number, and that $f(x) = 2^x$ exists for all real numbers x.

Exponential Functions

For $b > 0$, $b \neq 1$, and all real numbers x,

$$f(x) = b^x$$

defines the base b exponential function.

Limiting b to positive values ensures that outputs will be real numbers, and the restriction $b \neq 1$ is needed since $y = 1^x$ is a constant function (1 raised to *any* power is still 1). Specifically note the domain of an exponential function is *all real numbers,* and that all of the familiar properties of exponents still hold. A summary of these properties follows. For a complete review, see Section R.2.

Exponential Properties

For real numbers a, b, m, and n, with $a, b > 0$,

$$b^m \cdot b^n = b^{m+n} \qquad \frac{b^m}{b^n} = b^{m-n} \qquad (b^m)^n = b^{mn}$$

$$(ab)^n = a^n \cdot b^n \qquad b^{-n} = \frac{1}{b^n} \qquad \left(\frac{b}{a}\right)^{-n} = \left(\frac{a}{b}\right)^n$$

EXAMPLE 1 ▶ **Evaluating Exponential Functions**

Evaluate each exponential function for $x = 2$, $x = -1$, $x = \frac{1}{2}$, and $x = \pi$. Use a calculator for $x = \pi$, rounding to five decimal places.

a. $f(x) = 4^x$ **b.** $g(x) = \left(\frac{4}{9}\right)^x$

Solution ▶ **a.** For $f(x) = 4^x$,

$$f(2) = 4^2 = 16$$

$$f(-1) = 4^{-1} = \frac{1}{4}$$

$$f\left(\frac{1}{2}\right) = 4^{\frac{1}{2}} = \sqrt{4} = 2$$

$$f(\pi) = 4^\pi \approx 77.88023$$

b. For $g(x) = \left(\frac{4}{9}\right)^x$,

$$g(2) = \left(\frac{4}{9}\right)^2 = \frac{16}{81}$$

$$g(-1) = \left(\frac{4}{9}\right)^{-1} = \frac{9}{4}$$

$$g\left(\frac{1}{2}\right) = \left(\frac{4}{9}\right)^{\frac{1}{2}} = \sqrt{\frac{4}{9}} = \frac{2}{3}$$

$$g(\pi) = \left(\frac{4}{9}\right)^\pi \approx 0.07827$$

☑ **A.** You've just seen how we can evaluate an exponential function

Now try Exercises 7 through 10 ▶

B. Graphing Exponential Functions

To gain a better understanding of exponential functions, we'll graph examples of $y = b^x$ and note some of the characteristic features. Since $b \neq 1$, it seems reasonable that we graph one exponential function where $b > 1$ and one where $0 < b < 1$.

EXAMPLE 2 ▶ **Graphing Exponential Functions with $b > 1$**

Graph $y = 2^x$ using a table of values.

Solution ▶ To get an idea of the graph's shape we'll use integer values from -3 to 3 in our table, then draw the graph as a continuous curve, since the function is defined for all real numbers.

x	$y = 2^x$
-3	$2^{-3} = \frac{1}{8}$
-2	$2^{-2} = \frac{1}{4}$
-1	$2^{-1} = \frac{1}{2}$
0	$2^0 = 1$
1	$2^1 = 2$
2	$2^2 = 4$
3	$2^3 = 8$

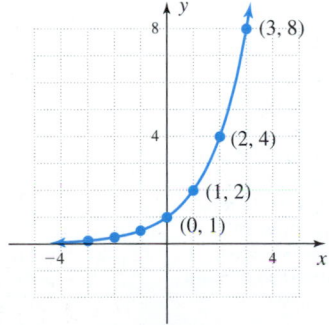

Now try Exercises 11 and 12 ▶

WORTHY OF NOTE

As in Example 2, functions that are increasing for all $x \in D$ are said to be **monotonically increasing** or simply **monotonic functions.** The function in Example 3 is monotonically decreasing.

Several important observations can now be made. First note the x-axis (the line $y = 0$) is a horizontal asymptote for the function, because as $x \to -\infty$, $y \to 0$. Second, the function is increasing over its entire domain, giving the function a range of $y \in (0, \infty)$.

EXAMPLE 3 ▶   Graphing Exponential Functions with $0 < b < 1$

Graph $y = \left(\frac{1}{2}\right)^x$ using a table of values.

Solution ▶   Using properties of exponents, we can write $\left(\frac{1}{2}\right)^x$ as $\left(\frac{2}{1}\right)^{-x} = 2^{-x}$. Again using integers from -3 to 3, we plot the ordered pairs and draw a continuous curve.

x	$y = 2^{-x}$
-3	$2^{-(-3)} = 2^3 = 8$
-2	$2^{-(-2)} = 2^2 = 4$
-1	$2^{-(-1)} = 2^1 = 2$
0	$2^0 = 1$
1	$2^{-1} = \frac{1}{2}$
2	$2^{-2} = \frac{1}{4}$
3	$2^{-3} = \frac{1}{8}$

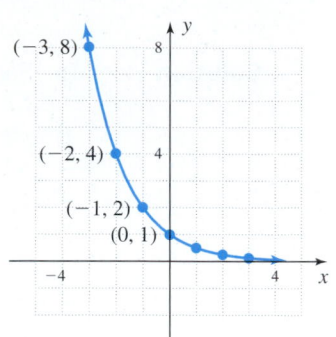

Now try Exercises 13 and 14 ▶

We note this graph is also asymptotic to the x-axis, but *decreasing on its domain.* In addition, both $y = 2^x$ and $y = 2^{-x} = \left(\frac{1}{2}\right)^x$ have a y-intercept of $(0, 1)$ and both are one-to-one, which suggests that an inverse function can be found. Finally, observe that $y = b^{-x}$ is *a reflection of $y = b^x$ across the y-axis,* a property that indicates these basic graphs might also be transformed in other ways, as were the toolbox functions. The characteristics of exponential functions are summarized here:

$f(x) = b^x$, $b > 0$ and $b \neq 1$

- one-to-one function
- domain: $x \in \mathbb{R}$
- increasing if $b > 1$

- y-intercept $(0, 1)$
- range: $y \in (0, \infty)$
- decreasing if $0 < b < 1$

- asymptotic to the x-axis (the line $y = 0$)

Figure 5.18   Figure 5.19

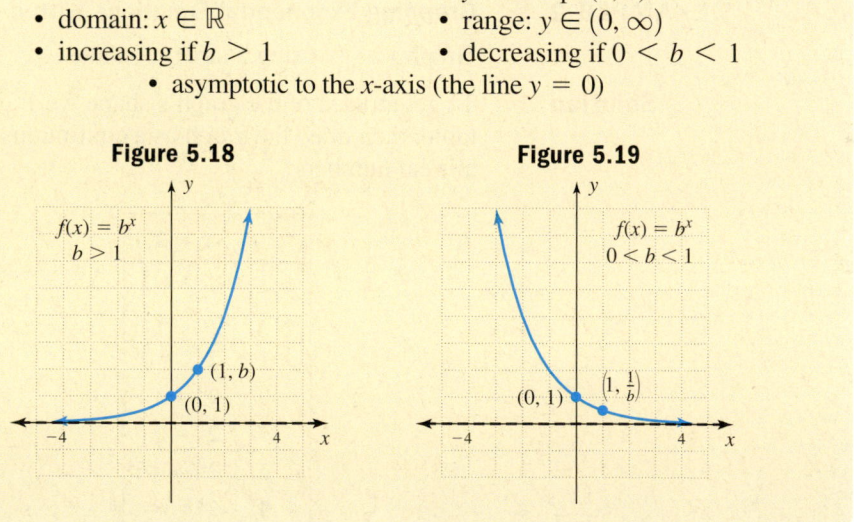

WORTHY OF NOTE

When an exponential function is increasing, it can be referred to as a "growth function." When decreasing, it is often called a "decay function." Each of the graphs shown in Figures 5.18 and 5.19 should now be added to your repertoire of basic functions, to be sketched from memory and analyzed or used as needed.

Just as the graph of a quadratic function maintains its parabolic shape regardless of the transformations applied, exponential functions will also maintain their general shape and features. Any sum or difference applied to the basic function ($y = b^x \pm k$ vs. $y = b^x$) will cause a vertical shift in the same direction as the sign, and any change to input values ($y = b^{x+h}$ vs. $y = b^x$) will cause a horizontal shift in a direction opposite the sign. For cases where multiple transformations are to be applied, refer to the sequence outlined on page 209.

EXAMPLE 4 ▶ **Graphing Exponential Functions Using Transformations**

Graph $F(x) = 2^{x-1} + 2$ using transformations of
the basic function $f(x) = 2^x$ (not by simply plotting
points). Clearly state what transformations are
applied.

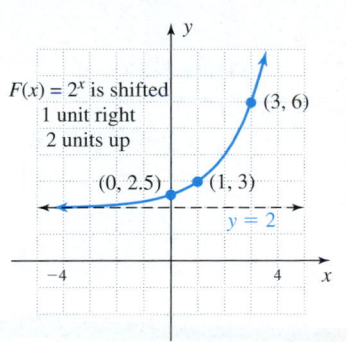

Solution ▶ The graph of F is that of the basic function
$f(x) = 2^x$ with a horizontal shift 1 unit right and
a vertical shift 2 units up. With this in mind the
horizontal asymptote also shifts from $y = 0$ to
$y = 2$ and $(0, 1)$ shifts to $(1, 3)$. The y-intercept of
F is at $(0, 2.5)$:

$$F(0) = 2^{(0)-1} + 2$$
$$= 2^{-1} + 2$$
$$= \frac{1}{2} + 2$$
$$= 2.5$$

☑ **B. You've just seen how
we can graph general
exponential functions**

To help sketch a more accurate graph, the point $(3, 6)$ can be used: $F(3) = 6$.

Now try Exercises 15 through 30 ▶

C. The Base-e Exponential Function: $f(x) = e^x$

In nature, exponential growth occurs when the rate of change in a population's growth
is in constant proportion to its current size. Using the rate of change notation,
$\frac{\Delta P}{\Delta t} = kP$, where k is a constant. For the city of Goldsboro, we know the population at
time t is given by $P(t) = 1000 \cdot 2^t$, but have no information on this value of k (**see
Exercise 90**). We can actually rewrite this function, and other exponential functions,
using a base that gives the value of k directly and without having to apply the difference quotient. This new base is an irrational number, symbolized by the letter e. In
Section 5.6 we'll develop the number e in the context of compound interest, while
making numerous references to our discussion here, where we define e as follows.

WORTHY OF NOTE

Just as the ratio of a circle's
circumference to its diameter is an
irrational number symbolized by π,
the irrational number that results
from $\left(1 + \dfrac{1}{x}\right)^x$ for infinitely large
x is symbolized by e. Writing
exponential functions in terms of e
simplifies many calculations in
advanced courses, and offers
additional advantages in applications
of exponential functions.

The Number e

For $x > 0$,

$$\text{as } x \to \infty, \left(1 + \frac{1}{x}\right)^x \to e$$

In words, e is the number that $\left(1 + \dfrac{1}{x}\right)^x$ approaches as x becomes infinitely large.

It has been proven that as x grows without bound, $\left(1 + \dfrac{1}{x}\right)^x$ indeed approaches the
unique, irrational number that we have named e. Table 5.5 gives approximate values of
the expression for selected values of x, and shows $e \approx 2.71828$ to five decimal places.

Figure 5.20

```
e^(2)
      7.389056099
```

The result is the base-e **exponential function:** $f(x) = e^x$, also called the **natural exponential function.** Instead of having to enter a decimal approximation when computing with e, most calculators have an "e^x" key, usually as the ⨀2nd function for the key marked ⨀LN. To find the value of e^2, use the keystrokes ⨀2nd ⨀LN 2 ⨀) ⨀ENTER, and the calculator display should read 7.389056099. Note the calculator supplies the left parenthesis for the exponent, and you must supply the right. See Figure 5.20.

Table 5.5

x	Approximate Value $(1 + \frac{1}{x})^x$
10	2.59
100	2.705
1000	2.7169
10,000	2.71815
100,000	2.718268
1,000,000	2.7182805
10,000,000	2.71828169

EXAMPLE 5 ▶ Evaluating the Natural Exponential Function

Use a calculator to evaluate $f(x) = e^x$ for the values of x given. Round to six decimal places as needed.

a. $f(3)$ b. $f(1)$ c. $f(0)$ d. $f(\frac{1}{2})$

Solution ▶ a. $f(3) = e^3 \approx 20.085537$ b. $f(1) = e^1 \approx 2.718282$
c. $f(0) = e^0 = 1$ (exactly)   d. $f(\frac{1}{2}) = e^{\frac{1}{2}} \approx 1.648721$

Now try Exercises 31 through 36 ▶

Although e is an irrational number, the graph of $y = e^x$ behaves in exactly the same way and has the same characteristics as other exponential graphs. Figure 5.21 shows this graph on the same grid as $y = 2^x$ and $y = 3^x$. As we might expect, all three graphs are increasing, have an asymptote at $y = 0$, and contain the point $(0, 1)$, with the graph of $y = e^x$ "between" the other two. The domain for all three functions, as with all basic exponential functions, is $x \in (-\infty, \infty)$ with range $y \in (0, \infty)$. The same transformations applied earlier can also be applied to the graph of $y = e^x$. **See Exercises 37 through 42.**

Figure 5.21

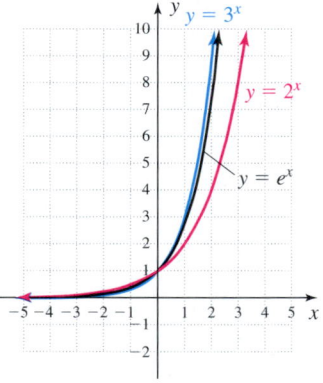

☑ **C.** You've just seen how we can graph base-e exponential functions

D. Solving Exponential Equations Using the Uniqueness Property

Since exponential functions are one-to-one, we can solve equations where each side is an exponential term with the identical base. This is because one-to-oneness guarantees a unique solution to the equation.

WORTHY OF NOTE

Exponential functions are very different from the power functions studied earlier. For power functions, the base is variable and the exponent is constant: $y = x^b$, while for exponential functions the *exponent is a variable* and the *base is constant*: $y = b^x$.

Exponential Equations and the Uniqueness Property

For all real numbers m, n, and b, where $b > 0$ and $b \neq 1$,

1. If $b^m = b^n$,
    then $m = n$.

2. If $m \neq n$,
    then $b^m \neq b^n$.

Equal bases imply exponents are equal.

The equation $2^x = 32$ can be written as $2^x = 2^5$, and we note $x = 5$ is a solution. Although $3^x = 32$ can be written as $3^x = 2^5$, the bases are not alike and the solution to this equation must wait until additional tools are developed in Section 5.5.

EXAMPLE 6 ▶ **Solving Exponential Equations**

Solve the exponential equations using the uniqueness property.

a. $3^{2x-1} = 81$ b. $\left(\frac{1}{6}\right)^{-3x-2} = 36^{x+1}$ c. $e^x e^2 = \dfrac{e^4}{e^{x+1}}$

Solution ▶

a.

$$3^{2x-1} = 81 \qquad \text{given}$$
$$3^{2x-1} = 3^4 \qquad \text{rewrite using base 3}$$
$$\Rightarrow 2x - 1 = 4 \qquad \text{uniqueness property}$$
$$x = \frac{5}{2} \qquad \text{solve for } x$$

Check ▶

$$3^{2x-1} = 81 \qquad \text{given}$$
$$3^{2\left(\frac{5}{2}\right)-1} = 81 \qquad \text{substitute } \tfrac{5}{2} \text{ for } x$$
$$3^{5-1} = 81 \qquad \text{simplify}$$
$$3^4 = 81 \qquad \text{result checks}$$
$$81 = 81$$

The remaining checks are left to the student.

b.

$$\left(\frac{1}{6}\right)^{-3x-2} = 36^{x+1} \qquad \text{given}$$
$$(6^{-1})^{-3x-2} = (6^2)^{x+1} \qquad \text{rewrite using base 6}$$
$$6^{3x+2} = 6^{2x+2} \qquad \text{power property of exponents}$$
$$\Rightarrow 3x + 2 = 2x + 2 \qquad \text{uniqueness property}$$
$$x = 0 \qquad \text{solve for } x$$

c.

$$e^x e^2 = \frac{e^4}{e^{x+1}} \qquad \text{given}$$
$$e^{x+2} = e^{4-(x+1)} \qquad \text{product property; quotient property}$$
$$e^{x+2} = e^{3-x} \qquad \text{simplify}$$
$$\Rightarrow x + 2 = 3 - x \qquad \text{uniqueness property}$$
$$2x = 1 \qquad \text{add } x, \text{ subtract 2}$$
$$x = \frac{1}{2} \qquad \text{solve for } x$$

Now try Exercises 43 through 62 ▶

Earlier in this section, we showed the exponential function $f(x) = b^x$ was defined for all real numbers and was a one-to-one function. This is important because it establishes that equations like $2^x = 7$ must have a solution, even if x is not rational. In fact, since $2^2 = 4$ and $2^3 = 8$, the following inequalities indicate the solution must be between 2 and 3

$$4 < 7 < 8 \qquad \text{7 is between 4 and 8}$$
$$2^2 < 2^x < 2^3 \qquad \text{replace 4 with } 2^2, \text{ 8 with } 2^3$$
$$\Rightarrow 2 < x < 3 \qquad x \text{ must be between 2 and 3}$$

Until we develop an inverse for exponential functions, we are unable to solve many of these equations in exact form. We can, however, get a very close approximation using a graphing calculator. For the equation $2^x = 7$, enter $Y_1 = 2^X$ and $Y_2 = 7$ on the ⬤Y= screen. Then press ⬤ZOOM 6 to graph both functions (see Figure 5.22). To find the point of intersection, press ⬤2nd ⬤TRACE **(CALC)** and select option **5:intersect,** then press ⬤ENTER *three* times (to identify the intersecting functions and bypass "Guess"). The x- and y-coordinates of the point of intersection will appear at the bottom of the screen, with the x-coordinate being the solution. As you can see, x is indeed between 2 and 3. **See Exercises 63 through 66.**

Two common applications of exponential functions involve appreciation (as when an item grows in value over time), and depreciation (as when tools and equipment decrease in value over time).

Figure 5.22

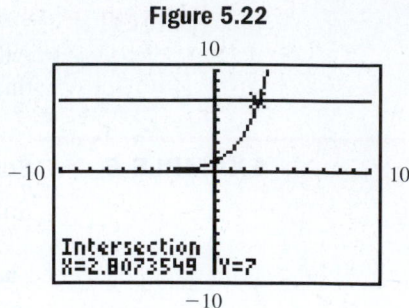

EXAMPLE 7 ▶ **Applications of Exponential Functions—Depreciation**

For insurance purposes, it is estimated that large household appliances lose $\frac{1}{5}$ of their value each year. The current value can then be modeled by the function $V(t) = V_0(\frac{4}{5})^t$, where V_0 is the initial value and $V(t)$ represents the value after t years. How many years does it take a washing machine that cost \$625 new, to depreciate to a value of \$256?

Solution ▶ For this exercise, $V_0 = \$625$ and $V(t) = \$256$. The formula yields

$$V(t) = V_0\left(\frac{4}{5}\right)^t \qquad \text{given}$$

$$256 = 625\left(\frac{4}{5}\right)^t \qquad \text{substitute known values}$$

$$\frac{256}{625} = \left(\frac{4}{5}\right)^t \qquad \text{divide by 625}$$

$$\left(\frac{4}{5}\right)^4 = \left(\frac{4}{5}\right)^t \qquad \text{equate bases } \frac{256}{625} = \left(\frac{4}{5}\right)^4$$

$$\Rightarrow 4 = t \qquad \text{Uniqueness Property}$$

After 4 yr, the washing machine's value has dropped to \$256.

Now try Exercises 69 through 74 ▶

Another very practical application of the natural exponential function involves **Newton's law of cooling.** This law or formula models the temperature of an object as it cools down, as when a pizza is removed from the oven and placed on the kitchen counter. The function model is

$$T(x) = T_R + (T_0 - T_R)e^{kx}, k < 0$$

where T_0 represents the initial temperature of the object, T_R represents the temperature of the room or surrounding medium, $T(x)$ is the temperature of the object x min later, and k is the cooling rate as determined by the nature and physical properties of the object.

EXAMPLE 8 ▶ **Applying an Exponential Function—Newton's Law of Cooling**

A pizza is taken from a 425°F oven and placed on the counter to cool. If the temperature in the kitchen is 75°F, and the cooling rate for this type of pizza is $k = -0.35$,

  a. What is the temperature (to the nearest degree) of the pizza 2 min later?

  b. To the nearest minute, how long until the pizza has cooled to a temperature below 90°F?

  c. If Zack and Raef like to eat their pizza at a temperature of about 110°F, how many minutes should they wait to "dig in"?

Solution ▶ Begin by substituting the given values to obtain the equation model:

$$T(x) = T_R + (T_0 - T_R)e^{kx} \quad \text{general equation model}$$
$$= 75 + (425 - 75)e^{-0.35x} \quad \text{substitute 75 for } T_R, 425 \text{ for } T_0, \text{ and } -0.35 \text{ for } k$$
$$= 75 + 350e^{-0.35x} \quad \text{simplify}$$

For part (a) we simply find $T(2)$:

a. $T(2) = 75 + 350e^{-0.35(2)}$ substitute 2 for x
$$\approx 249 \quad \text{result}$$

Two minutes later, the temperature of the pizza is near 249°.

b. Using the **TABLE** feature of a graphing calculator shows the pizza reaches a temperature of just under 90° after 9 min: $T(9) \approx 90°F$.

c. We elect to use the intersection-of-graphs method. After setting an appropriate window, we enter $Y_1 = 75 + 350e^{-0.35^x}$ and $Y_2 = 110$, then press **2nd** **CALC** option **5:intersect.** After pressing **ENTER** three times, the coordinates of the point of intersection appear at the bottom of the screen: $x \approx 6.6$, $y = 110$. It appears the boys should wait about $6\frac{1}{2}$ min for the pizza to cool.

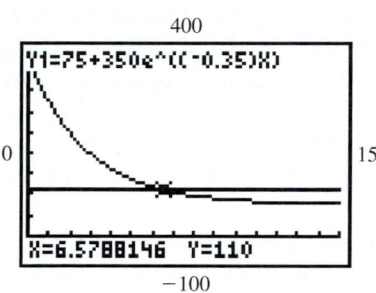

✓ D. You've just seen how we can solve exponential equations and applications

Now try Exercises 75 and 76 ▶

There are a number of additional applications of exponential functions in the Exercise Set. **See Exercises 77 through 84.**

5.2 EXERCISES

▶ CONCEPTS AND VOCABULARY

Fill in each blank with the appropriate word or phrase. Carefully reread the section if needed.

1. An exponential function is one of the form $y = \underline{\quad b^x \quad}$, where $\underline{\quad b \quad} > 0$, $\underline{\quad b \quad} \neq 1$, and $\underline{\quad x \quad}$ is any real number.

2. The domain of $y = b^x$ is all $\underline{\quad \text{real} \quad}$ $\underline{\quad \text{numbers} \quad}$, and the range is $y \in \underline{\quad (0, \infty) \quad}$. Further, as $x \to -\infty$, $y \underline{\quad \to 0 \quad}$.

3. For exponential functions of the form $y = ab^x$, the y-intercept is $(0, \underline{\quad a \quad})$, since $b^0 = \underline{\quad 1 \quad}$ for any real number b.

4. If each side of an equation can be written as an exponential term with the same base, the equation can be solved using the $\underline{\quad \text{uniqueness} \quad}$ $\underline{\quad \text{property} \quad}$.

5. State true or false and explain why: $y = b^x$ is always increasing if $0 < b < 1$.   False; for $|b| < 1$ and $x_2 > x_1$, $b^{x_2} < b^{x_1}$, so function is decreasing

6. Discuss/Explain the statement, "For $k > 0$, the y-intercept of $y = ab^x + k$ is $(0, a + k)$."   Answers will vary.

HOMEWORK SELECTION GUIDE

Core: 7–23 every other odd, 25–30, 31–63 every other odd, 65, 68, 69–79 odd, 89, 94 (30 Exercises)

Standard: 1–4, 7–23 every other odd, 25–30, 31–63 every other odd, 65, 68, 69–79 odd, 85, 86, 89, 92, 94 (37 Exercises)

Extended: 1–4, 7–23 every other odd, 25–30, 31–63 every other odd, 65, 67, 68, 69–79 odd, 85–88, 92–94 (40 Exercises)

In Depth: 1–6, 7–23 every other odd, 25–30, 31–63 every other odd, 65, 67, 68, 69–83 odd, 85–94 (47 Exercises)

▶ DEVELOPING YOUR SKILLS

Evaluate each function as indicated. Use a calculator only as needed, rounding results to thousandths.

7. $P(t) = 4^t$;
$t = 2, t = \frac{1}{2}, t = \frac{3}{2},$
$t = \sqrt{3}$
16, 2, 8, 11.036

8. $Q(t) = 8^t$;
$t = 2, t = \frac{1}{3}, t = \frac{5}{3},$
$t = \sqrt{5}$
64, 2, 32, 104.561

9. $V(n) = (\frac{1}{8})^n$;
$n = 0, n = 2, n = \frac{2}{3},$
$n = -2$
$1, \frac{1}{64}, \frac{1}{4}, 64$

10. $W(m) = (\frac{4}{9})^m$;
$m = 0, m = 3, m = \frac{3}{2},$
$m = -2$
$1, \frac{64}{729}, \frac{8}{27}, \frac{81}{16}$

Graph each function using a table of values and integer inputs between −3 and 3. Clearly label the y-intercept and one additional point, then state whether the function is increasing or decreasing.

11. $y = 3^x$

12. $y = 4^x$

13. $y = (\frac{1}{3})^x$

14. $y = (\frac{1}{4})^x$

Graph each of the following functions by *translating the basic function* $y = b^x$, sketching the asymptote, and strategically plotting a few points to round out the graph. Clearly state the basic function and what shifts are applied.

15. $y = 3^x + 2$

16. $y = 3^x - 3$

17. $y = 3^{x+3}$

18. $y = 3^{x-2}$

19. $y = 3^{-x}$

20. $y = 3^{-x} - 2$

21. $y = (\frac{1}{3})^x + 1$

22. $y = (\frac{1}{3})^x - 4$

23. $y = (\frac{1}{3})^{x-2}$

24. $y = (\frac{1}{3})^{x+2}$

Match each exponential equation to the correct graph.

25. $y = 5^{-x}$ e

26. $y = 4^{-x}$ c

27. $y = 3^{-x+1}$ a

28. $y = 3^{-x} + 1$ f

29. $y = 2^{x+1} - 2$ b

30. $y = 2^{x+2} - 1$ d

a.

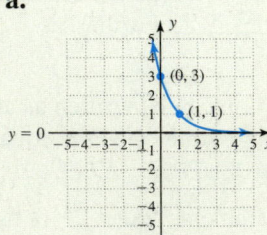

b.

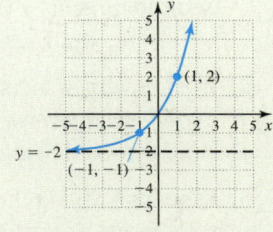

c.

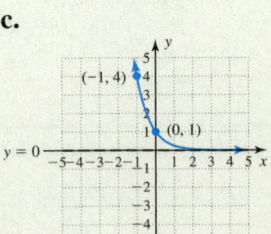

d.

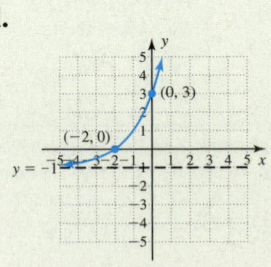

e.

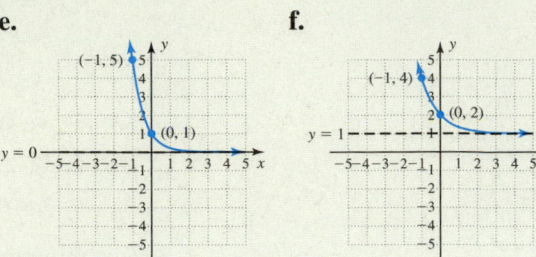

f.

Use a calculator to evaluate each expression, rounded to six decimal places.

31. e^1 2.718282

32. e^0 1

33. e^2 7.389056

34. $e^{-3.2}$ 0.040762

35. $e^{\sqrt{2}}$ 4.113250

36. e^π 23.140693

Graph each exponential function.

37. $f(x) = e^{x+3} - 2$

38. $g(x) = e^{x-2} + 1$

39. $r(t) = -e^t + 2$

40. $s(t) = -e^{t+2}$

41. $p(x) = e^{-x+2} - 1$

42. $q(x) = e^{-x-1} + 2$

Solve each exponential equation and check your answer by substituting into the original equation.

43. $10^x = 1000$ 3

44. $144 = 12^x$ 2

45. $25^x = 125$ $\frac{3}{2}$

46. $81 = 27^x$ $\frac{4}{3}$

47. $8^{x+2} = 32$ $-\frac{1}{3}$

48. $9^{x-1} = 27$ $\frac{5}{2}$

49. $32^x = 16^{x+1}$ 4

50. $100^{x+2} = 1000^x$ 4

51. $(\frac{1}{5})^x = 125$ −3

52. $(\frac{1}{4})^x = 64$ −3

53. $(\frac{1}{3})^{2x} = 9^{x-6}$ 3

54. $(\frac{1}{2})^{3x} = 8^{x-2}$ 1

55. $(\frac{1}{9})^{x-5} = 3^{3x}$ 2

56. $2^{-2x} = (\frac{1}{32})^{x-3}$ 5

57. $25^{3x} = 125^{x-2}$ −2

58. $27^{2x+4} = 9^{4x}$ 6

59. $\dfrac{e^4}{e^{2-x}} = e^3 e$ 2

60. $e^x(e^x + e) = \dfrac{e^x + e^{3x}}{e^{-x}}$ $\frac{1}{3}$

61. $(e^{2x-4})^3 = \dfrac{e^{x+5}}{e^2}$ 3

62. $e^x e^{x+3} = (e^{x+2})^3$ −3

Solve the following equations. First estimate the answer by bounding it between two integers, then solve the equation graphically. Adjust the viewing window as needed.

63. $3^x = 22$ $x \approx 2.8$

64. $2^x = 2.125$ $x \approx 1.17$

65. $e^{x-1} = 9$ $x \approx 3.2$

66. $e^{0.5x} = 8$
$x \approx 4.16$

Additional answers can be found in the Instructor Answer Appendix.

▶ WORKING WITH FORMULAS

67. The growth of bacteria: $P(t) = 1000 \cdot 3^t$

If the initial population of a common bacterium is 1000 and the population triples every day, its population is given by the formula shown, where $P(t)$ is the total population after t days. (a) Find the total population 12 hr, 1 day, $1\frac{1}{2}$ days, and 2 days later. (b) Do the outputs show the population is tripling every 24 hr (1 day)? (c) Explain why this is an increasing function. (d) Graph the function using an appropriate scale.

68. Spinners with numbers 1 to 4: $P(x) = (\frac{1}{4})^x$

Games that involve moving pieces around a board using a fair spinner are fairly common. If the spinner has the numbers 1 through 4, the probability that any one number is spun repeatedly is given by the formula shown, where x represents the number of spins and $P(x)$ represents the probability the same number results x times. (a) What is the probability that the first player spins a 2? (b) What is the probability that all four players spin a 2? (c) Explain why this is a decreasing function.

▶ APPLICATIONS

69. Depreciation: The financial analyst for a large construction firm estimates that its heavy equipment loses one-fifth of its value each year.

The current value of the equipment is then modeled by the function $V(t) = V_0(\frac{4}{5})^t$, where V_0 represents the initial value, t is in years, and $V(t)$ represents the value after t years. (a) How much is a large earthmover worth after 1 yr if it cost $125 thousand new? (b) How many years does it take for the earthmover to depreciate to a value of $64 thousand? **a.** $100,000 **b.** 3 yr

70. Depreciation: Photocopiers have become a critical part of the operation of many businesses, and due to their heavy use they can depreciate in value very quickly. If a copier loses $\frac{3}{8}$ of its value each year, the current value of the copier can be modeled by the function $V(t) = V_0(\frac{5}{8})^t$, where V_0 represents initial value, t is in years, and $V(t)$ represents the value after t yr. (a) How much is this copier worth after one year if it cost $64 thousand new? (b) How many years does it take for the copier to depreciate to a value of $25 thousand? **a.** $40,000 **b.** 2 yr

71. Depreciation: Margaret Madison, DDS, estimates that her dental equipment loses one-sixth of its value each year. (a) Determine the value of an x-ray machine after 5 yr if it cost $216 thousand new, and (b) determine how long until the machine is worth less than $125 thousand. **a.** ≈ $86,806 **b.** 3 yr

72. Exponential decay: The groundskeeper of a local high school estimates that due to heavy usage by the baseball and softball teams, the pitcher's

mound loses one-fifth of its height every month. Use this information to find an equation that models this information and then determine: (a) If the mound was 25 cm to begin, how long until the height becomes less than 16 cm high (meaning it

must be entirely rebuilt)? (b) If the mound were allowed to deteriorate further, find the height after 3 months. **a.** 2 mo **b.** 12.8 cm

73. Exponential growth: Similar to a small town doubling in size after a discovery of gold, a business that develops a product in high demand has the potential for doubling its revenue each year for a number of years. The revenue would be modeled by the function $R(t) = R_0 2^t$, where R_0 represents the initial revenue, and $R(t)$ represents the revenue after t years. (a) How much revenue is being generated after 4 yr, if the company's initial revenue was $2.5 million? (b) How many years does it take for the business to be generating $320 million in revenue? **a.** $40 million **b.** 7 yr

74. Exponential growth: If a company's revenue grows at a rate of 150% per year (rather than doubling as in Exercise 73), the revenue would be modeled by the function $R(t) = R_0(\frac{3}{2})^t$, where R_0 represents the initial revenue, and $R(t)$ represents the revenue after t years. (a) How much revenue is being generated after 3 yr, if the company's initial revenue was $256 thousand? (b) How long until the business is generating $1944 thousand in revenue? (*Hint:* Reduce the fraction.) **a.** $864 thousand **b.** 5 yr

Use Newton's law of cooling to complete Exercises 75 and 76: $T(x) = T_R + (T_0 - T_R)e^{kx}$.

75. Cold party drinks: Janae was late getting ready for the party, and the liters of soft drinks she bought were still at room temperature (73°F) with guests due to arrive in 15 min. If she puts these in her freezer at −10°F, will the drinks be cold enough (35°F) for her guests? Assume $k \approx -0.031$.
no, they will have to wait about 5 min

76. Warm party drinks: Newton's law of cooling applies equally well if the "cooling is negative," meaning the object is taken from a colder medium and placed in a warmer one. If a can of soft drink is taken from a 35°F cooler and placed in a room where the temperature is 75°F, how long will it take the drink to warm to 65°F? Assume $k \approx -0.031$.
about 45 min

Photochromatic sunglasses: Sunglasses that darken in sunlight (photochromatic sunglasses) contain millions of molecules of a substance known as *silver halide*. The molecules are transparent indoors in the absence of ultraviolet (UV) light. Outdoors, UV light from the sun causes the molecules to change shape, darkening the lenses in response to the intensity of the UV light. For certain lenses, the function $T(x) = 0.85^x$ models the transparency of the lenses (as a percentage) based on a UV index x. Find the transparency (to the nearest percent), if the lenses are exposed to

77. sunlight with a UV index of 7 (a high exposure).
32% transparent

78. sunlight with a UV index of 5.5 (a moderate exposure).   41% transparent

79. Given that a UV index of 11 is very high and most individuals should stay indoors, what is the minimum transparency percentage for these lenses?
17% transparent

80. Use a trial-and-error process and a graphing calculator to determine the UV index when the lenses are 50% transparent.
UV index of about 4.25

Modeling inflation: Assuming the rate of inflation is 5% per year, the predicted price of an item can be modeled by the function $P(t) = P_0(1.05)^t$, where P_0 represents the initial price of the item and t is in years. Use this information to solve Exercises 81 and 82.

81. What will the price of a new car be in the year 2015, if it cost $20,000 in the year 2010?   ≈$25,526

82. What will the price of a gallon of milk be in the year 2015, if it cost $3.95 in the year 2010? Round to the nearest cent.   ≈$5.04

Modeling radioactive decay: The half-life of a radioactive substance is the time required for half an initial amount of the substance to disappear through decay. The amount of the substance remaining is given by the formula $Q(t) = Q_0(\frac{1}{2})^{\frac{t}{h}}$, where h is the half-life, t represents the elapsed time, and $Q(t)$ represents the amount that remains (t and h must have the same unit of time). Use this information to solve Exercises 83 and 84.

83. Some isotopes of the substance known as thorium have a half-life of only 8 min. (a) If 64 grams are initially present, how many grams (g) of the substance remain after 24 min? (b) How many minutes until only 1 gram (g) of the substance remains?   a. 8 g   b. 48 min

84. Some isotopes of sodium have a half-life of about 16 hr. (a) If 128 g are initially present, how many grams of the substance remain after 2 days (48 hr)? (b) How many hours until only 1 g of the substance remains?   a. 16 g   b. 112 hr

▶ **EXTENDING THE CONCEPT**

85. If $10^{2x} = 25$, what is the value of 10^{-x}? $\frac{1}{5}$

86. If $5^{3x} = 27$, what is the value of 5^{2x}?   9

87. If $3^{0.5x} = 5$, what is the value of 3^{x+1}?   75

88. If $\left(\frac{1}{2}\right)^{x+1} = \frac{1}{3}$, what is the value of $\left(\frac{1}{2}\right)^{-x}$? $\frac{3}{2}$

89. The formula $f(x) = (\frac{1}{2})^x$ gives the probability that "x" number of flips result in heads (or tails). First determine the probability that 20 flips results in *20 heads in a row*. Then use the Internet or some other resource to determine the probability of winning a state lottery (expressed as a decimal). Which has the greater probability? Were you surprised?
9.5×10^{-7}; answers will vary

The growth rate constant that governs an exponential function was introduced on page 495.

90. In later sections, we will easily be able to find the growth constant k for Goldsboro, where $P(t) = 1000 \cdot 2^t$. For now we'll approximate its value using the rate of change formula on a very small interval of the domain. From the definition of an exponential function, $\frac{\Delta P}{\Delta t} = kP(t)$. Since k is constant, we can choose any value of t, say $t = 4$. For $h = 0.0001$, we have $\dfrac{1000 \cdot 2^{4+0.0001} - 1000 \cdot 2^4}{0.0001} = k \cdot P(4)$. (a) Use the equation shown to solve for k (round to thousandths). (b) Show that k is constant by completing the same exercise for $t = 2$ and $t = 6$. (c) Verify that $P(t) = 1000 \cdot 2^t$ and $P(t) = 1000e^{kt}$ give approximately the same results.
a. $k \approx 0.693$ b. verified   c. verified

► **MAINTAINING YOUR SKILLS**

91. (1.3) Given $f(x) = 2x^2 - 3x$, determine:
$f(-1), f(\frac{1}{3}), f(a), f(a + h)$

93. (R.6) Solve the following equations:

    a. $-2\sqrt{x - 3} + 7 = 21$ no solution

    b. $\dfrac{9}{x + 3} + 3 = \dfrac{12}{x - 3}$ $\{-5, 6\}$

Additional answers can be found in the Instructor Answer Appendix.

92. (2.2) Graph $g(x) = \sqrt{x + 2} - 1$ using a shift of a basic function. Then state the domain and range of g.
$x \in [-2, \infty), y \in [-1, \infty)$

94. (R.3) Identify each formula:

    a. $\frac{4}{3}\pi r^3$ volume of a sphere

    b. $\frac{1}{2}bh$ area of a triangle

    c. lwh volume of a rectangular prism

    d. $a^2 + b^2 = c^2$ Pythagorean theorem

5.3   Logarithms and Logarithmic Functions

LEARNING OBJECTIVES

In Section 5.3 you will see how we can:

☐ **A.** Write exponential equations in logarithmic form

☐ **B.** Find common logarithms and natural logarithms

☐ **C.** Graph logarithmic functions

☐ **D.** Find the domain of a logarithmic function

☐ **E.** Solve applications of logarithmic functions

A **transcendental function** is one whose solutions are beyond or *transcend* the methods applied to polynomial functions. The exponential function and its inverse, called the logarithmic function, are transcendental functions. In this section, we'll use the concept of an inverse to develop an understanding of the logarithmic function, which has numerous applications that include measuring pH levels, sound and earthquake intensities, barometric pressure, and other natural phenomena.

A. Exponential Equations and Logarithmic Form

While exponential functions have a large number of significant applications, we can't appreciate their full value until we develop the inverse function. Without it, we're unable to solve all but the simplest equations, of the type encountered in Section 5.2. Using the fact that $f(x) = b^x$ is one-to-one, we have the following:

  1. The function $f^{-1}(x)$ must exist.
  2. We can graph $f^{-1}(x)$ by interchanging the x- and y-coordinates of points from $f(x)$.
  3. The domain of $f(x)$ will become the range of $f^{-1}(x)$.
  4. The range of $f(x)$ will become the domain of $f^{-1}(x)$.
  5. The graph of $f^{-1}(x)$ will be a reflection of $f(x)$ across the line $y = x$.

Table 5.6 contains selected values for $f(x) = 2^x$. The values for $f^{-1}(x)$ in Table 5.7 were found by interchanging x- and y-coordinates. Both functions were then graphed using these values.

Table 5.6
$f(x): y = 2^x$

x	y
-3	$\frac{1}{8}$
-2	$\frac{1}{4}$
-1	$\frac{1}{2}$
0	1
1	2
2	4
3	8

Table 5.7
$f^{-1}(x): x = 2^y$

x	$y = f^{-1}(x)$
$\frac{1}{8}$	-3
$\frac{1}{4}$	-2
$\frac{1}{2}$	-1
1	0
2	1
4	2
8	3

The interchange of x and y and the graphs in Figure 5.23 show that $f^{-1}(x)$ has an x-intercept of $(1, 0)$, a vertical asymptote at $x = 0$, a domain of $x \in (0, \infty)$, and a range of $y \in (-\infty, \infty)$. To find *an equation* for $f^{-1}(x)$, we'll attempt to use the algebraic approach employed previously. For $f(x) = 2^x$,

1. use y instead of $f(x)$: $y = 2^x$.   **2.** interchange x and y: $x = 2^y$.

WORTHY OF NOTE

The word *logarithm* was coined by John Napier in 1614, and loosely translated from its Greek origins means "to reason with numbers."

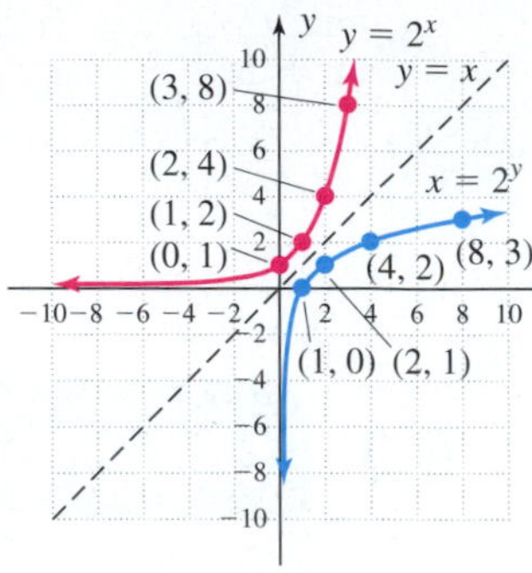

At this point we have an *implicit* equation for the inverse function, but no algebraic operations that enable us to solve *explicitly* for y in terms of x. Instead, we write $x = 2^y$ in function form by noting that "y is the exponent that goes on base 2 to obtain x." In the language of mathematics, this phrase is represented by $y = \log_2 x$ and is called a **logarithmic function** with base 2. For example, from Table 5.7 we have: -3 is the exponent that goes on base 2 to get $\frac{1}{8}$, and this is written $-3 = \log_2 \frac{1}{8}$ since $2^{-3} = \frac{1}{8}$ ✓. For $y = b^x, x = b^y \rightarrow y = \log_b x$ is the inverse function, and is read, "y is the logarithm base b of x." For this new function, we must always keep in mind what y represents—y is an exponent. In fact, y is the exponent that goes on base b to obtain x: $y = \log_b x$.

Figure 5.23

Logarithmic Functions

For positive numbers x and b, with $b \neq 1$,

$$y = \log_b x \text{ if and only if } x = b^y$$

The function $f(x) = \log_b x$ is a logarithmic function with base b. The expression $\log_b x$ is simply called a logarithm, and represents the exponent on b that yields x.

WORTHY OF NOTE

Since base-10 logarithms occur so frequently, we usually use only log x to represent $\log_{10} x$. We do something similar with square roots. Technically, the "square root of x" should be written $\sqrt[2]{x}$. However, square roots are so common we often leave off the two, assuming that if no index is written, an index of two is intended.

Finally, note the equations $x = b^y$ and $y = \log_b x$ are equivalent. We say that $x = b^y$ is the **exponential form** of the equation, whereas $y = \log_b x$ is written in **logarithmic form.** Of all possible bases for $\log_b x$, the most common are base 10 (likely due to our base-10 number system), and base e (due to the advantages it offers in advanced courses). The expression $\log_{10} x$ is called a **common logarithm,** and we simply write log x for $\log_{10} x$. The expression $\log_e x$ is called a **natural logarithm,** and is written in abbreviated form as ln x.

EXAMPLE 1 ▶   **Converting from Logarithmic Form to Exponential Form**

Write each equation in words, then in exponential form.

a. $3 = \log_2 8$ **b.** $1 = \log 10$ **c.** $0 = \ln 1$ **d.** $-2 = \log_3\left(\frac{1}{9}\right)$

Solution ▶   **a.** $3 = \log_2 8 \rightarrow 3$ is the exponent on base 2 for 8: $2^3 = 8$.

b. $1 = \log 10 \rightarrow 1$ is the exponent on base 10 for 10: $10^1 = 10$.

c. $0 = \ln 1 \rightarrow 0$ is the exponent on base e for 1: $e^0 = 1$.

d. $-2 = \log_3\left(\frac{1}{9}\right) \rightarrow -2$ is the exponent on base 3 for $\frac{1}{9}$: $3^{-2} = \frac{1}{9}$.

Now try Exercises 7 through 22 ▶

To convert from exponential form to logarithmic form, note the exponent on the base and read from there. For $5^3 = 125$, "3 is the exponent that goes on base 5 for 125," or *3 is the logarithm base 5 of 125*: $3 = \log_5 125$.

EXAMPLE 2 ▶ **Converting from Exponential Form to Logarithmic Form**

Write each equation in words, then in logarithmic form.

  a. $10^3 = 1000$ **b.** $2^{-1} = \frac{1}{2}$ **c.** $e^2 \approx 7.389$ **d.** $9^{\frac{3}{2}} = 27$

Solution ▶ **a.** $10^3 = 1000 \rightarrow 3$ is the exponent on base 10 for 1000, or
          3 is the logarithm base 10 of 1000: $3 = \log 1000$.

  b. $2^{-1} = \frac{1}{2} \rightarrow -1$ is the exponent on base 2 for $\frac{1}{2}$, or
          -1 is the logarithm base 2 of $\frac{1}{2}$: $-1 = \log_2(\frac{1}{2})$.

  c. $e^2 \approx 7.389 \rightarrow 2$ is the exponent on base e for 7.389, or
          2 is the logarithm base e of 7.389: $2 \approx \ln 7.389$.

  d. $9^{\frac{3}{2}} = 27 \rightarrow \frac{3}{2}$ is the exponent on base 9 for 27, or
          $\frac{3}{2}$ is the logarithm base 9 of 27: $\frac{3}{2} = \log_9 27$.

✓ **A.** You've just seen how we can write exponential equations in logarithmic form

Now try Exercises 23 through 38 ▶

B. Finding Common Logarithms and Natural Logarithms

Some logarithms are easy to evaluate. For example, $\log 100 = 2$ since $10^2 = 100$, and $\log \frac{1}{100} = -2$ since $10^{-2} = \frac{1}{100}$. But what about the expressions $\log 850$ and $\ln 4$? Because logarithmic functions are continuous on their domains, a value exists for $\log 850$ and the equation $10^x = 850$ must have a solution. Further, the inequalities

$$\log 100 < \log 850 < \log 1000$$
$$2 < \log 850 < 3$$

tell us that $\log 850$ must be between 2 and 3. Fortunately, modern calculators can compute base-10 and base-e logarithms instantly, often with nine-decimal-place accuracy. For $\log 850$, press **LOG**, then input 850 and press **ENTER**. The display should read 2.929418926. We can also use the calculator to verify $10^{2.929418926} = 850$ (see Figure 5.24). For $\ln 4$, press the **LN** key, then input 4 and press **ENTER** to obtain 1.386294361. Figure 5.25 verifies that $e^{1.386294361} = 4$.

Figure 5.24

```
log(850)
          2.929418926
10^Ans
                   850
```

Figure 5.25

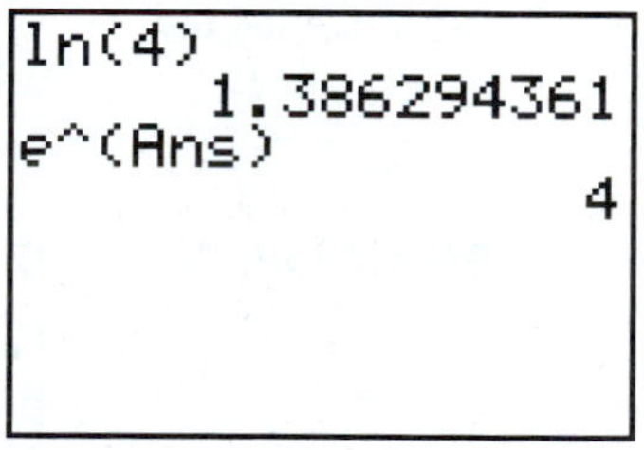

```
ln(4)
          1.386294361
e^(Ans)
                     4
```

EXAMPLE 3 ▶ **Finding the Value of a Logarithm**

Determine the value of each logarithm without using a calculator:

  a. $\log_2 8$ **b.** $\log_5(\frac{1}{25})$ **c.** $\ln e$ **d.** $\log \sqrt{10}$

Solution ▶ **a.** $\log_2 8$ represents the exponent on 2 for 8: $\log_2 8 = 3$, since $2^3 = 8$.

  b. $\log_5(\frac{1}{25})$ represents the exponent on 5 for $\frac{1}{25}$: $\log_5 \frac{1}{25} = -2$, since $5^{-2} = \frac{1}{25}$.

  c. $\ln e$ represents the exponent on e for e: $\ln e = 1$, since $e^1 = e$.

  d. $\log \sqrt{10}$ represents the exponent on 10 for $\sqrt{10}$: $\log \sqrt{10} = \frac{1}{2}$, since $10^{\frac{1}{2}} = \sqrt{10}$.

Now try Exercises 39 through 50 ▶

EXAMPLE 4 ▶ **Using a Calculator to Find Logarithms**

Use a calculator to evaluate each logarithmic expression. Verify the result.

a. log 1857   **b.** log 0.258   **c.** ln 3.592

Solution ▶
a. $\log 1857 = 3.268811904$,   **c.** $\ln 3.592 \approx 1.27870915$
$10^{3.268811904} = 1857 \checkmark$ $e^{1.27870915} \approx 3.592 \checkmark$

b. $\log 0.258 = -0.588380294$,
$10^{-0.588380294} = 0.258 \checkmark$

☑ **B. You've just seen how we can find common logarithms and natural logarithms**

Now try Exercises 51 through 58 ▶

Figure 5.26

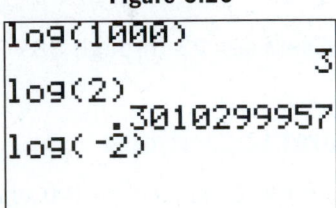

Finally, note that if $x > 10$, the value of log x is greater than 1, but if $0 < x < 10$ the value of log x is less than 1. Also, if $x < 0$, the expression log x does not represent a real number (the domain of $y = \log_b x$ does not include negative numbers). See Figures 5.26 and 5.27.

Figure 5.27

ERR:NONREAL ANS
1∎Quit
2:Goto

C. Graphing Logarithmic Functions

For convenience and ease of calculation, our first examples of logarithmic graphs are done using base-2 logarithms. However, the basic shape of a logarithmic graph remains unchanged regardless of the base used, and transformations can be applied to $y = \log_b(x)$ for any value of b. For $y = a \log(x \pm h) \pm k$, a continues to govern stretches, compressions, and vertical reflections, the graph will shift horizontally h units opposite the sign, and shift k units vertically in the same direction as the sign. Our earlier graph of $y = \log_2 x$ was completed using $x = 2^y$ as the inverse function for $y = 2^x$ (Figure 5.23). For reference, the graph is repeated in Figure 5.28.

WORTHY OF NOTE

As with the basic graphs we studied in Section 2.2, logarithmic graphs maintain the same characteristics when transformations are applied, and these graphs *should be added to your collection of basic functions,* ready for recall or analysis as the situation requires.

Figure 5.28

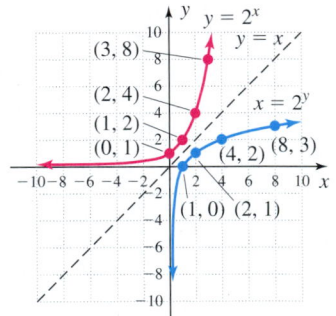

EXAMPLE 5 ▶ **Graphing Logarithmic Functions Using Transformations**

Graph $f(x) = \log_2(x - 3) + 1$ using transformations of $y = \log_2 x$ (not by simply plotting points). Clearly state what transformations are applied.

Solution ▶ The graph of f is the same as that of $y = \log_2 x$, shifted 3 units right and 1 unit up. The vertical asymptote will be at $x = 3$ and the point $(1, 0)$ from the basic graph becomes $(1 + 3, 0 + 1) = (4, 1)$. Knowing the graph's basic shape, we compute one additional point using $x = 7$:

$$f(7) = \log_2(7 - 3) + 1$$
$$= \log_2 4 + 1$$
$$= 2 + 1$$
$$= 3$$

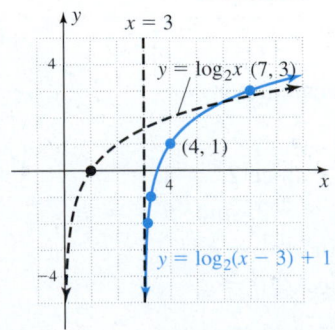

The point $(7, 3)$ is on the graph, shown in the figure.

Now try Exercises 59 through 72 ▶

As with the exponential functions, much can be learned from graphs of logarithmic functions and a summary of important characteristics is given here.

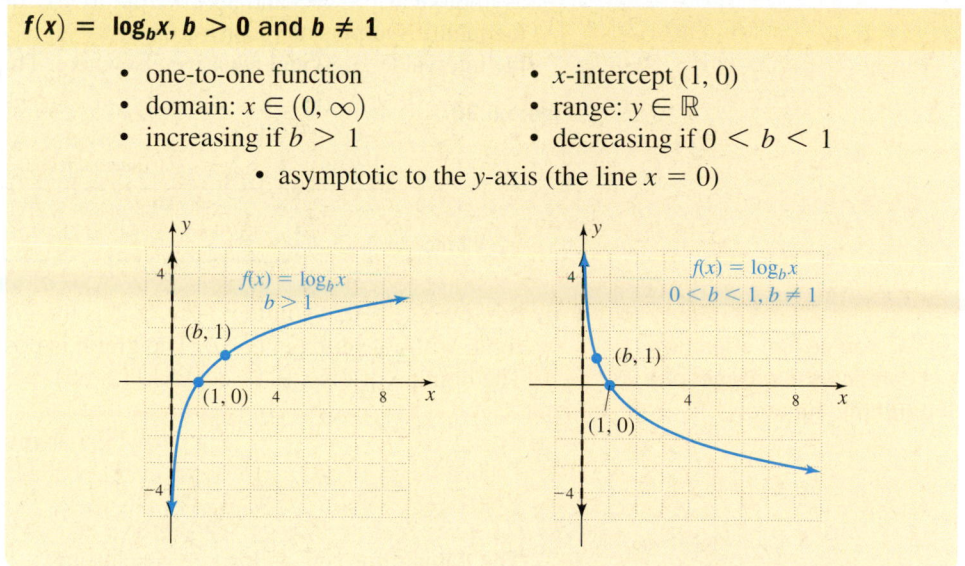

$f(x) = \log_b x, b > 0$ and $b \neq 1$

- one-to-one function
- domain: $x \in (0, \infty)$
- increasing if $b > 1$

- x-intercept $(1, 0)$
- range: $y \in \mathbb{R}$
- decreasing if $0 < b < 1$

- asymptotic to the y-axis (the line $x = 0$)

☑ **C. You've just seen how we can graph logarithmic functions**

D. Finding the Domain of a Logarithmic Function

Examples 5 and 6 illustrate how the domain of a logarithmic function can change when certain transformations are applied. Since the domain consists of *positive* real numbers, the argument of a logarithmic function must be greater than zero. This means finding the domain often consists of solving various inequalities, which can be done using the skills acquired in Sections 3.2 and 4.6.

EXAMPLE 6 ▶ **Finding the Domain of a Logarithmic Function**

Determine the domain of each function.

a. $p(x) = \log_2(2x + 3)$ **b.** $q(x) = \log_5(x^2 - 2x)$

c. $r(x) = \log\left(\dfrac{3 - x}{x + 3}\right)$ **d.** $f(x) = \ln|x - 2|$

Solution ▶ Begin by writing the argument of each function as a "greater than" inequality.

a. Solving $2x + 3 > 0$ for x gives $x > -\frac{3}{2}$, and the domain of p is $x \in (-\frac{3}{2}, \infty)$.

b. For $x^2 - 2x > 0$, we note $y = x^2 - 2x$ is a parabola, opening upward, with zeroes at $x = 0$ and $x = 2$ (see Figure 5.29). This means $x^2 - 2x$ will be positive for $x < 0$ and $x > 2$. The domain of q is $x \in (-\infty, 0) \cup (2, \infty)$.

Figure 5.29

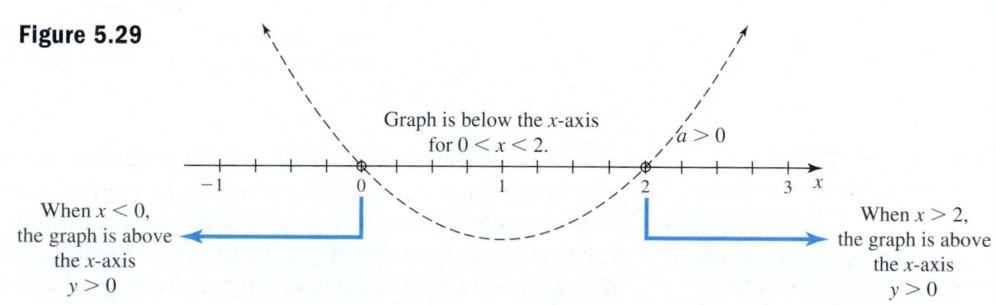

c. For $\dfrac{3-x}{x+3} > 0$, we note $y = \dfrac{3-x}{x+3}$ has a zero at $x = 3$, with a vertical asymptote at $x = -3$ and here we opt to use the interval test method to solve the inequality. Outputs are positive when $x = 0$ (see Figure 5.30), so y is positive in the interval $(-3, 3)$ and negative elsewhere. The domain of r is $x \in (-3, 3)$.

Figure 5.30

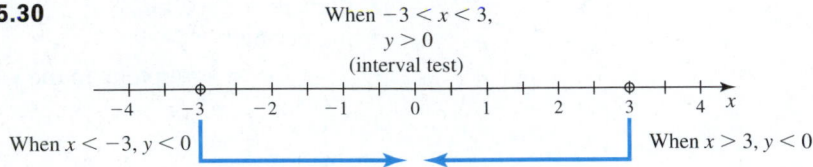

When $-3 < x < 3$,
$y > 0$
(interval test)

When $x < -3, y < 0$

When $x > 3, y < 0$

d. For $|x - 2| > 0$, we note $y = |x - 2|$ is the graph of $y = |x|$ shifted 2 units right, with its vertex at $(2, 0)$. The graph is positive for all x, except at $x = 2$. The domain of f is $x \in (-\infty, 2) \cup (2, \infty)$.

☑ **D.** You've just seen how we can find the domain of a logarithmic function

Now try Exercises 73 through 78 ▶

The domain for $r(x) = \log\left(\dfrac{3-x}{x+3}\right)$ from Example 7c can also be confirmed using the ⎡LOG⎤ key on a graphing calculator. Use this key to enter the equation as Y_1 on the ⎡Y=⎤ screen, then graph the function using the ⎡ZOOM⎤ **4:ZDecimal** option. Both the graph (Figure 5.31) and TABLE feature help to confirm the domain is $x \in (-3, 3)$.

Figure 5.31

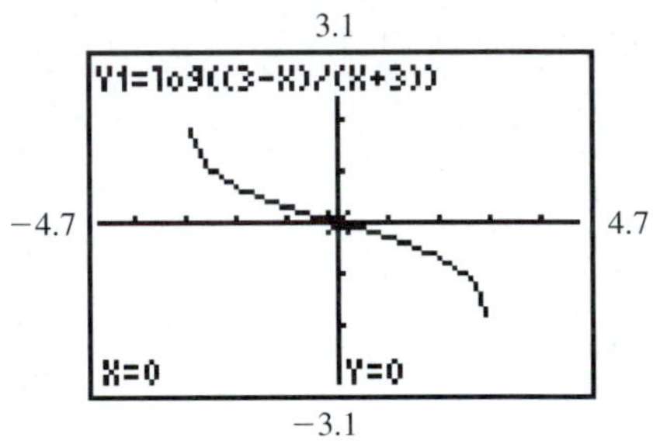

3.1

Y1=log((3-X)/(X+3))

-4.7   4.7

X=0   Y=0

-3.1

E. Applications of Logarithms

The use of logarithmic scales as a tool of measurement is primarily due to the range of values for the phenomenon being measured. For instance, time is generally measured on a linear scale, and for short periods a linear scale is appropriate. For the time line in Figure 5.32, each tick-mark *represents 1 unit,* and the time line can display a period of 10 yr. However, the scale would be useless in a study of geology or the age of the universe. If we scale the number line logarithmically, each tick-mark *represents a power of 10* (Figure 5.33) and a scale of the same length can now display a time period of 10 billion years.

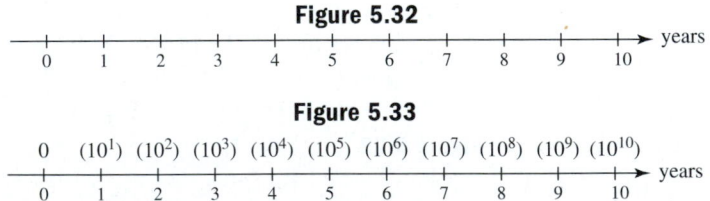

Figure 5.32

years

Figure 5.33

0   (10^1) (10^2) (10^3) (10^4) (10^5) (10^6) (10^7) (10^8) (10^9) (10^{10})

years

WORTHY OF NOTE

The **decibel** (dB) is the reference unit for sound, and is based on the faintest sound a person can hear, called the **threshold of audibility.** It is a base-10 logarithmic scale, meaning a sound 10 times more intense is one bel louder.

In much the same way, logarithmic measures are needed in a study of sound and earthquake intensity, as the scream of a jet engine is over 1 billion times more intense than the threshold of hearing, and the most destructive earthquakes are billions of times stronger than the slightest earth movement that can be felt. Similar ranges exist in the measurement of light, acidity, and voltage. Figures 5.34 and 5.35 show logarithmic scales for measuring sound in decibels (1 bel = 10 decibels) and earthquake intensity in Richter values (or magnitudes).

Figure 5.34

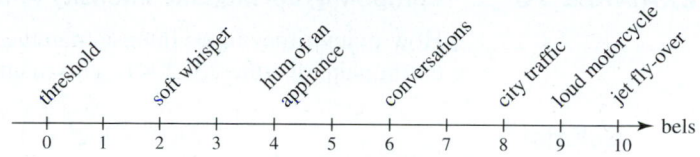

Figure 5.35

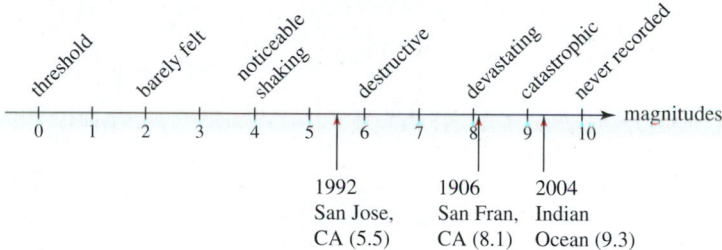

The slightest earth movement perceptible is called the **reference intensity** I_0, with the intensity I of stronger earthquakes expressed as a multiple of I_0. The earthquake that struck Haiti in January of 2010 was measured at over 10,500,000 times this reference intensity, or $I = 10,5000,000I_0$. To find the Richter value (magnitude) of this earthquake, we simply take the base-10 logarithm of the ratio $\dfrac{I}{I_0}$ to express these values on a logarithmic scale. In function form, $M(I) = \log\left(\dfrac{I}{I_0}\right)$, and we find that the Haitian earthquake had a magnitude of just over 7.0: $\log\left(\dfrac{10,500,000I_0}{I_0}\right) = \log(10,500,000) \approx 7.0$.

EXAMPLE 7A ▶ **Finding the Magnitude of an Earthquake**

Find the magnitude of the earthquakes (rounded to hundredths) with the intensities given.

   a. Eureka earthquake; January 9, 2010, near Humboldt county, California: $I = 3{,}162{,}000I_0$.

   b. Sumatra-Andaman earthquake; December 26, 2004, near the west coast of Sumatra, Indonesia: $I = 1{,}995{,}260{,}000I_0$.

Solution ▶   **a.**

$$M(I) = \log\left(\frac{I}{I_0}\right) \qquad \text{magnitude equation}$$

$$M(3{,}162{,}000I_0) = \log\left(\frac{3{,}162{,}000I_0}{I_0}\right) \qquad \text{substitute } 3{,}162{,}000I_0 \text{ for } I$$

$$= \log 3{,}162{,}000 \qquad \text{simplify}$$

$$\approx 6.5 \qquad \text{result}$$

The earthquake had a magnitude of about 6.5.

  b.

$$M(I) = \log\left(\frac{I}{I_0}\right) \qquad \text{magnitude equation}$$

$$M(1{,}995{,}260{,}000I_0) = \log\left(\frac{1{,}995{,}260{,}000I_0}{I_0}\right) \qquad \text{substitute } 1{,}995{,}260{,}000I_0 \text{ for } I$$

$$= \log 1{,}995{,}260{,}000 \qquad \text{simplify}$$

$$\approx 9.3 \qquad \text{result}$$

The earthquake had a magnitude of about 9.3.

EXAMPLE 7B ▶   **Comparing Earthquake Intensity to the Reference Intensity**

How many times more intense than the reference intensity I_0 was the Peruvian earthquake of June 23, 2001, with magnitude 8.4.

Solution ▶

$$M(I) = \log\left(\frac{I}{I_0}\right)$$ magnitude equation

$$8.4 = \log\left(\frac{I}{I_0}\right)$$ substitute 8.4 for $M(I)$

$$10^{8.4} = \left(\frac{I}{I_0}\right)$$ exponential form

$$I = 10^{8.4}I_0$$ solve for I

$$I \approx 251{,}188{,}643I_0$$ $10^{8.4} \approx 251{,}188{,}643$

The earthquake was over 251 million times more intense than the reference intensity.

EXAMPLE 7C ▶   **Comparing Earthquake Intensities**

Referring to Example 7A, how many times more intense was the Sumatra earthquake as compared to the Eureka earthquake?

Solution ▶   The Sumatra quake had a Richter value of 9.3, with an intensity of $10^{9.3}$. Similarly, the Eureka quake was measured at 6.5 on the Richter scale, with an intensity of $10^{6.5}$. Using these intensities, we find that the Sumatra quake was $\frac{10^{9.3}}{10^{6.5}} = 10^{2.8}$ or about 631 times more intense than the Eureka quake.

Now try Exercises 81 through 94 ▶

A second application of logarithmic functions involves the relationship between altitude and barometric pressure. The altitude or height above sea level can be determined by the formula $H = (30T + 8000) \ln\left(\frac{P_0}{P}\right)$, where H is the altitude in meters for a temperature T in degrees Celsius, P is the barometric pressure at a given altitude in units called **centimeters of mercury** (cmHg), and P_0 is the barometric pressure at sea level: 76 cmHg.

EXAMPLE 8 ▶   **Using Logarithms to Determine Altitude**

Hikers at the summit of Mt. Shasta in northern California take a pressure reading of 45.1 cmHg at a temperature of 9°C. How high is Mt. Shasta?

Solution ▶   For this exercise, $P_0 = 76$, $P = 45.1$, and $T = 9$. The formula yields

$$H = (30T + 8000) \ln\left(\frac{P_0}{P}\right)$$ given formula

$$= [30(9) + 8000] \ln\left(\frac{76}{45.1}\right)$$ substitute given values

$$= 8270 \ln\left(\frac{76}{45.1}\right)$$ simplify

$$\approx 4316$$ result

Mt. Shasta is about 4316 m high.

Now try Exercises 95 through 98 ▶

Our final application shows the versatility of logarithmic functions, and their value as a real-world model. Large advertising agencies are well aware that after a new ad campaign, sales will increase rapidly as more people become aware of the product. Continued advertising will give the new product additional market share, but once the "newness" wears off and the competition begins responding, sales tend to taper off—regardless of any additional amount spent on ads. This phenomenon can be modeled by the function

$$S(d) = k + a \ln d$$

where $S(d)$ is the number of expected sales after d dollars are spent, and a and k are constants related to product type and market size.

EXAMPLE 9 ▶ **Using Logarithms for Marketing Strategies**

Market research has shown that sales of the MusicMaster, a new system for downloading and playing music, can be approximated by the equation $S(d) = 2500 + 250 \ln d$, where $S(d)$ is the number of sales after d thousand dollars is spent on advertising. The graph of $y = S(d)$ is shown.

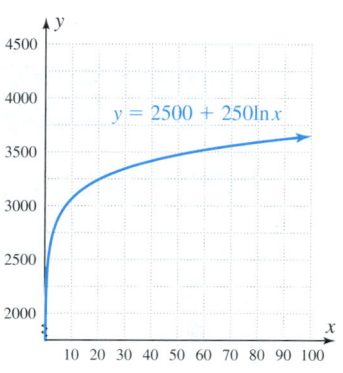

a. What sales volume is expected if the advertising budget is $40,000?

b. If the company needs to sell 3500 units to begin making a profit, how much should be spent on advertising?

Solution ▶ **a.** For sales volume, we simply evaluate the function for $d = 40$ (d in thousands):

$$
\begin{aligned}
S(d) &= 2500 + 250 \ln d &&\textcolor{red}{\text{given equation}} \\
S(40) &= 2500 + 250 \ln 40 &&\textcolor{red}{\text{substitute 40 for } d} \\
&\approx 2500 + 922 &&\textcolor{red}{250 \ln 40 \approx 922} \\
&= 3422
\end{aligned}
$$

Spending $40,000 on advertising will generate approximately 3422 sales.

b. To find the advertising budget needed, we substitute number of sales and solve for d.

$$
\begin{aligned}
S(d) &= 2500 + 250 \ln d &&\textcolor{red}{\text{given equation}} \\
3500 &= 2500 + 250 \ln d &&\textcolor{red}{\text{substitute 3500 for } S(d)} \\
1000 &= 250 \ln d &&\textcolor{red}{\text{subtract 2500}} \\
4 &= \ln d &&\textcolor{red}{\text{divide by 250}} \\
e^4 &= d &&\textcolor{red}{\text{exponential form}} \\
54.598 &\approx d &&\textcolor{red}{e^4 \approx 54.598}
\end{aligned}
$$

✓ **E.** You've just seen how we can solve applications of logarithmic functions

About $54,600 should be spent in order to sell 3500 units.

Now try Exercises 99 and 100 ▶

From the graph of $S(d)$ given in Example 10, it is apparent that while the number of sales continues to grow as more money is spent, the rate of growth slows considerably beyond $50,000. In cases like this a *numeric* view of what's happening can be more meaningful. Here we'll use the **TABLE** feature in an entirely new way to investigate the number of sales gained for each additional $1000 spent. For $Y_1 = 2500 + 250 \ln x$, we'll enter $Y_2 = Y_1(x) - Y_1(x - 1)$, which will automatically have the calculator find the difference between the current number of sales (spending x thousand dollars), and the number of sales made when $1000 less is spent ($x - 1$ dollars). Figure 5.36 shows that initially each additional $1000 spent results in a substantial sales increase, while Figure 5.37 shows very minor increases for a like amount spent.

Figure 5.36

X	Y1	Y2
3	2774.7	101.37
4	2846.6	71.921
5	2902.4	55.786
6	2947.9	45.58
7	2986.5	38.538
8	3019.9	33.383
9	3049.3	29.446

X=3

Figure 5.37

X	Y1	Y2
80	3595.5	3.1447
81	3598.6	3.1056
82	3601.7	3.0675
83	3604.7	3.0303
84	3607.7	2.994
85	3610.7	2.9586
86	3613.6	2.924

X=80

There are a number of other interesting applications of logarithmic functions in the Exercise set. **See Exercises 101 through 106.**

5.3 EXERCISES

▶ CONCEPTS AND VOCABULARY

Fill in each blank with the appropriate word or phrase. Carefully reread the section if needed.

1. A logarithmic function is of the form $y = \underline{\log_b x}$, where $\underline{b} > 0$, $\underline{b} \neq 1$ and inputs are $\underline{\text{greater}}$ than zero.

2. The range of $y = \log_b x$ is all $\underline{\text{real}}$ $\underline{\text{numbers}}$, and the domain is $x \in \underline{(0, \infty)}$. Further, as $x \to 0$, $y \to \underline{-\infty}$.

3. For logarithmic functions of the form $y = \log_b x$, the x-intercept is $\underline{(1, 0)}$, since $\log_b 1 = \underline{0}$.

4. The function $y = \log_b x$ is an increasing function if $\underline{b > 1}$, and a decreasing function if $\underline{0 < b < 1}$.

5. What number does the expression $\log_2 32$ represent? Discuss/Explain how $\log_2 32 = \log_2 2^5$ justifies this fact.   5; answers will vary

6. Explain how the graph of $Y = \log_b(x - 3)$ can be obtained from $y = \log_b x$. Where is the "new" x-intercept? Where is the new asymptote?

   shifts right 3; (4, 0); $x = 3$

▶ DEVELOPING YOUR SKILLS

Write each equation in exponential form.

7. $3 = \log_2 8$ $2^3 = 8$ 8. $2 = \log_3 9$ $3^2 = 9$

9. $-1 = \log_7 \frac{1}{7}$ $7^{-1} = \frac{1}{7}$ 10. $-3 = \ln \frac{1}{e^3}$ $e^{-3} = \frac{1}{e^3}$

11. $0 = \log_9 1$ $9^0 = 1$ 12. $0 = \ln 1$ $e^0 = 1$

13. $\frac{1}{3} = \log_8 2$ $8^{\frac{1}{3}} = 2$ 14. $\frac{1}{2} = \log_{81} 9$ $81^{\frac{1}{2}} = 9$

15. $1 = \log_2 2$ $2^1 = 2$ 16. $1 = \ln e$ $e^1 = e$

17. $\log_7 49 = 2$ $7^2 = 49$ 18. $\log_4 16 = 2$ $4^2 = 16$

19. $\log 100 = 2$ $10^2 = 100$ 20. $\log 10{,}000 = 4$ $10^4 = 10{,}000$

21. $\ln (54.598) \approx 4$ $e^4 \approx 54.598$ 22. $\log 0.001 = -3$ $10^{-3} = 0.001$

Write each equation in logarithmic form.

23. $4^3 = 64$ $\log_4 64 = 3$ 24. $e^3 \approx 20.086$ $3 \approx \ln 20.086$

25. $3^{-2} = \frac{1}{9}$ $\log_3 \frac{1}{9} = -2$ 26. $2^{-3} = \frac{1}{8}$ $\log_2 \frac{1}{8} = -3$

27. $e^0 = 1$ $0 = \ln 1$ 28. $8^0 = 1$ $\log_8 1 = 0$

29. $\left(\frac{1}{3}\right)^{-3} = 27$ $\log_{\frac{1}{3}} 27 = -3$ 30. $\left(\frac{1}{5}\right)^{-2} = 25$ $\log_{\frac{1}{5}}(25) = -2$

31. $10^3 = 1000$ $\log 1000 = 3$ 32. $e^1 = e$ $1 = \ln e$

33. $10^{-2} = \frac{1}{100}$ 34. $10^{-5} = \frac{1}{100{,}000}$

35. $4^{\frac{3}{2}} = 8$ 36. $e^{\frac{3}{4}} \approx 2.117$

37. $4^{-\frac{3}{2}} = \frac{1}{8}$ $\log_4 \frac{1}{8} = \frac{-3}{2}$ 38. $27^{-\frac{2}{3}} = \frac{1}{9}$ $\log_{27} \frac{1}{9} = \frac{-2}{3}$

Additional answers can be found in the Instructor Answer Appendix.

HOMEWORK SELECTION GUIDE

Core: 7–75 every other odd, 81–86, 91–94, 101, 102, 112 (31 Exercises)
Standard: 1–4, 7–75 every other odd, 79, 81–86, 91–94, 101–104, 108, 112, 114 (40 Exercises)

Extended: 1–4, 7–75 every other odd, 79–86, 91–94, 101–104, 108, 109, 112–114 (43 Exercises)
In Depth: 1–6, 7–75 every other odd, 79–86, 91–94, 101–104, 108–114 (48 Exercises)

Determine the value of each logarithm without using a calculator.

39. $\log_4 4$ 1

40. $\log_9 9$ 1

41. $\log_{11} 121$ 2

42. $\log_{12} 144$ 2

43. $\ln e$ 1

44. $\ln e^2$ 2

45. $\log_4 2$ $\frac{1}{2}$

46. $\log_{81} 9$ $\frac{1}{2}$

47. $\log_7 \frac{1}{49}$ -2

48. $\log_9 \frac{1}{81}$ -2

49. $\ln \frac{1}{e^2}$ -2

50. $\ln \frac{1}{\sqrt{e}}$ $-\frac{1}{2}$

Use a calculator to evaluate each expression, rounded to four decimal places.

51. $\log 50$ 1.6990

52. $\log 47$ 1.6721

53. $\ln 1.6$ 0.4700

54. $\ln 0.75$ -0.2877

55. $\ln 225$ 5.4161

56. $\ln 381$ 5.9428

57. $\log \sqrt{37}$ 0.7841

58. $\log 4\pi$ 1.0992

Graph each function *using transformations* of $y = \log_b x$ and strategically plotting a few points. Clearly state the transformations applied.

59. $f(x) = \log_2 x + 3$

60. $g(x) = \log_2(x - 2)$

61. $h(x) = \log_2(x - 2) + 3$

62. $p(x) = \log_3 x - 2$

63. $q(x) = \ln(x + 1)$

64. $r(x) = \ln(x + 1) - 2$

65. $Y_1 = -\ln(x + 1)$

66. $Y_2 = -\ln x + 2$

Use the transformation equation $y = af(x \pm h) \pm k$ and the asymptotes and intercept(s) of the parent function to match each equation to one of the graphs given. Assume $b > 1$.

67. $y = \log_b(x + 2)$ II

68. $y = 2\log_b x$ III

69. $y = 1 - \log_b x$ VI

70. $y = \log_b x - 1$ I

71. $y = \log_b x + 2$ V

72. $y = -\log_b x$ IV

I.

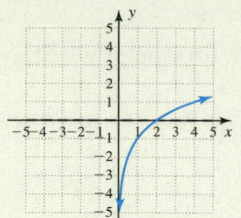

II.

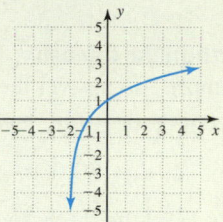

III.

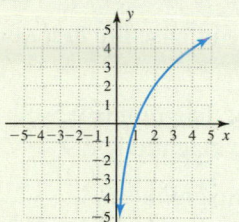

IV.

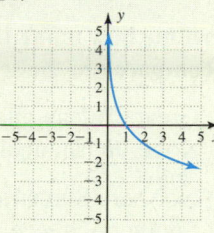

V.

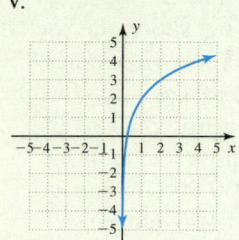

VI.

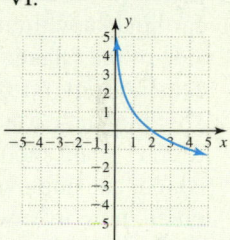

Determine the domain of the following functions.

73. $y = \log_6\left(\dfrac{x + 1}{x - 3}\right)$

74. $y = \ln\left(\dfrac{x - 2}{x + 3}\right)$

75. $y = \log_5 \sqrt{2x - 3}$

76. $y = \ln\sqrt{5 - 3x}$

77. $y = \log(9 - x^2)$

78. $y = \ln(9x - x^2)$

▶ **WORKING THE FORMULAS**

79. pH level: $f(x) = -\log x$

The pH level of a solution indicates the concentration of hydrogen (H^+) ions in a unit called *moles per liter*. The pH level $f(x)$ is given by the formula shown (often written as $pH = -\log[H^+]$), where x is the ion concentration (given in scientific notation). A solution with pH < 7 is called an acid (lemon juice: pH ≈ 2), and a solution with pH > 7 is called a base (household ammonia: pH ≈ 11). Use the formula to determine the pH level of tomato juice if $x = 7.94 \times 10^{-5}$ moles per liter. Is this an acid or base solution? pH ≈ 4.1; acid

80. Time required for an investment to double:

$$T(r) = \frac{\log 2}{\log(1 + r)}$$

The time required for an investment to double in value is given by the formula shown, where $T(r)$ represents the time required for an investment to double if invested at interest rate r (expressed as a decimal). How long would it take an investment to double if the interest rate were (a) 5%, (b) 8%, (c) 12%? **a.** ≈ 14.2 yr **b.** ≈ 9.0 yr **c.** ≈ 6.1 yr

Additional answers can be found in the Instructor Answer Appendix.

▶ APPLICATIONS

 Earthquake intensity: Use the information provided in Example 8 to answer the following.

81. Find the value of $M(I)$ given

    a. $I = 50,000I_0$ **b.** $I = 75,000,000\ I_0$.
      ≈ 4.7 ≈ 7.9

82. Find the intensity I of the earthquake given

    a. $M(I) = 3.2$ **b.** $M(I) = 8.1$.
      $\approx 1584.9I_0$ $\approx 125{,}892{,}541.2I_0$

Determine how many times more intense the first quake was compared to the second.

83. Great Chilean quake (1960): magnitude 9.5
    Kobe, Japan, quake (1995): magnitude 6.9
    about 398 times

84. Northern Sumatra (2004): magnitude 9.1
    Southern Greece (2008): magnitude 4.5
    about 39,811 times

85. Earthquake intensity: On June 25, 1989, an earthquake with magnitude 6.2 shook the southeast side of the Island of Hawaii (near Kalapana), causing some $1,000,000 in damage. On October 15, 2006, an earthquake measuring 6.7 on the Richter scale shook the northwest side of the island, causing over $100,000,000 in damage. How much more intense was the 2006 quake?
    about 3.2 times

86. Earthquake intensity: The most intense earthquake of the modern era occurred in Chile on May 22, 1960, and measured 9.5 on the Richter scale. How many times more intense was this earthquake, than the quake that hit Northern Sumatra (Indonesia) on March 28, 2005, and measured 8.7?   about 6.3 times

Brightness of a star: The brightness or intensity I of a star as perceived by the naked eye is measured in units called *magnitudes*. The brightest stars have magnitude 1 $[M(I) = 1]$ and the dimmest have magnitude 6 $[M(I) = 6]$. The magnitude of a star is given by the equation $M(I) = 6 - 2.5 \cdot \log\left(\dfrac{I}{I_0}\right)$, where I is the actual intensity of light from the star and I_0 is the faintest light visible to the human eye, called the reference intensity. The intensity I is often given as a multiple of this reference intensity.

87. Find the value of $M(I)$ given

    a. $I = 27I_0$ and   **b.** $I = 85I_0$.
      ≈ 2.4 ≈ 1.2

88. Find the intensity I of a star given

    a. $M(I) = 1.6$ and   **b.** $M(I) = 5.2$.
      $\approx 57.5I_0$ $\approx 2.1I_0$

Intensity of sound: The intensity of sound as perceived by the human ear is measured in units called decibels (dB). The loudest sounds that can be withstood without damage to the eardrum are in the 120- to 130-dB range, while a whisper may measure in the 15- to 20-dB range. Decibel measure is given by the equation $D(I) = 10 \log\left(\dfrac{I}{I_0}\right)$, where I is the actual intensity of the sound and I_0 is the faintest sound perceptible by the human ear—called the reference intensity. The intensity I is often given as a multiple of this reference intensity, with the constant 10^{-16} (watts per cm^2; W/cm^2) used as the threshold of audibility.

89. Find the value of $D(I)$ given

    a. $I = 10^{-14}$ and   **b.** $I = 10^{-4}$.
      20 dB   120 dB

90. Find the intensity I of the sound given

    a. $D(I) = 83$ and   **b.** $D(I) = 125$.
      199,526,231.5I_0 $\approx 3.2 \times 10^{12}I_0$

Determine how many times more intense the first sound is compared to the second.

91. pneumatic hammer: 11.2 bels
    heavy lawn mower: 8.5 bels   about 501 times

92. train horn: 7.5 bels
    soft music: 3.4 bels   about 12,589 times

93. Sound intensity of a hair dryer: Every morning (it seems), Jose is awakened by the mind-jarring, ear-jamming sound of his daughter's hair dryer (75 dB). He knew he was exaggerating, but told her (many times) of how it reminded him of his railroad days, when the air compressor for the pneumatic tools was running (110 dB). In fact, how many times more intense was the sound of the air compressor compared to the sound of the hair dryer?   about 3162 times

94. Sound intensity of a busy street: The decibel level of noisy, downtown traffic has been estimated at 87 dB, while the laughter and banter at a loud party might be in the 60 dB range. How many times more intense is the sound of the downtown traffic?
    about 501 times

The *barometric equation* $H = (30T + 8000) \ln\left(\dfrac{P_0}{P}\right)$ was discussed in Example 9.

95. Temperature and atmospheric pressure: Determine the height of Mount McKinley (Alaska), if the temperature at the summit is $-10°C$, with a barometric reading of 34 cmHg.   6194 m

Additional answers can be found in the Instructor Answer Appendix.

96. Temperature and atmospheric pressure: A large passenger plane is flying cross-country. The instruments on board show an air temperature of 3°C, with a barometric pressure of 22 cmHg. What is the altitude of the plane?   10,029 m

97. Altitude and atmospheric pressure: By definition, a mountain pass is a low point between two mountains. Passes may be very short with steep slopes, or as large as a valley between two peaks. Perhaps the highest drivable pass in the world is the Semo La pass in central Tibet. At its highest elevation, a temperature reading of 8°C was taken, along with a barometer reading of 39.3 cmHg. (a) Approximately how high is the Semo La pass? (b) While traveling up to this pass, an elevation marker is seen. If the barometer reading was 47.1 cmHg at a temperature of 12°C, what height did the marker give?
a. about 5434 m   **b.** 4000 m

98. Altitude and atmospheric pressure: Hikers on Mt. Everest take successive readings of 35 cmHg at 5°C and 30 cmHg at −10°C. (a) How far up the mountain are they at each reading? (b) Approximate the height of Mt. Everest if the temperature at the summit is −27°C and the barometric pressure is 22.2 cmHg.
a. 6319.4 m, 7157.4 m   **b.** 8848.3 m (true height: 8850 m)

99. Marketing budgets: An advertising agency has determined the number of items sold by a certain client is modeled by the equation $N(A) = 1500 + 315 \ln A$, where $N(A)$ represents the number of sales after spending A thousands of dollars on advertising. Determine the approximate number of items sold on an advertising budget of (a) $10,000; (b) $50,000. (c) Use the TABLE feature of a calculator to estimate how large a budget is needed (to the nearest $500 dollars) to sell 3000 items.   **a.** 2225 items   **b.** 2732 items   **c.** $117,000

100. Sports promotions: The accountants for a major boxing promoter have determined that the number of pay-per-view subscriptions sold to their championship bouts can be modeled by the function $N(d) = 15,000 + 5850 \ln d$, where $N(d)$ represents the number of subscriptions sold after spending d thousand dollars on promotional activities. Determine the number of subscriptions sold if (a) $50,000 and (b) $100,000 is spent. (c) Determine how much should be spent (to the nearest $1000 dollars) to sell over 50,000 subscriptions by simplifying the logarithmic equation and writing the result in exponential form.
a. 37,885 subscriptions   **b.** 41,940 subscriptions   **c.** $397,000

101. Home ventilation: In the construction of new housing, there is considerable emphasis placed on correct ventilation. If too little outdoor air enters a home, pollutants can sometimes accumulate to levels that pose a health risk. For homes of various sizes, ventilation requirements have been established and are based on floor area and the number of bedrooms. For a three-bedroom home, the relationship can be modeled by the function $C(x) = 42 \ln x - 270$, where $C(x)$ represents the number of cubic feet of air per minute (cfm) that should be exchanged with outside air in a home with floor area x (in square feet). (a) How many cfm of exchanged air are needed for a three-bedroom home with a floor area of 2500 ft^2? (b) If a three-bedroom home is being mechanically ventilated by a system with 40 cfm capacity, what is the square footage of the home, assuming it is built to code?
a. about 58.6 cfm   **b.** about 1605 ft^2

102. Runway takeoff distance: Many will remember the August 27, 2006, crash of a commuter jet at Lexington's Blue Grass Airport, that

was mistakenly trying to take off on a runway that was just too short. Forty-nine lives were lost. The minimum required length of a runway depends on the maximum allowable takeoff weight (mtw) of a specific plane. This relationship can be approximated by the function $L(x) = 2085 \ln x - 14,900$, where $L(x)$ represents the required length of a runway in feet, for a plane with x mtw in pounds.

a. The Airbus-320 has a 169,750 lb mtw. What minimum runway length is required for takeoff?   about 10,208 ft

b. By simplifying the logarithmic equation that results and writing the equation in exponential form, determine the mtw of a Learjet 30, which requires a runway of 5550 ft to takeoff safely.   about 18,181 lb

Memory retention: Under certain conditions, a person's retention of random facts can be modeled by the equation $P(x) = 95 - 14 \log_2 x$, where $P(x)$ is the percentage of those facts retained after x number of days. Find the percentage of facts a person might retain after x days for the values given. Note that many of the values given are powers of 2. Use the change-of-base formula those that are not.

103. a. 1 day   **b.** 4 days   **c.** 16 days
　　　95%　　　67%　　　39%
104. a. 32 days   **b.** 64 days   **c.** 78 days
　　　25%　　　11%　　　≈7%

105. pH level: Use the formula given in Exercise 79 to determine the pH level of black coffee if $x = 5.1 \times 10^{-5}$ moles per liter. Is black coffee considered an acid or base solution?   ≈4.3; acid

106. Tripling time: The length of time required for an amount of money to *triple* is given by the formula
$$T(r) = \frac{\log 3}{\log(1 + r)} \text{ (see Exercise 80).}$$ Use the TABLE feature of a graphing calculator to help estimate what interest rate is needed for an investment to triple in nine years.   ≈13%

▶ **EXTENDING THE CONCEPT**

107. Many texts and reference books give estimates of the noise level (in decibels dB) of common sounds. Through reading and research, try to locate or approximate where the following sounds would fall along this scale. In addition, determine at what point pain or ear damage begins to occur.   Answers will vary.

    a. threshold of audibility   **b.** lawn mower   90 dB
        0 dB
    c. whisper   15 dB   **d.** loud rock concert
                       120 dB
    e. lively party   100   **f.** jet engine   140 dB

108. Determine the value of x that makes the equation true: $\log_3[\log_3(\log_3 x)] = 0.$ 27

109. Find the value of each expression without using a calculator.

    a. $\log_{64}\frac{1}{16}$ $\frac{-2}{3}$ **b.** $\log_{\frac{4}{9}}\frac{27}{8}$ $\frac{-3}{2}$ **c.** $\log_{0.25}32$ $\frac{-5}{2}$

110. Suppose you and I represent two different numbers. Is the following cryptogram true or false? *The log of me base me is one and the log of you base you is one, but the log of you base me is equal to the log of me base you turned upside down.*

▶ **MAINTAINING YOUR SKILLS**

111. (2.2) Graph $g(x) = \sqrt[3]{x + 2} - 1$ by shifting the parent function. Then state the domain and range of g.

112. (R.4) Factor the following expressions:

    a. $x^3 - 8$ **b.** $a^2 - 49$
    c. $n^2 - 10n + 25$ **d.** $2b^2 - 7b + 6$

113. (4.2/4.6) For the graph shown, write the solution set for $f(x) < 0$. Then write the equation of the graph in factored form and in polynomial form.

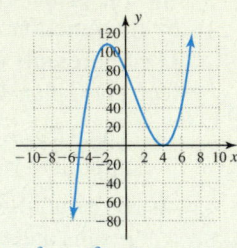

$x \in (-\infty, -5); f(x) = (x + 5)(x - 4)^2 = x^3 - 3x^2 - 24x + 80$

114. (2.1) A function $f(x)$ is defined by the ordered pairs shown in the table. Is the function (a) linear? (b) increasing? Justify your answers.

    a. No, rate of change is not constant;
    b. No, outputs are decreasing.

x	y
-10	0
-9	-2
-8	-8
-6	-18
-5	-50
-4	-72

Additional answers can be found in the Instructor Answer Appendix.

5.4   Properties of Logarithms

LEARNING OBJECTIVES

In Section 5.4 you will see how we can:

❑ **A.** Solve logarithmic equations using the fundamental properties of logarithms

❑ **B.** Apply the product, quotient, and power properties of logarithms

❑ **C.** Apply the change-of-base formula

❑ **D.** Solve applications using properties of logarithms

Logarithmic and exponential expressions have several fundamental properties that enable us to solve some basic equations. In this section, we'll learn how to use these relationships effectively, and introduce additional properties that enable us to simplify more complex equations before relying on the same fundamental properties to complete the solution.

A.   Solving Equations Using the Fundamental Properties of Logarithms

In Section 5.3, we converted expressions from exponential form to logarithmic form using the basic definition: $x = b^y \Leftrightarrow y = \log_b x$. This relationship reveals the following fundamental properties:

Fundamental Properties of Logarithms

For any base $b > 0$, $b \neq 1$,

I. $\log_b b = 1$, since $b^1 = b$ **III.** $\log_b b^x = x$, since $b^x = b^x$

II. $\log_b 1 = 0$, since $b^0 = 1$ **IV.** $b^{\log_b x} = x$ $(x > 0)$, since $\log_b x = \log_b x$

To see the verification of Property IV more clearly, again note that for $y = \log_b x$, $b^y = x$ is the exponential form, and substituting $\log_b x$ for y yields $b^{\log_b x} = x$. Also note that Properties III and IV demonstrate that $y = \log_b x$ and $y = b^x$ are inverse functions. In common language, "a base-b logarithm *undoes* a base-b exponential," and "a base-b exponential *undoes* a base-b logarithm." For $f(x) = \log_b x$ and $f^{-1}(x) = b^x$, using a composition verifies the inverse relationship just as in Example 5 from Section 5.1:

$$(f \circ f^{-1})(x) = f[f^{-1}(x)] \qquad (f^{-1} \circ f)(x) = f^{-1}[f(x)]$$
$$= \log_b b^x \qquad\qquad\qquad = b^{\log_b x}$$
$$= x \qquad\qquad\qquad\qquad = x$$

These properties can be used to solve basic equations involving logarithms and exponentials. From the uniqueness property for exponents (page 496), note that if $\log_b x = k$, then $b^{\log_b x} = b^k$, and we say that we have *exponentiated* both sides.

EXAMPLE 1 ▶ **Solving Basic Logarithmic Equations**

Solve each equation by applying fundamental properties. Answer in exact form and approximate form using a calculator (round to 1000ths).

a. $\ln x = 2$ **b.** $-0.52 = \log x$

Solution ▶

a. $\ln x = 2$ given

$e^{\ln x} = e^2$ exponentiate both sides

$x = e^2$ Property IV, exact form

≈ 7.389 approximate form

b. $-0.52 = \log x$ given

$10^{-0.52} = 10^{\log x}$ exponentiate both sides

$10^{-0.52} = x$ Property IV, exact form

$0.302 \approx x$ approximate form

Now try Exercises 7 through 10 ▶

Note that checking the exact solutions by substitution is a direct application of Property III (Figure 5.38).

Also, we observe that exponentiating both sides of the equation produces the same result as simply writing the original equation in exponential form, and the process can be viewed in terms of either approach.

Figure 5.38

```
ln(e^(2))
                    2
log(10^(-0.52))
                 -.52
```

EXAMPLE 2 ▶ **Solving Basic Exponential Equations**

Solve each equation by applying fundamental properties. Answer in exact form and approximate form using a calculator (round to 1000ths).

a. $e^x = 167$ **b.** $10^x = 8.223$

Solution ▶

a. $e^x = 167$ given

$\ln e^x = \ln 167$ use natural log

$x = \ln 167$ Property III, exact form

$x \approx 5.118$ approximate form

b. $10^x = 8.223$ given

$\log 10^x = \log 8.223$ use common log

$x = \log 8.223$ Property III, exact form

$x \approx 0.915$ approximate form

Now try Exercises 11 through 14 ▶

Similar to our observations from Example 1, taking the logarithm of both sides produced the same result as writing the equation in logarithmic form, and the process can be viewed in terms of either approach. Also note that here, checking the exact solution by substitution is a direct application of Property IV (Figure 5.39).

If an equation has a single logarithmic or exponential term (base 10 or base e), the equation can be solved by isolating this term and applying one of the fundamental properties as in Examples 1 and 2.

Figure 5.39

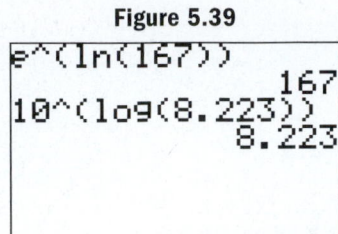

EXAMPLE 3 ▶ **Solving Exponential Equations**

Solve each equation. Write answers in exact form and approximate form to four decimal places.

    a. $10^x - 29 = 51$ **b.** $3e^{x+1} - 5 = 7$

Solution ▶   **a.** $10^x - 29 = 51$ given

              $10^x = 80$ add 29

Since the left-hand side is base 10, we apply the common logarithm.

$$\log 10^x = \log 80 \quad \text{take the common log of both sides}$$
$$x = \log 80 \quad \text{Property III (exact form)}$$
$$\approx 1.9031 \quad \text{approximate form}$$

b. $3e^{x+1} - 5 = 7$ given

    $3e^{x+1} = 12$ add 5

     $e^{x+1} = 4$ divide by 3

Since the left-hand side is base e, we apply the natural logarithm.

$$\ln e^{x+1} = \ln 4 \quad \text{take the natural log of both sides}$$
$$x + 1 = \ln 4 \quad \text{Property III}$$
$$x = \ln 4 - 1 \quad \text{solve for } x \text{ (exact form)}$$
$$\approx 0.3863 \quad \text{approximate form}$$

Now try Exercises 15 through 20 ▶

As an alternative to using the exact form to check solutions, we can **STO▸** (store) the exact result in storage location **X,T,θ,n** (the function variable x) and simply enter the original equation on the home screen. The verification for Example 3b is shown in Figure 5.40.

Figure 5.40

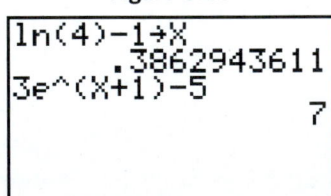

EXAMPLE 4 ▶ **Solving Logarithmic Equations**

Solve each equation. Write answers in exact form and approximate form to four decimal places.

    a. $2 \log (7x) + 1 = 4$ **b.** $-4 \ln (x + 1) - 5 = 7$

Solution ▶   **a.** $2 \log (7x) + 1 = 4$ given

$$2 \log (7x) = 3 \qquad \text{subtract 1}$$

$$\log (7x) = \frac{3}{2} \qquad \text{divide by 2}$$

$$7x = 10^{\frac{3}{2}} \qquad \text{exponential form}$$

$$x = \frac{10^{\frac{3}{2}}}{7} \qquad \text{divide by 7 (exact form)}$$

$$\approx 4.5175 \qquad \text{approximate form}$$

    b. $-4 \ln (x + 1) - 5 = 7$ given

$$-4 \ln (x + 1) = 12 \qquad \text{add 5}$$

$$\ln (x + 1) = -3 \qquad \text{divide by } -4$$

$$x + 1 = e^{-3} \qquad \text{exponential form}$$

$$x = e^{-3} - 1 \qquad \text{subtract 1 (exact form)}$$

$$\approx -0.9502 \qquad \text{approximate form}$$

Now try Exercises 21 through 26 ▶

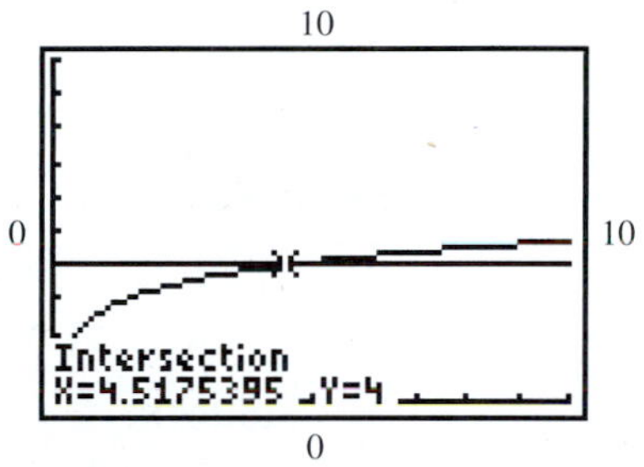

As with other kinds of equations, solutions to log-arithmic and exponential equations can be found using the intersection-of-graphs method or the zeroes method. For Example 4a and the intersection method, enter $Y_1 = 2 \log(7X) + 1$ and $Y_2 = 4$ on the (Y=) screen (from the domain of the function and the expected result, we know to set a window that includes only Quadrant I). Using the (2nd) (TRACE) **(CALC) 5:intersect** option, we press (ENTER) three times to identify each curve and bypass the "Guess" option. The calculator then finds the point of intersection and prints it at the bottom of the screen, and verifies our calculated result.

Figure 5.41

☑ **A. You've just seen how we can solve logarithmic equations using the fundamental properties of logarithms**

B. The Product, Quotient, and Power Properties of Logarithms

Generally speaking, equation solving involves simplifying the equation, isolating a variable term on one side, and applying an inverse to solve for the unknown. For loga-rithmic equations such as $\log x + \log (x + 3) = 1$, we must find a way to combine the terms on the left, before we can work toward a solution. This requires a further explo-ration of logarithmic properties.

Due to the close connection between exponents and logarithms, their properties are very similar. To illustrate, we'll use terms that can all be written in the form 2^x, and write the equations $8 \cdot 4 = 32$, $\frac{8}{4} = 2$, and $8^2 = 64$ in both exponential form and loga-rithmic form.

The exponents from a product are added:	exponential form:	$2^3 \cdot 2^2 = 2^{3+2}$
	logarithmic form:	$\log_2(8 \cdot 4) = \log_2 8 + \log_2 4$
The exponents from a quotient are subtracted:	exponential form:	$\dfrac{2^3}{2^2} = 2^{3-2}$
	logarithmic form:	$\log_2\left(\dfrac{8}{4}\right) = \log_2 8 - \log_2 4$
The exponents from a power are multiplied:	exponential form:	$(2^3)^2 = 2^{3 \cdot 2}$
	logarithmic form:	$\log_2 8^2 = 2 \cdot \log_2 8$

Each illustration can be generalized and applied with any base b.

WORTHY OF NOTE

For a more detailed verification of these properties, see Appendix VI.

Properties of Logarithms

Given M, N, and $b \neq 1$ are *positive* real numbers, and *any* real number p.

Product Property	**Quotient Property**	**Power Property**
$\log_b(MN) = \log_b M + \log_b N$	$\log_b\left(\dfrac{M}{N}\right) = \log_b M - \log_b N$	$\log_b M^p = p \log_b M$
The log of a product is a sum of logarithms.	The log of a quotient is a difference of logarithms.	The log of a quantity to a power is the power times the log of the quantity.

⚠ **CAUTION** ▶ It's very important that you read and understand these properties correctly. In particular:

(1) $\log(M + N) \neq \log M + \log N$, (2) $\log\left(\dfrac{M}{N}\right) \neq \dfrac{\log M}{\log N}$, and (3) $\dfrac{\log M}{\log N} \neq \log M - \log N$.

For $M = 100$ and $N = 10$, statement (1) would indicate $\log 110 = 2 + 1$✗, (2) would indicate that $1 = \frac{2}{1}$✗, and (3) would indicate that $\frac{2}{1} = 2 - 1$✗.

In the statement of these properties, it's worth reminding ourselves that the equal sign "works both ways," and we have $\log_b M + \log_b N = \log_b(MN)$. These properties are often used to write a sum or difference of logarithmic terms as a single expression.

EXAMPLE 5 ▶ **Rewriting Expressions Using Logarithmic Properties**

Use the properties of logarithms to write each expression as a single term.

a. $\log_2 7 + \log_2 5$ **b.** $2 \ln x + \ln(x + 6)$ **c.** $\log(x + 2) - \log x$

Solution ▶   **a.** $\log_2 7 + \log_2 5 = \log_2(7 \cdot 5)$ product property

$= \log_2 35$ simplify

b. $2 \ln x + \ln(x + 6) = \ln x^2 + \ln(x + 6)$ power property

$= \ln[x^2(x + 6)]$ product property

$= \ln[x^3 + 6x^2]$ simplify

c. $\log(x + 2) - \log x = \log\left(\dfrac{x + 2}{x}\right)$ quotient property

Now try Exercises 27 through 42 ▶

We can verify that these properties produce equivalent results by entering the original equation as Y_1, the result after applying the properties as Y_2, and viewing the results of various inputs on a **TABLE** screen. Results for Example 5b (power and product properties) and 5c (quotient property) are shown in Figures 5.42 and 5.43 respectively.

Figure 5.42

$2 \ln x + \ln(x + 6) = \ln(x^3 + 6x^2)$

X	Y₁	Y₂
.5	.48551	.48551
1	1.9459	1.9459
2.7183	4.1654	4.1654
3.1416	4.5023	4.5023
10	7.3778	7.3778
152	15.11	15.11
5000	25.553	25.553

X=5000

Figure 5.43

$\log(x + 2) - \log x = \log\left(\frac{x + 2}{x}\right)$

X	Y₁	Y₂
.1	1.3222	1.3222
.70711	.58302	.58302
.86603	.51975	.51975
1.5708	.35665	.35665
7.3891	.10403	.10403
10	.07918	.07918
123	.007	.007

X=123

EXAMPLE 6 ▶   **Rewriting Logarithmic Expressions Using the Power Property**

Use the power property of logarithms to rewrite each term as a product.

  a. $\ln 5^x$ **b.** $\log 32^{x+2}$ **c.** $\log \sqrt{x}$

Solution ▶   **a.** $\ln 5^x = x \ln 5$ power property

  b. $\log 32^{x+2} = (x + 2) \cdot \log 32$ power property

  c. $\log \sqrt{x} = \log x^{\frac{1}{2}}$ write radical using a rational exponent

  $\qquad\quad = \dfrac{1}{2} \log x$ power property

Now try Exercises 43 through 50 ▶

For examples of how these properties are used in context, **see Exercises 73 and 74.**

⚠ CAUTION ▶   Note from Example 6b that parentheses *must be used* whenever the exponent is a sum or difference. There is a huge difference between $(x + 2) \log 32$ and $x + 2 \log 32$.

Examples 5 and 6 illustrate how the properties of logarithms are used to *consolidate* logarithmic terms, primarily in preparation for equation solving. In other cases, the properties are used to rewrite or *expand* logarithmic expressions, so that certain other procedures can be applied more easily. Example 7 actually lays the foundation for more advanced mathematical work.

EXAMPLE 7 ▶   **Rewriting Expressions Using Logarithmic Properties**

Use the properties of logarithms to write the following expressions as sums or differences of simple logarithmic terms.

  a. $\log(x^2 z)$ **b.** $\ln \sqrt{\dfrac{x}{x + 5}}$

Solution ▶   **a.** $\log (x^2 z) = \log x^2 + \log z$ product property

  $\qquad\qquad\quad = 2 \log x + \log z$ power property

  b. $\ln \sqrt{\dfrac{x}{x + 5}} = \ln \left(\dfrac{x}{x + 5}\right)^{\frac{1}{2}}$ write radical using a rational exponent

  $\qquad\qquad\quad = \dfrac{1}{2} \ln\left(\dfrac{x}{x + 5}\right)$ power property

  $\qquad\qquad\quad = \dfrac{1}{2}[\ln x - \ln(x + 5)]$ quotient property

Now try Exercises 51 through 58 ▶

As you begin working with applications of logarithmic properties, it may help to have them written on a separate note card. This will enable you to compare each step and property as they are applied, remembering that "M" and "N" can represent any positive number or any positive, real-valued expression (see *Reinforcing Basic Concepts* on page 527). For Example 7(a) we then have

 B. You've just seen how we can apply the product, quotient, and power properties of logarithms

$$\log(x^2 z) = \log x^2 + \log z$$

C. The Change-of-Base Formula

Although base-10 and base-e logarithms dominate the mathematical landscape, there are many practical applications of other bases. Fortunately, a formula exists that will convert any given base into either base 10 or base e. It's called the **change-of-base formula.**

Change-of-Base Formula

For the positive real numbers M, a, and b, with $a, b \neq 1$,

$$\log_b M = \frac{\log M}{\log b} \qquad\qquad \log_b M = \frac{\ln M}{\ln b} \qquad\qquad \log_b M = \frac{\log_a M}{\log_a b}$$

$$\text{base 10} \qquad\qquad\qquad\qquad \text{base } e \qquad\qquad\qquad\qquad \text{arbitrary base } a$$

Proof of the Change-of-Base Formula

For $y = \log_b M$, we have $b^y = M$ in exponential form. It follows that

$$\log_a(b^y) = \log_a M \qquad \text{take base-}a \text{ logarithm of both sides}$$

$$y \log_a b = \log_a M \qquad \text{power property of logarithms}$$

$$y = \frac{\log_a M}{\log_a b} \qquad \text{divide by } \log_a b$$

$$\log_b M = \frac{\log_a M}{\log_a b} \qquad \text{substitute } \log_b M \text{ for } y$$

EXAMPLE 8 ▶ **Using the Change-of-Base Formula to Evaluate Expressions**

Find the value of each expression using the change-of-base formula. Answer in exact form and approximate form using nine digits, then *verify the result* using the original base. Note that either base 10 or base e can be used.

a. $\log_3 29$ **b.** $\log_5 3.6$

Solution ▶   **a.** $\log_3 29 = \dfrac{\log 29}{\log 3}$ **b.** $\log_5 3.6 = \dfrac{\ln 3.6}{\ln 5}$

≈ 3.065044752 ≈ 0.795888947

Check: $3^{3.065044752} = 29$ ✓   **Check:** $5^{0.795888947} = 3.6$ ✓

Now try Exercises 59 through 66 ▶

The change-of-base formula can also be used to study and graph logarithmic functions of *any* base. For $y = \log_b x$, the right-hand expression is simply rewritten using the formula and the equivalent function is $y = \dfrac{\log x}{\log b}$. The new function can then be evaluated as in Example 8, or used to study the graph of $y = \log_b x$ for any base b. **See Exercises 67 through 70.**

✓ **C. You've just seen how we can apply the change-of-base formula**

D. Solving Applications of Logarithms

We end this section with one additional application of logarithms. For all living things, the concentration of hydrogen ions in a solution plays an important role as their presence or absence alters the environment of other molecules in the solution. This can dramatically affect the functionality of the solution, or the ability of an organism to survive. The concentration of hydrogen ions (in moles per liter) is commonly expressed in terms of what is called the **pH scale.** A low pH number corresponds to high hydrogen ion concentration, and the solution is said to be acidic. A high pH corresponds to low hydrogen ion concentration, and the solution is said to be basic. Since the range of values is so large, the pH scale is logarithmic, meaning each unit change in the pH scale represents a 10-fold increase or decrease in the concentration of hydrogen ions. For example, tomato juice, with a pH level of 2, is 10 times more acidic than orange juice, with a pH level of 3, and 100 times more acidic than grape juice (pH = 4; the lower the pH number, the higher the ion concentration). The pH values range from 0 to 14, with pure water at pH = 7 being deemed "neutral" (neither basic nor acidic). Measuring pH levels plays an important role in biology, chemistry, food science, environmental science, medicine, oceanography, personal care products, and many other areas. The number of hydrogen atoms is usually represented by the term H^+, with the pH number defined as pH = $-\log[H^+]$

EXAMPLE 9A ▶ **The Concentration of Hydrogen Atoms in Ocean Water**

Ocean water has a pH number of near 7.9. What is the concentration of hydrogen ions? Write the result in scientific notation.

Solution ▶ Begin with the basic formula and work from there.

$$
\begin{array}{lll}
\text{pH} = -\log[H^+] & & \text{pH formula} \\
7.9 = -\log[H^+] & & \text{substitute 7.9 for pH} \\
-7.9 = \log[H^+] & & \text{multiply by } -1 \\
10^{-7.9} = H^+ & & \text{exponential form} \\
1.26 \times 10^{-8} = H^+ & & \text{result}
\end{array}
$$

The hydrogen ion concentration in ocean water is 1.26×10^{-8} moles/liter.

EXAMPLE 9B ▶ **Finding the pH Level of an Apple**

The concentration of hydrogen ions in an everyday apple is very near 7.94×10^{-4}. What is the pH level of an apple?

Solution ▶

$$
\begin{array}{lll}
\text{pH} = -\log[H^+] & & \text{pH formula} \\
= -\log[7.94 \times 10^{-4}] & & \text{substitute } 7.94 \times 10^{-4} \text{ for H}^+ \\
\approx 3.1 & & \text{result}
\end{array}
$$

☑ **D.** You've just seen how we can solve applications of logarithms

An apple has a pH level near 3.1.

Now try Exercises 73 through 78 ▶

5.4 EXERCISES

▶ **CONCEPTS AND VOCABULARY**

Fill in each blank with the appropriate word or phrase. Carefully reread the section if needed.

1. For $e^{-0.02x+1} = 10$, the solution process is most efficient if we apply a base __e__ logarithm to both sides.

2. To solve $\ln 2x - \ln(x + 3) = 0$, we can combine terms using the __quotient__ property, or add $\ln(x + 3)$ to both sides and use the __uniqueness__ property.

3. Since logarithmic functions are not defined for all real numbers, we should check all "solutions" for __extraneous__ roots.

4. The statement $\log_e 10 = \dfrac{\log 10}{\log e}$ is an example of the __change__-of-__base__ formula.

5. Use all factor pairs of 36 to illustrate the product property of logarithms. For example, since $36 = 4 \cdot 9$, is $\log(4 \cdot 9) = \log 4 + \log 9$?
   Answers will vary; Yes, 1.5663025 = 1.5663025

6. Use integer divisors of 24 to illustrate the quotient property of logarithms. For example, since $12 = \frac{24}{2}$, is $\log 12 = \log\left(\frac{24}{2}\right) = \log 24 - \log 2$?
   Answers will vary; Yes, 1.0791812 = 1.0791812

HOMEWORK SELECTION GUIDE

Core: 7–67 every other odd, 69, 73–77 odd, 82, 83 (22 Exercises)
Standard: 1–4, 7–67 every other odd, 69, 71, 73–80, 82, 84 (32 Exercises)

Extended: 1–4, 7–67 every other odd, 69, 71–80, 82, 84 (34 Exercises)
In Depth: 1–6, 7–67 every other odd, 69, 71–84 (37 Exercises)

▶ DEVELOPING YOUR SKILLS

Solve each equation by applying fundamental properties. Round to thousandths.

7. $\ln x = 3.4$ $x \approx 29.964$ **8.** $\ln x = \frac{1}{2}$ $x \approx 1.649$

9. $\log x = \frac{1}{4}$ $x \approx 1.778$ **10.** $\log x = 1.6$ $x \approx 39.811$

11. $e^x = 9.025$ $x \approx 2.200$ **12.** $e^x = 0.343$ $x \approx -1.070$

13. $10^x = 18.197$ $x \approx 1.260$ **14.** $10^x = 0.024$ $x \approx -1.620$

Solve each exponential equation. Write answers in exact form and in approximate form rounded to four decimal places.

15. $4e^{x-2} + 5 = 70$ **16.** $2 - 3e^{0.4x} = -7$

17. $10^{x+5} - 228 = -150$ **18.** $10^{2x} + 27 = 190$

19. $-150 = 290.8 - 190e^{-0.75x}$

20. $250e^{0.05x+1} + 175 = 1175$

Solve each logarithmic equation. Write answers in exact form and in approximate form rounded to four decimal places.

21. $3 \ln(x + 4) - 5 = 3$ **22.** $-15 = -8 \ln(3x) + 7$

23. $-1.5 = 2 \log(5 - x) - 4$

24. $-4 \log(2x) + 9 = 3.6$

25. $\frac{1}{2} \ln(2x + 5) + 3 = 3.2$

26. $\frac{3}{4} \ln(4x) - 6.9 = -5.1$

Use properties of logarithms to write each expression as a single term.

27. $\ln(2x) + \ln(x - 7)$ **28.** $\ln(x + 2) + \ln(3x)$

29. $\log(x + 1) + \log(x - 1)$

30. $\log(x - 3) + \log(x + 3)$

31. $\log_3 28 - \log_3 7$ **32.** $\log_6 30 - \log_6 10$

33. $\log x - \log(x + 1)$ **34.** $\log(x - 2) - \log x$

35. $\ln(x - 5) - \ln x$ **36.** $\ln(x + 3) - \ln(x - 1)$

37. $\ln(x^2 - 4) - \ln(x + 2)$

38. $\ln(x^2 - 25) - \ln(x + 5)$

39. $\log_2 7 + \log_2 6$ **40.** $\log_9 2 + \log_9 15$

41. $\log_5(x^2 - 2x) + \log_5 x^{-1}$

42. $\log_3(3x^2 + 5x) - \log_3 x$

Use the power property of logarithms to rewrite each term as the product of some quantity times a logarithmic term.

43. $\log 8^{x+2}$ $(x + 2) \log 8$ **44.** $\log 15^{x-3}$ $(x - 3) \log 15$

45. $\ln 5^{2x-1}$ $(2x - 1) \ln 5$ **46.** $\ln 10^{3x+2}$ $(3x + 2) \ln 10$

47. $\log \sqrt{22}$ $\frac{1}{2} \log 22$ **48.** $\log \sqrt[3]{34}$ $\frac{1}{3} \log 34$

49. $\log_5 81$ $4 \log_5 3$ **50.** $\log_7 121$ $2 \log_7 11$

Use the properties of logarithms to write the following expressions as sums or differences of simple logarithmic terms.

51. $\log(a^3 b)$ $3 \log a + \log b$ **52.** $\log(m^2 n)$ $2 \log m + \log n$

53. $\ln(x\sqrt[4]{y})$ $\ln x + \frac{1}{4} \ln y$ **54.** $\ln(\sqrt[3]{pq})$ $\frac{1}{3} \ln p + \ln q$

55. $\ln\left(\dfrac{x^2}{y}\right)$ $2 \ln x - \ln y$ **56.** $\ln\left(\dfrac{m^2}{n^3}\right)$ $2 \ln m - 3 \ln n$

57. $\log\left(\sqrt{\dfrac{x - 2}{x}}\right)$ **58.** $\log\left(\sqrt[3]{\dfrac{3 - v}{2v}}\right)$

Evaluate each expression using the change-of-base formula and either base 10 or base e. Answer in exact form and in approximate form using nine decimal places, then verify the result using the original base.

59. $\log_7 60$ **60.** $\log_8 92$

61. $\log_5 152$ **62.** $\log_6 200$

63. $\log_3 1.73205$ **64.** $\log_2 1.41421$

65. $\log_{0.5} 0.125$ **66.** $\log_{0.2} 0.008$

Use the change-of-base formula to write an equivalent function, then evaluate the function as indicated (round to six decimal places). Investigate and discuss any patterns you notice in the output values, then determine the next input that will continue the pattern.

67. $f(x) = \log_3 x; f(5), f(15), f(45)$

68. $g(x) = \log_2 x; g(5), g(10), g(20)$

69. $h(x) = \log_9 x; h(2), h(4), h(8)$

70. $H(x) = \log_\pi x; H(\sqrt{2}), H(2), H(\sqrt{2^3})$

▶ WORKING WITH FORMULAS

71. $\log_b M = \dfrac{1}{\log_M b}$

Use the change-of-base formula to verify the "formula" shown.   verified

72. $\log_B A \cdot \log_C B \cdot \log_D C = \log_D A$

Use the change-of-base formula to verify the "formula" shown.   verified

Additional answers can be found in the Instructor Answer Appendix.

▶ **APPLICATIONS**

73. **Pareto's 80/20 principle:** After observing that 80% of the land in his native Italy was owned by 20% of the population, Italian economist Vilfredo Pareto (1848–1923) noted this disparity in many other areas (20% of the workers produce 80% of the output, 20% of the customers create 80% of the revenue, etc.) and developed a mathematical model for this phenomenon, called **Pareto's law.** If N represents the number of people with incomes greater than X, then $\log N = \log A - m \log X$, where A and m are predetermined constants. (a) Solve the equation for N and (b) given $m = 1.5$ and $A = 9900$, find the number of people earning over $200,000. Assume X is in hundreds of thousands of dollars.
    a. $N = AX^{-m}$ **b.** ≈ 3500 people

74. **The species/area relationship:**

    The study of what is now known as *island biogeography* originated with Robert McArthur and Edward O. Wilson in the 1960s. In general, they found that the relationship between island area and the number of species present could be modeled by the equation $\log S = \log C + k \log A$, where S represents the total number of species, A represents the area of the island, while C and k are predetermined constants that depend on the size and proximity of other land masses as well as other factors. This makes it possible to predict the number of species on an island, when little other information is available. (a) Solve the equation for S and (b) given $k = 0.81$ and $C = 8$, find the predicted number of species an island with area of $A = 2000 \text{ km}^2$.   **a.** $S = CA^k$ **b.** ≈ 3775 species

75. **Blood plasma pH levels:** To be safe and usable, the blood plasma held by blood banks must have a pH level between 7.35 and 7.45. Blood outside of this normal range can cause disorientation,
    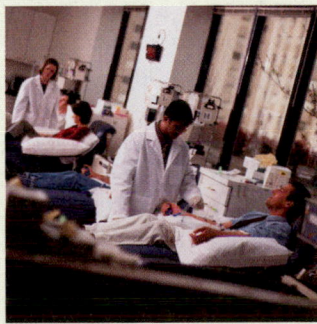
    behavioral changes, or even death. Using ion-sensitive electrodes, a sample of blood plasma is known to have a concentration of $H^+ = 4.786 \times 10^{-8}$. Is the plasma usable?
    no, pH $\approx$ 7.32

76. **Fresh milk:** As milk begins to sour, there is a corresponding decrease in pH level. Fresh milk has a pH level of near 6.5. After transport from farm to market, a sample of milk is tested using ion-sensitive electrodes and is found to have a concentration of $H^+ = 3.981 \times 10^{-5}$. Is this shipment of milk still suitable for market?
    no, pH $\approx$ 4.4 and the milk soured

77. **Soil acidity:** Throughout many parts of the Midwest, surface soils are neutral to slightly alkaline. While a majority of crops might prefer a pH neutral soil (pH = 7), some crops thrive in more acidic soils (potatoes, strawberries, others). For these crops, elemental sulfur is applied to help decrease the pH level (the optimum pH level for potato crops is near 5.2). Measurements of the soil on a certain midwestern farm indicate a hydrogen ion concentration of $H^+ = 1.259 \times 10^{-6}$. Is the soil ready for a potato crop to be planted?
    no, pH $\approx$ 5.9 and the soil must be treated further

78. **Acidity of gastric juices:** The normal pH value of human gastric juice can vary from 1 to 3, depending on genetics, diet, and other factors. The acidity is designed to control various harmful microorganisms that a person may ingest as they eat. Drinking large quantities of water before a meal can have a dramatic effect on this pH value, sometimes raising it
    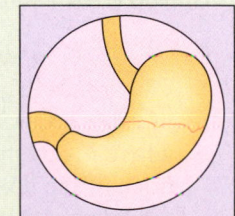
    beyond the normal range to as high as 4 or 5, making it possible for some harmful bacteria to survive. If a hospital patient's stomach fluid has a hydrogen ion concentration of $H^+ = 3.981 \times 10^{-3}$, is the pH level within a normal range?   yes, pH $\approx$ 2.4, within normal range

▶ **EXTENDING THE CONCEPT**

79. Logarithmic properties can also be used to compare the magnitude of very large numbers, numbers too large for a handheld calculator to manage. Use the power property of logarithms to compare the numbers 600^{601} and 601^{600}. Which number is larger? 600^{601}

80. Logarithmic properties can also be used to compare the magnitude of very small numbers, again numbers too small for a handheld calculator to manage. Use a negative exponent and the power property of logarithms to compare the numbers $\dfrac{1}{99^{100}}$ and $\dfrac{1}{100^{99}}$. Which number is smaller? $\dfrac{1}{99^{100}}$

▶ **MAINTAINING YOUR SKILLS**

81. (4.4/4.5) State the zeroes of f and the equation of any horizontal or vertical asymptotes given
$$f(x) = \frac{x^2 - x - 6}{x^2 - 1}.$$
zeroes at $x = 3$ and $x = -2$; HA: $y = 1$, VA $x = -1$, $x = 1$

82. (1.4) A sports shop can stock 36 cans of tennis balls on shelves that are 9 in. deep, and 54 cans on 12-in. shelves. Assuming the relationship is linear, (a) find the equation relating shelf size to number of cans, and (b) use it to determine what size shelf should be used to stock a full shipment of 72 cans.
a. $S = C/6 + 3$ **b.** 15-in. shelves

83. (R.4) Find all values of x that make the equation true: $2x(x - 3) + 4 = (3x + 5)(x - 1)$.
$x = 1$ or $x = -9$

84. (2.2) Determine the equation of the function shown in (a) shifted form and (b) standard form
a. $y = (x - 4)^2 - 1$
b. $y = x^2 - 8x + 15$

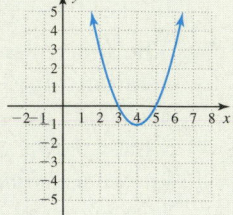

MID-CHAPTER CHECK

1. Write the following in logarithmic form.
   a. $27^{\frac{2}{3}} = 9$ $\frac{2}{3} = \log_{27}9$ **b.** $81^{\frac{5}{4}} = 243$ $\frac{5}{4} = \log_{81}243$

2. Write the following in exponential form.
   a. $\log_8 32 = \frac{5}{3}$ $8^{\frac{5}{3}} = 32$ **b.** $\log_{1296}6 = 0.25$
   $1296^{0.25} = 6$

3. Solve each equation for the unknown:
   a. $4^{2x} = 32^{x-1}$ $x = 5$ **b.** $\left(\frac{1}{3}\right)^{4b} = 9^{2b-5}$ $b = \frac{5}{4}$

4. Solve each equation for the unknown:
   a. $\log_{27}x = \frac{1}{3}$ $x = 3$ **b.** $\log_b 125 = 3$ $b = 5$

5. The homes in a popular neighborhood are growing in value according to the formula $V(t) = V_0(\frac{9}{8})^t$, where t is the time in years, V_0 is the purchase price of the home, and $V(t)$ is the current value of the home. (a) In 3 yr, how much will a \$50,000 home be worth? (b) Use the TABLE feature of your calculator to estimate how many years (to the nearest year) until the home doubles in value.
   a. \$71,191.41   **b.** 6 yr

6. The graph of the function $f(x) = 5^x$ has been shifted right 3 units, up 2 units, and stretched by a factor of 4. What is the equation of the resulting function? $F(x) = 4 \cdot 5^{x-3} + 2$

7. State the domain and range for $f(x) = \sqrt{x - 3} + 1$, then find $f^{-1}(x)$ and state its domain and range. Verify the inverse relationship using composition.
$f^{-1}(x) = (x - 1)^2 + 3$, $D: x \in [1, \infty)$; $R: y \in [3, \infty)$; verified

8. Write the following equations in logarithmic form, then verify the result on a calculator.
   a. $81 = 3^4$ **b.** $e^4 \approx 54.598$
   $4 = \log_3 81$, verified $4 \approx \ln 54.598$, verified

9. Write the following equations in exponential form, then verify the result on a calculator.
   a. $\dfrac{2}{3} = \log_{27}9$ **b.** $1.4 \approx \ln 4.0552$
   $27^{\frac{2}{3}} = 9$, verified $e^{1.4} \approx 4.0552$, verified

10. On August 15, 2007, an earthquake measuring 8.0 on the Richter scale struck coastal Peru. On October 17, 1989, right before Game 3 of the World Series between the Oakland A's and the San Francisco Giants, the Loma Prieta earthquake, measuring 7.1 on the Richter scale, struck the San Francisco Bay area. How much more intense was the Peruvian earthquake?   ≈7.9 times more intense

REINFORCING BASIC CONCEPTS

Understanding Properties of Logarithms

To effectively use the properties of logarithms as a mathematical tool, a student must attain some degree of comfort and fluency in their application. Otherwise we are resigned to using them as a template or formula, leaving little room for growth or insight. This feature is designed to promote an understanding of the product and quotient properties of logarithms, which play a role in the solution of logarithmic and exponential equations.

We begin by looking at some logarithmic expressions that are obviously true:

$$\log_2 2 = 1 \qquad \log_2 4 = 2 \qquad \log_2 8 = 3$$
$$\log_2 16 = 4 \qquad \log_2 32 = 5 \qquad \log_2 64 = 6$$

Next, we view the same expressions with their value *understood mentally,* illustrated by the numbers in the background, rather than expressly written.

$$\log_2 2 \qquad \log_2 4 \qquad \log_2 8 \qquad \log_2 16 \qquad \log_2 32 \qquad \log_2 64$$

This will make the product and quotient properties of equality much easier to "see." Recall the product property states: $\log_b M + \log_b N = \log_b(MN)$ and the quotient property states: $\log_b M - \log_b N = \log_b\left(\dfrac{M}{N}\right)$. Consider the following.

$$\log_2 4 + \log_2 8 = \log_2 32 \qquad\qquad \log_2 64 - \log_2 32 = \log_2 2$$

which is the same as saying $\qquad\qquad$ which is the same as saying

$$\log_2 4 + \log_2 8 = \log_2(4\cdot 8) \qquad\qquad \log_2 64 - \log_2 32 = \log_2\left(\tfrac{64}{32}\right)$$
$$\text{(since } 4\cdot 8 = 32) \qquad\qquad\qquad\qquad \text{(since } \tfrac{64}{32} = 2)$$

$$\log_b M + \log_b N = \log_b(MN) \qquad\qquad \log_b M - \log_b N = \log_b\left(\dfrac{M}{N}\right)$$

Exercise 1: Repeat this exercise using logarithms of base 3 and various sums and differences. Answers will vary.

Exercise 2: Use the basic concept behind these exercises to combine these expressions: (a) $\log(x) + \log(x + 3)$, (b) $\ln(x + 2) + \ln(x - 2)$, and (c) $\log(x) - \log(x + 3)$. **a.** $\log(x^2 + 3x)$ **b.** $\ln(x^2 - 4)$ **c.** $\log\frac{x}{x + 3}$

5.5 | Solving Exponential and Logarithmic Equations

LEARNING OBJECTIVES

In Section 5.5 you will see how we can:

☐ **A.** Solve general logarithmic and exponential equations

☐ **B.** Solve applications involving logistic, exponential, and logarithmic functions

In this section, we'll develop the ability to solve more general logarithmic and exponential equations. A logarithmic equation has at least one term that involves the logarithm of a variable. Likewise, an exponential equation is one that involves a variable exponent on some base. In the same way that we might square both sides or divide both sides of an equation in the solution process, we'll show that we can also exponentiate both sides or take logarithms of both sides to help obtain a solution.

A. Solving Logarithmic and Exponential Equations

One of the most common mistakes in solving exponential and logarithmic equations is to apply the inverse function too early—before the equation has been simplified. Just as we would naturally try to combine like terms for the equation $2\sqrt{x} + 7\sqrt{x} = 69$ (prior to squaring both sides), the logarithmic terms in $\log x + \log(x + 3) = 1$ must be combined prior to applying the exponential form. In addition, since the domain of $y = \log_b x$ is $x > 0$, logarithmic equations can sometimes produce **extraneous roots,** and checking all answers is a good practice. We'll illustrate by solving the equation $\log x + \log(x + 3) = 1$.

EXAMPLE 1 ▶ **Solving a Logarithmic Equation**

Solve for x and check your answer: $\log x + \log(x + 3) = 1$.

▼ **Algebraic Solution**

$$\log x + \log(x + 3) = 1 \qquad \text{original equation}$$
$$\log[x(x + 3)] = 1 \qquad \text{product property}$$
$$x^2 + 3x = 10^1 \qquad \text{exponential form,}$$
$$\text{distribute } x$$
$$x^2 + 3x - 10 = 0 \qquad \text{set equal to 0}$$
$$(x + 5)(x - 2) = 0 \qquad \text{factor}$$
$$x = -5 \text{ or } x = 2 \qquad \text{result}$$

▼ **Graphical Solution**

Using the intersection-of-graphs method, we enter $Y_1 = \log X + \log(X + 3)$ and $Y_2 = 1$. From the domain we know $x > 0$, indicating the solution will occur in QI. After graphing both functions using the window shown, the intersection method shows the only solution is $x = 2$.

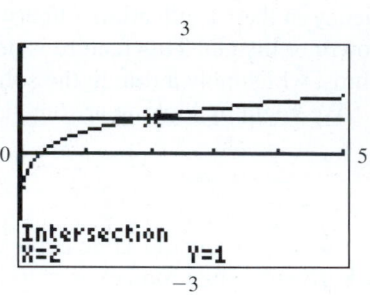

Check: The "solution" $x = -5$ is outside the domain and is ignored. For $x = 2$,

$$\log x + \log(x + 3) = 1 \qquad \text{original equation}$$
$$\log 2 + \log(2 + 3) = 1 \qquad \text{substitute 2 for } x$$
$$\log 2 + \log 5 = 1 \qquad \text{simplify}$$
$$\log(2 \cdot 5) = 1 \qquad \text{product property}$$
$$\log 10 = 1 \qquad \text{Property I}$$

You could also use a calculator to verify $\log 2 + \log 5 = 1$ directly.

Now try Exercises 7 through 14 ▶

If the simplified form of an equation yields a logarithmic term on both sides, the **uniqueness property of logarithms** provides an efficient way to work toward a solution. Since logarithmic functions are one-to-one, we have

> **The Uniqueness Property of Logarithms**
>
> For positive real numbers m, n, and $b \neq 1$,
>
> **1.** If $\log_b m = \log_b n$,   **2.** If $m \neq n$,
>
>   then $m = n$ then $\log_b m \neq \log_b n$
>
> Equal bases imply equal arguments.

EXAMPLE 2 ▶ **Solving Logarithmic Equations Using the Uniqueness Property**

Solve each equation using the uniqueness property.

     a. $\log(x + 2) = \log 7 + \log x$ **b.** $\ln 87 - \ln x = \ln 29$

▼ **Algebraic Solution**

a. $\log(x + 2) = \log 7 + \log x$
$$\log(x + 2) = \log 7x \qquad \text{properties of logarithms}$$
$$x + 2 = 7x \qquad \text{uniqueness property}$$
$$2 = 6x \qquad \text{solve for } x$$
$$\frac{1}{3} = x \qquad \text{result}$$

▼ **Graphical Solution**

Deciding which method to use (intersection or zeroes) can depend on the simplicity or complexity of the equation, how the equation is given, and/or which method gives the clearest view of the point(s) of intersection. Here, we opt to use the zeroes method.

a. After setting the equation equal to zero, we enter $Y_1 = \log(X + 2) - \log 7 - \log X$ and locate the x-intercept (if it exists) using **2nd** **TRACE** **(CALC) 2:Zero.** The graph reveals an x-intercept between 0 and 1, and using these values as bounds we locate the zero at $x = \dfrac{1}{3}$ (Figure 5.44).

Figure 5.44

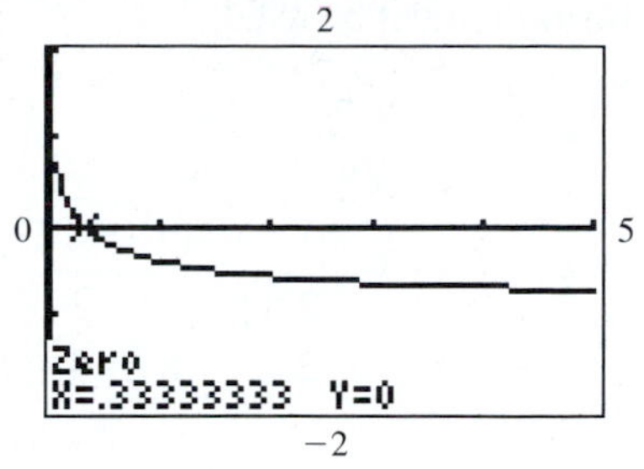

b. $\ln 87 - \ln x = \ln 29$

$\ln\left(\dfrac{87}{x}\right) = \ln 29$　　quotient property

$\dfrac{87}{x} = 29$　　uniqueness property

$87 = 29x$　　clear denominator

$3 = x$　　result

b. Once again setting the equation equal to zero, we enter $Y_1 = \ln 87 - \ln X - \ln 29$ to locate the x-intercept (if it exists). The graph shows an x-intercept near 3, with the calculator indicating the zero is exactly $x = 3$ (Figure 5.45).

Figure 5.45

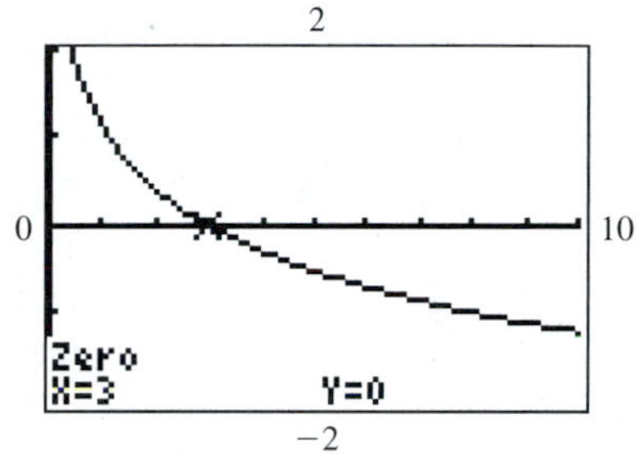

Now try Exercises 15 through 20 ▶

WORTHY OF NOTE

The uniqueness property can also be viewed as exponentiating both sides using the appropriate base, then applying Property IV.

Often the solution may depend on using a variety of algebraic skills in addition to logarithmic or exponential properties.

EXAMPLE 3 ▶　**Solving Logarithmic Equations**

Solve the equation and check your answer.

$$\log(x + 12) - \log x = \log(x + 9)$$

▼ Algebraic Solution

$\log(x + 12) - \log x = \log(x + 9)$　　given equation

$\log\left(\dfrac{x + 12}{x}\right) = \log(x + 9)$　　quotient property

$\dfrac{x + 12}{x} = x + 9$　　uniqueness property

$x + 12 = x^2 + 9x$　　clear denominator

$0 = x^2 + 8x - 12$　　set equal to 0

▼ Graphical Solution

Using the intersection-of-graphs method, we enter $\log(X + 12) - \log x$ as Y_1 and $\log(X + 9)$ as Y_2 on the **Y=** screen. Using **2nd** **TRACE** **(CALC) 5:intersect,** we find the graphs intersect at $x = 1.2915026$, and that *this is the only solution* (knowing the graphs' basic shape, we conclude they cannot intersect again).

The equation is not factorable, and the quadratic formula must be used.

$$x = \frac{-b \pm \sqrt{b^2 - 4ac}}{2a}$$ quadratic formula

$$= \frac{-8 \pm \sqrt{(8)^2 - 4(1)(-12)}}{2(1)}$$ substitute 1 for a, 8 for b, -12 for c

$$= \frac{-8 \pm \sqrt{112}}{2} = \frac{-8 \pm 4\sqrt{7}}{2}$$ simplify

$$= -4 \pm 2\sqrt{7}$$ result

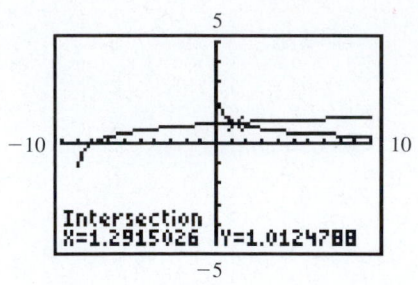

Substitution shows $x = -4 + 2\sqrt{7}$ ($x \approx 1.29150$) checks, but substituting $-4 - 2\sqrt{7}$ for x gives $\log(2.7085) - \log(-9.2915) = \log(-0.2915)$ and two of the three terms do not represent real numbers ($x = -4 - 2\sqrt{7}$ is an extraneous root).

Now try Exercises 21 through 36 ▶

⚠ **CAUTION ▶** Be careful not to dismiss or discard a possible solution simply because it's negative. For the equation $\log(-6 - x) = 1$, $x = -16$ is the solution (the domain here allows negative numbers: $-6 - x > 0$ yields $x < -6$ as the domain). In general, when a logarithmic equation has multiple solutions, all solutions should be checked.

Solving an exponential equation likewise involves isolating an exponential term on one side, or writing the equation where exponential terms of like base occur on each side. The latter case can be solved using the uniqueness property. If the exponential base is neither 10 nor e, logarithms of base b can be used along with the change-of-base formula to solve the equation.

EXAMPLE 4 ▶   **Solving an Exponential Equation Using Base b**

Solve the exponential equation. Answer in both exact form, and approximate form to four decimal places: $4^{3x} - 1 = 8$

Solution ▶ $4^{3x} - 1 = 8$ given equation

$4^{3x} = 9$ add 1

The left-hand side is neither base 10 or base e, so here we chose base 4 to solve.

$$\log_4 4^{3x} = \log_4 9$$ logarithms base 4

$$3x = \frac{\log 9}{\log 4}$$ Property III; change-of-base property

$$x = \frac{\log 9}{3 \log 4}$$ multiply by $\frac{1}{3}$ (exact form)

$$x \approx 0.5283$$ approximate form

A calculator check is shown here.

```
log(9)/(3*log(4)
)→X
         .5283208336
4^(3X)-1
              8
```

Now try Exercises 37 through 40 ▶

In some cases, two exponential terms with *unlike* bases may be involved. In this case, either common logs or natural logs can be used, but be sure to distinguish between constant terms like *ln 5* and variable terms like *x ln 5*. As with all equations, the goal is to isolate the *variable terms* on one side.

EXAMPLE 5 ▶ **Solving an Exponential Equation with Unlike Bases**

Solve the exponential equation $5^{x+1} = 6^{2x}$.

Algebraic Solution ▶ $5^{x+1} = 6^{2x}$ original equation

Begin by taking the natural log of both sides:

$$\ln(5^{x+1}) = \ln(6^{2x}) \qquad \text{\color{red}apply base-e logarithms}$$
$$(x+1)\ln 5 = 2x \ln 6 \qquad \text{\color{red}power property}$$
$$x \ln 5 + \ln 5 = 2x \ln 6 \qquad \text{\color{red}distribute}$$
$$\ln 5 = 2x \ln 6 - x \ln 5 \qquad \text{\color{red}variable terms to one side}$$
$$\ln 5 = x(2 \ln 6 - \ln 5) \qquad \text{\color{red}factor out x}$$
$$\frac{\ln 5}{2 \ln 6 - \ln 5} = x \qquad \text{\color{red}solve for x (exact form)}$$
$$0.8153 \approx x \qquad \text{\color{red}approximate form}$$

Graphical Solution ▶ In many cases, the quality of a graphical solution depends on the ability to determine an appropriate window size. Most often, this is accomplished using the domain of the functions involved, the context of the application, or a few test values. For $Y_1 = 5^{X+1}$ and $Y_2 = 6^{2X}$, we begin by observing that for $x = 0$, $Y_1 > Y_2$. However for $x = 2$, $Y_2 > Y_1$ (see Figure 5.46). This indicates that function values will be equal $[Y_1(X) = Y_2(X)]$ for some value of x between 0 and 1, and the window size for x must be set accordingly. These test values also show that y need not be greater than 36, though we may elect to use a higher value for both x and y to obtain a good window "frame." Using the window size indicated in Figure 5.47 reveals the solution to the equation (where the graphs intersect) is $x \approx 0.81528463$.

Figure 5.46

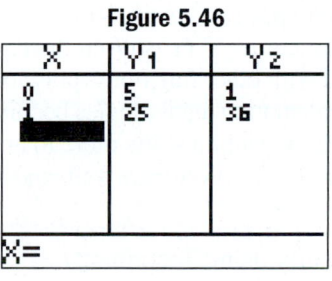

Figure 5.47

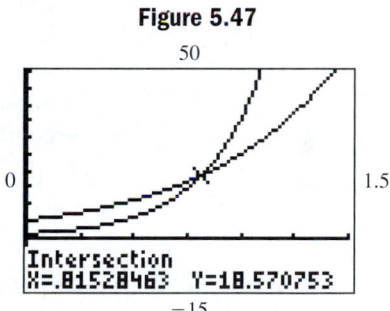

                                                Now try Exercises 41 through 44 ▶

As an alternative to taking the natural log of both sides directly, we can use the properties of exponents to simplify and combine the exponential terms as follows.

$$5^{x+1} = 6^{2x} \qquad \text{\color{red}original equation}$$
$$5^x 5^1 = (6^2)^x \qquad \text{\color{red}product and power properties}$$
$$5 \cdot 5^x = 36^x \qquad \text{\color{red}rewrite factors; simplify}$$
$$5 = \frac{36^x}{5^x} = \left(\frac{36}{5}\right)^x \qquad \text{\color{red}divide by 5^x, apply power property}$$
$$\ln 5 = x \ln 7.2 \qquad \text{\color{red}take natural logs ($\frac{36}{5} = 7.2$)}$$
$$\frac{\ln 5}{\ln 7.2} = x \qquad \text{\color{red}solve for x}$$

The result is equivalent to our original solution: $x \approx 0.81528463$.

Logarithmic equations come in many different forms, and the following ideas summarize the basic approaches used in solving them. For this summary, recall that for any positive real number k, $\log_b k$ *is a constant*. Assume M, N, and X represent algebraic expressions in x.

1. If the equation can be written in the form $\log_b M$ = constant, use the exponential form and algebra to solve: $b^{\text{constant}} = M$.

2. If the equation can be written in the form $\log_b M = \log_b N$, use the uniqueness property and algebra to solve: $M = N$.

3. If the equation has an additional constant term as in $\log_b M = \log_b N$ + constant, move all logarithmic terms to one side and consolidate using the product or quotient properties, then use the exponential form and algebra to solve:

$$\log_b M = \log_b N + \text{constant}$$

$$\log_b M - \log_b N = \text{constant}$$

$$\log_b\left(\frac{M}{N}\right) = \text{constant}$$

$$\frac{M}{N} = b^{\text{constant}}$$

4. If the equation has multiple logarithmic terms as in $\log_b X = \log_b M + \log_b N$, consolidate logarithmic terms using the product or quotient properties, then use the uniqueness property and algebra to solve:

$$\log_b X = \log_b M + \log_b N$$

$$\log_b X = \log_b(MN)$$

$$X = MN$$

Many other forms and varieties are possible.

In advanced applications, the equations used are sometimes impossible to solve using inverse functions. This is often the case when logarithmic or exponential functions are mixed with other functions (polynomial, radical, rational, etc.). In these cases graphing and calculating technologies become indispensible tools, and the emphasis in working towards a solution shifts more to an understanding of the domain, the graphical attributes of the functions involved, and setting an appropriate window.

EXAMPLE 6 ▶ **Solving Equations Using Technology**

Find all solutions to $e^{\sqrt[3]{x-8}} = \sqrt{x+9}$.

Solution ▶ Begin by noting that the domain of the function on the left, call it Y_1, is all real numbers, since this is the domain of both $y = e^x$ and $y = \sqrt[3]{x}$. However, the domain of the function on the right (call it Y_2) is $x \geq -9$, and we expect that any solution(s) to the equation must occur to the right of -9, so we opt for a standard viewing window to begin (Figure 5.48). At first it appears the graphs do not intersect to the left, but our knowledge of the domain

Figure 5.48

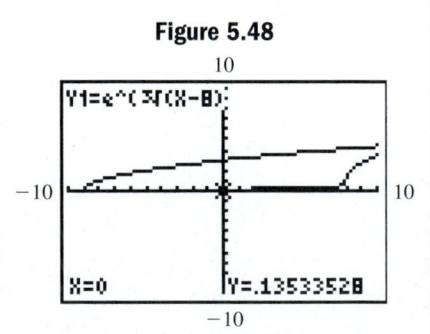

indicates they must intersect near $x = -9$, since Y_1 and Y_2 are both positive. Using the intersection-of-graphs method, we find one solution is $x \approx -8.994$. To the right, no point of intersection is initially visible so we extend our window to explore whether the graphs intersect again. Using Xmax = 20 we note the graphs indeed intersect (Figure 5.49) and locate a second solution at $x \approx 11.433$.

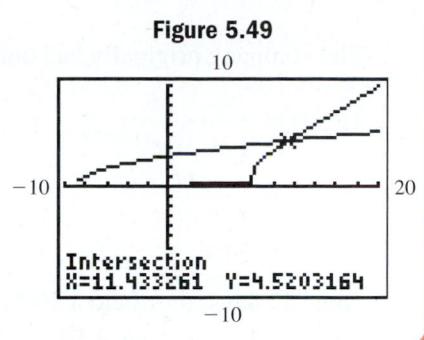

Figure 5.49

Intersection
X=11.433261 Y=4.5203164

☑ **A.** You've just seen how we can solve general logarithmic and exponential equations

Now try Exercises 45 through 48 ▶

B. Applications of Logistic, Exponential, and Logarithmic Functions

Applications of exponential and logarithmic functions take many different forms and it would be impossible to illustrate them all. As you work through the exercises, try to adopt a "big picture" approach, applying the general principles illustrated here to other applications. Some may have been introduced in previous sections. The difference here is that we can now *solve for the independent variable*, instead of simply evaluating the relationships.

In applications involving the **logistic growth** of animal populations, the initial stage of growth is virtually exponential, but due to limitations on food, space, or other resources, growth slows and at some point it reaches a limit. In business, the same principle applies to the logistic growth of sales or profits, due to market saturation. In these cases, the exponential term appears in the denominator of a quotient, and we "clear denominators" to begin the solution process.

EXAMPLE 7 ▶ **Solving a Logistic Equation**

A small business makes a new discovery and begins an aggressive advertising campaign, confident they can capture 66% of the market in a short period of time. They anticipate their market share will be modeled by the function

$M(t) = \dfrac{66}{1 + 10e^{-0.05t}}$, where $M(t)$ represents the percentage after t days. Use this function to answer the following.

a. What was the company's initial market share ($t = 0$)? What was their market share 30 days later?

b. How long will it take the company to reach a 60% market share?

▼ **Algebraic Solution**

a. $M(t) = \dfrac{66}{1 + 10e^{-0.05t}}$ given

$M(0) = \dfrac{66}{1 + 10e^{-0.05(0)}}$ substitute 0 for t

$= \dfrac{66}{11}$ simplify ($e^0 = 1$)

$= 6$ result

▼ **Graphical Solution**

a. After entering the function as Y_1, we can find $Y_1(0)$ and $Y_1(30)$ directly on the home screen (Figure 5.50). The company began the ad campaign with a 6% market share, and 30 days later they had secured over a 20.4% market share.

Figure 5.50

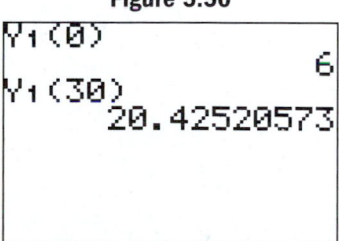

Y₁(0)
                    6
Y₁(30)
        20.42520573

The company originally had only a 6% market share.

$$M(30) = \frac{66}{1 + 10e^{-0.05(30)}}$$ substitute 30 for t

$$= \frac{66}{1 + 10e^{-1.5}}$$ simplify

$$\approx 20.4$$ result

After 30 days, they held a 20.4% market share.

b. For part (b), we replace $M(t)$ with 60 and solve for t.

$$60 = \frac{66}{1 + 10e^{-0.05t}}$$ given

$$60(1 + 10e^{-0.05t}) = 66$$ multiply by $1 + 10e^{-0.05t}$

$$1 + 10e^{-0.05t} = 1.1$$ divide by 60

$$10e^{-0.05t} = 0.1$$ subtract 1

$$e^{-0.05t} = 0.01$$ divide by 10

$$\ln e^{-0.05t} = \ln 0.01$$ apply base-e logarithms

$$-0.05t = \ln 0.01$$ Property III

$$t = \frac{\ln 0.01}{-0.05}$$ solve for t (exact form)

$$\approx 92$$ approximate form

b. Using the intersection-of-graphs method, we graph Y_1 with $Y_2 = 60$ to find any point(s) of intersection. For the window size, we reason that after 30 days, there is only a 20.4% market share (x must be much greater than 30), and a 60% market share is being explored (y must be greater than 60). Using the window indicated in Figure 5.51 reveals that a 60% market share will be attained shortly after the 92nd day.

Figure 5.51

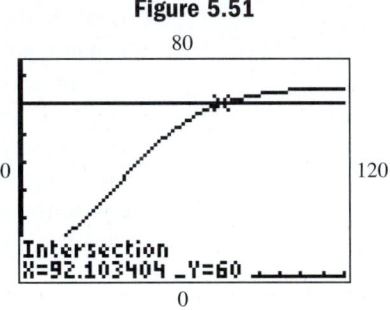

The company will reach a 60% market share in about 92 days.

Now try Exercises 49 and 50 ▶

Earlier we used the barometric equation $H = (30T + 8000) \ln\left(\dfrac{P_0}{P}\right)$ to find an altitude H, given a temperature and the atmospheric (barometric) pressure in centimeters of mercury (cmHg). Using the tools from this section, we are now able to find the atmospheric pressure for a given altitude and temperature.

EXAMPLE 8 ▶   Using Logarithms to Determine Atmospheric Pressure

Suppose a group of climbers has just scaled Mt. Rainier, the highest mountain of the Cascade Range in western Washington State. If the mountain is about 4395 m high and the temperature at the summit is $-22.5°C$, what is the atmospheric pressure at this altitude? The pressure at sea level is $P_0 = 76$ cmHg.

▼ Algebraic Solution

$$H = (30T + 8000) \ln\left(\frac{P_0}{P}\right)$$ given

$$4395 = [30(-22.5) + 8000] \ln\left(\frac{76}{P}\right)$$ substitute 4395 for H, 76 for P_0, and -22.5 for T

$$4395 = 7325 \ln\left(\frac{76}{P}\right)$$ simplify

$$0.6 = \ln\left(\frac{76}{P}\right)$$ divide by 7325

$$e^{0.6} = \frac{76}{P}$$ exponential form

$$Pe^{0.6} = 76$$ multiply by P

$$P = \frac{76}{e^{0.6}}$$ divide by $e^{0.6}$ (exact form)

$$\approx 41.7$$ approximate form

 **B.** You've just seen how we can solve applications involving logistic, exponential, and logarithmic functions

▼ Graphical Solution

To ensure that no algebraic errors are introduced, we'll enter the function as it appears *after the substitutions are made:*

$$Y_1 = [30(-22.5) + 8000] \ln\left(\frac{76}{x}\right).$$

For the window size, we reason that since x must be between 0 and 76, and y is equal to 4395, we only need the first Quadrant and a "frame" around the window that allows a clear view of the intersection point. Using the window indicated shows that at an altitude of 4395 m and a temperature of $-22.5°C$, the atmospheric pressure is about 41.7 cmHg.

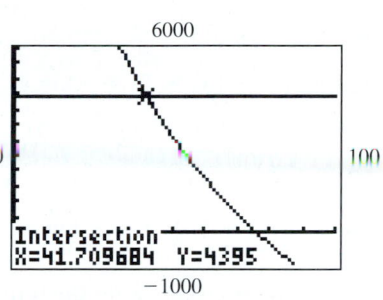

Now try Exercises 53 and 54 ▶

Additional applications involving appreciation/depreciation, Newton's law of cooling, space ship velocities and more, can be found in the Exercise set. **See Exercises 55 through 66.**

5.5 EXERCISES

▶ CONCEPTS AND VOCABULARY

Fill in each blank with the appropriate word or phrase. Carefully reread the section if needed.

1. The expression $\log_2 x$ represents a <u>variable</u> term, while the expression $\log_2 9$ represents a <u>constant</u> term.

2. To solve the equation $\ln(x + 3) - \ln x = 7$, we <u>combine</u> like terms using logarithmic properties, prior to writing the equation in <u>exponential</u> form.

3. If certain conditions are met, we know if $\log_b M = \log_b N$, then $M = N$. This is a statement of the <u>uniqueness</u> property, which is valid since logarithmic functions are <u>one</u>-to-<u>one</u>.

4. Since the domain of $y = \log_b x$ is <u>$x > 0$</u>. solving logarithmic equations will sometimes produce <u>extraneous</u> roots. Checking all solutions to logarithmic equations is a necessary step.

5. Answer true or false and explain your response:

$$\log_b(M + N) = \log_b(M) + \log_b(N)$$
False; answers will vary.

6. Answer true or false and explain your response:

$$\log_b\left(\frac{M}{N}\right) = \frac{\log_b M}{\log_b N}$$
False; answers will vary.

HOMEWORK SELECTION GUIDE

Core: 7–45 odd, 49–61 odd, 69, 70 (29 Exercises)
Standard: 1–4, 7–61 odd, 65, 69, 70, 71 (36 Exercises)

Extended: 1–4, 7–47 odd, 48, 49–61 odd, 65, 67, 69, 70–72 (39 Exercises)
In Depth: 1–6, 7–47 odd, 48, 49–69 odd, 70–73 (45 Exercises)

► DEVELOPING YOUR SKILLS

Solve each equation and check your answers.

7. $\log 4 + \log(x - 7) = 2$ $\quad x = 32$

8. $\log 5 + \log(x - 9) = 1$ $\quad x = 11$

9. $\log(2x - 5) - \log 78 = -1$ $\quad x = 6.4$

10. $\log(4 - 3x) - \log 145 = -2$ $\quad x = 0.85$

11. $\log(x - 15) - 2 = -\log x$ $\quad x = 20, -5$ is extraneous

12. $\log x - 1 = -\log(x - 9)$ $\quad x = 10, -1$ is extraneous

13. $\log(2x + 1) = 1 - \log x$ $\quad x = 2, -\frac{5}{2}$ is extraneous

14. $\log(3x - 13) = 2 - \log x$ $\quad x = \frac{25}{3}, -4$ is extraneous

Solve each equation using the uniqueness property.

15. $\log(5x + 2) = \log 2$ $\quad x = 0$

16. $\log(2x - 3) = \log 3$ $\quad x = 3$

17. $\log_4(x + 2) - \log_4 3 = \log_4(x - 1)$ $\quad x = \frac{5}{2}$

18. $\log_3(x + 6) - \log_3 x = \log_3 5$ $\quad x = \frac{3}{2}$

19. $\ln(8x - 4) = \ln 2 + \ln x$ $\quad x = \frac{2}{3}$

20. $\ln(x - 1) + \ln 6 = \ln(3x)$ $\quad x = 2$

Solve each equation using any appropriate method. State solutions in both exact form and in approximate form rounded to four decimal places. Clearly identify any extraneous roots. If there are no solutions, so state.

21. $\log(2x - 1) + \log 5 = 1$ $\quad x = \frac{3}{2}$

22. $\log(x - 7) + \log 3 = 2$ $\quad x = \frac{121}{3}$

23. $\log_2(9) + \log_2(x + 3) = 3$ $\quad x = \frac{-19}{9}$

24. $\log_3(x - 4) + \log_3(7) = 2$ $\quad x = \frac{37}{7}$

25. $\ln(x + 7) + \ln 9 = 2$ $\quad x = \frac{e^2 - 63}{9}; x \approx -6.1790$

26. $\ln 5 + \ln(x - 2) = 1$ $\quad x = \frac{e + 10}{5}; x \approx 2.5437$

27. $\log(x + 8) + \log x = \log(x + 18)$ $\quad x = 2; -9$ is extraneous

28. $\log(x + 14) - \log x = \log(x + 6)$ $\quad x = 2; -7$ is extraneous

29. $\ln(2x + 1) = 3 + \ln 6$ $\quad x = 3e^3 - \frac{1}{2}; x \approx 59.7566$

30. $\ln 21 = 1 + \ln(x - 2)$ $\quad x = 21e^{-1} + 2; x \approx 9.7255$

31. $\log(-x - 1) = \log(5x) + \log x$ $\quad$ no solution

32. $\log(1 - x) + \log x = \log(x + 4)$ $\quad$ no solution

33. $\log(x - 1) - \log x = \log(x - 3)$ $\quad x = 2 + \sqrt{3}; 2 - \sqrt{3}$ is extraneous

34. $\ln x + \ln(x - 2) = \ln 4$ $\quad x = 1 + \sqrt{5}, 1 - \sqrt{5}$ is extraneous

35. $7^x = 231$ $\qquad$ **36.** $6^x = 3589$

37. $5^{3x} - 2 = 128,965$ $\quad$ **38.** $9^{3x} - 3 = 78,462$

39. $2^{x+1} = 3^x$ $\qquad$ **40.** $7^x = 4^{2x-1}$

Solve each equation using the zeroes method or the intersection-of-graphs method. Round approximate solutions to three decimal places.

41. $\sqrt[3]{x} = \ln(x + 5)$ $\quad x \approx -4.815, x \approx 102.084$

42. $\dfrac{x^2 - 25}{x^2 - 9} = -\ln(x + 9) + 6$
$\quad x \approx -1.935, x \approx 1.707, x \approx 139.535$

43. $2^{x^2 - x - 6} = x^2 + x - 6$ $\quad x \approx 2.013, x \approx 3.608$

44. $x^3 - 9x = \dfrac{1}{2}e^x$
$\quad x \approx -2.999, x \approx -0.053, x \approx 3.941, x \approx 5.338$

45. $\dfrac{250}{1 + 4e^{-0.06x}} = 200$ $\quad x \approx 46.210$

46. $\dfrac{80}{1 + 15e^{-0.06x}} = 50$ $\quad x \approx 53.648$

► WORKING WITH FORMULAS

47. Logistic growth: $P(t) = \dfrac{C}{1 + ae^{-kt}}$

For populations that exhibit logistic growth, the population at time t is modeled by the function shown, where C is the carrying capacity of the population (the maximum population that can be supported over a long period of time), k is the growth constant, and $a = \frac{C - P(0)}{P(0)}$. Solve the formula for t, then use the result to find the value of t given $C = 450$, $a = 8$, $P = 400$, and $k = 0.075$.

$t = \dfrac{\ln\left(\dfrac{\frac{C}{P} - 1}{a}\right)}{-k}, t \approx 55.45$

Additional answers can be found in the Instructor Answer Appendix.

48. Estimating time of death: $h = -3.9 \cdot \ln\left(\dfrac{T - T_R}{T_0 - T_R}\right)$

Using the formula shown, a forensic expert can compute the approximate time of death for a person found recently expired, where T is the body temperature when it was found, T_R is the (constant) temperature of the room, T_0 is the body temperature at the time of death ($T_0 = 98.6°F$), and h is the number of hours since death. If the body was discovered at 9:00 A.M. with a temperature of 86.2°F, in a room at 73°F, at approximately what time did the person expire? (Note this formula is a version of Newton's law of cooling.) 6:25 AM

▶ **APPLICATIONS**

Use the *barometric equation* $H = (30T + 8000) \ln\left(\dfrac{P_0}{P}\right)$ for exercises 49 and 50. Recall that $P_0 = 76$ cmHg.

49. Altitude and temperature: A sophisticated spy plane is cruising at an altitude of 18,250 m. If the temperature at this altitude is $-75°C$, what is the barometric pressure?   about 3.2 cmHg

50. Altitude and temperature: A large weather balloon is released and takes altitude, pressure, and temperature readings as it climbs, and radios the information back to Earth. What is the pressure reading at an altitude of 5000 m, given the temperature is $-18°C$?   about 39 cmHg

51. Stocking a lake: A farmer wants to stock a private lake on his property with catfish. A specialist studies the area and depth of the lake, along with other factors, and determines it can support a maximum population of around 750 fish, with growth modeled by the function $P(t) = \dfrac{750}{1 + 24e^{-0.075t}}$, where $P(t)$ gives the current population after t months. (a) How many catfish did the farmer initially put in the lake? (b) How many months until the population reaches 300 fish?   a. 30 fish   b. about 37 months

52. Increasing sales: After expanding their area of operations, a manufacturer of small storage buildings believes the larger area can support sales of 40 units per month. After increasing the advertising budget and enlarging the sales force, sales are expected to grow according to the model $S(t) = \dfrac{40}{1 + 1.5e^{-0.08t}}$, where $S(t)$ is the expected number of sales after t months. (a) How many sales were being made each month, prior to the expansion? (b) How many months until sales reach 25 units per month?   a. 16 units   b. about 11.5 months

Use *Newton's law of cooling* $T = T_R + (T_0 - T_R)e^{kh}$ to complete Exercises 57 and 58. Recall that water freezes at 32°F and use $k = -0.012$. Refer to Section 5.2, page 498 as needed.

53. Making popsicles: On a hot summer day, Sean and his friends mix some Kool-Aid® and decide to freeze it in an ice tray to make popsicles. If the water used for the Kool-Aid® was 75°F and the freezer has a temperature of $-20°F$, how long will they have to wait to enjoy the treat?   about 50.2 min

54. Freezing time: Suppose the current temperature in Esconabe, Michigan, was 47°F when a 5°F arctic cold front moved over the state. How long would it take a puddle of water to freeze over?   about 36.8 min

Depreciation/appreciation: As time passes, the value of certain items decrease (appliances, automobiles, etc.), while the value of other items increase (collectibles, real estate, etc.). The time T in years for an item to reach a future value can be modeled by the formula $T = k \ln\left(\dfrac{V_n}{V_f}\right)$, **where V_n is the purchase price when new, V_f is its future value, and k is a constant that depends on the item.**

55. Automobile depreciation: If a new car is purchased for $28,500, find its value 3 yr later if $k = 5$. $15,641

56. Home appreciation: If a new home in an "upscale" neighborhood is purchased for $130,000, find its value 12 yr later if $k = -16$. $275, 210

Drug absorption: The time required for a certain percentage of a drug to be *absorbed* by the body after injection depends on the drug's absorption rate. This can be modeled by the function $T(p) = \dfrac{-\ln p}{k}$, **where p represents the percent of the drug that *remains unabsorbed* (expressed as a decimal), k is the absorption rate of the drug, and $T(p)$ represents the elapsed time.**

57. For a drug with an absorption rate of 7.2%, (a) find the time required (to the nearest hour) for the body to *absorb* 35% of the drug, and (b) find the percent of this drug (to the nearest half percent) that remains unabsorbed after 24 hr.   a. 6 hr   b. 18.0%

58. For a drug with an absorption rate of 5.7%, (a) find the time required (to the nearest hour) for the body to *absorb* 50% of the drug, and (b) find the percent of this drug (to the nearest half percent) that remains unabsorbed after 24 hr.   a. 12 hr   b. 25.5%

Spaceship velocity: In space travel, the change in the velocity of a spaceship V_s (in km/sec) depends on the mass of the ship M_s (in tons), the mass of the fuel which has been burned M_f (in tons) and the escape velocity of the exhaust V_e (in km/sec). Disregarding frictional forces, these are related by the equation

$$V_s = V_e \ln\left(\dfrac{M_s}{M_s - M_f}\right).$$

59. For the Jupiter VII rocket, find the mass of the fuel M_f that has been burned if $V_s = 6$ km/sec when $V_e = 8$ km/sec, and the ship's mass is 100 tons. $M_f = 52.76$ tons

60. For the Neptune X satellite booster, find the mass of the ship M_s if $M_f = 75$ tons of fuel has been burned when $V_s = 8$ km/sec and $V_e = 10$ km/sec. $M_s = 136.20$ tons

Learning curve: The job performance of a new employee when learning a repetitive task (as on an assembly line) improves very quickly at first, then grows more slowly over time. This can be modeled by the function $P(t) = a + b \ln t$, where a and b are constants that depend on the type of task and the training of the employee.

61. The number of toy planes an employee can assemble from its component parts depends on the length of time the employee has been working. This output is modeled by $P(t) = 5.9 + 12.6 \ln t$, where $P(t)$ is the number of planes assembled daily after working t days. (a) How many planes is an employee making after 5 days on the job? (b) How many days until the employee is able to assemble 34 planes per day?   **a.** 26 planes   **b.** 9 days

62. The number of circuit boards an associate can assemble from its component parts depends on the length of time the associate has been working. This output is modeled by $B(t) = 1 + 2.3 \ln t$, where $B(t)$ is the number of boards assembled daily after working t days. (a) How many boards is an employee completing after 9 days on the job? (b) How long will it take until the employee is able to complete 10 boards per day?   **a.** 6 boards   **b.** 50 days

▶ EXTENDING THE CONCEPT

Solve the following equations. Note that equations Exercises 63 and 64 are in quadratic form.

63. $2e^{2x} - 7e^x = 15$ $x = 1.609438$

64. $3e^{2x} - 4e^x - 7 = -3$ $x = 0.69314718$

65. Use the algebraic method to find the inverse function.
a. $f(x) = 2^{x+1}$ **b.** $y = 2 \ln(x - 3)$

66. Show that $g(x) = f^{-1}(x)$ by composing the functions.
a. $f(x) = 3^{x-2}$; $g(x) = \log_3 x + 2$
b. $f(x) = e^{x-1}$; $g(x) = \ln x + 1$

67. Use properties of logarithms and/or exponents to show
a. $y = 2^x$ is equivalent to $y = e^{x \ln 2}$.
b. $y = b^x$ is equivalent to $y = e^{rx}$, where $r = \ln b$.

68. Use test values for p and q to demonstrate that the following relationships are *false*.
a. $\ln(pq) = \ln p \ln q$ **b.** $\ln p + \ln q = \ln(p + q)$
c. $\ln\left(\dfrac{p}{q}\right) = \dfrac{\ln p}{\ln q}$ Answers will vary.

69. Match each equation with the most appropriate solution strategy, and justify/discuss why.
a. $e^{x+1} = 25$ _____d_____ apply base-10 logarithm to both sides
b. $\log(2x + 3) = \log 53$ _____e_____ rewrite and apply uniqueness property for exponentials
c. $\log(x^2 - 3x) = 2$ _____b_____ apply uniqueness property for logarithms
d. $10^{2x} = 97$ _____f_____ apply either base-10 or base-e logarithm
e. $2^{5x-3} = 32$ _____a_____ apply base-e logarithm
f. $7^{x+2} = 23$ _____c_____ write in exponential form

▶ MAINTAINING YOUR SKILLS

70. **(3.3)** Match the graph shown with its correct equation, without actually graphing the function.   b
a. $y = x^2 + 4x - 5$
b. $y = -x^2 - 4x + 5$
c. $y = -x^2 + 4x + 5$
d. $y = x^2 - 4x - 5$

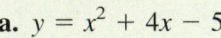

71. **(2.3/2.4)** State the domain and range of the functions.
a. $y = \sqrt{2x + 3}$ **b.** $y = |x + 2| - 3$
$x \in [-\frac{3}{2}, \infty), y \in [0, \infty)$ $x \in (-\infty, \infty), y \in [-3, \infty)$

72. **(4.5)** Graph the function $r(x) = \dfrac{x^2 - 4}{x - 1}$. Label all intercepts and asymptotes.

73. **(2.6)** Suppose the maximum load (in tons) that can be supported by a cylindrical post varies directly with its diameter raised to the fourth power and inversely as the square of its height. A post 8 ft high and 2 ft in diameter can support 6 tons. How many tons can be supported by a post 12 ft high and 3 ft in diameter?   13.5 tons

Additional answers can be found in the Instructor Answer Appendix.

5.6 Applications from Business, Finance, and Science

LEARNING OBJECTIVES

In Section 5.6 you will see how we can:

☐ **A.** Calculate simple interest and compound interest

☐ **B.** Calculate interest compounded continuously

☐ **C.** Solve applications of annuities and amortization

☐ **D.** Solve applications of exponential growth and decay

Would you pay $750,000 for a home worth only $250,000? Surprisingly, when a conventional mortgage is repaid over 30 years, this is not at all rare. Over time, the accumulated interest on the mortgage is easily more than two or three times the original value of the house. In this section we explore how interest is paid or charged, and look at other applications of exponential and logarithmic functions from business, finance, as well as the physical and social sciences.

A. Simple and Compound Interest

Simple interest is an amount of interest that is computed only once during the lifetime of an investment (or loan). In the world of finance, the initial deposit or base amount is referred to as the **principal p**, the **interest rate r** is given as a percentage and stated as an annual rate, with the term of the investment or loan most often given as *time t* in years. Simple interest is merely an application of the basic percent equation, with the additional element of time coming into play: *interest = principal × rate × time,* or $I = prt$. To find the total amount A that has accumulated (for deposits) or is due (for loans) after t years, we merely add the accumulated interest to the initial principal: $A = p + prt$.

WORTHY OF NOTE

If a loan is kept for only a certain number of months, weeks, or days, the time t should be stated as a fractional part of a year so the time period for the rate (years) matches the time period over which the loan is repaid.

Simple Interest Formula

If principal p is deposited or borrowed at interest rate r for a period of t years, the simple interest on this account will be

$$I = prt$$

The total amount A accumulated or due after this period will be

$$A = p + prt \quad \text{or} \quad A = p(1 + rt)$$

EXAMPLE 1 ▶ **Solving an Application of Simple Interest**

Many finance companies offer what have become known as *PayDay Loans*—a small $50 loan to help people get by until payday, usually no longer than 2 weeks. If the cost of this service is $12.50, determine the annual rate of interest charged by these companies.

Solution ▶ The interest charge is $12.50, the initial principal is $50.00, and the time period is 2 weeks or $\frac{2}{52} = \frac{1}{26}$ of a year. The simple interest formula yields

$$I = prt \qquad \text{simple interest formula}$$

$$12.50 = 50r\left(\frac{1}{26}\right) \qquad \text{substitute \$12.50 for } I, \text{ \$50.00 for } p, \text{ and } \tfrac{1}{26} \text{ for } t$$

$$6.5 = r \qquad \text{solve for } r$$

The annual interest rate on these loans is a whopping 650%!

Now try Exercises 7 through 16 ▶

Compound Interest

Many financial institutions pay **compound interest** on deposits they receive, which is interest paid on previously accumulated interest. The most common compounding periods are yearly, semiannually (two times per year), quarterly (four times per year), monthly (12 times per year), and daily (365 times per year). Applications of compound interest typically involve exponential functions. For convenience, consider $1000 in principal, deposited at 8% for 3 yr. The simple interest calculation shows $240 in interest is earned and there will be $1240 in the account: $A = 1000[1 + (0.08)(3)] = \$1240$. If the interest is *compounded each year* $(t = 1)$ instead of once at the start of the 3-yr period, the interest calculation shows

$$A_1 = 1000(1 + 0.08) = 1080 \text{ in the account at the end of year 1,}$$
$$A_2 = 1080(1 + 0.08) = 1166.40 \text{ in the account at the end of year 2,}$$
$$A_3 = 1166.40(1 + 0.08) \approx 1259.71 \text{ in the account at the end of year 3.}$$

The account has earned an additional $19.71 interest. More importantly, notice that we're multiplying by $(1 + 0.08)$ each compounding period, meaning results can be computed more efficiently by simply applying the factor $(1 + 0.08)^t$ to the initial principal p. For example,

$$A_3 = 1000(1 + 0.08)^3 \approx \$1259.71.$$

In general, for interest compounded yearly the **accumulated value** is $A = p(1 + r)^t$. Notice that solving this equation for p will tell us the amount we need to deposit *now*, in order to accumulate A dollars in t years: $p = \frac{A}{(1 + r)^t}$. This is called the **present value equation.**

Interest Compounded Annually

If a principal p is deposited at interest rate r and compounded yearly for a period of t yr, the *accumulated value* is

$$A = p(1 + r)^t$$

If an accumulated value A is desired after t yr, and the money is deposited at interest rate r and compounded yearly, the *present value* is

$$p = \frac{A}{(1 + r)^t}$$

EXAMPLE 2 ▶ **Finding the Doubling Time for Interest Compounded Yearly**

An initial deposit of $1000 is made into an account paying 6% compounded yearly. How long will it take for the money to double?

Solution ▶ Using the formula for interest compounded yearly we have

$$A = p(1 + r)^t \qquad \text{given}$$
$$2000 = 1000(1 + 0.06)^t \qquad \text{substitute 2000 for } A, \text{ 1000 for } p, \text{ and 0.06 for } r$$
$$2 = 1.06^t \qquad \text{isolate variable term}$$
$$\ln 2 = t \ln 1.06 \qquad \text{apply base-}e \text{ logarithms; power property}$$
$$\frac{\ln 2}{\ln 1.06} = t \qquad \text{solve for } t$$
$$11.9 \approx t \qquad \text{approximate form}$$

The money will double in just under 12 yr.

Now try Exercises 17 through 22 ▶

If interest is compounded monthly (12 times each year), the bank will divide the interest rate by 12 (the number of compoundings), but then pay you interest 12 times per year (interest is *compounded*). The net effect is an increased gain in the interest you earn, and the final compound interest formula takes this form:

$$\text{total amount} = \text{principal}\left(1 + \frac{\text{interest rate}}{\text{compoundings per year}}\right)^{(\text{years} \times \text{compoundings per year})}$$

Compounded Interest Formula

If principal p is deposited at interest rate r and compounded n times per year for a period of t yr, the *accumulated value* will be:

$$A = p\left(1 + \frac{r}{n}\right)^{nt}$$

EXAMPLE 3 ▶ **Solving an Application of Compound Interest**

Macalyn won $150,000 in the Missouri lottery and decides to invest the money for retirement in 20 yr. Of all the options available here, which one will produce the most money for retirement?

a. A certificate of deposit paying 5.4% compounded yearly.
b. A money market certificate paying 5.35% compounded semiannually.
c. A bank account paying 5.25% compounded quarterly.
d. A bond issue paying 5.2% compounded daily.

Solution ▶ **a.** $A = \$150{,}000\left(1 + \dfrac{0.054}{1}\right)^{(20 \times 1)}$ **c.** $A = \$150{,}000\left(1 + \dfrac{0.0525}{4}\right)^{(20 \times 4)}$

$\approx \$429{,}440.97$ $\approx \$425{,}729.59$

b. $A = \$150{,}000\left(1 + \dfrac{0.0535}{2}\right)^{(20 \times 2)}$ **d.** $A = \$150{,}000\left(1 + \dfrac{0.052}{365}\right)^{(20 \times 365)}$

$\approx \$431{,}200.96$ $\approx \$424{,}351.12$

✓ **A.** You've just seen how we can calculate simple interest and compound interest

The best choice is (b), semiannual compounding at 5.35% for 20 yr.

Now try Exercises 23 through 30 ▶

B. Interest Compounded Continuously

It seems natural to wonder what happens to the interest accumulation as n (the number of compounding periods) becomes very large. It appears the interest rate becomes very small (because we're dividing it by n), but the exponent becomes very large (since we're multiplying it by n). To see the result of this interplay more clearly, it will help to rewrite the compound interest formula $A = p(1 + \frac{r}{n})^{nt}$ using the substitution $n = xr$. This gives $\frac{r}{n} = \frac{1}{x}$, and by direct substitution (xr for n and $\frac{1}{x}$ for $\frac{r}{n}$) we obtain the form

$$A = p\left[\left(1 + \frac{1}{x}\right)^x\right]^{rt}$$

by regrouping. This allows for a more careful study of the "denominator versus exponent" relationship using $(1 + \frac{1}{x})^x$, *the same expression we used in Section 5.2 to define the number e* (also **see Section 5.2 Exercise 101**). Once again, note what happens as $x \to \infty$ (meaning the number of compounding periods increase without bound).

x	1	10	100	1000	10,000	100,000	1,000,000
$\left(1 + \dfrac{1}{x}\right)^x$	2	2.59374	2.70481	2.71692	2.71815	2.71827	2.71828

As before, as $x \to \infty$, $(1 + \frac{1}{x})^x \to e$. The net result of this investigation is a formula for **interest compounded continuously,** derived by replacing $(1 + \frac{1}{x})^x$ with the number e in the formula for compound interest, where

$$A = p\left[\left(1 + \frac{1}{x}\right)^x\right]^{rt} = pe^{rt}$$

Interest Compounded Continuously

If a principal p is deposited at interest rate r and compounded continuously for a period of t years, the *accumulated value* will be

$$A = pe^{rt}$$

EXAMPLE 4 ▶ **Solving an Application of Interest Compounded Continuously**

Jaimin has \$10,000 to invest and wants to have at least \$25,000 in the account in 10 yr for his daughter's college education fund. If the account pays interest compounded continuously, what interest rate is required?

Solution ▶ In this case, $P = \$10,000$, $A = \$25,000$, and $t = 10$.

▼ Algebraic Solution

$A = pe^{rt}$	given
$25,000 = 10,000e^{10r}$	substitute given values
$2.5 = e^{10r}$	isolate variable term
$\ln 2.5 = 10r \ln e$	use natural logs; power property
$\dfrac{\ln 2.5}{10} = r$	solve for r ($\ln e = 1$)
$0.092 \approx r$	approximate form

Jaimin will need an interest rate of about 9.2% to meet his goal.

▼ Graphical Solution

Using $Y_1 = 10,000e^{10X}$ and $Y_2 = 25,000$, we look for their point of intersection. For the window size, since 25,000 is the goal, $y \in [0, 30{,}000]$ seems reasonable for y. Although 12% interest ($x = 0.12$) is too good to be true, $x \in [0, 0.12]$ will create a nice frame for the x-values. The point of intersection shows an interest rate of about 9.2% is required.

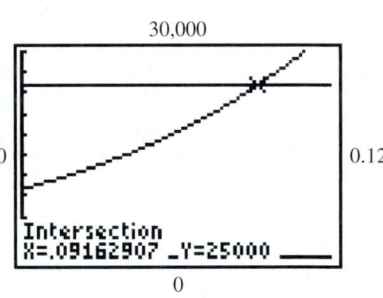

Now try Exercises 31 through 40 ▶

☑ **B. You've just seen how we can calculate interest compounded continuously**

C. Applications Involving Annuities and Amortization

Our previous calculations for simple and compound interest involved a single (lump) deposit (the principal) that accumulated interest over time. Many savings and investment plans involve a regular schedule of deposits (monthly, quarterly, or annual deposits) over the life of the investment. Such an investment plan is called an **annuity.**

Suppose that for 4 yr, \$100 is deposited annually into an account paying 8% compounded yearly. Using the compound interest formula we can track the accumulated value A in the account:

$$A = 100 + 100(1.08)^1 + 100(1.08)^2 + 100(1.08)^3$$

To develop an annuity formula, we multiply the annuity equation by 1.08, then subtract the original equation. This leaves only the first and last terms, since the other (interior) terms add to zero:

$$1.08A = 100(1.08) + 100(1.08)^2 + 100(1.08)^3 + 100(1.08)^4 \qquad \text{multiply by 1.08}$$

$$-A = -[100 + 100(1.08)^1 + 100(1.08)^2 + 100(1.08)^3] \qquad \text{original equation}$$

$$1.08A - A = 100(1.08)^4 - 100 \qquad \text{subtract (“interior terms” sum to zero)}$$

$$0.08A = 100[(1.08)^4 - 1] \qquad \text{factor out 100}$$

$$A = \frac{100[(1.08)^4 - 1]}{0.08} \qquad \text{solve for } A$$

This result can be generalized for any periodic payment $\mathcal{P}$, interest rate r, number of compounding periods n, and number of years t. This would give

$$A = \frac{\mathcal{P}\left[\left(1 + \dfrac{r}{n}\right)^{nt} - 1\right]}{\dfrac{r}{n}}$$

The formula can be made less formidable using $R = \frac{r}{n}$, where R is the interest rate per compounding period.

Accumulated Value of an Annuity

If a periodic payment $\mathcal{P}$ is deposited n times per year at an *annual interest rate r* with interest compounded n times per year for t years, the accumulated value is given by

$$A = \frac{\mathcal{P}}{R}[(1 + R)^{nt} - 1], \text{ where } R = \frac{r}{n}$$

This is also referred to as the **future value** of the account.

EXAMPLE 5 ▶   **Solving an Application of Annuities**

Since he was a young child, Fitisemanu's parents have been depositing $50 each month into an annuity that pays 6% annually and is compounded monthly. If the account is now worth $9875, how long has it been open?

Solution ▶   In this case $\mathcal{P} = 50$, $r = 0.06$, $n = 12$, $R = 0.005$, and $A = 9875$.

Algebraic Solution ▶

$$A = \frac{\mathcal{P}}{R}[(1 + R)^{nt} - 1] \qquad \text{future value formula}$$

$$9875 = \frac{50}{0.005}[(1.005)^{(12)(t)} - 1] \qquad \text{substitute given values}$$

$$1.9875 = 1.005^{12t} \qquad \text{simplify and isolate variable term}$$

$$\ln(1.9875) = 12t(\ln 1.005) \qquad \text{apply base-}e\text{ logarithms; power property}$$

$$\frac{\ln(1.9875)}{12\ln(1.005)} = t \qquad \text{solve for } t \text{ (exact form)}$$

$$11.5 \approx t \qquad \text{approximate form}$$

The account has been open approximately 11.5 yr.

Graphical Solution ▶ Here we'll use the intersection-of-graphs method. Entering

$$Y_1 = \left(\frac{50}{0.005}\right)(1.005^{12X} - 1) \text{ and}$$

$Y_2 = 9875$, we must next determine an appropriate window size. Since the goal is $9875, we'll use [0, 15,000] for y, leaving a large frame around the window. If no interest were paid, it would take

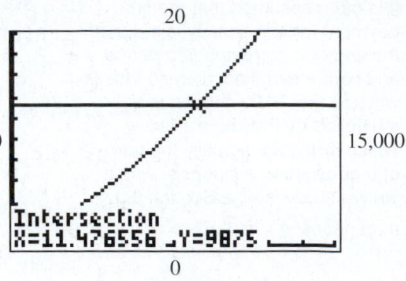

$$\frac{9875}{50(12)} \approx 16.5 \text{ yr to save 9875, so } [0, 20]$$

will also give a window size with plenty of room. The result is shown in the figure, and indicates the account has been open for about 11.5 yr.

Now try Exercises 41 through 44 ▶

The periodic payment required to meet a future goal or obligation can be computed by solving for P in the future value formula: $P = \dfrac{AR}{\left[(1 + R)^{nt} - 1\right]}$. In this form, P is referred to as a **sinking fund.**

EXAMPLE 6 ▶ **Solving an Application of Sinking Funds**

Sheila is determined to stay out of debt and decides to save $20,000 to pay cash for a new car in 4 yr. The best investment vehicle she can find pays 9% compounded monthly. If $300 is the most she can invest each month, can she meet her "4-yr" goal?

Solution ▶ Here we have $P = 300$, $A = 20,000$, $r = 0.09$, $n = 12$, and $R = 0.0075$. The sinking fund formula gives

$$P = \frac{AR}{\left[(1 + R)^{nt} - 1\right]} \qquad \text{sinking fund}$$

$$300 = \frac{(20,000)(0.0075)}{(1.0075)^{12t} - 1} \qquad \text{substitute 300 for } P, 20,000 \text{ for } A, 0.0075 \text{ for } R, \text{ and } 12 \text{ for } n$$

$$300(1.0075^{12t} - 1) = 150 \qquad \text{multiply in numerator, clear denominators}$$

$$1.0075^{12t} = 1.5 \qquad \text{isolate variable term}$$

$$12t \ln(1.0075) = \ln 1.5 \qquad \text{apply base-}e \text{ logarithms; power property}$$

$$t = \frac{\ln(1.5)}{12 \ln(1.0075)} \qquad \text{solve for } t \text{ (exact form)}$$

$$\approx 4.5 \qquad \text{approximate form}$$

☑ **C.** You've just seen how we can solve applications of annuities and amortization

No. She is close, but misses her original 4-yr goal.

Now try Exercises 45 and 46 ▶

For Example 6, we could have substituted 4 for t while leaving P and A unknown, to see if a payment of $300 per month would be sufficient. Using

$$Y_1 = \frac{x(0.0075)}{1.0075^{48} - 1} \text{ and the } \textbf{TABLE} \text{ feature of a}$$

calculator shows that just over $17,000 would be saved for monthly deposits of $300 (Figure 5.52), and that deposits of $347.70 would be required to save $20,000.

Figure 5.52

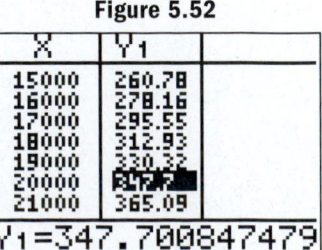

For additional practice with the formulas for interest earned or paid, the *Working with Formulas* portion of this Exercise Set has been expanded. **See Exercises 47 through 54.**

D. Applications Involving Exponential Growth and Decay

Closely related to interest compounded continuously are applications of **exponential growth** and **exponential decay.** If Q (quantity) and t (time) are variables, then Q grows exponentially as a function of t if $Q(t) = Q_0e^{rt}$ for positive constants Q_0 and r. Careful studies have shown that population growth, whether it be humans, bats, or bacteria, can be modeled by these "base-e" exponential growth functions. If $Q(t) = Q_0e^{-rt}$, then we say Q decreases or **decays exponentially** over time. The constant r determines how rapidly a quantity grows or decays and is known as the **growth rate** or **decay rate** constant.

EXAMPLE 7 ▶   Solving an Application of Exponential Growth

Because fruit flies multiply very quickly, they are often used in studies of genetics. Given the necessary space and food supply, a certain population of fruit flies is known to double every 12 days. If there were 100 flies to begin, find (a) the growth rate r and (b) the number of days until the population reaches 2000 flies.

▼ **Algebraic Solution**

a. Using the formula for exponential growth with $Q_0 = 100$, $t = 12$, and $Q(t) = 200$, we can solve for the growth rate r.

$$Q(t) = Q_0e^{rt} \quad \text{\color{magenta}exponential growth function}$$
$$200 = 100e^{12r} \quad \text{\color{magenta}substitute 200 for $Q(t)$ 100 for Q_0, and 12 for t}$$
$$2 = e^{12r} \quad \text{\color{magenta}isolate variable term}$$
$$\ln 2 = 12r \ln e \quad \text{\color{magenta}apply base-e logarithms; power property}$$
$$\frac{\ln 2}{12} = r \quad \text{\color{magenta}solve for r (exact form)}$$
$$0.05776 \approx r \quad \text{\color{magenta}approximate form}$$

The growth rate is approximately 5.78%.

b. To find the number of days until the fly population reaches 2000, we substitute 0.05776 for r in the exponential growth function.

$$Q(t) = Q_0e^{rt} \quad \text{\color{magenta}exponential growth function}$$
$$2000 = 100e^{0.05776t} \quad \text{\color{magenta}substitute 2000 for $Q(t)$, 100 for Q_0, and 0.05776 for r}$$
$$20 = e^{0.05776t} \quad \text{\color{magenta}isolate variable term}$$
$$\ln 20 = 0.05776t \ln e \quad \text{\color{magenta}apply base-e logarithms; power property}$$
$$\frac{\ln 20}{0.05776} = t \quad \text{\color{magenta}solve for t (exact form)}$$
$$51.87 \approx t \quad \text{\color{magenta}approximate form}$$

The fruit fly population will reach 2000 on day 51.

▼ **Graphical Solution**

a. After substituting the values given, we input $Y_1 = 100e^{12x}$ and $Y_2 = 200$ to use the intersection-of-graphs method. While growth rates vary widely for animal populations, we might expect the growth rate to be a decimal between and 0 and 0.2 (0% to 20%), but can adjust the window afterward if needed. Since a population of 200 is the target, we can use [0, 300] for y and [0, 0.2] for x. As seen in Figure 5.53, the graphs intersect at $x \approx 0.05776$.

Figure 5.53

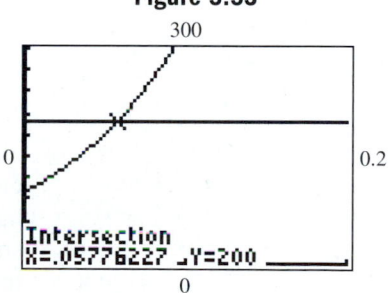

b. In part (a) we solved for the growth rate r. Here we'll use this growth rate and the intersection-of-graphs method to find the time t required for an initial population of 100 flies to grow to 2000. Begin by entering $100e^{0.05776X}$ as Y_1 and 2000 as Y_2. Setting the window for y is no problem, as we have a target population of 2000. For the x-values, consider the approximation $100e^{0.06t}$, and note that $t = 10$ gives too small a value ($100e^{0.6} \approx 182$), while $t = 100$ gives too large a value ($100e^{6.0} \approx 40{,}343$). Using the window $x \in [0, 100]$ and $y \in [0, 3000]$, we find the graphs intersect at $x \approx 51.87$, and the population of flies will reach 2000 in just less than 52 days. See Figure 5.54.

Figure 5.54

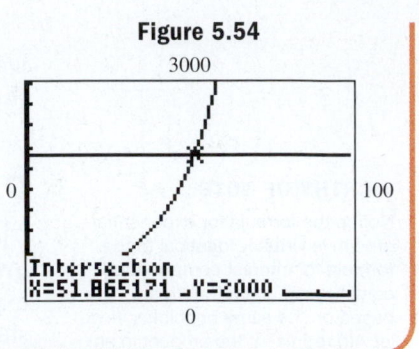

Now try Exercises 55 and 56 ▶

Perhaps the best-known examples of exponential decay involve radioactivity. Ever since the end of World War II, common citizens have been aware of the existence of **radioactive elements** and the power of atomic energy. Today, hundreds of additional applications have been found for these materials, from areas as diverse as biological research, radiology, medicine, and archeology. Radioactive elements decay of their own accord by emitting radiation. The rate of decay is measured using the **half-life** of the substance, which is the time required for a mass of radioactive material to decay until only one-half of its original mass remains. This half-life is used to find the rate of decay r, first mentioned in Section 5.5. In general, if h represents the half-life of the substance, one-half the initial amount remains when $t = h$.

$$Q(t) = Q_0 e^{-rt} \qquad \text{exponential decay function}$$

$$\frac{1}{2}Q_0 = Q_0 e^{-rh} \qquad \text{substitute } \tfrac{1}{2} Q_0 \text{ for } Q(t), h \text{ for } t$$

$$\frac{1}{2} = \frac{1}{e^{rh}} \qquad \text{divide by } Q_0; \text{ rewrite expression}$$

$$2 = e^{rh} \qquad \text{property of ratios}$$

$$\ln 2 = rh \ln e \qquad \text{apply base-}e \text{ logarithms; power property}$$

$$\frac{\ln 2}{h} = r \qquad \text{solve for } r \,(\ln e = 1)$$

Radioactive Rate of Decay

If h represents the half-life of a radioactive substance per unit time, the nominal rate of decay per a like unit of time is given by

$$r = \frac{\ln 2}{h}$$

The rate of decay for known radioactive elements varies greatly. For example, the element carbon-14 has a half-life of about 5730 yr, while the element lead-211 has a half-life of only about 3.5 min. Radioactive elements can be detected in extremely small amounts. If a drug is "labeled" (mixed with) a radioactive element and injected into a living organism, its passage through the organism can be traced and information on the health of internal organs can be obtained.

EXAMPLE 8 ▶ Solving a Radioactive Rate of Decay Application

The radioactive element potassium-42 is often used in biological experiments, since it has a half-life of only about 12.4 hr.

    a. How much of a 5-g sample will remain after 18 hr and 45 min?

    b. Use the intersection-of-graphs method to find the number of hours until only 0.5 g remain.

Solution ▶ **a.** To begin we find the nominal rate of decay r and use this value in the exponential decay function.

$$r = \frac{\ln 2}{h} \quad \text{radioactive rate of decay}$$

$$r = \frac{\ln 2}{12.4} \quad \text{substitute 12.4 for } h$$

$$r \approx 0.055899 \quad \text{result}$$

To determine how much of the sample remains after 18.75 hr, we use $r = 0.055899$ in the decay function and evaluate it at $t = 18.75$.

Figure 5.55

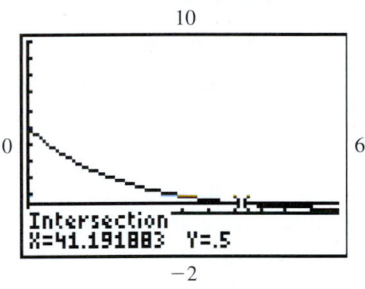

Intersection
X=41.191883   Y=.5

$$Q(t) = Q_0 e^{-rt} \quad \text{exponential decay function}$$

$$Q(18.75) = 5e^{(-0.055899)(18.75)} \quad \text{substitute 5 for } Q_0, 0.055899 \text{ for } r, \text{ and } 18.75 \text{ for } t$$

$$Q(18.75) \approx 1.75 \quad \text{evaluate}$$

After 18 hr and 45 min, only 1.75 g of potassium-42 will remain.

b. With $r = 0.055899$, the decay function becomes $Q(t) = 5e^{0.055899t}$. For the intersection-of-graphs method, setting the window for y poses no challenge as we have a target of 0.5 g. For the x-values, reason that starting with 5 g and a half-life of about 12 hr, 2.5 g remain after 12 hr, 1.25 g remain after 24 hr, 0.625 g after 36 hr, and that the time required must be greater than (but close to) 36 hr. Using $[0, 60]$ for x and $[-2, 10]$ for y, we find the graphs intersect near $x \approx 41.19$ (Figure 5.55). There will be only 0.5 g remaining shortly after 41 hr.

☑ **D.** You've just seen how we can solve applications of exponential growth and decay

Now try Exercises 57 through 62 ▶

5.6 EXERCISES

▶ **CONCEPTS AND VOCABULARY**

Fill in each blank with the appropriate word or phrase. Carefully reread the section if needed.

1. __Compound__ interest is interest paid to you on previously accumulated interest.

2. The formula for interest compounded __continuously__ is $A = pe^{rt}$, where e is approximately __2.72__.

3. Given constants Q_0 and r, and that Q decays exponentially as a function of t, the equation model is $Q(t) =$ ___$Q_0 e^{-rt}$___.

4. Investment plans calling for regularly scheduled deposits are called __annuities__. The annuity formula gives the __future__ value of the account.

5. Explain/Describe the difference between the future value and present value of an annuity. Include an example.   Answers will vary.

6. Describe/Explain how you would find the rate of growth r, given that a population of ants grew from 250 to 3000 in 6 weeks.   Answers will vary.

HOMEWORK SELECTION GUIDE

Core: 7–63 odd, 66 (30 Exercises)
Standard: 1–4, 7–65 odd, 66, 68 (36 Exercises)

Extended: 1–4, 7–51 odd, 52, 53–61 odd, 62–68 (40 Exercises)
In Depth: 1–6, 7–51 odd, 52, 53–61 odd, 62–69 (43 Exercises)

▶ DEVELOPING YOUR SKILLS

For simple interest accounts, the interest earned or due depends on the principal p, interest rate r, and the time t in years according to the formula $I = prt$.

7. Find p given $I = \$229.50$, $r = 6.25\%$, and $t = 9$ months. $4896

8. Find r given $I = \$1928.75$, $p = \$8500$, and $t = 3.75$ yr.   6.05%

9. Larry came up a little short one month at bill-paying time and had to take out a title loan on his car at Check Casher's, Inc. He borrowed \$260, and 3 weeks later he paid off the note for \$297.50. What was the annual interest rate on this title loan? (*Hint:* How much *interest* was charged?)   250%

10. Angela has \$750 in a passbook savings account that pays 2.5% simple interest. How long will it take the account balance to hit the \$1000 mark at this rate of interest, if she makes no further deposits? (*Hint:* How much *interest* will be paid?)   13.3 yr

For simple interest accounts, the amount A accumulated or due depends on the principal p, interest rate r, and the time t in years according to the formula $A = p(1 + rt)$.

11. Find p given $A = \$2500$, $r = 6.25\%$, and $t = 31$ months. $2152.47

12. Find r given $A = \$15,800$, $p = \$10,000$, and $t = 3.75$ yr.   15.47%

13. Olivette Custom Auto Service borrowed \$120,000 at 4.75% simple interest to expand their facility from three service bays to four. If they repaid \$149,925, what was the term of the loan?   5.25 yr

14. Healthy U sells nutritional supplements and borrows \$50,000 to expand their product line. When the note is due 3 yr later, they repay the lender \$62,500. If it was a simple interest note, what was the annual interest rate?   8.33%

15. **Simple interest:** The owner of Paul's Pawn Shop loans Larry \$200.00 using his Toro riding mower as collateral. Thirteen weeks later Larry comes back to get his mower out of pawn and pays Paul \$240.00. What was the annual simple interest rate on this loan?   80%

16. **Simple interest:** To open business in a new strip mall, Laurie's Custom Card Shoppe borrows \$50,000 from a group of investors at 4.55% simple interest. Business booms and blossoms, enabling Laurie to repay the loan fairly quickly. If Laurie repays \$62,500, how long did it take? $5\frac{1}{2}$ yr

Additional answers can be found in the Instructor Answer Appendix.

For accounts where interest is compounded annually, the amount A accumulated or due depends on the principal p, interest rate r, and the time t in years according to the formula $A = p(1 + r)^t$.

17. Find t given $A = \$48,428$, $p = \$38,000$, and $r = 6.25\%$.   4 yr

18. Find p given $A = \$30,146$, $r = 5.3\%$, and $t = 7$ yr. $21,000.57

19. How long would it take \$1525 to triple if invested at 7.1%?   16 yr

20. What interest rate will ensure a \$747.26 deposit will be worth \$1000 in 5 yr?   6%

For accounts where interest is compounded annually, the principal P needed to ensure an amount A has been accumulated in the time period t when deposited at interest rate r is given by the formula $P = \frac{A}{(1 + r)^t}$.

21. The Stringers need to make a \$10,000 balloon payment in 5 yr. How much should be invested now at 5.75%, so that the money will be available? $7561.33

22. Morgan is 8 yr old. If her mother wants to have \$25,000 for Morgan's first year of college (in 10 yr), how much should be invested now if the account pays a 6.375% fixed rate? $13,475.48

For compound interest accounts, the amount A accumulated or due depends on the principal p, interest rate r, number of compoundings per year n, and the time t in years according to the formula $A = p(1 + \frac{r}{n})^{nt}$.

23. Find t given $A = \$129,500$, $p = \$90,000$, and $r = 7.125\%$ compounded weekly.   about 5 yr

24. Find r given $A = \$95,375$, $p = \$65,750$, and $t = 15$ yr with interest compounded monthly.   2.48%

25. How long would it take a \$5000 deposit to double, if invested at a 9.25% rate and compounded daily?   7.5 yr

26. What principal should be deposited at 8.375% compounded monthly to ensure the account will be worth \$20,000 in 10 yr? $8681.04

27. **Compound interest:** As a curiosity, David decides to invest \$10 in an account paying 10% interest compounded 10 times per year for 10 yr. Is that enough time for the \$10 to triple in value?   no

28. **Compound interest:** As a follow-up experiment (see Exercise 27), David invests \$10 in an account paying 12% interest compounded 10 times per year for 10 yr, and another \$10 in an account paying 10% interest compounded 12 times per year for 10 yr. Which produces the better investment—more compounding periods or a higher interest rate?   higher interest rate

29. Compound interest: Due to demand, Donovan's Dairy (Wisconsin, USA) plans to double its size in 4 yr and will need $250,000 to begin development. If they invest $175,000 in an account that pays 8.75% compounded semiannually, (a) will there be sufficient funds to break ground in 4 yr? (b) If not, find the *minimum interest rate* that will enable the dairy to meet its 4-yr goal.　**a.** no　**b.** 9.12%

30. Compound interest: To celebrate the birth of a new daughter, Helyn invests 6000 Swiss francs in a college savings plan to pay for her daughter's first year of college in 18 yr. She estimates that 25,000 francs will be needed. If the account pays 7.2% compounded daily, (a) will she meet her investment goal? (b) If not, find the *minimum rate of interest* that will enable her to meet this 18-yr goal.　**a.** no　**b.** 7.93%

For accounts where interest is compounded continuously, the amount A accumulated or due depends on the principal p, interest rate r, and the time t in years according to the formula $A = pe^{rt}$.

31. Find t given $A = \$2500$, $p = \$1750$, and $r = 4.5\%$.　7.9 yr

32. Find r given $A = \$325,000$, $p = \$250,000$, and $t = 10$ yr.　2.62%

33. How long would it take $5000 to double if it is invested at 9.25%? Compare the result to Exercise 25.　7.5 yr

34. What principal should be deposited at 8.375% to ensure the account will be worth $20,000 in 10 yr? Compare the result to Exercise 26.　$8655.82

35. Interest compounded continuously: Valance wants to build an addition to his home outside Madrid (Spain) so he can watch over and help his parents in their old age. He hopes to have 20,000 euros put aside for this purpose within 5 yr. If he invests 12,500 euros in an account paying 8.6% interest compounded continuously, (a) will he meet his investment goal? (b) If not, find the *minimum rate of interest* that will enable him to meet this 5-yr goal.　**a.** no　**b.** 9.4%

36. Interest compounded continuously: Minh-Ho just inherited her father's farm near Mito (Japan), which badly needs a new barn. The estimated cost of the barn is 8,465,000 yen and she would like to begin construction in 4 yr. If she invests 6,250,000 yen in an account paying 6.5% interest compounded continuously, (a) will she meet her investment goal? (b) If not, find the *minimum rate of interest* that will enable her to meet this 4-yr goal.　**a.** no　**b.** 7.58%

37. Interest compounded continuously: William and Mary buy a small cottage in Dovershire (England), where they hope to move after retiring in 7 yr. The cottage needs about 20,000 euros worth of improvements to make it the retirement home they desire. If they invest 12,000 euros in an account paying 5.5% interest compounded continuously, (a) will they have enough to make the repairs? (b) If not, find the *minimum amount they need to deposit* that will enable them to meet this goal in 7 yr.　**a.** no　**b.** approx 13,609 euros

38. Interest compounded continuously: After living in Oslo (Norway) for 20 years, Kjell and Torill decide to move inland to help operate the family ski resort. They hope to make the move in 6 yr, after they have put aside 140,000 kroner. If they invest 85,000 kroner in an account paying 6.9% interest compounded continuously, (a) will they meet their 140,000 kroner goal? (b) If not, find the *minimum amount they need to deposit* that will enable them to meet this goal in 6 yr.　**a.** no　**b.** approx 92,540 kroner

The length of time T (in years) required for an initial principal P to grow to an amount A at a given interest rate r is given by $T = \frac{1}{r}\ln(\frac{A}{P})$.

39. Investment growth: A small business is planning to build a new $350,000 facility in 8 yr. If they deposit $200,000 in an account that pays 5% interest compounded continuously, will they have enough for the new facility in 8 yr? If not, what amount should be invested on these terms to meet the goal?　No; $234,612.02

40. Investment growth: After the twins were born, Sasan deposited $25,000 in an account paying 7.5% compounded continuously, with the goal of having $120,000 available for their college education 20 yr later. Will Sasan meet the 20-yr goal? If not, what amount should be invested on these terms to meet the goal?　No; $26,775.62

Ordinary annuities: If a periodic payment $\mathcal{P}$ is deposited n times per year, with annual interest rate r also compounded n times per year for t years, the future value of the account is given by $A = \frac{\mathcal{P}[(1 + R)^{nt} - 1]}{R}$, where $R = \frac{r}{n}$ (if the rate is 9% compounded monthly, $R = \frac{0.09}{12} = 0.0075$).

41. Saving for a rainy day: How long would it take Jasmine to save $10,000 if she deposits $90/month at an annual rate of 7.5 compounded monthly?　about 7 yr

42. Saving for a sunny day: What quarterly investment amount is required to ensure that Larry can save $4700 in 4 yr at an annual rate of 8.5% compounded quarterly?　$249.72

43. Saving for college: At the birth of their first child, Latasha and Terrance opened an annuity account and have been depositing $50 per month in the account ever since. If the account is now worth $30,000 and the interest on the account is 6.6% compounded monthly, how old is the child?   22 yr

44. Saving for a bequest: When Cherie (Brandon's first granddaughter) was born, he purchased an annuity account for her and stipulated that she should receive the funds (in trust, if necessary) upon his death. The quarterly annuity payments were $250 and interest on the account was 7.6% compounded quarterly. The account balance of $17,500 was recently given to Cherie. How much longer did Brandon live?   11 yr

45. Saving for a down payment: Tae-Hon is tired of renting and decides that within the next 5 yr he

must save $22,500 for the down payment on a home. He finds an investment company that offers 9% interest compounded monthly and begins depositing $250 each month in the account. (a) Is this monthly amount sufficient to help him meet his 5 yr goal? (b) If not, find the *minimum amount he needs to deposit each month* that will enable him to meet his goal in 5 yr.   a. no   b. $298.31

46. Saving to open a business: Madeline feels trapped in her current job and decides to save $75,000 over the next 7 yr to open up a Harley-Davidson franchise. To this end, she invests $145 every week in an account paying $7\frac{1}{2}$% interest compounded weekly. (a) Is this weekly amount sufficient to help her meet the seven-year goal? (b) If not, find the *minimum amount she needs to deposit each week* that will enable her to meet this goal in 7 yr.   a. no   b. $156.81

▶ **WORKING WITH FORMULAS**

Solve for the indicated unknowns.

47. $A = p + prt$
 a. solve for t
 b. solve for p

48. $A = p(1 + r)^t$
 a. solve for t
 b. solve for r

49. $A = p\left(1 + \dfrac{r}{n}\right)^{nt}$
 a. solve for r
 b. solve for t

50. $A = pe^{rt}$
 a. solve for p
 b. solve for r

51. $Q(t) = Q_0 e^{rt}$
 a. solve for Q_0
 b. solve for t

52. $p = \dfrac{AR}{[(1 + R)^{nt} - 1]}$
 a. solve for A
 b. solve for n

53. Amount of a mortgage payment: $P = \dfrac{AR}{1 - (1 + R)^{-nt}}$

The mortgage payment required to pay off (or amortize) a loan is given by the formula shown, where P is the payment amount, A is the original amount of the loan, t is the time in years, r is the annual interest rate, n is the number of payments per year, and $R = \frac{r}{n}$. Find the *monthly payment* required to amortize a $125,000 home, if the interest rate is 5.5%/year and the home is financed over 30 yr. $709.74

54. Time required to amortize a mortgage: $t = 16.71 \ln\left(\dfrac{x}{x - 1000}\right)$, $x > 1000$.

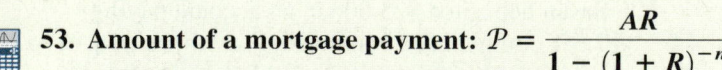

The number of years needed to amortize (pay off) a mortgage, depends on the amount of the regular monthly payment. The formula shown approximates the years t required to pay off a $200,000 mortgage at 6% interest, based on a monthly payment of x dollars.

 a. Use a **TABLE** to find the payment required to pay off this mortgage in 30 yr, and the amount of interest paid ($30 \times 12 = 360$ payments). $1199.14; $231,690.40 in interest

 b. Use the intersection-of-graphs method to find the payment required to pay off this mortgage in 20 yr, and the amount of interest that would be paid. How much interest was saved by making a higher payment? $1432.94, $143,905.60, $87,784.80

 c. Repeat part (b) for a complete payoff in 15 yr.   payment: $1687.82, interest paid: $103,807.60, interest saved: $127,882.80

▶ APPLICATIONS

55. Bacterial growth: As part of a lab experiment, Luamata needs to grow a culture of 200,000 bacteria, which are known to double in number in 12 hr. If he begins with 1000 bacteria, (a) find the growth rate r and (b) find how many hours it takes for the culture to produce the 200,000 bacteria.   **a.** 5.78% **b.** 91.67 hr

56. Rabbit populations: After the wolf population was decimated due to overhunting, the rabbit population in the Boluhti Game Reserve began to double every 6 months. If there were an estimated 120 rabbits to begin, (a) find the growth rate r and (b) find the number of months required for the population to reach 2500.   **a.** 11.55% **b.** 26.29 mo

For Exercises 57–60, (a) solve by finding the growth rate and using the decay formula $Q(t) = Q_0e^{-rt}$.

57. Iodine-131, radioactive decay: The radioactive element iodine-131 has a half-life of 8 days and is often used to help diagnose patients with thyroid problems. If a certain thyroid procedure requires 0.5 g and is scheduled to take place in 3 days, what is the minimum amount that must be on hand now (to the nearest hundredth of a gram)?   0.65 g

58. Sodium-24, radioactive decay: The radioactive element sodium-24 has a half-life of 15 hr and is used to help locate obstructions in blood flow. If the procedure requires 0.75 g and is scheduled to take place in 2 days (48 hr), what minimum amount must be on hand *now* (to the nearest hundredth of a gram)?   6.89 g

59. Americium-241, radioactive decay: The radioactive element americium-241 has a half-life of 432 yr and although extremely small amounts are used (about 0.0002 g), it is the most vital component of standard household smoke detectors. How many years will it take a 10-g mass of americium-241 to decay to 2.7 g?   about 816 yr

60. Carbon-14, radioactive decay: Carbon-14 is a radioactive compound that occurs naturally in all living organisms, with the amount in the organism constantly renewed. After death, no new carbon-14 is acquired and the amount in the organism begins to decay exponentially. If the half-life of carbon-14 is 5730 yr, how old is a mummy having only 45% of the normal amount of carbon-14?   about 6600 yr

Carbon-14 dating: If the percentage p of carbon-14 that remains in a fossil can be determined, the formula $T = -8267 \ln p$ can be used to estimate the number of years T since the organism died.

61. Dating the Lascaux Cave dwellers: Bits of charcoal from Lascaux Cave (home of the prehistoric Lascaux Cave Paintings) were used to estimate that the fire had burned some 17,255 yr ago. What percent of the original amount of carbon-14 remained in the bits of charcoal?   about 12.4%

62. Dating Stonehenge: Using organic fragments found near Stonehenge (England), scientists were able to determine that the organism that produced the fragments lived about 3925 yr ago. What percent of the original amount of carbon-14 remained in the organism?   about 62.2%

▶ EXTENDING THE CONCEPT

63. Many claim that inheritance taxes were put in place simply to prevent a massive accumulation of wealth by a select few. Suppose that in 1890, your great-grandfather deposited $10,000 in an account paying 6.2% compounded continuously. If the account were to pass to you untaxed, what would it be worth in 2010? Do some research on the inheritance tax laws in your state. In particular, what amounts can be inherited untaxed (i.e., before the inheritance tax kicks in)? $17,027,502.21

64. If you have not already completed Exercise 30, please do so. For *this* exercise, *solve the compound interest equation for r* to find the exact rate of interest that will enable Helyn to meet her 18-yr goal.   7.93%

65. If you have not already completed Exercise 43, please do so. Suppose the final balance of the account was $35,100 with interest again being compounded monthly. For *this* exercise, use a graphing calculator to find r, the exact rate of interest the account would have been earning.   7.2%

▶ MAINTAINING YOUR SKILLS

66. (1.1) In an effort to boost tourism, a trolley car is being built to carry sightseers from a strip mall to the top of Mt. Vernon, 1580-m high. Approximately how long will the trolley cables be?   2548.8 m

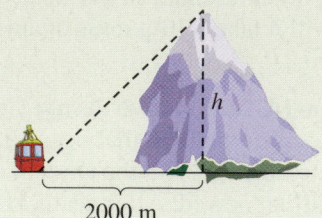

2000 m

67. (2.2/2.4) Name the toolbox functions that are (a) one-to-one, (b) even, (c) increasing for $x \in R$, and (d) asymptotic.

Additional answers can be found in the Instructor Answer Appendix.

68. (1.3) Is the following relation a function? If not, state how the definition of a function is violated.   yes

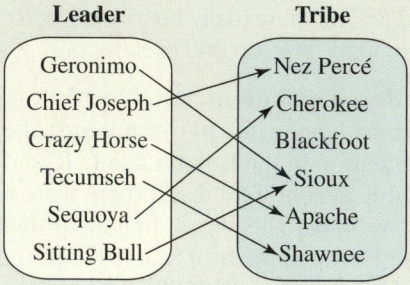

69. (4.2) A polynomial with real coefficients is known to have the zeroes $x = 3$, $x = -1$, and $x = 1 + 2i$. Find the equation of the polynomial, given it has degree 4 and a y-intercept of $(0, -15)$.
$P(x) = x^4 - 4x^3 + 6x^2 - 4x - 15$

5.7   Exponential, Logarithmic, and Logistic Equation Models

LEARNING OBJECTIVES

In Section 5.7 you will see how we can

- ☐ **A.** Choose an appropriate form of regression for a set of data
- ☐ **B.** Use a calculator to obtain exponential and logarithmic regression models
- ☐ **C.** Determine when a logistic model is appropriate and apply a logistic model to a set of data
- ☐ **D.** Use a regression model to answer questions and solve applications

WORTHY OF NOTE

For more information on the use of residuals, see the *Calculator Exploration and Discovery* feature on Residuals at the end of Chapter 3.

The basic concepts involved in calculating a regression equation were presented in Section 1.6 and 3.4. In this section, we extend these concepts to data sets that are best modeled by power, exponential, logarithmic, or logistic functions. All data sets, while contextual and accurate, *have been carefully chosen* to provide a maximum focus on regression fundamentals and related mathematical concepts. In reality, data sets are often not so "well-behaved" and many require sophisticated statistical tests before any conclusions can be drawn.

A.   Choosing an Appropriate Form of Regression

Most graphing calculators have the ability to perform several forms of regression, and selecting which of these to use is a critical issue. When various forms are applied to a given data set, some are easily discounted due to a poor fit. Others may fit very well for only a portion of the data, while still others may compete for being the "best-fit" equation. In a statistical study of regression, an in-depth look at the correlation coefficient (r), the coefficient of determination (r^2 or R^2), and a study of **residuals** are used to help make an appropriate choice. For our purposes, the correct or best choice will generally depend on two things: (1) how well the graph appears to fit the scatterplot, and (2) the context or situation that generated the data, coupled with a dose of common sense.

As we've noted previously, the final choice of regression can rarely be based on the scatterplot alone, although relying on the basic characteristics and end-behavior of certain graphs can be helpful **(see Exercise 67).** With an awareness of the toolbox functions, polynomial graphs, and applications of exponential and logarithmic functions, the context of the data can aid a decision.

EXAMPLE 1 ▶   **Choosing an Appropriate Form of Regression**

Suppose a set of data is generated from each context given. Use common sense, previous experience, or your own knowledge base to state whether a linear, quadratic, logarithmic, exponential, or power regression might be most appropriate. Justify your answers.

    a. population growth of the United States since 1800

    b. the distance covered by a jogger running at a constant speed

    c. height of a baseball t seconds after it's thrown

    d. the time it takes for a cup of hot coffee to cool to room temperature

Solution ▶   **a.** From examples in Section 5.6 and elsewhere, we've seen that animal and human populations tend to grow exponentially over time. Here, an exponential model is likely most appropriate.

    b. Since the jogger is moving at a constant speed, the rate-of-change $\dfrac{\Delta \text{distance}}{\Delta \text{time}}$ is constant and a linear model would be most appropriate.

    c. As seen in numerous places throughout the text, the height of a projectile is modeled by the equation $h(t) = -16t^2 + vt + k$, where $h(t)$ is the height in feet after t seconds. Here, a quadratic model would be most appropriate.

    d. Many have had the experience of pouring a cup of hot chocolate, coffee, or tea, only to leave it on the counter as they turn their attention to other things. The hot drink seems to cool quickly at first, then slowly approach room temperature. This experience, perhaps coupled with our awareness of *Newton's law of cooling,* shows a logarithmic or exponential model might be appropriate here.

☑ A. You've just seen how we can choose an appropriate form of regression for a set of data

Now try Exercises 7 through 20 ▶

B. Exponential and Logarithmic Regression Models

We now focus our attention on regression models that involve exponential and logarithmic functions. Recall the process of developing a regression equation involves these five stages: (1) clearing old data, (2) entering new data, (3) displaying the data, (4) calculating the regression equation, and (5) displaying and using the regression graph and equation.

EXAMPLE 2 ▶   **Calculating an Exponential Regression Model**

The number of centenarians (people who are 100 yr of age or older) has been climbing steadily over the last half century. The table shows the number of centenarians (per million population) for selected years. Use the data and a graphing calculator to draw the scatterplot, then use the scatterplot and context to decide on an appropriate form of regression.

Source: Data from 2004 *Statistical Abstract of the United States,* Table 14; various other years

Year "t" (1950 → 0)	Number "N" (per million)
0	16
10	18
20	25
30	74
40	115
50	262

Solution ▶ After clearing any existing data in the data lists, enter the input values (years since 1950) in L1 and the output values (number of centenarians per million population) in L2 (Figure 5.56). For the viewing window, scale the x-axis (years since 1950) from −10 to 70 and the y-axis (number per million) from −50 to 300 to comfortably fit the data and allow room for the coordinates to be shown at the bottom of the screen (Figure 5.57). The scatterplot rules out a linear model. While a quadratic model may fit the data, we expect that the correct model should exhibit asymptotic behavior since extremely few people lived to be 100 yr of age prior to dramatic advances in hygiene, diet, and medical care. This would lead us toward an exponential equation model. The keystrokes **STAT** ▷ brings up the **CALC** menu, with **ExpReg** (exponential regression) being option "0." The option can be selected by simply pressing "0," or by using the up arrow ⌃ or down arrow ⌄ to scroll to **0:ExpReg** then pressing **ENTER**. The exponential model seems to fit the data very well (Figures 5.58 and 5.59). To four decimal places the equation model is $y = (11.5090)1.0607^x$.

Figure 5.56

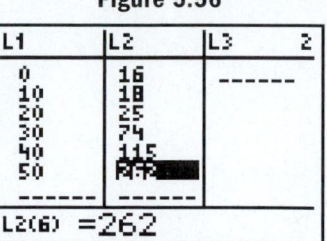

Figure 5.57

Figure 5.58

```
ExpReg
 y=a*b^x
 a=11.50896332
 b=1.060707503
```

Figure 5.59

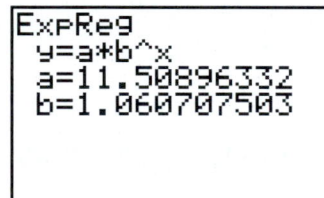

Now try Exercises 21 and 22 ▶

EXAMPLE 3 ▶ Calculating a Logarithmic Regression Model

One measure used in studies related to infant growth, nutrition, and development, is the relation between the circumference of a child's head and their age. The table to the right shows the average circumference of a female child's head for ages 0 to 36 months. Use the data and a graphing calculator to draw the scatterplot, then use the scatterplot and context to decide on an appropriate form of regression.

Source: *National Center for Health Statistics*

Age *a* (months)	Circumference *C* (cm)
0	34.8
6	43.0
12	45.2
18	46.5
24	47.5
30	48.2
36	48.6

Solution ▶ After clearing any existing data, enter the child's age (in months) as L1 and the circumference of the head (in cm) as L2. For the viewing window, scale the x-axis from −5 to 50 and the y-axis from 25 to 60 to comfortably fit the data (Figure 5.60). The scatterplot again rules out a linear model, and the context rules out a polynomial model due to end-behavior. As we expect the circumference of the head to continue

Figure 5.60

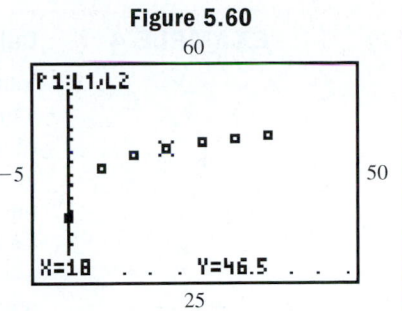

increasing slightly for many more months, it appears a logarithmic model may be the best fit. Note that since ln(0) is undefined, $a = 0.1$ was used to represent the age at birth (rather than $a = 0$), prior to running the regression. The **LnReg** (logarithmic regression) option is option 9, and the keystrokes ⬤STAT ▷ (**CALC**) **9** ⬤ENTER gives the equation shown in Figure 5.61, which fits the data very well (Figure 5.62).

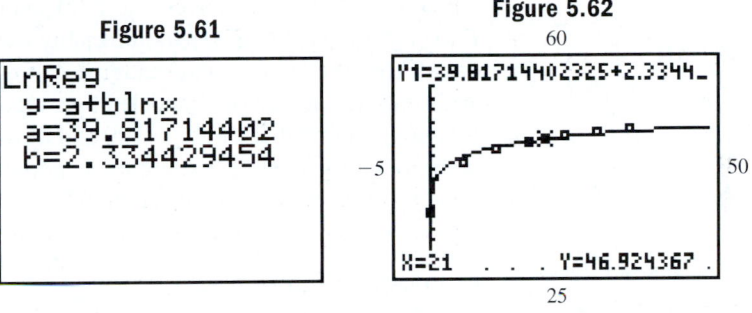

Figure 5.61   Figure 5.62

> **WORTHY OF NOTE**
>
> For applications involving exponential growth and logarithmic functions, it helps to remember that while both basic functions are increasing, a logarithmic function increases at a much slower rate.

☑ **B.** You've just seen how we can use a calculator to obtain exponential and logarithmic regression models

Now try Exercises 23 and 24 ▶

C. Logistic Equations and Regression Models

Many population growth models assume an unlimited supply of resources, nutrients, and room for growth, resulting in an exponential growth model. When resources become scarce or room for further expansion is limited, the result is often a **logistic growth model**. At first, growth is very rapid (like an exponential function), but this growth begins to taper off and slow down as nutrients are used up, living space becomes restricted, or due to other factors. Surprisingly, this type of growth can take many forms, including population growth, the spread of a disease, the growth of a tumor, or the spread of a stain in fabric. Specific logistic equations were encountered in Section 5.5. The general equation model for logistic growth is

> **Logistic Growth**
>
> Given constants a, b, and c, the logistic growth $P(t)$ of a population depends on time t according to the model
>
> $$P(t) = \frac{c}{1 + ae^{-bt}}$$

The constant c is called the **carrying capacity** of the population, in that as $t \rightarrow \infty$, $P(t) \rightarrow c$. In words, as the elapsed time becomes very large, the population will approach (but not exceed) c.

EXAMPLE 4 ▶ **Calculating a Logistic Regression Model**

Yeast cultures have a number of applications that are a great benefit to civilization and have been an object of study for centuries. A certain strain of yeast is grown in a lab, with its population checked at 2-hr intervals, and the data gathered are given in the table. Use the data and a graphing calculator to draw a scatterplot, and decide on an appropriate form of regression. If a logistic regression is the best model, attempt to estimate the capacity coefficient c prior to using your calculator to find the regression equation. How close were you to the actual value?

Elapsed Time (hours)	Population (100s)
2	20
4	50
6	122
8	260
10	450
12	570
14	630
16	650

Solution ▶ After clearing the data lists, enter the input values (elapsed time) in L1 and the output values (population) in L2. For the viewing window, scale the t-axis from -1 to 20 and the P-axis from -100 to 700 to comfortably fit the data. From the context and scatterplot, it's apparent the data are best modeled by a logistic function. Noting that Ymax = 700 and the data seem to level off near the top of the window, a good estimate for c would be about 675. Using logistic regression on the home screen (option **B:Logistic**), we obtain the equation $Y_1 = \dfrac{663}{1 + 123.9e^{-0.553X}}$ (rounded).

☑ **C.** You've just seen how we can determine when a logistic model is appropriate, and how to apply a logistic model to a set of data

Now try Exercises 25 and 26 ▶

When a regression equation is used to gather information, many of the equation solving skills from prior sections are employed. **Exercises 27 through 34** offer a variety of these equations for practice and warm-up.

D. Applications of Regression

Once the equation model for a data set has been obtained, it can be used to **interpolate** or approximate values that might occur *between* those given in the data set. It can also be used to **extrapolate** or predict future values. In this case, the investigation extends *beyond* the values from the data set, and is based on the assumption that projected trends will continue for an extended period of time.

Regardless of the regression applied, interpolation and extrapolation involve substituting a given or known value, then solving for the remaining unknown. We'll demonstrate here using the regression model from Example 3. The exercise set offers a large variety of regression applications, including some power regressions and additional applications of linear and quadratic regression.

EXAMPLE 5 ▶ **Using a Regression Equation to Interpolate or Extrapolate Information**

Use the regression equation from Example 3 to answer the following questions:
 a. What is the average circumference of a female child's head, if the child is 21 months old?
 b. According to the equation model, what will the average circumference be when the child turns $3\frac{1}{2}$ years old?
 c. If the circumference of the child's head is 44 cm, about how old is the child?

Solution ▶

Figure 5.63

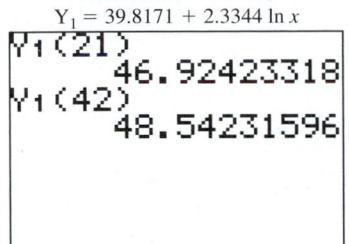
$Y_1 = 39.8171 + 2.3344 \ln x$

Figure 5.64

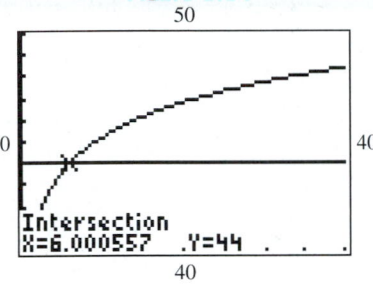

a. Using function notation we have $C(a) \approx 39.8171 + 2.3344 \ln(a)$. Substituting 21 for a gives:

$$C(21) \approx 39.8171 + 2.3344 \ln(21) \quad \text{substitute 21 for } a$$
$$\approx 46.9 \quad \text{result}$$

The circumference is approximately 46.9 cm. See Figure 5.63.

b. Substituting 3.5 yr $\times$ 12 = 42 months for a gives:

$$C(42) \approx 39.8171 + 2.3344 \ln(42) \quad \text{substitute 42 for } a$$
$$\approx 48.5 \quad \text{result}$$

The circumference will be approximately 48.5 cm. See Figure 5.63.

c. For part (c) we're given the circumference C and are asked to find the age a in which this circumference (44) occurs. Substituting 44 for $C(a)$ gives the equation 44 = 39.8171 + 2.3344 ln X, so we set $Y_1 = 39.8171 + 2.3344$ ln X and $Y_2 = 44$. For the window size, we know the formula is valid for female infants from 0 to 36 months, and from parts (a) and (b) a good range for the circumference will be from 40 to 50 cm. This indicates an appropriate window might be [0, 40] for x and [40, 50] for y. Using this window, we find the graphs intersect at about (6, 44), showing that a female child with a cranial circumference of 44 cm must be about 6 months old. See Figure 5.64.

Now try Exercises 37 through 44 ▶

☑ **D. You've just seen how we can use a regression model to answer questions and solve applications**

When extrapolating from a set of data, care and common sense must be used or results can be very misleading. For example, while the Olympic record for the 100-m dash has been steadily declining since the first Olympic Games, it would be foolish to think it will ever be run in 0 sec. There is a large variety of additional applications in the Exercise Set. **See Exercises 45 through 62.**

5.7 EXERCISES

▶ CONCEPTS AND VOCABULARY

Fill in each blank with the appropriate word or phrase. Carefully reread the section if needed.

1. The type of regression used often depends on (a) whether a particular graph appears to fit the __scatterplot__ and (b) the __context__ or __situation__ that generated the data.

2. The final choice of regression can rarely be based on the __scatterplot__ alone. Relying on the basic __characteristics__ and __end-behavior__ of certain graphs can be helpful.

3. To extrapolate means to use the data to predict values __beyond__ the given data.

4. To interpolate means to use the data to predict values __between__ the given data.

5. List the five steps used to find a regression equation using a calculator. Discuss possible errors that can occur if the first step is skipped. After the new data have been entered, what precautionary step should always be included? (1) clear out old data, (2) enter new data, (3) display the data, (4) calculate the regression equation, (5) display and use the results; Answers will vary.

6. Consider the eight toolbox functions and the exponential and logarithmic functions. How many of these satisfy the condition as $x \to \infty$, $y \to \infty$? For those that satisfy this condition, discuss/explain how you would choose between them judging from the scatterplot alone.
   $y = mx \, (m > 0), y = ax^2 \, (a > 0), y = a\sqrt{x} \, (a > 0), y = a|x|, (a > 0),$
   $y = ax^3 \, (a > 0), y = a\sqrt[3]{x} \, (a > 0), y = b^x \, (x > 0); y = a \log_b x \, (a > 0);$
   Answers will vary.

HOMEWORK SELECTION GUIDE

Core: 7–35 odd, 37–65 every other odd (23 Exercises)
Standard: 1–4, 7–35 odd, 37–65 every other odd, 66 (28 Exercises)

Extended: 1–4, 7–35 odd, 36, 37–61 every other odd, 62, 63, 65, 66 (31 Exercises)
In Depth: 1–6, 7–35 odd, 36, 37–61 every other odd, 62, 63–66 (34 Exercises)

▶ DEVELOPING YOUR SKILLS

Match each scatterplot given with one of the following: (a) likely linear, (b) likely quadratic, (c) likely exponential, (d) likely logarithmic, (e) likely logistic, or (f) none of these.

7.   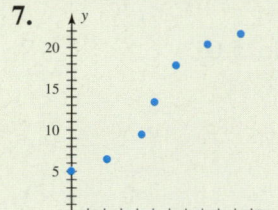   e

8.   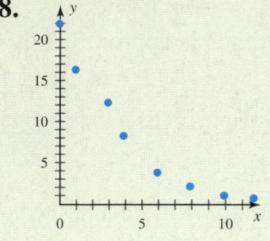   c

9.   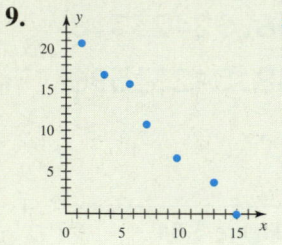   a

10.   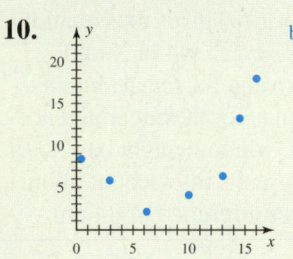   b

11.   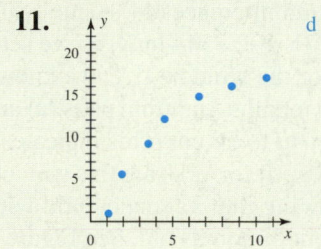   d

12.   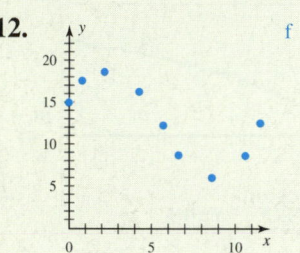   f

For Exercises 13 to 20, suppose a set of data is generated from the context indicated. Use common sense, previous experience, or your own knowledge base to state whether a linear, quadratic, logarithmic, exponential, power, or logistic regression might be most appropriate. Justify your answers.

13. total revenue and number of units sold   linear

14. page count in a book and total number of words   linear

15. years on the job and annual salary   exponential

16. time and population growth with unlimited resources   exponential

17. time and population growth with limited resources   logistic

18. elapsed time and the height of a projectile   quadratic

19. the cost of a gallon of milk over time   exponential

20. elapsed time and radioactive decay   exponential

Graph the data sets, then discuss why an exponential model could be an appropriate form of regression. Then find the regression equation.

21. Radioactive Studies

Time in Hours	Grams of Material
0.1	1.0
1	0.6
2	0.3
3	0.2
4	0.1
5	0.06

22. Rabbit Population

Month	Population (in hundreds)
0	2.5
3	5.0
6	6.1
9	12.3
12	17.8
15	30.2

Graph the data sets, then discuss why a logarithmic model could be an appropriate form of regression. Then find the regression equation.

23. Total number of sales compared to the amount spent on advertising

Advertising Costs ($1000s)	Total Number of Sales
1	125
5	437
10	652
15	710
20	770
25	848
30	858
35	864

24. Cumulative weight of diamonds extracted from a diamond mine

Time (months)	Weight (carats)
1	500
3	1748
6	2263
9	2610
12	3158
15	3501
18	3689
21	3810

Additional answers can be found in the Instructor Answer Appendix.

25. Spread of disease: Estimates of the cumulative number of SARS (sudden acute respiratory syndrome) cases reported in Hong-Kong during the spring of 2003 are shown in the table, with day 0 corresponding to February 20, (a) Use the data to draw a scatterplot, then use the context and scatterplot to decide on the best form of regression. (b) If a logistic model seems best, attempt to estimate the carrying capacity c, then (c) use your calculator to find the regression equation.

Days After Outbreak	Cumulative Total
0	100
14	560
21	870
35	1390
56	1660
70	1710
84	1750

Source: Center for Disease Control @ www.cdc.gov/ncidod/EID/vol9no12.

26. Cable television subscribers: The percentage of American households having cable television is given in the table for select years from 1976 to 2004. (a) Use the data to draw a scatterplot, then use the context and scatterplot to decide on the best form of regression. (b) If a logistic model seems best, attempt to estimate the carrying capacity c, then (c) use your calculator to find the regression equation (use 1976 $\rightarrow$ 0).

Source: Data pooled from the 2001 *New York Times Almanac*, p. 393; 2004 *Statistical Abstract of the United States*, Table 1120; various other years.

Year 1976 → 0	Percentage with Cable TV
0	16
4	22.6
8	43.7
12	53.8
16	61.5
20	66.7
24	68
28	70

The applications in this section require solving equations similar to those that follow. Solve each equation algebraically and graphically.

27. $96.35 = (9.4)1.6^x$ 4.95   **28.** $(3.7)2.9^x = 1253.93$ 5.47

29. $4.8x^{2.5} = 468.75$ 6.25   **30.** $4375 = 1.4x^{-1.25}$ 0.0016

31. $52 = 63.9 - 6.8 \ln x$ 5.75

32. $498.53 + 18.2 \ln x = 595.9$ 210.61

33. $52 = \dfrac{67}{1 + 20e^{-0.62x}}$ 6.84   **34.** $\dfrac{975}{1 + 82.3e^{-0.423x}} = 890$ 15.98

▶ **WORKING WITH FORMULAS**

35. Learning curve: $C(t) = 4.1 + 9.5 \ln t$

The number of circuit boards a newly hired employee can assemble from its component parts, depends on the experience of the employee as measured by the length of employment. This relationship is modeled by the formula shown, where $C(t)$ represents the number of circuit boards assembled per day, t days after employment. (a) How many boards are being assembled after 5 days on the job? (b) How many days until the employee is able to assemble 30 boards per day?

a. about 19 boards   **b.** about 15 days

36. Bicycle sales since 1920: $N(t) = 0.325(1.057)^t$

Despite the common use of automobiles and motorcycles, bicycle sales have continued to grow as a means of transportation as well as a form of recreation. The number of bicycles sold each year (in millions) can be approximated by the formula shown, where t is the number of years after 1920 (1920 → 0). According to this model, in what year did bicycle sales exceed 10 million?   1981

Source: 1976/1992 *Statistical Abstract of the United States*, Tables 406/395; various other years

▶ **APPLICATIONS**

Answer the questions using the given data and the related regression equation. All extrapolations assume the mathematical model will continue to represent future trends.

37. Weight loss: Harold needed to lose weight and started on a new diet and exercise regimen. The number of pounds he's lost since the diet began is given in the table. Draw the scatterplot, decide on an appropriate form of regression, and find an equation that models the data.

   a. What was Harold's total weight loss after 15 days?

   b. Approximately how many days did it take to lose a total of 18 pounds?

   c. According to the model, what is the projected weight loss for 100 days?

Time (days)	Pounds Lost
10	2
20	14
30	20
40	23
50	25.5
60	27.6
70	29.2
80	30.7

38. Depletion of resources: The longer an area is mined for gold, the more difficult and expensive it gets to obtain. The cumulative total of the ounces produced by a particular mine is shown in the table. Draw the scatterplot, use the scatterplot and context to determine whether an exponential or logarithmic model is more appropriate, then find an equation that models the data.

Time (months)	Ounces Mined
5	275
10	1890
15	2610
20	3158
25	3501
30	3789
35	4109
40	4309

a. What was the total number of ounces mined after 18 months?

b. About how many months did it take to mine a total of 4000 oz?

c. According to the model, what is the projected total after 50 months?

39. Number of U.S. post offices: Due in large part to the ease of travel and increased use of telephones, e-mail and instant messaging, the number of post offices in the United States has been on the decline since the twentieth century. The data given

Year (1900 → 0)	Offices (1000s)
1	77
20	52
40	43
60	37
80	32
100	28

show number of post offices (in thousands) for selected years. Use the data to draw a scatterplot, then use the context and scatterplot to find the regression equation (use 1900 → 0).

Source: Statistical Abstract of the United States; The First Measured Century

a. Approximately how many post offices were there in 1915?

b. In what year did the number of post offices drop below 34,000?

c. According to the model, how many post offices will there be in the year 2015?

40. Telephone use: The number of telephone calls per capita has been rising dramatically since the invention of the telephone in 1876. The table shows the number of phone calls per capita per year for selected years. Use the

Year (1900 → 0)	Number (per capita/ per year)
0	38
20	180
40	260
60	590
80	1250
97	2325

data to draw a scatterplot, then use the context and scatterplot to find the regression equation.

Source: The First Measured Century by Theodore Caplow, Louis Hicks, and Ben J. Wattenberg, The AEI Press, Washington, D.C., 2001.

a. What was the approximate number of calls per capita in 1970?

b. Approximately how many calls per capita will there be in 2015?

c. In what year did the number of calls per capita exceed 4000?

41. Milk production: Since 1980, the number of family farms with milk cows for commercial production has been decreasing. Use the data from the table given to draw a scatterplot, then use the context and scatterplot to find the regression equation.

Year (1980 → 0)	Number (in 1000s)
0	334
5	269
10	193
15	140
17	124
18	117
19	111

Source: Statistical Abstract of the United States, 2000.

a. What was the approximate number of farms with milk cows in 1993?

b. Approximately how many farms will have milk cows in 2010?

c. In what year will this number of farms drop below 45 thousand?

42. Froth height—carbonated beverages: The height of the froth on carbonated drinks and other beverages can be manipulated by the ingredients used in making the beverage and lends itself very well to the modeling process. The data in the table given show the froth height of a certain

Time (seconds)	Height of Froth (in.)
0	0.90
2	0.65
4	0.40
6	0.21
8	0.15
10	0.12
12	0.08

beverage as a function of time, after the froth has reached a maximum height. Use the data to draw a scatterplot, then use the context and scatterplot to find the regression equation.

a. What was the approximate height of the froth after 6.5 sec?

b. How long does it take for the height of the froth to reach one-half of its maximum height?

c. According to the model, how many seconds until the froth height is 0.02 in.?

43. Chicken production: In 1980, the production of chickens in the United States was about 392 million. In the next decade, the demand for chicken first dropped, then rose dramatically. The number of chickens produced is given in the table to the right for selected years. Use the data to draw a scatterplot, then use the context and scatterplot to find the regression equation.

Year (1980 → 0)	Number (millions)
0	392
5	370
9	356
14	386
16	393
17	410
18	424

Source: Statistical Abstract of the United States, 2000.

a. What was the approximate number of chickens produced in 1987?

b. Approximately how many chickens will be produced in 2004?

c. According to the model, for what years was the production of chickens below 365 million?

44. Veterans in civilian life: The number of military veterans in civilian life fluctuates with the number of persons inducted into the military (higher in times of war) and the passing of time. The number of living veterans is given in the table for selected years from 1950 to 1999. Use the data to draw a scatterplot, then use the context and scatterplot to find the regression equation.

Year (1950 → 0)	Number (millions)
0	19.1
10	22.5
20	27.6
30	28.6
40	27
48	25.1
49	24.6

Source: Statistical Abstract of the United States, 2000.

a. What was the approximate number of living military veterans in 1995?

b. Approximately how many living veterans will there be in 2015?

c. According to the model, in what years did the number of veterans exceed 26 million?

45. Use of debit cards: Since 1990, the use of debit cards to obtain cash and pay for purchases has become very common. The number of debit cards nationwide is given in the table for selected

Year (1990 → 0)	Number of Cards (millions)
0	164
5	201
8	217
10	230

years. Use the data to draw a scatterplot, then use the context and scatterplot to find the regression equation.

Source: Statistical Abstract of the United States, 2000.

a. Approximately how many debit cards were there in 1999?

b. Approximately how many debit cards will there be in 2015?

c. In what year did the number of debit cards exceed 300 million?

46. Quiz grade versus study time: To determine the value of doing homework, a student in college algebra records the time spent by classmates in preparation for a quiz the next day. Then she records their scores, which are shown in the table. Use the data to draw a scatterplot, then use the context and scatterplot to find the regression equation. According to the model, what grade can I expect if I study for 120 min?

x (min study)	y (score)
45	70
30	63
10	59
20	67
60	73
70	85
90	82
75	90

47. Population of coastal areas: The percentage of the U.S. population that can be categorized as living in *Pacific coastal areas* (minimum of 15% of the state's land area is a coastal watershed) has been growing steadily for decades, as indicated by the data given for selected years. Use the data to draw a scatterplot, then use the context and scatterplot to find the regression equation. According to the model, what is the predicted percentage of the population living in Pacific coastal areas in 2005, 2010 and 2015?

Year (1970 → 0)	Percentage
0	22.8
10	27.0
20	33.2
25	35.2
30	37.8
31	38.5
32	38.9
33	39.4

Source: 2004 Statistical Abstract of the United States, Table 23.

Additional answers can be found in the Instructor Answer Appendix.

48. Water depth and pressure: As anyone who's been swimming knows, the deeper you dive, the more pressure you feel on your body and eardrums. This pressure (in pounds per square inch or psi) is shown in the table for selected depths. Use the data to draw a scatterplot, then use the context and scatterplot to find the regression equation. According to the model, what pressure can be expected at a depth of 100 ft?

Depth (ft)	Pressure (psi)
15	6.94
25	11.85
35	15.64
45	19.58
55	24.35
65	28.27
75	32.68

49. Musical notes: The table shown gives the frequency (vibrations per second for each of the twelve notes in a selected octave) from the standard chromatic scale. Use the data to draw a scatterplot, then use the context and scatterplot to find the regression equation.

#	Note	Frequency
1	A	110.00
2	A#	116.54
3	B	123.48
4	C	130.82
5	C#	138.60
6	D	146.84
7	D#	155.56
8	E	164.82
9	F	174.62
10	F#	185.00
11	G	196.00
12	G#	207.66

a. What is the frequency of the "A" note that is an octave higher than the one shown? [*Hint:* The names repeat every 12 notes (one octave), so this would be the 13th note in this sequence.]

b. If the frequency is 370.00 what note is being played?

c. What pattern do you notice for the F#'s in each octave (the 10th, 22nd, 34th, and 46th notes in sequence)? Does the pattern hold for all notes?

50. Basketball salaries: In 1970, the average player salary for a professional basketball player was about $43,000. Since that time player salaries have risen dramatically. The average player salary for a professional player is given in the table shown for selected years. Use the data to draw a scatterplot,

Year (1970 → 0)	Salary ($1000s)
0	43
10	260
15	325
20	750
25	1900
27	2200
28	2600

then use the context and scatterplot to find the regression equation.

Source: Wall Street Journal Almanac.

a. What was the approximate salary for a player in 1993?

b. Approximately how much will the average salary be in 2005?

c. In what year did the average salary exceed $5,000,000?

51. Cost of cable service: The average monthly cost of cable TV has been rising steadily since it became very popular in the early 1980s. The data given shows the average monthly rate for selected years (1980 → 0). Use the data to draw a scatterplot, then use the context and scatterplot to find the regression equation. According to the model, what will be the cost of cable service in 2010? 2015?

Year (1980 → 0)	Monthly Charge
0	$7.69
5	$9.73
10	$16.78
20	$23.07
25	$30.70

Source: 2004–2005 Statistical Abstract of the United States, page 725, Table 1138.

52. Research and development expenditures: The development of new products, improved health care, greater scientific achievement, and other advances is fueled by huge investments in research and development (R & D). Since 1960, total R & D expenditures in the United States have shown a distinct pattern of growth, and the data

Year (1960 → 0)	R & D (billion $)
0	13.7
5	20.3
10	26.3
15	35.7
20	63.3
25	114.7
30	152.0
35	183.2
39	247.0

are given in the table for selected years from 1960 to 1999. Use the data to draw a scatterplot, then use the context and scatterplot to find the regression equation. According to the model, what was spent on R & D in 1992? In what year did expenditures for R & D exceed 450 billion?

53. Business start-up costs: As many new businesses open, they experience a period where little or no profit is realized due to start-up expenses, equipment purchases, and so on. The data given shows the profit of a new company for the first 6 months of business. Use the data to draw a scatterplot, then use the context and scatterplot to find the regression equation. According to the model, what is the first month that a profit will be earned?

Month	Profit ($1000s)
1	−5
2	−13
3	−18
4	−20
5	−21
6	−19

54. Low birth weight: For many years, the association between low birth weight (less than 2500 g or about 5.5 lb) and a mother's age has been well documented. The data given are grouped by age and give the percent of total births with low birth weight.

Source: National Vital Statistics Report, Vol. 50, No. 5, February 12, 2002.

Ages	Percent
15–19	8.5
20–24	6.5
25–29	5.2
30–34	5
35–39	6
40–44	8
45–54	10

a. Using the data and the median age of each group, draw a scatterplot and decide on an appropriate form of regression.

b. Find a regression equation that models the data. According to the model, what percent of births will have a low birth weight if the mother was 58 years old?

55. Growth of cell phone use: The tremendous surge in cell phone use that began in the early nineties has continued unabated into the new century. The total number of subscriptions is shown in the table for selected years, with 1990 → 0 and the number of subscriptions in millions. Use the data to draw a scatterplot. Does the data seem to follow an exponential or logistic pattern? Find the regression equation. According to the model, how many subscriptions were there in 1997? How many subscriptions does your model project for 2005? 2010? In what year will the subscriptions exceed 220 million?

Source: 2000/2004 Statistical Abstracts of the United States, Tables 919/1144.

Year (1990 → 0)	Subscriptions (millions)
0	5.3
3	16.0
6	44.0
8	69.2
12	140.0
13	158.7

56. Absorption rates of fabric: Using time lapse photography, the spread of a liquid is tracked in one-fifth of a second intervals, as a small amount of liquid is dropped on a piece of fabric. Use the data to draw a scatterplot, then use the context and scatterplot to find the regression equation. To the nearest hundredth of a second, how long did it take the stain to reach a size of 15 mm?

Time (sec)	Size (mm)
0.2	0.39
0.4	1.27
0.6	3.90
0.8	10.60
1.0	21.50
1.2	31.30
1.4	36.30
1.6	38.10
1.8	39.00

57. Planetary orbits: The table shown gives the time required for the first five planets to make one complete revolution around the Sun (in years), along with the average orbital radius of the planet in astronomical units (1 AU = 92.96 million miles). Use a graphing calculator to draw the scatterplot, then use the scatterplot, the context, and any previous experience to decide whether a polynomial, exponential, logarithmic, or power regression is most appropriate. Then (a) find the regression equation and use it to estimate the average orbital radius of Saturn, given it orbits the Sun every 29.46 yr, and (b) estimate how many years it takes Uranus to orbit the Sun, given it has an average orbital radius of 19.2 AU.

Planet	Years	Radius
Mercury	0.24	0.39
Venus	0.62	0.72
Earth	1.00	1.00
Mars	1.88	1.52
Jupiter	11.86	5.20

58. Ocean temperatures: The temperature of ocean water depends on several factors, including salinity, latitude, depth, and density. However, between depths of 125 m and 2000 m, ocean temperatures are relatively predictable, as indicated by the data shown for tropical oceans in the table. Use a graphing calculator to draw the scatterplot, then use the scatterplot, the context, and any previous experience to decide whether a polynomial, exponential, logarithmic, or power regression is

Depth (meters)	Temp (°C)
125	13.0
250	9.0
500	6.0
750	5.0
1000	4.4
1250	3.8
1500	3.1
1750	2.8
2000	2.5

most appropriate (end-behavior rules out linear and quadratic models as possibilities).

Source: UCLA at www.msc.ucla.oceanglobe/pdf/ thermo_plot_lab

a. Find the regression equation and use it to estimate the water temperature at a depth of 2850 m.

b. If the model were still valid at greater depths, what is the ocean temperature at the bottom of the Marianas Trench, some 10,900 m below sea level?

59. **Predater/prey model:** In the wild, some rodent populations vary inversely with the number of predators in the area. Over a period of time, a conservation team does an extensive study on this relationship and gathers the data shown. Draw a scatterplot of the data and (a) find a regression equation that models the data. According to the model, (b) if there are 150 predators in the area, what is the rodent population? (c) How many predators are in the area if studies show a rodent population of 3000 animals?

Predators	Rodents
10	5100
20	2500
30	1600
40	1200
50	950
60	775
70	660
80	575
90	500
100	450

60. **Children and AIDS:** Largely due to research, education, prevention, and better health care, estimates of the number of AIDS (acquired immune deficiency syndrome) cases diagnosed in children less than 13 yr of age have been declining. Data for the years 1995 through 2002 is given in the table.

Source: National Center for Disease Control and Prevention.

Years Since 1990	Cases
5	686
6	518
7	328
8	238
9	183
10	118
11	110
12	92

a. Use the data to draw a scatterplot and decide on an appropriate form of regression.

b. Find a regression equation that models the data. According to the model, how many cases of AIDS in children are projected for 2010?

c. In what year did the number of cases fall below 50?

61. **Growth rates of children:** After reading a report from The National Center for Health Statistics regarding the growth of children from age 0 to 36 months, Maryann decides to track the relationships (length in inches, weight in pounds) and (age in months, circumference of head in centimeters) for her newborn child, a beautiful baby girl—Morgan.

a. Use the (length, weight) data to draw a scatterplot, then use the context and scatterplot to find the regression equation. According to the model, how much will Morgan weigh when she reaches a height (length) of 39 in.? What will her length be when she weighs 28 lb?

b. Use the (age, circumference) data to draw a scatterplot, then use the context and scatterplot to find the regression equation. According to the model, what is the circumference of Morgan's head when she is 27 months old? How old will she be when the circumference of her head is 50 cm?

Exercise 61a		Exercise 61b	
Length (in.)	Weight (lb)	Age (months)	Circumference (cm)
17.5	5.50	1	38.0
21	10.75	6	44.0
25.5	16.25	12	46.5
28.5	19.00	18	48.0
33	25.25	21	48.3

62. **Correlation coefficients:** Although correlation coefficients can be very helpful, other factors must also be considered when selecting the most appropriate equation model for a set of data. To see why, use the data given to (a) find a linear regression equation and note its correlation coefficient, and (b) find an exponential regression equation and note its correlation coefficient. What do you notice? Without knowing the context of the data, would you be able to tell which model might be more suitable? (c) Use your calculator to graph the scatterplot and both functions. Which function appears to be a better fit?

▶ MAINTAINING YOUR SKILLS

63. **(4.4)** State the domain of the function, then write it in lowest terms:

$$h(x) = \frac{x^2 - 6x + 5}{x^3 - 4x^2 - 7x + 10}$$

$D: x \in (-\infty, -2) \cup (-2, 1) \cup (1, 5) \cup (5, \infty), \frac{1}{x+2}$

64. **(2.5)** Find a linear function that will make $p(x)$ continuous. $y = -\frac{3}{2}x + 7; 2 \le x < 4$

$$p(x) = \begin{cases} x^2 & -2 \le x < 2 \\ ?? & ? \le x < ? \\ \sqrt{x-4} + 1 & x \ge 4 \end{cases}$$

65. **(2.1)** For the graph of $f(x)$ given, estimate max/min values to the nearest tenth and state intervals where $f(x)\uparrow$ and $f(x)\downarrow$.

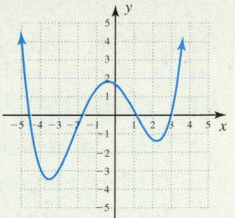

66. **(2.2)** The graph of $f(x) = x^{\frac{2}{3}}$ is given. Use it to sketch the graph of $F(x) = (x-2)^{\frac{2}{3}} + 3$, and use the graph to state the domain and range of F.

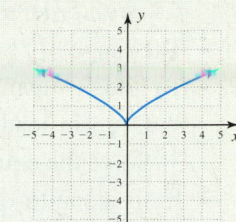

Additional answers can be found in the Instructor Answer Appendix.

MAKING CONNECTIONS

Making Connections: Graphically, Symbollically, Numerically, and Verbally

Eight graphs (a) through (h) are given. Match the characteristics or equations shown in 1 through 16 to one of the eight graphs.

(a)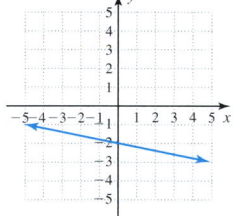

(b)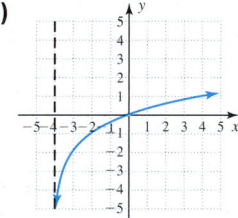

(c)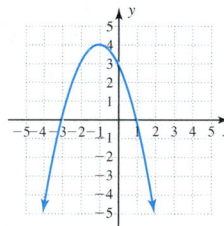

(d)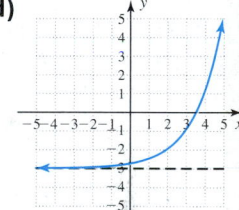

(e)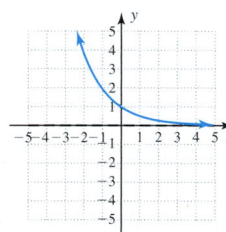

(f)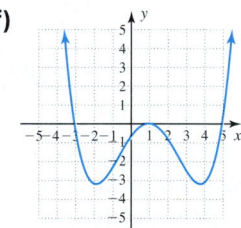

(g)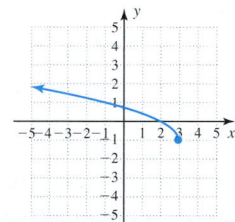

(h)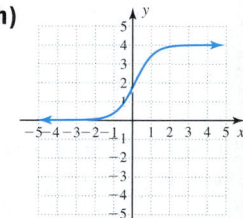

1. __a__ $y = -\frac{1}{5}x - 2$

2. __g__ domain: $x \in (-\infty, 3]$

3. __e__ as $x \to \infty, y \to 0$

4. __b__ $y = \log_2(x + 4) - 2$

5. __c__ $y = -(x + 1)^2 + 4$

6. __h__ as $x \to \infty, y \to 4$

7. __e__ $f(0) = 1, f(-2) = 4$

8. __d__ $f(x)\uparrow$ for $x \in (-\infty, \infty)$

9. __b__ range: $y \in (-\infty, \infty), f(-2) = -1$

10. __h__ $y = \frac{4}{1 + 1.5e^{-2x}}$

11. __g__ $y = \sqrt{3 - x} - 1$

12. __f__ $y = \frac{1}{20}(x + 3)(x - 1)^2(x - 5)$

13. __c__ axis of symmetry $x = -1$

14. __e__ $y = 2^{-x}$

15. __d__ $y = 2^{x-2} - 3$

16. __f__ $f(x) \le 0$ for $x \in [-3, 5]$

SUMMARY AND CONCEPT REVIEW

SECTION 5.1   One-to-One and Inverse Functions

KEY CONCEPTS

- A function is one-to-one if each element of the range corresponds to a unique element of the domain.
- If every horizontal line intersects the graph of a function in at most one point, the function is one-to-one.
- If f is a one-to-one function with ordered pairs (a, b), then the inverse of f exists and is that one-to-one function f^{-1} with ordered pairs of the form (b, a).
- The range of f becomes the domain of f^{-1}, and the domain of f becomes the range of f^{-1}.
- To find f^{-1} using the algebraic method:
  1. Use y instead of $f(x)$.   2. Interchange x and y.
  3. Solve the equation for y.   4. Substitute $f^{-1}(x)$ for y.
- If f is a one-to-one function, the inverse f^{-1} exists, where $(f \circ f^{-1})(x) = x$ and $(f^{-1} \circ f)(x) = x$.
- The graphs of f and f^{-1} are symmetric to the identity function $y = x$.

EXERCISES

Determine whether the functions given are one-to-one by noting the function family to which each belongs and mentally picturing the shape of the graph.

1. $h(x) = -|x - 2| + 3$ no   **2.** $p(x) = 2x^2 + 7$ no   **3.** $s(x) = \sqrt{x - 1} + 5$ yes

Find the inverse of each function given. Then show using composition that your inverse function is correct. State any necessary restrictions.

4. $f(x) = -3x + 2$ $f^{-1}(x) = \dfrac{x-2}{-3}$ **5.** $f(x) = x^2 - 2, x \geq 0$ **6.** $f(x) = \sqrt{x - 1}$ $f^{-1}(x) = x^2 + 1; x \geq 0$
$f^{-1}(x) = \sqrt{x + 2}$

Determine the domain and range for each function whose graph is given, and use this information to state the domain and range of the inverse function. Then use the line $y = x$ to estimate the location of three points on the graph, and use these to graph $f^{-1}(x)$ on the same grid.

7.
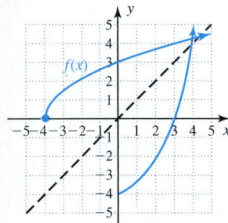

8.
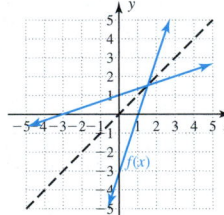

9.
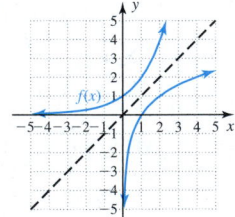

10. Fines for overdue material: Some libraries have set fees and penalties to discourage patrons from holding borrowed materials for an extended period. Suppose the fine for overdue DVDs is given by the function $f(t) = 0.15t + 2$, where $f(t)$ is the amount of the fine t days after it is due. (a) What is the fine for keeping a DVD seven (7) extra days? (b) Find $f^{-1}(t)$, then input your answer from part (a) and comment on the result. (c) If a fine of $3.80 was assessed, how many days was the DVD overdue?   **a.** \$3.05   **b.** $f^{-1}(t) = \frac{t-2}{0.15}, f^{-1}(3.05) = 7$ **c.** 12 days

SECTION 5.2   Exponential Functions

KEY CONCEPTS

- An exponential function is defined as $f(x) = b^x$, where $b > 0$, $b \neq 1$, and b, x are real numbers.
- The natural exponential function is $f(x) = e^x$, where $e \approx 2.71828182846$.
- For exponential functions, we have

- one-to-one function   • y-intercept $(0, 1)$ • domain: $x \in \mathbb{R}$
- range: $y \in (0, \infty)$ • increasing if $b > 1$ • decreasing if $0 < b < 1$
- asymptotic to x-axis

Additional answers can be found in the Instructor Answer Appendix.

- The graph of $y = b^{x \pm h} \pm k$ is a translation of the basic graph of $y = b^x$, horizontally h units opposite the sign and vertically k units in the same direction as the sign.
- If an equation can be written with like bases on each side, we solve it using the uniqueness property: If $b^m = b^n$, then $m = n$ (equal bases imply equal exponents).
- All previous properties of exponents also apply to exponential functions.

EXERCISES

Graph each function using *transformations of the basic function,* then strategically plot a few points to check your work and round out the graph. Draw and label the asymptote.

11. $y = 2^x + 3$ **12.** $y = 2^{-x} - 1$ **13.** $y = -e^{x+1} - 2$

Solve using the uniqueness property.

14. $3^{2x-1} = 27$ 2   **15.** $4^x = \frac{1}{16}$ -2 **16.** $e^x \cdot e^{x+1} = e^6$ $\frac{5}{2}$

17. A ballast machine is purchased new for \$142,000 by the AT & SF Railroad. The machine loses 15% of its value each year and must be replaced when its value drops below \$20,000. How many years will the machine be in service?   12.1 yr

SECTION 5.3 Logarithms and Logarithmic Functions

KEY CONCEPTS

- A logarithm is an exponent. For $x, b > 0$, and $b \neq 1$, the expression $\log_b x$ represents the exponent that goes on base b to obtain x: If $y = \log_b x$, then $b^y = x \Rightarrow b^{\log_b x} = x$ (by substitution).
- The equations $x = b^y$ and $y = \log_b x$ are equivalent. We say $x = b^y$ is the *exponential* form and $y = \log_b x$ is the *logarithmic* form of the equation.
- The value of $\log_b x$ can sometimes be determined by writing the expression in exponential form. If $b = 10$ or $b = e$, the value of $\log_b x$ can be found directly using a calculator.
- A logarithmic *function* is defined as $f(x) = \log_b x$, where $x, b > 0$, and $b \neq 1$.
  - $y = \log_{10} x = \log x$ is called the *common* logarithmic function.
  - $y = \log_e x = \ln x$ is called the *natural* logarithmic function.
- For $f(x) = \log_b x$ as defined we have
  - one-to-one function x-intercept $(1, 0)$ domain: $x \in (0, \infty)$ range: $y \in \mathbb{R}$
  - increasing if $b > 1$ decreasing if $0 < b < 1$ asymptotic to y-axis
- The graph of $y = \log_b(x \pm h) \pm k$ is a translation of the graph of $y = \log_b x$, horizontally h units opposite the sign and vertically k units in the same direction as the sign.

EXERCISES

Write each expression in *exponential* form.

18. $\log_3 9 = 2$ $3^2 = 9$ **19.** $\log_5 \frac{1}{125} = -3$ $5^{-3} = \frac{1}{125}$ **20.** $\ln 43 \approx 3.7612$ $e^{3.7612} \approx 43$

Write each expression in *logarithmic* form.

21. $5^2 = 25$ $\log_5 25 = 2$ **22.** $e^{-0.25} \approx 0.7788$ $\ln 0.7788 \approx -0.25$ **23.** $3^4 = 81$ $\log_3 81 = 4$

Find the value of each expression without using a calculator.

24. $\log_2 32$ 5   **25.** $\ln\left(\frac{1}{e}\right)$ -1 **26.** $\log_9 3$ $\frac{1}{2}$

Graph each function using *transformations of the basic function,* then strategically plot a few points to check your work and round out the graph. Draw and label the asymptote.

27. $f(x) = \log_2 x$ **28.** $f(x) = \log_2(x + 3)$ **29.** $f(x) = 2 + \ln(x - 1)$

Find the domain of the following functions.

30. $g(x) = \log \sqrt{2x + 3}$ $x \in \left(-\frac{3}{2}, \infty\right)$ **31.** $f(x) = \ln(x^2 - 6x)$ $x \in (-\infty, 0) \cup (6, \infty)$

Additional answers can be found in the Instructor Answer Appendix.

32. The magnitude of an earthquake is given by $M(I) = \log\dfrac{I}{I_0}$, where I is the intensity and I_0 is the reference intensity.

(a) Find $M(I)$ given $I = 62,000I_0$ and (b) find the intensity I given $M(I) = 7.3$.   **a.** 4.79 **b.** $10^{7.3}I_0$

SECTION 5.4   Properties of Logarithms

KEY CONCEPTS

- The basic definition of a logarithm gives rise to the following properties: For any base $b > 0$, $b \neq 1$,

  1. $\log_b b = 1$ (since $b^1 = b$)   **2.** $\log_b 1 = 0$ (since $b^0 = 1$)

  3. $\log_b b^x = x$ (since $b^x = b^x$)   **4.** $b^{\log_b x} = x$

- Since a logarithm is an exponent, they have properties that parallel those of exponents.

Product Property	**Quotient Property**	**Power Property**
like base and multiplication, add exponents:	like base and division, subtract exponents:	exponent raised to a power, multiply exponents:
$\log_b(MN) = \log_b M + \log_b N$	$\log_b\left(\dfrac{M}{N}\right) = \log_b M - \log_b N$	$\log_b M^p = p\log_b M$

- The logarithmic properties can be used to expand an expression: $\log(2x) = \log 2 + \log x$.

- The logarithmic properties can be used to contract an expression: $\ln(2x) - \ln(x + 3) = \ln\left(\dfrac{2x}{x + 3}\right)$.

- To evaluate logarithms with bases other than 10 or e, use the change-of-base formula:

$$\log_b M = \frac{\log M}{\log b} = \frac{\ln M}{\ln b}$$

- If an equation can be written with like bases on each side, we solve it using the uniqueness property: if $\log_b m = \log_b n$, then $m = n$ (equal bases imply equal arguments).
- If a single exponential or logarithmic term can be isolated on one side, then for any base b:

$$\text{If } b^x = k, \text{ then } x = \frac{\log k}{\log b} \qquad\qquad \text{If } \log_b x = k, \text{ then } x = b^k.$$

EXERCISES

33. Solve each equation by applying fundamental properties.

a. $\ln x = 32$ $x = e^{32}$ **b.** $\log x = 2.38$ $x = 10^{2.38}$ **c.** $e^x = 9.8$ $x = \ln 9.8$ **d.** $10^x = \sqrt{7}$ $x = \dfrac{1}{2}\log 7$

34. Solve each equation. Write answers in exact form and in approximate form to four decimal places.

a. $15 = 7 + 2e^{0.5x}$ **b.** $10^{0.2x} = 19$ **c.** $-2\log(3x) + 1 = -5$ **d.** $-2\ln x + 1 = 6.5$

35. Use the product or quotient property of logarithms to write each sum or difference as a single term.

a. $\ln 7 + \ln 6$ $\ln 42$ **b.** $\log_9 2 + \log_9 15$ $\log_9 30$ **c.** $\ln(x + 3) - \ln(x - 1)$ $\ln\left(\frac{x+3}{x-1}\right)$ **d.** $\log x + \log(x + 1)$ $\log(x^2 + x)$

36. Use the power property of logarithms to rewrite each term as a product.

a. $\log_5 9^2$ $2\log_5 9$ **b.** $\log_7 4^2$ $2\log_7 4$ **c.** $\ln 5^{2x-1}$ $(2x - 1)\ln 5$ **d.** $\ln 10^{3x+2}$ $(3x + 2)\ln 10$

37. Use the properties of logarithms to write the following expressions as sums or differences of simple logarithmic terms.

a. $\ln(x\sqrt[4]{y})$ $\ln x + \frac{1}{4}\ln y$ **b.** $\ln(\sqrt[3]{pq})$ $\frac{1}{3}\ln p + \frac{1}{3}\ln q$ **c.** $\log\left(\dfrac{\sqrt[3]{x^5 \cdot y^4}}{\sqrt{x^5 y^3}}\right)$ $\frac{5}{3}\log x + \frac{4}{3}\log y - \frac{5}{2}\log x - \frac{3}{2}\log y$ **d.** $\log\left(\dfrac{4\sqrt[3]{p^5 q^4}}{\sqrt{p^3 q^2}}\right)$ $\log 4 + \frac{5}{3}\log p + \frac{4}{3}\log q - \frac{3}{2}\log p - \log q$

38. Evaluate using a change-of-base formula. Answer in exact form and approximate form to thousandths.

a. $\log_6 45$ $\dfrac{\log 45}{\log 6} \approx 2.215$ **b.** $\log_3 128$ $\dfrac{\log 128}{\log 3} \approx 4.417$ **c.** $\log_2 124$ $\dfrac{\ln 124}{\ln 2} \approx 6.954$ **d.** $\log_5 0.42$ $\dfrac{\ln 0.42}{\ln 5} \approx -0.539$

SECTION 5.5 Solving Exponential and Logarithmic Equations

KEY CONCEPTS

- If an exponential equation uses base 10, isolate the exponential term, apply the common logarithm, then solve for x using algebra.
- If an exponential equation uses base e, isolate the exponential term, apply the natural logarithm, then solve for x using algebra.
- For a general base b, isolate the exponential term, apply the base-b logarithm, then solve for x using algebra and the change-of-base formula.
- If a logarithmic equation has a constant term as in $\log_b M = \log_b N + $ constant, move all logarithmic terms to one side and consolidate using the product or quotient properties, then use the exponential form and algebra to solve.
- If a logarithmic equation has multiple logarithmic terms as in $\log_b X = \log_b M + \log_b N$, consolidate logarithmic terms using the product or quotient properties, then use the uniqueness property and algebra to solve.

EXERCISES

Solve each equation. Answer in both exact form and approximate form.

39. $2^x = 7$ $x = \frac{\ln 7}{\ln 2}, x \approx 2.8074$

40. $3^{x+1} = 5$ $x \approx \frac{\ln 5}{\ln 3} - 1, x \approx 0.4650$

41. $e^{x-2} = 3^{-x}$ $\frac{2}{1 + \ln 3}, x \approx 0.9530$

42. $\ln(x + 1) = 2$ $x = e^2 - 1, x \approx 6.3891$

43. $\log x + \log(x - 3) = 1$ $x = 5; -2$ is extraneous

44. $\log_{25}(x + 2) - \log_{25}(x - 3) = \frac{1}{2}$ $x = 4.25$

45. The rate of decay for radioactive material is related to its half-life by the formula $R(h) = \frac{\ln 2}{h}$, where h represents the half-life of the material and $R(h)$ is the rate of decay expressed as a decimal. The element radon-222 has a half-life of approximately 3.9 days. (a) Find its rate of decay to the nearest hundredth of a percent. (b) Find the half-life of thorium-234 if its rate of decay is 2.89% per day.   **a.** 17.77%   **b.** 23.98 days

46. The *barometric equation* $H = (30T + 8000) \ln\left(\frac{P_0}{P}\right)$ relates the altitude H to atmospheric pressure P, where $P_0 = 76$ cmHg. Find the atmospheric pressure at the summit of Mount Pico de Orizaba (Mexico), whose summit is at 5657 m. Assume the temperature at the summit is $T = 12°C$.   38.6 cmHg

SECTION 5.6 Applications from Business, Finance, and Science

KEY CONCEPTS

- Simple interest: $I = prt$; p is the initial principal, r is the interest rate per year, and t is the time in years.
- Amount in an account after t years: $A = p + prt$ or $A = p(1 + rt)$.
- Interest compounded n times per year: $A = p\left(1 + \dfrac{r}{n}\right)^{nt}$; p is the initial principal, r is the interest rate per year, t is the time in years, and n is the times per year interest is compounded.
- Interest compounded continuously: $A = pe^{rt}$; p is the initial principal, r is the interest rate per year, and t is the time in years.
- If a loan or savings plan calls for a regular schedule of deposits, the plan is called an annuity.
- For periodic payment $\mathcal{P}$, deposited or paid n times per year, at annual interest rate r, with interest compounded or calculated n times per year for t years, and $R = \dfrac{r}{n}$:

  - The accumulated value of the account is $A = \dfrac{p}{R}[(1 + R)^{nt} - 1]$.

  - The payment required to meet a future goal is $\mathcal{P} = \dfrac{AR}{[(1 + R)^{nt} - 1]}$

  - The payment required to amortize an amount A is $\mathcal{P} = \dfrac{AR}{1 - (1 + R)^{-nt}}$.

  - The general formulas for exponential growth and decay are $Q(t) = Q_0 e^{rt}$ and $Q(t) = Q_0 e^{-rt}$, respectively.

EXERCISES

Solve each application.

47. Jeffery borrows $600.00 from his dad, who decides it's best to charge him interest. Three months later Jeff repays the loan plus interest, a total of $627.75. What was the annual interest rate on the loan?   18.5%

48. To save money for her first car, Cheryl invests the $7500 she inherited in an account paying 7.8% interest compounded monthly. She hopes to buy the car in 6 yr and needs $12,000. Is this possible?   Almost, she needs $42.15 more.

49. To save up for the vacation of a lifetime, Al-Harwi decides to save $15,000 over the next 4 yr. For this purpose he invests $260 every month in an account paying $7\frac{1}{2}$% interest compounded monthly. (a) Is this monthly amount sufficient to meet the four-year goal? (b) If not, find the *minimum amount he needs to deposit each month* that will enable him to meet this goal in 4 yr.   **a.** no   **b.** $268.93

50. Eighty prairie dogs are released in a wilderness area in an effort to repopulate the species. Five years later a statistical survey reveals the population has reached 1250 dogs. Assuming the growth was exponential, approximate the growth rate to the nearest tenth of a percent.   55.0%

SECTION 5.7   Exponential, Logarithmic, and Logistic Equation Models

KEY CONCEPTS

- The choice of regression models generally depends on: (a) whether the graph appears to fit the data, (b) the context or situation that generated the data, and (c) certain tests applied to the data.
- The regression equation can be used to *extrapolate* or predict future values or occurrences. When using extrapolation, values are projected *beyond* the given set of data.
- The regression equation can be used to *interpolate* or approximate intermediate values. When using interpolation, the values occur *between* those given in the data set.

EXERCISES

51. Vehicle fuel economy: While the average fuel economy of light trucks rose significantly in the late 1970s and early 1980s, this growth seemed to slow significantly thereafter, and increased at a much slower rate through the 1990s and beyond. The fuel economy for selected years is shown in the table (1975 → 0). Draw a scatterplot of the data then complete the following.

Source: Data from the Pew Charitable Trusts.

a. Decide on an appropriate form of regression and find a regression equation.

b. Use the regression equation to estimate the average fuel economy for light trucks in the year 1988.   16.9 mi/gal

c. If current trends continue, in what year will average fuel economy surpass 18.5 mi/gal?   the year 2011

Year (1975 → 0)	Average Miles per Gal
1	12.5
5	15.5
10	16.8
15	17.5
20	17.6
25	17.6
30	17.8

52. Biological studies—cell division: The relationship between cell size and cell division (mitosis) is a common object of biological research. In many cells, a natural mechanism exists that triggers cell division only after the cell has reached a certain size. If the supply of nutrients is scarce, cell growth slows after an initial spurt, and the time between cell divisions is lengthened. Suppose the normal size of a certain cell is 4 microns (four millionths of a meter), and the data in the table track the growth of this cell over time. Draw a scatterplot for the data and complete the following.

a. Decide on an appropriate form of regression and find a regression equation.

b. Use the regression equation to estimate the size of the cell after 2.5 hr.   ≈ 6.8 microns

c. If mitosis (cell division) begins when the cell has grown to 196% of its original size, how long until mitosis occurs?   (4 × 1.96) = 7.84; 5.56 hr (about 5 hr 34 min)

Time t (hr)	Cell Growth G (microns)
0	4
1	5.3
2	6.4
3	7.1
4	7.5
5	7.8
6	7.9

Additional answers can be found in the Instructor Answer Appendix.

MID-CHAPTER CHECK

1. Solve by graphing the system on a graphing calculator. State whether the system is consistent, inconsistent, or dependent.

$$\begin{cases} x - 3y = -2 \\ 2x + y = 3 \end{cases}$$ (1, 1) consistent

2. Solve the system using elimination. State whether the system is consistent, inconsistent, or dependent.

$$\begin{cases} x - 3y = -4 \\ 2x + y = 13 \end{cases}$$ (5, 3) consistent

3. Solve using a system of linear equations and any method you choose: How many ounces of a 40% acid should be mixed with 10 oz of a 64% acid to obtain a 48% acid solution?   20 oz

4. Determine whether the ordered triple is a solution to the system. If not, identify which equation(s) are not satisfied.   No; R2, R3

$$\begin{cases} 5x + 2y - 4z = 22 \\ 2x - 3y + z = -1 \quad (2, 0, -3) \\ 3x - 6y + z = 2 \end{cases}$$

5. The system given is a dependent system. Without solving, state why.   2R1 = R2

$$\begin{cases} x + 2y - 3z = 3 \\ 2x + 4y - 6z = 6 \\ x - 2y + 5z = -1 \end{cases}$$

Solve each system of equations: (1, 2, 3)

6. $$\begin{cases} x + 2y - 3z = -4 \\ 2y + z = 7 \\ 5y - 2z = 4 \end{cases}$$

Additional answers can be found in the Instructor Answer Appendix.

7. $$\begin{cases} 2x + 3y - 4z = -4 \\ x - 2y + z = 0 \quad (1, 2, 3) \\ -3x - 2y + 2z = -1 \end{cases}$$

8. Solve the following system and write the solution as an ordered triple in terms of the parameter p.

$$\begin{cases} 2x - y + z = 1 \\ -5x + 2y - 3z = 2 \end{cases}$$ $(p, p - 5, -p - 4)$

9. If you add Mozart's age when he wrote his first symphony, with the age of American chess player Paul Morphy when he began dominating the international chess scene, and the age of Blaise Pascal when he formulated his well-known *Essai pour les coniques* (Essay on Conics), the sum is 37. At the time of each event, Paul Morphy's age was 3 yr less than twice Mozart's, and Pascal was 3 yr older than Morphy. Set up a system of equations and find the age of each.   Morphy: 13, Mozart: 8, Pascal: 16

10. The *William Tell Overture* (Gioachino Rossini, 1829) is one of the most famous, and best-loved overtures known. It is played in four movements: a prelude, the storm (often used in animations with great clashes of thunder and a driving rain), the sunrise (actually, *A call to the dairy cows* . . .), and the finale (better known as the Lone Ranger theme song). The prelude takes 2.75 min. Depending on how fast the finale is played, the total playing time is about 11 min. The playing time for the prelude and finale is 1 min longer than the playing time of the storm and the sunrise. Also, the playtime of the storm plus twice the playtime of the sunrise is 1 min longer than twice the finale. Find the playtime for each movement.   prelude: 2.75 min, storm: 2.5 min, sunrise: 2.5 min, finale: 3.25 min

REINFORCING BASIC CONCEPTS

Window Size and Graphing Technology

Since most substantial applications involve noninteger values, technology can play an important role in applying mathematical models. However, with its use comes a heavy responsibility to use it carefully. A very real effort must be made to determine the best approach and to secure a reasonable estimate. This is the only way to guard against (the inevitable) keystroke errors, or ensure a window size that properly displays the results.

Rationale

On October 1, 1999, the newspaper *USA TODAY* ran an article titled, "Bad Math added up to Doomed Mars Craft." The article told of how a $125,000,000.00 spacecraft was lost, apparently because the team of scientists that *plotted the course* for the craft used U.S. units of measurement, while the team of scientists *guiding* the craft were using metric units. NASA's space chief was later quoted, "The problem here was not the error, it was the failure of . . . the checks and balances in our process to detect the error."

No matter how powerful the technology, always try to begin your problem-solving efforts with an estimate. Begin by exploring the **context** of the problem, asking questions about the range of possibilities: How fast can a human run? How much does a new car cost? What is a reasonable price for a ticket? What is the total available to invest? There is no calculating involved in these estimates, they simply rely on "horse sense" and human experience. In many applied problems, the input and output values must be positive—which means the solution will appear in the first quadrant, narrowing the possibilities considerably. This information will be used to set the viewing window of your graphing calculator, in preparation for solving the problem using a system and graphing technology.

Illustration 1 ▶ Erin just filled both her boat and Blazer with gas, at a total cost of $211.14. She purchased 35.7 gallons of premium for her boat and 15.3 gal of regular for her Blazer. Premium gasoline cost $0.10 per gallon more than regular. What was the cost per gallon of each grade of gasoline?

Solution ▶ Asking how much *you* paid for gas the last time you filled up should serve as a fair estimate. Certainly (in 2011) a cost of $6.00 or more per gallon in the United States is too high, and a cost of $2.50 per gallon or less would be too low. Also, we can estimate a solution by assuming that both kinds of gasoline cost the same. This would mean 51 gal were purchased for about $211, and a quick division would place the estimate at near $\frac{211}{51} \approx$ $4.14 per gallon. A good viewing window would be restricted to the first quadrant (since cost > 0) with maximum values of Xmax $= 6$ and Ymax $= 6$.

```
WINDOW
 Xmin=0
 Xmax=6
 Xscl=.5
 Ymin=0
 Ymax=6
 Yscl=.5
 Xres=1
```

Premium: $4.17/gal,   Regular: $4.07/gal

$$\begin{cases} 15.3R + 35.7P = 211.14 \\ P = R + 0.10 \end{cases}$$

Exercise 1: Solve Illustration 1 by writing a system of equations and graphing the system.

Exercise 2: Re-solve Exercises 63 and 64 from Section 6.1 using graphing technology. Verify results are identical.

Verified

6.3 Nonlinear Systems of Equations and Inequalities

LEARNING OBJECTIVES

In Section 6.3 you will see how we can:

☐ **A.** Visualize possible solutions

☐ **B.** Solve nonlinear systems using substitution

☐ **C.** Solve nonlinear systems using elimination

☐ **D.** Solve nonlinear systems of inequalities

☐ **E.** Solve applications of nonlinear systems

Equations where the variables have exponents other than 1 or that are transcendental (like logarithmic and exponential equations), are all *nonlinear* equations. A nonlinear system of equations has at least one nonlinear equation, and these systems occur in a great variety.

A. Possible Solutions for a Nonlinear System

When solving nonlinear systems, it is often helpful to *visualize* the graphs of each equation in the system. This can help determine the number of possible intersections and further assist the solution process.

EXAMPLE 1 ▶ Sketching Graphs to Visualize the Number of Possible Solutions

Identify each equation in the system as that of a line, parabola, circle, or one of the toolbox functions. Then determine the number of solutions possible by considering the different ways the graphs might intersect: $\begin{cases} x^2 + y^2 = 25 \\ x - y = 1 \end{cases}$. Finally, solve the system by graphing.

Solution ▶ The first equation contains a sum of second-degree terms with equal coefficients, which we recognize as the equation of a circle. The second equation is obviously linear. This means the system may have no solution, one solution, or two solutions, as shown in Figure 6.25. The graph of the system is shown in Figure 6.26 and consists of a line with slope $m = 1$ and y-intercept $(0, -1)$, with a circle of radius $r = 5$, centered at $(0, 0)$. The two points of intersection appear to be $(-3, -4)$ and $(4, 3)$. After checking these in the original equations we find that both are solutions to the system.

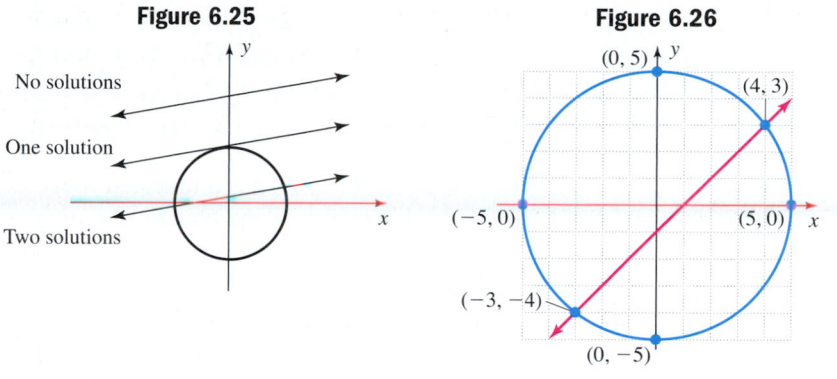

Figure 6.25

Figure 6.26

Now try Exercises 7 through 12 ▶

Solutions to nonlinear systems can also be found using a graphing calculator. From our work in Section 1.1, we know that graphing a relation often involves rewriting its equation in two "pieces," each of which is a function. To solve the system from Example 1, we begin by solving for y in each equation.

$$x^2 + y^2 = 25 \qquad \text{equation of a circle}$$
$$y^2 = 25 - x^2 \qquad \text{subtract } x^2$$
$$y = \sqrt{25 - x^2} \quad \text{or} \quad y = -\sqrt{25 - x^2} \qquad \text{square root property}$$

Figure 6.27

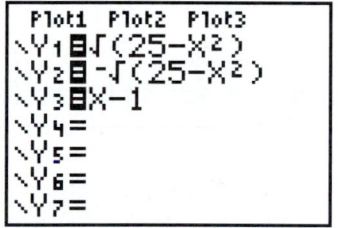

☑ **A. You've just seen how we can visualize possible solutions**

These two equations form the upper and lower halves of the circle (respectively), and are entered as Y_1 and Y_2. After solving the second equation for y, we have $y = x - 1$, which we enter as Y_3 (Figure 6.27).

We then find points of intersection, noting that we must find the intersection of Y_1 with Y_3 and Y_2 with Y_3. The intersection of Y_1 with Y_3 is shown in Figure 6.28 (using a "square" window) and verifies that $(4, 3)$ is a solution.

Figure 6.28

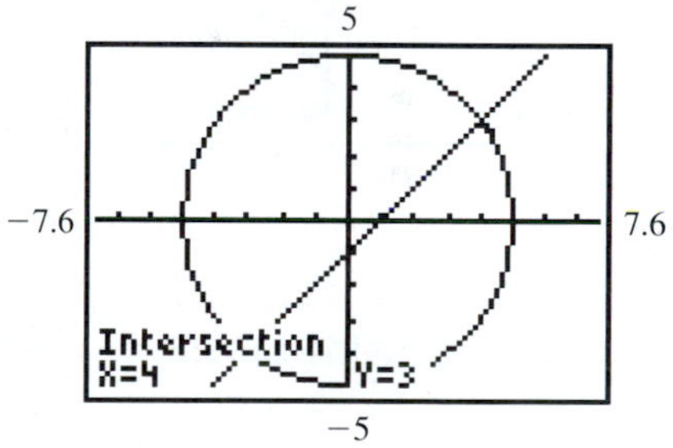

B. Solving Nonlinear Systems by Substitution

Since manually graphing nonlinear systems at best offers an *estimate* for the solution (points of intersection may not have rational values), we more often turn to algebraic methods or the use of a graphing calculator. Recall the substitution method involves solving one of the equations for a variable or expression that can be substituted in the other equation, to eliminate one of the variables. As with our study of linear systems, be aware that some nonlinear systems have no solutions.

EXAMPLE 2A ▶ **Solving a Nonlinear System Using Substitution and Graphing Technology**

Solve the system $\begin{cases} y = x^2 - 2x - 3 \\ 2x - y = 7 \end{cases}$ using

a. Substitution.   **b.** A graphing calculator.

Solution ▶   **a.** Substitution: The first equation is that of a parabola. The second equation is linear. Since the first equation is already written with y in terms of x, we can substitute $x^2 - 2x - 3$ for y in the second equation to solve.

$$
\begin{array}{ll}
2x - y = 7 & \text{second equation} \\
2x - (x^2 - 2x - 3) = 7 & \text{substitute } x^2 - 2x - 3 \text{ for } y \\
2x - x^2 + 2x + 3 = 7 & \text{distribute} \\
-x^2 + 4x + 3 = 7 & \text{combine like terms} \\
x^2 - 4x + 4 = 0 & \text{set equal to zero} \\
(x - 2)^2 = 0 & \text{factor}
\end{array}
$$

We find that $x = 2$ is a repeated root.

Since the second equation is simpler than the first, we substitute 2 for x in this equation and find $y = -3$. The ordered pair $(2, -3)$ checks in both equations and the system has only one (repeated) solution at $(2, -3)$.

b. Graphing calculator: The first equation is already in function form (y is written in terms of x), and we can immediately enter $y = x^2 - 2x - 3$ as Y_1 on the ⬭Y=⬭ screen. After solving $2x - y = 7$ for y, we obtain $y = 2x - 7$, which we enter as Y_2. Using the standard window (⬭ZOOM⬭ **6:ZStandard**), we obtain the graphs shown in Figure 6.29. Using the keystrokes ⬭2nd⬭ ⬭TRACE⬭ (CALC) **5:intersect**, we obtain the solution shown in Figure 6.30. We suspect that the calculator is trying to display $x = 2$ and $y = -3$ as the point of intersection, and for confirmation we use the **TABLE** feature (Figure 6.31), which shows this is indeed the case.

Figure 6.29

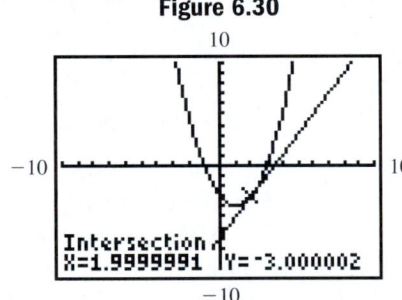

Figure 6.30

Figure 6.31

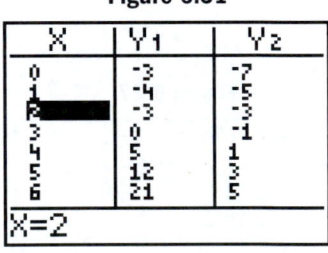

EXAMPLE 2B ▶   **Solving a System Using Substitution and Graphing Technology**

Solve the following system $\begin{cases} y = x^2 + 4x + 4 \\ -(x - 2)^2 = y \end{cases}$ using

a. Substitution.   **b.** A graphing calculator.

Solution ▶   **a.** Substitution: Expanding the binomial square in the second equation gives $y = -(x^2 - 4x + 4)$, then $y = -x^2 + 4x - 4$ after simplification.

The system then becomes: $\begin{cases} y = x^2 + 4x + 4 \\ y = -x^2 + 4x - 4 \end{cases}$. Substituting $-x^2 + 4x - 4$ for y in the first equation gives the following:

$$
\begin{array}{ll}
y = x^2 + 4x + 4 & \text{first equation} \\
-x^2 + 4x - 4 = x^2 + 4x + 4 & \text{substitute } -x^2 + 4x - 4 \text{ for } y \\
0 = 2x^2 + 8 & \text{add } x^2 \text{ and 4; subtract } 4x \\
-8 = 2x^2 & \text{isolate squared term} \\
-4 = x^2 & \text{divide by 2} \\
x = \sqrt{-4} \quad \text{or} \quad x = -\sqrt{-4} & \text{square root property} \\
x = 2i \quad \text{or} \quad x = -2i & \text{result}
\end{array}
$$

From this result we see that the system has no real solutions, but we continue working to provide the complete complex-number solution. Substituting $2i$ for x in the first equation gives

$$y = x^2 + 4x + 4 \qquad \text{first equation}$$
$$= (2i)^2 + 4(2i) + 4 \qquad \text{substitute } 2i \text{ for } x$$
$$= 4(-1) + 8i + 4 \qquad \text{simplify}$$
$$= 8i \qquad \text{result}$$

One solution is $(2i, 8i)$. Substituting $-2i$ for x shows the second solution is $(-2i, -8i)$. Both solutions can be verified using back-substitution.

b. Graphing calculator: Both equations can be entered directly on the ⬭Y=⬭ screen (both have y in terms of x) as Y_1 and Y_2, respectively. Using the standard window, we can easily see that the graphs do not intersect (see figure), indicating there are no real-number solutions.

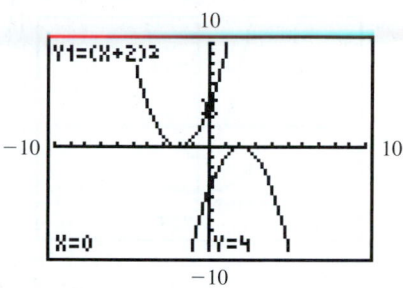

☑ **B.** You've just seen how we can solve nonlinear systems using substitution

Now try Exercises 13 through 20 ▶

C. Solving Nonlinear Systems by Elimination

When both equations in the system have second-degree terms with like variables, it is generally easier to use the elimination method, rather than substitution.

EXAMPLE 3 ▶ **Solving a Nonlinear System Using Elimination and Graphing Technology**

Solve the system $\begin{cases} y - \frac{1}{2}x^2 = -3 \\ x^2 + y^2 = 41 \end{cases}$ using

a. Elimination.   **b.** A graphing calculator.

Solution ▶ **a.** Elimination: The first equation can be rewritten as $y = \frac{1}{2}x^2 - 3$ and is a parabola opening upward with vertex $(0, -3)$. The second equation represents a circle with center at $(0, 0)$ and radius $r = \sqrt{41} \approx 6.4$. Mentally visualizing these graphs indicates there will be two solutions (see Figure 6.32). After writing the system with x- and y-terms in the same order, we find that using $2R1 + R2$ will eliminate the variable x. For $2(-\frac{1}{2}x^2 + y) = 2(-3)$ we have

Figure 6.32

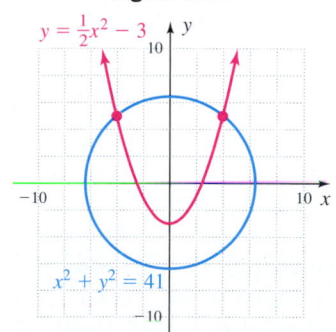

$$\begin{matrix} 2R1 \\ + \\ R1 \end{matrix} \begin{cases} -x^2 + 2y = -6 \qquad \text{rewrite first equation; multiply by 2} \\ \underline{x^2 + y^2 = 41} \qquad \text{second equation} \\ y^2 + 2y = 35 \qquad \text{add} \end{cases}$$

To find solutions, we can set the equation equal to zero and factor, or use the quadratic formula if needed.

$$y^2 + 2y - 35 = 0 \qquad \text{standard form}$$
$$(y + 7)(y - 5) = 0 \qquad \text{factored form}$$
$$y = -7 \text{ or } y = 5 \qquad \text{result}$$

Due to the radius of the circle, the solution $y = -7$ leads to nonreal solutions.

$$x^2 + y^2 = 41 \qquad \text{equation 2}$$
$$x^2 + (-7)^2 = 41 \qquad \text{substitute } -7 \text{ for } y$$
$$x^2 + 49 = 41 \qquad (-7)^2 = 49$$
$$x^2 = -8 \qquad \text{subtract 49}$$
$$x = \pm 2i\sqrt{2} \qquad \text{square root property}$$

The nonreal solutions are $(2i\sqrt{2}, -7)$ and $(-2i\sqrt{2}, -7)$. Using $y = 5$ in the second equation gives the following:

$$x^2 + y^2 = 41 \quad \text{equation 2}$$
$$x^2 + (5)^2 = 41 \quad \text{substitute 5 for } y$$
$$x^2 + 25 = 41 \quad 5^2 = 25$$
$$x^2 = 16 \quad \text{subtract 25}$$
$$x = \pm 4 \quad \text{square root property}$$

The real number solutions are $(-4, 5)$ and $(4, 5)$.

Figure 6.33

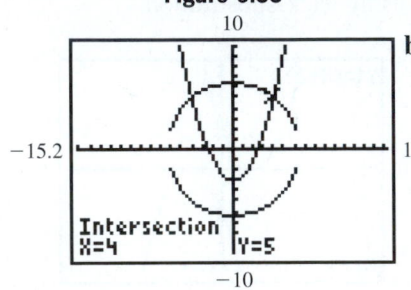

b. Graphing calculator: Enter the equation $y = \dfrac{1}{2}x^2 - 3$ as Y_1 on the [Y=] screen.

Solving $x^2 + y^2 = 41$ for y gives the equations $y = \sqrt{41 - x^2}$ and $y = -\sqrt{41 - x^2}$, which we enter as Y_2 and Y_3. Using a square window and the [2nd] [TRACE] (**CALC**) **5:intersect** feature as before, we obtain the graph shown in Figure 6.33, which shows the parabola and the circle intersect at $(4, 5)$. Since both the parabola and circle are symmetric to the y-axis, a second solution must be at $(-4, 5)$. These solutions check in both equations.

Now try Exercises 21 through 26 ▶

Nonlinear systems may involve other relations as well, including power, polynomial, logarithmic, or exponential functions. These are solved using the same methods.

EXAMPLE 4 ▶ **Solving a System of Logarithmic Equations**

Solve the system using the method of your choice: $\begin{cases} y = -\log(x + 7) + 2 \\ y = \log(x + 4) + 1 \end{cases}$.

Verify your solution using a graphing calculator.

Solution ▶ Since both equations have y written in terms of x, substitution appears to be the most convenient choice. The result is a logarithmic equation, which we can solve using the techniques from Chapter 5.

$$\log(x + 4) + 1 = -\log(x + 7) + 2 \quad \text{substitute } \log(x + 4) + 1 \text{ for } y \text{ in first equation}$$
$$\log(x + 4) + \log(x + 7) = 1 \quad \text{add } \log(x + 7); \text{ subtract 1}$$
$$\log(x + 4)(x + 7) = 1 \quad \text{product property of logarithms}$$
$$(x + 4)(x + 7) = 10^1 \quad \text{exponential form}$$
$$x^2 + 11x + 18 = 0 \quad \text{eliminate parentheses and set equal to zero}$$
$$(x + 9)(x + 2) = 0 \quad \text{factor}$$
$$x + 9 = 0 \quad \text{or} \quad x + 2 = 0 \quad \text{zero factor theorem}$$
$$x = -9 \quad \text{or} \quad x = -2 \quad \text{possible solutions}$$

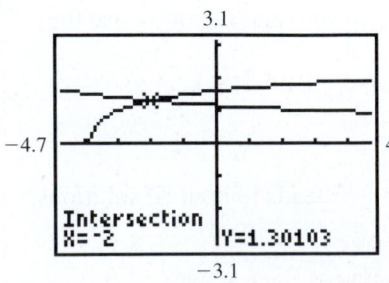

By inspection, we see that $x = -9$ is extraneous, since $\log(-9 + 4)$ and $-\log(-9 + 7)$ are not real numbers. Substituting -2 for x in the second equation we find one form of the (exact) solution is $(-2, \log 2 + 1)$. If we substitute -2 for x in the first equation the exact solution is $(-2, -\log 5 + 2)$. Using a calculator we can verify the ordered pairs are equivalent and approximately equal to $(-2, 1.3)$.

Using the [ZOOM] **4:ZDecimal** feature of graphing calculator produces the screen shown in the figure, which confirms the solution $(-2, \log 2 + 1)$ obtained by substitution.

Now try Exercises 27 through 38 ▶

☑ **C. You've just seen how we can solve nonlinear systems using elimination**

Figure 6.34

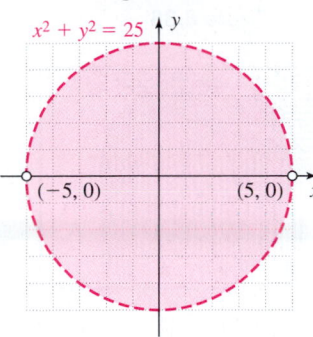

For practice solving more complex systems using a graphing calculator, **see Exercises 39 to 44.**

D. Solving Systems of Nonlinear Inequalities

Nonlinear *inequalities* can be solved by graphing the boundary given by the related equation, and checking the regions that result using a test point. For example, the inequality $x^2 + y^2 < 25$ is solved by first graphing $x^2 + y^2 = 25$, which is a circle with radius 5 centered at $(0, 0)$. We then decide whether the boundary is included or excluded (in this case it is not included). We then use a test point from either "outside" or "inside" the region formed. The test point $(0, 0)$ results in a true statement since $(0)^2 + (0)^2 < 25$, so the inside of the circle is shaded to indicate the solution region (Figure 6.34). For a *system* of nonlinear inequalities, we identify regions where the solution sets for both inequalities overlap, paying special attention to points of intersection.

EXAMPLE 5 ▶ **Solving Systems of Nonlinear Inequalities**

Solve the system $\begin{cases} x^2 + y^2 < 25 \\ 2y - x \geq 5 \end{cases}$ by graphing. Verify results using a graphing calculator.

Solution ▶ We recognize the first inequality from the system as a circle with radius 5, with the solution region in the interior. The second inequality is linear and after solving for x we'll use substitution to find points of intersection (if they exist). From $2y - x = 5$, we obtain $x = 2y - 5$.

$$
\begin{array}{ll}
x^2 + y^2 = 25 & \text{given} \\
(2y - 5)^2 + y^2 = 25 & \text{substitute } 2y - 5 \text{ for } x \\
4y^2 - 20y + 25 + y^2 = 25 & \text{expand} \\
5y^2 - 20y + 25 = 25 & \text{simplify} \\
y^2 - 4y = 0 & \text{subtract 25; divide by 5} \\
y(y - 4) = 0 & \text{factor} \\
y = 0 \quad \text{or} \quad y = 4 & \text{result}
\end{array}
$$

Back-substitution shows the graphs intersect at $(-5, 0)$ and $(3, 4)$. Graphing a line through these points and using $(0, 0)$ as a test point shows the upper half plane is the solution region for the linear inequality $[2(0) - 0 \geq 5$ is *false*]. The overlapping (solution) region for *both* inequalities is the circular section shown in purple. Note the points of intersection are graphed using "open dots" (see Figure 6.35), since points on the graph of the circle are excluded from the solution set.

Our manual sketch indicates the boundaries of the solution region are formed by the line and the upper half of the circle. For $x^2 + y^2 = 25$, the equation for the upper half of the circle is $y = \sqrt{25 - x^2}$, which we enter as Y_1. Solving the second equation for y we obtain $y = \dfrac{x + 5}{2}$ and enter this as Y_2.

Figure 6.35

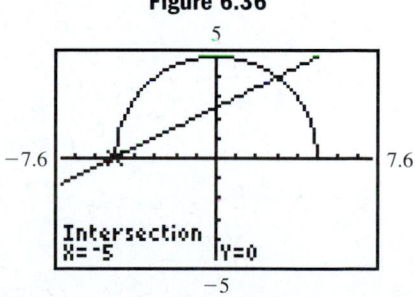

The resulting graph is shown in Figure 6.36 using a "squared" window (from the standard window, use 〔ZOOM〕 **5:ZSquare**, then 〔ZOOM〕 **2:Zoom In**). As the test point indicated, the region below the circle's circumference and above the line forms the solution region. Using the shading capability of the graphing

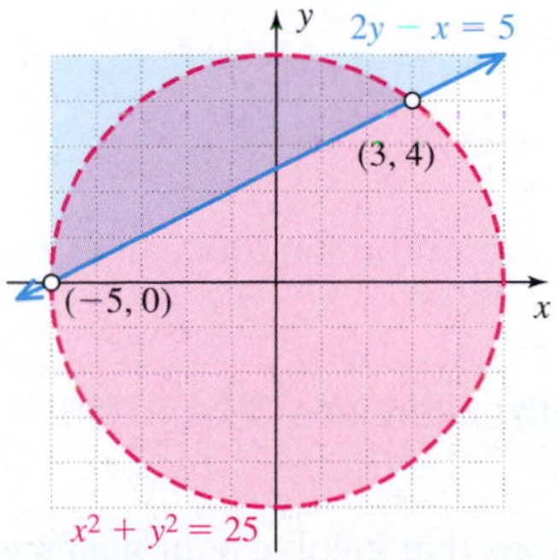

Figure 6.36

calculator found in the leftmost margin of the ⬭Y=⬭ screen, we use the "◣" marker to shade below the circle, and the "◥" marker to shade above the line (Figure 6.37). The resulting graph is shown in Figure 6.38, and appears to support the graphical solution completed by hand.

Figure 6.37

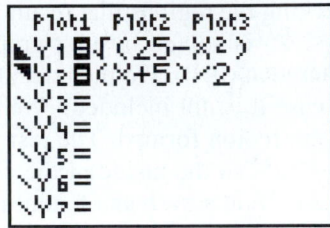

Figure 6.38

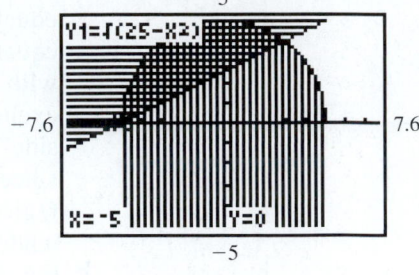

☑ **D.** You've just seen how we can solve nonlinear systems of inequalities

Now try Exercises 45 through 52 ▶

E. Applications of Nonlinear Systems

In the business world, a fast growing company can often reduce the average price of its products using what are called the **economies of scale.** These would include the ability to buy necessary materials in larger quantities, integrating new technology into the production process, and other means. However, there are also countering forces called the **diseconomies of scale,** which may include the need to hire additional employees, rent more production space, and the like. Companies often use what is called a **break-even analysis,** to determine the production level at which these forces stabilize.

EXAMPLE 6 ▶   **Solving an Application of Nonlinear Systems**

Suppose the cost to produce a new and inexpensive shoe made from molded plastic is modeled by the function $C(x) = x^2 - 5x + 18$, where $C(x)$ represents the cost to produce x thousand of these shoes. Similarly, revenue from the sales of these shoes is modeled by $R(x) = -x^2 + 10x - 4$. Use a break-even analysis to find the quantity of sales that will cause the company to break even.

Solution ▶   Essentially we are asked to solve the system formed by the two equations:
$$\begin{cases} C(x) = x^2 - 5x + 18 \\ R(x) = -x^2 + 10x - 4 \end{cases}$$. Since we want to know the point where the company breaks even, we set $C(x) = R(x)$ and solve.

$$C(x) = R(x)$$
$$x^2 - 5x + 18 = -x^2 + 10x - 4 \qquad \text{\color{magenta}substitute for } C(x) \text{ and } R(x)$$
$$2x^2 - 15x + 22 = 0 \qquad \text{\color{magenta}set equal to zero}$$
$$(2x - 11)(x - 2) = 0 \qquad \text{\color{magenta}factored form}$$
$$x = \frac{11}{2} \text{ or } x = 2 \qquad \text{\color{magenta}result}$$

With x in thousands, it appears the company will break even if either 2000 shoes or 5500 shoes are made and sold.

Now try Exercises 55 and 56 ▶

There are actually a large number of significant applications that involve nonlinear systems. Solutions to many of these are difficult to solve manually, and it's here that a skillful use of technology displays its ultimate value. One such application involves the manufacture of cardboard boxes (a billion dollar industry) used to ship goods and commodities all over the globe. Designing a box with the desired volume and needed dimensions often involves the use of a nonlinear system.

EXAMPLE 7 ▶ **Manufacturing Cardboard Boxes**

In order to transport seedlings from the nursery to the market, a manufacturing engineer designs an open box with a square bottom and four sides of equal height. For the most efficient fit, the volume of the box must be 900 in^3 and use 465 in^2 of cardboard stock. Find the dimensions of the box.

Figure 6.39

Solution ▶ After carefully drawing a diagram (see Figure 6.39), we let x represent the length and width of the square bottom, and y represent the height. The volume of the box is then represented by $x^2y = 900$ in^3 ($V = LWH$), and the surface area by $x^2 + 4xy = 465$

($SA = S^2 + 4\ SH$). The resulting system is $\begin{cases} x^2y & = 900 \\ x^2 + 4xy & = 465 \end{cases}$. Solving for y in

the first equation, we obtain $y = \dfrac{900}{x^2}$ and can now substitute this result into the

second equation to obtain an equation in x alone:

$$x^2 + 4xy = 465 \qquad \text{second equation}$$

$$x^2 + 4x\left(\frac{900}{x^2}\right) = 465 \qquad \text{substitute } \frac{900}{x^2} \text{ for } y$$

$$x^2 + \frac{3600}{x} = 465 \qquad \text{simplify}$$

$$x^3 + 3600 = 465x \qquad \text{multiply by } x \text{ (clear denominator)}$$

$$x^3 - 465x + 3600 = 0 \qquad \text{set equal to zero}$$

Figure 6.40

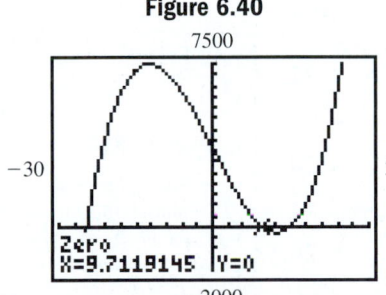

The zeroes of this cubic equation will give the dimensions of the base. A comprehensive graph of this function (using window size $[-30, 30]$ for x and $[-2000, 7500]$ for y) shows there are three zeroes, but one of them is negative and is discounted in this context (Figure 6.40).

Using the **2nd** **TRACE** (CALC) **2:zero** option, we find the positive zeroes are $x \approx 9.7$ (shown) and $x = 15$, and there turns out to be *two* solutions. Substituting 9.7 for x in the first equation,

$$x^2y = 900 \qquad \text{first equation}$$

$$(9.7)^2y = 900 \qquad \text{substitute 9.7 for } x$$

$$(94.09)y = 900 \qquad \text{square 9.7}$$

$$y \approx 9.6 \qquad \text{divide}$$

yields $y \approx 9.6$ in. for the height of the box, which does not seem practical for transporting seedlings. Substituting 15 for x gives $y = 4$ in. as the height, which seems more reasonable. The dimensions of the box will most likely be 15 in. × 15 in. × 4 in.

☑ **E.** You've just seen how we can solve applications of nonlinear systems

Now try Exercises 57 and 58 ▶

6.3 EXERCISES

▶ **CONCEPTS AND VOCABULARY**

Fill in the blanks with the appropriate word or phrase. Carefully reread the section if needed.

1. The solution to a system of nonlinear inequalities is a(n) ___region___ of the plane where the ___solutions___ for each individual inequality overlap.

2. When both equations in the system have at least one ___second___ -degree term, it is generally easier to use the ___elimination___ method to find a solution.

3. Suppose a nonlinear system contained a central hyperbola and an exponential function. Are three solutions possible? Are four solutions possible? Explain/Discuss.   Answers will vary.

4. Solve the system twice, once using elimination, then again using substitution. Compare/contrast each process and comment on which is more efficient in this case: $\begin{cases} x^2 + y^2 = 25 \\ x^2 + y = 5 \end{cases}$.
   Answers will vary.

5. Draw sketches showing the different ways each pair of relations can intersect and give one, two, three, and/or four points of intersection. If a given number of intersections is not possible, so state.

   a. circle and line
   b. parabola and line
   c. circle and parabola
   d. circle and absolute value function
   e. absolute value function and line
   f. absolute value function and parabola

6. By inspection only, identify the systems having *no solutions* and justify your choices.

   a. $\begin{cases} y = |x| - 6 \\ x^2 + y^2 = 9 \end{cases}$ **b.** $\begin{cases} y = x^2 + 4 \\ x^2 + y^2 = 4 \end{cases}$ **c.** $\begin{cases} y = x + 1 \\ x^2 + y^2 = 12 \end{cases}$

   a and b; circle between absolute value function and parabola above circle

▶ **DEVELOPING YOUR SKILLS**

Identify each equation in the system as that of a line, parabola, circle, or absolute value function, then solve the system by graphing.

7. $\begin{cases} x^2 + y = 6 \\ x + y = 4 \end{cases}$ 8. $\begin{cases} -x + y = 4 \\ x^2 + y^2 = 16 \end{cases}$

9. $\begin{cases} y^2 + x^2 = 100 \\ y = |x - 2| \end{cases}$ 10. $\begin{cases} x^2 + y^2 = 25 \\ x^2 + y = 13 \end{cases}$

11. $\begin{cases} -(x - 1)^2 + 2 = y \\ y - x^2 = -3 \end{cases}$ 12. $\begin{cases} y - 4 = -x^2 \\ y = -|x - 1| + 3 \end{cases}$

Solve using substitution and verify results (if possible) on a graphing calculator. Find both real and complex solutions (if they exist). In Exercises 17 and 18, solve for x^2 or y^2 and use the result as a substitution.

13. $\begin{cases} x^2 + y^2 = 25 \\ y - x = 1 \end{cases}$
   $(-4, -3), (3, 4)$

14. $\begin{cases} x + 7y = 50 \\ x^2 + y^2 = 100 \end{cases}$
   $(8, 6), (-6, 8)$

15. $\begin{cases} x^2 + y = 9 \\ -2x + y = 1 \end{cases}$
   $(2, 5), (-4, -7)$

16. $\begin{cases} x^2 - y = 8 \\ x + y = 4 \end{cases}$
   $(3, 1), (-4, 8)$

17. $\begin{cases} x^2 + y = 13 \\ x^2 + y^2 = 25 \end{cases}$
   $(-3, 4), (-4, -3), (3, 4), (4, -3)$

18. $\begin{cases} y^2 + (x - 3)^2 = 25 \\ y^2 + (x + 1)^2 = 9 \end{cases}$
   $(-1, 3), (-1, -3)$

19. $\begin{cases} y + x^2 = 6x \\ y - 23 = (x + 3)^2 \end{cases}$
   $(4i, 16 + 24i), (-4i, 16 - 24i)$

20. $\begin{cases} x^2 + y^2 = 16 \\ y - x = -6 \end{cases}$
   $(3 + i, -3 + 1), (3 - i, -3 - i)$

Solve each system using elimination and verify results (if possible) on a graphing calculator. Find both real and complex solutions (if they exist).

21. $\begin{cases} x^2 + y^2 = 25 \\ \frac{1}{4}x^2 + y = 1 \end{cases}$
   $(4, -3), (-4, -3), (2i\sqrt{6}, 7), (-2i\sqrt{6}, 7)$

22. $\begin{cases} y - \frac{1}{2}x^2 = -1 \\ x^2 + y^2 = 65 \end{cases}$
   $(-4, 7), (4, 7), (4i, -9), (-4i, -9)$

23. $\begin{cases} x^2 + y^2 = 6 \\ y = x - 4 \end{cases}$
   $(2 + i, -2 + i), (2 - i, -2 - i)$

24. $\begin{cases} y + x^2 = 8x \\ y - 34 = (x + 4)^2 \end{cases}$
   $(5i, 25 + 40i), (-5i, 24 - 40i)$

25. $\begin{cases} x^2 + y^2 = 65 \\ y = 3x + 25 \end{cases}$
   $(-8, 1), (-7, 4)$

26. $\begin{cases} y - 2x = 5 \\ x^2 + y^2 = 85 \end{cases}$
   $(2, 9), (-6, -7)$

Solve using the method of your choice and verify results (if possible) on a graphing calculator.

27. $\begin{cases} y - 5 = \log x \\ y = 6 - \log(x - 3) \end{cases}$
   $(5, \log 5 + 5)$

28. $\begin{cases} y = \log(x + 4) + 1 \\ y - 2 = -\log(x + 7) \end{cases}$
   $(-2, \log 2 + 1)$

29. $\begin{cases} y = \ln(x^2) + 1 \\ y - 1 = \ln(x + 12) \end{cases}$
   $(-3, \ln 9 + 1), (4, \ln 16 + 1)$

Additional answers can be found in the Instructor Answer Appendix.

HOMEWORK SELECTION GUIDE

Core: 7–51 every other odd, 55–65 odd, 70, 71 (20 Exercises)
Standard: 1–4, 7–51 every other odd, 54, 55–65 odd, 69, 70, 73 (26 Exercises)

Extended: 1–4, 7–51 every other odd, 54–58, 59–65 odd, 69, 70, 71, 73 (29 Exercises)
In Depth: 1–6, 7–51 every other odd, 53–58, 59–69 odd, 70–73 (34 Exercises)

30. $\begin{cases} \log(x + 1.1) = y + 3 \\ y + 4 = \log(x^2) \end{cases}$
$(-1, -4), (11, \log 12.1 - 3)$

31. $\begin{cases} y - 9 = e^{2x} \\ 3 = y - 7e^x \end{cases}$
$(0, 10), (\ln 6, 45)$

41. $\begin{cases} y = 2^x - 3 \\ y + 2x^2 = 9 \end{cases}$
$(-2.43, -2.81), (2, 1)$

42. $\begin{cases} y = -2\log(x + 8) \\ y + x^3 = 4x - 2 \end{cases}$
$(0.05, -1.81), (2, -2), (-2.05, -1.55)$

32. $\begin{cases} y - 2e^{2x} = 5 \\ y - 1 = 6e^x \end{cases}$
$(0, 7), (\ln 2, 13)$

33. $\begin{cases} y = 4^{x+3} \\ y - 2^{x^2+3x} = 0 \end{cases}$
$(-3, 1), (2, 1024)$

43. $\begin{cases} y = \dfrac{1}{(x - 3)^2} + 2 \\ (x - 3)^2 + y^2 = 10 \end{cases}$
$(0.72, 2.19), (2, 3), (4, 3), (5.28, 2.19)$

44. $\begin{cases} y^2 + x^2 = 5 \\ y = \dfrac{1}{x - 1} - 2 \end{cases}$
$(2, -1), (1.26, 1.85)$

34. $\begin{cases} y - 3^{x^2+2x} = 0 \\ y = 9^{x+2} \end{cases}$
$(2, 6561), (-2, 1)$

35. $\begin{cases} x^3 - y = 2x \\ y - 5x = -6 \end{cases}$
$(-3, -21), (1, -1), (2, 4)$

Solve each system of inequalities. Verify results using a graphing calculator.

36. $\begin{cases} y - x^3 = -2 \\ y + 4 = 3x \end{cases}$
$(-2, -10), (1, -1)$

37. $\begin{cases} x^2 - 6x = y - 4 \\ y - 2x = -8 \end{cases}$
$(2, -4), (6, 4)$

45. $\begin{cases} y - x^2 \geq 1 \\ x + y \leq 3 \end{cases}$

46. $\begin{cases} x^2 + y^2 \leq 25 \\ x + 2y \leq 5 \end{cases}$

38. $\begin{cases} y + x = -2 \\ y + 4x = x^2 \end{cases}$
$(1, -3), (2, -4)$

47. $\begin{cases} x^2 + y^2 > 16 \\ x^2 + y^2 \leq 64 \end{cases}$

48. $\begin{cases} y + 4 \geq x^2 \\ x^2 + y^2 \leq 34 \end{cases}$

 Solve each system using a graphing calculator. Round solutions to hundredths (as needed).

49. $\begin{cases} y - x^2 \leq -16 \\ y^2 + x^2 < 9 \end{cases}$

50. $\begin{cases} x^2 + y^2 \leq 16 \\ x + 2y > 10 \end{cases}$

39. $\begin{cases} x^2 + y^2 = 34 \\ y^2 + (x - 3)^2 = 25 \end{cases}$
$(3, 5), (3, -5)$

40. $\begin{cases} 5x^2 + 5y^2 = 40 \\ y + 2x = x^2 - 6 \end{cases}$
$(-2, 2), (-1.10, -2.61)$

51. $\begin{cases} y^2 + x^2 \leq 25 \\ |x| - 1 > -y \end{cases}$

52. $\begin{cases} y^2 + x^2 \leq 4 \\ x + y < 4 \end{cases}$

▶ WORKING WITH FORMULAS

53. Tunnel clearance: $h = \sqrt{r^2 - d^2}$

The maximum rectangular clearance allowed by a circular tunnel can be found using the formula shown, where $x^2 + y^2 = r^2$ models the tunnel's circular cross section and h is the height of the tunnel at a distance d from the center. If $r = 50$ ft, find the maximum clearance at distances of $d = 20, 30$, and 40 ft from center. $h \approx 45.8$ ft; $h = 40$ ft; $h = 30$ ft

54. Manufacturing cylindrical vents: $\begin{cases} A = 2\pi rh \\ V = \pi r^2 h \end{cases}$

In the manufacture of cylindrical vents, a rectangular piece of sheet metal is rolled, riveted, and sealed to form the vent. The radius and height required to form a vent with a specified volume, using a piece of sheet metal with a given area, can be found by solving the system shown. Use the system to find the radius and height if the volume required is 4071 cm³ and the area of the rectangular piece is 2714 cm². $r = 3$ cm, $h \approx 143.98$ cm

▶ APPLICATIONS

Solve the following applications of economies of scale.

55. World's most inexpensive car: Early in 2008, the Tata Company (India) unveiled the new Tata Nano, the world's most inexpensive car. With its low price and 54 miles per gallon, the car may prove to be very popular. *Assume* the cost to produce these cars is modeled by the function $C(x) = 2.5x^2 - 120x + 3500$, where $C(x)$ represents the cost to produce x-thousand cars. Suppose the revenue from the sale of these cars is modeled by $R(x) = -2x^2 + 180x - 500$. Use a break-even analysis to find the quantity of sales (to the nearest hundred) that will cause the company to break even.

The company breaks even if either 18,400 or 48,200 cars are sold.

56. Document reproduction: In a world of technology, document reproduction has become a billion dollar business. With very stiff competition, the price of a single black and white copy has varied greatly in recent years. Suppose the cost to produce these copies is modeled by the function $C(x) = 0.1x^2 - 1.2x + 7$, where $C(x)$ represents the cost to produce x hundred thousand copies. If the revenue from the sale of these copies is modeled by $R(x) = -0.1x^2 + 1.8x - 2$, use a break-even analysis to find the number of copies (to the nearest thousand) that will cause the company to break even.

The company breaks even if either 415,000 or 1,085,000 copies are sold.

Additional answers can be found in the Instructor Answer Appendix.

Build the system of nonlinear equations needed to solve the following manufacturing applications, then solve.

57. Dimensions of a pool: A homeowner wants to build a new swimming pool in her backyard. Due to size limitations, she decides to build a square pool with a flat bottom. If the volume of the pool must be 2000 ft³ and the tile surface for the sides and bottom will be 800 ft², what will be the final dimensions of the pool?

58. Box manufacturing: A cardboard box manufacturer receives an order for a large number of boxes for wrapping gifts during the Christmas season. The boxes are to have a square top and bottom, with a volume of 6750 cm³ and a surface area of 2700 cm². Find the dimensions of the box fitting these requirements.

Solve using a system of nonlinear equations.

59. Dimensions of a flag: A large American flag has an area of 85 m² and a perimeter of 37 m. Find the dimensions of the flag.   8.5 m × 10 m

60. Dimensions of a sail: The sail on a boat is a right triangle with a perimeter of 36 ft and a hypotenuse of 15 ft. Find the height and width of the sail.   9 ft, 12 ft

61. Dimensions of a tract: The area of a rectangular tract of land is 45 km². The length of a diagonal is $\sqrt{106}$ km. Find the dimensions of the tract.

5 km × 9 km

62. Dimensions of a deck: A rectangular deck has an area of 192 ft² and the length of the diagonal is 20 ft. Find the dimensions of the deck.   12 ft × 16 ft

63. Dimensions of a trailer: The surface area of a closed rectangular trailer with square ends is 928 ft². If the sum of all edges of the trailer is 164 ft, find its dimensions.   8 ft × 8 ft × 25 ft

64. Dimensions of a cylindrical tank: The surface area of a closed cylindrical tank is 192π m². Find the dimensions of the tank if the volume is 320π m³ and the radius is as small as possible. $r = 4$ m, $h = 20$ m

65. Supply and demand: Suppose the monthly market demand D (in ten-thousands of gallons) for a new synthetic oil is related to the price P in dollars by the equation $10P^2 + 6D = 144$. For the market price P, assume the amount D that manufacturers are willing to supply is modeled by $8P^2 - 8P - 4D = 12$. (a) What is the minimum price at which manufacturers are willing to begin supplying the oil? (b) Use this information to create a system of nonlinear equations, then solve the system to find the market equilibrium price (per gallon) and the quantity of oil supplied and sold at this price. $1.83; $3   90,000 gal $\begin{cases} 10P^2 + 6D = 144 \\ 8P^2 - 8P - 4D = 12 \end{cases}$

66. Supply and demand: The weekly demand D for organically grown carrots (in thousands of pounds) is related to the price per pound P by the equation $8P^2 + 4D = 84$. At this market price, the amount that growers are willing to supply is modeled by the equation $8P^2 + 6P - 2D = 48$. (a) What is the minimum price at which growers are willing to supply the organically grown carrots? (b) Use this information to create a system of nonlinear equations, then solve the system to find the market equilibrium price (per pound) and the quantity of carrots supplied and sold at this price. $2.11 $2.50   8500 lb $\begin{cases} 8P^2 + 4D = 84 \\ 8P^2 + 6P - 2D = 48 \end{cases}$

▶ **EXTENDING THE CONCEPT**

67. The area of a vertical parabolic segment is given by $A = \frac{2}{3}BH$, where B is the length of the horizontal base of the segment and H is the height from the base to the vertex. Investigate how this formula can be used to find the *area* of the solution region for the general system of inequalities shown.
$\begin{cases} y \geq x^2 - bx + c \\ y \leq c + bx - x^2 \end{cases}$ Answers will vary.

68. Find the area of the trapezoid formed by joining the points where the parabola $y = \frac{1}{2}x^2 - 26$ and the circle $x^2 + y^2 = 100$ intersect.   196 units²

69. A rectangular fish tank has a bottom and four sides made out of glass. Use a system of equations to help find the dimensions of the tank if the height is 18 in., surface area is 4806 in², the tank must hold 108 gal (1 gal = 231 in³), and all three dimensions are integers.   18 in. by 18 in. by 77 in.

Additional answers can be found in the Instructor Answer Appendix.

▶ MAINTAINING YOUR SKILLS

70. (R.4) Solve by factoring:

    a. $2x^2 + 5x - 63 = 0$ $x = -7, x = \frac{9}{2}$

    b. $4x^2 - 121 = 0$ $x = \frac{-11}{2}, x = \frac{11}{2}$

    c. $2x^3 - 3x^2 - 8x + 12 = 0$ $x = 2, x = -2, x = \frac{3}{2}$

71. (R.5/R.6) Solve each equation:

    a. $3x^2 + 4x - 12 = 0$ $x = \frac{-2 \pm 2\sqrt{10}}{3}$

    b. $\sqrt{3x + 1} - \sqrt{2x} = 1$ $x = 0, x = 8$

    c. $\dfrac{1}{x + 2} + \dfrac{3}{x^2 + 5x + 6} = \dfrac{2}{x + 3}$ $x = 2$

72. (6.2) Solve using any method. As an investment for retirement, Donovan bought three properties for a total of $250,000. Ten years later, the first property had doubled in value, the second property had tripled in value, and the third property was worth $10,000 less than when he bought it, for a current value of $485,000. Find the original purchase price if he paid $20,000 more for the first property than he did for the second. $95,000, $75,000, $80,000

73. (1.2) In 2005, a small business purchased a copier for $4500. In 2008, the value of the copier had decreased to $3300. Assuming the depreciation is linear: (a) find the rate of change $m = \dfrac{\Delta \text{value}}{\Delta \text{time}}$ and discuss its meaning in this context; (b) find the depreciation equation; and (c) use the equation to predict the copier's value in 2012. (d) If the copier is traded in for a new model when its value is less than $700, how long will the company use this copier?   **a.** $m = \frac{-400}{1}$, the copier depreciates by $400 a year   **b.** $y = -400x + 4500$ **c.** $1700   **d.** 9.5 yr

---

6.4   Systems of Inequalities and Linear Programming

LEARNING OBJECTIVES

In Section 6.4 you will see how we can:

☐ **A.** Solve a linear inequality in two variables

☐ **B.** Solve a system of linear inequalities

☐ **C.** Solve applications using a system of linear inequalities

☐ **D.** Solve applications using linear programming

In this section, we'll build on many of the ideas from Section 6.3, with a more direct focus on systems of linear inequalities. While systems of linear equations have an unlimited number of applications, there are many situations that can only be modeled using linear inequalities. For example, decisions in business and industry are often based on a large number of limitations or constraints, with many different ways these constraints can be satisfied.

A. Linear Inequalities in Two Variables

A linear equation in two variables is any equation that can be written in the form $Ax + By = C$, where A and B are real numbers, not simultaneously equal to zero. A **linear inequality** in two variables is similarly defined, with the " $=$ " sign replaced by the " $<$," " $>$," " $\leq$," or " $\geq$ " symbol:

$$Ax + By < C \qquad Ax + By > C$$
$$Ax + By \leq C \qquad Ax + By \geq C$$

Solving a linear inequality in two variables has many similarities with the one variable case. For one variable, we graph a *boundary point* on a number line, decide whether the endpoint is *included* or *excluded,* and *shade the appropriate half line.* For $x + 1 \leq 3$, we have the solution $x \leq 2$ with the endpoint (boundary point) included and the line shaded to the left (Figure 6.41):

Figure 6.42

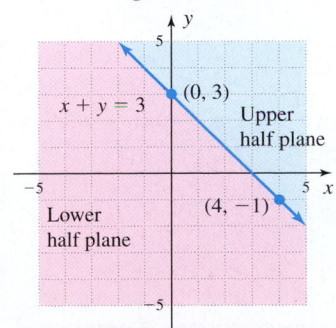

Figure 6.41

$-\infty$ ∞

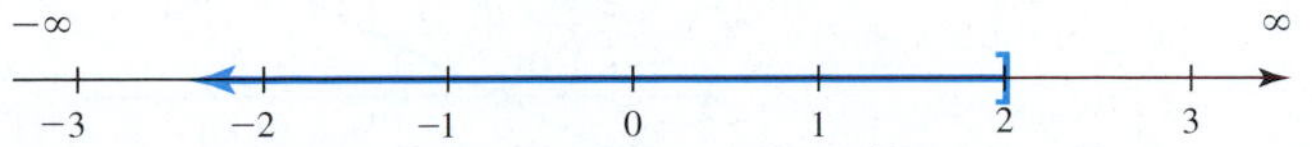

    -3 -2 -1 0 1 2 3

Interval notation: $x \in (-\infty, 2]$

For linear inequalities in two variables, we graph a *boundary line,* decide whether the boundary line is *included* or *excluded,* and *shade the appropriate half plane.* For $x + y \leq 3$, the boundary line $x + y = 3$ is graphed in Figure 6.42. Note it divides the coordinate plane into two regions called **half planes,** and it forms the **boundary** between the two regions. If the boundary is **included** in the solution set, we graph it using a *solid line*. If the boundary is **excluded,** a *dashed line* is used. Recall that solutions to a linear equation are ordered pairs that make the equation true. We use a similar idea to find or verify solutions to linear inequalities. If any one point in a half plane makes the inequality true, all points in that half plane will satisfy the inequality.

EXAMPLE 1 ▶   **Checking Solutions to an Inequality in Two Variables**

Determine whether the given ordered pairs are solutions to $-x + 2y \leq 2$:

    a. $(4, -3)$ **b.** $(-2, 1)$ **c.** $(-4, -1)$

Solution ▶   **a.** Substitute 4 for x and -3 for y: $-(4) + 2(-3) \leq 2$ substitute 4 for x, -3 for y

                                                  $-10 \leq 2$ true

      $(4, -3)$ is a solution.

    b. Substitute -2 for x and 1 for y: $-(-2) + 2(1) \leq 2$ substitute -2 for x, 1 for y

                                                      $4 \leq 2$ false

      $(-2, 1)$ is not a solution.

    c. Substitute -4 for x and -1 for y: $-(-4) + 2(-1) \leq 2$ substitute -4 for x, -1 for y

                                                     $2 \leq 2$ true

      $(-4, -1)$ is a solution.

Now try Exercises 7 through 10 ▶

WORTHY OF NOTE

This relationship is often called the **trichotomy axiom** or the *"three-part truth."* Given any two quantities, they are either equal to each other, or the first is less than the second, or the first is greater than the second.

Earlier we graphed linear equations by plotting a small number of ordered pairs or by solving for y and using the slope-intercept method. The line represented all ordered pairs that made the equation true, meaning *the left-hand expression was equal to the right-hand expression.* To graph linear inequalities, we reason that if the line represents all ordered pairs that make the expressions *equal,* then any point *not on that line* must make the expressions *unequal*—either greater than or less than. These ordered pair solutions must lie in one of the half planes formed by the line, which we shade to indicate the **solution region.** Note this implies the boundary line for any inequality *is determined by the related equation,* created by temporarily replacing the inequality symbol with an "=" sign.

EXAMPLE 2 ▶   **Solving an Inequality in Two Variables**

Solve the inequality $-x + 2y \leq 2$.

Solution ▶   The related equation and boundary line is $-x + 2y = 2$. Since the inequality is inclusive (less than *or equal to*), we graph a solid line. Using the intercepts, we graph the line through $(0, 1)$ and $(-2, 0)$ shown in Figure 6.43. To determine the solution region and which side to shade, we select $(0, 0)$ as a test point, which results in a true statement: $-(0) + 2(0) \leq 2$✓. Since $(0, 0)$ is in the "lower" half plane, we shade this side of the boundary (see Figure 6.44).

Figure 6.43

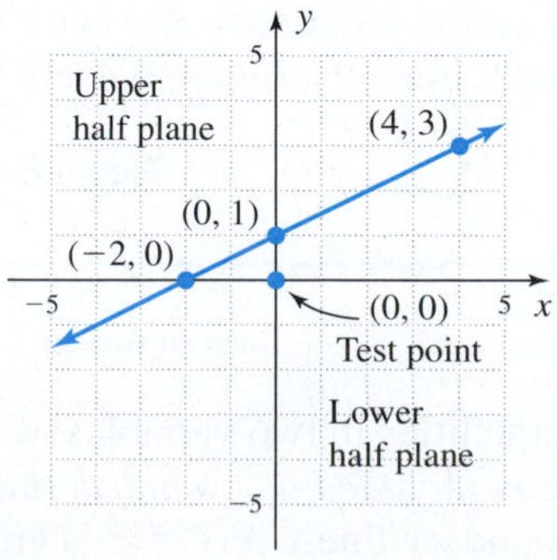

Figure 6.44

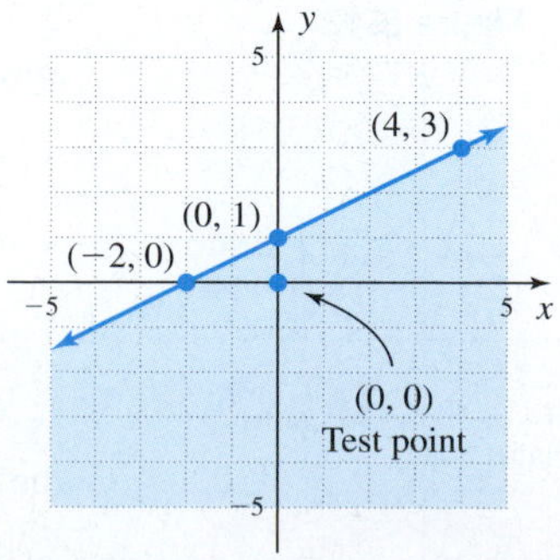

Now try Exercises 11 through 14 ▶

The same solution would be obtained if we first solved for y and graphed the boundary line using the slope-intercept method. However, using the slope-intercept method offers a distinct advantage—test points are no longer necessary since solutions to "less than" inequalities will always appear *below* the boundary line and solutions to "greater than" inequalities appear *above* the line. Written in slope-intercept form, the inequality from Example 2 is $y \leq \frac{1}{2}x + 1$. Note that $(0, 0)$ still results in a true statement, but the "less than or equal to" symbol now indicates directly that solutions will be found in the lower half plane. These observations lead to our general approach for solving linear inequalities:

> **Solving a Linear Inequality**
>
> **1.** Graph the boundary line by solving for y and using the slope-intercept form.
>   - Use a solid line if the boundary is included in the solution set.
>   - Use a dashed line if the boundary is excluded from the solution set.
> **2.** For "greater than" inequalities shade the upper half plane. For "less than" inequalities shade the lower half plane.

EXAMPLE 3 ▶ **Solving linear Inequalities in Two Variables Using Technology**

Solve the inequality $3x + 5y \leq 10$ using a graphing calculator.

Solution ▶ We begin by solving the inequality for y, so we can enter the equation on the $\boxed{Y=}$ screen.

$$3x + 5y \leq 10 \qquad \text{given inequality}$$
$$5y \leq -3x + 10 \qquad \text{subtract } 3x \text{ (isolate } y\text{-term)}$$
$$y \leq \frac{-3}{5}x + 2 \qquad \text{divide by 5}$$

Entering $Y_1 = \frac{-3}{5}X + 2$ on the $\boxed{Y=}$ screen, and graphing the boundary line using $\boxed{\text{ZOOM}}$ **6:ZStandard** produces the graph shown in Figure 6.45. The "less than" inequality indicates the region below the line should be shaded, so we return to the $\boxed{Y=}$ screen and move the cursor to the far left. From the default "connected line" setting, pressing $\boxed{\text{ENTER}}$ three times brings the "shade below the graph" marker "◣" into view (Figure 6.46), and pressing $\boxed{\text{GRAPH}}$ gives the solution shown in Figure 6.47. As a check, note that $(0, 0)$ is in the solution region, and is a solution to $3x + 5y \leq 10$.

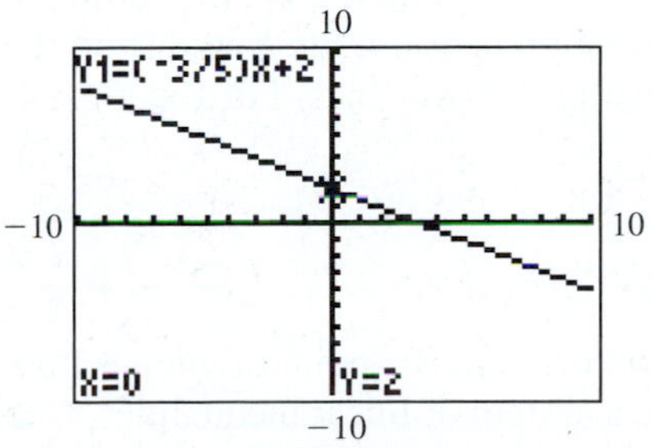

Figure 6.45

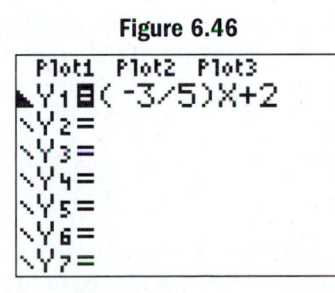

Figure 6.46

Figure 6.47

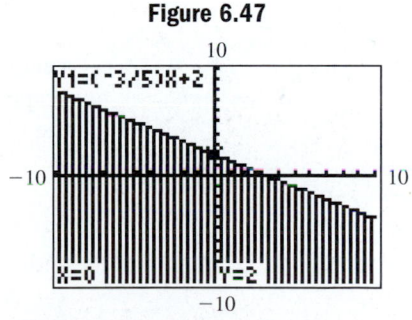

Now try Exercises 15 through 18 ▶

☑ **A.** You've just seen how we can solve a linear Inequality in two variables

B. Solving Systems of Linear Inequalities

To solve a **system of inequalities,** we apply the procedure outlined above to all inequalities in the system, and note the ordered pairs that satisfy *all inequalities simultaneously.* In other words, we find *the intersection of all solution regions* (where they overlap), which then represents the solution for the system. In the case of vertical boundary lines, the designations *"above"* or *"below" the line* cannot be applied, and instead we simply note that for any vertical line $x = k$, points with x-coordinates larger than k will occur to the right.

EXAMPLE 4 ▶   **Solving a System of Linear Inequalities**

Solve the system of inequalities: $\begin{cases} 2x + y \geq 4 \\ x - y < 2 \end{cases}$.

Solution ▶   Solving for y, we obtain $y \geq -2x + 4$ and $y > x - 2$. The line $y = -2x + 4$ will be a solid boundary line (included), while $y = x - 2$ will be dashed (not included). Both inequalities are "greater than" and so we shade the upper half plane for each. The regions overlap and form the solution region (the lavender region shown). This sequence of events is illustrated here:

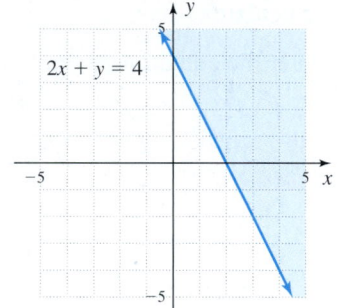

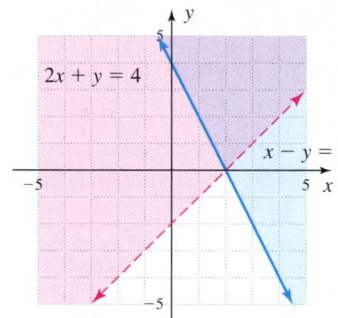

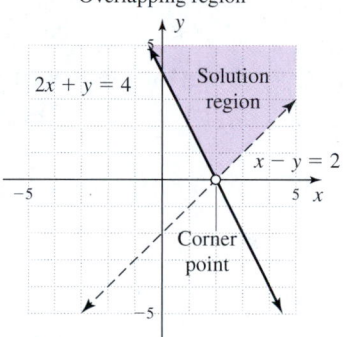

The solutions are all ordered pairs found in this region and its included boundaries. To verify the result, test the point $(2, 3)$ from inside the region, and $(5, -2)$ from outside the region [the point $(2, 0)$ is not a solution since it does not satisfy $x - y < 2$].

Now try Exercises 19 through 36 ▶

For future reference, the point of intersection $(2, 0)$ is called a **corner point** or **vertex** of the solution region. If the point of intersection is not easily found from the graph, we can find it by solving a linear system using the two lines. For Example 4, the system is

$$\begin{cases} 2x + y = 4 \\ x - y = 2 \end{cases}$$

and solving by elimination gives $3x = 6$, $x = 2$, and $(2, 0)$ as the point of intersection.

A graphing calculator can also be used to solve a system of linear inequalities. One method (there are several) involves these three steps, which are performed on each equation:

1. Solve for y and enter the results on the ⟨Y=⟩ screen to create the boundary lines.
2. Graph each line and test the related half plane.
3. Shade the appropriate half plane.

This process is illustrated in Example 5. Since many real-world applications of linear inequalities do not use negative numbers, we often include $x \geq 0$ and $y \geq 0$ as part of the system, and set **Xmin = 0** and **Ymin = 0** as part of the (WINDOW) size. The value of Xmax and Ymax will depend on equations or context given.

EXAMPLE 5 ▶ Solving a System of Inequalities Using Technology

Use a graphing calculator to solve the system: $\begin{cases} 3x + 2y < 14 \\ x + 2y < 8 \\ x \geq 0 \\ y \geq 0 \end{cases}$

Solution ▶ Following the steps outlined above, we have

1. *Enter the related equations.* For $3x + 2y = 14$, we have $y = -1.5x + 7$. For $x + 2y = 8$, we have $y = -0.5x + 4$. Enter these as Y_1 and Y_2 on the (Y=) screen.

2. *Graph the boundary lines.* Note the x- and y-intercepts of both lines are less than 10, so we can graph them using a "friendly window" where $x \in [0, 9.4]$ and $y \in [0, 6.2]$. After setting the window, press (GRAPH) to graph the lines.

3. *Shade the appropriate half plane.* Since both equations are in slope-intercept form and solving for y resulted in two "less than" inequalities, we shade *below* both lines, using the "◤" feature located to the far left of Y_1 and Y_2. Simply overlay the diagonal line and press (ENTER) repeatedly until the symbol appears (Figure 6.48). After pressing the GRAPH key, the calculator draws both lines and shades the appropriate regions (Figure 6.49). Note the calculator uses two different kinds of shading. This makes it easy to identify the solution region—it will be the "checker-board area" where the horizontal and vertical lines cross.

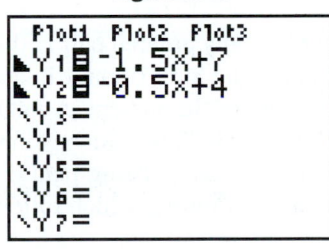

Figure 6.48

Figure 6.49

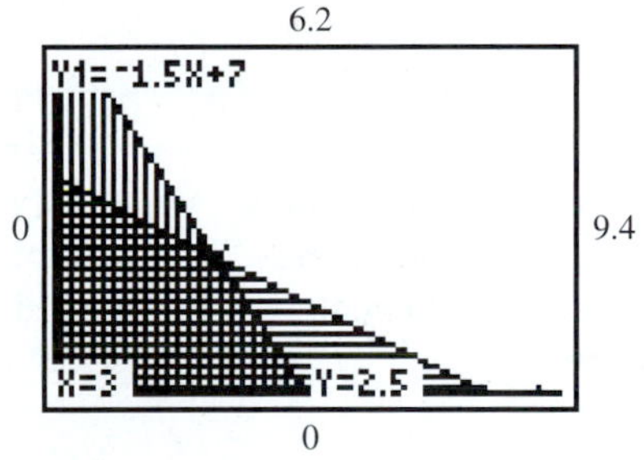

As a final check, we can navigate the position marker into the solution region and test a few points in both inequalities. Using the test point $(2, 2)$ from within the region yields

$3x + 2y < 14$	first inequality	$x + 2y < 8$	second inequality
$3(2) + 2(2) \leq 14$	substitute 2 for x, 2 for y	$(2) + 2(2) \leq 8$	substitute 2 for x, 2 for y
$10 \leq 14$	true	$6 \leq 8$	true

☑ **B.** You've just seen how we can solve a system of linear inequalities

Now try Exercises 37 through 50 ▶

C. Applications of Systems of Linear Inequalities

Systems of inequalities give us a way to model the decision-making process when certain **constraints** must be satisfied. A constraint is a fact or consideration that somehow limits or governs possible solutions, like the number of acres a farmer plants—which may be limited by time, size of land, government regulations, and so on.

EXAMPLE 6 ▶ **Solving Applications of Linear Inequalities**

As part of their retirement planning, James and Lily decide to invest up to $30,000 in two separate investment vehicles. The first is a bond issue paying 9% and the second is a money market certificate paying 5%. A financial adviser suggests they invest at least $10,000 in the certificate and not more than $15,000 in bonds. What various amounts can be invested in each?

Solution ▶ Consider the ordered pairs (B, C) where B represents the money invested in bonds and C the money invested in the certificate. Since they plan to invest no more than $30,000, the investment constraint would be $B + C \leq 30$ (in thousands). Following the adviser's recommendations, the constraints on each investment would be $B \leq 15$ and $C \geq 10$. Since they cannot invest less than zero dollars, the last two constraints are $B \geq 0$ and $C \geq 0$.

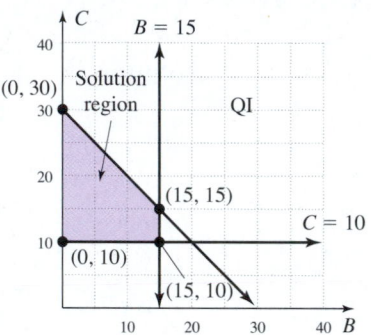

$$\begin{cases} B + C \leq 30 \\ B \leq 15 \\ C \geq 10 \\ B \geq 0 \\ C \geq 0 \end{cases}$$

The resulting system is shown in the figure, and indicates solutions will be in the first quadrant.

There is a vertical boundary line at $B = 15$ with shading to the left (less than) and a horizontal boundary line at $C = 10$ with shading above (greater than). After graphing $C = 30 - B$, we see the solution region is a quadrilateral with vertices at $(0, 10)$, $(0, 30)$, $(15, 10)$, and $(15, 15)$, as shown.

Now try Exercises 61 and 62 ▶

✓ **C. You've just seen how we can solve applications using a system of linear inequalities**

From Example 6, any ordered pair in this region or on its boundaries would represent an investment of the form (money in bonds, money in CDs) → (B, C), and would satisfy all constraints in the system. A natural follow-up question would be—What combination of (money in bonds, money in CDs) would offer the greatest return? This would depend on the interest being paid on each investment, and introduces us to a study of **linear programming,** which follows soon.

D. Linear Programming

To become as profitable as possible, corporations look for ways to maximize their revenue and minimize their costs, while keeping up with delivery schedules and product demand. To operate at peak efficiency, plant managers must find ways to maximize productivity, while minimizing related costs and considering employee welfare, union agreements, and other factors. Problems where the goal is to **maximize** or **minimize** the value of a given quantity under certain **constraints** or restrictions are called programming problems. The quantity we seek to maximize or minimize is called the **objective function.** For situations where *linear* programming is used, the objective function is given as a linear function in two variables and is denoted $f(x, y)$. A function in two variables is evaluated in much the same way as a single variable function. To evaluate $f(x, y) = 2x + 3y$ at the point $(4, 5)$, we substitute 4 for x and 5 for y: $f(4, 5) = 2(4) + 3(5) = 23$.

EXAMPLE 7 ▶ Determining Maximum Values

Determine which of the following ordered pairs maximizes the value of $f(x, y) = 5x + 4y$: $(0, 6)$, $(5, 0)$, $(0, 0)$, or $(4, 2)$.

Solution ▶ Organizing our work in table form gives

Given Point	Evaluate $f(x, y) = 5x + 4y$
$(0, 6)$	$f(0, 6) = 5(0) + 4(6) = 24$
$(5, 0)$	$f(5, 0) = 5(5) + 4(0) = 25$
$(0, 0)$	$f(0, 0) = 5(0) + 4(0) = 0$
$(4, 2)$	$f(4, 2) = 5(4) + 4(2) = 28$

The function $f(x, y) = 5x + 4y$ is maximized at $(4, 2)$.

Now try Exercises 51 through 54 ▶

Figure 6.50

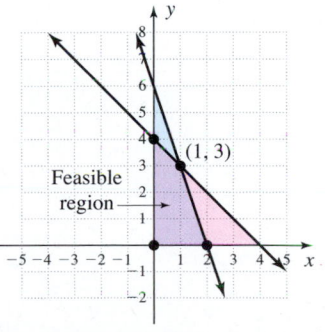

Convex   Not convex

When the objective is stated as a linear function in two variables and the constraints are expressed as a system of linear inequalities, we have what is called a **linear programming** problem. The systems of inequalities solved earlier produced solution regions that were either **bounded** (as in Example 6) or **unbounded** (as in Example 4). We interpret the word *bounded* to mean we can enclose the solution region within a circle of appropriate size. If we cannot draw a circle around the region because it extends indefinitely in some direction, the region is said to be *unbounded*. In this study, we will consider only situations that produce bounded solution regions, meaning the regions will have three or more vertices. The regions we study will also be **convex,** meaning that for any two points in the enclosed region, the line segment between them is also in the region (Figure 6.50). Under these conditions, it can be shown that the maximum or minimum values *must occur at one of the corner points of the solution region,* also called the **feasible region.**

EXAMPLE 8 ▶ Finding the Maximum of an Objective Function

Find the maximum value of the objective function $f(x, y) = 2x + y$ given the

constraints shown: $\begin{cases} x + y \leq 4 \\ 3x + y \leq 6 \\ x \geq 0 \\ y \geq 0 \end{cases}$.

Solution ▶ Begin by noting that the solutions must be in QI, since $x \geq 0$ and $y \geq 0$. Graph the boundary lines $y = -x + 4$ and $y = -3x + 6$, shading the lower half plane in each case since they are "less than" inequalities. This produces the feasible region shown in lavender. There are four corner points to this region: $(0, 0)$, $(0, 4)$, $(2, 0)$, and $(1, 3)$. Three of these points are intercepts and can be found quickly. The point $(1, 3)$ was found by solving the system $\begin{cases} x + y = 4 \\ 3x + y = 6 \end{cases}$. Knowing that the objective function will be maximized at one of the corner points, we test them in the objective function, using a table to organize our work.

Corner Point	Objective Function $f(x, y) = 2x + y$
$(0, 0)$	$f(0, 0) = 2(0) + (0) = 0$
$(0, 4)$	$f(0, 4) = 2(0) + (4) = 4$
$(2, 0)$	$f(2, 0) = 2(2) + (0) = 4$
$(1, 3)$	$f(1, 3) = 2(1) + (3) = 5$

The objective function $f(x, y) = 2x + y$ is maximized at $(1, 3)$.

Now try Exercises 55 through 58 ▶

To help understand why solutions must occur at a vertex, note the objective function $f(x, y)$ is maximized using only (x, y) ordered pairs from the feasible region. If we let K represent this maximum value, the function from Example 8 becomes $K = 2x + y$ or $y = -2x + K$, which is a line with slope -2 and y-intercept K. The table in Example 8 suggests that K should range from 0 to 5 and graphing $y = -2x + K$ for $K = 1$, $K = 3$, and $K = 5$ produces the family of parallel lines shown in Figure 6.51. Note that values of K larger than 5 will cause the line to miss the solution region, and the maximum value of 5 occurs where the line intersects the feasible region at the vertex $(1, 3)$. These observations lead to the following principles, which we offer without a formal proof.

Figure 6.51

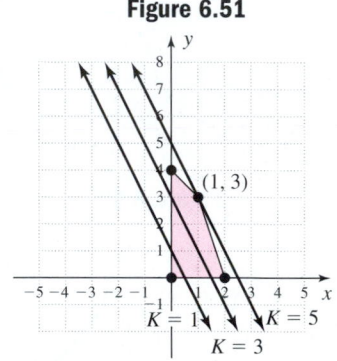

Linear Programming Solutions

1. If the feasible region is convex and bounded, a maximum and a minimum value exist.
2. If a unique solution exists, it will occur at a vertex of the feasible region.
3. If more than one solution exists, at least one of them occurs at a vertex of the feasible region with others on a boundary line.
4. If the feasible region is unbounded, a linear programming problem may have no solutions.

Solving linear programming problems depends in large part on two things: (1) identifying the **objective** and the **decision variables** (what each variable represents in context), and (2) using the decision variables to write the *objective function* and **constraint inequalities.** This brings us to our five-step approach for solving linear programming applications.

Solving Linear Programming Applications

1. Identify the main objective and the decision variables (descriptive variables may help) and write the objective function in terms of these variables.
2. Organize all information in a table, with the *decision variables* and *constraints* heading up the columns, and their *components* leading each row.
3. Complete the table using the information given, and write the constraint inequalities using the decision variables, constraints, and the domain.
4. Graph the constraint inequalities, determine the feasible region, and identify all corner points.
5. Test these points in the objective function to determine the optimal solution(s).

EXAMPLE 9 ▶   **Solving an Application of Linear Programming**

The owner of a snack food business wants to create two nut mixes for the holiday season. The regular mix will have 14 oz of peanuts and 4 oz of cashews, while the deluxe mix will have 12 oz of peanuts and 6 oz of cashews. The owner estimates he will make a profit of $3 on the regular mixes and $4 on the deluxe mixes. How many of each should be made in order to maximize profit, if only 840 oz of peanuts and 348 oz of cashews are available?

Solution ▶   Our *objective* is to maximize profit, and the *decision variables* could be r to represent the regular mixes sold, and d for the number of deluxe mixes. This gives $P(r, d) = \$3r + \$4d$ as our *objective function*. The information is organized in Table 6.1, using the variables r, d, and the constraints to head each column. Since the mixes are composed of peanuts and cashews, these lead the rows in the table.

Table 6.1

$$P(r, d) = \$3r \quad + \quad \$4d$$
$$\downarrow \qquad\qquad \downarrow$$

	Regular r	Deluxe d	Constraints: Total Ounces Available
Peanuts	14	12	840
Cashews	4	6	348

After filling in the appropriate values, reading the table from left to right along the "peanut" row and the "cashew" row, gives the constraint inequalities $14r + 12d \le 840$ and $4r + 6d \le 348$. Realizing we won't be making negative numbers of mixes, the remaining constraints are $r \ge 0$ and $d \ge 0$. The complete system is

$$\begin{cases} 14r + 12d \le 840 \\ 4r + 6d \le 348 \\ r \ge 0 \\ d \ge 0 \end{cases}$$

Note once again that the solutions must be in QI, since $r \ge 0$ and $d \ge 0$. Graphing the first two inequalities using slope-intercept form gives $d \le -\frac{7}{6}r + 70$ and $d \le -\frac{2}{3}r + 58$ producing the feasible region shown in lavender. The four corner points are $(0, 0)$, $(60, 0)$, $(0, 58)$, and $(24, 42)$. Three of these points are intercepts and can be read from a table of values or the graph itself. The point $(24, 42)$ was found by solving the system $\begin{cases} 14r + 12d = 840 \\ 4r + 6d = 348 \end{cases}$. Knowing the solution must occur at one of these points, we test them in the objective function (Table 6.2).

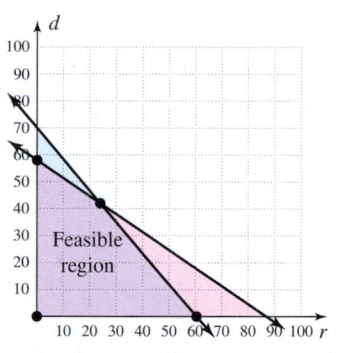

Table 6.2

Corner Point	Objective Function $P(r, d) = \$3r + \$4d$
$(0, 0)$	$P(0, 0) = \$3(0) + \$4(0) = \$0$
$(60, 0)$	$P(60, 0) = \$3(60) + \$4(0) = \$180$
$(0, 58)$	$P(0, 58) = \$3(0) + \$4(58) = \$232$
$(24, 42)$	$P(24, 42) = \$3(24) + \$4(42) = \$240$

Profit will be maximized if 24 boxes of the regular mix and 42 boxes of the deluxe mix are made and sold.

Now try Exercises 63 through 68 ▶

Linear programming can also be used to minimize an objective function, as in Example 10.

EXAMPLE 10 ▶ **Minimizing Costs Using Linear Programming**

A beverage producer needs to minimize shipping costs from its two primary plants in Kansas City (KC) and St. Louis (STL). All wholesale orders within the state are shipped from one of these plants. An outlet in Macon orders 200 cases of soft drinks on the same day an order for 240 cases comes from Springfield. The plant in KC has 300 cases ready to ship and the plant in STL has 200 cases. The cost of shipping each case to Macon is $0.50 from KC, and $0.70 from STL. The cost of shipping each case to Springfield is $0.60 from KC, and $0.65 from STL. How many cases should be shipped from each warehouse to minimize costs?

Solution ▶ Our *objective* is to minimize costs, which depends on the number of cases shipped from each plant. To begin we use the following assignments:

$$A \rightarrow \text{cases shipped from KC to Macon}$$
$$B \rightarrow \text{cases shipped from KC to Springfield}$$
$$C \rightarrow \text{cases shipped from STL to Macon}$$
$$D \rightarrow \text{cases shipped from STL to Springfield}$$

From this information, the equation for total cost T is

$$T = 0.5A + 0.6B + 0.7C + 0.65D,$$

an equation in *four* variables. To make the cost equation more manageable, note since Macon ordered 200 cases, $A + C = 200$. Similarly, Springfield ordered 240 cases, so $B + D = 240$. After solving for C and D, respectively, these equations enable us to substitute for C and D, resulting in an equation with just two variables. For $C = 200 - A$ and $D = 240 - B$ we have

$$T(A, B) = 0.5A + 0.6B + 0.7(200 - A) + 0.65(240 - B)$$
$$= 0.5A + 0.6B + 140 - 0.7A + 156 - 0.65B$$
$$= 296 - 0.2A - 0.05B$$

The constraints involving the KC plant are $A + B \le 300$ with $A \ge 0, B \ge 0$. The constraints for the STL plant are $C + D \le 200$ with $C \ge 0, D \ge 0$. Since we want a system in terms of A and B only, we again substitute $C = 200 - A$ and $D = 240 - B$ in all the STL inequalities:

$C + D \le 200$	STL inequalities	$C \ge 0$	$D \ge 0$
$(200 - A) + (240 - B) \le 200$	substitute $200 - A$ for C, $240 - B$ for D	$200 - A \ge 0$	$240 - B \ge 0$
		$200 \ge A$	$240 \ge B$
$440 - A - B \le 200$	simplify		
$240 \le A + B$	result		

Combining the new STL constraints with those from KC produces the following system and solution. All points of intersection were read from the graph or located using the related system of equations.

$$\begin{cases} A + B \le 300 \\ A + B \ge 240 \\ A \le 200 \\ B \le 240 \\ A \ge 0 \\ B \ge 0 \end{cases}$$

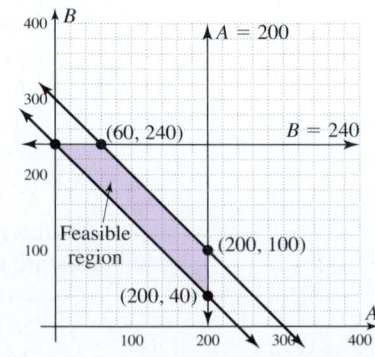

To find the minimum cost, we check each vertex in the objective function.

Vertices	Objective Function $T(A, B) = 296 - 0.2A - 0.05B$
(0, 240)	$P(0, 240) = 296 - 0.2(0) - 0.05(240) = \$284$
(60, 240)	$P(60, 240) = 296 - 0.2(60) - 0.05(240) = \$272$
(200, 100)	$P(200, 100) = 296 - 0.2(200) - 0.05(100) = \$251$
(200, 40)	$P(200, 40) = 296 - 0.2(200) - 0.05(40) = \$254$

The minimum cost occurs when $A = 200$ and $B = 100$, meaning the producer should ship the following quantities:

$A \rightarrow$ cases shipped from KC to Macon $= 200$
$B \rightarrow$ cases shipped from KC to Springfield $= 100$
$C \rightarrow$ cases shipped from STL to Macon $= 0$
$D \rightarrow$ cases shipped from STL to Springfield $= 140$

☑ **D.** You've just seen how we can solve applications using linear programming

Now try Exercises 69 and 70 ▶

6.4 EXERCISES

▶ CONCEPTS AND VOCABULARY

Fill in the blank with the appropriate word or phrase. Carefully reread the section if needed.

1. Any line $y = mx + b$ drawn in the coordinate plane divides the plane into two regions called ___half___ ___planes___.

2. For the line $y = mx + b$ drawn in the coordinate plane, solutions to $y > mx + b$ are found in the region ___above___ the line.

3. The overlapping region of two or more linear inequalities in a system is called the ___solution___ region.

4. If a linear programming problem has a unique solution (x, y), it must be a ___vertex___ of the feasible region.

5. Suppose two boundary lines in a system of linear inequalities intersect, but the point of intersection is not a vertex of the feasible region. Describe how this is possible.

6. Describe the conditions necessary for a linear programming problem to have multiple solutions. (*Hint:* Consider the diagram in Figure 6.51, and the slope of the line from the objective function.)

▶ DEVELOPING YOUR SKILLS

Determine whether the ordered pairs given are solutions.

7. $2x + y > 3$; $(0, 0), (3, -5), (-3, -4), (-3, 9)$

8. $3x - y > 5$; $(0, 0), (4, -1), (-1, -5), (1, -2)$

9. $4x - 2y \leq -8$; $(0, 0), (-3, 5), (-3, -2), (-1, 1)$

10. $3x + 5y \geq 15$; $(0, 0), (3, 5), (-1, 6), (7, -3)$

Solve the linear inequalities by shading the appropriate half plane. Verify your answer using a graphing calculator.

11. $x + 2y < 8$

12. $x - 3y > 6$

13. $2x - 3y \geq 9$

14. $4x + 5y \geq 15$

HOMEWORK SELECTION GUIDE   Additional answers can be found in the Instructor Answer Appendix.

Core: 7–23 odd, 25–45 every other odd, 47–57 odd, 61–69 odd, 73 (27 Exercises)
Standard: 1–4, 7–23 odd, 25–45 every other odd, 47–57 odd, 59, 61–69 odd, 71, 73, 75 (34 Exercises)

Extended: 1–4, 7–23 odd, 25–45 every other odd, 47–57 odd, 59, 60, 61–69 odd, 71–73, 75, 76 (37 Exercises)
In Depth: 1–6, 7–23 odd, 25–45 every other odd, 47–57 odd, 59, 60, 61–69 odd, 71–76 (40 Exercises)

Solve the following inequalities using a graphing calculator. Your answer should include a screen shot or facsimile, and comments regarding a test point.

15. $3x + 2y \leq 8$

16. $2x + 5y \geq 10$

17. $4x - 5y > -20$

18. $6x - 3y < 18$

Determine whether the ordered pairs given are solutions to the accompanying system.

19. $\begin{cases} 5y - x \geq 10 \\ 5y + 2x \leq -5 \end{cases}$;
$(-2, 1), (-5, -4), (-6, 2), (-8, 2.2)$ No, No, No, Yes

20. $\begin{cases} 8y + 7x \geq 56 \\ 3y - 4x \geq -12 \\ y \geq 4 \end{cases}$; $(1, 5), (4, 6), (8, 5), (5, 3)$ No, Yes, No, No

Solve each system of inequalities by graphing the solution region. Verify the solution using a test point.

21. $\begin{cases} x + 2y \geq 1 \\ 2x - y \leq -2 \end{cases}$

22. $\begin{cases} -x + 5y < 5 \\ x + 2y \geq 1 \end{cases}$

23. $\begin{cases} 3x + y > 4 \\ x > 2y \end{cases}$

24. $\begin{cases} 3x \leq 2y \\ y \geq 4x + 3 \end{cases}$

25. $\begin{cases} 2x + y < 4 \\ 2y > 3x + 6 \end{cases}$

26. $\begin{cases} x - 2y < -7 \\ 2x + y > 5 \end{cases}$

27. $\begin{cases} x > -3y - 2 \\ x + 3y \leq 6 \end{cases}$

28. $\begin{cases} 2x - 5y < 15 \\ 3x - 2y > 6 \end{cases}$

29. $\begin{cases} 5x + 4y \geq 20 \\ x - 1 \geq y \end{cases}$

30. $\begin{cases} 10x - 4y \leq 20 \\ 5x - 2y > -1 \end{cases}$

31. $\begin{cases} 0.2x > -0.3y - 1 \\ 0.3x + 0.5y \leq 0.6 \end{cases}$

32. $\begin{cases} x > -0.4y - 2.2 \\ x + 0.9y \leq -1.2 \end{cases}$

33. $\begin{cases} y \leq \dfrac{3}{2}x \\ 4y \geq 6x - 12 \end{cases}$

34. $\begin{cases} 3x + 4y > 12 \\ y < \dfrac{2}{3}x \end{cases}$

35. $\begin{cases} \dfrac{-2}{3}x + \dfrac{3}{4}y \leq 1 \\ \dfrac{1}{2}x + 2y \geq 3 \end{cases}$

36. $\begin{cases} \dfrac{1}{2}x + \dfrac{2}{5}y \leq 5 \\ \dfrac{5}{6}x - 2y \geq -5 \end{cases}$

37. $\begin{cases} x - y \geq -4 \\ 2x + y \leq 4 \\ x \geq 0, y \geq 0 \end{cases}$

38. $\begin{cases} 2x - y \leq 5 \\ x + 3y \leq 6 \\ x \geq 0, y \geq 0 \end{cases}$

39. $\begin{cases} y \leq x + 3 \\ x + 2y \leq 4 \\ x \geq 0, y \geq 0 \end{cases}$

40. $\begin{cases} 4y < 3x + 12 \\ y \leq x + 1 \\ x \geq 0, y \geq 0 \end{cases}$

41. $\begin{cases} 2x + 3y \leq 18 \\ 2x + y \leq 10 \\ x \geq 0, y \geq 0 \end{cases}$

42. $\begin{cases} 8x + 5y \leq 40 \\ x + y \leq 7 \\ x \geq 0, y \geq 0 \end{cases}$

 Use a graphing calculator to find the solution region for each system of linear inequalities. Your answer should include a screen shot or facsimile, and the location of any points of intersection.

43. $\begin{cases} y + 2x < 8 \\ y + x < 6 \\ x \geq 0 \\ y \geq 0 \end{cases}$

44. $\begin{cases} x + 2y < 10 \\ x + y < 7 \\ x \geq 0 \\ y \geq 0 \end{cases}$

45. $\begin{cases} -2x - y > -8 \\ -x - 2y < -7 \\ x \geq 0 \\ y \geq 0 \end{cases}$

46. $\begin{cases} y + 2x \geq 10 \\ 2y + x \leq 11 \\ x \geq 0 \\ y \geq 0 \end{cases}$

Use the equations given to write the system of linear inequalities represented by each graph.

47.
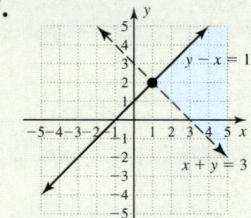

48.
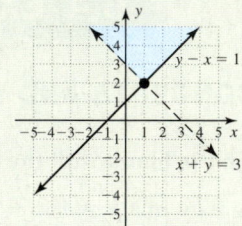

49.
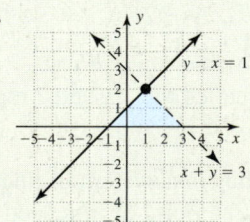

50.
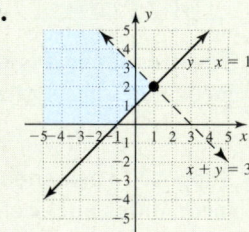

Determine which of the ordered pairs given produces the maximum value of $f(x, y)$.

51. $f(x, y) = 12x + 10y$; $(0, 0), (0, 8.5), (7, 0), (5, 3)$ (5, 3)

52. $f(x, y) = 50x + 45y$; $(0, 0), (0, 21), (15, 0),$
$(7.5, 12.5)$ (0, 21)

Determine which of the ordered pairs given produces the minimum value of $f(x, y)$.

53. $f(x, y) = 8x + 15y$; $(0, 20), (35, 0), (5, 15),$
$(12, 11)$ (12, 11)

54. $f(x, y) = 75x + 80y$; $(0, 9), (10, 0), (4, 5), (5, 4)$
   (5, 4)

Additional answers can be found in the Instructor Answer Appendix.

For Exercises 55 and 56, find the *maximum* value of the objective function $f(x, y) = 8x + 5y$ and where this value occurs, given the constraints shown.

55. $\begin{cases} x + 2y \leq 6 \\ 3x + y \leq 8 \\ x \geq 0 \\ y \geq 0 \end{cases}$ 26 at (2, 2) **56.** $\begin{cases} 2x + y \leq 7 \\ x + 2y \leq 5 \\ x \geq 0 \\ y \geq 0 \end{cases}$ 35 at (3, 1)

For Exercises 57 and 58, find the *minimum* value of the objective function $f(x, y) = 36x + 40y$ and where this value occurs, given the constraints shown.

57. $\begin{cases} 3x + 2y \geq 18 \\ 3x + 4y \geq 24 \\ x \geq 0 \\ y \geq 0 \end{cases}$ 264 at (4, 3) **58.** $\begin{cases} 2x + y \geq 10 \\ x + 4y \geq 3 \\ x \geq 2 \\ y \geq 0 \end{cases}$ 180 at (5, 0)

▶ WORKING WITH FORMULAS

Area Formulas

59. The area of a triangle is usually given as $A = \frac{1}{2}BH$, where B and H represent the base and height, respectively. The area of a rectangle can be stated as $A = BH$. If the base of both a triangle and rectangle is equal to 20 in., what are the possible values for H if the triangle must have an area *greater than* 50 in^2 and the rectangle must have an area *less than* 200 in^2? $5 < H < 10$

Volume Formulas

60. The volume of a cone is $V = \frac{1}{3}\pi r^2 h$, where r is the radius of the base and h is the height. The volume of a cylinder is $V = \pi r^2 h$. If the radius of both a cone and cylinder is equal to 10 cm, what are the possible values for h if the cone must have a volume *greater than* 200 cm^3 and the volume of the cylinder must be *less than* 850 cm^3? $\frac{6}{\pi} < h < \frac{17}{2\pi}$

▶ APPLICATIONS

Write a system of linear inequalities that models the information given, then solve. Verify the solution region using a graphing calculator.

61. Gifts to grandchildren: Grandpa Augustus is considering how to divide a $50,000 gift between his two grandchildren, Julius and Anthony. After weighing their respective positions in life and family responsibilities, he decides he must bequeath at least $20,000 to Julius, but no more than $25,000 to Anthony. Determine the possible ways that Grandpa can divide the $50,000.

62. Guns versus butter: Every year, governments around the world have to make the decision as to how much of their revenue must be spent on national defense and domestic improvements (guns versus butter). Suppose total revenue for these two needs was $120 billion, and a government decides they need to spend at least $42 billion on butter and no more than $80 billion on defense. Determine the possible amounts that can go toward each need.

Solve the following applications of linear programming.

63. Land/crop allocation: A farmer has 500 acres of land to plant corn and soybeans. During the last few years, market prices have been stable and the farmer anticipates a profit of $900 per acre on the corn harvest and $800 per acre on the soybeans. The farmer must take into account the time it takes

to plant and harvest each crop, which is 3 hr/acre for corn and 2 hr/acre for soybeans. If the farmer has at most 1300 hr to plant, care for, and harvest each crop, how many acres of each crop should be planted in order to maximize profits?

64. Coffee blends: The owner of a coffee shop has decided to introduce two new blends of coffee in order to attract new customers—a *Deluxe Blend* and a *Savory Blend.* Each pound of the deluxe blend contains 30% Colombian and 20% Arabian coffee, while each pound of the savory blend contains 35% Colombian and 15% Arabian coffee (the remainder of each is made up of cheap and plentiful domestic varieties). The profit on the deluxe blend will be $1.25 per pound, while the profit on the savory blend will be $1.40 per pound. How many pounds of each should the owner make in order to maximize profit, if only 455 lb of Colombian coffee and 250 lb of Arabian coffee are currently available?

65. Manufacturing screws: A machine shop manufactures two types of screws—sheet metal screws and wood screws, using three different machines. Machine Moe can make a sheet metal screw in 20 sec and a wood screw in 5 sec. Machine Larry can make a sheet metal screw in 5 sec and a wood screw in 20 sec. Machine Curly, the newest machine (nyuk, nyuk) can make a sheet metal screw in 15 sec and a wood screw in 15 sec.

(Shemp couldn't get a job because he failed the math portion of the employment exam.) Each machine can operate for only 3 hr each day before shutting down for maintenance. If sheet metal screws sell for 10 cents and wood screws sell for 12 cents, how many of each type should the machines be programmed to make in order to maximize revenue? (*Hint:* Standardize time units.)

66. **Hauling hazardous waste:** A waste disposal company is contracted to haul away some hazardous waste material. A full container of liquid waste weighs 800 lb and has a volume of 20 ft^3. A full container of solid waste weighs 600 lb and has a volume of 30 ft^3. The trucks used can carry at most 10 tons (20,000 lb) and have a carrying volume of 800 ft^3. If the trucking company makes $300 for disposing of liquid waste and $400 for disposing of solid waste, what is the maximum revenue per truck that can be generated?

67. **Maximizing profit—food service:** P. Barrett & Justin, Inc., is starting up a fast-food restaurant specializing in peanut butter and jelly sandwiches. Some of the peanut butter varieties are smooth, crunchy, reduced fat, and reduced sugar. The jellies will include those expected and common, as well as some exotic varieties such as kiwi and mango. Independent research has determined the two most popular sandwiches will be the traditional P&J (smooth peanut butter and grape jelly), and the Double-T (three slices of bread). A traditional P&J uses 2 oz of peanut butter and 3 oz of jelly. The Double-T uses 4 oz of peanut butter and 5 oz of jelly. The traditional sandwich will be priced at $2.00, and a Double-T at $3.50. If the restaurant has 250 oz of smooth peanut butter and 345 oz of grape jelly on hand for opening day, how many of each should they make and sell to maximize revenue?   65 traditionals, 30 Double-T's

68. **Maximizing profit—construction materials:** Mooney and Sons produces and sells two varieties of concrete mixes. The mixes are packaged in 50-lb bags. Type A is appropriate for finish work, and contains 20 lb of cement and 30 lb of sand. Type B is appropriate for foundation and footing work, and contains 10 lb of cement and 20 lb of sand. The remaining weight comes from gravel aggregate. The profit on type A is $1.20/bag, while the profit on type B is $0.90/bag. How many bags of each should the company make to maximize profit, if 2750 lb of cement and 4500 lb of sand are currently available?   225 bags of type B, 0 bags of type A

69. **Minimizing transportation costs:** Robert's Las Vegas Tours needs to drive 375 people and 19,450 lb of luggage from Salt Lake City, Utah, to Las Vegas, Nevada, and can charter buses from two companies. The buses from company X carry 45 passengers and 2750 lb of luggage at a cost of $1250 per trip. Company Y offers buses that carry 60 passengers and 2800 lb of luggage at a cost of $1350 per trip. How many buses should be chartered from each company in order for Robert to minimize the cost?

70. **Minimizing shipping costs:** An oil company is trying to minimize shipping costs from its two primary refineries in Tulsa, Oklahoma, and Houston, Texas. All orders within the region are shipped from one of these two refineries. An order for 220,000 gal comes in from a location in Colorado, and another for 250,000 gal from a location in Mississippi. The Tulsa refinery has 320,000 gal ready to ship, while the Houston refinery has 240,000 gal. The cost of transporting each gallon to Colorado is $0.05 from Tulsa and $0.075 from Houston. The cost of transporting each gallon to Mississippi is $0.06 from Tulsa and $0.065 from Houston. How many gallons should be distributed from each refinery to minimize the cost of filling both orders?

▶ **EXTENDING THE CONCEPT**

71. Graph the feasible region formed by the system
$$\begin{cases} x \geq 0 \\ y \geq 0 \\ y \leq 3 \\ x \leq 3 \end{cases}$$ (a) How would you describe this region?
(b) Select random points within the region or on any boundary line and evaluate the objective function $f(x, y) = 4.5x + 7.2y$. At what point (x, y) will this function be maximized? (c) How does this relate to optimal solutions to a linear programing problem?

72. Find the maximum value of the objective function $f(x, y) = 22x + 15y$ given the constraints
$$\begin{cases} 2x + 5y \leq 24 \\ 3x + 4y \leq 29 \\ x + 6y \leq 26 \\ x \geq 0 \\ y \geq 0 \end{cases}$$ max: 212.$\overline{6}$ at $\left(\frac{29}{3}, 0\right)$

Additional answers can be found in the Instructor Answer Appendix.

► **MAINTAINING YOUR SKILLS**

73. **(R.4)** Find all solutions (real and complex) by factoring: $x^3 - 5x^2 + 3x - 15 = 0$. $x = 5, \pm i\sqrt{3}$

74. **(4.6)** Solve the rational inequality. Write your answer in interval notation. $\dfrac{x + 2}{x^2 - 9} > 0$
    $x \in (-3, -2) \cup (3, \infty)$

75. **(2.6)** The resistance to current flow in copper wire varies directly as its length and inversely as the square of its diameter. A wire 8 m long with a 0.004-m diameter has a resistance of 1500 Ω. Find the resistance in a wire of like material that is 2.7 m long with a 0.005-m diameter.   324 Ω

76. **(5.5)** Solve for x: $-350 = 211e^{-0.025x} - 450$.
    $x \approx 29.87$

MAKING CONNECTIONS

Making Connections: Graphically, Symbollically, Numerically, and Verbally

Eight graphs (a) through (h) are given. Match the characteristics or equations shown in 1 through 16 to one of the eight graphs

(a) 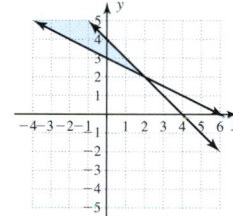   **(b)** 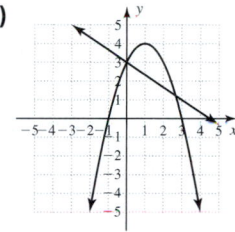   **(c)** 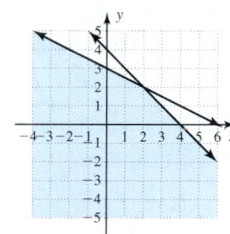   **(d)**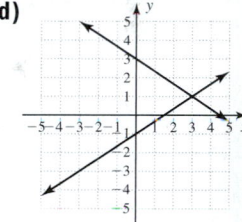

(e) 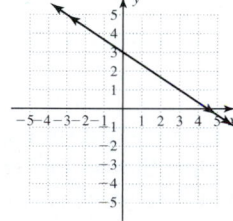   **(f)** 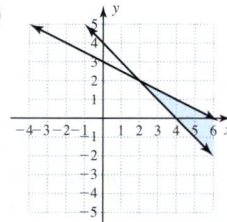   **(g)** 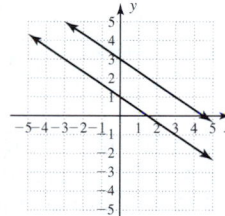   **(h)**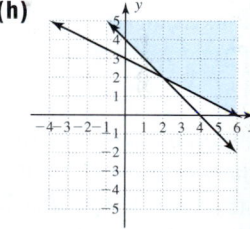

1. __c__ $\begin{cases} x + y \le 4 \\ x + 2y \le 6 \end{cases}$

2. __e__ $\begin{cases} 2x + 3y = 9 \\ x + \dfrac{3}{2}y = \dfrac{9}{2} \end{cases}$

3. __g__ $m_1 = m_2,\ b_1 \ne b_2$

4. __c__ $(0, 0)$ is a solution

5. __d__ $\begin{cases} 2x + 3y = 9 \\ -2x + 3y = -3 \end{cases}$

6. __a__ $\begin{cases} x + y \le 4 \\ x + 2y \ge 6 \end{cases}$

7. __a__ $(-1, 4)$ is a solution

8. __b__ nonlinear system

9. __g__ $\begin{cases} 2x + 3y = 9 \\ 2x + 3y = 3 \end{cases}$

10. __h__ $\begin{cases} x + y \ge 4 \\ x + 2y \ge 6 \end{cases}$

11. __e__ consistent, dependent

12. __g__ inconsistent system

13. __f__ $\begin{cases} x + y \ge 4 \\ x + 2y \le 6 \end{cases}$

14. __b__ $\begin{cases} 2x + 3y = 9 \\ x^2 + y = 2x + 3 \end{cases}$

15. __b__ exactly two solutions

16. __d__ consistent, independent, linear system

SUMMARY AND CONCEPT REVIEW

SECTION 6.1   Linear Systems in Two Variables with Applications

KEY CONCEPTS

- A *solution* to a linear system in two variables is an ordered pair (x, y) that makes all equations in the system true.
- Since every point on the graph of a line satisfies the equation of that line, a point where two lines intersect must satisfy both equations and is a solution of the system.
- A system with at least one solution is called a *consistent system*.
- If the lines have different slopes, there is a unique solution to the system (they intersect at a single point). The system is called a *consistent* and *independent system*.
- If the lines have equal slopes and the same y-intercept, they form identical or *coincident* lines. Since one line is right atop the other, they intersect at all points with an infinite number of solutions. The system is called a *consistent* and *dependent system*.
- If the lines have equal slopes but different y-intercepts, they will never intersect. The system has no solution and is called an *inconsistent system*.

EXERCISES

Solve each system by graphing manually. Verify your answer by graphing on a graphing calculator. If the system is inconsistent or dependent, so state.

1. $\begin{cases} 3x - 2y = 4 \\ -x + 3y = 8 \end{cases}$ **2.** $\begin{cases} 0.2x + 0.5y = -1.4 \\ x - 0.3y = 1.4 \end{cases}$ **3.** $\begin{cases} 2x + y = 2 \\ x - 2y = 4 \end{cases}$

Solve using substitution. Indicate whether each system is consistent, inconsistent, or dependent. Write unique solutions as an ordered pair. Check your answer using a graphing calculator.

4. $\begin{cases} y = 5 - x \\ 2x + 2y = 13 \end{cases}$ **5.** $\begin{cases} x + y = 4 \\ 0.4x + 0.3y = 1.7 \end{cases}$ **6.** $\begin{cases} x - 2y = 3 \\ x - 4y = -1 \end{cases}$
    no solution; inconsistent $(5, -1)$; consistent $(7, 2)$; consistent

Solve using elimination. Indicate whether each system is consistent, inconsistent, or dependent. Write unique solutions as an ordered pair. Check your answer using a graphing calculator.

7. $\begin{cases} 2x - 4y = 10 \\ 3x + 4y = 5 \end{cases}$ $(3, -1)$; consistent **8.** $\begin{cases} 2x = 3y + 6 \\ 2.4x + 3.6y = 6 \end{cases}$ $\left(\frac{11}{4}, \frac{-1}{6}\right)$; consistent

9. When it was first constructed in 1968, the John Hancock building in Chicago, Illinois, was the tallest structure in the world. In 1974, the Willis Tower in Chicago (formerly known as the Sears Tower) became the world's tallest structure. The Willis Tower is 323 ft taller than the John Hancock Building, and the sum of their heights is 2577 ft. How tall is each structure?
Willis Tower is 1450 ft; Hancock Building is 1127 ft.

10. The manufacturer of a revolutionary automobile spark plug, finds that demand for the plug can be modeled by the function $D(p) = -0.8p + 110$, where $D(p)$ represents the number of plugs bought (demanded, in tens of thousands) at price p in cents. The supply of these spark plugs is modeled by $S(p) = 0.24p - 14.8$, where $S(p)$ represents the number of plugs manufactured/supplied (in tens of thousands) at price p. Find the price for market equilibrium using a graphing calculator. $1.20

SECTION 6.2   Linear Systems in Three Variables with Applications

KEY CONCEPTS

- The graph of a linear equation in three variables is a *plane*.
- Systems in three variables can be solved using substitution and elimination.
- A linear system in three variables has the following possible solution sets:
  - If the planes intersect at a point, the system has one *unique solution* (x, y, z).
  - If the planes intersect at a line, the system has *linear dependence* and the solution (x, y, z) can be written as linear combinations of a single variable (a *parameter*).

Additional answers can be found in the Instructor Answer Appendix.

- If the planes are *coincident,* the equations in the system differ by a constant multiple, meaning they are all "disguised forms" of the *same equation.* The solutions have *coincident dependence,* and the solution set can be represented by any one of the equations.
- In all other cases, the system has *no solutions* and is an inconsistent system.

EXERCISES

Solve using elimination. If a system is inconsistent or dependent, so state. For systems with linear dependence, give the answer as an ordered triple using a parameter. Verify solutions on the home screen using the ⓐ키ⓗᴬ keys.

11. $\begin{cases} x + y - 2z = -1 \\ 4x - y + 3z = 3 \\ 3x + 2y - z = 4 \end{cases}$ (0, 3, 2) **12.** $\begin{cases} -x + y + 2z = 2 \\ x + y - z = 1 \\ 2x + y + z = 4 \end{cases}$ (1, 1, 1) **13.** $\begin{cases} 3x + y + 2z = 3 \\ x - 2y + 3z = 1 \\ 4x - 8y + 12z = 7 \end{cases}$ no solution, inconsistent

Solve using a system of three equations in three variables.

14. In one version of the card game Gin Rummy, numbered cards (N) 2 through 9 are worth 5 points, the 10s and all face cards (F) are worth 10 points, and aces (A) are worth 20 points. At the moment his opponent said "Gin!" Kenan had 12 cards in his hand, worth a total value of 125 points. If the value of his aces and face cards was equal to four times the value of his numbered cards, how many aces, face cards, and numbered cards was he holding?
3 aces, 4 face cards, 5 numbered cards

15. A vending machine accepts nickels, dimes, and quarters. At the end of a week, there is a total of $536 in the machine. The number of nickels and dimes combined is 360 more than the number of quarters. The number of quarters is 110 more than twice the number of nickels. How many of each type of coin are in the machine?
1530 quarters, 1180 dimes, 710 nickels

SECTION 6.3   Nonlinear Systems of Equations and Inequalities

KEY CONCEPTS

- Nonlinear systems of equations can be solved using substitution or elimination.
- First identify the graphs of the equations in the system to help determine the possible number of real solutions.
- For nonlinear systems of inequalities, graph the related equation for each inequality given, then use a test point to decide what region to shade as the solution.
- The solution for a system of inequalities is the overlapping region (if it exists) created by solutions to the individual inequalities.
- If the boundary is included, graph it using a solid line; if the boundary is not included use a dashed line.

EXERCISES

Find all solutions (real and complex) using substitution or elimination. Identify the graph of each relation before you begin. Verify your answers using a graphing calculator if possible.

16. $\begin{cases} x^2 + y^2 = 25 \\ y - x = -1 \end{cases}$ circle, line, (4, 3), (−3, −4) **17.** $\begin{cases} x = y^2 - 1 \\ x + 4y = -5 \end{cases}$ parabola, line, (3, −2) **18.** $\begin{cases} -x^2 + y = -1 \\ x^2 + y^2 = 7 \end{cases}$ parabola, circle, $(\sqrt{3}, 2), (-\sqrt{3}, 2),$ $(i\sqrt{2}, -2),$ $(-i\sqrt{2}, -3)$

19. $\begin{cases} x^2 + y^2 = 10 \\ y - 3x^2 = 0 \end{cases}$ **20.** $\begin{cases} y \leq x^2 - 2 \\ x^2 + y^2 \leq 16 \end{cases}$ **21.** $\begin{cases} x^2 + y^2 > 9 \\ x^2 + y \leq -3 \end{cases}$

circle, parabola, (1, 3), (−1, 3), $\left(\frac{i\sqrt{10}}{3}, \frac{-10}{3}\right), \left(\frac{i\sqrt{10}}{3}, \frac{10}{3}\right)$

SECTION 6.4   Systems of Linear Inequalities and Linear Programming

KEY CONCEPTS

- As in Section 6.3, to solve a *system of linear inequalities,* we find the intersecting or *overlapping areas* of the solution regions from the individual inequalities. The common area is the solution region for the system.
- The process known as *linear programming* seeks to *maximize* or *minimize* the value of a given quantity under certain *constraints* or restrictions.
- The quantity we attempt to maximize or minimize is called the *objective function.*
- The solution(s) to a linear programming problem *occur at one of the corner points of the feasible region.*

- The process of solving a linear programming application contains these six steps:
  - Identify the main objective and the decision variables.
  - Write the objective function in terms of these variables.
  - Organize all information in a table, using the decision variables and constraints.
  - Fill in the table with the information given and write the constraint inequalities.
  - Graph the constraint inequalities and determine the feasible region.
  - Identify all corner points of the feasible region and test these points in the objective function.

EXERCISES

Graph the solution region for each system of linear inequalities by first solving each equation for y. Verify each solution using a test point. Solve Exercise 24 using a graphing calculator.

22. $\begin{cases} -x - y > -2 \\ -x + y < -4 \end{cases}$

23. $\begin{cases} x - 4y \le 5 \\ -x + 2y \le 0 \end{cases}$

 **24.** $\begin{cases} x + 2y \ge 1 \\ 2x - y \le -2 \end{cases}$

25. Carefully graph the feasible region for the system of inequalities shown, then maximize the objective function: $f(x, y) = 30x + 45y$ $\begin{cases} 2x + y \le 10 \\ 2x + 3y \le 18 \\ x \ge 0, y \ge 0 \end{cases}$

26. After retiring, Oliver and Lisa Douglas buy and work a small farm (near Hooterville) that consists mostly of milk cows and egg-laying chickens. Although the price of a commodity is rarely stable, suppose that milk sales bring in an average of $85 per cow and egg sales an average of $50 per chicken over a period of time. During this time period, the new ranchers estimate that care and feeding of the animals took about 3 hr per cow and 2 hr per chicken, while maintaining the related equipment took 2 hr per cow and 1 hr per chicken. How many animals of each type should be maintained in order to maximize profits, if at most 1000 hr can be spent on care and feeding, and at most 525 hr on equipment maintenance? 50 cows, 425 chickens

PRACTICE TEST

Solve each system and state whether the system is consistent, inconsistent, or dependent. Verify solutions using a graphing calculator.

1. Solve graphically:

$\begin{cases} 3x + 2y = 12 \\ -x + 4y = 10 \end{cases}$

2. Solve using substitution:

$\begin{cases} 3x - y = 2 \\ -7x + 4y = -6 \end{cases}$

3. Solve using elimination:

$\begin{cases} 5x + 8y = 1 \\ 3x + 7y = 5 \end{cases}$

4. Solve using elimination:

$\begin{cases} x + 2y - z = -4 \\ 2x - 3y + 5z = 27 \\ -5x + y - 4z = -27 \end{cases}$

5. Solve using elimination:

$\begin{cases} 2x - y + z = 4 \\ -x + 2z = 1 \\ x - 2y + 8z = 11 \end{cases}$ $\{(x, y, z)\,|\,x = 2z - 1, y = 5z - 6, z \in \mathbb{R}\}$

6. Find values of a and b such that $(2, -1)$ is a solution of the system.

$\begin{cases} ax - by = 12 \\ bx + ay = -1 \end{cases}$ $a = 5, b = 2$

Create a system of equations to model each exercise, then solve using the method of your choice.

7. The perimeter of a "legal-size" paper is 114.3 cm. The length of the paper is 7.62 cm less than twice the width. Find the dimensions of a legal-size sheet of paper. 21.59 cm by 35.56 cm

8. The island nations of Tahiti and Tonga have a combined land area of 692 mi². Tahiti's land area is 112 mi² more than Tonga's. What is the land area of each island group? Tahiti, 402 mi²; Tonga, 290 mi²

9. Many years ago, two cans of corn (C), 3 cans of green beans (B), and 1 can of peas (P) cost $1.39. Three cans of C, 2 of B, and 2 of P cost $1.73. One can of C, 4 of B, and 3 of P cost $1.92. What was the price of a single can of C, B, and P? Corn 25¢, Beans 20¢, Peas 29¢

10. After inheriting $30,000 from a rich aunt, David decides to place the money in three different investments: a savings account paying 5%, a bond account paying 7%, and a stock account paying 9%. After 1 yr he earned $2080 in interest. Find how much was invested at each rate if $8000 less was invested at 9% than at 7%. $15,000 at 7%, $8000 at 5%, $7000 at 9%

11. Solve the system of inequalities by graphing.

$\begin{cases} x - y \le 2 \\ x + 2y \ge 8 \end{cases}$

12. Maximize the objective function: $P = 50x - 12y$, given the constraints shown.

$\begin{cases} x + 2y \le 8 \\ 8x + 5y \ge 40 \\ x \ge 0, y \ge 0 \end{cases}$

Additional answers can be found in the Instructor Answer Appendix.

Solve the linear programming problem.

13. A company manufactures two types of T-shirts, a plain T-shirt and a deluxe monogrammed T-shirt. To produce a plain shirt requires 1 hr of working time on machine A and 2 hr on machine B. To produce a deluxe shirt requires 1 hr on machine A and 3 hr on machine B. Machine A is available for at most 50 hr/week, while machine B is available for at most 120 hr/week. If a plain shirt can be sold at a profit of $4.25 each and a deluxe shirt can be sold at a profit of $5.00 each, how many of each should be manufactured to maximize the profit?

Solve each nonlinear system using the technique of your choice.

14. $\begin{cases} x^2 + y^2 = 16 \\ y - x = 2 \end{cases}$ **15.** $\begin{cases} 4y - x^2 = 1 \\ y^2 + x^2 = 4 \end{cases}$

$(-1 - \sqrt{7}, 1 - \sqrt{7}), (-1 + \sqrt{7}, 1 + \sqrt{7})$

16. A support bracket on the frame of a large ship is a steel right triangle with a hypotenuse of 25 ft and a perimeter of 60 ft. Find the lengths of the other sides using a system of nonlinear equations.   15 ft, 20 ft

17. Solve $\begin{cases} x^2 - y \leq 2 \\ y \leq \sqrt{9 - x^2} \end{cases}$ using a graphing calculator.

18. The company from Exercise 10 of the Summary and Concept Review, hired a market research firm to collect data about the next-generation spark plug that will increase fuel efficiency by over 20%. The data they collected are shown in the table. Enter the data into a graphing calculator, calculate the regression line for each set of data, then (a) display the lines and scatterplot on a single screen, and (b) use the lines to find the estimated price and quantity at which market equilibrium is achieved.

Price (cents)	Demand (10,000s)	Supply (10,000s)
135	2.9	19.7
90	37.2	8.0
62	59.6	0.45
112	19.7	12.7
75	46.0	3.69
121	12.4	15.9

19. Solve the system of inequalities.

$\begin{cases} 2x - y \leq -1 \\ 3x + 2y \geq 2 \end{cases}$

20. Write a system of four inequalities that describes the location of the dart on the dartboard shown.

Exercise 20

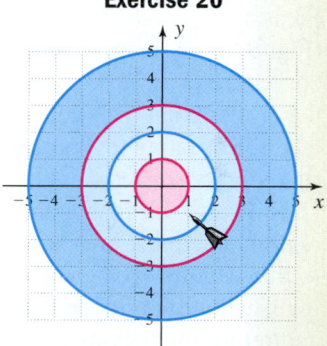

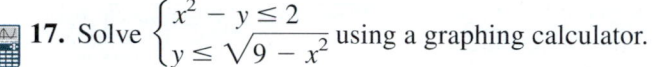

CALCULATOR EXPLORATION AND DISCOVERY

Optimal Solutions and Linear Programming

In this exercise, we'll use a graphing calculator to explore various areas of the feasible region, repeatedly evaluating the objective function to see where the maximal values (optimal solutions) seem to "congregate." If all goes as expected, ordered pairs nearest to a vertex should give relatively larger values. To demonstrate, we'll use Example 8 from Section 6.4, stated below.

Example 8 ▶ Find the maximum value of the objective function $f(x, y) = 2x + y$ given the

constraints shown: $\begin{cases} x + y \leq 4 \\ 3x + y \leq 6 \\ x \geq 0 \\ y \geq 0 \end{cases}$

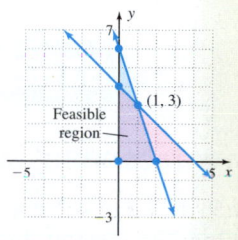

Solution ▶ The feasible region is shown in lavender. There are four corner points to this region: (0, 0), (0, 4), (2, 0), and (1, 3), and we found $f(x, y)$ was maximized at (1, 3): $f(1, 3) = 5$.

To explore this feasible region in terms of the objective function $f(x, y) = 2x + y$, enter the boundary lines $Y_1 = -X + 4$ and $Y_2 = -3X + 6$ on the (Y=) screen. However, instead of shading below the lines to show the feasible region (using the ◣ feature to the extreme left), we shade *above* both lines (using the ◤ feature) so that the feasible region remains clear. Setting the window size at $x \in [0, 3]$ and $y \in [-1.5, 4]$ produces Figure 6.52. Using $Y\text{min} = -1.5$ will leave a blank area just below QI that enables you to explore the feasible region as the *x*- and *y*-values are displayed. Next we place the calculator in "split-screen" mode so that we can view the graph and the home screen simultaneously. Press the **MODE** key and notice the line that reads **Full Horiz G-T.** The **Full** (screen) mode is the default operating mode. The **Horiz** mode splits the screen horizontally, placing the graph directly above a shorter home screen. Highlight **Horiz,** then press **ENTER** and **GRAPH** to have the calculator reset the screen in this mode. Most graphing calculators have a free-moving cursor that is brought into view by pressing the left **◁** or right **▷** arrow (Figure 6.53). A useful feature of this cursor is that it automatically stores the current X value as the variable X (**X,T,θ,n** or **ALPHA** **STO▸**) and the current Y value as the variable Y (**ALPHA** 1), which enables us to evaluate the objective function $f(x, y) = 2x + y$ right on the home screen. To access the graph and free-moving cursor you must press **GRAPH** each time, and to access the home screen you must press **2nd** **MODE** (**QUIT**) each time. Begin by moving the cursor to the upper-left corner of the region, near the *y*-intercept [we stopped at (~0.0957, 3.26̄)]. Once you have the cursor "tucked up into the corner," press **2nd** **MODE** (**QUIT**) to get to the home screen, then enter the objective function: 2X + Y. Pressing **ENTER** evaluates the function for the values indicated by the cursor's location (Figure 6.54). It appears the value of the objective function for points (x, y) in this corner are close to 4, and it's no accident that at the corner point (0, 4) the maximum value is in fact 4. Repeating this procedure for the lower-right corner suggests the maximum value near (2, 0) is also 4. Finally, press **GRAPH** to explore the region, where the lines intersect. Move the cursor to this vicinity, locate it very near the point of intersection [we stopped at (~0.957, 2.716̄)] and return to the home screen and evaluate (Figure 6.55). The value of the objective function is near 5 in this corner of the region, and at the corner point (1, 3) the maximum value is 5.

Figure 6.52

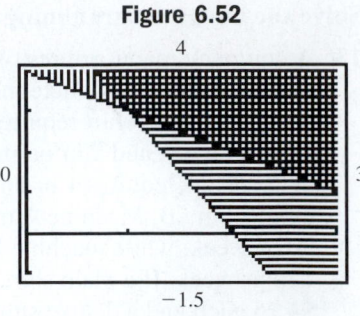

Figure 6.53

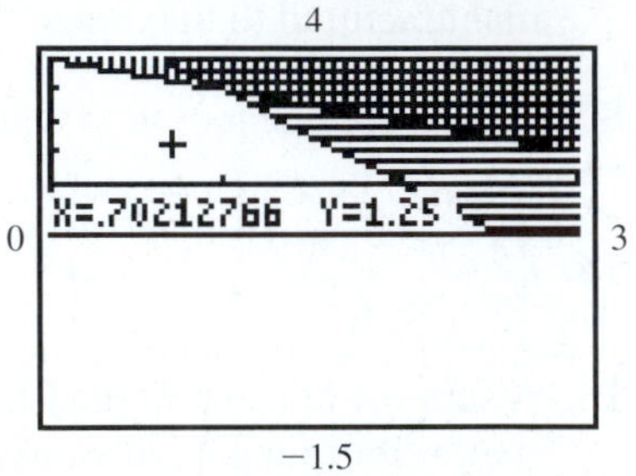

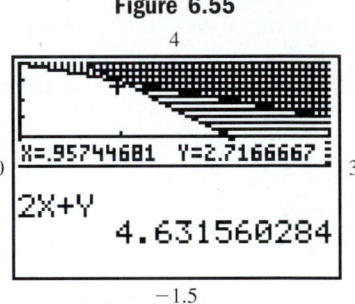

Figure 6.54

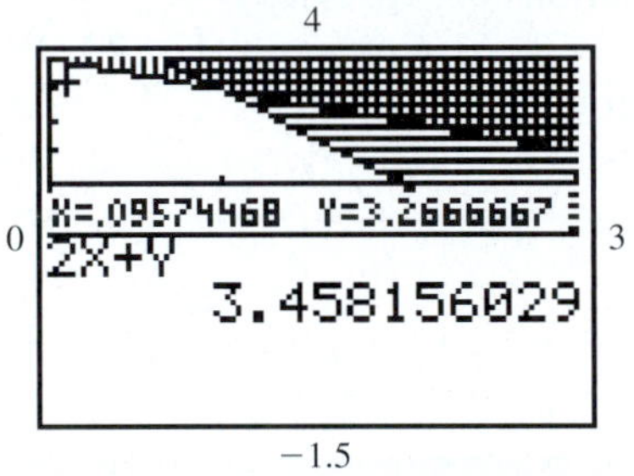

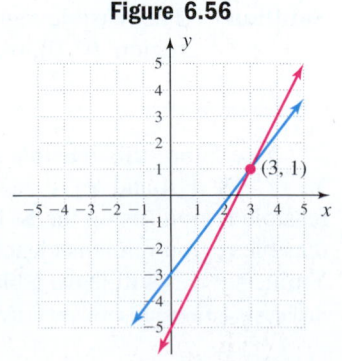

Figure 6.55

Exercise 1: The feasible region for the system given has four corner points. Use the ideas here to explore the area near each corner point of the feasible region to determine which point is the likely candidate to produce the *minimum* value of the objective function $f(x, y) = 2x + 4y$. Then solve the linear programming exercise to verify your guess.

$$\begin{cases} 2x + 2y \leq 15 \\ x + y \geq 6 \\ x + 4y \geq 9 \\ x \geq 0, y \geq 0 \end{cases}$$

(5, 1)

STRENGTHENING CORE SKILLS

Understanding Why Elimination and Substitution "Work"

When asked to solve a system of two equations in two variables, we first select an appropriate method. In Section 6.1, we learned three basic techniques: graphing, substitution, and elimination. In this feature, we'll explore how these methods are related using Example 2 from Section 6.1 where we were asked to solve the

system $\begin{cases} 4x - 3y = 9 \\ -2x + y = -5 \end{cases}$ by graphing. The resulting graph, shown here in

Figure 6.56, clearly indicates the solution is (3, 1).

As for the elimination method, either *x* or *y* can be easily eliminated. If the second equation is multiplied by 2, the *x*-coefficients will be additive inverses, and the sum results in an equation with *y* as the only unknown.

Figure 6.56

PRACTICE TEST

1. Write the expression $\log_3 81 = 4$ in exponential form. $3^4 = 81$

2. Write the expression $25^{1/2} = 5$ in logarithmic form. $\log_{25} 5 = \frac{1}{2}$

3. Write the expression $\log_b\left(\dfrac{\sqrt{x^5 y^3}}{z}\right)$ as a sum or difference of logarithmic terms. $\frac{5}{2}\log_b x + 3\log_b y - \log_b z$

4. Write the expression $\log_b m + \left(\frac{3}{2}\right)\log_b n - \frac{1}{2}\log_b p$ as a single logarithm. $\log_b \dfrac{m\sqrt{n^3}}{\sqrt{p}}$

Solve for x using the uniqueness property.

5. $5^{x-7} = 125$ $x = 10$

6. $2 \cdot 4^{3x} = \dfrac{8^x}{16}$ $x = \dfrac{-5}{3}$

Given $\log_a 3 \approx 0.48$ and $\log_a 5 \approx 1.72$, evaluate the following without the use of a calculator:

7. $\log_a 45$ 2.68

8. $\log_a 0.6$ -1.24

Graph using transformations of the parent function. Verify answers using a graphing calculator.

9. $g(x) = -2^{x-1} + 3$

10. $h(x) = \log_2(x - 2) + 1$

11. Use the change-of-base formula to evaluate. Verify results using a calculator.

   a. $\log_3 100$ 4.19 b. $\log_6 0.235$ -0.81

12. State the domain and range of $f(x) = (x - 2)^2 - 3$ and determine if f is a one-to-one function. If so, find its inverse. If not, restrict the domain of f to create a one-to-one function, then find the inverse of this new function, including the domain and range.

Solve each equation.

13. $3^{x-1} = 89$ $x = 1 + \frac{\ln 89}{\ln 3}, x \approx 5.0857$

14. $\log_5 x + \log_5(x + 4) = 1$ $x = 1, x = -5$ is extraneous

15. A copier is purchased new for $8000. The machine loses 18% of its value each year and must be replaced when its value drops below $3000. How many years will the machine be in service? ≈ 5 yr

16. In 1957, scientist Stanley Stevens proposed a mathematical model that attempted to compare the actual strength of a physical stimulus

with the human perception (dead-reckoning) of its strength. The model has been widely applied in comparisons of weight, sound, pressure, and other areas. If M represents the measured strength of the stimulus, and P the human perception of its strength, **Stevens' law** can be written as $\log P = \log k + \alpha \log M$, where α and k are constants determined by the type of stimulation applied. In a controlled experiment, subjects were given a known amount of weight to lift, then asked to select an unmarked weight they felt was equal to half the known weight. (a) Solve the equation for P and (b) use the result to determine what was perceived to be half of a 40-lb weight (assume $k = 0.89$ and $\alpha = 0.95$).   a. $P = kM^\alpha$ b. $P \approx 15.3$ lb $(M = 20)$

17. The number of ounces of unrefined platinum drawn from a mine is modeled by $Q(t) = -2600 + 1900 \ln(t)$, where $Q(t)$ represents the number of ounces mined in t months. How many months did it take for the number of ounces mined to exceed 3000?   19.1 months

18. Jacob decides to save $4000 over the next 5 yr so that he can present his wife with a new diamond ring for their 20th anniversary. He invests $50 every month in an account paying $8\frac{1}{4}\%$ interest compounded monthly. (a) Is this amount sufficient to meet the 5-yr goal? (b) If not, find the *minimum amount he needs to save monthly* that will enable him to meet this goal.   a. no   b. $54.09

19. Chaucer is a typical Welsh Corgi puppy. During his first year of life, his weight very closely follows the model $W(t) = 6.79 \ln t - 11.97$, where $W(t)$ is his weight in pounds after t weeks and $8 \le t \le 52$.

   a. How much will Chaucer weigh when he is 26 weeks old (to the nearest one-tenth pound)?   10.2 lb

   b. To the nearest week, how old is Chaucer when he weighs 12 lb?   34 weeks

20. Using time-lapse photography, the growth of a stain is tracked in 0.2 second intervals, as a small amount of liquid is dropped on various fabrics. Use the data given to draw a scatterplot and decide on an appropriate regression model. How long, to the nearest hundredth of a second, did it take the stain to reach a size of 15 mm?

Exercise 20

Time (sec)	Size (mm)
0.2	0.39
0.4	1.27
0.6	3.90
0.8	10.60
1.0	21.50
1.2	31.30
1.4	36.30
1.6	38.10
1.8	39.00

Additional answers can be found in the Instructor Answer Appendix.

CALCULATOR EXPLORATION AND DISCOVERY

Investigating Logistic Equations

As we saw in Section 5.5, logistic models have the form $P(t) = \dfrac{c}{1 + ae^{-bt}}$, where a, b, and c are constants and $P(t)$ represents the population at time t. For populations modeled by a logistic curve (sometimes called an "S" curve), growth is very rapid at first (like an exponential function), but this growth begins to slow down and level off due to various factors. This *Calculator Exploration and Discovery* is designed to investigate the effects that a and c have on the resulting graph.

I. Investigating a: From our earlier observation, as t becomes larger and larger, the term ae^{-bt} becomes smaller and smaller (approaching 0) because it is a decreasing function: as $t \to \infty$, $ae^{-bt} \to 0$. If we allow that the term eventually becomes so small it can be disregarded, what remains is $P(t) = \dfrac{c}{1}$ or c. This is why c is called the capacity constant and the population can get no larger than c. In Figure 5.65, the graph of $P(t) = \dfrac{1000}{1 + 50e^{-1x}}$ ($a = 50$, $b = 1$, and $c = 1000$) is shown using a lighter line, while the graph of $P(t) = \dfrac{750}{1 + 50e^{-1x}}$ ($a = 50$, $b = 1$, and $c = 750$), is given in bold.

Figure 5.65

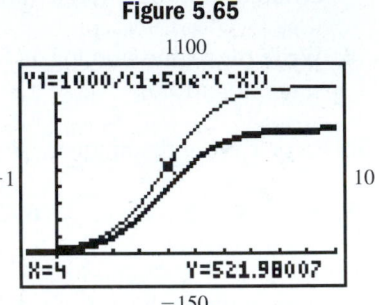

Also note that if a is held constant, smaller values of c cause the "interior" of the S curve to grow at a slower rate than larger values, a concept studied in some detail in a Calculus I class.

II. Investigating c: If $t = 0$, $ae^{-bt} = ae^0 = a$, and we note the ratio $P(0) = \dfrac{c}{1 + a}$ represents the *initial population*. This also means for constant values of c, larger values of a make the ratio $\dfrac{c}{1 + a}$ smaller; while smaller values of a make the ratio $\dfrac{c}{1 + a}$ larger. From this we conclude that a primarily affects the initial population. In Figures 5.66 and 5.67 shown next, $P(t) = \dfrac{1000}{1 + 50e^{-1x}}$ (from I) is graphed using a lighter line, while the graph of $P(t) = \dfrac{1000}{1 + 5e^{-1x}}$ ($a = 5$) and $P(t) = \dfrac{1000}{1 + 500e^{-1x}}$ ($a = 500$) are shown in bold.

Figure 5.66

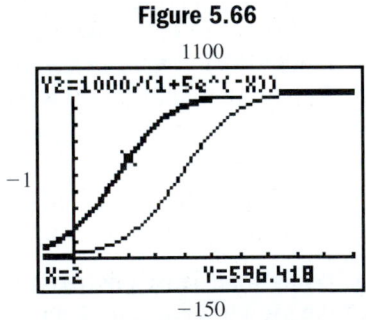

Figure 5.67

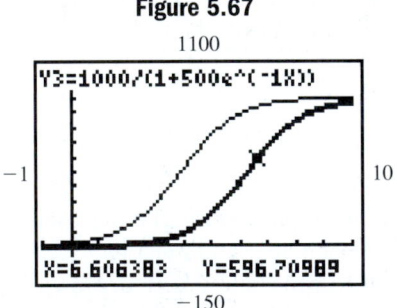

Note that changes in a appear to have no effect on the rate of growth in the interior of the S curve. The following exercises are based on the population of an ant colony, modeled by the logistic function $P(t) = \dfrac{2500}{1 + 25e^{-0.5x}}$. Respond to Exercises 1 through 6 without the use of a calculator.

Exercise 1: Identify the values of a, b, and c for this logistics curve. $a = 25$ $b = 0.5$ $c = 2500$

Exercise 2: What was the approximate initial population of the colony? $\frac{2500}{26} \approx 96$ ants

Exercise 3: Which gives a larger initial population:(a) $c = 2500$ and $a = 25$ or (b) $c = 3000$ and $a = 15$?   b

Exercise 4: What is the maximum population capacity for this colony?   2500

Exercise 5: Which causes a slower population growth: (a) $c = 2000$ and $a = 25$ or (b) $c = 3000$ and $a = 25$?   a

 Exercise 6: Verify your responses to Exercises 2 through 6 using a graphing calculator.   verified

STRENGTHENING CORE SKILLS

The HerdBurn Scale—What's Hot and What's Not

The human mouth can easily distinguish between heat levels (the burning sensation) when eating foods "spiced" with various peppers. The level of "heat" is generally given in *Scoville units*, which is a measure of the element capsaicin that causes the burn. Sweet bell peppers and others have no capsaicin and a Scoville rating of 0, while red habanero peppers have a Scoville rating of near 500,000. Although inedible, laboratory grades of capsaicin can have a Scoville rating of near 16,000,000! This range of values makes a unit scale impractical for common use, and a logarithmic scale once again becomes more desirable. Using the newly developed *HerdBurn scale (hb)*, we have the following measures of "heat" for well known peppers of various types.

The HerdBurn Scale

HerdBurn Units (hb)	Chili Pepper	General Sensation	Caustic Power
0	sweet banana peppers	not sensed	none
1	cherry peppers	delicate warmth	none
2	pepperoncini peppers	strong warmth	none
2.5	Sonora peppers	slight burn	some reaction
3	ancho peppers	moderate burn	fanning the mouth
3.5	jalapeno peppers	strong burn	eyes water
4	hidalgo peppers	sizzling burn	pain threshold
4.5	cayenne peppers	scorching burn	painful
5	Bahamian peppers	blistering burn	very painful
5.5	habanero peppers	ruthless burn	intense pain
6	naga jolokia peppers	merciless burn	debilitating
6.5	military grade pepper spray	*inedible*	incapacitating
7	laboratory grade capsaicin	*inedible*	ruinous
7.2	pure capsaicin	*inedible*	deadly

Similar to working with decibel levels or the Richter scale, we compare how many times hotter one pepper is than another by recognizing the values given are powers of 10. For example, a red habanero (5.7 hb), is about two times as hot as an orange habanero (5.4 hb): $\frac{10^{5.7}}{10^{5.4}} = 10^{0.3} \approx 2$, but nearly 100 times hotter than a red jalapeno pepper (3.7 hb): $\frac{10^{5.7}}{10^{3.7}} = 10^2$. Use this information to complete the following exercises.

Exercise 1: The "heat" in a rocotillo pepper measures about 3.4 on the HerdBurn scale, while a Jamaican hot pepper measures near 5.5. How many times hotter is the Jamaican pepper?   about 126 times hotter

Exercise 2: A naga jolokia (6.0 hb) pepper is about 63 times as hot as a serrano pepper. What is the HerdBurn number for a serrano pepper?   about 4.2 hb

CUMULATIVE REVIEW CHAPTERS R–5

Use the quadratic formula to solve for x.

1. $x^2 - 4x + 53 = 0$ $x = 2 \pm 7i$

2. $6x^2 + 19x = 36$ $x = \frac{4}{3}, x = \frac{-9}{2}$

3. Use substitution to show that $4 + 5i$ is a zero of
$f(x) = x^2 - 8x + 41$. $(4 + 5i)^2 - 8(4 + 5i) + 41 = 0$
$-9 + 40i - 32 - 40i + 41 = 0, 0 = 0\checkmark$

4. Graph using transformations of a basic function:
$y = 2\sqrt{x + 2} - 3$.

$f(g(x)) = x; g(f(x)) = x;$ Since $(f \circ g)(x) = (g \circ f)(x)$, they are inverse functions.
5. Find $(f \circ g)(x)$ and $(g \circ f)(x)$ and comment on what
you notice: $f(x) = x^3 - 2; g(x) = \sqrt[3]{x + 2}$.

6. State the domain of $h(x)$ in interval notation:

$h(x) = \dfrac{\sqrt{x + 3}}{x^2 + 6x + 8}$. $x \in [-3, -2) \cup (-2, \infty)$

7. According to the 2002 *National Vital Statistics Report* (Vol. 50, No. 5, page 19) there were 3100 sets of triplets born in the United States in 1991, and 6740 sets of triplets born in 1999. Assuming the relationship (year, sets of triplets) is linear: (a) find an equation of the line, (b) explain the meaning of the slope in this context, and (c) use the equation to estimate the number of sets born in 1996, and to project the number of sets that will be born in 2007 if this trend continues.

8. State the following geometric formulas:

a. area of a circle $A = \pi r^2$ **c.** perimeter of a rectangle
$P = 2L + 2W$
b. Pythagorean theorem **d.** area of a trapezoid
$a^2 + b^2 = c^2$ $A = \frac{h}{2}(B + b)$
9. Graph the following piecewise-defined function and state its domain, range, and intervals where it is increasing and decreasing.

$$h(x) = \begin{cases} -4 & -10 \le x < -2 \\ -x^2 & -2 \le x < 3 \\ 3x - 18 & x \ge 3 \end{cases}$$

10. Solve the inequality and write the solution in
interval notation: $\dfrac{2x + 1}{x - 3} \ge 0$. $x \in (-\infty, -\frac{1}{2}] \cup (3, +\infty)$

11. Use the rational roots theorem to find all zeroes of
$f(x) = x^4 - 3x^3 - 12x^2 + 52x - 48$.
$x = 3, x = 2$ (multiplicity 2); $x = -4$

12. Given $f(c) = \dfrac{9}{5}c + 32$, find k, where $k = f(25)$.

Then find the inverse function using the algebraic method, and verify that $f^{-1}(k) = 25$.

13. Solve the formula $V = \dfrac{1}{2}\pi b^2 a$ (the
volume of a paraboloid) for the
variable b. $\sqrt{\frac{2V}{\pi a}} = b$

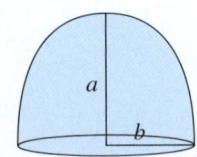

14. Use the *Guidelines for Graphing* to graph
a. $p(x) = x^3 - 4x^2 + x + 6$. **b.** $r(x) = \dfrac{5x^2}{x^2 + 4}$.

15. For $f(x) = \dfrac{2x + 3}{5}$, (a) find f^{-1}, (b) graph both functions and verify they are symmetric to the line $y = x$, and (c) show they are inverses using composition.

16. Solve for x: $10 = -2e^{-0.05x} + 25$. $x \approx -40.298$

17. Solve for x: $\ln(x + 3) + \ln(x - 2) = \ln(24)$.

18. Once in orbit, satellites are often powered by radioactive isotopes. From the natural process of radioactive decay, the power output declines over a period of time. For an initial amount of 50 g, suppose the power output is modeled by the function $p(t) = 50e^{-0.002t}$, where $p(t)$ is the power output in watts, t days after the satellite has been put into service. (a) Approximately how much power remains 6 *months* later (assume 1 mo. ≈ 30.5 days)? (b) How many *years* until only one-fourth of the original power remains? ≈ 30.5

19. Simon and Christine own a sport wagon and a minivan. The sport wagon has a power curve that is closely modeled by $H(r) = 123 \ln r - 897$, where $H(r)$ is the horsepower at r rpm, with $2200 \le r \le 5600$. The power curve for the minivan is $h(r) = 193 \ln r - 1464$, for $2600 < r \le 5800$.

a. How much horsepower is generated by each engine at 3000 rpm?
about 88 hp for sport wagon, about 81 hp for minivan
b. At what rpm are the engines generating the same horsepower? ≈ 3294 rpm

c. If Christine wants the maximum horsepower available, which vehicle should she drive? What is the maximum horsepower?
minivan, 208 hp at 5800 rpm
20. Wilson's disease is a hereditary disease that causes the body to retain copper. Radioactive copper, ^{64}Cu, has been used extensively to study and understand this disease. ^{64}Cu has a relatively short half-life of 12.7 hr. How many hours will it take for a 5-g mass of ^{64}Cu to decay to 1 g?   about 29.5 hr

Solve using a graphing calculator.

21. $e^{\frac{x}{2}} - 9.2 = 5 - \ln x$ $x \approx 5.064$

22. $5.2 - e^{0.25x} = 4 - \ln(x + 6)$ $x \approx -5.615, k \approx 5.135$

23. $e^{\sqrt[3]{x}} = \sqrt{x + 5}$ $x \approx 0.649, x \approx -4.967$

24. $\dfrac{x^2 - 25}{x^2 - 9} = \ln(x + 9) + 33$
$x \approx -8.920, x \approx -2.279, x \approx 2.370$

25. $2^{x^2 - x - 6} = x^2 + x - 6$ $x \approx 2.013, x \approx 3.608$

Additional answers can be found in the Instructor Answer Appendix.

Systems of Equations and Inequalities

CHAPTER OUTLINE

CHAPTER CONNECTIONS

At the turn of the century, there was an explosion in the number of handheld electronic devices available to consumers. One device in particular, the popular MP3 player, experienced a phenomenal growth in demand. With high demand and a large market, competition between manufacturers and suppliers is often fierce, with each fighting to earn and hold a share of the market. One significant factor in who gains the largest share is the price charged for the player, with suppliers willing to supply more at a greater price, and consumers willing to buy more at a lesser price. Determining where the price will stabilize is an important component in the economics of *"supply and demand."* This application occurs as Exercise 74 in Section 6.1.

Check out these other real-world connections:

▶ Appropriate Measurements in Dietetics (Section 6.1, Exercise 64)

▶ Allocating Winnings to Different Investments (Section 6.2, Exercise 53)

▶ Market Pricing for Organic Produce (Section 6.3, Exercise 66)

▶ Minimizing Shipping Costs (Section 6.4, Exercise 69)

Linear Systems in Two Variables with Applications

LEARNING OBJECTIVES

In Section 6.1 you will see how we can:

- ☐ **A.** Verify ordered pair solutions
- ☐ **B.** Solve linear systems by graphing
- ☐ **C.** Solve linear systems by substitution
- ☐ **D.** Solve linear systems by elimination
- ☐ **E.** Recognize inconsistent systems and dependent systems
- ☐ **F.** Use a system of equations to model and solve applications

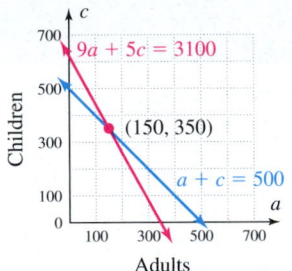

In earlier chapters, we used linear equations in two variables to model a number of real-world situations. Graphing these equations gave us a visual image of how the variables were related, and helped us better understand this relationship. In many applications, two different measures of the independent variable must be considered simultaneously, leading to a **system of two linear equations in two unknowns.** Here, a graphical presentation once again supports a better understanding, as we explore systems and their many applications.

A. Solutions to a System of Equations

A **system of equations** is a set of two or more equations for which a common solution is sought. Systems are widely used to model and solve applications when the information given enables the relationship between variables to be stated in different ways. For example, consider an amusement park that brought in $3100 in revenue by charging $9.00 for adults and $5.00 for children, while selling 500 tickets. Using a for adult and c for children, we could write one equation modeling the number of tickets sold: $a + c = 500$, and a second modeling the amount of revenue brought in: $9a + 5c = 3100$. To show that we're considering both equations simultaneously, a large "left brace" is used and the result is called a **system of two equations in two variables:**

$$\begin{cases} a + c = 500 & \text{number of tickets} \\ 9a + 5c = 3100 & \text{amount of revenue} \end{cases}$$

We note that both equations are linear and will have different slope values, so their graphs must intersect at some point. Since every point on a line satisfies the equation of that line, this point of intersection must satisfy *both* equations simultaneously and is the solution to the system. The figure that accompanies Example 1 shows the point of intersecion for this system is (150, 350).

EXAMPLE 1 ▶ **Verifying Solutions to a System**

Verify that (150, 350) is a solution to $\begin{cases} a + c = 500 \\ 9a + 5c = 3100 \end{cases}$.

Solution ▶ Substitute the 150 for a and 350 for c in each equation.

$a + c = 500$ first equation $9a + 5c = 3100$ second equation

$(150) + (350) = 500$ $9(150) + 5(350) = 3100$

$500 = 500 ✓$ $3100 = 3100 ✓$

Since (150, 350) satisfies both equations, it is the solution to the system and we find the park sold 150 adult tickets and 350 tickets for children.

Now try Exercises 7 through 18 ▶

To check the solution to Example 1 on a graphing calculator, recall that we can store values in any of the ⒜ALPHA⒝ characters using the ⒮STO•⒯ key. The home screen shown in Figure 6.1 shows that we've stored a value of 150 in alpha location **A**, and 350 in location **C**. The solution to a system can then be checked using these alpha keys and the operations indicated, as shown in Figure 6. 2.

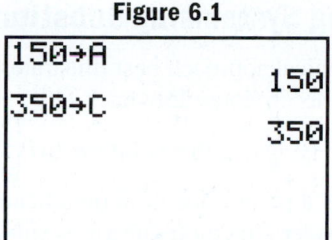

Figure 6.1

Figure 6.2

☑ **A.** You've just seen how we can verify ordered pair solutions

B. Solving Systems Graphically

To **solve a system of equations** means we apply various methods in an attempt to find ordered pair solutions. As Example 1 suggests, one method for finding solutions is to graph the system. Any method for graphing the lines can be employed, but to keep important concepts fresh, the slope-intercept method is used here.

EXAMPLE 2 ▶ **Solving a System Graphically**

Solve the system by graphing: $\begin{cases} 4x - 3y = 9 \\ -2x + y = -5 \end{cases}$.

Solution ▶ First write each equation in slope-intercept form (solve for y):

$$\begin{cases} 4x - 3y = 9 \\ -2x + y = -5 \end{cases} \rightarrow \begin{cases} y = \dfrac{4}{3}x - 3 \\ y = 2x - 5 \end{cases}$$

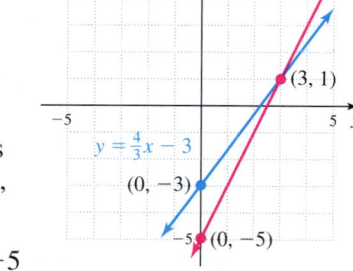

For the first line, $\frac{\Delta y}{\Delta x} = \frac{4}{3}$ with y-intercept $(0, -3)$. The second equation yields $\frac{\Delta y}{\Delta x} = \frac{2}{1}$ with $(0, -5)$ as the y-intercept. Both are then graphed on the grid as shown. The point of intersection appears to be $(3, 1)$, and checking this point in both equations gives

$$4x - 3y = 9 \qquad\qquad\qquad -2x + y = -5$$
$$4(3) - 3(1) = 9 \quad \substack{\text{substitute 3} \\ \text{for } x \text{ and 1 for } y} \quad -2(3) + (1) = -5$$
$$9 = 9 ✓ \qquad\qquad\qquad -5 = -5 ✓$$

This verifies that $(3, 1)$ is the solution to the system.

Now try Exercises 19 through 22 ▶

Graphical solutions to a system of equations can be found using a graphing calculator and the *intersection-of-graphs* method seen earlier. After solving for y and entering the equations on the ⬭ **Y=** screen (Figure 6.3), use the keystrokes ⬭ **2nd** ⬭ **TRACE** (**CALC**) **5:Intersect** to determine the point of intersection, if it exists (Figure 6.4).

☑ **B.** You've just seen how we can solve linear systems by graphing

Figure 6.3

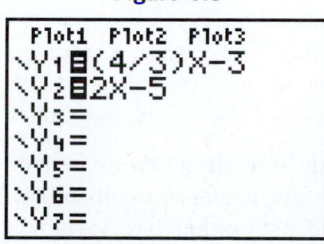

Figure 6.4

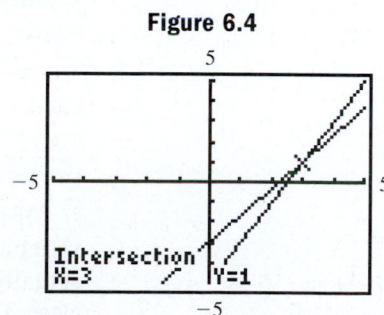

C. Solving Systems by Substitution

While a graphical approach best illustrates *why* the solution must be an ordered pair, it does have one obvious drawback—noninteger solutions are difficult to spot. The ordered pair $\left(\frac{2}{5}, \frac{12}{5}\right)$ is the solution to $\begin{cases} 4x + y = 4 \\ y = x + 2 \end{cases}$, but this would be difficult to "pinpoint" as a precise location on a hand-drawn graph. To overcome this limitation, we next consider a method known as **substitution.** The method involves converting a system of two equations in two variables into a single equation in one variable by using an appropriate substitution. For $\begin{cases} 4x + y = 4 \\ y = x + 2 \end{cases}$, the second equation says "*y* is two more than *x*." We reason that *all* points on this line are related this way, *including the point where this line intersects the other.* For this reason, we can substitute $x + 2$ for *y* in the first equation, obtaining a single equation in *x*.

EXAMPLE 3 ▶ **Solving a System Using Substitution**

Solve using substitution: $\begin{cases} 4x + y = 4 \\ y = x + 2 \end{cases}$.

Solution ▶ Since $y = x + 2$, we can replace *y* with $x + 2$ in the first equation.

$$4x + y = 4 \qquad \text{first equation}$$
$$4x + (x + 2) = 4 \qquad \text{substitute } x + 2 \text{ for } y$$
$$5x + 2 = 4 \qquad \text{simplify}$$
$$x = \frac{2}{5} \qquad \text{result}$$

The *x*-coordinate is $\frac{2}{5}$. To find the *y*-coordinate, we substitute $\frac{2}{5}$ for *x* into either of the original equations, a process known as **back-substitution.** Substituting in the second equation gives

$$y = x + 2 \qquad \text{second equation}$$
$$= \frac{2}{5} + 2 \qquad \text{substitute } \frac{2}{5} \text{ for } x$$
$$= \frac{12}{5} \qquad \frac{2}{1} = \frac{10}{5}, \frac{2}{5} + \frac{10}{5} = \frac{12}{5}$$

Figure 6.5

$Y_1 = -4X + 4, Y_2 = X + 2$

The solution to the system is $\left(\frac{2}{5}, \frac{12}{5}\right)$, which can be verified using the intersection-of-graphs method, and noting $\frac{2}{5} = 0.4$ and $\frac{12}{5} = 2.4$. See Figure 6.5. Recall that the keystrokes MATH 1:▶FRAC ENTER can be used to convert decimal numbers to fractions.

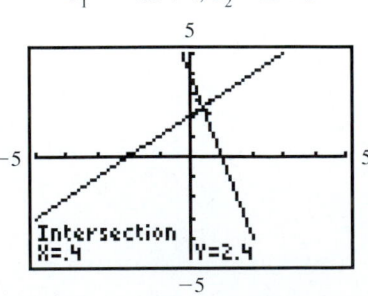

Now try Exercises 23 through 32 ▶

If neither equation allows an immediate substitution, we first solve for one of the variables, either *x* or *y*, and *then* substitute. The method is summarized here, and can actually be used with either like variables or like variable *expressions.* **See Exercises 33 to 36.**

☑ **C.** You've just seen how we can solve linear systems by substitution

> **Solving Systems Using Substitution**
>
> 1. Solve one of the equations for x in terms of y or y in terms of x.
> 2. Substitute for the appropriate variable in the *other* equation and solve for the variable that remains.
> 3. Substitute the value from step 2 into either of the original equations and solve for the other unknown.
> 4. Write the answer as an ordered pair and check the solution in both original equations.

D. Solving Systems Using Elimination

Now consider the system $\begin{cases} -2x + 5y = 13 \\ 2x - 3y = -7 \end{cases}$, where solving for any one of the variables will result in fractional values. The substitution method can still be used, but often the **elimination method** is more efficient. The method takes its name from what happens when you add certain equations in a system (by adding the like terms from each). If the coefficients of either x or y are additive inverses—they sum to zero and are *eliminated*. For the system shown, "adding the equations" produces $2y = 6$, giving $y = 3$, then $x = 1$ using back-substitution (verify).

If neither of the like-variable terms sum to zero, we can multiply one or both equations by a nonzero constant to "match up" the coefficients, so an elimination will take place. In doing so, we create an **equivalent system of equations,** meaning one that has the same solution as the original system. For $\begin{cases} 7x - 4y = 16 \\ -3x + 2y = -6 \end{cases}$, multiplying the second equation by 2 produces $\begin{cases} 7x - 4y = 16 \\ -6x + 4y = -12 \end{cases}$, and after adding the equations, we see that $x = 4$. Note the three systems produced are equivalent, and all have $(4, 3)$ as a solution ($y = 3$ was found using back-substitution).

1. $\begin{cases} 7x - 4y = 16 \\ -3x + 2y = -6 \end{cases}$ **2.** $\begin{cases} 7x - 4y = 16 \\ -6x + 4y = -12 \end{cases}$ **3.** $\begin{cases} 7x - 4y = 16 \\ x = 4 \end{cases}$

In summary,

> **Operations that Produce an Equivalent System**
>
> 1. Changing the order of the equations.
> 2. Replacing an equation by a nonzero constant multiple of that equation.
> 3. Replacing an equation with the sum of two equations from the system.

Before beginning a solution using elimination, check to make sure the equations are written in the **standard form** $Ax + By = C$, so that like terms will appear above/below each other. Throughout this chapter, we will use R1 to represent the equation in *row 1* of the system, R2 to represent the equation in *row 2*, and so on. These designations are used to help describe and document the steps being used to solve a system, as in Example 4 where 2R1 + R2 indicates the first equation has been multiplied by two, with the result added to the second equation.

EXAMPLE 4 ▶   **Solving a System by Elimination**

Solve using elimination: $\begin{cases} 2x - 3y = 7 \\ 6y + 5x = 4 \end{cases}$

Solution ▶ The second equation is not in standard form, so we rewrite the system as $\begin{cases} 2x - 3y = 7 \\ 5x + 6y = 4 \end{cases}$. If we "added the equations" now, we would get $7x + 3y = 11$, with neither variable eliminated. However, if we multiply *both sides* of the first equation by 2, the y-coefficients will become additive inverses. The sum then results in an equation with x as the only unknown.

$$Y_1 = \frac{7 - 2X}{-3}, \; Y_2 = \frac{4 - 5X}{6}$$

$$\begin{array}{r} 2R1 \\ + \\ R2 \\ \hline \text{sum} \end{array} \begin{cases} 4x - 6y = 14 \\ 5x + 6y = 4 \\ \hline 9x + 0y = 18 \end{cases} \quad \text{add}$$

$$9x = 18$$
$$x = 2 \quad \text{solve for } x$$

Substituting 2 for x back into either of the original equations yields $y = -1$. The ordered pair solution is $(2, -1)$. A graphical check is shown in the figure.

Now try Exercises 37 through 42 ▶

The elimination method is summarized here. If either equation has fraction or decimal coefficients, we can "clear" them using an appropriate constant multiplier.

> ### Solving Systems Using Elimination
>
> 1. Write each equation in standard form: $Ax + By = C$.
> 2. Multiply one or both equations by a constant that will create coefficients of x (or y) that are additive inverses.
> 3. Combine the two equations using vertical addition and solve for the variable that remains.
> 4. Substitute the value from step 3 into either of the original equations and solve for the other unknown.
> 5. Write the answer as an ordered pair and check the solution in both original equations.

WORTHY OF NOTE

As the elimination method involves adding two equations, it is sometimes referred to as the *addition method* for solving systems.

EXAMPLE 5 ▶ Solving a System Using Elimination

Solve using elimination: $\begin{cases} \frac{5}{8}x - \frac{3}{4}y = \frac{1}{4} \\ \frac{1}{2}x - \frac{2}{3}y = 1 \end{cases}$.

Solution ▶ Multiplying the first equation by 8 (8R1) and the second equation by 6 (6R2) will clear the fractions from each.

$$\begin{array}{c} 8R1 \\ 6R2 \end{array} \begin{cases} \frac{8}{1}(\frac{5}{8})x - \frac{8}{1}(\frac{3}{4})y = \frac{8}{1}(\frac{1}{4}) \\ \frac{6}{1}(\frac{1}{2})x - \frac{6}{1}(\frac{2}{3})y = 6(1) \end{cases} \rightarrow \begin{cases} 5x - 6y = 2 \\ 3x - 4y = 6 \end{cases}$$

The x-terms can now be eliminated if we use $3R1 + (-5R2)$.

$$\begin{array}{r} 3R1 \\ + \\ -5R2 \\ \hline \text{sum} \end{array} \begin{cases} 15x - 18y = 6 \\ -15x + 20y = -30 \\ \hline 0x + 2y = -24 \end{cases} \quad \text{add}$$

$$y = -12 \quad \text{solve for } y$$

Substituting $y = -12$ in either of the original equations yields $x = -14$, and the solution is $(-14, -12)$. We can check this solution by graphing, as we did for the solution to Example 4, or by substituting the *x*- and *y*-values into the original equations, as we did for Example 1. The check using substitution is shown in the figure. Note that -14  X has scrolled out of view.

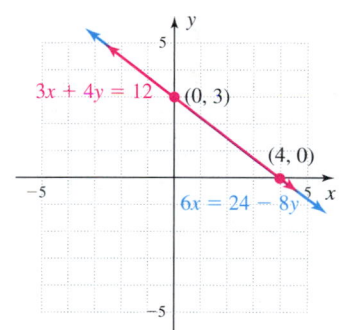

✓ **D.** You've just seen how we can solve linear systems by elimination

---

Now try Exercises 43 through 48 ▶

⚠ **CAUTION** ▶ Be sure to multiply *all* terms (on both sides) of the equation when using a constant multiplier. Also, note that for Example 5, we could have eliminated the *y*-terms using 2R1 with -3R2.

E. Inconsistent and Dependent Systems

A system having *at least one* solution is called a **consistent system.** As seen in Example 2, if the lines have different slopes, they intersect at a single point and the system has exactly one solution. Here, the lines are *independent* of each other and the system is called an **independent system.** If the lines have equal slopes *and* the same *y*-intercept, they are identical or **coincident lines.** Since one is right atop the other, they *intersect at all points,* and the system has an infinite number of solutions. Here, one line *depends* on the other and the system is called a **dependent system.** Using substitution or elimination on a dependent system results in the elimination of all variable terms and leaves a statement that is *always true,* such as $0 = 0$ or some other simple identity.

EXAMPLE 6 ▶ **Solving a Dependent System**

Solve using elimination: $\begin{cases} 3x + 4y = 12 \\ 6x = 24 - 8y \end{cases}$.

Solution ▶ Writing the system in standard form gives $\begin{cases} 3x + 4y = 12 \\ 6x + 8y = 24 \end{cases}$. By applying -2R1, we can eliminate the variable *x*:

$$\begin{array}{r} -2\text{R1} \\ + \\ \underline{\text{R2}} \\ \text{sum} \end{array} \begin{cases} -6x - 8y = -24 \\ 6x + 8y = 24 \end{cases}$$

add

$0x + 0y = 0$ variables are eliminated

$0 = 0$ true statement

Although we didn't expect it, both variables were eliminated and the final statement is true ($0 = 0$). This indicates the system is dependent, which the graph verifies (the lines are coincident). Writing both equations in slope-intercept form shows they represent the same line.

$$\begin{cases} 3x + 4y = 12 \\ 6x + 8y = 24 \end{cases} \longrightarrow \begin{cases} 4y = -3x + 12 \\ 8y = -6x + 24 \end{cases} \longrightarrow \begin{cases} y = -\dfrac{3}{4}x + 3 \\ y = -\dfrac{3}{4}x + 3 \end{cases}$$

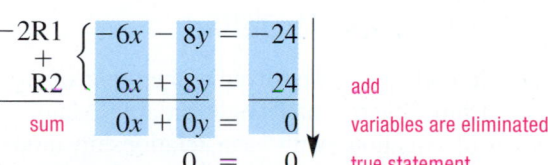

The solutions of a dependent system are often written in set notation as the set of ordered pairs (x, y), where y is a specified function of x. Here the solution would be $\{(x, y) | y = -\frac{3}{4}x + 3\}$. Using an ordered pair with an arbitrary variable, called a **parameter,** is also common: $\left(p, \dfrac{-3p}{4} + 3 \right)$.

Now try Exercises 49 through 60 ▶

Figure 6.7

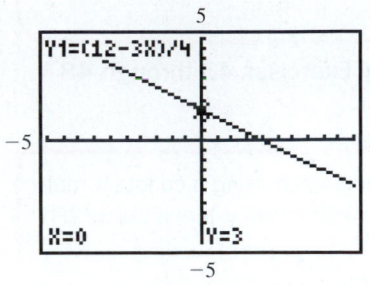

Figure 6.8

If we had attempted to solve the system in Example 6 by graphing (Figure 6.6), we could be mislead into thinking something is wrong—because only one line is visible (Figure 6.7). In this case, using the **TABLE** feature of the calculator would help verify that the system is dependent. Since the ordered pair solutions are identical (try scrolling through positive and negative values), the equations must be dependent (Figure 6.8).

Figure 6.6

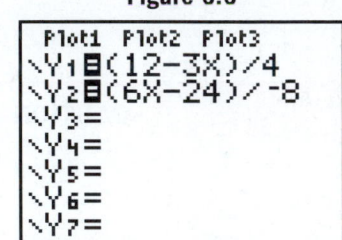

Finally, if the lines have equal slopes and *different y-intercepts,* they are parallel and the system will have no solution. A system with no solutions is called an **inconsistent system.** An "inconsistent system" produces an "inconsistent answer," such as $12 = 0$ or some other false statement when substitution or elimination is applied. In other words, all variable terms are once again eliminated, but the remaining statement is *false.* A summary of the three possibilities is shown in Figure 6.9 for arbitrary slope m and y-intercept $(0, b)$.

Figure 6.9

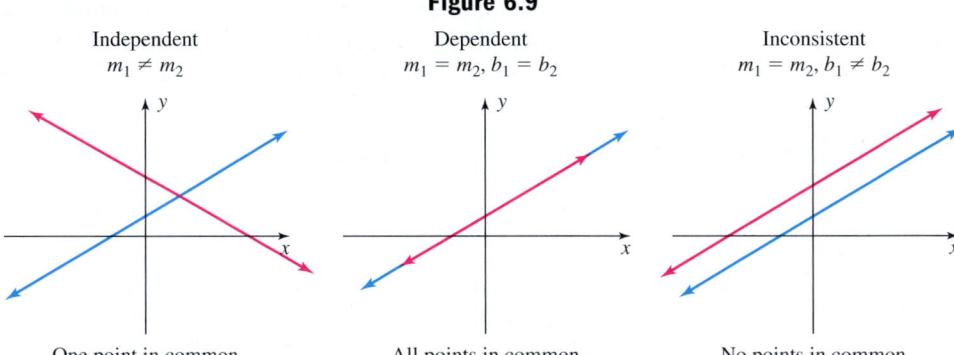

Independent
$m_1 \neq m_2$

Dependent
$m_1 = m_2, b_1 = b_2$

Inconsistent
$m_1 = m_2, b_1 \neq b_2$

One point in common   All points in common   No points in common

✅ **E.** You've just seen how we can recognize inconsistent systems and dependent systems

F. Systems and Modeling

In previous chapters, we solved numerous real-world applications by writing all given relationships in terms of a single variable. Many situations are easier to model using a system of equations with each relationship modeled independently using *two* variables. We begin here with a **mixture** application. Although they appear in many different forms (coin problems, metal alloys, investments, merchandising, and so on), mixture problems all have a similar theme. Generally one equation is related to *quantity* (how much of each item is being combined) and one equation is related to *value* (what is the value of each item being combined).

EXAMPLE 7 ▶   **Solving a Mixture Application**

A jeweler is commissioned to create a piece of artwork that will weigh 14 oz and consist of 75% gold. She has on hand two alloys that are 60% and 80% gold, respectively. How much of each should she use?

Solution ▶ Let x represent ounces of the 60% alloy and y represent ounces of the 80% alloy. The first equation must be $x + y = 14$, since the piece of art must weigh exactly 14 oz (this is the *quantity* equation). The x ounces are 60% gold, the y ounces are 80% gold, and the 14 oz will be 75% gold. This gives the *value* equation:

WORTHY OF NOTE

As an estimation tool, note that if equal amounts of the 60% and 80% alloys were used (7 oz each), the result would be a 70% alloy (halfway in between). Since a 75% alloy is needed, more of the 80% gold will be used.

$0.6x + 0.8y = 0.75(14)$. The system is $\begin{cases} x + y = 14 \\ 6x + 8y = 105 \end{cases}$ (after clearing decimals).

Solving for y in the first equation gives $y = 14 - x$. Substituting $14 - x$ for y in the second equation gives

$$
\begin{aligned}
6x + 8y &= 105 &&\text{second equation} \\
6x + 8(14 - x) &= 105 &&\text{substitute } 14 - x \text{ for } y \\
6x + 112 - 8x &= 105 &&\text{distribute} \\
-2x + 112 &= 105 &&\text{simplify} \\
x &= \frac{7}{2} &&\text{solve for } x
\end{aligned}
$$

$Y_1 = 14 - X,\ Y_2 = \dfrac{105 - 6X}{8}$

Intersection
X=3.5 ▬   Y=10.5

Substituting $\frac{7}{2}$ for x in the first equation gives $y = \frac{21}{2}$. She should use 3.5 oz of the 60% alloy and 10.5 oz of the 80% alloy. A graphical check is shown in the figure.

Now try Exercises 63 through 70 ▶

A second example involves an application of uniform motion (distance = rate · time), and explores concepts of great importance to the navigation of ships and airplanes. As a simple illustration, if you've ever walked at your normal rate r on the "moving walkways" at an airport, you likely noticed an increase in your total speed. This is because the resulting speed combines your walking rate r with the speed w of the walkway: *total speed* = $r + w$. If you walk in the opposite direction of the walkway, your total speed is much slower, as now *total speed* = $r - w$.

This same phenomenon is observed when an airplane is flying with or against the wind, or a ship is sailing with or against the current.

EXAMPLE 8 ▶ **Solving an Application of Systems—Uniform Motion**

An airplane flying due south from St. Louis, Missouri, to Baton Rouge, Louisiana, uses a strong, steady tailwind to complete the trip in only 2.5 hr. On the return trip, the same wind slows the flight and it takes 3 hr to get back. If the flight distance between these cities is 912 km, what is the cruising speed of the airplane (speed with no wind)? How fast is the wind blowing?

Solution ▶ Let r represent the rate of the plane and w the rate of the wind. Since $D = RT$, the flight to Baton Rouge can be modeled by $912 = (r + w)(2.5)$, and the return flight by $912 = (r - w)(3)$. This produces the system $\begin{cases} 912 = 2.5r + 2.5w \\ 912 = 3r - 3w \end{cases}$.

▼ **Algebraic Solution**

Dividing R1 by 2.5 and R2 by 3 produces the following sequence:

$\begin{array}{l} \dfrac{R1}{2.5} \\ \dfrac{R2}{3} \end{array} \begin{cases} 912 = 2.5r + 2.5w \\ 912 = 3r - 3w \end{cases} \rightarrow \begin{cases} 364.8 = r + w \\ 304.0 = r - w \end{cases}$

▼ **Graphical Solution**

Using x for w and y for r, we solve each equation for y and obtain:

$$Y_1 = \frac{912 - 2.5X}{2.5} \qquad Y_2 = \frac{912 + 3X}{3}$$

Using R1 + R2 gives $668.8 = 2r$, showing $334.4 = r$. The speed of the plane is 334.4 kph. Substituting 334.4 for r in the second equation, we have:

$912 = 3r - 3w$	equation
$912 = 3(334.4) - 3w$	substitute
$912 = 1003.2 - 3w$	multiply
$-91.2 = -3w$	subtract 1003.2
$30.4 = w$	divide by -3

The speed of the wind is 30.4 kph.

We then set an appropriate window and graph these equations to find the point of intersection.

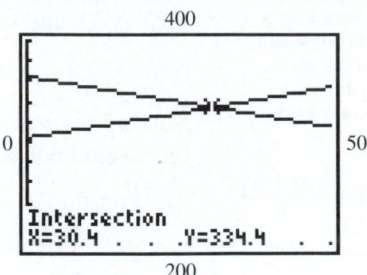

The speed of the wind (x) is 30.4 kph, and the speed of the plane (y) is 334.4 kph.

Now try Exercises 71 through 74 ▶

Systems of equations also play a significant role in *cost-based pricing* in the business world. The costs involved in running a business can broadly be understood as either a **fixed cost k** or a **variable cost v.** Fixed costs might include the monthly rent paid for facilities, which remains the same regardless of how many items are produced and sold. Variable costs would include the cost of materials needed to produce the item, which depends on the number of items made. The total cost can then be modeled by $C(x) = vx + k$ for x number of items. Once a **selling price p** has been determined, the revenue equation is simply $R(x) = px$ (price times number of items sold). We can now set up and solve a system of equations that will determine how many items must be sold to break even, performing what is called a **break-even analysis** where $C(x) = R(x)$.

EXAMPLE 9 ▶   Solving an Application of Systems: Break-Even Analysis

In home businesses that produce items to sell on Ebay®, fixed costs are easily determined by rent and utilities, and variable costs by the price of materials needed to produce the item. Karen's home business makes large decorative candles for all occasions. The cost of materials is $3.50 per candle, and her rent and utilities average $900 per month. If her candles sell for $9.50, how many candles must be sold each month to break even?

Solution ▶   Let x represent the number of candles sold. Her total cost is $C(x) = 3.5x + 900$ (variable cost plus fixed cost), and projected revenue is $R(x) = 9.5x$. This gives the

system $\begin{cases} C(x) = 3.5x + 900 \\ R(x) = 9.5x \end{cases}$. To break even, *Cost = Revenue* which gives

$$9.5x = 3.5x + 900$$
$$6x = 900$$
$$x = 150$$

The analysis shows that Karen must sell 150 candles each month to break even.

Now try Exercises 75 through 78 ▶

In a "free-market" economy, also referred to as a "supply-and-demand" economy, there are naturally occurring forces that invariably come into play if no outside forces act on the producers (suppliers) and consumers (demanders). Generally speaking, the higher the price of an item, the lower the demand. A good advertising campaign can increase the demand, but the increasing demand brings an increase in price, which moderates the demand—and so it goes until a balance is reached. These free-market forces ebb and flow until **market equilibrium** occurs, at the specific price where the supply and demand are equal.

In **Exercises 75 to 78,** the equation models were artificially constructed to yield a "nice" solution. In actual practice, the equations and coefficients are not so "well behaved" and are based on the collection and interpretation of real data. While market analysts have sophisticated programs and numerous models to help develop these equations, here we'll use our experience with regression to develop the supply and demand curves.

EXAMPLE 10 ▶ **Using Technology to Find Market Equilibrium**

A manufacturer of MP3 players has hired a consulting firm to do market research on their "next-generation" player. Over a 10-week period, the firm collected the data shown for the MP3 player market (data includes MP3 players sold and expected to sell).

Price (dollars)	Demand	Supply (Inventory)
107.10	6900	12,200
85.50	7900	9900
64.80	13,200	8000
52.20	13,500	7900
108.00	6700	14,000
91.80	7600	12,000
77.40	9200	9400
46.80	13,800	6100
74.70	10,600	8800
68.40	12,800	8600

a. Use a graphing calculator to simultaneously display the demand and supply scatterplots.

b. Calculate a line of best fit for each and graph them with the scatterplots (identify each curve).

c. Find the equilibrium point.

Solution ▶ **a.** Begin by clearing all lists. This can be done manually, or by pressing [2nd] [+] (**MEM**) and selecting option **4:ClrAllLists** (the command appears on the home screen). Pressing [ENTER] will execute the command, and the word DONE will appear. Carefully input price in L1, demand in L2, and supply in L3 (see Figure 6.10). With the window settings given in Figure 6.11, pressing [GRAPH] will display the price/demand and price/supply scatterplots shown. If this is not the case, use [2nd] [Y=] (**STAT PLOT**) to be sure that "On" is highlighted in Plot1 and Plot2, and that Plot1 uses L1 and L2, while Plot2 uses L1 and L3 (Figure 6.12). Note we've chosen a different mark to indicate the data points in Plot2.

b. Calculate the linear regression equation for L1 and L2 (demand), and paste it in Y_1: **LinReg (ax + b) L1, L2, Y₁** [ENTER]. Next, calculate the linear regression for L1 and L3 (supply) and paste it in Y_2: **LinReg (ax + b) L1, L3, Y₂** [ENTER] (recall that Y_1 and Y_2 are accessed using the [VARS] key). The resulting equations and graphs are shown in Figures 6.13 and 6.14.

c. Once again we use [2nd] [TRACE] (**CALC**) **5:intersect** to find the equilibrium point, which is approximately (80, 9931). Supply and demand for this MP3 player model are approximately equal at a price of about $80, with 9931 MP3 players bought and sold.

Figure 6.10

L1	L2	L3	1
107.1	6900	12200	
85.5	7900	9900	
64.8	13200	8000	
52.2	13500	7900	
108	6700	14000	
91.8	7600	12000	
77.4	9200	9400	

L1(1)=107.1

Figure 6.12

STAT PLOTS
1:Plot1...On
　L1　L2　□
2:Plot2...On
　L1　L3　+
3:Plot3...Off
　L1　L5　□
4↓PlotsOff

Figure 6.11

16,000

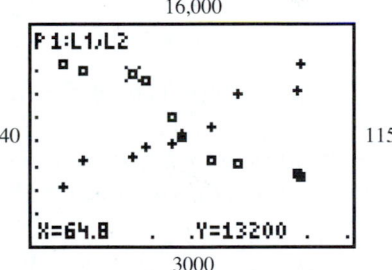

P1:L1,L2

40　　　　　　　　　　115

X=64.8　　.Y=13200.

3000

Figure 6.13

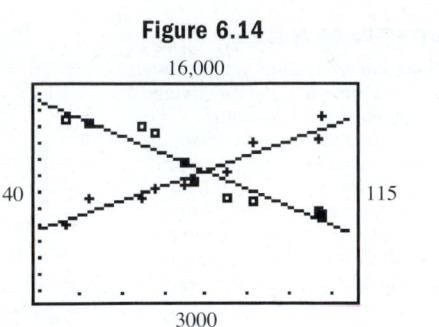

Figure 6.14

Now try Exercises 79 through 82 ▶

☑ **F.** You've just seen how we can use a system of equations to model and solve applications

Other interesting applications can be found in the Exercise set. **See Exercises 83 through 88.**

6.1 EXERCISES

▶ CONCEPTS AND VOCABULARY

Fill in the blank with the appropriate word or phrase. Carefully reread the section if needed.

1. Systems that have no solution are called _inconsistent_ systems.

2. Systems having at least one solution are called _consistent_ systems.

3. If the lines in a system intersect at a single point, the system is said to be _consistent_ and _independent_.

4. If the lines in a system are coincident, the system is referred to as _consistent_ and _dependent_.

5. The given systems are equivalent. How do we obtain the second system from the first?
$$\begin{cases} \frac{2}{3}x + \frac{1}{2}y = \frac{5}{3} \\ 0.2x + 0.4y = 1 \end{cases} \quad \begin{cases} 4x + 3y = 10 \\ 2x + 4y = 10 \end{cases}$$
Multiply the first equation by 6 and the second equation by 10.

6. For $\begin{cases} 2x + 5y = 8 \\ 3x + 4y = 5' \end{cases}$
which solution method would be more efficient, substitution or elimination? Discuss/Explain why.
Answers will vary.

▶ DEVELOPING YOUR SKILLS

Show the lines in each system would intersect in a single point by writing the equations in slope-intercept form.

7. $\begin{cases} 7x - 4y = 24 \\ 4x + 3y = 15 \end{cases}$
$y = \frac{7}{4}x - 6, y = \frac{-4}{3}x + 5$

8. $\begin{cases} 0.3x - 0.4y = 2 \\ 0.5x + 0.2y = -4 \end{cases}$
$y = \frac{3}{4}x - 5, y = \frac{-5}{2}x - 20$

An ordered pair is a solution to an equation if it makes the equation true. Given the graph shown here, determine which equation(s) have the indicated point as a solution. If the point satisfies more than one equation, write the system for which it is a solution.

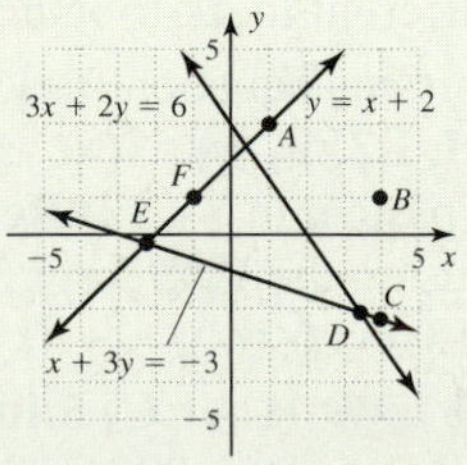

9. *A* $y = x + 2$

10. *B*   none

11. *C* $x + 3y = -3$

12. *D* $\begin{aligned} x + 3y &= -3, \\ 3x + 2y &= 6 \end{aligned}$

13. *E* $y = x + 2, x + 3y = -3$

14. *F* $y = x + 2$

Substitute the x- and y-values indicated by the ordered pair to determine if it is a solution to the system. Also check using the ⟨ALPHA⟩ keys on the home screen of a graphing calculator.

15. $\begin{cases} 3x + y = 11 \\ -5x + y = -13; \end{cases} (3, 2)$

16. $\begin{cases} 3x + 7y = -4 \\ 7x + 8y = -21; \end{cases} (-6, 2)$

17. $\begin{cases} 8x - 24y = -17 \\ 12x + 30y = 2; \end{cases} \left(-\frac{7}{8}, \frac{5}{12}\right)$

Additional answers can be found in the Instructor Answer Appendix.

HOMEWORK SELECTION GUIDE

Core: 7–14, 15–35 odd, 37–57 every other odd, 63–87 every other odd (32 Exercises)
Standard: 1–4, 7–14, 15–35 odd, 37–57 every other odd, 61, 63–87 every other odd 89, 93 (32 Exercises)

Extended: 1–4, 7–14, 15–35 odd, 37–57 every other odd, 61, 62, 63–87 every other odd 89, 91, 94 (32 Exercises)
In Depth: 1–6, 7–14, 15–35 odd, 37–57 every other odd, 61, 62, 63–87 every other odd 89, 90, 91–94 (32 Exercises)

18. $\begin{cases} 4x + 15y = 7 \\ 8x + 21y = 11; \end{cases} \left(\dfrac{1}{2}, \dfrac{1}{3}\right)$

Solve each system by *graphing* manually. Check results by graphing the system on a graphing calculator, and locating any points of intersection.

19. $\begin{cases} 3x + 2y = 12 \\ x - y = 9 \end{cases}$ **20.** $\begin{cases} 5x + 2y = -2 \\ -3x + y = 10 \end{cases}$

21. $\begin{cases} 5x - 2y = 5 \\ x + 3y = -16 \end{cases}$ **22.** $\begin{cases} 3x + y = -2 \\ 5x + 3y = 2 \end{cases}$

Solve each system using *substitution*. Write solutions as an ordered pair, and verify solutions using a graphing calculator.

23. $\begin{cases} x = 5y - 9 \\ x - 2y = -6 \end{cases}$
$(-4, 1)$ **24.** $\begin{cases} 4x - 5y = 7 \\ 2x - 5 = y \end{cases}$ $(3, 1)$

25. $\begin{cases} y = \frac{2}{3}x - 7 \\ 3x - 2y = 19 \end{cases}$
$(3, -5)$ **26.** $\begin{cases} 2x - y = 6 \\ y = \frac{3}{4}x - 1 \end{cases}$ $(4, 2)$

Identify the equation and variable that makes the substitution method easiest to use, then solve the system. Verify solutions using a graphing calculator.

27. $\begin{cases} 3x - 4y = 24 \\ 5x + y = 17 \end{cases}$
second equation, y, $(4, -3)$ **28.** $\begin{cases} 3x + 2y = 19 \\ x - 4y = -3 \end{cases}$
second equation, x, $(5, 2)$

29. $\begin{cases} 0.7x + 2y = 5 \\ x - 1.4y = 11.4 \end{cases}$
second equation, x, $(10, -1)$ **30.** $\begin{cases} 0.8x + y = 7.4 \\ 0.6x + 1.5y = 9.3 \end{cases}$
first equation, y, $(3, 5)$

31. $\begin{cases} 5x - 6y = 2 \\ x + 2y = 6 \end{cases}$
second equation, x, $\left(\frac{5}{2}, \frac{7}{4}\right)$ **32.** $\begin{cases} 2x + 5y = 5 \\ 8x - y = 6 \end{cases}$
second equation, y, $\left(\frac{5}{6}, \frac{2}{3}\right)$

The substitution method can be used for *like variables* or for *like expressions*. Solve the following systems, *using the expression* common to both equations (do not solve for x or y alone).

33. $\begin{cases} 2x + 4y = 6 \\ x + 12 = 4y \end{cases}$ $\left(-2, \frac{5}{2}\right)$ **34.** $\begin{cases} 8x = 3y + 24 \\ 8x - 5y = 36 \end{cases}$ $\left(\frac{3}{4}, -6\right)$

35. $\begin{cases} 5x - 11y = 21 \\ 11y = 5 - 8x \end{cases}$
$(2, -1)$ **36.** $\begin{cases} -6x = 5y - 16 \\ 5y - 6x = 4 \end{cases}$
$(1, 2)$

Solve using *elimination*. In some cases, the system must first be written in standard form. Verify solutions using a graphing calculator.

37. $\begin{cases} 2x - 4y = 10 \\ 3x + 4y = 5 \end{cases}$
$(3, -1)$ **38.** $\begin{cases} -x + 5y = 8 \\ x + 2y = 6 \end{cases}$
$(2, 2)$

39. $\begin{cases} 4x - 3y = 1 \\ 3y = -5x - 19 \end{cases}$
$(-2, -3)$ **40.** $\begin{cases} 5y - 3x = -5 \\ 3x + 2y = 19 \end{cases}$ $(5, 2)$

41. $\begin{cases} 2x = -3y + 17 \\ 4x - 5y = 12 \end{cases}$
$\left(\frac{11}{2}, 2\right)$ **42.** $\begin{cases} 2y = 5x + 2 \\ -4x = 17 - 6y \end{cases}$
$\left(1, \frac{7}{2}\right)$

43. $\begin{cases} 0.5x + 0.4y = 0.2 \\ 0.3y = 1.3 + 0.2x \end{cases}$
$(-2, 3)$ **44.** $\begin{cases} 0.2x + 0.3y = 0.8 \\ 0.3x + 0.4y = 1.3 \end{cases}$
$(7, -2)$

45. $\begin{cases} 0.32m - 0.12n = -1.44 \\ -0.24m + 0.08n = 1.04 \end{cases}$ $(-3, 4)$

46. $\begin{cases} 0.06g - 0.35h = -0.67 \\ -0.12g + 0.25h = 0.44 \end{cases}$ $(0.5, 2)$

47. $\begin{cases} -\frac{1}{6}u + \frac{1}{4}v = 4 \\ \frac{1}{2}u - \frac{2}{3}v = -11 \end{cases}$
$(-6, 12)$ **48.** $\begin{cases} \frac{3}{4}x + \frac{1}{3}y = -2 \\ \frac{3}{2}x + \frac{1}{5}y = 3 \end{cases}$
$(4, -15)$

Solve using any method and identify the system as consistent, inconsistent, or dependent. Verify solutions using a graphing calculator.

49. $\begin{cases} 4x + \frac{3}{4}y = 14 \\ -9x + \frac{5}{8}y = -13 \end{cases}$
$(2, 8)$; consistent/independent   **50.** $\begin{cases} \frac{2}{3}x + y = 2 \\ 2y = \frac{5}{6}x - 9 \end{cases}$
$(6, -2)$; consistent/independent

51. $\begin{cases} 0.2y = 0.3x + 4 \\ 0.6x - 0.4y = -1 \end{cases}$
$\varnothing$; inconsistent   **52.** $\begin{cases} 1.2x + 0.4y = 5 \\ 0.5y = -1.5x + 2 \end{cases}$
$\varnothing$; inconsistent

53. $\begin{cases} 6x - 22 = -y \\ 3x + \frac{1}{2}y = 11 \end{cases}$
$\{(x, y) | 6x + y = 22\}$; consistent/dependent   **54.** $\begin{cases} 15 - 5y = -9x \\ -3x + \frac{5}{3}y = 5 \end{cases}$

55. $\begin{cases} -10x + 35y = -5 \\ y = 0.25x \end{cases}$
$(4, 1)$; consistent/independent   **56.** $\begin{cases} 2x + 3y = 4 \\ x = -2.5y \end{cases}$
$(5, -2)$; consistent/independent

57. $\begin{cases} 7a + b = -25 \\ 2a - 5b = 14 \end{cases}$
$(-3, -4)$; consistent/independent   **58.** $\begin{cases} -2m + 3n = -1 \\ 5m - 6n = 4 \end{cases}$
$(2, 1)$; consistent/independent

59. $\begin{cases} 4a = 2 - 3b \\ 6b + 2a = 7 \end{cases}$
$\left(\frac{-1}{2}, \frac{4}{3}\right)$; consistent/independent   **60.** $\begin{cases} 3p - 2q = 4 \\ 9p + 4q = -3 \end{cases}$
$\left(\frac{1}{3}, \frac{-3}{2}\right)$; consistent/independent

▶ WORKING WITH FORMULAS

61. Uniform motion with current: $\begin{cases} (R + C)T_1 = D_1 \\ (R - C)T_2 = D_2 \end{cases}$

The formula shown can be used to solve uniform motion problems involving a *current,* where D represents distance traveled, R is the rate of the object with no current, C is the speed of the current, and T is the time. Chan-Li rows 9 mi up river (against the current) in 3 hr. It only took him 1 hr to row 5 mi downstream (with the current). How fast was the current? How fast can he row in still water?   1 mph, 4 mph

Additional answers can be found in the Instructor Answer Appendix.

62. Fahrenheit and Celsius temperatures: $\begin{cases} y = \frac{9}{5}x + 32 & °F \\ y = \frac{5}{9}(x - 32) & °C \end{cases}$

Many people are familiar with temperature measurement in degrees Celsius and degrees Fahrenheit, but few realize that the equations are linear and there is one temperature at which the two scales agree. Solve the system using the method of your choice and find this temperature. $-40°C = -40°F$

▶ APPLICATIONS

Solve each application by modeling the situation with a linear system. Be sure to clearly indicate what each variable represents. Check answers using a graphing calculator and the method of your choice.

Mixture

63. Theater productions: At a recent production of *A Comedy of Errors,* the Community Theater brought in a total of $30,495 in revenue. If adult tickets were $9 and children's tickets were $6.50, how many tickets of each type were sold if 3800 tickets in all were sold?   2318 adult tickets; 1482 child tickets

64. Milkfat requirements: A dietician needs to mix 10 gal of milk that is $2\frac{1}{2}$% milkfat for the day's rounds. He has some milk that is 4% milkfat and some that is $1\frac{1}{2}$% milkfat. How much of each should be used?   6 gallons of $1\frac{1}{2}$% milkfat; 4 gallons of 4% milkfat

65. Filling the family cars: Cherokee just filled both of the family vehicles at a service station. The total cost for 20 gal of regular unleaded and 17 gal of premium unleaded was $144.89. The premium gas was $0.10 more per gallon than the regular gas. Find the price per gallon for each type of gasoline.   premium: $3.97, regular: $3.87

66. Household cleaners: As a cleaning agent, a solution that is 24% vinegar is often used. How much pure (100%) vinegar and 5% vinegar must be mixed to obtain 50 oz of a 24% solution?   40 oz of 5%, 10 oz of pure

67. Alumni contributions: A wealthy alumnus donated $10,000 to his alma mater. The college used the funds to make a loan to a science major at 7% interest and a loan to a nursing student at 6% interest. That year the college earned $635 in interest. How much was loaned to each student?   nursing student $6500; science major $3500

68. Investing in bonds: A total of $12,000 is invested in two municipal bonds, one paying 10.5% and the other 12% simple interest. Last year the annual interest earned on the two investments was $1335. How much was invested at each rate? $7000 invested at 10.5%; $5000 invested at 12%

69. Saving money: Bryan has been doing odd jobs around the house, trying to earn enough money to buy a new Dirt-Surfer©. He saves all quarters and dimes in his piggy bank, while he places all nickels and pennies in a drawer to spend. So far, he has 225 coins

in the piggy bank, worth a total of $45.00. How many of the coins are quarters? How many are dimes?   150 quarters, 75 dimes

70. Coin investments: In 1990, Molly attended a coin auction and purchased some rare "Flowing Hair" fifty-cent pieces, and a number of very rare two-cent pieces from the Civil War Era. If she bought 47 coins with a face value of $10.06, how many of each denomination did she buy?   19 fifty-cent pieces, 28 two-cent pieces

Uniform Motion

71. Canoeing on a stream: On a recent camping trip, it took Molly and Sharon 2 hr to row 4 mi upstream from the drop in point to the campsite. After a leisurely weekend of camping, fishing, and relaxation, they rowed back downstream to the drop in point in just 30 min. Use this information to find (a) the speed of the current and (b) the speed Sharon and Molly would be rowing in still water.   **a.** 3 mph **b.** 5 mph

72. Taking a luxury cruise: A luxury ship is taking a Caribbean cruise from Caracas, Venezuela, to just off the coast of Belize City on the Yucatan Peninsula, a distance of 1435 mi. En route they encounter the Caribbean Current, which flows to the northwest, parallel to the coastline. From Caracas to the Belize coast, the trip took 70 hr. After a few days of fun in the sun, the ship leaves for Caracas, with the return trip taking 82 hr. Use this information to find (a) the speed of the Caribbean Current and (b) the cruising speed of the ship.   **a.** 1.5 mph **b.** 19 mph

73. Airport walkways: As part of an algebra field trip, Jason takes his class to the airport to use their moving walkways for a demonstration. The class measures the longest walkway, which turns out to be 256 ft long. Using a stop watch, Jason shows it takes him just 32 sec to complete the walk going in the same direction as the walkway. Walking in a direction opposite the walkway, it takes him 320 sec—10 times as long! The next day in class, Jason hands out a two-question quiz: (1) What was the speed of the walkway in feet per second? (2) What is my (Jason's) normal walking speed? Create the answer key for this quiz.   **a.** 3.6 ft/sec **b.** 4.4 ft/sec

74. Racing pigeons: The American Racing Pigeon Union often sponsors opportunities for owners to fly their birds in friendly competitions. During a recent competition, Steve's birds were liberated in Topeka, Kansas, and headed almost due north to their loft in Sioux Falls, South Dakota, a distance of 308 mi. During the flight, they encountered a steady wind from the north and the trip took 4.4 hr. The next month, Steve took his birds to a competition in Grand Forks, North Dakota, with the birds heading almost due south to home, also a distance of 308 mi. This time the birds were aided by the same wind from the north, and the trip took only 3.5 hr. Use this information to (a) find the racing speed of Steve's birds and (b) find the speed of the wind.   **a.** 79 mph   **b.** 9 mph

75. Lawn service: Dave and his sons run a lawn service, which includes mowing, edging, trimming, and aerating lawns. His fixed cost includes insurance, his salary, and monthly payments on equipment, and amounts to $4000/mo. The variable costs include gas, oil, hourly wages for his employees, and miscellaneous expenses, which run about $75 per lawn. The average charge for full-service lawn care is $115 per visit. Do a break-even analysis to (a) determine how many lawns Dave must service each month to break even and (b) the revenue required to break even.
a. 100 lawns/mo   **b.** $11,500/mo

76. Production of mini-microwave ovens: Due to high market demand, a manufacturer decides to introduce a new line of mini-microwave ovens for personal and office use. By using existing factory space and retraining some employees, fixed costs are estimated at $8400/mo. The components to assemble and test each microwave are expected to run $45 per unit. If market research shows consumers are willing to pay at least $69 for this product, find (a) how many units must be made and sold each month to break even and (b) the revenue required to break even.   **a.** 350 units/mo **b.** $24,150/mo

77. Farm commodities: One area where the law of supply and demand is clearly at work is farm commodities. Both growers and consumers watch this relationship closely, and use data collected by government agencies to track the relationship and make adjustments, as when a farmer decides to convert a large portion of her farmland from corn to soybeans to improve profits. Suppose that for x billion bushels of soybeans, supply is modeled by $y = 1.5x + 3$, where y is the current market price (in dollars per bushel). The related demand equation might be $y = -2.20x + 12$. (a) How many billion bushels will be supplied at a market price of $5.40? What will the demand be at this

price? Is supply less than demand? (b) How many billion bushels will be supplied at a market price of $7.05? What will the demand be at this price? Is demand less than supply? (c) To the nearest cent, at what price does the market reach equilibrium? How many bushels are being supplied/demanded?   **a.** 1.6 billion bu, 3 billion bu, yes   **b.** 2.7 billion bu, 2.25 billion bu, yes **c.** $6.65, 2.43 billion bu

78. Digital media: Market research has indicated that by 2015, sales of MP3 players and similar products will mushroom into a $70 billion dollar market. With a market this large, competition is often fierce—with suppliers fighting to earn and hold market shares. For x million MP3 players sold, supply is modeled by $y = 10.5x + 25$, where y is the current market price (in dollars). The related demand equation might be $y = -5.20x + 140$. (a) How many million MP3 players will be supplied at a market price of $88? What will the demand be at this price? Is supply less than demand? (b) How many million MP3 players will be supplied at a market price of $114? What will the demand be at this price? Is demand less than supply? (c) To the nearest cent, at what price does the market reach equilibrium? How many units are being supplied/demanded?   **a.** 6 million, 10 million, yes   **b.** about 8.5 million, 5 million, yes   **c.** $101.91, 7.32 million

 79. Pricing wakeboards: A water sports company that manufactures high-end wakeboards has hired an outside consulting firm to do some market research on their wakeboard. This consulting firm collected the following supply and demand data for this and comparable wakeboards over a 10-week period. Find the equilibrium point. Round your answer to the nearest integer and dollar.

Average Price (in U.S. dollars)	Quantity Demanded	Available Inventory
424.85	175	232
445.25	166	247
389.55	291	215
349.98	391	201
402.22	218	226
413.87	200	222
481.73	139	251
419.45	177	235
397.05	220	219
361.90	317	212

about 227 boards at $410 a piece

80. Pricing pet care products: A metal shop that manufactures pens for pet rabbits has collected some data on sales and production over the past 8 weeks. The following table shows the supply and demand data for these pens. Find the equilibrium point (round to the nearest cent and whole cage).

Average Price (in U.S. dollars)	Quantity Sold (Demand)	Production (Supply)
22.99	12	7
21.49	14	6
23.99	11	7
26.99	9	11
25.99	8	10
27.99	8	13
24.49	10	9
26.49	9	11

about 10 cages at $25.49 per cage

 81. Tracking supply and demand—oil products: The U.S. Bureau of Labor and Statistics tracks important data from many different markets. In May 2008, it collected the following supply-and-demand data for refined gasoline. Data were collected every Tuesday and Friday. Find the equilibrium point, rounding your answer to the nearest hundred thousand gallons and whole cent.

Average Price (in U.S. dollars)	Quantity Demanded (1×10^7 gal)	Available Inventory (1×10^7 gal)
3.17	8.82	9.10
3.12	8.87	9.05
3.04	9.08	8.97
2.84	9.22	8.91
3.11	8.92	9.02
3.15	8.76	9.08
3.10	9.01	8.99
3.11	8.94	9.01
2.93	9.13	8.93

about 90,000,000 gal at $3.04 gal

 82. Tracking supply and demand—energy efficient lightbulbs: The U.S. Bureau of Labor and Statistics has collected the following supply and demand data for the energy-efficient fluorescent lightbulbs sold each month for the past year. Find the equilibrium point, rounding your answer to the nearest ten thousand lightbulbs and whole cent. What is the yearly demand at the equilibrium point?

Average Price (in U.S. dollars)	Quantity Demanded (in millions)	Available Inventory (in millions)
9.40	0.84	1.23
8.51	1.17	0.95
8.78	1.05	1.11
10.82	0.68	1.29
6.77	1.47	0.77
9.33	0.91	1.21
8.34	1.25	0.88
10.37	0.76	1.27
8.62	1.09	1.02
8.44	1.21	0.92
8.58	1.18	0.97
8.96	1.01	1.17

about 1,060,000 lightbulbs at $8.87 per bulb. Yearly demand is 12,720,000 at equilibrium point.

Descriptive Translation

83. Important dates in U.S. history: If you sum the year that the Declaration of Independence was signed and the year that the Civil War ended, you get 3641. There are 89 yr that separate the two events. What year was the Declaration signed? What year did the Civil War end?   1776; 1865

84. Architectural wonders: When it was first constructed in 1889, the Eiffel Tower in Paris, France, was the tallest structure in the world. In 1975, the CN  Tower in Toronto, Canada, became the world's tallest structure. The CN Tower is 153 ft less than twice the height of the Eiffel Tower, and the sum of their heights is 2799 ft. How tall is each tower?
Eiffel Tower is 984 ft; CN Tower is 1815 ft

85. Pacific islands land area: In the South Pacific, the island nations of Tahiti and Tonga have a combined land area of 692 mi^2. Tahiti's land area is 112 mi^2 more than Tonga's. What is the land area of each island group?   Tahiti: 402 mi^2, Tonga: 290 mi^2

86. Card games: On a cold winter night, in the lobby of a beautiful hotel in Sante Fe, New Mexico, Marc and Klay just barely beat John and Steve in a close game of Trumps. If the sum of the team scores was 990 points, and there was a 12-point margin of victory, what was the final score?
Marc and Klay: 501 points, John and Steve: 489 points

Given any two points, the equation of a line through these points can be found using a system of equations. While there are certainly more efficient methods, using a system here will show how we can find equations for polynomials of higher degree. The key is to note that each point will yield an equation of the form $y = mx + b$. For instance, the

points $(3, 6)$ and $(-2, -4)$ yield the system $\begin{cases} 6 = 3m + b \\ -4 = -2m + b \end{cases}$.

87. Use a system of equations to find the equation of the line containing the points $(2, 7)$ and $(-4, -5)$. $y = 2x + 3$

88. Use a system of equations to find the equation of the line containing the points $(9, -1)$ and $(-3, 7)$. $y = -\dfrac{2}{3}x + 5$

▶ EXTENDING THE CONCEPT

89. Federal income tax reform has been a hot political topic for many years. Suppose tax plan A calls for a flat tax of 20% tax on all income (no deductions or loopholes). Tax plan B requires taxpayers to pay $5000 plus 10% of all income. For what income level do both plans require the same tax? $50,000

90. Suppose a certain amount of money was invested at 6% per year, and another amount at 8.5% per year, with a total return of $1250. If the amounts invested at each rate were switched, the yearly income would have been $1375. To the nearest whole dollar, how much was invested at each rate? $6552 at 8.5%; $11,551 at 6%

▶ MAINTAINING YOUR SKILLS

91. **(4.2)** Use the rational zeroes theorem to write the polynomial in completely factored form:
$3x^4 - 19x^3 + 15x^2 + 27x - 10$.
$(x - 5)(x - 2)(x + 1)(3x - 1)$

92. **(2.2)** Given the tool box function $f(x) = |x|$, sketch the graph of $F(x) = -|x + 3| - 2$.

93. **(3.2)** Graph $y = x^2 - 6x - 16$ and state the interval where $f(x) \leq 0$.

94. **(5.5)** Solve for x (rounded to the nearest thousandth): $33 = 77.5e^{-0.0052x} - 8.37$.
$x \approx 120.716$

Additional answers can be found in the Instructor Answer Appendix.

6.2 Linear Systems in Three Variables with Applications

LEARNING OBJECTIVES

In Section 6.2 you will see how we can:

☐ **A.** Visualize a solution in three dimensions

☐ **B.** Check ordered triple solutions

☐ **C.** Solve linear systems in three variables

☐ **D.** Recognize inconsistent and dependent systems

☐ **E.** Use a system of three equations in three variables to solve applications

The transition to systems of three equations in three variables requires a fair amount of "visual gymnastics" along with good organizational skills. Although the techniques used are identical and similar results are obtained, the third equation and variable give us more to track, and we must work more carefully toward the solution.

A. Visualizing Solutions in Three Dimensions

The solution to an equation in one variable is the single number that satisfies the equation. For $x + 1 = 3$, the solution is $x = 2$ and its graph is a single *point* on the number line, a **one-dimensional graph.** The solution to an equation in two variables, such as $x + y = 3$, is an ordered pair (x, y) that satisfies the equation. When we graph this solution set, the result is a *line* on the xy-coordinate grid, a **two-dimensional graph.** The solutions to an equation in three variables, such as $x + y + z = 6$, are the **ordered triples** (x, y, z) that satisfy the equation. When we graph this solution set, the result is a **plane in space**, a *graph in three dimensions*. Recall a plane is a flat surface having infinite length and width, but no depth. We can graph this plane using the intercept method and the result is shown in Figure 6.15. For graphs in three dimensions, the xy-plane is parallel to the ground (the y-axis points to the right) and z is the **vertical axis.** To find an additional point on this plane, we use any three numbers whose sum is 6, such as $(2, 3, 1)$. Move 2 units along the x-axis, 3 units parallel to the y-axis, and 1 unit parallel to the z-axis, as shown in Figure 6.16.

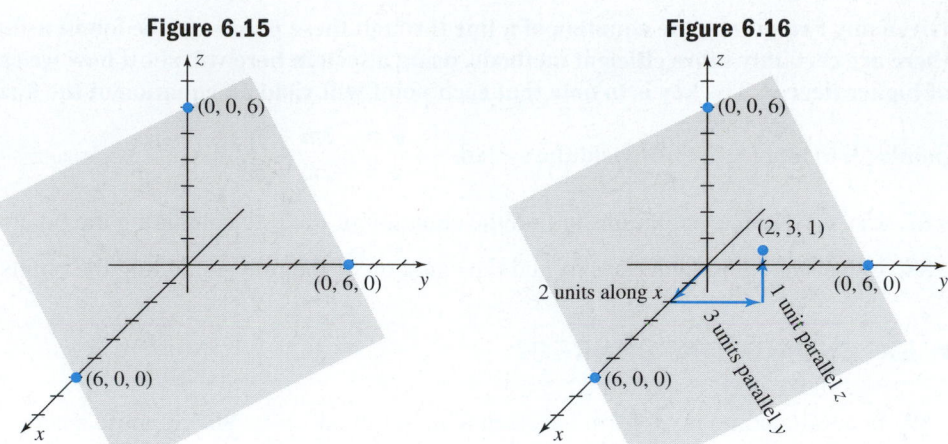

Figure 6.15   Figure 6.16

EXAMPLE 1 ▶   **Finding Solutions to an Equation in Three Variables**

Use a guess-and-check method to find four additional points on the plane determined by $x + y + z = 6$.

Solution ▶   We can begin by letting $x = 0$, then use any combination of y and z that sum to 6. Two examples are (0, 2, 4) and (0, 5, 1). We could also select any two values for x and y, then determine a value for z that results in a sum of 6. Two examples are $(-2, 9, -1)$ and $(8, -3, 1)$.

☑ **A.** You've just seen how we can visualize a solution in three dimensions

Now try Exercises 7 through 10 ▶

B. Solutions to a System of Three Equations in Three Variables

When solving a system of three equations in three variables, remember each equation represents a plane in space. These planes can intersect in various ways, creating different possibilities for a solution set (see Figures 6.17 to 6.20). The system could have a **unique solution** (a, b, c), if the planes intersect at a single point (Figure 6.17) (the point satisfies all three equations simultaneously). If the planes intersect in a line (Figure 6.18), the system is **linearly dependent** and there is an infinite number of solutions. Unlike the two-dimensional case, the equation of a line in three dimensions is somewhat complex, and the coordinates of all points on this line are usually *represented* by a specialized ordered triple, which we use to state the solution set. If the planes intersect at all points, the system has **coincident dependence** (see Figure 6.18). This indicates the equations of the system differ by only a constant multiple—they are all "disguised forms" of the *same equation*. The solution set is any ordered triple (a, b, c) satisfying this equation. Finally, the system may have no solutions. This can happen a number of different ways, most notably if the planes either intersect or are parallel, as shown in Figure 6.20 (other possibilities are discussed in the exercises). In the case of "no solutions," an ordered triple may satisfy none of the equations, only one of the equations, only two of the equations, but not all three equations.

Figure 6.17   Figure 6.18   Figure 6.19   Figure 6.20

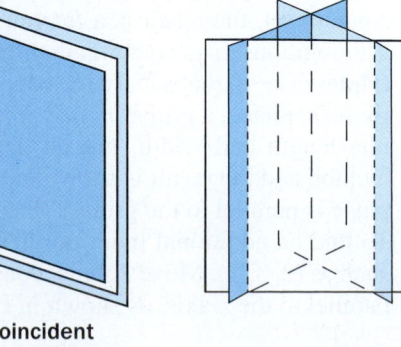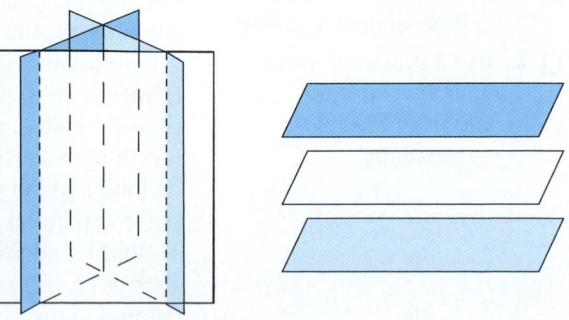

Unique solution   Linear   Coincident   No solutions
                    dependence   dependence

EXAMPLE 2 ▶ **Determining If an Ordered Triple Is a Solution**

Determine if the ordered triple $(1, -2, 3)$ is a solution to the systems shown.

a. $\begin{cases} x + 4y - z = -10 \\ 2x + 5y + 8z = 4 \\ x - 2y - 3z = -4 \end{cases}$ **b.** $\begin{cases} 3x + 2y - z = -4 \\ 2x - 3y - 2z = 2 \\ x - y + 2z = 9 \end{cases}$

Solution ▶ Substitute 1 for x, -2 for y, and 3 for z in the first system.

a. $\begin{cases} x + 4y - z = -10 \\ 2x + 5y + 8z = 4 \\ x - 2y - 3z = -4 \end{cases} \rightarrow \begin{cases} (1) + 4(-2) - (3) = -10 \\ 2(1) + 5(-2) + 8(3) = 4 \\ (1) - 2(-2) - 3(3) = -4 \end{cases} \rightarrow \begin{cases} -10 = -10 \text{ true} \\ 16 = 4 \text{ false} \\ -4 = -4 \text{ true} \end{cases}$

No, the ordered triple $(1, -2, 3)$ is not a solution to the first system. Now use the same substitutions in the second system.

b. $\begin{cases} 3x + 2y - z = -4 \\ 2x - 3y - 2z = 2 \\ x - y + 2z = 9 \end{cases} \rightarrow \begin{cases} 3(1) + 2(-2) - (3) = -4 \\ 2(1) - 3(-2) - 2(3) = 2 \\ (1) - (-2) + 2(3) = 9 \end{cases} \rightarrow \begin{cases} -4 = -4 \text{ true} \\ 2 = 2 \text{ true} \\ 9 = 9 \text{ true} \end{cases}$

The ordered triple $(1, -2, 3)$ is a solution to the second system only.

Now try Exercises 11 and 12 ▶

As with systems of two equations in two variables, solutions to larger systems can also be checked by storing values using the (ALPHA) keys, and entering each equation on the home screen of a graphing calculator. The screens shown in Figures 6.21 and 6.22 illustrate this process for Example 2(b).

Figure 6.21

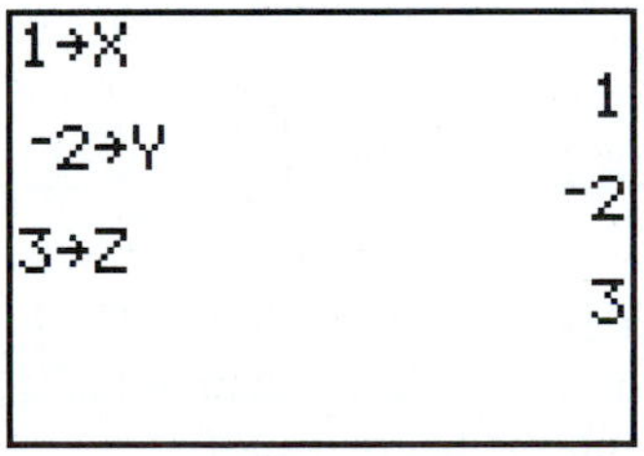

```
1→X
                1
-2→Y
               -2
3→Z
                3
```

Figure 6.22

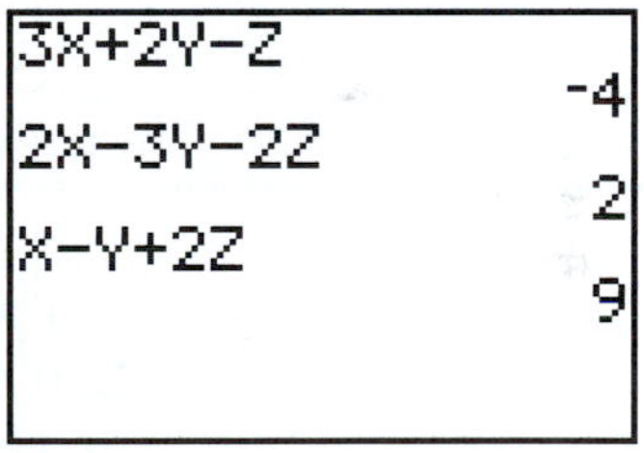

```
3X+2Y-Z
               -4
2X-3Y-2Z
                2
X-Y+2Z
                9
```

☑ **B.** You've just seen how we can check ordered triple solutions

C. Solving Systems of Three Equations in Three Variables Using Elimination

From Section 6.1, we know that two systems of equations are **equivalent** if they have the same solution set. The systems

$$\begin{cases} 2x + y - 2z = -7 \\ x + y + z = -1 \\ -2y - z = -3 \end{cases} \quad \text{and} \quad \begin{cases} 2x + y - 2z = -7 \\ y + 4z = 5 \\ z = 1 \end{cases}$$

are equivalent, as both have the unique solution $(-3, 1, 1)$. In addition, it is evident that the second system can be solved more easily, since R2 and R3 have fewer variables than the first system. In the simpler system, mentally substituting 1 for z into R2 immediately gives $y = 1$, and these values can be back-substituted into the first equation to find that $x = -3$. This observation guides us to a general approach for solving larger systems—we would like to *eliminate variables in the second and third equations, until we obtain an equivalent system that can easily be solved by back-substitution.* To begin, let's review the three operations that "transform" a given system, and produce an equivalent system.

Operations That Produce an Equivalent System

1. Changing the order of the equations.
2. Replacing an equation by a nonzero constant multiple of that equation.
3. Replacing an equation with the sum of two equations from the system.

Building on the ideas from Section 6.1, we develop the following approach for solving a system of three equations in three variables.

Solving a System of Three Equations in Three Variables

1. Write each equation in standard form: $Ax + By + Cz = D$.
2. If the "x" term in any equation has a coefficient of 1, interchange equations (if necessary) so this equation becomes R1.
3. Use the x-term in R1 to eliminate the x-terms from R2 and R3. The original R1, with the new R2 and R3, form an equivalent system that contains a smaller "subsystem" of two equations in two variables.
4. Solve the subsystem for either x or y and keep the result as the new R3. The result is an equivalent system that can be solved using back-substitution.

We'll begin by solving the system $\begin{cases} 2x + y - 2z = -7 \\ x + y + z = -1 \\ -2y - z = -3 \end{cases}$ using the elimination method and the procedure outlined. In Example 3, the notation $-2R1 + R2 \rightarrow R2$ indicates the equation in row 1 has been multiplied by -2 and added to the equation in row 2, with the result placed in the system as the new row 2.

EXAMPLE 3 ▶ **Solving a System of Three Equations in Three Variables**

Solve using elimination: $\begin{cases} 2x + y - 2z = -7 \\ x + y + z = -1 \\ -2y - z = -3 \end{cases}$.

Solution ▶ 1. The system is in standard form.
2. If the x-term in any equation has a coefficient of 1, interchange equations so this equation becomes R1.

$$\begin{cases} 2x + y - 2z = -7 \\ x + y + z = -1 \\ -2y - z = -3 \end{cases} \xrightarrow{\text{R2} \leftrightarrow \text{R1}} \begin{cases} x + y + z = -1 \\ 2x + y - 2z = -7 \\ -2y - z = -3 \end{cases}$$

3. Use R1 to eliminate the x-term in R2 and R3. Since R3 has no x-term, the only elimination needed is the x-term from R2. Using $-2R1 + R2$ will eliminate this term:

$$\begin{array}{rl} -2R1 & -2x - 2y - 2z = 2 \\ + & \\ \underline{R2} & \underline{2x + y - 2z = -7} \\ & 0x - 1y - 4z = -5 \quad \text{sum} \\ & y + 4z = 5 \quad \text{simplify} \end{array}$$

The new R2 is $y + 4z = 5$. The original R1 and R3, along with the new R2 form an equivalent system that contains a smaller **subsystem**

$$\begin{cases} x + y + z = -1 \\ 2x + y - 2z = -7 \\ -2y - z = -3 \end{cases} \xrightarrow[\text{R3} \rightarrow \text{R3}]{-2R1 + R2 \rightarrow R2} \begin{cases} x + y + z = -1 \\ y + 4z = 5 \\ -2y - z = -3 \end{cases} \begin{array}{l} \text{new} \\ \text{equivalent} \\ \text{system} \end{array}$$

4. Solve the subsystem for either y or z, and keep the result as a *new* R3. We choose to eliminate y using 2R2 + R3:

$$
\begin{array}{llrl}
\text{2R2} & & 2y + 8z = & 10 \\
+ & & & \\
\text{R3} & & -2y - z = & -3 \\
\hline
& & 0y + 7z = & 7 \quad \text{\textcolor{red}{sum}} \\
& & z = & 1 \quad \text{\textcolor{red}{simplify}}
\end{array}
$$

The new R3 is $z = 1$.

$$
\begin{cases} x + y + z = -1 \\ y + 4z = 5 \\ -2y - z = -3 \end{cases} \xrightarrow{\ 2R2 + R3 \to R3\ } \begin{cases} x + y + z = -1 \\ y + 4z = 5 \\ z = 1 \end{cases}
\begin{array}{l} \textcolor{red}{\text{new}} \\ \textcolor{red}{\text{equivalent}} \\ \textcolor{red}{\text{system}} \end{array}
$$

The new R3, along with the original R1 and R2 from step 3, form an equivalent system that can be solved using back-substitution. Substituting 1 for z in R2 yields $y = 1$. Substituting 1 for z and 1 for y in R1 yields $x = -3$. The solution is $(-3, 1, 1)$. Check this result using the (ALPHA) keys, as illustrated following Example 2.

Now try Exercises 13 through 16 ▶

While not absolutely needed for the elimination process, there are two reasons for wanting the coefficient of x to be "1" in R1. First, it makes the elimination method more efficient since we can more easily see what to use as a multiplier. Second, it lays the foundation for developing other methods of solving larger systems. If no equation has an x-coefficient of 1, we simply use the y- or z-variable instead (see Example 7). Since solutions to larger systems generally are worked out in stages, we will sometimes track the transformations used by writing them *between* the original system and the equivalent system, rather than to the left as we did in Section 6.1.

Here is an additional example illustrating the elimination process, but in *abbreviated form*. Verify the calculations indicated using a separate sheet.

EXAMPLE 4 ▶   **Solving a System of Three Equations in Three Variables**

Solve using elimination: $\begin{cases} -5y + 2x - z = -8 \\ -x + 3z + 2y = 13 \\ -z + 3y + x = 5 \end{cases}$.

Solution ▶   **1.** Write the equations in standard form: $\begin{cases} 2x - 5y - z = -8 \\ -x + 2y + 3z = 13 \\ x + 3y - z = 5 \end{cases}$

2. $\begin{cases} 2x - 5y - z = -8 \\ -x + 2y + 3z = 13 \\ x + 3y - z = 5 \end{cases} \xrightarrow{\ R3 \leftrightarrow R1\ } \begin{cases} x + 3y - z = 5 \\ -x + 2y + 3z = 13 \\ 2x - 5y - z = -8 \end{cases}
\begin{array}{l} \textcolor{red}{\text{equivalent}} \\ \textcolor{red}{\text{system}} \end{array}$

3. Using R1 + R2 will eliminate the x-term from R2, yielding $5y + 2z = 18$.
Using -2R1 + R3 eliminates the x-term from R3, yielding $-11y + z = -18$.

$\begin{cases} x + 3y - z = 5 \\ -x + 2y + 3z = 13 \\ 2x - 5y - z = -8 \end{cases} \xrightarrow[\ -2R1 + R3 \to R3\ ]{\ R1 + R2 \to R2\ } \begin{cases} x + 3y - z = 5 \\ 5y + 2z = 18 \\ -11y + z = -18 \end{cases}
\begin{array}{l} \textcolor{red}{\text{equivalent}} \\ \textcolor{red}{\text{system}} \end{array}$

4. Using -2R3 + R2 will eliminate z from the subsystem, leaving $27y = 54$.

$\begin{cases} x + 3y - z = 5 \\ 5y + 2z = 18 \\ -11y + z = -18 \end{cases} \xrightarrow{\ -2R3 + R2 \to R3\ } \begin{cases} x + 3y - z = 5 \\ 5y + 2z = 18 \\ 27y = 54 \end{cases}
\begin{array}{l} \textcolor{red}{\text{equivalent}} \\ \textcolor{red}{\text{system}} \end{array}$

Solving for y in R3 yields: $\dfrac{27y}{27} = \dfrac{54}{27}$, showing $y = 2$. Substituting 2 for y in R2 gives,

$$5(2) + 2z = 18 \quad \text{substitute 2 for } y$$
$$10 + 2z = 18 \quad \text{simplify}$$
$$2z = 8 \quad \text{subtract 10}$$
$$z = 4 \quad \text{divide by 2}$$

Substituting 2 for y and 4 for z in R1 gives,

$$x + 3(2) - 4 = 5 \quad \text{substitute 2 for } y, \text{ 4 for } z$$
$$x + 2 = 5 \quad \text{simplify}$$
$$x = 3 \quad \text{subtract 2}$$

☑ **C. You've just seen how we can solve linear systems in three variables**

The solution is $(3, 2, 4)$.

Now try Exercises 17 through 20 ▶

D. Inconsistent and Dependent Systems

As mentioned, it is possible for larger systems to have no solutions or an infinite number of solutions. As with our work in Section 6.1, an inconsistent system (no solutions) will produce inconsistent results, ending with a statement such as $0 = -3$ or some other contradiction.

EXAMPLE 5 ▶ **Attempting to Solve an Inconsistent System**

Solve using elimination: $\begin{cases} 2x + y - 3z = -3 \\ 3x - 2y + 4z = 2 \\ 4x + 2y - 6z = -7 \end{cases}$.

Solution ▶

1. This system has no equation where the coefficient of x is 1.
2. We can still use R1 to begin the solution process, but this time we'll use the variable y since it *does* have coefficient 1.

   Using $2R1 + R2$ eliminates the y-term from R2, leaving $7x - 2z = -4$. But using $-2R1 + R3$ to eliminate the y-term from R3 results in a contradiction:

2R1	$4x + 2y - 6z = -6$		$-2R1$	$-4x - 2y + 6z = 6$
+			+	
R2	$3x - 2y + 4z = 2$		R3	$4x + 2y - 6z = -7$
	$7x \quad\;\; - 2z = -4$			$0x + 0y + 0z = -1$

   $$0 = -1 \quad \text{contradiction}$$

   We conclude the system is inconsistent. The answer is the empty set $\varnothing$, and we need work no further.

Now try Exercises 21 and 22 ▶

Unlike our work with systems having only two variables, systems in three variables can have two forms of dependence—*linear dependence* (Figure 6.18) or *coincident dependence* (Figure 6.19). To help understand linear dependence, consider a system of two equations in three variables: $\begin{cases} -2x + 3y - z = 5 \\ x - 3y + 2z = -1 \end{cases}$. Each of these equations represents a plane, and unless the planes are parallel, their intersection will be a line. As in Section 6.1, we can state solutions to a dependent system using set notation with two of the variables written in terms of the third, or as an ordered triple using a parameter. The relationships named can then be used to generate specific solutions to the system.

Systems with two equations and two variables or three equations and three variables are called **square systems,** meaning there are exactly as many equations as there are variables. A system of linear equations cannot have a unique solution unless there are at least as many equations as there are variables in the system.

EXAMPLE 6 ▶ Solving a Dependent System

Solve using elimination: $\begin{cases} 2x + 5y - 3z = -5 \\ -x - 5y + 2z = 1 \end{cases}$.

Solution ▶ We immediately note that R1 + R2 eliminates the y-term from R2, yielding $x - z = -4$ and the new system $\begin{cases} 2x + 5y - 3z = -5 \\ x \quad\quad - z = -4 \end{cases}$. This means (x, y, z) will satisfy both equations only when $x = z - 4$ (the x-coordinate must be 4 less than the z-coordinate). Since x is written in terms of z, we substitute $z - 4$ for x in *either equation* to find how y is related to z. Using R2 in the original system we have: $-(z - 4) - 5y + 2z = 1$, which yields $y = \frac{1}{5}z + \frac{3}{5}$ (verify). This means the y-coordinate of the solution must be $\frac{3}{5}$ more than $\frac{1}{5}z$. In set notation, the solution is $\{(x, y, z,) \mid x = z - 4, y = \frac{1}{5}z + \frac{3}{5}, z \in \mathbb{R}\}$. Randomly choosing $z = -3, 2$, and 7, the solutions would be $(-7, 0, -3)$, $(-2, 1, 2)$, and $(3, 2, 7)$, respectively. Verify that these satisfy both equations. Using p as a parameter, the solution could be written $(p - 4, \frac{1}{5}p + \frac{3}{5}, p)$ in parameterized form.

Now try Exercises 23 through 26 ▶

The system in Example 6 was nonsquare, and we knew ahead of time the system would be dependent. The system in Example 7 *is* square, but only by applying the elimination process can we determine the nature of its solution(s).

EXAMPLE 7 ▶ Solving a Dependent System

Solve using elimination: $\begin{cases} 3x - 2y + z = -1 \\ 2x + y - z = 5 \\ 10x - 2y \quad = 8 \end{cases}$.

Solution ▶ This system has no equation where the coefficient of x is 1. We will still use R1, noting that R1 + R2 eliminates the z-term from R2, yielding $5x - y = 4$ (there is no z-term in R3).

$$\begin{cases} 3x - 2y + z = -1 \\ 2x + y - z = 5 \\ 10x - 2y \quad = 8 \end{cases} \xrightarrow[R3 \rightarrow R3]{R1 + R2 \rightarrow R2} \begin{cases} 3x - 2y + z = -1 \\ 5x - y \quad = 4 \\ 10x - 2y \quad = 8 \end{cases}$$

We next solve the subsystem. Using $-2R2 + R3$ eliminates the y-term in R3, but also all other terms:

$$\begin{array}{rl} -2R2 & -10x + 2y = -8 \\ + & \\ \underline{R3} & \underline{10x - 2y = 8} \\ & 0x + 0y = 0 \quad \text{sum} \\ & \quad\quad 0 = 0 \quad \text{result} \end{array}$$

Since R3 is the same as 2R2, the system is linearly dependent and equivalent to $\begin{cases} 3x - 2y + z = -1 \\ 5x - y \quad = 4 \end{cases}$. We can solve for y in R2 to write y in terms of x: $y = 5x - 4$. Substituting $5x - 4$ for y in R1 enables us to also write z in terms of x:

$$3x - 2y \qquad + z = -1 \qquad \text{R1}$$
$$3x - 2(5x - 4) + z = -1 \qquad \text{substitute } 5x - 4 \text{ for } y$$
$$3x - 10x + 8 \ + z = -1 \qquad \text{distribute}$$
$$-7x + z = -9 \qquad \text{simplify}$$
$$z = 7x - 9 \quad \text{solve for } z$$

The solution set is $\{(x, y, z)|x \in \mathbb{R}, y = 5x - 4, z = 7x - 9\}$. Three of the infinite number of solutions are $(0, -4, -9)$ for $x = 0$, $(2, 6, 5)$ for $x = 2$, and $(-1, -9, -16)$ for $x = -1$. Verify these triples satisfy all three equations. Again using the parameter p, the solution could be written as $(p, 5p - 4, 7p - 9)$.

Now try Exercises 27 through 30 ▶

☑ D. You've just seen how we can recognize inconsistent and dependent systems

For **coincident dependence** the equations in a system differ by only a constant multiple. After applying the elimination process—all variables are eliminated from the other equations, leaving statements that are always true (such as $2 = 2$ or some other). For additional practice solving various kinds of systems, **see Exercises 31 to 44.**

E. Applications

Applications of larger systems are simply an extension of our work with systems of two equations in two variables. Once again, the applications come in a variety of forms and from many fields. In the world of business and finance, systems can be used to diversify investments or spread out liabilities, a financial strategy hinted at in Example 8.

EXAMPLE 8 ▶   **Modeling the Finances of a Business**

A small business borrowed \$225,000 from three different lenders to expand their product line. The interest rates were 5%, 6%, and 7%. Find how much was borrowed at each rate if the annual interest came to \$13,000 and twice as much was borrowed at the 5% rate than was borrowed at the 7% rate.

Solution ▶   Let x, y, and z represent the amounts borrowed at 5%, 6%, and 7%, respectively. This means our first equation is $x + y + z = 225$ (in thousands). The second equation is determined by the total interest paid, which was \$13,000: $0.05x + 0.06y + 0.07z = 13$. The third is found by carefully reading the problem.

"twice as much was borrowed at the 5% rate than was borrowed at the 7% rate," or $x = 2z$.

These equations form the system: $\begin{cases} x + y + z = 225 \\ 0.05x + 0.06y + 0.07z = 13 \\ x = 2z \end{cases}$. The x-term of

the first equation has a coefficient of 1. Written in standard form we have:

$$\begin{cases} x + \ y + \ z = \ 225 \quad \text{R1} \\ 5x + 6y + 7z = 1300 \quad \text{R2} \quad \text{\color{red}multiply R2 by 100} \\ x - \qquad 2z = \quad 0 \quad \text{R3} \end{cases}$$

Using $-5R1 + R2$ will eliminate the x term in R2, while $-R1 + R3$ will eliminate the x-term in R3.

$$\begin{array}{rl} -5\text{R1} & -5x - 5y - 5z = -1125 \\ + & \\ \text{R2} & \underline{5x + 6y + 7z = \quad 1300} \\ & y + 2z = \quad 175 \end{array} \qquad \begin{array}{rl} -\text{R1} & -x - y - \ z = -225 \\ + & \\ \text{R3} & \underline{x \qquad - 2z = \qquad 0} \\ & -y - 3z = -225 \end{array}$$

The new R2 is $y + 2z = 175$, and the new R3 (after multiplying by -1) is

$y + 3z = 225$, yielding the equivalent system $\begin{cases} x + y + z = 225 \\ \phantom{x + {}} y + 2z = 175. \\ \phantom{x + {}} y + 3z = 225 \end{cases}$

Solving the 2×2 subsystem using $-R2 + R3$ yields $z = 50$. Back-substitution shows $y = 75$ and $x = 100$, yielding the solution $(100, 75, 50)$. This means \$50,000 was borrowed at the 7% rate, \$75,000 was borrowed at 6%, and \$100,000 at 5%.

A "substitution check" on a graphing calculator verifies this solution is correct (see Figures 6.23 and 6.24).

Figure 6.23

```
100→X
            100
75→Y
             75
50→Z
             50
```

Figure 6.24

```
X+Y+Z
            225
.05X+.06Y+.07Z
             13
X-2Z
              0
```

☑ **E.** You've just seen how we can use a system of three equations in three variables to solve applications

Now try Exercises 47 through 56 ▶

6.2 EXERCISES

▶ **CONCEPTS AND VOCABULARY**

Fill in the blank with the appropriate word or phrase. Carefully reread the section if needed.

1. The solution to an equation in three variables is an ordered ___triple___.

2. The graph of the solutions to an equation in three variables is a(n) ___plane___.

3. Systems that have the same solution set are called ___equivalent___ ___systems___.

4. If a 3×3 system is linearly dependent, the ordered triple solutions can be written in terms of a single variable called a(n) ___parameter___.

5. Find a value of z that makes the ordered triple $(2, -5, z)$ a solution to $2x + y + z = 4$. Discuss/Explain how this is accomplished. $z = 5$

6. Explain the difference between linear dependence and coincident dependence, and describe how the equations are related. Answers will vary.

▶ **DEVELOPING YOUR SKILLS**

Find any four ordered triples that satisfy the equation given.

7. $x + 2y + z = 9$ Answers will vary.

8. $3x + y - z = 8$ Answers will vary.

9. $-x + y + 2z = -6$ Answers will vary.

10. $2x - y + 3z = -12$ Answers will vary.

Use the ⟨ALPHA⟩ keys on a graphing calculator to determine if the given ordered triples are solutions to the system. If not a solution, identify which equation(s) are not satisfied.

11. $\begin{cases} x + y - 2z = -1 \\ 4x - y + 3z = 3 \\ 3x + 2y - z = 4 \end{cases}$; $\begin{array}{l} (0, 3, 2) \\ (-3, 4, 1) \end{array}$ yes   no; R2, R3

12. $\begin{cases} 2x + 3y + z = 9 \\ 5x - 2y - z = -32; \\ x - y - 2z = -13 \end{cases}$ $\begin{array}{l} (-4, 5, 2) \\ (5, -4, 11) \end{array}$ yes   no; R2

HOMEWORK SELECTION GUIDE

Core: 7–43 odd, 47–57 odd (25 Exercises)
Standard: 1–4, 7–43 odd, 45, 47–57 odd, 59, 63, 64 (33 Exercises)

Extended: 1–4, 7–43 odd, 45, 46, 47–57 odd, 59, 60, 63, 64 (35 Exercises)
In Depth: 1–6, 7–43 odd, 45, 46, 47–55 odd, 56–64 (41 Exercises)

Solve each system using elimination and back-substitution.

13. $\begin{cases} x - y - 2z = -10 \\ x - - z = 1 \\ z = 4 \end{cases}$
$(5, 7, 4)$

14. $\begin{cases} x + y + 2z = -1 \\ 4x - y = 3 \\ 3x = 6 \end{cases}$
$(2, 5, -4)$

15. $\begin{cases} x + 3y + 2z = 16 \\ -2y + 3z = 1 \\ 8y - 13z = -7 \end{cases}$
$(-2, 4, 3)$

16. $\begin{cases} -x + y + 5z = 1 \\ 4x + y = 1 \\ -3x - 2y = 8 \end{cases}$
$(2, -7, 2)$

17. $\begin{cases} -x + y + 2z = -10 \\ x + y - z = 7 \\ 2x + y + z = 5 \end{cases}$
$(4, 0, -3)$

18. $\begin{cases} x + y - 2z = -1 \\ 4x - y + 3z = 3 \\ 3x + 2y - z = 4 \end{cases}$
$(0, 3, 2)$

19. $\begin{cases} 2x - 3y + 2z = 0 \\ 3x - 4y + z = -20 \\ x + 2y - z = 16 \end{cases}$
$(5, 12, 13)$

20. $\begin{cases} 3x - y + z = 6 \\ 2x + 2y - z = 5 \\ 2x - y + z = 5 \end{cases}$
$(1, 6, 9)$

Solve using the elimination method. If a system is inconsistent or dependent, so state. For systems with linear dependence, write solutions in set notation and as an ordered triple in terms of a parameter.

21. $\begin{cases} 3x + y + 2z = 3 \\ x - 2y + 3z = 1 \\ 4x - 8y + 12z = 7 \end{cases}$

22. $\begin{cases} 2x - y + 3z = 8 \\ 3x - 4y + z = 4 \\ -4x + 2y - 6z = 5 \end{cases}$

23. $\begin{cases} 4x + y + 3z = 8 \\ x - 2y + 3z = 2 \end{cases}$

24. $\begin{cases} 4x - y + 2z = 9 \\ 3x + y + 5z = 5 \end{cases}$

25. $\begin{cases} 6x - 3y + 7z = 2 \\ 3x - 4y + z = 6 \end{cases}$

26. $\begin{cases} 2x - 4y + 5z = -2 \\ 3x - 2y + 3z = 7 \end{cases}$

Solve using elimination. If the system is linearly dependent, state the general solution in terms of a parameter (different forms of the solution are possible). Then find four specific ordered triples that satisfy the system (answers will vary).

27. $\begin{cases} 3x - 4y + 5z = 5 \\ -x + 2y - 3z = -3 \\ 3x - 2y + z = 1 \end{cases}$ $(p, 2p, p + 1)$

28. $\begin{cases} 5x - 3y + 2z = 4 \\ -9x + 5y - 4z = -12 \\ -3x + y - 2z = -12 \end{cases}$ $(p, p + 4, 8 - p)$

29. $\begin{cases} x + 2y - 3z = 1 \\ 3x + 5y - 8z = 7 \\ x + y - 2z = 5 \end{cases}$ $(p + 9, p - 4, p)$

30. $\begin{cases} -2x + 3y - 5z = 3 \\ 5x - 7y + 12z = -8 \\ x - y + 2z = -2 \end{cases}$ $(-p - 3, p - 1, p)$

Solve using the elimination method. If a system is inconsistent or dependent, so state. For systems with linear dependence, write the answer in terms of a parameter. For coincident dependence, state the solution in set notation.

31. $\begin{cases} 4x + 2y - 8z = 24 \\ -x - 0.5y + 2z = -6 \\ 2x + y - 4z = 12 \end{cases}$ $\left\{ (x, y, z) \mid x + \frac{1}{2}y - 2z = 6 \right\}$

32. $\begin{cases} 2x - 5y - 4z = 6 \\ x - 2.5y - 2z = 3 \\ -3x + 7.5y + 6z = -9 \end{cases}$ $\left\{ (x, y, z) \mid x - \frac{5}{2}y - 2z = 3 \right\}$

33. $\begin{cases} x - 2y + 2z = 6 \\ 2x - 6y + 3z = 13 \\ 3x + 4y - z = -11 \end{cases}$
$(-1, \frac{-3}{2}, 2)$

34. $\begin{cases} 4x - 5y - 6z = 5 \\ 2x - 3y + 3z = 0 \\ x + 2y - 3z = 5 \end{cases}$
$(2, 1, \frac{-1}{3})$

35. $\begin{cases} x - 5y - 4z = 3 \\ 2x - 9y - 7z = 2 \\ 3x - 14y - 11z = 5 \end{cases}$
$(-p - 17, -p - 4, p)$

36. $\begin{cases} 2x + 3y - 5z = 4 \\ x + y - 2z = 3 \\ x + 3y - 4z = -1 \end{cases}$
$(p + 5, p - 2, p)$

37. $\begin{cases} \frac{1}{6}x + \frac{1}{3}y - \frac{1}{2}z = 2 \\ \frac{3}{4}x - \frac{1}{3}y + \frac{1}{2}z = 9 \\ \frac{1}{2}x - y + \frac{1}{2}z = 2 \end{cases}$
$(12, 6, 4)$

38. $\begin{cases} \frac{x}{2} + \frac{y}{3} - \frac{z}{2} = 2 \\ \frac{2x}{3} - y - z = 8 \\ \frac{x}{6} + 2y + \frac{3z}{2} = 6 \end{cases}$
$(18, -6, 10)$

Some applications of systems lead to systems similar to those that follow. Solve using elimination.

39. $\begin{cases} -2A - B - 3C = 21 \\ B - C = 1 \\ A + B = -4 \end{cases}$ $(1, -5, -6)$

40. $\begin{cases} -A + 3B + 2C = 11 \\ 2B + C = 9 \\ B + 2C = 8 \end{cases}$ $\left(\frac{11}{3}, \frac{10}{3}, \frac{7}{3} \right)$

41. $\begin{cases} A + 2C = 7 \\ 2A - 3B = 8 \\ 3A + 6B - 8C = -33 \end{cases}$ $(1, -2, 3)$

42. $\begin{cases} A - 2B = 5 \\ B + 3C = 7 \\ 2A - B - C = 1 \end{cases}$ $(1, -2, 3)$

43. $\begin{cases} C = -2 \\ 5A - 2C = 5 \\ -4B - 9C = 16 \end{cases}$ $\left(\frac{1}{5}, \frac{1}{2}, -2 \right)$

44. $\begin{cases} C = 3 \\ 2A + 3C = 10 \\ 3B - 4C = -11 \end{cases}$ $\left(\frac{1}{2}, \frac{1}{3}, 3 \right)$

▶ **WORKING WITH FORMULAS**

45. Dimensions of a rectangular solid:

$$\begin{cases} 2w + 2h = P_1 \\ 2l + 2w = P_2 \\ 2l + 2h = P_3 \end{cases}$$

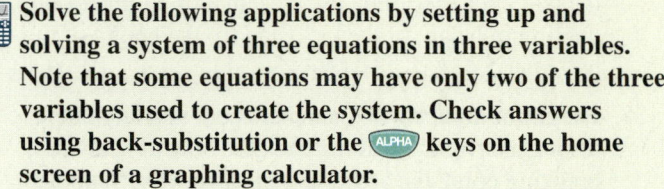

$P_2 = 16$ cm (top)
$P_3 = 18$ cm (large side)
$P_1 = 14$ cm (small side)

Using the formula shown, the dimensions of a rectangular solid can be found if the perimeters of the three distinct faces are known. Find the dimensions of the solid shown.   (5 cm, 3 cm, 4 cm)

46. Distance from a point (x, y, z) to the plane

$$Ax + By + Cz = D: \left| \frac{Ax + By + Cz - D}{\sqrt{A^2 + B^2 + C^2}} \right|$$

The perpendicular distance from a given point (x, y, z) to the plane defined by $Ax + By + Cz = D$ is given by the formula shown. Consider the plane given in Figure 6.15 ($x + y + z = 6$). What is the distance from this plane to the point $(3, 4, 5)$?
≈ 3.464 units

▶ **APPLICATIONS**

Solve the following applications by setting up and solving a system of three equations in three variables. Note that some equations may have only two of the three variables used to create the system. Check answers using back-substitution or the (ALPHA) keys on the home screen of a graphing calculator.

Investment/Finance and Simple Interest Problems

47. Investing the winnings: After winning $280,000 in the lottery, Maurika decided to place the money in three different investments: a certificate of deposit paying 4%, a money market certificate paying 5%, and some Aa bonds paying 7%. After 1 yr she earned $15,400 in interest. Find how much was invested at each rate if $20,000 more was invested at 7% than at 5%.
$80,000 at 4%; $90,000 at 5%; $110,000 at 7%

48. Purchase at auction: At an auction, a wealthy collector paid $7,000,000 for three paintings: a Monet, a Picasso, and a van Gogh. The Monet cost $800,000 more than the Picasso. The price of the van Gogh was $200,000 more than twice the price of the Monet. What was the price of each painting?
Monet $1,900,000; Picasso $1,100,000; van Gogh $4,000,000

Descriptive Translation

49. Major wars: The United States has fought three major wars in modern times: World War II, the Korean War, and the Vietnam War. If you sum the years that each conflict ended, the result is 5871. The Vietnam War ended 20 years after the Korean War and 28 years after World War II. In what year did each end?
World War II, 1945; Korean, 1953; Vietnam, 1973

50. Animal gestation periods: The average gestation period (in days) of an elephant, rhinoceros, and camel sum to 1520 days. The gestation period of a rhino is 58 days longer than that of a camel. Twice the camel's gestation period decreased by 162 gives the gestation period of an elephant. What is the gestation period of each?
elephant, 650 days; rhino, 464 days; camel, 406 days

51. Moments in U.S. history: If you sum the year the Declaration of Independence was signed, the year the 13th Amendment to the Constitution abolished slavery, and the year the Civil Rights Act was signed, the total would be 5605. Ninety-nine years separate the 13th Amendment and the Civil Rights Act. The Civil Rights Act was signed 188 years after the Declaration of Independence. What year was each signed? Declaration of Independence, 1776; 13th Amendment, 1865; Civil Rights Act, 1964

52. Aviary wingspan: If you combine the wingspan of the California Condor, the Wandering Albatross (see photo), and the prehistoric Quetzalcoatlus, you get an astonishing 18.6 m (over 60 ft). If

the wingspan of the Quetzalcoatlus is equal to five times that of the Wandering Albatross minus twice that of the California Condor, and six times the wingspan of the Condor is equal to five times the wingspan of the Albatross, what is the wingspan of each?
Albatross: 3.6 m, Condor: 3.0 m, Quetzalcoatlus: 12.0 m

Mixtures

53. Chemical mixtures: A chemist mixes three different solutions with concentrations of 20%, 30%, and 45% glucose to obtain 10 L of a 38% glucose solution. If the amount of 30% solution used is 1 L more than twice the amount of 20% solution used, find the amount of each solution used. 1 L 20% solution; 3 L 30% solution; 6 L 45% solution

54. Value of gold coins: As part of a promotion, a local bank invites its customers to view a large sack full of $5, $10, and $20 gold pieces, promising to give the sack to the first person able to state the number of coins for each

denomination. Customers are told there are exactly 250 coins, with a total face value of $1875. If there are also seven times as many $5 gold pieces as $20 gold pieces, how many of each denomination are there?
175 $5 gold pieces; 50 $10 gold pieces; 25 $20 gold pieces

Nutrition

55. **Industrial food production:** Acampana Soups is creating a new sausage and shrimp gumbo that contains three different types of fat: saturated, monounsaturated, and polyunsaturated. As a new member of the "Heart Healthy" menu, this soup must contain only 2.8 g of total fat per serving. The head chef Yev Kasem demands that the recipe provides exactly twice as much saturated fat as polyunsaturated fat. At the same time, the lead nutrition expert Florencia requires the amount of saturated fat in a serving to be 0.4 g less than the *combined* amount of unsaturated fats. How many grams of each type of fat will a serving of this soup contain?
saturated: 1.2 g, monounsaturated: 1.0 g, polyunsaturated: 0.6 g

56. **Geriatric nutrition:** The dietician at McKnight Place must create a balanced diet for Fred, consisting of a daily total of 1600 calories. For his successful rehabilitation, exact quantities of complex carbohydrates, fat, and protein must provide these calories. In this diet, carbohydrates provide 160 more calories than fat and protein together, while fat provides 1.25 times more calories than protein alone. How many calories should each nutrient provide on a daily basis?
carbohydrates: 880 cal, fat: 400 cal, protein: 320 cal

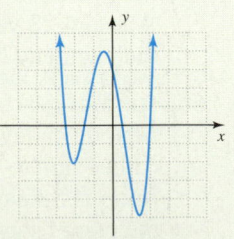 In Exercises 57 and 58 from Section 6.1, we found the equation of a line through two points by creating a system of two equations of the form $y = mx + b$. Just as any two points determine a unique line, three noncollinear points determine a unique parabola (as long as no two have the same first coordinate). Similar to the method used there, each point gives an equation of the form $y = ax^2 + bx + c$, and we create a 3 × 3 system that can be solved using elimination. For instance, the point (2, 41) gives the equation $41 = a(2)^2 + 2b + c$.

57. **Height of a soccer ball:** One second after being kicked ($t = 1$), a soccer ball is 26 ft high. After 2 sec ($t = 2$), the ball is 41 ft high, and after 6 sec the ball is 1 ft above the ground. Use the ordered pairs (time, height) to find a function $h(t)$ modeling the height of the ball after t sec, then use the equation to find (a) the maximum height of the kick, and (b) the height of the ball after 5.4 sec. $h(t) = -5t^2 + 30t + 1$; **a.** 46 ft **b.** 17.2 ft

58. **Height of an arrow:** An archer is out in a large field testing a new bow. Pulling back the bow to near its breaking point, the archer lets the arrow fly. Suppose that 1 sec after release ($t = 1$) the arrow was 184 ft high, four sec later ($t = 5$) it was 600 ft high, and after 12 sec, the arrow was 96 ft high. Use the ordered pairs (time, height) to find a function $h(t)$ modeling the height of the arrow after t sec, then use the equation to (a) find the maximum height of the shot and (b) determine how long the arrow was airborne. $h(t) = -16t^2 + 200t$; **a.** 625ft **b.** 12.5 sec

▶ EXTENDING THE CONCEPT

59. The system $\begin{cases} x - 2y - z = 2 \\ x - 2y + kz = 5 \\ 2x - 4y + 4z = 10 \end{cases}$ is dependent if $k =$ __2__, and inconsistent otherwise.

60. One form of the equation of a circle is $x^2 + y^2 + Dx + Ey + F = 0$. Use a system to find the equation of the circle through the points $(2, -1), (4, -3)$, and $(2, -5)$.
$x^2 + y^2 - 4x + 6y + 9 = 0$

▶ MAINTAINING YOUR SKILLS

61. **(4.6)** If $p(x) = 2x^2 - x - 3$, in what intervals is $p(x) \leq 0$? $[-1, \frac{3}{2}]$

62. **(4.3)** Graph the polynomial defined by $f(x) = x^4 - 5x^2 + 4$.

63. **(5.4)** Solve the logarithmic equation: $\log(x + 2) + \log x = \log 3$ $x = 1$

64. **(3.4)** Analyze the graph of g shown. Clearly state the domain and range, the zeroes of g, intervals where $g(x) > 0$, intervals where $g(x) < 0$, local maximums or minimums, and intervals where the function is increasing or decreasing. Assume each tick mark is one unit and estimate endpoints to the nearest tenths.

Additional answers can be found in the Instructor Answer Appendix.

R1 $4x - 3y = 9$

+

2R2 $-4x + 2y = -10$

sum $-y = -1$

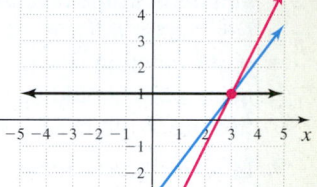

Figure 6.57

The result is $y = 1$ but remember, this is a system of *linear equations,* and $y = 1$ is still the equation of a (horizontal) line. Since the system $\begin{cases} 4x - 3y = 9 \\ \quad\quad y = 1 \end{cases}$ is equivalent to the original, it will have the same solution set. In Figure 6.57, we note the point of intersection for the new system is still $(3, 1)$. If we eliminate the y-terms instead (using R1 + 3R2), the result is $x = 3$, which is also the equation of a (vertical) line. Creating another equivalent system using this line produces $\begin{cases} x = 3 \\ y = 1 \end{cases}$ and the graph shown in Figure 6.58, where the vertical and horizontal lines intersect at $(3, 1)$, making the solution trivial.

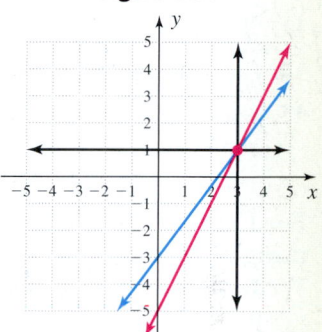

Figure 6.58

Note: Here we see a close connection to solving general equations, in that the goal is to write a series of equivalent yet simpler equations, continuing until the solution is obvious.

As for the substitution method, consider the second equation written as $y = 2x - 5$. This equation represents every point (x, y) on its graph, meaning the relationship for the ordered pair solutions can also be written $(x, 2x - 5)$. The same thing can be said for the line $4x - 3y = 9$, with its ordered pair solutions represented by $(x, \frac{4}{3}x - 3)$. At the point of intersection the y-coordinates must be identical, giving $2x - 5 = \frac{4}{3}x - 3$. Since these are equal, we can substitute $2x - 5$ for y in the first equation, or $\frac{4}{3}x - 3$ for y in the second equation, with both yielding the correct solution. Substituting $2x - 5$ for y in the first equation gives

$$
\begin{aligned}
4x - 3(2x - 5) &= 9 \qquad \text{substitute } 2x - 5 \text{ for } y \\
4x - 6x + 15 &= 9 \qquad \text{expand} \\
-2x &= -6 \qquad \text{simplify} \\
x &= 3
\end{aligned}
$$

and the solution $(x, 2x - 5)$ becomes $(3, 2(3) - 5)$ or $(3, 1)$.

All three methods will produce the same solution, and the best method to use at the time often depends on the nature of the system given, or even personal preference.

Exercise 1: Solve the system by (a) graphing, (b) elimination, and (c) substitution. Which method was most efficient for solving this system? $\begin{cases} 2x + y = 2 \\ 4x + 3y = 8 \end{cases}$ $(-1, 4)$, elimination

---

CUMULATIVE REVIEW CHAPTERS R–6

Graph each of the following. Include x- and y-intercepts and other important features of each graph.

1. $y = \frac{2}{3}x + 2$

2. $f(x) = |x - 2| + 3$

3. $g(x) = \sqrt{x - 3} + 1$

4. $h(x) = \dfrac{1}{x - 1} + 2$

5. $g(x) = (x - 3)(x + 1)(x + 4)$

6. $y = 2^x + 3$

7. Determine the following for the graph shown. Write all answers in interval notation:

   a. domain

   b. range

   c. interval(s) where $f(x)$ is increasing or decreasing

   d. interval(s) where $f(x)$ is constant

   e. location of any maximum or minimum value(s)

   f. interval(s) where $f(x)$ is positive or negative

   g. the average rate of change using $(-4, 0)$ and $(-2, 3.5)$. $\dfrac{\Delta y}{\Delta x} = \dfrac{7}{4}$

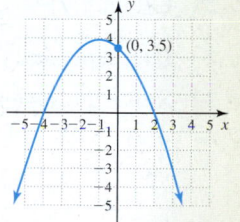

Additional answers can be found in the Instructor Answer Appendix.

8. Suppose the cost of making a rubber ball is given by $C(x) = 3x + 10$, where x is the number of balls in hundreds. If the revenue from the sale of these balls is given by $R(x) = -x^2 + 123x - 1990$, find the profit function (Profit = Revenue − Cost). How many balls should be produced and sold to obtain the maximum profit? What is this maximum profit?

9. Find all zeroes (real or complex):
$g(v) = v^3 - 9v^2 + 2v - 18$. $\quad x = 9,\ \pm i\sqrt{2}$

10. A polynomial has roots $x = 2$ and $x = 2 \pm 3i$. Find the polynomial and write it in standard form.
$P(x) = x^3 - 6x^2 + 21x - 26$

Given $f(x) = 2x - 5$ and $g(x) = 3x^2 + 2x$ find:

11. $(g - f)(x)$ $3x^2 + 5$ **12.** $(fg)(-2)$ -72 **13.** $(g \circ f)(2)$ 1

14. Calculate the difference quotient for $f(x) = x^2 - 3x$.
$2x + h - 3$

15. Use the rational roots theorem to factor the polynomial completely: $x^4 - 6x^3 - 13x^2 + 24x + 36$.
$(x - 2)(x + 2)(x - 3 - 3\sqrt{2})(x - 3 + 3\sqrt{2})$

For Exercises 16–25, all answers may be checked using a graphing calculator. Solve each inequality. Write your answer using interval notation.

16. $x^2 - 3x - 10 < 10$ **17.** $\dfrac{x - 2}{x + 3} \le 3$
$x \in (-2, 5)$ $x \in (-\infty, \frac{-11}{2}] \cup (-3, \infty)$

18. Solve each equation.

a. $\sqrt{x} - 2 = \sqrt{3x + 4}$

b. $x^{\frac{3}{2}} + 8 = 0$ **c.** $2|n + 4| + 3 = 13$

d. $x^2 - 6x + 13 = 0$ **e.** $x^{-2} - 3x^{-1} - 40 = 0$

f. $4 \cdot 2^{x+1} = \dfrac{1}{8}$ **g.** $3^{x-2} = 7$

h. $\log_3 81 = x$ **i.** $\log_3 x + \log_3(x - 2) = 1$

19. If a person invests \$5000 at 9% compounded continuously, how long until the account grows to \$12,000? ≈ 9.7 yr

20. Graph the piecewise function shown and state its domain and range.
$$f(x) = \begin{cases} |x + 2| & -5 \le x \le 0 \\ 2(x - 1)^2 & 0 < x \le 5 \end{cases}$$

21. Graph $h(x) = \dfrac{9 - x^2}{x^2 - 4}$. Give the coordinates of all intercepts and the equations of all asymptotes.

22. Solve the system of equations graphically.
$$\begin{cases} x - 3y = 1 \\ 2x + y = -5 \end{cases} \quad (-2, -1)$$

23. A truck is delivering a shipment of arrows, bowling balls, and cricket bats to a sporting goods store. There is a total of 120 items in the shipment. There are twice as many arrows as balls and bats combined. There are 10 more balls than bats. How many of each item are there?   80 arrows, 25 balls, 15 bats

24. Sketch the solution region.
$$\begin{cases} y < 2^x \\ x + 2y > 0 \\ 3x + y < 5 \end{cases}$$

25. Solve the system of equations.
$$\begin{cases} y = \log x + 4 \\ y = 5 - \log(x - 3) \end{cases} \quad x = 5, y \approx 4.7$$

Exercises 26 through 30 require the use of a graphing calculator.

26. For $Y_1 = \sqrt{x - 2}$ and $Y_2 = 3x + 6$, use the GRAPH function of a graphing calculator to determine the domain of

a. $Y_1 \circ Y_2$ $[\frac{-4}{3}, \infty)$ **b.** $Y_2 \circ Y_1$ $[2, \infty)$

27. Given $f(x) = x^3 + 8$,

a. Use the **DrawInv** feature of your graphing calculator to draw the graph of $f^{-1}(x)$.

b. Find the equation for $f^{-1}(x)$ and verify its graph matches the graph generated by the **DrawInv** feature.

28. Use the intersection-of-graphs method to solve the given equation. Round answer(s) to two decimal places: $\log(x^2 + 2.2) = 3.6 - \ln|x|$
$x \approx -6.79, x \approx 6.79$

29. Use the zeroes method to solve the given equation. Round your answer(s) to two decimal places.
$$\dfrac{75}{1 + 13e^{-0.09x}} = 55 \quad x \approx 39.74$$

30. Solve the system by graphing on a graphing calculator and locating any points of intersection. Check your answer using the alpha keys on the home screen of the calculator.
$$\begin{cases} -23x + 3y = -628.2 \\ 51x + 11y = 2162.6 \end{cases}$$

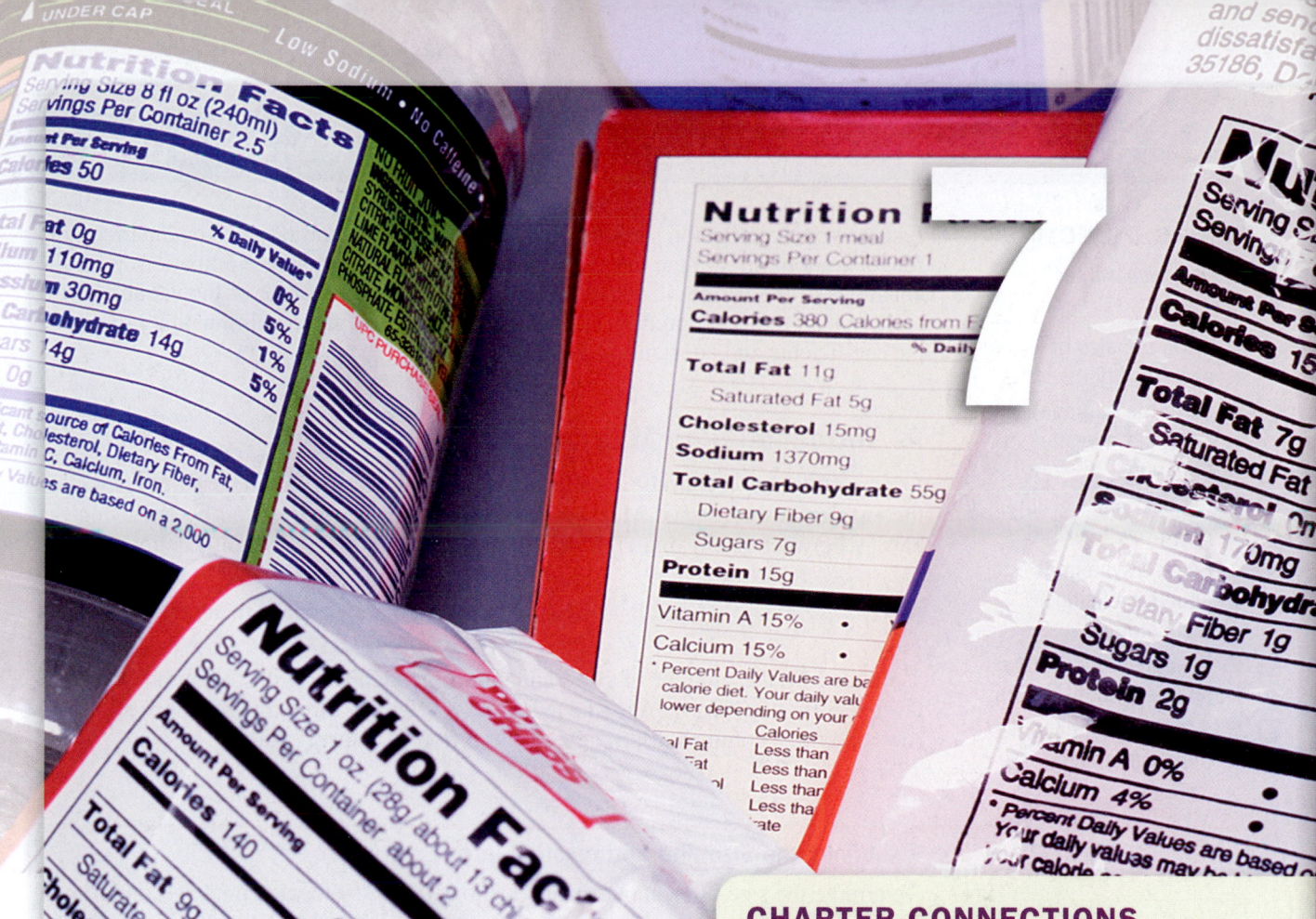

Matrices and Matrix Applications

CHAPTER OUTLINE

CHAPTER CONNECTIONS

From pediatric and geriatric care, to the training of a modern athlete, dietetic applications have become increasingly effective. In the latter case, athletes generally need high levels of carbohydrates and protein, but only moderate levels of fat. Suppose a physical trainer wants to supply one of her clients with 24 g of fat, 244 g of "carbs," and 40 g of protein for the noontime meal. Knowing the amounts of these nutrients contained in certain foods, the trainer can recommend a variety of foods and the amount of each that should be eaten. The matrix operations in this chapter demonstrate how to do this effectively. This application occurs as Exercise 84 in Section 7.3.

Check out these other real-world connections:

LEARNING OBJECTIVES

In Section 7.1 you will see how we can:

- ☐ **A.** State the size of a matrix and identify its entries
- ☐ **B.** Form the augmented matrix of a system of equations
- ☐ **C.** Solve a system of equations using row operations
- ☐ **D.** Solve a system of equations using technology
- ☐ **E.** Recognize inconsistent and dependent systems
- ☐ **F.** Solve applications using matrix methods

Just as synthetic division streamlines the process of polynomial division, matrices and row operations streamline the process of solving systems using elimination. With the equations of the system in standard form, the location and order of the variable terms and constant terms are predetermined, and we simply apply the elimination process on the coefficients and constants alone.

A. Introduction to Matrices

In general terms, a **matrix** is simply a rectangular arrangement of numbers, called the **entries** of the matrix. **Matrices** (plural of matrix) are denoted by enclosing the entries between a left and right bracket, and named using a capital letter, such as

$$A = \begin{bmatrix} 1 & -3 & 2 \\ 5 & 1 & -1 \end{bmatrix} \text{ and } B = \begin{bmatrix} 2 & -1 & 3 \\ 4 & 6 & -2 \\ 1 & 0 & -1 \end{bmatrix}. \text{ They occur in many different sizes}$$

as defined by the number of **rows** and **columns** each has, with the number of rows always given first. Matrix A is said to be a 2 × 3 (two by three) matrix, since it has two rows and three columns. Matrix B is a 3 × 3 (three by three) matrix.

EXAMPLE 1A ▶ **Identifying the Size and Entries of a Matrix**

Determine the size of each matrix and identify the entry located in the second row and first column.

$$\textbf{a. } C = \begin{bmatrix} 3 & -2 \\ 1 & 5 \\ -4 & 3 \end{bmatrix} \qquad \textbf{b. } D = \begin{bmatrix} 0.2 & -0.5 & 0.7 & 3.3 \\ -0.4 & 0.3 & 1 & 2 \\ 2.1 & -0.1 & 0.6 & 4.1 \end{bmatrix}$$

Solution ▶ **a.** Matrix C is 3 × 2. The row 2, column 1 entry is 1.

b. Matrix D is 3 × 4. The row 2, column 1 entry is −0.4.

If a matrix has the same number of rows and columns, it's called a **square matrix.** Matrix B above is a square matrix, while matrix A is not. For square matrices, the values on a diagonal line *from the upper left to the lower right* are called the **diagonal entries** and are said to be **on the diagonal** of the matrix. When solving systems using matrices, much of our focus is on these diagonal entries.

EXAMPLE 1B ▶ **Identifying the Diagonal Entries of a Square Matrix**

Name the diagonal entries of each matrix.

$$\textbf{a. } E = \begin{bmatrix} 1 & 4 \\ -2 & -3 \end{bmatrix} \qquad \textbf{b. } F = \begin{bmatrix} 0.2 & -0.5 & 0.7 \\ -0.4 & 0.3 & 1 \\ 2.1 & -0.1 & 0.6 \end{bmatrix}$$

Solution ▶ **a.** The diagonal entries of matrix E are 1 and −3.

b. For matrix F, the diagonal entries are 0.2, 0.3, and 0.6.

☑ **A.** You've just seen how we can state the size of a matrix and identify its entries

Now try Exercises 7 through 9 ▶

B. The Augmented Matrix of a System of Equations

A system of equations can be written in matrix form by augmenting or joining the **coefficient matrix,** formed by the variable coefficients, with the **matrix of constants.**

The coefficient matrix for the system $\begin{cases} 2x + 3y - z = 1 \\ x + z = 2 \\ x - 3y + 4z = 5 \end{cases}$ is $\begin{bmatrix} 2 & 3 & -1 \\ 1 & 0 & 1 \\ 1 & -3 & 4 \end{bmatrix}$ with

column 1 for the coefficients of x, column 2 for the coefficients of y, and so on. The

matrix of constants is $\begin{bmatrix} 1 \\ 2 \\ 5 \end{bmatrix}$. These two are joined to form the **augmented matrix,**

with a dotted line often used to separate the two as shown here: $\left[\begin{array}{ccc:c} 2 & 3 & -1 & 1 \\ 1 & 0 & 1 & 2 \\ 1 & -3 & 4 & 5 \end{array}\right]$.

It's important to note the use of a zero placeholder for the y-variable in the second row of the matrix, signifying there is no y-variable in the corresponding equation.

EXAMPLE 2 ▶ **Forming Augmented Matrices**

Form the augmented matrix for each system, and name the diagonal entries of each coefficient matrix.

a. $\begin{cases} 2x + y = 11 \\ -x + 3y = -2 \end{cases}$ **b.** $\begin{cases} \frac{1}{2}x + \phantom{\frac{2}{3}}y \phantom{+\frac{5}{6}z} = 5 \\ x + \frac{2}{3}y + \frac{5}{6}z = 11 \\ \phantom{x+\frac{2}{3}y+} z = 6 \end{cases}$

Solution ▶ **a.** $\begin{cases} 2x + y = 11 \\ -x + 3y = -2 \end{cases} \longrightarrow \left[\begin{array}{cc:c} 2 & 1 & 11 \\ -1 & 3 & -2 \end{array}\right]$

Diagonal entries: 2 and 3.

b. $\begin{cases} \frac{1}{2}x + \phantom{\frac{2}{3}}y \phantom{+\frac{5}{6}z} = 5 \\ x + \frac{2}{3}y + \frac{5}{6}z = 11 \\ \phantom{x+\frac{2}{3}y+} z = 6 \end{cases} \longrightarrow \left[\begin{array}{ccc:c} \frac{1}{2} & 1 & 0 & 5 \\ 1 & \frac{2}{3} & \frac{5}{6} & 11 \\ 0 & 0 & 1 & 6 \end{array}\right]$

Diagonal entries: $\frac{1}{2}$, $\frac{2}{3}$, and 1.

Notice that in the third row of part (b), the fact that $z = 6$ is indicated by zeroes in the x- and y-columns, 1 in the z-column, and a 6 in the column of constants.

Now try Exercises 10 through 12 ▶

This process can easily be reversed to write a system of equations from a given augmented matrix.

EXAMPLE 3 ▶ **Writing the System Corresponding to an Augmented Matrix**

Write the system of equations corresponding to each matrix, then solve the system using back-substitution.

a. $\left[\begin{array}{cc:c} 3 & -5 & -14 \\ 0 & 1 & 4 \end{array}\right]$ **b.** $\left[\begin{array}{ccc:c} 1 & 4 & -1 & -10 \\ 0 & -3 & 10 & 7 \\ 0 & 0 & 1 & 1 \end{array}\right]$

Solution ▶ **a.** $\left[\begin{array}{cc:c} 3 & -5 & -14 \\ 0 & 1 & 4 \end{array}\right] \longrightarrow \begin{cases} 3x - 5y = -14 \\ 1y = 4 \end{cases}$

With $y = 4$, we have

$3x - 5(4) = -14$ substitute 4 for y

$3x - 20 = -14$ multiply

$3x = 6$ add 20

$x = 2$ divide by 3

The solution is $(2, 4)$

b. $\begin{bmatrix} 1 & 4 & -1 & \vdots & -10 \\ 0 & -3 & 10 & \vdots & 7 \\ 0 & 0 & 1 & \vdots & 1 \end{bmatrix} \longrightarrow \begin{cases} 1x + 4y - 1z = -10 \\ \quad\quad -3y + 10z = 7 \\ \quad\quad\quad\quad 1z = 1 \end{cases}$

With $z = 1$, we first substitute 1 for z in the second equation.

$-3y + 10(1) = 7$ substitute 1 for z

$\quad -3y + 10 = 7$ multiply

$\quad\quad\quad -3y = -3$ subtract 10

$\quad\quad\quad\quad\quad y = 1$ divide by -3

We then substitute 1 for z and 1 for y in the first equation.

$x + 4(1) - 1 = 10$ substitute 1 for y and 1 for z

$\quad\quad x + 3 = 10$ simplify

$\quad\quad\quad\quad x = -7$ subtract 3

The solution is $(-7, 1, 1)$.

☑ **B.** You've just seen how we can form the augmented matrix of a system of equations

Now try Exercises 13 through 18 ▶

C. Solving a System Using Matrices

When a system of equations is written in augmented matrix form, we can solve the system by applying the same operations to each row of the matrix, that would be applied to the equations in the system. In this context, the operations are referred to as **elementary row operations.**

> **Elementary Row Operations**
>
> 1. Any two rows in a matrix can be interchanged.
> 2. The elements of any row can be multiplied by a nonzero constant.
> 3. Any two rows can be added together, and the sum used to replace one of the rows.

In this section, we'll use these operations to **triangularize the augmented matrix,** employing a solution method known as **Gaussian elimination.** A square matrix is said to be in **triangular form** when all of the entries below the diagonal are zero. For example,

the matrix $\begin{bmatrix} 1 & 4 & -1 & \vdots & -10 \\ 0 & -3 & 10 & \vdots & 7 \\ 0 & 0 & 1 & \vdots & 1 \end{bmatrix}$ is in triangular form: $\begin{bmatrix} 1 & 4 & -1 & \vdots & -10 \\ 0 & -3 & 10 & \vdots & 7 \\ 0 & 0 & 1 & \vdots & 1 \end{bmatrix}$.

In system form we have $\begin{cases} x + 4y - z = -10 \\ \quad -3y + 10z = 7 \\ \quad\quad\quad\quad z = 1 \end{cases}$, meaning a matrix written in triangular

form can be used to solve the system using back-substitution. We'll illustrate by

solving $\begin{cases} 1x + 4y - 1z = 4 \\ 2x + 5y + 8z = 15, \text{ using elimination to the left, and } \textit{row operations on the} \\ 1x + 3y - 3z = 1 \end{cases}$

augmented matrix to the right. As before, R1 represents the first equation in the system and the first row of the matrix, R2 represents equation 2 and row 2, and so on. The calculations involved are shown for the first stage only and are designed to offer a careful comparison. In actual practice, the format shown in Example 4 is used.

Elimination	**Row Operations**

Elimination
(System of Equations)
                      Row Operations
                      (Augmented Matrix)

$$\begin{cases} 1x + 4y - 1z = 4 \\ 2x + 5y + 8z = 15 \\ 1x + 3y - 3z = 1 \end{cases} \qquad \left[\begin{array}{ccc|c} 1 & 4 & -1 & 4 \\ 2 & 5 & 8 & 15 \\ 1 & 3 & -3 & 1 \end{array}\right]$$

To eliminate the x-term in R2, we use $-2\text{R1} + \text{R2} \rightarrow \text{R2}$. Identical operations are performed on the matrix, which begins the process of triangularizing the matrix.

System Form   **Matrix Form**

$$\begin{array}{lr} -2\text{R1} & -2x - 8y + 2z = -8 \\ + & \\ \underline{\text{R2}} & \underline{2x + 5y + 8z = 15} \\ \text{New R2} & -3y + 10z = 7 \end{array} \qquad \begin{array}{lrrrr} -2\text{R1} & -2 & -8 & 2 & -8 \\ + & \\ \underline{\text{R2}} & \underline{2} & \underline{5} & \underline{8} & \underline{15} \\ \text{New R2} & 0 & -3 & 10 & 7 \end{array}$$

For R3, the operations are $-1\text{R1} + \text{R3} \rightarrow \text{R3}$.

$$\begin{array}{lr} -1\text{R1} & -1x - 4y + 1z = -4 \\ + & \\ \underline{\text{R3}} & \underline{1x + 3y - 3z = 1} \\ \text{New R3} & -1y - 2z = -3 \end{array} \qquad \begin{array}{lrrrr} -1\text{R1} & -1 & -4 & 1 & -4 \\ + & \\ \underline{\text{R3}} & \underline{1} & \underline{3} & \underline{-3} & \underline{1} \\ \text{New R3} & 0 & -1 & -2 & -3 \end{array}$$

As always, we should look for opportunities to simplify any equation in the system (and any row in the matrix). Note that -1R3 will make the coefficients and related matrix entries positive. Here is the new system and matrix.

New System   **New Matrix**

$$\begin{cases} 1x + 4y - 1z = 4 \\ -3y + 10z = 7 \\ 1y + 2z = 3 \end{cases} \qquad \left[\begin{array}{ccc|c} 1 & 4 & -1 & 4 \\ 0 & -3 & 10 & 7 \\ 0 & 1 & 2 & 3 \end{array}\right]$$

On the left, we would finish by solving the 2×2 subsystem using $\text{R2} + 3\text{R3} \rightarrow \text{R3}$. In matrix form, we eliminate the corresponding entry (third row, second column) to triangularize the matrix.

$$\left[\begin{array}{ccc|c} 1 & 4 & -1 & 4 \\ 0 & -3 & 10 & 7 \\ 0 & 1 & 2 & 3 \end{array}\right] \xrightarrow{\text{R2} + 3\text{R3} \rightarrow \text{R3}} \left[\begin{array}{ccc|c} 1 & 4 & -1 & 4 \\ 0 & -3 & 10 & 7 \\ 0 & 0 & 16 & 16 \end{array}\right]$$

Dividing R3 by 16 gives $z = 1$ in the system, and entries of 0 0 1 1 in the augmented matrix. Completing the solution by back-substitution in the system gives the ordered triple (1, 1, 1). For practice using these row operations, **see Exercises 19 through 27.**

The general solution process is summarized here.

WORTHY OF NOTE

The procedure outlined for solving systems using matrices is virtually identical to that for solving systems by elimination. Using a 3×3 system for illustration, the "zeroes below the first diagonal entry" indicates we've eliminated the x-term from R2 and R3, the "zeroes below the second entry" indicates we've eliminated the y-term from the subsystem, and the division "to obtain a '1' in the final entry" indicates we have just solved for z.

Solving Systems by Triangularizing the Augmented Matrix

1. Write the system as an augmented matrix.
2. Use row operations to obtain zeroes below the first diagonal entry.
3. Use row operations to obtain zeroes below the second diagonal entry.
4. Continue until the matrix is triangularized (entries below diagonal are zero).
5. Divide to obtain a "1" in the last diagonal entry (if it is nonzero), then convert to equation form and solve using back-substitution.

Note: At each stage, look for opportunities to simplify row entries using multiplication or division. Also, to begin the process any equation with an x-coefficient of 1 can be made R1 by interchanging the equations.

EXAMPLE 4 ▶ **Solving Systems Using the Augmented Matrix**

Solve by triangularizing the augmented matrix: $\begin{cases} 2x + y - 2z = -7 \\ x + y + z = -1 \\ -2y - z = -3 \end{cases}$

Solution ▶ $\begin{cases} 2x + y - 2z = -7 \\ x + y + z = -1 \\ -2y - z = -3 \end{cases}$ matrix form → $\begin{bmatrix} 2 & 1 & -2 & \vdots & -7 \\ 1 & 1 & 1 & \vdots & -1 \\ 0 & -2 & -1 & \vdots & -3 \end{bmatrix}$ R1 ↔ R2 $\begin{bmatrix} 1 & 1 & 1 & \vdots & -1 \\ 2 & 1 & -2 & \vdots & -7 \\ 0 & -2 & -1 & \vdots & -3 \end{bmatrix}$

$\begin{bmatrix} 1 & 1 & 1 & \vdots & -1 \\ 2 & 1 & -2 & \vdots & -7 \\ 0 & -2 & -1 & \vdots & -3 \end{bmatrix}$ −2R1 + R2 → R2 $\begin{bmatrix} 1 & 1 & 1 & \vdots & -1 \\ 0 & -1 & -4 & \vdots & -5 \\ 0 & -2 & -1 & \vdots & -3 \end{bmatrix}$ −1R2 → R2 $\begin{bmatrix} 1 & 1 & 1 & \vdots & -1 \\ 0 & 1 & 4 & \vdots & 5 \\ 0 & -2 & -1 & \vdots & -3 \end{bmatrix}$

$\begin{bmatrix} 1 & 1 & 1 & \vdots & -1 \\ 0 & 1 & 4 & \vdots & 5 \\ 0 & -2 & -1 & \vdots & -3 \end{bmatrix}$ 2R2 + R3 → R3 $\begin{bmatrix} 1 & 1 & 1 & \vdots & -1 \\ 0 & 1 & 4 & \vdots & 5 \\ 0 & 0 & 7 & \vdots & 7 \end{bmatrix}$ $\frac{R3}{7}$ → R3 $\begin{bmatrix} 1 & 1 & 1 & \vdots & -1 \\ 0 & 1 & 4 & \vdots & 5 \\ 0 & 0 & 1 & \vdots & 1 \end{bmatrix}$

Converting the augmented matrix back into equation form yields $z = 1$. Back-substituting 1 for z in the second equation gives $y + 4(1) = 5$, with $y + 4 = 5$ showing $y = 1$. Back-substituting 1 for z and 1 for y in the first equation gives $x + 1 + 1 = -1$, or $x + 2 = -1$. Subtracting 2 from both sides shows $x = -3$. The solution is $(-3, 1, 1)$.

Now try Exercises 28 through 32 ▶

As mentioned, the process used in Example 4 is called **Gaussian elimination** (Carl Friedrich Gauss, 1777–1855), with the last matrix written in **row-echelon form.** It's also possible to solve a system entirely using only the augmented matrix, by continuing to use row operations until the diagonal entries are 1's, with 0's for all other entries in the coefficient matrix: $\begin{bmatrix} 1 & 0 & 0 & \vdots & a \\ 0 & 1 & 0 & \vdots & b \\ 0 & 0 & 1 & \vdots & c \end{bmatrix}$. The process is then called **Gauss-Jordan elimination** (Wilhelm Jordan, 1842–1899), with the final matrix written in *reduced row-echelon form* (see Appendix IV).

Note that with Gauss-Jordan elimination, our *initial* focus is less on getting 1's along the diagonal, and more on obtaining zeroes for all entries *other than* the diagonal entries. This will enable us to work with integer values in the solution process, as shown in Example 5.

EXAMPLE 5 ▶ **Solving a System Using Gauss-Jordan Elimination**

Solve using Gauss-Jordan elimination $\begin{cases} 2x + 5z - 15 = 2y \\ 2x + 3y = -1 + z. \\ 4y + z = -7 \end{cases}$

Solution ▶ standard form

$\begin{cases} 2x - 2y + 5z = 15 \\ 2x + 3y - 1z = -1 \\ 0x + 4y + 1z = -7 \end{cases}$ matrix form → $\begin{bmatrix} 2 & -2 & 5 & \vdots & 15 \\ 2 & 3 & -1 & \vdots & -1 \\ 0 & 4 & 1 & \vdots & -7 \end{bmatrix}$ −R1 + R2 → R2 $\begin{bmatrix} 2 & -2 & 5 & \vdots & 15 \\ 0 & 5 & -6 & \vdots & -16 \\ 0 & 4 & 1 & \vdots & -7 \end{bmatrix}$

$\begin{bmatrix} 2 & -2 & 5 & \vdots & 15 \\ 0 & 5 & -6 & \vdots & -16 \\ 0 & 4 & 1 & \vdots & -7 \end{bmatrix}$ 2R2 + 5R1 → R1, −4R2 + 5R3 → R3 $\begin{bmatrix} 10 & 0 & 13 & \vdots & 43 \\ 0 & 5 & -6 & \vdots & -16 \\ 0 & 0 & 29 & \vdots & 29 \end{bmatrix}$ $\frac{R3}{29}$ → R3 $\begin{bmatrix} 10 & 0 & 13 & \vdots & 43 \\ 0 & 5 & -6 & \vdots & -16 \\ 0 & 0 & 1 & \vdots & 1 \end{bmatrix}$

$$\begin{bmatrix} 10 & 0 & 13 & \vdots & 43 \\ 0 & 5 & -6 & \vdots & -16 \\ 0 & 0 & 1 & \vdots & 1 \end{bmatrix} \begin{array}{c} -13R3 + R1 \to R1 \\ \\ 6R3 + R2 \to R2 \end{array} \begin{bmatrix} 10 & 0 & 0 & \vdots & 30 \\ 0 & 5 & 0 & \vdots & -10 \\ 0 & 0 & 1 & \vdots & 1 \end{bmatrix} \begin{array}{c} \frac{R1}{10} \to R1 \\ \frac{R2}{5} \to R2 \end{array} \begin{bmatrix} 1 & 0 & 0 & \vdots & 3 \\ 0 & 1 & 0 & \vdots & -2 \\ 0 & 0 & 1 & \vdots & 1 \end{bmatrix}$$

The final matrix shows the solution is $x = 3$, $y = -2$, and $z = 1$, or $(3, -2, 1)$.

☑ **C.** You've just seen how we can solve a system of equations using row operations

Now try Exercises 33 through 36 ▶

D. Solving Systems of Equations Using Technology

Graphing calculators offer a very efficient way to solve systems of equations using matrices. Once the system has been written in matrix form, it can easily be entered into a calculator and solved by having the calculator instantly perform the row operations needed to produce the diagonal of 1's (with zeroes for all other coefficient entries). On many calculators, pressing [2nd] [x⁻¹] **(MATRIX)** gives a screen similar to the one shown in Figure 7.1, where we begin by selecting the **EDIT** option (push the right arrow twice). Pressing [ENTER] places you on a screen where you can **EDIT** matrix A, changing the size as needed, then input the entries of the matrix. To use the 3×4 matrix from Example 4, we press 3 [ENTER] and then 4 [ENTER], giving the screen shown in Figure 7.2. The dash marks to the right indicate that there is a fourth column that cannot be seen, but that comes into view as you enter the elements of the matrix. Begin entering the first row of the matrix, which is $\{2, 1, -2, -7\}$. Press [ENTER] after each entry and the cursor automatically goes to the next position in the matrix (note the calculator automatically shifts left and right to allow all four columns to be entered). After entering the second row $\{1, 1, 1, -1\}$ and the third row $\{0, -2, -1, -3\}$ the completed matrix should look like the one shown in Figure 7.3 (the matrix is currently shifted to the right, showing the fourth column).

Figure 7.1

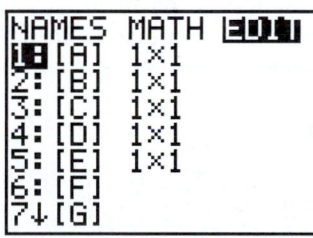

Figure 7.2

Figure 7.3

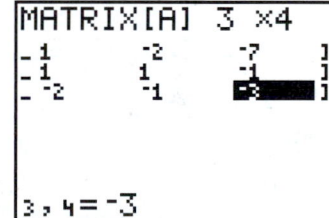

Most graphing calculators have an **rref** function, which is short for **reduced row echelon form.** To access this function, press [2nd] [x⁻¹] **(MATRIX)** and select the **MATH** option, then scroll upward (or downward) until you get to the **B:rref** option. Pressing [ENTER] places this function on the home screen, where we must tell it to perform the **rref** operation on matrix [A]. Pressing the [2nd] [x⁻¹] once again returns us to the matrix options, enabling us to select matrix **NAMES** (notice that **NAMES** is automatically highlighted). With the cursor over **1:[A] 3×4** we press [ENTER] and matrix [A] is placed on the home screen as the object of the **rref** function. After pressing [ )] and [ENTER] the calculator quickly computes the reduced row echelon form and displays it on the screen, as in Figure 7.4. The solution is easily read as $x = -3$, $y = 1$, and $z = 1$, just as we found in Example 4.

Figure 7.4

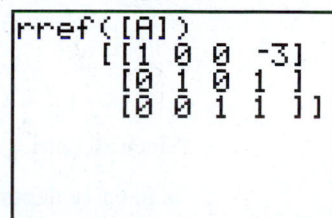

EXAMPLE 6 ▶ **Solving a System of Equations Using Technology**

Solve the following system using a graphing calculator.

$$\begin{cases} 2x + 3y - z = -4 \\ x - 3y + 2z = 1 \\ -3x + y - 4z = -5 \end{cases}$$

Solution ▶ Entering the system as matrix [A] produces the screen shown in Figure 7.5. Using the **rref([A])** command as shown above produces the screen in Figure 7.6. The solution is the ordered triple $(-2, 1, 3)$.

Check by substituting -2 for x, 1 for y, and 3 for z in the original equations.

Figure 7.5

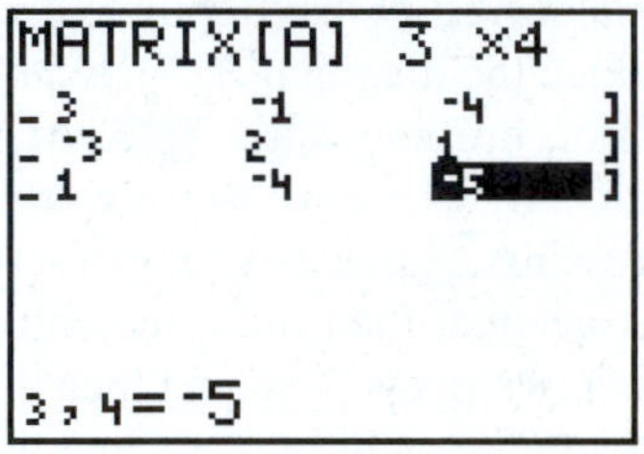

```
MATRIX[A] 3 ×4
 2     3    -1   -4  ]
 1    -3     2    1  ]
-3     1    -4   -5  ]



3,4=-5
```

Figure 7.6

```
rref([A])
   [[1 0 0 -2]
    [0 1 0 1 ]
    [0 0 1 3 ]]
```

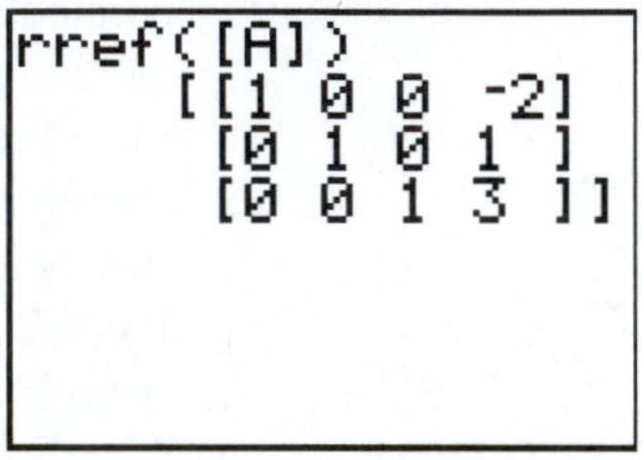

☑ **D.** You've just seen how we can solve a system of equations using technology

Now try Exercises 37 through 42 ▶

E. Inconsistent and Dependent Systems

Due to the strong link between a linear system and its augmented matrix, inconsistent and dependent systems can be recognized just as in Sections 6.1 and 6.2. An inconsistent system will yield an inconsistent or contradictory statement such as $0 = -12$, meaning all entries in a row of the matrix of coefficients are zero, but the constant is not. A linearly dependent system will yield an identity statement such as $0 = 0$, meaning all entries in *one* row of the matrix are zero. If the system has coincident dependence, *all* rows will be zero except one.

EXAMPLE 7 ▶ **Solving a Dependent System**

Solve the system using Gauss-Jordan elimination: $\begin{cases} x + y - 5z = 3 \\ -x + 2z = -1 \\ 2x - y - z = 0 \end{cases}$

Solution ▶

$$\begin{cases} x + y - 5z = 3 \\ -x + 2z = -1 \\ 2x - y - z = 0 \end{cases} \xrightarrow{\text{standard form}} \begin{cases} x + y - 5z = 3 \\ -x + 0y + 2z = -1 \\ 2x - y - z = 0 \end{cases} \xrightarrow{\text{matrix form}} \begin{bmatrix} 1 & 1 & -5 & | & 3 \\ -1 & 0 & 2 & | & -1 \\ 2 & -1 & -1 & | & 0 \end{bmatrix}$$

$$\begin{bmatrix} 1 & 1 & -5 & | & 3 \\ -1 & 0 & 2 & | & -1 \\ 2 & -1 & -1 & | & 0 \end{bmatrix} \begin{array}{l} \text{R1 + R2} \rightarrow \text{R2} \\ \\ -2\text{R1 + R3} \rightarrow \text{R3} \end{array} \begin{bmatrix} 1 & 1 & -5 & | & 3 \\ 0 & 1 & -3 & | & 2 \\ 0 & -3 & 9 & | & -6 \end{bmatrix} \begin{array}{l} -1\text{R2 + R1} \rightarrow \text{R1} \\ \\ 3\text{R2 + R3} \rightarrow \text{R3} \end{array} \begin{bmatrix} 1 & 0 & -2 & | & 1 \\ 0 & 1 & -3 & | & 2 \\ 0 & 0 & 0 & | & 0 \end{bmatrix}$$

Since all entries in the last row are zeroes and it's the only row of zeroes, we conclude the system is linearly dependent and equivalent to $\begin{cases} x - 2z = 1 \\ y - 3z = 2 \end{cases}$ (a calculator solution is shown in the figure). As in Chapter 6, we demonstrate this dependence by writing the (x, y, z) solution in terms of a parameter. Solving for y in R2 gives y in terms of z: $y = 3z + 2$. Solving for x in R1 gives x in terms of z: $x = 2z + 1$. As written, the solutions all depend on z: $x = 2z + 1$, $y = 3z + 2$, and $z = z$. Selecting p as the parameter (or some other "neutral" variable), we write the solution as $(2p + 1, 3p + 2, p)$. Two of the infinite number of solutions would be $(1, 2, 0)$ for $p = 0$, and $(-1, -1, -1)$ for $p = -1$. Test these triples in the original equations.

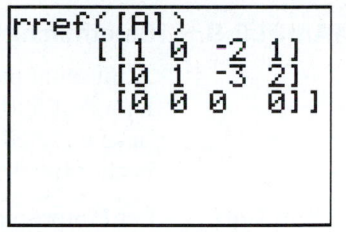

Now try Exercises 43 through 52 ▶

Since there was only one row of zeroes in the system from Example 7, we knew the system was *linearly* dependent. As mentioned previously, for coincident dependence the system will have only one *nonzero* row, with all other rows consisting entirely of zeroes. Note that if we change the constant in the third equation of the original system (the entry for row 3, column 4 was changed to a 5), an inconsistent system results as one equation is no longer dependent on the others (see Figures 7.7 and 7.8).

Figure 7.7   **Figure 7.8**

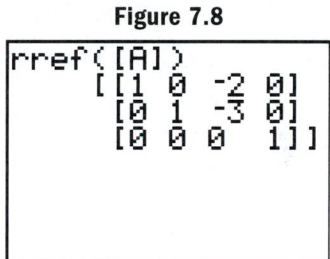

When the number of variables and equations in a system increases, solutions using technology can be particularly effective. In addition, some applications generate systems where the number of equations and number of variables are unequal, which the technology still handles efficiently.

EXAMPLE 8 ▶   **Solving a System of Equations Using Technology**

Solve the following system using a graphing calculator.

$$\begin{cases} x + 3y - z = 4 \\ -2x + y + 3z = 7 \\ -2y + 4z = 0 \\ -x + 4y + 2z = 11 \end{cases}$$

Solution ▶   After entering the system as matrix [A], using the **rref([A])** command produces the screen shown in the figure. Note that in this case, while the solution contains a row of zeroes, the system is *not linearly dependent* as there are more equations than variables, and unique values have been found for x, y, and z. The solution is $(-1, 2, 1)$.

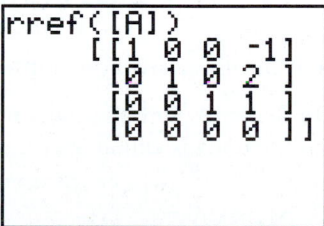

☑ **E.** You've just seen how we can recognize inconsistent and dependent systems

Now try Exercises 53 through 58 ▶

F. Solving Applications Using Matrices

As in other areas, solving applications using systems relies heavily on the ability to mathematically model information given verbally or in context. As you work through the exercises, read each problem carefully. Look for relationships that yield a system of two equations in two variables, three equations in three variables, and so on.

EXAMPLE 9 ▶ **Determining the Original Value of Collector's Items**

A museum purchases a famous painting, a ruby tiara, and a rare coin for its collection, spending a total of $30,000. One year later, the painting has tripled in value, while the tiara and the coin have doubled in value. The items now have a total value of $75,000. Find the purchase price of each if the original price of the painting was $1000 more than twice the coin.

Solution ▶ Let P represent the price of the painting, T the tiara, and C the coin.

Total spent was $30,000: $\rightarrow P + T + C = 30,000$
One year later: $\rightarrow 3P + 2T + 2C = 75,000$
Value of painting versus coin: $\rightarrow P = 2C + 1000$

$$\begin{cases} P + T + C = 30,000 \\ 3P + 2T + 2C = 75,000 \\ P = 2C + 1000 \end{cases} \text{standard form} \rightarrow \begin{cases} 1P + 1T + 1C = 30,000 \\ 3P + 2T + 2C = 75,000 \\ 1P + 0T - 2C = 1000 \end{cases} \text{matrix form} \rightarrow \begin{bmatrix} 1 & 1 & 1 & | & 30000 \\ 3 & 2 & 2 & | & 75000 \\ 1 & 0 & -2 & | & 1000 \end{bmatrix}$$

$$\begin{bmatrix} 1 & 1 & 1 & | & 30000 \\ 3 & 2 & 2 & | & 75000 \\ 1 & 0 & -2 & | & 1000 \end{bmatrix} \begin{matrix} -3R1 + R2 \rightarrow R2 \\ \\ -1R1 + R3 \rightarrow R3 \end{matrix} \begin{bmatrix} 1 & 1 & 1 & | & 30000 \\ 0 & -1 & -1 & | & -15000 \\ 0 & -1 & -3 & | & -29000 \end{bmatrix} \begin{matrix} -1R2 \rightarrow R2 \\ \\ -1R3 \rightarrow R3 \end{matrix} \begin{bmatrix} 1 & 1 & 1 & | & 30000 \\ 0 & 1 & 1 & | & 15000 \\ 0 & 1 & 3 & | & 29000 \end{bmatrix}$$

$$\begin{bmatrix} 1 & 1 & 1 & | & 30000 \\ 0 & 1 & 1 & | & 15000 \\ 0 & 1 & 3 & | & 29000 \end{bmatrix} -1R2 + R3 \rightarrow R3 \begin{bmatrix} 1 & 1 & 1 & | & 30000 \\ 0 & 1 & 1 & | & 15000 \\ 0 & 0 & 2 & | & 14000 \end{bmatrix} \frac{R3}{2} \rightarrow R3 \begin{bmatrix} 1 & 1 & 1 & | & 30000 \\ 0 & 1 & 1 & | & 15000 \\ 0 & 0 & 1 & | & 7000 \end{bmatrix}$$

From R3 of the triangularized form, $C = \$7000$ directly. Since R2 represents $T + C = 15,000$, we find the tiara was purchased for $T = \$8000$. Substituting these values into the first equation shows the painting was purchased for $15,000. The solution is (15,000, 8,000, 7,000). The solution found using a graphing calculator is also shown.

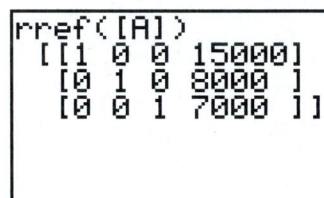

```
rref([A])
[[1 0 0 15000]
 [0 1 0 8000 ]
 [0 0 1 7000 ]]
```

☑ **F.** You've just seen how we can solve applications using matrix methods

Now try Exercises 61 through 68 ▶

7.1 EXERCISES

▶ CONCEPTS AND VOCABULARY

Fill in the blank with the appropriate word or phrase. Carefully reread the section if needed.

1. A matrix with the same number of rows and columns is called a(n) __square__ matrix.

2. When the coefficient matrix is used with the matrix of constants, the result is a(n) __augmented__ matrix.

3. Matrix $A = \begin{bmatrix} 2 & 4 & -3 \\ 1 & -2 & 1 \end{bmatrix}$ is a __2__ by __3__ matrix. The entry in the second row and third column is __1__.

4. Given matrix B shown here, the diagonal entries are __1__, __5__, and __1__. $B = \begin{bmatrix} 1 & 4 & 3 \\ -1 & 5 & 2 \\ 3 & -2 & 1 \end{bmatrix}$

5. The notation $-2R1 + R2 \rightarrow R2$ indicates that an equivalent matrix is formed by performing what operations/replacements? Multiply R1 by -2 and add that result to R2. This sum will be the new R2.

6. Describe how to tell an inconsistent system apart from a dependent system when solving using matrix methods (row reduction). If all entries in a row are zero, we conclude the system is dependent. If all entries in a row are zero except for the constant, the system is inconsistent.

HOMEWORK SELECTION GUIDE

Core: 7–31 every other odd, 37–49 odd, 53, 55, 57, 61, 63, 64, 65, 67 (22 Exercises)
Standard: 1–4, 7–31 every other odd, 37–49 odd, 53, 55, 57, 61, 63, 64, 65, 67, 69, 73 (28 Exercises)
Extended: 1–4, 7–31 every other odd, 37–49 odd, 53–63 odd, 64–70, 72, 73 (33 Exercises)
In Depth: 1–6, 7–31 every other odd, 37–49 odd, 53–63 odd, 64–74 (37 Exercises)

▶ DEVELOPING YOUR SKILLS

Determine the size of each matrix and identify the third row and second column entry. If the matrix given is a square matrix, identify the diagonal entries.

7. $\begin{bmatrix} 1 & 0 \\ 2.1 & 1 \\ -3 & 5.8 \end{bmatrix}$ $3 \times 2, 5.8$ **8.** $\begin{bmatrix} 1 & 0 & 4 \\ 1 & 3 & -7 \\ 5 & -1 & 2 \end{bmatrix}$
$3 \times 3, -1, 1, 3, 2$

9. $\begin{bmatrix} 1 & 0 & 4 \\ 1 & 3 & -7 \\ 5 & -1 & 2 \\ 2 & -3 & 9 \end{bmatrix}$ $4 \times 3, -1$

Form the augmented matrix, then name the diagonal entries of the coefficient matrix.

10. $\begin{cases} 2x - 3y - 2z = 7 \\ x - y + 2z = -5 \\ 3x + 2y - z = 11 \end{cases}$ $\begin{bmatrix} 2 & -3 & -2 & 7 \\ 1 & -1 & 2 & -5 \\ 3 & 2 & -1 & 11 \end{bmatrix}$;
diagonal entries 2, −1, −1

11. $\begin{cases} x + 2y - z = 1 \\ x + z = 3 \\ 2x - y + z = 3 \end{cases}$ $\begin{bmatrix} 1 & 2 & -1 & 1 \\ 1 & 0 & 1 & 3 \\ 2 & -1 & 1 & 3 \end{bmatrix}$;
diagonal entries 1, 0, 1

12. $\begin{cases} 2x + 3y + z = 5 \\ 2y - z = 7 \\ x - y - 2z = 5 \end{cases}$ $\begin{bmatrix} 2 & 3 & 1 & 5 \\ 0 & 2 & -1 & 7 \\ 1 & -1 & -2 & 5 \end{bmatrix}$;
diagonal entries 2, 2, −2

Write the system of equations for each matrix. Then use back-substitution to find its solution.

13. $\begin{bmatrix} 1 & 4 & 5 \\ 0 & 1 & \frac{1}{2} \end{bmatrix}$ $\begin{cases} x + 4y = 5 \\ y = \frac{1}{2} \end{cases} \rightarrow (3, \frac{1}{2})$

14. $\begin{bmatrix} 1 & -5 & -15 \\ 0 & -1 & -2 \end{bmatrix}$ $\begin{cases} x - 5y = -15 \\ -y = -2 \end{cases} \rightarrow (-5, 2)$

15. $\begin{bmatrix} 1 & 2 & -1 & 0 \\ 0 & 1 & 2 & 2 \\ 0 & 0 & 1 & 3 \end{bmatrix}$ $\begin{cases} x + 2y - z = 0 \\ y + 2z = 2 \rightarrow (11, -4, 3) \\ z = 3 \end{cases}$

16. $\begin{bmatrix} 1 & 0 & 7 & -5 \\ 0 & 1 & -5 & 15 \\ 0 & 0 & 1 & -26 \end{bmatrix}$ $\begin{cases} x + 7z = -5 \\ y - 5z = 15 \rightarrow (177, -115, -26) \\ z = -26 \end{cases}$

17. $\begin{bmatrix} 1 & 3 & -4 & 29 \\ 0 & 1 & -\frac{3}{2} & \frac{21}{2} \\ 0 & 0 & 1 & 3 \end{bmatrix}$ $\begin{cases} x + 3y - 4z = 29 \\ y - \frac{3}{2}z = \frac{21}{2} \rightarrow (-4, 15, 3) \\ z = 3 \end{cases}$

18. $\begin{bmatrix} 1 & 2 & -1 & 3 \\ 0 & 1 & \frac{1}{6} & \frac{2}{3} \\ 0 & 0 & 1 & \frac{22}{7} \end{bmatrix}$ $\begin{cases} x + 2y - z = 3 \\ y + \frac{1}{6}z = \frac{2}{3} \rightarrow (\frac{41}{7}, \frac{1}{7}, \frac{22}{7}) \\ z = \frac{22}{7} \end{cases}$

Perform the indicated row operation(s) in the order given and write the new matrix.

19. $\begin{bmatrix} \frac{1}{2} & -3 & -1 \\ -5 & 2 & 4 \end{bmatrix}$ $\begin{matrix} 2R1 \rightarrow R1, \\ 5R1 + R2 \rightarrow R2 \end{matrix}$

20. $\begin{bmatrix} 7 & 4 & 3 \\ 4 & -8 & 12 \end{bmatrix}$ $\begin{matrix} \frac{1}{4}R2 \rightarrow R2, \\ R1 \leftrightarrow R2 \end{matrix}$

21. $\begin{bmatrix} -2 & 1 & 0 & 4 \\ 5 & 8 & 3 & -5 \\ 1 & -3 & 3 & 2 \end{bmatrix}$ $\begin{matrix} R1 \leftrightarrow R3, \\ -5R1 + R2 \rightarrow R2 \end{matrix}$

22. $\begin{bmatrix} -3 & 2 & 0 & 0 \\ 1 & 1 & 2 & 6 \\ 4 & 1 & -3 & 2 \end{bmatrix}$ $\begin{matrix} R1 \leftrightarrow R2, \\ -4R1 + R3 \rightarrow R3 \end{matrix}$

23. $\begin{bmatrix} 3 & 1 & 1 & 8 \\ 6 & -1 & -1 & 10 \\ 4 & -2 & -3 & 22 \end{bmatrix}$ $\begin{matrix} -2R1 + R2 \rightarrow R2, \\ -4R1 + 3R3 \rightarrow R3 \end{matrix}$

24. $\begin{bmatrix} 2 & 1 & -1 & -3 \\ 3 & 1 & 1 & 0 \\ 4 & 3 & 2 & 3 \end{bmatrix}$ $\begin{matrix} -3R1 + 2R2 \rightarrow R2, \\ -2R1 + R3 \rightarrow R3 \end{matrix}$

What row operations would produce zeroes beneath the first entry in the diagonal?

25. $\begin{bmatrix} 1 & 3 & 0 & 2 \\ -2 & 4 & 1 & 1 \\ 3 & -1 & -2 & 9 \end{bmatrix}$ $\begin{matrix} 2R_1 + R_2 \rightarrow R_2 \\ -3R_1 + R_3 \rightarrow R_3 \end{matrix}$

26. $\begin{bmatrix} 1 & 1 & -4 & -3 \\ 3 & 0 & 1 & 5 \\ -5 & 3 & 2 & 3 \end{bmatrix}$ $\begin{matrix} -3R_1 + R_2 \rightarrow R_2 \\ 5R_1 + R_3 \rightarrow R_3 \end{matrix}$

27. $\begin{bmatrix} 1 & 2 & 0 & 10 \\ 5 & 1 & 2 & 6 \\ -4 & 3 & -3 & 2 \end{bmatrix}$ $\begin{matrix} -5R_1 + R_2 \rightarrow R_2 \\ 4R_1 + R_3 \rightarrow R_3 \end{matrix}$

Solve each system by triangularizing the augmented matrix and using back-substitution. Simplify by clearing fractions or decimals before beginning.

28. $\begin{cases} 2y = 5x + 4 \\ -5x = 2 - 4y \end{cases}$ $\left(\frac{-6}{5}, -1\right)$

29. $\begin{cases} 0.15g - 0.35h = -0.5 \\ -0.12g + 0.25h = 0.1 \end{cases}$ $(20, 10)$

30. $\begin{cases} -\frac{1}{5}u + \frac{1}{4}v = 1 \\ \frac{1}{10}u + \frac{1}{2}v = 7 \end{cases}$ $(10, 12)$

31. $\begin{cases} x - 2y + 2z = 7 \\ 2x + 2y - z = 5 \\ 3x - y + z = 6 \end{cases}$ $(1, 6, 9)$

32. $\begin{cases} 2x - 3y - 2z = 7 \\ x - y + 2z = -5 \\ 3x + 2y - z = 11 \end{cases}$ $(2, 1, -3)$

Additional answers can be found in the Instructor Answer Appendix.

Solve using Gauss-Jordan elimination.

33. $\begin{cases} x + 2y - z = 1 \\ x + z = 3 \\ 2x - y + z = 3 \end{cases}$
(1, 1, 2)

34. $\begin{cases} 2x + 3y + z = 5 \\ 2y - z = 7 \\ x - y - 2z = 5 \end{cases}$
(1, 2, −3)

35. $\begin{cases} -x + y + 2z = 2 \\ x + y - z = 1 \\ 2x + y + z = 4 \end{cases}$
(1, 1, 1)

36. $\begin{cases} x + y - 2z = -1 \\ 4x - y + 3z = 3 \\ 3x + 2y - z = 4 \end{cases}$
(0, 3, 2)

Solve each system of equations using a graphing calculator. Verify each solution on the home screen using the ⒶⓁⓅⒽⒶ **keys, as in Chapter 6.**

37. $\begin{cases} x + 2y + 3z = 9 \\ -2x + y - 5z = -20 \\ 3x - 5y + z = 14 \end{cases}$
(2, −1, 3)

38. $\begin{cases} 3x + 5y + z = 6 \\ -4x + 7y - 2z = 24 \\ x - 10y + 3z = -40 \end{cases}$
(1, 2, −7)

39. $\begin{cases} 0.2x + 0.5y + z = 13 \\ -x + 0.7y - 0.4z = -9 \\ -0.5x - y + 0.8z = -4.6 \end{cases}$
(10, 6, 8)

40. $\begin{cases} 0.1x + 2y + 0.6z = -36.4 \\ -3x + 0.8y - 0.2z = -35 \\ 0.5x - y + 0.4z = 25 \end{cases}$
(6, −20, 5)

41. $\begin{cases} \frac{1}{2}x + \frac{3}{4}y + \phantom{\frac{3}{2}}z = \frac{3}{4} \\ -x + \frac{1}{4}y + \frac{3}{2}z = -\frac{5}{8} \\ \frac{5}{4}x - \phantom{\frac{3}{2}}y + \frac{1}{2}z = -\frac{1}{2} \end{cases}$
$\left(\frac{1}{2}, 1, -\frac{1}{4}\right)$

42. $\begin{cases} \frac{5}{4}x + \frac{1}{2}y + \phantom{\frac{7}{8}}z = -\frac{1}{4} \\ -x + \frac{3}{2}y - \frac{7}{8}z = -\frac{3}{5} \\ \frac{1}{4}x - \phantom{\frac{7}{8}}y + \frac{1}{4}z = \frac{2}{5} \end{cases}$
$\left(\frac{8}{5}, -\frac{1}{2}, -2\right)$

Solve each system by triangularizing the augmented matrix and using back-substitution. If the system is linearly dependent, give the solution in terms of a parameter. If the system has coincident dependence, answer in set notation as in Chapter 6.

43. $\begin{cases} 4x - 8y + 8z = 24 \\ 2x - 6y + 3z = 13 \\ 3x + 4y - z = -11 \end{cases}$
$\left(-1, \frac{-3}{2}, 2\right)$

44. $\begin{cases} 3x + y + z = -2 \\ x - 2y + 3z = 1 \\ 2x - 3y + 5z = 3 \end{cases}$
(−9, 13, 12)

45. $\begin{cases} x + 3y + 5z = 20 \\ 2x + 3y + 4z = 16 \\ x + 2y + 3z = 12 \end{cases}$
linear dependence
$(p - 4, -2p + 8, p)$

46. $\begin{cases} -x + 2y + 3z = -6 \\ x - y + 2z = -4 \\ 3x - 6y - 9z = 18 \end{cases}$
linear dependence
$(-7p - 14, -5p - 10, p)$

47. $\begin{cases} 3x - 4y + 2z = -2 \\ \frac{3}{2}x - 2y + z = -1 \\ -6x + 8y - 4z = 4 \end{cases}$
coincident dependence
$\{(x, y, z)|3x - 4y + 2z = -2\}$

48. $\begin{cases} 2x - y + 3z = 1 \\ 4x - 2y + 6z = 2 \\ 10x - 5y + 15z = 5 \end{cases}$
coincident dependence
$\{(x, y, z)|2x - y + 3z = 1\}$

49. $\begin{cases} 2x - y + 3z = 1 \\ 2y + 6z = 2 \\ x - \frac{1}{2}y + \frac{3}{2}z = 5 \end{cases}$
inconsistent, no solution

50. $\begin{cases} x - 2y + 3z = 2 \\ 3x + 4y - z = 6 \\ 4x + 2y + 2z = 7 \end{cases}$
inconsistent, no solution

51. $\begin{cases} x + 2y + z = 4 \\ 3x - 4y + z = 4 \\ 6x - 8y + 2z = 8 \end{cases}$
linear dependence, $(p, \frac{1}{3}p, 4 - \frac{5}{3}p)$

52. $\begin{cases} -2x + 4y - 3z = 4 \\ 5x - 6y + 7z = -12 \\ x + 2y + z = -4 \end{cases}$
linear dependence,
$(-\frac{5}{4}p - 3, \frac{1}{8}p - \frac{1}{2}, p)$

Solve each system of equations using a graphing calculator. Verify each solution on the home screen using the ⒶⓁⓅⒽⒶ **keys, as in Chapter 6.**

53. $\begin{cases} 2x - 3y + z = 0 \\ x + 2y - z = 5 \\ 3x + 2z = 4 \\ x + 3y + z = 4 \end{cases}$
(2, 1, −1)

54. $\begin{cases} -2x + 5y = -4 \\ 3x - 4y + z = 0 \\ x + y - 3z = -8 \\ 5x - 9y + z = 4 \end{cases}$
(−3, −2, 1)

55. $\begin{cases} x + 3y + z = 1 \\ x - 2z = 7 \\ -2y + 3z = -6 \\ 2x + 3y - z = 8 \end{cases}$
(3, 0, −2)

56. $\begin{cases} -2x + 3y = 4 \\ - 2y + 5z = -19 \\ x - 4z = 13 \\ -x + 3y - 4z = 17 \end{cases}$
(1, 2, −3)

57. $\begin{cases} x + 2y - 3z + w = -13 \\ -2x + 3y + z - 2w = -3 \\ -x - y + 2z + 3w = 4 \\ 3x - 2y - z - w = 5 \end{cases}$ $(1, -2, 3, -1)$

58. $\begin{cases} 2x + 3y - 4z + w = -10 \\ x - 2y - 3w = 6 \\ 3x + 5z + 2w = 1 \\ -x + y + 2z - 5w = 21 \end{cases}$ $(-1, 1, 2, -3)$

▶ WORKING WITH FORMULAS

Area of a triangle in the plane: $A = \pm\dfrac{1}{2}(x_1y_2 - x_2y_1 + x_2y_3 - x_3y_2 + x_3y_1 - x_1y_3)$

The area of a triangle in the plane is given by the formula shown, where the vertices of the triangle are located at the points (x_1, y_1), (x_2, y_2), **and** (x_3, y_3), **and the sign is chosen to ensure a positive value.**

59. Find the area of a triangle whose vertices are $(-1, -3)$, $(5, 2)$, and $(1, 8)$. 28 units²

60. Find the area of a triangle whose vertices are $(6, -2)$, $(-5, 4)$, and $(-1, 7)$. 28.5 units²

▶ APPLICATIONS

Model each problem using a system of linear equations. Then solve using the augmented matrix.

Descriptive Translation

61. The distance (via air travel) from Los Angeles (LA), California, to Saint Louis (STL), Missouri, to Cincinnati (CIN), Ohio, to New York City (NYC), New York, is approximately 2480 mi. Find the length of each flight if the distance from LA to STL is 50 mi more than five times the distance between STL and CIN, and 110 mi less than three times the distance from CIN to NYC.
LA to STL, 1600 mi; STL to CIN, 310 mi; CIN to NY, 570 mi

62. In the 2006 NBA Championship Series, Dwayne Wade of the Miami Heat carried his team to the title after the first two games were lost to the Dallas Mavericks. If 187 points were scored in the title game and the Heat won by 3 points, what was the final score? Heat: 95, Mavericks: 92

63. Moe is lecturing Larry and Curly once again (Moe, Larry, and Curly of *The Three Stooges* fame) claiming he is twice as smart as Larry and three times as smart as Curly. If he is correct and the sum of their IQs is 165, what is the IQ of each stooge? Moe 90, Larry 45, Curly 30

64. A collector of rare books buys a handwritten, autographed copy of Edgar Allan Poe's *Annabel Lee,* an original advance copy of L. Frank Baum's *The Wonderful Wizard of Oz,* and a first print copy of *The Caine Mutiny* by Herman Wouk, paying a total of $100,000. Find the cost of each one, given that the cost of *Annabel Lee* and twice the cost of *The Caine Mutiny* sum to the price paid for *The Wonderful Wizard of Oz,* and *The Caine Mutiny* cost twice as much as *Annabel Lee.*
Poe, $12,500; Baum, $62,500; Wouk, $25,000

Geometry

65. A right triangle has a hypotenuse of 39 m. If the perimeter is 90 m, and the longer leg is 6 m longer than twice the shorter leg, find the dimensions of the triangle. 15 m, 36 m, 39 m

66. In triangle *ABC*, the sum of angles *A* and *C* is equal to three times angle *B*. Angle *C* is 10 degrees more than twice angle *B*. Find the measure of each angle.
$A = 35°$, $B = 45°$, $C = 100°$

Investment and Finance

67. Suppose $10,000 is invested in three different investment vehicles paying 5%, 7%, and 9% annual interest. Find the amount invested at each rate if the interest earned after 1 yr is $760 and the amount invested at 9% is equal to the sum of the amounts invested at 5% and 7%.
$2000 at 5%; $3000 at 7%; $5000 at 9%

68. The trustee of a union's pension fund has invested funds in three ways: a savings fund paying 4% annual interest, a money market fund paying 7%, and government bonds paying 8%. Find the amount invested in each if the interest earned after one year is $0.178 million and the amount in government bonds is $0.3 million more than twice the amount in money market funds. The total amount invested is $2.5 million dollars. $0.4 million at 4%; $0.6 million at 7%; $1.5 million at 8%

► **EXTENDING THE CONCEPT**

69. Given the drawing shown, use a system of equations and the matrix method to find the measure of the angles labeled as x and y. Recall that vertical angles are equal and that the sum of the angles in a triangle is $180°$. $x = 84°; y = 25°$

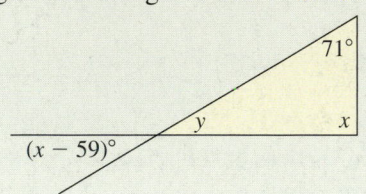

$(x - 59)°$

70. The system given here has a solution of $(1, -2, 3)$. Find the value of a and b.

$$\begin{bmatrix} 1 & a & b & | & 1 \\ 2b & 2a & 5 & | & 13 \\ 2a & 7 & 3b & | & -8 \end{bmatrix} \quad a = \frac{3}{4}, b = \frac{1}{2}$$

► **MAINTAINING YOUR SKILLS**

71. (3.5) Given $f(x) = x^3 - 8$ and $g(x) = x - 2$, find $f + g, f - g, fg,$ and $\dfrac{f}{g}$.

$x^3 + x - 10; x^3 - x - 6; x^4 - 2x^3 - 8x + 16; x^2 + 2x + 4$

72. (4.2) Given $x = 2$ is a zero of $h(x) = x^4 - x^2 - 12$, find all zeroes of h, real and complex.
$x = \pm 2; x = \pm i\sqrt{3}$

73. (5.6) Since 2005, cable installations for an Internet company have been modeled by the function $C(t) = 15 \ln (t + 1)$, where $C(t)$ represents cable installations in thousands, t yr after 2005. In what year will the number of installations be greater than 30,000? $C > 30,000$ in the year 2011 $(t \approx 6.39)$

74. (5.6) If a set amount of money p is deposited regularly (daily, weekly, monthly, etc.) n times per year at a fixed interest rate r, the amount of money A accumulated in t years is given by the formula shown. If a parent deposited \$250 per month for 18 yr at 4.6% beginning when her first child was born, how much has been accumulated to help pay for college expenses? \$83,811.95

$$A = \dfrac{p\left[\left(1 + \dfrac{r}{n}\right)^{nt} - 1\right]}{\dfrac{r}{n}}$$

7.2 The Algebra of Matrices

LEARNING OBJECTIVES

In Section 7.2 you will see how we can:

☐ **A.** Determine if two matrices are equal

☐ **B.** Add and subtract matrices

☐ **C.** Compute the product of two matrices

Matrices serve a much wider purpose than just a convenient method for solving systems. To understand their broader application, we need to know more about matrix theory, the various ways matrices can be combined, and some of their more practical uses. The common operations of addition, subtraction, multiplication, and division are all defined for matrices, as are other operations. Practical applications of matrix theory can be found in the social sciences, inventory management, genetics, operations research, engineering, and many other fields.

A. Equality of Matrices

To effectively study matrix algebra, we first give matrices a more general definition. For the *general* matrix A, all entries will be denoted using the lowercase letter "a," with their position in the matrix designated by the dual subscript a_{ij}. The letter "i" gives the *row* and the letter "j" gives the *column* of the entry's location. The general $m \times n$ matrix A is written

$$
\begin{array}{c}
\quad\quad\quad\ \text{col 1}\quad\ \text{col 2}\quad\ \text{col 3}\quad\quad\quad\ \text{col } j\quad\quad\ \text{col } n \\
\begin{array}{l}
\text{row 1} \to \\
\text{row 2} \to \\
\text{row 3} \to \\
\\
\text{row } i \to \\
\\
\text{row } m \to
\end{array}
\begin{bmatrix}
a_{11} & a_{12} & a_{13} & \cdots & a_{1j} & \cdots & a_{1n} \\
a_{21} & a_{22} & a_{23} & \cdots & a_{2j} & \cdots & a_{2n} \\
a_{31} & a_{32} & a_{33} & \cdots & a_{3j} & \cdots & a_{3n} \\
\vdots & \vdots & \vdots & & \vdots & & \vdots \\
a_{i1} & a_{i2} & a_{i3} & \cdots & \mathbf{a_{ij}} & \cdots & a_{in} \\
\vdots & \vdots & \vdots & & \vdots & & \vdots \\
a_{m1} & a_{m2} & a_{m3} & \cdots & a_{mj} & \cdots & a_{mn}
\end{bmatrix}
\end{array}
$$

a_{ij} is a general matrix element

The size of a matrix is also referred to as its **order,** and we say the order of general matrix A is $m \times n$. Note that diagonal entries have the same row and column number, a_{ij}, where $i = j$. Also, where the general entry of matrix A is a_{ij}, the general entry of matrix B is b_{ij}, of matrix C is c_{ij}, and so on.

EXAMPLE 1 ▶ **Identifying the Order and Entries of a Matrix**

State the order of each matrix and name the entries corresponding to a_{22}, a_{31}; b_{22}, b_{31}; and c_{22}, c_{31}.

$$
\textbf{a. } A = \begin{bmatrix} 1 & 4 \\ -2 & -3 \end{bmatrix}
\quad
\textbf{b. } B = \begin{bmatrix} 3 & -2 \\ 1 & 5 \\ -4 & 3 \end{bmatrix}
\quad
\textbf{c. } C = \begin{bmatrix} 0.2 & -0.5 & 0.7 \\ -1 & 0.3 & 1 \\ 2.1 & -0.1 & 0.6 \end{bmatrix}
$$

Solution ▶   **a.** matrix A: order 2×2. Entry $a_{22} = -3$ (the row 2, column 2 entry is -3). There is no a_{31} entry (A is only 2×2).

   b. matrix B: order 3×2. Entry $b_{22} = 5$, entry $b_{31} = -4$.

   c. matrix C: order 3×3. Entry $c_{22} = 0.3$, entry $c_{31} = 2.1$.

Now try Exercises 7 through 12 ▶

Equality of Matrices

Two matrices are equal if they have the same order and their corresponding entries are equal. In symbols, this means that $A = B$ if $a_{ij} = b_{ij}$ for all i and j.

EXAMPLE 2 ▶ **Determining If Two Matrices Are Equal**

Determine whether the following statements are true, false, or conditional. If false, explain why. If conditional, find values that will make the statement true.

$$
\textbf{a. } \begin{bmatrix} 1 & 4 \\ -2 & -3 \end{bmatrix} = \begin{bmatrix} -3 & -2 \\ 4 & 1 \end{bmatrix}
\quad\quad
\textbf{b. } \begin{bmatrix} 3 & -2 \\ 1 & 5 \\ -4 & 3 \end{bmatrix} = \begin{bmatrix} 3 & -2 & 1 \\ 5 & -4 & 3 \end{bmatrix}
$$

$$
\textbf{c. } \begin{bmatrix} 1 & 4 \\ -2 & -3 \end{bmatrix} = \begin{bmatrix} a - 2 & 2b \\ c & -3 \end{bmatrix}
$$

Solution ▶   **a.** $\begin{bmatrix} 1 & 4 \\ -2 & -3 \end{bmatrix} = \begin{bmatrix} -3 & -2 \\ 4 & 1 \end{bmatrix}$ is false. The matrices have the same order and entries, but corresponding entries are not equal.

   b. $\begin{bmatrix} 3 & -2 \\ 1 & 5 \\ -4 & 3 \end{bmatrix} = \begin{bmatrix} 3 & -2 & 1 \\ 5 & -4 & 3 \end{bmatrix}$ is false. Their orders are not equal.

   c. $\begin{bmatrix} 1 & 4 \\ -2 & -3 \end{bmatrix} = \begin{bmatrix} a - 2 & 2b \\ c & -3 \end{bmatrix}$ is conditional. The statement is true when

☑ **A.** You've just seen how we can determine if two matrices are equal

$a - 2 = 1$ ($a = 3$), $2b = 4$ ($b = 2$), $c = -2$, and is false otherwise.

Now try Exercises 13 through 16 ▶

B. Addition and Subtraction of Matrices

A sum or difference of matrices is found by combining the corresponding entries. This limits the operations to matrices of like orders, so that every entry in one matrix has a "corresponding entry" in the other. This also means the result is a new matrix of *like order,* whose entries are the corresponding sums or differences. If you attempt to add matrices of unlike size on a graphing calculator, an error message is displayed: **ERR: DIM MISMATCH.** Note that since a_{ij} represents a general entry of matrix A, $[a_{ij}]$ represents the entire matrix.

> **Addition and Subtraction of Matrices**
>
> Given matrices A, B, and C having like orders.
>
> The sum $A + B = C$,   The difference $A - B = C$,
> where $[a_{ij} + b_{ij}] = [c_{ij}]$.   where $[a_{ij} - b_{ij}] = [c_{ij}]$.

EXAMPLE 3 ▶ **Adding and Subtracting Matrices**

Compute the sum or difference of the matrices indicated.

$$A = \begin{bmatrix} 2 & 6 \\ 1 & 0 \\ 1 & -3 \end{bmatrix} \qquad B = \begin{bmatrix} -3 & 2 & -1 \\ -5 & 4 & 3 \end{bmatrix} \qquad C = \begin{bmatrix} 3 & -2 \\ 1 & 5 \\ -4 & 3 \end{bmatrix}$$

a. $A + C$ **b.** $A + B$ **c.** $C - A$

Solution ▶ **a.** $A + C = \begin{bmatrix} 2 & 6 \\ 1 & 0 \\ 1 & -3 \end{bmatrix} + \begin{bmatrix} 3 & -2 \\ 1 & 5 \\ -4 & 3 \end{bmatrix}$ sum of *A* and *C*

$$= \begin{bmatrix} 2+3 & 6+(-2) \\ 1+1 & 0+5 \\ 1+(-4) & -3+3 \end{bmatrix} = \begin{bmatrix} 5 & 4 \\ 2 & 5 \\ -3 & 0 \end{bmatrix}$$ add corresponding entries

b. $A + B = \begin{bmatrix} 2 & 6 \\ 1 & 0 \\ 1 & -3 \end{bmatrix} + \begin{bmatrix} -3 & 2 & -1 \\ -5 & 4 & 3 \end{bmatrix}$ Addition and subtraction are not defined for matrices of unlike order.

c. $C - A = \begin{bmatrix} 3 & -2 \\ 1 & 5 \\ -4 & 3 \end{bmatrix} - \begin{bmatrix} 2 & 6 \\ 1 & 0 \\ 1 & -3 \end{bmatrix}$ difference of *C* and *A*

$$= \begin{bmatrix} 3-2 & -2-6 \\ 1-1 & 5-0 \\ -4-1 & 3-(-3) \end{bmatrix} = \begin{bmatrix} 1 & -8 \\ 0 & 5 \\ -5 & 6 \end{bmatrix}$$ subtract corresponding entries

Now try Exercises 17 through 20 ▶

Operations on matrices can be very laborious for larger matrices and for matrices with noninteger or large entries. For these, we can turn to available technology for assistance. This shifts our focus from a meticulous computation of entries, to carefully entering each matrix into the calculator, double-checking each entry, and appraising results to see if they're reasonable.

EXAMPLE 4 ▶ **Using Technology for Matrix Operations**

Use a calculator to compute the difference $A - B$ for the matrices given.

$$A = \begin{bmatrix} \frac{2}{11} & -0.5 & \frac{6}{5} \\ 0.9 & \frac{3}{4} & -4 \\ 0 & 6 & -\frac{5}{12} \end{bmatrix} \quad B = \begin{bmatrix} \frac{1}{6} & \frac{-7}{10} & 0.75 \\ \frac{11}{25} & 0 & -5 \\ -4 & \frac{-5}{9} & \frac{-5}{12} \end{bmatrix}$$

Solution ▶ The entries for matrix A are shown in Figure 7.9. After entering matrix B, exit to the home screen [2nd MODE (**QUIT**)], call up matrix A, press the ⟨ – ⟩ (subtract) key, then call up matrix B and press ENTER. The calculator quickly finds the difference and displays the results shown in Figure 7.10. The last line on the screen shows the result can be stored for future use in a new matrix C by pressing the STO▶ key, calling up matrix C, and pressing ENTER.

Figure 7.9

```
MATRIX[A]  3 ×3
[ .1B1B2   -.5      1.2      ]
[ .9        .75    -4        ]
[ 0         6      ■■■■■■■    ]



3,3=-.416666666…
```

Figure 7.10

```
[A]-[B]
[[.0151515152  .…
 [.46
 [4              6.…
Ans→[C]
```

Now try Exercises 21 through 24 ▶

Figure 7.11

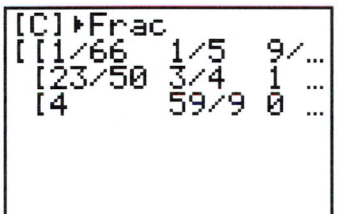

```
[C]▶Frac
[[1/66    1/5    9/…
 [23/50   3/4    1  …
 [4       59/9   0  …
```

In Figure 7.10 the dots to the right on the calculator screen indicate there are additional digits or matrix columns that can't fit on the display, as often happens with larger matrices or decimal numbers. Sometimes, converting entries to fraction form will provide a display that's easier to read. Here, this is done by calling up the matrix C, and using the MATH **1: ▶ Frac** option. After pressing ENTER, all entries are converted to fractions in simplest form (where possible), as in Figure 7.11. The third column can be viewed by pressing the right arrow.

Since the addition of two matrices is defined as the sum of corresponding entries, we find the properties of matrix addition closely resemble those of real number addition. Similar to standard algebraic properties, $-A$ represents the product $-1 \cdot A$ and any subtraction can be rewritten as an algebraic sum: $A - B = A + (-B)$. As noted in the properties box, for any matrix A, the sum $A + (-A)$ will yield the **zero matrix Z,** a matrix of like size whose entries are all zeroes. Also note that matrix $-A$ is the **additive inverse** for A, while Z is the **additive identity.**

✓ **B. You've just seen how we can add and subtract matrices**

Properties of Matrix Addition

Given matrices A, B, C, and Z are $m \times n$ matrices, with Z the zero matrix. Then,

I. $A + B = B + A$ ⟶ matrix addition is commutative

II. $(A + B) + C = A + (B + C)$ ⟶ matrix addition is associative

III. $A + Z = Z + A = A$ ⟶ Z is the additive identity

IV. $A + (-A) = (-A) + A = Z$ ⟶ $-A$ is the additive inverse of A

C. Matrices and Multiplication

The algebraic terms $2a$ and ab have counterparts in matrix algebra. The product $2A$ represents a constant times a matrix and is called **scalar multiplication.** The product AB represents the product of two matrices.

Scalar Multiplication

Scalar multiplication is defined by taking the product of the constant with *each entry* in the matrix, forming a new matrix of like size. In symbols, for any real number k and matrix A, $kA = [ka_{ij}]$.

EXAMPLE 5 ▶ **Computing Operations on Matrices**

Given $A = \begin{bmatrix} 4 & 3 \\ \frac{1}{2} & 1 \\ 0 & -3 \end{bmatrix}$ and $B = \begin{bmatrix} 3 & -2 \\ 0 & 6 \\ -4 & 0.4 \end{bmatrix}$, compute the following:

a. $\frac{1}{2}B$ **b.** $-4A - \frac{1}{2}B$

Solution ▶ **a.** $\frac{1}{2}B = \left(\frac{1}{2}\right)\begin{bmatrix} 3 & -2 \\ 0 & 6 \\ -4 & 0.4 \end{bmatrix}$

$$= \begin{bmatrix} (\frac{1}{2})(3) & (\frac{1}{2})(-2) \\ (\frac{1}{2})(0) & (\frac{1}{2})(6) \\ (\frac{1}{2})(-4) & (\frac{1}{2})(0.4) \end{bmatrix} = \begin{bmatrix} \frac{3}{2} & -1 \\ 0 & 3 \\ -2 & 0.2 \end{bmatrix}$$

b. $-4A - \dfrac{1}{2}B = -4A + \left(-\dfrac{1}{2}\right)B$ rewrite using algebraic addition

$$= \begin{bmatrix} (-4)(4) & (-4)(3) \\ (-4)(\frac{1}{2}) & (-4)(1) \\ (-4)(0) & (-4)(-3) \end{bmatrix} + \begin{bmatrix} (-\frac{1}{2})(3) & (-\frac{1}{2})(-2) \\ (-\frac{1}{2})(0) & (-\frac{1}{2})(6) \\ (-\frac{1}{2})(-4) & (-\frac{1}{2})(0.4) \end{bmatrix}$$

$$= \begin{bmatrix} -16 & -12 \\ -2 & -4 \\ 0 & 12 \end{bmatrix} + \begin{bmatrix} -\frac{3}{2} & 1 \\ 0 & -3 \\ 2 & -0.2 \end{bmatrix}$$ simplify

$$= \begin{bmatrix} -16 + (-\frac{3}{2}) & -12 + 1 \\ -2 + 0 & -4 + (-3) \\ 0 + 2 & 12 + (-0.2) \end{bmatrix} = \begin{bmatrix} -\frac{35}{2} & -11 \\ -2 & -7 \\ 2 & 11.8 \end{bmatrix}$$ result

Now try Exercises 25 through 28 ▶

Matrix Multiplication

Consider a cable company offering three different levels of Internet service: Bronze—fast, Silver—very fast, and Gold—lightning fast. Table 7.1 shows the number and types of programs sold to households and businesses for the week. Each program has an incentive package consisting of a rebate and a certain number of free weeks, as shown in Table 7.2.

Table 7.1 Matrix A

	Bronze	Silver	Gold
Homes	40	20	25
Businesses	10	15	45

Table 7.2 Matrix B

	Rebate	Free Weeks
Bronze	$15	2
Silver	$25	4
Gold	$35	6

To compute the amount of rebate money the cable company paid to households for the week, we would take the first row (R1) in Table 7.1 and multiply by the corresponding entries (bronze with bronze, silver with silver, and so on) in the first column (C1) of Table 7.2 and add these products. In matrix form, we have

$$[40 \ \ 20 \ \ \mathbf{25}] \cdot \begin{bmatrix} 15 \\ 25 \\ \mathbf{35} \end{bmatrix} = 40 \cdot 15 + 20 \cdot 25 + \mathbf{25} \cdot \mathbf{35} = \$1975. \text{ Using R1 of Table 7.1}$$

with C2 from Table 7.2 gives the number of free weeks awarded to homes:

$$[40 \ \ 20 \ \ \mathbf{25}] \cdot \begin{bmatrix} 2 \\ 4 \\ \mathbf{6} \end{bmatrix} = 40 \cdot 2 + 20 \cdot 4 + \mathbf{25} \cdot \mathbf{6} = 310. \text{ Using the second row (R2) of}$$

Table 7.1 with the two columns from Table 7.2 will give the amount of rebate money and the number of free weeks, respectively, awarded to business customers. When all computations are complete, the result is a product matrix P with order 2×2. This is because the product of R1 from matrix A, with C1 from matrix B, *gives the entry in*

$$position \ P_{11} \ of \ the \ product \ matrix: \begin{bmatrix} 40 & 20 & \mathbf{25} \\ 10 & 15 & 45 \end{bmatrix} \cdot \begin{bmatrix} 15 & 2 \\ 25 & 4 \\ \mathbf{35} & 6 \end{bmatrix} = \begin{bmatrix} 1975 & 310 \\ 2100 & 350 \end{bmatrix}.$$

Likewise, the product $R1 \cdot C2$ will give entry P_{12} (310), the product of R2 with C1 will give P_{21} (2100), and so on. This "row × column" multiplication can be generalized, and leads to the following. Given $m \times n$ matrix A and $s \times t$ matrix B,

In more formal terms, we have the following definition of matrix multiplication.

Matrix Multiplication

Given the $m \times n$ matrix $A = [a_{ij}]$ and the $s \times t$ matrix $B = [b_{ij}]$. If $n = s$, then matrix multiplication is possible and the product AB is an $m \times t$ matrix $P = [p_{ij}]$, where p_{ij} is product of the ith row of A with the jth column of B.

In less formal terms, matrix multiplication involves multiplying the row entries of the first matrix with the corresponding column entries of the second, and adding them together. In Example 6, two of the matrix products [parts (a) and (b)] are shown in full detail, with the first entry of the product matrix color-coded.

EXAMPLE 6 ▶   Multiplying Matrices

Given the matrices A through E shown here, compute the following products:

    a. AB **b.** CD **c.** DC **d.** AE **e.** EA

$$A = \begin{bmatrix} -2 & 1 \\ 3 & 4 \end{bmatrix} \quad B = \begin{bmatrix} 4 & 3 \\ 6 & 1 \end{bmatrix} \quad C = \begin{bmatrix} -2 & 1 & 3 \\ 1 & 0 & 2 \\ 4 & 1 & -1 \end{bmatrix} \quad D = \begin{bmatrix} 2 & 5 & 1 \\ 4 & -1 & 1 \\ 0 & 3 & -2 \end{bmatrix} \quad E = \begin{bmatrix} -2 & -1 \\ 3 & 0 \\ 1 & 2 \end{bmatrix}$$

Solution ▶ **a.** $AB = \begin{bmatrix} -2 & 1 \\ 3 & 4 \end{bmatrix}\begin{bmatrix} 4 & 3 \\ 6 & 1 \end{bmatrix} = \begin{bmatrix} -2 & -5 \\ 36 & 13 \end{bmatrix}$

Computation: $\begin{bmatrix} (-2)(4) + (1)(6) & (-2)(3) + (1)(1) \\ (3)(4) + (4)(6) & (3)(3) + (4)(1) \end{bmatrix}$

> A B
> (2×2) (2×2)
> multiplication is possible since $2 = 2$
>
> A B
> (2×2) (2×2)
> result will be a 2×2 matrix

b. $CD = \begin{bmatrix} -2 & 1 & 3 \\ 1 & 0 & 2 \\ 4 & 1 & -1 \end{bmatrix}\begin{bmatrix} 2 & 5 & 1 \\ 4 & -1 & 1 \\ 0 & 3 & -2 \end{bmatrix} = \begin{bmatrix} 0 & -2 & -7 \\ 2 & 11 & -3 \\ 12 & 16 & 7 \end{bmatrix}$

> C D
> (3×3) (3×3)
> multiplication is possible since $3 = 3$
>
> C D
> (3×3) (3×3)
> result will be a 3×3 matrix

Computation: $\begin{bmatrix} (-2)(2) + (1)(4) + (3)(0) & (-2)(5) + (1)(-1) + (3)(3) & (-2)(1) + (1)(1) + (3)(-2) \\ (1)(2) + (0)(4) + (2)(0) & (1)(5) + (0)(-1) + (2)(3) & (1)(1) + (0)(1) + (2)(-2) \\ (4)(2) + (1)(4) + (-1)(0) & (4)(5) + (1)(-1) + (-1)(3) & (4)(1) + (1)(1) + (-1)(-2) \end{bmatrix}$

c. $DC = \begin{bmatrix} 2 & 5 & 1 \\ 4 & -1 & 1 \\ 0 & 3 & -2 \end{bmatrix}\begin{bmatrix} -2 & 1 & 3 \\ 1 & 0 & 2 \\ 4 & 1 & -1 \end{bmatrix} = \begin{bmatrix} 5 & 3 & 15 \\ -5 & 5 & 9 \\ -5 & -2 & 8 \end{bmatrix}$

> D C
> (3×3) (3×3)
> multiplication is possible since $3 = 3$
>
> D C
> (3×3) (3×3)
> result will be a 3×3 matrix

d. $AE = \begin{bmatrix} -2 & 1 \\ 3 & 4 \end{bmatrix}\begin{bmatrix} -2 & -1 \\ 3 & 0 \\ 1 & 2 \end{bmatrix}$

> A E
> (2×2) (3×2)
> multiplication is not possible since $2 \neq 3$

e. $EA = \begin{bmatrix} -2 & -1 \\ 3 & 0 \\ 1 & 2 \end{bmatrix}\begin{bmatrix} -2 & 1 \\ 3 & 4 \end{bmatrix} = \begin{bmatrix} 1 & -6 \\ -6 & 3 \\ 4 & 9 \end{bmatrix}$

> E A
> (3×2) (2×2)
> multiplication is possible since $2 = 2$
>
> E A
> (3×2) (2×2)
> result will be a 3×2 matrix

Now try Exercises 29 through 40 ▶

Example 6 shows that in general, matrix multiplication is not commutative. Parts (b) and (c) show $CD \neq DC$ since we get different results, and parts (d) and (e) show $AE \neq EA$, since AE is not defined while EA is.

As with the addition and subtraction of matrices, matrix multiplication becomes cumbersome and time consuming for larger matrices, and we will often turn to the technology available in such cases.

EXAMPLE 7 ▶ **Using Technology for Matrix Operations**

Use a calculator to compute the product AB.

$$A = \begin{bmatrix} 2 & -3 & 0 \\ -1 & 5 & 4 \\ 6 & 0 & 2 \\ 3 & 2 & -1 \end{bmatrix} \quad B = \begin{bmatrix} \frac{1}{2} & -0.7 & 1 \\ 0.5 & 3.2 & -3 \\ -2 & \frac{3}{4} & 4 \end{bmatrix}$$

Solution ▶ Carefully enter matrices A and B into the calculator, then press 2nd MODE (**QUIT**) to get to the home screen. Use [A][B] ENTER, and the calculator finds the product shown in the figure. Just for "fun," we'll

> A B
> (4×3) (3×3)
> multiplication is possible since $3 = 3$
>
> A B
> (4×3) (3×3)
> result will be a 4×3 matrix

double check the first entry of the product matrix:
$(2)(1/2) + (-3)(0.5) + (0)(-2) = -0.5.$ ✔

$$AB = \begin{bmatrix} 2 & -3 & 0 \\ -1 & 5 & 4 \\ 6 & 0 & 2 \\ 3 & 2 & -1 \end{bmatrix} \begin{bmatrix} \frac{1}{2} & -0.7 & 1 \\ 0.5 & 3.2 & -3 \\ -2 & \frac{3}{4} & 4 \end{bmatrix}$$

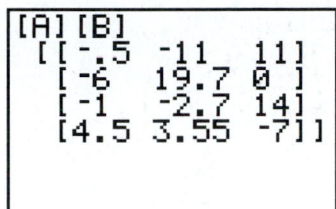

```
[A][B]
[[-.5  -11   11]
 [-6   19.7  0 ]
 [-1   -2.7  14]
 [4.5  3.55  -7]]
```

Now try Exercises 41 through 52 ▶

Properties of Matrix Multiplication

Earlier, Example 6 demonstrated that matrix multiplication is not commutative. Here is a group of properties that *do* hold for matrices. You are asked to check these properties in the Exercise Set using various matrices. **See Exercises 53 through 56.**

> **Properties of Matrix Multiplication**
>
> Given matrices A, B, and C for which the products are defined:
>
> **I.** $A(BC) = (AB)C$ ⟶ matrix multiplication is associative
> **II.** $A(B + C) = AB + AC$ → matrix multiplication is distributive from the left
> **III.** $(B + C)A = BA + CA$ → matrix multiplication is distributive from the right
> **IV.** $k(A + B) = kA + kB$ ⟶ a constant k can be distributed over addition

We close this section with an application of matrix multiplication. There are many other interesting applications in the Exercise Set.

EXAMPLE 8 ▶ **Using Matrix Multiplication to Track Volunteer Enlistments**

In a certain country, the number of males and females that will join the military depends on their age. This information is stored in matrix A (Table 7.3). The likelihood a volunteer will join a *particular branch* of the military also depends on their age, with this information stored in matrix B (Table 7.4). (a) Compute the product $P = AB$ and discuss/interpret what is indicated by the entries P_{11}, P_{13}, and P_{24} of the product matrix. (b) How many males are expected to join the Navy this year?

Table 7.3 Matrix A

A	Age Group		
Sex	**18–19**	**20–21**	**22–23**
Female	1000	1500	500
Male	2500	3000	2000

Table 7.4 Matrix B

B	Likelihood of Joining			
Age Group	**Army**	**Navy**	**Air Force**	**Marines**
18–19	0.42	0.28	0.17	0.13
20–21	0.38	0.26	0.27	0.09
22–23	0.33	0.25	0.35	0.07

Solution ▶ **a.** Matrix A has order 2×3 and matrix B has order 3×4. The product matrix P can be found and is a 2×4 matrix. Carefully enter the matrices in your calculator. Figure 7.12 shows the entries of matrix B. Using $[A][B]$ [ENTER], the calculator finds the product matrix shown in Figure 7.13. Pressing the right arrow shows the complete product matrix is

$$P = \begin{bmatrix} 1155 & 795 & 750 & 300 \\ 2850 & 1980 & 1935 & 735 \end{bmatrix}.$$

The entry P_{11} is the product of R1 from A and C1 from B, and indicates that for the year, 1155 females are projected to join the Army. In like manner, entry P_{13} shows that 750 females are projected to join the Air Force. Entry P_{24} indicates that 735 males are projected to join the Marines.

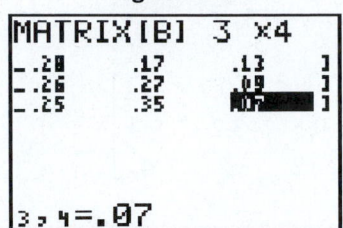

Figure 7.12

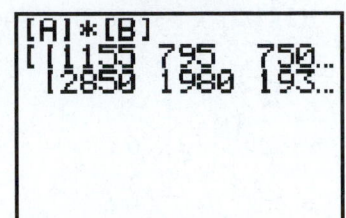

Figure 7.13

b. The product R2 (males) · C2 (Navy) gives P_{22} = 1980, meaning 1980 males are expected to join the Navy.

✓ **C.** You've just seen how we can compute the product of two matrices

Now try Exercise 59 through 66 ▶

7.2 EXERCISES

▶ CONCEPTS AND VOCABULARY

Fill in the blank with the appropriate word or phrase. Carefully reread the section if needed.

1. Two matrices are equal if they are like size and the corresponding entries are equal. In symbols, $A = B$ if $\underline{a_{ij}}$ = $\underline{b_{ij}}$.

2. The sum of two matrices (of like size) is found by adding the corresponding entries. In symbols, $A + B = \underline{[a_{ij} + b_{ij}]}$.

3. The product of a constant times a matrix is called $\underline{scalar}$ multiplication.

4. The size of a matrix is also referred to as its $\underline{order}$. The order of $A = \begin{bmatrix} 1 & 2 & 3 \\ 4 & 5 & 6 \end{bmatrix}$ is $\underline{2 \times 3}$.

5. Give two reasons why matrix multiplication is generally not commutative. Include several examples using matrices of various sizes. Answers will vary.

6. Discuss the conditions under which matrix multiplication is defined. Include several examples using matrices of various sizes.   Answers will vary.

▶ DEVELOPING YOUR SKILLS

State the order of each matrix and name the entries in positions a_{12} and a_{23} if they exist. Then name the position a_{ij} of the "5" in each.

7. $\begin{bmatrix} 1 & -3 \\ 5 & -7 \end{bmatrix}$

8. $\begin{bmatrix} 19 \\ -11 \\ 5 \end{bmatrix}$

9. $\begin{bmatrix} 2 & -3 & 0.5 \\ 0 & 5 & 6 \end{bmatrix}$

10. $\begin{bmatrix} 2 & 0.4 \\ -0.1 & 5 \\ 0.3 & -3 \end{bmatrix}$

11. $\begin{bmatrix} -2 & 1 & -7 \\ 0 & 8 & 1 \\ 5 & -1 & 4 \end{bmatrix}$

12. $\begin{bmatrix} 89 & 55 & 34 & 21 \\ 13 & 8 & 5 & 3 \\ 2 & 1 & 1 & 0 \end{bmatrix}$

Determine if the following statements are true, false, or conditional. If false, explain why. If conditional, find values of $a, b, c, p, q,$ and r that will make the statement true.

13. $\begin{bmatrix} \sqrt{1} & \sqrt{4} & \sqrt{8} \\ \sqrt{16} & \sqrt{32} & \sqrt{64} \end{bmatrix} = \begin{bmatrix} 1 & 2 & 2\sqrt{2} \\ 4 & 4\sqrt{2} & 8 \end{bmatrix}$

14. $\begin{bmatrix} \frac{3}{2} & \frac{-7}{5} & \frac{13}{10} \\ \frac{-1}{2} & \frac{-2}{5} & \frac{1}{3} \end{bmatrix} = \begin{bmatrix} 1.5 & -1.4 & 1.3 \\ -0.5 & -0.4 & 0.\overline{3} \end{bmatrix}$

15. $\begin{bmatrix} -2 & 3 & a \\ 2b & -5 & 4 \\ 0 & -9 & 3c \end{bmatrix} = \begin{bmatrix} c & 3 & -4 \\ 6 & -5 & -a \\ 0 & -3b & -6 \end{bmatrix}$

16. $\begin{bmatrix} 2p + 1 & -5 & 9 \\ 1 & 12 & 0 \\ q + 5 & 9 & -2r \end{bmatrix} = \begin{bmatrix} 7 & -5 & 2 - q \\ 1 & 3r & 0 \\ -2 & 3p & -8 \end{bmatrix}$

Additional answers can be found in the Instructor Answer Appendix.

HOMEWORK SELECTION GUIDE

Core: 7, 11, 13, 15, 17–49 every other odd, 53, 55, 57, 59–65 odd, 70 (21 Exercises)
Standard: 1–4, 7, 11, 13, 15, 17–49 every other odd, 53, 55, 57, 59–65 odd, 67, 70, 71 (27 Exercises)

Extended: 1–4, 7, 11, 13, 15, 17–49 every other odd, 53, 55, 57, 58, 59–63, 65, 67, 68, 70–72 (32 Exercises)
In Depth: 1–6, 7, 11, 13, 15, 17–49 every other odd, 53, 55, 57, 58, 59–63, 65, 67–72 (35 Exercises)

For matrices A through J as given, perform the indicated operation(s), if possible. If an operation cannot be completed, state why. Use a calculator only for those exercises designated by an icon.

$$A = \begin{bmatrix} 2 & 3 \\ 5 & 8 \end{bmatrix} \qquad B = \begin{bmatrix} 2 \\ 1 \\ -3 \end{bmatrix}$$

$$C = \begin{bmatrix} 2 & 0.5 \\ 0.2 & 5 \\ -1 & 3 \end{bmatrix} \qquad D = \begin{bmatrix} 1 & 0 & 0 \\ 0 & 1 & 0 \\ 0 & 0 & 1 \end{bmatrix}$$

$$E = \begin{bmatrix} 1 & -2 & 0 \\ 0 & -1 & 2 \\ 4 & 3 & -6 \end{bmatrix} \qquad F = \begin{bmatrix} 6 & -3 & 9 \\ 12 & 0 & -6 \end{bmatrix}$$

$$G = \begin{bmatrix} -1 & 2 & 0 \\ 0 & 1 & -2 \\ -4 & -3 & 6 \end{bmatrix} \qquad H = \begin{bmatrix} 8 & -3 \\ -5 & 2 \end{bmatrix}$$

$$I = \begin{bmatrix} \frac{1}{2} & -\frac{3}{8} & \frac{1}{4} \\ \frac{1}{4} & \frac{3}{2} & -\frac{5}{8} \\ \frac{1}{8} & \frac{3}{4} & -\frac{5}{2} \end{bmatrix} \qquad J = \begin{bmatrix} \frac{7}{32} & \frac{5}{8} \\ -\frac{5}{16} & \frac{3}{16} \end{bmatrix}$$

17. $A + H$ **18.** $E + G$

19. $F + H$ **20.** $G + D$

 **21.** $H + J$ **22.** $A - J$

23. $G + I$ **24.** $I - G$

25. $3H - 2A$ **26.** $2E + 3G$

27. $\frac{1}{2}E - 3D$ **28.** $F - \frac{2}{3}F$

29. ED **30.** DE

31. AH **32.** HA

33. CB **34.** FH

35. HF **36.** EB

37. H^2 **38.** F^2

39. FE **40.** EF

For matrices A through H as given, use a calculator to perform the indicated operation(s), if possible. If an operation cannot be completed, state why.

$$A = \begin{bmatrix} -5 & 4 \\ 3 & 9 \end{bmatrix} \qquad B = \begin{bmatrix} 1 & 0 \\ 0 & 1 \end{bmatrix}$$

$$C = \begin{bmatrix} \frac{\sqrt{3}}{2} & \frac{\sqrt{3}}{3} \\ \sqrt{3} & 2\sqrt{3} \end{bmatrix} \qquad D = \begin{bmatrix} 1 & 0 & 0 \\ 0 & 1 & 0 \\ 0 & 0 & 1 \end{bmatrix}$$

$$E = \begin{bmatrix} 1 & -2 & 0 \\ 0 & -1 & 2 \\ 4 & 3 & -6 \end{bmatrix} \qquad F = \begin{bmatrix} 12 & -8 & 32 \\ 4 & 8 & 16 \end{bmatrix}$$

$$G = \begin{bmatrix} 0 & \frac{3}{4} & \frac{1}{4} \\ -\frac{1}{2} & \frac{3}{8} & \frac{1}{8} \\ -\frac{1}{4} & \frac{11}{16} & \frac{1}{16} \end{bmatrix} \qquad H = \begin{bmatrix} -\frac{3}{19} & \frac{4}{57} \\ \frac{1}{19} & \frac{5}{57} \end{bmatrix}$$

41. AH **42.** HA

43. EG **44.** GE

45. HB **46.** BH

47. DG **48.** GD

49. C^2 **50.** E^2

51. FG **52.** AF

For Exercises 53 through 56, use a calculator and matrices A, B, and C to verify each statement.

$$A = \begin{bmatrix} -1 & 3 & 5 \\ 2 & 7 & -1 \\ 4 & 0 & 6 \end{bmatrix} \qquad B = \begin{bmatrix} 0.3 & -0.4 & 1.2 \\ -2.5 & 2 & 0.9 \\ 1 & -0.5 & 0.2 \end{bmatrix}$$

$$C = \begin{bmatrix} 45 & -1 & 3 \\ -6 & 10 & -15 \\ 21 & -28 & 36 \end{bmatrix}$$

53. Matrix multiplication is not generally commutative: (a) $AB \neq BA$, (b) $AC \neq CA$, and (c) $BC \neq CB$.   verified

54. Matrix multiplication is distributive from the left: $A(B + C) = AB + AC$.   verified

55. Matrix multiplication is distributive from the right: $(B + C)A = BA + CA$.   verified

56. Matrix multiplication is associative: $(AB)C = A(BC)$.   verified

▶ WORKING WITH FORMULAS

$$\begin{bmatrix} 2 & 2 \\ W & 0 \end{bmatrix} \cdot \begin{bmatrix} L \\ W \end{bmatrix} = \begin{bmatrix} \text{Perimeter} \\ \text{Area} \end{bmatrix}$$

The perimeter and area of a rectangle can be simultaneously calculated using the matrix formula shown, where L represents the length and W represents the width of the rectangle. Use the matrix formula and your calculator to find the perimeter and area of the rectangles shown, then check the results using $P = 2L + 2W$ and $A = LW$.

57.

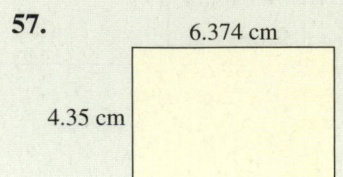

6.374 cm

4.35 cm

58.

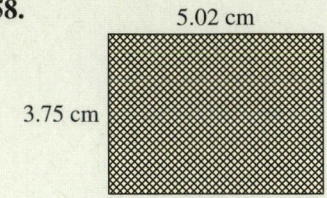

5.02 cm

3.75 cm

▶ APPLICATIONS

59. Custom T's designs and sells specialty T-shirts and sweatshirts, with factories in Verdi and Minsk. The company offers this apparel in three quality levels: standard, deluxe, and premium. Last fall the Verdi plant produced 3820 standard, 2460 deluxe, and 1540 premium T-shirts, along with 1960 standard, 1240 deluxe, and 920 premium sweatshirts. The Minsk plant produced 4220 standard, 2960 deluxe, and 1640 premium T-shirts, along with 2960 standard, 3240 deluxe, and 820 premium sweatshirts in the same time period.

  a. Write a 3 × 2 "production matrix" for each plant [$V \to$ Verdi, $M \to$ Minsk], with a *T-shirt* column, a *sweatshirt* column, and three rows showing how many of the different types of apparel were manufactured.

  b. Use the matrices from part (a) to determine how many more or fewer articles of clothing were produced by Minsk than Verdi.

  c. Use scalar multiplication to find how many shirts of each type will be made at Verdi and Minsk next fall, if each is expecting a 4% increase in business.

  d. Write a matrix that shows Custom T's total production next fall (from both plants), for each type of apparel.

60. Terry's Tire Store sells automobile and truck tires through three retail outlets. Sales at the Cahokia store for the months of January, February, and March broke down as follows: 350, 420, and 530 auto tires and 220, 180, and 140 truck tires. The Shady Oak branch sold 430, 560, and 690 auto tires and 280, 320, and 220 truck tires during the same 3 months. Sales figures for the downtown store were 864, 980, and 1236 auto tires and 535, 542, and 332 truck tires.

  a. Write a 2 × 3 "sales matrix" for each store [$C \to$ Cahokia, $S \to$ Shady Oak, $D \to$ Downtown], with *January, February,* and *March* columns, and two rows showing the sales of auto and truck tires respectively.

  b. Use the matrices from part (a) to determine how many more or fewer tires of each type the downtown store sold (each month) over the other two stores combined.

  c. Market trends indicate that for the same three months in the following year, the Cahokia store will likely experience a 10% increase in sales, the Shady Oak store a 3% decrease, with sales at the downtown store remaining level (no change). Write a matrix that shows the combined monthly sales from all three stores next year, for each type of tire.

61. Home improvements: Dream-Makers Home Improvements specializes in replacement windows, replacement doors, and new siding. During the peak season, the number of contracts that came from various parts of the city (North, South, East, and West) are shown in matrix C. The average profit per contract is shown in matrix P. Compute the product PC and discuss what each entry of the product matrix represents.

$$\begin{array}{c} \\ \text{Windows} \\ \text{Doors} \\ \text{Siding} \end{array} \begin{array}{cccc} N & S & E & W \\ \begin{bmatrix} 9 & 6 & 5 & 4 \\ 7 & 5 & 7 & 6 \\ 2 & 3 & 5 & 2 \end{bmatrix} \end{array} = C$$

$$\begin{array}{ccc} \text{Windows} & \text{Doors} & \text{Siding} \\ \begin{bmatrix} 1500 & 500 & 2500 \end{bmatrix} \end{array} = P$$

62. Classical music: Station 90.7—*The Home of Classical Music*—is having their annual fund drive. Being a loyal listener, Mitchell decides that for the next 3 days he will donate money according to his favorite composers, by the number of times their music comes on the air: $3 for every piece by Mozart (*M*), $2.50 for every piece by Beethoven (*B*), and $2 for every piece by Vivaldi (*V*).

This information is displayed in matrix D. The number of pieces he heard from each composer is displayed in matrix C. Compute the product DC and discuss what each entry of the product matrix represents.

$$\begin{array}{c} \text{Mon. Tue. Wed.} \\ \begin{array}{c} M \\ B \\ V \end{array} \begin{bmatrix} 4 & 3 & 5 \\ 3 & 2 & 4 \\ 2 & 3 & 3 \end{bmatrix} = C \end{array}$$

$$\begin{array}{ccc} M & B & V \\ [3 & 2.5 & 2] = D \end{array}$$

63. Pizza and salad: The science department and math department of a local college are at a pre-semester retreat, and decide to have pizza, salads, and soft drinks for lunch. The quantity of food ordered by each department is shown in matrix Q. The cost of the food item at each restaurant is shown in matrix C using the published prices from three popular restaurants: Pizza Home (PH), Papa Jeff's (PJ), and Dynamos (D).

  a. What is the total cost to the math department if the food is ordered from Pizza Home?

  b. What is the total cost to the science department if the food is ordered from Papa Jeff's?

  c. Compute the product QC and discuss the meaning of each entry in the product matrix.

$$\begin{array}{c} \quad\quad\quad \text{Pizza} \quad \text{Salad} \quad \text{Drink} \\ \begin{array}{c} \text{Science} \\ \text{Math} \end{array} \begin{bmatrix} 8 & 12 & 20 \\ 10 & 8 & 18 \end{bmatrix} = Q \end{array}$$

$$\begin{array}{c} \quad\quad \text{PH} \quad \text{PJ} \quad \text{D} \\ \begin{array}{c} \text{Pizza} \\ \text{Salad} \\ \text{Drink} \end{array} \begin{bmatrix} 8 & 7.5 & 10 \\ 1.5 & 1.75 & 2 \\ 0.90 & 1 & 0.75 \end{bmatrix} = C \end{array}$$

64. Manufacturing pool tables: Cue Ball Incorporated makes three types of pool tables, for homes, commercial use, and professional use. The amount of time required to pack, load, and install each is summarized in matrix T, with all times in hours. The cost of these components in dollars per hour, is summarized in matrix C for two of its warehouses, one on the west coast and the other in the midwest.

  a. What is the cost to package, load, and install a commercial pool table from the coastal warehouse?

  b. What is the cost to package, load, and install a commercial pool table from the warehouse in the midwest?

  c. Compute the product TC and discuss the meaning of each entry in the product matrix.

Additional answers can be found in the Instructor Answer Appendix.

$$\begin{array}{c} \quad\quad\quad \text{Pack} \quad \text{Load} \quad \text{Install} \\ \begin{array}{c} \text{Home} \\ \text{Comm} \\ \text{Prof} \end{array} \begin{bmatrix} 1 & 0.2 & 1.5 \\ 1.5 & 0.5 & 2.2 \\ 1.75 & 0.75 & 2.5 \end{bmatrix} = T \end{array}$$

$$\begin{array}{c} \quad\quad\quad \text{Coast} \quad \text{Midwest} \\ \begin{array}{c} \text{Pack} \\ \text{Load} \\ \text{Install} \end{array} \begin{bmatrix} 10 & 8 \\ 12 & 10.5 \\ 13.5 & 12.5 \end{bmatrix} = C \end{array}$$

65. Joining a club: Each school year, among the students planning to join a club, the likelihood a student joins a particular club depends on their class standing. This information is stored in matrix C. The number of males and females from each class that are projected to join a club each year is stored in matrix J. Compute the product JC and use the result to answer the following:

  a. Approximately how many females joined the chess club?

  b. Approximately how many males joined the writing club?

  c. What does the entry p_{13} of the product matrix tells us?

$$\begin{array}{c} \quad\quad\quad \text{Fresh} \quad \text{Soph} \quad \text{Junior} \\ \begin{array}{c} \text{Female} \\ \text{Male} \end{array} \begin{bmatrix} 25 & 18 & 21 \\ 22 & 19 & 18 \end{bmatrix} = J \end{array}$$

$$\begin{array}{c} \quad\quad\quad \text{Spanish} \quad \text{Chess} \quad \text{Writing} \\ \begin{array}{c} \text{Fresh} \\ \text{Soph} \\ \text{Junior} \end{array} \begin{bmatrix} 0.6 & 0.1 & 0.3 \\ 0.5 & 0.2 & 0.3 \\ 0.4 & 0.2 & 0.4 \end{bmatrix} = C \end{array}$$

66. Designer shirts: The SweatShirt Shoppe sells three types of designs on its products: stenciled (S), embossed (E), and applique (A). The quantity of each size sold is shown in matrix Q. The retail price of each sweatshirt depends on its size and whether it was finished by hand or machine. Retail prices are shown in matrix C. Assuming all stock is sold, compute the product QC and use the result to answer the following.

  a. How much revenue was generated by the large sweatshirts?

  b. How much revenue was generated by the extra-large sweatshirts?

  c. What does the entry p_{11} of the product matrix QC tell us?

$$\begin{array}{c} \quad\quad\quad \text{S} \quad \text{E} \quad \text{A} \\ \begin{array}{c} \text{med} \\ \text{large} \\ \text{x-large} \end{array} \begin{bmatrix} 30 & 30 & 15 \\ 60 & 50 & 20 \\ 50 & 40 & 30 \end{bmatrix} = Q \end{array}$$

$$\begin{array}{c} \quad\quad\quad \text{Hand} \quad \text{Machine} \\ \begin{array}{c} \text{S} \\ \text{E} \\ \text{A} \end{array} \begin{bmatrix} 40 & 25 \\ 60 & 40 \\ 90 & 60 \end{bmatrix} = C \end{array}$$

▶ **EXTENDING THE CONCEPT**

67. For the matrix A shown, use your calculator to compute A^2, A^3, A^4, and A^5. Do you notice a pattern? Try to write a "matrix formula" for A^n, where n is a positive integer, then use your formula to find A^6. Check results using a calculator.

$$A = \begin{bmatrix} 1 & 0 & 1 \\ 1 & 1 & 1 \\ 1 & 0 & 1 \end{bmatrix}$$

68. The matrix $M = \begin{bmatrix} 2 & 1 \\ -3 & -2 \end{bmatrix}$ has some very interesting properties. Compute the powers M^2, M^3, M^4, and M^5, then discuss what you find. Try to find/create another 2×2 matrix that has similar properties.

▶ **MAINTAINING YOUR SKILLS**

69. (6.2) Solve the system using elimination.

$$\begin{cases} x + 2y - z = 3 \\ -2x - y + 3z = -5 \\ 5x + 3y - 2z = 2 \end{cases}$$

70. (2.1/4.6) Given $f(x) = x^4 + 10x^2 - 9$, solve $f(x) \geq 0$.

71. (5.4) Evaluate using the change-of-base formula, then check using exponentiation.

$$\log_2 21$$

72. (4.1) Find the quotient using synthetic division, then check using multiplication.

$$\frac{x^3 - 9x + 10}{x - 2}$$

MID-CHAPTER CHECK

State the size of each matrix and identify the entry in second row, third column.

1. $A = \begin{bmatrix} 0.4 & 1.1 & 0.2 \\ -0.2 & 0.1 & -0.9 \\ 0.7 & 0.4 & 0.8 \end{bmatrix}$ $3 \times 3, -0.9$

2. $B = \begin{bmatrix} -2 & 1 & \frac{1}{2} & 5 \\ 4 & \frac{3}{4} & 0 & -3 \end{bmatrix}$ $2 \times 4, 0$

Write each system in matrix form and solve using row operations to triangularize the matrix. If the system is linearly dependent, write the solution using a parameter.

3. $\begin{cases} 2x + 3y = -5 \\ -5x - 4y = 2 \end{cases}$ $(2, -3)$

4. $\begin{cases} -x + y - 5z = 23 \\ 2x + 4y - z = 9 \\ 3x - 5y + z = 1 \end{cases}$ $(2, 0, -5)$

5. $\begin{cases} x + y - 3z = -11 \\ 4x - y - 2z = -4 \\ 3x - 2y + z = 7 \end{cases}$ $(p - 3, 2p - 8, p)$

6. For matrices A and B given, compute:

$$A = \begin{bmatrix} -3 & -2 \\ 5 & 4 \end{bmatrix} \quad B = \begin{bmatrix} 10 & 15 \\ -30 & -5 \end{bmatrix}$$

a. $A - B$ **b.** $\frac{2}{5}B$ **c.** $5A + B$

$\begin{bmatrix} -13 & -17 \\ 35 & 9 \end{bmatrix}$ $\begin{bmatrix} 4 & 6 \\ -12 & -2 \end{bmatrix}$ $\begin{bmatrix} -5 & 5 \\ -5 & 15 \end{bmatrix}$

Additional answers can be found in the Instructor Answer Appendix.

 7. For matrices C and D given, use a calculator to find:

$$C = \begin{bmatrix} -0.2 & 0 & 0.2 \\ 0.4 & 0.8 & 0 \\ 0.1 & -0.2 & -0.1 \end{bmatrix} D = \begin{bmatrix} 5 & 2.5 & 10 \\ -2.5 & 0 & -5 \\ 10 & 2.5 & 10 \end{bmatrix}$$

a. $C + \frac{1}{5}D$ **b.** $-0.6D$ **c.** CD

8. For the matrices A, B, C, and D given, compute the products indicated (if possible):

$$A = \begin{bmatrix} 4 & -1 \\ 0 & -5 \end{bmatrix} \qquad B = \begin{bmatrix} 6 & -2 \\ 0 & 1 \\ 4 & 7 \end{bmatrix}$$

$$C = \begin{bmatrix} 4 & -8 & -3 \\ -1 & 0 & 1 \end{bmatrix} \quad D = \begin{bmatrix} 2 & 0 & -6 \\ -1 & -3 & 0 \\ 1 & 5 & -4 \end{bmatrix}$$

a. AC **b.** $-2CD$ **c.** BA **d.** $CB - 4A$

9. Create a system of equations to model this exercise, then write the system in matrix form and solve. The campus bookstore offers both new and used texts to students. In a recent biology class with 24 students, 14 bought used texts and 10 bought new texts, with the class as a whole paying $2370. Of the 6 premed students in class, 2 bought used texts, and 4 bought new texts, with the group paying a total of $660. How much does a used text cost? How much does a new text cost?   used: $80, new: $125

10. Table A shown gives the number and type of extended warranties sold to individual car owners and to business fleets. Table B shows the promotions offered to those making the purchase. Write the entries of each table in matrix form and compute the product matrix $P = AB$, then state what each entry of the product matrix represents.

Table A

Extended Warranties	80,000 mi	100,000 mi	120,000 mi
Individuals	30	25	10
Businesses	20	12	5

Table B

Promotions	Rebate	Free AAA Membership
80,000 mi	$50	1 yr
100,000 mi	$75	2 yr
120,000 mi	$100	3 yr

REINFORCING BASIC CONCEPTS

More on Matrix Multiplication

To help understand and master the concept of matrix multiplication, it helps to take a closer look at the entries of the product matrix. Recall for the product $AB = P$, the entry p_{11} in the product matrix is the result of multiplying the 1st row of A with the 1st column of B, the entry p_{12} is the result of multiplying 1st row of A, with the 2nd column of B, and so on.

$$P = \begin{bmatrix} 1 & 2 \\ 6 & -2 \\ 7 & 7 \end{bmatrix}$$

Exercise 1: The product of the 3rd row of A with the 2nd column of B, gives what entry in P? p_{32}

Exercise 2: The entry p_{13} is the result of what product? The entry p_{22} is the result of what product?

Exercise 3: If p_{33} is the last entry of the product matrix, what are the possible sizes of A and B?

Exercise 4: Of the eight matrices shown here, only two produce the product matrix P shown. Use the ideas highlighted above to determine which two.

$$A = \begin{bmatrix} 3 & -1 \\ 2 & 1 \\ 4 & 1 \end{bmatrix} \qquad B = \begin{bmatrix} 1 & 1 \\ 2 & 0 \end{bmatrix} \qquad C = \begin{bmatrix} 2 & 0 & -6 \\ -1 & -3 & 0 \\ 1 & 5 & -4 \end{bmatrix} \qquad D = \begin{bmatrix} 2 & 4 \\ -1 & 0 \\ 3 & 5 \end{bmatrix}$$

$$E = \begin{bmatrix} 1 & 2 \\ 3 & -1 \end{bmatrix} \qquad F = \begin{bmatrix} 1 & 0 \\ 0 & 2 \\ 4 & 1 \end{bmatrix} \qquad G = \begin{bmatrix} 1 \\ 4 \\ 6 \end{bmatrix} \qquad H = \begin{bmatrix} 1 & 3 \end{bmatrix} \quad \text{only } FE = P$$

Additional answers can be found in the Instructor Answer Appendix.

7.3 Solving Linear Systems Using Matrix Equations

LEARNING OBJECTIVES

In Section 7.3 you will see how we can:

☐ **A.** Recognize the identity matrix for multiplication

☐ **B.** Find the inverse of a square matrix

☐ **C.** Solve systems using matrix equations

☐ **D.** Use determinants to find whether a matrix is invertible

While using matrices and row operations offers a degree of efficiency in solving systems, we are still required to solve for each variable *individually*. Using matrix multiplication we can actually rewrite a given system as a single *matrix equation*, in which the solutions are computed *simultaneously*. As with other kinds of equations, the use of identities and inverses are involved, which we now develop in the context of matrices.

A. Multiplication and Identity Matrices

From the properties of real numbers, 1 is the identity for multiplication since $n \cdot 1 = 1 \cdot n = n$. A similar identity exists for matrix multiplication. Consider the 2×2 matrix $A = \begin{bmatrix} 1 & 4 \\ -2 & 3 \end{bmatrix}$. While matrix multiplication is not *generally* commutative, if we can find a matrix B where $AB = BA = A$, then B is a prime candidate for the identity matrix, which is denoted I. For the products AB and BA to be possible and have the same order as A, we note B must also be a 2×2 matrix. Using the arbitrary matrix $B = \begin{bmatrix} a & b \\ c & d \end{bmatrix}$, we have the following.

EXAMPLE 1A ▶ **Solving $AB = A$ to Find the Identity Matrix**

For $\begin{bmatrix} 1 & 4 \\ -2 & 3 \end{bmatrix} \begin{bmatrix} a & b \\ c & d \end{bmatrix} = \begin{bmatrix} 1 & 4 \\ -2 & 3 \end{bmatrix}$, use matrix multiplication, the equality of matrices, and systems of equations to find the value of a, b, c, and d.

Solution ▶ The product on the left gives $\begin{bmatrix} a + 4c & b + 4d \\ -2a + 3c & -2b + 3d \end{bmatrix} = \begin{bmatrix} 1 & 4 \\ -2 & 3 \end{bmatrix}$.

Since corresponding entries must be equal (shown by matching colors), we can find a, b, c, and d by solving the systems $\begin{cases} a + 4c = 1 \\ -2a + 3c = -2 \end{cases}$ and $\begin{cases} b + 4d = 4 \\ -2b + 3d = 3 \end{cases}$. For the first system, 2R1 + R2 shows $a = 1$ and $c = 0$. Using 2R1 + R2 for the second shows $b = 0$ and $d = 1$. It appears $\begin{bmatrix} 1 & 0 \\ 0 & 1 \end{bmatrix}$ is a candidate for the identity matrix.

Before we name B as the identity matrix, we must show that $AB = BA = A$.

EXAMPLE 1B ▶ **Verifying $AB = BA = A$**

Given $A = \begin{bmatrix} 1 & 4 \\ -2 & 3 \end{bmatrix}$ and $B = \begin{bmatrix} 1 & 0 \\ 0 & 1 \end{bmatrix}$, determine if $AB = A$ and $BA = A$.

Solution ▶ $AB = \begin{bmatrix} 1 & 4 \\ -2 & 3 \end{bmatrix} \begin{bmatrix} 1 & 0 \\ 0 & 1 \end{bmatrix}$ $BA = \begin{bmatrix} 1 & 0 \\ 0 & 1 \end{bmatrix} \begin{bmatrix} 1 & 4 \\ -2 & 3 \end{bmatrix}$

$= \begin{bmatrix} 1(1) + 4(0) & 1(0) + 4(1) \\ -2(1) + 3(0) & -2(0) + 3(1) \end{bmatrix}$ $= \begin{bmatrix} 1(1) + 0(-2) & 1(4) + 0(3) \\ 0(1) + 1(-2) & 0(4) + 1(3) \end{bmatrix}$

$= \begin{bmatrix} 1 & 4 \\ -2 & 3 \end{bmatrix} = A ✓$ $= \begin{bmatrix} 1 & 4 \\ -2 & 3 \end{bmatrix} = A ✓$

Since $AB = A = BA$, B is the identity matrix I.

Now try Exercises 7 through 10 ▶

By replacing the entries of $A = \begin{bmatrix} 1 & 4 \\ -2 & -3 \end{bmatrix}$ with those of the general matrix $\begin{bmatrix} a_{11} & a_{12} \\ a_{21} & a_{22} \end{bmatrix}$, we can show that $I = \begin{bmatrix} 1 & 0 \\ 0 & 1 \end{bmatrix}$ is the identity for *all* 2×2 matrices. In considering the identity for larger matrices, we find that only *square matrices* have inverses, since $AI = IA$ is the primary requirement (the multiplication must be possible in both directions). This is commonly referred to as *multiplication from the right* and *multiplication from the left*. Using the same procedure as before we can show $\begin{bmatrix} 1 & 0 & 0 \\ 0 & 1 & 0 \\ 0 & 0 & 1 \end{bmatrix}$ is the identity for 3×3 matrices (denoted I_3). The $n \times n$ identity matrix I_n consists of 1's down the main diagonal and 0's for all other entries. Also, the identity I_n for a square matrix is unique.

As in Section 7.2, a graphing calculator can be used to investigate operations on matrices and matrix properties. For the 3×3 matrix

$$A = \begin{bmatrix} 2 & 5 & 1 \\ 4 & -1 & 1 \\ 0 & 3 & -2 \end{bmatrix} \text{ and } I_3 = \begin{bmatrix} 1 & 0 & 0 \\ 0 & 1 & 0 \\ 0 & 0 & 1 \end{bmatrix}, \text{ a}$$

calculator will confirm that $AI_3 = A = I_3A$. Carefully enter A into your calculator as matrix A, and I_3 as matrix B. Figure 7.14 shows $AB = A$ and $BA = A$. **See Exercises 11 through 14.**

Figure 7.14

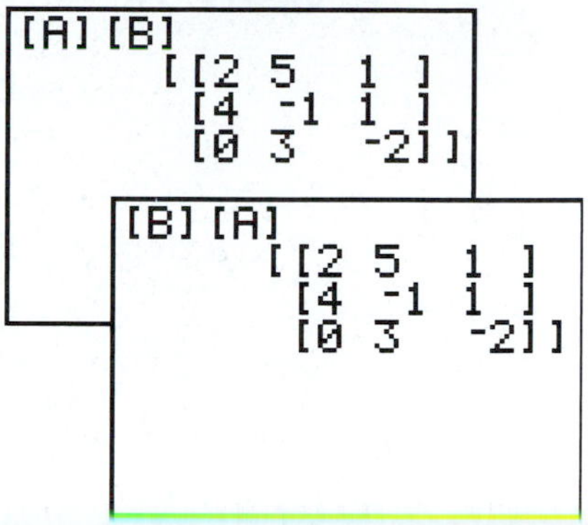

```
[A] [B]
     [[2 5  1 ]
      [4 -1 1 ]
      [0 3  -2] ]
[B] [A]
     [[2 5  1 ]
      [4 -1 1 ]
      [0 3  -2] ]
```

☑ **A.** You've just seen how we can recognize the identity matrix for multiplication

B. The Inverse of a Matrix

Again from the properties of real numbers, we know the multiplicative inverse for a is $a^{-1} = \dfrac{1}{a}$ $(a \neq 0)$, since the products $a \cdot a^{-1}$ and $a^{-1} \cdot a$ yield the identity 1. To show that a similar inverse exists for matrices, consider the square matrix $A = \begin{bmatrix} 6 & 5 \\ 2 & 2 \end{bmatrix}$ and an arbitrary matrix $B = \begin{bmatrix} a & b \\ c & d \end{bmatrix}$. If we can find a matrix B, where $AB = BA = I$, then B is a prime candidate for the inverse matrix of A, which is denoted A^{-1}. In an attempt to find such a matrix B, we proceed as in Examples 1A and 1B.

EXAMPLE 2A ▶ **Solving $AB = I$ to find A^{-1}**

For $\begin{bmatrix} 6 & 5 \\ 2 & 2 \end{bmatrix} \begin{bmatrix} a & b \\ c & d \end{bmatrix} = \begin{bmatrix} 1 & 0 \\ 0 & 1 \end{bmatrix}$, use matrix multiplication, the equality of matrices, and systems of equations to find the entries of B.

Solution ▶ The product on the left gives $\begin{bmatrix} 6a + 5c & 6b + 5d \\ 2a + 2c & 2b + 2d \end{bmatrix} = \begin{bmatrix} 1 & 0 \\ 0 & 1 \end{bmatrix}$. Since corresponding entries must be equal (shown by matching colors), we find the values of a, b, c, and d by solving the systems $\begin{cases} 6a + 5c = 1 \\ 2a + 2c = 0 \end{cases}$ and $\begin{cases} 6b + 5d = 0 \\ 2b + 2d = 1 \end{cases}$. Using $-3R2 + R1$ for the first system shows $a = 1$ and $c = -1$, while $-3R2 + R1$ for the second system shows $b = -2.5$ and $d = 3$. Matrix $B = \begin{bmatrix} a & b \\ c & d \end{bmatrix} = \begin{bmatrix} 1 & -2.5 \\ -1 & 3 \end{bmatrix}$ is the prime candidate for A^{-1}.

It may have occurred to you that the systems used in Example 2A have *exactly the same coefficients*. In matrix form, these systems are $\begin{bmatrix} 6 & 5 & | & 1 \\ 2 & 2 & | & 0 \end{bmatrix}$ and $\begin{bmatrix} 6 & 5 & | & 0 \\ 2 & 2 & | & 1 \end{bmatrix}$. Instead of solving these two systems separately, we can solve them *simultaneously* by simply using the coefficient matrix augmented with the identity matrix: $\begin{bmatrix} 6 & 5 & | & 1 & 0 \\ 2 & 2 & | & 0 & 1 \end{bmatrix}$. Row operations are then used to transform the given matrix into the identity: $\begin{bmatrix} 1 & 0 & | & e & f \\ 0 & 1 & | & g & h \end{bmatrix}$, in a sense "preserving" the inverse operations needed. The inverse matrix is then $\begin{bmatrix} e & f \\ g & h \end{bmatrix}$.

EXAMPLE 2B ▶   Finding a 2 × 2 Inverse Using the Augmented Matrix

Use the augmented matrix method to find A^{-1} for $A = \begin{bmatrix} 6 & 5 \\ 2 & 2 \end{bmatrix}$. Then use a calculator to verify $A^{-1}A = AA^{-1} = I$.

Solution ▶   First we augment the given matrix with the identity: $\begin{bmatrix} 6 & 5 & | & 1 & 0 \\ 2 & 2 & | & 0 & 1 \end{bmatrix}$. Then we use row operations to transform A into the identity matrix.

$$\begin{bmatrix} 6 & 5 & | & 1 & 0 \\ 2 & 2 & | & 0 & 1 \end{bmatrix} \quad -3R2 + R1 \rightarrow R2 \quad \begin{bmatrix} 6 & 5 & | & 1 & 0 \\ 0 & -1 & | & 1 & -3 \end{bmatrix}$$

$$\begin{bmatrix} 6 & 5 & | & 1 & 0 \\ 0 & -1 & | & 1 & -3 \end{bmatrix} \quad 5R2 + R1 \rightarrow R1 \quad \begin{bmatrix} 6 & 0 & | & 6 & -15 \\ 0 & -1 & | & 1 & -3 \end{bmatrix}$$

$$\begin{bmatrix} 6 & 0 & | & 6 & -15 \\ 0 & -1 & | & 1 & -3 \end{bmatrix} \quad \begin{matrix} \frac{R1}{6} \rightarrow R1 \\ -1R2 \rightarrow R2 \end{matrix} \quad \begin{bmatrix} 1 & 0 & | & 1 & -2.5 \\ 0 & 1 & | & -1 & 3 \end{bmatrix}$$

```
[B][A]
     [[1 0]
      [0 1]]
[A][B]
     [[1 0]
      [0 1]]
```

As in Example 2A, a prime candidate for A^{-1} is $\begin{bmatrix} 1 & -2.5 \\ -1 & 3 \end{bmatrix}$. Using a calculator with $A = [A]$ and $A^{-1} = [B]$ confirms that $A^{-1}A = AA^{-1} = I$ (see figure).

Now try Exercises 15 through 22 ▶

These observations guide us to the following definition of an inverse matrix.

The Inverse of a Matrix

Given an $n \times n$ matrix A, if there exists an $n \times n$ matrix A^{-1} such that $AA^{-1} = A^{-1}A = I_n$, then A^{-1} is the inverse of matrix A.

We will soon discover that while only square matrices have inverses, not every square matrix has an inverse. If an inverse exists, the matrix is said to be **invertible.** For 2×2 matrices that are invertible, a simple formula exists for computing the inverse. The formula is derived in the *Strengthening Core Skills* feature at the end of Chapter 7.

The Inverse of a 2 × 2 Matrix

If $A = \begin{bmatrix} a & b \\ c & d \end{bmatrix}$, then $A^{-1} = \dfrac{1}{ad - bc}\begin{bmatrix} d & -b \\ -c & a \end{bmatrix}$ provided $ad - bc \neq 0$

To "test" the formula, again consider the matrix $A = \begin{bmatrix} 6 & 5 \\ 2 & 2 \end{bmatrix}$, where $a = 6$, $b = 5$, $c = 2$, and $d = 2$:

$$A^{-1} = \frac{1}{(6)(2) - (5)(2)}\begin{bmatrix} 2 & -5 \\ -2 & 6 \end{bmatrix}$$

$$= \frac{1}{2}\begin{bmatrix} 2 & -5 \\ -2 & 6 \end{bmatrix} = \begin{bmatrix} 1 & -2.5 \\ -1 & 3 \end{bmatrix} \checkmark$$

See Exercises 63 through 66 for more practice with this formula.

Almost without exception, real-world applications involve much larger matrices, with entries that are not integer-valued. Although the augmented matrix method from Example 2 can be extended to find the inverse of larger matrices, the process becomes very tedious and too time consuming to be useful. The process for a 3×3 matrix is discussed in the *Strengthening Core Skills* feature at the end of Chapter 7. But for practical reasons, we will rely on a calculator to produce these larger inverse matrices. This is done by (1) carefully entering a square matrix A into the calculator, (2) returning to the home screen, and (3) calling up matrix A and pressing the ⬛ x^{-1} key and ⬛ to find A^{-1}. In the context of matrices, calculators are programmed to compute an inverse matrix using this key, rather than to somehow find a reciprocal. **See Exercises 23 through 26.**

☑ **B.** You've just seen how we can find the inverse of a square matrix

C. Solving Systems Using Matrix Equations

One reason matrix multiplication has its row $\times$ column definition is to assist in writing a linear system of equations as a single matrix equation. The equation consists of the matrix of constants B on the right, and a product of the coefficient matrix A with the matrix of variables X on the left: $AX = B$. For $\begin{cases} x + 4y - z = 10 \\ 2x + 5y - 3z = 7, \text{ the matrix} \\ 8x + y - 2z = 11 \end{cases}$

equation is $\begin{bmatrix} 1 & 4 & -1 \\ 2 & 5 & -3 \\ 8 & 1 & -2 \end{bmatrix} \begin{bmatrix} x \\ y \\ z \end{bmatrix} = \begin{bmatrix} 10 \\ 7 \\ 11 \end{bmatrix}$. Note that computing the product on the left will yield the original system.

Once written as a matrix equation, the system can be solved using an inverse matrix and the following sequence. If A represents the matrix of coefficients, X the matrix of variables, B the matrix of constants, and I the appropriate identity, the sequence is

(1) $AX = B$ matrix equation

(2) $A^{-1}(AX) = A^{-1}B$ multiply from the left by the inverse of A

(3) $(A^{-1}A)X = A^{-1}B$ associative property

(4) $IX = A^{-1}B$ $A^{-1}A = I$

(5) $X = A^{-1}B$ $IX = X$

Lines 1 through 5 illustrate the steps that make the method work. In actual practice, after carefully entering the matrices, only step 5 is used when solving matrix equations using technology. Once matrix A is entered, the calculator will automatically *find* and *use* A^{-1} as we enter $A^{-1}B$.

EXAMPLE 3 ▶   **Using Technology to Solve a Matrix Equation**

Use a calculator and a matrix equation to solve the system

$\begin{cases} x + 4y - z = 10 \\ 2x + 5y - 3z = 7. \\ 8x + y - 2z = 11 \end{cases}$

Solution ▶   As before, the matrix equation is $\begin{bmatrix} 1 & 4 & -1 \\ 2 & 5 & -3 \\ 8 & 1 & -2 \end{bmatrix} \begin{bmatrix} x \\ y \\ z \end{bmatrix} = \begin{bmatrix} 10 \\ 7 \\ 11 \end{bmatrix}$.

Carefully enter (and double-check) the matrix of coefficients as matrix A in your calculator, and the matrix of constants as matrix B (see Figure 7.15). The product $A^{-1}B$ shows the solution is $x = 2$, $y = 3$, $z = 4$ (see Figure 7.16). Verify by substitution.

Figure 7.15

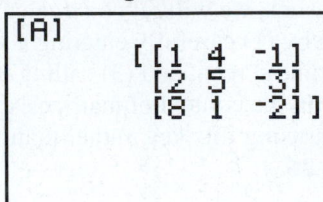

Figure 7.16

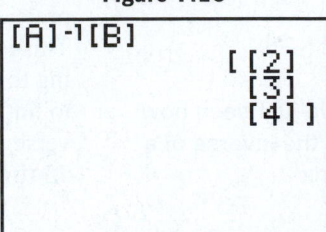

Now try Exercises 27 through 44 ▶

Figure 7.17

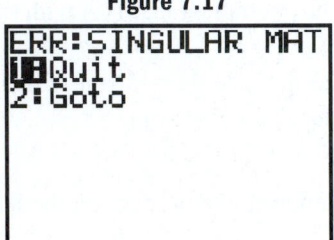

The matrix equation method does have a few shortcomings. Consider the system whose corresponding matrix equation is $\begin{bmatrix} 4 & -10 \\ -2 & 5 \end{bmatrix}\begin{bmatrix} x \\ y \end{bmatrix} = \begin{bmatrix} -8 \\ 13 \end{bmatrix}$. After entering the matrix of coefficients A and matrix of constants B, attempting to compute $A^{-1}B$ results in the error message shown in Figure 7.17. The calculator is unable to return a solution due to something called a **"singular matrix."** To investigate further, we attempt to find A^{-1} for $\begin{bmatrix} 4 & -10 \\ -2 & 5 \end{bmatrix}$ using the formula for a 2×2 matrix. With $a = 4$, $b = -10$, $c = -2$, and $d = 5$, we have

$$A^{-1} = \frac{1}{ad - bc}\begin{bmatrix} d & -b \\ -c & a \end{bmatrix} = \frac{1}{(4)(5) - (-10)(-2)}\begin{bmatrix} 5 & 10 \\ 2 & 4 \end{bmatrix} = \frac{1}{0}\begin{bmatrix} 5 & 10 \\ 2 & 4 \end{bmatrix}$$

☑ **C. You've just seen how we can solve systems using matrix equations**

Since division by zero is undefined, we conclude that matrix A has no inverse. A matrix having no inverse is said to be **singular** or **noninvertible.** Solving systems using matrix equations is only possible when the matrix of coefficients is **nonsingular.**

D. Determinants and Singular Matrices

As a practical matter, it becomes important to know ahead of time whether a particular matrix has an inverse. To help with this, we introduce one additional operation on a square matrix, that of calculating its **determinant.** For a 1×1 matrix the determinant is the entry itself. For a 2×2 matrix $A = \begin{bmatrix} a_{11} & a_{12} \\ a_{21} & a_{22} \end{bmatrix}$, the determinant of A, written as $\det(A)$ or denoted with vertical bars as $|A|$, is computed as *a difference of diagonal products* beginning with the upper-left entry:

2nd diagonal product

$$\det(A) = \begin{vmatrix} a_{11} & a_{12} \\ a_{21} & a_{22} \end{vmatrix} = a_{11}a_{22} - a_{21}a_{12}$$

1st diagonal product

The Determinant of a 2 × 2 Matrix

Given any 2×2 matrix $A = \begin{bmatrix} a_{11} & a_{12} \\ a_{21} & a_{22} \end{bmatrix}$,

$$\det(A) = |A| = a_{11}a_{22} - a_{21}a_{12}$$

EXAMPLE 4 ▶   **Calculating Determinants**

Compute the determinant of each matrix.

a. $B = \begin{bmatrix} 3 & 2 \\ 1 & -6 \end{bmatrix}$ **b.** $C = \begin{bmatrix} 5 & 2 & 1 \\ -1 & -3 & 4 \end{bmatrix}$ **c.** $D = \begin{bmatrix} 4 & -10 \\ -2 & 5 \end{bmatrix}$

Solution ▶   **a.** $\det(B) = \begin{vmatrix} 3 & 2 \\ 1 & -6 \end{vmatrix} = (3)(-6) - (1)(2) = -20$

b. Determinants are only defined for square matrices (see figure).

c. $\det(D) = \begin{vmatrix} 4 & -10 \\ -2 & 5 \end{vmatrix} = (4)(5) - (-2)(-10) = 20 - 20 = 0$

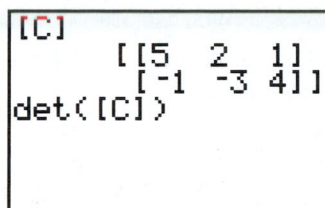 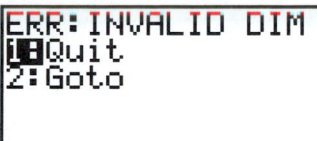

Now try Exercises 45 through 48 ▶

Notice from Example 4(c), the determinant of $\begin{bmatrix} 4 & -10 \\ -2 & 5 \end{bmatrix}$ is zero, and this is the same matrix we earlier found had no inverse. This observation can be extended to larger matrices and offers the connection we seek between a given matrix, its inverse, and matrix equations.

> **Singular Matrices**
>
> If A is a square matrix and $\det(A) = 0$, the inverse matrix *does not exist* and A is said to be *singular* or *noninvertible*.

In summary, inverses exist only for square matrices, but not every square matrix has an inverse. If the determinant of a square matrix is zero, an inverse does not exist and the method of matrix equations cannot be used to solve the system.

WORTHY OF NOTE

For the determinant of a general $n \times n$ matrix using cofactors, see Appendix IV.

To use the determinant test for a 3×3 system, we need to compute a 3×3 determinant. At first glance, our experience with 2×2 determinants appears to be of little help. However, every entry in a 3×3 matrix is associated with a smaller 2×2 matrix, formed by *deleting the row and column* of that entry and using the entries that remain. These 2×2's are called the **associated minor matrices** or simply the **minors.** Using a general matrix of coefficients, we'll identify the minors associated with the entries in the first row.

$$\begin{bmatrix} \textcircled{a_{11}} & a_{12} & a_{13} \\ a_{21} & a_{22} & a_{23} \\ a_{31} & a_{32} & a_{33} \end{bmatrix} \qquad \begin{bmatrix} a_{11} & \textcircled{a_{12}} & a_{13} \\ a_{21} & a_{22} & a_{23} \\ a_{31} & a_{32} & a_{33} \end{bmatrix} \qquad \begin{bmatrix} a_{11} & a_{12} & \textcircled{a_{13}} \\ a_{21} & a_{22} & a_{23} \\ a_{31} & a_{32} & a_{33} \end{bmatrix}$$

$$\begin{array}{ccc} \textbf{Entry: } a_{11} & \textbf{Entry: } a_{12} & \textbf{Entry: } a_{13} \\ \textbf{associated minor} & \textbf{associated minor} & \textbf{associated minor} \end{array}$$

$$\begin{bmatrix} a_{22} & a_{23} \\ a_{32} & a_{33} \end{bmatrix} \qquad\qquad \begin{bmatrix} a_{21} & a_{23} \\ a_{31} & a_{33} \end{bmatrix} \qquad\qquad \begin{bmatrix} a_{21} & a_{22} \\ a_{31} & a_{32} \end{bmatrix}$$

To illustrate, consider the system shown, and (1) form the matrix of coefficients, (2) identify the minor matrices associated with the entries in the first row, and (3) compute the determinant of each *minor*.

$$\begin{cases} 2x + 3y - z = 1 \\ x - 4y + 2z = -3 \\ 3x + y = -1 \end{cases} \quad \text{(1) Matrix of coefficients} \quad \begin{bmatrix} 2 & 3 & -1 \\ 1 & -4 & 2 \\ 3 & 1 & 0 \end{bmatrix}$$

(2)
$$\begin{bmatrix} \textcircled{2} & 3 & 1 \\ 1 & -4 & 2 \\ 3 & 1 & 0 \end{bmatrix} \qquad \begin{bmatrix} 2 & \textcircled{3} & -1 \\ 1 & -4 & 2 \\ 3 & 1 & 0 \end{bmatrix} \qquad \begin{bmatrix} 2 & 3 & \textcircled{-1} \\ 1 & -4 & 2 \\ 3 & 1 & 0 \end{bmatrix}$$

Entry a_{11}: 2	**Entry a_{12}: 3**	**Entry a_{13}: -1**
associated minor	associated minor	associated minor
$\begin{bmatrix} -4 & 2 \\ 1 & 0 \end{bmatrix}$	$\begin{bmatrix} 1 & 2 \\ 3 & 0 \end{bmatrix}$	$\begin{bmatrix} 1 & -4 \\ 3 & 1 \end{bmatrix}$
(3) Determinant of minor	**Determinant of minor**	**Determinant of minor**
$(-4)(0) - (1)(2) = -2$	$(1)(0) - (3)(2) = -6$	$(1)(1) - (3)(-4) = 13$

For computing a 3×3 determinant, we illustrate a technique called **expansion by minors.**

The Determinant of a 3 × 3 Matrix—Expansion by Minors

For the matrix M shown, $\det(M)$ is the unique number computed as follows:

matrix M

$$\begin{bmatrix} a_{11} & a_{12} & a_{13} \\ a_{21} & a_{22} & a_{23} \\ a_{31} & a_{32} & a_{33} \end{bmatrix}$$

1. Select any row or column and form the product of each entry with its minor matrix. The illustration here uses the entries in row 1:

$$\det(M) = +a_{11}\begin{vmatrix} a_{22} & a_{23} \\ a_{32} & a_{33} \end{vmatrix} - a_{12}\begin{vmatrix} a_{21} & a_{23} \\ a_{31} & a_{33} \end{vmatrix} + a_{13}\begin{vmatrix} a_{21} & a_{22} \\ a_{31} & a_{32} \end{vmatrix}$$

2. The *signs used between terms* of the expansion depends on the row or column chosen, according to the *sign chart* shown.

Sign Chart

$$\begin{bmatrix} + & - & + \\ - & + & - \\ + & - & + \end{bmatrix}$$

The determinant of a matrix is unique and *any* row or column can be used. For this reason, it's helpful to select the row or column having the most zero, positive, and/or smaller entries.

EXAMPLE 5 ▶ **Calculating a 3 × 3 Determinant**

Compute the determinant of $M = \begin{bmatrix} 2 & 1 & -3 \\ 1 & -1 & 0 \\ -2 & 1 & 4 \end{bmatrix}$.

Solution ▶ Since the second row has the "smallest" entries as well as a zero entry, we compute the determinant using this row. According to the sign chart, the signs of the terms will be negative–positive–negative, giving

$$\det(M) = -(1)\begin{vmatrix} 1 & -3 \\ 1 & 4 \end{vmatrix} + (-1)\begin{vmatrix} 2 & -3 \\ -2 & 4 \end{vmatrix} - (0)\begin{vmatrix} 2 & 1 \\ -2 & 1 \end{vmatrix}$$

$$= -1(4 + 3) + (-1)(8 - 6) - (0)(2 + 2)$$

$$= \quad -7 \quad + \quad (-2) \quad - \quad 0$$

$$= -9$$

The value of $\det(M)$ is -9.

Now try Exercises 49 through 52 ▶

Try computing the determinant of M two more times, using a different row or column each time. Since the determinant is unique, you should obtain the same result.

There are actually other alternatives for computing a 3×3 determinant. The first is called **determinants by column rotation,** and takes advantage of patterns generated from the expansion of minors. This method is applied to the matrix shown, which uses alphabetical entries for simplicity.

$$\det \begin{bmatrix} a & b & c \\ d & e & f \\ g & h & i \end{bmatrix} \begin{array}{l} = a(ei - fh) - b(di - fg) + c(dh - eg) \quad \text{\color{magenta}expansion using R1} \\ = aei - afh - bdi + bfg + cdh - ceg \quad \text{\color{magenta}distribute} \\ = aei + bfg + cdh - afh - bdi - ceg \quad \text{\color{magenta}rewrite result} \end{array}$$

Although history is unsure of who should be credited, notice that if you repeat the first two columns to the right of the given matrix ("rotation of columns"), identical products are obtained using the six diagonals formed—three in the downward direction using addition, three in the upward direction using subtraction.

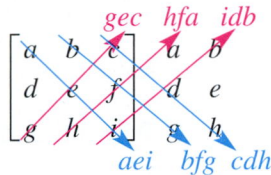

Adding the products in **blue** (regardless of sign) and subtracting the products in **red** (regardless of sign) gives the determinant. This method is more efficient than expansion by minors, *but can only be used for 3×3 matrices!*

EXAMPLE 6 ▶ **Calculating det(A) Using Column Rotation**

Use the column rotation method to find the determinant of $A = \begin{bmatrix} 1 & 5 & 3 \\ -2 & -8 & 0 \\ -3 & -11 & 1 \end{bmatrix}$.

Solution ▶ Rotate columns 1 and 2 to the right, and compute the diagonal products.

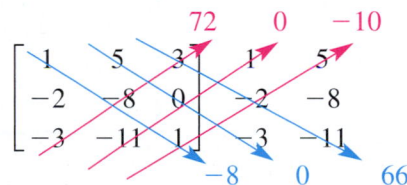

Adding the products in **blue** (regardless of sign) and subtracting the products in **red** (regardless of sign) shows $\det(A) = -4$:

$$-8 + 0 + 66 - 72 - 0 - (-10) = -4.$$

Now try Exercises 53 through 56 ▶

The final method is presented in the *Extending the Concept* feature of the Exercise Set, and shows that if certain conditions are met, the determinant of a matrix can be found using its triangularized form.

As with the operations studied in Section 7.2, the process of computing a determinant becomes very cumbersome for larger matrices, or those with rational or radical entries. Most graphing calculators are programmed to handle these computations easily. After accessing the matrix menu (2nd x⁻¹), calculating a determinant is the first option under the **MATH** submenu (Figure 7.18). The calculator results for det([A]) and det([B]) as defined are shown in Figures 7.19 and 7.20. **See Exercises 57 and 58.**

Figure 7.18   **Figure 7.19**   **Figure 7.20**

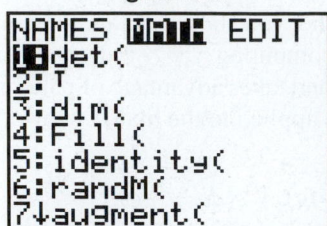   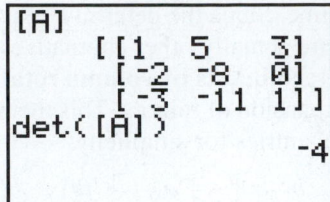   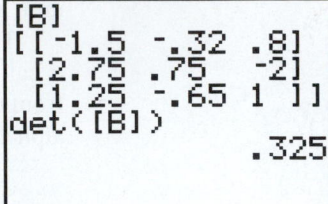

EXAMPLE 7 ▶ **Solving a System after Verifying A is Invertible**

Given the system shown here, (1) form the matrix equation $AX = B$; (2) compute the determinant of the coefficient matrix and determine if you can proceed; and (3) if so, solve the system using a matrix equation.

$$\begin{cases} 2x + 1y - 3z = 11 \\ 1x - 1y \qquad = 1 \\ -2x + 1y + 4z = -8 \end{cases}$$

Solution ▶   **1.** Form the matrix equation $AX = B$:

$$\begin{bmatrix} 2 & 1 & -3 \\ 1 & -1 & 0 \\ -2 & 1 & 4 \end{bmatrix} \begin{bmatrix} x \\ y \\ z \end{bmatrix} = \begin{bmatrix} 11 \\ 1 \\ -8 \end{bmatrix}$$

2. Enter the matrices A and B into the calculator. Since $\det(A)$ is nonzero (from Example 5 and Figure 7.21), we can proceed.

3. Enter $A^{-1}B$ on the home screen and press ⏎ (Figure 7.22).

Figure 7.21   **Figure 7.22**

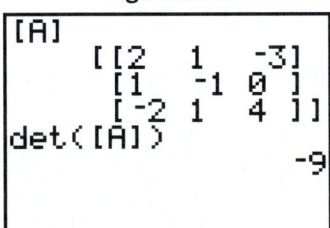   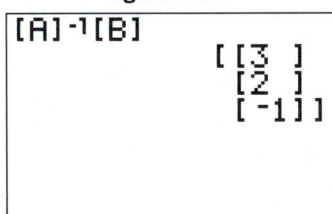

The solution is the ordered triple $(3, 2, -1)$.

Now try Exercises 59 through 62 ▶

We close this section with an application involving a 4×4 system. There is a large variety of additional applications in the Exercise Set.

EXAMPLE 8 ▶ **Solving an Application Using Technology and Matrix Equations**

A local theater sells four sizes of soft drinks: 32 oz @ $2.25; 24 oz @ $1.90; 16 oz @ $1.50; and 12 oz @ $1.20/each. As part of a "free guest pass" promotion, the manager asks employees to try and determine the number of each size sold, given the following information: (1) the total revenue from soft drinks was $719.80; (2) there were 9096 oz of soft drink sold; (3) there was a total of 394 soft drinks sold; and (4) the number of 24-oz and 12-oz drinks sold was 12 more than the number of 32-oz and 16-oz drinks sold. Write a system of equations that models this information, then solve the system using a matrix equation.

Solution ▶ If we let x, l, m, and s represent the number of 32-oz, 24-oz, 16-oz, and 12-oz soft drinks sold, the following system is produced:

$$
\begin{array}{rl}
\text{revenue:} & \\
\text{ounces sold:} & \\
\text{quantity sold:} & \\
\text{amounts sold:} &
\end{array}
\begin{cases}
2.25x + 1.90l + 1.50m + 1.20s = 719.8 \\
32x + 24l + 16m + 12s = 9096 \\
x + l + m + s = 394 \\
l + s = x + m + 12
\end{cases}
$$

When written as a matrix equation the system becomes:

$$
\begin{bmatrix}
2.25 & 1.9 & 1.5 & 1.2 \\
32 & 24 & 16 & 12 \\
1 & 1 & 1 & 1 \\
-1 & 1 & -1 & 1
\end{bmatrix}
\begin{bmatrix}
x \\ l \\ m \\ s
\end{bmatrix}
=
\begin{bmatrix}
719.8 \\ 9096 \\ 394 \\ 12
\end{bmatrix}
$$

To solve, carefully enter the matrix of coefficients as matrix A (see Figure 7.23), and the matrix of constants as matrix B, then compute $A^{-1}B = X$ [since $\det(A) \neq 0$]. This gives a solution of $(x, l, m, s) = (112, 151, 79, 52)$ (Figure 7.24).

Figure 7.23

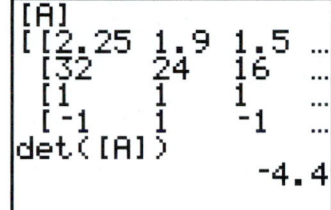

Figure 7.24

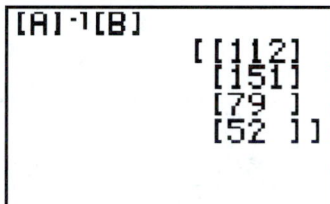

☑ **D.** You've just seen how we can use determinants to find whether a matrix is invertible

Now try Exercises 67 through 78 ▶

7.3 EXERCISES

▶ CONCEPTS AND VOCABULARY

Fill in the blank with the appropriate word or phrase. Carefully reread the section if needed.

1. The $n \times n$ identity matrix I_n consists of 1's down the ___diagonal___ and ___zeroes___ for all other entries.

2. The product of a square matrix A and its inverse A^{-1} yields the ___identity___ matrix.

3. Given square matrices A and B of like size, B is the inverse of A if ___AB___ = ___BA___ = ___I___. Notationally we write $B = $ ___A^{-1}___.

4. If the determinant of a matrix is zero, the matrix is said to be ___singular___ or ___noninvertible___, meaning no inverse exists.

5. Explain why inverses exist only for square matrices, then discuss why some square matrices do not have an inverse. Illustrate each point with an example.   Answers will vary.

6. What is the connection between the determinant of a 2×2 matrix and the formula for finding its inverse? Use the connection to create a 2×2 matrix that is invertible, and another that is not.
   Answers will vary.

HOMEWORK SELECTION GUIDE

Core: 7, 11–19 odd, 23, 27, 29, 33, 39, 43–49 odd, 57–71 odd, 75, 83, 89 (24 Exercises)
Standard: 1–4, 11–19 odd, 23, 27, 29, 33, 39, 43–49 odd, 57–75 odd, 83, 86, 90, 91 (33 Exercises)

Extended: 1–4, 7, 11–19 odd, 23, 27, 29, 33, 43–49 odd, 57–83 odd, 85, 86, 90–92 (38 Exercises)
In Depth: 1–6, 11–19 odd, 23, 27, 29, 33, 39, 39, 43–49 odd, 57–83 odd, 85, 86, 89–92 (41 Exercises)

▶ DEVELOPING YOUR SKILLS

Use matrix multiplication, equality of matrices, and the arbitrary matrix given to show that $\begin{bmatrix} a & b \\ c & d \end{bmatrix} = \begin{bmatrix} 1 & 0 \\ 0 & 1 \end{bmatrix}$.

7. $A = \begin{bmatrix} 2 & 5 \\ -3 & -7 \end{bmatrix} \begin{bmatrix} a & b \\ c & d \end{bmatrix} = \begin{bmatrix} 2 & 5 \\ -3 & -7 \end{bmatrix}$ *verified*

8. $A = \begin{bmatrix} 9 & -7 \\ -5 & 4 \end{bmatrix} \begin{bmatrix} a & b \\ c & d \end{bmatrix} = \begin{bmatrix} 9 & -7 \\ -5 & 4 \end{bmatrix}$ *verified*

9. $A = \begin{bmatrix} 0.4 & 0.6 \\ 0.3 & 0.2 \end{bmatrix} \begin{bmatrix} a & b \\ c & d \end{bmatrix} = \begin{bmatrix} 0.4 & 0.6 \\ 0.3 & 0.2 \end{bmatrix}$ *verified*

10. $A = \begin{bmatrix} \frac{1}{2} & \frac{1}{4} \\ \frac{1}{3} & \frac{1}{8} \end{bmatrix} \begin{bmatrix} a & b \\ c & d \end{bmatrix} = \begin{bmatrix} \frac{1}{2} & \frac{1}{4} \\ \frac{1}{3} & \frac{1}{8} \end{bmatrix}$ *verified*

For $I_2 = \begin{bmatrix} 1 & 0 \\ 0 & 1 \end{bmatrix}$, $I_3 = \begin{bmatrix} 1 & 0 & 0 \\ 0 & 1 & 0 \\ 0 & 0 & 1 \end{bmatrix}$, and

$I_4 = \begin{bmatrix} 1 & 0 & 0 & 0 \\ 0 & 1 & 0 & 0 \\ 0 & 0 & 1 & 0 \\ 0 & 0 & 0 & 1 \end{bmatrix}$, show $AI = IA = A$ for the

matrices of like size. Use a calculator for Exercise 14.

11. $\begin{bmatrix} -3 & 8 \\ -4 & 10 \end{bmatrix}$ *verified*

12. $\begin{bmatrix} 0.5 & -0.2 \\ -0.7 & 0.3 \end{bmatrix}$ *verified*

13. $\begin{bmatrix} -4 & 1 & 6 \\ 9 & 5 & 3 \\ 0 & -2 & 1 \end{bmatrix}$ *verified*

14. $\begin{bmatrix} 9 & 1 & 3 & -1 \\ 2 & 0 & -5 & 3 \\ 4 & 6 & 1 & 0 \\ 0 & -2 & 4 & 1 \end{bmatrix}$ *verified*

Find the inverse of each 2 × 2 matrix using matrix multiplication, equality of matrices, and a system of equations.

15. $\begin{bmatrix} 5 & -4 \\ 2 & 2 \end{bmatrix}$ $\begin{bmatrix} \frac{1}{9} & \frac{2}{9} \\ \frac{-1}{9} & \frac{5}{18} \end{bmatrix}$

16. $\begin{bmatrix} 1 & -5 \\ 0 & -4 \end{bmatrix}$ $\begin{bmatrix} 1 & \frac{-5}{4} \\ 0 & \frac{-1}{4} \end{bmatrix}$

Find the inverse of each matrix by augmenting of the the identity matrix and using row operations.

17. $\begin{bmatrix} 1 & -3 \\ 4 & -10 \end{bmatrix}$ $\begin{bmatrix} -5 & 1.5 \\ -2 & 0.5 \end{bmatrix}$

18. $\begin{bmatrix} -2 & 0.4 \\ 1 & 0.8 \end{bmatrix}$ $\begin{bmatrix} -0.4 & 0.2 \\ 0.5 & 1 \end{bmatrix}$

Demonstrate that $B = A^{-1}$, by showing $AB = BA = I$. Do not use a calculator.

19. $A = \begin{bmatrix} 1 & 5 \\ -2 & -9 \end{bmatrix}$

$B = \begin{bmatrix} -9 & -5 \\ 2 & 1 \end{bmatrix}$ *verified*

20. $A = \begin{bmatrix} -2 & -6 \\ 4 & 11 \end{bmatrix}$

$B = \begin{bmatrix} 5.5 & 3 \\ -2 & -1 \end{bmatrix}$ *verified*

21. $A = \begin{bmatrix} 4 & -5 \\ 0 & 2 \end{bmatrix}$

$B = \begin{bmatrix} \frac{1}{4} & \frac{5}{8} \\ 0 & \frac{1}{2} \end{bmatrix}$ *verified*

22. $A = \begin{bmatrix} -2 & 5 \\ 3 & -4 \end{bmatrix}$

$B = \begin{bmatrix} \frac{4}{7} & \frac{5}{7} \\ \frac{3}{7} & \frac{2}{7} \end{bmatrix}$ *verified*

Use a calculator to find $A^{-1} = B$, then confirm the inverse by showing $AB = BA = I$.

23. $A = \begin{bmatrix} -2 & 3 & 1 \\ 5 & 2 & 4 \\ 2 & 0 & -1 \end{bmatrix}$ $\begin{bmatrix} \frac{-2}{39} & \frac{1}{13} & \frac{10}{39} \\ \frac{1}{3} & 0 & \frac{1}{3} \\ \frac{-4}{39} & \frac{2}{13} & \frac{-19}{39} \end{bmatrix}$

24. $A = \begin{bmatrix} 0.5 & 0.2 & 0.1 \\ 0 & 0.3 & 0.6 \\ 1 & 0.4 & -0.3 \end{bmatrix}$ $\begin{bmatrix} \frac{22}{5} & \frac{-4}{3} & \frac{-6}{5} \\ -8 & \frac{10}{3} & 4 \\ 4 & 0 & -2 \end{bmatrix}$

25. $A = \begin{bmatrix} -7 & 5 & -3 \\ 1 & 9 & 0 \\ 2 & -2 & -5 \end{bmatrix}$ $\begin{bmatrix} \frac{-9}{80} & \frac{31}{400} & \frac{27}{400} \\ \frac{1}{80} & \frac{41}{400} & \frac{-3}{400} \\ \frac{-1}{20} & \frac{-1}{100} & \frac{-17}{100} \end{bmatrix}$

26. $A = \frac{1}{12}\begin{bmatrix} 12 & -6 & 3 & 0 \\ 0 & -4 & 8 & -12 \\ 12 & -12 & 0 & 0 \\ 0 & 12 & 0 & -12 \end{bmatrix}$ $\begin{bmatrix} 1 & \frac{-3}{8} & 0 & \frac{3}{8} \\ 1 & \frac{-3}{8} & -1 & \frac{3}{8} \\ 2 & \frac{3}{4} & -2 & \frac{-3}{4} \\ 1 & \frac{-3}{8} & -1 & \frac{-5}{8} \end{bmatrix}$

Write each system in the form of a matrix equation. Do not solve.

27. $\begin{cases} 2x - 3y = 9 \\ -5x + 7y = 8 \end{cases}$ $\begin{bmatrix} 2 & -3 \\ -5 & 7 \end{bmatrix}\begin{bmatrix} x \\ y \end{bmatrix} = \begin{bmatrix} 9 \\ 8 \end{bmatrix}$

28. $\begin{cases} 0.5x - 0.6y = 0.6 \\ -0.7x + 0.4y = -0.375 \end{cases}$ $\begin{bmatrix} 0.5 & -0.6 \\ -0.7 & 0.4 \end{bmatrix}\begin{bmatrix} x \\ y \end{bmatrix} = \begin{bmatrix} 0.6 \\ -0.375 \end{bmatrix}$

29. $\begin{cases} x + 2y - z = 1 \\ x + z = 3 \\ 2x - y + z = 3 \end{cases}$ $\begin{bmatrix} 1 & 2 & -1 \\ 1 & 0 & 1 \\ 2 & -1 & 1 \end{bmatrix}\begin{bmatrix} x \\ y \\ z \end{bmatrix} = \begin{bmatrix} 1 \\ 3 \\ 3 \end{bmatrix}$

30. $\begin{cases} 2x - 3y - 2z = 4 \\ \frac{1}{4}x - \frac{2}{5}y + \frac{3}{4}z = \frac{-1}{3} \\ -2x + 1.3y - 3z = 5 \end{cases}$ $\begin{bmatrix} 2 & -3 & -2 \\ \frac{1}{4} & \frac{-2}{5} & \frac{3}{4} \\ -2 & 1.3 & -3 \end{bmatrix}\begin{bmatrix} x \\ y \\ z \end{bmatrix} = \begin{bmatrix} 4 \\ \frac{-1}{3} \\ 5 \end{bmatrix}$

31. $\begin{cases} -2w + x - 4y + 5z = -3 \\ 2w - 5x + y - 3z = 4 \\ -3w + x + 6y + z = 1 \\ w + 4x - 5y + z = -9 \end{cases}$

32. $\begin{cases} 1.5w + 2.1x - 0.4y + z = 1 \\ 0.2w - 2.6x + y = 5.8 \\ 3.2x + z = 2.7 \\ 1.6w + 4x - 5y + 2.6z = -1.8 \end{cases}$

Additional answers can be found in the Instructor Answer Appendix.

Write each system as a matrix equation and solve (if possible) using inverse matrices and your calculator. If the coefficient matrix is singular, write *no solution*.

33. $\begin{cases} 0.05x - 3.2y = -15.8 \\ 0.02x + 2.4y = 12.08 \end{cases}$ (4, 5)

34. $\begin{cases} 0.3x + 1.1y = 3.5 \\ -0.5x - 2.9y = -10.1 \end{cases}$ (−3, 4)

35. $\begin{cases} \frac{-1}{6}u + \frac{1}{4}v = 1 \\ \frac{1}{2}u - \frac{2}{3}v = -2 \end{cases}$ (12, 12)

36. $\begin{cases} \sqrt{2}a + \sqrt{3}b = 2\sqrt{6} \\ \sqrt{6}a + b = 4\sqrt{2} \end{cases}$ ($\sqrt{3}$, $\sqrt{2}$)

37. $\begin{cases} \frac{-5}{2}a + \frac{3}{5}b = -\frac{3}{10} \\ \frac{5}{16}a - \frac{3}{2}b = -\frac{7}{16} \end{cases}$ $(\frac{1}{5}, \frac{1}{3})$

38. $\begin{cases} 3\sqrt{2}a + 2\sqrt{3}b = 12 \\ 5\sqrt{2}a - 3\sqrt{3}b = 1 \end{cases}$ ($\sqrt{2}$, $\sqrt{3}$)

39. $\begin{cases} 0.2x - 1.6y + 2z = -1.9 \\ -0.4x - y + 0.6z = -1 \\ 0.8x + 3.2y - 0.4z = 0.2 \end{cases}$ (1.5, −0.5, −1.5)

40. $\begin{cases} 1.7x + 2.3y - 2z = 41.5 \\ 1.4x - 0.9y + 1.6z = -10 \\ -0.8x + 1.8y - 0.5z = 16.5 \end{cases}$ (5, 10, −5)

41. $\begin{cases} x - 2y + 2z = 9 \\ 2x - 1.5y + 1.8z = 12 \\ \frac{-2}{3}x + \frac{1}{2}y - \frac{3}{5}z = -4 \end{cases}$ (3, 2, 5)

42. $\begin{cases} 4x - 5y - 6z = 33 \\ \frac{1}{8}x - \frac{3}{5}y + \frac{5}{4}z = 9 \\ -0.5x + 2.4y - 4z = -32 \end{cases}$ (8, −5, 4)

43. $\begin{cases} -2w + 3x - 4y + 5z = -3 \\ 0.2w - 2.6x + y - 0.4z = 2.4 \\ -3w + 3.2x + 2.8y + z = 6.1 \\ 1.6w + 4x - 5y + 2.6z = -9.8 \end{cases}$ (−1, −0.5, 1.5, 0.5)

44. $\begin{cases} 2w - 5x + 3y - 4z = 7 \\ 1.6w + 4.2y - 1.8z = 5.4 \\ 3w + 6.7x - 9y + 4z = -8.5 \\ 0.7x - 0.9z = 0.9 \end{cases}$ $(0.5, 0, \frac{2}{3}, -1)$

Compute the determinant of each matrix and state whether an inverse matrix exists. Do not use a calculator.

45. $\begin{bmatrix} 4 & -7 \\ 3 & -5 \end{bmatrix}$ 1, yes   **46.** $\begin{bmatrix} 0.6 & 0.3 \\ 0.4 & 0.5 \end{bmatrix}$ 0.18, yes

47. $\begin{bmatrix} 1.2 & -0.8 \\ 0.3 & -0.2 \end{bmatrix}$ 0, no   **48.** $\begin{bmatrix} -2 & 6 \\ -3 & 9 \end{bmatrix}$ 0, no

Additional answers can be found in the Instructor Answer Appendix.

Compute the determinant of each matrix without using a calculator. If the determinant is zero, write *singular matrix*.

49. $A = \begin{bmatrix} 1 & 0 & -2 \\ 0 & -1 & -1 \\ 2 & 1 & -4 \end{bmatrix}$ 1   **50.** $B = \begin{bmatrix} -2 & 2 & 1 \\ 0 & -1 & 2 \\ 4 & -4 & 0 \end{bmatrix}$ 4

51. $C = \begin{bmatrix} -2 & 3 & 4 \\ 0 & 6 & 2 \\ 1 & -1.5 & -2 \end{bmatrix}$ singular matrix

52. $D = \begin{bmatrix} 1 & 2 & -0.8 \\ 2.5 & 5 & -2 \\ 3 & 0 & -2.5 \end{bmatrix}$ singular matrix

Compute the determinant of each matrix using the column rotation method.

53. $\begin{bmatrix} 2 & -3 & 1 \\ 4 & -1 & 5 \\ 1 & 0 & -2 \end{bmatrix}$ −34   **54.** $\begin{bmatrix} -3 & 2 & 4 \\ -1 & -2 & 0 \\ 3 & 1 & 5 \end{bmatrix}$ 60

55. $\begin{bmatrix} 1 & -1 & 2 \\ 3 & -2 & 4 \\ 4 & 3 & 1 \end{bmatrix}$ 7   **56.** $\begin{bmatrix} 5 & 6 & 2 \\ -2 & 1 & -2 \\ 3 & 4 & -1 \end{bmatrix}$ −35

Use a calculator to compute the determinant of each matrix. If the determinant is zero, write *singular matrix*. If the determinant is nonzero, find A^{-1} and store the result as matrix B (STO▸ 2nd x^{-1} 2: [B] ENTER). Then verify the inverse by showing $AB = BA = I$.

57. $A = \begin{bmatrix} 1 & 0 & 3 & -4 \\ 2 & 5 & 0 & 1 \\ 8 & 15 & 6 & -5 \\ 0 & 8 & -4 & 1 \end{bmatrix}$ singular matrix

58. $M = \begin{bmatrix} 1 & 2 & 1 & 1 \\ 0 & 1 & -3 & 2 \\ -1 & 0 & 2 & -3 \\ 2 & -1 & 1 & 4 \end{bmatrix}$ $\begin{bmatrix} \frac{1}{2} & \frac{-9}{4} & -3 & \frac{-5}{4} \\ \frac{1}{3} & \frac{1}{3} & \frac{1}{3} & 0 \\ 0 & \frac{1}{2} & 1 & \frac{1}{2} \\ \frac{-1}{6} & \frac{13}{12} & \frac{4}{3} & \frac{3}{4} \end{bmatrix}$

For each system shown, form the matrix equation $AX = B$; compute the determinant of the coefficient matrix and determine if you can proceed; and if possible, solve the system using the matrix equation.

59. $\begin{cases} x - 2y + 2z = 7 \\ 2x + 2y - z = 5 \\ 3x - y + z = 6 \end{cases}$ **60.** $\begin{cases} 2x - 3y - 2z = 7 \\ x - y + 2z = -5 \\ 3x + 2y - z = 11 \end{cases}$

det(A) = −5;
(1, 6, 9)

61. $\begin{cases} x - 3y + 4z = -1 \\ 4x - y + 5z = 7 \\ 3x + 2y + z = -3 \end{cases}$ **62.** $\begin{cases} 5x - 2y + z = 1 \\ 3x - 4y + 9z = -2 \\ 4x - 3y + 5z = 6 \end{cases}$

det(A) = 0   det(A) = 0

▶ WORKING WITH FORMULAS

The inverse of a 2 × 2 matrix: $A = \begin{bmatrix} a & b \\ c & d \end{bmatrix} \rightarrow A^{-1} = \dfrac{1}{ad-bc} \cdot \begin{bmatrix} d & -b \\ -c & a \end{bmatrix}$

The inverse of a 2 × 2 matrix can be found using the formula shown, as long as $ad - bc \neq 0$. Use the formula to find inverses for the matrices here (if possible), then verify by showing $A \cdot A^{-1} = A \cdot A^{-1} = I$.

63. $A = \begin{bmatrix} 3 & -5 \\ 2 & 1 \end{bmatrix}$ **64.** $B = \begin{bmatrix} 2 & 3 \\ -5 & -4 \end{bmatrix}$ **65.** $C = \begin{bmatrix} 0.3 & -0.4 \\ -0.6 & 0.8 \end{bmatrix}$ **66.** $\begin{bmatrix} 0.2 & 0.3 \\ -0.4 & -0.6 \end{bmatrix}$

singular   singular

▶ APPLICATIONS

Solve each application using a matrix equation.

 Descriptive Translation

67. Convenience store sales: The local Moto-Mart sells four different sizes of Slushies—behemoth, 60 oz @ $2.59; gargantuan, 48 oz @ $2.29; mammoth, 36 oz @ $1.99; and jumbo, 24 oz @ $1.59. As part of a promotion, the owner offers free gas to any customer who can tell how many of each size were sold last week, given the following information: (1) The total revenue for the Slushies was $402.29; (2) 7884 ounces were sold; (3) a total of 191 Slushies were sold; and (4) the number of behemoth Slushies sold was one more than the number of jumbo. How many of each size were sold?
31 behemoth, 52 gargantuan, 78 mammoth, 30 jumbo

68. Cartoon characters: In America, four of the most beloved cartoon characters are Foghorn Leghorn, Elmer Fudd, Bugs Bunny, and Tweety Bird. Suppose that Bugs Bunny is four times as tall as Tweety Bird. Elmer Fudd is as tall as the combined height of Bugs Bunny and Tweety Bird. Foghorn Leghorn is 20 cm taller than the combined height of Elmer Fudd and Tweety Bird. The combined height of all four characters is 500 cm. How tall is each one?
Foghorn 200 cm; Elmer 150 cm; Bugs 120 cm; Tweety 30 cm

69. Rolling Stones music: One of the most prolific and popular rock-and-roll bands of all time is the Rolling Stones. Four of their many great hits include: *Jumpin' Jack Flash, Tumbling Dice, You Can't Always Get What You Want,* and *Wild Horses.* The total playing time of all four songs is 20.75 min. The combined playing time of *Jumpin' Jack Flash* and *Tumbling Dice* equals that of *You Can't Always Get What You Want. Wild Horses* is 2 min longer than *Jumpin' Jack Flash,* and *You Can't Always Get What You Want* is twice as long as *Tumbling Dice.* Find the playing time of each song.

70. Mozart's arias: Mozart wrote some of vocal music's most memorable arias in his operas, including *Tamino's Aria, Papageno's Aria,* the *Champagne Aria,* and the *Catalogue Aria.* The total playing time of all four arias is 14.3 min. *Papageno's Aria* is 3 min shorter than the *Catalogue Aria.* The *Champagne Aria* is 2.7 min shorter than *Tamino's Aria.* The combined time of *Tamino's Aria* and *Papageno's Aria* is five times that of the *Champagne Aria.* Find the playing time of all four arias.

Manufacturing

71. Resource allocation: Time Pieces Inc. manufactures four different types of grandfather clocks. Each clock requires these four stages: (1) assembly, (2) installing the clockworks, (3) inspection and testing, and (4) packaging for delivery. The time required for each stage is shown in the table, for each of the four clock types. At the end of a busy week, the owner determines that personnel on the assembly line worked for 262 hr, the installation crews for 160 hr, the testing department for 29 hr, and the packaging department for 68 hr. How many clocks of each type were made?

Dept.	Clock A	Clock B	Clock C	Clock D
Assemble	2.2	2.5	2.75	3
Install	1.2	1.4	1.8	2
Test	0.2	0.25	0.3	0.5
Pack	0.5	0.55	0.75	1.0

72. Resource allocation: Figurines Inc. makes and sells four sizes of metal figurines, mostly historical figures and celebrities. Each figurine goes through four stages of development: (1) casting, (2) trimming, (3) polishing, and (4) painting. The time required for each stage is shown in the table, for each of the four sizes. At the end of a busy week, the manager finds that the casting department put in 62 hr, and the trimming department worked for 93.5 hr, with the polishing and painting departments logging 138 hr and 358 hr, respectively. How many figurines of each type were made?

Dept.	Small	Medium	Large	X-Large
Casting	0.5	0.6	0.75	1
Trimming	0.8	0.9	1.1	1.5
Polishing	1.2	1.4	1.7	2
Painting	2.5	3.5	4.5	6

73. **Thermal conductivity:** In lab experiments designed to measure the heat conductivity of a square metal plate of uniform density, the edges are held at four different (constant) temperatures. The *mean-value principle* from physics tells us that the temperature at a given point p_i on the plate is equal to the average temperature of nearby points. Use this information to form a system of four equations in four variables, and determine the temperature at interior points p_1, p_2, p_3, and p_4 on the plate shown. (*Hint:* Use the temperature of the four points closest to each.)

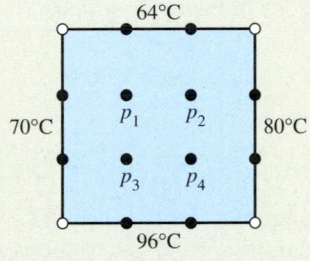

74. **Thermal conductivity:** Repeat Exercise 73 if (a) the temperatures at the top and bottom of the plate were *increased* by 10°, with the temperatures at the left and right edges *decreased* by 10° (what do you notice?); (b) the temperature at the top and the temperature to the left were *decreased* by 10°, with the temperatures at the bottom and right held at their original temperature.

Curve Fitting

75. **Quadratic fit:** Use a matrix equation to find a quadratic function of the form $y = ax^2 + bx + c$ such that $(-4, -5)$, $(0, -5)$, and $(2, 7)$ are on the graph of the function. $y = x^2 + 4x - 5$

76. **Quadratic fit:** Use a matrix equation to find a quadratic function of the form $y = ax^2 + bx + c$ such that $(-4, -0)$, $(1, 5)$ and, $(2, -6)$ are on the graph of the function. $y = -2x^2 - 5x + 12$

77. **Cubic fit:** Use a matrix equation to find a cubic function of the form $y = ax^3 + bx^2 + cx + d$ such that $(-4, -6)$, $(-1, 0)$, $(1, -16)$, and $(3, 8)$ are on the graph of the function.

78. **Cubic fit:** Use a matrix equation to find a cubic function of the form $y = ax^3 + bx^2 + cx + d$ such that $(-2, 5)$, $(0, 1)$, $(2, -3)$, and $(3, 25)$ are on the graph of the function.

Additional answers can be found in the Instructor Answer Appendix.

Investing

79. **Wise investing:** Morgan received an $800 gift from her grandfather, and showing wisdom beyond her years, decided to place the money in a certificate of deposit (CD) and a money market fund (MM). At the time, CDs were earning 3.5% and MMs were earning 2.5%. At the end of 1 yr, she cashed both in and received a total of $824.50. How much was deposited in each?
$450 in the CD, $350 in the MM

80. **Baseball cards:** Gary has a passion for baseball, which includes a collection of rare baseball cards. His most prized cards feature Willie Mays (1953 Topps) and Mickey Mantle (1959 Topps). The Willie Mays card has appreciated 28% and the Mickey Mantle card 25% since he purchased them, and together they are now worth $17,100. If he paid a total of $13,507.50 at auction for both cards, what was the original price of each?
Willie Mays: $7187.50, Mickey Mantle: $6320.00

81. **Retirement planning:** Using payroll deduction, Jeanette was able to put aside $4800 per month last year for her impending retirement. Last year, her company retirement fund paid 4.2% and her mutual funds returned 5.75%, but her stock fund actually decreased 2.5% in value. If her net gain for the year was $104.50 and $300 more was placed in stocks than in mutual funds, how much was placed in each investment vehicle?
$1500 in retirement fund, $1500 in mutual fund, $1800 in stock fund

82. **Charitable giving:** The hyperbolic funnels seen at many shopping malls are primarily used by nonprofit organizations to raise funds for worthy causes. A coin is launched down a ramp into the funnel and seemingly makes endless circuits before finally disappearing down a "black hole" (the collection bin). During one such collection, the bin was found to hold $112.89, and 1450 coins consisting of pennies, nickels, dimes, and quarters. How many of each denominator were there, if the number of quarters and dimes was equal to the number of nickels, and the number of pennies was twice the number of quarters.   207 quarters, 311 dimes, 518 nickels, 414 pennies

Nutrition

83. **Animal diets:** A zoo dietician needs to create a specialized diet that regulates an animal's intake of fat, carbohydrates, and protein during a meal. The table given shows three different foods and the amount of these nutrients (in grams) that each ounce of food provides. How many ounces of each should the dietician recommend to supply 20 g of fat, 30 g of carbohydrates, and 44 g of protein?

Nutrient	Food I	Food II	Food III
Fat	2	4	3
Carb.	4	2	5
Protein	5	6	7

84. **Training diet:** A physical trainer is designing a workout diet for one of her clients, and wants to supply him with 24 g of fat, 244 g of carbohydrates, and 40 g of protein for the noontime meal. The table given shows three different foods and the amount of these nutrients (in grams) that each ounce of food provides. How many ounces of each should the trainer recommend?

Nutrient	Food I	Food II	Food III
Fat	2	5	0
Carb.	10	15	18
Protein	2	10	0.75

▶ EXTENDING THE CONCEPT

85. Some matrix applications require that you solve a matrix equation of the form $AX + B = C$, where A, B, and C are matrices with the appropriate number of rows and columns and A^{-1} exists. Investigate the solution process for such equations using $A = \begin{bmatrix} 2 & 3 \\ -5 & -4 \end{bmatrix}$, $B = \begin{bmatrix} 4 \\ 9 \end{bmatrix}$, $C = \begin{bmatrix} 12 \\ -4 \end{bmatrix}$, and $X = \begin{bmatrix} x \\ y \end{bmatrix}$, then solve $AX + B = C$ for X *symbolically* (using A^{-1}, I, and so on).   Answers will vary.

86. Another alternative for finding determinants uses the triangularized form of a matrix and is offered without proof: *If nonsingular matrix A is written in triangularized form using standard row operations but without exchanging any rows and without using the operation kR_i to replace any row (k a constant), then det(A) is equal to the product of resulting diagonal entries.* Compute the determinant of each matrix using this method. Be careful not to interchange rows and do not replace any row by a multiple of that row in the process.

a. $\begin{bmatrix} 1 & -2 & 3 \\ -4 & 5 & -6 \\ 2 & 5 & 3 \end{bmatrix}$ b. $\begin{bmatrix} 2 & 5 & -1 \\ -2 & -3 & 4 \\ 4 & 6 & 5 \end{bmatrix}$ c. $\begin{bmatrix} -2 & 4 & 1 \\ 5 & 7 & -2 \\ 3 & -8 & -1 \end{bmatrix}$ d. $\begin{bmatrix} 3 & -1 & 4 \\ 0 & -2 & 6 \\ -2 & 1 & -3 \end{bmatrix}$
   −45   52   −19   −4

▶ MAINTAINING YOUR SKILLS

87. **(4.2)** Solve using the rational zeroes theorem:
   $x^3 - 7x^2 = -36$ $x = 6, x = 3, x = -2$

88. **(2.2/5.3)** Match each equation to its related graph. Justify your answers.
   shifts right 2   shifts down 2
   b $y = \log_2(x - 2)$ a $y = \log_2 x - 2$

   a. 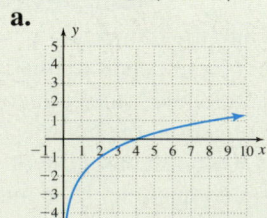   b.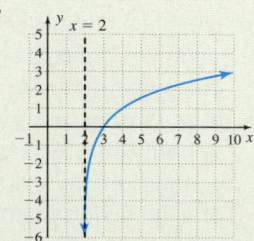

89. **(2.3)** Solve the absolute value inequality:
   $-3|2x + 5| - 7 \le -19$. $x \in \left(-\infty, -\frac{9}{2}\right] \cup \left[-\frac{1}{2}, \infty\right)$

90. **(2.6)** A coin collector believes that the value of a coin varies inversely as the number of coins still in circulation. If 4 million coins are in circulation, the coin has a value of $25. (a) Find the variation equation and (b) determine how many coins are in circulation if the value of the coin is $6.25.

   a. $V = 100\left(\frac{1}{C}\right)$, C in millions   b. 16 Million

Applications of Matrices and Determinants: Cramer's Rule, Partial Fractions, and More

In addition to solving systems, matrices can be used to accomplish such diverse things as finding the volume of a three-dimensional solid or establishing certain geometrical relationships in the coordinate plane. Numerous uses are also found in higher mathematics, such as checking whether solutions to a differential equation are linearly independent.

A. Solving Systems Using Determinants and Cramer's Rule

In addition to identifying singular matrices, determinants can actually be used to *develop a formula* for the solution of a system. Consider the following solution to a *general* 2×2 system, which parallels the solution to a specific 2×2 system. With a view toward a solution involving determinants, the coefficients of x are written as a_{11} and a_{21} in the general system, and the coefficients of y are a_{12} and a_{22}.

Specific System

$$\begin{cases} 2x + 5y = 9 \\ 3x + 4y = 10 \end{cases}$$

eliminate the x-term

$-3R1 + 2R2$

sums to zero

$$\begin{cases} \boxed{-3 \cdot 2x} - 3 \cdot 5y = -3 \cdot 9 \\ \boxed{2 \cdot 3x} + 2 \cdot 4y = 2 \cdot 10 \end{cases}$$

$$2 \cdot 4y - 3 \cdot 5y = 2 \cdot 10 - 3 \cdot 9$$

General System

$$\begin{cases} a_{11}x + a_{12}y = c_1 \\ a_{21}x + a_{22}y = c_2 \end{cases}$$

eliminate the x-term

$-a_{21}R1 + a_{11}R2$

sums to zero

$$\begin{cases} \boxed{-a_{21}a_{11}x} - a_{21}a_{12}y = -a_{21}c_1 \\ \boxed{a_{11}a_{21}x} + a_{11}a_{22}y = a_{11}c_2 \end{cases}$$

$$a_{11}a_{22}y - a_{21}a_{12}y = a_{11}c_2 - a_{21}c_1$$

Notice the x-terms sum to zero in both systems. We are deliberately leaving the solution on the left unsimplified to show the pattern developing on the right. Next we solve for y.

Factor Out y

$$(2 \cdot 4 - 3 \cdot 5)y = 2 \cdot 10 - 3 \cdot 9$$

$$y = \frac{2 \cdot 10 - 3 \cdot 9}{2 \cdot 4 - 3 \cdot 5} \text{ divide}$$

Factor Out y

$$(a_{11}a_{22} - a_{21}a_{12})y = a_{11}c_2 - a_{21}c_1$$

$$\text{divide} \quad y = \frac{a_{11}c_2 - a_{21}c_1}{a_{11}a_{22} - a_{21}a_{12}}$$

On the left we find $y = \frac{-7}{-7} = 1$ and back-substitution shows $x = 2$. But more important, on the right we obtain a formula for the y-value of *any* 2×2 system: $y = \dfrac{a_{11}c_2 - a_{21}c_1}{a_{11}a_{22} - a_{21}a_{12}}$. If we had chosen to solve for x, the solution would be $x = \dfrac{a_{22}c_1 - a_{12}c_2}{a_{11}a_{22} - a_{21}a_{12}}$. Note these formulas are defined only if $a_{11}a_{22} - a_{21}a_{12} \neq 0$. You may have already noticed, but this denominator is the *determinant of the matrix of coefficients* $\begin{bmatrix} a_{11} & a_{12} \\ a_{21} & a_{22} \end{bmatrix}$ from the previous section! Since the numerator is also a difference of two products, we investigate the possibility that it too can be expressed as a determinant. Working backward, we're able to reconstruct the numerator for x in determinant form as $\begin{bmatrix} c_1 & a_{12} \\ c_2 & a_{22} \end{bmatrix}$, where it is apparent this matrix was formed by *replacing the coefficients of the x-variables with the constant terms.*

(removed)

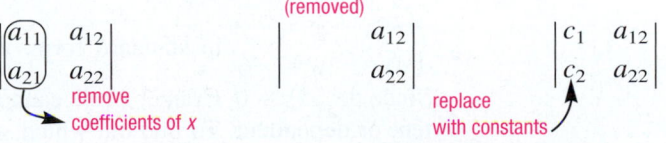

It is also apparent the numerator for y can be also written in determinant form as $\begin{vmatrix} a_{11} & c_1 \\ a_{21} & c_2 \end{vmatrix}$, or the determinant formed by *replacing the coefficients of the y-variables with the constant terms*:

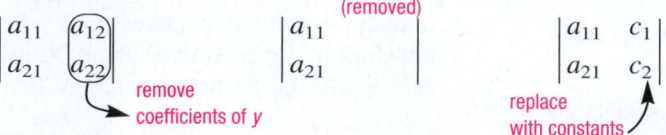

(removed)

remove coefficients of y

replace with constants

If we use the notation D_y for this determinant, D_x for the determinant where x coefficients were replaced by the constants, and D as the determinant for the matrix of coefficients—the solutions can be written as shown next, with the result known as **Cramer's rule.**

Cramer's Rule for 2 × 2 Systems

Given a 2 × 2 system of linear equations

$$\begin{cases} a_{11}x + a_{12}y = c_1 \\ a_{21}x + a_{22}y = c_2 \end{cases}$$

the solution (if one exists) is the ordered pair (x, y), where

$$x = \frac{D_x}{D} = \frac{\begin{vmatrix} c_1 & a_{12} \\ c_2 & a_{22} \end{vmatrix}}{\begin{vmatrix} a_{11} & a_{12} \\ a_{21} & a_{22} \end{vmatrix}} \quad \text{and} \quad y = \frac{D_y}{D} = \frac{\begin{vmatrix} a_{11} & c_1 \\ a_{21} & c_2 \end{vmatrix}}{\begin{vmatrix} a_{11} & a_{12} \\ a_{21} & a_{22} \end{vmatrix}}$$

provided $D \neq 0$.

EXAMPLE 1 ▶ **Solving a System Using Cramer's Rule**

Use Cramer's rule to solve the system $\begin{cases} 2x - 5y = 9 \\ -3x + 4y = -10 \end{cases}$.

Solution ▶ Begin by finding the value of D, D_x, and D_y.

$$D = \begin{vmatrix} 2 & -5 \\ -3 & 4 \end{vmatrix} \qquad D_x = \begin{vmatrix} 9 & -5 \\ -10 & 4 \end{vmatrix} \qquad D_y = \begin{vmatrix} 2 & 9 \\ -3 & -10 \end{vmatrix}$$

$$(2)(4) - (-3)(-5) \qquad (9)(4) - (-10)(-5) \qquad (2)(-10) - (-3)(9)$$

$$= -7 \qquad\qquad\qquad = -14 \qquad\qquad\qquad = 7$$

This gives $x = \dfrac{D_x}{D} = \dfrac{-14}{-7} = 2$ and $y = \dfrac{D_y}{D} = \dfrac{7}{-7} = -1$. The solution is $(2, -1)$. Check by substituting these values into the original equations.

Now try Exercises 7 through 14 ▶

Regardless of the method used to solve a system, always be aware that a consistent, inconsistent, or dependent system is possible. The system $\begin{cases} y - 2x = -3 \\ 4x + 6 = 2y \end{cases}$ yields $\begin{cases} -2x + y = -3 \\ 4x - 2y = -6 \end{cases}$ in standard form, with $D = \begin{vmatrix} -2 & 1 \\ 4 & -2 \end{vmatrix} = (-2)(-2) - (4)(1) = 0$. Since $\det(D) = 0$, Cramer's rule cannot be applied, and the system is either inconsistent or dependent. To find out which, we write the equations in function form (solve for y). The result is $\begin{cases} y = 2x - 3 \\ y = 2x + 3 \end{cases}$, showing the system consists of two parallel lines and has no solutions.

Cramer's Rule for 3 × 3 Systems

Cramer's rule can be extended to a 3 × 3 system of linear equations, using the same pattern as for 2 × 2 systems. Given the general 3 × 3 system

$$\begin{cases} a_{11}x + a_{12}y + a_{13}z = c_1 \\ a_{21}x + a_{22}y + a_{23}z = c_2 \\ a_{31}x + a_{32}y + a_{33}z = c_3 \end{cases}$$

the solutions are $x = \dfrac{D_x}{D}$, $y = \dfrac{D_y}{D}$, and $z = \dfrac{D_z}{D}$, where D_x, D_y, and D_z are again formed by replacing the coefficients of the indicated variable with the constants, and D is the determinant of the matrix of coefficients ($D \neq 0$).

Cramer's Rule Applied to 3 × 3 Systems

Given a 3 × 3 system of linear equations

$$\begin{cases} a_{11}x + a_{12}y + a_{13}z = c_1 \\ a_{21}x + a_{22}y + a_{23}z = c_2 \\ a_{31}x + a_{32}y + a_{33}z = c_3 \end{cases}$$

The solution (if one exists) is an ordered triple (x, y, z), where

$$x = \frac{\begin{vmatrix} c_1 & a_{12} & a_{13} \\ c_2 & a_{22} & a_{23} \\ c_3 & a_{32} & a_{33} \end{vmatrix}}{\begin{vmatrix} a_{11} & a_{12} & a_{13} \\ a_{21} & a_{22} & a_{23} \\ a_{31} & a_{32} & a_{33} \end{vmatrix}} \qquad y = \frac{\begin{vmatrix} a_{11} & c_1 & a_{13} \\ a_{21} & c_2 & a_{23} \\ a_{31} & c_3 & a_{33} \end{vmatrix}}{\begin{vmatrix} a_{11} & a_{12} & a_{13} \\ a_{21} & a_{22} & a_{23} \\ a_{31} & a_{32} & a_{33} \end{vmatrix}} \qquad z = \frac{\begin{vmatrix} a_{11} & a_{12} & c_1 \\ a_{21} & a_{22} & c_2 \\ a_{31} & a_{32} & c_3 \end{vmatrix}}{\begin{vmatrix} a_{11} & a_{12} & a_{13} \\ a_{21} & a_{22} & a_{23} \\ a_{31} & a_{32} & a_{33} \end{vmatrix}},$$

provided $D \neq 0$.

EXAMPLE 2 ▶   **Solving a 3 × 3 System Using Cramer's Rule**

Solve using Cramer's rule: $\begin{cases} x - 2y + 3z = -1 \\ -2x + y - 5z = 1 \\ 3x + 3y + 4z = 2 \end{cases}$

Solution ▶   Begin by computing the determinant of the matrix of coefficients, to ensure that Cramer's rule can be applied. Using the third row, we have

$$D = \begin{vmatrix} 1 & -2 & 3 \\ -2 & 1 & -5 \\ 3 & 3 & 4 \end{vmatrix} = +3\begin{vmatrix} -2 & 3 \\ 1 & -5 \end{vmatrix} - 3\begin{vmatrix} 1 & 3 \\ -2 & -5 \end{vmatrix} + 4\begin{vmatrix} 1 & -2 \\ -2 & 1 \end{vmatrix}$$

$$= 3(7) - 3(1) + 4(-3) = 6$$

Since $D \neq 0$ we continue, electing to compute the remaining determinants using a calculator (see figures).

$$D_x = \begin{vmatrix} -1 & -2 & 3 \\ 1 & 1 & -5 \\ 2 & 3 & 4 \end{vmatrix} = 12 \quad D_y = \begin{vmatrix} 1 & -1 & 3 \\ -2 & 1 & -5 \\ 3 & 2 & 4 \end{vmatrix} = 0 \quad D_z = \begin{vmatrix} 1 & -2 & -1 \\ -2 & 1 & 1 \\ 3 & 3 & 2 \end{vmatrix} = -6$$

```
[A]
    [[-1  -2   3 ]
     [1    1  -5]
     [2    3   4 ]]
det([A])
                12
```

```
[B]
    [[1   -1   3 ]
     [-2   1  -5]
     [3    2   4 ]]
det([B])
                0
```

```
[C]
    [[1   -2  -1]
     [-2   1   1 ]
     [3    3   2 ]]
det([C])
               -6
```

As the final screen in this example shows, the

solution is $x = \dfrac{D_x}{D} = \dfrac{12}{6} = 2$, $y = \dfrac{D_y}{D} = \dfrac{0}{6} = 0$, and

$z = \dfrac{D_z}{D} = \dfrac{-6}{6} = -1$, or $(2, 0, -1)$ in triple form. Check this

solution in the original equations.

```
det([A])/6
               2
det([B])/6
               0
det([C])/6
              -1
```

☑ **A. You've just seen how we can solve a system using determinants and Cramer's rule**

Now try Exercises 15 through 22 ▶

B. Rational Expressions and Partial Fractions

Recall that a rational expression is one of the form $\dfrac{P(x)}{Q(x)}$, where P and Q are polynomials and $Q(x) \neq 0$. The addition of rational expressions is widely taught in courses prior to college algebra, and involves combining two rational expressions into a single term using a common denominator. In some applications of higher mathematics, we seek to reverse this process and *decompose* a rational expression into a sum of its **partial fractions**. To begin, we make the following observations:

1. Consider the sum $\dfrac{7}{x + 2} + \dfrac{5}{x - 3}$, noting both terms are proper fractions (the degree of the numerator is less than the degree of the denominator) and have distinct linear denominators.

$$\frac{7}{x + 2} + \frac{5}{x - 3} = \frac{7(x - 3)}{(x + 2)(x - 3)} + \frac{5(x + 2)}{(x - 3)(x + 2)} \qquad \text{common denominator}$$

$$= \frac{7(x - 3) + 5(x + 2)}{(x + 2)(x - 3)} \qquad \text{combine numerators}$$

$$= \frac{12x - 11}{(x + 2)(x - 3)} \qquad \text{result}$$

Assuming we didn't have the original sum to look at, reversing the process would require us to begin with a **decomposition template** such as

$$\frac{12x - 11}{(x + 2)(x - 3)} = \frac{A}{x + 2} + \frac{B}{x - 3}$$

and solve for the *constants A and B*. We know the numerators must be constant, else the fraction(s) would be improper while the original expression is not.

2. Consider the sum $\dfrac{3}{x - 1} + \dfrac{5}{x^2 - 2x + 1}$, again noting both terms are proper fractions.

$$\frac{3}{x-1} + \frac{5}{x^2-2x+1} = \frac{3}{x-1} + \frac{5}{(x-1)(x-1)} \qquad \text{factor denominators}$$

$$= \frac{3(x-1)}{(x-1)(x-1)} + \frac{5}{(x-1)(x-1)} \qquad \text{common denominator}$$

$$= \frac{(3x-3)+5}{(x-1)(x-1)} \qquad \text{combine numerators}$$

$$= \frac{3x+2}{(x-1)^2} \qquad \text{result}$$

Note that while the new denominator is the repeated factor $(x-1)^2$, *both* $(x-1)$ and $(x-1)^2$ were denominators in the original sum. Assuming we didn't know the original sum, reversing the process would require us to begin with the template

$$\frac{3x+2}{(x-1)^2} = \frac{A}{x-1} + \frac{B}{(x-1)^2}$$

and solve for the constants A and B. As with observation 1, we know the numerator of the first term must be constant. While the second term would still be a proper fraction if the numerator were linear (degree 1), the denominator is a *repeated* linear factor and using a single constant in the numerator of *all such fractions* will ensure we obtain unique values for A and B. In the end, for any repeated linear factor $(ax+b)^n$ in the original denominator, terms of the form $\dfrac{A_1}{ax+b} + \dfrac{A_2}{(ax+b)^2} + \cdots + \dfrac{A_{n-1}}{(ax+b)^{n-1}} +$

$\dfrac{A_n}{(ax+b)^n}$ must appear in the decomposition template, although some of these numerators may turn out to be zero.

EXAMPLE 3 ▶ **Writing the Decomposition Template for Unique and Repeated Linear Factors**

Write the decomposition template for

a. $\dfrac{x-8}{2x^2+5x+3}$ \qquad b. $\dfrac{x+1}{x^2-6x+9}$

Solution ▶ a. Factoring the denominator gives $\dfrac{x-8}{(2x+3)(x+1)}$. With two distinct linear factors in the denominator, the decomposition template is

$$\frac{x-8}{(2x+3)(x+1)} = \frac{A}{2x+3} + \frac{B}{x+1} \qquad \text{decomposition template}$$

b. After factoring we have $\dfrac{x+1}{(x-3)^2}$, and the denominator is a repeated linear factor. Using our previous observations the template would be

$$\frac{x+1}{(x-3)^2} = \frac{A}{x-3} + \frac{B}{(x-3)^2} \qquad \text{decomposition template}$$

Now try Exercises 23 through 28 ▶

When both distinct and repeated linear factors are present in the denominator, the decomposition template maintains the elements illustrated in both observations 1 and 2.

EXAMPLE 4 ▶ **Writing the Decomposition Template for Unique and Repeated Linear Factors**

Write the decomposition template for $\dfrac{x^2 - 4x - 15}{x^3 - 2x^2 + x}$.

Solution ▶ Factoring the denominator gives $\dfrac{x^2 - 4x - 15}{x(x^2 - 2x + 1)}$ or $\dfrac{x^2 - 4x - 15}{x(x - 1)^2}$ after factoring

completely. With a distinct linear factor of x, and the repeated linear factor $(x - 1)^2$, the decomposition template becomes

$$\frac{x^2 - 4x - 15}{x(x - 1)^2} = \frac{A}{x} + \frac{B}{x - 1} + \frac{C}{(x - 1)^2} \qquad \text{decomposition template}$$

Now try Exercises 29 and 30 ▶

To continue our observations,

3. Consider the sum $\dfrac{4}{x} + \dfrac{2x + 3}{x^2 + 1}$, noting the denominator of the first term is linear, while the denominator of the second is an irreducible quadratic.

$$\frac{4}{x} + \frac{2x + 3}{x^2 + 1} = \frac{4(x^2 + 1)}{x(x^2 + 1)} + \frac{(2x + 3)x}{(x^2 + 1)x} \qquad \text{find common denominator}$$

$$= \frac{(4x^2 + 4) + (2x^2 + 3x)}{x(x^2 + 1)} \qquad \text{combine numerators}$$

$$= \frac{6x^2 + 3x + 4}{x(x^2 + 1)} \qquad \text{result}$$

Here, reversing the process would require us to begin with the template

$$\frac{6x^2 + 3x + 4}{x(x^2 + 1)} = \frac{A}{x} + \frac{Bx + C}{x^2 + 1},$$

allowing that the numerator of the second term might be linear since the denominator is quadratic but *not due to a repeated linear factor*.

4. Finally, consider the sum $\dfrac{1}{x^2 + 3} + \dfrac{x - 2}{(x^2 + 3)^2}$, where the denominator of the first term is an irreducible quadratic, with the second being *the same factor* with multiplicity two.

$$\frac{1}{x^2 + 3} + \frac{x - 2}{(x^2 + 3)^2} = \frac{1(x^2 + 3)}{(x^2 + 3)(x^2 + 3)} + \frac{x - 2}{(x^2 + 3)(x^2 + 3)} \qquad \text{common denominator}$$

$$= \frac{(x^2 + 3) + (x - 2)}{(x^2 + 3)(x^2 + 3)} \qquad \text{combine numerators}$$

$$= \frac{x^2 + x + 1}{(x^2 + 3)^2} \qquad \text{result after simplifying}$$

Reversing the process would require us to begin with the template

$$\frac{x^2 + x + 1}{(x^2 + 3)^2} = \frac{Ax + B}{x^2 + 3} + \frac{Cx + D}{(x^2 + 3)^2}$$

allowing that the numerator of either term might be nonconstant for the reasons in observation 3. Similar to our reasoning in observation 2, all powers of a repeated quadratic factor must be present in the template.

When both distinct and repeated factors are present in the denominator, the decomposition template maintains the essential elements determined by observations 1 through 4. Using these observations, we can formulate a general approach to the decomposition template.

> **Decomposition Template for Rational Expressions**
>
> For the rational expression $\dfrac{P(x)}{Q(x)}$ in lowest terms . . .
>
> 1. Factor Q completely into linear factors and irreducible quadratic factors.
> 2. For the linear factors, each distinct linear factor and each power of a repeated linear factor must appear in the decomposition template with a constant numerator.
> 3. For the irreducible quadratic factors, each distinct quadratic factor and each power of a repeated quadratic factor must appear in the decomposition template with a linear numerator.
> 4. If the degree of P is greater than or equal to the degree of Q, find the quotient and remainder using polynomial division. Only the remainder portion need be decomposed into partial fractions.

EXAMPLE 5 ▶ **Writing the Decomposition Template for Linear and Quadratic Factors**

Write the decomposition template for

a. $\dfrac{x^2 + 10x + 1}{(x + 1)(x^2 + 3x + 1)}$ b. $\dfrac{x^2}{(x^2 + 2)^3}$

Solution ▶ a. One factor of the denominator is a distinct linear factor, and the other is an irreducible quadratic. The decomposition template is

$$\frac{x^2 + 10x + 1}{(x + 1)(x^2 + 3x + 1)} = \frac{A}{x + 1} + \frac{Bx + C}{x^2 + 3x + 1} \qquad \text{decomposition template}$$

b. The denominator consists of a repeated, irreducible quadratic factor. Using our previous observations the template would be

$$\frac{x^2}{(x^2 + 2)^3} = \frac{Ax + B}{x^2 + 2} + \frac{Cx + D}{(x^2 + 2)^2} + \frac{Ex + F}{(x^2 + 2)^3} \qquad \text{decomposition template}$$

Now try Exercises 31 and 32 ▶

Once the template is obtained, we multiply both sides of the equation by the factored form of the original denominator and simplify. The resulting equation is an identity—a true statement for all real numbers x, and in many cases the constants A, B, C, and so on can be identified using a choice of **convenient values** for x, as in Example 6.

EXAMPLE 6 ▶ **Decomposing a Rational Expression with Linear Factors**

Decompose the expression $\dfrac{4x + 11}{x^2 + 7x + 10}$ into partial fractions.

Solution ▶ Factoring the denominator gives $\dfrac{4x + 11}{(x + 5)(x + 2)}$, with two distinct linear factors in the denominator. The required template is

$$\frac{4x + 11}{(x + 5)(x + 2)} = \frac{A}{x + 5} + \frac{B}{x + 2} \qquad \text{decomposition template}$$

Multiplying both sides by $(x + 5)(x + 2)$ clears all denominators and yields

$$4x + 11 = A(x + 2) + B(x + 5) \qquad \text{clear denominators}$$

Since the equation must be true for all x, using $x = -5$ will *conveniently* eliminate the term with B, and enable us to solve for A directly:

$$4(-5) + 11 = A(-5 + 2) + B(-5 + 5) \qquad \text{substitute } -5 \text{ for } x$$
$$-20 + 11 = -3A + B(0) \qquad \text{simplify}$$
$$-9 = -3A \qquad \text{term with } B \text{ is eliminated}$$
$$3 = A \qquad \text{solve for } A$$

To find B, we repeat this procedure, using an x-value that *conveniently* eliminates the term with A, namely, $x = -2$.

$$4x + 11 = A(x + 2) + B(x + 5) \qquad \text{original equation}$$
$$4(-2) + 11 = A(-2 + 2) + B(-2 + 5) \qquad \text{substitute } -2 \text{ for } x$$
$$-8 + 11 = A(0) + 3B \qquad \text{simplify}$$
$$3 = 3B \qquad \text{term with } A \text{ is eliminated}$$
$$1 = B \qquad \text{solve for } B$$

With $A = 3$ and $B = 1$, the complete decomposition is

$$\frac{4x + 11}{(x + 5)(x + 2)} = \frac{3}{x + 5} + \frac{1}{x + 2}$$

which can be checked by adding the rational expressions on the right.

Now try Exercises 33 through 38 ▶

EXAMPLE 7 ▶ **Decomposing a Rational Expression with Repeated Linear Factors**

Decompose the expression $\dfrac{9}{(x + 5)(x^2 + 7x + 10)}$ into partial fractions.

Solution ▶ Factoring the denominator gives $\dfrac{9}{(x + 5)(x + 2)(x + 5)} = \dfrac{9}{(x + 2)(x + 5)^2}$
(one distinct linear factor, one repeated linear factor). The decomposition template
is $\dfrac{9}{(x + 2)(x + 5)^2} = \dfrac{A}{x + 2} + \dfrac{B}{x + 5} + \dfrac{C}{(x + 5)^2}$. Multiplying both sides by
$(x + 2)(x + 5)^2$ clears all denominators and yields

$$9 = A(x + 5)^2 + B(x + 2)(x + 5) + C(x + 2).$$

Using $x = -5$ will eliminate the terms with A and B, giving

$$9 = A(-5 + 5)^2 + B(-5 + 2)(-5 + 5) + C(-5 + 2) \qquad \text{substitute } -5 \text{ for } x$$
$$9 = A(0) + B(-3)(0) - 3C \qquad \text{simplify}$$
$$9 = -3C \qquad \text{terms with } A \text{ and } B \text{ are eliminated}$$
$$-3 = C \qquad \text{solve for } C$$

Using $x = -2$ will eliminate the terms with B and C, and we have

$$9 = A(x + 5)^2 + B(x + 2)(x + 5) + C(x + 2) \qquad \text{original equation}$$
$$9 = A(-2 + 5)^2 + B(-2 + 2)(-2 + 5) + C(-2 + 2) \qquad \text{substitute } -2 \text{ for } x$$
$$9 = A(3)^2 + B(0)(3) + C(0) \qquad \text{simplify}$$
$$9 = 9A \qquad \text{terms with } B \text{ and } C \text{ are eliminated}$$
$$1 = A \qquad \text{solve for } A$$

To find B, we substitute $A = 1$ and $C = -3$ into the original equation, *with any value of x that does not eliminate B.* For efficiency, we'll often use $x = 0$ or $x = 1$ for this purpose (if possible).

$$9 = A(x + 5)^2 + B(x + 2)(x + 5) + C(x + 2)$$ original equation

$$9 = 1(0 + 5)^2 + B(0 + 2)(0 + 5) - 3(0 + 2)$$ substitute 1 for A, -3 for C, 0 for x

$$9 = 25 + 10B - 6$$ simplify

$$-1 = B$$ solve for B

With $A = 1$, $B = -1$, and $C = -3$ the complete decomposition is

$$\frac{9}{(x + 2)(x + 5)^2} = \frac{1}{x + 2} + \frac{-1}{x + 5} + \frac{-3}{(x + 5)^2}$$

$$= \frac{1}{x + 2} - \frac{1}{x + 5} - \frac{3}{(x + 5)^2}$$

Now try Exercises 39 and 40 ▶

As an alternative to using convenient values, a system of equations can be set up by multiplying out the right-hand side (after clearing fractions) and equating coefficients of the terms with like degrees.

EXAMPLE 8 ▶   **Decomposing a Rational Expression with Linear and Quadratic Factors**

Decompose the given expression into partial fractions: $\dfrac{3x^2 - x - 11}{x^3 - 3x^2 + 4x - 12}$.

Solution ▶   A careful inspection indicates the denominator will factor by grouping, giving $x^3 - 3x^2 + 4x - 12 = x^2(x - 3) + 4(x - 3) = (x - 3)(x^2 + 4)$. With one linear factor and one irreducible quadratic factor, the required template is

$$\frac{3x^2 - x - 11}{(x - 3)(x^2 + 4)} = \frac{A}{x - 3} + \frac{Bx + C}{x^2 + 4}$$ decomposition template

$$3x^2 - x - 11 = A(x^2 + 4) + (Bx + C)(x - 3)$$ multiply by $(x - 3)(x^2 + 4)$ (clear denominators)

$$= Ax^2 + 4A + Bx^2 - 3Bx + Cx - 3C$$ distribute/F-O-I-L

$$= (A + B)x^2 + (C - 3B)x + 4A - 3C$$ group and factor

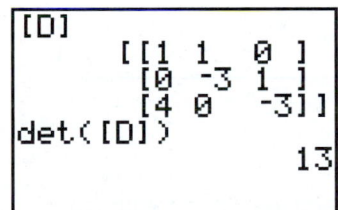

```
[D]
      [[1   1    0 ]
       [0  -3    1 ]
       [4   0   -3]]
det([D])
                 13
```

For the left side to equal the right, we must equate coefficients of terms with like degree: $A + B = 3$, $C - 3B = -1$, and $4A - 3C = -11$. This gives the 3×3 system $\begin{bmatrix} 1 & 1 & 0 & | & 3 \\ 0 & -3 & 1 & | & -1 \\ 4 & 0 & -3 & | & -11 \end{bmatrix}$ in matrix form (verify this). Using the matrix of coefficients we find that $D = 13$ (see figure), and we complete the solution using Cramer's rule.

$$D_A = \begin{vmatrix} 3 & 1 & 0 \\ -1 & -3 & 1 \\ -11 & 0 & -3 \end{vmatrix} = 13 \quad D_B = \begin{vmatrix} 1 & 3 & 0 \\ 0 & -1 & 1 \\ 4 & -11 & -3 \end{vmatrix} = 26 \quad D_C = \begin{vmatrix} 1 & 1 & 3 \\ 0 & -3 & -1 \\ 4 & 0 & -11 \end{vmatrix} = 65$$

The result is $A = \frac{13}{13} = 1$, $B = \frac{26}{13} = 2$, and $C = \frac{65}{13} = 5$, giving the decomposition

$$\frac{3x^2 - x - 11}{(x - 3)(x^2 + 4)} = \frac{1}{x - 3} + \frac{2x + 5}{x^2 + 4}.$$

Now try Exercises 41 through 46 ▶

In some cases, the "convenient values" method cannot be applied and a system of equations is our *only* option. Also, if the decomposition template produces a large or cumbersome system, a graphing calculator can assist the solution process using a matrix equation or the **"rref("** feature. **See Exercises 47 and 48.**

As a final reminder, if the degree of the numerator is *greater than* the degree of the denominator, divide using long division and apply the preceding methods to the remainder

B. You've just seen how we can decompose a rational expression into partial fractions

polynomial. For instance, you can check that $\dfrac{3x^3 + 6x^2 + 5x - 7}{x^2 + 2x + 1} = 3x + \dfrac{2x - 7}{(x + 1)^2}$, and

decomposing the remainder polynomial gives a final result of $3x + \dfrac{2}{x + 1} - \dfrac{9}{(x + 1)^2}$.

C. Determinants, Geometry, and the Coordinate Plane

As mentioned in the introduction, the use of determinants extends far beyond solving systems of equations. Here, we'll demonstrate how determinants can be used to find the area of a triangle whose vertices are given as three points in the coordinate plane.

The Area of a Triangle in the *xy*-Plane

Given a triangle with vertices at (x_1, y_1), (x_2, y_2), and (x_3, y_3),

$$\text{Area} = \left| \frac{\det(T)}{2} \right| \quad \text{where } T = \begin{bmatrix} x_1 & y_1 & 1 \\ x_2 & y_2 & 1 \\ x_3 & y_3 & 1 \end{bmatrix}$$

EXAMPLE 9 ▶ **Finding the Area of a Triangle Using Determinants**

Find the area of a triangle with vertices at $(3, 1)$, $(-2, 3)$, and $(1, 7)$ (see Figure 7.25).

Solution ▶ Begin by forming matrix T and computing $\det(T)$ (see Figure 7.26):

Figure 7.25

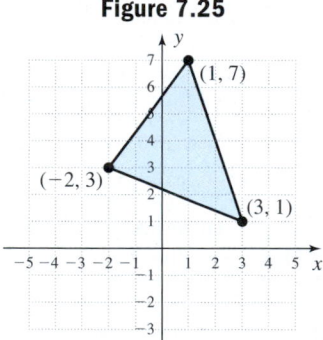

$$\det(T) = \begin{vmatrix} x_1 & y_1 & 1 \\ x_2 & y_2 & 1 \\ x_3 & y_3 & 1 \end{vmatrix} = \begin{vmatrix} 3 & 1 & 1 \\ -2 & 3 & 1 \\ 1 & 7 & 1 \end{vmatrix}$$

$$= 3(3 - 7) - 1(-2 - 1) + 1(-14 - 3)$$

$$= -12 + 3 + (-17) = -26$$

Compute the area: $A = \left| \dfrac{\det(T)}{2} \right| = \left| \dfrac{-26}{2} \right|$

$$= 13$$

The area of this triangle is 13 units^2.

Figure 7.26
T = [A]

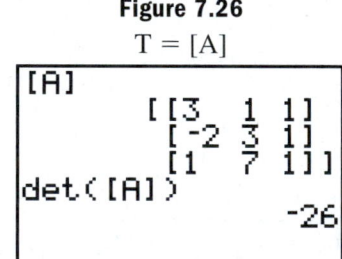

Now try Exercises 51 through 56 ▶

As an extension of this formula, what if the three points were collinear? After a moment, it may occur to you that the formula would give an area of 0 units^2, since no triangle could be formed. This gives rise to a **test for collinear points.**

Test for Collinear Points

Three points (x_1, y_1), (x_2, y_2), and (x_3, y_3) are collinear if

$$\det(A) = \begin{vmatrix} x_1 & y_1 & 1 \\ x_2 & y_2 & 1 \\ x_3 & y_3 & 1 \end{vmatrix} = 0.$$

C. You've just seen how we can use determinants in applications involving geometry in the coordinate plane

See Exercises 57 through 62. There are a variety of additional applications in the Exercise Set. See Exercises 63 through 68.

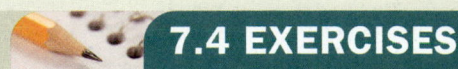

7.4 EXERCISES

▶ CONCEPTS AND VOCABULARY

Fill in the blank with the appropriate word or phrase. Carefully reread the section if needed.

1. The determinant $\begin{vmatrix} a_{11} & a_{12} \\ a_{21} & a_{22} \end{vmatrix}$ is evaluated

 as: $\underline{a_{11}a_{22} - a_{21}a_{12}}$.

2. $\underline{\text{Cramer's}}$ rule uses a ratio of determinants to solve for the unknowns in a system.

3. Given the matrix of coefficients D, the matrix D_x is formed by replacing the coefficients of x with the $\underline{\text{constant}}$ terms.

4. The three points (x_1, y_1), (x_2, y_2), and (x_3, y_3) are collinear if $|T| = \begin{vmatrix} x_1 & y_1 & 1 \\ x_2 & y_2 & 1 \\ x_3 & y_3 & 1 \end{vmatrix}$ has a value of $\underline{0}$.

5. Discuss/Explain the process of writing $\dfrac{8x - 3}{x^2 - x}$ as a sum of partial fractions.   Answers will vary.

6. Discuss/Explain why Cramer's rule cannot be applied if $D = 0$. Use an example to illustrate.   Answers will vary.

▶ DEVELOPING YOUR SKILLS

Write the determinants D, D_x, and D_y for the systems given, but do not solve.

7. $\begin{cases} 2x + 5y = 7 \\ -3x + 4y = 1 \end{cases}$

8. $\begin{cases} -x + 5y = 12 \\ 3x - 2y = -8 \end{cases}$

11. $\begin{cases} \dfrac{x}{8} + \dfrac{y}{4} = 1 \\ \dfrac{y}{5} = \dfrac{x}{2} + 6 \end{cases}$ $\left(\dfrac{-26}{3}, \dfrac{25}{3}\right)$ 12. $\begin{cases} \dfrac{2}{3}x - \dfrac{3}{8}y = \dfrac{7}{5} \\ \dfrac{5}{6}x + \dfrac{3}{4}y = \dfrac{11}{10} \end{cases}$ $\left(\dfrac{9}{5}, \dfrac{-8}{15}\right)$

Solve each system of equations using Cramer's rule, if possible. Do not use a calculator.

9. $\begin{cases} 4x + y = -11 \\ 3x - 5y = -60 \end{cases}$ $(-5, 9)$

10. $\begin{cases} x = -2y - 11 \\ y = 2x - 13 \end{cases}$ $(3, -7)$

13. $\begin{cases} 0.6x - 0.3y = 8 \\ 0.8x - 0.4y = -3 \end{cases}$ no solution

14. $\begin{cases} -2.5x + 6y = -1.5 \\ 0.5x - 1.2y = 3.6 \end{cases}$ no solution

The two systems given in Exercises 15 and 16 are identical except for the third equation. For the first system given, (a) write the determinants D, D_x, D_y, and D_z then (b) determine if a solution using Cramer's rule is possible by computing $|D|$ without the use of a calculator (do not solve the system). Then (c) compute $|D|$ for the second system and try to determine how the equations in the second system are related.

15. $\begin{cases} 4x - y + 2z = -5 \\ -3x + 2y - z = 8 \\ x - 5y + 3z = -3 \end{cases}$, $\begin{cases} 4x - y + 2z = -5 \\ -3x + 2y - z = 8 \\ x + y + z = 3 \end{cases}$

16. $\begin{cases} 2x + 3z = -2 \\ -x + 5y + z = 12 \\ 3x - 2y + z = -8 \end{cases}$, $\begin{cases} 2x + 3z = -2 \\ -x + 5y + z = 12 \\ x + 5y + 4z = 10 \end{cases}$

Use Cramer's rule to solve each system of equations. Verify computations using a graphing calculator.

17. $\begin{cases} x + 2y + 5z = 10 \\ 3x + 4y - z = 10 \\ x - y - z = -2 \end{cases}$ $(1, 2, 1)$

18. $\begin{cases} x + 3y + 5z = 6 \\ 2x - 4y + 6z = 14 \\ 9x - 6y + 3z = 3 \end{cases}$ $(-1, -1, 2)$

19. $\begin{cases} y + 2z = 1 \\ 4x - 5y + 8z = -8 \\ 8x - 9z = 9 \end{cases}$

20. $\begin{cases} x + 2y + 5z = 10 \\ 3x - z = 8 \\ -y - z = -3 \end{cases}$ $\left(\dfrac{14}{5}, \dfrac{13}{5}, \dfrac{2}{5}\right)$

21. $\begin{cases} w + 2x - 3y = -8 \\ x - 3y + 5z = -22 \\ 4w - 5x = 5 \\ -y + 3z = -11 \end{cases}$ $(0, -1, 2, -3)$

22. $\begin{cases} w - 2x + 3y - z = 11 \\ 3w - 2y + 6z = -13 \\ 2x + 4y - 5z = 16 \\ 3x - 4z = 5 \end{cases}$ $(1, -1, 2, -2)$

Additional answers can be found in the Instructor Answer Appendix.

HOMEWORK SELECTION GUIDE

Core: 7–15 odd, 23, 27, 31, 33, 35, 49, 51,53, 55, 63, 65, 67, 74 (18 Exercises)
Standard: 1–4, 7–15 odd, 23, 27, 31, 33, 35, 37, 41, 49, 51, 53, 55, 63, 65–67, 73, 74 (26 Exercises)

Extended: 1–4, 7–15 odd, 23, 27, 31, 33, 35, 37, 41, 43, 45, 49, 51, 53, 55, 63, 65–68, 70, 72–74 (31 Exercises)
In Depth: 1–6, 7–15 odd, 23, 27, 31, 33, 35, 37, 41, 43, 45, 49, 51, 53, 55, 63, 65–68, 70–74 (34 Exercises)

▶ DECOMPOSITION OF RATIONAL EXPRESSIONS

Exercises 23 through 32 are designed solely to reinforce the various possibilities for decomposing a rational expression. All are proper fractions whose denominators are completely factored. Set up the decomposition template using appropriate numerators, but *do not solve*.

23. $\dfrac{3x + 2}{(x + 3)(x - 2)}$

24. $\dfrac{-4x + 1}{(x - 2)(x - 5)}$

25. $\dfrac{3x^2 - 2x + 5}{(x - 1)(x + 2)(x - 3)}$

26. $\dfrac{-2x^2 + 3x - 4}{(x + 3)(x + 1)(x - 2)}$

27. $\dfrac{x^2 + 5}{x(x - 3)(x + 1)}$

28. $\dfrac{x^2 - 7}{(x + 4)(x - 2)x}$

29. $\dfrac{x^2 + x - 1}{x^2(x + 2)}$

30. $\dfrac{x^2 - 3x + 5}{(x - 3)(x + 2)^2}$

31. $\dfrac{x^3 + 3x - 2}{(x + 1)(x^2 + 2)^2}$

32. $\dfrac{2x^3 + 3x^2 - 4x + 1}{x(x^2 + 3)^2}$

Decompose each rational expression into partial fractions.

33. $\dfrac{4 - x}{x^2 + x}$ $\dfrac{4}{x} - \dfrac{5}{x + 1}$

34. $\dfrac{3x + 13}{x^2 + 5x + 6}$ $\dfrac{7}{x + 2} - \dfrac{4}{x + 3}$

35. $\dfrac{2x - 27}{2x^2 + x - 15}$ $\dfrac{-4}{2x - 5} + \dfrac{3}{x + 3}$

36. $\dfrac{-11x + 6}{5x^2 - 4x - 12}$ $\dfrac{5x + 6}{} - \dfrac{1}{x - 2}$

37. $\dfrac{8x^2 - 3x - 7}{x^3 - x}$

38. $\dfrac{x^2 + 24x - 12}{x^3 - 4x}$

39. $\dfrac{3x^2 + 7x - 1}{x^3 + 2x^2 + x}$

40. $\dfrac{-2x^2 - 7x + 28}{x^3 - 4x^2 + 4x}$

41. $\dfrac{3x^2 + 10x + 4}{8 - x^3}$

42. $\dfrac{3x^2 + 4x - 1}{x^3 - 1}$

43. $\dfrac{6x^2 + x + 13}{x^3 + 2x^2 + 3x + 6}$

44. $\dfrac{2x^2 - 14x - 7}{x^3 - 2x^2 + 5x - 10}$

45. $\dfrac{x^4 - x^2 - 2x + 1}{x^5 + 2x^3 + x}$ $\dfrac{1}{x} - \dfrac{3x + 2}{(x^2 + 1)^2}$

46. $\dfrac{-3x^4 + 13x^2 + x - 12}{x^5 + 4x^3 + 4x}$ $\dfrac{-3}{x} + \dfrac{25x + 1}{(x^2 + 2)^2}$

47. $\dfrac{x^3 - 17x^2 + 76x - 98}{(x^2 - 6x + 9)(x^2 - 2x - 3)}$ $\dfrac{3}{x + 1} - \dfrac{2}{x - 3} + \dfrac{1}{(x - 3)^3}$

48. $\dfrac{16x^3 - 66x^2 + 98x - 54}{(2x^2 - 3x)(4x^2 - 12x + 9)}$ $\dfrac{2}{x} + \dfrac{3}{(2x - 3)^2} - \dfrac{1}{(2x - 3)^3}$

▶ WORKING WITH FORMULAS

Area of a Norman window: $A = \begin{vmatrix} L & r^2 \\ -\dfrac{\pi}{2} & W \end{vmatrix}$. The determinant shown can be used to find the area of a Norman

window (rectangle + half-circle) with length L, width W, and radius $r = \dfrac{W}{2}$. Find the area of the following windows.

49.

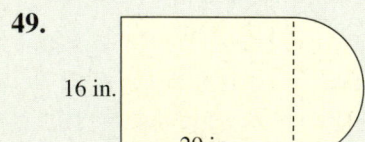

16 in.

20 in.

$320 + 32\pi \approx 420.5$ in^2

50.

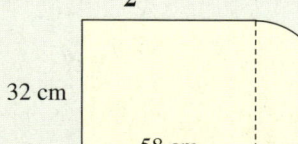

32 cm

58 cm

$1856 + 128\pi \approx 2258.1$ cm^2

▶ APPLICATIONS

Geometric Applications

Find the area of the triangle with the vertices given. Assume units are in centimeters.

51. $(2, 1)$, $(3, 7)$, and $(5, 3)$ 8 cm^2

52. $(-2, 3)$, $(-3, -4)$, and $(-6, 1)$ 13 cm^2

Find the area of the parallelogram with vertices given. Assume units are in feet.

53. $(-4, 2)$, $(-6, -1)$, $(3, -1)$, and $(5, 2)$ 27 ft^2

54. $(-5, -6)$, $(5, 0)$, $(5, 4)$, and $(-5, -2)$ 40 ft^2

Additional answers can be found in the Instructor Answer Appendix.

The volume of a triangular pyramid is given by the formula $V = \frac{1}{3}Bh$, where B represents the area of the triangular base and h is the height of the pyramid. **Find the volume of a triangular pyramid whose height is given and whose base has the coordinates shown. Assume units are in meters.**

55. $h = 6$ m; vertices $(3, 5)$, $(-4, 2)$, and $(-1, 6)$ 19 m³   **56.** $h = 7.5$ m; vertices $(-2, 3)$, $(-3, -4)$, and $(-6, 1)$

32.5 m³

Determine if the following sets of points are collinear.

57. $(1, 5)$, $(-2, -1)$, and $(4, 11)$ yes   **58.** $(1, 1)$, $(3, -5)$, and $(-2, 9)$ no

59. $(-2.5, 5.2)$, $(1.2, -5.6)$, and $(2.2, -8.5)$ no   **60.** $(-0.5, 2.55)$, $(-2.8, 1.63)$, and $(3, 3.95)$ yes

For each linear equation given, substitute the first two points to verify they are solutions. Then use the test for collinear points to determine if the third point is also a solution.

61. $2x - 3y = 7$; $(2, -1)$, $(-1.3, -3.2)$, $(-3.1, -4.4)$ **62.** $5x + 2y = 4$; $(2, -3)$, $(3.5, -6.75)$, $(-2.7, 8.75)$

yes, yes, yes   yes, yes, yes

Write a linear system that models each application. Then solve using Cramer's rule.

63. Return on investments: If $15,000 is invested at a certain interest rate and $25,000 is invested at another interest rate, the total return was $2900. If the investments were reversed the return would be $2700. What was the interest rate paid on each investment?

64. Cost of fruit: Many years ago, two pounds of apples, 2 lb of kiwi, and 10 lb of pears cost $3.26. Three pounds of apples, 2 lb of kiwi, and 7 lb of pears cost $2.98. Two pounds of apples, 3 lb of kiwi, and 6 lb of pears cost $2.89. Find the cost of a pound of each fruit.

65. Forces on trusses of a roof: Triangular trusses have been used for decades in the construction of homes, bridges, tower supports, and other projects. If we consider a very simple truss in the form of an equilateral triangle, the forces exerted along the rafters of the truss by a weight at the apex can be modeled by a 2 × 2 system of linear equations. If a 180-lb carpenter is working at the center of this truss, the forces along each rafter can be modeled by the system shown. Find the force along each rafter.

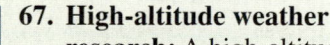

$60°$

F_1 F_2

$60°$ $60°$

$$\begin{cases} \dfrac{\sqrt{3}}{2}(F_1 + F_2) = 180 \\ F_1 - F_2 = 0 \end{cases}$$

$F_1 = F_2 = 60\sqrt{3}$
≈ 103.9 lb

66. Dietary research for pets: As part of a research project, a college student is mixing a special diet for pet mice from two available sources. The diet must offer exactly 22.8 g of protein and 5 g of fat. Given the protein and fat values for the food sources shown, how much of each should be used?   source 1: 40 g, source 2: 65 g

	Source 1	Source 2
protein value	0.18	0.24
fat value	0.06	0.04

67. High-altitude weather research: A high-altitude weather balloon carrying a heavy payload has suddenly ruptured and is plummeting back to Earth. Using an onboard altimeter, the payload radios its height in feet every 2 sec after rupture. For data of the form (time in seconds, height in feet), three of the readings are $(5, 9600)$, $(10, 8400)$, and $(15, 6400)$. (a) Use these data to find an equation of the form $h = at^2 + bt + c$ that models the height of the balloon at any time t. (b) At what height did the balloon rupture? (c) What is the altitude of the balloon after 20 sec? (d) How many seconds until the payload hits the ground?

a. $h = 10{,}000 - 16t^2$ b. 10,000 ft   c. 3600 ft   d. 25 sec

68. Manufacturing surfboards: Australian Waterglide is a manufacturer of custom surfboards for beginners, recreational

surfers, and surfers participating in international competitions. For each board, production is handled in three stages: forming, fiberglass, and finishing.

The number of hours required for each stage are given in the table. If the company has 80 labor hours per week available for forming, 152 hr available for fiberglass, and 145 hr available for finishing, how many boards of each type should be made?

7 beginner, 11 recreational, 3 competition

	Beginner	Recreational	Competition
forming	3	4	5
fiberglass	4.5	8.5	9
finishing	5.5	7.5	8

▶ EXTENDING THE CONCEPT

69. Find the area of the pentagon whose vertices are: $(-5, -5)$, $(5, -5)$, $(8, 6)$, $(-8, 6)$, and $(0, 12.5)$.

$A = 195$ units^2

70. The polynomial form for the equation of a circle is $x^2 + y^2 + Dx + Ey + F = 0$. Find the equation of the circle that contains the points $(-1, 7)$, $(2, 8)$, and $(5, -1)$. $x^2 + y^2 - 4x - 6y - 12 = 0$

▶ MAINTAINING YOUR SKILLS

71. (4.3) Graph the polynomial using information about end-behavior, y-intercept, x-intercept(s), and midinterval points: $f(x) = x^3 - 2x^2 - 7x + 6$.

72. (2.2) Which is the graph (left or right) of $g(x) = -|x + 1| + 3$? Justify your answer.

left graph; graph shifts left 1

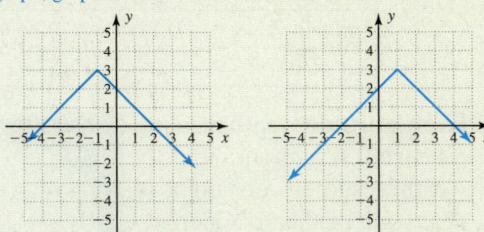

73. (5.3/5.5) Solve the equation $3^{2x-1} = 9^{2-x}$ two ways. First using logarithms, then by equating the bases and using properties of equality.

$3^{2x-1} = 3^{4-2x}$; $x = 1.25$

74. (4.3) Which is the graph (left or right) of a degree 3 polynomial? Justify your answer.

left graph; right shows four roots

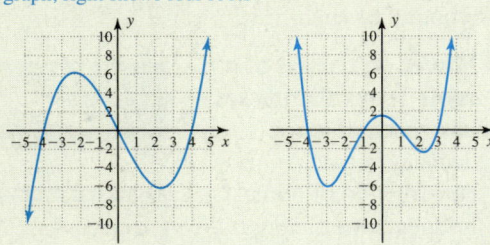

Additional answers can be found in the Instructor Answer Appendix.

7.5   Matrix Applications and Technology Use

LEARNING OBJECTIVES

In Section 7.5 you will see how we can:

☐ **A.** Use matrix equations to solve static systems

☐ **B.** Use matrices for encryption/decryption

Most of the skills needed for this study have been presented in previous sections. Here we'll use various types of regression, combined with systems of equations, to solve practical applications from business and industry.

A. Solving Static Systems with Varying Constraints

When the considerations of a business or industry involve more than two variables, solutions using matrix methods have a distinct advantage over other methods. Companies often have to perform calculations using basic systems weekly, daily, or even hourly, to keep up with trends, market changes, changes in cost of raw materials, and so on. In many situations, the basic requirements remain the same, but the frequently changing inputs require a recalculation each time they change.

EXAMPLE 1 ▶ **Determining Supply Inventories Using Matrices**

BNN Soft Drinks receives new orders daily for its most popular drink, Saratoga Cola. It can deliver the carbonated beverage in a twelve-pack of 12-ounce (oz) cans, a six-pack of 20-oz bottles, or in a 2-L bottle. The ingredients required to produce a twelve-pack include 1 gallon (gal) of carbonated water, 1.25 pounds (lb) of sugar, 2 cups (c) of flavoring, and 0.5 grams (g) of caffeine. For the six-pack, 0.8 gal of carbonated water, 1 lb of sugar, 1.6 c of flavoring, and 0.4 g of caffeine are needed. The 2-L bottle contains 0.47 gal of carbonated water, 0.59 lb of sugar, 0.94 c of flavoring, and 0.24 g of caffeine. How much of each ingredient must be on hand for Monday's order of 300 twelve-packs, 200 six-packs, and 500 2-L bottles? What quantities must be on hand for Tuesday's order: 410 twelve-packs, 320 six-packs, and 275 2-L bottles?

Solution ▶ Begin by setting up a general system of equations, letting x represent the number of twelve-packs, y the number of six-packs, and z the number of 2-L bottles:

$$\begin{cases} 1x + 0.8y + 0.47z = \text{gallons of carbonated water} \\ 1.25x + 1y + 0.59z = \text{pounds of sugar} \\ 2x + 1.6y + 0.94z = \text{cups of flavoring} \\ 0.5x + 0.4y + 0.24z = \text{grams of caffeine} \end{cases}$$

As a matrix equation we have

$$\begin{bmatrix} 1 & 0.8 & 0.47 \\ 1.25 & 1 & 0.59 \\ 2 & 1.6 & 0.94 \\ 0.5 & 0.4 & 0.24 \end{bmatrix} \begin{bmatrix} x \\ y \\ z \end{bmatrix} = \begin{bmatrix} w \\ s \\ f \\ c \end{bmatrix}$$

Enter the 4×3 matrix as matrix A, and the size of the order as matrix as B. Using a calculator, we find

$$AB = \begin{bmatrix} 1 & 0.8 & 0.47 \\ 1.25 & 1 & 0.59 \\ 2 & 1.6 & 0.94 \\ 0.5 & 0.4 & 0.24 \end{bmatrix} \begin{bmatrix} 300 \\ 200 \\ 500 \end{bmatrix} = \begin{bmatrix} 695 \\ 870 \\ 1390 \\ 350 \end{bmatrix},$$

```
[A] [B]
       [[695 ]
        [870 ]
        [1390]
        [350 ]]
```

and BNN Soft Drinks will need 695 gal of carbonated water, 870 lb of sugar, 1390 c of flavoring, and 350 g of caffeine for Monday's order. After entering $C = \begin{bmatrix} 410 \\ 320 \\ 275 \end{bmatrix}$

for Tuesday's orders, computing the product AC shows 795.25 gal of carbonated water, 994.75 lb of sugar, 1590.5 c of flavoring, and 399 g of caffeine are needed for Tuesday.

Now try Exercises 1 through 6 ▶

Example 1 showed how the creation of a static matrix can help track and control inventory requirements. In Example 2, we use a static matrix to solve a system that will identify the amount of data traffic used by a company during various hours of the day.

EXAMPLE 2 ▶ **Identifying the Source of Data Traffic Using Matrices**

Mariño Imports is a medium-size company that is considering upgrading from a 1.544 megabytes per sec (Mbps) T1 Internet line to a fractional T3 line with a bandwidth of 7.72 Mbps. They currently use their bandwidth for phone traffic, office data, and Internet commerce. The IT (Information Technology) director devises a plan to monitor how much data traffic each resource uses on an hourly basis. Because of the physical arrangement of the hardware, she cannot monitor each resource individually. The table shows the information she collected for the first 3 hr. Determine how many gigabytes (GB) each resource used individually during these 3 hr.

	Phone, Data, and Commerce	Phone and Data	Data and Commerce
9–10:00 A.M.	5.4 GB	4.0 GB	4.2 GB
10–11:00 A.M.	5.3 GB	3.8 GB	4.2 GB
11–12:00 P.M.	5.1 GB	3.5 GB	3.6 GB

Solution ▶ Using p to represent the phone traffic, d for office data, and c for Internet commerce, we create the system shown, which models data use for the 9:00 o'clock hour. Since we actually need to solve two more systems whose only difference is the constant terms (for the 10 and 11 o'clock hours), using a matrix equation to solve the system (Section 7.3) will be most convenient. Begin by writing the related matrix equation for this system:

$$\begin{cases} p + d + c = 5.4 \\ p + d = 4.0 \\ d + c = 4.2 \end{cases}$$

$$AX = B: \begin{bmatrix} 1 & 1 & 1 \\ 1 & 1 & 0 \\ 0 & 1 & 1 \end{bmatrix} \begin{bmatrix} p \\ d \\ c \end{bmatrix} = \begin{bmatrix} 5.4 \\ 4.0 \\ 4.2 \end{bmatrix}$$

[A]⁻¹[B]
[[1.2]
[2.8]
[1.4]]

Using $X = A^{-1}B$ to solve the system (see figure), we find there was 1.2 GB of phone traffic, 2.8 GB of office data, and 1.4 GB of Internet commerce during this hour. Note that the IT director may make this calculation 10 or more times a day (once for every hour of business). While we *could* solve for the 10 and 11 o'clock hours using $C = \begin{bmatrix} 5.3 \\ 3.8 \\ 4.2 \end{bmatrix}$ and $D = \begin{bmatrix} \mathbf{5.1} \\ \mathbf{3.5} \\ \mathbf{3.6} \end{bmatrix}$, then calculate $A^{-1}C$ and $A^{-1}D$, these calculations *can all be performed simultaneously* by combining the matrices B, C, and D into one 3 × 3 matrix and multiplying by A^{-1}. Due to the properties of matrix multiplication, each column of the product will represent the data information for a given hour, as shown here:

$$[A^{-1}] \begin{bmatrix} 5.4 & 5.3 & \mathbf{5.1} \\ 4.0 & 3.8 & \mathbf{3.5} \\ 4.2 & 4.2 & \mathbf{3.6} \end{bmatrix} = \begin{bmatrix} 1.2 & 1.1 & 1.5 \\ 2.8 & 2.7 & 2.0 \\ 1.4 & 1.5 & 1.6 \end{bmatrix}$$

In the second hour, there was 1.1 GB of phone traffic, 2.7 GB of office data, and 1.5 GB of Internet commerce. In the third hour, 1.5 GB of phone traffic, 2.0 GB of office data, and 1.6 GB of Internet commerce bandwidth was used.

✓ **A.** You've just seen how we can use matrix equations to solve static systems

Now try Exercises 7 through 10 ▶

B. Using Matrices to Encrypt Messages

In *The Gold-Bug*, by Edgar Allan Poe, the narrator deciphers the secret message on a treasure map by accounting for the frequencies of specific letters and words. The coding used was a simple substitution cipher, which today can be broken with relative ease. In modern times where information wields great power, a simple cipher like the one used there will not suffice. When you pay your tuition, register for classes, make on-line purchases, and so on, you are publicly transmitting very private data, which needs to be protected. While there are many complex encryption methods available (including the now famous symmetric-key and public-key techniques), we will use a matrix-based technique. The nature of matrix multiplication makes it very difficult to determine the exact matrices that yield a given product, and we'll use this fact to our advantage. Beginning with a fixed, invertible matrix A, we will develop a matrix B such that the product AB is possible, and our secret message is encrypted in AB. At the receiving end, they will need to know A^{-1} to decipher the message, since $A^{-1}(AB) = (A^{-1}A)B = B$, which is the original message. Note that in case an intruder were to find matrix A (perhaps purchasing the information from a disgruntled employee), we must be able to change it easily. This means we should develop a method for generating a matrix A, with integer entries, where A is invertible and A^{-1} also has integer entries.

EXAMPLE 3 ▶ **Finding an Invertible Matrix A Where Both A and A^{-1} Have Integer Entries**

Find an invertible 3×3 matrix A as just described, and its inverse A^{-1}.

Solution ▶ Begin with any 3×3 matrix that has only 1s or -1s on its main diagonal, and 0s below the diagonal. The upper triangle can consist of any integer values you choose, as in

$$\begin{bmatrix} -1 & 5 & -1 \\ 0 & 1 & 8 \\ 0 & 0 & 1 \end{bmatrix}.$$

WORTHY OF NOTE

Performing row operations is explained in more detail in the graphing calculator manual accompanying this text.

Now, use **any** of the elementary row operations to make the matrix more complex. For instance, we'll use a calculator to create a new matrix by (1) using R1 + R2 → R2 to create matrix [C], then (2) using R1 + R3 → R3 to create matrix [D], then (3) using R3 + R2 → R2 to create matrix [E], and finally (4) −2 R1 + R3 → R3 to obtain our final matrix [A]. To begin, enter the initial matrix as matrix [B]. For (1) R1 + R2 → R2, go to the (MATRIX) **MATH** submenu, select option **D:row+(** and press ⏎ to bring this option to the home screen. This feature requires us to name the matrix we're using, and to indicate what rows to add, so we enter **D:row+([B], 1, 2).** The screen shown in Figure 7.27 indicates we've placed the result in matrix [C]. For R1 + R3 → R3, recall **D:row+([B], 1, 2)** using 2nd ⏎ and change it to **D:row+([C], 1, 3)** STO▸ [D] (Figure 7.28). Repeat this process for (3) R3 + R2 → R2 to create matrix [E]: **D:row+([D], 3, 2)** STO▸ [E] (Figure 7.29). Finally, we compute (4) −2R1 + R3 → R3 using the (new) option **F:*row+(−2, [E], 1, 3)** STO▸[A] and the process is complete (Figure 7.30). The entries of matrix A are all integers, and A^{-1} exists and also has integer entries (Figure 7.31). This will always be the case for matrices created in this way.

Figure 7.27

```
row+([B],1,2)→[C
]
         [[-1 5 -1]
          [-1 6 7 ]
          [0  0 1 ]]
```

Figure 7.28

```
row+([C],1,3)→[D
]
         [[-1 5 -1]
          [-1 6 7 ]
          [-1 5 0 ]]
```

Figure 7.29

```
row+([D],3,2)→[E
]
         [[-1 5  -1]
          [-2 11 7 ]
          [-1 5  0 ]]
```

Figure 7.30

```
*row+(-2,[E],1,3
)→[A]
         [[-1 5  -1]
          [-2 11 7 ]
          [1  -5 2 ]]
```

Figure 7.31

```
[A]⁻¹
         [[-57 5 -46]
          [-11 1 -9 ]
          [1   0 1  ]]
```

Now try Exercises 11 through 16 ▶

EXAMPLE 4 ▶ **Using Matrices to Encrypt Messages**

Set up a substitution cipher to encode the message MATH IS SWEET, and then use the matrix A from Example 3 to encrypt it.

Solution ▶ For the cipher, we will associate a unique number to every letter in the alphabet. This can be done randomly or using a systematic approach. Here we choose to associate 0 with a blank space, and assign 1 to A, -1 to B, 2 to C, -2 to D, and so on.

Blank	A	B	C	D	E	F	G	H	I	J	K	L	M
0	1	-1	2	-2	3	-3	4	-4	5	-5	6	-6	7

N	O	P	Q	R	S	T	U	V	W	X	Y	Z
-7	8	-8	9	-9	10	-10	11	-11	12	-12	13	-13

Now encode the secret message as shown:

M	A	T	H		I	S		S	W	E	E	T
7	1	-10	-4	0	5	10	0	10	12	3	3	-10

We next enter the coded message into a new matrix B, by entering it letter by letter *into the columns* of B. Note that since the encrypting matrix A is 3×3, B must have 3 rows for multiplication to be possible. The result is

$$B = \begin{bmatrix} M & H & S & W & T \\ A & * & * & E & * \\ T & I & S & E & * \end{bmatrix} = \begin{bmatrix} 7 & -4 & 10 & 12 & -10 \\ 1 & 0 & 0 & 3 & 0 \\ -10 & 5 & 10 & 3 & 0 \end{bmatrix}$$

Since the message is too short to fill matrix B, we use blank spaces to complete the final column. Computing the product AB encrypts the message, and only someone with access to A^{-1} will be able to read it:

$$AB = \begin{bmatrix} -1 & 5 & -1 \\ -2 & 11 & 7 \\ 1 & -5 & 2 \end{bmatrix} \begin{bmatrix} 7 & -4 & 10 & 12 & -10 \\ 1 & 0 & 0 & 3 & 0 \\ -10 & 5 & 10 & 3 & 0 \end{bmatrix}$$

$$= \begin{bmatrix} 8 & -1 & -20 & 0 & 10 \\ -73 & 43 & 50 & 30 & 20 \\ -18 & 6 & 30 & 3 & -10 \end{bmatrix}$$

The encrypted message is 8, -73, -18, -1, 43, 6, -20, 50, 30, 0, 30, 3, 10, 20, -10.

Now try Exercises 17 through 22 ▶

EXAMPLE 5 ▶ **Deciphering Encrypted Messages Using an Inverse Matrix**

Decipher the encrypted message from Example 4 using A^{-1} from Example 3.

Solution ▶ The received message is 8, -73, -18, -1, 43, 6, -20, 50, 30, 0, 30, 3, 10, 20, -10, and is the result of the product AB. To find matrix B, we apply A^{-1} since $A^{-1}(AB) = (A^{-1}A)B = B$. Writing the received message in matrix form we have

$$AB = \begin{bmatrix} 8 & -1 & -20 & 0 & 10 \\ -73 & 43 & 50 & 30 & 20 \\ -18 & 6 & 30 & 3 & -10 \end{bmatrix}$$

Next multiply AB by A^{-1} on the left, to determine matrix B:

$$A^{-1}(AB) = \begin{bmatrix} -57 & 5 & -46 \\ -11 & 1 & -9 \\ 1 & 0 & 1 \end{bmatrix} \begin{bmatrix} 8 & -1 & -20 & 0 & 10 \\ -73 & 43 & 50 & 30 & 20 \\ -18 & 6 & 30 & 3 & -10 \end{bmatrix}$$

$$= \begin{bmatrix} 7 & -4 & 10 & 12 & -10 \\ 1 & 0 & 0 & 3 & 0 \\ -10 & 5 & 10 & 3 & 0 \end{bmatrix} = B$$

Writing matrix B in sentence form gives 7, 1, -10, -4, 0, 5, 10, 0, 10, 12, 3, 3, -10, 0, 0, and using the substitution cipher to replace numbers with letters, reveals the message MATH IS SWEET.

 **B.** You've just seen how we can use matrices for encryption/decryption

Now try Exercises 23 through 28 ▶

7.5 EXERCISES

1. Slammin' Drums manufactures several different types of drums. Its most popular drums are the 22″ bass drum, the 12″ tom, and the 14″ snare drum. The 22″ bass drum requires 7 ft² of skin, 8.5 ft² of wood veneer, 8 tension rods, and 11.5 ft of hoop. The 12″ tom requires 2 ft² of skin, 3 ft² of wood veneer, 6 tension rods, and 6.5 ft of hoop. The 14″ snare requires 2.5 ft² of skin, 1.5 ft² of wood veneer, 10 tension rods, and 7 ft of hoop. In February, Slammin' Drums received orders for 15 bass drums, 21 toms, and 27 snares. Use your calculator and a matrix equation to determine how much of each raw material they need to have on hand to fill these orders.
214.5 ft² of skin, 231.0 ft² of wood veneer, 516 tension rods, and 498 ft of hoop

2. In March, Slammin' Drums' orders consisted of 19 bass drums, 19 toms, and 25 snares. Use your calculator and a matrix equation to determine how much of each raw material they need to have on hand to fill their orders. (See Exercise 1.)
233.5 ft² of skin, 256 ft² of wood veneer, 516 tension rods, and 517 ft of hoop

3. The following table represents Slammin's orders for the months of April through July. Use your calculator and a matrix equation to determine how much of each raw material they need to have on hand to fill these orders. (See Exercise 1.) (*Hint:* Using a clever 4 × 3 and 3 × 1 matrix can reduce this problem to a single step.)

	April	May	June	July
Bass drum	23	21	17	14
Tom	20	18	15	17
Snare drum	29	35	27	25

955 ft² of skin, 1021.5 ft² of wood veneer, 2180 tension rods, and 2129.5 ft of hoop

4. The following table represents Slammin's orders for the months of August through November. Use your calculator and a matrix equation to determine how much of each raw material they need to have on hand to fill their orders. (See Exercise 1.) (*Hint:* Using a clever 4 × 3 and 3 × 1 matrix can reduce this problem to a single step.)

	August	September	October	November
Bass drum	17	22	16	12
Tom	15	14	13	11
Snare drum	32	28	27	21

845 ft² of skin, 890.5 ft² of wood veneer, 1934 tension rods, and 1871 ft of hoop

5. Midwest Petroleum (MP) produces three types of combustibles using common refined gasoline and vegetable products. The first is E10 (also known as gasohol), the second is E85, and the third is biodiesel. One gallon of E10 requires 0.90 gal of gasoline, 2 lb of corn, 1 oz of yeast, and 0.5 gal of water. One gallon of E85 requires 0.15 gal of gasoline, 17 lb of corn, 8.5 oz of yeast, and 4.25gal of water. One gallon of biodiesel requires 20 lb of corn and 3 gal of water. One week's production at MP consisted of 100,000 gal of E10, 15,000 gal of E85, and 7000 gal of biodiesel. Use your calculator and a matrix equation to determine how much of each raw material they used to fill their orders.
92,250 gal gasoline, 595,000 lb corn, 227,500 oz yeast, and 134,750 gal water

HOMEWORK SELECTION GUIDE

Core: 1–7 odd, 13, 19 (6 Exercises)
Standard: 1, 2, 3, 5, 7, 13, 19 (7 Exercises)

Extended: 1–5, 7, 13, 19 (8 Exercises)
In Depth: 1–7, 13, 19 (9 Exercises)

6. The following table represents Midwest Petroleum's production for the next 3 weeks. Use your calculator and a matrix equation to determine the total amount of raw material they need to fill their orders (see Exercise 5). 287,250 gal gasoline, 2,035,000 lb corn, 777,500 oz yeast, and 460,750 gal water

	Week 2	Week 3	Week 4
E10	110,000	95,000	105,000
E85	17,000	18,000	20,000
Biodiesel	6,000	8,000	10,000

7. Roll-X Watches makes some of the finest wristwatches in the world. Their most popular model is the Clam. It comes in three versions: Silver, Gold, and Platinum. Management thinks there might be a thief in the production line, so they decide to closely monitor the precious metal consumption. A Silver Clam contains 1.2 oz of silver and 0.2 oz of gold. A Gold Clam contains 0.5 oz of silver, 0.8 oz of gold, and 0.1 oz of platinum. A Platinum Clam contains 0.2 oz of silver, 0.5 oz of gold, and 0.7 oz of platinum. During the first week of monitoring, the production team used 10.9 oz of silver, 9.2 oz of gold, and 2.3 oz of platinum. Use your calculator and a matrix equation to determine the number of each type of watch that should have been produced. 5 Silver, 9 Gold, and 2 Platinum

8. The following table contains the precious metal consumption of the Roll-X Watch production line during the next five weeks (see Exercise 7). Use your graphing calculator to determine the number of each type of watch that should have been produced each week. For which week does the data seem to indicate a possible theft of precious metal?

Ounces	Week 1	Week 2	Week 3	Week 4	Week 5
Silver	13.1	9	12.9	11.9	11.2
Gold	11	7.7	8.6	8.4	9.5
Platinum	2.5	1.5	0.9	2.8	1.7

9. There are three classes of grain, of which three bundles from the first class, two from the second, and one from the third make 39 measures. Two of the first, three of the second, and one of the third make 34 measures. And one of the first, two of the second, and three of the third make 26 measures. How many measures of grain are contained in one bundle of each class? (*This is the historic problem from the Chiu chang suan shu.*)

10. During a given week, the measures of grain that make up the bundles in Exercise 9 can vary slightly. Three local Chinese bakeries always buy the same numbers of bundles, as outlined in Exercise 9. That is to say, bakery 1 buys three bundles of the first class, two of the second, and one of the third. Bakery 2 buys two of the first, three of the second, and one of the third. And finally, bakery 3 buys one of the first, two of the second, and three of the third. The following table outlines how many measures of grain each bakery received each day. How many measures of grain were contained in one bundle of each class, on each day?

	Mon	Tues	Wed	Thurs	Fri
Bakery 1 (measures)	39	38	38	37.75	39.75
Bakery 2 (measures)	34	33	33.5	32.5	35
Bakery 3 (measures)	26	26	27	26.25	27.25

For Exercises 11–16, use the criteria indicated to find 3×3 matrices A and A^{-1}, where the entries of both are all integers.

11. The lower triangle is all zeroes.   Answers will vary.

12. The upper triangle is all zeroes.   Answers will vary.

13. $a_{2,1} = 5$ Answers will vary.

14. $a_{3,2} = -2$ Answers will vary.

15. $a_{3,1} = 1$ and $a_{2,3} = 2$ Answers will vary.

16. $a_{2,1} = -3$ and $a_{1,3} = 1$ Answers will vary.

17. Use the matrix A you created in Exercise 11 and the substitution cipher from Example 4 to encrypt your full name.   Answers will vary.

18. Use the matrix A you created in Exercise 12 and the substitution cipher from Example 4 to encrypt your school's name.   Answers will vary.

19. Design your own substitution cipher. Then use it and the matrix A you created in Exercise 13 to encrypt the title of your favorite movie. Answers will vary.

20. Design your own substitution cipher. Then use it and the matrix A you created in Exercise 14 to encrypt the title of your favorite snack food. Answers will vary.

21. Design your own substitution cipher. Then use it and the matrix A you created in Exercise 15 to encrypt the White House switchboard phone number, 202-456-1414.   Answers will vary.

22. Design your own substitution cipher. Then use it and the matrix A you created in Exercise 16 to encrypt the Casa Rosada switchboard phone number 54-11-4344-3600. The Casa Rosada, or Pink House, consists of the offices of the president of Argentina.   Answers will vary.

23. Use the matrix A^{-1} from Exercise 11, and the appropriate substitution cipher to decrypt the message from Exercise 17.   Answers will vary.

24. Use the matrix A^{-1} from Exercise 12, and the appropriate substitution cipher to decrypt the message from Exercise 18.   Answers will vary.

25. Use the matrix A^{-1} from Exercise 13, and the appropriate substitution cipher to decrypt the message from Exercise 19.   Answers will vary.

26. Use the matrix A^{-1} from Exercise 14, and the appropriate substitution cipher to decrypt the message from Exercise 20.   Answers will vary.

27. Use the matrix A^{-1} from Exercise 15, and the appropriate substitution cipher to decrypt the message from Exercise 21.   Answers will vary.

28. Use the matrix A^{-1} from Exercise 16, and the appropriate substitution cipher to decrypt the message from Exercise 22.   Answers will vary.

MAKING CONNECTIONS

Making Connections: Graphically, Symbolically, Numerically, and Verbally

Eight matrices A through H are given. Use a graphing calculator to help match the characteristics or operations indicated in 1 through 16 to one of the eight matrices. In some cases, the response requires two matrices.

$$A = \begin{bmatrix} 3 & -2 \\ 1 & 4 \end{bmatrix} \qquad B = \begin{bmatrix} -2 & 3 \\ 2 & -4 \end{bmatrix} \qquad C = \begin{bmatrix} 4 & 0 & -2 \\ 1 & -3 & 5 \end{bmatrix} \qquad D = \begin{bmatrix} -2 & 5 \\ 1 & 4 \\ 0 & 3 \end{bmatrix}$$

$$E = \begin{bmatrix} 0 & 3 & 2 \\ -2 & 4 & -1 \\ 1 & 5 & -6 \end{bmatrix} \qquad F = \begin{bmatrix} 1 & 0 & 0 \\ 0 & 1 & 0 \\ 0 & 0 & 1 \end{bmatrix} \qquad G = \begin{bmatrix} 1 & 3 & 0 \\ -3 & 4 & -1 \\ -2 & 7 & -1 \end{bmatrix} \qquad H = \begin{bmatrix} 3 & -1 & | & 1 & 0 \\ -5 & 2 & | & 0 & 1 \end{bmatrix}$$

1. __G__ 3 × 3, noninvertible

2. __F__ determinant is 1

3. __D__ entry $a_{3,2}$ is 3

4. __A, B__ the sum is $\begin{bmatrix} 1 & 1 \\ 3 & 0 \end{bmatrix}$

5. __A__ determinant is 14

6. __B__ matrix squared is $\begin{bmatrix} 10 & -18 \\ -12 & 22 \end{bmatrix}$

7. __B__ matrix inverse is $\begin{bmatrix} -2 & -1.5 \\ -1 & -1 \end{bmatrix}$

8. __G__ entry $a_{3,1}$ is −2

9. __CD__ the product is $\begin{bmatrix} -8 & 14 \\ -5 & 8 \end{bmatrix}$

10. __DC__ the product is $\begin{bmatrix} -3 & -15 & 29 \\ 8 & -12 & 18 \\ 3 & -9 & 15 \end{bmatrix}$

11. __E__ determinant is −67

12. __G__ determinant is 0

13. __D__ 3 × 2 matrix

14. __C__ 2 × 3 matrix

15. __H__ augmented matrix

16. __F__ identity matrix

SUMMARY AND CONCEPT REVIEW

SECTION 7.1   Solving Linear Systems Using Matrices and Row Operations

KEY CONCEPTS

- A *matrix* is a rectangular arrangement of numbers. An $m \times n$ matrix has m rows and n columns.
- An *augmented matrix* is derived from a system of linear equations by augmenting the *coefficient matrix* (formed by the variable coefficients) with the *matrix of constants*.
- One matrix method for solving systems of equations is by triangularizing the augmented matrix.
- An inconsistent system with no solutions will yield a contradictory statement such as $0 = 1$. A dependent system with infinitely many solutions will yield an identity statement such as $0 = 0$.

EXERCISES

1. Write an example of the following matrices:

   a. 2×3 **b.** 3×2 **c.** 3×3, in triangular form   Answers will vary.

Solve by triangularizing the augmented matrix. If the system is linearly dependent, state the answer using a parameter. Use a calculator for Exercise 5.

2. $\begin{cases} x - 2y = 6 \\ 4x - 3y = 4 \end{cases}$
$(-2, -4)$

3. $\begin{cases} x - 2y + 2z = 7 \\ 2x + 2y - z = 5 \\ 3x - y + z = 6 \end{cases}$
$(1, 6, 9)$

4. $\begin{cases} 2x - y + 2z = -1 \\ x + 2y + 2z = -3 \\ 3x - 4y + 2z = 1 \end{cases}$
$(-\frac{6}{5}p - 1, -\frac{2}{5}p - 1, p)$

5. $\begin{cases} 2w + x + 2y - 3z = -19 \\ w - 2x - y + 4z = 15 \\ x + 2y - z = 1 \\ 3w - 2x - 5z = -60 \end{cases}$
$(-2, 7, 1, 8)$

SECTION 7.2   The Algebra of Matrices

KEY CONCEPTS

- The entries of a matrix are denoted a_{ij}, where i gives the row and j gives the column of its location.
- Two matrices A and B of equal size (or *order*) are equal if corresponding entries are equal.
- The sum or difference of two matrices of equal order is found by combining corresponding entries: $A + B = [a_{ij} + b_{ij}]$
- The *identity matrix for addition* is an $m \times n$ matrix whose entries are all zeroes.
- To perform *scalar multiplication,* take the product of the constant with each entry in the matrix, forming a new matrix of like size. For matrix A: $kA = [ka_{ij}]$.
- *Matrix multiplication* is performed as row entry $\times$ column entry. For an $m \times n$ matrix $A = [a_{ij}]$ and an $s \times t$ matrix $B = [b_{ij}]$, AB is possible if $n = s$. The result will be an $m \times t$ matrix $P = [p_{ij}]$, where p_{ij} is the product of the ith row of A with the jth column of B.
- When technology is used to perform operations on matrices, carefully enter each matrix into the calculator. Then double check that each entry is correct and appraise the results to see if they are reasonable.

EXERCISES

Compute the operations indicated below (if possible), using the following matrices.

$$A = \begin{bmatrix} \frac{-1}{4} & \frac{-3}{4} \\ \frac{-1}{8} & \frac{-7}{8} \end{bmatrix} \qquad B = \begin{bmatrix} -7 & 6 \\ 1 & -2 \end{bmatrix} \qquad C = \begin{bmatrix} -1 & 3 & 4 \\ 5 & -2 & 0 \\ 6 & -3 & 2 \end{bmatrix} \qquad D = \begin{bmatrix} 2 & -3 & 0 \\ 0.5 & 1 & -1 \\ 4 & 0.1 & 5 \end{bmatrix}$$

6. $A + B$ **7.** $B - A$ **8.** $C - B$ **9.** $8A$ **10.** BA

11. $C + D$ **12.** $D - C$ **13.** BC **14.** $-4D$ **15.** CD

Additional answers can be found in the Instructor Answer Appendix.

SECTION 7.3 Solving Linear Systems Using Matrix Equations

KEY CONCEPTS

- The *identity matrix for multiplication I*, has 1's on the main diagonal and 0's for all other entries. For any $n \times n$ matrix A, the identity matrix is also an $n \times n$ matrix I_n, where $AI_n = I_n A = A$.
- For an $n \times n$ (square) matrix A, the *inverse matrix* for multiplication is a matrix B such that $AB = BA = I_n$. For matrix A the inverse is denoted A^{-1}. Only square matrices have inverses.
- Any $n \times n$ system of equations can be written as a matrix equation and solved (if a unique solution exists) using an inverse matrix. The system

$$\begin{cases} 2x + 3y = 7 \\ x - 4y = -2 \end{cases} \text{ is written as } \begin{bmatrix} 2 & 3 \\ 1 & -4 \end{bmatrix} \begin{bmatrix} x \\ y \end{bmatrix} = \begin{bmatrix} 7 \\ -2 \end{bmatrix}.$$

- Every square matrix has a real number associated with it, called its *determinant*. For 2×2 matrix $A = \begin{bmatrix} a_{11} & a_{12} \\ a_{21} & a_{22} \end{bmatrix}$, $\det(A) = a_{11}a_{22} - a_{21}a_{12}$.
- If the determinant of a matrix is zero, the matrix is said to be *singular* or *noninvertible*. If the coefficient matrix of a matrix equation is noninvertible, the system is either inconsistent or dependent.

EXERCISES

Complete Exercises 16 through 18 using the following matrices:

$$A = \begin{bmatrix} 1 & 0 \\ 0 & 1 \end{bmatrix} \qquad B = \begin{bmatrix} 0.2 & 0.2 \\ -0.6 & 0.4 \end{bmatrix} \qquad C = \begin{bmatrix} 2 & -1 \\ 3 & 1 \end{bmatrix} \qquad D = \begin{bmatrix} 10 & -6 \\ -15 & 9 \end{bmatrix}$$

16. Exactly one of the matrices given is singular. Compute each determinant to identify it.   *D*

17. Show that $AB = BA = B$. What can you conclude about matrix A?   It's an identity matrix.

18. Show that $BC = CB = I$. What can you conclude about matrix C?   It's the inverse of *B*.

Use a graphing calculator to complete Exercises 19 through 21, using the matrices given:

$$E = \begin{bmatrix} 1 & -2 & 3 \\ -2 & 1 & -5 \\ -1 & -1 & -2 \end{bmatrix} \qquad F = \begin{bmatrix} 1 & -1 & 1 \\ 0 & 1 & 0 \\ -2 & 1 & -1 \end{bmatrix} \qquad G = \begin{bmatrix} -1 & 0 & -1 \\ 0 & 1 & 0 \\ 2 & 1 & 1 \end{bmatrix}$$

19. Exactly one of the matrices is singular. Determine which one.   *E*

20. Compute the products FG and GF. What can you conclude about matrix G?   It is the inverse of *F*.

21. Verify that $EG \neq GE$ and $EF \neq FE$. What can you conclude?   Matrix multiplication is not generally commutative.

Solve manually using a matrix equation.

22. $\begin{cases} 2x - 5y = 14 \\ -3y + 4x = -14 \end{cases}$ $(-8, -6)$

Solve using a matrix equation and your calculator.

23. $\begin{cases} 0.5x - 2.2y + 3z = -8 \\ -0.6x - y + 2z = -7.2 \\ x + 1.5y - 0.2z = 2.6 \end{cases}$ $(2, 0, -3)$

SECTION 7.4 Applications of Matrices and Determinants: Cramer's Rule, Partial Fractions, and More

KEY CONCEPTS

- Cramer's rule uses a ratio of determinants to solve systems of equations (if solutions exist).
- The determinant of the 2×2 matrix $\begin{vmatrix} a_{11} & a_{12} \\ a_{21} & a_{22} \end{vmatrix}$ is $a_{11}a_{22} - a_{21}a_{12}$.
- To compute the value of 3×3 and larger determinants, a calculator is generally used.
- Determinants can be used to find the area of a triangle in the plane if the vertices of the triangle are known, and as a test to see if three points are collinear.
- A system of equations can be used to write a rational expression as a sum of its partial fractions.

EXERCISES

Solve using Cramer's rule. Use a graphing calculator for Exercise 26.

24. $\begin{cases} 5x + 6y = 8 \\ 10x - 2y = -9 \end{cases}$ $\left(\dfrac{-19}{35}, \dfrac{25}{14}\right)$ **25.** $\begin{cases} 2x + y - z = -1 \\ x - 2y + z = 5 \\ 3x - y + 2z = 8 \end{cases}$ $(1, -1, 2)$  **26.** $\begin{cases} 2x + y = -2 \\ -x + y + 5z = 12 \\ 3x - 2y + z = -8 \end{cases}$ $\left(\dfrac{-37}{19}, \dfrac{36}{19}, \dfrac{31}{19}\right)$

27. Find the area of a triangle whose vertices have the coordinates $(6, 1)$, $(-1, -6)$, and $(-6, 2)$. $\dfrac{91}{2}$ units^2

28. Find the partial fraction decomposition for $\dfrac{7x^2 - 5x + 17}{x^3 - 2x^2 + 3x - 6}$. $\dfrac{5}{x - 2} + \dfrac{2x - 1}{x^2 + 3}$

SECTION 7.5   Matrix Applications and Technology Use

KEY CONCEPTS

- In studies involving numerous constraints, many variables, and/or large amounts of data, matrix methods have a distinct advantage over other solution methods.
- Square matrices with integer coefficients rarely have an inverse that also has integer coefficients. However these can be carefully constructed.
- A matrix [A] can be used to encode a written message, with the matrix $[A]^{-1}$ being used to decode it.

 **29.** J.P. Sailing and Co. hand builds two types of classic sailing boats made entirely of wood. Both the 13-foot Laser class dinghy and 25-foot Bermuda sloop yacht are constructed primarily with three types of wood: Douglas fir, Brazilian jequitiba, and, of course, Asian teak. Each dingy is built using 120 ft^2 of fir, 132 ft^2 of jequitiba, and 50 ft^2 of teak. The construction process for the yacht uses 270 ft^2 of fir, 260 ft^2 of jequitiba, and 108 ft^2 of teak. In their first year of production, J.P. Sailing built 11 dinghies and 5 yachts. For the following 4 yr, their dingy production increased by 3 per year, while the yacht production increased by 2 per year. Use your calculator and a matrix equation to determine how much of each type of wood they used during each of their first five years of production as well as the 5-yr totals.

30. In addition to being one of the world's first celebrity chefs, Marie-Antoine Carême (1784–1833) was a master pastry chef. Had he used matrices to encode his favorite culinary math secret, perhaps his message would have been:

$$11, 10, 25, -25, -6, -59, 33, 13, 76, 34, 43, 69, -3, 0, -6$$

If Carême used the cipher on page 696 (Example 4) and the matrix $A = \begin{bmatrix} 1 & 2 & 3 \\ 1 & 3 & 1 \\ 2 & 4 & 7 \end{bmatrix}$ for the encoding process, decode the message to find the unconventional culinary secret that is actually familiar to students of mathematics.   PIE ARE SQUARE

PRACTICE TEST

Solve each system by triangularizing the augmented matrix and using back-substitution.

1. $\begin{cases} 3x + 8y = -5 \\ x + 10y = 2 \end{cases}$ $\left(-3, \dfrac{1}{2}\right)$

2. $\begin{cases} 3x - y + 5z = 1 \\ 3x + y + 4z = 4 \\ x + y + z = \frac{7}{3} \end{cases}$ $\left(-3p + \dfrac{16}{3}, p, 2p - 3\right)$

3. $\begin{cases} 4x - 5y - 6z = 5 \\ 2x - 3y + 3z = 0 \\ x + 2y - 3z = 5 \end{cases}$ $\left(2, 1, \dfrac{-1}{3}\right)$

4. Given matrices A and B, compute:

   a. $A - B$ **b.** $\dfrac{2}{5}B$ **c.** AB **d.** A^{-1} **e.** $|A|$

$$A = \begin{bmatrix} -3 & -2 \\ 5 & 4 \end{bmatrix} \qquad B = \begin{bmatrix} 3 & 3 \\ -3 & -5 \end{bmatrix}$$

5. Given matrices C and D, use a calculator to find:
  a. $C - D$ **b.** $-0.6D$ **c.** DC **d.** D^{-1} **e.** $|D|$

$$C = \begin{bmatrix} 0.5 & 0 & 0.2 \\ 0.4 & -0.5 & 0 \\ 0.1 & -0.4 & -0.1 \end{bmatrix}$$

$$D = \begin{bmatrix} 0.5 & 0.1 & 0.2 \\ -0.1 & 0.1 & 0 \\ 0.3 & 0.4 & 0.8 \end{bmatrix}$$

6. Use matrices to find three different solutions of the dependent system:

$$\begin{cases} 2x - y + z = 4 \\ 3x - 2y + 4z = 9 \\ x - 2y + 8z = 11 \end{cases}$$

$(-1, -6, 0), (1, -1, 1), (3, 4, 2),$ answers vary as $(2p - 1, 5p - 6, p)$

7. Solve using Cramer's rule: $\begin{cases} 2x - 3y = 2 \\ x - 6y = -2 \end{cases}$ $\left(2, \dfrac{2}{3}\right)$

8. Solve using a calculator and Cramer's rule:

$$\begin{cases} 2x + 3y + z = 3 \\ x - 2y - z = 4 \\ x - y - 2z = -1 \end{cases}$$ $(3, -2, 3)$

9. Solve using a matrix equation and your calculator:

$$\begin{cases} 2x - 5y = 11 \\ 4x + 7y = 4 \end{cases}$$ $\left(\dfrac{97}{34}, \dfrac{-18}{17}\right)$

10. Solve using a matrix equation and your calculator:

$$\begin{cases} x - 2y + 2z = 7 \\ 2x + 2y - z = 5 \\ 3x - y + z = 6 \end{cases}$$ $(1, 6, 9)$

11. Use the equality of matrices to write and solve a system that gives values of x, y, and z so that $A = B$.

$$A = \begin{bmatrix} 2x + y & 3 \\ x + z & 3x + 2z \end{bmatrix} \text{ and }$$

$$B = \begin{bmatrix} z - 1 & 3 \\ 2y + 5 & y + 8 \end{bmatrix}$$ $x = 1, y = -1, z = 2$

12. Given matrix X is a solution to $AX = B$ for the matrix A given, find matrix B.

$$X = \begin{bmatrix} -1 \\ \frac{-3}{2} \\ 2 \end{bmatrix} \quad A = \begin{bmatrix} 1 & -2 & 2 \\ 2 & -6 & 3 \\ 3 & 4 & -1 \end{bmatrix} \quad B = \begin{bmatrix} 6 \\ 13 \\ -11 \end{bmatrix}$$

13. Use matrices and a graphing calculator to determine which three of the following four points are collinear: $(-1, 4), (1, 3), (2, 1), (4, -1)$ $(-1, 4), (2, 1), (4, -1)$

14. A farmer plants a triangular field with wheat. The first vertex of the triangular field is 1 mi east and 1 mi north of his house. The second vertex is 3 mi east and 1 mi south of his house. The third vertex is 1 mi west and 2 mi south of his house. What is the area of the field? 5 mi^2

15. For $A = \begin{bmatrix} r & 2 \\ 3 & s \end{bmatrix}$ and $A^2 = \begin{bmatrix} 10 & -2 \\ -3 & 7 \end{bmatrix}$ given, find r and s. $r = -2, s = 1$

Create a system of equations to model each exercise, then solve using any matrix method.

16. Dr. Brown and Dr. Stamper graduate from medical school with $155,000 worth of student loans. Due to her state's tuition reimbursement plan, Dr. Brown owes one fourth of what Dr. Stamper owes. How much does each doctor owe?

17. Justin is rehabbing two old houses simultaneously. He calculates that last week he spent 23 hr working on these houses. If he spent 8 more hours on one of the houses, how many hours did he spend on each house? 7.5 hr, 15.5 hr

18. In his first month as assistant principal of Washington High School, Mr. Johnson gave out 20 detentions. They were either for 1 day, 2 days, or 5 days. He recorded a total of 38 days of detention served. He also noted that there were twice as many 2-day detentions as 5-day detentions. How many of each type of detention did Mr. Johnson give out?

19. The city of Cherrywood has approved a $1,800,000 plan to renovate its historic commercial district. The money will be coming from three separate sources. The first is a federal program that charges a low 2% interest annually. The second is a municipal bond offering that will cost 5% annually. The third is a standard loan from a neighborhood bank, but it will cost 8.5% annually. In the first year, the city will not make any repayment on these loans and will accrue $94,500 more debt. The federal program and bank loan together are responsible for $29,500 of this interest. How much money was originally provided by each source?

20. Decompose the expression into partial fractions:

$$\frac{4x^2 - 4x + 3}{x^3 - 27}$$ $\dfrac{1}{x - 3} + \dfrac{3x + 2}{x^2 + 3x + 9}$

CALCULATOR EXPLORATION AND DISCOVERY

Cramer's Rule

In Section 7.4, we saw that one interesting application of matrices is Cramer's rule. You may have noticed that when technology is used with Cramer's rule, the chances of making an error are fairly high, as we need to input the entries for numerous matrices. However, as we mentioned in the chapter introduction, one of the advantages of matrices is that they *are easily programmable,* and we can actually write a very simple program that will make Cramer's rule a more efficient method.

To begin, press the **PRGM** key, and then the right arrow ▷ twice and **ENTER** to create a name for our program. At the prompt, we'll enter **CRAMER2.** As we write the program, note that the needed commands (**ClrHome, Disp, Pause, Prompt, Stop**) are all located in the submenus of the **PRGM** key, and the = sign is found under the **TEST** menu, accessed using the **2nd** **MATH** keys (the arrows "→" are used to indicate the **STO▸** key.

The following program takes the coefficients and constants of a 2 × 2 linear system, and returns the ordered pair solution in the form of $x = h$ and $y = k$ (for constants h and k). Even with minimal programming experience, reading through the program will help you identify that Cramer's rule is being used.

ClrHome	**(CE–BF)/(AE–BD)→X**
Disp "2×2 SYSTEMS"	**(AF–DC)/(AE–BD)→Y**
Pause	**ClrHome**
Disp "AX+BY = C"	**Disp "THE SOLUTION IS"**
Disp "DX+EY = F"	**Disp ""**
Disp ""	**Disp "X="**
Disp "ENTER THE VALUES"	**Disp X**
Disp "FOR A,B,C,D,E,F"	**Disp "Y="**
Disp ""	**Disp Y**
Prompt A,B,C,D,E,F	**Stop**

Exercise 1: Use the program to check the answers to Exercises 1 and 7 of the Practice Test. $\left(3, \frac{1}{2}\right), \left(2, \frac{2}{3}\right)$

Exercise 2: Create 2 × 2 systems of your own that are (a) consistent, (b) inconsistent, and (c) dependent. Then verify results using the program.   Answers will vary.

Exercise 3: Use the box on page 681 of Section 7.4 to write a similar program for 3 × 3 systems. Call the program CRAMER3, and repeat parts (a), (b), and (c) from Exercise 2.   Answers will vary.

STRENGTHENING CORE SKILLS

Augmented Matrices and Matrix Inverses

The formula for finding the inverse of a 2 × 2 matrix has its roots in the more general method of computing the inverse of an $n \times n$ matrix. This involves augmenting a square matrix M with its corresponding identity I_n on the right (forming an $n \times 2n$ matrix), and using row operations to *transform M into the identity.* In some sense, as the original matrix is transformed, the "identity part" keeps track of the operations we used to convert M and we can use the results to "get back home," so to speak. We'll illustrate with the 2 × 2 matrix from Section 7.3, Example 2B, where we found that $\begin{bmatrix} 1 & -2.5 \\ -1 & 3 \end{bmatrix}$ was the inverse matrix for $\begin{bmatrix} 6 & 5 \\ 2 & 2 \end{bmatrix}$. We begin by augmenting $\begin{bmatrix} 6 & 5 \\ 2 & 2 \end{bmatrix}$ with the 2 × 2 identity matrix.

$$\begin{bmatrix} 6 & 5 & | & 1 & 0 \\ 2 & 2 & | & 0 & 1 \end{bmatrix} \xrightarrow{-3R2 + R1 \to R2} \begin{bmatrix} 6 & 5 & | & 1 & 0 \\ 0 & -1 & | & 1 & -3 \end{bmatrix} \xrightarrow{-1R2 \to R2} \begin{bmatrix} 6 & 5 & | & 1 & 0 \\ 0 & 1 & | & -1 & 3 \end{bmatrix}$$

$$\begin{bmatrix} 6 & 5 & | & 1 & 0 \\ 0 & 1 & | & -1 & 3 \end{bmatrix} \xrightarrow{-5R2 + R1 \to R1} \begin{bmatrix} 6 & 0 & | & 6 & -15 \\ 0 & 1 & | & -1 & 3 \end{bmatrix} \xrightarrow{\frac{R1}{6} \to R1} \begin{bmatrix} 1 & 0 & | & 1 & -2.5 \\ 0 & 1 & | & -1 & 3 \end{bmatrix}$$

As you can see, the identity is automatically transformed into the inverse matrix when this method is applied.

Performing similar row operations on the general matrix $\begin{bmatrix} a & b \\ c & d \end{bmatrix}$ results in the formula given earlier.

As you might imagine, attempting this on a general 3×3 matrix is problematic at best, and instead we simply apply the augmented matrix method to find A^{-1} for the 3×3 matrix shown in blue.

$$
\begin{bmatrix}
2 & 1 & 0 & | & 1 & 0 & 0 \\
-1 & 3 & -2 & | & 0 & 1 & 0 \\
3 & -1 & 2 & | & 0 & 0 & 1
\end{bmatrix}
\xrightarrow[-3R1 + 2R3 \to R3]{R1 + 2R2 \to R2}
\begin{bmatrix}
2 & 1 & 0 & | & 1 & 0 & 0 \\
0 & 7 & -4 & | & 1 & 2 & 0 \\
0 & -5 & 4 & | & -3 & 0 & 2
\end{bmatrix}
$$

$$
\xrightarrow[5R2 + 7R3 \to R3]{R2 - 7R1 \to R1}
\begin{bmatrix}
-14 & 0 & -4 & | & -6 & 2 & 0 \\
0 & 7 & -4 & | & 1 & 2 & 0 \\
0 & 0 & 8 & | & -16 & 10 & 14
\end{bmatrix}
\xrightarrow{\dfrac{R3}{8} \to R3}
\begin{bmatrix}
-14 & 0 & -4 & | & -6 & 2 & 0 \\
0 & 7 & -4 & | & 1 & 2 & 0 \\
0 & 0 & 1 & | & -2 & 1.25 & 1.75
\end{bmatrix}
$$

$$
\xrightarrow[4R3 + R1 \to R1]{4R3 + R2 \to R2}
\begin{bmatrix}
-14 & 0 & 0 & | & -14 & 7 & 7 \\
0 & 7 & 0 & | & -7 & 7 & 7 \\
0 & 0 & 1 & | & -2 & 1.25 & 1.75
\end{bmatrix}
\xrightarrow[\dfrac{R2}{7} \to R2]{\dfrac{R1}{-14} \to R1}
\begin{bmatrix}
1 & 0 & 0 & | & 1 & -0.5 & -0.5 \\
0 & 1 & 0 & | & -1 & 1 & 1 \\
0 & 0 & 1 & | & -2 & 1.25 & 1.75
\end{bmatrix}.
$$

To verify, we show $AA^{-1} = I$: $\begin{bmatrix} 2 & 1 & 0 \\ -1 & 3 & -2 \\ 3 & -1 & 2 \end{bmatrix}\begin{bmatrix} 1 & -0.5 & -0.5 \\ -1 & 1 & 1 \\ -2 & 1.25 & 1.75 \end{bmatrix} = \begin{bmatrix} 1 & 0 & 0 \\ 0 & 1 & 0 \\ 0 & 0 & 1 \end{bmatrix}$ ✓ ($A^{-1}A = I$ also checks).

Exercise 1: Use the preceding inverse and a matrix equation to solve the system

$$
\begin{cases}
2x + y = -2 \\
-x + 3y - 2z = -15. \quad (1, -4, 1) \\
3x - y + 2z = 9
\end{cases}
$$

CUMULATIVE REVIEW CHAPTERS R–7

1. Perform the operations indicated.
  a. $(3 - 2i)(2 + i)$ $8 - i$
  b. $(5 - 3i)^2$ $16 - 30i$
  c. $\dfrac{8 - i}{2 + i}$ $3 - 2i$
  d. i^{49} i

2. Solve $S = 2\pi rh + 2\pi r^2$ for h.

Solve the following equations. Verify solutions using a graphing calculator.

3. $2x - 4(3x + 1) = 5 - 4x$ $x = \dfrac{-3}{2}$

4. $\dfrac{x + 6}{x + 2} - \dfrac{1}{x} = \dfrac{12}{x^2 + 2x}$ $x = -7, x = 2$

5. $9x^2 + 1 = 6x$ $x = \dfrac{1}{3}$

6. $\sqrt{2x + 11} - x = 6$ $x = -5$

7. Find an equation of the line that passes through the points $(2, -2)$ and $(-3, 5)$.

8. Find an equation of the line perpendicular to the line shown, with the same y-intercept.

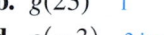

9. Given $f(x) = 3 - 4x - x^2$ and $g(x) = |4 - \sqrt{x + 2}|$, find

  a. $f(3)$ -18 **b.** $g(23)$ 1
  c. $f(-2)$ 7 **d.** $g(-3)$ -3 is not in the domain

Graph the following by using transformations of the parent function.

10. $y = (x - 3)^2$ **11.** $y = \dfrac{1}{2}|x| - 3$

12. $y = -(x + 2)^3$ **13.** $y = \sqrt{-x} - 2$

Solve the following inequalities. Express your answer in interval notation.

14. $2(x - 2) + 3 \leq 8$ **15.** $x^2 + 5 > 6x$

16. A long-distance calling card advertises one of the lowest rates available. It claims to charge 3¢ per minute for a long distance call, after a 15¢ connection fee. The cost of a call is modeled by the ceiling function $c(m) = 9\lceil \frac{1}{3}m \rceil + 15$, where m is the length of the call in minutes and c is the cost of the call in cents. Graph this function and explain why this card may not be as inexpensive as advertised.   It bills in 3-min increments.

Name interval(s) where the following functions are increasing, decreasing, or constant. Write answers using interval notation.

17. $y = f(x)$

18. $g(x) = \dfrac{1}{(x + 3)} + 3$

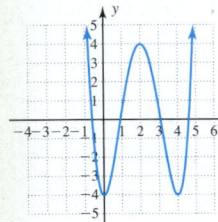
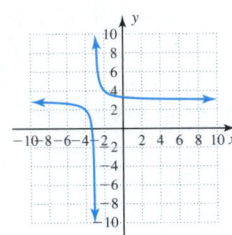

19. Compute the quotient $(4y^2 - 3y + 10) \div (2y + 1)$.

20. Use the remainder theorem to find $P(-3)$ for $P(x) = x^3 + 4x^2 - 15$.

21. Find a cubic polynomial with real coefficients having the roots $x = -2$ and $x = 3 - i$. Assume a lead coefficient of 1.

22. State the end-behavior and y-intercept of the following functions.
   a. $f(x) = 3x^3 - 16$
   b. $g(x) = -5x^6 + 2x^3 - 3x + 7$
   c. $h(x) = 10 - 2x^5$

23. Solve the system of equations using any method.
$$\begin{cases} 5x + 6y = -30 \\ -3x + 2y = 4 \end{cases}$$

24. Solve the system of inequalities by graphing. Verify your solution using a graphing calculator.
$$\begin{cases} y > x^2 - 1 \\ x^2 + y^2 \le 13 \end{cases}$$

25. Simon Legré, a notorious French criminal, is planning to embezzle money from the Prefect of Police, and believes he can remove 12% of their

$5,000,000 budget without being detected. He plans to use his ill-gotten gain to buy a legitimate business, and purchase a condominium and a delivery truck for the business. The business will cost $20,000 more than twice the delivery truck, and the condominium will cost $200,000 more than the truck and business combined. How much will each cost?   condominium: $400,000; delivery truck: $60,000; business: $140,000

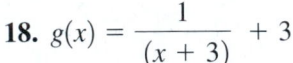 **For Exercises 26 through 30, use of a graphing calculator to solve.**

26. Solve the system using a graphing calculator. Round your answers to two decimal places.
$$\begin{cases} x + y^2 = 10 \\ 3x^2 + 2y^2 = 30 \end{cases}$$ (−1.52, 3.39), (2.19, 2.79), (−1.52, −3.39), (2.19, −2.79)

27. Use a calculator to compute the determinant of each matrix. If the determinant is zero, write *singular matrix*.

a. $\begin{bmatrix} 1 & -5 & 3 & 2 \\ 0 & 3 & 2 & -3 \\ 1 & 0 & 1 & 1 \\ 1 & 0 & 5 & 7 \end{bmatrix}$ −144

b. $\begin{bmatrix} 0 & 1 & -1 & 5 \\ 4 & -1 & 1 & -4 \\ 6 & -2 & 4 & 9 \\ -9 & 7 & 6 & 3 \end{bmatrix}$ 1148

28. In Cuivre River State Park, there are two dominant squirrel species: gray-tailed and red-tailed. The function $g(t) = \ln|t^5 - 3t^4 + 2t^3 - 5t^2 + t - 3|$ models the population of gray-tailed squirrels from 1990 to 2010, while $r(t) = \ln|-2t^3 + 5t^2 - 3t + 1| + 3$ models the red-tailed population during this same 20-yr period. These two functions give the respective number of squirrels (in hundreds) living in this park during year t, where $t = 0$ corresponds to 1990. Using integer-valued years and the **TABLE** feature of your calculator, determine when these two populations were closest in size. What was the difference in populations that year?   1992, four squirrels

29. Determine the intervals over which the function $f(x) = 4x^3 - 4x^2 - 32x + 17$ is increasing. $(-\infty, \frac{-4}{3}] \cup [2, \infty)$

30. Find all solutions (if they exist) using a graphing calculator.
$$\sqrt{\dfrac{x^2}{x + 1}} = e^{\frac{x^2}{x+1}}$$ no solution

8

Analytic Geometry and the Conic Sections

CHAPTER OUTLINE

CHAPTER CONNECTIONS

One of the most breathtaking, daredevil stunts performed at air shows is the *power dive*. In some cases, as the pilot dives toward the ground and pulls out of the dive just in time, the plane flys along a path that can be modeled by a hyperpola (the third member of the family of conic sections). If we consider a given point on the ground and under the grandstand as the origin (0, 0), we can use the equation that models the hyperbolic path to determine its minimum altitude as it passes over the stands. This application appears as Exercise 81 in Section 8.3.

Check out these other real-world connections:

▶ Designing an Elliptical Garden with Fountains (Section 8.2, Exercise 66)

▶ The Design of a Lithotripter for Treating Kidney Stones with Shockwaves (Section 8.2, Exercise 68)

▶ Locating a Ship Using Radar (Section 8.3, Exercise 85)

▶ Parabolic Shape of a Solar Furnace (Section 8.4, Exercise 94)

LEARNING OBJECTIVES

In Section 8.1 you will see how we can:

☐ **A.** Verify theorems from basic geometry involving the distance between two points

☐ **B.** Verify that points (x, y) are an equal distance from a given point and a given line

☐ **C.** Use the defining characteristics of a conic section to find its equation

Generally speaking, **analytical geometry** is a study of geometry using the tools of algebra and a coordinate system. These tools include the midpoint and distance formulas; the algebra of parallel, perpendicular, and intersecting lines; and other tools that help establish geometric concepts. In this section, we'll use these tools to verify certain relationships, then use these relationships to introduce a family of curves known as the **conic sections.**

A. Verifying Relationships from Plane Geometry

For the most part, the algebraic tools used in this study were introduced in previous chapters. As the midpoint and distance formulas play a central role, they are restated here for convenience.

Algebraic Tools Used in Analytical Geometry

Given two points $P_1 = (x_1, y_1)$ and $P_2 = (x_2, y_2)$ in the xy-plane.

Midpoint Formula	**Distance Formula**
The midpoint of line segment P_1P_2 is	The distance from P_1 to P_2 is
$(x, y) = \left(\dfrac{x_1 + x_2}{2}, \dfrac{y_1 + y_2}{2} \right)$	$d = \sqrt{(x_2 - x_1)^2 + (y_2 - y_1)^2}$

These formulas can be used to verify the conclusion of many theorems from Euclidean geometry, while providing important links to an understanding of the conic sections.

EXAMPLE 1 ▶ **Verifying a Theorem from Basic Geometry**

A theorem from basic geometry states: *The midpoint of the hypotenuse of a right triangle is an equal distance from all three vertices.* Verify this statement for the right triangle formed by $(-4, -2)$, $(4, -2)$, and $(4, 4)$.

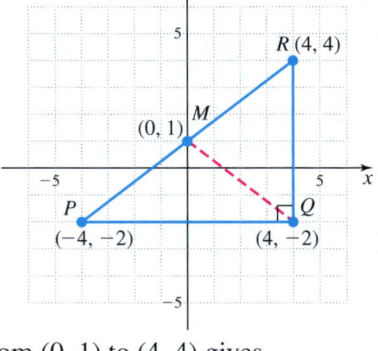

Solution ▶ After the plotting points and drawing a triangle, we note the hypotenuse has endpoints $(-4, -2)$ and $(4, 4)$, with midpoint

$$\left(\frac{4 + (-4)}{2}, \frac{4 + (-2)}{2} \right) = (0, 1).$$

Using the distance formula to find the distance from $(0, 1)$ to $(4, 4)$ gives

$$\begin{aligned} d &= \sqrt{(4 - 0)^2 + (4 - 1)^2} \\ &= \sqrt{(4)^2 + (3)^2} \\ &= \sqrt{25} \\ &= 5 \end{aligned}$$

From the definition of midpoint, $(0, 1)$ is also 5 units from $(-4, -2)$.

Checking the distance from $(0, 1)$ to the vertex $(4, -2)$ gives

$$\begin{aligned} d &= \sqrt{(4 - 0)^2 + (-2 - 1)^2} \\ &= \sqrt{4^2 + (-3)^2} \\ &= \sqrt{25} \\ &= 5 \end{aligned}$$

The midpoint of the hypotenuse *is* an equal distance from all three vertices (see the figure).

Now try Exercises 7 through 12 ▶

Recall from Section 1.1 that a circle is the set of all points that are an equal distance (called the radius) from a given point (called the center). If all three vertices of a triangle lie on the circumference of a circle, we say the circle **circumscribes** the triangle. Based on our earlier work, it appears we could also state the theorem in Example 1 as *For any circle in the xy-plane whose center (h, k) is the midpoint of the hypotenuse L of a right triangle, the circle defined by* $(x - h)^2 + (y - k)^2 = (\frac{L}{2})^2$, *circumscribes the triangle.* The circle and triangle from Example 1 illustrate this theorem in Figure 8.1, where the equation of the circle is $(x - 0)^2 + (y - 1)^2 = (\frac{10}{2})^2$. **See Exercises 13 through 20.**

Figure 8.1

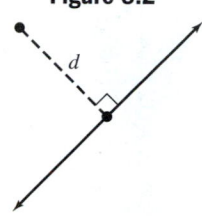

☑ **A.** You've just seen how we can verify theorems from basic geometry involving the distance between two points

Figure 8.2

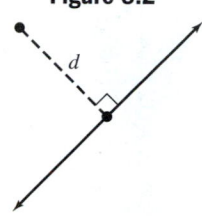

B. The Distance between a Point and a Line

In a study of analytical geometry, we are also interested in the distance d between a point and a *line*. This is always defined as the **perpendicular distance,** or the length of a line segment perpendicular to the given line, with the given point and the point of intersection as endpoints (see Figure 8.2).

EXAMPLE 2 ▶ **Locating Points That Are an Equal Distance from a Given Point and Line**

In Figure 8.3, the origin (0, 0) is seen to be an equal distance from the point (0, 2) and the line $y = -2$. Show that the following points are also an equal distance from (0, 2) and $y = -2$:

 a. $(2, \frac{1}{2})$ **b.** (4, 2)   **c.** (8, 8)

Solution ▶ Since the given line is horizontal, the perpendicular distance from the line to each point can be found by vertically counting the units. It remains to show that this is also the distance from the given point to (0, 2) (see Figure 8.4).

a. The distance from $(2, \frac{1}{2})$ to $y = -2$ is **2.5 units**. The distance from $(2, \frac{1}{2})$ to (0, 2) is

Figure 8.3

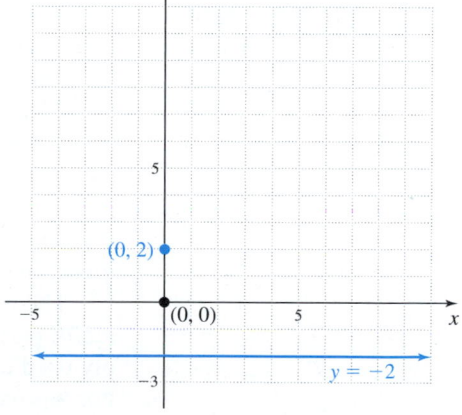

$$d = \sqrt{(0 - 2)^2 + (2 - 0.5)^2}$$
$$= \sqrt{(-2)^2 + (1.5)^2}$$
$$= \sqrt{6.25}$$
$$= 2.5 ✓$$

Figure 8.4

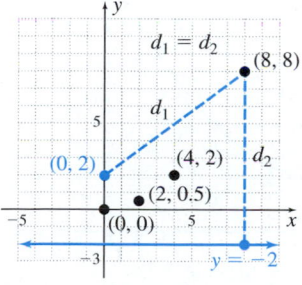

b. The distance from (4, 2) to $y = -2$ is **4 units**. The distance from (4, 2) to (0, 2) is

$$d = \sqrt{(0 - 4)^2 + (2 - 2)^2}$$
$$= \sqrt{(-4)^2 + (0)^2}$$
$$= \sqrt{16}$$
$$= 4 ✓$$

c. The distance d_2 from (8, 8) to $y = -2$ is **10 units**. The distance d_1 from (8, 8) to (0, 2) is

$$d_1 = \sqrt{(0 - 8)^2 + (2 - 8)^2}$$
$$= \sqrt{(-8)^2 + (-6)^2}$$
$$= \sqrt{100}$$
$$= 10 ✓$$

☑ **B.** You've just seen how we can verify that points (x, y) are an equal distance from a given point and a given line

Now try Exercises 23 through 26 ▶

Figure 8.5

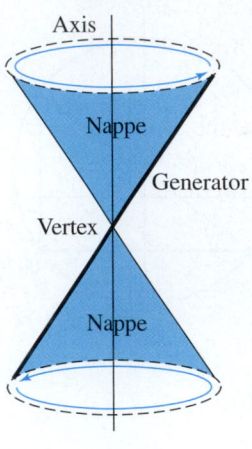

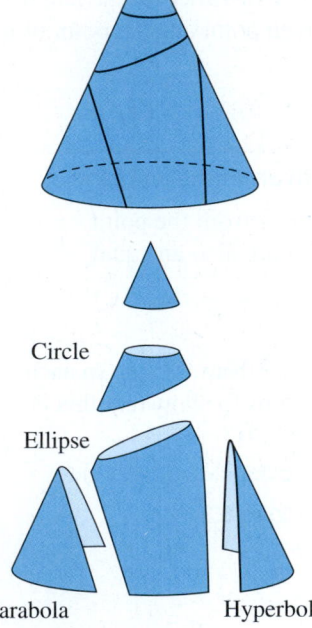

Circle

Ellipse

Parabola   Hyperbola

C. Characteristics of the Conic Sections

Examples 1 and 2 bring us one step closer to the wider application of these ideas in a study of the conic sections. But before the connection is clearly made, we'll introduce some background on this family of curves. In common use, a cone might bring to mind the conical paper cups found at a water cooler. The point of the cone is called the **vertex** and the sheet of paper forming the sides is called a **nappe.** In mathematical terms, a cone has two nappes, formed by rotating a nonvertical line (called the generator), about a vertical line (called the axis), at their point of intersection—the vertex (see Figure 8.5). The conic sections are so named because all curves in the family can be formed by a *section* of the *cone,* or more precisely the intersection of a plane and a cone. Figure 8.6 shows that if the plane does not go through the vertex, the intersection will produce a circle, ellipse, parabola, or hyperbola.

Figure 8.6

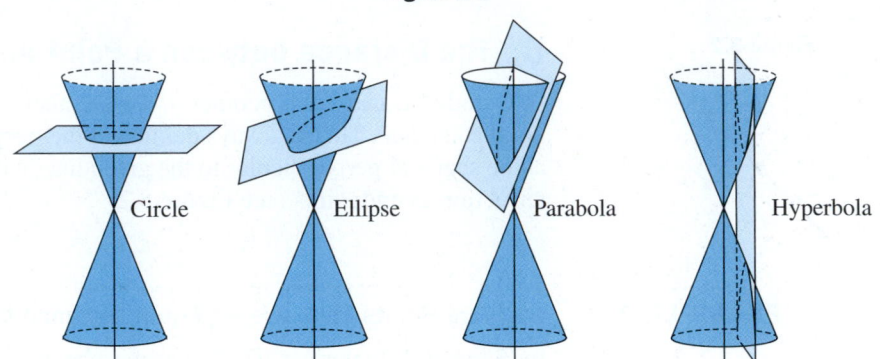

Circle   Ellipse   Parabola   Hyperbola

If the plane *does* go through the vertex, the result is a single point, a single line (if the plane contains the generator), or a pair of intersecting lines (if the plane contains the axis).

The connection we seek to make is that each conic section can be defined in terms of the distance between points in the plane, as in Example 1, or the distance between a given point and a line, as in Example 2. In Example 1, we noted the points $(-4, -2)$, $(4, -2)$, and $(4, 4)$ were all on a circle of radius 5 with center $(0, 1)$, in line with the analytic definition of a circle: *A circle is the set of all points that are an equal distance (called the radius) from a given point (called the center).*

In Example 2, you may have noticed that the points seemed to form the right branch of a parabola (see Figure 8.7), and in fact, this example illustrates the analytic definition of a parabola: *A parabola is the set of all points that are an equal distance from a given point (called the **focus**), and a given line (called the **directrix**).*

The focus and directrix are not actually part of the graph, they are simply used to locate points on the graph. For this reason all foci (plural of focus) will be represented by a "✳" symbol rather than a point.

Figure 8.7

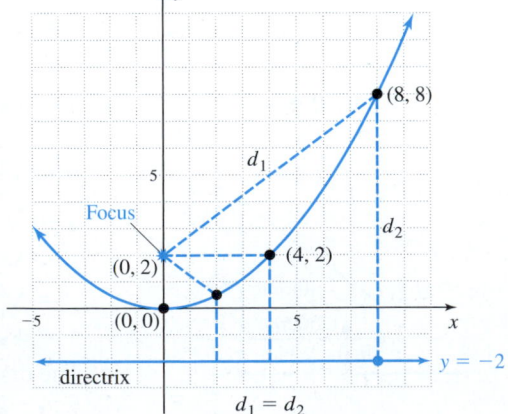

EXAMPLE 3 ▶ **Finding an Equation for All Points That Form a Certain Parabola**

With Example 2 as a pattern, use the analytic definition to find a formula (equation) for the set of all points that form the parabola.

Solution ▶ Use the ordered pair (x, y) to represent an arbitrary point on the parabola. Since any point on the line $y = -2$ has coordinates $(x, -2)$, we set the distance from $(x, -2)$ to (x, y) equal to the distance from $(0, 2)$ to (x, y). The result is

$Y_1 = \frac{1}{8}X^2,\ Y_2 = -2$

$$\sqrt{(x - x)^2 + [y - (-2)]^2} = \sqrt{(x - 0)^2 + (y - 2)^2} \quad \text{distances are equal}$$
$$\sqrt{(y + 2)^2} = \sqrt{x^2 + (y - 2)^2} \quad \text{simplify}$$
$$(y + 2)^2 = x^2 + (y - 2)^2 \quad \text{power property}$$
$$y^2 + 4y + 4 = x^2 + y^2 - 4y + 4 \quad \text{expand binomials}$$
$$8y = x^2 \quad \text{simplify}$$
$$y = \frac{1}{8}x^2 \quad \text{result}$$

All points satisfying these conditions are on the parabola defined by $y = \frac{1}{8}x^2$. See the figure.

Now try Exercises 27 and 28 ▶

At this point, it seems reasonable to ask what happens when the distance from the focus to (x, y) is *less than* the distance from the directrix to (x, y). For example, what if the distance is only two-thirds as long? As you might guess, the result is one of the other conic sections, in this case an ellipse. If the distance from the focus to a point (x, y) is *greater than* the distance from the directrix to (x, y), one branch of a hyperbola is formed. While we will defer a development of their general equations until later in the chapter, the following diagrams serve to illustrate this relationship for the ellipse, and show why we refer to the conic sections as a *family of curves*. In Figure 8.8, the line segment from the focus to each point on the graph (shown in blue), is exactly two-thirds the length of the line segment from the directrix to the same point (shown in red). Note the graph of these points forms the right half of an ellipse. In Figure 8.9, the lines and points forming the first half are removed to more clearly show the remaining points that form the complete graph.

Figure 8.8   **Figure 8.9**

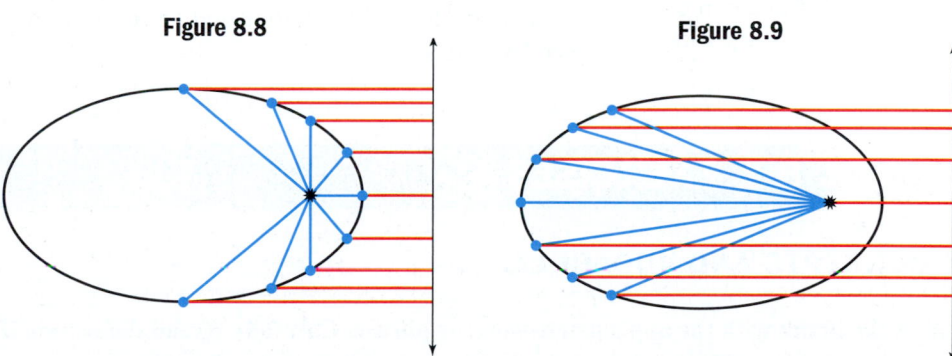

EXAMPLE 4 ▶ **Finding an Equation for All Points That Form a Certain Ellipse**

Suppose we arbitrarily select the point $(1, 0)$ as a focus and the (vertical) line $x = 4$ as the directrix. Use these to find an equation for the set of all points where the distance from the focus to a point (x, y) is $\frac{1}{2}$ the distance from the directrix to (x, y).

Solution ▶ Since any point on the line $x = 4$ has coordinates $(4, y)$, we have:

Distance from $(1, 0)$ to $(x, y) = \frac{1}{2}$ [distance from $(4, y)$ to (x, y)]   in words

$$\sqrt{(x - 1)^2 + [y - (0)]^2} = \frac{1}{2}\sqrt{(x - 4)^2 + (y - y)^2} \quad \text{resulting equation}$$

$$\sqrt{(x - 1)^2 + y^2} = \frac{1}{2}\sqrt{(x - 4)^2} \quad \text{simplify}$$

$$(x - 1)^2 + y^2 = \frac{1}{4}(x - 4)^2 \quad \text{power property}$$

$$x^2 - 2x + 1 + y^2 = \frac{1}{4}(x^2 - 8x + 16) \quad \text{expand binomials}$$

$$x^2 - 2x + 1 + y^2 = \frac{1}{4}x^2 - 2x + 4 \quad \text{distribute}$$

$$\frac{3}{4}x^2 + y^2 = 3 \quad \text{simplify: } 1x^2 - \frac{1}{4}x^2 = \frac{3}{4}x^2$$

$$3x^2 + 4y^2 = 12 \quad \text{polynomial form}$$

All points satisfying these conditions are on the ellipse defined by $3x^2 + 4y^2 = 12$.

Now try Exercises 29 and 30 ▶

Figure 8.10

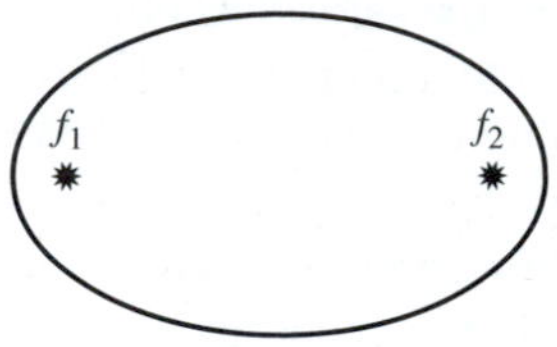

f_1 f_2

Actually, any given ellipse has two foci (see Figure 8.10) and the equation from Example 4 could also have been developed using the left focus (with the directrix also on the left). This symmetrical relationship leads us to an *alternative definition* for the ellipse, which we will explore further in Section 8.2:

> *For foci f_1 and f_2, an ellipse is the set of all points (x, y) where the sum of the distances from f_1 to (x, y) and f_2 to (x, y) is constant.*

See Figure 8.11 and **Exercises 31 and 32.** Both the focus/directrix definition and the two foci definition have merit, and simply tend to call out different characteristics and applications of the ellipse. The hyperbola also has a focus/directrix definition and a two foci definition. **See Exercises 33 and 34.**

☑ **C.** You've just seen how we can use the defining characteristics of a conic section to find its equation

Figure 8.11

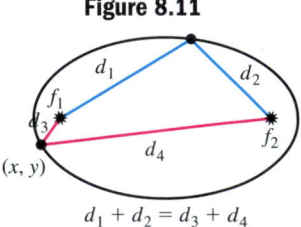

$d_1 + d_2 = d_3 + d_4$

![pencil icon] **8.1 EXERCISES**

▶ **CONCEPTS AND VOCABULARY**

Fill in the blank with the appropriate word or phrase. Carefully reread the section if needed.

1. Analytical geometry is a study of __geometry__ using the tools of __algebra__.

2. The distance formula is $d = \sqrt{(x_2 - x_1)^2 + (y_2 - y_1)^2}$; the midpoint formula is $M = \left(\frac{x_1 + x_2}{2}, \frac{y_1 + y_2}{2}\right)$.

3. The distance between a point and a line always refers to the __perpendicular__ distance.

4. The conic sections are formed by the intersection of a __plane__ and a __cone__.

5. If a plane intersects a cone at its vertex, the result is a __point__, a line, or a pair of __intersecting__ lines.

6. A circle is defined relative to an equal distance between two __points__. A parabola is defined relative to an equal distance between a __point__ and a __line__.

HOMEWORK SELECTION GUIDE

Core: 7–33 odd, 40 (14 Exercises)
Standard: 1–4, 7–33 odd, 39, 41 (19 Exercises)

Extended: 1–4, 7–21 odd, 22, 23–33 odd, 35, 39, 40, 41 (22 Exercises)
In Depth: 1–6, 7–21 odd, 22, 23–33 odd, 35, 36, 37–40 (26 Exercises)

▶ DEVELOPING YOUR SKILLS

The three points given form a right triangle. Find the midpoint of the hypotenuse and verify that the midpoint is an equal distance from all three vertices.

7. $P_1 = (-5, 2)$
 $P_2 = (1, 2)$
 $P_3 = (-5, -6)$

8. $P_1 = (3, 2)$
 $P_2 = (3, 14)$
 $P_3 = (8, 2)$

9. $P_1 = (-2, 1)$
 $P_2 = (6, -5)$
 $P_3 = (2, -7)$

10. $P_1 = (0, -5)$
 $P_2 = (-6, 4)$
 $P_3 = (6, -1)$

11. $P_1 = (10, -21)$
 $P_2 = (-6, -9)$
 $P_3 = (3, 3)$

12. $P_1 = (6, -6)$
 $P_2 = (-12, 18)$
 $P_3 = (20, 42)$

13. Find an equation of the circle that circumscribes the triangle in Exercise 7. $\quad (x + 2)^2 + (y + 2)^2 = 5^2$

14. Find an equation of the circle that circumscribes the triangle in Exercise 8. $\quad \left(x - \frac{11}{2}\right)^2 + (y - 8)^2 = \left(\frac{13}{2}\right)^2$

15. Find an equation of the circle that circumscribes the triangle in Exercise 9. $\quad (x - 2)^2 + (y + 2)^2 = 5^2$

16. Find an equation of the circle that circumscribes the triangle in Exercise 10. $\quad x^2 + \left(y - \frac{3}{2}\right)^2 = \left(\frac{13}{2}\right)^2$

17. Find an equation of the circle that circumscribes the triangle in Exercise 11. $\quad \left(x - \frac{13}{2}\right)^2 + (y + 9)^2 = \left(\frac{25}{2}\right)^2$

18. Find an equation of the circle that circumscribes the triangle in Exercise 12. $\quad (x - 13)^2 + (y - 18)^2 = 25^2$

19. Of the following six points, four are an equal distance from the point $A(2, 3)$ and two are not. (a) Identify which four, and (b) find any two additional points that are this same (nonvertical, nonhorizontal) distance from $(2, 3)$:

 $B(7, 15) \quad C(-10, 8) \quad D(9, 14) \quad E(-3, -9)$

 $F(5, 4 + 3\sqrt{10}) \quad G(2 - 2\sqrt{30}, 10)$
 **a.** $d = 13; B, C, E, G$ **b.** Answers will vary.

20. Of the following six points, four are an equal distance from the point $P(-1, 4)$ and two are not. (a) Identify which four, and (b) find any two additional points that are the same (nonvertical, nonhorizontal) distance from $(-1, 4)$.

 $Q(-9, 10) \quad R(5, 12) \quad S(-7, 11) \quad T(4, 4 + 5\sqrt{3})$

 $U(-1 + 4\sqrt{6}, 6) \quad V(-7, 4 + \sqrt{51})$
 **a.** $d = 10; Q, R, T, U$ **b.** Answers will vary.

▶ WORKING WITH FORMULAS

The Perpendicular Distance from a Point to a Line: $d = \dfrac{|Ax_1 + By_1 + C|}{\sqrt{A^2 + B^2}}$. **The perpendicular distance from a point (x_1, y_1) to a given line can be found using the formula shown, where $Ax + By + C = 0$ is the equation of the line in standard form (A, B, and C are integers).**

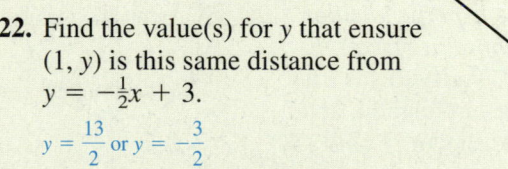

21. Use the formula to verify that $P(-6, 2)$ and $Q(6, 4)$ are an equal distance from the line $y = -\frac{1}{2}x + 3$.
 Verified, $d = \dfrac{8\sqrt{5}}{5}$

22. Find the value(s) for y that ensure $(1, y)$ is this same distance from $y = -\frac{1}{2}x + 3$.
 $y = \dfrac{13}{2}$ or $y = -\dfrac{3}{2}$

▶ APPLICATIONS

23. Of the following four points, three are an equal distance from the point $A(0, 1)$ and the line $y = -1$. (a) Identify which three, and (b) find any two additional points that satisfy these conditions.

 $B(-6, 9) \quad C(4, 4) \quad D(-2\sqrt{2}, 6) \quad E(4\sqrt{2}, 8)$
 **a.** B, C, E **b.** Answers will vary.

24. Of the following four points, three are an equal distance from the point $P(2, 4)$ and the line $y = -4$. (a) Identify which three, and (b) find any two additional points that satisfy these conditions.

 $Q(-10, 9) \quad R(2 + 4\sqrt{2}, 3) \quad S(10, 4)$
 $T(2 - 4\sqrt{5}, 5)$ **a.** Q, S, T **b.** Answers will vary.

25. Consider the fixed *point* $(0, -4)$ and the fixed *line* $y = 4$. Verify that the distance from each point given to $(0, -4)$, is equal to the distance from the point to the line $y = 4$.

 $A(4, -1) \quad B\left(10, -\dfrac{25}{4}\right) \quad C(4\sqrt{2}, -2)$
 $D(8\sqrt{5}, -20)$ Verified

26. Consider the fixed *point* $(0, -2)$ and the fixed *line* $y = 2$. Verify that the distance from each point given to $(0, -2)$, is equal to the distance from the point to the line $y = 2$.

 $P(12, -18) \quad Q\left(6, -\dfrac{9}{2}\right) \quad R(4\sqrt{5}, -10)$
 $S(4\sqrt{6}, -12)$ Verified

27. The points from Exercise 25 are on the graph of a parabola. Find an equation of the parabola. $y = -\frac{1}{16}x^2$

28. The points from Exercise 26 are on the graph of a parabola. Find an equation of the parabola. $y = -\frac{1}{8}x^2$

29. Using $(0, -2)$ as the focus and the horizontal line $y = -8$ as the directrix, find an equation for the set of all points (x, y) where the distance from the focus to (x, y) is one-half the distance from the directrix to (x, y). $4x^2 + 3y^2 = 48$

30. Using $(4, 0)$ as the focus and the vertical line $x = 9$ as the directrix, find an equation for the set of all points (x, y) where the distance from the focus to (x, y) is two-thirds the distance from the directrix to (x, y).
$5x^2 + 9y^2 = 180$

31. From Exercise 29, verify the points $(-3, 2)$ and $(\sqrt{12}, 0)$ are on the ellipse defined by $4x^2 + 3y^2 = 48$. Then verify that $d_1 + d_2 = d_3 + d_4$.
Verified, verified

Exercise 31

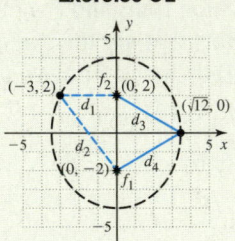

32. From Exercise 30, verify the points $(4, \frac{10}{3})$ and $(-3, -\sqrt{15})$ are on the ellipse defined by $5x^2 + 9y^2 = 180$. Then verify that $d_1 + d_2 = d_3 + d_4$.
Verified, verified

Exercise 32

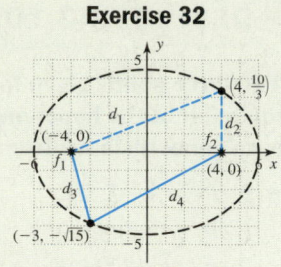

33. From the focus/directrix definition of a hyperbola: *If the distance from the focus to a point (x, y) is **greater than** the distance from the directrix to (x, y), one branch of a hyperbola is formed.* Using $(2, 0)$ as the focus and the vertical line $x = \frac{1}{2}$ as the directrix, find an equation for the set of all points (x, y) where the distance from the focus to (x, y), is twice the distance from the directrix to (x, y).
$3x^2 - y^2 = 3$

34. From the two foci definition of a hyperbola: *For foci f_1 and f_2, a hyperbola is the set of all points (x, y) where the difference of the distances from f_1 to (x, y) and f_2 to (x, y) is constant.* Verify the points $(2, 3)$ and $(-3, -2\sqrt{6})$ are on the graph of the hyperbola from Exercise 33. Then verify $d_1 - d_2 = d_3 - d_4$.   Verified, verified

▶ **EXTENDING THE CONCEPT**

35. **Properties of a circle:** A theorem from elementary geometry states: *If a radius is perpendicular to a chord, it bisects the chord.* Verify this is true for the circle, radii, and chords shown.   Verified

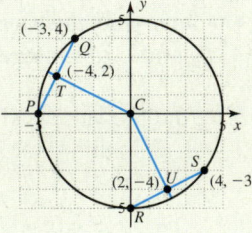

36. Verify that points $C(-2, 3)$ and $D(2\sqrt{2}, \sqrt{6})$ are points on the ellipse with foci at $A(-2, 0)$ and $B(2, 0)$, by verifying $d(AC) + d(BC) = d(AD) + d(BD)$. The expression that results has the form $\sqrt{U + V} + \sqrt{U - V}$, which prior to the common use of technology, had to be simplified using the formula $\sqrt{U + V} + \sqrt{U - V} = \sqrt{a} + \sqrt{b}$, where $a = 2U$ and $b = 4(U^2 - V^2)$. Use this relationship to simplify the equation above.
Verified (both add to 8)

▶ **MAINTAINING YOUR SKILLS**

37. **(5.6)** $5000 is deposited at 4% compounded continuously. How many years will it take for the account to exceed $8000?   about 12 yr

38. **(5.5)** Solve for x in both exact and approximate form:

a. $5 = \dfrac{10}{1 + 9e^{-0.5x}}$ **b.** $345 = 5e^{0.4x} + 75$

$x = 2 \ln 9, x \approx 4.39$ $x = \dfrac{5}{2} \ln 54, x \approx 9.97$

39. **(4.3)** Use the rational zeroes theorem and other tools to factor $f(x)$ and sketch its graph: $f(x) = x^4 - 3x^3 - 3x^2 + 11x - 6$.

40. **(4.4)** Sketch a complete graph of $h(x) = \dfrac{x^2 - 9}{x^2 - 4}$.

Clearly label all intercepts and asymptotes.

Additional answers can be found in the Instructor Answer Appendix.

8.2 The Circle and the Ellipse

LEARNING OBJECTIVES

In Section 8.2 you will see how we can:

- ☐ **A.** Use the characteristics of a circle and its graph to understand the equation of an ellipse
- ☐ **B.** Use the equation of an ellipse to graph central and noncentral ellipses
- ☐ **C.** Locate the foci of an ellipse and use the foci and other features to write the equation
- ☐ **D.** Solve applications involving the foci

In Section 8.1, we introduced the equation of an ellipse using analytical geometry and the focus-directrix definition. Here we'll take a different approach, and use the equation of a circle to demonstrate that a circle is simply a special ellipse. In doing so, we'll establish a relationship between the foci and vertices of the ellipse, that enables us to apply these characteristics in context.

A. The Equation and Graph of a Circle

Recall that the equation of a circle with radius r and center at (h, k) is

$$(x - h)^2 + (y - k)^2 = r^2.$$

As in Section 1.1, the standard form can be used to construct the equation of the circle given the center and radius as in Example 1, or to graph the circle as in Example 2.

EXAMPLE 1 ▶ **Determining the Equation of a Circle Given Its Center and Radius**

Find an equation of the circle with radius 5 and center at $(2, -1)$, then graph the relation on a calculator.

Solution ▶ With a center of $(2, -1)$, we have $h = 2$, $k = -1$, and $r = 5$. Making the corresponding substitutions into the standard form we obtain

$$(x - h)^2 + (y - k)^2 = r^2 \quad \text{standard form}$$
$$(x - 2)^2 + [y - (-1)]^2 = 5^2 \quad \text{substitute 2 for } h, -1 \text{ for } k, \text{ and 5 for } r$$
$$(x - 2)^2 + (y + 1)^2 = 25 \quad \text{simplify}$$

The equation of this circle is $(x - 2)^2 + (y + 1)^2 = 25$.

Recall from Section 1.1 that circles (and other relations) can be graphed by solving for y, then graphing the upper and lower halves of the circle.

$$(x - 2)^2 + (y + 1)^2 = 25 \quad \text{original equation}$$
$$(y + 1)^2 = 25 - (x - 2)^2 \quad \text{isolate term containing } y$$
$$y + 1 = \pm\sqrt{25 - (x - 2)^2} \quad \text{take square roots}$$
$$y = \pm\sqrt{25 - (x - 2)^2} - 1 \quad \text{subtract 1}$$
$$Y_1 = +\sqrt{25 - (X - 2)^2} - 1, \quad Y_2 = -\sqrt{25 - (X - 2)^2} - 1$$

The graph is shown in the figure using a square window. Note the point $(5, 3)$ satisfies the original equation and is a point on the graph and that $(5, -5)$, $(-1, -5)$, and $(-1, 3)$ must also be on the graph due to symmetry.

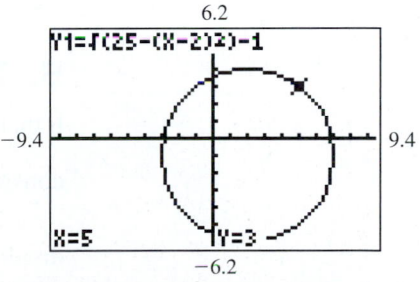

Now try Exercises 7 through 12 ▶

If the equation is given in polynomial form, recall that we first complete the square in x and y to identify the center and radius.

EXAMPLE 2 ▶ **Completing the Square to Graph a Circle**

Find the center and radius of the circle whose equation is given, then sketch its graph: $x^2 + y^2 - 6x + 4y - 3 = 0$.

Solution ▶ Begin by completing the square in both x and y.

$$(x^2 - 6x + \underline{}) + (y^2 + 4y + \underline{}) = 3 \qquad \text{group } x\text{- and } y\text{-terms; add 3}$$

$$(x^2 - 6x + 9) + (y^2 + 4y + 4) = 3 + 9 + 4 \qquad \text{complete the square}$$

adds 9 to left side   adds 4 to left side   add 9 + 4 to right side

$$(x - 3)^2 + (y + 2)^2 = 16 \qquad \text{factor and simplify}$$

The center is at $(3, -2)$, with radius $r = \sqrt{16} = 4$.

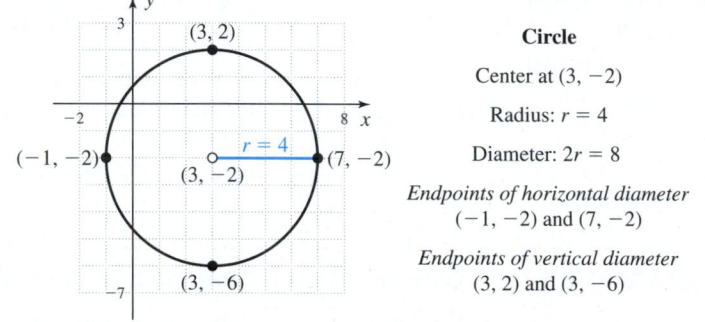

Circle

Center at $(3, -2)$

Radius: $r = 4$

Diameter: $2r = 8$

Endpoints of horizontal diameter
$(-1, -2)$ and $(7, -2)$

Endpoints of vertical diameter
$(3, 2)$ and $(3, -6)$

Now try Exercises 13 through 18 ▶

The equation of a circle in **standard form** provides a useful link to some of the other conic sections, and is obtained by *setting the equation equal to 1*. In the case of a circle, this means we simply divide by r^2.

$$(x - h)^2 + (y - k)^2 = r^2 \qquad \text{standard form}$$

$$\frac{(x - h)^2}{r^2} + \frac{(y - k)^2}{r^2} = 1 \qquad \text{divide by } r^2$$

✓ **A.** You've just seen how we can use the characteristics of a circle and its graph to understand the equation of an ellipse

In this form, the value of r in each denominator gives the *horizontal* and *vertical* distances, respectively, from the center to the graph. This is not so important in the case of a circle, since this distance is the same in *any* direction. But for other conics, these horizontal and vertical distances are *not* the same, making the new form a valuable tool for graphing. To distinguish the horizontal from the vertical distance, r^2 is replaced by a^2 in the "x-term" (horizontal distance), and by b^2 in the "y-term" (vertical distance). This distinction leads us directly into our study of the ellipse.

B. The Equation of an Ellipse

It then seems reasonable to ask, "What happens to the graph when $a \neq b$?" To answer, consider the equation from Example 2. We have $\dfrac{(x - 3)^2}{4^2} + \dfrac{(y + 2)^2}{4^2} = 1$ (after dividing by 16), which we now compare to $\dfrac{(x - 3)^2}{4^2} + \dfrac{(y + 2)^2}{3^2} = 1$, where $a = 4$ and $b = 3$. The center of the graph is still at $(3, -2)$, since $h = 3$ and $k = -2$ remain unchanged. Substituting $y = -2$ to find additional points, eliminates the y-term and gives two values for x:

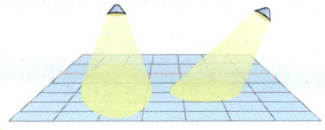
$$\frac{(x - 3)^2}{4^2} + \frac{(-2 + 2)^2}{3^2} = 1 \qquad \text{substitute } -2 \text{ for } y$$

$$\frac{(x - 3)^2}{4^2} + 0 = 1 \qquad \text{simplify}$$

$$(x - 3)^2 = 16 \qquad \text{multiply by } 4^2 = 16$$

$$x - 3 = \pm 4 \qquad \text{property of square roots}$$

$$x = 3 \pm 4 \qquad \text{add 3}$$

$$x = 7 \text{ and } x = -1$$

This shows the horizontal distance from the center to the graph is still $a = 4$, and the points $(-1, -2)$ and $(7, -2)$ are on the graph (see Figure 8.12). Similarly, for $x = 3$ we have $(y + 2)^2 = 9$, giving $y = -5$ and $y = 1$, and showing the vertical distance from the center to the graph is now $b = 3$, with points $(3, 1)$ and $(3, -5)$ on the graph. Using this information to sketch the curve reveals the "circle" is elongated and has become a **horizontal** ellipse.

For this ellipse, the line segment through the center, parallel the x-axis, and with endpoints on the ellipse is called the **major axis,** with the endpoints of the major axis called **vertices.** The segment perpendicular to and bisecting the major axis (with its endpoints on the ellipse) is called the **minor axis,** as shown in Figure 8.13.

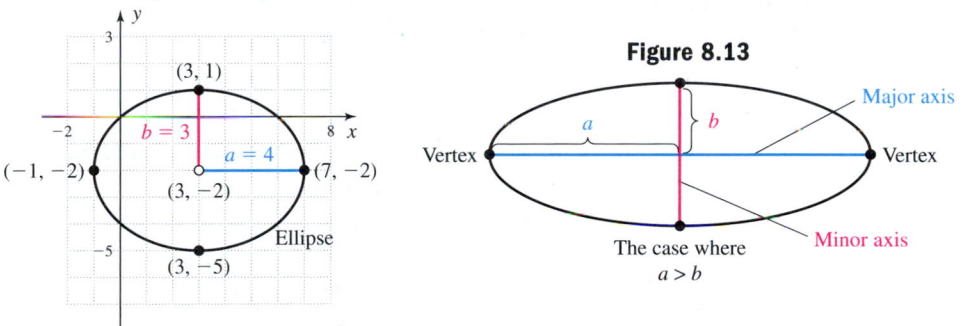

Figure 8.12

Figure 8.13

The case where $a > b$

- If $a^2 > b^2$, the major axis is horizontal (parallel to the x-axis) with length $2a$, and the minor axis is vertical with length $2b$ (see Example 3).
- If $a^2 < b^2$ the major axis is vertical (parallel to the y-axis) with length $2b$, and the minor axis is horizontal with length $2a$ (see Example 4).

Generalizing this observation we obtain the equation of an ellipse in standard form.

The Equation of an Ellipse in Standard Form

Given $\dfrac{(x - h)^2}{a^2} + \dfrac{(y - k)^2}{b^2} = 1.$

If $a \neq b$ the equation represents the graph of an ellipse with center at (h, k).

- $|a|$ gives the horizontal distance from center to graph.
- $|b|$ gives the vertical distance from center to graph.

Finally, note the line segment from center to vertex is called the **semimajor axis,** with the perpendicular line segment from center to graph called the **semiminor axis.**

EXAMPLE 3 ▶ **Graphing a Horizontal Ellipse**

Sketch the graph of the ellipse defined by
$$\frac{(x - 2)^2}{25} + \frac{(y + 1)^2}{9} = 1.$$

Solution ▶

Noting $a \neq b$, we have an ellipse with
center $(h, k) = (2, -1)$. The horizontal
distance from the center to the graph is
$a = 5$, and the vertical distance from the
center to the graph is $b = 3$. After plotting
the corresponding points and connecting
them with a smooth curve, we obtain the
graph shown.

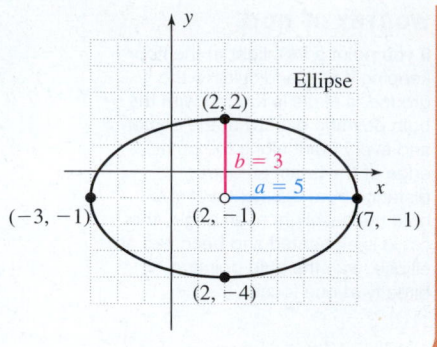

Now try Exercises 19 through 24 ▶

WORTHY OF NOTE

In general, for the equation
$Ax^2 + By^2 = F$ $(A, B, F > 0)$, the
equation represents a circle if
$A = B$, and an ellipse if $A \neq B$.

As with the circle, the equation of an ellipse can be given in polynomial form, and
here our knowledge of circles is helpful. For the equation $25x^2 + 4y^2 = 100$, we
know the graph cannot be a circle since the coefficients are unequal, and the center of
the graph must be at the origin since $h = k = 0$. To actually draw the graph, we convert the equation to standard form. Note that a circle whose center is at $(0, 0)$ is called
a **central circle,** and an ellipse with center at $(0, 0)$ is called a **central ellipse.**

EXAMPLE 4 ▶

Graphing a Vertical Ellipse

For $25x^2 + 4y^2 = 100$,

a. Write the equation in standard form and identify the center and the values of a
   and b.

b. Identify the major and minor axes and name the vertices.

c. Sketch the graph.

d. Graph the relation on a graphing calculator using a "friendly" window, then
   use the ⟨TRACE⟩ feature to find four additional points on the graph whose
   coordinates are rational.

Solution ▶

The coefficients of x^2 and y^2 are unequal, and 25, 4, and 100 have like signs. The
equation represents an ellipse with center at $(0, 0)$. To obtain standard form:

a. $25x^2 + 4y^2 = 100$ given equation

$$\frac{25x^2}{100} + \frac{4y^2}{100} = 1$$ divide by 100

$$\frac{x^2}{4} + \frac{y^2}{25} = 1$$ standard form

$$\frac{x^2}{2^2} + \frac{y^2}{5^2} = 1$$ write denominators in squared form; $a = 2$, $b = 5$

b. The result shows $a = 2$ and $b = 5$, indicating the major axis will be vertical and
   the minor axis will be horizontal. With the center at the origin, the x-intercepts will
   be $(-2, 0)$ and $(2, 0)$,
   with the vertices (and
   y-intercepts) at
   $(0, -5)$ and $(0, 5)$.

c. Plotting these
   intercepts and
   sketching the ellipse
   results in the graph
   shown in
   Figure 8.14.

Figure 8.14

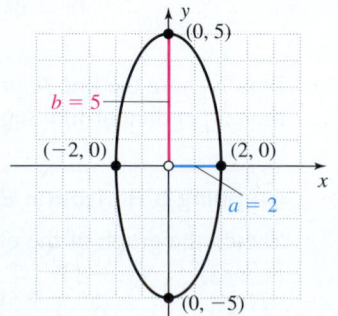

Vertical ellipse

Center at $(0, 0)$

Endpoints of major axis (vertices)
$(0, -5)$ and $(0, 5)$

Endpoints of minor axis
$(-2, 0)$ and $(2, 0)$

Length of major axis $2b$: $2(5) = 10$
Length of minor axis $2a$: $2(2) = 4$

d. As with the circle, we begin by solving for y.

$$25x^2 + 4y^2 = 100 \qquad \text{original equation}$$
$$4y^2 = 100 - 25x^2 \qquad \text{isolate term containing } y$$
$$y^2 = \frac{100 - 25x^2}{4} \qquad \text{divide by 4}$$
$$y = \pm\sqrt{\frac{100 - 25x^2}{4}} \qquad \text{take square roots}$$
$$Y_1 = +\sqrt{\frac{100 - 25X^2}{4}}, \; Y_2 = -\sqrt{\frac{100 - 25X^2}{4}}$$

Figure 8.15

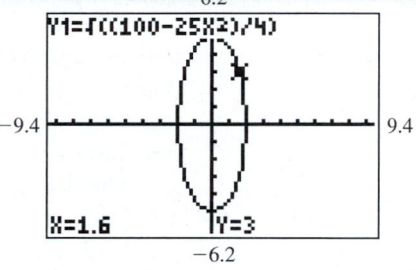

The graph is shown in Figure 8.15, where we note that $(1.6, 3)$ is a point on the graph. Due to the symmetry of the ellipse, $(-1.6, 3)$, $(-1.6, -3)$, and $(1.6, -3)$ are also on the graph.

Now try Exercises 25 through 36 ▶

WORTHY OF NOTE

After writing the equation in standard form, it is possible to end up with a constant that is zero or negative. In the first case, the graph is a single point. In the second case, no graph is possible since roots of the equation will be complex numbers. These are called *degenerate cases*. **See Exercise 84.**

If the center of the ellipse is not at the origin, the polynomial form has additional linear terms and we must first complete the square in x and y, then write the equation in standard form to sketch the graph (see the Reinforcing Basic Concepts feature for more on completing the square). Figure 8.16 illustrates how the central ellipse and the shifted ellipse are related.

Figure 8.16

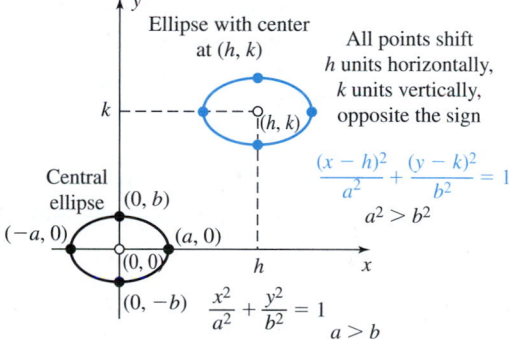

EXAMPLE 5 ▶ **Completing the Square to Graph an Ellipse**

Sketch the graph of $25x^2 + 4y^2 + 150x - 16y + 141 = 0$, then state the domain and range of the relation.

Solution ▶ The coefficients of x^2 and y^2 are unequal and have like signs, and we assume the equation represents an ellipse but wait until we have the factored form to be certain (it could be a degenerate ellipse).

$$25x^2 + 4y^2 + 150x - 16y + 141 = 0 \qquad \text{given equation (polynomial form)}$$
$$25x^2 + 150x + 4y^2 - 16y = -141 \qquad \text{group like terms; subtract 141}$$
$$25(x^2 + 6x + \underline{}) + 4(y^2 - 4y + \underline{}) = -141 \qquad \text{factor out leading coefficient from each group}$$
$$25(x^2 + 6x + 9) + 4(y^2 - 4y + 4) = -141 + 225 + 16 \qquad \text{complete the square}$$

add $225 + 16$ to right

adds $25(9) = 225$ \qquad adds $4(4) = 16$

$$25(x + 3)^2 + 4(y - 2)^2 = 100 \qquad \text{factor}$$

$$\frac{25(x + 3)^2}{100} + \frac{4(y - 2)^2}{100} = \frac{100}{100} \qquad \text{divide both sides by 100}$$

$$\frac{(x + 3)^2}{4} + \frac{(y - 2)^2}{25} = 1 \qquad \text{simplify (standard form)}$$

$$\frac{(x + 3)^2}{2^2} + \frac{(y - 2)^2}{5^2} = 1 \qquad \text{write denominators in squared form}$$

The result is a vertical ellipse with center at $(-3, 2)$, with $a = 2$ and $b = 5$. The vertices are a vertical distance of 5 units from center, and the endpoints of the minor axis are a horizontal distance of 2 units from center. Note this is the same ellipse as in Example 4, but shifted 3 units left and 2 up.

The domain of this relation is $x \in [-5, -1]$, and the range is $y \in [-3, 7]$.

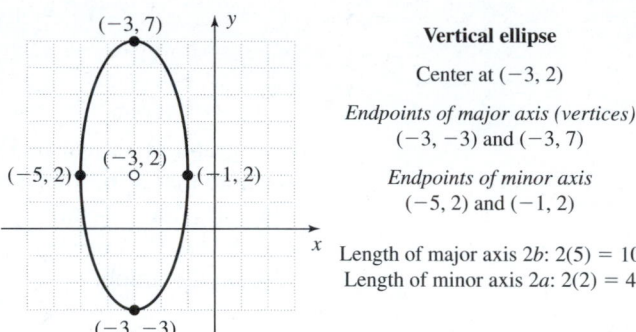

Vertical ellipse

Center at $(-3, 2)$

Endpoints of major axis (vertices)
$(-3, -3)$ and $(-3, 7)$

Endpoints of minor axis
$(-5, 2)$ and $(-1, 2)$

Length of major axis $2b$: $2(5) = 10$
Length of minor axis $2a$: $2(2) = 4$

✅ **B.** You've just seen how we can use the equation of an ellipse to graph central and noncentral ellipses

Now try Exercises 37 through 44 ▶

C. The Foci of an Ellipse

In Section 8.1, we noted that an ellipse could also be defined in terms of two special points called the **foci**. The Museum of Science and Industry in Chicago, Illinois (http://www.msichicago.org), has a permanent exhibit called the *Whispering Gallery.* The construction of the room is based on some of the reflective properties of an ellipse. If two people stand at designated points in the room and one of them whispers very softly, the other person can hear the whisper quite clearly—even though they are over 40 ft apart! The point where each person stands is a **focus** of an ellipse. This reflective property also applies to light and radiation, giving the ellipse some powerful applications in science, medicine, acoustics, and other areas. To understand and appreciate these applications, we introduce the analytic definition of an ellipse.

WORTHY OF NOTE

You can easily draw an ellipse that satisfies the definition. Press two pushpins (these form the foci of the ellipse) halfway down into a piece of heavy cardboard about 6 in. apart. Take an 8-in. piece of string and loop each end around the pins. Use a pencil to draw the string taut and keep it taut as you move the pencil in a circular motion—and the result is an ellipse! A different length of string or a different distance between the foci will produce a different ellipse.

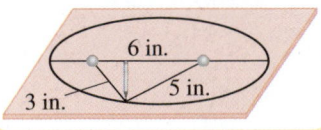

Definition of an Ellipse

Given two fixed points f_1 and f_2 in a plane, an ellipse is the set of all points (x, y) where the distance from f_1 to (x, y) added to the distance from f_2 to (x, y) remains constant.

$$d_1 + d_2 = k$$

The fixed points f_1 and f_2 are called the *foci* of the ellipse, and the points $P(x, y)$ are on the graph of the ellipse.

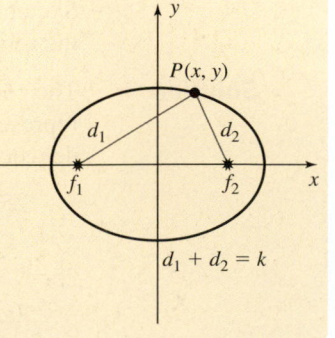

To find the equation of an ellipse in terms of a and b we combine the definition just given with the distance formula. Consider the ellipse shown in Figure 8.17 (for

Figure 8.17

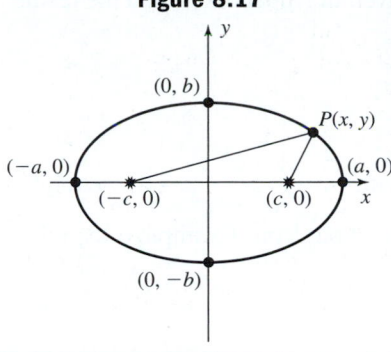

calculating ease we use a central ellipse). Note the vertices have coordinates $(-a, 0)$ and $(a, 0)$, and the endpoints of the minor axis have coordinates $(0, -b)$ and $(0, b)$ as before. It is customary to assign foci the coordinates $f_1 \rightarrow (-c, 0)$ and $f_2 \rightarrow (c, 0)$. We can calculate the distance between $(c, 0)$ and any point $P(x, y)$ on the ellipse using the distance formula:

$$\sqrt{(x - c)^2 + (y - 0)^2}$$

Likewise the distance between $(-c, 0)$ and any point (x, y) is

$$\sqrt{(x + c)^2 + (y - 0)^2}$$

According to the definition, the sum must be constant:

$$\sqrt{(x - c)^2 + y^2} + \sqrt{(x + c)^2 + y^2} = k$$

EXAMPLE 6 ▶ **Finding the Value of k from the Definition of an Ellipse**

Use the definition of an ellipse and the diagram given to determine the constant k used for this ellipse (also see the following *Worthy of Note*). Note that $a = 5, b = 3,$ and $c = 4$.

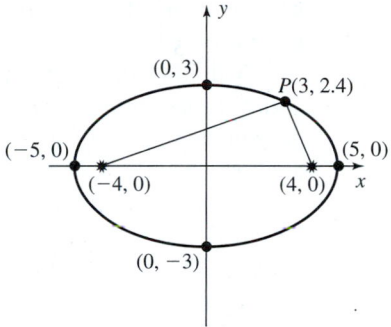

Solution ▶

$$\sqrt{(x - c)^2 + (y - 0)^2} + \sqrt{(x + c)^2 + (y - 0)^2} = k \quad \text{given}$$
$$\sqrt{(3 - 4)^2 + (2.4 - 0)^2} + \sqrt{(3 + 4)^2 + (2.4 - 0)^2} = k \quad \text{substitute}$$
$$\sqrt{(-1)^2 + 2.4^2} + \sqrt{7^2 + 2.4^2} = k \quad \text{add}$$
$$\sqrt{6.76} + \sqrt{54.76} = k \quad \text{simplify radicals}$$
$$2.6 + 7.4 = k \quad \text{compute square roots}$$
$$10 = k \quad \text{result}$$

The constant value for this ellipse is 10 units.

WORTHY OF NOTE

Note that if the foci are coincident (both at the origin) the "ellipse" will actually be a circle with radius $\dfrac{k}{2}$;
$\sqrt{x^2 + y^2} + \sqrt{x^2 + y^2} = k$ leads to $x^2 + y^2 = \dfrac{k^2}{4}$. In Example 6 we found $k = 10$, giving $\dfrac{10}{2} = 5$, and if we used the "string" to draw the circle, the pencil would be 5 units from the center, creating a circle of radius 5.

Now try Exercises 45 through 48 ▶

In Example 6, the sum of the distances could also be found by moving the point (x, y) to the location of a vertex $(a, 0)$, then using the symmetry of the ellipse. The sum is identical to the length of the major axis, since the overlapping part of the string from $(c, 0)$ to $(a, 0)$ is the same length as from $(-a, 0)$ to $(-c, 0)$ (see Figure 8.18). This shows the constant k is equal to 2a *regardless of the distance between foci.*

As we noted, the result is

Figure 8.18

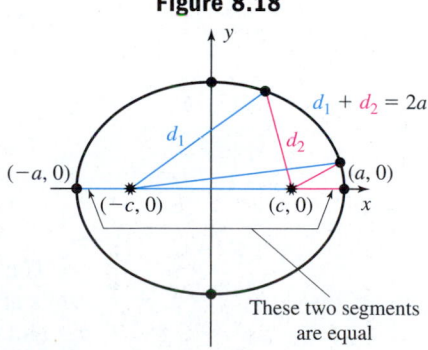

$$\sqrt{(x - c)^2 + y^2} + \sqrt{(x + c)^2 + y^2} = 2a \quad \text{substitute 2a for k}$$

The details for simplifying this expression are given in Appendix V, and the result is very close to the standard form seen previously:

$$\frac{x^2}{a^2} + \frac{y^2}{a^2 - c^2} = 1$$

By comparing the standard form $\frac{x^2}{a^2} + \frac{y^2}{b^2} = 1$ with $\frac{x^2}{a^2} + \frac{y^2}{a^2 - c^2} = 1$, we might suspect that $b^2 = a^2 - c^2$, and this is indeed the case. Note from Example 6 the relationship yields

$$b^2 = a^2 - c^2$$
$$3^2 = 5^2 - 4^2$$
$$9 = 25 - 16$$

Additionally, when we consider that $(0, b)$ is a point on the ellipse, the distance from $(0, b)$ to $(c, 0)$ must be equal to a due to symmetry (the "constant distance" used to form the ellipse is always $2a$). We then see in Figure 8.19, that $b^2 + c^2 = a^2$ (Pythagorean Theorem), yielding $b^2 = a^2 - c^2$ as above.

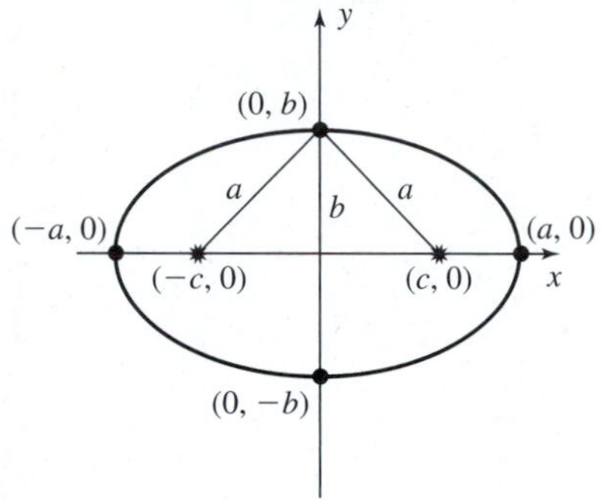

With this development, we now have the ability to *locate the foci of any ellipse*—an important step toward using the ellipse in practical applications. Because we're often asked to find the location of the foci, it's best to rewrite the relationship in terms of c^2, using absolute value bars to allow for a major axis that is vertical: $c^2 = |a^2 - b^2|$.

Figure 8.19

EXAMPLE 7 ▶ **Completing the Square to Graph an Ellipse and Locate the Foci**

For the ellipse defined by $25x^2 + 9y^2 - 100x - 54y - 44 = 0$, find the coordinates of the center, vertices, foci, and endpoints of the minor axis. Then sketch the graph.

Solution ▶

$$25x^2 + 9y^2 - 100x - 54y - 44 = 0 \qquad \text{given}$$
$$25x^2 - 100x + 9y^2 - 54y = 44 \qquad \text{group terms; add 44}$$
$$25(x^2 - 4x + \underline{}) + 9(y^2 - 6y + \underline{}) = 44 \qquad \text{factor out lead coefficients}$$
$$25(x^2 - 4x + 4) + 9(y^2 - 6y + 9) = 44 + 100 + 81 \qquad \text{add } 100 + 81 \text{ to right-hand side}$$

adds $25(4) = 100$ adds $9(9) = 81$

$$25(x - 2)^2 + 9(y - 3)^2 = 225 \qquad \text{factored form}$$
$$\frac{25(x - 2)^2}{225} + \frac{9(y - 3)^2}{225} = \frac{225}{225} \qquad \text{divide by 225}$$
$$\frac{(x - 2)^2}{9} + \frac{(y - 3)^2}{25} = 1 \qquad \text{simplify (standard form)}$$
$$\frac{(x - 2)^2}{3^2} + \frac{(y - 3)^2}{5^2} = 1 \qquad \text{write denominators in squared form}$$

The result shows a vertical ellipse with $a = 3$ and $b = 5$. The center of the ellipse is at $(2, 3)$. The vertices are a vertical distance of $b = 5$ units from center at $(2, 8)$ and $(2, -2)$. The endpoints of the minor axis are a horizontal distance of $a = 3$ units from center at $(-1, 3)$ and $(5, 3)$. To locate the foci, we use the foci formula

for an ellipse: $c^2 = |a^2 - b^2|$, giving $c^2 = |3^2 - 5^2| = 16$. This shows the foci "✳"
are located a vertical distance of 4 units from center at $(2, 7)$ and $(2, -1)$.

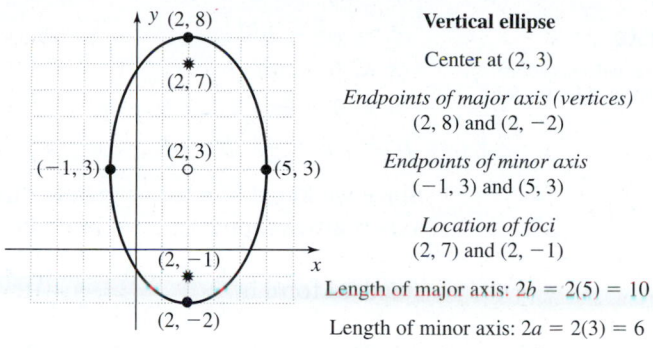

Vertical ellipse

Center at $(2, 3)$

Endpoints of major axis (vertices)
$(2, 8)$ and $(2, -2)$

Endpoints of minor axis
$(-1, 3)$ and $(5, 3)$

Location of foci
$(2, 7)$ and $(2, -1)$

Length of major axis: $2b = 2(5) = 10$

Length of minor axis: $2a = 2(3) = 6$

Now try Exercises 49 through 54 ▶

For an ellipse, a **focal chord** is a line segment perpendicular to the major axis, through a focus and with endpoints on the ellipse. In the Exercise Set, you are asked to verify that the focal chord of an ellipse has length $L = \dfrac{2m^2}{n}$, where m is the length of the semiminor axis and n is the length of the semimajor axis. This means the distance from the foci to the graph (along a focal chord) is $\dfrac{m^2}{n}$, a fact can often be used to help graph an ellipse. For Example 7, $m = 3$ and $n = 5$, so the horizontal distance from focus to graph (in either direction) is $\dfrac{3^2}{5} = \dfrac{9}{5}$. From the upper focus $(2, 7)$, we can now graph the additional points $(2 - 1.8, 7) = (0.2, 7)$ and $(2 + 1.8, 7) = (3.8, 7)$, and from the lower focus $(2, -1)$ we obtain $(0.2, -1)$ and $(3.8, -1)$ without having to evaluate the original equation. Graphical verification is provided in Figure 8.20. Also **see Exercises 83 and 85.**

Figure 8.20

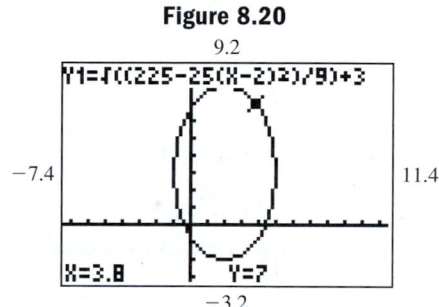

For future reference, remember the foci of an ellipse always occur on the major axis, with $a > c$ and $a^2 > c^2$ for a horizontal ellipse, with $b > c$ and $b^2 > c^2$ for a vertical ellipse. This makes it easier to remember the **foci formula** for ellipses: $c^2 = |a^2 - b^2|$. If any two of the values for a, b, and c are known, the relationship between them can be used to construct the equation of the ellipse.

EXAMPLE 8 ▶   **Finding the Equation of an Ellipse**

Find the equation of the ellipse (in standard form) that has foci at $(0, -2)$ and $(0, 2)$, with a minor axis 6 units in length. Then graph the ellipse

a. By hand.

b. On a graphing calculator.

c. Find the distance from foci to graph along a focal chord $\left(\text{using } \dfrac{m^2}{n}\right)$, and use the result to verify that the endpoints of both focal chords are all on the graph.

Solution ▶ Since the foci are on the y-axis and an equal distance from (0, 0), we know this is a vertical and central ellipse with $c = 2$ and $c^2 = 4$. The minor axis has a length of $2a = 6$ units, meaning $a = 3$ and $a^2 = 9$. To find b^2, use the foci equation and solve.

LOOKING AHEAD

For the hyperbola, we'll find that $c > a$, and the formula for the foci of a hyperbola will be $c^2 = a^2 + b^2$.

$$c^2 = |a^2 - b^2| \quad \text{foci equation (ellipse)}$$
$$4 = |9 - b^2| \quad \text{substitute}$$
$$-4 = 9 - b^2 \qquad 4 = 9 - b^2 \quad \text{solve the absolute value equation}$$
$$b^2 = 13 \qquad b^2 = 5 \quad \text{result}$$

Figure 8.21

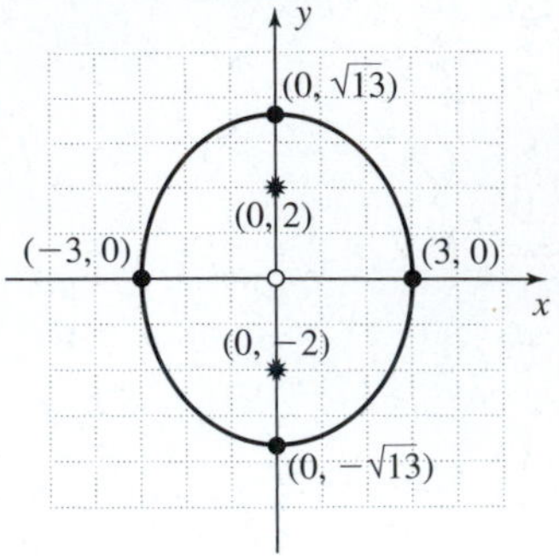

Since we know b^2 must be greater than a^2 (the major axis is always longer), $b^2 = 5$ can be discarded. The standard form is $\frac{x^2}{3^2} + \frac{y^2}{(\sqrt{13})^2} = 1$.

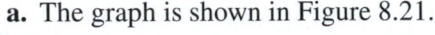

a. The graph is shown in Figure 8.21.

b. For a calculator generated graph, begin by solving for y.

$$\frac{x^2}{9} + \frac{y^2}{13} = 1 \quad \text{original equation}$$
$$13x^2 + 9y^2 = 117 \quad \text{clear denominators}$$
$$9y^2 = 117 - 13x^2 \quad \text{isolate } y\text{-term}$$
$$y^2 = \frac{117 - 13x^2}{9} \quad \text{divide by 9}$$
$$y = \pm\sqrt{\frac{117 - 13x^2}{9}} \quad \text{take square roots}$$
$$Y_1 = +\sqrt{\frac{117 - 13X^2}{9}},\ Y_2 = -\sqrt{\frac{117 - 13X^2}{9}}$$

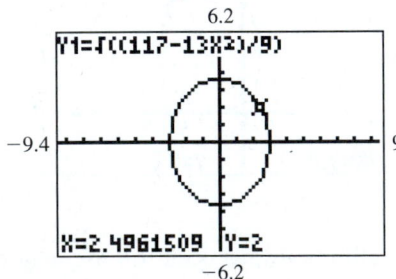

The graph is shown in Figure 8.22.

Figure 8.22

Figure 8.23

c. From the discussion prior to Example 8, the horizontal distance from foci to graph must be $\frac{m^2}{n} = \frac{9}{\sqrt{13}}$. Using the TRACE feature and entering $x = 9/\sqrt{13}$ verifies that $\left(\frac{9}{\sqrt{13}}, 2\right)$ is a point on the graph (Figure 8.23), and that $\left(-\frac{9}{\sqrt{13}}, 2\right)$, $\left(-\frac{9}{\sqrt{13}}, -2\right)$, and $\left(\frac{9}{\sqrt{13}}, -2\right)$ must also be on the graph due to symmetry.

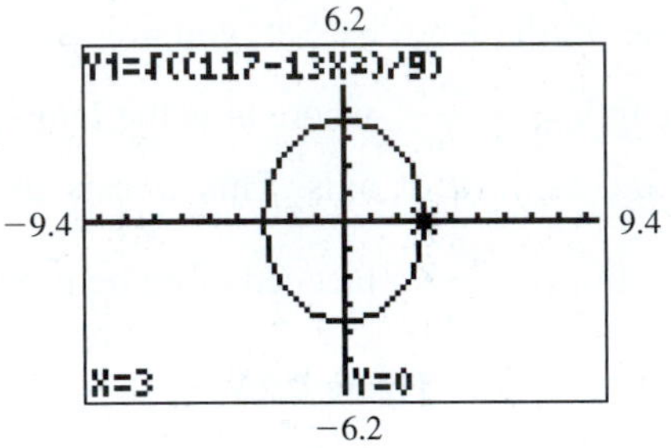

Now try Exercises 55 through 62 ▶

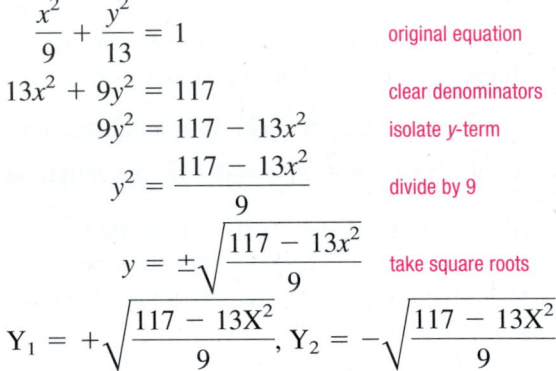

☑ **C.** You've just seen how we can locate the foci of an ellipse and use the foci and other features to write the equation

D. Applications Involving Foci

Applications involving the foci of a conic section can take various forms. In many cases, only partial information about the conic section is available and the ideas from Example 8 must be used to "fill in the gaps." In other applications, we must rewrite a known or given equation to find information related to the values of a, b, and c.

EXAMPLE 9 ▶ Solving Applications Using the Characteristics of an Ellipse

In Washington, D.C., there is a park called the *Ellipse* located between the White House and the Washington Monument. The park is surrounded by a path that forms an ellipse with the length of the major axis being about 1502 ft and the minor axis having a length of 1280 ft. Suppose the park manager wants to install water fountains at the location of the foci. Find the distance between the fountains rounded to the nearest foot.

Solution ▶ Since the major axis has length $2a = 1502$, we know $a = 751$ and $a^2 = 564{,}001$. The minor axis has length $2b = 1280$, meaning $b = 640$ and $b^2 = 409{,}600$. To find c, use the foci equation:

$$
\begin{aligned}
c^2 &= a^2 - b^2 & \text{since we know } a > b \\
&= 564{,}001 - 409{,}600 & \text{substitute} \\
&= 154{,}401 & \text{subtract} \\
c &\approx -393 \text{ and } c \approx 393 & \text{square root property}
\end{aligned}
$$

✓ **D.** You've just seen how we can solve applications involving the foci

The distance between the water fountains would be $2(393) = 786$ ft.

Now try Exercises 65 through 80 ▶

8.2 EXERCISES

▶ CONCEPTS AND VOCABULARY

Fill in the blank with the appropriate word or phrase. Carefully reread the section if needed.

1. For an ellipse, the relationship between a, b, and c is given by the foci equation $\underline{c^2 = |a^2 - b^2|}$, since $c < a$ or $c < b$.

2. The greatest distance across an ellipse is called the $\underline{\text{major}}$ $\underline{\text{axis}}$ and the endpoints are called $\underline{\text{vertices}}$.

3. For a vertical ellipse, the length of the minor axis is $\underline{2a}$ and the length of the major axis is $\underline{2b}$.

4. To write the equation $2x^2 + y^2 - 6x = 7$ in standard form, $\underline{\text{complete}}$ the $\underline{\text{square}}$ in x.

5. Explain/Discuss how the relations $a > b$, $a = b$ and $a < b$ affect the graph of a conic section with equation $\dfrac{(x - h)^2}{a^2} + \dfrac{(y - k)^2}{b^2} = 1$. Answers will vary.

6. Suppose foci are located at $(-3, 2)$ and $(5, 2)$. Discuss/Explain the conditions necessary for the graph to be an ellipse. Answers will vary.

▶ DEVELOPING YOUR SKILLS

Find an equation of the circle satisfying the conditions given, then graph the result on a graphing calculator and locate two additional points on the graph.

7. center $(0, 0)$, radius 7

8. center $(0, 0)$, radius 9

9. center $(5, 0)$, radius $\sqrt{3}$

10. center $(0, 4)$, radius $\sqrt{5}$

11. diameter has endpoints $(4, 9)$ and $(-2, 1)$

12. diameter has endpoints $(-2, -3)$, and $(3, 9)$

Write each equation in standard form to identify the center and radius of the circle, then sketch its graph.

13. $x^2 + y^2 - 12x - 10y + 52 = 0$

14. $x^2 + y^2 + 8x - 6y - 11 = 0$

15. $x^2 + y^2 - 4x + 10y + 4 = 0$

16. $x^2 + y^2 + 4x + 6y - 3 = 0$

17. $x^2 + y^2 + 6x - 5 = 0$

18. $x^2 + y^2 - 8y - 5 = 0$

Sketch the graph of each ellipse.

19. $\dfrac{(x - 1)^2}{9} + \dfrac{(y - 2)^2}{16} = 1$

20. $\dfrac{(x - 3)^2}{4} + \dfrac{(y - 1)^2}{25} = 1$

21. $\dfrac{(x - 2)^2}{25} + \dfrac{(y + 3)^2}{4} = 1$

22. $\dfrac{(x + 5)^2}{1} + \dfrac{(y - 2)^2}{16} = 1$

HOMEWORK SELECTION GUIDE

Core: 7–59 every other odd, 63, 65–81 every other odd (20 Exercises)
Standard: 1–4, 7–59 every other odd, 63, 65–81 every other odd, 83, 85, 88 (27 Exercises)

Extended: 1–4, 7–59 every other odd, 61, 63–65, 69, 71, 72, 73, 75, 77, 81, 83, 85, 87, 88 (33 Exercises)
In Depth: 1–6, 7–59 every other odd, 61, 63–65, 69, 71, 72, 73, 75, 77, 81, 83–88 (37 Exercises)

23. $\dfrac{(x+1)^2}{16} + \dfrac{(y+2)^2}{9} = 1$

24. $\dfrac{(x+1)^2}{36} + \dfrac{(y+3)^2}{9} = 1$

For each exercise, (a) write the equation in standard form, then identify the center and the values of *a* and *b*, (b) state the coordinates of the vertices and the coordinates of the endpoints of the minor axis, (c) sketch the graph, and (d) for 25–28 (only) graph the relations on a graphing calculator and identify four additional points on the graph whose coordinates are rational.

25. $x^2 + 4y^2 = 16$ **26.** $9x^2 + y^2 = 36$

27. $16x^2 + 9y^2 = 144$ **28.** $25x^2 + 9y^2 = 225$

29. $2x^2 + 5y^2 = 10$ **30.** $3x^2 + 7y^2 = 21$

Identify each equation as that of an ellipse or circle, then sketch its graph.

31. $(x+1)^2 + 4(y-2)^2 = 16$

32. $9(x-2)^2 + (y+3)^2 = 36$

33. $2(x-2)^2 + 2(y+4)^2 = 18$

34. $(x-6)^2 + y^2 = 49$

35. $4(x-1)^2 + 9(y-4)^2 = 36$

36. $25(x-3)^2 + 4(y+2)^2 = 100$

Complete the square in both *x* and *y* to write each equation in standard form. Then draw a complete graph of the relation and identify all important features, including the domain and range.

37. $4x^2 + y^2 + 6y + 5 = 0$

38. $x^2 + 3y^2 + 8x + 7 = 0$

39. $x^2 + 4y^2 - 8y + 4x - 8 = 0$

40. $3x^2 + y^2 - 8y + 12x - 8 = 0$

41. $5x^2 + 2y^2 + 20y - 30x + 75 = 0$

42. $4x^2 + 9y^2 - 16x + 18y - 11 = 0$

43. $2x^2 + 5y^2 - 12x + 20y - 12 = 0$

44. $6x^2 + 3y^2 - 24x + 18y - 3 = 0$

Use the definition of an ellipse to find the constant *k* for each ellipse (figures are not drawn to scale).

45. $k = 20$ **46.** $k = 30$

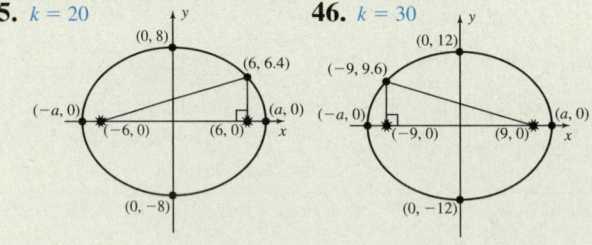

47. $k = 20$ **48.** $k = 200$

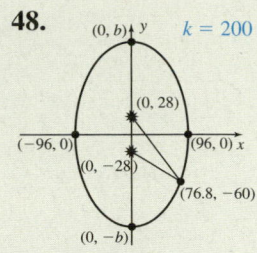

Find the coordinates of the (a) center, (b) vertices, (c) foci, and (d) endpoints of the minor axis. Then (e) sketch the graph.

49. $4x^2 + 25y^2 - 16x - 50y - 59 = 0$

50. $9x^2 + 16y^2 - 54x - 64y + 1 = 0$

51. $25x^2 + 16y^2 - 200x + 96y + 144 = 0$

52. $49x^2 + 4y^2 + 196x - 40y + 100 = 0$

53. $6x^2 + 24x + 9y^2 + 36y + 6 = 0$

54. $5x^2 - 50x + 2y^2 - 12y + 93 = 0$

Find the equation of the ellipse (in standard form) that satisfies the following conditions. Then (a) graph the ellipse by hand, (b) confirm your graph by graphing the ellipse on a graphing calculator, and (c) find the length of the focal chords and verify the endpoints of the chords are on the graph.

55. vertices at $(-6, 0)$ and $(6, 0)$; foci at $(-4, 0)$ and $(4, 0)$

56. vertices at $(-8, 0)$ and $(8, 0)$; foci at $(-5, 0)$ and $(5, 0)$

57. foci at $(3, -6)$ and $(3, 2)$; length of minor axis: 6 units

58. foci at $(-4, -3)$ and $(8, -3)$; length of minor axis: 8 units

Use the characteristics of an ellipse and the graph given to write the related equation and find the location of the foci.

59.   **60.**

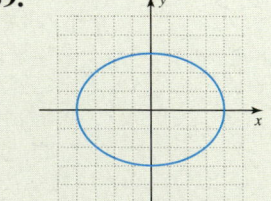

$\dfrac{x^2}{16} + \dfrac{y^2}{9} = 1, (\pm\sqrt{7}, 0)$

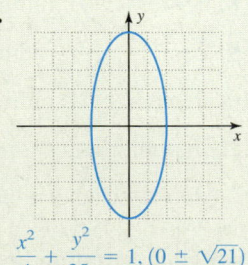

$\dfrac{x^2}{4} + \dfrac{y^2}{25} = 1, (0 \pm \sqrt{21})$

61.   **62.**

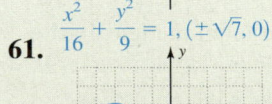

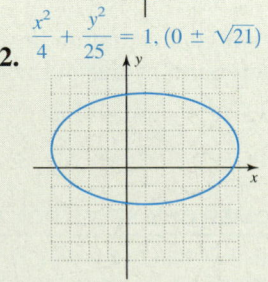

▶ WORKING WITH FORMULAS

63. Area of an Ellipse: $A = \pi ab$

The area of an ellipse is given by the formula shown, where a is the distance from the center to the graph in the horizontal direction and b is the distance from center to graph in the vertical direction. Find the area of the ellipse defined by $16x^2 + 9y^2 = 144$. $A = 12\pi$ units^2

64. The Perimeter of an Ellipse: $P = 2\pi\sqrt{\dfrac{a^2 + b^2}{2}}$

The perimeter of an ellipse can be *approximated* by the formula shown, where a represents the length of the semimajor axis and b represents the length of the semiminor axis. Find the perimeter of the ellipse defined by the equation $\dfrac{x^2}{49} + \dfrac{y^2}{4} = 1$.

$\sqrt{106}\,\pi \approx 32.34$ units

▶ APPLICATIONS

65. Decorative fireplaces: A bricklayer intends to build an elliptical fireplace 3 ft high and 8 ft wide, with two glass doors that open at the middle. The hinges to these doors are to be screwed onto a spine that is perpendicular to the hearth and goes through the foci of the ellipse. How far from center will the spines be located? How tall will each spine be?

$\sqrt{7} \approx 2.65$ ft
2.25 ft

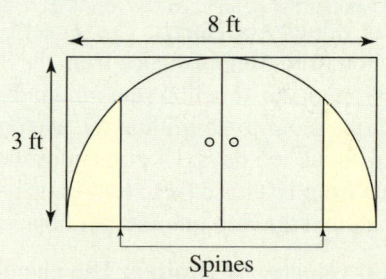

Spines

66. Decorative gardens: A retired math teacher decides to present her husband with a beautiful elliptical garden to help celebrate their 50th anniversary. The ellipse is to be 8 m long and 5 m across, with decorative fountains located at the foci. How far from the center of the ellipse should the fountains be located (round to the nearest 100th of a meter)? How far apart are the fountains?   3.12 m
6.24 ft

67. Attracting attention to art: As part of an art show, a gallery owner asks a student from the local university to design a unique exhibit that will highlight one of the more significant pieces in the collection, an ancient sculpture. The student decides to create an elliptical showroom with reflective walls, with a rotating laser light on a stand at one focus, and the sculpture placed at the other focus on a stand of equal height. The laser light then points continually at the sculpture as it rotates. If the elliptical room is 24 ft long and 16 ft wide, how far from the center of the ellipse should the stands be located (round to the nearest 10th of a foot)? How far apart are the stands?   8.9 ft
17.9 ft

Additional answers can be found in the Instructor Answer Appendix.

68. Medical procedures: The medical procedure called *lithotripsy* is a noninvasive medical procedure that is used to break up kidney and bladder stones in the body. A machine called a *lithotripter* uses its three-dimensional semielliptical shape and the foci properties of an ellipse to concentrate shock waves generated at one focus, on a kidney stone located at the other focus (see diagram—not drawn to scale). If the lithotripter has a length (semimajor axis) of 16 cm and a radius (semiminor axis) of 10 cm, how far from the vertex should a kidney stone be located for the best result? Round to the nearest hundredth.   28.49 cm

Exercise 68

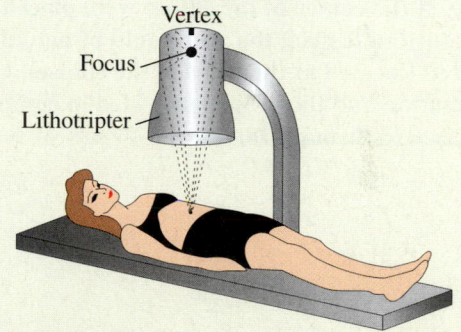

69. Elliptical arches: In some situations, bridges are built using uniform elliptical archways as shown in the figure given. Find the equation of the ellipse forming each arch if it has a total width of 30 ft and a maximum center height (above level ground) of 8 ft. What is the height of a point 9 ft to the right of the center of each arch?

Exercise 69

8 ft

60 ft

70. Elliptical arches: An elliptical arch bridge is built across a one-lane highway. The arch is 20 ft across and has a maximum center height of 12 ft. Will a farm truck hauling a load 10 ft wide with a clearance height of 11 ft be able to go under the bridge without damage? (*Hint:* See Exercise 69.)   no

71. Plumbing: By allowing the free flow of air, a properly vented home enables water to run freely throughout its plumbing system, while helping to prevent sewage gases from entering the home. Find the equation of the elliptical hole cut in a roof in order to allow a 3-in. vent pipe to exit, if the roof has a slope of $\frac{4}{3}$. $\dfrac{x^2}{6.25} + \dfrac{y^2}{2.25} = 1$

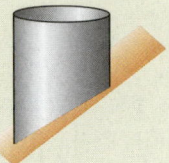

72. Light projection: Standing a short distance from a wall, Kymani's flashlight projects a circle of radius 30 cm. When holding the flashlight at an angle, a vertical ellipse 50 cm long is formed, with the focus 10 cm from the vertex (see *Worthy of Note*, page 717). Find the equation of the circle and ellipse, and the area of the wall that each illuminates. circle: $x^2 + y^2 = 30^2$, $A = 900\pi$ cm^2; ellipse: $\dfrac{x^2}{25^2} + \dfrac{y^2}{20^2} = 1$, $A = 500\pi$ cm^2

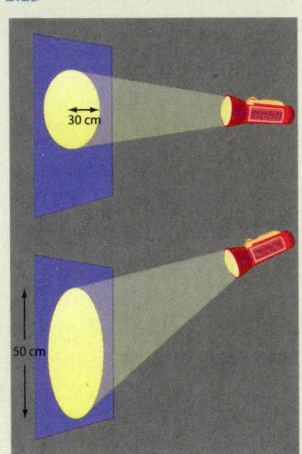

As a planet orbits around the Sun, it traces out an ellipse. If the center of the ellipse were placed at **(0, 0)** on a coordinate grid, the Sun would be actually off-centered (located at the *focus* of the ellipse). Use this information and the graphs provided to complete Exercises 73 through 78.

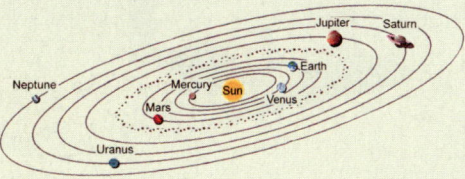

Exercise 73

73. Orbit of Mercury: The approximate orbit of the planet Mercury is shown in the figure given. Find an equation that models this orbit. $\dfrac{x^2}{36^2} + \dfrac{y^2}{(35.25)^2} = 1$

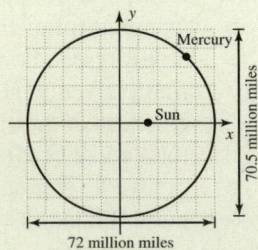

Exercise 74

74. Orbit of Pluto: The approximate orbit of the Kuiper object formerly known as Pluto is shown in the figure given. Find an equation that models this orbit. $\dfrac{x^2}{1825^2} + \dfrac{y^2}{1770^2} = 1$

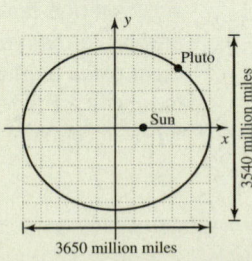

75. Planetary orbits: Except for small variations, a planet's orbit around the Sun is elliptical with the Sun at one focus. The aphelion (maximum distance from the Sun) of the planet Mars is approximately 156 million miles, while the perihelion (minimum distance from the Sun) of Mars is about 128 million miles. Use this information to find the lengths of the semimajor and semiminor axes, rounded to the nearest million. If Mars has an orbital velocity of 54,000 miles per hour (1.296 million miles per day), how many days does it take Mars to orbit the Sun? (*Hint:* Use the formula from Exercise 64.) $a \approx 142$ million miles, $b \approx 141$ million miles, orbit time ≈ 686 days

76. Planetary orbits: The aphelion (maximum distance from the Sun) of the planet Saturn is approximately 940 million miles, while the perihelion (minimum distance from the Sun) of Saturn is about 840 million miles. Use this information to find the lengths of the semimajor and semiminor axes, rounded to the nearest million. If Saturn has an orbital velocity of 21,650 miles per hour (about 0.52 million miles per day), how many days does it take Saturn to orbit the Sun? How many years? $a \approx 890$ million miles, $b \approx 889$ million miles, orbit time $\approx 10{,}748$ days; 29.4 yr

77. Orbital velocity of Earth: The planet Earth has a perihelion (minimum distance from the Sun) of about 91 million mi, an aphelion (maximum distance from the Sun) of close to 95 million mi, and completes one orbit in about 365 days. Use this information and the formula from Exercise 64 to find Earth's orbital speed around the Sun in miles per hour. about 66,697 mph

78. Orbital velocity of Jupiter: The planet Jupiter has a perihelion of 460 million mi, an aphelion of 508 million mi, and completes one orbit in about 4329 days. Use this information and the formula from Exercise 64 to find Jupiter's orbital speed around the Sun in miles per hour. about 29,252 mph

79. Area of a race track: Suppose the *Toronado 500* is a car race that is run on an elliptical track. The track is bounded by two ellipses with equations of $4x^2 + 9y^2 = 900$ and $9x^2 + 25y^2 = 900$, where x and y are in hundreds of yards. Use the formula given in Exercise 63 to find the area of the race track. 90,000 π yd^2

Exercise 80

80. Area of a border: The tablecloth for a large oval table is elliptical in shape. It is designed with two concentric ellipses (one within the other) as shown in the figure. The equation of the outer ellipse is $9x^2 + 25y^2 = 225$, and the equation of the inner ellipse is $4x^2 + 16y^2 = 64$ with x and y in feet. Use the formula given in Exercise 63 to find the area of the border of the tablecloth. 7π ft^2

Additional answers can be found in the Instructor Answer Appendix.

81. Whispering galleries: Due to their unique properties, ellipses are used in the construction of *whispering galleries* like those in St. Paul's Cathedral (London) and Statuary Hall in the U.S. Capitol. Regarding the latter, it is known that John Quincy Adams (1767–1848), while a member of the House of Representatives, situated his desk at a focal point of the elliptical ceiling, easily eavesdropping on the private conversations of other House members located near the other focal point. Suppose a whispering gallery was built using the equation

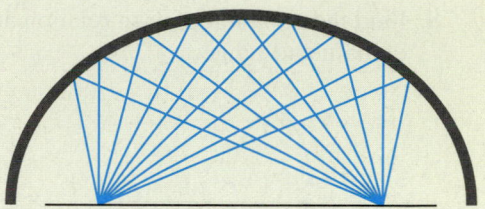

$\dfrac{x^2}{2809} + \dfrac{y^2}{2025} = 1$, with the dimensions in feet. (a) How tall is the ceiling at its highest point? (b) How wide is the gallery vertex to vertex? (c) How far from the base of the doors at either end, should a young couple stand so that one can clearly hear the other whispering, "I love you."?   **a.** 45 ft   **b.** 106 ft   **c.** 25 ft

82. While an elliptical billiard table has little practical value, it offers an excellent illustration of elliptical properties. A ball placed at one focus and hit with the cue stick from any angle, will hit the cushion and immediately rebound to the other focus and continue through each focus until coming to rest. Suppose one such table was constructed using the equation $\dfrac{x^2}{9} + \dfrac{y^2}{4} = 1$ as a model, with the dimensions in feet. (a) How far apart are the vertices? (b) How far apart are the foci? As a side note, Lewis Carroll (1832–1898) did invent a game of circular billiards, complete with rules.   **a.** 6 ft   **b.** about 4 ft $5\dfrac{2}{3}$ in.

▶ **EXTENDING THE CONCEPT**

83. For $6x^2 + 36x + 3y^2 - 24y + 74 = -28$, does the equation appear to be that of a circle, ellipse, or parabola? Write the equation in factored form. What do you notice? What can you say about the graph of this equation?
ellipse, since squared terms are positive and $A \neq B$; $6(x + 3)^2 + 3(y - 4)^2 = 0$, the constant term becomes zero; the graph is the single point $(-3, 4)$

84. Algebraically verify that for the ellipse $\dfrac{x^2}{a^2} + \dfrac{y^2}{b^2} = 1$ with $b > a$, the length of the focal chord is still $\dfrac{2a^2}{b}$.   Verified

▶ **MAINTAINING YOUR SKILLS**

85. (5.4) Evaluate the expression using the change-of-base formula: $\log_3 20$. $\dfrac{\log 20}{\log 3} \approx 2.73$

86. (3.1) Compute the product $z_1 z_2$ and quotient $\dfrac{z_1}{z_2}$ of: $z_1 = 2\sqrt{3} + 2i\sqrt{3}$; $z_2 = 5\sqrt{3} - 5i$

87. (2.3) Solve the absolute value inequality (a) graphically and (b) analytically: $-2|x - 3| + 10 > 4$.

88. (2.6) The resistance R to current flow in an electrical wire varies directly as the length L of the wire and inversely as the square of its diameter d. (a) Write the equation of variation; (b) find the constant of variation if a wire 2 m long with diameter $d = 0.005$ m has a resistance of 240 ohms (Ω); and (c) find the resistance in a similar wire 3 m long and 0.006 m in diameter.
  a. $R = \dfrac{kL}{d^2}$ **b.** $k = 0.003$ **c.** 250Ω

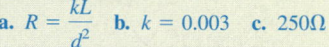

MID-CHAPTER CHECK

Sketch the graph of each conic section.

1. $(x - 4)^2 + (y + 3)^2 = 9$

2. $x^2 + y^2 - 10x + 4y + 4 = 0$

3. $\dfrac{(x - 2)^2}{16} + \dfrac{(y + 3)^2}{1} = 1$

4. $9x^2 + 4y^2 + 18x - 24y + 9 = 0$

Additional answers can be found in the Instructor Answer Appendix.

5. $\dfrac{(x + 3)^2}{9} + \dfrac{(y - 4)^2}{4} = 1$

6. $9x^2 + 16y^2 - 36x + 96y + 36 = 0$

7. Find the equation for all points located an equal distance from the point $(0, 3)$ and the line $y = -3$. $y = \dfrac{1}{12}x^2$

8. Find the equation of each relation and state its domain and range.

a.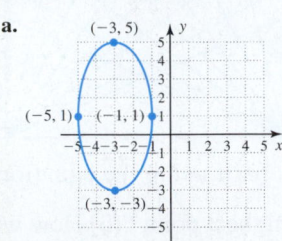

b.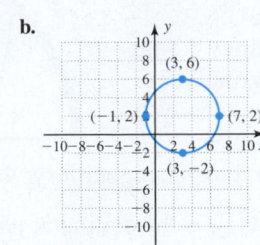

9. Find the equation of the ellipse having foci at $(0, 13)$ and $(0, -13)$, with a minor axis of length 10 units. $\dfrac{x^2}{25} + \dfrac{y^2}{194} = 1$

10. Find the equation of the ellipse (in standard form) if the vertices are $(-4, 0)$ and $(4, 0)$ and the distance between the foci is $4\sqrt{3}$ units. $\dfrac{x^2}{16} + \dfrac{y^2}{4} = 1$

REINFORCING BASIC CONCEPTS

More on Completing the Square

From our work so far in Chapter 8, we realize the process of *completing the square* has much greater use than simply as a tool for working with quadratic equations. It is a valuable tool in the application of the conic sections, as well as other areas. The purpose of this *Reinforcing Basic Concepts* is to strengthen the ability and confidence needed to apply the process correctly. This is important because in some cases the values of a and b are rational or irrational numbers. No matter what the context,

1. **The process begins** *with a coefficient of 1.* For $20x^2 + 120x + 27y^2 - 54y + 192 = 0$, we recognize the equation of an ellipse, since the coefficients of the squared terms are positive and unequal. To study or graph this ellipse, we'll use the standard form to identify the values of a, b, and c. Grouping the like-variable terms gives

$$(20x^2 + 120x \quad) + (27y^2 - 54y \quad) + 192 = 0$$

and to complete the square, we factor out the lead coefficient of each group (to get a coefficient of 1):

$$20(x^2 + 6x \quad) + 27(y^2 - 2y \quad) + 192 = 0$$

Subtracting 192 from both sides brings us to the fundamental step for completing the square.

2. **The quantity** $\left(\dfrac{1}{2} \cdot \textit{linear cofficient}\right)^2$ **will complete a trinomial square.** For this example we obtain

$$\left(\frac{1}{2} \cdot 6\right)^2 = 9 \text{ for } x, \text{ and } \left(\frac{1}{2} \cdot -2\right)^2 = 1 \text{ for } y, \text{ with these numbers inserted in the appropriate group:}$$

$$20(x^2 + 6x + 9) + 27(y^2 - 2y + 1) = -192 \quad \text{\textcolor{red}{complete the square}}$$

Due to the distributive property, we have in effect added $20 \cdot 9 = 180$ and $27 \cdot 1 = 27$ (for a total of 207) to the left side of the equation:

$$20(x^2 + 6x + 9) + 27(y^2 - 2y + 1) = -192$$

$$\underset{\text{to left side}}{\text{adds } 20 \cdot 9 = 180} \qquad \underset{\text{to left side}}{\text{adds } 27 \cdot 1 = 27}$$

This brings us to the final step.

3. **Keep the equation in balance.** Since the left side was increased by 207, we also increase the right side by 207.

$$20(x^2 + 6x + 9) + 27(y^2 - 2y + 1) = -192 + 207$$

$$\underset{\text{to left side}}{\text{adds } 20 \cdot 9 = 180} \qquad \underset{\text{to left side}}{\text{adds } 27 \cdot 1 = 27} \qquad \underset{\text{to right side}}{\text{add } 180 + 27 = 207}$$

The quantities in parentheses factor, giving $20(x + 3)^2 + 27(y - 1)^2 = 15$. We then divide by 15 and simplify, obtaining the standard form $\dfrac{4(x + 3)^2}{3} + \dfrac{9(y - 1)^2}{5} = 1$. Note the coefficient of each binomial square is not 1, even after setting the equation equal to 1. In the *Strengthening Core Skills* feature of this chapter, we'll look at how to write equations of this type in standard form to obtain the values of a and b. For now, practice completing the square using these exercises.

Exercise 1: $100x^2 - 400x + 18y^2 - 108y + 554 = 0 \quad \dfrac{25(x - 2)^2}{2} + \dfrac{9(y - 3)^2}{4} = 1$

Exercise 2: $28x^2 - 56x + 48y^2 + 192y + 195 = 0 \quad \dfrac{28(x - 1)^2}{25} + \dfrac{48(y + 2)^2}{25} = 1$

8.3 The Hyperbola

LEARNING OBJECTIVES

In Section 8.3 you will see how we can:

☐ **A.** Use the equation of a hyperbola to graph central and noncentral hyperbolas

☐ **B.** Distinguish between the equations of circles, ellipses, and hyperbolas

☐ **C.** Locate the foci of a hyperbola and use the foci and other features to write its equation

☐ **D.** Solve applications involving foci

As seen in Section 8.1 (see Figure 8.24), a hyperbola is a conic section formed by a plane that cuts both nappes of a right circular cone. A hyperbola has two symmetric parts called **branches,** which open in opposite directions. Although the branches appear to resemble parabolas, we will soon discover they are actually a very different curve.

Figure 8.24

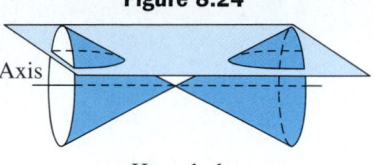

A. The Equation of a Hyperbola

In Section 8.2, we noted that for the equation $Ax^2 + By^2 = F$, if $A = B$, the equation is that of a circle, if $A \neq B$, the equation represents an ellipse. Both cases contain a *sum* of second-degree terms. Perhaps driven by curiosity, we might wonder what happens if the equation has a *difference* of second-degree terms. Consider the equation $9x^2 - 16y^2 = 144$. It appears the graph will be centered at $(0, 0)$ since no shifts are applied (h and k are both zero). Using the intercept method to graph this equation reveals an entirely new curve, called a *hyperbola*.

EXAMPLE 1 ▶ **Graphing a Central Hyperbola**

Graph the equation $9x^2 - 16y^2 = 144$ using intercepts and additional points as needed.

Solution ▶

$$9x^2 - 16y^2 = 144 \quad \text{given}$$
$$9(0)^2 - 16y^2 = 144 \quad \text{substitute 0 for } x$$
$$-16y^2 = 144 \quad \text{simplify}$$
$$y^2 = -9 \quad \text{divide by } -16$$

Since y^2 can never be negative, we conclude that the graph has *no y-intercepts.* Substituting $y = 0$ to find the x-intercepts gives

$$9x^2 - 16y^2 = 144 \quad \text{given}$$
$$9x^2 - 16(0)^2 = 144 \quad \text{substitute 0 for } y$$
$$9x^2 = 144 \quad \text{simplify}$$
$$x^2 = 16 \quad \text{divide by 9}$$
$$x = \sqrt{16} \quad \text{and} \quad x = -\sqrt{16} \quad \text{square root property}$$
$$x = 4 \quad \text{and} \quad x = -4 \quad \text{simplify}$$
$$(4, 0) \quad \text{and} \quad (-4, 0) \quad \text{x-intercepts}$$

Knowing the graph has no y-intercepts, we select inputs greater than 4 and less than -4 to help sketch the graph. Using $x = 5$ and $x = -5$ yields

$9x^2 - 16y^2 = 144$	given	$9x^2 - 16y^2 = 144$
$9(5)^2 - 16y^2 = 144$	substitute for x	$9(-5)^2 - 16y^2 = 144$
$9(25) - 16y^2 = 144$	$5^2 = (-5)^2 = 25$	$9(25) - 16y^2 = 144$
$225 - 16y^2 = 144$	simplify	$225 - 16y^2 = 144$
$-16y^2 = -81$	subtract 225	$-16y^2 = -81$
$y^2 = \dfrac{81}{16}$	divide by -16	$y^2 = \dfrac{81}{16}$
$y = \dfrac{9}{4} \quad y = -\dfrac{9}{4}$	square root property	$y = \dfrac{9}{4} \quad y = -\dfrac{9}{4}$
$y = 2.25 \quad y = -2.25$	decimal form	$y = 2.25 \quad y = -2.25$
$(5, 2.25) \quad (5, -2.25)$	ordered pairs	$(-5, 2.25) \quad (-5, -2.25)$

Plotting these points and connecting them with a smooth curve, while *knowing there are no y-intercepts,* produces the graph in the figure. The point at the origin (in blue) is not a part of the graph, and is given only to indicate the "center" of the hyperbola. The points $(-4, 0)$ and $(4, 0)$ are called **vertices,** and the **center** of the hyperbola is always the point halfway between them.

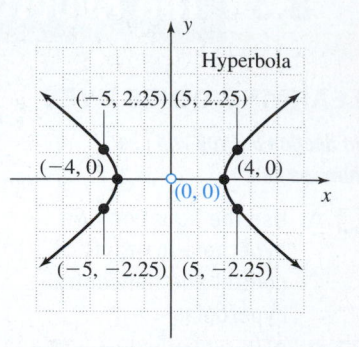

Now try Exercises 7 through 22 ▶

As with the circle and ellipse, the hyperbola fails the vertical line test and we must graph the relation on a calculator by writing its equation in two parts, each of which is a function. For the hyperbola in Example 1, this gives

$9x^2 - 16y^2 = 144$ original equation

$-16y^2 = 144 - 9x^2$ isolate *y*-term

$16y^2 = 9x^2 - 144$ multiply by -1

$y^2 = \dfrac{9x^2 - 144}{16}$ divide by 16

$y = \pm\sqrt{\dfrac{9x^2 - 144}{16}}$ take square roots

$Y_1 = +\sqrt{\dfrac{9X^2 - 144}{16}}, \quad Y_2 = -\sqrt{\dfrac{9X^2 - 144}{16}}$

Figure 8.25

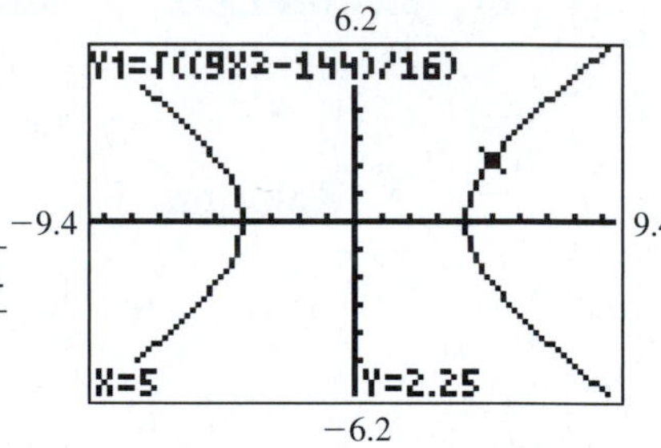

The graph is shown in Figure 8.25.

Since the hyperbola in Example 1 crosses a horizontal line of symmetry, it is referred to as a **horizontal hyperbola.** If the center is at the origin, we have a **central hyperbola.** The line passing through the center and both vertices is called the **transverse axis** (vertices are always on the transverse axis), and the line passing through the center and perpendicular to this axis is called the **conjugate axis** (see Figure 8.26).

In Example 1, the coefficient of x^2 was positive and we were subtracting $16y^2$: $9x^2 - 16y^2 = 144$. The result was a horizontal hyperbola. If the y^2-term is positive and we subtract the term containing x^2, the result is a **vertical hyperbola** (Figure 8.27).

Figure 8.26

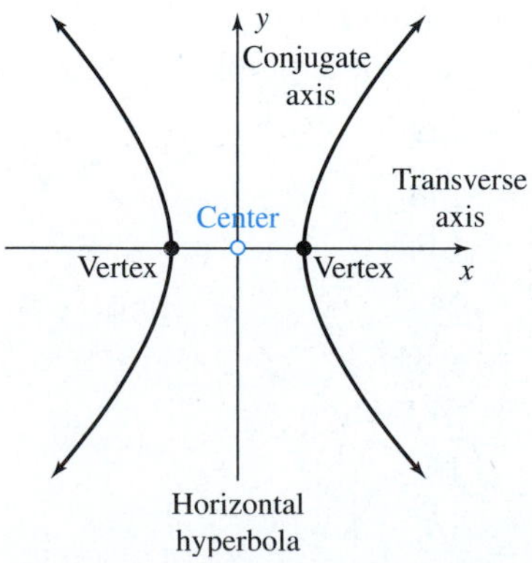

Horizontal hyperbola

Figure 8.27

Vertical hyperbola

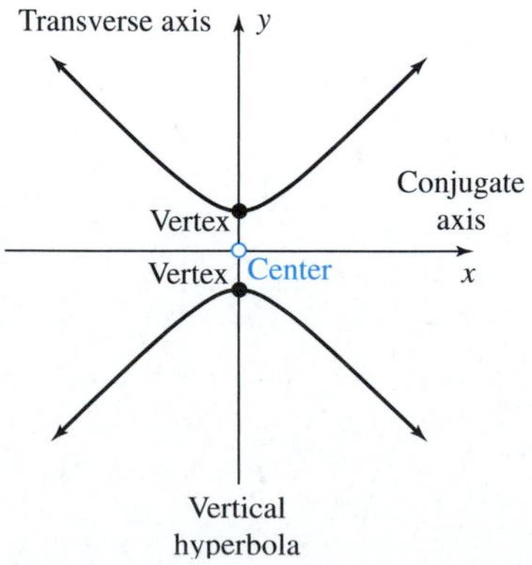

EXAMPLE 2 ▶ **Identifying the Axes, Vertices, and Center of a Hyperbola from Its Graph**

For the hyperbola shown, state the location of the vertices and the equation of the transverse axis. Then identify the location of the center and the equation of the conjugate axis.

Solution ▶ By inspection we locate the vertices at $(0, 0)$ and $(0, 4)$. The equation of the transverse axis is $x = 0$. The center is halfway between the vertices at $(0, 2)$, meaning the equation of the conjugate axis is $y = 2$.

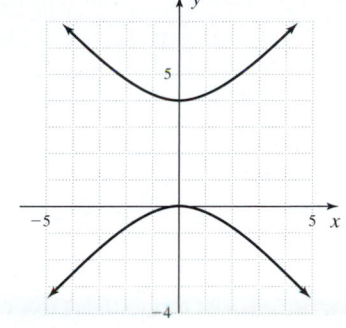

Now try Exercises 23 through 26 ▶

Standard Form

As with the ellipse, the polynomial form of the equation is helpful for *identifying* hyperbolas, but not very helpful when it comes to *graphing* a hyperbola (since we still must go through the laborious process of finding additional points). For graphing, standard form is once again preferred. Consider the hyperbola $9x^2 - 16y^2 = 144$ from Example 1. To write the equation in standard form, we divide by 144 and obtain $\dfrac{x^2}{4^2} - \dfrac{y^2}{3^2} = 1$. By comparing the standard form to the graph, we note $a = 4$ represents the distance from center to vertices, similar to the way we used a previously. But since the graph has no y-intercepts, what could $b = 3$ represent? The answer lies in the fact that branches of a hyperbola are **asymptotic,** meaning they will approach and become very close to imaginary lines that can be used to sketch the graph. For a central hyperbola, the slopes of the asymptotic lines are given by the ratios $\dfrac{b}{a}$ and $-\dfrac{b}{a}$, with the related equations being $y = \dfrac{b}{a}x$ and $y = -\dfrac{b}{a}x$. The graph from Example 1 is repeated in Figure 8.28, with the asymptotes drawn. For a clearer understanding of how the equations for the asymptotes were determined, **see Exercise 87.**

A second method of drawing the asymptotes involves drawing a **central rectangle** with dimensions $2a$ by $2b$, as shown in Figure 8.29. The asymptotes will be the *extended diagonals* of this rectangle. This brings us to the equation of a hyperbola in standard form.

Figure 8.28

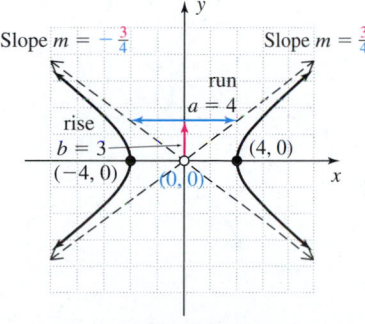

Slope method

Figure 8.29

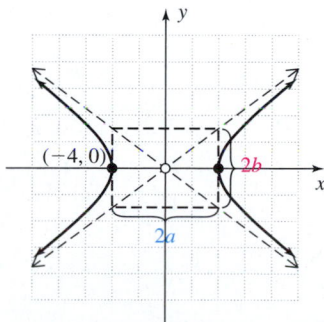

Central rectangle method

The Equation of a Hyperbola in Standard Form

The equation

$$\frac{(x - h)^2}{a^2} - \frac{(y - k)^2}{b^2} = 1$$

represents a *horizontal* hyperbola with center (h, k)
- *transverse* axis $y = k$
- *conjugate* axis $x = h$
- $|a|$ gives the distance from center to vertices.

The equation

$$\frac{(y - k)^2}{b^2} - \frac{(x - h)^2}{a^2} = 1$$

represents a *vertical* hyperbola with center (h, k)
- *transverse* axis $x = h$
- *conjugate* axis $y = k$
- $|b|$ gives the distance from center to vertices.

- Asymptotes can be drawn by starting at (h, k) and using slopes $m = \pm\dfrac{b}{a}$.

EXAMPLE 3 ▶ **Graphing a Hyperbola Using Its Equation in Standard Form**

Sketch the graph of $16(x - 2)^2 - 9(y - 1)^2 = 144$, and label the center, vertices, and asymptotes.

Solution ▶ Begin by noting a difference of the second-degree terms, with the x^2-term occurring first. This means we'll be graphing a horizontal hyperbola whose center is at $(2, 1)$. Continue by writing the equation in standard form.

$$16(x - 2)^2 - 9(y - 1)^2 = 144 \qquad \text{given equation}$$

$$\frac{16(x - 2)^2}{144} - \frac{9(y - 1)^2}{144} = \frac{144}{144} \qquad \text{divide by 144}$$

$$\frac{(x - 2)^2}{9} - \frac{(y - 1)^2}{16} = 1 \qquad \text{simplify}$$

$$\frac{(x - 2)^2}{3^2} - \frac{(y - 1)^2}{4^2} = 1 \qquad \text{write denominators in squared form}$$

Since $a = 3$ the vertices are a horizontal distance of 3 units from the center $(2, 1)$, giving $(2 + 3, 1) \rightarrow (5, 1)$ and $(2 - 3, 1) \rightarrow (-1, 1)$. After plotting the center and vertices, we can begin at the center and count off slopes of $m = \pm\dfrac{b}{a} = \pm\dfrac{4}{3}$, or draw a rectangle centered at $(2, 1)$ with dimensions $2(3) = 6$ (horizontal dimension) by $2(4) = 8$ (vertical dimension) to sketch the asymptotes. The complete graph is shown here.

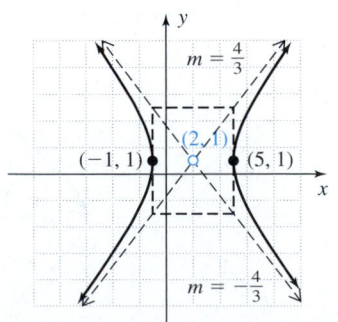

Horizontal hyperbola

Center at $(2, 1)$

Vertices at $(-1, 1)$ and $(5, 1)$

Transverse axis: $y = 1$
Conjugate axis: $x = 2$

Width of rectangle
$\begin{pmatrix} \text{horizontal dimension and} \\ \text{distance between vertices} \end{pmatrix}$
$2a = 2(3) = 6$

Length of rectangle
(vertical dimension)
$2b = 2(4) = 8$

Now try Exercises 27 through 44 ▶

If the hyperbola in Example 3 were a central hyperbola, the equations of the asymptotes would be $y = \frac{4}{3}x$ and $y = -\frac{4}{3}x$. But the center of this graph has been shifted 2 units right and 1 unit up. Using our knowledge of shifts and translations, the equations for the asymptotes of the shifted hyperbola must be

1. $(y - 1) = +\frac{4}{3}(x - 2)$, or $y = \frac{4}{3}x - \frac{5}{3}$ in simplified form, and

2. $(y - 1) = -\frac{4}{3}(x - 2)$ or $y = -\frac{4}{3}x + \frac{11}{3}$.

Using $Y_1 = +\sqrt{\dfrac{16(X - 2)^2 - 144}{9}} + 1$, and $Y_2 = -\sqrt{\dfrac{16(X - 2)^2 - 144}{9}} + 1$

(obtained by solving for y in the original equation), a calculator generated graph of the hyperbola and its asymptotes is shown here (Figures 8.30 and 8.31).

Figure 8.30

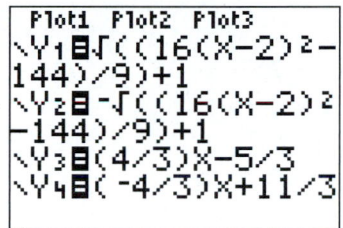

Figure 8.31

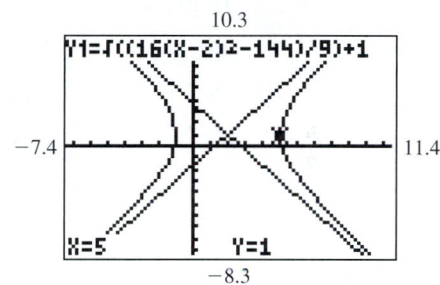

Polynomial Form

If the equation is given as a polynomial in expanded form, complete the square in x and y, then write the equation in standard form.

EXAMPLE 4 ▶   Graphing a Hyperbola by Completing the Square

Graph the equation $9y^2 - x^2 + 54y + 4x + 68 = 0$ by completing the square. Label the center and vertices and sketch the asymptotes. Then graph the hyperbola on a graphing calculator and use the ⬭TRACE⬭ feature with a "friendly" window to locate four additional points whose coordinates are rational.

Solution ▶ Since the y^2-term occurs first, we assume the equation represents a vertical hyperbola, but wait for the factored form to be sure (**see Exercise 91**).

$$9y^2 - x^2 + 54y + 4x + 68 = 0 \qquad \text{given}$$

$$9y^2 + 54y - x^2 + 4x = -68 \qquad \text{collect like-variable terms; subtract 68}$$

$$9(y^2 + 6y + \underline{\quad}) - 1(x^2 - 4x + \underline{\quad}) = -68 \qquad \text{factor out 9 from } y\text{-terms and } -1 \text{ from } x\text{-terms}$$

$$9(y^2 + 6y + 9) - 1(x^2 - 4x + 4) = -68 + 81 + (-4) \qquad \text{complete the square}$$
$$\underset{\text{adds } 9(9) = 81}{\underbrace{\qquad\qquad}} \qquad \underset{\text{adds } -1(4) = -4}{\underbrace{\qquad\qquad}} \qquad\qquad\qquad \text{add } 81 + (-4) \text{ to right}$$

$$9(y + 3)^2 - 1(x - 2)^2 = 9 \qquad \text{factor} \rightarrow \text{vertical hyperbola}$$

$$\frac{(y + 3)^2}{1} - \frac{(x - 2)^2}{9} = 1 \qquad \text{divide by 9 (standard form)}$$

$$\frac{(y + 3)^2}{1^2} - \frac{(x - 2)^2}{3^2} = 1 \qquad \text{write denominators in squared form}$$

The center of the hyperbola is $(2, -3)$ with $a = 3$, $b = 1$, and a transverse axis of $x = 2$. The vertices are at $(2, -3 + 1)$ and $(2, -3 - 1) \rightarrow (2, -2)$ and $(2, -4)$. After plotting the center and vertices, we draw a rectangle centered at $(2, -3)$ with a horizontal "width" of $2(3) = 6$ and a vertical "length" of $2(1) = 2$ to sketch the asymptotes. The completed graph is given in Figure 8.32.

Figure 8.32

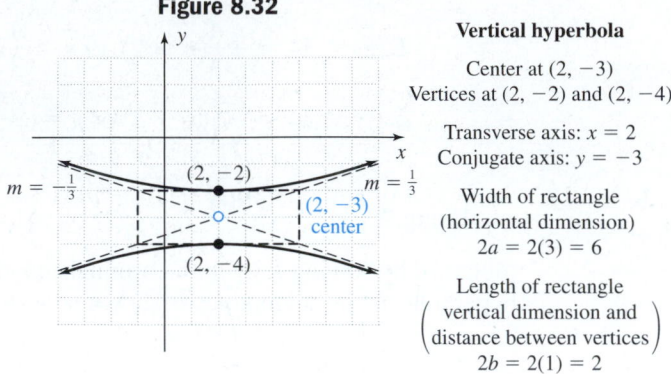

Vertical hyperbola

Center at $(2, -3)$
Vertices at $(2, -2)$ and $(2, -4)$

Transverse axis: $x = 2$
Conjugate axis: $y = -3$

Width of rectangle
(horizontal dimension)
$2a = 2(3) = 6$

Length of rectangle
$\begin{pmatrix} \text{vertical dimension and} \\ \text{distance between vertices} \end{pmatrix}$
$2b = 2(1) = 2$

To graph the hyperbola on a calculator, we again solve for y.

$$9(y + 3)^2 - (x - 2)^2 = 9 \qquad \text{factored form}$$

$$9(y + 3)^2 = 9 + (x - 2)^2 \qquad \text{isolate term containing } y$$

$$(y + 3)^2 = \frac{9 + (x - 2)^2}{9} \qquad \text{divide by 9}$$

$$y + 3 = \pm\sqrt{\frac{9 + (x - 2)^2}{9}} \qquad \text{take square roots}$$

$$y = \pm\sqrt{\frac{9 + (x - 2)^2}{9}} - 3 \qquad \text{subtract 3}$$

$$Y_1 = +\sqrt{\frac{9 + (X - 2)^2}{9}} - 3, \quad Y_2 = -\sqrt{\frac{9 + (X - 2)^2}{9}} - 3$$

Figure 8.33

Using the friendly window shown, we can use the arrow keys to TRACE though x-values and find the points $(-5.2, -0.4)$ and $(9.2, -0.4)$ on the upper branch, with $(-5.2, -5.6)$ and $(9.2, -5.6)$ on the lower branch. Note how these points show that a hyperbola is symmetric to its center, as well as the horizontal line and vertical line through its center. The graph is shown in Figure 8.33.

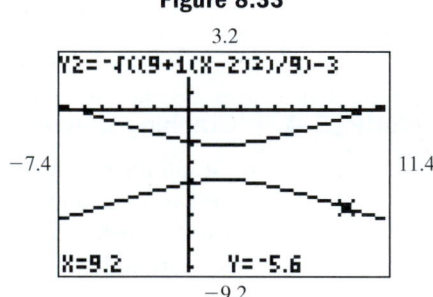

✓ **A.** You've just seen how we can use the equation of a hyperbola to graph central and noncentral hyperbolas

Now try Exercises 45 through 48 ▶

B. Distinguishing between the Equations of Circles, Ellipses, and Hyperbolas

So far we've explored numerous graphs of circles, ellipses, and hyperbolas. In Example 5 we'll attempt to identify a given conic section from its equation alone (without graphing the equation). As you've seen, the corresponding equations have unique characteristics that can help distinguish one from the other.

EXAMPLE 5 ▶ **Identifying a Conic Section from Its Equation**

Identify each equation as that of a circle, ellipse, or hyperbola. Justify your choice and name the center, but do not draw the graphs.

  a. $y^2 = 36 + 9x^2$ **b.** $4x^2 = 16 - 4y^2$
  c. $x^2 = 225 - 25y^2$ **d.** $25x^2 = 100 + 4y^2$
  e. $3(x - 2)^2 + 4(y + 3)^2 = 12$ **f.** $4(x + 5)^2 = 36 + 9(y - 4)^2$

Solution ▶ **a.** Writing the equation as $y^2 - 9x^2 = 36$ shows $h = 0$ and $k = 0$. Since the equation contains a difference of second-degree terms, it is the equation of a (vertical) hyperbola (A and B have opposite signs). The center is at $(0, 0)$.

b. Rewriting the equation as $4x^2 + 4y^2 = 16$ and dividing by 4 gives $x^2 + y^2 = 4$. The equation represents a circle of radius 2 ($A = B$), with the center at $(0, 0)$.

c. Writing the equation as $x^2 + 25y^2 = 225$ we note a sum of second-degree terms with unequal coefficients. The equation is that of an ellipse ($A \neq B$), with the center at $(0, 0)$.

d. Rewriting the equation as $25x^2 - 4y^2 = 100$ we note the equation contains a difference of second-degree terms. The equation represents a central (horizontal) hyperbola (A and B have opposite signs), whose center is at $(0, 0)$.

e. The equation is in factored form and contains a sum of second-degree terms with unequal coefficients. This is the equation of an ellipse ($A \neq B$) with the center at $(2, -3)$.

f. Rewriting the equation as $4(x + 5)^2 - 9(y - 4)^2 = 36$ we note a difference of second-degree terms. The equation represents a horizontal hyperbola (A and B have opposite signs) with center $(-5, 4)$.

☑ **B. You've just seen how we can distinguish between the equations of circles, ellipses, and hyperbolas**

Now try Exercises 49 through 60 ▶

C. The Foci of a Hyperbola

Like the ellipse, the foci of a hyperbola play an important part in their application. A long distance radio navigation system (called LORAN for short), can be used to determine the location of ships and airplanes and is based on the characteristics of a hyperbola (**see Exercises 85 and 86**). Hyperbolic mirrors are also used in some telescopes, and have the property that a beam of light directed at one focus will be reflected to the second focus. To understand and appreciate these applications, we use the analytic definition of a hyperbola:

Definition of a Hyperbola

Given two fixed points f_1 and f_2 in a plane, a hyperbola is the set of all points (x, y) such that the distance d_1 from f_1 to (x, y) and the distance d_2 from f_2 to (x, y), satisfy the equation

$$|d_1 - d_2| = k.$$

In other words, the difference of these two distances is a positive constant.

  The fixed points f_1 and f_2 are called the foci of the hyperbola, and all such points (x, y) are on the graph of the hyperbola.

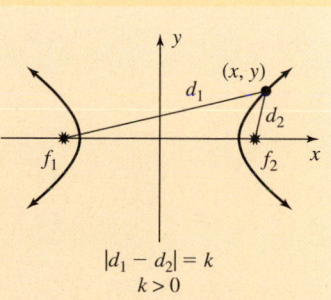

Figure 8.34

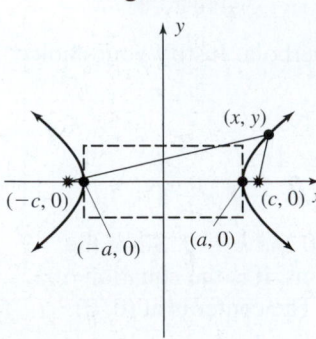

As with the analytic definition of the ellipse, it can be shown that the constant k is again equal to $2a$ (for horizontal hyperbolas). To find the equation of a hyperbola in terms of a and b, we use an approach similar to that of the ellipse (see Appendix V), and the result is identical to that seen earlier: $\dfrac{x^2}{a^2} - \dfrac{y^2}{b^2} = 1$ where $b^2 = c^2 - a^2$ (see Figure 8.34).

We now have the ability to *find the foci of any hyperbola*—and can use this information in many significant applications. Since the location of the foci play such an important role, it is best to remember the relationship as $c^2 = a^2 + b^2$ (called the **foci formula** for hyperbolas), noting that for a hyperbola, $c > a$ and $c^2 > a^2$ (also $c > b$ and $c^2 > b^2$). Be sure to note that for ellipses, the foci formula is $c^2 = |a^2 - b^2|$ since $a > c$ (horizontal ellipses) or $b > c$ (vertical ellipses).

EXAMPLE 6 ▶ **Graphing a Hyperbola and Identifying Its Foci by Completing the Square**

For the hyperbola defined by $7x^2 - 9y^2 - 14x + 72y - 200 = 0$, find the coordinates of the center, vertices, foci, and the dimensions of the central rectangle. Then sketch the graph, including the asymptotes.

Solution ▶

$$7x^2 - 9y^2 - 14x + 72y - 200 = 0 \qquad \text{given}$$
$$7x^2 - 14x - 9y^2 + 72y = 200 \qquad \text{group terms; add 200}$$
$$7(x^2 - 2x + \underline{\quad}) - 9(y^2 - 8y + \underline{\quad}) = 200 \qquad \text{factor out leading coefficients}$$
$$7(x^2 - 2x + 1) - 9(y^2 - 8y + 16) = 200 + 7 + (-144) \qquad \text{complete the square}$$
$$\to \text{add } 7 + (-144)$$

adds $7(1) = 7$ · · · adds $-9(16) = -144$ · · · to right-hand side

$$7(x - 1)^2 - 9(y - 4)^2 = 63 \qquad \text{factored form}$$
$$\frac{(x - 1)^2}{9} - \frac{(y - 4)^2}{7} = 1 \qquad \text{divide by 63 and simplify}$$
$$\frac{(x - 1)^2}{3^2} - \frac{(y - 4)^2}{(\sqrt{7})^2} = 1 \qquad \text{write denominators in squared form}$$

This is a horizontal hyperbola with $a = 3$ ($a^2 = 9$) and $b = \sqrt{7}$ ($b^2 = 7$). The center is at $(1, 4)$, with vertices $(-2, 4)$ and $(4, 4)$. Using the foci formula $c^2 = a^2 + b^2$ yields $c^2 = 9 + 7 = 16$, showing the foci are $(-3, 4)$ and $(5, 4)$ (4 units from center). The central rectangle is $2(3) = 6$ by $2\sqrt{7} \approx 5.29$. Drawing the rectangle and sketching the asymptotes results in the graph shown.

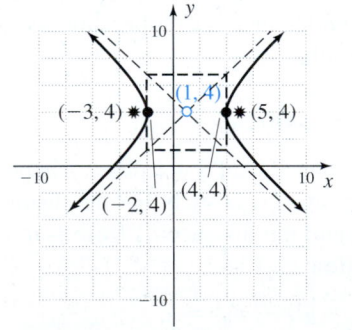

Horizontal hyperbola

Center at $(1, 4)$
Vertices at $(-2, 4)$ and $(4, 4)$

Transverse axis: $y = 4$
Conjugate axis: $x = 1$
Location of foci: $(-3, 4)$ and $(5, 4)$

Width of rectangle
$\begin{pmatrix} \text{horizontal dimension and} \\ \text{distance between vertices} \end{pmatrix}$
$2a = 2(3) = 6$

Length of rectangle
(vertical dimension)
$2b = 2(\sqrt{7}) \approx 5.29$

Now try Exercises 61 through 70 ▶

The focal chord for a horizontal hyperbola is a vertical line segment through the focus with endpoints on the hyperbola. Similar to the focal chord of an ellipse, we can use its length to find additional points on the graph of the hyperbola. The total length is once again $L = \dfrac{2b^2}{a}$ (for a horizontal hyperbola), meaning the distance from the foci to the graph (along the focal chord) is $\dfrac{b^2}{a}$.

For Example 6, $a = 3$ and $b^2 = 7$, so the vertical distance from focus to graph (in either direction) is $\dfrac{7}{3} = 2.\overline{3}$. From the left focus $(-3, 4)$, we can now graph the additional points $(-3, 4 + 2.\overline{3}) = (-3, 6.\overline{3})$, and $(-3, 4 - 2.\overline{3}) = (-3, 1.\overline{6})$. From the right focus $(5, 4)$, we obtain $(5, 6.\overline{3})$ and $(5, 1.\overline{6})$. Graphical verification is provided in Figure 8.35. Also **see Exercise 80.**

Figure 8.35

☑ **C.** You've just seen how we can locate the foci of a hyperbola and use the foci and other features to write its equation

As with the ellipse, if any two of the values for a, b, and c are known, the relationship between them can be used to construct the equation of the hyperbola. **See Exercises 71 through 78.**

D. Applications Involving Foci

Applications involving the foci of a conic section can take many forms. As before, only partial information about the hyperbola may be available, and we'll determine a solution by manipulating a given equation or constructing an equation from given facts.

EXAMPLE 7 ▶ **Applying the Properties of a Hyperbola—The Path of a Comet**

Comets with a high velocity cannot be captured by the Sun's gravity, and are slung around the Sun in a hyperbolic path with the Sun at one focus. If the path illustrated by the graph shown is modeled by the equation $2116x^2 - 400y^2 = 846{,}400$, how close did the comet get to the Sun? Assume units are in millions of miles and round to the nearest million.

Solution ▶ We are essentially asked to find the distance between a vertex and focus. Begin by writing the equation in standard form:

$2116x^2 - 400y^2 = 846{,}400$ given

$\dfrac{x^2}{400} - \dfrac{y^2}{2116} = 1$ divide by 846,400

$\dfrac{x^2}{20^2} - \dfrac{y^2}{46^2} = 1$ write denominators in squared form

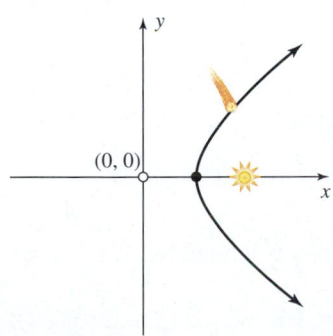

This is a horizontal hyperbola with $a = 20$ ($a^2 = 400$) and $b = 46$ ($b^2 = 2116$). Use the foci formula to find c^2 and c.

$$c^2 = a^2 + b^2$$
$$c^2 = 400 + 2116$$
$$c^2 = 2516$$
$$c \approx -50 \text{ and } c \approx 50$$

Since $a = 20$ and $|c| \approx 50$, the comet came within about $50 - 20 = 30$ million miles of the Sun.

Now try Exercises 81 through 84 ▶

EXAMPLE 8 ▶ **Applying the Properties of a Hyperbola—The Location of a Storm**

Two amateur meteorologists, living 4 km apart (4000 m), see a storm approaching. The one farthest from the storm hears a loud clap of thunder 9 sec after the one nearest. Assuming the speed of sound is 340 m/sec, determine an equation that models possible locations for the storm at this time.

Solution ▶ Let M_1 represent the meteorologist nearest the storm and M_2 the farthest. Since M_2 heard the thunder 9 sec after M_1, M_2 must be $9 \cdot 340 = 3060$ m farther away from the storm S. In other words, from our definition of a hyperbola, we have $|d_1 - d_2| = 3060$. The set of all points that satisfy this equation will be on the graph of a hyperbola, and we'll use this fact to develop an equation model for possible locations of the storm. Let's place the information on a coordinate grid. For convenience, we'll use the straight line distance between M_1 and M_2 as the x-axis, with the origin an equal distance from each. With the constant difference equal to 3060, we have $2a = 3060$, $a = 1530$ from the definition of a hyperbola, giving $\dfrac{x^2}{1530^2} - \dfrac{y^2}{b^2} = 1$. With $c = 2000$ m (the distance from the origin to M_1 or M_2), we find the value of b using the equation $c^2 = a^2 + b^2$: $2000^2 = 1530^2 + b^2$ or $b^2 = (2000)^2 - (1530)^2 = 1{,}659{,}100 \approx 1288^2$. The equation that models possible locations of the storm is $\dfrac{x^2}{1530^2} - \dfrac{y^2}{1288^2} \approx 1$.

☑ **D. You've just seen how we can solve applications involving foci**

Now try Exercises 85 and 86 ▶

8.3 EXERCISES

▶ CONCEPTS AND VOCABULARY

Fill in the blank with the appropriate word or phrase. Carefully reread the section if needed.

1. The line that passes through the vertices of a hyperbola is called the __transverse__ axis.

2. The center of a hyperbola is located __midway__ between the vertices.

3. The conjugate axis is __perpendicular__ to the __transverse__ axis and contains the __center__ of the hyperbola.

4. The center of the hyperbola defined by $\dfrac{(x-2)^2}{4^2} - \dfrac{(y-3)^2}{5^2} = 1$ is at __(2, 3)__.

HOMEWORK SELECTION GUIDE

Core: 7–75 every other odd, 81, 83, 85 (21 Exercises)
Standard: 1–4, 7–75 every other odd, 79, 81, 83, 84, 85, 89, 91, 92 (30 Exercises)

Extended: 1–4, 7–75 every other odd, 79–85, 87, 89, 90–92 (34 Exercises)
In Depth: 1–6, 7–75 every other odd, 79–93 (39 Exercises)

5. Compare/Contrast the two methods used to find the asymptotes of a hyperbola. Include an example illustrating both methods. *Answers will vary.*

6. Explore/Explain why $A(x - h)^2 - B(y - k)^2 = F$, $(A, B > 0)$ results in a hyperbola regardless of whether $A = B$ or $A \neq B$. Illustrate with an example. *Answers will vary.*

▶ DEVELOPING YOUR SKILLS

Graph each hyperbola. Label the center, vertices, and any additional points used.

7. $\dfrac{x^2}{9} - \dfrac{y^2}{4} = 1$

8. $\dfrac{x^2}{16} - \dfrac{y^2}{9} = 1$

9. $\dfrac{x^2}{4} - \dfrac{y^2}{9} = 1$

10. $\dfrac{x^2}{25} - \dfrac{y^2}{16} = 1$

11. $\dfrac{x^2}{49} - \dfrac{y^2}{16} = 1$

12. $\dfrac{x^2}{25} - \dfrac{y^2}{9} = 1$

13. $\dfrac{x^2}{36} - \dfrac{y^2}{16} = 1$

14. $\dfrac{x^2}{81} - \dfrac{y^2}{16} = 1$

15. $\dfrac{y^2}{9} - \dfrac{x^2}{1} = 1$

16. $\dfrac{y^2}{1} - \dfrac{x^2}{4} = 1$

17. $\dfrac{y^2}{12} - \dfrac{x^2}{4} = 1$

18. $\dfrac{y^2}{9} - \dfrac{x^2}{18} = 1$

19. $\dfrac{y^2}{9} - \dfrac{x^2}{9} = 1$

20. $\dfrac{y^2}{4} - \dfrac{x^2}{4} = 1$

21. $\dfrac{y^2}{36} - \dfrac{x^2}{25} = 1$

22. $\dfrac{y^2}{16} - \dfrac{x^2}{4} = 1$

For the graphs given, state the location of the vertices and the equation of the transverse axis. Then identify the location of the center and the equation of the conjugate axis. Note the scale used on each axis.

23.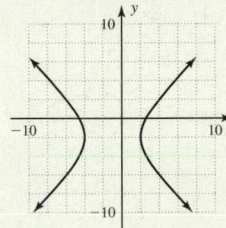

24.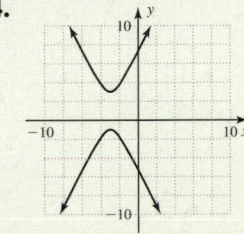

25.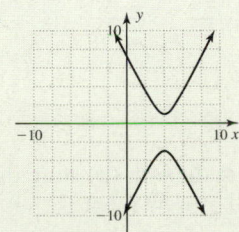

26.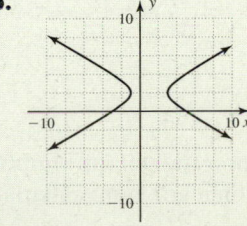

Additional answers can be found in the Instructor Answer Appendix.

Sketch a complete graph of each equation, including the asymptotes. Be sure to identify the center and vertices.

27. $\dfrac{(y + 1)^2}{4} - \dfrac{x^2}{25} = 1$

28. $\dfrac{y^2}{4} - \dfrac{(x - 2)^2}{9} = 1$

29. $\dfrac{(x - 3)^2}{36} - \dfrac{(y + 2)^2}{49} = 1$

30. $\dfrac{(x - 2)^2}{9} - \dfrac{(y - 1)^2}{4} = 1$

31. $\dfrac{(y + 1)^2}{7} - \dfrac{(x + 5)^2}{9} = 1$

32. $\dfrac{(y - 3)^2}{16} - \dfrac{(x + 2)^2}{5} = 1$

33. $(x - 2)^2 - 4(y + 1)^2 = 16$

34. $9(x + 1)^2 - (y - 3)^2 = 81$

35. $2(y + 3)^2 - 5(x - 1)^2 = 50$

36. $9(y - 4)^2 - 5(x - 3)^2 = 45$

37. $12(x - 4)^2 - 5(y - 3)^2 = 60$

38. $8(x - 4)^2 - 3(y - 3)^2 = 24$

39. $16x^2 - 9y^2 = 144$ **40.** $16x^2 - 25y^2 = 400$

41. $9y^2 - 4x^2 = 36$ **42.** $25y^2 - 4x^2 = 100$

43. $12x^2 - 9y^2 = 72$ **44.** $36x^2 - 20y^2 = 180$

Graph each hyperbola by writing the equation in standard form. Label the center and vertices, and sketch the asymptotes. Then use a graphing calculator to graph each relation and locate four additional points whose coordinates are rational.

45. $4x^2 - y^2 + 40x - 4y + 60 = 0$

46. $x^2 - 4y^2 - 12x - 16y + 16 = 0$

47. $4y^2 - x^2 - 24y - 4x + 28 = 0$

48. $-9x^2 + 4y^2 - 18x - 24y - 9 = 0$

Classify each equation as that of a circle, ellipse, or hyperbola. Justify your response (assume all are nondegenerate).

49. $-4x^2 - 4y^2 = -24$ *circle; $A = B$*

50. $9y^2 = -4x^2 + 36$ *ellipse; $A \neq B$*

51. $x^2 + y^2 = 2x + 4y + 4$ circle; $A = B$

52. $x^2 = y^2 + 6y - 7$ hyperbola; A, B opposite signs

53. $2x^2 - 4y^2 = 8$ hyperbola; A, B opposite signs

54. $36x^2 + 25y^2 = 900$ ellipse; $A \neq B$

55. $x^2 + 5 = 2y^2$ hyperbola; A, B opposite signs

56. $x + y^2 = 3x^2 + 9$ hyperbola; A, B opposite signs

57. $2x^2 = -2y^2 + x + 20$ circle; $A = B$

58. $2y^2 + 3 = 6x^2 + 8$ hyperbola; A, B opposite signs

59. $16x^2 + 5y^2 - 3x + 4y = 538$ ellipse; $A \neq B$

60. $9x^2 + 9y^2 - 9x + 12y + 4 = 0$ circle; $A = B$

Use the definition of a hyperbola to find the distance between the vertices and the dimensions of the rectangle centered at (h, k). Figures are not drawn to scale. Note that Exercises 63 and 64 are *vertical hyperbolas*.

61.

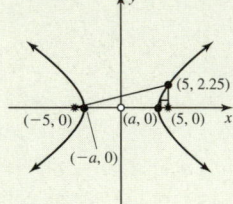

$8, 2a = 8, 2b = 6$

62.

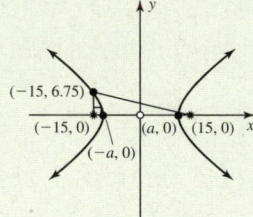

$24, 2a = 24, 2b = 18$

63.

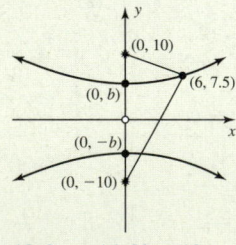

$12, 2a = 16, 2b = 12$

64.

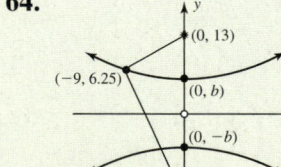

$10, 2a = 24, 2b = 10$

Write each equation in standard form to find and list the coordinates of the (a) center, (b) vertices, (c) foci, and (d) dimensions of the central rectangle. Then (e) sketch the graph, including the asymptotes.

65. $4x^2 - 9y^2 - 24x + 72y - 144 = 0$

66. $4x^2 - 36y^2 - 40x + 144y - 188 = 0$

67. $4y^2 - 16x^2 - 24y - 28 = 0$

68. $4y^2 - 81x^2 - 162x - 405 = 0$

69. $9x^2 - 3y^2 - 54x - 12y + 33 = 0$

70. $10x^2 + 60x - 5y^2 + 20y - 20 = 0$

Find the equation of the hyperbola (in standard form) that satisfies the following conditions:

71. vertices at $(-6, 0)$ and $(6, 0)$; foci at $(-8, 0)$ and $(8, 0)$ $\dfrac{x^2}{36} - \dfrac{y^2}{28} = 1$

72. vertices at $(-4, 0)$ and $(4, 0)$; foci at $(-6, 0)$ and $(6, 0)$ $\dfrac{x^2}{16} - \dfrac{y^2}{20} = 1$

73. foci at $\left(-2, -3\sqrt{2}\right)$ and $\left(-2, 3\sqrt{2}\right)$; length of conjugate axis: 6 units $\dfrac{y^2}{9} - \dfrac{(x+2)^2}{9} = 1$

74. foci at $(-5, 2)$ and $(7, 2)$; length of conjugate axis: 8 units $\dfrac{(x-1)^2}{20} - \dfrac{(y-2)^2}{16} = 1$

Use the characteristics of a hyperbola and the graph given to write the related equation and state the location of the foci (75 and 76) or the dimensions of the central rectangle (77 and 78).

75.

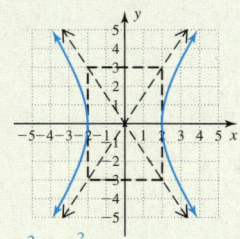

$\dfrac{x^2}{4} - \dfrac{y^2}{9} = 1, (\pm\sqrt{13}, 0)$

76.

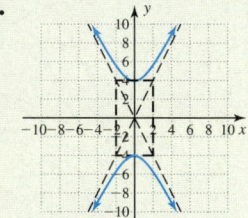

$\dfrac{y^2}{16} - \dfrac{x^2}{4} = 1, (0, \pm 2\sqrt{5})$

77.

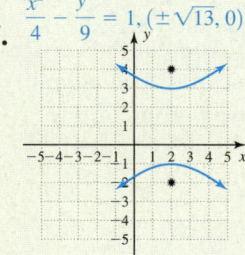

$\dfrac{(y-1)^2}{4} - \dfrac{(x-2)^2}{5} = 1, 4 \text{ by } 2\sqrt{5}$

78.

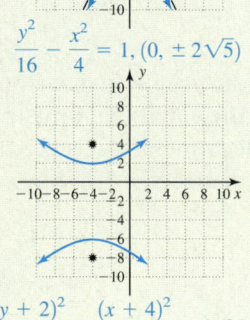

$\dfrac{(y+2)^2}{16} - \dfrac{(x+4)^2}{20} = 1, 8 \text{ by } 4\sqrt{5}$

▶ WORKING WITH FORMULAS

 79. Equation of a semi-hyperbola: $y = \sqrt{\dfrac{36 - 4x^2}{-9}}$

The "upper half" of a certain hyperbola is given by the equation shown. (a) Simplify the radicand, (b) state the domain of the expression, and (c) enter the expression as Y_1 on a graphing calculator and graph. What is the equation for the "lower half" of this hyperbola?

80. **Focal chord of a hyperbola:** $L = \dfrac{2m^2}{n}$

The focal chords of a hyperbola are line segments parallel to the conjugate axis with endpoints on the hyperbola, and containing points f_1 and f_2 (see grid). The length of the chord is given by the formula shown, where n is the distance from center to vertex and m is the distance from center to one side of the central rectangle. Use the formula to find the length of the focal chord for the hyperbola indicated, then compare the calculated value with the length estimated from the given graph:

$$\frac{(x-2)^2}{4} - \frac{(y-1)^2}{5} = 1. \quad 5$$

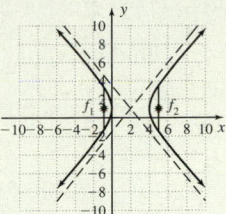

▶ APPLICATIONS

81. **Stunt pilots:** At an air show, a stunt plane dives along a hyperbolic path whose vertex is directly over the grandstands. If the plane's flight path can be modeled by the hyperbola $25y^2 - 1600x^2 = 40{,}000$, what is the minimum altitude of the plane as it passes over the stands? Assume x and y are in yards.   40 yd

82. **Flying clubs:** To test their skill as pilots, the members of a flight club attempt to drop sandbags on a target placed in an open field, by diving along a hyperbolic path whose vertex is directly over the target area. If the flight path of the plane flown by the club's president is modeled by $9y^2 - 16x^2 = 14{,}400$, what is the minimum altitude of her plane as it passes over the target? Assume x and y are in feet.   40 ft

83. **Charged particles:** It has been shown that when like particles with a common charge are hurled at each other, they deflect and travel along paths that are hyperbolic. Suppose the paths of two such particles is modeled by the hyperbola $x^2 - 9y^2 = 36$. What is the minimum distance between the particles as they approach each other? Assume x and y are in microns.   12 microns

84. **Nuclear cooling towers:** The natural draft cooling towers for nuclear power stations are called *hyperboloids of one sheet.* The perpendicular cross sections of these hyperboloids form two branches of a hyperbola. Suppose the central cross section of one such tower is modeled by the hyperbola $1600x^2 - 400(y - 50)^2 = 640{,}000$. What is the minimum distance between the sides of the tower? Assume x and y are in feet.   40 ft

85. **Locating a ship using radar:** Under certain conditions, the properties of a hyperbola can be used to help locate the position of a ship. Suppose two radio stations are located 100 km apart along a straight shoreline. A ship is sailing parallel to the shore and is 60 km out to sea. The ship sends out a distress call that is picked up by the closer station in 0.4 milliseconds (msec—one-thousandth of a second), while it takes 0.5 msec to reach the station that is farther away. Radio waves travel at a speed of approximately 300 km/msec. Use this information to find the equation of a hyperbola that will help you find the location of the ship, then find the coordinates of the ship. (*Hint:* Draw the hyperbola on a coordinate system with the radio stations on the x-axis at the foci, then use the definition of a hyperbola.)
$\dfrac{x^2}{225} - \dfrac{y^2}{2275} = 1$, about $(24.1, 60)$ or $(-24.1, 60)$

86. **Locating a plane using radar:** Two radio stations are located 80 km apart along a straight shoreline, when a "mayday" call (a plea for immediate help) is received from a plane that is about to ditch in the ocean (attempt a water landing). The plane was flying at low altitude, parallel to the shoreline, and 20 km out when it ran into trouble. The plane's distress call is picked up by the closer station in 0.1 msec, while it takes 0.3 msec to reach the other. Use this information to construct the equation of a hyperbola that will help you find the location of the ditched plane, then find the coordinates of the plane. Also see Exercise 85.
$\dfrac{x^2}{900} - \dfrac{y^2}{700} = 1$, about $(-37.6, 20)$ or $(37.6, 20)$

Additional answers can be found in the Instructor Answer Appendix.

▶ EXTENDING THE CONCEPT

87. For a greater understanding as to *why* the branches of a hyperbola are asymptotic, solve $\dfrac{x^2}{a^2} - \dfrac{y^2}{b^2} = 1$ for y, then consider what happens as $x \to \infty$ (note that $x^2 - k \approx x^2$ for large x).

88. Which has a greater area: (a) The central rectangle of the hyperbola given by $(x - 5)^2 - (y + 4)^2 = 57$, (b) the circle given by $(x - 5)^2 + (y + 4)^2 = 57$, or (c) the ellipse given by $9(x - 5)^2 + 10(y + 4)^2 = 570$?   a

89. It is possible for the plane to intersect only the vertex of the cone or to be tangent to the sides. These are called **degenerate cases** of a conic section. Many times we're unable to tell if the equation represents a degenerate case until it's written in standard form. Write the following equations in standard form and comment.

    a. $4x^2 - 32x - y^2 + 4y + 60 = 0$ **b.** $x^2 - 4x + 5y^2 - 40y + 84 = 0$

$$\frac{(x - 4)^2}{\frac{1}{4}} - (y - 2)^2 = 0 \qquad\qquad (x - 2)^2 + \frac{(y - 4)^2}{\frac{1}{5}} = 0$$

▶ MAINTAINING YOUR SKILLS

90. (2.5) Graph the piecewise-defined function:

$$f(x) = \begin{cases} 4 - x^2 & -2 \le x < 3 \\ 5 & x \ge 3 \end{cases}$$

91. (4.1) Use synthetic division and the remainder theorem to determine if $x = 2$ is a zero of $g(x) = x^5 - 5x^4 + 4x^3 + 16x^2 - 32x + 16$. If yes, find its multiplicity.   yes, 3

92. (4.2) The number $z = 1 + i\sqrt{2}$ is a solution to two out of the three equations given. Which two?   b and c

    a. $x^4 + 4 = 0$

    b. $x^3 - 6x^2 + 11x - 12 = 0$

    c. $x^2 - 2x + 3 = 0$

93. (6.4) A government-approved company is licensed to haul toxic waste. Each container of solid waste weighs 800 lb and has a volume of 100 ft³. Each container of liquid waste weighs 1000 lb and is 60 ft³ in volume. The revenue from hauling solid waste is $300 per container, while the revenue from liquid waste is $350 per container. The truck used by this company has a weight capacity of 39.8 tons and a volume capacity of 6960 ft³. What combination of solid and liquid waste containers will produce the maximum revenue?   42 solid, 46 liquid

Additional answers can be found in the Instructor Answer Appendix.

8.4   The Analytic Parabola; More on Nonlinear Systems

LEARNING OBJECTIVES

In Section 8.4 you will see how we can:

☐ **A.** Graph parabolas with a horizontal axis of symmetry

☐ **B.** Identify and use the focus-directrix form of the equation of a parabola

☐ **C.** Solve nonlinear systems involving the conic sections

☐ **D.** Solve applications of the analytic parabola

In previous coursework, you likely learned that the graph of a quadratic function was a parabola. Parabolas are actually the fourth and final member of the family of conic sections, and as we saw in Section 8.1, the graph can be obtained by observing the intersection of a plane and a cone. If the plane is parallel to the generator of the cone (shown as a dark line in Figure 8.36), the intersection of the plane with one nappe forms a parabola. In this section we develop the general equation of a parabola from its analytic definition, opening a new realm of applications that extends far beyond those involving only zeroes and extreme values.

Figure 8.36

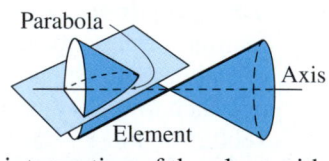

A. Parabolas with a Horizontal Axis

An introductory study of parabolas generally involves those with a vertical axis, defined by the equation $y = ax^2 + bx + c$. Unlike the previous conic sections, this equation has *only one second-degree (squared) term in x* and defines a function. As a

Figure 8.37

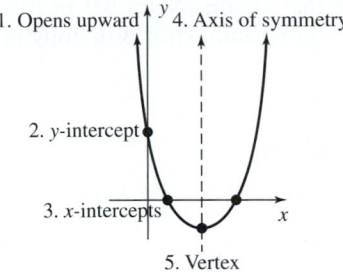

1. Opens upward
2. y-intercept
3. x-intercepts
4. Axis of symmetry
5. Vertex

review, the primary characteristics are listed here and illustrated in Figure 8.37. See **Exercises 7 through 12.**

Vertical Parabolas

For a second-degree equation of the form $y = ax^2 + bx + c$, the graph is a vertical parabola with these characteristics:

1. opens upward if $a > 0$, downward if $a < 0$.
2. y-intercept: $(0, c)$ (substitute 0 for x)
3. x-intercept(s): substitute 0 for y and solve.
4. axis of symmetry: $x = \dfrac{-b}{2a}$
5. vertex: $(h, k) = \left(\dfrac{-b}{2a}, \dfrac{4ac - b^2}{4a} \right)$

Horizontal Parabolas

Similar to our study of horizontal and vertical hyperbolas, the graph of a parabola can open *to the right or left,* as well as up or down. After interchanging the variables x and y in the standard equation, we obtain the parabola $x = ay^2 + by + c$, noting the resulting graph will be a reflection about the line $y = x$. Here, the axis of symmetry is a horizontal line and factoring or the quadratic formula is used to find the *y-intercepts* (if they exist). Note that although the graph is still a parabola—*it is not the graph of a function.*

Horizontal Parabolas

For a second-degree equation of the form $x = ay^2 + by + c$, the graph is a horizontal parabola with these characteristics:

1. opens right if $a > 0$, left if $a < 0$.
2. x-intercept: $(c, 0)$ (substitute 0 for y)
3. y-intercepts(s): substitute 0 for x and solve.
4. axis of symmetry: $y = \dfrac{-b}{2a}$
5. vertex: $\left(\dfrac{4ac - b^2}{4a}, \dfrac{-b}{2a} \right)$

EXAMPLE 1 ▶   **Graphing a Horizontal Parabola**

Graph the relation whose equation is $x = y^2 + 3y - 4$, then state the domain and range of the relation.

Solution ▶   Since the equation has a single squared term in y, the graph will be a horizontal parabola. With $a > 0$ $(a = 1)$, the parabola opens to the right. The x-intercept is $(-4, 0)$. Factoring shows the y-intercepts are $y = -4$ and $y = 1$. The axis of symmetry is $y = \dfrac{-3}{2} = -1.5$, and substituting this value into the original equation gives $x = -6.25$. The coordinates of the vertex are $(-6.25, -1.5)$. Using horizontal and vertical boundary lines we find the domain for this relation is $x \in [-6.25, \infty)$ and the range is $y \in (-\infty, \infty)$. The graph is shown.

Points on graph: $(-4, 0)$, $(0, 1)$, $(-6.25, -1.5)$, $(0, -4)$, with axis of symmetry $y = -1.5$.

Now try Exercises 13 through 18 ▶

As with the vertical parabola, the equation of a horizontal parabola can be written as a transformation: $x = a(y \pm k)^2 \pm h$ by completing the square. Note that in this case, the vertical shift is k units *opposite the sign,* with a horizontal shift of h units in the same direction as the sign.

EXAMPLE 2 ▶   **Graphing a Horizontal Parabola by Completing the Square**

Graph by completing the square: $x = -2y^2 - 8y - 9$, then state the domain and range.

Solution ▶   Using the original equation, we note the graph will be a horizontal parabola opening to the left ($a = -2$) and have an x-intercept of $(-9, 0)$. Completing the square gives $x = -2(y^2 + 4y + 4) - 9 + 8$, so the equation of this parabola in shifted form is

$$x = -2(y + 2)^2 - 1$$

The vertex is at $(-1, -2)$ and $y = -2$ is the axis of symmetry. This means there are no y-intercepts, a fact that comes to light when we attempt to solve the equation after substituting 0 for x:

$$-2(y + 2)^2 - 1 = 0 \qquad \text{substitute 0 for } x$$

$$(y + 2)^2 = -\frac{1}{2} \qquad \text{no real roots}$$

Using symmetry, the point $(-9, -4)$ is also on the graph. After plotting these points we obtain the graph shown. From the graph, the domain is $x \in (-\infty, -1]$ and the range is $y \in \mathbb{R}$.

Now try Exercises 19 through 36 ▶

As with the other relations graphed in this chapter, a horizontal parabola also fails the vertical line test and we must graph the relation in two pieces. Using the equation from Example 2 we have

$$x = -2(y + 2)^2 - 1 \qquad \text{shifted form}$$

$$x + 1 = -2(y + 2)^2 \qquad \text{isolate } y\text{-term}$$

$$\frac{x + 1}{-2} = (y + 2)^2 \qquad \text{divide by } -2$$

$$\pm\sqrt{\frac{x + 1}{-2}} = y + 2 \qquad \text{take square roots}$$

$$-2 \pm \sqrt{\frac{x + 1}{-2}} = y \qquad \text{solve for } y$$

$$Y_1 = -2 + \sqrt{\frac{X + 1}{-2}}, Y_2 = -2 - \sqrt{\frac{X + 1}{-2}}$$

The graph is given in Figure 8.38, and shows that $(-3, -1)$ is also a point on the graph.

Figure 8.38

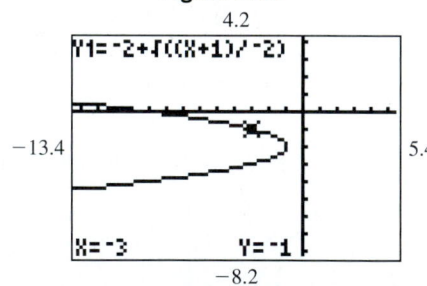

✅ **A.** You've just seen how we can graph parabolas with a horizontal axis of symmetry

B. The Focus-Directrix Form of the Equation of a Parabola

As with the ellipse and hyperbola, many significant applications of the parabola rely on its analytical definition rather than its algebraic form. From the construction of radio telescopes to the manufacture of flashlights, the location of the focus of a parabola is critical. To understand these and other applications, we use the analytic definition of a parabola first introduced in Section 8.1.

Definition of a Parabola

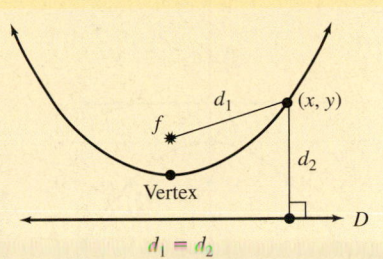

Given a fixed point f and fixed line D in the plane, a parabola is the set of all points (x, y) such that the distance from f to (x, y) is equal to the distance from line D to (x, y). The fixed point f is the **focus** of the parabola, and the fixed line is the **directrix.**

$d_1 = d_2$

WORTHY OF NOTE

For the analytic parabola, we use p to designate the focus since c is so commonly used as the constant term in $y = ax^2 + bx + c$.

The general equation of a parabola can be obtained by combining this definition with the distance formula. With no loss of generality, we can assume the parabola shown in the definition box is oriented in the plane with the vertex at $(0, 0)$ and the focus at $(0, p)$. As the diagram in Figure 8.39 indicates, this gives the directrix an equation of $y = -p$ with all points on D having coordinates of $(x, -p)$. Using $d_1 = d_2$ the distance formula yields

Figure 8.39

$$\sqrt{(x - 0)^2 + (y - p)^2} = \sqrt{(x - x)^2 + (y + p)^2} \quad \text{from the definition}$$

$$(x - 0)^2 + (y - p)^2 = (x - x)^2 + (y + p)^2 \quad \text{square both sides}$$

$$x^2 + y^2 - 2py + p^2 = 0 + y^2 + 2py + p^2 \quad \text{simplify; expand binomials}$$

$$x^2 - 2py = 2py \quad \text{subtract } p^2 \text{ and } y^2$$

$$x^2 = 4py \quad \text{isolate } x^2$$

The resulting equation is called the **focus-directrix form** of a *vertical parabola* with center at $(0, 0)$. If we had begun by orienting the parabola so it opened to the right, we would have obtained the equation of a *horizontal parabola* with center $(0, 0)$: $y^2 = 4px$.

The Equation of a Parabola in Focus-Directrix Form

Vertical Parabola

$$x^2 = 4py$$

focus $(0, p)$, directrix: $y = -p$
If $p > 0$, opens upward.
If $p < 0$, opens downward.

Horizontal Parabola

$$y^2 = 4px$$

focus at $(p, 0)$, directrix: $x = -p$
If $p > 0$, opens to the right.
If $p < 0$, opens to the left.

For a parabola, note there is only one second-degree term.

EXAMPLE 3 ▶ **Locating the Focus and Directrix of a Parabola**

Find the vertex, focus, and directrix for the parabola defined by $x^2 = -12y$. Then sketch the graph, including the focus and directrix.

Solution ▶ Since the x-term is squared and no shifts have been applied, the graph will be a vertical parabola with a vertex of $(0, 0)$. Use a direct comparison between the given equation and the focus-directrix form to determine the value of p:

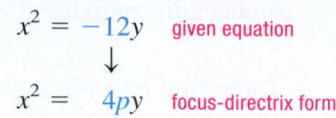

$$x^2 = -12y \quad \text{given equation}$$
$$\downarrow$$
$$x^2 = 4py \quad \text{focus-directrix form}$$

This shows:

$$4p = -12$$
$$p = -3$$

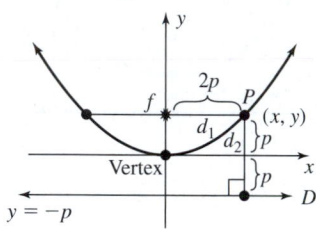

Since $p = -3$ ($p < 0$), the parabola opens downward, with the focus at $(0, -3)$ and directrix $y = 3$. To complete the graph we need a few additional points. Since 36 is divisible by 12, we can use inputs of $x = 6$ and $x = -6$ $[(-6)^2 = 6^2 = 36]$, giving the points $(6, -3)$ and $(-6, -3)$. Note the axis of symmetry is $x = 0$. The graph is shown.

Now try Exercises 37 through 48 ▶

Figure 8.40

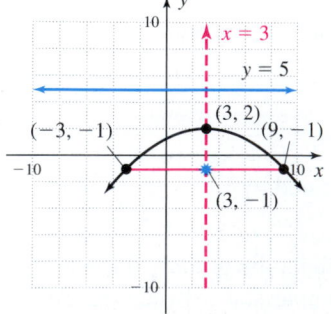

As an alternative to calculating additional points to sketch the graph, we can use what is called the **focal chord** of the parabola. Similar to the ellipse and hyperbola, the focal chord is the line segment that contains the focus, is parallel to the directrix, and has its endpoints on the graph. Using the definition of a parabola and the diagram in Figure 8.40, we note the vertical distance from (x, y) to the directrix $y = -p$ is $2p$. Since $d_1 = d_2$ a line segment parallel to the directrix from the focus to the graph will also have a length of $|2p|$, and the focal chord of any parabola has a total length of $|4p|$. Note that in Example 3, the points we happened to choose were actually the endpoints of the focal chord.

Finally, if the vertex of a vertical parabola is shifted to (h, k), the equation will have the form $(x \pm h)^2 = 4p(y \pm k)$. As with the other conic sections, both the horizontal and vertical shifts are "opposite the sign."

EXAMPLE 4 ▶   **Locating the Focus and Directrix of a Parabola**

Find the vertex, focus, and directrix for the parabola whose equation is given, then sketch the graph, including the focus, focal chord, and directrix: $x^2 - 6x + 12y - 15 = 0$.

Solution ▶   Since only the x-term is squared, the graph will be a vertical parabola. To find the end-behavior, vertex, focus, and directrix, we complete the square in x and use a direct comparison between the shifted form and the focus-directrix form:

$$x^2 - 6x + 12y - 15 = 0 \qquad \text{given equation}$$
$$x^2 - 6x + \underline{} = -12y + 15 \qquad \text{complete the square in } x$$
$$x^2 - 6x + 9 = -12y + 24 \qquad \text{add 9}$$
$$(x - 3)^2 = -12(y - 2) \qquad \text{factor}$$

Notice the parabola has been shifted 3 units right and 2 up, so *all features of the parabola will likewise be shifted*. Since we have $4p = -12$ (the coefficient of the linear term), we know $p = -3$ ($p < 0$) and the parabola opens downward. If the parabola were in standard position, the vertex would be at $(0, 0)$, the focus at $(0, -3)$ and the directrix a horizontal line at $y = 3$. But since the parabola is shifted 3 right and 2 up, we add 3 to all x-values and 2 to all y-values to locate the features of the shifted parabola. The vertex is at $(0 + 3, 0 + 2) = (3, 2)$. The focus is $(0 + 3, -3 + 2) = (3, -1)$ and the directrix is $y = 3 + 2 = 5$. Finally, the horizontal distance from the focus to the graph is $|2p| = 6$ units (since $|4p| = 12$), giving us the additional points $(-3, -1)$ and $(9, -1)$ as endpoints of the focal chord. The graph is shown.

Now try Exercises 49 through 60 ▶

In many cases, we need to construct the equation of the parabola when only partial information in known, as illustrated in Example 5.

EXAMPLE 5 ▶ **Constructing the Equation of a Parabola**

Find the equation of the parabola with vertex (4, 4) and focus (1, 4). Then graph the parabola using the equation and focal chord.

Solution ▶ As the vertex and focus are on a horizontal line, we have a horizontal parabola with general equation $(y \pm k)^2 = 4p(x \pm h)$. The distance p from vertex to focus is 3 units, and with the focus to the left of the vertex, the parabola opens left so $p = -3$. Using the focal chord, the vertical distance from (1, 4) to the graph is $|2p| = |2(-3)| = 6$, giving points (1, 10) and (1, -2). The vertex is shifted 4 units right and 4 units up from (0, 0), showing $h = 4$ and $k = 4$, and the equation of the parabola must be $(y - 4)^2 = -12(x - 4)$, with directrix $x = 7$. The graph is shown.

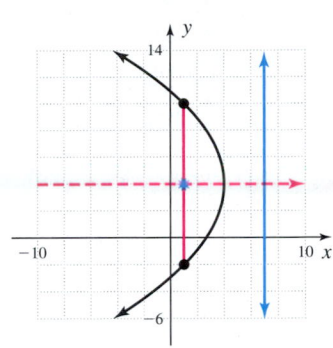

✓ **B.** You've just seen how we can identify and use the focus-directrix form of the equation of a parabola

Now try Exercises 61 through 76 ▶

C. Nonlinear Systems and the Conic Sections

Similar to our work with nonlinear systems in Section 6.3, the graphing, substitution, or elimination method can still be used when the system involves a conic section. When both equations in the system have at least one second degree term, it is generally easier to use the elimination method.

EXAMPLE 6 ▶ **Solving a System of Nonlinear Equations**

Solve the system using elimination: $\begin{cases} 2y^2 - 5x^2 = 13 \\ 3x^2 + 4y^2 = 39 \end{cases}$

Solution ▶ The first equation represents a vertical and central hyperbola, while the second represents a horizontal and central ellipse. After writing the system with the x- and y-terms in the same order, we obtain $\begin{cases} -5x^2 + 2y^2 = 13 \\ 3x^2 + 4y^2 = 39 \end{cases}$. Using $-2R1 + R2$ will eliminate the y-term.

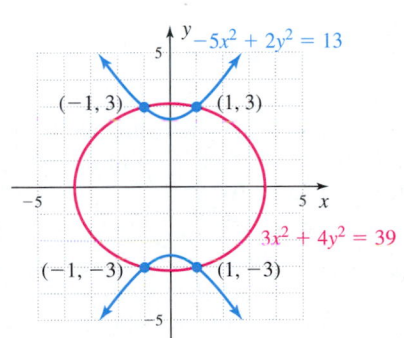

$$\begin{cases} 10x^2 - 4y^2 = -26 \quad \text{—2R1} \\ \underline{3x^2 + 4y^2 = \quad 39} \quad \text{+R2} \\ 13x^2 + 0 = \quad 13 \quad \text{sum} \end{cases}$$

$$x^2 = 1 \qquad \text{divide by 13}$$

$$x = -1 \quad \text{or} \quad x = 1 \qquad \text{square root property}$$

Substituting $x = 1$ and $x = -1$ into the second equation we obtain

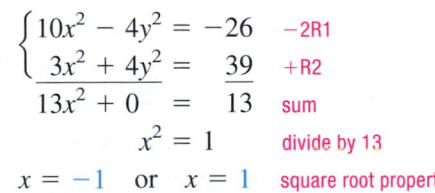

$3(1)^2 + 4y^2 = 39$	$3(-1)^2 + 4y^2 = 39$
$3 + 4y^2 = 39$	$3 + 4y^2 = 39$
$4y^2 = 36$	$4y^2 = 36$
$y^2 = 9$	$y^2 = 9$
$y = -3 \quad \text{or} \quad y = 3$	$y = -3 \quad \text{or} \quad y = 3$

Since -1 and 1 each generated *two outputs,* there are a total of four ordered pair solutions: (1, -3), (1, 3), (-1, -3), and (-1, 3). The graph is shown and supports our results.

Now try Exercises 77 through 82 ▶

Nonlinear systems like the one in Example 6 can also be solved by graphing the system on a graphing calculator and looking for points of intersection. Solving each equation for y yields the following results.

$$2y^2 - 5x^2 = 13 \qquad \text{original equation} \qquad 3x^2 + 4y^2 = 39$$

$$2y^2 = 13 + 5x^2 \qquad \text{isolate } y\text{-term} \qquad 4y^2 = 39 - 3x^2$$

$$y^2 = \frac{13 + 5x^2}{2} \qquad \text{divide by coefficient} \qquad y^2 = \frac{39 - 3x^2}{4}$$

$$y = \pm\sqrt{\frac{13 + 5x^2}{2}} \qquad \text{take square roots} \qquad y = \pm\sqrt{\frac{39 - 3x^2}{4}}$$

$$Y_1 = \sqrt{\frac{13 + 5X^2}{2}}, Y_2 = -\sqrt{\frac{13 + 5X^2}{2}} \qquad Y_3 = \sqrt{\frac{39 - 3X^2}{4}}, Y_4 = -\sqrt{\frac{39 - 3X^2}{4}}$$

The equations and graphs are shown in Figures 8.41 and 8.42, and verify that the point $(1, 3)$ is a solution to the system. Using symmetry, the other solutions are $(-1, 3)$, $(-1, -3)$, and $(1, -3)$. **See Exercises 83 through 86.**

Figure 8.41

```
Plot1  Plot2  Plot3
\Y1■√((13+5X²)/2
)
\Y2■-Y1
\Y3■√((39-3X²)/4
)
\Y4■-Y3
\Y5=
```

Figure 8.42

☑ **C.** You've just seen how we can solve nonlinear systems involving the conic sections

D. Application of the Analytic Parabola

Here is just one of the many ways the analytic definition of a parabola can be applied. There are several others in the Exercise Set. Many applications use the parabolic property that light or sound coming in parallel to the axis of a parabola will be reflected to the focus.

EXAMPLE 7 ▶   Locating the Focus of a Parabolic Receiver

The diagram shows the cross section of a radio antenna dish. Engineers have located a point on the cross section that is 0.75 m above and 6 m to the right of the vertex. At what coordinates should the engineers build the focus of the antenna?

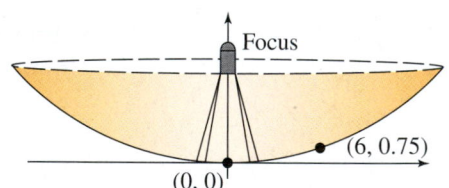

Solution ▶   By inspection we see this is a vertical parabola with center at $(0, 0)$. This means its equation must be of the form $x^2 = 4py$. Because we know $(6, 0.75)$ is a point on this graph, we can substitute $(6, 0.75)$ in this equation and solve for p:

$$x^2 = 4py \qquad \text{equation for vertical parabola, vertex at (0, 0)}$$
$$(6)^2 = 4p(0.75) \qquad \text{substitute 6 for } x \text{ and 0.75 for } y$$
$$36 = 3p \qquad \text{simplify}$$
$$p = 12 \qquad \text{result}$$

With $p = 12$, we see that the focus must be located at $(0, 12)$, or 12 m directly above the vertex.

Now try Exercises 89 through 96 ▶

☑ **D.** You've just seen how we can solve applications of the analytic parabola

Note that in many cases, the focus of a parabolic dish may be above the rim of the dish.

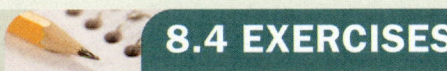

8.4 EXERCISES

▶ CONCEPTS AND VOCABULARY

Fill in the blank with the appropriate word or phrase. Carefully reread the section if needed.

1. The equation $x = ay^2 + by + c$ is that of a(n) _horizontal_ parabola, opening to the ___right___ if $a > 0$ and to the left if _$a < 0$_ .

2. If point P is on the graph of a parabola with directrix D, the distance from P to line D is equal to the distance between P and the ___focus___ .

3. Given $y^2 = 4px$, the focus is at __$(p, 0)$__ and the equation of the directrix is __$x = -p$__ .

4. Given $x^2 = -16y$, the value of p is ___-4___ and the coordinates of the focus are ___$(0, -4)$___ .

5. Discuss/Explain how to find the vertex, directrix, and focus from the equation $(x - h)^2 = 4p(y - k)$.
   Answers will vary.

6. If a horizontal parabola has a vertex of $(2, -3)$ with $a > 0$, what can you say about the y-intercepts? Will the graph always have an x-intercept? Explain.   Answers will vary.

▶ DEVELOPING YOUR SKILLS

Find the x- and y-intercepts (if they exist) and the vertex of the parabola. Then sketch the graph by using symmetry and a few additional points or completing the square and shifting a parent function. Scale the axes as needed to comfortably fit the graph and state the domain and range.

7. $y = x^2 - 2x - 3$ 8. $y = x^2 + 6x + 5$

9. $y = -2x^2 + 8x + 10$ 10. $y = -3x^2 - 12x + 15$

11. $y = 2x^2 + 5x - 7$ 12. $y = 2x^2 - 7x + 3$

Find the x- and y-intercepts (if they exist) and the vertex of the graph. Then sketch the graph using symmetry and a few additional points (scale the axes as needed). Finally, state the domain and range of the relation.

13. $x = y^2 - 2y - 3$ 14. $x = y^2 - 4y - 12$

15. $x = -y^2 + 6y + 7$ 16. $x = -y^2 + 8y - 12$

17. $x = -y^2 + 8y - 16$ 18. $x = -y^2 + 6y - 9$

Sketch by completing the square and using symmetry and shifts of a basic function. Be sure to find the x- and y-intercepts (if they exist) and the vertex of the graph, then state the domain and range of the relation.

19. $x = y^2 - 6y$ 20. $x = y^2 - 8y$

21. $x = y^2 - 4$ 22. $x = y^2 - 9$

23. $x = -y^2 + 2y - 1$ 24. $x = -y^2 + 4y - 4$

25. $x = y^2 + y - 6$ 26. $x = y^2 + 4y - 5$

27. $x = y^2 - 10y + 4$ 28. $x = y^2 + 12y - 5$

29. $x = 3 - 8y - 2y^2$ 30. $x = 2 - 12y + 3y^2$

31. $y = (x - 2)^2 + 3$ 32. $y = (x + 2)^2 - 4$

33. $x = (y - 3)^2 + 2$ 34. $x = (y + 1)^2 - 4$

35. $x = 2(y - 3)^2 + 1$ 36. $x = -2(y + 3)^2 - 5$

Find the vertex, focus, and directrix for the parabolas defined by the equations given, then use this information to sketch a complete graph (illustrate and name these features). For Exercises 49 to 60, also include the focal chord.

37. $x^2 = 8y$ 38. $x^2 = 16y$

39. $x^2 = -24y$ 40. $x^2 = -20y$

41. $x^2 = 6y$ 42. $x^2 = 18y$

43. $y^2 = -4x$ 44. $y^2 = -12x$

45. $y^2 = 18x$ 46. $y^2 = 20x$

47. $y^2 = -10x$ 48. $y^2 = -14x$

49. $x^2 - 8x - 8y + 16 = 0$

50. $x^2 - 10x - 12y + 25 = 0$

51. $x^2 - 14x - 24y + 1 = 0$

52. $x^2 - 10x - 12y + 1 = 0$

53. $3x^2 - 24x - 12y + 12 = 0$

54. $2x^2 - 8x - 16y - 24 = 0$

55. $y^2 - 12y - 20x + 36 = 0$

56. $y^2 - 6y - 16x + 9 = 0$

57. $y^2 - 6y + 4x + 1 = 0$

HOMEWORK SELECTION GUIDE

Additional answers can be found in the Instructor Answer Appendix.

Core: 7–83 every other odd, 89–95 odd (24 Exercises)
Standard: 1–4, 7–83 every other odd, 89, 90, 91–99 odd, 100 (32 Exercises)

Extended: 1–4, 7–83 every other odd, 85, 87, 89–93, 95, 97–101 (35 Exercises)
In Depth: 1–6, 7–83 every other odd, 85, 87–95, 97–102 (42 Exercises)

58. $y^2 - 2y + 8x + 9 = 0$

59. $2y^2 - 20y + 8x + 2 = 0$

60. $3y^2 - 18y + 12x + 3 = 0$

For Exercises 61–72, find the equation of the parabola in standard form that satisfies the conditions given.

61. focus: $(0, 2)$
directrix: $y = -2$
$x^2 = 8y$

62. focus: $(0, -3)$
directrix: $y = 3$
$x^2 = -12y$

63. focus: $(4, 0)$
directrix: $x = -4$
$y^2 = 16x$

64. focus: $(-3, 0)$
directrix: $x = 3$
$y^2 = -12x$

65. focus: $(0, -5)$
directrix: $y = 5$
$x^2 = -20y$

66. focus: $(5, 0)$
directrix: $x = -5$
$y^2 = 20x$

67. vertex: $(2, -2)$
focus: $(-1, -2)$
$(y + 2)^2 = -12(x - 2)$

68. vertex: $(4, 1)$
focus: $(1, 1)$
$(y - 1)^2 = -12(x - 4)$

69. vertex: $(4, -7)$
focus: $(4, -4)$
$(x - 4)^2 = 12(y + 7)$

70. vertex: $(-3, -4)$
focus: $(-3, -1)$
$(x + 3)^2 = 12(y + 4)$

71. focus: $(3, 4)$
directrix: $y = 0$
$(x - 3)^2 = 8(y - 2)$

72. focus: $(-1, 2)$
directrix: $x = -5$
$(y - 2)^2 = 8(x + 3)$

For the graphs in Exercises 73–76, only two of the following four features are displayed: vertex, focus, directrix, and endpoints of the focal chord. Find the remaining two features and the equation of the parabola.

73.

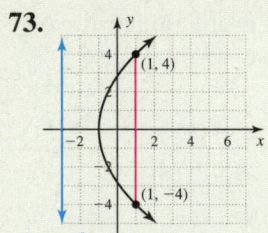

$y^2 = 8(x + 1)$; vertex $(-1, 0)$;
focus $(1, 0)$

74.

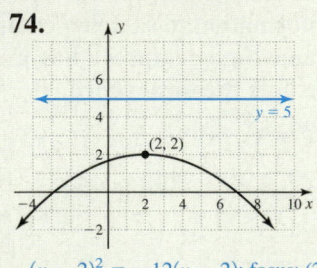

$(x - 2)^2 = -12(y - 2)$; focus: $(2, -1)$;
endpoints $(-4, -1)$ and $(8, -1)$

75.

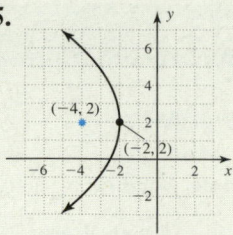

$(y - 2)^2 = -8(x + 2)$; directrix: $x = 0$;
endpoints $(-4, 6)$ and $(-4, -2)$

76.

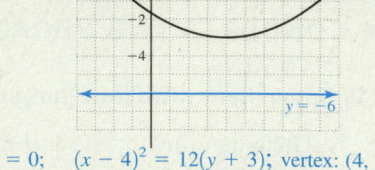
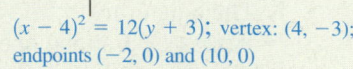

$(x - 4)^2 = 12(y + 3)$; vertex: $(4, -3)$;
endpoints $(-2, 0)$ and $(10, 0)$

 Solve using substitution or elimination, then verify your solutions by graphing the system on a graphing calculator.

77. $\begin{cases} x^2 + y^2 = 25 \\ 2x^2 - 3y^2 = 5 \end{cases}$

78. $\begin{cases} y^2 - x^2 = 12 \\ x^2 + y^2 = 20 \end{cases}$

79. $\begin{cases} x^2 - y = 4 \\ y^2 - x^2 = 16 \end{cases}$

80. $\begin{cases} 2x^2 - 3y^2 = 38 \\ x^2 + 5y = 35 \end{cases}$

81. $\begin{cases} 5x^2 - 2y^2 = 75 \\ 2x^2 + 3y^2 = 125 \end{cases}$

82. $\begin{cases} 3x^2 - 7y^2 = 20 \\ 4x^2 + 9y^2 = 45 \end{cases}$

Solve the following systems using a graphing calculator. Round approximate solutions to three decimal places.

83. $\begin{cases} (x - 2)^2 + y^2 = 20 \\ \dfrac{x^2}{4} + y = 8 \end{cases}$ $(4, 4), (6.187, -1.571)$

84. $\begin{cases} 4x^2 - (y - 12)^2 = 441 \\ x^2 + (y - 12)^2 = 1764 \end{cases}$ $(\pm 21, \pm 48.373)$

85. $\begin{cases} (x - 10)^2 + (y - 10)^2 = 144 \\ (x + 4)^2 + y^2 = 144 \end{cases}$ $(-1.863, 11.808), (7.863, -1.808)$

86. $\begin{cases} 3(x + 24)^2 + y^2 = 196 \\ 4x + 4y = -31 \end{cases}$ no solutions

▶ WORKING WITH FORMULAS

87. The area of a right parabolic segment: $A = \frac{2}{3}ab$

A *right parabolic segment* is that part of a parabola formed by a line perpendicular to its axis, which cuts the parabola. The area of this segment is given by the formula shown, where b is the length of the chord cutting the parabola and a is the perpendicular distance from the vertex to this chord. What is the area of the parabolic segment shown in the figure?
16 units^2

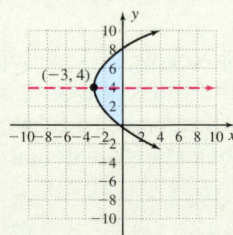

 88. The arc length of a right parabolic segment:

$$\frac{1}{2}\sqrt{b^2 + 16a^2} + \frac{b^2}{8a}\ln\left(\frac{4a + \sqrt{b^2 + 16a^2}}{b}\right)$$

Although a fairly simple concept, finding the length of the parabolic arc traversed by a projectile requires a good deal of computation. To find the length of the arc ABC shown, we use the formula given where a is the maximum height attained by the projectile, b is the horizontal distance it traveled, and "ln" represents the natural log function. Suppose a baseball thrown from centerfield reaches a maximum height of 20 ft and traverses an arc length of 340 ft. Will the ball reach the catcher 310 ft away without bouncing? yes

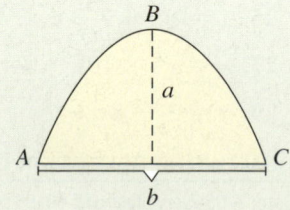

Additional answers can be found in the Instructor Answer Appendix.

▶ APPLICATIONS

89. Parabolic car headlights: The cross section of a typical car headlight can be modeled by an equation similar to $25x = 16y^2$, where x and y are in inches and $x \in [0, 4]$. Use this information to graph the relation for the indicated domain.

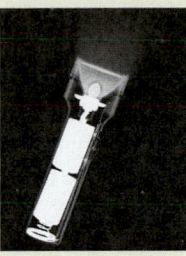

90. Parabolic flashlights: The cross section of a typical flashlight reflector can be modeled by an equation similar to $4x = y^2$, where x and y are in centimeters and $x \in [0, 2.25]$. Use this information to graph the relation for the indicated domain.

91. Parabolic sound receivers: Sound technicians at professional sports events often use parabolic receivers as they move along the sidelines. If a two-dimensional cross section of the receiver is modeled by the equation $y^2 = 54x$, and is 36 in. in *diameter,* how deep is the parabolic receiver? What is the location of the focus? [*Hint:* Graph the parabola on the coordinate grid (scale the axes).]
6 in.; (13.5, 0)

Exercise 91

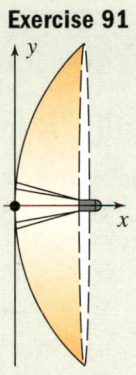

92. Parabolic sound receivers: Private investigators will often use a smaller and less expensive parabolic receiver (see Exercise 91) to gather information for their clients. If a two-dimensional cross section of the receiver is modeled by the equation $y^2 = 24x$, and the receiver is 12 in. in *diameter,* how deep is the parabolic dish? What is the location of the focus?   1.5 in.; (6, 0)

93. Parabolic radio wave receivers: The program known as S.E.T.I. (Search for Extra-Terrestrial Intelligence) involves a group of scientists using radio telescopes to look for radio signals from possible intelligent species in outer space. The radio telescopes are actually parabolic dishes that vary in size from a few feet to hundreds of feet in diameter. If a particular radio telescope is 100 ft in diameter

and has a cross section modeled by the equation $x^2 = 167y$, how deep is the parabolic dish? What is the location of the focus? [*Hint:* Graph the parabola on the coordinate grid (scale the axes).]
≈14.97 ft, (0, 41.75)

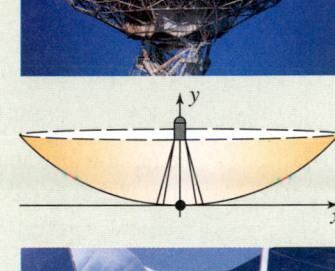

94. Solar furnace: Another form of technology that uses a parabolic dish is called a solar furnace. In general, the rays of the Sun are reflected by the dish and concentrated at the focus, producing extremely high temperatures. Suppose the dish of one of these parabolic reflectors has a 30-ft diameter and a cross section modeled by the equation $x^2 = 50y$. How deep is the parabolic dish? What is the location of the focus?   4.5 ft, (0, 12.5)

95. Commercial flashlights: The reflector of a large, commercial flashlight has the shape of a parabolic dish, with a diameter of 10 cm and a depth of 5 cm. What equation will the engineers and technicians use for the manufacture of the dish? How far from the vertex (the lowest point of the dish) will the bulb be placed? (*Hint:* Analyze the information using a coordinate system.)
$y^2 = 5x$ or $x^2 = 5y$, 1.25 cm

96. Industrial spotlights: The reflector of an industrial spotlight has the shape of a parabolic dish with a diameter of 120 cm. What is the depth of the dish if the correct placement of the bulb is 11.25 cm above the vertex (the lowest point of the dish)? What equation will the engineers and technicians use for the manufacture of the dish? (*Hint:* Analyze the information using a coordinate system.)
80 cm, $y^2 = 45x$ or $x^2 = 45y$

▶ EXTENDING THE CONCEPT

97. In a study of quadratic graphs from the equation $y = ax^2 + bx + c$, no mention is made of a parabola's focus and directrix. Generally, when $a \geq 1$, the focus of a parabola is very near its vertex. Complete the square of the function $y = 2x^2 - 8x$ and write the result in the form $(x - h)^2 = 4p(y - k)$. What is the value of p? What are the coordinates of the vertex?
$(x - 2)^2 = \frac{1}{2}(y + 8); p = \frac{1}{8}; (2, -8)$

98. Like the ellipse and hyperbola, the focal chord of a parabola (also called the **latus rectum**) can be used to help sketch its graph. From our earlier work, we know the endpoints of the focal chord are $2p$ units from the focus. Write the equation $-12y + 15 = x^2 - 6x$ in the form $4p(y \pm k) = (x \pm h)^2$, and use the endpoints of the focal chord to help graph the parabola.

Additional answers can be found in the Instructor Answer Appendix.

▶ **MAINTAINING YOUR SKILLS**

99. (6.2) Construct a system of three equations in three variables using the equation $y = ax^2 + bx + c$ and the points $(-3, 3)$, $(0, 6)$, and $(1, -1)$. Then use a matrix equation to find the equation of the parabola containing these points. $y = -2x^2 - 5x + 6$

100. (3.1/4.2) Find all roots (real and complex) of the equation $x^6 - 64 = 0$. (*Hint:* Begin by factoring as the difference of two perfect squares.)
$-2, 2, 1 + i\sqrt{3}, 1 - i\sqrt{3}, -1 + i\sqrt{3}, -1 - i\sqrt{3}$

101. (2.1) What are the characteristics of an *even* function? What are the characteristics of an *odd* function?
symmetric to the y-axis, $f(-x) = f(x)$; symmetric to the origin, $f(-x) = -f(x)$

102. (4.2/4.3) Use the function
$f(x) = x^5 + 2x^4 + 17x^3 + 34x^2 - 18x - 36$
to comment and give illustrations of the tools available for working with polynomials:
(a) synthetic division, (b) rational roots theorem, (c) the remainder and factor theorems, (d) the tests for $x = -1$ and $x = 1$, (e) the upper/lower bounds property, (f) Descartes' rule of signs, and (g) roots of multiplicity (bounces, crosses, alternating intervals).   Answers will vary.

MAKING CONNECTIONS

Making Connections: Graphically, Symbolically, Numerically, and Verbally

Eight graphs (a) through (h) are given. Match the characteristics shown in 1 through 16 to one of the eight graphs.

(a) 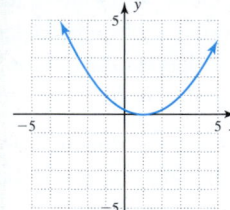 **(b)** 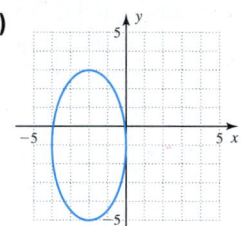 **(c)** 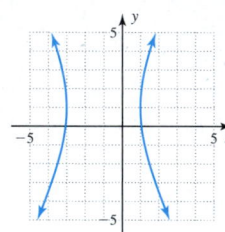 **(d)**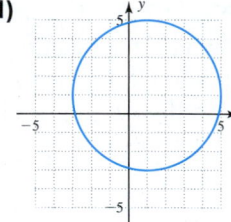

(e) 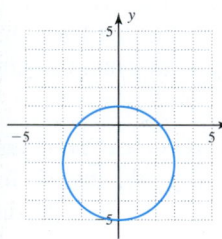 **(f)** 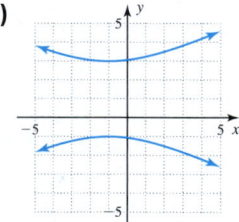 **(g)** 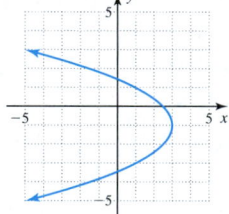 **(h)**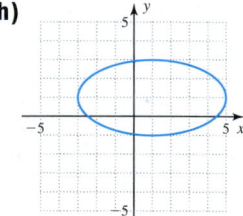

1. __d__ $(x - 1)^2 + (y - 1)^2 = 16$

2. __a__ $y = \dfrac{1}{4}(x - 1)^2$

3. __f__ foci at $(-1, 1 \pm \sqrt{5})$

4. __c__ transverse axis $y = 1$

5. __g__ $x = -\dfrac{1}{2}(y + 1)^2 + 3$

6. __d__ domain: $x \in [-3, 5]$, range: $y \in [-3, 5]$

7. __c__ $4(x + 1)^2 - (y - 1)^2 = 16$

8. __e__ $x^2 + (y - 2)^2 = 9$

9. __h__ vertices at $(-3, 1)$ and $(5, 1)$

10. __f__ $4(y - 1)^2 - (x + 1)^2 = 16$

11. __e__ center at $(0, -2)$

12. __a__ focus at $(1, 1)$

13. __b__ $4(x + 2)^2 + (y + 1)^2 = 16$

14. __h__ $(x - 1)^2 + 4(y - 1)^2 = 16$

15. __g__ axis of symmetry: $y = -1$

16. __b__ domain: $x \in [-4, 0]$, range: $y \in [-5, 3]$

SUMMARY AND CONCEPT REVIEW

SECTION 8.1 A Brief Introduction to Analytical Geometry

KEY CONCEPTS

- The midpoint and distance formulas play important roles in the study of analytical geometry:

$$\text{midpoint: } (x, y) = \left(\frac{x_2 + x_1}{2}, \frac{y_2 + y_1}{2} \right) \quad \text{distance: } d = \sqrt{(x_2 - x_1)^2 + (y_2 - y_1)^2}$$

- The perpendicular distance from a point to a line is the length of the line segment perpendicular to the given line with the given point and the point of intersection as endpoints.
- Using these tools, we can verify or construct relationships between points, lines, and curves in the plane; verify properties of geometric figures; prove theorems from Euclidean geometry; and construct relationships that define the conic sections.

EXERCISES

1. Verify the closed figure with vertices $(-3, -4)$, $(-5, 4)$, $(3, 6)$, and $(5, -2)$ is a square.
   verified (segments are perpendicular and equal length)
2. Find the equation of the circle that circumscribes the square in Exercise 1. $x^2 + (y - 1)^2 = 34$

3. A theorem from Euclidean geometry states: *If any two points are equidistant from the endpoints of a line segment, they are on the perpendicular bisector of the segment.* Determine if the line through $(-3, 6)$ and $(6, -9)$ is the perpendicular bisector of the segment through $(-5, -2)$ and $(5, 4)$.   yes

4. Four points are given here. Verify that the distance from each point to the line $y = -1$ is the same as the distance from the given point to the fixed point $(0, 1)$: $(-6, 9)$, $(-2, 1)$, $(4, 4)$, and $(8, 16)$.   verified

SECTION 8.2 The Circle and the Ellipse

KEY CONCEPTS

- The equation of a circle centered at (h, k) with radius r is $(x - h)^2 + (y - k)^2 = r^2$.

- Dividing both sides by r^2, we obtain the standard form $\dfrac{(x - h)^2}{r^2} + \dfrac{(y - k)^2}{r^2} = 1$, showing the horizontal and vertical distance from center to graph is r.

- The equation of an ellipse in standard form is $\dfrac{(x - h)^2}{a^2} + \dfrac{(y - k)^2}{b^2} = 1$. The center of the ellipse is (h, k), with horizontal distance a and vertical distance b from center to graph.

- Given two fixed points f_1 and f_2 in a plane (called the foci), an ellipse is the set of all points (x, y) such that the distance from the first focus to (x, y), plus the distance from the second focus to (x, y), remains constant.

- For an ellipse, the distance from center to vertex is *greater than* the distance c from center to one focus.

- To find the foci of a horizontal ellipse, use: $a^2 = b^2 + c^2$ (since $a > c$), or $c^2 = |a^2 - b^2|$.

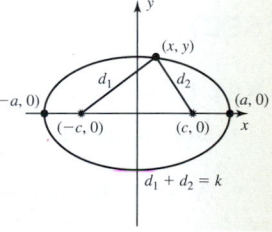

EXERCISES

Sketch the graph of each equation in Exercises 5 through 9.

5. $x^2 + y^2 = 16$

6. $x^2 + 4y^2 = 36$

7. $9x^2 + y^2 - 18x - 27 = 0$

8. $x^2 + y^2 + 6x + 4y + 12 = 0$

9. $\dfrac{(x + 3)^2}{16} + \dfrac{(y - 2)^2}{9} = 1$

10. Find the equation of the ellipse with minor axis of length 6 and foci at $(-4, 0)$ and $(4, 0)$. Then graph the equation on a graphing calculator using a "friendly" window and use the ⬭TRACE feature to locate four additional points on the graph with coordinates that are rational.

Additional answers can be found in the Instructor Answer Appendix.

11. Find the equation of the ellipse with vertices at (a) $(-13, 0)$ and $(13, 0)$, foci at $(-12, 0)$ and $(12, 0)$; (b) foci at $(0, -16)$ and $(0, 16)$, major axis: 40 units. **a.** $\dfrac{x^2}{169} + \dfrac{y^2}{25} = 1$ **b.** $\dfrac{x^2}{144} + \dfrac{y^2}{400} = 1$

12. Write the equation in standard form and sketch the graph, noting all of the characteristic features of the ellipse.
$4x^2 + 25y^2 - 16x - 50y - 59 = 0$

SECTION 8.3 The Hyperbola

KEY CONCEPTS

- The equation of a *horizontal* hyperbola in standard form is $\dfrac{(x - h)^2}{a^2} - \dfrac{(y - k)^2}{b^2} = 1$. The center of the hyperbola

  is (h, k) with horizontal distance a from center to vertices, and vertical distance b from center to the midpoint of the sides of the central rectangle.

- Given two fixed points f_1 and f_2 in a plane (called the foci), a hyperbola is the set of all points (x, y) such that the distance from one focus to point (x, y), less the distance from the other focus to (x, y), remains a positive constant: $|d_1 - d_2| = k$.

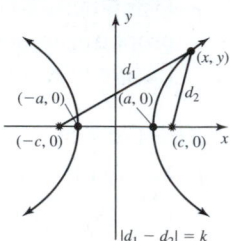

- For a hyperbola, the distance from center to one vertex is *less than* the distance from center to the focus c.

- To find the foci of a hyperbola: $c^2 = a^2 + b^2$ (since $c > a$).

EXERCISES

Sketch the graph of each equation in Exercises 13 through 17, indicating the center, vertices, and asymptotes.

13. $4y^2 - 25x^2 = 100$

14. $\dfrac{(y - 3)^2}{16} - \dfrac{(x + 2)^2}{9} = 1$

15. $\dfrac{(x + 2)^2}{9} - \dfrac{(y - 1)^2}{4} = 1$

16. $9y^2 - x^2 - 18y - 72 = 0$ **17.** $x^2 - 4y^2 - 12x - 8y + 16 = 0$

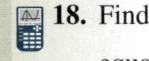 **18.** Find the equation of the hyperbola with vertices at $(-3, 0)$ and $(3, 0)$, and asymptotes of $y = \pm\dfrac{4}{3}x$. Then graph the

equation on a graphing calculator using a "friendly" window and use the TRACE feature to locate two additional points with rational coordinates.

19. Find the equation of the hyperbola with (a) vertices at $(\pm15, 0)$, foci at $(\pm17, 0)$, and (b) foci at $(0, \pm5)$ with vertical dimension of central rectangle 8 units. **a.** $\dfrac{x^2}{225} - \dfrac{y^2}{64} = 1$ **b.** $\dfrac{y^2}{16} - \dfrac{x^2}{9} = 1$

20. Write the equation in standard form and sketch the graph, noting all of the characteristic features of the hyperbola.
$4x^2 - 9y^2 - 40x + 36y + 28 = 0$

SECTION 8.4 The Analytic Parabola; More on Nonlinear Systems

KEY CONCEPTS

- Horizontal parabolas have equations of the form $x = ay^2 + by + c;\ a \neq 0$.

- A horizontal parabola will open to the right if $a > 0$, and to the left if $a < 0$. The axis of symmetry is $y = \dfrac{-b}{2a}$,

  with the vertex (h, k) found by evaluating at $y = \dfrac{-b}{2a}$ or by completing the square and writing

  the equation in shifted form: $(x - h) = a(y - k)^2$.

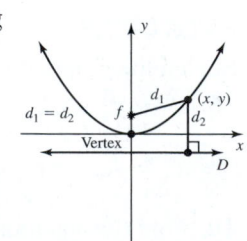

- Given a fixed point f (called the focus) and fixed line D (called the directrix) in the plane, a parabola is the set of all points (x, y) such that the distance from f to (x, y) is equal to the distance from (x, y) to line D.

- The equation $x^2 = 4py$ describes a vertical parabola, opening upward if $p > 0$, and opening downward if $p < 0$.

- The equation $y^2 = 4px$ describes a horizontal parabola, opening to the right if $p > 0$, and opening to the left if $p < 0$.

- p is the distance from the vertex to the focus (or from the vertex to the directrix).
- The focal chord of a parabola is a line segment that contains the focus and is parallel the directrix, with its endpoints on the graph. It has a total length of $|4p|$, meaning the distance from the focus to a point on the graph (as described) is $|2p|$. It is commonly used to assist in drawing a graph of the parabola.

EXERCISES

For Exercises 21 and 22, find the vertex and x- and y-intercepts if they exist. Then sketch the graph using symmetry and a few points or by completing the square and shifting a parent function.

21. $x = y^2 - 4$ **22.** $x = y^2 + y - 6$

For Exercises 23 and 24, find the vertex, focus, and directrix for each parabola. Then sketch the graph using this information and the focal chord. Also graph the directrix.

23. $x^2 = -20y$ **24.** $x^2 - 8x - 8y + 16 = 0$

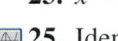 **25.** Identify the conic sections in the system, then solve. Check solutions using the intersection-of-graphs method and a graphing calculator.

$$\begin{cases} (x + 7)^2 + 4^2 = 20 \\ y^2 - 7 = x \end{cases}$$

circle, parabola; $(-3, 2), (-3, -2)$

PRACTICE TEST

By inspection only (no graphing, completing the square, etc.), match each equation to its correct description.

1. $x^2 + y^2 - 6x + 4y + 9 = 0$ _____c_____

2. $4y^2 + x^2 - 4x + 8y + 20 = 0$ _____d_____

3. $y - x^2 - 4x + 20 = 0$ _____a_____

4. $x^2 - 4y^2 - 4x + 12y + 20 = 0$ _____b_____

    a. Parabola **b.** Hyperbola **c.** Circle **d.** Ellipse

Graph each conic section, and label the center, vertices, foci, focal chords, asymptotes, and other important features where applicable.

5. $(x - 4)^2 + (y + 3)^2 = 9$

6. $\dfrac{(x - 2)^2}{16} + \dfrac{(y + 3)^2}{1} = 1$

7. $\dfrac{(x + 3)^2}{9} - \dfrac{(y - 4)^2}{4} = 1$

8. $x^2 + y^2 - 10x + 4y + 4 = 0$

9. $9x^2 + 4y^2 + 18x - 24y + 9 = 0$

10. $9x^2 - 4y^2 + 18x - 24y - 63 = 0$

11. $x = (y + 3)^2 - 2$

12. $y^2 - 6y - 12x - 15 = 0$

Solve each nonlinear system using any method.

13. a. $\begin{cases} 4x^2 - y^2 = 16 \\ y - x = 2 \end{cases}$ **b.** $\begin{cases} 2y^2 - x^2 = 4 \\ x^2 + y^2 = 8 \end{cases}$

14. A support bracket on the frame of a large ship is a steel right triangle with a hypotenuse of 25 ft and a perimeter of 60 ft. Find the lengths of the other sides using a system of nonlinear equations.   15 ft, 20 ft

15. Find an equation for the circle whose center is at $(-2, 5)$ and whose graph goes through the point $(0, 3)$. $(x + 2)^2 + (y - 5)^2 = 8$

16. Find the equation of the ellipse (in standard form) with vertices at $(-4, 0)$ and $(4, 0)$ with foci located at $(-2, 0)$ and $(2, 0)$. Then use a graphing calculator to determine where this ellipse and the circle $x^2 + y^2 = 13$ intersect.

17. The orbit of Mars around the Sun is elliptical, with the Sun at one focus. When the orbit is expressed as a central ellipse on the coordinate grid, its equation is $\dfrac{x^2}{(141.65)^2} + \dfrac{y^2}{(141.03)^2} = 1$, with a and b in millions of miles. Use this information to find the *aphelion* of Mars (distance from the Sun at its farthest point), and the *perihelion* of Mars (distance from the Sun at its closest point).
154.89 million miles; 128.41 million miles

Determine the equation of each relation and state its domain and range. For the parabola and the ellipse, also give the location of the foci.

18.

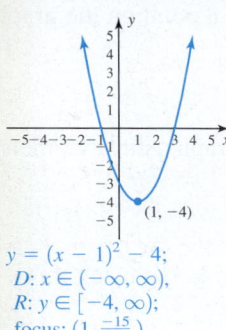

$y = (x - 1)^2 - 4$;
$D: x \in (-\infty, \infty)$,
$R: y \in [-4, \infty)$;
focus: $(1, \frac{-15}{4})$

19.

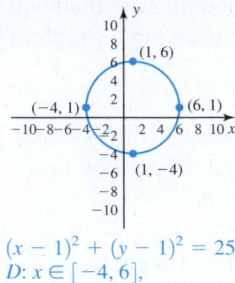

$(x - 1)^2 + (y - 1)^2 = 25$;
$D: x \in [-4, 6]$,
$R: y \in [-4, 6]$

20.

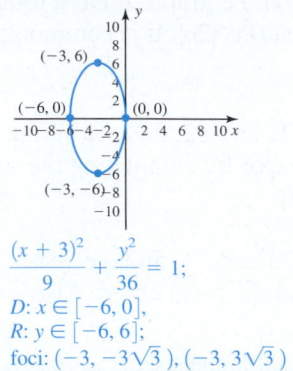

$\dfrac{(x + 3)^2}{9} + \dfrac{y^2}{36} = 1$;
$D: x \in [-6, 0]$,
$R: y \in [-6, 6]$;
foci: $(-3, -3\sqrt{3}), (-3, 3\sqrt{3})$

CALCULATOR EXPLORATION AND DISCOVERY

Elongation and Eccentricity

Technically speaking, a circle is an ellipse with both foci at the center. As the distance between foci increases, the ellipse becomes more elongated. We saw other instances of elongation in stretches and compressions of parabolic graphs, and in hyperbolic graphs where the asymptotic slopes varied depending on the values a and b. The measure used to quantify this elongation is called the *eccentricity e*, and is determined by the ratio $e = \dfrac{c}{a}$. For this *Exploration and Discovery*, we'll use the **repeat graph** feature of a graphing calculator to explore the eccentricity of the graph of a conic. The "repeat graph" feature enables you to graph a family of curves by enclosing changes in a parameter in braces "{ }." For instance, entering $\{-2, -1, 0, 1, 2\}X + 3$ as Y_1 on the (Y=) screen will automatically graph these five lines:

$$y = -2x + 3, \ y = -x + 3, \ y = 3,$$
$$y = x + 3, \text{ and } y = 2x + 3.$$

We'll use this feature to graph a family of ellipses, observing the result and calculating the eccentricity for each curve in the family. The standard form is $\dfrac{x^2}{a^2} + \dfrac{y^2}{b^2} = 1$, which we'll solve for y and enter as Y_1 and Y_2. After simplification the result is $y = \pm b\sqrt{1 - \dfrac{x^2}{a^2}}$, but for this investigation we'll use the constant $b = 2$ and vary the parameter a using the values $a = 2, 4, 6,$ and 8. The result is $y = 2\sqrt{1 - \dfrac{x^2}{\{4, 16, 36, 64\}}}$. Note from Figure 8.43 that we've set $Y_2 = -Y_1$ to graph the lower half of the ellipse. Using the "friendly window" shown (Figure 8.44) gives the result shown in Figure 8.45, where we see the ellipse is increasingly elongated in the horizontal direction (note when $a = 2$ the result is a circle since $a = b$).

Figure 8.43

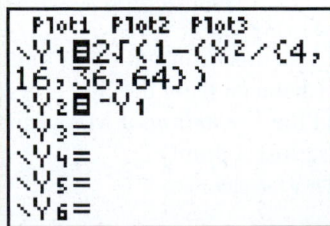

Figure 8.44

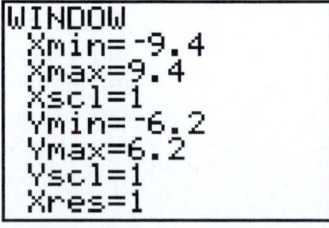

Figure 8.45

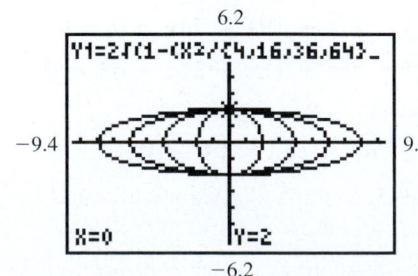

Using $a = 2, 4, 6,$ and 8 with $b = 2$ in the foci formula $c = \sqrt{a^2 - b^2}$ gives $c = 0, 2\sqrt{3}, 4\sqrt{2},$ and $2\sqrt{15}$, respectively, with these eccentricities:

$e = \dfrac{0}{2}, \dfrac{2\sqrt{3}}{4}, \dfrac{4\sqrt{2}}{6}$, and $\dfrac{2\sqrt{15}}{8}$. While difficult to see in radical form, we find that the eccentricity of an ellipse always satisfies the inequality $0 < e < 1$ (excluding the circle = ellipse case). To two decimal places, the values are $e = 0$, 0.87, 0.94, and 0.97, respectively.

As a final note, it's interesting how the $e = \dfrac{c}{a}$ definition of eccentricity relates to our everyday use of the word "eccentric." A normal or "noneccentric" person is thought to be well-rounded, and sure enough $e = 0$ produces a well-rounded figure—a circle. A person who is highly eccentric is thought to be far from the norm, deviating greatly from the center, and greater values of e produce very elongated ellipses.

Exercise 1: Perform a similar exploration using a family of *hyperbolas*. What do you notice about the eccentricity? $e > 1$

Exercise 2: Perform a similar exploration using a family of *parabolas*. What do you notice about the eccentricity? $e = 1$

STRENGTHENING CORE SKILLS

Ellipses and Hyperbolas with Rational/Irrational Values of *a* and *b*

Using the process known as completing the square, we were able to convert from the polynomial form of a conic section to the standard form. However, for some equations, values of a and b are somewhat difficult to identify, since the coefficients are not factors. Consider the equation $20x^2 + 120x + 27y^2 - 54y + 192 = 0$, the equation of an ellipse.

$$20x^2 + 120x + 27y^2 - 54y + 192 = 0 \qquad \text{original equation}$$
$$20(x^2 + 6x + \underline{\quad}) + 27(y^2 - 2y + \underline{\quad}) = -192 \qquad \text{subtract 192}$$
$$20(x^2 + 6x + 9) + 27(y^2 - 2y + 1) = -192 + 27 + 180 \qquad \text{complete the square in } x \text{ and } y$$
$$20(x + 3)^2 + 27(y - 1)^2 = 15 \qquad \text{factor and simplify}$$
$$\dfrac{4(x + 3)^2}{3} + \dfrac{9(y - 1)^2}{5} = 1 \qquad \text{standard form}$$

Unfortunately, we cannot easily identify the values of a and b, since the coefficients of each binomial square are not "1." In these cases, we can write the equation in standard form by using a simple property of fractions—the numerator and denominator of any fraction can be divided by the same quantity to obtain an equivalent fraction. Although the result may look odd, it can nevertheless be applied here, giving a result of $\dfrac{(x + 3)^2}{\frac{3}{4}} + \dfrac{(y - 1)^2}{\frac{5}{9}} = 1$.

We can now identify a and b by writing these denominators in squared form, which gives the following expression: $\dfrac{(x + 3)^2}{\left(\frac{\sqrt{3}}{2}\right)^2} + \dfrac{(y - 1)^2}{\left(\frac{\sqrt{5}}{3}\right)^2} = 1$. The values of a and b are now easily seen as $a = \dfrac{\sqrt{3}}{2} \approx 0.866$ and $b = \dfrac{\sqrt{5}}{3} \approx 0.745$.

Use this idea to complete the following exercises.

Exercise 1: Write the equation in standard form, then identify the values of a and b and use them to graph the ellipse.

$$\dfrac{4(x + 3)^2}{49} + \dfrac{25(y - 1)^2}{36} = 1$$

Exercise 2: Write the equation in standard form, then identify the values of a and b and use them to graph the hyperbola.

$$\dfrac{9(x + 3)^2}{80} - \dfrac{4(y - 1)^2}{81} = 1$$

Exercise 3: Identify the values of a and b by writing the equation $100x^2 - 400x - 18y^2 - 108y + 230 = 0$ in standard form.

Exercise 4: Identify the values of a and b by writing the equation $28x^2 - 56x + 48y^2 + 192y + 195 = 0$ in standard form.

Additional answers can be found in the Instructor Answer Appendix.

CUMULATIVE REVIEW CHAPTERS R–8

Solve each equation.

1. $x^3 - 2x^2 + 4x - 8 = 0$ $x = 2, x = \pm 2i$

2. $2|n + 4| + 3 = 13$ $n = 1, n = -9$

3. $\sqrt{x - 3} + 5 = x$ $x = 7; x = 4$ is extraneous

4. $x^{\frac{3}{2}} + 8 = 0$ no solution

5. $x^2 - 6x + 13 = 0$ $x = 3 \pm 2i$

6. $4 \cdot 2^{x+1} = \dfrac{1}{8}$ $x = -6$

7. $3^{x-2} = 7$ $x = 2 + \dfrac{\ln 7}{\ln 3}$

8. $\ln x = 2$ $x = e^2$

9. $\log x + \log (x - 3) = 1$ $x = 5; x = -2$ is extraneous

Graph each relation. Include vertices, x- and y-intercepts, asymptotes, and other features.

10. $y = \dfrac{2}{3}x + 2$ **11.** $y = |x - 2| + 3$

12. $y = \dfrac{1}{x - 1} + 2$ **13.** $y = \sqrt{x - 3} + 1$

14. a. $g(x) = (x - 3)(x + 1)(x + 4)$
 b. $f(x) = x^4 + x^3 - 13x^2 - x + 12$

15. $h(x) = \dfrac{x - 2}{x^2 - 9}$ **16.** $q(x) = 2^x + 3$

17. $f(x) = \log_2(x + 1)$ **18.** $x = y^2 + 4y + 7$

19. $x^2 + y^2 + 10x - 4y + 20 = 0$

20. $4(x - 1)^2 - 36(y + 2)^2 = 144$

21. Determine the following for the indicated graph (write all answers in interval notation): (a) the domain, (b) the range, (c) interval(s) where $f(x)$ is increasing or decreasing, (d) interval(s) where $f(x)$ is constant, (e) location of any maximum or minimum value(s), (f) interval(s) where $f(x)$ is positive, and (g) interval(s) where $f(x)$ is negative.

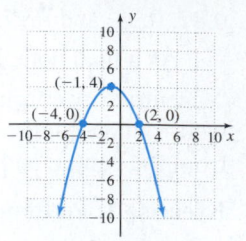

a. $x \in (-\infty, \infty)$ **b.** $y \in (-\infty, 4]$
c. $f(x)\uparrow: x \in (-\infty, -1), f(x)\downarrow: x \in (-1, \infty)$ **d.** none   **e.** max: $(-1, 4)$
f. $f(x) > 0: x \in (-4, 2)$ **g.** $f(x) < 0: x \in (-\infty, -4) \cup (2, \infty)$

Solve each system of equations with a graphing calculator. Use a matrix equation for Exercise 22, and the intersection-of-graphs method for Exercise 23.

22. $\begin{cases} 4x + 3y = 13 \\ -9y + 5z = 19 \\ x - 4z = -4 \end{cases}$ **23.** $\begin{cases} x^2 + y^2 = 25 \\ 64x^2 + 12y^2 = 768 \end{cases}$

$(4, -1, 2)$ $(3, 4), (3, -4), (-3, 4), (-3, -4)$

24. If a person invests $5000 at 9% compounded quarterly, how long would it take for the money to grow to $12,000? ≈ 9.8 yr

25. A radiator contains 10 L of liquid that is 40% antifreeze. How much should be drained off and replaced with pure antifreeze for a 60% mixture? $3\frac{1}{3}$ L

Solve each equation using a graphing calculator.

26. $\dfrac{1}{8}x^3 - 4x + 3 = \dfrac{1}{4}x^2 - 5$ $x = 2, x \approx \pm 5.657$

27. $|x + 4| = 8 - |x|$ $x = -6, x = 2$

28. $e^{2x} - 3e^x = 4$ $x \approx 1.386\ (x = \ln 4)$

29. $3^{x-1} = 2^{2-x}$ $x \approx 1.387 \left(x = \dfrac{2\ln 2 + \ln 3}{\ln 2 + \ln 3} \text{ or } \dfrac{\ln 12}{\ln 6} \right)$

30. $\dfrac{x + 3}{x - 4} \geq 3$ $x \in (4, 7.5]$

Additional Topics in Algebra

CHAPTER OUTLINE

CHAPTER CONNECTIONS

For a corporation of any size, decisions made by upper management often depend on a large number of factors, with the desired outcome attainable in many different ways. For instance, consider a legal firm that specializes in family law, with a support staff of 15 employees—6 paralegals and 9 legal assistants. Due to recent changes in the law, the firm wants to send some combination of five support staff to a conference dedicated to the new changes. In Chapter 9, we'll see how counting techniques and probability can be used to determine the various ways such a group can be randomly formed, even if certain constraints are imposed. This application appears as Exercise 34 in Section 9.6.

Check out these other real-world connections:

▶ Determining the Effects of Inflation (Section 9.1, Exercise 86)

▶ Calculating Possible Movements of a Computer Animation (Section 9.2, Exercise 77)

▶ Counting the Number of Possible Area Codes and Phone Numbers (Section 9.5, Exercises 84 and 85)

▶ Tracking and Improving Customer Service Using Probability (Section 9.6, Exercise 53)

LEARNING OBJECTIVES

In Section 9.1 you will see how we can:

☐ **A.** Write out the terms of a sequence given the general or *n*th term

☐ **B.** Work with recursive sequences and sequences involving a factorial

☐ **C.** Find the partial sum of a series

☐ **D.** Use summation notation to write and evaluate series

☐ **E.** Use sequences to solve applications

A *sequence* can be thought of as a pattern of numbers listed in a prescribed order. A *series* is the sum of the numbers in a sequence. Sequences and series come in countless varieties, and we'll introduce some general forms here. In following sections we'll focus on two special types: arithmetic and geometric sequences. These are used in a number of different fields, with a wide variety of significant applications.

A. Finding the Terms of a Sequence Given the General Term

Suppose a person had $10,000 to invest, and decided to place the money in government bonds that guarantee an annual return of 7%. From our work in Chapter 5, we know the amount of money in the account after x years can be modeled by the function $f(x) = 10,000(1.07)^x$. If you reinvest your earnings each year, the amount in the account would be (rounded to the nearest dollar):

Year:	$f(1)$	$f(2)$	$f(3)$	$f(4)$	$f(5)\dots$
	↓	↓	↓	↓	↓
Value:	$10,700	$11,449	$12,250	$13,108	$14,026\dots

Note the relationship (year, value) is a function that pairs 1 with $10,700, 2 with $11,449, 3 with $12,250, and so on. This is an example of a **sequence.** To distinguish sequences from other algebraic functions, we commonly name the functions a instead of f, use the variable n instead of x, and employ a subscript notation. The function $f(x) = 10,000(1.07)^x$ would then be written $a_n = 10,000(1.07)^n$. Using this notation $a_1 = 10,700$, $a_2 = 11,449$, and so on.

The values $a_1, a_2, a_3, a_4, \dots$ are called the **terms** of the sequence. If the account were closed after a certain number of years (for example, after the fifth year) we have a **finite sequence.** If we let the investment grow indefinitely, the result is called an **infinite sequence.** The expression a_n that defines the sequence is called the **general** or **n**th **term** and the terms immediately preceding it are called the $(n - 1)$st term, the $(n - 2)$nd term, and so on.

Sequences

A *finite sequence* is a function a_n whose domain is the set of natural numbers from 1 to n. The terms of the sequence are labeled

$$a_1, a_2, a_3, \dots, a_k, a_{k+1}, \dots, a_{n-1}, a_n$$

where a_k represents an arbitrary "interior" term and a_n also represents the last term of the sequence.

An *infinite sequence* is a function a_n whose domain is the set of <u>all</u> natural numbers.

EXAMPLE 1A ▶ Computing Specified Terms of a Sequence

For $a_n = \dfrac{n + 1}{n^2}$, find $a_1, a_3, a_6,$ and a_7.

Solution ▶

$$a_1 = \frac{1 + 1}{1^2} = 2 \qquad\qquad a_3 = \frac{3 + 1}{3^2} = \frac{4}{9}$$

$$a_6 = \frac{6 + 1}{6^2} = \frac{7}{36} \qquad\qquad a_7 = \frac{7 + 1}{7^2} = \frac{8}{49}$$

EXAMPLE 1B ▶ Computing the First *k* Terms of a Sequence

Find the first four terms of the sequence $a_n = (-1)^n 2^n$. Write the terms of the sequence as a list.

Solution ▶

$$a_1 = (-1)^1 2^1 = -2 \qquad\qquad a_2 = (-1)^2 2^2 = 4$$
$$a_3 = (-1)^3 2^3 = -8 \qquad\qquad a_4 = (-1)^4 2^4 = 16$$

The sequence can be written $-2, 4, -8, 16, \ldots$, or more generally as $-2, 4, -8,$ $16, \ldots, (-1)^n 2^n, \ldots$ to show how each term was generated.

WORTHY OF NOTE

When the terms of a sequence *alternate in sign* as in Example 1B, we call it an **alternating sequence**.

Now try Exercises 7 through 22 ▶

Much of the beauty and power of studying sequences comes from patterns detected within the sequence, or the ability to find a particular term quickly. Here, the calculator becomes an invaluable tool, aiding computations to be sure … but also enabling explorations not generally possible with paper and pencil alone. Most graphing calculators offer a **"seq("** feature (often in a **STAT** or **LIST** menu), where the left parenthesis indicates we need to enter the following four items of information: the nth term formula defining the sequence, the variable in use, the starting value, and the ending value. The sequence generated can be seen on the home screen, or stored in a list for future use. In Figure 9.1, this feature was used to generate the first four terms of the sequence defined in Example 1(B) using the keystrokes [2nd] [STAT] (**LIST**) [▶] (**OPS**) **5:seq(**.

In addition to the **seq(** feature, most calculators have a sequence [MODE], which is especially useful when working with several sequences simultaneously. In sequence [MODE] (Figure 9.2A), the [Y=] screen presents a very different look (Figure 9.2B), first of all showing that the minimum value of n is 1, then naming the functions u(n), v(n), and w(n) instead of Y_1, Y_2, and Y_3. The entries following each, for example "u(nMin) =," are used in a study of the recursive sequences, which are covered later. Note that function names for the sequences (u, v, and w) are the [2nd] function for the numbers 7, 8, and 9, respectively.

Figure 9.1

```
seq((-1)^(X)*2^(
X),X,1,4)
     {-2 4 -8 16}
```

Figure 9.2B

```
Plot1 Plot2 Plot3
 nMin=1
\u(n)=
 u(nMin)=
\v(n)=
 v(nMin)=
\w(n)=
 w(nMin)=
```

Figure 9.2A

```
NORMAL  SCI  ENG
FLOAT  0123456789
RADIAN  DEGREE
FUNC  PAR  POL  SEQ
CONNECTED  DOT
SEQUENTIAL  SIMUL
REAL  a+bi  re^θi
FULL  HORIZ  G-T
SETCLOCK08/07/09 1:13PM
```

EXAMPLE 2 ▶

Finding the Terms of a Sequence Using Technology

Use a calculator in sequence [MODE] to

a. Find the sixth term (as a fraction) for the sequence defined in Example 1A.

b. Generate the first five terms for the sequence defined in Example 1B and store the results in a list.

Solution ▶

a. On the [Y=] screen, define the sequence u(n) as $u(n) = \dfrac{n+1}{n^2}$ (Figure 9.3), leaving the second line blank. Then go to the home screen and access the **"seq("** feature as before, and supply the information required. Since we only want the sixth term (a_6), we enter **seq(u(n), n, 6, 6) ▶Frac**, and after pressing [ENTER], we obtain the expected result $\dfrac{7}{36}$ (Figure 9.4).

Figure 9.3

```
Plot1 Plot2 Plot3
 nMin=1
\u(n)=(n+1)/n²
 u(nMin)=
\v(n)=
 v(nMin)=
\w(n)=
 w(nMin)=
```

Figure 9.4

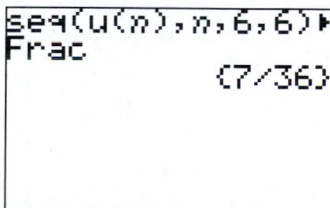

```
seq(u(n),n,6,6)▶
Frac
          {7/36}
```

Figure 9.6

L1	L2	L3	1
-2	-----	-----	
4			
-8			
16			
-32			

L1(1) = -2

☑ **A.** You've just seen how we can write out the terms of a sequence given the general or nth term

b. For the first five terms, we change $u(n)$ to $u(n) = (-1)^n \cdot 2^n$ then go to the home screen and use the "**5:seq(**" command once again, this time for terms one through five. Knowing that we want the results stored in a list, we follow the command with (STO•) (2nd) (1) (**L1**) (Figure 9.5). The list is shown in Figure 9.6.

Figure 9.5

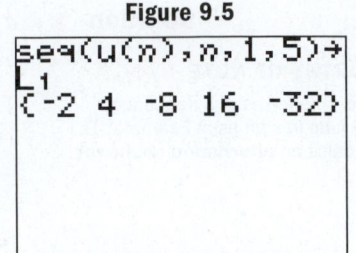
```
seq(u(n),n,1,5)→
L1
{-2 4 -8 16 -32}
```

Now try Exercises 23 through 32 ▶

B. Recursive Sequences and Factorial Notation

Sometimes the formula defining a sequence uses the preceding term or terms to generate those that follow. These are called **recursive sequences** and are particularly useful in writing computer programs. Because of how they are defined, recursive sequences must give an inaugural term or **seed element(s),** to begin the recursion process.

Perhaps the most famous recursive sequence is associated with the work of Leonardo of Pisa (A.D. 1180–1250), better known to history as *Fibonacci*. In the Fibonacci sequence, each successive term is the sum of the previous two, beginning with 1, 1, ...

EXAMPLE 3 ▶ **Computing the Terms of a Recursive Sequence**

Write out the first eight terms of the Fibonacci sequence, which is defined by $c_1 = 1$, $c_2 = 1$, and $c_n = c_{n-1} + c_{n-2}$.

Solution ▶ The first two terms are given, so we begin with $n = 3$.

$$c_3 = c_{3-1} + c_{3-2} \qquad c_4 = c_{4-1} + c_{4-2} \qquad c_5 = c_{5-1} + c_{5-2}$$
$$= c_2 + c_1 \qquad\qquad = c_3 + c_2 \qquad\qquad = c_4 + c_3$$
$$= 1 + 1 \qquad\qquad = 2 + 1 \qquad\qquad = 3 + 2$$
$$= 2 \qquad\qquad\quad = 3 \qquad\qquad\quad = 5$$

WORTHY OF NOTE

One application of the Fibonacci sequence involves the Fibonacci spiral, found in the growth of many ferns and the spiral shell of many mollusks.

At this point we can simply use the fact that each successive term is simply the sum of the preceding two, and find that $c_6 = 3 + 5 = 8$, $c_7 = 5 + 8 = 13$, and $c_8 = 8 + 13 = 21$. The first eight terms are 1, 1, 2, 3, 5, 8, 13, and 21.

Now try Exercises 33 through 38 ▶

Since a recursive sequence is defined using a preceding term or terms, the *first term(s)* must be given in order to determine those that follow. To generate the sequence using technology, we enter these initial terms as "u(nMin) =" on the (Y=) screen, enclosed in braces. For the Fibonacci sequence from Example 3, we would enter u(nMin) = {1, 1}" as shown in Figure 9.7, with the result shown in Figure 9.8 (use the right arrow to view any remaining terms).

Some sequences may involve the computation of a **factorial,** which is the product of a given natural number with all those that precede it. The expression **5!** is read, "five factorial," and is evaluated as: $5! = 5 \cdot 4 \cdot 3 \cdot 2 \cdot 1 = 120$.

Figure 9.8

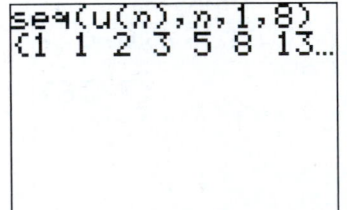
```
seq(u(n),n,1,8)
{1 1 2 3 5 8 13...
```

Figure 9.7

```
Plot1 Plot2 Plot3
nMin=1
\u(n)☐u(n-1)+u(n
-2)
u(nMin)☐{1,1}
\v(n)=
v(nMin)=
\w(n)=
```

Factorials

For any natural number n,

$$n! = n \cdot (n-1) \cdot (n-2) \cdot \ldots \cdot 3 \cdot 2 \cdot 1$$

Rewriting a factorial in equivalent forms often makes it easier to simplify certain expressions. For example, we can rewrite 5! as $5 \cdot 4!$ or $5! = 5 \cdot 4 \cdot 3!$, and so on. Consider Example 4.

EXAMPLE 4 ▶ **Simplifying Expressions Using Factorial Notation**

Simplify by writing the numerator in an equivalent form.

a. $\dfrac{9!}{7!}$　　　　b. $\dfrac{11!}{8!2!}$　　　　c. $\dfrac{6!}{3!5!}$

Solution ▶

a. $\dfrac{9!}{7!} = \dfrac{9 \cdot 8 \cdot 7!}{7!}$　　b. $\dfrac{11!}{8!2!} = \dfrac{11 \cdot 10 \cdot 9 \cdot 8!}{8!2!}$　　c. $\dfrac{6!}{3!5!} = \dfrac{6 \cdot 5!}{3!5!}$

$= 9 \cdot 8$　　　　　　$= \dfrac{990}{2}$　　　　　　　$= \dfrac{6}{6}$

$= 72$　　　　　　　$= 495$　　　　　　　$= 1$

Now try Exercises 39 through 44 ▶

Most calculators have a factorial option or key. On many calculator models it is located on a submenu of the **MATH** key: **MATH** ◀ **PRB 4: !**.

EXAMPLE 5 ▶ **Computing a Specified Term from a Sequence Defined Using Factorials**

Find and simplify the third term of each sequence.

a. $a_n = \dfrac{n!}{2^n}$　　　　　　b. $c_n = \dfrac{(-1)^n(2n-1)!}{n!}$

Algebraic Solution ▶

a. $a_3 = \dfrac{3!}{2^3}$　　　　b. $c_3 = \dfrac{(-1)^3[2(3)-1]!}{3!}$

$= \dfrac{6}{8} = \dfrac{3}{4}$　　　　$= \dfrac{(-1)(5!)}{3!} = \dfrac{(-1)[5 \cdot 4 \cdot 3!]}{3!}$

$= -20$

Technology Solution ▶

For this exercise we enter $\dfrac{n!}{2^n}$ as u(n), and

$\dfrac{(-1)^n(2n-1)!}{n!}$ as v(n). Once these functions have been defined on the **Y=** screen, we can evaluate them on the home screen as with other functions (Figure 9.9), or use the **TABLE** feature (Figure 9.10).

Figure 9.9, 9.10

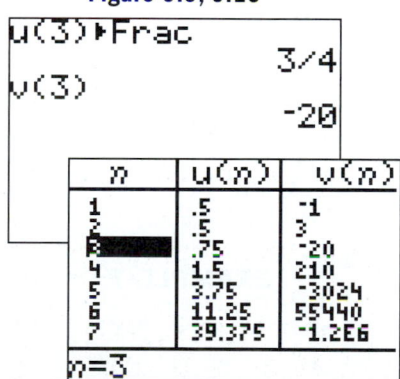

☑ **B.** You've just seen how we can work with recursive sequences and sequences involving a factorial

Now try Exercises 45 through 50 ▶

Figure 9.11

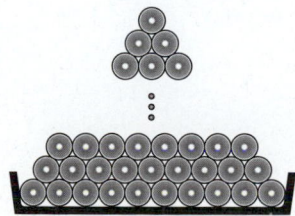

C. Series and Partial Sums

Sometimes the terms of a sequence are dictated by context rather than a formula. Consider the stacking of large pipes in a storage yard. If there are 10 pipes in the bottom row, then 9 pipes, then 8 (see Figure 9.11), how many pipes are in the stack if there is a single pipe at the top? The sequence generated is 10, 9, 8, ... , 3, 2, 1 and to answer the question we would have to *compute the sum of all terms in the sequence*. When the terms of a finite sequence are added, the result is called a **finite series**.

Finite Series

Given the sequence $a_1, a_2, a_3, a_4, \ldots, a_n$, the sum of the terms is called a **finite series** or **partial sum** and is denoted S_n:

$$S_n = a_1 + a_2 + a_3 + \ldots + a_{n-1} + a_n$$

EXAMPLE 6 ▶   Computing a Partial Sum

Given $a_n = 2n$, find the value of   **a.** S_4 **b.** S_7.

Solution ▶ Since we eventually need the sum of the first seven terms, begin by writing out these terms: 2, 4, 6, 8, 10, 12, and 14.

a. $S_4 = a_1 + a_2 + a_3 + a_4$ **b.** $S_7 = a_1 + a_2 + a_3 + a_4 + a_5 + a_6 + a_7$
$\qquad = 2 + 4 + 6 + 8$ $\qquad\qquad\qquad = 2 + 4 + 6 + 8 + 10 + 12 + 14$
$\qquad = 20$ $\qquad\qquad\qquad\qquad\qquad = 56$

Now try Exercises 51 through 56 ▶

Figure 9.12

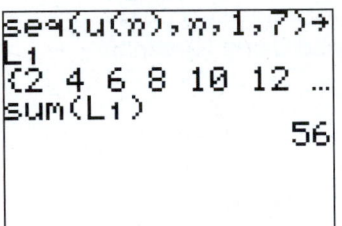

☑ **C.** You've just seen how we can find the partial sum of a series

There are several ways of computing a partial sum using technology, with the most common being (1) storing the sequence in a list then computing the sum of the list elements, or (2) computing the sum of a sequence directly on the home screen using the appropriate commands. On many calculators, the **"sum("** option is in a **MATH** submenu, accessed using ⊙2nd⊙ ⊙STAT⊙ (LIST) ◁ (MATH) 5:sum(. For the sum in Example 6(b), we have $u(n) = 2n$, with option (1) demonstrated in Figure 9.12 and option (2) in Figure 9.13.

Figure 9.13

```
sum(seq(u(n),n,1
,7))
                 56
```

D. Summation Notation

When the general term of a sequence is known, the Greek letter *sigma* Σ can be used to write the related series as a formula. For instance, to indicate the sum of the first four term of $a_n = 3n + 2$, we write $\sum_{i=1}^{4}(3i + 2)$ with this notation indicating we are to compute the sum of all terms generated as i cycles from 1 through 4. This result is called **summation** or **sigma notation** and the letter i is called the **index of summation**. The letters $j, k, l,$ and m are also used as index numbers, and the summation need not start at 1.

EXAMPLE 7 ▶   Computing a Partial Sum

Compute each sum:

a. $\displaystyle\sum_{i=1}^{4}(3i + 2)$ **b.** $\displaystyle\sum_{j=1}^{5}\frac{1}{j}$ **c.** $\displaystyle\sum_{k=3}^{6}(-1)^k k^2$

d. Check each sum using a graphing calculator.

Solution ▶ **a.** $\displaystyle\sum_{i=1}^{4}(3i + 2) = (3 \cdot 1 + 2) + (3 \cdot 2 + 2) + (3 \cdot 3 + 2) + (3 \cdot 4 + 2)$
$\qquad\qquad\qquad\qquad = 5 + 8 + 11 + 14 = 38$

b. $\displaystyle\sum_{j=1}^{5}\frac{1}{j} = \frac{1}{1} + \frac{1}{2} + \frac{1}{3} + \frac{1}{4} + \frac{1}{5}$
$\qquad\quad = \frac{60}{60} + \frac{30}{60} + \frac{20}{60} + \frac{15}{60} + \frac{12}{60} = \frac{137}{60}$

c. $\displaystyle\sum_{k=3}^{6}(-1)^k k^2 = (-1)^3 \cdot 3^2 + (-1)^4 \cdot 4^2 + (-1)^5 \cdot 5^2 + (-1)^6 \cdot 6^2$

$$= -9 + 16 + (-25) + 36 = 18$$

d. Begin by entering the functions in parts (a), (b), and (c) as u(n), v(n), and w(n) respectively on the [Y=] screen (Figure 9.14). Note that while different indices are used in this example, all are entered into the calculator using the variable n. The results are shown in Figures 9.15 and 9.16.

Figure 9.14

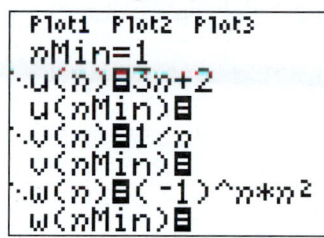

Figure 9.15

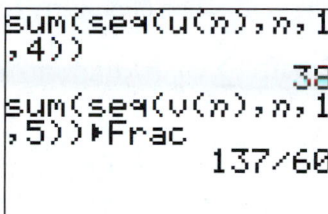

Figure 9.16

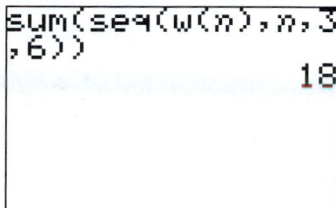

Now try Exercises 57 through 68 ▶

If a definite pattern is noted in a given series expansion, this process can be reversed, with the expanded form being expressed in summation notation using the nth term.

EXAMPLE 8 ▶   **Writing a Sum in Sigma Notation**

Write each of the following sums in summation (sigma) notation.

a. $1 + 3 + 5 + 7 + 9$ b. $6 + 9 + 12 + 15 + \cdots$

Solution ▶

a. The series has five terms and each term is an odd number, or 1 less than a multiple of 2. The general term is $a_n = 2n - 1$, and the series is $\displaystyle\sum_{n=1}^{5}(2n - 1)$.

b. The raised ellipsis "$\cdots$" indicates the sum continues infinitely. Since the terms are multiples of 3, we identify the general term as $a_n = 3n$, while noting the series starts at $n = 2$ (instead of $n = 1$). Since the sum continues indefinitely, we use the infinity symbol ∞ as the "ending" value in sigma notation. The series is $\displaystyle\sum_{n=2}^{\infty} 3n$.

WORTHY OF NOTE

By varying the function given and/or where the sum begins, more than one acceptable form is possible.

For Example 8(b) $\displaystyle\sum_{k=1}^{\infty}(3 + 3k)$ also works.

Now try Exercises 69 through 78 ▶

Since the commutative and associative laws hold for the addition of real numbers, summations have the following properties:

Properties of Summation

Given any real number c and natural number n,

(I) $\displaystyle\sum_{i=1}^{n} c = cn$

If you add a constant c "n" times the result is cn.

(II) $\displaystyle\sum_{i=1}^{n} ca_i = c\sum_{i=1}^{n} a_i$

A constant can be factored out of a sum.

(III) $\displaystyle\sum_{i=1}^{n}(a_i \pm b_i) = \sum_{i=1}^{n}a_i \pm \sum_{i=1}^{n}b_i$

A summation can be distributed to two (or more) sequences.

(IV) $\displaystyle\sum_{i=1}^{m}a_i + \sum_{i=m+1}^{n}a_i = \sum_{i=1}^{n}a_i;\ 1 \le m < n$

A summation is cumulative and can be written as a sum of smaller parts.

The verification of property II depends solely on the distributive property.

Proof: $\displaystyle\sum_{i=1}^{n}ca_i = ca_1 + ca_2 + ca_3 + \cdots + ca_n$ expand sum

$= c(a_1 + a_2 + a_3 + \cdots + a_n)$ factor out c

$= c\displaystyle\sum_{i=1}^{n}a_i$ write series in summation form

The verifications of properties III and IV simply use the commutative and associative properties. You are asked to prove property III in **Exercise 94.**

EXAMPLE 9 ▶ **Computing a Sum Using Summation Properties**

Recompute the sum $\displaystyle\sum_{i=1}^{4}(3i + 2)$ from Example 7(a) using summation properties.

Solution ▶ $\displaystyle\sum_{i=1}^{4}(3i + 2) = \sum_{i=1}^{4}3i + \sum_{i=1}^{4}2$ property III

$= 3\displaystyle\sum_{i=1}^{4}i + \sum_{i=1}^{4}2$ property II

$= 3(10) + 2(4)$ $1 + 2 + 3 + 4 = 10$; property I

$= 38$ result

☑ **D.** You've just seen how we can use summation notation to write and evaluate series

Now try Exercises 79 through 82 ▶

E. Applications of Sequences

To solve applications of sequences, (1) identify where the sequence begins (the initial term), (2) write out the first few terms to help identify the nth term, and (3) decide on an appropriate approach or strategy.

EXAMPLE 10 ▶ **Solving an Application—Accumulation of Stock**

Hydra already owned 1420 shares of stock when her company began offering employees the opportunity to purchase 175 discounted shares per year. If she made no purchases other than these discounted shares each year, how many shares will she have 9 yr later? If this continued for the 25 yr she will work for the company, how many shares will she have at retirement?

Solution ▶ To begin, it helps to simply write out the first few terms of the sequence. Since she already had 1420 shares before the company made this offer, we let $a_0 = 1420$ be the inaugural element, showing $a_1 = 1595$ (after 1 yr, she owns $1420 + 175 = 1595$ shares). The first few terms are 1595, 1770, 1945, 2120, and so on. This supports a general term of $a_n = 1595 + 175(n - 1)$.

After 9 years	**After 25 years**
$a_9 = 1595 + 175(8)$	$a_{25} = 1595 + 175(24)$
$= 2995$	$= 5795$

After 9 yr she would have 2995 shares. Upon retirement she would own 5795 shares of company stock.

Now try Exercises 85 through 90 ▶

Surprisingly, sequences and series have a number of other interesting properties, applications, and mathematical connections. For instance, some of the most celebrated numbers in mathematics can be approximated using a series, as demonstrated in Example 11.

EXAMPLE 11 ▶ Use a calculator to find the partial sums S_4, S_8, and S_{12} for the sequences given. If any sum seems to be approaching a fixed number, name that number.

a. $a_n = \dfrac{1}{n!}$ **b.** $a_n = \dfrac{1}{3^n}$

Solution ▶ **a.** Begin by entering $\dfrac{1}{n!}$ as u(n) on the ⬭Y= screen. Using the "**sum(**" and "**seq(**" commands as before produces the results shown in Figures 9.17 through 9.19, where it appears the sum becomes very close to $e - 1$ for larger values of n.

Figure 9.17	**Figure 9.18**	**Figure 9.19**
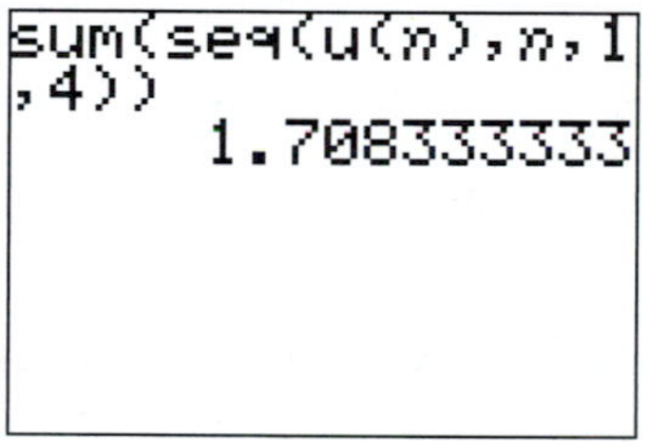 sum(seq(u(n),n,1,4)) 1.708333333	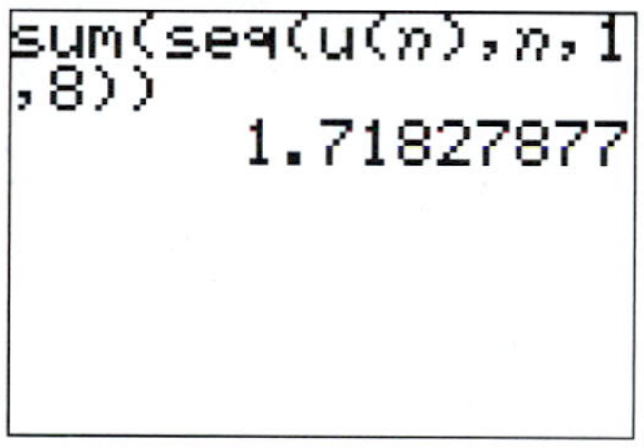 sum(seq(u(n),n,1,8)) 1.71827877	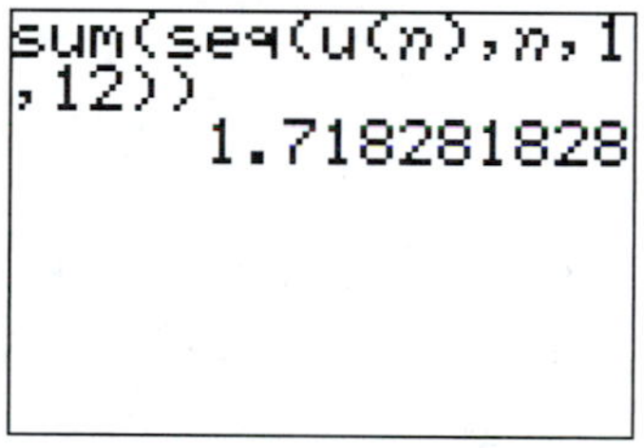 sum(seq(u(n),n,1,12)) 1.718281828

b. After entering $\dfrac{1}{3^n}$ as v(n) on the ⬭Y= screen, we once again compute the sums indicated as shown in Figures 9.20 through 9.22. It appears the sum becomes very close to $\dfrac{1}{2}$ for larger values of n.

Figure 9.20	**Figure 9.21**	**Figure 9.22**
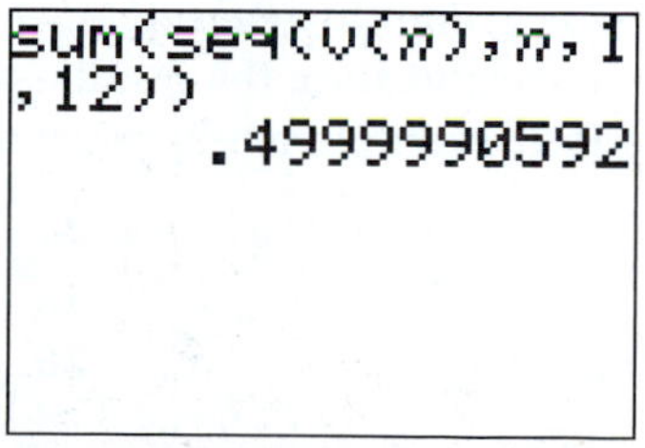 sum(seq(v(n),n,1,4)) .4938271605	sum(seq(v(n),n,1,8)) .4999237921	sum(seq(v(n),n,1,12)) .4999990592

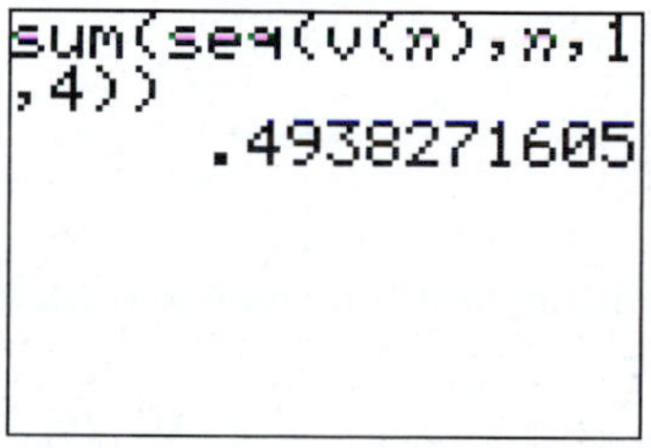

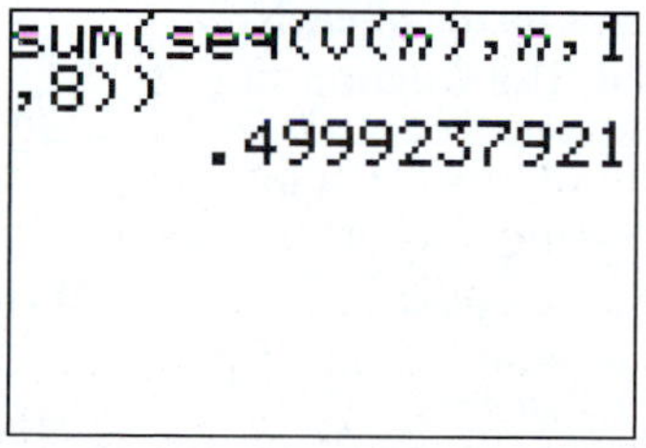

☑ **E.** You've just seen how we can use sequences to solve applications

Now try Exercises 91 and 92 ▶

9.1 EXERCISES

▶ **CONCEPTS AND VOCABULARY**

Fill in the blank with the appropriate word or phrase. Carefully reread the section if needed.

1. A sequence is a(n) __pattern__ of numbers listed in a specific __order__.

2. A series is the __sum__ of the numbers from a given sequence.

3. A sequence that uses the preceding term(s) to generate those that follow is called a __recursive__ sequence.

4. The notation $n!$ represents the __product__ of the natural number n, with all those __preceding__ n.

5. Describe the characteristics of a recursive sequence and give one example.
   formula defining the sequence uses the preceding term(s);
   answers will vary.

6. Describe the characteristics of an alternating sequence and give one example.
   terms alternate in sign; answers will vary.

▶ **DEVELOPING YOUR SKILLS**

Find the first four terms, then find the 8th and 12th term for each nth term given.

7. $a_n = 2n - 1$
   $1, 3, 5, 7; a_8 = 15; a_{12} = 23$

8. $a_n = 2n + 3$
   $5, 7, 9, 11; a_8 = 19; a_{12} = 27$

9. $a_n = 3n^2 - 3$
   $0, 9, 24, 45; a_8 = 189; a_{12} = 429$

10. $a_n = 2n^3 - 12$ $-10, 4, 42,$
    $116; a_8 = 1012; a_{12} = 3444$

11. $a_n = (-1)^n n$
    $-1, 2, -3, 4; a_8 = 8; a_{12} = 12$

12. $a_n = \dfrac{(-1)^n}{n}$
    $-1, \frac{1}{2}, -\frac{1}{3}, \frac{1}{4}; a_8 = \frac{1}{8}; a_{12} = \frac{1}{12}$

13. $a_n = \dfrac{n}{n + 1}$

14. $a_n = \left(1 + \dfrac{1}{n}\right)^n$

15. $a_n = \left(\dfrac{1}{2}\right)^n$

16. $a_n = \left(\dfrac{2}{3}\right)^n$

17. $a_n = \dfrac{1}{n}$

18. $a_n = \dfrac{1}{n^2}$

19. $a_n = \dfrac{(-1)^n}{n(n + 1)}$

20. $a_n = \dfrac{(-1)^{n+1}}{2n^2 - 1}$

21. $a_n = (-1)^n 2^n$

22. $a_n = (-1)^n 2^{-n}$

Use a calculator to: (a) find the indicated term for each sequence, and (b) generate the first five terms of each sequence and store the results in a list. Use fractions if possible; round to tenths when necessary.

23. $a_n = n^2 - 2; a_9$
    a. 79 **b.** $\{-1, 2, 7, 14, 23\}$

24. $a_n = (n - 2)^2; a_9$
    a. 49 **b.** $\{1, 0, 1, 4, 9\}$

25. $a_n = \dfrac{(-1)^{n+1}}{n}; a_5$
    a. $\frac{1}{5}$ **b.** $\{1, \frac{-1}{2}, \frac{1}{3}, \frac{-1}{4}, \frac{1}{5}\}$

26. $a_n = \dfrac{(-1)^{n+1}}{2n - 1}; a_5$
    a. $\frac{1}{9}$ **b.** $\{1, \frac{-1}{3}, \frac{1}{5}, \frac{-1}{7}, \frac{1}{9}\}$

27. $a_n = 2\left(\dfrac{1}{2}\right)^{n-1}; a_7$
    a. $\frac{1}{32}$ **b.** $\{2, 1, \frac{1}{2}, \frac{1}{4}, \frac{1}{8}\}$

28. $a_n = 3\left(\dfrac{1}{3}\right)^{n-1}; a_7$
    a. $\frac{1}{243}$ **b.** $\{3, 1, \frac{1}{3}, \frac{1}{9}, \frac{1}{27}\}$

29. $a_n = \left(1 + \dfrac{1}{n}\right)^n; a_{10}$
    a. approx. 2.6 **b.** $\{2, \frac{9}{4}, \frac{64}{27}, \frac{625}{256}, \frac{7776}{3125}\}$

30. $a_n = \left(n + \dfrac{1}{n}\right)^n; a_9$
    a. approx. 432,655,358.1
    b. $\{2, \frac{25}{4}, \frac{1000}{27}, \frac{83321}{256}, \frac{11,881,376}{3125}\}$

31. $a_n = \dfrac{1}{n(2n + 1)}; a_4$ **a.** $\frac{1}{36}$ **b.** $\{\frac{1}{3}, \frac{1}{10}, \frac{1}{21}, \frac{1}{36}, \frac{1}{55}\}$

32. $a_n = \dfrac{1}{(2n - 1)(2n + 1)}; a_5$ **a.** $\frac{1}{99}$ **b.** $\{\frac{1}{3}, \frac{1}{15}, \frac{1}{35}, \frac{1}{63}, \frac{1}{99}\}$

Find the first five terms of each recursive sequence.

33. $\begin{cases} a_1 = 2 \\ a_n = 5a_{n-1} - 3 \end{cases}$
    $2, 7, 32, 157, 782$

34. $\begin{cases} a_1 = 3 \\ a_n = 2a_{n-1} - 3 \end{cases}$
    $3, 3, 3, 3, 3$

35. $\begin{cases} a_1 = -1 \\ a_n = (a_{n-1})^2 + 3 \end{cases}$
    $-1, 4, 19, 364, 132,499$

36. $\begin{cases} a_1 = -2 \\ a_n = a_{n-1} - 16 \end{cases}$
    $-2, -18, -34, -50, -66$

37. $\begin{cases} c_1 = 64, c_2 = 32 \\ c_n = \dfrac{c_{n-2} - c_{n-1}}{2} \end{cases}$
    $64, 32, 16, 8, 4$

38. $\begin{cases} c_1 = 1, c_2 = 2 \\ c_n = c_{n-1} + (c_{n-2})^2 \end{cases}$
    $1, 2, 3, 7, 16$

Simplify each factorial expression.

39. $\dfrac{8!}{5!}$ 336

40. $\dfrac{12!}{10!}$ 132

41. $\dfrac{9!}{7!2!}$ 36

42. $\dfrac{6!}{3!3!}$ 20

43. $\dfrac{8!}{2!6!}$ 28

44. $\dfrac{10!}{3!7!}$ 120

Write out the first four terms in each sequence.

45. $a_n = \dfrac{n!}{(n + 1)!}$ $\frac{1}{2}, \frac{1}{3}, \frac{1}{4}, \frac{1}{5}$

46. $a_n = \dfrac{n!}{(n + 3)!}$ $\frac{1}{24}, \frac{1}{60}, \frac{1}{120}, \frac{1}{210}$

Additional answers can be found in the Instructor Answer Appendix.

HOMEWORK SELECTION GUIDE

Core: 7–79 every other odd, 85, 87, 89, 91 (23 Exercises)
Standard: 1–4, 7–79 every other odd, 83, 85, 87, 88, 89, 91, 95, 97, 100 (31 Exercises)

Extended: 1–4, 7–79 every other odd, 83, 84, 85, 87, 88, 89, 91, 92, 95, 97, 100 (34 Exercises)
In Depth: 1–6, 7–79 every other odd, 83, 84, 85–92, 95–100 (40 Exercises)

47. $a_n = \dfrac{(n + 1)!}{(3n)!}$ **48.** $a_n = \dfrac{(n + 3)!}{(2n)!}$ $12, 5, 1, \frac{1}{8}$

$\frac{1}{3}, \frac{1}{120}, \frac{1}{15,120}, \frac{1}{3,991,680}$

49. $a_n = \dfrac{n^n}{n!}$ $1, 2, \frac{9}{2}, \frac{32}{3}$ **50.** $a_n = \dfrac{2^n}{n!}$ $2, 2, \frac{4}{3}, \frac{2}{3}$

Find the indicated partial sum for each sequence.

51. $a_n = n$; S_5 15 **52.** $a_n = n^2$; S_7 140

53. $a_n = 2n - 1$; S_8 64 **54.** $a_n = 3n - 1$; S_6 57

55. $a_n = \dfrac{1}{n}$; S_5 $\frac{137}{60}$ **56.** $a_n = \dfrac{n}{n + 1}$; S_4 $\frac{163}{60}$

Expand and evaluate each series. Verify results using a graphing calculator.

57. $\displaystyle\sum_{i=1}^{4}(3i - 5)$ **58.** $\displaystyle\sum_{i=1}^{5}(2i - 3)$ **59.** $\displaystyle\sum_{k=1}^{5}(2k^2 - 3)$

60. $\displaystyle\sum_{k=1}^{5}(k^2 + 1)$ **61.** $\displaystyle\sum_{k=1}^{7}(-1)^k k$ **62.** $\displaystyle\sum_{k=1}^{5}(-1)^k 2^k$

63. $\displaystyle\sum_{i=1}^{4}\dfrac{i^2}{2}$ **64.** $\displaystyle\sum_{i=2}^{4} i^2$ **65.** $\displaystyle\sum_{j=3}^{7} 2j$

66. $\displaystyle\sum_{j=3}^{7}\dfrac{j}{2^j}$ **67.** $\displaystyle\sum_{k=3}^{8}\dfrac{(-1)^k}{k(k - 2)}$ **68.** $\displaystyle\sum_{k=2}^{6}\dfrac{(-1)^{k+1}}{k^2 - 1}$

Write each sum using sigma notation. Answers are not necessarily unique.

69. a. $4 + 8 + 12 + 16 + 20$ $\displaystyle\sum_{n=1}^{5} 4n$

 b. $5 + 10 + 15 + 20 + 25$ $\displaystyle\sum_{n=1}^{5} 5n$

70. a. $-1 + 4 - 9 + 16 - 25 + 36$ $\displaystyle\sum_{n=1}^{6}(-1)^n n^2$

 b. $1 - 8 + 27 - 64 + 125 - 216$ $\displaystyle\sum_{n=1}^{6}(-1)^{n+1} n^3$

71. a. $1 + 3 + 5 + 7 + 9 + 11 + \cdots$ $\displaystyle\sum_{k=1}^{\infty}(2k - 1)$

 b. $1 + \dfrac{1}{2} + \dfrac{1}{4} + \dfrac{1}{8} + \dfrac{1}{16} + \dfrac{1}{32} + \cdots$ $\displaystyle\sum_{k=1}^{\infty}\dfrac{1}{k^2}$

72. a. $0.1 + 0.01 + 0.001 + 0.0001 + \cdots$ $\displaystyle\sum_{k=1}^{\infty}\dfrac{1}{10^k}$

 b. $1 + \dfrac{1}{2} + \dfrac{1}{6} + \dfrac{1}{24} + \dfrac{1}{120} + \dfrac{1}{720} + \cdots$ $\displaystyle\sum_{k=1}^{\infty}\dfrac{1}{k!}$

For the given general term a_n, write the indicated sum using sigma notation.

73. $a_n = n + 3$; S_5 $\displaystyle\sum_{n=1}^{5}(n + 3)$

74. $a_n = \dfrac{n^2 + 1}{n + 1}$; S_4 $\displaystyle\sum_{n=1}^{4}\dfrac{n^2 + 1}{n + 1}$

75. $a_n = \dfrac{n^2}{3}$; third partial sum $\displaystyle\sum_{n=1}^{3}\dfrac{n^2}{3}$

76. $a_n = 2n - 1$; sixth partial sum $\displaystyle\sum_{n=1}^{6}(2n - 1)$

77. $a_n = \dfrac{n}{2^n}$; sum for $n = 3$ to 7 $\displaystyle\sum_{n=3}^{7}\dfrac{n}{2^n}$

78. $a_n = n^2$; sum for $n = 2$ to 6 $\displaystyle\sum_{n=2}^{6} n^2$

Compute each sum by applying properties of summation.

79. $\displaystyle\sum_{i=1}^{5}(4i - 5)$ 35 **80.** $\displaystyle\sum_{i=1}^{6}(3 + 2i)$ 60

81. $\displaystyle\sum_{k=1}^{4}(3k^2 + k)$ 100 **82.** $\displaystyle\sum_{k=1}^{4}(2k^3 + 5)$ 220

▶ WORKING WITH FORMULAS

83. Sum of $a_n = 3n - 2$: $S_n = \dfrac{n(3n - 1)}{2}$

The sum of the first n terms of the sequence defined by $a_n = 3n - 2 = 1, 4, 7, 10, \ldots$, $(3n - 2), \ldots$ is given by the formula shown. Find S_5 using the formula, then verify by direct calculation. 35, verified

84. Sum of $a_n = 3n - 1$: $S_n = \dfrac{n(3n + 1)}{2}$

The sum of the first n terms of the sequence defined by $a_n = 3n - 1 = 2, 5, 8, 11, \ldots, (3n - 1), \ldots$ is given by the formula shown. Find S_8 using the formula, then verify by direct calculation. Observing the formulas from Exercises 83 and 84, can you now state the sum formula for $a_n = 3n - 0$? 100, verified, $S_n = \dfrac{n(3n + 3)}{2}$

▶ APPLICATIONS

Use the information given in each exercise to determine the nth term a_n for the sequence described. Then use the nth term to list the specified number of terms.

85. Wage increases: Latisha gets \$7.25 an hour for filling candy machines for Archtown Vending. Each year she receives a \$0.50 hourly raise. List Latisha's hourly wage for the first 5 yr. How much will she make in the fifth year if she works 8 hr per day for 240 working days? \$7.25, \$7.75, \$8.25, \$8.75, \$9.25; \$17,760

86. Average birth weight: The average birth weight of a certain animal species is 900 g, with the baby gaining 125 g each day for the first 10 days. List the infant's weight for the first 10 days. How much does the infant weigh on the 10th day? 900, 1025, 1150, 1275, 1400, 1525, 1650, 1775, 1900, 2025; 2025 g

Additional answers can be found in the Instructor Answer Appendix.

87. **Blue-book value:** Steve's car has a blue-book value of \$6000. Each year it loses 20% of its value (its value each year is 80% of the year before). List the value of Steve's car for the next 5 yr. (*Hint:* $a_0 = 6000$.)
$a_n = 6000(0.8)^{n-1}$; 6000, 4800, 3840, 3072, 2457.60, 1966.08

88. **Effects of inflation:** Suppose inflation (an increase in value) will average 4% for the next 5 yr. List the growing cost (year by year) of a DVD that costs \$15 right now. (*Hint:* $a_0 = 15$.)
$a_n = 15(1.04)^{n-1}$; 15, 15.60, 16.22, 16.87, 17.55, 18.25

89. **Stocking a lake:** A local fishery stocks a large lake with 1500 bass and then adds an additional 100 mature bass per month until the lake nears maximum capacity. If the bass population grows at a rate of 5% per month through natural reproduction, the number of bass in the pond after n months is given by the recursive sequence $b_0 = 1500$, $b_n = 1.05b_{n-1} + 100$. How many bass will be in the lake after 6 months?   ≈2690

90. **Species preservation:** The Interior Department introduces 50 wolves (male and female) into a large wildlife area in an effort to preserve the species. Each year about 12 additional adult wolves are added from capture and relocation programs. If the wolf population grows at a rate of 10% per year through natural reproduction, the number of wolves in the area after n years is given by the recursive sequence $w_0 = 50$, $w_n = 1.10w_{n-1} + 12$. How many wolves are in the wildlife area after 6 years?   ≈181

Use your calculator to find the partial sums for $n = 4$, $n = 8$, and $n = 12$ for the summations given, and attempt to name the number the summation approximates:

91. $\displaystyle\sum_{k=1}^{n} \frac{1}{2^k}$ approaches 1

92. $\displaystyle\sum_{k=1}^{n} \frac{2^k + 3^k}{6^k}$ $\frac{3}{2}$

▶ **EXTENDING THE CONCEPT**

93. Verify that a constant can be factored out of a sum. That is, verify that the following statement is true:
$$\sum_{j=1}^{n} ca_j = c\sum_{j=1}^{n} a_j$$

94. Verify that a summation may be distributed to two (or more) sequences. That is, verify that the following statement is true:
$$\sum_{i=1}^{n} (a_i \pm b_i) = \sum_{i=1}^{n} a_i \pm \sum_{i=1}^{n} b_i.$$ verified

Regarding Exercises 91 and 92, sometimes a series will approach a fixed number very slowly, and many more terms must be added before this value is recognized. Use your graphing calculator to compute the sums S_{10}, S_{25}, and S_{50} for the following sequences to see if you can recognize the number. Add more terms if necessary.

95. $a_n = \dfrac{1}{n(n+1)(n+2)}$ $\frac{1}{4}$ or 0.25

96. $a_n = \dfrac{1}{(2n-1)(2n+1)}$ $\frac{1}{2}$ or 0.5

▶ **MAINTAINING YOUR SKILLS**

97. **(5.3)** Write $\log_3 \frac{1}{81} = -x$ in exponential form, then solve by equating bases. $3^{-x} = \frac{1}{81}$; $x = 4$

98. **(3.6)** Set up the difference quotient for $f(x) = \sqrt{x}$, then rationalize the numerator.
$\dfrac{\sqrt{x+h} - \sqrt{x}}{h}$; $\dfrac{1}{\sqrt{x+h} + \sqrt{x}}$

99. **(8.4)** Solve the nonlinear system. $\begin{cases} x^2 + y^2 = 9 \\ 9y^2 - 4x^2 = 16 \end{cases}$
$(\sqrt{5}, -2), (\sqrt{5}, 2), (-\sqrt{5}, -2), (-\sqrt{5}, 2)$

100. **(7.3)** Solve the system using a matrix equation.
$\begin{cases} 25x + y - 2z = -14 \\ 2x - y + z = 40 \\ -7x + 3y - z = -13 \end{cases}$ (3, 21, 55)

Additional answers can be found in the Instructor Answer Appendix.

LEARNING OBJECTIVES

In Section 9.2 you will see how we can:

☐ **A.** Identify an arithmetic sequence and its common difference

☐ **B.** Find the nth term of an arithmetic sequence

☐ **C.** Find the nth partial sum of an arithmetic sequence

☐ **D.** Solve applications involving arithmetic sequences

Similar to the way polynomials fall into certain groups or families (linear, quadratic, cubic, etc.), sequences and series with common characteristics are likewise grouped. In this section, we focus on sequences where each successive term is generated by adding a constant value, as in the sequence 1, 8, 15, 22, 29, ... , where 7 is added to a given term in order to produce the next term.

A. Identifying an Arithmetic Sequence and Finding the Common Difference

An **arithmetic sequence** is one where each successive term is found by adding a fixed constant to the preceding term. For instance 3, 7, 11, 15, ... is an arithmetic sequence, since adding 4 to any given term produces the next term. This also means if you take the difference of any two consecutive terms, the result will be 4 and in fact, 4 is called the **common difference** d for this sequence. Using the notation developed earlier, we can write $d = a_{k+1} - a_k$, where a_k represents any term of the sequence and a_{k+1} represents the term that follows a_k.

> **Arithmetic Sequences**
>
> Given a sequence $a_1, a_2, a_3, \ldots, a_k, a_{k+1}, \ldots, a_n$, where $k, n \in \mathbb{N}$ and $k < n$, if there exists a common difference d such that $a_{k+1} - a_k = d$ for all k, then the sequence is an *arithmetic sequence*.

The difference of successive terms can be rewritten as $a_{k+1} = a_k + d$ (for $k \geq 1$) to highlight that each following term is found by adding d to the previous term.

EXAMPLE 1 ▶ **Identifying an Arithmetic Sequence**

Determine if the given sequence is arithmetic. If yes, name the common difference. If not, try to determine the pattern that forms the sequence.

    a. 2, 5, 8, 11, ...   **b.** $\frac{1}{2}, \frac{5}{6}, \frac{13}{12}, \frac{77}{60}, \frac{29}{20}, \ldots$

Solution ▶ **a.** Begin by looking for a common difference $d = a_{k+1} - a_k$. Checking each pair of consecutive terms we have

$$5 - 2 = 3 \qquad 8 - 5 = 3 \qquad 11 - 8 = 3 \quad \text{and so on.}$$

This is an arithmetic sequence with common difference $d = 3$.

b. Checking each pair of consecutive terms yields

$$\frac{5}{6} - \frac{1}{2} = \frac{5}{6} - \frac{3}{6} \qquad \frac{13}{12} - \frac{5}{6} = \frac{13}{12} - \frac{10}{12} \qquad \frac{77}{60} - \frac{13}{12} = \frac{77}{60} - \frac{65}{60}$$

$$= \frac{2}{6} = \frac{1}{3} \qquad\qquad = \frac{3}{12} = \frac{1}{4} \qquad\qquad = \frac{12}{60} = \frac{1}{5}$$

Since the difference is not constant, this is not an arithmetic sequence. It appears the sequence is formed by adding $\frac{1}{k}$ to each previous term, for natural numbers k.

Now try Exercises 7 through 18 ▶

EXAMPLE 2 ▶ **Writing the First *k* Terms of an Arithmetic Sequence**

Write the first five terms of the arithmetic sequence, given the first term a_1 and the common difference *d*.

    a. $a_1 = 12$ and $d = -4$ **b.** $a_1 = \frac{1}{2}$ and $d = \frac{1}{3}$

Solution ▶ **a.** $a_1 = 12$ and $d = -4$. Starting at $a_1 = 12$, add -4 to each new term to generate the sequence: 12, 8, 4, 0, -4.

    b. $a_1 = \frac{1}{2}$ and $d = \frac{1}{3}$. Starting at $a_1 = \frac{1}{2}$ and adding $\frac{1}{3}$ to each new term will generate the sequence: $\frac{1}{2}, \frac{5}{6}, \frac{7}{6}, \frac{3}{2}, \frac{11}{6}$. Note that since the common denominator is 6, terms of the sequence can quickly be found by adding $\frac{1}{3} = \frac{2}{6}$ to the previous term and reducing if possible.

☑ A. You've just seen how we can identify an arithmetic sequence and its common difference

Now try Exercises 19 through 30 ▶

B. Finding the *n*th Term of an Arithmetic Sequence

If the values a_1 and *d* from an arithmetic sequence are known, we could generate the terms of the sequence by adding *multiples of d to the first term,* instead of adding *d* to each new term. For example, we can generate the sequence 3, 8, 13, 18, 23 by adding multiples of 5 to the first term $a_1 = 3$:

$$
\begin{aligned}
3 &= 3 + (0)5 & a_1 &= a_1 + 0d \\
8 &= 3 + (1)5 & a_2 &= a_1 + 1d \\
13 &= 3 + (2)5 & a_3 &= a_1 + 2d \\
18 &= 3 + (3)5 & a_4 &= a_1 + 3d \\
23 &= 3 + (4)5 & a_5 &= a_1 + 4d
\end{aligned}
$$

current term ⟶ initial term ⟷ coefficient of common difference

It's helpful to note the coefficient of *d* is 1 less than the subscript of the current term (as shown): $5 - 1 = 4$. This observation leads us to a formula for the *n*th term.

The *n*th Term of an Arithmetic Sequence

The *n*th term of an *arithmetic sequence* is given by

$$a_n = a_1 + (n - 1)d$$

where *d* is the common difference.

EXAMPLE 3 ▶ **Finding a Specified Term in an Arithmetic Sequence**

Find the 24th term of the sequence 0.1, 0.4, 0.7, 1, ...

Solution ▶ Instead of creating all terms up to the 24th, we determine the constant *d* and use the *n*th term formula. By inspection we note $a_1 = 0.1$ and $d = 0.3$.

$$
\begin{aligned}
a_n &= a_1 + (n - 1)d & &\text{\textit{n}th term formula} \\
&= 0.1 + (n - 1)0.3 & &\text{substitute 0.1 for } a_1 \text{ and 0.3 for } d \\
&= 0.1 + 0.3n - 0.3 & &\text{eliminate parentheses} \\
&= 0.3n - 0.2 & &\text{simplify}
\end{aligned}
$$

To find the 24th term we substitute 24 for *n*:

$$
\begin{aligned}
a_{24} &= 0.3(24) - 0.2 & &\text{substitute 24 for } n \\
&= 7.0 & &\text{result}
\end{aligned}
$$

Now try Exercises 31 through 42 ▶

EXAMPLE 4 ▶ **Determining the Number of Terms in an Arithmetic Sequence**

Find the number of terms in the arithmetic sequence $2, -5, -12, -19, \ldots, -411$.

Solution ▶ By inspection we see that $a_1 = 2$ and $d = -7$. As before,

$$
\begin{aligned}
a_n &= a_1 + (n-1)d && \text{\textit{n}th term formula} \\
&= 2 + (n-1)(-7) && \text{substitute 2 for } a_1 \text{ and } -7 \text{ for } d \\
&= 2 - 7n + 7 && \text{distribute } -7 \\
&= -7n + 9 && \text{simplify}
\end{aligned}
$$

Although we don't know the number of terms in the sequence, we *do* know the last or *n*th term is -411. Substituting -411 for a_n gives

$$
\begin{aligned}
-411 &= -7n + 9 && \text{substitute } -411 \text{ for } a_n \\
60 &= n && \text{solve for } n
\end{aligned}
$$

There are 60 terms in this sequence.

Now try Exercises 43 through 50 ▶

Note that in both Examples 3 and 4, the *n*th term had the form of a linear equation ($y = mx + b$) after simplifying: $a_n = 0.3n - 0.2$ and $a_n = -7n + 9$. This is a characteristic of arithmetic sequences, with the common difference d corresponding to the slope m. This means the graph of an arithmetic sequence will always be a set of discrete points that lie on a straight line. After entering $u(n) = 0.3n - 0.2$ (from Example 3), Figure 9.23 shows the table of values for $a_n = 0.3n - 0.2$, with the graph in Figure 9.24A.

Figure 9.23

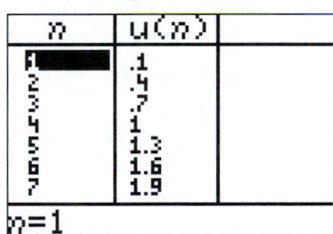

Figure 9.24A

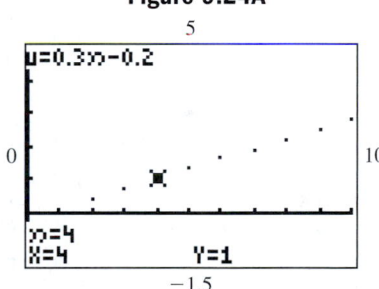

In sequence **MODE**, we still set the size of the viewing window as before, but we can also stipulate the range of values of n to be used (nMin and nMax), which term we want to plot as a beginning (PlotStart), and whether we want all following terms graphed (PlotStep = 1), every second term graphed (PlotStep = 2), and so on. The graph in Figure 9.24A was generated using the values shown in Figure 9.24B (Ymin $= -1.5$, Ymax $= 5$, and Yscl $= 1$ cannot be seen).

Figure 9.24B

```
WINDOW
 nMin=1
 nMax=10
 PlotStart=1
 PlotStep=1
 Xmin=0
 Xmax=10
↓Xscl=1
```

To see the graph of a sequence more distinctly, we can enter the natural numbers 1 through 10 in L1, then define L2 as u(L1) (Figures 9.25 and 9.26) and plot these points (Figure 9.27).

Figure 9.25

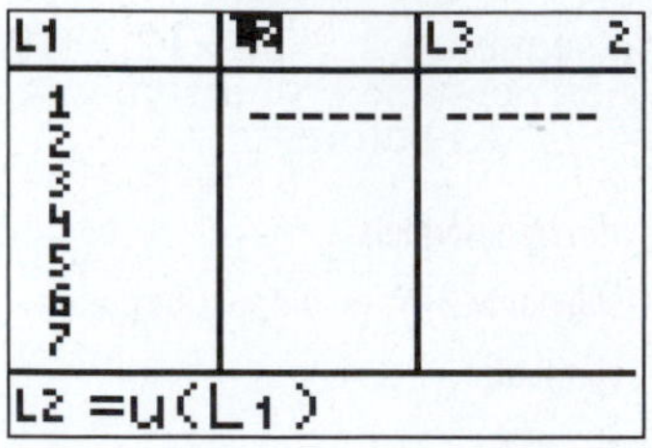

Figure 9.26

Figure 9.27

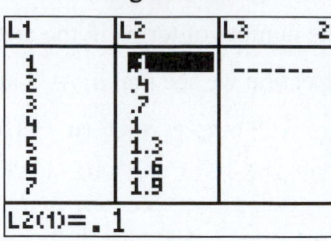

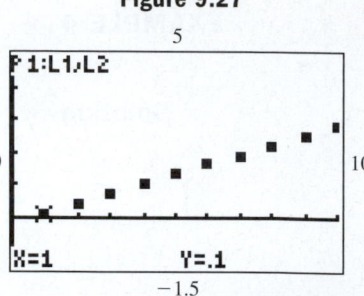

One additional advantage of this approach is that we can easily verify the common difference is d using the "**ΔList(**" feature, which automatically computes the difference between each successive term in a specified list. This option is located in the same submenu as the "**seq(**" option, accessed using (2nd) (STAT) (LIST) ▷ (OPS) 7:ΔList(L2) (ENTER). See Figures 9.28 and 9.29.

Figure 9.28

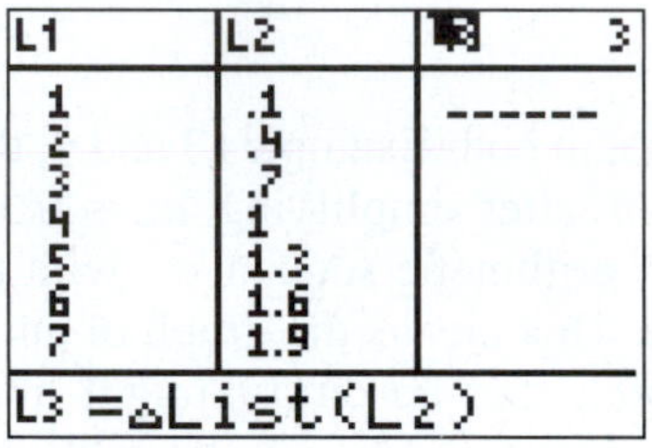

Figure 9.29

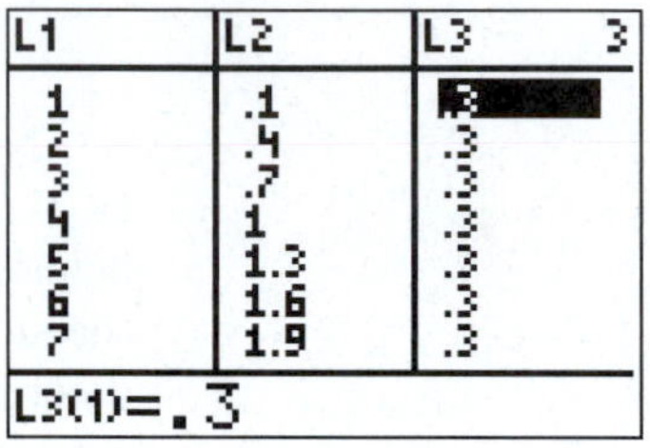

EXAMPLE 5 ▶ **Graphing Arithmetic Sequences**

Enter the natural numbers 1 through 6 in L1, and the terms of the sequence −0.45, −0.1, 0.25, 0.6, 0.95, 1.3 in L2. Then verify the sequence in L2 is arithmetic by

  a. Graphing the related points to see if they appear linear.

  b. Using the **ΔList(** feature.

  c. Finding the nth term that defines the sequence and graph the sequence.

Solution ▶ **a.** The plotted points are shown in Figure 9.30 and appear to be linear.

  b. The **ΔList(** feature shows there is a common difference of 0.35 (Figure 9.31).

  c. With $a_1 = -0.45$ and $d = 0.35$, the nth term must be

$$a_n = a_1 + (n - 1)d \qquad \text{\textcolor{magenta}{nth term formula}}$$
$$= -0.45 + (n - 1)(0.35) \qquad \text{\textcolor{magenta}{substitute for a_1 and d}}$$
$$= 0.35n - 0.8 \qquad \text{\textcolor{magenta}{simplify}}$$

The nth term for this sequence is $a_n = 0.35n - 0.8$. The graph is shown in Figure 9.32.

Figure 9.30

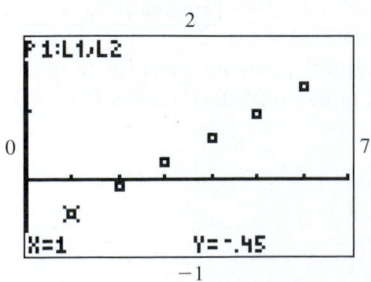

Figure 9.31

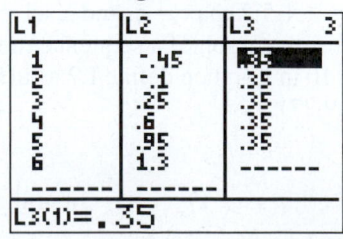

Figure 9.32

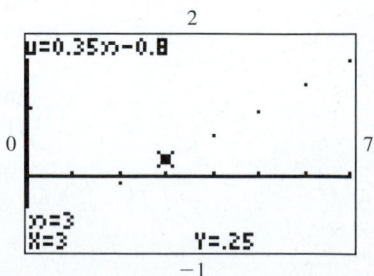

Now try Exercises 51 through 54 ▶

If the term a_1 is unknown but a term a_k is given, the nth term can be written

$$a_n = a_k + (n - k)d$$

(the subscript of the term a_k and coefficient of d sum to n).

EXAMPLE 6 ▶   **Finding the First Term of an Arithmetic Sequence**

Given an arithmetic sequence where $a_6 = 0.55$ and $a_{13} = 0.9$, find the common difference d and the value of a_1.

Solution ▶   At first it seems that not enough information is given, but recall we can express a_{13} as the sum of any earlier term and the appropriate multiple of d. Since a_6 is known, we write $a_{13} = a_6 + 7d$ (note $13 = 6 + 7$ as required).

$$a_{13} = a_6 + 7d \qquad \text{a_1 is unknown}$$
$$0.9 = 0.55 + 7d \qquad \text{substitute 0.9 for a_{13} and 0.55 for a_6}$$
$$0.35 = 7d \qquad \text{subtract 0.55}$$
$$d = 0.05 \qquad \text{solve for d}$$

Having found d, we can now solve for a_1.

$$a_{13} = a_1 + 12d \qquad \text{nth term formula for $n = 13$}$$
$$0.9 = a_1 + 12(0.05) \qquad \text{substitute 0.9 for a_{13} and 0.05 for d}$$
$$0.9 = a_1 + 0.6 \qquad \text{simplify}$$
$$a_1 = 0.3 \qquad \text{solve for a_1}$$

☑ **B. You've just seen how we can find the nth term of an arithmetic sequence**

The first term is $a_1 = 0.3$ and the common difference is $d = 0.05$.

Now try Exercises 55 through 60 ▶

C. Finding the nth Partial Sum of an Arithmetic Sequence

Using sequences and series to solve applications often requires computing the sum of a given number of terms. To develop and understand the approach used, consider the sum of the first 10 natural numbers. Using S_{10} to represent this sum, we have $S_{10} = 1 + 2 + 3 + 4 + 5 + 6 + 7 + 8 + 9 + 10$. We could just use brute force, but if we rewrite the sum a second time *but in reverse order,* then add it to the first, we find that each column adds to 11.

S_{10}	=	1	+	2	+	3	+	4	+	5	+	6	+	7	+	8	+	9	+	10
S_{10}	=	10	+	9	+	8	+	7	+	6	+	5	+	4	+	3	+	2	+	1
$2S_{10}$	=	11	+	11	+	11	+	11	+	11	+	11	+	11	+	11	+	11	+	11

Since there are 10 columns, the total is $11 \times 10 = 110$ but this is twice the actual sum and we find that $2S_{10} = 110$, so $S_{10} = 55$.

Now consider the sequence $a_1, a_2, a_3, a_4, \ldots, a_n$ with common difference d. Use S_n to represent the sum of the first n terms and write the original series, then the series in reverse order underneath. Since one row increases at the same rate the other decreases, the sum of each column remains constant, and for simplicity's sake we choose $a_1 + a_n$ to represent this sum.

$$
\begin{array}{lcccccccccccc}
S_n = & a_1 & + & a_2 & + & a_3 & + \cdots + & a_{n-2} & + & a_{n-1} & + & a_n & \quad \text{add} \\
S_n = & a_n & + & a_{n-1} & + & a_{n-2} & + \cdots + & a_3 & + & a_2 & + & a_1 & \quad \text{columns} \\
\hline
2S_n = & (a_1 + a_n) & + & (a_1 + a_n) & + & (a_1 + a_n) & + \cdots + & (a_1 + a_n) & + & (a_1 + a_n) & + & (a_1 + a_n) & \quad \text{vertically}
\end{array}
$$

To understand why each column adds to $a_1 + a_n$, consider the sum in the second column: $a_2 + a_{n-1}$. From $a_2 = a_1 + d$ and $a_{n-1} = a_n - d$, we obtain $a_2 + a_{n-1} = (a_1 + d) + (a_n - d)$ by adding the equations, which gives a result of $a_1 + a_n$. Since there are n columns, we end up with $2S_n = n(a_1 + a_n)$, and solving for S_n gives the formula for the first n terms of an arithmetic sequence.

The *n*th Partial Sum of an Arithmetic Sequence

Given an arithmetic sequence with first term a_1, the *n*th partial sum is given by

$$S_n = \frac{n}{2}(a_1 + a_n).$$

In words: The sum of an arithmetic sequence is one-half the number of terms times the sum of the first and last term.

EXAMPLE 7 ▶ **Computing the Sum of an Arithmetic Sequence**

Use the summation formula to find the sum of the first 75 positive odd integers: $\sum_{n=1}^{75} (2n - 1)$. Verify the result using a graphing calculator.

Solution ▶ The initial terms of the sequence are 1, 3, 5, ... and we note $a_1 = 1$, $d = 2$, and $n = 75$. To use the sum formula, we need the value of a_{75}: $2(75) - 1 = 149$. formula shows $a_{75} = a_1 + 74d = 1 + 74(2)$, so $a_{75} = 149$.

$$S_n = \frac{n}{2}(a_1 + a_n) \qquad \text{\color{magenta}sum formula}$$

$$S_{75} = \frac{75}{2}(a_1 + a_{75}) \qquad \text{\color{magenta}substitute 75 for } n$$

$$= \frac{75}{2}(1 + 149) \qquad \text{\color{magenta}substitute 1 for } a_1, \text{ 149 for } a_{75}$$

$$= 5625 \qquad \text{\color{magenta}result}$$

The sum of the first 75 positive odd integers is 5625.

To verify, we enter $u(n) = 2n - 1$ on the 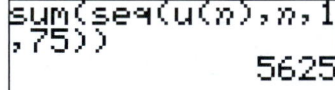 screen, and find the sum of the first 75 terms of the sequence on the home screen as before. See figure.

```
sum(seq(u(n),n,1
,75))
                5625
```

Now try Exercises 61 through 66 ▶

By substituting the *n*th term formula directly into the formula for partial sums, we're able to find a partial sum without actually having to find the *n*th term:

$$S_n = \frac{n}{2}(a_1 + a_n) \qquad \text{\color{magenta}sum formula}$$

✓ C. You've just seen how we can find the *n*th partial sum of an arithmetic sequence

$$= \frac{n}{2}(a_1 + [a_1 + (n - 1)d]) \qquad \text{\color{magenta}substitute } a_1 + (n-1)d \text{ for } a_n$$

$$= \frac{n}{2}[2a_1 + (n - 1)d] \qquad \text{\color{magenta}alternative formula for the } n\text{th partial sum}$$

See Exercises 67 through 72 for more on this alternative formula.

D. Applications

Figure 9.33

spiral fern

In the evolution of certain plants and shelled animals, sequences and series seem to have been one of nature's favorite tools. The sprials found on many ferns and other plants are excellent examples of sequences in nature, as are the size of the chambers in nautilus shells (see Figures 9.33 and 9.34). Sequences and series also provide a good mathematical model for a variety of other situations as well.

Figure 9.34

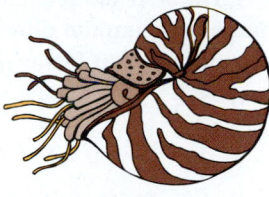

nautilus

EXAMPLE 8 ▶ **Solving an Application of Arithmetic Sequences: Seating Capacity**

Cox Auditorium is an amphitheater that has 40 seats in the first row, 42 seats in the second row, 44 in the third, and so on. If there are 75 rows in the auditorium, what is the auditorium's seating capacity?

Solution ▶ The number of seats in each row gives the terms of an arithmetic sequence with $a_1 = 40$, $d = 2$, and $n = 75$. To find the seating capacity, we need to find the total number of seats, which is the sum of this arithmetic sequence. Since the value of a_{75} is unknown, we opt for the alternative formula $S_n = \dfrac{n}{2}[2a_1 + (n-1)d]$.

$$S_n = \frac{n}{2}[2a_1 + (n-1)d] \qquad \text{sum formula}$$

$$S_{75} = \frac{75}{2}[2(40) + (75-1)(2)] \qquad \text{substitute 40 for } a_1, 2 \text{ for } d, \text{ and 75 for } n$$

$$= \frac{75}{2}(228) \qquad \text{simplify}$$

$$= 8550 \qquad \text{result}$$

☑ **D. You've just seen how we can solve applications involving arithmetic sequences**

The seating capacity of Cox Auditorium is 8550.

Now try Exercises 75 through 80 ▶

9.2 EXERCISES

▶ **CONCEPTS AND VOCABULARY**

Fill in the blank with the appropriate word or phrase. Carefully reread the section if needed.

1. Consecutive terms in an arithmetic sequence differ by a constant called the __common__ __difference__.

2. The sum of the first n terms of an arithmetic sequence is called the nth __partial__ __sum__.

3. The formula for the nth partial sum of an arithmetic sequence is $S_n = \dfrac{n(a_1 + a_n)}{2}$, where a_n is the __nth__ term.

4. The nth term formula for an arithmetic sequence is $a_n = \dfrac{a_1 + (n-1)d}{}$, where a_1 is the __first__ term and d is the __common__ __difference__.

5. Discuss how the terms of an arithmetic sequence can be written in various ways using the relationship $a_n = a_k + (n-k)d$. Answers will vary.

6. Describe how the formula for the nth partial sum was derived, and illustrate its application using a sequence from the exercise set. Answers will vary.

HOMEWORK SELECTION GUIDE

Core: 7–71 every other odd, 75, 77, 79 (22 Exercises)
Standard: 1–4, 7–51 every other odd, 52, 53, 55–71 every other odd, 73, 75, 76, 77–83 odd (30 Exercises)

Extended: 1–4, 7–51 every other odd, 52, 53, 55–71 every other odd, 73, 75–79, 81–83, 85 (33 Exercises)
In Depth: 1–6, 7–51 every other odd, 52, 53, 55–71 every other odd, 73–86 (39 Exercises)

▶ DEVELOPING YOUR SKILLS

Determine if the sequence given is arithmetic. If yes, name the common difference. If not, try to determine the pattern that forms the sequence.

7. $-5, -2, 1, 4, 7, 10, \ldots$ arithmetic; $d = 3$

8. $1, -2, -5, -8, -11, -14, \ldots$ arithmetic; $d = -3$

9. $0.5, 3, 5.5, 8, 10.5, \ldots$ arithmetic; $d = 2.5$

10. $1.2, 3.5, 5.8, 8.1, 10.4, \ldots$ arithmetic; $d = 2.3$

11. $2, 3, 5, 7, 11, 13, 17, \ldots$ not arithmetic; all prime

12. $1, 4, 8, 13, 19, 26, 34, \ldots$
   not arithmetic; each difference increases by 1

13. $\frac{1}{24}, \frac{1}{12}, \frac{1}{8}, \frac{1}{6}, \frac{5}{24}, \ldots$ arithmetic; $d = \frac{1}{24}$

14. $\frac{1}{12}, \frac{1}{15}, \frac{1}{20}, \frac{1}{30}, \frac{1}{60}, \ldots$ arithmetic; $d = -\frac{1}{60}$

15. $1, 4, 9, 16, 25, 36, \ldots$ not arithmetic; $a_n = n^2$

16. $-125, -64, -27, -8, -1, \ldots$
   not arithmetic; $a_n = (n-6)^3$

17. $\pi, \dfrac{5\pi}{6}, \dfrac{2\pi}{3}, \dfrac{\pi}{2}, \dfrac{\pi}{3}, \dfrac{\pi}{6}, \ldots$ arithmetic; $d = \frac{-\pi}{6}$

18. $\pi, \dfrac{7\pi}{8}, \dfrac{3\pi}{4}, \dfrac{5\pi}{8}, \dfrac{\pi}{2}, \ldots$ arithmetic; $d = \frac{-\pi}{8}$

Write the first four terms of the arithmetic sequence with the given first term and common difference.

19. $a_1 = 2, d = 3$ 2, 5, 8, 11   **20.** $a_1 = 8, d = 3$
                                                                  8, 11, 14, 17

21. $a_1 = 7, d = -2$ 7, 5, 3, 1   **22.** $a_1 = 60, d = -12$
                                                              60, 48, 36, 24

23. $a_1 = 0.3, d = 0.03$ **24.** $a_1 = 0.5, d = 0.25$
    0.3, 0.33, 0.36, 0.39   0.5, 0.75, 1, 1.25

25. $a_1 = \frac{3}{2}, d = \frac{1}{2}$ $\frac{3}{2}, 2, \frac{5}{2}, 3$ **26.** $a_1 = \frac{1}{5}, d = \frac{1}{10}$ $\frac{1}{5}, \frac{3}{10}, \frac{2}{5}, \frac{1}{2}$

27. $a_1 = \frac{3}{4}, d = -\frac{1}{8}$ $\frac{3}{4}, \frac{5}{8}, \frac{1}{2}, \frac{3}{8}$ **28.** $a_1 = \frac{1}{6}, d = -\frac{1}{3}$
                                                            $\frac{1}{6}, \frac{-1}{6}, \frac{-1}{2}, \frac{-5}{6}$

29. $a_1 = -2, d = -3$ **30.** $a_1 = -4, d = -4$
    $-2, -5, -8, -11$ $-4, -8, -12, -16$

Identify the first term and the common difference, then write the expression for the general term a_n and use it to find the 6th, 10th, and 12th terms of the sequence.

31. $2, 7, 12, 17, \ldots$ **32.** $7, 4, 1, -2, -5, \ldots$

33. $5.10, 5.25, 5.40, \ldots$ **34.** $9.75, 9.40, 9.05, \ldots$

35. $\frac{3}{2}, \frac{9}{4}, 3, \frac{15}{4}, \ldots$ **36.** $\frac{5}{7}, \frac{3}{14}, -\frac{2}{7}, -\frac{11}{14}, \ldots$

Find the indicated term using the information given.

37. $a_1 = 5, d = 4$; find a_{15} 61

38. $a_1 = 9, d = -2$; find a_{17} -23

39. $a_1 = \frac{3}{2}, d = -\frac{1}{12}$; find a_7 1

40. $a_1 = \frac{12}{25}, d = -\frac{1}{10}$; find a_9 $\frac{-8}{25}$

41. $a_1 = -0.025, d = 0.05$; find a_{50} 2.425

42. $a_1 = 3.125, d = -0.25$; find a_{20} -1.625

Find the number of terms in each sequence.

43. $a_1 = 2, a_n = -22, d = -3$ 9

44. $a_1 = 4, a_n = 42, d = 2$ 20

45. $a_1 = 0.4, a_n = 10.9, d = 0.25$ 43

46. $a_1 = -0.3, a_n = -36, d = -2.1$ 18

47. $-3, -0.5, 2, 4.5, 7, \ldots, 47$ 21

48. $-3.4, -1.1, 1.2, 3.5, \ldots, 38$ 19

49. $\frac{1}{12}, \frac{1}{8}, \frac{1}{6}, \frac{5}{24}, \frac{1}{4}, \ldots, \frac{9}{8}$ 26   **50.** $\frac{1}{12}, \frac{1}{15}, \frac{1}{20}, \frac{1}{30}, \ldots, -\frac{1}{4}$
                                                                  21

📇 **For Exercises 51 through 54, enter the natural numbers 1 through 6 in L1 on a graphing calculator, and the terms of the given sequence in L2. Then determine if the sequence is arithmetic by (a) graphing the related points to see if they appear linear, and (b) using the ΔList(feature. If an arithmetic sequence, (c) find the nth term and graph the sequence.**

51. $1.5, 2.25, 3, 3.75, 4.5, 5.25, \ldots$

52. $\dfrac{47}{18}, \dfrac{19}{9}, \dfrac{29}{18}, \dfrac{10}{9}, \dfrac{11}{18}, \dfrac{1}{9}, \ldots$

53. $9, 8, 6, 3, -1, -6, \ldots$ **54.** $\dfrac{1}{1}, \dfrac{1}{2}, \dfrac{1}{3}, \dfrac{1}{4}, \dfrac{1}{5}, \dfrac{1}{6}, \ldots$
   a. appears nonlinear   **a.** appears nonlinear
   b. no common difference   **b.** no common difference

Find the common difference d and the value of a_1 using the information given.

55. $a_3 = 7, a_7 = 19$ **56.** $a_5 = -17, a_{11} = -2$
       $d = 3, a_1 = 1$ $d = \frac{5}{2}, a_1 = -27$

57. $a_2 = 1.025, a_{26} = 10.025$ $d = 0.375, a_1 = 0.65$

58. $a_6 = -12.9, a_{30} = 1.5$ $d = 0.6, a_1 = -15.9$

59. $a_{10} = \frac{13}{18}, a_{24} = \frac{27}{2}$ **60.** $a_4 = \frac{5}{4}, a_8 = \frac{9}{4}$
         $d = \frac{115}{126}, a_1 = \frac{-472}{63}$ $d = \frac{1}{4}, a_1 = \frac{1}{2}$

Evaluate each sum. For Exercises 65 and 66, use the summation properties from Section 9.1. Verify all results on a graphing calculator.

61. $\displaystyle\sum_{n=1}^{30} (3n - 4)$ **62.** $\displaystyle\sum_{n=1}^{29} (4n - 1)$

63. $\displaystyle\sum_{n=1}^{37} \left(\frac{3}{4}n + 2\right)$ **64.** $\displaystyle\sum_{n=1}^{20} \left(\frac{5}{2}n - 3\right)$

65. $\displaystyle\sum_{n=4}^{15} (3 - 5n)$ **66.** $\displaystyle\sum_{n=7}^{20} (7 - 2n)$

Additional answers can be found in the Instructor Answer Appendix.

Use the alternative formula for the nth partial sum to compute the sums indicated.

67. The sum S_{15} for the sequence
$-12 + (-9.5) + (-7) + (-4.5) + \cdots$ 82.5

68. The sum S_{20} for the sequence $\frac{9}{2} + \frac{7}{2} + \frac{5}{2} + \frac{3}{2} + \cdots$
   -100

69. The sum S_{30} for the sequence
$0.003 + 0.173 + 0.343 + 0.513 + \cdots$ 74.04

70. The sum S_{50} for the sequence
$(-2) + (-7) + (-12) + (-17) + \cdots$ -6225

71. The sum S_{20} for the sequence
$\sqrt{2} + 2\sqrt{2} + 3\sqrt{2} + 4\sqrt{2} + \cdots$ $210\sqrt{2}$

72. The sum S_{10} for the sequence
$12\sqrt{3} + 10\sqrt{3} + 8\sqrt{3} + 6\sqrt{3} + \cdots$ $30\sqrt{3}$

▶ WORKING WITH FORMULAS

73. Sum of the first n natural numbers: $S_n = \dfrac{n(n+1)}{2}$

The sum of the first n natural numbers can be found using the formula shown, where n represents the number of terms in the sum. Verify the formula by adding the first six natural numbers by hand, and then evaluating S_6. Then find the sum of the first 75 natural numbers. $S_6 = 21$; $S_{75} = 2850$

74. Sum of the squares of the first n natural numbers: $S_n = \dfrac{n(n+1)(2n+1)}{6}$

If the first n natural numbers are squared, the sum of these squares can be found using the formula shown, where n represents the number of terms in the sum. Verify the formula by computing the sum of the squares of the first six natural numbers by hand, and then evaluating S_6. Then find the sum of the squares of the first 20 natural numbers: $(1^2 + 2^2 + 3^2 + \cdots + 20^2)$. $S_6 = 91$; $S_{20} = 2870$

▶ APPLICATIONS

75. Temperature fluctuation: At 5 P.M. in Coldwater, the temperature was a chilly 36°F. If the temperature decreased by 3°F every half-hour for the next 7 hr, at what time did the temperature hit 0°F?   at 11 P.M.

76. Arc of a baby swing: When Mackenzie's baby swing is started, the first swing (one way) is a 30-in. arc. As the swing slows down, each successive arc is $\frac{3}{2}$ in. less than the previous one. Find (a) the length of the tenth swing and (b) how far Mackenzie has traveled during the 10 swings.   16.5 in.; 232.5 in.

77. Computer animations: The animation on a new computer game initially allows the hero of the game to jump a (screen) distance of 10 in. over booby traps and obstacles. Each successive jump is limited to $\frac{3}{4}$ in. less than the previous one. Find (a) the length of the seventh jump and (b) the total distance covered after seven jumps.   5.5 in.; 54.25 in.

78. Seating capacity: The Fox Theater creates a "theater in the round" when it shows any of Shakespeare's plays. The first row has 80 seats, the second row has 88,

the third row has 96, and so on. How many seats are in the 10th row? If there is room for 25 rows, how many chairs will be needed to set up the theater?   152; 4400

79. Sales goals: *At the time that I was newly hired, 100 sales per month was what I required. Each following month—the last plus 20 more, as I work for the goal of top sales award. When 2500 sales are thusly made, it's Tahiti, Hawaii, and piña coladas in the shade.* How many sales were made by this person in the seventh month? What were the total sales after the 12th month? Was the goal of 2500 total sales met? $a_7 = 220$; $a_{12} = 2520$; yes

80. Bequests to charity: *At the time our mother left this Earth, she gave \$9000 to her children of birth. This we kept and each year added \$3000 more, as a lasting memorial from the children she bore. When \$42,000 is thusly attained, all goes to charity that her memory be maintained.* What was the balance in the sixth year? In what year was the goal of \$42,000 met? $a_6 = 27{,}000$; In the 11th year: $a_{11} = 42{,}000$

Additional answers can be found in the Instructor Answer Appendix.

▶ EXTENDING THE CONCEPT

81. From a study of numerical analysis, a function is known to be linear if its "first differences" (differences between successive outputs) are constant. Likewise, a function is known to be quadratic if its "first differences" form an *arithmetic sequence.* Use this information to determine if the following sets of output come from a linear or quadratic function:

   a. 19, 11.8, 4.6, −2.6, −9.8, −17, −24.2, … linear function
   b. −10.31, −10.94, −11.99, −13.46, −15.35, … quadratic

82. From elementary geometry it is known that the interior angles of a triangle sum to 180°, the interior angles of a quadrilateral sum to 360°, the interior angles of a pentagon sum to 540°, and so on. Use the pattern created by the relationship between the number of sides and the number of angles to develop a formula for the sum of the interior angles of an n-sided polygon. The interior angles of a decagon (10 sides) sum to how many degrees? $180(n − 2)$, 1440°

▶ MAINTAINING YOUR SKILLS

83. (5.5) Solve for t: $2530 = 500e^{0.45t}$ $t \approx 3.6$

84. (3.2) Graph by completing the square. Label all important features: $y = x^2 − 2x − 3$.

85. (1.3) In 2000, the deer population was 972. By 2005 it had grown to 1217. Assuming the growth is linear, find the function that models this data and use it to estimate the deer population in 2008.
   $f(x) = 49x + 972$; 1364

86. (2.6) Given y varies inversely with x and directly with w. If $y = 14$ when $x = 15$ and $w = 52.5$, find the value of y when $x = 32$ and $w = 208$. $y = 26$

Additional answers can be found in the Instructor Answer Appendix.

9.3   Geometric Sequences

LEARNING OBJECTIVES

In Section 9.3 you will see how we can:

☐ **A.** Identify a geometric sequence and its common ratio

☐ **B.** Find the nth term of a geometric sequence

☐ **C.** Find the nth partial sum of a geometric sequence

☐ **D.** Find the sum of an infinite geometric series

☐ **E.** Solve application problems involving geometric sequences and series

Recall that arithmetic sequences are those where each term is found by *adding* a constant value to the preceding term. In this section, we consider **geometric sequences,** where each term is found by *multiplying* the preceding term by a constant value. Geometric sequences have many interesting applications, as do **geometric series.**

A. Geometric Sequences

A geometric sequence is one where each successive term is found by multiplying the preceding term by a fixed constant. Consider growth of a bacteria population, where a single cell splits in two every hour over a 24-hr period. Beginning with a single bacterium ($a_0 = 1$), after 1 hr there are 2, after 2 hr there are 4, and so on. Writing the number of bacteria as a sequence we have:

hours:	a_1	a_2	a_3	a_4	a_5	…
	↓	↓	↓	↓	↓	
bacteria:	2	4	8	16	32	…

The sequence 2, 4, 8, 16, 32, … is a geometric sequence since each term is found by multiplying the previous term by the constant factor 2. This also means that the ratio of any two consecutive terms must be 2 and in fact, 2 is called the **common ratio r** for this sequence. Using the notation from Section 9.1 we can write $r = \dfrac{a_{k+1}}{a_k}$, where a_k represents any term of the sequence and a_{k+1} represents the term that follows a_k.

> ### Geometric Sequences
>
> Given a sequence $a_1, a_2, a_3, \ldots, a_k, a_{k+1}, \ldots, a_n$, where $k, n \in \mathbb{N}$ and $k < n$,
>
> if there exists a common ratio r such that $\dfrac{a_{k+1}}{a_k} = r$ for all k,
>
> then the sequence is a *geometric sequence*.

The ratio of successive terms can be rewritten as $a_{k+1} = a_k r$ (for $k \geq 1$) to highlight that each term is found by multiplying the preceding term by r.

EXAMPLE 1 ▶ **Testing a Sequence for a Common Ratio**

Determine if the given sequence is geometric. If not geometric, try to determine the pattern that forms the sequence.

    a. $1, 0.5, 0.25, 0.125, \ldots$ **b.** $\frac{1}{7}, \frac{2}{7}, \frac{6}{7}, \frac{24}{7}, \frac{120}{7}, \ldots$

Solution ▶ Apply the definition to check for a common ratio $r = \dfrac{a_{k+1}}{a_k}$.

    a. For $1, 0.5, 0.25, 0.125, \ldots$, the ratio of consecutive terms gives

$$\frac{0.5}{1} = 0.5, \qquad \frac{0.25}{0.5} = 0.5, \qquad \frac{0.125}{0.25} = 0.5, \qquad \text{and so on.}$$

This is a geometric sequence with common ratio $r = 0.5$.

    b. For $\frac{1}{7}, \frac{2}{7}, \frac{6}{7}, \frac{24}{7}, \frac{120}{7}, \ldots$, we have:

$$\frac{2}{7} \div \frac{1}{7} = \frac{2}{7} \cdot \frac{7}{1} \qquad \frac{6}{7} \div \frac{2}{7} = \frac{6}{7} \cdot \frac{7}{2} \qquad \frac{24}{7} \div \frac{6}{7} = \frac{24}{7} \cdot \frac{7}{6} \quad \text{and so on.}$$

$$= 2 \qquad\qquad\qquad = 3 \qquad\qquad\qquad = 4$$

Since the ratio is not constant, this is not a geometric sequence. The sequence appears to be formed by dividing $n!$ by 7: $a_n = \dfrac{n!}{7}$.

Now try Exercises 7 through 24 ▶

EXAMPLE 2 ▶ **Writing the Terms of a Geometric Sequence**

Write the first five terms of the geometric sequence, given the first term $a_1 = -16$ and the common ratio $r = 0.25$.

Solution ▶ Given $a_1 = -16$ and $r = 0.25$. Starting at $a_1 = -16$, multiply each term by 0.25 to generate the sequence.

$$a_2 = -16 \cdot 0.25 = -4 \qquad a_3 = -4 \cdot 0.25 = -1$$
$$a_4 = -1 \cdot 0.25 = -0.25 \qquad a_5 = -0.25 \cdot 0.25 = -0.0625$$

The first five terms of this sequence are $-16, -4, -1, -0.25,$ and -0.0625.

☑ **A.** You've just seen how we can identify a geometric sequence and its common ratio

Now try Exercises 25 through 32 ▶

B. Find the *n*th Term of a Geometric Sequence

If the values a_1 and r from a geometric sequence are known, we could generate the terms of the sequence by applying *additional factors of r to the first term,* instead of multiplying each new term by r. If $a_1 = 3$ and $r = 2$, we simply begin at a_1, and continue applying additional factors of r for each successive term.

$$3 = 3 \cdot 2^0 \qquad a_1 = a_1 r^0$$
$$6 = 3 \cdot 2^1 \qquad a_2 = a_1 r^1$$
$$12 = 3 \cdot 2^2 \qquad a_3 = a_1 r^2$$
$$24 = 3 \cdot 2^3 \qquad a_4 = a_1 r^3$$
$$48 = 3 \cdot 2^4 \qquad a_5 = a_1 r^4$$

current term — initial term — exponent on common ratio

From this pattern, we note the exponent on r is always 1 less than the subscript of the current term: $5 - 1 = 4$, which leads us to the formula for the nth term of a geometric sequence.

> **The nth Term of a Geometric Sequence**
>
> The nth term of a *geometric sequence* is given by
>
> $$a_n = a_1 r^{n-1}$$
>
> where r is the common ratio.

EXAMPLE 3 ▶ **Finding a Specific Term in a Sequence**

Identify the common ratio r, and use it to write the expression for the nth term. Then find the 10th term of the sequence: $3, -6, 12, -24, \ldots$.

Solution ▶ By inspection we note that $a_1 = 3$ and $r = -2$. This gives

$$a_n = a_1 r^{n-1} \qquad \text{\textit{n}th term formula}$$
$$= 3(-2)^{n-1} \qquad \text{substitute 3 for } a_1 \text{ and } -2 \text{ for } r$$

To find the 10th term we substitute $n = 10$:

$$a_{10} = 3(-2)^{10-1} \qquad \text{substitute 10 for } n$$
$$= 3(-2)^9 = -1536 \qquad \text{simplify}$$

Now try Exercises 33 through 46 ▶

EXAMPLE 4 ▶ **Determining the Number of Terms in a Geometric Sequence**

Find the number of terms in the geometric sequence $4, 2, 1, \ldots, \frac{1}{64}$.

Solution ▶ Observing that $a_1 = 4$ and $r = \frac{1}{2}$, we have

$$a_n = a_1 r^{n-1} \qquad \text{\textit{n}th term formula}$$
$$= 4\left(\frac{1}{2}\right)^{n-1} \qquad \text{substitute 4 for } a_1 \text{ and } \frac{1}{2} \text{ for } r$$

Although we don't know the number of terms in the sequence, we *do* know the last or nth term is $\frac{1}{64}$. Substituting $a_n = \frac{1}{64}$ gives

$$\frac{1}{64} = 4\left(\frac{1}{2}\right)^{n-1} \qquad \text{substitute } \frac{1}{64} \text{ for } a_n$$
$$\frac{1}{256} = \left(\frac{1}{2}\right)^{n-1} \qquad \text{divide by 4} \left(\text{multiply by } \frac{1}{4}\right)$$

From our work in Chapter 5, we attempt to write both sides as exponentials with a like base, or apply logarithms. Since $256 = 2^8$, we equate bases.

$$\left(\frac{1}{2}\right)^8 = \left(\frac{1}{2}\right)^{n-1} \qquad \text{write } \frac{1}{256} \text{ as } \left(\frac{1}{2}\right)^8$$

$$\rightarrow 8 = n - 1 \qquad \text{like bases imply exponents must be equal}$$

$$9 = n \qquad \text{solve for } n$$

This shows there are nine terms in the sequence.

> **Now try Exercises 47 through 58** ▶

Note that in both Examples 3 and 4, the *n*th term had the form of an exponential equation $(y = a \cdot b^n)$ after simplifying: $a_n = 3(-2)^{n-1}$ and $a_n = 4\left(\frac{1}{2}\right)^{n-1}$. This is in fact a characteristic of geometric sequences, with the common ratio r corresponding to the base b. This means the graph of a geometric sequence will always be a set of discrete points that lie on an exponential graph (see *Worthy of Note*). After entering

$$u(n) = 4\left(\frac{1}{2}\right)^{n-1}$$

(from Example 4), Figure 9.35 shows the table of values for this sequence, with the graph in Figure 9.36.

As before, we can see the graph of the sequence more distinctly by entering the natural numbers 1 through 8 in L1, then defining L2 as u(L1). The resulting graph is shown in Figure 9.37.

Figure 9.35

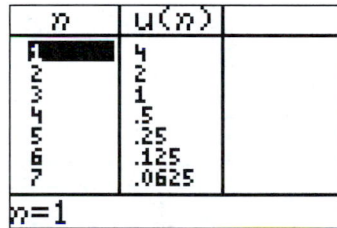

Figure 9.36

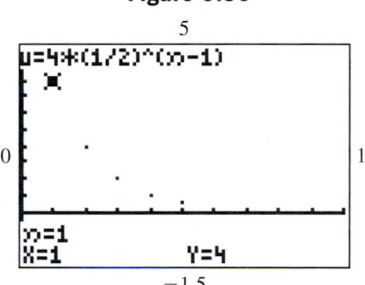

Figure 9.37

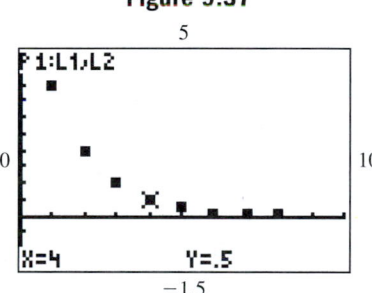

One additional advantage of using a list is that we can easily verify whether or not a common ratio r exists. We do this by duplicating L2 in L3 (define L3 = L2), then deleting *the first entry of L3* and *the last entry of L2* (to keep the same number of terms in each list). See Figure 9.38. This enables us to find the ratio $\frac{a_{k+1}}{a_k}$ for the entire list by defining L4 as the ratio L3/L2 (which automatically computes the ratio of successive terms). See Figures 9.39 and 9.40.

Figure 9.38

L1	L2	L3	3
1	4	2	
2	2	1	
3	1	.5	
4	.5	.25	
5	.25	.125	
6	.125	.0625	
7	.0625	.03125	

L3 = {2, 1, .5, .25, ...}

Figure 9.39

L2	L3	L4	4
4	2	------	
2	1		
1	.5		
.5	.25		
.25	.125		
.125	.0625		
.0625	.03125		

L4 = L3/L2

Figure 9.40

L2	L3	L4	4
4	2	.5	
2	1	.5	
1	.5	.5	
.5	.25	.5	
.25	.125	.5	
.125	.0625	.5	
.0625	.03125	.5	

L4(1) = .5

EXAMPLE 5 ▶ **Graphing Geometric Sequences**

Enter the natural numbers 1 through 6 in L1, and the terms of the sequence 0.25, 1.25, 6.25, 31.25, 156.25, 781.25 in L2. Then determine if the sequence is geometric by

a. Graphing the related points to see if they appear to form along an exponential graph.

b. Finding the successive ratios between terms.

c. If the sequence is geometric, find the nth term and graph the sequence.

Solution ▶ **a.** The plotted points are shown in Figure 9.41 and appear to lie along an exponential graph.

Figure 9.41

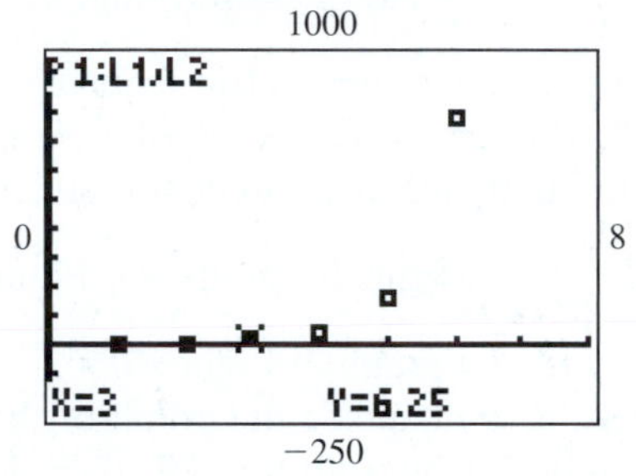

Figure 9.42

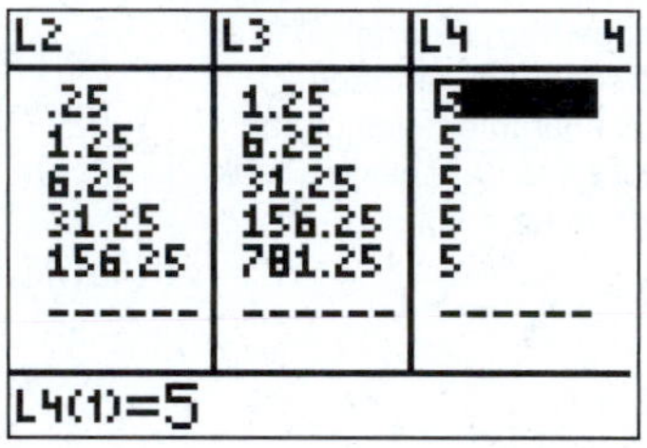

b. Using the approach described prior to Example 5, we find there is a common ratio of $r = 5$ (Figure 9.42).

Figure 9.43

c. With $a_1 = 0.25$ and $r = 5$, the nth term must be

$$a_n = a_1 r^{n-1} \qquad \text{\textit{n}th term formula}$$
$$= 0.25(5)^{n-1} \qquad \text{substitute for } a_1 \text{ and } r$$

The nth term for this sequence is $a_n = 0.25(5)^{n-1}$. The graph is shown in Figure 9.43.

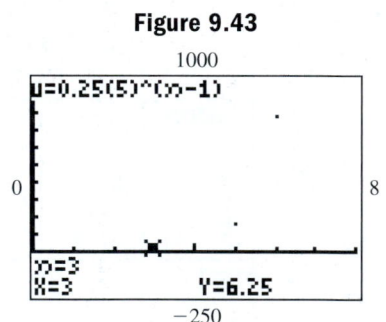

Now try Exercises 59 through 62 ▶

If the term a_1 is unknown but a term a_k is given, the nth term can be written

$$a_n = a_k r^{n-k},$$

(the subscript on the term a_k and the exponent on r sum to n).

EXAMPLE 6 ▶ **Finding the First Term of a Geometric Sequence**

Given a geometric sequence where $a_4 = 0.075$ and $a_7 = 0.009375$, find the common ratio r and the value of a_1.

Solution ▶ Since a_1 is not known, we express a_7 as the product of a known term and the appropriate number of common ratios: $a_7 = a_4 r^3$ ($7 = 4 + 3$, as required).

$$a_7 = a_4 \cdot r^3 \qquad a_1 \text{ is unknown}$$
$$0.009375 = 0.075 r^3 \qquad \text{substitute 0.009375 for } a_7 \text{ and 0.075 for } a_4$$
$$0.125 = r^3 \qquad \text{divide by 0.075}$$
$$r = 0.5 \qquad \text{solve for } r$$

Having found r, we can now solve for a_1

$$a_7 = a_1 r^6 \qquad \textit{nth term formula}$$
$$0.009375 = a_1(0.5)^6 \qquad \textit{substitute 0.009375 for } a_7 \textit{ and 0.5 for } r$$
$$0.009375 = a_1(0.015625) \qquad \textit{simplify}$$
$$a_1 = 0.6 \qquad \textit{solve for } a_1$$

☑ **B.** You've just seen how we can find the *n*th term of a geometric sequence

The first term is $a_1 = 0.6$ and the common ratio is $r = 0.5$.

Now try Exercises 63 through 68 ▶

C. Find the *n*th Partial Sum of a Geometric Sequence

As with arithmetic series, applications of geometric series often involve computing sums of consecutive terms. We can adapt the method for finding the sum of an arithmetic sequence to develop a formula for adding the first *n* terms of a geometric sequence.

For the *n*th term $a_n = a_1 r^{n-1}$, we have $S_n = a_1 + a_1 r + a_1 r^2 + a_1 r^3 + \cdots + a_1 r^{n-1}$. If we multiply S_n by $-r$ then add the original series, the "interior terms" sum to zero.

$$-rS_n = -a_1 r + (-a_1 r^2) + (-a_1 r^3) + \cdots + (-a_1 r^{n-1}) + (-a_1 r^n)$$
$$+ \ S_n = \ a_1 + a_1 r + a_1 r^2 + \cdots + a_1 r^{n-2} + a_1 r^{n-1}$$
$$\overline{S_n - rS_n = \ a_1 + \ 0 \ + \ 0 \ + 0 + \ 0 \ + \ 0 \ + (-a_1 r^n)}$$

We then have $S_n - rS_n = a_1 - a_1 r^n$, and can now solve for S_n:

$$S_n(1 - r) = a_1 - a_1 r^n \qquad \textit{factor out } S_n$$
$$S_n = \frac{a_1 - a_1 r^n}{1 - r} \qquad \textit{solve for } S_n \textit{ (divide by } 1 - r)$$

The result is a formula for the *n*th partial sum of a geometric sequence.

The *n*th Partial Sum of a Geometric Sequence

Given a geometric sequence with first term a_1 and common ratio r, the *n*th partial sum (the sum of the first *n* terms) is

$$S_n = \frac{a_1 - a_1 r^n}{1 - r} = \frac{a_1(1 - r^n)}{1 - r}, r \neq 1$$

In words: The sum of the first *n* terms of a geometric sequence is the difference of the first and $(n + 1)$st term, divided by 1 minus the common ratio.

EXAMPLE 7 ▶ **Computing a Partial Sum**

Use the preceding summation formula to find the sum: $\sum_{i=1}^{9} 3^i$ (the first nine powers of 3). Verify the result on a graphing calculator.

Solution ▶ The initial terms of this series are $3 + 9 + 27 + \cdots$, and we note $a_1 = 3, r = 3$, and $n = 9$. We could find the first nine terms and add, but using the partial sum formula is much faster and gives

$$S_n = \frac{a_1(1 - r^n)}{1 - r} \qquad \textit{sum formula}$$
$$S_9 = \frac{3(1 - 3^9)}{1 - 3} \qquad \textit{substitute 3 for } a_1, \textit{9 for } n, \textit{and 3 for } r$$

$$= \frac{3(-19{,}682)}{-2} \qquad \text{simplify}$$

$$= 29{,}523 \qquad \text{result}$$

To verify, we enter $u(n) = 3^n$ on the ⟨Y=⟩ screen, and find the sum of the first nine terms of the sequence on the home screen as before. See figure.

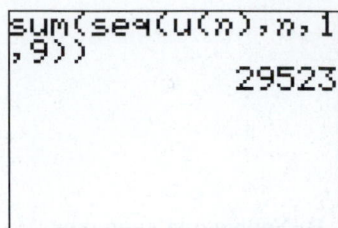

☑ **C. You've just seen how we can find the *n*th partial sum of a geometric sequence**

Now try Exercises 69 through 92 ▶

D. The Sum of an Infinite Geometric Series

To this point we've considered only partial sums of a geometric series. While it is impossible to add an infinite number of these terms, some of these "infinite sums" appear to have what is called a **limiting value.** The sum appears to get ever closer to this value but never exceeds it—much like the asymptotic behavior of some graphs. We will define the sum of this **infinite geometric series** to be this limiting value, if it exists. Consider the illustration in Figure 9.44, where a standard sheet of typing paper is cut in half. One of the halves is again cut in half and the process is continued indefinitely, as shown. Notice the "halves" create an infinite sequence $\frac{1}{2}, \frac{1}{4}, \frac{1}{8}, \frac{1}{16}, \frac{1}{32}, \cdots$ with $a_1 = \frac{1}{2}$ and $r = \frac{1}{2}$. The corresponding infinite series is $\frac{1}{2} + \frac{1}{4} + \frac{1}{8} + \frac{1}{16} + \frac{1}{32} + \cdots + \frac{1}{2^n} + \cdots$.

Figure 9.44

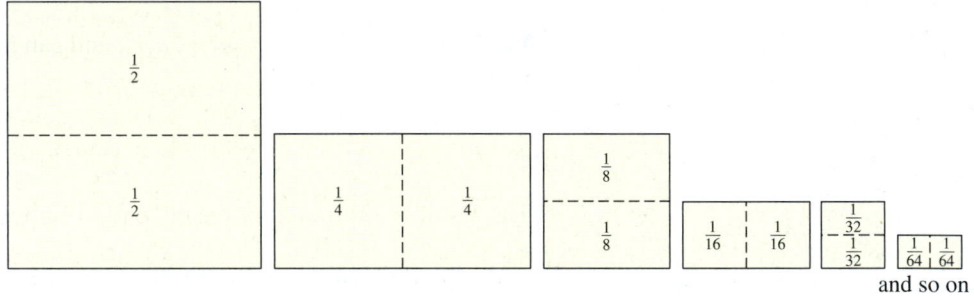

and so on

Figure 9.45

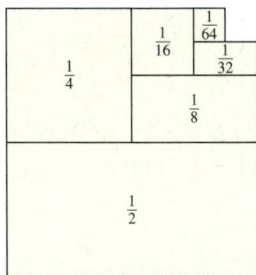

If we arrange one of the halves from each stage as shown in Figure 9.45, we would be rebuilding the original sheet of paper. As we add more and more of these halves together, we get closer and closer to the size of the original sheet. We gain an intuitive sense that this series must add to 1, because the *pieces* of the original sheet of paper must add to 1 whole sheet. To explore this idea further, consider what happens to $\left(\frac{1}{2}\right)^n$ as n becomes large.

$$n = 4{:} \left(\frac{1}{2}\right)^4 = 0.0625 \qquad n = 8{:} \left(\frac{1}{2}\right)^8 \approx 0.004 \qquad n = 12{:} \left(\frac{1}{2}\right)^{12} \approx 0.0002$$

Further exploration with a calculator seems to support the idea that as $n \to \infty$, $\left(\frac{1}{2}\right)^n \to 0$, although a definitive proof is left for a future course. In fact, it can be shown that for any $|r| < 1$, r^n becomes very close to zero as n becomes large. In symbols: as $n \to \infty$, $r^n \to 0$. For $S_n = \dfrac{a_1 - a_1 r^n}{1 - r} = \dfrac{a_1}{1 - r} - \dfrac{a_1 r^n}{1 - r}$, note that if $|r| < 1$ and "we sum an infinite number of terms," the second term becomes zero, leaving only the first term. In other words, the limiting value (represented by S_∞) is $S_\infty = \dfrac{a_1}{1 - r}$.

WORTHY OF NOTE

The formula for the sum of an infinite geometric series can also be derived by noting that $S_\infty = a_1 + a_1 r + a_1 r^2 + a_1 r^3 + \cdots$ can be rewritten as $S_\infty = a_1 + r(a_1 + a_1 r + a_1 r^2 + a_1 r^3 + \cdots) = a_1 + rS_\infty$.

$$S_\infty - rS_\infty = a_1$$
$$S_\infty(1 - r) = a_1$$
$$S_\infty = \frac{a_1}{1 - r}.$$

Infinite Geometric Series

Given a geometric sequence with first term a_1 and $|r| < 1$, the sum of the related infinite series is given by

$$S_\infty = \frac{a_1}{1 - r}; r \neq 1$$

If $|r| > 1$, no finite sum exists.

EXAMPLE 8 ▶ **Computing an Infinite Sum**

Find the limiting value of each infinite geometric series (if it exists).

a. $1 + 2 + 4 + 8 + \cdots$ **b.** $3 + 2 + \frac{4}{3} + \frac{8}{9} + \cdots$

c. $0.185 + 0.000185 + 0.000000185 + \cdots$

Solution ▶ Begin by determining if the infinite series is geometric with $|r| < 1$. If so, use $S_\infty = \dfrac{a_1}{1 - r}$.

a. Since $r = 2$ (by inspection), a finite sum does not exist.

b. Using the ratio of consecutive terms we find $r = \frac{2}{3}$ and the infinite sum exists. With $a_1 = 3$, we have

$$S_\infty = \frac{3}{1 - \frac{2}{3}} = \frac{3}{\frac{1}{3}} = 9$$

c. This series is equivalent to the repeating decimal $0.185185185\ldots = 0.\overline{185}$. The common ratio is $r = \frac{0.000185}{0.185} = 0.001$ and the infinite sum exists:

$$S_\infty = \frac{0.185}{1 - 0.001} = \frac{5}{27}$$

Now try Exercises 93 through 108 ▶

While it is impossible for us to sum an infinite number of terms, we can often see a pattern that strongly suggests a limiting value exists. As in Section 9.2, we do this by increasing the number terms we sum, until such a limiting value seems apparent. But be very careful to note that such a calculator approach *is far from an actual proof,* and a more definitive means must be used to actually prove a limiting value exists. For the sequence in Example 8(b), we enter $3\left(\dfrac{2}{3}\right)^{n-1}$ as u(n), and compute the sums S_{10}, S_{20}, S_{30}, and S_{40} (Figures 9.46 and 9.47), where it indeed appears the limiting value is 9.

Figure 9.46

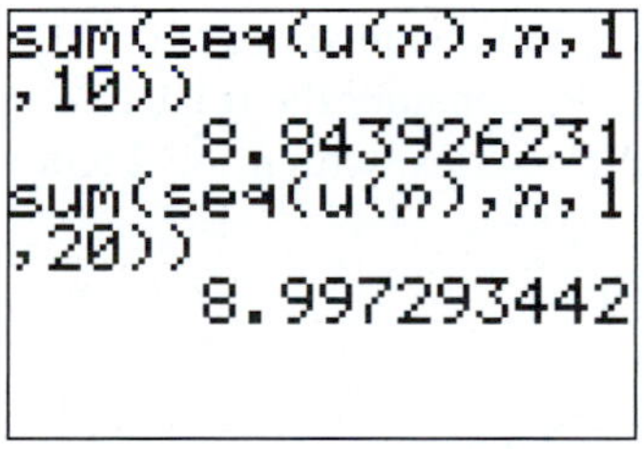

```
sum(seq(u(n),n,1
,10))
          8.843926231
sum(seq(u(n),n,1
,20))
          8.997293442
```

Figure 9.47

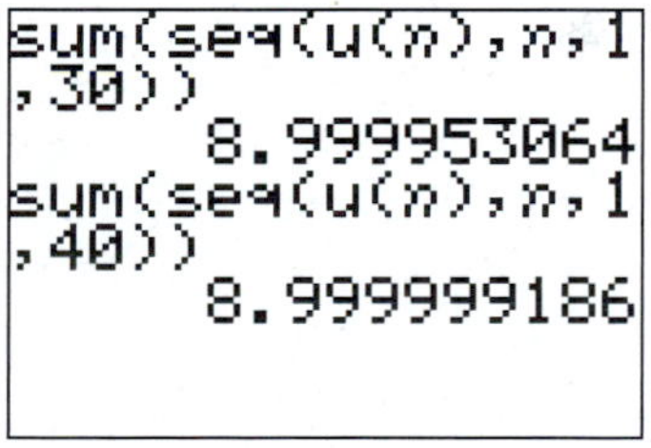

```
sum(seq(u(n),n,1
,30))
          8.999953064
sum(seq(u(n),n,1
,40))
          8.999999186
```

☑ **D.** You've just seen how we can find the sum of an infinite geometric series

E. Applications Involving Geometric Sequences and Series

Here are a few of the ways these ideas can be put to use.

EXAMPLE 9 ▶ **Solving an Application of Geometric Sequences: Pendulums**

A pendulum is any object attached to a fixed point and allowed to swing freely under the influence of gravity. Suppose each swing is 0.9 the length of the previous one. Gradually the swings become shorter and shorter and at some point the pendulum will appear to have stopped (although *theoretically* it never does).

a. How far does the pendulum travel on its eighth swing, if the first was 2 m?

b. What is the total distance traveled by the pendulum for these eight swings?

c. How many swings until the length of each swing falls below 0.5 m?

d. What total distance does the pendulum travel before coming to rest?

Verify your response to parts (a) through (c) using a graphing calculator.

Solution ▶

a. The lengths of each swing form the terms of a geometric sequence with $a_1 = 2$ and $r = 0.9$. The first few terms are 2, 1.8, 1.62, 1.458, and so on. For the 8th term we have:

$a_n = a_1 r^{n-1}$ *n*th term formula

$a_8 = 2(0.9)^{8-1}$ substitute 8 for *n*, 2 for a_1, and 0.9 for *r*

≈ 0.957

The pendulum travels about 0.957 m on its 8th swing. See Figure 9.48.

Figure 9.48

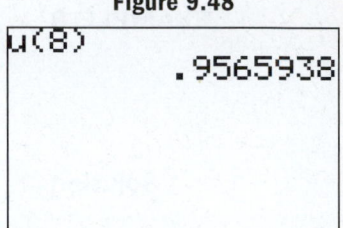

b. For the total distance traveled after eight swings, we compute the value of S_8.

$S_n = \dfrac{a_1(1 - r^n)}{1 - r}$ *n*th partial sum formula

$S_8 = \dfrac{2(1 - 0.9^8)}{1 - 0.9}$ substitute 2 for a_1, 0.9 for *r*, and 8 for *n*

≈ 11.4

The pendulum has traveled about 11.4 m by the end of the 8th swing. See Figure 9.49.

Figure 9.49

sum(seq(u(n),n,1
,8))
              11.3906558

c. To find the number of swings until the length of each swing is less than 0.5 m, we solve for *n* in the equation $0.5 = 2(0.9)^{n-1}$. This yields

$0.25 = (0.9)^{n-1}$ divide by 2

$\ln 0.25 = (n - 1)\ln 0.9$ take the natural log, apply power property

$\dfrac{\ln 0.25}{\ln 0.9} + 1 = n$ solve for *n* (exact form)

$14.16 \approx n$ solve for *n* (approximate form)

After the 14th swing, each successive swing will be less than 0.5 m. See Figure 9.50.

Figure 9.50

n	$u(n)$
9	.86093
10	.77484
11	.69736
12	.62762
13	.56486
14	.50837
15	.45754

$n=15$

d. For the total distance traveled before coming to rest, we consider the related infinite geometric series, with $a_1 = 2$ and $r = 0.9$.

$S_\infty = \dfrac{a_1}{1 - r}$ infinite sum formula

$S_\infty = \dfrac{2}{1 - 0.9}$ substitute 2 for a_1 and 0.9 for *r*

$= 20$ result

The pendulum would travel 20 m before coming to rest. Note that summing a larger number of terms on a calculator takes an increasing amount of time. The values for S_{15} and S_{150} are shown in Figure 9.51.

Figure 9.51

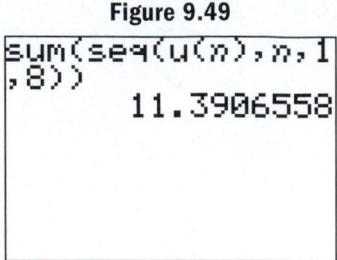

Now try Exercises 111 and 112 ▶

As mentioned in Section 9.1, sometimes the sequence or series for a particular application will use the preliminary or inaugural term a_0, as when an initial amount of money is deposited before any interest is earned, or the efficiency of a new machine after purchase—prior to any wear and tear.

EXAMPLE 10 ▶ **Equipment Efficiency—Furniture Manufacturing**

The manufacturing of mass-produced furniture requires robotic machines to drill numerous holes for the bolts used in the assembly process. When new, the drill bits are capable of drilling through hardwood at a rate of 6 cm/sec. As the bit becomes worn, it loses 4% of its drilling speed per day.

    a. How many cm/sec can the bit drill through after a 5-day workweek?

    b. When the drilling speed falls below 3.6 cm/sec, the bit must be replaced. After how many days must the bit be replaced?

Solution ▶ The efficiency of a new drill bit (prior to use) is given as $a_0 = 6$ cm/sec. Since the bit *loses* $4\% = 0.04$ of its efficiency per day, it maintains $96\% = 0.96$ of its efficiency, showing that after 1 day of use $a_1 = 0.96(6) = 5.76$. This means the nth term formula will be $a_n = 5.76(0.96)^{n-1}$.

    a. At the end of day 5 we have

$$a_5 = 5.76(0.96)^{5-1}$$
$$= 5.76(0.96)^4$$
$$\approx 4.9$$

After 5 days, the bit can drill through the hardwood at about 4.9 cm/sec.

    b. To find the number of days until the efficiency falls below 3.6 cm/sec, we replace a_n with 3.6 and solve for n.

$3.6 = 5.76(0.96)^{n-1}$	substitute 3.6 for a_n
$0.625 = 0.96^{n-1}$	divide
$\ln 0.625 = \ln 0.96^{n-1}$	take the natural log of both sides
$\ln 0.625 = (n-1)\ln 0.96$	power property
$\dfrac{\ln 0.625}{\ln 0.96} = n - 1$	divide
$\dfrac{\ln 0.625}{\ln 0.96} + 1 = n$	solve for n (exact form)
$12.5 \approx n$	solution (approximate form)

The drill bit must be replaced after 12 full days of use.

 **E.** You've just seen how we can solve application problems involving geometric sequences and series

Now try Exercises 113 through 126 ▶

9.3 EXERCISES

▶ **CONCEPTS AND VOCABULARY**

Fill in the blank with the appropriate word or phrase. Carefully reread the section if needed.

1. In a geometric sequence, each successive term is found by <u>multiplying</u> the preceding term by a fixed value r.

2. In a geometric sequence, the common ratio r can be found by computing the <u>quotient</u> of any two consecutive terms.

3. The nth term of a geometric sequence is given by $a_n = \underline{a_1 r^{n-1}}$, for any $n \geq 1$.

4. For the general sequence $a_1, a_2, a_3, \ldots, a_k, \ldots$, the fifth partial sum is given by $S_5 = \underline{\hspace{1cm}}$. $\dfrac{a_1(1-r^5)}{1-r}$

5. Describe/Discuss how the formula for the nth partial sum is related to the formula for the sum of an infinite geometric series. Answers will vary.

6. Describe the difference(s) between an arithmetic and a geometric sequence. How can a student prevent confusion between the formulas? Answers will vary.

HOMEWORK SELECTION GUIDE

Core: 7–107 every other odd, 109, 110, 111–121 odd (34 Exercises)
Standard: 1–4, 7–107 every other odd, 109, 110, 111–123 odd, 127, 129, 132, 133 (43 Exercises)

Extended: 1–4, 7–107 every other odd, 109, 110, 111–125 odd, 127–129, 132–134 (46 Exercises)
In Depth: 1–6, 7–107 every other odd, 109, 110, 111–125 odd, 127–130, 132–135 (50 Exercises)

▶ DEVELOPING YOUR SKILLS

Determine if the sequence given is geometric. If yes, name the common ratio. If not, try to determine the pattern that forms the sequence.

7. $4, 8, 16, 32, \ldots$ $r = 2$

8. $2, 6, 18, 54, 162, \ldots$ $r = 3$

9. $3, -6, 12, -24, 48, \ldots$ $r = -2$

10. $128, -32, 8, -2, \ldots$ $r = \frac{-1}{4}$

11. $2, 5, 10, 17, 26, \ldots$ not geometric; $a_n = n^2 + 1$

12. $-13, -9, -5, -1, 3, \ldots$ not geometric; $a_n = 4n - 17$

13. $3, 0.3, 0.03, 0.003, \ldots$ $r = 0.1$

14. $12, 0.12, 0.0012, 0.000012, \ldots$ $r = 0.01$

15. $-1, 3, -12, 60, -360, \ldots$

16. $-\frac{2}{3}, 2, -8, 40, -240, \ldots$

17. $25, 10, 4, \frac{8}{5}, \ldots$ $r = \frac{2}{5}$

18. $-36, 24, -16, \frac{32}{3}, \ldots$

19. $\frac{1}{2}, \frac{1}{4}, \frac{1}{8}, \frac{1}{16}, \ldots$ $r = \frac{1}{2}$

20. $\frac{2}{3}, \frac{4}{9}, \frac{8}{27}, \frac{16}{81}, \ldots$ $r = \frac{2}{3}$

21. $3, \dfrac{12}{x}, \dfrac{48}{x^2}, \dfrac{192}{x^3}, \ldots$ $r = \frac{4}{x}$

22. $5, \dfrac{10}{a}, \dfrac{20}{a^2}, \dfrac{40}{a^3}, \ldots$ $r = \frac{2}{a}$

23. $240, 120, 40, 10, 2, \ldots$

24. $-120, -60, -20, -5, -1, \ldots$

Write the first four terms of the sequence, given a_1 and r.

25. $a_1 = 5, r = 2$

26. $a_1 = 2, r = -4$

27. $a_1 = -6, r = -\frac{1}{2}$

28. $a_1 = \frac{2}{3}, r = \frac{1}{5}$

29. $a_1 = 4, r = \sqrt{3}$

30. $a_1 = \sqrt{5}, r = \sqrt{5}$

31. $a_1 = 0.1, r = 0.1$

32. $a_1 = 0.024, r = 0.01$

Write the expression for the nth term, then find the indicated term for each sequence.

33. $a_1 = -24, r = \frac{1}{2}$; find a_7

34. $a_1 = 48, r = -\frac{1}{3}$; find a_6

35. $a_1 = -\frac{1}{20}, r = -5$; find a_4

36. $a_1 = \frac{3}{20}, r = 4$; find a_5

37. $a_1 = 2, r = \sqrt{2}$; find a_7

38. $a_1 = \sqrt{3}, r = \sqrt{3}$; find a_8

Additional answers can be found in the Instructor Answer Appendix.

Identify a_1 and r, then write the expression for the nth term $a_n = a_1 r^{n-1}$ and use it to find a_6, a_{10}, and a_{12}.

39. $\frac{1}{27}, -\frac{1}{9}, \frac{1}{3}, -1, 3, \ldots$ 40. $-\frac{7}{8}, \frac{7}{4}, -\frac{7}{2}, 7, -14, \ldots$

41. $729, 243, 81, 27, 9, \ldots$ 42. $625, 125, 25, 5, 1, \ldots$

43. $\frac{1}{2}, \frac{\sqrt{2}}{2}, 1, \sqrt{2}, 2, \ldots$

44. $36\sqrt{3}, 36, 12\sqrt{3}, 12, 4\sqrt{3}, \ldots$

45. $0.2, 0.08, 0.032, 0.0128, \ldots$

46. $0.5, -0.35, 0.245, -0.1715, \ldots$

Find the number of terms in each sequence.

47. $a_1 = 9, a_n = 729, r = 3$ 5

48. $a_1 = 1, a_n = -128, r = -2$ 8

49. $a_1 = 16, a_n = \frac{1}{64}, r = \frac{1}{2}$ 11

50. $a_1 = 4, a_n = \frac{1}{512}, r = \frac{1}{2}$ 12

51. $a_1 = -1, a_n = -1296, r = \sqrt{6}$ 9

52. $a_1 = 2, a_n = 1458, r = -\sqrt{3}$ 13

53. $2, -6, 18, -54, \ldots, -4374$ 8

54. $3, -6, 12, -24, \ldots, -6144$ 12

55. $64, 32\sqrt{2}, 32, 16\sqrt{2}, \ldots, 1$ 13

56. $243, 81\sqrt{3}, 81, 27\sqrt{3}, \ldots, 1$ 11

57. $\frac{3}{8}, -\frac{3}{4}, \frac{3}{2}, -3, \ldots, 96$ 9

58. $-\frac{5}{27}, \frac{5}{9}, -\frac{5}{3}, -5, \ldots, -135$ 7

For Exercises 59 through 62, enter the natural numbers 1 through 6 in L1 on a graphing calculator, and the terms of the given sequence in L2. Then determine if the sequence is geometric by (a) graphing the related points to see if they appear to lie on an exponential curve, and (b) computing the successive ratios of all terms. If a geometric sequence, find the nth term and graph the sequence.

59. $131.25, 26.25, 5.25, 1.05, 0.21, 0.042, \ldots$

60. $2, 2\sqrt{5}, 10, 10\sqrt{5}, 50, 50\sqrt{5}, \ldots$

61. $20, 16, 12, 8, 4, 0, \ldots$

62. $\dfrac{1}{6}, \dfrac{1}{3}, \dfrac{1}{2}, \dfrac{2}{3}, \dfrac{5}{6}, 1, \ldots$

Find the common ratio r and the value of a_1 using the information given (assume $r > 0$).

63. $a_3 = 324, a_7 = 64$ $r = \frac{2}{3}, a_1 = 729$

64. $a_5 = 6, a_9 = 486$ $r = 3, a_1 = \frac{2}{27}$

65. $a_4 = \frac{4}{9}, a_8 = \frac{4}{4}$ $r = \frac{3}{2}, a_1 = \frac{32}{243}$

66. $a_2 = \frac{16}{81}, a_5 = \frac{32}{3}$ $r = \frac{3}{2}, a_{12} = \frac{32}{243}$

67. $a_4 = \frac{32}{3}, a_8 = 54$ $r = \frac{3}{2}, a_1 = \frac{256}{81}$

68. $a_3 = \frac{16}{25}, a_7 = 25$ $r = \frac{5}{2}, a_1 = \frac{64}{625}$

Find the partial sum indicated.

69. $a_1 = 8$, $r = -2$; find S_{12} $-10{,}920$

70. $a_1 = 2$, $r = -3$; find S_8 -3280

71. $a_1 = 96$, $r = \frac{1}{3}$; find S_5 $\frac{3872}{27} \approx 143.41$

72. $a_1 = 12$, $r = \frac{1}{2}$; find S_8 $\frac{765}{32} \approx 23.91$

73. $a_1 = 8$, $r = \frac{3}{2}$; find S_7 $\frac{2059}{8} = 257.375$

74. $a_1 = -1$, $r = -\frac{3}{2}$; find S_{10} $\frac{11{,}605}{512} \approx 22.67$

75. $2 + 6 + 18 + \cdots$; find S_6 728

76. $2 + 8 + 32 + \cdots$; find S_7 $10{,}922$

77. $16 - 8 + 4 - \cdots$; find S_8 $\frac{85}{8} = 10.625$

78. $4 - 12 + 36 - \cdots$; find S_8 -6560

79. $\frac{4}{3} + \frac{2}{9} + \frac{1}{27} + \cdots$; find S_9 ≈ 1.60

80. $\frac{1}{18} - \frac{1}{6} + \frac{1}{2} - \cdots$; find S_7 $\frac{547}{18} \approx 30.39$

Find the partial sum indicated, and verify the result using a graphing calculator. For Exercises 85 and 86, use the summation properties from Section 9.1.

81. $\displaystyle\sum_{j=1}^{5} 4^j$

82. $\displaystyle\sum_{k=1}^{10} 2^k$

83. $\displaystyle\sum_{k=1}^{8} 5\left(\frac{2}{3}\right)^{k-1}$

84. $\displaystyle\sum_{j=1}^{7} 3\left(\frac{1}{5}\right)^{j-1}$

85. $\displaystyle\sum_{i=4}^{10} 9\left(-\frac{1}{2}\right)^{i-1}$

86. $\displaystyle\sum_{i=3}^{8} 5\left(-\frac{1}{4}\right)^{i-1}$

Find the indicated partial sum using the information given. Write all results in simplest form.

87. $a_2 = -5$, $a_5 = \frac{1}{25}$; find S_5 $\frac{521}{25}$

88. $a_3 = 1$, $a_6 = -27$; find S_6 $\frac{-182}{9}$

89. $a_3 = \frac{4}{9}$, $a_7 = \frac{9}{64}$; find S_6 $\frac{3367}{1296}$

90. $a_2 = \frac{16}{81}$, $a_5 = \frac{2}{3}$; find S_8 $\frac{6305}{972}$

91. $a_3 = 2\sqrt{2}$, $a_6 = 8$; find S_7 $14 + 15\sqrt{2}$

92. $a_2 = 3$, $a_5 = 9\sqrt{3}$; find S_7 $39 + 40\sqrt{3}$

Determine whether the infinite geometric series has a finite sum. If so, find the limiting value.

93. $9 + 3 + 1 + \cdots$ $\frac{27}{2}$

94. $36 + 24 + 16 + \cdots$ 108

95. $3 + 6 + 12 + 24 + \cdots$ no

96. $4 + 8 + 16 + 32 + \cdots$ no

97. $25 + 10 + 4 + \frac{8}{5} + \cdots$ $\frac{125}{3}$

98. $10 + 2 + \frac{2}{5} + \frac{2}{25} + \cdots$ $\frac{25}{2}$

99. $6 + 3 + \frac{3}{2} + \frac{3}{4} + \cdots$ 12

100. $-49 + (-7) + \left(-\frac{1}{7}\right) + \cdots$ $\frac{-343}{6}$

101. $6 - 3 + \frac{3}{2} - \frac{3}{4} + \cdots$ 4

102. $10 - 5 + \frac{5}{2} - \frac{5}{4} + \cdots$ $\frac{20}{3}$

103. $0.3 + 0.03 + 0.003 + \cdots$ $\frac{1}{3}$

104. $0.63 + 0.0063 + 0.000063 + \cdots$ $\frac{7}{11}$

105. $\displaystyle\sum_{k=1}^{\infty} \frac{3}{4}\left(\frac{2}{3}\right)^k$ $\frac{3}{2}$

106. $\displaystyle\sum_{i=1}^{\infty} 5\left(\frac{1}{2}\right)^i$ 5

107. $\displaystyle\sum_{j=1}^{\infty} 9\left(-\frac{5}{4}\right)^j$

No finite sum exists.

108. $\displaystyle\sum_{k=1}^{\infty} 12\left(\frac{4}{3}\right)^k$

No finite sum exists.

▶ **WORKING WITH FORMULAS**

109. **Sum of the cubes of the first n natural numbers:**

$$S_n = \frac{n^2(n + 1)^2}{4}$$

Compute $1^3 + 2^3 + 3^3 + \cdots + 8^3$ using the formula given. Then confirm the result by direct calculation. 1296

110. **Student loan payment: $A_n = P(1 + r)^n$**

If P dollars is borrowed at an annual interest rate r with interest compounded annually, the amount of money to be paid back after n years is given by the indicated formula. Find the total amount of money that the student must repay to clear the loan, if $8000 is borrowed at 4.5% interest and the loan is paid back in 10 yr. $12,423.76

Additional answers can be found in the Instructor Answer Appendix.

▶ **APPLICATIONS**

Write the nth term formula for each application, then solve.

111. Pendulum movement: On each swing, a pendulum travels only 80% as far as it did on the previous swing. If the first swing is 24 ft, how far does the pendulum travel on the 7th swing? What total distance is traveled before the pendulum comes to rest?
$a_n = 24(0.8)^{n-1}, a_7 \approx 6.3$ ft, $S_\infty = 120$ ft

112. Tire swings: Ernesto is swinging to and fro on his backyard tire swing. Using his legs and body, he pumps each swing until reaching a maximum height, then suddenly relaxes until the swing comes to a stop. With each swing, Ernesto travels 75% as far as he did on the previous swing. If the first arc (or swing) is 30 ft, find the distance Ernesto travels on the 5th arc. What total distance will he travel before coming to rest?
$a_n = 30(0.75)^{n-1}, a_5 \approx 9.5$ ft, $S_\infty = 120$ ft

Identify the inaugural term and write the nth term formula for each application, then solve.

113. Depreciation—automobiles: A certain new SUV depreciates in value about 20% per year (meaning it holds 80% of its value each year). If the SUV is purchased for $46,000, how much is it worth 4 yr later? How many years until its value is less than $5000?
$a_0 = 46{,}000; a_n = 36{,}800(0.8)^{n-1}, a_4 = \$18{,}841.60, 10$ yr

114. Depreciation—business equipment: A new photocopier under heavy use will depreciate about 25% per year (meaning it holds 75% of its value each year). If the copier is purchased for $7000, how much is it worth 4 yr later? How many years until its value is less than $1246?
$a_0 = 7{,}000; a_n = 5250(0.75)^{n-1}, a_4 \approx \$2215.00, 6$ yr

115. Equipment aging—industrial oil pumps: Tests have shown that the pumping power of a heavy-duty oil pump decreases by 3% per month. If the pump can move 160 gallons per minute (gpm) new, how many gpm can the pump move 8 months later? If the pumping rate falls below 118 gpm, the pump must be replaced. How many months until this pump is replaced?
$a_0 = 160; a_n = 155.2(0.97)^{n-1}, a_8 \approx 125.4$ gpm, 10 mo

Additional answers can be found in the Instructor Answer Appendix.

116. Equipment aging—lumber production: At the local mill, a certain type of saw blade can saw approximately 2 log-feet/sec when it is new. As time goes on, the blade becomes worn, and loses 6% of its cutting speed each week. How many log-feet/sec can the saw blade cut after 6 weeks? If the cutting speed falls below 1.2 log-feet/sec, the blade must be replaced. During what week of operation will this blade be replaced?
$a_0 = 2; a_n = 1.88(0.94)^{n-1}, a_6 \approx 1.38$ log-ft/sec, 8th week

117. Population growth—United States: At the beginning of the year 2000, the population of the United States was approximately 277 million. If the population is growing at a rate of 2.3% per year, what was the population in 2010, 10 yr later?
$a_0 = 277; a_n = 283.37(1.023)^{n-1}, a_{10} \approx 347.7$ million

118. Population growth—space colony: The population of the Zeta Colony on Mars is 1000 people. Determine the population of the Colony 20 yr from now, if the population is growing at a constant rate of 5% per year.
$a_0 = 1000; a_n = 1050(1.05)^{n-1}, a_{20} \approx 2653$ people

119. Creating a vacuum: To create a vacuum, a hand pump is used to remove the air from an air-tight cube with a volume of 462 in^3. With each stroke of the pump, two-fifths of the air that remains in the cube is removed. How much air remains inside after the 5th stroke? How many strokes are required to remove all but 12.9 in^3 of the air?

120. Atmospheric pressure: In 1654, scientist Otto Von Guericke performed his famous demonstration of atmospheric pressure and the strength of a vacuum in front of Emperor Ferdinand III of Hungary. After joining two hemispheres with mating rims, he used a vacuum pump to remove all of the air from the sphere formed. He then attached a team of 15 horses to each hemisphere and despite their efforts, they could not pull the hemispheres apart. If the sphere held a volume of 4200 in^3 of air and one-tenth of the remaining air was removed with each stroke of the pump, how much air was still in the sphere after the 11th stroke? How many strokes were required to remove 85% of the air?

121. Treating swimming pools: In preparation for the summer swim season, chlorine is added to swimming pools to control algae and bacteria. However, careful measurements must be taken as levels above 5 ppm (parts per million) can be highly irritating to the eyes and throat, while levels below 1 ppm will be ineffective (3.0 to 3.5 ppm is ideal). In addition, the water must be treated daily since within a 24-hr period, about 25% of the chlorine will dissipate into the air. If the chlorine level in a swimming pool is 8 ppm after its initial treatment, how many days should the County Pool Supervisor wait before opening it up to the public? If left untreated, how many days until the chlorine level drops below 1 ppm?

122. **Venting landfill gases:** The gases created from the decomposition of waste in landfills must be carefully managed, as their release can cause terrible odors, harm the landfill structure, damage vegetation, or even cause an explosion. Suppose the accumulated volume of gas is 50,000 ft^3, and civil engineers are able to vent 2.5% of this gas into the atmosphere daily. What volume of gas remains after 21 days? How many days until the volume of gas drops below 10,000 ft^3?

123. **Population growth—bacteria:** A biologist finds that the population of a certain type of bacteria doubles *each half-hour*. If an initial culture has 50 bacteria, what is the population after 5 hr? How long will it take for the number of bacteria to reach 204,800?

124. **Population growth—boom towns:** Suppose the population of a "boom town" in the old west doubled *every 2 months* after gold was discovered. If the initial population was 219, what was the population 8 months later? How many months until the population exceeded 28,000?

125. **Elastic rebound—super balls:** Megan discovers that a rubber ball dropped from a height of 2 m rebounds four-fifths of the distance it has previously fallen. How high does it rebound on the 7th bounce? How far does the ball travel before coming to rest?

126. **Elastic rebound—computer animation:** The screen saver on my computer is programmed to send a colored ball vertically down the middle of the screen so that it rebounds 95% of the distance it last traversed. If the ball always begins at the top and the screen is 36 cm tall, how high does the ball bounce on its 8th rebound? How far does the ball travel before coming to rest (and a new screen saver starts)?

▶ **EXTENDING THE CONCEPT**

127. A standard piece of typing paper is approximately 0.001 in. thick. Suppose you were able to fold this piece of paper in half 26 times. How thick would the result be? As tall as a hare, as tall as a hen, as tall as a horse, as tall as a house, or over 1 mi high? Find the actual height by computing the 27th term of a geometric sequence. Discuss what you find.
about 67,109 in. This is almost 1.06 mi.

128. As part of a science experiment, identical rubber balls are dropped from a given height onto these surfaces: slate, cement, and asphalt. When dropped onto slate, the ball rebounds 80% of the height from which it last fell. Onto cement the figure is 75% and onto asphalt the figure is 70%. The ball is dropped from 130 m onto the slate, 175 m onto the cement, and 200 m onto the asphalt. Which ball has traveled the shortest total distance at the time of the fourth bounce? Which ball will travel farthest before coming to rest?
dropped onto slate; dropped onto cement

129. Consider the following situation. A person is hired at a salary of $40,000 per year, with a guaranteed raise of $1750 per year. At the same time, inflation is running about 4% per year. How many years until this person's salary is overtaken and eaten up by the actual cost of living?
$40,000 + 1750x$ vs. $40,000(1.04)^x$; 6 yr

130. Find an alternative formula for the sum

$$S_n = \sum_{k=1}^{n} \log k,$$ that does not use the sigma notation.
$S_n = \log n!$

131. Verify the following statements:

a. If $a_1, a_2, a_3, \ldots, a_n$ is a geometric sequence with r and a_1 greater than zero, then $\log a_1, \log a_2, \log a_3, \ldots, \log a_n$ is an arithmetic sequence.

b. If $a_1, a_2, a_3, \ldots, a_n$ is an arithmetic sequence, then $10^{a_1}, 10^{a_2}, \ldots, 10^{a_n}$, is a geometric sequence.

▶ **MAINTAINING YOUR SKILLS**

132. **(3.2)** Find the zeroes of f using the quadratic formula: $f(x) = x^2 + 5x + 9$.
$x = \dfrac{-5}{2} \pm \dfrac{\sqrt{11}}{2}i$

133. **(R.5)** Solve for x: $\dfrac{3}{x^2 - 3x - 10} - \dfrac{4}{x - 5} = \dfrac{1}{x + 2}$
$x = 0$

134. **(4.5)** Graph the rational function: $h(x) = \dfrac{x^2}{x - 1}$

135. **(5.6)** Given the logistic function shown, find $p(50)$, $p(75)$, $p(100)$, and $p(150)$:

$$p(t) = \dfrac{4200}{1 + 10e^{-0.055t}}$$

$p(50) \approx 2562.1$, $p(75) \approx 3615.6$, $p(100) \approx 4035.1$, $p(150) \approx 4189.1$

Additional answers can be found in the Instructor Answer Appendix.

LEARNING OBJECTIVES

In Section 9.4 you will see how we can:

- ☐ **A.** Use subscript notation to evaluate and compose functions
- ☐ **B.** Apply the principle of mathematical induction to sum formulas involving natural numbers
- ☐ **C.** Apply the principle of mathematical induction to general statements involving natural numbers

Since middle school (or even before) we have accepted that, "The product of two negative numbers is a positive number." But have you ever been asked to *prove* it? It's not as easy as it seems. We may think of several patterns that yield the result, analogies that indicate its truth, or even number line illustrations that lead us to believe the statement. But most of us have never seen a *proof* (see www.mhhe.com/coburn). In this section, we introduce one of mathematics' most powerful tools for proving a statement, called **proof by induction.**

A. Subscript Notation and Function Notation

One of the challenges in understanding a proof by induction is working with the notation. Earlier in the chapter, we introduced subscript notation as an alternative to function notation, since it is more commonly used when the functions are defined by a sequence. But regardless of the notation used, the functions can still be simplified, evaluated, composed, and even graphed. Consider the function $f(x) = 3x^2 - 1$ and the sequence defined by $a_n = 3n^2 - 1$. Both can be evaluated and graphed, with the only difference being that $f(x)$ is continuous with domain $x \in \mathbb{R}$, while a_n is discrete (made up of distinct points) with domain $n \in \mathbb{N}$.

EXAMPLE 1 ▶ **Using Subscript Notation for a Composition**

For $f(x) = 3x^2 - 1$ and $a_n = 3n^2 - 1$, find $f(k + 1)$ and a_{k+1}.

Solution ▶

$$\begin{aligned} f(k + 1) &= 3(k + 1)^2 - 1 \\ &= 3(k^2 + 2k + 1) - 1 \\ &= 3k^2 + 6k + 2 \end{aligned} \qquad \begin{aligned} a_{k+1} &= 3(k + 1)^2 - 1 \\ &= 3(k^2 + 2k + 1) - 1 \\ &= 3k^2 + 6k + 2 \end{aligned}$$

Now try Exercises 7 through 18 ▶

☑ **A. You've just seen how we can use subscript notation to evaluate and compose functions**

No matter which notation is used, every occurrence of the input variable is replaced by the new value or expression indicated by the composition.

B. Mathematical Induction Applied to Sums

Consider the sum of odd numbers $1 + 3 + 5 + 7 + 9 + 11 + 13 + \cdots$. The sum of the first four terms is $1 + 3 + 5 + 7 = 16$, or $S_4 = 16$. If we now add a_5 (the next term in line), would we get the same answer as if we had simply computed S_5? Common sense would say, "Yes!" since $S_5 = 1 + 3 + 5 + 7 + 9 = 25$ and $S_4 + a_5 = 16 + 9 = 25$✓. In diagram form, we have

add next term $a_5 = 9$ to S_4

$$1 + 3 + 5 + 7 + 9 + 11 + 13 + 15 + \cdots$$

S_4 —— sum of 4 terms

S_5 —— sum of 5 terms

Our goal is to develop this same degree of clarity in the *notational scheme* of things. For a given series, if we find the kth partial sum S_k (shown next) and then add the next term a_{k+1}, would we get the same answer if we had simply computed S_{k+1}? In other words, is $S_k + a_{k+1} = S_{k+1}$ true?

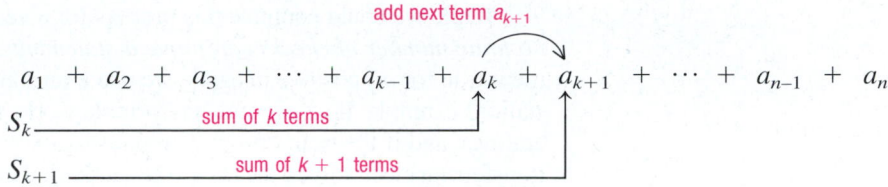

Now, let's return to the sum $1 + 3 + 5 + 7 + \cdots + (2n - 1)$. This is an arithmetic series with $a_1 = 1$, $d = 2$, and nth term $a_n = 2n - 1$. Using the sum formula for an arithmetic sequence, an alternative formula for *this sum* can be established.

$$S_n = \frac{n(a_1 + a_n)}{2} \qquad \text{summation formula for an arithmetic sequence}$$

$$= \frac{n(1 + 2n - 1)}{2} \qquad \text{substitute 1 for } a_1 \text{ and } 2n-1 \text{ for } a_n$$

$$= \frac{n(2n)}{2} \qquad \text{simplify}$$

$$= n^2 \qquad \text{result}$$

This shows that the sum of the first n positive odd integers is given by $S_n = n^2$. As a check we compute $S_5 = 1 + 3 + 5 + 7 + 9 = 25$ and compare to $S_5 = 5^2 = 25$✓. We also note $S_6 = 6^2 = 36$, and $S_5 + a_6 = 25 + 11 = 36$, showing $S_6 = S_5 + a_6$. For more on this relationship, **see Exercises 19 through 24.** While it may seem simplistic now, showing $S_5 + a_6 = S_6$ and $S_k + a_{k+1} = S_{k+1}$ (in general) is a critical component of a proof by induction.

Unfortunately, general summation formulas for many sequences cannot be established from known formulas. In addition, just because a formula works for the first few values of n, we cannot assume that it will hold true for *all* values of n (there are infinitely many). As an illustration, the formula $a_n = n^2 - n + 41$ yields a prime number for *every natural number n from 1 to 40,* but fails to yield a prime for $n = 41$. This helps demonstrate the need for a more conclusive proof, particularly when a relationship appears to be true, and can be "verified" in a finite number of cases, but whether it is true in *all* cases remains in doubt.

Proof by induction is based on a relatively simple idea. To help understand how it works, consider n relay stations that are used to transport electricity from a generating plant to a distant city. If we know the generating plant is operating, and if we assume that the kth relay station (any station in the series) is making the transfer to the $(k + 1)$st station (the

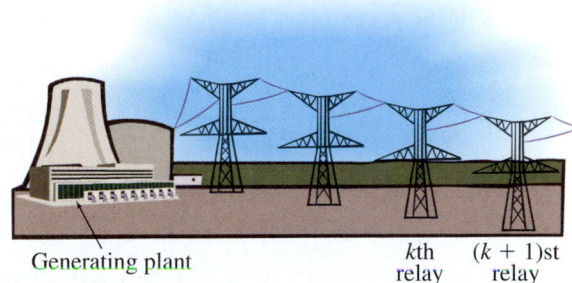

Generating plant kth $(k + 1)$st
                                    relay   relay

next station in the series), then we're sure the city will have electricity. This idea can be applied mathematically as follows. Consider the statement, "The sum of the first n positive even integers is $n^2 + n$." In other words, $2 + 4 + 6 + 8 + \cdots + 2n = n^2 + n$. We can certainly verify the statement for the first few even numbers:

The first even number is 2 and ... $(1)^2 + 1 = 2$

The sum of the first *two* even numbers is $2 + 4 = 6$ and ... $(2)^2 + 2 = 6$

The sum of the first *three* even numbers is
$2 + 4 + 6 = 12$ and ... $(3)^2 + 3 = 12$

The sum of the first *four* even numbers is
$2 + 4 + 6 + 8 = 20$ and ... $(4)^2 + 4 = 20$

While we could continue this process for a very long time (or even use a computer), *no finite number of checks can prove a statement is universally true.* To prove the statement true for *all* positive integers, we use a reasoning similar to that applied in the relay stations example. If we are sure the formula works for $n = 1$ (the generating station is operating), and if the truth of $n = k$ implies that $n = k + 1$ is true [the kth relay station is transferring electricity to the $(k + 1)$st station], then the statement is true for all n (the city will get its electricity). The case where $n = 1$ is called the **base case** of an inductive proof, and the assumption that the formula is true for $n = k$ is called the **induction hypothesis.** When the induction hypothesis is applied to a sum formula, we attempt to show that $S_k + a_{k+1} = S_{k+1}$. Since k and $k + 1$ are arbitrary, the statement must be true for all n.

Mathematical Induction Applied to Sums

Let S_n be a sum formula involving positive integers.

If   **1.** S_1 is true, and

   2. the truth of S_k implies that S_{k+1} is true,

then S_n must be true for all positive integers n.

WORTHY OF NOTE
To satisfy our finite minds, it might help to show that S_n is true for the first few cases, prior to extending the ideas to the infinite case.

Both parts 1 and 2 must be verified for the proof to be complete. Since the process requires the terms S_k, a_{k+1}, and S_{k+1}, we will usually compute these first.

EXAMPLE 2 ▶   **Proving a Statement Using Mathematical Induction**

Use induction to prove that *the sum of the first n perfect squares is given by*

$$1 + 4 + 9 + 16 + 25 + \cdots + n^2 = \frac{n(n + 1)(2n + 1)}{6}.$$

Solution ▶   Given $a_n = n^2$ and $S_n = \dfrac{n(n + 1)(2n + 1)}{6}$, the needed components are ...

For $a_n = n^2$: $a_k = k^2$ and $a_{k+1} = (k + 1)^2$

For $S_n = \dfrac{n(n + 1)(2n + 1)}{6}$: $S_k = \dfrac{k(k + 1)(2k + 1)}{6}$ and $S_{k+1} = \dfrac{(k + 1)(k + 2)(2k + 3)}{6}$

1. Show S_n is true for $n = 1$.

$$S_n = \frac{n(n + 1)(2n + 1)}{6} \qquad \text{sum formula}$$

$$S_1 = \frac{1(2)(3)}{6} \qquad \text{base case: } n = 1$$

$$= 1\checkmark \qquad \text{result checks, the first term is 1}$$

2. Assume S_k is true,

$$1 + 4 + 9 + 16 + \cdots + k^2 = \frac{k(k + 1)(2k + 1)}{6} \qquad \text{induction hypothesis: } S_k \text{ is true}$$

and use it to show the truth of S_{k+1} follows. That is,

$$\underbrace{1 + 4 + 9 + 16 + \cdots + k^2}_{S_k} + \underbrace{(k + 1)^2}_{a_{k+1}} = \underbrace{\frac{(k + 1)(k + 2)(2k + 3)}{6}}_{S_{k+1}}$$

Working with the left-hand side, we have

$$\underbrace{1 + 4 + 9 + 16 + \cdots + k^2} + (k + 1)^2$$

$$= \frac{k(k + 1)(2k + 1)}{6} + (k + 1)^2 \qquad \text{induction hypothesis: substitute } \frac{k(k + 1)(2k + 1)}{6}$$
$$\text{for } 1 + 4 + 9 + 16 + 25 + \cdots + k^2$$

$$= \frac{k(k + 1)(2k + 1) + 6(k + 1)^2}{6} \qquad \text{common denominator}$$

$$= \frac{(k+1)[k(2k+1)+6(k+1)]}{6} \quad \text{factor out } k+1$$

$$= \frac{(k+1)[2k^2+7k+6]}{6} \quad \text{multiply and combine terms}$$

$$= \frac{(k+1)(k+2)(2k+3)}{6} \quad \text{factor the trinomial, result is } S_{k+1}$$

☑ **B.** You've just seen how we can apply the principle of mathematical induction to sum formulas involving natural numbers

Since the truth of S_{k+1} follows from S_k, the formula is true for all n.

Now try Exercises 27 through 38 ▶

C. The General Principle of Mathematical Induction

Proof by induction can be used to verify many other kinds of relationships involving a natural number n. In this regard, the basic principles remain the same but are stated more broadly. Rather than using S_n to represent a sum, we will use P_n to represent *any proposed statement or relationship* we might wish to verify. This broadens the scope of the proof and makes it more widely applicable, while maintaining its connection to the sum formulas verified earlier.

The General Principle of Mathematical Induction

Let P_n be a statement involving natural numbers.

If   **1.** P_1 is true, and

    2. the truth of P_k implies that P_{k+1} is also true

then P_n must be true for all natural numbers n.

EXAMPLE 3 ▶ **Proving a Statement Using the General Principle of Mathematical Induction**

Use the general principle of mathematical induction to show the statement P_n is true for all natural numbers n. P_n: $2^n \geq n+1$

Solution ▶ The statement P_n is defined as $2^n \geq n+1$. This means that P_k is represented by $2^k \geq k+1$ and P_{k+1} by $2^{k+1} \geq k+2$.

1. Show P_n is true for $n=1$:

$$P_n: \quad 2^n \geq n+1 \quad \text{given statement}$$
$$P_1: \quad 2^1 \geq 1+1 \quad \text{base case: } n=1$$
$$2 \geq 2 ✓ \quad \text{true}$$

Although not a part of the formal proof, a table of values can help to illustrate the relationship we're trying to establish. It *appears* that the statement is true.

n	1	2	3	4	5
2^n	2	4	8	16	32
$n+1$	2	3	4	5	6

2. Assume that P_k is true.

$$P_k: \quad 2^k \geq k+1 \quad \text{induction hypothesis}$$

and use it to show the truth of P_{k+1}. That is,

$$P_{k+1}: \quad 2^{k+1} \geq (k+1)+1$$
$$\geq k+2$$

Begin by working with the left-hand side of the inequality, 2^{k+1}.

$$2^{k+1} = 2(\mathbf{2^k}) \qquad \text{properties of exponents}$$

$$\geq 2(\mathbf{k+1}) \qquad \textit{induction hypothesis:} \text{ substitute } k+1 \text{ for } 2^k$$

$$\text{(symbol changes since } k+1 \text{ is less than or equal to } 2^k)$$

$$\geq 2k + 2 \qquad \text{distribute}$$

Since k is a positive integer, $2^{k+1} \geq 2k + 2 \geq k + 2$,

showing $2^{k+1} \geq k + 2$.

WORTHY OF NOTE

Note there is no reference to a_n, a_k, or a_{k+1} in the statement of the general principle of mathematical induction.

Since the truth of P_{k+1} follows from P_k, the formula is true for all n.

Now try Exercises 39 through 42 ▶

EXAMPLE 4 ▶   **Proving Divisibility Using Mathematical Induction**

Let P_n be the statement, "$4^n - 1$ *is divisible by* 3 *for all positive integers n.*" Use mathematical induction to prove that P_n is true.

Solution ▶   If a number is evenly divisible by three, it can be written as the product of 3 and some positive integer we will call p.

1. Show P_n is true for $n = 1$:

$$P_n: \quad 4^n - 1 = 3p \qquad \text{given statement, } p \in \mathbb{Z}$$
$$P_1: \quad 4^{(1)} - 1 = 3p \qquad \text{substitute 1 for } n$$
$$3 = 3p \checkmark \qquad \text{statement is true for } n = 1$$

2. Assume that P_k is true.

$$P_k: \quad 4^k - 1 = 3p \qquad \text{induction hypothesis}$$
$$4^k = 3p + 1 \qquad \text{isolate } 4^k$$

and use it to show the truth of P_{k+1}. That is,

$$P_{k+1}: \quad 4^{k+1} - 1 = 3q \text{ for } q \in \mathbb{Z} \text{ is also true.}$$

Beginning with the left-hand side we have:

$$4^{k+1} - 1 = 4 \cdot \mathbf{4^k} - 1 \qquad \text{properties of exponents}$$
$$= 4 \cdot (\mathbf{3p+1}) - 1 \qquad \text{induction hypothesis: substitute } 3p+1 \text{ for } 4^k$$
$$= 12p + 3 \qquad \text{distribute and simplify}$$
$$= 3(4p + 1) = 3q \qquad \text{factor}$$

The last step shows $4^{k+1} - 1$ is divisible by 3. Since the original statement is true for $n = 1$, and the truth of P_k implies the truth of P_{k+1}, the statement, "$4^n - 1$ *is divisible by* 3" is true for all positive integers n.

Now try Exercises 43 through 47 ▶

We close this section with some final notes. Although the base step of a proof by induction seems trivial, both the base step and the induction hypothesis are necessary parts of the proof. For example, the statement $\dfrac{1}{3^n} < \dfrac{1}{3n}$ is false for $n = 1$, but true for

☑ **C.** You've just seen how we can apply the principle of mathematical induction to general statements involving natural numbers

all other positive integers. Finally, for a fixed natural number p, some statements are false for all $n < p$, but true for all $n \geq p$. By modifying the base case to begin at p, we can use the induction hypothesis to prove the statement is true for all n greater than p. For example, $n < \frac{1}{3}n^2$ is false for $n < 4$, but true for all $n \geq 4$.

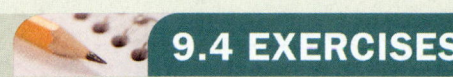

9.4 EXERCISES

▶ CONCEPTS AND VOCABULARY

Fill in the blank with the appropriate word or phrase. Carefully reread the section if needed.

1. No ___finite___ number of verifications can prove a statement ___universally___ true.

2. Showing a statement is true for $n = 1$ is called the ___base___ ___case___ of an inductive proof.

3. Assuming that a statement/formula is true for $n = k$ is called the ___induction___ ___hypothesis___.

4. The graph of a sequence is ___discrete___, meaning it is made up of distinct points.

5. Explain the equation $S_k + a_{k+1} = S_{k+1}$. Begin by saying, "Since the kth term is arbitrary ..." (continue from here).   Answers will vary.

6. Discuss the similarities and differences between mathematical induction applied to sums and the general principle of mathematical induction. Answers will vary.

▶ DEVELOPING YOUR SKILLS

For the given nth term a_n, find $a_4, a_5, a_k,$ and a_{k+1}.

7. $a_n = 10n - 6$ 8. $a_n = 6n - 4$

9. $a_n = n$ 10. $a_n = 7n$

11. $a_n = 2^{n-1}$ 12. $a_n = 2(3^{n-1})$

For the given sum formula S_n, find $S_4, S_5, S_k,$ and S_{k+1}.

13. $S_n = n(5n - 1)$ 14. $S_n = n(3n - 1)$

15. $S_n = \dfrac{n(n + 1)}{2}$ 16. $S_n = \dfrac{7n(n + 1)}{2}$

17. $S_n = 2^n - 1$ 18. $S_n = 3^n - 1$

Verify that $S_4 + a_5 = S_5$ for each exercise. Note that each S_n is identical to those in Exercises 13 through 18.

19. $a_n = 10n - 6; S_n = n(5n - 1)$

20. $a_n = 6n - 4; S_n = n(3n - 1)$

21. $a_n = n; S_n = \dfrac{n(n + 1)}{2}$

22. $a_n = 7n; S_n = \dfrac{7n(n + 1)}{2}$

23. $a_n = 2^{n-1}; S_n = 2^n - 1$

24. $a_n = 2(3^{n-1}); S_n = 3^n - 1$

▶ WORKING WITH FORMULAS

25. **Sum of the first n cubes (alternative form): $(1 + 2 + 3 + 4 + \cdots + n)^2$**

   Earlier we noted the formula for the sum of the first n cubes was $\dfrac{n^2(n + 1)^2}{4}$. An alternative is given by the formula shown.

   a. Verify the formula for $n = 1, 5,$ and 9.

   b. Verify the formula using

   $$1 + 2 + 3 + \cdots + n = \dfrac{n(n + 1)}{2}.$$

26. **Powers of the imaginary unit: $i^{n+4} = i^n$, where $i = \sqrt{-1}$**

   Use a proof by induction to prove that powers of the imaginary unit are cyclic. That is, that they cycle through the numbers $i, -1, -i,$ and 1 for consecutive powers.

▶ APPLICATIONS

Use mathematical induction to prove the indicated sum formula is true for all natural numbers n.

27. $2 + 4 + 6 + 8 + 10 + \cdots + 2n;$
   $a_n = 2n, S_n = n(n + 1)$

28. $3 + 7 + 11 + 15 + 19 + \cdots + (4n - 1);$
   $a_n = 4n - 1, S_n = n(2n + 1)$

29. $5 + 10 + 15 + 20 + 25 + \cdots + 5n;$
   $a_n = 5n, S_n = \dfrac{5n(n + 1)}{2}$

30. $1 + 4 + 7 + 10 + 13 + \cdots + (3n - 2);$
   $a_n = 3n - 2, S_n = \dfrac{n(3n - 1)}{2}$

Additional answers can be found in the Instructor Answer Appendix.

HOMEWORK SELECTION GUIDE

Core: 1–4, 7–23 odd, 27, 31, 33, 39 (17 Exercises)
Standard: 1–4, 7–23 odd, 26, 27, 31, 33, 37, 39, 45, 48, 52, 53 (23 Exercises)

Extended: 1–4, 7–23 odd, 25–27, 31, 33, 37, 39, 45, 48, 50, 52–54 (26 Exercises)
In Depth: 1–6, 7–23 odd, 25–27, 31, 33, 35, 37, 39, 45, 48, 50–54 (30 Exercises)

31. $5 + 9 + 13 + 17 + \cdots + (4n + 1)$;
$a_n = 4n + 1, S_n = n(2n + 3)$

32. $4 + 12 + 20 + 28 + 36 + \cdots + (8n - 4)$;
$a_n = 8n - 4, S_n = 4n^2$

33. $3 + 9 + 27 + 81 + 243 + \cdots + 3^n$;
$a_n = 3^n, S_n = \dfrac{3(3^n - 1)}{2}$

34. $5 + 25 + 125 + 625 + \cdots + 5^n$;
$a_n = 5^n, S_n = \dfrac{5(5^n - 1)}{4}$

35. $2 + 4 + 8 + 16 + 32 + 64 + \cdots + 2^n$;
$a_n = 2^n, S_n = 2^{n+1} - 2$

36. $1 + 8 + 27 + 64 + 125 + 216 + \cdots + n^3$;
$a_n = n^3, S_n = \dfrac{n^2(n + 1)^2}{4}$

37. $\dfrac{1}{1(3)} + \dfrac{1}{3(5)} + \dfrac{1}{5(7)} + \cdots + \dfrac{1}{(2n - 1)(2n + 1)}$;
$a_n = \dfrac{1}{(2n - 1)(2n + 1)}, S_n = \dfrac{n}{2n + 1}$

38. $\dfrac{1}{1(2)} + \dfrac{1}{2(3)} + \dfrac{1}{3(4)} + \cdots + \dfrac{1}{n(n + 1)}$;
$a_n = \dfrac{1}{n(n + 1)}, S_n = \dfrac{n}{n + 1}$

Use the principle of mathematical induction to prove that each statement is true for all natural numbers n.

39. $3^n \geq 2n + 1$ **40.** $2^n \geq n + 1$

41. $3 \cdot 4^{n-1} \leq 4^n - 1$ **42.** $4 \cdot 5^{n-1} \leq 5^n - 1$

43. $n^2 - 7n$ is divisible by 2

44. $n^3 - n + 3$ is divisible by 3

45. $n^3 + 3n^2 + 2n$ is divisible by 3

46. $5^n - 1$ is divisible by 4

47. $6^n - 1$ is divisible by 5

▶ EXTENDING THE CONCEPT

48. You may have noticed that the sum formula for the first n integers was *quadratic,* and the formula for the first n integer squares was *cubic.* Is the formula for the first n integer cubes, if it exists, a quartic (degree four) function? Use your calculator to run a quartic regression on the first five perfect cubes (enter 1 through 5 in L1 and the cumulative sums in L2). What did you find? How is this exercise related to Exercise 36?
$Y_1 = \frac{1}{4}X^4 + \frac{1}{2}X^3 + \frac{1}{4}X^2$, the formula obtained in Exercise 48 can be rewritten to match the conclusion of Exercise 36.

49. Use mathematical induction to prove that $\dfrac{x^n - 1}{x - 1} = (1 + x + x^2 + x^3 + \cdots + x^{n-1})$. verified

50. Use mathematical induction to prove that for $1^4 + 2^4 + 3^4 + \cdots + n^4$, where $a_n = n^4$,
$S_n = \dfrac{n(n + 1)(2n + 1)(3n^2 + 3n - 1)}{30}$. verified

▶ MAINTAINING YOUR SKILLS

51. (7.2) Given the matrices $A = \begin{bmatrix} -1 & 2 \\ 3 & 1 \end{bmatrix}$ and
$B = \begin{bmatrix} 2 & -1 \\ 4 & 3 \end{bmatrix}$, find $A + B, A - B, 2A - 3B$,
AB, BA, and B^{-1}.

52. (2.5) State the domain and range of the piecewise function shown here.
$D: x \in (-\infty, \infty)$
$R: y \in [-2, \infty)$

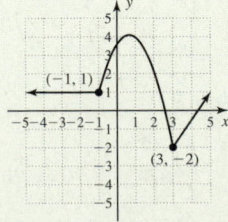

53. (1.1) State the equation of the circle whose graph is shown here. $(x - 4)^2 + (y - 3)^2 = 25$

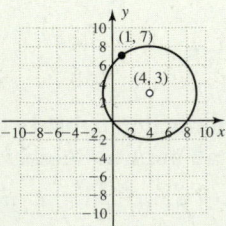

54. (5.5) Solve: $3e^{2x-1} + 5 = 17$. Answer in exact form. $x = \dfrac{1 + \ln 4}{2}$

Additional answers can be found in the Instructor Answer Appendix.

MID-CHAPTER CHECK

In Exercises 1 through 3, the nth term is given. Write the first three terms of each sequence and find a_9.

1. $a_n = 7n - 4$ 3, 10, 17, $a_9 = 59$

2. $a_n = n^2 + 3$ 4, 7, 12, $a_9 = 84$

3. $a_n = (-1)^n(2n - 1)$ $-1, 3, -5, a_9 = -17$

4. Evaluate the sum $\displaystyle\sum_{n=1}^{4} 3^{n+1}$ 360

5. Rewrite using sigma notation.
$$1 + 4 + 7 + 10 + 13 + 16 \qquad \sum_{k=1}^{6}(3k-2)$$

Match each formula to its correct description.

6. $S_n = \dfrac{n(a_1 + a_n)}{2}$ d   **7.** $a_n = a_1 r^{n-1}$ e

8. $S_\infty = \dfrac{a_1}{1 - r}$ a   **9.** $a_n = a_1 + (n-1)d$ b

10. $S_n = \dfrac{a_1(1 - r^n)}{1 - r}$ c

    a. sum of an infinite geometric series

    b. nth term formula for an arithmetic series

    c. sum of a finite geometric series

    d. summation formula for an arithmetic series

    e. nth term formula for a geometric series

11. Identify a_1 and the common difference d. Then find an expression for the general term a_n.

    a. $2, 5, 8, 11, \ldots$ $a_1 = 2, d = 3, a_n = 3n - 1$

    b. $\frac{3}{2}, \frac{9}{4}, 3, \frac{15}{4}, \ldots$ $a_1 = \frac{3}{2}, d = \frac{3}{4}, a_n = \frac{3}{4}n + \frac{3}{4}$

Find the number of terms in each series and then find the sum. Verify results on a graphing calculator.

12. $2 + 5 + 8 + 11 + \cdots + 74$ $n = 25, S_{25} = 950$

13. $\frac{1}{2} + \frac{3}{2} + \frac{5}{2} + \frac{7}{2} + \cdots + \frac{31}{2}$ $n = 16, S_{16} = 128$

14. For an arithmetic series, $a_3 = -8$ and $a_7 = 4$. Find S_{10}. $S_{10} = -5$

15. For a geometric series, $a_3 = -81$ and $a_6 = 3$. Find S_{10}. $S_{10} = \dfrac{-14{,}762}{27}$

16. Identify a_1 and the common ratio r. Then find an expression for the general term a_n.

    a. $2, 6, 18, 54, \ldots$ $a_1 = 2, r = 3, a_n = 2(3)^{n-1}$

    b. $\frac{1}{2}, \frac{1}{4}, \frac{1}{8}, \frac{1}{16}, \ldots$ $a_1 = \frac{1}{2}, r = \frac{1}{2}, a_n = (\frac{1}{2})^n$

17. Find the number of terms in the series then compute the sum. $\frac{1}{54} + \frac{1}{18} + \frac{1}{6} + \cdots + \frac{81}{2}$ $n = 8, S_8 = \frac{1640}{27}$

18. Find the infinite sum (if it exists).
$-49 + (-7) + (-1) + (-\frac{1}{7}) + \cdots$ $\frac{-343}{6}$

19. Barrels of toxic waste are stacked at a storage facility in pyramid form, with 60 barrels in the first row, 59 in the second row, and so on, until there are 10 barrels in the top row. How many barrels are in the storage facility? Verify results using a graphing calculator.   1785

20. As part of a conditioning regimen, a drill sergeant orders her platoon to do 25 continuous standing broad jumps. The best of these recruits was able to jump 96% of the distance from the previous jump, with a first jump distance of 8 ft. Use a sequence/series to determine the distance the recruit jumped on the 15th try, and the total distance traveled by the recruit after all 25 jumps. Verify results using a graphing calculator. ≈ 4.5 ft; ≈ 127.9 ft

REINFORCING BASIC CONCEPTS

Applications of Summation

 The properties of summation play a large role in the development of key ideas in a first semester calculus course, and the following summation formulas are an integral part of these ideas. The first three formulas were verified in Section 9.4, while proof of the fourth was part of Exercise 48 on page 802.

(1) $\displaystyle\sum_{i=1}^{n} c = cn$ **(2)** $\displaystyle\sum_{i=1}^{n} i = \frac{n(n+1)}{2}$ **(3)** $\displaystyle\sum_{i=1}^{n} i^2 = \frac{n(n+1)(2n+1)}{6}$ **(4)** $\displaystyle\sum_{i=1}^{n} i^3 = \frac{n^2(n+1)^2}{4}$

To see the various ways they can be applied consider the following.

Illustration 1 ▶ Over several years, the owner of Morgan's LawnCare has noticed that the company's monthly profits (in thousands) can be approximated by the sequence $a_n = 0.0625n^3 - 1.25n^2 + 6n$, with the points plotted in Figure 9.52 (the continuous graph is shown for effect only). Find the company's approximate annual profit.

Solution ▶ The most obvious approach would be to simply compute terms a_1 through a_{12} (January through December) and find their sum: **sum(seq(Y1, X, 1, 12)**, which gives a result of 35.75 or \$35,750.

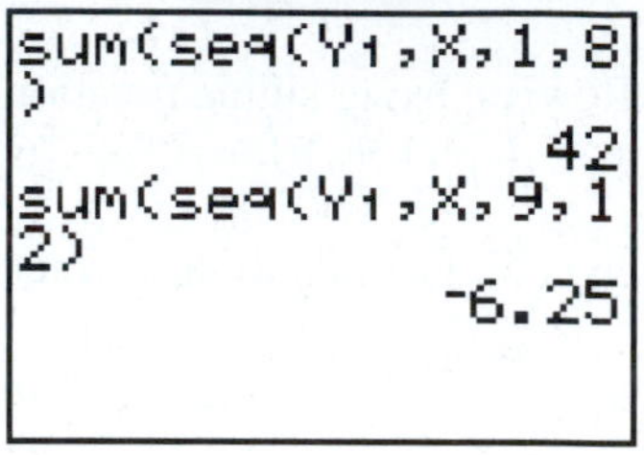

As an alternative, we could add the amount of profit earned by the company in the first 8 months, then add the amount the company lost (or broke even) during the last 4 months. In other words, we could apply summation property IV:

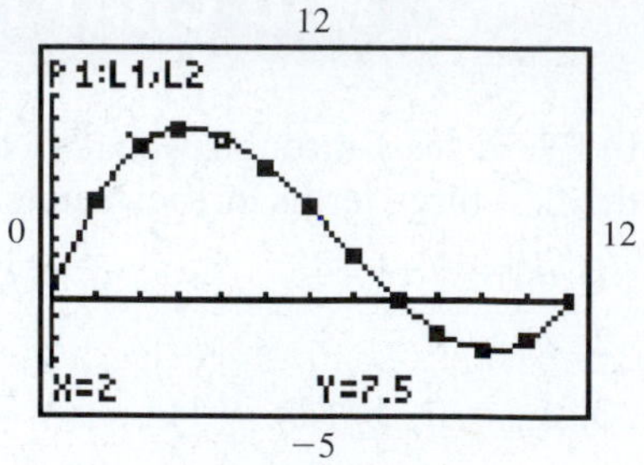

$$\sum_{i=1}^{12} a_n = \sum_{i=1}^{8} a_n + \sum_{i=9}^{12} a_n \text{ [(see Figure 9.53), which gives the same result:}$$

$$42 + (-6.25) = 35.75 \text{ or } \$35,750].$$

As a third option, we could use summation properties along with the appropriate summation formulas, and compute the result manually. Note the function is now written in terms of "i." *Distribute summation and factor out constants* (properties II and III):

$$\sum_{i=1}^{12} (0.0625i^3 - 1.25i^2 + 6i) = 0.0625 \sum_{i=1}^{12} i^3 - 1.25 \sum_{i=1}^{12} i^2 + 6 \sum_{i=1}^{12} i$$

Figure 9.52

Figure 9.53

Replace each summation with the appropriate summation formula, then substitute 12 for n:

$$= 0.0625\left[\frac{n^2(n+1)^2}{4}\right] - 1.25\left[\frac{n(n+1)(2n+1)}{6}\right] + 6\left[\frac{n(n+1)}{2}\right]$$

$$= 0.0625\left[\frac{(12)^2(13)^2}{4}\right] - 1.25\left[\frac{(12)(13)(25)}{6}\right] + 6\left[\frac{(12)(13)}{2}\right]$$

$$= 0.0625(6084) - 1.25(650) + 6(78)$$

$$= 35.75$$

As we expected, the result shows profit was \$35,750. While some approaches seem "easier" than others, all have great value, are applied in different ways at different times, and are necessary to adequately develop key concepts in future classes.

Exercise 1: Repeat Illustration 1 if the profit sequence is $a_n = 0.125x^3 - 2.5x^2 + 12x$.   \$71,500

9.5 Counting Techniques

LEARNING OBJECTIVES

In Section 9.5 you will see how we can:

☐ **A.** Count possibilities using lists and tree diagrams

☐ **B.** Count possibilities using the fundamental principle of counting

☐ **C.** Quick-count distinguishable permutations

☐ **D.** Quick-count nondistinguishable permutations

☐ **E.** Quick-count using combinations

How long would it take to estimate the number of fans sitting shoulder-to-shoulder at a sold-out basketball game? Well, it depends. You could actually begin counting 1, 2, 3, 4, 5, ..., which would take a very long time, or you could try to simplify the process by counting the number of fans in the first row and multiplying by the number of rows. Techniques for "quick-counting" the objects in a set or various subsets of a large set play an important role in a study of probability.

A. Counting by Listing and Tree Diagrams

Consider the simple spinner shown in Figure 9.54, which is divided into three equal parts. What are the different possible outcomes for two spins, spin 1 followed by spin 2? We might begin by organizing the possibilities using a **tree diagram.** As the name implies, each choice or possibility appears as the branch of a tree, with the total possibilities being equal to the number of (unique)

Figure 9.54

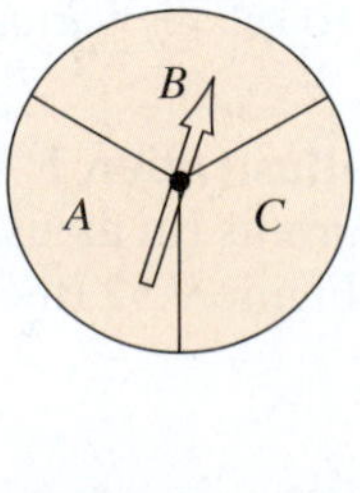

paths from the beginning point to the end of a branch. Figure 9.55 shows how the spinner exercise would appear (possibilities for two spins). Moving from top to bottom we can trace nine possible paths: *AA*, *AB*, *AC*, *BA*, *BB*, *BC*, *CA*, *CB*, and *CC*.

Figure 9.55

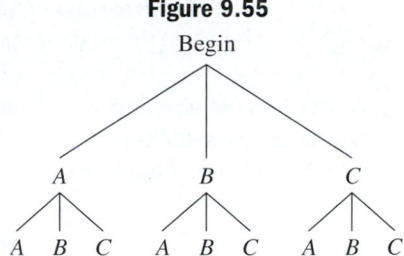

EXAMPLE 1 ▶ **Listing Possibilities Using a Tree Diagram**

A basketball player is fouled and awarded three free throws. Let H represent the possibility of a hit (basket is made), and M the possibility of a miss. Determine the possible outcomes for the three shots using a tree diagram.

Solution ▶ Each shot has two possibilities, hit (H) or miss (M), so the tree will branch in two directions at each level. As illustrated in the figure, there are a total of eight possibilities: HHH, HHM, HMH, HMM, MHH, MHM, MMH, and MMM.

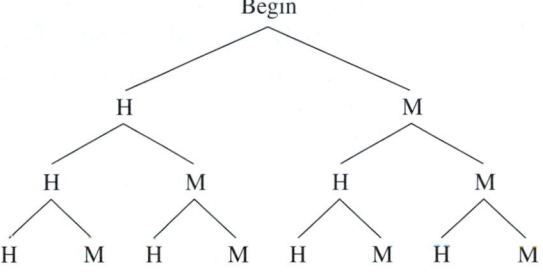

Now try Exercises 7 through 10 ▶

WORTHY OF NOTE

Sample spaces may vary depending on how we define the experiment, and for simplicity's sake we consider only those experiments having outcomes that are equally likely.

To assist our discussion, an **experiment** is any task that can be done repeatedly and has a well-defined set of possible outcomes. Each repetition of the experiment is called a **trial**. A **sample outcome** is any potential outcome of a trial, and a **sample space** is a set of all possible outcomes.

In our first illustration, the *experiment* was spinning a spinner, there were *three sample outcomes* (*A*, *B*, or *C*), the experiment had *two trials* (spin 1 and spin 2), and there were *nine* elements in the *sample space*. Note that after the first trial, each of the three sample outcomes will again have three possibilities (*A*, *B*, and *C*). For two trials we have $3^2 = 9$ possibilities, while three trials would yield a sample space with $3^3 = 27$ possibilities. In general, for *N* equally likely outcomes we have

> **A "Quick-Counting" Formula for a Sample Space**
>
> If an experiment has *N* sample outcomes that are equally likely and the experiment is repeated *t* times, the number of elements in the sample space is
> $$N^t.$$

EXAMPLE 2 ▶ **Counting the Outcomes in a Sample Space**

Many combination locks have the digits 0 through 39 arranged along a circular dial. Opening the lock requires stopping at a sequence of three numbers within this range, going counterclockwise to the first number, clockwise to the second, and counterclockwise to the third. How many three-number combinations are possible?

Solution ▶ There are 40 sample outcomes ($N = 40$) in this experiment, and three trials ($t = 3$). The number of possible combinations is identical to the number of elements in the sample space. The quick-counting formula gives $40^3 = 64,000$ possible combinations.

☑ **A.** You've just seen how we can count possibilities using lists and tree diagrams

Now try Exercises 11 and 12 ▶

B. Fundamental Principle of Counting

The number of possible outcomes may differ depending on how the event is defined. For example, some security systems, license plates, and telephone numbers exclude certain numbers. For example, phone numbers cannot begin with 0 or 1 because these are reserved for operator assistance, long distance, and international calls. Constructing a three-digit area code is like filling in three blanks $\underset{\text{digit}}{\underline{\quad}}\ \underset{\text{digit}}{\underline{\quad}}\ \underset{\text{digit}}{\underline{\quad}}$ with three digits. Since the area code must start with a number between 2 and 9, there are eight choices for the first blank. Since there are 10 choices for the second digit and 10 choices for the third, there are $8 \cdot 10 \cdot 10 = 800$ possibilities in the sample space.

EXAMPLE 3 ▶ **Counting Possibilities for a Four-Digit Security Code**

A digital security system requires that you enter a four-digit PIN (personal identification number), using only the digits 1 through 9. How many codes are possible if

  a. Repetition of digits is allowed?
  b. Repetition is not allowed?
  c. The first digit must be even and repetitions are not allowed?

Solution ▶ **a.** Consider filling in the four blanks $\underset{\text{digit}}{\underline{\quad}}\ \underset{\text{digit}}{\underline{\quad}}\ \underset{\text{digit}}{\underline{\quad}}\ \underset{\text{digit}}{\underline{\quad}}$ with the number of ways the digit can be chosen. If repetition is allowed, the experiment is similar to that of Example 2 and there are $N^t = 9^4 = 6561$ possible PINs.

  b. If repetition is not allowed, there are only eight possible choices for the second digit of the PIN, then seven for the third, and six for the fourth. The number of possible PIN numbers decreases to $9 \cdot 8 \cdot 7 \cdot 6 = 3024$.

  c. There are four choices for the first digit (2, 4, 6, 8). Once this choice has been made there are eight choices for the second digit, seven for the third, and six for the last: $4 \cdot 8 \cdot 7 \cdot 6 = 1344$ possible codes.

Now try Exercises 13 through 16 ▶

Given *any* experiment involving a sequence of tasks, if the first task can be completed in p possible ways, the second task has q possibilities, and the third task has r possibilities, a tree diagram will show that the number of possibilities in the sample space for task$_1$–task$_2$–task$_3$ is $p \cdot q \cdot r$. This situation is simply a variation of the previous quick-counting formula. Even though the examples we've considered to this point have varied a great deal, this idea was fundamental to counting all possibilities in a sample space and is, in fact, known as the **fundamental principle of counting** (FPC).

Fundamental Principle of Counting (Applied to Three Tasks)

Given any experiment with three defined tasks, if there are p possibilities for the first task, q possibilities for the second, and r possibilities for the third, the total number of ways the experiment can be completed is

$$p \cdot q \cdot r.$$

This fundamental principle can be extended to include any number of tasks, and can be applied in many different ways. **See Exercises 17 through 20.**

EXAMPLE 4 ▶ **Counting Possibilities for Seating Arrangements**

Adrienne, Bob, Carol, Dax, Earlene, and Fabian bought tickets to see *The Marriage of Figaro*. Assuming they sat together in a row of six seats, how many different seating arrangements are possible if

   a. Bob and Carol are sweethearts and must sit together?

   b. Bob and Carol are enemies and must not sit together?

Solution ▶ **a.** Since a restriction has been placed on the seating arrangement, it will help to divide the experiment into a sequence of tasks: *task 1:* they sit together; *task 2:* either Bob is on the left or Bob is on the right; and *task 3:* the other four are seated. Bob and Carol can sit together in five different ways, as shown in Figure 9.56, so there are five possibilities for task 1. There are two ways they can be side-by-side: Bob on the left and Carol on the right, as shown, or Carol on the left and Bob on the right. The remaining four people can be seated randomly, so task 3 has 4! = 24 possibilities. Under these conditions they can be seated 5 · 2 · 4! = 240 ways.

Figure 9.56

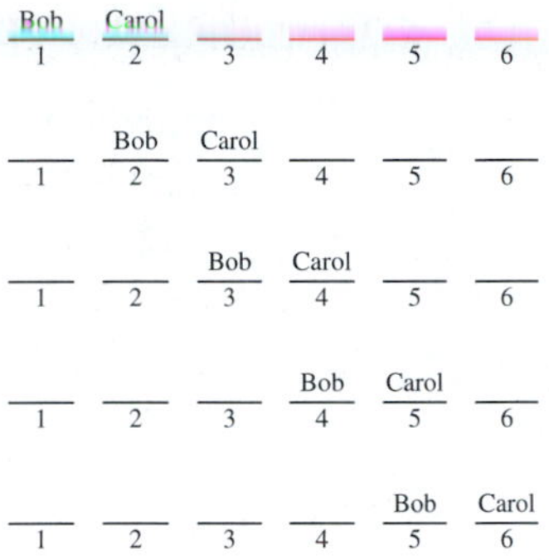

   b. This is similar to part (a), but now we have to count the number of ways they can be separated by *at least one seat: task 1:* Bob and Carol are in nonadjacent seats; *task 2:* either Bob is on the left or Bob is on the right; and *task 3:* the other four are seated. For tasks 1 and 2, be careful to note there is no multiplication involved, just a simple counting. If Bob sits in seat 1 (to the left of Carol), there are four nonadjacent seats on the right. If Bob sits in seat 2, there are three nonadjacent seats on the right. With Bob in seat 3, there are two nonadjacent seats to his right. Similar reasoning for the remaining seats shows there are 10 · 2 = 20 possibilities for Bob and Carol not sitting together (by symmetry, Bob could also sit to the right of Carol). Multiplying by the number of ways the other four can be seated task 3 gives 20 · 4! = 480 possible seating arrangements. We could also reason that since there are 6! = 720 random seating arrangements and 240 of them consist of Bob and Carol sitting together, the remaining 720 − 240 = 480 must consist of Bob and Carol *not* sitting together. More will be said about this type of reasoning in Section 9.6.

☑ **B.** You've just seen how we can count possibilities using the fundamental principle of counting

Now try Exercises 21 through 28 ▶

C. Distinguishable Permutations

In the game of Scrabble® (Milton Bradley), players attempt to form words by rearranging letters. Suppose a player has the letters P, S, T, and O at the end of the game. These letters could be rearranged or *permuted* to form the words POTS, SPOT, TOPS, OPTS, POST, or STOP. These arrangements are called permutations of the four letters. A permutation is any new arrangement, listing, or sequence of objects obtained by changing an existing order. A **distinguishable permutation** is a permutation that produces a result different from the original. For example, a distinguishable permutation of the digits in the number 1989 is 8199.

Example 4 considered six people, six seats, and the various ways they could be seated. But what if there were fewer seats than people? By the FPC, with six people and four seats there could be 6 · 5 · 4 · 3 = 360 different arrangements, with six people and three seats there are 6 · 5 · 4 = 120 different arrangements, and so on. These rearrangements are called distinguishable permutations. You may have noticed that for six people and six seats, we will use all six factors of 6!, while for six people and four seats we used the first four, six people and three seats required only the first three, and so on. Generally, for n people and r seats, the first r factors of $n!$ will be used. The notation and formula for *distinguishable permutations of n objects taken r at a time* is

$$_nP_r = \frac{n!}{(n-r)!}.$$ By defining 0! = 1, the formula includes the case where all n objects

are selected, which of course results in $_nP_n = \dfrac{n!}{(n-n)!} = \dfrac{n!}{0!} = \dfrac{n!}{1} = n!.$

> ### Distinguishable Permutations: Unique Elements
>
> If r objects are selected from a set containing n unique elements ($r \leq n$) and placed in an ordered arrangement, the number of distinguishable permutations is
>
> $$_nP_r = \frac{n!}{(n-r)!} \quad \text{or} \quad _nP_r = n(n-1)(n-2)\cdots(n-r+1)$$

EXAMPLE 5 ▶ **Computing a Permutation**

Compute each value of $_nP_r$ using the methods described previously.

    a. $_7P_4$ **b.** $_{10}P_3$

Solution ▶ Begin by evaluating each expression using the formula $_nP_r = \dfrac{n!}{(n-r)!}$, noting the third line (in bold) gives the first r factors of $n!$.

    a. $_7P_4 = \dfrac{7!}{(7-4)!}$ **b.** $_{10}P_3 = \dfrac{10!}{(10-3)!}$

                $= \dfrac{7 \cdot 6 \cdot 5 \cdot 4 \cdot 3!}{3!}$ $= \dfrac{10 \cdot 9 \cdot 8 \cdot 7!}{7!}$

                $= \mathbf{7 \cdot 6 \cdot 5 \cdot 4}$ $= \mathbf{10 \cdot 9 \cdot 8}$

                $= 840$ $= 720$

Now try Exercises 29 through 36 ▶

Figure 9.57

When the number of objects is very large, the formula for permutations can become somewhat unwieldy and the computed result is often a very large number. When needed, most graphing calculators have the ability to compute permutations, with this option accessed using **MATH** ◀ **(PRB) 2:nPr.** Figure 9.57 verifies the computation for Example 5(b), and also shows that if there were 15 people and 7 chairs, the number of possible seating arrangements exceeds 32 million! Note that the value of n is entered first, followed by the **nPr** command and the value of r.

```
10 nPr 3
             720
15 nPr 7
       32432400
```

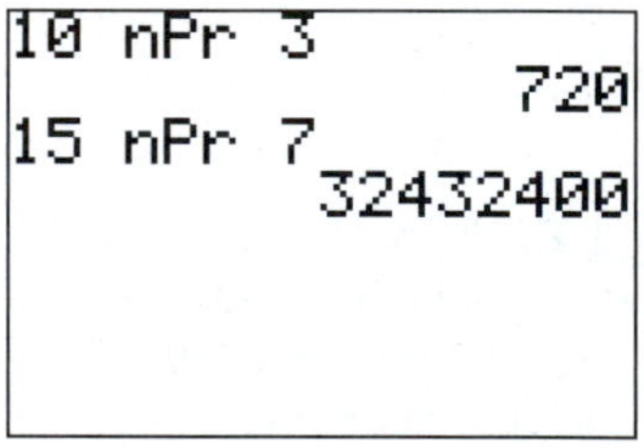

EXAMPLE 6 ▶ **Counting the Possibilities for Finishing a Race**

As part of a sorority's initiation process, the nine new inductees must participate in a 1-mi race. Assuming there are no ties, how many first- through fifth-place finishes are possible if it is well known that Mediocre Mary will finish fifth and Lightning Louise will finish first?

Solution ▶ To help understand the situation, we can diagram the possibilities for finishing first through fifth. Since Louise will finish first, this slot can be filled in only one way, by Louise herself. The same goes for Mary and her fifth-place finish:

Louise				Mary
1st	2nd	3rd	4th	5th

The remaining three slots can be filled in $_7P_3 = 7 \cdot 6 \cdot 5$ different ways, indicating that under these conditions, there are $1 \cdot 7 \cdot 6 \cdot 5 \cdot 1 = 210$ different ways to finish.

☑ **C.** You've just seen how we can quick-count distinguishable permutations

Now try Exercises 37 through 42 ▶

D. Nondistinguishable Permutations

As the name implies, certain permutations are nondistinguishable, meaning you cannot tell one apart from another. Such is the case when the original set contains elements or sample outcomes that are identical. Consider a family with four children, Lyddell, Morgan, Michael, and Mitchell, who are at the photo studio for a family picture. Michael and Mitchell are identical twins and cannot be told apart. In how many ways can they be lined up for the picture? Since this is an ordered arrangement of four children taken from a group of four, there are $_4P_4 = 24$ ways to line them up. A few of them are

Lyddell	Morgan	Michael	Mitchell	Lyddell	Morgan	Mitchell	Michael
Lyddell	Michael	Morgan	Mitchell	Lyddell	Mitchell	Morgan	Michael
Michael	Lyddell	Morgan	Mitchell	Mitchell	Lyddell	Morgan	Michael

But of these six arrangements, half will appear to be the same picture, since the difference between Michael and Mitchell cannot be distinguished. In fact, of the 24 total permutations, every picture where Michael and Mitchell have switched places will be nondistinguishable. To find the *distinguishable* permutations, we need to take the total permutations ($_4P_4$) *and divide by* 2!, *the number of ways the twins can be permuted:* $\dfrac{_4P_4}{(2)!} = \dfrac{24}{2} = 12$ distinguishable pictures.

These ideas can be generalized and stated in the following way.

Nondistinguishable Permutations: Nonunique Elements

In a set containing n elements where one element is repeated p times, another is repeated q times, and another is repeated r times ($p + q + r = n$), the number of nondistinguishable permutations is

$$\frac{_nP_n}{p!q!r!} = \frac{n!}{p!q!r!}$$

The idea can be extended to include any number of repeated elements.

EXAMPLE 7 ▶ **Counting Nondistinguishable Permutations**

A Scrabble player starts the game with the seven letters S, A, O, O, T, T, and T in her rack. How many distinguishable arrangements can be formed as she attempts to play a word?

Solution ▶ Essentially the exercise asks for the number of distinguishable permutations of the seven letters, given T is repeated three times and O is repeated twice (for S and A, 1! = 1). There are $\dfrac{_7P_7}{3!2!} = 420$ distinguishable permutations.

☑ **D.** You've just seen how we can quick-count nondistinguishable permutations

Now try Exercises 43 through 54 ▶

E. Combinations

Similar to nondistinguishable permutations, there are other times the total number of permutations must be reduced to quick-count the elements of a desired subset. Consider a vending machine that offers a variety of 40¢ candies. If you have a quarter (Q), dime (D), and nickel (N), the machine wouldn't care about the order the coins are deposited. Even though QDN, QND, DQN, DNQ, NQD, and NDQ give the $_3P_3 = 6$ possible permutations, the machine considers them as equal and will vend your snack. Using sets, this is similar to saying the set $A = \{X, Y, Z\}$ has only one subset with three elements, since $\{X, Z, Y\}$, $\{Y, X, Z\}$, $\{Y, Z, X\}$, and so on, all represent the same set. Similarly, there are six two-letter permutations of X, Y, and Z ($_3P_2 = 6$): XY, XZ, YX,

WORTHY OF NOTE

In Example 7, if a Scrabble player is able to play all seven letters in one turn, he or she "bingos" and is awarded 50 extra points. The player in Example 7 did just that. Can you determine what word was played?

YZ, ZX, and *ZY,* but only three two-letter subsets: $\{X, Y\}$, $\{X, Z\}$ and $\{Y, Z\}$. When permutations having the same elements are considered identical, the result is the number of possible **combinations** and is denoted $_nC_r$. Since the r objects can be selected in $r!$ ways, we divide $_nP_r$ by $r!$ to "quick-count" the number of possibilities: $_nC_r = \dfrac{_nP_r}{r!}$, which can be thought of as *the first r factors of n!, divided by r!.* By substituting $\dfrac{n!}{(n-r)!}$ for $_nP_r$ in this formula, we find an alternative method for computing $_nC_r$ is $\dfrac{n!}{r!(n-r)!}$. Take special note that when r objects are selected from a set with n elements and the order they're listed is unimportant (because you end up with the same subset), the result is a *combination,* not a permutation.

Combinations

The number of combinations of n objects taken r at a time is given by

$$_nC_r = \frac{_nP_r}{r!} \qquad \text{or} \qquad _nC_r = \frac{n!}{r!(n-r)!}$$

EXAMPLE 8 ▶ **Computing Combinations Using a Formula**

Compute each value of $_nC_r$ given.

a. $_7C_4$ **b.** $_8C_3$ **c.** $_5C_2$

Solution ▶   **a.** $_7C_4 = \dfrac{7 \cdot 6 \cdot 5 \cdot 4}{4!}$ **b.** $_8C_3 = \dfrac{8 \cdot 7 \cdot 6}{3!}$ **c.** $_5C_2 = \dfrac{5 \cdot 4}{2!}$

           $= 35$ $= 56$ $= 10$

Now try Exercises 55 through 64 ▶

As with permutations, when the number of objects is very large, the formula for combinations can also become somewhat cumbersome. Most graphing calculators have the ability to compute combinations, with this option accessed on the same submenu as **nPr:** (MATH) ◀ (PRB) **3:nCr.** Figure 9.58 verifies the computation from Example 8(b), and also shows that in a Political Science class with 30 students, 5 can be picked at random to attend a seminar in the nation's capitol 142,506 ways.

Figure 9.58

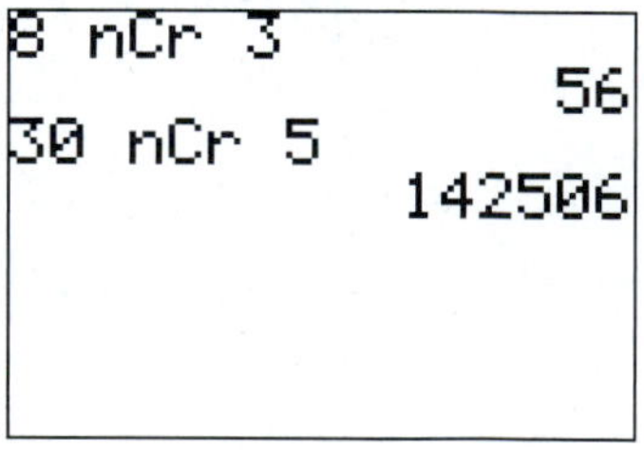

```
8 nCr 3
                56
30 nCr 5
            142506
```

EXAMPLE 9 ▶ **Applications of Combinations-Lottery Results**

A small city is getting ready to draw five Ping-Pong balls of the nine they have numbered 1 through 9 to determine the winner(s) for its annual raffle. If a ticket holder has the same five numbers, they win. In how many ways can the winning numbers be drawn?

Solution ▶ Since the winning numbers can be drawn in any order, we have a combination of 9 things taken 5 at a time. The five numbers can be

drawn in $_9C_5 = \dfrac{9 \cdot 8 \cdot 7 \cdot 6 \cdot 5}{5!} = 126$ ways.

Now try Exercises 65 and 66 ▶

Somewhat surprisingly, there are many situations where the order things are listed is not important. Such situations include

- The formation of committees, since the order people volunteer is unimportant
- Card games with a standard deck, since the order cards are dealt is unimportant
- Playing BINGO, since the order the winning numbers are called is unimportant

When the order in which people or objects are selected from a group is unimportant, the number of possibilities is a *combination,* not a permutation.

Another way to tell the difference between permutations and combinations is the following memory device: *Permutations* have *Priority* or *Precedence;* in other words, the *Position* of each element matters. By contrast, a *Combination* is like a *Committee* of *Colleagues* or *Collection* of *Commoners;* all members have equal rank. For permutations, *a-b-c* is different from *b-a-c.* For combinations, *a-b-c* is the same as *b-a-c.*

EXAMPLE 10 ▶ **Applications of Quick-Counting—Committees and Governance**

The Sociology Department of Lakeside Community College has 12 dedicated faculty members. (a) In how many ways can a three-member textbook selection committee be formed? (b) If the department is in need of a Department Chair, Curriculum Chair, and Technology Chair, in how many ways can the positions be filled?

Solution ▶ a. Since textbook selection depends on a *Committee* of *Colleagues,* the order members are chosen is not important. This is a *Combination* of 12 people taken 3 at a time, and there are $_{12}C_3 = 220$ ways the committee can be formed.

   b. Since those selected will have *Position* or *Priority,* this is a *Permutation* of 12 people taken 3 at a time, giving $_{12}P_3 = 1320$ ways the positions can be filled.

Now try Exercises 67 through 78 ▶

☑ **E. You've just seen how we can quick-count using combinations**

The Exercise Set contains a wide variety of additional applications. **See Exercises 81 through 107.**

9.5 EXERCISES

▶ **CONCEPTS AND VOCABULARY**

Fill in the blank with the appropriate word or phrase. Carefully reread the section if needed.

1. A(n) ___experiment___ is any task that can be repeated and has a(n) ___well-defined___ set of possible outcomes.

2. When unique elements of a set are rearranged, the result is called a(n) ___distinguishable___ permutation.

3. If an experiment has N equally likely outcomes and is repeated t times, the number of elements in the sample space is given by ___N^t___ .

4. If some elements of a group are identical, certain rearrangements are identical and the result is a(n) ___nondistinguishable___ permutation.

5. A three-digit number is formed from digits 1 to 9. Explain how forming the number with repetition differs from forming it without repetition.
   Answers will vary.

6. Discuss/Explain the difference between a permutation and a combination. Try to think of new ways to help remember the distinction.
   Answers will vary.

HOMEWORK SELECTION GUIDE

Core: 7–75 every other odd, 77, 81, 85, 89, 93, 97, 99, 103, 105 (27 Exercises)
Standard: 1–4, 7–75 every other odd, 77, 81, 85, 89, 93, 96, 97, 99, 103, 105, 110, 111 (34 Exercises)

Extended: 1–4, 7–75 every other odd, 77, 80, 81, 85, 89, 93, 96, 97, 99, 103, 105, 107, 108, 110–112 (38 Exercises)
In Depth: 1–6, 7–75 every other odd, 77, 79, 80, 81, 85, 89, 93, 96, 97, 99, 100, 103, 105, 107, 108, 110–113 (43 Exercises)

▶ DEVELOPING YOUR SKILLS

7. For the spinner shown here, (a) draw a tree diagram illustrating all possible outcomes for two spins and (b) create an ordered list showing all possible outcomes for two spins.

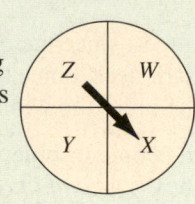

8. For the fair coin shown here, (a) draw a tree diagram illustrating all possible outcomes for four flips and (b) create an ordered list showing the possible outcomes for four flips.

9. A fair coin is flipped five times. If you extend the tree diagram from Exercise 8, how many possibilities are there?   32

10. A spinner has the two equally likely outcomes A or B and is spun four times. How is this experiment related to the one in Exercise 8? How many possibilities are there?   16

11. An inexpensive lock uses the numbers 0 to 24 for a three-number combination. How many different combinations are possible?   15,625

12. Grades at a local college consist of A, B, C, D, F, and W. If four classes are taken, how many different report cards are possible?   1296

License plates. In a certain (English-speaking) country, license plates for automobiles consist of two letters followed by one of four symbols (■, ◆, ○, or ●), followed by three digits. How many license plates are possible if

13. Repetition is allowed?   2,704,000

14. Repetition is not allowed?   1,872,000

15. A remote access door opener requires a five-digit (1–9) sequence. How many sequences are possible if (a) repetition is allowed? (b) repetition is not allowed?   a. 59,049   b. 15,120

16. An instructor is qualified to teach Math 020, 030, 140, and 160. How many different four-course schedules are possible if (a) repetition is allowed? (b) repetition is not allowed?   a. 256   b. 24

Use the fundamental principle of counting and other quick-counting techniques to respond.

17. **Menu items:** At Joe's Diner, the manager is offering a dinner special that consists of one choice of entree (chicken, beef, soy meat, or pork), two vegetable servings (corn, carrots, green beans, peas, broccoli, or okra), and one choice of pasta, rice, or potatoes. How many different meals are possible?   360 if double veggies are not allowed, 432 if double veggies are allowed.

18. **Getting dressed:** A frugal businessman has five shirts, seven ties, four pairs of dress pants, and three pairs of dress shoes. Assuming that all possible arrangements are appealing, how many different shirt-tie-pants-shoes outfits are possible?   420

19. **Number combinations:** How many four-digit numbers can be formed using the even digits 0, 2, 4, 6, 8, if (a) no repetitions are allowed; (b) repetitions are allowed; (c) repetitions are not allowed and the number must be less than 6000 and divisible by 10.   a. 120   b. 625   c. 12

20. **Number combinations:** If I was born in March, April, or May, after the 19th but before the 30th, and after 1949 but before 1981, how many different MM–DD–YYYY dates are possible for my birthday?   930

Seating arrangements: William, Xayden, York, and Zelda decide to sit together at the movies. How many ways can they be seated if

21. They sit in random order?   24

22. York must sit next to Zelda?   12

23. York and Zelda must be on the outside?   4

24. William must have the aisle seat?   6

Course schedule: A college student is trying to set her schedule for the next semester and is planning to take five classes: English, art, math, fitness, and science. How many different schedules are possible if

25. The classes can be taken in any order.   120

26. She wants her science class to immediately follow her math class.   24

27. She wants her English class to be first and her fitness class to be last.   6

28. She can't decide on the best order and simply takes the classes in alphabetical order.   1

Find the value of $_nP_r$ in two ways: (a) compute r factors of $n!$ and (b) use the formula $_nP_r = \dfrac{n!}{(n-r)!}$.

29. $_{10}P_3$ 720

30. $_{12}P_2$ 132

31. $_9P_4$ 3024

32. $_5P_3$ 60

33. $_8P_7$ 40,320

34. $_8P_1$ 8

Additional answers can be found in the Instructor Answer Appendix.

Determine the number of three-letter permutations of the letters given, then use an organized list to write them all out. How many of them are actually words or common names?

35. T, R, and A   6; 3   **36.** P, M, and A   6; 3

37. The regional manager for an office supply store needs to replace the manager and assistant manager at the downtown store. In how many ways can this be done if she selects the personnel from a group of 10 qualified applicants?   90

38. The local chapter of Mu Alpha Theta will soon be electing a president, vice-president, and treasurer. In how many ways can the positions be filled if the chapter has 15 members?   2730

39. The local school board is going to select a principal, vice-principal, and assistant vice-principal from a pool of eight qualified candidates. In how many ways can this be done?   336

40. From a pool of 32 applicants, a board of directors must select a president, vice-president, labor relations liaison, and a director of personnel for the company's day-to-day operations. Assuming all applicants are qualified and willing to take on any of these positions, how many ways can this be done?   863,040

41. A hugely popular chess tournament now has six finalists. Assuming there are no ties, (a) in how many ways can the finalists place in the final round? (b) In how many ways can they finish first, second, and third? (c) In how many ways can they finish if it's sure that Roberta Fischer is going to win the tournament and that Geraldine Kasparov will come in sixth?   a. 720   b. 120   c. 24

42. A field of 10 horses has just left the paddock area and is heading for the gate. Assuming there are no ties in the big race, (a) in how many ways can the horses place in the race? (b) In how many ways can they finish in the win, place, or show positions? (c) In how many ways can they finish if it's sure that John Henry III is going to win, Seattle Slew III will come in second (place), and either Dumb Luck II or Calamity Jane I will come in tenth?   a. 3,628,800   b. 720   c. 10,080

Assuming all multiple births are identical and the children cannot be told apart, how many distinguishable photographs can be taken of a family of six, if they stand in a single row and there is

43. one set of twins   360

44. one set of triplets   120

45. one set of twins and one set of triplets   60

46. one set of quadruplets   30

47. How many distinguishable numbers can be made by rearranging the digits of 105,001?   60

48. How many distinguishable numbers can be made by rearranging the digits in the palindrome 1,234,321?   630

How many distinguishable permutations can be formed from the letters of the given word?

49. logic   120   **50.** leave   60

51. lotto   30   **52.** levee   20

A Scrabble player (see Example 7) has the six letters shown remaining in her rack. How many distinguishable, six-letter permutations can be formed? (If all six letters are played, what was the word?)

53. A, A, A, N, N, B   60, BANANA

54. D, D, D, N, A, E   120, ADDEND

Find the value of $_nC_r$: (a) using $_nC_r = \dfrac{_nP_r}{r!}$ (r factors of $n!$ over $r!$) and (b) using $_nC_r = \dfrac{n!}{r!(n-r)!}$.

55. $_9C_4$ 126   **56.** $_{10}C_3$ 120   **57.** $_8C_5$ 56

58. $_6C_3$ 20   **59.** $_6C_6$ 1   **60.** $_6C_0$ 1

Use a calculator to verify that each pair of combinations is equal.

61. $_9C_4$, $_9C_5$ verified   **62.** $_{10}C_3$, $_{10}C_7$ verified

63. $_8C_5$, $_8C_3$ verified   **64.** $_7C_2$, $_7C_5$ verified

65. A platoon leader needs to send four soldiers to do some reconnaissance work. There are 12 soldiers in the platoon and each soldier is assigned a number between 1 and 12. The numbers 1 through 12 are placed in a helmet and drawn randomly. If a soldier's number is drawn, then that soldier goes on the mission. In how many ways can the reconnaissance team be chosen?   495

66. Seven colored balls (red, indigo, violet, yellow, green, blue, and orange) are placed in a bag and three are then withdrawn. In how many ways can the three colored balls be drawn?   35

67. When the company's switchboard operators went on strike, the company president asked for three volunteers from among the managerial ranks to temporarily take their place. In how many ways can the three volunteers "step forward," if there are 14 managers and assistant managers in all?   364

68. Becky has identified 12 books she wants to read this year and decides to take four with her to read while on vacation. She chooses *Pastwatch* by Orson Scott Card for sure, then decides to randomly choose any three of the remaining books. In how many ways can she select the four books she'll end up taking?   165

69. A new garage band has built up their repertoire to 10 excellent songs that really rock. Next month they'll be playing in a *Battle of the Bands* contest, with the winner getting some guaranteed gigs at the city's most popular hot spots. In how many ways can the band select 5 of their 10 songs to play at the contest?   252

70. Pierre de Guirré is an award-winning chef and has just developed 12 delectable, new main-course recipes for his restaurant. In how many ways can he select three of the recipes to be entered in an international culinary competition?   220

For each exercise, determine whether a permutation, a combination, counting principles, or a determination of the number of subsets is the most appropriate tool for obtaining a solution, then solve. Some exercises can be completed using more than one method.

71. In how many ways can eight second-grade children line up for lunch? $8! = 40,320$

72. If you flip a fair coin five times, how many different outcomes are possible? $2^5 = 32$

73. Eight sprinters are competing for the gold, silver, and bronze medals. In how many ways can the medals be awarded? $_8P_3 = 336$

74. Motorcycle license plates are made using two letters followed by three numbers. How many plates can be made if repetition of letters (only) is allowed? $26^2 \cdot {}_{10}P_3 = 486,720$

75. A committee of five students is chosen from a class of 20 to attend a seminar. How many different ways can this be done? $_{20}C_5 = 15,504$

76. If onions, cheese, pickles, and tomatoes are available to dress a hamburger, how many different hamburgers can be made? $\sum_{x=0}^{4} {}_4C_x = 16$

77. A caterer offers eight kinds of fruit to make various fruit trays. How many different trays can be made using four different fruits? $_8C_4 = 70$

78. Eighteen females try out for the basketball team, but the coach can only place 15 on her roster. How many different teams can be formed? $_{18}C_{15} = 816$

▶ WORKING WITH FORMULAS

79. Stirling's Formula: $n! \approx \sqrt{2\pi} \cdot (n^{n+0.5}) \cdot e^{-n}$

Values of $n!$ grow very quickly as n gets larger (13! is already in the billions). For some applications, scientists find it useful to use the approximation for $n!$ shown, called Stirling's Formula.

a. Compute the value of 7! on your calculator, then use Stirling's Formula with $n = 7$. By what percent does the approximate value differ from the true value? $\approx 1.2\%$

b. Compute the value of 10! on your calculator, then use Stirling's Formula with $n = 10$. By what percent does the approximate value differ from the true value? $\approx 0.83\%$

80. Factorial formulas: For $n, k \in \mathbb{W}$, where $n > k$, $\dfrac{n!}{(n-k)!} = n(n-1)(n-2) \cdots (n-k+1)$

a. Verify the formula for $n = 7$ and $k = 5$.   verified

b. Verify the formula for $n = 9$ and $k = 6$.   verified

▶ APPLICATIONS

81. Yahtzee: In the game of "Yahtzee"® (Milton Bradley) five dice are rolled simultaneously on the first turn in an attempt to obtain various arrangements (worth various point values). How many different arrangements are possible?   7776

82. Twister: In the game of "Twister"® (Milton Bradley) a simple spinner is divided into four quadrants designated Left Foot (LF), Right Hand (RH), Right Foot (RF), and Left Hand (LH), with four different color possibilities in each quadrant (red, green, yellow, blue). Determine the number of possible outcomes for three spins.   4096

83. Clue: In the game of "Clue"® (Parker Brothers) a crime is committed in one of nine rooms, with one of six implements, by one of six people. In how many different ways can the crime be committed?   324

Phone numbers in North America have 10 digits: a three-digit area code, a three-digit exchange number, and the four final digits that make each phone number unique. Neither area codes nor exchange numbers can start with 0 or 1. Prior to 1994 the second digit of the area code *had to be* a 0 or 1. Sixteen area codes are reserved for special services (such as 911 and 411). In 1994, the last area code was used up and the rules were changed to allow the digits 2 through 9 as the middle digit in area codes.

84. How many different area codes were possible prior to 1994?   144

85. How many different exchange numbers were possible prior to 1994?   800

86. How many different phone numbers were possible *prior to* 1994?   1,152,000,000

87. How many different phone numbers were possible *after* 1994?   6,272,000,000

Aircraft N-Numbers: In the United States, private aircraft are identified by an "N-Number," which is generally the letter "N" followed by five characters and includes these restrictions: (1) the N-Number can consist of five digits, four digits followed by one letter, or three digits followed by two letters; (2) the first digit cannot be a zero; (3) to avoid confusion with the numbers zero and one, the letters O and I cannot be used; and (4) repetition of digits and letters is allowed. How many unique N-Numbers can be formed

88. that have four digits and one letter?   216,000

89. that have three digits and two letters?   518,400

90. that have five digits?   90,000

91. that have three digits, two letters with no repetitions of any kind allowed?   357,696

Seating arrangements: Eight people would like to be seated. Assuming some will have to stand, in how many ways can the seats be filled if the number of seats available is

92. eight   40,320

93. five   6720

94. three   336

95. one   8

Seating arrangements: In how many different ways can eight people (six students and two teachers) sit in a row of eight seats if

96. the teachers must sit on the ends   1440

97. the teachers must sit together   10,080

Television station programming: A television station needs to fill eight half-hour slots for its Tuesday evening schedule with eight programs. In how many ways can this be done if

98. there are no constraints   40,320

99. *Seinfeld* must have the 8:00 P.M. slot   5040

100. *Seinfeld* must have the 8:00 P.M. slot and *The Drew Carey Show* must be shown at 6:00 P.M.   720

101. *Friends* can be aired at 7:00 or 9:00 P.M. and *Everybody Loves Raymond* can be aired at 6:00 or 8:00 P.M.   2880

Scholarship awards: Fifteen students at Roosevelt Community College have applied for six available scholarship awards. How many ways can the awards be given if

102. there are six different awards given to six different students   3,603,600

103. there are six identical awards given to six different students   5005

Committee composition: The local city council has 10 members and is trying to decide if they want to be governed by a committee of three people or by a president, vice-president, and secretary.

104. If they are to be governed by committee, how many unique committees can be formed?   120

105. How many different president, vice-president, and secretary possibilities are there?   720

106. Team rosters: A soccer team has three goalies, eight defensive players, and eight forwards on its roster. How many different starting line-ups can be formed (one goalie, three defensive players, and three forwards)?   9408

107. e-mail addresses: A business wants to standardize the e-mail addresses of its employees. To make them easier to remember and use, they consist of two letters and two digits (followed by @esmtb.com), with zero being excluded from use as the first digit and no repetition of letters or digits allowed. Will this provide enough unique addresses for their 53,000 employees worldwide?   52,650, no

▶ **EXTENDING THE CONCEPT**

Tic-Tac-Toe: In the game *Tic-Tac-Toe,* players alternately write an "X" or an "O" in one of nine squares on a 3 × 3 grid. If either player gets three in a row horizontally, vertically, or diagonally, that player wins. If all nine squares are played with neither person winning, the game is a draw. Assuming "X" always goes first,

108. How many different "ending boards" are possible if the game ends after five plays? $8 \cdot (_6\,_nC_r\,2) = 120$

109. How many different "ending boards" are possible if the game ends after six plays?
$6 \cdot (_6\,_nC_r\,3 - 2) + 2 \cdot 20 = 148$

▶ **MAINTAINING YOUR SKILLS**

110. **(6.4)** Solve the given system of linear inequalities by graphing. Shade the feasible region.

$$\begin{cases} 2x + y < 6 \\ x + 2y < 6 \\ x \geq 0 \\ y \geq 0 \end{cases}$$

111. **(9.2)** For the series $1 + 5 + 9 + 13 + \cdots + 197$, state the nth term formula then find the 35th term and the sum of the first 35 terms. $a_n = 1 + (n-1)4; 137; 2415$

112. **(7.2/7.3)** Given matrices A and B shown, use a calculator to find $A + B, AB$, and A^{-1}.

$$A = \begin{bmatrix} 1 & 0 & 3 \\ -2 & 5 & 1 \\ 2 & 1 & 4 \end{bmatrix} \qquad B = \begin{bmatrix} 0.5 & 0.2 & -7 \\ -9 & 0.1 & 8 \\ 1.2 & 0 & 6 \end{bmatrix}$$

113. **(8.3)** Graph the hyperbola that is defined by

$$\frac{(x-2)^2}{4} - \frac{(y+3)^2}{9} = 1.$$

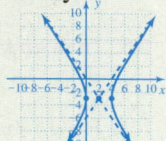

Additional answers can be found in the Instructor Answer Appendix.

9.6   Introduction to Probability

LEARNING OBJECTIVES

In Section 9.6 you will see how we can:

☐ **A.** Define an event on a sample space
☐ **B.** Compute elementary probabilities
☐ **C.** Use certain properties of probability
☐ **D.** Compute probabilities using quick-counting techniques
☐ **E.** Compute probabilities involving nonexclusive events

There are few areas of mathematics that give us a better view of the world than **probability** and **statistics.** Unlike statistics, which seeks to analyze and interpret data, probability (for our purposes) attempts to use observations and data to make statements concerning the likelihood of future events. Such predictions of what *might* happen have found widespread application in such diverse fields as politics, manufacturing, gambling, opinion polls, product life, and many others. In this section, we develop the basic elements of probability.

A. Defining an Event

In Section 9.5 we defined the following terms: experiment and sample outcome. Flipping a coin twice in succession is an *experiment,* and two sample outcomes are HH and HT. An **event E** is *any designated set of sample outcomes,* and is a subset of the sample space. One event might be E_1: (two heads occur), another possibility is E_2: (at least one tail occurs).

> **EXAMPLE 1** ▶ **Stating a Sample Space and Defining an Event**
>
> Consider the experiment of rolling one standard, six-sided die (plural is dice). State the sample space S and define any two events relative to S.
>
> **Solution** ▶ S is the set of all possible outcomes, so $S = \{1, 2, 3, 4, 5, 6\}$. Two possible events are E_1: (a 5 is rolled) and E_2: (an even number is rolled).

☑ **A.** You've just seen how we can define an event on a sample space

Now try Exercises 7 through 10 ▶

B. Elementary Probability

When rolling a die, we know the result can be any of the six equally likely outcomes in the sample space, so the chance of E_1:(a five is rolled) is $\frac{1}{6}$. Since three of the elements in S are even numbers, the chance of E_2:(an even number is rolled) is $\frac{3}{6} = \frac{1}{2}$. This suggests the following definition.

The Probability of an Event E

Given S is a sample space of equally likely events and E is an event relative to S, the probability of E, written $P(E)$, is computed as

$$P(E) = \frac{n(E)}{n(S)}$$

where $n(E)$ represents the number of elements in E, and $n(S)$ represents the number of elements in S.

WORTHY OF NOTE

Our study of probability will involve only those sample spaces with events that are equally likely.

A standard deck of playing cards consists of 52 cards divided in four groups or *suits*. There are 13 hearts (♥), 13 diamonds (♦), 13 spades (♠), and 13 clubs (♣). As you can see in Figure 9.59, each of the 13 cards in a suit is labeled A, 2, 3, 4, 5, 6, 7, 8, 9, 10, J, Q, and K. Also notice that 26 of the cards are red (hearts and diamonds), 26 are black (spades and clubs), and 12 of the cards are "face cards" (J, Q, K of each suit).

Figure 9.59

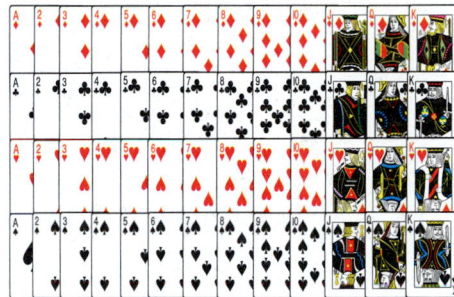

EXAMPLE 2 ▶ **Stating a Sample Space and the Probability of a Single Outcome**

A single card is drawn from a well-shuffled deck. Define S and state the probability of any single outcome. Then define E as *a King is drawn* and find $P(E)$.

Solution ▶ Sample space: $S = \{$the 52 cards$\}$. There are 52 equally likely outcomes, so the probability of any one outcome is $\frac{1}{52}$. Since S has four Kings,

$$P(E) = \frac{n(E)}{n(S)} = \frac{4}{52} \text{ or about } 0.077.$$

Now try Exercises 11 through 14 ▶

EXAMPLE 3 ▶ **Stating a Sample Space and the Probability of a Single Outcome**

A family of five has two girls and three boys named Sophie, Maria, Albert, Isaac, and Pythagoras. Their ages are 21, 19, 15, 13, and 9, respectively. One is to be selected randomly. Find the probability a teenager is chosen.

Solution ▶ The sample space is $S = \{9, 13, 15, 19, 21\}$. Three of the five are teenagers, meaning the probability is $\frac{3}{5}$, 0.6, or 60%.

✓ **B.** You've just seen how we can compute elementary probabilities

Now try Exercises 15 and 16 ▶

C. Properties of Probability

A study of probability necessarily includes recognizing some basic and fundamental properties. For example, when a fair die is rolled, what is $P(E)$ if E is defined as a *1, 2, 3, 4, 5, or 6* is rolled? The event E will occur 100% of the time, since 1, 2, 3, 4, 5, 6 are the only possibilities. In symbols we write $P(\text{outcome is in the sample space})$ or simply $P(S) = 1$ (100%).

What percent of the time will a result *not* in the sample space occur? Since the die has only the six sides numbered 1 through 6, the probability of rolling something else is zero. In symbols, $P(\text{outcome is not in sample space}) = 0$ or simply $P(\sim S) = 0$.

> **Properties of Probability**
>
> Given sample space S and any event E defined relative to S.
>
> **1.** $P(S) = 1$ **2.** $P(\sim S) = 0$ **3.** $0 \leq P(E) \leq 1$

EXAMPLE 4 ▶ **Determining the Probability of an Event**

A game is played using a spinner like the one shown. Determine the probability of the following events:

E_1: A nine is spun. E_2: An integer greater than 0 and less than 9 is spun.

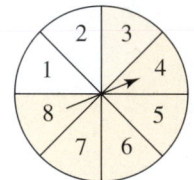

Solution ▶ The sample space consists of eight equally likely outcomes.

$$P(E_1) = \frac{0}{8} = 0 \qquad P(E_2) = \frac{8}{8} = 1.$$

Technically, E_1: A nine is spun is not an "event," since it is not in the sample space and cannot occur, while E_2 contains the entire sample space and must occur.

Now try Exercises 17 and 18 ▶

Because we know $P(S) = 1$ and all sample outcomes are equally likely, the probabilities of all single events defined on the sample space must sum to 1. For the experiment of rolling a fair die, the sample space has six outcomes that are equally likely. Note that $P(1) = P(2) = P(3) = P(4) = P(5) = P(6) = \frac{1}{6}$, and $\frac{1}{6} + \frac{1}{6} + \frac{1}{6} + \frac{1}{6} + \frac{1}{6} + \frac{1}{6} = 1$.

> **Probability and Sample Outcomes**
>
> Given a sample space S with n equally likely sample outcomes $s_1, s_2, s_3, \ldots, s_n$.
>
> $$\sum_{i=1}^{n} P(s_i) = P(s_1) + P(s_2) + P(s_3) + \cdots + P(s_n) = 1$$

The **complement** of an event E is the set of sample outcomes in S not contained in E. Symbolically, $\sim E$ is the complement of E.

> **Probability and Complementary Events**
>
> Given sample space S and any event E defined relative to S, the complement of E, written $\sim E$, is the set of all outcomes not in E and:
>
> **1.** $P(E) = 1 - P(\sim E)$ **2.** $P(E) + P(\sim E) = 1$

EXAMPLE 5 ▶ **Stating a Probability Using Complements**

Use complementary events to answer the following questions:

  a. A single card is drawn from a well-shuffled deck. What is the probability that it is not a diamond?

  b. A single letter is picked at random from the letters in the word "divisibility." What is the probability it is not an "i"?

Solution ▶ **a.** Since there are 13 diamonds in a standard 52-card deck:
$P(\sim D) = 1 - P(D) = 1 - \frac{13}{52} = \frac{39}{52} = 0.75$.

WORTHY OF NOTE

Probabilities can be written in fraction form, decimal form, or as a percent. For $P(\sim D)$ from Example 5(a), the probability could be written $\frac{3}{4}$, 0.75, or 75%.

  b. Of the 12 letters in d-i-v-i-s-i-b-i-l-i-t-y, 5 are "i's." This means $P(\sim i) = 1 - P(i)$, or $1 - \frac{5}{12} = \frac{7}{12}$. The probability of choosing a letter other than i is $0.58\overline{3}$.

Now try Exercises 19 through 22 ▶

EXAMPLE 6 ▶ **Stating a Probability Using Complements**

Inter-Island Waterways has just opened hydrofoil service between several islands. The hydrofoil is powered by two engines, one forward and one aft, and will operate if either of its two engines is functioning. Due to testing and past experience, the company knows the probability of the aft engine failing is $P(\text{aft engine fails}) = 0.05$, the probability of the forward engine failing is $P(\text{forward engine fails}) = 0.03$, and the probability that both fail is $P(\text{both engines simultaneously fail}) = 0.012$. What is the probability the hydrofoil completes its next trip?

Solution ▶ Although the answer may *seem* complicated, note that $P(\text{trip is completed})$ and $P(\text{both engines simultaneously fail})$ are complements.

$$P(\text{trip is completed}) = 1 - P(\text{both engines simultaneously fail})$$
$$= 1 - 0.012$$
$$= 0.988$$

There is close to a 99% probability the trip will be completed.

Now try Exercises 23 and 24 ▶

The chart in Figure 9.60 shows all 36 possible outcomes (the sample space) from the experiment of rolling two fair dice.

Figure 9.60

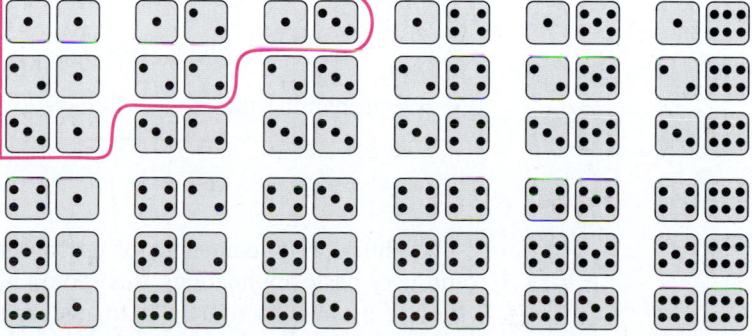

EXAMPLE 7 ▶ **Stating a Probability Using Complements**

Two fair dice are rolled. What is the probability the sum of both dice is greater than or equal to 5, $P(\text{sum} \geq 5)$?

Solution ▶   See Figure 9.60. For $P(\text{sum} \geq 5)$ it may be easier to use complements as there are far fewer possibilities: $P(\text{sum} \geq 5) = 1 - P(\text{sum} < 5)$, which gives

$$1 - \frac{6}{36} = 1 - \frac{1}{6} = \frac{5}{6} = 0.8\overline{3}.$$

☑ **C.** You've just seen how we can use certain properties of probability

Now try Exercises 25 and 26 ▶

D. Probability and Quick-Counting

Quick-counting techniques were introduced earlier to help count the number of elements in a large or more complex sample space, and the number of sample outcomes in an event.

EXAMPLE 8A ▶   **Stating a Probability Using Combinations**

Five cards are drawn from a shuffled 52-card deck. Calculate the probability of E_1:(*all five cards are face cards*) or E_2:(*all five cards are hearts*).

Solution ▶   The sample space for both events consists of all five-card groups that can be formed from the 52 cards or $_{52}C_5$. For E_1 we are to select five face cards from the 12 that are available (three from each suit), or $_{12}C_5$. The probability of five face cards is $\dfrac{n(E)}{n(S)} = \dfrac{_{12}C_5}{_{52}C_5}$, which gives $\dfrac{792}{2,598,960} \approx 0.0003$. For E_2 we are to select five hearts from the 13 available, or $_{13}C_5$. The probability of five hearts is $\dfrac{n(E)}{n(S)} = \dfrac{_{13}C_5}{_{52}C_5}$, which is $\dfrac{1287}{2,598,960} \approx 0.0005$.

WORTHY OF NOTE

It seems reasonable that the probability of 5 hearts is slightly higher, as 13 of the 52 cards are hearts, while only 12 are face cards.

EXAMPLE 8B ▶   **Stating a Probability Using Combinations and the Fundamental Principle of Counting**

Of the 42 seniors at Jacoby High School, 23 are female and 19 are male. A group of five students is to be selected at random to attend a conference in Reno, Nevada. What is the probability the group will have exactly three females?

Solution ▶   The sample space consists of all five-person groups that can be formed from the 42 seniors or $_{42}C_5$. The event consists of selecting 3 females from the 23 available $(_{23}C_3)$ and 2 males from the 19 available $(_{19}C_2)$. Using the fundamental principle of counting $n(E) = {_{23}C_3} \cdot {_{19}C_2}$ and the probability the group has 3 females is $\dfrac{n(E)}{n(S)} = \dfrac{_{23}C_3 \cdot {_{19}C_2}}{_{42}C_5}$, which gives $\dfrac{302,841}{850,668} \approx 0.356$. There is approximately a 35.6% probability the group will have exactly 3 females.

Now try Exercises 27 through 34 ▶

While the fundamentals of probability are usually introduced using dice, cards, and very basic applications, this should not take away from its true power and utility. We use probability to help us to analyze things quantitatively (with numbers) so that important decisions can be made. In manufacturing, the probability a product becomes defective during its warranty period is crucial to the company's financial stability. Health organizations rely heavily on probability to plan against the anticipated spread of a symptom through a given population. In these and many other situations, the sample spaces are very large and the defined events likewise large, making the use of technology an integral part of probability studies. The computed result of Example 8A is shown in Figure 9.61 (in scientific notation—shift the decimal four places to the left). The computed result of Example 8B appears in Figure 9.62.

Figure 9.61

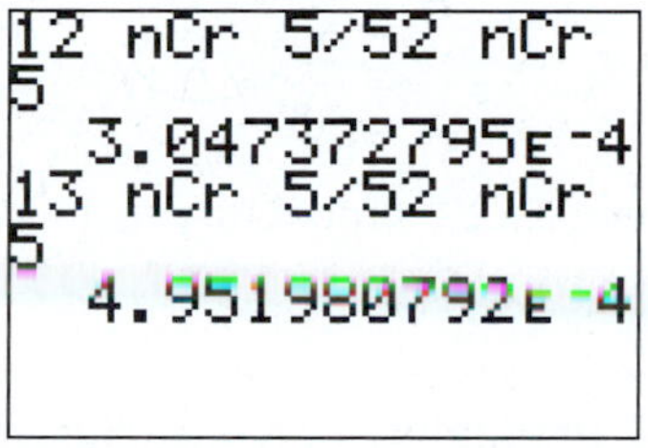

```
12 nCr 5/52 nCr
5
   3.047372795E-4
13 nCr 5/52 nCr
5
   4.951980792E-4
```

Figure 9.62

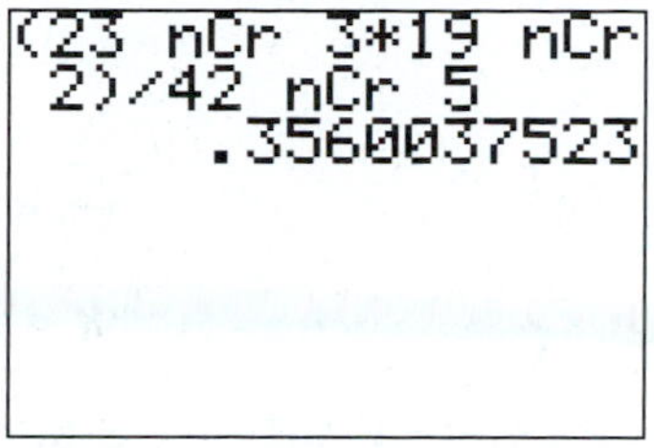

```
(23 nCr 3*19 nCr
2)/42 nCr 5
           .3560037523
```

☑ **D.** You've just seen how we can compute probabilities using quick-counting techniques

E. Probability and Nonexclusive Events

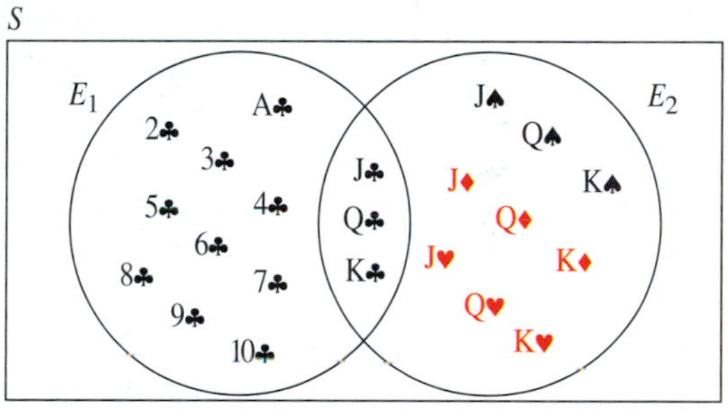

Sometimes the way events are defined causes them to share sample outcomes. Using a standard deck of playing cards once again, if we define the events E_1:(a club is drawn) and E_2:(a face card is drawn), they share the outcomes J♣, Q♣, and K♣ as shown in Figure 9.63. This overlapping region is the intersection of the events, or $E_1 \cap E_2$. If we compute $n(E_1 \cup E_2)$

Figure 9.63

as $n(E_1) + n(E_2)$ as before, this intersecting region gets counted *twice*!

In cases where the events are **nonexclusive** (not mutually exclusive), we maintain the correct count by subtracting one of the two intersections, obtaining $n(E_1 \cup E_2) = n(E_1) + n(E_2) - n(E_1 \cap E_2)$. This leads to the following calculation for the probability of nonexclusive events:

$$P(E_1 \cup E_2) = \frac{n(E_1) + n(E_2) - n(E_1 \cap E_2)}{n(S)} \quad \text{definition of probability}$$

$$= \frac{n(E_1)}{n(S)} + \frac{n(E_1)}{n(S)} - \frac{n(E_1 \cap E_2)}{n(S)} \quad \text{property of rational expressions}$$

$$= P(E_1) + P(E_2) - P(E_1 \cap E_2) \quad \text{definition of probability}$$

WORTHY OF NOTE

This can be verified by simply counting the elements involved: $n(E_1) = 13$ and $n(E_2) = 12$ so $n(E_1) + n(E_2) = 25$. However, there are only 22 possibilities in $E_1 \cup E_2$—the J♣, Q♣, and K♣ got counted twice.

> **Probability and Nonexclusive Events**
>
> Given sample space S and *nonexclusive events* E_1 and E_2 defined relative to S, the probability of E_1 *or* E_2 is given by
>
> $$P(E_1 \cup E_2) = P(E_1) + P(E_2) - P(E_1 \cap E_2)$$

EXAMPLE 9A ▶ **Stating the Probability of Nonexclusive Events**

What is the probability that a club or a face card is drawn from a standard deck of 52 well-shuffled cards?

Solution ▶ As before, define the events E_1:(a club is drawn) and E_2:(a face card is drawn). Since there are 13 clubs and 12 face cards, $P(E_1) = \frac{13}{52}$ and $P(E_2) = \frac{12}{52}$. But three of the face cards are clubs, so $P(E_1 \cap E_2) = \frac{3}{52}$. This leads to

$$P(E_1 \cup E_2) = P(E_1) + P(E_2) - P(E_1 \cap E_2) \quad \text{nonexclusive events}$$

$$= \frac{13}{52} + \frac{12}{52} - \frac{3}{52} \quad \text{substitute}$$

$$= \frac{22}{52} \approx 0.423 \quad \text{combine terms}$$

There is about a 42% probability that a club or face card is drawn.

EXAMPLE 9B ▶ **Stating the Probability of Nonexclusive Events**

A survey of 100 voters was taken to gather information on critical issues and the demographic information collected is shown in the table. One out of the 100 voters is to be drawn at random to be interviewed on the 5 P.M. News. What is the probability the person is a woman (W) or a Republican (R)?

	Women	Men	Totals
Republican	17	20	37
Democrat	22	17	39
Independent	8	7	15
Green Party	4	1	5
Tax Reform	2	2	4
Totals	53	47	100

Solution ▶ Since there are 53 women and 37 Republicans, $P(W) = 0.53$ and $P(R) = 0.37$. The table shows 17 people are both female and Republican so $P(W \cap R) = 0.17$.

$$P(W \cup R) = P(W) + P(R) - P(W \cap R) \quad \text{nonexclusive events}$$

$$= 0.53 + 0.37 - 0.17 \quad \text{substitute}$$

$$= 0.73 \quad \text{combine}$$

There is a 73% probability the person is a woman or a Republican.

Now try Exercises 35 through 48 ▶

Two events that have no common outcomes are called **mutually exclusive** events (one excludes the other and vice versa). For example, in rolling one die, E_1:(a 2 is rolled) and E_2:(an odd number is rolled) are mutually exclusive, since 2 is not an odd number. For the probability of E_3:(a 2 is rolled or an odd number is rolled), we note that $n(E_1 \cap E_2) = 0$ and the previous formula simply reduces to $P(E_1) + P(E_2)$. **See Exercises 49 and 50.** There is a large variety of additional applications in the Exercise Set. **See Exercises 53 through 68.**

When probability calculations require a repeated use of permutations or combinations, tables can make the work more efficient and help to explore the patterns they generate. For instance, to choose r children from a group of six children ($n = 6$), we can set **TBLSET** to **AUTO,** then press [Y=] and enter $6\,_nC_r\,X$ as Y_1 (Figure 9.64). Access the **TABLE** ([2nd] [GRAPH]) and note that the calculator has automatically computed the value of $_6C_0, _6C_1, _6C_2, \ldots, _6C_6$ (Figure 9.65) and the pattern of outputs is symmetric. For calculations like those required in Example 8B ($_{23}C_3 \cdot _{19}C_2$), we can enter $Y_1 = 23\,_nC_r\,X$, $Y_2 = 19\,_nC_r\,(5 - X)$, and $Y_3 = Y_1 \cdot Y_2$ to further explore the number of ways groups with different numbers of females can be chosen. Also **see Exercise 70.**

Figure 9.64

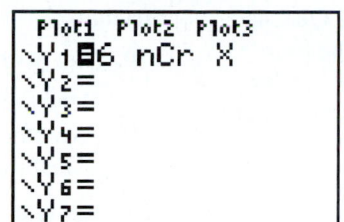

Figure 9.65

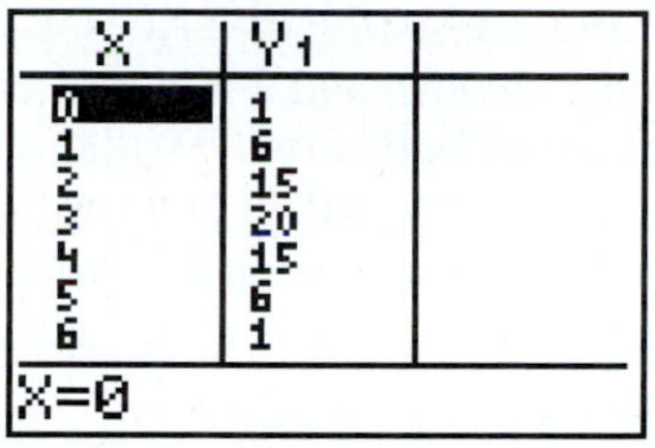

✅ **E.** You've just seen how we can compute probabilities involving nonexclusive events

9.6 EXERCISES

► CONCEPTS AND VOCABULARY

Fill in the blank with the appropriate word or phrase. Carefully reread the section if needed.

1. Given a sample space S and an event E defined relative to S, $P(E) = \dfrac{n(E)}{n(S)}$.

2. In elementary probability, we consider all events in the sample space to be ___equally___ likely.

3. Given a sample space S and an event E defined relative to S: __0__ $\leq P(E) \leq$ __1__, $P(S) =$ __1__, and $P(\sim S) =$ __0__.

4. The ___complement___ of an event E is the set of sample outcomes in S which are not contained in E.

5. Discuss/Explain the difference between mutually exclusive events and nonexclusive events. Give an example of each.   Answers will vary.

6. A single die is rolled. With no calculations, explain why the probability of rolling an even number is greater than rolling a number greater than four. Answers will vary.

► DEVELOPING YOUR SKILLS

State the sample space S and the probability of a single outcome. Then define any two events E relative to S (many answers possible).

Exercise 8

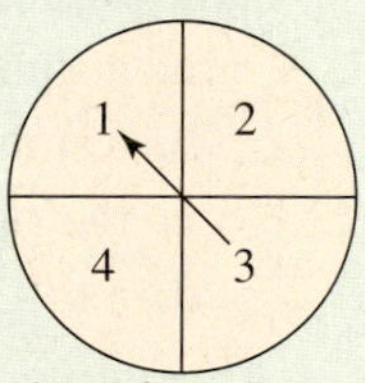

7. Two fair coins (heads and tails) are flipped. $S = \{HH, HT, TH, TT\}, \frac{1}{4}$

8. The simple spinner shown is spun. $S = \{1, 2, 3, 4,\}, \frac{1}{4}$

9. The head coaches for six little league teams (the Patriots, Cougars, Angels, Sharks, Eagles, and Stars) have gathered to discuss new changes in the rule book. One of them is randomly chosen to ask the first question.
$S = \{$coach of Patriots, Cougars, Angels, Sharks, Eagles, Stars$\}, \frac{1}{6}$

10. Experts on the planets Mercury, Venus, Mars, Jupiter, Saturn, Uranus, Neptune, and the Kuiper object formerly known as the planet Pluto have gathered at a space exploration conference. One group of experts is selected at random to speak first. $S = \{$experts for Mercury, Venus, Mars, Jupiter, Saturn, Uranus, Neptune, Pluto$\}, \frac{1}{8}$

Find $P(E)$ for the events defined.

11. Nine index cards numbered 1 through 9 are shuffled and placed in an envelope, then one of the cards is randomly drawn. Define event E as *the number drawn is even.* $P(E) = \frac{4}{9}$

12. Eight flash cards used for studying basic geometric shapes are shuffled and one of the cards is drawn at random. The eight cards include information on circles, squares, rectangles, kites, trapezoids, parallelograms, pentagons, and triangles. Define event E as *a quadrilateral is drawn.* $P(\text{quad}) = \frac{5}{8}$

13. One card is drawn at random from a standard deck of 52 cards. What is the probability of
    a. drawing a Jack $\frac{1}{13}$
    b. drawing a spade $\frac{1}{4}$
    c. drawing a black card $\frac{1}{2}$
    d. drawing a red three $\frac{1}{26}$

HOMEWORK SELECTION GUIDE

Core: 7–63 odd, 69 (30 Exercises)
Standard: 1–4, 7–65 odd, 69, 72, 73 (37 Exercises)

Extended: 1–4, 7–51 odd, 52, 53–69 odd, 70–73 (41 Exercises)
In Depth: 1–6, 7–51 odd, 52, 53–57 odd, 58, 59–69 odd, 70–74 (45 Exercises)

14. Pinochle is a card game played with a deck of 48 cards consisting of 2 Aces, 2 Kings, 2 Queens, 2 Jacks, 2 Tens, and 2 Nines in each of the four standard suits [hearts (♥), diamonds (♦), spades (♠), and clubs (♣)]. If one card is drawn at random from this deck, what is the probability of

a. drawing an Ace $\frac{1}{6}$ **b.** drawing a club $\frac{1}{4}$

c. drawing a red card $\frac{1}{2}$

d. drawing a face card (Jack, Queen, King) $\frac{1}{2}$

15. A group of finalists on a game show consists of three males and five females. Hank has a score of 520 points, with Harry and Hester having 490 and 475 points, respectively. Madeline has 532 points, with Mackenzie, Morgan, Maggie, and Melanie having 495, 480, 472, and 470 points, respectively. One of the contestants is randomly selected to start the final round. Define E_1 as *Hester is chosen,* E_2 as *a female is chosen,* and E_3 as *a contestant with fewer than 500 points is chosen.* Find the probability of each event. $P(E_1) = \frac{1}{8}, P(E_2) = \frac{5}{8}, P(E_3) = \frac{3}{4}$

16. Soccer coach Maddox needs to fill the last spot on his starting roster for the opening day of the season and has to choose between three forwards and five defenders. The forwards have jersey numbers 5, 12, and 17, while the defenders have jersey numbers 7, 10, 11, 14, and 18. Define E_1 as *a forward is chosen,* E_2 as *a defender is chosen,* and E_3 as *a player whose jersey number is greater than 10 is chosen.* Find the probability of each event. $P(E_1) = \frac{3}{8}, P(E_2) = \frac{5}{8}, P(E_3) = \frac{5}{8}$

17. A game is played using a spinner like the one shown. For each spin,

a. What is the probability the arrow lands in a shaded region? $\frac{3}{4}$

b. What is the probability your spin is less than 5? 1

c. What is the probability you spin a 2? $\frac{1}{4}$

d. What is the probability the arrow points to prime number? $\frac{1}{2}$

18. A game is played using a spinner like the one shown here. For each spin,

a. What is the probability the arrow lands in a lightly shaded region? $\frac{1}{3}$

b. What is the probability your spin is greater than 2? $\frac{2}{3}$

c. What is the probability the arrow lands in a shaded region? $\frac{5}{6}$

d. What is the probability you spin a 5? $\frac{1}{6}$

Use the complementary events to complete Exercises 19 through 22.

19. One card is drawn from a standard deck of 52. What is the probability it is not a club? $\frac{3}{4}$

20. Four standard dice are rolled. What is the probability the sum is less than 24? $\frac{1295}{1296}$

21. A single digit is randomly selected from among the digits of 10!. What is the probability the digit is not a 2? $\frac{6}{7}$

22. A corporation will be moving its offices to Los Angeles, Miami, Atlanta, Dallas, or Phoenix. If the site is randomly selected, what is the probability Dallas is not chosen? $\frac{4}{5}$

23. A large manufacturing plant can remain at full production as long as one of its two generators is functioning. Due to past experience and the age difference between the systems, the plant manager estimates the probability of the main generator failing is 0.05, the probability of the secondary generator failing is 0.01, and the probability of both failing is 0.009. What is the probability the plant remains in full production today? 0.991

24. A fire station gets an emergency call from a shopping mall in the mid-afternoon. From a study of traffic patterns, Chief Nozawa knows the probability the most direct route is clogged with traffic is 0.07, while the probability of the secondary route being clogged is 0.05. The probability both are clogged is 0.02. What is the probability they can respond to the call unimpeded using one of these routes? 0.98

25. Two fair dice are rolled (see Figure 9.60). What is the probability of

a. a sum less than four $\frac{1}{12}$

b. a sum less than eleven $\frac{11}{12}$

c. the sum is not nine $\frac{8}{9}$

d. a roll is not a "double" (both dice the same) $\frac{5}{6}$

"Double-six" dominos is a game played with the 28 numbered tiles shown in the diagram.

26. The 28 dominos are placed in a bag, shuffled, and then one domino is randomly drawn. What is the probability the total number of dots on the domino

a. is three or less $\frac{3}{14}$

b. is greater than three $\frac{11}{14}$

c. does not have a blank half $\frac{3}{4}$

d. is not a "double" (both sides the same) $\frac{3}{4}$

Find $P(E)$ given the values for $n(E)$ and $n(S)$ shown.

27. $n(E) = {}_6C_3 \cdot {}_4C_2$; $n(S) = {}_{10}C_5$ $\frac{10}{21}$

28. $n(E) = {}_{12}C_9 \cdot {}_8C_7$; $n(S) = {}_{20}C_{16}$ $\frac{352}{969}$

29. $n(E) = {}_9C_6 \cdot {}_5C_3$; $n(S) = {}_{14}C_9$ $\frac{60}{143}$

30. $n(E) = {}_7C_6 \cdot {}_3C_2$; $n(S) = {}_{10}C_8$ $\frac{7}{15}$

31. Five cards are drawn from a well-shuffled, standard deck of 52 cards. Which has the greater probability: (a) all five cards are red or (b) all five cards are numbered cards? How much greater?   b, about 12%

32. Five cards are drawn from a well-shuffled pinochle deck of 48 cards (see Exercise 14). Which has the greater probability, (a) all five cards are face cards (King, Queen, or Jack) or (b) all five cards are black? How much greater?   probabilities are equal, 0

33. A dietetics class has 24 students. Of these, 9 are vegetarians and 15 are not. The instructor receives enough funding to send six students to a conference. If the students are selected randomly, what is the probability the group will have

    a. exactly two vegetarians   0.3651

    b. exactly four nonvegetarians   0.3651

    c. at least three vegetarians   0.3969

34. A large law firm has a support staff of 15 employees: six paralegals and nine legal assistants. Due to recent changes in the law, the firm wants to send five of them to a forum on the new changes. If the selection is done randomly, what is the probability the group will have

    a. exactly three paralegals   0.2398

    b. exactly two legal assistants   0.2398

    c. at least two paralegals   0.7063

Find the probability indicated using the information given.

35. Given $P(E_1) = 0.7$, $P(E_2) = 0.5$, and $P(E_1 \cap E_2) = 0.3$, compute $P(E_1 \cup E_2)$.   0.9

36. Given $P(E_1) = 0.6$, $P(E_2) = 0.3$, and $P(E_1 \cap E_2) = 0.2$, compute $P(E_1 \cup E_2)$.   0.7

37. Given $P(E_1) = \frac{3}{8}$, $P(E_2) = \frac{3}{4}$, and $P(E_1 \cup E_2) = \frac{5}{6}$; compute $P(E_1 \cap E_2)$. $\frac{7}{24}$

38. Given $P(E_1) = \frac{1}{2}$, $P(E_2) = \frac{3}{5}$, and $P(E_1 \cup E_2) = \frac{17}{20}$; compute $P(E_1 \cap E_2)$. $\frac{1}{4}$

39. Given $P(E_1 \cup E_2) = 0.72$, $P(E_2) = 0.56$, and $P(E_1 \cap E_2) = 0.43$; compute $P(E_1)$.   0.59

40. Given $P(E_1 \cup E_2) = 0.85$, $P(E_1) = 0.4$, and $P(E_1 \cap E_2) = 0.21$; compute $P(E_2)$.   0.66

41. Two fair dice are rolled. What is the probability the sum of the dice is

    a. a multiple of 3 and an odd number $\frac{1}{6}$

    b. a sum greater than 5 and a 3 on one die $\frac{7}{36}$

    c. an even number and a number greater than 9 $\frac{1}{9}$

    d. an odd number and a number less than 10 $\frac{4}{9}$

42. *Eight Ball* is a game played on a pool table with 15 balls numbered 1 through 15 and a cue ball that is solid white. Of the 15 numbered balls, 8 are a solid (nonwhite) color and numbered 1 through 8, and seven are striped balls numbered 9 through 15. The fifteen numbered pool balls (no cueball) are placed in a large bowl and mixed, then one is drawn out. What is the probability of drawing

    a. the eight ball $\frac{1}{15}$

    b. a number greater than fifteen   0

    c. an even number $\frac{7}{15}$

    d. a multiple of three $\frac{1}{3}$

    e. a solid color and an even number $\frac{4}{15}$

    f. a striped ball and an odd number $\frac{4}{15}$

    g. an even number and a number divisible by 3 $\frac{2}{15}$

    h. an odd number and a number divisible by 4   0

43. A survey of 50 veterans was taken to gather information on their service career and what life is like out of the military. A breakdown of those surveyed is shown in the table. One out of the 50 will be selected at random for an interview and a biographical sketch. What is the probability the person chosen is

	Women	Men	Totals
Private	6	9	15
Corporal	10	8	18
Sergeant	4	5	9
Lieutenant	2	1	3
Captain	2	3	5
Totals	24	26	50

a. a woman and a sergeant $\frac{2}{25}$

b. a man and a private $\frac{9}{50}$

c. a private and a sergeant 0

d. a woman and an officer $\frac{2}{25}$

e. a person in the military 1

44. Referring to Exercise 43, what is the probability the person chosen is

a. a woman or a sergeant $\frac{29}{50}$

b. a man or a private $\frac{16}{25}$

c. a woman or a man 1

d. a woman or an officer $\frac{14}{25}$

e. a captain or a lieutenant $\frac{4}{25}$

A computer is asked to randomly generate a three-digit number. What is the probability the

45. ten's digit is odd or the one's digit is even $\frac{3}{4}$

46. first digit is prime and the number is a multiple of 10 $\frac{2}{45}$

A computer is asked to randomly generate a four-digit number. What is the probability the number is

47. at least 4000 or a multiple of 5 $\frac{11}{15}$

48. less than 7000 and an odd number $\frac{1}{3}$

49. Two fair dice are rolled. What is the probability of

a. boxcars (sum of 12) or snake eyes (sum of 2) $\frac{1}{18}$

b. a sum of 7 or a sum of 11 $\frac{2}{9}$

c. an even-numbered sum or a prime sum $\frac{8}{9}$

d. an odd-numbered sum or a sum that is a multiple of 4 $\frac{3}{4}$

e. a sum of 15 or a multiple of 12 $\frac{1}{36}$

f. a sum that is a prime number $\frac{5}{12}$

50. Suppose all 16 balls from a game of pool (see Exercise 42) are placed in a large leather bag and mixed, then one is drawn out. Consider the cue ball as "0." What is the probability of drawing

a. a striped ball $\frac{7}{16}$

b. a solid-colored ball $\frac{9}{16}$

c. a polka-dotted ball 0

d. the cue ball $\frac{1}{16}$

e. the cue ball or the eight ball $\frac{1}{8}$

f. a striped ball or a number less than five $\frac{3}{4}$

g. a solid color or a number greater than 12 $\frac{3}{4}$

h. an odd number or a number divisible by 4 $\frac{3}{4}$

▶ WORKING WITH FORMULAS

51. **Games involving a fair spinner (with numbers 1 through 4): $P(n) = \left(\frac{1}{4}\right)^n$**

Games that involve moving pieces around a board using a fair spinner are fairly common. If a fair spinner has the numbers 1 through 4, the probability that any one number is spun n times in succession is given by the formula shown, where n represents the number of spins. What is the probability (a) the first player spins a two? (b) all four players spin a two? (c) Discuss the graph of $P(n)$ and explain the connection between the graph and the probability of consistently spinning a two. $\frac{1}{4}$; $\frac{1}{256}$; answers will vary.

52. **Games involving a fair coin (heads and tails): $P(n) = \left(\frac{1}{2}\right)^n$**

When a fair coin is flipped, the probability that heads (or tails) is flipped n times in a row is given by the formula shown, where n represents the number of flips. What is the probability (a) the first flip is heads? (b) the first four flips are heads? (c) Discuss the graph of $P(n)$ and explain the connection between the graph and the probability of consistently flipping heads. $\frac{1}{2}$; $\frac{1}{16}$; answers will vary.

▶ APPLICATIONS

53. To improve customer service, a company tracks the number of minutes a caller is "on hold" and waiting for a customer service representative. The table shows the probability that a caller will wait m minutes. Based on the table, what is the probability a caller waits

a. at least 2 minutes   0.33

b. less than 2 minutes   0.67

c. 4 minutes or less   1

d. over 4 minutes   0

e. less than 2 or more than 4 minutes   0.67

f. 3 or more minutes   0.08

Wait Time (minutes m)	Probability
0	0.07
$0 < m < 1$	0.28
$1 \leq m < 2$	0.32
$2 \leq m < 3$	0.25
$3 \leq m < 4$	0.08

54. To study the impact of technology on American families, a researcher first determines the probability that a family has *n* computers at home. Based on the table, what is the probability a home

   a. has at least one computer   91%

   b. has two or more computers   40%

   c. has fewer than four computers   97%

   d. has five computers   0%

   e. has one, two, or three computers   88%

   f. does not have two computers   72%

Number of Computers	Probability
0	9%
1	51%
2	28%
3	9%
4	3%

55. Jolene is an experienced markswoman and is able to hit a 10 in. by 20 in. target 100% of the time at a range of 100 yd. Assuming the probability she hits a target is related to its area, what is the probability she hits the shaded portions shown?

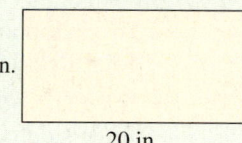

10 in.   20 in.

   a. isosceles triangle $\frac{1}{2}$ **b.** right triangle $\frac{1}{2}$

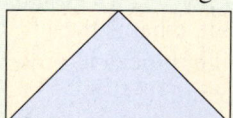 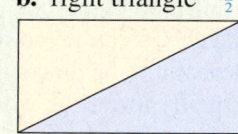

   c. isosceles right triangle $\frac{1}{8}$

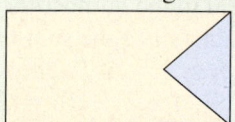

56. a. square $\frac{1}{2}$ **b.** circle $\frac{\pi}{8}$

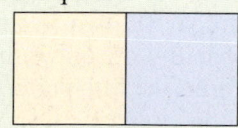 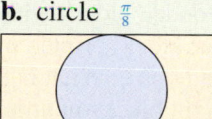

   c. isosceles trapezoid with $b = \frac{B}{2}$ $\frac{3}{4}$

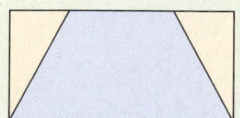

57. A circular dartboard has a total radius of 8 in., with circular bands that are 2 in. wide, as shown. You are skilled enough to hit this board 100% of the time so you always score at least two points each time you throw a dart. Assuming the probabilities are related to area, on the next dart that you throw what is the probability you

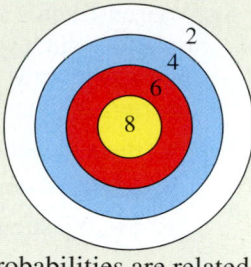

   a. score at least a 4? $\frac{9}{16}$ **b.** score at least a 6? $\frac{1}{4}$

   c. hit the bull's-eye? $\frac{1}{16}$ **d.** score exactly 4 points? $\frac{5}{16}$

58. Three red balls, six blue balls, and four white balls are placed in a bag. What is the probability the first ball you draw out is

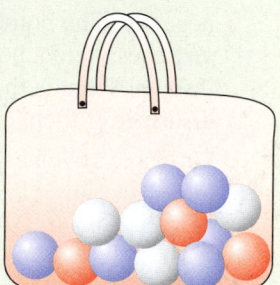

   a. red $\frac{3}{13}$ **b.** blue $\frac{6}{13}$

   c. not white $\frac{9}{13}$

   d. purple 0

   e. red or white $\frac{7}{13}$

   f. red and white 0

59. Three red balls, six blue balls, and four white balls are placed in a bag, then two are drawn out and placed in a rack. What is the probability the balls drawn are

   a. first red, second blue $\frac{3}{26}$

   b. first blue, second red $\frac{3}{26}$

   c. both white $\frac{1}{13}$

   d. first blue, second not red $\frac{9}{26}$

   e. first white, second not blue $\frac{2}{13}$

   f. first not red, second not blue $\frac{11}{26}$

60. Consider the 210 discrete points found in the first and second quadrants where $-10 \leq x \leq 10$, $1 \leq y \leq 10$, and x and y are integers. The coordinates of each point is written on a slip of paper and placed in a box. One of the slips is then randomly drawn. What is the probability the point (x, y) drawn

   a. is on the graph of $y = |x|$ $\frac{2}{21}$

   b. is on the graph of $y = 2|x|$ $\frac{1}{21}$

   c. is on the graph of $y = 0.5|x|$ $\frac{1}{21}$

   d. has coordinates $(x, y > -2)$ 1

   e. has coordinates $(x \leq 5, y > -2)$ $\frac{16}{21}$

   f. is between the branches of $y = x^2$ $\frac{1}{5}$

61. Your instructor surprises you with a True/False quiz for which you are totally unprepared and must guess randomly. What is the probability you pass the quiz with an 80% or better if there are

   a. three questions $\frac{1}{8}$

   b. four questions $\frac{1}{16}$

   c. five questions $\frac{3}{16}$

62. A robot is sent out to disarm a timed explosive device by randomly changing some switches from a neutral position to a *positive flow* or *negative flow* position. The problem is, the switches are independent and unmarked, and it is unknown which direction is positive and which direction is negative. The bomb is harmless if a majority of the switches yield a positive flow. All switches must be thrown. What is the probability the device is disarmed if there are

   a. three switches $\frac{1}{2}$

   b. four switches $\frac{5}{16}$

   c. five switches $\frac{1}{2}$

63. A survey of 100 retirees was taken to gather information concerning how they viewed the Vietnam War back in the early 1970s. A breakdown of those surveyed is shown in the table. One out of the hundred will be selected at random for a personal, taped interview. What is the probability the person chosen had a

   a. career of any kind and opposed the war $\frac{47}{100}$

   b. medical career and supported the war $\frac{2}{25}$

   c. military career and opposed the war $\frac{3}{100}$

   d. legal or business career and opposed the war $\frac{9}{50}$

   e. academic or medical career and supported the war $\frac{11}{100}$

Career	Support	Opposed	Total
Military	9	3	12
Medical	8	16	24
Legal	15	12	27
Business	18	6	24
Academics	3	10	13
Totals	53	47	100

64. Referring to Exercise 63, what is the probability the person chosen

   a. had a career of any kind or opposed the war 1

   b. had a medical career or supported the war $\frac{69}{100}$

   c. supported the war or had a military career $\frac{14}{25}$

   d. had a medical or a legal career $\frac{51}{100}$

   e. supported or opposed the war 1

65. The Board of Directors for a large hospital has 15 members. There are six doctors of nephrology (kidneys), five doctors of gastroenterology (stomach and intestines), and four doctors of endocrinology (hormones and glands). Eight of them will be selected to visit the nation's premier hospitals on a 3-week, expenses-paid tour. What is the probability the group of eight selected consists of exactly

   a. four nephrologists and four gastroenterologists $\frac{5}{429}$

   b. three endocrinologists and five nephrologists $\frac{8}{2145}$

66. A support group for hodophobics (an irrational fear of travel) has 32 members. There are 15 aviophobics (fear of air travel), eight siderodrophobics (fear of train travel), and nine thalassophobics (fear of ocean travel) in the group. Twelve of them will be randomly selected to participate in a new therapy. What is the probability the group of 12 selected consists of exactly

   a. two aviophobics, six siderodrophobics, and four thalassophobics ≈ 0.002

   b. five thalassophobics, four aviophobics, and three siderodrophobics ≈ 0.043

67. A trained chimpanzee is given a box containing eight wooden cubes with the letters p, a, r, a, l, l, e, l printed on them (one letter per block). Assuming the chimp can't read or spell, what is the probability he draws the eight blocks in order and actually forms the word "parallel"? $\frac{1}{3360}$

68. A number is called a "perfect number" if the sum of its proper factors is equal to the number itself. Six is the first perfect number since the sum of its proper factors is six: $1 + 2 + 3 = 6$. Twenty-eight is the second since: $1 + 2 + 4 + 7 + 14 = 28$. A young child is given a box containing eight wooden blocks with the following numbers (one per block) printed on them: four 3's, two 5's, one 0, and one 6. What is the probability she draws the eight blocks in order and forms the fifth perfect number: 33,550,336? $\frac{1}{840}$

▶ **EXTENDING THE CONCEPT**

69. The function $f(x) = (\frac{1}{2})^x$ gives the probability that x number of flips will all result in heads (or tails). Compute the probability that 20 flips results in *20 heads in a row,* then use the Internet or some other resource to find the probability of winning a state lottery. Which is more likely to happen (which has the greater probability)? Were you surprised?
$\frac{1}{1,048,576}$; answers will vary; 20 heads in a row.

 70. Recall that a function is a relation in which each element of the domain is paired with only one element from the range. Is the relation defined by $C(x) = {}_nC_x$ (n is a constant) a function? To

investigate, plot the points generated by $C(x) = {}_8C_x$ for $x = 0$ to $x = 8$ using a "friendly" window ($x \in [0, 9.4]$, $y \in [0, 93]$) and answer the following questions:

  a. Is the resulting graph continuous or discrete (made up of distinct points)?   discrete

  b. Does the resulting graph pass the vertical line test?   yes

  c. Discuss the features of the relation and its graph, including the domain, range, maximum or minimum values, and symmetries observed. Answers will vary.

▶ **MAINTAINING YOUR SKILLS**

71. (7.1) Solve the system using matrices and row reduction: $\begin{cases} x - 2y + 3z = 10 \\ 2x + y - z = 18 \\ 3x - 2y + z = 26 \end{cases}$ (9, 1, 1)

72. (5.4) Complete the following logarithmic properties:

$\log_b b = \underline{\ 1\ }$ $\log_b 1 = \underline{\ 0\ }$

$\log_b b^n = \underline{\ n\ }$ $b^{\log_b n} = \underline{\ n\ }$

73. (4.6) Solve the inequality by graphing the function and labeling the appropriate interval(s):

$\dfrac{x^2 - 1}{x} \geq 0.$ $x \in [-1, 0) \cup [1, \infty)$

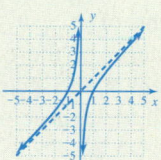

74. (9.3) A rubber ball is dropped from a height of 25 ft onto a hard surface. With each bounce, it rebounds 60% of the height from which it last fell. Use sequences/series to find (a) the height of the sixth bounce, (b) the total distance traveled up to the sixth bounce, and (c) the distance the ball will travel before coming to rest.
a. ≈ 1.17 ft   **b.** ≈ 94.17 ft   **c.** 100 ft

9.7   The Binomial Theorem

LEARNING OBJECTIVES

In Section 9.7 you will see how we can:

☐ **A.** Use Pascal's triangle to find $(a + b)^n$

☐ **B.** Find binomial coefficients using $\binom{n}{k}$ notation

☐ **C.** Use the binomial theorem to find $(a + b)^n$

☐ **D.** Find a specific term of a binomial expansion

☐ **E.** Solve applications of binomial powers

Strictly speaking, a binomial is a polynomial with two terms. This limits us to terms with real number coefficients and whole number powers on variables. In this section, we will loosely regard a binomial as the sum or difference of *any* two terms. Hence

$3x^2 - y^4$, $\sqrt{x} + 4$, $x + \dfrac{1}{x}$, and $-\dfrac{1}{2} + \dfrac{\sqrt{3}}{2}i$ are all "binomials." Our goal is to develop

an ability to raise a binomial to any natural number power, with the results having important applications in genetics, probability, polynomial theory, and other areas. The tool used for this purpose is called the *binomial theorem.*

A. Binomial Powers and Pascal's Triangle

Much of our mathematical understanding comes from a study of patterns. One area where the study of patterns has been particularly fruitful is **Pascal's triangle** (Figure 9.66), named after the French scientist Blaise Pascal (although the triangle was well known before his time). It begins with a "1" at the vertex of the triangle, with 1's extending diagonally downward to the left and right as shown. The entries on the interior of the triangle are found by adding the two entries directly above and to the left and right of each new position.

Figure 9.66

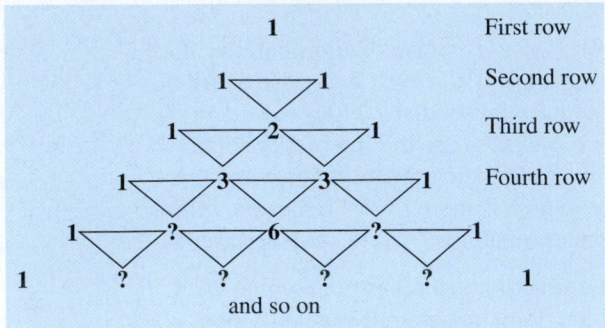

There are a variety of patterns hidden within the triangle. In this section, we'll use the *horizontal rows* of the triangle to help us raise a binomial to various powers. To begin, recall that $(a + b)^0 = 1$ and $(a + b)^1 = 1a + 1b$ (unit coefficients are included for emphasis). In our earlier work, we saw that a binomial square (a binomial raised to the second power) always followed the pattern $(a + b)^2 = 1a^2 + 2ab + 1b^2$. Observe the overall pattern that is developing as we include $(a + b)^3$:

$$
\begin{array}{lcr}
(a + b)^0 & 1 & \text{row 1} \\
(a + b)^1 & 1a + 1b & \text{row 2} \\
(a + b)^2 & 1a^2 + 2ab + 1b^2 & \text{row 3} \\
(a + b)^3 & 1a^3 + 3a^2b + 3ab^2 + 1b^3 & \text{row 4}
\end{array}
$$

Apparently the coefficients of $(a + b)^n$ will occur in row $n + 1$ of Pascal's triangle. Also observe that in each term of the expansion, the exponent of the first term a *decreases by 1* as the exponent on the second term b *increases by 1*, keeping the degree of each term constant (recall the degree of a term with more than one variable is the sum of the exponents).

$$1a^3b^0 + 3a^2b^1 + 3a^1b^2 + 1a^0b^3$$

| $3 + 0$ | $2 + 1$ | $1 + 2$ | $0 + 3$ |
| degree 3 | degree 3 | degree 3 | degree 3 |

These observations help us to quickly expand a binomial power.

EXAMPLE 1 ▶ **Expanding a Binomial Using Pascal's Triangle**

Use Pascal's triangle and the patterns noted to expand $(x + \frac{1}{2})^4$.

Solution ▶ Working step-by-step we have

1. The coefficients will be in the fifth row of Pascal's triangle.

$$1 \quad\quad 4 \quad\quad 6 \quad\quad 4 \quad\quad 1$$

2. The exponents on x begin at 4 and *decrease,* while the exponents on $\frac{1}{2}$ begin at 0 and *increase.*

$$1x^4\left(\frac{1}{2}\right)^0 + 4x^3\left(\frac{1}{2}\right)^1 + 6x^2\left(\frac{1}{2}\right)^2 + 4x^1\left(\frac{1}{2}\right)^3 + 1x^0\left(\frac{1}{2}\right)^4$$

3. Simplify each term.

The result is $x^4 + 2x^3 + \frac{3}{2}x^2 + \frac{1}{2}x + \frac{1}{16}$.

Now try Exercises 7 through 10 ▶

If the exercise involves a difference rather than a sum, we simply rewrite the expression using algebraic addition and proceed as before.

EXAMPLE 2 ▶ **Raising a Complex Number to a Power Using Pascal's Triangle**

Use Pascal's triangle and the patterns noted to compute $(3 - 2i)^5$.

Solution ▶ Begin by rewriting $(3 - 2i)^5$ as $[3 + (-2i)]^5$.

1. The coefficients will be in the sixth row of Pascal's triangle.

$$1 \qquad 5 \qquad 10 \qquad 10 \qquad 5 \qquad 1$$

2. The exponents on 3 begin at 5 and *decrease,* while the exponents on $(-2i)$ begin at 0 and *increase.*

$$\mathbf{1}(3^5)(-2i)^0 + \mathbf{5}(3^4)(-2i)^1 + \mathbf{10}(3^3)(-2i)^2 + \mathbf{10}(3^2)(-2i)^3 + \mathbf{5}(3^1)(-2i)^4 + \mathbf{1}(3^0)(-2i)^5$$

3. Simplify each term.

$$243 - 810i - 1080 + 720i + 240 - 32i$$

The result is $-597 - 122i$.

Now try Exercises 11 and 12 ▶

Expanding Binomial Powers $(a + b)^n$

1. The coefficients will be in row $n + 1$ of Pascal's triangle.
2. The exponents on *the first term* begin at n and *decrease,* while the exponents on *the second term* begin at 0 and *increase.*
3. For any binomial difference $(a - b)^n$, rewrite the base as $[a + (-b)]^n$ using algebraic addition and proceed as before, then simplify each term.

☑ **A.** You've just seen how we can use Pascal's triangle to find $(a + b)^n$

B. Binomial Coefficients and Factorials

Pascal's triangle can easily be used to find the coefficients of $(a + b)^n$, as long as the exponent is relatively small. If we needed to expand $(a + b)^{25}$, writing out the first 26 rows of the triangle would be rather tedious. To overcome this limitation, we introduce a *formula* that enables us to find the coefficients of any binomial expansion.

The Binomial Coefficients

For natural numbers n and r where $n \geq r$, the expression $\binom{n}{r}$, read "n choose r," is called the **binomial coefficient** and evaluated as:

$$\binom{n}{r} = \frac{n!}{r!(n - r)!}$$

Notice the formula for determining binomial coefficients is identical to that for $_nC_r$, and it turns out the coefficients are actually found using a combination, with the new notation used primarily as a convenience. In Example 1, we found the coefficients of $(a + b)^4$ using the fifth or $(n + 1)$st row of Pascal's triangle. In Example 3, these coefficients are found using the formula for binomial coefficients.

EXAMPLE 3 ▶ **Computing Binomial Coefficients**

Evaluate $\binom{n}{r} = \dfrac{n!}{r!(n - r)!}$ as indicated:

a. $\binom{4}{1}$ **b.** $\binom{4}{2}$ **c.** $\binom{4}{3}$

Solution ▶ **a.** $\dbinom{4}{1} = \dfrac{4!}{1!(4-1)!} = \dfrac{4 \cdot 3!}{1!3!} = 4$

b. $\dbinom{4}{2} = \dfrac{4!}{2!(4-2)!} = \dfrac{4 \cdot 3 \cdot 2!}{2!2!} = \dfrac{4 \cdot 3}{2} = 6$

c. $\dbinom{4}{3} = \dfrac{4!}{3!(4-3)!} = \dfrac{4 \cdot 3!}{3!1!} = 4$

Now try Exercises 13 through 20 ▶

Note $\dbinom{4}{1} = 4$, $\dbinom{4}{2} = 6$, and $\dbinom{4}{3} = 4$ give the *interior entries* in the fifth row of Pascal's triangle: 1 4 6 4 1. For consistency and symmetry, we define $0! = 1$, which enables the formula to generate all entries of the triangle, including the "1's."

$\dbinom{4}{0} = \dfrac{4!}{0!(4-0)!}$ *apply formula* $\qquad \dbinom{4}{4} = \dfrac{4!}{4!(4-4)!} = \dfrac{4!}{4! \cdot 0!}$ *apply formula*

$= \dfrac{4!}{1 \cdot 4!} = 1$ $0! = 1$ $\qquad\qquad\qquad = \dfrac{4!}{4! \cdot 1} = 1$ $0! = 1$

The formula for $\dbinom{n}{r}$ with $0 \le r \le n$ now gives all coefficients in the $(n+1)$st row. For $n = 5$, we have

$$\dbinom{5}{0} \qquad \dbinom{5}{1} \qquad \dbinom{5}{2} \qquad \dbinom{5}{3} \qquad \dbinom{5}{4} \qquad \dbinom{5}{5}$$
$$1 \qquad\quad 5 \qquad\quad 10 \qquad\quad 10 \qquad\quad 5 \qquad\quad 1$$

EXAMPLE 4 ▶ **Computing Binomial Coefficients**

Compute the binomial coefficients:

a. $\dbinom{9}{0}$ **b.** $\dbinom{9}{1}$ **c.** $\dbinom{6}{5}$ **d.** $\dbinom{6}{6}$

Solution ▶ **a.** $\dbinom{9}{0} = \dfrac{9!}{0!(9-0)!}$ **b.** $\dbinom{9}{1} = \dfrac{9!}{1!(9-1)!}$

$\qquad = \dfrac{9!}{9!} = 1$ $\qquad\qquad\qquad = \dfrac{9!}{8!} = 9$

c. $\dbinom{6}{5} = \dfrac{6!}{5!(6-5)!}$ **d.** $\dbinom{6}{6} = \dfrac{6!}{6!(6-6)!}$

$\qquad = \dfrac{6!}{5!} = 6$ $\qquad\qquad\qquad = \dfrac{6!}{6!} = 1$

Now try Exercises 21 through 24 ▶

✓ **B.** You've just seen how we can find binomial coefficients using $\dbinom{n}{k}$ notation

As mentioned, the formulas for $\dbinom{n}{r}$ and $_nC_r$ yield like results for given values of n and r. For future use, it will help to commit the general results from Example 4 to memory: $\dbinom{n}{0} = 1$, $\dbinom{n}{1} = n$, $\dbinom{n}{n-1} = n$, and $\dbinom{n}{n} = 1$.

C. The Binomial Theorem

Using $\begin{pmatrix} n \\ r \end{pmatrix}$ notation and the observations made regarding binomial powers, we can now state the **binomial theorem.**

Binomial Theorem

For any binomial $(a + b)$ and natural number n,

$$(a + b)^n = \binom{n}{0}a^n b^0 + \binom{n}{1}a^{n-1}b^1 + \binom{n}{2}a^{n-2}b^2 + \cdots$$

$$+ \binom{n}{n-1}a^1 b^{n-1} + \binom{n}{n}a^0 b^n$$

The theorem can also be stated in summation form as

$$(a + b)^n = \sum_{r=0}^{n} \binom{n}{r}a^{n-r}b^r$$

The expansion actually looks overly impressive in this form, and it helps to summarize the process in words, as we did earlier. The exponents on the first term a begin at n and decrease, while the exponents on the second term b begin at 0 and increase, keeping the degree of each term constant. The $\begin{pmatrix} n \\ r \end{pmatrix}$ notation simply gives the coefficients of each term. As a final note, observe that the r in $\begin{pmatrix} n \\ r \end{pmatrix}$ gives the exponent on b.

EXAMPLE 5 ▶ **Expanding a Binomial Using the Binomial Theorem**

Expand $(a + b)^6$ using the binomial theorem.

Solution ▶ $(a + b)^6 = \binom{6}{0}a^6 b^0 + \binom{6}{1}a^5 b^1 + \binom{6}{2}a^4 b^2 + \binom{6}{3}a^3 b^3 + \binom{6}{4}a^2 b^4 + \binom{6}{5}a^1 b^5 + \binom{6}{6}a^0 b^6$

$= \dfrac{6!}{0!6!}a^6 + \dfrac{6!}{1!5!}a^5 b^1 + \dfrac{6!}{2!4!}a^4 b^2 + \dfrac{6!}{3!3!}a^3 b^3 + \dfrac{6!}{4!2!}a^2 b^4 + \dfrac{6!}{5!1!}a^1 b^5 + \dfrac{6!}{6!0!}b^6$

$= 1a^6 + 6a^5 b + 15a^4 b^2 + 20a^3 b^3 + 15a^2 b^4 + 6ab^5 + 1b^6$

Now try Exercises 25 through 32 ▶

EXAMPLE 6 ▶ **Using the Binomial Theorem to Find the Initial Terms of an Expansion**

Find the first three terms of $(2x + y^2)^{10}$.

Solution ▶ Use the binomial theorem with $a = 2x$, $b = y^2$, and $n = 10$.

$(2x + y^2)^{10} = \binom{10}{0}(2x)^{10}(y^2)^0 + \binom{10}{1}(2x)^9(y^2)^1 + \binom{10}{2}(2x)^8(y^2)^2 + \cdots$ first three terms

$= (1)1024x^{10} + (10)512x^9 y^2 + \dfrac{10!}{2!8!}256x^8 y^4 + \cdots$ $\binom{10}{0} = 1, \binom{10}{1} = 10$

$= 1024x^{10} + 5120x^9 y^2 + (45)256x^8 y^4 + \cdots$ $\dfrac{10!}{2!8!} = 45$

$= 1024x^{10} + 5120x^9 y^2 + 11,520x^8 y^4 + \cdots$ result

☑ **C.** You've just seen how we can use the binomial theorem to find $(a + b)^n$

Now try Exercises 33 through 36 ▶

D. Finding a Specific Term of the Binomial Expansion

In some applications of the binomial theorem, our main interest is a *specific term* of the expansion, rather than the expansion as a whole. To find a specified term, it helps to consider that the expansion of $(a + b)^n$ has $n + 1$ terms: $(a + b)^0$ has one term, $(a + b)^1$ has two terms, $(a + b)^2$ has three terms, and so on. Because the notation $\binom{n}{r}$ always begins at $r = 0$ for the first term, the value of r will be *1 less than the term we are seeking*. In other words, for the seventh term of $(a + b)^9$, we use $r = 6$.

The kth Term of a Binomial Expansion

For the binomial expansion $(a + b)^n$, the kth term is given by

$$\binom{n}{r}a^{n-r}b^r, \text{ where } r = k - 1.$$

EXAMPLE 7 ▶

Finding a Specific Term of a Binomial Expansion

Find the eighth term in the expansion of $(x + 2y)^{12}$.

Solution ▶

By comparing $(x + 2y)^{12}$ to $(a + b)^n$ we have $a = x$, $b = 2y$, and $n = 12$. Since we want the eighth term, $k = 8$ and $r = 7$. The eighth term of the expansion is

$$\binom{12}{7}x^5(2y)^7 = \frac{12!}{7!5!}128x^5y^7 \qquad \textcolor{red}{2^7 = 128}$$
$$= (792)(128x^5y^7) \qquad \textcolor{red}{\binom{12}{7} = 792}$$
$$= 101{,}376x^5y^7 \qquad \textcolor{red}{\text{result}}$$

☑ **D.** You've just seen how we can find a specific term of a binomial expansion

Now try Exercises 37 through 42 ▶

E. Applications

One application of the binomial theorem involves a **binomial experiment** and **binomial probability.** For binomial probabilities, the following must be true: (1) The experiment must have only two possible outcomes, typically called success and failure, and (2) if the experiment has n trials, the probability of success must be constant for all n trials. If the probability of success for each trial is p, the formula $\binom{n}{k}(1 - p)^{n-k}p^k$ gives the probability that exactly k trials will be successful.

Binomial Probability

Given a binomial experiment with n trials, where the probability for success in each trial is p. The probability that exactly k trials are successful is given by

$$\binom{n}{k}(1 - p)^{n-k}p^k.$$

EXAMPLE 8 ▶

Applying the Binomial Theorem—Binomial Probability

Paula Rodrigues has a free-throw shooting average of 85%. On the last play of the game, with her team behind by three points, she is fouled at the three-point line, and is awarded two additional free throws via technical fouls on the opposing coach (a total of five free-throws). What is the probability she makes *at least three* (meaning they at least tie the game)?

Solution ▶ Here we have $p = 0.85$, $1 - p = 0.15$, and $n = 5$. The key idea is to recognize the phrase *at least three* means "3 or 4 or 5." So $P(\text{at least } 3) = P(3 \cup 4 \cup 5)$.

$$P(\text{at least } 3) = P(3 \cup 4 \cup 5) \qquad \text{"or" implies a union}$$
$$= P(3) + P(4) + P(5) \qquad \text{sum of probabilities (mutually exclusive events)}$$
$$= \binom{5}{3}(0.15)^2(0.85)^3 + \binom{5}{4}(0.15)^1(0.85)^4 + \binom{5}{5}(0.15)^0(0.85)^5$$
$$\approx 0.1382 + 0.3915 + 0.4437$$
$$= 0.9734$$

Paula's team has an excellent chance ($\approx 97.3\%$) of at least tying the game.

Now try Exercises 45 and 46 ▶

As you can see, calculations involving binomial probabilities can become quite extensive. Here again, a conceptual understanding of what the numbers mean can be combined with the use of technology to solve significant applications of the idea. Most graphing calculators provide a *binomial probability distribution function*, abbreviated **"binompdf("** and accessed using [2nd] [VARS] **(DISTR) 0:binompdf(**. The function requires three inputs: the number of trials n, the probability of success p for each trial, and the value of k. As with the evaluation of other functions, k can be a single value or a list of values enclosed in braces: "{ }." The resulting calculation for Example 8 is shown in Figure 9.67, and verifies each of the individual probabilities, although we must use the right arrow to see them all. To find the sum of these probabilities, we simply precede the **"binompdf("** command with the **"sum("** feature used previously. The final result is shown in Figure 9.68, and verifies our earlier calculation. **See Exercises 47 and 48.**

Figure 9.67

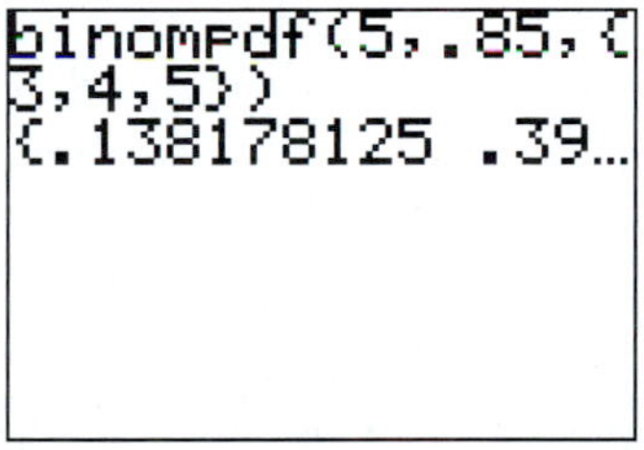

```
binompdf(5,.85,{
3,4,5})
{.138178125 .39…
```

Figure 9.68

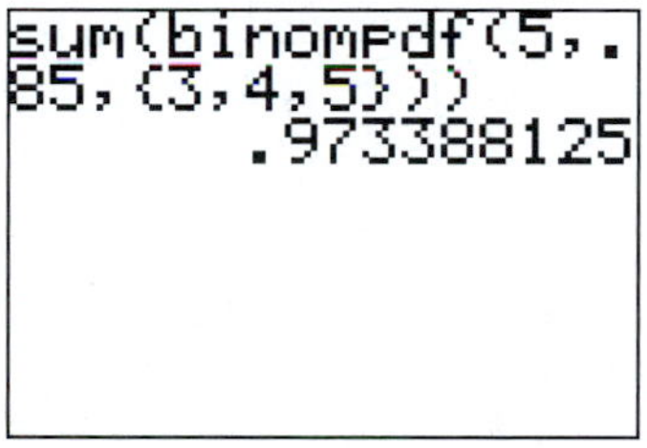

```
sum(binompdf(5,.
85,{3,4,5}))
           .973388125
```

☑ **E.** You've just seen how we can solve applications of binomial powers

9.7 EXERCISES

▶ CONCEPTS AND VOCABULARY

Fill in the blank with the appropriate word or phrase. Carefully reread the section if needed.

1. In any binomial expansion, there is always ___one___ more term than the power applied.

2. In all terms in the expanded form of $(a + b)^n$, the exponents on a and b must sum to ___n___.

3. To expand a binomial *difference* such as $(a - 2b)^5$, we rewrite the binomial as $(a + (-2b))^5$ and proceed as before.

4. In a binomial experiment with n trials, the probability there are exactly k successes is given by the formula _____. $\binom{n}{k}(1 - p)^{n-k}p^k$

5. Discuss why the expansion of $(a + b)^n$ has $n + 1$ terms.   Answers will vary.

6. For any defined binomial experiment, discuss the relationships between the phrases, "exactly k success," and "at least k successes."   Answers will vary.

HOMEWORK SELECTION GUIDE

Core: 7–43 odd, 45–49 (24 Exercises)
Standard: 1–4, 7–43 odd, 45–49, 51, 52 (30 Exercises)

Extended: 1–4, 7–43 odd, 44–53 (33 Exercises)
In Depth: 1–6, 7–43 odd, 44–54 (36 Exercises)

► **DEVELOPING YOUR SKILLS**

Use *Pascal's triangle* and the patterns explored to write each expansion.

7. $(x + y)^5$ **8.** $(a + b)^6$ **9.** $(2x + 3)^4$

10. $(x^2 + \frac{1}{3})^3$ **11.** $(1 - 2i)^5$ **12.** $(2 - 5i)^4$

Evaluate each of the following

13. $\binom{7}{4}$ 35   **14.** $\binom{8}{2}$ 28   **15.** $\binom{5}{3}$ 10

16. $\binom{9}{5}$ 126   **17.** $\binom{20}{17}$ 1140   **18.** $\binom{30}{26}$ 27,405

19. $\binom{40}{3}$ 9880   **20.** $\binom{45}{3}$ 14,190   **21.** $\binom{6}{0}$ 1

22. $\binom{5}{0}$ 1   **23.** $\binom{15}{15}$ 1   **24.** $\binom{10}{10}$ 1

Use the *binomial theorem* to expand each expression. Write the general form first, then simplify.

25. $(c + d)^5$ **26.** $(v + w)^4$ **27.** $(a - b)^6$

28. $(x - y)^7$ **29.** $(2x - 3)^4$ **30.** $(a - 2b)^5$

31. $(1 - 2i)^3$ **32.** $(2 + i\sqrt{3})^5$

Use the *binomial theorem* to write the first three terms.

33. $(x + 2y)^9$ **34.** $(3p + q)^8$ **35.** $(v^2 - \frac{1}{2}w)^{12}$

36. $(\frac{1}{2}a - b^2)^{10}$

Find the indicated term for each binomial expansion.

37. $(x + y)^7$; 4th term   **38.** $(m + n)^6$; 5th term

39. $(p - 2)^8$; 7th term   **40.** $(a - 3)^{14}$; 10th term

41. $(2x + y)^{12}$; 11th term   **42.** $(3n + m)^9$; 6th term

► **WORKING WITH FORMULAS**

43. Binomial probability: $P(k) = \binom{n}{k}\left(\frac{1}{2}\right)^k\left(\frac{1}{2}\right)^{n-k}$

The theoretical probability of getting exactly k heads in n flips of a fair coin is given by the formula above. What is the probability that you would get exactly 5 heads in 10 flips of the coin? ≈0.25

44. Binomial probability: $P(k) = \binom{n}{k}\left(\frac{1}{5}\right)^k\left(\frac{4}{5}\right)^{n-k}$

A multiple choice test has five options per question. The probability of guessing correctly k times out of n questions is found using the formula shown. What is the probability a person scores a 70% by guessing randomly (7 out of 10 questions correct)? ≈0.0008

► **APPLICATIONS**

45. Batting averages: Tony Gwynn (San Diego Padres) had a lifetime batting average of 0.347, ranking him as one of the greatest hitters of all time. Suppose he came to bat five times in any given game.

   a. What is the probability that he will get exactly three hits? ≈17.8%

   b. What is the probability that he will get at least three hits? ≈23.0%

46. Pollution testing: Erin suspects that a nearby iron smelter is contaminating the drinking water over a large area. A statistical study reveals that 83% of the wells in this area are likely contaminated. If the figure is accurate, find the probability that if another 10 wells are tested

   a. exactly 8 are contaminated ≈29.3%

   b. at least 8 are contaminated ≈76.6%

47. Late rental returns: The manager of Victor's DVD Rentals knows that 6% of all DVDs rented are returned *late*. Of the eight videos rented in the last hour, what is the probability that

   a. exactly five are returned on time ≈0.89%

   b. exactly six are returned on time ≈7.0%

   c. at least six are returned on time ≈99.0%

   d. none of them will be returned late ≈61.0%

48. Opinion polls: From past experience, a research firm knows that 20% of telephone respondents will agree to answer an opinion poll. If 20 people are contacted by phone, what is the probability that

   a. exactly 18 refuse to be polled ≈13.7%

   b. exactly 19 refuse to be polled ≈5.8%

   c. at least 18 refuse to be polled ≈20.7%

   d. none of them agree to be polled ≈1.2%

▶ **EXTENDING THE CONCEPT**

49. If you sum the entries in each row of Pascal's triangle, a pattern emerges. Find a formula that generalizes the result for any row of the triangle, and use it to find the sum of the entries in the 12th row of the triangle. 2^{n-1}, 2048

50. The *derived polynomial* of $f(x)$ is $f(x + h)$ or the original polynomial evaluated at $x + h$. Use Pascal's triangle or the binomial theorem to find the derived polynomial for $f(x) = x^3 + 3x^2 + 5x - 11$. Simplify the result completely.
$x^3 + 3x^2h + 3xh^2 + h^3 + 3x^2 + 6xh + 3h^2 + 5x + 5h - 11$

▶ **MAINTAINING YOUR SKILLS**

51. (2.5) Graph the function shown and find $f(3)$:

$$f(x) = \begin{cases} x + 2 & x \le 2 \\ (x - 4)^2 & x > 2 \end{cases}$$

52. (3.1) Show that $x = -1 + i$ is a solution to $x^4 + 2x^3 - x^2 - 6x - 6 = 0$.   verified

Additional answers can be found in the Instructor Answer Appendix.

53. (4.3/4.6) Graph the function $g(x) = x^3 - x^2 - 6x$. Clearly indicate all intercepts and intervals where $g(x) > 0$.

54. (5.6) If \$2500 is deposited at 6% compounded continuously, how much would be in the account 10 years later?   \$4555.30

MAKING CONNECTIONS

Making Connections: Graphically, Symbolically, Numerically, and Verbally

Eight situations are described in (a) through (h) below. Match the characteristics, formulas, operations, or results indicated in 1 through 16 to one of the eight situations.

(a) $\displaystyle\sum_{i=1}^{7} \frac{3^{i-1}}{18}$

(b) $-2 + 0.5 + 3 + 5.5 + 8 + 10.5 + \cdots + 33$

(c)

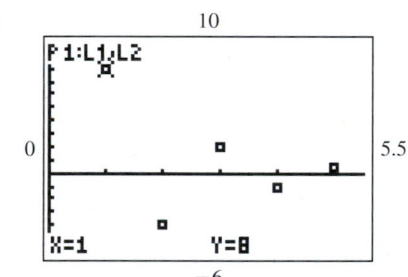

(d)

L1	⬛	L3 2
1	2.9	------
2	2.2	
3	1.5	
4	.8	
5	.1	
6	-.6	
7	-1.3	

$L2 = \{2.9, 2.2, 1.5\ldots$

(e) $16a^4 - 32a^3b + 24a^2b^2 - 8ab^3 + b^4$

(f) $-29, -23, -17, -11, \ldots$

(g) $1, 1, 2, 3, 5, 8, 13, \ldots$

(h) $a^4 + 8a^3b + 24a^2b^2 + 32ab^3 + 16b^4$

1. __c__ alternating sequence

2. __g__ Fibonacci sequence

3. __b__ 232.5

4. __d__ $d = -0.7$

5. __d__ $a_n = 3.6 - 0.7n$

6. __f__ $S_{39} = 3315$

7. __b__ arithmetic series

8. __h__ $(a + 2b)^4$

9. __d__ $a_{22} = -11.8$

10. __a__ geometric series

11. __a__ $r = 3$

12. __c__ $S_\infty = \dfrac{16}{3}$

13. __e__ $(2a - b)^4$

14. __f__ $a_n = 6n - 35$

15. __g__ recursively defined

16. __a__ $\dfrac{1093}{18}$

SUMMARY AND CONCEPT REVIEW

SECTION 9.1   Sequences and Series

KEY CONCEPTS

- A *finite sequence* is a function a_n whose domain is the set of natural numbers from 1 to n.
- The terms of the sequence are labeled $a_1, a_2, a_3, \ldots, a_{k-1}, a_k, a_{k+1}, \ldots, a_{n-2}, a_{n-1}, a_n$.
- The expression a_n, which defines the sequence (generates the terms in order), is called the nth term.
- An *infinite sequence* is a function whose domain is the set of natural numbers.
- When each term of a sequence is larger than the preceding term, it is called an increasing sequence.
- When each term of a sequence is smaller than the preceding term, it is called a decreasing sequence.
- When successive terms of a sequence alternate in sign, it is called an alternating sequence.
- When the terms of a sequence are generated using previous term(s), it is called a recursive sequence.
- Sequences are sometimes defined using factorials, which are the product of a given natural number with all natural numbers that precede it: $n! = n \cdot (n-1) \cdot (n-2) \cdot \cdots \cdot 3 \cdot 2 \cdot 1$.
- Given the sequence $a_1, a_2, a_3, a_4, \ldots, a_n$ the sum is called a finite series and is denoted S_n.
- $S_n = a_1 + a_2 + a_3 + a_4 + \cdots + a_n$. The sum of the first n terms is called a partial sum.
- In sigma notation, the expression $\displaystyle\sum_{i=1}^{k} a_i = a_1 + a_2 + \cdots + a_k$ represents a finite series,

  and the letter "i" is called the index of summation.

EXERCISES

Write the first four terms that are defined and the value of a_{10}.

1. $a_n = 5n - 4$ 1, 6, 11, 16; $a_{10} = 46$

2. $a_n = \dfrac{n+1}{n^2+1}$ $1, \frac{3}{5}, \frac{2}{5}, \frac{5}{17}; a_{10} = \frac{11}{101}$

Find the general term a_n for each sequence, and the value of a_6.

3. $1, 16, 81, 256, \ldots$ $a_n = n^4; a_6 = 1296$

4. $-17, -14, -11, -8, \ldots$ $a_n = -17 + (n-1)(3); a_6 = -2$

Find the eighth partial sum (S_8).

5. $\frac{1}{2}, \frac{1}{4}, \frac{1}{8}, \ldots$ $\frac{255}{256}$

6. $-21, -19, -17, \ldots$ -112

Evaluate each sum.

7. $\displaystyle\sum_{n=1}^{7} n^2$ 140

8. $\displaystyle\sum_{n=1}^{5} (3n - 2)$ 35

Write the first five terms that are defined.

9. $a_n = \dfrac{n!}{(n-2)!}$ 2, 6, 12, 20, 30

10. $\begin{cases} a_1 = \frac{1}{2} \\ a_{n+1} = 2a_n - \frac{1}{4} \end{cases}$ $\frac{1}{2}, \frac{3}{4}, \frac{5}{4}, \frac{9}{4}, \frac{17}{4}$

Write as a single summation and evaluate.

11. $\displaystyle\sum_{i=1}^{7} i^2 + \sum_{i=1}^{7} (3i - 2)$ $\displaystyle\sum_{i=1}^{7} (i^2 + 3i - 2); 210$

12. A large wildlife preserve brings in 40 rare hawks (male and female) in an effort to repopulate the species. Each year they are able to add an average of 10 additional hawks in cooperation with other wildlife areas. If the population of hawks grows at a rate of 12% through natural reproduction, the number of hawks in the preserve after x yr is given by the recursive sequence $h_0 = 40$, $h_n = 1.12 h_{n-1} + 10$. (a) How many hawks are in the wildlife preserve after 5 yr? (b) How many years before the number of hawks exceeds 200?   **a.** about 134 hawks   **b.** 8 yr

SECTION 9.2   Arithmetic Sequences

KEY CONCEPTS

- In an arithmetic sequence, successive terms are found by adding a fixed constant to the preceding term.
- In a sequence, if there exists a number d, called the common difference, such that $a_{k+1} - a_k = d$, then the sequence is arithmetic. Alternatively, $a_{k+1} = a_k + d$ for $k \geq 1$.
- The nth term n of an arithmetic sequence is given by $a_n = a_1 + (n - 1)d$, where a_1 is the first term and d is the common difference.
- If the initial term is unknown or is not a_1 the nth term can be written $a_n = a_k + (n - k)d$, where the subscript of the term a_k and the coefficient of d sum to n.
- For an arithmetic sequence with first term a_1, the nth partial sum (the sum of the first n terms) is given by

$$S_n = \frac{n(a_1 + a_n)}{2}.$$

EXERCISES

Find the general term (a_n) for each arithmetic sequence. Then find the indicated term.

13. $2, 5, 8, 11, \ldots$; find a_{40} $a_n = 2 + 3(n - 1)$; 119

14. $3, 1, -1, -3, \ldots$; find a_{35} $a_n = 3 + (-2)(n - 1)$; -65

Find the sum of each series.

15. $-1 + 3 + 7 + 11 + \cdots + 75$ 740

16. $1 + 4 + 7 + 10 + \cdots + 88$ 1335

17. $3 + 6 + 9 + 12 + \cdots$; S_{20} 630

18. $1 + \frac{3}{4} + \frac{1}{2} + \frac{1}{4} + \cdots$; S_{15} -11.25

19. $\displaystyle\sum_{n=1}^{25} (3n - 4)$ 875

20. From a point just behind the cockpit, the width of a modern fighter plane's swept-back wings is 1.25 m. The width of the wings, measured in equal increments, increases according to the pattern $1.25, 2.15, 3.05, 3.95, \ldots$. Find the width of the wings on the eighth measurement.   7.55 m

SECTION 9.3   Geometric Sequences

KEY CONCEPTS

- In a geometric sequence, successive terms are found by multiplying the preceding term by a nonzero constant.

- In other words, if there exists a number r, called the common ratio, such that $\dfrac{a_{k+1}}{a_k} = r$, then the sequence is geometric. Alternatively, we can write $a_{k+1} = a_k r$ for $k \geq 1$.

- The nth term a_n of a geometric sequence is given by $a_n = a_1 r^{n-1}$, where a_1 is the first term and a_n represents the general term of a finite sequence.

- If the initial term is unknown or is not a_1, the nth term can be written $a_n = a_k r^{n-k}$, where the subscript of the term a_k and the exponent on r sum to n.

- The nth partial sum of a geometric sequence is $S_n = \dfrac{a_1(1 - r^n)}{1 - r}$.

- If $|r| < 1$, the sum of an infinite geometric series is $S_\infty = \dfrac{a_1}{1 - r}$.

EXERCISES

Find the indicated term for each geometric sequence.

21. $a_1 = 5, r = 3$; find a_7 3645

22. $a_1 = 4, r = \sqrt{2}$; find a_7 32

23. $a_1 = \sqrt{7}, r = \sqrt{7}$; find a_8 2401

Find the indicated sum, if it exists.

24. $16 - 8 + 4 - \cdots$; find S_7 10.75

25. $2 + 6 + 18 + \cdots$; find S_8 6560

26. $\frac{4}{5} + \frac{2}{5} + \frac{1}{5} + \frac{1}{10} + \cdots$; find S_{12} $\frac{819}{512}$

27. $4 + 8 + 16 + 32 + \cdots$ does not exist

28. $5 + 0.5 + 0.05 + 0.005 + \cdots$ $\frac{50}{9}$

29. $6 - 3 + \frac{3}{2} - \frac{3}{4} + \cdots$ 4

30. $\sum_{k=1}^{8} 5\left(\frac{2}{3}\right)^k$ $\frac{63,050}{6561}$ **31.** $\sum_{k=1}^{\infty} 12\left(\frac{4}{3}\right)^k$ does not exist   **32.** $\sum_{k=1}^{\infty} 5\left(\frac{1}{2}\right)^k$ 5

33. Sumpter reservoir contains 121,500 ft^3 of water and is being drained in the following way. Each day one-third of the water is *drained* (and not replaced). Use a sequence/series to compute how much water *remains in the pond* after 7 days. $a_0 = 121,500, a_1 = 81,000, a_n = 81,000\left(\frac{2}{3}\right)^{n-1}, a_7 \approx 7111 \text{ ft}^3$

34. Credit-hours taught at Cody Community College have been increasing at 7% per year since it opened in 2001 and taught 1225 credit-hours. For the new faculty, the college needs to predict the number of credit-hours that will be taught in 2015. Use a sequence/series to compute the credit-hours for 2015 and to find the total number of credit hours taught through the 2015 school year. $a_0 = 1225, a_1 \approx 1311, a_n = 1311(1.07)^{n-1}, a_{15} \approx 3380, S_{15} \approx 32,944$

SECTION 9.4   Mathematical Induction

KEY CONCEPTS

- Functions written in subscript notation can be evaluated, graphed, and composed with other functions.
- A sum formula involving only natural numbers n as inputs can be proven valid using a proof by induction. Given that S_n represents a sum formula involving natural numbers, if (1) S_1 is true and (2) $S_k + a_{k+1} = S_{k+1}$, then S_n must be true for all natural numbers.
- Proof by induction can also be used to validate other relationships, using a more general statement of the principle. Let P_n be a statement involving the natural numbers n. If (1) P_1 is true (P_n for $n = 1$) and (2) the truth of P_k implies that P_{k+1} is also true, then P_n must be true for all natural numbers n.

EXERCISES

Use the principle of mathematical induction to prove the indicated sum formula is true for all natural numbers n.

35. $1 + 2 + 3 + 4 + 5 + \cdots + n$;

$a_n = n$ and $S_n = \dfrac{n(n+1)}{2}$.

36. $1 + 4 + 9 + 16 + 25 + 36 + \cdots + n^2$;

$a_n = n^2$ and $S_n = \dfrac{n(n+1)(2n+1)}{6}$.

Use the principle of mathematical induction to prove that each statement is true for all natural numbers n.

37. $4^n \geq 3n + 1$ **38.** $6 \cdot 7^{n-1} \leq 7^n - 1$ **39.** $3^n - 1$ is divisible by 2

SECTION 9.5   Counting Techniques

KEY CONCEPTS

- An experiment is any task that can be repeated and has a well-defined set of possible outcomes.
- Each repetition of an experiment is called a trial.
- Any potential outcome of an experiment is called a sample outcome.
- The set of all sample outcomes is called the sample space.
- An experiment with N (equally likely) sample outcomes that is repeated t times, has a sample space with N^t elements.
- If a sample outcome can be used more than once, the counting is said to be with repetition. If a sample outcome can be used only once, the counting is said to be without repetition.
- The fundamental principle of counting states: If there are p possibilities for a first task, q possibilities for the second, and r possibilities for the third, the total number of ways the experiment can be completed is pqr. This fundamental principle can be extended to include any number of tasks.
- If the elements of a sample space have precedence or priority (order or rank is important), the number of elements is counted using a permutation, denoted $_nP_r$ and read, "the distinguishable permutations of n objects taken r at a time."
- To expand $_nP_r$, we can write out the first r factors of $n!$ or use the formula $_nP_r = \dfrac{n!}{(n-r)!}$.

- If any of the sample outcomes are identical, certain permutations will be nondistinguishable. In a set containing n elements where one element is repeated p times, another is repeated q times, and another r times ($p + q + r = n$), the number of distinguishable permutations is given by $\dfrac{_nP_n}{p!q!r!} = \dfrac{n!}{p!q!r!}$.

- If the elements of a set have no rank, order, or precedence (as in a committee of colleagues) permutations with the same elements are considered identical. The result is the number of combinations, $_nC_r = \dfrac{n!}{r!(n-r)!}$.

EXERCISES

40. Three slips of paper with the letters A, B, and C are placed in a box and randomly drawn one at a time. Show all possible ways they can be drawn using a tree diagram.

41. The combination for a certain bicycle lock consists of three digits. How many combinations are possible if (a) repetition of digits is not allowed and (b) repetition of digits is allowed.   720; 1000

42. Jethro has three work shirts, four pairs of work pants, and two pairs of work shoes. How many different ways can he dress himself (shirt, pants, shoes) for a day's work?   24

43. From a field of 12 contestants in a pet show, three cats are chosen at random to be photographed for a publicity poster. In how many different ways can the cats be chosen?   220

44. Compute the following values by hand, showing all work:

     a. $7!$ 5040   **b.** $_7P_4$ 840   **c.** $_7C_4$ 35

45. Six horses are competing in a race at the McClintock Ranch. Assuming there are no ties, (a) how many different ways can the horses finish the race? (b) How many different ways can the horses finish first, second, and third place? (c) How many finishes are possible if it is well known that Nellie-the-Nag will finish last and Sea Biscuit will finish first?   **a.** 720   **b.** 120   **c.** 24

46. How many distinguishable permutations can be formed from the letters in the word "tomorrow"?   3360

47. Quality Construction Company has 12 equally talented employees. (a) How many ways can a three-member crew be formed to complete a small job? (b) If the company is in need of a Foreman, Assistant Foreman, and Crew Chief, in how many ways can the positions be filled?   **a.** 220   **b.** 1320

SECTION 9.6   Introduction to Probability

KEY CONCEPTS

- An event E is any designated set of sample outcomes.
- Given S is a sample space of equally likely sample outcomes and E is an event relative to S, the probability of E, written $P(E)$, is computed as $P(E) = \dfrac{n(E)}{n(S)}$, where $n(E)$ represents the number of elements in E, and $n(S)$ represents the number of elements in S.
- The complement of an event E is the set of sample outcomes in S, but not in E and is denoted $\sim E$.
- Given sample space S and any event E defined relative to S:
  (1) $P(\sim S) = 0$,   (2) $0 \le P(E) \le 1$,   (3) $P(S) = 1$,   (4) $P(E) = 1 - P(\sim E)$, and
  (5) $P(E) + P(\sim E) = 1$.
- Two events that have no outcomes in common are said to be mutually exclusive.
- If two events are not mutually exclusive, $P(E_1 \text{ or } E_2) \to P(E_1 \cup E_2) = P(E_1) + P(E_2) - P(E_1 \cap E_2)$.
- If two events are mutually exclusive, $P(E_1 \text{ or } E_2) \to P(E_1 \cup E_2) = P(E_1) + P(E_2)$.

EXERCISES

48. One card is drawn from a standard deck. What is the probability the card is a ten or a face card? $\frac{4}{13}$

49. One card is drawn from a standard deck. What is the probability the card is a Queen or a face card? $\frac{3}{13}$

50. One die is rolled. What is the probability the result is not a three? $\frac{5}{6}$

51. Given $P(E_1) = \frac{3}{8}$, $P(E_2) = \frac{3}{4}$, and $P(E_1 \cup E_2) = \frac{5}{6}$, compute $P(E_1 \cap E_2)$. $\frac{7}{24}$

52. Find $P(E)$ given that $n(E) = {}_7C_4 \cdot {}_5C_3$ and $n(S) = {}_{12}C_7$. $\frac{175}{396}$

53. To determine if more physicians should be hired, a medical clinic tracks the number of days between a patient's request for an appointment and the actual appointment date. The table given shows the probability that a patient must wait d days. Based on the table, what is the probability a patient must wait

Wait (days d)	Probability
0	0.002
$0 < d < 10$	0.07
$10 \leq d < 20$	0.32
$20 \leq d < 30$	0.43
$30 \leq d < 40$	0.178

a. at least 20 days 0.608
b. less than 20 days 0.392
c. 40 days or less 1
d. over 40 days 0
e. less than 40 and more than 10 days 0.928
f. 30 or more days 0.178

SECTION 9.7 The Binomial Theorem

KEY CONCEPTS

- To expand $(a + b)^n$ for n of "moderate size," we can use Pascal's triangle and observed patterns.

- For any natural numbers n and r, where $n \geq r$, the expression $\binom{n}{r}$ (read "n choose r") is called the *binomial coefficient* and evaluated as $\binom{n}{r} = \dfrac{n!}{r!(n-r)!}$.

- If n is large, it is more efficient to expand using the binomial coefficients and binomial theorem.

- The following binomial coefficients are useful/common and should be committed to memory:

$$\binom{n}{0} = 1 \qquad \binom{n}{1} = n \qquad \binom{n}{n-1} = n \qquad \binom{n}{n} = 1$$

- We define $0! = 1$; for example $\binom{n}{n} = \dfrac{n!}{n!(n-n)!} = \dfrac{1}{0!} = \dfrac{1}{1} = 1$.

- The binomial theorem: $(a + b)^n = \binom{n}{0}a^n b^0 + \binom{n}{1}a^{n-1}b^1 + \binom{n}{2}a^{n-2}b^2 + \cdots + \binom{n}{n-1}a^1 b^{n-1} + \binom{n}{n}a^0 b^n$.

- The kth term of $(a + b)^n$ can be found using the formula $\binom{n}{r}a^{n-r}b^r$, where $r = k - 1$.

EXERCISES

54. Evaluate each of the following:

a. $\binom{7}{5}$ 21
b. $\binom{8}{3}$ 56

55. Use Pascal's triangle to expand the expressions:

a. $(x - y)^4$
$x^4 - 4x^3y + 6x^2y^2 - 4xy^3 + y^4$
b. $(1 + 2i)^5$ $41 - 38i$

Use the binomial theorem to:

56. Write the first four terms of

a. $(a + \sqrt{3})^8$
b. $(5a + 2b)^7$

57. Find the indicated term of each expansion.

a. $(x + 2y)^7$; fourth $280x^4y^3$
b. $(2a - b)^{14}$; 10th $-64{,}064a^5b^9$

58. Mark Leland is a professional bowler who is able to roll a strike (knocking down all 10 pins on the first ball) 91% of the time. (a) What is the probability he rolls at least four strikes in the first five frames? (b) What is the probability he rolls five strikes (and scares the competition)? **a.** about 93.3% **b.** about 62.4%

PRACTICE TEST

1. The general term of a sequence is given. Find the first four terms and the 8th term.

a. $a_n = \dfrac{2n}{n + 3}$
b. $a_n = \dfrac{(n + 2)!}{n!}$

c. $a_n = \begin{cases} a_1 = 3 \\ a_{n+1} = \sqrt{(a_n)^2 - 1} \end{cases}$

2. Expand each series and evaluate.

a. $\displaystyle\sum_{k=2}^{6} (2k^2 - 3)$ 165
b. $\displaystyle\sum_{j=2}^{6} (-1)^j \left(\dfrac{j}{j + 1}\right)$ $\frac{311}{420}$
c. $\displaystyle\sum_{j=1}^{5} (-2)\left(\dfrac{3}{4}\right)^j$ $\frac{-2343}{512}$
d. $\displaystyle\sum_{k=1}^{\infty} 7\left(\dfrac{1}{2}\right)^k$ 7

Additional answers can be found in the Instructor Answer Appendix.

3. Identify the first term and the common difference or common ratio. Then find the general term a_n.

    a. $7, 4, 1, -2, \ldots$ **b.** $-8, -6, -4, -2, \ldots$

    c. $4, -8, 16, -32, \ldots$ **d.** $10, 4, \frac{8}{5}, \frac{16}{25}, \ldots$

4. Find the indicated value for each sequence.

    a. $a_1 = 4, d = 5$; find a_{40} 199

    b. $a_1 = 2, a_n = -22, d = -3$; find n 9

    c. $a_1 = 24, r = \frac{1}{2}$; find a_6 $\frac{3}{4}$

    d. $a_1 = -2, a_n = 486, r = -3$; find n 6

5. Find the sum of each series.

    a. $7 + 10 + 13 + \cdots + 100$ 1712

    b. $\displaystyle\sum_{k=1}^{37}(3k + 2)$ 2183

    c. For $4 - 12 + 36 - 108 + \cdots$, find S_7 2188

    d. $6 + 3 + \frac{3}{2} + \frac{3}{4} + \cdots$ 12

6. Each swing of a pendulum (in one direction) is 95% of the previous one. If the first swing is 12 ft, (a) find the length of the seventh swing and (b) determine the distance traveled by the pendulum for the first seven swings.   **a.** ≈ 8.82 ft   **b.** ≈ 72.4 ft

7. A rare coin that cost $3000 appreciates in value 7% per year. Find the value after 12 yr. $6756.57

8. A car that costs $50,000 decreases in value by 15% per year. Find the value of the car after 5 yr. $22,185.27

9. Use mathematical induction to verify that for $a_n = 5n - 3$, the sum formula $S_n = \dfrac{5n^2 - n}{2}$ is true for all natural numbers n.

10. Use the principle of mathematical induction to verify that P_n: $2 \cdot 3^{n-1} \leq 3^n - 1$ is true for all natural numbers n.

11. Three colored balls (aqua, brown, and creme) are to be drawn without replacement from a bag. List all possible ways they can be drawn using (a) a tree diagram and (b) an organized list.

12. Suppose that license plates for motorcycles must consist of three numbers followed by two letters. How many license plates are possible if zero and "Z" cannot be used and no repetition is allowed?   302,400

13. If one icon is randomly chosen from the following set, find the probability a mailbox is not chosen: $\{$📫, 📪, 📬, 📭, 📧, 📨$\}$. $\frac{1}{3}$

14. Compute the following values by hand, showing all work: (a) $6!$ (b) $_6P_3$ (c) $_6C_3$ **a.** 720   **b.** 120   **c.** 20

15. An English major has built a collection of rare books that includes two identical copies of *The Canterbury Tales* (Chaucer), three identical copies of *Romeo and Juliet* (Shakespeare), four identical copies of *Faustus* (Marlowe), and four identical copies of *The Faerie Queen* (Spenser). If these books are to be arranged on a shelf, how many distinguishable permutations are possible?   900,900

16. A company specializes in marketing various *cornucopia* (traditionally a curved horn overflowing with fruit, vegetables, gourds, and ears of grain) for Thanksgiving table settings. The company has seven fruit, six vegetable, five gourd, and four grain varieties available. If two from each group (without repetition) are used to fill the horn, how many different cornucopia are possible?   302,400

17. Use Pascal's triangle to expand/simplify:

    a. $(x - 2y)^4$ **b.** $(1 + i)^4$

18. Use the binomial theorem to write the first three terms of (a) $(x + \sqrt{2})^{10}$ and (b) $(a - 2b^3)^8$.

19. Michael and Mitchell are attempting to make a nonstop, 100-mi trip on a tandem bicycle. The probability that Michael cannot continue pedaling for the entire trip is 0.02. The probability that Mitchell cannot continue pedaling for the entire trip is 0.018. The probability that neither one can pedal the entire trip is 0.011. What is the probability that they complete the trip?   0.989

20. The spinner shown is spun once. What is the probability of spinning

    a. a striped wedge $\frac{1}{4}$

    b. a shaded wedge $\frac{5}{12}$

    c. a clear wedge $\frac{1}{3}$

    d. an even number $\frac{1}{2}$

    e. a two or an odd number $\frac{7}{12}$

    f. a number greater than nine $\frac{1}{4}$

    g. a shaded wedge *or* a number greater than 12 $\frac{5}{12}$

    h. a shaded wedge *and* a number greater than 12   0

21. To improve customer service, a cable company tracks the number of days a customer must wait until their cable service is installed. The table shows the probability that a customer must wait d days. Based on the table, what is the probability a customer waits

Wait (days d)	Probability
0	0.02
$0 < d < 1$	0.30
$1 \leq d < 2$	0.60
$2 \leq d < 3$	0.05
$3 \leq d < 4$	0.03

    a. at least 2 days   0.08   **b.** less than 2 days   0.92

    c. 4 days or less   1   **d.** over 4 days   0

    e. less than 2 or at least 3 days   0.95

    f. three or more days   0.03

Additional answers can be found in the Instructor Answer Appendix.

22. An experienced archer can hit the rectangular target shown 100% of the time at a range of 75 m. Assuming the probability the target is hit is related to its area, what is the probability the archer hits within the

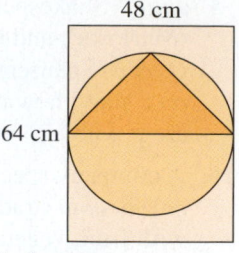

48 cm

64 cm

a. triangle   0.1875   **b.** circle   0.589

c. circle but outside the triangle   0.4015

d. lower half-circle   0.2945

e. rectangle but outside the circle   0.4110

f. lower half-rectangle, outside the circle   0.2055

23. A survey of 100 union workers was taken to register concerns to be raised at the next bargaining session. A breakdown of those surveyed is shown in the table

Expertise Level	Women	Men	Total
Apprentice	16	18	34
Technician	15	13	28
Craftsman	9	9	18
Journeyman	7	6	13
Contractor	3	4	7
Totals	50	50	100

in the right column. One out of the hundred will be selected at random for a personal interview. What is the probability the person chosen is a

a. woman or a craftsman $\frac{59}{100}$

b. man or a contractor $\frac{53}{100}$

c. man and a technician $\frac{13}{100}$

d. journeyman or an apprentice $\frac{47}{100}$

24. Cheddar is a 12-year-old male box turtle. Provolone is an 8-year-old female box turtle. The probability that Cheddar will live another 8 yr is 0.85. The probability that Provolone will live another 8 yr is 0.95. Find the probability that

a. both turtles live for another 8 yr   0.8075

b. neither turtle lives for another 8 yr   0.0075

c. at least one of them will live another 8 yr   0.9925

25. The quality control department at a lightbulb factory has determined that the company is losing money because their manufacturing process produces a defective bulb 12% of the time. If a random sample of 10 bulbs is tested, (a) what is the probability that none are defective? (b) What is the probability that no more than 3 bulbs are defective?
a. about 27.9%   **b.** about 97.6%

CALCULATOR EXPLORATION AND DISCOVERY

Infinite Series, Finite Results

Although there were many earlier flirtations with infinite processes, it may have been the paradoxes of Zeno of Elea (~450 B.C.) that crystallized certain questions that simultaneously frustrated and fascinated early mathematicians. The first paradox, called the dichotomy paradox, can be summarized by the following question: How can one ever finish a race, seeing that one-half the distance must first be traversed, then one-half the remaining distance, then one-half the distance that then remains, and so on an infinite number of times? Although we easily accept that races can be finished, the subtleties involved in this question stymied mathematicians for centuries and were not satisfactorily resolved until the eighteenth century. In modern notation, Zeno's first paradox says $\frac{1}{2} + \frac{1}{4} + \frac{1}{8} + \frac{1}{16} + \cdots < 1$. This is a geometric series with $a_1 = \frac{1}{2}$ and $r = \frac{1}{2}$.

Illustration 1 ▶ For the geometric sequence with $a_1 = \dfrac{1}{2}$ and $r = \dfrac{1}{2}$, the nth term is $a_n = \dfrac{1}{2^n}$. Use the "sum(" and "seq(" features of your calculator to compute S_5, S_{10}, and S_{15} (see Section 9.1). Does the sum appear to be approaching some "limiting value?" If so, what is this value? Now compute S_{20}, S_{25}, and S_{30}. Does there still appear to be a limit to the sum? What happens when you have the calculator compute S_{35}?   The calculator returns a "1."

Figure 9.69

```
sum(seq(0.5^X,X,
1,5))
              .96875
sum(seq(0.5^X,X,
1,10))
       .9990234375
sum(seq(0.5^X,X,
1,15))
```

Solution ▶ `CLEAR` the calculator and enter sum(seq (0.5^X, X, 1, 5)) on the home screen. Pressing `ENTER` gives $S_5 = 0.96875$ (Figure 9.69). Press `2nd` `ENTER` to recall the expression and overwrite the 5, changing it to a 10. Pressing `ENTER` shows $S_{10} = 0.9990234375$. Repeating these steps gives $S_{15} = 0.9999694824$, and it seems that "1" may be a limiting value. Our conjecture receives further support as S_{20}, S_{25}, and S_{30} are closer and closer to 1, but do not exceed it.

Note that the sum of additional terms will create a longer string of 9's. That the sum of an infinite number of these terms *is* **1** can be understood by converting the repeating decimal $0.\overline{9}$ to its fractional form (as shown). For $x = 0.\overline{9}$, $10x = 9.\overline{9}$ and it follows that

$$
\begin{aligned}
10x &= 9.\overline{9} \\
-x &= -0.\overline{9} \\
\hline
9x &= 9 \\
x &= 1
\end{aligned}
$$

For a geometric sequence, the result of an infinite sum can be verified using $S_\infty = \dfrac{a_1}{1 - r}$. However, there are many nongeometric, infinite series that also have a limiting value. In some cases these require many, many more terms before the limiting value can be observed.

Use a calculator to write the first five terms and to find S_5, S_{10}, and S_{15}. Decide if the sum appears to be approaching some limiting value, then compute S_{20} and S_{25}. Do these sums support your conjecture?

Exercise 1: $a_1 = \frac{1}{3}$ and $r = \frac{1}{3}$ $\frac{1}{2}$ **Exercise 2:** $a_1 = 0.2$ and $r = 0.2$ $\frac{1}{4}$ **Exercise 3:** $a_n = \dfrac{1}{(n - 1)!}$ e

Additional Insight: Zeno's first paradox can also be "resolved" by observing that the "half-steps" needed to complete the race require increasingly shorter (infinitesimally short) amounts of time. Eventually the race is complete.

STRENGTHENING CORE SKILLS

Probability, Quick-Counting, and Card Games

The card game known as *Five Card Stud* is often played for fun and relaxation, using toothpicks, beans, or pocket change as players attempt to develop a winning "hand" from the five cards dealt. The various "hands" are given here with the higher value hands listed first (e.g., a full house is a better/higher hand than a flush).

Five Card Hand	Description	Probability of Being Dealt
royal flush	five cards of the same suit in sequence from 10 to Ace	0.000 001 540
straight flush	any five cards of the same suit in sequence (exclude royal)	0.000 013 900
four of a kind	four cards of the same rank, any fifth card	
full house	three cards of the same rank, with one pair	
flush	five cards of the same suit, no sequence required	0.001 970
straight	five cards in sequence, regardless of suit	
three of a kind	three cards of the same rank, any two other cards	
two pairs	two cards of the one rank, two of another rank, one other card	0.047 500
one pair	two cards of the same rank, any three others	0.422 600

For this study, we will consider the hands that are based on suit (the flushes) and the sample space to be five cards dealt from a deck of 52, or $_{52}C_5$.

A flush consists of five cards in the same suit, a straight consists of five cards in sequence. Let's consider that an Ace can be used as either a high card (as in 10, J, Q, K, A) or a low card (as in A, 2, 3, 4, 5). Since the dominant characteristic of a flush is its *suit,* we first consider choosing one suit of the four, then the number of ways that the straight can be formed (if needed).

Illustration 1 ▶ What is the probability of being dealt a royal flush?

Solution ▶ For a royal flush, all cards must be of one suit. Since there are four suits, it can be chosen in $_4C_1$ ways. A royal flush must have the cards A, K, Q, J, and 10 and once the suit has been decided, it can happen in only this (one) way or $_1C_1$. This means $P \text{ (royal flush)} = \dfrac{_4C_1 \cdot _1C_1}{_{52}C_5} \approx 0.000\,001\,540.$

Illustration 2 ▶ What is the probability of being dealt a straight flush?

Solution ▶ Once again all cards must be of one suit, which can be chosen in $_4C_1$ ways. A straight flush is any five cards in sequence and once the suit has been decided, this can happen in 10 ways (Ace on down, King on down, Queen on down, and so on). By the FCP, there are $_4C_1 \cdot _{10}C_1 = 40$ ways this can happen, but *four of these will be royal flushes that are of a higher value* and must be subtracted from this total. So in the intended context we have

$$P \text{ (straight flush)} = \dfrac{_4C_1 \cdot _{10}C_1 - 4}{_{52}C_5} \approx 0.000\,013\,900$$

Using these examples, determine the probability of being dealt

Exercise 1: a simple flush (no royal or straight flushes)

Exercise 2: three cards of the same suit and any two other (nonsuit) cards

Exercise 3: four cards of the same suit and any one other (nonsuit) card

Exercise 4: a flush having no face cards

CUMULATIVE REVIEW CHAPTERS R–9

1. Robot Moe is assembling memory cards for computers. At 9:00 A.M., 52 cards had been assembled. At 11:00 A.M., a total of 98 had been made. Assuming the production rate is linear
   a. Find the slope of this line and explain what it means in this context.
   b. Find a linear equation model for this data.
   c. Determine how many cards Moe can assemble in an eight-hour day.
   d. Determine the approximate time that Moe began work this morning.

2. Verify by direct substitution that $x = 2 + i$ is a solution to $x^2 - 4x + 5 = 0$.

3. Solve using the quadratic formula: $3x^2 + 5x - 7 = 0$. State your answer in exact and approximate form.

4. Sketch the graph of $y = \sqrt{x + 4} - 3$ using transformations of a tool box function. Plot the x- and y-intercepts.

5. Write a variation equation and find the value of k: Y varies inversely with X and jointly with V and W. Y is equal to 10 when $X = 9$, $V = 5$, and $W = 12$.

6. Graph the piecewise-defined function and state the domain and range.
$$y = \begin{cases} -2 & -3 \le x \le -1 \\ x & -1 < x < 2 \\ x^2 & 2 \le x \le 3 \end{cases}$$

7. Verify that $f(x) = x^3 - 5$ and $g(x) = \sqrt[3]{x + 5}$ are inverse functions. How are the graphs of f and g related?

8. For the graph of $g(x)$ shown, state where
   a. $g(x) = 0$ b. $g(x) < 0$
   c. $g(x) > 0$ d. $g(x)\uparrow$
   e. $g(x)\downarrow$ f. local max
   g. local min   h. $g(x) = 2$
   i. $g(4)$ j. $g(-1)$
   k. as $x \to -1^+$, $g(x) \to$ _____
   l. as $x \to \infty$, $g(x) \to$ _____
   m. the domain of $g(x)$

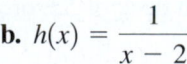

9. Compute the difference quotient for each function given.

   a. $f(x) = 2x^2 - 3x$ b. $h(x) = \dfrac{1}{x - 2}$

10. Graph the polynomial function given. Clearly indicate all intercepts. $f(x) = x^3 + x^2 - 4x - 4$

11. Graph the rational function $h(x) = \dfrac{2x^2 - 8}{x^2 - 1}$. Clearly indicate all asymptotes and intercepts.

12. Write each expression in logarithmic form:

   a. $x = 10^y$ **b.** $\dfrac{1}{81} = 3^{-4}$

13. Write each expression in exponential form:

   a. $3 = \log_x(125)$ **b.** $\ln(2x - 1) = 5$

14. What interest rate is required to ensure that $2000 will double in 10 yr if interest is compounded continuously?

15. Solve for x.

   a. $e^{2x-1} = 217$ **b.** $\log(3x - 2) + 1 = 4$

16. Solve using matrices and row reduction:

$$\begin{cases} 2a + 3b - 6c = 15 \\ 4a - 6b + 5c = 35 \\ 3a + 2b - 5c = 24 \end{cases}$$

17. Solve using a calculator and inverse matrices.

$$\begin{cases} 0.7x + 1.2y - 3.2z = -32.5 \\ 1.5x - 2.7y + 0.8z = -7.5 \\ 2.8x + 1.9y - 2.1z = 1.5 \end{cases}$$

18. Find the equation of the hyperbola with foci at $(-6, 0)$ and $(6, 0)$ and vertices at $(-4, 0)$ and $(4, 0)$.

19. Identify the center, vertices, and foci of the conic section defined by $x^2 + 4y^2 - 24y + 6x + 29 = 0$.

20. Use properties of sequences to determine a_{20} and S_{20}.

   a. $262{,}144,\ 65{,}536,\ 16{,}384,\ 4096, \ldots$

   b. $\dfrac{7}{8}, \dfrac{27}{40}, \dfrac{19}{40}, \dfrac{11}{40}, \ldots$

21. Empty 55-gal drums are stacked at a storage facility in the form of a pyramid with 52 barrels in the first row, 51 barrels in the second row, and so on, until there are 10 barrels in the top row. Use properties of sequences to determine how many barrels are in this stack.   1333

22. Three $20 bills, six $10 bills, and four $5 bills are placed in a box, then two bills are drawn out and placed in a savings account. What is the probability the bills drawn are

   a. first $20, second $10 $\frac{3}{26}$

   b. first $10, second $20 $\frac{3}{26}$

   c. both $5 $\frac{1}{13}$

   d. first $5, second not $20 $\frac{3}{13}$

   e. first $5, second not $10 $\frac{2}{13}$

   f. first not $20, second $20 $\frac{5}{26}$

Additional answers can be found in the Instructor Answer Appendix.

23. The manager of Tom's Tool and Equipment Rentals knows that 4% of all tools rented are returned late. Of the 12 tools rented in the last hour, what is the probability that

   a. exactly ten will be returned on time $\approx 7.0\%$

   b. at least eleven will be returned on time $\approx 91.9\%$

   c. at least ten will be returned on time $\approx 98.9\%$

   d. none of them will be returned on time $\binom{12}{0}(0.04)^{12}(0.96)^0$; virtually nil

24. Use a proof by induction to verify $3 + 7 + 11 + 15 + \cdots + (4n - 1) = n(2n + 1)$ for all natural numbers n.   verified

25. A park ranger tracks the number of campers at a popular park from March ($m = 3$) to September ($m = 9$) and collects the following data (month, number of campers): $(3, 56)$, $(5, 126)$, and $(9, 98)$. Assuming the data is quadratic, draw a scatterplot and construct a 3×3 system of equations and solve to obtain a parabolic equation model. (a) What month had the maximum number of campers? (b) What was this maximum number? (c) How many campers might be expected in April? (d) Based on your model, what month(s) is the park apparently closed to campers (number of campers is zero or negative)?

Exercises 26 through 30 require the use of a graphing calculator.

26. Solve the system. For this system x, y, and z are the variables, with π and e the well-known constants. Round your answer to two decimal places.

$$\begin{cases} \pi x + \sqrt{2}y + ez = 5 \\ \sqrt{2}x + ey + \pi z = 5 \\ ex + \pi y + \sqrt{2}z = 5 \end{cases}$$ (0.69, 0.69, 0.69)

27. Find the solution region for the system of linear inequalities. Your answer should include a screen shot or facsimile, and the location of any points of intersection, rounded to four decimal places.

$$\begin{cases} 112x + 39y > 438 \\ 57x - 64y < 101 \\ x \geq 0 \\ y \geq 0 \end{cases}$$

28. A triangle has vertices at $(112.3, 98.5)$, $(67.7, -39)$, and $(-27\pi, 21.5)$. Use the determinant formula to determine its area, rounded to the nearest tenth. 11,835.1 units^2

29. A recursive sequence is defined by $a_1 = 0.3$ and $a_{k+1} = a_k^2 - 0.4a_k$. Find S_{40}, rounded to four decimal places.   0.2794

30. Solve the equation. Round your answer to two decimal places. $\log x - 2\ln x + 3\log_3 x = 6$. $x \approx 172.46$

Appendix I

The Language, Notation, and Numbers of Mathematics

The most fundamental requirement for learning algebra is mastering the words, symbols, and numbers used to express mathematical ideas. "Words are the symbols of knowledge, the keys to accurate learning" (Norman Lewis in *Word Power Made Easy*, Penguin Books).

A. Sets of Numbers, Graphing Real Numbers, and Set Notation

To effectively use mathematics as a problem-solving tool, we must first be familiar with the **sets of numbers** used to quantify (give a numeric value to) the things we investigate. Only then can we make comparisons and develop equations that lead to informed decisions.

Natural Numbers

The most basic numbers are those used to count physical objects: 1, 2, 3, 4, and so on. These are called **natural numbers** and are represented by the capital letter $\mathbb{N}$, often written in the special font shown. We use **set notation** to list or describe a set of numbers. Braces { } are used to group **members** or **elements** of the set, commas separate each member, and three dots (called an *ellipsis*) are used to indicate a pattern that continues indefinitely. The notation $\mathbb{N} = \{1, 2, 3, 4, 5, \ldots\}$ is read, "$\mathbb{N}$ is the set of numbers 1, 2, 3, 4, 5, and so on." To show membership in a set, the symbol $\in$ is used. It is read "is an element of" or "belongs to." The statements $6 \in \mathbb{N}$ (6 is an element of $\mathbb{N}$) and $0 \notin \mathbb{N}$ (0 is not an element of $\mathbb{N}$) are true statements. A set having no elements is called the **empty** or **null set,** and is designated by empty braces { } or the symbol $\varnothing$.

EXAMPLE 1 ▶ **Writing Sets of Numbers Using Set Notation**

List the set of natural numbers that are

   a. greater than 100   **b.** negative

   c. greater than or equal to 5 and less than 12

Solution ▶   **a.** $\{101, 102, 103, 104, \ldots\}$

   b. { }; all natural numbers are positive.

   c. $\{5, 6, 7, 8, 9, 10, 11\}$

Now try Exercises 7 and 8 ▶

Whole Numbers

Combining zero with the natural numbers produces a new set called the **whole numbers** $\mathbb{W} = \{0, 1, 2, 3, 4, \ldots\}$. We say that the natural numbers are a **proper subset** of the whole numbers, denoted $\mathbb{N} \subset \mathbb{W}$, since every natural number is also a whole number. The symbol $\subset$ means "is a proper subset of."

EXAMPLE 2 ▶ Determining Membership in a Set

Given $A = \{1, 2, 3, 4, 5, 6\}$, $B = \{2, 4\}$, and $C = \{0, 1, 2, 3, 5, 8\}$, determine whether the following statements are true or false. Justify your response.

a. $B \subset A$ **b.** $B \subset C$ **c.** $C \subset \mathbb{W}$

d. $C \subset \mathbb{N}$ **e.** $104 \in \mathbb{W}$ **f.** $0 \in \mathbb{N}$

g. $2 \notin \mathbb{W}$

Solution ▶ **a.** True: Every element of B is in A. **b.** False: $4 \notin C$, so $B \nsubseteq C$.

c. True: All elements are whole numbers. **d.** False: $0 \notin \mathbb{N}$, so $C \nsubseteq \mathbb{N}$.

e. True: 104 is a whole number. **f.** False: $0 \notin \mathbb{N}$.

g. False: 2 *is* a whole number.

Now try Exercises 9 through 14 ▶

Integers

Numbers greater than zero are **positive numbers.** Every positive number has an *opposite* that is a **negative number** (a number less than zero). Combining zero and the natural numbers with their opposites produces the set of **integers** $\mathbb{Z} = \{\ldots, -3, -2, -1, 0, 1, 2, 3, \ldots\}$. We can illustrate the location and magnitude of a number (in relation to other numbers) using a **number line** (see Figure AI.1).

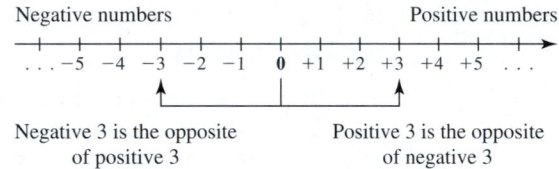

Figure AI.1

The number that corresponds to a given point on the number line is called the **coordinate** of that point. When we want to note a specific location on the line, a bold dot "•" is used and we have then **graphed** the number. Since we need only one coordinate to denote a location on the number line, it is referred to as a **one-dimensional graph.**

Rational Numbers

WORTHY OF NOTE

The integers are a subset of the rational numbers: $\mathbb{Z} \subset \mathbb{Q}$, since any integer can be written as a fraction using a denominator of one: $-2 = \frac{-2}{1}$ and $0 = \frac{0}{1}$, etc.

Fractions and mixed numbers are part of a set called the **rational numbers** $\mathbb{Q}$. A rational number is one that can be written as a fraction with an integer numerator and an integer denominator other than zero. In set notation we write $\mathbb{Q} = \{\frac{p}{q} | p, q \in \mathbb{Z}; q \neq 0\}$. The vertical bar "|" is read "such that" and indicates that a description follows. In words, we say, "$\mathbb{Q}$ is the set of numbers of the form p over q, such that p and q are integers and q is not equal to zero."

EXAMPLE 3 ▶ Graphing Rational Numbers

Graph the fractions by converting to decimal form and estimating their location between two integers:

a. $-2\frac{1}{3}$ **b.** $\frac{7}{2}$

Solution ▶ **a.** $-2\frac{1}{3} = -2.3333333\ldots$ or $-2.\overline{3}$ **b.** $\frac{7}{2} = 3.5$

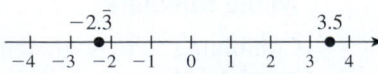

Now try Exercises 15 through 18 ▶

Since the division $\frac{7}{2}$ **terminated,** the result is called a **terminating decimal.** The decimal form of $-2\frac{1}{3}$ is called **repeating** and **nonterminating** (note that $-2.3 \neq -2.\overline{3}$). Recall that a repeating decimal is written with a horizontal bar over the first block of digit(s) that repeat. For instance $\frac{119}{55} = 2.1454545\ldots = 2.1\overline{45}$.

When using a calculator for computations involving repeating decimals, you must either use the rational form or "fill the display" with the digits that repeat. As an exploration, suppose that you are to inherit $\frac{1}{3} = 0.\overline{3}$ of a \$90,000 estate. How many "repeating threes" (times 90,000) are needed until the calculator returns an answer of \$30,000? **See Exercises 19 and 20.**

Irrational Numbers

Although any fraction can be written in decimal form, not all decimal numbers can be written as a fraction. One example is the number represented by the Greek letter π (pi), frequently seen in a study of circles. Although we often approximate π using 3.14, its true value has a **nonrepeating** and *nonterminating* decimal form. Other numbers of this type include $2.101001000100001\ldots$ (there is no block of digits that repeat—the number of zeroes between each "1" is increasing), and $\sqrt{5} \approx 2.2360679\ldots$ (the decimal form never terminates). Numbers with a nonrepeating and nonterminating decimal form belong to the set of irrational numbers $\mathbb{H}$.

EXAMPLE 4 ▶ **Approximating Irrational Numbers**

Use a calculator as needed to approximate the value of each number given (round to 100ths), then graph them on the number line:

   a. $\sqrt{3}$ **b.** π **c.** $\sqrt{19}$ **d.** $-\frac{\sqrt{2}}{2}$

Solution ▶ **a.** $\sqrt{3} \approx 1.73$ **b.** $\pi \approx 3.14$ **c.** $\sqrt{19} \approx 4.36$ **d.** $-\frac{\sqrt{2}}{2} \approx -0.71$

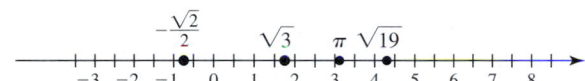

Now try Exercises 21 through 24 ▶

WORTHY OF NOTE

Checking the approximation for $\sqrt{5}$ shown, we obtain $2.2360679^2 = 4.999999653$. While we can find better approximations by using more and more decimal places, we never obtain five *exactly* (although some calculators will say the result is 5 due to limitations in programming).

```
2.2360679²
          4.999999653
```

Real Numbers

The set of rational numbers combined with the set of irrational numbers produces the set of **real numbers** $\mathbb{R}$. Figure AI.2 illustrates the relationship between the sets of numbers we've discussed so far. Notice how each subset appears "nested" in a larger set.

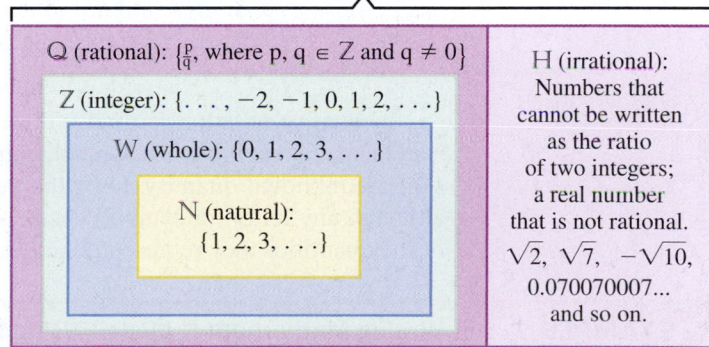

Figure AI.2

EXAMPLE 5 ▶ **Identifying Members of a Number Set**

List the numbers in set $A = \{-2, 0, 5, \sqrt{7}, 12, \frac{2}{3}, 4.5, \sqrt{21}, \pi, -0.75\}$ that belong to

    a. $\mathbb{Q}$ **b.** $\mathbb{H}$ **c.** $\mathbb{W}$ **d.** $\mathbb{Z}$

Solution ▶   **a.** $-2, 0, 5, 12, \frac{2}{3}, 4.5, -0.75 \in \mathbb{Q}$ **b.** $\sqrt{7}, \sqrt{21}, \pi \in \mathbb{H}$

      c. $0, 5, 12 \in \mathbb{W}$ **d.** $-2, 0, 5, 12 \in \mathbb{Z}$

Now try Exercises 25 through 28 ▶

EXAMPLE 6 ▶ **Evaluating Statements about Sets of Numbers**

Determine whether the statements are true or false. Justify your response.

    a. $\mathbb{N} \subset \mathbb{Q}$ **b.** $\mathbb{H} \subset \mathbb{Q}$ **c.** $\mathbb{W} \subset \mathbb{Z}$ **d.** $\mathbb{Z} \subset \mathbb{R}$

Solution ▶   **a.** True: All natural numbers can be written as fractions over 1.

      b. False: No irrational number can be written in fraction form.

      c. True: All whole numbers are integers.

      d. True: Every integer is a real number.

☑ **A.** You've just reviewed sets of numbers, graphing real numbers, and set notation

Now try Exercises 29 through 40 ▶

B. Inequality Symbols and Order Relations

We compare numbers of different size using **inequality notation,** known as the **greater than** ($>$) and **less than** ($<$) symbols. Note that $-4 < 3$ is the same as saying -4 is to the left of 3 on the number line. In fact, on a number line, any given number is smaller than any number to the right of it (see Figure AI.3).

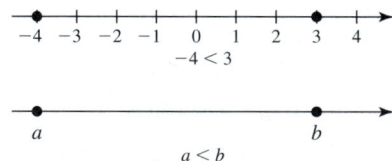

Figure AI.3

Order Property of Real Numbers

Given any two real numbers a and b.

    1. $a < b$ if a is to the left of b on the number line.

    2. $a > b$ if a is to the right of b on the number line.

Inequality notation is used with numbers and variables to write mathematical statements. A **variable** is a symbol, commonly a letter of the alphabet, used to represent an unknown quantity. Over the years x, y, and n have become most common, although any letter (or symbol) can be used. Often we'll use variables that remind us of the quantities they represent, like L for length, and D for distance.

EXAMPLE 7 ▶ **Writing Mathematical Models Using Inequalities**

Use a variable and an inequality symbol to represent the statement: "To hit a home run out of Jacobi Park, the ball must travel over three hundred twenty-five feet."

Solution ▶ Let D represent distance: $D > 325$ ft.

Now try Exercises 41 through 44 ▶

In Example 7, note the number 325 itself is not a possible value for D. If the ball traveled *exactly* 325 ft, it would hit the fence and stay in play. Numbers that mark the limit or boundary of an inequality are called **endpoints.** If the endpoint(s) are *not* included, the less than ($<$) or greater than ($>$) symbols are used. When the endpoints *are* included, the *less than or equal to symbol* ($\leq$) or the *greater than or equal to symbol* ($\geq$) is used. The decision to *include* or *exclude* an endpoint is often an important one, and many mathematical decisions (and real-life decisions) depend on a clear understanding of the distinction. **See Exercises 45 through 50.**

☑ **B.** You've just reviewed inequality symbols and order relations

C. The Absolute Value of a Real Number

Any nonzero real number "n" is either a positive number or a negative number. But in some applications, our primary interest is simply the *size* of n, rather than its sign. This is called the **absolute value** of n, denoted $|n|$, and can be thought of as its *distance from zero on the number line*, regardless of the direction (see Figure AI.4). Since distance is always positive or zero, $|n| \geq 0$.

Figure AI.4

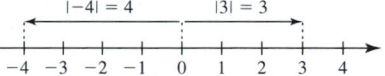

EXAMPLE 8 ▶ **Absolute Value Reading and Reasoning**

In the table shown, the absolute value of a number is given in column 1. Complete the remaining columns.

Solution ▶

Column 1 (In Symbols)	Column 2 (Spoken)	Column 3 (Result)	Column 4 (Reason)
$\lvert 7.5 \rvert$	"the absolute value of seven and five-tenths"	7.5	the distance between 7.5 and 0 is 7.5 units
$\lvert -2 \rvert$	"the absolute value of negative two"	2	the distance between -2 and 0 is 2 units
$-\lvert -6 \rvert$	"the opposite of the absolute value of negative six"	-6	the distance between -6 and 0 is 6 units, the opposite of 6 is -6

Now try Exercises 51 through 58 ▶

Example 8 illustrates that the absolute value of a positive number is the number itself, while the absolute value of a negative number is the *opposite of that number* (recall that $-n$ is positive if n itself is negative). For this reason the formal definition of absolute value is stated as follows.

Absolute Value

For any real number n,

$$|n| = \begin{cases} n & \text{if} \quad n \geq 0 \\ -n & \text{if} \quad n < 0 \end{cases}$$

The concept of absolute value can actually be used to find the distance between *any* two numbers on a number line. For instance, we know the distance between 2 and 8 is 6 (by counting). Using absolute values, we can write $|8 - 2| = |6| = 6$, or $|2 - 8| = |-6| = 6$. Generally, if a and b are two numbers on the real number line, the distance between them is $|a - b|$, which is identical to $|b - a|$.

EXAMPLE 9 ▶ Using Absolute Value to Find the Distance between Points

Find the distance between -5 and 3 on the number line.

Solution ▶ Substituting -5 for a and 3 for b in the formula shown gives

$$|-5 - 3| = |-8| = 8 \quad \text{or} \quad |3 - (-5)| = |8| = 8.$$

☑ **C.** You've just reviewed the absolute value of a real number

Now try Exercises 59 through 66 ▶

TECHNOLOGY SUPPORT

On many calculators, the absolute value function is accessed by pressing the **MATH** key, then navigating to "**NUM**" (number options) and selecting option **1: abs(**. The calculator provides the left parenthesis, you must supply the right.

```
abs(8-2)
              6
abs(2-8)
              6
```

D. The Order of Operations

The operations of addition, subtraction, multiplication, and division are defined for the set of real numbers, and the concept of absolute value plays an important role. Prior to our study of the order of operations, we will review fundamental concepts related to division and zero, exponential notation, and square roots/cube roots.

Division and Zero

The quotient $\frac{36}{9} = 4$ can be checked using the related multiplication: $4 \cdot 9 = 36\checkmark$. A similar check can be used to understand quotients involving zero.

EXAMPLE 10 ▶ Understanding Division with Zero by Writing the Related Product

Rewrite each quotient *using the related product*.

    a. $0 \div 8 = p$ **b.** $\frac{16}{0} = q$ **c.** $\frac{0}{12} = n$

Solution ▶   **a.** $0 \div 8 = p$, if $p \cdot 8 = 0$. **b.** $\frac{16}{0} = q$, if $q \cdot 0 = 16$.   **c.** $\frac{0}{12} = n$, if $n \cdot 12 = 0$.
        This shows $p = 0$.   There is no such number q.   This shows $n = 0$.

Now try Exercises 67 through 70 ▶

WORTHY OF NOTE

When a pizza is delivered to your home, it often has "8 parts to the whole," and in fraction form we have $\frac{8}{8}$. When all 8 pieces are eaten, 0 pieces remain and the fraction form becomes $\frac{0}{8} = 0$. However, the expression $\frac{8}{0}$ is meaningless (undefined), since it would indicate a pizza that has "0 parts to the whole (??)."

In Example 10(a), a dividend of 0 and a divisor of 8 means we are going to divide zero into eight groups. The related multiplication shows there will be zero in each group. As in Example 10(b), an expression with a divisor of 0 *cannot be computed or checked.* Although it seems trivial, division by zero has many implications in a study of mathematics, so make an effort to know the facts: The quotient of zero and any nonzero number is zero $\left(\frac{0}{n} = 0\right)$ but division *by* zero is undefined $\left(\frac{n}{0}\right.$ is undefined$\left.\right)$. The special case of $\frac{0}{0}$ is said to be **indeterminate,** as $\frac{0}{0} = n$ *appears* to be true for all real numbers n (since the check gives $n \cdot 0 = 0\checkmark$). The expression $\frac{0}{0}$ is studied in greater detail in more advanced classes.

Division and Zero

The quotient of zero and any real number n is zero ($n \neq 0$):

$$0 \div n = 0 \qquad\qquad \frac{0}{n} = 0.$$

The expressions $n \div 0$ and $\dfrac{n}{0}$ are undefined.

TECHNOLOGY SUPPORT

On a graphing calculator, we can evaluate the same expression simultaneously for multiple values using braces "{ }". Note that when the divisor is any nonzero number x, the calculator shows that $\frac{0}{x} = 0$ (see Figures AI.5 and AI.6), while $\frac{x}{0}$ returns an error message (Figure AI.7).

Figure AI.5

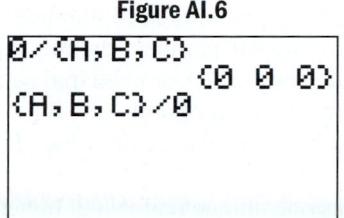

```
-279→A
            -279
1572→B
            1572
√(2)→C
      1.414213562
```

Figure AI.6

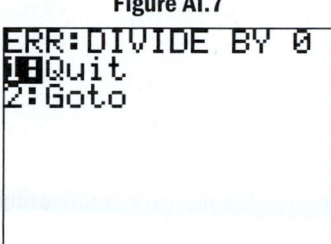

```
0/{A,B,C}
            {0 0 0}
{A,B,C}/0
```

Figure AI.7

```
ERR:DIVIDE BY 0
1:Quit
2:Goto
```

Squares, Cubes, and Exponential Form

When a number is repeatedly multiplied by itself as in $(10)(10)(10)(10)$, we write it using **exponential notation** as 10^4. The number used for repeated multiplication (in this case 10) is called the **base,** and the superscript number is called an **exponent.** The exponent tells how many times the base occurs as a factor, and we say 10^4 is written in **exponential form.** Numbers that result from squaring an integer are called **perfect squares,** while numbers that result from cubing an integer are called **perfect cubes.** These are often collected into a table, such as Table AI.1, and students are strongly encouraged to memorize these values to help complete many common calculations mentally. Only the square and cube of selected positive integers are shown.

Table AI.1

Perfect Squares				Perfect Cubes	
N	N^2	N	N^2	N	N^3
1	1	7	49	1	1
2	4	8	64	2	8
3	9	9	81	3	27
4	16	10	100	4	64
5	25	11	121	5	125
6	36	12	144	6	216

EXAMPLE 11 ▶ **Evaluating Numbers in Exponential Form**

Write each exponential in expanded form, then determine its value.

    a. 4^3 **b.** $(-6)^2$ **c.** -6^2 **d.** $\left(\frac{2}{3}\right)^3$

Solution ▶   **a.** $4^3 = 4 \cdot 4 \cdot 4 = 64$ **b.** $(-6)^2 = (-6) \cdot (-6) = 36$

                c. $-6^2 = -(6 \cdot 6) = -36$ **d.** $\left(\frac{2}{3}\right)^3 = \frac{2}{3} \cdot \frac{2}{3} \cdot \frac{2}{3} = \frac{8}{27}$

Now try Exercises 71 and 72 ▶

    Examples 11(b) and 11(c) illustrate an important distinction. The expression $(-6)^2$ gives a single operation, "the square of negative six" and the negative sign is included in both factors. The expression -6^2 gives two operations, "six is squared, and the result is made negative." The square of six is calculated first, with the negative sign applied afterward.

Square Roots and Cube Roots

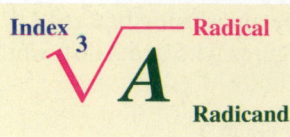

Index — Radical
$$\sqrt[3]{A}$$
Radicand

For the square root operation, either the $\sqrt{}$ or $\sqrt[2]{}$ notation can be used. The $\sqrt{}$ symbol is called a **radical**, the number under the radical is called the **radicand**, and the small case number used is called the **index** (see figure). The index tells how many factors are needed to obtain the radicand. For example, $\sqrt{25} = 5$, since $5 \cdot 5 = 5^2 = 25$ (when the $\sqrt{}$ symbol is used, the index is understood to be 2). In general, $\sqrt{a} = b$ only if $b^2 = a$. All numbers greater than zero have one positive and one negative square root. The *positive* or **principal square root** of 49 is 7 $(\sqrt{49} = 7)$ since $7^2 = 49$. The *negative* square root of 49 is -7 $(-\sqrt{49} = -7)$.

The cube root of a number has the form $\sqrt[3]{a} = b$, where $b^3 = a$. This means $\sqrt[3]{27} = 3$ since $3^3 = 27$, and $\sqrt[3]{-8} = -2$ since $(-2)^3 = -8$. The cube root of a real number has one unique real value. In general, we have the following:

Square Roots	Cube Roots
$\sqrt{a} = b$ if $b^2 = a$	$\sqrt[3]{a} = b$ if $b^3 = a$
$(a \geq 0)$	$(a \in \mathbb{R})$
This indicates that	This indicates that
$\sqrt{a} \cdot \sqrt{a} = a$	$\sqrt[3]{a} \cdot \sqrt[3]{a} \cdot \sqrt[3]{a} = a$
or $(\sqrt{a})^2 = a$	or $(\sqrt[3]{a})^3 = a$

WORTHY OF NOTE

It is helpful to note that both 0 and 1 are their own square root, cube root, and *n*th root. That is, $\sqrt{0} = 0$, $\sqrt[3]{0} = 0, \dots, \sqrt[n]{0} = 0$; and $\sqrt{1} = 1$, $\sqrt[3]{1} = 1, \dots, \sqrt[n]{1} = 1$.

EXAMPLE 12 ▶ **Evaluating Square Roots and Cube Roots**

Determine the value of each expression.

a. $\sqrt{49}$ **b.** $\sqrt[3]{125}$ **c.** $\sqrt{\frac{9}{16}}$ **d.** $-\sqrt{16}$ **e.** $\sqrt{-25}$

Solution ▶   **a.** $\sqrt{49} = 7$ since $7 \cdot 7 = 49$ **b.** $\sqrt[3]{125} = 5$ since $5 \cdot 5 \cdot 5 = 125$

c. $\sqrt{\frac{9}{16}} = \frac{3}{4}$ since $\frac{3}{4} \cdot \frac{3}{4} = \frac{9}{16}$ **d.** $-\sqrt{16} = -4$ since $\sqrt{16} = 4$

e. $\sqrt{-25}$ is not a real number [note that $5 \cdot 5 = (-5)(-5) = 25$]

Now try Exercises 73 through 78 ▶

TECHNOLOGY SUPPORT

On many graphing calculators, the cube root function is accessed by pressing the **MATH** key and using option **4**: $\sqrt[3]{(}$. The properties stated in the tan box for cube roots are illustrated here using the cube root of 7.

```
³√(7)
          1.912931183
Ans^3
                    7
³√(7)³√(7)³√(7)
                    7
```

WORTHY OF NOTE

Sometimes the acronym **PEMDAS** is used as a more concise way to recall the order of operations: **P**arentheses, **E**xponents, **M**ultiplication, **D**ivision, **A**ddition, and **S**ubtraction. The idea has merit, so long as you remember that multiplication and division *have an equal rank,* as do addition and subtraction, and these must be computed in the order they occur (from left to right).

For square roots, if the radicand is a perfect square or has perfect squares in both the numerator and denominator, the result is a rational number as in Examples 12(a) and 12(c). If the radicand is not a perfect square, the result is an irrational number. Similar statements can be made regarding cube roots [Example 12(b)].

The Order of Operations

When basic operations are combined into a larger mathematical expression, we use a specified **priority** or **order of operations** to evaluate them.

The Order of Operations

1. Simplify within grouping symbols (parentheses, brackets, braces, etc.). If there are "nested" symbols of grouping, begin with the innermost group. If a fraction bar is used, simplify the numerator and denominator separately.
2. Evaluate all exponents and roots.
3. Compute all multiplications or divisions *in the order they occur from left to right.*
4. Compute all additions or subtractions *in the order they occur from left to right.*

EXAMPLE 13 ▶ **Evaluating Expressions Using the Order of Operations**

Simplify using the order of operations:

a. $5 + 2 \cdot 3$

b. $8 + 36 \div 4(12 - 3^2)$

c. $\dfrac{-4.5(8) - 3}{\sqrt[3]{125} + 2^3}$

d. $7500\left(1 + \dfrac{0.075}{12}\right)^{12 \cdot 15}$

Solution ▶ **a.** $5 + 2 \cdot 3 = 5 + 6$ multiplication before addition

$= 11$ result

WORTHY OF NOTE

Many common tendencies are hard to overcome. For instance, let's evaluate the expressions $3 + 4 \cdot 5$ and $24 \div 6 \cdot 2$. For the first, the correct result is 23 (multiplication before addition), though some will get 35 by adding first. For the second, the correct result is 8 (multiplication or division *in order*), though some will get 2 by multiplying first.

b. $8 + 36 \div 4(12 - 3^2)$

$= 8 + 36 \div 4(12 - 9)$ simplify within parentheses

$= 8 + 36 \div 4(3)$ $12 - 9 = 3$

$= 8 + 9(3)$ division before multiplication

$= 8 + 27$ multiply

$= 35$ result

c. $\dfrac{-4.5(8) - 3}{\sqrt[3]{125} + 2^3}$ original expression

$= \dfrac{-36 - 3}{5 + 8}$ simplify terms in the numerator and denominator

$= \dfrac{-39}{13}$ combine terms

$= -3$ result

d. $7500\left(1 + \dfrac{0.075}{12}\right)^{12 \cdot 15}$ original expression

$= 7500(1.00625)^{12 \cdot 15}$ simplify within the parenthesis (division before addition)

$= 7500(1.00625)^{180}$ simplify the exponent so it can be applied

$\approx 7500(3.069451727)$ exponents before multiplication

$\approx 23{,}020.88795$ result

☑ **D.** You've just reviewed the order of operations

Now try Exercises 79 through 104 ▶

TECHNOLOGY SUPPORT

While graphing and calculating technology give us enormous power and convenience, great care must still be taken in its use and application. In the case of Example 13(d), note that the product in the exponent must be parenthesized so that the calculator executes the operations as intended.

```
7500(1+.075/12)^
12*15
           121233.6674
7500(1+.075/12)^
(12*15)
         23020.88795
```

A.I EXERCISES

▶ CONCEPTS AND VOCABULARY

Fill in each blank with the appropriate word or phrase. Carefully reread the section, if necessary.

1. The symbol $\subset$ means: is a __proper__ __subset__ of and the symbol $\in$ means: is an __element__ of.

2. A number corresponding to a point on the number line is called the __coordinate__ of that point.

3. Every positive number has two square roots, one __positive__ and one __negative__. The two square roots of 49 are __7__; and __−7__ $\sqrt{49}$ represents the __principal__ square root of 49.

4. The decimal form of $\sqrt{7}$ contains an infinite number of non __repeating__ and non __terminating__ digits. This means that $\sqrt{7}$ is a(n) __irrational__ number.

5. Discuss/Explain why the value of $12 \cdot \frac{1}{3} + \frac{2}{3}$ is $4\frac{2}{3}$ and not 12.   Order of operations requires multiplication before addition.

6. Discuss/Explain (a) why $(-5)^2 = 25$, while $-5^2 = -25$; and (b) why $-5^3 = (-5)^3 = -125$.
   a. $(-5)^2 = (-5)(-5) = 25$ b. $-5^3 = -(5)(5)(5)$
      $-5^2 = -(5 \cdot 5) = -25$ $= -125$
      $(-5)^3 = (-5)(-5)(-5)$
      $= -125$

▶ DEVELOPING YOUR SKILLS

7. List the natural numbers that are
   a. less than 6. $\{1, 2, 3, 4, 5\}$
   b. less than 1. $\{\ \}$

8. List the natural numbers that are
   a. between 0 and 1. $\{\ \}$
   b. greater than 50. $\{51, 52, 53, 54, \ldots\}$

Identify each of the following statements as either true or false. If false, give an example that shows why.

9. $\mathbb{N} \subset \mathbb{W}$ True

10. $\mathbb{W} \not\subset \mathbb{N}$ True

11. $\{33, 35, 37, 39\} \subset \mathbb{W}$ True

12. $\{2.2, 2.3, 2.4, 2.5\} \subset \mathbb{W}$ False; 2.2 is not a whole number.

13. $6 \in \{0, 1, 2, 3, \ldots\}$ True

14. $1297 \notin \{0, 1, 2, 3, \ldots\}$ False; 1297 is a whole number.

Convert to decimal form and graph by estimating the number's location between two integers.

15. $\frac{4}{3}$ 16. $-\frac{7}{8}$ 17. $2\frac{5}{9}$ 18. $-1\frac{5}{6}$

19. A Texas rancher has 120,000 acres of range land, and wants to use two-thirds of it for cattle and the remaining $\frac{1}{3} = 0.\overline{3}$ for sheep. (a) Using the fraction $\frac{1}{3}$, how many acres will be set aside for sheep? (b) Using a calculator, determine the number of "repeating 3's" that are required (0.3, 0.33, 0.333, etc.) before the correct answer is returned.
   a. 40,000 acres   b. ten 3's (0.333 333 333 3, may vary by calculator model)

20. An architect is reviewing the floor plan for a new office building that offers 36,000 ft² of office space on the first floor. On this floor, $\frac{13}{18} = 0.7\overline{2}$ is considered premium frontage space. (a) Using the fraction $\frac{13}{18}$, how many square feet is considered "premium?" (b) Using a calculator, determine the number of "repeating 2's" that are required (0.72, 0.722, 0.7222, etc) before the correct answer is returned.   a. 26,000 ft²   b. nine 2's (0.722 222 222 2, may vary by calculator model)

Use a calculator to approximate the value of each number (round to hundredths as needed). Then graph each number by estimating its location between two integers.

21. 7   22. $\frac{75}{4}$ 23. 3   24. $\frac{25\pi}{2}$

For the sets in Exercises 25 through 28:

a. List all numbers that are elements of (i) $\mathbb{N}$, (ii) $\mathbb{W}$, (iii) $\mathbb{Z}$, (iv) $\mathbb{Q}$, (v) $\mathbb{H}$, and (vi) $\mathbb{R}$.

b. Reorder the elements of each set from smallest to largest.

c. Graph the elements of each set on a number line.

25. $\{-1, 8, 0.75, \frac{9}{2}, 5.\overline{6}, 7, \frac{3}{5}, 6\}$

26. $\{-7, 2.\overline{1}, 5.73, -3\frac{5}{6}, 0, -1.12, \frac{7}{8}\}$

27. $\{-5, \sqrt{49}, 2, -3, 6, -1, \sqrt{3}, 0, 4, \pi\}$

28. $\{-8, 5, -2\frac{3}{5}, 1.75, -\sqrt{2}, -0.6, \pi, \frac{7}{2}, \sqrt{64}\}$

State true or false. If false, state why.

29. $\mathbb{R} \subset \mathbb{H}$ False; not all real numbers are irrational.   30. $\mathbb{N} \subset \mathbb{R}$ True

31. $\mathbb{Q} \subset \mathbb{Z}$ False; not all rational numbers are integers.   32. $\mathbb{Z} \subset \mathbb{Q}$ True

33. $\sqrt{25} \in \mathbb{H}$ False; $\sqrt{25} = 5$ is not irrational.   34. $\sqrt{19} \in \mathbb{H}$ True

Match each set with its correct symbol and description/illustration.

35. _c_ IV Irrational numbers   a. $\mathbb{R}$ I. $\{1, 2, 3, 4, \ldots\}$

36. _f_ V Integers   b. $\mathbb{Q}$ II. $\{\frac{a}{b}, | a, b \in \mathbb{Z}, b \neq 0\}$

37. _a_ VI Real numbers   c. $\mathbb{H}$ III. $\{0, 1, 2, 3, 4, \ldots\}$

38. _b_ II Rational numbers   d. $\mathbb{W}$ IV. $\{\pi, \sqrt{7}, -\sqrt{13}, \text{etc.}\}$

39. _d_ III Whole numbers   e. $\mathbb{N}$ V. $\{\ldots -3, -2, -1, 0, 1, 2, 3, \ldots\}$

40. _e_ I Natural numbers   f. $\mathbb{Z}$ VI. $\mathbb{N}, \mathbb{W}, \mathbb{Z}, \mathbb{Q}, \mathbb{H}$

Use a descriptive variable or the variable given with an inequality symbol ($<$, $>$, $\leq$, $\geq$) to write a model for each statement.

41. To spend the night at a friend's house, Kylie must be at least 6 years old.
Let a represent Kylie's age: $a \geq 6$ years.

42. Monty can spend at most $2500 on the purchase of a used automobile.
Let a represent the amount: $a \leq 2500$ dollars.

43. If Jerod gets no more than two words incorrect on his spelling test he can play in the soccer game this weekend.
Let n represent the number of incorrect words: $n \leq 2$ incorrect.

44. Andy must weigh less than 112 lb to be allowed to wrestle in his weight class at the meet.
Let w represent Andy's weight: $w < 112$ lb.

45. In order for the expression $\sqrt{2x - 3}$ to represent a real number, x must be greater than or equal to $\frac{3}{2}$. $x \geq \frac{3}{2}$

46. In order for the expression $\sqrt{5 - 4x}$ to represent a real number, x must be less than or equal to $\frac{5}{4}$. $x \leq \frac{5}{4}$

47. In order for the expression $\dfrac{1}{\sqrt{2 - x}}$ to represent a real number, x must be less than 2. $x < 2$

48. In order for the expression $\dfrac{1}{\sqrt{x - 7}}$ to represent a real number, x must be greater than 7. $x > 7$

49. In order for a weight sensor to function properly, an item must weigh at least 5 grams, but less than 32 grams. $5 \leq w < 32$

50. To warn against trespassers, a new motion detector is installed. The detector's range is from 2 m, to no more than 20 m. $2 \leq r \leq 20$

Evaluate/simplify each expression.

51. $|-2.75|$ 2.75   52. $|-7.24|$ 7.24

53. $-|-4|$ -4 54. $-|-6|$ -6

55. $\left|\dfrac{1}{2}\right|$ $\dfrac{1}{2}$ 56. $\left|\dfrac{2}{5}\right|$ $\dfrac{2}{5}$

57. $\left|-\dfrac{3}{4}\right|$ $\dfrac{3}{4}$ 58. $\left|-\dfrac{3}{7}\right|$ $\dfrac{3}{7}$

Use the concept of absolute value to complete Exercises 59 to 68.

59. Write the statement two ways, then simplify. "The distance between -7.5 and 2.5 is..."
$|-7.5 - 2.5|, |2.5 - (-7.5)|, 10$

60. Write the statement two ways, then simplify. "The distance between $13\frac{2}{5}$ and $-2\frac{3}{5}$ is..."
$|-13\frac{2}{5} - (-2\frac{3}{5})|, |-2\frac{3}{5} - 13\frac{2}{5}|, 16$

61. What two numbers on the number line are five units from negative three? $-8, 2$

62. What two numbers on the number line are three units from two? $5, -1$

63. If n is positive, then $-n$ is ___negative___.

64. If n is negative, then $-n$ is ___positive___.

65. If $n < 0$, then $|n| = $ ___$-n$___.

66. If $n > 0$, then $|n| = $ ___n___.

Determine which expressions are equal to zero and which are undefined. Justify your responses by writing the related multiplication.

67. $12 \div 0$ undefined, since $12 \div 0 = k$ implies $k \cdot 0 = 12$ 68. $0 \div 12$ $0 \div 12 = 0$, since $0 \cdot 12 = 0$

69. $\dfrac{7}{0}$ undefined, since $7 \div 0 = k$ implies $k \cdot 0 = 7$ 70. $\dfrac{0}{7}$ $0 \div 7 = 0$, since $0 \cdot 7 = 0$

Without computing the actual answer, state whether the result will be positive or negative. Be careful to note what power is used and whether the negative sign is included in parentheses.

71. a. $(-7)^2$ Positive   b. -7^2 negative
    c. $(-7)^5$ negative   d. -7^5 negative

72. a. $(-7)^3$ negative   b. -7^3 negative
    c. $(-7)^4$ positive   d. -7^4 negative

Evaluate without the aid of a calculator.

73. $-\sqrt{\dfrac{121}{36}}$ $-\frac{11}{6}$

74. $-\sqrt{\dfrac{25}{49}}$ $-\frac{5}{7}$

75. $\sqrt[3]{-8}$ -2

76. $\sqrt[3]{-64}$ -4

77. What perfect square is closest to 78? $9^2 = 81$ is closest

78. What perfect cube is closest to -71?
$(-4)^3 = -64$ is closest

Perform the operation indicated without the aid of a calculator.

79. $-24 - (-31)$ 7

80. $-45 - (-54)$ 9

81. $7.045 - 9.23$ -2.185

82. $0.0762 - 0.9034$ -0.8272

83. $4\frac{5}{6} + \left(-\frac{1}{2}\right)$ $4\frac{1}{3}$

84. $1\frac{1}{8} + \left(-\frac{3}{4}\right)$ $\frac{3}{8}$

85. $\left(-\frac{2}{3}\right)\left(3\frac{5}{8}\right)$ $-\frac{29}{12}$ or $-2\frac{5}{12}$

86. $(-8)\left(2\frac{1}{4}\right)$ -18

87. $(12)(-3)(0)$ 0

88. $(-1)(0)(-5)$ 0

89. $-60 \div 12$ -5

90. $75 \div (-15)$ -5

91. $\frac{4}{5} \div (-8)$ $-\frac{1}{10}$

92. $-15 \div \frac{1}{2}$ -30

93. $-\frac{2}{3} \div \frac{16}{21}$ $-\frac{7}{8}$

94. $-\frac{3}{4} \div \frac{7}{8}$ $-\frac{6}{7}$

Evaluate without a calculator, using the order of operations.

95. $12 - 10 \div 2 \times 5 + (-3)^2$ -4

96. $(5 - 2)^2 - 16 \div 4 \cdot 2 - 1$ 0

97. $\sqrt{\dfrac{9}{16}} - \dfrac{3}{5} \cdot \left(\dfrac{5}{3}\right)^2$ $-\frac{11}{12}$

98. $\left(\dfrac{3}{2}\right)^2 \div \left(\dfrac{9}{4}\right) - \sqrt{\dfrac{25}{64}}$ $\frac{3}{8}$

99. $\dfrac{4(-7) - 6^2}{6 - \sqrt{49}}$ 64

100. $\dfrac{5(-6) - 3^2}{9 - \sqrt{64}}$ -39

Evaluate using a calculator (round to hundredths).

101. $2475\left(1 + \dfrac{0.06}{4}\right)^{4 \cdot 10}$ 4489.70

102. $5100\left(1 + \dfrac{0.078}{52}\right)^{52 \cdot 20}$ $24{,}241.64$

103. $\dfrac{-4 + \sqrt{(-4)^2 - 4(3)(-39)}}{2(3)}$ 3

104. $\dfrac{-12 - \sqrt{(-12)^2 - 4(-2)(32)}}{2(-2)}$ 8

▶ WORKING WITH FORMULAS

105. **Pitch diameter:** $D = \dfrac{d \cdot n}{n + 2}$

Mesh gears are used to transfer rotary motion and power from one shaft to another. The *pitch diameter D* of a drive gear is given by the formula shown, where *d* is the outer diameter of the gear and *n* is the number of teeth on the gear. Find the pitch diameter of a gear with 12 teeth and an outer diameter of 5 cm. $D \approx 4.3$ cm

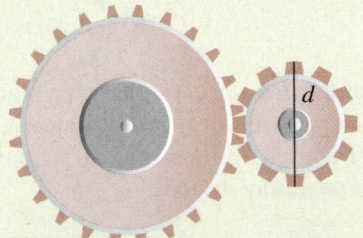

106. **Pediatric dosages and Clark's rule:** $D_C = \dfrac{D_A \cdot W}{150}$

The amount of medication prescribed for young children depends on their weight, height, age, body surface area and other factors. **Clark's rule** is a formula that helps estimate the correct child's dose D_C based on the adult dose D_A and the weight W of the child (an average adult weight of 150 lb is assumed). Compute a child's dose if the adult dose is 50 mg and the child weighs 30 lb. $D_C = 10$ mg

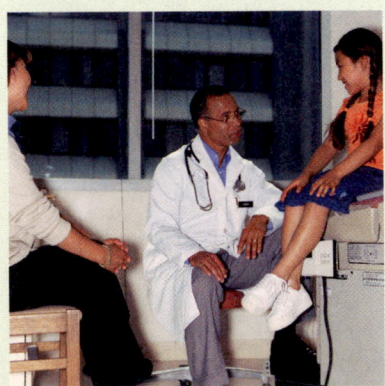

▶ APPLICATIONS

Use positive and negative numbers to model the situation, then compute.

107. Temperature changes: At 6:00 P.M., the temperature was 50°F. A cold front moves through that causes the temperature to *drop* 3°F each hour until midnight. What is the temperature at midnight?   32°F

108. Air conditioning: Most air conditioning systems are designed to create a 2° *drop* in the air temperature each hour. How long would it take to reduce the air temperature from 86° to 71°? $7\frac{1}{2}$ hr

109. Record temperatures: The state of California holds the record for the greatest temperature swing between a record high and a record low. The record high was 134°F and the record low was −45°F. How many degrees *difference* are there between the record high and the record low?   179°F

110. Cold fronts: In Juneau, Alaska, the temperature was 17°F early one morning. A cold front later moved in and the temperature *dropped* 32°F by lunchtime. What was the temperature at lunchtime?   −15°F

Use a calculator and the rational/irrational numbers given to compute.

111. An insect crawls $15\sqrt{2}$ cm along the diagonal of a square, $\frac{31}{2}$ cm along the length of a line segment, and 10π cm around the circumference of a circle. What is the total distance traveled (to the nearest 100th of a cm)?

112. Find the distance between points A and B (rounded to the nearest 10th cm), given that the square and equilateral triangle have sides of length 10 cm, the circle has a *circumference* of 22 cm, and the line segment is $\frac{35}{2}$ cm long.

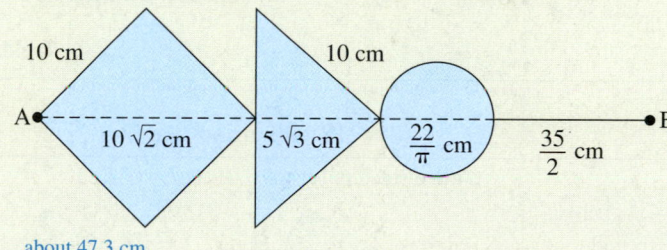

about 68.13 cm

about 47.3 cm

▶ EXTENDING THE CONCEPT

113. Here are some historical approximations for π. Which one is closest to the true value?

Archimedes: $3\frac{1}{7}$ Tsu Ch'ung-chih: $\frac{355}{113}$

Aryabhata: $\frac{62,832}{20,000}$ Brahmagupta: $\sqrt{10}$

Tsu Ch'ung-chih: $\frac{355}{113}$

114. If $A > 0$ and $B < 0$, is the product $A \cdot (-B)$ positive or negative?   positive

115. If $A < 0$ and $B < 0$, is the quotient $-(A \div B)$ positive or negative?   negative

Appendix II

Geometry Review with Unit Conversions

Developing the ability to use mathematics as a descriptive tool is a major goal of this text. Without a solid understanding of basic geometry, this goal would be difficult to achieve—as many of the tasks we perform daily are based on decisions regarding size, measurement, configuration, and the like.

A. Perimeter and Area Formulas

Basic geometry plays an important role in the application of mathematics. For your convenience, the most common formulas are collected in Table A.II.1, and a focused effort should be made to commit them to memory. Note that some of the formulas use **subscripted variables,** or a variable with a small case number to the lower right (s_1, s_2, and so on). To help understand the table, we quickly review some fundamental terms

Table A.II.1

	Definition and Diagram	Perimeter Formula (linear units or *units*)	Area Formula (square units or *units²*)
triangle	a three sided polygon	$P = s_1 + s_2 + s_3$	$A = \dfrac{1}{2}bh$
rectangle	a quadrilateral with four right angles and opposite sides parallel	$P = 2L + 2W$	$A = LW$
square	a rectangle with four equal sides	$P = 4s$	$A = s^2$
trapezoid	a quadrilateral with one pair of parallel sides (called bases b_1 and b_2)	sum of all sides $P = s_1 + s_2 + s_3 + s_4$	$A = \dfrac{h}{2}(b_1 + b_2)$
circle	the set of all points lying in a plane that are an equal distance (called the radius r) from a given point (called the center C).	$C = 2\pi r$ or $C = \pi d$	$A = \pi r^2$

and their meaning. A **plane** is the infinite extension of length and width along a flat surface. **Perimeter** is the distance around a two-dimensional figure, or a closed figure that lies in a plane. Many times these figures are **polygons,** or closed figures composed of line segments. The general name for a four-sided polygon is a **quadrilateral. A right angle** is an angle measuring 90°. A quadrilateral with four right angles is called a **rectangle. Area** is a measure of the amount of surface covered by a plane figure, with the measurement given in **square units.**

The formulas $C = \pi d$ and $C = 2\pi r$ both use the symbol "π," which represents the ratio of a circle's circumference to its diameter. We will use a two decimal approximation in calculations done by hand: $\pi \approx 3.14$. On many calculators, π is the [2nd] function to the [^] key and produces a much better approximation (see Figure A.II.1). When using a calculator, we most often use all displayed digits and round only the answer to the desired level of accuracy.

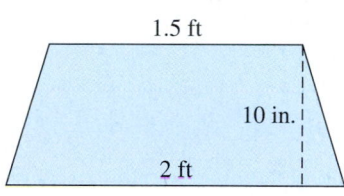

If a problem or application uses a formula, begin by stating the formula rather than by immediately making any substitutions. This will help to prevent many careless errors. For Example 1, recall that a trapezoid is a quadrilateral with two parallel sides.

EXAMPLE 1 ▶ **Finding the Area of a Window**

A basement window is shaped like an isosceles trapezoid (base angles equal, opposite sides equal in length), with a height of 10 in. and bases of 1.5 ft and 2 ft. What is the area of the glass in the window?

Solution ▶ Before applying the area formula, all measures must use the same unit. In inches we have 1.5 ft = 18 in. and 2 ft = 24 in.

$$A = \frac{h}{2}(b_1 + b_2) \qquad \text{given formula}$$

$$= \frac{10 \text{ in.}}{2}(18 \text{ in.} + 24 \text{ in.}) \qquad \text{substitute 10 in. for } h, \text{ 18 in. for } b_1 \text{ and 24 in. for } b_2$$

$$= (5 \text{ in.})(42 \text{ in.}) \qquad \text{simplify}$$

$$= 210 \text{ in}^2 \qquad \text{result}$$

The area of the glass in the window is 210 in².

Now try Exercises 7 through 18 ▶

WORTHY OF NOTE

In actual practice, most calculations are done without using the units of measure, with the correct units supplied in the final answer. When like units occur in an exercise, they are treated just as the numeric factors. If they are part of a product, we write the units with an appropriate exponent as in Example 1. If the like units occur in the numerator and denominator, they "cancel" as we will see in Example 5.

Composite Figures

The largest part of geometric applications, whether in art, construction, or architecture, involve **composite figures,** or figures that combine basic shapes. In many cases we are able to **partition** or break the figure into more common shapes using an **auxiliary line,** or a dashed line drawn to highlight certain features of the diagram. When computing a perimeter, we use only the exposed, outer edges, much as a soldier would guard the base camp by marching along the outer edge—the perimeter. For composite figures, it's helpful to verbally describe the situation given, creating an English language model that can easily be translated into an equation model.

EXAMPLE 2 ▶   **Determining the Perimeter and Area of a Composite Figure**

Find the perimeter and area of the composite
Figure A.II.2. Use $\pi \approx 3.14$.

Figure A.II.2

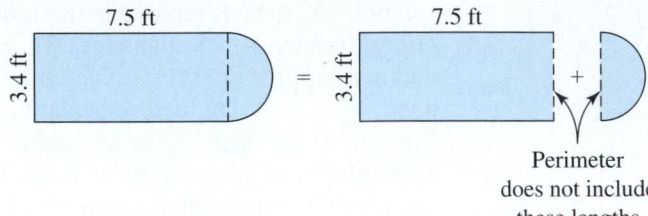

7.5 ft

3.4 ft

Solution ▶   To compute the perimeter, use only the exposed
(outer) edges as shown in Figure A.II.3.

Figure A.II.3

7.5 ft   7.5 ft

3.4 ft   =   3.4 ft   +

Perimeter
does not include
these lengths

- Perimeter = three sides of rectangle + one-half circle   verbal model

$$P = 2L + W + \frac{\pi d}{2}$$ formula model

$$\approx 2(7.5) + 3.4 + \frac{(3.14)(3.4)}{2}$$ substitute 7.5 for L, 3.4 for W, and 3.14 for π

$$\approx 15 + 3.4 + 5.338$$ simplify

$$\approx 23.738$$ result

The perimeter of the figure is about 23.7 ft.

- Total Area = area of rectangle + one-half area of circle   verbal model

$$A = LW + \frac{\pi r^2}{2}$$ formula model

$$\approx (7.5)(3.4) + \frac{(3.14)(1.7)^2}{2}$$ substitute 1.7 for r (d/2)

$$\approx 25.5 + 4.5373$$ simplify

$$\approx 30.0373$$ result

The area of the figure is about 30.0 ft².

☑ **A.** You've just reviewed
how to find the perimeter and
area of common geometric
figures.

Now try Exercises 19 through 30 ▶

B. Volume

Volume is a measure of the amount of space occupied by a three-dimensional object
and is measured in **cubic units.** Some of the more common formulas are given in
Table AII.2.

Table A.II.2

	Illustration		Volume Formula (cubic units or units^3)
rectangular solid	a six-sided solid figure with opposite faces congruent and adjacent faces meeting at right angles		$V = LWH$
cube	a rectangular solid with six congruent, square faces		$V = s^3$
sphere	the set of all points an equal distance (called the radius) from a given point (called the center)		$V = \dfrac{4}{3}\pi r^3$
right circular cylinder	union of all line segments connecting two congruent circles in parallel planes, meeting each at a right angle		$V = \pi r^2 h$
right circular cone	union of all line segments connecting a given point (vertex) to a given circle (base) and whose altitude meets the center of the base at a right angle		$V = \dfrac{1}{3}\pi r^2 h$
right square pyramid	union of all line segments connecting a given point (vertex) to a given polygon (base) and whose altitude meets the center of the base at a right angle		$V = \dfrac{1}{3}s^2 h$

 EXAMPLE 3 ▶ **Determining the Volume of a Composite Figure**

Sand at a cement factory is being dumped from a conveyor belt into a pile shaped like a right circular cone atop a right circular cylinder (see figure). How many cubic feet of sand are there at the moment the cone is 6 ft high with a diameter of 10 ft?

WORTHY OF NOTE

It is again worth noting that units of measure are treated as though they were numeric factors. For the cylinder in Example 3:
$\pi \cdot (5 \text{ ft})^2 \cdot (3 \text{ ft}) = \pi \cdot 25 \text{ ft}^2 \cdot 3 \text{ ft} = 75\pi \text{ ft}^3$. This concept is an important part of the unit conversions often used in the application of mathematics.

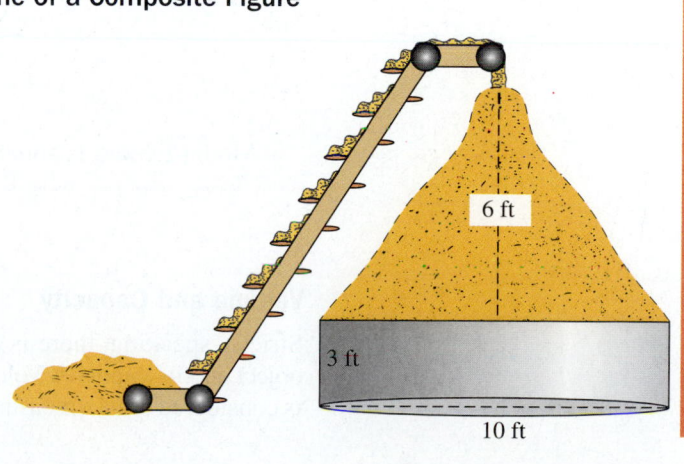

Solution ▶   Total Volume = volume of cylinder + volume of cone   verbal model

$$V = \pi r^2 h_1 + \frac{1}{3}\pi r^2 h_2$$ formula model

$$= \pi(5)^2(3) + \frac{1}{3}\pi(5)^2(6)$$ substitute 5 for r, 3 for h_1, and 6 for h_2

$$= 75\pi + 50\pi$$ simplify

$$= 125\pi$$ result (exact form)

There are about 392.7 ft^3 of sand in the pile.

✓ **B.** You've just reviewed how to compute the volume of common geometric solids.

Now try Exercises 31 through 36 ▶

C. Unit Conversion Factors

Unit conversion factors are used to convert from one unit of measure to a related unit, like ounces to pounds. The procedure used is based on the fact that multiplying any quantity by the equivalent of 1 does not change its value. Several examples of unit conversion factors are shown here.

a. $\dfrac{12 \text{ inches}}{1 \text{ foot}}$ **b.** $\dfrac{2000 \text{ pounds}}{1 \text{ ton}}$ **c.** $\dfrac{4 \text{ quarts}}{1 \text{ gallon}}$

OR   OR   OR

$\dfrac{1 \text{ foot}}{12 \text{ inches}}$ $\dfrac{1 \text{ ton}}{2000 \text{ pounds}}$ $\dfrac{1 \text{ gallon}}{4 \text{ quarts}}$

A unit factor is set up so that the units you are converting *from* cancel out, leaving only the units you are converting *to*. When doing these conversions, it is helpful to set up the units first and then write the numeric values with their related unit.

EXAMPLE 4 ▶   **Unit Conversions—Feet to Miles**

Mt. Everest is 29,035 ft high and is the highest mountain in the world when measured from see level. Convert 29,035 ft to miles.

Solution ▶   **a.** Set up the units of the conversion factor so that "feet" will cancel.

$$\frac{29,035 \text{ ft}}{1}\left(\frac{\text{mi}}{\text{ft}}\right)$$

b. Recall that 5280 ft = 1 mi, and write the numeric values with their related unit: "5280" with feet and "1" with mile. Then simplify the result and state your answer.

$$\frac{29,035 \text{ ft}}{1}\left(\frac{1 \text{ mi}}{5280 \text{ ft}}\right) = \frac{29,035}{5280}\text{ mi}$$

$$\approx 5.49905303$$

Mount Everest is about 5.5 mi high.

Now try Exercises 37 and 38 ▶

Volume and Capacity

Strictly speaking, there is a distinction between the **volume** of a three dimensional object and its **capacity.** Volume is the amount of space occupied by the object, capacity is considered to be a measure of the object's potential for holding or storing something.

Some of the common relationships between units of volume and capacity are given in the following table.

Volume and Capacity
$1 \text{ gal} = 231 \text{ in}^3$ $1 \text{ L} = 1000 \text{ cm}^3 = 1 \text{ dm}^3$ $1 \text{ mL} = 1 \text{ cm}^3$

EXAMPLE 5 ▶ **Unit Conversions—Cubic Inches to gallons**

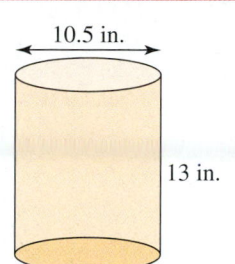
10.5 in.
13 in.

While taking inventory of a warehouse, workers find a large cylindrical paint can with no label. The cylinder measures 13 in. high and has a diameter of 10.5 in. (see figure). How many *gallons* of paint does the container hold?

Solution ▶ **a.** Compute the volume in cubic inches.

$$V = \pi r^2 h \qquad \text{volume of a cylinder}$$
$$\approx (3.14)(5.25)^2(13) \qquad \text{substitute } \frac{10.5}{2} = 5.25 \text{ for } r, 13 \text{ for } h, \text{ and } 3.14 \text{ for } \pi$$
$$\approx 1125 \text{ in}^3 \qquad \text{simplify}$$

b. Convert cubic inches to gallons using a unit conversion factor.

$$1125 \text{ in}^3 = \frac{1125 \text{ in}^3}{1} \left(\frac{1 \text{ gal}}{231 \text{ in}^3} \right) \qquad \text{set up units}$$

$$\approx 4.87 \text{ gallons} \qquad \text{result}$$

There are just under 4.9 gal of paint in the container.

Now try Exercises 39 and 40 ▶

U.S. Units and Metric Units

Since U.S. Customary units and metric units were developed independently of each other, there is no "direct link" between the two, and we usually look up equivalent values in a reference book or appendix. Since two equal quantities have a ratio of 1, any conversion factor needed can be formed as before. The length/distance conversions shown in Table AII.3 are rounded to three decimal places.

Table A.II.3

U.S. to Metric	Metric to U.S.
1 inch = 2.54 centimeters	1 centimeter ≈ 0.3937 inch
1 foot = 0.3048 meter	1 meter ≈ 3.2808 feet
1 yard = 0.9144 meter	1 meter ≈ 1.0936 yards
1 mile ≈ 1.6093 kilometers	1 kilometer ≈ 0.6214 mile

Using the left half of the table when doing U.S. to metric conversions and the right-half when doing metric to U.S. conversions helps to simplify the calculation.

EXAMPLE 6 ▶ **Unit Conversions—Kilometers to Miles**

The flight distance between New York City and London is 5536 kilometers. Approximately how many *miles* is the related flight?

Solution ▶ Set up the conversion factor, noting from the table that

$$1 \text{ km} \approx 0.6214 \text{ mi} \qquad \text{\color{red}conversion factors}$$

$$5536 \text{ km} = \frac{5536 \text{ \cancel{km}}}{1}\left(\frac{0.6214 \text{ miles}}{1 \text{ \cancel{km}}}\right) \qquad \text{\color{red}set up units}$$

$$\approx 3440 \text{ mi} \qquad \text{\color{red}result}$$

The flight distance is approximately 3440 mi.

Now try Exercises 41 through 44 ▶

Conversion factors for weight, volume, and capacity can be found online or in reference books and applied in the same manner.

Similar Triangles

Another important geometric relationship **is that of similar triangles.** Two triangles are similar if corresponding angles are equal. The angles are usually named with capital letters and the side opposite each angle is named used the related lower case letter (see Figure A.II.4). Similar triangles have the following useful properties:

Figure A.II.4

Similar Triangles

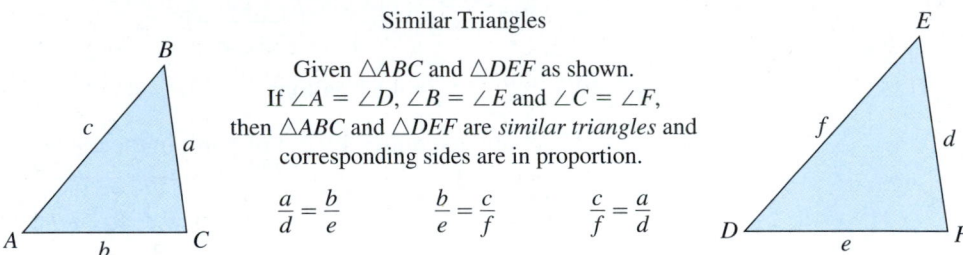

Given $\triangle ABC$ and $\triangle DEF$ as shown.
If $\angle A = \angle D$, $\angle B = \angle E$ and $\angle C = \angle F$,
then $\triangle ABC$ and $\triangle DEF$ are *similar triangles* and
corresponding sides are in proportion.

$$\frac{a}{d} = \frac{b}{e} \qquad \frac{b}{e} = \frac{c}{f} \qquad \frac{c}{f} = \frac{a}{d}$$

The phrase *corresponding sides* means sides that are in the same relative position in each triangle. This property enables us to find the length of a missing side by setting up and solving a proportion. One important application of similar triangles involves ramps or other inclined planes.

EXAMPLE 7 ▶ **Determining the Height of a Ramp**

The ramp leading up to a loading dock has a base of 8 m with vertical braces placed every 2 m. If the first vertical support is 1.5 m high, how high is the dock?

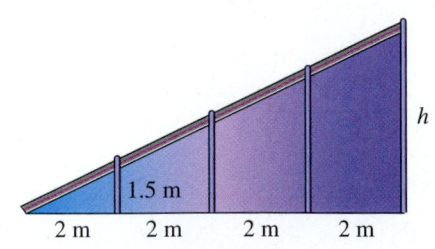

Solution ▶ Since the two triangles in question are similar, we set up and solve a proportion. Let h represent the height of the dock (see figure).

$$\frac{\text{brace height}}{\text{horizontal length}} = \frac{\text{dock height}}{\text{dock length}} \qquad \text{\color{red}corresponding sides are in proportion}$$

$$\frac{1.5}{2} = \frac{h}{8} \qquad \text{\color{red}substitute known values}$$

$$12 = 2h \qquad \text{\color{red}clear denominators}$$

$$6 = h \qquad \text{\color{red}solve for } h$$

The dock is 6 ft high.

Now try Exercises 47 and 48 ▶

Additional Applications

Cubes and rectangular boxes are part of a larger family of solids called **prisms**. A prism **is any solid** figure with bases of the same size and shape (the top and bottom are both called **bases**). A **right prism** has sides which are perpendicular to the bases, with all cross sections congruent. The volume of a right prism is the area of its base times its height. This is easily seen in the case of a rectangular box: $V = LWH = (LW)H = (\text{area of base}) \cdot H$.

Volume of a Right Prism

triangular prism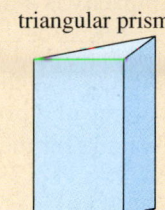

$V = Bh$,
where B represents the area of the base
and h represents the height.

trapezoidal prism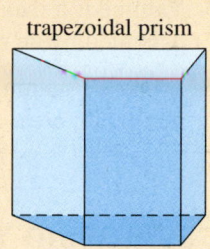

EXAMPLE 8 ▶   **Determining the Volume of a Right Prism**

The feeding trough at a large cattle ranch is a right prism with trapezoidal bases (see figure). The trough is 1.4 ft deep and 38 ft long. If the bases of the trapezoid are 2 ft and 3 ft, how many loads of feed are needed to fill the trough if each load is 54 ft^3?

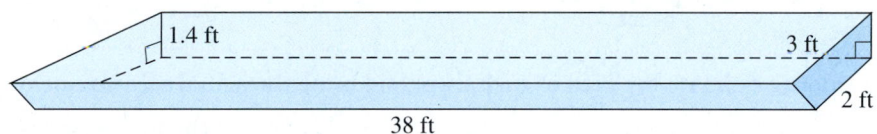

Solution ▶   Notice the trough is a right trapezoidal prism.

a. Find the area of the trapezoidal bases:

$$A = \frac{h}{2}(b_1 + b_2) \quad \text{area of a trapezoid}$$

$$= \frac{1.4}{2}(3 + 2) \quad \text{substitute 1.4 for } h \text{, 3 for } b_1 \text{, and 2 for } b_2$$

$$= 3.5 \text{ ft}^2 \quad \text{result}$$

The area of each base is 3.5 ft^2.

b. Compute the volume of the trough (the right prism).

$$V = Bh \qquad \text{volume = area } \cdot \text{ height}$$

$$= (3.5)(38) \quad \text{substitute 3.5 for } B \text{, 38 for } h$$

$$= 133 \text{ ft}^3 \quad \text{result}$$

The volume of the trough is 133 ft^3.

c. Find how many 54-ft^3 loads are needed for 133 ft^3 (divide).

$$\text{loads} \approx \frac{133 \text{ ft}^3}{54 \text{ ft}^3}$$

$$\approx 2.463$$

Approximately 2.5 loads of feed are needed to fill the trough.

☑ **C.** You've just reviewed how to use unit conversion factors effectively.

Now try Exercises 49 through 66 ▶

A.II EXERCISES

▶ CONCEPTS AND VOCABULARY

Fill in each blank with the appropriate word or phrase. Carefully reread the section, if necessary.

1. A geometric figure composed of more than one basic shape is a __composite__ figure.

2. A trapezoid is any __quadrilateral__ that has exactly two __parallel__ sides.

3. Volume is a measure of the amount of __space__ occupied by an object.

4. Capacity is a measure of an object's potential to __hold__ or __store__ something.

5. Discuss/Explain the methods available to find the perimeter, area, and volume of composite figures.
   Answers will vary.

6. Discuss/Explain the formula for computing the volume of a prism. Does the concept apply to the formula for the volume of a cylinder? Explain.
   Answers will vary.

▶ DEVELOPING YOUR SKILLS

Compute the area of each trapezoid using the dimensions given.

7.

88 mm
3 cm
64 mm
$A = 2280$ mm^2 or 22.8 cm^2

8.
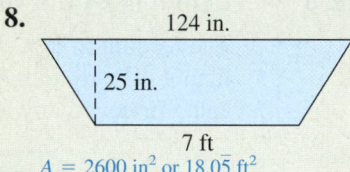
124 in.
25 in.
7 ft
$A = 2600$ in^2 or $18.0\overline{5}$ ft^2

Use a ruler to estimate the area of each trapezoid using the actual measurements.

9. In square millimeters,
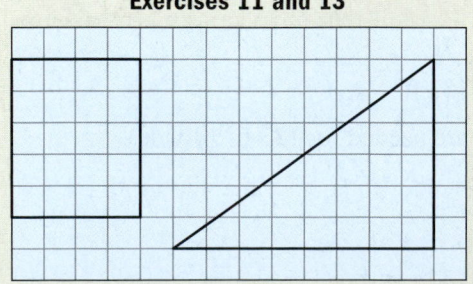
$A \approx 950$ mm^2

10. In square inches,
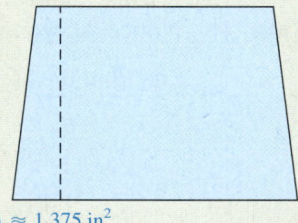
$A \approx 1.375$ in^2

Graph paper can be an excellent aid in understanding why perimeter must be measured in linear units, while area is measured in square units. Suppose the graph paper shown has squares which are 1 cm by 1 cm. Use the grid to answer each question by counting and by computation.

11. What are the perimeter and area of the rectangle shown?

Exercises 11 and 13

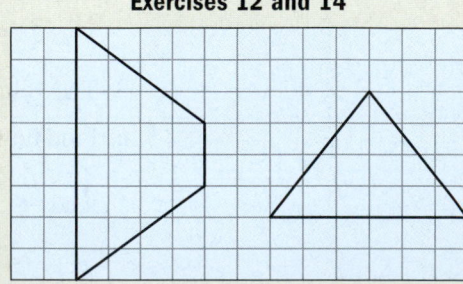

12. What are the perimeter and area of the trapezoid shown?

Exercises 12 and 14

13. What are the perimeter and area of the triangle shown?

14. What are the perimeter and area of the triangle shown?

Additional answers can be found in the Instructor Answer Appendix.

15. What is the outer circumference of a circular gear with a radius (center to teeth) of 2.5 cm? Round to tenths. $C \approx 15.7$ cm

16. If the inner radius of the gear from Exercise 15 (center to base of teeth) is 2.2 cm, what is the working depth of the gear (height of gear teeth)?
0.3 cm or 3 mm

17. Find the missing length, then compute the area and perimeter.

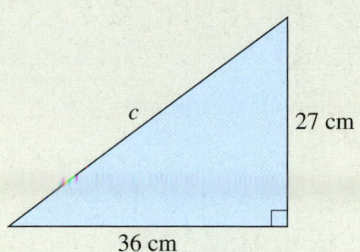

18. Find the missing length, then compute the area and perimeter.

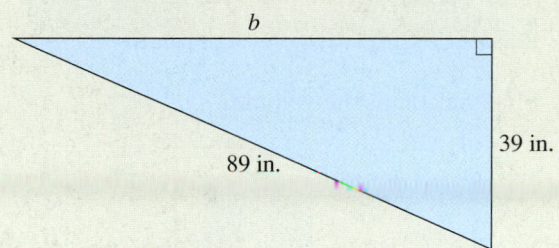

In Exercises 19–26, a composite figure is shown. Construct an appropriate formula and use it to find the perimeter, area, or volume as indicated.

19. Perimeter

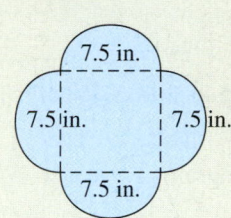

$P = 2(\pi d)$ $15\pi \approx 47.1$ in.

20. Perimeter
$P = 2s + \dfrac{\pi d}{2}$ $24 + 6\pi \approx 42.8$ cm

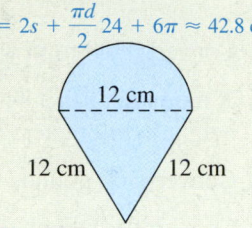

21. Area

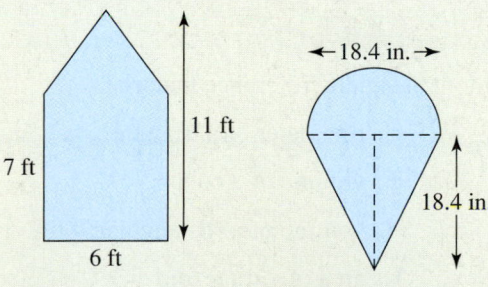

22. Area

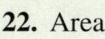

23. Area

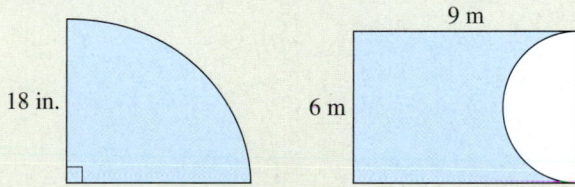

24. Area

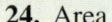

25. Determine the outer perimeter and total area of the track and field shown.

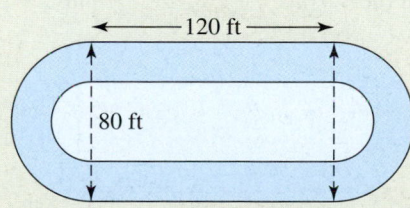

26. Find the perimeter and area of the skirt pattern shown.

27. How many square inches of paper are needed to cover the kite?   504 in²

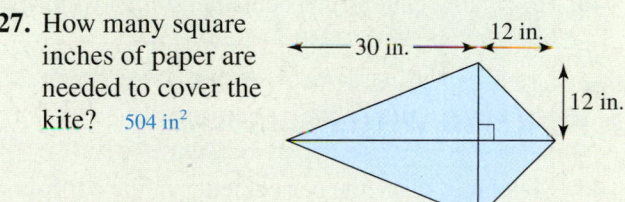

28. Find the area of the mainsail and the area of the jib sail of the sailboat.   Main: 36 ft² Jib: 18 ft²

29. Find the length of the missing side. $W = 12$ cm

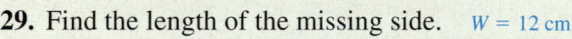
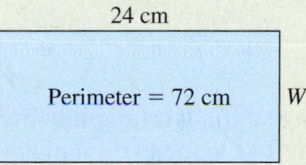

30. Find the height of the trapezoid. $h = 24$ in.

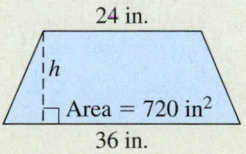

Additional answers can be found in the Instructor Answer Appendix.

For each figure, (a) draw auxiliary lines that partition the figure into basic shapes and label the sides as needed.
(b) Give a verbal model and a formula model that could be used to complete the exercise.

31. Area   **32.** Volume   **33.** Volume   **34.** Perimeter

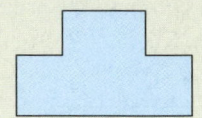   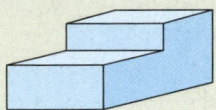   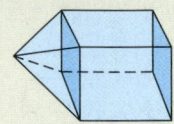   

35. Find the composite volume.   **36.** Find the composite volume.

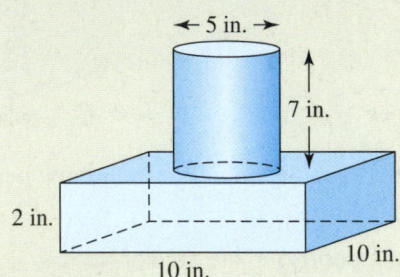   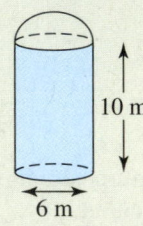

37. Convert 625 quarts to gallons.   156.25 gal   **41.** Convert 6500 m to yards.   7108.4 yd

38. Convert 52,600 pounds to tons.   26.3 tons   **42.** A mountain has a height of 18,300 ft. What is the
                                                          height in kilometers?   about 5.58 km tall
39. Convert 2887.5 in^3 to gallons.   12.5 gal
                                                         43. Convert 300 pounds to kilograms.   661.5 kg
40. How many gallons in a container with a volume of
    623.7 in^3?   about 2.7 gal   **44.** Convert 95 gallons to liters.   about 359.6 L

▶ WORKING WITH FORMULAS

45. Matching: Place the correct letter in the corresponding blank without using any reference material.

A. perimeter of a square   **B.** perimeter of a rectangle   **C.** perimeter of a triangle

D. area of a triangle   **E.** circumference of a circle   **F.** volume of a cube

G. area of a square   **H.** perimeter of a trapezoid   **I.** volume of a rectangular solid

J. area of a circle   **K.** area of a rectangle   **L.** area of a trapezoid

  __K__ LW __C__ $S_1 + S_2 + S_3$ __J__ πr^2

  __A__ $4S$ __I__ LWH __D__ $\dfrac{1}{2}BH$

  __H__ $S_1 + S_2 + S_3 + S_4$ __E__ $2\pi r$ __G__ S^2

  __B__ $2L + 2W$ __F__ S^3 __L__ $\dfrac{h(b_1 + b_2)}{2}$

46. **Surface Area of a Rectangular Box: SA = 2(LW + LH + WH)** The surface area of a   **Exercise 46**
    rectangular box is found by summing the areas of all six sides. Find the surface area of a
    box 15 in. tall, 8 in. wide, and 3 in. deep.   378 in^2

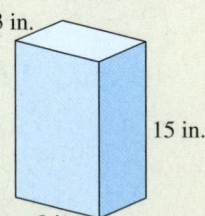

▶ APPLICATIONS (ROUND TO TENTHS AS NEEDED)

47. Similar triangles: To estimate the height of a flagpole, Mitchell reasons his shadow must be proportional to the shadow of the flagpole due to similar triangles. Mitchell is 4 feet tall and his shadow measures 7.5 ft. The shadow of the flagpole measures 60 ft. How high is the pole? 32 ft

Exercise 47

48. Similar triangles: A triangular image measuring 8 in. × 9 in. × 10 in. is projected on a screen using an overhead projector. If the smallest side of the projected image is 2 feet, what are the other dimensions of the projected image? 27 in., 30 in.

49. Volume of a prism: Using a sheet of canvas that is 10 ft by 16 ft, a simple tent is made using a long pole through the middle and pegging the sides into the ground as shown. If the tent is 4 feet high at the apex, what is the volume of space contained in the tent? 192 ft³

Exercise 49

16 ft — 4 ft

50. Volume of a prism: The feeding trough at a pig farm is a right prism with bases that are isosceles triangles (two equal sides). The trough is 3 ft wide, 2 ft deep, and 12 ft long. What is its volume? If each load is 7 ft³, how many loads of slop are needed to fill the trough? 36 ft³, six trips

Exercise 50

3 ft — 2 ft — 12 ft

51. Length of a cable: A radio tower is secured by cables which are clamped 21.5 m up the tower and anchored in the ground 9 m from its base. If 30 cm lengths are needed to secure the cable at each end, how long are the cables? Round to hundredths of a meter. 23.91 m

Exercise 51

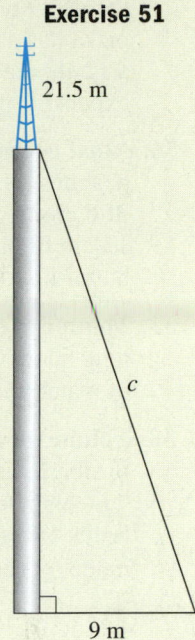

21.5 m

c

9 m

52. Height of a kite: Benjamin Franklin is flying his kite in a storm once again … and has let out 200 m of string. John Adams has walked to a position directly under the kite and is 150 m from Ben. How high is the kite to the nearest meter? 132 m

Exercise 52

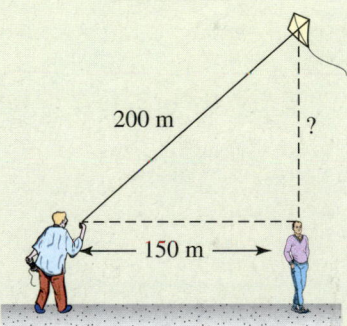

200 m   ?

150 m

53. Unit conversions: A U.S. citizen living in Brazil wants a Brazilian carpenter to build a wooden chest 3 ft high, 2.5 ft wide, and 4 ft long. If the carpenter knows only the metric system, what dimensions should the carpenter be given? 91.44 cm × 76.2 cm × 121.92 cm

54. Unit conversions: A baseball pitcher at the major-league level can throw the ball around 145 ft/sec. What is the speed of the ball in kilometers per hour?

Exercise 54

about 159.1 Km/hr

55. Most economical purchase: Missy's Famous Pizza Emporium is running a special—one large for $8.99 or two mediums for $13.49. If a large pizza has a 14-in. diameter and a medium pizza has a 12-in. diameter, which is the better buy (least expensive per square inch)?   one large at about 5.8¢/in²

56. Most economical purchase: A large can of peaches costs 79¢ and is 15 cm high with a radius of 6 cm. A small can of peaches costs 39¢ and is 15 cm high with a radius of 4 cm. Which is the better buy (least expensive per cubic cm)?   large can at about 4.7¢/cm³

57. Volume of a jacuzzi: A certain Jacuzzi tub is 84 in. long 60 in. wide and 30 in. deep. How many gallons of water will it take to fill this tub?   about 654.5 gal

58. Volume of a birdbath: I am trying to fill an outdoor birdbath, and all I can find is a plastic dish pan 7 in. high, with a 12 in. by 15 in. base. If the birdbath holds 15 gal, how many trips will have to be made?   three trips will be required

Exercise 58

59. Paving a walkway: Current plans call for building a circular fountain 6 m in diameter, surrounded by a circular walkway 1.5 m wide. (a) What is the approximate area of the walkway? (b) If the concrete for the walkway is 6 cm deep, what volume of cement must be used? (c) If the cement costs $125/m³ with a 7% sales tax, what is the cost of the materials for this walkway?   **(a)** 35.3 m²   **(b)** 2.1 m³   **(c)** about $280.88

Exercise 59

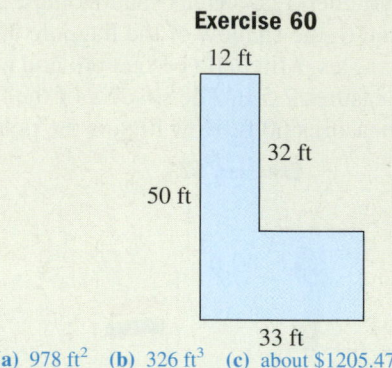

6 m

1.5 m

60. Paving a driveway: A driveway and turn-about has the dimensions shown in the figure. (a) What is the area of the driveway? (b) If the concrete was poured to a depth of 4 in., what volume of cement was used to construct the driveway? (c) If the

cement cost $3.50 ft³ and a 5.65% tax was paid, what was the total cost of the materials used to complete the driveway?

Exercise 60

12 ft

32 ft

50 ft

33 ft

(a) 978 ft²   **(b)** 326 ft³   **(c)** about $1205.47

61. Cost of drywall: After the studs are up, the wall shown in the figure must be covered in drywall. (a) How many square feet of drywall are needed? (b) If drywall is sold only in 4 ft by 8 ft sheets, about how many sheets are required? (c) If drywall costs $8.25 per sheet and a 5.75% tax must be paid, what is the total cost of the material?

Exercise 61

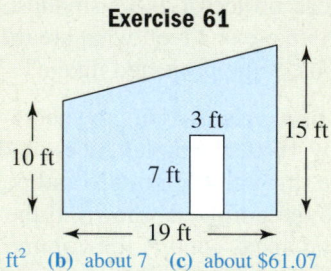

10 ft   3 ft   15 ft

7 ft

19 ft

(a) 216.5 ft²   **(b)** about 7   **(c)** about $61.07

62. Cost of baseboards: The dimensions for the living room/dining room of a home are shown, (a) How many feet and inches of molding are needed for the baseboards around the perimeter of the room? (b) If the molding is only sold in 8-ft lengths, how many are needed? (c) If the molding costs $1.74 per foot and sales tax is 5.75%, what is the total cost of the baseboards?

Exercise 62

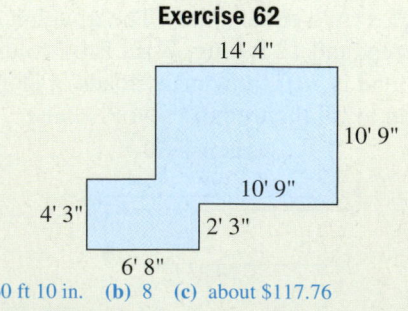

14' 4"

10' 9"

10' 9"

4' 3"

2' 3"

6' 8"

(a) 60 ft 10 in.   **(b)** 8   **(c)** about $117.76

63. Dimensions of an index card: A popular-sized index card has a length that is 1 in. less than twice its width. If the card has an area of 15 in², find the length and width.   5 in. × 3 in.

64. Dimensions of a ruler: The plastic ruler that Albert uses for graphing lines has a length that is 1 cm more than seven times its width. If the ruler has an area of 30 cm², find its length and width.   15 cm × 2 cm

65. Tracking an oil leak: At an oil storage facility, one of the tanks has a slow leak. If the tank shown was full to begin with, how many gallons have been lost at the moment the height of the oil in the tank is 24 ft?

Exercise 65

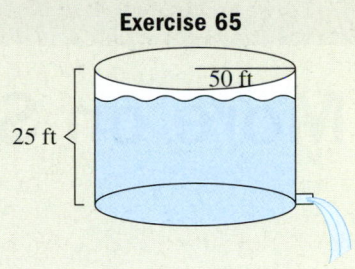

66. Tracking a water leak: A cylindrical water tank has developed a slow leak. If the tank is standing vertically and is 4 ft tall with a radius of 9 in., how many gallons have leaked out at the moment the height of the water in the tank is 3 ft?

▶ **EXTENDING THE CONCEPT**

67. The area (in cm²) of the region shown in the figure is

  a. $35 - xy$ **b.** $35 + xy$ **c.** $35 + 7y - 5x$

  d. $35 + xy - 7y$ **e.** $35 - xy + 5x$ **f.** none of these

Exercise 67

x cm

y cm

5 cm

7 cm

68. A 2-quart saucepan is 75% full of leftover soup. Which plastic container will best store the leftover soup (i.e., with minimum wasted volume)?

  a. a hemispheric bowl with radius 3.5 in.

  b. a cylinder with radius 3 in. and height 5 in.

  c. a rectangular container 4 in. × 6 in. × 3 in.

  d. None of these will hold the soup.

Additional answers can be found in the Instructor Answer Appendix.

Appendix III

More on Synthetic Division

As the name implies, synthetic division simulates the long division process, but in a condensed and more efficient form. It's based on a few simple observations of long division, as noted in the division $(x^3 - 2x^2 - 13x - 17) \div (x - 5)$ shown in Figure AIII.1.

Figure AIII.1

$$
\begin{array}{r}
x^2 + 3x + 2 \\
x - 5 \overline{\smash{)}\, x^3 - 2x^2 - 13x - 17} \\
\underline{-\,(x^3 - 5x^2)} \\
3x^2 - 13x \\
\underline{-\,(3x^2 - 15x)} \\
2x - 17 \\
\underline{-\,(2x - 10)} \\
-7 \quad \text{remainder}
\end{array}
$$

Figure AIII.2

$$
\begin{array}{r}
1 \quad\; 3 \quad\;\; 2 \\
x - 5 \overline{\smash{)}\, 1 \quad -2 \quad -13 \quad -17} \\
5 \\
\underline{} \\
3 \\
15 \\
\underline{} \\
2 \\
10 \\
\underline{} \\
-7 \quad \text{remainder}
\end{array}
$$

A careful observation reveals a great deal of repetition, as any term in red is a duplicate of the term above it. In addition, since the dividend and divisor must be written in decreasing order of degree, the variable part of each term is unnecessary as we can let the *position of each coefficient* indicate the degree of the term. In other words, we'll agree that

$$1 \quad -2 \quad -13 \quad -17 \quad \text{represents the polynomial} \quad 1x^3 - 2x^2 - 13x - 17.$$

Finally, we know in advance that we'll be subtracting each partial product, so we can "distribute the negative," shown at each stage. Removing the repeated terms and variable factors, then distributing the negative to the remaining terms produces Figure AIII.2. The entire process can now be condensed by vertically compressing the rows of the division so that a minimum of space is used (Figure AIII.3).

Figure AIII.3

$$
\begin{array}{r}
1 \quad\;\; 3 \quad\;\;\; 2 \quad \text{quotient} \\
x - 5 \overline{\smash{)}\, 1 \quad -2 \quad -13 \quad -17} \quad \text{dividend} \\
5 \quad\;\; 15 \quad\;\; 10 \quad \text{products} \\
\underline{} \\
3 \quad\;\;\; 2 \quad\; -7 \quad \text{sums}
\end{array}
$$

Figure AIII.4

$$
\begin{array}{r}
1 \quad\;\; 3 \quad\;\;\; 2 \\
x - 5 \overline{\smash{)}\, 1 \quad -2 \quad -13 \quad -17} \quad \text{dividend} \\
5 \quad\;\; 15 \quad\;\; 10 \quad \text{products} \\
\underline{} \\
1 \quad\;\;\; 3 \quad\;\;\; 2 \quad\; -7 \quad \text{remainder} \\
\text{quotient}
\end{array}
$$

Further, if we include the lead coefficient in the bottom row (Figure AIII.4), the coefficients in the top row (in **blue**) are duplicated and no longer necessary, since the quotient and remainder now appear in the last row. Finally, note all entries in the product row (in **red**) are five times the sum from the prior column. There is a direct connection between this multiplication by 5 and the divisor $x - 5$, and in fact, it is the *zero of the divisor* that is used in synthetic division ($x = 5$ from $x - 5 = 0$). A simple change in format makes this method of division easier to use, and highlights the location of the divisor and remainder (the **blue** brackets in Figure AIII.5). Note the process begins by "dropping the lead coefficient into place" (shown in **bold**). The full process of synthetic division is shown in Figure AIII.6 for the same exercise.

Figure AIII. 5

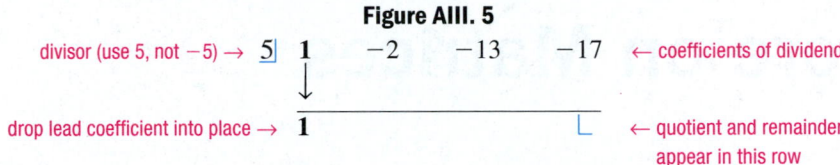

We then multiply this coefficient by the "divisor," place the result in the next column and add. In a sense, we "multiply in the diagonal direction," and "add in the vertical direction." Continue the process until the division is complete.

Figure AIII. 6

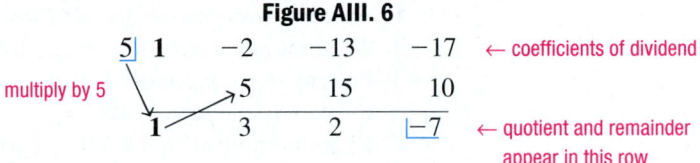

The result is $x^2 + 3x + 2 + \dfrac{-7}{x - 5}$, read from the last row.

Appendix IV

More on Matrices

Reduced Row-Echelon Form

A matrix is in reduced row-echelon form if it satisfies the following conditions:

1. All null rows (zeroes for all entries) occur at the bottom of the matrix.
2. The first non-zero entry of any row must be a 1.
3. For any two consecutive, nonzero rows, the leading 1 in the higher row is to the left of the 1 in the lower row.
4. Every column with a leading 1 has zeroes for all other entries in the column.

Matrices A through D are in reduced row-echelon form.

$$A = \begin{bmatrix} 0 & 1 & 0 & 0 & 0 & 5 \\ 0 & 0 & 0 & 1 & 0 & 3 \\ 0 & 0 & 0 & 0 & 1 & 2 \end{bmatrix} \quad B = \begin{bmatrix} 1 & 0 & 0 & 5 \\ 0 & 1 & 0 & 3 \\ 0 & 0 & 0 & 0 \end{bmatrix} \quad C = \begin{bmatrix} 1 & 0 & 0 & 5 \\ 0 & 1 & 3 & -2 \\ 0 & 0 & 0 & 0 \end{bmatrix} \quad D = \begin{bmatrix} 1 & 0 & 5 & 0 \\ 0 & 1 & 2 & 0 \\ 0 & 0 & 0 & 1 \end{bmatrix}$$

Where *Gaussian elimination* places a matrix in *row-echelon form* (satisfying the first three conditions), *Gauss-Jordan elimination* places a matrix in *reduced row-echelon form*. To obtain this form, continue applying row operations to the matrix until the fourth condition above is also satisfied. For a 3×3 system having a unique solution, the diagonal entries of the coefficient matrix will be 1's, with 0's for all other entries. To illustrate, we'll extend Example 4 from Section 7.1 until reduced row-echelon form is obtained.

EXAMPLE 4 ▶ **Solving Systems Using the Augmented Matrix and Gauss-Jordan Elimination**

Solve using Gauss-Jordan elimination: $\begin{cases} 2x + y - 2z = -7 \\ x + y + z = -1 \\ -2y - z = -3 \end{cases}$

$\begin{cases} 2x + y - 2z = -7 \\ x + y + z = -1 \\ -2y - z = -3 \end{cases}$ matrix form → $\begin{bmatrix} 2 & 1 & -2 & -7 \\ 1 & 1 & 1 & -1 \\ 0 & -2 & -1 & -3 \end{bmatrix}$ R1 ↔ R2 $\begin{bmatrix} 1 & 1 & 1 & -1 \\ 2 & 1 & -2 & -7 \\ 0 & -2 & -1 & -3 \end{bmatrix}$

$\begin{bmatrix} 1 & 1 & 1 & -1 \\ 2 & 1 & -2 & -7 \\ 0 & -2 & -1 & -3 \end{bmatrix}$ −2R1 + R2 → R2 $\begin{bmatrix} 1 & 1 & 1 & -1 \\ 0 & -1 & -4 & -5 \\ 0 & -2 & -1 & -3 \end{bmatrix}$ −1R2 $\begin{bmatrix} 1 & 1 & 1 & -1 \\ 0 & 1 & 4 & 5 \\ 0 & -2 & -1 & -3 \end{bmatrix}$

$\begin{bmatrix} 1 & 1 & 1 & -1 \\ 0 & 1 & 4 & 5 \\ 0 & -2 & -1 & -3 \end{bmatrix}$ 2R2 + R3 → R3 $\begin{bmatrix} 1 & 1 & 1 & -1 \\ 0 & 1 & 4 & 5 \\ 0 & 0 & 7 & 7 \end{bmatrix}$ $\dfrac{R3}{7} \to R3$ $\begin{bmatrix} 1 & 1 & 1 & -1 \\ 0 & 1 & 4 & 5 \\ 0 & 0 & 1 & 1 \end{bmatrix}$

$\begin{bmatrix} 1 & 1 & 1 & -1 \\ 0 & 1 & 4 & 5 \\ 0 & 0 & 1 & 1 \end{bmatrix}$ −R2 + R1 → R1 $\begin{bmatrix} 1 & 0 & -3 & -6 \\ 0 & 1 & 4 & 5 \\ 0 & 0 & 1 & 1 \end{bmatrix}$ 3R3 + R1 → R1   −4R3 + R2 → R2 $\begin{bmatrix} 1 & 0 & 0 & -3 \\ 0 & 1 & 0 & 1 \\ 0 & 0 & 1 & 1 \end{bmatrix}$

The final matrix is in reduced row-echelon form with solution $(-3, 1, 1)$ just as in Section 7.1.

The Determinant of a General Matrix

To compute the determinant of a general square matrix, we introduce the idea of a **cofactor.** For an $n \times n$ matrix A, $A_{ij} = (-1)^{i+j}|M_{ij}|$ is the cofactor of matrix element a_{ij}, where $|M_{ij}|$ represents the determinant of the corresponding minor matrix. Note that $i + j$ is the sum of the row and column of the entry, and if this sum is even, $(-1)^{i+j} = 1$, while if the sum is odd, $(-1)^{i+j} = -1$ (this is how the sign table for a 3×3 determinant was generated). To compute the determinant of an $n \times n$ matrix, multiply each element in any row or column by its cofactor and add. The result is a tier-like process in which the determinant of a larger matrix requires computing the determinant of smaller matrices. In the case of a 4×4 matrix, each of the minor matrices will be size 3×3, whose determinant then requires the computation of other 2×2 determinants. In the following illustration, two of the entries in the first row are zero for convenience. For

$$A = \begin{bmatrix} -2 & 0 & 3 & 0 \\ 1 & 2 & 0 & -2 \\ 3 & -1 & 4 & 1 \\ 0 & -3 & 2 & 1 \end{bmatrix},$$

we have: $\det(A) = -2 \cdot (-1)^{1+1} \begin{vmatrix} 2 & 0 & -2 \\ -1 & 4 & 1 \\ -3 & 2 & 1 \end{vmatrix} + (3) \cdot (-1)^{1+3} \begin{vmatrix} 1 & 2 & -2 \\ 3 & -1 & 1 \\ 0 & -3 & 1 \end{vmatrix}$

Computing the first 3×3 determinant gives -16, the second 3×3 determinant is 14. This gives:

$$\det(A) = -2(-16) + 3(14)$$
$$= 74$$

Appendix V

Deriving the Equation of a Conic

The Equation of an Ellipse

In Section 8.2, the equation $\sqrt{(x+c)^2 + y^2} + \sqrt{(x-c)^2 + y^2} = 2a$ was developed using the distance formula and the definition of an ellipse. To find the standard form of the equation, we treat this result as a radical equation, isolating one of the radicals and squaring both sides.

$$\sqrt{(x+c)^2 + y^2} = 2a - \sqrt{(x-c)^2 + y^2} \qquad \text{isolate one radical}$$
$$(x+c)^2 + y^2 = 4a^2 - 4a\sqrt{(x-c)^2 + y^2} + (x-c)^2 + y^2 \qquad \text{square both sides}$$

We continue by simplifying the equation, isolating the remaining radical, and squaring again.

$$x^2 + 2cx + c^2 + y^2 = 4a^2 - 4a\sqrt{(x-c)^2 + y^2} + x^2 - 2cx + c^2 + y^2 \quad \substack{\text{expand} \\ \text{binomials}}$$
$$4cx = 4a^2 - 4a\sqrt{(x-c)^2 + y^2} \qquad \text{simplify}$$
$$a\sqrt{(x-c)^2 + y^2} = a^2 - cx \qquad \text{isolate radical; divide by 4}$$
$$a^2[(x-c)^2 + y^2] = a^4 - 2a^2cx + c^2x^2 \qquad \text{square both sides}$$
$$a^2x^2 - 2a^2cx + a^2c^2 + a^2y^2 = a^4 - 2a^2cx + c^2x^2 \qquad \text{expand and distribute } a^2 \text{ on left}$$
$$a^2x^2 - c^2x^2 + a^2y^2 = a^4 - a^2c^2 \qquad \text{add } 2a^2cx \text{ and rewrite equation}$$
$$x^2(a^2 - c^2) + a^2y^2 = a^2(a^2 - c^2) \qquad \text{factor}$$
$$\frac{x^2}{a^2} + \frac{y^2}{a^2 - c^2} = 1 \qquad \text{divide by } a^2(a^2 - c^2)$$

Since $a > c$, we know $a^2 > c^2$ and $a^2 - c^2 > 0$. For convenience, let $b^2 = a^2 - c^2$ (it also follows that $a^2 > b^2$ and $a > b$, since $c > 0$). Substituting b^2 for $a^2 - c^2$ we obtain the standard form of the equation of an ellipse (major axis horizontal, since we stipulated $a > b$): $\frac{x^2}{a^2} + \frac{y^2}{b^2} = 1$. Note once again the x-intercepts are $(\pm a, 0)$, while the y-intercepts are $(0, \pm b)$. For the foci, $c^2 = |a^2 - b^2|$ to allow for a major axis that may be vertical.

The Equation of a Hyperbola

In Section 8.3, the equation $\sqrt{(x+c)^2 + y^2} - \sqrt{(x-c)^2 + y^2} = 2a$ was developed using the distance formula and the definition of a hyperbola. To find the standard form of this equation, we apply the same procedures as before.

$$\sqrt{(x+c)^2 + y^2} = 2a + \sqrt{(x-c)^2 + y^2} \qquad \text{isolate one radical}$$
$$(x+c)^2 + y^2 = 4a^2 + 4a\sqrt{(x-c)^2 + y^2} + (x-c)^2 + y^2 \qquad \text{square both sides}$$
$$x^2 + 2cx + c^2 + y^2 = 4a^2 + 4a\sqrt{(x-c)^2 + y^2} + x^2 - 2cx + c^2 + y^2 \quad \substack{\text{expand} \\ \text{binomials}}$$
$$4cx = 4a^2 + 4a\sqrt{(x-c)^2 + y^2} \qquad \text{simplify}$$
$$cx - a^2 = a\sqrt{(x-c)^2 + y^2} \qquad \text{isolate radical; divide by 4}$$
$$c^2x^2 - 2a^2cx + a^4 = a^2[(x-c)^2 + y^2] \qquad \text{square both sides}$$
$$c^2x^2 - 2a^2cx + a^4 = a^2x^2 - 2a^2cx + a^2c^2 + a^2y^2 \qquad \text{expand and distribute } a^2 \text{ on the right}$$
$$c^2x^2 - a^2x^2 - a^2y^2 = a^2c^2 - a^4 \qquad \text{add } 2a^2cx \text{ and rewrite equation}$$
$$x^2(c^2 - a^2) - a^2y^2 = a^2(c^2 - a^2) \qquad \text{factor}$$
$$\frac{x^2}{a^2} - \frac{y^2}{c^2 - a^2} = 1 \qquad \text{divide by } a^2(c^2 - a^2)$$

From the definition of a hyperbola we have $0 < a < c$, showing $c^2 > a^2$ and $c^2 - a^2 > 0$. For convenience, let $b^2 = c^2 - a^2$ and substitute to obtain the standard form of the equation of a hyperbola (transverse axis horizontal): $\dfrac{x^2}{a^2} - \dfrac{y^2}{b^2} = 1$.

Note the x-intercepts are $(0, \pm a)$ and there are no y-intercepts. For the foci, $c^2 = a^2 + b^2$.

The Asymptotes of a Central Hyperbola

From our work in Section 8.3, a central hyperbola with a horizontal axis will have asymptotes at $y = \pm \dfrac{b}{a}x$. To understand why, recall that for asymptotic behavior we investigate what happens to the relation for large values of x, meaning as $|x| \to \infty$. Starting with $\dfrac{x^2}{a^2} - \dfrac{y^2}{b^2} = 1$, we have

$$b^2x^2 - a^2y^2 = a^2b^2 \qquad \text{clear denominators}$$

$$a^2y^2 = b^2x^2 - a^2b^2 \qquad \text{isolate term with } y$$

$$a^2y^2 = b^2x^2\left(1 - \frac{a^2}{x^2}\right) \qquad \text{factor out } b^2x^2 \text{ from right side}$$

$$y^2 = \frac{b^2}{a^2}x^2\left(1 - \frac{a^2}{x^2}\right) \qquad \text{divide by } a^2$$

$$y = \pm\frac{b}{a}x\sqrt{1 - \frac{a^2}{x^2}} \qquad \text{take square roots of both sides}$$

As $|x| \to \infty$, $\dfrac{a^2}{x^2} \to 0$, and we find that for large values of x, $y \approx \pm\dfrac{b}{a}x$.

Appendix VI

Proof Positive—A Selection of Proofs from College Algebra

Proofs from Chapter 4

The Remainder Theorem

If a polynomial $p(x)$ is divided by $(x - c)$ using synthetic division, the remainder is equal to $p(c)$.

Proof of the Remainder Theorem

From our previous work, any number c used in synthetic division will occur as the factor $(x - c)$ when written as (quotient)(divisor) + remainder: $p(x) = (x - c)q(x) + r$. Here, $q(x)$ represents the quotient polynomial and r is a constant. Evaluating $p(c)$ gives

$$p(x) = (x - c)q(x) + r$$
$$p(c) = (c - c)q(c) + r$$
$$= 0 \cdot q(c) + r$$
$$= r \checkmark$$

The Factor Theorem

Given a polynomial $p(x)$,
(1) if $p(c) = 0$, then $x - c$ is a factor of $p(x)$, and
(2) if $x - c$ is a factor of $p(x)$, then $p(c) = 0$.

Proof of the Factor Theorem

1. Consider a polynomial p written in the form $p(x) = (x - c)q(x) + r$. From the remainder theorem we know $p(c) = r$, and substituting $p(c)$ for r in the equation shown gives

$$p(x) = (x - c)q(x) + p(c)$$

and $x - c$ is a factor of $p(x)$, if $p(c) = 0$,

$$p(x) = (x - c)q(x) \checkmark$$

2. The steps from part 1 can be reversed, since any factor $(x - c)$ of $p(x)$, can be written in the form $p(x) = (x - c)q(x)$. Evaluating at $x = c$ produces a result of zero:

$$p(c) = (c - c)q(x)$$
$$= 0 \checkmark$$

Complex Conjugates Theorem

Given $p(x)$ is a polynomial with real number coefficients, complex solutions must occur in conjugate pairs. If $a + bi$, $b \neq 0$ is a solution, then $a - bi$ must also be a solution.

To prove this for polynomials of degree $n > 2$, we let $z_1 = a + bi$ and $z_2 = c + di$ be complex numbers, and let $\bar{z}_1 = a - bi$, and $\bar{z}_2 = c - di$ represent their conjugates, and observe the following properties:

1. The conjugate of a sum is equal to the sum of the conjugates.

sum: $z_1 + z_2$ sum of conjugates: $\bar{z}_1 + \bar{z}_2$

$(a + bi) + (c + di)$ $(a - bi) + (c - di)$

$(a + c) + (b + d)i$ → conjugate of sum → $(a + c) - (b + d)i$ ✓

2. The conjugate of a product is equal to the product of the conjugates.

product: $z_1 \cdot z_2$ product of conjugates: $\bar{z}_1 \cdot \bar{z}_2$

$(a + bi) \cdot (c + di)$ $(a - bi) \cdot (c - di)$

$ac + adi + bci + bdi^2$ $ac - adi - cbi + bdi^2$

$(ac - bd) + (ad + bc)i$ → conjugate of product → $(ac - bd) - (ad + bc)i$ ✓

Since polynomials involve only sums and products, and the complex conjugate of any real number is the number itself, we have the following:

Proof of the Complex Conjugates Theorem

Given polynomial $p(x) = a_n x^n + a_{n-1} x^{n-1} + \cdots + a_1 x^1 + a_0$, where $a_n, a_{n-1}, \cdots, a_1, a_0$ are real numbers and $z = a + bi$ is a zero of p, we must show that $\bar{z} = a - bi$ is also a zero.

$$a_n z^n + a_{n-1} z^{n-1} + \cdots + a_1 z^1 + a_0 = p(z) \qquad \text{evaluate } p(x) \text{ at } z$$

$$a_n z^n + a_{n-1} z^{n-1} + \cdots + a_1 z^1 + a_0 = 0 \qquad p(z) = 0 \text{ given}$$

$$\overline{a_n z^n + a_{n-1} z^{n-1} + \cdots + a_1 z^1 + a_0} = \overline{0} \qquad \text{conjugate both sides}$$

$$\overline{a_n z^n} + \overline{a_{n-1} z^{n-1}} + \cdots + \overline{a_1 z^1} + \overline{a_0} = \overline{0} \qquad \text{property 1}$$

$$\bar{a}_n(\bar{z}^n) + \bar{a}_{n-1}(\bar{z}^{n-1}) + \cdots + \bar{a}_1(\bar{z}^1) + \bar{a}_0 = \overline{0} \qquad \text{property 2}$$

$$a_n(\bar{z}^n) + a_{n-1}(\bar{z}^{n-1}) + \cdots + a_1(\bar{z}^1) + a_0 = 0 \qquad \text{conjugate of a real number is the number}$$

$$p(\bar{z}) = 0 \quad \checkmark \text{ result}$$

An immediate and useful result of this theorem is that any polynomial of odd degree must have at least one real root.

Linear Factorization Theorem

If $p(x)$ is a complex polynomial of degree $n \geq 1$, then p has exactly n linear factors and can be written in the form $p(x) = a_n(x - c_1)(x - c_2) \cdot \cdots \cdot (x - c_n)$, where $a_n \neq 0$ and $c_1, c_2, \ldots, c_n$ are complex numbers. Some factors may have multiplicities greater than 1 ($c_1, c_2, \ldots, c_n$ are not necessarily distinct).

Proof of the Linear Factorization Theorem

Given $p(x) = a_n x^n + a_{n-1} x^{n-1} + \cdots + a_1 x + a_0$ is a complex polynomial, the Fundamental Theorem of Algebra establishes that $p(x)$ has a least one complex zero, call it c_1. The factor theorem stipulates $(x - c_1)$ must be a factor of P, giving

$$p(x) = (x - c_1)q_1(x)$$

where $q_1(x)$ is a complex polynomial of degree $n - 1$.

Since $q_1(x)$ is a complex polynomial in its own right, it too must also have a complex zero, call it c_2. Then $(x - c_2)$ must be a factor of $q_1(x)$, giving

$$p(x) = (x - c_1)(x - c_2)q_2(x)$$

where $q_2(x)$ is a complex polynomial of degree $n - 2$.

Repeating this rationale n times will cause $p(x)$ to be rewritten in the form

$$p(x) = (x - c_1)(x - c_2) \cdot \cdots \cdot (x - c_n)q_n(x)$$

where $q_n(x)$ has a degree of $n - n = 0$, a nonzero constant typically called a_n.

The result is $p(x) = a_n(x - c_1)(x - c_2) \cdot \cdots \cdot (x - c_n)$, and the proof is complete.

Proofs from Chapter 5

The Product Property of Logarithms

Given M, N, and b ≠ 1 are positive real numbers,
$\log_b(MN) = \log_b M + \log_b N.$

Proof of the Product Property

For $P = \log_b M$ and $Q = \log_b N$, we have $b^P = M$ and $b^Q = N$ in exponential form. It follows that

$$\log_b(MN) = \log_b(b^P b^Q) \qquad \text{substitute } b^P \text{ for } M \text{ and } b^Q \text{ for } N$$
$$= \log_b(b^{P+Q}) \qquad \text{properties of exponents}$$
$$= P + Q \qquad \text{log property 3}$$
$$= \log_b M + \log_b N \qquad \text{substitute } \log_b M \text{ for } P \text{ and } \log_b N \text{ for } Q$$

The Quotient Property of Logarithms

Given M, N, and b ≠ 1 are positive real numbers,
$\log_b\left(\dfrac{M}{N}\right) = \log_b M - \log_b N.$

Proof of the Quotient Property

For $P = \log_b M$ and $Q = \log_b N$, we have $b^P = M$ and $b^Q = N$ in exponential form. It follows that

$$\log_b\left(\frac{M}{N}\right) = \log_b\left(\frac{b^P}{b^Q}\right) \qquad \text{substitute } b^P \text{ for } M \text{ and } b^Q \text{ for } N$$
$$= \log_b(b^{P-Q}) \qquad \text{properties of exponents}$$
$$= P - Q \qquad \text{log property 3}$$
$$= \log_b M - \log_b N \qquad \text{substitute } \log_b M \text{ for } P \text{ and } \log_b N \text{ for } Q$$

The Power Property of Logarithms

Given M, N, and b ≠ 1 are positive real numbers and any real number x,
$\log_b M^x = x \log_b M.$

Proof of the Power Property

For $P = \log_b M$, we have $b^P = M$ in exponential form. It follows that

$$\log_b(M)^x = \log_b(b^P)^x \qquad \text{substitute } b^P \text{ for } M$$
$$= \log_b(b^{Px}) \qquad \text{properties of exponents}$$
$$= Px \qquad \text{log property 3}$$
$$= (\log_b M)x \qquad \text{substitute } \log_b M \text{ for } P$$
$$= x \log_b M \qquad \text{rewrite factors}$$

Instructor Answer Appendix

CHAPTER R
Exercises R.1, pp. 8–10

29. Let w represent the width in meters. Then $2w$ represents twice the width and $2w - 3$ represents three meters less than twice the width.

30. Let b represent the base in centimeters. Then $3b$ represents three times the base and $3b - 6$ represents six centimeters less than three times the base.

31. Let b represent the speed of the bus. Then $b + 15$ represents 15 mph more than the speed of the bus.

32. Let t represent the time for Remus to finish the race. Then $t + 3$ represents three minutes more time than Remus.

61.

x	Output
-3	14
-2	6
-1	0
0	-4
1	-6
2	-6
3	-4

-1 gives an output of 0.

62.

x	Output
-3	12
-2	5
-1	0
0	-3
1	-4
2	-3
3	0

-1 and 3 give outputs of 0.

63.

x	Output
-3	-18
-2	-15
-1	-12
0	-9
1	-6
2	-3
3	0

3 gives an output of 0.

64.

x	Output
-3	20
-2	15
-1	10
0	5
1	0
2	-5
3	-10

1 gives an output of 0.

65.

x	Output
-3	-5
-2	8
-1	9
0	4
1	-1
2	0
3	13

2 gives an output of 0.

66.

x	Output
-3	-24
-2	0
-1	12
0	18
1	24
2	36
3	60

-2 gives an output of 0.

Exercises R.2, pp. 21–24

73. polynomial, none of these, degree 3   **74.** polynomial, none of these, degree 3   **75.** nonpolynomial because exponents are not whole numbers, NA, NA   **76.** nonpolynomial, variable in a denominator, NA, NA   **77.** polynomial, binomial, degree 3   **78.** nonpolynomial, exponents are not whole numbers, NA, NA

101. $21v^2 - 47v + 20$ **102.** $12w^2 + 28w - 5$ **103.** $9 - m^2$
104. $25 - n^2$ **105.** $p^2 + 1.1p - 9$ **106.** $q^2 - 3.7q - 5.88$
107. $x^2 + \frac{3}{4}x + \frac{1}{8}$ **108.** $z^2 + \frac{7}{6}z + \frac{5}{18}$ **109.** $m^2 - \frac{9}{16}$ **110.** $n^2 - \frac{4}{25}$
111. $6x^2 + 11xy - 10y^2$ **112.** $6a^2 + 19ab + 3b^2$
113. $12c^2 + 23cd + 5d^2$ **114.** $10x^2 - 9xy - 9y^2$ **115.** $2x^4 - x^2 - 15$
116. $6y^4 - y^2 - 2$

142. d.

t	$S(t)$
1	9
2	16
3	21
4	24
5	25
6	24
7	21
8	16
9	9
10	0

Exercises R.3, pp. 35–39

41. $\{a | a \geq 2\}$; ; $a \in [2, \infty)$

42. $\{n | n < \frac{17}{6}\}$; ; $n \in (-\infty, \frac{17}{6})$

43. $\{n | n \geq 1\}$; ; $n \in [1, \infty)$

44. $\{x | x > -3\}$; ; $x \in (-3, \infty)$

45. $\{x | x < \frac{-32}{5}\}$; ; $x \in (-\infty, \frac{-32}{5})$

46. $\{y | y < -4\}$; ; $y \in (-\infty, -4)$

59. $x \in (-\infty, -2) \cup (1, \infty)$;

60. $x \in (-\infty, -5) \cup (5, \infty)$;

61. $x \in [-2, 5)$;

62. $x \in [-4, 3)$;

65. $x \in (-\infty, \infty)$;

66. $x \in (-\infty, 2]$;

67. $x \in [-5, 0]$;

68. $x \in [-4, 0]$;

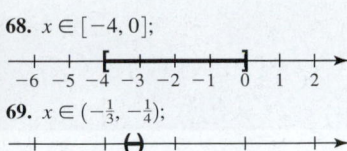

69. $x \in (-\frac{1}{3}, -\frac{1}{4})$;

70. $x \in (\frac{2}{3}, \frac{5}{4}]$;

71. $x \in (-\infty, \infty)$;

72. $x \in (-\infty, -4) \cup (5, \infty)$;

73. $x \in [-4, 1)$;

74. $x \in (2, \frac{23}{3}]$;

75. $x \in [-1.4, 0.8]$;

76. $x \in (2.3, 9.6)$;

77. $x \in [-16, 8)$;

78. $x \in (3, 45]$;

Exercises R.4, pp. 49–53

104. $V = \frac{4}{3}\pi(R^3 - r^3) = \frac{4}{3}\pi(R - r)(R^2 + Rr + r^2)$;
3.276π cm³; 10.3 cm³   **105.** $V = x(x + 5)(x + 3)$ **a.** 3 in.
b. 5 in.   **c.** $V = 24(29)(27) = 18{,}792$ in³   **106.** $B = x(x - 6)(x - 7)$
a. 7 fewer   **b.** 6 fewer   **c.** $B = 10(4)(3) = 120$ books per box

107. $L = L_0\sqrt{\left(1 + \dfrac{v}{c}\right)\left(1 - \dfrac{v}{c}\right)}$, $L = 12\sqrt{(1 + 0.75)(1 - 0.75)}$
$= 3\sqrt{7}$ in. ≈ 7.94 in.

108. $v = \dfrac{G}{4\eta}(R + r)(R - r)$ $v = \dfrac{15}{4(0.25)}(0.8)(0.2) = 2.4$

Exercises R.5, pp. 61–64

89. Price rises rapidly for first four days, then begins a gradual decrease. Yes, on the 35th day of trading.

Day	Price
0	10
1	16.67
2	32.76
3	47.40
4	53.51
5	52.86
6	49.25
7	44.91
8	40.75
9	37.03
10	33.81

CHAPTER 1
Exercises 1.1, pp. 98–102

7.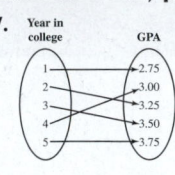

8.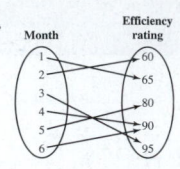

$D = \{1, 2, 3, 4, 5\}$
$R = \{2.75, 3.00, 3.25, 3.50, 3.75\}$

$D = \{1, 2, 3, 4, 5, 6\}$
$R = \{60, 65, 80, 90, 95\}$

13. 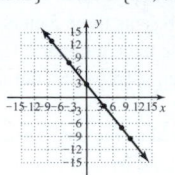   **14.** 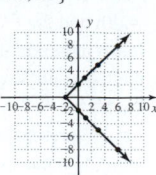   **15.**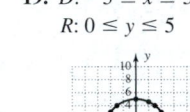

16.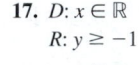

17. $D: x \in \mathbb{R}$
$R: y \geq -1$
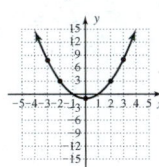

18. $D: x \in \mathbb{R}$
$R: y \leq 3$
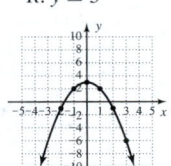

19. $D: -5 \leq x \leq 5$
$R: 0 \leq y \leq 5$

20. $D: -13 \leq x \leq 13$
$R: 0 \leq y \leq 13$
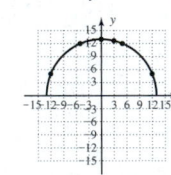

21. $D: x \geq 1$
$R: y \in \mathbb{R}$
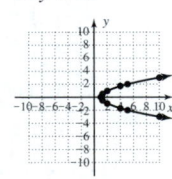

22. $D: x \geq -2$
$R: y \in \mathbb{R}$
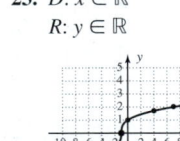

23. $D: x \in \mathbb{R}$
$R: y \in \mathbb{R}$
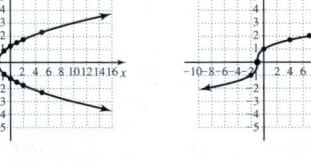

24. $D: x \in \mathbb{R}$
$R: y \in \mathbb{R}$
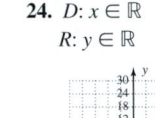
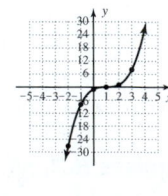

25. $(0, 2), (-5, 0)$

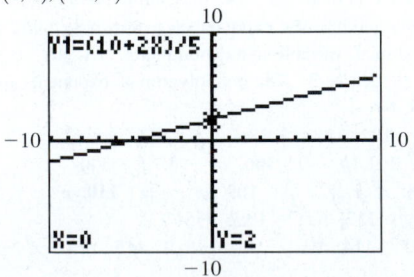

26. $(0, 2), (4, 0)$

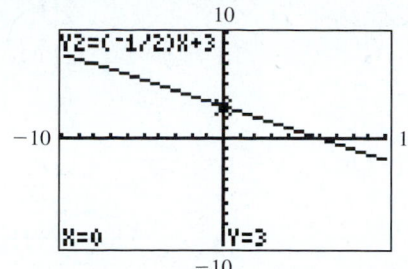

27. $(0, 0), (0, 4)$

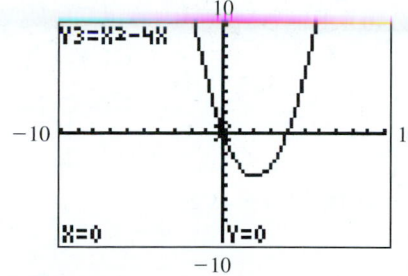

28. $(0, 3), (-1, 0), (3, 0)$

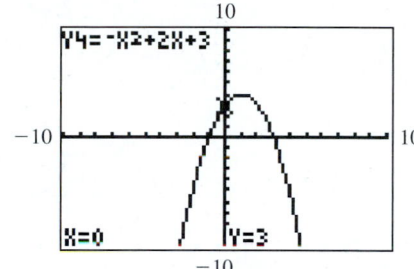

49. $x^2 + y^2 = 9$ **50.** $x^2 + y^2 = 36$ **51.** $(x - 5)^2 + y^2 = 3$

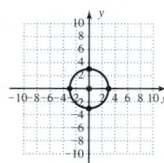 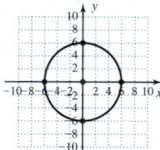 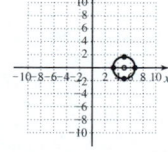

52. $x^2 + (y - 4)^2 = 5$ **53.** $(x - 4)^2 + (y + 3)^2 = 4$

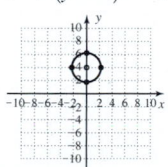 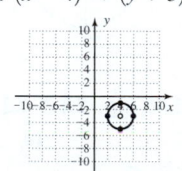

54. $(x - 3)^2 + (y + 8)^2 = 81$ **55.** $(x + 7)^2 + (y + 4)^2 = 7$

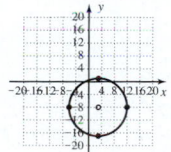 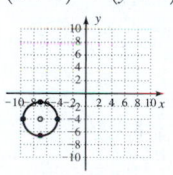

56. $(x + 2)^2 + (y + 5)^2 = 6$ **57.** $(x - 1)^2 + (y + 2)^2 = 9$

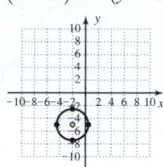 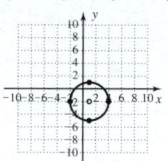

58. $(x + 2)^2 + (y - 3)^2 = 25$ **59.** $(x - 4)^2 + (y - 5)^2 = 12$

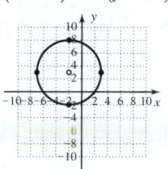 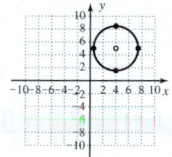

60. $(x - 5)^2 + (y - 1)^2 = 20$ **61.** $(x - 7)^2 + (y - 1)^2 = 100$

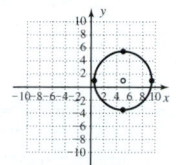 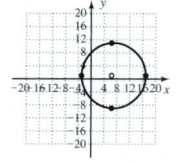

62. $(x + 8)^2 + (y - 3)^2 = 169$ **63.** $(x - 3)^2 + (y - 4)^2 = 41$

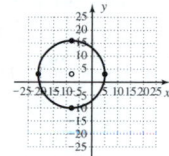 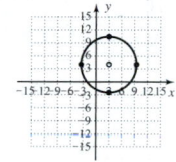

64. $(x + 5)^2 + (y - 2)^2 = 17$ **65.** $(x - 5)^2 + (y - 4)^2 = 9$

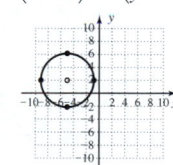 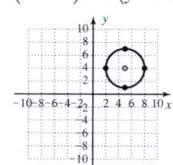

66. $(x - 5)^2 + (y - 3)^2 = 9$ **67.** $(2, 3), r = 2, x \in [0, 4], y \in [1, 5]$

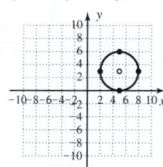 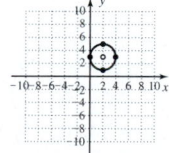

68. $(5, 1), r = 3, x \in [2, 8], y \in [-2, 4]$

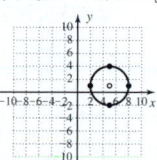

69. $(-1, 2), r = 2\sqrt{3}, x \in [-1 - 2\sqrt{3}, -1 + 2\sqrt{3}],$
$y \in [2 - 2\sqrt{3}, 2 + 2\sqrt{3}]$

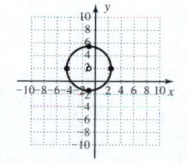

70. $(7, -4), r = 2\sqrt{5}, x \in [7 - 2\sqrt{5}, 7 + 2\sqrt{5}],$
$y \in [-4 - 2\sqrt{5}, -4 + 2\sqrt{5}]$

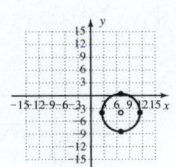

71. $(-4, 0), r = 9, x \in [-13, 5], y \in [-9, 9]$

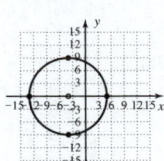

72. $(0, 3), r = 7, x \in [-7, 7], y \in [-4, 10]$

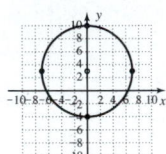

73. $(x - 5)^2 + (y - 6)^2 = 57, (5, 6), r = \sqrt{57}$

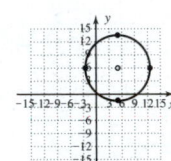

74. $(x + 3)^2 + (y - 4)^2 = 31, (-3, 4), r = \sqrt{31}$

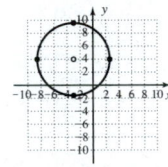

75. $(x - 5)^2 + (y + 2)^2 = 25, (5, -2), r = 5$

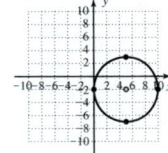

76. $(x + 3)^2 + (y + 2)^2 = 1, (-3, -2), r = 1$

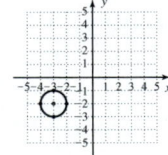

77. $x^2 + (y + 3)^2 = 14, (0, -3), r = \sqrt{14}$

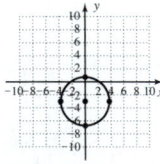

78. $(x - 4)^2 + y^2 = 4, (4, 0), r = 2$

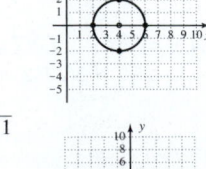

79. $(x + 2)^2 + (y + 5)^2 = 11, (-2, -5), r = \sqrt{11}$

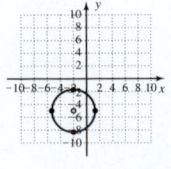

80. $(x - 4)^2 + (y - 7)^2 = 112, (4, 7), r = 4\sqrt{7}$

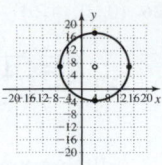

81. $(x + 7)^2 + y^2 = 37, (-7, 0), r = \sqrt{37}$

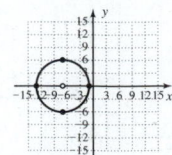

82. $x^2 + (y - 11)^2 = 126, (0, 11), r = 3\sqrt{14}$

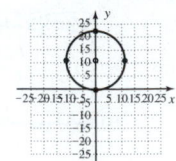

83. $(x - 3)^2 + (y + 5)^2 = 32, (3, -5), r = 4\sqrt{2}$

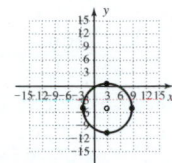

84. $(x - 4)^2 + (y + 3)^2 = 24, (4, -3), r = 2\sqrt{6}$

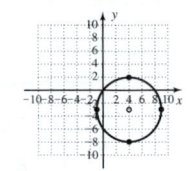

93. d.

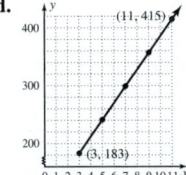

99.

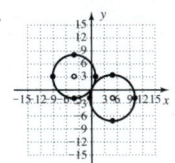

No, distance between centers is less than sum of radii.

100. ≈ 66 mi

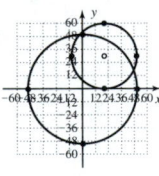

Exercises 1.2, pp. 112–116

8.

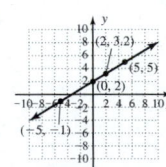

10.

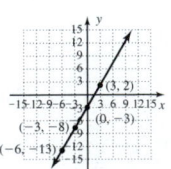

11. $-0.5 = \frac{3}{2}(-3) + 4$
$-0.5 = -\frac{9}{2} + 4$
$-0.5 = -0.5\checkmark$
$\frac{19}{4} = \frac{3}{2}(\frac{1}{2}) + 4$
$\frac{19}{4} = \frac{3}{4} + 4$
$\frac{19}{4} = \frac{19}{4}\checkmark$

12. $-5.5 = \frac{5}{3}(-1.5) - 3$
$-5.5 = -2.5 - 3$
$-5.5 = -5.5\checkmark$
$\frac{37}{6} = \frac{5}{3}(\frac{11}{2}) - 3$
$\frac{37}{6} = \frac{55}{6} - 3$
$\frac{37}{6} = \frac{37}{6}\checkmark$

13.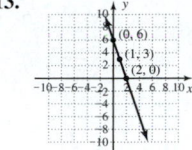

14.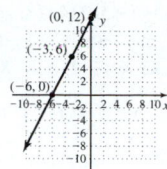

15.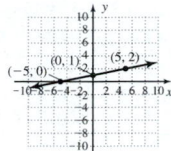

16.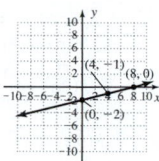

17.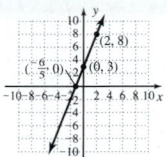

18.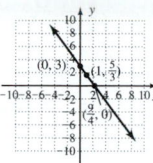

19.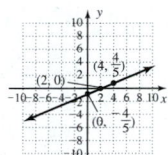

20.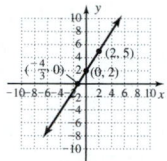

21.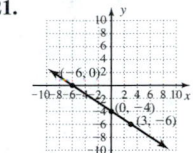

22.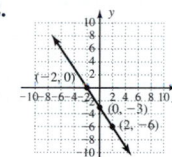

23.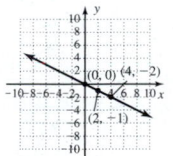

24.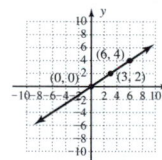

25.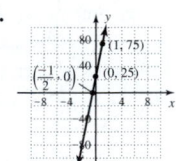

26.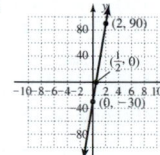

27.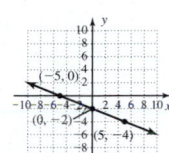

28.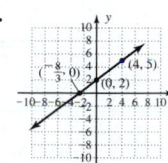

29.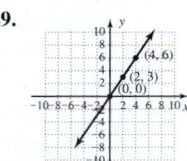

30.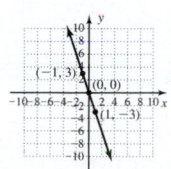

31.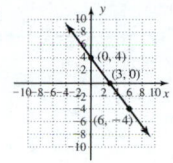

32.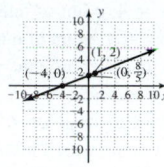

33. $m = 1$;
(2, 4) and (1, 3)

34. $m = \frac{5}{7}$;
(−9, −2) and (12, 13)

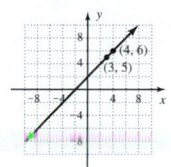

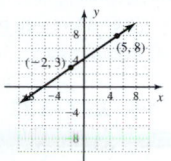

35. $m = \frac{4}{3}$;
(7, −1) and (1, −9)

36. $m = \frac{8}{3}$;
(−6, −9) and (3, 15)

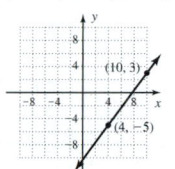

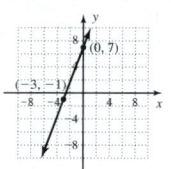

37. $m = \frac{-15}{4}$;
(5, −23) and (−7, 22)

38. $m = -2$;
(−3, 1) and (−7, 9)

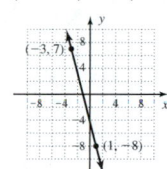

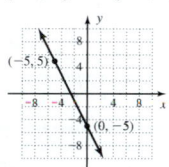

39. $m = \frac{-4}{7}$;
(−10, 10) and (11, −2)

40. $m = -3$;
(−4, 2) and (−1, −7)

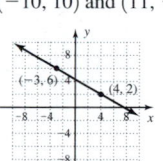

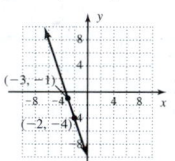

46. a. $m = 2200$; the plane climbs 2200 ft/min   **b.** 6 min

47. In inches: (0, −6) and (576, −18): $m = \frac{-1}{48}$. The sewer line is 1 in.
deeper for each 48 in. in length.

48. In feet: (0, 4) and (12, 0): $m = \frac{-1}{3}$. The roof decreases 1 ft in height
for every 3 ft in horizontal distance from the ridge.

49.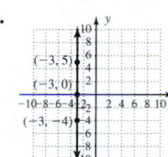

50.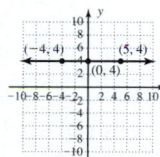

51.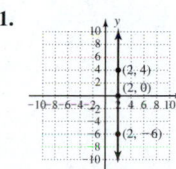

52.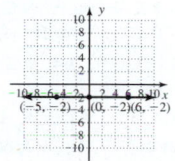

53. L_1: $x = 2$; L_2: $y = 4$; point of intersection (2, 4)
54. L_1: $x = -3$; L_2: $y = 1$; point of intersection $(-3, 1)$
55. a. For any two points chosen $m = 0$, indicating there has been no increase or decrease in the number of supreme court justices.   **b.** For any two points chosen $m = \frac{1}{10}$, which indicates that over the last 5 decades, one nonwhite or nonfemale justice has been added to the court every 10 yr.

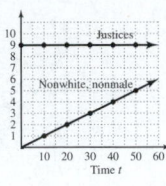

56. For any two points chosen, $m = \frac{-9}{5000}$ and the boiling point of water decreases by 9°F for each increase of 5000 ft in altitude.

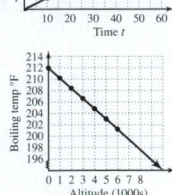

Exercises 1.3, pp. 127–131

31. 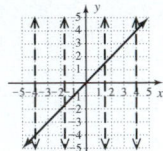 **32.**

function   function

33.   **34.**

function   nonfunction

39. 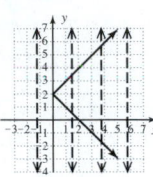 $p \in (-\infty, 3)$

40. $x \in (-2, \infty)$

41. $m \in (-\infty, 5]$

42. $n \in [-4, \infty)$

43. $x \in (-\infty, 1) \cup (1, \infty)$

44. $x \in (-\infty, -3) \cup (-3, \infty)$

45. $x \in (2, 5)$

46. 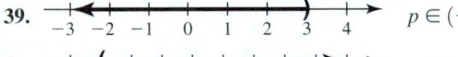 $p \in (-3, 4]$

51. function, $x \in [-4, 5]$, $y \in [-2, 3]$
52. function, $x \in [-4, \infty)$, $y \in (-\infty, 5]$
53. function, $x \in [-4, \infty)$, $y \in [-4, \infty)$
54. not a function, $x \in [0, 3]$, $y \in [-4, 4]$
55. function, $x \in [-4, 4]$, $y \in [-5, -1]$
56. function, $x \in (-\infty, \infty)$, $y \in (-\infty, \infty)$
57. function, $x \in (-\infty, \infty)$, $y \in (-\infty, \infty)$
58. not a function, $x \in [-3, 4]$, $y \in [-3, 4]$
59. not a function, $x \in [-3, 5]$, $y \in [-3, 3]$
60. function, $x \in (-\infty, \infty)$, $y \in [-3, \infty)$
61. not a function, $x \in (-\infty, 3]$, $y \in (-\infty, \infty)$
62. function, $x \in [-4, \infty)$, $y \in [0, \infty)$
63. $x \in (-\infty, 5) \cup (5, \infty)$ **64.** $x \in (-\infty, -3) \cup (-3, \infty)$
65. $a \in [\frac{-5}{3}, \infty)$ **66.** $a \in [\frac{2}{5}, \infty)$
67. $x \in (-\infty, -5) \cup (-5, 5) \cup (5, \infty)$
68. $x \in (-\infty, -7) \cup (-7, 7) \cup (7, \infty)$
69. $v \in (-\infty, -3\sqrt{2}) \cup (-3\sqrt{2}, 3\sqrt{2}) \cup (3\sqrt{2}, \infty)$
70. $q \in (-\infty, -2\sqrt{3}) \cup (-2\sqrt{3}, 2\sqrt{3}) \cup (2\sqrt{3}, \infty)$
71. $x \in (-\infty, \infty)$ **72.** $x \in (-\infty, \infty)$ **73.** $n \in (-\infty, \infty)$
74. $t \in (-\infty, \infty)$ **75.** $x \in (-\infty, \infty)$ **76.** $x \in (-\infty, \infty)$
77. $x \in (-\infty, -2) \cup (-2, 5) \cup (5, \infty)$

78. $x \in (-\infty, -5) \cup (-5, 3) \cup (3, \infty)$
79. $x \in [2, \frac{5}{2}) \cup (\frac{5}{2}, \infty)$ **80.** $x \in [-1, \frac{-2}{3}) \cup (\frac{-2}{3}, \infty)$
81. $x \in (-4, \infty)$ **82.** $x \in (2, \infty)$ **83.** $x \in (3, \infty)$ **84.** $x \in (-\infty, 5)$
87. $f(-6) = 0, f(\frac{3}{2}) = \frac{15}{4}, f(2c) = c + 3, f(c + 1) = \frac{1}{2}c + \frac{7}{2}$
88. $f(-6) = -9, f(\frac{3}{2}) = -4, f(2c) = \frac{4c}{3} - 5, f(c + 1) = \frac{2}{3}c - \frac{13}{3}$
89. $f(-6) = 132, f(\frac{3}{2}) = \frac{3}{4}, f(2c) = 12c^2 - 8c, f(c + 1) = 3c^2 + 2c - 1$
90. $f(-6) = 54, f(\frac{3}{2}) = 9, f(2c) = 8c^2 + 6c, f(c + 1) = 2c^2 + 7c + 5$
91. $h(3) = 1, h(\frac{-2}{3}) = \frac{-9}{2}, h(3a) = \frac{1}{a}, h(a - 2) = \frac{3}{a - 2}$
92. $h(3) = \frac{2}{9}, h(\frac{-2}{3}) = \frac{9}{2}, h(3a) = \frac{2}{9a^2}, h(a - 2) = \frac{2}{a^2 - 4a + 4}$
93. $h(3) = 5, h(\frac{-2}{3}) = -5, h(3a) = -5$ if $a < 0$ or 5 if $a > 0$, $h(a - 2) = 5$ if $a > 2$ or -5 if $a < 2$
94. $h(3) = 4, h\left(\frac{-2}{3}\right) = -4, h(3a) = -4$ if $a < 0$ or 4 if $a > 0, h(a - 2) = 4$ if $a > 2$ or -4 if $a < 2$
95. $g(4) = 8\pi, g\left(\frac{3}{2}\right) = 3\pi, g(2c) = 4\pi c, g(c + 3) = 2\pi(c + 3)$
96. $g(4) = 8\pi h, g\left(\frac{3}{2}\right) = 3\pi h, g(2c) = 4\pi ch, g(c + 3) = 2\pi h(c + 3)$
97. $g(4) = 16\pi, g\left(\frac{3}{2}\right) = \frac{9}{4}\pi, g(2c) = 4\pi c^2, g(c + 3) = (c^2 + 6c + 9)\pi$
98. $g(4) = 16\pi h, g\left(\frac{3}{2}\right) = \frac{9}{4}\pi h, g(2c) = 4\pi c^2 h, g(c + 3) = (c^2 + 6c + 9)\pi h$
99. $p(5) = \sqrt{13}, p\left(\frac{3}{2}\right) = \sqrt{6}, p(3a) = \sqrt{6a + 3}, p(a - 1) = \sqrt{2a + 1}$
100. $p(5) = \sqrt{19}, p\left(\frac{3}{2}\right) = \sqrt{5}, p(3a) = \sqrt{12a - 1}, p(a - 1) = \sqrt{4a - 5}$
101. $p(5) = \frac{14}{5}, p\left(\frac{3}{2}\right) = \frac{7}{9}, p(3a) = \frac{27a^2 - 5}{9a^2}, p(a - 1) = \frac{3a^2 - 6a - 2}{a^2 - 2a + 1}$
102. $p(5) = \frac{53}{25}, p\left(\frac{3}{2}\right) = \frac{10}{3}, p(3a) = \frac{6a^2 + 1}{3a^2}, p(a - 1) = \frac{2a^2 - 4a + 5}{a^2 - 2a + 1}$
103. a. D: $\{-1, 0, 1, 2, 3, 4, 5\}$ **b.** R: $\{-2, -1, 0, 1, 2, 3, 4\}$ **c.** 1
d. -1 **104. a.** D: $\{-5, -4, -3, -2, -1, 0, 1, 2, 3, 4, 5\}$
b. R: $\{-1, 0, 1, 2, 3, 4, 5\}$ **c.** 0 **d.** $-3, 5$ **105. a.** D: $[-5, 5]$
b. $y \in [-3, 4]$ **c.** -2 **d.** -4 and 0 **106. a.** D: $[-3, 5]$
b. $y \in [-4, 5]$ **c.** -4 **d.** 1 and 3 **107. a.** D: $[-3, \infty)$ **b.** $y \in (-\infty, 4]$
c. 2 **d.** -2 and 2 **108. a.** D: $[-5, \infty)$ **b.** $y \in [-2, \infty)$ **c.** -2
d. 1 and 3 **117. a.** $c(t) = 42.50t + 50$ **b.** $156.25   **c.** 5 hr
d. $t \in [0, 10.6]$; $c \in [0, 500]$ **118. a.** Yes. Each x is paired with exactly one y.   **b.** 9 P.M.   **c.** $3\frac{1}{2}$ m   **d.** 5 P.M. and 1 A.M.   **119. a.** Yes. Each x is paired with exactly one y.   **b.** 10 P.M.   **c.** 0.9 m   **d.** 7 P.M. and 1 A.M.
121. negative outputs become positive

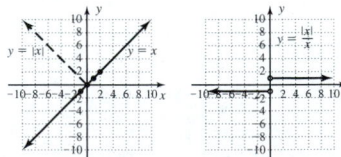

122. negative outputs become positive

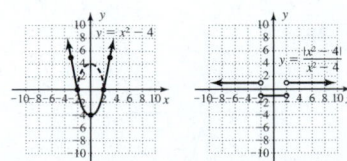

123. a. $x \in (-\infty, -2) \cup (2, \infty)$; $x = \frac{2y + 3}{1 - y}$; $y \in (-\infty, 1) \cup (1, \infty)$
b. $x \in \mathbb{R}$ $x = \pm\sqrt{y + 3}$; $y \in [-3, \infty)$

124. $(x - 4)^2 + (y + 1)^2 = 25$

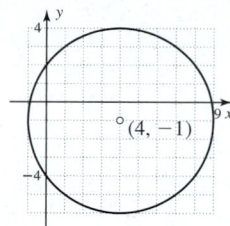

Mid-Chapter Check, p. 132

1.
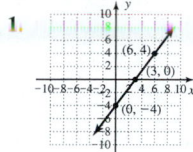

3. positive, loss is decreasing (profit is increasing); $m = \frac{3}{2}$, yes; $\frac{1.5}{1}$, each year Data.com's loss decreases by 1.5 million.

4. a. $E(x) = 7.5x + 950$ **b.** \$1100, \$1175, \$1250
   c. $x \in [0, 75, 5]$ and $y \in [-200, 2000, 200]$
   d. 47 snowboards

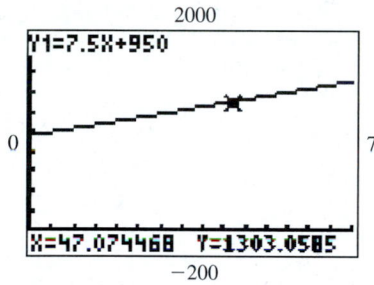

5. $x = -3$; no; input -3 is paired with more than one output.
6. $y = 2$; yes **8.** from $x = 1$ to $x = 2$; steeper line $\rightarrow$ greater slope
9. $\frac{\Delta F}{\Delta p} = \frac{3}{4}$; For each increase of 4000 pheasants, the fox population increases by 300; 1100 foxes. **10. a.** $x \in \{-3, -2, -1, 0, 1, 2, 3, 4\}$, $y \in \{-3, -2, -1, 0, 1, 2, 3, 4\}$ **b.** $x \in [-3, 4]$, $y \in [-3, 4]$
   c. $x \in (-\infty, \infty)$, $y \in (-\infty, \infty)$

Exercises 1.4, pp. 143–147

7. $y = \frac{-4}{5}x + 2$

x	y
-5	6
-2	$\frac{18}{5}$
0	2
1	$\frac{6}{5}$
3	$\frac{-2}{5}$

8. $y = \frac{2}{3}x + 3$

x	y
-5	$\frac{-1}{3}$
-2	$\frac{5}{3}$
0	3
1	$\frac{11}{3}$
3	5

9. $y = 2x + 7$

x	y
-5	-3
-2	3
0	7
1	9
3	13

10. $y = \frac{2}{7}x - 3$

x	y
-5	$\frac{-31}{7}$
-2	$\frac{-25}{7}$
0	-3
1	$\frac{-19}{7}$
3	$\frac{-15}{7}$

11. $y = \frac{-5}{3}x - 5$

x	y
-5	$\frac{10}{3}$
-2	$\frac{-5}{3}$
0	-5
1	$\frac{-20}{3}$
3	-10

12. $y = \frac{7}{3}x + 14$

x	y
-5	$\frac{7}{3}$
-2	$\frac{28}{3}$
0	14
1	$\frac{49}{3}$
3	21

19.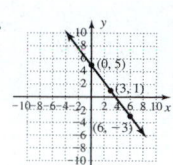
20.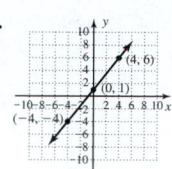
21.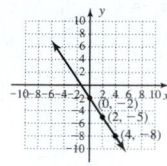

22.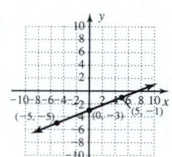
23.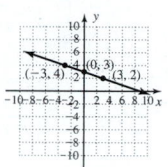
24.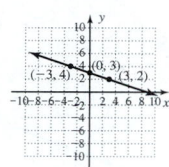

25. a. $\frac{-3}{4}$ **b.** $y = \frac{-3}{4}x + 3$ **c.** The coeff. of x is the slope and the constant is the y-intercept. **26. a.** $\frac{2}{3}$ **b.** $y = \frac{2}{3}x - 2$ **c.** The coeff. of x is the slope and the constant is the y-intercept. **27. a.** $\frac{2}{5}$ **b.** $y = \frac{2}{5}x - 2$
c. The coeff. of x is the slope and the constant is the y-intercept.
28. a. $\frac{-2}{3}$ **b.** $y = \frac{-2}{3}x + 3$ **c.** The coeff. of x is the slope and the constant is the y-intercept. **29. a.** $\frac{4}{5}$ **b.** $y = \frac{4}{5}x + 3$ **c.** The coeff. of x is the slope and the constant is the y-intercept. **30. a.** $\frac{-6}{5}$
b. $y = \frac{-6}{5}x - 5$ **c.** The coeff. of x is the slope and the constant is the y-intercept. **31.** $y = \frac{-2}{3}x + 2, f(x) = \frac{-2}{3}x + 2, m = \frac{-2}{3}$, y-intercept $(0, 2)$
32. $y = \frac{3}{4}x + 3, f(x) = \frac{3}{4}x + 3, m = \frac{3}{4}$, y-intercept $(0, 3)$
33. $y = \frac{-5}{4}x + 5, f(x) = \frac{-5}{4}x + 5, m = \frac{-5}{4}$, y-intercept $(0, 5)$
34. $y = -2x + 4, f(x) = -2x + 4, m = -2$, y-intercept $(0, 4)$
35. $y = \frac{1}{3}x, f(x) = \frac{1}{3}x, m = \frac{1}{3}$, y-intercept $(0, 0)$ **36.** $y = \frac{-2}{5}x, f(x) = \frac{-2}{5}x$, $m = \frac{-2}{5}$, y-intercept $(0, 0)$ **37.** $y = \frac{-3}{4}x + 3, f(x) = \frac{-3}{4}x + 3, m = \frac{-3}{4}$, y-intercept $(0, 3)$ **38.** $y = \frac{3}{5}x - 4, f(x) = \frac{3}{5}x - 4, m = \frac{3}{5}$, y-intercept $(0, -4)$

51. $y = -\frac{3}{5}x + 4$ **52.** $y = \frac{1}{2}x + 2$ **53.** $y = \frac{2}{3}x - 5$

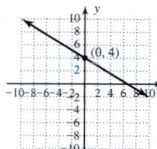   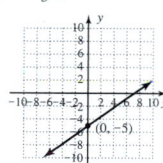

54. $y = \frac{3}{2}x + 2$ **55.**   **56.**

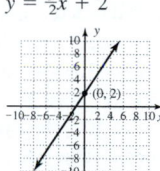   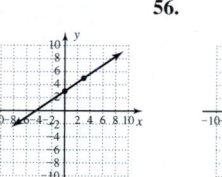

57. **58.** **59.**

60. **61.** **62.**

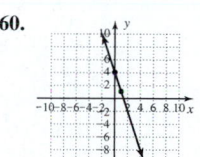   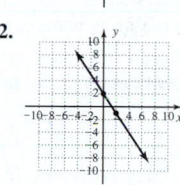

77. a. $y = \frac{-3}{4}x - \frac{5}{2}$ **b.** $y = \frac{4}{3}x - \frac{20}{3}$ **78. a.** $y = \frac{-2}{5}x + \frac{17}{5}$
b. $y = \frac{5}{2}x + \frac{1}{2}$ **79. a.** $y = \frac{4}{9}x + \frac{31}{9}$ **b.** $y = \frac{-9}{4}x + \frac{3}{4}$
80. a. $y = \frac{-2}{3}x - \frac{11}{6}$ **b.** $y = \frac{3}{2}x - 4$ **81. a.** $y = \frac{-1}{2}x - 2$
b. $y = 2x - 2$ **82. a.** $y = -x + 4$ **b.** $y = x + 2$

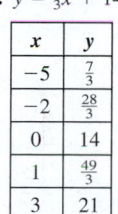

83.

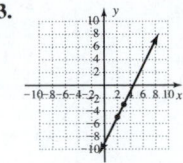

$$y + 5 = 2(x - 2)$$

84.

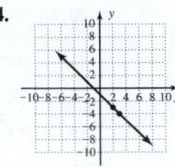

$$y + 3 = -1(x - 2)$$

85.

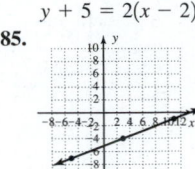

$$y + 4 = \frac{3}{8}(x - 3)$$

86.

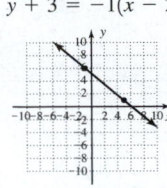

$$y - 6 = \frac{-5}{6}(x + 1)$$

87.

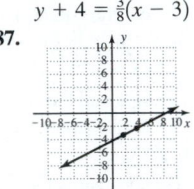

$$y + 3.1 = 0.5(x - 1.8)$$

88.

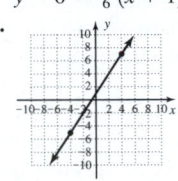

$$y + 0.125 = 1.5(x + 0.75)$$

89. $y - 2 = \frac{6}{5}(x - 4)$; For each 5000 additional sales, income rises $6000.   **90.** $y - 9 = \frac{-3}{2}(x - 3)$; Every two years, 30,000 typewriters are no longer in service.   **91.** $y - 100 = \frac{-20}{1}(x - 0.5)$; For every hour of television, a student's final grade falls 20%.   **92.** $y - 3 = \frac{3}{7}(x - 2)$; Every 7000 investors increases the number of online brokerage houses by 3.
93. $y - 10 = \frac{35}{2}(x - \frac{1}{2})$; Every 2 in. of rainfall increases the number of cattle raised per acre by 35.   **94.** $y - 2 = \frac{2}{5}(x - 60)$; For every 5°F rise in temperature there are 2 additional eggs per hen per week.   **103.** $m = \frac{-a}{b}$, y-intercept $= \frac{c}{b}$ **a.** $m = \frac{-3}{4}$, y-intercept $(0, 2)$ **b.** $m = \frac{-2}{5}$, y-intercept $(0, -3)$ **c.** $m = \frac{5}{6}$, y-intercept $(0, 2)$ **d.** $m = \frac{5}{3}$, y-intercept $(0, 3)$

104. a. $(5, 0), (0, 2)$ **b.** $(-4, 0), (0, 3)$ **c.** $\left(\frac{8}{5}, 0\right)$, $(0, 2)$ Slope m is always equal to $\frac{-k}{h}$.   **105. a.** As the temperature increases 5°C, the velocity of sound waves increases 3 m/s. At a temperature of 0°C, the velocity is 331 m/s.   **b.** 343 m/s   **c.** 50°C   **106. a.** Every 5 seconds the velocity is increasing 26 ft/sec. The initial velocity is 60 ft/sec.   **b.** 108.88 ft/sec   **c.** $t \approx 7.7$ sec   **107. a.** $V = \frac{20}{3}t + 150$ **b.** Every 3 yr the value of the coin increases by $20; the initial value was $150.   **c.** $223.33
d. 15 years, in 2013   **e.** 3 yr   **108. a.** $V = -3500t + 18,500$ **b.** Every 1 yr the equipment decreases in value by $3500; the initial value was $18,500.   **c.** $4500   **d.** 3.6 yr   **e.** 5 yr   **109. a.** $N = 7t + 9$ **b.** Every 1 yr the number of homes with Internet access increases by 7 million.
c. 1993   **d.** 86 million   **e.** 13 yr   **f.** 2010   **110. a.** $S = 14.8t + 72$
b. Every 1 yr sales increase by $14.8 billion dollars.   **c.** 2007
d. $220 billion   **e.** 14 yr   **f.** 2010   **111. a.** $P = 58,000t + 740,000$
b. Each year, the prison population increases by 58,000.   **c.** 1,900,000
112. a. $M = 2.7t + 143$ **b.** Each year, the number of restaurant meals increases by about 3 meals.   **c.** about 197 meals/yr

Exercises 1.5, pp. 159–162

7.

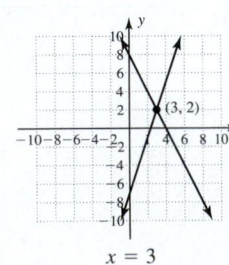

$$x = 3$$

8.

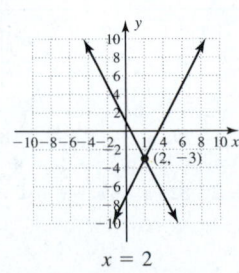

$$x = 2$$

Exercises 1.6, pp. 171–177

7. a.

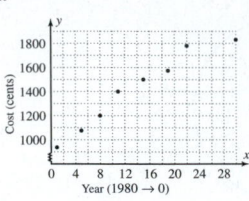

8. a.

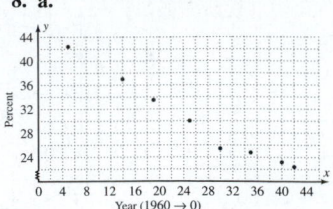

9. a.

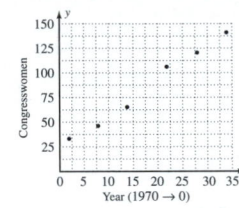

10. a.

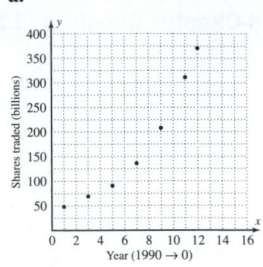

11. a.

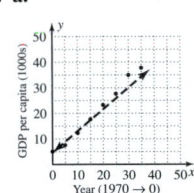

12. a.

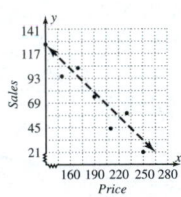

13. a. (A) (D) (C) (B)
b.

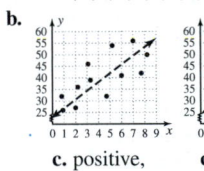

c. positive,   **c.** positive,   **c.** negative,   **c.** negative,
d. $m \approx 3.8$;   **d.** $m \approx 4.2$;   **d.** $m \approx -2.4$;   **d.** $m \approx -4.6$

14. a. (A) (D) (B) (C)
b.

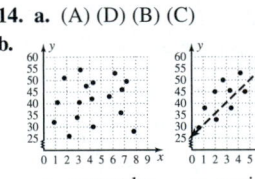

c. cannot be   **c.** positive,   **c.** negative,   **c.** positive,
determined,   **d.** $m \approx 4.8$;   **d.** $m \approx -5.0$;   **d.** $m \approx 2.8$
d. not possible;

21. a.

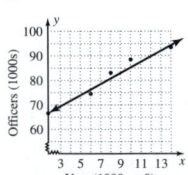

22. a.

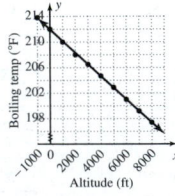

25. a.

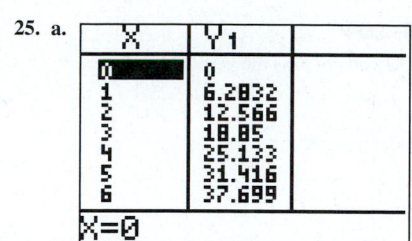

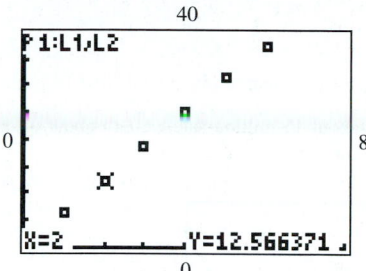

26. a.

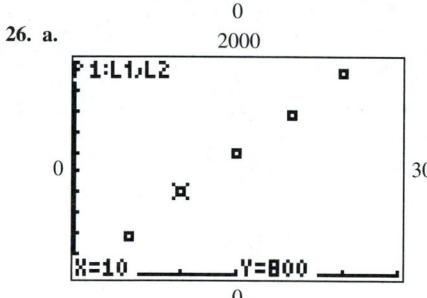

27. a.

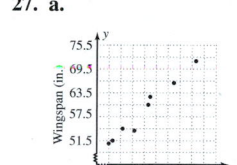

28. a.

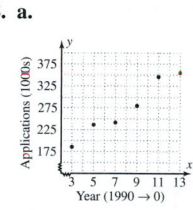

29. a.

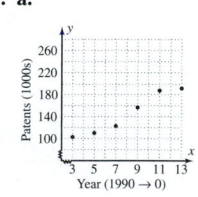

30. a.

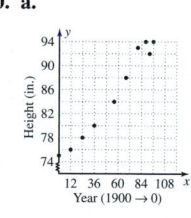

31. a.

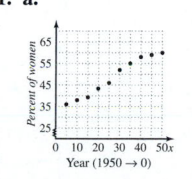 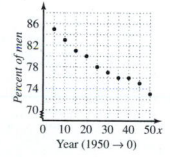

32. a.

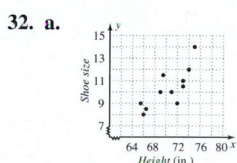

33. a.

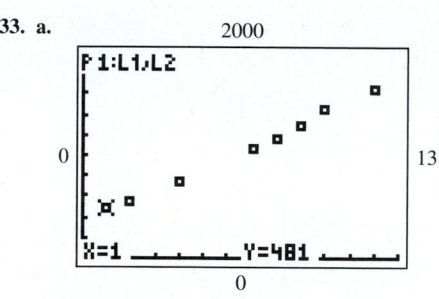

34. b. $f(x) = 47.2x - 62.4$

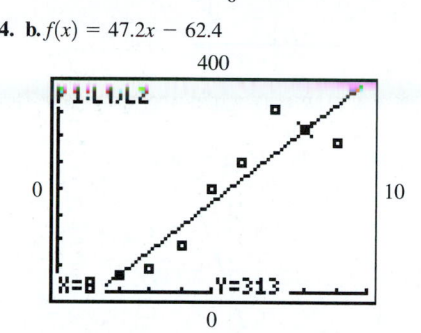

Summary and Concept Review, pp. 178–182

1. $x \in \{-7, -4, 0, 3, 5\}$ $y \in \{-2, 0, 1, 3, 8\}$

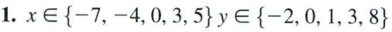

3.

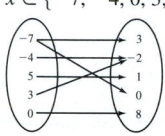

$(0, -5), (-3, 0)$

6.

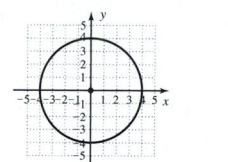

7.

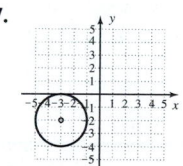

9. a.

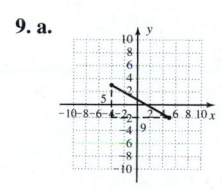

$\frac{-5}{9}, (14, -7)$

b.

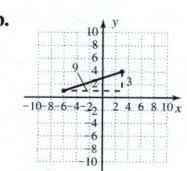

$\frac{1}{3}, (0, 3)$

11. a.

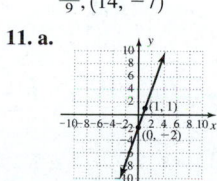

b.

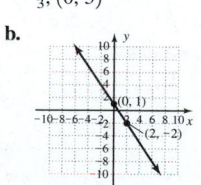

12. a. 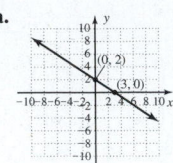 **b.**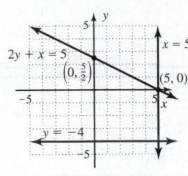

13. a. vertical
b. horizontal
c. neither

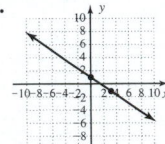

19. I. a. $D = \{-1, 0, 1, 2, 3, 4, 5\}, R = \{-2, 1, 0, 1, 2, 3, 4\}$ **b.** 1 **c.** 2
II. a. $x \in (-\infty, \infty), y \in (-\infty, \infty)$ **b.** -1 **c.** 3
III. a. $x \in [-3, \infty), y \in [-4, \infty)$ **b.** -1 **c.** -3 or 3

21. a. 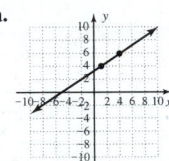 **b.**

22. a. **b.**
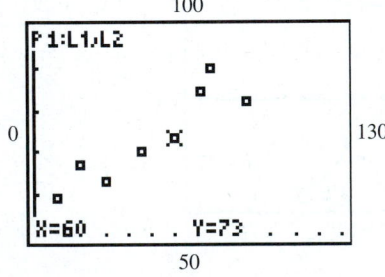

24. $y = \frac{-3}{4}x + \frac{11}{4}$ **27. a.** $(y - 90) = \frac{-15}{2}(x - 2)$ **b.** $(14, 0), (0, 105)$
c. $f(x) = \frac{-15}{2}x + 105$ **d.** $f(20) = -45, x = 12$

38. a.
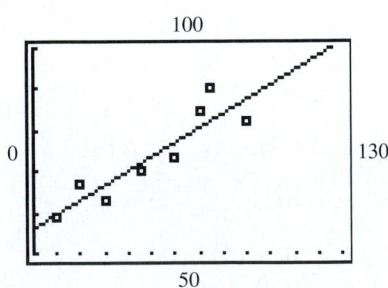

b. linear **c.** positive
39. a. $f(x) = 0.35x + 56.10$
b.

c. strong
40. $f(120) = 98.1.$ over 98%

Practice Test, pp. 183–184

5. a and c are nonfunctions, they do not pass the vertical line test
7. 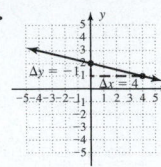 **8.** 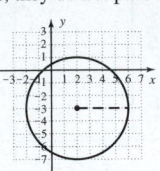 $(2, -3); r = 4$

13. a. $x \in \{-4, -2, 0, 2, 4, 6\}$ $y \in \{-2, -1, 0, 1, 2, 3\}$
b. $x \in [-2, 6]$ $y \in [1, 4]$

16. a. $\dfrac{\Delta \text{sales}}{\Delta \text{time}} = \dfrac{13.5}{1}$ **b.** sales are increasing at a rate of 13.5 million

phones per year **c.** 2008: about 15 million sales, 2010: about 42 million
sales, 2011: about 55.5 million sales
19. a. **b.** linear **c.** positive

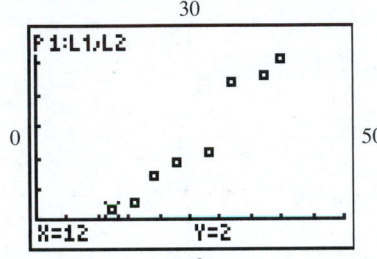

Strengthening Core Skills, pp. 184–185

1. a. $\frac{1}{3}$, increasing **b.** $y - 5 = \frac{1}{3}(x - 0)$,
$y = \frac{1}{3}x + 5$
c. $(0, 5), (-15, 0)$

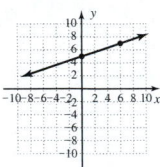

2. a. $\frac{-7}{3}$, decreasing **b.** $y - 9 = \frac{-7}{3}(x - 0)$,
$y = \frac{-7}{3}x + 9$
c. $(0, 9), (\frac{22}{7}, 0)$

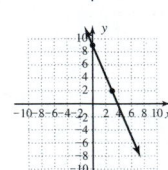

3. a. $\frac{1}{2}$, increasing **b.** $y - 2 = \frac{1}{2}(x - 3)$,
$y = \frac{1}{2}x + \frac{1}{2}$
c. $(0, \frac{1}{2}), (-1, 0)$

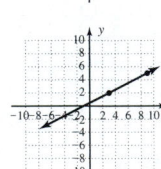

4. a. $\frac{3}{4}$, increasing **b.** $y + 4 = \frac{3}{4}(x + 5)$,
$y = \frac{3}{4}x - \frac{1}{4}$
c. $(0, \frac{-1}{4}), (\frac{1}{3}, 0)$

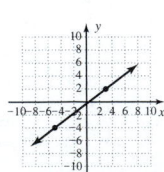

5. a. $\frac{-3}{4}$, decreasing **b.** $y - 5 = \frac{-3}{4}(x + 2)$,
$y = \frac{-3}{4}x + \frac{7}{2}$
c. $(0, \frac{7}{2}), (\frac{14}{3}, 0)$

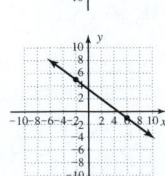

6. a. $\frac{-1}{2}$, decreasing **b.** $y + 7 = \frac{-1}{2}(x - 2)$,
$y = \frac{-1}{2}x - 6$
c. $(0, -6), (-12, 0)$

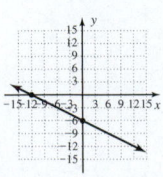

CHAPTER 2

Exercises 2.1, pp. 196–201

33. $V(x)\uparrow$: $x \in (-3, 1) \cup (4, 6)$; $V(x)\downarrow$: $x \in (-\infty, -3) \cup (1, 4)$; constant: none

34. $H(x)\uparrow$: $x \in (-2, 0) \cup (3, 5)$; $H(x)\downarrow$: $x \in (-\infty, -2)$; $H(x)$ constant: $x \in (0, 3)$

35. $f(x)\uparrow$: $x \in (1, 4)$; $f(x)\downarrow$: $x \in (-2, 1) \cup (4, \infty)$; constant: $x \in (-\infty, -2)$

36. $g(x)\uparrow$: $x \in (0, 3) \cup (5, 9)$; $g(x)\downarrow$: $x \in (3, 5) \cup (9, \infty)$; constant: none

37. a. $p(x)\uparrow$: $x \in (-\infty, \infty)$; $p(x)\downarrow$: none **b.** down, up

38. a. $q(x)\uparrow$: none; $q(x)\downarrow$: $x \in (-\infty, \infty)$ **b.** up, down

39. a. $f(x)\uparrow$: $x \in (-3, 0) \cup (3, \infty)$; $f(x)\downarrow$: $x \in (-\infty, -3) \cup (0, 3)$

b. up, up **40. a.** $g(x)\uparrow$: $x \in (-\infty, -2) \cup (1, 6) \cup (8, \infty)$; $g(x)\downarrow$: $x \in (-2, 1) \cup (6, 8)$ **b.** down, up

41. a. $x \in (-\infty, \infty)$; $y \in (-\infty, 5]$ **b.** $x = 1, 3$

c. $H(x) \geq 0$: $x \in [1, 3]$; $H(x) \leq 0$: $x \in (-\infty, 1] \cup [3, \infty)$

d. $H(x)\uparrow$: $x \in (-\infty, 2)$; $H(x)\downarrow$: $x \in (2, \infty)$ **e.** local max: $y = 5$ at $(2, 5)$

42. a. $x \in (-\infty, \infty)$; $y \in (-\infty, \infty)$ **b.** $x = -3.5, 0, 3.5$

c. $f(x) \geq 0$: $x \in [-3.5, 0] \cup [3.5, \infty)$; $f(x) \leq 0$: $x \in (-\infty, -3.5] \cup [0, 3.5]$ **d.** $f(x)\uparrow$: $x \in (-\infty, -2) \cup (2, \infty)$; $f(x)\downarrow$: $x \in (-2, 2)$ **e.** local max: $y = 3$ at $(-2, 3)$; local min: $y = -3$ at $(2, -3)$ **43. a.** $x \in (-\infty, \infty)$; $y \in (-\infty, \infty)$ **b.** $x = -1, 5$

c. $g(x) \geq 0$: $x \in [-1, \infty)$; $g(x) \leq 0$: $x \in (-\infty, -1] \cup \{3.5\}$

d. $g(x)\uparrow$: $x \in (-\infty, 1) \cup (5, \infty)$; $g(x)\downarrow$: $x \in (1, 5)$ **e.** local max: $y = 6$ at $(1, 6)$; local min: $y = 0$ at $(5, 0)$

44. a. $x \in (-\infty, 5]$; $y \in (-\infty, 5]$

b. $x = -5, -3, 1$ **c.** $h(x) \geq 0$: $x \in [-5, -3] \cup [1, 5]$; $h(x) \leq 0$: $x \in (-\infty, 5] \cup [-3, 1]$

d. $h(x)\uparrow$: $x \in (-\infty, -4) \cup (-1, 2) \cup (4, 5)$; $h(x)\downarrow$: $x \in (-4, -1) \cup (2, 4)$

e. local max: $y = 3$ at $(-4, 3)$, $y = 4$ at $(2, 4)$; local min: $y = -4$ at $(-1, -4)$, $y = 1$ at $(4, 1)$; endpoint max $y = 5$ at $(5, 5)$

45. a. $x \in [-4, \infty)$; $y \in (-\infty, 3]$

b. $x = -4, 2$ **c.** $Y_1 \geq 0$: $x \in [-4, 2]$; $Y_1 \leq 0$: $x \in [2, \infty)$

d. $Y_1\uparrow$: $x \in (-4, -2)$; $Y_1\downarrow$: $x \in (-2, \infty)$ **e.** local max: $y = 3$ at $(-2, 3)$; endpoint min $y = 0$ at $(-4, 0)$

46. a. $x \in (-\infty, \infty)$, $y \in (-\infty, 3]$ **b.** $x = 0, 2$ **c.** $Y_2 \geq 0$: $x \in [0, 2]$; $Y_2 \leq 0$: $x \in (-\infty, 0] \cup [2, \infty)$ **d.** $Y_2\uparrow$: $x \in (-\infty, 1)$; $Y_2\downarrow$: $x \in (1, \infty)$

e. local max: $y = 3$ at $(1, 3)$ **47. a.** $x \in (-\infty, \infty)$, $y \in (-\infty, \infty)$

b. $x = -4$ **c.** $p(x) \geq 0$: $x \in [-4, \infty)$; $p(x) \leq 0$: $x \in (-\infty, -4]$

d. $p(x)\uparrow$: $x \in (-\infty, -3) \cup (-3, \infty)$; $p(x)\downarrow$: never decreasing **e.** local max: none; local min: none

48. a. $x \in (-\infty, \infty)$, $y \in [3, \infty)$ **b.** none

c. $q(x) \geq 0$: $x \in (-\infty, \infty)$; $q(x) \leq 0$: $q(x)$ is always positive

d. $q(x)\uparrow$: $x \in (5, \infty)$; $q(x)\downarrow$: $x \in (-\infty, 5)$ **e.** local max: none; local min: $y = 3$ at $(5, 3)$ **49.** max: $y \approx 1.58$ at $x \approx 0.78$; min: $y \approx -0.47$ at $x \approx 2.55$ **50.** max: $y \approx 2.54$ at $x \approx -2.22$; min: $y \approx -0.76$ at $x \approx -0.45$ **51.** max: $y \approx 1.54$ at $x \approx -6.21$, $y \approx 3.28$ at $x \approx 2.55$; min: $y \approx -3.28$ at $x \approx -2.55$, $y \approx -1.54$ at $x \approx 6.21$

52. max: $y \approx 0$ at $x \approx 0$, $y \approx 1.57$ at $x \approx 4.27$; min: $y \approx -10.38$ at $x \approx -3.74$, $y \approx -0.85$ at $x \approx 1.88$

53. max: $y \approx 3.08$ at $x = \frac{8}{3}$; min: $y = 0$ at $x = 4$ (endpoint)

54. max: $y \approx 2.46$ at $x = -2.4$; min: $y = -2$ at $x = -3$ (endpoint), $y = -2$ at $x = 0$

55. a. $x \in (-\infty, -3] \cup [3, \infty)$; $y \in [0, \infty)$ **b.** $(-3, 0), (3, 0)$

c. $f(x)\uparrow$: $x \in (3, \infty)$; $f(x)\downarrow$: $x \in (-\infty, -3)$ **d.** even

e. $x = \pm\sqrt{\dfrac{9y^2 + 36}{2}}$ **56.** $y = \sin x$: **a.** $y \in [-1, 1]$

b. $(-180, 0), (0, 0), (180, 0), (360, 0)$

c. $f(x)\uparrow$: $x \in (-360, -270) \cup (-90, 90) \cup (270, 360)$; $f(x)\downarrow$: $x \in (-270, -90) \cup (90, 270)$

d. min: $(-90, -1)$ and $(270, -1)$, max: $(-270, 1)$ and $(90, 1)$

e. odd; $y = \cos x$: **a.** $y \in [-1, 1]$

b. $(-270, 0), (-90, 0), (90, 0), (270, 0)$

c. $f(x)\uparrow$: $x \in (-180, 0) \cup (180, 360)$; $f(x)\downarrow$: $x \in (-360, -180) \cup (0, 180)$

d. min: $(-180, -1)$ and $(180, -1)$, max: $(-360, 1)$, $(0, 1)$ and $(360, 1)$

e. even **61. a.** D: $t \in [1983, 2009]$, R: $I \in [5, 14]$; **b.** $I(t)\uparrow$: $t \in (1983, 1984) \cup (1986, 1987) \cup (1993, 1994) \cup (1998, 2000) \cup (2005, 2006)$; $I(t)\downarrow$: $t \in (1984, 1986) \cup (1989, 1993) \cup (1994, 1998) \cup (2000, 2003) \cup (2006, 2009)$; $I(t)$ constant: $(1987, 1989) \cup (2003, 2005)$; **c.** global max: $I = 14$ in 1984, global min: $I = 5$ in 2009 (also an endpoint min);

d. greatest increase: (1993, 1994), greatest decrease: (1985, 1986)

62. a. D: $t \in [1980, 2008]$, R: $S \in [-410, 240]$; **b.** $S(t)\uparrow$: $t \in (1983, 1984) \cup (1986, 1987) \cup (1992, 2000) \cup (2004, 2007)$; $S(t)\downarrow$: $t \in (1981, 1983) \cup (1984, 1985) \cup (1987, 1992) \cup (2000, 2004) \cup (2007, 2008)$; $S(t)$ constant: $(1980, 1981) \cup (1985, 1986)$; **c.** global max: $S = 240$ in 2000, global min: $S = -410$ in 2004; **d.** greatest increase: (1999, 2000), greatest decrease: (2001, 2002)

63. zeroes: $(-8, 0), (-4, 0), (0, 0), (4, 0)$; min: $(-10, -6), (-2, -1), (4, 0)$; max: $(-6, 2), (2, 2)$

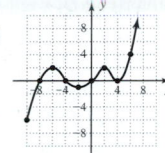

64. zeroes: $(-9, 0), (-3, 0), (6, 0)$; min: $(-6, -6), (6, 0)$; max: $(3, 6)$

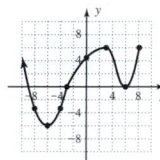

71. a. $\dfrac{12}{4 - x^2}$ **b.** $\dfrac{9}{4 - x^2}$

Exercises 2.2, pp. 212–217

7. a. quadratic; **b.** up/up, $(-2, -4)$, $x = -2$, $(-4, 0)$, $(0, 0)$, $(0, 0)$; **c.** D: $x \in \mathbb{R}$, R: $y \in [-4, \infty)$

8. a. quadratic; **b.** down/down, $(1, 1)$, $x = 1$, $(0, 0)$, $(2, 0)$, $(0, 0)$; **c.** D: $x \in \mathbb{R}$, R: $y \in (-\infty, 1]$

9. a. quadratic; **b.** up/up, $(1, -4)$, $x = 1$, $(-1, 0)$, $(3, 0)$, $(0, -3)$; **c.** D: $x \in \mathbb{R}$, R: $y \in [-4, \infty)$

10. a. quadratic; **b.** down/down, $(1, 9)$, $x = 1$, $(-2, 0)$, $(4, 0)$, $(0, 8)$; **c.** D: $x \in \mathbb{R}$, R: $y \in (-\infty, 9]$

11. a. quadratic; **b.** up/up, $(2, -9)$, $x = 2$, $(-1, 0)$, $(5, 0)$, $(0, -5)$; **c.** D: $x \in \mathbb{R}$, R: $y \in [-9, \infty)$

12. a. quadratic; **b.** up/up, $(-3, -4)$, $x = -3$, $(-5, 0)$, $(-1, 0)$, $(0, 5)$; **c.** D: $x \in \mathbb{R}$, R: $y \in [-4, \infty)$

13. a. square root; **b.** up to the right, $(-4, -2)$, $(-3, 0)$, $(0, 2)$; **c.** D: $x \in [-4, \infty)$, R: $y \in [-2, \infty)$

14. a. square root; **b.** down to the right, $(-4, 2)$, $(-3, 0)$, $(0, -2)$; **c.** D: $x \in [-4, \infty)$, R: $y \in (-\infty, 2]$

15. a. square root; **b.** down to the left, $(4, 3)$, $(3, 0)$, $(0, -3)$; **c.** D: $x \in (-\infty, 4]$, R: $y \in (-\infty, 3]$

16. a. square root; **b.** up to the right, $(-1, -4)$, $(3, 0)$, $(0, -2)$; **c.** D: $x \in [-1, \infty)$, R: $y \in [-4, \infty)$

17. a. square root; **b.** up to the left, $(4, 0)$, $(4, 0)$, $(0, 4)$; **c.** D: $x \in (-\infty, 4]$, R: $y \in [0, \infty)$

18. a. square root; **b.** down to the right, $(-1, 4)$, $(3, 0)$, $(0, 2)$; **c.** D: $x \in [-1, \infty)$, R: $y \in (-\infty, 4]$

19. a. absolute value; **b.** up/up, $(-1, -4)$, $x = -1$, $(-3, 0)$, $(1, 0)$, $(0, -2)$; **c.** D: $x \in \mathbb{R}$, R: $y \in [-4, \infty)$

20. a. absolute value; **b.** down/down, $(2, 3)$, $x = 2$, $(1, 0)$, $(3, 0)$, $(0, -3)$; **c.** D: $x \in \mathbb{R}$, R: $y \in (-\infty, 3]$

21. a. absolute value; **b.** down/down, $(-1, 6)$, $x = -1$, $(-4, 0)$, $(2, 0)$, $(0, 4)$; **c.** D: $x \in \mathbb{R}$, R: $y \in (-\infty, 6]$

22. a. absolute value; **b.** up/up, $(2, -6)$, $x = 2$, $(0, 0)$, $(4, 0)$, $(0, 0)$; **c.** D: $x \in \mathbb{R}$, R: $y \in [-6, \infty)$

23. a. absolute value;   **b.** down/down, (0, 6), $x = 0$, (−2, 0), (2, 0), (0, 6);
c. $D: x \in \mathbb{R}, R: y \in (-\infty, 6]$
24. a. absolute value;   **b.** up/up, (−1, 0), $x = -1$, (−1, 0), (0, 2);
c. $D: x \in \mathbb{R}, R: y \in [0, \infty)$
25. a. cubic;   **b.** up/down, (1, 0), (1, 0), (0, 1);   **c.** $D: x \in \mathbb{R}, R: y \in \mathbb{R}$
26. a. cubic;   **b.** down/up, (−1, 0), (−1, 0), (0, 1);   **c.** $D: x \in \mathbb{R}, R: y \in \mathbb{R}$
27. a. cubic;   **b.** down/up, (0, 1), (−1, 0), (0, 1);   **c.** $D: x \in \mathbb{R}, R: y \in \mathbb{R}$
28. a. cube root;   **b.** up/down, (0, 1), (1, 0), (0, 1);   **c.** $D: x \in \mathbb{R}, R: y \in \mathbb{R}$
29. a. cube root;   **b.** down/up, (1, −1), (2, 0), (0, −2);
c. $D: x \in \mathbb{R}, R: y \in \mathbb{R}$
30. a. cube root;   **b.** up/down, (−1, −1), (−2, 0), (0, −2);
c. $D: x \in \mathbb{R}, R: y \in \mathbb{R}$
31. square root function; y-int (0, 2); x-int (−3, 0); initial point (−4, −2);
up on right; $D: x \in [-4, \infty), R: y \in [-2, \infty)$
32. quadratic function; x-int (−3, 0), (1, 0); y-int (0, 3); vertex (−1, 4);
down, down; $D: x \in \mathbb{R}, R: y \in (-\infty, 4]$
33. cubic function; y-int (0, −2); x-int (−2, 0); inflection point (−1, −1);
up, down; $D: x \in \mathbb{R}, R: y \in \mathbb{R}$
34. absolute value; x-int (−1, 0), (3, 0); y-int (0, −2); vertex (1, −4);
up, up; $D: x \in \mathbb{R}, R: y \in [-4, \infty)$

35.

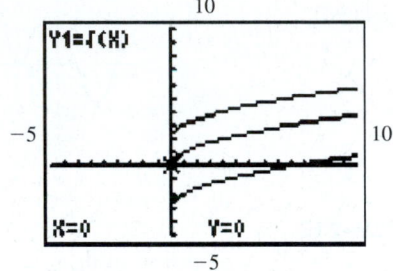

the graph of g is f shifted up 2 units; the graph
of h is h shifted down 3 units

36.

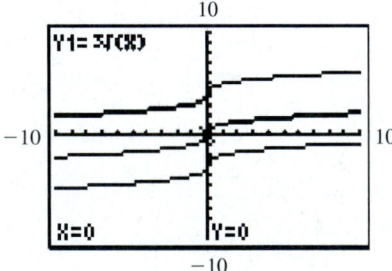

the graph of g is f shifted down 3 units;
the graph of h is h shifted up 4 units

37.

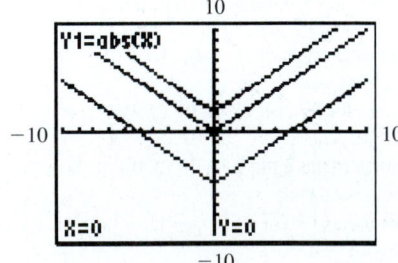

the graph of q is p shifted down 5 units;
the graph of r is p shifted up 2 units

38.

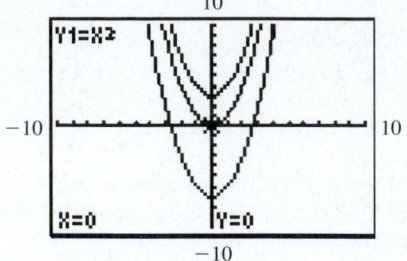

the graph of q is p shifted down 7 units;
the graph of r is p shifted up 3 units

39. 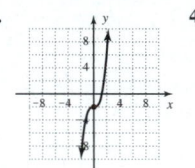   **40.** 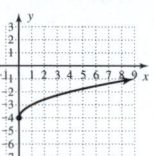   **41.**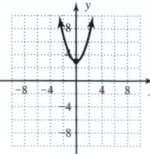

42.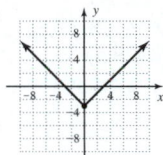

43.

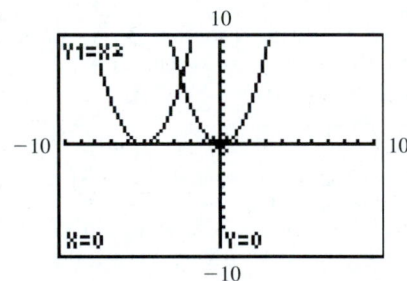

the graph of q is p shifted left 5 units

44.

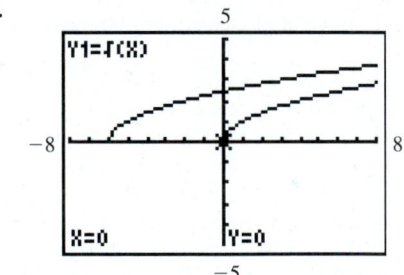

the graph of g is f shifted left 4 units

45.

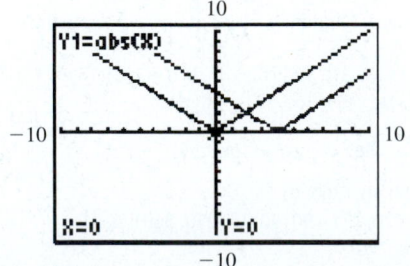

the graph of Y_2 is Y_1 shifted right 4 units

46.

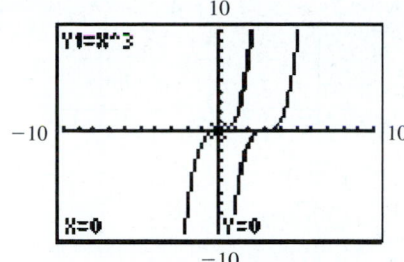

the graph of *H* is *h* shifted right 4 units

47. 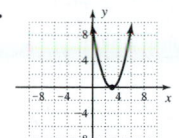   **48.** 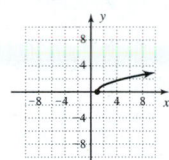   **49.**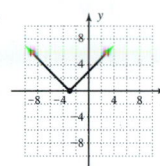

50. 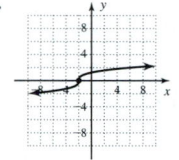   **51.** 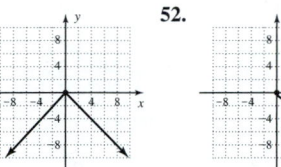   **52.**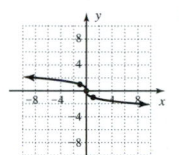

53. 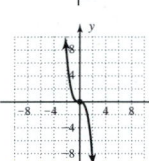   **54.**

55.

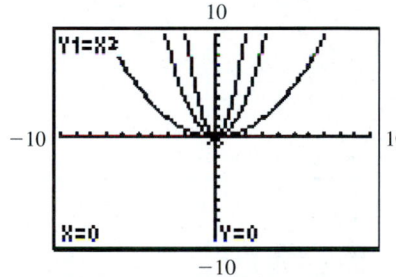

the graph of *q* is *p* vertically stretched;
the graph of *r* is *p* vertically compressed

56.

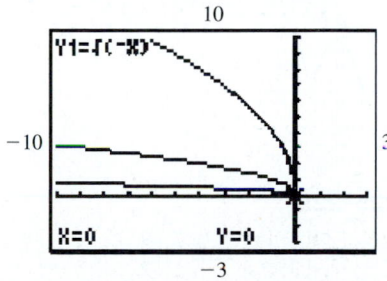

the graph of *g* is *f* vertically stretched;
the graph of *h* is *f* vertically compressed

57.

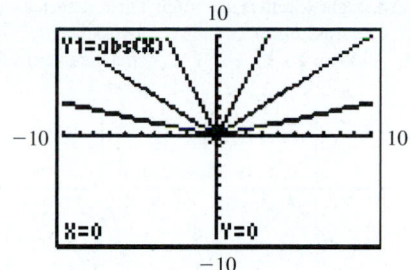

the graph of Y$_2$ is Y$_1$ vertically stretched;
the graph of Y$_3$ is Y$_1$ vertically compressed

58.

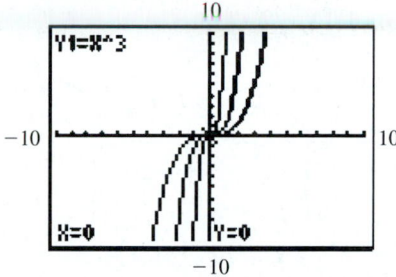

the graph of *v* is *u* vertically stretched;
the graph of *w* is *u* vertically compressed

59. 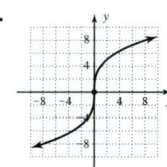   **60.** 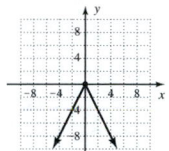   **61.**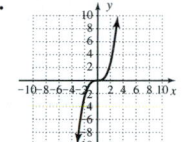

62.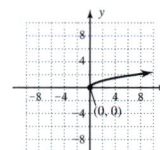

75. left 2, down 1   **76.** right 3, up 2

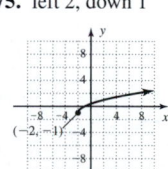   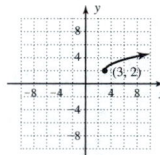

77. left 3, reflected across *x*-axis, down 2   **78.** right 2, reflected across *x*-axis, up 5   **79.** left 3, down 1

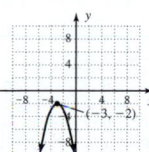   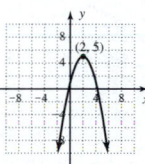   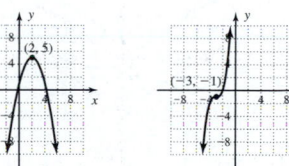

80. right 2, up 1   **81.** left 1, down 2   **82.** right 3, up 1

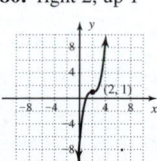   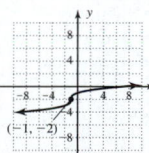   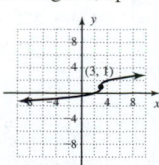

83. left 3, reflected across *x*-axis, down 2

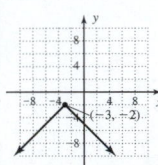

84. right 4, reflected across *x*-axis, down 2

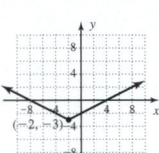

85. left 1, reflected across *x*-axis, stretched vertically, down 3

86. left 2, compressed vertically, down 3

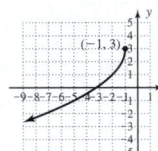

87. left 2, reflected across *x*-axis, compressed vertically, down 1

88. left 1, stretched vertically, up 2

89. left 1, reflected across *x* = −1, reflected across *x*-axis, stretched vertically, up 3

90. right 2, reflected across *x* = 2, stretched vertically, down 1

91. right 3, compressed vertically, up 1

92. right 3, reflected across *x*-axis, stretched vertically, up 4

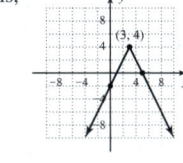

93. a.

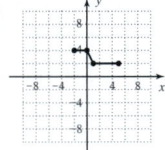

b.

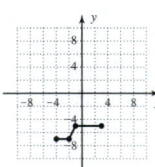

c.

d.

94. a.

b.

c.

d.

95. a.

b.

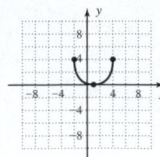

c.

d.

96. a.

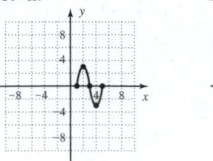

b.

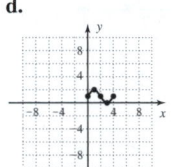

c.

d.

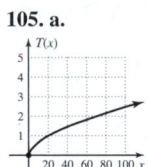

103. a.

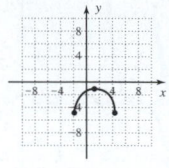

104. a.

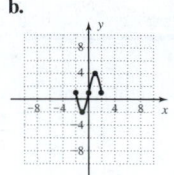

105. a.

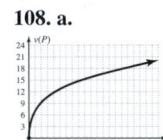

106. a.

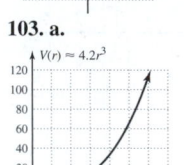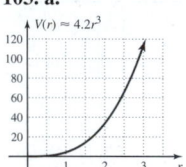

107. a.

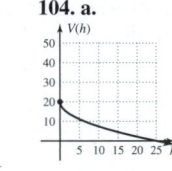

108. a.

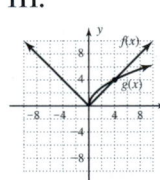

109. a.

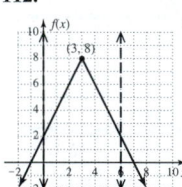

110. a.

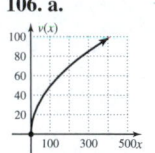

111.

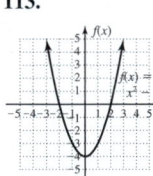

112.

113.

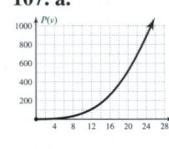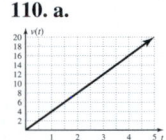

Exercises 2.3, pp. 225–228

27. $(-5, -3)$ **28.** $[-1, 5]$ **31.** $\left[\dfrac{8}{3}, \dfrac{14}{3}\right]$ **32.** $\left(-\dfrac{9}{2}, \dfrac{3}{2}\right)$

33. { } **34.** { } **35.** $\left(-1, \dfrac{3}{5}\right)$ **36.** $\left(-\dfrac{2}{3}, 2\right)$ **37.** $\left[-\dfrac{7}{4}, 0\right]$

38. $\left[\dfrac{3}{8}, \dfrac{33}{8}\right]$ **43.** $\left(-\infty, -\dfrac{7}{3}\right] \cup \left[\dfrac{7}{3}, \infty\right)$

44. $\left(-\infty, -\dfrac{15}{4}\right] \cup \left[\dfrac{15}{4}, \infty\right)$ **45.** $\left(-\infty, \dfrac{3}{7}\right] \cup [1, \infty)$

46. $\left(-\infty, -\dfrac{9}{2}\right] \cup \left[-\dfrac{5}{2}, \infty\right)$ **66. a.** $|v - 726| \le 235$; **b.** $[491, 961]$

Mid Chapter Check, pp. 228–229

5. a. cubic **b.** up on the left, down on the right; inflection point: (2, 2);
x-int: (4, 0); *y*-int: (0, 5) **c.** *D*: $x \in (-\infty, \infty)$; *R*: $y \in (-\infty, \infty)$
d. $k = 1$

6.

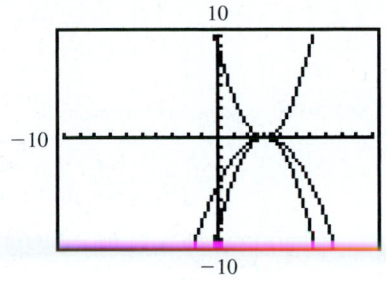

$q(x)$ is a reflection of
$p(x)$ across the *x*-axis,
and $r(x)$ is the same as
$q(x)$, but compressed by
a factor of $\frac{1}{2}$

Exercises 2.4, pp. 241–245

7. a. as $x \to -\infty, y \to 2$
as $x \to \infty, y \to 2$
b. as $x \to 1^{-}, y \to -\infty$
as $x \to 1^{+}, y \to \infty$

8. a. as $x \to -\infty, y \to -2$
as $x \to \infty, y \to -2$
b. as $x \to 1^{-}, y \to -\infty$
as $x \to 1^{+}, y \to \infty$

9. a. as $x \to -\infty, y \to 1$
as $x \to \infty, y \to 1$
b. $y = 1$
c. as $x \to -2^{-}, y \to \infty$
as $x \to -2^{+}, y \to \infty$

10. a. as $x \to -\infty, y \to 2$
as $x \to \infty, y \to 2$
b. $y = 2$
c. as $x \to -2^{-}, y \to -\infty$
as $x \to -2^{+}, y \to -\infty$

11. down 1, $x \in (-\infty, 0) \cup (0, \infty), y \in (-\infty, -1) \cup (-1, \infty)$

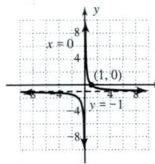

12. up 2, $x \in (-\infty, 0) \cup (0, \infty), y \in (-\infty, 2) \cup (2, \infty)$

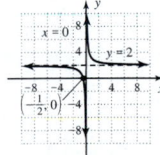

13. left 2, $x \in (-\infty, -2) \cup (-2, \infty), y \in (-\infty, 0) \cup (0, \infty)$

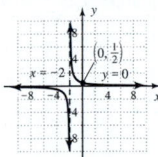

14. right 3, $x \in (-\infty, 3) \cup (3, \infty), y \in (-\infty, 0) \cup (0, \infty)$

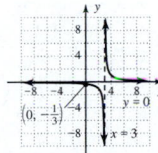

15. right 2, reflected across *x*-axis,
$x \in (-\infty, 2) \cup (2, \infty), y \in (-\infty, 0) \cup (0, \infty)$

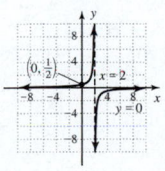

16. reflected across *x*-axis, down 2,
$x \in (-\infty, 0) \cup (0, \infty), y \in (-\infty, -2) \cup (-2, \infty)$

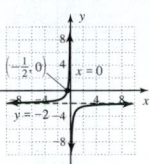

17. left 2, down 1,
$x \in (-\infty, -2) \cup (-2, \infty), y \in (-\infty, -1) \cup (-1, \infty)$

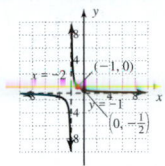
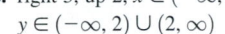

18. right 3, up 2, $x \in (-\infty, 3) \cup (3, \infty)$,
$y \in (-\infty, 2) \cup (2, \infty)$

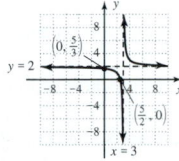

19. right 1, $x \in (-\infty, 1) \cup (1, \infty), y \in (0, \infty)$

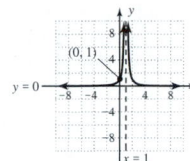

20. left 5, $x \in (-\infty, -5) \cup (-5, \infty), y \in (0, \infty)$

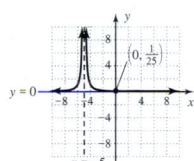

21. left 2, reflected across *x*-axis,
$x \in (-\infty, -2) \cup (-2, \infty), y \in (-\infty, 0)$

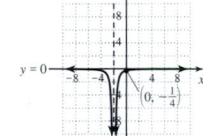

22. reflected across *x*-axis, down 2,
$x \in (-\infty, 0) \cup (0, \infty)$,
$y \in (-\infty, -2)$

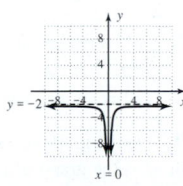

23. down 2, $x \in (-\infty, 0) \cup (0, \infty), y \in (-2, \infty)$

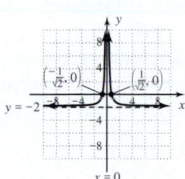

24. up 3, $x \in (-\infty, 0) \cup (0, \infty), y \in (3, \infty)$

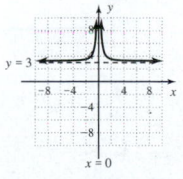

25. left 2, up 1, $x \in (-\infty, -2) \cup (-2, \infty)$, $y \in (1, \infty)$

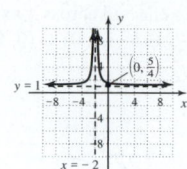

26. right 1, down 2, $x \in (-\infty, 1) \cup (1, \infty)$, $y \in (-2, \infty)$

$$\left(\frac{2 - \sqrt{2}}{2}, 0\right), \left(\frac{2 + \sqrt{2}}{2}, 0\right)$$

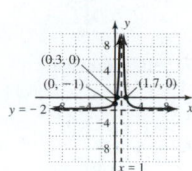

27. reciprocal quadratic, $S(x) = \dfrac{1}{(x + 1)^2} - 2$

28. reciprocal quadratic, $s(x) = \dfrac{-1}{(x + 2)^2} + 2$

29. reciprocal function, $Q(x) = \dfrac{1}{x + 1} - 2$

30. reciprocal function, $q(x) = \dfrac{1}{x - 1} - 2$

31. reciprocal quadratic, $v(x) = \dfrac{1}{(x + 2)^2} - 5$

32. reciprocal function, $w(x) = \dfrac{1}{x + 2} - 3$

39. $g(x)$ increases faster $(3 > 2)$ **a.** $x = 0$ and 1,
  b. $(-\infty, 0) \cup (0, 1)$,   **c.** $(1, \infty)$

40. $g(x)$ increases faster $(5 > 4)$,   **a.** $x = 0$ and 1,
  b. $(-\infty, 0) \cup (0, 1)$,   **c.** $(1, \infty)$

41. $f(x)$ increases faster $(4 > 2)$ **a.** $x = -1, 0,$ and 1,
  b. $(-1, 0) \cup (0, 1)$,   **c.** $(-\infty, -1) \cup (1, \infty)$

42. $g(x)$ increases faster $(5 > 3)$,   **a.** $x = -1, 0,$ and 1,
  b. $(-\infty, -1) \cup (0, 1)$,   **c.** $(-1, 0) \cup (1, \infty)$

43. $g(x)$ increases faster $\left(\dfrac{4}{5} > \dfrac{2}{3}\right)$ **a.** $x = -1, 0,$ and 1,
  b. $(-1, 0) \cup (0, 1)$,   **c.** $(-\infty, -1) \cup (1, \infty)$

44. $f(x)$ increases faster $\left(\dfrac{7}{4} > \dfrac{3}{2}\right)$ **a.** $x = 0$ and 1,   **b.** $(0, 1)$,   **c.** $(1, \infty)$

45. $g(x)$ increases faster $\left(\dfrac{1}{3} > \dfrac{1}{6}\right)$ **a.** $x = 0$ and 1,   **b.** $(0, 1)$,   **c.** $(1, \infty)$

46. $g(x)$ increases faster $\left(\dfrac{1}{4} > \dfrac{1}{5}\right)$ **a.** $x = 0$ and 1,   **b.** $(0, 1)$,   **c.** $(1, \infty)$

47. $g(x)$ increases faster $\left(\dfrac{5}{4} > \dfrac{2}{3}\right)$ **a.** $x = 0$ and 1,   **b.** $(0, 1)$,   **c.** $(1, \infty)$

48. $f(x)$ increases faster $\left(\dfrac{3}{2} > \dfrac{3}{4}\right)$ **a.** $x = 0$ and 1,   **b.** $(0, 1)$,   **c.** $(1, \infty)$

59. F is the graph of f shifted left 1 unit and down 2; verified
60. G is the graph of g shifted right 3 units and up 2; verified
61. P is the graph of p shifted right 2 units and reflected across the x-axis; verified
62. Q is the graph of q stretched vertically by a factor of 2 and shifted down 5 units; verified

65. c.

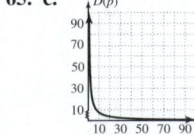

66. c.

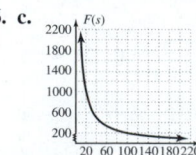

67. c.

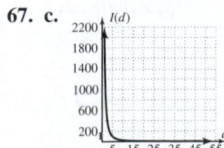

68. c.

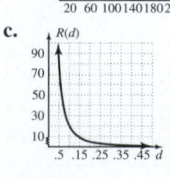

69. a. \$20,000, \$80,000, \$320,000; cost increases dramatically
  b.

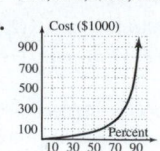

  c. as $p \to 100, C \to \infty$

70. a. \$733,333, \$2,200,000, \$6,600,000; cost increases dramatically
  b.

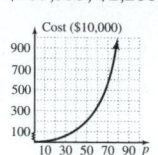

  c. as $p \to 100^-, C \to \infty$

75. a.

76. a.

77. a.

78. a.

81. $y = \frac{-2}{3}x + 5$,

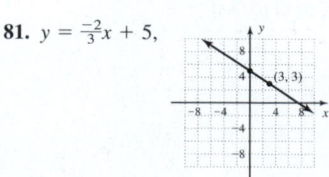

Exercises 2.5, pp. 254–258

13. $D: x \in [-6, \infty); R: y \in [-4, \infty)$

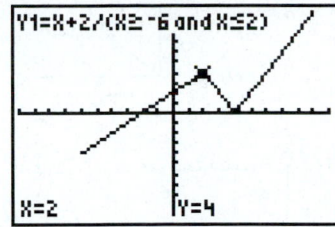

14. $D: x \in [-4, 7]; R: y \in [0, 5]$

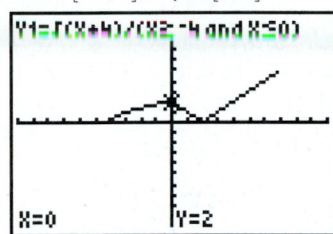

15. $D: x \in [-2, \infty); R: y \in [-4, \infty)$

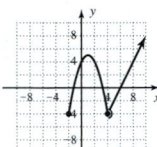

16. $D: x \in (-\infty, 5]; R: y \in (-\infty, 6]$

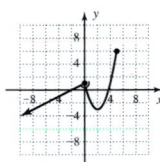

17. $D: x \in (-\infty, 9); R: y \in [2, \infty)$

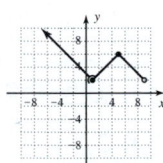

18. $D: x \in (-\infty, 6]; R: y \in (-\infty, 9]$

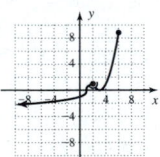

19. $D: x \in (-\infty, \infty); R: y \in [0, \infty)$

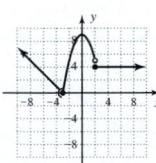

20. $D: x \in (-\infty, \infty); R: y \in [0, \infty)$

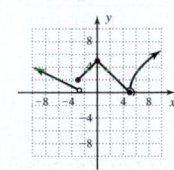

21. $D: x \in (-\infty, \infty); R: y \in (-\infty, 3) \cup (3, \infty)$

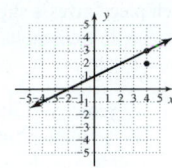

22. $D: x \in (-\infty, \infty); R: y \in (-\infty, 3) \cup (3, \infty)$

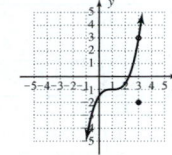

23. discontinuity at $x = -3$, redefine $f(x) = -6$ at $x = -3; c = -6$

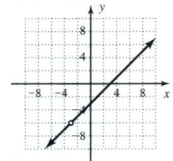

24. discontinuity at $x = 5$, redefine $f(x) = 7$ at $x = 5; c = 7$

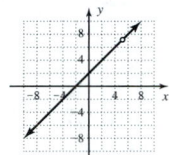

25. discontinuity at $x = 1$, redefine $f(x) = 3$ at $x = 1; c = 3$

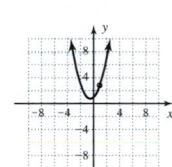

26. discontinuity at $x = -2$, redefine $f(x) = -8$ at $x = -2; c = -8$

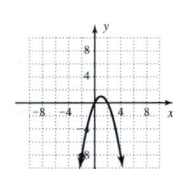

31. Graph is discontinuous at $x = 0; f(x) = 1$ for $x > 0; f(x) = -1$ for $x < 0$.

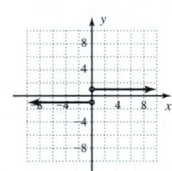

33. a. $S(t) = \begin{cases} -t^2 + 6t & 0 \le t \le 5 \\ 5 & t > 5 \end{cases}$ **b.** $S(t) \in [0, 9]$

34. a. $f(t) = \begin{cases} -0.13t^2 + 8.1t + 208 & 4 \le t \le 38 \\ -5.75|t - 46| + 374 & 38 < t < 54 \\ -2.45t + 460 & t \ge 54 \end{cases}$ **b.** $f(t) \in [0, 374]$

35. a.

Year (0 → 1950)	Percent
5	7.33
15	14.13
25	14.93
35	22.65
45	41.55
55	60.45
65	79.35

b. Each piece gives a slightly different value due to rounding of coefficients in each model. At $t = 30$, we use the "first" piece: $P(30) = 13.08$.

36. a.

Year $(0 \rightarrow 1980)$	Barrels (billions)
3	1.183
9	2.115
15	2.68
25	4.01
35	5.34

b. About 1.1 billion barrels

37. $C(h) = \begin{cases} 0.09h & 0 \le h \le 1000 \\ 0.18h - 90 & h > 1000 \end{cases}$

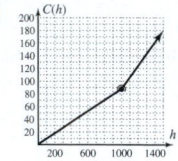

$C(1200) = \$126$

38. $C(w) = \begin{cases} 0.05w & 0 \le w \le 5000 \\ 0.10w - 250 & w > 5000 \end{cases}$

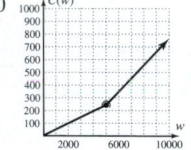

$C(9500) = \$700$

39. $C(t) = \begin{cases} 0.75t & 0 \le t \le 25 \\ 1.5t - 18.75 & t > 25 \end{cases}$

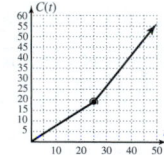

$C(45) = \$48.75$

40. $T(x) = \begin{cases} -0.21x^2 + 6.1x + 52 & 5 \le x \le 15 \\ 4.53x + 28.3 & x > 15 \end{cases}$

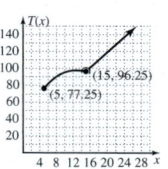

92,000, 119,000, 142,000; 164,000

41. $S(t) = \begin{cases} -1.35t^2 + 31.9t + 152 & 0 \le t \le 12 \\ 2.5t^2 - 80.6t + 950 & 12 < t \le 22 \end{cases}$

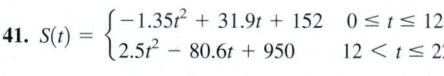

$498 billion, $653 billion, $931 billion

42. $T(x) = \begin{cases} \frac{x}{10} & 0 \le x \le 200 \\ 0.001x^2 - 0.3x + 40 & x > 200 \end{cases}$

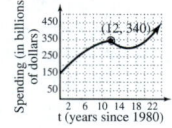

75 tickets

43. $c(m) = \begin{cases} 3.3m & 0 \le m \le 30 \\ 7m - 111 & m > 30 \end{cases}$

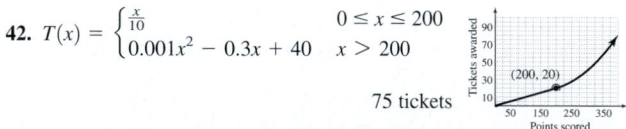

$2.11

44. $w(h) = \begin{cases} 9.50h & 0 \le h \le 40 \\ 14.25h - 190 & 40 < h \le 48 \\ 19h - 418 & 48 < h \le 84 \end{cases}$

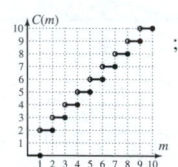

$608

45. $C(a) = \begin{cases} 0 & a < 2 \\ 2 & 2 \le a < 13 \\ 5 & 13 \le a < 20 \\ 7 & 20 \le a < 65 \\ 5 & a \ge 65 \end{cases}$

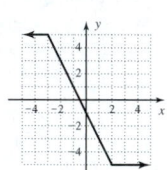

$38

46. a. $A(t) = \lfloor t \rfloor$, **b.** $0 \le t < 123$, **c.** 36 yr, **d.** 36 yr, **e.** 37 yr, **f.** 37 yr **47. a.** $C(w - 1) = 17\lceil w - 1 \rceil + 88$ **b.** $0 < w \le 13$ **c.** 88¢ **d.** 173¢ **e.** 173¢ **f.** 173¢ **g.** 190¢

48. a. $C(m) = \begin{cases} 0 & 0 < m \le 1 \\ \lceil m \rceil & m > 1 \end{cases}$; **b.**

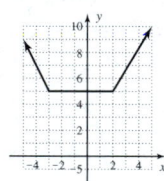

c. $3 + 14 + 2 + 4 + 9 = 32$, yes; **d.** 28 min, 55 sec

49. yes; $h(x) = \begin{cases} 5 & x \le -3 \\ -2x - 1 & -3 < x < 2 \\ -5 & x \ge 2 \end{cases}$

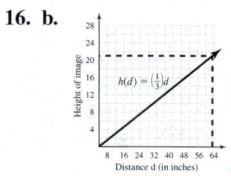

50. yes; $H(x) = \begin{cases} -2x - 1 & x < -3 \\ 5 & -3 \le x \le 2 \\ 2x + 1 & x > 2 \end{cases}$

Exercises 2.6, pp. 265–269

15. b.

16. b.

23. a. Area varies directly as a side squared. **b.** $A = ks^2$ **c.**

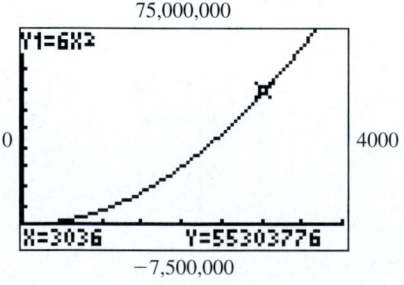

d.

X	Y₁	
0	0	
5	150	
10	600	
15	1350	
20	2400	
25	3750	
30	5400	

X=0

e. $k = 6$; $A = 6s^2$; 55,303,776 m²
24. a. Area varies directly as a side squared.   **b.** $A = ks^2$
c.

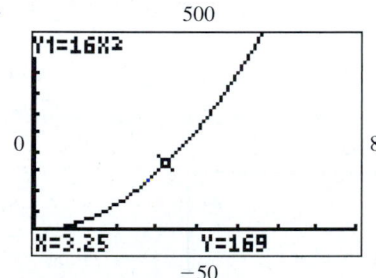

80,000

Y1=0.433X²

0   500

X=400   Y=69280

−8000

d.

X	Y₁	
0	0	
10	43.3	
20	173.2	
30	389.7	
40	692.8	
50	1082.5	
60	1558.8	

X=0

e. $k = 0.433$; $A = 0.433s^2$; 69,280 mi²
25. a. Distance varies directly as time squared.   **b.** $D = kt^2$
c.

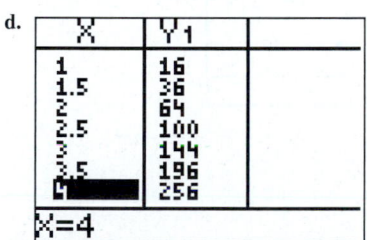

500

Y1=16X²

0   8

X=3.25   Y=169

−50

d.

X	Y₁	
1	16	
1.5	36	
2	64	
2.5	100	
3	144	
3.5	196	
4	256	

X=4

e. $k = 16$; $d = 16t^2$; about 3.5 sec; 121 ft
26. a. Area varies directly as radius squared.   **b.** $A = kr^2$
c.

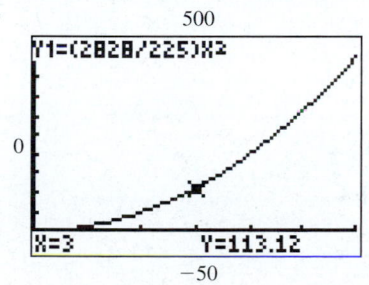

500

Y1=(2828/225)X²

0   6

X=3   Y=113.12

−50

d.

X	Y₁	
0	0	
.5	3.1422	
1	12.569	
1.5	28.28	
2	50.276	
2.5	78.556	
3	113.12	

X=0

e. $k = \dfrac{2828}{225}$; $A = \dfrac{2828}{225}r^2$; about 3 in.; about 28 in²

66.

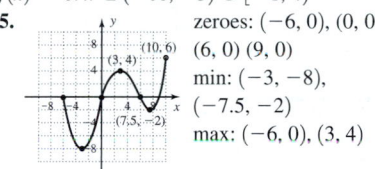

Summary and Concept Review, pp. 271–275

1. $D: x \in (-\infty, \infty)$, $R: y \in [-5, \infty)$, $f(x)\uparrow: x \in (2, \infty)$,
$f(x)\downarrow: x \in (-\infty, 2)$, $f(x) > 0: x \in (-\infty, -1) \cup (5, \infty)$,
$f(x) < 0: x \in (-1, 5)$
2. $D: x \in [-3, \infty)$, $R: y \in (-\infty, 0]$, $f(x)\uparrow:$ none, $f(x)\downarrow: x \in (-3, \infty)$,
$f(x) > 0:$ none, $f(x) < 0: x \in (-3, \infty)$
3. $D: x \in (-\infty, \infty)$, $R: y \in (-\infty, \infty)$, $f(x)\uparrow: x \in (-\infty, -3) \cup (1, \infty)$,
$f(x)\downarrow: x \in (-3, 1)$, $f(x) > 0: x \in (-5, -1) \cup (4, \infty)$,
$f(x) < 0: x \in (-\infty, -5) \cup [-1, 4)$
5. zeroes: $(-6, 0)$, $(0, 0)$, $(6, 0)$ $(9, 0)$
min: $(-3, -8)$, $(-7.5, -2)$
max: $(-6, 0)$, $(3, 4)$

7. squaring function   **a.** up on left/up on the right;   **b.** x-intercept: $(-4, 0)$, $(0, 0)$; y-intercept: $(0, 0)$ **c.** vertex $(-2, -4)$
d. $x \in (-\infty, \infty)$, $y \in [-4, \infty)$
8. square root function   **a.** down on the right;   **b.** x-intercept: $(0, 0)$; y-intercept: $(0, 0)$ **c.** initial point $(-1, 2)$;   **d.** $x \in [-1, \infty)$, $y \in (-\infty, 2]$
9. cubing function   **a.** down on left/up on the right   **b.** x-intercepts: $(2, 0)$; y-intercept: $(0, -2)$ **c.** inflection point: $(1, -1)$
d. $x \in (-\infty, \infty)$, $y \in (-\infty, \infty)$
10. absolute value function   **a.** down on left/down on the right
b. x-intercepts: $(-1, 0)$, $(3, 0)$; y-intercept: $(0, 1)$ **c.** vertex: $(1, 2)$;
d. $x \in (-\infty, \infty)$, $y \in (-\infty, 2]$
11. cube root function   **a.** up on left, down on right   **b.** x-intercept: $(1, 0)$; y-intercept: $(0, 1)$ **c.** inflection point: $(1, 0)$ **d.** $x \in (-\infty, \infty)$, $y \in (-\infty, \infty)$

12. quadratic

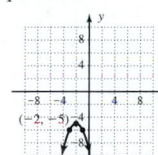

13. absolute value

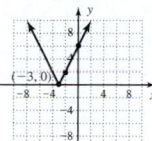

14. cubic

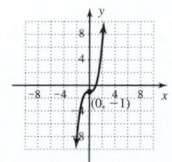

15. square root

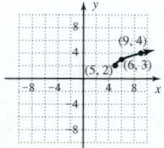

16. cube root

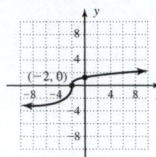

17. a.

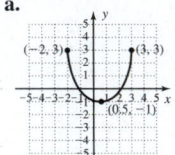

b.

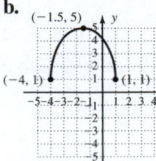

c.

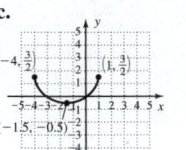

30.

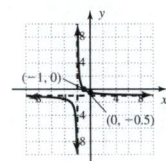

31.

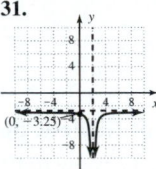

32.   a. ≈$32,143; $75,000; $175,000; $675,000; cost increases dramatically
c. as $p \to 100$, $C \to \infty$

b.

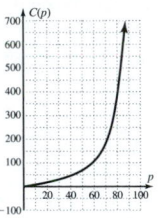

33.

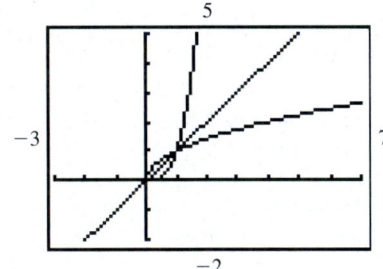

Domain of $f(x)$ is $(-\infty, \infty)$; Domain of $g(x)$ and $h(x)$ is $[0, \infty)$.

34. a. 88.4 hr,   **b.** 570 km

35. a. $f(x) = \begin{cases} 5 & x \le -4 \\ -x + 1 & -4 < x \le 3 \\ 3\sqrt{x - 3} - 1 & x > 3 \end{cases}$ **b.** $R: y \in [-2, \infty)$

36.

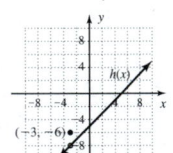

$D: x \in (-\infty, \infty)$,
$R: y \in (-\infty, -8) \cup (-8, \infty)$,
discontinuity at $x = -3$;
define $h(x) = -8$ at $x = -3$

38. $D: x \in (-\infty, \infty)$ $R: y \in [-4, \infty)$

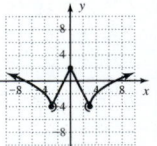

39. $\begin{cases} 20x & x \le 2 \\ 30x - 20 & 2 < x \le 4 \\ 40x - 60 & x > 4 \end{cases}$
For 5 hr the total cost is $140.

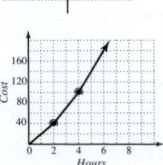

Practice Test, pp. 275–277

1. a. $D: x \in [-4, \infty)$; $R: y \in [-3, \infty)$ **b.** $f(-1) \approx 2.2$
c. $f(x) < 0: x \in (-4, -3)$; $f(x) > 0: x \in (-3, \infty)$
d. $f(x)\uparrow: x \in (-4, \infty)$; $f(x)\downarrow$: none   **e.** $f(x) = 3\sqrt{x + 4} - 3$

2.

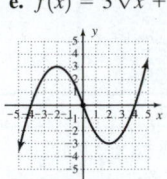

3.

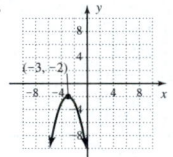

4. max: $y = 8$ at $x = -2$; min: $y = -7$ at $x \approx -5.87$ and $y = -7$ at $x \approx 1.87$

5. I. a. square root   **b.** $x \in [-4, \infty)$, $y \in [-3, \infty)$ **c.** $(-2, 0), (0, 1)$
d. up on right   **e.** $x \in (-2, \infty)$ **f.** $x \in [-4, -2)$
II. a. cubic   **b.** $x \in (-\infty, \infty)$, $y \in (-\infty, \infty)$ **c.** $(2, 0), (0, -1)$
d. down on left, up on right   **e.** $x \in (2, \infty)$ **f.** $x \in (-\infty, 2)$
III. a. absolute value   **b.** $x \in (-\infty, \infty)$, $y \in (-\infty, 4]$
c. $(-1, 0), (3, 0), (0, 2)$ **d.** down on left, down on right   **e.** $x \in (-1, 3)$
f. $x \in (-\infty, -1) \cup (3, \infty)$
IV. a. quadratic   **b.** $x \in (-\infty, \infty)$; $y \in [-5.5, \infty)$
c. $(0, 0), (5, 0), (0, 0)$ **d.** up on left, up on right
e. $x \in (-\infty, 0) \cup (5, \infty)$ **f.** $x \in (0, 5)$

6.

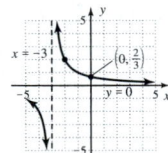

7.

8. $\left(-\infty, \frac{-20}{3}\right) \cup \left(\frac{22}{3}, \infty\right)$ **9.** $[-4, 0)$ **10.** $x = 0.75$
11.

15. a.

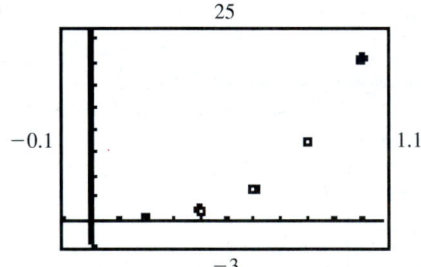

b. $S(t) = 17.27\, t^{2.50}$ **c.** 3.05 mm,   **d.** 0.95 sec
18. b.

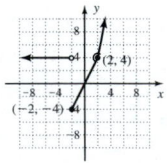

Cumulative Review Chapters R–2, pp. 279–280

4. $r = \dfrac{-\pi h \pm \sqrt{\pi^2 h^2 + 2\pi A}}{2\pi}$

8. a. 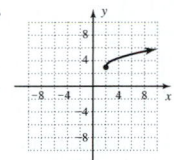 **b.**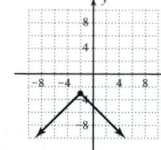

9. 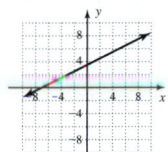 $y = \frac{1}{2}x + \frac{7}{2}$

11. 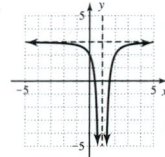 **12.**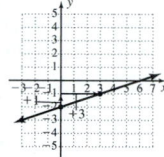

13. a. D: $x \in (-\infty, 8]$, R: $y \in [-4, \infty)$ **b.** $5, -3, -3, 1, 2$
c. $(-2, 0)$ **d.** $f(x) < 0$: $x \in (-2, 2)$; $f(x) > 0$: $x \in (-\infty, -2) \cup [2, 8]$
e. min: $(0, -4)$, max: $(8, 7)$ **f.** $f(x)\!\uparrow$: $x \in (0, 8)$; $f(x)\!\downarrow$: $x \in (-\infty, 0)$

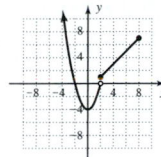

18. No, Raphael corresponds to both the School of Athens and Parnassus.
24. $m_1 = \frac{1}{2}$, $m_2 = -2 \Rightarrow y = -2x + 4$
25. $P = 15 + \sqrt{97}$ units ≈ 24.8 units. No, it is not a right triangle.
$5^2 + (\sqrt{97})^2 \neq 10^2$

27.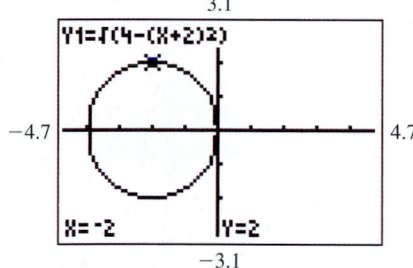

29.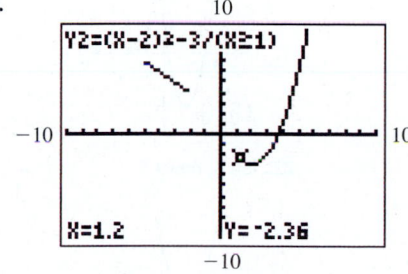

30. a. $x \in [0, 10]$, $y \in [0, 10]$, linear; **b.** $Y_1 = 0.21X + 4.36$;
$Y_2 = 0.56X + 5.33$, **c.** 2001 to 2007, $\dfrac{0.56}{0.21} = 2\frac{2}{3}$ times as fast.

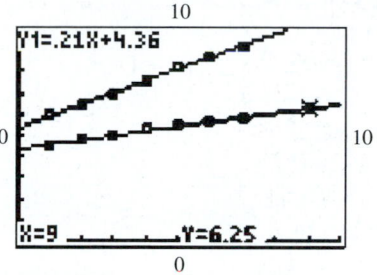

CHAPTER 3

Exercises 3.1, pp. 289–292

13. a. $1 + i$ **b.** $2 + i\sqrt{3}$ **14. a.** $8 - i\sqrt{2}$ **b.** $2 + 3i\sqrt{5}$

15. a. $4 + 2i$ **b.** $2 - i\sqrt{2}$ **16. a.** $\dfrac{3}{2} - \dfrac{3\sqrt{2}}{2}i$ **b.** $\dfrac{3}{2} + \dfrac{5\sqrt{2}}{4}i$

21. a. $4 + 5i\sqrt{2}$; $a = 4$, $b = 5\sqrt{2}$ **b.** $-5 + 3i\sqrt{3}$; $a = -5$, $b = 3\sqrt{3}$
22. a. $-2 + 4i\sqrt{3}$; $a = -2$, $b = 4\sqrt{3}$
b. $7 + 5i\sqrt{3}$; $a = 7$, $b = 5\sqrt{3}$

23. a. $\dfrac{7}{4} + \dfrac{7\sqrt{2}}{8}i$; $a = \dfrac{7}{4}$, $b = \dfrac{7\sqrt{2}}{8}$ **b.** $\dfrac{1}{2} + \dfrac{\sqrt{10}}{2}i$; $a = \dfrac{1}{2}$, $b = \dfrac{\sqrt{10}}{2}$

24. a. $\dfrac{7}{4} + \dfrac{\sqrt{7}}{4}i$; $a = \dfrac{7}{4}$, $b = \dfrac{\sqrt{7}}{4}$ **b.** $\dfrac{4}{3} + \dfrac{\sqrt{3}}{2}i$; $a = \dfrac{4}{3}$, $b = \dfrac{\sqrt{3}}{2}$

Exercises 3.2, pp. 308–313

21. $m = \pm 4$ **22.** $p = \pm 7$
23. $y = \pm 2\sqrt{7}$; $y \approx \pm 5.29$ **24.** $m = \pm 2\sqrt{5}$; $m \approx \pm 4.47$
25. no real solutions **26.** no real solutions **27.** $x = \pm\dfrac{\sqrt{21}}{4}$; $x \approx \pm 1.15$
28. $y = \pm\dfrac{\sqrt{13}}{3}$; $y \approx \pm 1.20$ **29.** $n = 9$; $n = -3$ **30.** $p = 2$; $p = -12$
31. $w = -5 \pm \sqrt{3}$; $w \approx -3.27$ or $w \approx -6.73$
32. $m = 4 \pm \sqrt{5}$; $m \approx 6.24$ or $m \approx 1.76$ **33.** no real solutions
34. no real solutions **35.** $m = 2 \pm \dfrac{3\sqrt{2}}{7}$; $m \approx 2.61$ or $m \approx 1.39$
36. $x = 5 \pm \dfrac{2\sqrt{3}}{5}$; $x \approx 5.69$ or $x \approx 4.31$
45. $p = 3 \pm \sqrt{6}$; $p \approx 5.45$ or $p \approx 0.55$
46. $n = 2 \pm \sqrt{14}$; $n \approx 5.74$ or $n \approx -1.74$
47. $p = -3 \pm \sqrt{5}$; $p \approx -0.76$ or $p \approx -5.24$
48. $x = 4 \pm \sqrt{17}$; $x \approx 8.12$ or $x \approx -0.12$
49. $m = \dfrac{-3}{2} \pm \dfrac{\sqrt{13}}{2}$; $m \approx 0.30$ or $m \approx -3.30$
50. $n = \dfrac{-5}{2} \pm \dfrac{\sqrt{33}}{2}$; $n \approx 0.37$ or $n \approx -5.37$
51. $n = \dfrac{5}{2} \pm \dfrac{3\sqrt{5}}{2}$; $n \approx 5.85$ or $n \approx -0.85$
52. $w = \dfrac{7}{2} \pm \dfrac{\sqrt{37}}{2}$; $w \approx 6.54$ or $w \approx 0.46$
53. a. $x = \frac{1}{2}$ or $x = -4$ **b.** $x = 0.5$ or $x = -4$
54. a. $w = 2$ or $w = \frac{2}{3}$ **b.** $w = 2$ or $w \approx 0.67$
55. a. $n = 3$ or $n = \dfrac{-3}{2}$ **b.** $n = 3$ or $n = -1.5$
56. a. $p = \dfrac{5}{4} \pm \dfrac{\sqrt{33}}{4}$ **b.** $p \approx 2.69$ or $p \approx -0.19$
57. a. $p = \dfrac{3}{8} \pm \dfrac{\sqrt{41}}{8}$ **b.** $p \approx 1.18$ or $p \approx -0.43$
58. a. $x = \dfrac{-5}{6} \pm \dfrac{\sqrt{97}}{6}$ **b.** $x \approx 0.81$ or $x \approx -2.47$
59. a. $m = \dfrac{7}{2} \pm \dfrac{\sqrt{33}}{2}$ **b.** $m \approx 6.37$ or $m \approx 0.63$
60. a. $a = 2 \pm \sqrt{19}$ **b.** $a \approx 6.36$ or $a \approx -2.36$
61. $x = 6$ or $x = -3$
62. $w = -3 \pm \sqrt{10}$; $w \approx 0.16$ or $w \approx -6.16$ **63.** $m = \pm\dfrac{5}{2}$
64. $a = \dfrac{1}{2} \pm \dfrac{\sqrt{2}}{2}$; $a \approx 1.21$ or $a \approx -0.21$
65. $n = 1 \pm \dfrac{\sqrt{5}}{2}$; $n \approx 2.12$ or $n \approx -0.12$

66. $x = 1 \pm \frac{\sqrt{6}}{2}i; x \approx 1 \pm 1.22i$

67. $w = \frac{2}{3}$ or $w = \frac{-1}{2}$

68. $a = \frac{5}{6} \pm \frac{\sqrt{47}}{6}i; a \approx 0.8\overline{3} \pm 1.14i$

69. $m = \frac{3}{2} \pm \frac{\sqrt{6}}{2}i; m \approx 1.5 \pm 1.12i$

70. $p = 0$ or $p = \frac{-1}{3}$ **71.** $n = \pm\frac{3}{2}$ **72.** $x = \frac{-3}{4}$ or $x = 1$

73. $w = \frac{-4}{5}$ or $w = 2$ **74.** $m = \frac{-2}{3}$ or $m = 3$

75. $a = \frac{1}{6} \pm \frac{\sqrt{23}}{6}i; a \approx 0.1\overline{6} \pm 0.80i$

76. $n = \frac{1}{3} \pm \frac{\sqrt{10}}{3}; n \approx 1.39$ or $n \approx -0.72$

77. $p = \frac{3}{5} \pm \frac{2\sqrt{6}}{5}; p \approx 1.58$ or $p \approx -0.38$

78. $x = -\frac{1}{4} \pm \frac{\sqrt{23}}{4}i; x \approx -0.25 \pm 1.20i$

79. $w = \frac{1}{10} \pm \frac{\sqrt{21}}{10}; w \approx 0.56$ or $w \approx -0.36$

80. $m = -\frac{1}{3}$ or $m = 2$ **81.** $a = \frac{3}{4} \pm \frac{\sqrt{31}}{4}i; a \approx 0.75 \pm 1.39i$

82. $n = -2 \pm 2\sqrt{3}; n \approx 1.46$ or $n \approx -5.46$

83. $p = 1 \pm \frac{3\sqrt{2}}{2}i; p \approx 1 \pm 2.12i$

84. $x = \frac{5}{16} \pm \frac{\sqrt{57}}{16}; x \approx 0.78$ or $x \approx -0.16$

85. $w = \frac{-1}{3} \pm \frac{\sqrt{2}}{3}; w \approx 0.14$ or $w \approx -0.80$

86. $m = \frac{16}{15} \pm \frac{\sqrt{226}}{15}; m \approx 2.07$ or $m \approx 0.06$

87. $a = -3 \pm \frac{3\sqrt{2}}{2}; a \approx -0.88$ or $a \approx -5.12$

88. $n = \frac{3}{4} \pm \frac{\sqrt{321}}{12}; n \approx 2.24$ or $n \approx -0.74$

89. $p = \frac{2}{3} \pm \frac{\sqrt{394}}{6}; p \approx 3.97$ or $p \approx -2.64$

90. $x = \frac{24}{25} \pm \frac{3\sqrt{1006}}{50}; x \approx 2.86$ or $x \approx -0.94$

91. two rational; factorable   **92.** two rational; factorable

93. two nonreal   **94.** two nonreal   **95.** two rational; factorable

96. two rational; factorable   **97.** two nonreal   **98.** two nonreal

99. two irrational   **100.** two irrational   **101.** one repeated; factorable

102. one repeated; factorable   **103.** $x = \frac{3}{2} \pm \frac{1}{2}i$ **104.** $x = \frac{5}{2} \pm \frac{3}{2}i$

105. $x = -\frac{1}{2} \pm \frac{\sqrt{3}}{2}i$ **106.** $x = -1 \pm 3i\sqrt{2}$ **107.** $x = \frac{5}{4} \pm \frac{3\sqrt{7}}{4}i$

108. $x = \frac{2}{5} \pm \frac{\sqrt{11}}{5}i$ **111.** $x \in (-\infty, -5] \cup [1, \infty)$

112. $x \in (-\infty, -2] \cup [5, \infty)$ **113.** $x \in (-1, \frac{7}{2})$

114. $x \in (-\frac{5}{2}, 1)$ **115.** $x \in [-\sqrt{7}, \sqrt{7}]$

116. $x \in [-\sqrt{13}, \sqrt{13}]$ **117.** $x \in [-\frac{3}{2} - \frac{\sqrt{33}}{2}, -\frac{3}{2} + \frac{\sqrt{33}}{2}]$

118. $x \in [\frac{5}{2} - \frac{\sqrt{33}}{2}, \frac{5}{2} + \frac{\sqrt{33}}{2}]$ **119.** $x \in (-\infty, -\frac{5}{3}] \cup [1, \infty)$

120. $x \in (-\infty, -1] \cup [\frac{7}{4}, \infty)$ **125.** $x \in (-\infty, 5) \cup (5, \infty)$

126. $x \in (-\infty, 7) \cup (7, \infty)$ **129.** $x \in (-\infty, \infty)$

130. $x \in (-\infty, \infty)$ **131.** $x \in (-\infty, \infty)$ **132.** $x \in (-\infty, \infty)$

155. $t = \dfrac{v \pm \sqrt{v^2 - 64h}}{32}$ **156.** $r = \dfrac{-2\pi h \pm \sqrt{4\pi^2 h^2 + 8\pi A}}{4\pi}$

157. $t = 3 + \frac{\sqrt{138}}{2}$ sec, $t \approx 8.87$ sec

158. $t = \frac{15}{4} - \frac{\sqrt{119}}{4}$ sec, $t \approx 1.02$ sec

169. a. $7x^2 + 6x - 16 = 0$ **b.** $6x^2 + 5x - 14 = 0$
c. $5x^2 - x - 6 = 0$

170. a. $c > 4$ **b.** $c = 4$ **c.** $c < 4$ **d.** $c = -5, 3$, others

171. $z = -2i; z = 5i$ **172.** $z = -2i; z = 11i$

173. $z = \dfrac{-3}{4}i; z = 2i$ **174.** $z = \dfrac{-13i}{2}; z = 2i$

175. $z = -1 - i; z = -13 - i$

176. $z = 1 + 3i; z = -9 + 3i$

Exercises 3.3, pp. 322-326

7. left 2, down 9

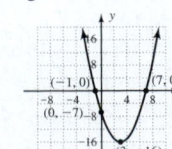

8. right 3, down 16

9. right 1, reflected across x-axis, up 4

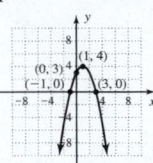

10. right 4, reflected across x-axis, up 9

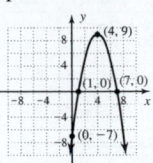

11. left 1, stretched vertically, down 8

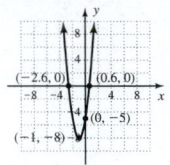

12. right 3, stretched vertically, down 21

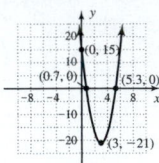

13. right 2, reflected across x-axis, stretched vertically, up 15

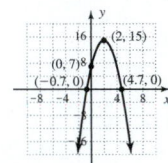

14. right 2, reflected across x-axis, stretched vertically, up 5

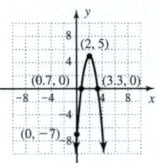

15. right $\frac{7}{4}$, stretched vertically, down $\frac{25}{8}$

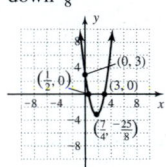

16. right $\frac{9}{8}$, stretched vertically, down $\frac{49}{16}$

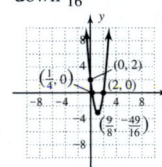

17. left $\frac{7}{6}$, reflected across x-axis, stretched vertically, up $\frac{121}{12}$

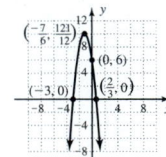

18. right $\frac{9}{4}$, reflected across x-axis, stretched vertically, up $\frac{25}{8}$

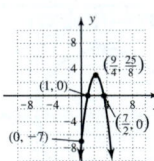

19. right $\frac{5}{2}$, down $\frac{17}{4}$

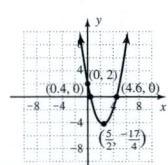

20. left $\frac{7}{2}$, down $\frac{33}{4}$

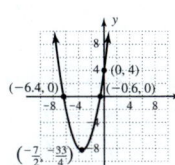

21. left 1, down 7

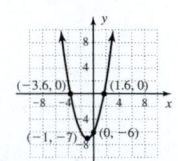

22. left 4, down 5

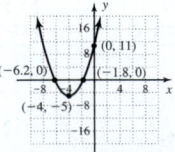

23. right 2, reflected across x-axis, up 6

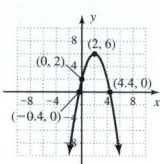

24. right 5, reflected across x-axis, up 6

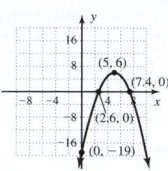

25. left 3, compressed vertically up $\frac{5}{2}$

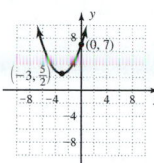

26. right 5, compressed vertically up 3

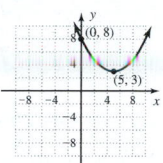

27. right $\frac{5}{2}$, reflected across x-axis, stretched vertically, up $\frac{11}{2}$

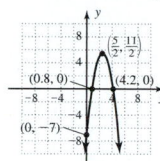

28. right 2, reflected across x-axis, stretched vertically, up 5

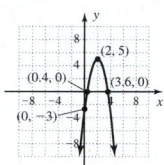

29. right $\frac{3}{2}$, stretched vertically, down 6

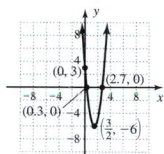

30. left 2, stretched vertically, down 7

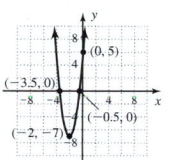

31. left 3, compressed vertically, down $\frac{19}{2}$

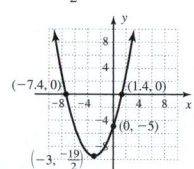

32. right 3, compressed vertically, down 7

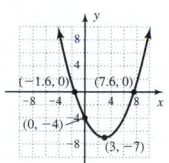

48. b.

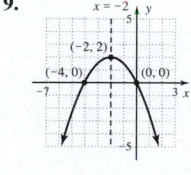

Mid-Chapter Check, p. 327

3.
$$(1 + 2i)^2 - 2(1 + 2i) + 5 = 0$$
$$1 + 4i + 4i^2 - 2 - 4i + 5 = 0$$
$$1 + (-4) - 2 + 5 = 0$$
$$0 = 0\checkmark \text{ Yes.}$$

9.

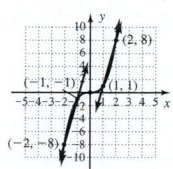

Reinforcing Basic Concepts, pp. 327-328

Exercise 1: $\frac{7}{2} + (-1) = \frac{5}{2} = -\frac{b}{a}\checkmark$ $\frac{7}{2} \cdot (-1) = \frac{-7}{2} = \frac{c}{a}\checkmark$

Exercise 2: $\frac{2 + 3\sqrt{2}}{2} + \frac{2 - 3\sqrt{2}}{2} = \frac{4}{2} = \frac{-b}{a}\checkmark$

$\frac{2 + 3\sqrt{2}}{2} \cdot \frac{2 - 3\sqrt{2}}{2} = \frac{-14}{4} = \frac{-7}{2} = \frac{c}{a}\checkmark$

Exercise 3: $(5 + 2\sqrt{3}i) + (5 - 2\sqrt{3}i) = 10 = \frac{-b}{a}\checkmark$

$(5 + 2\sqrt{3}i)(5 - 2\sqrt{3}i) = 25 + 12 = 37 = \frac{c}{a}\checkmark$

Exercise 4: $x_1 + x_2 = -\frac{b}{a}$ since radical terms sum to 0, and $2\left(\frac{-b}{2a}\right) = \frac{-b}{a}$;

$x_1 x_2 = \frac{c}{a}$ since $\left(\frac{-b}{2a}\right)^2 - \left(\frac{\sqrt{b^2 - 4ac}}{2a}\right)^2 = \frac{b^2}{4a^2} - \frac{b^2 - 4ac}{4a^2} = \frac{4ac}{4a^2} = \frac{c}{a}$

Exercises 3.4, pp. 335-339

7. a. $y = 1.45x^2 - 2.59x - 4.55$ **b.** 4.1475 **c.** $x = 6.446$
8. a. $y = 2.36x^2 - 0.65x + 2.58$ **b.** 29.215 **c.** $x = 4.069$
9. a. $y = -0.851x^2 + 3.153x + 64.428$ **b.** 65.037 **c.** $x = 7.623$
10. a. $y = -2.746x^2 - 17.482x + 11.935$ **b.** -15.615 **c.** $x = 7.654$
11. a. $y = 3.485x^2 - 26.424x - 60.505$ **b.** 10.422
c. $x = 1.746$ or 5.836   **12. a.** $y = 9.647x^2 - 73.724x + 63.871$
b. -76.986 **c.** $x = 0.570$ or 7.073
13. a. $y = 0.113x^2 - 5.796x + 61.583$ **b.** -12.882
c. $x = 7.368$ or 44.019   **14. a.** $y = -0.030x^2 + 3.032x - 2.168$
b. 74.179   **c.** $x = 9.940$ or 90.766
43. a. 14,570 ft **b.** 14,174 ft **c.** -198 ft/sec
d. $h(10) = 13,340$ ft; $h(12) = 12,624$ ft; average rate of change $=$
-358 ft/sec, almost twice as fast **44. a.** 6.1 **b.** 6.1, They are the same.
45. a. $R = -43.07t^2 + 976.53t - 126.8$ **b.** 4598
c. $t \approx 8.26$ days (early in the ninth day) **d.** about 5408 participants
46. a. $T = -960w^2 + 17,375w + 1091$ **b.** 63,966 **c.** $w \approx 5.87$ weeks
(late in the sixth week) **d.** about 9708 fans
47. a. $T = 5.92n^2 - 83.13n + 349.86$ **b.** about 82 sec
c. 16 tourists **d.** about 58 sec with seven tourists
48. a. $T = -0.0016w^2 + 0.5898w - 24.7717$ **b.** 28 tomatoes
c. 89 gal or 276 gal **d.** 29 tomatoes **49. a.** $\frac{\Delta\text{weight}}{\Delta\text{time}} = \frac{50}{1}$, positive, 50 g
are gained each week **b.** 25th to 29th: $\frac{\Delta w}{\Delta t} = \frac{50}{1}$; 32nd to 36th: $\frac{\Delta w}{\Delta t} = \frac{250}{1}$; the
weight gain is five time greater in the later weeks. **50. a.** $\frac{\Delta\text{fertility}}{\Delta\text{time}} = \frac{-1}{20}$,
negative, fertility is decreasing by one child every 20 yr **b.** 1940 to 1950;
$\frac{\Delta f}{\Delta t} = \frac{0.8}{10}$; positive, fertility is increasing by less than one child every 10 yr
c. 1940 to 1950: $\frac{\Delta f}{\Delta t} = \frac{0.8}{10}$; 1980 to 1990; $\frac{\Delta f}{\Delta t} = \frac{0.2}{10}$, the fertility rate was
increasing four times as fast from 1940 to 1950.
51. a. 7 **b.** 7 **c.** They are the same.
d. Slopes are equal.

57. c. $\frac{\Delta r}{\Delta t} = 0.74$ in./sec when $t \in [0, 1]$ and 0.04 in./sec when $t \in [18, 19]$;
$\frac{\Delta V}{\Delta r} \approx 64.745$ in³/in. when $t \in [0, 1]$ and 22.602 when $t \in [18, 19]$.
Answers will vary.

58. a. $D(x) = \frac{f(x + h) - f(x)}{(x + h) - x} = \frac{f(x + h) - f(x)}{h}$

60.

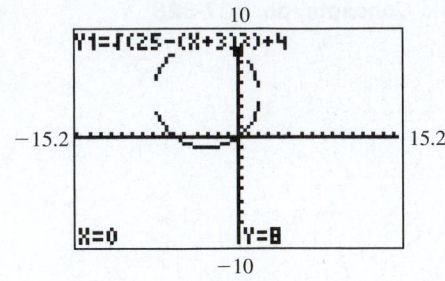

$(x + 3)^2 + (y - 4)^2 = 25$; $C = (-3, 4)$, $r = 5$,
$Y_1 = \sqrt{25 - (X + 3)^2} + 4$, $Y_2 = -\sqrt{25 + (X + 3)^2} + 4$

Exercises 3.5, pp. 346–352

7. a. $x \in \mathbb{R}$ **b.** $f(-2) - g(-2) = 13$

8. a. $x \in \mathbb{R}$ **b.** $f(5) - g(5) = 10$

9. a. $h(x) = x^2 - 6x - 3$ **b.** $h(-2) = 13$ **c.** they are identical

10. a. $h(x) = 2x^2 - 3x - 25$ **b.** $h(5) = 10$ **c.** they are identical

11. a. $x \in [3, \infty)$ **b.** $h(x) = \sqrt{x - 3} + 2x^3 - 54$

c. $h(4) = 75$, 2 is not in the domain of h.

12. a. $x \in [\frac{5}{2}, \infty)$ **b.** $h(x) = 4x^2 - 2x + 3 - \sqrt{2x - 5}$

c. $h(7) = 182$, 2 is not in the domain of h.

13. a. $x \in [-5, 3]$ **b.** $r(x) = \sqrt{x + 5} + \sqrt{3 - x}$

c. $r(2) = \sqrt{7} + 1$, 4 is not in the domain of r.

14. a. $x \in [-2, 6]$ **b.** $r(x) = \sqrt{6 - x} - \sqrt{x + 2}$

c. -3 is not in the domain of r, $r(2) = 0$

15. a. $x \in [-4, \infty)$ **b.** $h(x) = \sqrt{x + 4}(2x + 3)$

c. $h(-4) = 0$, $h(21) = 225$

16. a. $x \in [7, \infty)$ **b.** $h(x) = (-3x + 5)\sqrt{x - 7}$

c. $h(8) = -19$, $h(11) = -56$

17. a. $x \in [-1, 7]$ **b.** $r(x) = \sqrt{-x^2 + 6x + 7}$

c. 15 is not in the domain of r, $r(3) = 4$

18. a. $x \in [-4, 4]$ **b.** $r(x) = \sqrt{16 - x^2}$

c. -5 is not in the domain of r, $r(-3) = \sqrt{7}$

19. a. $x \in (-\infty, -4) \cup (-4, \infty)$ **b.** $h(x) = x - 4, x \neq -4$

20. a. $x \in (-\infty, 7) \cup (7, \infty)$ **b.** $h(x) = x + 7, x \neq 7$

21. a. $x \in (-\infty, -4) \cup (-4, \infty)$ **b.** $h(x) = x^2 - 2, x \neq -4$

22. a. $x \in (-\infty, 5) \cup (5, \infty)$ **b.** $h(x) = x^2 + 2, x \neq 5$

23. a. $x \in (-\infty, 1) \cup (1, \infty)$ **b.** $h(x) = x^2 - 6x, x \neq 1$

24. a. $x \in (-\infty, 1) \cup (1, \infty)$ **b.** $h(x) = x^2 + x + 1, x \neq 1$

25. a. $x \in (-\infty, 5) \cup (5, \infty)$ **b.** $h(x) = \dfrac{x + 1}{x - 5}, x \neq 5$

26. a. $x \in (-\infty, 7) \cup (7, \infty)$ **b.** $h(x) = \dfrac{x + 3}{x - 7}, x \neq 7$

27. a. $x \in (-\infty, -2)$ **b.** $r(x) = \dfrac{2x - 3}{\sqrt{-2 - x}}$

c. 6 is not in the domain of r, $r(-6) = -\dfrac{15}{2}$

28. a. $x \in (-\infty, 3)$ **b.** $r(x) = \dfrac{1 - x}{\sqrt{3 - x}}$

c. 6 is not in the domain of r, $r(-6) = \dfrac{7}{3}$

29. a. $x \in (5, \infty)$ **b.** $r(x) = \dfrac{x - 5}{\sqrt{x - 5}}$

c. $r(6) = 1$, -6 is not in the domain of r.

30. a. $x \in (-3, \infty)$ **b.** $r(x) = \dfrac{x + 2}{\sqrt{x + 3}}$

c. $r(6) = \dfrac{8}{3}$, -6 is not in the domain of r.

31. a. $x \in \left(-\dfrac{13}{2}, \infty\right)$ **b.** $r(x) = \dfrac{x^2 - 36}{\sqrt{2x + 13}}$ **c.** $r(6) = 0$, $r(-6) = 0$

32. a. $x \in \left(-\dfrac{7}{3}, \infty\right)$ **b.** $r(x) = \dfrac{x^2 - 6x}{\sqrt{7 + 3x}}$

c. $r(6) = 0$, -6 is not in the domain of r.

33. a. $h(x) = \dfrac{2x + 4}{x - 3}$ **b.** $x \in (-\infty, 3) \cup (3, \infty)$ **c.** $x \neq -2, x \neq 0$

34. a. $h(x) = \dfrac{2x - 4}{x + 1}$ **b.** $x \in (-\infty, -1) \cup (-1, \infty)$

c. $x \neq 0, x \neq 2$

35. sum: $3x + 1, x \in (-\infty, \infty)$; difference: $x + 5, x \in (-\infty, \infty)$;
product: $2x^2 - x - 6, x \in (-\infty, \infty)$;
quotient: $\dfrac{2x + 3}{x - 2}, x \in (-\infty, 2) \cup (2, \infty)$

36. sum: $3x - 8, x \in (-\infty, \infty)$; difference: $-x - 2, x \in (-\infty, \infty)$;
product: $2x^2 - 13x + 15, x \in (-\infty, \infty)$;
quotient: $\dfrac{x - 5}{2x - 3}, x \in \left(-\infty, \dfrac{3}{2}\right) \cup \left(\dfrac{3}{2}, \infty\right)$

37. sum: $x^2 + 3x + 5, x \in (-\infty, \infty)$; difference: $x^2 - 3x + 9$,
$x \in (-\infty, \infty)$; product: $3x^3 - 2x^2 + 21x - 14, x \in (-\infty, \infty)$;
quotient: $\dfrac{x^2 + 7}{3x - 2}, x \in \left(-\infty, \dfrac{2}{3}\right) \cup \left(\dfrac{2}{3}, \infty\right)$

38. sum: $x^2 - 2x + 4, x \in (-\infty, \infty)$; difference: $x^2 - 4x - 4$,
$x \in (-\infty, \infty)$; product: $x^3 + x^2 - 12x, x \in (-\infty, \infty)$;
quotient: $\dfrac{x^2 - 3x}{x + 4}, x \in (-\infty, -4) \cup (-4, \infty)$

39. sum: $x^2 + 3x - 4, x \in (-\infty, \infty)$; difference: $x^2 + x - 2$,
$x \in (-\infty, \infty)$; product: $x^3 + x^2 - 5x + 3, x \in (-\infty, \infty)$;
quotient: $x + 3, x \in (-\infty, 1) \cup (1, \infty)$

40. sum: $x^2 - x - 12, x \in (-\infty, \infty)$; difference: $x^2 - 3x - 18$,
$x \in (-\infty, \infty)$; product: $x^3 + x^2 - 21x - 45, x \in (-\infty, \infty)$;
quotient: $x - 5, x \in (-\infty, -3) \cup (-3, \infty)$

41. sum: $3x + 1 + \sqrt{x - 3}, x \in [3, \infty)$; difference: $3x + 1 - \sqrt{x - 3}$,
$x \in [3, \infty)$; product: $(3x + 1)\sqrt{x - 3}, x \in [3, \infty)$;
quotient: $\dfrac{3x + 1}{\sqrt{x - 3}}, x \in (3, \infty)$

42. sum: $x + 2 + \sqrt{x + 6}, x \in [-6, \infty)$; difference: $x + 2 - \sqrt{x - 6}$,
$x \in [-6, \infty)$; product: $(x + 2)\sqrt{x + 6}, x \in [-6, \infty)$;
quotient: $\dfrac{x + 2}{\sqrt{x + 6}}, x \in (-6, \infty)$

43. sum: $2x^2 + \sqrt{x + 1}, x \in [-1, \infty)$; difference: $2x^2 - \sqrt{x + 1}$,
$x \in [-1, \infty)$; product: $2x^2\sqrt{x + 1}, x \in [-1, \infty)$;
quotient: $\dfrac{2x^2}{\sqrt{x + 1}}, x \in (-1, \infty)$

44. sum: $x^2 + 2 + \sqrt{x - 5}, x \in [5, \infty)$; difference: $x^2 + 2 - \sqrt{x - 5}$.
$x \in [5, \infty)$; product: $(x^2 + 2)\sqrt{x - 5}, x \in [5, \infty)$;
quotient: $\dfrac{x^2 + 2}{\sqrt{x - 5}}, x \in (5, \infty)$

45. sum: $\dfrac{7x - 11}{(x - 3)(x + 2)}, x \in (-\infty, -2) \cup (-2, 3) \cup (3, \infty)$;
difference: $\dfrac{-3x + 19}{(x - 3)(x + 2)}, x \in (-\infty, -2) \cup (-2, 3) \cup (3, \infty)$;
product: $\dfrac{10}{(x - 3)(x + 2)}, x \in (-\infty, -2) \cup (-2, 3) \cup (3, \infty)$;
quotient: $\dfrac{2x + 4}{5(x - 3)}, x \in (-\infty, -2) \cup (-2, 3) \cup (3, \infty)$

46. sum: $\dfrac{5x + 17}{(x - 3)(x + 5)}, x \in (-\infty, -5) \cup (-5, 3) \cup (3, \infty)$;
difference: $\dfrac{3x + 23}{(x - 3)(x + 5)}, x \in (-\infty, -5) \cup (-5, 3) \cup (3, \infty)$;
product: $\dfrac{4}{(x + 5)(x - 3)}, x \in (-\infty, -5) \cup (-5, 3) \cup (3, \infty)$;
quotient: $\dfrac{4x + 20}{x - 3}, x \in (-\infty, -5) \cup (-5, 3) \cup (3, \infty)$

49. a. $1 billion **b.** $5 billion **c.** 2003, 2007, 2010
d. $t \in (2000, 2003) \cup (2007, 2010)$ **e.** $t \in (2003, 2007)$
f. $R(5) - C(5)$; $4 billion
50. a. $6 billion **b.** $5 billion **c.** 2001, 2004, 2007
d. $t \in (2000, 2001) \cup (2004, 2007)$ **e.** $t \in (2001, 2004) \cup (2007, 2010)$
f. $D(t) + R(t)$; $9 billion

51. a. 4 **b.** 0 **c.** 2 **d.** 3 **e.** $-\dfrac{1}{3}$ **f.** 6 **g.** -3 **h.** 1 **i.** 1
j. undefined
52. a. 0 **b.** -12 **c.** 2 **d.** 2 **e.** $\dfrac{1}{6}$ **f.** -6 **g.** -20 **h.** -5 **i.** 3
j. undefined
69. a. $V = 400 - 400e^{-0.08t}$ **b.** $f(t) = 400, g(t) = 400e^{-0.08t}$,
$V(1) \approx 400 - 369 = 31$ ft/sec, $V(2) \approx 400 - 341 = 59$ ft/sec,
$V(20) \approx 400 - 81 = 319$ ft/sec **c.** 400 ft/sec
81. a. 1995 to 1996; 1999 to 2004 **b.** 30; 1995 **c.** 20 seats; 1997
d. The total number of seats in the senate (50); the number of additional
seats held by the majority

84.

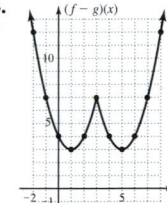

Exercises 3.6, pp. 365–369

7. $0, 0, 4a^2 - 10a - 14, a^2 - 9a$
8. $0, -10, 27t^3 - 27t, t^3 + 3t^2 - 6t - 8$
9. a. $h(x) = \sqrt{2x - 2}$ **b.** $H(x) = 2\sqrt{x + 3} - 5$
c. D of $h(x): x \in [1, \infty)$; D of $H(x): x \in [-3, \infty)$
10. a. $h(x) = \sqrt{9 - x^2} + 3$ **b.** $H(x) = \sqrt{-x^2 - 6x}$
c. D of $h(x): x \in [-3, 3]$; D of $H(x): x \in [-6, 0]$
11. a. $h(x) = \sqrt{3x + 1}$ **b.** $H(x) = 3\sqrt{x - 3} + 4$
c. D of $h(x): x \in \left[-\frac{1}{3}, \infty\right)$; D of $H(x): x \in [3, \infty)$
12. a. $h(x) = 2\sqrt{x + 1}$ **b.** $H(x) = 4\sqrt{x + 5} - 1$
c. D of $h(x): x \in [-1, \infty)$; D of $H(x): x \in [-5, \infty)$
13. a. $h(x) = x^2 + x - 2$ **b.** $H(x) = x^2 - 3x + 2$
c. D of $h(x): x \in (-\infty, \infty)$; D of $H(x): x \in (-\infty, \infty)$
14. a. $h(x) = 18x^2 + 24x + 7$ **b.** $H(x) = 6x^2 - 1$
c. D of $h(x): x \in (-\infty, \infty)$; D of $H(x): x \in (-\infty, \infty)$
15. a. $h(x) = x^2 + 7x + 8$ **b.** $H(x) = x^2 + x - 1$
c. D of $h(x): x \in (-\infty, \infty)$; D of $H(x): x \in (-\infty, \infty)$
16. a. $h(x) = x^2 - 8x + 14$ **b.** $H(x) = x^2 - 4x$
c. D of $h(x): x \in (-\infty, \infty)$; D of $H(x): x \in (-\infty, \infty)$
17. a. $h(x) = |-3x + 1| - 5$ **b.** $H(x) = -3|x| + 16$
c. D of $h(x): x \in (-\infty, \infty)$; D of $H(x): x \in (-\infty, \infty)$
18. a. $h(x) = |3x - 7|$ **b.** $H(x) = 3|x - 2| - 5$
c. D of $h(x): x \in (-\infty, \infty)$; D of $H(x): x \in (-\infty, \infty)$
19. $h(-3) = -6, h(\sqrt{2}) \approx 9.071$
$h(\frac{1}{2}) = 2.75, h(5) = 50$
20. $h(-3) = 50, h(\sqrt{2}) \approx 8.44$
$h(\frac{1}{2}) = 20.25, h(5) = -54$
21. $h(-3) = 1, h(\sqrt{2}) \approx 145.91$
$h(\frac{1}{2}) = 64, h(5) = $ ERR;
$x = 5$ is not in the domain of g
22. $h(-3) = 49, h(\sqrt{2}) \approx 3.343$
$h(\frac{1}{2}) = $ ERR, $h(5) = 81$;
$x = \frac{1}{2}$ is not in the domain of g
23. $h(-3) = 2, h(\sqrt{2}) = 5$
$h(\frac{1}{2}) = 5.5, h(5) = $ ERR;
$g(5)$ is not in the domain of f

24. $h(-3) = $ ERR, $h(\sqrt{2}) = -1$
$h(\frac{1}{2}) = -0.8, h(5) = -0.4375$;
$g(-3)$ is not in the domain of f
25. a. $(f \circ g)(x)$: For $g(x)$ to be defined, $x \neq 0$.

For $f[g(x)] = \dfrac{2g(x)}{g(x) + 3}$, $g(x) \neq -3$ so $x \neq -\dfrac{5}{3}$.

domain: all real numbers except $x = -3$ and $x = -\dfrac{5}{3}$
b. $(g \circ f)(x)$: For $f(x)$ to be defined, $x \neq -3$.

For $g[f(x)] = \dfrac{5}{f(x)}$, $f(x) \neq 0$ so $x \neq 0$.

domain: all real numbers except $x = -3$ and $x = 0$
c. $(f \circ g)(x) = \dfrac{10}{5 + 3x}$; $(g \circ f)(x) = \dfrac{5x + 15}{2x}$; the domain of a
composition cannot always be determined from the composed form
26. a. $(f \circ g)(x)$: For $g(x)$ to be defined, $x \neq 2$.

For $f[g(x)] = \dfrac{-3}{g(x)}$, $g(x) \neq 0$ so $x \neq 0$.

domain: $\{x | x \neq 0, x \neq 2\}$
b. $(g \circ f)(x)$: For $f(x)$ to be defined, $x \neq 0$.

For $g[f(x)] = \dfrac{f(x)}{f(x) - 2}$, $f(x) \neq 2$ so $x \neq -\dfrac{3}{2}$.

domain: all real numbers except $x = 0$ and $x = -\dfrac{3}{2}$
c. $(f \circ g)(x) = \dfrac{-3x + 6}{x}$; $(g \circ f)(x) = \dfrac{3}{2x + 3}$; the domain of a
composition cannot always be determined from the composed form
27. a. $(f \circ g)(x)$: For $g(x)$ to be defined, $x \neq 5$.

For $f[g(x)] = \dfrac{4}{g(x)}$, $g(x) \neq 0$ and $g(x)$ is never zero

domain: $\{x | x \neq 5\}$
b. $(g \circ f)(x)$: For $f(x)$ to be defined, $x \neq 0$.

For $g[f(x)] = \dfrac{1}{f(x) - 5}$, $f(x) \neq 5$ so $x \neq \dfrac{4}{5}$.

domain: all real numbers except $x = 0$ and $x = \dfrac{4}{5}$
c. $(f \circ g)(x) = 4x - 20$; $(g \circ f)(x) = \dfrac{x}{4 - 5x}$; the domain of a
composition cannot always be determined from the composed form
28. a. $(f \circ g)(x)$: For $g(x)$ to be defined, $x \neq 2$.

For $f[g(x)] = \dfrac{3}{g(x)}$, $g(x) \neq 0$ and $g(x)$ is never zero

domain: $\{x | x \neq 2\}$
b. $(g \circ f)(x)$: For $f(x)$ to be defined, $x \neq 0$.

For $g[f(x)] = \dfrac{1}{f(x) - 2}$, $f(x) \neq 2$ so $x \neq \dfrac{3}{2}$.

domain: all real numbers except $x = 0$ and $x = \dfrac{3}{2}$
c. $(f \circ g)(x) = 3x - 6$; $(g \circ f)(x) = \dfrac{x}{3 - 2x}$; the domain of a composition
cannot always be determined from the composed form
31. $g(x) = \sqrt{x - 2} + 1, f(x) = x^3 - 5$
32. $q(x) = x^2 - 5, p(x) = \sqrt[3]{x} + 2$
47. a. $\dfrac{\Delta g}{\Delta x} = 2x + 2 + h$ **b.** $\dfrac{\Delta g}{\Delta x} = -3.9$ **c.** $\dfrac{\Delta g}{\Delta x} = 3.01$
d.

The rates of change have opposite sign, with the
secant line to the left being slightly more steep.

48. a. $\frac{\Delta j}{\Delta x} = 2x - 6 + h$　**b.** $\frac{\Delta j}{\Delta x} = -2.1$　**c.** $\frac{\Delta j}{\Delta x} = 4.01$

d. 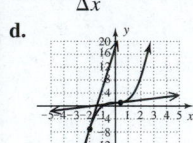　The rates of change have opposite sign, with the secant line on the right being more steep.

49. a. $\frac{\Delta g}{\Delta x} = 3x^2 + 3xh + h^2$　**b.** $\frac{\Delta g}{\Delta x} = 12.61$　**c.** $\frac{\Delta g}{\Delta x} \approx 0.49$

d. 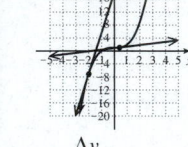　Both lines have a positive slope, but the line at $x = -2$ is much steeper.

50. a. $\frac{\Delta v}{\Delta x} = \frac{1}{\sqrt{x+h} + \sqrt{x}}$　**b.** $\frac{\Delta v}{\Delta x} \approx 0.49$　**c.** $\frac{\Delta v}{\Delta x} \approx 0.25$

d. 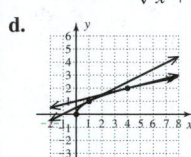　Both lines have a positive slope, but the line at $x = 4$ is less steep.

51. $\frac{\Delta j}{\Delta x} = \frac{-2x - h}{x^2(x+h)^2}$;

$[0.50, 0.51]: \frac{\Delta j}{\Delta x} \approx -15.5$;

$[1.50, 1.51]: \frac{\Delta j}{\Delta x} \approx -0.6$;

Answers will vary.

52. $\frac{\Delta f}{\Delta x} = 2x - 4 + h$;

$[0.00, 0.01]: \frac{\Delta f}{\Delta x} = -3.99$;

$[3.00, 3.01]: \frac{\Delta f}{\Delta x} = 2.01$;

Answers will vary.

53. $\frac{\Delta g}{\Delta x} = 3x^2 + 3xh + h^2$;

$[-2.01, -2.00]: \frac{\Delta g}{\Delta x} \approx 12.1$;

$[0.40, 0.41]: \frac{\Delta g}{\Delta x} \approx 0.5$;

Answers will vary.

54. $\frac{\Delta r}{\Delta x} = \frac{1}{\sqrt{x+h} + \sqrt{x}}$

$[1.00, 1.01]: \frac{\Delta r}{\Delta x} \approx 0.5$;

$[4.00, 4.01]: \frac{\Delta r}{\Delta x} \approx .025$;

Answers will vary.

55. a. $f[g(x)] = (x-2)^2 + 4(x-2) + 3$
$= x^2 - 4x + 4 + 4x - 8 + 3$
$= x^2 - 1 ✓$

b. verified

63. a. $L(0) = 500$ lions and $H(500) = 400$ hyenas
b. $H[L(x)] = 400 + 0.0075x$, $(H \circ L)(16,000) = 520$ hyenas　**c.** prior to an increase of 30,000

64. a. $p(0) = 750$ purple martins and $m(750) = 16,250$ mosquitoes
b. $m[p(r)] = 16,250 + 168.75r$, about 24,688 mosquitoes
c. an increase of 120 raccoons

65. a. $\frac{\Delta d}{\Delta h} \approx 0.2$　**b.** $\frac{\Delta d}{\Delta h} \approx 0.04$

c. 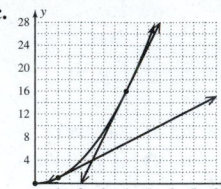　As height increases you can see farther, but the sight distance increases at a slower rate.

66. a. $\frac{\Delta P}{\Delta x} = 2.01$　**b.** $\frac{\Delta P}{\Delta x} = 8.01$

c. 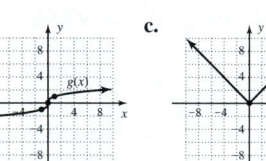　The projected image grows at a faster rate, the farther you move away from the screen.

67. a. March: $\frac{\Delta d}{\Delta t} \approx 15$, June: $\frac{\Delta d}{\Delta t} \approx 3$, 5 times faster

b. $t = 6.75$, late June　**c.** decreasing $\left(\frac{\Delta d}{\Delta t} < 0 \right)$; 5000 units/month

68. a. $s = 30$ mph: $\frac{\Delta m}{\Delta s} \approx 0.4$, $s = 45$ mph: $\frac{\Delta m}{\Delta s} \approx 0.1$, 4 times faster

b. $s = 50$ mph, 25 mpg　**c.** decreasing $\left(\frac{\Delta m}{\Delta s} < 0 \right)$; -0.4

70. a. $h(x) = \frac{1}{|x+1| - 4}$,　**b.** $x \neq 3, x \neq -5$; yes

c. $[-1, 3) \cup (3, \infty)$

71. a. For $y = \frac{1}{x}: \frac{\Delta y}{\Delta x} \approx -3.91$;　For $y = \frac{1}{x^2}: \frac{\Delta y}{\Delta x} \approx -15.53$

b. Less—decrease is more gradual; For $y = \frac{1}{x}: \frac{\Delta y}{\Delta x} \approx -1.54$;
For $y = \frac{1}{x^2}: \frac{\Delta y}{\Delta x} \approx -3.83$

73. a. 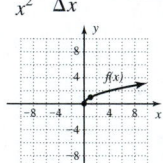　**b.** 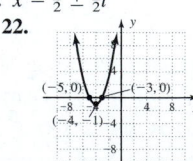　**c.**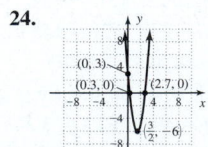

Summary and Concept Review, pp. 370–375

13. a. $x = 5$ or $x = -2$　**b.** $x = -5$ or $x = 5$　**c.** $x = -\frac{5}{3}$ or $x = 3$
d. $x = -2$ or $x = 2$ or $x = 3$　**14. a.** $x = \pm 3$　**b.** $x = 2 \pm \sqrt{5}$
c. $x = \pm i\sqrt{5}$　**d.** $x = \pm 5$　**15. a.** $x = 3$ or $x = -5$
b. $x = -8$ or $x = 2$　**c.** $x = 1 \pm \frac{\sqrt{10}}{2}$; $x \approx 2.58$ or $x \approx -0.58$
d. $x = 2$ or $x = \frac{1}{3}$　**16. a.** $x = 2 \pm i\sqrt{5}$; $x \approx 2 \pm 2.24i$
b. $x = \frac{3}{2} \pm \frac{\sqrt{2}}{2}$; $x \approx 2.21$ or $x \approx 0.79$　**c.** $x = \frac{3}{2} \pm \frac{1}{2}i$

21.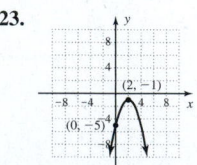

22.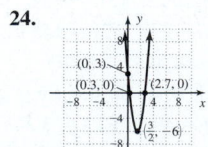

23.

24.

27. a.

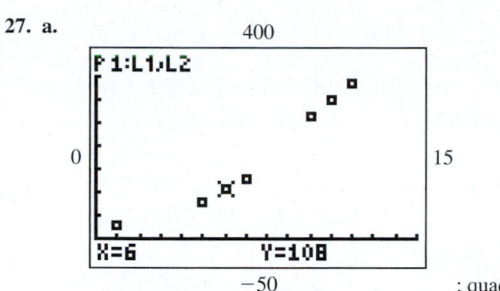

; quadratic

b. $A(x) = 2.144x^2 + 1.010x + 25.847$, about \$48.2 billion   **c.** about \$460.2 billion   **d.** year $16 \rightarrow 2016$

28. a.

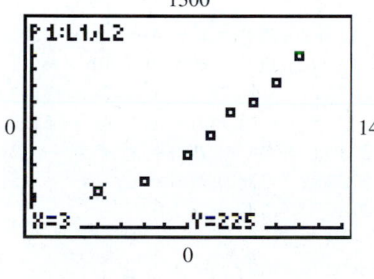

, quadratic

b. $F_d(x) = 0.35x^2 - 0.3x$, 1242 units   **c.** about 82 mph

29. a. $\dfrac{\Delta N}{\Delta t} = \dfrac{2.95}{1}$, 295 stores per year   **b.** for 2000 to 2002: $\dfrac{\Delta N}{\Delta t} \approx \dfrac{11.93}{1}$, 1193 stores per year; $295 \times 4 \approx 1180$ ✓   **c.** for 2002 to 2004: $\dfrac{\Delta N}{\Delta t} \approx \dfrac{13.42}{1}$, 1342 stores per year; for 2006 to 2008: $\dfrac{\Delta N}{\Delta t} \approx \dfrac{13.20}{1}$, 1320 stores per year, very close   **30. a.** 20 ft^3 **b.** approx. 18.251 ft^3

c. approx. -1.749 ft^3/sec   **d.** approx. -0.149 ft^3/sec   **e.** $t \approx 22.4$ sec

Practice Test, pp. 375–376

10.
$$(2 - 3i)^2 - 4(2 - 3i) + 13 = 0$$
$$4 - 12i - 9 - 8 + 12i + 13 = 0$$
$$-13 + 13 = 0$$
$$0 = 0 \checkmark$$

13. a. $f(x) = -(x - 5)^2 + 9$ **b.** $g(x) = \frac{1}{2}(x + 4)^2 + 8$

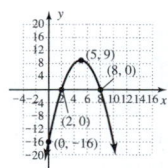   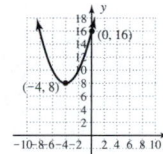

16. a.

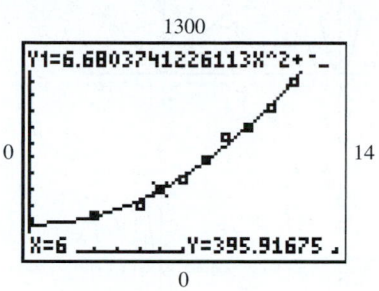

quadratic

b. $y \approx 6.68x^2 - 3.48x + 176.30$

c. about 396,000; about 4,264,000

17. a. 4750 books per year   **b.** approx. 11,267 books per year

c. -2400 books per year and 18,100 books per year

19. a. No, new company and sales should be growing   **b.** 15 for [4, 5]; 19 for [5, 6]   **c.** $\frac{\Delta S}{\Delta t} = 4t - 3 + 2h$. For small h, sales volume is approximately $\frac{37,000 \text{ unit}}{1 \text{ mo}}$ in month 10, $\frac{69,000 \text{ units}}{1 \text{ mo}}$ in month 18, and $\frac{93,000 \text{ units}}{1 \text{ mo}}$ in month 24

20. a. $V(t) = \frac{4}{3}\pi(\sqrt{t})^3$ **b.** 36π in^3

Calculator Exploration and Discovery, pp. 376–377

Exercise 1. a. linear: $r \approx 0.99$; residuals form a quadratic pattern (not random); quadratic: $r \approx 0.997$, residuals appear random, we do not expect time to begin decreasing; power: $r \approx 0.999$, residuals appear random, context suggests a power function.   **b.** 2.01 sec, 2.63 sec   **c.** 59 cm, 89 cm

Strengthening Core Skills, pp. 378–379

1.

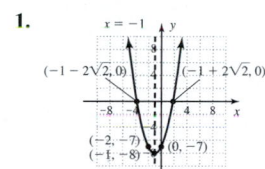

2.

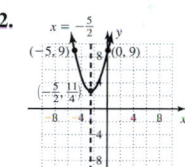

3.

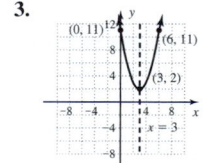

4.

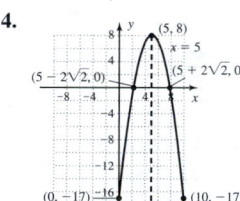

5.

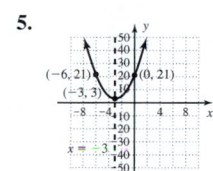

6.

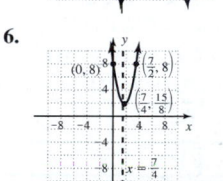

Cumulative Review Chapters R–3, pp. 379–380

3. a. $(x - 1)(x^2 + x + 1)$ **b.** $(x - 3)(x + 2)(x - 2)$

4. $x = \dfrac{-2 \pm \sqrt{2}}{2}$, $x \approx -0.29$, $x \approx -1.71$

6.

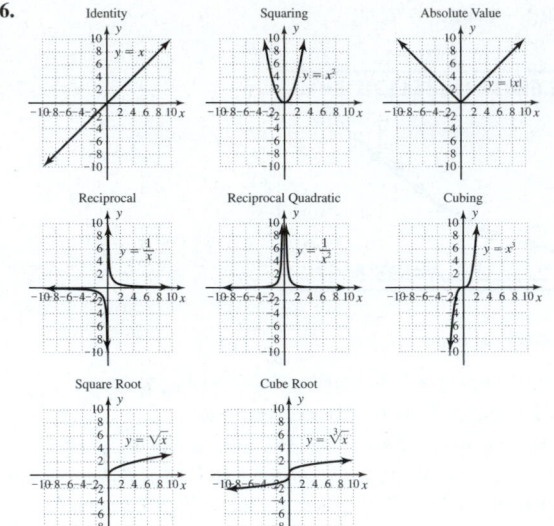

8. $(f \cdot g)(x) = 3x^3 - 12x^2 + 12x;\ \left(\dfrac{f}{g}\right)(x) = 3x,\ x \neq 2;\ (g \circ f)(-2) = 22$

9. $y = \dfrac{11}{60}x + \dfrac{1009}{60}$; 39 min, driving time increases 11 min every 60 days

12.

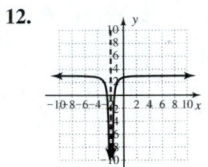

14. a. $x \in (-\infty, \infty)$
   b. $x \in (2, \infty)$

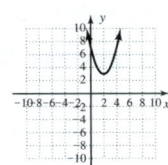

15.

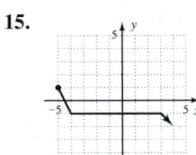

16.

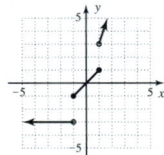

19. a. $f(4) = -1,\ g(2) = 4,\ (f \circ g)(2) = -1$
b. $g(4) = 0,\ f(8) = 4,\ (g \circ f)(8) = 0$
c. $(fg)(0) = (-2)(4) = -8,\ (\frac{g}{f})(0) = \frac{4}{-2} = -2$
d. $(f + g)(1) = -3 + 5 = 2,\ (g - f)(9) = 2 - 2 = 0$
21. $2x^2 - 9.2x + 14.5 = 0$ and $1.2x^2 = 5.52x - 8.7$
22. min: $y \approx -7.40$ at $x \approx -2.04$; max: $y \approx -0.60$ at $x \approx 2.04$
23. $y = 3x^2 - 1.5x + 7$ **24.** $x = 0$ and $x \approx 4.78$
25. $x \in (-\infty, -0.2) \cup (-0.2, 0.2) \cup (0.2, \infty)$

X	Y1
-.3	-40
-.2	ERR:
-.1	0
0	-25
.1	-66.67
.2	ERR:
.3	80
X= -.3	

CHAPTER 4

Exercises 4.1, pp. 390–394

13. a. $\dfrac{2x^2 - 5x - 3}{x - 3} = (2x + 1) + \dfrac{0}{x - 3}$

b. $2x^2 - 5x - 3 = (x - 3)(2x + 1) + 0$

14. a. $\dfrac{3x^2 + 13x - 10}{x + 5} = (3x - 2) + \dfrac{0}{x + 5}$

b. $3x^2 + 13x - 10 = (x + 5)(3x - 2) + 0$

15. a. $\dfrac{x^3 - 3x^2 - 14x - 8}{x + 2} = (x^2 - 5x - 4) + \dfrac{0}{x + 2}$

b. $x^3 - 3x^2 - 14x - 8 = (x + 2)(x^2 - 5x - 4) + 0$

16. a. $\dfrac{x^3 - 6x^2 - 24x - 17}{x + 1} = (x^2 - 7x - 17) + \dfrac{0}{x + 1}$

b. $x^3 - 6x^2 - 24x - 17 = (x + 1)(x^2 - 7x - 17) + 0$

17. a. $\dfrac{x^3 - 5x^2 - 4x + 23}{x - 2} = (x^2 - 3x - 10) + \dfrac{3}{x - 2}$

b. $x^3 - 5x^2 - 4x + 23 = (x - 2)(x^2 - 3x - 10) + 3$

18. a. $\dfrac{x^3 + 12x^2 + 34x - 9}{x + 7} = (x^2 + 5x - 1) + \dfrac{-2}{x + 7}$

b. $x^3 + 12x^2 + 34x - 9 = (x + 7)(x^2 + 5x - 1) - 2$

19. a. $\dfrac{2x^3 - 5x^2 - 11x - 17}{x - 4} = (2x^2 + 3x + 1) + \dfrac{-13}{x - 4}$

b. $2x^3 - 5x^2 - 11x - 17 = (x - 4)(2x^2 + 3x + 1) - 13$

20. a. $\dfrac{3x^3 - x^2 - 7x + 27}{x - 1} = (3x^2 + 2x - 5) + \dfrac{22}{x - 1}$

b. $3x^3 - x^2 - 7x + 27 = (x - 1)(3x^2 + 2x - 5) + 22$
25. $3x^3 - 8x + 12 = (x - 1)(3x^2 + 3x - 5) + 7$
26. $2x^3 + 7x - 81 = (x - 3)(2x^2 + 6x + 25) - 6$
27. $n^3 + 27 = (n + 3)(n^2 - 3n + 9) + 0$
28. $m^3 - 8 = (m - 2)(m^2 + 2m + 4) + 0$
29. $x^4 + 3x^3 - 16x - 8 = (x - 2)(x^3 + 5x^2 + 10x + 4) + 0$
30. $x^4 + 3x^2 + 29x - 21 = (x + 3)(x^3 - 3x^2 + 12x - 7) + 0$

31. $\dfrac{2x^3 + 7x^2 - x + 26}{x^2 + 3} = (2x + 7) + \dfrac{-7x + 5}{x^2 + 3}$

32. $\dfrac{x^4 + 3x^3 + 2x^2 - x - 5}{x^2 - 2} = (x^2 + 3x + 4) + \dfrac{5x + 3}{x^2 - 2}$

33. $\dfrac{x^4 - 5x^2 - 4x + 7}{x^2 - 1} = (x^2 - 4) + \dfrac{-4x + 3}{x^2 - 1}$

34. $\dfrac{x^4 + 2x^3 - 8x - 16}{x^2 + 5} = (x^2 + 2x - 5) + \dfrac{-18x + 9}{x^2 + 5}$

51.
-3	1	2	-5	-6
		-3	3	6
	1	-1	-2	0

52.
-6	1	3	-16	12
		-6	18	-12
	1	-3	2	0

53.
2	1	0	-7	6
		2	4	-6
	1	2	-3	0

54.
-4	1	0	-13	12
		-4	16	-12
	1	-4	3	0

55.
$\frac{2}{3}$	9	18	-4	-8
		6	16	8
	9	24	12	0

56.
$-\frac{3}{5}$	5	13	-9	-9
		-3	-6	9
	5	10	-15	0

57. $P(x) = (x + 2)(x - 3)(x + 5),\ P(x) = x^3 + 4x^2 - 11x - 30$
58. $P(x) = (x - 1)(x + 4)(x - 2),\ P(x) = x^3 + x^2 - 10x + 8$
59. $P(x) = (x + 2)(x - \sqrt{3})(x + \sqrt{3}),\ P(x) = x^3 + 2x^2 - 3x - 6$
60. $P(x) = (x - \sqrt{5})(x + \sqrt{5})(x - 4),\ P(x) = x^3 - 4x^2 - 5x + 20$
61. $P(x) = (x + 5)(x - 2\sqrt{3})(x + 2\sqrt{3}),\ P(x) = x^3 + 5x^2 - 12x - 60$
62. $P(x) = (x - 4)(x - 3\sqrt{2})(x + 3\sqrt{2}),\ P(x) = x^3 - 4x^2 - 18x + 72$
63. $P(x) = (x - 1)(x + 2)(x - \sqrt{10})(x + \sqrt{10}),$
$P(x) = x^4 + x^3 - 12x^2 - 10x + 20$
64. $P(x) = (x - \sqrt{7})(x + \sqrt{7})(x - 3)(x + 1),$
$P(x) = x^4 - 2x^3 - 10x^2 + 14x + 27$

Exercises 4.2, pp. 406–410

15. $(x - 5)^3(x + 9)^2$; $x = 5$, multiplicity 3; $x = -9$, multiplicity 2
16. $(x + 6)^3(x - 4)^2$; $x = -6$, multiplicity 3; $x = 4$, multiplicity 2
17. $(x - 7)^2(x + 2)^2(x + 7)$; $x = 7$, multiplicity 2; $x = -2$,
multiplicity 2; $x = -7$, multiplicity 1

18. $(x - 6)^2(x - 3)^2(x + 6)$; $x = 6$, multiplicity 2; $x = 3$, multiplicity 2; $x = -6$, multiplicity 1

33. $\{\pm 1, \pm 15, \pm 3, \pm 5, \pm \frac{1}{4}, \pm \frac{15}{4}, \pm \frac{3}{4}, \pm \frac{5}{4}, \pm \frac{1}{2}, \pm \frac{15}{2}, \pm \frac{3}{2}, \pm \frac{5}{2}\}$

34. $\{\pm 1, \pm 20, \pm 2, \pm 10, \pm 4, \pm 5, \pm \frac{1}{3}, \pm \frac{20}{3}, \pm \frac{2}{3}, \pm \frac{10}{3}, \pm \frac{4}{3}, \pm \frac{5}{3}\}$

35. $\{\pm 1, \pm 15, \pm 3, \pm 5, \pm \frac{1}{2}, \pm \frac{15}{2}, \pm \frac{3}{2}, \pm \frac{5}{2}\}$

36. $\{\pm 1, \pm 14, \pm 2, \pm 7, \pm \frac{1}{2}, \pm \frac{7}{2}\}$

37. $\{\pm 1, \pm 28, \pm 2, \pm 14, \pm 4, \pm 7, \pm \frac{1}{6}, \pm \frac{14}{3}, \pm \frac{1}{3}, \pm \frac{7}{3}, \pm \frac{2}{3}, \pm \frac{7}{6}, \pm \frac{1}{2}, \pm \frac{7}{2}, \pm \frac{28}{3}, \pm \frac{4}{3}\}$

38. $\{\pm 1, \pm 36, \pm 2, \pm 18, \pm 3, \pm 12, \pm 4, \pm 9, \pm 6, \pm \frac{1}{7}, \pm \frac{36}{7}, \pm \frac{2}{7}, \pm \frac{18}{7}, \pm \frac{3}{7},$
$\pm \frac{12}{7}, \pm \frac{4}{7}, \pm \frac{9}{7}, \pm \frac{6}{7}\}$

39. $\{\pm 1, \pm 3, \pm \frac{1}{32}, \pm \frac{1}{2}, \pm \frac{1}{16}, \pm \frac{1}{4}, \pm \frac{1}{8}, \pm \frac{3}{32}, \pm \frac{3}{2}, \pm \frac{3}{16}, \pm \frac{3}{4}, \pm \frac{3}{8}\}$

40. $\{\pm 1, \pm 6, \pm 2, \pm 3, \pm \frac{1}{24}, \pm \frac{1}{2}, \pm \frac{1}{12}, \pm \frac{1}{3}, \pm \frac{1}{8}, \pm \frac{1}{4}, \pm \frac{1}{6}, \pm \frac{3}{4}, \pm \frac{3}{2}, \pm \frac{2}{3}, \pm \frac{3}{8}\}$

83. a. possible roots: $\{\pm 1, \pm 8, \pm 2, \pm 4\}$;   **b.** neither -1 nor 1 is a root;   **c.** 3 or 1 positive roots, 1 negative root;   **d.** roots must lie between -2 and 2   **84. a.** possible roots: $\{\pm 1, \pm 6, \pm 2, \pm 3\}$;   **b.** neither -1 nor 1 is a root;   **c.** 1 positive root, 3 or 1 negative roots;   **d.** roots must lie between -3 and 2   **85. a.** possible roots: $\{\pm 1, \pm 2\}$;   **b.** -1 is a root;   **c.** 2 or 0 positive roots, 3 or 1 negative roots;   **d.** roots must lie between -3 and 2
86. a. possible roots: $\{\pm 1, \pm 4, \pm 2\}$;   **b.** $x = 1$ is a root;   **c.** 3 or 1 positive roots, 2 or 0 negative roots;   **d.** roots must lie between -2 and 1   **87. a.** possible roots: $\{\pm 1, \pm 12, \pm 2, \pm 6, \pm 3, \pm 4\}$;   **b.** $x = 1$ and $x = -1$ are roots;   **c.** 4, 2, or 0 positive roots, 1 negative root;   **d.** roots must lie between -1 and 4   **88. a.** possible roots: $\{\pm 1, \pm 14, \pm 2, \pm 7\}$;   **b.** $x = 1$ and $x = -1$ are roots;   **c.** 3 or 1 positive roots, 2 or 0 negative roots;   **d.** roots must lie between -3 and 4   **89. a.** possible roots: $\{\pm 1, \pm 20, \pm 2, \pm 10, \pm 4, \pm 5, \pm \frac{1}{2}, \pm \frac{5}{2}\}$;   **b.** $x = 1$ is a root;   **c.** 1 positive root, 1 negative root;   **d.** roots must lie between -2 and 1   **90. a.** possible roots: $\{\pm 1, \pm 24, \pm 2, \pm 12, \pm 3, \pm 8, \pm 4, \pm 6, \pm \frac{1}{3}, \pm \frac{2}{3}, \pm \frac{8}{3}, \pm \frac{4}{3}\}$;   **b.** $x = -1$ is a root; 1   **c.** 1 positive root, 1 negative root;   **d.** roots must lie between -1 and 4

116. 2, 1, 4

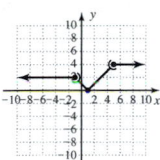

118. a. $f(x) \geq 0$: $x \in [-4, -1] \cup [1, \infty)$ **b.** max: $(-3, 4)$; min: $(0, -2)$ **c.** $f(x)\uparrow$: $x \in (-\infty, -3) \cup (0, \infty)$, $f(x)\downarrow$: $x \in (-3, 0)$

Exercises 4.3, pp. 423–428

23. a. even   **b.** -3 odd, -1 even, 3 odd   **c.** $f(x) = (x + 3)(x + 1)^2(x - 3)$, deg 4   **d.** $x \in \mathbb{R}, y \in [-9, \infty)$ **24. a.** odd   **b.** -3 odd, -1 even, 2 even   **c.** $f(x) = (x + 3)(x + 1)^2(x - 2)^2$, deg 5   **d.** $x \in \mathbb{R}, y \in \mathbb{R}$ **25. a.** even   **b.** -3 odd, -1 odd, 2 odd, 4 odd   **c.** $f(x) = -(x + 3)(x + 1)(x - 2)(x - 4)$, deg 4   **d.** $x \in \mathbb{R}, y \in (-\infty, 25]$ **26. a.** even   **b.** -3 odd, -2 even, 1 odd, 2 even   **c.** $f(x) = (x + 3)(x + 2)^2(x - 1)(x - 2)^2$, deg 6   **d.** $x \in \mathbb{R}, y \in [-52, \infty)$ **27. a.** odd   **b.** -1 even, 3 odd   **c.** $f(x) = -(x + 1)^2(x - 3)$, deg 3   **d.** $x \in \mathbb{R}, y \in \mathbb{R}$ **28. a.** odd   **b.** -3 odd, -1 even, 1 odd, 2 odd   **c.** $f(x) = -(x + 3)(x + 1)^2(x - 1)(x - 2)$, deg 5   **d.** $x \in \mathbb{R}, y \in \mathbb{R}$

43. 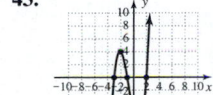   **44.** 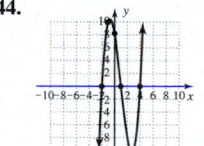   **45.**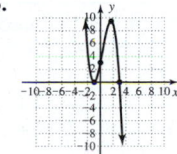

46. 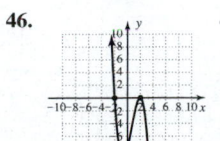   **47.** 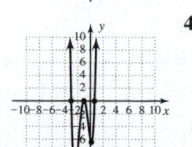   **48.**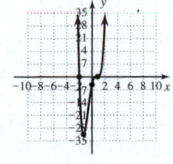

49. 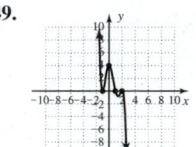   **50.** 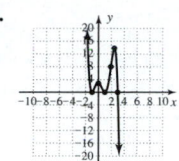   **51.**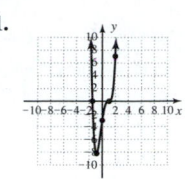

52. 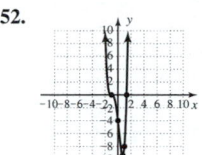   **53.** 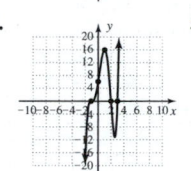   **54.**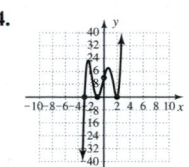

55. 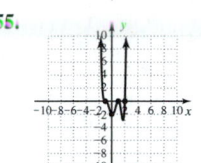   **56.** 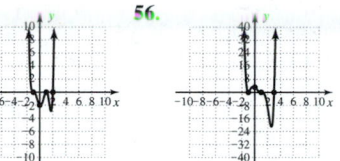   **57.**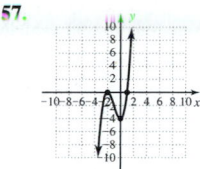

58. 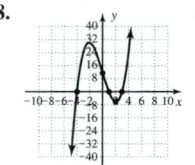   **59.** 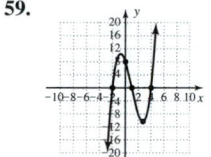   **60.**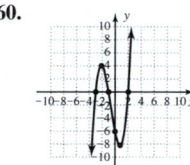

61. 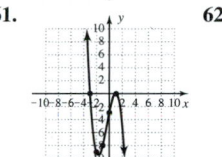   **62.** 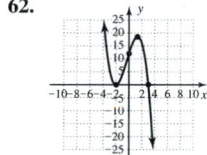   **63.**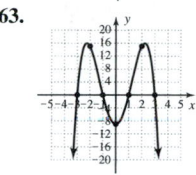

64. 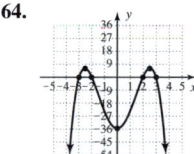   **65.** 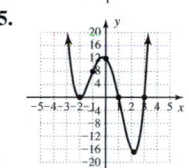   **66.**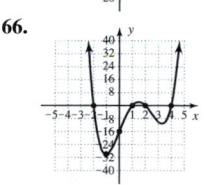

67. 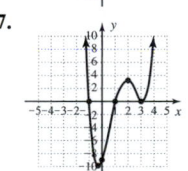   **68.** 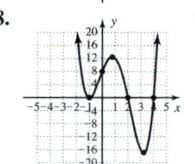   **69.**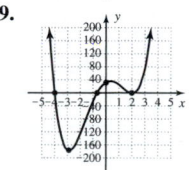

70. 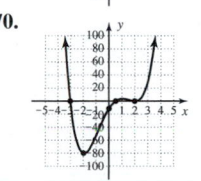   **71.** 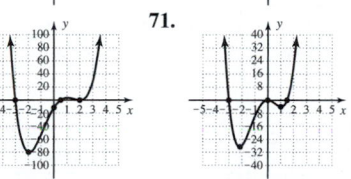   **72.**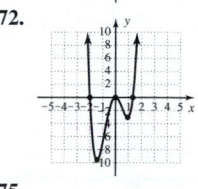

73. 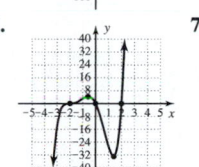   **74.** 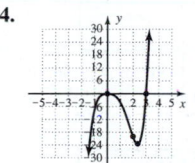   **75.**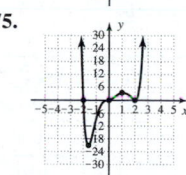

76.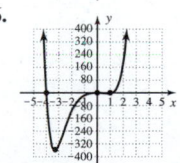

77. $h(x) = (x + 4)(x - \sqrt{3})(x + \sqrt{3})(x - i\sqrt{3})(x + i\sqrt{3})$
78. $H(x) = (x + 5)(x - \sqrt{2})(x + \sqrt{2})(x - i\sqrt{2})(x + i\sqrt{2})$

79. $f(x) = 2(x + \frac{5}{2})(x - \sqrt{2})(x + \sqrt{2})(x - \sqrt{3})(x + \sqrt{3})$

80. $g(x) = 3(x + \frac{2}{3})(x - \sqrt{2})(x + \sqrt{2})(x - \sqrt{6})(x + \sqrt{6})$

81. $P(x) = \frac{1}{6}(x + 4)(x - 1)(x - 3)$, $P(x) = \frac{1}{6}(x^3 - 13x + 12)$

82. $P(x) = \frac{1}{12}(x + 3)(x - 1)^2(x - 4)$,
$P(x) = \frac{1}{12}(x^4 - 3x^3 - 9x^2 + 23x - 12)$

83 a. $(-3)^2 + (-1)^2 + (2)^2 + (4)^2 = (-2)^2 - 2(-13)$
$$9 + 1 + 4 + 16 = 4 + 26$$
$$30 = 30 ✓$$
   b. $(x + 3)(x + 1)(x - 2)(x - 4) = x^4 - 2x^3 - 13x^2 + 14x + 24 ✓$

84. $P(x) = x^4 - 2x^3 + 2x^2 + 6x - 15$

85. a. 280 vehicles above average, 216 vehicles below average, 154 vehicles above average   **b.** 6:00 A.M. ($t = 0$), 10:00 A.M. ($t = 4$), 3:00 P.M. ($t = 9$), 6:00 P.M. ($t = 12$)

c. max: about 300 vehicles above average at 7:30 A.M.;
   min: about 220 vehicles below average at 12 noon

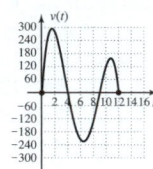

86. a. $p(3) = 432$ hundred thousand, $p(10) = 180$ hundred thousand
b. $m = 7$ (July), $m = 12$ (December)
c. max: about 50,000,000 insects in March;
   min: 0 (zero) insects in July

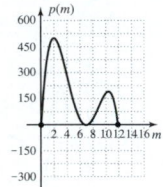

87. c. $B(x) = \frac{1}{4}x(x - 4)(x - 9)$, $-\$80,000$

88. c. $P(x) = \frac{3}{4}(x - 1)(x - 2)(x - 4)(x - 6)$, 4.5 in., -9 in.

89. a.

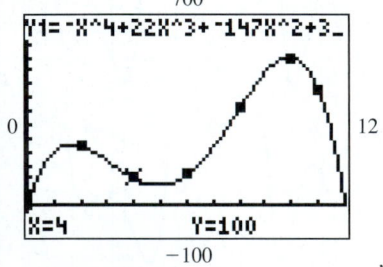

, quartic;

b. $t \approx 1.7$ (7:42 A.M.), 227 vehicles; $t \approx 9.9$ (3:54 P.M.), 551 vehicles
c. $t \approx 7.93$ (1:56 P.M.) and $t \approx 11.27$ (5:16 P.M.)

90. a.

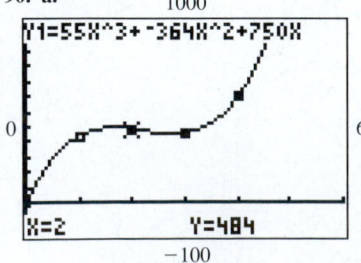

, cubic;

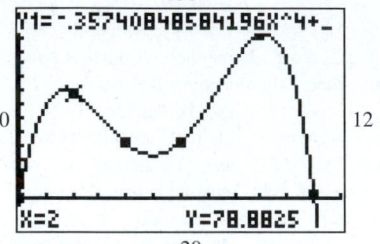

b. $t \approx 1.64$ sec, 493.6 ft/sec   **c.** $t \approx 2.77$ sec, 453.5 ft/sec   **d.** $t \approx 4.48$ sec

91. a.

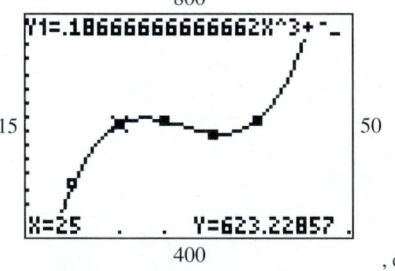

, quartic;

b. morning: $t \approx 1.72$ (11:43 A.M.); evening: $t \approx 9.11$, (7:07 P.M.)
c. $t \approx 4.98$ (2:59 P.M.), about 33 customers
d. $t \approx 7.92$ (5:55 P.M.) and $t \approx 10.02$ (8:01 P.M.)

92. a.

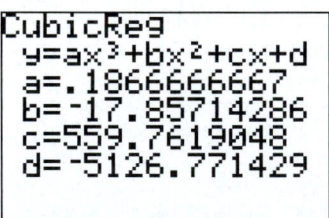

, cubic;

b. 644.4 W at 27.7 mph   **c.** 572.5 W   **d.** about 44.9 mph

94. verified; $P(x) = x^4 - 2x^3 - 13x^2 + 14x + 24$

98. $h(x) = \dfrac{1 - 2x}{x^2}$; $D : x \in \{x | x \neq 0\}$; $H(x) = \dfrac{1}{x^2 - 2x}$;
$D: x \in \{x | x \neq 0, x \neq 2\}$

Mid Chapter, p. 428

1. a. $x^3 + 8x^2 + 7x - 14 = (x^2 + 6x - 5)(x + 2) - 4$
   b. $\dfrac{x^3 + 8x^2 + 7x - 14}{x + 2} = x^2 + 6x - 5 - \dfrac{4}{x + 2}$
2. $f(x) = (2x + 3)(x + 1)(x - 1)(x - 2)$ **4.** $f(x) = x^3 - 2x + 4$
5. $g(2) = -8$ and $g(3) = 5$ have opposite signs
6. $f(x) = (x - 2)(x + 1)(x + 2)(x + 4)$

7. $x = -2, x = 1, x = -1 \pm 3i$

8. 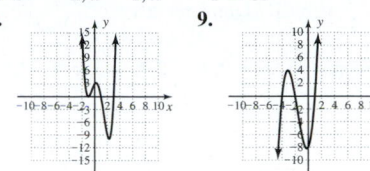 **9.**

10. a. degree 4; three turning points   **b.** 2 sec

c. $A(t) = (t - 1)^2(t - 3)(t - 5)$, $A(t) = t^4 - 10t^3 + 32t^2 - 38t + 15$

$A(2) = 3$; altitude is 300 ft above hard-deck, $A(4) = -9$; altitude is 900 ft below hard-deck

Exercises 4.4, pp. 440–445

7. $x = 3, x \in (-\infty, 3) \cup (3, \infty)$ **8.** $x = \frac{3}{2}, x \in (-\infty, \frac{3}{2}) \cup (\frac{3}{2}, \infty)$

9. $x = 3, x = -3, x \in (-\infty, -3) \cup (-3, 3) \cup (3, \infty)$

10. $x = \frac{2}{3}, x = \frac{-2}{3}, x \in (-\infty, -\frac{2}{3}) \cup (-\frac{2}{3}, \frac{2}{3}) \cup (\frac{2}{3}, \infty)$

11. $x = \frac{-5}{2}, x = 1, x \in (-\infty, -\frac{5}{2}) \cup (-\frac{5}{2}, 1) \cup (1, \infty)$

12. $x = \frac{3}{2}, x = -1, x \in (-\infty, -1) \cup (-1, \frac{3}{2}) \cup (\frac{3}{2}, \infty)$

13. No V.A., $x \in (-\infty, \infty)$ **14.** No V.A., $x \in (-\infty, \infty)$

15. $x = 3$, yes; $x = -2$, yes   **16.** $x = 5$, yes; $x = -4$, yes

17. $x = 3$, no   **18.** $x = 2$, no   **19.** $x = 2$, yes; $x = -2$, no

20. $x = -1$, no; $x = 1$, yes   **21.** $y = 0$, crosses at $(\frac{3}{2}, 0)$ **22.** $y = 0$,

crosses at $(-\frac{3}{4}, 0)$ **23.** $y = 4$, crosses at $(-\frac{21}{4}, 4)$ **24.** $y = 2$, crosses at

$(-20, 2)$ **25.** $y = 3$, does not cross   **26.** $y = 3$, does not cross

27. $q(x) = 0, r(x) = 8x$ directly; the graph will cross the horizontal asymptote at $x = 0$.

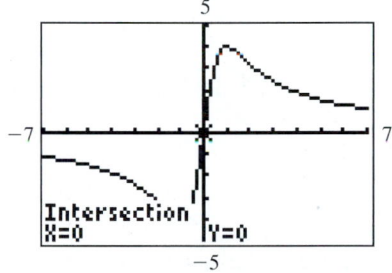

28. $q(x) = 0, r(x) = 4x + 8$ directly; the graph will cross the horizontal asymptote at $x = -2$.

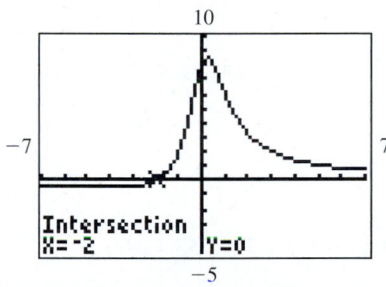

29. $q(x) = 2, r(x) = -8x - 8$; the graph will cross the horizontal asymptote at $x = 1$.

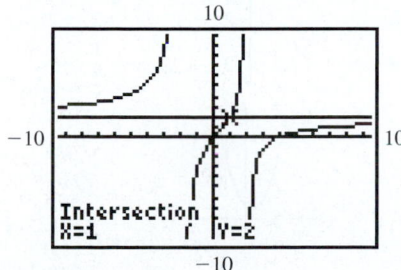

30. $q(x) = 1, r(x) = -x - 5$; the graph will cross the horizontal asymptote at $x = -5$.

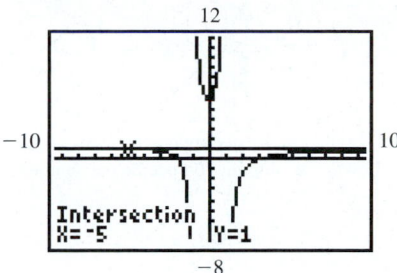

31. $(0, 0)$ cross, $(3, 0)$ cross   **32.** $(0, 0)$ cross, $(2, 0)$ cross   **33.** $(-4, 0)$ cross, $(0, 4)$ **34.** $(-6, 0)$ cross, $(-1, 0)$ cross, $(0, -3)$ **35.** $(0, 0)$ cross, $(3, 0)$ bounce   **36.** $(-2, 0)$ bounce, $(0, 0)$ cross

37. 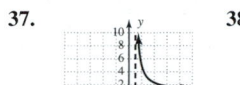   **38.** 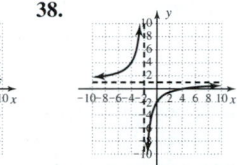   **39.**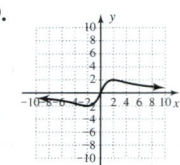

40. 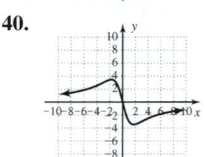   **41.** 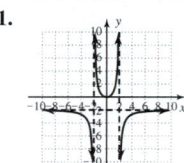   **42.**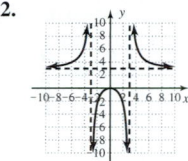

43. 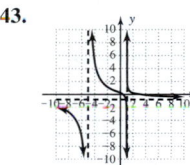   **44.** 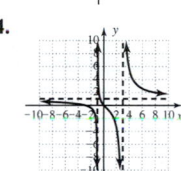   **45.**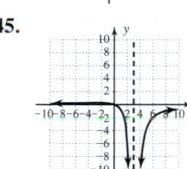

46. 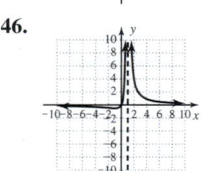   **47.** 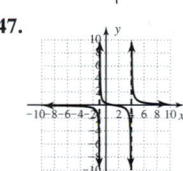   **48.**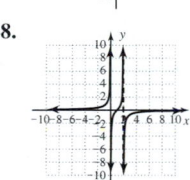

49. 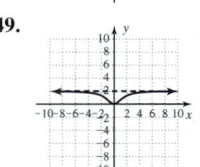   **50.** 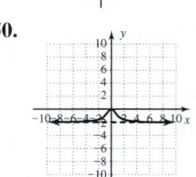   **51.**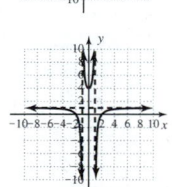

52. 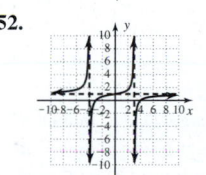   **53.** 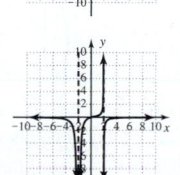   **54.**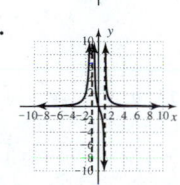

55. $f(x) = \dfrac{(x - 4)(x + 1)}{(x + 2)(x - 3)}$ **56.** $f(x) = \dfrac{5x}{(x + 3)^2(x - 3)}$

57. $f(x) = \dfrac{x^2 - 4}{9 - x^2}$ **58.** $f(x) = \dfrac{(x - 1)^2}{(x - 2)^2}$

59. 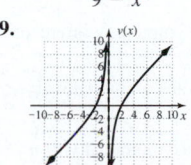   **60.**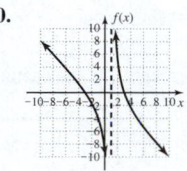

61. 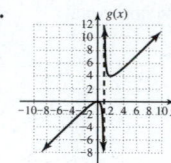   **62.**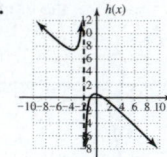

63. a. 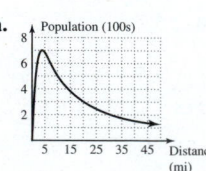   Population density approaches zero far from town.   **c.** 4.5 mi, 704 people per square mi

64. 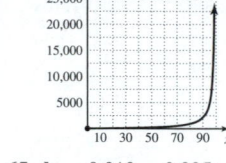   **a.** It is impossible to remove 100% of the pollutants.   **b.** $250 thousand, $277.8 thousand   **c.** 90%

65. b. −0.019, −0.005; As the number of hours increases, the rate of change decreases.   **66. a.** 12 days; about $102,000   **b.** 0.55, −0.06; rate of change is positive and large in first interval; negative and small in second   **c.** $t \to \infty, S \to 0^+$; horizontal asymptote
67. a. $20,000, $80,000, $320,000; cost increases dramatically
b. 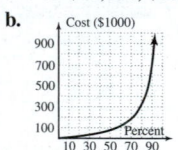   **c.** as $p \to 100^-, C \to \infty$

68. a. $733,333, $2,200,000, $6,600,000; cost increases dramatically
b. 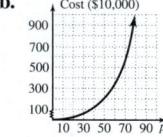   **c.** as $p \to 100^-, C \to \infty$

69. a. 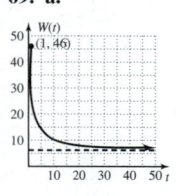   **70. a.** 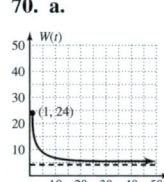   **71. a.**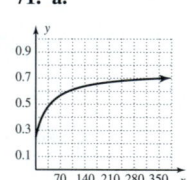

72. b. $C(6) = 0.12$ oz/gal;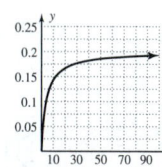
$C(28) = 0.175$ oz/gal;
320 gal

73. d. The horizontal asymptote at $y = 125$ means the average cost approaches $125 as monthly production gets very large. Due to limitations on production (maximum of 5000 heaters) the average cost will never fall below $A(5000) = 135$.   **74. d.** $y = 6$ means the average cost will never be less than $6; no; 20,000   **75. c.** The horizontal asymptote at $y = 95$ means her average grade will approach 95 as the number of tests taken increases; no   **76. c.** The horizontal asymptote at $y = 4$ means his GPA will approach 4 as the number of 4-point credit hours earned increases; no

78. a. horizontal asymptote at $y = \dfrac{a}{b}$; no vertical asymptote; y-intercept

at $\left(0, \dfrac{k}{h}\right)$; no x-intercepts   **b.** no change in asymptotes; y-intercept at

$\left(0, \dfrac{k}{h}\right)$; x-intercepts at $\left(\pm\sqrt{\dfrac{-k}{a}}, 0\right)$ **c.** horizontal asymptote at $y = a$

d. horizontal asymptote at $y = \dfrac{a}{b}$; vertical asymptotes at $x = \pm\sqrt{\dfrac{-h}{b}}$;

y-intercept at $\left(0, \dfrac{k}{h}\right)$; no x-intercepts   **e.** $a = 3, b = 2, h = 3, k = -12$

$$f(x) = \frac{3x^2 - 12}{2x^2 + 3}$$

79. $y = \dfrac{-4}{3}x - \dfrac{1}{3}$ **81.** $\frac{-16}{3}, \frac{3}{4}$; $(3x + 16)(4x - 3) = 0$
82. $\mathbb{C}: a + bi$, where $a, b \in \mathbb{R}, i = \sqrt{-1}$

$\mathbb{Q}: \left\{\dfrac{a}{b} \middle| a, b \in \mathbb{Z}; b \neq 0\right\}$

$\mathbb{Z}: \{\dots, -2, -1, 0, 1, 2, \dots\}$

Exercises 4.5, pp. 454–458

7. $F(x) = \begin{cases} \dfrac{x^2 - 4}{x + 2} & x \neq -2 \\ -4 & x = -2 \end{cases}$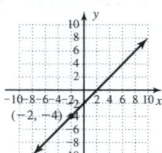

8. $F(x) = \begin{cases} \dfrac{x^2 - 9}{x + 3} & x \neq -3 \\ -6 & x = -3 \end{cases}$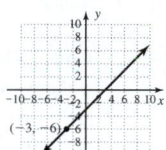

9. $G(x) = \begin{cases} \dfrac{x^2 - 2x - 3}{x + 1} & x \neq -1 \\ -4 & x = -1 \end{cases}$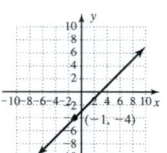

10. $G(x) = \begin{cases} \dfrac{x^2 - 3x - 10}{x - 5} & x \neq 5 \\ 7 & x = 5 \end{cases}$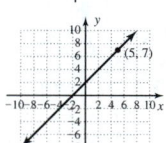

11. $H(x) = \begin{cases} \dfrac{3x - 2x^2}{2x - 3} & x \neq \dfrac{3}{2} \\ \dfrac{-3}{2} & x = \dfrac{3}{2} \end{cases}$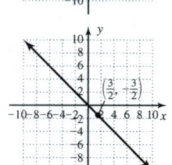

12. $H(x) = \begin{cases} \dfrac{4x - 5x^2}{5x - 4} & x \neq \dfrac{4}{5} \\ \dfrac{-4}{5} & x = \dfrac{4}{5} \end{cases}$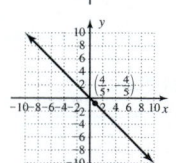

13. $P(x) = \begin{cases} \dfrac{x^3 - 8}{x - 2} & x \neq 2 \\ 12 & x = 2 \end{cases}$ **14.** $P(x) = \begin{cases} \dfrac{8x^3 - 1}{2x - 1} & x \neq \dfrac{1}{2} \\ 3 & x = \dfrac{1}{2} \end{cases}$

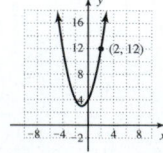   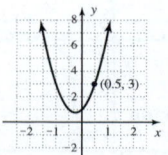

15. $Q(x) = \begin{cases} \dfrac{x^3 - 7x - 6}{x + 1} & x \neq -1 \\ -4 & x = -1 \end{cases}$

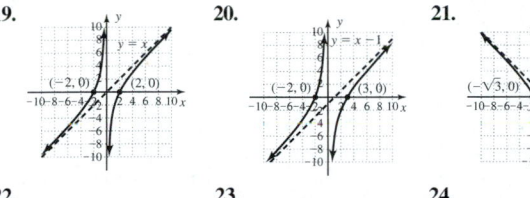

16. $Q(x) = \begin{cases} \dfrac{x^3 - 3x + 2}{x + 2} & x \neq -2 \\ 9 & x = -2 \end{cases}$

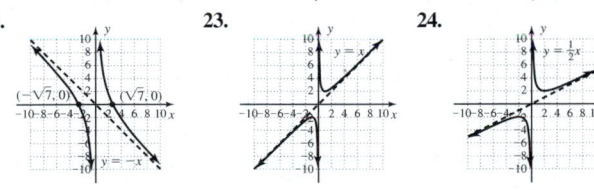

17. $R(x) = \begin{cases} \dfrac{x^3 + 3x^2 - x - 3}{x^2 + 2x - 3} & x \neq -3, x \neq 1 \\ -2 & x = -3 \\ 2 & x = 1 \end{cases}$

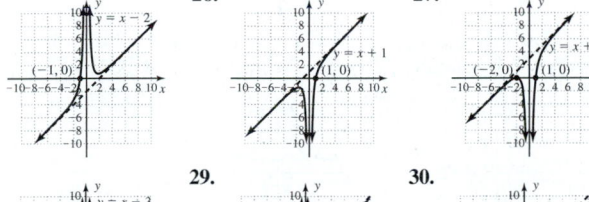

18. $R(x) = \begin{cases} \dfrac{x^3 + 2x^2 - 4x + 8}{x^2 - 4} & x \neq -2, x \neq 2 \\ -4 & x = -2 \\ 0 & x = 2 \end{cases}$

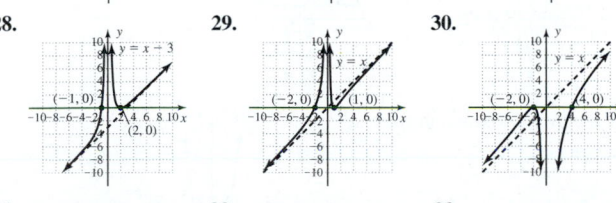

19. **20.** **21.**

22. **23.** **24.**

25. **26.** **27.**

28. **29.** **30.**

31. **32.** **33.**

34. **35.** **36.**

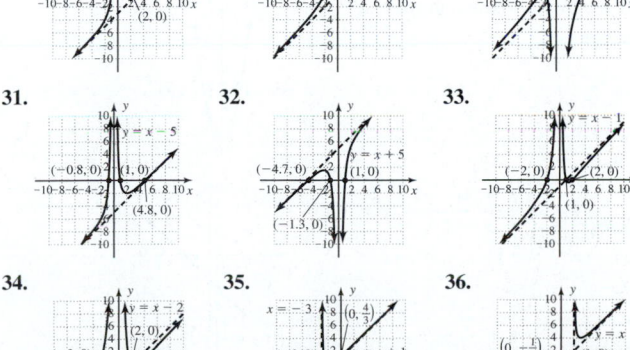

37. **38.** **39.**

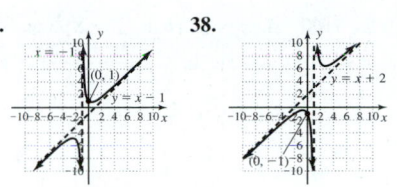

40. **41.** **42.**

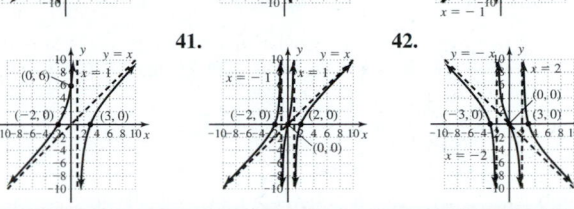

43. **44.** **45.**

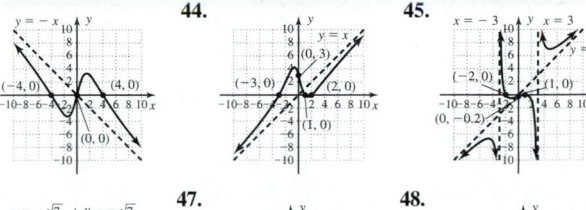

46. **47.** **48.**

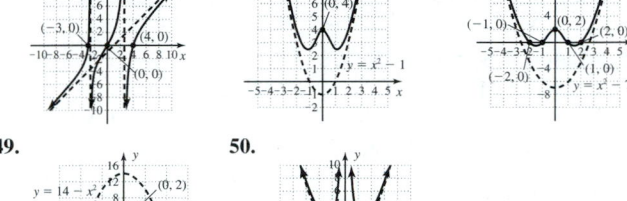

49. **50.**

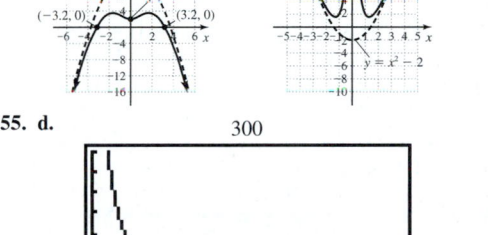

55. d.

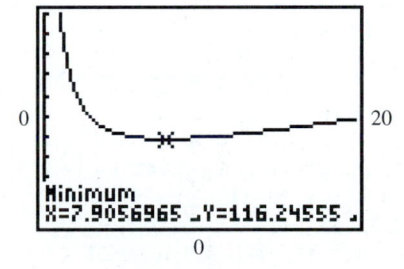

56. a. $A(x) = \dfrac{5x^2 + 94x + 576}{x}$; $x = 0$, $y = 5x + 94$

b. cost: \$675, \$784, \$903; avg. cost: \$675, \$392, \$301

c. 11; \$201.36; no   **d.**

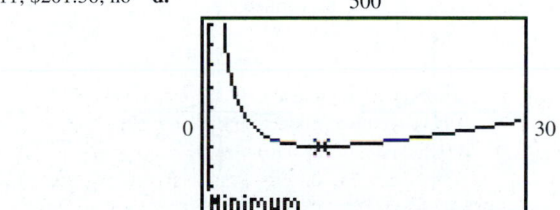

58. a. $S(x, y) = 2x(x + 2) + 2y(x + 2) + 2xy$; $V(x, y) = (x^2 + 2x)y$

b. $y = \dfrac{36}{x^2 + 2x}$; $S(x) = \dfrac{2x^4 + 8x^3 + 8x^2 + 144x + 144}{x(x + 2)}$

c. $S(x)$ is asymptotic to $y = 2x^2 + 4x$.   **d.** $x \approx 2$ ft 7 in.; $y \approx 3$ ft $\frac{1}{2}$ in.; 2 ft 7 in. $\times$ 4 ft 7 in. $\times$ 3 ft $\frac{1}{2}$ in.   **59. a.** $A(x, y) = xy$;

$R(x, y) = (x - 2.5)(y - 2)$ **b.** $y = \dfrac{2x + 55}{x - 2.5}$; $A(x) = \dfrac{2x^2 + 55x}{x - 2.5}$

c. $A(x)$ is asymptotic to $y = 2x + 60$ **d.** $x \approx 11.16$ in.; $y \approx 8.93$ in.

60. a. $A(x, y) = xy$; $R(x, y) = (x - 5)(y - 5)$ **b.** $y = \dfrac{5x + 475}{x - 5}$;

$A(x) = \dfrac{5x^2 + 475x}{x - 5}$ **c.** $A(x)$ is asymptotic to $y = 5x + 500$.

d. $x \approx 27.36$ in.; $y \approx 27.36$ in.   **61. a.** $h = \dfrac{V}{\pi r^2}$ **b.** $S = 2\pi r^2 + \dfrac{2V}{r}$

c. $S = \dfrac{2\pi r^3 + 2V}{r}$ **d.** $r \approx 5.76$ cm, $h \approx 11.51$ cm; $S \approx 625.13$ cm^2

62. a. $r = \dfrac{3V + \pi h^3}{3\pi h^2}$ **b.** $S = \dfrac{2(3V + \pi h^3)}{3h}$

c. $S(h) = \dfrac{2(1500 + \pi h^3)}{3h}$, $h \approx 6.20$ cm   **d.** $r \approx 6.20$ cm

64. a. $m = \dfrac{k}{h - a}$; $y = \dfrac{k}{h - a}(x - a)$ **b.** $\left(0, \dfrac{ka}{a - h}\right)$, $(a, 0)$

c. $A = \dfrac{1}{2}\left(\dfrac{ka^2}{a - h}\right)$ **d.** $A = \dfrac{1}{2}\left[\dfrac{4a^2}{a - 4}\right]$ has a minimum at $(8, 32)$,
32 units^2; the triangle is isosceles   **67.** $y = \frac{3}{4}x - 4$, $m = \frac{3}{4}$, $(0, -4)$
68. a. $b^2 - 4ac > 0$, with $b^2 - 4ac$ a perfect square   **b.** $b^2 - 4ac > 0$,
but not a perfect square   **c.** $b^2 - 4ac = 0$ **d.** not possible   **e.** not possible
f. $b^2 - 4ac < 0$ **69. a.** $P = 30$ cm,   **b.** $\overline{CD} = \frac{60}{13}$ cm,   **c.** 30 cm^2,
d. $A = \frac{750}{169}$ cm^2, and $A = \frac{4320}{169}$ cm^2

Exercises 4.6, pp. 465–469

7. $x \in (-3, 5)$ **8.** $x \in (-7, 2)$ **9.** $x \in [4, \infty) \cup \{-1\}$
10. $x \in (-\infty, -6] \cup \{1\}$ **11.** $x \in (-\infty, -2] \cup \{2\} \cup [4, \infty)$
12. $x \in [1, 3] \cup \{-2\}$ **13.** $x \in (-2 - \sqrt{3}, -2 + \sqrt{3})$
14. $x \in (-\infty, 3 - \sqrt{5}) \cup (3 + \sqrt{5}, \infty)$ **17.** $x \in (-3, 1) \cup (2, \infty)$
18. $x \in (-4, 1) \cup (3, \infty)$ **19.** $x \in (-\infty, -3) \cup (-1, 1) \cup (3, \infty)$
20. $x \in (-3, -2) \cup (2, 3)$ **21.** $x \in (-\infty, -2) \cup (-2, 1) \cup (3, \infty)$
22. $x \in (-\infty, -2) \cup (1, 2) \cup (4, \infty)$ **26.** $x \in (-\infty, -1) \cup (2, 4)$
27. $x \in (-\infty, -2] \cup (-1, 1) \cup [3, \infty)$ **28.** $x \in (-\infty, -3) \cup (3, 7)$
33. $x \in (-\infty, -2) \cup [2, 3)$ **34.** $x \in (-2, 1] \cup (4, \infty)$
35. $x \in (-\infty, -5) \cup (0, 1) \cup (2, \infty)$
36. $x \in (-\infty, -3) \cup (-1, 0) \cup (3, \infty)$
37. $x \in (-4, -2] \cup (1, 2] \cup (3, \infty)$ **38.** $x \in (-\infty, -3) \cup (-3, 1)$
41. $x \in (-\infty, -2] \cup (0, 2)$ **42.** $x \in (-3, 0) \cup [\frac{9}{2}, \infty)$
43. $x \in (-\infty, -17) \cup (-2, 1) \cup (7, \infty)$
44. $x \in (-\infty, -5) \cup (-3, 1) \cup (7, \infty)$ **45.** $x \in (-3, \frac{-7}{4}] \cup (2, \infty)$
46. $x \in (-\infty, -4) \cup [1, 6)$ **47.** $x \in (-2, \infty)$ **48.** $x \in (-\infty, 3)$
49. $x \in (-1, \infty)$ **50.** $x \in (-\infty, 2)$ **51.** $(-\infty, -3) \cup (3, \infty)$
52. $x \in (-4, -2) \cup (2, 5)$ **53.** $x \in (-\infty, -3] \cup [5, \infty)$
54. $x \in (-\infty, -6] \cup [3, \infty)$ **55.** $x \in [-3, 0] \cup [3, \infty)$
56. $x \in (-\infty, -2] \cup [0, 2]$ **57.** d   **58.** d   **59.** b   **60.** b
61. a. verified
b. $D = -4(p + \frac{3}{4})(p + 3)^2$; $p = -3$, $q = -2$; $p = \frac{-3}{4}$, $q = \frac{1}{4}$

c. $(-\infty, -3) \cup (-3, \frac{-3}{4})$ **d.** verified

62. a.

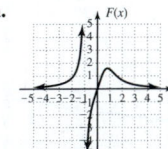 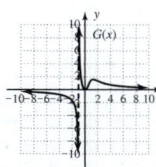

c. $(-\infty, -1) \cup (0, \infty)$
65. b. horizontal: $r_2 = 20$, as r_1 increases, r_2 decreases to maintain $R = 40$;
vertical: $r_1 = 20$, as r_1 decreases, r_2 increases to maintain $R = 40$
66. b. horizontal: $r_2 = 30$, as r_1 increases, r_2 decreases to maintain
$R = 60$; vertical: $r_1 = 30$, as r_1 decreases, r_2 increases to maintain $R = 60$

71. a. yes, $x^2 \geq 0$ **b.** yes, $\dfrac{x^2}{x^2 + 1} \geq 0$

72. a. $x \in [-3, 1] \cup [\frac{5}{2}, \infty)$ **b.** $x \in [-4, \frac{3}{2}] \cup [2, \infty)$
73. $x(x + 2)(x - 1)^2 > 0$; $\dfrac{x(x + 2)}{(x - 1)^2} > 0$
74. $F(x) > 0$ for $x \in (-2, 1) \cup (4, \infty)$
77.

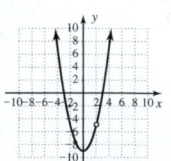

Summary and Concept Review, pp. 470–474

3. $\underline{-7|}$ 2 13 -6 9 14

		-14	7	-7	-14
	2	-1	1	2	$\underline{|0}$

Since $r = 0$, -7 is a root and $x + 7$ is a factor.

7. $\frac{1}{2}|$ 4 8 -3 -1

		2	5	1
	4	10	2	$\underline{|0}$

Since $r = 0$, $\frac{1}{2}$ is a root and $(x - \frac{1}{2})$ is a factor.

8. $3i|$ 1 -2 9 -18

			$3i$	$-9 - 6i$	18
	1	$-2 + 3i$	$-6i$		$\underline{|0}$

Since $r = 0$, $3i$ is a zero

9. $\underline{-7|}$ 1 9 13 -10

		-7	-14	7
	1	2	-1	$\underline{|-3}$

$h(-7) = -3$

13. $\{\pm 1, \pm 10, \pm 2, \pm 5, \pm\frac{1}{2}, \pm\frac{5}{2}, \pm\frac{1}{4}, \pm\frac{5}{4}\}$ **14.** $x = -\frac{1}{2}, 2, \frac{5}{2}$
15. $P(x) = (2x + 3)(x - 4)(x + 1)$ **16.** only possibilities are $\pm 1, \pm 3$,
none give a remainder of zero   **17.** $[1, 2], [4, 5]$; verified   **18.** one sign
change for $g(x) \rightarrow 1$ positive zero; three sign changes for $g(-x) \rightarrow 3$ or 1
negative zeroes; 1 positive, 3 negative, 0 complex, or 1 positive, 1 negative,
2 complex; 1 positive, 1 negative, 2 complex, verified

21.   **22.**   **23.**

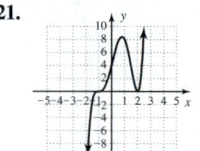 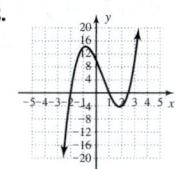 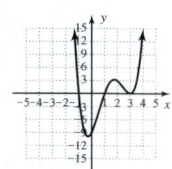

25. a. $\{x | x \in \mathbb{R}; x \neq -1, 4\}$ **b.** HA: $y = 1$; VA: $x = -1, x = 4$
c. $V(0) = \frac{9}{4}$ (y-intercept); $x = -3, 3$ (x-intercepts)   **d.** $V(1) = \frac{4}{3}$
27.   **28.**

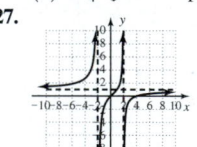 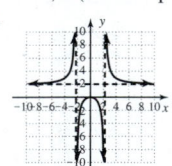

31. removable discontinuity at $(2, -5)$;

32. $H(x) = \begin{cases} \dfrac{x^2 - 3x - 4}{x + 1} & x \neq -1 \\ -5 & x = -1 \end{cases}$

33. 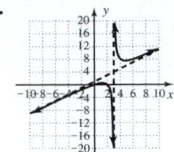   **34.**    **35. a.**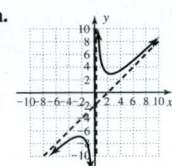

36. factored form $(x + 4)(x - 1)(x - 2) > 0$

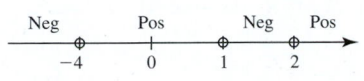

outputs are positive for $x \in (-4, 1) \cup (2, \infty)$

37. $\dfrac{x^2 - 3x - 10}{x - 2} = \dfrac{(x - 5)(x + 2)}{x - 2} \geq 0$

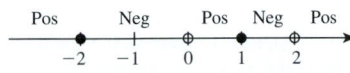

outputs are positive or zero for $x \in [-2, 2) \cup [5, \infty)$

38. $\dfrac{(x + 2)(x - 1)}{x(x - 2)} \leq 0$

outputs are negative or zero for $x \in [-2, 0) \cup [1, 2)$

Practice Test, pp. 474–475

1. a. $f(x) = -(x - 5)^2 + 9$ **b.** $g(x) = \frac{1}{2}(x + 4)^2 + 8$

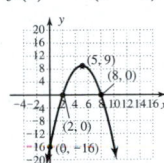   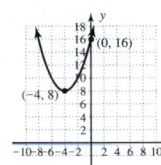

6. $\underline{-3}|$ 1   0 -15 -10 24

$\qquad\qquad -3$ 9   18 -24

$\overline{\qquad 1 \quad -3 \quad -6 \quad 8 \quad |\underline{0}}$ $r = 0$ ✓

10. a. $\pm 1, \pm 18, \pm 2, \pm 9, \pm 3, \pm 6$ **b.** 1 positive zero, 3 or 1 negative zeroes; 2 or 0 complex zeroes   **c.** $C(x) = (x + 2)(x - 1)(x - 3i)(x + 3i)$
11. a. 2002, 2004, 2008   **b.** 4 yr   **c.** deficit of $7.5 million
12.   **13.**   **14.**

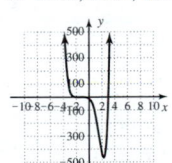   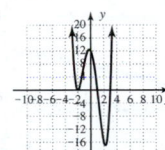   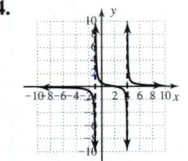

15. a. removal of 100% of the contaminants   **b.** $500,000; $3,000,000; dramatic increase   **c.** 88%
16. a.   **b.**

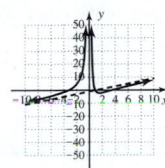   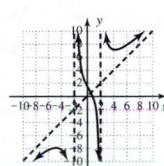

18. a. $x \in (-\infty, -3] \cup [-1, 4]$ **b.** $x \in (-\infty, -4) \cup (0, 2)$
19. a.

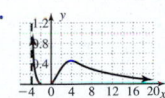

b. $h = -\sqrt[3]{55}$; no   **c.** 28.6%, 29.6%   **d.** ≈ 11.7 hr   **e.** 4 hr, 43.7%
f. The amount of the chemical in the bloodstream becomes neglible.

Cumulative Review Chapters R–4 pp. 477–478

3. a. $(x - 1)(x^2 + x + 1)$ **b.** $(x - 3)(x + 2)(x - 2)$
4. $x = \dfrac{-2 \pm \sqrt{2}}{2}$, $x \approx -0.29, -1.71$

6.

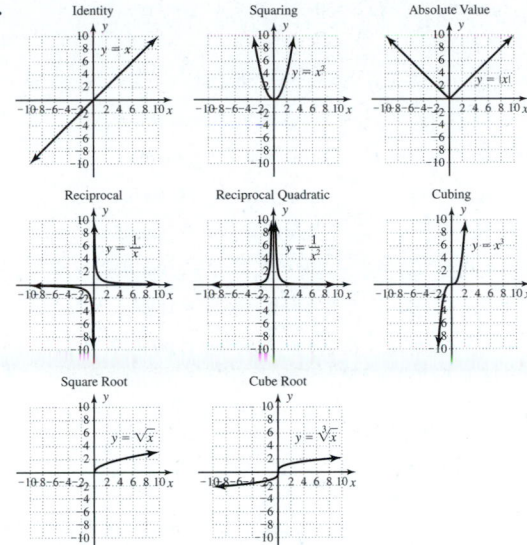

9. $y = \dfrac{11}{60}x + \dfrac{1009}{60}$; 39 min, driving time increases 11 min every 60 days

12.   **13.** $f^{-1}(x) = \dfrac{x^3 + 3}{2}$

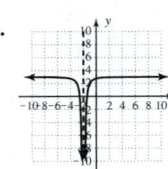

14. a. $x \in (-\infty, \infty)$
b. $x \in (2, \infty)$ **15.**

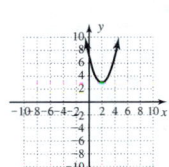   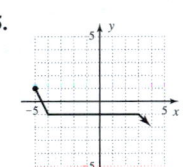

16.   **19.**   **20.** $x \in (-2, 1] \cup (2, \infty)$

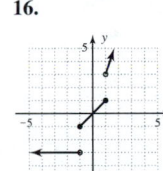   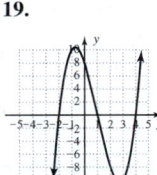   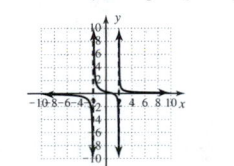

CHAPTER 5
Exercises 5.1, pp. 488–491

7. one-to-one   **8.** not one-to-one, fails horizontal line test, $x = 0$ and $x = -4$ are each paired with $y = 1$ **9.** one-to-one   **10.** one-to-one
11. not one-to-one, fails horizontal line test: $x = -3, x = -0.5$ and $x = 2$ are paired with $y = 0$ **12.** one-to-one   **13.** not a function   **14.** not one-to-one, fails horizontal line test: $x = -5, x = -1, x = 1$, and $x = 5$ are all paired with $y = 1$ **15.** one-to-one   **16.** not one-to-one, $y = 7$ is paired with $x = -2$ and $x = 2$ **17.** not one-to-one, $y = 1$ is paired with $x = -6$ and $x = 8$ **18.** one-to-one   **19.** one-to-one   **20.** one-to-one
21. not one-to-one; $h(x) < 3$, corresponds to two x-values
22. not one-to-one; $p(t) > 5$, corresponds to two t-values
23. one-to-one   **24.** one-to-one   **25.** not one-to-one; $y = 3$ corresponds to more than one x-value   **26.** one-to-one
27. $f^{-1}(x) = \{(1, -2), (4, -1), (5, 0), (9, 2), (15, 5)\}$
28. $g^{-1}(x) = \{(30, -2), (11, -1), (4, 0), (3, 1), (2, 2)\}$
29. $v^{-1}(x) = \{(3, -4), (2, -3), (1, 0), (0, 5), (-1, 12), (-2, 21), (-3, 32)\}$
30. $w^{-1}(x) = \{(4, -6), (2.5, -5), (-2, -2), (-5, 0), (-9.5, 3), (-11, 4), (-15.5, 7)\}$ **31.** $f^{-1}(x) = x - 5$ **32.** $g^{-1}(x) = x + 4$

33. $p^{-1}(x) = \dfrac{-5}{4}x$ **34.** $r^{-1}(x) = \dfrac{4}{3}x$ **35.** $f^{-1}(x) = \dfrac{x-3}{4}$

36. $g^{-1}(x) = \dfrac{x+2}{5}$ **37.** $t(x) = x^3 + 4$ **38.** $s(x) = x^3 - 2$

39. $x \in \mathbb{R}, y \in \mathbb{R}; f^{-1}(x) = x^3 + 2, x \in \mathbb{R}, y \in \mathbb{R}$; verified

40. $x \in \mathbb{R}, y \in \mathbb{R}; f^{-1}(x) = x^3 - 3, x \in \mathbb{R}, y \in \mathbb{R}$; verified

41. $x \in \mathbb{R}, y \in \mathbb{R}; f^{-1}(x) = \sqrt[3]{x - 1}, x \in \mathbb{R}, y \in \mathbb{R}$; verified

42. $x \in \mathbb{R}, y \in \mathbb{R}; f^{-1}(x) = \sqrt[3]{x + 2}, x \in \mathbb{R}, y \in \mathbb{R}$; verified

43. $x \neq -2, y \neq 0; f^{-1}(x) = \dfrac{8}{x} - 2, x \neq 0, y \neq -2$; verified

44. $x \neq 1, y \neq 0; f^{-1}(x) = \dfrac{12}{x} + 1, x \neq 0, y \neq 1$; verified

45. $x \neq -1, y \neq 1; f^{-1}(x) = \dfrac{x}{1-x}, x \neq 1, y \neq -1$; verified

46. $x \neq 1, y \neq -1; f^{-1}(x) = \dfrac{x-2}{x+1}, x \neq -1, y \neq 1$; verified

47. a. $x \geq -5, y \geq 0$ **b.** $f^{-1}(x) = \sqrt{x} - 5, x \geq 0, y \geq -5$

48. a. $x \geq 0, y \geq 3$ **b.** $g^{-1}(x) = \sqrt{x} - 3, x \geq 3, y \geq 0$

49. a. $x > 3, y > 0$ **b.** $v^{-1}(x) = \sqrt{\dfrac{8}{x} + 3}, x > 0, y > 3$

50. a. $x > 0, y > 2$ **b.** $V^{-1}(x) = \sqrt{\dfrac{4}{x-2}}, x > 2, y > 0$

51. a. $x \geq -4, y \geq -2$ **b.** $p^{-1}(x) = \sqrt{x + 2} - 4, x \geq -2, y \geq -4$

52. a. $x > 2, y > 1$ **b.** $q^{-1}(x) = \sqrt{\dfrac{4}{x-1}} + 2, x > 1, y > 2$

53. $(f \circ g)(x) = x, (g \circ f)(x) = x$ **54.** $(f \circ g)(x) = x, (g \circ f)(x) = x$

55. $(f \circ g)(x) = x, (g \circ f)(x) = x$ **56.** $(f \circ g)(x) = x, (g \circ f)(x) = x$

57. $(f \circ g)(x) = x, (g \circ f)(x) = x$ **58.** $(f \circ g)(x) = x, (g \circ f)(x) = x$

59. $(f \circ g)(x) = x, (g \circ f)(x) = x$ **60.** $(f \circ g)(x) = x, (g \circ f)(x) = x$

61. $f^{-1}(x) = \dfrac{x+5}{3}$ **62.** $f^{-1}(x) = \dfrac{x-4}{5}$ **63.** $f^{-1}(x) = 2x + 5$

64. $f^{-1}(x) = 3x - 4$ **65.** $f^{-1}(x) = 2x + 6$ **66.** $f^{-1}(x) = \frac{3}{2}(x - 1)$

67. $f^{-1}(x) = \sqrt[3]{x - 3}$ **68.** $f^{-1}(x) = \sqrt[3]{x + 4}$ **69.** $f^{-1}(x) = \dfrac{x^3 - 1}{2}$

70. $f^{-1}(x) = \dfrac{x^3 + 2}{3}$ **71.** $f^{-1}(x) = 2\sqrt[3]{x} + 1$

72. $f^{-1}(x) = -3\sqrt[3]{x} - 3$

73. $D: x \geq -\dfrac{2}{3}, R: y \geq 0; f^{-1}(x) = \dfrac{x^2 - 2}{3}, D: x \geq 0, R: y \geq -\dfrac{2}{3}$

74. $D: x \geq \dfrac{5}{2}, R: y \geq 0; g^{-1}(x) = \dfrac{x^2 + 5}{2}, D: x \geq 0, R: y \geq \dfrac{5}{2}$

75. $D: x \geq 3, R: y \geq 0; p^{-1}(x) = \dfrac{x^2}{4} + 3, D: x \geq 0, R: y \geq 3$

76. $D: x \geq -1, R: y \geq 0; q^{-1}(x) = \dfrac{x^2}{16} - 1, D: x \geq 0, R: y \geq -1$

77. $D: x \geq 0, R: y \geq 3; v^{-1}(x) = \sqrt{x - 3}, D: x \geq 3, R: y \geq 0$

78. $D: x \geq 0, R: y \geq -1; w^{-1}(x) = \sqrt{x + 1}, D: x \geq -1, R: y \geq 0$

79.

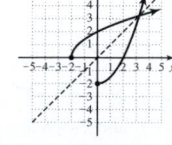

$D: x \in [0, \infty), R: y \in [-2, \infty);$
$D: x \in [-2, \infty), R: y \in [0, \infty)$

80.

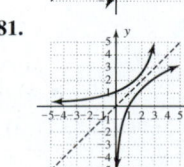

$D: x \in (-\infty, \infty), R: y \in (-\infty, \infty);$
$D: x \in (-\infty, \infty), R: y \in (-\infty, \infty)$

81.

$D: x \in (0, \infty), R: y \in (-\infty, \infty);$
$D: x \in (-\infty, \infty), R: y \in (0, \infty)$

82.

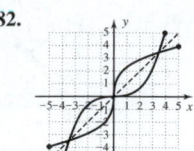

$D: x \in [-4, 4], R: y \in [-5, 5];$
$D: x \in [-5, 5], R: y \in [-4, 4]$

83.

$D: x \in (-\infty, 4], R: y \in (-\infty, 4];$
$D: x \in (-\infty, 4], R: y \in (-\infty, 4]$

84.

$D: x \in (-\infty, \infty), R: y \in (-\infty, \infty);$
$D: x \in (-\infty, \infty), R: y \in (-\infty, \infty)$

85. a. $f^{-1}(x) = \dfrac{x-1}{2}$ **b.** $(-3, -5), (0, 1),$ and $(1, 3)$ are on the graph of $f; (-5, -3), (1, 0),$ and $(3, 1)$ are on the graph of f^{-1} **c.** verified
d.

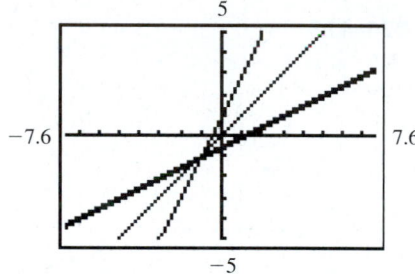

86. a. $g^{-1}(x) = \sqrt{x - 1}, x \geq 1$ **b.** $(0, 1), (2, 5),$ and $(3, 10)$ are on the graph of $g; (1, 0), (5, 2),$ and $(10, 3)$ are on the graph of g^{-1} **c.** verified
d.

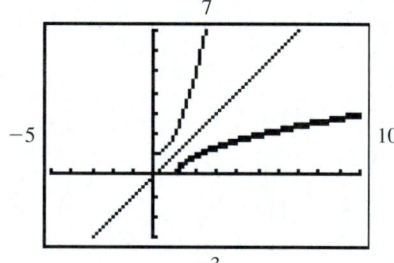

87. a. $h^{-1}(x) = \dfrac{x}{1-x}$ **b.** $(0, 0), (1, \frac{1}{2}),$ and $(2, \frac{2}{3})$ are on the graph of $h; (0, 0), (\frac{1}{2}, 1),$ and $(\frac{2}{3}, 2)$ are on the graph of h^{-1} **c.** verified
d.

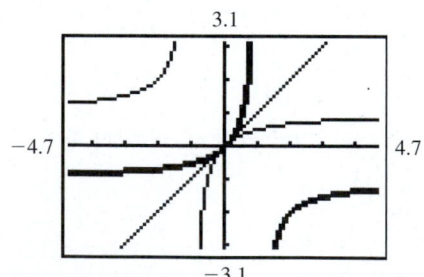

88. a. $j^{-1}(x) = (\frac{x+6}{2})^2 - 9, x \geq 6$ **b.** $(-8, -4), (-5, -2),$ and $(7, 2)$ are on the graph of $j; (-4, -8), (-2, -5),$ and $(2, 7)$ are on the graph of j^{-1} **c.** verified

d.

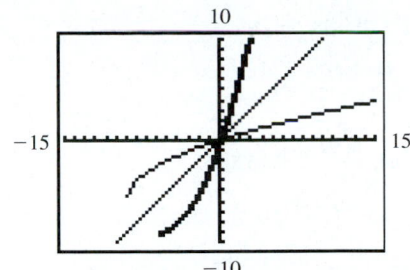

89. a. 31.5 cm   **b.** The result is 80 cm. It gives the distance of the projector from the screen.   **90. a.** 15 in.   **b.** The result is 14,130 in³. It gives the volume of the sphere.   **91. a.** −63.5°F   **b.** $f^{-1}(x) = \frac{-2}{7}(x - 59)$; independent: temperature, dependent: altitude   **c.** 22,000 ft

92. a. $220   **b.** $f^{-1}(x) = \frac{x + 560}{12}$; independent: fine, dependent: speed

c. 61 mph   **93. a.** 144 ft   **b.** $f^{-1}(x) = \frac{\sqrt{x}}{4}$, independent: distance fallen,

dependent: time fallen   **c.** 7 sec   **94. a.** 314 ft²

b. $f^{-1}(x) = \sqrt{\dfrac{x}{\pi}}$, independent: area, dependent: radius   **c.** 20 ft

95. a. 28,260 ft³   **b.** $f^{-1}(x) = \sqrt[3]{\dfrac{3x}{\pi}}$, independent: volume, dependent:

height   **c.** 9 ft   **96. a.** 10.8 hp   **b.** $f^{-1}(x) = \sqrt[3]{2500x}$, independent: horsepower, dependent: wind speed   **c.** 40 mph
98. d   **99. a.** $P = 2l + 2w$ **b.** $A = \pi r^2$ **c.** $V = \pi r^2 h$ **d.** $V = \frac{1}{3}\pi r^2 h$
e. $C = 2\pi r$ **f.** $A = \frac{1}{2}bh$ **g.** $A = \frac{1}{2}(b_1 + b_2)h$ **h.** $V = \frac{4}{3}\pi r^3$
i. $a^2 + b^2 = c^2$ **100.** $x \in [-1, 2]$ **101.** ≈0.472, ≈0.365; rate of change is greater in [1, 2] due to shape of the graph.
102. a. $x = 0, x = \pm\sqrt{5}$ **b.** $x = 2, x = -2, x = 7$
c. $x = 0, x = 3$ **d.** $x = 0, x = 4, x = -1$

Exercises 5.2, pp. 499–503

11.

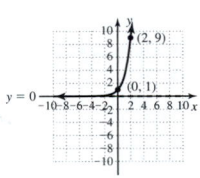

increasing

12.

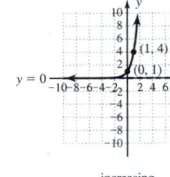

increasing

13.

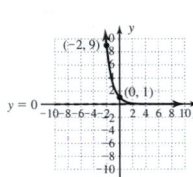

decreasing

14.

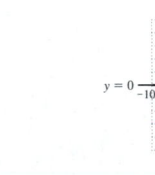

decreasing

15. $y = 3^x$; up 2

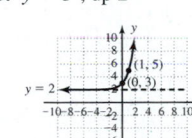

16. $y = 3^x$; down 3

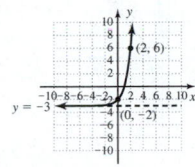

17. $y = 3^x$; left 3

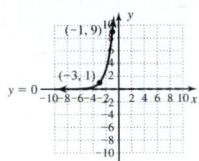

18. $y = 3^x$; right 2

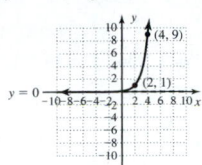

19. $y = 3^x$; reflect across y-axis

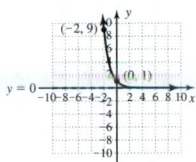

20. $y = 3^x$; reflect across y-axis, down 2

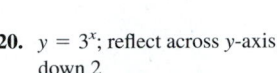

21. $y = (\frac{1}{3})^x$; up 1

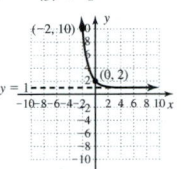

22. $y = (\frac{1}{3})^x$; down 4

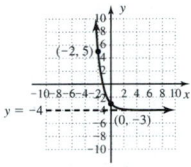

23. $y = (\frac{1}{3})^x$; right 2

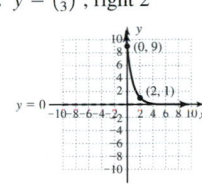

24. $y = (\frac{1}{3})^x$; left 2

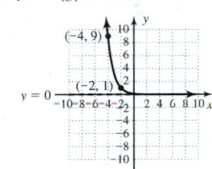

37.

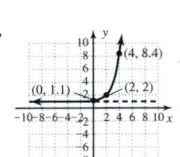

38.

39.

40.

41.

42.

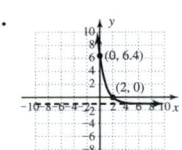

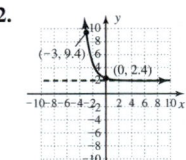

67. a. 1732, 3000, 5196, 9000   **b.** yes   **c.** as $t \to \infty, P \to \infty$
d.

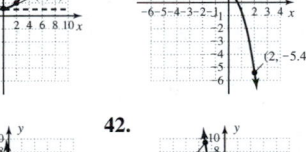

68. a. $\frac{1}{4}$ **b.** $\frac{1}{256}$ **c.** as $x \to \infty, P \to 0$
91. 5; $\dfrac{-7}{9}$; $2a^2 - 3a$; $2a^2 + 4ah + 2h^2 - 3a - 3h$

92.

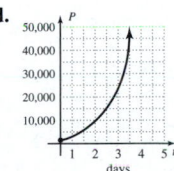

Exercises 5.3, pp. 512–516

33. $\log_{\frac{1}{100}} = -2$ **34.** $\log_{\frac{1}{100,000}} = -5$

35. $\log_4 8 = \frac{3}{2}$ **36.** $\frac{3}{4} = \ln 2.117$

59. shift up 3

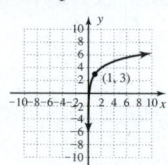

60. shift right 2

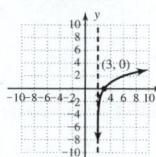

61. shift right 2, up 3

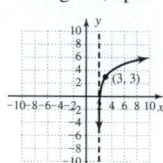

62. shift down 2

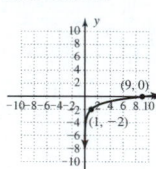

63. shift left 1

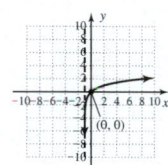

64. shift left 1, down 2

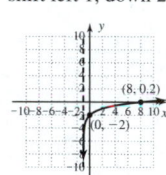

65. shift left 1, reflect across x-axis

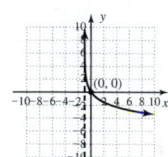

66. reflect across x-axis, shift up 2

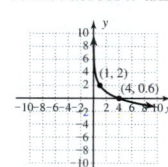

73. $x \in (-\infty, -1) \cup (3, \infty)$ **74.** $x \in (-\infty, -3) \cup (2, \infty)$
75. $x \in \left(\frac{3}{2}, \infty\right)$ **76.** $x \in \left(-\infty, \frac{5}{3}\right)$ **77.** $x \in (-3, 3)$ **78.** $x \in (0, 9)$
110. True **111.** D: $x \in \mathbb{R}$, R: $y \in \mathbb{R}$

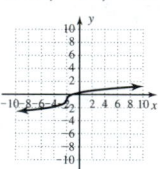

112. a. $(x - 2)(x^2 + 2x + 4)$ **b.** $(a + 7)(a - 7)$ **c.** $(n - 5)(n - 5)$
d. $(2b - 3)(b - 2)$

Exercises 5.4, pp. 523–526

15. $x = \ln\frac{65}{4} + 2, x \approx 4.7881$ **16.** $x = \frac{\ln 3}{0.4}, x \approx 2.7465$

17. $x = \log(78) - 5, x \approx -3.1079$ **18.** $x = \frac{\log 163}{2}, x \approx 1.1061$

19. $x = -\frac{\ln 2.32}{0.75}, x \approx -1.1221$ **20.** $x = \frac{\ln 4 - 1}{0.05}, x \approx 7.7259$

21. $x = e^{\frac{8}{3}} - 4, x \approx 10.3919$ **22.** $x = \frac{e^{2.75}}{3}, x \approx 5.2142$

23. $x = 5 - 10^{1.25}, x \approx -12.7828$ **24.** $x = \frac{10^{1.35}}{2}, x \approx 11.1936$

25. $x = \frac{e^{0.4} - 5}{2}, x \approx -1.7541$ **26.** $x = \frac{e^{2.4}}{4}, x \approx 2.7558$

27. $\ln(2x^2 - 14x)$ **28.** $\ln(3x^2 + 6x)$ **29.** $\log(x^2 - 1)$
30. $\log(x^2 - 9)$ **31.** $\log_3 4$ **32.** $\log_6 3$
33. $\log\left(\frac{x}{x + 1}\right)$ **34.** $\log\left(\frac{x - 2}{x}\right)$ **35.** $\ln\left(\frac{x - 5}{x}\right)$

36. $\ln\left(\frac{x + 3}{x - 1}\right)$ **37.** $\ln(x - 2)$ **38.** $\ln(x - 5)$

39. $\log_2 42$ **40.** $\log_9 30$ **41.** $\log_5(x - 2)$
42. $\log_3(3x + 5)$ **57.** $\frac{1}{2}[\log(x - 2) - \log x]$
58. $\frac{1}{3}[\log(3 - v) - \log 2 - \log v]$

59. $\frac{\ln 60}{\ln 7}$; 2.104076884 **60.** $\frac{\ln 92}{\ln 8}$; 2.174520652

61. $\frac{\ln 152}{\ln 5}$; 3.121512475 **62.** $\frac{\ln 200}{\ln 6}$; 2.957047225

63. $\frac{\log 1.73205}{\log 3}$; 0.499999576 **64.** $\frac{\log 1.41421}{\log 2}$; 0.499996366

65. $\frac{\log 0.125}{\log 0.5}$; 3 **66.** $\frac{\log 0.008}{\log 0.2}$; 3

67. $f(x) = \frac{\log(x)}{\log(3)}$; $f(5) \approx 1.4650$; $f(15) \approx 2.4650$; $f(45) \approx 3.4650$;
outputs increase by 1; $f(3^3 \cdot 5) \approx 4.4650$

68. $g(x) = \frac{\log(x)}{\log(2)}$; $g(5) \approx 2.3219$; $g(10) \approx 3.3219$; $g(20) \approx 4.3219$;
outputs increase by 1; $g(2^3 \cdot 5) \approx 5.3219$

69. $h(x) = \frac{\log(x)}{\log(9)}$; $h(2) \approx 0.3155$; $h(4) \approx 0.6309$; $h(8) \approx 0.9464$;
outputs are multiples of 0.3155; $h(2^4) \approx 4(0.3155) \approx 1.2619$

70. $H(x) = \frac{\log(x)}{\log(\pi)}$; $H(\sqrt 2) \approx 0.3028$; $H(2) \approx 0.6055$; $H(\sqrt{2^3}) \approx 0.9083$;
outputs are multiples of 0.3028; $H(\sqrt{2^4}) = H(4) \approx 1.2110$

Exercises 5.5, pp. 535–538

35. $x = \frac{\ln 231}{\ln 7}$; $x \approx 2.7968$ **36.** $x = \frac{\ln 3589}{\ln 6}$; $x \approx 4.5685$

37. $x = \frac{\ln 128,967}{3 \ln 5}$; $x \approx 2.4371$ **38.** $x = \frac{\ln 78,465}{3 \ln 9}$; $x \approx 1.7098$

39. $x = \frac{\ln 2}{\ln 3 - \ln 2}$; $x \approx 1.7095$ **40.** $x = \frac{\ln 4}{2 \ln 4 - \ln 7}$; $x \approx 1.6769$

65. a. $\begin{aligned} y &= 2^{x+1} \\ x &= 2^{y+1} \\ \ln x &= (y + 1)\ln 2 \\ \frac{\ln x}{\ln 2} &= y + 1 \\ \frac{\ln x}{\ln 2} - 1 &= y \end{aligned}$ **b.** $\begin{aligned} y &= 2\ln(x - 3) \\ x &= 2\ln(y - 3) \\ \frac{x}{2} &= \ln(y - 3) \\ e^{\frac{x}{2}} &= y - 3 \\ y &= e^{\frac{x}{2}} + 3 \end{aligned}$

66. a. $(f \circ g)(x) = 3^{(\log_3 x + 2) - 2} = 3^{\log_3 x} = x$;
$(g \circ f)(x) = \log_3(3^{x-2}) + 2 = x - 2 + 2 = x$
b. $(f \circ g)(x) = e^{(\ln x + 1) - 1} = e^{\ln x} = x$;
$(g \circ f)(x) = \ln e^{x-1} + 1 = x - 1 + 1 = x$
67. a. $y = e^{x \ln 2} = e^{\ln 2^x} = 2^x$;
$y = 2^x \Rightarrow \ln y = x \ln 2$, $e^{\ln y} = e^{x \ln 2} \Rightarrow y = e^{x \ln 2}$
b. $y = b^x$, $\ln y = x \ln b$, $e^{\ln y} = e^{x \ln b}$, $y = e^{xr}$ for $r = \ln b$
72.

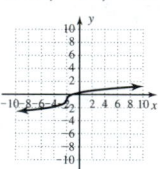

Exercises 5.6, pp. 547–552

47. a. $t = \frac{A - p}{pr}$ **b.** $p = \frac{A}{1 + rt}$ **48. a.** $t = \frac{\ln\left(\frac{A}{p}\right)}{\ln(1 + r)}$

b. $r = \sqrt[t]{\dfrac{A}{p}} - 1$ **49. a.** $r = n\left(\sqrt[nt]{\dfrac{A}{p}} - 1\right)$ **b.** $t = \dfrac{\ln\left(\dfrac{A}{p}\right)}{n \ln\left(1 + \dfrac{r}{n}\right)}$

50. a. $p = \dfrac{A}{e^{rt}}$ **b.** $r = \dfrac{\ln\left(\dfrac{A}{p}\right)}{t}$ **51. a.** $Q_0 = \dfrac{Q(t)}{e^{rt}}$ **b.** $t = \dfrac{\ln\left(\dfrac{Q(t)}{Q_0}\right)}{r}$

52. a. $A = \dfrac{p[(1 + R)^{nt} - 1]}{R}$ **b.** $n = \dfrac{\ln\left(\dfrac{AR}{p} + 1\right)}{t \ln(1 + R)}$

67. a. $f(x) = x^3, f(x) = x, f(x) = \sqrt{x}, f(x) = \sqrt[3]{x}, f(x) = \dfrac{1}{x}$

b. $f(x) = |x|, f(x) = x^2, f(x) = \dfrac{1}{x^2}$ **c.** $f(x) = x, f(x) = x^3, f(x) = \sqrt{x},$

$f(x) = \sqrt[3]{x}$ **d.** $f(x) = \dfrac{1}{x}, f(x) = \dfrac{1}{x^2}$

Exercises 5.7, pp. 557–565

21. As time increases, the amount of radioactive material decreases but will never truly reach 0 or become negative. Exponential with $b < 1$ and $k > 0$ is the best choice. $y \approx (1.042)0.5626^x$

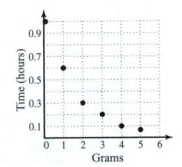

22. Populations usually grow exponentially or logistically. This growth does not appear to taper off. Exponential with $b > 1$ and $k > 0$ is the best choice. $y \approx (2.6550)1.1754^x$

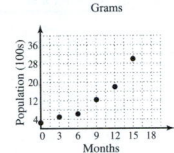

23. Sales will increase rapidly, then level off as the market is saturated with ads and advertising becomes less effective, possibly modeled by a logarithmic function. $y \approx 120.4938 + 217.2705 \ln(x)$

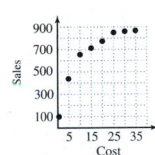

24. In a productive mine, we expect that initially, the diamonds may be nearer the surface and more plentiful, becoming more scarce and harder to find as time goes on. A logarithmic model seems to fit this description. $y \approx 454.7845 + 1087.8962 \ln(x)$

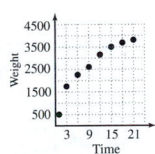

25. a.
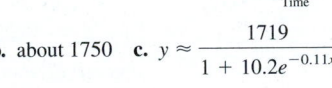
logistic **b.** about 1750 **c.** $y \approx \dfrac{1719}{1 + 10.2e^{-0.11x}}$

26. a.
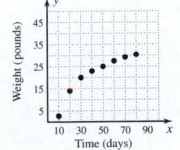
logistic **b.** about 72 **c.** $y \approx \dfrac{69.99}{1 + 4.00e^{-0.22x}}$

37. logarithmic, $y \approx -27.4 + 13.5 \ln x$
  a. 9.2 lb **b.** 29 days **c.** 34.8 lb

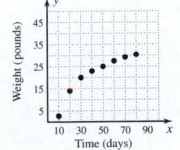

38. logarithmic, $y \approx -2635.6 + 1904.8 \ln x$
  a. 2870 oz **b.** 32.6 months **c.** 4816 oz

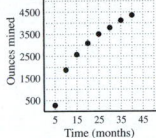

39. logarithmic, $y \approx 78.8 - 10.3 \ln x$
  a. 51,000 **b.** 1977 **c.** 29,900

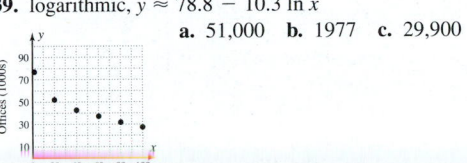

40. exponential, $y \approx 53.24(1.04)^x$
  a. 829 **b.** 4843 **c.** 2010

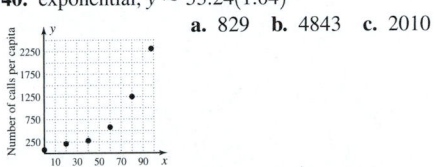

41. exponential, $y \approx 346.79(0.94)^x$
  a. 155,100 **b.** 54,200 **c.** 2013

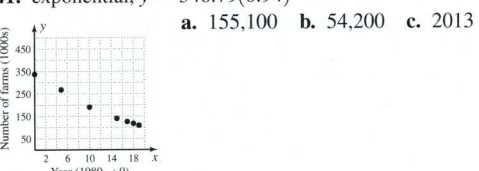

42. exponential, $y \approx 0.89(0.81)^x$
  a. 0.23 in. **b.** 3.2 sec **c.** 18 sec

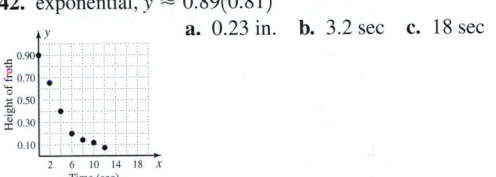

43. quadratic $y \approx 0.576x^2 - 8.879x + 394$
  a. 360 million **b.** about 513 million
  c. from 1984 to 1990

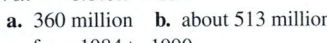
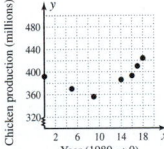

44. quadratic, $y \approx -0.010x^2 + 0.629x + 18.542$
  a. 26.6 million **b.** 17.2 million
  c. from 1965 to 1997

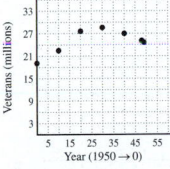

45. linear, $y \approx 6.555x + 165.308$
  a. 224.3 million **b.** 329.2 million **c.** 2010

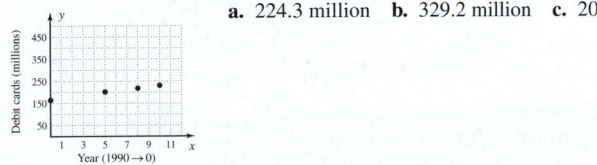

46. linear, $y \approx 0.35x + 56.10$; 98

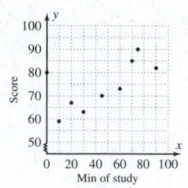

47. linear, $P(t) \approx 0.51t + 22.51$,
2005: 40.4%, 2010: 43.0%, 2015: 45.5%

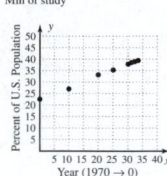

48. linear, $y \approx 0.42x + 0.81$, about 43 psi

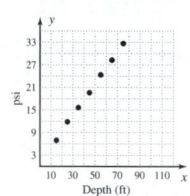

49. exponential, $y \approx 103.83\,(1.0595)^x$
    a. 220 **b.** The 22nd note, or F#
    c. frequency doubles, yes

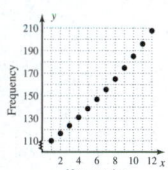

50. exponential, $y \approx 46.215\,(1.155)^x$
    a. \$1,271,000 **b.** \$7,164,000 **c.** 2002

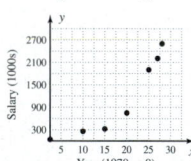

51. exponential, $y \approx 8.02\,(1.0564)^x$
\$41.59/mo, \$54.72/mo

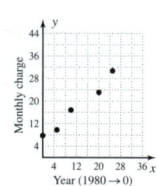

52. exponential, $y \approx 13.29\,(1.08^x)$;
about \$156 billion; 2005

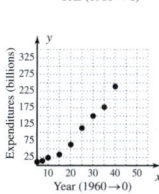

53. quadratic, $y \approx 1.18x^2 - 10.99x + 4.60$;
month 8

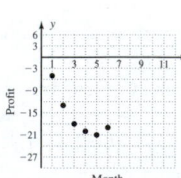

54. a. quadratic regression: **b.** about 16.1%;
$p(t) \approx 0.0148t^2 - 0.9175t + 19.5601$

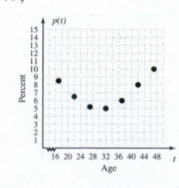

55. logistic, $y \approx \dfrac{222.133}{1 + 32.280e^{-0.336x}}$;

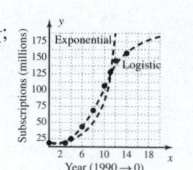

about 55 million, about 184 million, about 214 million; 2014

56. logistic, $y \approx \dfrac{39.116}{1 + 314.662e^{-5.948x}}$;

about 0.89 sec

57.

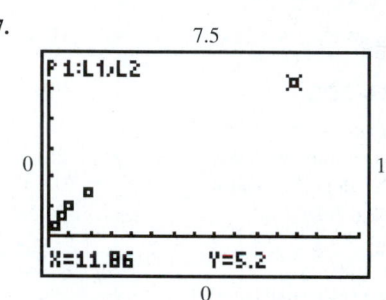

power regression,
a. $y \approx x^{0.665}$, 9.5 AU;
b. 84.8 yr

58.

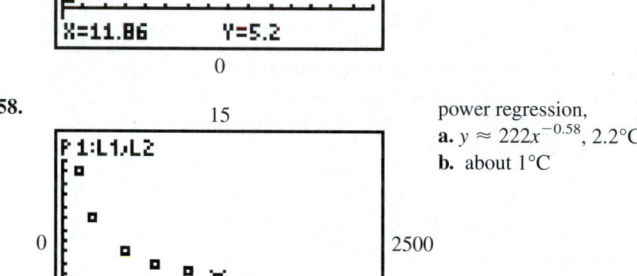

power regression,
a. $y \approx 222x^{-0.58}$, 2.2°C
b. about 1°C

59.

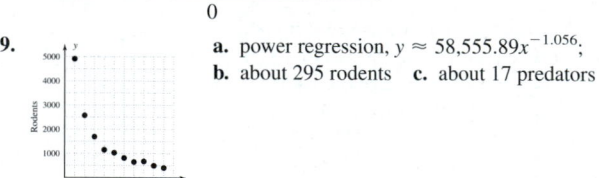

a. power regression, $y \approx 58,555.89x^{-1.056}$;
b. about 295 rodents **c.** about 17 predators

60. a.

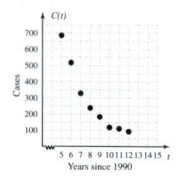

power regression;
b. $C(t) \approx 36,579x^{-2.428}$; about 25 cases;
c. 2005

61. a.

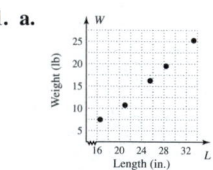

linear, $W \approx 1.24L - 15.83$, 32.5 lb, 35.3 in.

61. b.

logarithmic, $C(a) \approx 37.9694 + 3.4229 \ln(a)$,
about 49.3 cm, about 34 mo

62. a. $y \approx 3.332x - 1.168$, $r \approx 0.968$ **b.** $y = (1.140)1.695^x$, $r \approx 0.969$
The correlation coefficients are nearly equal. No.
c. exponential

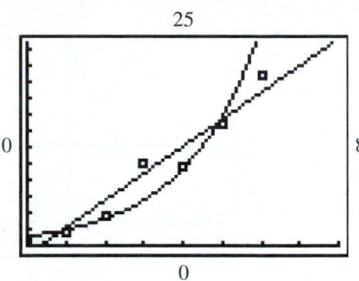

65. max: $(-0.4, 1.8)$
min: $(-3.5, -3.5)$, $(2.3, -1.4)$
$f(x)\uparrow$: $x \in (-3.5, -0.4) \cup (2.3, \infty)$
$f(x)\downarrow$: $x \in (-\infty, -3.5) \cup (-0.4, 2.3)$
66. D: $x \in (-\infty, \infty)$, R: $y \in [3, \infty)$

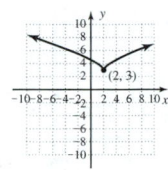

Summary and Concept Review, pp. 566–570

7. $f(x)$: D: $x \in [-4, \infty)$, R: $y \in [0, \infty)$; $f^{-1}(x)$: D: $x \in [0, \infty)$,
R: $y \in [-4, \infty)$ **8.** $f(x)$: D: $x \in (-\infty, \infty)$, R: $y \in (-\infty, \infty)$;
$f^{-1}(x)$: D: $(-\infty, \infty)$, R: $y \in (-\infty, \infty)$ **9.** $f(x)$: D: $x \in (-\infty, \infty)$,
R: $y \in (0, \infty)$; $f^{-1}(x)$: D: $x \in (0, \infty)$, R: $y \in (-\infty, \infty)$

11. 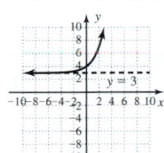   **12.** 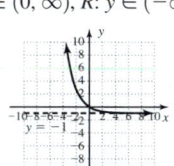   **13.**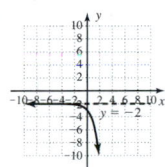

27. 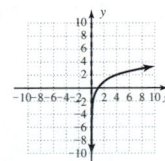   **28.** 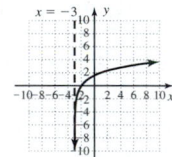   **29.**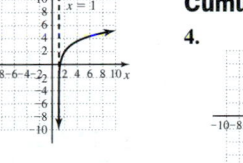

34. a. $x = \dfrac{\ln 4}{0.5}$, $x \approx 2.7726$ **b.** $x = \dfrac{\log 19}{0.2}$, $x \approx 6.3938$

c. $x = \dfrac{10^3}{3}$, $x \approx 333.3333$ **d.** $x = e^{-2.75}$, $x \approx 0.0639$

51.

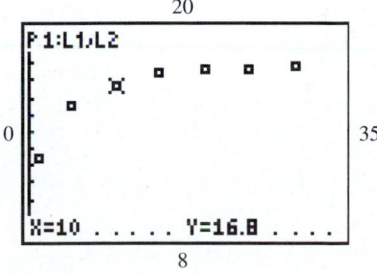

a. logarithmic, $y \approx 12.772 + 1.595 \ln x$

52.

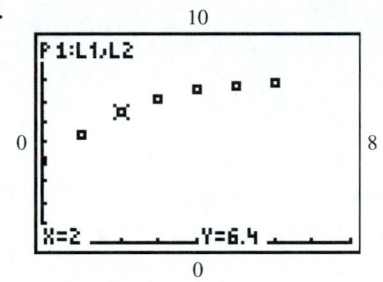

a. logistic, $g(t) = \dfrac{8.0}{1 + 1.0e^{-0.7t}}$

Practice Test, p. 571

9. 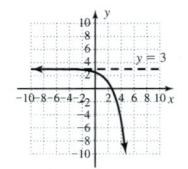   **10.**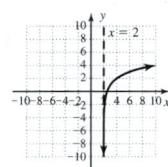

12. f is a parabola (hence not one-to-one), $x \in \mathbb{R}$, $y \in [-3, \infty)$; vertex is
at $(2, -3)$, so restricted domain could be $x \in [2, \infty)$ to create a one-to-one
function; $f^{-1}(x) = \sqrt{x + 3} + 2$, $x \in [-3, \infty)$, $y \in [2, \infty)$.

20. logistic; $y = \dfrac{39.1156}{1 + 314.6617e^{-5.9483x}}$; 0.89 sec

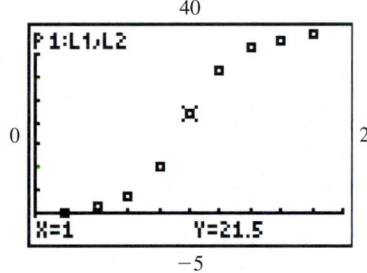

Cumulative Review Chapters R–5, p. 574

4.

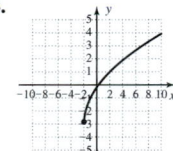

7. a. $T(t) = 455t + 2645$ (1991 → year 1)   **b.** $\dfrac{\Delta T}{\Delta t} = \dfrac{455}{1}$, triple births
increase by 455 each year   **c.** $T(6) = 5375$ sets of triplets,
$T(17) = 10{,}380$ sets of triplets

9. D: $x \in [-10, \infty)$, R: $y \in [-9, \infty)$
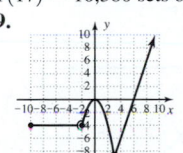 $h(x)\uparrow$: $x \in (-2, 0) \cup (3, \infty)$ $h(x)\downarrow$: $x \in (0, 3)$

12. $k = 77$, $f^{-1}(c) = \frac{5}{9}(c - 32)$, $f^{-1}(77) = \frac{5}{9}(77 - 32) = 25$ ✓
14. a. 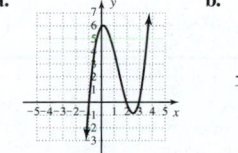   **b.**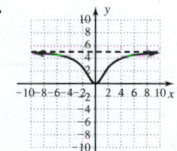

15. a. $f^{-1}(x) = \dfrac{5x - 3}{2}$ **b.** 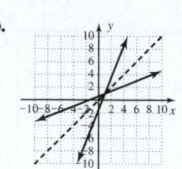   **c.** $f^{-1}(f(x)) = x$

17. $x = 5, x = -6$ is an extraneous root   **18. a.** $p(183) \approx 34.7\ W$

b. $p(t) = \dfrac{50}{4}$; $t \approx 693$ days Approx. 1 yr 11 months

CHAPTER 6

Exercises 6.1, pp. 586–591

15. 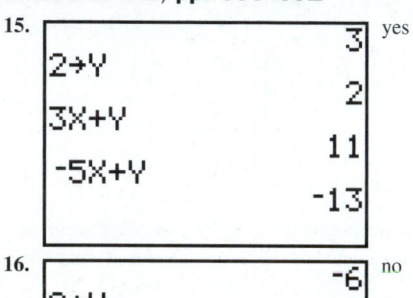 yes

16. 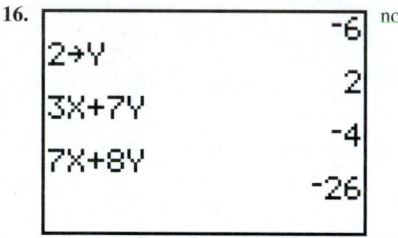 no

17. 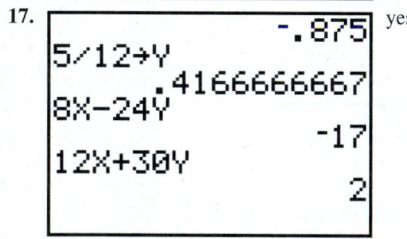 yes

18. 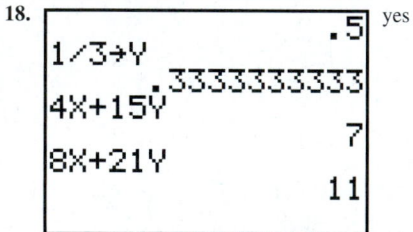 yes

19. 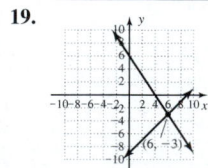 $Y_1 = \dfrac{12 - 3X}{2}$, $Y_2 = X - 9$

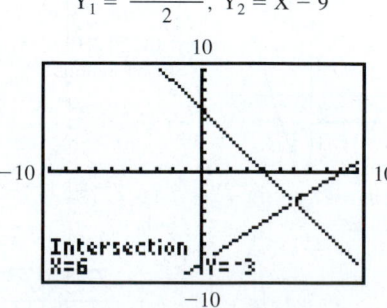

20. 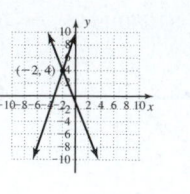 $Y_1 = \dfrac{-2 - 5X}{2}$, $Y_2 = 3X + 10$

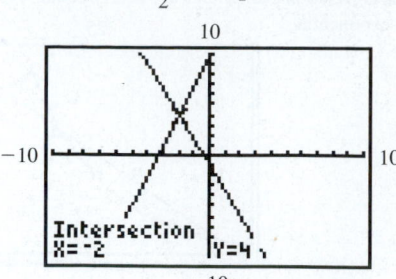

21. 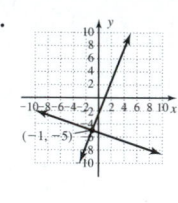 $Y_1 = \dfrac{5 - 5X}{-2}$, $Y_2 = \dfrac{-16 - X}{3}$

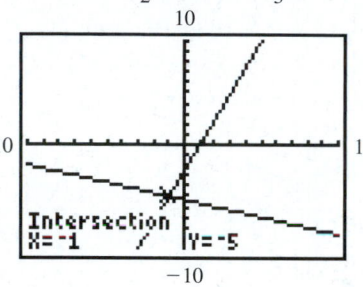

22. 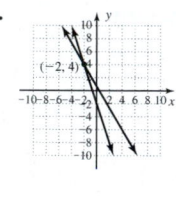 $Y_1 = -2 - 3X$, $Y_2 = \dfrac{2 - 5X}{3}$

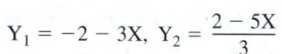

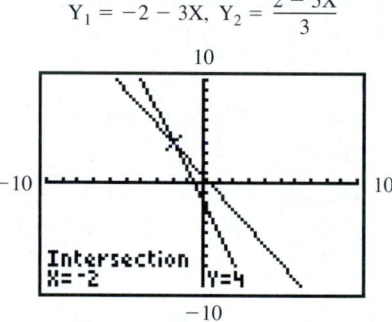

54. $\{(x, y)\,|\,9x - 5y = -15\}$; consistent/dependent

92. 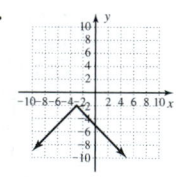   **93.** $x \in [-2, 8]$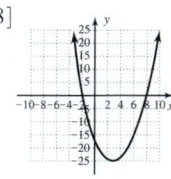

Exercises 6.2, pp. 599–602

21. no solution, inconsistent
22. no solution, inconsistent
23. $\{(x, y, z)\,|\,x \in \mathbb{R}, y = 2 - x, z = 2 - x\}$; $(p, 2 - p, 2 - p)$,
other solutions possible
24. $\{(x, y, z)\,|\,x = -z + 2, y = -2z - 1, z \in \mathbb{R}\}$; $(-p + 2, -2p - 1, p)$,
other solutions possible
25. $\left\{(x, y, z)\,\middle|\,x = -\dfrac{5}{3}z - \dfrac{2}{3}, y = -z - 2, z \in \mathbb{R}\right\}$; $\left(-\dfrac{5}{3}p - \dfrac{2}{3}, -p - 2, p\right)$,
other solutions possible
26. $\left\{(x, y, z)\,\middle|\,x = -\dfrac{1}{4}z + 4, y = \dfrac{9}{8}z + \dfrac{5}{2}, z \in \mathbb{R}\right\}$; $\left(-\dfrac{1}{4}p + 4, \dfrac{9}{8}p + \dfrac{5}{2}, p\right)$,
other solutions possible
62.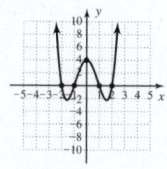

64.
$D: x \in (-\infty, \infty)$
$R: y \in [-5, \infty)$
zeroes: $x = -2.5, x = -1.5, x = 0.5, x = 2$
$g(x) > 0: x \in (-\infty, -2.5) \cup (-1.5, 0.5) \cup (2, \infty)$
$g(x) < 0: x \in (-2.5, -1.5) \cup (0.5, 2)$
max: $(-0.5, 4)$
min: $(-2, -2), (1.5, -5)$
$f(x)\uparrow: x \in (-2, -0.5) \cup (1.5, \infty)$
$f(x)\downarrow: x \in (-\infty, -2) \cup (-0.5, 1.5)$

Exercises 6.3, pp. 612–615

5. a. 3 or 4 not possible

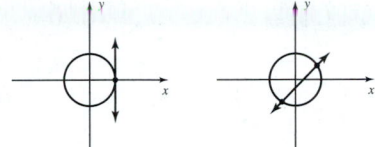

b. 3 or 4 not possible

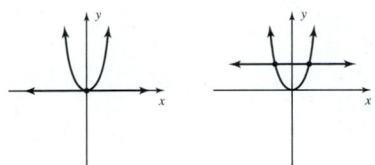

c.

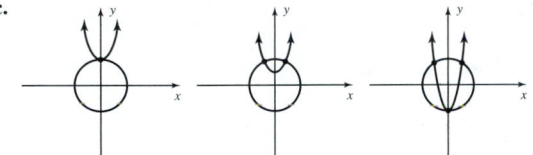

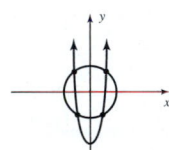

d.

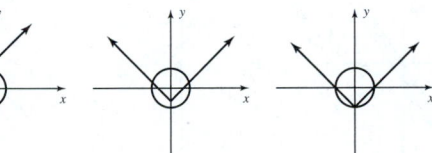

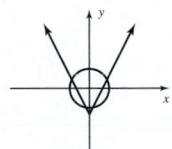

e. 3 or 4 solutions not possible

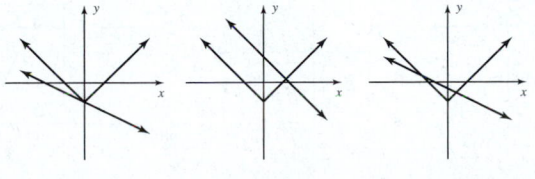

f.

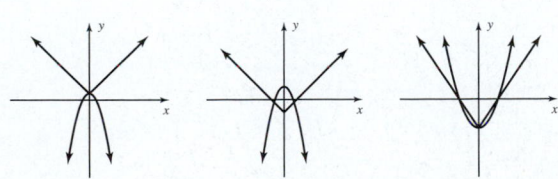

7. 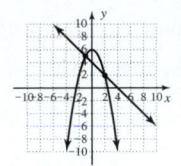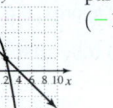 parabola, line;
$(-1, 5), (2, 2)$

8. 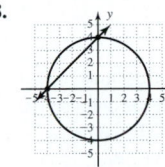 line, circle;
$(-4, 0), (0, 4)$

9. 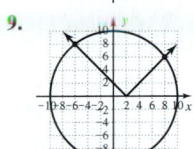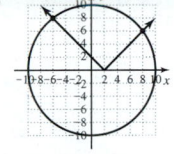 circle, absolute value;
$(-6, 8), (8, 6)$

10. 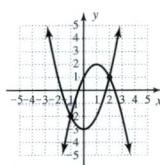 circle, parabola;
$(3, 4), (-3, 4),$
$(-4, -3), (4, -3)$

11. 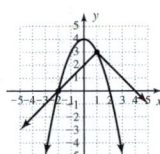 parabola, parabola;
$(-1, -2), (2, 1)$

12. 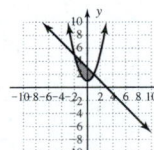 parabola, absolute value;
$(-2, 0), (1, 3)$

45. 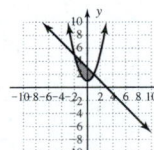   **46.**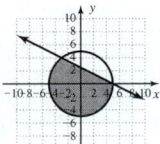

47. 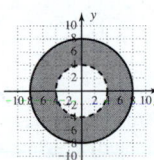   **48.**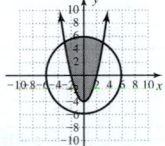

49. no solution   **50.** no solution

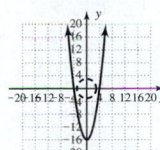   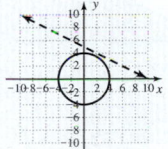

51. 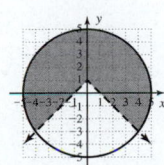   **52.**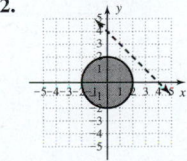

57. $\begin{cases} x^2y = 2000 \\ x^2 + 4xy = 800 \end{cases}$; approx. (12.4, 13) or (20, 5); The pool will likely have the dimensions 20 ft by 20 ft by 5 ft.

58. $\begin{cases} x^2y = 6750 \\ 2x^2 + 4xy = 2700 \end{cases}$; approx. (11, 55.8) or (30, 7.5); The box will likely have the dimensions 30 cm by 30 cm by 7.5 cm.

Exercises 6.4, pp. 625–629

5. The feasible region may be bordered by three or more oblique lines, with two of them intersecting outside and away from the feasible region.
6. Objective function and one of the boundary lines may be collinear.
7. No, No, No, No **8.** No, Yes, No, No **9.** No, Yes, Yes, No
10. No, Yes, Yes, No

11. 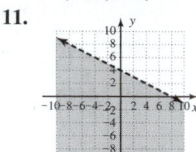   **12.** 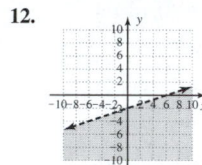   **13.**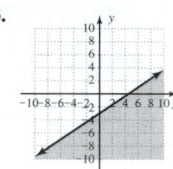

14.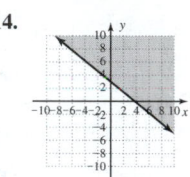

15. 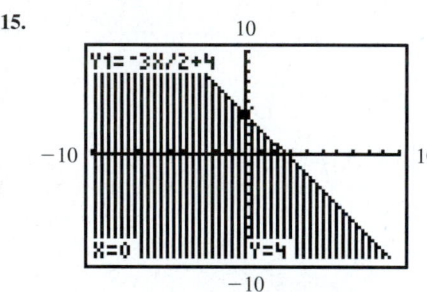   (0, 0) is in the solution region, and $3(0) + 2(0) \le 8$ is true.

16. 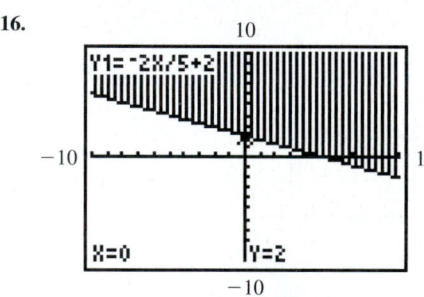   (0, 0) is not in the solution region, and $2(0) + 5(0) \ge 10$ is false.

17. 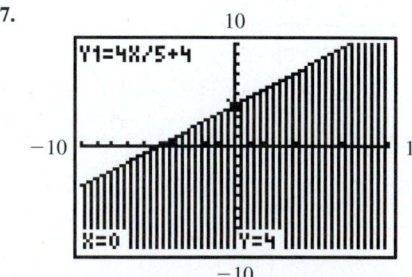   (0, 0) is in the solution region, and $4(0) - 5(0) > -20$ is true.

18. 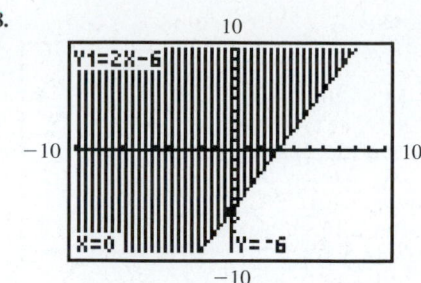   (0, 0) is in the solution region, and $6(0) - 3(0) < 18$ is true.

21. 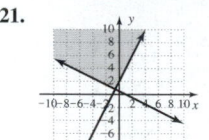   **22.** 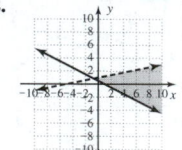   **23.**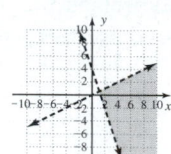

24. 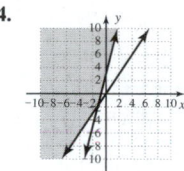   **25.** 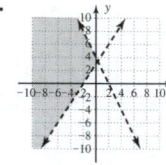   **26.**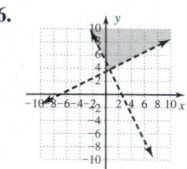

27. 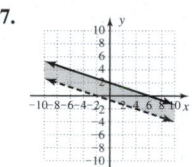   **28.** 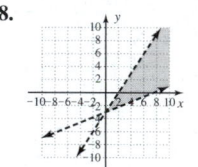   **29.**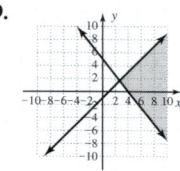

30. 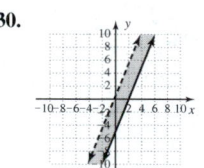   **31.** 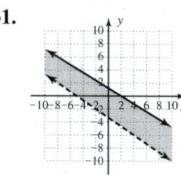   **32.**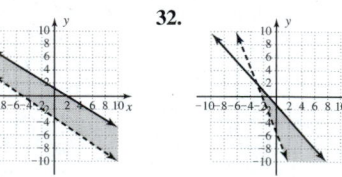

33. 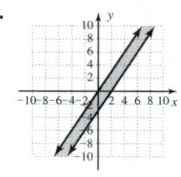   **34.** 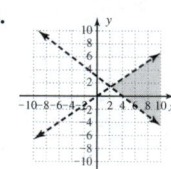   **35.**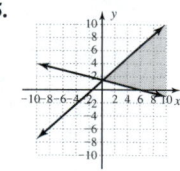

36. 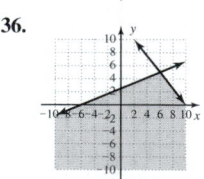   **37.** 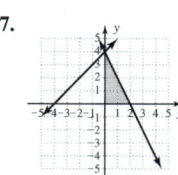   **38.**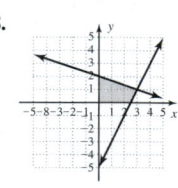

39. 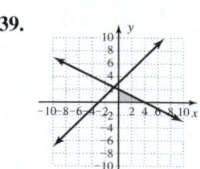   **40.** 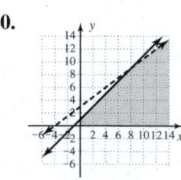   **41.**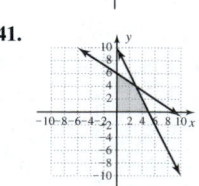

42. 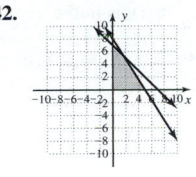   **43.**

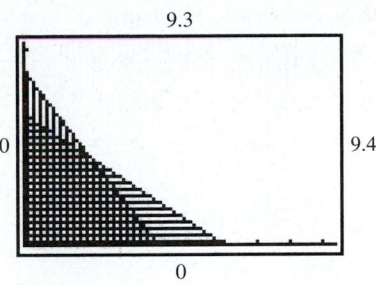

(2, 4)

44.

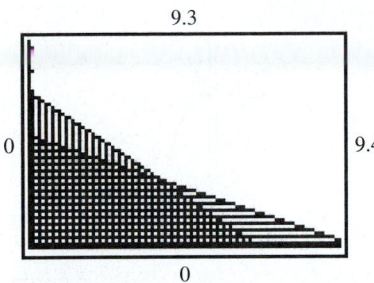

(4, 3)

45.

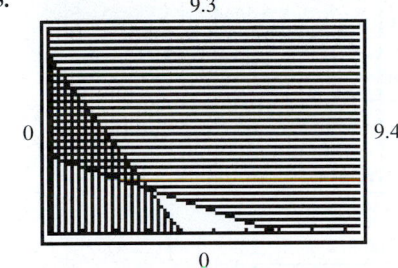

(3, 2)

46.

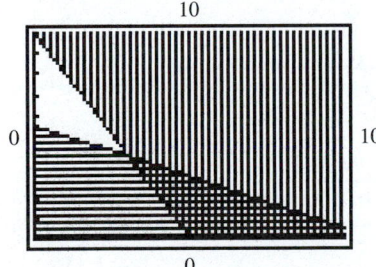

(3, 4)

47. $\begin{cases} y - x \le 1 \\ x + y > 3 \end{cases}$ **48.** $\begin{cases} y - x \ge 1 \\ x + y > 3 \end{cases}$

49. $\begin{cases} y - x \le 1 \\ x + y < 3 \\ y \ge 0 \end{cases}$ **50.** $\begin{cases} y - x \ge 1 \\ x + y < 3 \\ y \ge 0 \end{cases}$

61. 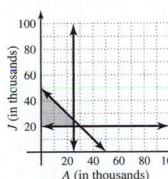   **62.**

$J + A \le 50{,}000$ $G + B \le 120$

$J \ge 20{,}000$ $B \ge 42$

$A \le 25{,}000$ $G \le 80$

63. 300 acres of corn; 200 acres of soybeans
64. 770 lb of deluxe and 640 lb of savory
65. 240 sheet metal screws; 480 wood screws   **66.** $11,000
69. 3 buses from company X; 4 buses from company Y

70. 220,000 gallons from Tulsa to Colorado; 100,000 gal from Tulsa to Mississippi; 0 gal from Houston to Colorado; 150,000 gal from Houston to Mississippi

71. 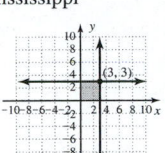   **a.** the region is a square
                     b. Max is 35.1 at (3, 3)
                     c. optimal solutions occur at vertices

Summary and Concept Review, pp. 630–632

1. 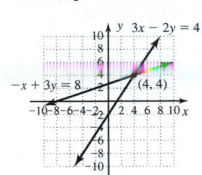   **2.**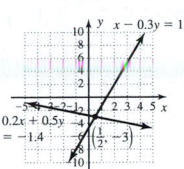

3.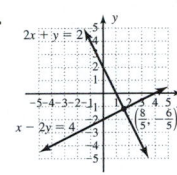

20. 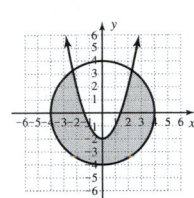   **21.**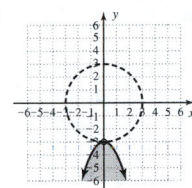

parabola, circle   note the open circle showing noninclusion at $(0, -3)$; circle, parabola

22. 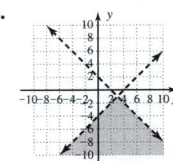   **23.**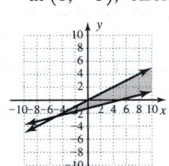

24.

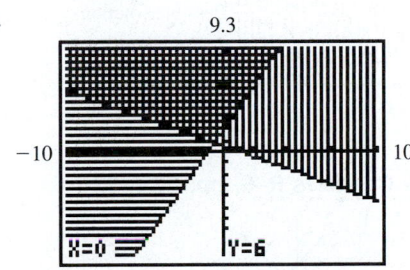

25. 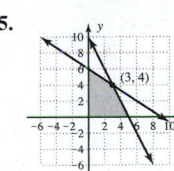   Maximum of 270 occurs at both (0, 6) and (3, 4) on the line $y = \dfrac{-2}{3} x + 6$, and all points on this line in between these endpoints.

Practice Test, pp. 632–633

1.

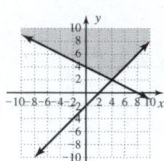

(2, 3)   **2.** $\left(\frac{2}{5}, \frac{-4}{5}\right)$ **3.** $(-3, 2)$ **4.** $(2, -1, 4)$

11.

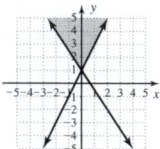

12. Maximum of $P = 400$ at $(8, 0)$

13. 30 plain; 20 deluxe

15. $(\sqrt{3}, 1), (-\sqrt{3}, 1), (i\sqrt{21}, -5), (-i\sqrt{21}, -5)$

17.

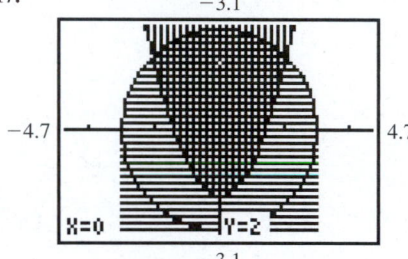

18. a.

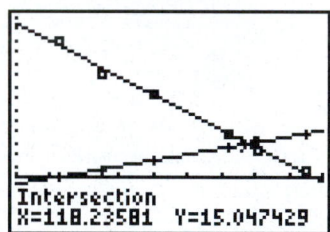

b. equilibrium is achieved when approx 150,000 plugs are sold at a price near $1.18

19.

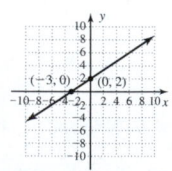

20. Answers may vary. Possible solution:
$$\begin{cases} x^2 + y^2 > 1 \\ x^2 + y^2 < 4 \\ x > 0, y < 0 \end{cases}$$

Cumulative Review Chapters R–6, pp. 635–636

1.

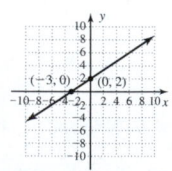

2.

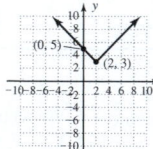

3.

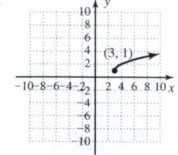

4.

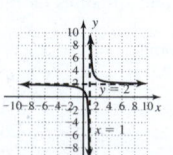

5.

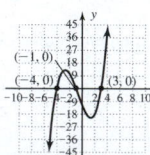

6.

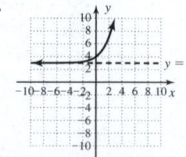

7. a. $D: x \in (-\infty, \infty)$ **b.** $R: y \in (-\infty, 4]$ **c.** $f(x)\uparrow: x \in (-\infty, -1)$ $f(x)\downarrow: x \in (-1, \infty)$ **d.** n/a   **e.** max: $(-1, 4)$ **f.** $f(x) > 0: x \in (-4, 2)$ $f(x) < 0: x \in (-\infty, -4) \cup (2, \infty)$ **8.** $P(x) = -x^2 + 120x - 2000$; 6000; $1600

18. a. no solution   **b.** no solution   **c.** $n = 1, n = -9$ **d.** $x = 3 \pm 2i$ **e.** $x = \frac{1}{8}, x = \frac{-1}{5}$ **f.** $x = -6$ **g.** $x = \frac{\ln 7}{\ln 3} + 2$ **h.** $x = 4$ **i.** $x = 3$

20.

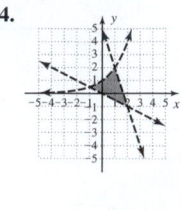

21.

$D: x \in [-5, 5]$
$R: y \in [0, 32]$

22.

24.

27. a.

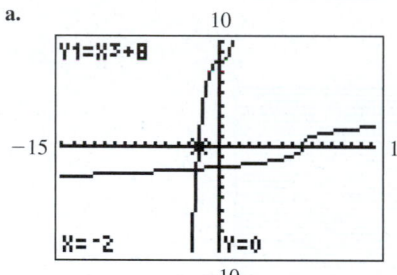

b. $y = \sqrt[3]{x} - 8$, verified

30.

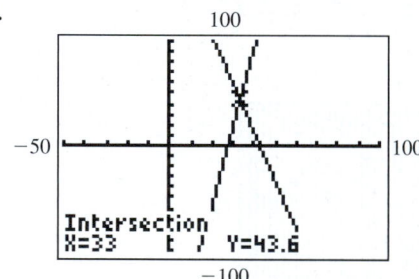

CHAPTER 7

Exercises 7.1, pp. 646–650

19. $\begin{bmatrix} 1 & -6 & | & -2 \\ 0 & -28 & | & -6 \end{bmatrix}$ **20.** $\begin{bmatrix} 1 & -2 & | & 3 \\ 7 & 4 & | & 3 \end{bmatrix}$ **21.** $\begin{bmatrix} 1 & -3 & 3 & | & 2 \\ 0 & 23 & -12 & | & -15 \\ -2 & 1 & 0 & | & 4 \end{bmatrix}$

22. $\begin{bmatrix} 1 & 1 & 2 & | & 6 \\ -3 & 2 & 0 & | & 0 \\ 0 & -3 & -11 & | & -22 \end{bmatrix}$ **23.** $\begin{bmatrix} 3 & 1 & 1 & | & 8 \\ 0 & -3 & -3 & | & -6 \\ 0 & -10 & -13 & | & 34 \end{bmatrix}$

24. $\begin{bmatrix} 2 & 1 & -1 & | & -3 \\ 0 & -1 & 5 & | & 9 \\ 0 & 1 & 4 & | & 9 \end{bmatrix}$

Exercises 7.2, pp. 658–662

7. $2 \times 2, a_{12} = -3, a_{21} = 5$ **8.** $3 \times 1, a_{31} = 5$

9. $2 \times 3, a_{12} = -3, a_{23} = 6, a_{22} = 5$ **10.** $3 \times 2, a_{12} = 0.4, a_{22} = 5$

11. $3 \times 3, a_{12} = 1, a_{23} = 1, a_{31} = 5$ **12.** $3 \times 4, a_{12} = 55, a_{23} = 5$

13. true   **14.** true   **15.** conditional, $c = -2, a = -4, b = 3$

16. conditional, $p = 3, q = -7, r = 4$ **17.** $\begin{bmatrix} 10 & 0 \\ 0 & 10 \end{bmatrix}$

18. $\begin{bmatrix} 0 & 0 & 0 \\ 0 & 0 & 0 \\ 0 & 0 & 0 \end{bmatrix}$ **19.** different orders, sum not possible

20. $\begin{bmatrix} 0 & 2 & 0 \\ 0 & 2 & -2 \\ -4 & -3 & 7 \end{bmatrix}$ **21.** $\begin{bmatrix} \frac{263}{32} & -\frac{19}{8} \\ -\frac{85}{16} & \frac{35}{16} \end{bmatrix}$ **22.** $\begin{bmatrix} \frac{57}{32} & \frac{19}{8} \\ \frac{85}{16} & \frac{125}{16} \end{bmatrix}$

23. $\begin{bmatrix} -\frac{1}{2} & \frac{13}{8} & -\frac{1}{4} \\ \frac{1}{4} & \frac{5}{2} & -\frac{21}{8} \\ -\frac{31}{8} & -\frac{9}{4} & \frac{7}{2} \end{bmatrix}$ **24.** $\begin{bmatrix} \frac{3}{2} & -\frac{19}{8} & -\frac{1}{4} \\ \frac{1}{4} & \frac{1}{2} & \frac{11}{8} \\ \frac{33}{8} & \frac{15}{4} & -\frac{17}{2} \end{bmatrix}$ **25.** $\begin{bmatrix} 20 & -15 \\ -25 & -10 \end{bmatrix}$

26. $\begin{bmatrix} -1 & 2 & 0 \\ 0 & 1 & -2 \\ -4 & -3 & 6 \end{bmatrix}$ **27.** $\begin{bmatrix} -\frac{5}{2} & -1 & 0 \\ 0 & -\frac{7}{2} & 1 \\ 2 & \frac{3}{2} & -6 \end{bmatrix}$ **28.** $\begin{bmatrix} 2 & -1 & 3 \\ 4 & 0 & -2 \end{bmatrix}$

29. $\begin{bmatrix} 1 & -2 & 0 \\ 0 & -1 & 2 \\ 4 & 3 & -6 \end{bmatrix}$ **30.** $\begin{bmatrix} 1 & -2 & 0 \\ 0 & -1 & 2 \\ 4 & 3 & -6 \end{bmatrix}$ **31.** $\begin{bmatrix} 1 & 0 \\ 0 & 1 \end{bmatrix}$ **32.** $\begin{bmatrix} 1 & 0 \\ 0 & 1 \end{bmatrix}$

33. matrix mult. not possible **34.** matrix mult. not possible

35. $\begin{bmatrix} 12 & -24 & 90 \\ -6 & 15 & -57 \end{bmatrix}$ **36.** $\begin{bmatrix} 0 \\ -7 \\ 29 \end{bmatrix}$ **37.** $\begin{bmatrix} 79 & -30 \\ -50 & 19 \end{bmatrix}$

38. matrix mult. not possible **39.** $\begin{bmatrix} 42 & 18 & -60 \\ -12 & -42 & 36 \end{bmatrix}$

40. matrix mult. not possible **41.** $\begin{bmatrix} 1 & 0 \\ 0 & 1 \end{bmatrix}$ **42.** $\begin{bmatrix} 1 & 0 \\ 0 & 1 \end{bmatrix}$

43. $\begin{bmatrix} 1 & 0 & 0 \\ 0 & 1 & 0 \\ 0 & 0 & 1 \end{bmatrix}$ **44.** $\begin{bmatrix} 1 & 0 & 0 \\ 0 & 1 & 0 \\ 0 & 0 & 1 \end{bmatrix}$ **45.** $\begin{bmatrix} \frac{-3}{19} & \frac{4}{57} \\ \frac{1}{19} & \frac{5}{57} \end{bmatrix}$ **46.** $\begin{bmatrix} \frac{-3}{19} & \frac{4}{57} \\ \frac{1}{19} & \frac{5}{57} \end{bmatrix}$

47. $\begin{bmatrix} 0 & \frac{3}{4} & \frac{1}{4} \\ \frac{-1}{2} & \frac{3}{8} & \frac{1}{8} \\ \frac{-1}{4} & \frac{11}{16} & \frac{1}{16} \end{bmatrix}$ **48.** $\begin{bmatrix} 0 & \frac{3}{4} & \frac{1}{4} \\ \frac{-1}{2} & \frac{3}{8} & \frac{1}{8} \\ \frac{-1}{4} & \frac{11}{16} & \frac{1}{16} \end{bmatrix}$ **49.** $\begin{bmatrix} 1.75 & 2.5 \\ 7.5 & 13 \end{bmatrix}$

50. $\begin{bmatrix} 1 & 0 & -4 \\ 8 & 7 & -14 \\ -20 & -29 & 42 \end{bmatrix}$ **51.** $\begin{bmatrix} -4 & 28 & 4 \\ -8 & 17 & 3 \end{bmatrix}$ **52.** $\begin{bmatrix} -44 & 72 & -96 \\ 72 & 48 & 240 \end{bmatrix}$

57. $P = 21.448$ cm; $A = 27.7269$ cm²
58. $P = 17.54$ cm; $A = 18.825$ cm²
59. a.

$$V = \begin{array}{c} \\ S \\ D \\ P \end{array} \begin{matrix} T & S \\ \begin{bmatrix} 3820 & 1960 \\ 2460 & 1240 \\ 1540 & 920 \end{bmatrix} \end{matrix} \qquad M = \begin{array}{c} \\ S \\ D \\ P \end{array} \begin{matrix} T & S \\ \begin{bmatrix} 4220 & 2960 \\ 2960 & 3240 \\ 1640 & 820 \end{bmatrix} \end{matrix}$$

b. 3900 more by Minsk

c. $V = \begin{bmatrix} 3972.8 & 2038.4 \\ 2558.4 & 1289.6 \\ 1601.6 & 956.8 \end{bmatrix}$ **d.** $\begin{bmatrix} 8361.6 & 5116.8 \\ 5636.8 & 4659.2 \\ 3307.2 & 1809.6 \end{bmatrix}$

$M = \begin{bmatrix} 4388.8 & 3078.4 \\ 3078.4 & 3369.6 \\ 1705.6 & 852.8 \end{bmatrix}$

60. a.

$$C = \begin{array}{c} \\ A \\ T \end{array} \begin{matrix} J & F & M \\ \begin{bmatrix} 350 & 420 & 530 \\ 220 & 180 & 140 \end{bmatrix} \end{matrix}$$

b. $\begin{bmatrix} 84 & 0 & 16 \\ 35 & 42 & -28 \end{bmatrix}$

$$S = \begin{array}{c} \\ A \\ T \end{array} \begin{matrix} J & F & M \\ \begin{bmatrix} 430 & 560 & 690 \\ 280 & 320 & 220 \end{bmatrix} \end{matrix}$$

$$D = \begin{array}{c} \\ A \\ T \end{array} \begin{matrix} J & F & M \\ \begin{bmatrix} 864 & 980 & 1236 \\ 535 & 542 & 332 \end{bmatrix} \end{matrix}$$

c. $\begin{bmatrix} 1666.1 & 1985.2 & 2488.3 \\ 1048.6 & 1050.4 & 699.4 \end{bmatrix}$

61. [22,000 19,000 23,500 14,000]; Total Profit N: $22,000, S: $19,000, E: $23,500, W: $14,000 **62.** [23.5 20 31]; Donations M: $23.50, T: $20, W: $31

63. a. $108.20 **b.** $101

c. $\begin{bmatrix} 100 & 101 & 119 \\ 108.2 & 107 & 129.5 \end{bmatrix}$ First row, total cost for science from each restaurant; second row, total cost for math from each restaurant.

64. a. $50.70 **b.** $44.75
c. $\begin{bmatrix} 32.65 & 28.85 \\ 50.7 & 44.75 \\ 60.25 & 53.125 \end{bmatrix}$ First row: total cost for home tables from each location; second row, total cost for commercial tables from each location; third row, total cost for professional tables from each location

65. $\begin{bmatrix} 32.4 & 10.3 & 21.3 \\ 29.9 & 9.6 & 19.5 \end{bmatrix}$ **a.** 10 **b.** 20

c. p_{13} gives the approximate number of females expected to join the writing club

66. $\begin{bmatrix} 4350 & 2850 \\ 7200 & 4700 \\ 7100 & 4650 \end{bmatrix}$ **a.** $11,900 **b.** $11,750

c. p_{11} gives the revenue from the medium, hand-finished sweatshirts

67. $\begin{bmatrix} 2^{n-1} & 0 & 2^{n-1} \\ 2^n - 1 & 1 & 2^n - 1 \\ 2^{n-1} & 0 & 2^{n-1} \end{bmatrix}$

68. all odd powers generate the original matrix
69. $(-1, 1, -2)$ **70.** $x \in [-3, -1] \cup [1, 3]$ **71.** ≈ 4.39
72. $x^2 + 2x - 5$

Mid-Chapter Check pp. 662–663

7. a. $\begin{bmatrix} 0.8 & 0.5 & 2.2 \\ -0.1 & 0.8 & -1 \\ 2.1 & 0.3 & 1.9 \end{bmatrix}$ **b.** $\begin{bmatrix} -3 & -1.5 & -6 \\ 1.5 & 0 & 3 \\ -6 & -1.5 & -6 \end{bmatrix}$ **c.** $\begin{bmatrix} 1 & 0 & 0 \\ 0 & 1 & 0 \\ 0 & 0 & 1 \end{bmatrix}$

8. a. $\begin{bmatrix} 17 & -32 & -13 \\ 5 & 0 & -5 \end{bmatrix}$ **b.** $\begin{bmatrix} -26 & -18 & 24 \\ 2 & -10 & -4 \end{bmatrix}$ **c.** $\begin{bmatrix} 24 & 4 \\ 0 & -5 \\ 16 & -39 \end{bmatrix}$

d. $\begin{bmatrix} -4 & -33 \\ -2 & 29 \end{bmatrix}$

10. $\begin{bmatrix} 4375 & 110 \\ 2400 & 59 \end{bmatrix}$,

p_{11}: total rebates paid to individuals, p_{21}: total rebates paid to business, p_{12}: free AAA years given to individuals, p_{22}: free AAA years given to business

Reinforcing Basic Concepts, p. 663
Exercise 2:
1st row of A with 3rd column of B
2nd row of A with 2nd column of B

Exercise 3:
$[A] \to 3 \times 1$; $[B] \to 1 \times 3$
$[A] \to 3 \times 2$; $[B] \to 2 \times 3$
$[A] \to 3 \times 3$; $[B] \to 3 \times 3$
$[A] \to 3 \times n$; $[B] \to n \times 3$; $n \in \mathbb{N}$

Exercises 7.3, pp. 673–678

31. $\begin{bmatrix} -2 & 1 & -4 & 5 \\ 2 & -5 & 1 & -3 \\ -3 & 1 & 6 & 1 \\ 1 & 4 & -5 & 1 \end{bmatrix} \begin{bmatrix} w \\ x \\ y \\ z \end{bmatrix} = \begin{bmatrix} -3 \\ 4 \\ 1 \\ -9 \end{bmatrix}$

32. $\begin{bmatrix} 1.5 & 2.1 & -0.4 & 1 \\ 0.2 & -2.6 & 1 & 0 \\ 0 & 3.2 & 0 & 1 \\ 1.6 & 4 & -5 & 2.6 \end{bmatrix} \begin{bmatrix} w \\ x \\ y \\ z \end{bmatrix} = \begin{bmatrix} 1 \\ 5.8 \\ 2.7 \\ -1.8 \end{bmatrix}$

60. $\det(A) = -37$; $(2, 1, -3)$

63. $A^{-1} = \begin{bmatrix} \frac{1}{13} & \frac{5}{13} \\ \frac{-2}{13} & \frac{3}{13} \end{bmatrix}$ **64.** $B^{-1} = \begin{bmatrix} \frac{-4}{7} & \frac{-3}{7} \\ \frac{5}{7} & \frac{2}{7} \end{bmatrix}$

verified   verified

69. Jumpin' Jack Flash: 3.75 min
Tumbling Dice: 3.75 min
You Can't Always Get: 7.5 min
Wild Horses: 5.75 min

70. Tamino's Aria: 4.1
Papageno's Aria: 2.9
Champagne Aria: 1.4
Catalogue Aria: 5.9
71. 30 of clock A; 20 of clock B; 40 of clock C; 12 of clock D
72. 25 small; 15 medium; 30 large; 18 x-large
73. $p_1 = 72.25°, p_2 = 74.75°, p_3 = 80.25°, p_4 = 82.75°$
74. a. $p_1 = 72.25°, p_2 = 74.75°, p_3 = 80.25°,$
     $p_4 = 82.75°$; no change in temperatures
  b. $p_1 = 64.75°, p_2 = 69.75°, p_3 = 75.25°, p_4 = 80.25°$
77. $y = x^3 + 2x^2 - 9x - 10$ **78.** $y = 2x^3 - 10x + 1$
83. 2 oz Food I, 1 oz Food II, 4 oz Food III
84. 7 oz Food I; 2 oz Food II; 8 oz Food III

Exercises 7.4 pp. 689–692

7. $D = \begin{vmatrix} 2 & 5 \\ -3 & 4 \end{vmatrix}; D_x = \begin{vmatrix} 7 & 5 \\ 1 & 4 \end{vmatrix}; D_y = \begin{vmatrix} 2 & 7 \\ -3 & 1 \end{vmatrix}$

8. $D = \begin{vmatrix} -1 & 5 \\ 3 & -2 \end{vmatrix}; D_x = \begin{vmatrix} 12 & 5 \\ -8 & -2 \end{vmatrix}; D_y = \begin{vmatrix} -1 & 12 \\ 3 & -8 \end{vmatrix}$

15. a. $D = \begin{vmatrix} 4 & -1 & 2 \\ -3 & 2 & -1 \\ 1 & -5 & 3 \end{vmatrix} \quad D_x = \begin{vmatrix} -5 & -1 & 2 \\ 8 & 2 & -1 \\ -3 & -5 & 3 \end{vmatrix}$

$D_y = \begin{vmatrix} 4 & -5 & 2 \\ -3 & 8 & -1 \\ 1 & -3 & 3 \end{vmatrix} \quad D_z = \begin{vmatrix} 4 & -1 & -5 \\ -3 & 2 & 8 \\ 1 & -5 & -3 \end{vmatrix},$

b. $|D| = 22$, solutions possible
c. $|D| = 0$, Cramer's rule cannot be used: coefficients $R_1 + R_2 = R_3$

16. a. $D = \begin{vmatrix} 2 & 0 & 3 \\ -1 & 5 & 1 \\ 3 & -2 & 1 \end{vmatrix}, \quad D_x = \begin{vmatrix} -2 & 0 & 3 \\ 12 & 5 & 1 \\ -8 & -2 & 1 \end{vmatrix},$

$D_y = \begin{vmatrix} 2 & -2 & 3 \\ -1 & 12 & 1 \\ 3 & -8 & 1 \end{vmatrix}, \quad D_z = \begin{vmatrix} 2 & 0 & -2 \\ -1 & 5 & 12 \\ 3 & -2 & -8 \end{vmatrix}$

b. $|D| = -25$, solutions possible
c. $|D| = 0$, Cramer's rule cannot be used: coefficients $R_1 + R_2 = R_3$

19. $\left(\frac{3}{4}, \frac{5}{3}, \frac{-1}{3}\right)$ **23.** $\frac{A}{x+3} + \frac{B}{x-2}$ **24.** $\frac{A}{x-2} + \frac{B}{x-5}$

25. $\frac{A}{x-1} + \frac{B}{x+2} + \frac{C}{x-3}$ **26.** $\frac{A}{x+3} + \frac{B}{x+1} + \frac{C}{x-2}$

27. $\frac{A}{x} + \frac{B}{x-3} + \frac{C}{x+1}$ **28.** $\frac{A}{x+4} + \frac{B}{x-2} + \frac{C}{x}$

29. $\frac{A}{x} + \frac{B}{x^2} + \frac{C}{x+2}$ **30.** $\frac{A}{x-3} + \frac{B}{x+2} + \frac{C}{(x+2)^2}$

31. $\frac{A}{x+1} + \frac{Bx+C}{x^2+2} + \frac{Dx+E}{(x^2+2)^2}$ **32.** $\frac{A}{x} + \frac{Bx+C}{x^2+3} + \frac{Dx+E}{(x^2+3)^2}$

37. $\frac{7}{x} + \frac{2}{x+1} - \frac{1}{x-1}$ **38.** $\frac{3}{x} - \frac{7}{x+2} + \frac{5}{x-2}$

39. $\frac{-1}{x} + \frac{4}{x+1} + \frac{5}{(x+1)^2}$ **40.** $\frac{7}{x} - \frac{9}{x-2} + \frac{3}{(x-2)^2}$

41. $\frac{3}{2-x} - \frac{4}{x^2+2x+4}$ **42.** $\frac{2}{x-1} + \frac{x+3}{x^2+x+1}$

43. $\frac{5}{x+2} + \frac{x-1}{x^2+3}$ **44.** $\frac{-3}{x-2} + \frac{5x-4}{x^2+5}$

63. $\begin{cases} 15{,}000x + 25{,}000y = 2900 \\ 25{,}000x + 15{,}000y = 2700 \end{cases}$; 6%, 8%

64. $\begin{cases} 2x + 2y + 10z = 3.26 \\ 3x + 2y + 7z = 2.98 \\ 2x + 3y + 6z = 2.89 \end{cases}$;

apples, 29¢/lb
kiwi, 39¢/lb
pears, 19¢/lb

71.

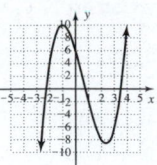

Exercises 7.5, pp. 697–699

8.

Watch Model	Week 1	Week 2	Week 3	Week 4	Week 5
Silver	6	4	7	6.5	5
Gold	11	8	9	7	10
Platinum	2	1	0	3	1

It looks like precious metal was stolen in week 4.
9. one bundle of first class = 9.25 measures of grain;
one bundle of second class = 4.25 measures of grain;
one bundle of third class = 2.75 measures of grain

10.

(measures/ bundle)	Mon	Tues	Wed	Thur	Fri
First Class	9.25	9	8.75	9	9.25
Second Class	4.25	4	4.25	3.75	4.5
Third Class	2.75	3	3.25	3.25	3

Summary and Concept Review, pp. 700–702

6. $\begin{bmatrix} -7.25 & 5.25 \\ 0.875 & -2.875 \end{bmatrix}$ **7.** $\begin{bmatrix} -6.75 & 6.75 \\ 1.125 & -1.125 \end{bmatrix}$ **8.** not possible

9. $\begin{bmatrix} -2 & -6 \\ -1 & -7 \end{bmatrix}$ **10.** $\begin{bmatrix} 1 & 0 \\ 0 & 1 \end{bmatrix}$ **11.** $\begin{bmatrix} 1 & 0 & 4 \\ 5.5 & -1 & -1 \\ 10 & -2.9 & 7 \end{bmatrix}$

12. $\begin{bmatrix} 3 & -6 & -4 \\ -4.5 & 3 & -1 \\ -2 & 3.1 & 3 \end{bmatrix}$ **13.** not possible **14.** $\begin{bmatrix} -8 & 12 & 0 \\ -2 & -4 & 4 \\ -16 & -0.4 & -20 \end{bmatrix}$

15. $\begin{bmatrix} 15.5 & 6.4 & 17 \\ 9 & -17 & 2 \\ 18.5 & 20.8 & 13 \end{bmatrix}$

29. Year 1: fir = 2670 ft², jequitba = 2752 ft², teak = 1090 ft²;
Year 2: fir = 3570 ft², jequitba = 3668 ft², teak = 1456 ft²;
Year 3: fir = 4470 ft², jequitba = 4584 ft², teak = 1822 ft²;
Year 4: fir = 5370 ft², jequitba = 5500 ft², teak = 2188 ft²;
Year 5: fir = 6270 ft², jequitba = 6416 ft², teak = 2554 ft²;
Total: fir = 22,350 ft², jequitba = 22,920 ft², teak = 9110 ft²

Practice Test, pp. 702–703

4. a. $\begin{bmatrix} -6 & -5 \\ 8 & 9 \end{bmatrix}$ **b.** $\begin{bmatrix} 1.2 & 1.2 \\ -1.2 & -2 \end{bmatrix}$ **c.** $\begin{bmatrix} -3 & 1 \\ 3 & -5 \end{bmatrix}$ **d.** $\begin{bmatrix} -2 & -1 \\ 2.5 & 1.5 \end{bmatrix}$

e. -2

5. a. $\begin{bmatrix} 0 & -0.1 & 0 \\ 0.5 & -0.6 & 0 \\ -0.2 & -0.8 & -0.9 \end{bmatrix}$ **b.** $\begin{bmatrix} -0.3 & -0.06 & -0.12 \\ 0.06 & -0.06 & 0 \\ -0.18 & -0.24 & -0.48 \end{bmatrix}$

c. $\begin{bmatrix} 0.31 & -0.13 & 0.08 \\ -0.01 & -0.05 & -0.02 \\ 0.39 & -0.52 & -0.02 \end{bmatrix}$ **d.** $\begin{bmatrix} \frac{40}{17} & 0 & \frac{-10}{17} \\ \frac{40}{17} & 10 & \frac{-10}{17} \\ \frac{-35}{17} & -5 & \frac{30}{17} \end{bmatrix}$

e. $\frac{17}{500}$ **16.** Dr. Brown owes $31,000; Dr. Stamper owes $124,000
18. 11 one-day, 6 two-day, 3 five-day **19.** federal program: $200,000;
municipal bonds: $1,300,000; bank loan: $300,000

Cumulative Review Chapters R–7, pp. 705–706

2. $h = \dfrac{S - 2\pi r^2}{2\pi r}$ **7.** $y = \dfrac{-7}{5}x + \dfrac{4}{5}$ **8.** $y = -\dfrac{3}{2}x + 3$

10. 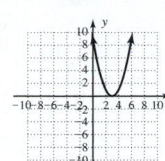   **11.** 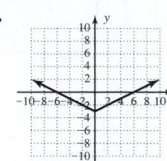   **12.**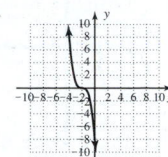

13. 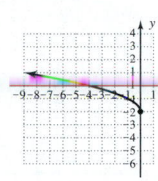   **14.** $\left(-\infty, \dfrac{9}{2}\right]$ **15.** $(-\infty, 1) \cup (5, \infty)$

17. $f(x)\!\uparrow: x \in (0, 2) \cup (4, \infty), f(x)\!\downarrow: x \in (-\infty, 0) \cup (2, 4)$, constant: none
18. $g(x)\!\uparrow:$ none, $g(x)\!\downarrow: x \in (-\infty, -3) \cup (-3, \infty)$, constant: none

19. $2y - \dfrac{5}{2} + \dfrac{25}{2(2y + 1)}$

20. $-3 \big| \begin{array}{rrrr} 1 & 4 & 0 & -15 \\ & -3 & -3 & 9 \\ \hline 1 & 1 & -3 & \underline{|-6} \end{array}$

  $P(-3) = -6$
21. $x^3 - 4x^2 - 2x + 20$
22. a. down/up, -16 **b.** down/down, 7   **c.** up/down, 10

23. $\left(-3, \dfrac{-5}{2}\right)$ **24.**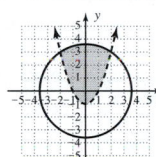

CHAPTER 8

Exercises 8.1, pp. 712–714

7. $(-2, -2)$; verified   **8.** $\left(\dfrac{11}{2}, 8\right)$; verified   **9.** $(2, -2)$; verified
10. $\left(0, \dfrac{3}{2}\right)$; verified   **11.** $\left(\dfrac{13}{2}, -9\right)$; verified   **12.** $(13, 18)$; verified
39. $f(x) = (x + 2)(x - 1)^2(x - 3)$ **40.**

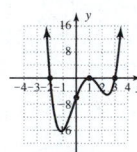   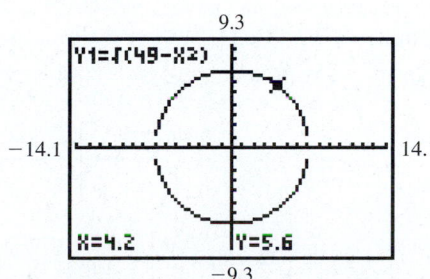

Exercises 8.2, pp. 725–729

7. $x^2 + y^2 = 49$

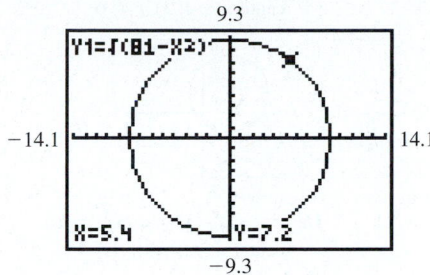

$(4.2, 5.6), (4.2, -5.6)$

8. $x^2 + y^2 = 81$

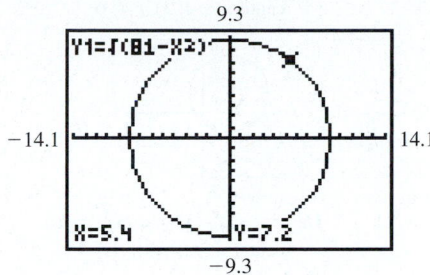

$(5.4, 7.2), (5.4, -7.2)$
9. $(x - 5)^2 + y^2 = 3$

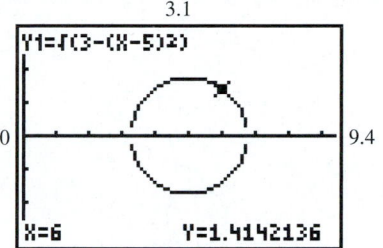

$(6, \sqrt{2}), (6, -\sqrt{2})$
10. $x^2 + (y - 4)^2 = 5$

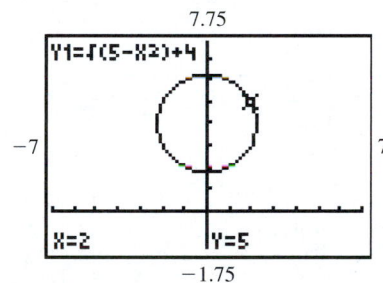

$(2, 5), (2, 3)$
11. $(x - 1)^2 + (y - 5)^2 = 25$

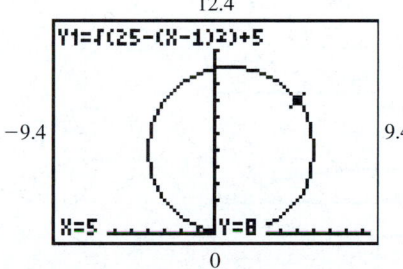

$(5, 8), (5, 2)$
12. $\left(x - \dfrac{1}{2}\right)^2 + (y - 3)^2 = \dfrac{169}{4}$

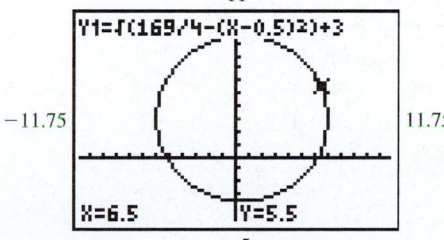

$(6.5, 5.5), (6.5, 0.5)$

13. $(x - 6)^2 + (y - 5)^2 = 9$
center: $(6, 5)$, $r = 3$

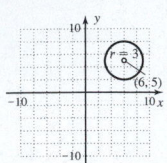

14. $(x + 4)^2 + (y - 3)^2 = 36$
center: $(-4, 3)$, $r = 6$

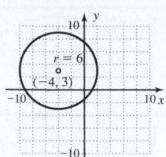

15. $(x - 2)^2 + (y + 5)^2 = 25$
center: $(2, -5)$, $r = 5$

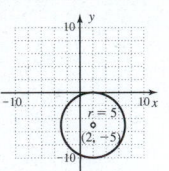

16. $(x + 2)^2 + (y + 3)^2 = 16$
center: $(-2, -3)$, $r = 4$

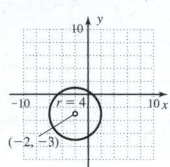

17. $(x + 3)^2 + y^2 = 14$
center: $(-3, 0)$, $r = \sqrt{14}$

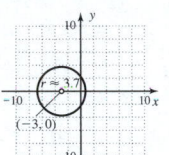

18. $x^2 + (y - 4)^2 = 21$
center: $(0, 4)$, $r = \sqrt{21}$

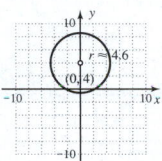

19.

20.

21.

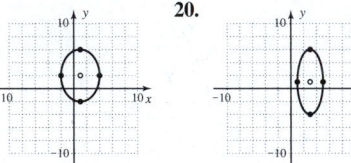
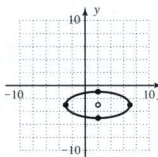

22.

23.

24.

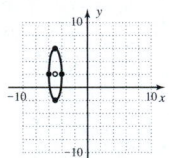
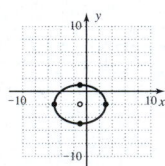
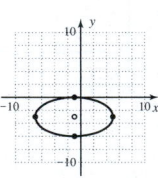

25. a. $\dfrac{x^2}{16} + \dfrac{y^2}{4} = 1$, $(0, 0)$, $a = 4$, $b = 2$
b. $(-4, 0)$, $(4, 0)$, $(0, -2)$, $(0, 2)$ **c.**
d.

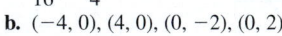

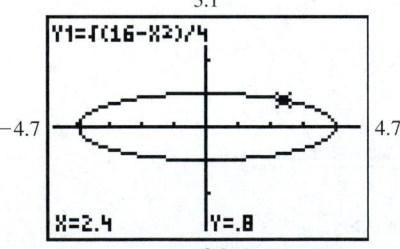

$(2.4, 0.8)$, $(2.4, -0.8)$, $(-2.4, 0.8)$, $(-2.4, -0.8)$

26. a. $\dfrac{x^2}{4} + \dfrac{y^2}{36} = 1$, $(0, 0)$, $a = 2$, $b = 6$
b. $(0, -6)$, $(0, 6)$, $(-2, 0)$, $(2, 0)$
c.

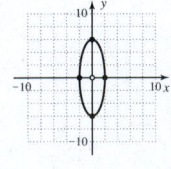

d.

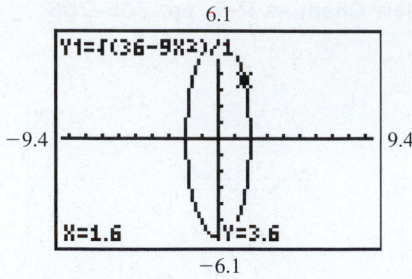

$(1.6, 3.6)$, $(1.6, -3.6)$, $(-1.6, 3.6)$, $(-1.6, -3.6)$,

27. a. $\dfrac{x^2}{9} + \dfrac{y^2}{16} = 1$, $(0, 0)$, $a = 3$, $b = 4$
b. $(0, -4)$, $(0, 4)$, $(-3, 0)$, $(3, 0)$ **c.**

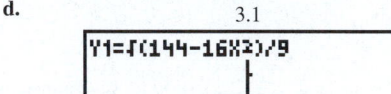

d.

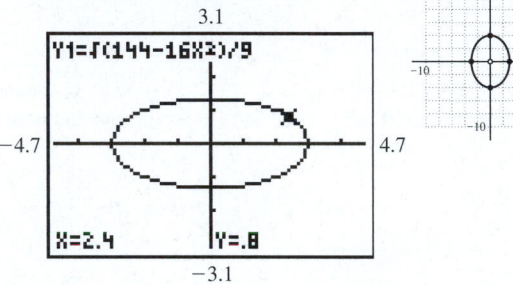

$(2.4, 0.8)$, $(2.4, -0.8)$, $(-2.4, 0.8)$, $(-2.4, -0.8)$,

28. a. $\dfrac{x^2}{9} + \dfrac{y^2}{25} = 1$, $(0, 0)$, $a = 3$, $b = 5$
b. $(0, -5)$, $(0, 5)$, $(-3, 0)$, $(3, 0)$ **c.**

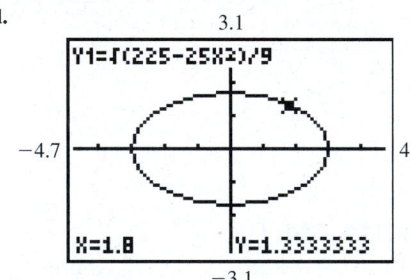

d.

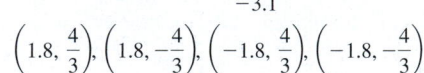

$\left(1.8, \dfrac{4}{3}\right), \left(1.8, -\dfrac{4}{3}\right), \left(-1.8, \dfrac{4}{3}\right), \left(-1.8, -\dfrac{4}{3}\right)$

29. a. $\dfrac{x^2}{5} + \dfrac{y^2}{2} = 1$, $(0, 0)$, $a = \sqrt{5}$, $b = \sqrt{2}$
b. $(-\sqrt{5}, 0)$, $(\sqrt{5}, 0)$, $(0, -\sqrt{2})$, $(0, \sqrt{2})$ **c.**

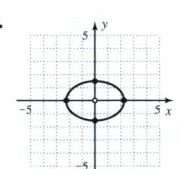

30. a. $\dfrac{x^2}{7} + \dfrac{y^2}{3} = 1$, $(0, 0)$, $a = \sqrt{7}$, $b = \sqrt{3}$
b. $(-\sqrt{7}, 0)$, $(\sqrt{7}, 0)$, $(0, -\sqrt{3})$, $(0, \sqrt{3})$ **c.**

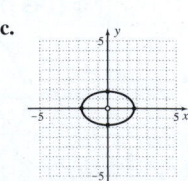

31. ellipse

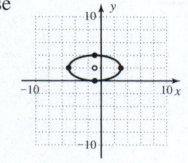

32. ellipse

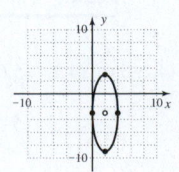

33. circle

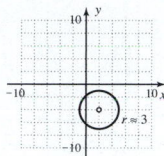

34. circle

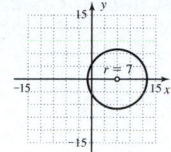

35. ellipse

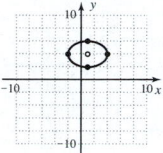

36. ellipse

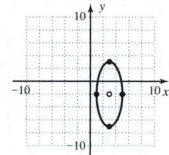

37. $x^2 + \dfrac{(y+3)^2}{4} = 1$

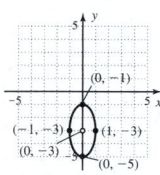

$D: x \in [-1, 1], R: y \in [-5, -1]$

38. $\dfrac{(x+4)^2}{9} + \dfrac{y^2}{3} = 1$

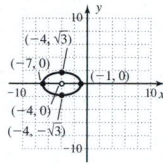

$D: x \in [-7, -1], R: y \in [-\sqrt{3}, \sqrt{3}]$

39. $\dfrac{(x+2)^2}{16} + \dfrac{(y-1)^2}{4} = 1$

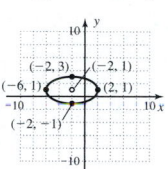

$D: x \in [-6, 2], R: y \in [-1, 3]$

40. $\dfrac{(x+2)^2}{12} + \dfrac{(y-4)^2}{36} = 1$

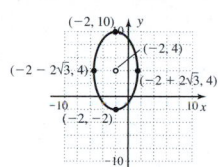

$D: x \in [-2 - 2\sqrt{3}, -2 + 2\sqrt{3}],$
$R: y \in [-2, 10]$

41. $\dfrac{(x-3)^2}{4} + \dfrac{(y+5)^2}{10} = 1$

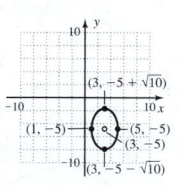

$D: x \in [1, 5],$
$R: y \in [-5 - \sqrt{10}, -5 + \sqrt{10}]$

42. $\dfrac{(x-2)^2}{9} + \dfrac{(y+1)^2}{4} = 1$

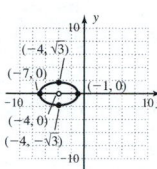

$D: x \in [-1, 5], R: y \in [-3, 1]$

43. $\dfrac{(x-3)^2}{25} + \dfrac{(y+2)^2}{10} = 1$

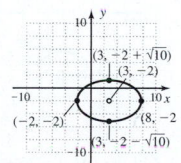

$D: x \in [-2, 8],$
$R: y \in [-2 - \sqrt{10}, -2 + \sqrt{10}]$

44. $\dfrac{(x-2)^2}{9} + \dfrac{(y+3)^2}{18} = 1$

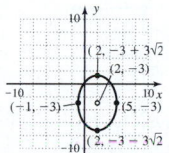

$D: x \in [-1, 5],$
$R: y \in [-3 - 3\sqrt{2}, -3 + 3\sqrt{2}]$

49. a. (2, 1) **b.** (−3, 1) and (7, 1) **c.** $(2 - \sqrt{21}, 1)$ and $(2 + \sqrt{21}, 1)$
d. (2, 3) and (2, −1) **e.**

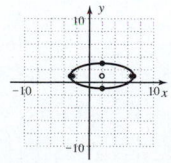

50. a. (3, 2) **b.** (−1, 2) and (7, 2) **c.** $(3 - \sqrt{7}, 2)$ and $(3 + \sqrt{7}, 2)$
d. (3, 5) and (3, −1) **e.**

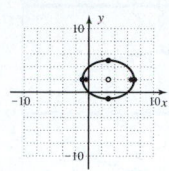

51. a. (4, −3) **b.** (4, 2) and (4, −8) **c.** (4, 0) and (4, −6) **d.** (0, −3)
and (8, −3) **e.**

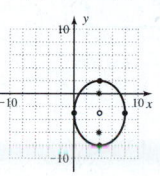

52. a. (−2, 5) **b.** (−2, 12) and (−2, −2) **c.** $(-2, 5 + 3\sqrt{5})$ and
$(-2, 5 - 3\sqrt{5})$ **d.** (−4, 5) and (0, 5) **e.**

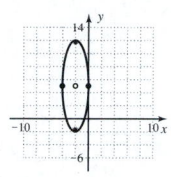

53. a. (−2, −2) **b.** (−5, −2) and (1, −2)
c. $(-2 + \sqrt{3}, -2)$ and $(-2 - \sqrt{3}, -2)$ **d.** $(-2, -2 + \sqrt{6})$ and
$(-2, -2 - \sqrt{6})$ **e.**

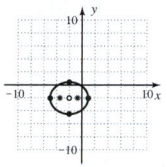

54. a. (5, 3) **b.** (5, 8) and (5, −2) **c.** $(5, 3 + \sqrt{15})$ and $(5, 3 - \sqrt{15})$
d. $(5 - \sqrt{10}, 3)$ and $(5 + \sqrt{10}, 3)$ **e.**

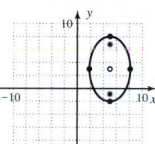

55. $\dfrac{x^2}{36} + \dfrac{y^2}{20} = 1$ **a.**

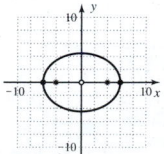

b.

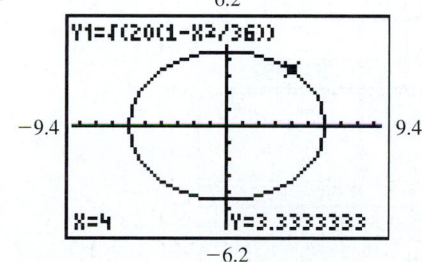

c. $L = 6.\overline{6}$ **d.** verified

56. $\dfrac{x^2}{64} + \dfrac{y^2}{39} = 1$ **a.**

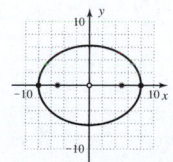

b.

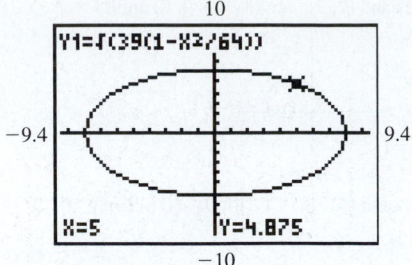

c. $L = 9.75$ **d.** verified

57. $\dfrac{(x-3)^2}{9} + \dfrac{(y+2)^2}{25} = 1$ **a.**

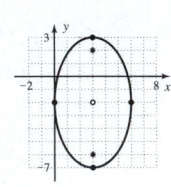

b.

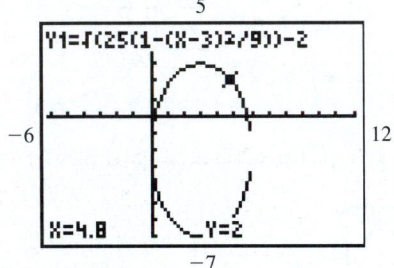

c. $L = 3.6$ **d.** verified

58. $\dfrac{(x-2)^2}{52} + \dfrac{(y+3)^2}{16} = 1$ **a.**

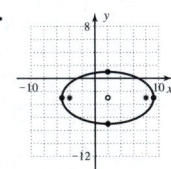

b.

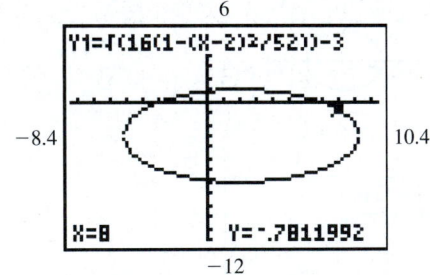

c. $L \approx 4.4$ **d.** verified

61. $\dfrac{(x+3)^2}{4} + \dfrac{(y+1)^2}{16} = 1,\ (-3,\, -1 \pm 2\sqrt{3})$

62. $\dfrac{(x-1)^2}{25} + \dfrac{(y-1)^2}{9} = 1,\ (5,1),\ (-3,1)$

69. $\dfrac{x^2}{15^2} + \dfrac{y^2}{8^2} = 1;\ 6.4$ ft

86. $(30 + 10\sqrt{3}) + (30 - 10\sqrt{3})i;\ \dfrac{3 - \sqrt{3}}{10} + \dfrac{3 + \sqrt{3}}{10}i$

87. a. $x \in (0, 6)$

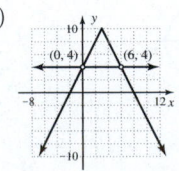

b. $-2|x - 3| + 10 > 4$
$|x - 3| < 3$
$x - 3 > -3$ or $x - 3 < 3$
$x > 0$ or $x < 6$
$0 < x < 6$

Mid-Chapter Check, pp. 729–730

1.   **2.**   **3.**

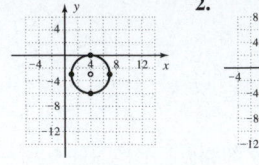   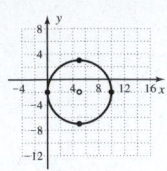   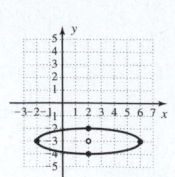

4.   **5.**   **6.**

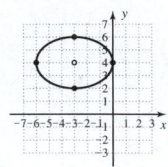   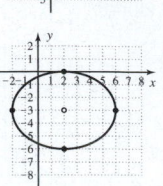

8. a. $\dfrac{(x+3)^2}{4} + \dfrac{(y-1)^2}{16} = 1;\ D: x \in [-5, -1],\ R: y \in [-3, 5]$

b. $(x - 3)^2 + (y - 2)^2 = 16;\ D: x \in [-1, 7],\ R: y \in [-2, 6]$

Exercises 8.3, pp. 740–744

7.   **8.**   **9.**

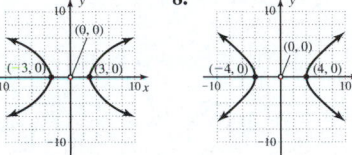   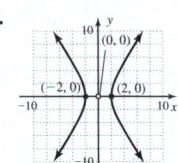

10.   **11.**   **12.**

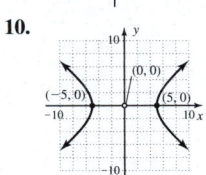   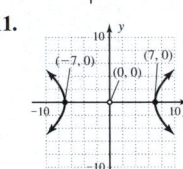   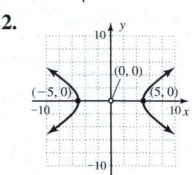

13.   **14.**   **15.**

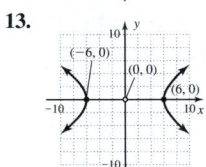   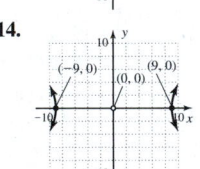   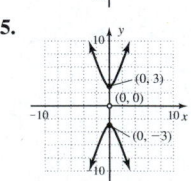

16.   **17.**   **18.**

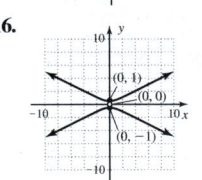   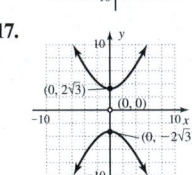   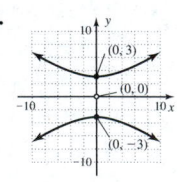

19.   **20.**   **21.**

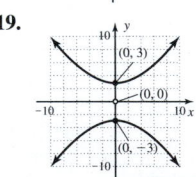   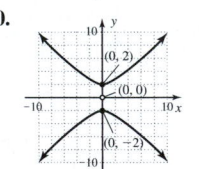   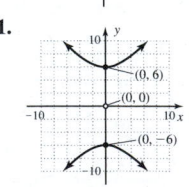

22.

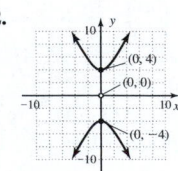

23. $(-4, -2), (2, -2), y = -2, (-1, -2), x = -1$
24. $(-3, 3), (-3, -1), x = -3, (-3, 1), y = 1$
25. $(4, 1), (4, -3), x = 4, (4, -1), y = -1$
26. $(-1, 2), (3, 2), y = 2, (1, 2), x = 1$

27.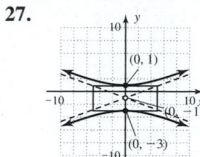

28.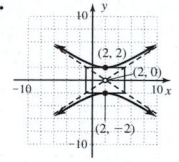

29.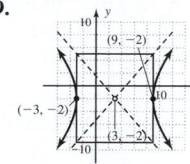

30.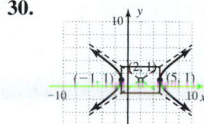

31.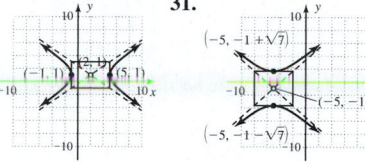

32.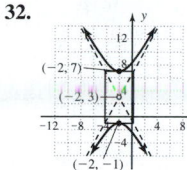

33.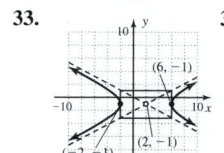

34.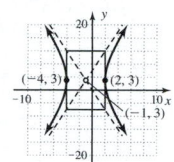

35.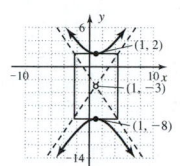

36.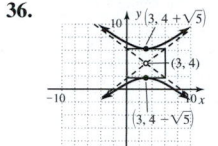

37.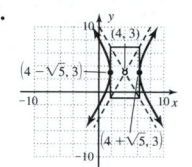

38.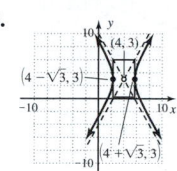

39.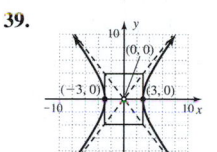

40.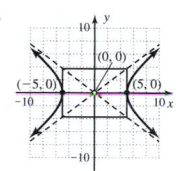

41.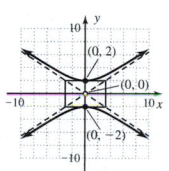

42.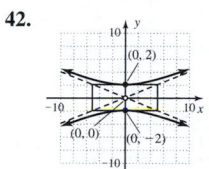

43.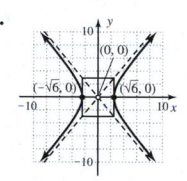

44.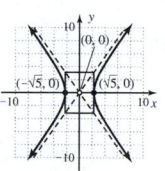

45. $\dfrac{(x + 5)^2}{9} - \dfrac{(y + 2)^2}{36} = 1$

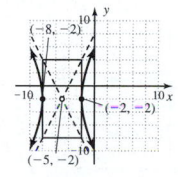

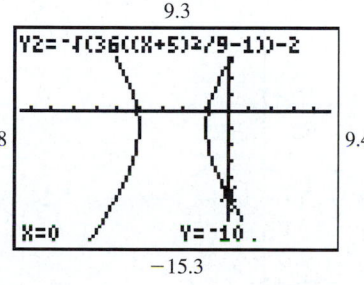

$(0, -10), (0, 6), (-10, -10), (-10, 6)$

46. $\dfrac{(x - 6)^2}{4} - \dfrac{(y + 2)^2}{1} = 1$

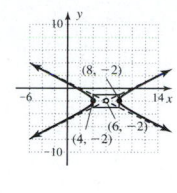

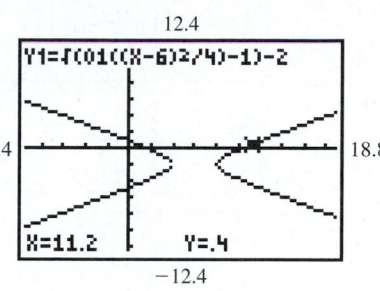

$(11.2, 0.4), (11.2, -4.4), (0.8, -4.4), (0.8, 0.4)$

47. $\dfrac{(y - 3)^2}{1} - \dfrac{(x + 2)^2}{4} = 1$

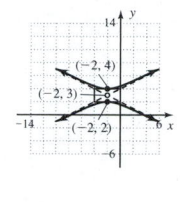

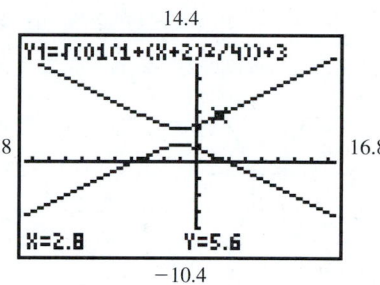

$(2.8, 5.6), (2.8, 0.4), (-6.8, 0.4), (-6.8, 5.6)$

48. $\dfrac{(y - 3)^2}{9} - \dfrac{(x + 1)^2}{4} = 1$

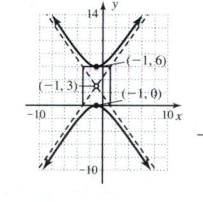

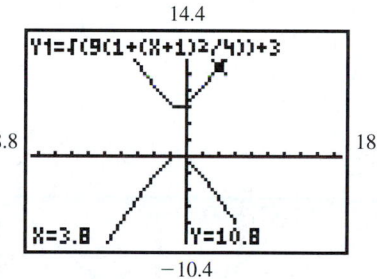

$(3.8, 10.8), (3.8, -4.8), (-5.8, -4.8), (-5.8, 10.8)$

65. $\dfrac{(x - 3)^2}{9} - \dfrac{(y - 4)^2}{4} = 1$ **a.** $(3, 4)$ **b.** $(0, 4)$ and $(6, 4)$
c. $(3 - \sqrt{13}, 4)$ and $(3 + \sqrt{13}, 4)$ **d.** $2a = 6, 2b = 4$
e.

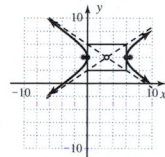

66. $\dfrac{(x - 5)^2}{36} - \dfrac{(y - 2)^2}{4} = 1$ **a.** $(5, 2)$ **b.** $(-1, 2)$ and $(11, 2)$
c. $(5 - 2\sqrt{10}, 2)$ and $(5 + 2\sqrt{10}, 2)$ **d.** $2a = 12, 2b = 4$
e.

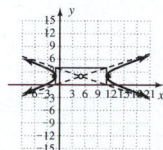

67. $\dfrac{(y - 3)^2}{16} - \dfrac{x^2}{4} = 1$ **a.** $(0, 3)$ **b.** $(0, 7), (0, -1)$
c. $(0, 3 + 2\sqrt{5}), (0, 3 - 2\sqrt{5})$ **d.** $2a = 4, 2b = 8$

e. $y = 2x + 3, y = -2x + 3$

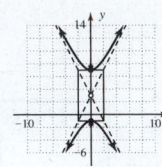

68. $\dfrac{y^2}{81} - \dfrac{(x+1)^2}{4} = 1$ **a.** $(-1, 0)$ **b.** $(-1, 9), (-1, -9)$
c. $(-1, \sqrt{85}), (-1, -\sqrt{85})$,   **d.** $2a = 4, 2b = 18$
e. $y = 4.5x + 4.5, y = -4.5x - 4.5$

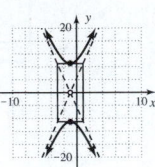

69. $\dfrac{(x-3)^2}{4} - \dfrac{(y+2)^2}{12} = 1$ **a.** $(3, -2)$ **b.** $(1, -2)$ and $(5, -2)$
c. $(-1, -2)$ and $(7, -2)$ **d.** $2a = 4, 2b = 4\sqrt{3}$
e.

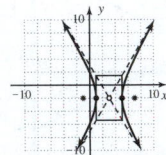

70. $\dfrac{(x+3)^2}{9} - \dfrac{(y-2)^2}{18} = 1$ **a.** $(-3, 2)$ **b.** $(-6, 2)$ and $(0, 2)$
c. $(-3 - 3\sqrt{3}, 2)$ and $(-3 + 3\sqrt{3}, 2)$ **d.** $2a = 6, 2b = 6\sqrt{2}$
e.

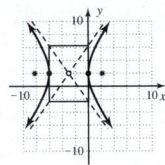

79. a. $y = \frac{2}{3}\sqrt{x^2 - 9}$ **b.** $x \in (-\infty, -3] \cup [3, \infty)$
c.

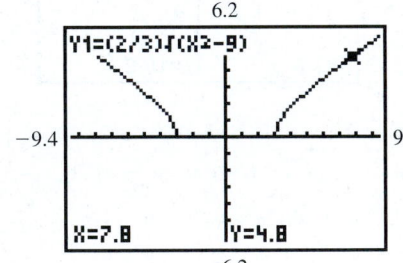

$y = \dfrac{-2}{3}\sqrt{x^2 - 9}$

87. $y = \pm\sqrt{\dfrac{b^2}{a^2}x^2 - b^2}$, as $x \to \infty, y \to \pm\sqrt{\dfrac{b^2}{a^2}x^2} = \pm\dfrac{b}{a}x$

90.

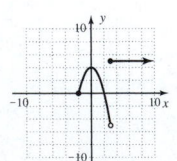

Exercises 8.4, pp. 751–754

7. $x \in (-\infty, \infty), y \in [-4, \infty)$

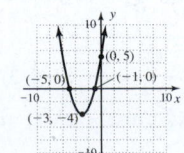

8. $x \in (-\infty, \infty), y \in [-4, \infty)$

9. $x \in (-\infty, \infty), y \in (-\infty, 18]$

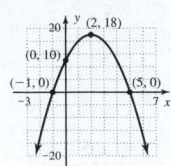

10. $x \in (-\infty, \infty), y \in (-\infty, 27]$

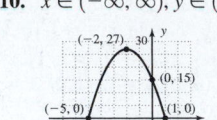

11. $x \in (-\infty, \infty),$
$y \in [-10.125, \infty)$

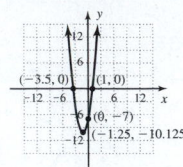

12. $x \in (-\infty, \infty),$
$y \in [-3.125, \infty)$

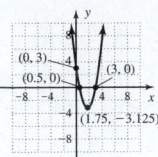

13. $x \in [-4, \infty), y \in (-\infty, \infty)$

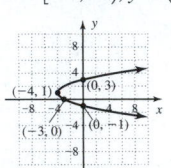

14. $x \in [-16, \infty), y \in (-\infty, \infty)$

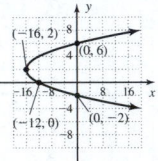

15. $x \in (-\infty, 16], y \in (-\infty, \infty)$

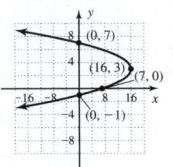

16. $x \in (-\infty, 4], y \in (-\infty, \infty)$

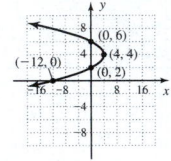

17. $x \in (-\infty, 0], y \in (-\infty, \infty)$

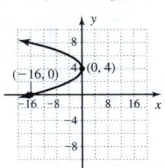

18. $x \in (-\infty, 0], y \in (-\infty, \infty)$

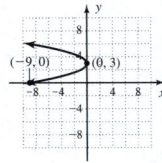

19. $x \in [-9, \infty), y \in (-\infty, \infty)$

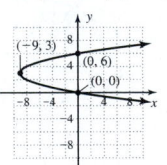

20. $x \in [-16, \infty), y \in (-\infty, \infty)$

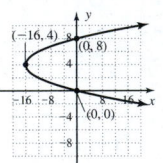

21. $x \in [-4, \infty), y \in (-\infty, \infty)$

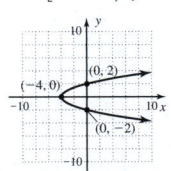

22. $x \in [-9, \infty), y \in (-\infty, \infty)$

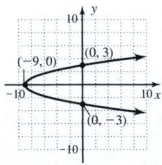

23. $x \in (-\infty, 0], y \in (-\infty, \infty)$

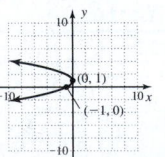

24. $x \in (-\infty, 0], y \in (-\infty, \infty)$

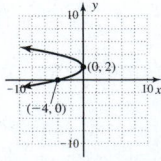

25. $x \in [-6.25, \infty), y \in (-\infty, \infty)$

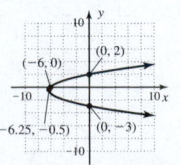

26. $x \in [-9, \infty), y \in (-\infty, \infty)$

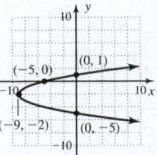

27. $x \in [-21, \infty), y \in (-\infty, \infty)$

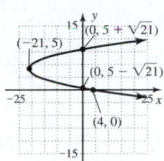

28. $x \in [-41, \infty), y \in (-\infty, \infty)$

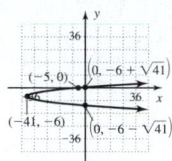

29. $x \in (-\infty, 11], y \in (-\infty, \infty)$

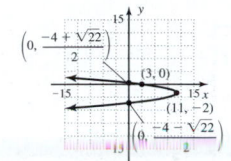

30. $x \in [-10, \infty), y \in (-\infty, \infty)$

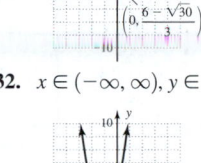

31. $x \in (-\infty, \infty), y \in [3, \infty)$

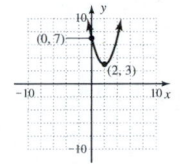

32. $x \in (-\infty, \infty), y \in [-4, \infty)$

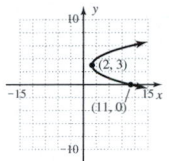

33. $x \in [2, \infty), y \in (-\infty, \infty)$

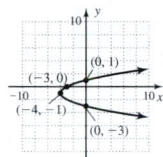

34. $x \in [-4, \infty), y \in (-\infty, \infty)$

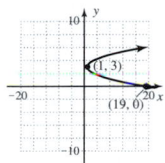

35. $x \in [1, \infty), y \in (-\infty, \infty)$

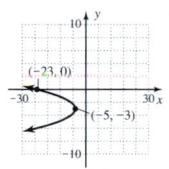

36. $x \in (-\infty, -5], y \in (-\infty, \infty)$

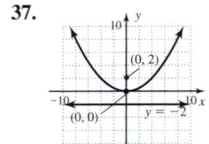

37.

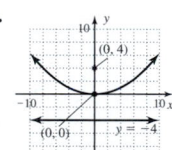

38.

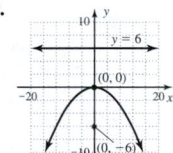

39.

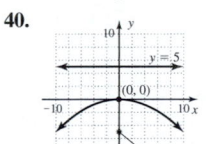

40.

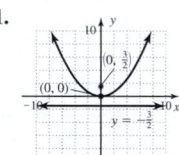

41.

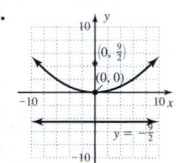

42.

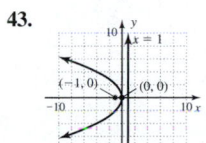

43.

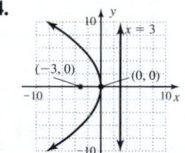

44.

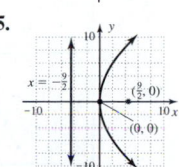

45.

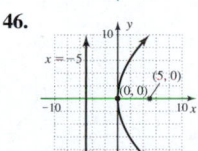

46.

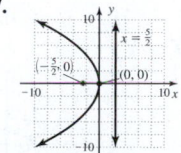

47.

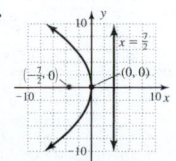

48.

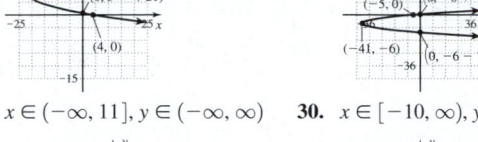

49.

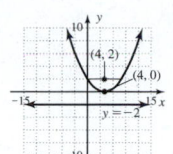

50.

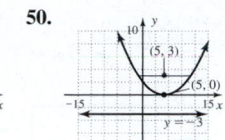

51.

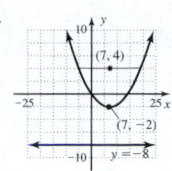

52.

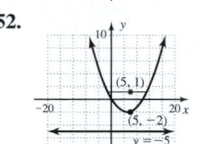

53.

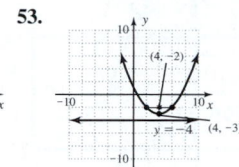

54.

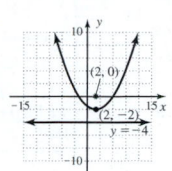

55.

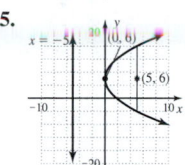

56.

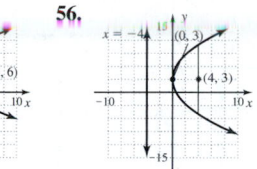

57.

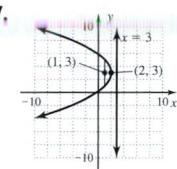

58.

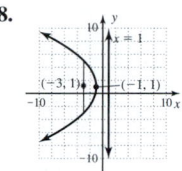

59.

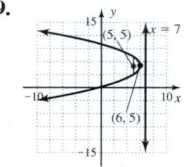

60.

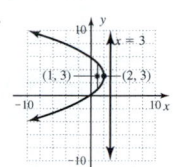

77. $(4, 3), (4, -3), (-4, 3), (-4, -3)$

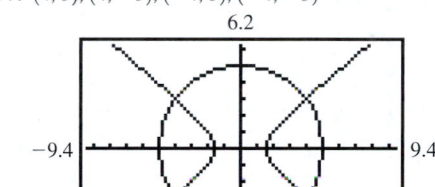

78. $(2, 4), (2, -4), (-2, 4), (-2, -4)$

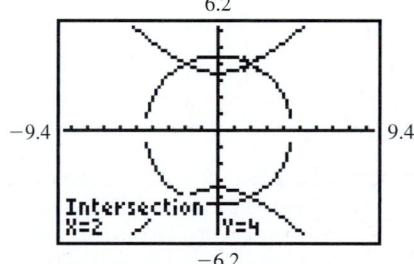

79. $(3, 5), (-3, 5), (0, -4)$

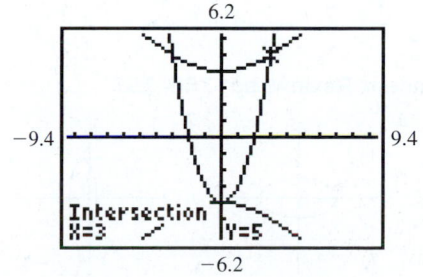

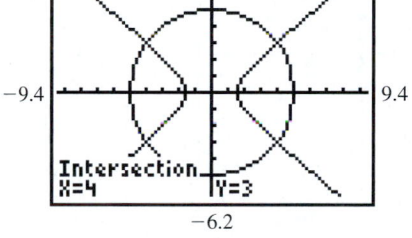
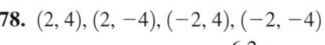
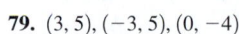

80. $(5, 2), (-5, 2), \left(\dfrac{\sqrt{555}}{3}, -\dfrac{16}{3}\right), \left(-\dfrac{\sqrt{555}}{3}, -\dfrac{16}{3}\right)$

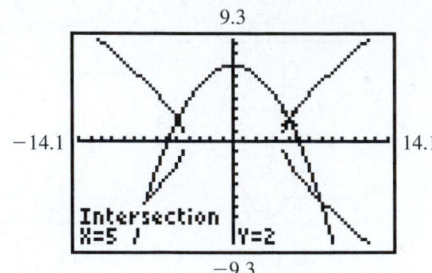

81. $(5, 5), (5, -5), (-5, 5), (-5, -5)$

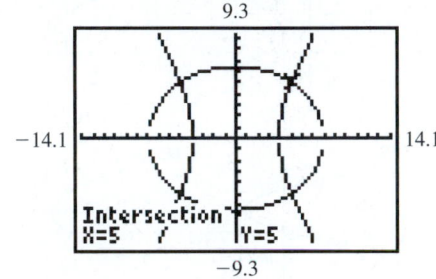

82. $(3, 1), (3, -1), (-3, 1), (-3, -1)$

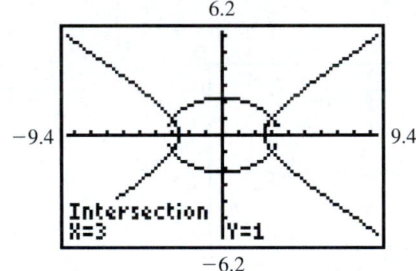

89. 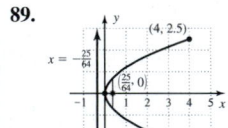 **90.**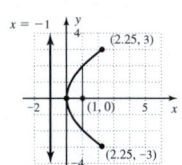

98.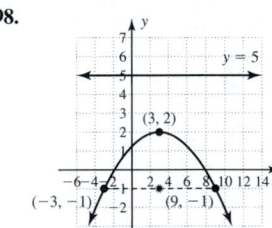

$-12(y - 2) = (x - 3)^2$; vertex $(3, 2)$;
$|p| = 3$, focus $(3, -1)$, directrix $y = 5$;
endpoints of focal chord $(3 - 6, -1)$,
$(3 + 6, -1)$ or $(-3, -1)$ and $(9, -1)$

Summary and Concept Review, pp. 755–757

5. 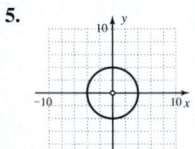 **6.** 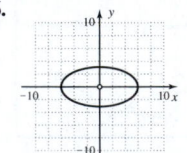 **7.**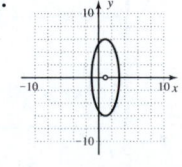

8. **9.**
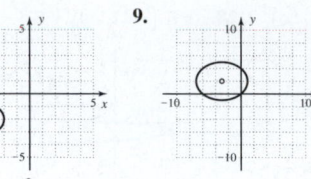

10. $\dfrac{x^2}{25} + \dfrac{y^2}{9} = 1$

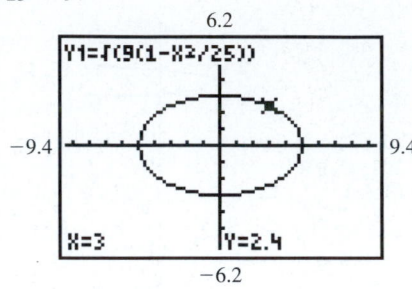

one possibility is $(3, 2.4), (-3, 2.4), (3, -2.4), (-3, -2.4)$

12. $\dfrac{(x - 2)^2}{25} + \dfrac{(y - 1)^2}{4} = 1$ **13.**

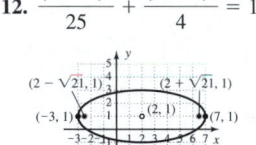 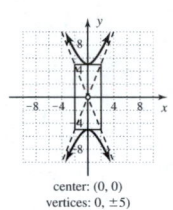

center: $(0, 0)$
vertices: $0, \pm 5$

14. **15.** **16.**
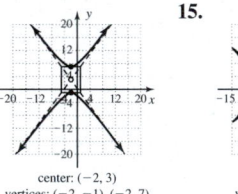 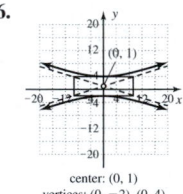

center: $(-2, 3)$
vertices: $(-2, -1), (-2, 7)$

center: $(-2, 1)$
vertices: $(-5, 1), (1, 1)$

center: $(0, 1)$
vertices: $(0, -2), (0, 4)$

17.
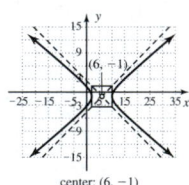

center: $(6, -1)$
vertices: $(2, -1), (10, -1)$

18. $\dfrac{x^2}{9} - \dfrac{y^2}{16} = 1$

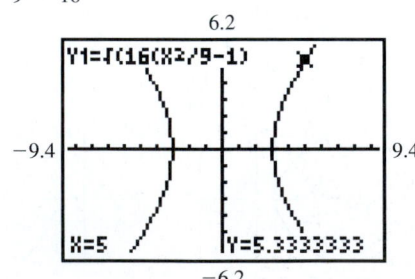

$(5, 5.\overline{3}), (-5, 5.\overline{3}), (5, -5.\overline{3}), (-5, -5.\overline{3}),$

20. $\dfrac{(x - 5)^2}{9} - \dfrac{(y - 2)^2}{4} = 1$ **21.**

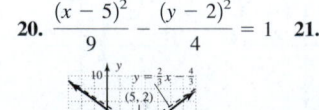 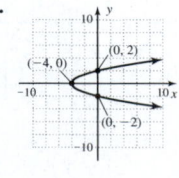

22.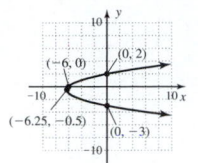

23.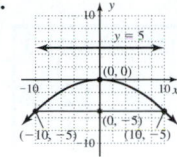

24.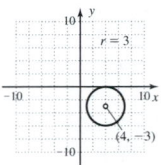

3. $\dfrac{(x+3)^2}{\left(\frac{7}{2}\right)^2} + \dfrac{(y-1)^2}{\left(\frac{6}{5}\right)^2} = 1$; $a = \dfrac{7}{2}, b = \dfrac{6}{5}$

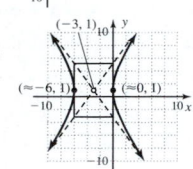

4. $\dfrac{(x+3)^2}{\left(\frac{4\sqrt{5}}{3}\right)^2} - \dfrac{(y-1)^2}{\left(\frac{9}{2}\right)^2} = 1$; $a = \dfrac{4\sqrt{5}}{3} \approx 3, b = \dfrac{9}{2}$

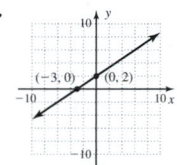

Practice Test, pp. 757–758

5.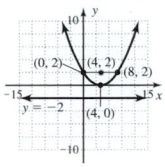

6. $C(2,-3)$
$F_1(2-\sqrt{15}, -3)$
$F_2(2+\sqrt{15}, -3)$

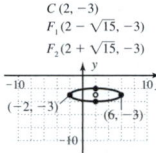

7.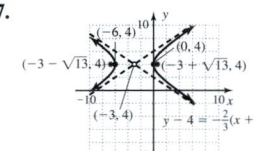

8.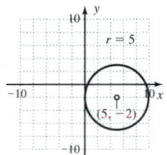

9.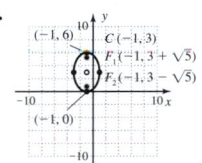

10.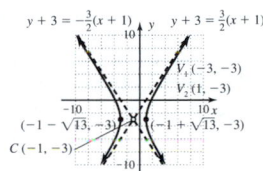

11.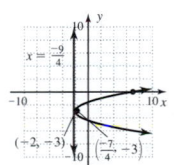

12.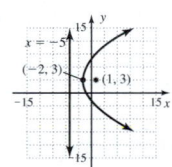

13. a. $\left(\dfrac{10}{3}, \dfrac{16}{3}\right), (-2, 0)$ **b.** $(2,2),(-2,2),(-2,-2),(2,-2)$

16. $\dfrac{x^2}{16} + \dfrac{y^2}{12} = 1$

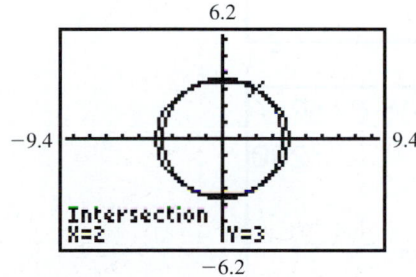

$(2,3), (2,-3), (-2,-3), (-2,3)$

Strengthening Core Skills, p. 759

1. $\dfrac{(x-2)^2}{\left(\frac{\sqrt{2}}{5}\right)^2} - \dfrac{(y+3)^2}{\left(\frac{2}{3}\right)^2} = 1$

$a = \dfrac{\sqrt{2}}{5}, b = \dfrac{2}{3}$

2. $\dfrac{(x-1)^2}{\left(\frac{5\sqrt{7}}{14}\right)^2} + \dfrac{(y+2)^2}{\left(\frac{5\sqrt{3}}{12}\right)^2} = 1$

$a = \dfrac{5\sqrt{7}}{14}, b = \dfrac{5\sqrt{3}}{12}$

Cumulative Review Chapters R–8, p. 760

10.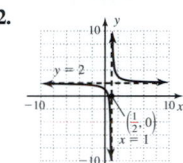

11.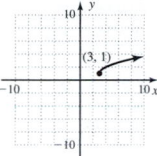

12.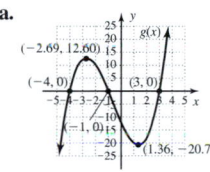

13.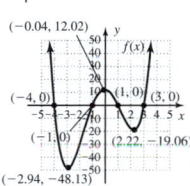

14. a. 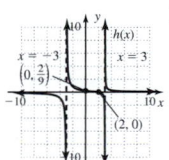 **b.**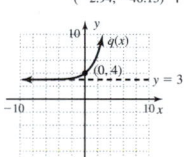

15.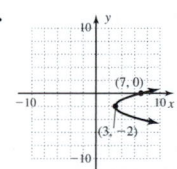

16.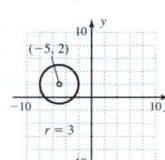

17.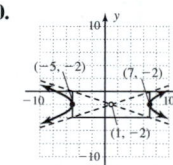

18.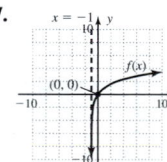

19.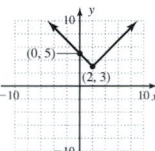

20. (bottom graph)

CHAPTER 9
Exercises 9.1, pp. 770–772

13. $\dfrac{1}{2}, \dfrac{2}{3}, \dfrac{3}{4}, \dfrac{4}{5}; a_8 = \dfrac{8}{9}; a_{12} = \dfrac{12}{13}$

14. $2, \dfrac{9}{4}, \dfrac{64}{27}, \dfrac{625}{256}; a_8 = \left(\dfrac{9}{8}\right)^8; a_{12} = \left(\dfrac{13}{12}\right)^{12}$

15. $\dfrac{1}{2}, \dfrac{1}{4}, \dfrac{1}{8}, \dfrac{1}{16}; a_8 = \dfrac{1}{256}; a_{12} = \dfrac{1}{4096}$

16. $\dfrac{2}{3}, \dfrac{4}{9}, \dfrac{8}{27}, \dfrac{16}{81}; a_8 = \left(\dfrac{2}{3}\right)^8; a_{12} = \left(\dfrac{2}{3}\right)^{12}$

17. $1, \dfrac{1}{2}, \dfrac{1}{3}, \dfrac{1}{4}; a_8 = \dfrac{1}{8}; a_{12} = \dfrac{1}{12}$ **18.** $1, \dfrac{1}{4}, \dfrac{1}{9}, \dfrac{1}{16}; a_8 = \dfrac{1}{64}; a_{12} = \dfrac{1}{144}$

19. $\dfrac{-1}{2}, \dfrac{1}{6}, \dfrac{-1}{12}, \dfrac{1}{20}; a_8 = \dfrac{1}{72}; a_{12} = \dfrac{1}{156}$

20. $1, \dfrac{-1}{7}, \dfrac{1}{17}, \dfrac{-1}{31}; a_8 = \dfrac{-1}{127}; a_{12} = \dfrac{-1}{287}$

21. $-2, 4, -8, 16; a_8 = 256; a_{12} = 4096$

22. $\frac{-1}{2}, \frac{1}{4}, \frac{-1}{8}, \frac{1}{16}; a_8 = \frac{1}{256}; a_{12} = \frac{1}{4096}$

57. $-2 + 1 + 4 + 7 = 10$ **58.** $-1 + 1 + 3 + 5 + 7 = 15$

59. $-1 + 5 + 15 + 29 + 47 = 95$ **60.** $2 + 5 + 10 + 17 + 26 = 60$

61. $-1 + 2 + (-3) + 4 + (-5) + 6 + (-7) = -4$

62. $-2 + 4 + (-8) + 16 + (-32) = -22$

63. $0.5 + 2 + 4.5 + 8 = 15$ **64.** $4 + 9 + 16 = 29$

65. $6 + 8 + 10 + 12 + 14 = 50$

66. $\frac{3}{8} + \frac{1}{4} + \frac{5}{32} + \frac{3}{32} + \frac{7}{128} = \frac{119}{128}$

67. $-\frac{1}{3} + \frac{1}{8} + \left(-\frac{1}{15}\right) + \frac{1}{24} + \left(-\frac{1}{35}\right) + \frac{1}{48} = -\frac{27}{112}$

68. $-\frac{1}{3} + \frac{1}{8} + \left(-\frac{1}{15}\right) + \frac{1}{24} + \left(-\frac{1}{35}\right) = -\frac{11}{42}$

93. $\sum_{j=1}^{n} ca_j = ca_1 + ca_2 + ca_3 + \cdots + ca_{n-1} + ca_n$

$\qquad = c(a_1 + a_2 + a_3 + \cdots + a_{n-1} + a_n)$

$\qquad = c\sum_{j=1}^{n} a_j$

Exercises 9.2, pp. 779–782

31. $a_1 = 2, d = 5, a_n = 5n - 3, a_6 = 27, a_{10} = 47, a_{12} = 57$

32. $a_1 = 7, d = -3, a_n = -3n + 10,$
$a_6 = -8, a_{10} = -20, a_{12} = -26$

33. $a_1 = 5.10, d = 0.15,$
$a_n = 0.15n + 4.95, a_6 = 5.85,$
$a_{10} = 6.45, a_{12} = 6.75$

34. $a_1 = 9.75, d = -0.35,$
$a_n = -0.35n + 10.10$
$a_6 = 8.00, a_{10} = 6.60, a_{12} = 5.90$

35. $a_1 = \frac{3}{2}, d = \frac{3}{4}, a_n = \frac{3}{4}n + \frac{3}{4},$
$a_6 = \frac{21}{4}, a_{10} = \frac{33}{4}, a_{12} = \frac{39}{4}$

36. $a_1 = \frac{5}{7}, d = -\frac{1}{2}; a_n = -\frac{1}{2}n + \frac{17}{14}; a_6 = \frac{-25}{14}; a_{10} = \frac{-53}{14}; a_{12} = \frac{-67}{14}$

51. **a.** appears linear   **b.** $d = 0.75$ **c.** $a_n = 0.75n + 0.75$

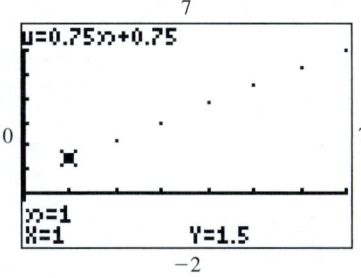

52. **a.** appears linear   **b.** $d = -\dfrac{1}{2}$ **c.** $a_n = -\dfrac{1}{2}n + \dfrac{28}{9}$

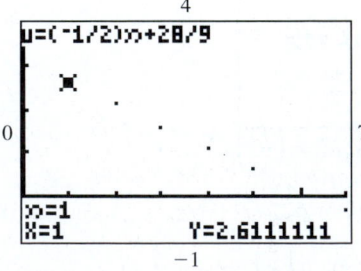

61. 1275

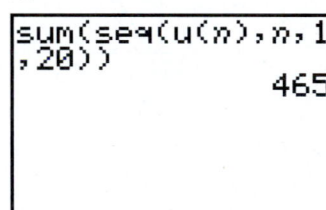

62. 1711

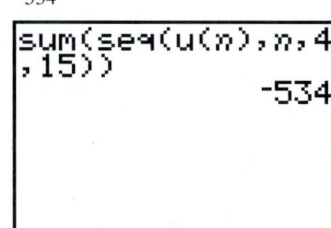

63. 601.25

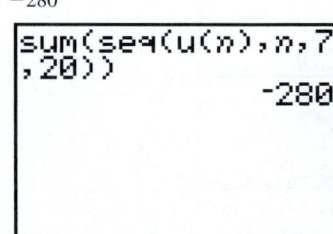

64. 465

65. -534

66. -280

84.

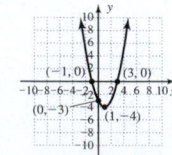

Exercises 9.3, pp. 791–795

15. not geometric; ratio of terms decreases by 1

16. not geometric; ratio of terms decreases by 1

18. $r = \frac{-2}{3}$ **23.** not geometric; $a_n = \frac{240}{n!}$ **24.** not geometric, $a_n = \frac{-120}{n!}$

25. 5, 10, 20, 40 **26.** 2, −8, 32, −128 **27.** −6, 3, $\frac{-3}{2}$, $\frac{3}{4}$

28. $\frac{2}{3}, \frac{2}{15}, \frac{2}{75}, \frac{2}{375}$ **29.** 4, $4\sqrt{3}$, 12, $12\sqrt{3}$ **30.** $\sqrt{5}$, 5, $5\sqrt{5}$, 25

31. 0.1, 0.01, 0.001, 0.0001

32. 0.024, 0.00024, 0.0000024, 0.000000024

33. $a_n = -24\left(\frac{1}{2}\right)^{n-1}$; $a_7 = -\frac{3}{8}$ **34.** $a_n = 48\left(-\frac{1}{3}\right)^{n-1}$; $a_6 = -\frac{16}{81}$

35. $a_n = -\frac{1}{20}(-5)^{n-1}$; $a_4 = \frac{25}{4}$ **36.** $a_n = \frac{3}{20}(4)^{n-1}$; $a_5 = \frac{192}{5}$

37. $a_n = 2(\sqrt{2})^{n-1} = (\sqrt{2})^{n+1}$; $a_7 = 16$

38. $a_n = \sqrt{3}(\sqrt{3})^{n-1} = (\sqrt{3})^n$; $a_8 = 81$

39. $a_1 = \frac{1}{27}, r = -3, a_n = \frac{1}{27}(-3)^{n-1}, a_6 = -9, a_{10} = -729$, $a_{12} = -6561$

40. $a_1 = -\frac{7}{8}; r = -2, a_n = -\frac{7}{8}(-2)^{n-1}; a_6 = 28; a_{10} = 448; a_{12} = 1792$

41. $a_1 = 729, r = \frac{1}{3}, a_n = 729(\frac{1}{3})^{n-1}, a_6 = 3, a_{10} = \frac{1}{27}, a_{12} = \frac{1}{243}$

42. $a_1 = 625, r = \frac{1}{5}, a_n = 625(\frac{1}{5})^{n-1}, a_6 = \frac{1}{5}, a_{10} = \frac{1}{3125}, a_{12} = \frac{1}{78,125}$

43. $a_1 = \frac{1}{2}, r = \sqrt{2}, a_n = \frac{1}{2}(\sqrt{2})^{n-1}, a_6 = 2\sqrt{2}, a_{10} = 8\sqrt{2}$, $a_{12} = 16\sqrt{2}$

44. $a_1 = 36\sqrt{3}, r = \frac{1}{\sqrt{3}}, a_n = 36\sqrt{3}\left(\frac{1}{\sqrt{3}}\right)^{n-1}, a_6 = 4$, $a_{10} = \frac{4}{9}, a_{12} = \frac{4}{27}$ **45.** $a_1 = 0.2, r = 0.4, a_n = 0.2(0.4)^{n-1}$ $a_6 = 0.002048, a_{10} = 0.0000524288, a_{12} = 0.000008388608$

46. $a_1 = 0.5, r = -0.7, a_n = 0.5(-0.7)^{n-1}, a_6 = -0.084035$, $a_{10} = -0.0201768035, a_{12} \approx -0.0098866337$

59. a. appears exponential **b.** $r = 0.2$ **c.** $a_n = 131.25(0.2)^{n-1}$

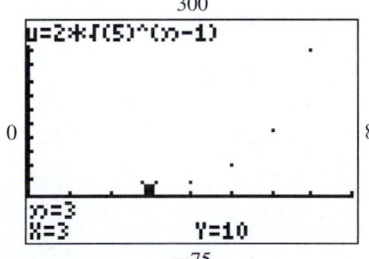

60. a. appears exponential **b.** $r = \sqrt{5}$ **c.** $a_n = 2(\sqrt{5})^{n-1}$

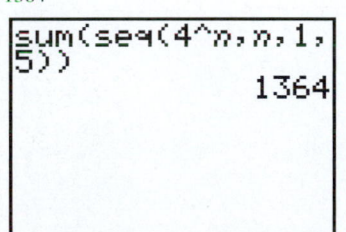

61. a. appears nonexponential **b.** no common ratio

62. a. appears nonexponential **b.** no common ratio

81. 1364

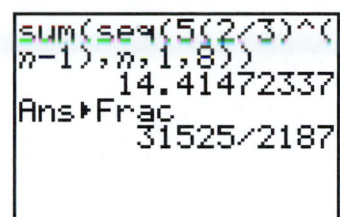

82. 2046

83. $\frac{31,525}{2187}$

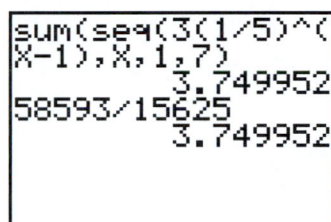

84. $\frac{58,593}{15,625}$

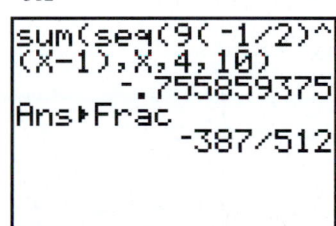

85. $-\frac{387}{512}$

86. $\frac{4095}{16,384}$

119. $a_0 = 462; a_n = 277.2\left(\frac{3}{5}\right)^{n-1}$, $a_5 \approx 35.9$ in^3, 7 strokes

120. $a_0 = 4200, a_n = 3780(0.9)^{n-1}$, about 1318 in^3, 0.15(4200) = 630, about 18 strokes **121.** $a_0 = 8, a_n = 6(0.75)^{n-1}$, 2 days, 8 days

122. $a_0 = 50,000, a_n = 48,750(0.975)^{n-1}$, about 29,381 ft^3, 64 days

123. $a_0 = 50; a_n = 100(2)^{n-1}, a_{10} = 51,200$ bacterium, 12 half-hours (6 hr)

124. $a_0 = 219; a_n = 438(2)^{n-1}, a_4 = 3504$ people, 7 periods (14 mo)

125. $a_0 = 2$ m; $a_n = 1.6\left(\frac{4}{5}\right)^{n-1}$, $a_7 \approx 0.42$ m,

total distance $= a_0 + 2S_\infty = 18$ m

126. $a_0 = 36$ cm; $a_n = 34.2(0.95)^{n-1}$, $a_8 \approx 23.9$ cm,
total distance $= a_0 + 2S_\infty = 1404$ cm

131. a. For an arithmetic sequence, the difference $d = a_k - a_{k-1}$ must be constant. For $a_k = a_1 r^{k-1}$ and $a_{k-1} = a_1 r^{k-2}$, we have
$d = \log(a_1 r^{k-1}) - \log(a_1 r^{k-2})$
$= \log(a_1) + \log(r^{k-1}) - [\log(a_1) + \log(r^{k-2})]$
$= \log(a_1) + (k-1)\log(r) - \log(a_1) - (k-2)\log(r)$
$= \log(r)[(k-1) - (k-2)]$
$= \log(r)\checkmark$

b. For a geometric sequence, the ratio $r = \dfrac{a_k}{a_{k-1}}$ must be constant.
For $a_k = a_1 + (k-1)d$ and $a_{k-1} = a_1 + (k-2)d$ we have
$r = \dfrac{10^{a_1 + (k-1)d}}{10^{a_1 + (k-2)d}}$
$= \dfrac{10^{a_1} 10^{(k-1)d}}{10^{a_1} 10^{(k-2)d}}$
$= \dfrac{10^{(k-1)d}}{10^{(k-2)d}}$
$= 10^{(k-1)d - (k-2)d}$
$= 10^{d[(k-1) - (k-2)]}$
$= 10^d \checkmark$

134.

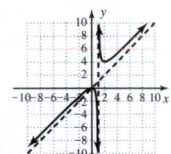

Exercises 9.4, pp. 801–802

7. $a_n = 10n - 6$
$a_4 = 10(4) - 6 = 40 - 6 = 34$;
$a_5 = 10(5) - 6 = 50 - 6 = 44$;
$a_k = 10k - 6$;
$a_{k+1} = 10(k+1) - 6 = 10k + 10 - 6 = 10k + 4$

8. $a_n = 6n - 4$
$a_4 = 6(4) - 4 = 24 - 4 = 20$;
$a_5 = 6(5) - 4 = 30 - 4 = 26$;
$a_k = 6k - 4$;
$a_{k+1} = 6(k+1) - 4 = 6k + 6 - 4 = 6k + 2$

9. $a_n = n$
$a_4 = 4$;
$a_5 = 5$;
$a_k = k$;
$a_{k+1} = k + 1$

10. $a_n = 7n$
$a_4 = 7(4) = 28$;
$a_5 = 7(5) = 35$;
$a_k = 7k$;
$a_{k+1} = 7(k+1) = 7k + 7$

11. $a_n = 2^{n-1}$
$a_4 = 2^{4-1} = 2^3 = 8$;
$a_5 = 2^{5-1} = 2^4 = 16$;
$a_k = 2^{k-1}$;
$a_{k+1} = 2^{k+1-1} = 2^k$

12. $a_n = 2(3^{n-1})$
$a_4 = 2(3^{4-1}) = 2(3^3) = 2(27) = 54$;
$a_5 = 2(3^{5-1}) = 2(3^4) = 2(81) = 162$;
$a_k = 2(3^{k-1})$;
$a_{k+1} = 2(3^{k+1-1}) = 2(3^k)$

13. $S_n = n(5n - 1)$
$S_4 = 4(5(4) - 1) = 4(20 - 1) = 4(19) = 76$;
$S_5 = 5(5(5) - 1) = 5(25 - 1) = 5(24) = 120$;
$S_k = k(5k - 1)$;
$S_{k+1} = (k+1)(5(k+1) - 1) = (k+1)(5k + 5 - 1) =$
$(k+1)(5k + 4)$

14. $S_n = n(3n - 1)$
$S_4 = 4(3(4) - 1) = 4(12 - 1) = 4(11) = 44$;
$S_5 = 5(3(5) - 1) = 5(15 - 1) = 5(14) = 70$;
$S_k = k(3k - 1)$;
$S_{k+1} = (k+1)(3(k+1) - 1) = (k+1)(3k + 3 - 1) =$
$(k+1)(3k + 2)$

15. $S_n = \dfrac{n(n+1)}{2}$
$S_4 = \dfrac{4(4+1)}{2} = \dfrac{4(5)}{2} = 10$;
$S_5 = \dfrac{5(5+1)}{2} = \dfrac{5(6)}{2} = 15$;
$S_k = \dfrac{k(k+1)}{2}$;
$S_{k+1} = \dfrac{(k+1)(k+1+1)}{2} = \dfrac{(k+1)(k+2)}{2}$

16. $S_n = \dfrac{7n(n+1)}{2}$
$S_4 = \dfrac{7(4)(4+1)}{2} = \dfrac{28(5)}{2} = 70$;
$S_5 = \dfrac{7(5)(5+1)}{2} = \dfrac{35(6)}{2} = 105$;
$S_k = \dfrac{7k(k+1)}{2}$;
$S_{k+1} = \dfrac{7(k+1)(k+1+1)}{2} = \dfrac{7(k+1)(k+2)}{2}$

17. $S_n = 2^n - 1$
$S_4 = 2^4 - 1 = 16 - 1 = 15$;
$S_5 = 2^5 - 1 = 32 - 1 = 31$;
$S_k = 2^k - 1$;
$S_{k+1} = 2^{k+1} - 1$

18. $S_n = 3^n - 1$
$S_4 = 3^4 - 1 = 81 - 1 = 80$;
$S_5 = 3^5 - 1 = 243 - 1 = 242$;
$S_k = 3^k - 1$;
$S_{k+1} = 3^{k+1} - 1$

19. $a_n = 10n - 6$; $S_n = n(5n - 1)$
$S_4 = 4(5(4) - 1) = 4(20 - 1) = 4(19) = 76$;
$a_5 = 10(5) - 6 = 50 - 6 = 44$;
$S_5 = 5(5(5) - 1) = 5(25 - 1) = 5(24) = 120$;
$S_4 + a_5 = S_5$
$76 + 44 = 120$
$120 = 120$
Verified

20. $a_n = 6n - 4$; $S_n = n(3n - 1)$
$S_4 = 4(3(4) - 1) = 4(12 - 1) = 4(11) = 44$;
$a_5 = 6(5) - 4 = 30 - 4 = 26$;
$S_5 = 5(3(5) - 1) = 5(15 - 1) = 5(14) = 70$;
$S_4 + a_5 = S_5$
$44 + 26 = 70$
$70 = 70$
Verified

21. $a_n = n$; $S_n = \dfrac{n(n+1)}{2}$
$S_4 = \dfrac{4(4+1)}{2} = \dfrac{4(5)}{2} = 10$;
$a_5 = 5$;
$S_5 = \dfrac{5(5+1)}{2} = \dfrac{5(6)}{2} = 15$;
$S_4 + a_5 = S_5$
$10 + 5 = 15$
$15 = 15$
Verified

22. $a_n = 7n$; $S_n = \dfrac{7n(n+1)}{2}$

$S_4 = \dfrac{7(4)(4+1)}{2} = \dfrac{28(5)}{2} = 70$;

$a_5 = 7(5) = 35$;

$S_5 = \dfrac{7(5)(5+1)}{2} = \dfrac{35(6)}{2} = 105$;

$S_4 + a_5 = S_5$

$70 + 35 = 105$

$\qquad 105 = 105$

Verified

23. $a_n = 2^{n-1}$; $S_n = 2^n - 1$

$S_4 = 2^4 - 1 = 16 - 1 = 15$;

$a_5 = 2^{5-1} = 2^4 = 16$;

$S_5 = 2^5 - 1 = 32 - 1 = 31$;

$S_4 + a_5 = S_5$

$15 + 16 = 31$

$\qquad 31 = 31$

Verified

24. $a_n = 2(3^{n-1})$; $S_n = 3^n - 1$

$S_4 = 3^4 - 1 = 81 - 1 = 80$;

$a_5 = 2(3^{5-1}) = 2(3^4) = 2(81) = 162$;

$S_5 = 3^5 - 1 = 243 - 1 = 242$;

$S_4 + a_5 = S_5$

$80 + 162 = 242$

$\qquad 242 = 242$

Verified

25. a. $a_n = n^3$; $S_n = (1 + 2 + 3 + 4 + \cdots + n)^2$

$S_1 = 1^2 = 1^3$

$S_5 = (1 + 2 + 3 + 4 + 5)^2$

$\quad = 15^2$

$\quad = 225$

$1 + 8 + 27 + 64 + 125 = 225$

$S_9 = (1 + 2 + \cdots + 9)^2$

$\quad = 45^2$

$\quad = 2025$

$1 + 8 + \cdots + 729 = 2025$

b. $\left[\dfrac{n(n+1)}{2}\right]^2 = \dfrac{n^2(n+1)^2}{4}$

26. P_n: $i^{n+4} = i^n$

1. Show P_n is true for $n = 1$.

$i^{1+4} = i \cdot i^4 = i \cdot 1 = i^1$

2. Assume true for P_k

$i^{k+4} = i^k$

and use it to show the truth of P_{k+1} follows. That is: $i^{(k+1)+4} = i^{k+1}$.

Working with the left hand side:

i^{k+1+4}

$= i^{k+4+1}$

$= i^{k+4} i^1$

$= i^k i^1$

$= i^{k+1}$

Since the truth of P_{k+1} follows from P_k, the formula is true for all n.

27. 1. Show S_n is true for $n = 1$.

$S_1 = 1(1+1) = 1(2) = 2$

Verified

2. Assume S_k is true: $2 + 4 + 6 + 8 + 10 + \cdots + 2k = k(k+1)$

and use it to show the truth of S_{k+1} follows. That is:

$2 + 4 + 6 + \cdots + 2k + 2(k+1) = (k+1)(k+1+1)$

$S_k + a_{k+1} = S_{k+1}$

Working with the left hand side:

$2 + 4 + 6 + \cdots + 2k + 2(k+1)$

$= k(k+1) + 2(k+1)$

$= (k+1)(k+2)$

$= S_{k+1}$

Since the truth of S_{k+1} follows from S_k, the formula is true for all n.

28. 1. Show S_n is true for $n = 1$.

$S_1 = 1(2(1) + 1) = 1(2 + 1) = 1(3) = 3$

Verified

2. Assume S_k is true:

$3 + 7 + 11 + \cdots + (4k - 1) = k(2k + 1)$ and use it to show the truth of S_{k+1} follows. That is:

$3 + 7 + 11 + \cdots + (4k - 1) + [4(k+1) - 1]$

$\quad = (k+1)[2(k+1) + 1]$

$S_k + a_{k+1} = S_{k+1}$

Working with the left hand side:

$3 + 7 + 11 + \cdots + (4k - 1) + (4k + 3)$

$= k(2k + 1) + 4k + 3$

$= 2k^2 + k + 4k + 3$

$= 2k^2 + 5k + 3$

$= (k+1)(2k + 3)$

$= S_{k+1}$

Since the truth of S_{k+1} follows from S_k, the formula is true for all n.

29. 1. Show S_n is true for $n = 1$.

$S_1 = \dfrac{5(1)(1+1)}{2} = \dfrac{5(2)}{2} = 5$

Verified

2. Assume S_k is true:

$5 + 10 + 15 + \cdots + 5k = \dfrac{5k(k+1)}{2}$

and use it to show the truth of S_{k+1} follows. That is:

$5 + 10 + 15 + \cdots + 5k + 5(k+1) = \dfrac{5(k+1)(k+1+1)}{2}$

$S_k + a_{k+1} = S_{k+1}$

Working with the left hand side:

$5 + 10 + 15 + \cdots + 5k + 5(k+1)$

$= \dfrac{5k(k+1)}{2} + 5(k+1)$

$= \dfrac{5k(k+1) + 10(k+1)}{2}$

$= \dfrac{(k+1)(5k + 10)}{2}$

$= \dfrac{5(k+1)(k+2)}{2}$

$= S_{k+1}$

Since the truth of S_{k+1} follows from S_k, the formula is true for all n.

30. 1. Show S_n is true for $n = 1$.

$S_1 = \dfrac{1(3(1) - 1)}{2} = \dfrac{3 - 1}{2} = \dfrac{2}{2} = 1$

Verified

2. Assume S_k is true:

$1 + 4 + 7 + \cdots + (3k - 2) = \dfrac{k(3k - 1)}{2}$

and use it to show the truth of S_{k+1} follows. That is:

$1 + 4 + 7 + \cdots + (3k - 2) + [3(k+1) - 2]$

$\quad = \dfrac{(k+1)[3(k+1) - 1]}{2}$

$S_k + a_{k+1} = S_{k+1}$

Working with the left hand side:

$1 + 4 + 7 + \cdots + (3k - 2) + [3(k+1) - 2]$

$= \dfrac{k(3k - 1)}{2} + 3k + 1$

$= \dfrac{k(3k - 1) + 2(3k + 1)}{2}$

$= \dfrac{3k^2 - k + 6k + 2}{2}$

$= \dfrac{3k^2 + 5k + 2}{3}$

$= \dfrac{(k+1)(3k + 2)}{2}$

$= S_{k+1}$

Since the truth of S_{k+1} follows from S_k, the formula is true for all n.

31. 1. Show S_n is true for $n = 1$.
$S_1 = 1(2(1) + 3) = 5$
Verified

2. Assume S_k is true:
$5 + 9 + 13 + 17 + \cdots + (4k + 1) = k(2k + 3)$
and use it to show the truth of S_{k+1} follows. That is:
$5 + 9 + 13 + 17 + \cdots + (4k + 1) + [4(k + 1) + 1]$
$= (k + 1)(2(k + 1) + 3)$
$S_k + a_{k+1} = S_{k+1}$
Working with the left hand side:
$5 + 9 + 13 + 17 + \cdots + (4k + 1) + [4(k + 1) + 1]$
$= k(2k + 3) + 4k + 5$
$= 2k^2 + 3k + 4k + 5$
$= 2k^2 + 7k + 5$
$= (k + 1)(2k + 5)$
$= S_{k+1}$
Since the truth of S_{k+1} follows from S_k, the formula is true for all n.

32. 1. Show S_n is true for $n = 1$.
$S_1 = 4(1)^2 = 4(1) = 4$
Verified

2. Assume S_k is true:
$4 + 12 + 20 + \cdots + (8k - 4) = 4k^2$
and use it to show the truth of S_{k+1} follows. That is:
$4 + 12 + 20 + \cdots + (8k - 4) + [8(k + 1) - 4] = 4(k + 1)^2$
$S_k + a_{k+1} = S_{k+1}$
Working with the left hand side:
$4 + 12 + 20 + \cdots + (8k - 4) + [8(k + 1) - 4]$
$= 4k^2 + 8k + 4$
$= 4(k^2 + 2k + 1)$
$= 4(k + 1)^2$
$= S_{k+1}$
Since the truth of S_{k+1} follows from S_k, the formula is true for all n.

33. 1. Show S_n is true for $n = 1$.

$S_1 = \dfrac{3(3^1 - 1)}{2} = \dfrac{3(3 - 1)}{2} = \dfrac{3(2)}{2} = 3$

Verified

2. Assume S_k is true:

$3 + 9 + 27 + \cdots + 3^k = \dfrac{3(3^k - 1)}{2}$

and use it to show the truth of S_{k+1} follows. That is:

$3 + 9 + 27 + \cdots + 3^k + 3^{k+1} = \dfrac{3(3^{k+1} - 1)}{2}$

$S_k + a_{k+1} = S_{k+1}$
Working with the left hand side:
$3 + 9 + 27 + \cdots + 3^k + 3^{k+1}$

$= \dfrac{3(3^k - 1)}{2} + 3^{k+1}$

$= \dfrac{3(3^k - 1) + 2(3^{k+1})}{2}$

$= \dfrac{3^{k+1} - 3 + 2(3^{k+1})}{2}$

$= \dfrac{3(3^{k+1}) - 3}{2}$

$= \dfrac{3(3^{k+1} - 1)}{2}$

$= S_{k+1}$
Since the truth of S_{k+1} follows from S_k, the formula is true for all n.

34. 1. Show S_n is true for $n = 1$.

$S_1 = \dfrac{5(5^1 - 1)}{4} = \dfrac{5(5 - 1)}{4} = \dfrac{5(4)}{4} = 5$

Verified

2. Assume S_k is true:

$5 + 25 + 125 + \cdots + 5^k = \dfrac{5(5^k - 1)}{4}$

and use it to show the truth of S_{k+1} follows. That is:

$5 + 25 + 125 + \cdots + 5^k + 5^{k+1} = \dfrac{5(5^{k+1} - 1)}{4}$

$S_k + a_{k+1} = S_{k+1}$
Working with the left hand side:
$5 + 25 + 125 + \cdots + 5^k + 5^{k+1}$

$= \dfrac{5(5^k - 1)}{4} + 5^{k+1}$

$= \dfrac{5(5^k - 1) + 4(5)^{k+1}}{4}$

$= \dfrac{5^{k+1} - 5 + 4(5)^{k+1}}{4}$

$= \dfrac{5(5)^{k+1} - 5}{4}$

$= \dfrac{5(5^{k+1} - 1)}{4}$

$= S_{k+1}$
Since the truth of S_{k+1} follows from S_k, the formula is true for all n.

35. 1. Show S_n is true for $n = 1$.
$S_1 = 2^{1+1} - 2 = 2^2 - 2 = 4 - 2 = 2$
Verified

2. Assume S_k is true:
$2 + 4 + 8 + \cdots + 2^k = 2^{k+1} - 2$
and use it to show the truth of S_{k+1} follows. That is:
$2 + 4 + 8 + \cdots + 2^k + 2^{k+1} = 2^{k+2} - 2$
$S_k + a_{k+1} = S_{k+1}$
Working with the left hand side:
$2 + 4 + 8 + \cdots + 2^k + 2^{k+1}$
$= 2^{k+1} - 2 + 2^{k+1}$
$= 2(2^{k+1}) - 2$
$= 2^{k+2} - 2$
$= S_{k+1}$
Since the truth of S_{k+1} follows from S_k, the formula is true for all n.

36. 1. Show S_n is true for $n = 1$.

$S_1 = \dfrac{(1)^2(1 + 1)^2}{4} = \dfrac{1(2)^2}{4} = \dfrac{4}{4} = 1$

Verified

2. Assume S_k is true:

$1 + 8 + 27 + \cdots + k^3 = \dfrac{k^2(k + 1)^2}{4}$

and use it to show the truth of S_{k+1} follows. That is:
$1 + 8 + 27 + \cdots + k^3 + (k + 1)^3$

$= \dfrac{(k + 1)^2(k + 1 + 1)^2}{4}$

$S_k + a_{k+1} = S_{k+1}$
Working with the left hand side:
$1 + 8 + 27 + \cdots + k^3 + (k + 1)^3$

$= \dfrac{k^2(k + 1)^2}{4} + (k + 1)^3$

$= \dfrac{k^2(k + 1)^2 + 4(k + 1)^3}{4}$

$= \dfrac{(k + 1)^2(k^2 + 4(k + 1))}{4}$

$= \dfrac{(k + 1)^2(k^2 + 4k + 4)}{4}$

$= \dfrac{(k + 1)^2(k + 2)^2}{4}$

$= S_{k+1}$
Since the truth of S_{k+1} follows from S_k, the formula is true for all n.

37. 1. Show S_n is true for $n = 1$.

$$S_1 = \frac{1}{2(1) + 1} = \frac{1}{2 + 1} = \frac{1}{3}$$

Verified

2. Assume S_k is true:

$$\frac{1}{3} + \frac{1}{15} + \frac{1}{35} + \cdots + \frac{1}{(2k - 1)(2k + 1)} = \frac{k}{2k + 1}$$

and use it to show the truth of S_{k+1} follows. That is:

$$\frac{1}{3} + \frac{1}{15} + \frac{1}{35} + \cdots + \frac{1}{(2k - 1)(2k + 1)}$$

$$+ \frac{1}{(2(k + 1) - 1)(2(k + 1) + 1)} = \frac{k + 1}{2(k + 1) + 1}$$

$S_k + a_{k+1} = S_{k+1}$

Working with the left hand side:

$$\frac{1}{3} + \frac{1}{15} + \frac{1}{35} + \cdots + \frac{1}{(2k - 1)(2k + 1)} + \frac{1}{(2k + 1)(2k + 3)}$$

$$= \frac{k}{2k + 1} + \frac{1}{(2k + 1)(2k + 3)}$$

$$= \frac{k(2k + 3) + 1}{(2k + 1)(2k + 3)}$$

$$= \frac{2k^2 + 3k + 1}{(2k + 1)(2k + 3)}$$

$$= \frac{(2k + 1)(k + 1)}{(2k + 1)(2k + 3)}$$

$$= \frac{k + 1}{2k + 3}$$

$$= S_{k+1}$$

Since the truth of S_{k+1} follows from S_k, the formula is true for all n.

38. 1. Show S_n is true for $n = 1$.

$$S_1 = \frac{1}{1 + 1} = \frac{1}{2}$$

Verified

2. Assume S_k is true:

$$\frac{1}{2} + \frac{1}{6} + \frac{1}{12} + \cdots + \frac{1}{k(k + 1)} = \frac{k}{k + 1}$$

and use it to show the truth of S_{k+1} follows. That is:

$$\frac{1}{2} + \frac{1}{6} + \frac{1}{12} + \cdots + \frac{1}{k(k + 1)} + \frac{1}{(k + 1)(k + 1 + 1)}$$

$$= \frac{k + 1}{k + 1 + 1}$$

$S_k + a_{k+1} = S_{k+1}$

Working with the left hand side:

$$\frac{1}{2} + \frac{1}{6} + \frac{1}{12} + \cdots + \frac{1}{k(k + 1)} + \frac{1}{(k + 1)(k + 2)}$$

$$= \frac{k}{k + 1} + \frac{1}{(k + 1)(k + 2)}$$

$$= \frac{k(k + 2) + 1}{(k + 1)(k + 2)}$$

$$= \frac{k^2 + 2k + 1}{(k + 1)(k + 2)}$$

$$= \frac{(k + 1)^2}{(k + 1)(k + 2)}$$

$$= \frac{k + 1}{k + 2}$$

$$= S_{k+1}$$

Since the truth of S_{k+1} follows from S_k, the formula is true for all n.

39. 1. Show P_n is true for $n = 1$.

P_1:

$$3^1 \geq 2(1) + 1$$
$$3 \geq 2 + 1$$
$$3 \geq 3$$

Verified

2. Assume P_k: $3^k \geq 2k + 1$ is true and use it to show the truth of P_{k+1} follows. That is: $3^{k+1} \geq 2(k + 1) + 1$.

Working with the left hand side:

$$3^{k+1} = 3(3^k)$$
$$\geq 3(2k + 1)$$
$$\geq 6k + 3$$

Since k is a positive integer,

$$6k + 3 \geq 2k + 3$$

Showing P_{k+1}: $3^{k+1} \geq 2k + 3$

Since the truth of P_{k+1} follows from P_k, the statement is true for all n.

40. 1. Show P_n is true for $n = 1$.

P_1:

$$2^1 \geq 1 + 1$$
$$2 \geq 2$$

Verified

2. Assume P_k: $2^k \geq k + 1$ is true and use it to show the truth of P_{k+1} follows. That is: $2^{k+1} \geq k + 1 + 1$.

Working with the left hand side:

$$2^{k+1} = 2(2^k)$$
$$\geq 2(k + 1)$$
$$\geq 2k + 2$$

Since k is a positive integer,

$$2k + 2 \geq k + 2$$

Showing P_{k+1}: $2^{k+1} \geq k + 2$

Since the truth of P_{k+1} follows from P_k, the statement is true for all n.

41. 1. Show P_n is true for $n = 1$.

P_1:

$$3 \cdot 4^{1-1} \leq 4^1 - 1$$
$$3 \cdot 4^0 \leq 4 - 1$$
$$3 \cdot 1 \leq 3$$
$$3 \leq 3$$

Verified

2. Assume P_k: $3 \cdot 4^{k-1} \leq 4^k - 1$ is true and use it to show the truth of P_{k+1} follows. That is: $3 \cdot 4^{k+1-1} \leq 4^{k+1} - 1$.

Working with the left hand side:

$$3 \cdot 4^k = 3 \cdot 4(4^{k-1})$$
$$= 4 \cdot 3(4^{k-1})$$
$$\leq 4(4^k - 1)$$
$$\leq 4^{k+1} - 4$$

Since k is a positive integer, $4^{k+1} - 4 \leq 4^{k+1} - 1$

Showing P_{k+1}: $3 \cdot 4^k \leq 4^{k+1} - 1$

Since the truth of P_{k+1} follows from P_k, the statement is true for all n.

42. 1. Show P_n is true for $n = 1$.

P_1:

$$4 \cdot 5^{1-1} \leq 5^1 - 1$$
$$4 \cdot 5^0 \leq 5 - 1$$
$$4 \cdot 1 \leq 4$$
$$4 \leq 4$$

Verified

2. Assume P_k: $4 \cdot 5^{k-1} \leq 5^k - 1$ is true and use it to show the truth of P_{k+1} follows. That is: $4 \cdot 5^{k+1-1} \leq 5^{k+1} - 1$.

Working with the left hand side:

$$4 \cdot 5^k = 4 \cdot 5(5^{k-1})$$
$$= 5 \cdot 4(5^{k-1})$$
$$\leq 5(5^k - 1)$$
$$\leq 5^{k+1} - 5$$

Since k is a positive integer, $5^{k+1} - 5 \leq 5^{k+1} - 1$

Showing P_{k+1}: $4 \cdot 5^k \leq 5^{k+1} - 1$

Since the truth of P_{k+1} follows from P_k, the statement is true for all n.

43. $n^2 - 7n$ is divisible by 2

1. Show P_n is true for $n = 1$.

P_n: $n^2 - 7n = 2m$

P_1:

$$(1)^2 - 7(1) = 2m$$
$$1 - 7 = 2m$$
$$-6 = 2m \quad \text{Verified}$$

2. Assume P_k: $k^2 - 7k = 2m$ for $m \in \mathbb{Z}$ and use it to show the truth of P_{k+1} follows. That is: $(k + 1)^2 - 7(k + 1) = 2p$ for $p \in \mathbb{Z}$.
Working with the left hand side:
$(k + 1)^2 - 7(k + 1)$
$= k^2 + 2k + 1 - 7k - 7$
$= k^2 - 7k + 2k - 6$
$= 2m + 2k - 6$
$= 2(m + k - 3)$
is divisible by 2.
Since the truth of P_{k+1} follows from P_k, the statement is true for all n.

44. $n^3 - n + 3$ is divisible by 3
1. Show P_n is true for $n = 1$. P_n: $n^3 - n + 3 = 3m$
P_1:
$(1)^3 - 1 + 3 = 3m$
$1 - 1 + 3 = 3m$
$3 = 3m$ Verified
2. Assume P_k: $k^3 - k + 3 = 3m$ for $m \in \mathbb{Z}$ and use it to show the truth of P_{k+1} follows.
That is: $(k + 1)^3 - (k + 1) + 3 = 3p$ for $p \in \mathbb{Z}$.
Working with the left hand side:
$(k + 1)^3 - (k + 1) + 3$
$= k^3 + 3k^2 + 3k + 1 - k - 1 + 3$
$= k^3 + 3k^2 + 3k - k + 3$
$= (k^3 - k + 3) + 3(k^2 + k)$
$= 3m + 3(k^2 + k)$ is divisible by 3.
Since the truth of P_{k+1} follows from P_k, the statement is true for all n.

45. $n^3 + 3n^2 + 2n$ is divisible by 3
1. Show P_n is true for $n = 1$. P_n: $n^3 + 3n^2 + 2n = 3m$
P_1:
$(1)^3 + 3(1)^2 + 2(1) = 3m$
$1 + 3 + 2 = 3m$
$6 = 3m$
$2 = m$
Verified
2. Assume P_k: $k^3 + 3k^2 + 2k = 3m$ for $m \in \mathbb{Z}$ and use it to show the truth of P_{k+1} follows.
That is: $(k + 1)^3 + 3(k + 1)^2 + 2(k + 1) = 3p$ for $p \in \mathbb{Z}$.
Working with the left hand side:
$(k + 1)^3 + 3(k + 1)^2 + 2(k + 1)$
$= k^3 + 3k^2 + 3k + 1 + 3(k^2 + 2k + 1) + 2k + 2$
$= k^3 + 3k^2 + 2k + 3(k^2 + 2k + 1) + 3k + 3$
$= k^3 + 3k^2 + 2k + 3(k^2 + 2k + 1) + 3(k + 1)$
$= 3m + 3(k^2 + 2k + 1) + 3(k + 1)$ is divisible by 3.
Since the truth of P_{k+1} follows from P_k, the statement is true for all n.

46. $5^n - 1$ is divisible by 4
1. Show P_n is true for $n = 1$. P_n: $5^n - 1 = 4m$
P_1:
$5^1 - 1 = 4m$
$5 - 1 = 4m$
$4 = 4m$
$1 = m$
Verified
2. Assume P_k: $5^k - 1 = 4m$ or $5^k = 4m + 1$ for $m \in \mathbb{Z}$ and use it to show the truth of P_{k+1} follows.
That is: $5^{k+1} - 1 = 4p$ for $p \in \mathbb{Z}$.
Working with the left hand side:
$5^{k+1} - 1$
$= 5(5^k) - 1$
$= 5(4m + 1) - 1$
$= 20m + 5 - 1$
$= 20m + 4$
$= 4(5m + 1)$
is divisible by 4.
Since the truth of P_{k+1} follows from P_k, the statement is true for all n.

47. $6^n - 1$ is divisible by 5
1. Show P_n is true for $n = 1$. P_n: $6^n - 1 = 5m$
P_1:
$6^1 - 1 = 5m$
$6 - 1 = 5m$
$5 = 5m$
$1 = m$
Verified
2. Assume P_k: $6^k - 1 = 5m$ or $6^k = 5m + 1$ for $m \in \mathbb{Z}$ and use it to show the truth of P_{k+1} follows.
That is: $6^{k+1} - 1 = 5p$ for $p \in \mathbb{Z}$.
Working with the left hand side:
$6^k - 1$
$= 6(6^k) - 1$
$= 6(5m + 1) - 1$
$= 30m + 6 - 1$
$= 30m + 5$
$= 5(6m + 1)$
is divisible by 5.
Since the truth of P_{k+1} follows from P_k, the statement is true for all n.

51. $A + B = \begin{bmatrix} 1 & 1 \\ 7 & 4 \end{bmatrix}$; $A - B = \begin{bmatrix} -3 & 3 \\ -1 & -2 \end{bmatrix}$; $2A - 3B = \begin{bmatrix} -8 & 7 \\ -6 & -7 \end{bmatrix}$;
$AB = \begin{bmatrix} 6 & 7 \\ 10 & 0 \end{bmatrix}$; $BA = \begin{bmatrix} -5 & 3 \\ 5 & 11 \end{bmatrix}$; $B^{-1} = \begin{bmatrix} 0.3 & 0.1 \\ -0.4 & 0.2 \end{bmatrix}$

Exercises 9.5, pp. 811–816

7. a. 16 possible

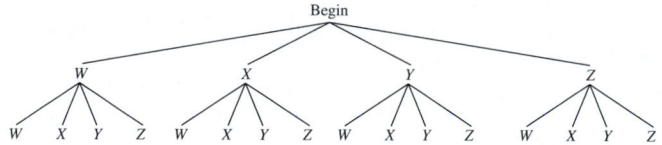

b. WW, WX, WY, WZ, XW, XX, XY, XZ, YW, YX, YY, YZ, ZW, ZX, ZY, ZZ
8. a. 16 possible

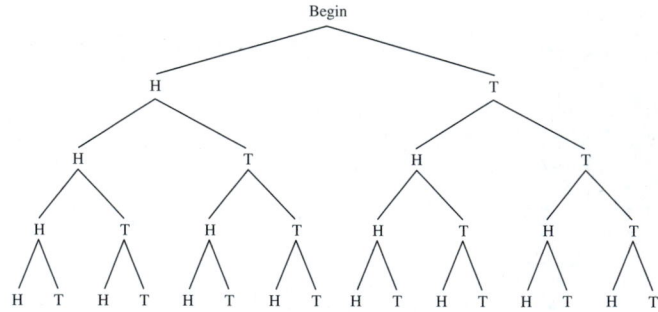

b. HHHH, HHHT, HHTH, HHTT, HTHH, HTHT, HTTH, HTTT, THHH, THHT, THTH, THTT, TTHH, TTHT, TTTH, TTTT

112. $A + B = \begin{bmatrix} 1.5 & 0.2 & -4 \\ -11 & 5.1 & 9 \\ 3.2 & 1 & 10 \end{bmatrix}$, $AB = \begin{bmatrix} 4.1 & 0.2 & 11 \\ -44.8 & 0.1 & 60 \\ -3.2 & 0.5 & 18 \end{bmatrix}$,

$A^{-1} = \begin{bmatrix} \frac{-19}{17} & \frac{-3}{17} & \frac{15}{17} \\ \frac{-10}{17} & \frac{2}{17} & \frac{7}{17} \\ \frac{12}{17} & \frac{1}{17} & \frac{-5}{17} \end{bmatrix}$

Exercises 9.7, pp. 835–837

7. $x^5 + 5x^4y + 10x^3y^2 + 10x^2y^3 + 5xy^4 + y^5$
8. $a^6 + 6a^5b + 15a^4b^2 + 20a^3b^3 + 15a^2b^4 + 6ab^5 + b^6$
9. $16x^4 + 96x^3 + 216x^2 + 216x + 81$
10. $x^6 + x^4 + \frac{1}{3}x^2 + \frac{1}{27}$ **11.** $41 + 38i$ **12.** $41 + 840i$
25. $c^5 + 5c^4d + 10c^3d^2 + 10c^2d^3 + 5cd^4 + d^5$
26. $v^4 + 4v^3w + 6v^2w^2 + 4vw^3 + w^4$

27. $a^6 - 6a^5b + 15a^4b^2 - 20a^3b^3 + 15a^2b^4 - 6ab^5 + b^6$
28. $x^7 - 7x^6y + 21x^5y^2 - 35x^4y^3 + 35x^3y^4 - 21x^2y^5 + 7xy^6 - y^7$
29. $16x^4 - 96x^3 + 216x^2 - 216x + 81$
30. $a^5 - 10a^4b + 40a^3b^2 - 80a^2b^3 + 80ab^4 - 32b^5$
31. $-11 + 2i$ **32.** $-118 - 31i\sqrt{3}$ **33.** $x^9 + 18x^8y + 144x^7y^2$
34. $6561p^8 + 17{,}496p^7q + 20{,}412p^6q^2$ **35.** $v^{24} - 6v^{22}w + \frac{33}{2}v^{20}w^2$
36. $\frac{1}{1024}a^{10} - \frac{5}{256}a^9b^2 + \frac{45}{256}a^8b^4$ **37.** $35x^4y^3$ **38.** $15m^2n^4$
39. $1792p^2$ **40.** $-39{,}405{,}366a^5$ **41.** $264x^2y^{10}$ **42.** $10{,}206n^4m^5$
51. 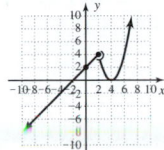 $f(3) = 1$

53. $g(x) > 0: x \in (-2, 0) \cup (3, \infty)$

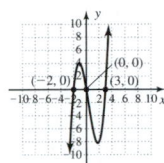

Summary and Concept Review, pp. 838–842

35. (1) Show S_n is true for $n = 1$: $S_1 = \dfrac{1(1+1)}{2} = 1$✓

(2) Assume S_k is true: $1 + 2 + 3 + \cdots + k = \dfrac{k(k+1)}{2}$

Use it to show the truth of S_{k+1}:

$$1 + 2 + 3 + \cdots + k + (k+1) = \frac{(k+1)(k+1+1)}{2}$$

left hand side: $1 + 2 + 3 + \cdots + k + (k+1)$

$$= \frac{k(k+1)}{2} + (k+1) = \frac{k(k+1) + 2(k+1)}{2}$$

$$= \frac{(k+1)(k+2)}{2}$$

36. (1) Show S_n is true for $n = 1$: $S_1 = \dfrac{1(1+1)[2(1)+1]}{6} = 1$✓

(2) Assume S_k is true: $1 + 4 + 9 + \cdots + k^2 = \dfrac{k(k+1)(2k+1)}{6}$

Use it to show the truth of S_{k+1}:

$$1 + 4 + 9 + \cdots + k^2 + (k+1)^2 = \frac{(k+1)(k+1+1)[2(k+1)+1]}{6}$$

left hand side: $1 + 4 + 9 + \cdots + k^2 + (k+1)^2$

$$= \frac{k(k+1)(2k+1)}{6} + \frac{6(k+1)^2}{6} = \frac{(k+1)[(2k^2 + k + 6k + 6]}{6}$$

$$= \frac{(k+1)(2k^2 + 7k + 6)}{6} = \frac{(k+1)(k+2)(2k+3)}{6}$$

37. (1) Show P_n is true for $n = 1$: P_1: $4^1 \geq 3(1) + 1$✓
(2) Assume P_k is true: $4^k \geq 3k + 1$
Use it to show the truth of P_{k+1}:
$4^{k+1} \geq 3(k+1) + 1 = 3k + 4$
left hand side: $4^{k+1} = 4(4^k)$
$$\geq 4(3k+1) = 12k + 4$$
Since k is a positive integer, $12k + 4 \geq 3k + 4$ showing
$4^{k+1} \geq 3k + 4$

38. (1) Show P_n is true for $n = 1$: P_1: $6 \cdot 7^{1-1} \leq 7^1 - 1$✓
(2) Assume P_k is true: $6 \cdot 7^{k-1} \leq 7^k - 1$
Use it to show the truth of P_{k+1}:
$6 \cdot 7^k \leq 7^{k+1} - 1$
left hand side: $6 \cdot 7^k = 7 \cdot 6 \cdot 7^{k-1}$
$$\leq 7 \cdot 7^k - 1$$
$$\leq 7^{k+1} - 7$$
Since k is a positive integer, $7^{k+1} - 7 \leq 7^{k+1} - 1$.

39. (1) Show P_n is true for $n = 1$: P_1: $3^1 - 1 = 2$ or $2(1)$✓
(2) Assume P_k is true: $3^k - 1 = 2p$ or $3^k = 2p + 1$ for $p \in \mathbb{Z}$
Use it to show the truth of P_{k+1}:
$3^{k+1} - 1 = 2q$ for $q \in \mathbb{Z}$
left hand side: $3^{k+1} - 1 = 3 \cdot 3^k - 1$
$$= 3 \cdot (2p + 1) - 1$$
$$= 3 \cdot 2p + 3 - 1$$
$$= 3 \cdot 2p + 2$$
$$= 2(3p + 1)$$
$$= 2q \text{ is divisible by 2}$$

40. six ways

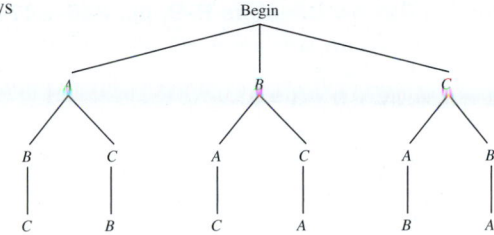

56. a. $a^8 + 8\sqrt{3}a^7 + 84a^6 + 168\sqrt{3}a^5$
b. $78{,}125a^7 + 218{,}750a^6b + 262{,}500a^5b^2 + 175{,}000a^4b^3$

Practice Test, pp. 842–844

1. a. $\frac{1}{2}, \frac{4}{5}, 1, \frac{8}{7}; a_8 = \frac{16}{11}$ **b.** $6, 12, 20, 30; a_8 = 90$
c. $3, 2\sqrt{2}, \sqrt{7}, \sqrt{6}; a_8 = \sqrt{2}$
3. a. $a_1 = 7, d = -3, a_n = 10 - 3n$
b. $a_1 = -8, d = 2, a_n = 2n - 10$ **c.** $a_1 = 4, r = -2, a_n = 4(-2)^{n-1}$
d. $a_1 = 10, r = \frac{2}{5}, a_n = 10(\frac{2}{5})^{n-1}$
9. $a_k = 5k - 3, a_{k+1} = 5(k+1) - 3,$

$$S_k = \frac{5k^2 - k}{2}, S_{k+1} = \frac{5(k+1)^2 - (k+1)}{2};$$

For $n = 1$: $S_1 = \dfrac{5(1)^2 - 1}{2} = 2$✓

Assume: $S_k = \dfrac{5k^2 - k}{2}$ is true,

Prove: $\dfrac{5k^2 - k}{2} + 5(k+1) - 3 = \dfrac{5(k+1)^2 - (k+1)}{2}$

$$\frac{5k^2 - k}{2} + \frac{10(k+1) - 6}{2} = \frac{(k+1)[5(k+1) - 1]}{2}$$

$$\frac{5k^2 + 9k + 4}{2} = \frac{(k+1)(5k+4)}{2}$$

$$\frac{(5k+4)(k+1)}{2} = \frac{(5k+4)(k+1)}{2}$$ ✓

10. For $n = 1$: $2 \cdot 3^{1-1} \leq 3^1 - 1$
$$2(1) \leq 2$$ ✓
Assume: $2 \cdot 3^{k-1} \leq 3^k - 1$
Prove: $2 \cdot 3^{(k+1)-1} \leq 3^{k+1} - 1$
$$2 \cdot 3^{(k+1)-1} = 2 \cdot 3^{(k-1)+1}$$
$$= 2 \cdot 3^{(k-1)} \cdot 3$$
$$\leq (3^k - 1) \cdot 3$$
$(3^k - 1) \cdot 3 = 3^{k+1} - 3 \leq 3^{k+1} - 1$ ✓

11. a.

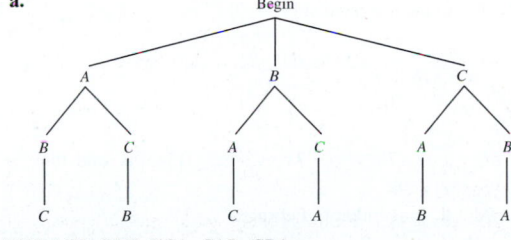

b. $ABC, ACB, BAC, BCA, CAB, CBA$
17. a. $x^4 - 8x^3y + 24x^2y^2 - 32xy^3 + 16y^4$ **b.** -4
18. a. $x^{10} + 10\sqrt{2}\,x^9 + 90x^8$ **b.** $a^8 - 16a^7b^3 + 112a^6b^6$

Strengthening Core Skills, pp. 845–846

Exercise 1. $\dfrac{_4C_1 \cdot _{13}C_5 - 40}{_{52}C_5} \approx 0.001\,970$

Exercise 2. $\dfrac{4 \cdot _{13}C_3 \cdot _{39}C_2}{_{52}C_5} \approx 0.326\,170$

Exercise 3. $\dfrac{4 \cdot _{13}C_4 \cdot _{39}C_1}{_{52}C_5} \approx 0.042\,917$

Exercise 4. $\dfrac{4 \cdot _{10}C_5}{_{52}C_5} \approx 0.000\,388$

Cumulative Review Chapters R–9, pp. 846–847

1. a. 23 cards are assembled each hour.   **b.** $y = 23x - 155$
c. 184 cards   **d.** $\approx 6{:}45$ A.M.
2. $(2 + i)^2 - 4(2 + i) + 5 = 0$
$4 + 4i + i^2 - 8 - 4i + 5 = 0$
$4 + (-1) - 8 + 5 = 0\,\checkmark$
3. $x = \dfrac{-5 \pm \sqrt{109}}{6}$; $x \approx 0.91$; $x \approx -2.57$

4.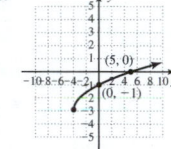

5. $Y = \dfrac{kVW}{X}$; $Y = \dfrac{3VW}{2X}$

6. 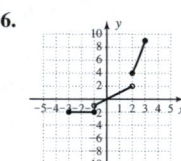 $D: x \in [-3, 3], R: y \in \{-2\} \cup (-1, 2) \cup [4, 9]$

7. verified; reflections across $y = x$ **8. a.** $x = 0$ **b.** $x \in (-1, 0)$
c. $x \in (-\infty, -1) \cup (0, \infty)$ **d.** $x \in (-\infty, -1) \cup (-1, 1)$
e. $x \in (1, \infty)$ **f.** $y = 3$ at $(1, 3)$ **g.** none   **h.** $x \approx -2.3, 0.4, 2$
i. $g(4) \approx 0.25$ **j.** does not exist   **k.** $-\infty$ **l.** 0
m. $x \in (-\infty, -1) \cup (-1, \infty)$

9. a. $4x + 2h - 3$ **b.** $\dfrac{-1}{(x + h - 2)(x - 2)}$

10. 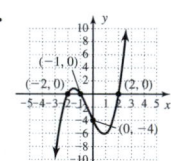   **11.**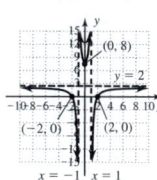

12. a. $\log_{10} x = y$ **b.** $\log_3 \frac{1}{81} = -4$ **13. a.** $x^3 = 125$
b. $e^5 = 2x - 1$ **14.** 6.93%   **15. a.** $x \approx 3.19$ **b.** $x = 334$
16. $(9, 1, 1)$ **17.** $(5, 10, 15)$
18. $\dfrac{x^2}{16} - \dfrac{y^2}{20} = 1$
19. $\dfrac{(x + 3)^2}{16} + \dfrac{(y - 3)^2}{4} = 1$; $(-3, 3)$; $(-7, 3)$, $(1, 3)$, $(-3 - 2\sqrt{3}, 3)$,
$(-3 + 2\sqrt{3}, 3)$

20. a. $a_{20} = \dfrac{1}{1{,}048{,}576} \approx 0.00000095$; $S_{20} = 349{,}525.\overline{3}$
b. $a_{20} = -\frac{117}{40}$; $S_{20} = -20.5$
25. $\begin{cases} 9a + 3b + c = 56 \\ 25a + 5b + c = 126; \\ 81a + 9b + c = 98 \end{cases}$ $y = -7x^2 + 91x - 154$ **a.** mid-June
b. ≈ 142 **c.** 98   **d.** November to February

27.

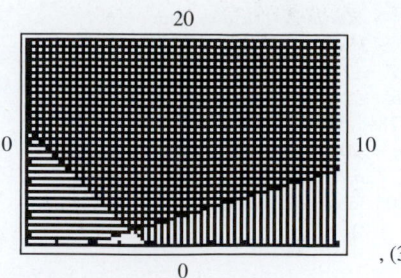

, $(3.4044, 1.4539)$

CHAPTER APPENDIX

APPENDIX A.I pp. A-10–A-13

15. $1.\overline{3}$
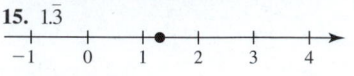

16. -20.875

17. $2.\overline{5}$

18. $-1.\overline{83}$

21. ≈ 2.65

22. ≈ 4.33

23. ≈ 1.73

24. ≈ 6.27

25. a. i. $\{8, 7, 6\}$ **ii.** $\{8, 7, 6\}$ **iii.** $(-1, 8, 7, 6)$
iv. $\{-1, 8, 0.75, \frac{9}{2}, 5.\overline{6}, 7, \frac{3}{5}, 6\}$ **v.** $\{\ \}$
vi. $\{-1, 8, 0.75, \frac{9}{2}, 5.\overline{6}, 7, \frac{3}{5}, 6\}$ **b.** $\{-1, \frac{3}{5}, 0.75, \frac{9}{2}, 5.\overline{6}, 6, 7, 8\}$
c.

26. a. i. $\{\ \}$ **ii.** $\{0\}$ **iii.** $\{-7, 0\}$
iv. $\{-7, 2.\overline{1}, 5.73, -3\frac{5}{6}, 0, -1.12, \frac{7}{8}\}$ **v.** $\{\ \}$
vi. $\{-7, 2.\overline{1}, 5.73, -3\frac{5}{6}, 0, -1.12, \frac{7}{8}\}$
b. $\{-7, -3\frac{5}{6}, -1.12, 0, \frac{7}{8}, 2.\overline{1}, 5.73\}$
c.

27. a. i. $\{\sqrt{49}, 2, 6, 4\}$ **ii.** $\{\sqrt{49}, 2, 6, 0, 4\}$
iii. $\{-5, \sqrt{49}, 2, -3, 6, -1, 0, 4\}$ **iv.** $\{-5, \sqrt{49}, 2, -3, 6, -1, 0, 4\}$
v. $\{\sqrt{3}, \pi\}$ **vi.** $\{-5, \sqrt{49}, 2, -3, 6, -1, \sqrt{3}, 0, 4, \pi\}$
b. $\{-5, -3, -1, 0, \sqrt{3}, 2, \pi, 4, 6, \sqrt{49}\}$
c.

28. a. i. $\{5, \sqrt{64}\}$ **ii.** $\{5, \sqrt{64}\}$ **iii.** $\{-8, 5, \sqrt{64}\}$
iv. $\{-8, 5, -2\frac{3}{5}, 1.75, -0.6, \frac{7}{2}, \sqrt{64}\}$ **v.** $\{-\sqrt{2}, \pi\}$
vi. $\{-8, 5, -2\frac{3}{5}, 1.75, -\sqrt{2}, -0.6, \pi, \frac{7}{2}, \sqrt{64}\}$
b. $\{-8, -2\frac{3}{5}, -\sqrt{2}, -0.6, 1.75, \pi, \frac{7}{2}, 5, \sqrt{64}\}$
c.

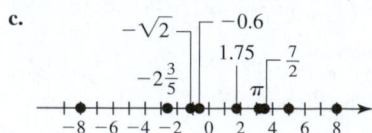

APPENDIX A.II pp. A-22–A-27

11. $P = 18$ cm
$A = 20$ cm^2

12. $P = 20$ cm
$A = 20$ cm^2

13. $P = 24$ cm
$A = 24$ cm^2

14. $P = 16$ cm
$A = 12$ cm^2

17. $c = 45$ cm
$A = 486$ cm^2
$P = 108$ cm

18. $b = 80$ in.
$A = 1560$ in^2
$P = 208$ in.

21. $A = \dfrac{bh}{2} + LW$
$A = 54$ ft^2

22. $A = \dfrac{bh}{2} + \dfrac{\pi r^2}{2}$
$A = 169.28 + 42.32\pi \approx 302.2$ in^2

23. $A = \dfrac{\pi r^2}{4}$
$A = 81\pi \approx 254.5$ in^2

24. $A = LW - \dfrac{\pi r^2}{2}$
$A = 54 - \dfrac{9\pi}{2} \approx 39.9$ m^2

25. $P = 2L + \pi d$
$P = 240 + 80\pi \approx 491.3$ ft;
$A = LW + \pi r^2$
$A = 9600 + 1600\pi \approx 14{,}626.5$ ft^2

26. $P = s_1 + s_2 + \dfrac{\pi R}{2} + \dfrac{\pi r}{2}$
$P = 54 + 22.5\pi \approx 124.7$ in.
$A = \dfrac{\pi R^2}{4} - \dfrac{\pi r^2}{4}$
$A = 303.75\pi \approx 954.3$ in^2

31. a.

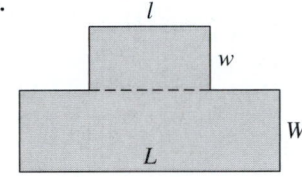

b. Total Area = area of larger rectangle plus area of smaller rectangle
c. $A = LW + Iw$

32. a.

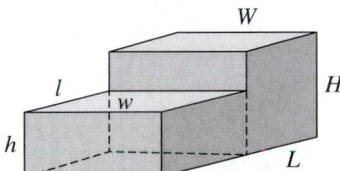

b. Total Volume = volume of larger rectangular solid plus volume of smaller rectangular solid
c. $V = LWH + Iwh$

33. a.

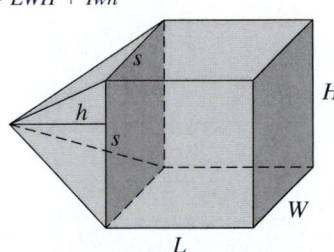

b. Total volume = volume of rectangular solid plus volume of pyramid
c. $V = LWH + LWh$ or $V = LWH + s^2h$

34. a.

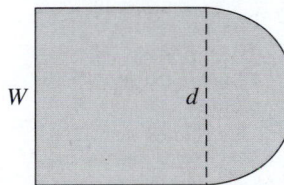

b. Perimeter = three sides of rectangle plus one half the circumference of the circle.
c. $P = 2L + W + \pi\dfrac{W}{2}$ or $P = 2L + W + \dfrac{\pi d}{2}$

35. $V = LWH + \pi r^2 h$
$V = 200 + 43.75\pi \approx 337.4$ in^3

36. $V = \pi r^2 h + \dfrac{2}{3}\pi r^3$
$V = 108\pi \approx 339.3$ m^3

65. $2500\pi \approx 7854$ ft^3
about 58,752 gal

66. $0.5625\pi \approx 1.77$ ft^3
about 13.24 gal

67. d. $35 + xy - 7y$

68. a. the hemispheric bowl with radius 3.5 in.

Photo Credits

Chapter R

p. 1: © Royalty-Free/CORBIS; p. 8: © Photodisc/Getty Images/RF; p. 24: © Royalty-Free/CORBIS; p. 79: © Glen Allison/Getty Images/RF.

Chapter 1

Opener/p. 87: © Neil Beer/Getty Images/RF; p. 88 left: © PhotoAlto/RF; p. 88 middle: © Brand X Pictures/PunchStock/RF; p. 88 right: © Lars Niki/RF; p. 134: © Royalty-Free/CORBIS; p. 163: NASA/RF; p. 176: © Tom Grill/Corbis/RF; p. 186: © Lourens Smak/Alamy/RF.

Chapter 2

Opener/p. 189: © Getty Images, Inc./RF; p. 190: © Siede Preis/Getty Images/RF; p. 191: © The McGraw-Hill Companies, Inc./Ken Cavanagh Photographer; p. 229: © PhotoLink/Getty Images/RF; p. 248: © Alan and Sandy Carey/Getty Images/RF; p. 259: Courtesy John Coburn; p. 266: © Royalty-Free/CORBIS.

Chapter 3

Opener/p. 281: © Christian Pondella Photography; p. 311: © Photodisc Collection/Getty Images/RF; p. 329: © Photodisc Collection/Getty Images/RF; p. 330: Courtesy NASA; p. 338: © Steve Cole/Getty Images/RF; p. 352: © U.S. Fish & Wildlife Service/Tracy Brooks/RF; p. 360: © Royalty-Free/CORBIS; p. 361: © The McGraw-Hill Companies, Inc./Barry Barker, photographer; p. 362: © Royalty-Free/CORBIS; p. 367/top: © Patrick Clark/Getty Images/RF; p. 367/bottom: © Digital Vision/PunchStock/RF; p. 368 left: © Goodshoot/PunchStock/RF; p. 368 right: © Royalty-Free/CORBIS; p. 369: © Charles Smith/CORBIS/RF.

Chapter 4

Opener/p. 381: © Royalty-Free/CORBIS; p. 390: © Adalberto Rios Szalay/Sexto Sol/Getty Images/RF; p. 393, p. 409, p. 444, p. 468: © Royalty-Free/CORBIS.

Chapter 5

Opener/p. 479: © Comstock Images/RF; p. 501 left: © Geostock/Getty Images/RF; p. 501 right: © Lawrence M. Sawyer/Getty Images/RF; p. 510: © U.S. Geological Survey; p. 511: © Lars Niki/RF; p. 515: © Medioimages/Superstock/RF; p. 525 top-left: © Ingram Publishing/age Fotostock/RF; p. 525 bottom-left: © Keith Brofsky/Getty Images/RF; p. 525 top-right: © The McGraw-Hill Companies, Inc./Andrew Resek, photographer; p. 525 bottom-right: © McGraw-Hill Higher Education/Carlyn Iverson, photographer; p. 537: © Stock Trek/Getty Images/RF; p. 554: Courtesy Dawn Bercier; p. 571: © CMCD/Getty Images/RF; p. 573: © John A. Rizzo/Getty Images/RF.

Chapter 6

Opener/p. 575: © Stockbyte/Getty Images/RF; p. 584: © I. Rozenbaum/F. Cirou/Photo Alto/RF; p. 589: © 2009 Jupterimages Corporation/RF; p. 590: © Royalty-Free/CORBIS; p. 601: © Creatas/Punchstock/RF; p. 611: © Ryan McVay/Getty Images/RF; p. 614: © The McGraw-Hill Companies, Inc./Jill Braaten, photographer; p. 630: © F. Shussler/PhotoLink/Getty Images/RF.

Chapter 7

Opener/p. 637: © C. Sherburne/PhotoLink/Getty Images/RF; p. 691: © Ingram Publishing/SuperStock/RF; p. 692: © Stockbyte/Getty Images/RF; p. 694: © Steve Cole/Getty Images/RF; p. 702: © Getty Images/RF.

Chapter 8

Opener/p. 707: © Royalty-Free/CORBIS; p. 739: © Brand X Pictures/Punchstock/RF; p. 740: © Digital Vision/Getty Images/RF; p. 743: © H. Wiesenhofer/PhotoLink/Getty Images/RF; p. 753 left: © Jim Wehtje/Getty Images/RF; p. 753 top: © Edmond Van Hoorick/Getty Images/RF; p. 753/bottom: © Creatas/PunchStock/RF.

Chapter 9

Opener/p. 761: © Digital Vision/RF; p. 781: © Royalty-Free/CORBIS; p. 794: © Anderson Ross/Getty Images/RF; p. 825: © Royalty-Free/CORBIS.

Appendices I

p. A-12: © Ryan McVay/Getty Images/RF.

Appendices II

p. A-25: © Getty Images/Digital Vision/RF; p. A-26: © Sandra Ivany/Brand X Pictures/Getty Images/RF.

Index